Handbook of Solvent Extraction

Handbook of Solvent Extraction

Edited by

Teh C. Lo
Hoffmann-La Roche Inc.
Nutley, New Jersey
USA

Malcolm H. I. Baird
McMaster University
Hamilton, Ontario
Canada

Carl Hanson
University of Bradford
Bradford, West Yorkshire
United Kingdom

A Wiley-Interscience Publication

JOHN WILEY & SONS

New York • Chichester • Brisbane • Toronto • Singapore

Library of Congress Cataloging in Publication Data:

Main entry under title:

Handbook of solvent extraction.

"A Wiley-Interscience publication."
Includes index.
1. Extraction (Chemistry)–Handbooks, manuals, etc.
I. Lo, Teh C. (Teh Cheng) II. Baird, Malcolm H. I.
III. Hanson, Carl.

TP156.E8H26 1982 660.2′84248 82-15957
ISBN 0-471-04164-5

Printed in the United States of America

10 9 8 7 6 5 4 3 2 1

CONTRIBUTORS

Sven O. S. Andersson, Director, MX-Processer Reinhardt and Company AB, Sweden.

Allan W. Ashbrook, Director of Corporate Technology, Eldorado Nuclear Ltd., Canada.

Neil F. Ashton, Lecturer, Department of Chemical Engineering, University of Birmingham, United Kingdom.

Philip J. Bailes, Senior Lecturer, School of Chemical Engineering, University of Bradford, United Kingdom.

Malcolm H. I. Baird, Chairman and Professor, Department of Chemical Engineering, McMaster University, Canada.

Eli Barnea, Formerly Deputy Managing Director, IMI Institute of Research and Development, Israel; now Group Technology Manager, Edward L. Bateman Ltd., South Africa.

Harold H. Bieber, Director, Engineering Design Department, Hoffmann–La Roche Inc., United States.

Ruth Blumberg, Senior Principal Scientific Investigator, Miles Israel Ltd., Israel; and Miles Laboratories, Inc., Industrial Products Group, United States.

William J. Bowers, Manager, Safety Section, Safety and Technical Services, British Nuclear Fuels Ltd., United Kingdom.

A. Brench, Unilever Research, United Kingdom.

Ulrich Bühlmann, Head of Research and Development, Kühni Ltd., Switzerland.

Stuart D. Cavers, Professor, Department of Chemical Engineering, The University of British Columbia, Canada.

John Coleby, Formerly of R. Graesser Ltd., United Kingdom; now Part-Time Consultant, North Wales, United Kingdom.

Michael Cox, Principal Lecturer, School of Natural Sciences, Hatfield Polytechnic, United Kingdom.

T. E. Degaleesan, Formerly of University of Madras, India; now Professor, Department of Chemical Engineering, A. C. College of Technology, Perarignar Anna University of Technology, Madras, India.

Rod I. Edwards, Formerly of National Institute for Metallurgy, South Africa; and Research and Development Manager, Engelhard Industries Ltd., United Kingdom; now Council for Mineral Technology, Chemical Engineering Research Group, Chemical Engineering Department, University of Natal, Durban, South Africa.

Jim F. C. Fisher, Formerly Manager, Zambian Mining Industry Technical Services, Zambia; now senior Divisional Metallurgist, Anglo American Corporation, Republic of South Africa.

Douglas S. Flett, Head, Metals Extraction Division, Department of Industry, Warren Spring Laboratory, United Kingdom.

J. C. Godfrey, Senior Research Fellow, Schools of Chemical Engineering, University of Bradford, United Kingdom.

David A. Gudelis, Director of Lubes and Specialties Laboratory, Products Research Division, Exxon Research and Engineering Company, United States.

M. Hafez, Formerly Assistant Professor, Chemical Engineering Department, McMaster University, Canada; now Group Leader, Lube Process Division, Research Department, Imperial Oil, Canada.

Wolf Hamm, Unilever Research, United Kingdom.

Carl Hanson, Professor and Chairman, School of Chemical Engineering, University of Bradford, United Kingdom.

Stanley Hartland, Professor, Department of Chemical and Industrial Engineering, Swiss Federal Institute of Technology, Switzerland.

N. F. Haasen, DSM, Central Laboratory (TEC), The Netherlands.

John Ingham, Senior Lecturer, School of Chemical Engineering, University of Bradford, United Kingdom.

Robert Kern, Manager, Concept Layout and Piping Design, Hoffmann–La Roche Inc., United States.

Tai K. Kim, Engineering Specialist, Chemical and Metallurgical Division, GTE Products Corporation, United States.

C. Judson King, Professor and Chairman, Department of Chemical Engineering, University of California, Berkeley, United States.

Walter C. G. Kosters, Group Leader, Physical Separations, Shell Internationale Petroleum Maatschappij B. V., The Netherlands.

G. S. Laddha, Formerly of University of Madras, India; now Director of Chemical Engineering, A. C. College of Technology, Perarignar Anna University of Technology, Madras, India.

Vaikuntam I. Lakshmanan, Formerly of Eldorado Nuclear Ltd., Canada; now Manager, Mineral Processing and Hydrometallurgy Groups.

Philip J. D. Lloyd, Director, Metallurgy Laboratory, Chamber of Mines of South Africa Research Organisation, South Africa.

Teh C. Lo, Engineering Fellow, Technical Development Department, Hoffmann-La Roche Inc., United States.

D. H. Logsdail, Principal Scientific Officer, Separation Processes Services, AERE Harwell, United Kingdom.

Leonard Lowes, Formerly of Research and Development Department, British Nuclear Fuels Ltd., Windscale Works, United Kingdom; now retired, Seascale, United Kingdom.

Martin B. MacInnis, Engineering Manager, Chemicals, Chemical and Metallurgical Division, GTE Products Corporation, United States.

Donald Mackay, Professor, Department of Chemical Engineering and Applied Chemistry, University of Toronto, Canada.

Jan Marek, Manager, Computations and Information, Luwa AG, Switzerland.

Colin McDermott, Lecturer, Department of Chemical Engineering, University of Birmingham, United Kingdom.

Magda Medir, Professor, Facultat De Quimica, Universitat De Barcelona, Spain.

John Melling, Senior Scientific Officer, Department of Industry, Warren Spring Laboratory, United Kingdom.

David Meyer, Manager, Process Design Department, IMI Institute for Research and Development, Israel.

Alfred L. Mills, Chemical Technology Division, Atomic Energy Research Establishment, United Kingdom.

Thomas Míšek, Head, Research Department, Research Institute of Chemical Equipment, Czechoslovakia.

Albert Mögli, Consulting Engineer, Chemical Engineering and Manufacturing, Kühni Ltd., Switzerland.

Eckart Müller, Lurgi Kohle und Mineralöltechnik GmbH, West Germany.

Arthur Naylor, Senior Group Manager (Chemistry), Research and Development Department, British Nuclear Fuels Ltd., United Kingdom.

Cor W. Notebaart, Formerly of Nchanga Consolidated Copper Mines Ltd. and Roan Consolidated Mines Ltd., Research and Development, Zambia; now Section Leader, Billiton Research B. V., The Netherlands.

James Y. Oldshue, Vice President, Mixing Technology, Mixing Equipment Company, United States.

Donald F. Othmer, Distinguished Professor, Polytechnic Institute of New York, United States.

Seymour P. Plonsky, Manager, Electrical and Instrumentation Engineering, Hoffmann-La Roche Inc., United States.

Henry R. Clive Pratt, Visiting Professor, Department of Chemical Engineering, University of Melbourne, Australia.

Michael W. T. Pratt, Senior Lecturer, School of Chemical Engineering, University of Bradford, United Kingdom.

Jaroslav Prochazka, Leading Scientist, Professor, Institute of Chemical Process Fundamentals, Czechoslovak Academy of Sciences, Czechoslovakia.

Hans Reinhardt, Managing Director, MX-Processer Reinhardt and Company AB, Sweden.

Nevill M. Rice, Senior Lecturer, Department of Mining and Mineral Engineering, University of Leeds, United Kingdom.

Kenneth Ridgway, Senior Lecturer in Pharmaceutical Engineering, School of Pharmacy, University of London, United Kingdom.

Gordon M. Ritcey, Head, Solution Treatment Section, Extractive Metallurgy Laboratory, Department of Energy, Mines and Resources, CANMET, Canada.

George A. Rowden, Research Project Manager, Davy McKee Ltd., United Kingdom.

Jan H. A. Rydberg, Professor, Department of Nuclear Chemistry, Chalmers University of Technology, Sweden.

Bruce M. Sankey, Senior Research Associate, Imperial Oil Research Department, Canada.

Edward G. Scheibel, Formerly Consultant, Suntech, Inc., United States; now President, E. G. Scheibel, Inc., United States.

James B. Scuffham, Manager, Research and Development, Davy McKee Ltd., United Kingdom.

Man Mohan Sharma, Professor, Department of Chemical Technology, University of Bombay, India.

L. Sherrington, Formerly of RTZ Ltd., United Kingdom; now retired, Wokingham, United Kingdom.

Anton J. F. Simons, DSM, Central Laboratory (TEC), The Netherlands.

Oksana A. Sinegribova, Senior Reader, The Mendeleev Institute of Chemical Technology, U.S.S.R.

Michael J. Slater, Senior Lecturer, School of Chemical Engineering, University of Bradford, United Kingdom.

Ladislav Steiner, Senior Assistant, Department of Chemical and Industrial Engineering, Swiss Federal Institute of Technology, Switzerland.

H. M. Stönner, Lurgi Kohle and Mineralöltechnik GmbH, West Germany.

W. A. M. te Riele, Council for Mineral Technology, Chemical Engineering Department, University of Natal, Durban, South Africa.

Edward E. Thorpe, Senior Industrial Engineer, Beecham Pharmaceuticals Ltd., United Kingdom.

Gunnar Thorsen, Laboratory of Chemical Engineering, University of Trondheim, Norway.

Tamotsu Ueno, Manager, Chemical Research Division, Niigata Research Laboratory, Mitsubishi Gas Chemical Company, Japan.

José A. Vidueira, Senior Chemical Engineer, Union Carbide Corporation, United States.

George C. I. Warwick (deceased 7/24/81), formerly of Davy McKee (Oil and Chemicals) Ltd., United Kingdom and British Nuclear Fuels Ltd., United Kingdom.

William L. Wilkinson, formerly Professor of Chemical Engineering, University of Bradford, United Kingdom; now Deputy Director, Reprocessing Division, British Nuclear Fuels Ltd., United Kingdom.

John Arnold Williams, Head of Safety and Technical Services, British Nuclear Fuels Ltd., United Kingdom.

Peter D. Wilson, Assistant Group Manager (Chemistry), Research and Development Department, British Nuclear Fuels Ltd., United Kingdom.

Donald R. Woods, Professor, Department of Chemical Engineering, McMaster University, Canada.

Gennady A. Yagodin, Correspondent Member, Soviet Academy of Science; Professor and Rector, The Mendeleev Institute of Chemical Technology, U.S.S.R.

PREFACE

There is often a need in the process industries to separate the components of homogeneous liquid mixtures. Several approaches are available, most depending on the addition or creation of a second phase followed by differential distribution of the components between the resultant two phases. Liquid–liquid extraction, commonly known as *solvent extraction*, is a technique for separating the components of a solution by distribution between two liquid phases. In many ways it is complementary to distillation. The distribution of components between liquid and vapor phases depends largely on molecular size, whereas a distribution between two conjugate liquid phases is more dependent on chemical type. The term "solvent extraction" is also used for certain solid–liquid operations, in particular the recovery of essential oils by treatment of seeds and so on with an organic solvent. This operation is more akin in its characteristics to *leaching*, although it uses an organic solvent rather than an inorganic reagent. Solid–liquid operations are not covered by this handbook, for the purposes of which *solvent extraction* is interpreted as being *liquid–liquid* extraction.

With the enormous growth of research and the industrial applications of solvent extraction over the past few decades, the need has arisen for a single reference work presenting the state of knowledge in all areas of this subject. This need was in the minds of the editors when they met several times during 1975–1977 and devised the basic scheme of chapters used in this handbook. It was agreed that four distinct parts should be created that would correspond to different special needs of the potential readership, with sufficient cross-referencing to provide a coherent overall picture of the subject. Starting in late 1977, world authorities in various areas of extraction were asked to submit chapters on their field; these were written and subjected to review and editing during the period to September 1981.

Part I of this handbook deals with *general principles*. It covers not only the standard approaches that will be familiar to most chemical engineers, but also specialized chemical and physical effects that have assumed increasing importance in recent years, such as metals extraction chemistry and axial mixing. Chapter 8 summarizes special aspects that have not yet been studied in great detail but could be of importance in the future.

Part II, concerning *industrial extraction equipment*, is of interest to all who design and operate extractors. The emphasis is on the physical characteristics, selection, and operation of the very large number of types of extractor available. Wherever possible, authors with industrial experience in operating this equipment have been asked to contribute these chapters.

Industrial processes using extraction are described in Part III. Here the emphasis is on applied chemistry and flowsheets, with the equipment characteristics being only one factor in the overall design picture. This section gives many examples of the reasons for selecting extraction rather than other separation techniques and the selection criteria for various types of equipment.

Part IV, on *cost and engineering*, deals with other practical aspects of extraction processes, such as control, costing, and safety. These must be considered before any extraction process can be fully designed.

More than 70 authors from many countries, with different experience and academic training, have contributed to this handbook. Because of this and the very rapid rate of advance of knowledge of extraction, the different chapters do not always show the same approach and outlook. The editors have made every reasonable effort to achieve consistency and provide cross-references, but, in general, no attempt has been made to enforce a particular viewpoint. For example, there may be sincere disagreement between authors of different chapters about the mechanism of a particular industrially important metal extraction process or the best method of compensating for mixing effects, and so on. In such cases each author's view has been allowed to stand under that author's name, this being the "state of the art" as the handbook goes to press. The alternative would be to delete all reference to such controversial material, but this would greatly reduce the scope of the handbook.

This handbook should be helpful to people involved in processing, equipment selection and design, and research and academic study of solvent extraction. It is hoped that it will be readable (at least in part!) by anyone with a background in chemical or mechanical engineering, chemistry, or process metallurgy.

In presenting this handbook, the editors would like to pay tribute to the late Professor Robert E. Treybal, whose book *Liquid Extraction* (now out of print) was for many years the standard text in this field, providing an invaluable teaching source during the formative years of the technology.

The initial proposal by the editors for this handbook was appraised for the publishers by Dr. Donald F. Othmer, Distinguished Professor of the Polytechnic Institute of New York, who continued to maintain an active interest in its development. The editors are extremely grateful for his interest and encouragement and have invited him to write the Foreword in which some of the historical background of solvent extraction is reviewed along with his memories of some of his own work in the field.

The editors wish to express their appreciation to their respective employers for their tolerance and understanding during the preparation of this handbook. Special thanks are due to Mrs. Patricia A. Hargreaves and the many other secretaries who helped with the heavy volume of correspondence and to Ms. Anne Toto for her editorial assistance.

We also wish to thank the extraction experts who acted as external reviewers but who are too numerous to be listed individually and Mr. Patrick Wong of Hoffmann–La Roche Inc. for his work on the index.

Last but not least, our thanks go to the many authors who have given unsparingly of their time and expertise. Such has been their enthusiasm that, in many cases, the submitted contributions at first exceeded the required length! However, the authors have always shown great cooperation in response to editorial requests for changes in length and style, and it is our hope that this handbook will faithfully reflect the current state of the art in solvent extraction.

TEH C. LO
Nutley, New Jersey
USA

MALCOLM H. I. BAIRD
Hamilton, Ontario
Canada

CARL HANSON
Bradford, West Yorkshire
United Kingdom

December 1982

RANDOM REMINISCENCES OF PROBLEMS IN SOLVENT EXTRACTION

DONALD F. OTHMER
Distinguished Professor
Polytechnic Institute of New York
Brooklyn, New York

Solvent extraction to obtain from an initial liquid a dissolved material such as another liquid or a solid is a chemical engineering separation operation that has many variations and many applications in the process industries, and uses many types of equipment. It may be one of a long series of sequential operations making possible (1) the miracles of the modern process industries that have grown and increased mightily from prehistory and (2) the simple changes of the chemistry of materials immediately available to humans. Now there are substantial reasons why the use of solvent extraction will continue to develop and to expand in benefit to the distant reaches of the future.

The major development and use of solvent extraction, as the term is used here, have taken place in only a little more than the last 50 years. Of extreme interest and good fortune to the writer, who has been honored in the request to write this Foreword in the form of some of his personal experiences and reminiscences of the field, is the fact that this growth and maturing of what has become such an important tool of the chemical engineering profession in the design of the processes and plants of the chemical industry has coincided with the span of his own professional life.

In addition, he regards himself as most fortunate, since he has no biologic children, to have been blessed with so many brilliant academic children whom he can recognize with almost paternal pride. Thus he so regards one editor of this monumental treatise, for another substantial accomplishment that Teh C. Lo has achieved, with his international partners, Malcolm Baird and Carl Hanson, who are also leaders because of their vast contributions to the subject field.

The solvent extraction of solids from solids is an older, possibly simpler art—not the subject of this handbook, although certainly its parent. The leaching or lixiviation of alkaki from wood ashes allowed the preparation of the first crude soap and later glass. This early practice of washing of solids was extended later to that of cloth and clothing and also to some simple examples of wet metallurgy. It was, at first, not countercurrent and no more scientific than the modern television commercial "Ours is better!" for another brand of aspirin or detergent.

One of the most elusive historical questions of extraction in metallurgy is that of the amalgamation by mercury of the precious metals. Particularly, no one is sure whether silver was first so obtained from other solids, as its ores, almost 450 years ago in Europe or by metallurgists with the conquistadors in Mexico thereafter. An operation with an equal degree of antiquity is also related to extraction with chemical reaction. This is the parting of a melted alloy of gold and silver by sulfur, antimonium sulfide, or other inorganic liquids that dissolve the liquid silver.

The advantage of countercurrent operation has been known for centuries but not mathematically modeled as a first scientific approach to this art of washing a solid with a solvent. Usually, however, the solvent was that of a running stream of water of relatively infinite amount, and as the extracted material was seldom of value, it was discarded in the *solvent layer.* It has been only in our century that a large number of liquids have been available industrially, as the first solvent for a solute or as the second solvent for extraction of solute from the first. Thus the field of this book, solvent extraction, is confined to what has long been known also as *liquid–liquid extraction.*

The classic method of separation of two liquids from their solution has been by distillation. Even with modern processing and equipment, this may require considerable thermal energy. The thermodynamic minimum heat cost of the separation by distillation, as by extraction, is the supply of the mutual heat of solution of the two liquids; and extraction, by itself, often effects the separation with much less additional energy. However, the stripping of the solvent from the extract and raffinate layers may require much heat in the distilling columns usually required as accessories.

Simultaneously with many liquids becoming available in or by the petroleum refining and the chemical industries, many problems have arisen in regard to separating these liquids, as well as many uses for liquids as solvents for separating solutes in other liquids, often inorganic and metallic in nature. It was fortunate that when the petroleum industry became the processor of the largest value and diversity of fluids by some orders of magnitude, many of the tools of the chemical engineer already had been designed and forged. They merely had to be shaped to do the fluid separations, particular to the liquids in the new petroleum industries.

Physicochemical Relations

The design of any extracting equipment requires equilibrium data, usually of mutual solubilities of the components on a ternary diagram, and distribution data, that is, of amount of the solute in one liquid layer versus that in the other. Of the large amount of such data now available for design of processes, much of that early determined for ternary liquid systems was published by Othmer and his students.

In addition to many such data, Othmer, et al. [1] indicated a very simple but precise method of the determinations of solubilities of three liquids. Also, a useful means of expressing such data, particularly in the extraction by a solvent of a solute liquid from water, was the construction of effective concentration curves. Here the value of the amount of solute divided by the sum of that of the solute plus that of water was plotted for the solvent layer against that for the water layer. This gives a plot of the effective concentration in the two layers without reference to the solvent present in each and helps in screening liquids for a particular solvent extraction.

Othmer and Tobias [2] showed how to (1) apply the lever rule graphically in determining tie lines (i.e., lines on a ternary-phase diagram connecting compositions of the two liquid phases in equilibrium with each other), and (2) reduce to a new linear equation based on work of Nernst, and to graph as a straight line, data for mutual solubility and thus to be able readily to correlate and predict such data, and (3) develop and use an equation and graphs to allow the interrelation and prediction of liquid mutual solubility data from data of vapor pressures or of vapor compositions of boiling binary solutions. The reverse program was also demonstrated, allowing the translation of liquid–liquid solubility data to give vapor pressures or vapor compositions.

A simple method for calculation from known mutual solubility data of the number of transfer units required in an extractor for a particular design was also presented by Scheibel and Othmer [3]. Although solvent extractions are usually carried out at or near ambient temperatures, a linear method for the interrelation of distribution coefficients to temperature was developed to allow calculations and extrapolations at other temperatures [4]. Data for other ternary systems important in the design of separating processes for particular industries were also presented by Othmer et al. in 1945 [5] and later by Othmer and Ku [6].

Extraction Equipment

Although chemists have used solvent extraction on the laboratory bench for much longer, it has been used in industrial plants for only about 100 years. One reason has been that there had not been available either the necessary solvents or large standardized extracting equipment. Also, process designers had to design and build their own extractor, sometimes as simple as one with a spray of droplets of the liquid to be extracted settling down through a tall vessel of solvent.

From such early solvent extractions came the spray- or droplet-type column—depending simply on droplets of light liquid introduced at the bottom and settling upward through a heavy liquid in the lower part of the column to an interface with a continuous phase of light liquid in the upper part. At the top, droplets of the heavy liquid are introduced to settle down through the light liquid to the interface. The sum of the heights of light and heavy liquids have been sufficient to give as much as 80 ft of countercurrent contact. Always empirically designed from experimental pilot plants, many large plants have been very successful in such separations.

Often these "open" extractors were filled with packing shapes, as used in distillation columns. These prevent convection or other currents in the continuous liquid phases from interfering with the countercurrent flows of the respective droplets in the upper and lower continuous phases. The same droplet-type equipment has been used [7, 8, 9] as inexpensive, direct contact liquid–liquid heat exchangers without metallic heat-transfer surfaces. In such columns, for example, droplets of a heavy, hot liquid descend through and are cooled by a rising stream of warmer, lighter liquid, which, of course, must be quite immiscible, as water and a petroleum oil. In some cases it has been possible to include a solvent extraction function with the heat exchange function in a single operation and column [9].

Inexpensive as such spray- or droplet-type extractors were to build, although they do have a large holdup and inventory of solvent that may be expensive, other types of extractor have often superseded them.

Simultaneously with the development of droplet extractors in high vertical columns was that of the mixer–settler type. The first were batch units with a single tank for mixing, then settling, and finally withdrawing separately the solvent and the raffinate layers for countercurrent treatment of other batches waiting in auxiliary vessels. This assemblage of tanks could then be made to give satisfactory extraction; and it was usually mounted in a more or less horizontal plane, like a row of separating funnels in a line on a laboratory bench. There was successive (1) mixing (shaking) of the two liquids, (2) settling, (3) decanting, and then (4) moving (in large units, pumping) of the two phases, respectively, to adjacent vessels right and left.

From the batch operation it was only a step to a continuous mixer–settler operation. Many variations have been developed over the past 50 years and are still being described in journal articles and patents. These vary in size from the row of separatory funnels to very large units, with each tank having a capacity of thousands of gallons in metallurgical and petroleum refineries. At first, operation was always in batch units and manually controlled and later it was continuous and automatic.

Compared to the simplicity of design and construction of the open-column extractors, with bubbles of one liquid rising and/or falling through a continuous phase of another liquid, or of a single tank alternately used as a mixer and then as a settler, was the complexity of piping and fittings of another extractor used in the 1920s that was developed first for extracting acetic acid from its solutions. This had a "staircase" of 20 or more units each: mixing tanks, settling tanks, and air lifts or other pumps to lift the heavier layer of one settler to the next higher mixer, whereas the settled lower layer decanted to overflow in an inverted syphon down to the next-lower mixer [10].

The late 1920s and early 1930s brought columns that also had as many as 20 or more alternate sections for mixing and for settling, but superimposed above each other, so that the agitators in the mixing sections could have a single drive shaft. Now both vertical liquid flows—up or down—of the two layers could proceed automatically as a result of gravity alone. In one of these extractors [11] it was found that, as constructed and operated, a mixer–settler pair might have the efficiency of somewhat more than one equilibrium unit as a result of the additional countercurrent extracting action in each settler. This extractor had all the agitators of the mixer sections on a single axial drive shaft for units of up to 4 ft in diameter; larger units had individual, externally driven agitators between the baffled settlers. Somewhat later, other columnar mixer–settler extractors were developed, patented, named after, and commercialized by Othmer's former students, Scheibel and York, and then Karr. These also have been improved and/or operated and reported by Teh C. Lo. All these, and others, are reported later in this book because of their industrial importance.

Two Common Liquid Solutions Difficult to Separate

As stated in the rule in solubility relations, "like likes like," acetic (and even more so, formic) acid, with the ionizing hydrogen atom, prefers to remain with the water that is so similar in this respect, rather than being separated by dissolving in any other liquid in contact. This is true, however much the "organic part" of the acetic acid molecule may resemble or be attracted to the organic nature of a solvent, and for formic acid, the "organic part" has only half as many carbon atoms as does acetic acid. Under the same rule, solutions of alcohol and water are difficult to separate because of the common hydroxyl group.

However, the dilute aqueous solutions of both alcohol and acetic acid satisfied human needs for thousands of years before it was realized, in attempts to separate these two solutions, that they were solutions of individual liquids. These liquids came naturally, but they were also manufactured in two of the first processing industries—the beer and wine by fermenting carbohydrate foods with the loss of carbon dioxide, and the vinegar from subsequent fermenting with simultaneous oxidation of the alcohol in beer or wine. These two solutions are among the very few generally known during all of human history. After long use as solutions, much of the development work in the theory, process design, and equipment for separating liquids was concerned with them. For many hundreds of years distillation practice was devoted almost totally to separating alcohol–water solutions and for the last hundred years has been equally concerned with separating water–acetic acid solutions. Both solutions are now applied in many processes in many different industries.

For at least the last 600 years, human ingenuity in chemical engineering process development has been exercised in separating water from alcohol solutions made by fermentation so as to make a beverage that intoxicated one more quickly. Distillation has been thc usual unit operation used because of its much greater simplicity for this system and the product desired, but solvent extraction has also been shown to separate the alcohol [12, 13]. The current interest in alcohol as a motor fuel has renewed interest in every possible means of separation of this pair of liquids, so much alike as to make difficult the driving of a wedge between them by any separation process, including extraction. However, many extraction processes are now proposed to take advantage of possibly lower thermal cost in the 1980s. None, so far as is known, has resulted in a large installation.

As industry has grown, there has been an increasing awareness of the difficulty and large cost in separating what has amounted to millions of tons of water from solutions of acetic acid of various concentrations. Besides oxidative fermentation of alcohol to yield a vinegar of about 6% concentration, condensation of the vapors from the distillation of wood gives pyroligneous acid of about the same strength. But acetic acid is difficult to separate from water, even though, unlike every other volatile organic acid, it has no constant boiling mixture—at any pressure—with water [14]. Traditionally in its production acetic acid was won chemically from its dilute solutions by (1) neutralizing with lime to give calcium acetate, (2) evaporation of the water and drying of this salt, and (3) springing of the acetic with sulfuric acid and evaporating off an acid of about 70% concentration, which was distilled and rectified into fractions of 28% and 56%, and finally some part came as glacial, 99.5%.

An early example of the industrial use of solvent extraction was that of acetic acid from pyroligneous acid by a solvent, ethyl acetate, patented by Theodor Goering [15]. Although this development was 100 years ago, his countercurrent process is still in industrial use in one plant today, with some modifications of the distillation system for recovery of ethyl acetate, which has some disadvantages as a solvent.

Simpler and less expensive solvent extraction processes have been developed since by many others and described in many patents. Those by Othmer have separated billions of pounds of acetic acid from solutions coming in the original production or in its recovery from use as a solvent or reagent, or as a by-product, in the manufacture of—among other tonnage materials—aspirin, cellulose acetate, aromatic and paraffinic petroleum fractions, and cyclonite or RDX, the powerful chemical explosive of World War II.

Two variations of countercurrent solvent extraction of one component of a fluid from another, as well as the basic liquid–liquid extraction, also were first used as operations to remove the water from acetic acid. The first was *extractive distillation.* (Suida built several plants in Europe, and Albin two in this country during 1925–1935.) Here, in a distilling column, a descending stream of a quite high boiling solvent extracts acetic acid from the rising vapors containing also water. The vapors go overhead substantially free of acid, and the extract layer—acetic acid, water, and the very high boiling solvent—go out the bottom to be distilled separately to give acid-free solvent and acetic acid with about 25% water. This is then fractionated by usual rectification.

In another extractive distillation developed at about the same time, a solvent was used that had at least a semichemical affinity for one of the components in the distillation of the vapors of methanol and acetone, which has a constant boiling mixture. A brine of aqueous calcium chloride descending from the top of the distillation column allows the acetone to go overhead while discharging the methanol in the brine at the base for separate distillation therefrom [16]. A somewhat higher temperature is necessary in this distillation because the calcium chloride lowers the vapor pressure of methanol, but not of acetone.

Azeotropic distillation, in its best embodiment, is also a method of solvent extraction; it uses a solvent having a boiling point somewhat lower than that used in the true extractive distillation. This descending liquid also extracts from a rising stream of vapors in a distilling column the acetic acid to be discharged completely anhydrous and free of solvent at the base. The solvent simultaneously steam distills the water, free of acid, over the top (Othmer, numerous journal articles and patents 1930–1970, also ref. [17]). The process has been used in many plants throughout the world, and the largest installation ever has just been engineered for early installation.

Usually, a solvent extractor discharges two liquid streams: (1) the extract layer, containing most of the original solute, along with some of its original solvent and most of the second or added solvent; and (2) the raffinate layer, containing a minimum of both the original solute and the added solvent, and most of the original solvent. Each of the two liquid streams must be distilled in at least one column for separation of the added solvent contained, which is returned to the extractor. Often these auxiliary columns use azeotropic distillation. Since each layer contains three components, two distillation columns might be required for each. The distillations use much more energy than the extractor does, and their processing must be minimized and designed carefully. In one process, however, only one extractor is used with four azeotropic distillations in one method of minimizing these heat costs [18].

Extraction of Some Other Mixtures

Quite the contrary, anhydrous acetic acid itself may be used as an excellent solvent to extract undesired aromatic and naphthenic compounds from the very much less soluble paraffin components of a crude petroleum oil or distillate. The acetic acid in the raffinate from this first extraction is then extracted by a small amount of water, whereas the solvent layer is extracted with weaker acid from subsequent washings. This dilutes the acetic acid therein and causes the precipitation of the oil containing the aromatics and naphthenes. The water added is minimized by the use of another extraction of this precipitated oil. Thus the aromatics and naphthenes come out in a new phase separation.

By this combination of extractions, a very small amount of water as dilute acid is used successively in respective extraction columns. The two layers from each of these are each subsequently extracted with dilute acetic acid to cause phase separation. This results in four auxiliary extractors, which with the one main one, total five extractors. Obviously, the four auxiliary extractors use the dilute acetic acid, which comes as solvent layer from each in countercurrent flow. This builds its strength up to 90%.

A single azeotropic distilling column using an earlier Othmer process distills the water from this 90% acetic acid in concentrating it back to 99.5% for recycle to the first extractor of the series of five. Truly this process and plant use a minimum of energy with the five extractors and only one main azeotropic column. A very much smaller distilling column strips the solvent for the azeotropic distillation from the small amount of water, which is reused successively for its several countercurrent extracting functions.

This thermally efficient system of separating aromatics, such as benzene, toluene, xylene, or naphthenes from the paraffins of petroleum fractions to be used as lubricants, burning oils, motor fuels, and so on uses as an extracting liquid, acetic acid, whose solvent power for the aromatics and naphthenes is markedly changed by the addition of a very small amount of an added liquid—water in this case. This is analogous but opposite to the principle used by many generations of chemists—seldom countercurrently—in the "salting out" of a solute into one layer by a heavy addition of another solid solute—often a salt—very soluble in the other liquid.

Seldom would such additions of a huge amount of a solid solute be economical in the large volumes of liquids to be extracted industrially. However, sometimes industrial solutions are very concentrated with solid solutes and do have relatively smaller amounts of other values that classically are separated by expensive processes or completely discarded to recover the principal value of the solution.

Two examples of the separation of such smaller amounts of materials from highly concentrated solutions by what were found to be extraordinarily efficient solvent extractions are described by Othmer [19]. These take advantage of the inordinately large extracting power of a solvent for the particular minor amount of a solute or solutes when present in such a concentrated solution—usually aqueous. The solvent used may be miscible in all proportions with the liquid itself (e.g., water) when free of the solutes. Thus a minor amount of a solute in an aqueous solution concentrated with a principal solute may be extracted with ethanol, methanol, acetone, or acetic acid, any one of which is completely miscible with water by itself.

One example is in refining of raw sugar, where it has long been known that the large heat expense of the usual melting (dissolving) of the raw sugar crystals in water and recrystallization time after time from increasingly impure mother liquor—finally molasses—can be greatly reduced. Meanwhile a purer product is obtained by simply washing the film of impurities off the surface with methanol or anhydrous acetic acid. Solvent extraction with these solvents, acetone, or other water-soluble solvents will separate the same impurities from sugar solutions or heavy syrups of 50% or more concentration (e.g., molasses) because the water-soluble solvent does not dissolve in the syrup. It cannot compete with sugar and its many hydroxyl groups for solubility in the water [20].

Equally interesting as a study in mutual miscibilities and the separation processes that may be derived therefrom is the relation of water-soluble solvents—particularly acetone—to the acetic acid in the black liquors from the kraft or the neutral sulfate semichemical processes of wood pulping. After evaporation of these liquors, as usual, but before burning to recover the soda values, the acetic acid and formic acid are sprung with sulfuric acid. Then in the presence of the high concentration of salts and other solids a solvent such as acetone, normally completely miscible with water, does not dissolve in this very concentrated solution. Instead it extracts the acetic acid, for which it has a distribution coefficient several times as favorable as that of any more usual organic solvent in the ternary system with water and acetic acid alone [18, 21].

During the long sleep of science, the alchemists dreamed of a wonderful solvent that would dissolve anything and everything. They woke up when they tried to imagine a possible container that would not, by itself, be dissolved by such a liquid. Conversely, there is water—our most unusual but most common liquid—with the widest solvency ability of any liquid. So, what is an effective solvent extractant for water from its dilute solution, seawater, the largest volume of any material on the surface of earth? Such a solvent would be valuable indeed in developing an economical process for desalination.

Water, the most common liquid in chemical industry, biology, botany, and elsewhere, is most often the extracting solvent, or the liquid whose solute is being extracted. The solvent extraction of a part of the water as freshwater from seawater and its salt is an interesting exercise that many have tried with huge ratios of common petroleum fractions and with lesser ratios of comparatively uncommon liquids. Usually it has been attempted to take advantage of (1) the natural repugnance to salt of many organic solvents having more—although often very little—solubility for water and/or (2) the change of the amount of this solubility of water in these organic solvents with temperature.

Not only in desalination has there been used a change in solubilities with temperature, but also in solvent extractions of other materials. In the simplest case a solvent is brought to equilibrium with a solution in another liquid at one temperature, the temperature is changed to one where the solubility is less, and much of the other liquid goes out of solution to form a second layer at the new temperature.

The change may be to either a higher or lower temperature. For use in desalination, some liquids have greater solubility for water at higher temperatures whereas others dissolve more at lower temperatures. Thus processes have been designed to use both types of solvent. After solvent extraction of the seawater at one temperature, high or low (whichever temperature gives most solubility of the water in the particular solvent), the temperature of the extract layer is increased or lowered, as the case may be. Then the lesser solubility of the water at that temperature causes it to form a separate layer, which is decanted. Thus water separates from different solvents at a temperature higher or lower than that of the extraction; and processes of both types have been described [9] wherein butanol is exemplified as a liquid dissolving more water when hot than when cold.

Accessory equipment for heat transfer, distillation, and particularly multistage flash evaporation is required. By the use of such an extraction, much of the water is obtained without requiring vaporization; thus the overall heat requirement of a multistage flash evaporation system may be reduced very much by the initial solvent extraction, followed by ancillary multistage flash evaporation steps, which can accomplish the usual solvent stripping from the raffinate and the extract streams.

And the Future?

The coming years will see a vast development of process engineering for the expanding chemical and metallurgical industries—particularly as different materials and ores of lower value must be used as feedstocks. Even more challenging, as to complexity and size of equipment and of its throughput, will be the task of the chemical engineer to *manufacture fuels*. Hence solvent extraction, as all the tools of our profession, will likewise become more versatile and more sophisticated in its increasing uses.

If we can extrapolate to the future through an analysis of the past, we can estimate how far we should go in this great field by reading the following pages, and also by considering an incident of the middle 1920s when this writer simply *had* to have an efficient extractor of many equivalent stages to make possible an essential process. The company asked him to pose the problem to the most capable chemical engineering theoretician and consultant of the time, who, however, had had no occasion to do any extraction. The dogmatic response as to how to obtain the necessary countercurrent liquid contacting was to use a bubble cap distilling column!

We have gone a long way since then in understanding the mechanism and practice of solvent extraction, and the information given in this handbook will be the conveyance for a longer journey for the rest of the years of this century!

REFERENCES

1. D. F. Othmer, R. E. White, and E. Truegar, *Ind. Eng. Chem.* **33**, 1240 (1941).
2. D. F. Othmer and P. E. Tobias, *Ind. Eng. Chem.* **34**, 690, 693, 696 (1942).
3. E. G. Scheibel and D. F. Othmer, *Transact. AIChE* **38**, 339, 883 (1942).
4. D. F. Othmer and M. S. Thakar, *Ind. Eng. Chem.* **44**, 1654 (1952).
5. D. F. Othmer, W. S. Bergen, N. Schlechter, and P. F. Bruins, *Ind. Eng. Chem.* **37**, 890 (1945).
6. D. F. Othmer and P. L. Ku, *J. Chem. Eng. Data*, **4**, 42 (1959).
7. D. F. Othmer, *Kirk-Othmer Encyclopedia of Chemical Technology*, 2nd ed., Vol. 22, p. 44, Wiley-Interscience, New York, 1970.
8. D. F. Othmer, *Chem. Proc. Eng.* **43**, 566 (1962).
9. D. F. Othmer, U.S. Patent 3,692,634 (September 19, 1972).
10. E. Ricard and H. M. Guinot, German Patent 598,595 (June 15, 1934).
11. D. F. Othmer, U.S. Patent 2,000,606 (May 7, 1935).
12. D. F. Othmer and E. Trueger, *Transact. AIChE* **37**, 597 (1941).
13. D. F. Othmer and R. L. Ratcliffe, Jr., *Ind. Eng. Chem.* **35**, 798 (1943).
14. D. F. Othmer, S. J. Silvis, and A. Spiel, *Ind. Eng. Chem.* **44**, 1864 (1952).
15. T. Goering, German Patent 28064 (December 18, 1883).
16. D. F. Othmer, U.S. Patent application (1930).
17. D. F. Othmer, "Azeotropic and Extractive Distillation," in *Kirk-Othmer Encyclopedia of Chemical Technology*, 3rd ed., Vol. 3, Wiley-Interscience, New York, 1978, p. 352.
18. D. F. Othmer, *Chem. Eng. Progr.* **54**, p. 48 (1958).
19. D. F. Othmer, "Extraction of Concentrated Solutions," in *Thermodynamic Behaviour of Electrolytes in Mixed Solvents II*, Advances in Chemistry Series, W. F. Furter, Ed., American Chemical Society, 1979, No. 177, p. 1.
20. D. F. Othmer, U.S. Patent No. 4,116,712 (September 26, 1978).
21. D. F. Othmer, U.S. Patent 2,878,283 (March 17, 1959).

CONTENTS

INTERNATIONAL SYSTEM OF UNITS (SI) AND CONVERSION FACTORS

All industrial nations have converted, or are in the process of converting, to SI units. This unit system is sometimes loosely termed the metric system, but it refers to a specific set of metric base units and prefixes (Tables 1 and 2) that must not be confused with other metric systems such as the obsolescent centimeter, gram, second (CGS) system, which still survives in the common unit for surface tension.

The base units can be used to define SI units for derived quantities such as force (newton, N), energy (joule, J), pressure (pascal, Pa) and power (watt, W). The unit of volume m^3 is inconveniently large for many chemical applications, and so the liter is permitted as an alternative. For the same reason, "mol" denotes the molecular weight in grams, not kilograms. The molecular weight in kilograms is denoted kmol in SI units. Generally speaking, the use of prefixes (Table 2) allows convenient adjustment of the unit sizes.

TABLE 1 SI BASE UNITS

Quantity	Base Unit	Symbol
Length	metre*	m
Mass	kilogram	kg
Time	second	s
Electric current	ampere	A
Thermodynamic temperature	kelvin	K
Amount of substance	mole	mol
Luminous intensity	candela	cd

*Note that the U.S. spelling "meter" is used in this handbook.

TABLE 2 PREFIXES

Multiple	SI Prefix	Symbol
10^{18}	exa	E
10^{15}	peta	P
10^{12}	tera	T
10^{9}	giga	G
10^{6}	mega	M
10^{3}	kilo	k
10^{2}	hecto	h
10	deka	da
10^{-1}	deci	d
10^{-2}	centi	c
10^{-3}	milli	m
10^{-6}	micro	μ
10^{-9}	nano	n
10^{-12}	pico	p
10^{-15}	femto	f
10^{-18}	atto	a

The use of SI units is preferred in this handbook, but it has not been mandatorily imposed on authors. Many industries, particularly in the United States, still operate wholly or partially in non-SI units, and until recently much of the extraction data published in industrially oriented journals have been expressed in non-SI units.

For these reasons, Table 3 is provided, giving some of the more frequently used terms in extraction in non-SI and SI units, with appropriate conversion factors. (The conversion factor is the number by which a non-SI quantity must be multiplied to give the corresponding SI quantity.)

TABLE 3 CONVERSION FACTORS TO SI FOR SOME COMMONLY USED QUANTITIES

Measurement	Non-SI	SI	Conversion Factor
Acceleration	ft/s^2	m s^{-2}	0.3048
Absolute temperature	°R or °F abs	K	0.5556
Area	ft^2	m^2	0.0929
Capacity	gallon (U.S.)	m^3	0.003785
	gallon (U.S.)	liter	3.785
	gallon (Imperial)	liter	4.536
	ft^3	m^3	0.02832
Concentration (molar)	lb · mol/ft^3	mole liter^{-1} or kmol m^{-3}	16.02
Density (or mass concentration)	lb$_m$/ft^3	kg m^{-3}	16.02
Energy	ft · lb$_f$	J	1.356
Force	lb$_f$	N	4.448
Gas constant	ft ⟩ lb$_f$ lb mol^{-1} °R^{-1} [1546]	J mol^{-1} K^{-1} [8.314]	
	cal mol^{-1} K^{-1} [1.987]		
Gravity correction factor (g_c)	lb$_m$ · ft s^{-2} lb$_f^{-1}$ [32.17]	kg m^{-1} s^{-2} N^{-1} [1.000]	
Heat	BTU	J	1055
Heat of reaction	BTU/lb · mole	J mol^{-1}	2.326
Length	ft	m	0.3048
Mass	lb$_m$	kg	0.4536
	short ton (2000 lb)	mg (metric ton)	0.9072
Mass flow per unit area	lb$_m$/ft^{-2} s^{-1}	kg m^{-2} s^{-1}	4.883
Moles	lb · mole	mol	453.6
Molecular diffusivity	cm^2 s^{-1}	m^2 s^{-1}	10^{-4}
Power	hp (horsepower)	W	746
Pressure	lb$_f$/in^2	kPa	6.90
	atm	kPa	101.35
	mm Hg	kPa	0.1334
Specific heat	BTU lb$_m^{-1}$ °R^{-1}	J kg^{-1} K^{-1}	4187
	BTU lb^{-1} mol^{-1} °R^{-1}	J mol^{-1} K^{-1}	4.187
Specific interfacial area	ft^2/ft^3	m^2 m^{-3}	3.281
Surface or interfacial tension	dynes/cm	mN m^{-1}	1.000
Velocity	ft/s	m s^{-1}	0.3048
Viscosity	poise	Pa s	0.1000
Volume (see Capacity)			

Handbook of Solvent Extraction

Part I

General Principles

1

CHEMISTRY OF EXTRACTION OF NONREACTING SOLUTES

N. F. Ashton and C. McDermott
University of Birmingham
United Kingdom

A. Brench
Unilever Ltd.,
United Kingdom

1. PHYSICAL CHEMISTRY OF NONREACTING SOLUTES

The questions of whether two liquid phases can form in a given system, and, if so, how the components will distribute themselves between the phases, can be answered only by a study of the relevant thermodynamics. The appearance of phase separation can be explained by stability criteria analogous to those for mechanical systems, and this approach can also be used in regard to the distribution of components between phases.

1.1. Free Energy and the Stability Condition

The separation of a liquid of specified composition into two liquid phases in thermodynamic equilibrium may be conveniently explained by the dependence of the energy of the mixture as a function of its composition. The measure of the total energy of the system for our purposes is taken as being the Gibbs free energy G defined by

$$G = H - TS \quad (1)$$

The change in molar Gibbs free energy g^m on the formation of 1 mole of the mixture from n pure components at a given temperature and pressure is given by

$$g^m = \sum_{i=1}^{n} RTx_i \ln a_i \quad (2)$$

In an ideal system the activity a_i is equal to the mole fraction x_i. In nonideal systems the activity is related to the mole fraction by the activity coefficient γ_i as follows:

$$a_i = x_i\gamma_i \quad (3)$$

It follows then that the energy of mixing g^m of any given mixture consists of an ideal component g^{id} and the "excess" Gibbs energy g^E thus

$$g^m = g^{\mathrm{id}} + g^E \quad (4)$$

where

$$g^{\mathrm{id}} = \sum_{i=1}^{n} RT \ln x_i \quad (5)$$

and

$$g^E = \sum_{i=1}^{n} RT \ln \gamma_i \quad (6)$$

1.1.1. Binary Systems

Returning to the composition dependence of g^m we see that ideal systems, that is, systems with zero values of g^E, may be represented as in Fig. 1. Here the free-energy curve is everywhere concave upward, which means that at any composition it is not possible to replace that liquid by an equivalent liquid pair whose combined energy is less than that of the single liquid phase. Geomet-

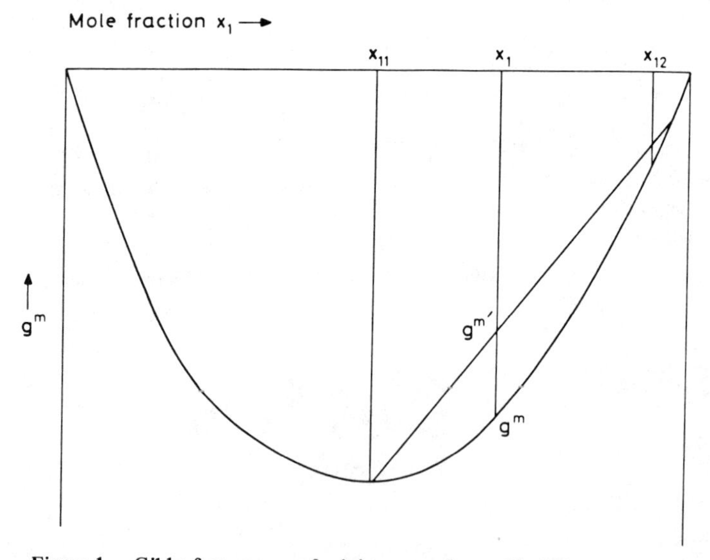

Figure 1. Gibbs free energy of mixing curve for an ideal binary system.

rically, this means that no two compositions x_{11} and x_{12} can be found such that the chord joining them on the free-energy curve intersects the ordinate at x_1 at a value $g^{m'}$ that is less than g^m. $g^{m'}$ is the weighted average of the mixing free energies at x_{11} and x_{12}, which is the total mixing free energy of two separate liquid phases with these compositions and a total effective composition equal to x_1. Supposing then that, under these conditions, we have n_1 moles of liquid with composition x_{11} and n_2 moles with composition x_{12}, we then have

$$(n_1 + n_2) x_1 = n_1 x_{11} + n_2 x_{12} \tag{7}$$

and n_1 and n_2 are inversely proportional to $x_{12} - x_1$ and $x_{11} - x_1$, respectively. In the case where no such $g^{m'}$ can be found there is no possibility that the liquid will separate into more than one liquid phase. We may conclude, therefore, that phase separation does not occur in ideal systems.

Turning now to systems exhibiting nonideal behavior, it is evident that the presence of a significant g^E contribution will have the effect of modifying the shape and, therefore, the concavity of the g^m curve. In most cases the effect of this is to produce a curve where the symmetry has been destroyed, but at no point along the curve has the second derivative changed sign; in other words, the curvature has remained everywhere of the same sign and the curve is still everywhere upward concave. The stability characteristics of such moderately nonideal systems are then the same as those of ideal systems; that is, no liquid phase separation may be expected.

The behavior of highly nonideal systems is typified by the curve depicted in Fig. 2. Here the excess free energy has produced a change of sign of the second derivative between points P and Q so that in this range the curve is upward convex. The result of this is that any two liquid compositions in the immediate neighborhood of x_1 whose equivalent total composition is the same as x_1 must have a combined free energy of mixing that is less than that of the single liquid of composition x_1. Such a separation will take place spontaneously as a result of normal molecular activity in the liquid. Once started, the separation will continue to increase until the point is reached where further separation will only serve to increase the effective combined free energy of mixing g^m. At this stage the chord to the free-energy curve between points with compositions x_{11} and x_{12} becomes the common tangent at these points.

Compositions x_{11} and x_{12} are the compositions of the two liquid phases to which any liquid with an overall composition between these values will normally separate. Compositions x_{11} and x_{12} then represent the miscibility gap for the given binary system. A liquid with composition lying outside this range will remain as a single liquid phase.

Within the miscibility gap two different modes of behavior can be distinguished and the

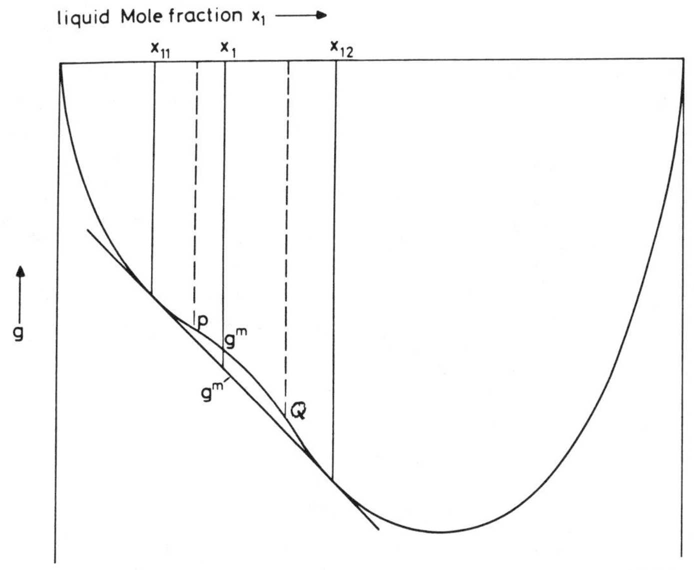

Figure 2. Gibbs free energy of mixing curve for a highly nonideal system exhibiting liquid-phase separation.

form of the liquid instability depends on in which part of the range composition x_1 lies. At composition x_1 we have seen that when the second derivative is negative the result of any, even infinitesimal, disturbance is to produce a phase separation that increases until the equilibrium compositions for the two phases are achieved. Points P and Q in the diagram are points of inflection, that is, points where the second derivative changes sign. Between these two points the second derivative is negative and outside this range it is positive. Range PQ thus represents a range of complete instability where a single liquid phase cannot exist. The sections of the curve lying within the miscibility gap but outside range PQ have positive second derivatives and so are stable but, provided a sufficiently large perturbation is applied, are capable of separating to form two liquid phases. Such regions of the curve are described as metastable by analogy with the similar situation that arises in the study of mechanical stability.

The preceding considerations may be summarized by the following condition of stability; that is, when

$$\frac{d^2g}{dx_1^2} > 0 \qquad (8)$$

the system is either totally stable or metastable. Otherwise, the system is unstable and the boundary between metastability and instability occurs where

$$\frac{d^2g}{dx_1^2} = 0 \qquad (9)$$

These conclusions, first arrived at by Gibbs [1], are more fully explained by Prigogine and Defay [2]. They are also to be found in some texts on phase equilibria and chemical thermodynamics, such as that by Rowlinson [3].

1.1.2. Ternary and Multicomponent Systems

A full discussion of the stability of multicomponent systems is provided by Prigogine and Defay [2]. Here we limit the treatment to a brief review of results corresponding to those already given for binary systems.

To facilitate the description of multicomponent systems, it is convenient to make use of the chemical potential μ_i of component i defined by

$$\mu_i = \frac{\partial G}{\partial n_i} \qquad (10)$$

and its derivative

$$\mu_{ij} = \frac{\partial^2 G}{\partial n_i \partial n_j} \qquad (11)$$

where n_i and n_j denote the number of moles of components i and j, respectively. With the use of this notation it can be shown that an n component mixture is stable or metastable provided the $n \times n$ matrix, A, whose i, jth element is μ_{ij} is positive definite. This means, in effect, that any expression of the form

$$\sum_{i,j=1}^{n} \mu_{ij} z_i z_j \qquad (12)$$

can take only positive values irrespective of the meaning or values, positive or otherwise, of z_i and z_j. Finally, by appeal to the Gibbs–Duhem equation, it is possible to reduce by one the order of matrix A in the statement of the stability condition. The final form of the stability condition states that provided the matrix A_k, formed by deleting from A the kth row and kth column, is positive definite, the system is stable. It can be shown to be immaterial which row (and corresponding column) is deleted for, since provided one such matrix A_k is positive definite, so are all the others.

A simple test for the positive definiteness of matrix A_k is available in the form of Sylvester's criterion, which states that A_k is positive definite if and only if all the leading principal minors of A_k are positive. The stability condition thus becomes

$$\mu_{11}, \begin{vmatrix} \mu_{11} & \mu_{12} \\ \mu_{12} & \mu_{22} \end{vmatrix}, \ldots, |A_k| > 0 \qquad (13)$$

The stability boundary is the curve joining points on which the determinant $|A_k|$ is zero.

It is easily shown that the preceding stability condition is consistent with that presented for binary systems by substituting

$$x_i = \frac{n_i}{n_1 + n_2}$$

into the preceding stability condition [Eq. (13)] which, for a binary system, reduces to

$$\frac{\partial^2 G}{\partial n_1^2} > 0 \qquad (14)$$

First we observe that

$$G(n_1, n_2) = G(x_1, N) = Ng(x_1) \tag{15}$$

using the rule for the derivative of a function of a function [4], we have

$$\mu_1 = \frac{\partial G}{\partial n_1} = \sum_{i=1}^{2} \frac{\partial G}{\partial x_i} \cdot \frac{\partial x_i}{\partial n_1} + \frac{\partial G}{\partial N} \cdot \frac{\partial N}{\partial n_1} \tag{16}$$

which leads to

$$\mu_1 = g + (1 - x_1) \frac{dg}{dx_1} \tag{17}$$

which is a form of the Gibbs–Duhem equation to be considered again in the next section. A further application of Eq. (16) gives

$$\frac{\partial^2 G}{\partial n_1^2} = \frac{x_2^2}{N} \cdot \frac{d^2 g}{dx_1^2} \tag{18}$$

Equation (18) shows that conditions (8) and (14) are equivalent.

In the case of a ternary system the stability condition (13) leads to the following requirement for stability

$$\mu_{11} > 0 \tag{19}$$

and

$$\mu_{11}\mu_{22} - \mu_{12}^2 \geqslant 0 \tag{20}$$

Now by reuse of (10) and (11) and definition of the mole fractions by

$$x_i = \frac{n_i}{n_1 + n_2 + n_3} \quad (i = 1, 2, 3) \tag{21}$$

an argument similar to that given previously for binary systems leads to the equivalent stability conditions

$$\frac{\partial^2 g}{\partial x_1^2} > 0 \tag{22}$$

and

$$\frac{\partial^2 g}{\partial x_1^2} \frac{\partial^2 g}{\partial x_2^2} - \left(\frac{\partial^2 g}{\partial x_1 \partial x_2}\right)^2 \geqslant 0 \tag{23}$$

The equations of the stability boundary, the curve separating the stable and unstable regions, is

$$\mu_{11}\mu_{22} - \mu_{12}^2 = 0 \tag{24}$$

which may also be expressed in the form of (23), where the left-hand side is equated to zero. The stability boundary or spinodal curve of a ternary system is illustrated in Fig. 3. The region enclosed by the spinodal curve and the 1–2 binary axis is the unstable section of the diagram, and the remainder is either metastable or completely stable.

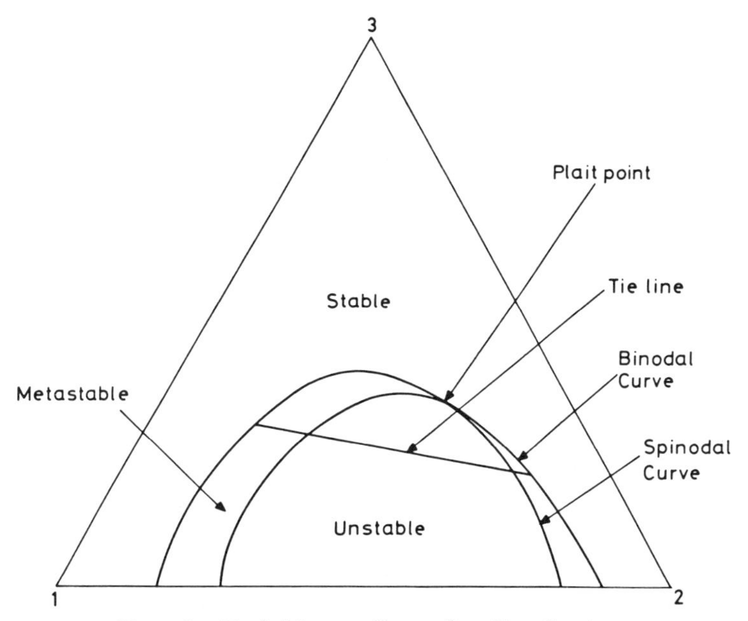

Figure 3. Typical ternary diagram for a Type I system.

Also shown on Fig. 3 is the binodal curve, which is that curve joining all points representing compositions of coexisting equilibrium liquid phases. Any liquid with composition lying within the binodal curve may split to form two liquid phases with compositions corresponding to the ends of the tie line. The tie-line passes through the feed or overall composition and ends on the binodal curve. The spinodal curve ends on the 1–2 binary at the points corresponding to the points of inflection on the binary free energy of mixing curve and the binodal curve ends at the points of contact of the common tangent in Fig. 2. The spinodal curve may, in principle, be located without difficulty because we have an equation that must be satisfied by points lying along it. The same cannot be said of the binodal curve as the "common tangents" requirement has not been formulated as an equation. The next section deals with the conditions satisfied by points on this curve.

1.2. Activities and the Gibbs–Duhem Equation

The Gibbs–Duhem equation, a form of which has already appeared as Eq (17), is central to any treatment of phase equilibrium. King [5] shows that any partial molar quantity satisfies the Gibbs–Duhem equation and states its various forms as applied to the chemical potential and activity coefficient. The particular versions in which we are interested are as follows.

1. In terms of the total number of moles present and chemical potentials

$$\sum_{i=1}^{n} n_i d\mu_i = V\,dP - S\,dT \tag{25}$$

with a corresponding isothermal isobaric form with zero right-hand side.

2. In terms of mole fractions and activity coefficients

$$\sum_{i=1}^{n} x_i\,d(\ln \gamma_i) = \frac{v^m\,dP}{RT} - \frac{h^m\,dT}{RT^2} \tag{26}$$

again with a corresponding isothermal isobaric form. The differentials on the left-hand sides of Eqs. (25) and (26) may be divided by the differential of an independent variable to give the derivative form of the equation. For example, division of Eq. (26) by dx_1, the isothermal isobaric form for chemical potential, yields

$$x_i \frac{d\mu_1}{dx_1} + x_2 \frac{d\mu_2}{dx_1} = 0 \tag{27}$$

It is then a simple matter to combine the corresponding form of Eq. (2) for the chemical potential

$$g = x_1\mu_1 + x_2\mu_2 \tag{28}$$

with Eq. (27) to obtain Eq. (17).

A geometric interpretation of Eq. (17) is given in Fig. 4, where we see that the tangent to the free-

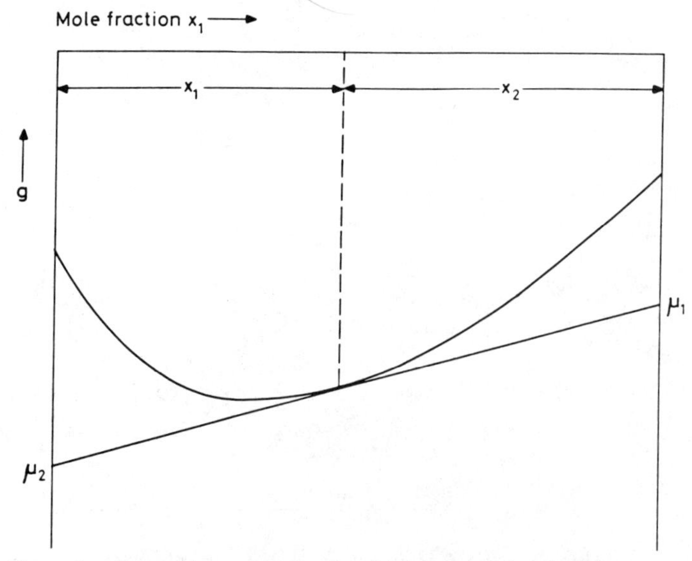

Figure 4. Geometric construction for obtaining partial molar free energy.

energy curve intersects ordinates at the right-hand end of the composition range with value μ_1. The other end of the tangent gives the value of μ_2. From Fig. 2 we can now see that the two points with the common tangent must also have the same values of μ_1 and μ_2. The condition for two liquid phases to coexist in equilibrium then may be written

$$\mu_i' = \mu_i'' \; (i = 1, \ldots, n) \tag{29}$$

where superscripts $'$ and $''$ denote the individual liquid phases. Although only justified here for binary systems, Eq. (29) is also valid for multicomponent systems.

A more convenient form of Eq. (29) may be obtained by making use of the relationship given by Prigogine and Defay [2]

$$\mu_i = \mu_i^0 + RT \ln x_i \gamma_i \tag{30}$$

which, in view of Eq. (3), enables the equilibrium condition

$$a_i' = a_i'' \tag{31}$$

to be used.

The problem of calculating points on the binodal curve is essentially that of solving Eqs. (31) for the x_i subject to the specification of the overall composition.

1.3. The Critical Point

1.3.1. Binary Systems

The nonideality of most systems tends to decrease as the temperature increases. The result of this is that, if we consider a system such as that shown in Fig. 2, increasing the temperature has the effect of reducing the convex portion of the curve. At some temperature points P and Q will become coincident, and at this point the two liquid phases become indistinguishable. This we term the *critical point*, and the conditions obtaining at the critical point are

$$\frac{d^2 g}{dx_1^2} = 0 \tag{32}$$

and

$$\frac{d^3 g}{dx_1^3} = 0 \tag{33}$$

For a more detailed discussion, reference may be made to King [5], page 69.

1.3.2. Ternary Systems

Reference to Fig. 3 would indicate that, as the temperature is increased, the binodal and spinodal curves will collapse toward the 1–2 binary. Thus the ternary critical point corresponding to that described previously for the binary will be the critical point for the 1–2 binary.

Again referring to Fig. 3, we see that critical behavior also occurs at the point of contact of the binodal and spinodal curves. Here we have a tie line of infinitesimal length and to correspond to it, two liquid phases of virtually the same composition. The first condition to be satisfied at this point (the "plait point") is that it lies on the stability boundary, and so Eq. (24) must be satisfied. The second condition may be obtained by considering the extension of the infinitesimal tie line at the plait point into the stable region. Prigogine and Defay [2] show that as we pass from the stable to unstable regions, condition (23) breaks down before (22) does. Therefore, as we move along the infinitesimal tie line, the left-hand side of (23) becomes positive, negative, and positive again in a very short interval. This means that the first derivative of this function is zero at the plait point, and this is our required second condition.

A precise formulation of the conditions obtaining at the plait point is presented by Gibbs [1] and in a more usable form by Prigogine and Defay [2]. Further elaboration is inappropriate here as, once a few tie lines are available, the location of the plait point is straightforward by the graphical methods described in the next section.

2. MEANS OF PRESENTATION OF EQUILIBRIUM DATA

Only systems that exhibit a region of partial miscibility are of interest. From a practical viewpoint, some systems may be considered to be completely immiscible over wide temperature ranges. However, in reality all liquids are soluble in each other to a limited extent.

2.1. Binary Systems

The limited miscibility of a binary pair of components is referred to as the *mutual solubility*. The effect of temperature is usually of the form shown in Fig. 5(a), which is a plot of the compositions of the saturated liquid phases at equilibrium for components A and B. Curve RNP represents the composition of a saturated solution of A in B as a function of temperature. Similarly,

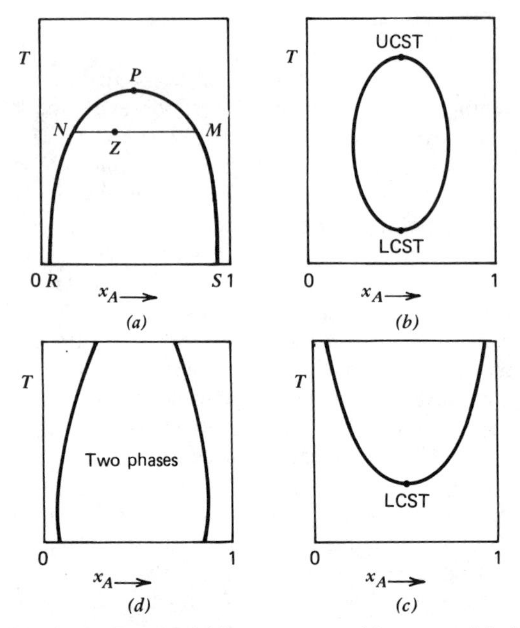

Figure 5. Types of binary system: (*a*) upper critical solution temperature; (*b*) upper and lower critical solution temperatures; (*c*) lower critical solution temperature; (*d*) no critical solution temperature.

curve *SMP* represents the solubility of *B* in *A* as a function of temperature. The area enclosed by the curve represents the two-phase region, and the area outside the curve represents mixtures that are completely miscible.

For *Z* moles of a mixture of overall composition *x* represented by point *Z* at system temperature *T*, separation into two liquid phases is represented by points *N* and *M*, and line *NZM* is the *tie line*. Points *N* and *M* now represent the number of moles of the two liquid phases and x_N and x_M, the corresponding compositions. The relative amounts of the two liquid phases may be calculated from a material balance

$$Z = N + M \tag{34}$$

$$Zx_Z = Nx_N + Mx_M \tag{35}$$

$$\frac{N}{M} = \frac{x_M - x_Z}{x_Z - x_N} \tag{36}$$

This important equation is known as the *lever-arm rule*, [see Chapter 5, Eq. (8)].

For this assumed mixture, as the system temperature is raised, the solubility of each component in the other increases and the length of the tie line decreases. Eventually at point *P* the tie line disappears and a single liquid phase results. This is defined as the *critical solution temperature* (see Section 1.3). Generally, this critical point is not at the midpoint of composition, nor are the solubility curves symmetric.

With a sufficient change in temperature, systems may exhibit both upper and lower critical solution temperatures (the latter occurring at a minimum temperature) as indicated in Fig. 5*b*. At normal pressures boiling may occur below the upper critical solution temperature with the resulting diagram of the form shown in Fig. 5*c*. Finally, both upper and lower critical solution temperatures may lie outside the range of interest or outside the temperature range in which the system remains liquid so that the system is everywhere two phase as shown in Fig. 5*d*.

2.2. Ternary Systems

Most ternary liquid–liquid equilibrium data sets are obtained under isothermal conditions, each set covering a range of ternary compositions. Such data sets are conveniently represented on triangular diagrams by use of either equilateral or right-angle triangles as a basis.

2.2.1. Triangular Diagrams

It is convenient to represent ternary systems by using equilateral triangular coordinates as shown in Fig. 6*a*. A property of an equilateral triangle is that the sum of the perpendicular distances from any point within the triangle to the three sides is equal to the altitude of the triangle. Hence, by letting the altitude represent 100% composition, any ternary composition may be represented by a point within the triangle.

Referring to Fig. 6*a* each corner of the triangle represents a pure component defined as *A*, *B*, or *C*. For a point within the triangle *M*, the perpendicular distance from *M* to base *AB* represents the amount of component *C* in the mixture. Similarly, the distance from *M* to side *AC* represents the amount of component *B* and the distance from *M* to side *BC* represents the amount of component *A*.

Any point on a side of the triangle represents a binary mixture.

Curve *NPR* shown within the triangle in Fig. 6*a* represents the boundary of the two-phase region for this type of system. System types are discussed fully later, but it is convenient to use this here as an example. This is defined as a Type I (or closed) system as it has only one pair of components that are partially miscible. The side of the triangle *AB* represents a binary mixture of components *A* and *B*. Points *N* and *R* represent

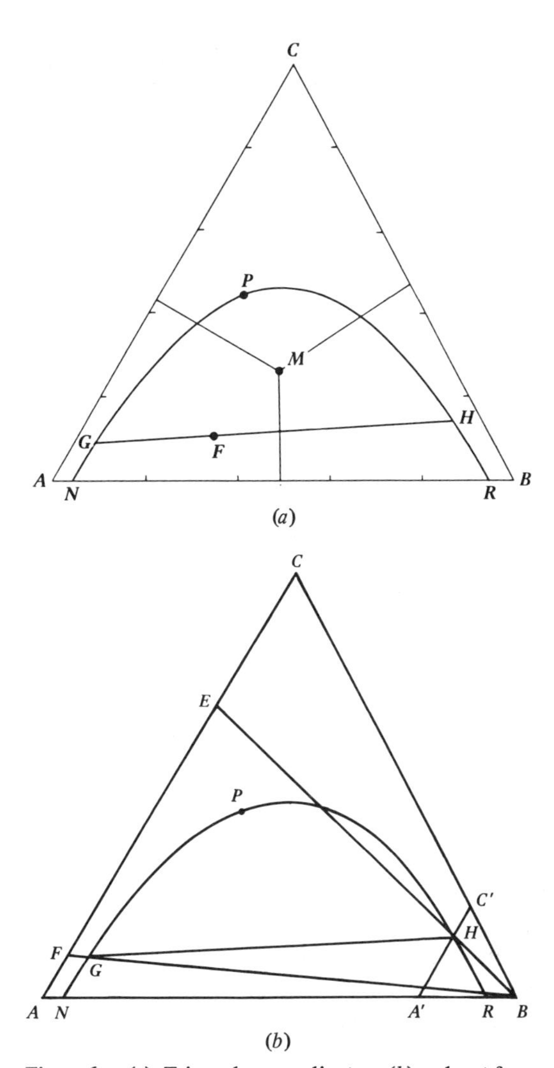

Figure 6. (*a*) Triangular coordinates; (*b*) solvent-free basis.

the mutual solubilities of B in A and of A in B at the temperature fixed for the diagram. The ternary solubility curve NPR is defined as a binodal curve and is specific to one defined temperature. Pressure has little effect on liquid–liquid equilibrium.

A line such as GH joining points on the binodal curve is defined as a tie line and joins points representing the liquid phases existing in equilibrium.

As discussed in Section 1, the limit of immiscibility is represented by point P, defined as the "plait point." As more component C is added to the mixture, the length of the tie lines decreases and the compositions of the equilibrium phases move closer together. As the plait point is approached, the length of the tie line diminishes to zero and the system becomes completely miscible.

It is frequently useful to specify liquid compositions on a "solvent-free" basis as demonstrated in Fig. 6*b*. Regarding component B as the solvent, the mole fraction of C at point H in the solvent-free pseudobinary $A'C'$ is the same as that of point E on the AC binary. Solvent-free compositions are important as they are closely related to product compositions after solvent recovery. The solvent-free composition for a given liquid is obtained by producing the line through the composition point and the solvent corner B to its intersection with the AC binary.

2.2.2. *Rectangular Coordinates*

An alternative to the equilateral triangle is to use rectangular coordinates as illustrated in Fig. 7. The composition of point M is defined in the same way as for the equilateral triangle, but it should be noted that the scales are not equal. For example, the binary mixture AB changes from 0 to 1.0 over the length AB while the binary mixture BC changes from 0 to 1.0 over the length CB, which is equal to $(\sqrt{2})\, AB$. A further alternative using rectangular co-ordinates is the use of unequal scales for A and B to obtain clarity on the diagram for certain types of system. The third component C is obtained by difference.

A coordinate system devised by Janecke [6] is illustrated in Fig. 8. Here, the ratio $B/(A + C)$ is plotted as ordinate against the fraction of C on a B(solvent)-free basis as abscissa for points on the binodal curve. On this type of diagram, the scales may be extended at will and it has advantages in the simplification of some process calculations.

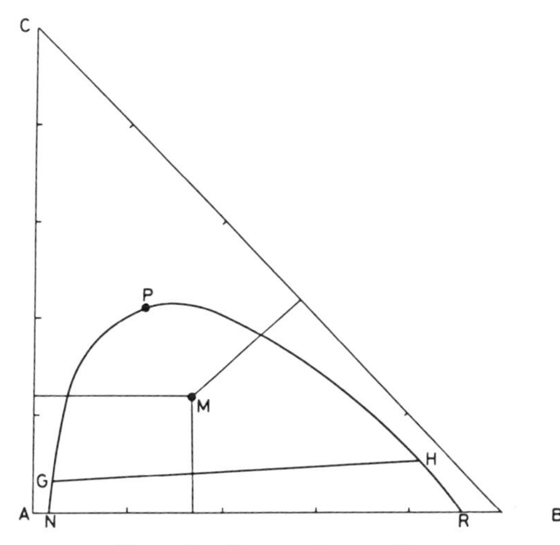

Figure 7. Rectangular coordinates.

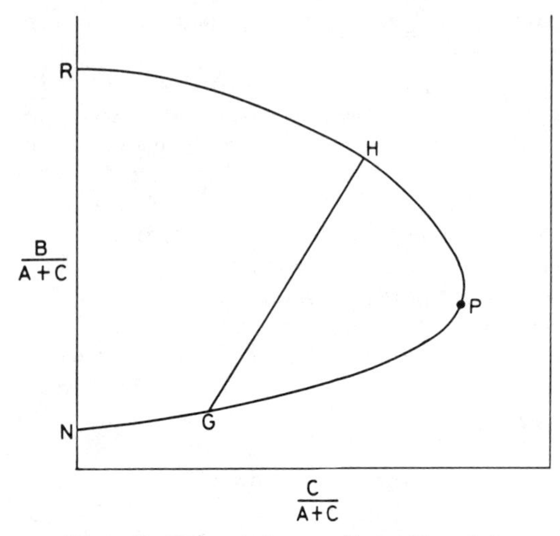

Figure 8. Rectangular coordinates (Janecke).

2.2.3. *Distribution Curves*

The distribution coefficient (or partition coefficient) is defined as the ratio of the mole fraction of solute in the solvent-rich phase to that in the A-rich phase. With reference to Fig. 6, for the tie line represented by GH, the distribution coefficient for component C may be defined as

$$m_C = \frac{x_{CB}}{x_{CA}} \tag{37}$$

where x_{CA} is the concentration of component C in the A-rich liquid phase (point G) and x_{CB} is the concentration of component C in the B-rich liquid phase (point H).

Plotting with use of rectilinear arithmetic coordinates results in a curve such as that shown in Fig. 9. For a typical Type I system, the curve

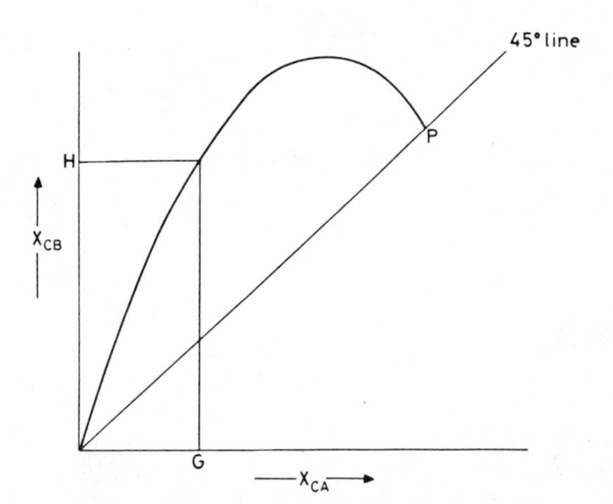

Figure 9. Distribution curve.

may pass through a maximum and return to the 45° line at a point that corresponds with the plait point where the region of complete miscibility begins.

2.2.4. *Selectivity*

Considering the system represented by Fig. 6, the component B may be considered to be the solvent used to separate a mixture of A and C. The effectiveness of the solvent is expressed by comparing the ratio of component C to component A in the B-rich phase to that in the A-rich phase at equilibrium. This is defined as the "selectivity," β, of the solvent. For the tie line GH,

$$\beta = \frac{x_{CH}/x_{AH}}{x_{CG}/x_{AG}} \tag{38}$$

Referring to Fig. 6b, we see that

$$\frac{x_{CH}}{x_{AH}} = \frac{A'H}{C'H} = \frac{AE}{CE}$$

so that

$$\beta = \frac{AE \cdot CF}{AF \cdot CE}$$

and it is clear that the more widely separated are the solvent-free concentrations E and F, the more selective is solvent B for solute C. A solvent that has a high selectivity for the solute will have tie lines that slope away from rather than toward the solvent corner.

2.3. Multicomponent Systems

Multicomponent systems are difficult to represent graphically. Basically two types may be considered: (1) two solutes distributed between two solvents; and (2) a multicomponent mixture to be separated by a single solvent. Case 1 may be represented by constructing a separate ternary diagram for each solvent, but it is often impractical to attempt this. Case 2 may often be simplified to a pseudoternary mixture and a single ternary diagram constructed.

In general, for all multicomponent systems, it is more straightforward to use the prediction and correlation methods described in Section 3 to generate the phase equilibrium data in terms of component activity coefficients and phase compositions. These methods are usually combined with stage-to-stage (stagewise) calculation algorithms that are best solved by computer methods.

2.4. System Types

The basic properties of binary systems have been discussed in Section 2.1. No further elaboration is required beyond emphasizing that binary mutual solubilities are dependent on temperature but are almost independent of pressure. Complete miscibility of the binary pair may be achieved for most systems by either raising the temperature above the upper critical value or lowering it below the lower critical value.

As stated by Treybal [7], if all three components of a ternary mixture mix in all proportions to form homogeneous solutions, the system is of no importance in liquid extraction. Treybal suggested the following classification of partially miscible ternary systems:

Type I—Contains one partially miscible binary pair. Also referred to as a *closed system.* An example is shown in Fig. 10*a*.

Type II—Contains two partially miscible binary pairs. Also referred to as an open system. Examples of this type are shown in Fig. 10(*b–d*).

Type III—Contains three partially miscible binary pairs, as shown in Fig. 10*f*.

Type IV—Contains a solid phase.

Types III and IV—Of little significant interest and are not discussed further.

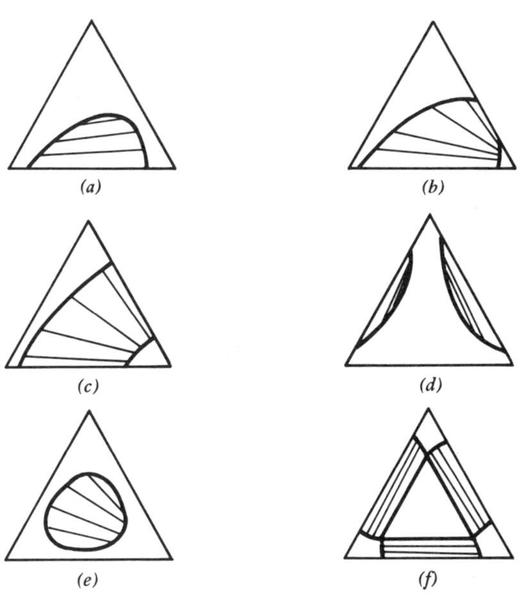

Figure 10. (*a*) Type I system. (*b*) Type II system (transitional between I and II); (*c*) Type II system; (*d*) Type II system (consisting of two Type I regions); (*e*) island curve; (*f*) three coexisting liquid phases.

2.4.1. *Type I Systems*

The properties of Type I systems have been described in Section 2.2.1 under the discussion of graphical methods for representation of data. Type I systems, sometimes known as *closed* systems, are the most important for the purposes of liquid–liquid extraction and include about 75% of ternary systems on which experimental data are available.

The tie-lines shown in Fig. 10*a* are merely a selection of the infinitely many tie lines making up any two-liquid-phase region. For most systems, the orientation of the tie-lines changes steadily in the same direction so that the trend followed by the distribution ratio remains the same everywhere. A few systems exhibit the behavior shown in Fig. 11, where the slope of the tie-lines changes sign together with a corresponding change in the distribution ratio through a value of unity. Such systems are termed *solu-*

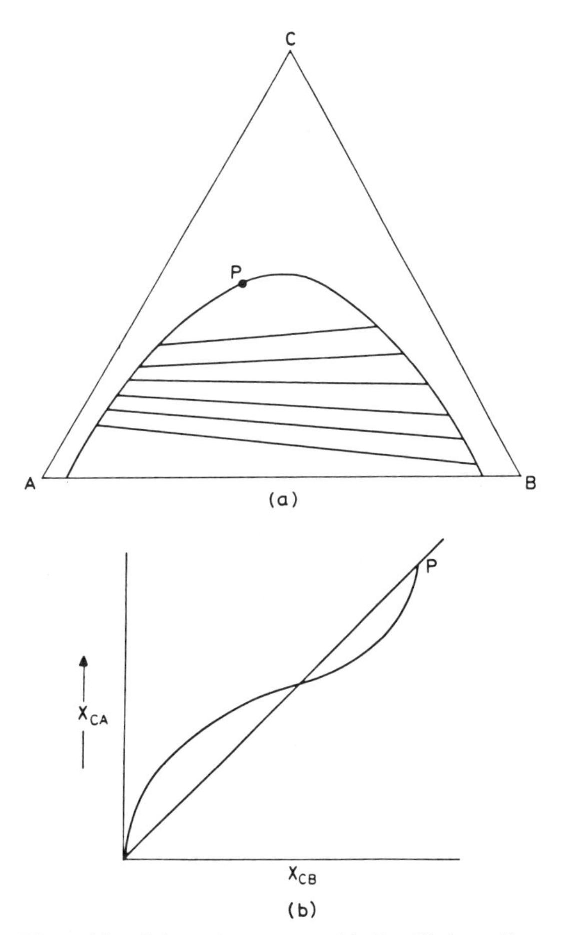

Figure 11. Solutropic system. (*a*) Equilibrium diagram; (*b*) distribution curve.

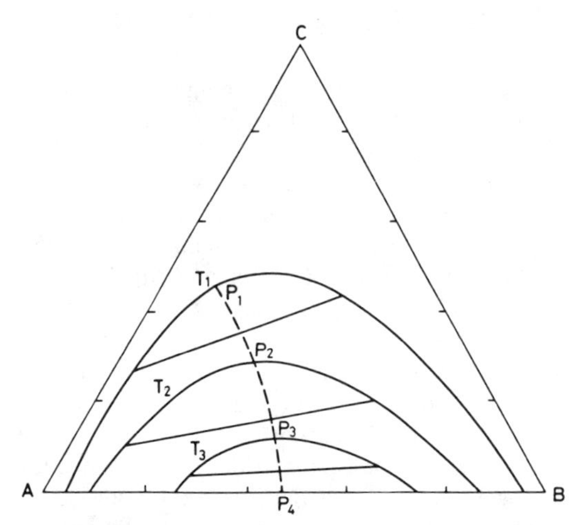

Figure 12. Effect of temperature for a Type I system. Note: $T_1 < T_2 < T_3$.

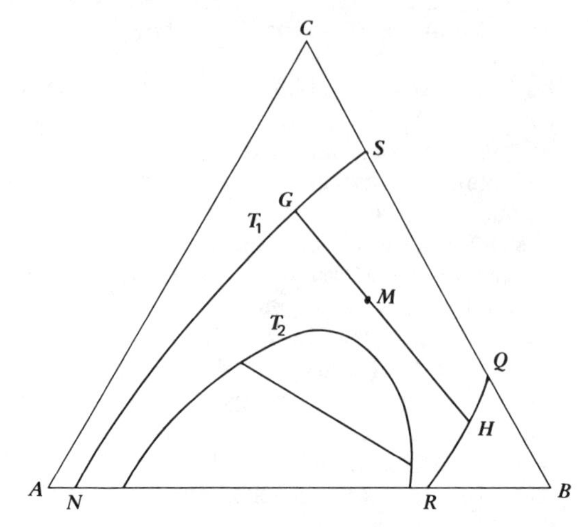

Figure 13. Effect of temperature for a Type II system. Note: $T_1 < T_2$.

tropic by analogy with azeotropic systems in distillation and may cause similar problems in the design of staged separations. The solutropy may disappear with a change of units (e.g., weight fraction to mole fraction) and does not necessarily pose a problem, although difficulties are more likely to arise with such systems.

The presence of an isopycnic tie-line, where the densities of the coexisting liquid phases are equal, is important from the point of view of the physical separation of the two phases. The isopycnic tie line is usually indicated as a dashed line on the phase diagram and may be encountered in binary and multicomponent systems of any type.

(a) Effect of Temperature. Generally, the effect of increasing system temperature is to increase the solubilities of the two liquid phases. This has a significant effect on the size and shape of the binodal curve as illustrated in Fig. 12. Not only does the area of immiscibility change, but the slopes of the tie lines may also change. The most simple situation is presented in Fig. 12 and a full discussion of the possible temperature effects is presented by Treybal [7].

2.4.2. Type II Systems

The Type II system is composed of two partially miscible binary pairs as illustrated in Fig. 13, where components A and B are partially miscible and components B and C are also partially miscible, at the system temperature.

As defined for the Type I system, points N and R represent the mutual solubilities of the AB binary pair and points S and Q represent the mutual solubilities of the BC binary pair. Curves NGS and RHQ represent the boundaries of this area of immiscibility. A point such as M within the region of immiscibility represents a composition that will generate two equilibrium liquid phases, represented by points G and H that are joined by the tie line.

(a) Effect of Temperature. Increase in system temperature generally increases the mutual solubilities of the two liquid phases and may also affect the slope of the tie-lines (see Fig. 13). It can be seen that it is possible to change from a Type II system to a Type I system by a suitable change of system temperature.

2.5. Tie-Line Correlations

Generally with published equilibrium data, the number of tie-lines presented is limited, and for process calculations it is necessary to interpolate and extrapolate to cover the complete region of interest. Estimation "by eye" on suitable diagrams is possible with fair accuracy, but several methods have been proposed that are claimed to give reliable results.

2.5.1. Binary Systems

For a binary system of the type represented on Fig. 5, King [5] has presented an application of the "law of rectilinear diameters." Basically, this states that the midpoints of tie-lines will lie on a straight line. Thus if two tie-lines are known, a straight line can be constructed that joins their midpoints. In theory, extrapolation of this line will allow an accurate determination of the critical point. There is no rigorous justification for this procedure, although when more than two tie lines are available, the curve connecting their midpoints may still be similarly extrapolated.

2.5.2. *Ternary Systems*

As discussed in Section 2.2, ternary systems are generally presented in the form of a plot on triangular coordinates. Obviously, it is possible to interpolate "by eye" between known tie lines when sufficient data are available. Also, the distribution curve can be used as an aid to interpolation.

Tie-line correlation or conjugation curves may be constructed as shown in Fig. 14. Lines parallel to one of the sides of the triangle are drawn from the tie-line ends G_1, G_2, and so on, and lines parallel to a different side are drawn from ends H_1, H_2, and so on. The points of intersection of these lines lie on the conjugation curve. The conjugation curve generally curves gently and can be

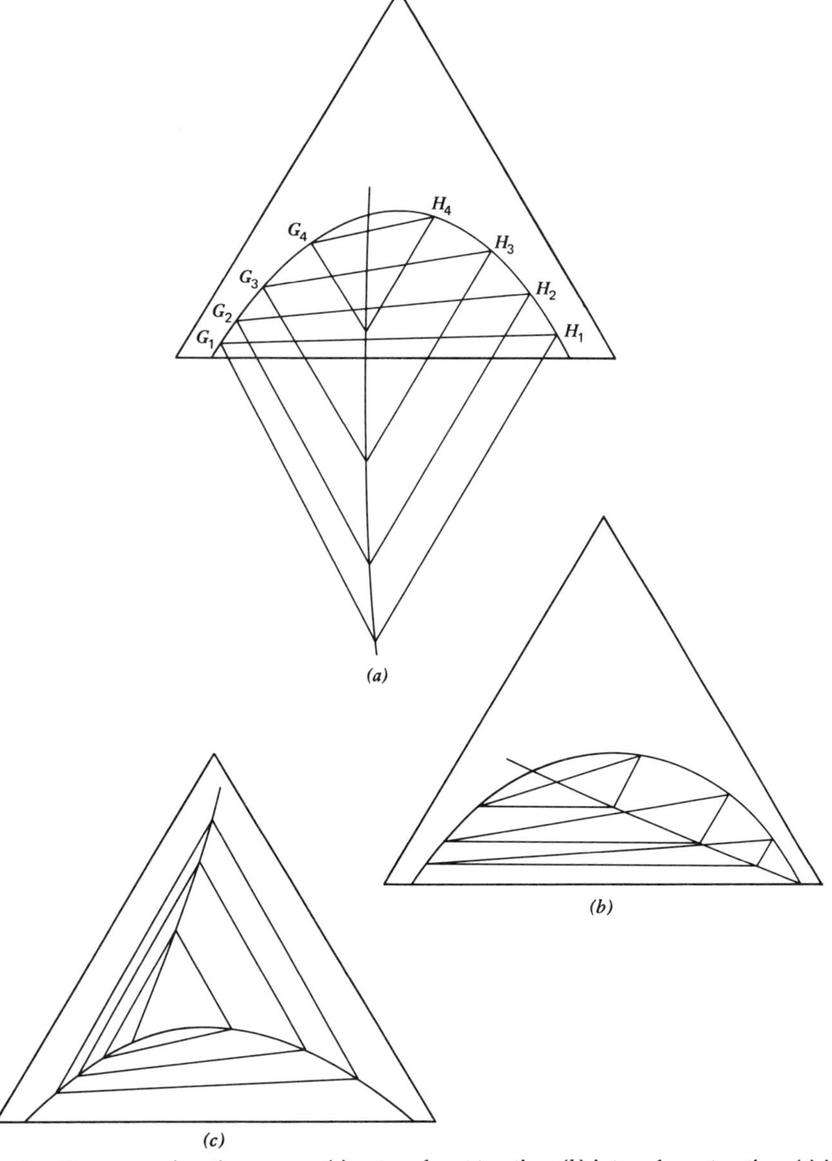

Figure 14. Ternary conjugation curves: (*a*) external construction; (*b*) internal construction; (*c*) internal construction.

extrapolated to meet the solubility curve at the plait point. Interpolation of tie lines from an available conjugation curve is a simple reversal of the procedure of construction. The method is satisfactory when several tie-lines are available for the construction of the conjugation curve, but caution is needed when only two are used because of the curvature of the conjugation curve. The plait point can only be reliably located when tie-lines in its vicinity are available.

It can be seen from Fig. 14 that the conjugation curve may lie either outside or inside the triangular diagram. When the lines drawn from one end of a set of tie lines intersect the other side of the binodal curve before crossing a side of the triangle, an internal conjugation curve will be obtained as in Figs. 14*b* and 14*c*. When this applies to neither set of lines, an external curve or construction will result as in Fig. 14*a*. It is usually more convenient to aim for the internal construction so that the constructed curve and triangular diagram may remain on the same sheet.

When one set of lines is parallel to the partially miscible binary as in Fig. 14*b*, the conjugation curve will pass through one or the other of the binary mutual solubilities depending on the orientation of the tie lines. Including this point is an aid in the construction of this particular conjugation curve.

For some nonsolutropic systems, the tie lines may be produced to meet the base AB of the triangle at a single point as shown in Fig. 15. For those systems that behave in this way, simple interpolation is possible. The method also allows rough extrapolation when only one or two tie lines are available. When two tie lines intersect on the extended base, it is likely that the others will also pass through the same point. Further tie lines and the plait point may then be easily estimated. For a single tie line, the assumption that the other tie lines also pass through its point of intersection with the base is unlikely to be worse than any other.

To generate further tie lines by the preceding graphical methods, it is necessary that the binodal or solubility curve be available. As is shown in Section 6, the solubility curve is more easily determined than are the tie lines, so this is frequently the case.

An elegant plot of equilibrium composition data is that proposed by Hand [8] where the ratio of mole fraction of solute to solvent in the solvent-rich phase x_{CB}/x_{BB} is plotted against the ratio of mole fraction of solute to diluent in the diluent-rich phase x_{CA}/x_{AA} on logarithmic coordinates. According to Hand, the result should be reasonably approximated by a straight line. An interesting modification of the Hand plot, suggested by Treybal et al. [9], is to add a plot of the solubility curve in the form of x_C/x_B versus x_C/x_A as indicated in Fig. 16. The extrapolation of the Hand plot meets the solubility curve at the plait point. As before, for a reliable estimate of the location of the plait point, tie lines in its vicinity are needed.

The shape of the solubility curve plot as shown in Fig. 16 depends on the shape of the two-phase region but will always tend to the point (0, 0), that is, the line at 45° to the axes as the AB binary is approached.

Othmer and Tobias [10] also suggest that a plot of $(1 - x_{AA})/x_{AA}$ versus $(1 - x_{BB})/x_{BB}$ on logarithmic coordinates produces a linear relationship for the equilibrium compositions. This has also enjoyed wide application and no doubt could be modified to include the solubility curve for plait point location.

The validity of the assumptions underlying the linearity of the Hand and Othmer–Tobias

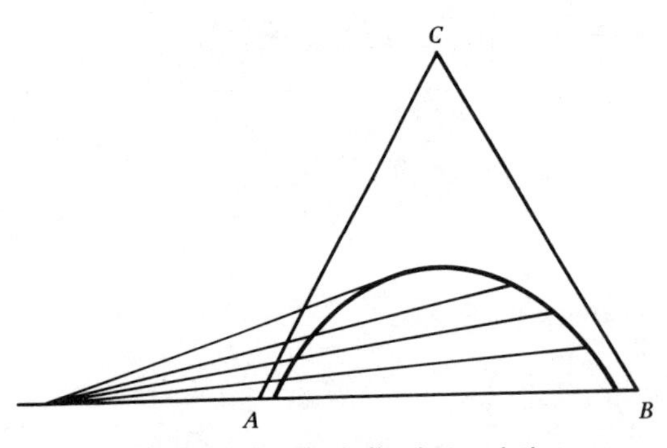

Figure 15. Simple tie-line interpolation.

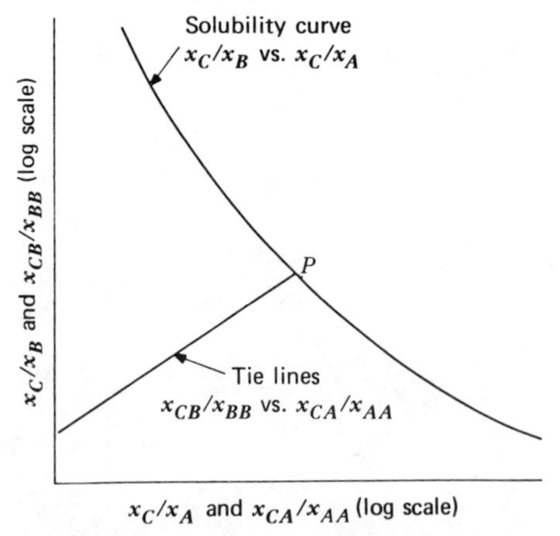

Figure 16. Hand plot for tie-line correlation.

plots has been questioned by Carniti et al. [11], although their conclusions do not affect the usefulness of the graphical representation.

3. CORRELATING EQUATIONS

The condition of equality of activities of a component in two liquid phases in equilibrium [Eq. (31)] suggests that analytic representation of the activity coefficient may be used to correlate liquid equilibrium data. Although almost any of the great range of correlating equations available for the activity coefficient could be used for this purpose, most of the simple forms fail to describe adequately the highly nonideal behavior of systems exhibiting phase separation. A further difficulty is that there are very few cases where the same set of parameters can be used to describe both the liquid–liquid and liquid–vapor equilibrium for a particular system. Although this is not a problem as far as the representation of liquid–liquid equilibria alone is concerned, it must be remembered that most of published correlations were obtained from liquid–vapor equilibria and thus cannot be expected to be used directly for the corresponding liquid-phase equilibria. The fact that the thermodynamic basis of both kinds of phase equilibrium lies with the activity coefficient causes the preceding limitations to cast some doubt on the validity of the assumptions made or models used to generate the correlating equations. This also becomes important when equilibrium data are to be extrapolated or used to predict the behavior of systems with added components. Representation of closed-curve Type I systems is particularly difficult with the use of activity coefficient or free-energy correlations. A slight error in the value of free energy in a range near to the plait point can lead to large errors in the location of the binodal curve.

The situation has improved with the use of some of the newer local composition equations, and discussion is confined to these. A more detailed review of most of the available correlating equations has been provided by Sorensen et al. [12].

The activity coefficient is most conveniently represented by making use of the excess function Q, defined by

$$Q = \sum x_i \ln \gamma_i = \frac{g^E}{RT} \tag{39}$$

From Eqs. (16) and (30) we then have

$$\ln \gamma_i = \frac{\partial Q}{\partial n_i} \tag{40}$$

The difficulties of correlating Q are small compared with those of the direct correlation of the activity coefficient. This is due to the relatively simple functional dependence of free energy on composition as opposed to that of the activity coefficient. The function Q, like g^E, takes the value zero at the ends of the composition range.

3.1. Wilson Equation

Historically the generation of correlating equations that have been most successful in representing liquid–liquid equilibria, that is, equations based on the local composition concept, began with the Wilson equation. In its original form the Wilson equation is unable to handle the case of liquid phase splitting, although its success has been very marked in the correlation of vapor–liquid equilibria. The equation, as developed by Wilson [13], depends on the two-liquid theory due to Scott [14]. Scott's theory regards the mixture as being made up of hypothetical fluids, one for each component. The fluid consists of cells each having a molecule of the corresponding component at its center. The local composition x_{ij} within a cell is defined as the composition of molecules of component j surrounding a molecule of component i. Wilson proposed that the local and overall composition in terms of mole fractions are related by

$$\frac{x_{ij}}{x_{ii}} = \frac{x_j \exp(-\lambda_{ij}/RT)}{x_i \exp(-\lambda_{ii}/RT)} \tag{41}$$

where λ_{ij} is the interaction energy between a molecule of component i and one of component j. Each mole fraction is then weighted by a suitable Boltzmann factor, which introduces a temperature effect into the model.

Wilson then used this expression to calculate the effective volume fraction around a molecule of each type and used in the Flory–Huggins equations for athermal ($h^E = 0$) mixtures

$$\frac{g^E}{RT} = x_1 \ln \frac{\phi_1}{x_1} + x_2 \ln \frac{\phi_2}{x_2} \tag{42}$$

This gives

$$\frac{g^E}{RT} = -x_1 \ln (x_1 + \Lambda_{12} x_2) - x_2 \ln (\Lambda_{21} x_1 + x_2) \qquad (43)$$

where

$$\Lambda_{12} = \frac{v_2}{v_1} \exp \left[\frac{-(\lambda_{12} - \lambda_{11})}{RT}\right] \qquad (44)$$

$$\Lambda_{21} = \frac{v_1}{v_2} \exp \left[\frac{-(\lambda_{12} - \lambda_{22})}{RT}\right] \qquad (45)$$

The corresponding activity coefficient expression is

$$\ln \gamma_1 = -\ln (x_1 + \Lambda_{12}) - x_2 \left[\frac{\Lambda_{21}}{x_2 + x_1 \Lambda_{21}} - \frac{\Lambda_{12}}{x_1 + x_2 \Lambda_{12}}\right] \qquad (46)$$

with the expression for $\ln \gamma_2$ obtained by interchanging subscripts.

The multicomponent form of Wilson's equation is

$$\ln \gamma_k = -\ln \left[\sum_{j=1}^{n} x_j \Lambda_{kj}\right] + 1 - \sum_{i=1}^{n} \frac{x_i \Lambda_{ik}}{\sum_{j=1}^{n} x_j \Lambda_{ij}} \qquad (47)$$

and it has the important property that only binary parameters are used.

The Wilson equation was introduced in the chemical engineering literature by Orye and Prausnitz [15] and has been shown by them to give a much better correlation and prediction of vapor–liquid equilibria than those equations based on series expansions of g^E. In spite of its excellent properties with regard to the correlations and predictions of excess free energies, the Wilson equation suffers from the serious disadvantage that no parameter set can be found for which condition (8) does not hold. Thus, as stated earlier, the Wilson equation is totally unable to predict phase splitting. This was demonstrated both by Wilson himself and by Renon and Prausnitz [16].

Wilson recognized the failure of his equation to predict partial miscibility regardless of the values given in the parameters. To overcome this he suggested simply the multiplication of the excess free-energy expression [Eq. (43)] by a constant. No physical significance or theoretical justification was offered at the time, but later Renon and Prausnitz [17] provided a derivation based on the integration of an excess enthalpy expression.

The three-parameter form of the Wilson equation is identical to (43) apart from the multiplication constant c. By virtue of Eq. (40), the same constant is applied to the $\ln \gamma$, and so multiplication of the right-hand side of Eq. (46) by c will give the required correlating equation. The third parameter presents problems in prediction of multicomponent activity coefficients, because, in order to use the equation, the same value of c must be used for each constituent binary. The effect of different values of c on the quality of correlation for different types of system has been investigated by Hiranuma [18], who suggests specific values to use in each case. The value seems to be influenced by the degree of hydrogen bonding for systems with association and/or solvation, but dispersion forces appear to produce little effect.

An important point made by Kayatama [19] is that the three-parameter Wilson equation for excess free energy can have at most two points of inflection in the binary range, whereas the NRTL equation, presented in the next section, can produce up to four. This means that the miscibility gap predicted by the Wilson equation is unique, which is an advantage when the gap is being used to generate equilibrium data.

Other modifications of the Wilson equation have been proposed and these are reviewed by Sorensen et al. [12]. None seems to enjoy any particular advantage over the nonrandom two-liquid (NRTL) and universal quasichemical (UNIQUAC) equations that follow.

3.2. NRTL Equation

The Wilson and other local composition equations were critically examined by Renon and Prausnitz [16], who sought to modify two steps common to their derivation. These steps were the relationship between local and overall mole fractions as given in Eq. (41) and the use of local mole fractions to give local volume fractions for inclusion in Eq. (42).

Renon and Prausnitz proposed that Eq. (41) should be replaced by

$$\frac{x_{21}}{x_{11}} = \frac{x_2}{x_1} \frac{\exp(-\alpha_{12}\lambda_{21}/RT)}{\exp(-\alpha_{12}\lambda_{11}/RT)} \qquad (48)$$

where the single constant α_{12} is a measure of the nonrandomness of the mixture of components 1 and 2. Second, instead of the Flory–Huggins athermal expression [Eq. (42)], Renon related the free energy directly to the molecular interactions, giving

$$g^E = x_1 x_{12}(\lambda_{21} - \lambda_{11}) + x_2 x_{12}(\lambda_{12} - \lambda_{22}) \tag{49}$$

The final form of the NRTL equation is

$$\frac{g^E}{RT} = x_1 x_2 \left(\frac{\tau_{21} G_{21}}{x_1 + x_2 G_{21}} + \frac{\tau_{12} G_{21}}{x_2 + x_1 G_{21}} \right) \tag{50}$$

with the activity coefficient given by

$$\ln \gamma_1 = x_2^2 \left(\frac{\tau_{21} G_{21}^2}{(x_1 + x_2 G_{21})^2} + \frac{\tau_{12} G_{12}}{(x_2 + x_1 G_{12})^2} \right) \tag{51}$$

where τ_{12} and τ_{21} are given by

$$\tau_{ij} = \frac{\lambda_{ij} - \lambda_{jj}}{RT}$$

$$G_{12} = \exp(-\alpha_{12}\tau_{12})$$

$$G_{21} = \exp(-\alpha_{21}\tau_{21})$$

and

$$\alpha_{12} = \alpha_{21} \tag{52}$$

The multicomponent forms of the preceding equations are given by Renon and Prausnitz [16]. An important property of the NRTL equation is that, like the Wilson equation, only binary parameters are required for multicomponent prediction. The Wilson equation involves only two parameters per binary whereas the NRTL equation involves three parameters, and one might expect that the increased flexibility would improve the representation of highly nonideal systems.

The binary miscibility gap can be represented by using only two of the three parameters. The third parameter, α, may be specified according to rules provided by Renon or may be adjusted to improve the representation of the vapor–liquid equilibria. The value of α that, according to Renon, provides a measure of the nonrandomness of the mixture, can be expected to lie between about 0.2 and 0.5. The lower values of α are better for representing systems with phase splitting, and Renon showed that with values of α greater than 0.426, phase splitting cannot occur at all.

This figure was later modified to 0.42773 by Heidemann and Mandhane [20], who made a detailed investigation into the effect of increased α values. It was found that instead of eliminating phase splitting, α values greater than 0.42773 can lead to multiple miscibility gaps and must be treated with caution. Further detailed investigation of α and associated values of τ_{12} and τ_{21} that produce phase splitting has been carried out by Tassios [21], who provides a chart showing ranges of permitted values of τ_{12} and τ_{21} corresponding to a given value of α. Mattelin and Verhoeye [22], as well as discussing in detail the circumstances in which phase splitting occurs, provide a series of curves from which parameter values may be obtained to correspond to a known miscibility gap.

3.3. UNIQUAC Equation

The UNIQUAC equation, proposed by Abrams and Prausnitz (23), has had success similar to that of the NRTL in the correlation and prediction of liquid–liquid equilibria. The equation is derived by introducing local area fractions of the molecules as the primary concentration variables into the quasichemical analysis by Guggenheim [24] with molecular size and shape parameters obtained from pure component data. The resulting equations are

$$\begin{aligned}\frac{g^E}{RT} &= x_1 \ln \frac{\phi_1}{x_1} + x_2 \ln \frac{\phi_2}{x_2} \\ &+ \frac{Z}{2}\left(q_1 x_1 \ln \frac{\theta_1}{\phi_1} + q_2 x_2 \ln \frac{\theta_2}{\phi_2} \right) \\ &- q_1 x_1 \ln(\theta_1 + \theta_2 \tau_{21}) \\ &- q_2 x_2 \ln(\theta_2 + \theta_1 \tau_{12})\end{aligned} \tag{53}$$

and

$$\begin{aligned}\ln \gamma_i = \ln \frac{\phi_1}{x_1} &+ \frac{Z}{2} q_1 \ln \frac{\theta_1}{\phi_1} + \phi_2 \left(l_1 - \frac{r_1}{r_2} l_2 \right) \\ &- q_1 \ln(\theta_1 + \theta_2 \tau_{21}) \\ &+ \theta_2 q_1 \left(\frac{\tau_{21}}{\theta_1 + \theta_2 \tau_{21}} - \frac{\tau_{12}}{\theta_2 + \theta_1 \tau_{21}} \right)\end{aligned} \tag{54}$$

where

$$l_1 = \frac{Z}{2}(r_1 - q_1) - (r_1 - 1)$$

$$\phi_1 = \frac{x_1 r_1}{x_1 r_1 + x_2 r_2}$$

$$\theta_1 = \frac{x_1 q_1}{x_1 q_1 + x_2 q_2}$$

$$\tau_{12} = \exp\left(-\frac{\lambda_{12} - \lambda_{22}}{RT}\right) \tag{55}$$

with subscripts reversed for component 2.

The adjustable parameters $(\lambda_{21} - \lambda_{11})$ and $(\lambda_{12} - \lambda_{22})$ have the same significance as in the Wilson and NRTL equations. Pure component parameters r_1 and q_1 are, respectively, measures of molecular van der Waals volumes and molecular surface areas. They are calculated as the sum of the group–volume and group–area parameters R_k and Q_k:

$$r_1 = \sum_k \nu_k^{(1)} R_k \quad \text{and} \quad q_1 = \sum_k \nu_k^{(1)} Q_k \tag{56}$$

where $\nu_k^{(1)}$, always an integer, is the number of groups of type k in molecule 1. Group parameters R_k and Q_k are obtained from the van der Waals group volume and surface areas V_k and A_k given by Bondi [25]:

$$R_k = \frac{V_k}{15.17} \quad \text{and} \quad Q_k = \frac{A_k}{2.5 \times 10^9} \tag{57}$$

Here Z is the coordination number of the system, and, whereas the values of the adjustable parameters are dependent on Z, the quality of the final prediction is independent of Z provided $6 \leqslant Z \leqslant 12$. (Abrams recommends $Z = 10$.)

The generation of the pure component parameters r and q is identical to the corresponding method for the UNIQUAC functional group activity coefficient (UNIFAC) equation described in Section 5.

The local composition equations generally have parameters given in the form of Eqs. (55) and so provide, at least in part, some temperature dependence of the activity coefficients. However, there is no reason to assume that the $(\lambda_{12} - \lambda_{22})$ and $(\lambda_{21} - \lambda_{11})$ are themselves independent of temperature.

Cukor and Prausnitz [26] suggested a linear temperature dependence of the above parameters but found it gave no improvement in the representation of isobaric data. Other authors, including Asselineau and Renon [27], seem to reach the same conclusion, although Stroud [28], performing bubble-point calculations for partially miscible systems, noted an improvement with use of the temperature parameters given by Renon.

To summarize, for most correlations including isobaric, the built-in temperature dependence in Eq. (55) is sufficient. For certain special requirements or when a range of temperatures is to be covered, a linear dependence of the energy parameters may be justified.

3.4. Discussion and Comparison

The preceding correlating equations may be used either for representation of known data or for the prediction of multicomponent data from that of component systems. From the remarks introducing this section, it is clear that the predicted ternary and multicomponent data should be treated with caution. It is essential that, where such data are to be used for design purposes, some experimental verification is obtained. The use of a particular correlating equation then requires that a set of parameters be determined to represent the data.

For a binary system, the UNIQUAC parameters τ_{12} and τ_{21} are completely determined by the equality of activities of the two components at the ends of the miscibility gap. Similarly, the same condition serves to fix two of the three adjustable parameters in the NRTL equation. The third parameter may be specified according to the rules mentioned earlier or adjusted to give the best representation of either the vapor–liquid equilibria of the system or the ternary binodal curve. A value of 0.2 for α has been found satisfactory for many cases of partial miscibility.

For ternary and multicomponent systems, the number of adjustable parameters is dependent on whether the binary τ_{ij} are regarded as independent. When they are taken to be independent, six parameters are available with use of the UNIQUAC equation to represent a ternary binodal curve. The mutual solubilities for the partially miscible binary may be used to determine two of these, leaving four to provide a fit of the rest of the binodal curve. A choice may be made to use all six parameters to improve the fit of the curve at the plait point at the expense of the exact representation of the binary mutual solubilities.

Alternatively, when the parameters are taken to be defined by equations of the form of Eq. (55) only five may be specified as a result of the existence of only five differences within the six parameters λ_{ij}. Thus, according to Eq. (55), we would have

$$\tau_{12}\tau_{23}\tau_{31} = \tau_{13}\tau_{32}\tau_{21} \quad (58)$$

on the reasonable assumption that

$$\lambda_{ij} = \lambda_{ji} \quad (59)$$

As may be expected, the overall fit of the binodal curve improves as the number of parameters is increased at the expense of an increase in computer time. The most acceptable alternative, according to Sorensen et al. [12], is to use four independent adjustable parameters (two fixed by the partially miscible binary) to represent the binodal curve for a Type I system. Sorensen et al. also reasonably conclude that the objective function used for the fit should be based on the composition residuals at the ends of the tie-lines rather than activity differences. This reflects the necessity to accurately reproduce tie-line compositions.

Similar remarks also apply to the parameters in the NRTL equation, but in this case, where the parameters are defined by Eq. (52), the τ_{ij} are related by

$$\tau_{12} + \tau_{23} + \tau_{31} = \tau_{13} + \tau_{32} + \tau_{21} \quad (60)$$

The NRTL equation also has the parameter α for each binary that can assist in the representation of difficult systems. However, with a large number of adjustable parameters, computational problems become severe.

The UNIQUAC equation, with only two adjustable parameters per binary and some predictive capacity based on pure component properties, seems to have the advantage in the representation of liquid–liquid equilibria.

3.5. Equations of State

An alternative approach to the problem without the use of activity coefficients has been suggested by Heidemann [29] and Peng and Robinson [30]. The equilibrium relationships are expressed in terms of an equation of state for the mixture. In this case the equation used is the version of the Redlich Kwong equation proposed by Wilson [31], but any equation of state may be used. The equation of state, once a composition and pressure are specified, may be used to calculate the appropriate volume. Generally there are three possible roots to the equation, one corresponding to the liquid, one to the vapor, and the third without physical significance.

In the proposed method three initial compositions are chosen (two liquid and one vapor) and the equation of state is used to calculate the individual phase volumes and, by a related equation, the corresponding fugacities. The fugacities are closely related to the activities, permitting calculation of the free energy of mixing. A minimization procedure is then used to modify the guessed compositions so that the calculated free energy is reduced. The method described is designed to locate the minimum rapidly without the possibility of producing any negative mole fractions.

The success of the method is that it can predict the qualitative behavior of the system, but its accuracy seems to be poorer than with use of methods based on the calculation of activity coefficients. Knowledge of mixing rules to provide multicomponent parameters for equations of state is less advanced than that of prediction of multicomponent activity coefficients, and although it is an interesting development at present, it cannot provide a realistic alternative to the well-established methods described earlier.

4. DATA PREDICTION

In Section 3 methods are presented for the correlation of liquid–liquid equilibrium data. However, experimental data are seldom available for a particular system, and so a method of prediction is required. Modern prediction methods lead to activity coefficients at the required temperature over the complete composition range. These are then used to calculate the compositions of the coexisting liquid phases.

The complexity of the interactions occurring in the liquid phase for systems that exhibit partial miscibility has proved to be an insurmountable problem for prediction from basic thermodynamics. However, particularly in the preliminary stages of a process design, some form of prediction is useful for solvent selection.

The prediction methods discussed in this section enable some estimation of activity coefficients to be made. The solubility parameter method, based on the regular solution theory of Hildebrand and Scott [32], makes use of the heat of vaporization of the pure components. The method of Pierotti et al. [33] correlates infinite dilution activity coefficients with the molecular group structure of the components involved. Other, more recent group contribution methods allow the prediction of the complete binodal curve and tie-lines. In some cases these make use of the "local composition" correlating equations for which coefficients are generated. In all cases the aim is to identify group interaction parameters from the reduction of large data sets. The parameters are then used to predict new systems of components already represented within the data set or, exceptionally, to generate data for components not included in the original data.

4.1. Regular Solution Theory

Hildebrand and Scott [32, 34] and Scatchard [35] made the following basic assumptions in developing their equation: (1) the mutual energy of two molecules depends only on the distance between them; (2) the molecules in the mixture are randomly located; and (3) there is no change of volume on mixing. On this basis, the change of energy on mixing is given by, for a binary mixture,

$$\Delta E^m = (x_1 V_1 + x_2 V_2) \phi_1 \phi_2 (C_{11} + C_{22} - 2C_{12}) \tag{61}$$

where

$$\phi_i = \frac{x_i V_i}{x_1 V_1 + x_2 V_2}$$

Following Scatchard [35] and taking C_{12} as the geometric mean of C_{11} and C_{22}, the introduction of solubility parameters yields

$$\Delta E^m = (x_1 V_1 + x_2 V_2) \phi_1 \phi_2 (\delta_1 - \delta_2)^2 \tag{62}$$

where $\delta_i = (-E_i/V_i)^{1/2}$, the cohesive energy density of component i.

Since it was initially assumed that there was no volume change, this is equivalent to the enthalpy change on mixing.

For systems where the solvent differs substantially in molar volume from the other system components, a partial molar entropy of mixing term can be derived [32, 36, 37] to allow for this:

$$\Delta s_1^m = -R \left[\ln \phi_1 + \phi_2 \left(1 - \frac{V_1}{V_2}\right) \right] \tag{63}$$

Combination of the two preceding expressions to give the partial molal pure energy of mixing leads to one form of the Scatchard–Hildebrand equation:

$$\ln a_1 = \ln \phi_1 + \phi_2 \left(1 - \frac{V_1}{V_2}\right) + V_2 \phi_2^2 \frac{(\delta_1 - \delta_2)^2}{RT} \tag{64}$$

Reference to the derivation of this equation [32] and simple algebraic manipulation [38] can be shown to give a multicomponent form of the equation such that

$$\ln a_i = \ln \phi_i + \left(1 - \frac{V_i}{\bar{V}}\right) + V_i \frac{\delta_i - \bar{\delta}}{RT} \tag{65}$$

where

$$\phi_i = \frac{x_i V_i}{\sum_{j=i}^{n} x_j V_j} \tag{66a}$$

$$\bar{\delta} = \sum_{j=1}^{n} \phi_j \delta_j \tag{66b}$$

and

$$\bar{V} = \sum_{j=1}^{n} x_j V_j \tag{66c}$$

4.1.1. Application of Regular Solution Theory to Real Solutions

The regular solution theory is found to apply to certain systems involving hydrocarbons, fluorocarbons, and so on [34], but its region of applicability is quite limited. Analysis of the situations in which it does apply, indicates that once permanent dipole–dipole interactions and hydrogen bonding occur, the theory fails to remain accurate.

Burrell [39] first attempted an application of the regular solution theory to practical polar systems by introducing a solvent classification system. He grouped solvents into those groups of high, medium, and low hydrogen bonding capacities and with the solubility parameter produced a technique for solvent selection in the paint industry. This approach has been refined by use of a quantified hydrogen bonding term, and eventually a model using three terms was developed [40, 41]. In this case the three parameters are related to the original conception of the solubility parameter, the dipole movement, and the hydrogen bonding term (based on the shift in the micron band for methyl deuteroxide), respectively.

Other approaches [42–44], have been used to cope with the polar terms, basically using a two-parameter model to cover nonpolar forces. This concept uses homomorphs, which are nonpolar molecules of the same size and shape as the polar one of interest, to generate the nonpolar term and hence the polar contribution.

Experience [38, 42] in the paint industry has suggested that the three solubility parameter approach is viable, provided the weightings attached to the three parameters are not equal. Numerical analysis of a range of liquid–liquid and liquid–vapor equilibrium data sets has indicated that the following equation can be used to accurately reduce data with a failure rate of 4% (i.e., 8 sets of 201 studied [38]):

$$\ln a_i = \ln \phi_i + \left(1 - \frac{V_i}{\bar{V}}\right) + \frac{V_i}{RT}\,[(\delta_{D,i} - \bar{\delta}_D)^2 + A_i((\delta_{p,i} - \bar{\delta}_p)^2 + (\delta_{H,i} - \bar{\delta}_H)^2)] \tag{67}$$

where $\delta_{D,i}$ = the dispersion forces solubility parameter
$\delta_{p,i}$ = that for dipole interactions
$\delta_{H,i}$ = that for hydrogen bonding interactions
A_i = the weighting factor

The field of chromatography has further refined this concept [46] and split the hydrogen bonding interactions into two parts relating to proton donor properties δ_a and proton acceptor prop-

TABLE 1a PREDICTION OF OVERALL SOLUBILITY PARAMETERS[a]

Group	$V\delta$ [(cal · cm^3)$^{1/2}$/mol]	Group	$V\delta$ [(cal · cm^3)$^{1/2}$/mol]
$-CH_3$	148.3	—S—	209.42
$-CH_2-$	131.5	Cl_2	342.67
>CH—	85.99	Cl (primary)	205.06
>C<	32.03	Cl (secondary)	208.27
$CH_2=$	216.54	Cl (aromatic)	161.0
—CH=	121.53	Br	257.88
>C=	84.51	Br (aromatic)	205.60
—CH= (aromatic)	117.12	F	41.33
—C= (aromatic)	98.12		
—O— (ether, acetal)	114.98	*Structural Features*	
—O— (epoxide)	176.20	Conjugation	23.26
—CHO	292.64	Cis	-7.13
$-(CO)_2O$	567.29	Trans	-13.50
—OH	225.84	Four-membered ring	77.76
—H	-50.47	Five-membered ring	20.99
—OH (aromatic)	170.99	Six-membered ring	-23.44
$-NH_2$	226.56	Ortho substitution	9.69
—NH—	180.03	Meta substitution	6.6
—N—	61.08	Para substitution	40.33
—C≡N	354.66		22.56
—COO—	326.58		62.5
>C=O	262.96		
—NCO	358.66	Base value	135.1

[a]*Overall solubility parameter:* $\delta = \sum V\delta / V$, where V is the molar volume of the compound and $\sum V\delta$ is the sum of groups contributions for the compound plus the basic value.

erties δ_b. This approach has been suggested for use in the field of solvent selection [47] and in this case involves, instead of the difference in solubility parameters—that is, $(\delta_1 - \delta_2)^2$—the use of products—that is, the terms $(\delta_a)_1(\delta_b)_S$, $(\delta_a)_S(\delta_b)_1$, $(\delta_a)_1(\delta_b)_1$, and $(\delta_a)_S(\delta_b)_S$. The object of the approach is to maximize the first two terms and minimize the last two. In practice, most compounds have zero δ_a values and only alcohols have significant δ_a values. On this basis the approach is an attempt to deal with the difficulties encountered with such compounds and aqueous solutions in chromatography and does not offer a significant improvement over the three-parameter model in the solvent extraction field.

4.1.2. *Estimation of Solubility Parameters*

A knowledge of the value of the solubility parameters is a necessary prerequisite for the use of the modified Scatchard-Hildebrand equation, and a number of experimental procedures are detailed in the literature [34]. For solvent selection purposes, however, work by Small [48] and Hoy [49] showed that the overall solubility parameter δ defined by

$$\delta^2 = \delta_D^2 + \delta_p^2 + \delta_H^2 \tag{68}$$

can be estimated on a group additive basis. Similar additive rules have been found [38] for the polar and hydrogen bonding parameters, and the group contributions and additive basis are given in Table 1.

This approach is very similar in many respects to those reported for infinite dilution activity coefficients [50] and the analytical solution of groups (ASOG), UNIQUAC, and UNIFAC techniques [51]. Hansen [45] has presented a very useful summary of solubility parameters and their estimation and use.

The weighting parameter A used in the modified equation has been found to be a pure component property and Table 2 summarizes a number of experimentally established values. These

TABLE 1b PREDICTION OF POLAR AND HYDROGEN BONDING SOLUBILITY PARAMETERS[a,b,c]

Group	Polar Contribution $V\delta_p$ [(cal · cm^3)$^{1/2}$/mol]	Hydrogen Bonding Contribution $V\delta_H^2$ (cal/mol)	
		Aliphatic	Aromatic
—F	225 ± 25	0	0
—C_1	300 ± 100	100 ± 20	100 ± 20
>C_{1_2}	175 ± 25	165 ± 10	180 ± 10
—Br	300 ± 25	500 ± 100	500 ± 100
—I	325 ± 25	1000 ± 200	
—O—	200 ± 50	1150 ± 300	1250 ± 300
>C=O	390 ± 15	800 ± 250	400 ± 125
—COO—	250 ± 25	1250 ± 150	800 ± 150
—CN	525 ± 50	500 ± 200	550 ± 200
—NO_2	500 ± 50	400 ± 50	400 ± 50
—NH_2	300 ± 100	1350 ± 200	2250 ± 200
>NH	100 ± 15	750 ± 200	
—OH	250 ± 30	4650 ± 400	4650 ± 50
(—OH)$_n$	n(170 ± 25)	n(4650 ± 400)	n(4650 ± 400)
—COOH	220 ± 10	2750 ± 250	2250 ± 250

[a]*Polar solubility parameter:* $\delta_p = \sum V \sum \delta_p / V$, where V is the molar volume of the compound and $\sum V\delta_p$ is the sum of group contribution.
[b]*Hydrogen bonding solubility parameter:* $\delta_H = (\sum V\delta_H^2)/V^{1/2}$, where V is the molar volume of the compound and $\sum V\delta_H^2$ the sum of group contributions.
[c]*Dispersion solubility parameter:* $\delta_D = (\delta^2 - \delta_p^2 - \delta_H^2)^{1/2}$.

TABLE 2 SUMMARY OF BEST WEIGHTING PARAMETER VALUES, *A*

Component	*A*	Component	*A*
Triolein	0.159	*N*-Butanol	0.238
Trilinolein	0.166	*N*-Propanol	0.241
Olive oil	0.176	Isopropanol	0.200
Diolein	0.195	Ethanol	0.200
Oleic acid	0.175	Methyl acetate	0.160
Methanol	0.262	Ethylene diamine	0.290
Acetone	0.140	Ethylene glycol	0.306
Methyl ethyl ketone	0.144	*N*-Butyl acetate	0.185
Benzene	0.154	Ethyl benzene	0.117
1-Pentanol	0.192	Styrene	0.129
Dioxane	0.099	*S*-Octanol	0.233
Glycerol	0.202	*N*-Hexanol	0.276
Furfural	0.259	Propylene carbonate	0.413
N-Heptane	0.196	*N*-Hexane	0.223
Ethyl acetate	0.142	Toluene	0.064
Isobutanol	0.233	Pyridine	0.108
Propyl acetate	0.164	Acetic acid	0.118
Amyl acetate	0.193	Methyl isobutyl ketone	0.133
Ethyl propionate	0.198	Chloroform	0.217
Ethyl butyrate	0.185	Tetrachloroethylene	0.241
Isooctane	0.167	Trichloroethylene	0.198
Cyclohexane	0.194	Carbon tetrachloride	0.125
sec-Butanol	0.194	1,1,2-Trichloroethane	0.176
t-Butanol	0.202	Water	0.319

Source: Based on data from Brench [38].

appear to be a function of the homologous series and can be interpreted physically on the basis that the dispersion forces represent a "billiard ball" effect over the whole molecule and the polar and hydrogen bonding terms are localized on groups and thus not uniformly distributed.

4.2. Prediction from Group Contribution Methods

4.2.1. *Correlations by Pierotti et al.*

In systems where mutual solubilities are small, good estimates of distributions may be made by using infinite dilution activity coefficients.

The distribution coefficient m is given by Eq. (38) as

$$m = \frac{x_{BS}}{x_{BA}} = \frac{\gamma_{BA}}{\gamma_{BS}} \tag{69}$$

representing the relative distribution of component B between the solvent-rich phase S and the A-rich phase. At infinite dilution, the mole fraction of solute B approaches zero, so that

$$m^0 = \frac{\gamma_{BA}^0}{\gamma_{BS}^0} \tag{70}$$

This approach neglects the solubility of component A in the solvent and in such systems will allow a good estimation of the distribution coefficient at low concentrations of B.

Pierotti et al. [33] correlated the infinite dilution activity coefficients of structurally related systems to the number of carbon atoms in the solute N_A and the solvent N_B. The equation is of the form

$$\log \gamma_A^0 = \alpha + \frac{\zeta^0}{N_A} + \frac{\theta}{N_B} + \epsilon \frac{N_A}{N_B} + \eta(N_A - N_B)^2 \tag{71}$$

where the constant θ is a function of the solvent series B, ζ of the solute series A, ϵ and η are functions of both, and α is independent of both.

Constants for a number of solute–solvent series have been presented by Treybal [7].

4.2.2. *UNIFAC Method*

The UNIFAC model was developed for the prediction of Vapor–Liquid equilibrium and has been defined by Fredenslund [51, 52]. The form of the equations has been developed from the UNIQUAC correlating equation;

$$\ln \gamma_i = \ln \gamma_i^S + \ln \gamma_i^G \tag{72}$$

with

$$\ln \gamma_i^S = \ln \frac{\phi_i}{x_i} + 5q_i \ln \frac{\theta_i}{\phi_i} + l_i - \frac{\phi_i}{x_i} \sum x_{jj} \tag{73}$$

where

$$l_i = 5(r_i - q_i) - (r_i - 1)$$

$$\theta_i = \frac{x_i q_i}{\sum_j x_j q_j}$$

$$\phi_i = \frac{x_i r_i}{\sum_j x_j r_j}$$

$$r_i = \sum_k v_k^{(i)} R_k$$

R_k = van der Waals volume

$$q_i = \sum_k \nu_k^{(i)} Q_k$$

Q_k = surface area of group k

$\nu_k^{(i)}$ = groups of type k in component i

and $\ln \gamma_i^G$ is calculated from the relationship

$$\ln \gamma_i^G = \sum_k \nu_k^{(i)} [\ln r_k - \ln r_k^{(i)}] \tag{74}$$

where r_k and $r_k^{(i)}$ represent the activity coefficients of group k in solution and in pure i and r is expressed as a function of the group fraction X, defined as

$$X_k = \frac{\sum_i x_i \nu_k^{(i)}}{\sum_k \sum_j x_j \nu_k^{(j)}} \tag{75}$$

the summation extending over all components.

The terms $\ln r_k$ and $\ln r_k^{(i)}$ are given by the following expressions:

$$\ln r_k = Q_k \left[1 - \ln \left(\sum \lambda_m \psi_{mk} \right) - \sum_m \frac{\lambda_m \psi_{km}}{\sum_n \lambda_n \psi_{nm}} \right] \tag{76}$$

with

$$\lambda_m = \frac{Q_m X_m}{\sum_n Q_n X_n} \quad \text{and} \quad \psi_{mn} = \exp\left[- \left(\frac{a_{mn}}{T} \right) \right]$$

where a_{mn} are binary group parameters.

Parameters for vapor–liquid equilibrium data predictions are well developed and available from Fredenslund [52] and Skjold-Jorgensen et al. [53].

Considering the prediction of liquid–liquid equilibrium data, Magnussen et al. [54] have presented a comparison of the prediction of data using group contribution parameters from various sources. They conclude that predicting liquid–liquid equilibrium data from group contribution parameters based on vapor–liquid equilibrium data will yield results that are in qualitative agreement with experimental data. A new set of group contribution parameters is being developed by Magnussen et al. [54] based on the liquid–liquid equilibrium data bank due to Sorenson and Arlt [55].

The use of this set of parameters has resulted in good agreement with experimental data, and Magnussen concludes that this demonstrates the ability of the group contribution method to predict liquid–liquid equilibrium data of "semiquantitative" reliability but emphasizes that although the binodal curve may be predicted well, there are likely to be serious errors in the prediction of distribution ratio.

4.2.3. *ASOG Method*

Derr and Deal [56] developed the method of predicting activity coefficients by group contributions by use of a form similar to that presented in Section 4.2.2 for UNIFAC. Primarily, it was developed for the prediction of vapor–liquid equilibria, and a large number of group interaction parameters have been presented by Kojima and Tochigi [57].

Little work has been published on the use of the ASOG method for liquid–liquid equilibrium data. Tochigi and Kojima [58] used parameters obtained from vapor–liquid data [59] to predict liquid–liquid data for nine ternary systems comprised of the groups CH_2, OH, and CO as alcohols, ketones, *n*-paraffins, and water. The results were in qualitative agreement with experimental data, but particular differences were obtained with closed systems. Sugi and Katayama [60] attempted to predict liquid–liquid data for three ternary systems by using parameters calculated from the binary mutual solubilities. The predicted data showed differences from the experimental data.

5. SOLVENT SELECTION

The success of a liquid–liquid extraction process is strongly dependent on the selection of the most appropriate solvent. The final choice of solvent will be a compromise between solvent selectivity, solvent capacity, toxicological constraints, system physical properties, and solvent recovery difficulties to give the optimum process design.

The design of the solvent recovery operation must normally consider the influence of temperature changes on product quality and removal of trace levels of solvent to meet safety requirements while maintaining a high energy economy. With natural feedstocks, contamination of the solvent with water is also likely, and hence solvent drying facilities may be needed. In view of the preceding comments, the selection of the most appropriate solvent will thus be dominated by the balance between solvent selectivity and capacity.

The obvious route to solvent selection is to determine equilibrium data for the separation of interest with a variety of solvents and select the most appropriate of those evaluated. Since the acquisition of the equilibrium data is essential for a process design, this is a logical procedure, but it suffers from a number of disadvantages. From a strictly scientific viewpoint, it is not possible to state that the optimum solvent has been found, unless every possible solvent has been evaluated. This is normally a tedious and expensive exercise, and with large molecules and difficult analytical techniques it can be prohibitively expensive. Thus a means of highlighting possible solvent groups is very desirable.

Any classification system for solvents will fall into two categories: empirical origin and thermodynamic origin. The empirical selection techniques are based on practical observations and evaluations on a large range of systems, whereas the thermodynamic techniques stem from phase equilibrium considerations.

5.1. Empirical Selection Techniques

A search of the literature shows that a variety of properties have been used to assess the selectivity of solvents. Francis [61, 62] has shown that critical solution temperatures can provide a good guide to solvent selectivity, but the study has been limited to petroleum industry products. Drew and Hixson [63] have indicated similar behavior with fatty acids and propane. Ewell et al. [64] proposed a system of classifying liquids on the basis of hydrogen bonding into one of five classes.

Developing from this approach, a number of references are to be found in the literature where attempts have been made to rank miscibility. Rothmund [65] arranged a series of liquids in increasing order of miscibility with aliphatic hydrocarbons or decreasing miscibility with water and suggested that this could lead to a numerical value depending on the position of a given liquid in the series. Francis [61] used the concept in an eight-place system of interrelationships but could not produce a self-consistent system. To date the most successful version of this approach is due to Godfrey [66], who has developed a diagram covering 31 classes of liquid. For the purposes of the diagram, the term "borderline miscibility" is defined by initial solution temperature between 25 and 75°C. The system suffers from the disadvantage that the occasional compound has a miscibility number in excess of 31 and some compounds have dual-class memberships. However, the technique can be very rapidly used to identify possible liquid–liquid extraction solvents and to indicate which component is preferably extracted. The approach does work on complex systems but can suffer from a "size effect" with large molecules that distorts the picture.

In summary, the empirical screening techniques give at best a preliminary estimate of where solvents may lie but cannot specifically identify good solvents and in general do not contribute to any understanding of the phenomena.

5.2. Solvent Selection Analysis of Modified Regular Solution Model

In considering the suitability of a solvent for a given separation, it is necessary to consider two characteristics of the solvent: (1) the selectivity of the solvent for the component to be extracted and (2) the partition coefficient of the component to be extracted in the solvent. These concepts can be developed further by considering the case where component B is extracted from component A by a solvent, component S with the partition coefficient m_B given by Eq. (69).

Equation (67) can be rearranged to give a form for the activity coefficient as follows:

$$\ln \gamma_i = \ln V_i - \ln \bar{V} + \left(1 - \frac{V_i}{\bar{V}}\right) + \frac{V_i}{RT} \{(\delta_{D,i} - \bar{\delta}_D)^2 + A_i [(\delta_{p,i} - \bar{\delta}_p)^2 + (\delta_{H,i} - \bar{\delta}_H)^2]\} \quad (77)$$

which leads to the following expression for the partition coefficient:

$$\ln m_B = -\ln \bar{V} - \frac{V_B}{\bar{V}} + \frac{V_B}{RT} \{(\delta_{D,B} - \bar{\delta}_D)^2 + A_B [(\delta_{p,B} - \bar{\delta}_p)^2 + (\delta_{H,B} - \bar{\delta}_H)^2]\} + \ln \bar{V}^s + \frac{V_B}{\bar{V}^s} - \frac{V_B}{RT} \{(\delta_{D,B} - \bar{\delta}_D^s)^2 + A_B [(\delta_{p,B} - \bar{\delta}_p^s)^2 + (\delta_{H,B} - \bar{\delta}_H^s)^2]\} \quad (78)$$

Similarly, the solvent selectivity for component B, β_B can be defined as

$$\beta_B = \frac{m_B}{m_A} \quad (79)$$

where m_A is the partition coefficient for component A. Hence

$$\ln \beta_B = -\frac{V_B}{\bar{V}} + \frac{V_B}{RT} \{(\delta_{D,B} - \bar{\delta}_D)^2 + A_B [(\delta_{p,B} - \bar{\delta}_H)^2]\} + \frac{V_B}{\bar{V}^s} - \frac{V_B}{RT} \{(\delta_{D,B} - \bar{\delta}_D)^2 + A_B [(\delta_{p,B} - \bar{\delta}_p^s)^2 + (\delta_{H,B} - \bar{\delta}_H^s)^2]\} + \frac{V_A}{\bar{V}} - \frac{V_A}{RT} \{(\delta_{D,A} - \bar{\delta}_D)^2 + A_A [(\delta_{p,A} - \bar{\delta}_p)^2 + (\delta_{H,A} - \bar{\delta}_H)^2]\} - \frac{V_A}{\bar{V}^s} + \frac{V_A}{RT} \{(\delta_{D,A} - \bar{\delta}_D^s)^2 + A_A [(\delta_{p,A} - \bar{\delta}_p^s)^2 + (\delta_{H,A} - \bar{\delta}_H)^2]\} \quad (80)$$

The preceding equations can be changed to the infinite dilution expressions for partition coefficient and selectivity, by substituting as follows:

$$\bar{V} = V_A$$
$$\bar{V}^s = V_s$$
$$\bar{\delta} = \delta_A$$
$$\bar{\delta}^s = \delta_s$$

In other words, the solvent phase is pure solvent and the other phase pure component A.

In analyzing the preceding equations for solvent selection, the equations can be divided into two subsections, one containing the solubility parameter terms and the other the molar volume terms.

Considering the solubility parameter terms first, analysis of "real" liquid systems [38] indicates that, in general, the difference between the dispersion forces solubility parameters is much smaller than those between the polar and hydrogen bonding solubility parameters. Thus this difference can frequently be ignored. Similarly, since components B and A are initially miscible, their solubility parameters are likely to be nearer to each other than to those of the solvent. Hence

$$(\delta_B - \bar{\delta})^2 \ll (\delta_B - \bar{\delta}^s)^2$$
$$(\delta_A - \bar{\delta})^2 \ll (\delta_A - \bar{\delta}^s)^2$$

With these approximations, the expressions for the partition coefficient and selectivity become

$$m_B \propto [(\delta_{p,B} - \bar{\delta}_p^s)^2 + (\delta_{H,B} - \bar{\delta}_H^s)^2]^{-1}$$
$$\beta_B \propto \frac{(\delta_{p,A} - \bar{\delta}_p^s)^2 + (\delta_{H,A} - \bar{\delta}_H^s)^2}{(\delta_{p,B} - \bar{\delta}_p^s)^2 + (\delta_{H,B} - \bar{\delta}_H^s)^2} \quad (81)$$

If a δ_p/δ_H plane is considered, it becomes possible to plot the ternary equilibrium situation on this plane and hence relate the selectivity and partition coefficient to the distance between points on the plane, that is,

$$m_B \propto r_{B,S}^{-2}$$
$$\beta_B \propto \left(\frac{r_{A,S}}{r_{B,S}}\right)^2 \quad (82)$$

where $r_{i,j}$ is the distance between points i and j as shown in Fig. 17. Hence the maximum selectivity is obtained with the maximum separation of components A and S and the minimum separation of components B and S; that is, the best sol-

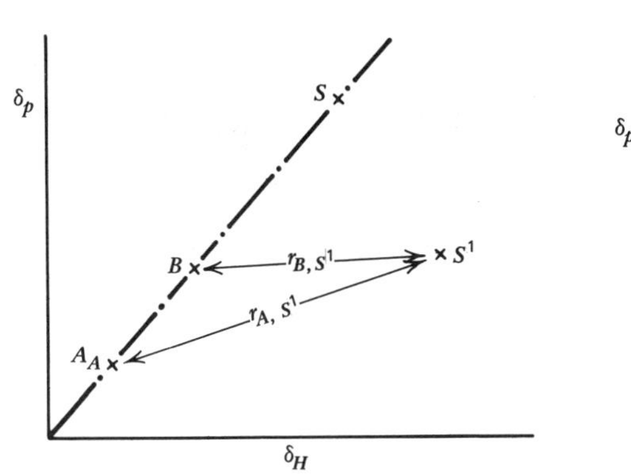

Figure 17. Solvent selection diagram.

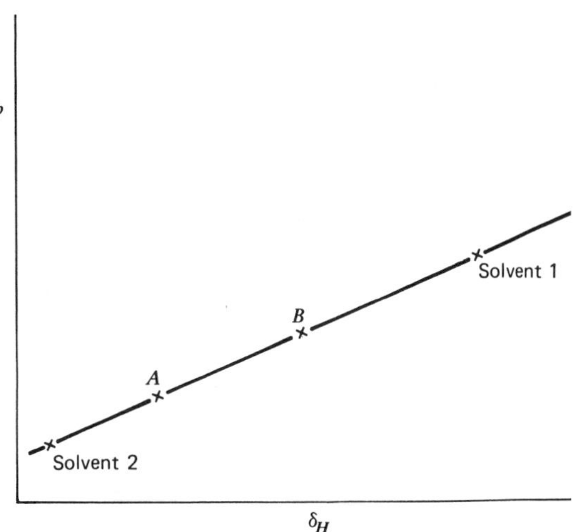

Figure 18. Two-solvent systems.

vent lies on the line joining components B and A, closer to component B than to component A (see Fig. 17). This observation conflicts with the partition coefficient in that, as the distance between component B and A is increased, its value falls, indicating possible low solvent capacity. Thus the best solvent lies on the line through components B and A as indicated previously, sufficiently far from component B to generate two liquid phases, but no farther than is absolutely necessary to ensure adequate solvent capacity and low solvent usage.

Considering Fig. 17, if components B and A are interchanged, it becomes very difficult to specify a solvent to extract component B because there is insufficient space between component B and the axis to locate a solvent and generate two liquid phases. Hence a solvent to extract component A in this case is the only possible single-solvent option, although, in general, it is assumed that component B is present in lower concentration in the feedstock to cause its nomination as component B. Thus the model suggests that this will be a difficult process to operate and a two solvent scheme could be attractive. The model can deal with this option since, returning to the original notation, it is obvious that the second solvent should again lie on the line joining components B and A, but this time closer to component B than component A, since it should have the greater selectivity for component B in this case (Fig. 18).

Having defined a route to indicate the best solvent in terms of solubility parameters, it is obvious that since there are only a finite number of solvents, one will seldom fall in the desired area. One possible solution to this problem is to consider a mixed solvent, because as shown in Fig. 19, by mixture of two solvents the "solubility parameters" of the mixture will line up on a line joining the two components of the mixed solvent, and hence the "ideal" solvent" can be achieved. In practice, a number of solvent pairs may be capable of giving the ideal, but with varying practical difficulties. Referring to Fig. 19, we see that solvent pairs S_3 and S_4 are capable of giving the desired mixed solvent, but, by considering components A and B as a pseudosolvent, it is evident that this pseudosolvent will selectively extract component S_3 and thus demix the solvent so that the ideal mixture is not effectively used. Hence solvent components S_1 and S_2 offer a better potential ideal solvent since demixing is less likely. Extending this concept, it is possible to plot a large number of solvents on the δ_p/δ_H plane and to indicate regions in which solvents cannot be found either as a single component or by mixing. Figure 20 summarizes these data and indicates the lack of solvents with large hydrogen bonding characteristics and low polar interaction.

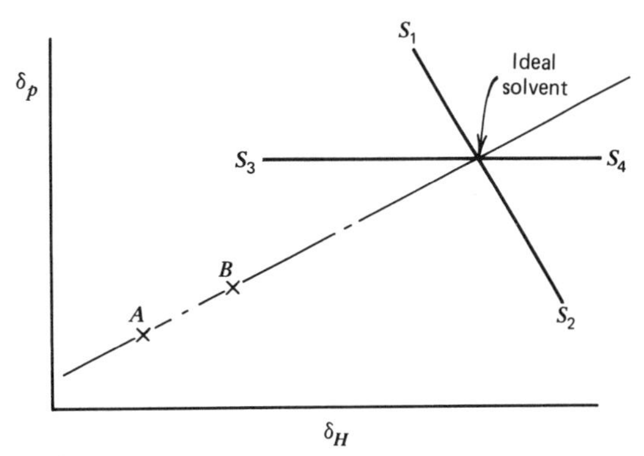

Figure 19. Mixed solvents.

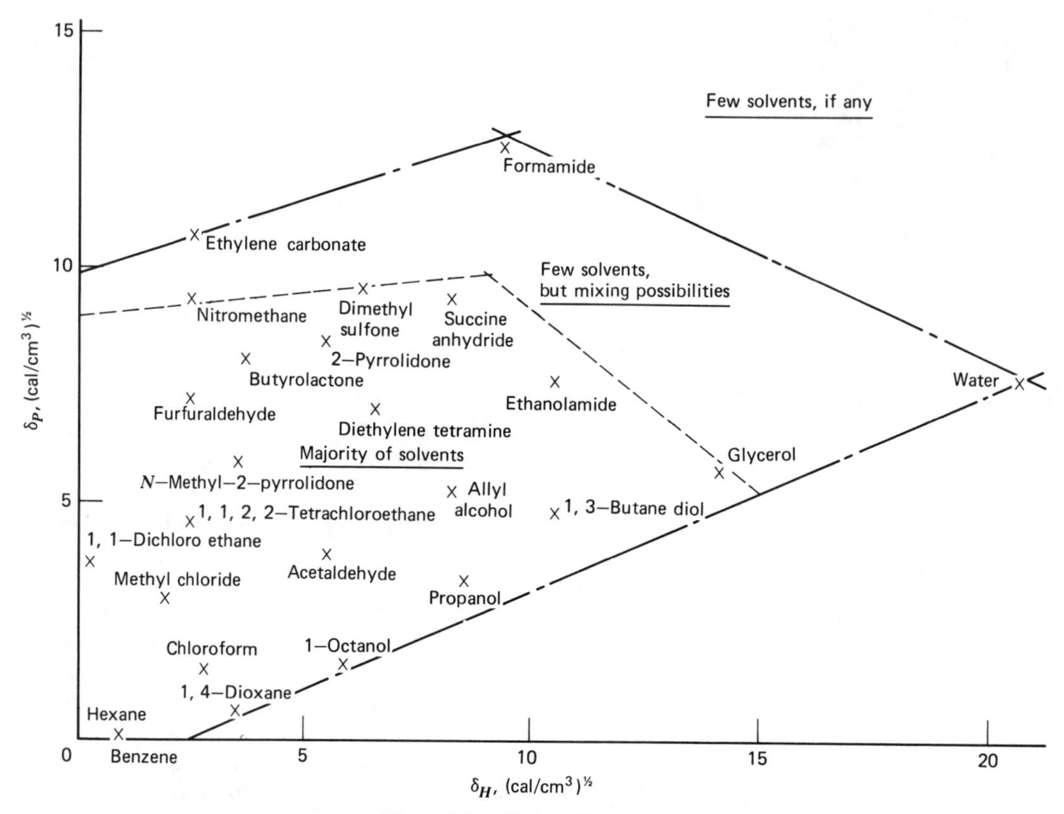

Figure 20. "Solvent" locations.

In considering the terms involving molar volumes, it is very difficult to identify a selection procedure and in practice frequently it is impossible, because once the solubility parameters of the solvent have been selected, there is very little freedom of selection on molar volumes because of the limited number of solvents available. Hence, where large differences in molecular size are encountered, the molar volume term must be considered as an unavoidable modification to the selection procedure and its influence must be allowed for, possibly by offsetting the final solvent choice from the solvent selection line as compensation.

A separate aspect of the solvent molar volume is that it can indicate whether more than one phase is possible [38]. From a mathematical viewpoint, the activity equation must exhibit a region where a single activity value can be obtained from two different compositions. Hence,

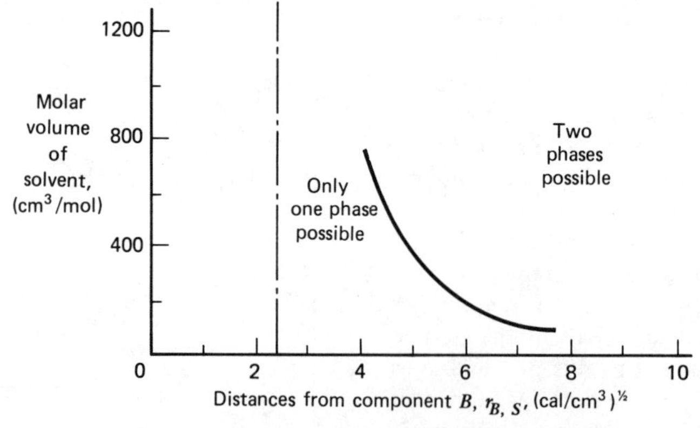

Figure 21. Influence of solvent molar volume.

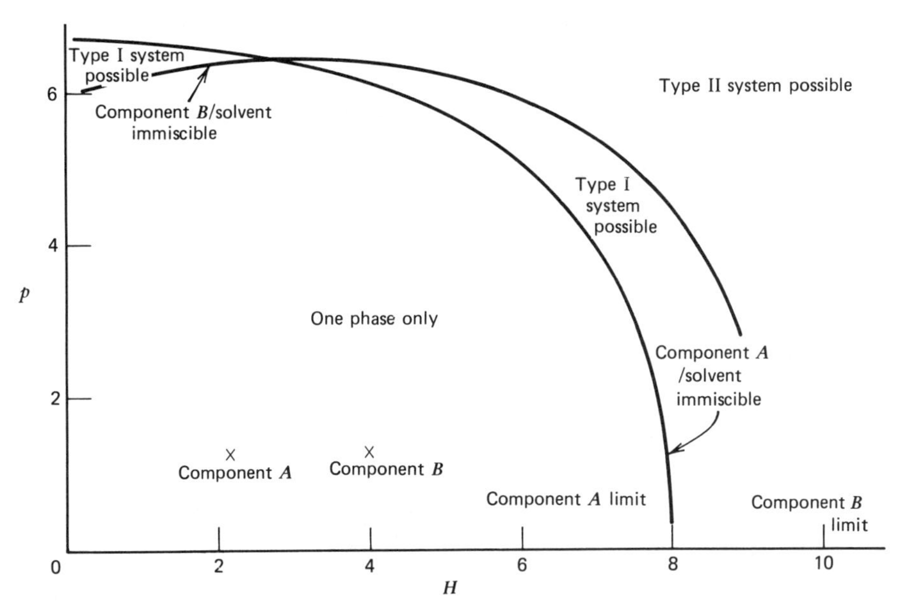

Figure 22. Ternary system behavior (with fixed solvent molar volume).

if two liquid phases are present, the curve must exhibit at least one maximum and minimum and in the limit will be on the point of potentially forming two liquid phases if an inflection occurs in the curve. Analysis of the activity equation on this basis produces the result that (1) if $V_s < V_{\text{lim}}$, only one phase is possible; and (2) if $V_s > V_{\text{lim}}$, where $V_{\text{lim}} = V_B/[1 + 2C - (8C)^{1/2}]$ and $C = A_B V_B r_{Bs}^2/RT$, two phases are possible but may not exist.

The analysis also indicates that below a limiting value of $r_{B,s}$ given by $r_{B,\text{lim}} = (0.5\ A_B V_B/RT)^{1/2}$, this volume dependence does not occur. Figure 21 summarizes this behavior so that solvents located below the line do not form two liquid phases whereas those above potentially can. By considering solvent interactions with components B and A, it becomes possible by this technique, to divide the δ_p/δ_H plane into regions offering potentially different liquid–liquid equilibrium behavior for a given solvent size. Figure 22 summarizes a typical diagram [7] in terms of Types I and II behavior.

6. DATA SOURCES

6.1. Experimental Determination of Data

The measurement of liquid–liquid equilibrium data is basically simple and requires little in the way of sophisticated apparatus. Garner and Bourne [67] have briefly reviewed published apparatus and emphasize the importance of (1) purity of materials, (2) temperature control, (3) attainment of true equilibrium, (4) sampling, and (5) analysis.

With the development of technology over the last 20 years, many of the problems have been resolved. Modern electronics allow accurate temperature control; magnetic stirrers enable equilibrium to be attained in a relatively short time and the availability of GLC, as a general laboratory tool, enables accurate analysis to be made.

6.1.1. *Solubility Determination*

Many authors [7, 68, 69] suggest that it is advantageous to determine the binodal (solubility) curve before attempting equilibrium determinations. This is usually done by titration of one component into a vessel containing a known mixture of the other two components. Referring to Fig. 23, we see that a known mixture of components A and B gives point M to which component C is added by titration until the mixture becomes turbid. This establishes point 1 on the binodal curve. A known amount of component B may then be added to yield point 2 in the miscible region. To this new mixture, component C is again added by titration until the mixture again becomes turbid locating point 3 on the binodal curve. The composition of the mixture is always known as measured amounts of the com-

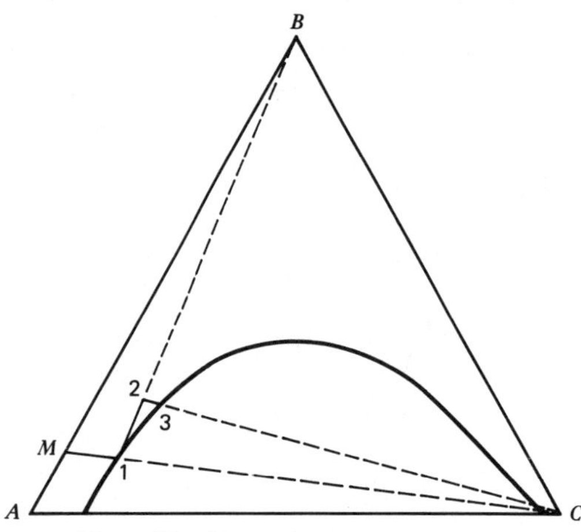

Figure 23. Determination of solubility.

ponents are added. Thus the complete binodal curve may be determined.

6.1.2. Tie-Line Determination

A simple cell is illustrated in Fig. 24. A small total volume (about 20 ml) is advantageous and temperature control is by circulation of constant temperature fluid. A magnetic stirrer enables intimate contact of the phases to be achieved in order to attain equilibrium. After settling, both phases may be sampled using a syringe for direct injection into a chromatograph.

This procedure will yield directly the full equilibrium data in terms of tie-lines. By suitable choice of the overall composition charged to the cell, the full composition range may be covered.

A more sophisticated device for obtaining equilibrium data is the AKUFVE (Swedish abbreviation for "apparatus for continuous measurement of partition factors in solvent extraction") apparatus [70]. Davis et al. [71] have recently suggested a modification for high phase ratios. Details are given in Chapter 17.2.

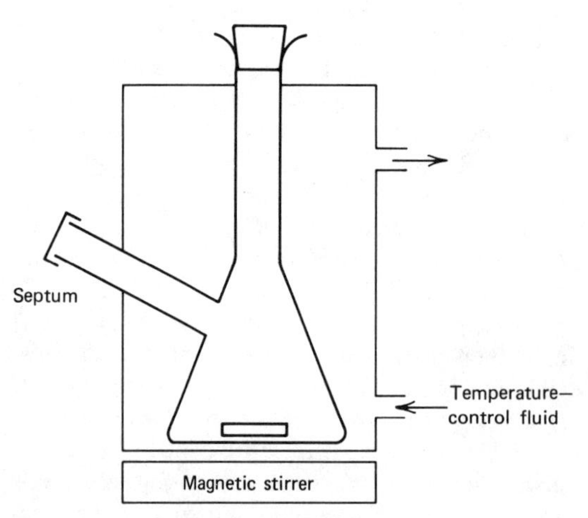

Figure 24. Equilibrium cell.

6.2. Published Data

6.2.1. Literature Sources

No compilation (bibliography) of published data has existed until recently. Garner and Bourne [67] commented that data are widely scattered throughout the literature and presented a useful tabulation of sources of data. More recently, Sorensen et al. [12] made similar comments and recommended reference to sources by Francis [69], Stephen and Stephen [72], Seidel [73], and Landolt-Bornstein [74]. Sorensen and Arlt in collaboration with the University of Dortmund and DECHEMA, have collected, correlated, and predicted all published liquid–liquid equilibrium data to form a computerized data bank [55]. A computer-aided bibliography by Wisniak and Tamir [75] has recently appeared.

6.2.2. Data Banks

Many commercial organizations are considering entering the data bank field, for instance, PPDS and ESDU. The most advanced is that prepared by Sorensen et al. in collaboration with the University of Dortmund and published in the DECHEMA Chemistry Data series [55]. The aim of this effort is to establish a comprehensive computerized data bank for liquid–liquid equilibrium data and to correlate these data by using the UNIQUAC and NRTL equations. Also, it is intended to use this data bank to establish group contribution parameters for the UNIFAC prediction method for liquid–liquid equilibrium data.

Full details of the data bank are given by Sorensen and Arlt [55] and Sorensen et al. [12]. Briefly, the components are water and organic, nonpolymeric compounds that must contain two or more of the atoms C, H, O, N, F, Cl, Br, and S. All components have a normal boiling point above 0°C and the temperature range of the data is 0–150°C, with pressures ranging up to 20 atm, although pressure has little effect on liquid–liquid data.

The data bank contains experimental binary, ternary, and quaternary liquid–liquid equilibrium data giving the compositions in each of the coexisting phases, that is, tie-line data. In 1979 Sorensen et al. stated that the data bank contained 884 binary data sets, 772 ternary and 23 quaternary and commented that water appeared in 55% of the binary and 78% of the ternary mixtures and that 75% of the ternary data were within 10°C of ambient temperature.

NOTATION†

A	$n \times n$ matrix
A_i	weighting parameter for component i
A_k	matrix with kth row and column deleted *or* van der Waals group surface area [Eq. (57)]
a_i	activity of component i
G	Gibbs free energy, J
G_{ij}	functions defined in Eq. (51)
g^m	molar Gibbs free energy, J/mol
g^{id}	see Eq. (5)
g^E	see Eq. (6)
H	enthalpy, J
h^m	molar heat of mixing, J mol^{-1}
l_i	see Eq. (73)
l_1	parameter defined in Eq. (55)
m	distribution coefficient
m^0	distribution coefficient at infinite dilution
M	total number of moles present in one phase
N	total number of moles present in another phase
n	number of components
n_i, n_j	number of moles of i, j
P	Pressure, Pa
Q	excess function [see Eq. (39)]
Q_k	van der Waals group area parameter
q_i	sum of group area contributions [see Eq. (56)]
R	gas constant, J K^{-1} mol^{-1}
R_k	van der Waals group volume parameter [see Eq. (57)]
r_i	sum of group volume contributions [see Eq. (56)]
$r_{i,j}$	distance between points i and j in Fig. 17
r_k	activity coefficient of group k in solution
$r_k^{(i)}$	activity coefficient of group k in pure i
S	entropy, J K^{-1}
T	temperature, K
V	molar volume, m^3 mol^{-1}
$\bar{V}$	average molar volume [see Eq. (66c)], m^3 mol^{-1}
V_k	van der Waals group volume for type k
v^m	molar volume of mixing, m^3 mol^{-1}
v_i	liquid molar volume of component i, m^3 mol^{-1}
X_k	see Eq. (75)
x_i	mole fraction of component i
x_{ij}	local composition of component j, with respect to component i [see Eq. (41)]
Z	coordination number

†Note: Unit symbols other than SI (International System of Units) are still widely used, as noted in some of the tables and figures.

Greek Symbols

α	term in Eq. (71)
α_{ij}	measure of nonrandomness of mixture of components i, j
β	selectivity [see Eq. (79)]
γ_i	activity coefficient, component i
ΔE^m	change in energy on mixing, J mol^{-1}
Δs_1^m	partial molar entropy of mixing, [see Eq. (63)], J mol^{-1} K^{-1}
δ_i	cohesive energy density of component i, $[J\ m^{-3}]^{0.5}$
$\bar{\delta}$	weighted average value of δ_i [see Eq. (66b)]
ϵ	see Eq. (71)
Λ	see eqs. (44) and (45)
λ_{ij}	interaction energy between components i and j, J mol^{-1}
λ_m	see Eq. (76)
μ_i	chemical potential of component i, J mol^{-1}
μ_{ij}	chemical potential derivative, J mol^{-2}
ν_k	number of type k groups in molecule
ϕ_i	volume fraction of component i
Ψ_{mn}	see Eq. (76)
τ_{ij}	see Eq. (52)
ϕ_i	see Eq. (55)
d	constant in Eq. (71)
η	see Eq. (71)

REFERENCES

1. J. W. Gibbs, *Collected Works*, Yale University Press, New Haven, 1928.
2. I. Prigogine and R. Defay, *Chemical Thermodynamics*, Longmans, London, 1954.

3. J. S. Rowlinson, *Liquids and Liquid Mixtures*, 2nd ed., Butterworths, London, 1969.
4. J. H. Perry, Ed., *Chemical Engineers Handbook*, 4th ed., McGraw-Hill, New York, 1963, p. 2-25.
5. M. B. King, *Phase Equilibrium in Mixtures*, Pergamon, Oxford, 1969.
6. E. Janecke, *Z. Anorg. Chem.* **51**, 132 (1906).
7. R. E. Treybal, *Liquid Extraction*, McGraw-Hill, New York, 1963.
8. D. B. Hand, *J. Phys. Chem.* **34**, 1961 (1930).
9. R. E. Treybal, L. D. Weber and J. F. Daley, *Ind. Eng. Chem.* **38**, 817 (1946).
10. D. F. Othmer and P. E. Tobias, *Ind. Eng. Chem.* **34**, 693 (1942).
11. P. Carniti, L. Cori, and V. Ragaini, *Fluid Phase Equilibria* **2**(1), 39 (1978).
12. J. M. Sorensen, T. Magnussen, P. Rasmussen, and A. Fredenslund, *Fluid Phase Equilibria* **3**(1), 47 (1979).
13. G. M. Wilson, *J. Am. Chem. Soc.* **86**, 127 (1964).
14. R. L. Scott, *J. Chem. Phys.* **25**, 193 (1956).
15. R. V. Orye and J. M. Prausnitz, *Ind. Eng. Chem.* **57**(5), 18 (1965).
16. H. Renon and J. M. Prausnitz, *J. AIChE* **14**, 135 (1968).
17. H. Renon and J. M. Prausnitz, *Ind. Eng. Chem. Proc. Des. Devel.* **8**, 413 (1969).
18. M. Hiranuma, *J. Chem. Eng. Jap.* **8**, 162 (1975).
19. T. Kayatama, *J. Chem. Eng. Jap.* **6**, 357 (1973).
20. R. A. Heidemann and J. M. Mandhane, *Chem. Eng. Sci.* **28**, 1213 (1973).
21. D. Tassios, *Ind. Eng. Chem. Proc. Des. Devel.* **15**, 574 (1976).
22. A. C. Mattelin and L. A. J. Verhoeye, *Chem. Eng. Sci.* **30**, 193 (1975).
23. D. S. Abrams and J. M. Prausnitz, *J. AIChE* **21**, 62 (1975).
24. E. A. Guggenheim, *Mixtures*, Oxford University Press, 1952.
25. A. Bondi, *Physical Properties of Molecular Crystals, Liquids and Gases*, Wiley, New York, 1968.
26. P. M. Cukor and J. M. Prausnitz, *Inst. Chem. Eng. Symp. Ser.* **32**(3), 88 (1969).
27. L. Asselineau and H. Renon, *Chem. Eng. Sci.* **25**, 1211 (1970).
28. S. Stroud, Ph.D. thesis, University of Birmingham, England, 1976.
29. R. A. Heidemann, *J. AIChE* **20**, 847 (1974).
30. D. Y. Peng and D. A. Robinson, *Can. J. Chem. Eng.* **54**, 595 (1976).
31. G. M. Wilson, *Adv. Cryogen. Eng.* **9**, 168 (1964).
32. J. H. Hildebrand and R. L. Scott, *Solubility of Non-Electrolytes*, 3rd ed., Dover, New York, 1963.
33. G. J. Pierotti, C. H. Deal, and E. L. Derr, *Ind. Eng. Chem.* **51**, 95 (1959).
34. J. H. Hildebrand and R. L. Scott, *Regular Solutions*, Prentice-Hall, Englewood Cliffs, NJ, 1962.
35. G. Scatchard, *Chem. Res.* **8**, 321 (1931).
36. G. Scatchard, *J. Am. Chem. Soc.* **56**, 995 (1934).
37. G. Scatchard, *Transact. Faraday Soc.* **33**, 160 (1937).
38. A. Brench, *Soc. Chem. Ind. Symp.*, unpublished presentation (1978).
39. H. Burrell, *Am. Chem. Soc. Org. Coat. Plast. Chem.* **28**(1), 682 (1968).
40. A. E. Van Arkel, *Transact. Faraday Soc.* **42B**, 81 (1946).
41. Kirk-Othmer *Encyclopedia of Chemical Technology*, 2nd ed. suppl. vol., Wiley, New York, 1971, pp. 889-909.
42. R. F. Blanks and J. M. Prausnitz, *Ind. Eng. Chem. Fund.* **3**, 1 (1964).
43. R. F. Weimer and J. M. Prausnitz, *Hydrocarbon Process.* **44**, 237 (1965).
44. J. G. Helpinstill and M. Van Winkle, *Ind. Eng. Chem.* **7**, 213 (1968).
45. C. M. Hansen, *Ind. Eng. Chem. Process Des. Devel.* **8**, 2 (1969).
46. B. L. Karger, L. Snyder, R. Lloyd, and C. Horvath, *An Introduction to Separation Science*, Wiley, New York, 1973.
47. L. Snyder, *Chem. Technol.* **9**, 750 (December (1979).
48. J. Small, *Appl. Chem.* **3**, 71 (1953).
49. K. L. Hoy, *J. Paint Technol.* **42**(541), 76 (1970).
50. H. R. Null and D. A. Palmer, *Chem. Eng. Progr.* **65**, 47 (1969).
51. A. Fredenslund, R. L. Jones, and J. M. Prausnitz, *J. AIChE* **21**, 1086 (1975).
52. A. Fredenslund, J. Gmehling, and P. Rasmussen, *Vapor-Liquid Equilibria using UNIFAC*, Elsevier, Amsterdam, 1977.
53. S. Skjold-Jorgensen, B. Kolbe, J. Gmehling, and P. Rasmussen, *Ind. Eng. Chem. Proc. Des. Devel.* **18**, 714 (1979).
54. T. Magnussen, J. Sorensen, P. Rasmussen, and A. Fredenslund, *Fluid Phase Equilibria* **4**, 151 (1980).
55. J. Sorensen and W. Arlt, *Liquid-Liquid Equilibrium Data Collection*, DECHEMA Chemistry Data Series, Frankfurt, Part 1, 1979; Parts 2 and 3, 1980.
56. E. L. Derr and C. H. Deal, *Inst. Chem. Eng. Symp. Ser.* **32**(3), 37 (1969).
57. K. Kojima and K. Tochigi, *Prediction of Vapour/Liquid Equilibria by Using the ASOG Method*, Elsevier, Amsterdam, 1979.
58. K. Tochigi and K. Kojima, *J. Chem. Eng. Jap.* **10**(1), 61 (1977).
59. K. Tochigi and K. Kojima, *J. Chem. Eng. Jap.* **9**(4), 207 (1976).

60. H. Sugi and T. Katayama, *J. Chem. Eng. Jap.* **10,** 400 (1977).
61. A. W. Francis, *Ind. Eng. Chem.* **36,** 764 (1944).
62. A. W. Francis, *Critical Solution Temperatures*, American Chemical Society, Washington DC, 1961.
63. D. A. Drew and A. N. Hixson, *Transact. Am. Inst. Chem. Eng.* **40,** 675 (1944).
64. R. H. Ewell, J. M. Harrison, and L. Berg, *Ind. Eng. Chem.* **36,** 871 (1944).
65. V. Rothmund, *Z. Phys. Chem.* **26,** 433 (1898).
66. N. B. Godfrey, *Chem. Technol.* **2,** 359 (1972).
67. F. H. Garner and J. R. Bourne, in *Chemical Engineering Practice*, Vol. 5, H. W. Cremer and T. Davies, Eds., Butterworths, London, 1958, p. 368.
68. L. Alders, *Liquid-Liquid Extraction*, Elsevier, Amsterdam, 1959.
69. A. W. Francis, *Liquid-Liquid Equilibriums*, Interscience, New York, 1963.
70. H. Reinhardt and J. Rydberg, *Acta Chem. Scand.* **23,** 2773 (1969).
71. S. S. Davis, G. Elson, E. Tomlinson, G. Harrison, and J. C. Dearden, *Chem. Ind.* 677 (1976).
72. H. Stephen and T. Stephen, *The Solubilities of Inorganic and Organic Compounds*, Pergamon, Oxford, 1963.
73. A. Seidel, *Solubilities of Organic Compounds*, Van Nostrand, Princeton, NJ, 1941.
74. H. H. Landolt and R. Bornstein, *Zahlenwerte und Funktionen*, Springer-Verlag, Berlin, 1959.
75. J. Wisniak and A. Tamir, *Liquid-Liquid Equilibrium and Extraction*, Elsevier, Amsterdam, 1980.

2.1

EXTRACTION WITH REACTION

M. M. Sharma
University of Bombay
India

1. INTRODUCTION

Extraction accompanied by chemical reaction of a sparingly soluble solute is of great industrial importance. A variety of such cases can be cited, and Table 1 gives some of the more important examples. The two-phase reaction may be a part of the process (e.g., item 3, 9, or 13 in Table 1) or may be concerned with removal and/or recovery of a solute (e.g., item 4 or 12 in Table 1). The extent of reaction or the degree of removal of solute may vary, sometimes exceeding the 99% level of conversion or removal. In many respects the theoretical aspects of extraction with chemical reaction resemble those of gas absorption accompanied by chemical reaction. However, in extraction accompanied by chemical reaction the true mass-transfer coefficients are

TABLE 1 LIQUID–LIQUID REACTIONS: SOME EXAMPLES OF INDUSTRIAL IMPORTANCE

Number	System
1	Hydrolysis and/or saponification of a variety of esters
2	Hydrolysis of organic chloro compounds (C_5 chlorides, nitrochlorobenzenes, polychlorobenzenes, etc.)
3	Nitration of aromatic compounds (benzene, chlorobenzene, toluene, etc.)
4	Removal of COS from liquefied C_3/C_4 fractions by treatment with aqueous NaOH and alkanolamine solutions
5	Manufacture of dithiocarbamates by reaction between aqueous solutions of amines and CS_2
6	Alkylation of organic compounds (alkylation of benzene/toluene with olefins; alkylation of phenols with isobutylene; toluene with isobutylene and butenes; isobutane with butenes and isobutylene, etc.)
7	Oligomerization of 1-butene in sulfuric acid; dimerization of isobutylene and isoamylene
8	Conversion of propylene, butenes, and amylenes into corresponding alcohols by extraction into sulfuric acid; extraction of isoamylene from loaded H_2SO_4 solutions into *n*-heptane
9	Manufacture of oximes (cyclohexanone oxime by reaction between cyclohexanone and aqueous solutions of hydroxylamine sulfate/phosphate)
10	Reaction between ethylene dichloride and aqueous ammonia for production of ethylenediamine
11	Reduction of a variety of nitroaromatic compounds with aqueous Na_2S_x
12	Recovery of formic and acetic acid from dilute aqueous streams, in solvents such as tributyl phosphate (TBP), trioctylphosphine oxide (TOPO), etc. containing trioctyl or trinonyl amine
13	Extraction of metals, such as, uranium, copper, nickel, cobalt, and vanadium from aqueous solutions
14	Pyrometallurgical operations involving melts and molten slag
15	Interfacial polycondensation reactions

generally lower than in gas–liquid contactors, and, in principle, the reaction can occur in both the phases.

Despite its considerable industrial importance, extraction with reaction has received limited attention. Sharma [1] and Hanson [2] have reviewed some aspects of this subject, and there is a more recent survey by Doraiswamy and Sharma [3]. Table 2 gives a general classification of reactions involving two liquid phases.

This chapter deals mainly with the effects of irreversible reactions within one phase on rates of interfacial mass transfer. These effects are considered in relation to the mass-transfer coefficient due to diffusion (or *true* mass-transfer coefficient; see Chapter 3). Such reactions can enhance the rates of interfacial mass transfer, relative to the rates predicted from diffusion

TABLE 2 CLASSIFICATION OF REACTIONS

Number	System
1	Irreversible reactions of general order
2	Reversible reactions of general order
3	Two-phase reactions involving one of the phases as aqueous but with reaction occurring in the organic phase
4	Consecutive reactions
5	Consecutive/parallel reactions
6	Extraction of a solute accompanied by reaction with two reactants
7	Simultaneous extraction with reaction of two solutes
8	Extraction with reaction in both phases

theory. Reversible reactions are covered only briefly in this chapter, with a much more detailed treatment given in Chapter 2.2 in regard to extraction of metals. The ability of a slow, reversible reaction at an interface to increase the overall resistance to mass transfer is also discussed in Section 3.1 of Chapter 3.

2. CONTROLLING REGIMES

Consider the following irreversible reaction, which is *m*th order with respect to the dissolved solute A and *n*th order with respect to reactive component B:

$$A + ZB \longrightarrow \text{Products}$$

Component B and the products are assumed to be insoluble in phase A′, which contains solute A. Solute A is sparingly soluble in phase B′. The reaction occurs only in phase B′, which may consist of pure component B, or it may be dissolved in an inert solvent; it is assumed that there is no resistance to mass transfer in the A′ phase. We further assume that isothermal conditions prevail in the system. The rate of homogeneous reaction per unit volume of the B′ phase is given by $k_{mn}[A]^m[B]^n$. As in the case of gas–liquid reactions, it is assumed that the B′ phase is saturated with component A "at the interface," that is, in the B′ phase within only a few molecular diameters from the interface.

Different theories of diffusional mass transfer, such as the two-film theory, Higbie's penetration theory, and Danckwerts' surface renewal theory (see Chapter 3), have been propounded, but their predictions in regard to the effect of chemical reaction on rates of mass transfer do not differ substantially, particularly when the diffusivities of components A and B are comparable [4]. The two-film theory is relatively simple to adopt and is used here, unless the differences between the predictions based on different theories are substantial.

Depending on the relative rates of diffusion and chemical reaction the system may conform to one of the following well-defined controlling regimes:

Regime 1. Very slow reaction.
Regime 2. Slow reaction.
Regime 3. Very fast reaction.
Regime 4. Instantaneous reaction.

Intermediate cases can also be considered, and for practically all cases of industrial importance, there are analytical expressions to calculate the specific rate of extraction.

Rate expressions are given in Tables 3a and 3b and the practical implications of the different regimes are summarized in the following paragraphs. The corresponding concentration profiles for A and B are given in Fig. 1.

2.1. Regime 1: Very Slow Reaction

The reaction rate is so slow that phase B′ is essentially saturated with component A. Diffusional factors are unimportant; thus increase in the interfacial area or overall mass-transfer coefficient will have no effect on the rate of reaction per unit volume of the B′ phase. Thus, if a mechanically agitated contactor is used, a saving in energy consumption can be made by properly selecting the speed of agitation. However, the agitation should not be reduced to the point where the regime condition (see Table 3a) begins to be violated. Within this regime, scale-up requires only a knowledge of kinetics of the reaction.

2.2. Regime 2: Slow Reaction

The reaction is sufficiently fast that the concentration of A is zero in phase B′ except in the mass transfer film [Fig. 1(*a*)].

Here kinetic factors are unimportant, and so is the concentration of B. Since diffusional factors control the rate, the effect of temperature on the rate of reaction will be relatively small compared to the case of regime 1. In this regime the scale-up requires only a knowledge of the true mass-transfer coefficient k_L and the interfacial area.

2.3. Regime 3: Very Fast Reaction

Component A is depleted completely within the mass-transfer film, but the concentration of B is constant [Fig. 1(*b*)].

In this regime, in the special case of reactions that are first order in A, R_A is proportional to $[A^*]$, whereas for reactions that are zero order in A, the specific rate of extraction will be proportional to $[A^*]^{0.5}$.

The value of R_A is proportional to $k_{mn}^{0.5}$, and hence when a choice of reactive species is available for extraction, the value of R_A can be enhanced substantially by using a faster-reacting

TABLE 3a SALIENT FEATURES OF VARIOUS CONTROLLING REGIMES FOR IRREVERSIBLE REACTION[a]

Regime	Conditions	Rate Expression	Distinctive Features
1	$k_L a[A^*] >> l k_{mn}[A^*]^m[B_0]^n$	$R'_A = k_{mn}[A^*]^m[B_0]^n$	Diffusional factors unimportant; rate governed by kinetic factors
2	$k_L a[A^*] << l k_{mn}[A^*]^m[B_0]^n$	$R_A = k_L[A^*]$	Rate governed by diffusional factors and independent of concentration of B (concentration profile shown in Fig. 1*a*)
	$\frac{\sqrt{2/(m+1) D_A k_{mn}[A^*]^{m-1}[B_0]^n}}{k_L} = \sqrt{M} << 1$		
3	$\frac{\sqrt{2/(m+1) D_A k_{mn}[A^*]^{m-1}[B_0]^n}}{k_L} >> 1$	$R_A = [A^*]\sqrt{2/(m+1) D_A[A^*]^{m-1}[B_0]^n}$	Specific rate of mass transfer a unique function of physicochemical properties and independent of hydrodynamic factors (concentration profile shown in Fig. 1*b*)
	$\frac{\sqrt{2/(m+1) D_A k_{mn}[A^*]^{m-1}[B_0]^n}}{k_L} << \frac{[B_0]}{Z[A^*]}\sqrt{\frac{D_B}{D_A}}$		
4	$\frac{\sqrt{2/(m+1) D_A k_{mn}[A^*]^{m-1}[B_0]^n}}{k_L} >> \frac{[B_0]}{Z[A^*]}\sqrt{\frac{D_B}{D_A}}$	$R_A = k_L[A^*]\left(1 + \frac{[B_0]}{Z[A^*]}\frac{D_B}{D_A}\right)$ (film theory)	Rate governed by diffusional factors; kinetics unimportant (concentration profile shown in Fig. 1*c*)
	$\frac{[B_0]}{Z[A^*]} >> 1$	$R_A = k_L \frac{[B_0]}{Z}\frac{D_B}{D_A}$	Rate independent of concentration of A
		$R_A \approx k_L[A^*]\left(\sqrt{\frac{D_A}{D_B}} + \frac{[B_0]}{Z[A^*]}\sqrt{\frac{D_B}{D_A}}\right)$ (surface renewal theory)	
	$\frac{[B_0]}{Z[A^*]} >> 1$	$R_A = \frac{k_L[B_0]}{Z}\sqrt{\frac{D_B}{D_A}}$	

Source: Data from Doraiswamy and Sharma [3], Danckwerts [4], and Sharma and Nanda [14].

[a] $A + ZB \rightarrow$ Products; Reaction *m*th Order in [A] and *n*th Order in [B].

TABLE 3b SALIENT FEATURES OF INTERMEDIATE CONTROLLING REGIMES[a]

Intermediate Regime	Conditions	Rate Expression	Distinctive Features
1–2	$k_L a[A^*] \simeq l k_{mn}[A^*]^m[B_0]^n$	$\frac{[A^*]}{R_A a} = \frac{1}{k_L a} + \frac{1}{l k_{1n}[B_0]^n}$ (for $m = 1$)	Both diffusional and kinetic factors important (concentration profile shown in Fig. 1*d*)
2–3	$\frac{\sqrt{2/(m+1) D_A k_{mn}[A^*]^{m-1}[B_0]^n}}{k_L} \simeq 1$	$R_A \simeq \frac{k_L[A^*]\sqrt{M}}{\tanh\sqrt{M}}$	Both physicochemical data and hydrodynamics important (concentration profile shown in Fig. 1*e*)
	$\frac{\sqrt{2/(m+1) D_A k_{mn}[A^*]^{m-1}[B_0]^n}}{k_L} << \frac{[B_0]}{Z[A^*]}\sqrt{\frac{D_B}{D_A}}$	$R_A = [A^*]\sqrt{D_A k_{1n}[B_0]^n + k_L^2}$ (for $m = 1$; Danckwerts' surface renewal theory)	
3–4	$\frac{\sqrt{2/(m+1) D_A k_{mn}[A^*]^{m-1}[B_0]^n}}{k_L} \simeq \frac{[B_0]}{Z[A^*]}\sqrt{\frac{D_B}{D_A}}$ $\frac{[B_0]}{Z[A^*]}\sqrt{\frac{D_B}{D_A}} >> 3$	$\phi = \sqrt{M}\left(\frac{\phi_a - \phi}{\phi_a - 1}\right)^{n/2}$ where $\phi = \frac{k_{LR}}{k_L} = \frac{R_A}{[A^*]k_L}$ and $\phi_a = 1 + \frac{[B_0]}{Z[A^*]}\frac{D_B}{D_A}$ (film theory) $\phi_a \simeq \sqrt{\frac{D_A}{D_B}} + \frac{[B_0]}{Z[A^*]}\sqrt{\frac{D_B}{D_A}}$ (surface renewal theory)	Both physicochemical data and hydrodynamics important (concentration profile shown in Fig. 1*f*)

Source: Data from Doraiswamy and Sharma [3], Danckwerts [4], and Sharma and Nanda [14].
[a] A + ZB → Products; reaction *m*th Order in [A] and *n*th Order in [B].

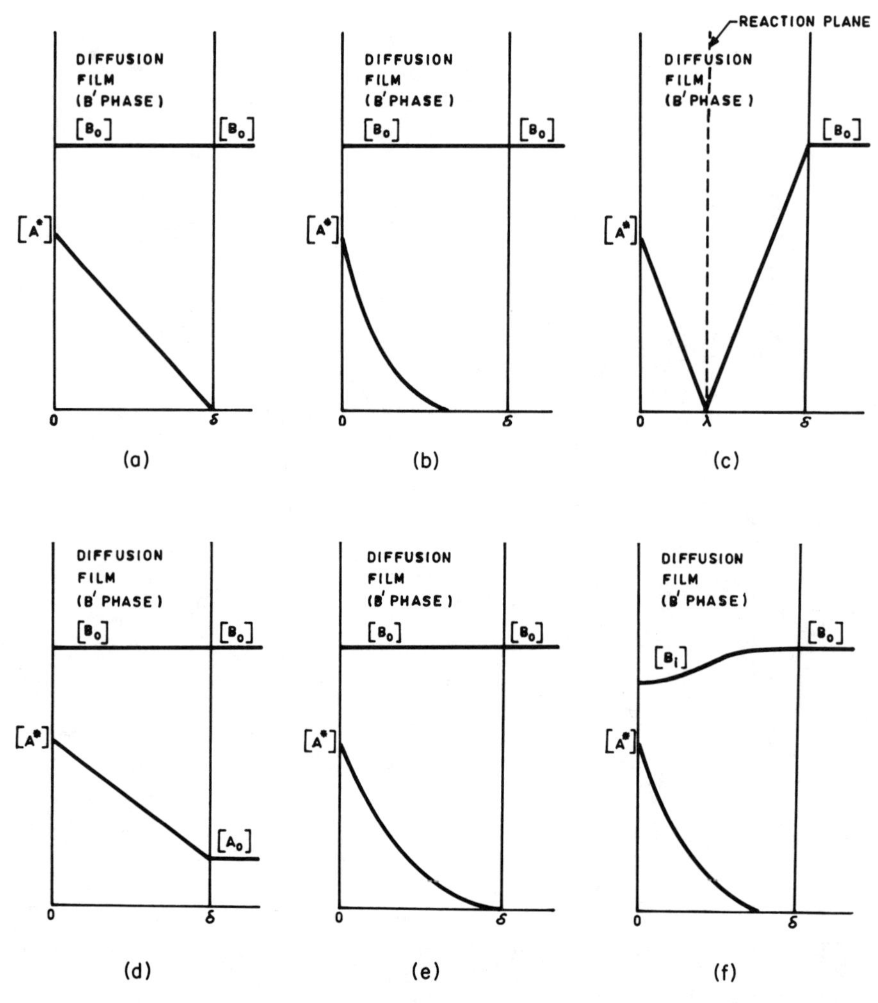

Figure 1. Concentration profiles for various controlling regimes: (*a*) regime 2 (slow reaction); (*b*) regime 3 (very fast reaction); (*c*) regime 4 (instantaneous reaction); (*d*) regime between 1 and 2; (*e*) regime between 2 and 3; (*f*) regime between 3 and 4.

species (see the example of removal of COS in Section 4.3).

In this regime the various forms of surface renewal theories lead to the expression shown in Table 3a, and there are no differences between predictions based on different theories. The measurement of R_A, coupled with the knowledge of $[A^*]$, D_A, m, and n, will allow the determination of k_{mn}. Within this regime the scale-up requires a knowledge of the specific rate of mass transfer R_A and effective interfacial area.

2.4. Regime 4: Instantaneous Reaction

The reaction is so fast that it can be assumed to occur in an infinitesimally thin zone (plane) where both reactants approach zero concentration [see Fig. 1(*c*)].

We have a unique situation as R_A becomes independent of the concentration of A (provided $[B_0] >> [A^*]$). In this regime the scale-up requires a knowledge of the overall mass-transfer coefficient. Since the mass-transfer resistance in phase B′ is very low, a check should be made to see whether the resistance has shifted to phase A′ (see Section 2.7).

2.5. Intermediate Regimes

Table 3b indicates the conditions and rate expressions for the three possible intermediate regimes, and the concentration profiles are shown in Figs. 1 (*d–f*). The various mass-transfer models give slightly different rate expressions in these cases.

2.6. Reversible Reaction

Reversible reactions are encountered commonly in gas–liquid systems. Reversible reactions are also found in liquid–liquid systems, notably in the extraction of metals (see Chapter 2.2). In most cases the bulk liquid can be assumed to be in equilibrium. When the reaction is instantaneous, it can also be assumed that equilibrium exists throughout the liquid, including the interface, whereas for fast reactions, the equilibrium is not reached at every point. It is necessary to treat these two cases separately [4, 5].

2.7. Shifting of Resistance

In regimes 3 and 4 (Table 3a) the enhancement of the B′-phase transfer coefficient (k_{LR}) may result in a situation where the rate of reaction is controlled almost completely by the diffusional resistance in the A′ phase. In such a case the kinetics of the reaction becomes unimportant. This situation can be expressed on the basis of the addition of resistances:

$$\frac{1}{K_0} = \frac{1}{k_{LA(A'\,\text{phase})}} + \frac{1}{m_A k_{LR}} \tag{1}$$

If the value of $m_A k_{LR}$ is far greater than $k_{LA(A'\,\text{phase})}$, the resistance will be confined to the A′ phase. This situation can also arise because of moderate to high solubility of A in the B′ phase, which will result in a relatively higher value of m_A [see also Chapter 3, Eq. (58)].

2.8. Importance of Interfacial Turbulence

Interfacial turbulence changes the local value of true mass-transfer coefficient (see Chapter 3), and hence the controlling regime can change. For instance, a system in the regime between 3 and 4, that is, with depletion of B at the interface, may be moved to regime 3 by the onset of interfacial turbulence that overcomes the depletion of B. However, there is very limited information in the literature on the role of interfacial turbulence in extraction with reaction.

3. EXPERIMENTAL IDENTIFICATION OF THE CONTROLLING REGIME

A variety of model contactors can be used to discern the controlling mechanism. To determine whether diffusional factors are important, a standard, fully baffled, mechanically agitated contactor provided with a four- or six-blade turbine impeller is recommended. This contactor can be provided with a jacket and internal heating coil to operate at a specified temperature (Fig. 2). For laboratory purposes, a contactor internal diameter of 10–15 cm should suffice, and the turbine diameter can be taken as 40% of the diameter of the contactor. The contactor should preferably be of borosilicate glass to permit visual observation, but if glass is not suitable, plastics or any alloy steel may be used. The impeller and the shaft can be made from an appropriate plastic material or an alloy steel. In such a contactor, values of the effective interfacial area and overall mass-transfer coefficient vary approximately linearly with the speed of agitation over the speed range 500–2500 rpm [6–8]. If the conversion in phase B′ is less than 20% in the first 10 min, it is desirable to operate the contactor batchwise. With very high rates of reaction, it would be desirable to operate the contactor continuously, with the system in backmixed flow conditions at relatively high speeds of agitation.

In batchwise operation we can conveniently study the effect of the following variables on conversion to discern whether diffusional factors are important: (1) speed of agitation; (2) dispersed-phase holdup; (3) reversal of phases; and (4) use of an emulsifying agent. For studying the effect of speed of agitation, the A′ phase is dispersed and its holdup is set at about 10%. It is desirable to use pure A. If it is found that the speed of agitation has a nominal effect on the rate of reaction per unit volume of the B′ phase up to a speed of agitation of (say) 1000 rpm and thereafter has no effect up to a speed of agitation of (say) 2500 rpm, it is a clear indication that diffusional factors have been eliminated beyond a speed of agitation of 1000 rpm. A further check can be provided by varying the dispersed-phase holdup. It is known that in the range of holdup of 5–30% the specific interfacial area of the dispersion varies approximately linearly with the dispersed-phase holdup. Thus if at 1500–2000 rpm, there is no effect of the variation in holdup on the rate of reaction per unit volume of the B′ phase, there is a further proof that diffusional factors have been eliminated. Additional

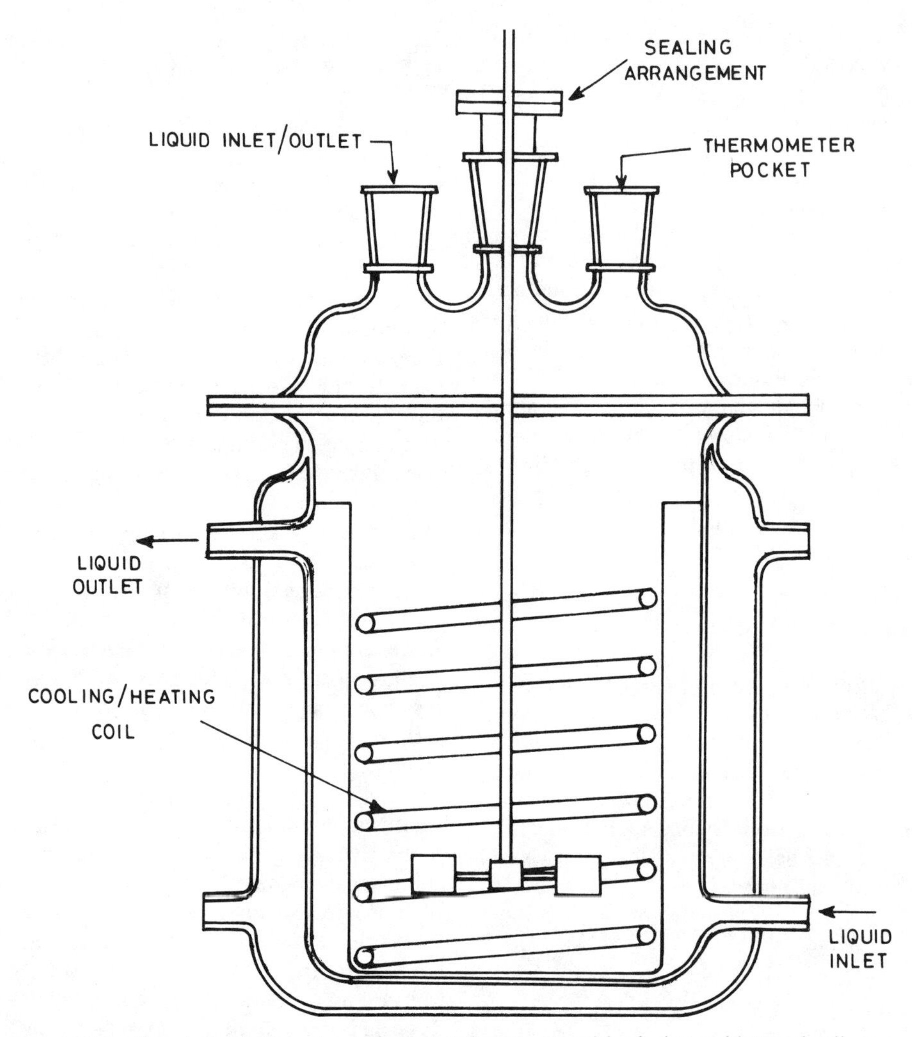

Figure 2. A standard mechanically agitated contactor with a jacket and internal coil.

proof can be provided by inverting the phases, that is, converting the dispersed phase into the continuous phase. It is known that the interfacial area per unit volume of dispersion changes drastically with this reversal of phases. We can also study the effect of an emulsifying agent that will increase the interfacial area enormously under otherwise uniform conditions. If there is no effect of emulsifying agent on the rate per unit volume of B′ phase under otherwise uniform conditions, diffusional factors would appear to be unimportant.

For the special case of reactions that are zero order in A, once diffusional factors are eliminated, the rate of reaction (under otherwise uniform conditions) will be independent of the concentration of A in the A′ phase.

This procedure has been used in studying a variety of reactions. Typical examples are (1) alkylation of aromatics with propylene, butylenes, and higher olefins [9, 10], (2) nitration of chlorobenzene [11], and (3) sulfonation of aromatic substances [12]. This procedure can also be conveniently adopted for the continuous mode of operation of the model contactor.

Once it is established that diffusional factors are unimportant, the kinetics of the two-phase reaction and even the effect of temperature can be studied. Thus scale-up data can be obtained with confidence, subject to the assurance that for the most demanding situation, the diffusional factors are unimportant. While studying the effect of the variation in the interfacial concentration [A*] in the B′ phase, it should be ensured that there is no resistance to mass transfer in the A′ phase. This can be checked by simply doubling the dispersed phase (A′ phase) holdup under a specified set of conditions, and if this change does not affect the rate of reaction per unit volume of the B′ phase, there is probably no resistance to mass transfer in the A′ phase.

For industrial purposes, a contactor other than the type shown in Fig. 2 may be chosen, or it may not be considered desirable to operate a mechanically agitated contactor at speed of agitation sufficient to eliminate diffusional resistance. In such cases it is necessary to assess the importance of diffusional factors on the basis of the knowledge of the kinetics of the reaction and values of overall mass-transfer coefficient and interfacial area.

If it is found that diffusional factors are important, as indicated by the pronounced effect of the speed of agitation on the rate of reaction per unit volume of the B′ phase, it is essential to use a model contactor with a known interfacial area for ascertaining whether we are in regime 2, 3, or 4.

3.1. Model Contactors with Known Interfacial Area

In principle, we can use model contactors with known interfacial area and known hydrodynamics, such as the wetted wall column, the wetted sphere column, and the laminar jet apparatus (see Chapter 3). However, these apparatuses must be operated continuously and are not very adaptable. It is far more convenient to use batch contactors of simpler design where the minimum condition of known interfacial area is satisfied. The oldest design is that proposed by Lewis, which is commonly referred to as a *Lewis stirred cell* (see Chapter 3, Fig. 10). Simpler designs using just beakers and a cruciform-type stirrer can also be used [13, 14]. Recently further modifications in the design of the Lewis type of cell have been suggested [15, 16]. Blakeley et al. [17] have suggested that vertically opposed, submerged jets, one in each phase, can be used to provide agitation in both the phases and yet a plane interface can be maintained. The simple design due to Lewis or the more sophisticated design due to Landau and Chin [16] is recommended.

In these contactors the values of the true mass-transfer coefficient k_L for the transfer of A into the B′ phase are known to be affected by the speed of stirring. Hence any effect of the speed of agitation beyond a certain minimum speed required to ensure good mixing in both the phases, on the specific rate of mass transfer R_A would indicate the controlling regime being 2 or 4; in the case of regime 3, R_A is independent of k_L (see Table 3a). The concentration of B has no effect on R_A in regime 2, but in the case of regime 4, R_A is proportional to $[B_0]$. Thus we can clearly distinguish between the different controlling regimes. An additional check can be provided by varying the concentration of A, using an inert diluent in the A′ phase. If there is no effect of the variation in mole fraction of A in the A′ phase on R_A, and provided there is no resistance to mass transfer in the A′ phase, regime 4 rather than regime 2 is concluded to be the pertinent case. However, the properties of the B′ phase should not change significantly when the concentration of species B is varied. If necessary, an inert substance may be deliberately used in the B′ phase to ensure that physical properties (e.g., viscosity) do not change significantly with the variation in the concentration of species B.

4. EXAMPLES

4.1. Alkaline Hydrolysis of Organic Compounds [13, 18–20]

Alkaline hydrolysis of a variety of organic compounds, such as esters of acetic acid, formic acid, and so on or benzyl chloride, nitrochlorobenzenes, alkyl chlorides/bromides, and so on, is of considerable industrial importance. Depending on the values of the rate constant and mass-transfer coefficient, we could be in regimes 1, 2, or 3 or intermediate regimes. Thus the alkaline hydrolysis of esters of acetic acid, benzyl chloride, and so on may well be in the kinetically controlled slow reaction situation of regime 1. On the other hand, hydrolysis of nitrochlorobenzenes may be in the diffusion controlled regime 2. For faster-reacting esters such as the higher esters of formic acid and chloracetic acid, the kinetics may conform to the fast reaction regime (regime 3). The following typical values illustrate the preceding points:

$$m = 1$$

$$n = 1$$

$$k_2 = 50, 1000, 40{,}000 \text{ ml mol}^{-1} \text{ s}^{-1}$$

$$[A^*] = 1 \times 10^{-5} \text{ mol/ml}$$

$$[B_0] = 2 \times 10^{-3} \text{ mol/ml}$$

$$k_L = 2 \times 10^{-3} \text{ cm/sec}$$

$$D_B = D_A$$

$$a = 30 \text{ cm}^2/\text{cm}^3$$

$$l = 0.8.$$

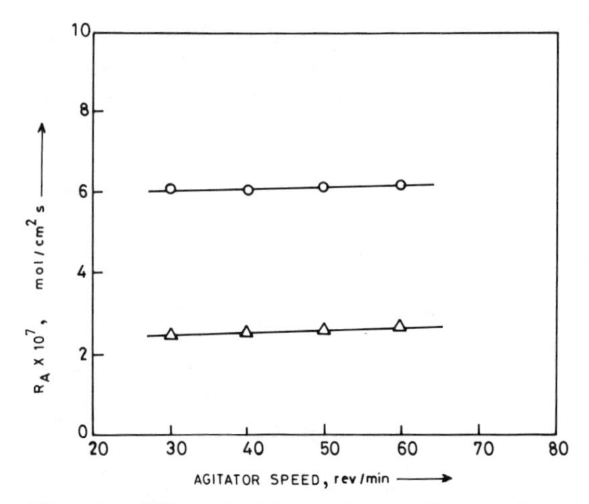

Figure 3. Effect of stirring on the specific rate of extraction of *n*-butyl formate in aqueous sodium hydroxide solutions in stirred cells: o–*n*-butyl formate–2.04 *M* NaOH, interfacial area 33 cm^2; Δ–*n*-butyl formate–4.05 *M* NaOH, interfacial area 47 cm^2. (Adopted from Nanda and Sharma [13].)

For the lowest value of k_2, we will be in regime 1; for the intermediate value of k_2, we will be in regime 2; and for the highest value of k_2, we will be in regime 3. (See regime conditions in Tables 3a and 3b).

Figure 3 shows a typical system in regime 3; there is no effect of speed of stirring on the specific rate of extraction of *n*-butyl formate into an aqueous solution of sodium hydroxide in a stirred cell [13].

4.2. Nitration of Aromatic Substances [11, 21-24]

This process has been studied for more than three decades, but only in the last decade have the engineering aspects of the reactions been clearly established. The nitration reaction involves two steps:

1. Formation of nitronium ions

$$HONO_2 + H^+ \rightleftharpoons H_2O + NO_2^+$$

2. Reaction between dissolved aromatic substance, ArH, and nitronium ions

$$ArH + NO_2^+ \rightleftharpoons ArNO_2 + H^+$$

In most cases of practical interest the second step is rate controlling. It is also believed that practically all the reaction occurs in the aqueous phase. Depending on the nature of the aromatic substance and the concentrations of sulfuric acid and nitric acid, we could be in regime 1, 2, or 3 (or intermediate regimes). For substances such as chloro- or dichlorobenzenes, nitrobenzene, etc., we may be in kinetically controlled regime 1, whereas for substances such as benzene or toluene, we may be in regimes 2 or 3.

A unique feature in the nitration of toluene has been brought out in the light of the two-step mechanism given in the preceding paragraph [22]. At higher concentrations of toluene, the reaction may well be controlled by the first step (formation of nitronium ions), and intrinsically the reaction will be zero order in toluene; at lower concentrations of toluene, with toluene diluted in an inert material or nitrotoluenes, the reaction may be controlled by the second step (reaction between aromatic substance and nitronium ions) and will be intrinsically first order in toluene. Thus, if we are in the fast-reaction regime (regime 3), at higher concentrations of toluene, the specific rate of reaction will be proportional to $[A^*]^{0.5}$.

A recent compilation of papers [23] on industrial and laboratory nitration covers various aspects of two-phase nitrations.

4.3. Removal of COS From Liquefied Petroleum (C_3/C_4) Fractions and Reactions of CS_2 with Amines

4.3.1. Removal of COS

Dissolved COS is usually removed from liquefied petroleum (C_3/C_4) fractions by treatment with aqueous solutions of sodium hydroxide or alkanolamines. In the case of alkanolamines, the rate constant varies over a wide range depending on the basicity and structure of the amine. For instance, the ratio of second-order rate constants for ethylaminoethanol and diethanolamine is 25, and in regime 3 a changeover to a faster reacting amine will result in increasing the value of R_A by a factor of 5 as the values of $[A^*]$ and D_A are comparable in the two cases. Thus the contactor volume can be reduced by a factor of 5 [25].

4.3.2. Reaction Between CS_2 and Amines

The reaction between CS_2 and amines such as methylamine, dimethylamine, and cyclohexylamine is of industrial importance for the manufacture of pesticides and rubber chemicals [26]. Depending on circumstances, the system may fall in regime 3 or in between regimes 2 and 3.

4.4. Alkylation of Organic Compounds with Olefins

Aklylation of benzene (or toluene) with a variety of olefins in the presence of different acidic catalysts (e.g., H_2SO_4) is of considerable industrial importance. It appears that with sulfuric acid in a mechanically agitated contactor, the reaction between benzene/toluene and olefins such as propylene and butenes falls in the kinetically controlled regime [9, 10, 27]. This is illustrated in Fig. 4, which shows data from a 9-cm-inner-diameter laboratory contactor for benzene and 1- or 2-butene.

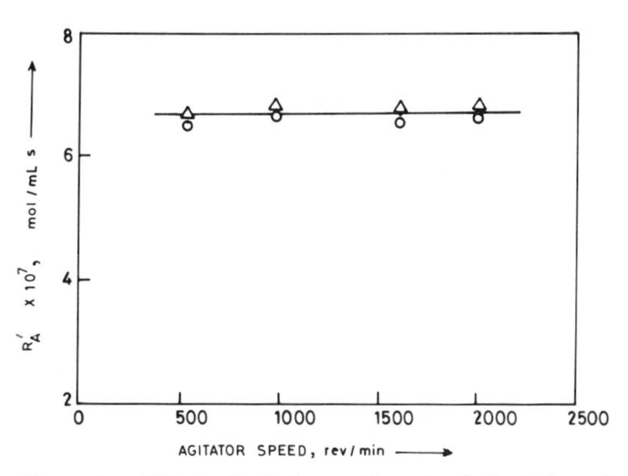

Figure 4. Effect of stirring on the rate of alkylation of benzene with 1-butene and 2-butene in a 9-cm-i.d. fully baffled mechanically agitated contactor: ○–1-butene-benzene-aqueous H_2SO_4 (16.8 mol/liter); Δ–2-butene-benzene-aqueous H_2SO_4 (16.8 mol/liter). (Adopted from Tiwari and Sharma [9].)

The alkylation of isobutane with butenes and isobutylene is of great practical value for the production of high-octane motor fuel. Here H_2SO_4, HF, and so on are used as catalysts. With H_2SO_4 as a catalyst, the system behaves according to regime 3 [28].

4.5. Reduction of Aromatic Nitro Compounds to Corresponding Aromatic Amines with Aqueous Na_2S/Na_2S_x

Bhave and Sharma [29] have studied the reduction of a variety of aromatic nitrocompounds, such as nitrotoluenes, *m*-nitrochlorobenzene, *m*-dinitrobenzene, and *p*-nitroaniline, with aqueous Na_2S/Na_2S_2 at 60°C in a 10-cm-i.d. mechanically agitated contactor. The effect of speed of agitation, dispersed-phase holdup, reversal of phases, and so on on the rate of reduction per unit volume of the aqueous phase was studied. Figure 5 shows the effect of the speed of agitation and dispersed-phase fraction on the rate of reduction for the *m*-nitrochlorobenzene–aqueous Na_2S system. There was no effect of the speed of agitation beyond 600 rpm, and the dispersed (organic) phase holdup in the range of 0.05–0.25 on the rate of reduction per unit volume of the

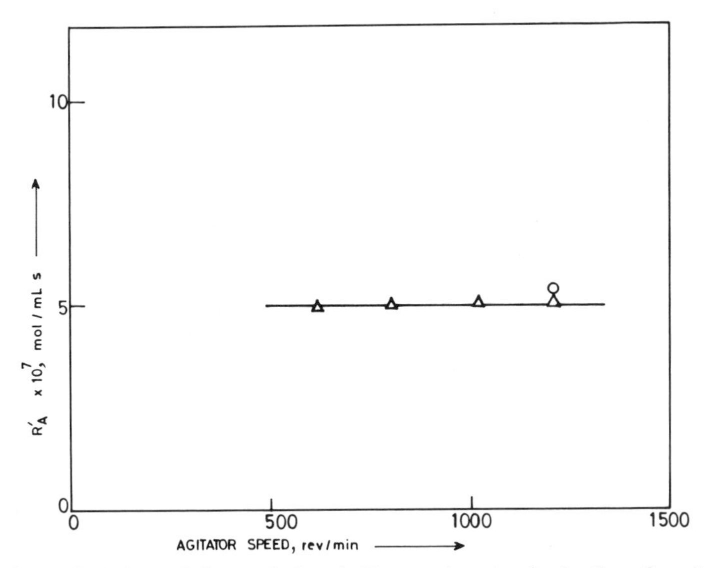

Figure 5. Effect of stirring and dispersed-phase holdup on the rate of reduction of *m*-nitrochlorobenzene with aqueous sodium sulfide in a 10-cm-i.d., fully baffled, mechanically agitated contactor. Temperature, 60°C; ionic strength, 3 g ion/liter; $[B_0]_{av}$, 0.155 mol/liter; continuous phase–aqueous Na_2S; dispersed phase–*m*-nitrochlorobenzene, fractional holdup of dispersed phase ○–0.25; Δ–0.05. ([Adopted from Bhave and Sharma [29].)

aqueous phase. Furthermore, there was no effect of the reversal of phases on the rate of reduction per unit volume of the aqueous phase. From the foregoing it is clear that reaction conforms to the very slow reaction regime 1 and that diffusional factors are unimportant.

4.6. Solvent Extraction of Metals

The solvent extraction of metal salts from aqueous solutions has acquired great commercial importance, particularly for the recovery of copper, nickel, cobalt, uranium, vanadium, and other metals from their lean ores. The reactions involved are mainly reversible, but some of the principles described in this chapter are applicable. More details are given in Chapter 2.2.

5. MEASUREMENT OF INTERFACIAL AREA AND TRUE MASS-TRANSFER COEFFICIENT [3, 30]

It is necessary to know the values of a and k_L separately, to discern the controlling mechanism. In regime 3 we need a knowledge of a to design large reactors.

The interfacial area a may be obtained from drop size data (Chapter 4), or the theory of extraction with fast chemical reaction (regime 3) can be adopted for the measurement of effective interfacial area. Nanda and Sharma [13] had first introduced this method, and it was suggested that alkaline hydrolysis of formic acid esters (n-butyl formate, n- or isoamyl formate, etc.) could be used. This method, usually referred to as the *chemical method of measuring effective interfacial area*, is now widely used and some aspects have been discussed by Sharma and Danckwerts [30]. A number of new industrially important systems have been suggested that offer some special advantages. For instance, the extraction of tertiary olefins, such as isoamylene, from inert organic solvents by aqueous solutions of sulfuric acid can be employed and the extent of reaction can be followed by the analysis of the organic phase. Extraction of pinenes from an inert organic solvent into 30–45% H_2SO_4 can also be employed for the measurement of a [8]. Figure 6 shows values of a obtained by using the β-pinene–aqueous H_2SO_4 system, as a function of the speed of stirring in mechanically agitated contactors. The theory of desorption preceded by a fast pseudo-mth-order reaction [31] can also be employed in the measurement of a in liquid–liquid systems. This has been done by extracting isoamylene (0.1–0.2 mol fraction in sulfuric acid) into an inert solvent such as n-heptane [32].

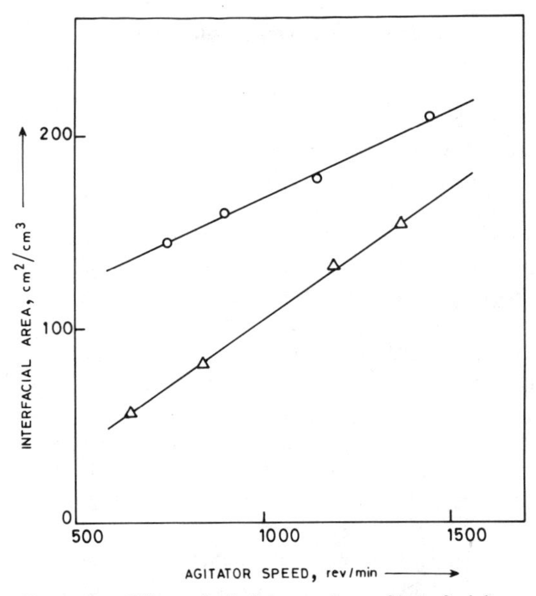

Figure 6. Effect of stirring on values of interfacial area in fully baffled mechanically agitated contactors. Systems–β-pinene–aqueous H_2SO_4; continuous phase–42% w/w aqueous H_2SO_4; dispersed phase–0.0975 mole fraction β-pinene in monochlorobenzene.

	Contactor i.d., cm	Fractional Holdup of Dispersed Phase
△	9.8	0.25
○	16.5	0.192

(Adopted from Laddha and Sharma [8].)

The technique of interfacial polycondensation can also be used for measurement of a (see Section 6 of this chapter).

The theory of regime 2 can be employed for the measurement of k_La. A convenient system is alkaline hydrolysis of esters of acetic acid (n-butyl, n- or isoamyl, etc.). The theory of regime 4 can also be used for measurement of k_La. A convenient system in this case is extraction of pure acid such as n-heptanoic acid or of 2-ethylhexoic acid into aqueous sodium hydroxide solutions.

For the regime between 2 and 3, we can make plots $[R_Aa/[A^*]]^2$ against $[B_0]^n$ and obtain values of a and k_L from the slope and intercept (see Table 3b). Such plots are sometimes referred to as *Danckwerts' plots;* it must be ensured that the physical properties of the B′ phase do not change significantly when the concentration of species B is varied. Extraction of formic acid esters into sodium hydroxide solutions can be utilized for making a Danckwerts' plot. Under certain conditions the nitration of toluene also lends itself to measurement of a from regime 3 and k_L and a from the Danckwerts' plot.

Simultaneous measurement of a employing the theory of regime 3 and at the same time the physical mass-transfer coefficient of an inert substance into the B′ phase (or mass transfer of an inert substance dissolved in the B′ phase into the A′ phase) will allow the calculation of true k_L. It must be ensured that for physical mass-transfer, the resistance is positively in the B′ phase.

Typical values of $k_L a$ and a in different liquid-liquid systems have been reported [3, 30].

6. REACTIONS IN ORGANIC PHASE, INCLUDING INTERFACIAL POLYCONDENSATION

6.1. Reactions in Organic Phase

It is often tacitly assumed that the reaction between a sparingly soluble organic compound and component B (dissolved in an aqueous phase) takes place in the aqueous phase. Although this is generally true, there are some cases where the reaction occurs exclusively in the organic phase. A classic example is that of hydrolysis of triglycerides of C_{12}, C_{14}, C_{16}, and C_{18} acids (natural fats). Here under conditions of industrial importance the solubility of water in the organic phase is very high compared to that of fat in water and the reaction occurs in the organic phase. Another example is that of nitration of higher olefins (e.g., 1-dodecene) with aqueous nitric acid. Here it was found that the rates of nitration of 1-dodecene and 1-octadecene, under otherwise identical conditions, were practically the same despite very large differences in the value of [A*] for the preceding olefins in the aqueous phase. This and some other arguments give a clear indication that the reaction occurs in the organic phase [33].

6.1.1. *Phase-Transfer Catalysis*

Many two-phase reactions involving an aqueous phase are inconveniently slow. Attempts have been made to catalyze these reactions by using quaternary ammonium and phosphonium compounds. These catalysts are referred to as *phase-transfer catalysts* as they allow the reactive ions in the aqueous phase to be transferred to the organic phase, where the reaction occurs. For example, the conversion of 1-chlorooctane to 1-cyanooctane with aqueous sodium cyanide [34] does not occur significantly even after 2 weeks under refluxing conditions. However, the use of a phase-transfer catalyst such as 1.3 mol% of tributyl (hexadecyl) phosphonium bromide allows the reaction to occur to an extent of 99% in less than 2 hr. This is truly remarkable, and as the name suggests, we are able to take the cyanide ions to the organic phase.

It is known that the solubility of ionic materials in organic solvent increases rapidly when a large "onium" cation is used in place of the alkali metal cation. In the preceding case different alkyl chlorides, with vastly different solubilities in the aqueous phase, gave practically the same rate of reaction, indicating that the locale of the reaction is the organic phase. It was established by Starks and Owens [34] that diffusional factors were unimportant.

The catalysts employed do not operate through an increase in effective interfacial area, and the reaction occurs in the organic phase, and not at the interface or in the micelles. It has also been reported that the reaction in the organic phase is first order with respect to the concentration of phase-transfer catalyst. A schematic representation of the reaction scheme is as follows:

$$NaCN + Q^+Cl^- \rightleftharpoons Q^+CN^- + NaCl \qquad \text{Aqueous phase}$$

$$\text{-----}\ \upharpoonleft\downharpoonright\ \text{-----}\ \upharpoonleft\downharpoonright\ \text{---------} \qquad \text{Interface}$$

$$RCN + Q^+Cl^- \rightleftharpoons Q^+CN^- + RCl \qquad \text{Organic phase}$$

where Q^+ represents the quaternary ammonium, phosphonium cation, or a crown ether–alkali metal cation complex.

The phase-transfer catalysis is potentially very attractive for conducting a variety of reactions of industrial importance [35]. In many cases of single-phase reactions, which require scrupulously anhydrous conditions, it may be attractive to employ a two-phase reaction system with phase-transfer catalysis. Recently, the use of solid phase-transfer catalysts has been suggested [36]. For the preceding case of the conversion of 1-chlorooctane to 1-cyanooctane, a polystyrene resin, cross-linked with 2% divinylbenzene, bearing quaternary ammonium or phosphonium groups or crownethers, has been found to be effective. This type of catalyst system is referred to as *triphase catalysis* and appears to be very attractive.

6.2. Interfacial Polycondensation

This is based on the reaction between diacid halides (usually chloride), dissolved in an organic solvent insoluble in water, such as xylenes and CCl_4; and diamines or diols, dissolved in water. Alkali is added to neutralize the liberated acid. This reaction can be conducted under ambient conditions and conforms to regime 4, the case of extraction accompanied by instantaneous reaction. A film of polymer is formed on the organic phase droplets dispersed in water, and thus droplets are encapsulated and interfacial area measurements can be made. This reaction occurs in the organic phase in view of a relatively high distribution coefficient of (say) diamines in the organic phase compared to that of diacyl chlorides in the aqueous phase.

7. REACTION IN BOTH PHASES

In principle, if there is finite solubility of B in the A′ phase, we can have a situation where the reaction occurs in both the phases. Mathematical analyses of this problem, where there are different controlling regimes, have been made by Rod [37] and Mhaskar and Sharma [38]. The occurrence of the reaction in the A′ phase may have a beneficial effect on the yield of the reactor provided no undesirable by-products are formed. It is conceivable that an undesirable reaction may occur in one of the phases, and this should be avoided if possible. In the case of an instantaneous reaction (regime 4), conditions have been suggested that will allow the reaction to occur in the desired phase, and thus improved yields may be realized.

An industrial example of occurrence of reaction in both the phases is the oximation of cyclohexanone, an important step in the manufacture of caprolactam (monomer for Nylon-6) [39]. Industrial reactor design has been approached from first principles in this case.

8. EXTRACTION ACCOMPANIED BY COMPLEX REACTIONS

In the field of gas absorption with reaction, many complex reaction sequences have been studied theoretically and experimentally. Typical cases are (1) consecutive reactions or consecutive/parallel reactions, (2) one gas and two reactants, (3) simultaneous absorption with reaction of two gases (reaction with species B or reaction between dissolved gases), and so on [3]. These analyses can easily be extended to liquid–liquid reactions. In the case of consecutive reactions, theory can be used to determine whether diffusional factors affect the selectivity. Similarly, with two solutes undergoing a reaction with dissolved B, we can find not only the specific rates of mass transfer for both solutes, but also the conditions that will favor selectivity with respect to the desired solute.

Data from such complex systems should be analyzed with care. For instance, competitive rates of nitration are often considered for finding the rate data for the unknown substance. Such a study would give the desired rate constant provided *both* the solutes react and conform to either regime 1 or regime 3. In other cases we will not obtain the rate constant for the unknown substance, as shown by Hanson and Ismail [40] for nitration of a mixture of benzene and toluene.

9. SEPARATION OF CLOSE BOILING ACIDIC AND BASIC MIXTURES BY DISSOCIATION EXTRACTION

Dissociation extraction is concerned with separation of mixtures of organic acidic–basic compounds that are difficult to separate by common methods of separation such as distillation or crystallization. Typical examples of such mixures are *m*-cresol-*p*-cresol-2,6-xylenol, *o*-cresol-6-chloro-*o*-cresol, 2,4-xylidine-2,6-xylidine, 3-picoline-4-picoline, and *N*-alkylanilines. The technique of dissociation extraction exploits the difference between the dissociation constants, as well as the difference between the distribution coefficients of the components of the mixture in a two-phase system. This technique is discussed in more detail in Section 3.1 of Chapter 21, where a theoretical expression [41] for the binary separation factor is given. Recently, the theory has been extended to multicomponent systems [42, 43].

Example: Separation of *m*-Cresol and 2,6-Xylenol

m-Cresol and 2,6-xylenol have respective boiling points (at 760 mm Hg) of 202 and 203°C. Their dissociation constants are, however, 7.94×10^{-11} and 2.398×10^{-11} g · ion/liter, respectively, and their separation by dissociation extraction can be attempted. The distribution coefficient of *m*-cresol between benzene and water is 11.2, and that of 2,6-xylenol is 73.0. Consider a case where single-stage dissociation extraction with 1 *N* NaOH gives an organic phase whose total molarity is 0.5 g · mol/liter and composition is 80% 2,6-xylenol and 20% *m*-cresol. (These are the conditions of the equilibriated organic phase.) Knowing these data, we can predict the separation factor from the relationship given in Section 3.1 of Chapter 21, and this works out to be 21.07. The experimental value [44] is 22.1. Thus the value of the separation factor is extremely good, and the technique of dissociation extraction appears to be attractive for the separation of *m*-cresol and 2,6-xylenol.

In conventional dissociation extraction, the consumption of the extractant alkali or acid makes the process economically less attractive. To overcome this difficulty, Anwar et al. [45] have developed a modified dissociation extraction in which instead of strong acid or alkali a weakly acidic extractant is used. In the case of some very sparingly soluble substances, it may be advantageous to carry out dissociation extraction without a solvent [46]. Further details are given in Chapter 21.

10. CONCLUSIONS

In many respects the basic features of extraction with reaction are akin to those for gas absorption with chemical reaction. The latter area has received considerable attention. However, there are special features in liquid–liquid systems regarding procedures for discerning the controlling mechanism and the possibilities of reaction occurring in the A′ phase or both the phases. In most cases of practical importance a systematic approach can probably be made to discern the controlling mechanism and arrive at procedures for the design of large-scale reactors.

NOTATION

A	solute
A′	phase containing A
$[A_0]$	concentration of dissolved A in bulk B′ phase, mol/ml
$[A^*]$	concentration of A at liquid–liquid interface (interfacial concentration of A), mol/ml
a	interfacial area per unit volume of dispersion, cm^2/cm^3
B	dissolved reactive species present in the B′ phase
B′	phase containing B
$[B_i]$	concentration of species B at interface, mol/ml
$[B_0]$	concentration of dissolved B in bulk B′ phase, mol/ml
D	diffusion coefficient of dissolved species indicated in subscript, cm^2/sec
k_L	true mass-transfer coefficient of A with respect to phase B′, cm/sec (defined by $R_A = k_L([A^*] - [A_0])$)
$k_{LA(A'\,phase)}$	true mass-transfer coefficient of A in A′ phase, cm/sec
k_{LR}	mass-transfer coefficient with chemical reaction, cm/sec (defined by $R_A = k_{LR}[A^*]$)
K_0	overall mass-transfer coefficient, cm/sec
k_{mn}	rate constant for a general-order reaction, $(ml/mol)^{m+n-1}$ sec^{-1}
l	fraction holdup of B′ phase in contactor
$\sqrt{M}$	$\dfrac{\left\{\dfrac{2}{m+1} D_A k_{mn} [A^*]^{m-1} [B_0]^n\right\}^{1/2}}{k_L}$ for a general-order reaction
m	order of reaction with respect to A
m_A	distribution coefficient for A in B′ phase
n	order of reaction with respect to B
R_A	specific rate of extraction of species A, mol cm^{-2} s^{-1}
R'_A	rate of homogeneous reaction per unit volume of B′ phase, mol s^{-1} ml^{-1} s^{-1}
Z	stoichiometric coefficient of reaction, that is, number of moles of B reactive with 1 mol of A

Greek Symbols

δ	thickness of diffusion film in B′ phase, cm
ϕ	enhancement factor
ϕ_a	asymptotic value of the enhancement factor

Subscripts

a	asymptotic value

REFERENCES

1. M. M. Sharma, *Chem. Age India* **17**, 445 (1966).
2. C. Hanson, *Recent Advances in Liquid-Liquid Extraction*, 1st ed., Pergamon Press, Oxford, 1971, p. 429.
3. L. K. Doraiswamy and M. M. Sharma, *Heterogeneous Reactions: Analysis, Examples and Reactor Design*, Wiley, New York, in press.
4. P. V. Danckwerts, *Gas-Liquid Reactions*, McGraw-Hill, London, 1970.
5. R. M. Secor and J. A. Beutler, *J. AIChE* **13**, 365 (1967).
6. J. B. Fernandes and M. M. Sharma, *Chem. Eng. Sci.* **22**, 1267 (1967).
7. D. S. Sankholkar and M. M. Sharma, *Chem. Eng. Sci.* **28**, 2089 (1973).
8. S. S. Laddha and M. M. Sharma, *Chem. Eng. Sci.* **31**, 843 (1976).
9. R. K. Tiwari and M. M. Sharma, *Chem. Eng. Sci.* **32**, 1253 (1977).
10. I. Komasawa, T. Inoue, and T. Otake, *J. Chem. Eng. Jap.* **5**, 34 (1972).
11. P. R. Cox and A. N. Strachan, *Chem. Eng. Sci.* **27**, 457 (1972).
12. M. Sohrabi, T. Kaghazchi, and C. Hanson, *J. Appl. Chem. Biotechnol.* **27**, 453 (1977).
13. A. K. Nanda and M. M. Sharma, *Chem. Eng. Sci.* **21**, 707 (1966).
14. M. M. Sharma and A. K. Nanda, *Transact. Inst. Chem. Eng.* **46**, T44 (1968).
15. J. Bulicka and J. Prochazka, *Chem. Eng. Sci.* **31**, 137 (1976).
16. J. Landau and M. Chin, *Can. J. Chem. Eng.* **55**, 161 (1977).
17. D. Blakeley, J. T. Davies, W. J. McManamey, and S. K. S. Multani, *Chem. Eng. Sci.* **32**, 1457 (1977).
18. A. K. Nanda and M. M. Sharma, *Chem. Eng. Sci.* **22**, 769 (1967).
19. R. C. Sharma and M. M. Sharma, *J. Appl. Chem. Biotechnol.* **19**, 162 (1969).
20. R. C. Sharma and M. M. Sharma, *Bull. Chem. Soc. Jap.* **43**, 642 (1970).
21. L. F. Albright and C. Hanson, *Loss Prevention* (CEP Technical Manual) **3**, 26 (1969).
22. J. W. Chapman and A. N. Strachan, *J. Chem. Soc. Chem. Commun.* 293 (1974).
23. L. F. Albright and C. Hanson, Eds. *Industrial and Laboratory Nitration*, Vol. 22, American Chemical Society Symposium Series, 1975.
24. C. Hanson and H. A. M. Ismail, *J. Appl. Chem. Biotechnol.* **26**, 111 (1976).
25. M. M. Sharma, Kinetics of Gas Absorption, Ph.D. thesis, University of Cambridge, 1964; *Transact. Faraday Soc.* **61**, 681 (1965).
26. P. J. Kothari and M. M. Sharma, *Chem. Eng. Sci.* **21**, 391 (1966).
27. J. A. Richardson and H. F. Rase, *Ind. Eng. Chem. Prod. Res. Devel.* **17**, 287 (1978).
28. L. M. Lee and P. Harriot, *Ind. Eng. Chem. Prod. Res. Devel.* **16**, 282 (1977).
29. R. R. Bhave and M. M. Sharma, *J. Chem. Technol. Biotechnol.* **31**, 93 (1981).
30. M. M. Sharma and P. V. Danckwerts, *Brt. Chem. Eng.* **15**, 522 (1970).
31. Y. T. Shah and M. M. Sharma, *Transact. Inst. Chem. Eng.* **54**, 1 (1976).
32. D. S. Sankholkar and M. M. Sharma, *Chem. Eng. Sci.* **30**, 729 (1975).
33. D. P. Gregory, R. J. Martens, C. E. Stubbs, and J. D. Wagner, *J. Appl. Chem. Biotechnol.* **26**, 623 (1976).
34. C. M. Starks and R. M. Owens, *J. Am. Chem. Soc.* **95**, 3613 (1973).
35. E. V. Dehmlow, *Angew. Chem. Internatl. Ed.* **13**, 170 (1974).
36. S. L. Regen, *Angew. Chem. Internatl. Ed.* **18**, 421 (1979).
37. V. Rod, *Chem. Eng. J.* **7**, 137 (1974).
38. R. D. Mhaskar and M. M. Sharma, *Chem. Eng. Sci.* **30**, 811 (1975).
39. V. Rod, Chemical Reaction Engineering, in *Proceedings of the 4th International/6th European Symposium*, Heidelberg, West Germany, 1976, p. 275.
40. C. Hanson and H. A. M. Ismail, *Chem. Eng. Sci.* **32**, 775 (1977).
41. M. M. Anwar, C. Hanson, and M. W. T. Pratt, *Proceedings of International Solvent Extraction Conference* (Soc. Chem. Ind., London), 1971, p. 119.
42. V. V. Kafarov, V. G. Vygon, A. I. Chulok, and V. A. Kostyuk, *Russ. J. Phys. Chem.* **50**, 1618 (1976).
43. V. V. Wadekar and M. M. Sharma, *J. Sep. Proc. Technol.* **2**, 28 (1981).
44. V. V. Wadekar and M. M. Sharma, *J. Chem. Technol. Biotechnol.* **31**, 279 (1981).
45. M. M. Anwar, C. Hanson, and M. W. T. Pratt, *Transact. Inst. Eng.* **49**, 93 (1971).
46. S. S. Laddha and M. M. Sharma, *J. Appl. Chem. Biotechnol.* **28**, 69 (1978).

2.2

METAL EXTRACTANT CHEMISTRY

Michael Cox
Hatfield Polytechnic
United Kingdom

Douglas S. Flett
Warren Spring Laboratory
United Kingdom

1. INTRODUCTION

The solvent extraction of metals has become increasingly important in connection with hydrometallurgical extraction processes (see Chapter 25). The heavy organic chemical extraction processes generally rely on irreversible chemical reactions; however, the hydrometallurgical and allied solvent extraction processes involve equilibrium chemical reactions. The former processes are treated in Chapter 2.1, whereas this chapter considers the basic chemistry associated with solvent extraction through reversible chemical reactions.

The main areas of application of solvent extraction with reversible chemical reaction are hydrometallurgy and analytical chemistry. Cations and anions are extracted from an aqueous phase into an organic phase, and because such extraction is energetically very unfavorable, it is necessary to neutralize the ionic charge prior to extraction. This neutralization is achieved either by forming organic-soluble neutral complexes in the aqueous phase between cations and anions or by direct reaction between the ionic species of interest and an appropriate organic compound to form a neutral species soluble in the organic phase.

In hydrometallurgical applications there is

TABLE 1 CLASSIFICATION OF EXTRACTION REAGENTS

Extractant Type	What They Do	What They Are
Organic acids, such as carboxylic, sulfonic, phosphoric, phosphonic, phosphinic acids; and acidic chelating agents	Extraction by compound formation	Cation extractants
Polyphenylmetalloid type, polyalkylsulfonium type, polyalkylammonium type, and salts of high-molecular-weight aliphatic amines	Extraction by ion-pair formation	Anion exchangers
Carbon-, sulfur-, or phosphorus-bonded oxygen-bearing extractants; alkylsulfides; and so on	Extraction by solvation	Solvating agents

usually a low solubility of the organic extraction reagent in the aqueous phase, a very low solubility of the aqueous ions in the organic phase, and a high solubility of the extractant–ion complex in the organic phase. Thus the extent of extraction achieved under given reaction conditions can be directly described by a *heterogeneous chemical reaction equilibrium*. This overall expression is made up of a variety of interactive equilibria relating bulk-phase concentrations of the reactants on either side of the aqueous/organic interface to the final equilibrium concentrations of the reaction products. In analytical chemical applications, however, the situation is more relaxed, permitting the use of reagents with quite appreciable water solubility.

These factors have profound consequences when the rate of mass transfer is considered. Thus solvent extraction with chemical reaction can be taken to mean either (1) bulk chemical reaction in one phase accompanied by mass transfer of reactant(s) to or from another phase or (2) a transfer process involving a chemical change at the interface. These extreme conditions are related by the concept of a reaction zone that can extend into both phases [1]. Mass transfer may occur either under *chemical kinetic* control or *diffusional* control, and the actual location of the chemical reaction, that is, in the aqueous- or organic-phase reaction zones or at the interface itself, is of considerable importance. In this context the phrase "at the interface" denotes a region within a few molecule diameters of the interface. This region is several orders of magnitude smaller than the diffusional boundary layer.

Extractants almost by definition must be interfacially active, and without exception the interfacial tension is lowered as extractant concentrations increase. Thus the interfacial physical chemistry of these systems is important in respect to the mechanism of extraction, particularly where interfacial chemical processes occur. These factors are discussed in detail in this chapter. There are several classifications of extraction reagents available depending on whether reagents are listed by what they do or what they are (Table 1). Over the years, many papers have appeared on the equilibrium chemistry of extraction in the areas of cation exchange, anion exchange, and extraction by solvation. This chapter cannot review this subject in full, but it highlights the salient features and governing principles as they are understood at present.

2. EQUILIBRIUM EXTRACTION CHEMISTRY

Chapter 1 of this handbook discussed the physical chemistry of partition of a solute between two immiscible or partially miscible phases. The concepts and definitions of the distribution coefficient, the partition coefficient, and so on, will already be familiar. Here, however, the situation is more complicated as the partition of a solute between two phases is accompanied by the reaction between ions or neutral complexes and organic compounds usually dissolved in an organic solvent (*diluent*) immiscible in the aqueous phase. These organic compounds, called *reagents* or *extractants*, can obviously partition between the organic and aqueous phases.

This partition will be influenced by the physical chemistry of the extractant in the organic phase, which in turn will relate to the nature and effect of the diluent and also to the state of the reagent in the aqueous phase. The behavior of the extractant in both organic and aqueous phases will thus affect the chemistry of metal extraction. For a full description of solvent extraction involving equilibrium chemical reactions, therefore, the nature of the extractant species in the organic phase and their partition behavior must be discussed prior to any consideration of the chemistry of metal extraction. The extractants are dealt with in three sections under the headings given in Table 1.

2.1. Acids and Chelating Acidic Extractants

2.1.1. Properties of Extractants Alone

For a simple monobasic acid extractant RH dissolved in an aqueous-immiscible diluent, partitioning occurs according to the following equilibria:

$$\overline{RH} \underset{}{\overset{p_{RH}}{\rightleftharpoons}} RH \underset{}{\overset{K_D}{\rightleftharpoons}} R^- + H^+ \quad (1)$$

where the overbar denotes species in the organic phase (see Notation at end of this chapter). From Eq. (1) it can be shown that

$$\log[\overline{RH}] = \log(p_{RH}K_D^{-1}) + \log[R^-][H^+] \quad (2)$$

thus a plot of $\log[\overline{RH}]$ against $\log[R^-][H^+]$ will yield a straight line of unit slope with an intercept equal to $\log(p_{RH}K_D^{-1})$.

The acid distribution coefficient D is given by

$$D = \frac{[\overline{RH}]}{[RH] + [R^-]} \quad (3)$$

and

$$\log D = \log(p_{RH}) - \log(1 + K_D[H^+]^{-1}) \quad (4)$$

A plot of $\log D$ against pH will thus give a smooth curve that is of the normalized form $\log y = \log(x + 1)$. The horizontal asymptote $\log y = 0$ yields $\log p_{RH}$, whereas the point of intersection of this asymptote with the asymptote $\log y = \log x$ gives pK_D = pH. This approach is demonstrated for the partition of diisopropyl salicylic acid between benzene and 1 *M* potassium nitrate aqueous solution, in Figs. 1 and 2, where data representing Eqs. (2) and (4), respectively, are plotted. The figures yield the values of 5.39 and 1.51 for pK_D and $\log p_{RH}$, respectively.

For a simple system, therefore, the distribution of the extractant will depend on the value of the thermodynamic partition constant, the *pK* value of the acidic group, and the aqueous pH. The partition constant will depend on the relative solubilities of the extractant in the organic and aqueous phases, and in general p_{RH} will increase with increasing molecular weight in any homologous series and will also depend on the nature of the diluent. This simple picture can be complicated by (1) polymerization or aggregation of the extractant in the organic phase, (2) direct interaction or adduct formation between the extractant and the diluent, (3) formation of acid hydrates soluble in the aqueous phase, (4) amphoteric behavior of the extractant, mostly found for chelating extractants such as 8-quinolinol, and (5) polymerization or aggregation of the extractant and the ionized extractant anion in the aqueous phase.

For carboxylic acids, the major complications are dimerization of the acids in the organic phase and the presence of singly charged anion molecule dimers, as well as doubly charged dimeric species in the aqueous phase. The mathematical treatment of such effects on the distribution of carboxylic acids has been fully discussed by Flett and Jaycock [2], as has the aqueous and organic phase chemistry of importance in solvent extraction systems. Quantitative information on distribution and dimerization constants of carboxylic acids is given by Sekine and Hasegawa [3].

The corresponding behavior of acidic organophosphorus compounds has been reviewed by Marcus and Kertes [4] up to about 1967. Here again a number of equilibria relating to polymerization, hydrate formation, and so on, have to be assumed in the distribution of these acids between the organic and aqueous phases. The degree of polymerization again depends on the nature of the diluent, and it is found that the monoalkyl dihydrogen phosphoric acids $ROPO(OH)_2$ are more highly polymerized than the dialkyl monohydrogen phosphoric acids. The acids are also amphoteric, extracting mineral acids under certain conditions, and thus the equilibria associated with this behavior have to be included. Hydrogen-bonded interactions between the diluent and other donor compounds such as tri-*n*-butylphosphate and alkylamines [5] are also possible. Hydrogen-bonding interactions are,

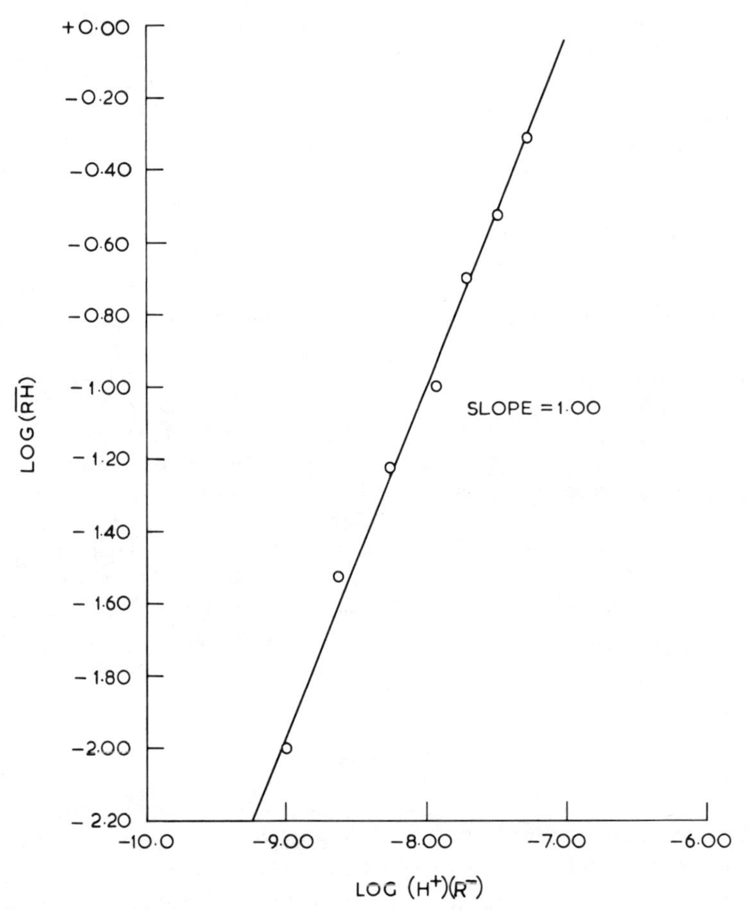

Figure 1. Acidity in aqueous phase (1 *M* KNO_3) as a function of the concentration of 3,5-diisopropyl salicylic acid in benzene.

in fact, important in all these systems and most of the interactions are likely to be competitive. The combination of extractant with a neutral donor or another extractant can be of considerable significance in metal ion extraction and can give rise to the phenomena of *synergism* and *antagonism*, which are discussed later. Data on distribution, dissociation, dimerization constants, and other parameters are available from several sources [3, 4, 6].

Sulfonic acids are even more highly aggregated than the alkylphosphoric acids and tend to form micelles even at very low organic-phase concentrations. The presence of micelles thus dominates the whole extraction chemistry of sulfonic acids. The solvent extraction behavior of these acids has been reviewed [4, 7], and a most useful up-to-date bibliography has been published by King Industries Inc. [8]. The physical chemistry of aggregation and micellization of carboxylic acids, sulfonic acids, and their salts in organic solvents has also been reviewed recently [9]. Great care must be taken when examining the literature for values of dimerization constants, critical micelle concentrations (CMC), and similar parameters. Anhydrous organic solution data must be treated cautiously, for it is certain that in a solvent extraction situation the presence of water will produce hydrates and, through competitive hydrogen-bonding with the extractants, materially affect the degree of polymerization and observed CMC values.

The chelating extractants tend to behave in a more regular fashion than the organic acids. Nevertheless, hydroxyoximes dimerize in the organic phase to an extent that depends not only on the aromaticity of the diluent, but also on the hydroxyoxime type [10]. The presence of higher aggregates has recently been established [11], and the degree of aggregation has considerable

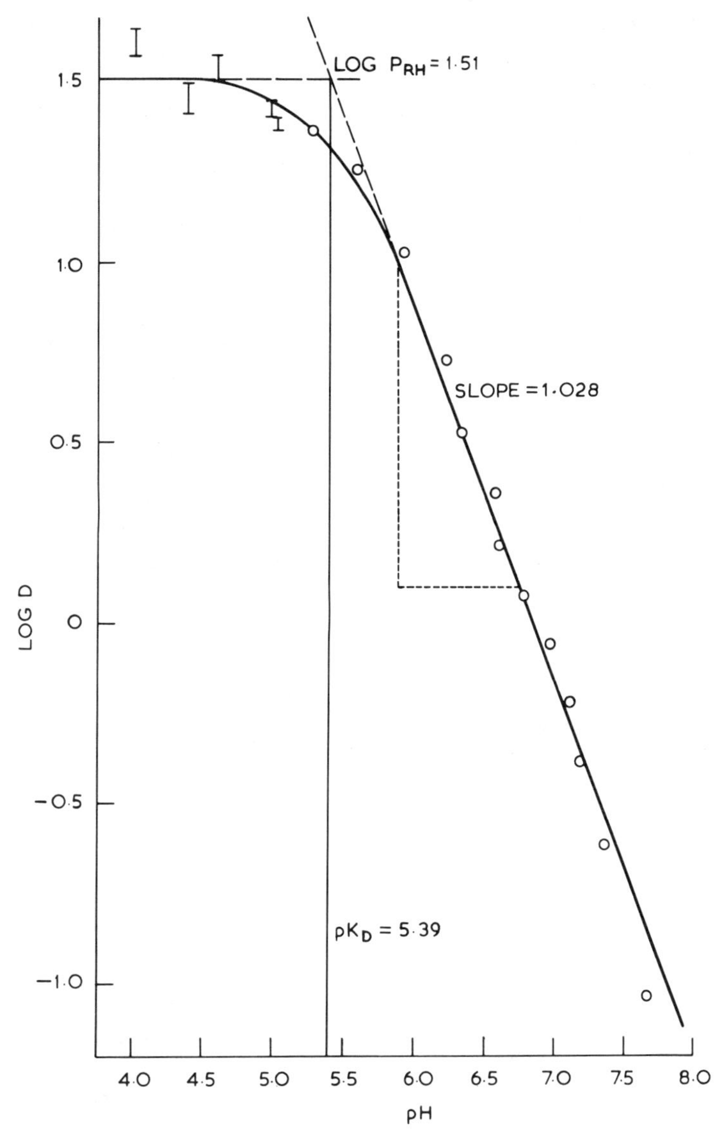

Figure 2. Distribution of 3,5-diisopropyl salicylic acid as a function of the pH in aqueous phase (1 *M* KNO_3).

significance in the interpretation of kinetic data [12]. A full description of the physical properties and extraction data for proprietary hydroxyoxime reagents has been given by Ashbrook [13, 14] and others [15–20].

The β-diketones, probably the most widely studied group of chelating extractants, undergo enolization, which is an additional factor affecting extractant behavior and distribution. Dissociation constants and distribution constants for a variety of diketones between carbon tetrachloride and 0.1 *M* sodium perchlorate solution have been provided by Sekine and Hasegawa [3], who have also reviewed the extensive literature on this class of extractant.

Amphoteric behavior is shown by some of these chelating acid extractants; thus 8-quinolinol-based reagents and α-hydroxyoximes are known to extract acids and anionic or neutral metal complexes from aqueous solution. This behavior is related to the presence of weakly basic atoms in the extractant molecule, the chemistry of which is similar to that of anion exchangers and solvating reagents. Such behavior must be borne in mind when metal extraction over a wide range of acidities is considered.

2.1.2. *Extraction of Metal Complexes*

Metal cations can react with organic acids and acidic chelating agents to form neutral complexes that are preferentially dissolved by the organic phase:

$$M^{n+} + n\overline{RH} \rightleftharpoons \overline{MR_n} + nH^+ \tag{5}$$

where M^{n+} is an n-valent metal cation.

Equation (5) describes a cation exchange reaction wherein hydrogen ions are exchanged for the metal cation, so the degree of extraction of the metal ions depends on the pH of the aqueous phase. The degree of extraction will also vary with the nature of the metal; in the absence of steric effects, an increasing charge to radius ratio of the metal cation will generally parallel increased extraction. The rules of coordination chemistry and ligand field theory are obeyed.

Metal extraction by acidic reagents is conventionally shown as a stepwise process whereby the reagent partitions between the organic and aqueous phases, ionizes and then the acid anion reacts stepwise with the metal ion until the neutral complex is formed that then itself partitions between the two phases:

$$\overline{RH} \rightleftharpoons RH \qquad p_{RH} = \frac{[RH]}{[\overline{RH}]} \tag{6}$$

$$RH \rightleftharpoons R^- + H^+ \qquad K_D = \frac{[R^-]\,[H^+]}{[RH]} \tag{7}$$

$$M^{n+} + R^- \rightleftharpoons MR^{(n-1)+}$$

$$K_f = \frac{[MR_n]}{[M^+]\,[R^-]^n} \tag{8}$$

$$(MR_{n-1})^+ + R^- \rightleftharpoons MR_n$$

$$MR_n \rightleftharpoons \overline{MR_n} \qquad p_{MR_n} = \frac{[MR_n]}{[\overline{MR_n}]} \tag{9}$$

This conventional description clearly presupposes an extraction mechanism that is wholly dependent on *chemical reactions in the aqueous phase*, requiring at least a reasonable water solubility of the extractant. This concept represents one extreme of a spectrum of conditions, the other extreme being the *wholly interfacial reactions* when the extractants are highly water insoluble. In this case Eqs. (6) and (9) should be replaced by adsorption isotherms relating interfacial concentration of the extractant and metal complex to the bulk-phase concentrations. Depending on the acidity of the extractant and the pH range being considered, it may sometimes be appropriate to consider direct reaction between the molecular form of the extractant and the metal cation rather than the extractant anion. Use of the scheme given previously, however, has considerable advantages in describing the chemistry of metal extraction by acid reagents.

From the definition of the distribution coefficient in Eq. (3), the application of the law of mass action to Eq. (5) yields

$$\log D = \log K + n\,\mathrm{pH} + n \log[\overline{RH}] \tag{10}$$

where $K = p_{RH}^{-2} K_D^2 K_f p_{MR}$ and $[\overline{MR_n}]$ and $[M^{n+}]$ are assumed to represent the total metal concentrations in each phase. The metal ion distribution coefficient thus increases with increasing aqueous solubility and increasing acidity of the reagent, increasing pH and also increasing reagent concentration. The value of the formation constant K_f clearly will affect the value of D, reflecting the specific interactions related to chelating extractants and also other factors such as steric effects from extractant molecular geometry.

(a) Qualitative Considerations. The degree of extraction of metal ions by organic acids depends on the metal ion basicity and thus in general follows the lyotropic series (see Figs. 3 and 4) in the absence of specific extractant metal–ion interactions. The pH of 50% extraction, when $\log D = 0$, is often used to rank metal ions for a fixed reagent concentration. Several such selectivity series have been published for carboxylic acids [2] and for di-2-ethylhexylphosphoric acid [21]. Differences between series for individual acids may be due to specific interactions, steric factors in relation to extractant geometry, or experimental error as the pH range covering some of the metals given in these series is very small.

The variation of degree of extraction with the aqueous solubility of the extractant [Eq. (10)] is partly offset by the solubility of the resultant metal complex, so it is probably more sensible to consider the variation in the value of $p_{MR_n} \cdot p_{RH}^{-2}$. Since K_f and K_D also change in any homologous series of organic acids, this relationship cannot be quantified. Variation of the aqueous phase anion can affect the selectivity order due to the metal–anion interactions in the aqueous phase. Provided no metal–inorganic-anion complexes are extracted, the distribution coefficient will be de-

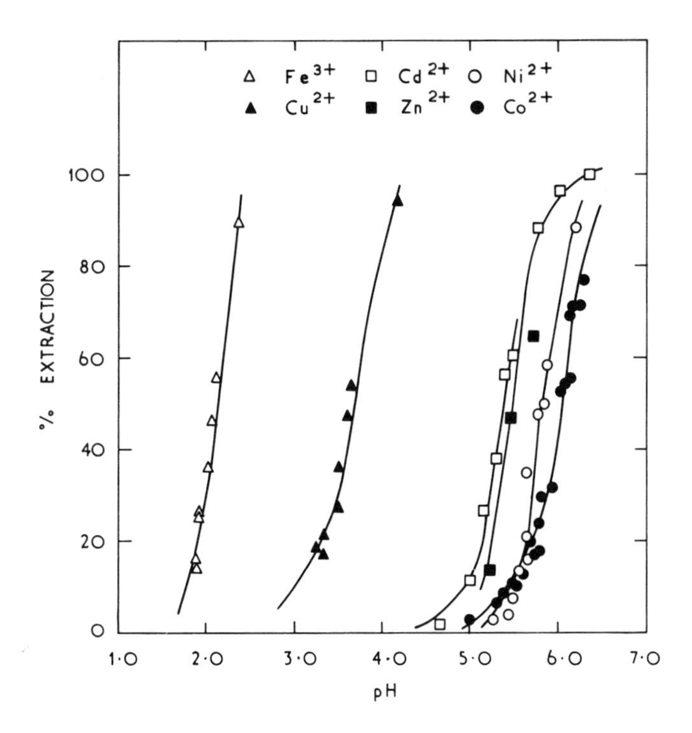

Figure 3. Extraction of Fe^{3+}, Cu^{2+}, Cd^{2+}, Zn^{2+}, Ni^{2+}, and Co^{2+} with naphthenic acid.

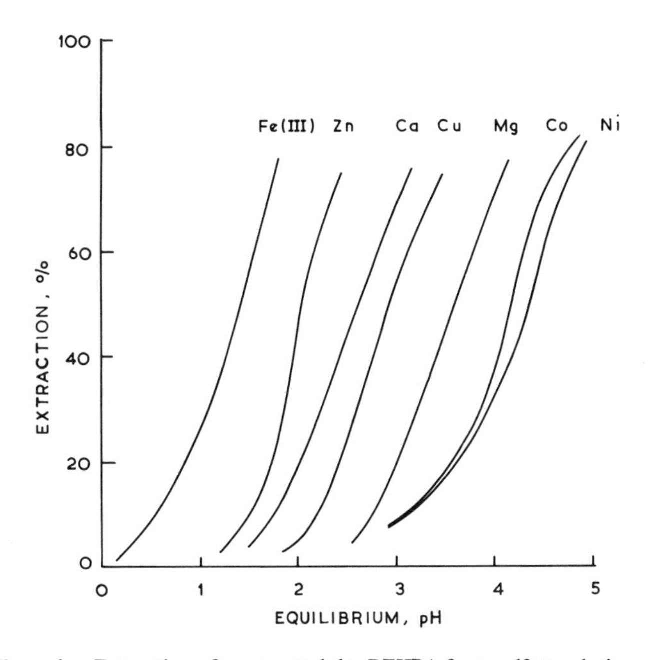

Figure 4. Extraction of some metals by DEHPA from sulfate solution.

creased by such interactions as follows:

$$\log D = \log K + n\text{pH} + n \log[\overline{\text{RH}}] - \log \sum_{0}^{i} \beta_i[\text{ML}_i] \qquad (11)$$

In general, the aqueous interaction would be expected to increase in the order $ClO_4^- < NO_3^- < Cl^- < CO_3^{2-} < SO_4^{2-}$. The effect of anion change on selectivity order is shown by the reversal of the order of extraction of zinc and cadmium by a carboxylic acid when the anion is changed from chloride to sulfate (Fig. 5 [22]). Equation (11) also applies to a neutral ligand when nonextractable complexes such as ammonia are formed. The degree of extraction increases with increasing acidity of the extractant is in the following general order; carboxylic acids < alkylphosphoric acids < sulfonic acids.

Obviously, chelating extractants cannot be ranked in this series as the K_f value, arising from specific interactions with these compounds, dominates the system. However, the effect of increasing acid strength within such classes of reagents can be demonstrated for hydroxyoximes where chlorination of 4-nonyl-2-hydroxybenzophenone oxime in the 3 position changes the pH_{50} for copper from 0.96 [23] to 0.00 [24], whereas the acid dissociation constants for the antiisomers are 11.5 and 10.5, respectively [14]. A similar effect, of course, exists within each class of simple acids, and pH_{50} for copper with lauric acid is reduced from 4.1 to 3.4 when α-bromolauric acid is used [59].

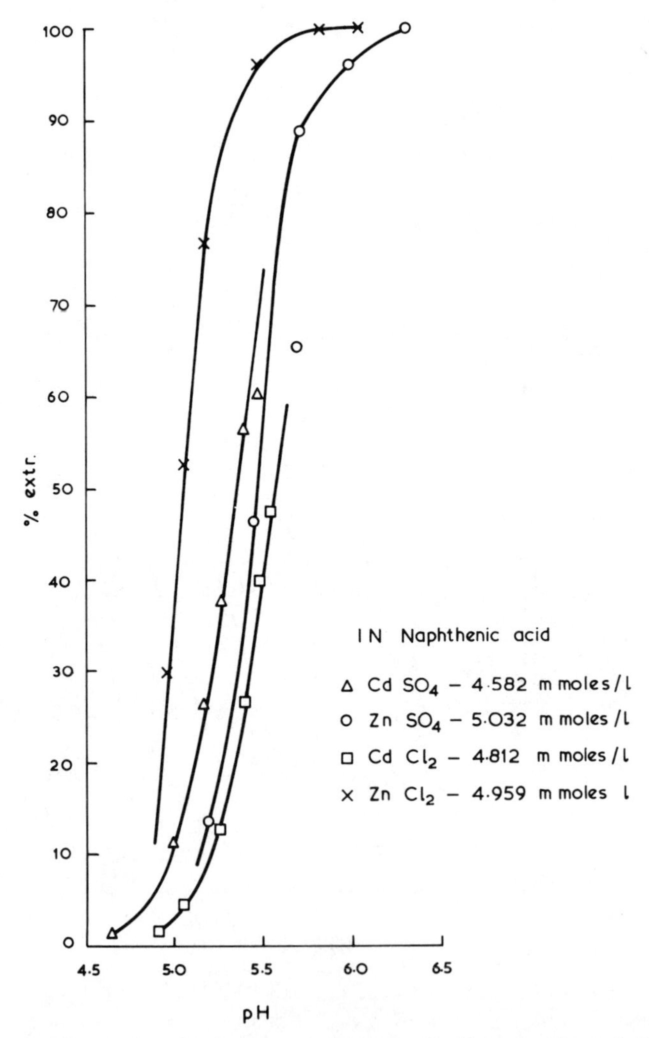

Figure 5. Extraction of cadmium and zinc from chloride and sulfate solutions.

(b) Stoichiometry and Structure of Extracted Species. By far the most popular way to study the stoichiometry of the extracted species has been the use of *slope analysis.* This relies on plots of log D versus pH at constant extractant concentration and also log D versus log[RH] or log[R^-] at constant pH. However, the use of Eq. (10) for this purpose is based on the following simplifying assumptions:

1. No polymeric species are formed in either phase.
2. Activity factors are essentially constant, so equilibrium constants based on concentration can be used.
3. Only uncharged species are extracted.
4. The formation of intermediate nonextractable complexes can be neglected.
5. No adduct formation between complexes and undissociated extractant molecules or the organic diluent or modifier takes place.
6. All hydrolysis reactions are insignificant.

It is thus very unusual for the simple extraction equation [Eq. (5)] and the derived equation [Eq. (10)] to account completely for metal extraction by acidic extractants. Assumption 2, however, can be complied with as nearly as possible by using constant ionic strength media and low metal concentrations. Thus the activity coefficients are constant and the free organic extractant concentration will also remain essentially constant, provided $[\mathrm{RH}]/n\colon [M^{n+}] \gg 1$. Assumption 3 is normally valid. Assumption 1 is seldom valid, as it has already been noted that many of the extractants dimerise to an extent dependent on the nature of the organic diluent. Monoalkylphosphoric acids polymerize and sulfonic acids form micelles, usually with a micelle number of 7. The extraction equation has to reflect this behavior if correct interpretations are to be drawn from slope analysis. The formation of intermediate nonextractable complexes with the extractant as in assumption 4 [$MR^{(n-1)+}$, $MR_2^{(n-2)+}$, etc.] will reduce the distribution coefficient and give rise to a curved plot of log D versus log[RH] or log D versus [R^-]. The plot of log D versus pH will not be affected unless charged species are extracted. The effect of adduct formation and hydrolysis on the form of the extraction plots can be seen in Figs. 7 and 8 in comparison with the ideal case in Fig. 6. In the latter the log D–pH plots show an increase in D at increasing $[\overline{\mathrm{RH}}]$ but have a common plateau that is numerically equal to p_{MR}. The log D–log[R^-] plots for all RH concentrations coincide. When adducts [i.e., $\overline{MR_n \cdot (RH)_m}$] are formed, log $D_{\max}$ increases with increasing $[\overline{\mathrm{RH}}]$ and the log D–log[R^-] plots are no longer coincident (Fig. 7). Adduct formation usually occurs when the coordination requirements of the extracted metal remain unsatisfied after formation of the neutral MR_n species; thus the water molecules in the remaining coordination positions are replaced by extractant molecules through direct coordination to the central metal ion. Addition of diluent molecules is also possible when sufficiently polar diluents such as alcohols and ether are used. When extractants such as carboxylic acid and dialkylphosphoric acids are involved, the adduct formation is also believed to involve ring formation through the hydrogen-bonded adduct.

Behavior when hydrolytic species are involved in the reaction is illustrated in Fig. 8. This is also representative of cases where nonextractable complexes are formed in the aqueous phase with ligands other than the extractant. The maximum distribution coefficient will depend on the ex-

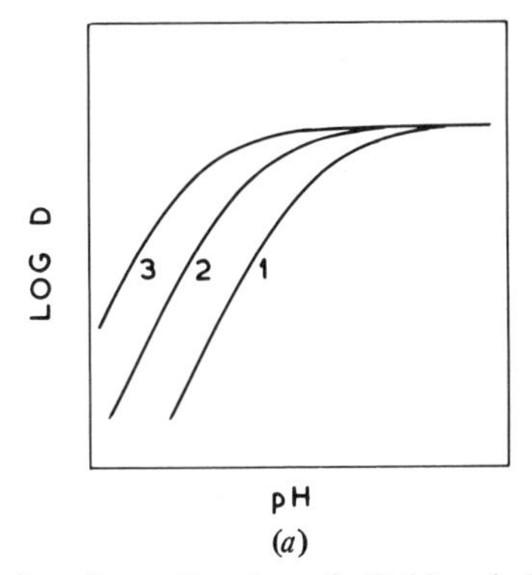

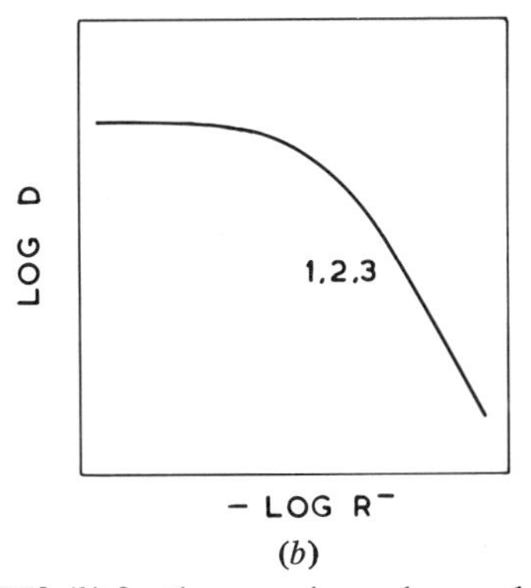

Figure 6. Log D as a function of pH (*a*) and –log [R^-] (*b*) for the case when only complexes of the form MR_n are formed: 1, 2, and 3 denote different increasing $\overline{\mathrm{RH}}$ concentrations.

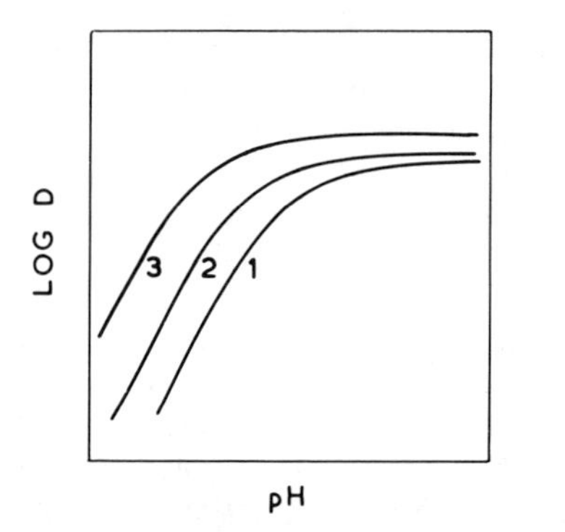

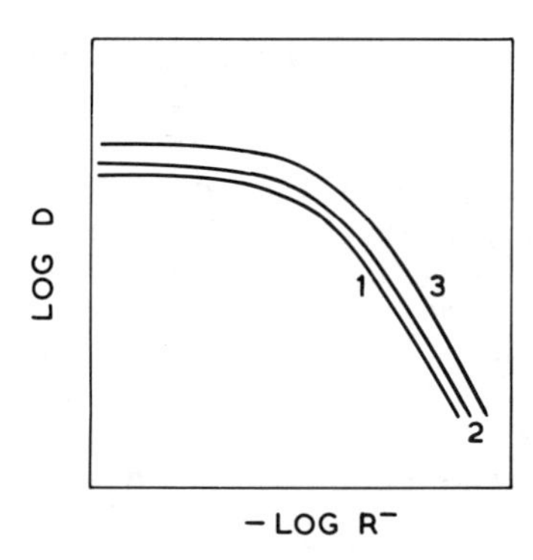

Figure 7. Log D as a function of pH and −log $[R^-]$ for the case when both complexes of the form MR_n and $MR_n\,(RH)_m$ are formed and extracted: 1, 2, and 3 denote different increasing $\overline{RH}$ concentrations.

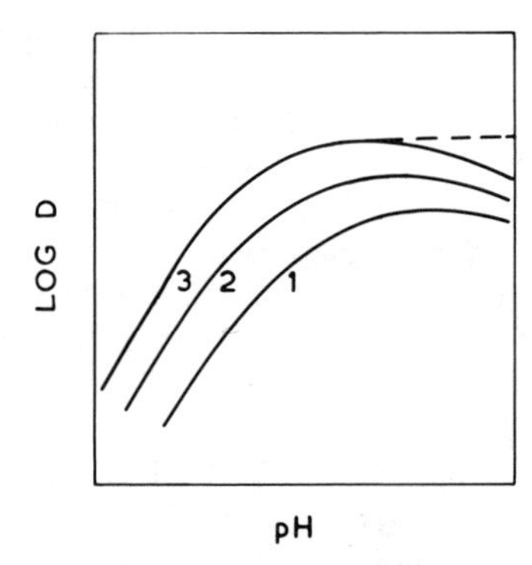

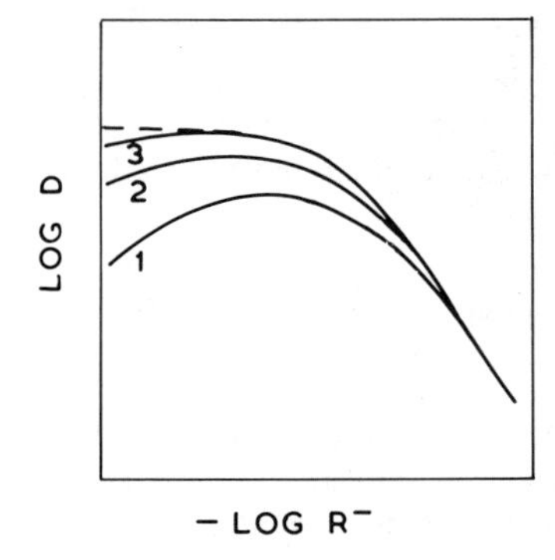

Figure 8. Log D as a function of pH and −log $[R^-]$ for the case when extractable complexes of the form MR_n and nonextractable complexes of the form $M(OH)_p$ are formed: 1, 2, and 3 denote different increasing $\overline{RH}$ concentrations.

tractant concentration and increases with increasing concentration until formation of the nonextractable complexes can be neglected. Polymerization of the extracted metal complexes is also fairly common with carboxylic acids [2] and organophosphoric acids [4] but less so with chelating extractants. This phenomenon results in an increase in log D with increasing metal ion concentration.

Variation of temperature can also affect the extraction chemistry, as instanced by the cobalt–nickel extraction system with di-2-ethylhexylphosphoric acid (D2EHPA). As temperature increases, the extracted cobalt species changes from a hexa coordinated dihydrate to a tetrahedral anhydrous species with a consequent increase in the distribution coefficient (Fig. 9), whereas increasing cobalt concentration at constant temperature causes polymerization of the organic-phase species with consequent increase of distribution coefficient (Fig. 10). The nickel complex does not show this behavior at all, and this difference is the basis of a separation of cobalt from nickel in sulfate solution by D2EHPA [25].

Various attempts have been made to model the effects of all these variables, as in the case of carboxylic acids [2]. Computer programs have been written on this basis to analyze extraction data and determine the likeliest species present in the organic phase. However, it is quite clear from the foregoing discussion that slope analysis in such complex situations is fraught with difficulty and should be used only in conjunction with physical measurements (e.g., infrared, ultraviolet, and Raman spectroscopy) to obtain an unambiguous description of the extraction system. An excellent critique of the slope analysis technique has been produced by Marcus [26].

Stoichiometric studies have been carried out for all types of acidic extractant. Generally the chelating extractants form readily defined species, whereas for the acidic nonchelating extractants, a variety of species tend to coexist in the

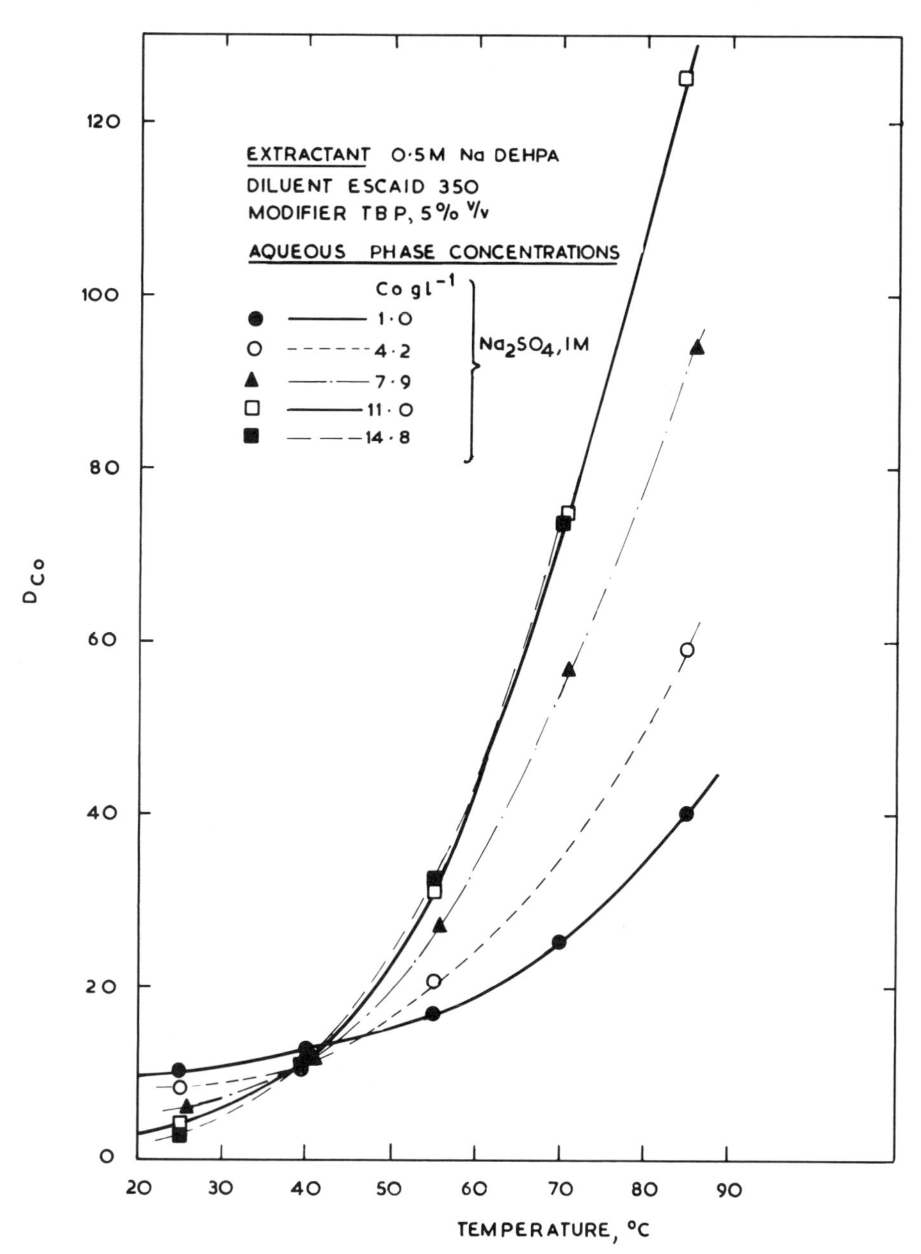

Figure 9. Effect of temperature on extraction of Co^{2+} from SO_4^{2-} solution by the sodium salt of DEHPA.

organic phase, with their relative importance depending on the system variables. Extensive data on specific systems is available [3, 4, 27, 28]. Thermodynamic data, formation constants, and so on, for metal complexes with carboxylic acids; alkylphosphoric, phosphonic, phosphinic acids; and sulfonic acids have been published by Marcus et al. [6, 29]. Because the acidic chelating extractants behave in a relatively simple manner, considerable interest has been shown in the application of regular solution theory to predict their distribution behavior in solvent extraction systems. Considerable success in this regard has been achieved for the β-diketones, and an excellent review of the use of theory in various systems, including β-diketones, 8-quinolinol, carboxylic acids, and similar reagents has been published by Irving [30].

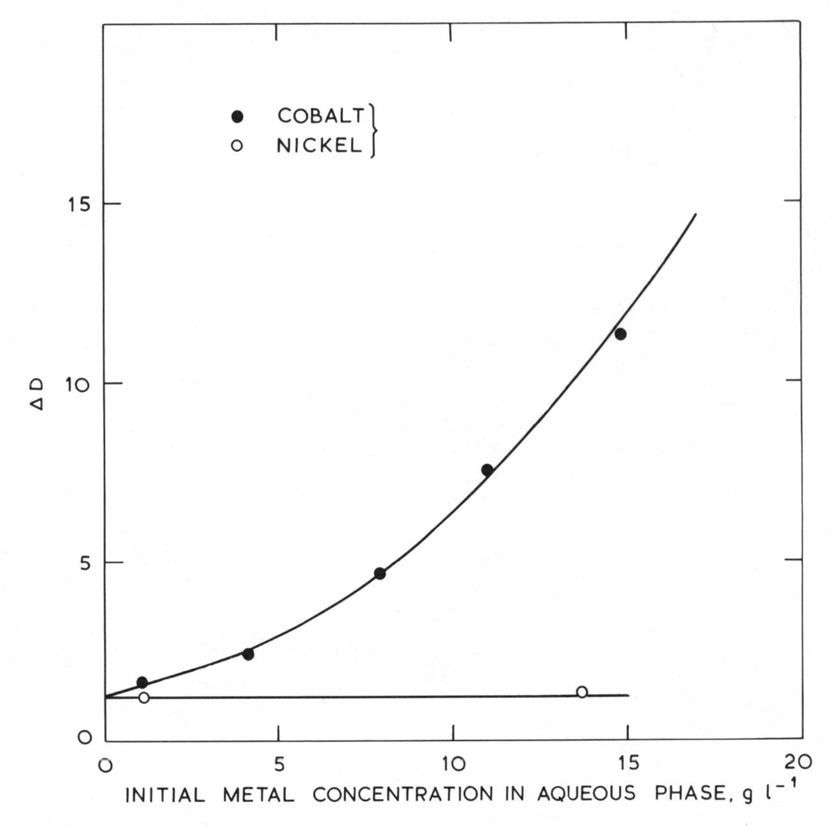

Figure 10. Increase in distribution coefficient of cobalt and nickel with metal ion concentration at 55°C.

2.2. Anion Exchangers

2.2.1. Properties of Extractants Alone

As shown in Table 1, anion exchangers in solvent extraction are generally protonated forms of primary, secondary, and tertiary high-molecular-weight amines and quaternary compounds. As already shown, the form of the extractant in both phases and its distribution behavior must be known in order to understand an extraction process. For anion exchangers, it is essential to elucidate the formation and nature of the quaternary salt and its relationship with the free base. In the following discussions examples are taken mainly from well-known quaternary ammonium compounds and amine salts that are the most widely used in commercial processing.

Quaternary ammonium compounds have been reviewed by Marcus and Kertes [4]. A feature of these compounds is their ionic character, which is retained under both acidic and basic conditions. They fall into two distinguishable groups according to their solubility, surface activity, and tendency for molecular association. Those compounds with 24 or more carbon atoms per molecule are readily soluble in organic solvents and give a high distribution coefficient when distributed between an aqueous electrolyte and an immiscible organic phase. Quaternary ammonium compounds that contain five or less carbon atoms are readily water soluble. High-molecular-weight salts are only sparingly soluble in water and are assumed to behave as strong electrolytes.

At low concentrations, short-chain alkyl quaternary ammonium salts are present in aqueous solution as monomers. However, as the concentration is increased, micelle formation takes place and the critical micelle concentration is dependent on the nature of the solute molecule, the length of the hydrocarbon chains, the temperature, and the presence and concentration of other electrolytes [4].

In organic solution quaternary ammonium

salts tend to aggregate, increasingly with increases in the molecular weight of the solute. Normally the high-molecular-weight salts are readily soluble in nonpolar aromatic hydrocarbons and the more polar diluents such as nitrobenzene and ethylene dichloride but show only sparing solubility in aliphatic hydrocarbons.

Quaternary onium compounds exchange anions by means of the following reaction formulated for simplicity for quaternary ammonium salts:

$$\overline{R_4N^+A^-} + B^- \rightleftharpoons \overline{R_4N^+B^-} + A^- \qquad (12)$$

In contrast, primary, secondary, and tertiary amines require protonation before they can act as anion exchangers and thus also have acid extracting properties in the free-base form:

$$\overline{R_3N} + H^+ + \overline{A} \overset{K}{\rightleftharpoons} \overline{R_3NH^+A^-} + B^- \rightleftharpoons \overline{R_3NH^+B^-} + A^- \qquad (13)$$

which gives a mass action constant K:

$$K = \frac{[\overline{R_3NH^+A^-}]}{[\overline{R_3N}]\,[H^+]\,[A^-]} \qquad (14)$$

The base strengths of aliphatic amines as measured by the values of K vary widely according to the diluent used for the measurements. Grinstead and Davis [31] found that, although increasing substitution of the nitrogen by alkyl groups results in an increase in base strength from primary to tertiary amines at least in the gas phase [32], in solution the increasing steric hindrance by the bulky alkyl groups impedes solvation of the cation–anion ion pair and thus reduces the stability of the salt relative to the free amine. Many workers have also found that the value of K increases with increasing concentration of the amine salt, and this is attributed to aggregation of the salt in the organic phase:

$$\overline{R_3NH^+A^-} \rightleftharpoons \overline{(R_3NH^+A^-)_2} \rightleftharpoons \overline{(R_3NH^+A^-)_n} \rightleftharpoons \overline{(R_3NH^+A^-)_{micelle}} \qquad (15)$$

The degree of aggregation depends on the properties of the diluent and the nature of the ammonium cation and anion. In toluene, alkyl ammonium chloride salts increase in degree of aggregation in the following order [31]: tertiary alkyl ammonium < secondary < primary.

The effect of the anion has been reported [33] for hydrohalic acid extraction by laurylamine. The ammonium salts usually exist as ion pairs at low concentrations in solutions of low dielectric constant; however, with increasing concentration, higher ion association aggregates are formed. This aggregation decreases with increasing polarity or dielectric constant of the diluent and increases in the order $I^- < Br^- < Cl^-$. The mass action constant for acid extraction, on the other hand, increases in the opposite order and also according to the following order of diluents: cyclohexane < carbon tetrachloride < benzene < chloroform < nitrobenzene. Danesi et al [34] similarly find an aggregation order for trilaurylamine salts: $ClO_4^- < NO_3^- < Br^- < Cl^-$, whereas the mass action constants for acid extraction are again in the reverse order.

Water uptake by the organic phase is proportional to the amine salt concentration and decreases with increasing anion size. The large amines do not seem to be hydrated very much in common diluents. The degree of hydration of the free base increases in the following order: tertiary < secondary < primary [31]. It would appear that the amount of water coextracted with the anion on amine salt formation is dependent on the ionic character of the amine salt and on the nature of the diluent. Thus the tertiary amine hydrochlorides are monohydrates in benzene and toluene, but less water is coextracted into chloroform or carbon tetrachloride, whereas cyclohexane shows the greatest amount of coextracted water. This obviously relates to the relative ability of the diluent to solvate the amine-salt-ion pairs. Roddy and Coleman [35] have examined the extraction of water by tri-*n*-octylamine and several of its salts in two diluents. In benzene the amount of water extracted by the amine salt always exceeded that extracted by the free base and increased in the following order: acetate < perchlorate < nitrate < chloride < bisulfate < sulfate.

The effect of coextracted water on the degree of extraction is not fully clear. Although Marcus and Kertes [4] conclude that water plays only a minor role, Muhammed et al. [36] have shown that water can have a significant effect on the aggregation of free tri-*n*-octylamine in benzene. Although the degree of aggregation of tertiary amine sulfates is considerably less than that of tertiary amine bisulfates, it is not clear whether this is due to the greater amount of water associated with the sulfate form or the increased opportunities for hydrogen bonding with the bisulfate system. Certainly, the greater degree of aggregation found for secondary amine sulfate salts compared with tertiary amine sulfate salts has been directly attributed to opportunities for extensive hydrogen bonding in the former, which is not possible in the case of the latter [37].

The order of extraction of common acids by Alamine 336 (tricaprylamine, General Mills, Inc.) has been given by Agers et al. [38]. For kerosene diluent, the order of extraction is as follows: $HNO_3 > HBr > HCl > H_2SO_4 > H_3PO_4$. For benzene and chloroform, the order for HCl and H_2SO_4 is reversed. The order of acid extraction power decreases in the order primary to secondary to tertiary. Marcus et al. [39] have ordered the affinities of anions for Alamine 336 as follows: $H_2PO_4^- < HSO_4^- < Cl^- < Br^-$ for toluene diluent, in agreement with the earlier work of Agers et al. The principal species present in the organic phase on extraction of phosphoric acid by Alamine 336 have been shown to be $\overline{R_3NH \cdot H_2PO_4}$ and $\overline{R_3NH \cdot H_2PO_4 \cdot H_3PO_4}$. Excess acid extraction as shown by the second species is also found for other acids, such as nitric acid. Organic-phase properties can change during acid extraction; for example, the dielectric constant can change markedly with increasing acid loading when the amine comprises a significant proportion of the organic phase.

These effects have been studied by use of a variety of techniques, such as light scattering, vapor-phase osmometry, vapor-pressure lowering, cryoscopy, infrared spectroscopy, and the two-phase titration technique. Computer analysis of distribution data has also been widely used, as exemplified by the Swedish program Letagropvrid. A general compilation of equilibrium constants of solvent extraction reactions involving alkyl ammonium salt extractants has been given [40]. An up-to-date discussion of the salt-forming reactions of amines and the aggregation of the salts together with a critical evaluation of the tabulated equilibrium constants [40] has been given by Kertes [41]. A titrimetric technique for sequential determination of amine and quaternary ammonium ion in the organic phase has been described recently [42].

In addition to the alkyl ammonium compounds, several other polyalkyl or polyphenyl onium compounds, such as quaternary arsonium, phosphonium and sulfonium salts, have been studied. The properties of these reagents are very similar to those already discussed and are not further described here. Further information can be found elsewhere in the literature [3, 4, 43].

2.2.2. *Extraction of Metal Complexes*

The extraction of metal complexes can proceed by either an anion exchange reaction, or an addition reaction as follows:

$$\overline{R_4N^+X^-} + MY_{n+1}^- \rightleftharpoons \overline{R_4N^+MY_{n+1}^-} + X^- \tag{16}$$

or

$$\overline{R_4N^+X^-} + MX_n \rightleftharpoons \overline{R_4N^+MX_{n+1}^-} \tag{17}$$

There is little doubt that both extraction mechanisms take place even when the anion in the aqueous and organic phase is the same. The controlling factor is the free concentration of X^- in the aqueous phase and the dominant species present in the aqueous phase [$MX^{(n-1)+}$, $MX_2^{(n-2)+}$, ..., MX_n, MX_{n+1}^-, etc.]; thus a knowledge of the aqueous-phase chemistry of these systems is of prime importance. At high X^- concentration, Eq. (17) will be the dominant reaction whereas at low X^- concentration, reaction (16) will predominate.

The interaction of amine salts and quaternary ammonium ions with metal anionic complexes is mainly electrostatic; thus, as the degree of extraction is due to the extent of ion-pair formation, selectivity will depend on charge, ion size, and extent of complex formation in the aqueous phase. Selectivity is generally achieved by varying the ligand concentration in the aqueous phase.

Stoichiometry for acidic extractants has traditionally been determined by slope analysis techniques, plotting the distribution coefficient of the metal of interest as a function of the aqueous concentration and extractant concentration. Interpretation of these results can be difficult because of hydrolytic polymerization of aqueous-phase complexes when oxyanions are concerned and the degree of aggregation of the amine salt and amine salt metal complex in the organic phase. Extensive computer studies of distribution data have been carried out, notably with use of the Letagropvrid routine [44]. Again, the use of slope analysis or numerical analysis alone for determination of stoichiometry must be done cautiously unless there are other appropriate physical chemical data.

Third-phase formation, that is, the splitting of the organic phase into two parts, is common in these systems and occurs through solubility problems relating to aggregation. This almost always results in a light organic phase rich in diluent and lean in amine and a dense organic phase rich in amine salt–metal complex. Third-phase formation is eliminated by addition of diluent *modifiers* such as long-chain aliphatic alcohols that carry out specific solvation of the amine salts and metal complexes through either dipole–dipole interactions or hydrogen bonding.

Extractable anionic species are found with several different types of ligand. For example, some metals occur as oxyanions that are extractable by amines, such as ReO_4^-, TcO_4^-, and MnO_4^-. The Group VI metals chromium, molybdenum, and tungsten also occur as oxyanions, but here the system is more complex in that aggregation and proton association with the anionic complexes can also occur. The anions derived from mineral acids are also known to form complexes with metals. The extent of complexation of metals by these anions, as before, is as follows: $ClO_4^- < NO_3^- < Cl^- < SO_4^{2-}$; thus the extraction of metals by quaternary ammonium ions follows this ligand order.

The nitrate ion tends to form only weak complexes with metal ions in aqueous solution, but a few elements can be extracted in the form of nitrate complexes with amine salts. These elements tend to be tetra- and hexavalent actinides and tervalent lanthanides, the heavier transition elements of Groups IV and V, and some of the platinum elements [45, 46].

The order of extraction of anionic metal halides seems independent of the class of amine, although in general the order is quaternary > tertiary > secondary > primary. This order is the reverse of that for hydrochloric acid extraction, and an explanation of this in terms of back donation between chlorine atoms and the amine protons has been proposed by Diamond [32]. The polyvalent metals exhibit the highest distribution ratio, followed by the divalent transition metals [4]. Frequently, extraction from alkali chloride media is preferred to mineral acid, presumably because of competition in the latter case by extraction of the mineral acid itself. The selectivity order of metals follows the tendency to form stable anionic chlorospecies thus: Au(III) > Ga(III) > Fe(III) > In(III). However, the stoichiometry of the extracted complex often indicates the formation of ionic aggregates greater than the simple 1 : 1 species.

The sulfuric acid system is complicated by the presence of the sulfate–bisulfate equilibrium. Thus the extraction of metal sulfate complexes by amine salts is strongly pH dependent since the bisulfate ion (which predominates at low pH) will compete for extractant with the metal-sulfate complex. The distribution data for metal sulfates exhibit a maximum at acidities of about 0.1 *M*, above which the extraction rapidly decreases as the bisulfate concentration increases. Extraction also depends strongly on the class of amine with the order primary to secondary to tertiary being followed. This order is a reversal of that found for both nitrato and chloro species but the same as that for sulfuric acid extraction. The elements that can be extracted are very similar to those given for the nitrate system [45].

Some other systems that form anionic metal complexes have been studied, for example, thiocyanate; carboxylates, in particular acetate, formate, and oxalate; and phosphate [3].

2.3. Solvating Extractants

2.3.1. Properties of Extractants Alone

Extraction by solvation requires the transfer of a formally neutral species from the aqueous to the organic phase by solvation of the hydrogen ion in the case of acid extraction or the hydrogen ion of a complex metal acid species (e.g., $HFeCl_4$) or by solvation of the metal ion of a neutral salt species. From Table 1 it is seen that all oxygen-bearing organic solvents will act in this way, as will organophosphorus compounds such as the triesters of phosphoric, phosphonic, and phosphinic acids; alkyl phosphites; amine oxides; organic sulfoxides; and organic sulfides. However, the solvating power depends strongly on the basicity of the reagent. As Marcus and Kertes [4] point out, oxygen-containing solvents are very weakly basic and in general tend not to form oxonium salts. In the carbon-bonded oxygen donor systems (e.g., ethers and ketones), water usually forms an essential part of the complex, forming a hydrogen-bonded bridge between the solvating agent and the solute. However, in the organophosphorus systems, water is often eliminated from the organic phase. Whereas ethers, ketones, and so on are all electron-donating compounds, alcohols are amphoteric and exhibit both donor and acceptor properties. The relevant physical chemistry of carbon-bonded oxygen donors and the extraction of acids with these reagents have been reviewed by Marcus and Kertes [4].

The extraction of acids can be represented by the equation

$$m\mathrm{HX} + n\bar{\mathrm{S}} + x\mathrm{H_2O} \rightleftharpoons \overline{(\mathrm{HX})_m \mathrm{S}_n(\mathrm{H_2O})_x} \tag{18}$$

and specific hydrates such as $H_3O^+X^-$ are important features in acid extraction. The extensive literature data on the mutual solubility of the aqueous and organic phases [3, 4, 28] is important in meaningful interpretation of distribution data. A number of the extractants are appreciably miscible with water; hence ternary phase diagrams are required to understand the extraction process. As the molecular weight of extractant increases, this solubility decreases.

The pitfalls in the use of slope analysis to determine the stoichiometry of extracted species without regard to ionic hydration and solvation have been discussed by Widmer [47]. This author also shows how solvent extraction can be used to determine association numbers and association orders of the extracting species. A recent compilation of equilibrium constants [29] indicates the large amount of available data on the extraction of acids by carbon-bonded oxygen donor solvents, although the only commercially important acid in this respect is phosphoric acid.

Extraction by phosphorus-bonded oxygen donor solvents has been extensively studied and is of particular interest in the atomic energy field. The most well known and most used organophosphorus ester is tri-*n*-butylphosphate (TBP). With regard to the physical chemistry of this type of solvent in organic solution, nonideal behavior is the rule. Hydrogen bonding type and dipole–dipole interactions are common, and both self-association and extractant–diluent association complexes are well known for this class of extractant. Equilibrium data for such interactions have been tabulated [6].

Data on aqueous solubilities and hydrates of phosphorus-bonded oxygen donor solvents is abundant, particularly for TBP. Because of the increasing polar nature of the phosphoryl group, the solubility of neutral organophosphorus compounds in water decreases in the following order: phosphine oxides > phosphinates > phosphonates > phosphates. Solubility also decreases with increasing alkyl chain length. Like the carbon oxygen donors, these phosphorus oxygen compounds extract mineral acids from aqueous solution. The order of extractability depends on a number of factors, such as mineral acid strength, size, and hydration of the anion as well as acid concentration in the aqueous phase. The nature of the extracted species also varies with the mineral acid and can be related to a competition for the proton between the extractant, the water, and the anion. Thus strong acids are usually extracted with the proton retaining its primary hydration shell, which is then solvated by the extractant through hydrogen bonds between the water and phosphoryl oxygen atoms. The amount of water coextracted also varies with the acid, so a complicated series of equilibria exist. In anhydrous systems, of course, the acids are present as molecular adducts through the phosphoryl oxygen atom. In general, extraction power increases with increase in the number of carbon–phosphorus bonds. Excellent bibliographies are available [3, 4, 28], and equilibrium constant data are tabulated [6], including data on the extraction of water and the hydration of the extractants.

Other oxygen-containing neutral extractants that have received some attention recently are amine oxides and sulfoxides; much of this work has been carried out in the Soviet Union, and a good empirical description of the chemistry of these compounds, together with the chemistry of organic sulfides as extractants has been given by Mikhailov [48].

2.3.2. *Extraction of Metal Complexes*

The extent of extraction by solvating extractants depends on the extent of complex formation in the aqueous phase, as in the case of the basic extractants. Extraction is by solvation of either the central metal atom of the complex or the proton in the case of formation of a complex acid species:

$$MX_n + y\bar{S} \rightleftharpoons \overline{MX_nS_y} \tag{19}$$

or

$$HMX_{n+1} + x\bar{S} \rightleftharpoons \overline{(HS_x)^+(MX_{n+1})^-} \tag{20}$$

In the extraction of these metal complexes, as for acid extraction, the solvating agent will replace primary and/or secondary waters of hydration, thus rendering the complex soluble in the organic phase.

The degree of extraction of a metal by a solvating extractant should depend on a number of factors, including the nature and concentration of the anionic coordinating ligand X, which, in turn, influences the type of metal complex formed; the degree of hydration of these aqueous metal complexes; and the relative strength of the water–metal and extractant–metal bonds, leading to a competition between water and the extractant for the solvation sites. Marcus and Kertes [4] have attempted to relate these factors to the electropositivity of the metal. The electrovalency requirements of strongly electropositive metals can be satisfied by the strongly polar organophosphorus reagents and hence are readily extractable in an unhydrated form. A less strongly electropositive metal is usually extracted by the same reagent in a partially hydrated form, a similar situation to the extraction of a strongly electropositive metal with the less polar carbon–oxygen extractants. This is explained by the persisting attraction of the metal for water. Similarly, it would be expected that mixed water–extractant complexes would be formed by extraction of the less electropositive metals by carbon–oxygen extractants. Here it is postulated that the extractant cannot compete successfully with water for the primary coordination sites and that extraction occurs by replacement of the secondary hydration sphere by extractant molecules hydrogen bonding to water in the primary sphere.

The use of carbon–oxygen-bonded extractants for metals is well established, especially in halide and nitrate systems. Details of the various systems can be found in later sections of this chapter and in other sources [3, 4, 28, 29, 49]. Generally it is found that ketones are better extractants for metal species than ethers because of their greater electron donor properties and thus are more widely used. Among these ketones, methyl isobutyl ketone has been used in the laboratory, but its low flash point has limited its commercial use to separations such as niobium and tantalum, where only a small-scale plant is needed to satisfy demand.

Most of the work with phosphorus–oxygen-donor extractants has been concerned with halide and nitrate systems. In contrast to the extraction of acids by these reagents, the neutral metal salts are usually extracted into the organic phase in an anhydrous form. Once again, very extensive studies have been made of these reagents in the extraction of metal salts [3, 4, 6, 28], including physicochemical studies of the nature of extracted species in both solution and the solid state. A few general conclusions can be made concerning these systems. First, the nature of the diluent used with the reagent will affect the extraction behavior. Dilution itself affects the physical properties of the organic phase and also the water content and the activity of the reagent. Moreover, extraction decreases with increasing diluent polarity, probably as a result of competition for the phosphoryl oxygen by the various potential hydrogen-bonding molecules in the system. However, in most cases the distribution ratios cannot be correlated with physical properties of the diluents. Obviously, the ability of the diluent to solvate the extracted metal complex is important, and in these systems third-phase formation is fairly common in particular. Chemical stability of the reagents is another important factor; ester linkages can undergo hydrolysis, leading to production of acidic species. This, of course, changes the nature of the extractant and may produce a system completely different (i.e., extraction by an acidic reagent) from that originally proposed. Purification of organic phases and avoidance of degradation are thus important in these systems.

2.4. Mixed-Complex Formation

Mixed complexes can often be formed in the solvent extraction of metal ions. This is the basis of *synergism*, wherein the degree of extraction of metal by the mixed system exceeds the combined effects of the individual components, that is, a stability constant effect associated with the mixed complex. It was formerly thought that an opposite effect called *antisynergism* existed, but it is now clear that the destruction of synergism is due merely to mass-action effects or changes in physical properties of the organic phase, associated with the predominance of one of the reactants in the organic phase. Thus it is expected that mixed-complex systems will obey the same rules as single-extractant systems. The compounds added to the organic phase to produce mixed complexes will also partition between both phases just like the primary extractant and may also interact with all species present in both phases. Organic-phase interactions between the two organic reactants are of extreme importance as they will materially affect the metal extraction behavior of the system. The thermodynamics of mixed-complex formation is of special interest, and mixed-complex phenomena are discussed in the following paragraphs under the individual extractant class headings.

2.4.1. Acid Extractants

Mixed-complex formation in acid extractant systems can occur in several ways. By far the most common is the addition to the systems of neutral donor molecules such as alcohols, organophosphorus compounds, nitrogen containing ligands such as amines, and acids where the compound remains undissociated. Mixed complexes are also formed between two acidic extractants with both exchanging protons in the extraction reaction. Thus a wide variety of complexes and behavior arise, and it is common to find the same extractant acting as an acid toward one metal ion in a complex and as an un-ionized donor molecule toward another. The determination of formation constants and a general algebraic description of mixed-complex formation in acidic systems has been discussed by Dyrssen [50].

(a) Acid Extractants and Neutral Donor Molecules. The formation of mixed complexes in this system can be described in two ways:

$$\overline{MR_n} + x\bar{S} \rightleftharpoons \overline{MR_nS_x} \tag{21}$$

and

$$\overline{MR_n(RH)_x} + y\bar{S} \rightleftharpoons \overline{MR_n(RH)_{x-y}S_y} + \overline{(RH)_y} \tag{22}$$

Equation (21) describes an addition mechanism, whereas Eq. (22) describes a substitution mechanism. Clearly, the addition mechanism will most likely occur when the central metal atom in the complex MR_n is coordinately unsaturated. However, this is not always the case, and an increase in coordination number of the central metal atom to accommodate the neutral ligand is also found. It would seem, therefore, that in all cases the neutral ligand bonds directly to the central metal atom. Kertes [28] has listed some general rules for the stability of mixed complexes: (1) stability increases with increasing base strength of the neutral ligand, in the absence of steric factors; (2) change of the diluent can materially affect complex stability, although not stoichiometry; and (3) the stronger the original complex, the less will be the tendency to form a mixed-complex species.

Formation of solvated species in extraction systems can be considered as a special case of mixed-complex formation, where the neutral donor molecule is the undissociated acid. The formation of other mixed complexes from the solvated species by substitution can be readily understood. Substitution is also obvious in the replacement of water in hydrated chelated molecules by neutral donor compounds. It is less easy to understand the formation of species such as $MR_n(RH)_x$ in the extraction of metals with dimeric extractants such as carboxylic acids or dialkyl monohydrogen phosphoric acids. The acid dimer may interact in the organic phase with the neutral ligand according to the equation

$$\overline{(RH)_2} + 2\bar{S} \rightleftharpoons 2\overline{(RH \cdot S)} \tag{23}$$

but no proof exists that (RH · S) species actually take part in the extraction process. Indeed, this would be highly unlikely as the extraction reactions generally take place at the interface, and for acidic systems, there is considerable evidence that the interfacial species are monomeric. It would thus be unlikely for the species RH · S to be present at the interface and thus be active in the extraction reaction. Nevertheless, interaction between neutral ligands and the extractant can have a profound effect on system behavior; for example, as the concentration of neutral ligand is increased, the associated mass-action effects can very from synergistic to a considerable reduction in the degree of metal ion extraction. A corresponding effect on reaction kinetics will occur, as discussed later.

In metal chelate extraction by far the most studied reagents are the β-diketones, which can form mixed complexes with a wide variety of neutral ligands. The lower the stability of the metal diketones and the greater the basicity of the donor atom, the greater is the stability of the adduct formed. Large exothermic enthalpy changes usually characterize these systems [51]. Metal β-diketonates also form mixed complexes with alkylamines and alkylamine hydrochlorides. The inclusion of the amine hydrochloride in the complex is unusual; Newman and Klotz [52] have postulated the formation of AmT_3R_3NHCl, AMT_3R_3NHT, and $AmT_3NHClHT$ in the extraction of americium by a mixture of tri-*n*-octylamine(R_3N) and thenoyltrifluroacetone (HT). Ke and Li [53] have also found adduct formation with copper β-diketonates and alkylamine and alkylamine hydrochlorides. Visible and infrared spectra suggest that the alkylamines bond directly to the copper through the nitrogen atom while the amine hydrochloride binds directly through the chlorine atom. Full discussions of these systems are available [3, 4, 28, 54].

Of particular interest is the recent discovery that metal complexes of alkylated 8-quinolinol and also hydroxyoximes can form adducts with neutral donor molecules, for although this type of complex with 8-quinolinol is recorded [55, 56], little was known about the behavior of hydroxyoximes in this situation. Addition of carboxylic and dialkylphosphoric acids to metal-hydroxyoxime systems produced mixed complexes and, for some metals, a considerable amount of synergism; thus synergistic extraction by mixtures of an aliphatic α-hydroxyoxime (LIX63, Henkel Corporation) and various carboxylic and phosphoric acids has been widely reported [57–59]. Investigations of nickel extraction showed that addition of the carboxylic acid to the four-coordinate orange $(LIX63)_2Ni$ complex produced a green six-coordinate mixed complex [60]. Spectral studies showed that the hydroxyoxime chelate structure remained intact and the acids were acting as neutral donor molecules. It is also shown that no interaction between the two types of donor molecule occurred. An interesting feature of this mixed system was that the rate of extraction of nickel was very slow [61]. Other systems showing similar mixed complexes are aromatic β-hydroxyoximes and carboxylic [62]; dialkylphosphoric [63]; and aromatic sulfonic acids [63] for copper, nickel, and cobalt; and alkylated 8-quinolinol and carboxylic acids for nickel [62], cobalt [64], zinc [65], and cadmium [65]. In this latter system, interaction between the carboxylic acid and the 8-quinolinol has been found [66] that materially affects the interfacial properties of the system and thus permits the extraction of cobalt without simultaneous oxidation to cobalt(III). Although addition of nonyl phenol to hydroxyoxime systems does not seem to cause mixed-complex formation, the two reagents do interact to cause a decrease in copper extraction with increasing nonyl phenol concentration and a decrease in the rate of extraction [67, 68]. There is some evidence for a similar effect with nonyl phenol and alkylated 8-quinolinol mixtures, but these reagents do show a small synergistic effect accompanying mixed complex formation in the extraction of copper [69]. Obviously, this is a case where the observed effect will depend critically on the relative amounts of chelating agent and neutral ligand in the organic phase.

In the carboxylic acid extraction system, mixed complexes form on addition of amines to the system [2] and for tri-*n*-butylphosphate salicylic acid mixtures [2]. Little synergism, however, is found with either system. Synergism was found in the extraction of cesium with mixtures of 4-phenylvaleric acid and 4-*sec*-butyl-2-(α-methylbenzyl) phenol [70]: here the phenol behaves as a neutral solvating agent at pH values below 11. Some data on mixed complex stability constants and adduct formation with neutral ligands have been compiled [29].

Mixed-complex formation and synergism in dialkylphosphoric acid systems have been reported extensively, with particular interest in the addition of neutral organophosphorus compounds. Once again the stability of the mixed complex increases with increasing donor properties of the molecule; thus for neutral organophosphorus compounds the series, in order of effectiveness, is $(RO)_3PO < R(RO)_2PO < R_2(RO)PO < R_3PO$. Interaction between the extractant and the neutral ligand is again important in the determination of the chemistry of these systems. Mixed extractant systems containing alkylphosphoric acid have also been studied with other neutral donors such as alkylamines and aliphatic and aromatic alcohols. Dialkylphosphoric acid alkylamine systems are unlikely to form mixed complexes as mixtures of these reagents evolve heat during the formation of the amine alkylphosphate salts. A large decrease in the extraction of metals can be expected in such systems. However, if the pH of the aqueous phase is so high that the amine is in the free-base form, adducts similar to those reported for the carboxylic acids could be formed. Similarly, when the alkylphosphoric acid is in a salt form, for example, the sodium salt, adduct formation with amines should be possible. A full description of these systems may be found elsewhere [3, 4, 28], together with a useful compilation of stability constant data [6].

Mixed complexes in sulfonic acid, neutral ligand systems have been reported by Wang and Li [71] for the extraction of zinc by dinonylnaphthalene sulfonic acid (DNNSA) and its sodium salt with trioctylphosphine oxide (TOPO), tri-*n*-butylphosphate (TBP), or 2-ethylhexanol. No mixed complexes were formed with the systems containing the free acid, and the observed decrease in zinc extraction on addition of the neutral ligands can be attributed to interaction between DNNSA and the ligands. As expected, TOPO interacts more strongly with DNNSA than does TBP. A mixed complex was formed between zinc, the sodium salt of DNNSA, and TOPO, and synergism occurred, but neither TBP or 2-ethylhexanol had any effect.

(b) Two Acidic Extractants. In some mixed β-diketone systems both diketones combine in satisfying the charge and coordination requirements of the central metal ion. This type of mixed-complex formation does not necessarily lead to synergism. Differentiation must be made between systems wherein mixed-complex formation is produced through adduct formation [72] and by mixed-chelate formation. It has been suggested [73] that the stability of the mixed-chelate species is achieved only by a combination of a weak and strong β-diketone. Similar phenomena for iridium and europium complexes with a β-diketone and β-isopropyltropolone have been reported [74]. Recently the formation of mixed complexes of lanthanide and actinide cations with dinonylnaphthalene sulfonic acid (DNNSA) and di-2-ethylhexylphosphoric acid

(D2EHPA) has been reported [75]. The degree of complexation with D2EHPA and the dimensions of the complexed ion are shown to be important factors for their inclusion in the DNNSA micelles. The core volume of the DNNSA micelles is about the size of the trivalent metal ion D2EHPA complex, but inclusion of the neutral D2EHPA metal complex does not occur. Instead, only positively charged species are included in the DNNSA micelle, and the charge neutralization is achieved by loss of protons. Interfacial tension changes noted in these systems correlate with bulk changes in metal ion extraction behavior. Yet again, the importance of interfacial phenomena in solvent extraction systems is emphasized.

2.4.2. *Anion Exchangers*

Little is known of the formation of mixed complexes in amine extraction systems. However, Deptula and Minc [76] have shown that addition of alkylphosphoric acids to tri-*n*-octylamine depressed the extraction of sulfuric acid through the strong interaction between the organic acid and base. Their conclusion that extraction of other inorganic species such as UO_2SO_4 and $PtCl_6^{2-}$ by alkylamines should be depressed by addition of alkylphosphoric acids was supported by a study of the extraction of $PtCl_6^{2-}$ with tri-iso-octylamine in the presence of di-*n*-butylphosphoric acid [77]. Similarly, at low sulfuric acid concentration the extraction of uranium from UO_2SO_4 by a mixture of tri-*n*-octylamine and di-2-ethylhexylphosphoric acid was depressed by formation of the alkylphosphate amine salt. However, as the sulfuric acid concentration was increased, synergistic extraction of uranium was found [78] that was attributed to the formation of a mixed complex $(R_3'NH)^+(UO_2SO_4HR_2)^-$. A similar behavior was discovered by Liem and Sinegribova [79] in the extraction of hafnium by di-2-ethylhexylphosphoric acid and tri-*n*-octylamine. In this case the mixed complex had the structure $(R_3NH^+)_2[Hf(SO_4)_2R]^{2-}$. Deptula [80] showed that similar mixed complexes were formed in the extraction of uranyl sulfate by amine–monoalkylphosphoric acid mixtures. However, the decrease in extraction found in earlier studies was absent in this system because of the weaker acid strength of the monoalkylphosphoric acids reducing the interaction with the amine in the organic phase. It is clear, therefore, that amines can act in mixed-complex systems both as adducts and extractants similar to the acid extractants such as carboxylic and alkylphosphoric acids.

Mixtures of two amines have been shown to produce mixed amine complexes in the extraction of uranium from sulfuric acid [81], with a synergistic effect. This behavior is analogous to the formation of mixed chelates with acidic extractants. No information on mixed-complex formation of metals with amine salts and neutral donors is available.

2.4.3. *Solvating Extractants*

The role of solvating extractants and neutral ligands as donor compounds in acidic extraction systems has already been discussed. Systems with two neutral ligands do exist, although not much information is available; Kertes [28] has given a short account. Clearly, hydrogen bonding is of prime importance as also is nonideality in these mixtures. Synergism seems to occur in these systems when one of the extractants has a much higher dielectric constant than the other, or when one is a donor capable of hydrogen bonding and the other is an acceptor molecule. Diluent extraction interaction may also be important with organophosphorus reagents, and Nishimura et al. [82] have shown that hydrogen bonding of chloroform to tri-*n*-butylphosphate and tri-*n*-octylphosphine oxide appreciably depressed the extraction power of these extractants. The study showed a greater hydrogen acceptor strength for the phosphine oxide than the alkyl phosphate. A similar effect was found for alkylamines, but they were much poorer hydrogen acceptors in spite of their greater proton affinity.

3. INTERFACIAL PROPERTIES OF EXTRACTANTS

As has been stated earlier, the chemical reactions involved in metal extraction can take place anywhere between the limits of the bulk phases including the aqueous–organic interface itself. Regardless of the mechanism and the location of the chemical reaction of kinetic significance, mass transfer must occur across the interface. Thus it is important to have information about the nature of the interface and the species adsorbed thereon; in the particular case where chemical reactions take place at the interface itself, the nature and concentration of the species involved should be established. Such information is of great value in assisting mechanistic studies of solvent extraction systems. Solvent extraction reagents all exhibit interfacial activity to some extent and thus are amenable to study by standard interfacial physicochemical techniques. The most common interfacial properties studied in this context are interfacial tension, interfacial potential, and interfacial viscosity.

3.1. Interfacial Tension

Interfacial tension measurements have hitherto provided most of the interfacial property data. The lowering of interfacial tension by addition of an extractant is the result of penetration of the interfacial surface by hydrophilic groups. The mutual repulsion of these polar groups causes the contractile tendency of the interface to be reduced. The interfacial tension reduction depends on reagent type, bulk-phase concentration, diluent and also the water solubility of the reagent. Various techniques for the measurement of interfacial tension are well documented [83, 84] and will not be considered further here.

Values of the interfacial concentration, so important for the interpretation of heterogeneous kinetics, can be obtained from interfacial tension data by use of the Gibbs isotherm [84]. However, this precise thermodynamic statement should be used only when reduction of the general equation to its single solute form can be justified. Also, where the adsorbed species changes its molecular complexity from that in the bulk phase, the appropriate concentrations must be known. Because of these limitations, caution must be taken in using the Gibbs adsorption isotherm to obtain quantitative data for adsorbed species in solvent extraction systems.

Even under the most suitable conditions, the Gibbs isotherm only permits indirect determination of interfacial area or concentration of the adsorbed species. Direct determinations at an oil/water interface can be made by means of a modified Langmuir trough [85], but this technique is unsuitable for routine work. Also because many solvent extraction reagents are only weakly surface active, movement of the barrier to obtain the force–area measurements may disturb the adsorbed molecular film, resulting in a change of interfacial concentration by desorption into the bulk phase. Methods involving radiotracers [86] also have limitations arising from the availability of suitable labeled compounds. Thus the use of interfacial tension data is generally restricted to qualitative rather than quantitative interpretation. A further restriction arises from the extreme purity required of the reagents being studied, as the presence of small amounts of surface-active impurities can cause large errors.

3.1.1. Acidic Extractants

The carboxylic acids have been widely studied and the results show that, as expected, the interfacial tension decreases with increasing pH as the acid ionizes. The interfacial tension also decreases with increasing reagent concentration reaching a limiting value at relatively low molar concentration implying surface saturation. Calculation of molecular areas from these data indicate that in benzene–water [87] and heptane–water [88] systems the molecules are parallel to the interface in the dimeric hydrogen bonded form. The plot of interfacial tension against concentration should be linear, but in some cases deviations occur that have been attributed to micelle formation [89], although application of the Gibbs isotherm to this situation merely indicates a change in the adsorbed species from that in the bulk phase. In some cases a discontinuity occurs in this plot at a concentration that equals the critical micelle concentration for the reagent.

It is difficult to generalize on the effect of added metal ions on the interfacial tension, as this depends on a number of factors including the pH of the system and the relative interfacial activities of the extractant, its ionized form, and the metal complex. Thus addition of metal ions may either increase or decrease the observed interfacial tension, depending on the extractant used.

The interfacial tensions of chelating acids, mainly the α- and β-hydroxyoximes and their copper derivatives have also been studied. However, some of these data should be interpreted very carefully as the commercial reagents often contain other surface-active compounds. Thus the calculations of interfacial area per molecule as carried out by several workers [17] cannot be related to the configurations of the molecule at the interface with any certainty. As expected from the very weakly acidic nature of these extractants, no variation of interfacial tension with pH is found below pH 9 [16, 90], and the results [19] on commercial samples are most probably associated with impurities in the system. Cox and Flett [17] have reported the copper complex to be much less interfacially active than the parent hydroxyoxime, although Hughes [91] recently has presented data that suggest that the difference is marginal. The nature of the interfacial species in hydroxyoxime systems also remains to be determined. The existence of dimers of the aromatic β-hydroxyoximes at the interface has been invoked [92, 93], but it is more likely that considerable dissociation of the extractant aggregates occurs at the interface as a result of competitive hydrogen bonding with water molecules.

3.1.2. Anion Exchangers

The interfacial activity of amines increases with decreasing pH, that is, as protonation of the free base increases. This phenomenon has been used to establish the stoichiometry of interfacial complexes, a discontinuity occurring in the interfacial tension plot at the requisite concentration [94]. The nature of the counterion is also important and follows the order established from extraction data; thus for tri-*n*-octylamine–toluene–water, the interfacial pressures are in the following order: $Cl^- > NO_3^- > ClO_4^-$ [95]. These data can be interpreted by consideration of the interfacially absorbed species. The interface is positively charged and expansion of the interface is a function of the number of adsorbed cations per unit area. This electrostatic repulsion can be reduced by interaction between anions in the aqueous phase and the adsorbed cations; however, the effect of these anions on the water structure must also be considered [96]. The combination of these two effects leads to the observed order. When more water-soluble amine salts such as tetrabutylammonium halides are studied, the observed order of interfacial pressure is as follows: $I^- > Br^- > Cl^-$ [97]. This can be explained by the reduction of charge on the adsorbed monolayer by anion interaction, allowing more ion pairs to enter the interface from the aqueous phase. Thus a greater cation–anion interaction will promote a larger interfacial pressure. Therefore, the observed interfacial tension parameters of amines and amine salts depend on the nature of the counterion and the water solubility of the species involved and their hydration or water structure breaking properties.

3.1.3. Solvating Extractants

Few studies have been made of this class of extractant. However, the complexes between tri-*n*-butylphosphate and ionic species are more interfacially active than the reagent alone and the stoichiometry of the adsorbed species has been formulated from discontinuities in the interfacial tension–concentration plots [98, 99]. A relationship between equilibrium extraction parameters and interfacial tension has been demonstrated [100] for the extraction of metal thiocyanates by long-chain alcohols in both toluene–water and heptane–water systems.

3.1.4. Mixed-Extractant Systems

It is unfortunate that few interfacial studies of mixed extractants or diluent modifier–extractant systems have been carried out, since much useful information about these systems could be so obtained. Indeed, whenever hydrogen-bonding interactions occur between extractants and between extractants and diluent modifiers, these interactions are reflected in changes in extraction equilibria, kinetics, and naturally in the interfacial physical chemistry. Aggregation phenomena are thus seen as merely particular examples of a more general situation. The limited available data on mixed systems illustrate the general principles given previously. Thus addition of tri-*n*-butylphosphate to the system uranium(VI) (UO^{2+})–di-2-ethylhexylphosphoric acid increases the interfacial tension [101], and although considerable enhancement of equilibrium extraction occurs, the rate of uranium extraction is reduced [102].

The interaction between a carboxylic acid and an alkylated 8-quinolinol was inferred from the increase in interfacial tension with addition of acid until it reflected that of the carboxylic acid itself. This anomalous behavior has been explained in terms of hydrogen-bonding interaction between the two extractants to form a less interfacially active species; thus as the carboxylic acid concentration was increased the 8-quinolinol was replaced at the interface by carboxylic acid molecules. The interaction was confirmed by proton magnetic resonance studies [66]. This result is significant when the chemistry of metal extraction by such mixed systems is considered. For example, the oxidation of cobalt on extraction with 8-quinolinol to the inert cobalt(III) species is an interfacial reaction as addition of excess carboxylic acid effectively prevents oxidation and permits extraction of the cobalt(II) species as a mixed complex. Similar behavior may be anticipated when carboxylic acids are replaced by alkyl phosphoric acids. However, addition of sulfonic acids that are highly interfacially active effectively displaces all other molecules from the interface and has a considerable effect on solvent extraction kinetics [61, 62].

The effects of addition of nonyl phenol to hydroxyoxime and 8-quinolinol-based extraction systems have already been discussed. Interaction between the nonyl phenol and an alkylated 8-quinolinol has been inferred from the increase in the interfacial tension of the system with increasing nonyl phenol concentration [69]. This observation agrees with a reduction in extraction rate for copper, but mixed-complex formation appears to take place as the equilibrium extraction of copper is enhanced. Similar interfacial behavior is anticipated for hydroxyoxime–nonyl phenol mixtures, as nonyl phenol addition reduces copper extraction rate, but in this context no mixed-complex formation takes place and the degree of copper extraction is decreased.

It is of interest that the α-hydroxyoxime LIX63 (Henkel Corporation) that is added to the β-hydroxybenzophenone oxime (LIX65N) as an accelerator for copper extraction is less interfacially active than LIX65N [16, 17]. The accelerator effect in this case does not seem to involve interfacial physical chemistry, in contrast to that for sulfonic acids and similar compounds. Such phenomena are discussed more fully in the section on kinetics.

3.2. Interfacial Potential [84]

The penetration of hydrophilic groups of molecules adsorbed at an oil/water interface into the aqueous phase will cause orientation of nearby water molecule dipoles, resulting in a potential difference across the interphase volume (Fig. 11).

Water Oil

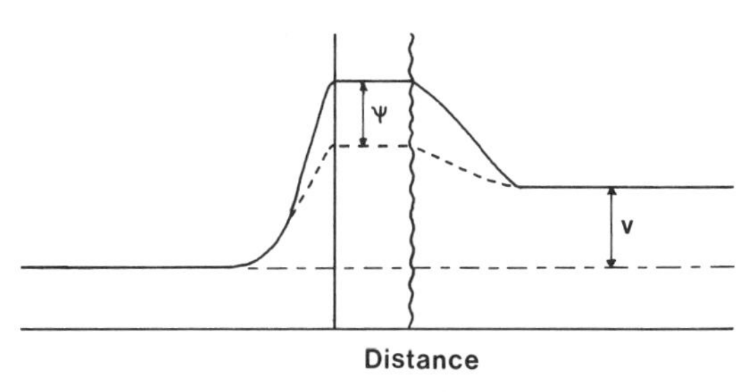

Figure 11. Schematic of interfacial potential caused by adsorption of molecules containing hydrophilic groups at an oil/water interface.

The size of this potential difference Ψ will depend on the system and also parameters such as pH, temperature, and ionic strength. If the organic phase is sufficiently polar to permit appreciable ionization of extracted complexes, a further distribution potential V must be included [84], but this is negligible for the nonpolar diluents normally used in metal extraction. Results obtained from these measurements complement those obtained from interfacial tension studies as they are dependent on molecular orientation of the adsorbed molecules and interactions with metal ions in the aqueous phase.

Because of experimental difficulties associated with the two liquid phases, few data are available on metal extraction related systems. Much more data are available for the surface potential at an air–water interface that is closely related to that at a liquid–liquid interface. For precise information on molecular geometry at the interface, the potential must be broken down into its individual dipole and electrostatic components. This has been achieved for surface potential measurements [103, 104] and for insoluble films at liquid–liquid interfaces [103]. Although the interfacial potential is related simply to the number of adsorbed molecules, the relationship to standard dipole moments of the adsorbed molecules is unknown and probably complicated [84].

The penetration of polar head groups into the aqueous phase will cause a distribution of charge within the interphase volume that, in turn, influences the concentration of ions in this region. This can be demonstrated by the adsorption of dyes; for example, an acid dye adsorbed at a benzene/water interface will indicate a pH more acidic than that of the bulk aqueous phase [105]. Also the interfacial tension of adsorbed acids reaches the value indicative of half-ionization at a pH three to four units on the alkaline side when compared to bulk aqueous-phase data [106]. If this concentration gradient of ions follows a Boltzmann-type distribution of the form

$$C_s = C_b \exp\left(-\frac{\epsilon\Psi}{kT}\right) \tag{24}$$

Then

$$\mathrm{pH}_s = \mathrm{pH}_b + \frac{\epsilon C}{2.303kT} \tag{25}$$

Thus only when $\Psi = 0$ will $\mathrm{pH}_s = \mathrm{pH}_b$, and if Ψ is negative, then $\mathrm{pH}_s < \mathrm{pH}_b$. A value of Ψ of the order of 200 mV will give a difference between surface and bulk pH values of three to four units.

Similar results can be obtained from measurement of the zeta potential. The following relationship was found between interfacial pH and the zeta potential for lauric acid soaps for a number of liquid–liquid systems [107]:

$$pH_s = pH_b - \frac{\zeta}{59} \tag{26}$$

Much more data on these systems are required before interfacial potentials can be fully utilized in the interpretation of interfacial reactions in solvent extraction.

3.3. Interfacial Viscosity [84]

Adsorption at a liquid/liquid interface will produce at surface saturation a coherent monolayer, which will behave similarly to monolayers produced at, for example, an air/water interface by the spreading of carboxylic acids. However, because of the experimental difficulties in obtaining results on liquid–liquid systems, data appropriate to solvent extraction is very scarce. These difficulties arise mainly from the tendency of the indicating device to disrupt weakly adsorbed surface films. Special viscometers [108] have been developed for much lower surface viscosities, but they have some limitations [17]. In spite of the experimental difficulties, more data on interfacial viscosity would be valuable for the interpretation of solvent extraction mechanisms. For example, it would be useful to know whether a particular adsorbed film is of a gaseous, liquid, or solid type, and this information is not readily obtainable from other measurements.

One of the most important effects of interfacial viscosity is on the bulk phases on either side of the monolayer. Thus when a monolayer flows along a liquid surface, some of the underlying liquid is carried along with it, and conversely, a moving bulk phase will drag a uniform monolayer along until the stress caused by the viscous traction is balanced by the backspreading pressure of the monolayer. This phenomenon can reduce or prevent circulation in moving liquid drops and is thus of great importance in the consideration of liquid–liquid extraction with surface-active reagents (Fig. 12). The effect of surfactants on mass transfer in nonreacting systems is discussed in Chapter 3 (see Fig. 12 of that chapter).

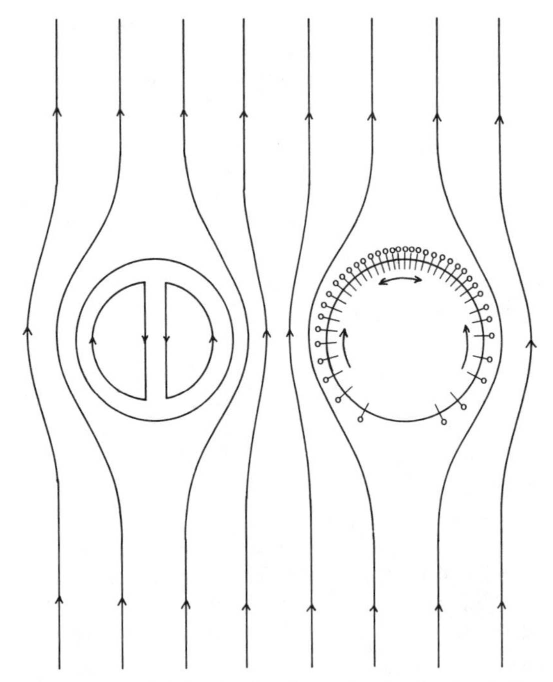

Figure 12. Schematic showing reduction in circulation in moving liquid drops caused by monolayer formation on the drop surface.

4. KINETICS

The kinetics of solvent extraction of reactive solutes are complex, involving mass transfer with chemical reaction in a heterogeneous system. If extraction rate is diffusion controlled, it will depend on the interfacial area and the concentration of the slow-diffusing species. When the extraction rate is chemically controlled it is important to ascertain the location of the rate-controlling chemical reaction or reactions, that is, within a bulk phase or at the interface or in a thin zone adjacent to the interface. For a bulk-phase rate controlling chemical reaction (see Chapter 2.1, regime 1), the important parameters will be solubility of reactants, their distribution coefficients (which will vary with diluent choice and ionic strength of the aqueous phase), ionization constants if appropriate, and phase volume. For an interfacial chemical reaction under diffusion controlled kinetics (see Chapter 3, Section 3.3.1) the composition of the interface will correspond to the concentration of species as given by the equilibrium expression for the interfacial reaction. When interfacial chemical reactions are rate controlling, the important parameters are interfacial area, interfacial activity of reacting species, and molecular geometry with respect to preferential molecular orientation at the interface. Under such rate-controlling conditions the composition of the interface will be that of reactants only and interfacial physical chemical measurements will be very useful as an aid to data interpretation.

Interfacial tension measurements are not sufficient for elucidation of the rate-controlling steps in a given system. Fundamental kinetic studies are necessary, and three main methods are used:

1. In the *single-drop technique,* drops of known size and hence interfacial area are passed through the continuous phase in a vertical column length. See also Section 4.4 of Chapter 3. Mass transfer and diffusion are strongly enhanced when there is internal circulation within the drops; however, as shown in the previous section, this is unlikely to be present in several metal extraction systems of commercial interest.
2. The *Lewis cell* avoids the problem of boundary-layer renewal encountered in the single-drop system. This apparatus consists of a vessel containing the phases that are stirred simultaneously to achieve mixing within each phase without causing interfacial rippling. A more detailed description is given in Section 4.3 of Chapter 3. Interfacial area is controlled by the vessel geometry and the rate of interfacial mass transfer per unit interfacial area can thus be measured accurately.
3. Complete dispersion of the phases can be achieved in *shaken or stirred contactors,* such as that described in Section 3 of Chapter 2.1. A contactor that has been extremely useful in studying metal extraction is the AKUFVE equipment, which essentially consists of a mixer and a centrifuge for rapid and complete phase separation and recycle [109] (see also Chapter 17.2). As this is a totally enclosed system, on-line sensing can be used in the recycle loops to yield kinetic data, which provides an advantage over a simple stirred mixer or shaken flask. However, the AKUFVE technique suffers from the drawback common to all dispersion contactors that the interfacial area cannot be measured. This is a disadvantage when interfacial rate-controlling processes are concerned, particularly since the interfacial area is likely to vary with the level of extraction due to interfacial property changes during a kinetic run.

Obviously, each of these three methods employs quite different hydrodynamic conditions and this should be borne in mind when comparisons are made of kinetic data obtained by these techniques. Hydrodynamic conditions are of major importance in the study of rate processes in solvent extraction. An extensive review of the basic kinetic models and their application to experimental systems has been published by Danesi and Chiarizia [169] that extends this section.

4.1. Acidic Extractants

If the complex forming reactions are homogeneous in the aqueous phase, the reaction scheme is that given in Eqs. (6)–(9). Several interpretations of kinetic data for relatively water soluble chelating extractants have been made [3, 110], giving a rate expression of the form

$$r = \frac{k_f[M^{n+}]^x[E^-]^y}{[H^+]^z} \tag{27}$$

This expression represents the forward extraction rate at a point far removed from equilibrium, and a similar expression can be written for the reverse or stripping reaction. At equilibrium the forward and backward rates are equal and the two rate equations combine to form the mass action expression as defined by the stoichiometry of the equilibrium extraction reaction.

However, when extractants are relatively insoluble, interfacial chemical reactions must be considered and a model must be developed to account for both hydrodynamic and chemical kinetic factors. Such a model has been developed by Danesi et al. [111] and Ortiz et al. [112]. By analogy from Eq. (5), the forward reaction rate for extraction of M^{n+} is given in terms of interfacial concentrations:

$$r' = \frac{k'_f[M^{n+}]_i^x[E^-]_i^y}{[H^+]_i^z} \tag{28}$$

The interfacial and bulk-phase concentrations are related by the equations of mass transfer, and, for simplicity, it is assumed that the system is in steady state, both bulk phases are well mixed and that the aqueous-phase- and organic-phase-soluble species are insoluble in the other phase. It thus follows that transfer of M^{n+} from the aqueous to the organic phase can take place only by chemical reaction; thus the mass-transfer rate of M^{n+} from the bulk aqueous phase to the interface is equal to the rate of chemical reaction and the rates of transfer of other species are linked to the transfer of M^{n+} by the stoichiometry of Eq. (5); hence

$$r' = \frac{k_f' \{[M^{n+}] - r'/K_{M^{n+}}\}^x \{[RH] - 2r'/K_{\overline{RH}}\}^y}{\{[H^+] + 2r'/K_{H^+}\}^z} \tag{29}$$

or

$$r' = k_f' \frac{[M^{n+}]^x [\overline{RH}]^y}{[H^+]^z} F \tag{30}$$

where

$$F = \frac{\{1 - r'/K_{M^{n+}}[M^{n+}]\}^x \{1 - nr'/K_{\overline{RH}}[\overline{RH}]\}^y}{\{1 + nr'/K_{H^+}[H^+]\}^z} \tag{31}$$

Thus from Eq. (30) the kinetic expression for a heterogeneous reaction differs from that for a homogeneous reaction by a factor of F that is less than unity. It is shown in Section 3.1 of Chapter 3 that from the mass-transfer point of view, a slow interfacial reaction is responsible for additional resistance; that is, the mass-transfer rate is slower than it would be if the interfacial equilibrium were reached rapidly. Factor F in Eq. (31) depends on the hydrodynamic conditions of the system by way of the mass-transfer coefficients of the species involved in the reaction. However, as may be seen, F is affected not only by the hydrodynamic conditions of the system, but also by the concentration of the reactants and the rate of chemical reaction itself. Thus values of the apparent reaction orders x, y, and z obtained from experimental rate data will depend on the concentrations of reactants and the hydrodynamic conditions, and unless due consideration is given to the relationship between chemical reaction rates and mass transfer in the concentration ranges of interest, extraction mechanisms deduced from apparent reaction orders may be in error. Relatively few kinetic studies have been reported compared with the large amount of data on equilibrium studies.

4.1.1. Acidic Nonchelating Extractants

No definitive kinetic studies have been reported for carboxylic acid extractants, but several studies have been carried out for dialkyl phosphoric acids and sulfonic acids. Table 2 shows that all these systems have interfacial rate-controlling chemical reactions. The variety of rate-controlling reactions shown cannot at present be rationalized with respect to the extraction system, that is extractant and diluent, metal ion, and rate-controlling reaction. The effects of extractant

TABLE 2 KINETIC STUDIES OF NONCHELATING ACID EXTRACTANTS[a]

Metal	Anion	Diluent	Technique	Reaction Site	Mechanism	Reference
					Extraction with Organophosphoric Acids	
Cu^{2+}	SO_4^{2-}		Q	I	$M^{2+} + (H_2A_2)_i \rightleftharpoons (MA_2H^+)_i + H^+$	113
Co^{2+}	SO_4^{2-}		Q	I		114
Ca^{2+}	NO_3^-	Dodecane	Q	I	$Ca^{2+} + HA_i \rightleftharpoons (CaA_2)_i + 2H^+$ $(CaA_2)_i + 2(H_2A_2) \rightleftharpoons \overline{CaA_2 \cdot 4HA}$	115
Fe^{3+}	ClO_4^-	*n*-Octane	Q, V	I	$Fe^{3+} + A^- \rightleftharpoons FeOH^{2+} + HA \rightleftharpoons FeA^{2+}$ (low HA) $FeA^{2+} + A^- \rightleftharpoons FeA^{2+} \cdot HA$ (high HA)	116
Fe^{3+}	ClO_4^-	Benzene	Q	I		117
Tm^{3+}	ClO_4^-	Toluene	DC	I	Diffusion model employed	118
					Extraction with Alkyl Sulfonic Acids	
Fe^{3+}	ClO_4^-	Toluene	Q	I	$M^{3+} + 2(HS)_i \rightleftharpoons (MS_2^+)_i + 2H^+$	111, 119
Eu^{3+}	ClO_4^-	Toluene	Q	I	$(MS_2^+)_i + \overline{(HS)_m} \rightleftharpoons \overline{(MH_{3-m}S_m)} + 2(HS)_i + H^+$	119, 120
Tm^{3+}	ClO_4^-	Toluene	DC	I	Diffusion model employed	118

[a]Abbreviations: Q = quiescent interface; V = vigorous stirring; DC = diffusion cell; I = interface.

molecular geometry and acid strength have not been studied in detail, and further rationalization awaits such studies. It is of interest, however, that chloride ions accelerate the extraction of ferric iron through displacement of the less labile water of hydration by the more labile chloride ligands to give a more kinetically active species [116]. Other organic and inorganic ligands achieve the same effect [116]. It should also be noticed that a mixed diffusional–chemical reaction kinetic regime was found in several of the studies.

4.1.2. Acidic Chelating Extractants

This subgroup of extractants has provided the largest amount of kinetic data, which are summarized in Table 3. For some of the data, rate-controlling reactions occur in a bulk phase and relate generally to the extraction mechanisms given in Eqs. (6) to (9), whereas in the other cases the chemical reactions of kinetic significance take place at the interface. It would be of considerable interest to have data for a homologous series of chelating agents with a wide range of aqueous solubilities to show whether these two mechanisms are distinct, or, as is more likely, that as the aqueous concentration of the extractant increases the reaction zone moves from the interface right into the bulk aqueous phase.

Table 3 shows that a variety of mechanisms have been postulated for dithizone extractions. The earlier studies were carried out by using shaken flasks, and thus there would be difficulty in differentiating between interfacial or homogeneous aqueous-phase reactions. The recent study by Nitsch and Kruis [133] showed clearly by use of a Lewis cell technique that the kinetically important chemical reactions were interfacial. The rate-controlling step varied with the zinc concentration. At high zinc concentration, diffusion of dithizone to the interface was rate controlling, but the chemical reaction between zinc and dithizone anions at the interface was rate controlling at low metal concentration.

No interfacial chemical reactions have been postulated in the kinetic studies with β-diketones in shaken flasks. Reactions with diketonate anions are found to be rate controlling, except at high acidities when molecular diketone is the reactive species. In the only study using a quiescent interface cell, diffusional processes were found to be rate controlling, although the proposed mechanistic model was dependent on pH. Further studies with higher-molecular-weight diketones of lower aqueous solubility such as are now being developed for commercial application in hydrometallurgy will be interesting.

Copper extraction with hydroxyoximes has been of particular interest recently because of its commercial significance (see Chapter 25.1). There is no disagreement that the kinetically important reactions occur at the interface. However, all three experimental techniques mentioned earlier (items 1–3 in the list at the beginning of Section 4) have been used in the studies, and with different extraction and hydrodynamic conditions, it is not surprising that the reaction orders assigned to the reactants (Table 4) differ considerably. Critical reviews of the kinetic studies on this system have been given by Flett [10] and Hummelstedt [12]. The problems encountered with impurities in these commercial extractants have been emphasized, and it has been shown that the extraction rate is reduced in the presence of nonyl phenol impurity, used in the synthesis of the aromatic β-hydroxyoximes [67]. Thus whereas the rate changes with respect to reactant concentrations in the same system should be of some mechanistic significance, it is not possible to make quantitative comparisons between commercial samples of different hydroxyoximes as impurity concentrations will certainly differ. However, there is no doubt that the β-hydroxyaldoxime, P1 (Acorga Ltd.), extracts copper much faster than does the β-hydroxyketoximes. This is assumed to be due to its superior geometry with respect to molecular conformation at the organic/aqueous interface. Indeed, a relationship between extraction rate and extractant interfacial properties has been demonstrated for different diluents [10, 139] (Fig. 13), but this has not so far been quantified in relation to extractant molecular geometry.

Aggregation of hydroxyoximes in the bulk organic phase is of considerable kinetic importance, and the neglect of this has given rise to a variety of apparent reaction orders for these systems, depending on the concentration ranges and the diluents used. Hummelstedt [18] has shown that variation in the degree of aggregation of the oximes with concentration and diluent type would give rise to the variation in reaction orders found in the various studies and also explain the effect of diluent on activation energy. He also concluded that hydrodynamic effects render the single-drop experimental technique unreliable at pH values greater than 2, which explains the conflicting results concerning pH dependence of the rate in Table 4. Ortiz et al. [112] have shown how neglect of hydrodynamic effects can also

TABLE 3 KINETIC STUDIES OF CHELATING ACID EXTRACTANTS[a]

Metal	Anion	Diluent	Technique	Reaction Site	Mechanism	References
			Extraction with β-Diketones			
Fe^{3+}	ClO_4^-	$CHCl_3$, CCl_4	SF	A	$Fe^{3+} + HA \rightleftharpoons FeA^{2+} + H^+$ $FeOH^{2+} + HA \rightleftharpoons FeA^{2+} + H_2O$	121
	Cl^-, NO_3^-	MIBK	SF	O	$FeX_3 \rightleftharpoons \overline{FeX_3}; \overline{FeX_3} + A^- \quad \overline{FeX_2A}$	122, 123
Fe^{3+}	ClO_4^-	Benzene	V	A	$Fe^{3+} + HA \rightleftharpoons FeA^{2+} + H^+$ $FeOH^{2+} + HA \rightleftharpoons FeA^{2+} + H_2O$	124
Fe^{3+}	ClO_4^-	Benzene	SF	A	$Fe^{3+} + HA \rightleftharpoons FeA^{2+} + H^+$	126
	SCN^-	Benzene	SF	A	$Fe^{3+} + SCN^- \rightleftharpoons FeSCN^{2+}$	
Be^{2+}	Cl^-, ClO_4^-, NO_3^-	CCl_4	SF	A	$Be^{2+} + HA \rightleftharpoons BeA^+ + H^+$	125
	Cl^-, ClO_4^-, NO_3^-	MIBK	SF	O	$BeX_2 \rightleftharpoons \overline{BeX_2}; \overline{BeX_2} + A^- \rightleftharpoons \overline{BeXA}$	
Ga^{3+}	ClO_4^-	$CHCl_3$	SF	A	$GaOH^{2+} + A^- \rightleftharpoons GaOHA^+$	127
Cu^{2+}	NO_3^-	Benzene	Q	A	Inclusion of diffusion term; change of mechanism with pH	128
			Extraction with Thiocarbazones			
Zn^{2+}, Ni^{2+}, Co^{2+}, Cd^{2+}	ClO_4^-	$CHCl_3$, CCl_4	SF	A	Loss of hydrating water from cation	110, 129, 130
Cu^{2+}		CCl_4	SF	I	$Cu^{2+} + HT_i \longrightarrow$	131
Zn^{2+}	Cl^-	$CHCl_3$, CCl_4	Q, D	I	Mechanism dependent on zinc concentration	132, 133

[a]Abbreviations: Q = quiescent interface; SF = shake flask; D = single drop; A = aqueous phase; O = organic phase; V = vigorous stirring; I = interface; MIBK = isobutyl methyl ketone; X = anion.

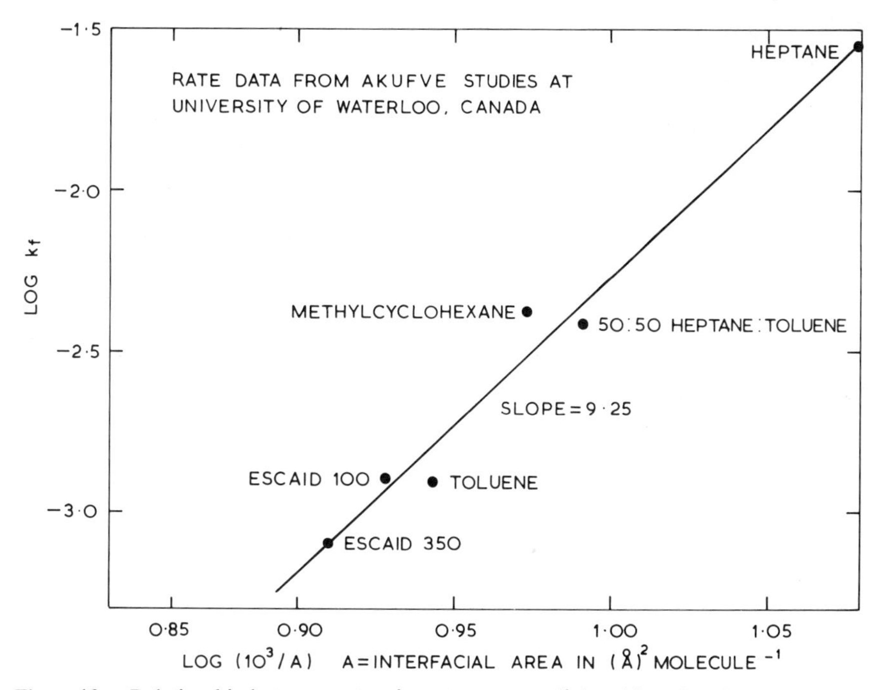

Figure 13. Relationship between extraction rate constant (k_f) and interfacial concentration.

lead to incorrect conclusions concerning reaction order and demonstrate this by calculations according to equation [29]. Thus, although rate controlling interfacial chemical reactions involving unionised extractants are not disputed, the actual rate controlling reaction is still in doubt. Hummelstedt [18] concludes that the rate of water exchange on copper ions is rate controlling, on the basis that the aquocupric cation is the kinetically active species. This is indirectly supported by the observation that the extraction of copper is faster from chloride than from sulfate solution [140]. Kinetically active copper-chloro complexes with a faster ligand exchange rate than the aquoion would account for this observation. Little work has been carried out with other metal ions, and no information on the effect of metal ion on kinetics and mechanisms is available at present.

The kinetic studies of metal extraction with

TABLE 4 SUMMARY OF APPARENT ORDERS OF REACTION FOR COPPER EXTRACTION WITH HYDROXYOXIMES

Experimental Method	Extractant	Order of Reaction			Reference
		H^+	RH	Cu^{2+}	
AKUFVE	LIX65N	-0.9	1.01	1	134
Lewis cell	LIX65N	-0.6	1.10	1	16
Single drop	LIX65N	ND	0.5[a]	1	135
AKUFVE	LIX64N	-1.0	1.59[b]	1	134
Single drop	LIX64N	0	0.5[c]	1	136
Single drop	LIX65N	-1.0[d]	–[e]	1	137
Lewis cell	LIX65N	-2.0–0	2.0–1.0	1.0–0[f]	138

[a]Average value from published data.
[b]Value obtained from varying LIX64N concentration.
[c]Value obtained by varying LIX65N concentration at constant LIX63 concentration.
[d]Only below a rate of 0.1 mmol m^{-2} s^{-1}.
[e]No chemical order ascribable to RH concentration dependence.
[f]Order found to change between these limits with extent of reaction.

8-quinolinol derivatives show a variety of behavior. The compound Kelex 100 (Ashland Chemical Company), produced for commercial use in copper extraction, has been studied using both a fully agitated system [141] and a quiescent interface technique [16]. There is agreement on an interfacial rate-controlling chemical reaction, a first-order rate dependence on copper concentration and pH, but no agreement on the reaction order for the extractant concentration. It seems likely that the reaction takes place with unionized quinoline molecules and thus addition of a second ligand would probably be the rate-controlling step. However, in the extraction of copper by 8-quinolinol itself at a quiescent interface it has been shown that the greater aqueous solubility of this unsubstituted reagent leads to a reaction site in the bulk aqueous phase. Also, the reaction between the extractant and metal cation is fast so that diffusion of the 8-quinolinol to the aqueous phase now becomes the rate-controlling step [142].

4.2. Anion Exchangers

As shown earlier, the extraction of metals by amines involves formation of the amine salt, followed by substitution by a complex metal containing anionic species for the simple anion. Thus the rate of extraction is concerned largely with the substitution process but may depend also on the rate of formation of the amine salt. Both the amines and their salts are surface active, although the salts are more so, and it has been found [95] from interfacial property measurements that the anions are partially associated with the amine cation at the interface. Thus kinetic expreessions relating interfacial chemical reaction with the associated hydrodynamic factors can be derived as for the interfacial reactions in metal cation extraction [143].

As discussed earlier, amines can extract acids as well as metals from aqueous solution, and the kinetics of both processes have been studied under quiescent interface conditions in a stirred cell. In the first case it has been shown [143] in the extraction of hydrochloric acid by trilaurylamine in toluene that the kinetics are best explained by a combination of both interfacial chemical reaction and diffusion of the amine salt away from the interface. The overall scheme is

$$H^+ + (TLA)_i \leftrightharpoons (TLAH)_i^+ \text{ (slow)} \tag{32}$$

$$Cl^- + (TLAH)_i^+ \leftrightharpoons (TLAHCl)_i \text{ (fast)} \tag{33}$$

$$(TLAHCl)_i + \overline{TLA} \leftrightharpoons \overline{TLAHCl} + (TLA)_i \text{ (slow)} \tag{34}$$

The slow interfacial protonation reaction is not indicated by two earlier studies of the extraction of hydrochloric acid by Amberlite LA-2 in various diluents by use of a rising-drop technique [144], and the extraction of both hydrochloric and nitric acids by trilaurylamine in toluene [145]. In both these cases the rate-determining step was concluded to be the rate of diffusion of the ion pairs away from the interface. However, in these studies and others using a diffusion cell [146] and laminar jet [147] techniques, the design of apparatus may emphasize the diffusional nature of the reaction. Thus a recent examination of the extraction of hydrochloric acid by Amberlite LA-2 in *n*-hexane by use of a stirred cell [148] that confirms the earlier indication of diffusion control was carried out at amine concentrations and stirrer speeds at which the diffusional part of the overall rate expression [143] would predominate.

Several kinetic studies have been made on the extraction of metals by amine salts, including the following systems: UO_2^{2+}–SO_4^{2-}–di-*n*-decylamine [149]; Fe^{3+}–Cl^-–trilaurylamine [150]; and Pu^{4+}–NO_3^-–trilaurylamine [151]. Considering the known strong interfacial activity of the amine salts, it is expected that interfacial reactions will predominate in these systems as well, especially when studied at a quiescent interface. The mechanisms proposed by Danesi and Chiarizia [150] for iron(III) and plutonium(IV) for extraction and stripping are as follows:

$$FeCl_3 + q(TLAHCl)_i \rightleftharpoons [FeCl_3(TLAHCl)_q]_i \tag{35}$$

$$FeCl_3(TLAHCl)_q)_i + \overline{TLAHCl} \rightleftharpoons \overline{FeCl_3(TLAHCl)} + q(TLAHCl)_i \tag{36}$$

$$\overline{(FeCl_3(TLAHCl)_q)_i} + 3\overline{TLAHCl} \rightleftharpoons \overline{FeCl_3(TLAHCl)_3} + q(TLAHCl)_i \tag{37}$$

The plutonium mechanism is similar, except the polymeric species in the organic phase involves the decamer of $TLAHNO_3$, which would be the favored species.

McDowell and Coleman [149] studied the extraction of uranium(VI) in sulfate solution in a quiescent interface cell, measuring the transfer of both uranium and sulfate. The extraction mechanism was found to depend on the aqueous sulfate concentration. At low concentration a neutral transfer process was found:

$$3\overline{(R_2NH_2)_2SO_4} + UO_2SO_4 \rightleftharpoons \overline{(R_2NH_2)_6UO_2(SO_4)_4} \quad (39)$$

whereas anionic exchange occurred at high sulfate concentrations:

$$\overline{3(R_2NH_2)_2SO_4} + UO_2(SO_4)_2^{2-} \rightleftharpoons \overline{(R_2NH_2)_6UO_2(SO_4)_4} + SO_4^{2-} \quad (40)$$

The location of the reaction site was taken as the organic/aqueous interface, as was also shown in the extraction of copper from chloride media by a quaternary amine salt [152].

In all these processes it is recognized that the extractant molecules will be associated in the bulk organic phase and that these aggregated species will be in equilibrium with each other and with the monomer. However, if these aggregation equilibrium reactions are fast, the reaction of the monomer will be kinetically indistinguishable from that of the aggregate.

4.3. Solvating Extractants

In this class of reagents most of the kinetic work has been concerned with the organophosphorus reagents. These reagents are capable of extracting both acids and metallic species, and kinetic studies of both types of process have been reported. Yagodin and Tarasov [153, 154] have stripped acids from loaded organic phases by using a cell that allows measurement at very short contact times. Initially the mass-transfer process is controlled by an interfacial chemical reaction, but as the reaction proceeds, the process becomes diffusion controlled. In the latter case the rate-determining step is diffusion of the solute from the bulk organic phase to the interface. This behavior was shown for the following systems: HNO_3–TBP; HSCN–TBP; HNO_3–TOPO; and HSCN–TOPO, being stripped into water. For stripping of $HClO_4$–TBP, the mass-transfer process seems to be diffusion controlled at all times. Introduction of excess reagent or surfactants slowed the mass-transfer process, probably by reducing interfacial mobility. Diffusion control of the extraction of nitric acid by tri-*n*-butylphosphate has also been recorded [155].

The extraction and stripping of uranium(VI)/organophosphorus reagent systems has been studied using diffusion cell [156], single drop [157, 158], short contact time [159], and Lewis cell equipment [160]. These studies have all recently been summarized by Yagodin [159], who has suggested a mechanism for the stripping process. Similar studies have been carried out on the systems plutonium(IV)–tri-*n*-butylphosphate–nitrate [157] and yttrium(III)–tri-octylphosphine oxide–nitrate [158] systems by use of the single-drop technique.

The increasing interest in the "crown ethers" is reflected in the kinetic study of the extraction of potassium by dibenzo-18-crown-6 in a stirred cell [161]. Analysis of the rate of mass transfer indicates that extraction occurs by an interfacial chemical reaction, but the mechanism of transfer of the interfacial complex into the bulk phase cannot be deduced from the data. A further interesting development is the separation of palladium(II) from platinum(II) by use of the faster extraction rate of palladium with dialkylsulfides [162]. No detailed kinetic work is available yet on this system.

4.4. Mixed-Extractant Systems

Thus far consideration has been given only to simple organic-phase systems consisting of an extractant and a diluent. Addition of other compounds such as modifiers to the organic phase is likely to affect not only phase properties, but also the kinetics, due to interaction with the extractant through hydrogen bonding, competition for interfacial sites and so on. More recently, the commercial introduction of accelerator compounds has led to considerations of how such compounds act to increase the extraction rate. Thus the extraction kinetics of mixed-extractant systems are of considerable importance.

Although the addition of surfactants causes an increase in interfacial area, it commonly reduces the rate of mass transfer per unit area in extraction systems without chemical reaction. For extraction systems with chemical reaction, surfactants may retard or enhance the extraction rate, and the effect is dependent on the nature of the reaction (i.e., cation or anion extraction) and the nature of the charge developed at the interface by adsorption of the surfactant. Thus adsorption of a positively charged species at the interface retards the rate of extraction of cations, whereas adsorption of a negatively charged species would enhance the extraction rate. For anion extraction, the opposite arguments would apply. Thus Yagodin [159] has shown that addition of a cationic surfactant reduced the stripping rate of $UO_2(NO_3)_2(TOPO)_2$ whereas an anionic surfactant increased the rate. The retardation of extraction rate is simply explained by invoking a blocking mechanism that either prevents the chemical reaction taking place and/or drastically reduces the phase boundary transfer rate of reactants and products of chemical reaction. Enhancement is achieved by interaction between the surfactant ions and the metal ion or anionic complex of interest, whereby it is rapidly transferred into the organic phase to undergo rapid ligand exchange reactions yielding the extracted species expected in the absence of the surfactant. This latter mechanism is well known in organic chemistry, where it is termed *phase transfer catalysis* (see also Section 6.1 of Chapter 2.1). Several specific examples exist in metal cation extraction that involve the formation of negatively charged interfacially active species capable of ionic interaction with the metal cation. Thus addition of dinonylnaphthalene sulfonic acid [162, 163], dioctylsulfosuccinate [163], di-2-ethylhexylphosphoric acid [57, 164], and a carboxylic acid [165, 166] all enhance the rate of copper extraction by hydroxyoxime extractants. The carboxylic acid example in particular demonstrates the importance of acid ionization in the rate enhancement as no enhancement is observed at pH values below 3.0, in which pH range no ionization of the carboxylic acid would be expected.

Modifier compounds, like diluents, can interact with the extractant through hydrogen bonding to affect the rate of extraction. As shown earlier, this effect is related directly to a modification of the equilibrium of the chemical reaction, in distinction from the effect of added surfactants. It is axiomatic that any change in the overall thermodynamic equilibrium of the system will affect the rate of chemical reaction. Thus the use of modifiers such as tri-n-butylphosphate, isodecanol, and nonyl phenol, which can interact with the extractant and in some cases the extracted complex (through hydrogen bonding), inevitably results in a change in the kinetics. Thus far reductions in extraction rate by modifiers have always been found, probably because of reduction in activity (concentration) of one of the active species in the rate equation (extractant concentration) through formation of kinetically inactive modifier–extractant adduct species. Examples are the reduction in copper extraction rate on addition of nonyl phenol to hydroxyoxime systems [67, 68] and to 8-quinolinol systems [69], where adduct formation between the extractant and modifier is known to occur.

The so-called accelerator compounds used or proposed for use with the β-hydroxyoxime extractants belong to a different class of additive. The best known of these is 5,8-diethyl-7-hydroxydodeca-6-oxime, commonly known as LIX63 (registered trademark, Henkel Corporation), whereas other hydroxyoximes, dioximes, other nitrogen containing compounds, have been shown also to possess accelerator properties [166]. Here the accelerator compound does not affect the equilibrium properties of the system, so a catalytic mechanism must be sought. Various mechanisms have been proposed [10, 12, 13, 16, 135, 136, 167] but have also been subject to criticism [18, 168]. It is likely that the answer lies in the suggestion made by Hummelstedt [18] that addition of the α-hydroxyoxime enhances the rate of water exchange and dehydration of the copper ion, which he suggests is the rate-controlling factor in copper extraction. This suggestion can be extended to yield the following hypothesis.

The key to the kinetic enigmas associated with the hydroxyoxime systems (and others) may well be concerned with the exact nature of the kinetically active species in the aqueous phase. If this is the hydrated copper cation, the rate of water exchange may well be the rate-controlling process as proposed by Hummelstedt [18]. However, if one or more water molecules have already been removed (as suggested by the kinetic difference between chloride and sulfate solutions) with the formation of a new kinetically active species capable of more rapid ligand exchange, the requirements for acceleration would be satisfied. Of crucial importance here will be the aqueous-phase interaction between hydrated copper ions and the various hydroxyoxime- or nitrogen-containing species present to form copper–nitrogen adducts that are kinetically active. Nitrogen basicity, donor power, and aqueous solubility all require correlation, but it is believed that through such concepts the outstanding questions relating to acceleration of extraction reactions by essentially interfacially inactive species will be answered.

In concluding this section, it is worth noting that erroneous rate expressions in such systems have been published for the forward extraction. These have the form

$$r = k_B \frac{[M^{n+}]^a[\overline{RH}]^b}{[H^+]^c} \cdot [\bar{B}]^p \qquad (41)$$

where B is the accelerator and a, b, c, and p are reaction orders. It has rightly been pointed out [18, 135, 138] that this expression must also include the rate of extraction in the absence of accelerator:

$$r = k_f \frac{[M^{n+}]^x[\overline{RH}]^y}{[H^+]^z} + k_B \frac{[M^{n+}]^a[\overline{RH}]^b}{[H^+]^c} \cdot [\bar{B}]^p \tag{42}$$

As techniques have been developed for experimental study of mass-transfer kinetics involving chemical reactions, the results have increasingly indicated the importance of the interface. Thus reinvestigation of earlier kinetic work under defined interfacial conditions is leading to a reassessment of the location of the reaction site [128, 133]. However, it is clear that the choice of equipment and experimental parameters such as concentration ranges of reactants can predetermine the nature of the results obtained [112]. Thus in the rising- and falling-drop technique, a change in kinetic regime may occur between (1) drop formation and coalescence and (2) steady state rising and falling; the Lewis cell, if operated under conditions where the rate is independent of stirring speed, reduces the overall effect of diffusion processes. Some examples of these effects for nonmetal reaction systems with comparatively simple chemistry have been given in Chapter 2.1, Section 3. Unfortunately, it is impossible to choose a single apparatus sufficiently flexible to allow study of all possible systems. The best that can be achieved is for the investigators to realize the limitations of their data and consider these in their interpretations.

NOTATION

a, b, c	reaction orders [Eqs. (41) and (42)]
A^-	anion
B^-	anion
B	accelerator
C	concentration of ions
D	distribution coefficient [see Eq. (3)]
E	reacting species of extractant [Eq. (27)], that is, RH or R^-
F	correlation factor for heterogeneous reaction [see Eq. (31)]
K	mass-action constant [see Eq. (10)]
K_D	acid dissociation constant [see Eq. (7)]
K_F	formation constant of metal complex [see Eq. (8)]
$K_{M^{n+}}, K_{\overline{RH}}, K_{H^+}$	mass-transfer coefficients for species as indicated by subscripts, cm/s
k	Boltzmann constant, J/K
k_B	rate constant for forward reaction accelerated by B [Eq. (41)], liter mol^{-1} s^{-1}
k_f	rate constant of forward reaction [see Eq. (27)], liter mol^{-1} s^{-1}
k_f'	rate constant of forward interfacial reaction [see Eq. (28)], liter mol^{-1} s^{-1}
L	inorganic ligand
M^{n+}	n-valent metal cation
MR_n	metal complex of monobasic acid or chelating agent
p	reaction order for accelerator [Eqs. (41) and (42)]
p_{MR_n}	thermodynamic partition constant of metal complex [see Eq. (9)]
p_{RH}	thermodynamic partition constant of monobasic acid or chelating agent [see Eq. (6)]
RH	monobasic acid or chelating agent
R^-	anion of RH
r	rate of extraction [see Eq. (27)], $liter^{(1+z-x-y)}$ $mol^{(x+y-z-1)}/s$
r'	rate of extraction, see eq. (28), $liter^{(1+z-x-y)}$ $mol^{(x+y-z-1)}/s$
S	solvating agent
T	temperature, Kelvins
TLA	trilaurylamine
x, y, z	reaction orders
X^-	anion
Y^-	anion

Greek Letters

β_i	overall formation constant of ith species
ϵ	electronic charge, C (coulombs)
Ψ	interfacial potential, V
ζ	zeta potential, V

Subscripts and Parentheses

$[-]$	concentration of species in aqueous phase, mol/liter
$\overline{[-]}$	concentration of species in organic phase, mol/liter
—	overbar denoting species in organic phase
b	subscript denoting bulk phase

i subscript denoting interfacial species
micelle subscript denoting micelle species
s subscript denoting surface

REFERENCES

1. C. Hanson, M. A. Hughes, and J. G. Marsland, *Proceedings of the International Solvent Extraction Conference* (ISEC74), Vol. 3 (Soc. Chem. Ind., London), 1974, p. 2401.
2. D. S. Flett and M. J. Jaycock, in J. A. Marinsky and Y. Marcus, Eds., *Ion Exchange and Solvent Extraction*, Vol. 3, Marcel Dekker, New York, 1973, p. 1.
3. T. Sekine and Y. Hasegawa, *Solvent Extraction Chemistry*, Marcel Dekker, New York, 1977.
4. Y. Marcus and A. S. Kertes, *Ion Exchange and Solvent Extraction of Metal Complexes*, Interscience, New York, 1969.
5. D. H. Liem, *Acta Chem. Scand.* **26**, 191 (1972).
6. Y. Marcus, A. S. Kertes, and E. Yanir, *Equilibrium Constants of Liquid-Liquid Distribution Reactions, Introduction and Part 1: Organophosphorus Extractants*, IUPAC Additional Publication, Butterworths, London, 1974.
7. G. Y. Markovits, and G. R. Choppin, in J. A. Marinsky and Y. Marcus, Eds., *Ion Exchange and Solvent Extraction*, Vol. 3, Marcel Dekker, New York, 1973, p. 51.
8. King Industries, Inc., Norwalk, Connecticut, *SYNEX Liquid Ion Exchange Reagents, Product Bulletin and Bibliography.*
9. A. S. Kertes and H. Gutmann, in E. Matijevic, Ed., *Surface and Colloid Science*, Vol. 8, Interscience, New York, 1975, p. 193.
10. D. S. Flett, *Acc. Chem. Res.* **10**, 99 (1977).
11. C. Hanson, M. A. Hughes, and R. J. Whewell, *J. Appl. Chem. Biotechnol.* **28**, 426 (1978).
12. L. Hummelstedt, T. Tammi, E. Paetero, H. Andresen, and J. Karjaluoto, *Proceedings of 4th International Congress in Scandinavia on Chemical Engineering*, 1977, p. 123.
13. A. W. Ashbrook, *Coord. Chem. Rev.* **16**, 285 (1975).
14. A. W. Ashbrook, *Hydrometallurgy* **1**, 5 (1975).
15. S. Dobson and A. J. van der Zeeuw, *Chem. Ind.* **1975**, 175 (1975).
16. C. Fleming, National Institute for Metallurgy, Report No 1793, Johannesburg, South Africa, 1976.
17. M. Cox and D. S. Flett, *Proceedings of the International Solvent Extraction Conference* (ISEC77) Vol. 1 (Can. Inst. Min. Met., Montreal), 1979, p. 63.
18. L. Hummelstedt, *Proceedings of the International Solvent Extraction Conference* (ISEC77), Vol. 1, (Can. Inst. Min. Met., Montreal), 1979, p. 86.
19. T. A. B. Al-Diwan, M. A. Hughes, and R. J. Whewell, *J. Inorg. Nucl. Chem.* **39**, 1419 (1977).
20. H. J. Foakes, J. S. Preston, and R. J. Whewell, *Anal. Chim. Acta*, **97**, 349 (1978).
21. F. Krasovic and C. Klofutar, *J. Inorg. Nucl. Chem.* **27**, 2437, (1965).
22. D. S. Flett, *Nature* **200**, 465 (1963).
23. J. C. Carpenter, P. G. Christie, V. I. Lakshmanan, G. J. Lawson, and P. S. Nyholm, in J. C. Yannopoulos and J. C. Agarwal, Eds., *Extractive Metallurgy of Copper*, Vol. 2, AIME, New York, 1976, p. 1025.
24. V. I. Lakshmanan and G. J. Lawson, *J. Inorg. Nucl. Chem.* **37**, 207 (1975).
25. D. S. Flett and D. W. West, *Complex Metallurgy '78*, IMM, London, 1978, p. 49.
26. Y. Marcus, *Pure Appl. Chem.* **20**, 85 (1969).
27. Yu. A. Zolotov, *Extraction of Chelate Compounds*, Ann Arbor-Humphrey Science, London, 1970.
28. A. S. Kertes, in C. Hanson, Ed., *Recent Advances in Liquid-Liquid Extraction*, Pergamon Press, London, 1971, p. 15.
29. Y. Marcus, E. Yanir, and A. S. Kertes, *Equilibrium Constants of Liquid-Liquid Distribution Reactions, Part III: Compound Forming Extractants, Solvating Solvents and Inert Solvents.*, IUPAC, Pergamon Press, London, 1977.
30. H. M. N. H. Irving, in J. A. Marinsky and Y. Marcus, Eds., *Ion Exchange and Solvent Extraction*, Vol. 6, Marcel Dekker, New York, 1974, p. 139.
31. R. R. Grinstead and J. C. Davis, *J. Phys. Chem.* **72**, 1630 (1968).
32. R. M. Diamond, in D. Dyrssen, J. O. Liljensin, and J. Rydberg, Eds., *Solvent Extraction Chemistry*, North-Holland, Amsterdam, 1967, p. 349.
33. W. Muller and R. M. Diamond, *J. Phys. Chem.* **70**, 3469, (1966).
34. P. R. Danesi, F. Orlandini, and G. Scibona, *J. Inorg. Nucl. Chem.* **30**, 2513 (1968).
35. J. W. Roddy and C. F. Coleman, *J. Inorg. Nucl. Chem.* **31**, 3599, (1969).
36. M. Muhammed, J. Szabon, and E. Hogfeldt, *Chem. Scripta* **6**, 61 (1974).
37. R. W. Cattrell and S. J. E. Slater, *J. Inorg. Nucl. Chem.* **36**, 947 (1974).
38. D. W. Agers, J. E. House, and J. L. Drobnick, Paper presented at Annual Meeting of AIME, Dallas, Texas, February 1963.
39. Y. Marcus, L. E. Asher, and H. Barak, *J. Inorg. Nucl. Chem.* **40**, 325, (1978).
40. A. S. Kertes, Y. Marcus, and E. Yamir, *Equilibrium Constants of Liquid-Liquid Distribution Reactions, Part II: Alkylammonium Salt Extrac-*

tants, IUPAC Additional Publication, Butterworths, London, 1974.

41. A. S. Kertes, *Critical Evaluation of Some Equilibrium Constants Involving Alkylammonium Extractants*, IUPAC, Commission on Equilibrium Data, Pergamon Press, New York, 1977.
42. St. J. H. Blakeley and V. J. Zatka, *Anal. Chim. Acta* **74,** 139 (1975).
43. E. L. T. M. Spitzer, *Het Ingenieursblad* **41,** 418 (1972).
44. L. G. Sillen, *Acta Chem. Scand.* **18,** 1085 (1964).
45. T. Ishimori and E. Nakamura, Japan Atomic Energy Research Institute Reports, JAERI 1047 (1963); JAERI 1062, (1964–1965).
46. F. G. Seeley and D. J. Crouse, *J. Chem. Eng. Data* **16,** 393 (1971).
47. H. M. Widmer, *Proceedings of the International Solvent Extraction Conference* (ISEC71), Vol. 1 (Soc. Chem. Ind., London), 1971, p. 37.
48. V. A. Mikhailov, *Proceedings of the International Solvent Extraction Conference* (ISEC77), Vol. 1 (Can. Inst. Min. Met., Montreal), 1979, p. 52.
49. T. Ishimori, E. Akatsu, K. Tsukeuchi, T. Kobune, Y. Usuba, K. Kimura, G. Onawa, and H. Uchiyama, Japan Atomic Energy Research Institute Report, JAERI 1106 (1966).
50. D. Dyrssen, *Proc. Symp. Coord. Chem. Tihany 1964*, Akademini Kiado, Budapest, 1966, p. 707.
51. E. F. Kassierer and A. S. Kertes, *J. Inorg. Nucl. Chem.* **34,** 3209 (1972).
52. L. Newman and P. Klotz, *Inorg. Chem.* **5,** 461 (1966).
53. C. H. Ke and N. C. Li, *J. Inorg. Nucl. Chem.* **31,** 1383 (1969).
54. G. Duyckaerts and J. F. Desreux, *Proceedings of the International Solvent Extraction Conference* (ISEC77), Vol. 1 (Can. Inst. Min. Met., Montreal), 1979, p. 73.
55. G. K. Schweitzer, R. B. Neil, and F. R. Clifford, *Anal. Chim. Acta* **33,** 514 (1965).
56. J. C. Spigorelli and C. F. Meloan, *Sep. Sci.* **9,** 79, (1974).
57. E. G. Joe, G. M. Ritcey, and A. W. Ashbrook, *J. Metals NY* **18,** 18 (1966).
58. D. S. Flett and D. W. West, *Proceedings of the International Solvent Extraction Conference* (ISEC71), Vol. 1 (Soc. Chem. Ind., London), 1971, p. 214.
59. M. Cox and D. S. Flett, *Proceedings of the International Solvent Extraction Conference* (ISEC71), Vol. 1 (Soc. Chem. Ind., London), 1971, p. 204.
60. D. S. Flett, M. Cox, and J. D. G. Heels, *Proceedings of the International Solvent Extraction Conference* (ISEC74), Vol. 3 (Soc. Chem. Ind., London), 1974, p. 2559.
61. D. S. Flett, M. Cox, and J. D. G. Heels, *J. Inorg. Nucl. Chem.* **37,** 2533 (1975).
62. B. G. Nyman and L. Hummelstedt, *Proceedings of the International Solvent Extraction Conference* (ISEC74), Vol. 1 (Soc. Chem. Ind., London), 1974, p. 669.
63. R. H. Moore and J. A. Partridge, Battelle Pacific Northwest Laboratories Report No. BNWL-SA-4476 (1972).
64. V. I. Lakshmanan and G. J. Lawson. *J. Inorg. Nucl. Chem.* **35,** 4285 (1973).
65. G. Harrison, V. I. Lakshmanan, and G. J. Lawson, *Hydrometallurgy* **1,** 339 (1976).
66. D. S. Flett, M. Cox, and J. D. G. Heels, *J. Inorg. Nucl. Chem.* **37,** 2197 (1975).
67. C. Hanson, M. A. Hughes, J. S. Preston, and R. J. Whewell, *J. Inorg. Nucl. Chem.* **38,** 2306 (1976).
68. R. F. Dalton, *Proceedings of the International Solvent Extraction Conference* (ISEC77), Vol. 1, (Can. Inst. Min. Met., Montreal), 1979, p. 40.
69. D. S. Flett and D. R. Spink, *Proceedings of the International Solvent Extraction Conference* (ISEC77), Vol. 2, (Can. Inst. Min. Met., Montreal), 1979, p. 496.
70. R. A. Zingaro and C. F. Coleman, *J. Inorg. Nucl. Chem.* **29,** 1287 (1967).
71. S. M. Wang and N. C. Li, *J. Inorg. Nucl. Chem.* **28,** 1091 (1966).
72. C. Woo, W. F. Wagner, and D. E. Sands, *J. Inorg. Nucl. Chem.* **34,** 307, (1972).
73. L. Newman and P. Klotz, in D. Dyrssen, J. O. Liljensin, and J. Rydberg, Eds., *Solvent Extraction Chemistry*, North-Holland, Amsterdam, 1967, p. 128.
74. T. Sekine and D. Dyrssen, *J. Inorg. Nucl. Chem.* **29,** 1489 (1967).
75. A. van Dalen, J. Wijkstra, and K. W. Gerritsma, *J. Inorg. Nucl. Chem.* **40,** 875 (1978).
76. C. Deptula and S. Minc, *Nukleonika* **10,** 421 (1965).
77. C. Deptula, *J. Inorg. Nucl. Chem.* **29,** 1097 (1967).
78. C. Deptula and S. Minc, *J. Inorg. Nucl. Chem.* **29,** 159 (1967).
79. D. H. Liem and O. A. Sinegribova, *Acta Chem. Scand.* **25,** 301 (1971).
80. C. Deptula, *J. Inorg. Nucl. Chem.* **32,** 277 (1970).
81. G. Rinelli and C. Abbruzzese, in M. J. Jones, Ed., *Geology, Mining and Extractive Processing of Uranium*, IMM, London, 1977, p. 20.
82. S. Nishimura, C. H. Ke, and N. C. Li, *J. Phys. Chem.* **72,** 1297 (1968).
83. P. P. Pugechevich, in B. Vodar and B. Le Neindre, Eds., *Experimental Thermodynamics*, Vol. 2, IUPAC, Butterworths, London, 1974, p. 991.
84. J. T. Davies and E. K. Rideal, *Interfacial Phe-*

nomena, 2nd ed., Academic Press, New York, 1963.

85. F. A. Askew and J. F. Danielli, *Transact. Faraday Soc.* **36,** 785 (1940).

86. D. E. Graham, L. Chatergoon, and M. C. Phillips, *J. Phys. E* **8,** 696 (1975).

87. E. Hutchinson, *J. Colloid Sci.* **3,** 235 (1948).

88. F. Seelich, *Monatsch. Chem.* **79,** 348 (1948).

89. R. Chiarizia, P. R. Danesi, G. D'Alessandro, and B. Scuppa, *J. Inorg. Nucl. Chem.* **38,** 1367 (1976).

90. C. G. Hirons, Ph.D. thesis, Council of National Academic Awards (Hatfield Polytechnic) 1981.

91. M. A. Hughes, E. Mistry, and R. J. Whewell, *J. Inorg. Nucl. Chem.* **40,** 1694 (1978).

92. R. Price and J. A. Tumilty, in G. A. Davies and J. B. Scuffham, Eds., *Inst. Chem. E. Symp. Ser. No. 42, Hydrometallurgy*, 1975, paper 18.

93. R. F. Dalton, F. Hauxwell, and J. A. Tumilty, *Chem. Ind.* **1976,** 181 (1976).

94. V. V. Tarasov, G. A. Yagodin, E. V. Yurtov, and I. N. Gritsko, *Tr. Mosk. Khim.-Technol. Inst.* **81,** 73 (1974).

95. G. Scibona, P. R. Danesi, A. Conte, and B. Scuppa, *J. Colloid Interface Sci.* **35,** 631 (1971).

96. R. M. Diamond, *J. Phys. Chem.* **67,** 2513 (1963).

97. K. Tamaki, *Bull. Chem. Soc. Jap.* **40,** 38 (1967).

98. E. Chifu, Z. Andrei, and M. Tomoaia, *Ann. Chim. (Rome)* **64,** 869 (1974).

99. Z. Andrei and E. Chifu, *Stud. Univ. Babes-Bolyai Ser. Chem.* **21,** 10 (1976).

100. G. I. Starobinets, V. L. Lomako, and E. R. Mazovka, *Dokl. Acad. Nauk BSSR* **18,** 817 (1974).

101. M. E. J. Birch, M.Sc. thesis, Loughborough University of Technology, 1969.

102. A. A. North, private communication cited in Ref. 100.

103. J. T. Davies and E. K. Rideal, *Can. J. Chem.* **33,** 947 (1955).

104. J. T. Davies, *Biochem. Biophys. Acta* **11,** 165 (1953).

105. G. S. Hartley and J. W. Roe, *Transact. Faraday Soc.* **36,** 101 (1940).

106. R. A. Peters, *Proc. Roy. Soc. A* **133,** 147 (1931).

107. C. J. Cante, J. E. McDermott, F. Z. Saleeb, and H. L. Rosano, *J. Colloid Interface Sci.* **50,** 1 (1975).

108. J. T. Davies and G. R. A. Mayers, *Transact. Faraday Soc.* **56,** 690 (1960).

109. J. Rydberg, H. Reinhardt, and J. O. Liljensin, in J. A. Marinsky and Y. Marcus, Eds., *Ion Exchange and Solvent Extraction*, Vol. 3, Marcel Dekker, New York, p. 111.

110. J. S. Oh and H. Freiser, *Anal. Chem.* **39,** 295 (1967).

111. P. R. Danesi, R. Chiarizia, and A. Santelli, *J. Inorg. Nucl. Chem.* **38,** 1687 (1976).

112. E. S. Perez de Ortiz, M. Cox, and D. S. Flett, *Proceedings of the International Solvent Extraction Conference* (ISEC77), Vol. 1 (Can. Inst. Min. Met., Montreal), 1979, p. 198.

113. M. L. Brisk and W. J. McManamey, *J. Appl. Chem.* **19,** 109 (1969).

114. M. L. Brisk and W. J. McManamey, *J. Appl. Chem.* **19,** 103 (1969).

115. G. F. Vandergrift and E. P. Horwitz, *J. Inorg. Nucl. Chem.* **42,** 119 (1980).

116. J. W. Roddy, C. F. Coleman, and S. Arai, *J. Inorg. Nucl. Chem.* **33,** 1099 (1971).

117. Yu. B. Kletenik and V. A. Navrotskaya, *Russ. J. Inorg. Chem.* **12,** 1648 (1967).

118. A. T. Kandil and G. R. Choppin, *J. Inorg. Nucl. Chem.* **37,** 1787 (1975).

119. R. Chiarizia and P. R. Danesi, *J. Inorg. Nucl. Chem.* **39,** 525 (1977).

120. P. R. Danesi, R. Chiarizia, and W. A. A. Sanad, *J. Inorg. Nucl. Chem.* **39,** 519 (1977).

121. T. Sekine, J. Yumikura, and Yu. Komatsu, *Bull. Chem. Soc. Jap.* **46,** 2356 (1973).

122. T. Sekine and Yu. Komatsu, *J. Inorg. Nucl. Chem.* **37,** 185 (1975).

123. Yu. Komatsu, H. Honda, and T. Sekine, *J. Inorg. Nucl. Chem.* **38,** 1861 (1976).

124. A. Adin and L. Newman, *J. Inorg. Nucl. Chem.* **32,** 3321 (1970).

125. T. Sekine, Y. Koike, and Yu. Komatsu, *Bull. Chem. Soc. Jap.* **44,** 2903 (1971).

126. H. L. Finston and Y. Inoue, *J. Inorg. Nucl. Chem.* **29,** 199 (1967).

127. T. Sekine, Y. Komatsu, and J. Yumikura, *J. Inorg. Nucl. Chem.* **35,** 3891 (1973).

128. K. Kondo, S. Takakashi, T. Tsuneyuki, and F. Nakashio, *J. Chem. Eng. Jap.* **11,** 193 (1978).

129. B. E. McClellan and H. Freiser, *Anal. Chem.* **36,** 2262 (1964).

130. P. R. Subbaraman, Sr. M. Cordes, and H. Freiser, *Anal. Chem.* **41,** 1878 (1969).

131. R. W. Geiger and E. B. Sandell, *Anal. Chim. Acta* **8,** 197 (1953).

132. W. Nitsch and K. Hillekamp, *Chem. Zeit.* **96,** 254 (1972).

133. W. Nitsch and B. Kruis, *J. Inorg. Nucl. Chem.* **40,** 857 (1978).

134. D. S. Flett, D. N. Okuhara, and D. R. Spink, *J. Inorg. Nucl. Chem.* **35,** 2471 (1973).

135. R. J. Whewell, M. A. Hughes, and C. Hanson, *J. Inorg. Nucl. Chem.* **38,** 2071 (1976).

136. R. L. Attwood, D. N. Thatcher, and J. D. Miller, *Met. Transact. B* **6B,** 465 (1975).

137. R. J. Whewell, M. A. Hughes, and C. Hanson, *J. Inorg. Nucl. Chem.* **37,** 2303 (1975).

138. C. A. Fleming, M. J. Nicol, R. D. Hancock, and N. P. Finkelstein, *J. Appl. Chem. Biotechnol*, **28,** 443 (1978).

139. R. J. Whewell, M. A. Hughes, and C. Hanson, in M. J. Jones, Ed., *Advances in Extractive Metallurgy 1977*, IMM, London, 1977, p. 21.

140. R. J. Whewell, M. A. Hughes, and C. Hanson, *Proceedings ISEC77*, Vol. 1, CIMM, Montreal, 1979, p. 185.

141. D. S. Flett and D. R. Spink, *J. Inorg. Nucl. Chem.* **37,** 1967 (1975).

142. V. Rod and L. Rychnovsky, *Internatl. Congr. Chem. Eng. Chem. Equip. Des. Autom.* (*Proc.*) (5th) **S-1,** 1.7 (1975).

143. P. R. Danesi, R. Chiarizia, and M. Muhammad, *J. Inorg. Nucl. Chem.* **40,** 1581 (1978).

144. F. Nakashio, T. Tsuneguki, K. Inoue, and W. Sakai, *Proceedings of the International Solvent Extraction Conference* (ISEC71), Vol. 2 (Soc. Chem. Ind., London), 1971, p. 831.

145. L. D. Fenton, Ph.D. thesis, Massachusetts Institute of Technology, Cambridge, MA, 1959, cited in Ref. 4, p. 433.

146. M. Harada, H. Araki, and W. Eguchi, *Kagaku Kogaku* **35,** 1136 (1971) through Chem. Abstr. **76,** 5230*d* (1972).

147. T. Kataoka, T. Nishiki, T. Nakaya, and K. Ueyama, *Kagaku Kogaku* **35,** 1157 (1971); through *Chem. Abstr.* **76,** 47677*k* (1972).

148. T. Tsuneguki, K. Kondo, Y. Kawano, and F. Nakashio, *J. Chem. Eng. Jap.* **11,** 198 (1978).

149. W. J. McDowell and C. F. Coleman, *J. Inorg. Nucl. Chem.* **29,** 1325 (1967).

150. R. Chiarizia and P. R. Danesi, *J. Inorg. Nucl. Chem.* **40,** 1811 (1978).

151. R. Chiarizia, P. R. Danesi, and C. Domenichini, *J. Inorg. Nucl. Chem.* **40,** 1409 (1978).

152. M. C. Fuerstenau, M. R. Elmore, and B. R. Palmer, 12th International Mineral Processing Congress Sao Paulo, Brazil, 1977, Meeting 9, Paper 5.

153. G. A. Yagodin and V. V. Tarasov, *Proceedings of the International Solvent Extraction Conference* (ISEC71), Vol. 2 (Soc. Chem. Ind., London), 1971, p. 888.

154. G. A. Yagodin, V. V. Tarasov, and N. F. Kizim, *Proceedings ISEC74*, Vol. 3, SCI, London, 1974, p. 2541.

155. D. R. Olander and M. Benedict, *Nucl. Sci. Eng.* **15,** 354 (1963).

156. H. T. Hahn, *J. Am. Chem. Soc.* **79,** 4625 (1957).

157. F. Baumgartner and I. Finsterwalder, *J. Phys. Chem.* **74,** 108 (1970).

158. L. Farbu, H. A. C. McKay, and A. G. Wain, *Proceedings of the International Solvent Extraction Conference* (ISEC74), Vol. 3 (Soc. Chem. Ind., London), 1974, p. 2427.

159. G. A. Yagodin, V. V. Tarasov, A. B. Ivanov, and N. E. Kruchinina, *Izvest Vyssh. Uchebu, Zaved., Khim. Technol.* **20,** 1496, (1977).

160. V. V. Tarasov, E. V. Turtov, and G. A. Yagodin, *J. Phys. Chem.* **52,** 596 (1978).

161. P. R. Danesi, R. Chiarizia, M. Pizzichini, and A. Saltelli, *J. Inorg. Nucl. Chem.* **40,** 1119 (1978).

162. R. I. Edwards, *Proceedings of the International Solvent Extraction Conference* (ISEC77), Vol. 1 (Can. Inst. Min. Met., Montreal), 1979, p. 24.

163. E. A. Morin and H. D. Peterson, U.S.Patent 3878286 (1975).

164. W. C. Hazen and E. L. Coltrinari, U.S. Patent 3872209 (1975).

165. M. B. Goren and E. L. Coltrinari, U.S. Patent 3927169 (1975).

166. A. J. van der Zeeuw and R. Kok, *Proc. Internatl. Solv. Extr. Conf.* (ISEC77) **1,** 17, 203 (Can. Inst. Min. Met., Montreal) (1979).

167. J. D. Miller and R. L. Attwood, *J. Inorg. Nucl. Chem.* **37,** 2539 (1975).

168. D. S. Flett, J. Melling, and D. R. Spink, *J. Inorg. Nucl. Chem.* **39,** 700 (1977).

169. P. R. Danesi and R. Chiarizia, *CRC Crit. Rev. Anal. Chem.* 1980 (November), 1.

3

INTERPHASE MASS TRANSFER

H. R. C. Pratt
University of Melbourne
Australia

1. INTRODUCTION

A knowledge of the rate of interphase mass transfer is necessary for the design of extraction equipment once the equilibrium relationship has been established. Such rates are dependent on the *interfacial area* of contact of the phases; the *mass-transfer coefficient*, which is a measure of the specific rate of transport of mass across the interface; and the concentration difference, that is, the *driving force*, which effects the transfer.

In most types of equipment one of the liquid phases is dispersed intimately as droplets within the other in order to provide a large area of contact and hence a high performance. Unfortunately, the resulting complex hydrodynamical pattern makes exact analysis of performance difficult, and various simplified forms of laboratory apparatus that provide a known contact area have thus been devised with a view to identification of the major controlling variables. Although there is now a reasonable understanding of mass transfer under such idealized conditions, the situation is still far from satisfactory for the dispersions present in commercial equipment. Thus contactor efficiencies predicted from both theoretical models and experimental data for single droplets have been found to differ widely from measured values [1]. However, better agreement has since been reported for the specific case of packed columns [2], and it may be expected that this will eventually become true of other types of contactor.

This chapter describes the theoretical basis of liquid–liquid mass transfer, and the experimental methods used in laboratory studies. Consideration is also given to certain interfacial phenomena (including heterogeneous reaction) that accompany mass transfer.

2. DIFFUSION AND MASS TRANSFER IN BINARY SYSTEMS

Binary mass transfer occurs in simple extraction processes involving the transfer of a solute between two effectively immiscible solvents. The theory of binary mass transfer is reasonably well established in comparison to that for multicomponent transfer, and the analytical treatment is considerably simpler. As a further simplification, only one-dimensional transport is considered. For a more detailed treatment the books by Sherwood et al. [3] and others [4–6] should be consulted.

2.1. Mechanisms of Transfer

The rate of transfer of solute within a given phase, between the bulk and the interface, is governed by two independent mechanisms. The first, termed *molecular diffusion*, arises from the random thermal motion of the individual molecules. Although these move with very high velocities, their path lengths in liquids are extremely short, so that their rate of migration is slow. Molecular diffusion can arise from gradients of concentration, temperature, pressure, or an external potential such as an electric field, but only concentration diffusion is of importance in liquid extraction.

The second mechanism of transfer, known as *eddy* (or *turbulent*) *diffusion*, arises from the random movement of small packets of fluid. These transport the solute from one region to another, after which they blend with the surrounding fluid, with molecular diffusion playing a part in finally dispersing the solute. As might be expected, eddies tend to be damped near a solid surface or a liquid/liquid interface, so that molecular diffusion is controlling in this region. However, within a very short distance of the interface, the eddying motion increases rapidly and eddy diffusion becomes predominant in the bulk phase. In liquid extraction processes both mechanisms must generally be taken into account, and several mass-transfer models have been proposed on this basis.

2.2. Molecular Diffusion

2.2.1. Laws of Diffusion

(a) Fick's First Law. The diffusional rate of transfer of component 1 of a binary mixture is given by Fick's first law:

$$J_1 = -D_{12} \frac{dc_1}{dz} \tag{1}$$

where J_1 is the flux, dc_1/dz is the concentration gradient that provides the driving force, and D_{12} is the *diffusion coefficient.* A similar expression describes the transfer of component 2. The units of D_{12} are those of flux divided by concentration gradient, m^2/s, irrespective of whether mass or mole units are used.

Equation (1) bears an obvious resemblance to Fourier's law of heat conduction, but there is an important difference: the transfer of mass, unlike that of heat, involves the bodily movement of the fluid as a whole, except in the special circumstance in which the components move at equal rates in opposite directions. In general, the total flux $\hat{N}_1$ differs from the diffusional flux J_1 because of the presence of superimposed *convective* (i.e., *bulk*) flow, and the two must be related by specifying a suitable moving *reference frame*.

The choice of reference frame may be based on stationary coordinates or on the volume average, mass average, or molar average velocities. These velocities are defined as follows:

$$\mathbf{v}_r = a_1 \mathbf{v}_1 + a_2 \mathbf{v}_2 \tag{2}$$

where a_i is the appropriate *weighting function*, defined in the following ways:

1. **Volume Average Velocity.** $a_i = c_i \bar{V}_i$, where c_i is the concentration of component i in mass or molar units and $\bar{V}_i$ is the corresponding partial specific volume or partial molar volume.
2. **Mass Average Velocity.** $a_i = w_i$, the mass fraction.
3. **Molar Average Velocity.** $a_i = \bar{x}_i$, the mole fraction.

The various reference velocities are vector quantities that differ in both direction and magnitude in the generalized three-dimensional case [7]. The diffusion flux J_i of component i is related to these velocities by expressions of the type

$$J_i = c_i(\mathbf{v}_i - \mathbf{v}_r) \tag{3}$$

where $\mathbf{v}_i$ is its diffusion velocity relative to stationary coordinates, and J_i, c_i are expressed in consistent units, molar or mass. The definition of a flux is not complete unless *both* the units and the reference frame are specified.

The most commonly used reference frame for diffusion calculations is now that of *no net volume flux* [3, 8]. If it can be assumed that the mass density for liquid systems is constant with composition, then

$$\bar{V}_i = \left(\frac{\partial V}{\partial m_i}\right)_{m_{i \neq j}} = \frac{1}{\rho} \tag{4}$$

so that the partial specific volumes $\bar{V}_i$ are all equal, and

$$c_i \bar{V}_i = \frac{c_i}{\rho} = w_i \tag{5}$$

Consequently, the mass average velocity is equal to the volume average velocity and can be used as a reference basis. Similarly, the molar average velocity can be used if the molar density is constant. In liquid extraction mass or molar concentrations are always used, with the implicit assumption of the use of the corresponding mass or molar average velocity rather than the less convenient volume average value. However, the partial specific rather than the partial molar volume is more often found to be approximately constant; thus the use of mass units is generally more appropriate. The treatment that follows is applicable in either system of units.

In the one-dimensional case the vector quantities in Eq. (3) can be replaced by scalars. Combination with Eq. (1) gives, therefore, for each component

$$J_1 = -D_{12} \frac{dc_1}{dz} = c_1(v_1 - v_r) \tag{6}$$

$$J_2 = -D_{21} \frac{dc_2}{dz} = c_2(v_2 - v_r) \tag{7}$$

The equations of continuity for the individual components, expressed with reference to stationary coordinates, reduce as follows for steady-state diffusion [4]

$$\frac{d\hat{N}_1}{dz} = \frac{d\hat{N}_2}{dz} = 0 \tag{8}$$

and integration gives $\hat{N}_1$ = const and $\hat{N}_2$ = const. Combination of this result with Eqs. (6) and (7), noting that $c_i v_i = N_i$ and, from Eq. (2), $c_i v_r = x_i \Sigma \hat{N}_i$, finally gives

$$\hat{N}_1 = x_1(\hat{N}_1 + \hat{N}_2) - c_{tx} D \frac{dx_1}{dz} = \text{const} \tag{9}$$

$$\hat{N}_2 = x_2(\hat{N}_1 + \hat{N}_2) - c_{tx} D \frac{dx_2}{dz} = \text{const} \tag{10}$$

where c_{tx} is the total concentration, that is, the mass or molar density. In these expressions the subscripts on D are dropped since on addition, noting that $x_1 + x_2 = 1.0$, they reduce to $D_{12} = D_{21}$.

Equations (9) and (10) lead to the important result that the total flux of each component, given by the algebraic sum of the diffusion flux and the bulk flow, is constant over the diffusion path. On the other hand, x_i varies, and hence the diffusional fluxes $J_i = -c_{tx} D\, dx_i/dz$ are not constant except in the limiting case of zero transport, when $(\hat{N}_1 + \hat{N}_2) = 0$; however, addition of Eqs. (9) and (10) shows that $(J_1 + J_2) = 0$. In other words, the *sum* of the diffusional fluxes is zero throughout the diffusion path.

Further discussion of this subject is deferred to Section 3.1.1.

(b) Fick's Second Law. Steady-state diffusion is a special case of the more general one of transient diffusion in which the concentrations and fluxes vary with time. Unsteady-state diffusion is relevant to some experimental methods of determining diffusion coefficients and to the "penetration" models of the mass-transfer process described later.

The differential equation for transient diffusion is obtained by combining Fick's first law expression with the equation of continuity. For the one-dimensional case, the latter is obtained by equating the rate of accumulation of component 1 in a slice of liquid of thickness dz in a direction normal to the direction of diffusion, to the corresponding change in flux, giving

$$\frac{\partial c_1}{\partial t} + \frac{\partial \hat{N}_1}{dz} = 0 \tag{11}$$

Combination of Eqs. (1) and (11), assuming zero net transport (i.e., $\hat{N}_1 = J_1$), gives Fick's second law:

$$\frac{\partial c_1}{\partial t} = \frac{\partial}{\partial z}\left(D \frac{\partial c_1}{\partial z}\right) \tag{12}$$

Solutions of Eq. (12) for a variety of geometric shapes and boundary conditions are given in books by Crank [9] and by Carslaw and Jaeger [10], the latter for related problems in heat conduction.

2.2.2. *Effect of Concentration on Diffusivity*

(a) Differential and Integral Diffusion Coefficients. Liquid-phase diffusivities differ from gas-phase diffusivities in two important respects: (1) their values are much smaller, falling mostly within the range 0.1-5 × 10^{-9} m^2/s at normal temperatures; and (2) they exhibit a marked variation with solute concentration, often showing a sharp minimum or maximum value.

Early workers measured liquid diffusivities by means of various forms of transient diffusion cell, using an appropriate solution of Eq. (12) to obtain the diffusion coefficient. Such methods, however, yield values of an *integral diffusion coefficient,* which is meaningful only in relation to the actual range of concentrations used, a fact that unfortunately was not recognized at the time.

Some workers also developed optical interference methods for the determination of local values of the concentration gradient. Clack [11] made an important advance by using a steady-state diffusion cell, which enabled the diffusion flux to be measured directly, in conjunction with an optical method of this type. By this means he obtained the first reliable measurements of the local, that is, the *differential* diffusional coefficient for several salt–water systems. The relationship between the differential and integral diffusion coefficients, D and $\bar{D}$, respectively, is

$$\bar{D} = \frac{1}{c' - c''} \int_{c_0}^{c_t} D \, dc \tag{13}$$

where c' and c'' are the concentrations at each end of the diffusion path. More detailed accounts of this subject are available elsewhere [12, 13, 23].

(b) Thermodynamic Correction Factor. It was first suggested by Hartley [14] that the true driving force causing diffusion is the gradient of chemical potential. On this basis, Eq. (1) can be written as

$$J_1 = \frac{-\mathcal{D}_{12} c_t x_1}{RT} \frac{d\mu_i}{dz} \tag{14}$$

Since μ_i is related to the thermodynamic activity a_i as follows,

$$\mu_i = \mu_i^0 + RT \ln a_i \tag{15}$$

combination with Eq. (14) gives

$$J_1 = -\mathcal{D}_{12}\left(\frac{d \ln a_1}{d \ln x_1}\right)\frac{dc_1}{dz} \tag{16}$$

$$= -\mathcal{D}_{12}\left(1 + \frac{d \ln \gamma_1}{d \ln x_1}\right)\frac{dc_1}{dz} \tag{17}$$

where γ_1 is the corresponding activity coefficient. A similar expression is obtained for component 2. Because of the logarithmic differentiation, any convenient scale of units can be used for the activity [13], and also for concentration, provided c_t can be considered constant.

The term in brackets in Eqs. (16) and (17), known as the *thermodynamic factor*, reduces to unity for ideal solutions, that is, for γ_i = 1.0, giving the usual Fick's law expression. The experimentally measured diffusivity D_{12} is thus related to a corrected value $\mathcal{D}_{12}$ as follows:

$$D_{12} = \mathcal{D}_{12}\,\alpha \tag{18}$$

where

$$\alpha = 1 + \frac{d \ln \gamma_1}{d \ln x_1} = 1 + \frac{d \ln \gamma_2}{d \ln x_2} \tag{19}$$

The corrected differential diffusion coefficients for nonelectrolytes are found in most cases to exhibit a more regular variation with composition than the original experimental data, and Vignes [15] proposed the following empirical relationship between the two:

$$\frac{D_{12}}{\alpha} = \mathcal{D}_{12} = (D_{12}^0)^{\bar{x}_2} (D_{21}^0)^{\bar{x}_1} \tag{20}$$

where D_{12}^0 and D_{21}^0 are the limiting values of D_{12} for $\bar{x}_1 \to 0$ and $\bar{x}_2 \to 0$, respectively.

Cullinan [16] later gave a derivation of the Vignes expression in terms of absolute rate theory, and Dullien [17] in a statistical test found errors of up to nearly 20% in some cases. A modified expression taking viscosity into account was proposed by Leffler and Cullinan [18] as follows:

$$\frac{D_{12}\mu}{\alpha} = (D_{12}^0 \mu_2)^{\bar{x}_2} (D_{21}^0 \mu_1)^{\bar{x}_1} \qquad (21)$$

This gave a somewhat better representation of the data in most cases, with the notable exception of mixtures of n-alkanes. Sanchez and Clifton [19] found the following empirical relationship to fit the data well, even for systems far from ideal:

$$D_{12} = (\bar{x}_1 D_{12}^0 + \bar{x}_2 D_{21}^0)(1 - m + m\alpha) \qquad (22)$$

where m is an empirical constant. At least one experimental point in the middle of the concentration range is necessary to permit evaluation of m.

Diffusivity and activity coefficient data are available from several sources [15, 17, 20–23].

2.2.3. *Measurement of Diffusivity*

The diaphragm cell is by far the most convenient measurement method for routine purposes. This employs a porous diaphragm between two well-agitated solutions so that the concentration gradient is set up entirely in the diaphragm and is not upset by convection. A useful account of the early development of the method has been given by Gordon [24]. Two later forms of apparatus [25, 26], using magnetic stirring [27], are shown in Fig. 1.

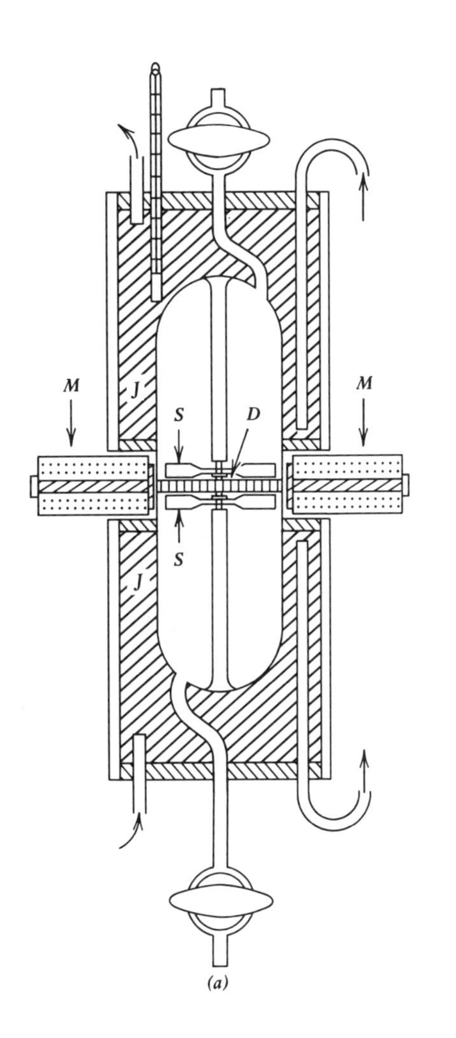

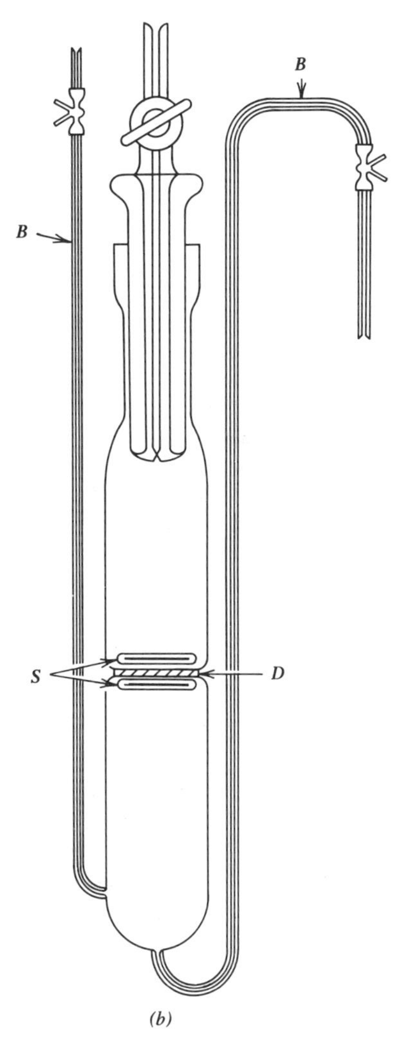

Figure 1. Diaphragm cells for measurement of diffusivity in liquids: (*a*) Lewis cell [25]; (*b*) Dullien-Shemilt cell [26] [*B*, capillaries for charging; *D*, diaphragm; *J*, water jacket; *S*, stirrer; *M*, solenoids (eight in all)].

The method consists of filling the compartments with the two solutions, with careful elimination of air, establishing the concentration gradient in the diaphragm and recharging the compartments. Diffusion is then allowed to proceed normally for at least 3 days, after which the contents of the compartments are analyzed. The integral diffusion coefficient is calculated from the quasi-steady-state relation

$$\bar{D} = \frac{1}{\beta t} \ln \frac{\Delta c_0}{\Delta c_t} \tag{23}$$

where Δc_0 and Δc_t are respectively the initial and final differences in concentration between the compartments and β is the cell constant; this is defined formally as

$$\beta = \frac{A}{L}\left(\frac{1}{V_1} + \frac{1}{V_2}\right) \tag{24}$$

where A is the effective area for diffusion, L is the length of diffusion path, and V_1 and V_2 are the compartment volumes. In practice, β is obtained from calibration runs by using 0.1 N KCl and water at 20°C for which the diffusivity values are well established [28]. Care must be taken to operate at a stirrer speed sufficiently high to avoid effect on β.

Gordon [24] devised a numerical method for the calculation of the differential diffusion coefficient from the measured integral values. Lewis [25] proposed an alternative method in which zero concentration was maintained in one compartment by continuous flushing with solvent and the concentration monitored in the other compartment. This leads to simple relations for both differential and integral diffusivities but involves restrictions on the analytical methods that can be used.

2.2.4. *Prediction of Diffusivity*

A satisfactory method for the prediction of diffusivity in liquids is much needed in view of the tedious nature of the experimental method. However, the kinetic theory of the liquid state is not sufficiently well developed to provide the basis for a satisfactory predictive method, with the single exception of electrolyte solutions at infinite dilution. On the other hand, it is fortunate that uncertainty in diffusivity values is partly offset by the fact that mass-transfer rates are proportional to the diffusivity raised to an exponent of 0.50–0.67 rather than unity (see Section 3.1).

(a) Nonelectrolytes. Many of the proposed methods are based on the early Stokes–Einstein equation that relates the drag force on a solute molecule in a continuum of solvent to the diffusion velocity, giving

$$D_{12} = \frac{kT}{6\pi r_1 \mu_2} \tag{25}$$

where r_1 is the radius of the solute molecule, μ_2 the viscosity of the solvent, and k the Boltzmann constant. This was modified by Wilke and Chang [29] to give the following *dimensional* equation

$$D_{12}^0 = 7.4 \times 10^{-12}\left(\frac{(\phi M_2)^{0.5}T}{\mu_2 V_1^{0.6}}\right) \tag{26}$$

where D_{12}^0 = diffusivity of component 1 at very low concentration in component 2, that is, the solvent, m²/s

μ_2 = viscosity of the solvent, centipoise

T = absolute temperature, Kelvin

M_2 = molecular weight of solvent

V_1 = molecular volume of solute at normal boiling point, ml/mol

ϕ = "association parameter"

Values of V_1 are obtained by summing the Le Bas atomic and group volumes listed in Table 1; values of ϕ are given in Table 2. Sherwood et al. [3] have pointed out that the values given by Eq. (26) may be expected to approach asymptotically, but not exceed those given by Eq. (25), as shown in Fig. 2.

Reddy and Doriswamy [30] proposed an alternative equation that did not involve an association parameter, and a modified form was given by Laddha and Smith [31], who found an average deviation of only ±13% from the experimental data on which it was based.

Physical properties such as the latent heat of vaporization have also been introduced to account for abnormal behavior. A summary of many of the equations proposed is given by Laddha and Degaleesan [32].

The various relations given previously are suitable only for predicting the diffusivity at infinite dilution, and two such values can be obtained by treating each of the components in turn as solute. These can then be used in conjunction with the thermodynamic correction

TABLE 1 ADDITIVE-VOLUME[a] INCREMENTS OF LE BAS FOR CALCULATING MOLAL VOLUMES AT THE NORMAL BOILING POINT

Atom	Increment, $cm^3/g.$ mole
Carbon	14.8
Hydrogen	3.7
Oxygen (except as below)	7.4
In methyl esters and ethers	9.1
In ethyl esters and ethers	11.0
In acids	12.0
Joined to S, P, N	8.3
Nitrogen	
Doubly bonded	15.6
In primary amines	10.5
In secondary amines	12.0
Fluorine	8.7
Chlorine	24.6
Bromine	27.0
Iodine	37.0
Sulfur	25.6
Ring:	
Three membered	-6.0
Four membered	-8.5
Five membered	-11.5
Six membered	-15.0
Naphthalene	-30.0
Anthracene	-47.5

[a]The additive-volume procedure must not be used for simple molecules. In such cases the following values should be used: H_2O, 18.9; NH_3, 25.8; Cl_2, 48.4; Br_2, 53.2; and I_2, 71.5.

TABLE 2 ASSOCIATION PARAMETERS FOR SOLVENTS

Solvent	ϕ
Water	2.6
Methanol	1.9
Ethanol	1.5
Benzene	1.0
Ether	1.0
Heptane	1.0
Other unassociated solvents	1.0

Source: Wilke and Chang [29].

factor described in Section 2.2.2(b), to obtain the diffusity at any desired concentration.

(b) Electrolytes. In an electrolyte solution the diffusing dissociated ions have different mobilities according to their size. However, in the absence of an applied electrical potential the attraction between ions of opposite charge causes them to diffuse at the same rate, apparently as a single molecule.

This "molecular" diffusion coefficient of a single salt can be predicted very accurately at infinite dilution by means of the Nernst–Haskell equation as follows:

$$D_{12}^0 \; (m^2/s) = 10^{-4} \, \frac{RT}{F^2} \, \frac{(1/n_+ + 1/n_-)}{(1/\lambda_+^0 + 1/\lambda_-^0)} \tag{27}$$

where λ_+^0, λ_-^0 = limiting (i.e., zero concentration) ionic conductances of cation and anion, respectively, mhos (gram equivalent)$^{-1}$

n_+, n_- = valencies of cation and anion, respectively

F = Faraday constant = 96,488 C (gram equivalent)$^{-1}$.

R = gas constant = 8.315 J K^{-1} (mol)$^{-1}$

T = temperature, Kelvin

Values of the limiting conductances are available in several sources [13, 33], including Perry's *Chemical Engineers' Handbook.*

As the salt concentration is increased, D_{12} first decreases somewhat and then increases slowly, often passing through a maximum. Gordon [34] has recommended the following semiempirical equation for electrolyte solutions up to a concentration of about 2 N:

$$D_{12} = D_{12}^0 \left(\frac{\mu_2}{\mu c_2 \bar{V}_2} \right) \left(1 + m \, \frac{\delta \ln \gamma^{\pm}}{\delta m} \right) \tag{28}$$

where c_2 = concentration of water, mol/ml

m = molality of solution, mol/kg of solvent

$\bar{V}_2$ = partial molal volume of water, ml/mol

$\gamma^{\pm}$ = mean ionic activity coefficient based on molality

μ, μ_2 = viscosity of solution and water, respectively

Values of $\gamma^{\pm}$ for a number of aqueous solutions have been given by Harned and Owen [33].

Vinograd and McBain [35] have given expressions for simultaneous diffusion in solutions of

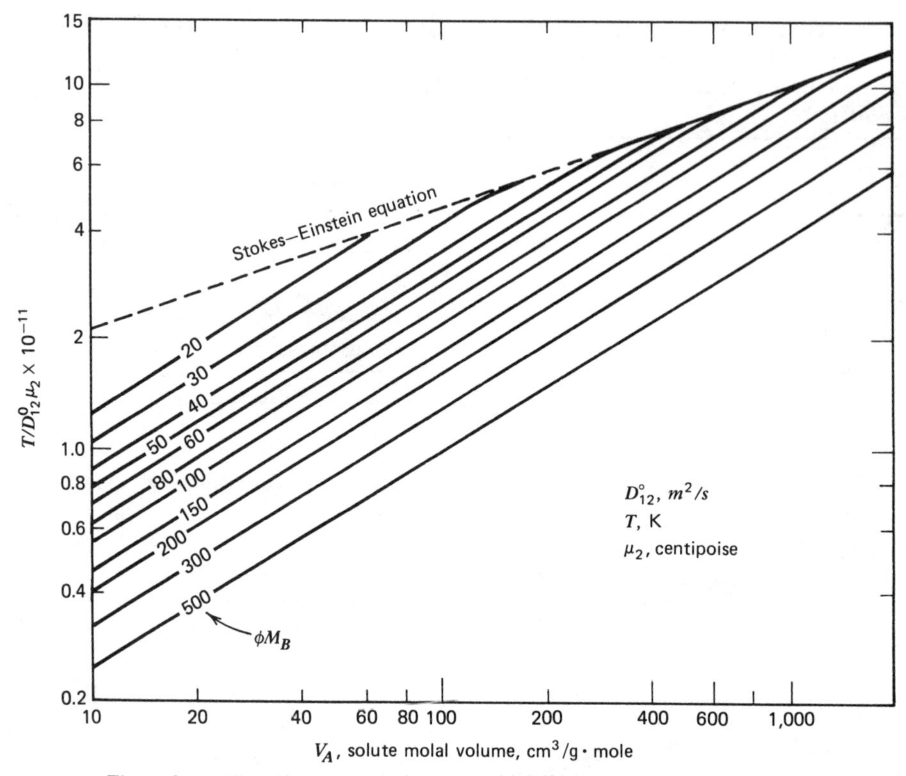

Figure 2. Wilke–Chang correlation of liquid diffusivity at low concentration.

mixed electrolytes. No great error is involved in using the "molecular" diffusion coefficients for the species involved, provided ions of greatly differing conductances, particularly H^+ and OH^-, are not involved [3].

2.3. Eddy Diffusion

2.3.1. Nature of Turbulent Flow

In turbulent flow, rapidly fluctuating velocities are superimposed on the main flow. These vary in both magnitude and direction in an irregular manner, but the time-averaged value $\bar{u}_j'$ of a fluctuating component in the j direction is zero, that is,

$$\bar{u}_j' = \frac{1}{t}\int_0^t u_j'\,dt = 0 \qquad (29)$$

where j refers to the coordinate directions x, y, or z. The instantaneous velocity at any point is thus the sum of a time-averaged value and an instantaneous *fluctuating* (or *deviating*) velocity; thus for unidirectional flow in the x direction, the three components are

$$u_x = \bar{u}_x + u_x', \qquad u_y = u_y', \qquad u_z = u_z' \qquad (30)$$

It is apparent from considerations of continuity that the transport of fluid in a given direction by a velocity fluctuation must be compensated by a return flow in the opposite direction and that for an incompressible fluid, this must be immediate. This leads to the concept of turbulence as an eddying motion, with continuous random formation, mixing, breakup, disappearance, and reformation of eddies of a wide spectrum of sizes throughout the fluid.

2.3.2. Transport Rates

In laminar flow the shear stress acting on a plane parallel to the direction of flow arises from the thermal motions of molecules that transfer their momentum between "layers" moving at different velocities. This is expressed quantitatively in terms of Newton's law of viscosity as follows:

$$\tau_l = -\nu\,\frac{d(\rho u_x)}{dz} \qquad (31)$$

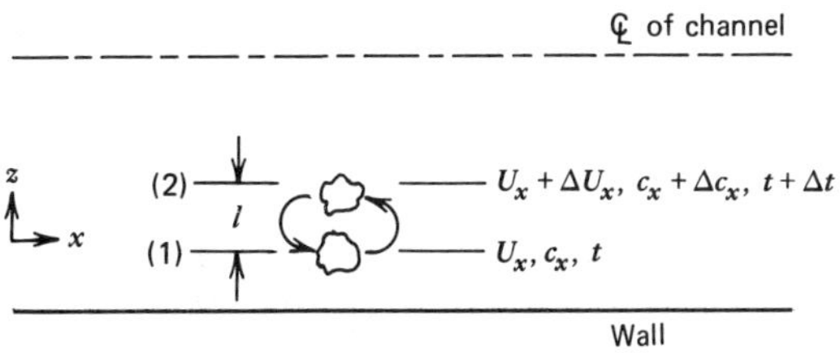

Figure 3. Eddy diffusion of momentum, mass, and heat.

where the laminar shear stress τ_l has units of momentum flux and, $\nu = \mu/\rho$, the *kinematic viscosity*, is the coefficient of momentum transfer. In this expression the gradient of ρu_x, the momentum per unit volume, is the driving force which effects the transport of momentum.

A similar situation arises in turbulent flow, with the molecular motion replaced by that of "packets" of fluid that transfer their momentum to adjacent layers normal to the main flow with a deviating velocity u_z'. This is illustrated in Fig. 3, in which the mean velocity in the x direction at location 2 is greater than that at 1 by an amount

$$\Delta \bar{U}_x = |\bar{u}_x'| = -l \left| \frac{d\bar{u}_x}{dz} \right| \tag{32}$$

where l, the *Prandtl mixing length*, is somewhat analogous to the mean free path of molecules in kinetic theory. A packet of fluid with a transverse momentum per unit volume of $\rho u_z'$ gives rise to a momentum flux of $\rho |u_x'| u_z'$ on arrival at its new location; this is equal to the turbulent shear stress so that, taking temporal averages and assuming that $|\bar{u}_x'| \simeq |\bar{u}_z'|$,

$$\tau_t = \rho |\bar{u}_x'| \bar{u}_z' \simeq \rho l^2 \left| \frac{d\bar{u}_x}{dz} \right| \frac{d\bar{u}_x}{dz} \tag{33}$$

Hence an *eddy kinematic viscosity* E_ν can be defined by

$$\tau_t = -E_\nu \frac{d(\rho \bar{u}_x)}{dz} \tag{34}$$

where

$$E_\nu = l^2 \left| \frac{d\bar{u}_x}{dz} \right| \tag{35}$$

The total shear, including the molecular component, is thus

$$\tau = -(\nu + E_\nu) \frac{d(\rho \bar{u}_x)}{dz} \tag{36}$$

The eddies also bring about a similar transfer of dissolved solute or of heat. It can be shown by analogy that the total mass flux due to both molecular and eddy transport is

$$J = -(D + E_D) \frac{dc}{dz} \tag{37}$$

where the eddy diffusivity E_D is given by the right-hand side of Eq. (35).

A similar procedure leads to the following expression for the heat flux due to molecular and eddy transport:

$$Q = -(\alpha_H + E_H) \frac{d(\rho c_p \theta)}{dz} \tag{38}$$

where $\alpha_H = \kappa/\rho c_p$ is the molecular thermal diffusivity, and E_H, again given by Eq. (35), is the eddy thermal diffusivity.

Assuming ρ to be constant, Eqs. (36) and (37) can be combined as follows:

$$\frac{\rho(\nu + E_\nu)\, du_x}{\tau} = \frac{(D + E_D)\, dc}{J} \tag{39}$$

The derivations of E_ν and E_D suggest that they have the same value; if it is also assumed that the *Schmidt number* ν/D is unity or that $E_\nu \gg \nu$ and $E_D \gg D$, there is a simple analogy between momentum and mass transfer. It can be shown that

$$\mathrm{St}_M = \frac{k_j}{\bar{U}} = \frac{\tau_w}{\rho \bar{U}^2} = \frac{f}{2} \tag{40}$$

where St_M is the mass-transfer Stanton number, and the *mass-transfer coefficient* k_j is defined as follows in terms of the wall flux J_w and the concentration difference between the fluid at the wall and the bulk fluid:

$$J_w = k_j(c_b - c_w) \tag{41}$$

A similar derivation for heat transfer assuming that the Prandtl number $c_p \mu/\kappa$ is unity or that $E_\nu \gg \nu$ and $E_H \gg \alpha_H$ gives

$$\mathrm{St}_H = \frac{h}{c_p \rho \bar{U}} = \frac{f}{2} \tag{42}$$

These simple expressions, which correspond to the well-known *Reynolds analogy*, agree fairly well with experimental data on heat and mass transfer to gases, for which the Prandtl and Schmidt groups are close to unity. However, in normal liquids these groups have much higher values, so allowance must be made for the laminar regions near the wall. The Reynolds analogy is also not applicable to more complex geometries such as spherical droplets, where pressure loss arises from form drag as well as from skin friction (see Section 4 of this chapter).

The preceding treatment of turbulent flow theory is introductory, and more comprehensive accounts are available [3, 5, 36–38].

3. INTERPHASE MASS TRANSFER

3.1. Binary Mass-Transfer Models

The concept of the mass-transfer coefficient k_j has already been introduced in relation to eddy transport to a tube wall [Eq. (41)]. In practice, the mechanism is more complex since molecular transport near the wall must also be taken into account; further, in liquid extraction two liquid phases are involved, with transfer taking place through an interface that is usually *free*, that is, unconstrained by the presence of a solid boundary.

A number of theoretical mass-transfer models have been proposed; of these the simple two-film theory, described first in the following paragraphs, provides the basic relationships. In the following account the terms "X" and "Y" phases relate to the *feed* and *extractant* phases respectively, regardless of which is dispersed. However, the alternative designations "dispersed" and "continuous" phases are also used in describing mass transfer between droplets and surrounding phase.

3.1.1. *Stagnant-Film Theory*

The two-film theory proposed by Whitman [39] involves two postulates: (1) the resistance to transfer resides in two stagnant films, one on each side of the interface, through which transfer of solute occurs by *molecular* diffusion; and (2) the phases are in equilibrium at the interface itself. This is illustrated in Fig. 4*a*, in which the Y-phase concentration is expressed as c_y/K, where the distribution coefficient K is assumed constant. Postulate 1 requires that the concentration falls rapidly in the X-phase film from the constant value c_{xb} in the bulk to the interfacial value c_{xi} and in the Y-phase film from c_{yi}/K at the interface to c_{yb}/K in the bulk; according to postulate 2, the points c_{xi} and c_{yi}/K are coincident at the interface, as shown. In practice, the concentration profiles vary smoothly from bulk phase to interface, as indicated by the broken lines in Fig. 4*a*.

The rate of transfer through the X-phase film is obtained from Eq. (9), which is expressed as follows assuming a film thickness of δ_x and putting $\eta = z/\delta_x$;

$$\hat{N}_1 = x(\hat{N}_1 + \hat{N}_2) - F_x c_{tx} \frac{dx}{d\eta} \tag{43}$$

where

$$F_x = \frac{D_x}{\delta_x} \tag{44}$$

Since there is no accumulation in the film, $\hat{N}_1$ and $\hat{N}_2$ are constant and Eq. (43) can be integrated with the boundary conditions $x = x_b$ (the

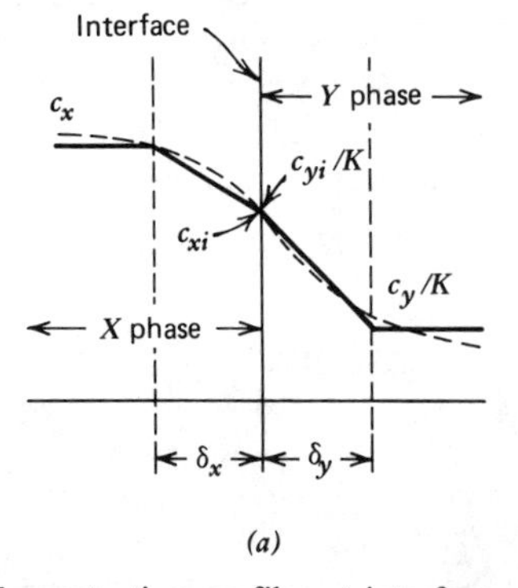

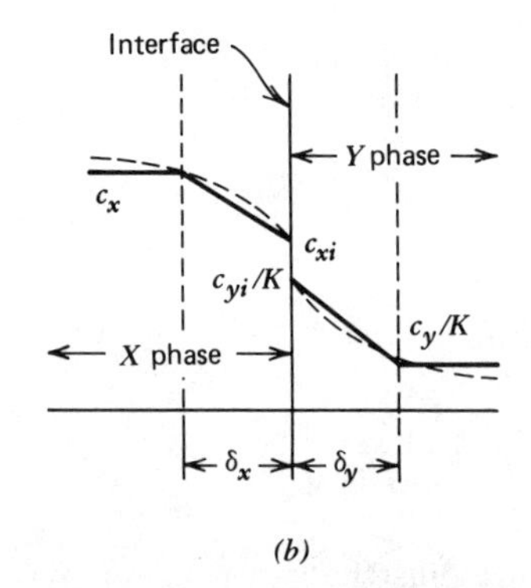

Figure 4. Concentration profiles at interface on basis of film theory: (*a*) with equilibrium at interface; (*b*) with slow heterogeneous reaction. The broken lines show more realistic forms of the profiles.

bulk-phase composition) at $\eta = 0$ and $x = x_i$ (the interface composition) at $\eta = 1$, giving

$$\frac{1}{F_x c_{tx}} = \frac{1}{\hat{N}_1 + \hat{N}_2} \ln \left[\frac{x_i(\hat{N}_1 + \hat{N}_2) - \hat{N}_1}{x_b(\hat{N}_1 + \hat{N}_2) - \hat{N}_1} \right]$$

This can be put into the form

$$\hat{N}_1 = F_x z_x c_{tx} \ln \left[\frac{z_x - x_i}{z_x - x_b} \right] \quad (45)$$

where $z_x = \hat{N}_1/(\hat{N}_1 + \hat{N}_2)$, with the $\hat{N}_i$ given the appropriate signs; hence z_x is the relative rate of transport of the components through the film. Equation (45) can be expressed more conveniently as

$$\hat{N}_1 = k_x c_{tx}(x_b - x_i) \quad (46)$$

$$= k_x(c_{xb} - c_{xi}) \quad (47)$$

where

$$k_x = \frac{F_x}{x_D} \quad (48)$$

and

$$x_D = \frac{(z_x - x)_{\text{lm}}}{z_x} \quad (49a)$$

$$= \frac{1}{z_x} \left(z_x - \frac{c_x}{c_{tx}} \right)_{\text{lm}} \quad (49b)$$

and $(z_x - x)_{\text{lm}}$ refers to the logarithmic mean of $(z_x - x_i)$ and $(z_x - x_b)$.

The term x_D is sometimes called the *drift factor*, since it allows for bulk flow or "drift" in the film. In liquid extraction z_x normally has a value of unity since, with effectively immiscible solvents, the solute diffuses through stagnant solvent. The drift factor is thus equal to the logarithmic mean of the *solvent* concentrations in the bulk phase and at the interface, that is, to $(1 - x)_{\text{lm}}$.

The transfer in the Y phase takes place from the interface into the bulk and can be treated in an identical manner, giving

$$\hat{N}_1 = k_y c_{ty}(y_i - y_b) \quad (50)$$

$$= k_y(c_{yi} - c_{yb}) \quad (51)$$

where

$$k_y = \frac{F_y}{y_D} \quad (52)$$

and

$$y_D = \frac{(z_y - y)_{\text{lm}}}{z_y} \quad (53)$$

Again, in general, $z_y = 1.0$ for liquid extraction. It must be emphasized that the flux $\hat{N}_1$ is the same on both sides of the interface since there is no accumulation in the films.

The mass-transfer coefficient k_j is the *practical* or *as-measured* value since it is equal to the ratio of *total* flux to driving force, as shown by Eqs. (47) and (51). The coefficient F_j, first used by Colburn and Drew [7, 40], corresponds to the value that would be obtained in the absence of bulk flow, as may be seen by putting $(\hat{N}_1 + \hat{N}_2) = 0$ in Eq. (43), and it is thus termed the *coefficient of equimass* or *equimolar counterflow* or, more simply, the *zero-transport* coefficient. It is important when comparing values of the mass-transfer coefficients for different conditions or when using the analogy to heat or momentum transport, to convert the k_j values to the F_j form by multiplying by the drift factor, as indicated by Eqs. (48) and (52), in order to express them with respect to a stationary reference frame.

The foregoing relationships apply regardless of whether the concentrations are expressed in mass or molar units provided that they are consistent. As pointed out in Section 2.2.1(a), however, the use of zero mass average velocity as reference frame usually gives the better approximation to the more correct zero-volume average velocity in liquid diffusion calculations and obviates the need for using inconvenient volume fraction units. The use of mass units is thus generally to be preferred, even though molar units have customarily been used in the past. As may be expected, the numerical difference between the values of k_j in the two systems of units arises entirely from a difference in the drift terms, since the zero transport F_j coefficients are the same in both systems.

A physical interpretation of the film coefficients is shown in Fig. 5*a*, where *ABC* represents the equilibrium relationship, not necessarily linear, and the operating point E corresponds to bulk-phase concentrations c_{xb} and c_{yb}. From Eqs. (47) and (51), since $\hat{N}_1$ is the same for both films, it follows that

$$\hat{N}_1 = k_x(c_{xb} - c_{xi}) = k_y(c_{yi} - c_{yb}) \quad (54)$$

so that

$$-\frac{k_x}{k_y} = \frac{c_{yi} - c_{yb}}{c_{xi} - c_{xb}} \quad (55)$$

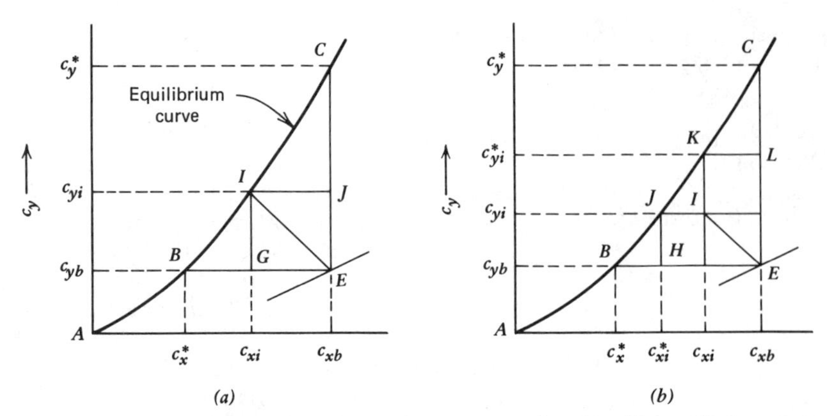

Figure 5. Relationship between driving forces for mass transfer: (*a*) with equilibrium at interface; (*b*) with slow heterogeneous reaction.

A line of slope $-k_x/k_y$ through E thus intersects the equilibrium curve at I with coordinates (c_{xi}, c_{yi}), and the driving forces in the X and Y phases are represented by IJ and IG, respectively.

It is also convenient to define overall mass-transfer coefficients k_{ox} and k_{oy} as follows:

$$\hat{N}_1 = k_{ox}(c_{xb} - c_x^*) = k_{oy}(c_y^* - c_{yb}) \tag{56}$$

where c_x^* and c_y^* are the compositions in equilibrium with c_{yb} and c_{xb}, respectively, as shown in Fig. 5*a*. The driving force $(c_{yi} - c_{yb})$ in Eq. (54) can be expressed as $m'_y(c_{xi} - c_x^*)$, where m'_y is the slope of the cord IB, so that Eq. (54) becomes

$$\hat{N}_1\left[\frac{1}{k_x} + \frac{1}{m'_y k_y}\right] = (c_{xb} - c_{xi}) + (c_{xi} - c_x^*)$$

$$= (c_{xb} - c_x^*) \tag{57}$$

The overall driving force on an X-phase basis is thus $IJ + GB = EB$, and Eqs. (56) and (57) give

$$\frac{1}{k_{ox}} = \frac{1}{k_x} + \frac{1}{m'_y k_y} \tag{58}$$

Similarly, on a Y-phase basis the overall driving force is $IG + JC = EC$, and

$$\frac{1}{k_{oy}} = \frac{m'_x}{k_x} + \frac{1}{k_y} \tag{59}$$

where m'_x is the slope of chord IC. The values of m'_x and m'_y in these expressions can be replaced by K, the distribution coefficient, when this is constant.

The individual terms in Eqs. (58) and (59) correspond to *mass-transfer resistances*, so that the overall resistance is equal to the sum of the separate phase resistances. It is apparent that the term $1/m'_y k_y$ in Eq. (58) diminishes in importance with increase in distribution coefficient so that $k_{ox} \to k_x$, and if less than about 10% of the overall resistance resides in the Y-phase the system is said to be *X-phase controlled.* Similarly, the term m'_x/k_x in Eq. (59) diminishes in importance with decrease in distribution coefficient so that $k_{oy} \to k_y$, and the system is considered to be *Y-phase controlled* if more than 90% of the overall resistance resides in this phase.

An alternative definition [4] of mass-transfer coefficient $k_x^{\bullet}$ is based on the *net diffusional* flux, but its use in the present context is not recommended [41].

3.1.2. *Boundary-Layer Theories*

Simple film theory is inadequate in that it fails to take into account eddy transport within the bulk phases. On the other hand, Eqs. (40) and (42), derived on the basis of the Reynolds analogy for transfer to a tube wall, also fail in most instances since they make no allowance for molecular transport near the wall. Prandtl [42] and Taylor [43] proposed a model that combined molecular transport in a thermal (or diffusional) boundary layer adjacent to the wall with the Reynolds analogy in the central core. In an alternative approach, Murphree [44] integrated Eqs. (36) and (38) on the assumption that E_v and E_H vary as the cube of the distance from the wall and obtained an expression for heat transfer that agreed well with experimental data over a limited range of Pr.

An important advance was made by von Karman [45], who used the "universal velocity distribution" for pipe flow to evaluate E_v and hence E_H. This distribution is expressed in terms of parameters u^+ and z^+, defined by

$$u^+ = \frac{\bar{U}}{u^*}; \quad z^+ = \frac{zu^*}{\nu} \tag{60}$$

where z is the distance from the wall and $u^* = \sqrt{\tau_w/\rho} = \bar{U}\sqrt{f/2}$ is the *friction velocity*. Although the experimental velocity distribution is continuous, it was found convenient to divide it into three regions as follows:

Laminar sublayer

$$u^+ = z^+ \qquad 0 < z^+ < 5$$

Buffer layer

$$u^+ = -3.05 + 5.0 \ln z^+ \qquad 5 < z^+ < 30$$

Turbulent core

$$u^+ = 5.5 + 2.5 \ln z^+ \qquad z^+ > 30$$

The procedure was to differentiate the expressions for the buffer layer and turbulent core and to substitute the results in Eq. (36) to obtain E_v/ν as a function of r/r_w. Equation (37) or (38) was then solved, together with the expression for the laminar sublayer, assuming that E_v/E_M or E_v/E_H = constants β_M or β_H, the *turbulent Schmidt* or *turbulent Prandtl* numbers, respectively, to obtain the temperature or concentration distribution, from which the mass- or heat-transfer coefficient can be obtained. The result for mass transfer is as follows [46, 47]:

$$\mathrm{St}_M = \frac{k_x x_D}{U_x}$$

$$= \frac{f/2}{\beta_M + 5\beta_M\sqrt{f/2}\,\{\psi - 1 + \ln[(1 + 5\psi)/6]\}} \tag{61}$$

where $\psi = \mathrm{Sc}/\beta_M$. The corresponding result for heat transfer with β_H taken as unity was found [45] to agree well with experimental data for turbulent flow in the range $0.73 < \mathrm{Pr} < 35$; in this connection it must be emphasized that no heat- or mass-transfer data are involved in the derivation of Eq. (61).

Various workers have attempted to improve on Eq. (61) by semitheoretical or other means [3]. Much of the difficulty lies in a lack of understanding of conditions very close to the wall, in the laminar sublayer region. In addition, two problems arise in connection with the treatment of eddy diffusion: (1) the analogy between the transport properties is limited by the fact that momentum is a vector, whereas heat and mass are scalar quantities; and (2) momentum transported by eddies is likely to be absorbed almost immediately in a new location, whereas heat and solute are absorbed more slowly, by breakdown of the packets of fluid into smaller units and ultimately by molecular transport. The turbulent Schmidt (or Prandtl) group β was introduced with a view to correcting for the second of these factors; experimental data have shown it to have a value of about 0.7, irrespective of the physical properties of the system, for processes other than heat transfer to liquid metals [47].

Chilton and Colburn [48] proposed an alternative, more empirical set of relationships that represent the data at least as well as the more fundamental forms:

$$\frac{k_x x_D}{U_x}\left(\frac{\mu}{\rho D}\right)^{2/3} = j_M = \frac{h}{c_p \rho U}\left(\frac{c_p \mu}{\kappa}\right)^{2/3}$$

$$= j_H = \frac{f}{2} \tag{62}$$

where f is usually expressed as a simple power function of Reynolds number, for example, $f = 0.023\ \mathrm{Re}^{-0.20}$ for turbulent flow in smooth tubes. When Sc or Pr = 1.0, Eqs. (61) and (62) reduce to the simple Reynolds analogy forms, that is, Eqs. (40) and (42).

In liquid extraction simple geometries such as tubes and flat surfaces are seldom used. Even when they are, as with the wetted-wall column (Section 4.1), the flow pattern is affected by drag between the phases so that the coefficients deviate widely from the predicted values. Most types of extraction equipment in fact operate with one of the phases dispersed as droplets, giving rise to a "free" interface, as stated earlier. For design purposes, therefore, it is necessary to obtain experimental values of the individual mass-transfer coefficients for each phase and to correlate them by means of empirical relationships of the type

$$\frac{k_j j_D}{U_j} = \text{const } \mathrm{Re}_j^{-m}\,\mathrm{Sc}_j^{-n} \tag{63}$$

where $n = 0.67$ on the basis of boundary-layer theory and j_D is the drift factor for phase j. The corresponding expression for heat transfer is

$$\frac{h}{c_p \rho U} = \text{const } \mathrm{Re}^{-m}\,\mathrm{Pr}^{-n} \tag{64}$$

The analogy between these equations can be used to predict heat-transfer rates for extractors from mass-transfer data and vice versa.

These expressions can be put into alternative forms by multiplying both sides by the product of Re and Sc or Pr, giving, in the case of Eq. (63)

$$\mathrm{Sh} = \frac{k_j j_D d}{D_j} = \text{const } \mathrm{Re}_j^{1-m}\ \mathrm{Sc}_j^{1-n} \tag{65}$$

From Eqs. (44) and (48), $D_j/k_j j_D = \delta_j$, the thickness of the hypothetical stagnant film, so that the left-hand side of Eq. (65) is equal to d/δ_j. Hence this equation is in fact a correlation of the ratio of the film thickness to some characteristic dimension d of the equipment.

3.1.3. Penetration Theories

Penetration theories differ from film-type theories in that unsteady-state diffusion is postulated, from interface into bulk phase, for a limited time after which the surface is renewed by some means and the process repeats itself. The diffusion is thus of a transient nature with the rate decreasing with time, and is described by Fick's second law [Eq. (12)]. This becomes, assuming the diffusion coefficient to be constant over the range of concentration involved,

$$\frac{\partial c}{\partial t} = D\frac{\partial^2 c}{\partial z^2} \tag{66}$$

Higbie [49] was first to propose a model of this type, specifically for gas absorption from bubbles into liquid or into a liquid film on a packing surface. A small element of liquid of uniform concentration c_b is contacted with the second phase for a time t, during which its interfacial composition is maintained constant at c_i. Transfer of solute takes place by molecular diffusion and, provided the element is of effectively infinite thickness (i.e., $z \gg \sqrt{Dt}$), the initial and boundary conditions are

$$\begin{aligned} c &= c_b, & z &> 0, & t &= 0 \\ c &= c_i, & z &= 0, & t &> 0 \\ c &= c_b, & z &= \infty, & t &> 0 \end{aligned} \tag{67}$$

and the solution of Eq. (66) is the time-dependent concentration profile:

$$c - c_b = (c_i - c_b)\left[1 - \mathrm{erf}\,\frac{z}{2\sqrt{Dt}}\right] \tag{68}$$

The instantaneous flux of solute into the element is given by Eq. (6), and combination with Eq. (68) gives

$$J = -D\left(\frac{dc}{dz}\right)_{z=0} = (c_i - c_b)\sqrt{\frac{D}{\pi t}} \tag{69}$$

The flux is thus infinite when the fresh surface is first brought into contact with the second phase and decreases with exposure time. Taking the time mean of Eq. (69) and rearranging, one obtains the mass-transfer coefficient as follows, in terms of phase Y:

$$F_y = \frac{\bar{J}_y}{(c_{yi} - c_{yb})} = 2\sqrt{\frac{D_y}{\pi t_e}} \tag{70}$$

where t_e is the mean exposure time of the element. Higbie [49] suggested that for bubble-type gas absorbers t_e would be approximately equal to the time required for a bubble to rise a distance equal to its own diameter, and a similar interpretation would appear to apply to droplet contactors used in liquid extraction.

An extension of the penetration theory was proposed by Danckwerts [50], who replaced the fixed exposure time proposed by Higbie by a wide spectrum of times and averaged the varying degrees of penetration. It was assumed that the chance of a surface element being replaced by fresh liquid from the bulk by turbulent eddies is independent of its exposure time. On this basis, the surface age distribution function is given by

$$\phi = se^{-st} \tag{71}$$

where s is the fraction of surface replaced in unit time, assumed constant, and ϕ is the probability that a surface element will be exposed for a time t before replacement. The mean steady-state flux of solute is then

$$\begin{aligned} \bar{J}_y &= (c_{yi} - c_{yb})\left(\frac{D_y}{\pi}\right)^{1/2}\int_0^\infty \frac{se^{-st}}{t^{1/2}}\,dt \\ &= (c_{yi} - c_{yb})\sqrt{D_y s} \end{aligned} \tag{72}$$

so that

$$F_y = k_y y_D = \sqrt{D s_y} \tag{73}$$

Equations (70) and (73) have both been derived for the case of solute transfer *into* the Y phase, from interface to bulk phase. The mechanism is less obvious for the reverse direction of transfer, in which case the diffusion path can no longer be considered infinite. However, since the total diffusional flux at a point is always zero, the mechanism is plausible on the basis that the *solvent* diffuses *away* from the interface.

Both forms [49, 50] of the penetration theory are similar in that they lead to the relationship $k_j \propto D_j^{0.50}$. This is in marked contrast to the simple film theory, which requires an exponent of unity on diffusivity, and the boundary-layer theory, for which an empirical exponent of 0.67 is found to be most appropriate for single-phase transfer in a tube.

The penetration theories introduce new parameters t_e or s in place of the effective film thickness δ of the earlier theories, and again their values must be obtained from experimental data. However, Eq. (73) can be brought to the same form as Eq. (63) by assuming that s is proportional to a velocity gradient U_j/d (with U_j taken as the impeller tip speed for rotary-agitated contactors):

$$\frac{k_j j_D}{U_j} = \text{const} \left(\frac{dU_j}{\nu_j}\right)^{-0.5} \left(\frac{\nu_j}{D_j}\right)^{-0.5} \qquad (74)$$

Similarly, the Higbie expression [Eq. (70)] can be modified by assuming that $t_e = d_d/v_d$, leading to an equation in terms of droplet Reynolds number:

$$\frac{k_j j_D}{v_d} = \text{const} \left(\frac{d_d v_d}{\nu_c}\right)^{-0.5} \left(\frac{\nu_j}{D_j}\right)^{-0.5} \left(\frac{\nu_j}{\nu_c}\right)^{0.5} \qquad (75)$$

These simplified interpretations for s and t_e both lead to an exponent of −0.5 on Re, but somewhat different exponents may be observed in practice.

Kishinevsky and Pamfilov [51] proposed a form of penetration theory in which both eddy and molecular diffusion were assumed to be responsible for the transfer. Experiments with a stirred-cell gas absorber and a bubble contactor showed that under conditions of high agitation the rate of absorption was independent of the nature of the gas, with eddy diffusion predominating. A similar mechanism was once considered to apply to liquid extraction in stirred cells, but this now appears unlikely (Section 4.3).

3.1.4. *Combined Film–Penetration Theories*

Toor and Marchello [52] proposed a combined model according to which solute diffuses from an interface into elements of a surface film of thickness δ; beyond this the concentration is maintained constant at the bulk value c_b by turbulent eddies that at times enter the surface region. This concept is introduced into the penetration model by replacing the third boundary condition of Eq. (67) by $c = c_b$ at $z = \delta$ (instead of at $z = \infty$). The resulting solution to Eq. (66) for the instantaneous flux takes different forms for short and long times ($t \ll$ or $\gg \delta^2/D$, respectively), which are averaged by using either the Higbie or the Danckwerts distribution function. The expressions thus obtained for the mean flux reduce to the penetration theory [Eq. (70) or (73)] for short times and to the stagnant-film theory [Eq. (44)] for long times.

Modified theories of this type have been described by Marchello and Toor [53], Harriott [54], and others, mostly with application to transfer at a solid wall; a concise review of these has been given by Sherwood et al. [3]. Such models all involve parameters that must be determined experimentally for the particular form of extraction equipment to be used. These, in turn, must be correlated empirically by expressions that may be expected to take the general form of Eq. (63), (74), or (75) and to be relatively insensitive to the particular type of model assumed.

3.2. Significance of Interfacial Area

In the foregoing treatment the mass-transfer rate has been expressed as the mass (or molar) flux $\hat{N}$ normal to the interface between the phases. The definition is applicable directly to equipment, usually of the laboratory type, in which there is a single extended boundary between the phases. With most industrial equipment, however, one phase is dispersed in the other phase as small droplets so that a relatively large interfacial area can be provided in a given volume. In such cases it is necessary to express the flux in terms of the cross section of the extractor, and a parameter a, defined as the superficial area of contact of the phases per unit volume of dispersion, is used (see Chapter 4).

On this basis Eqs. (47), (51), and (56) are expressed in differential form, for example, in terms of k_{ox}, as follows:

$$dN_1 = k_{ox} a (c_{xb} - c_x^*)\, dz \qquad (76)$$

where N_1 is the flux based on unit cross section of the contactor and z is the distance from one end. Methods for the integration of this equation, to determine the size of contactor required for a specified duty, are described in Chapter 5.

Most of the early published data on mass transfer was expressed in terms of $k_{oj}a$, the *overall volume coefficient*, or the equivalent *height of a transfer unit* (discussed in Chapter 5). The considerable variations observed in $k_{oj}a$ values are due almost entirely to changes in the interfacial area, which is determined by the holdup and droplet size of the dispersed phase (see Chapter 4), whereas the mass-transfer coefficients themselves are relatively constant for a given system, as is discussed later.

The need to treat the volume coefficient as the product of two quite distinct variables, instead of as a "lumped parameter," has become recognized in recent years. In particular, extensive studies have been made of the interfacial area in packed extraction columns, and the mass-transfer coefficients have been thus calculated from mass-transfer data [2]. Limited data on interfacial area are also available for other contactor types and for liquid–liquid mixers, but there is an urgent need for more extensive data to enable contactor design to be put on a more reliable basis. Further complication in this regard often arises from the effect of interfacial tension gradients, as described later.

3.3. Influence of Interface

It has been assumed in the foregoing treatment that equilibrium is established at the interface itself so that it offers no resistance to transfer; this does appear to be the case with many simple physical systems. However, even in the case of a "clean" interface (i.e., uncontaminated by surface films), there can be a departure from equilibrium as a result of a slow heterogeneous reaction at the interface, leading to a *reduction* in mass-transfer rate. Conversely, an *enhancement* in mass-transfer rate can be caused by spontaneous interfacial turbulence generated by interfacial tension gradients (Marangoni effect).

3.3.1. Heterogeneous Reaction

A solute may exist in different chemical forms in the two phases, causing it to undergo a heterogeneous reaction on transferring from one phase to the other. This is found to occur in the extraction of carboxylic acids from water into *nonpolar* solvents such as benzene, in which they exist as dimers; thus for benzoic acid

$$2C_6H_5COOH\,(aq) \rightleftharpoons (C_6H_5COOH)_2\,(solvent)$$

A further example occurs in the extraction of amines such as diethylamine from water into organic solvents with the loss of their hydration water.

Inorganic solvent extraction systems provide many examples of complex formation between solute and solvent. A typical example is the extraction of uranyl nitrate with oxygenated solvents such as methyl isobutyl ketone; thus

$$UO_2(NO_3)_2 \cdot 6H_2O + 2S \rightleftharpoons UO_2(NO_3)_2 \cdot 2S \cdot 4H_2O + 2H_2O$$

where S represents solvent. Slow reactions frequently occur with chelating solvents, such as in the extraction of copper ions with oxime-type solvents, where the reaction is often a rate-limiting step (see Chapter 2.2).

The effect of heterogeneous reaction is illustrated in terms of the two-film theory in Fig. 4*b*, where it is seen that the points representing the interfacial concentrations c_{xi} and c_{yi}/K are no longer coincident. This can be explained quantitatively by considering the case of an mth-order forward reaction and nth-order reverse reaction, with rate constants k_m^+ and k_n^-, respectively. The net rate of reaction, which is equal to the flux of solute through the two interfacial films, is then

$$N_1 = k_m^+ c_{xi}^m - k_n^- c_{yi}^n \tag{77}$$

$$= k_m^+ \left(c_{xi}^m - \frac{k_n^-}{k_m^+} c_{yi}^n\right) \tag{78}$$

At equilibrium the reaction rate is zero, so that

$$\frac{k_n^-}{k_m^+} = \frac{c_{xi}^{*m}}{c_{yi}^n} \tag{79}$$

where c_{xi}^* is the concentration on the X-phase side of the interface in equilibrium with c_{yi} on the Y side. These expressions can be combined with Eqs. (54) and (55) to give the relationship between the resistances, as follows:

$$\frac{1}{k_{ox}} = \frac{1}{k_x} + \frac{1}{m_y' k_y} + \frac{1}{k_m^+}\left[\frac{c_{xi} - c_{xi}^*}{c_{xi}^m - c_{xi}^{*m}}\right] \tag{80}$$

Similarly

$$\frac{1}{k_{oy}} = \frac{m_x'}{k_x} + \frac{1}{k_y} + \frac{1}{k_n^-}\left[\frac{c_{yi}^* - c_{yi}}{c_{yi}^{*n} - c_{yi}^n}\right] \tag{81}$$

The right-hand terms in these equations represent the effect of a slow reaction on the transfer rate, specifically, the *chemical reaction resistance;* this is concentration independent only for first-order reactions, when m or $n = 1$. This relationship is illustrated in Fig. 5*b*, in which point I representing conditions at the interface no longer terminates on the equilibrium curve. The driving forces for the X phase, the chemical reaction, and the Y phase are respectively represented on an X-phase basis by KL, JI, and BH and on a Y-phase basis by CL, KI, and JH; the overall driving forces are BE and CE, respectively, as in the absence of reaction. The values of m_x' and m_y' are now given by the slopes of the chords CK and JB.

The reaction velocity constants k_m^+ and k_n^- can be expressed in terms of either the classical Arrhenius equation or the activated complex theory [55]. It is not possible to predict these constants theoretically, however, because of a lack of knowledge of the necessary liquid-phase partition functions. The reactions must thus be treated in a less fundamental manner by measuring the rates experimentally and using the theoretical relations to evaluate the activation energy, or the enthalpy and entropy of activation.

In cases where a homogeneous reaction involving the solute occurs in the extractant phase, an *enhancement* of mass transfer results. These cases are reviewed in Chapter 2.1 of this handbook.

3.3.2. Spontaneous Turbulence

The dramatic effect of interfacial tension gradients in causing motion of liquid surfaces was noted as early as 1855 by Thomson [56] (quoted in Ref. 57), who observed on placing a small quantity of alcohol carefully on a water surface that "a rapid rushing of the surface is found to occur outwards from the place where the spirit is introduced." Marangoni in 1865 described the spreading of drops of one liquid on the surface of another liquid, and similar observations were reported by other workers shortly afterward [57]. The efficacy of capryl and other higher alcohols as foam breaking agents has been known for a long time in laboratory and industrial practice.

The spreading effect is due to the *gradient* of interfacial tension and consequent *opposite* gradient of surface pressure produced by the additive [58], (see Fig. 6). As a result of viscous drag, a thin layer of the underlying liquid, up to 10 μm thick, is carried with the surface, thus destroying the liquid laminae between adjacent gas bubbles or liquid droplets and breaking the foam or emulsion [59].

The performance of liquid–liquid contactors is strongly influenced in several ways by interfacial effects of this type. For example, drop–drop coalescence is generally enhanced when transfer occurs from organic drops to the aqueous continuous phase and is retarded in the opposite direction of transfer [60]. More details on this are given in Chapter 4 (see Figure 6 of that chapter).

Mass transfer to and from single drops suspended in a second liquid phase has been found to give rise to violent kicks and oscillations with many solutes, particularly organic ones [61]. This results from random fluctuations of solute concentration produced by eddies; thus a high concentration at some point on the surface results in a lowering of interfacial tension, a consequent increase in surface pressure, and a rapid motion of the drop surface in an attempt to restore stability. The effect can also be demonstrated by squirting a pulse of acetone from a fine syringe on to the "equator" of a water drop suspended in toluene, causing the drop to give a sharp kick toward the jet as soon as the acetone arrives [62]. Such effects lead to an increase in mass flux across the interface.

Similar disturbances and eruptions can be observed on plane interfaces, again leading to enhanced mass-transfer rates (Section 4.3). The mechanism in this case is best visualized in terms of the random surface renewal model, leading to random local changes in interfacial tension. Haydon [62] proposed a simple theory on this basis, suggesting that spontaneous turbulence should also occur for reverse transfer, that is, from water into solvent phase, at high solute con-

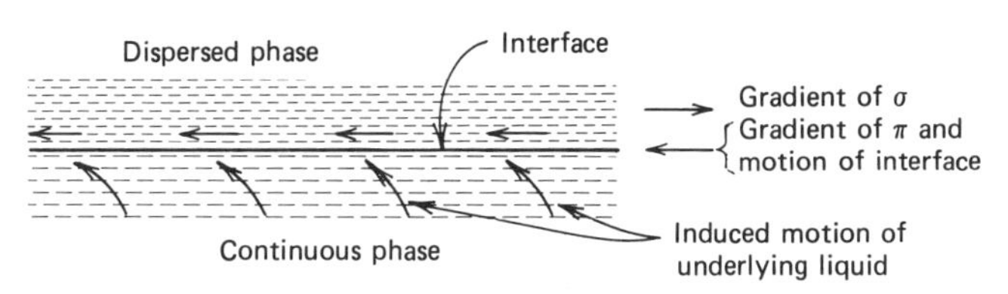

Figure 6. Motion of liquid surface due to interfacial tension gradient.

centrations; this has been observed experimentally [63]. Sternling and Scriven [64] showed on the basis of hydrodynamic stability theory that amplification of interfacial disturbances can occur, leading inter alia to the formation of both stationary and propagating *roll cells* (Fig. 7). Various workers have extended this theory and have attempted to relate the onset of instability to a critical value of the *Marangoni number*, defined as follows:

$$\mathrm{Ma} = \frac{d\sigma}{dc}\,\frac{(c_i - c)\,\delta}{\mu D} \tag{82}$$

where δ is the boundary-layer thickness or penetration depth. Brian et al. [65, 66] have shown that solute storage in the Gibbs adsorption layer can have an important stabilizing effect.

In addition to the preceding theoretical mass-transfer studies, much qualitative work on both droplets and plane interfaces has been carried out by using schlieren and other flow visualization techniques. An excellent review of this topic up to 1971 has been given by Sawistowski [57]; see also Davies [67]. The effect of these phenomena on the performance of extractors is complex and is only briefly reviewed here.

In *mixer–settler cascades*, transfer from organic to aqueous phase leads to an increase in both droplet size and area mass-transfer coefficient as compared with the opposite direction of transfer, but the overall volumetric mass-transfer rate may be lowered; the coalescence rate in the settler will be higher, causing the throughput to be increased. Performance may be improved by increasing the agitation rate, especially at the high concentration (i.e., feed) end.

In *columns*, the droplet size is increased and the dispersed-phase holdup reduced when the transfer takes place from organic to aqueous phase; both effects combine to reduce the interfacial area so that the mass-transfer rate is likely to be lowered in spite of an increased mass-transfer coefficient. The lower holdup leads to an increased flooding rate, causing the throughput to be increased. The performance can usually be improved by increasing the agitation rate with mechanically agitated columns, but care must be taken that this does not lead to flooding at the low-concentration end, where interfacial turbulence effects are at a minimum or absent.

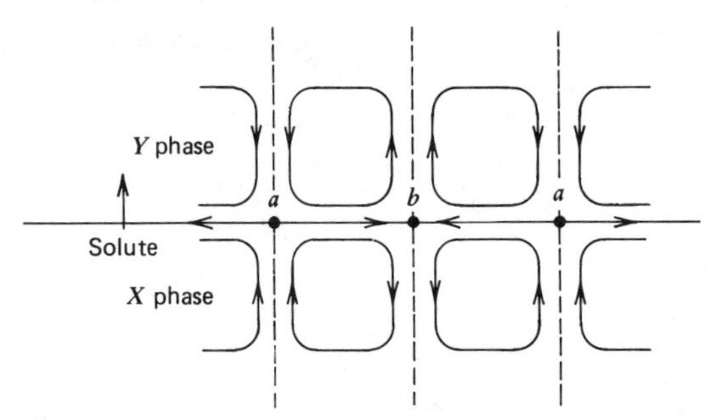

Figure 7. Formation of convection (i.e., roll) cells. The concentration gradients between *a* and *b* are maintained by the circulation established by interfacial tension gradients.

3.4. Multicomponent Mass Transfer

Binary mass-transfer theory is limited in use to ternary systems with immiscible solvents. Multicomponent theory must be used when the solvent miscibility in ternary systems is appreciable and with quaternary and higher systems when two or more solutes are present. These two cases differ in that, with the former, a countertransfer of the solvents is superimposed on the transfer of a single solute, whereas with quaternary and higher systems, the solutes transfer in the same direction through stagnant solvent.

The theory is less developed for multicomponent than for binary transfer, and, although it can be presented in generalized matrix form, it becomes increasingly difficult to use as the number of components is increased beyond 3. The basic theory, particularly in linearized form, is well described by Cussler [8]; however, an exact, nonlinear approach pioneered by Krishna and Standart [68] has since become available and is presented in a modified form in the paragraphs that follow.

3.4.1. Diffusion in Multicomponent Systems

Two alternative methods are used to express multicomponent diffusion. Of these the generalized Fick's law formulation, which has come into use in recent years, takes the following form for unidirectional transport

$$J_i = -c_{tx} \sum_{k=1}^{n-1} D_{ik} \frac{dx_k}{dz} \qquad i = 1, 2, \ldots, n-1 \tag{83}$$

The diffusional flux J_i is related as follows to the total component flux $\hat{N}_i$ relative to a stationary reference frame [cf. Eq. (9)]

$$J_i = \hat{N}_i - x_i \sum_{i=1}^{n} \hat{N}_i \qquad i = 1, 2, \ldots, n \tag{84}$$

The D_{ik} are "practical" multicomponent diffusion coefficients in the sense that measurable concentration gradients appear as driving forces; thus Eq. (83) becomes, for component 1

$$J_1 = \hat{N}_1 - x_1 \hat{N}_T = -c_{tx}\left[D_{11}\frac{dx_1}{dz} + D_{12}\frac{dx_2}{dz} + \cdots\right] \quad (85)$$

For an n-component system, Eq. (83) contains $(n - 1)^2$ diffusion coefficients; however, these are subject to constraints imposed by the Onsager reciprocal relations of irreversible thermodynamics [8] so that the number is reduced to $n(n - 1)/2$. Many workers have measured values of the D_{ik} for ternary systems (see Ref. 8 for a bibliography).

In multicomponent diffusion, kinetic interactions of the components are important and can lead to the phenomena of *osmotic* and *reverse* diffusion, in which the flux is nonzero with zero and negative concentration gradients, respectively, and *diffusion barrier*, with zero flux under a positive gradient [69]. These effects are determined by the magnitude of the cross-coefficients $D_{ik}\,(i \neq k)$; however, the effect of thermodynamic interactions is also contained in the coefficients so that their values are subject to considerable variation.

The other, more fundamental formulation of multicomponent diffusion consists of the generalized Stefan–Maxwell equations:

$$\frac{x_i}{RT}\frac{d\mu_i}{dz} = \sum_{\substack{k=1\\k\neq i}}^{n} \frac{x_i\hat{N}_k - x_k\hat{N}_i}{c_{tx}Ð_{xik}} \qquad i = 1, 2, \ldots, n \quad (86)$$

where the $Ð_{ik}$ are the diffusion coefficients for the pair ik in the mixture; because of the symmetry relationship $Ð_{ik} = Ð_{ki}$, these number $n(n - 1)/2$ as before. Only $(n - 1)$ of the chemical potential gradients are independent in view of the Gibbs–Duhem relationship

$$\sum_{i=1}^{n} x_i \frac{d\mu_i}{dz} = 0 \quad (87)$$

so that an additional determinancy condition is required. Evaluation of the chemical potential gradients in Eq. (86), as in deriviation of Eq. (17), results in

$$\sum_{k=1}^{n-1} \Gamma_{x,ik}\frac{dx_k}{dz} = \sum_{\substack{k=1\\k\neq i}}^{n} \frac{x_i\hat{N}_k - x_k\hat{N}_i}{c_{tx}Ð_{xik}} \qquad i = 1, 2, \ldots, n - 1 \quad (88)$$

where the thermodynamic factor can be expressed as a matrix with the elements

$$\Gamma_{x,ik} = \delta_{ik} + \frac{x_i}{x_k}\frac{\delta \ln \gamma_i}{\delta \ln x_k} \qquad i, k = 1, 2, \ldots, n - 1 \quad (89)$$

in which δ_{ik} is the Kronecker delta.

The diffusivities $Ð_{ik}$ incorporate the effects of the kinetic, but not the thermodynamic, interactions of the molecular species, so that they are subject to less variation than the Fick's law coefficients. Further, their values are approximately predictable for binary systems over the entire concentration range by the methods described in Section 2.2.2(b), although they are likely to be affected by the presence of the other components. Use of these diffusivities in conjunction with the Fick-type formulation is made possible by an approximate linearized matrix theory [70, 71]. However, Krishna and Standart [68] have described an exact solution of Eq. (88) based on the two-film model, which is more convenient, and Krishna [72] has extended this specifically to the liquid phase. A modified form of this method is described in the following paragraphs.

3.4.2. Matrix Solution in Terms of Mass-Transfer Coefficients

The binary mass-transfer coefficients F_{xik} can be introduced into Eq. (88) by substitution of $\eta = z/\delta$, the dimensionless distance within the film, and putting

$$F_{xik} = \frac{Ð_{xik}}{\delta_x} \quad (90)$$

By defining a matrix $[\phi_x]$ with the elements

$$\phi_{x,ii} = \frac{\hat{N}_i}{F_{xin}} + \sum_{\substack{k=1\\k\neq i}}^{n} \frac{\hat{N}_k}{F_{xik}} \qquad i = 1, 2, \ldots, n - 1 \quad (91)$$

$$\phi_{x,ik} = -\hat{N}_i\left(\frac{1}{F_{xik}} - \frac{1}{F_{xin}}\right) \qquad i, k = 1, 2, \ldots, n - 1;\ i \neq k \quad (92)$$

and a column vector (ζ) with elements

$$\zeta_{x,i} = \frac{-\hat{N}_i}{F_{xin}} \qquad i = 1, 2, \ldots, n-1 \tag{93}$$

Eq. (88) can be written as a $(n-1)$-dimensional first-order matrix equation

$$c_{tx}[\Gamma_x] \frac{d(x)}{d\eta} = [\phi_x](x) + (\zeta_x) \tag{94}$$

To solve Eq. (94), it is necessary to assume that both the D_{ik} (and hence the F_{ik}) and the activity coefficients are constant over the range of concentration in the film. On this basis $[\phi_x]$ and $[\Gamma_x]$ are constant, so that integration with the boundary conditions

$$\eta = 0 \text{ (bulk phase)}, \qquad (x) = (x_b)$$

$$\eta = 1 \text{ (interface)}, \qquad (x) = (x_I) \tag{95}$$

gives the concentration profiles in the film as

$$(x - x_b) = \{\exp[\theta_x]\,\eta - \ulcorner I \lrcorner\} \cdot \{\exp[\theta_x] - \ulcorner I \lrcorner\}^{-1}(x_I - x_b) \tag{96}$$

where

$$[\theta_x] = [\Gamma_x]^{-1}[\phi_x]\,c_{tx}^{-1} \tag{97}$$

The composition gradient on the bulk liquid side of the film is obtained by differentiating Eq. (96), giving

$$\left.\frac{d(x)}{d\eta}\right|_{\eta=0} = -[\theta_x]\{\exp[\theta_x] - \ulcorner I \lrcorner\}^{-1}(x_b - x_I) \tag{98}$$

Equation (88) can be expressed in an alternative matrix form by defining a parameter z_ν as follows [73]:

$$z_\nu = \frac{-\hat{N}_n}{\Sigma_{k=1}^{n-1} \hat{N}_k} \tag{99}$$

so that $z_\nu = 1$ for zero transport (i.e., equimass or equimolar countertransfer) and $z_\nu = 0$ for diffusion through stagnant solvent (treating the latter as component n). Hence Eq. (88) becomes [74]

$$c_{tx}[\Gamma_x] \frac{d(x)}{d\eta} = -[B_x](\hat{N}) \tag{100}$$

where the elements of $[B_x]$ are

$$B_{x,ii} = \frac{z_\nu x_i}{F_{xin}} + \sum_{\substack{k=1 \\ k \neq i}}^{n} \frac{x_k}{F_{xik}} \qquad i = 1, 2, \ldots, n-1 \tag{101}$$

$$B_{x,ik} = -x_i\left(\frac{1}{F_{xik}} - \frac{z_\nu}{F_{xin}}\right) \qquad i = 1, 2, \ldots, n-1; i \neq k \tag{102}$$

The mass-transfer rates can be defined as follows in terms of the total flux $(\hat{N})$ and a mass-transfer coefficient matrix $[k_x]$:

$$(\hat{N}) = -c_{tx}[k_x] \left.\frac{d(x)}{d\eta}\right|_{\eta=0} \tag{103}$$

A comparison with eq. (100) for $\eta = 0$, that is, for the bulk phase, shows that

$$[k_x] = [B_{xb}]^{-1}[\Gamma_x] \tag{104}$$

and combination of Eqs. (98) and (103) gives

$$(\hat{N}) = c_{tx}[k_x][\theta_x]\{\exp[\theta_x] - \ulcorner I \lrcorner\}^{-1}(x_b - x_I) \tag{105}$$

This can be expressed more concisely by defining a correction factor matrix $[\Xi_x]$, as follows:

$$[\Xi_x] = [\theta_x]\{\exp[\theta_x] - \ulcorner I \lrcorner\}^{-1} \tag{106}$$

so that

$$(\hat{N}) = c_{tx}[k_x^c](x_b - x_I) \tag{107}$$

where

$$[k_x^c] = [k_x][\Xi_x] \tag{108}$$

The mass-transfer flux, for instance, for component 1, is thus

$$\hat{N}_1 = k_{x11}^c c_{tx}(x_{1b} - x_{1I}) + k_{x12}^c c_{tx}(x_{2b} - x_{2I}) + \cdots \tag{109}$$

Equations (108) and (109) reduce to Eq. (47) when $n = 2$, that is, for binary systems.

Krishna's solution [72], in terms of the *net diffusional* flux (J), can be obtained by putting $z_\nu = 1.0$ in Eqs. (100)–(102), giving a "finite-flux" coefficient matrix $[k_x^\bullet]$. This suffers from the same disadvantages as the corresponding binary coefficient $k_x^\bullet$ (Section 3.1.1).

3.4.3. *Computational Procedure*

The solution of Eq. (107) for the individual component fluxes requires the calculation of the coefficients F_{xik} from Eq. (90) with use of values of δ, the film thickness, obtained from a suitable correlation in terms of Re and Sc [e.g., Eq. (65)]. The correction factor matrix $[\Xi_x]$ can be evaluated from $[\theta_x]$ by means of Sylvester's theorem [75]; explicit solutions are available for ternary systems [68]. However, since the fluxes are contained in the elements of $[\theta_x]$, an iterative procedure is required for the calculation.

In liquid extraction, changes in solvent miscibility can usually be disregarded and the multicomponent process treated as one in which two or more solutes diffuse through stagnant solvent (i.e., $z_v = 0$). The procedure for this case is as follows:

1. Calculate the elements of $[k_x]$ from Eqs. (89), (101), (102), and (104), taking $z_v = 0$.
2. Assume $[\Xi_x]$ to be equal to the unit matrix $\ulcorner I \lrcorner$ and calculate the $(n - 1)$ total fluxes $(\hat{N})$ from Eq. (107); the nth flux is given by the determinancy condition $\hat{N}_n = 0$.
3. Using these values of the fluxes, calculate the elements of $[\theta_x]$ from Eqs. (89), (91), (92), and (97), and hence obtain $[\Xi_x]$ from Eq. (106).
4. Calculate $[k_x^c]$ from Eq. (108) and use this to obtain revised values of the fluxes $(\hat{N})$ from Eq. (107).
5. Repeat steps 3 and 4 until convergence is obtained for each individual $\hat{N}_i$.

If the solvent miscibility cannot be neglected, a similar procedure applies for the special case where $z_v = 1.0$, that is, zero net transport. A more complex treatment is required for the general case in which $z_v \neq 0$ or 1.0, based on individual component balances around the interface in terms of their distribution coefficients (see Refs. 74 and 76 for alternative treatments of the comparable case of nonequimolar distillation).

3.4.4. *Overall Coefficient Matrices*

It is apparent that an expression analogous to Eq. (107) can be obtained for the Y phase, as follows:

$$(\hat{N}) = c_{ty}[k_y^c]\,(y_I - y_b) \tag{110}$$

where

$$[k_y^c] = [k_y]\,[\Xi_y] \tag{111}$$

The fluxes $(\hat{N})$ are identical to those for the X phase since there is no accumulation in the films. It is also possible to define *overall* mass-transfer coefficient matrices, as follows [cf. Eq. (56)]:

$$(\hat{N}) = c_{tx}[k_{ox}^c]\,(x_b - x^*) = c_{ty}[k_{oy}^c]\,(y^* - y_b) \tag{112}$$

where (x^*) and (y^*) are the concentration vectors corresponding to equilibrium with (y_b) and (x_b), respectively.

Relationships between overall and individual film coefficients can be obtained in a manner analogous to those for binary systems. Thus the driving force vector $(x_b - x_I)$ in Eq. (107) is expressed in terms of the Y-phase concentrations by premultiplication of both sides by $[m_x'] \; [k_x^c]^{-1}$, where $[m_x']$ is the equilibrium matrix with elements given by

$$m_{xik}' = \frac{c_{ty}(y_i^* - y_{Ii})}{c_{tx}(x_{bk} - x_{Ik})} \qquad i, k = 1, 2, \ldots, n-1$$

The interfacial concentrations (x_I) are then eliminated by using Eq. (110) and the resulting expression combined with Eq. (112); noting that the flux vector can also be eliminated, one finds

$$[k_{oy}^c]^{-1} = [k_y^c]^{-1} + [m_x']\,[k_x^c]^{-1} \tag{113}$$

Similarly

$$[k_{ox}^c]^{-1} = [k_x^c]^{-1} + [m_y']^{-1}\,[k_y^c]^{-1} \tag{114}$$

These expressions reduce to Eqs. (59) and (58), respectively, for binary systems.

The matrices $[m_x']$ and $[m_y']$ both reduce to the matrix $[M]$, with elements $(c_{ty}/c_{tx})(\delta y_i/\delta x_i)$, if the equilibria can be linearized over the films [77]; this becomes identical with the matrix of distribution coefficients $[K]$ if these are constant. If the equilibria of the components are noninteractive, that is, are unaffected by the presence of the other components, the cross-terms in these matrices are zero and they assume the diagonal forms $\ulcorner m_x' \lrcorner$, $\ulcorner m_y' \lrcorner$ and $\ulcorner M \lrcorner$, respectively.

4. MASS TRANSFER FOR SIMPLE GEOMETRIES

Mass-transfer coefficients for a single liquid at a solid surface correlate closely with those for gases and, by the Chilton–Colburn analogy, with the corresponding heat-transfer rates. However, the situation is much more complex with liquid extraction, because of the effects of interfacial drag on the velocity profiles and of droplet coalescence and redispersion.

Many experimental studies of liquid–liquid mass transfer have been carried out in the laboratory by using an apparatus that provides a known interfacial area and relatively simple hydrodynamical behavior. The principal objectives of these studies were to (1) determine which type of model (e.g., film or penetration) is applicable, (2) ascertain whether the mass-transfer resistances are in fact additive, as suggested by Eqs. (58) and (59), (3) explore for possible interfacial effects, including an interfacial resistance, and (4) study the kinetics of extraction accompanied by chemical reaction (see Section 3.3.1 and Chapter 2.1). Several forms of apparatus that have been devised are briefly reviewed in thc following paragraphs.

4.1. Wetted-Wall Contactors

Wetted-wall (falling-film) columns (Fig. 8) provide a known contact area for the phases, apart from a small discrepancy due to film rippling. Studies of gas absorption, vaporization, and distillation in such columns have given mass-transfer coefficient values for the gaseous core in reasonable agreement with the Chilton–Colburn analogy [i.e., Eq. (62)]. However, the analogy does not hold for a liquid core because of distortion of the velocity profiles in both core and wall liquids as a result of drag. Thus, although the Reynolds number values for the core fall well below the usual transition value of 2100, turbulence is induced by the ripples on the surface of the wall film. Results for the extraction of phenol [78] and acetic acid [79], each at two temperatures, and for uranyl nitrate extraction with two solvents [80] are shown in Fig. 9. These cover a range of Sc values of nearly 100 and indicate an exponent on this group of − 0.67, in conformity with the boundary-layer model. The line shown is represented by

$$\frac{k_c}{U_c} = 9.0\left(\frac{dU_c\,\rho_c}{\mu_c}\right)^{-0.40}\left(\frac{\mu_c}{\rho_c D_c}\right)^{-0.67} \tag{115}$$

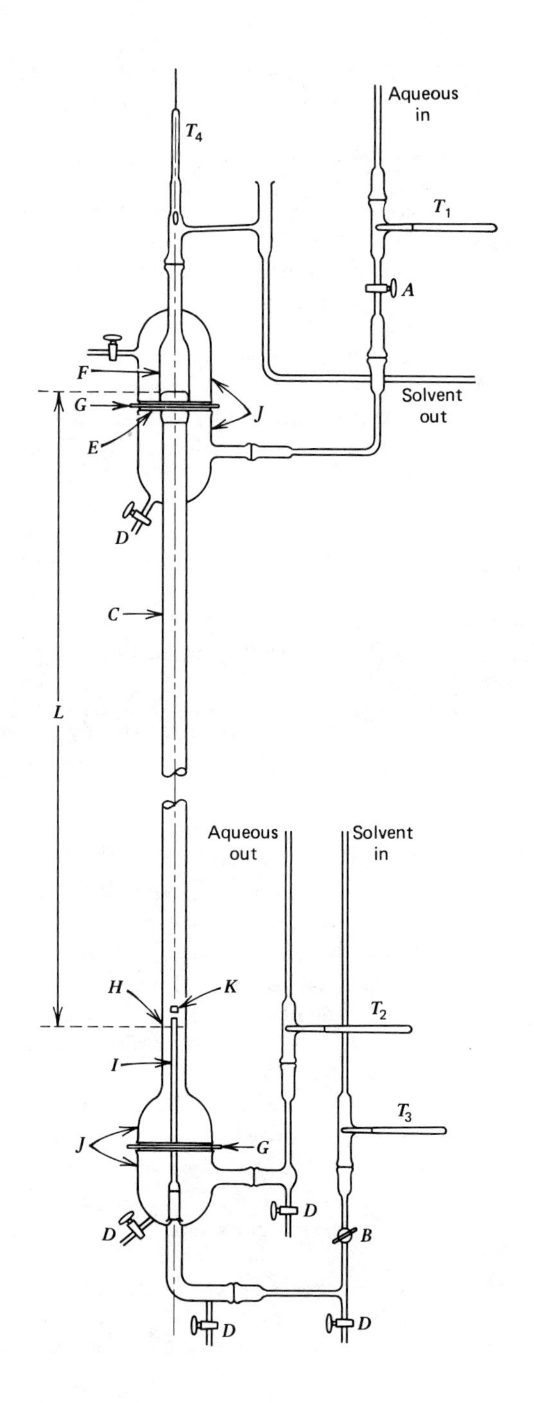

Key

A—Aqueous inlet tap
B—Solvent inlet tap
C—Column—25 mm i.d.
D—Draincocks
E—Weir
F—Cap
G—Gaskets: rubber or PVC
H—Interface level
I—Solvent inlet tube
J—Expanded ends 3—in. i.d.
K—Solvent inlet cap
L—Column height 124.6 cm
T—Thermometers

Figure 8. Wetted-wall column for liquid–liquid systems [80].

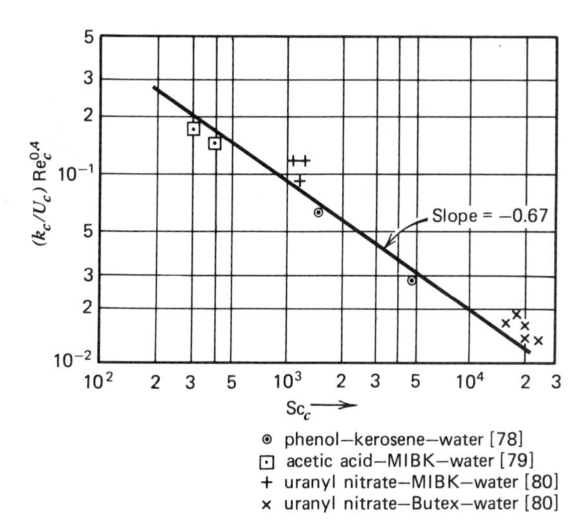

Figure 9. Correlation of core fluid mass-transfer data for wetted-wall column.

indicating, as expected, an exponent on Re different from the value of about −0.2 obtained for gas–liquid systems.

Two results of particular interest emerged from the work on uranyl nitrate: (1) there were indications of an appreciable interfacial resistance due to slow heterogeneous reaction (Section 3.3.1); and (2) the mass transfer in the aqueous phase was apparently a third-order process. Since diffusion is necessarily of first order, the latter result was interpreted as resulting from stripping of the solute, assumed to be un-ionized salt, from the surface layer and replenishment from the ions by a rate-controlling step. Some confirmation of this theory is provided by the fact that a similar phenomenon did not occur in the extraction of uranyl nitrate in a packed column [81], in which appreciable depletion of the diffusing component would not be expected due to the rapid renewal of the droplet surface. On allowing for this effect, the results for the wall film were in approximate agreement with those obtained for the acetic acid system [79],

$$\frac{k_w}{D_w}\left(\frac{\mu_w}{\rho_w^2 g}\right)^{1/3} = 0.00135\left(\frac{4\Gamma}{\mu_w}\right) \cdot \left(\frac{\Delta\rho}{\rho_w}\right)\left(\frac{\mu_w}{\rho_w D_w}\right)^{0.62} \tag{116}$$

where $4\Gamma/\mu_w$ is the Reynolds number for film flow.

Two alternative types of film contactor have also been described, one of which [82], used to investigate spontaneous interfacial turbulence (Section 3.3.2) was of annular cross section with the film on the inner surface. The other consisted of a single film-wetted sphere [83], which was recommended for the determination of the diffusivity of sparingly soluble liquids.

4.2. Laminar Jets

Two types of laminar jet apparatus have been described by Quinn et al. [84, 85], who also summarized the theory. The later of these, capable of giving contact times as short as 0.01 s, was used to determine the diffusivities of benzene and toluene in water. Attempts to measure the interfacial resistance in the transfer inter alia of acetic and benzoic acids between benzene and water indicated that this was negligibly small. A similar result was obtained by Kimura and Miyauchi [86] in a study of the transfer of diethylamine and butyric acid between the same solvents in a similar apparatus.

4.3. Stirred Cells

The transfer cell with plane interface and individually stirred phases, first introduced by Lewis [87], has been widely adopted for the study of liquid–liquid and, more recently, gas–liquid mass transfer [88, 89]. The Lewis cell, (Fig. 10*a*) was provided with horizontal baffles to give an annular interface; the stirrers were mounted on separate shafts, concentric with the support for the central baffle, and were driven at accurately controllable speeds. Most cells of this type have been operated on a transient basis, although a continuous flow type has also been described [90].

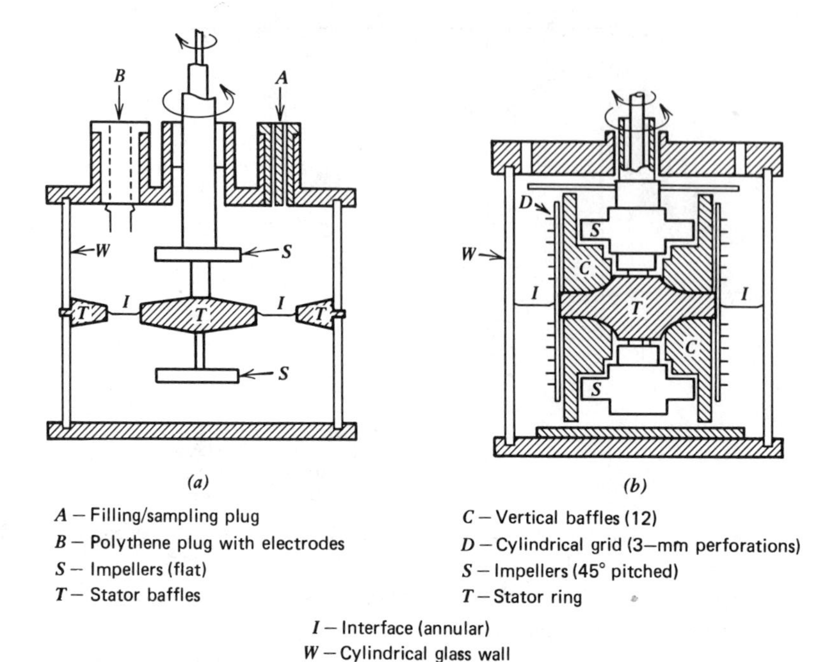

Figure 10. Stirred mass-transfer cells: (*a*) original Lewis design [87]; (*b*) modified design [98].

The operating procedure consists in charging the cell, starting the stirrers, and withdrawing samples of each phase at regular intervals for analysis. The mass-transfer flux for phase j is given by the mass balance, and combination with the rate equation gives.

$$N_j = \frac{V_j}{A}\frac{dc_j}{dt} = k_j(c_{j\infty} - c_j) \tag{117}$$

where V_j is the compartment volume, A is the area of interface, and $c_{j\infty}$ is the final (i.e., equilibrium) concentration. Integration gives

$$k_j = \frac{V_j}{At}\ln(c_{j\infty} - c_j) \tag{118}$$

so that k_j is obtainable from the slope of a plot of $\log(c_{j\infty} - c_j)$ against time.

Lewis first calibrated his cell by measurement of the individual phase resistances by using the Colburn-Welsh method [91], which consists of determining the approach to saturation of pairs of partially miscible liquids (e.g., ethyl acetate–water). The results were correlated in terms of the Reynolds numbers of each phase, but without any Schmidt number terms. The only anomaly was the transfer of water into furfural (but not the reverse), which was very fast; this has since been attributed to the high heat of mixing [92]. Overall mass-transfer coefficients were then measured for the transfer of a solute between the phases and compared with those predicted from the binary data by using Eq. (58). In some cases agreement was good, but in other cases the experimental values were low or high as a result of interfacial resistance or interfacial turbulence, respectively.

Lewis [93] also studied the transfer of uranyl nitrate between water and three solvents and found that the interfacial resistance, which was often zero initially, increased with time of contact; this could have resulted from interfacial turbulence, which would enhance the coefficient initially but would decrease as equilibrium was approached. McManamey [94] in similar experiments observed an interfacial resistance of 40–60% of the overall in the extraction of copper, cobalt, and nickel nitrates into *n*-butanol; interfacial turbulence, leading to an increased mass-transfer rate, was observed for the opposite direction of transfer.

The effect of interfacial films was also studied by Lewis [87], who found that rigid films inhibit the transfer of turbulence across the interface. A similar result was obtained by Davies and Mayers [95], who considered that the mechanism is one of inhibiting surface renewal.

Several workers [94, 96, 97] have queried Lewis' finding [87] that the transfer rate is independent of molecular diffusivity. This matter appears to have been finally settled by Bulicka and Prochazka [98], who modified the original Lewis cell as shown in Fig. 10*b*, in order to obtain greater turbulence together with increased stability of the interface. The results with this cell clearly indicated an exponent on D of 0.50, in accordance with the surface renewal form of the penetration theory and with their resulting semitheoretical relationship

$$\mathrm{Sh}_j = \frac{k_j d}{D_j} = \text{const } \mathrm{Re}_j^{0.75}\, \mathrm{Sc}_j^{0.50}\, \psi \qquad (119)$$

where $\mathrm{Re}_j = d^2 N_j / \nu_j$ and ψ is an "interaction factor," expressing the effect of turbulence of one phase on the other.

A modified Lewis cell has also been used for a study of multicomponent transfer in the system acetone–glycerol–water, which has an appreciable solvent miscibility [99]. The results showed the importance of the cross (i.e., nondiagonal) terms in the mass-transfer coefficient matrix (Section 3.4.2).

It may be concluded that the transfer cell has proved to be an invaluable tool for the study of liquid–liquid mass transfer, and its potentiality has not yet been fully exploited.

4.4. Single-Droplet Contactors

In most practical contactors mass transfer takes place between droplet "swarms" and surrounding continuous phase, with drop coalescence and breakdown taking place. Empirical correlations for mass transfer and interfacial area are given in the appropriate chapters on equipment. The difficulty of modeling performance under such conditions has led to extensive studies with "single-file" columns, through which pass a steady flow of individual droplets (Fig. 11). These have been supplemented by theoretical studies of mass-transfer mechanisms in both droplet and surrounding continuous phases.

The measurement of the individual phase coefficients for droplets presents difficulty for the majority of systems, in the absence of a convenient method of varying the droplet velocity independently. However, these coefficients are obtainable directly with systems of very high or very low distribution coefficient, in which the transfer is controlled almost entirely by one or other phase. The Colburn–Welsh method (Section 4.3) is also used, but the results are sometimes suspect because of the possible occurrence of spontaneous turbulence [2, 100].

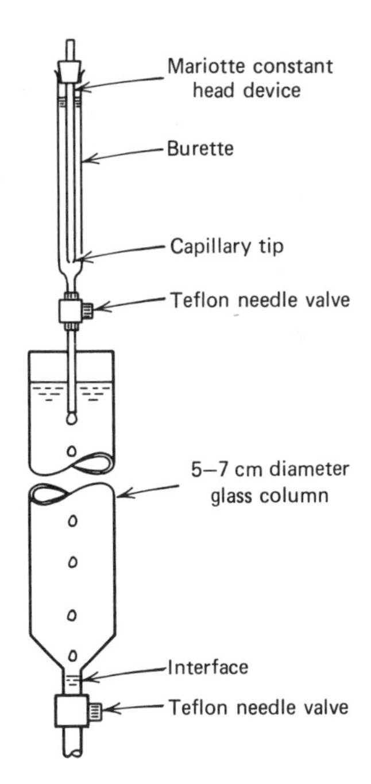

Figure 11. "Single-file" droplet contactor.

Mass transfer to or from drops occurs in three different stages: (1) during formation at a nozzle; (2) during free rise or fall; and (3) during coalescence at a interface. Methods have been devised to determine the relative magnitude of the three stages, and separate consideration is given to these in the following paragraphs.

4.4.1. Free-Moving Droplets: Continuous-Phase Resistance

(a) Solid Spheres. Much theoretical and experimental work has been done on forced-convection heat and mass transfer to solid spheres. Although circulating liquid drops behave somewhat differently, the solid-sphere results provide a useful basis for comparison. The theoretical treatment by Frossling and others [36] is based on a combination of the momentum balance, continuity, and energy or mass transport equations in the boundary layer with the assumption of potential flow outside this. Although numerical solution is required for specific values of Sc, experimental data can be represented closely by the relation (for mass transfer)

$$\mathrm{Sh}_c = \frac{k_c x_D d_d}{D_c} = A + B\, \mathrm{Re}_D^{0.5}\, \mathrm{Sc}_c^{0.33} \qquad (120)$$

An extensive review of the available data by Rowe et al. [101] indicated preferred values of

$A = 2$ and for water (and presumably other liquids) $B \doteq 0.79$. A summary of many of the proposed correlations [31] indicates that the preceding exponents of Re_D and Sc_c are the most appropriate, although there are individual differences in the values of A and B. The exponent on Sc_c, corresponding to $D^{0.67}$, is indicative that the boundary layer rather than the penetration model is applicable.

(b) Liquid Drops. These differ from solid spheres in two respects: (1) internal circulation occurs in all but the smallest sizes; and (2) distortion to ellipsoidal shape occurs with the larger sizes. Droplets of smaller sizes, therefore, have a lower drag coefficient and higher terminal velocity than solid spheres of the same density, whereas the opposite is true of the larger sizes ($\mathrm{Re}_D > 150$–200), which also exhibit cyclic shedding of toroidal vortices from alternate sides. With still larger droplets ($\mathrm{Re}_D > 800$), random wake shedding occurs together with form oscillation of the drops themselves [102, 103].

Griffith [104] gave a number of formulas for Sh_c for various conditions, of which those for $\mathrm{Re}_D > 1$ differed according to the degree of mobility assumed for the droplet surface. Experimental data were presented for the range $\mathrm{Re}_D =$ 2–120, which in most cases were best fitted by the expression for a rapidly moving surface; this was similar to Eq. (120), with $A = 2$ and $B = 1.13$, together with an additional term dependent on the viscosity ratio of the phases. Weber [105] corrected an error in an earlier theoretical analysis and recommended the following approximate form, applicable in the range $\mathrm{Re}_D < 120$, $\mu_r = \mu_d/\mu_c \leqslant 2$, $0 \leqslant \rho_d/\rho_c \leqslant 4$:

$$\mathrm{Sh}_c = \frac{2}{\sqrt{\pi}}\left(1 - \mathrm{Re}_D^{-1/2}(2.89 + 2.15\,\mu_r^{0.64})\right)^{1/2} \mathrm{Pe}^{1/2} \qquad (121)$$

where $\mathrm{Pe} = \mathrm{Re}_D \cdot \mathrm{Sc}_c$. This fitted Griffith's [104] data with an error of less than 5% in $\mathrm{Sh}/\mathrm{Pe}^{1/2}$, and probably represents the best available method for the prediction of Sh_c in this range.

Garner et al. [106] presented data for the range $\mathrm{Re}_D =$ 8-800, obtained by the Colburn-Welsh [91] method for several binary systems. The results were reasonably well correlated for all but one system by the relationship

$$\mathrm{Sh}_c = -126 + 1.8\mathrm{Re}_D^{0.50}\,\mathrm{Sc}_c^{0.42} \qquad (122)$$

Thorsen and Terjesen [107] used an alternative method to obtain continuous phase coefficients, based on the transfer of *o*-nitrophenol between water and various solvents; since these systems had distribution coefficients of 126–325, the organic (droplet)-phase resistance was negligible. Radioactive carbon tetrachloride was also transferred between the inactive form and water, a system for which the distribution coefficient is 1990. The data are shown in Fig. 12*a*, where the upper line is represented by

$$\mathrm{Sh}_c = -178 + 3.62\mathrm{Re}_D^{0.5}\,\mathrm{Sc}_c^{0.33} \qquad (123)$$

The lower line represents data obtained with a strong surface-active agent present and agrees with Eq. (120) with $B = 0.95$, recommended by Garner and Suckling [108] for solid spheres. Figure 12*b* shows a comparison of the data obtained in the absence of surface active agent with Eq. (122); it is apparent that the effect of the different exponent on Sc is small.

A number of other expressions have been proposed for the continuous-phase coefficient; these are summarized in Refs. 31 and 109.

4.4.2. Free-Moving Droplets: Dispersed-Phase Resistance

The limiting rate of transfer for a drop corresponds to that for unsteady molecular diffusion into a rigid sphere. The theoretical solution for this case expressed as the fraction extracted, that is, the extraction efficiency E_d, is as follows for a constant interface composition c_{di} [9, 10]:

$$E_d = \frac{c_d^0 - c_{dm}}{c_d^0 - c_{di}} = 1 - \frac{6}{\pi^2}\sum_{n=1}^{\infty}\frac{1}{n^2} \cdot \exp\left(\frac{-4\pi^2 n^2 D_d t}{d_d^2}\right) \qquad (124)$$

where c_d^0 is the initial composition and c_{dm} the mean value after time t. The Sherwood number may be derived in simple form from this expression by using only the first term of the series:

$$\mathrm{Sh}_d = \frac{k_d x_D d_d}{D_d} \simeq \frac{2\pi^2}{3} = 6.58 \qquad (125)$$

This underestimates the coefficient, and the actual value is 10% greater when $(4D_d t/d_d^2)$ has a value of 0.067, corresponding to an efficiency of 67.5%; the error continues to decrease for longer times.

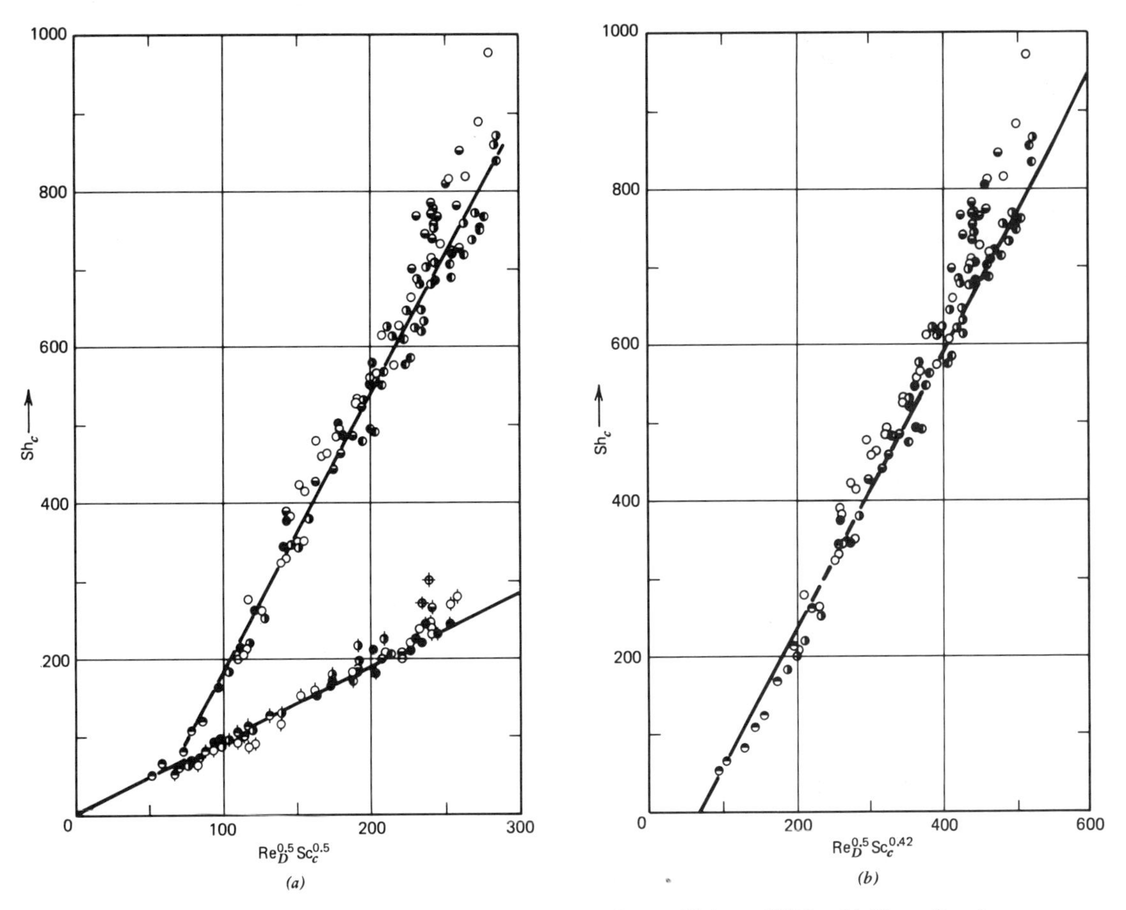

Figure 12. Correlation of continuous-phase mass-transfer coefficients [107]. (*a*) Upper line (pure solvents) given by Eq. (123). Lower line (surfactant present) given by solid sphere correlation [Eq. (120)] with $A = 2, B = 0.95$. (*b*) Data for pure solvents compared with Eq. (122).

In practice, mass-transfer rates are enhanced as a result of mixing within the droplets. Kronig and Brink [110] examined such a case, assuming a streamline circulatory flow of the Hadamard type. The solution to the resulting eigenvalue problem took the form of a series that can be approximated as before to give

$$\mathrm{Sh}_d \simeq 16.7 \tag{126}$$

The rate of transfer is thus enhanced by a factor of about 2.5 compared with the rigid-sphere case. Handlos and Baron [111] proposed an alternative model in which the Hadamard streamlines were replaced by a system of concentric circles giving complete mixing in a single circuit. Their resulting expression was

$$\frac{k_d}{v_d} = \frac{0.00375}{1 + \mu_d/\mu_c} \tag{127}$$

indicating that the rate of transfer is independent of diffusivity for this model. A more complete solution in terms of k_{od}, that is, allowing for the continuous-phase resistance, has been given [112].

Experimental evidence suggests that the Kronig–Brink theory is applicable for $\mathrm{Re}_D < 50$, but above this the coefficient values increase fairly rapidly [113]. Brunson and Wellek [114] have examined 12 models that have been proposed for the oscillating-droplet range (i.e., $\mathrm{Re}_D > 150$–200) and concluded that published data from several sources were best represented by an empirical equation obtained by Skelland and Wellek [115]:

$$\mathrm{Sh}_d = 0.320 \mathrm{Re}_d^{0.68} \left(\frac{\sigma^3 \rho_c^2}{g \mu_c^4 \Delta\rho} \right)^{0.10} \left(\frac{4 D_d t_c}{d_d^2} \right)^{-0.14} \tag{128}$$

where t_c is the contact time of the droplets. This expression gave an average absolute deviation of 15.6%; however, it was found, surprisingly, that the data were fitted nearly as well, and in fact better for high-interfacial-tension systems, by the simple Higbie penetration model [Eq. (70)].

Rozen and Bezzubova [116] have presented dispersed-phase coefficient data for nine systems, using mineral acids and acetic acid as solutes and taking great care to eliminate surface-active agents. Separate correlations were given for (1) medium-sized (spherical) and (2) large (oblate) droplets, as follows:

(1) $\mathrm{Re}_d = 10 - 10^3$:

$$\mathrm{Sh}_d = 0.32\mathrm{Re}_d^{0.63}\,\mathrm{Sc}_d^{0.50}\left(1 + \frac{\mu_d}{\mu_c}\right)^{-0.50} \quad (129)$$

(2) $\mathrm{Re}_c = 10^2 - 1.5 \times 10^3$:

$$\mathrm{Sh}_d = 7.5 \times 10^{-5}\,\mathrm{Re}_c^{2.0}\,\mathrm{Sc}_d^{0.56}\left(1 + \frac{\mu_d}{\mu_c}\right)^{-0.50} \quad (130)$$

Alternative correlations were also presented that included the effect of interfacial tension.

4.4.3. *Drop Formation and Coalescence*

The magnitude of the "end effects," that is, the proportion of the mass transfer that occurs during droplet formation and coalescence, has usually been determined by varying the height and extrapolating the amount of mass transfer to zero height; in some cases special provision has been made to minimize the effect of droplet coalescence. This method has been questioned on the grounds that the plots tend to be nonlinear [117], but this was later explained as having been due to the unsuspected presence of surface-active agents that inhibit circulation within the drops [116]. An alternative method has also been used in which drops are successively formed and withdrawn at the same nozzle, without release [118, 119].

The rate of mass transfer to a growing droplet can be expressed in terms of the Higbie penetration theory, allowing for the increase in area during growth. Thus from Eq. (69) the amount dM transferred during time dt is given by

$$dM = \Delta c A \left(\frac{D}{\pi}\right)^{1/2} t^{-1/2}\,dt \quad (131)$$

where Δc is the driving force. The area A, in terms of that of a full-grown droplet A_f, is given by $A_f(t/t_f)^{2/3}$, where t_f is the time of formation. Substitution of this value in Eq. (131) and integration between the limits $t = 0$ and $t = t_f$ gives the mass-transfer coefficient k_{df} as follows:

$$k_{df} = \frac{M}{A_f t_f \Delta c} = \frac{6}{7}\left(\frac{D}{\pi}\right)^{1/2} t_f^{-1/2} \quad (132)$$

Several alternative mechanisms have been proposed, all of which lead to similar expressions apart from the value of the numerical constant [118, 119], and Popovich et al. [119] found that their own data agreed best with a model due to Ilkovic, for which the constant had a value of 1.52. However, Skelland and Minhas [120], using an improved method of separating the effects of drop formation and coalescence, found that their data were considerably higher than predicted. This was attributed to internal circulation and oscillation of droplets, and the data were thus correlated empirically by dimensional analysis. The same workers also presented data for droplet coalescence at an interface, whereas Hemler [121, 122] obtained data for the continuous-phase coefficients during droplet formation and coalescence. Again all these data were correlated by dimensional analysis.

The hydrodynamics of drop formation and coalescence are quite complex (see Chapter 4). In general, the mass-transfer characteristics of a forming or coalescing dispersion (as opposed to a single drop) cannot be predicted without a careful examination of the hydrodynamic effects in the particular system under study.

NOTATION

		Dimensions	SI units
A_i	concentration of solute at interface	ML^{-3} †	kg m^{-3}
a	interfacial area per unit volume	L^{-1}	m^{-1}
a_i	thermodynamic activity	‡	‡

†Mole units are also applicable.
‡Any convenient units can be used [see Eq. (16) *et seq.* and Ref. 13].

		Dimensions	SI units
c_j	concentration in phase j	ML^{-3} †	kg m^{-3} †
c_j^*	concentration in phase j at equilibrium with other phase	ML^{-3} †	kg m^{-3} †
c_p	specific heat capacity	$L^2T^{-2}\theta^{-1}$	J kg^{-1} K^{-1}
c_{tj}	total concentration in phase j	ML^{-3} †	kg m^{-3} †
D	binary diffusivity	L^2T^{-1}	m^2 s^{-1}
D_{ik}^0	binary diffusivity at low concentration of component i	L^2T^{-1}	m^2 s^{-1}
$Đ_{ik}$	corrected diffusivity of pair ik	L^2T^{-1}	m^2 s^{-1}
d	characteristic dimension of equipment	L	m
d_d	diameter of droplet	L	m
E_D	eddy diffusivity	L^2T^{-1}	m^2 s^{-1}
E_H	eddy thermal diffusivity	L^2T^{-1}	m^2 s^{-1}
E_ν	eddy kinematic viscosity	L^2T^{-1}	m^2 s^{-1}
F_j	mass-transfer coefficient for zero net transport	LT^{-1}	m s^{-1}
f	friction factor (Fanning)		
g	gravitational acceleration	LT^{-2}	m s^{-2}
h	heat-transfer coefficient	$MT^{-3}\theta^{-1}$	W m^{-2} K^{-1}
J	diffusional flux	$ML^{-2}T^{-1}$ †	kg m^{-2} s^{-1} †
j_D	drift factor for component j		
K	distribution coefficient, c_y^*/c_x		
k_j	mass-transfer coefficient for phase j	LT^{-1}	m s^{-1}
k_{oj}	overall mass-transfer coefficient based on phase j	LT^{-1}	m s^{-1}
k_m^+	rate constant for mth-order forward reaction	$M^{(1-m)}L^{(3m-2)}T^{-1}$ †	$kg^{(1-m)}m^{(3m-2)}s^{-1}$
k_n^-	rate constant for nth-order reverse reaction	$M^{(1-n)}L^{(3n-2)}T^{-1}$ †	$kg^{(1-n)}m^{(3n-2)}s^{-1}$
l	Prandtl mixing length	L	m
m_i	total mass of species	M	kg
m_j'	slope of chord to equilibrium curve [Eqs. (57)-(59)]		
$\hat{N}_i$	total flux of component through interface	$ML^{-2}T^{-1}$ †	kg m^{-2} s^{-1} †
N_i	total flux of component based on contractor cross section	$ML^{-2}T^{-1}$ †	kg m^{-2} s^{-1} †
Pr	Prandtl number, $c_p\mu/\kappa$		
Q	heat flux	MT^{-3}	W m^{-2}
R	gas constant, 8315 J K^{-1} $(kmol)^{-1}$	$ML^2T^{-2}K^{-1}$ $(mol)^{-1}$	J K^{-1} $(kmol)^{-1}$
Re_j	Reynolds number, dU_j/ν_j		
Re_D	Reynolds number for droplets or solid spheres, $d_d v_d/\nu_c$		
r	radial distance from center of tube	L	m
r_w	radius of tube	L	m
Sc_j	Schmidt number, ν_j/D_j		
Sh_j	Sherwood number, $k_j j_D d/D_j$		

†Mole units are also applicable.

		Dimensions	SI units
St_H	Stanton heat-transfer number, $h/c_p \rho U$		
St_M	Stanton mass-transfer number, $k_j j_D / U_j$		
s	fractional rate of renewal of interface	T^{-1}	s^{-1}
t	time	T	s
t_e	time of exposure of interface	T	s
t_f	time of formation of droplet	T	s
U	velocity	LT^{-1}	$\mathrm{m\ s}^{-1}$
$\bar{U}$	mean velocity	LT^{-1}	$\mathrm{m\ s}^{-1}$
u_x, u_y, u_z	components of velocity	LT^{-1}	$\mathrm{m\ s}^{-1}$
u_x', u_y', u_z'	deviating velocities in x, y, z directions	LT^{-1}	$\mathrm{m\ s}^{-1}$
$\bar{u}_x$	mean velocity in x direction	LT^{-1}	$\mathrm{m\ s}^{-1}$
V	total volume	L^3	m^3
$\bar{V}_i$	partial specific volume of component i	$\mathrm{L}^3\mathrm{M}^{-1}$†	$\mathrm{m}^3\ \mathrm{kg}^{-1}$†
$\mathbf{v}_i$	vector velocity of component i	LT^{-1}	$\mathrm{m\ s}^{-1}$
$\mathbf{v}_r$	reference velocity [Eq. (2)]	LT^{-1}	$\mathrm{m\ s}^{-1}$
v_d	droplet terminal velocity	LT^{-1}	$\mathrm{m\ s}^{-1}$
w	mass fraction		
x, y, z	coordinate directions		
x	mass or mole fraction in X (i.e., feed) phase		
$\bar{x}$	mole fraction in X (i.e., feed) phase		
x^*	mass or mole fraction in X phase at equilibrium with Y phase		
x_D	drift factor for X phase		
y	mass or mole fraction in Y (i.e., extract) phase		
y^*	mass or mole fraction in Y phase at equilibrium with X phase		
y_D	drift factor for Y phase		
z	$= \hat{N}_1/(\hat{N}_1 + \hat{N}_2)$, relative velocity through film		
z_ν	defined by Eq. (99)		

Greek Symbols

α	thermodynamic factor [Eq. (19)]		
α_H	$= \kappa/\rho c_p$, thermal diffusivity	$\mathrm{L}^2\mathrm{T}^{-1}$	$\mathrm{m}^2\ \mathrm{s}^{-1}$
β	= turbulent Prandtl or Schmidt number, E_ν/E_H or E_ν/E_M, respectively		
Γ	peripheral liquid rate [Eq. (116)]	$\mathrm{ML}^{-1}\mathrm{T}^{-1}$	$\mathrm{kg\ m}^{-1}\ \mathrm{s}^{-1}$
γ_i	activity coefficient		
δ	film thickness	L	m
θ	temperature	θ	K
κ	thermal conductivity	$\mathrm{MLT}^{-3}\theta^{-1}$	$\mathrm{W\ m}^{-1}\ \mathrm{K}^{-1}$
μ	viscosity	$\mathrm{ML}^{-1}\ \mathrm{T}^{-1}$	$\mathrm{N\ s\ L}^{-2}$
μ_i	chemical potential of component i	$\mathrm{ML}^2\ \mathrm{T}^{-2}\ (\mathrm{mol})^{-1}$	$\mathrm{J\ (kmol)}^{-1}$

†Mole units are also applicable.

		Dimensions	SI units
$\Delta\pi$	surface pressure gradient	$ML^{-1}T^{-2}$	$N\ m^{-2}$
ν	kinematic viscosity	L^2T^{-1}	$m^2\ s^{-1}$
ρ	density	ML^{-3}	$kg\ m^{-3}$
$\Delta\rho$	density difference, $\rho_c - \rho_d$	ML^{-3}	$kg\ m^{-3}$
σ	interfacial tension	MT^{-2}	$N\ m^{-1}$
τ	shear stress	$ML^{-1}T^{-2}$	$kg\ m^{-1}\ s^{-2}$

Subscripts

c	continuous phase
c	core liquid phase (Section 4.1)
d	dispersed phase; df refers to k_d for droplet formation
H	heat transfer
I	at interface (Section 3.4 only)
i	number of component (Section 3.4 only)
i	at interface (except Section 3.4)
j	phase X or Y; phase c or d
k	dummy variable
l	laminar
lm	logarithmic mean
M	mass transfer
t	turbulent
w	wall phase (Section 4.1)
x	X (i.e., feed) phase
y	Y (i.e., extract) phase
$1,2,\ldots,n$	number of component

Matrix Symbols

[ ]	square matrix
( )	column vector
$\ulcorner\ \lrcorner$	diagonal matrix
$\ulcorner I \lrcorner$	unit matrix

REFERENCES

1. W. J. Korchinsky, *Can. J. Chem. Eng.* **52,** 468 (1974).
2. H. R. C. Pratt and W. J. Anderson, *Proceedings of the International Conference on Solvent Extraction* (ISEC77), Toronto, (Can. Inst. Min. Met., Montreal), 1979, p. 242.
3. T. K. Sherwood, R. L. Pigford, and C. R. Wilke, *Mass Transfer*, McGraw-Hill, New York, 1975.
4. R. B. Bird, W. E. Stewart, and E. N. Lightfoot, *Transport Phenomena*, McGraw-Hill, New York, 1960.
5. C. O. Bennett and J. E. Myers, *Momentum, Heat and Mass Transfer*, McGraw-Hill, New York, 1962.
6. R. E. Treybal, *Mass Transfer Operations*, 2nd ed., McGraw-Hill, New York, 1968.
7. C. H. Bedingfield and T. B. Drew, *Ind. Eng. Chem.* **42,** 1164 (1950).
8. E. L. Cussler, *Multicomponent Diffusion,* Elsevier, Amsterdam, 1976.
9. J. Crank, *The Mathematics of Diffusion*, Oxford University Press, London, 1956.
10. H. S. Carslaw and J. C. Jaeger, *Conduction of Heat in Solids*, 2nd ed., Oxford University Press, London, 1959.
11. B. W. Clack, *Proc. Phys. Soc.* (*Lond.*) **36,** 313 (1924).
12. L. G. Longsworth, *Ann. NY Acad. Sci.* **44,** 211 (1945).

13. R. A. Robinson and R. H. Stokes, *Electrolyte Solutions*, 2nd ed. (revised), Butterworths, London, 1965.
14. G. S. Hartley, *Phil. Mag.* **12,** 473 (1931).
15. A. Vignes, *Ind. Eng. Chem. Fund.* **5,** 189 (1966).
16. H. T. Cullinan, *Ind. Eng. Chem. Fund.* **5,** 281 (1966).
17. F. A. L. Dullien, *Ind. Eng. Chem. Fund.* **10,** 41 (1971).
18. J. Leffler and H. T. Cullinan, *IEC Fund.* **9,** 84 (1970).
19. V. Sanchez and M. Clifton, *Ind. Eng. Chem. Fund.* **16,** 318 (1975).
20. R. L. Robinson, W. C. Edmister, and F. A. L. Dullien, *Ind. Eng. Chem. Fund.* **5,** 75 (1966).
21. S. S. Rao and C. O. Bennett, *AIChE J.* **17,** 75 (1971).
22. J. L. Haluska and C. P. Colver, *Ind. Eng. Chem. Fund.* **10,** 610 (1971).
23. P. A. Johnson and A. L. Babb, *Chem. Rev.* **56,** 387 (1956).
24. A. R. Gordon, *Ann. NY Acad. Sci.* **44,** 285 (1945).
25. J. B. Lewis, *J. Appl. Chem.* **5,** 228 (1955).
26. F. A. L. Dullien and L. W. Shemilt, *Can. J. Chem. Eng.* **39,** 242 (1961).
27. R. H. Stokes, *J. Am. Chem. Soc.* **72,** 763, 2243 (1950).
28. R. H. Stokes, *J. Am. Chem. Soc.* **73,** 3527 (1951).
29. C. R. Wilke and P. C. Chang, *AIChE J.* **1,** 264 (1955).
30. K. A. Reddy and L. K. Doriswamy, *Ind. Eng. Chem. Fund.* **6,** 77 (1967).
31. G. S. Laddha and J. M. Smith, *Indian Chem. Eng.* **11,** Transact. 109 (October 1969).
32. G. S. Laddha and T. E. Degaleesan, *Transport Phenomena in Liquid Extraction*, Tata McGraw-Hill, New Delhi, 1976.
33. H. S. Harned and B. B. Owen, *The Physical Chemistry of Electrolytic Solutions*, American Chemical Society Monograph 95, Reinhold, New York, 1950.
34. A. R. Gordon, *J. Chem. Phys.* **5,** 522 (1937).
35. J. R. Vinograd and J. W. McBain, *J. Am. Chem. Soc.* **63,** 2008 (1941).
36. H. Schlichting, *Boundary Layer Theory*, McGraw-Hill, New York, 1960.
37. J. O. Hinze, *Turbulence*, McGraw-Hill, New York, 1959.
38. J. T. Davies, *Turbulence Phenomena*, Academic Press, New York, 1972.
39. W. G. Whitman, *Chem. Met. Eng.* **29,** 147 (1923); *Ind. Eng. Chem.* **16,** 1215 (1924).
40. A. P. Colburn and T. B. Drew, *Transact. Am. Inst. Chem. Eng.* **33,** 197 (1937).
41. H. R. C. Pratt and P. G. Tuohey, *Chem. Eng. J.* **18,** 251 (1979).
42. T. L. Prandtl, *Z. Physik* **11,** 1072 (1910); **29,** 487 (1928).
43. G. I. Taylor, *Rep. Mem. Br. Adv. Comm. Aeronaut*, **272,** 423 (1916).
44. E. V. Murphree, *Ind. Eng. Chem.* **24,** 726 (1932).
45. Th. von Karman, *Transact ASME* **61,** 705 (1939).
46. T. K. Sherwood, *Transact Am. Inst. Chem. Eng.* **36,** 817 (1940).
47. T. K. Sherwood, *Ind. Eng. Chem.* **42,** 2077 (1950).
48. T. H. Chilton and A. P. Colburn, *Ind. Eng. Chem.* **26,** 1183 (1934).
49. R. Higbie, *Transact. Am. Inst. Chem. Eng.* **31,** 365 (1935).
50. P. V. Danckwerts, *Ind. Eng. Chem.* **43,** 1460 (1951).
51. M. K. Kishinevsky and A. V. Pamfilov, *Zh. Prik. Khim.* **22,** 1173 (1949): M. K. Kishinevsky and V. T. Sebransky, ibid **29,** 17 (1956); M. K. Kishinevsky and L. A. Mochalova, ibid. **29,** 170 (1956).
52. H. L. Toor and J. M. Marchello, *AIChE J.* **4,** 97 (1958).
53. J. M. Marchello and H. L. Toor, *Ind. Eng. Chem. Fund.* **2,** 8 (1963).
54. P. Harriott, *Chem. Eng. Sci.* **17,** 149 (1962).
55. S. Glasstone, K. J. Laidler, and H. Eyring, *The Theory of Rate Processes*, McGraw-Hill, New York, 1941.
56. J. Thomson, *Phil. Mag.* **10** (4), 330 (1855).
57. H. Sawistowski, in C. Hanson, ed., *Recent Advances in Liquid–Liquid Extraction*, Pergamon Press, Oxford, 1971, Chapter 9.
58. J. T. Davies and E. K. Rideal, *Interfacial Phenomena*, 2nd ed., Academic Press, New York, 1963.
59. W. E. Ewers and K. L. Sutherland, *Aust. J. Sci. Res.* **5,** 697 (1952).
60. H. Groothuis and F. J. Zuiderweg, *Chem. Eng. Sci.* **12,** 288 (1960).
61. J. B. Lewis and H. R. C. Pratt, *Nature* **171,** 1155 (1953).
62. D. A. Haydon, *Nature* **176,** 839 (1955); *Proc. Roy. Soc.* **A243,** 483 (1958).
63. N. G. Maroudas and H. Sawistowski, *Nature* **188,** 1186 (1960); see also Ref. 80.
64. C. V. Sternling and L. E. Scriven, *AIChE J.* **5,** 514 (1959); *J. Fluid Mech.* **19,** 321 (1964).
65. P. L. T. Brian, *AIChE J.* **17,** 765 (1971).
66. P. L. T. Brian and K. A. Smith, *AIChE J.* **18,** 231 (1972); P. L. T. Brian and J. R. Ross, ibid., 583.
67. J. T. Davies, *Turbulence Phenomena*, Academic Press, New York, 1972.
68. R. Krishna and G. L. Standart, *AIChE J.* **22,** 383 (1976).
69. H. L. Toor, *AIChE J.* **3,** 198 (1957).
70. H. L. Toor, *AIChE J.* **10,** 448, 460 (1964).

71. W. E Stewart and R. Prober, *Ind. Eng. Chem. Fund.*, **3,** 224 (1964).

72. R. Krishna, *Lett. Heat Mass Transf.* **3,** 153 (1976).

73. E. N. Turevskii, I. A. Aleksandrov, and V. G. Gorechenkov, *Internatl. Chem. Eng.* **14,** 112 (1974).

74. P. G. Tuohey, *Binary and Ternary Mass Transfer in Distillation*, University of Melbourne Ph.D. thesis (1980).

75. N. R. Amundson, *Mathematical Methods in Chemical Engineering; Matrices and Their Application*, Prentice-Hall, Englewood Cliffs, NJ, 1966.

76. R. Krishna and G. L. Standart, *Lett. Heat Mass Transf.* **3,** 173 (1976); R. Krishna, *Chem. Eng. Sci.* **32,** 1197 (1977).

77. R. Krishna, *Lett. Heat Mass Transf.* **3,** 41 (1976).

78. R. Fallah, T. G. Hunter, and A. W. Nash, *J. Soc. Chem. Ind.* **54,** 49T (1935).

79. D. A. Brinsmade and H. Bliss, *Transact. Am. Inst. Chem. Eng.* **39,** 679 (1943).

80. R. Murdoch and H. R. C. Pratt, *Transact. Inst. Chem. Eng.* **31,** 307 (1953).

81. L. E. Smith, J. D. Thornton, and H. R. C. Pratt, *Transact. Inst. Chem. Eng.* **35,** 292 (1957).

82. N. G. Maroudas and H. Sawistowski, *Chem. Eng. Sci.* **19,** 919 (1964).

83. G. A. Ratcliff and K. J. Reid, *Transact. Inst. Chem. Eng.* **39,** 423 (1961).

84. J. A. Quinn and P. G. Jeannin, *Chem. Eng. Sci.* **15,** 243 (1961).

85. W. J. Ward and J. A. Quinn, *AIChE J.* **10,** 155 (1964); **11,** 1005 (1965).

86. M. S. Kimura and T. Miyauchi, *Chem. Eng. Sci.* **21,** 1057 (1966).

87. J. B. Lewis, *Chem. Eng. Sci.* **3,** 248, 260 (1954).

88. J. T. Davies, A. A. Kilner, and G. A. Ratcliff, *Chem. Eng. Sci.* **19,** 583 (1964).

89. O. Levenspiel and J. H. Godfrey, *Chem. Eng. Sci.* **29,** 1723 (1974).

90. D. R. Olander and L. B. Reddy, *Chem. Eng. Sci.* **19,** 67 (1964).

91. A. P. Colburn and D. G. Welsh, *Transact. Am. Inst. Chem. Eng.* **38,** 179 (1942).

92. G. A. Davies and J. D. Thornton, *Lett. Heat Mass Transf.* **4,** 287 (1977).

93. J. B. Lewis, *Nature* **178,** 274 (1956); *Chem. Eng. Sci.* **8,** 295 (1958).

94. W. J. McManamey, *Chem. Eng. Sci.* **15,** 210, 251 (1961).

95. J. T. Davies and G. R. A. Mayers, *Chem. Eng. Sci.* **16,** 55 (1961).

96. G. R. A. Mayers, *Chem. Eng. Sci.* **16,** 69 (1961).

97. W. J. McManamey, J. T. Davies, J. M. Woollen, and J. R. Coe, *Chem. Eng. Sci.* **28,** 1061 (1973).

98. J. Bulicka and J. Prochazka, *Chem. Eng. Sci.* **31,** 137 (1976).

99. G. L. Standart, H. T. Cullinan, A. Paybarah, and N. Louizos, *AIChE J.* **21,** 554 (1975).

100. L. J. Austin, W. E. Ying, and H. Sawistowski, *Chem. Eng. Sci.* **21,** 1109 (1966).

101. P. N. Rowe, K. T. Claxton, and J. B. Lewis, *Transact. Inst. Chem. Eng.* **43,** T14 (1965).

102. J. Yeheskel and E. Kehat, *Chem. Eng. Sci.* **26,** 1223 (1971).

103. W. J. Anderson and H. R. C. Pratt, *Chem. Eng. Sci.* **33,** 995 (1978).

104. R. M. Griffith, *Chem. Eng. Sci.* **12,** 198 (1960).

105. M. E. Weber, *Ind. Eng. Chem. Fund.* **14,** 165 (1975).

106. F. H. Garner, A. Foord, and M. Tayeban, *J. Appl. Chem.* **9,** 315 (1959).

107. G. Thorsen and S. G. Terjesen, *Chem. Eng. Sci.* **17,** 137 (1962).

108. F. A. Garner and R. D. Suckling, *AIChE J.* **4,** 114 (1958).

109. C. I. Pritchard and S. K. Biswas, *Br. Chem. Eng.* **12,** 879 (1967).

110. R. Kronig and J. C. Brink, *Appl. Sci. Res.* **A2,** 142 (1950).

111. A. E. Handlos and T. Baron, *AIChE J.* **3,** 127 (1957).

112. W. J. Korchinsky and J. J. C. Cruz-Pinto, *Chem. Eng. Sci.* **34,** 551 (1979).

113. A. I. Johnson and A. E. Hamielec, *AIChE J.* **6,** 145 (1960).

114. R. J. Brunson and R. M. Wellek, *Can. J. Chem. Eng.* **48,** 267 (1970).

115. A. H. P. Skelland and R. M. Wellek, *AIChE J.* **10,** 491 (1964).

116. A. M. Rozen and A. I. Bezzubova, *Theor. Found. Chem. Eng.* **2,** 715 (1968); translated from *Teor. Osnovy Khim. Tekh.* **2,** 850 (1968).

117. W. Licht and W. F. Pansing, *Ind. Eng. Chem.* **45,** 1885 (1953).

118. J. M. Coulson and S. J. Skinner, *Chem. Eng. Sci.* **1,** 197 (1952).

119. A. T. Popovitch, R. E. Jervis, and O. Trass, *Chem. Eng. Sci.* **19,** 357 (1964).

120. A. H. P. Skelland and S. S. Minhas, *AIChE J.* **17,** 1316 (1971).

121. C. L. Hemler, Ph.D. thesis, University of Notre Dame (1974); see ref. 122.

122. A. H. P. Skelland and W. L. Conger, *Ind. Eng. Chem. Proc. Des. Devel.* **12,** 445 (1973).

4

DISPERSION AND COALESCENCE

G. S. Laddha and T. E. Degaleesan
University of Madras
India†

1. INTRODUCTION

Phase dispersion and coalescence phenomena are important in liquid–liquid extraction with either stagewise or differential contact. Systems with high interfacial tension require applied mechanical agitation in order to achieve adequate dispersion, and the scale of turbulence is the criterion controlling drop size. Systems with low interfacial tension may be handled in nonagitated gravity columns whereas agitated dispersions usually require additional volume for phase separation as in mixer–settler units. For difficult separation systems with an emulsification tendency and low density difference between the

†Present address: Anna University, Madras, India.

phases, centrifugal extractors provide the force required for phase separation.

The breakup of liquid jets issuing from orifice distributors as in spray and sieve-plate columns results in a spread of drop sizes, rise velocities, and contact times for the drop population as the drops move through the continuous phase. The presence of column internals [as in packed and rotating-disk contactor (RDC) columns] and use of additional mechanical energy (as in RDC, Scheibel, Oldshue–Rushton columns, and the agitated vessel/mixer–settler) further affect the dispersion characteristics. The mean drop size in a contactor is controlled by coalescence–redispersion phenomena. Large drops entering the contactor emerge as smaller droplets, whereas small droplets often coalesce and emerge as large droplets of an equilibrium size. A dynamic equilibrium is established between the continuous breakup and coalescence processes occurring in the various contactors. The mean equilibrium drop size depends on the mode of operation of the contactor and the type and extent of agitation as well as the physical properties of the liquid phases. High agitation energies cause breakdown of drops, yielding large interfacial areas, but this advantage is greatly offset by the absence of oscillation or circulation in the smaller drops. In large drops, interfacial phenomena such as circulation, oscillation, interfacial turbulence and spontaneous emulsification considerably influence the coalescence rates. These effects are sensitive to trace contamination, and in commercial extraction equipment such contamination is often heavy. However, when polar organic solvents are used (of low interfacial tension against water) there is less sensitivity to the adsorption of impurities. Moreover, the drop breakup and coalescence process form clean interfaces while much of the mass transfer is occurring. High shear rates in agitated vessels may also largely overcome the surface tension gradients and may even collapse the surface films and sweep them away [1].

1.1. Measurement of Mean Drop Size

This can be done conveniently by photography and chemical reaction techniques [2–4]. The success of the photographic method depends on the proper choice of site or location within the vessel, near the wall. The distribution of the photographed drop images must truly represent the mean distribution of drops in the entire contractor after correcting for the distortion and enlargement when macrophotography is used [4]. For nonspherical drops, the major and minor axes of the drop images should be measured and reduced to the actual values d_1 and d_2, taking into account the enlargement factor. The equivalent size of the drop d_e can then be calculated by using the method proposed by Lewis et al. [3] as follows:

$$d_e = \sqrt[3]{d_1^2 d_2} \tag{1}$$

For spherical drops, the drop diameter is taken as d_e. On the basis of the number of drops (clear images) counted in the photographs and their d_e values, the mean drop size usually calculated as Sauter mean drop size d_{vs} can be obtained thus:

$$d_{vs} = \frac{\sum_1^n d_e^3}{\sum_1^n d_e^2} \tag{2}$$

This definition of mean drop size is useful in mass-transfer calculations (see Chapter 3) since the specific interfacial area a of the dispersion is given simply as $6x/d_{vs}$. The mean drop size may also be obtained from direct estimation of interfacial area by the chemical reaction technique if holdup is known (see also Chapter 2.1, Section 5).

The direct area measurement method involves a liquid–liquid chemical reaction such as the hydrolysis of an organic ester such as cyclohexyl formate by sodium hydroxide. The total reaction rate R_A' of such a fast pseudo-first-order reaction under dispersion conditions may be estimated experimentally on the basis of the fall in the normality of the alkali. The rate of reaction R_A per unit interfacial area is independent of hydrodynamic conditions and may be estimated from the reaction rate constant by using the appropriate rate equation (see Chapter 2.1, "regime 3" in Tables 3a and 3b). The effective interfacial area per unit volume is found from the values of R_A' and R_A as follows:

$$a = \frac{R_A'}{R_A V} \tag{3}$$

where V is the volume of the dispersion.

2. PHASE DISPERSION

2.1. Single-Drop Systems

Drops released at very low flow rates from a nonwetting nozzle in a quiescent liquid are uniform in size and have a certain ratio of equivalent drop size d_e to nozzle size d_N. The value of (d_e/d_N) is governed by complex factors and ranges from about 3 for very small nozzles (hypodermic needles) to near unity for large nozzle sizes (small tubes). The diameter of the largest nozzle that can be used for producing a steady stream of drops (without backflow within the nozzle tube) can be estimated approximately as $\pi(\gamma/\Delta\rho\, g)^{1/2}$, which also gives the size of maximum possible stable drop in a stationary quiescent medium.

Drop formation at a nozzle occurs in two hydrodynamic regions [5–10] depending on the terminal velocity range of drops. As the nozzle size is increased, the size of drop produced also increases, until a size is reached beyond which the drop terminal velocity remains almost independent of drop size. The following simple relationships give fairly useful estimates [10] of peak drop size d_p (above which drops are broken down and below which the drops could be either stable or coalescing) and peak terminal velocity u_p:

$$d_p = 1.38\left(\frac{\gamma}{\Delta\rho\, g}\right)^{1/2} \tag{4}$$

and

$$u_p = 1.59\left(\frac{\gamma\,\Delta\rho\, g}{\rho_c^2}\right)^{1/4} \tag{5}$$

The drop size and terminal velocity above and below this peak condition may be related in terms of a gravity group $G\ (g\,\Delta\rho\,\rho_c d_e^3/\mu_c^2)$ and a property group $P(\gamma^3\rho_c^2/\Delta\rho\, g\mu_c^4)$ in addition to the drop Reynolds number $\mathrm{Re}(d_e u_t \rho_c/\mu_c)$.

$$\frac{G^{0.54}}{P^{0.25}} = C\left[\frac{\mathrm{Re}}{P^{0.25}}\right]^n \tag{6}$$

where the constants C and n vary according to the region of operation, as follows:

For region 1 where $\dfrac{\mathrm{Re}}{P^{1/4}} < 0.22$,

$$n = 1.0 \quad \text{and} \quad C = 1.225$$

For region 2 where $\dfrac{\mathrm{Re}}{P^{1/4}} > 0.22$,

$$n = 1.33 \quad \text{and} \quad C = 0.94$$

The mean size of drops released at nozzle tips at flow rates up to low jetting velocities may be estimated by the following equation of Devotta [11]:

$$d_e\left(\frac{\gamma}{\Delta\rho\, g}\right)^{-1/2} = 2.3\left(\frac{d_N^2\,\Delta\rho\, g}{\gamma}\right)^{0.235} \cdot \left(\frac{u_N^2}{2g\, d_N}\right)^{0.022} \tag{7}$$

In Eq. (7), the Bond group representing the static forces $(d_N^2\,\Delta\rho\, g/\gamma)$ appears to be the dominating factor. No significant effect of viscosity was observed over a wide range for either phase.

2.2. Dispersion from Jetting Streams

The proper choice of nozzle size requires calculation of jet velocity u_j and diameter d_j of the maximum-length jet, which, according to Hixson and co-workers [12, 13], gives dispersions that have a maximum interfacial area. The use of jet velocities between 10 and 25 cm/s has been recommended [14]. The mean drop size of the dispersion is estimated [15, 16] to be twice the diameter of the jet d_j, which may be obtained by the following relationship [13]:

$$\frac{d_N}{d_j} = p + q(\phi)^n \tag{8}$$

where the factor ϕ is defined as $d_N/(\gamma/\Delta\rho\, g)^{0.5}$ and values of p, q, and n are given by

For $\phi < 0.785$ $\quad p = 1.0;\quad q = 0.485;\quad n = 2$

For $\phi > 0.785$ $\quad p = 0.12;\quad q = 1.51;\quad n = 1$

The recommended limits of nozzle size for obtaining proper liquid jets with known distributions of drop sizes on jet breakup are [2]

$$\frac{1}{2}\left(\frac{\gamma}{\Delta\rho\, g}\right)^{1/2} < d_N < \pi\left(\frac{\gamma}{\Delta\rho\, g}\right)^{1/2} \tag{9}$$

The lower limit of diameter may be preferred for systems with high interfacial tension and the upper limit, for low interfacial tension systems.

2.3. Dispersion in Spray Columns

Vedaiyan et al. [2] noted that the drop size distributions obtained from single and multiple nozzles in the size range given by Eq. (9) varied from near normal at low nozzle velocity u_N to highly skewed at high u_N and passed through a bimodal distribution stage at moderate values of u_N. Figure 1 shows typical distribution patterns reported for isoamyl alcohol dispersed into water using 0.1-cm nozzles. Values of d_{vs} in spray

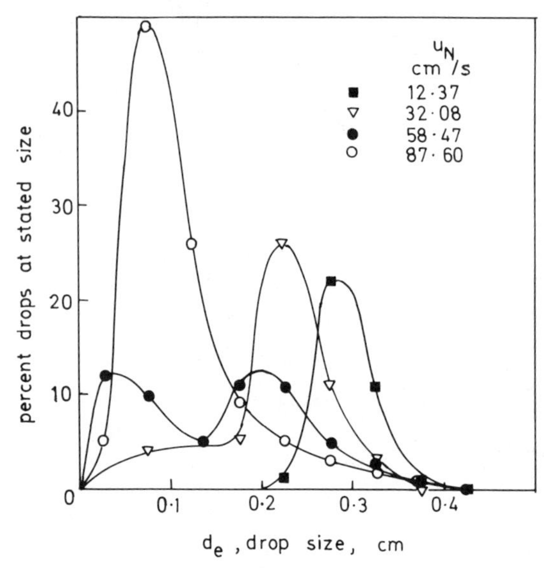

Figure 1. Drop size distribution versus nozzle velocity [2].

columns were correlated [2] for several systems covering a wide range of u_N by

$$\frac{d_{vs}}{(\gamma/\Delta\rho\, g)^{1/2}} = \alpha_1 \left(\frac{u_N^2}{2gd_N}\right)^{\beta_1} \tag{10}$$

Umesh et al. [17] noted that Eq. (10) is not significantly affected by variations in μ_c or μ_d although the u_N at which the onset of bimodality occurred decreased with increase of μ_c. Whereas the reported values of the constants α_1 and β_1 in Eq. (10) (viz., 1.592 and -0.067, respectively), were found to be valid both in the absence of mass transfer and for solute transfer from the continuous phase to dispersed phase ($c \rightarrow d$), these were affected for $d \rightarrow c$ transfer (viz., 2.1 and +0.1, respectively) [18]. In $d \rightarrow c$ transfer the occurrence of heavy coalescence of droplets due to Marangoni effects appeared to increase the mean size of droplets of the swarm with a consequent increase in the mean velocity of rise of droplets in the column [18] (see Sections 3.1 and 4.2.1).

2.3.1. Dispersion in Sieve-Plate Columns

If each compartment of a sieve-plate column behaved like a spray column, the mean drop size could be computed from Eq. (10). However, photographic analysis of drop size data covering a wide range of sieve-plate hole sizes indicates [19] that the mean drop size is independent of the nozzle Froude number but could be represented by a simple equation of the form given by Eq. (4), in which the constant varied with hole size d_N as shown in Fig. 2. The decrease of drop size is steep below a critical hole size d_{Nc} equal to about 0.25 cm. The drop size obtained for the sieve plate with critical hole size d_{Nc} corresponded to the value for peak drop size d_p given by Eq. (4). The effect of hole size on droplet characteristic velocity $\bar{u}_0$ shows a similar transition at d_{Nc} as illustrated in Fig. 3 (see also Section 3.1 and Table 2).

2.4. Dispersion in Packed Columns with Packings Preferentially Wetted by the Continuous Phase

Lewis et al. [3] observed that the exit droplet size was independent of packing size D_p if this

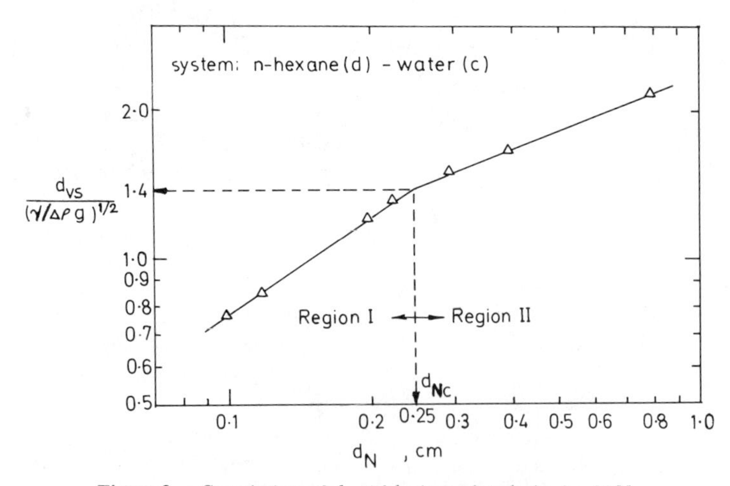

Figure 2. Correlation of d_{vs} with sieve-plate hole size [19].

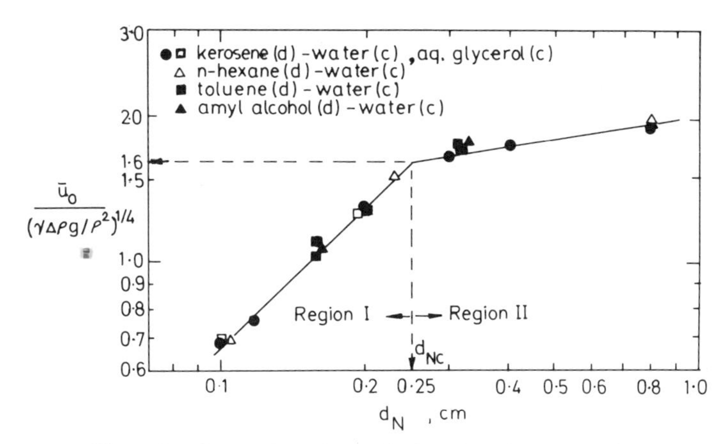

Figure 3. Correlation of $\bar{u}_0$ with sieve-plate hole size [19].

exceeded the critical size D_{pc}, given by

$$D_{pc} = 2.42 \left(\frac{\gamma}{\Delta \rho \, g} \right)^{1/2} \qquad (11)$$

The exit drop size d_{vs} for organic–aqueous systems could be estimated by the simplified correlation form proposed by Lewis et al. [3]:

$$d_{vs} = C' \left(\frac{\gamma}{\Delta \rho \, g} \right)^{1/2} \qquad (12)$$

with the value of the constant $C = 0.92$. This equilibrium drop size $(d_{vs})_{eq}$ is possibly achieved within the packed column at some height h' from the inlet of the packed section as represented by an equation of the following form [20]:

$$(d_{vs})_{eq} = A + B \exp{(Ch')} \qquad (13)$$

where the constants A, B, and C depend on the direction of solute transfer; $(d_{vs})_{eq}$ was found to be smaller than the critical drop size d_c, which varied with system properties [21] according to the simplified correlation

$$d_c = 1.32 \left(\frac{\gamma}{\Delta \rho \, g} \right)^{1/2} \qquad (14)$$

This may be compared with the peak drop size d_p given by Eq. (4) and the following equation given by Hu and Kintner [5, 23] for d_c in the absence of packing:

$$d_c = 1.476 \left(\frac{\gamma}{\Delta \rho \, g} \right)^{1/2} \qquad (15)$$

Very little information is available on the mean drop size of dispersion when packings below D_{pc} given by Eq. (11) are used, such as in bench-scale packed columns and wire-mesh packed beds that aid coalescence [22, 24].

2.5. Dispersion in Agitated Systems

The droplet breakup mechanism in a given hydrodynamic field may be considered in terms of energy dissipation [25, 26]. If the scale of the process (drop size) is greater than the minimum eddy size, the breakup rate depends mainly on the energy dissipation rate per unit mass of fluid.

2.5.1. Kolmogoroff-Hinze Equation

The dynamic forces due to turbulent eddies tend to break up the drop in opposition to interfacial forces. Balancing of these forces and assumption of homogeneous isotropic turbulence results in the following equation for the maximum stable drop sizes [25, 26]:

$$d_{max} \propto \frac{\gamma^{0.6}}{(P/v)^{0.4} \rho_c^{0.2}} \qquad (16)$$

Equation (16) has been modified by Hinze to give the following equation for d_{max} in terms of ϵ, the power dissipated per unit mass of fluid:

$$d_{max} = C_1 \left(\frac{g_c \gamma}{\rho_c} \right)^{0.6} (\epsilon)^{-0.4} \qquad (17)$$

where C_1 is a constant (reported to be 0.72 for a rotating-cylinder apparatus by Clay [27]) and is dependent on the system properties and the phase dispersed; the relationships for ϵ are summarized in Table 1 for some well-known types of contactor. It has been noted that the ratio of mean diameter to maximum diameter of drops in the swarm remains constant in the absence of mass transfer; the ratio has been reported to be 0.5 for drops in stirred vessels [28] and 0.7 for RDC contactors [53]. The mean drop size of agitated dispersions is affected by the solute-transfer process as it introduces Marangoni effects with consequent changes in droplet size distribution and interdroplet coalescence frequencies. Laddha et al. [29, 30] noted mass-transfer effects on drop size in RDC and other columns. Hence the above approach through Eq. (17) and Table 1 may be used for an approximate estimate of drop size in the absence of mass transfer.

2.5.2. *Drop Size in Agitated Vessels*

The Sauter mean drop size of a dilute dispersion for a noncoalescing system d_{vs}^0 was related to impeller Weber number by a modification of Eq. (17) by Shinnar and Church [31] as follows:

$$d_{vs}^0 = C_1'(\mathrm{We})^{-0.6}\, D_R \tag{18}$$

The value for C_1' at extremely low holdup ($x < 0.01$) has been found to be 0.053 by Chen and Middleman [32]. Thornton and Bouyatiotis [33] described the effect of holdup on d_{vs}^0 by the following relationship for impeller speeds above the minimum critical value:

$$d_{vs} = d_{vs}^0 + mx \tag{19}$$

Empirical relations between d_{vs}^0 and m under the experimental conditions were given [33]. However, Doulah [34] suggested the following relationship:

$$d_{vs} = d_{vs}^0\,(1 + \beta x) \tag{20}$$

and identified the term $(1 + \beta x)$ with the change in the ratio (ν_e/ν_c) of the continuous-phase viscosity under dispersing and nondispersing conditions. Since the experimentally measured value of β ranged from 0 to 9 depending on the holdup range covered in the individual studies [4, 32, 34, 40–44], Gnansundaram et al. [4] postulated that the ratio (ν_e/ν_c) is expected to increase exponentially with holdup in the region below a certain maximum critical speed (We $<$ 1000) beyond which the coalescence characteristics are offset by high turbulence. They recommended the following relationships for estimation of d_{vs} at any holdup in batch turbine-agitated vessels:

$$\text{For We} < 10{,}000 \qquad \frac{d_{vs}}{D_R} = 0.052\,(\mathrm{We})^{-0.6}\, e^{4x} \tag{21}$$

$$\text{For We} > 10{,}000 \qquad \frac{d_{vs}}{D_R} = 0.39\,(\mathrm{We})^{-0.6} \tag{22}$$

TABLE 1 RELATIONSHIPS FOR ϵ FOR SOME CONTACTORS

Contactor and Reference	Relationship	Equation Number	Remarks
Spray Column [35]	$\epsilon = u_d g \dfrac{\Delta\rho}{\rho_c}$	(1)	Assuming that entire energy is dissipated in the continuous phase
Plain disk RDC [36–38]	$\epsilon = N_{\mathrm{Po}}\left(\dfrac{4}{\pi}\,\dfrac{N^3 D_R^5}{n_C Z_C D_T^2}\right)$	(2)	$N_{\mathrm{Po}} \approx 0.03$ for Re $> 10^5$; C_1 = 0.4–0.6 (organic dispersed), C_1 = 0.8–1.2 (aqueous dispersed) in Eq. (17); $d_{vs} \sim 0.7\, d_{\max}$
Pulsed sieve-plate Columns [15]	$\epsilon = \dfrac{\pi^2 n_c[(S_t/S_N)^2 - 1]\,(Af)^3}{2 g_c c_N^2 Z_T}$	(3)	Predicts low ϵ for low S_t/S_N ratios ($<$0.35)
Oldshue–Rushton columns [15]	$N_{\mathrm{Po}} = 4.4$	(4)	In Eq. (2) above
Scheibel columns [39]	$N_{\mathrm{Po}} = 1.85$	(5)	In Eq. (2) above

Equation (21) reduces to Eq. (18) for $x < 0.01$ with $C_1' = 0.053$ as proposed by Chen and Middleman [32].

3. HOLDUP AND FLOODING

Normally the phase that preferentially wets the contactor/column internals will be selected as the continuous phase so that the dispersion consists of discrete droplets moving freely within the continuous phase; otherwise the wetting phase, when dispersed, flows either as rivulets or streams of uneven chunks of liquid, yielding a poor dispersion with unpredictable hydrodynamics. The volume of droplets in the contactor during steady operation is termed the *operational holdup* of the dispersed phase x, generally expressed as a fraction of the total dispersion volume or effective volume of the contactor. In any contactor some dispersed phase always remains stationary, blocked at the junction points and blind corners provided by the contactor internals; this "static" holdup contributes little to mass transfer, unlike the "operational" holdup [45].

In a countercurrent column-type contactor, steady operation is possible when the rate of arrival of droplets does not exceed the coalescence rate at the main interface; otherwise droplet buildup occurs at the interface, gradually extending over the entire column and leading to "flooding." In a column with no internals such as a spray column (or in a sieve-plate column compartment) the height of the droplet buildup can be related to the mean size of droplets arriving at the interface zone [46]. Letan and Kehat [47, 48] investigated three possible regimes of spray column operation—dispersed, restrained, and dense packing of drops, in the order of increasing holdup. The desired regime was obtained mainly by controlling the position and size of the coalescing interface.

In a sieve-plate column the static holdup comprising the coalesced layer formed under each plate is the result of three major effects attributable to continuous-phase flow rate, orifice resistance, and interfacial tension effects [49, 50]. The height of static holdup layer under the plate h_t is given by an equation [49] to represent the head due to these three effects:

$$h_t = \frac{4.5\, u_c^2 \rho_c}{2g\, \Delta\rho} + \frac{6\gamma g_c}{(d_e^0)\Delta\rho\, g} + \frac{u_N^2[1 - (S_N/S_t)^2]\, \rho_d}{2gC_N^2\, \Delta\rho} \tag{23}$$

where the orifice coefficient C_N may be taken as 0.67 for circular holes, although Major and Hertzog [50] suggested a Reynolds number dependence. A knowledge of the preceding height of coalesced layer (static holdup) is necessary for estimation by difference of the dispersion volume in the sieve-plate column compartments. The static holdup increases strongly with continuous-phase flow, affecting the main interface of the compartments and finally resulting in a condition approaching flooding. On the other hand, significant growth of the coalesced layer occurs with increase in dispersed phase rate only at high continuous-phase rates. This effect also depends on the free hole area in the plates.

In a plain RDC, the operational holdup x is affected by rotor speed N, stator ring opening D_S, and compartment height Z_C in addition to the phase flow rates [29, 51, 52]. Increase of N produces greater shear, causing drop breakup, and because of the lower settling velocities of the broken droplets and the increased tortuosity of the flow path of the droplets along the toroidal vortices [53], the drop population in the column increases. Similarly, increase of D_R at a given N results in increased peripheral speeds of the rotor disk, causing finer drop dispersions and larger holdup. However, increases of D_S and Z_C offer a less tortuous path for the drops, resulting in a decrease in holdup and allowing greater throughput of the liquid phases.

In a rotary annular column (see Chapter 14), the vortex height between successive spiral ring patterns of dispersed droplets should be minimised for increased holdup and contactor efficiency, and the rotor peripheral speed must be above a critical value [54]. The holdup also depends on the nature of the vortices and the flow pattern due to the different wetting characteristics of the stator and rotor surfaces [55].

In a cylindrical batch mixer the holdup of the dispersed phase is generally assumed to be equal to the fractional volume of that phase in the total volume of feed charged into the vessel when the impeller speed is above a certain minimum value [15, 56] at which the mixing index I_M becomes unity. The mixing index I_M is defined for batch mixing in terms of the droplet holdup x and the phase volumes v_d and v_c:

$$I_M = \frac{x}{v_d/(v_d + v_c)} \tag{24}$$

There is an uncertain region of operation in which it is difficult to predict whether the dispersion will be of water-in-oil (water–oil) or oil-in-water (oil–water) type, but at high solvent to aqueous feeds, the aqueous phase would be dispersed [57]. Laity and Treybal [58] found that the power requirements for two-phase dispersion could be obtained from Rushton's correlation for single-phase systems, when the mean density and viscosity properties of the dispersion were used. The power number remains almost constant in the region $\mathrm{Re}' > 10^4$, where I_M can be assumed to be unity. The minimum impeller speed for this condition at which complete dispersion of one of the phases occurs may be predicted by the following empirical correlation [59]:

$$\frac{D_R^2 N}{g} = C\left(\frac{D_T}{D_R}\right)^n \left(\frac{\mu_c}{\mu_d}\right)^{0.11} \left(\frac{\Delta\rho}{\rho_c}\right)^{0.25} \left(\frac{\gamma}{D_R^2 \rho_c g}\right)^{0.3} \tag{25}$$

where the constants C and n reported (for flat-blade turbines located at a height of $1.5D_R$ from the bottom of tank of diameter $D_T = 3D_R$) are 4.0 and 0.881, respectively. Radial flow impellers gave a lower minimum speed for complete dispersion than axial flow impellers (e.g., propeller) [59]. Weinstein and Treybal [60] found that holdup in continuous-flow-agitated tanks was represented by a slip velocity relationship at specific power inputs in excess of 340 W/m^3. At power inputs below this limit a different form of dimensionless correlation was obtained.

Steady operation of an agitated vessel is possible provided no phase reversal occurs. Generally, phase reversal limits contactor operation since beyond this point the interfacial area decreases as the holdup of the phase reversed (now dispersed) decreases. It is found [57] that water-oil emulsion settles faster than oil-water emulsion since the coalescence rates of water droplets in oil are higher than those of oil droplets in water. This is important for settler design, and the throughput capacity of the contacting equipment should be greater if the oil phase is made continuous and the water phase dispersed.

3.1. Characteristic Velocity and Operational Holdup

The slip velocity concept proposed by Lapidus and Elgin [61] and the relative velocity concept due to Pratt and co-workers [45, 62] (based on Steinour's analysis [63]) has been useful in correlating holdup with phase flow rate; for a countercurrent column, the slip or relative velocity u_s is given by

$$u_s \equiv \frac{u_c}{1-x} - \frac{u_d}{x} = \bar{u}_0(1-x) \tag{26}$$

where $\bar{u}_0$ is defined as the characteristic velocity and may be identified with the average terminal velocity $(u_t)_{\mathrm{av}}$ of the droplets [64]. One may predict holdup for any given set of phase flow rates provided a suitable correlation for estimation of $\bar{u}_0$ is available. Several investigators have suggested $\bar{u}_0$ correlations for various types of contactor (see Table 2). It is important to note that the direction of solute transfer affects the droplet coalescence characteristics as well as the mean drop size, holdup and $\bar{u}_0$ within the contactor due to Marangoni effects and other interfacial instabilities. Under the $d \rightarrow c$ transfer condition, the mean size of drops has been found to increase within the contactor due to coalescence, resulting in a decrease in holdup and increase in settling velocity or $\bar{u}_0$. Laddha et al. [29, 30, 67] have examined the design aspects of various contactors in the light of the effect of direction of solute transfer on the contactor hydrodynamics. These results supplement the general design data given in Chapter 13 of this handbook.

3.2. Holdup at Flooding

Flooding in extraction columns may be characterized by the following conditions:

$$\left(\frac{\delta u_c}{\delta x}\right)_f = \left(\frac{\delta u_d}{\delta x}\right)_f = 0 \tag{27}$$

Application of the preceding criteria [Eq. (26)] yields the following relationships at flooding:

$$u_{df} = 2\bar{u}_0 x_f^2 (1 - x_f) \tag{28}$$

$$u_{cf} = \bar{u}_0 (1 - x_f)^2 (1 - 2x_f) \tag{29}$$

and

$$x_f = \frac{[1 + 8(u_c/u_d)_f]^{0.5} - 3}{4(u_c/u_d)_f - 4} \tag{30}$$

These relationships are not restricted to any particular type of column and may be applied to calculate flooding conditions in any column provided estimates of $\bar{u}_0$ under actual transfer conditions are possible (see Table 2).

Holdup at flooding for a packed column may be estimated by the following correlation based on Eq. (28) and the $\bar{u}_0$ correlation given by Laddha et al. [65-68]:

$$x_f\sqrt{1 - x_f} = C_1\left(\frac{u_{df}^2 a}{g e^3} \cdot \frac{\rho_c}{\Delta\rho}\right)^n \tag{31}$$

TABLE 2 SOME IMPORTANT CORRELATIONS FOR CHARACTERISTIC VELOCITY

Column and Reference	Correlation	Remarks
Spray column [18, 64]	$\dfrac{\bar{u}_0}{(\gamma \Delta \rho g/\rho_c^2)^{1/4}} = \alpha_2 \left(\dfrac{u_N^2}{2gd_N}\right)^{\beta_2}$	$\begin{cases} \text{No mass transfer and } c \to d \text{ transfer:} & \alpha_2 = 1.088, \ \beta_2 = -0.082 \\ d \to c \text{ transfer:} & \alpha_2 = 1.42, \ \beta_2 = +0.125 \end{cases}$
Packed column [65–68]	$\bar{u}_0 = C\left(\dfrac{a}{e^3 g}\dfrac{\rho_c}{\Delta \rho}\right)^{-1/2}$ For packings above D_{pc} [Eq. (11)]	$\begin{cases} \text{No mass transfer:} & C = 0.683 \\ c \to d \text{ transfer:} & C = 0.637 \\ d \to c \text{ transfer:} & C = 0.820 \end{cases}$
Sieve–plate column [19]	$\bar{u}_0 = C'\left(\dfrac{\gamma \Delta \rho g}{\rho_c^2}\right)^{1/4}$	where C' varied with hole size d_N as shown in Fig. 3; the variation is steep below a critical hole size $d_{Nc} = 0.25$ cm at which $\bar{u}_0$ corresponds to the value obtained for peak terminal velocity u_p given by Eq. (5)
RDC [29]	$\dfrac{\bar{u}_0}{(\gamma \Delta \rho g/\rho_c^2)^{1/4}} \ G_f = C[\mathrm{Fr} \cdot \Psi^m]^n$	$\bar{u}_0$ varies steeply above a critical speed; analysis includes data due to Logsdail et al. [51] and Kung and Beckmann [52]

Operating Condition	C	n	m	
In absence of mass transfer:				
$\mathrm{Fr} \cdot \Psi$ >180	1.08	0.08	1.0	$G_f \equiv \left(\dfrac{Z_C}{D_R}\right)^{0.9} \left(\dfrac{D_S}{D_R}\right)^{2.1} \left(\dfrac{D_R}{D_T}\right)^{2.4}$
$\mathrm{Fr} \cdot \Psi$ <180	0.01	1.0	1.0	
For $c \to d$ transfer:				
$\mathrm{Fr} \cdot \Psi^{1/2}$ >16	1.40	0.08	0.5	$\Psi \equiv \left(\dfrac{\gamma^3 \rho_c}{\mu_c^4 g}\right)^{1/4} \left(\dfrac{\Delta \rho}{\rho_c}\right)^{3/5}$
$\mathrm{Fr} \cdot \Psi^{1/2}$ <16	0.08	1.0	0.5	
For $d \to c$ transfer:				
$\mathrm{Fr} \cdot \Psi^{1/2}$ >25	1.40	0.08	0.5	$\mathrm{Fr} \equiv \left(\dfrac{g}{D_R N^2}\right)$
$\mathrm{Fr} \cdot \Psi^{1/2}$ <25	0.11	1.0	0.5	

Column and Reference	Correlation	Remarks
RDC [69]	$\bar{u}_0 = \left(\dfrac{u_s}{1-x}\right) \exp\left[-\left(\dfrac{z}{\alpha} - 4.1x\right)\right]$	z—coalescence correction factor; α—back-mixing coefficient
Rotary annular columns [70]	$\dfrac{\bar{u}_0 \mu_c}{\gamma} = 0.028 \left(\dfrac{\Delta \rho}{\rho_c}\right)^{1.5} \left(\dfrac{D_T}{D_R}\right)^{1.2} (\mathrm{Fr})^{1.1} (\mathrm{Re})^{-0.18}$	Above critical speed, $\mathrm{Re} \equiv (D_R^2 N \rho_c/\mu c)$; $\mathrm{Fr} \equiv (g/D_R N^2)$; no solute transfer
Agitated vessel [33]	$\dfrac{u_0 \rho_c}{\mu_c g} = C\left(\dfrac{(P/v)^3 g_c^3}{\rho_c^2 \mu_c g^4}\right)^{-0.64} \left(\dfrac{\rho_c \gamma^3}{\mu_c^4 g}\right)^{0.14} \left(\dfrac{\Delta \rho}{\rho_c}\right)^{1.1}$	Concurrent flow is assumed; $C = 1.95\ (10^{-5})$ for no mass transfer

where C_1 and n in the absence of mass transfer are [74], respectively: Raschig rings, 0.52 and 0.083; Lessing rings, 0.47 and 0.074; Berl Saddles, 0.54 and 0.092; and spheres, 0.46 and 0.076.

In a pulsed sieve-plate column, pulsing action introduces a coalescence–redispersion cycle in the liquid phases. The phases flowing countercurrently undergo alternate dispersion and coalescence when the phase ratio u_c/u_d is in the order of unity; otherwise the phase flowing at a lesser rate generally forms the dispersed phase [75–77]. Of the various distinct regimes of operation of a pulsed column (see Chapter 11), the mixer-settler and emulsion regimes are preferable as they provide efficient operation because of intimate contact of the phases [78]. The dispersed-phase holdup could be correlated by the slip velocity equation (Eq. 26) with the characteristic velocity dependent on the column and pulse variables and the physical properties of the system [79]. Holdup at flooding is related to phase flow rates and characteristic velocity by Eqs. (28)–(30).

In pulsed packed columns, holdup and $\bar{u}_0$ are affected by pulse variables, geometry factors, and physical properties of the liquid system in addition to fluid wetting effects [77]. Data on holdup x_f at flooding (for nonwetting dispersed phase) have been correlated by Potnis [80] by the following relationship:

$$x_f = 0.98e^{0.59}\left(\frac{u_d^2 a}{ge^3}\right)^{0.30}\left(\frac{\rho_c}{\Delta\rho}\right)^{0.44}\left(\frac{Af\rho_c}{a\mu_c}\right)^{-0.14} \quad (32)$$

3.3. Holdup and Flooding in Coalescer Beds of Liquid–Liquid Separators

Various packings—namely, rings, knitted-mesh, or fiber beds—are used in the settling zones of liquid–liquid separators as coalescence promoters. The packing type and depth required depend on the characteristics of the dispersions to be separated. Thus ceramic or metallic packings smaller than the critical size D_{pc} [Eq. (11)] have been used for primary dispersions ($>100\,\mu$m) that are otherwise difficult to settle. The dispersed phase droplets coalesce at the packing surface to form large supported drops, several times the size of the packing through which they move. The equilibrium drop size is larger than in the case of packings above the critical size, but the holdup at and below flooding can still be predicted [81] by Eqs. (31) and (26), respectively. In Scheibel columns the flow capacity is limited by flooding at the packing, and according to Piper [82], the usual flooding correlations are not applicable in this case as the hydrodynamic phenomenon is complex. However, Honekamp and Burkhart [83] indicate that the behavior of a four-stage Scheibel extractor at constant stirrer speed with respect to drop size and holdup parallels that of a packed extraction column.

Flooding of knitted wire-mesh packed bed coalescers has been studied by Davies et al. [22, 84]. Secondary dispersions as well as primary dispersions that tend to build up a droplet dense phase beneath the interface are coalesced by use of such wire-mesh beds. Flooding of wire-mesh beds made of dispersed-phase wetting materials (consisting of either single or dual wire) occurs

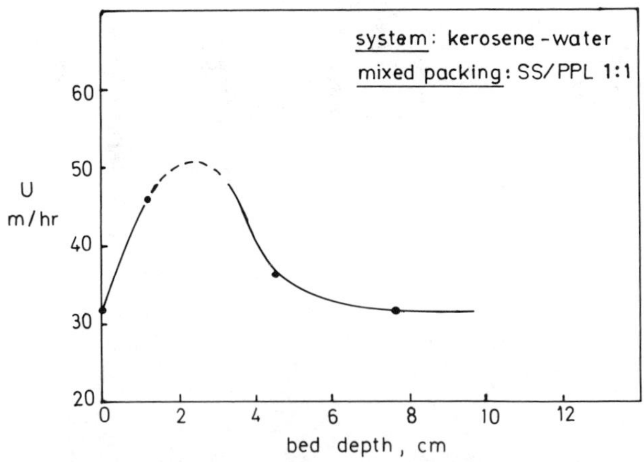

Figure 4. Flooding velocity versus coalescer bed depth [22].

when the backflow of the continuous phase released by coalescence attains a limiting value, beyond which the continuous phase is entrained as fine droplets by the coalesced dispersed phase leaving the bed. The onset of flooding is apparent from the asymptotic increase in the pressure drop across the bed. The bed depth at which the flooding velocity is a maximum is chosen in designing a coalescer. Figure 4 illustrates the typical variation of flooding velocity with depth of a bed of dual wire packing [stainless steel (SS)/polypropylene (PPL)] used in the settling zone of a spray column [22]. For coalescer beds with dispersed-phase nonwetting wire materials (either stainless steel or polypropylene alone), flooding has been observed at very low flow rates. The onset of flooding occurs as a buildup of dispersed droplet layers at the inlet face of the packing, as shown in Figs. 5*a* and 5*b* [84]. The drop buildup heights without packing are also shown for comparison. It is seen that for organic phase dispersed, coalescence is greatly inhibited when large packing heights are used but facilitated when small packing heights are used. However, for water phase dispersed, the trend appears to be reversed, particularly at high flow rates. In the case of very fine dispersions (drop size $< 10\ \mu m$), the limiting velocity of the emulsion phase may be defined as the velocity at which the tiny droplets adhering to the fiber surface on interception are reentrained [84]. At this condition adherence of the droplets to the fiber surface on interception is also inhibited (see also Sections 6.1 and 6.2).

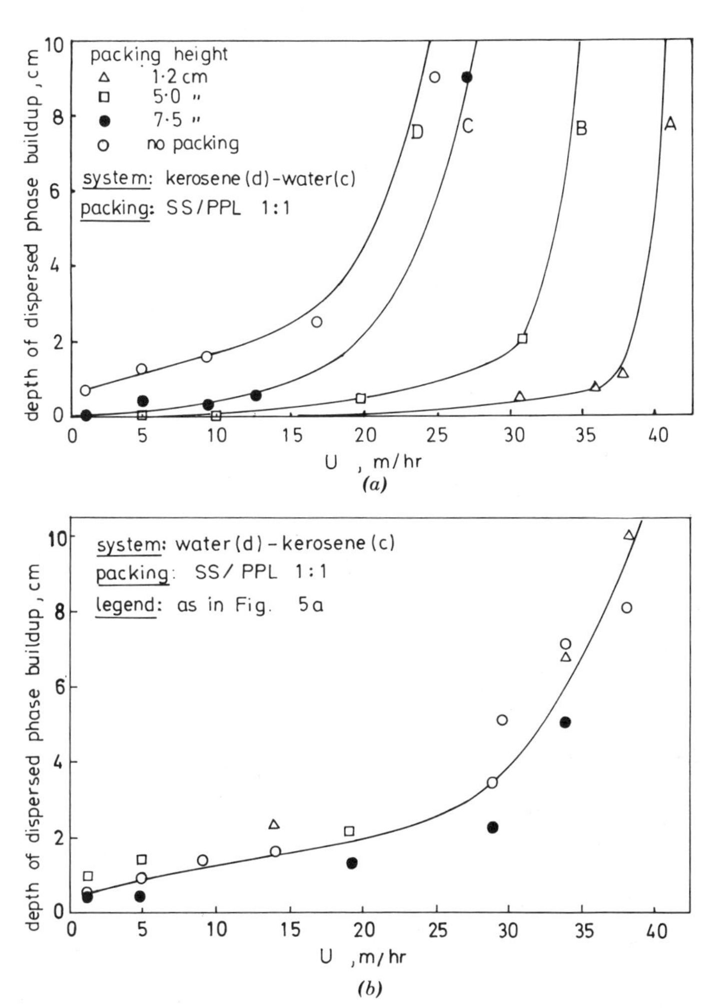

Figure 5. Depth of droplet buildup layer at coalescer inlet [22]: (*a*) organic phase dispersed; (*b*) aqueous phase dispersed.

4. SINGLE-DROP COALESCENCE

Coalescence is associated with the decrease in free energy of the liquid–liquid interface and is aided by the suppression of turbulence, which helps the droplets to aggregate to form a heterogeneous dense packed zone at the main interface between the bulk phases. The rate of migration of droplets to the coalescing main interface depends on the type of dispersion and the properties and interfacial characteristics of the system. The actual coalescence mechanisms are complex, involving the factors that govern the thinning of the continuous-phase film between the two coalescing interfaces [85]; depending on conditions, coalescence may occur either at the plane interface or at the drop/drop interface. Coalescence at the plane interface occurs at the exit end of a contactor after mass transfer is over, whereas drop–drop coalescence arises both within the droplet band awaiting coalescence and within the contactor. Detailed reviews of liquid–liquid coalescence phenomena have been made by Jeffreys and Davies [85] and Rietema [86].

4.1. Coalescence of Single Drops at Flat Interfaces

Studies with flat interfaces have helped in the understanding of coalescence without the complexities of drop–drop and multidrop interactions. Investigators have found that the time interval between the arrival of a droplet at an interface and its final coalescence is not constant but exhibits a distribution of time although the distribution is nearly Gaussian. The process may be considered to occur in successive stages, as follows [87–92]:

1. The drop as well as the interface are deformed as the drop approaches.
2. Oscillation of the drop at the interface is dampened and a film of continuous phase is held between the drop and its bulk phase.
3. The film thins by drainage and ruptures, initiating the drop coalescence process.
4. The rupture hole expands and the drop contents flow (either completely or partially) into its main phase.

The mean coalescence time $\bar{t}$, which is also termed the *rest time*, comprises the mean of several observations of the total time taken for stages 1–3. The time taken for stage 1 is termed the *predrainage time* and that for stages 2 and 3, the *drainage time*. The coalescence time denotes the sum of predrainage and drainage times. The time taken for film removal (after its rupture) and deposition of drop contents in stage 4 has been found to be almost negligible (of the order of 0.06–0.08 s). Most of the total coalescence time is occupied by drainage, during which the drop rests on the thinning film at the interface [85]. The rate of coalescence is thus controlled mainly by the drainage and rupture of the film (or disk) of continuous phase held between the drop and the interface; the diameter of this film d_f is related to the drop diameter d as follows [87]:

$$\frac{d_f}{d} = 0.707 \left(\frac{d^2\ \Delta\rho\ g}{\gamma}\right)^{1/2} \tag{33}$$

Rupture of the film occurs after a time t_0 at some minimum thickness h_0, which depends on the physical properties of the system. The probability for coalescence exists if time for coalescence t is greater than t_0, and Cockbain and McRoberts [87] have suggested plotting the ratio of N, the number of drops uncoalesced at time t, to N_0, the total number of drops studied, against time t. In absence of transfer of solute between the phases, the following relationship for t is proposed [88]:

$$\log\left(\frac{N}{N_0}\right) = -K(t - t_0)^{1.5} \tag{34}$$

where the coalescence constant K is related to the physical properties, the drop diameter, and the frequency of disturbance. The values of K and t_0 have been found to increase linearly with temperature and drop size. An alternate simpler relationship [89] may be valid for systems that show ready coalescence, as in the case of $d \rightarrow c$ transfer of solute:

$$\log\left(\frac{N}{N_0}\right) = -Ct^n \tag{35}$$

where the exponent n is 2 for concentrated electrolytes and 3 for other systems.

4.1.1. *Factors that Affect Coalescence*

The coalescence of a liquid drop at a flat interface is controlled mainly by the following factors, which affect the film thinning and rupture process [88]: (1) drop size; (2) distance of fall of drop to the interface; (3) curvature of the drop side interface; (4) density difference between phases; (5) viscosity ratio of the phases; (6) interfacial tension effects; (7) temperature effects; (8) vibration and electrical effects; and (9) presence of electrical double layers and solute-transfer effects.

The coalescence time increases with drop size. A large drop tends to flatten into an oblate spheroid when resting on the interface, thus increasing the area between the drop and interface. Consequently, the resistance impeding the film drainage increases and the film may thicken at the center, with the drop surface assuming a dimple shape whereby the draining time and consequently the coalescence time increase. This occurs when ionic components, surfactants, and solute-transfer effects are absent.

The distance of fall of the drop could either increase or decrease the stability of the drop, depending on the thermal or mechanical disturbances produced [93]. Generally the stability of the drop increases with increase in the distance of fall [94, 95]. The effect of drop curvature [96] at the resting point is to increase the drop stability if the curvature of the drop surface facing the interface is concave (dimple-shaped drop) as indicated previously.

Increase in the density difference between the drop liquid and the field phase liquid results in an increase in coalescence time since the drop is more deformed, whereas the hydrostatic force at the center causing the film drainage does not increase proportionately. Also, an increase in viscosity of the field phase relative to the drop phase increases the film drainage time and hence the coalescence time. High interfacial tension helps the drop to resist deformation, resulting in an increase of film thinning rate and a decrease of coalescence time. This effect outweighs the effect of increased interfacial tension in inhibiting the drainage flow of the film itself. The effect of temperature on the coalescence time is complex, as it affects all the physical properties. Generally, increase of temperature tends to reduce coalescence time [95–97] unless there is transfer of heat from one phase to the other, which could bring in further complexities due to Marangoni instability.

Induced electric fields accelerate the coalescence of oil–water emulsions [98]. For drops containing electrolytes, the coalescence time is lower as compared to nonelectrolytes as a result of formation of electrical double layers at the interfaces between the drop and the draining film and between the draining film and the bulk phases of the coalesced liquid. The increase in the coalescence time is caused by the double layers tending to retard the flow of the draining film across the force of attraction of the opposite charges at the interfaces and in the flowing film. This *electroviscous effect* is more pronounced when the thickness of the film is of a magnitude of the same order as that of the electrical double layer at which the apparent viscosity may be as high as five times the normal viscosity of the draining film [85, 89]. Other electrical effects are discussed in Chapter 8.

Particles at the interface tend to promote coalescence if they are preferentially wetted by the drop phase; they act as connecting links across the draining film, causing film rupture almost instantaneously [90]. Surfactants, on the other hand, tend to retard coalescence by slowing down the film drainage, possibly as a result of increase of interfacial viscosity as well as the flattening of the drop interface [96].

The presence of an undistributed solute, especially in the drop phase, causes mass-transfer-induced turbulence due to Marangoni effects, resulting in decrease of the coalescence time [91]. MacKay and Mason [91] measured the rate of film thinning beneath water drops of 650-μm diameter in cinnamaldehyde and gave the following relationship:

$$-\frac{dh}{dt} = kh^3 \tag{36}$$

Typically, the coalescence time $\bar{t}$ was 5.3 s when 1% acetone was present in the drop phase, whereas in the absence of the solute it was 8.1 for a value of $k = 3$. Brown and Hanson [98] defined k as equal to $(8F/3\pi\, d_f^4 \mu_c)$ in terms of the diameter of the film on contact d_f and the force F exerted by the drop $(\pi/6)d^3\ \Delta\rho\, g$.

The following useful correlation has been suggested by Jeffreys et al. [85] on the basis of dimensional analysis of the major controlling variables (i.e., mean rest time $\bar{t}$, drop size d, viscosity of the continuous phase μ_c, density difference of the phases $\Delta\rho$, interfacial tension γ, and distance of drop fall L):

$$\frac{\gamma \bar{t}}{d\mu_c} = 1.32(10^5)\left(\frac{L}{d}\right)^{0.18}\left(\frac{d^2\ \Delta\rho\, g}{\gamma}\right)^{0.32} \tag{37}$$

Equation (37) has been modified by Smith and Davies [99] for use when neighboring drops influence the drainage rate of continuous-phase film (as in the case of a dense drop buildup beneath the main interface of a spray column):

$$\frac{\gamma \bar{t}}{d_0 \mu_c} = 31(10^3)\left(\frac{d_0^2 \, \Delta\rho \, g}{\gamma}\right)^{-1.24}\left(\frac{\mu_d}{\mu_c}\right)^{1.03} \tag{38}$$

The coalescence time $\bar{t}$ given by Eq. (38) may then be used to obtain the depth of droplet buildup H of drops awaiting coalescence beneath the phase boundary (as in a spray column) by substituting for $\bar{t}$ in the following equation [99]:

$$\frac{H}{d_0} = 0.24 \left(\frac{u_d \, \bar{t}}{d_0}\right)^{1.1} \tag{39}$$

Various theoretical models have also been developed to predict film drainage time, which almost represents the coalescence time, as coalescence is imminent once the film has ruptured. These models are based on the concepts that (1) the mechanical work done is equal to the energy dissipated due to friction or viscous shear within the draining film [88] and (2) the shapes of the drop and the deformed interface coexist in equilibrium [95, 100–102].

4.2. Drop–Drop Coalescence

Few experimental coalescence studies with pairs of drops are reported, as it is difficult to carry out drainage measurements between two drops that remain close to each other. However, various mathematical models have been proposed [103–107]. The significant omission in most of these models is the effect of internal circulation of drops on the coalescence time or film thinning process. Neilson et al. [96] measured coalescence times of drops at flat interfaces, showing the progressive increase of coalescence time with surfactant concentration; specifically, as the drop circulation is dampened by the surfacant, the film drainage rate is also reduced. Attempts at modelling to include circulation within the drops have been reported recently [108] for the case of coalescence of two liquid drops approaching each other.

Partial or stepwise coalescence in the case of drop–drop systems has also been observed [90, 103, 109, 110] leading to the formation of a secondary droplet. This occurs when β, the diameter ratio (d_2/d_1) of the drops approaching each other, is greater than about 3.5. The drop coalescence behavior becomes identical to that [Fig. 6*b*], the film becomes depleted of solute and the higher interfacial tension has the effect observed for partial coalescence at a flat interface when $\beta > 12$, at which the surface of the larger drop behaved as if it were a flat surface [103]. The fractional decrease in the surface area due to coalescence is given by the following relationship [85]:

$$\frac{\Delta s}{s} = 1 - \frac{(1 + \beta^3)^{2/3}}{1 + \beta^2} \tag{40}$$

When β is unity, $\Delta s/s$ has a maximum value. The surface energy released at coalescence by rupture of the continuous phase film is available as kinetic energy, and the resulting pressure difference between the drops is given by

$$\Delta p = \frac{8\gamma}{d_2}(\beta - 1) \tag{41}$$

The smaller of the two drops will drain its liquid into the larger drop since its inside pressure will be higher because of its smaller radius of curvature.

It is difficult to obtain experimental data on film drainage as the drops would need mechanical support, but a simulated study [111] involving a free drop held on a convex liquid surface has been made. McAvoy et al. [107] investigated coalescence immediately following drainage of the continuous phase from the moment of rupture until the final coalesced drop was obtained. The results show that the mean velocity of expansion of the hole is a function of interfacial tension but insensitive to both density difference and the radius of the intervening continuous-phase film before its rupture.

4.2.1. Effect of Solute Transfer on Drop–Drop Coalescence

Groothuis and Zuiderweg [71] demonstrated the effect of solute transfer on the coalescence of pairs of drops held at nozzle tips opposite to each other within a tank holding the continuous phase. Coalescence was promoted by solute transfer from drop phase to continuous phase, even when only one of the drops in the pair contained the solute. However, when the solute diffused into the droplets from the continuous phase, coalescence was strongly inhibited. Treybal [15] has pointed out that interfacial tension is lowered by solute; therefore, for $d \rightarrow c$ transfer, there is a region of low interfacial tension in the film [Fig. 6*a*] leading to enhanced drainage and rapid coalescence. Conversely, for $c \rightarrow d$ transfer

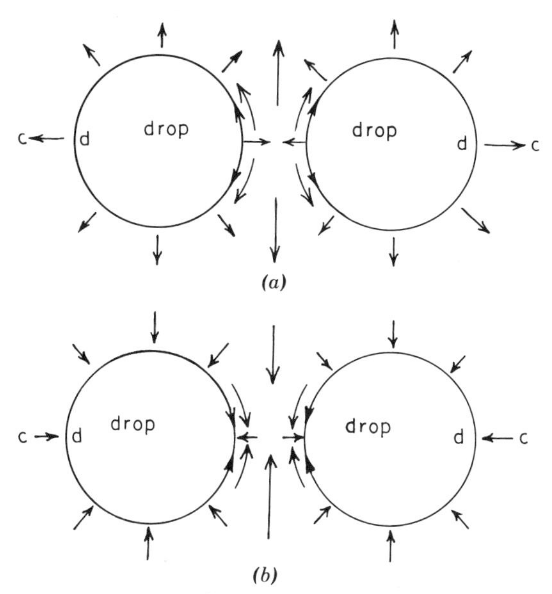

Figure 6. Effect of solute transfer on interdrop coalescence: (*a*) $d \rightarrow c$ direction of transfer; (*b*) $c \rightarrow d$ direction of transfer.

[Fig. 6*b*], the film becomes depleted of solute and the higher interfacial tension has the effect of drawing in the continuous phase and counteracting the tendancy to drainage of the film.

It is possible that factors other than the mass-transfer effect on interfacial tension, such as those caused by interfacial turbulence and variation in physical properties accompanying solute transfer (e.g., viscoelastic behavior of interface), could also contribute to the change in stability of the continuous-phase film between the drops [112]. However, the solute-transfer effects on coalescence illustrated in Fig. 6 have been widely observed [72, 91, 113].

Coalescence of droplets can affect the performance of extraction equipment; for instance, interdrop coalescence in the extraction zone of spray columns was found to reduce transfer efficiency [114]. An understanding of coalescence in high-shear fields is also important in the design of mechanically agitated contactors where equilibrium between coalescence and droplet breakup establishes the mean drop size. As the shear rate increases (even when droplets collide "head on"), the probability for coalescence decreases as the contact time becomes less than the film drainage time required to cause coalescence.

4.3. Drop-Drop Coalescence in Applied Electric Fields

Under applied electric fields, at any specific shear rate, there is a threshold field strength above which coalescence is enhanced. This enhancement arises from the increased rate of film thinning as a result of electrostatic attraction and enhancement of the electric field between the drops by induced charge. Also, the critical field strengths required to promote coalescence increase in shear rate as the film thinning rates are affected at high shear rates [98].

4.4. Coalescence of Drops at Solid Surfaces

Study of coalescence at flat solid surfaces is important in the understanding of drop coalescence at complex solid surfaces such as in packed and sieve-plate extraction columns, in wire-mesh cartridge packings of Scheibel column settlers, and at walls and baffles of other extractors and settlers. It is also necessary to study separately the factors affecting coalescence of drops of primary dispersions ($d > 100\ \mu m$) and those affecting secondary dispersions.

4.4.1. Coalescence of a Single Drop at a Flat Solid Surface

This is almost identical to coalescence of a single drop at a flat liquid interface; however, an important deviation is the behavior after film rupture. Depending on the properties of the system, especially the contact angle and the three-phase interfacial tension, the drop after film rupture may either spread over the surface or stay on it as a sessile drop assuming an equilibrium shape [115–117]. The final shape of the drop on a smooth surface will depend on the contact angle at equilibrium θ_{eq}, given here in terms of the surface energies of the liquid/liquid and liquid/solid interfacial tensions:

$$\cos \theta_{eq} = \frac{\gamma_{1s} - \gamma_{2s}}{\gamma_{12}} \qquad (42)$$

For a smooth surface and a wholly nonwetting drop, θ_{eq} will approach 180°. However, most liquid–liquid systems have partial wetting characteristics with θ_{eq} varying between 0 and 180°. For a drop liquid that does not preferentially wet the solid surface, $\theta_{eq} > 90°$ (Fig. 7).

The contact angle and coalescence may be affected by surface coatings (hydrophilic or hydrophobic) or by surfactants. Also, the contact angle may depend on the motion of the droplet on the surface, that is, advancing or receding. It is thus desirable to use the dynamic contact angle for characterising the system [117–119]. The importance of surface wetting has been indicated for sieve plates as well as packings [120]. When

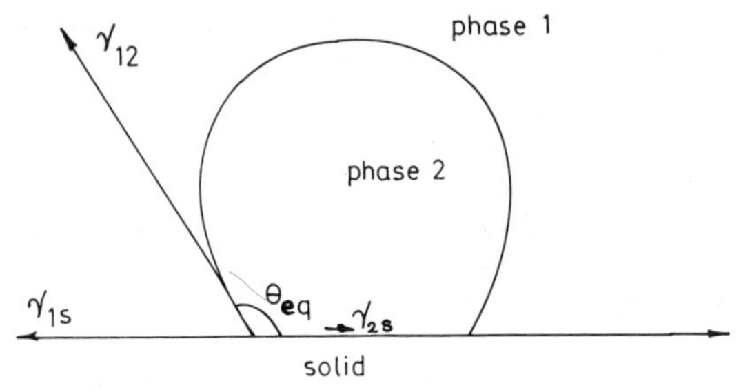

Figure 7. Schematic diagram of a sessile drop on a flat solid surface.

the kinetic energy of an approaching drop is less than the increase in surface energy required to deform the drop, droplets that wet the plate cannot pass through the sieve–plate perforations, whereas for a nonwetting plate, the droplets pass through the perforations without change in size, even when the hole size is 0.8 times the approaching drop size.

5. COALESCENCE OF DISPERSIONS

There are two main aspects of the coalescence of dispersions: (1) interdroplet coalescence within the dispersion during the process of dispersion and contacting; and (2) droplet coalescence at the phase boundary to effect a separation of the phases from the dispersion. These aspects are discussed in the following paragraphs with reference to various types of contactor.

5.1. Coalescence in Spray Column

A uniform droplet size distribution in a spray column can be obtained by careful design of distributors and choice of nozzles at the inlet. Variations in the distribution during passage through the column due to coalescence can be studied from the observation of the dispersion at the main interface at the exit end of the column [121]. The buildup of a dense bed of droplets beneath the main interface before coalescence with the main interface has been investigated by Smith and Davies [46]. Bed depth H can be predicted by the following relationship obtained from Eqs. (38) and (39) [99]:

$$\frac{H}{d_0} = 15.1(10^3)\left(\frac{u_d \mu_c}{\gamma}\right)^{1.10} \cdot \left(\frac{d_0^2 \, \Delta\rho \, g}{\gamma}\right)^{-1.38} \left(\frac{\mu_d}{\mu_c}\right)^{1.13} \tag{43}$$

Lee and Lewis [121], using carefully purified systems, have observed frequent interdroplet coalescence within the dense dispersion band with a wide range of drop size distribution. However, in the presence of small quantities of contaminants, such interdrop coalescence was found to be absent and only the coalescence between droplets and the main interface occurred. This resulted in *interfacial separation*, a situation in which the continuous phase released by coalescence flows back so rapidly that the droplet bed is expanded. Allak and Jeffreys [122], using a refractive-index matching technique, have observed that the dense droplet band at the phase boundary consists of three distinct segments or zones: (1) the interface coalescing zone with one or two layers of dodecahedra-shaped drops just below the main coalescing interface; (2) the main packed droplet zone having considerable interdroplet coalescence with the plateau borders of pentagonal dodecahedra drops forming the drainage network for the continuous phase; and (3) the flocculating zone, comprising a less dense packed bed of spherical droplets without much interdrop coalescence. The frequency of interdroplet coalescence ω in zone (2) has been shown to be related to the group $(\mu_c u_d/\gamma)$, the total depth of the droplet bed H, and the mean droplet size d_0:

$$\omega = 1 - 0.559\left(\frac{\gamma}{\mu_c u_d}\right)^{0.054} \left(\frac{d_0}{H}\right)^{0.19} \tag{44}$$

where ω is the frequency of coalescence expressed as a fraction of total number of drops of mean size d_0 that coalesce in a plane in the band at a depth H. The mean rest time for interfacial separation is influenced by the terminal velocity and the diameter of the droplet in the swarm and droplet bed voidage. Treybal and co-workers [123, 124] have reported interdroplet coalescence studies in a liquid–liquid fluidized bed with uniform-sized droplets. Maraschino and Treybal [123] proposed an interdroplet coalescence model, whereas Ramamoorthy and Treybal [124] suggested the following empirical correlation for the coalescence frequency ω:

$$\left(\frac{\omega d_e}{u_s}\right) 10^5 = 0.02\left(\frac{\gamma}{\mu_c u_s}\right)^{1.18} \left(\frac{\mu_d}{\mu_c}\right)^{0.45} \tag{45}$$

5.2. Coalescence in Packed Sections

Droplet coalescence in packings depends on the mobility of drops through the voids and on the preferential wetting characteristics of the packing material. At wetting packings, the droplets coalesce to form a film surrounding the packing and detach by a drip-point mechanism. However, at nonwetting packings, there are two different coalescence mechanisms, depending on whether the packing is above or below a critical value D_{pc} based on the physical properties of the system [Eq. 11)]. For packings above the critical size, the droplets move freely through the voids and coalescence appears to take place between droplets above the peak drop size d_p held against the packing structure; not every interdroplet collision results in coalescence unless the direction of solute transfer is from the dispersed phase to the continuous phase. Sometimes droplet breakup has also been observed. However, when the packing is smaller than the critical size ($D_p < D_{pc}$, usually packings of $\leqslant 10$ mm) for a nonwetting packing, the approaching droplets coalesce to form large chunks of fluid, sometimes enveloping a few packings that provide a scaffolding-type support for the chunks. These move slowly through the packing interstices, assuming various kinds of shapes while constantly rubbing against the packing surface wetted by the continuous-phase film [81, 125].

In the case of knitted packings used as coalescence promoters in the settling zones of Scheibel columns and settlers, the performance of packings depends on their wettability characteristics [120].

5.3. Coalescence and Redispersion in Mixers

In the operation of mixers, coalescence and redispersion phenomena are important as they influence the mass-transfer rates within the mixer. Coalescence of smaller drops produces liquid mixing, whereas new surface is formed during redispersion of large drops. In an agitated vessel the drops are broken at the tip of the impeller blades and are swept away radially to parts of the vessel away from the impeller turbulence, where they tend to coalesce. In general, liquids of low interfacial tension (e.g., organic esters) are most prone to fairly rapid coalescence and subsequent easy redispersion by the impeller blades. Such interdrop coalescence will be greater when the dispersed-phase holdup is high [1, 42, 44]. However, the repeated coalescence and redispersion of the dispersed phase has been difficult to characterize. In mixers, coalescence frequencies have been measured experimentally in terms of the fraction of all drops coalescing per second f and expressed as a function of impeller speed and dispersed-phase holdup [126, 127]:

$$f \propto N^p x^q \tag{46}$$

The exact values of exponents p and q depend on the system properties and the type of impeller but can be approximately taken as 2.5 and 0.7, respectively [1] for systems in which there is no solute transfer occurring between the phases. However, $d \rightarrow c$ transfer is reported to increase the coalescence frequency considerably [128, 129]. It has been found that the factors affecting coalescence frequency also affect the redispersion mechanism, indicating the importance of the coalescence–redispersion mechanism [42, 44, 130, 131].

5.4. Coalescence in Settlers

Two stages of batch settling may be identified [132]. In the first stage, called *primary break*, the larger drops settle faster, reaching the phase boundary (main interface), where they coalesce and mix with the bulk phase. In the second stage, defined as *secondary break*, very small droplets in the micrometer size range appear as a fine haze that settles very slowly. Such haze occurs during the coalescence process when dual emulsions are formed containing secondary droplets of either phase [110].

In the operation of a continuous settler attached to a single mixing unit wherein the droplets moved horizontally along the settler in plug flow and primary phase separation was completed before the settler outlet, Jeffreys et al. [133] noted that the emulsion (two-phase mixture) was wedge shaped (Fig. 8). The length of the coalescence wedge was found to increase with an increase in dispersed-phase flow rate per unit settler width and a decrease in the size of entering drops. They [133] modeled the settler, assuming the existence of interdrop and interfacial coalescence modes such that the drop residence time became equal to the coalescence time. They introduced the concept of a coalescence stage in that a stepwise form of wedge was assumed with a face-centred cubic packing structure for the drops. Wedge length was calculated by the equation

$$L = nv\tau \tag{47}$$

where n is the number of coalescence stages, τ is the coalescence time of an ideal stage, and v is the axial velocity within the wedge. In practice, since continuous coalescence takes place throughout the wedge (rather than in idealized stages)

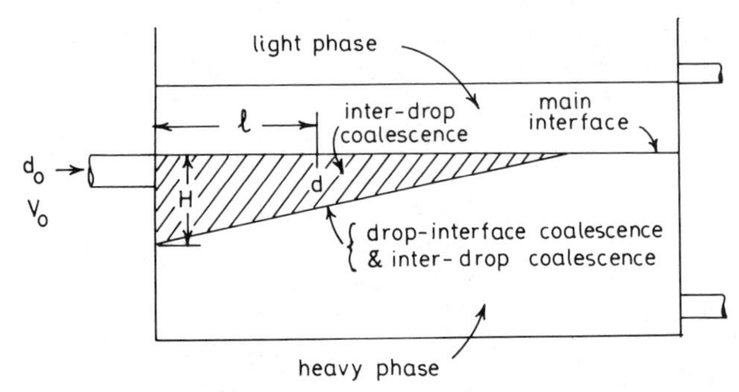

Figure 8. Coalescence wedge in a continuous settler (heavy phase dispersed).

and coalescence rates are found to vary, Davies and Jeffreys [133] gave an alternative approach in which the volume flow rate of dispersed phase entering the wedge is equated to the overall coalescence rate of the dispersed phase. The mean diameter d of drops at a distance l from the inlet of the settler is given by the following expression [133]:

$$d^{3/2} - d_0^{3/2} = \frac{l}{4\,vf} \qquad (48)$$

where v is the axial velocity of drops in the settler and f can be taken as the proportionality constant in Eq. (46). The final correlation relating to the volumetric flow rate of the dispersed phase per unit settler width V_0 and the length of the coalescence wedge L is given as [85, 133]

$$V_0 = 4\eta^* v \left[\left(d^{3/2} + \frac{L}{4\,vf} \right)^{2/3} - d_0 \right] \qquad (49)$$

where η^* is the surface packing efficiency of drops of size d at the phase boundary. A fair agreement between predicted and experimental results has been reported [22, 120]. The design and location of the entry and exit pipes should be such that backmixing and circulation patterns are avoided in the entrance region to facilitate the formation of a stable packing of drops.

6. PROMOTION AND INHIBITION OF COALESCENCE

Coalescence promotion may be needed with three kinds of dispersion: (1) primary dispersions; (2) dispersions involving fine droplets (>100 μm) nearly stabilized by inhibited coalescence; and (3) secondary dispersions (<10 μm) comprising micrometer and submicrometer size misty droplet particles. Also, in gravity settling of primary dispersions a large inflow of dispersed phase will lead to the formation of a dispersion band at the main interface (phase boundary). Separation within this band takes place by droplet coalescence. But a further increase in flow rate causes this dispersion band to grow until it fills the available settling space, giving rise to a flooding condition [121]. Hence attempts have been made to promote coalescence in order to increase the separation rate of the phases and thus achieve increased throughputs. Special packings or coalescence promoters for this purpose range from conventional ring or saddle packings of small sizes to knitted or woven wire-mesh packings.

6.1. Coalescence Promotion

6.1.1. Primary Dispersion in Packed Separators

Coalescence in packed separators is maximized when the dispersed phase preferentially wets the packing surface [22] and coalesces to form liquid films or drops that flow through its voids. Flooding conditions within the separator are governed by the superficial velocity of the released continuous phase flowing countercurrently through the voids. Under normal conditions (below flooding) the characteristics of the dispersion leaving the packing are determined by the conditions at the outlet, particularly the size of voids and the velocity of the coalesced phase through the voids. At low flow rates droplets are formed on the downstream face of the packing by a drip-point mechanism. The aperture size or wetted area of the packing at the drip point is important since the surface forces acting on the drop at detachment are in equilibrium with the bouyancy forces and kinetic forces. On this basis, Davies et al. [22] suggested the following Harkins–Brown [134] relationship to obtain the equivalent diameter d_e of the drop formed at the drip point at very low flow rates:

$$d_e = \left[6F\left(\frac{d_N\gamma}{\Delta\rho g}\right)\right]^{1/3} \qquad (50)$$

The effective wetted diameter of the drip point d_N may be assumed to be equal to the upper critical value indicated by Eq. (9). However, at such subjetting velocities at the drip point, drop size may be simply predicted by Eq. (7). The drop sizes so predicted compare well [11] with those given by the relationship proposed by Scheele and Meister [135]. The operation of packed separators at drip-point exit conditions was found to be satisfactory by Thomas and Mumford [136], although operation at such low velocities was considered to be below the economic range by Davies et al. [22]. The maximum capacity of a separator is indicated by a buildup of dispersed-phase drops at the inlet end, leading to flooding. At such high dispersed phase velocities the drip-point mechanism changes to a jetting mechanism that becomes disrupted by Rayleigh disturbances; the mean drop size of the dispersion produced by jet breakup may be obtained by using Eq. (8) (Section 2.1) or Eq. (10) given by Vedaiyan et al. [2], which predicts mean drop size at any velocity in the jetting range where the kinetic force predominates. The following condition of DeChazal and Ryan [137] may be used to locate the jetting regime:

$$\left(\frac{d_N u_N^2 \rho_d}{\gamma}\right)^{0.5} > \left[1.513 - 0.53\left(\frac{d_N^2 \Delta\rho g}{\gamma}\right)\right] \qquad (51)$$

6.1.2. Fine Dispersions in Knitted-Mesh Packed Sections

Knitted wire-mesh packings made of a single material are commonly used in Scheibel extractors as well as differential column contactors (e.g., Oldshue-Rushton, rotary disk, or annular columns) for coalescence of fine dispersions in the separation zone. The material may be stainless steel or a plastic and is selected so as to be wetted by the dispersed phase. Such packings effect partial coalescence of the droplets while allowing countercurrent flow of continuous and dispersed phases. On the other hand, the use of nonwetting mesh packing prevents the droplet phase from entering the packed section, allowing the droplets to accumulate at the upstream side of packing, resulting in premature flooding [22].

(a) Phase Inversion. In practice, liquid–liquid contactors must operate over a wide range of flow rates and phase ratios with frequent changes in the physical properties of the phases such as viscosity, density, and interfacial tension across the extractor. A combination of these factors can cause phase inversion [22]. For most systems, it is possible to disperse either of the contacting phases within a certain range of flow ratios, in which one might thus expect phase inversion to occur [138–140]. This is especially so in a Scheibel column with use of the conventional knitted wire mesh of a single material intended to coalesce droplets of the wetting liquid phase (see Fig. 9*a*). On phase inversion, the mesh is unable to coalesce a dispersion of droplets of the nonwetting liquid phase, whereas the incoming droplets continue to accumulate on the upstream

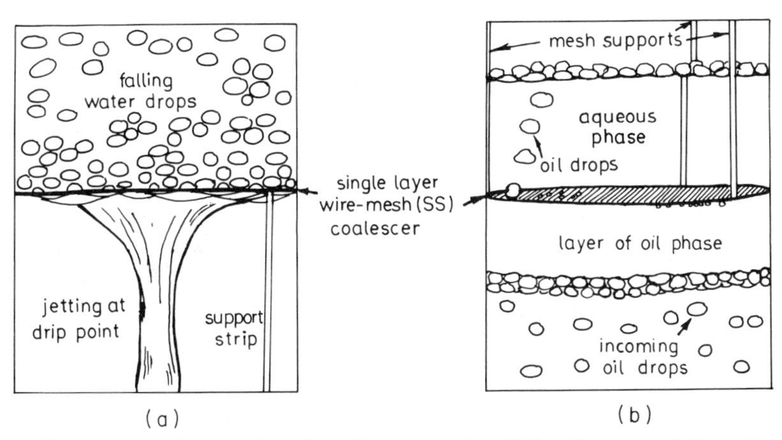

Figure 9. Sketch of stainless steel mesh packing operation [22]: (*a*) water–oil dispersion; (*b*) oil–water dispersion.

side of the packing, causing a premature flooding of the column (see Fig. 9*b*). To avoid such predicaments, Davies et al. [22] devised a new mixed packing by knitting alternately the elements of stainless steel (a high-surface-energy material) and polypropylene (a low-surface-energy material), as shown in Fig. 10. Higher throughputs and flooding velocities were obtained with mixed packings than with identical knitted packings of a single material. Thus both water–oil and oil–water dispersions can be efficiently handled by mixed packings, increasing the separator capacity. The stitch length, wire diameter, and crimp can be varied to give packings of the desired voidages. In addition, the proportion of each material used in the mixed packing can be varied to give a satisfactory performance with many desirable characteristics such as low pressure drop, low holdup, and maximum separation velocities (high allowable throughput). Mixed packings are recommended [22] for improving separator design and for Scheibel column internals. Woven material and sprayed and etched surfaces promote coalescence by the "junction effect" that can also be applied [141].

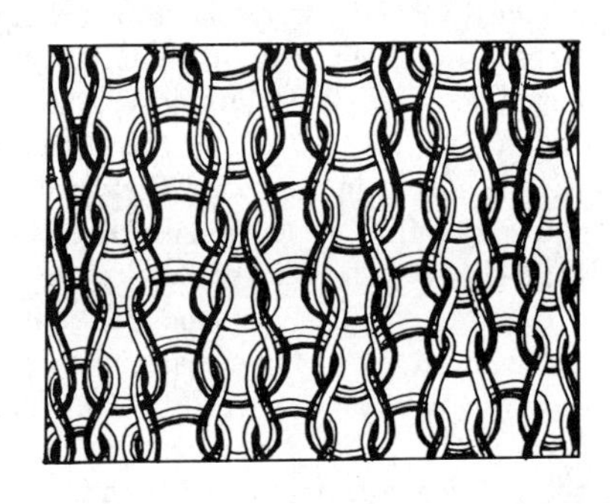

Figure 10. Sketch of mixed packing [22].

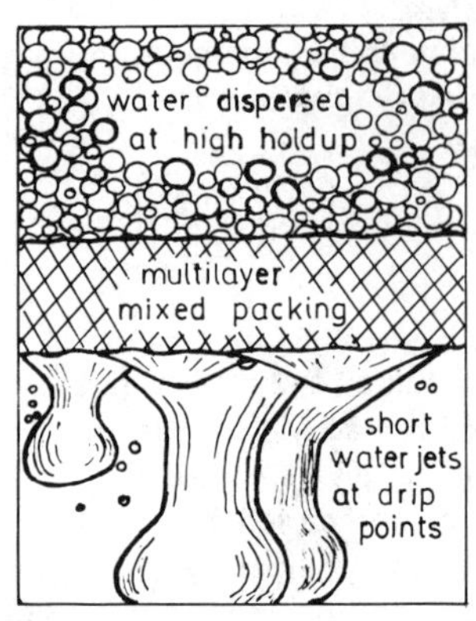

Figure 12. Operation of mixed packing at high loading [22].

Figures 9, 11, and 12 are based on photographs [22] and indicate the performance of the various knitted wire-mesh packings. Figure 9*a* shows the operation of stainless steel mesh packings for water–oil dispersions in which water drops neatly coalesce at the mesh, emerging as a thick jet of liquid; Fig. 9*b* shows the operation of the same packing element when the phase is reversed (i.e., indicating flooding). Figures 11*a* and 11*b* show the efficient coalescence of oil–water and water–oil dispersion, respectively, when the mixed packing is used. Figure 12 shows the performance of the mixed packing at high loading conditions, coalescing water drops in oil. In the event of a phase inversion the mixed packings can still function without allowing the column to flood since the junctions of high-energy/low-energy-surface materials in the packing promote

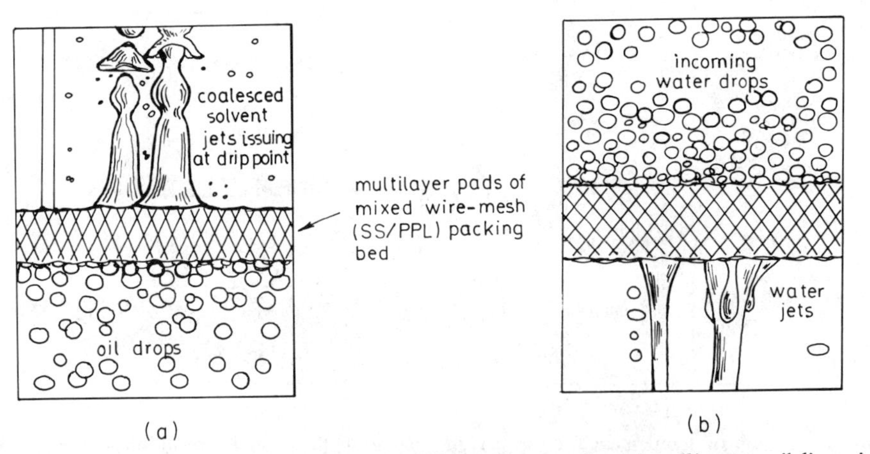

Figure 11. Sketch of mixed packing operation [22]: (*a*) oil–water dispersion; (*b*) water–oil dispersion.

and enhance coalescence of both types of dispersion, namely, oil–water or water–oil. High flooding velocities have also been realized for these packings (see Fig. 12). The use of mixed packings in a commercial column contactor of 1.8 m diameter in the final phase separation stage is reported to have reduced the entrainment of the dispersed phase in the outlet continuous phase from 10% to less than 0.1% when the column is operated at loading rates of up to 50 m/hr of dispersed phase [84].

6.2. Secondary Dispersions Through Fibrous Beds in Continuous Settlers

Secondary dispersions and nonstirred hazy emulsions made up of droplet particles of micrometer and submicrometer size are slow to settle and coalesce. Fibrous beds containing several pads of suitable fine-fiber packings (e.g., nylon, Teflon, cotton, and stainless steel) have been found effective in such cases, as they facilitate coalescence between droplets held by an interception mechanism until the resulting drops are sufficiently large to be carried away by the continuous phase and separated by gravity as a primary dispersion. The voids in such beds are very large in relation to the size of the droplets. Photographic studies [24, 142] have confirmed that droplets first attach individually to the surface of the packing fiber and then build up (possibly as a result of van der Waals forces [143]) before they finally coalesce to a bigger drop that then detaches and is swept away by the current of continuous phase fluid (see Fig. 13). High separation efficiency depends on the mechanism of approach of the first droplet and its adhesion to the packing for a sufficient length of time to facilitate adherence of further droplets and eventual coalescence. The probability of tiny droplets reaching a packing surface by direct interception depends on the relative sizes of fibers, void spaces, and droplets and on the local flow conditions. Submicrometer-sized droplets may also reach the fibre surface by brownian movement or by electrostatic effects. A major factor governing the droplet adhesion is considered to be surface roughness rather than the contact angle. Retention of the droplets also requires that the forces of adhesion must exceed the local shear forces on the droplets. There is a maximum separation velocity above which the droplet retention time on the surface is less than the time required for coalescence [24]. Studies with cotton and synthetic fiber beds indicate that the maximum separation velocity initially increases with bed depth, reaches a maximum, and then decreases steadily with further increase in bed depth. This may be because of increased viscous shear forces in the deep beds, preventing the droplets from adhering to the fiber surfaces. This is more obvious for the case of low-interfacial-tension systems and smoother-packing fibers such as glass, stainless steel, nylon, and Teflon packings [85, 120] for which viscous shear forces are believed to control adhesion. Separation is more efficient with high-interfacial-tension and high-density-difference systems, such as isooctane dispersed into water and kerosene dispersed into water [120]. With a proper choice of packing fiber having high-surface roughness and high-energy characteristics with minimum possible fiber thickness and bed depth, a maximum separation velocity of 1.5 cm/s can be reached. Rosenfeld and Wasan [142] have developed a model to describe the operation of a fibrous bed coalescer.

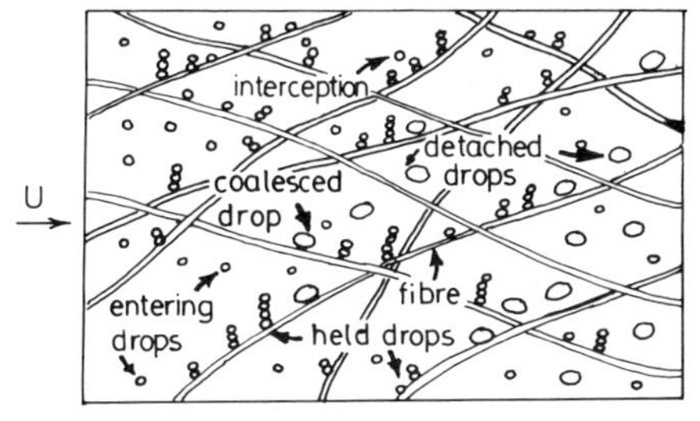

Figure 13. Schematic diagram of fibrous bed in operation.

NOTATION

a	superficial area of packings, m^{-1} [interfacial area of contact in Eq. (3), m^{-1}]
A	pulse amplitude (stroke), m
$c \to d$	solute transfer from continuous phase to dispersed phase
$d \to c$	solute transfer from dispersed phase to continuous phase
C	constant
C_N	orifice coefficient [Eq. (23) and Table 1]
d	drop diameter (Sections 4 and 5), m
d_c	critical drop size (Section 4.1), m
d_e	equivalent drop diameter, m
d_e^0	value of d_e at a linear velocity of 3 cm/s in the perforation, m

d_f	diameter of continuous-phase film in contact with coalescing drop, m
d_j	diameter of liquid jet issuing from nozzle [Eq. (8)], m
d_{max}	maximum stable drop size, m
d_N	nozzle or orifice diameter, m
d_{Nc}	critical nozzle size (Figs. 2 and 3), m
d_0	mean size of drop entering the settling zone (Sections 4 and 5), m
d_p	peak drop size [Eq. (4)], m
d_{vs}	Sauter mean volume–surface diameter of drops, m
d_{vs}^0	Sauter mean drop size at low holdup ($x \to 0$) [Eq. (18)]
D_R	diameter of impeller or disk, m
D_T	diameter of column or vessel, m
e	voidage of packing, fraction
f	pulse frequency [Eq. (32) and Table 1], s^{-1}
f	fraction of drops coalescing per second [Eq. (46)], s^{-1}
F	Harkins–Brown correction factor [Eq. (50)]
g	acceleration due to gravity, m/s^2
g_c	gravitational conversion factor, kg m s^{-2} N^{-1}
h	thickness of continuous-phase film [Eq. (36), Section 4], m
h_t	liquid head due to continuous phase, interfacial tension, and orifice effect, respectively [Eq. (23)], m
H	height of droplet dense bed beneath the main interface [Eqs. (39) and (43)], m
l	distance from inlet of settler [Eq. (48)], m
L	length of coalescence wedge in continuous settler [Eq. (47)]; distance of drop fall in Eq. (37), m
m	slope of plot of d_{vs} versus x [Eq. (19)]
n_C	number of compartments
N	rotational speed of rotor or impeller, rev/s
P/v	energy dissipated per unit volume of fluid phases [Eq. (16) and Table 2)], W m^{-3}
P_n	total power spent in agitation (Table 1)
R_A	specific reaction rate per unit interfacial area, mol/s · m^2
R_A'	total reaction rate, mol/s
$s, \Delta s$	surface area and difference in surface area of drops [Eq. (40)]
$S_t; S_N$	area of sieve plate and area of all holes in the plate, respectively [Table 1 and Eq. (23)], m^2
$\bar{t}$	mean coalescence time (Section 4), s
u_c, u_d	superficial velocity of the continuous and dispersed phases, respectively, based on column cross section, m/s
u_N	velocity at the nozzle or orifice, m/s
$\bar{u}_0$	mean characteristic velocity of droplets [Eq. (26) and Table 2], m/s
u_p	peak terminal velocity of drop [Eq. (5)], m/s
u_s	slip velocity [Eq. (26)], m/s
u_t	drop terminal velocity, m/s
U	superficial velocity or throughput of entering stream based on coalescer bed area, m/s or m/hr (Figs. 4 and 5)
v	axial velocity within coalescence wedge [Eq. (47)], m/s
V_0	volumetric flow rate of dispersed phase per unit settler width, m^2/s
We	Weber number ($D_R^3 N^2 \rho_c/\gamma$), dimensionless
x	operational holdup of dispersed phase, fraction
Z_T, Z_C	height of column and compartment, respectively, m
ϵ	power dissipated per unit mass of fluid phase [Eq. (17) and Table 1], W/kg
γ	interfacial tension, N/m
μ	viscosity, Pa s
π	3.1416
$\rho, \Delta\rho$	density and density difference, respectively, kg/m^3
θ	contact angle, radians

Subscripts

c, d	continuous and dispersed phases, respectively
cr	at critical condition
e	of emulsion
eq	at equilibrium
f	at flooding condition

REFERENCES

1. J. T. Davies, *Turbulence Phenomena*, Academic Press, New York, 1972, Chapter 10.
2. S. Vedaiyan, T. E. Degaleesan, G. S. Laddha, and H. E. Hoelscher, *AIChE J.* **18,** 161 (1972).
3. J. B. Lewis, I. Jones, and H. R. C. Pratt, *Transact. Inst. Chem. Eng.* **29,** 126 (1951).

4. S. Gnanasundaram, T. E. Degaleesan, and G. S. Laddha, *Can. J. Chem. Eng.* **57**, 141 (1979).
5. S. Hu and R. C. Kintner, *AIChE J.* **1**, 42 (1955).
6. W. Licht and G. S. R. Narasimhamurthy, *AIChE J.* **1**, 366 (1955).
7. A. J. Klee and R. E. Treybal, *AIChE J.* **2**, 444 (1956).
8. P. M. Krishna, D. Venkateswarlu, and G. S. R. Narasimhamurthy, *J. Chem. Eng. Data*, **4**, 336, 340 (1959).
9. S. P. Bhattacharya and D. Venkateswarlu, *Transact. Indian Inst. Chem. Eng.* **10**, 25 (1957–1958).
10. T. R. Krishnaswamy, S. Chandramouli, M. G. Subba Rau, and G. S. Laddha, *Indian Chem. Eng.* **9**, T59 (1967).
11. S. Devotta, Ph.D. thesis, 1978, University of Madras; K. Chandrasekharan and S. Devotta, paper No. H2.21-CHISA Conference 1978, Prague.
12. F. W. Keith and A. N. Hixson, *Ind. Eng. Chem.* **47**, 258 (1955).
13. R. M. Christiansen and A. N. Hixson, *Ind. Eng. Chem.* **49**, 1017 (1957).
14. F. D. Mayfield and W. L. Church, *Ind. Eng. Chem.* **44**, 2253 (1952).
15. R. E. Treybal, *Liquid Extraction*, 2nd ed., McGraw-Hill, New York, 1963.
16. G. A. Hughmark, *Ind. Eng. Chem. Fund.* **6**, 408 (1967).
17. K. R. Umesh, S. Vedaiyan, T. E. Degaleesan, and G. S. Laddha, *Indian J. Technol.* **16**, 303 (1978).
18. L. Satish, T. E. Degaleesan, and G. S. Laddha, *Indian Chem. Eng.* **16**, 36 (1974).
19. N. Prabhu, A. K. Agarwal, T. E. Degaleesan, and G. S. Laddha, *Indian J. Technol.* **14**, 55 (1976).
20. C. Ramshaw and J. D. Thornton, *Institute of Chemical Engineers Symposium on Liquid-Liquid Extraction*, April 1967, No. 26, p. 73.
21. J. D. Thornton, *Ind. Chem.* **39**, 632 (1963).
22. G. A. Davies, G. V. Jeffreys, and M. Azfal, *Br. Chem. Eng.* **17**, 709 (1972).
23. R. C. Kintner, *Advances in Chemical Engineering*, Vol. 4, Academic Press, New York, 1963.
24. S. S. Sareen, P. M. Rose, R. C. Gudesen, and R. C. Kintner, *AIChE J.* **12**, 1045 (1966).
25. A. N. Kolmogoroff, *Dokl. Akad. Nauk USSR* **66**, 825 (1949).
26. J. O. Hinze, *AIChE J.* **1**, 289 (1955).
27. P. H. Clay, *Ned. Akad. Van. Wetensk. Proc.* **43**, 852, 979 (1940).
28. J. W. van Heuven and W. J. Beek, *Proceedings of the International Solvent Extraction Conference, The Hague, 1971* (Soc. Chem. Ind., London), 1971, p. 70.
29. G. S. Laddha, R. Kannappan, and T. E. Degaleesan, *Can. J. Chem. Eng.* **56**, 137 (1978).
30. G. S. Laddha and T. E. Degaleesan, *Transport Phenomena in Liquid Extraction*, Tata McGraw-Hill, New Delhi, 1976.
31. R. Shinnar and J. M. Church, *Ind. Eng. Chem.* **52**, 253 (1960); **43**, 479 (1961).
32. H. T. Chen and S. Middleman, *AIChE J.* **13**, 989 (1967).
33. J. D. Thornton and B. A. Bouyatiotis, *Ind. Chem.* **39**, 298 (1963).
34. M. S. Doulah, *Ind. Eng. Chem. Fund.* **14**, 137 (1975).
35. M. H. I. Baird and R. G. Rice, *Chem. Eng. J.* **9**, 171 (1975).
36. G. H. Reman and R. B. Olney, *Chem. Eng. Progr.* **51**, 141 (1955).
37. G. H. Reman and J. G. Van de Vusse, *Pet. Ref.* **34**, 129 (1955).
38. G. H. Reman, *Joint Symposium on Scaling up of Chemical Plant and Processes*, Institute of Chemical Engineers, 1957, p. 26.
39. D. G. Jordan, *Chemical Process Development*, Part 2, Interscience, New York, 1968.
40. P. H. Calderbank, *Transact. Inst. Chem. Eng.* **36**, 443 (1958); **37**, 173 (1959); also V. W. Uhl and J. D. Gray, eds., *in Mixing Theory and Practice*, Academic Press, New York, 1969.
41. Y. Mlynek and W. Resnick, *AIChE J.* **18**, 122 (1972).
42. C. A. Coulaloglou and L. L. Tavlarides, *AIChE J.* **22**, 289 (1976).
43. D. E. Brown and K. Pitt, *Chem. Eng. Sci.* **27**, 577 (1972); **29**, 345 (1974).
44. M. A. Delichatsios and R. F. Probstein, *Ind. Eng. Chem. Fund.* **15**, 134 (1976).
45. R. Gayler and H. R. C. Pratt, *Transact. Inst. Chem. Eng.* **29**, 89 (1951).
46. D. V. Smith and G. A. Davies, *Can. J. Chem. Eng.* **48**, 628 (1970).
47. R. Letan and E. Kehat, *AIChE J.* **13**, 443 (1967).
48. E. Kehat and R. Letan, *Ind. Eng. Chem. Proc. Des. Devel.* **7**, 385 (1968).
49. R. Bussalori, S. Schiff, and R. E. Treybal, *Ind. Eng. Chem.* **45**, 2413 (1953).
50. C. J. Major and R. R. Hertzog, *Chem. Eng. Progr.* **51** (January), 17J–21J (1955).
51. D. H. Logsdail, J. D. Thornton, and H. R. C. Pratt, *Transact. Inst. Chem. Eng.* **35**, 302 (1957).
52. E. Y. Kung and R. B. Beckmann, *AIChE J.* **7**, 319 (1961).
53. C. P. Strand, R. B. Olney, and G. H. Ackerman, *AIChE J.* **8**, 252 (1962).
54. R. Spence and R. J. W. Streeton, *Atom. Energy Res. Estab. (UK)*, Report No. R4091, 33 (1962).
55. M. W. Davis and E. J. Weber, *Ind. Eng. Chem.* **52**, 929 (1960).
56. S. Nagata, N. Yoshioka, T. Yokoyama, and D. Teramoto, *Transact. Soc. Chem. Eng. (Jap.)* **8**, 43 (1950).
57. W. A. Rodger, V. G. Trice, and J. H. Rushton, Chem. Eng. Progr. **52**, 515 (1956).

58. D. S. Laity and R. E. Treybal, *AIChE J.* **3**, 1976 (1957).
59. A. H. P. Skelland and R. Seksaria, *Ind. Eng. Chem. Proc. Des. Devel.* **17**, 56 (1978).
60. B. Weinstein and R. E. Treybal, *AIChE J.* **19**, 851 (1973).
61. L. Lapidus and J. C. Elgin, *AIChE J.* **3**, 63 (1957).
62. R. Gayler, N. W. Roberts, and H. R. C. Pratt, *Transact. Inst. Chem. Eng.* **31**, 57 (1953).
63. H. H. Steinour, *Ind. Eng. Chem.* **36**, 618 (1944).
64. S. Vedaiyan, T. E. Degaleesan, and G. S. Laddha, *Indian J. Technol.* **12**, 135 (1974).
65. T. Sitaramayya and G. S. Laddha, *Chem. Eng. Sci.* **13**, 263 (1961).
66. P. Chandrasekharan and G. S. Laddha, *Chemical Process Design, A Symposium, 24-25 July 1961*, Council of Scientific and Industrial Research, New Delhi, 1963, p. 129.
67. T. E. Degaleesan and G. S. Laddha, *Indian J. Technol.* **3**, 137 (1965).
68. M. R. V. Krishman, T. E. Degaleesan, and G. S. Laddha, Paper No. HMT-9-71, First National Heat and Mass Transfer Conference, Indian Institute of Technology, Madras, December 1971.
69. T. Misek, *Rotating Disc Extractors and Their Calculation*, No. 13, Metody a Pockody Chemicke Technologie, State Publishing House of Technical Literature, Prague, 1964.
70. J. D. Thornton and H. R. C. Pratt, *Transact. Inst. Chem. Eng.* **31**, 289 (1953).
71. H. Groothuis and F. J. Zuiderweg, *Chem. Eng. Sci.* **12**, 288 (1960).
72. A. R. Smith, J. E. Caswell, P. P. Larsen, and S. D. Cavers, *Can. J. Chem. Eng.* **41**, 150 (1963).
73. R. Gayler and H. R. C. Pratt, *Transact. Inst. Chem. Eng.* **31**, 69 (1953).
74. P. Chandrasekaran and G. S. Laddha, *Proceedings of the First Australasian Conference on Hydraulics and Fluid Mechanics, 1962*, Pergamon Press, London, 1963, p. 243.
75. H. F. Wiegandt and R. L. Von Berg, *Chem. Eng.* **61**, 183 (1954).
76. C. Hanson, *Chem. Eng.* **75**, 76 (1968).
77. R. H. Sobotik and D. M. Himmelblau, *AIChE J.* **6**, 619 (1960).
78. G. Sege and F. M. Woodfield, *Chem. Eng. Progr.* **50**, (8), 396 (1954).
79. J. D. Thornton, *Transact. Inst. Chem. Eng.* **35**, 316 (1957).
80. G. V. Potnis, Studies in Pulsed Liquid-Liquid Extraction Column and the Effect of Physical Properties of Systems on its Flow Capacity, Ph.D. thesis, Bombay University, India, 1970.
81. L. Gurumurthi, M. R. V. Krishman, T. E. Degaleesan, and G. S. Laddha, *Indian J. Technol.* **16**, 257 (1978).
82. H. B. Piper, M.Sc. thesis, 1966, University of Manchester, UK.
83. J. R. Honekamp and L. E. Burkhart, *Ind. Eng. Chem. Proc. Des. Devel.* **1**, 177 (1962).
84. G. A. Davies, G. V. Jeffreys, and F. A. Ali, Institute of Chemical Engineers Symposium on Process Development, Birmingham, 1969; *Br. Chem. Eng. Proc. Technol.* **17**, 709 (1972).
85. G. V. Jeffreys and G. A. Davies, in C. Hanson, ed., *Recent Advances in Liquid-Liquid Extraction*, Pergamon Press, Oxford, 1971, Chapter 14.
86. K. Rietema, *Advances in Chemical Engineering*, Vol. 5, Academic Press, New York, 1964, Chapter 7.
87. E. G. Cockbain and T. S. McRoberts, *J. Colloid Sci.* **8**, 440 (1953).
88. T. Gillespie and E. K. Rideal, *Transact. Faraday Soc.* **52**, 173 (1956).
89. G. A. H. Elton and R. G. Picknett, *Proceedings of the Second International Congress on Surface Activity*, Vol. 1, Butterworths, London, 1957, p. 288.
90. G. E. Charles and S. G. Mason, *J. Colloid Sci.* **15**, 105, 235 (1960).
91. G. D. M. MacKay and S. G. Mason, *Nature* **191**, 488 (1961).
92. S. B. Lang, Lawrence Radiation Laboratory, UCRL Report 10097, University of California, Berkeley, CA, 1962.
93. S. B. Lang, Ph.D. thesis, University of California, Berkeley, 1962.
94. G. B. Lawson, Ph.D. thesis, University of Manchester, UK, 1967; *Chem. Proc. Eng.* (May), 45 (1967).
95. G. V. Jeffreys and J. L. Hawksley, *AIChE J.* **11**, 413, 418 (1965).
96. L. E. Neilsen, R. Wall, and G. Adams, *J. Coll. Sci.* **13**, 441 (1958).
97. S. Vijayan, M. Furrer, and A. B. Ponter, *Can. J. Chem. Eng.* **54**, 269 (1976).
98. A. H. Brown and C. Hanson, *Chem. Eng. Sci.* **23**, 841 (1968); *Nature* **214**, 76 (1967).
99. D. V. Smith and G. A. Davies, *AEChE Symp. Ser.* **68** (124), 1 (1972).
100. S. B. Lang and C. R. Wilke, *Ind. Eng. Chem.* **Fund. 10**, 329 (1971).
101. H. M. Princen and S. G. Mason, *J. Colloid Sci.* **20**, 156 (1965).
102. S. Hartland, *Chem. Eng. Sci.* **24**, 987 (1969).
103. G. D. M. MacKay and S. G. Mason, *Can. J. Chem. Eng.* **41**, 203 (1963).
104. R. M. McAvoy and R. C. Kintner, *J. Colloid Sci.* **20**, 188 (1965).
105. H. M. Princen, *J. Colloid Sci.* **18**, 178 (1963).
106. G. D. M. MacKay and S. G. Mason, *J. Colloid Sci.* **16**, 632 (1961).

107. R. M. McAvoy, W. A. Weigard, E. E. Tomkin, and R. C. Kintner, *AIChE-Inst. Chem. Eng. Joint Symposium*, reprinted in *Adv. Sep. Tech.* **1**, 16 (1965).
108. E. Rushton and G. A. Davies, quoted by G. V. Jeffreys and G. A. Davies, in C. Hanson, ed., *Recent Advances in Liquid-Liquid Extraction*, Pergamon Press, Oxford, 1971, Chapter 14.
109. A. H. Brown and C. Hanson, *Inst. Chem. Eng. Symp. Ser.* No. 26, 57 (1967).
110. G. A. Davies, G. V. Jeffreys, D. V. Smith, and F. A. Ali, *Can. J. Chem. Eng.* **48,** 329 (1970).
111. D. V. Smith, M.Sc. thesis, University of Manchester, 1966.
112. C. V. Sternling and L. E. Scriven, *AIChE J.* **5,** 514 (1959).
113. G. V. Jeffreys and G. B. Lawson, *Transact. Inst. Chem. Eng.* **43,** 294 (1965).
114. H. F. Johnson and H. Bliss, *Transact. Inst. Chem. Eng.* **42,** 331 (1946).
115. R. N. Wenzel, *J. Chem. Phys.* **53,** 1466 (1949).
116. J. F. Padday, *Proceedings of the Second International Congress Surface Activity*, Vol. 3, Butterworths, London, 1957.
117. L. I. Oxipow, *Surface Chemistry*, Reinhold, New York, 1962.
118. G. D. Yarnold and B. J. Mason, *Proc. Phys. Soc.* **B62,** 125 (1949).
119. G. E. D. Elliott and A. C. Riddiford, *J. Colloid Interface Sci.* **23,** 389 (1967).
120. G. A. Davies and G. V. Jeffreys, *Filtr. Sep.* **6,** 349 (1969).
121. J. C. Lee and G. Lewis, *Institute of Chemical Engineers Symposium on Liquid Extraction*, Newcastle, UK, 1967.
122. A. M. A. Allak and G. V. Jeffreys, *AIChE J.* **20,** 564 (1974).
123. M. J. Maraschino and R. E. Treybal, *AIChE J.* **17,** 1174 (1971).
124. P. Ramamoorthy and R. E. Treybal, *AIChE J.* **24,** 985 (1978).
125. I. Leibson and R. B. Beckmann, *Chem. Eng. Progr.* **49,** 405 (1953).
126. A. J. Madden and G. L. Damerell, *AIChE J.* **8,** 233 (1962).
127. H. D. Schindler and R. E. Treybal, *AIChE J.* **14,** 790 (1968).
128. J. T. Davies and E. K. Rideal, *Interfacial Phenomena*, Academic Press, New York, 1963.
129. H. Groothuis and F. J. Zuiderweg, *Chem. Eng. Sci.* **19,** 63 (1964).
130. W. J. Howrath, *Chem. Eng. Sci.* **19,** 33 (1964).
131. W. J. Howrath, *AIChE J.* **13,** 1007 (1967).
132. H. P. Meissner and B. Chertow, *Ind. Eng. Chem.* **38,** 856 (1946).
133. G. V. Jeffreys, G. A. Davies, and K. Pitt, *AIChE J.* **16,** 823, 827 (1970).
134. W. D. Harkins and F. E. Brown, *J. Am. Chem. Soc.* **41,** 499 (1919).
135. G. F. Scheele and B. J. Meister, *AIChE J.* **14,** 9 (1968).
136. R. J. Thomas and C. J. Mumford, *Proceedings of the International Solvent Extraction Conference, The Hague, 1971* (Soc. Chem. Ind., London), 1971, p. 400.
137. L. E. M. DeChazal and J. T. Ryan, *AIChE J.* **17,** 1226 (1971).
138. A. H. Selkar and C. A. Sleicher, *Can. J. Chem. Eng.* **43,** 298 (1965).
139. J. A. Quinn and D. B. Sigloh, *Can. J. Chem. Eng.* **41,** 15 (1963).
140. R. W. Luhning and H. Sawistowski, *Proceedings of the International Solvent Extraction Conference, The Hague, 1971* (Soc. Chem. Ind., London), 1971, p. 873.
141. British Patent No. 33444/71; G. A. Davies, G. V. Jeffreys, and M. Azfal, *Br. Chem. Eng.* **17,** 709 (1972).
142. J. I. Rosenfeld and D. T. Wasan, *Can. J. Chem. Eng.* **52,** 3 (1974); D. T. Wasan (discussion).
143. L. A. Spielman and S. L. Goren, *Ind. Eng. Chem. Fund.* **11,** 66, 73 (1972).

5

COMPUTATION OF STAGEWISE AND DIFFERENTIAL CONTACTORS: PLUG FLOW

H. R. C. Pratt
University of Melbourne
Australia

1. INTRODUCTION

1.1. General

Liquid–liquid extraction belongs to the class of *countercurrent diffusional separation processes*, in which it ranks second in importance to distillation. It is also sometimes classified with solid–liquid extraction (leaching) under the composite title of *solvent extraction*, or still more broadly with solvent dewaxing, crystallization, and so on, under the generic title of *solvent processing*. As a convenient simplification, it is referred to in this chapter as *liquid extraction* to distinguish it from solid extraction and other related processes.

Liquid extraction employs an added solvent as the *separating agent*, analogous to the use of heat in distillation. The solvent is immiscible with at least one of the feed components, so the minimum number of components involved is 3. In this respect the analogy with distillation is not exact, since distillation involves a minimum of only two components, leading to a difference in the phase rule relationship; there is, therefore, a closer analogy to fractional gas absorption and to sorption processes, in which the separating agent comprises a solvent, or an adsorbent or ion-exchange resin, respectively. However, there is a further type of liquid extraction process that is unique in that it uses *two* mutually immiscible solvents to separate a feed mixture, which may or may not be dissolved in one of the solvents. This process, known as *dual* (or *double*)-*solvent extraction*, thus involves a minimum of four components, and the additional degree of freedom results in marked differences from other types of separation process.

Liquid extraction can be carried out either continuously or batchwise; the former mode is by far the most important outside the laboratory. Conditions are generally isothermal, although a temperature gradient is sometimes imposed over the contactor in order to obtain an improved separation. However, auxiliary distillation or evaporation equipment is often necessary, both to recover solvent from the outgoing streams and, when used, to provide reflux. Alternatively, as in the case of metal extraction (see Chapter 2.2), the extract component is recovered by stripping the solvent extract with mineral acid or other reagent, thus using *chemical energy;* this is manifest in the form of the thermal, mechanical, and/or electrical energy required to produce or regenerate the reagent. In the final analysis, therefore, direct or indirect energy costs always represent an important component of the total cost of extraction processes.

The numerous types of contactor available fall into two main classes: (1) mixer–settler cascades, comprising a series of stages within each of which the phases are intimately mixed and then allowed to separate; and (2) column extractors covering the range from compartmental types with or without mechanical agitation and interstage settling, to true differential columns. Details of the various types of extractor are given in Chapters 9–17.

1.2. Theoretical Basis of Computational Methods

Liquid extraction theory has the objective of establishing quantitative relationships between system properties (particularly the phase equilibrium), equipment characteristics and relevant physical properties to enable extractor performance to be predicted over a wide range of operating conditions. The applications of the theory are twofold: (1) the *design* of new equipment to meet required performance specifications; and (2) the *performance analysis* of existing equipment, for purposes of scale-up or to satisfy changed operational requirements.

The theoretical treatment concerns two separate aspects of contactor behavior: the permissible *throughput* and the *mass-transfer performance.* Regarding the former aspect, the maximum throughput of a mixer–settler stage is controlled by the *coalescence rate* in the settler, and this, in turn, determines the stage volume and hence the solvent inventory; with column extractors, on the other hand, the limit on throughput is determined by the *flooding rate*, above which the excess flow of dispersed phase is rejected at the inlet. In both cases the limiting factors are hydrodynamic in nature, as discussed generally in Chapter 4 and more specifically in Chapters 9–17.

The mass-transfer performance of a contactor, which forms the subject of this chapter, is determined, on the other hand, by the number of stages or the length of column required to produce a given degree of separation or vice versa. The controlling variables in this case are the system properties, particularly the form of the equilibrium relationship and the rates of interphase mass transfer by a combination of molecular and eddy diffusion (Chapter 3). Although the factors controlling throughput and performance are independent in principle, interaction can occur in practice; thus an increase in agitation rate may enhance performance by increasing the interfacial area available for mass transfer and simultaneously reduce the coalescence or flooding rate and hence the throughput.

2. DESIGN PROCEDURES: CONTINUOUS PROCESSES

2.1. Process Selection

2.1.1. Process Classification

Continuous extraction processes can be carried out according to five different operating schemes, as outlined in the following paragraphs. Of these all but the last relate to single-solvent systems, that is, ternary or higher systems that utilize a single added solvent, pure or mixed, to effect the separation.

The simplest type of process is *single-contact extraction*, in which the feed and solvent pass at controlled rates into a mixing vessel followed by a phase separator and the extract and raffinate are withdrawn continuously. This corresponds to the use of a single mixer–settler stage, such as the left-hand-stage in Fig. 1*a*.

In *multistage crossflow extraction* (Fig. 1*a*) the feed is passed successively through a number of mixer–settler stages arranged in series with a separate feed of fresh solvent to each. With the use of a sufficient number of stages, the raffinate can be almost completely stripped of solute, but the solvent usage is high and the mixed extract concentration correspondingly low.

Continuous countercurrent extraction is analogous to gas absorption and stripping and can be carried out by use of either a mixer–settler cascade arranged as shown in Fig. 1*b* or a countercurrent column, as shown in Fig. 2*a*. In both cases feed and solvent are introduced at opposite ends, where the extract and the raffinate, respectively, are withdrawn. By this means the outgoing extract is brought into contact with entering feed containing the highest concentration of solute, ensuring a low solvent consumption and consequent high extract composition.

In countercurrent extraction the highest possible extract concentration corresponds to equilibrium with the entering feed. However, this can be increased by extending the contactor beyond the feed point and refluxing part of the extract to the last stage after removing the solvent by suitable means [1] (e.g., by fractional distillation or evaporation). This process is termed *single-solvent fractional extraction* and can be carried out in a column, as shown in Fig. 2*b* or in a mixer–settler cascade.

The solvent removal device serves to convert extract into the raffinate phase and is analogous to the condenser used in distillation to convert vapor into liquid. At the other end the introduction of fresh solvent is analogous to the direct injection of steam, so a device equivalent to a

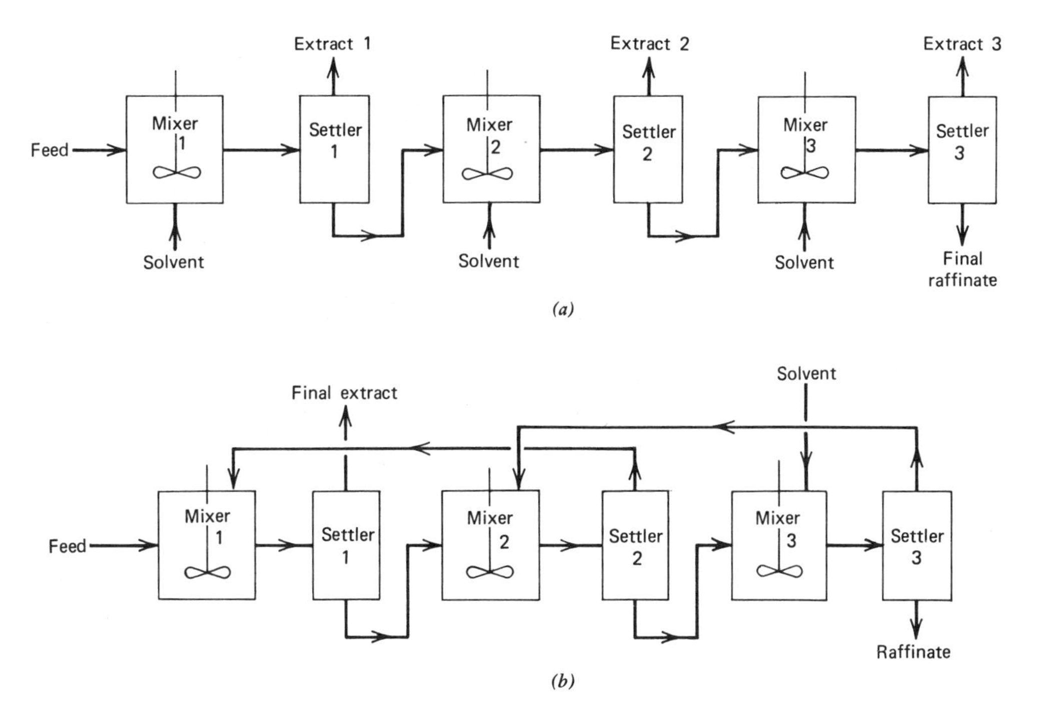

Figure 1. Continuous multistage extraction: (*a*) crossflow; (*b*) countercurrent.

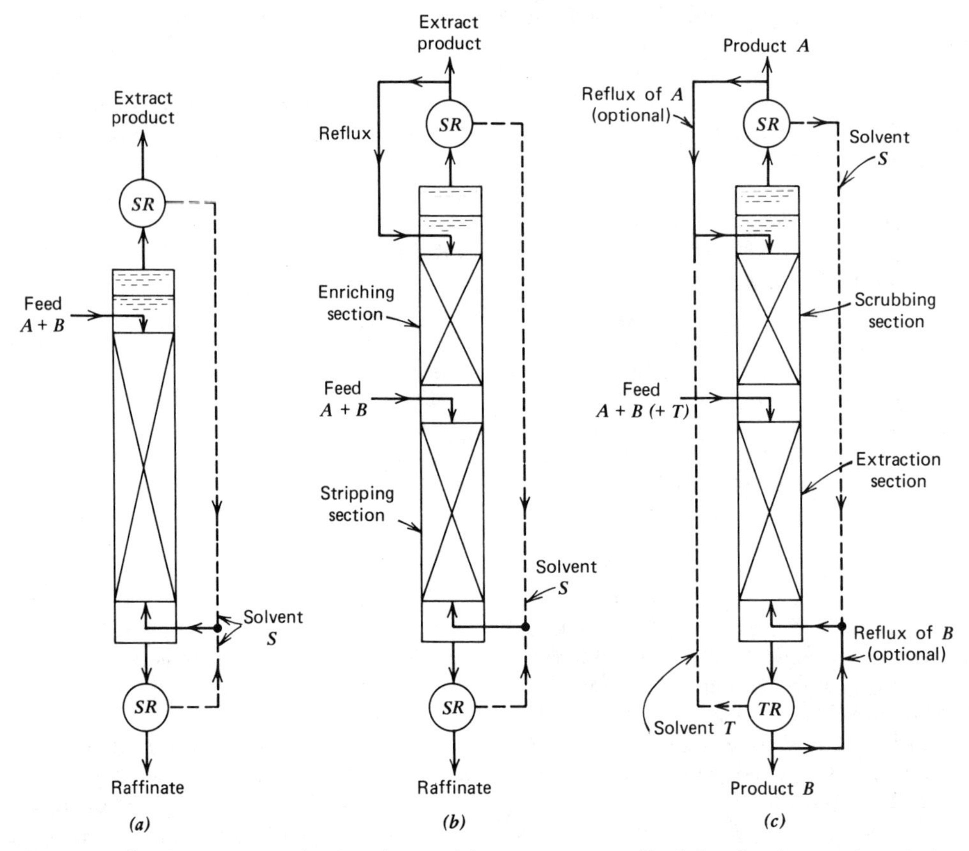

Figure 2. Continuous extraction in columns: (*a*) countercurrent; (*b*, *c*) fractional extraction, single and dual solvent, respectively (*SR*, *TR* = solvent *S*, *T* removal, respectively).

reboiler is not required. The use of extract reflux is subject to limitations not encountered in distillation; thus the reflux return must not result in complete miscibility at either end, and the density difference of the phases must be adequate to maintain countercurrent flow throughout the contactor.

In *dual-solvent fractional extraction* two immiscible solvents are used; these enter the extractor at opposite ends, and the feed, often—although not necessarily—dissolved in one of the solvents, is introduced at some intermediate point [2]. A minimum of four components (two solutes and two solvents) is thus involved. External reflux at one or both ends can be used if desired and is sometimes advantageous, although not essential. This process can also be carried out in either a column (Fig. 2*c*) or a mixer-settler cascade.

2.1.2. Choice of Method

Single-contact and multistage crossflow extraction are used only outside the laboratory in relatively small operations. Countercurrent extraction, on the other hand, uses the solvent much more efficiently and is widely used for both organic and inorganic separations.

Although single-solvent fractional extraction is capable of giving a more concentrated extract than simple countercurrent treatment, it is seldom justified with Type 3/1 systems (i.e., with *three* components and *one* immiscible pair of components), because of limitations imposed by the plait point composition. The term "Type I" is often applied to the present Type 3/1, and "Type II" to 3/2 systems with banded two-phase regions, that is, in which the two two-phase regions merge without forming a three-phase zone. With the comparatively rare banded Type 3/2 systems, it is possible in principle to obtain a complete separation of the feed components. No such constraint exists for dual-solvent fractional extraction, which is the most powerful extraction method available for large-scale use.

It should be noted that Type 4/1 systems can often be used in both single- and double-solvent processes. For example, a solution of ethanol and acetone in toluene can be extracted with water to recover the two solutes as a mixture; alternatively and more effectively, the ethanol and acetone can be separated from one another by dual-solvent extraction with water and toluene.

2.2. Basis of Design Methods

Extractor theory is based on a combination of the equilibrium relationship with the material balance and the mass-transfer rate expression. Both graphical and analytical methods can be used for computational purposes; the graphical method has advantages, despite the limited accuracy, when the solvent miscibility is appreciable. Numerical methods also have their place, especially with double-solvent and multicomponent extraction.

Contactor performance can be conveniently expressed in terms of either *theoretical (i.e., ideal) stages*, or of *transfer units;* the former concept is applicable strictly to mixer–settler type contractors and the latter, which is based on mass-transfer theory, to differential columns. However, agitated compartmental columns, of which many types exist in practice, form an intermediate class and can be treated in terms of either concept. It is shown later that there is a simple theoretical relationship between the two concepts.

The underlying principles of the various graphical methods of design for single contact extraction are described in Section 2.3. Similarly, the basic analytical relationships in terms of both theoretical stages and transfer units are described under countercurrent extraction in Section 2.5. The treatment of countercurrent processes throughout this chapter is restricted to the case where the phases are in *plug* (also termed *piston*) flow, that is, with negligible axial dispersion. The theory is considerably more complex when axial dispersion is taken into account (see Chapter 6).

2.3. Single-Contact Extraction

Miscibility boundary and equilibrium tie line data for ternary systems can be represented directly on both equilateral triangular and rectangular (i.e., right-angled triangular) coordinates, and these diagrams can also be used as a basis for graphical design procedures in terms of theoretical stages. The rectilinear solvent-free coordinate system of Janecke [3] can also be used and is particularly convenient for design purposes. As an alternative, the equilibrium distribution and selectivity diagrams may be used for the determination of theoretical stages by the McCabe–Thiele stepwise procedure. The application of these methods to single-contact extraction is described in the paragraphs that follow.

2.3.1. Triangular Diagrams

A comparison of the two triangular coordinate systems is shown in Fig. 3, in which curves *CRPEH* represent the miscibility boundary and *RE*, a typical equilibrium tie line. Concentrations are usually expressed in mass fraction units, although mole fractions can be used if desired. Point F represents the composition of the feed, which enters the stage at a rate F and contains mass fractions x_{AF}, x_{BF}, and x_{SF} of components A and B and of solvent S, respectively; similarly, S' represents the solvent entering and E and R

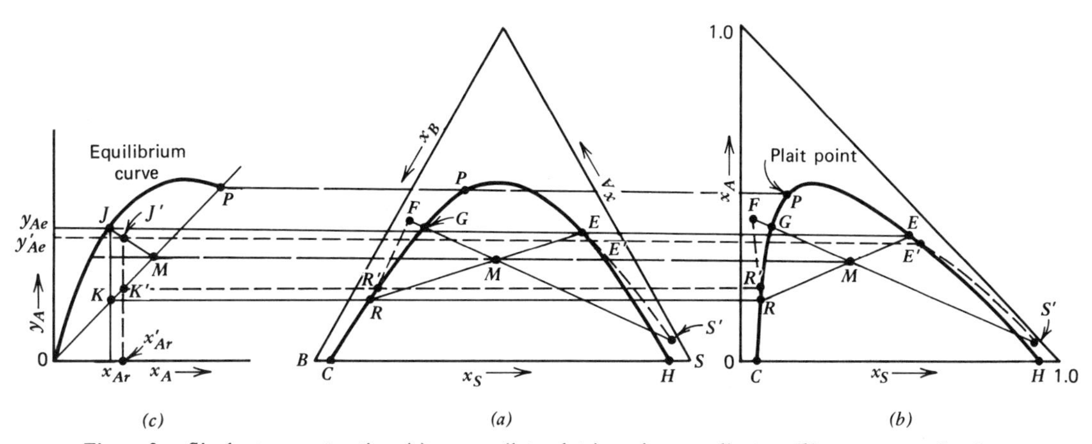

Figure 3. Single-stage extraction (*a*) on equilateral triangular coordinates; (*b*) on rectangular (i.e., right-angled triangular) coordinates; (*c*) projection onto equilibrium distribution diagram.

the extract and raffinate phases leaving the stage. Mass balances are then taken around the stage on total material, distributed component A, and solvent S, as follows:

Total material $A + B + S$:

$$F + S = R + E = M \tag{1}$$

Distributed component A:

$$Fx_{AF} + Sy_{As} = Rx_{Ar} + Ey_{Ae} = Mx_{Am} \tag{2}$$

Solvent S:

$$Fx_{SF} + Sy_{Ss} = Rx_{Sr} + Ey_{Se} = Mx_{Sm} \tag{3}$$

In these equations M, x_{Am}, and x_{Sm} refer to mixture M. Combination of the terms in F, S, and M gives

$$F + S - M = 0 \tag{4}$$

$$Fx_{AF} + Sy_{As} - Mx_{Am} = 0 \tag{5}$$

$$Fx_{SF} + Sy_{Ss} - Mx_{Sm} = 0 \tag{6}$$

Equations (4)–(6) comprise a set of linear homogeneous equations in F, S, and M, which have a nontrivial solution only if the determinant (see Notation at end of chapter) of the coefficients is zero; thus

$$\begin{vmatrix} 1 & 1 & 1 \\ x_{AF} & y_{As} & x_{Am} \\ x_{SF} & y_{Ss} & x_{Sm} \end{vmatrix} = 0 \tag{7}$$

Equation (7) expresses the condition that the three points (x_{AF}, x_{SF}), (y_{As}, y_{Ss}), and (x_{Am}, x_{Sm}) lie on a straight line, that is, that M lies on the line FS' in Fig. 3. The location of M is found by elimination of M from Eq. (4) and (5), giving

$$\frac{F}{S} = \frac{x_{Am} - y_{As}}{x_{AF} - x_{Am}} = \frac{MS'}{MF} \tag{8}$$

This is the *lever-arm rule*, applicable to all *conserved-property* diagrams such as the Janecke diagram described in the following paragraphs or the Ponchon–Savarit diagram (involving enthalpy) used in distillation.

The compositions and amounts of extract and raffinate products can be obtained by applying the same procedure to the terms in E, R, and M in Eqs. (1)–(3) or by direct use of the lever-arm rule. This shows that points E, R, and M are collinear, so that E and R can be located on the miscibility boundary by finding by trial the equilibrium tie line that passes through M. The quantity of extract produced is obtained by eliminating R from Eqs. (1) and (2) and solving for E, as follows:

$$E = \frac{M(x_{Am} - x_{Ar})}{y_{Ae} - x_{Ar}} \tag{9}$$

The amount of raffinate is then obtainable from Eq. (1). Similarly, elimination of M gives

$$\frac{R}{E} = \frac{y_{Ae} - x_{Am}}{x_{Am} - x_{Ar}} \tag{10}$$

so that $R/E = EM/MR$.

The equilibrium distribution diagram can be constructed by projection from the triangular diagrams as shown in Fig. 3*c*. The equilibrium curve is obtained by projection of equilibrium tie lines such as RE and terminates on the diagonal at the plait point P. The stage operating line MJ is obtained by projecting mixture point M onto the diagonal and drawing a line through this of slope $-(R/E)$, as indicated by Eq. (10). The tie line RE on the triangular diagrams can thus be located directly by projecting back from the distribution diagram, without the need for trial and error.

In a continuous flow extractor complete equilibrium is unlikely to be attained, even at slow flow rates, because of the nature of the residence time distribution curve; therefore, the phase compositions would pass along curves from F and S' such as those shown as broken lines in Fig. 3 to E' and R', respectively. These points can be determined by projection from the distribution diagram if the Hausen efficiency MJ'/MJ (Section 4.1.3) is known.

2.3.2. *Janecke Diagram*

The Janecke diagram is constructed by plotting the *solvent ratio*, $S/(A + B)$, that is, the ratio of solvent to total solute, against the extract and raffinate compositions on a solvent-free basis. A typical plot for a Type 3/1 system is shown in Fig. 4*a*, where $CRPEH$ represents the miscibility boundary, P the plait point, and RE the equilibrium tie line. If the compositions and flows are expressed on a solvent-free basis (denoted by the overbar), the material balances take the following form:

Total nonsolvent $A + B$:

$$\bar{F} + \bar{S} = \bar{R} + \bar{E} = \bar{M} \tag{11}$$

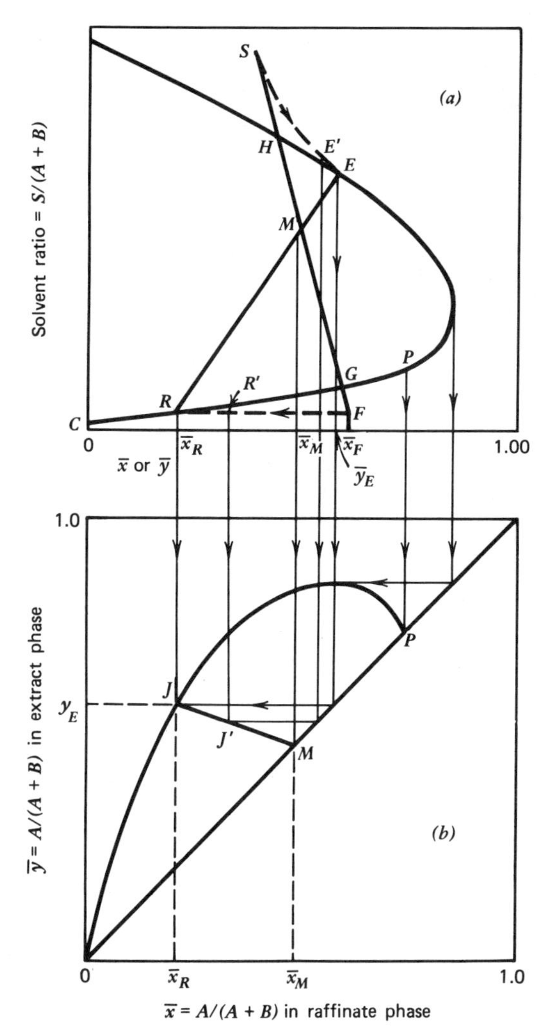

Figure 4. Single-stage extraction (*a*) on Janecke coordinates; (*b*) projection onto equilibrium selectivity diagram.

Distributed component A:

$$\bar{F}\bar{x}_F + \bar{S}\bar{y}_s = \bar{R}\bar{x}_r + \bar{E}\bar{y}_e = \bar{M}\bar{x}_m \qquad (12)$$

Solvent S:

$$\bar{F}s_F + \bar{S}s_s = \bar{R}s_r + \bar{E}s_e = \bar{M}s_m \qquad (13)$$

In these equations $\bar{S}$ refers to total *nonsolvent* in the solvent stream, that is, to $S(1 - y_s)$, where S is the flow of impure solvent and y_s its solvent content; s_s is the solvent ratio of the nonsolvent in the solvent stream, given by $y_s/(1 - y_s)$, and $\bar{y}_s$ is the mass fraction of distributed component (i.e., solute) in the nonsolvent; and $\bar{M}$ is the flow of mixture to the settler on a solvent-free basis, and $\bar{x}_m$ is its solute content and s_m, its solvent ratio.

The coefficients of $\bar{F}$, $\bar{S}$, and $\bar{M}$ in Eqs. (11)–(13) can be written in determinantal form as in Eq. (7), showing that points $(\bar{x}_F, s_F)$, $(\bar{y}_s, s_s)$, and $(\bar{x}_m, s_m)$, that is, F, S, and M in Fig. 4*a*, lie on a straight line. Similarly, on expressing the terms in $\bar{R}$, $\bar{E}$, and $\bar{M}$ in the same way, it is evident that the points R, E, and M are collinear, intersecting line FMS at M; this corresponds to the equilibrium tie line through M, which must be found by trial, or more conveniently by projection from the selectivity curve (Fig. 4*b*), as is now shown here. The coordinates of M are obtained as follows from eqs. (11)–(13):

$$\bar{x}_m = \frac{\bar{F}x_F + \bar{S}y_s}{\bar{F} + \bar{S}}; \qquad s_m = \frac{\bar{F}s_F + \bar{S}s_s}{\bar{F} + \bar{S}} \qquad (14\text{–}15)$$

If the entering solvent is pure, $\bar{S} = 0$; however, s_s is then infinite and the term $\bar{S}s_s$ in Eq. (15) is equal to S. The point S in Fig. 4*a* is therefore located at infinity and M lies vertically above $\bar{x}_F$ with the ordinate $(s_F + S/\bar{F})$, as shown by Eq. (15).

Definite limits exist to the possible range of solvent ratios $S/\bar{F}$ that can be used. Thus the minimum ratio is that for which point M coincides with G on Fig. 4*a* and corresponds to GF/SG. Similarly, the maximum solvent ratio, for which M coincides with H, is given by FH/HS. The ratio of solvent-free raffinate to extract product produced $\bar{R}/\bar{E}$ is given by EM/MR.

The equilibrium selectivity diagram can be constructed by projection from the Janecke diagram as shown in Fig. 4*b*. The equilibrium curve, obtained by projection of equilibrium tie lines such as RE, terminates at the plait point P. The stage operating line MJ is obtained by projecting mixture point M onto the diagonal and drawing a line through this of slope $-(\bar{R}/\bar{E})$, as may be shown by eliminating M from Eqs. (11) and (12) to give an expression analogous to Eq. (10). The tie line RE on the Janecke diagram can thus be located directly, without trial and error, by projection of point J to R on the miscibility curve and extending RM to E.

The effect of incomplete attainment of equilibrium is shown by the two broken lines $FR'R$ and $SE'E$ in Fig. 4*a*; these are analogous to the corresponding lines in Figs. 3*a* and 3*b*. If the actual exit compositions are assumed to be represented by points E' and R', the Hausen efficiency is defined on a selectivity diagram basis by MJ'/MJ (Fig. 4*b*).

2.4. Multistage Crossflow Extraction

2.4.1. *Graphical Methods*

(a) General Case. Since this process is an extension of single-contact extraction in which the raffinate from each stage is retreated with fresh solvent, the course of the extraction can be shown in a similar manner on the triangular and Janecke diagrams. An example of the use of the Janecke diagram is shown in Fig. 5, in which the successive raffinates are indicated by R_1, R_2, and so on. Since the raffinate from any stage provides the feed to the next, Eqs. (11)–(15) are written in terms of stage n, replacing $\bar{R}$ and $\bar{E}$ by $\bar{R}_n$ and $\bar{E}_n$, respectively, and similarly $\bar{F}$ and $\bar{S}$ by $\bar{R}_{n-1}$ and $\bar{S}_n$. In Fig. 5 the use of impure solvent of composition S is shown on the right and that of pure solvent, on the left. The Janecke diagram can be projected onto the selectivity diagram if desired, in the same way as for single-stage extraction.

The various extracts are usually combined, giving a mixture of composition

$$y_{av} = \frac{\sum_n \bar{E}_n \bar{y}_n}{\sum_n \bar{E}_n} \tag{16}$$

Recovery of solvent, if required, is normally carried out on the combined extracts and on the final raffinate. In general, unless the solvent is solute free, one of the tie lines on extension will pass through point S on Fig. 5; this corresponds to the maximum possible degree of extraction and would require an infinite number of stages.

(b) Immiscible Solvents. A simplified procedure can be used when the solvents are effectively immiscible over the range of solute concentrations involved. For this purpose the flow rates are expressed in terms of the pure solvents S and B as R_i and E_i, with the concentrations in corresponding mass (or mole) ratio units x' and y', equal to x_A/x_B and y_A/y_S, respectively. A mass balance on solute over any stage n then gives

$$E_i(y'_n - y'_s) = R_i(x'_{n-1} - x'_n) \tag{17}$$

where y'_s and x'_{n-1} are the solvent and raffinate compositions entering stage n and y'_n and x'_n are the exit compositions. On rearranging, this becomes

$$\frac{y'_s - y'_n}{x'_{n-1} - x'_n} = -\frac{R_i}{E_i} \tag{17a}$$

This relationship is shown on the distribution diagram in Fig. 6, in which $OLJG$ represents the equilibrium curve and F, the feed state inlet compositions (x'_F, y'_s). A straight line FG with a slope of $-(R_i/E_{i1})$ intersects the equilibrium curve at G, giving the compositions of the first-stage extract and raffinate y'_1 and x'_1, respectively, assuming equilibrium to have been attained. This raffinate forms the feed to the second stage, and a line of slope $-(R_i/E_{i2})$ drawn

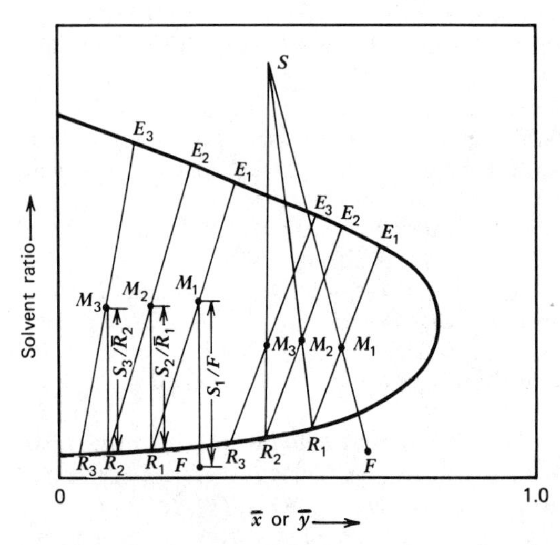

Figure 5. Multistage crossflow extraction on Janecke coordinates. Right-hand construction applies to impure solvent; left-hand, to pure solvent.

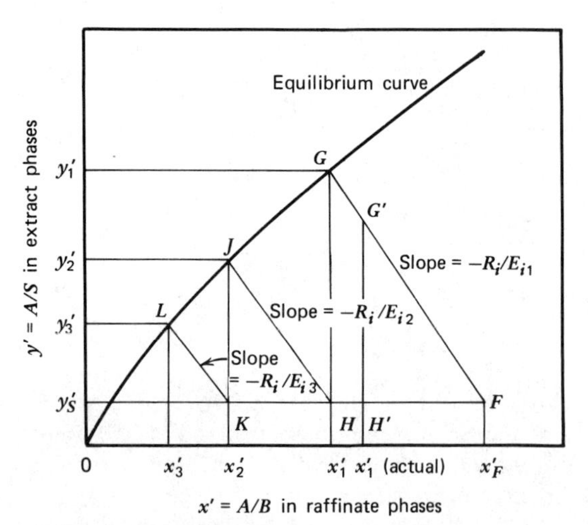

Figure 6. Multistage crossflow extraction with immiscible solvents.

through H gives the second-stage compositions y'_2, x'_2, and so on. The combined extract composition is then obtained from Eq. (16). If equilibrium is not reached, the lines FG, and so on, must be terminated at points such as G', where FG'/FG is the Hausen efficiency E_H (Section 4.1.3).

2.4.2. *Analytical Method*

A simple analytical solution is available for the case of effectively immiscible solvents if in addition both the distribution coefficient $K = y'_n/x'_n$ and the stage flow ratio R_i/E_i are constant; however, this equilibrium constraint can conveniently be relaxed somewhat by the use of the dimensionless concentration units described in Section 2.5.4, since on this basis the equilibrium relationship is approximated by a straight line of slope m not necessarily passing through the origin. For this purpose, the concentrations are first expressed in mass/volume units c_x, c_y and then converted to the dimensionless forms X and Y, using Eqs. (60) and (62) in Section 2.5.4, with the term $c_{y,N+1}$ replaced by c_s, the solute content of the entering solvent. In these units the initial feed concentration becomes $X_F = 1.0$ and the inlet solvent concentration, $Y_S = 0$.

The stage material balance [Eq. (17)] expressed in these units becomes

$$\mathbf{S}Y_n = (X_{n-1} - X_n) \tag{18}$$

where $\mathbf{S}$, the stripping ratio, is given by E/mR with E/R taken as the average value over the contactor. The equilibrium relationship in the same units is $Y_n^* = X_n$ so that, assuming equilibrium stages, Eq. (18) becomes

$$X_n(1 + \mathbf{S}) - X_{n-1} = 0 \tag{18a}$$

The solution to this equation, on putting $X_n = X_N$ when $n = N_S$, is

$$X_N = \frac{1}{(\mathbf{S} + 1)^{N_S}} \tag{19}$$

A convenient graphical solution to Eq. (19) is given in Fig. 7.

Equation (19) applies only to the case where the total solvent fed, E_{iT}, is equally distributed between the N_S stages. This distribution policy can be shown to be the most efficient for crosscurrent extraction. As N_S is increased to infinity,

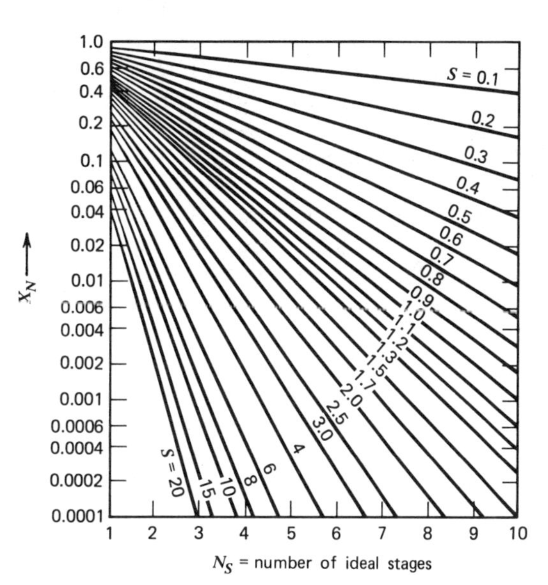

Figure 7. Multistage crossflow extraction: graphical solution for number of theoretical stages ($\mathbf{S}$ = stripping factor per stage).

the value of X_N for a given total solvent feed reaches a limit corresponding to batch differential extraction (Section 5.3).

Equation (19) also shows that for a given number of stages, the degree of extraction from the feed always increases (i.e., X_N decreases) as the solvent quantity is increased. However, the proportion removed by a given amount of solvent relative to that removed by the same amount by use of infinite stages always passes through a minimum as the solvent quantity is increased; this occurs at $\mathbf{S} = 1.6$–1.8, depending on the number of stages [4, 5]. The relative efficiency, defined in this way, does not increase very rapidly for $N_S > 4$, the minimum, for example, having a value of 94% for five stages.

It is assumed in the preceding treatment that the equilibrium relationship is linear, so that variations in the stripping factor $\mathbf{S}$ result only from changes in the flow ratio in different stages. Frequently, however, the equilibrium distribution is nonlinear, for instance, taking the form $y'^* = Kx'^r$. In such cases it has been shown [6] that although an infinite amount of solvent is required for complete extraction of solute when $r \geqslant 1.0$, for $r < 1.0$ with pure solvent and infinite stages, extraction is complete with an amount of solvent given by

$$E_{iT} = \frac{R_i x_F'^{(1-r)}}{K(1-r)} \tag{20}$$

2.5. Continuous Countercurrent Extraction

2.5.1. *Graphical Solution for Stages*

(a) General Solution. The two triangular and the Janecke coordinate systems described in Section 2.3 for single-contact extraction can all be extended to cover continuous countercurrent extraction in stagewise contactors. Janecke coordinates are the most convenient in a number of respects, and their use is described in the following paragraphs [7].

The necessary relationships are established by taking material balances between any two stages n and $n+1$ and around the solvent feed end in terms of solvent-free flows and composition as follows (cf. Fig. 11):

Nonsolvent $A + B$:

$$\bar{R}_n - \bar{E}_{n+1} = \bar{R}_p - \bar{S} = \bar{J} \tag{21}$$

Solute A:

$$\bar{R}_n \bar{x}_n - \bar{E}_{n+1} \bar{y}_{n+1} = \bar{R}_p \bar{x}_{pr} - \bar{S}\bar{y}_s = \bar{J}\bar{x}_j \tag{22}$$

Solvent S:

$$\bar{R}_n s_{x,n} - \bar{E}_{n+1} s_{y,n+1} = \bar{R}_p s_{pr} - \bar{S} s_s = \bar{J} s_j \tag{23}$$

The terms containing $\bar{J}$ refer to the *net flows*, which are constant throughout the extractor. A set of linear homogeneous equations is obtained on combining the terms in $\bar{R}$, $\bar{E}$, and $\bar{J}$, the nontrivial solution to which is given by

$$\begin{vmatrix} 1 & 1 & 1 \\ \bar{x}_n & \bar{y}_{n+1} & \bar{x}_j \\ s_{x,n} & s_{y,n+1} & s_j \end{vmatrix} = 0 \tag{24}$$

As with Eq. (7), Eq. (24) expresses the condition that the points $(\bar{x}_n, s_{x,n})$, $(\bar{y}_{n+1}, s_{y,n+1})$, and $(\bar{x}_j, s_j)$ are collinear; consequently, a line drawn through the net flow point J on the Janecke diagram (Fig. 8*a*) and the miscibility boundary will intersect the latter at points that satisfy the material balance equations. It is noted, on writing in determinant form the equations obtained on combining the terms in $\bar{R}_p$, $\bar{S}$, and $\bar{J}$ in Eqs. (21)–(23), that points $(\bar{x}_{pr}, s_{pr})$, $(\bar{y}_s, s_s)$, and $(\bar{x}_j, s_j)$ are also collinear. Hence J lies on the extension of SR_p, where R_p has the coordinates $(\bar{x}_{pr}, s_{pr})$; the coordinates of J are obtained as follows from Eqs. (21)–(23).

$$x_j = \frac{\bar{R}_p \bar{x}_{pr} - \bar{S}\bar{y}_s}{\bar{R}_p - \bar{S}}, \qquad s_j = \frac{\bar{R}_p s_{pr} - \bar{S} s_s}{\bar{R}_p - \bar{S}} \tag{25–26}$$

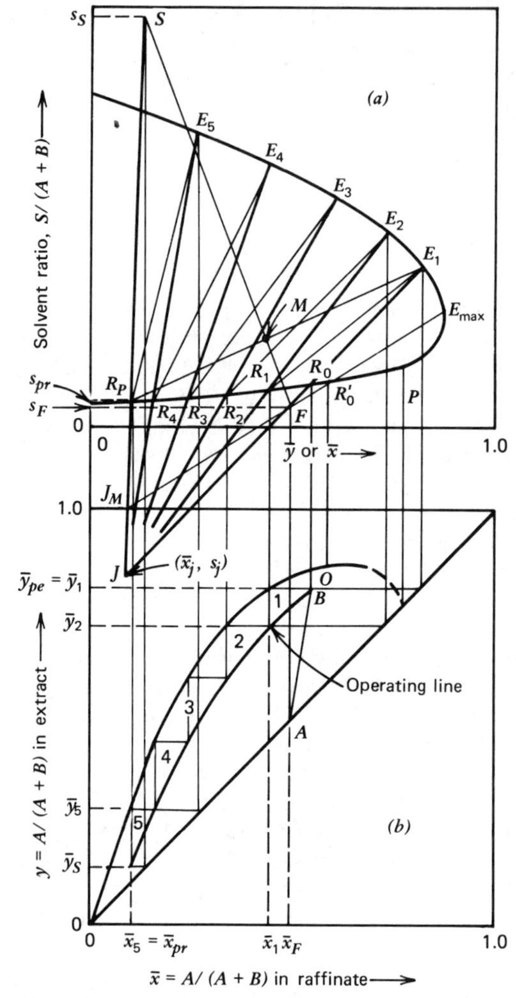

Figure 8. Continuous countercurrent extraction (*a*) on Janecke coordinates; (*b*) projection onto selectivity diagram.

If the entering solvent is solute free, $\bar{S} = 0$ and $\bar{x}_j = \bar{x}_{pr}$; also, s_s is infinite and $\bar{S}s_s = S$, the quantity of pure solvent fed. Hence $s_j = (s_{pr} - S/\bar{R}_p)$, so that J is located vertically below $\bar{R}_p$ at a distance $S/\bar{R}_p$, with S at infinity. This distance thus corresponds directly to the *solvent ratio*, that is, the ratio of solvent fed to raffinate product.

With the use of these results, the number of theoretical stages can be determined by using the construction shown in Fig. 8*a*. The net flow point is located as described earlier and is joined to feed point F, with coordinates $(\bar{x}_F, s_F)$. On extension, this line meets the miscibility boundary at E_1, giving the composition and solvent content of the extract product. Equilibrium tie line E_1R_1 is then drawn through E_1, and JR_1 is extended to meet the miscibility boundary at E_2, giving the second-stage extract composition.

Such alternate equilibrium tie lines and material balance lines are repeatedly drawn until the specified exit raffinate composition R_p is reached, that is, after five stages (Fig. 8*a*).

The quantities of solvent-free extract and raffinate products are obtained from the overall material balance:

$$\bar{R}_p + \bar{E}_1 = \bar{F} + \bar{S} \tag{27}$$

The overall mixing point M is located on the Janecke diagram at the intersection of FS and R_pE_1; hence, by the lever-arm rule, $\bar{E}_1/\bar{R}_p = R_pM/ME_1$, and substitution of this value in Eq. (27) gives R_p and E_1.

As the solvent rate is decreased, the net flow point J approaches R_p, and the required number of stages increases until finally a point JM is reached, at which the number becomes infinite. This limit, known as the *minimum solvent ratio*, usually occurs when an equilibrium tie line, on extension, passes through both F and J, as shown by line $E_{max}FJ_M$ in Fig. 8*a*; E_{max} then corresponds to the highest attainable extract product composition. On occasion, however, when there is an inflection in the slopes of the tie lines, other tie lines on extension may intersect the extension of SR_p below this, in which case J_M corresponds to the intersection most remote from R_p.

Although the number of theoretical stages decreases as the solvent ratio is increased, the contactor cross section and the cost of solvent recovery from the extract product are also both increased. In practice, a solvent ratio of about 20–40% above the minimum would be used, but the exact value must be determined at the design stage on economic grounds.

If desired, the number of stages can be determined by the McCabe–Thiele method after projection onto the selectivity diagram, as shown in Fig. 8*b*. For this purpose, the equilibrium curve is obtained by projection of the tie lines, and the operating line by drawing lines at random through J and projecting the points of intersection with the miscibility boundary. The stages are then stepped off in the usual way, as shown. Line AB shows the composition change on saturation of the feed with solvent in the first stage. At the minimum solvent ratio the operating line "pinches in" to the equilibrium curve at C, the projection of point R'_0, thus requiring infinite stages.

In practice, the stage efficiency would be below 100%, so that the number of actual stages required would be greater than that of theoretical stages. Allowance for this is best made in terms of the Murphree efficiency (see Section 4.1.2).

(b) Dual Feeds. It is sometimes necessary to operate an extractor with a second feed of lower strength introduced at an intermediate stage. In this case the extractor must be considered as two regions, one between the first and second feeds, the other between the second feed and the raffinate stage. Within each region, stagewise construction can be carried out as detailed in Section 2.5.1(a). However, the net flow points for the two regions are different. It can readily be shown that the two net flow points are collinear with the second feed composition, and the lever-arm rule is obeyed in regard to net flows and combined feed flow.

(c) Immiscible Solvents. When the solvents are effectively immiscible, the distribution diagram can be simplified by the use of mass ratio coordinates [8, 9], as described for crossflow extraction [Section 2.4.1(b)]. The extractor operating line, which relates the compositions between stages, is obtained by taking material balances on solute around the solvent inlet; this gives (see Fig. 11)

$$y'_{n+1} = \frac{R_i}{E_i}\, x'_n + \left[y'_{N+1} - \frac{R_i}{E_i}\, x'_N\right] \tag{28}$$

This expression, which represents a straight line of slope R_i/E_i, is shown in Fig. 9, together with the equilibrium curve. Stage operating lines of slope $-(R_i/E_i)$, obtained in the same manner as Eq. (17a), are also shown; these correspond to the diagonals of the rectangles constructed from the equilibrium steps. In practice, however, it is

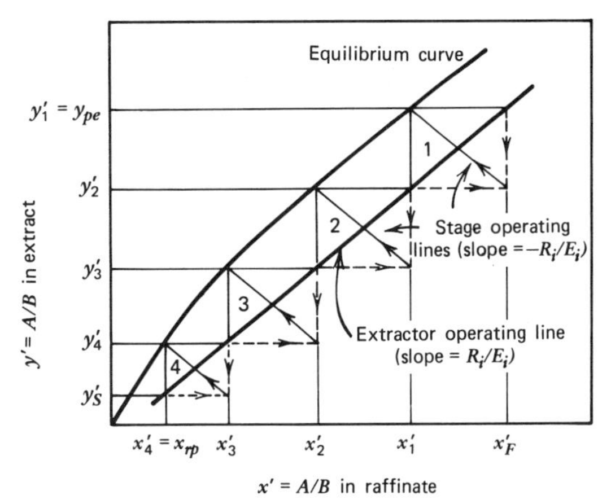

Figure 9. Countercurrent multistage extraction with immiscible solvents.

sufficient to determine the number of theoretical stages required by the McCabe–Thiele stepwise construction. Allowance for nonideal stages can again be made in terms of the Murphree efficiency (Section 4.1.2).

2.5.2. Analytical Methods: Stagewise Contactors

Analytical methods are more convenient and accurate than graphical solutions in a number of respects. However, in liquid extraction their use is limited by the complex nature of the equilibrium relationship and changes in solvent miscibility. Thus a method due to Colburn [10] for the calculation of theoretical stages is restricted to cases where the partition coefficient and solvent ratio are both constant, that is, to dilute systems; however, modified methods are also available that allow for some curvature of the equilibrium relationship [11, 12].

The following sections describe an analytical treatment that divides the extractor into sections, in each of which the equilibrium and operating lines are approximated by straight line segments. These are then expressed in terms of dimensionless concentrations, giving a result analogous to that described by Colburn [10]. Somewhat different notation and units are used in this treatment; this involves the designation of the feed (i.e., raffinate) and extract streams as the X and Y phases, with corresponding lowercase subscripts. Furthermore, the concentrations

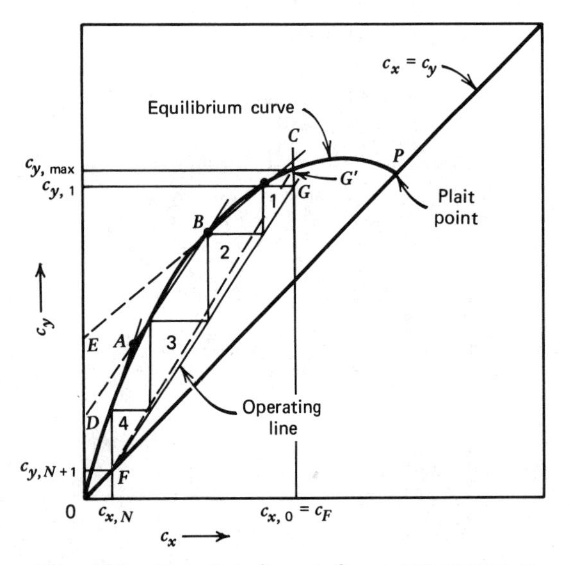

Figure 10. Operating diagram for countercurrent extraction on distribution coordinates.

c_j will be expressed in the more convenient units of mass (or moles) per unit volume, with the flows F_j and velocities U_j in volumetric units. These can also be put into terms of mass or mole fraction by noting that, for instance, $c_x = c_{tx}x$, where c_{tx} is the *total* mass or molar concentration (i.e., density); however, c_t generally varies with concentration, especially in molar units.

(a) Operating Diagram. In Fig. 10 an equilibrium curve is shown on distribution coordinates, together with its representation by straight line segments OA, AB, and BC. These segments are represented algebraically by the relationship

$$c^*_{y,n} = m'c_{x,n} + q' \tag{29}$$

or by

$$c^*_{x,n} = mc_{y,n} + q \tag{30}$$

where

$$m = \frac{1}{m'} \quad \text{and} \quad q = -mq' \tag{31}$$

The operating line equation is obtained by taking a solute balance around the solvent inlet end, as shown in Fig. 11; this gives

$$F_{y,n+1}c_{y,n+1} + F_{x,N}c_{x,N} = F_{y,N+1}c_{y,N+1} + F_{x,n}c_{x,n} \tag{32}$$

Solving for $c_{y,n+1}$, neglecting changes in F_x and F_y, yields

$$c_{y,n+1} = \frac{F_x}{F_y}c_{x,n} + \left(c_{y,N+1} - \frac{F_x}{F_y}c_{x,N}\right) \tag{33}$$

If average values can be used for F_x and F_y, Eq. (33) represents a straight line of slope F_x/F_y on Fig. 10. On the other hand, the transfer of solute from one phase to the other causes both F_x and F_y to decrease toward the raffinate outlet end, resulting in curvature of the operating line FG. The true operating line for this case can be constructed by using the corresponding solute balance in mass ratio units [i.e., Eq. (28)], and converting to volume concentration units, noting that, for example, $c_x = c_{tx}x'/(1 + x')$. Finally, if the solvent miscibility also changes, it is necessary to obtain the operating line coordinates by projection from the Janecke diagram (Fig. 8*b*) and conversion of the concentration units.

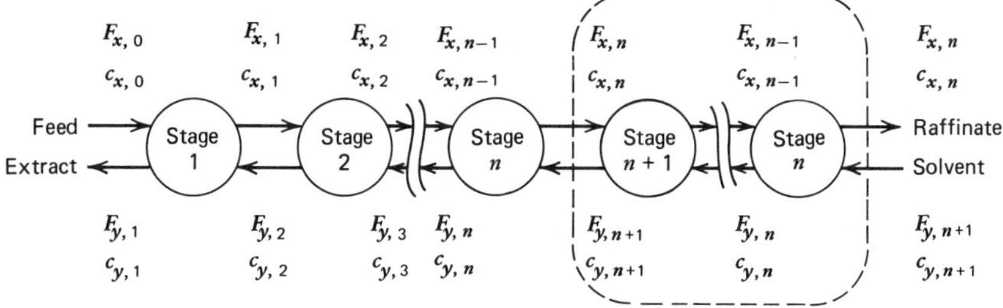

Figure 11. Material balance for continuous countercurrent extraction.

In all cases it is assumed that the operating line can be approximated, either as a whole or in sections, by expressions of the form

$$c_{y,n+1} = pc_{x,n} + b \tag{34}$$

If both operating and equilibrium lines are approximated by straight line segments, the corresponding intersections should be arranged to occur at the same c_x or c_y values as appropriate.

(b) Overall Material Balance and Solvent Ratio. The overall solute material balance, that is, around the entire contactor, is obtained by putting $n = 0$ in Eq. (33). The composition of the final extract can thus be calculated if the other compositions and the solvent ratio are known. Alternatively, it can be solved for the solvent ratio, assuming average values of the flows:

$$\left(\frac{F_y}{F_x}\right)_{av} = \frac{c_{x,0} - c_{x,N}}{c_{y,1} - c_{y,N+1}} \tag{35}$$

Again, this equation must be expressed in terms of mass ratio concentrations x' and y' and the corresponding flow rates of pure solvents R_i and E_i if the inlet solute concentration is high.

The minimum solvent ratio corresponds to operating line FG' in Fig. 10, which intersects the equilibrium curve at G' with abscissa c_F. The maximum possible extract composition is then $c_{y,\max}$, and substitution of this for $c_{y,1}$ in Eq. (35) gives the minimum solvent ratio. In practice, a somewhat higher solvent ratio must be used, for instance, corresponding to operating line FG, giving extract composition $c_{y,1}$.

(c) Number of Theoretical Stages. The general equation of the extractor in terms of c_x is obtained by eliminating $c_{y,n+1}$ between Eqs. (29) and (34), giving

$$c_{x,n+1} - \frac{p}{m'} c_{x,n} = \frac{b - q'}{m'} \tag{36}$$

This is a linear, first-order difference equation, the solution to which, with the boundary condition $n = 0$, $c_x = c_{x,0}$, gives the c_x profile as follows:

$$c_{x,n} = \left(c_{x,0} - \frac{1}{D_x}\right)\left(\frac{p}{m'}\right)^n + \frac{1}{D_x} \tag{37}$$

where

$$D_x = \frac{m' - p}{b - q'} \tag{38}$$

The number of theoretical stages required is obtained by putting $n = N_S$, $c_{x,n} = c_{x,N}$ and solving for N_S, giving

$$N_S = \frac{\log [(D_x c_{x,0} - 1)/(D_x c_{x,N} - 1)]}{\log m'/p} \tag{39}$$

Similar solutions in terms of c_y are obtained by elimination of $c_{x,n}$ from Eqs. (29) and (34). The results are included in Table 1, together with a summary of the foregoing equations.

(d) Numerical Methods. Numerical methods have been devised for the computation of extractors for ternary systems, using the equilibrium data either directly or in activity coefficient form. For convenience, these methods are considered in the treatment of multicomponent systems in Section 6.3.2.

2.5.3. *Differential Contactors*

In true differential contactors the compositions of the phases change continuously from one end

TABLE 1 SUMMARY OF DESIGN EQUATIONS

Equation Number	Equations for	Staged Contactors
		(a) Normal[a]
1	Equilibrium line	$c^*_{y,n} = m'c_{x,n} + q'$ or $c^*_{x,n} = mc_{y,n} + q$
2	Operating line	$c_{y,n+1} = pc_{x,n} + b$ where $p = \frac{F_x}{F_y}$ $b = (c_{y,N+1} - pc_{x,N})$
3	Overall material balance	$c_{y,1} = pc_{x,0} + b$
4	Concentration profiles	$c_{x,n} = \left[c_{x,0} - \frac{1}{D_x}\right]\left[\frac{p}{m'}\right]^n + \frac{1}{D_x}$ $c_{y,n} = \left[c_{y,1} - \frac{1}{D_y}\right]\left[\frac{p}{m'}\right]^{n-1} + \frac{1}{D_y}$ where $D_x = \frac{m' - p}{b - q'}$ $D_y = \frac{m' - p}{m'b - pq'}$
5	NTS or NTU	$N_S = \frac{\log [(D_x c_{x,0} - 1)/(D_x c_{x,N} - 1)]}{\log m'/p}$ $= \frac{\log [(D_y c_{y,1} - 1)/(D_y c_{y,N+1} - 1)]}{\log m'/p}$
6	Extractor efficiency	

[a]With concentrations expressed as kg/m^3 or $kmol/m^3$.

Staged Contactors	Differential Contactors	
(b) Dimensionless	(c) Normal[a]	(d) Dimensionless
$Y_n^* = X_n$	$c_y^* = m'c_x + q'$ or $c_x^* = mc_y + q$	$Y^* = X$
$Y_{n+1} = \mathbf{E}(X_n - X_N)$	$c_y = pc_x + b$ where $p = \dfrac{U_x}{U_y}$ $b = (c_{y,I} - pc_{x,I})$	$Y = \mathbf{E}(X - X_I)$
$Y_1 = \mathbf{E}(1 - X_N)$	$c_{y,0} = pc_{x,0} + b$	$Y_0 = \mathbf{E}(1 - X_I)$
$X_n = Y_n = \dfrac{\mathbf{E}^{N_S+1} - \mathbf{E}^n}{\mathbf{E}^{N_S+1} - 1}$ $(\mathbf{E} \neq 1.0)$ $X_n = Y_n = \dfrac{1 + N_S - n}{1 + N_S}$ $(\mathbf{E} = 1.0)$	$c_x = \left(c_{x0} - \dfrac{1}{D_x}\right) \exp(\mu) + \dfrac{1}{D_x}$ $c_y = \left(c_{y0} - \dfrac{1}{D_y}\right) \exp(\mu) + \dfrac{1}{D_y}$ where $\mu = N_{Tox}\left(\dfrac{m'}{p} - 1\right)$ or $\mu = N_{Toy}\left(1 - \dfrac{m'}{p}\right)$	$X = \dfrac{\mathbf{S} \exp[\lambda L(1 - Z)] - 1}{\mathbf{S} \exp(\lambda L) - 1}$ $Y = \dfrac{\exp[\lambda L(1 - Z)] - 1}{\mathbf{S} \exp(\lambda L) - 1}$ $(\mathbf{S} \neq 1.0)$ $X = \dfrac{L(1 - Z) + H_{ox}}{L + H_{ox}}$ $Y = \dfrac{L(1 - Z)}{L + H_{ox}}$ $(\mathbf{S} = 1.0)$ where $\lambda = \dfrac{\mathbf{S} - 1}{\mathbf{S}H_{Tox}}$ or $\dfrac{\mathbf{S} - 1}{H_{Toy}}$
$N_S = \dfrac{\log[(1 - 1/\mathbf{S})(1/X_N) + 1/\mathbf{S}]}{\log \mathbf{S}}$ $(\mathbf{S} \neq 1.0)$ $= \dfrac{1}{X_N} - 1$ $(\mathbf{S} = 1.0)$	$N_{Tox} = \dfrac{m'}{m' - p} \ln\left[\dfrac{D_x c_{x,0} - 1}{D_x c_{x,I} - 1}\right]$ $N_{Toy} = \dfrac{p}{m' - p} \ln\left[\dfrac{D_y c_{y,0} - 1}{D_y c_{y,I} - 1}\right]$	$N_{Tox} = \dfrac{\mathbf{S}}{\mathbf{S} - 1} \ln\left[\left(1 - \dfrac{1}{\mathbf{S}}\right)\dfrac{1}{X_I} + \dfrac{1}{\mathbf{S}}\right]$ $N_{Toy} = \dfrac{1}{\mathbf{S} - 1} \ln\left[\left(1 - \dfrac{1}{\mathbf{S}}\right)\dfrac{1}{X_I} + \dfrac{1}{\mathbf{S}}\right]$ $(\mathbf{S} \neq 1.0)$ $N_{Tox} = N_{Toy} = \dfrac{1}{X_I} - 1$ $(\mathbf{S} = 1.0)$
$E_{sx} = (1 - X_N)\mathbf{S}Y_1$ $= \dfrac{\mathbf{S}^{N_S+1} - \mathbf{S}}{\mathbf{S}^{N_S+1} - 1}$ $(\mathbf{S} \neq 1.0)$ $= \dfrac{N_S}{N_S + 1}$ $(\mathbf{S} = 1.0)$		E_{ox} or $E_{oy} = (1 - X_I)\mathbf{S}Y_0$ $= \dfrac{\mathbf{S} \exp(\nu) - \mathbf{S}}{\mathbf{S} \exp(\nu) - 1}$ $(\mathbf{S} \neq 1.0)$ $= \dfrac{N_T}{1 + N_T}$ $(\mathbf{S} = 1.0)$ where $\nu = \dfrac{N_{Tox}(\mathbf{S} - 1)}{\mathbf{S}}$ or $\nu = N_{Toy}(\mathbf{S} - 1)$ $N_T = N_{Tox}$ or N_{Toy}

to the other, rather than in steps. A simplified representation of a column contactor of this type is shown in Fig. 12, in which the two phases are depicted for simplicity as occupying vertical flow channels with the interface between them.

(a) Transfer Units. Considering first the X phase in Fig. 12, the flux of solute component A within the section of height δZ is as follows (see Chapter 3, Section 3.1.1):

$$dN_A = -U_x\, dc_x = k_x a(c_x - c_{xi})\, dz \quad (40)$$

where c_x represents the bulk-phase composition and c_{xi}, the interfacial composition. Rearrangement and integration between the limits $z = 0$, $c_x = c_{x0}$ and $z = L$, $c_x = c_{xI}$ gives

$$\frac{k_x aL}{U_x} = \int_{c_{xI}}^{c_{x0}} \frac{dc_x}{(c_x - c_{xi})} = N_{Tx} \quad (41)$$

The integral N_{Tx} in this expression is termed the *number of X-phase transfer units.* This can be expressed in the form

$$L = H_{Tx} N_{Tx} \quad (42)$$

where H_{Tx} is the height of an X-phase transfer unit, given by

$$H_{Tx} = \frac{U_x}{k_x a} \quad (43)$$

Equations (41) and (42) show that H_{Tx} can be identified as the height of column that produces a change in concentration Δc_x numerically equal to the mean driving force over the interval. Similarly, N_{Tx} can be considered as a dimensionless measure of the difficulty of a given separation.

A similar treatment can be accorded to the Y phase, giving

$$\frac{k_y aL}{U_y} = \int_{c_{yI}}^{c_{y0}} \frac{dc_y}{(c_{yi} - c_y)} = N_{Ty} \quad (44)$$

together with $L = H_{Ty} N_{Ty}$, where

$$H_{Ty} = \frac{U_y}{k_y a} \quad (45)$$

The use of Eqs. (41) and (44) requires a knowledge of the interfacial compositions c_{xi} and c_{yi}. In the absence of heterogeneous reaction

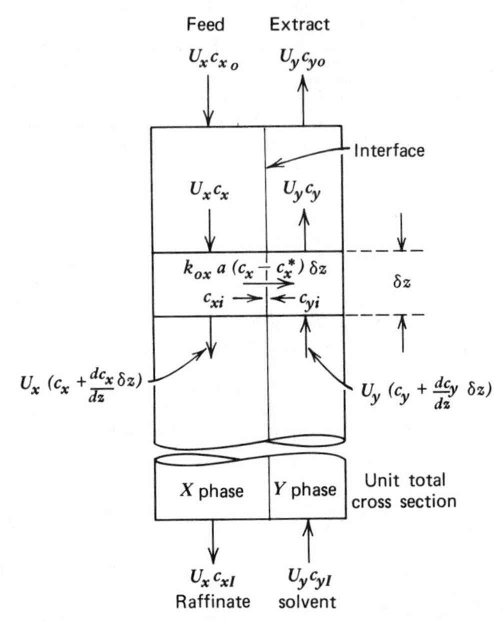

Figure 12. Shell balance on differential extractor over height δz. For simplicity, the phases are shown as occupying separate flow channels.

at the interface, these can be obtained as shown in Fig. 5*a* of Chapter 3 by drawing tie lines of slope $-k_x/k_y$ from points on the operating line to intersect the equilibrium curve at points (c_{xi}, c_{yi}). Equation (41) can be then integrated graphically by plotting $1/(c_x - c_{xi})$ against c_x and measuring the area under the curve between $c_x = c_{x0}$ and c_{xI} or by numerical means. The value of the integral in Eq. (44) can be obtained in a similar manner.

(b) Overall Transfer Units. Because it is difficult to obtain reliable values of the individual coefficients k_x and k_y, mass-transfer data are usually based on the overall coefficients k_{ox} and k_{oy} (Chapter 3, Section 3.1.1). In this form Eq. (40) becomes

$$dN_A = -U_x\, dc_x = k_{ox} a(c_x - c_x^*)\, dz \quad (46)$$

where c_x^* is the X-phase concentration in equilibrium with c_y. Integration as before then gives

$$N_{Tox} = \int_{c_{xI}}^{c_{x0}} \frac{dc_x}{(c_x - c_x^*)},$$

$$H_{Tox} = \frac{U_x}{k_{ox} a} \quad (47\text{–}48)$$

and similarly for the Y phase

$$N_{Toy} = \int_{c_{yI}}^{c_{y0}} \frac{dc_y}{(c_y^* - c_y)},$$

$$H_{Toy} = \frac{U_y}{k_{oy}a} \qquad (49\text{–}50)$$

These integrals are simple to evaluate since c_x^* and c_y^* are directly obtainable as the equilibrium values corresponding to c_y and c_x, respectively.

The individual and overall values of H_T are related by the summation of resistances given in Chapter 3, Eq. (58). Thus for the X phase, that equation is multiplied through by U_x/a to give

$$H_{Tox} = H_{Tx} + \mathbf{E}H_{Ty} \qquad (51)$$

where $\mathbf{E} = U_x/m_y'U_y$, the *extraction factor* (defined here to be analogous to the absorption factor in absorption). Similarly, for the Y phase, on multiplying through by U_y/a

$$H_{Toy} = H_{Ty} + \mathbf{S}H_{Tx} \qquad (52)$$

where $\mathbf{S} = m_x'U_y/U_x$, the *stripping factor.*

Although very convenient, the use of overall H_T values involves an important restriction; when the equilibrium line is curved, m_x' and m_y' are no longer constant, and consequently the overall coefficients and H_T values vary, even if the individual film values are constant.

Equation (51) suggests that a plot of H_{Tox} against $\mathbf{E}$ should give a straight line of slope H_{Ty} and intercept H_{Tx}, provided these are constant. However, although mass-transfer data plotted in this way often appear to give good straight lines, in some instances negative intercepts are obtained with droplet-type contactors. Since k_x and k_y are known to be approximately constant for given systems in such contactors (Chapter 3, Section 4.4), this result must be attributed to variations in specific interfacial area a. Thus a is known to be proportional to the holdup [13], and, in turn, to the dispersed phase flow [14]; hence if the Y phase is dispersed, H_{Ty} will be constant and H_{Tx} proportional to $(U_x/U_y)^n$. This has been observed for packed columns [15, 16], for which $n = 0.54\text{–}0.98$. It may be concluded, therefore, that Eqs. (51) and (52) do not provide a sound basis for obtaining individual H_{Tj} values.

(c) Alternative Expressions for Transfer Units. In the past, flux has often been expressed in molar units as follows, for example, on an overall X-phase basis:

$$dN_A = -d(V_x x) = k_{ox}ac_{tx}(x - x^*)\,dz \qquad (53)$$

Also

$$d(V_x x) = V_{ix}\,d\left(\frac{x}{1-x}\right) = \frac{V_x\,dx}{1-x} \qquad (54)$$

After rearranging, the variance in k_{ox} with concentration is removed prior to integration by multiplying both sides by the drift factor, $(1 - x)_{om}$, that is, the logarithmic mean of $(1 - x)$ and $(1 - x^*)$. The result is

$$\frac{k_{ox}a(1-x)_{om}L}{U_x} = \int_{x_I}^{x_0} \frac{(1-x)_{om}\,dx}{(1-x)(x-x^*)} = N_{Tox} \qquad (55)$$

where $V_x/c_{tx,av}$ is replaced by U_x, with $c_{tx,av}$ the mean value of the total molar concentration, averaged over the contactor.

Equation (55) can be solved as before by graphical integration. However, if $(1 - x)$ and $(1 - x^*)$ differ by a factor of not more than 2, the arithmetic mean of these can replace the logarithmic mean $(1 - x)_{om}$ with negligible error. The integral then becomes [17]

$$N_{Tox} = \int_{x_I}^{x_0} \frac{dx}{(x - x^*)} + \frac{1}{2}\ln\frac{1 - x_I}{1 - x_0} \qquad (56)$$

The integral here is identical with that in Eq. (47) provided the total concentration c_{tx} is constant or can be averaged; the second term is thus a correction mainly for drift. Hence it is more convenient to use Eq. (47) directly, taking k_{ox}, a and U_x in Eq. (48) as the average values within the contactor. In this form, c_x can equally well be expressed in terms of mass or moles per unit volume. However, appropriate units must be used when converting k_{ox} to the Colburn–Drew zero transport form F_{ox} by multiplying by the drift factor $(1 - x)_{om}$, for instance, as for correlation purposes. Thus, as shown in Chapter 3, Section 3.1.1, the volume average velocity is the correct reference velocity for diffusion in the absence of volume change on mixing, and this is usually approached more closely by the mass average rather than the molar average velocity. Hence in such cases the drift factor should be expressed in mass rather than mole fraction units, although this has seldom been the custom.

(d) Direct Solution for Contactor Length. If both equilibrium and operating lines are straight, or more generally if the driving force $(c_x - c_x^*)$ is linear in c_x, Eq. (46) can be integrated directly, giving

$$L = \frac{U_x(c_{x0} - c_{xI})}{k_{ox}a(c_x - c_x^*)_{\text{lm}}} \tag{57}$$

where $(c_x - c_x^*)_{\text{lm}}$ is the logarithmic mean of $(c_{x0} - c_{x0}^*)$ and $(c_{xI} - c_{xI}^*)$. This result can also be put into the form $L = H_{Tox}N_{Tox}$, where

$$N_{Tox} = \frac{c_{x0} - c_{xI}}{(c_x - c_x^*)_{\text{lm}}} \tag{58}$$

Similar expressions are obtained in terms of the Y phase.

(e) Analytical Solutions for Transfer Units. In preparing the operating diagram, the necessary relationships are similar to those for stagewise contactors [Eqs. (29)–(35)], except that the stage designations n and N are dropped and the F_j are replaced by U_j (see Table 1 for summary of equations).

To integrate Eq. (47), $(c_x - c_x^*)$ is first evaluated by eliminating c_y between Eqs. (29) and (34); the final result is

$$N_{Tox} = \frac{m'}{m' - p} \ln \left(\frac{D_x c_{x0} - 1}{D_x c_{xI} - 1} \right) \tag{59}$$

The counterpart in terms of N_{Toy} is obtained similarly from Eq. (44) and is given in Table 1, together with the concentration profiles.

2.5.4. *Use of Dimensionless Concentrations*

(a) Definitions. The foregoing results can be expressed more concisely in terms of dimensionless concentrations X and Y by means of a change of coordinates. This is shown in Fig. 13*a*, which represents a typical operating diagram on $c_x - c_y$ coordinates; since the X phase is the feed (i.e., raffinate) phase, the process is a stripping operation and the operating line lies below the equilibrium line.

The procedure is as follows: (1) the concentration $c_{x,N+1}^*$ in equilibrium with the inlet Y-phase composition $c_{y,N+1}$ is located and the coordinate origin is moved to point A, as shown; (2) the coordinate scales are expanded or contracted so that the inlet concentration $c_{x,0}$ and the equilibrium Y-phase composition $c_{y,0}^*$ both have values of 1.0. By this means the new coordinates of any point, say, P, are defined by $X = AD/AC$ and $Y = AE/AG$. The resulting modified operating diagram is shown in Fig. 13*b*, in which the equilibrium line is represented by $Y^* = X$.

The dimensionless concentrations are expressed algebraically as follows in terms of the measured concentrations c_x and c_y, using units of mass or moles per unit volume:

$$X = \frac{c_x - c_{x,N+1}^*}{c_{x,0} - c_{x,N+1}^*} = \frac{c_x - (mc_{y,N+1} + q)}{c_{x,0} - (mc_{y,N+1} + q)} \tag{60}$$

$$Y = \frac{c_y - c_{y,N+1}}{c_{y,0}^* - c_{y,N+1}} \tag{61}$$

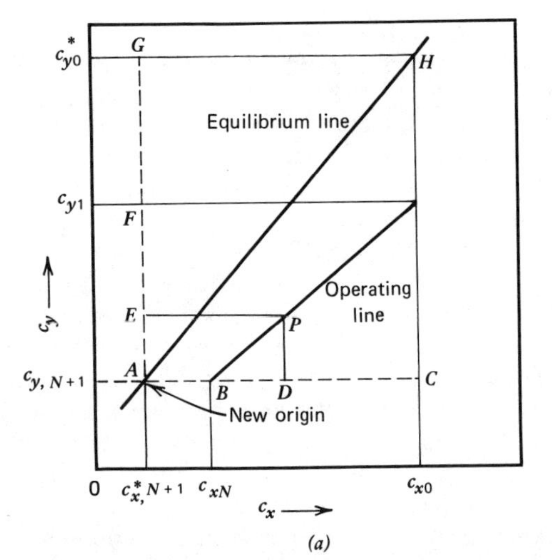

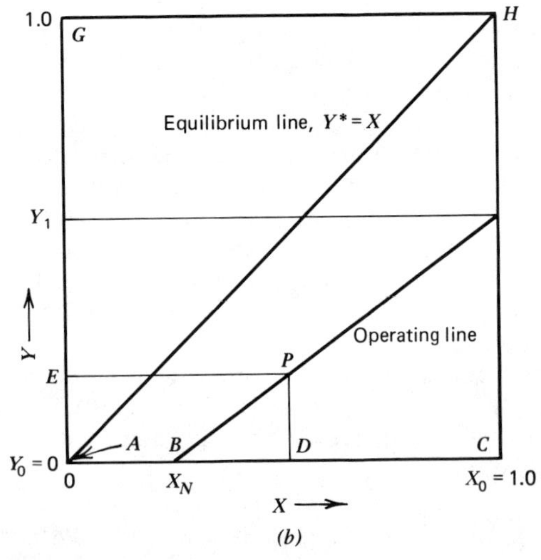

Figure 13. Operating diagrams for countercurrent extraction (*a*) on $c_x - c_y$ coordinates; (*b*) on dimensionless coordinates.

Since m is defined by

$$m = \frac{c_{x,0} - c^*_{x,N+1}}{c^*_{y,0} - c_{y,N+1}}$$

Eq. (61) becomes

$$Y = \frac{m(c_y - c_{y,N+1})}{c_{x,0} - (mc_{y,N+1} + q)} \tag{62}$$

The corresponding expressions for a differential contactor are obtained by replacing subscript $N+1$ by I. The dimensionless forms of Eqs. (29), (30), and (34), are included in Table 1.

(b) Applications. Substitution for c_x in Eq. (39) from Eq. (60) and insertion of the values of p and b gives

$$N_S = \frac{\log\left[(1 - 1/\mathbf{S})(1/X_N) + 1/\mathbf{S}\right]}{\log \mathbf{S}} \qquad (\mathbf{S} \neq 1.0) \tag{63}$$

$$= \frac{1}{X_N} - 1 \qquad (\mathbf{S} = 1.0) \tag{64}$$

This expression, which is useful for design, reduces to one given by Colburn [10] on putting $q = 0$ in Eq. (60) but is more general since it can be used for equilibrium curve segments that do not pass through the origin.

The foregoing equations can also be arranged in a form due to Kremser [18, 19] so as to give the *performance* directly as a function of the number of stages. Thus, on defining an *overall stripping efficiency* $E_{Sx} = (1 - X_N)\,\mathbf{S}Y_1$, Eq. (63) gives

$$E_{Sx} = \frac{\mathbf{S}^{N_S+1} - \mathbf{S}}{\mathbf{S}^{N_S+1} - 1} \qquad (\mathbf{S} \neq 1.0) \tag{65}$$

$$= \frac{N_S}{N_S + 1} \qquad (\mathbf{S} \neq 1.0) \tag{66}$$

Similar results are obtained in terms of transfer units (see Table 1). Convenient graphical solutions to the various equations are given in Figs. 14–17.

(c) Relationships Between Stages and Transfer Units. Combination of Eq. (63) with the two corresponding expressions for transfer units from

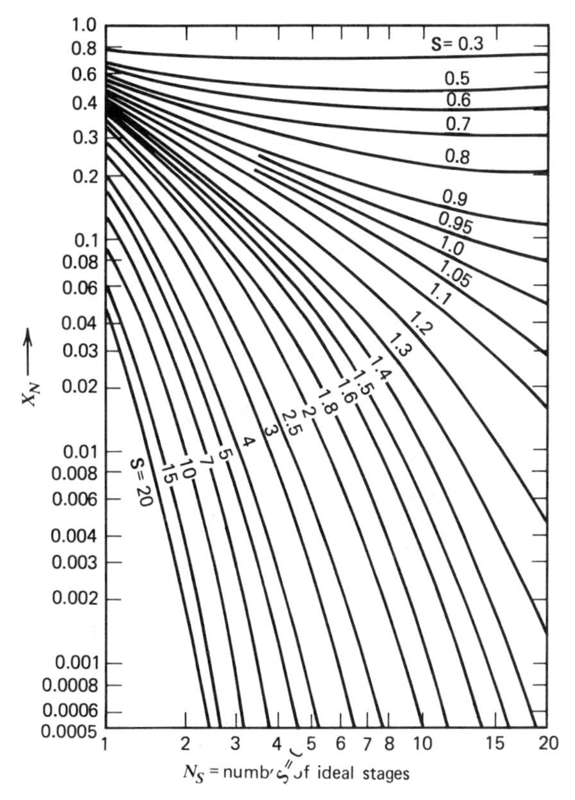

Figure 14. Countercurrent extraction: graphical solution for theoretical stages.

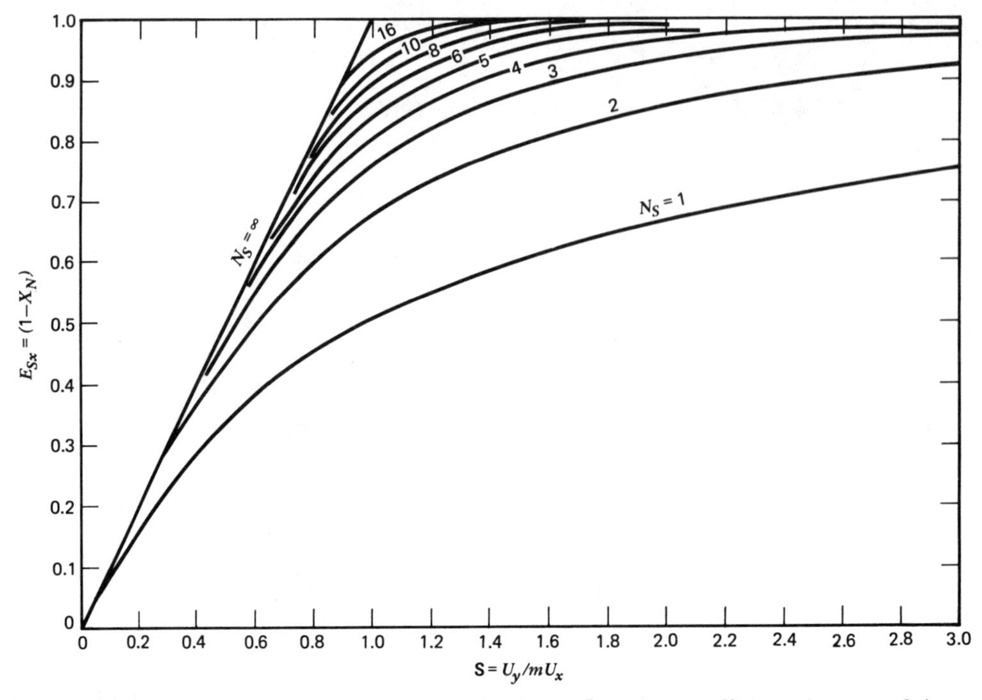

Figure 15. Countercurrent extraction: graphical solution for extractor efficiency in terms of theoretical stages.

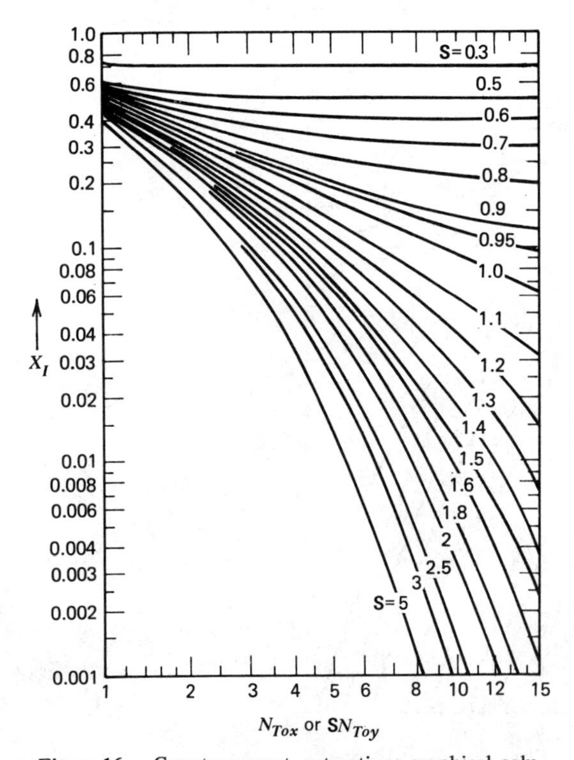

Figure 16. Countercurrent extraction: graphical solution for number of overall transfer units.

Table 1 gives

$$\frac{N_{Tox}}{N_S} = \frac{\mathbf{S} \ln \mathbf{S}}{S - 1} \qquad (\mathbf{S} \neq 1.0) \tag{67}$$

$$\frac{N_{Toy}}{N_S} = \frac{\ln \mathbf{S}}{\mathbf{S} - 1} \qquad (\mathbf{S} \neq 1.0) \tag{68}$$

$$\frac{N_{Tox}}{N_{Toy}} = \mathbf{S} \tag{69}$$

When $\mathbf{S} = 1.0$, $N_{Tox} = N_{Toy} = N_S$.

2.6. Fractional Extraction (Single Solvent)

In this process (Fig. 2*b*) the stripping section is a countercurrent extractor of the type considered in Section 2.5.1 and can be treated in the same manner on the Janecke diagram, as shown in Fig. 18. A banded Type 3/2 system is considered.

For the enriching section [7], material balances are taken between any two stages m and $m + 1$ and around the solvent separator, as follows:

Nonsolvent $A + B$:

$$\bar{E}_{m+1} - \bar{R}_m = \bar{P}_E + \bar{S}' = \bar{K} \tag{70}$$

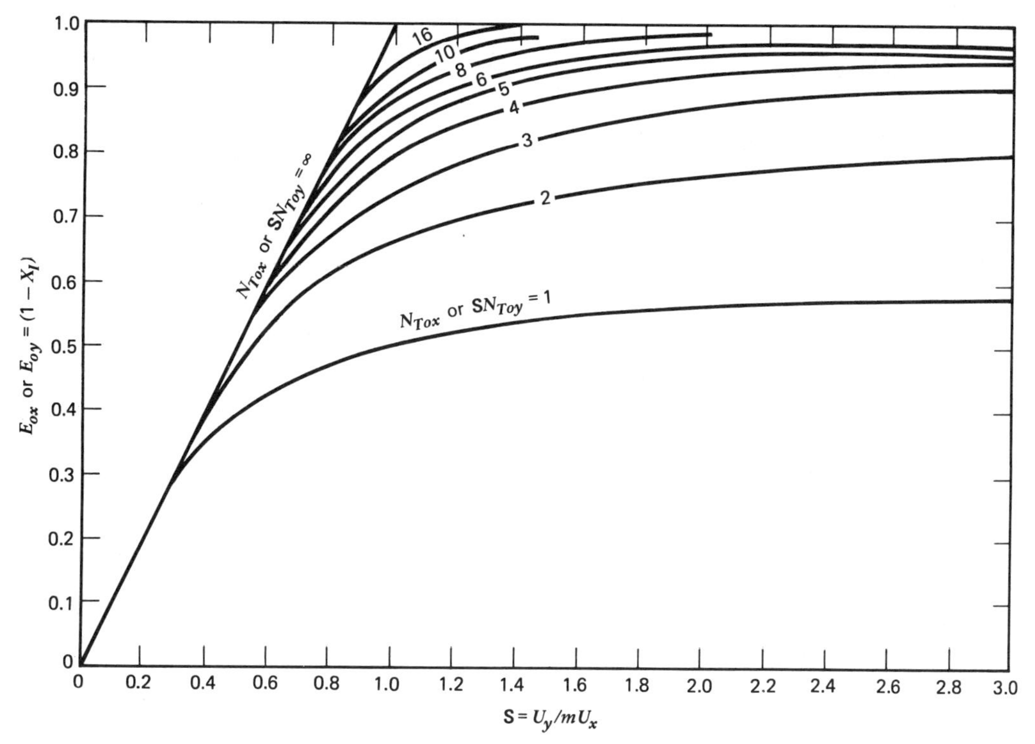

Figure 17. Countercurrent extraction: graphical solution for extractor efficiency in terms of transfer units.

Solute A:

$$\bar{E}_{m+1}\bar{y}_{m+1} - \bar{R}_m\bar{x}_m = \bar{P}_E\bar{y}_{pe} + \bar{S}'\bar{y}'_s = \overline{K}\bar{y}_k \quad (71)$$

Solvent S:

$$\bar{E}_{m+1}s_{y,m+1} - \bar{R}_m s_{x,m} = \bar{P}_E s_{pe} + \bar{S}'s'_s = \bar{K}s_k \quad (72)$$

The values of $\bar{S}'$, $\bar{y}'_s$, and s'_s are usually similar to $\bar{S}$, $\bar{y}_s$, and s_s in Eqs. (21)–(23) since the solvent is normally recycled. Elimination of $\bar{E}$, $\bar{R}$, and $\bar{K}$ by writing in determinantal form as usual shows that the points $(\bar{x}_m, s_{x,m})$, $(\bar{y}_{m+1}, s_{y,m+1})$, and $(\bar{y}_k, s_k)$ are collinear, with $\bar{y}_k$ and s_k given by

$$y_k = \frac{\bar{P}_E\bar{y}_{pe} + \bar{S}'\bar{y}'_s}{\bar{P}_E + \bar{S}'} \quad (73)$$

$$s_k = \frac{\bar{P}_E s_{pe} + \bar{S}'s'_s}{\bar{P}_E + \bar{S}'} \quad (74)$$

When the solute content of the solvent leaving the solvent separator is zero, these expressions become $\bar{y}_k = \bar{y}_{pe}$ and $s_k = (s_{pe} + S/\bar{P}_E)$.

Similar balances on $\bar{F}$, $\bar{R}_P$, and $\bar{P}_E$, that is, around the whole extractor, show that the points JFK are also collinear. The construction for the enriching section can now be completed on the Janecke diagram (Fig. 18) as follows. The extract product point P_E, with coordinates $(\bar{x}_{pe}, s_{pe})$, is joined to S', and K_M, corresponding to the *minimum extract reflux ratio*, is located by extending equilibrium tie lines on the right of F to cut $S'P_E$; the intersection most remote from P_E corresponds to K_M, which is given by $K_M E_1/E_1 P_E$. A somewhat higher reflux ratio, corresponding to K, for instance, is then chosen and KF is extended to meet the extension of the lines SR_P at J, thus satisfying the overall material balance. The usual stage construction is then carried out by drawing equilibrium tie lines (e.g., R_1E_1), followed by stage operating lines (e.g., KE_2R_1), alternately until the feed point is reached. The operating lines are then placed through the lower net flow point J and the procedure continued until the specified raffinate composition is reached. The equilibrium tie lines and stage operating lines can, if desired, be projected onto the selectivity diagram and the stages stepped off by the McCabe–Thiele method.

As the reflux ratio is increased, point K moves

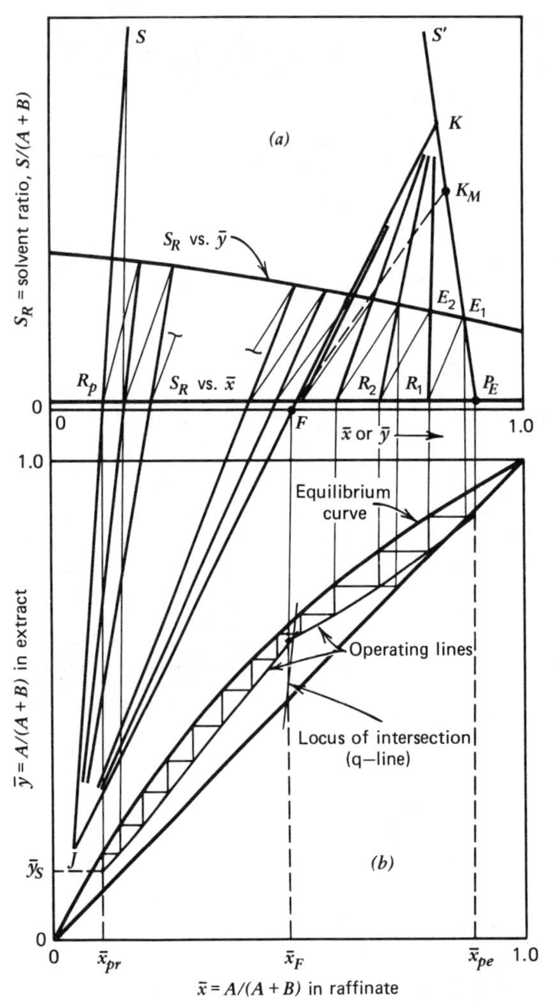

Figure 18. Single-solvent fractional extraction on (*a*) Janecke and (*b*) selectivity diagrams.

further from P_E until finally, with pure solvent, it approaches infinity and the stage operating lines become vertical. This corresponds to the *total reflux* condition, which would require infinite heat for solvent recovery, since no product is obtained. At the minimum reflux condition, on the other hand, the heat requirement per unit of product is a minimum but the number of stages, and hence the capital cost, would be infinite. Clearly, there will be an *optimum reflux ratio* that corresponds to a minimum in a graph of operating plus capital charges against reflux ratio, just as in distillation.

2.7. Fractional Extraction (Dual Solvent)

This process involves a minimum of four components, comprising two solutes, A and B, that are separated by fractional distribution between two immiscible solvents, S and T. In general, such systems are of Type 4/1, that is, with one immiscible pair (the two solvents), although this is not a necessary restriction.

Phase equilibria for quaternary systems can be represented in three-dimensional form on equilateral or right-angled tetrahedra or on the triangular prism. The miscibility boundary is a surface joining the separate boundaries for the two ternary systems AST and BST, and the tie-line equilibria occupy the three-dimensional space within this boundary. Much experimental data are thus required for such systems; moreover, three-dimensional models for general use are difficult to construct, and although methods of two-dimensional projection have been described [20], these are very tedious. Therefore, it is usually assumed in design that the solvents are immiscible throughout the extractor; this is normally reasonable for inorganic solutes, but less so for organic solutes, especially in the region of the feed point where solute concentrations are high, as is seen later.

2.7.1. *Graphical Solution*

The stagewise arrangement of the extractor is shown in Fig. 19 together with notation. The conventional designations of *scrubbing* and *extraction* sections arose from the need to treat the solvent extract with an aqueous "scrub"

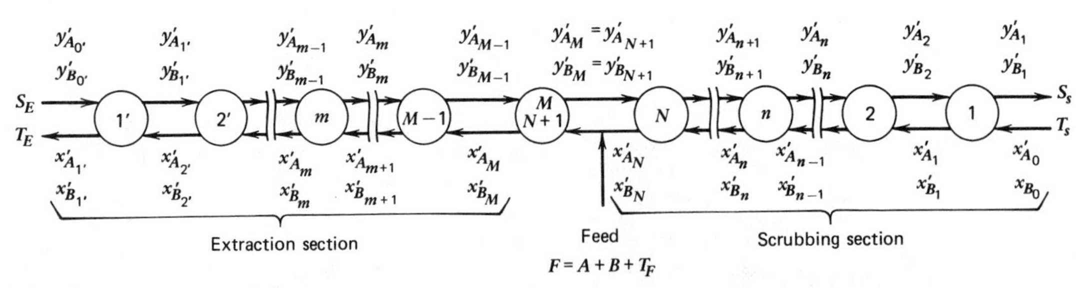

Figure 19. Dual-solvent fractional extraction in stagewise contactor.

solution in inorganic separations. However, when no solvent is present in the feed, as in some organic separations, these terms tend to lose their meaning and it is more appropriate to use the conventional terms *enriching* and *stripping* sections, respectively, in relation to the desired product.

A material balance taken between stages n and $(n + 1)$ of the scrubbing section and around the T inlet gives

$$S_S y'_{n+1} + T_S x'_0 = S_S y'_1 + T_S x'_n \qquad (75)$$

where the solvent flows S_S and T_S are expressed on a solute-free basis and x' and y' are the corresponding concentrations in weight or mole ratio units. The absences of subscripts A and B on the latter indicate that the expression applies to either solute. Solution for y'_{n+1} yields

$$y'_{n+1} = \frac{T_S}{S_S} x'_n + \left(y'_1 - \frac{T_S}{S_S} x'_0\right) \qquad (76)$$

This equation represents operating lines of slope T_S/S_S passing through points (x'_0, y'_1) on the $y' - x'$ diagrams for each component, as shown in Fig. 20. A similar balance for the extraction section gives

$$y'_m = \frac{T_E}{S_E} x'_{m+1} + \left(y'_{0'} - \frac{T_E}{S_E} x'_{1'}\right) \qquad (77)$$

This, similarly, represents lines of slope T_E/S_E on the same diagrams, passing through points $(x'_{1'}, y'_{0'})$; since one or other (but seldom both) solvents are often present in the feed, the flow rates in the two sections are related by

$$S_S = S_E + S_F; \qquad T_E = T_S + T_F \qquad (78)$$

The operating lines and the corresponding equilibrium curves are plotted separately for each component as shown in Fig. 20, and the theoretical stages are stepped off in the usual way. The feed stage compositions must now be matched, assuming for the present that the same number is required for each component. This is done by plotting the compositions against stage number for each component in each section on the same diagram and finding the feed stage by trial, corresponding to the rectangle shown in Fig. 21. If the compositions are expressed in terms of x', that is, for the T phase, the matching must be done at the Mth stage in the extraction and the $(N + 1)$th in the scrubbing section, where the total stage requirement is $(M + N)$.

It is apparent from Fig. 20 that the concentrations of the two components increase from minima at the ends of the contactor to maxima at the feed stage; these profiles are displaced so that the concentration of one component is greater at one end and that of the other is greater at the other end, thus effecting the separation.

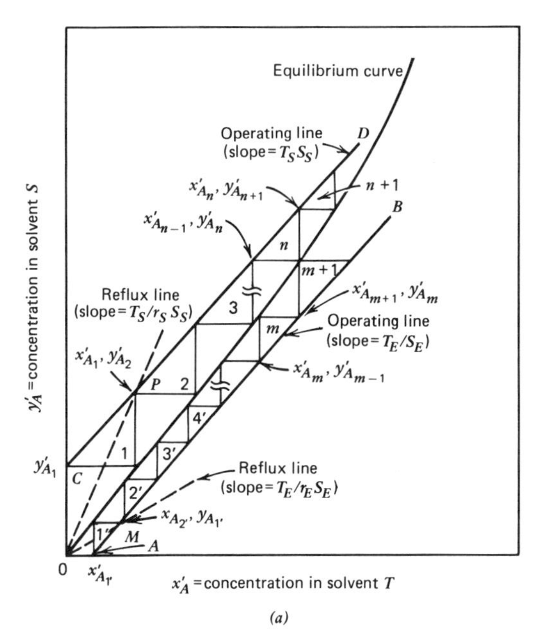

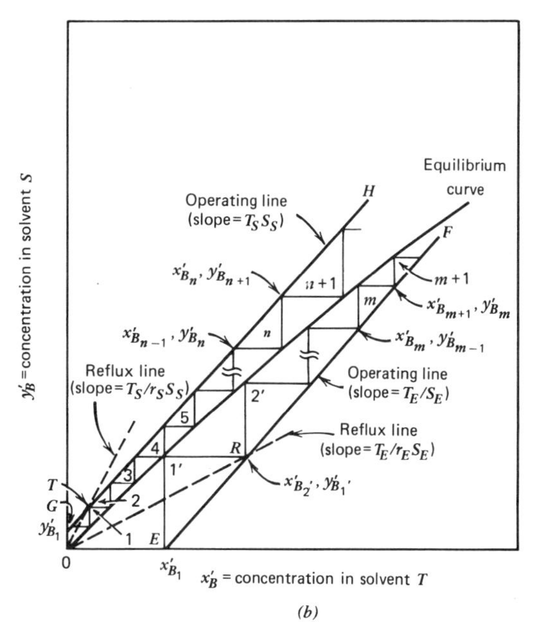

Figure 20. Operating diagrams for dual-solvent fractional extraction of feed components A and B.

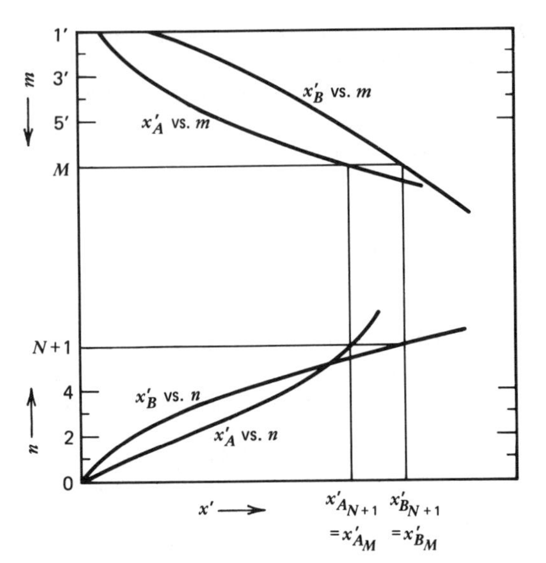

Figure 21. Dual-solvent extraction: matching of feed component concentrations in feed stage.

The effect of reflux on these profiles is described in Section 2.7.7.

2.7.2. *Stagewise Calculation*

The accuracy of the graphical method is limited, especially in "pinched-in" regions where the operating and equilibrium lines are close. To overcome this, stagewise (stage-to-stage) calculations may be carried out from each end of the contactor for each component; thus the equilibrium relationship $y'_k = K_k x'_k$ and the appropriate operating line equations are used alternately to calculate compositions within and between the stages, respectively. As an example, starting from the T inlet with the use of specified values of y'_{k1} for each component, x'_{k1} is given by y'_{k1}/K_k and y'_{k2} is then obtained by using Eq. (76). This procedure is continued for as many stages as required. Similar calculations are then made from the S inlet, calculating alternately $y'_{km} = K_k x'_{km}$ and substituting this in Eq. (77) to obtain $x'_{k,m+1}$.

Since the equilibria are seldom linear, the K_k values are functions of concentration. If the concentration of one component does not influence the K value for the other, it is sufficient to plot $K_k(x'_k)$ as a function of x'_k. If this is not so, the stage calculations must be carried out by use of an iterative procedure, first assuming the K_k values, calculating the corresponding equilibrium compositions, and using these to obtain adjusted values of the K_k.

Such calculations can conveniently be conducted by computer (see Section 2.7.6). If the partition coefficients are not constant, they are best expressed analytically as functions of the concentrations or in terms of activity coefficients, by use of a suitable multicomponent form of the Margules equation, for instance. Alternative methods of carrying out stagewise calculations involving matrix inversion have also been described (see Section 6.3.2).

2.7.3. *Design Parameters*

Before stage calculations can be started, it is necessary to know the concentrations of the exit solvent streams, which depend on the solvent rates chosen. In design these, in turn, depend on the following initial specification of the problem: (1) the solvent-free composition and flow rate of the feed and its solvent content (if any); and (2) the purities of the two products on a solvent-free basis or the purity of one of the products and the recovery of its major component. For such a specification there is a critical condition corresponding to fixed flow rates of each solvent, at which the separation just becomes possible with infinite stages. This corresponds to the minimum reflux condition in distillation.

At higher absolute solvent flows there is increasing flexibility in the permissible range of solvent ratio T_S/S_S (Section 2.7.5). However, a distinct optimum occurs, corresponding to a minimum total number of stages, and a relatively small deviation one way or the other from this leads to infinite stages being required in either scrubbing or extraction section, as shown in Fig. 22. Also, the absolute solvent flows must

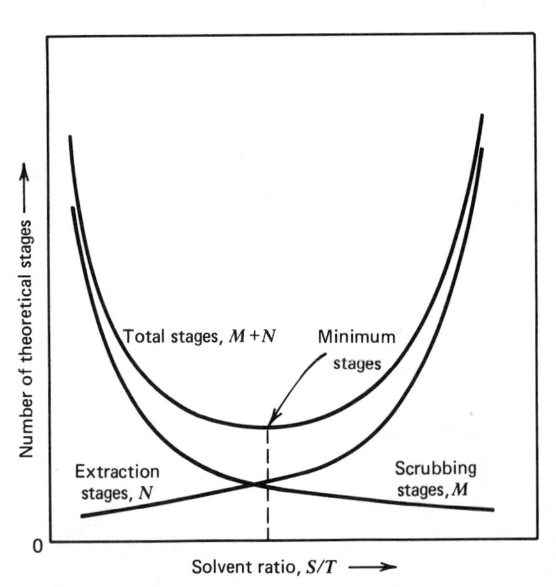

Figure 22. Dual-solvent extraction: effect of solvent ratio on stage number.

be sufficiently high to prevent separation of a solid phase (with inorganic systems) or phase miscibility (with organic systems) in the feed stage, where the total solute concentration is greatest.

Similar considerations apply to the prediction of the performance of a contactor with a given feedstock and number of stages, if the solvent flows and solvent ratio can be selected freely. If, however, these are specified in advance so that the operating lines have predetermined slopes, the problem becomes one of adjusting the intercepts to correspond to the given number of stages.

2.7.4. *Effect of Feed Solvent*

When no solvent is present in the feed, the operating lines for the extraction and scrubbing sections are parallel. The introduction of part of one of the solvents with the feed necessitates more stages in the corresponding scrubbing section; for instance, if solvent T is present in the feed, the slope T_S/S_S of the operating lines *CD* and *GH* in Fig. 20 is reduced and the enrichment per stage is consequently less.

A useful relationship for the coordinates of intersection of the operating lines (x_i', y_i') is obtained by solving Eqs. (76)-(78) for these compositions, as follows:

$$y_i' = \frac{T_S - T_E}{S_S - S_E} x_i' + \frac{S_S y_i' - S_E y_{0'}' - T_S x_0' + T_E x_{1'}'}{S_S - S_E} \quad (79)$$

A material balance on solute $K(= A \text{ or } B)$ around the entire extractor, where K_F is the rate of feed of that component, gives

$$K_F = S_S y_1' - S_E y_{0'}' - T_S x_0' + T_E x_{1'}' \quad (80)$$

A combination of Eqs. (78)-(80) finally gives

$$y_i' = -\frac{T_F}{S_F} x_i' + \frac{K_F}{S_F} \quad (81)$$

In the general case in which both solvents are present in the feed, the operating line locus for each component is a line of slope $-(T_F/S_F)$ with intercepts of x_F'/S_F on the y' axis and x_F'/T_F on the x' axis. Although both solvents are seldom present together in the feed, three special cases can occur in practice, as follows [21]:

Case 1. When $S_F = 0$, $S_S = S_E$ and $x_i' = x_F'$. The operating lines thus intersect on the ordinate $x' = x_F'$, as shown in Fig. 23*a*.

Case 2. When $T_F = 0$, $T_S = T_E$ and $y_i' = y_F$. The operating lines thus intersect on the ordinate $y' = y_F'$ (Fig. 23*b*).

Case 3. When $T_F = S_F = 0$, x_i' and $y_i' \rightarrow \infty$, so that the operating lines are parallel.

In the subsequent treatment case 1 is assumed, for which y_i' is obtained as follows from Eq. (77), assuming the entering solvents to be solute free:

$$y_i' = \frac{T_E}{S}(x_F' - x_1') \quad (82)$$

where the subscript on S has been dropped.

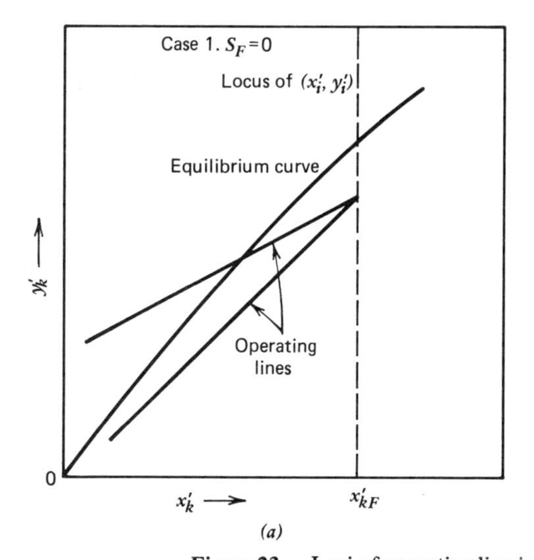

(*a*)

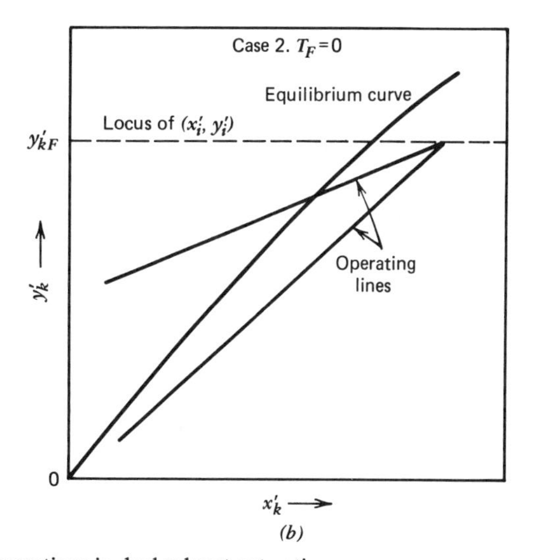

(*b*)

Figure 23. Loci of operating line intersections in dual-solvent extraction.

2.7.5. *Permissible Solvent Rates*

Prochazka and Jiricny [22] have obtained a solution to the design problem based on the following specification of variables:†

$$A = \frac{x'_{BF}}{x'_{AF}}, \qquad Z = \frac{T_E x'_{A1'}}{T_F x'_{AF}}, \qquad B = \frac{y'_{B1}}{y'_{A1}} \tag{83}$$

together with

$$P = \frac{T_S}{T_F}, \qquad 1 + P = \frac{T_E}{T_F}, \qquad R = \frac{S}{T_F} \tag{84}$$

The exit concentrations corresponding to these relations are

$$x'_{A1'} = \frac{x'_{AF} Z}{1 - P} \tag{85}$$

$$x'_{B1} = \frac{x'_{AF}[A - B(1 - Z)]}{1 + P} \tag{86}$$

$$y'_{A1} = \frac{x'_{AF}(1 - Z)}{R} \tag{87}$$

$$y'_{B1} = \frac{x'_{AF} B(1 - Z)}{R} \tag{88}$$

The selected values of A, B, and Z are subject to the following inequality, as may be seen from Eq. (86):

$$\frac{A}{B(1 - Z)} > 1 \tag{89}$$

By use of these definitions, Eq. (82) can be expressed as follows for the two components:

$$y'_{Ai} = \frac{x'_{AF}(1 + P - Z)}{R} \tag{90}$$

$$y'_{Bi} = \frac{x'_{AF}[AF + B(1 - Z)]}{R} \tag{91}$$

Special cases arise when the two operating lines for a component intersect on the corresponding equilibrium line. If this occurs for component A, the corresponding value of R_A is obtained by substituting $y'_{Ai} = K'_A x'_{AF}$ in Eq. (90), giving

$$R_A = \frac{1 + P - Z}{K_A} \tag{92}$$

†This notation applies only in this section (Section 2.7.5).

Similarly, if the intersection with the equilibrium line occurs for component B, substitution of $y'_{Bi} = K_B A x'_{AF}$ in Eq. (91) gives

$$R_B = \frac{AP + B(1 - Z)}{K_B A} \tag{93}$$

Equations (92) and (93) represent two straight lines on a plot of R versus P (Fig. 24); these have been termed the *limiting lines* by Prochazka and Jiricny [22]. These lines intersect at point (P_K, R_K), termed the *critical point*, at which both pairs of operating lines intersect simultaneously on the corresponding equilibrium lines.

A solution to the design problem that satisfies Eqs. (85)-(88) must also satisfy the following requirements:

$$y'_{AM} = y'_{A,N+1} \quad \text{and} \quad y'_{BM} = y_{B,N+1} \tag{94}$$

together with

$$\frac{N_B}{N_A} = \frac{M_B}{M_A} = k_r \tag{95}$$

where N_k and M_k are the numbers of *actual* stages for component k in the scrubbing and the extraction section, respectively. This relation allows for the fact that the stage efficiencies, or the heights of a theoretical stage, may differ for the two solutes.

Prochazka and Jiricny [22] have shown that a sufficient condition for the existence of a solu-

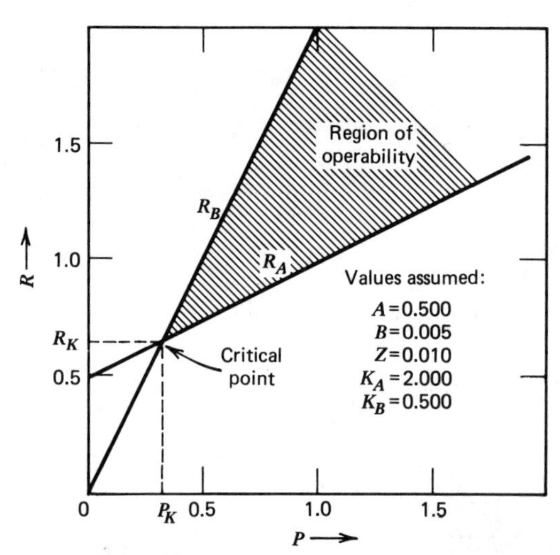

Figure 24. Plot of limiting lines in dual-solvent extraction.

tion is that

$$P > P_K; \qquad R_A \leqslant R \leqslant R_B \tag{96}$$

This corresponds to operation within the hatched area in Fig. 24. Operation at the critical point itself (P_K, R_K) corresponds to the overall minimum solvent rates, requiring infinite stages if $k_r = 1.0$, but this is apparently not necessarily the case for $k_r \neq 1.0$.

2.7.6. Solution by Computer

If the foregoing result is used to select suitable solvent rates, the stage requirements of an extractor can be determined by means of the computer program shown in Fig. 25. In this program y'_{BM} is the iteration variable and is matched against $y'_{B,N+1}$. Thus M_B is calculated for an assumed value of y'_{BM}, and N_B for $y'_{BM} = y'_{B,N+1}$, followed by $y'_{A,N+1}$ for $N_A = N_B/k_r$ and M_A for $y'_{AM} = y'_{A,N+1}$. The value of $(M_A - M_B/k_r)$ is then compared with the specified maximum error, and if this is too great, a new value of $y'_{B,N+1}$ is obtained and the computation is repeated.

Further details are given by Prochazka and Jiricny [22], and the results of one of their typical calculations are given below. The input data, in mass units, were as follows:

$$x'_{AF} = 0.233, \quad Z = 0.010, \quad P = 0.274$$
$$A = 0.098, \quad K_A = 3.645, \quad R = 0.406$$
$$B = 0.002, \quad K_B = 0.630, \quad k_r = 0.5$$

The chosen value of P satisfies the constraint $P > P_K = 0.132$, and that of R lies between $R_A = 0.347$ and $R_B = 0.467$, as required by Eq. (96). Assuming $\delta = 0.1$, the results after three iterations were

$$M_A = 19.17, \quad N_A = 16.18, \quad y'_{AM} = 0.696$$
$$M_B = 9.58, \quad N_B = 8.09, \quad y'_{BM} = 0.014$$

and the actual error was $\delta = 0.000$.

For nonlinear equilibria, the distribution coefficient must be computed for each stage, and if the values for the two components interact, iteration is necessary on each stage, as described in Section 2.7.2. Under such circumstances Eq. (96) can be taken as a guide only, and the equilibrium curves so that the concentration of or chords joining the origin to the intersections

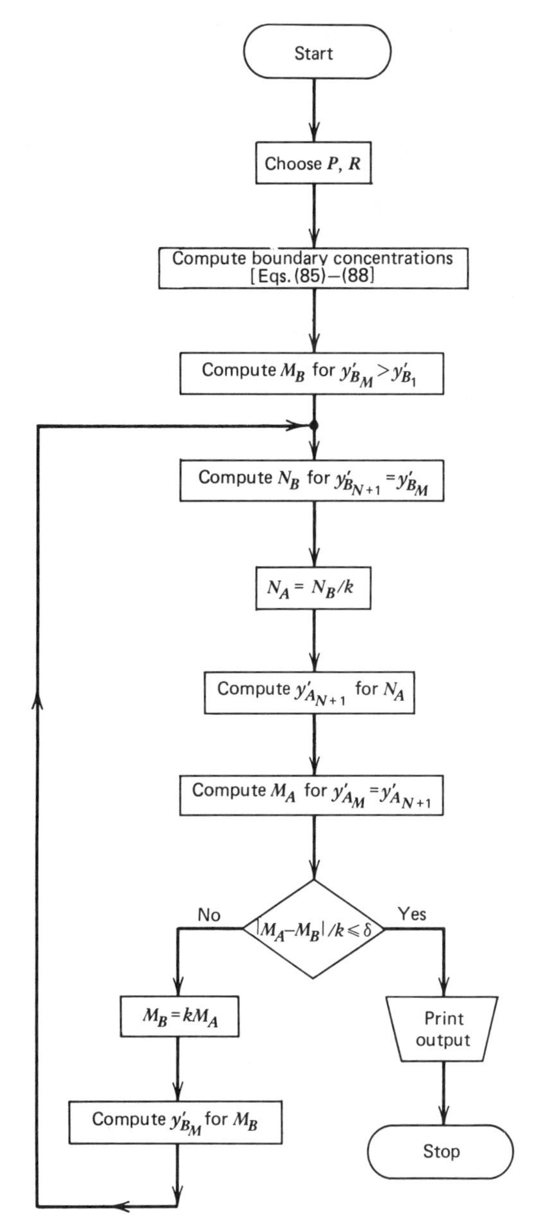

Figure 25. Flow chart for numerical solution of dual-solvent extraction [22].

with the line $x = x_F$ in order to obtain effective values of the K_k.

The solvent: feed flow ratios P and R used in the procedure just described can be varied over considerable limits. However, for a given value of P, there will be a well-defined value of R corresponding to the optimum solvent ratio for which the total number of stages is a minimum. Increase in P while P/R is maintained constant will further reduce the number of stages at the expense of producing more dilute products that require more heat for recovery of solvent; also, a larger cross section will be required to accommodate the increased flows. There will, therefore, be an optimum value of P for which the sum of the operating costs and capital charges is a minimum, and this must be determined by computation over a range of P values.

2.7.7. *Use of Reflux*

External reflux, shown in Figs. 2*c* and 26, can sometimes be beneficial as it increases the concentrations at the ends of the extractor. This may be shown by taking balances for each solute around the solvent removal unit; thus if r_S is the fraction of the solute that is refluxed at the end of the scrubbing section, corresponding to a reflux ratio of $R_S = r_S/(1 - r_S)$, then

$$T_S x'_0 = S r_S y'_1$$

$$\therefore y'_1 = \frac{T_S}{r_S S} x'_0 \qquad (97)$$

This represents straight lines through the origins, of slopes $T_S/r_S S$, in Figs. 20*a* and 20*b*, as shown by lines OP and OT. Similarly, for the end of the extraction section, where a fraction r_E is refluxed

$$y'_{0'} = \frac{r_E T_E}{S} x'_{1'} \qquad (98)$$

This represents the straight lines OM and OR of slopes $r_E T_E/S$ on the same diagrams. The operating lines terminate on these reflux lines, and the stages are determined from the intersection points.

When reflux is used, six cases can arise according to the relative slopes of the reflux and equilibrium lines, as follows [6, 21].

1. **Scrubbing Section**
   a. If the reflux line lies *above* the equilibrium curve, the exit solute concentration in solvent S will be *less* then in the feed stage.
   b. If the reflux line lies *below* the equilibrium curve, the exit solute concentration in solvent S will be *greater* than in the feed stage.
   c. If the reflux line and equilibrium curve have the *same slopes*, the solute concentration will remain constant throughout the section.
2. **Extraction Section**
   a. If the reflux line lies *above* the equilibrium curve, the exit solute concentration in solvent T will be *greater* than in the feed stage.
   b. If the reflux line lies *below* the equilibrium curve, the exit solute concentration in solvent T will be *less* than in the feed stage.
   c. If the reflux line and equilibrium curve have the *same slopes*, the solute concentration will remain constant throughout the section.

Hence, by suitable adjustment of the reflux ratios, a solute concentration can be made to increase progressively from one end of the extractor to the other, to pass through a maximum or a minimum at the feed stage or to remain constant in one or both sections. In the example shown in Fig. 27, the reflux lines lie between the equilibrium curves so that the concentration of component A increases steadily from the S to the T inlet, whereas the opposite occurs for component B.

At total reflux, the reflux and operating lines coincide, the latter passing through the origin when extended. The output of the extractor is

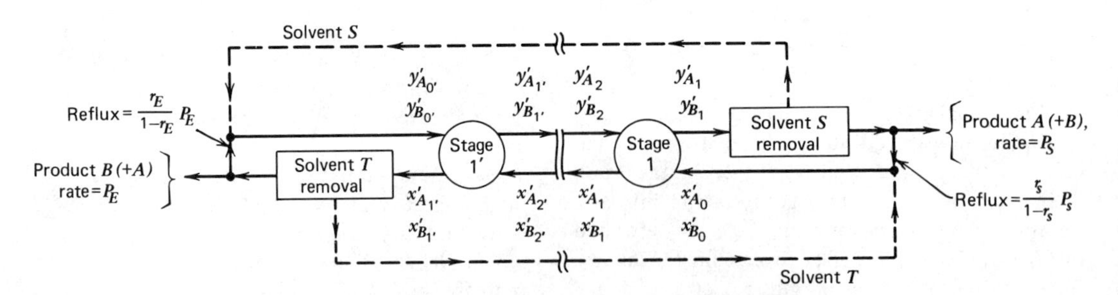

Figure 26. Dual-solvent fractional extraction with reflux.

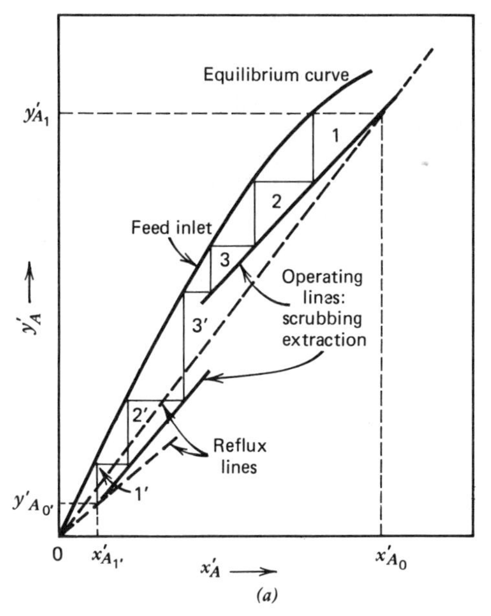

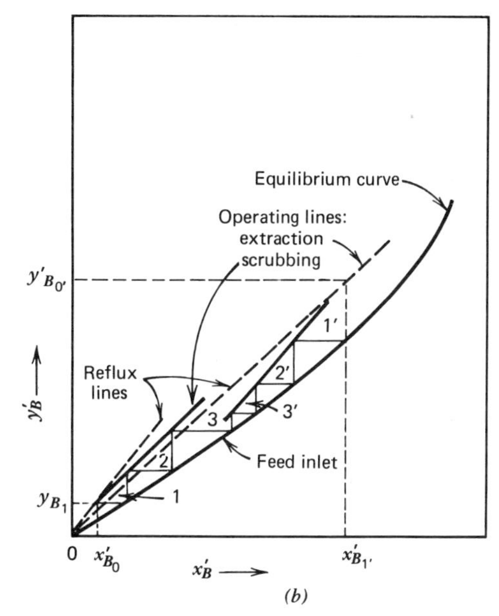

Figure 27. Dual-solvent extraction with reflux lines between equilibrium curves.

then zero and the number of theoretical stages is a minimum and is given as follows by the usual Underwood-Fenske equation:

$$(N+M)_{\min} = \frac{\log\left(\dfrac{y'_{A1}x'_{B1'}}{x'_{A1}y'_{B1}}\right)}{\log\beta} = \frac{\log(P_A/P_B)}{\log\beta} \tag{99}$$

where P_k is the ratio of the amounts of a solute recovered from solvents S and T, respectively (the *rejection ratio*) and $\beta = K_A/K_B$. This relationship is valid only when the partition coefficients are constant.

Although the use of reflux will generally decrease the number of theoretical stages required [2], it will increase the solvent requirements if the feed stage concentrations must be kept below some maximum value; thus the cross-section must be increased. In addition, the cost of solvent removal must be taken into account; hence the problem of optimization is very complex. In this regard the use of a computer program [22] is very helpful, but the assumption that Eq. (96) applies may not be valid when reflux is used.

2.7.8. *Analytical Methods*

When the partition coefficients are constant, the stage calculations can also be carried out analytically by using suitably modified forms of the expression for the extraction efficiency E_{Sx} in Table 1. Thus for the extraction section, this gives, on putting $q = 0$,

$$\frac{x'_M - x'_{1'}}{x'_M - y'_{0'}/K} = \frac{\mathbf{S}_E^M - \mathbf{S}_E}{\mathbf{S}_E^M - 1} \tag{100}$$

where $S_E = KS_E/T_F$. On substituting for $y'_{0'}$ from Eq. (98) and solving for x'_M, this becomes

$$x'_M = \frac{x'_{1'}[\mathbf{S}_E^M - 1 - r_E(\mathbf{S}_E^{M-1} - 1)]}{(\mathbf{S}_E - 1)} \tag{101}$$

A similar expression is obtained for the scrubbing section, as follows:

$$y'_{N+1} = \frac{y'_1[1 - \mathbf{S}_S^{N+1} - r_S\mathbf{S}_S(1 - \mathbf{S}_S^N)]}{\mathbf{S}_S^N(1 - \mathbf{S}_S)} \tag{102}$$

where $\mathbf{S}_S = KS_S/T_S$. At the feed stage

$$y'_{N+1} = Kx_M \tag{103}$$

Substitution of Eqs. (101) and (102) into Eq. (103) gives two equations, one for each component, which in principle can be solved for M and N. This is difficult, however, but Eqs. (101) and (102) can be used to calculate the stage compositions for matching graphically at the feed stage, as in Fig. 21. A number of special cases is discussed in the following paragraphs.

(a) Pure Solvents; No Reflux. Equations (101)–(103) give, on putting $r_E = r_S = 0$

$$M = \frac{\log Z_A}{\log \mathbf{S}_{E,A}} = \frac{\log Z_B}{\log \mathbf{S}_{E,B}} \tag{104}$$

where for $S_F = 0$

$$Z_k = \frac{P_k(\mathbf{S}_E - 1)(1 - \mathbf{S}_S^{N+1})}{\mathbf{S}_E \mathbf{S}_S^N (1 - \mathbf{S}_S)} + 1 \tag{105}$$

and for $T_F = 0$

$$Z_k = \frac{P_k(\mathbf{S}_E - 1)(1 - \mathbf{S}_S^{N+1})}{\mathbf{S}_S^{N+1}(1 - \mathbf{S}_E)} + 1 \tag{106}$$

In these expressions P_k, the *rejection ratio* for feed solute K, is given by

$$P_k = \frac{p_k}{1 - p_k} = \frac{S_S y'_{k1}}{T_E x'_{k1'}} \tag{107}$$

where p_k is the fraction of the feed solute that is withdrawn with the solvent S from the scrubbing section. Equation (104) with (105) or (106) provides two expressions, one for each solute, that can be solved by trial for N and M.

(b) With Reflux: Neither Solvent Present in Feed. In this case $\mathbf{S}_E = \mathbf{S}_S = \mathbf{S}$, and Eqs. (101)–(103) become

$$M = \frac{\log Z_A}{\log \mathbf{S}_A} = \frac{\log Z_B}{\log \mathbf{S}_B} \tag{108}$$

where

$$Z_k = \frac{(1 - r_E)}{(\mathbf{S} - r_E)} \cdot \left(\mathbf{S}(P_k + 1) + \frac{P_k(r_S \mathbf{S} - 1)}{\mathbf{S}^N (1 - r_S)} \right) \tag{109}$$

and P_k is now given by

$$P_k = \frac{p_k}{1 - p_k} = \frac{S y'_{k1}(1 - r_S)}{T x'_{k1'}(1 - r_E)} \tag{110}$$

The two expressions provided by Eq. (108) can be solved by trial for M and N.

(c) As (b), Without Reflux. A combination of Eqs. (108) and (109) gives, on setting $r_E = r_S = 0$ and solving for P_k [23, 24]

$$P_k = \frac{\mathbf{S}^{N+1}(\mathbf{S}^M - 1)}{\mathbf{S}^{N+1} - 1} \tag{111}$$

It can be shown from Eq. (111) that $P_k = M/(N + 1)$ for a solute for which $\mathbf{S} = 1$, so that in a center-fed contactor such a solute would be divided equally between the two products. On the other hand, when $\mathbf{S} \gg 1, P_k \simeq \mathbf{S}^M$, in which case most of the solute would leave with solvent $\mathbf{S}$ from the end of the scrubber; and when $S \ll 1$, $1/P_k \simeq \mathbf{S}^{N+1}$, in which case most of the solute would leave with solvent T from the end of the extraction section.

The total number of stages and the feed point location are strongly influenced by the solvent ratio S/T, as shown in Fig. 22. Klinkenberg et al. [23, 24] have described six cases that permit easy calculation of this diagram so that the minimum number of stages can be determined (see also Refs. 6 and 25). Of these, their case 4, for $M = N$ corresponding to $P_A = \mathbf{S}_A^N$ and $1/P_B = \mathbf{S}_B^N$, is of particular interest, giving

$$M + N = \frac{2\log (P_A/P_B)}{\log \beta} - 1 \tag{112}$$

This is twice the minimum number of stages at total reflux as given by Eq. (99), less one, and is identical with the optimum number of stages for an ideal tapered cascade of *separating elements* [26], such as gaseous diffusion cells and mixer-settler stages. A result identical to Eq. (112), but with a different feed point, is obtained for the case where $\mathbf{S}_A \mathbf{S}_B = 1$, corresponding to a symmetrical separation (i.e., for which $p_A = q_B$). This requires a somewhat higher T/S ratio, and the optimum generally lies between the two values.

The preceding six cases have been summarized by Treybal [6] and Alders [25], and Treybal has given six additional cases for equal reflux ratios, that is, for $r_S = r_E$. A graphical solution for determining the minimum total stages, feed location, and solvent ratio has been described by Klinkenberg [27], and Scheibel [28] has proposed an empirical expression for calculating the minimum number of stages.

3. HEAT TRANSFER IN LIQUID–LIQUID CONTACTORS

3.1. Applications

Liquid–liquid contactors can be used for the transfer of heat between two immiscible liquids.

This results in a considerable cost reduction, as compared with conventional heat exchangers, as a result of the replacement of fixed heat-transfer surface by interfacial surface. Woodward [29] proposed the use of spray towers in water desalting plants for heat exchange between feed water and the exit product and concentrated brine streams; two towers are required, with a water-insoluble heat-transfer liquid circulating between them, as shown in Fig. 28.

Woodward's original proposal involved the use of spray towers operating in the so-called dense-packed region, that is, at very high holdup, above the normal flood-point value; but an attempt to scale up from a laboratory size unit was later reported to have failed [30]. Other workers, particularly Letan and Kehat [31], have continued to study the heat-transfer characteristics of such contactors and have confirmed the major adverse effect on performance of backmixing [32] that has been observed by others for mass transfer (see Chapter 6).

The attraction of spray towers for this purpose probably arose as much from their nonfouling characteristics and ease of cleaning as from low capital cost. However, other types of liquid-liquid contactor are equally suitable for heat transfer, and compromise designs that combine

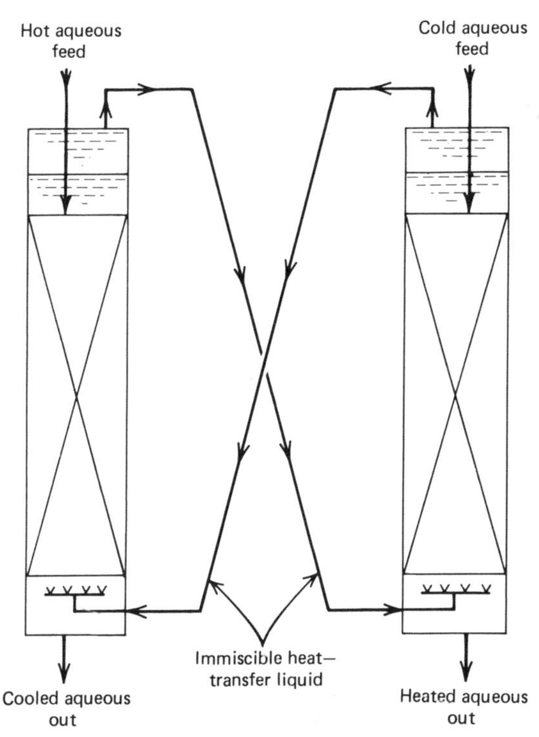

Figure 28. Arrangement of contactors for heat exchange between two aqueous streams.

a reasonable ease of cleaning and relatively low capital cost with greatly reduced backmixing, and hence an acceptable performance, are possible. The design of contactors for heat transfer closely parallels that for mass transfer and the various available methods are reviewed briefly below in this context, neglecting backmixing effects.

3.2. Design Methods for Heat Transfer

The equilibrium relation for heat transfer takes the simple form $t_x = t_y$ and is represented on a t_y-t_x plot by a line of slope unity through the origin. An operating line can be drawn on the same diagram to relate the temperatures of the two phases between stages or, for a differential column, at any cross section. This is obtained by taking a heat balance around one end of the contactor giving, for a differential contactor, on solving for t_y [cf. Eq. (34)]:

$$t_y = \frac{G_x c_{sx}}{G_y c_{sy}} t_x + \left(t_{yI} - \frac{G_x c_{sx}}{G_y c_{sy}} t_{xI}\right) \tag{113}$$

For stagewise contactors, t_{yI} is replaced by $t_{y,N+1}$ and t_{xI} by $t_{x,N}$. Since solute transfer does not occur, the flows G_j are constant unless the miscibility changes appreciably with temperature. Equation (113) thus represents a straight line on the t_y-t_x diagram, and equilibrium stages can be stepped off between this and the equilibrium line in the usual way.

For differential contactors, thermal transfer units can be defined in the same way as for mass transfer. Thus a heat balance across a height dz, assuming the X phase to enter at the higher temperature, gives

$$-G_x c_{sx}\, dt_x = h_o a(t_x - t_y)\, dz \tag{114}$$

where h_o is the overall heat-transfer coefficient. Integration between $z = 0$, $t_x = t_{x0}$, and $z = L$, $t_x = t_{xI}$ gives

$$N_{Tox}^t = \int_{txI}^{t_{x0}} \frac{dt_x}{t_x - t_y} = \frac{h_o aL}{G_x c_{sx}} \tag{115}$$

where N_{Tox}^t is the number of overall thermal transfer units on an X-phase basis.

Similarly, a height of an overall thermal transfer unit can be defined by

$$H_{Tox}^t = \frac{G_x c_{sx}}{h_o a} \tag{116}$$

Corresponding expressions are obtained in terms of the Y phase as follows:

$$N_{Toy}^t = \int_{t_{yI}}^{t_{yo}} \frac{dt_y}{t_x - t_y} = \frac{h_o a L}{G_y c_{sy}} \tag{117}$$

$$H_{Toy}^t = \frac{G_y c_{sy}}{h_o a} \tag{118}$$

The analytical solutions obtained earlier for mass transfer in terms of dimensionless concentrations, together with the corresponding graphical solutions, can be used directly for heat transfer by expressing the temperature in dimensionless form. For this purpose, again assuming the X phase to be the "feed" (i.e., higher-temperature) phase, the required definitions are as follows:

$$T_x = \frac{t_x - t_{yI}}{t_{x0} - t_{yI}}; \qquad T_y = \frac{t_y - t_{yI}}{t_{x0} - t_{yI}} \tag{119}$$

The thermal stripping factor $\mathbf{S}^t = G_y c_{sy}/G_x c_{sx}$, with T_{xN} or T_{xI}, can be used directly in place of $\mathbf{S}$ and X_N or X_I in the expressions for N_S and N_{Toj} in Table 1 and in Figs. 14 and 16 to obtain N_S^t and N_{Toj}^t, and in the expressions for extractor efficiency in the same Table and in Figs. 15 and 17 to obtain the corresponding thermal efficiencies.

Values of H_T^t can be predicted from those of H_T and vice versa by use of the analogy between heat- and mass-transfer coefficients described in Chapter 3. This analogy is used to relate the individual phase values of the coefficients, that is, h_x to k_x and h_y to k_y, after which the overall values are obtained in the usual way.

4. STAGE EFFICIENCY

4.1. Definitions

In batch extraction the theoretical stage concept is valid as long as the phases are contacted closely for an adequate time and an excessively slow reaction step is not involved. On the other hand, under continuous flow conditions the form of the residence time distribution is such that the departure from equilibrium must be appreciable at realistic flow rates. For this reason, it is necessary to define a *stage efficiency*, analogous to the plate efficiency in distillation, which relates the performance of a real stage to that of an ideal theoretical stage.

As pointed out by Standart [33] in regard to distillation columns, there are two separate aspects of efficiency. In terms of extraction, these are (1) its use to differentiate the behavior of a real extractor from that of an ideal one, and (2) the interpretation of the efficiency in terms of the actual mechanisms of mass transfer. Several definitions of stage efficiency have been proposed, as follows. Since liquid extraction is essentially an isothermal process, stage efficiency can be more precisely defined than in the case of distillation [33].

The present account is concerned primarily with the case where only one solute is present. When two or more solutes are present, the situation is greatly complicated by coupling between the motions of the diffusing species. The theory of multicomponent mass transfer is described in Chapter 3, Section 3.4; however, its application in terms of stage efficiency has yet to be fully resolved.

4.1.1. Overall Efficiency

The overall efficiency E_o of an extractor is defined in a simple manner as the ratio of the number of ideal to real stages required to accomplish the same duty, that is, the same concentration change with the given flows; thus

$$E_o = \frac{N_{S,\,\text{ideal}}}{N_{S,\,\text{real}}} \tag{120}$$

Although convenient for design purposes, this efficiency is meaningful only for a linear equilibrium relationship. Its relation to other efficiencies under such circumstances is given in Table 2.

4.1.2. Murphree Efficiencies

The Murphree efficiency [34], which can be expressed in terms of either the X or Y phase, is defined as the ratio of the actual concentration change of that phase within the stage to the change that would have occurred if equilibrium had been reached. This is illustrated in Fig. 29 by using mass (or mole) fraction units (volumetric units are also applicable); hence

$$E_{Mx} = \frac{x_{n-1} - x_n}{x_{n-1} - x_n^*} \tag{121}$$

and

$$E_{My} = \frac{y_n - y_{n+1}}{y_n^* - y_{n+1}} \tag{122}$$

TABLE 2 RELATIONSHIPS BETWEEN EFFICIENCIES

Efficiency Required	In Terms of	X Phase	Y Phase
		Individual Stage	
E_{Mx}, E_{My}	E_{My}, E_{Mx}	$E_{Mx} = \dfrac{\mathbf{S}E_{My}}{1 + E_{My}(\mathbf{S} - 1)}$	$E_{My} = \dfrac{\mathbf{E}E_{Mx}}{1 + E_{Mx}(\mathbf{E} - 1)}$
E_H	E_{Mx}, E_{My}	$E_H = \dfrac{E_{Mx}(1 + \mathbf{E})}{\mathbf{E}E_{Mx} + 1}$	$E_H = \dfrac{E_{My}(1 + \mathbf{S})}{\mathbf{S}E_{Mx} + 1}$
E_{Mx}, E_{My}	E_H	$E_{Mx} = \dfrac{E_H}{\mathbf{E}(1 - E_H) + 1}$	$E_{My} = \dfrac{E_H}{\mathbf{S}(1 - E_H) + 1}$
		Crossflow Extractors	
E_o	E_{Mx}, E_{My}	$E_o = \dfrac{\log \{(E_{Mx} + \mathbf{S})/[E_{Mx}(1 - \mathbf{S}) + \mathbf{S}]\}}{\log (1 + \mathbf{S})}$	$E_o = \dfrac{\log (\mathbf{S}E_{My} + 1)}{\log (1 + \mathbf{S})}$
E_o	E_H	$E_o = \dfrac{\log [1 + \mathbf{S}(1 - E_H)]}{\log (1 + \mathbf{S})}$	
		Countercurrent Extractors	
E_o	E_{Mx}, E_{My}	$E_o = \dfrac{\log [1 + E_{Mx}(\mathbf{E} - 1)]}{\log \mathbf{E}}$	$E_o = \dfrac{\log [1 + E_{My}(\mathbf{S} - 1)]}{\log \mathbf{S}}$
E_o	E_H	$E_o = \dfrac{\log \{(1 + \mathbf{S} - E_H)/[1 + \mathbf{S}(1 - E_H)]\}}{\log \mathbf{S}}$	

These efficiencies relate to overall stage values; that is, the concentrations x_{n-1} and y_{n+1} refer to those of the streams entering the stage and x_n, y_n, to those leaving. A pseudoequilibrium (i.e., efficiency) line can thus be drawn, using Eq. (121) or (122) to obtain the coordinates x_n, y_n, with the stages stepped off as shown.

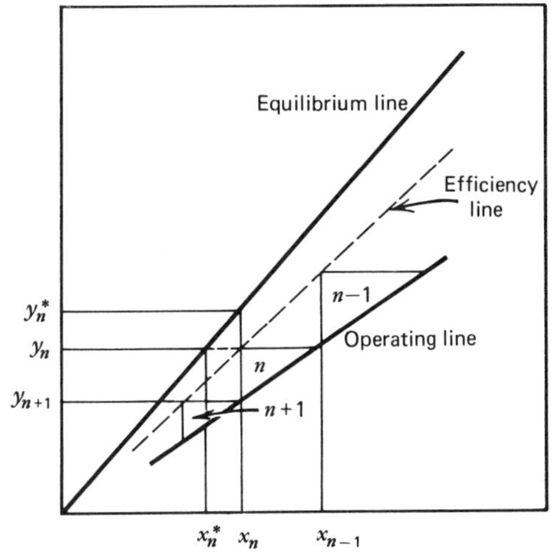

Figure 29. Murphree efficiency.

In most mixer–settlers the phases are either fully mixed or pass through the mixer concurrently; the end result is the same in both cases, and Eqs. (121) and (122) apply. However, it is recognized that on distillation plates an additional enrichment can occur as a result of the crossflow of the liquid with respect to the vapor. In such circumstances it is necessary to define *local* (or *point*) efficiencies as follows:

$$E_{Mx}^1 = \frac{x_{n+1} - x}{x_{n-1} - x^*} \tag{123}$$

$$E_{My}^1 = \frac{y - y_{n+1}}{y^* - y_{n+1}} \tag{124}$$

where x, y refer to concentrations at points along the liquid path and x^*, y^*, to the corresponding equilibrium values. Such an effect would be less commonly encountered in liquid extraction, although it could appear with unagitated perforated plate columns in which cross flow of the continuous phase occurs to some degree.

Lewis [35] derived theoretical relations between point and overall plate efficiencies for distillation that could be applied to extractors of this type. The results are available as plots of overall versus point efficiency with the stripping factor as parameter [35, 36].

4.1.3. *Hausen Efficiency*

Hausen [37] has proposed an alternative definition of plate efficiency in distillation, based on the individual plate operating line. In terms of liquid extraction, this is obtained by taking a balance on solute around the stage. If changes in flow are neglected, this gives

$$N_A = G_x(x_1 - x_2) = G_y(y_2 - y_1)$$

$$\therefore \frac{y_1 - y_2}{x_1 - x_2} = -\frac{G_x}{G_y} \qquad (125)$$

This equation is represented as line AB on Fig. 30. If the exit streams had come to equilibrium, they would have had solute concentrations x_e and y_e, corresponding to point C, with an amount N_{Ae} of solute transferred. The Hausen stage efficiency E_H is then defined as

$$E_H = \frac{N_A}{N_{Ae}} = \frac{x_1 - x_2}{x_1 - x_e}$$

$$= \frac{y_2 - y_1}{y_e - y_1} = \frac{AB}{AC} \qquad (126)$$

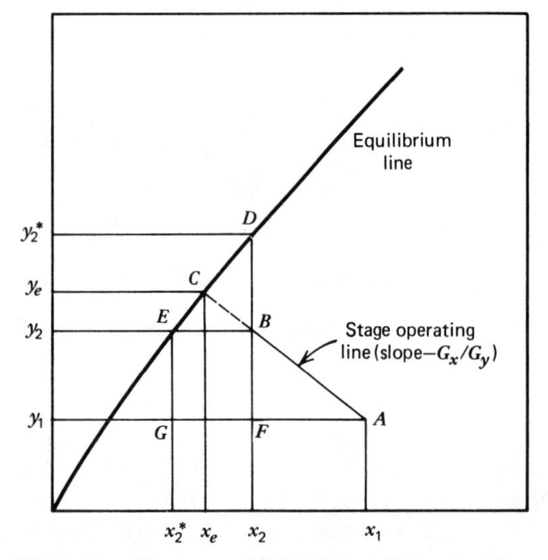

Figure 30. Hausen and Murphree efficiencies for a single stage.

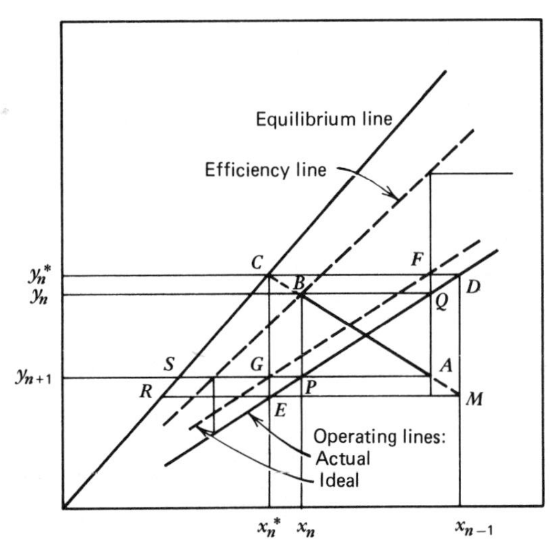

Figure 31. Hausen efficiency for countercurrent extraction.

The Hausen efficiency has the advantage of being independent of the phase on which it is based; that is, E_{Mx} is represented by AF/AG and E_{My}, by FB/FD.

Unfortunately, the Hausen efficiency suffers from a disadvantage when applied to a countercurrent cascade. This is shown in Fig. 31, where the streams x_{n-1} and y_{n+1} enter the real stage PBQ at A and leave at B. For retention of the same inlet concentrations, the corresponding equilibrium step must be displaced to GCF and the extractor operating line to FG, thus necessitating a change in the raffinate (or extract) product composition. Alternatively, if the true operating line is retained so that the equilibrium stage is represented by ECD, the stage efficiency becomes AB/MC; in this case the driving force at the stage inlet on an X-phase basis corresponds to MR instead of AS and is incorrect.

Therefore, although the Hausen efficiency is perfectly satisfactory for a single stage, it is somewhat unrealistic when applied to countercurrent multistage processes. Relationships between the Hausen and other efficiencies as given by Treybal [6] and Ho and Prince [38] are included in Table 2.

4.2. Transfer Units and Efficiency

The performance of a mixer–settler stage or a single compartment of a stagewise column can itself be expressed in terms of transfer units. The operating diagram for such a stage is shown in Fig. 32a, where the feed entering stage n at point D comprises X phase of composition c_{x1} from

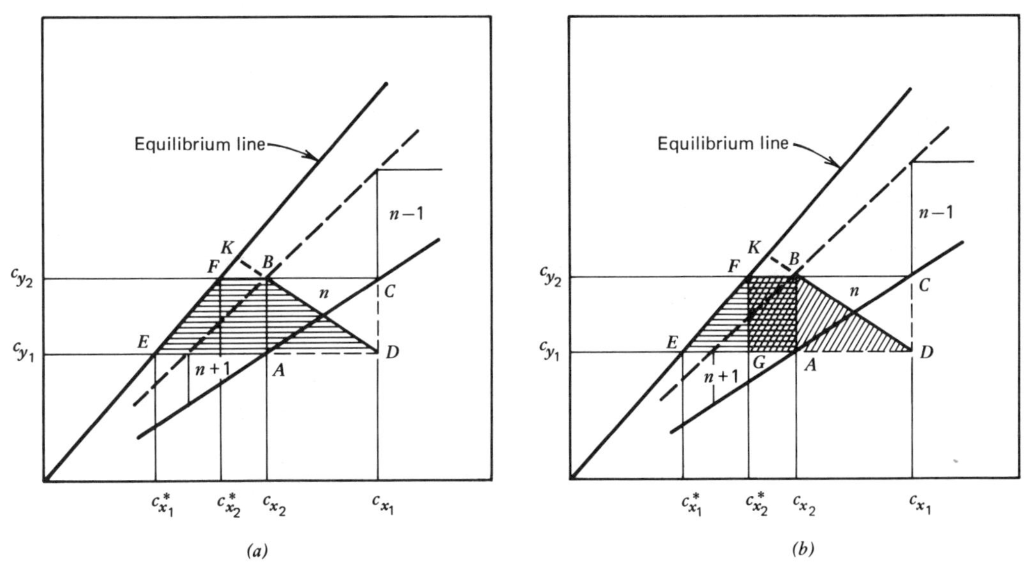

Figure 32. Range of driving force for (*a*) plug flow stage, (*b*) complete backmixing of X phase (area $AEFB$) and Y phase (area $DGFB$).

stage $(n - 1)$ and Y phase of composition c_{y1} from stage $(n + 1)$. The compositions of the phases leaving the stage lie on the stage operating line DK of slope $-(U_x/U_y)$ [see Eq. (125)]. The rate of mass transfer in a section of length dh is given by Eq. (46), neglecting changes in flow rate:

$$dN_A = -U_x\, dc_x = k_{ox}a(c_x - c_x^*)\, dh \quad (127)$$

If it is initially assumed that the phases pass through the stage in cocurrent plug flow, the compositions follow the operating line DB and the driving force changes progressively from DE to BF, as shown by the hatched area $EDBF$. Integration of Eq. (127) on this basis thus gives

$$N^1_{ox,P} = \int_{c_{x2}}^{c_{x1}} \frac{dc_x}{c_x - c_x^*} = \frac{c_{x1} - c_{x2}}{(c_x - c_x^*)_{\text{lm}}} \quad (128)$$

$$= \frac{k_{ox}ah_c}{U_x} \quad (129)$$

where $(c_x - c_x^*)_{\text{lm}}$ is the logarithmic mean of $(c_{x1} - c_{x1}^*)$ and $(c_{x2} - c_{x2}^*)$ and N^1_{ox} is the number of overall transfer units in the stage, with subscript P referring to plug flow.

4.2.1. *Effect of Mixing Within Stage*

In practice, considerable backmixing of either or of both phases is likely. If the Y phase is dispersed, backmixing is likely to be more severe in the X phase unless coalescence and redispersion of the Y phase is very rapid. If the backmixing of the X phase is complete, its composition will remain constant throughout the stage at c_{x2}, in which case integration of Eq. (127) gives

$$N^1_{ox,X} = \frac{(c_{x1} - c_{x2})}{(c_{x2} - c_x^*)_{\text{lm}}} = \frac{k_{ox}ah_c}{U_x} \quad (130)$$

where $(c_{x2} - c_x^*)_{\text{lm}}$ is the logarithmic mean of $(c_{x2} - c_{x1}^*)$ and $(c_{x2} - c_{x2}^*)$. The range of driving force corresponds to the hatched area $AEFB$ in Fig. 32*b*. This result can be put into another form by expressing c_x^* in Eq. (127) in terms of c_y by using the linear equilibrium relationship (Table 1) and noting that $dc_x = U_y/U_x\, dc_y$. On performing the integration and expressing c_{x2} in terms of c_{y2}^*, one obtains

$$N^1_{ox,X} = -\mathbf{S} \ln\left(\frac{c_{y2}^* - c_{y2}}{c_{y2}^* - c_{y1}}\right)$$

$$= -\mathbf{S} \ln\left(1 - \frac{c_{y2} - c_{y1}}{c_{y2}^* - c_{y1}}\right)$$

$$= -\mathbf{S} \ln(1 - E_{My}) \quad (131)$$

If, on the other hand, the X phase is dispersed and the Y phase is completely backmixed so that its composition remains constant at c_{y2}, c_x^* has the constant value c_{x2}^* throughout the stage. The range of driving force then corresponds to the area *DGFB* on Fig. 32*b*, giving an average value on integration of $(c_x - c_{x2}^*)_{\text{lm}}$, the logarithmic mean of $(c_{x1} - c_{x2}^*)$ and $(c_{x2} - c_{x2}^*)$.

Finally, if both phases are completely backmixed, the driving force is constant at $(c_{x2} - c_{x2}^*)$ throughout the stage, and

$$N_{ox,B}^1 = \frac{c_{x1} - c_{x2}}{c_{x2} - c_{x2}^*} = \frac{k_{ox}ah_c}{U_x} \tag{132}$$

$$= \frac{E_{Mx}}{1 - E_{Mx}} \tag{133}$$

When the Y-phase resistance is controlling, it is preferable to use the corresponding relationships to the preceding ones in terms of N_{oy}^1. These can be obtained in the same manner, and the results for all eight cases are summarized in Table 3. In Table 3 the expressions for N_{ox}^1 and N_{oy}^1 are given in terms of dimensionless concentrations, defined over the stage so that the inlet concentrations are $X_1 = 1.0$ and $Y_1 = 0$; solutions are also given for the stage exit concentrations X_2 or Y_2.

4.2.2. *Stage Effectiveness*

In all four cases discussed previously, N_{ox}^1 is obtained in two forms: (1) in terms of the specific transfer rate, that is, as $k_{ox}ah_c/U_x$, which is the *same* in all cases; and (2) as the ratio of the concentration change over the stage to the mean driving force, which *differs* for each case. The first form is thus a measure of the potential performance of the stage as determined mainly by the degree of dispersion of the phases,

TABLE 3 RELATIONSHIPS BETWEEN STAGE TRANSFER UNITS, EXIT COMPOSITIONS, AND EFFICIENCIES

Case Number and Description	X Phase	Y Phase
Case 1—Piston flow	$N_{oxP}^1 = \dfrac{-\ln[X_2(\mathbf{E}+1) - \mathbf{E}]}{\mathbf{E}+1}$	$N_{oyP}^1 = \dfrac{-\ln[1 - Y_2(\mathbf{S}+1)]}{\mathbf{S}+1}$
	$X_2 = \dfrac{\mathbf{E} + \exp[-N_{oxP}^1(\mathbf{E}+1)]}{\mathbf{E}+1}$	$Y_2 = \dfrac{1 - \exp[-N_{oyP}^1(\mathbf{S}+1)]}{\mathbf{S}+1}$
Case 2—X phase completely backmixed	$N_{oxX}^1 = \dfrac{1}{\mathbf{E}} \ln\left[\dfrac{X_2}{X_2(\mathbf{E}+1) - \mathbf{E}}\right]$	$N_{oyX}^1 = \ln\left[\dfrac{1 - \mathbf{S}Y_2}{1 - Y_2(\mathbf{S}+1)}\right]$
	$= -\mathbf{S}\ln(1 - E_{My})$	$= -\ln(1 - E_{My})$
	$X_2 = \dfrac{\mathbf{E}\exp(EN_{oxX}^1)}{(1+\mathbf{E})\exp(\mathbf{E}N_{oxX}^1) - 1}$	$Y_2 = \dfrac{\exp(N_{oyX}^1) - 1}{(\mathbf{S}+1)\exp(N_{oyX}^1) - \mathbf{S}}$
Case 3—Y phase completely backmixed	$N_{oxY}^1 = \ln\left[\dfrac{1 - \mathbf{E}(1 - X_2)}{X_2(\mathbf{E}+1) - \mathbf{E}}\right]$	$N_{oyY}^1 = \dfrac{1}{\mathbf{S}} \ln\left[\dfrac{1 - Y_2}{1 - Y_2(\mathbf{S}+1)}\right]$
	$= -\ln(1 - E_{mx})$	$= -\mathbf{E}\ln(1 - E_{Mx})$
	$X_2 = \dfrac{1 + \mathbf{E}[\exp(N_{oxY}^1) - 1]}{(1+\mathbf{E})\exp(N_{oxY}^1) - \mathbf{E}}$	$Y_2 = \dfrac{\exp(\mathbf{S}N_{oyY}^1) - 1}{(\mathbf{S}+1)\exp(\mathbf{S}N_{oyY}^1) - 1}$
Case 4—Both phases completely backmixed	$N_{oxB}^1 = \dfrac{1 - X_2}{X_2(\mathbf{E}+1) - \mathbf{E}}$	$N_{oyB}^1 = \dfrac{Y_2}{1 - Y_2(\mathbf{S}+1)}$
	$= \dfrac{E_{Mx}}{1 - E_{Mx}}$	$= \dfrac{E_{My}}{1 - E_{My}}$
	$X_2 = \dfrac{\mathbf{E}N_{oxB}^1 + 1}{N_{oxB}^1(\mathbf{E}+1) + 1}$	$Y_2 = \dfrac{N_{oyB}^1}{N_{oyB}^1(\mathbf{S}+1) + 1}$

whereas the second is based on the actual flow pattern that prevails.

The effect of flow pattern is illustrated in Fig. 33, in which the exit composition X_2 is plotted against N_{ox}^1, with E = 0.50, for each of the four modes of operation by using the relationships given in Table 3; it is clear that plug flow gives the best, and totally mixed flow the worst performance. However, the fraction extracted, given by $(1 - X_2)$, does not differ greatly for the various cases; thus comparison of the completely mixed and plug flow cases reveals the difference to reach only about 23% for $N_{ox}^1 = 1.0$, decreasing as N_{ox}^1 is further increased.

In formulating models of stagewise extraction, including the backflow model to be described in Chapter 6, it is usual to assume that both phases are completely mixed before separation so that the driving force is constant throughout the stage at the exit value of $(c_{x2} - c_{x2}^*)$. This implicitly assumes that the dispersed phase behaves as a second continuous phase, that is, that coalescence and redispersion are very rapid. Miyauchi and Vermeulen [39] have shown on theoretical grounds that this assumption is permissible provided the droplet size, holdup, and overall mass-transfer coefficient are all constant, and that a linear equilibrium relationship applies. Support for this result is provided by Fig. 33, which shows that differences in stage outlet concentrations for the four limiting cases are relatively small.

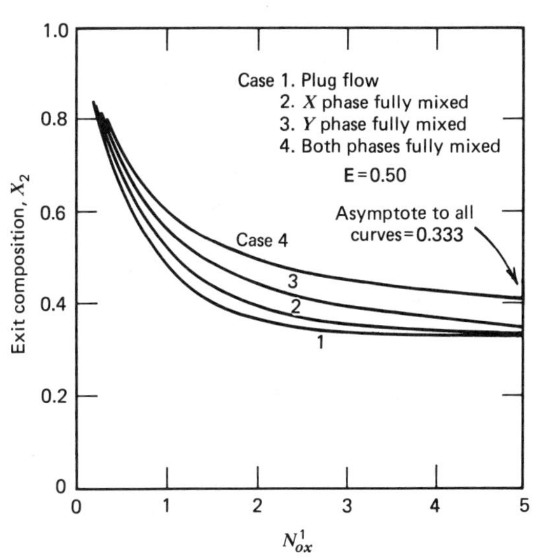

Figure 33. Dimensionless exit composition of X phase from a single stage as a function of number of transfer units and degree of backmixing.

Indices N_{ox}^1 and N_{oy}^1 provide useful measures of the effectiveness of a single stage, but since these can vary from zero up to infinity for an ideal theoretical stage, it is preferable to express them in terms of a fully mixed transfer unit efficiency E_{Tj}, which varies from zero to unity:

$$E_{Tj} = \frac{N_{oj,B}^1}{N_{oj,B}^1 + 1} \tag{134}$$

As shown in Table 3 under case 4, this efficiency is identical with the Murphree efficiency E_{Mj} if the stages are in fact completely mixed. If this is not the case, the measured performance will correspond to a somewhat smaller number of true transfer units, and this must be taken into account in deriving the mass-transfer coefficient from the results.

5. DESIGN PROCEDURES: BATCH PROCESSES

5.1. Types of Process

Batch extraction processes form a group of operations more diverse than the continuous processes but are restricted in use to the laboratory and to small-scale preparative work. A brief outline of the various processes is given in the following paragraphs prior to quantitative treatments.

5.1.1. Single- and Multiple-Contact Batch Extraction

This process, the basic laboratory method of extraction, consists in adding a solvent to a feed mixture contained in a separating funnel, shaking, and allowing the phases to disengage. After removal of extract phase, the treatment can be repeated with the addition of fresh solvent for as many times as desired. Special complex operating schemes are used for difficult separations (see Section 5.1.4). Other forms are used for the simulation of the various continuous processes in the laboratory and are described in Chapter 17.1.

5.1.2. Differential Batch Extraction

In this process, analogous to Rayleigh distillation, a charge is extracted repeatedly with infinitesimal quantities of solvent and the extract separated as formed. This condition can be approached in practice by means of an apparatus

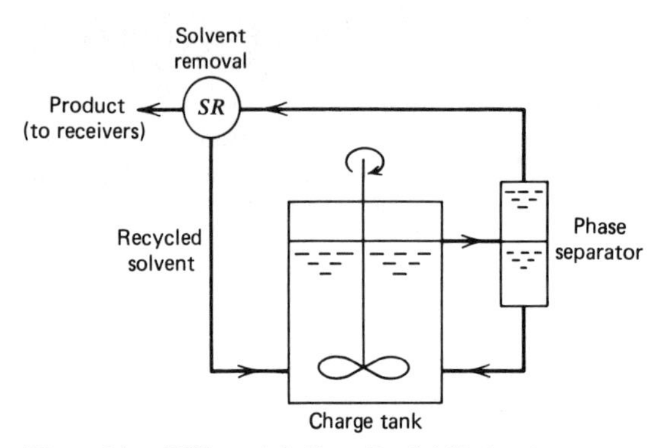

Figure 34. Differential (i.e., Rayleigh) batch extraction.

(Fig. 34) in which fresh solvent is added continuously to the charge in the mixing tank; the resulting mixture passes to a phase separator, from which the extract phase is withdrawn to the solvent removal unit and the raffinate is recycled back to the mixer. This process, like Rayleigh distillation, cannot give a high degree of separation unless the separation factor is very large.

5.1.3. Batch Fractional Extraction

The degree of separation obtainable in the process described in Section 5.1.2 can be increased by adding a countercurrent contactor between the charge tank and reflux unit and refluxing part of the separated extract product to the end of the contactor [40], as shown in Fig. 35. In

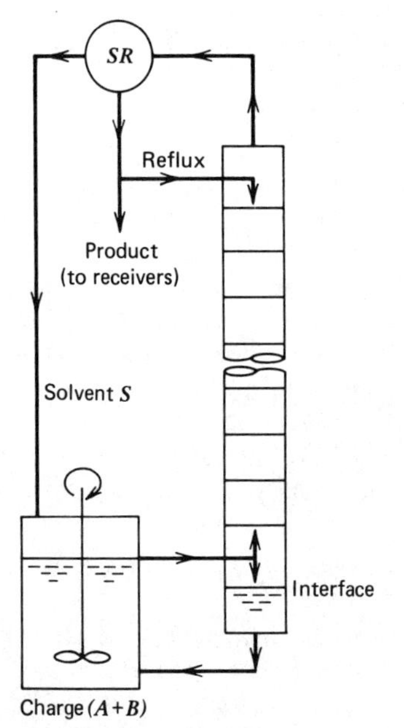

Figure 35. Normal single-solvent batch fractional extraction (SR = solvent S removal).

this form fractions of decreasing solubility are obtained and the process bears an obvious resemblance to normal batch distillation, which gives successive fractions of decreasing volatility. However, an alternative form, analogous to the lesser-known inverted batch distillation [26], can be used if it is desired to separate the least-soluble component first [40]. Dual-solvent forms of both normal and inverted type have also been proposed [40].

This process has been used in the laboratory to fractionate lubricating oil with acetone [41] and a petroleum fraction with phenol [42]. No larger-scale applications have been reported.

5.1.4. Countercurrent Multiple-Contact Batch Fractionation

This process, more commonly known as *countercurrent distribution*, is much used by chemists for the analysis and small-scale laboratory separation of complex materials, particularly of biological origin, by using separating funnels or an automatic multistage extractor such as that of Craig (see Chapter 17.1). In *transient batch countercurrent extraction* [43, 44], the feed is introduced as a single pulse into countercurrent flows of two solvents.

5.2. Single- and Multiple-Contact Extraction

Single-contact batch extraction can be represented on the triangular and Janecke diagrams (Figs. 3 and 4) in the same way as the corresponding continuous process by substitution of mass for flow rate. Equilibrium between the phases is likely with the batch process since all fluid elements are contacted for the same time. Multiple-contact extraction, involving repeated extraction of raffinate with fresh batches of solvent, can be treated in the same manner, for instance, by using the Janecke diagram as shown in Fig. 5; alternatively, the analytical method of Section 2.4.2 can be used.

5.3. Differential Batch Extraction

5.3.1. Immiscible Solvents

Study of Figure 34 reveals that the slow continuous flow of solvent through the stirred vessel leads to a progressive depletion of solute from the charge. Depletion can be calculated as a function of the amount of solvent added if it is assumed that the two solvents S and B are immiscible. A solute balance then gives

$$-R_i\,dx' = (y' - y'_s)\,dE_i \tag{135}$$

where y'_s is the solute content of the entering solvent. The total solvent used E_{iT} is obtained by integration between the limits of x_F and x_f, the initial and final charge compositions, respectively; thus

$$E_{iT} = \int_0^{E_{iT}} dE_i = R_i \int_{x_f}^{x_F} \frac{dx'}{y' - y'_s} \tag{136}$$

Equation (136) can be solved graphically by plotting $1/(y' - y'_s)$ against x', taking y' and x' from the equilibrium distribution diagram if the phases attain equilibrium; if they do not, it is necessary to allow for the efficiency on a Hausen or Murphree basis. For a given amount of pure solvent, a greater proportion of solute will be removed by differential extraction than by single-stage batch extraction.

5.3.2. *General Case*

The general case where solvent miscibility cannot be neglected can be treated using triangular [6] or Janecke coordinates. The material balance in the latter coordinates, assuming pure solvent, leads to a form of the Rayleigh equation as follows [cf. Eq. (136)]:

$$\int_{\bar{R}_f}^{\bar{R}_0} \frac{d\bar{R}}{\bar{R}} = \ln \frac{\bar{R}_0}{\bar{R}_f} = \int_{\bar{x}_f}^{\bar{x}_0} \frac{d\bar{x}}{(\bar{y} - \bar{x})} \tag{137}$$

Equation (137) can be integrated graphically by measuring the area under a plot of $1/(\bar{y} - \bar{x})$ against $\bar{x}$, using values of $\bar{y}$ corresponding to $\bar{x}$ obtained from the selectivity diagram if the phases are at equilibrium.

For some banded Type 3/2 systems, the selectivity ratio β, defined as

$$\beta = \frac{K_A}{K_B} = \frac{\bar{y}(1 - \bar{x})}{\bar{x}(1 - \bar{y})} \tag{138}$$

is constant in weight or mole fraction units. For such cases, it can be shown [26] that

$$\ln \frac{\bar{R}_0}{\bar{R}_f} = \frac{1}{(\beta - 1)} \left(\ln \frac{\bar{x}_0}{\bar{x}_f} + \beta \ln \frac{1 - \bar{x}_f}{1 - \bar{x}_0} \right) \tag{139}$$

5.4. Batch Fractional Extraction

In single solvent batch fractional extraction the single-stage contactor shown in Fig. 34 is replaced by a multistage unit, so that $(\bar{y} - \bar{x})$ in Eq. (137) becomes $(\bar{x}_{pr} - \bar{x}_b)$, where $\bar{x}_b$ is the residual charge composition. The Janecke diagram can be used to prepare a plot of $\bar{x}_b$ versus $\bar{x}_{pr}$ can be prepared for the given solvent ratio, starting from a series of values of $\bar{x}_{pr}$ and constructing the stages to obtain $\bar{x}_b$. In a similar manner, Eq. (136) could be applied to each component in the corresponding dual-solvent process, by using incremental material balances to obtain the successive charge compositions.

Integration of Eqs. (136) and (137) in this manner would enable *batch extraction curves*, analogous to batch distillation curves [26], to be prepared, showing the variation of product and residue concentrations with $(1 - R_f/R_0)$, the fraction extracted, at a given solvent ratio. However, when operating in this way it is necessary to take into account the holdup of the phases, with its consequent effect on residence time within the contactor. The effect of holdup has been found to be both complex and significant for distillation [45, 46] and is likely to be so for most liquid–liquid extraction contactors, necessitating detailed computation for determination of the batch curves.

5.5. Countercurrent Distribution

This process, much used by chemists for resolving complex mixtures, is suitable only for handling small quantities and is not amenable to large-scale operation. However, one form, involving double withdrawal of products with repeated feed [49], is the batchwise analogue of dual solvent fractional extraction, which provides a link between the laboratory process and practical application.

Reference should be made to Chapter 17.1 and to the reviews by Craig [47], Hecker [48], and Alders [25] for details of this process.

6. MULTICOMPONENT SEPARATIONS

6.1. Types of Operation

The processes considered thus far have all involved the separation of only two feed components, one of which is also a solvent in the case of single solvent operations. Multicomponent separations can be defined in the broadest sense as those in which either (1) the added solvent or solvents are mixtures, (2) the feed contains three or more components, or (3) both (1) and (2) apply simultaneously. However, the use of mixed solvents, although often important in practice, is relatively trivial in the sense that it is usually possible to treat these as single entities, except at high solute concentrations where the miscibility is appreciable. The case where the nondistributed feed component of effective single solvent systems is a mixture also falls into this category.

For the present purpose, multicomponent separations are defined as those in which three or more components are present in the feed, two of which are distributed in the case of single-solvent processes; solvent present in the feed to dual-solvent processes is not considered as a feed component. The systems involved thus comprise a minimum of four components in the single-solvent case and five with dual solvents. Although miscibility and tie-line equilibria for quaternary systems can be represented by means of solid models or by projection from these [20], this is inconvenient in practice; with five or more components, such representation is not possible at all, and in general multicomponent data must be presented in parametric form.

For systems of considerable complexity, it has been suggested that the compositions on the AB axis of the triangular diagram (Fig. 3) be replaced by a relevant physical property. By this means, with the use of a given feedstock and solvent, it is possible to determine the miscibility boundary and tie-line equilibria in the laboratory by measurements of the chosen physical property on the separated phases. For example, this method has been suggested for lubricating oil stocks, where the *viscosity–gravity constant* is used as a measure of aromaticity [50], and for glyceride oils where the *iodine number* is used to determine the degree of unsaturation. In the latter case it has also been proposed [51] to use the Janecke diagram with the iodine number on the horizontal axis.

This technique has, however, been shown [6, 52] to possess serious limitations, as may be demonstrated for a quaternary system with two distributed components A and D. In such cases projection of the miscibility boundary surface onto the base of a tetrahedral phase diagram for different proportions of A and D will lead to different ternary miscibility boundaries. Consequently, a phase diagram thus obtained will apply only to the particular feed mixture used and will not be applicable to repeated (i.e., crossflow) or countercurrent extraction. Therefore, the design of such an extractor must be based on a laboratory simulation of the entire process by use of separating funnels or tests with a small-scale continuous extractor (Chapter 17.1).

6.2. Single- and Multistage Crossflow

The solute distribution in single-stage extraction of Type 4/1 quaternary systems can be determined by a method involving orthogonal projection onto the boundary surfaces of the regular tetrahedron [20]. However, this procedure is very tedious, and with five or more components, such methods are not possible.

Simplified methods of the type used for ternary systems are applicable when the solvent miscibility can be neglected, especially if the distribution coefficients are unaffected by the concentrations of the other solutes. In such cases the graphical method illustrated in Fig. 6 can be applied directly, using a separate diagram for each solute and drawing stage operating lines of the same slope on each. This procedure is obviously capable of extension to multiple contact extraction of both batchwise and continuous crossflow types.

Under the same circumstances, the composition after a given number of stages can be obtained analytically by applying Eq. (19) in turn to each component, to allow calculation of the composition of the composite extract. For a constant distribution coefficient, the fraction q_n of solute remaining in the raffinate after n stages is [cf. Eq. (19)]

$$q_n = \frac{1}{(\mathbf{S}_j + 1)^n} \tag{140}$$

and p_n, the fraction in the nth stage extract, is $q_n S_j$; the degree of separation of any pair of components, say, A and B, after N_S stages is thus given by $(q_A - q_B)_{N_S}$. Treybal [6] has shown that this can be maximized with respect to either stage number or solvent ratio E_i/R_i as follows:

$$N_{S,\text{opt}} = \frac{\log\,[\log\,(\mathbf{S}_A + 1)/\log\,(\mathbf{S}_B + 1)]}{\log\,[(\mathbf{S}_A + 1)/(\mathbf{S}_B + 1)]} \tag{141}$$

$$\mathbf{S}_{A,\text{opt}} = \frac{K_A E_i}{R_i} = \frac{\beta[\beta^{1/(N_S+1)} - 1]}{\beta - \beta^{1/(N_S+1)}} \tag{142}$$

Equation (141) permits the optimum number of stages to be determined, given the solvent ratio and Eq. (142), the optimum solvent ratio for a given number of stages. Used together, these equations yield the result $N_S = \infty$ and $(E_i/R_i) = 0$ per stage, showing that differential extraction gives the best separation.

When the distribution coefficients are interdependent, an iterative procedure is required for the calculation of the distribution. This can be done graphically for a quaternary system by plotting a series of equilibrium curves on the diagram for solute A, each corresponding to a fixed concentration of solute B, and vice versa. The stage operating line terminations, such as G on GF in Fig. 6, are then located so that the equilibrium concentrations on each diagram correspond with one another. This procedure becomes increasingly complex as the number is increased beyond 2, since one diagram is required for each.

6.3. Single-Solvent Countercurrent Processes

6.3.1. *Short-Cut Methods*

In its simplest form, this type of process involves the simultaneous extraction of two or more solutes of constant partition coefficient in an end-fed extractor by use of effectively immiscible solvents, an operation analogous to multicomponent absorption of dilute gas mixtures. In such cases the number of theoretical stages required can be determined graphically by using the distribution diagram, as shown in Fig. 36, in which the equilibrium lines refer to components A, B, and C of high, medium, and low partition coefficient, respectively. As the operating lines for each component are parallel, with slopes R/E, they can be placed by trial so that the same number of stages is required for each component between the specified feed and the inlet solvent concentrations.

It is convenient in practice to designate one of the solutes, of intermediate partition coefficient, as the *key component*, for which a suitable solvent ratio is selected; this would generally correspond to a value of $\mathbf{S} = 0.5$–0.8, but the value assumed initially may need to be revised later to obtain the desired recoveries of the other components. On specifying the concentration of the key component in either raffinate or extract product, the operating line can be drawn and the stages stepped off in the usual way. Parallel operating lines are then drawn for each of the other solutes in turn and their positions adjusted to give the same number of stages.

This procedure is particularly simple with solutes of relatively high or relatively low partition coefficient. Thus in Fig. 36, component B is selected as key and A, with a high partition

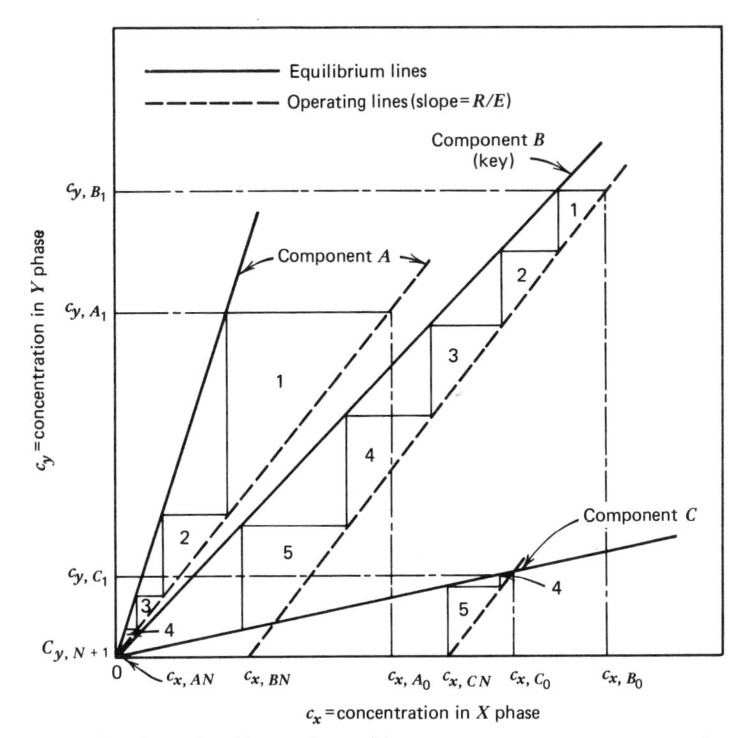

Figure 36. Operating diagram for multicomponent countercurrent extraction.

coefficient is seen to be virtually completely stripped from the raffinate so that its operating line passes effectively through the origin. Solute C, on the other hand, has a low partition and will be effectively at equilibrium at the feed inlet, so that its operating line pinches in to the equilibrium curve at that point, as shown.

A similar procedure can be carried out analytically, using the Colburn-type relationship [Eq. 5(b) in Table 1] to determine the number of stages required for the key component. By use of this result, the efficiency of extraction of the other components, and hence their concentrations in the exit raffinate, can be calculated, using the Kremser relationship [Eq. 6(b) in Table 1].

The graphical method can still be used with nonlinear equilibria, provided there is no undue change in solvent miscibility so that the operating lines are reasonably straight. When this is not the case, extensive equilibrium data are required for all components, including solvents, so that approximate numerical methods can be used. Smith [53] has described such a method, applicable also to the case where there is extract reflux, involving the use of averaged flow rates and effective stripping factors that are corrected in subsequent iterations. Such methods are subjective, however, and must be used with caution if large changes in flow can occur [53].

6.3.2. *Numerical Methods*

The computer calculation of single-solvent extraction is similar to that of gas absorption but is simpler because temperature changes due to heat of absorption do not occur. Sometimes an extractor is operated with a predetermined temperature profile in order to take advantage of the solubility behavior of the system, but this must be imposed by suitable heat-transfer arrangements and is unaffected by the mass transfer.

Two alternative procedures have been described for the computation of multistage extractors, one involving stagewise calculation from one end and the other an overall matrix formulation. Both methods require the initial specification of (1) the flow rates and compositions of each feed, (2) the number of theoretical stages and locations of the feed points, (3) the solvent ratio, (4) the extract reflux ratio, and (5) the temperature profile. The unknowns to be determined are then the raffinate and extract flows and compositions at each stage, and several different combinations of iteration variables are possible. In this form, which is most convenient for computational purposes, the results are directly applicable to the evaluation of the performance of a given extractor. For design purposes, however, calculations are required for various combinations of numbers of stages, solvent ratio, and so on, for determination of the optimum conditions, and here the short-cut methods are useful in providing a suitable starting point.

Smith and Brinkley [54] have described a method involving stagewise calculation, assuming as iteration variables (1) the extract product rate, (2) the composition, and (3) a linear extract rate profile. Their method was refined by Friday [55] to give perfect convergence; details are given in Smith's [53] book.

Tierney et al. [56] have described a procedure for the direct solution of the individual stage material balances in matrix form, with the equilibria expressed in terms of activity coefficients. Tierney [57] has also described a procedure, restricted to three components, in which the equilibrium phase compositions are used directly; a similar method has been given by Roche [58–60]. The original papers should be consulted for details.

6.4. Dual-Solvent Extraction

In principle, dual-solvent extraction can give a complete separation of each component of a multicomponent mixture. As with distillation, this will require $(C - 1)$ extractors for the separation of a C-component mixture; however, since the solvent effectiveness frequently depends on complex formation, especially with inorganic separations, the same solvent pair cannot always be used to separate each component. Very little work has been done on the theory of this process in multicomponent form, and the following account is confined to general principles.

In drawing up the specification for a given separation, it is necessary, as in distillation, to specify two *key components* that determine the sharpness of separation. Assuming the feed to be dissolved in solvent T, the *less soluble key* is that component that appears in very small but specifiable amount in the extract product, that is, in the exit solvent S from the scrubbing section. Similarly, the *more soluble key* is that component that is present in small but specifiable amount in the raffinate product, that is, in the solvent T leaving the extraction section.

If the two keys have adjacent solubilities, it is possible to draw up an overall material balance, assigning components of lower partition coefficient than the less soluble key to the raffinate product, and those of higher partition coefficient than the more soluble key to the extract product. However, with *split keys*, when there are one or more components with partition coefficients between those of the keys, the distribution of these between the two products cannot be specified independently and must be obtained by stagewise calculation.

If it can now be assumed as an approximation that the partition coefficients of the key components are unaffected by the concentrations of the non-keys, the extraction can be treated as a binary separation of the two keys. The overall minimum solvent to feed ratios corresponding to the critical point, that is, P_K and R_K, and the permissible range of P and R values can thus be obtained as described in Section 2.7.5. On specification of a pair of values within this range, stagewise calculation can be used to determine the number of stages required to give the desired separation of the keys.

With split keys, the distribution of the intermediate components must be determined by trial-and-error calculation. Thus a concentration is assumed in one of the products, and stagewise calculations are made to the other end through the feed point to obtain the concentration in the other product, assuming the same number of stages as for the keys. If these concentrations do not satisfy the overall material balance for that component, the calculation is repeated with a new assumed value.

This method should be applicable to nonlinear equilibria, but if the partition coefficient of a component is influenced by the concentrations of other components, excessive amounts of partition data are required with systems containing more than three or four feed components. For example, in nuclear fuel processing where more than 12 major components are present, extensive small-scale experimental work is essential for determination of the conditions necessary to obtain optimum performance. Batch simulation schemes with separating funnels can be used, but it is generally more expedient to use small-scale multistage extractors with capacities of a few millilitres per stage, which have been specially devised for this purpose (see Chapter 17.1).

NOTATION

		Dimensions	SI Units
a	specific interfacial area of contact of the phases	L^{-1}	m^{-1}
b	intercept of operating line [Eq. (34)]		
C	number of components in system		
c_j	concentration in phase j	ML^{-3} †	$kg\ m^{-3}$ †
c_{tj}	mean *total* concentration in phase j	ML^{-3} †	$kg\ m^{-3}$ †
c_{sj}	specific heat capacity of phase j	$L^2T^{-2}\theta^{-1}$	$J\ kg^{-1}\ K^{-1}$
D_x	defined in Table 1, Eq. 4(a)		
D_y	defined in Table 1, Eq. 4(a)		
E	extraction factor, mU_x/U_y		
E	extract rate	MT^{-1} †, ‡	$kg\ s^{-1}$ †, ‡
$\bar{E}$	extract rate on solvent-free basis	MT^{-1} †	$kg\ s^{-1}$ †
E_i	extract solvent rate on solute-free basis	MT^{-1} †	$kg\ s^{-1}$ †
E_H	Hausen efficiency		
E_{Mj}	Murphree efficiency based on phase j		
E_o	overall efficiency		
E_{oj}	extractor efficiency based on phase j (differential)		

†Molar units are also applicable.

‡M or moles (kg or kmol) for batch processes (Section 5.2).

		Dimensions	SI Units
E_{Sx}	extractor efficiency (stagewise)		
F	feed quantity or rate	MT^{-1} †, ‡	$kg\ s^{-1}$ †, ‡
$\bar{F}$	feed rate on solvent-free basis	MT^{-1} †	$kg\ s^{-1}$ †
F_j	total flow of phase j	L^3T^{-1}	$m^3 s^{-1}$ †
G_j	mass flow of phase j	$ML^{-2}T^{-1}$	$kg\ m^{-2}\ s^{-1}$
H_{Tj}	height of a transfer unit for phase j	L	m
H_{Toj}	height of an overall transfer unit based on phase j	L	m
h	height (or length) along stage	L	m
h_c	total height (or length) of single stage or compartment	L	m
h_j	heat-transfer coefficient for phase j	$MT^{-3}\theta^{-1}$	$W\ m^{-2}\ K^{-1}$
h_o	overall heat-transfer coefficient	$MT^{-3}\theta^{-1}$	$W\ m^{-2}\ K^{-1}$
$\bar{J}$	net flow in stripping section on solvent-free basis	MT^{-1} †	$kg\ s^{-1}$ †
$\bar{K}$	net flow in enriching section on solvent-free basis	MT^{-1} †	$kg\ s^{-1}$ †
K_k	partition coefficient of solute K $(= c_y^*/c_x)$		
k_j	mass-transfer coefficient for phase j	LT^{-1}	$m\ s^{-1}$
k_{oj}	overall mass-transfer coefficient based on phase j	LT^{-1}	$m\ s^{-1}$
L	height (or length) of differential extractor	L	m
M	quantity or rate of flow of mixed phases	MT^{-1} †, ‡	$kg\ s^{-1}$ †, ‡
M	total number of stages in extraction section (dual-solvent extraction)		
$\bar{M}$	flow of mixed phases on solvent-free basis	MT^{-1} †	$kg\ s^{-1}$ †
m	stage number within enriching section counted from extract reflux inlet (single-solvent operation), or within extraction section counted from solvent T exit (dual-solvent operation)		
m	reciprocal slope of equilibrium line [Eq. (30)]		
m'	slope of equilibrium line [Eq. (29)]		
N	total number of stages in stripping section (single-solvent operation) or scrubbing section (dual-solvent operation)		
N_A	flux of solute A	$ML^{-2}T^{-1}$ †	$kg\ m^{-2}\ s^{-1}$ †
N_S	number of theoretical stages		
N_{Tj}	number of transfer units based on phase j		
N_{Toj}	number of overall transfer units based on phase j		
N_{oj}^1	number of overall transfer units in single stage or compartment, based on phase j (with subscripts P, X, Y, or B, Section 4.2.1)		

†Molar units are also applicable.
‡M or moles (kg or kmol) for batch processes (Section 5.2).

		Dimensions	SI Units
n	stage number within stripping section counted from feed inlet (single-solvent operation) or within scrubbing section counted from solvent S exit (dual-solvent operation)		
P	product rate on solvent-free basis	MT^{-1} †	kg s^{-1} †
P_k	rejection ratio for solute k [Eqs. (107) and (110)]		
p	slope of operating line [Eq. (34)]		
p_k	fraction of feed solute K withdrawn with extraction solvent		
q	intercept of equilibrium line [Eq. (30)]		
q'	intercept of equilibrium line [Eq. (29)]		
q_k	fraction of feed solute K remaining in raffinate solvent		
R	raffinate rate	MT^{-1} †, ‡	kg s^{-1} †, ‡
$\bar{R}$	raffinate rate on solvent-free basis	MT^{-1} †	kg s^{-1} †
R_i	raffinate solvent rate on solute-free basis	MT^{-1} †	kg s^{-1}
r_E, r_S	fraction of solute refluxed at end of extraction or scrubbing section, respectively		
S	stripping factor, U_y/mU_x		
S	flow rate of solvent including solutes, if any (single-solvent operation)	MT^{-1} †	kg s^{-1} †
S	flow rate of solvent S on solute-free basis (dual-solvent operation); subscripts E, S, and F refer to extraction and scrubbing sections and feed, respectively	MT^{-1} †	kg s^{-1} †
$\bar{S}$	flow rate of nonsolvent in solvent	MT^{-1} †	kg s^{-1} †
s	solvent ratio, that is, ratio of solvent to total nonsolvent		
T	flow rate of solvent T on solute-free basis (dual-solvent operation); subscripts E, S, and F refer to extraction and scrubbing sections and feed, respectively	MT^{-1} †	kg s^{-1} †
T_j	dimensionless temperature of phase j [Eq. (119)]		
t_j	temperature of phase j	θ	K
U_j	superficial velocity of phase j in extractor	LT^{-1}	m s^{-1}
V_j	molar flow rate of phase j	mol $L^{-2}T^{-1}$	(kmol) m^{-2} s^{-1}
X	dimensionless concentration of solute in X phase [Eq. (60)]		
x	weight or mole fraction of component in X phase		
x'	weight or mole ratio of solute to solvent in X phase		
$\bar{x}$	weight or mole fraction on solvent-free basis in X phase		

†Molar units are also applicable.

‡M or moles (kg or kmol) for batch processes (Section 5.2).

		Dimensions	SI Units
Y	dimensionless concentration of solute in Y phase [Eq. (62)]		
y	weight or mole fraction of component in Y phase		
y'	weight or mole ratio of solute in Y phase		
$\bar{y}$	weight or mole fraction on solvent-free basis in Y phase		
Z	$= z/L$, dimensionless length		
z	height (or length) within extractor	L	m

Greek Symbols

β	$= K_A/K_B$, selectivity ratio		
λ	defined in Table 1, Eq. 4(d)		
ρ_j	density of phase j	ML^{-3} †	kg m^{-3} †

Subscripts

A	distributed component (solute) in feed
B	nondistributed feed component (single-solvent processes); second distributed component (dual-solvent processes)
b	residual charge in batch extraction
E	extraction section (Section 2.7)
E, e	extract phase
F	feed
f	final, that is, end of batch
I	solvent inlet/raffinate outlet end
i	solute-free rate
i	value at interface
i	intersection of operating lines (Section 2.7.4)
j	X or Y phase
j	net flow in stripping section
K	critical point (Section 2.7.5)
k	solute component K (dual-solvent operation)
k	net flow in enriching section
M	last (i.e., feed) stage of extraction section
m	stage within enriching or extraction section
N	last stage (i.e., adjacent to feed) of scrubbing section
n	stage within stripping or scrubbing section
o	feed inlet/extract outlet end of extractor

†Molar units are also applicable.

		Dimensions	SI Units
o	initial raffinate at start of batch		
p	product		
R, r	raffinate phase		
S	scrubbing section (Section 2.7)		
S	solvent (component of mixture)		
s	solvent (i.e., extract) phase		
1, 2	inlet and exit, respectively, of stage (Section 4.2)		

Superscripts

1	local value
t	thermal
*	equilibrium value

Determinant

$$\begin{vmatrix} x_1 & y_1 & 1 \\ x_2 & y_2 & 1 \\ x_3 & y_3 & 1 \end{vmatrix}$$

represents twice the area of a triangle with vertices (x_1, y_1), (x_2, y_2), and (x_3, y_3). If the three points are collinear, the determinant is zero.

REFERENCES

1. R. N. J. Saal and W. J. D. van Dijck, *Proceedings of the World Petroleum Congress (London)*, Vol. II, 1933, p. 352.
2. E. G. Scheibel, *Chem. Eng. Progr.* **44,** 681, 771 (1948).
3. E. Janecke, *Z. Anorg. Chem.* **51,** 132 (1906).
4. T. W. Evans, *Ind. Eng. Chem.* **26,** 438 (1934).
5. F. M. Tiller, *Chem. Eng. Progr.* **45,** 391 (1949).
6. R. E. Treybal, *Liquid Extraction*, 2nd ed., McGraw-Hill, New York, 1963.
7. J. O. Maloney and A. E. Schubert, *Transact. Am. Inst. Chem. Eng.* **36,** 741 (1940).
8. T. G. Hunter and A. W. Nash, *J. Soc. Chem. Ind.* **51,** 285T (1932).
9. T. W. Evans, *Ind. Eng. Chem.* **26,** 860 (1934).
10. A. P. Colburn, *Transact. Am. Inst. Chem. Eng.* **35,** 91 (1939).
11. A. P. Colburn, *Ind. Eng. Chem.* **33,** 459 (1941).
12. E. G. Scheibel and D. F. Othmer, *Transact. Am. Inst. Chem. Eng.* **38,** 339 (1942).
13. R. Gayler and H. R. C. Pratt, *Transact. Inst. Chem. Eng.* **31,** 69 (1953).
14. R. Gayler and H. R. C. Pratt, *Transact. Inst. Chem. Eng.* **29,** 110 (1951).
15. A. P. Colburn and D. G. Welsh, *Transact. Am. Inst. Chem. Eng.* **38,** 179 (1942).
16. G. S. Laddha and J. M. Smith, *Chem. Eng. Progr.* **46,** 195 (1950).
17. J. H. Wiegand, *Transact. Am. Inst. Chem. Eng.* **36,** 679 (1940).
18. A. Kremser, *Natl. Petrol. News* **22** (21), 42 (1930).
19. M. Souders and G. G. Brown, *Ind. Eng. Chem.* **24,** 519 (1932).
20. A. V. Brancker, T. G. Hunter, and A. W. Nash, *J. Phys. Chem.* **44,** 683 (1940); *Ind. Eng. Chem.* **33,** 880 (1941).
21. G. F. Asselin and E. W. Comings, *Ind. Eng. Chem.* **42,** 1198 (1950).
22. J. Prochazka and V. Jiricny, *Chem. Eng. Sci.* **31,** 179 (1976).
23. A. Klinkenberg, *Chem. Eng. Sci.* **1,** 86 (1951).
24. A. Klinkenberg, H. A. Lauwerier, and G. H. Reman, *Chem. Eng. Sci.* **1,** 93 (1951).
25. L. Alders, *Liquid-Liquid Extraction*, 2nd ed., Elsevier, Amsterdam, 1959, Chapter II.
26. H. R. C. Pratt, *Countercurrent Separation Processes*, Elsevier, Amsterdam, 1967.
27. A. Klinkenberg, *Ind. Eng. Chem.* **45,** 653 (1953).
28. E. G. Scheibel, *Ind. Eng. Chem.* **46,** 16 (1954).
29. T. Woodward, *Chem. Eng. Progr.* **57,** 52 (1961).

30. W. S. T. Thomson, T. Woodward, W. A. Shrode, E. D. Baird, and D. A. Oliver, *Off. Sal. Water Progr. Rep.* **63,** (1962); **78,** (1963).
31. R. Letan and E. Kehat, *AIChE J.* **13,** 443 (1967); **14,** 398 (1968); **16,** 955 (1970); J. Yeheskel and E. Kehat, *AIChE J.* **19,** 720 (1973).
32. W. J. Anderson and H. R. C. Pratt, *Chem. Eng. Sci.* **33,** 995 (1978).
33. G. Standart, *Chem. Eng. Sci.* **20,** 611 (1956).
34. E. V. Murphree, *Ind. Eng. Chem.* **17,** 747 (1925).
35. W. K. Lewis, *Ind. Eng. Chem.* **28,** 399 (1936).
36. R. H. Perry, ed., *Chemical Engineers Handbook*, 5th ed., Section 18, McGraw-Hill, New York, 1973.
37. H. Hausen, *Chem. Ing. Technol.* **25,** 595 (1953).
38. G. E. Ho and R. G. H. Prince, *Transact. Inst. Chem. Eng.* **48,** T.101 (1970).
39. T. Miyauchi and T. Vermeulen, *Ind. Eng. Chem. Fund.* **2,** 305 (1963).
40. H. R. C. Pratt, *Chem. Eng. Sci.* **3,** 189 (1954).
41. M. R. Cannon and M. R. Fenske, *Ind. Eng. Chem.* **28,** 1035 (1936).
42. A. A. Kondrat'ev, *Internatl. Chem. Eng.* **4,** 434 (1964).
43. R. E. Cornish, R. C. Archibald, E. A. Murphy, and H. M. Evans, *Ind. Eng. Chem.* **26,** 397 (1934).
44. A. J. P. Martin and R. L. M. Synge, *Biochem. J.* **35,** 91 (1941).
45. A. Rose, R. C. Johnson, and T. J. Williams, *Chem. Eng. Progr.* **48,** 549 (1952).
46. A. Rose and R. C. Johnson, *Chem. Eng. Progr.* **49,** 15 (1953).
47. L. C. Craig and D. Craig, A. Weissburger, Ed., in *Techniques of Organic Chemistry*, 2nd ed., Vol. III, Part 1, Interscience, New York, 1956, pp. 149–332.
48. E. Hecker, *Verteilungsverfahren im Laboratorium*, Verlag Chemie, Weinheim/Bergstr., 1955.
49. E. L. Compere and A. L. Ryland, *Ind. Eng. Chem.* **43,** 239 (1951); **46,** 24 (1954).
50. T. G. Hunter and A. W. Nash, *Ind. Eng. Chem.* **27,** 836 (1935).
51. R. F. Ruthruff and D. F. Wilcock, *Transact. Am. Inst. Chem. Eng.* **37,** 649 (1941).
52. L. Alders, *J. Inst. Petrol.* **42,** 228 (1956).
53. B. D. Smith, *Design of Equilibrium Stage Processes*, McGraw-Hill, New York, 1963.
54. B. D. Smith and W. K. Brinkley, *AIChE J.* **6,** 451 (1960).
55. J. R. Friday, Ph.D. research, Purdue University, Lafayette, IN, 1961, as quoted by Smith [53].
56. J. W. Tierney, J. L. Yanosik, J. A. Bruno, and A. J. Brainard, *Proceedings of the International Solvent Extraction Conference, 1971*, Vol. 2, Society of Chemical Industry, London, 1971, p. 1051.
57. J. W. Tierney, *Proceedings of the International Solvent Extraction Conference, 1977*, Toronto (ISEC77), Vol. 1, (Can. Inst. Min. Met.), 1979, p. 242.
58. E. C. Roche, in J. Christensen, Ed., *Computer Calculations for Chemical Engineering Education*, Vol. VI, *Stagewise Operations*, National Academy of Engineering (USA), 1973, p. 46.
59. E. C. Roche, *Br. Chem. Eng.* **14,** 1393 (1969).
60. E. C. Roche, *Br. Chem. Eng.* **16,** 821 (1971).

6

AXIAL DISPERSION

H. R. C. Pratt
University of Melbourne
Australia

M. H. I. Baird
McMaster University
Canada

1. INTRODUCTION

1.1. Nature of Axial Dispersion

The treatment of continuous countercurrent extraction in Chapter 5 relates strictly to the case of pure plug flow of the phases, which is attainable only with extractors in which phase separation between stages in virtually complete, such as in discrete-stage mixer–settlers and perforated-plate columns. The performance of other contactor types is influenced adversely by departures from the plug flow pattern, and a more advanced treatment is thus necessary. The factors that contribute to this reduction in performance are complex but can be identified broadly as follows:

1. Circulatory flow of continuous phase arising from the dissipation of the potential energy of dispersed-phase droplets [1] or films.
2. Transport and shedding of continuous phase in wakes attached to the rear of dispersed-phase droplets.
3. Molecular and turbulent diffusion of continuous phase in both axial and radial directions along concentration gradients.
4. Circulation of continuous phase and consequent entrainment of dispersed phase in mechanically agitated contactors.
5. Channeling and consequent maldistribution due to particular characteristics of the contactor geometry or, when appropriate, of the packing or internal fittings.
6. Nonuniform velocity profiles of one or both phases due to frictional drag of stationary surfaces.
7. Dispersion of droplet velocities as a result of the range of droplet diameters present ("forward mixing" [2]).

The first two factors result in pure backmixing of continuous phase; however, wake shedding (factor 2) occurs only at values of $Re_d > 150–200$ and is then only 5–10% of the circulatory flow [1, 3, 4]. The third and fourth factors also lead, *inter alia*, to a degree of backmixing; particularly the latter is also responsible for backmixing of the dispersed phase.

The sixth factor, which differs in nature from the previous factors, is a consequence of the velocity profile that exists in single-phase flow between a stationary surface, where the velocity is zero, and the center of the stream, where it is a maximum. Thus to an observer moving with the average velocity of the flow, some elements of the stream will appear to move backward, although to a stationary observer, there will be no backflow. This effect leads to a distribution of residence times of the fluid elements, which may also be expected to influence the performance of an extractor adversely. The combined result of the various effects is thus more accurately termed *axial dispersion.*

Of the various methods available for the measurement of axial dispersion (see Section 7), that involving continuous tracer injection near the phase outlet gives a measure of true backmixing and is interpreted in terms of a plug flow–backmix model. The transient tracer method, in which a pulse or step change is introduced near the phase inlet, gives a measure of the residence time distribution and hence of *total* axial dispersion. However, the results are normally interpreted by "force fitting" to the plug flow–backmix model, and the same is true of results obtained from measured concentration profiles. With use of the latter methods, it is apparent, therefore, that this model is in fact being used to cover all the first six factors just described. Although the seventh factor also influences the residence time distribution of the dispersed phase, it has in fact a different effect on performance and is thus discussed separately in Section 6.

1.2. Axial Dispersion Models

Two distinct types of model have been proposed, namely, the *diffusion model*, which assumes a turbulent backdiffusion of solute superimposed on plug flow of the phases [5–7], and the *backflow model*, with well-mixed nonideal stages between which backflow occurs [8]. These two models in fact represent limiting cases; thus the diffusion model is approached in practice by differential extractors such as packed and baffle-plate columns and the backflow model, by mixer–settlers of cocurrent settler type with heavy entrainment in the separated phases.

Between these two extremes there is a variety of multicompartment extractors of both the non-coalescing type (e.g., multiimpeller and pulsed-plate columns) and the countercurrent settler type (e.g., the Scheibel column), which do not conform closely to either model. However, a relationship exists between the two models that becomes closer as the number of compartments is increased, allowing the performance of such contactors to be expressed in terms of either model with reasonable accuracy [9].

2. TYPES OF MODEL

2.1. Diffusion Model

The assumptions inherent in this model are as follows [7]:

1. The backmixing of each phase can be characterized by a constant turbulent diffusion coefficient E_j.
2. The mean velocity and concentration of each phase is constant through every cross section.
3. The volume mass-transfer coefficient is constant or can be averaged over the column.
4. The solute concentration gradients in each phase are continuous (except at the phase inlets, as is seen later).
5. The solvent and raffinate phases are effectively immiscible or have a constant miscibility irrespective of solute concentration.
6. The volumetric flow rates of feed and solvent (i.e., X and Y) phases are constant throughout.
7. The equilibrium relation is linear or can be approximated by a straight line.

2.1.1. *Derivation of Equations*

Material balances over a differential length of contactor are shown in Fig. 1 for one-dimensional countercurrent flow (see Fig. 12 in Chapter 5 for the corresponding plug flow case). These give

$$E_x \frac{d^2 c_x}{dz^2} - U_x \frac{dc_x}{dz} - k_{ox} a(c_x - c_x^*) = 0 \quad (1)$$

$$E_y \frac{d^2 c_y}{dz^2} + U_y \frac{dc_y}{dz} + k_{ox} a(c_x - c_x^*) = 0 \quad (2)$$

With backmixing absent, $E_j = 0$ and these expressions reduce to those for plug flow, such as to Eq. (46) of Chapter 5 for the X phase. If a linear equilibrium relation given by

$$c_x^* = mc_y + q \quad (3)$$

is assumed, Eqs. (1) and (2) can be expressed in dimensionless form as follows:

$$\frac{d^2 X}{dZ^2} - P_x B \frac{dX}{dZ} - N_{ox} P_x B(X - Y) = 0 \quad (4)$$

$$\frac{d^2 Y}{dZ^2} + P_y B \frac{dY}{dZ} + \mathbf{E} N_{ox} P_y B(X - Y) = 0 \quad (5)$$

The dimensionless variables P_j, B, Z, E, and N_{ox} are defined in the Notation section at the end of this chapter; it is to be noted that N_{ox} refers to "true" transfer units and not to the plug flow value N_{oxP} (N_{Tox} in Chapter 5). For a reason that will become apparent later, the external concentrations at the ends of the extractor are designated by *superscripts* 0 and I at the feed and

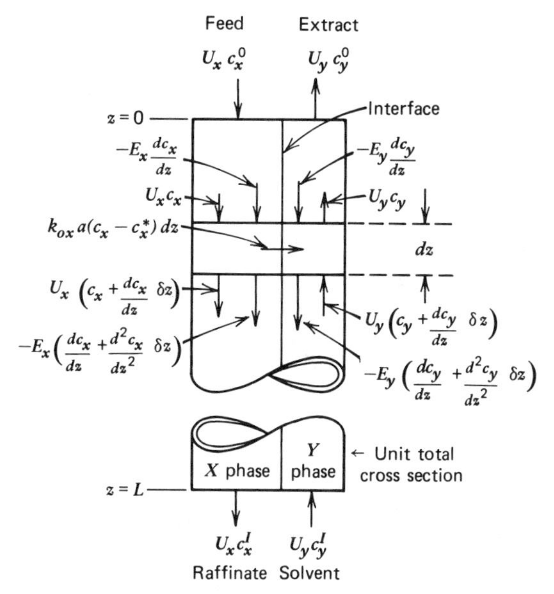

Figure 1. Diffusion model: material balance over differential section.

solvent phase inlets, respectively, and the corresponding internal concentrations (i.e., at the ends of the contacting section) by similar *subscripts.* Hence the dimensionless concentrations are defined as follows [cf. Eqs. (60) and (62) in Chapter 5]:

$$X = \frac{c_x - (mc_y^I + q)}{c_x^0 - (mc_y^I + q)} \tag{6}$$

$$Y = \frac{m(c_y - c_y^I)}{c_x^0 - (mc_y^I + q)} \tag{7}$$

For the backflow model described later, the superscript I is replaced by $N + 1$.

Elimination of Y between Eqs. (4) and (5) gives

$$\frac{d^4X}{dZ^4} - \alpha\frac{d^3X}{dZ^3} - \beta\frac{d^2X}{dZ^2} - \gamma\frac{dX}{dZ} = 0 \tag{8}$$

where

$$\alpha = B(P_x - P_y) \tag{9}$$

$$\beta = N_{ox}B(P_x + \mathbf{E}P_y) + P_xP_yB^2 \tag{10}$$

$$\gamma = N_{ox}P_xP_yB^2(1 - \mathbf{E}) \tag{11}$$

2.1.2. *Solutions for Profiles*

The solutions for the X and Y concentrations as a function of fractional height are obtained from Eqs. (8) and (4) as follows [5–7, 10]:

$$X = \sum_{i=1}^{4} A_i \exp(\lambda_i Z) \tag{12}$$

$$Y = \sum_{i=1}^{4} A_i a_i \exp(\lambda_i Z) \tag{13}$$

where $\lambda_1 = 0$ and the remaining λ_i are the roots of the characteristic equation

$$\lambda^3 - \alpha\lambda^2 - \beta\lambda - \lambda = 0 \tag{14}$$

The a_i are given by

$$a_i = 1 + \frac{\lambda_i}{N_{ox}} - \frac{\lambda_i^2}{N_{ox}P_xB} \tag{15}$$

$$= \frac{\mathbf{E}(1 - \lambda_i/P_xB)}{(1 + \lambda_i/P_yB)} \tag{16}$$

where the second, more convenient form is obtained by eliminating N_{ox} between Eqs. (14) and (15).

The roots λ_i of Eq. (14) are given by

$$\lambda_i = \frac{\alpha}{3} + 2p^{1/2}\cos\left(\frac{u}{3} + k\right) \tag{17}$$

with $k = 0°$, $120°$, and $240°$ for λ_2, λ_3, and λ_4, respectively, and

$$p = \left(\frac{\alpha}{3}\right)^2 + \frac{\beta}{3}$$

$$q = \left(\frac{\alpha}{3}\right)^3 + \frac{\alpha\beta}{6} + \frac{\gamma}{2}$$

$$u = \cos^{-1}\left(\frac{q}{p^{3/2}}\right)$$

where the angle u lies between $0°$ and $180°$. This solution applies only if

$$q^2 - p^3 < 0$$

which is always the case in the present application.

The complete solution requires the evaluation of the constants A_i from the boundary conditions. Before derivation of these, the foregoing results are put into an alternative form that is more convenient for design purposes.

2.1.3. *Modified Solution*

The preceding form of solution is satisfactory as it stands for evaluating the performance of an extractor of known length by putting $Z = 1.0$. It is less convenient for design purposes, when the length is unknown, since the roots λ_i are themselves functions of $B(=L/d_c)$, necessitating an iterative solution. This can be obviated by substitution of $\lambda_i'L$ for λ_i, L/d_c for B, and L/H_{ox} for N_{ox} in Eqs. (18)–(16). By this means L is eliminated from the characteristic equation, which becomes

$$\lambda'^3 - \alpha'\lambda'^2 - \beta'\lambda' - \gamma' = 0 \tag{18}$$

with

$$\alpha' = \frac{P_x - P_y}{d_c} \tag{19}$$

$$\beta' = \frac{P_y + \mathbf{E}P_y}{H_{ox}d_c} + \frac{P_x P_y}{d_c^2} \tag{20}$$

$$\gamma' = \frac{P_x P_y (1 - \mathbf{E})}{H_{ox} d_c^2} \tag{21}$$

This form of the characteristic equation is used in Section 3.1.1 in the development of simplified solutions for extractor length. However, for convenience, the full solutions are first presented in the original form.

2.1.4. *Boundary Conditions*

The four boundary conditions required for evaluation of the four constants A_i are obtained as follows [5–7, 11].

(a) X **Phase,** $Z = 0$. At the X-phase inlet the rate of solute transfer away from the boundary of the contacting section, that is, at $Z = 0$, where the concentration is $c_{x,0}$, is given by

$$-E_x \left[\frac{dc_x}{dz}\right]_0 + U_x c_{x,0}$$

The solute flow entering with the feed is $U_x c_x^0$; hence, equating the flows yields

$$-E_x \left[\frac{dc_x}{dz}\right]_0 = U_x (c_x^0 - c_{x,0}) \tag{22}$$

In dimensionless units this becomes

$$-\left[\frac{dX}{dZ}\right]_0 = P_x B(1 - X_0) \tag{23}$$

(b) X **Phase,** $Z = 1.0$. At the X-phase outlet where the dimensionless internal and external concentrations are X_I and X^I, respectively, similar reasoning leads to

$$-\left[\frac{dX}{dZ}\right]_I = P_x B(X^I - X_I)$$

Since $P_x B$ is always positive or zero, the gradient given by the left-hand term is of opposite sign to that of the right. The only condition possible is thus

$$X^I = X_I; \qquad \left[\frac{dX}{dZ}\right]_I = 0 \tag{24}$$

(c) Y **Phase,** $Z = 1.0$. Reasoning similar to that in Section 2.1.4(a) gives

$$-\left[\frac{dY}{dZ}\right]_I = P_y B Y_I \tag{25}$$

(d) Y **Phase,** $Z = 0$. A procedure similar to that described in Section 2.1.4(b) gives

$$Y^0 = Y_0; \qquad \left[\frac{dY}{dZ}\right]_0 = 0 \tag{26}$$

Substitution of the solutions for the profiles [Eqs. (12) and (13)] into Eqs. (23)–(26) gives

$$A_i + \sum_{i=1}^{4} A_i \left(\frac{1 - \lambda_i}{P_x B}\right) = 1 \tag{27}$$

$$\sum_{i=2}^{4} A_i \lambda_i e^{\lambda_i} = 0 \tag{28}$$

$$\sum_{i=2}^{4} A_i a_i \lambda_i = 0 \tag{29}$$

$$A_1 + \sum_{i=2}^{4} A_i a_i e^{\lambda_i} \left(1 + \frac{\lambda_i}{P_y B}\right) = 0 \tag{30}$$

Sleicher [7] solved the set of equations (12), (14), and (27–30) numerically to obtain the profiles, whereas Miyauchi and Vermeulen [6] obtained the analytical solution for the A_i in determinantal form. The complete solution in the latter form is summarized in Table 1.

2.1.5. *Solution for* $\mathbf{E} = 1.0$

When $\mathbf{E} = 1.0$, Eq. (11) shows that $\gamma = 0$ so that the characteristic equation reduces to a quadratic. The solution thus takes the modified form given in Table 1.

2.1.6. *Concentration Profiles*

Typical concentration profiles, as calculated by the preceding method, are shown in Fig. 2 together with the corresponding curves for the plug flow case. The effect of backmixing is apparent from the concentration "jumps" AB and CD at the X- and Y-phase inlets [12, 13], and the zero slopes at the exits, which result from the specified boundary conditions [6]. These both have

TABLE 1 GENERAL SOLUTIONS

	Diffusion Model	Backflow Model
Case 1	P_x, P_y, H_{ox}, L finite, $\mathbf{E} \neq 1.0$	$\alpha_x, \alpha_y, N_{ox}^1, N$ finite, $\mathbf{E} \neq 1.0$
Profiles	$X = A_1 + A_2 e^{\lambda_2 Z} + A_3 e^{\lambda_3 Z} + A_4 e^{\lambda_4 Z}$ $Y = A_1 + A_2 a_2^{\lambda_2 Z} + A_3 a_3 e^{\lambda_3 Z} + A_4 a_4 e^{\lambda_4 Z}$ where $a_i = \dfrac{\mathbf{E}(1 - \lambda_i/P_x B)}{(1 + \lambda_i/P_y B)}$	$X_n = A_1 + A_2 \mu_2^n + A_3 \mu_3^n + A_4 \mu_4^n$ $Y_n = A_1 + A_2 a_2 \mu_2^n + A_3 a_3 \mu_3^n + A_4 a_4 \mu_4^n$ where $a_i = \dfrac{\mathbf{E}[1 - \alpha_x(\mu_i - 1)]}{\mu_i + \alpha_y(\mu_i - 1)}$
Characteristic equation	$\lambda^3 - \alpha\lambda^2 - \beta\lambda - \gamma = 0$ where α, β, γ are given by Eqs. (9)–(11) and the solution by Eq. (17)	$(\mu - 1)^3 - \alpha(\mu - 1)^2 - \beta(\mu - 1) - \gamma = 0$ where α, β, and γ are given by Eqs. (41)–(43) and the solution by Eq. (17)
Coefficients A_i	$A_i = \dfrac{D_{A_i}}{D_{A_1} - D_A}$ where $D_A = \begin{vmatrix} 1 - \dfrac{\lambda_2}{P_x B} & 1 - \dfrac{\lambda_3}{P_x B} & 1 - \dfrac{\lambda_4}{P_x B} \\ a_2\lambda_2 & a_3\lambda_3 & a_4\lambda_4 \\ \lambda_2 e^{\lambda_2} & \lambda_3 e^{\lambda_2} & \lambda_4 e^{\lambda_4} \end{vmatrix}$ $D_{A_1} = \begin{vmatrix} a_2\lambda_2 & a_3\lambda_3 & a_4\lambda_4 \\ \lambda_2 e^{\lambda_2} & \lambda_3 e^{\lambda_3} & \lambda_4 e^{\lambda_4} \\ a_2 e^{\lambda_2} f(\lambda_2) & a_3 e^{\lambda_3} f(\lambda_3) & a_4 e^{\lambda_4} f(\lambda_4) \end{vmatrix}$ with $f(\lambda_i) = \left(1 + \dfrac{\lambda_i}{P_y B}\right)$ $D_{A_2} = \lambda_3\lambda_4(a_4 e^{\lambda_3} - a_3 e^{\lambda_4})$ $D_{A_3} = \lambda_2\lambda_4(a_2 e^{\lambda_4} - a_4 e^{\lambda_2})$ $D_{A_4} = \lambda_2\lambda_3(a_3 e^{\lambda_2} - a_2 e^{\lambda_3})$	$A_i = \dfrac{D_{A_i}}{D_{A_1} - D_A}$ where $D_A = \begin{vmatrix} [1 - \alpha_x(\mu_2 - 1)] & [1 - \alpha_x(\mu_3 - 1)] & [1 - \alpha_x(\mu_4 - 1)] \\ a_2(\mu_2 - 1) & a_3(\mu_3 - 1) & a_4(\mu_4 - 1) \\ \mu_2^N(\mu_2 - 1) & \mu_3^N(\mu_3 - 1) & \mu_4^N(\mu_4 - 1) \end{vmatrix}$ $D_{A_1} = \begin{vmatrix} a_2(\mu_2 - 1) & a_3(\mu_3 - 1) & a_4(\mu_4 - 1) \\ \mu_2^N(\mu_2 - 1) & \mu_3^N(\mu_3 - 1) & \mu_4^N(\mu_4 - 1) \\ a_2\mu_2^N f(\mu_2) & a_3\mu_3^N f(\mu_3) & a_4\mu_4^N f(\mu_4) \end{vmatrix}$ with $f(\mu_i) = [\mu_i + \alpha_y(\mu_i - 1)]$ $D_{A_2} = (\mu_3 - 1)(\mu_4 - 1)(a_4\mu_3^N - a_3\mu_4^N)$ $D_{A_3} = (\mu_2 - 1)(\mu_4 - 1)(a_2\mu_4^N - a_4\mu_2^N)$ $D_{A_4} = (\mu_2 - 1)(\mu_3 - 1)(a_3\mu_2^N - a_2\mu_3^N)$

Case 2

P_x, P_y, H_{ox}, L finite, $\mathbf{E} = 1.0$

Profiles

$$X = A_1 + A_2 e^{\lambda_2 Z} + A_3 e^{\lambda_3 Z} + A_4 Z$$

$$Y = A_1 + A_2 a_2 e^{\lambda_2 Z} + A_3 a_3 e^{\lambda_3 Z} + A_4 \left(Z + \frac{1}{N_{ox}}\right)$$

where a_i are given as above with $\mathbf{E} = 1.0$

Characteristic equation

$$\lambda^2 - \alpha\lambda - \beta = 0$$

where α and β are given as above with $\mathbf{E} = 1.0$

$$\lambda_2, \lambda_3 = \frac{\alpha}{2} \pm \sqrt{\left(\frac{\alpha}{2}\right)^2 + \beta} \quad (+ve \text{ for } \lambda_2)$$

Coefficients A_i

$$A_i = \frac{D_{A_i}}{D_{A_1} - D_A}$$

where

$$D_A = \begin{vmatrix} 1 - \frac{\lambda_2}{P_x B} & 1 - \frac{\lambda_3}{P_x B} & -\frac{1}{P_x B} \\ a_2\lambda_2 & a_3\lambda_3 & 1 \\ \lambda_2 e^{\lambda_2} & \lambda_3 e^{\lambda_3} & 1 \end{vmatrix}$$

$$D_{A_1} = \begin{vmatrix} a_2\lambda_2 & a_3\lambda_3 & 1 \\ \lambda_2 e^{\lambda_2} & \lambda_3 e^{\lambda_3} & 1 \\ a_2 e^{\lambda_2} f(\lambda_2) & a_3 e^{\lambda_3} f(\lambda_3) & \left(1 + \frac{1}{N_{ox}} + \frac{1}{P_y B}\right) \end{vmatrix}$$

$$D_{A_2} = \lambda_3(e^{\lambda_3} - a_3)$$

$$D_{A_3} = \lambda_2(a_2 - e^{\lambda_2})$$

$$D_{A_4} = \lambda_2\lambda_3(a_3 e^{\lambda_2} - a_2 e^{\lambda_3})$$

$\alpha_x, \alpha_y, N_{ox}^1, N$ finite, $\mathbf{E} = 1.0$

Profiles

$$X = A_1 + A_2\mu_2^n + A_3\mu_3^n + A_4 n$$

$$Y = A_1 + A_2 a_2 \mu_2^n + A_3 a_3 \mu_3^n + A_4 \left(n + \frac{1}{N_{ox}^1}\right)$$

where a_i are given as above with $\mathbf{E} = 1.0$

Characteristic equation

$$(\mu - 1)^2 - \alpha(\mu - 1) - \beta = 0$$

where α and β are given as above with $\mathbf{E} = 1.0$

$$(\mu_2 - 1)(\mu_3 - 1) = \frac{\alpha}{2} \pm \sqrt{\left(\frac{\alpha}{2}\right)^2 + \beta} \quad (+ve \text{ for } \mu_2)$$

Coefficients A_i

$$A_i = \frac{D_{A_i}}{D_{A_1} - D_A}$$

where

$$D_A = \begin{vmatrix} [1 - \alpha_x(\mu_2 - 1)] & [1 - \alpha_x(\mu_3 - 1)] & -\alpha_x \\ a_2(\mu_2 - 1) & a_3(\mu_3 - 1) & 1 \\ \mu_2^N(\mu_2 - 1) & \mu_3^N(\mu_3 - 1) & 1 \end{vmatrix}$$

$$D_{A_1} = \begin{vmatrix} a_2(\mu_2 - 1) & a_3(\mu_3 - 1) & 1 \\ \mu_2^N(\mu_2 - 1) & \mu_3^N(\mu_3 - 1) & 1 \\ a_2\mu_2^N f(\mu_2) & a_3\mu_3^N f(\mu_3) & \left[N + \frac{1}{N_{ox}^1} + (1 + \alpha_y)\right] \end{vmatrix}$$

$$D_{A_2} = (\mu_3 - 1)(\mu_3^N - a_3)$$

$$D_{A_3} = (\mu_2 - 1)(a_2 - \mu_2^N)$$

$$D_{A_4} = (\mu_2 - 1)(\mu_3 - 1)(a_3\mu_2^N - a_2\mu_3^N)$$

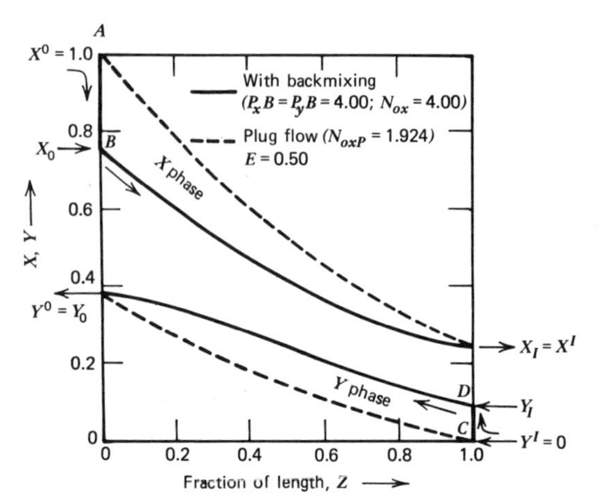

Figure 2. Effect of backmixing on concentration profiles.

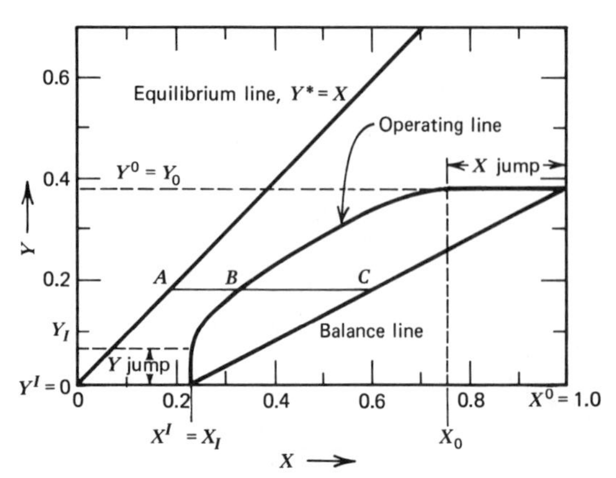

Figure 3. Operating diagram on X-Y coordinates. The parameter values are the same as those for Fig. 2.

the effect of reducing the concentration difference between the phases so that more transfer units are required when backmixing occurs.

The inlet concentration jumps are also apparent in the Y-X diagram shown in Fig. 3 for the same case. The true operating line suffers considerable displacement from the plug flow operating line, which is sometimes termed the *balance line* since it represents the overall material balance on the extractor. The driving force on an X-phase basis is typically reduced by backmixing to AB from the plug flow value AC.

With some extractor types, backmixing occurs only in the continuous phase, so that the dispersed phase profile does not show the inlet jump or the zero gradient at the exit. Such cases are discussed in Section 2.5.

2.1.7. *Relationship Between N_{ox} and N_{oxP}*

Integration of Eq. (1) gives the following expression for the true number of backmix transfer units:

$$N_{ox} = \int_{c_{xI}}^{c_{x0}} \frac{d[c_x - (E_x/U_x)(dc_x/dz)]}{(c_x - c_x^*)_B} = \frac{k_{ox}aL}{U_x} \tag{31}$$

where the driving force $(c_x - c_x^*)_B$ represents the "true" value equivalent to, say, AB in Fig. 3. This may be compared with the plug flow value given by

$$N_{oxP} = \int_{c_{xI}}^{c_{x0}} \frac{dc_x}{(c_x - c_x^*)_P} \tag{32}$$

It is apparent that the "true" N_{ox} differs from the plug flow value N_{oxP} in two respects: (1) the amount of solute transferred is effectively reduced by the term $(E_x/U_x)\, dc_x/dz$ due to remixing; and (2) the driving force is reduced, for example, from AC to AB in Fig. 3. Since dc_x/dz is negative, both effects increase the number of transfer units required for a given concentration range, as compared with plug flow. Some workers [12, 13] have employed a third definition, termed the *measured* or *interior apparent* value [6] N_{oxM}, using Eq. (32) with $(c_x - c_x^*)_P$ replaced by the true driving force $(c_x - c_x^*)_B$ obtained from experimentally determined concentration profiles, to evaluate extractor performance. Such values are intermediate between the true and plug flow values; however, there is no theoretical justification for the use of this definition. Each of these three definitions of the number of transfer units (NTU) is associated with a corresponding value of the height of a transfer unit (HTU), that is, $H_{ox} = L/N_{ox}$, and so on.

To simplify the design of backmixed extractors, it has been suggested that the effect of backmixing can be expressed in terms of a parameter N_{oxD}, the *number of overall dispersion transfer units* [6]; thus

$$\frac{1}{N_{oxP}} = \frac{1}{N_{ox}} + \frac{1}{N_{oxD}} \tag{33}$$

An empirical method was also proposed [6] for the correlation of N_{oxD} in terms of $P_x B$ and $P_y B$, based on the solution for "perfect" mass transfer (case 11 in Table 4, i.e., for $k_{ox}a \to \infty$, not $L \to \infty$ [14]). However, by combining eq. (4) with a material balance around one end of the column it is found that N_{oxD} is given by [15]

$$\frac{1}{N_{oxD}} = \frac{\mathbf{E}}{P_x B} - \frac{I_x'}{N_{ox} N_{oxP}^I P_x B} + \frac{I_y}{N_{oxP}^I P_x B} - \frac{1}{N_{oxP}^I} + \frac{1}{N_{oxP}} \tag{34}$$

where N_{oxP}^I is the value of N_{oxP} between $X^I (= 1.0)$ and X_I:

$$I_x' = \int_{(dX/dZ)_I}^{(dX/dZ)_0} \frac{d(dX/dZ)}{X(1 - \mathbf{E}) + \mathbf{E} X^I},$$

$$I_y = \int_Y^{Y^0} \frac{dY}{X(1 - \mathbf{E}) + \mathbf{E} X^I} \tag{35}$$

This relationship was confirmed for a number of cases from the calculated profiles [15].

The relationship between N_{ox} and N_{oxP} is thus clearly complex and cannot be expressed in terms of a simple parameter N_{oxD} as suggested by Eq. (33). Alternative simplified forms of the solutions given in Table 1 are presented in Section 3.1.

2.2. Backflow Model

This model assumes a series of stages interconnected as shown in Fig. 4; each stage contains a mixing device and may or may not include a settler in which partial or complete coalescence takes place. The inherent assumptions are as follows:

1. Each stage is well mixed, and backmixing occurs by mutual entrainment of the phases between stages, after coalescence if appropriate.
2. The backmixing is expressed in terms of

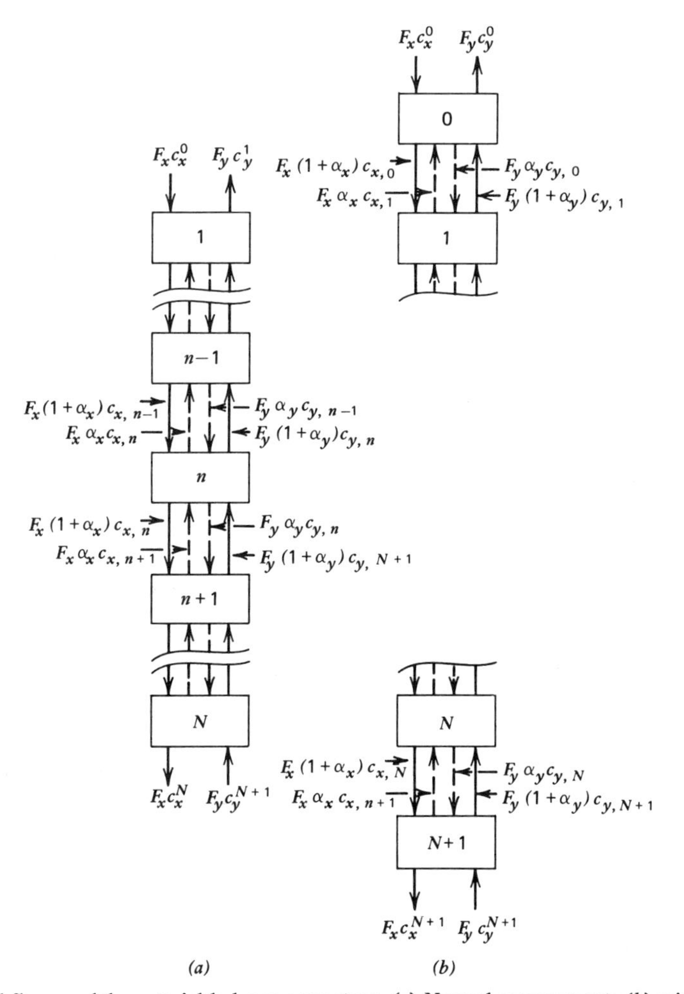

Figure 4. Backflow model: material balance over stage. (*a*) Normal arrangement; (*b*) with fictitious end stages.

the ratios α_j of backmixed to net forward interstage flow and is constant for all stages.

3. All mass transfer occurs in the mixer.
4. The value of $k_{ox}aV$, the product of volume mass-transfer coefficient and stage volume, is constant for each stage.

5-7. As (5)-(7) for the diffusion model (Section 2.1).

2.2.1. *Derivation of Equations* [8]

The material balances for the two phases around stage n are shown in Fig. 4. For the X-phase on collecting terms, this gives

$$(1+\alpha_x)\,c_{x,n-1} - (1+2\alpha_x)\,c_{x,n} + \alpha_x c_{x,n+1} = \frac{k_{ox}ash_c}{F_x}(c_{x,n} - c^*_{x,n}) \quad (36)$$

and similarly for the Y-phase, this gives

$$\alpha_y c_{y,n-1} - (1+2\alpha_y)\,c_{y,n} + (1+\alpha_y)\,c_{y,n+1} = -\frac{k_{ox}ash_c}{F_y}(c_{x,n} - c^*_{x,n}) \quad (37)$$

Assuming a linear equilibrium relationship given by Eq. (3), these balances can be put into dimensionless form as follows:

$$(1+\alpha_x)\,X_{n-1} - (1+2\alpha_x)\,X_n + \alpha_x X_{n+1} = N^1_{ox}(X_n - Y_n) \quad (38)$$

$$\alpha_y Y_{n-1} - (1+2\alpha_y)\,Y_n + (1+\alpha_y)\,Y_{n+1} = -N^1_{ox}\mathbf{E}(X_n - Y_n) \quad (39)$$

where $N^1_{ox} = k_{ox}aV/F_x = k_{ox}ah_c/U_x$, the number of perfectly mixed transfer units per stage [Chapter 5, Eq. (132)].

By expressing Eqs. (38) and (39) in E operator form (i.e., $EX_n = X_{n+1}$), Y can be eliminated to give

$$[(E-1)^4 - \alpha(E-1)^3 - \beta(E-1)^2 - \gamma(E-1)]\;X_n = 0 \quad (40)$$

where

$$\alpha = \frac{1+N^1_{ox}}{\alpha_x} - \frac{1-\mathbf{E}N^1_{ox}}{(1+\alpha_y)} \quad (41)$$

$$\beta = N^1_{ox}\left[\frac{1}{\alpha_x} + \frac{\mathbf{E}}{1+\alpha_y}\right] + \frac{1+N^1_{ox}(1-\mathbf{E})}{\alpha_x(1+\alpha_y)} \quad (42)$$

$$\gamma = \frac{N^1_{ox}(1-\mathbf{E})}{\alpha_x(1+\alpha_y)} \quad (43)$$

2.2.2. *Solutions for Profiles*

The solution for the X profile is obtained from Eq. (40) as follows:

$$X_n = \sum A_i \mu_i^n \quad (44)$$

Combination of this result with Eq. (38) gives

$$Y_n = \sum A_i a_i \mu_i^n \quad (45)$$

with

$$a_i = \frac{1}{\mu_i}\left[1 + \frac{1+N^1_{ox}}{N^1_{ox}}(\mu_i - 1) - \frac{\alpha_x}{N^1_{ox}}(\mu_i - 1)^2\right] \quad (46)$$

$$= \frac{\mathbf{E}[1 - \alpha_x(\mu_i - 1)]}{\mu_i + \alpha_y(\mu_i - 1)} \quad (47)$$

where the more convenient second form is obtained by eliminating N^1_{ox} between Eq. (46) and the characteristic equation [Eq. (48)].

The μ_i in Eqs. (44)-(47) are the roots of the characteristic equation, which on dividing through by $(\mu - 1)$ becomes

$$(\mu-1)^3 - \alpha(\mu-1)^2 - \beta(\mu-1) - \gamma = 0 \quad (48)$$

Hence $\mu_1 = 1.0$ and μ_2 to μ_4 are obtained by solving Eq. (48) by means of Eq. (17).

2.2.3. *Boundary Conditions*

Following Sleicher [8], it is assumed that fictitious after-settlers 0 and $(N+1)$, in which no mass transfer takes place, are provided at the ends of the contactor as shown in Fig. 4. The four boundary conditions required for the evaluation of the constants A_i in Eqs. (44) and (45) are then obtained as follows.

(a) ***X* phase, *n* = 0.** A balance on X phase around stage 0 gives, on cancelling F_x

$$c^0_x = c_{x,0} + \alpha_x(c_{x0} - c_{x,1})$$

In dimensionless concentrations this becomes

$$X_0 + \alpha_x(X_0 - X_1) = 1 \qquad (49)$$

(b) Y **Phase, $n = 0$.** A similar balance on Y phase gives

$$c_y^0 + \alpha_y c_{y,0} = (1 + \alpha_y)\, c_{y,1}$$

Since there is no mass transfer, $c_y^0 = c_{y,0}$ so that $c_y^0 = c_{y,1}$, that is,

$$Y^0 = Y_0 = Y_1 \qquad (50)$$

(c) Y **Phase, $n = N + 1$.** A balance on Y phase around stage $(N + 1)$ gives

$$Y_{N+1} - \alpha_y(Y_N - Y_{N+1}) = 0 \qquad (51)$$

(d) X **Phase, $n = N + 1$.** A procedure similar to that described in Section 2.2.3(b) gives

$$X^{N+1} = X_{N+1} = X_N \qquad (52)$$

Substitution of Eqs. (44) and (45) in (49) to (52) gives the following four equations that determine the A_i:

$$A_1 + \sum_{i=2}^{4} A_i[1 - \alpha_x(\mu_i - 1)] = 1 \qquad (53)$$

$$\sum_{i=2}^{4} A_i a_i(\mu_i - 1) = 0 \qquad (54)$$

$$A_1 + \sum_{i=2}^{4} A_i a_i \mu_i^N [\mu_i + \alpha_y(\mu_i - 1)] = 0 \qquad (55)$$

$$\sum_{i=2}^{4} A_i \mu_i^N (\mu_i - 1) = 0 \qquad (56)$$

These equations can be solved in determinantal form for the A_i, as for the diffusional model. The complete solution is summarized in Table 1.

2.2.4. *Solution for* $\mathbf{E} = 1.0$

As for the diffusion model, $\gamma = 0$ when $\mathbf{E} = 1.0$ so that the characteristic equation reduces to a quadratic. A modified solution thus applies and is given in Table 1.

2.3. Comparison of Models

2.3.1. *Parameters Involved*

Both models involve five parameters, as shown in Table 2. In the solution to the diffusion model, the parameters P_j and B always appear in combination as $P_j B = U_j L/E_j$, termed the *column Peclet numbers.* This appears to reduce the number of parameters from 5 to 4 and has led to a belief that the diffusion model is inherently simpler than the backmixing model, which involves five parameters [10]. However, this form of the Peclet number involves the loss of a degree of freedom since it is proportional to column length, whereas in fact the axial dispersion as defined by the *local* Peclet number is usually approximately constant. No such ambiguity arises with the modified form of solution referred to in Section 2.1.3 and used later in Section 3.1.

2.3.2. *Relationship between Models*

Miyauchi and Vermeulen [9] derived a number of working relationships between the diffusional and backflow models, of which the most direct is obtained by approximating the differential equations for the former in finite-difference form. Thus Eq. (4) becomes, on dividing through by $P_x B$ and replacing the differential terms by

TABLE 2 RELATIONSHIP BETWEEN MODEL PARAMETERS

Number and Description of Parameter	Diffusion Model		Backflow Model	
	Name	Symbol	Name	Symbol
1 Backmixing, X phase	Peclet no.	P_x	Backmixing ratio	α_x
2 Backmixing, Y phase	Peclet no.	P_y	Backmixing ratio	α_y
3 Extractor efficiency	HTU	H_{ox}	TU per actual stage	N_{ox}^1
4 Extractor size	Length	L^a	Number of actual stages	N
5 Flow ratio	Extraction factor	$\mathbf{E}$	Extraction factor	$\mathbf{E}$

[a] Or $B(= L/d_c)$.

lowest-order central differences

$$\frac{1}{P_x B}\left[\frac{X_{n+1} - 2X_n + X_{n+1}}{\Delta Z^2}\right] - \frac{(X_{n+1} - X_{n-1})}{2\Delta Z} = N_{ox}(X - Y) \quad (57)$$

For a total of N segments, the size of each is $\Delta Z = 1/N$. Substitution of this value in Eq. (57), putting $N_{ox}^1 = N_{ox}/N$, gives

$$\left[\frac{N}{P_x B} + \frac{1}{2}\right](X_{n-1} - X_n) - \left[\frac{N}{P_x B} - \frac{1}{2}\right] \cdot (X_n - X_{n+1}) = N_{ox}^1(X_n - Y_n) \quad (58)$$

Equation (38) for the backflow model can be arranged as follows:

$$(1 + \alpha_x)(X_{n-1} - X_n) - \alpha_x(X_n - X_{n-1}) = N_{ox}^1(X_n - Y_n) \quad (59)$$

A comparison of Eqs. (58) and (59) and of the corresponding expressions for the Y phase shows that these are essentially the same when the following relationships are satisfied:

$$\frac{1}{P_x B} = \frac{1}{N}\left(\alpha_j + \frac{1}{2}\right) \quad (60)$$

Hence

$$P_j B \simeq \frac{N}{\alpha_j} \quad (N \to \infty) \quad (61)$$

The boundary conditions for the diffusional model [Eqs. (23)–(26)] are obtained from those for the backflow model [Eqs. (49)–(52)] by substitution of Eq. (61) in the latter and letting $1/N = \Delta Z \to 0$.

The other relations [9] differ in relatively minor ways from Eq. (60), for example, with N replaced by $(N - 1)$. Consequently, either model should apply with reasonable accuracy to extractors of intermediate type, such as the rotary disk and pulsed-plate columns, provided the number of stages is large.

2.4. Effect of Entrance Sections

Wilburn [16] extended the diffusion model to allow for the stagnant zones at the ends of a differential contactor, beyond the feed and solvent inlets (i.e., for $Z < 0$ and > 1.0), obtaining a set of six differential equations and 12 boundary conditions. The solutions, which resembled Eqs. (12) and (13), contained 12 constants A_{ij} that were obtained by numerical inversion of a 12×12 matrix. Typical results showed that the inlet concentration jumps were increased by the presence of stagnant zones.

The method was later extended [17] to allow for variations in dispersed-phase holdup and for a nonlinear equilibrium. Corresponding solutions were obtained by McSwain and Durbin [18] in terms of the backflow model.

2.5. Reduced-Parameter Solutions

Under many circumstances the full range of five-parameter values does not apply in practice, in which case modified solutions result [6, 8, 10]. These are discussed briefly in the following paragraphs and the detailed solutions are summarized in Tables 3 and 4. In presenting the results, a consistent system of numbering the roots λ_i, μ_i has been adopted, noting that λ_2, μ_2 determine conditions near the X-phase entry, λ_3, μ_3 determine those near the Y-phase entry, and λ_4, μ_4 determine those in the central section [19(a)].

2.5.1. No Backmixing in One or Both Phases (Cases 3–8)

The results for these cases are summarized in Table 3. Of these, the first four, relating to backmixing in one or other phase alone, apply to a number of important contactor types that exhibit backmixing only in the continuous phase.

Cases 7 and 8, for no backmixing in either phase, correspond to plug flow, and in the case of the diffusional model the expressions for X reduce to those for N_{Tox} in Eq. (5) of Table 1, Chapter 5, on putting $Z = 1$. The stagewise forms are important since they relate to nonideal stages for which N_{ox}^1 is finite ($E_{Mj} < 1.0$).

2.5.2. Equilibrium Between Phases (Cases 9–12)

On letting $N_{ox}^1 \to \infty$, it follows that $\mu_4 \to E$; thus cases 7 and 8 for the backflow model reduce to cases 9 and 10; on putting $n = N$, these become identical with Eq. (6) for staged contactors in Table 1 of Chapter 5. If α_x and α_y are nonzero and $N_{ox}^1 \to \infty$, cases 11 and 12 (Table 4) are obtained; these can be used for approximate design purposes in terms of theoretical stages, using the overall stage efficiency (Chapter 5, Section 4.1.1).

These cases are unrealistic for the diffusional model since they correspond to equilibrium over

TABLE 3 SPECIAL SOLUTIONS: BACKMIXING IN ONE OR NEITHER PHASE[a]

Case Number		Diffusion Model		Backflow Model
3	$P_y \to \infty; P_x,$ H_{ox}, L finite $\mathbf{E} \neq 1.0$	$X = A_1 + A_2 e^{\lambda_2 Z} + A_4 e^{\lambda_4 Z}$ $Y = A_1 + A_2 a_2 e^{\lambda_2 Z} + A_4 a_4 e^{\lambda_4 Z}$ $\lambda_2, \lambda_4 = \frac{\beta}{2} \pm [(\beta/2)^2 + \gamma]^{1/2}$ (+ for λ_2) where $\beta = \mathbf{E} N_{ox} + P_x B$ $\gamma = N_{ox} P_x B(1 - \mathbf{E})$ $A_i = \frac{D_{A_i}}{D_{A_1} - D_A}$ where $D_A = \lambda_4 e^{\lambda_4}\left(1 - \frac{\lambda_2}{P_x B}\right) - \lambda_2 e^{\lambda_2}\left(1 - \frac{\lambda_3}{P_x B}\right)$ $D_{A_1} = e^{(\lambda_2 + \lambda_4)}(a_2 \lambda_4 - a_4 \lambda_2)$ $D_{A_2} = -\lambda_4 e^{\lambda_4}$ $D_{A_4} = \lambda_2 e^{\lambda_2}$	$\alpha_y = 0; \alpha_x,$ N_{ox}^1, N finite; $\mathbf{E} \neq 1.0$	$X_n = A_1 + A_2 \mu_2^n + A_4 \mu_4^n$ $Y_n = A_1 + A_2 a_2 \mu_2^n + A_4 a_4 \mu_4^n$ $(\mu_2 - 1), (\mu_4 - 1) = \frac{\beta}{2} \pm [(\beta/2)^2 + \gamma]^{1/2}$ (+ for μ_2) where $\beta = N_{ox}^1 \mathbf{E} + \frac{1 + N_{ox}^1}{\alpha_x}$ $\gamma = N_{ox}^1 \frac{1 - \mathbf{E}}{\alpha_x}$ $A_i = \frac{D_{A_i}}{D_{A_1} - D_A}$ where $D_A = \mu_4^N(\mu_4 - 1) - \mu_2^N(\mu_2 - 1)$ $+ \alpha_x(\mu_2 - 1)(\mu_4 - 1)(\mu_2^N - \mu_4^N)$ $D_{A_1} = \mathbf{E}\mu_2^N \mu_4^N(\mu_4 - \mu_2)$ $D_{A_2} = -\mu_4^N(\mu_4 - 1)$ $D_{A_4} = \mu_2^N(\mu_2 - 1)$
4	As (3) with $\mathbf{E} = 1.0$	$X = A_1 + A_2 e^{\lambda_2 Z} + A_4 Z$ $Y = A_1 + A_2 a_2 e^{\lambda_2 Z} + A_4 \left(Z + \frac{1}{N_{ox}}\right)$ $\lambda_2 = (N_{ox} + P_x B)$ $A_i = \frac{D_{A_i}}{D_{A_1} - D_A}$	As (3) with $\mathbf{E} = 1.0$	$Y_n = A_1 + A_2 \mu_2^n + A_4 n$ $Y_n = A_1 + A_2 a_2 \mu_2^n + A_4 \left(n + \frac{1}{N_{ox}^1}\right)$ $\mu_2 - 1 = \frac{1 + N_{ox}^1(1 + \alpha_x)}{\alpha_x}$ $A_i = \frac{D_{A_i}}{D_{A_1} - D_A}$

TABLE 3 (*Continued*)

Case Number		Diffusion Model		Backflow Model
		where $D_A = \dfrac{\lambda_2 e^{\lambda_2} - N_{ox}}{P_x B}$ $D_{A_1} = e^{\lambda_2}\left[a_2 - \lambda_2\left(1 + \dfrac{1}{N_{ox}}\right)\right]$ $D_{A_2} = -1.0; D_{A_4} = \lambda_2 e^{\lambda_2}$		where $D_A = [1 + \alpha_x(\mu_2 - 1)] + \alpha_x \mu_2^N(\mu_2 - 1)$ $D_{A_1} = \mu_2^N\left[a_2\mu_2 - (\mu_2 - 1)\left(N + 1 + \dfrac{1}{N_{ox}^1}\right)\right]$ $D_{A_2} = -1.0; D_{A_4} = \mu_2^N(\mu_2 - 1)$
5	$P_x \to \infty; P_y$, H_{ox}, L finite, $\mathbf{E} \neq 1.0$	$X = A_1 + A_3 e^{\lambda_3 Z} + A_4 e^{\lambda_4 Z}$ $Y = A_1 + A_3 a_3 e^{\lambda_3 Z} + A_4 a_4 e^{\lambda_4 Z}$ $\lambda_3, \lambda_4 = \dfrac{-\beta}{2} \pm [(\beta/2)^2 - \gamma]^{1/2}$ (+ for λ_4) where $\beta = N_{ox} + P_y B$ $\gamma = N_{ox} P_y B(1 - \mathbf{E})$ $A_i = \dfrac{D_{A_i}}{D_{A_1} + D_A}$ where $D_A = (a_4\lambda_4 - a_3\lambda_3)$ $D_{A_1} = \mathbf{E}[a_3\lambda_3 e^{\lambda_4} - a_4\lambda_4 e^{\lambda_3}]$ $D_{A_3} = a_4\lambda_4; D_{A_4} = -a_3\lambda_3$	$\alpha_x = 0; \alpha_y$, N_{ox}^1, N finite; $\mathbf{E} \neq 1.0$	$X_n = A_1 + A_3\mu_3^n + A_4\mu_4^n$ $Y_n = A_1 + A_3 a_3 \mu_3^n + A_4 a_4 \mu_4^n$ $(\mu_3 - 1), (\mu_4 - 1) = \dfrac{-\beta}{2} \pm [(\beta/2)^2 - \gamma]^{1/2}$ (+ for μ_4) where $\beta = \dfrac{1 + N_{ox}^1(2 - \mathbf{E} + \alpha_y)}{(1 + \alpha_y)(1 + N_{ox}^1)}$ $\gamma = \dfrac{N_{ox}^1(1 - \mathbf{E})}{(1 + \alpha_y)(1 + N_{ox}^1)}$ $A_i = \dfrac{D_{A_i}}{D_{A_1} + D_A}$ where $D_A = a_4(\mu_4 - 1) - a_3(\mu_3 - 1)$ $D_{A_1} = \mathbf{E}[a_3\mu_4^N(\mu_3 - 1) - a_4\mu_3^N(\mu_4 - 1)]$ $D_{A_3} = a_4(\mu_4 - 1); D_{A_4} = -a_3(\mu_3 - 1)$
6	As (5) with $\mathbf{E} = 1.0$	$X = A_1 + A_3 e^{\lambda_3 Z} + A_4 Z$ $Y = A_1 + A_3 a_3 e^{\lambda_3 Z} + A_4\left(Z + \dfrac{1}{N_{ox}}\right)$	As (5) with $\mathbf{E} = 1.0$	$X_n = A_1 + A_3\mu_3^n + A_4 n$ $Y_n = A_1 + A_3 a_3 \mu_3^n + A_4\left(n + \dfrac{1}{N_{ox}}\right)$

		$\lambda_3 = -(N_{ox} + P_y B)$ $A_i = \dfrac{D_{A_i}}{D_{A_1} + 1}$ where $D_{A_1} = a_3 \lambda_3 \left(1 + \dfrac{1}{N_{ox}} + \dfrac{1}{P_y B}\right) - e^{\lambda_3}$ $D_{A_3} = 1.0; D_{A_4} = -a_3 \lambda_3$		$\mu_3 - 1 = -\dfrac{1 + N_{ox}^1(1 + \alpha_y)}{(1 + N_{ox}^1)(1 + \alpha_y)}$ $A_i = \dfrac{D_{A_i}}{D_{A_1} + 1}$ where $D_{A_1} = a_3(\mu_3 - 1)\left(N + \dfrac{1}{N_{ox}^1} + \alpha_y + 1\right) - \mu_3^N$ $D_{A_3} = 1.0; D_{A_4} = -a_3(\mu_3 - 1)$
7	$P_x, P_y \to \infty$; H_{ox}, L finite; $\mathbf{E} \neq 1.0$	$X = \dfrac{\exp[\lambda_4(1 - Z)] - \mathbf{E}}{\exp(\lambda_4) - \mathbf{E}}$ $Y = \dfrac{\mathbf{E} \exp[\lambda_4(1 - Z)] - \mathbf{E}}{\exp(\lambda_4) - \mathbf{E}}$ where $\lambda_4 = N_{ox}(1 - \mathbf{E})$	$\alpha_x, \alpha_y = 0$; N_{ox}^1, N finite; $\mathbf{E} \neq 1.0$	$X_n = \dfrac{\mathbf{E}\mu_4^N - \mu_4^n}{\mathbf{E}\mu_4^N - 1}$ $Y_n = \dfrac{\mathbf{E}(\mu_4^N - \mu^{n-1})}{\mathbf{E}\mu_4^N - 1}$ where $\mu_4 = \dfrac{1 + N_{ox}^1 \mathbf{E}}{1 + N_{ox}^1}$
8	As (7) with $\mathbf{E} = 1.0$	$X = \dfrac{N_{ox}(1 - Z) + 1}{N_{ox} + 1}$ $Y = \dfrac{N_{ox}(1 - Z)}{N_{ox} + 1}$	As (7) with $\mathbf{E} = 1.0$	$X_n = \dfrac{N_{ox}^1(1 + N - n) + 1}{N_{ox}^1(1 + N) + 1}$ $Y_n = \dfrac{N_{ox}^1(1 + N - n)}{N_{ox}^1(1 + N) + 1}$
9	$P_x, P_y \to \infty$; $H_{ox} \to 0$; L finite; $\mathbf{E} \neq 1.0$	$X = Y = 1 \quad (\mathbf{E} > 1.0)$ $X = Y = 0 \quad (\mathbf{E} < 1.0)$	$\alpha_x, \alpha_y = 0$; $N_{ox}^1 \to \infty$; N finite; $\mathbf{E} \neq 1.0$	$X_n = Y_n = \dfrac{\mathbf{E}^{N+1} - \mathbf{E}^n}{\mathbf{E}^{N+1} - 1}$
10	As (9) with $\mathbf{E} = 1.0$	$X = Y = (1 - Z)$	As (9) with $\mathbf{E} = 1.0$	$X_n = Y_n = \dfrac{1 + N - n}{1 + N}$

[a]When required, a_i is obtained from Eq. (16) or (47).

TABLE 4 FURTHER SPECIAL SOLUTIONS

Case Number		Diffusion Model		Backflow Model
11	$H_{ox} \to 0$; P_x, P_y, L finite $\mathbf{E} \neq 1.0$	$X = Y = \dfrac{\mathbf{E}^2 - \mathbf{E} \exp [(1 - \mathbf{E}) P_{oy} B(1 - Z)]}{\mathbf{E}^2 - \exp [(1 - \mathbf{E}) P_{oy} B]}$ where $P_{oy} = \left[\dfrac{\mathbf{E}}{P_x} + \dfrac{1}{P_y}\right]^{-1}$	$N_{ox}^1 \to \infty$ α_x, α_y, N finite $\mathbf{E} \neq 1.0$	$X_n = Y_n = \dfrac{\mathbf{E}^2 \mu^N - \mathbf{E}\mu^n}{\mathbf{E}^2 \mu^N - \mu}$ where $\mu = \dfrac{\mathbf{E}(\alpha_x + 1) + \alpha_y}{\mathbf{E}\alpha_x + 1 + \alpha_y}$
12	As (11) with $\mathbf{E} = 1.0$	$X = Y = \dfrac{1 + P_{oy} B(1 - Z)}{2 + P_{oy} B}$	As (11) with $\mathbf{E} = 1.0$	$X_n = Y_n = \dfrac{\alpha_x + \alpha_y + 1 + N - n}{2(\alpha_x + \alpha_y) + 1 + N}$
13	$P_y = 0$; P_x, H_{ox}, L, $\mathbf{E}$ finite	$X = \dfrac{\mathbf{E}C + D + \lambda_3 \exp(\lambda_3 + \lambda_2 Z) - \lambda_2 \exp(\lambda_2 + \lambda_3 Z)}{(E + 1) C + D}$ $Y = \dfrac{\mathbf{E}C + D}{(\mathbf{E} + 1) C + D}$ $\lambda_2, \lambda_3 = \dfrac{\alpha}{2} \pm [(\alpha/2)^2 + \beta]^{1/2}$ (+ for λ_2) where $\alpha = P_x B; \beta = N_{ox} P_x B$ $C = e^{\lambda_3}(\lambda_3 + N_{ox}) - e^{\lambda_2}(\lambda_2 + N_{ox})$ $D = \mathbf{E}(\lambda_2 - \lambda_3) \exp(P_x B)$	$\alpha_y \to \infty$ α_x, N_{ox}^1, N, $\mathbf{E}$ finite	$X_n = \dfrac{\mathbf{E}C + D + \mu_3^N(\mu_3 - 1)\mu_2^n - \mu_2^N(\mu_2 - 1)\mu_3^n}{(\mathbf{E} + 1) C + D}$ $Y_n = \dfrac{\mathbf{E}C + D}{(\mathbf{E} + 1) C + D}$ $(\mu_2 - 1), (\mu_3 - 1) = \dfrac{\alpha}{2} \pm [(\alpha/2)^2 + \beta]^{1/2}$ (+ for μ_2) where $\alpha = \dfrac{1 + N_{ox}^1}{\alpha_x}; \beta = \dfrac{N_{ox}^1}{\alpha_x}$ $C = \mu_3^N [(\mu_3 - 1) + N_{ox}^1] - \mu_2^N [(\mu_3 - 1) + N_{ox}^1]$ $D = \mathbf{E}\mu_2^N \mu_3^N (\mu_2 - \mu_3)$

14	$P_x = 0$ $P_y, H_{ox}, L,$ $\mathbf{E}$ finite	$X = \dfrac{\mathbf{E}G}{(\lambda_2 - \lambda_3) + (\mathbf{E} + 1)\, G}$ $Y = \dfrac{\mathbf{E}\,[G - \lambda_3 e^{\lambda_2 Z} + \lambda_2 e^{\lambda_3 Z}]}{(\lambda_2 - \lambda_3) + (\mathbf{E} + 1)\, G}$ $\lambda_2, \lambda_3 = \dfrac{\alpha}{2} \pm [(\alpha/2)^2 + \beta]^{1/2}$ (+ for λ_2) where $\alpha = P_y B;\ \beta = \mathbf{E} N_{ox} P_y B$ $G = e^{\lambda_2}(\lambda_3 - \mathbf{E}N_{ox}) - e^{\lambda_3}(\lambda_2 - \mathbf{E}N_{ox})$	$\alpha_x \to \infty;$ $\alpha_y, N_{ox}^1, N,$ $\mathbf{E}$ finite	$X_n = \dfrac{\mathbf{E}G}{(\mu_2 - \mu_3) + (\mathbf{E} + 1)\, G}$ $Y_n = \dfrac{\mathbf{E}\,[G - (\mu_3 - 1)\, \mu_2^n + (\mu_2 - 1)\, \mu_3^n}{(\mu_2 - \mu_3) + (\mathbf{E} + 1)\, G}$ $(\mu_2 - 1), (\mu_3 - 1) = \dfrac{\alpha}{2} \pm [(\alpha/2)^2 + \beta]^{1/2}$ (+ for μ_2) where $\alpha = \dfrac{N_{ox}^1 \mathbf{E} - 1}{1 + \alpha_y};\ \beta = \dfrac{N_{ox}^1 \mathbf{E}}{1 + \alpha_y}$ $G = \mu_2^N (\mu_3 - 1)\, f(\mu_2) - \mu_3^N (\mu_2 - 1)\, f(\mu_3)$ $f(\mu_i) = [\mu_i + \alpha_y(\mu_i - 1)]$
15	As (13), with $P_x \to \infty$	$X = \dfrac{\mathbf{E}(1 - e^{\lambda_3}) + e^{\lambda_3 Z}}{\mathbf{E}(1 - e^{\lambda_3}) + 1}$ $Y = \dfrac{\mathbf{E}(1 - e^{\lambda_3})}{\mathbf{E}(1 - e^{\lambda_3}) + 1}$ $\lambda_3 = -N_{ox}$	As (13), with $\alpha_x = 0$	$Y_n = \dfrac{\mathbf{E}(1 - \mu_3^N) + \mu_3^n}{\mathbf{E}(1 - \mu_3^N) + 1}$ $Y_n = \dfrac{\mathbf{E}(1 - \mu_3^N)}{\mathbf{E}(1 - \mu_3^N) + 1}$ $\mu_3 = \dfrac{1}{N_{ox}^1 + 1}$
16	As (14), with $P_y \to \infty$	$X = \dfrac{\mathbf{E}}{1 + \mathbf{E} - e^{-\lambda_2}}$ $Y = \dfrac{\mathbf{E}(1 - e^{-\lambda_2(1 - Z)})}{1 + \mathbf{E} - e^{-\lambda_2}}$ $\lambda_2 = \mathbf{E} N_{ox}$	As (14), with $\alpha_y = 0$	$X_n = \dfrac{\mathbf{E}}{1 + \mathbf{E} - \mu_2^{-(N+1)}}$ $Y_n = \dfrac{\mathbf{E}(1 - \mu_2^{n - (N+1)})}{1 + \mathbf{E} - \mu_2^{-(N+1)}}$ $\mu_2 = \mathbf{E} N_{ox}^1 + 1$

TABLE 4 *(Continued)*

Case Number		Diffusion Model		Backflow Model
17	$P_x = P_y = 0$ $H_{ox}, L, \mathbf{E}$ finite	$X = \dfrac{1 + \mathbf{E} N_{ox}}{1 + (\mathbf{E} + 1) N_{ox}}$ $Y = \dfrac{\mathbf{E} N_{ox}}{1 + (\mathbf{E} + 1) N_{ox}}$	$\alpha_x, \alpha_y \to \infty$ $N_{ox}^1, N, \mathbf{E}$, finite	$X_n = \dfrac{1 + \mathbf{E} N_{ox}^1 N}{1 + (\mathbf{E} + 1) N_{ox}^1 N}$ $Y_n = \dfrac{\mathbf{E} N_{ox}^1 N}{1 + (\mathbf{E} + 1) N_{ox}^1 N}$
18	$\mathbf{E} = 0; P_x$, P_y, H_{ox}, L finite	$X = \dfrac{P_x B [\lambda_3 e^{-\lambda_2 (1 - Z)} - \lambda_2 e^{-\lambda_3 (1 - Z)}]}{\lambda_3^2 e^{-\lambda_2} - \lambda_2^2 e^{-\lambda_3}}$ $Y = 0^a$ $\lambda_2, \lambda_3 = \dfrac{\alpha}{2} \pm [(\alpha/2)^2 + \beta]^{1/2}$ (+ for λ_2) where $\alpha = P_x B; \beta = N_{ox} P_x B$	$\mathbf{E} = 0; \alpha_x$, α_y, N_{ox}^1, N finite	$X_n = \dfrac{(\mu_3 - 1) \mu_3^N \mu_2^n - (\mu_2 - 1) \mu_2^N \mu_3^n}{\mu_3^N (\mu_3 + N_{ox}^1 - 1) - \mu_2^N (\mu_2 + N_{ox}^1 - 1)}$ $Y_n = 0^a$ $(\mu_2 - 1), (\mu_3 - 1) = \dfrac{\alpha}{2} \pm [(\alpha/2)^2 + \beta]^{1/2}$ (+ for μ_2) where $\alpha = \dfrac{1 + N_{ox}^1}{\alpha_x}; \beta = \dfrac{N_{ox}^1}{\alpha_x}$

[a] In case 18, Y or $Y_n = 0$ refers to unreacted solute; the exit reacted solute concentration is obtained by substitution of X_I or X_N into the overall mass balance [Eq. (64)].

the whole concentration range, instead of at individual point values as with stagewise contactors, and cases 9 and 10, with no backmixing, are in fact trivial.

2.5.3. Perfect Mixing of One or Both Phases (Cases 13-17)

The results for these cases are listed in Table 4, the first two corresponding to perfect mixing in one phase, that is, to $P_j = 0$ or $\alpha_j \to \infty$, with normal backmixing in the other. A limit occurs with the second phase in plug flow, giving cases 15 and 16 and a further limit, case 17, with both phases perfectly mixed; the latter is identical with case 4 in Table 3 in Chapter 5.

2.5.4. Homogeneous Reaction (Case 18)

If an infinitely rapid irreversible first-order reaction occurs in the Y phase, then $E \to 0$, giving case 18. In practice, this could be applicable, for instance, to the removal of acid or base from an organic phase by scrubbing with aqueous alkali or acid, respectively.

3. SIMPLIFIED ANALYTICAL DESIGN METHODS

3.1. Linear Equilibria

The foregoing analytical relationships are of considerable complexity, and iteration procedures using a digital computer are required for solution of extractor length or number of stages. For the diffusional model, therefore, a number of empirical methods have been proposed relating the number of "true" transfer units N_{ox} to N_{oxP} and the Peclet numbers [6, 7, 21, 22]. However, the relationship between N_{ox} and N_{oxP} is itself complex, as shown in Section 2.1.7, and such methods are thus unreliable.

An alternative approach was adopted by Mecklenburgh and Hartland [19], who showed that the roots of the characteristic equation λ_i have properties such that a considerable simplification of the analytical solution is possible. Unfortunately, their method is restricted to cases where the extraction factor **E** is close to unity, but Pratt [15] devised an alternative method that gives a closed solution for length and is free from this restriction. A similar method was developed for staged contactors using the backflow model [23]. These methods, described in the following paragraphs, lead to relatively simple design procedures that do not require the use of a computer.

3.1.1. Diffusion Model

(a) Properties of Roots. An examination of the values of the roots λ_i of the characteristic equation [Eq. (14)] over a wide range of parameter values shows that they possess the following properties [19]: (1) λ_2 is large and positive, increasing with P_x; (2) λ_3 is negative and $|\lambda_3|$ is large, increasing with P_y; and (3) $\lambda_4 = 0$ when $\mathbf{E} = 1.0$, small and negative when $\mathbf{E} < 1.0$, and small and positive when $\mathbf{E} > 1.0$. Consequently, all terms containing $\exp(-\lambda_2)$ and $\exp(\lambda_3)$ can be disregarded in most circumstances because of properties (1) and (2).

(b) Derivation of Equations [15]. Considering first case 1 in Table 1, the expression for the Y profile is expanded by writing out the A_i in full. The numerator and denominator are then divided by $\exp(\lambda_2)$ and all terms containing $\exp(-\lambda_2)$ and $\exp(\lambda_3)$ disregarded; further, since λ_2 is large, the term containing $\exp[-\lambda_2(1 - Z)]$ can be neglected, except near the X-phase outlet, where $Z \to 1.0$. On putting $\lambda_i = \lambda_i' L$, the resulting expression becomes

$$1 - Y \simeq \frac{a_4\lambda_2'(\lambda_4' - \lambda_3') - \mathbf{E}a_4\lambda_2'(\lambda_4' e^{\lambda_3' LZ} - \lambda_3' e^{\lambda_4' LZ})}{\mathbf{E}^2\lambda_3'(\lambda_2' - \lambda_4')\, e^{\lambda_4' L} + a_4\lambda_2'(\lambda_4' - \lambda_3')} \tag{62}$$

After substitution of $Y = Y^0$ at $Z = 0$, this can be solved for the contactor length L, as follows:

$$L \simeq \frac{1}{\lambda_4'} \ln \frac{a_4\lambda_2'(\lambda_4' - \lambda_3')(\mathbf{E} - Y^0)}{\mathbf{E}^2\lambda_3'(\lambda_4' - \lambda_2')(1 - Y^0)} \tag{63}$$

A corresponding solution in terms of X^I can be obtained by substitution for Y^0 in Eq. (63) by means of the overall material balance (cf. Table 1 in Chapter 5):

$$Y^0 = \mathbf{E}(1 - X^I) \tag{64}$$

The calculation of extractor length for given values of P_x, P_y, H_{ox}, and **E** is now a simple matter. Thus the characteristic equation in the form given by Eqs. (18)-(21) is solved by means of Eq. (17) for the roots λ_i', and a_4 is calculated; substitution in Eq. (63) then gives the length directly. In this regard the values of the a_i remain unchanged when expressed in terms of the modified roots λ_i', and Eq. (16) becomes

$$a = \frac{\mathbf{E}(1 - \lambda_i' d_c/P_x)}{1 + \lambda_i' d_c/P_y} \tag{65}$$

The corresponding relationships for backmixing in one phase only, and for $\mathbf{E} = 1.0$, that is, for cases 2–6 in Tables 1 and 3, are obtained in a similar manner. The resulting expressions for L are summarized in Table 5, and the simplified profile equations in Table 6.

(c) Accuracy of Solutions. Equation (63) was tested for accuracy [15] by comparison of the results for a wide range of parameter values with the exact computed values taken from the extensive tabulation by McMullen et al. [24]. As an empirical rule, it was suggested [15] that the error would be well within 5% if the parameter $L_{calc}(\lambda_2' - \lambda_3')$ exceeds 7–8 with $N_{ox} \geqslant 2$. The accuracy is generally somewhat better when backmixing occurs in only one of the phases.

The use of the method is illustrated by the following worked example.

Example 1. Calculate the column height required to meet the specification $c_x^0 = 120.0$, $c_x^I = 34.34$, and $c_y^I = 8.00$, all in kg/m^3. The design parameters are $U_x/U_y = 0.800$, $P_x = 0.025$, $P_y = 0.100$, $d_c = 0.10$ ft, and $H_{ox} = 2.00$ ft; the equilibrium data are closely represented by

$$c_x^* = 0.625 c_y + 0.500$$

1. **Material Balance.** From the equilibrium relationship,

$$c_x^{I*} = 0.625 \times 8.00 + 0.500 = 5.50$$

Hence from Eq. (6)

$$X^I = \frac{34.43 - 5.50}{120.0 - 5.50} = 0.2527$$

Since $\mathbf{E} = 0.625 \times 0.800 = 0.500$, Eq. (64) gives

$$Y^0 = \mathbf{E}(1 - X^I) = 0.500\,(1 - 0.2527)$$
$$= 0.3736$$

2. **Characteristic Equation.** From Eqs. (19)–(21),

$$\alpha' = \frac{0.025 - 0.100}{0.100} = -0.750$$

$$\beta' = \frac{0.025 + 0.500 \times 0.10}{2.00 \times 0.10} + \frac{0.025 \times 0.10}{0.10^2} = 0.625$$

$$\gamma' = \frac{0.025 \times 0.100\,(1.0 - 0.500)}{2.00 \times 0.10^2}$$
$$= 0.0625$$

Substitution into Eq. (17) gives the roots of Eq. (18) as

$$\lambda_2' = +0.5614525; \qquad \lambda_3' = -1.2202247;$$
$$\lambda_4' = -0.0912278$$

From Eq. (65), $a_4 = 0.750964$.

3. **Height Required.** Substitution into Eq. (63) gives $L = 7.99$ ft.

Since $L(\lambda_2' - \lambda_3') = 14.25$ and $N_{ox} = 7.99/2.00 = 4.0$, this result should be of adequate accuracy, and in fact the exact result is 8.00 ft.

3.1.2. *Backflow Model*

(a) Properties of Roots. In this case the roots μ_i of the characteristic equation [Eq. (48)] possess the following properties [23]): (1) μ_2 is relatively large and positive, so that terms containing μ_2^{-N} can be disregarded if N is not too small; (2), $0 < \mu_3 < 1$, so that terms containing μ_3^N can be disregarded if N is not too small; and (3) $\mu_4 = 1.0$ when $\mathbf{E} = 1.0$; also, $(\mu_4 - 1)$ is small and positive when $\mathbf{E} > 1.0$ and is small and negative when $\mathbf{E} < 1.0$.

(b) Derivation of Equations. Following a procedure similar to that for the diffusional model, the equation for X_n or Y_n of case 1 in Table 2 is written out in full, with the numerator and denominator divided by μ_2^N and terms containing μ_2^{-N} and μ_3^N disregarded. On substitution of $Y = Y^0$, $n = 0$ in the simplified expression for Y_n, the resulting expression can be solved for the total number of stages N. The simplified solutions thus obtained for the number of stages and concentration profiles, together with those for cases 2–6, are included in Tables 5 and 6.

(c) Accuracy of Solutions. As for the diffusion model, the accuracy was tested by substituting exact computed exit concentrations for a given number of stages into the simplified equations [23]. It was apparent that the simplified method is of acceptable accuracy for design provided that N is not less than 4–6, and for increasing numbers of stages, the error rapidly becomes negligible. In all cases the error was found to be negative, so that the calculated number should be rounded off upward to the nearest integer.

Example 2. Calculate the number of contactor stages required to meet the specification $c_x^0 = 120.0$, $c_x^N = 53.24$, and $c_y^1 = 8.00$, all in kg/m^3. The contactor characteristics are $\alpha_x = 3.00$, $\alpha_y = 1.50$, $N_{ox}^1 = 0.160$, and $F_x/F_y = 1.040$. The equilibrium relationship is the same as for Example 1.

1. **Material Balance.** As in Example 1, $c_x^{N*} = 5.50$; also, $X^N = 0.4169$ and $\mathbf{E} = 0.650$, so that Eq. (64) gives $Y^1 = 0.3790$.
2. **Characteristic Equation.** The coefficients in Eq. (48) are given by Eqs. (41)–(43) as follows:

   $$\alpha = 0.02827, \qquad \beta = 0.23573,$$
   $$\gamma = 0.0074667$$

   Substitution into Eq. (17) gives the roots as follows:

   $$(\mu_2 - 1) = 0.514576;$$
   $$(\mu_3 - 1) = -0.454375;$$
   $$(\mu_4 - 1) = -0.031936$$

3. **Number of Stages.** Equation (47) gives $a_4 = 0.77407$. Hence substitution in the equation for case 1 (backflow model) in Table 5 gives

   $$N = 10.99 \text{ stages}$$

   Rounding off upward gives 11 stages which is in agreement with the exact value. Details of this calculation are given by Pratt [23].

3.2. Nonlinear Equilibria

The foregoing methods are not applicable when the equilibrium line is curved, and computer methods of the type described in Section 4.2 must be used to obtain an exact solution. The graphical methods of Section 4.1 are also applicable but, apart from limits on accuracy, are difficult to apply when backmixing occurs in both phases.

As an alternative, a solution can usually be obtained by subdividing the contactor into sections, within each of which the equilibrium relation can be approximated by straight line segments. A complex method of this type, requiring computer solution, has been described by Wilburn and Nicholson [17] (see Section 2.4). A simpler approach was adopted by Pratt [25], who proposed two alternative methods for the diffusion model based on the simplified solutions given previously. These are described in Section 3.2.1 together with an adaptation of the first method to the backflow model.

3.2.1. *Diffusion Model*

(a) First Method. In this method the equilibrium curve is approximated by two or three straight line segments; to simplify the presentation, two sections are assumed. The procedure is then as follows (see Fig. 5):

1. The jump concentration $c_{y,I}$ is assumed, and the equilibrium line is approximated by the two straight lines AB and BE, the latter terminating at $c_{y,I}$, that is, at E. The two lines are then extended to cover the full concentration range, giving ABC and $DEBG$.
2. The dimensionless concentrations Y^0 and Y_k, the latter corresponding to point B, are calculated; these will differ for the two sections since the two equilibrium lines have different slopes.
3. The appropriate equation from Table 5 is used to calculate the overall length L_f^0, assuming equilibrium line ABC, and the fraction Z_f of this required for the feed section, down to concentration Y_k, is obtained by using the appropriate equation in Table 7. The actual length of this section is then $L_f^0 Z_f$.
4. The values of L_s^0 and Z_s for the solvent inlet section are obtained in the same way by using equilibrium line DBG. The length of this section is then $L_s^0(1 - Z_s)$.
5. The jump concentrations $c_{x,0}$ and $c_{y,I}$ are calculated for each section by substituting $Z = 0$ and 1.0, respectively, in the X- and Y-profile equations in Table 6. These are used to relocate the straight line segments if necessary (particularly BE, in Fig. 5, noting that no mass transfer

TABLE 5 SIMPLIFIED DESIGN EQUATIONS (LINEAR EQUILIBRIUM RELATIONSHIP

Case Number		Diffusion Model		Backflow Model
1	P_x, P_y, H_{ox}, L finite, $\mathbf{E} \neq 1.0$	$L \simeq \frac{1}{\lambda'_4} \ln \left\{ \frac{\lambda'_2 a_4 (\lambda'_4 - \lambda'_3)(\mathbf{E} - Y^0)}{\mathbf{E}^2 \lambda'_3 (\lambda'_4 - \lambda'_2)(1 - Y^0)} \right\}$ λ'_i - given by Eqs. (17)–(21)	$\alpha_x, \alpha_y, N^1_{ox}$, N finite, $\mathbf{E} \neq 1.0$	$N \simeq \frac{\log \left\{ \frac{a_4(\mu_2 - 1)(\mu_4 - \mu_3)(\mathbf{E} - Y^0)}{\mathbf{E}^2(\mu_3 - 1)(\mu_4 - \mu_2)(1 - Y^0)} \right\}}{\log \mu_4}$ μ_i - given by Eqs. (17), (41)–(43), and (48)
2	As (1) with $\mathbf{E} = 1.0$	$L \simeq \frac{Y^0}{1 - Y^0} \left[H_{ox} + d_c \left(\frac{1}{P_x} + \frac{1}{P_y} \right) \right] - d_c \left[\frac{1}{P_x} + \frac{1}{P_y} \right] + \left[\frac{1}{\lambda'_2} - \frac{1}{\lambda'_3} \right]$ $\lambda'_2, \lambda'_3 = \frac{\alpha'}{2} \pm [(\alpha'/2)^2 + \beta']^{1/2}$ (+ for λ'_2) with α', β' given by Eqs. (19)–(20)	As (1) with $\mathbf{E} = 1.0$	$N \simeq \frac{Y^0}{1 - Y^0} \left[\frac{1}{N^1_{ox}} + \alpha_x + \alpha_y + 1 \right] - (\alpha_x + \alpha_y + 1) + \left[\frac{1}{\mu_2 - 1} - \frac{1}{\mu_3 - 1} \right]$ μ_2, μ_3—see Table 1, case 2
3	$P_y \to \infty; P_x$, H_{ox}, L finite; $\mathbf{E} \neq 1.0$	$L \simeq \frac{1}{\lambda'_4} \ln \left\{ \frac{\lambda'_2 a_4 (\mathbf{E} - Y^0)}{\mathbf{E}^2 (\lambda'_2 - \lambda'_4)(1 - Y^0)} \right\}$ $\lambda'_2, \lambda'_4 = \frac{\beta'}{2} \pm [(\beta'/2)^2 + \gamma]^{1/2}$ (+ for λ'_2)	$\alpha_y = 0; \alpha_x$, N^1_{ox}, N finite, $\mathbf{E} \neq 1.0$	$N \simeq \frac{\log \left\{ \frac{a_4 \mu_4 (\mu_2 - 1)(\mathbf{E} - Y^0)}{\mathbf{E}^2(\mu_2 - \mu_4)(1 - Y^0)} \right\}}{\log \mu_4}$ μ_2, μ_4—see Table 3, case 3

		with $\beta' = \dfrac{\mathbf{E}}{H_{ox}} + \dfrac{P_x}{d_c}$ $\gamma' = P_x \dfrac{1 - \mathbf{E}}{H_{ox} d_c}$		
5	$P_x \to \infty; P_y$, H_{ox}, L finite; $\mathbf{E} \neq 1.0$	$L \simeq \dfrac{1}{\lambda_4'} \ln \left\{ \dfrac{a_4(\lambda_3' - \lambda_4')(\mathbf{E} - Y^0)}{\mathbf{E}^2 \lambda_3'(1 - Y^0)} \right\}$ $\lambda_3', \lambda_4' = \dfrac{-\beta'}{2} \pm [(\beta'/2)^2 - \gamma]^{1/2}$ (+ for λ_4') with $\beta' = \dfrac{1}{H_{ox}} + \dfrac{P_y}{d_c}$ $\gamma' = P_y \dfrac{1 - \mathbf{E}}{H_{ox} d_c}$	$\alpha_x = 0; \alpha_y$, N_{ox}^1, N finite; $\mathbf{E} \neq 1.0$	$N \simeq \dfrac{\log \left\{ \dfrac{a_4(\mu_3 - \mu_4)(\mathbf{E} - Y^0)}{\mathbf{E}^2(\mu_3 - 1)(1 - Y^0)} \right\}}{\log \mu_4}$ μ_3, μ_4—see Table 3, case 5
4, 6	As (3) and (5), respectively, with $\mathbf{E} = 1.0$	$L \simeq \dfrac{Y^0 \left[2 + 2 \dfrac{d_c}{H_{ox} P_j} + \dfrac{H_{ox} P_j}{d_c} \right] - \dfrac{d_c}{H_{ox} P_j}}{(1 - Y^0) \left[\dfrac{1}{H_{ox}} + \dfrac{P_j}{d_c} \right]}$ (P_j refers to finite P_x or P_y)	As (3) and (5), respectively, with $\mathbf{E} = 1.0$	$N \simeq \dfrac{Y^0}{(1 - Y^0)} \left[\dfrac{1}{N_{ox}^1} + \alpha_j + 1 \right] - \dfrac{N_{ox}^1 \alpha_j (1 + \alpha_j)}{1 + N_{ox}^1 (1 + \alpha_j)}$ (α_j refers to nonzero α_x or α_y)

TABLE 6 SIMPLIFIED PROFILE EQUATIONS[a]

Case Number		Diffusion Model
1	P_x, P_y, H_{ox}, L finite, $\mathbf{E} \neq 1.0$	$X \simeq \dfrac{\mathbf{E}\{C_1 + C_2 e^{-\lambda_2' L(1-Z)} + \lambda_2'(a_4 \lambda_4' e^{\lambda_3' LZ} - a_3 \lambda_3' e^{\lambda_4' LZ})\}}{\mathbf{E}C_1 + a_3 a_4 \lambda_2'(\lambda_4' - \lambda_3')}$ where $C_1 = \mathbf{E} a_3 \lambda_3'(\lambda_2' - \lambda_4')\, e^{\lambda_4' L}$ $C_2 = a_3 \lambda_3' \lambda_4' e^{\lambda_4' L}$
2	As (1) with $\mathbf{E} = 1.0$	$X \simeq \dfrac{C_3 + a_3 \lambda_2' \lambda_3' L + a_3 \lambda_3' e^{-\lambda_2' L(1-Z)} + \lambda_2' e^{\lambda_3' LZ} - a_3 \lambda_2' \lambda_3' C_5}{C_3 + C_4 + a_3 \lambda_2' \lambda_3' L}$ where $C_3 = a_3 \lambda_3' \left[\lambda_2' \left(H_{ox} + \dfrac{d_c}{P_y}\right) - \left(1 - \dfrac{\lambda_2' d_c}{P_x}\right)\right]$ $C_4 = \lambda_2' \left[\left(1 - \dfrac{\lambda_3' d_c}{P_x}\right) + \left(\dfrac{a_3 \lambda_3' d_c}{P_x}\right)\right]$ $C_5 = LZ$ (X profile) $C_5 = (LZ + H_{ox})$ (Y profile)
3	$P_y \to \infty$; P_x, H_{ox}, L finite; $\mathbf{E} \neq 1.0$	$X \simeq \dfrac{\mathbf{E}\{\mathbf{E}(\lambda_2' - \lambda_4')\, e^{\lambda_4' L} + \lambda_4' e^{\lambda_4' L} e^{-\lambda_2' L(1-Z)} - \lambda_2' e^{\lambda_4' LZ}\}}{\mathbf{E}^2(\lambda_2' - \lambda_4')\, e^{\lambda_4' L} - a_4 \lambda_2'}$
4	As (3) with $\mathbf{E} = 1.0$	$X \text{ or } Y \simeq \dfrac{\lambda_2'(L + H_{ox}) + (d_c/H_{ox} P_x) + C_6}{\lambda_2'(L + H_{ox} + d_c/P_x) + d_c/H_{ox} P_x}$ where $C_6 = e^{-\lambda_2' L(1-Z)} - \lambda_2' LZ$ (X profile) $C_6 = -\left[\dfrac{d_c}{H_{ox} P_x}\right] e^{-\lambda_2' L(1-Z)} - \lambda_2'(LZ + H_{ox})$ (Y profile)
5	$P_x \to \infty$; P_y, H_{ox}, L finite; $\mathbf{E} \neq 1.0$	$X \simeq \dfrac{\mathbf{E}\{\mathbf{E} a_3 \lambda_3' e^{\lambda_4' L} + a_4 \lambda_4' e^{\lambda_3' LZ} - a_3 \lambda_3' e^{\lambda_4' LZ}\}}{\mathbf{E}^2 a_3 \lambda_3' e^{\lambda_4' L} + a_3 a_4(\lambda_4' - \lambda_3')}$
6	As (5) with $\mathbf{E} = 1.0$	$X \text{ or } Y \simeq \dfrac{-\lambda_3'(L + H_{ox} + d_c/P_y) + C_7}{-\lambda_3'(L + H_{ox} + d_c/P_y) + d_c/H_{ox} P_y}$ where $C_7 = \dfrac{d_c}{H_{ox} P_y}\, e^{\lambda_3' LZ} + \lambda_3' LZ$ (X profile) $C_7 = -e^{\lambda_3' LZ} + \lambda_3'(LZ + H_{ox})$ (Y profile)

[a]For Y profile in cases 1–3 and 5, multiply terms containing exp $[-\lambda_2' L(1 - Z)]$, exp $(\lambda_3' LZ)$, and exp $(\lambda_4' LZ)$ by corresponding a_i from Eq. (65) or those containing μ_2^{n-N}, μ_3^n, and μ_4^n by a_i from Eq. (47). For values of roots λ_i' or μ_i, see Table 5 for cases 1–3 and 5 and Table 7 for cases 4 and 6.

	Backflow Model
$\alpha_x, \alpha_y,$ $N^1_{ox},$ N finite, $\mathbf{E} \neq 1.0$	$X_n \simeq \frac{\mathbf{E}\{C_1 + C_2\mu_2^{n-N} + (\mu_2 - 1)[a_4(\mu_4 - 1)\mu_3^n - a_3(\mu_3 - 1)\mu_4^n]\}}{\mathbf{E}C_1 + a_3 a_4(\mu_2 - 1)(\mu_4 - \mu_3)}$ where $C_1 = \mathbf{E}a_3(\mu_3 - 1)(\mu_2 - \mu_4)\mu_4^N$ $C_2 = a_3(\mu_3 - 1)(\mu_4 - 1)\mu_4^N$
As (1) with $\mathbf{E} = 1.0$	$X_n \simeq \frac{C_3 + a_3(\mu_2 - 1)(\mu_3 - 1)N + a_3(\mu_3 - 1)\mu_2^{n-N} + (\mu_2 - 1)\mu_3^n - C_5}{C_3 + C_4 + a_3(\mu_2 - 1)(\mu_3 - 1)N}$ where $C_3 = a_3(\mu_3 - 1)\left\{(\mu_2 - 1)\left(\alpha_x + \alpha_y + 1 + \frac{1}{N^1_{ox}}\right) - 1\right\}$ $C_4 = (\mu_2 - 1)[1 - \alpha_x(\mu_3 - 1)(1 - a_3)]$ $C_5 = a_3(\mu_2 - 1)(\mu_3 - 1)n$ (X profile) $C_5 = a_3(\mu_2 - 1)(\mu_3 - 1)\left(n + \frac{1}{N^1_{ox}}\right)$ (Y profile)
$\alpha_y = 0;$ $\alpha_x, N^1_{ox},$ N finite; $\mathbf{E} \neq 1.0$	$X_n \simeq \frac{\mathbf{E}\{\mathbf{E}(\mu_2 - \mu_4)\mu_4^N + (\mu_4 - 1)\mu_4^N\mu_2^{n-N} - (\mu_2 - 1)\mu_4^n\}}{\mathbf{E}^2(\mu_2 - \mu_4)\mu_4^N - a_4(\mu_2 - 1)\mu_4}$
As (3) with $\mathbf{E} = 1.0$	X_n or $Y_n \simeq \frac{1 - (\mu_2 - 1)(\alpha_x + N + 1 + 1/N^1_{ox}) + C_6}{1 - (\mu_2 - 1)(2\alpha_x + N + 1 + 1/N^1_{ox})}$ where $C_6 = -\mu_2^{n-N} + (\mu_2 - 1)n$ (X profile) $C_6 = -[1 - \alpha_x(\mu_2 - 1)]\mu_2^{n-N-1} + (\mu_2 - 1)\left(n + \frac{1}{N^1_{ox}}\right)$ (Y profile)
$\alpha_x = 0;$ α_y, N^1_{ox}, N finite; $\mathbf{E} \neq 1.0$	$X_n \simeq \frac{\mathbf{E}\{\mathbf{E}a_3(\mu_3 - 1)\mu_4^N + a_4(\mu_4 - 1)\mu_3^n - a_3(\mu_3 - 1)\mu_4^n\}}{\mathbf{E}^2 a_3(\mu_3 - 1)\mu_4^N + a_3 a_4(\mu_4 - \mu_3)}$
As (5) with $\mathbf{E} = 1.0$	X_n or $Y_n \simeq \frac{(\mu_3 - 1)(\alpha_y + N + 1 + 1/N^1_{ox}) + C_7}{(\mu_3 - 1)(2\alpha_y + N + 1 + 1/N^1_{ox}) + \mu_3}$ where $C_7 = [\mu_3 + \alpha_y(\mu_3 - 1)]\mu_3^n - (\mu_3 - 1)n$ (X profile) $C_7 = \mu_3^n - (\mu_3 - 1)\left(n + \frac{1}{N^1_{ox}}\right)$ (Y profile)

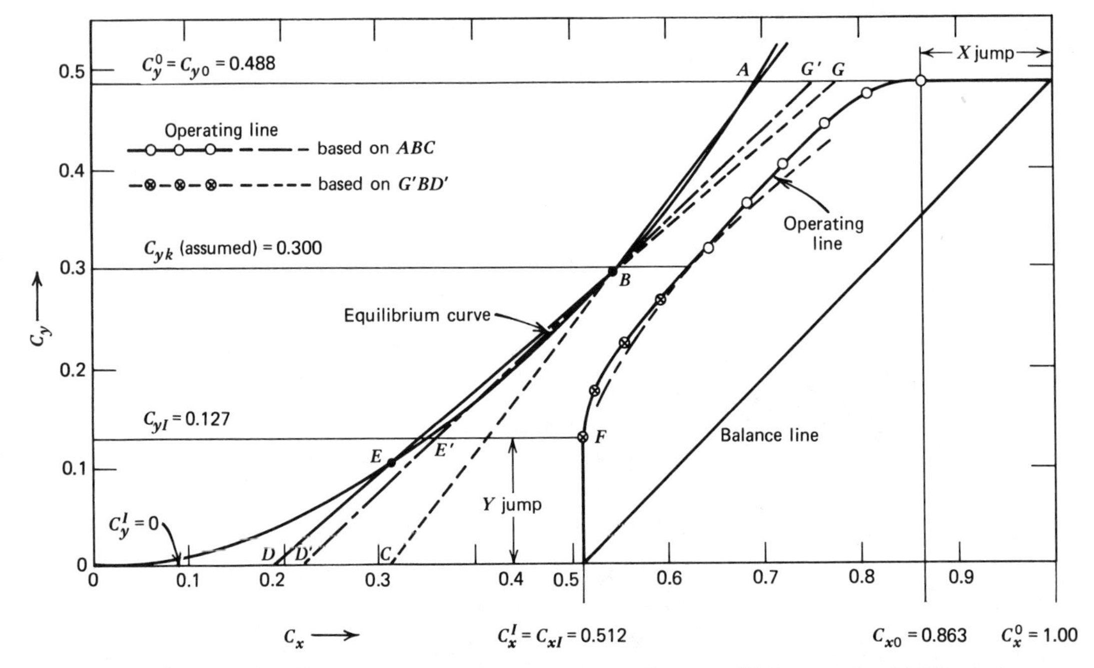

Figure 5. Operating diagram on c_y-c_x coordinates for nonlinear equilibrium relationship (Examples 3 and 4).

occurs at concentrations below $c_{y,I}$), in which case steps 3–5 are repeated.

6. The total height required is then $L = L_f^0 Z_f + L_s^0(1 - Z_s)$.
7. If a third, middle section is used, the values of Z_{m1} and Z_{m2} at the two ends must be calculated and $L_m^0(Z_{m2} - Z_{m1})$ added to the length in step 6.

The equations listed in Table 7, used in steps 3 and 4, are obtained from the simplified profile equations in Table 6 by omitting the terms $\exp[-\lambda_2' L^0(1 - Z)]$ and $\exp(\lambda_3' L^0 Z)$ and solving for Z. These are valid in the range $Z \simeq 0.35$–0.65, and if the calculated value lies outside this range, the location of B on AC or the line AC itself, must be adjusted. Alternatively, the calculated L^0 and Z values can be used in the appropriate Y-profile equation in Table 6 to obtain a more exact value of Y_k, which is then used to readjust the straight line segments.

Example 3. Calculate the extractor length required to meet the specification $c_x^0 = 1.0$, $c_x^I = 0.512$, and $c_y^I = 0.0$, all in kmol/m^3; other parameters are $U_x/U_y = 1.0$, $P_x = P_y = 0.040$, $H_{ox} = 2.00$ ft, and $d_c = 0.10$ ft. The equilibrium relationship is given by $c_y = c_x^2$.

1. **Equilibria.** The equilibrium curve is shown in Fig. 5, together with the balance line. The exit solvent phase composition, by overall material balance, is $c_y^0 = 1.00\ (1.000–0.512) = 0.488$. Since P_y is small, the Y jump will be considerable, so that the lower part of the equilibrium curve can be disregarded. As a first trial, therefore, the column will be divided at $c_{yk} = 0.300$ with equilibria represented by AB and BE.
2. **Feed Entry Section.** Extending AB to C, intercept $q_f = 0.312$. Also, $m_f = (0.6986 - 0.312)/0.488 = 0.7922 = \mathbf{E}_f$. Transformation of these coordinates over the *full* concentration range AC gives

$$Y_f^0 = \frac{0.7922(0.488 - 0)}{1.0 - 0.312} = 0.5619,$$

$$Y_{kf} = \frac{0.7922(0.300 - 0)}{1.0 - 0.312} = 0.3454$$

Solution of the characteristic equation, that is, Eqs. (18)–(21) with

$$P_x/d_c = P_y/d_c = 0.400$$

$$H_{ox} = 2.0 \qquad \text{and} \qquad \mathbf{E}_f = 0.7922$$

gives, with Eq. (65) for a_4

$$\lambda_2' = 0.735555, \qquad \lambda_3' = -0.703425,$$

$$\lambda_4' = -0.032129, \qquad a_4 = 0.930580$$

Hence from the equation for case 1 in Table 5, $L_f^0 = 10.540$ ft, so that $\exp(\lambda_4' L_f^0) = 0.71273$. Substitution of the latter in the corresponding equation in Table 7 gives $Z_f = 0.5480$, so that $L_f = L_f^0 Z_f = 5.776$ ft.

3. **Solvent Entry Section.** Line BD has intercept $q_s = 0.187$ and slope $m_s = \mathbf{E}_s = 1.210$. Hence

$$Y_s^0 = \frac{1.210(0.488 - 0)}{1.0 - 0.187} = 0.7263,$$

$$Y_{ks} = \frac{1.210(0.300 - 0)}{1.0 - 0.187} = 0.4465$$

The λ_i' and a_4 are

$$\lambda_2' = 0.761538, \qquad \lambda_3' = -0.789481,$$

$$\lambda_4' = 0.027943, \qquad a_4 = 1.051982$$

As before,

$$L_s^0 = 11.531 \text{ ft} \quad \text{and}$$

$$\exp(\lambda_4' L_s^0) = 1.364835,$$

giving $Z_s = 0.6370$. Hence $L_s = L_s^0(1 - Z_s) = 4.041$ ft.

4. **Jump Concentrations.** Substitution of $Z = 0$ and $Z = 1$ in the profile equations in Table 6 for the feed and solvent entry sections, respectively, gives $c_{x,0} = 0.8625$, which differs adequately from the corresponding equilibrium concentration A in Fig. 5, and $c_{y,I} = 0.1325$.
5. **Total Length.** $L = 5.776 + 4.041 = 9.82$ ft. This result can be refined by adjusting the lower equilibrium line to $D'E'G'$, where E' better corresponds to the Y-phase jump concentration at F, giving $L_s = 4.180$. Hence the final length is 9.96 ft, which compares well with the exact value of 10.0 ft [19(c)].

(b) Second Method. In practice, situations can arise where the preceding method fails due to an equilibrium line segment intersecting the operating line on extension, such as in the vicinity of $c_{x,0}$ in Fig. 5. An alternative method was thus devised [25] in which the length of each section is calculated in terms of its *actual* concentration range, treating it as an independent extractor, and applying special boundary conditions based on continuity of concentration gradients at the junctions between sections. Full details of the method and its application to the solution of Example 3 are given by Pratt [25].

3.2.2. Backflow Model

Of the procedures described previously, only the first is applicable to the backflow model since the stepwise nature of the concentration changes precludes the use of continuity conditions at the junction of two sections. The procedure using the first method is essentially the same as that for the differential model. Thus the total number of stages required over the full concentration range is calculated for each of the straight equilibrium line approximations by use of the simplified solutions given in Table 5. The numbers required over the ranges of approximate coincidence of the straight lines with the equilibrium curve are then calculated using the expressions for $n(Y_k)$ in Table 7 and added together to give the total number of stages; this is then rounded off to the next higher integer number.

The lower limit of concentration to be approximated by the straight line for the solvent entry section corresponds to $c_{y,N}$ (see $c_{y,I}$ in Fig. 5); this can be calculated by substitution of $n = N$ in the appropriate simplified profile equation in Table 6, noting that the term in μ_3^n can be disregarded. The value of the "jump" concentration $c_{y,N+1}$ can similarly be obtained by substitution of $n = (N + 1)$ in the same equation, but this is seldom required. The procedure is illustrated by means of the following worked example.

Example 4. Repeat Example 3, assuming the use of a stagewise contactor with the following characteristics: $\alpha_x = 3.00$; $\alpha_y = 1.50$; and Murphree stage efficiency $E_{Mx} = 60\%$.

1. **Equilibria.** As for Example No. 3, the equilibrium curve in Fig. 5 will be approximated by straight lines AB and BE, intersecting at $c_{yk} = 0.300$.
2. **Value of N_{ox}^1.** From Eq. (133) in Chapter 5, $N_{ox}^1 = 0.60/(1 - 0.60) = 1.50$ transfer units per stage.

TABLE 7 NONLINEAR EQUILIBRA: VALUES OF $Z(Y_k)$ AND $n(Y_k)$[a]

Case Number		Diffusion Model
1	P_x, P_y, H_{ox}, L finite, $\mathbf{E} \neq 1.0$	$Z \simeq \frac{1}{\lambda_4' L} \ln \left\{ \frac{\mathbf{E}^2 \lambda_3'(\lambda_2' - \lambda_4') e^{\lambda_4' L}(1 - Y_k) - a_4 \lambda_2'(\lambda_4' - \lambda_3') Y_k}{\mathbf{E} a_4 \lambda_2' \lambda_3'} \right\}$
2	As (1) with $\mathbf{E} = 1.0$	$Z \simeq \frac{K(1 - Y_k) - \lambda_2' \left[1 - \frac{\lambda_3' d_c}{P_x}(1 - a_3)\right] Y_k}{a_3 \lambda_2' \lambda_3' L} - \frac{H_{ox}}{L}$ where $K = a_3 \lambda_3' \left\{ \lambda_2' \left[L + H_{ox} + d_c \left(\frac{1}{P_x} + \frac{1}{P_y} \right) \right] - 1 \right\}$
3	$P_y \to \infty$; P_x, H_{ox}, L finite; $\mathbf{E} \neq 1.0$	$Z \simeq \frac{1}{\lambda_4' L} \ln \left\{ \frac{\mathbf{E}^2 (\lambda_2' - \lambda_4') e^{\lambda_4' L}(1 - Y_k) + a_4 \lambda_2' Y_k}{\mathbf{E} a_4 \lambda_2'} \right\}$
4	As (3) with $\mathbf{E} = 1.0$	$Z \simeq \frac{\left(\lambda_2' L + \frac{d_c}{H_{ox} P_x}\right)(1 - Y_k) - \lambda_2' \left(H_{ox} + \frac{d_c}{P_x}\right) Y_k}{\lambda_2' L}$ where $\lambda_2' = \frac{1}{H_{ox}} + \frac{P_x}{d_c}$
5	$P_x \to \infty$; P_y, H_{ox}, L finite, $\mathbf{E} \neq 1.0$	$Z \simeq \frac{1}{\lambda_4' L} \ln \left\{ \frac{\mathbf{E}^2 \lambda_3' e^{\lambda_4' L}(1 - Y_k) - a_4(\lambda_4' - \lambda_3') Y_k}{\mathbf{E} a_4 \lambda_3'} \right\}$
6	As (5) with $\mathbf{E} = 1.0$	$Z \simeq \frac{\left[\lambda_3' L + \frac{d_c}{P_y}\right](1 - Y_k) + \left[\frac{d_c}{H_{ox} P_y} - \lambda_3' H_{ox}\right] Y_k}{\lambda_3' L}$ where $\lambda_3' = -\left(\frac{1}{H_{ox}} + \frac{P_y}{d_c}\right)$

[a]For values of roots λ_j' or μ_i for other than cases 4 and 6, see Table 5.

	Backflow Model
$\alpha_x, \alpha_y, N_{ox}^1$, N finite, $\mathbf{E} \neq 1.0$	$$n \simeq \frac{\log \left\{ \dfrac{\mathbf{E}^2(\mu_3 - 1)(\mu_2 - \mu_4)\,\mu_4^N(1 - Y_k) - a_4(\mu_2 - 1)(\mu_4 - \mu_3)\,Y_k}{\mathbf{E}a_4(\mu_2 - 1)(\mu_3 - 1)} \right\}}{\log \mu_4}$$
As (1) with $\mathbf{E} = 1.0$	$$n \simeq \frac{K(1 - Y_k) - (\mu_2 - 1)\,[1 - \alpha_x(\mu_3 - 1)(1 - a_3)]\;Y_k}{a_3(\mu_2 - 1)(\mu_3 - 1)} - \frac{1}{N_{ox}^1}$$ where $$K = a_3(\mu_3 - 1)\left\{\left[(\mu_2 - 1)\left(\alpha_x + \alpha_y + 1 + \frac{1}{N_{ox}^1} + N\right)\right] - 1\right\}$$
$\alpha_y = 0$; α_x, N_{ox}^1, N finite, $\mathbf{E} \neq 1.0$	$$n \simeq \frac{\log \left\{ \dfrac{\mathbf{E}^2(\mu_2 - \mu_4)\,\mu_4^N(1 - Y_k) + a_4\mu_4(\mu_2 - 1)\,Y_k}{\mathbf{E}a_4(\mu_2 - 1)} \right\}}{\log \mu_4}$$
As (3) with $\mathbf{E} = 1.0$	$$n \simeq \left[\left(\alpha_x + 1 + N + \frac{1}{N_{ox}^1}\right) - \frac{N_{ox}^1(1 + \alpha_x)}{(\mu_2 - 1)}\right](1 - Y_k) - \alpha_x Y_k - \frac{1}{N_{ox}^1}$$ where $$(\mu_2 - 1) = \frac{1 + N_{ox}^1(1 + \alpha_x)}{\alpha_x}$$
$\alpha_x = 0$; α_y, N_{ox}^1, N finite; $\mathbf{E} \neq 1.0$	$$n \simeq \frac{\log \left\{ \dfrac{\mathbf{E}^2(\mu_3 - 1)\,\mu_4^N(1 - Y_k) - a_4(\mu_4 - \mu_3)\,Y_k}{\mathbf{E}a_4(\mu_3 - 1)} \right\}}{\log \mu_4}$$
As (5) with $\mathbf{E} = 1.0$	$$n \simeq \left(1 + \alpha_y + \frac{1}{N_{ox}^1} + N\right)(1 - Y_k) - \frac{Y_k}{a_3(\mu_3 - 1)} - \frac{1}{N_{ox}^1}$$ where $$(\mu_3 - 1) = -\frac{1 + N_{ox}^1(1 + \alpha_y)}{(1 + N_{ox}^1)(1 + \alpha_y)}$$

3. **Feed Entry Section.** As for Example 3, $Y_f^0 = 0.5619$; $Y_{k,f} = 0.3454$; $\mathbf{E}_f = 0.7922$. Solution of the characteristic equation, that is, Eq. (48) with (41)–(43), for the $(\mu_i - 1)$ and Eq. (47) for a_4, gives

$$(\mu_2 - 1) = 1.629964;$$
$$(\mu_3 - 1) = -0.684036;$$
$$(\mu_4 - 1) = -0.037275;$$
$$a_4 = 0.971301$$

Substitution in the equation for case 1 in Table 5 gives $N_f = 7.50$ stages. Substitution of this value in the corresponding equation in Table 7 then gives $n_f = 5.13$ stages.

4. **Solvent Entry Section.** As before, $Y_s^0 = 0.7263$; $Y_{k,s} = 0.4465$; $E_s = 1.210$. Also

$$(\mu_2 - 1) = 1.856665;$$
$$(\mu_3 - 1) = -0.728388;$$
$$(\mu_4 - 1) = 0.031057;$$
$$a_4 = 1.018210$$

Hence $N_s = 8.66$ and $n_s = 6.71$ stages. The number of stages required in the section is thus $N_s - n_s = 1.95$. Also, from the Y-profile equation (Table 6, case 1), $c_{y,N+1} = 0.119$ and $c_{y,N} = 0.198$.

5. **Total Stages.** $N_T = 5.13 + 1.95 = 7.08$ stages are required; in practice, eight stages would be used. Since the solvent inlet section covers the concentration range $c_y = 0.198$–0.300, the equilibrium curve would be better represented by the line $D'E'B$; however, recalculation on this basis causes N_T to increase by only 0.10 stage, so that the previous result of eight stages is unchanged.

4. DIRECT-SOLUTION METHODS

When applicable the foregoing analytical solutions, especially in simplified form, are the most convenient for use for both design and performance evaluation. However, in some cases with nonlinear equilibria it is necessary to resort to direct solution of the governing differential or difference equations. Both graphical and numerical methods have been described for this purpose and are, in fact, closely related. In the following account the graphical methods are described first since they serve not only to simplify the subsequent numerical treatment, but also to provide a clearer insight into the principles involved.

4.1. Graphical Solutions

Rod [26] has described an approximate graphical method of design for differential columns involving stepwise integration of the governing equations for the diffusional model. A similar method was later proposed for the backflow model [27]; both methods can also be adapted for the computation of the relevant mass transfer and backmixing parameters from measured concentration profiles [28].

As becomes apparent later, these methods are particularly convenient when the backmixing is confined to a single phase. When both phases are backmixed, on the other hand, a tedious trial-and-error procedure is required; this is discussed further in Section 4.2.

4.1.1. Diffusion Model

By use of Rod's method [26], the true operating line is constructed by stepwise integration of Eq. (1) in conjunction with a material balance on solute. The latter, taken through the upper boundary of the differential section shown in Fig. 1 and around the feed inlet end of the column, gives

$$J = U_x \left[c_x - \frac{E_x}{U_x} \frac{dc_x}{dz} \right] - U_y \left[c_y + \frac{E_y}{U_y} \frac{dc_y}{dz} \right] \tag{66}$$

where $J = U_x c_x^0 - U_y c_{y,0}$, the net solute flow per unit cross section. By introduction of new variables defined as

$$C_x = c_x - \frac{E_x}{U_x} \frac{dc_x}{dz} \tag{67}$$

$$C_y = c_y + \frac{E_y}{U_y} \frac{dc_y}{dz} \tag{68}$$

Eqs. (1) and (66) become

$$\frac{dC_x}{dz} = -\frac{k_{ox}a}{U_x} (c_x - c_x^*) \tag{69}$$

$$J = U_x C_x - U_y C_y \tag{70}$$

A plot of Eq. (70) on $C_x - C_y$ coordinates gives the plug flow operating line (also termed the *balance line*), shown in Figs. 6 and 7. The boundary conditions [Eqs. (23)-(26)] become in these units

$$z = 0: \quad C_{x,0} = c_x^0 \text{ and } C_{y,0} = c_{y,0} \tag{71}$$

$$z = L: \quad C_{x,I} = c_{x,I} \text{ and } C_{y,I} = c_y^I \tag{72}$$

An approximate solution to Eqs. (67)-(70) can be obtained by replacing the differentials dc_x, dc_y, and dz by the finite differences $\Delta c_{x,i} = c_{x,i+1} - c_{x,i}$, $\Delta c_{y,i} = c_{y,i+1} - c_{y,i}$, and Δz, respectively. They thus become

$$\Delta c_{x,i} = -\frac{U_x \Delta z}{E_x} (C_{x,i} - c_{x,i}) \tag{73}$$

$$\Delta c_{y,i} = -(c_{y,i} - C_{y,i+1}) \Big/ \left[1 + \frac{E_y}{U_y \Delta z}\right] \tag{74}$$

$$\Delta C_{x,i} = -\frac{k_{ox} a \Delta z}{U_x} (c_{x,i} - c^*_{x,i}) \tag{75}$$

$$C_{y,i+1} = \frac{U_x}{U_y} C_{x,i+1} - \frac{J}{U_y} \tag{76}$$

(a) Backmixing in One Phase. When backmixing occurs in only one phase, say, Y, the variables C_x and c_x are equal, so that Eqs. (75) and (76) become

$$\Delta c_{x,i} = -\frac{k_{ox} a \Delta z}{U_x} (c_{x,i} - c^*_{x,i}) \tag{77}$$

$$C_{y,i+1} = \frac{U_x}{U_y} c_{x,i+1} - \frac{J}{U_y} \tag{78}$$

Since the inlet jump concentrations are not known, the computation is started from the entry point of the phase in which there is no

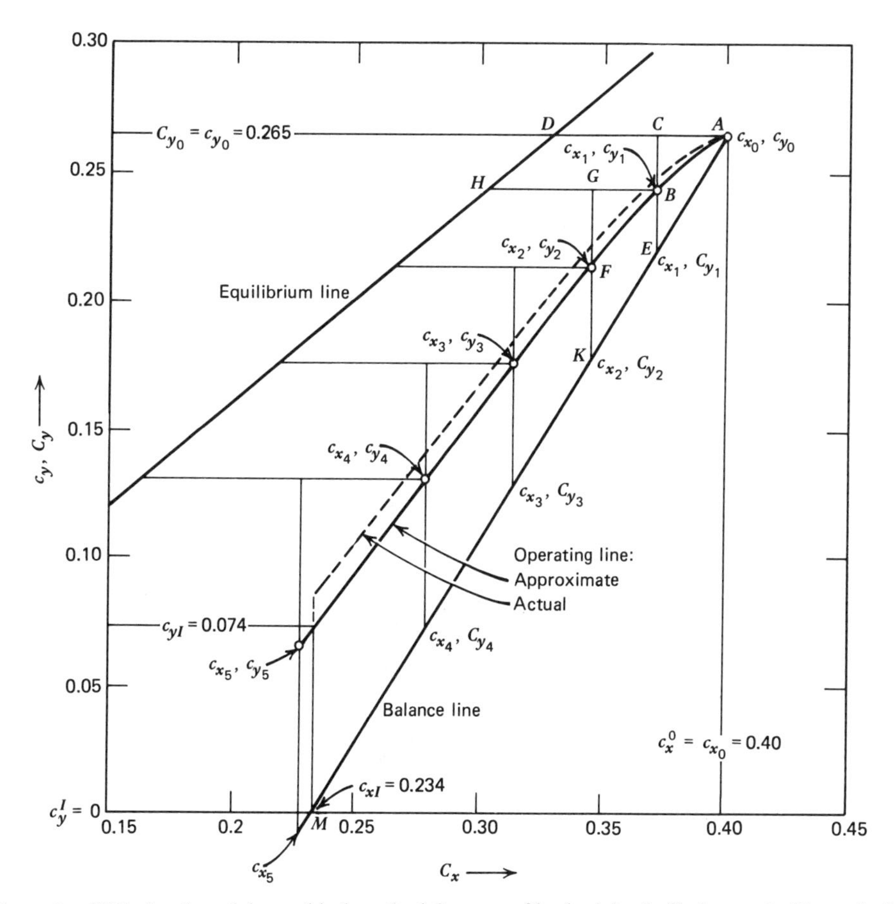

Figure 6. Diffusional model: graphical method for case of backmixing in Y phase only (Example 5). [2b]

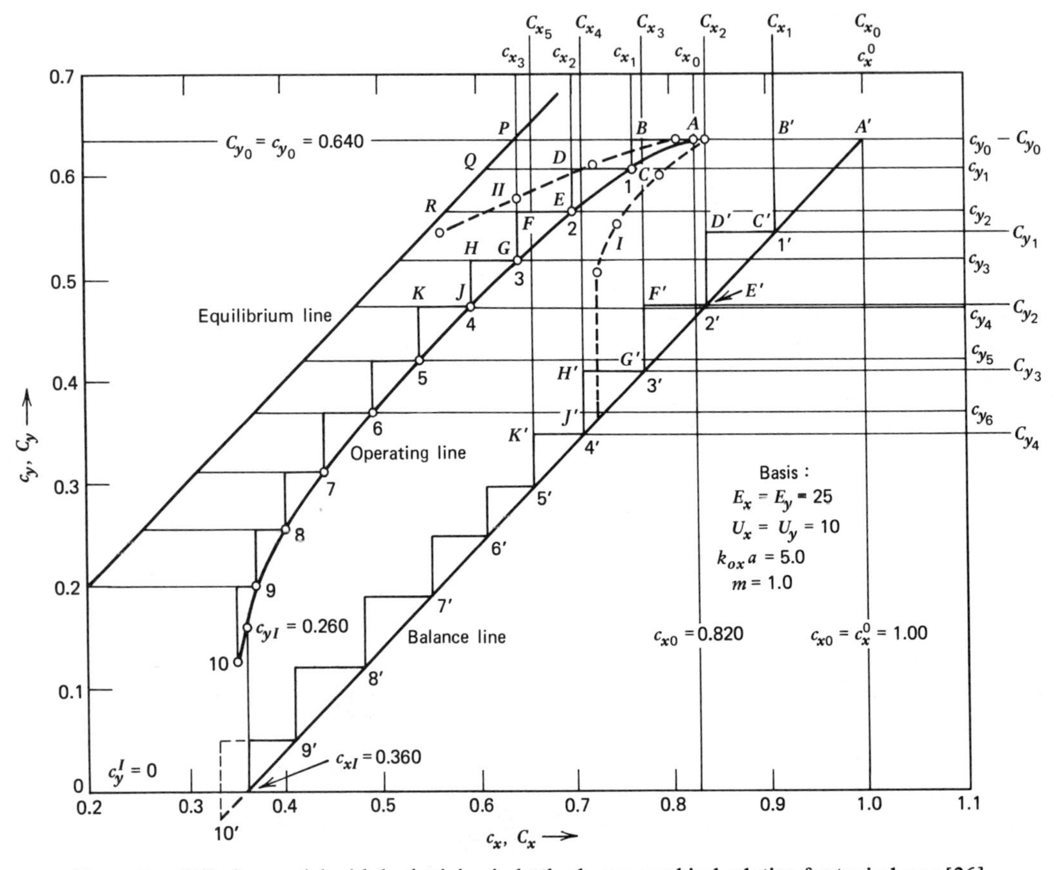

Figure 7. Diffusion model with backmixing in both phases: graphical solution for typical case [26]. Curves I and II show effect of assuming incorrect jump concentration c_{x_0}.

backmixing, that is, the X phase in the present case. Hence the terminal concentrations $c_x^0 = c_{x,0}$ and $c_{y,0} = C_{y,0}$ are located on the balance line at A as shown in Fig. 6. A suitable value of the incremental height Δz is selected so that the value of the parameter $k_{ox}a\Delta z/U_x$ lies between 0.1 and 0.5, after which the procedure is as follows:

1. Draw a horizontal line through A to meet the equilibrium curve at D, and divide AD at C so that $AC/AD = k_{ox}a\Delta z/U_x$, in accordance with Eq. (77).
2. Draw a vertical line through C to intersect the balance line at E, that is, at $C_{y,1}$ [from Eq. (78)].
3. Subdivide CE at B in the ratio $CB/CE = 1/(1 + E_y/U_y\Delta z)$, giving $c_{y,1}$ [from Eq. (74)].
4. Draw a horizontal line through B and repeat steps 1–3 until the specified exit composition $c_{x,I}$ is reached.
5. Multiply the number of height elements thus obtained by Δz to obtain the required overall column height.

This procedure is illustrated by the following example, taken from Rod [26], who assumed a linear equilibrium so that the result could be compared with the exact solution.

Example 5. Calculate the contacting height required to meet the following specification: $c_x^0 = 0.40$, $c_y^I = 0.0$, and $c_{y,0} = 0.265$ (all in kg mol/m^3); $U_x = 16$, $U_y = 10$ m/hr; $k_{ox}a = 3.2$ hr^{-1}; $E_x = 0$, $E_y = 25$ m^2/hr; and linear equilibrium relationship with $m = 1.25$, $q = 0$.

From the overall material balance, that is, $16\,(0.40 - c_{x,I}) = 10\,(0.265–0.0)$, the exit raffinate composition $c_{x,I} = 0.234$. The balance line is shown in Fig. 6 and point A, with coordinates (0.40, 0.265) is located on it. Selection of a value of $\Delta z = 2.0$ gives $k_{ox}a\Delta z/U = 0.40$, so that $\Delta c_{x,i}/(c_{x,i} - c_{x,i}^*) = -0.40$; also, from Eq. (74), $\Delta c_{y,i}/(c_{y,i} - C_{y,i+1}) = -0.44$.

With the use of these values, the procedure described previously is carried out as shown in Fig. 6, requiring 4.9 steps to reach the specified raffinate exit composition at N with the Y-phase inlet jump MN. The height required is thus $4.9 \times 2.0 = 9.8$ m.

This result is in reasonable agreement with the exact value of 10.0 m given by Rod [26]. The discrepancy would be even less if a smaller value of Δz were used, in which case the operating line would lie closer to the theoretical line, shown as a broken curve in Fig. 6.

When backmixing occurs only in the X phase the construction is started from the Y-phase entry using the following equations

$$\Delta c_{x,i'} = (C_{x,i'+1} - c_{x,i'}) \Big/ \left[1 + \frac{E_x}{U_x \Delta z}\right] \tag{79}$$

$$\Delta c_{y,i'} = \frac{k_{oy} a \Delta z}{U_y} (c^*_{y,i'} - c_{y,i'}) \tag{80}$$

$$C_{x,i'+1} = \frac{U_y}{U_x} c_{y,i'+1} + \frac{J}{U_x} \tag{81}$$

In these expressions i' refers to the number of the step counted from the Y-phase inlet, that is, from the point $(c_{x,I}, c_y^I)$, and K_{oy} is the overall mass-transfer coefficient based on the Y phase; otherwise the construction is similar to that for the preceding case.

(b) Backmixing in Both Phases. The solution is more complicated in this case since inlet concentration jumps occur at both phase entries, making trial-and-error calculation essential. Thus a value for $c_{x,0}$ must be assumed when calculating from the X-phase inlet and refined if the specified exit composition cannot be reached. The procedure, using Eq. (73)–(76), is described in the following list and illustrated in Fig. 7, where point A' on the balance line corresponds to the external concentrations $(c_x^0, c_{y,0})$, which are identical with $(C_{x,0}, C_{y,0})$.

1. Draw a horizontal line $A'P$ through A' to meet the equilibrium curve at P and assume a value for $c_{x,0}$, for example, at A.
2. Locate B' $(= C_{x,I})$ on $A'P$ from the relation $A'B'/AP = k_{ox} a \Delta z / U_x$ [Eq. (75)] and draw a vertical line through B' to meet the balance line at C' [$= C_{y,1}$ from Eq. (76)].
3. Locate B $(= c_{x,I})$ on $A'P$ so that $AB/A'A = U_x \Delta z / E_x$ [Eq. (73)].
4. Locate C $(= c_{y,1})$ vertically below B so that $BC/B'C = 1/(1 + E_y/U_y \Delta z)$ [Eq. (74)].
5. Draw a horizontal line through C to meet the equilibrium curve at Q and repeat from step 2; if the desired raffinate exit composition is reached after $i(=N_i)$ steps, the required height $L = N_i \Delta z$.
6. If the desired X-phase exit composition cannot be reached, repeat from step 1, using a new assumed value for $c_{x,0}$.

If the value of $c_{x,0}$ assumed in the first step were incorrect, the operating line would go to $+\infty$ or $-\infty$ without reaching the specified raffinate composition. This is illustrated by the broken lines I and II in Fig. 7, which correspond to high and low values, respectively, of $c_{x,0}$ [26]. It is apparent that the computation is very sensitive to small errors in $c_{x,0}$, and this becomes extreme as E_x (or E_y in the case of $c_{y,I}$, when starting from the solvent entry) is decreased, that is, with low backmixing in that phase (see Section 4.2.2 for further discussion).

4.1.2. *Backflow Model*

The auxiliary variables in this case [27] are

$$C_{x,n} = (1 + \alpha_x) c_{x,n} - \alpha_x c_{x,n+1} \tag{82}$$

$$C_{y,n} = (1 + \alpha_y) c_{y,n} - \alpha_y c_{y,n-1} \tag{83}$$

which transform the governing equation [Eq. (36)] and the material balance around the feed end [equivalent to Eq. (66)] to

$$C_{x,n-1} - C_{x,n} = N_{ox}^1 (c_{x,n} - c^*_{x,n}) \tag{84}$$

$$J' = F_x C_{x,n} - F_y C_{y,n+1} \tag{85}$$

The boundary conditions [Eqs. (49)–(52)] similarly transform to

$$n = 0: \quad C_{x,0} = c_x^0, \qquad C_{y,1} = c_{y,1} \tag{86}$$

$$n = N+1: \quad C_{x,N} = c_{x,N} \quad C_{y,N+1} = c_y^{N+1} \tag{87}$$

(a) Backmixing in One Phase. If backmixing occurs only in, say, the Y phase, the construction is started from the X-phase inlet end, putting $C_x = c_x$ throughout. Further, N_{ox}^1 can be related to the Murphree efficiency by Eq. (133) in

Chapter 5 giving, on combining with Eq. (84)

$$E_{Mx} = \frac{N_{ox}^1}{1 + N_{ox}^1} = \frac{c_{x,n} - c_{x,n+1}}{c_{x,n} - c_{x,n+1}^*} \tag{88}$$

Also, Eq. (83) can be expressed as follows:

$$c_{y,n+1} - C_{y,n+1} = \frac{\alpha_y}{1 + \alpha_y}(c_{y,n} - C_{y,n+1}) \tag{89}$$

The construction is shown in Fig. 8. After locating $(c_x^0, c_{y,1})$ on the balance line at A, the procedure is as follows:

1. Draw AD horizontally to meet the equilibrium curve at D and locate C so that $AC/AD = E_{Mx}$ [Eq. (88)].
2. Draw a vertical line through C to meet the balance line at E, that is, at $C_{y,2}$ [Eq. (85)].
3. Locate B on the operating line so that $BE/CE = \alpha_y/(1 + \alpha_y)$, that is, by using Eq. (89), thus obtaining $c_{y,2}$.
4. Draw a horizontal line BH and repeat the procedure from step 1 until the desired raffinate composition $c_{x,I}$ is reached at stage N.

In Fig. 8 the same concentrations and flows are assumed as in Example 5 together with $\alpha_y = 2.0$ and $N_{ox}^1 = 1.0$, that is, $E_{Mx} = 0.50$, requiring 4.8 actual stages. The Y-phase inlet jump appears as the step at the left-hand end of the operating line.

A similar procedure is used if backmixing occurs only in the X phase, starting from the Y-phase entry. In this case the mass transfer is expressed on a Y-phase basis, that is, as $N_{oy}^1 = k_{oy}ah_c/U_y$, with the corresponding efficiency E_{My}.

(b) Backmixing in Both Phases. The procedure is more complicated in this case, requiring trial-

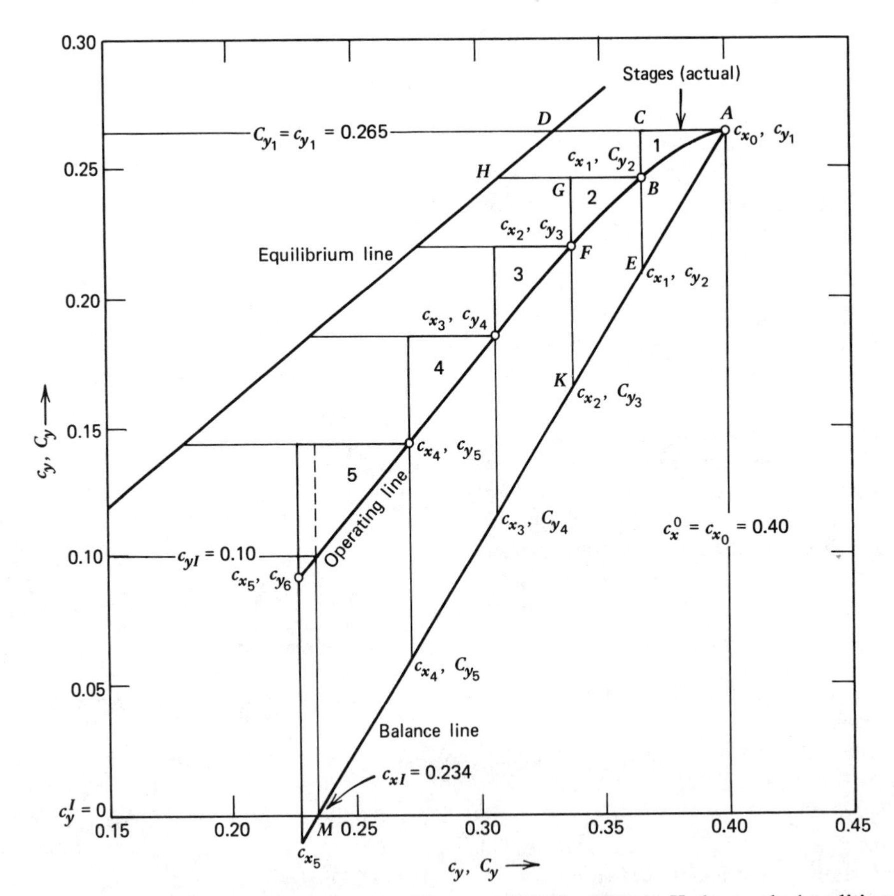

Figure 8. Backflow model: graphical method for case of backmixing in Y phase only (conditions as for Fig. 6 with $\alpha_y = 1.0; E_{Mx} = 0.75$).

and-error location of the inlet jump concentration $c_{x,0}$ as with the diffusional model. Equations (82), (84) and (85) are relevant to this case, together with Eq. (89). Equation (84) is retained as such since it cannot be expressed in terms of E_{Mx}.

The detailed construction is shown in Fig. 9; the procedure is as follows, starting from point A with coordinates $(c_x^0, c_{y,1})$, on the balance line:

1. Assume a value for $c_{x,0}$, for example, point B.
2. Locate point C, that is, $c_{x,1}$, from the relationship $BC = AB/\alpha_x$, as required by Eq. (82), and extend AC to the equilibrium curve at P.
3. Locate $D(C_{x,1})$ from $AD = N_{ox}^1 \cdot CP$ [Eq. (84)], and draw a vertical line through D to intersect the balance line at E, that is, $C_{y,2}$ [Eq. (85)].
4. Locate F $(= c_{y,2})$ from $FE/DE = \alpha_y/(1 + \alpha_y)$, that is, from Eq. (89), and draw horizontal line FQ.
5. Locate G $(= c_{x,2})$ from $HG = FH/\alpha_x$ [Eq. (82)].
6. Repeat from step 3 until [if the correct value were assumed in step (1) for $c_{x,0}$] the specified exit raffinate composition $c_{x,N}$ is reached.
7. If the operating line thus obtained fails to reach composition $c_{x,N}$, meeting either the equilibrium line or the ordinate c_y^{N+1}, the computation is repeated with a different assumed value of $c_{x,0}$.

As with the diffusional case, the calculation is sensitive to the assumed value of $c_{x,0}$, becoming extremely so with small values of the backmixing ratio α_x.

4.2. Numerical Solution

4.2.1. Basic Procedures

The design of differential column extractors can be carried out in terms of the diffusion model by

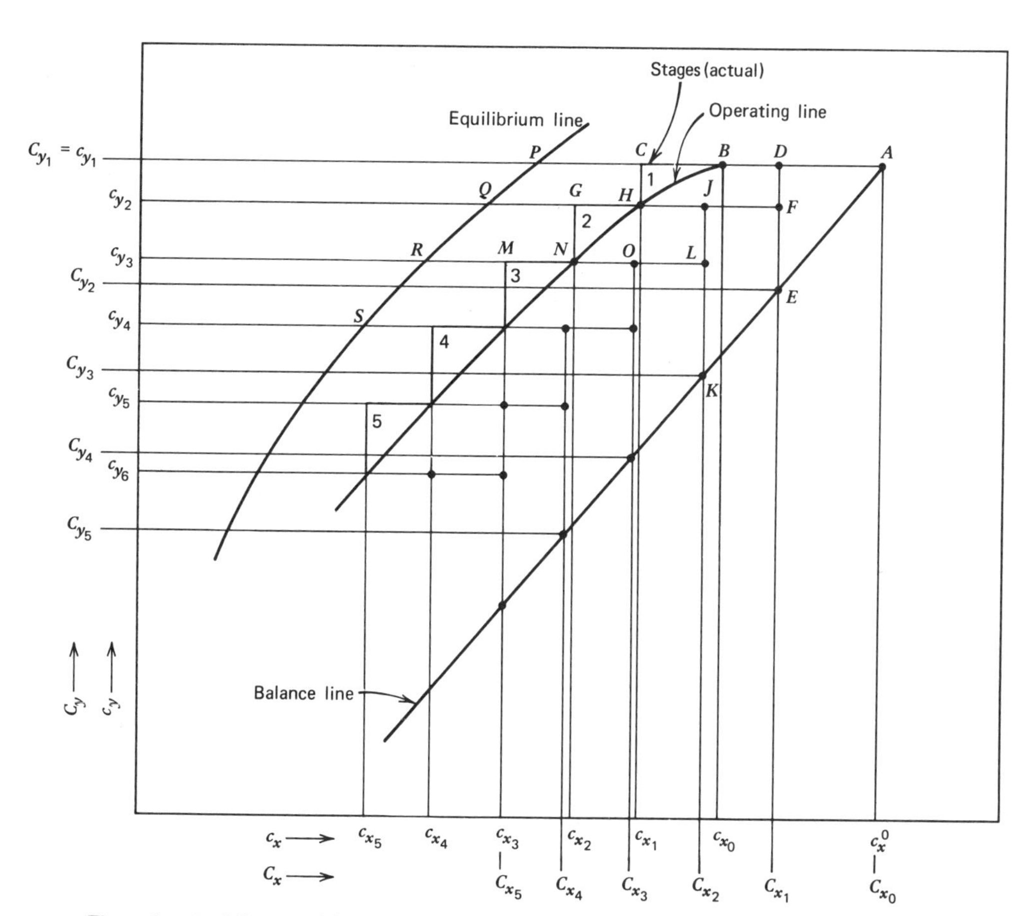

Figure 9. Backflow model: basis of graphical method for case of backmixing in both phases.

numerical integration of Eqs. (67)–(69) by using, for example, a fourth-order Runge–Kutta method. However, it is more expedient to employ a stepwise procedure based either on Rod's method [26] by use of Eqs. (73)–(76) or on the backflow model by use of Eqs. (82)–(85), together with the relationships [cf. Eq. (60)]

$$\alpha_j = \frac{N}{P_j B} - \frac{1}{2} \tag{90}$$

$$N_{ox}^1 = \frac{N_{ox}}{N} = \frac{L}{H_{ox} N} \tag{91}$$

Computer programs of either type can easily be written by analogy with the graphical procedures described in Section 4.1, or, for the backflow model, from flowsheets given by Rod [27].

When backmixing occurs in only one phase, simple direct solutions are obtainable, as with the graphical methods, by starting the calculation at the inlet of the phase that is not backmixed. With backmixing in both phases, however, iteration is required on one or other of the inlet "jump" concentrations, thus introducing difficulty in achieving accuracy if backmixing in the starting phase is low. Several alternative procedures have been devised to alleviate this problem, as described in Sections 4.2.2–4.2.4.

4.2.2. *Boundary Iteration Method*

In its original form this procedure is the exact analogue of Rod's graphical method [26]. Thus a value is assumed for one of the inlet jump concentrations, say, $c_{x,0}$, and the compositions are calculated in a stepwise manner toward the other terminal by using Eqs. (73)–(76). Unless the assumed value of $c_{x,0}$ were correct, the calculated c_y compositions will go to $\pm\infty$ and the calculation must be repeated by using a new value of $c_{x,0}$.

The procedure is similar for the backflow model, using Eqs. (82)–(85). It has been shown that the error with this method is of the order of $1/N^2$ as compared with $1/N$ for Rod's method, where N is the number of steps in each case [29]. This method, therefore, has the advantage of requiring fewer steps.

The boundary iteration method is extremely sensitive to the assumed value of $c_{x,0}$ (or $c_{y,I}$) irrespective of the model used. A starting value for this can be obtained by use of the simplified analytical solutions for cases 1 and 2 given in Table 6, after linearizing the equilibria. Alternatively, solutions obtained for one and two steps, which converge rapidly, can be used as a basis for increasing the number of steps progressively until the desired degree of accuracy is obtained [29].

An additional problem arises with this method when the backmixing is low; thus by use of a 39-bit computer it was found in particular cases with low backmixing that convergence was not obtainable with a 1-bit error in $c_{x,0}$, that is, with an accuracy of 1 in 10^{12} [19(a)]. A satisfactory solution is obtainable with low backmixing in only one phase, provided the calculation is started at the inlet of the other phase. With low backmixing in both phases, however, convergence is likely only with a small number of steps, even with use of the backflow model, so that the result is likely to be inaccurate. No difficulty arises with backmixing in only one of the phases, as described in the previous section.

4.2.3. *Direct Matrix Solution*

As an alternative to boundary iteration, the set of $2N$ equations obtained by putting $n = 1$ to N in the backflow model Eqs. (36) and (37) can be solved simultaneously in matrix form. Thus McSwain and Durbin [18] used a Gaussian elimination procedure to solve the resulting quindiagonal matrix equation, and also applied the Newton–Raphson iterative method to the case of nonlinear equilibria. Ricker et al. [30] further developed the method for multisolute operation with more complex equilibria, obtaining block tridiagonal matrices that were solved by the Newton–Raphson method.

4.2.4. *Unsteady-State Method*

An entirely different method of solution [29] consists in expressing Eqs. (84) and (85) in unsteady-state form and integrating until the profiles become steady. Thus these equations become

$$\frac{dc_{x,n}}{d\theta} = C_{x,n-1} - C_{x,n} - N_{ox}^1(c_{x,n} - c_{x,n}^*) \tag{92}$$

$$\frac{dc_{y,n}}{d\theta} = C_{y,n+1} - C_{y,n} + QN_{ox}^1(c_{x,n} - c_{x,n}^*) \tag{93}$$

where θ is a variable proportional to time. For differential contactors, the corresponding values of the α_j in Eqs. (82) and (83), and of N_{ox}^1, are obtained from the P_j and H_{ox}, respectively, using Eqs. (90) and (91).

The recommended procedure consists in first calculating the exit compositions for $N = 1$; these are then used as a starting point for the simultaneous integration of Eqs. (92) and (93) for a progressively increasing number of stages starting from $N = 2$, until the desired degree of accuracy is reached for the exit compositions. The resulting solution gives the contactor performance; for design, it is necessary to assume the length or number of stages and to iterate on these until the desired exit composition is obtained. Full details of the method are given in Appendix II of Mecklenburgh and Hartland [29]. Further discussions of unsteady-state methods are given in Chapters 7 and 27.2.

4.3. Analog Simulation

Equations (92) and (93) are also suitable for solution by analog computer, or by digital computer by use of an analog simulation language such as MIMIC or CSMP. Thus the transient response characteristics of the contactor and the final steady-state concentration profiles can be found in a simple and direct manner. An example of such a computation by using the MIMIC program has been given for the case of a five-stage extractor [31].

5. APPLICATION TO DUAL-SOLVENT EXTRACTION

The two axial dispersion models described previously apply specifically to the case of simple single-solvent extraction. In practice, the more important separations are of the dual-solvent type with two (or more) solutes, and it is of interest to consider how the models can be applied to such cases.

In Chapter 5, Section 2.7.1, it has been shown that in the absence of external reflux the concentrations of each solute increase from minima at the two ends of the extractor to maxima at the feed point. Thus, provided mixing of the feed at its entry is adequate, the extraction and scrubbing sections can be treated as separate contactors arranged end to end, with the T-solvent phase treated as feed to the extraction section and the S-solvent phase as feed to the scrubbing section. Furthermore, if backmixing occurs in both phases, there will be concentration jumps in the T phase between the feed stage and adjacent extraction section stage and in the S phase between the feed and adjacent scrubber stage, as well as at the external solvent inlets (Fig. 19 in Chapter 5).

The following design procedure, using the graphical method of matching the feed stage compositions shown in Fig. 21 of Chapter 5, is therefore applicable provided that the equilibria for the two solutes are noninteractive:

1. Specify the feed flow rate and composition (including solvent, if any), the product purities and solvent flow rates, using Eq. (83), (84), and (96) in Chapter 5; hence calculate the solvent exit concentrations using Eq. (85)–(88) in Chapter 5.
2. Assume a series of values of the concentration of solute A in solvent T at the feed stage (or feed point) and calculate the number of stages or length of column required in the extraction section for each, treating the T phase as feed. Repeat for solute B.
3. Repeat step 2 for the scrubbing section, treating the S phase as feed.
4. Using overall material balances around the extraction section, calculate the exit concentrations in solvent S corresponding to the inlet concentrations in solvent T assumed in step 2.
5. Plot the calculated number of stages or column length against solute concentration in solvent S for each section and match the feed stage (or feed point) compositions as shown in Fig. 21 in Chapter 5.

In steps 2 and 3 the number of stages or column length can be calculated by using the simplified analytical methods for linear or nonlinear equilibria described in Section 3; alternatively, the graphical or numerical methods of Section 4 can be used. As an alternative to graphical matching of the feed point compositions the numerical procedure illustrated in Fig. 25 in Chapter 5 can be used in conjunction with the present methods of calculating stages or column length and exit compositions.

In the general case when the partition coefficients of the solutes are interactive, numerical solution for stages or column length in steps 2 and 3 would appear essential. This would involve a complex interactive procedure, which thus far does not appear to have received any attention.

6 POLYDISPERSIVITY AND "FORWARD MIXING"

6.1. Polydisperse Systems

In practice, the dispersed phase in most contactors consists of drops of varying diameter, leading to a different form of axial dispersion as noted in Section 1.1. As a result, the performance is reduced in comparison with that of a monodispersion due to the differing specific areas and residence times of the various size fractions. This form of axial dispersion was first described by Olney [32], who formulated the mass-transfer equations in terms of the diffusion model without solving them. The effect has been named *forward mixing* [2], a rather inaccurate description, since the reduction in performance occurs even when all the droplets have the same velocity, as shown in the following paragraphs.

Rod [2] explained the mechanism in terms of a simplified system with only two droplet size fractions present, as shown in the operating diagram (Fig. 10). Thus the two fractions enter the column at $Z = 1.0$ with the same concentration c_y^I; as the mass-transfer rate is greater for the small-size fraction and the initial concentration gradient thus steeper, its operating line approaches the equilibrium curve more rapidly than that for the larger-size fraction. These gradients become similar further along the column, and the exit dispersed phase, after coalescence, has the same composition as for a monodispersion, although a greater contactor height is required. It is assumed in this model that the continuous-phase concentration remains constant at each cross section, although those of the two dispersed-phase fractions always differ.

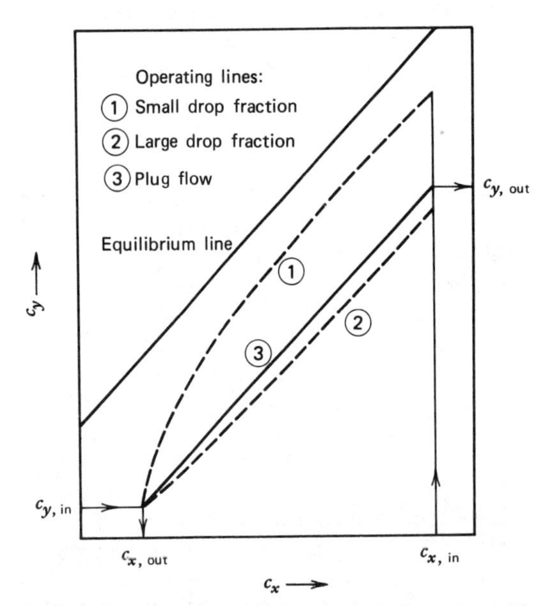

Figure 10. Operating diagram for polydispersion with two droplet size fractions and no backmixing (Based on Rod [2]).

This example illustrates an important difference between forward and backmixing, namely, that the former on its own does not lead to an inlet concentration jump. Also, forward mixing does not increase with increasing column diameter since its magnitude is determined solely by the characteristics of the dispersion.

6.2. Theoretical Treatment

6.2.1. *Differential Model*

The effect of forward mixing can be introduced into the diffusional model by taking separate solute balances for each size fraction i of the dispersed phase [32, 33]. Assuming the Y phase to be dispersed and not backmixed and expressing the mass-transfer coefficients in terms of k_{oy} gives

$$\frac{dY_i}{dZ} = -N_{oy,i}(X - Y_i) \tag{94}$$

where

$$N_{oy,i} = \frac{6k_{oy,i}\phi f_i L}{U_{y,i}d_i} = \frac{6k_{oy,i}L}{v_i d_i} \tag{95}$$

The corresponding expression for the X phase, assuming it to be backmixed, is

$$\frac{d^2X}{dZ^2} - P_xB\frac{dX}{dZ} = \frac{P_xBL}{\mathbf{E}U_y}\sum_{i=1}^{n}\frac{6k_{oy,i}\phi f_i}{d_i}(X - Y_i) \tag{96}$$

By introducing a new variable $\mathbf{X}$ defined as follows [cf. Eq. (67)],

$$\mathbf{X} = X - \frac{1}{P_xB}\frac{dX}{dZ} \tag{97}$$

Eq. (96) can be separated into two first-order differential equations:

$$\frac{dX}{dZ} = P_xB(X - \mathbf{X}) \tag{98}$$

$$\frac{d\mathbf{X}}{dZ} = -\frac{L}{\mathbf{E}U_y} \sum_{i=1}^{n} \frac{6k_{oy,i}\phi f_i}{d_i} (X - Y_i) \quad (99)$$

The boundary conditions for the X phase at the two ends are given by Eqs. (23) and (24); that for the inlet Y-phase, that is, at $Z = 1.0$, is given by $Y_i = Y^I = 0$ in the absence of backmixing in that phase and for the exit Y phase by

$$Z = 0: \quad c_y^I = c_{y,I} = \sum_{i=1}^{n} \frac{\phi f_i v_i c_{y,i}^I}{U_y} \quad (100)$$

Rod [2] obtained an approximate solution of these equations based on limiting values of the HTU at the two ends of the column. He also described a computer programme [27] that took into account backmixing in both phases.

An extensive computer study by Chartres and Korchinsky [33], assuming no dispersed phase backmixing, indicated a performance loss for polydispersed systems, as compared with monodispersed systems of the same d_{32}, the magnitude of which could not be adequately predicted by means of axial dispersion coefficients. It was also found that the effect of size dispersion is reduced by coalescence, which tends to equalize the dispersed-phase concentration laterally.

6.2.2. *Stagewise Model*

The effect of forward mixing has also been introduced into the backflow model [34, 35]. Computed concentration profiles and extraction efficiencies for polydispersions and the corresponding monodispersions again showed the magnitude of the forward mixing effect to be appreciable [35]. No allowance was made in this study for droplet coalescence, although this could easily be incorporated into the model.

6.3 Effect of Coalescence on Performance

The effect of forward dispersion on column performance is influenced to a considerable extent by droplet coalescence and redispersion [33]. Komasawa et al. [36] proposed a "collision model" for packed columns, which assumes that coalescence occurs between trapped and moving droplets, and they measured coalescence rates for such conditions by using a chemical method; however, this model applies only to packings below the "critical size" for which coalescence rates are high, and not to the more realistic case of larger packings with freely moving droplets [37, 38].

Coalescence rates have been measured for a packed column as a function of droplet size by Hamilton and Pratt [39], using a method in which two equal streams of size-equilibrated methyl isobutyl ketone droplets containing nickel xanthate (yellow) and dithizone (green) were allowed to mingle and passed into the packing. Coalescence of dissimilar colored droplets gave red droplets, which were identified by color photography after leaving the packing. From the results, coalescence and breakage rate constants were obtained and were used to predict the effects of coalescence and breakage on the extraction of 5% aqueous acetic acid with methyl isobutyl ketone, by use of a Monte Carlo procedure. The calculated column heights for 10% dispersed-phase holdup and 0.1% acetic acid in the raffinate were in the ratio 1.032:1.00:0.94 for zero, the measured, and infinite coalescence/breakage, respectively [40], the last corresponding to equalization of the droplet concentration at every cross section; the relative lowering of the mass-transfer rates is shown in Fig. 11. The results assuming a monodispersion of the same Sauter mean diameter were close to those for the infinite coalescence/breakage case. These effects were relatively somewhat less when allowance was made for backmixing in the continuous

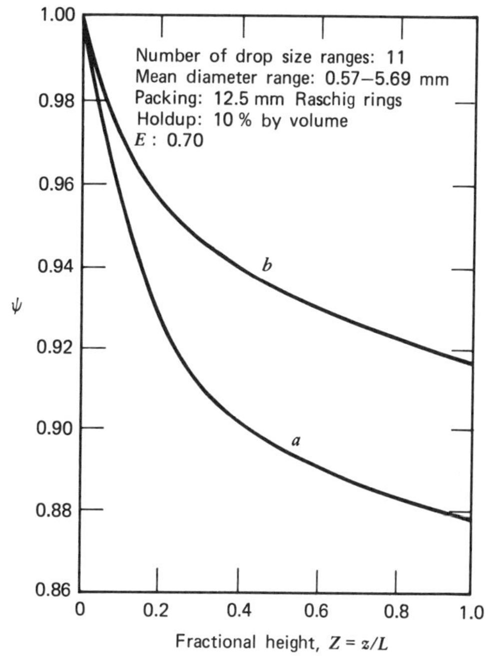

Figure 11. Relative mass-transfer rates ψ as a function of distance from dispersed-phase inlet (*a*) without and (*b*) with droplet coalescence and breakage; extraction of aqueous acetic acid from 5 to 1% w/v with methyl isobutyl ketone, without backmixing [40].

phase, due to the resulting increase in height; they would undoubtedly be greater if the velocities (and hence residence times) of the various size fractions differed.

Similar work with a pulsed perforated plate column [41] showed that droplet coalescence and breakage rates were considerably lower than in a packed column. This was apparently due to coalescence and breakage occurring only within, or adjacent to the plates, and not in the space between them. In addition, the droplets were considerably smaller, and the size range narrower than in the packed column. As a result, the predicted overall effect of polydispersivity, even in the absence of droplet coalescence and breakage, was very small.

It may be concluded that, in using the scale-up method given in Chapter 16 (Section 3.4.1), the effect of forward mixing can be assumed to be included in the measured H_{ox} or N_{ox}^1 values and to be unaffected by column diameter.

7. MEASUREMENT OF AXIAL DISPERSION

The most common methods for obtaining data on axial dispersion use tracer injection. The selected tracer should not transfer between phases, nor should it be adsorbed significantly at the liquid–liquid interface. The tracer may be a dye, the concentration of which can be measured rapidly by photometer. Inorganic salt solutions in conjunction with electrical conductivity measurement are useful in aqueous phases, but precautions may be necessary in regard to natural-convection effects. Radioactive tracers have the advantage of being detectable without the need of sampling devices.

7.1. Steady Tracer Injection

If a nontransferring tracer is injected steadily at a certain point in an extraction column, an upstream concentration profile of tracer is established as a result of backmixing in the phase in which the tracer dissolves. The governing steady-state equation according to the diffusion model is given by Eq. (1) with the mass-transfer term omitted, that is, by

$$E_j \frac{d^2 c_j}{dz^2} - U_j \frac{dc_j}{dz} = 0 \tag{101}$$

where subscript j refers to the phase to which the tracer is added and the sign convention is for z to increase in the direction of the flow of phase j. If $z = 0$ is the axial position at which tracer is injected, and $c_{j,0}$ the concentration of tracer produced by mixing the tracer flow with the bulk flow, it can easily be shown that a declining concentration profile of tracer is established by backmixing *upstream* of the injection point (i.e., with $z \leqslant 0$), as follows:

$$c_j = c_{j,0} \exp\left(\frac{U_j z}{E_j}\right) \tag{102}$$

The value of E_j is found by measuring the tracer concentration at several upstream locations and solving Eq. (102) [42]; the experimental technique is shown in Fig. 12*a*.

The steady-state tracer injection method has also been used in association with the stagewise backflow model [43]. It follows from the steady-state material balance in the absence of interphase mass transfer that the upstream tracer concentration profile is given by

$$c_{j,N'} = c_{j,0} \left[\frac{\alpha_j}{1 + \alpha_j}\right]^{N'} \tag{103}$$

where N' refers to the stage number counted upstream from the stage at which tracer is added. Equations (102) and (103) can be solved simultaneously to give the relationship between the effective axial dispersion coefficient (due to backmixing) and the backflow ratio:

$$E_j = \frac{U_j h_c}{\ln\left[(1 + \alpha)/\alpha\right]} \tag{104}$$

It must be emphasized that the values of E_j appearing in Eqs. (102) and (104) are due to *backmixing only* (see Section 1.1). Steady tracer injection techniques are thus useful only where backmixing is predominant, such as in well-agitated contactors or gravity columns. The steady injection method is not applicable to the dispersed phase with normal chemical tracers because of the difficulty of incorporating these into the drops.

In using the steady-state tracer injection method, (1) it is usually advisable to take samples at several different points in a column, and (2) sufficient time must be allowed before each measurement for the steady state to become established.

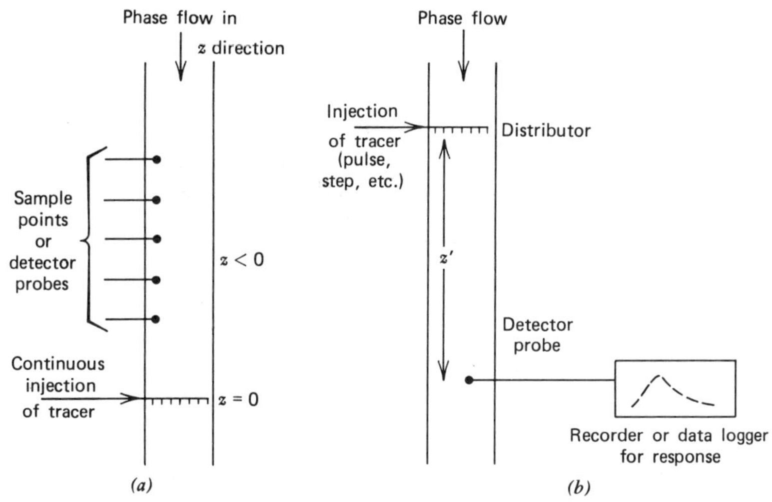

Figure 12. Tracer methods for measuring axial dispersion: (*a*) steady state; (*b*) unsteady state (single probe).

7.2. Unsteady Tracer Injection

Because of the factors mentioned in Section 7.1, unsteady-state tracer injection is used more frequently than steady-state injection. The one-dimensional conservation equation for tracer dispersion (diffusion model) is given by adding an accumulation term to Eq. (101):

$$E_j \frac{\partial^2 c_j}{dz^2} - U_j \frac{\partial c_j}{\partial z} - \phi_j \frac{\partial c_j}{\partial t} = 0 \qquad (105)$$

Tracer may be injected as a single sharp pulse (delta function) or a pulse of finite length, a step change, sinusoidal wave form, and so on (see also Chapter 7, Section 3.1). Mathematically, the simplest case is the delta function, where the injection time is very short in comparison to the time over which response occurs. Tracer concentration is measured continuously by means of a single detector probe situated downstream of the injection point, as illustrated in Fig. 12*b*.

Levenspiel and Smith [44] gave an analytical solution for the tracer concentration at the probe with delta function injection, for the case of single phase flow ($\epsilon_j = 1$). For two-phase flow, this solution becomes

$$c_j = \frac{m'}{2s\sqrt{\pi E_j t/\phi_j}} \exp\left[\frac{-(z' - U_j t/\phi_j)^2}{4E_j t/\phi_j}\right] \qquad (106)$$

where z' is the distance of the probe from the injection point and m' is the mass of tracer injected in the pulse at time $t = 0$. The value of E_j is found from the recorded curve by a fitting procedure or by calculation from the variance:

$$\frac{E_j}{U_j z'} = \frac{(8\sigma^2 + 1)^{1/2} - 1}{8} \qquad (107)$$

The values of E_j obtained in this case include both the backmixing and forward-mixing effects. A variation of this technique is the use of a tracer (e.g., sodium hydroxide) that undergoes an instantaneous reaction with a component (e.g., hydrochloric acid) in phase j. The change in color of an indicator is observed, and E_j can be calculated therefrom [45].

Although single probes are normally used, the results are sensitive to the experimental difficulty of producing a perfectly sharp pulse of tracer. The use of *two* probes at different axial positions allows for accurate measurement of axial dispersion regardless of the nature of the pulse of tracer, as discussed in Chapter 7, Section 3.1.

The response to pulse tracer injection can also be used to obtain the mixing parameters for the backflow model. If the number of stages N' between the injection point and the downstream probe is fairly large, the appropriate form of Eq. (60) applies. Since the distance z' is given as $N'h_c$, this equation can also be written

$$E_j = U_j h_c(\alpha_j + 0.5) \qquad (108)$$

In cases where backmixing predominates and $\alpha_j >> 1$, the steady and unsteady methods

[Eqs. (104) and (108), respectively] give the same relationship:

$$(E_j)_{\text{steady}} \simeq (E_j)_{\text{unsteady}} \simeq U_j h_c \alpha_j \quad (109)$$

However, with $\alpha_j = 0$, that is, with no interstage backflow,

$$(E_j)_{\text{unsteady}} = 0.5 U_j h_c \quad (110)$$

The steady-state tracer injection method is not capable of detecting any mixing between stages in this case, which corresponds to mixing only *within* compartments.

7.3. Measurement of Concentration Profiles of Transferring Solute

Results of tracer measurements of axial dispersion are open to the criticism that they may not be valid under mass-transfer conditions. For example, mass-transfer-induced effects such as interfacial turbulence (see Chapter 3, Section 3.3.2) could result in axial dispersion rates different from those obtained in the absence of mass transfer.

A potential solution to this problem consists of direct measurement of the concentration profiles of transferring solute in the column, followed by solution of the transport equations for the dispersion coefficient or the Peclet number [19(a), 28]. Experimental studies using this method have been reported [46, 47]. More recently, special sampling probes have been developed that minimize the interfacial mass transfer as the phases are withdrawn from the column [48].

The experimental difficulty of carrying out profile studies has discouraged extensive use of this method for determining axial dispersion coefficients. Also, in many cases the sensitivity of the profiles to axial dispersion is low. Data from tracer studies (Sections 7.1 and 7.2) are generally considered acceptable, unless there are known mass-transfer-induced hydrodynamic effects.

8. DATA ON AXIAL DISPERSION

As already noted in Section 1.1, axial dispersion in liquid–liquid contactors occurs by a number of different contributing mechanisms, depending on the equipment type and size and the system used. Consequently, there are many different forms of correlation available for axial dispersion data. This section gives only a brief survey of the main results, but further data are given in some of the chapters dealing with specific types of equipment. An extensive survey of axial dispersion data up to 1970 has been given by Ingham [49].

In general, it can be stated that axial dispersion coefficients E_j are seldom less than 10^{-4} m^2/s and may rise to 10^{-2} m^2/s or more in conditions where circulation is strong, compared with values in the order of 10^{-9} m^2/s for the molecular diffusivity of solutes.

8.1. Continuous Phase

8.1.1. Spray Columns

Axial mixing effects are much greater in spray columns than in baffled or packed columns. Not only is mixing carried out by the dispersed-phase droplet wakes, but gross circulation may be induced [1], and it is doubtful whether the diffusion model is applicable under these conditions. These circulation effects increase strongly with column diameter, and Wijffels and Rietema [50] report a fiftyfold increase in apparent E_c as column diameter increases from 150 to 1000 mm. Considerable information is available for small-diameter (≤100-mm) columns. The correlation by Zheleznyak and Landau [51] is recommended in Chapter 10 of this handbook, and a slightly simpler equation has since been developed by Laddha et al. [52].

8.1.2. Unagitated Columns with Internals

Axial dispersion data in such columns are generally correlated as Peclet numbers referred to the appropriate internal dimension. Vermeulen et al. [53] expressed the Peclet number in a *packed column* as $\epsilon U_c d_p / E_c$, where ϵ is the voidage of the packing. This Peclet number was found [53, 54] to decrease from the single-phase value of 0.5 as the dispersed-phase flow was increased. Pratt and Anderson [55] have found improved accuracy for Raschig ring packings with the correlating equation

$$\frac{\epsilon U_c d_p}{E_c} = 0.00786 \left(\frac{d_p U_c}{\nu_c}\right)^{0.47} \cdot \left(\frac{U_c}{U_d}\right)^{0.43} \left(\frac{d_c}{d_{32}}\right)^{1.0} \quad (111)$$

where d_{32} is the Sauter mean droplet diameter, which can be calculated from system properties. Further data are given in Chapter 10. There is some evidence [56] that large increases in E_c can occur as a result of circulation effects in large-diameter packed columns, although this conflicts with the observed absence of any significant change in column performance when column diameter increased from 75 to 300 mm [57].

In unagitated *tray columns* there is little back-flow between compartments, but the continuous phase can be assumed to be well mixed within compartments (see Chapter 10), and Eq. (110) is applicable here.

8.1.3. Pulsed Columns

Vermeulen et al. [53] found that axial dispersion in *packed* columns was increased *or* decreased by pulsation depending on the type of packing used. Rosen et al. [58] correlated data from laboratory pulsed packed columns by the relationship

$$E_c \simeq 0.5 d_p [U_c + 2a'f] \qquad (112)$$

For large-diameter pulsed packed columns, axial dispersion is enhanced by circulation, and Rosen and Krylov [56] have suggested that

$$E_c = 0.5 d_p [U_c + 2a'f] + \frac{k_1 D^2 (\Delta U)^2}{0.5 d_p [U_c + 2a'f]} \qquad (113)$$

In the last term, which gives the contribution due to circulation, ΔU represents the average deviation from uniform velocity distribution. It is seen later that the circulation contribution predominates at large column diameters D but that it is reduced by pulsation. The terms k_1 and ΔU are not known *a priori.*

In conventional *pulsed plate* extraction columns, the fractional open area of the plate is sufficiently small that circulation effects through the plates are small. Pulsation has the effect of increasing the backflow between compartments, but the effect of compartment height is less pronounced than indicated in Eq. (108). Correlations due to Miyauchi and Oya [59] and Rosen et al. [56] give contradictory effects of column diameter. Kagan et al. [60] found no significant difference between the mixing data for a 56-mm- and a 300-mm-diameter column. Their simple correlation is

$$E_c = 0.49 h_c^{0.76} (U_c + 2a'f) \qquad (114)$$

The constant of 0.49 applies when the quantities are given in centimeters and seconds. A similar lack of effect of column diameter was reported by Rouyer et al. [61], using columns of 45- and 600-mm diameter.

Recently Garg and Pratt [62] analyzed previously published mass-transfer data for pulsed columns for the "emulsion" region and found in contrast to the foregoing results that there was a significant effect of column diameter in the range $74 < D < 305$ mm. Their correlation for the continuous-phase backflow ratio is as follows:

$$\alpha_c = D^{0.8} (a'f)^{0.10} \left(0.170 + 0.302 \frac{U_d}{U_c}\right) \qquad (115)$$

This equation is applicable with the use of centimeter and second units. Support for the positive effect of column diameter on axial mixing is also provided by Woodfield and Sege [63], who found that the performance of a pulsed column deteriorated on scale-up from 75- to 750-mm diameter, unless special baffles were provided in the larger column size. It seems possible that the lack of effect of column diameter observed by some workers [60, 61] may have been due to operation of their columns in the "mixer–settler" rather than the "emulsion" regime.

8.1.4. Reciprocating-Plate Columns

Reciprocating-plate columns are hydrodynamically quite similar to pulsed-plate columns. Nemecek and Prochazka [64] found that for reciprocating plates having a relatively low open area (20%), there was approximate agreement with Kagan's correlation, namely, Eq. (114). Hafez et al. [65] worked with a 150-mm-diameter Karr-type reciprocating column with plate open area of 60% and observed that at low agitation levels the axial dispersion coefficient tended to increase as a result of circulation effects. This effect had not been noted in a previous study [45] of a 50-mm-diameter reciprocating-plate column. Hafez et al. [65] found that these circulation effects could be considerably reduced by appropriate changes in the plate design and arrangement.

8.1.5. Rotary-Agitated Columns

The early correlation by Stemerding et al. [66] for continuous-phase axial dispersion in rotating-disk contactors (*RDC*s) is considered acceptable for design purposes (see Chapter 13.1).

$$E_c = 0.5U_c h_c + 0.012 D_I n h_c \left(\frac{S}{D}\right)^2 \quad (116)$$

In common with Eq. (112), the right-hand side of this equation contains a first term corresponding to a "well-mixed stage" model and a second agitation-dependent term that relates mainly to backflow between stages.

Misek [67] showed that continuous-phase axial dispersion data in *RDC* and *ARD* contactors could be expressed in a general form

$$E_c = 0.5U_c h_c + k_2 D_I n h_c \left(\frac{S}{D}\right)^2 \cdot \left(\frac{D_I}{D}\right)^{2/3} N_p^{1/3} \left(\frac{l}{h_c}\right)^{1/3} \quad (117)$$

where the product $k_2 l^{1/3} N_p^{1/3}$ is approximately $0.04\ \text{cm}^{1/3}$.

The *Oldshue-Rushton* impeller-agitated column (see Chapter 13.4) has been studied by Ingham [43], who used continuous tracer injection and gave axial mixing correlations for the standard type of column and for a design in which a draft tube baffle was included. More recently, Komasawa and Ingham [68] measured backmixing in the presence of transferring solute. Under these conditions, continuous-phase axial mixing was found to be unaffected by mass transfer, but dispersed-phase mixing was affected.

The impeller-agitated *Kuhni* column was investigated by Ingham et al. [69], who used a steady-state tracer injection technique. The continuous-phase axial dispersion coefficient was found to be approximately proportional to $D_I n h_c$, indicating that backflow effects are the main cause of mixing in this case.

The *RTL* (*Graesser raining bucket*) horizontal-axis contactor was the subject of an axial mixing study by Sheikh et al. [70], who found that the axial mixing coefficient (from unsteady-state tracer response) was given by an equation of the form

$$E_c = 0.5U_c h_c + k_3 D_I n h_c U_c \quad (118)$$

where k_3 is a dimensional constant for the apparatus. The backmixing coefficient (from steady-state tracer profiles; see Section 7.1) was substantially *greater* than the total axial mixing coefficient, but it was concluded [70] from measured concentration profiles that backflow is not the predominant mixing mechanism in this type of equipment.

8.2. Dispersed Phase

True backmixing in the dispersed phase is generally less pronounced than in the continuous phase, and the most common cause is the circulatory flow induced between the compartments of rotary agitated columns. Thus Rod [71] found that uniformly sized suspended particles in a well-agitated RDC contactor had essentially the same degree of backmixing as the continuous phase and that it was represented by an expression similar to Eq. (116).

It seems clear that reported values of dispersed-phase axial mixing coefficients for some column types, such as packed [53] and pulsed-plate [59] columns are due mainly to forward mixing. In fact, the forward mixing mechanism (Section 6) is of such importance that the *total* measured axial dispersion coefficient of the dispersed phase can often exceed that of the continuous phase. This is shown in Fig. 13 for a 1.5-m-diameter pulsed-plate column [72]. Forward-mixing effects are greatest at low agitation levels and at low dispersed-phase holdup, that is, when the drop size distribution is least uniform. A similar trend has been found in pulsed packed columns of 0.1- and 0.7-m diam-

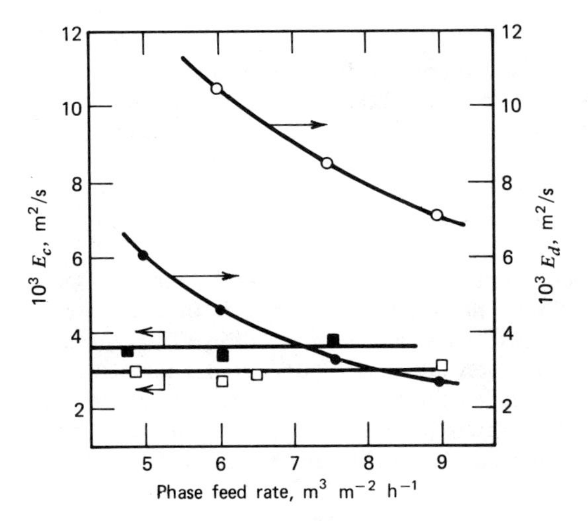

Symbol	Phase	$2a'f$, m/min
■	Continuous	1.3
□	Continuous	0.8
•	Dispersed	1.3
o	Dispersed	0.8

Figure 13. Dispersed and continuous-phase axial mixing in a 1.5-m pulsed plate column (redrawn from Karpacheva et al. [7]).

eter [73]. At low agitation, forward mixing due to droplet nonuniformity is important; this effect diminishes as agitation is increased, but backmixing becomes increasingly predominant, and the nonlinearity of these two effects results in a plot of E_d against agitation that shows a minimum value [73].

In spray columns, forward mixing is predominant in the dispersed phase at small diameters, as indicated by the work of Laddha et al. [52], who found a strong interrelation between dispersed-phase mixing and drop size distribution in a 50-mm column.

8.3. Conclusions

This short survey has shown that although much data have been collected on axial dispersion, particularly in the continuous phase and in small columns, there remain areas of uncertainty and even disagreement. The correlations quoted in this section should thus be used with caution if experimental conditions differ widely from those on which the correlations are based. In such cases data should preferably be obtained by pilot scale testing by using one of the procedures outlined in Section 7.

NOTATION

		Dimensions	SI Units
A_i	coefficients in Eqs. (12) and (44)		
a	superficial area of contact of the phases	L^{-1}	m^{-1}
a_i	coefficients in Eqs. (13) and (45) given by Eqs. (15) and (16) and (46) and (47), respectively		
a'	amplitude	L	m
B	$= L/d_c$		
C_j	variables defined by Eqs. (67) and (68) and (82–83)	ML^{-3} †	$kg\ m^{-3}$ ‡
c_j	concentration in phase j	ML^{-3} †	$kg\ m^{-3}$ ‡
D	diameter of contactor	L	m
D_I	diameter of impeller	L	m
d_c	characteristic dimension	L	m
d_i	diameter of droplet of size group i	L	m
d_{32}	mean volume–surface (i.e., Sauter mean) diameter of droplets	L	m
d_p	characteristic size of packing	L	m
E	extraction factor, mU_x/U_y		
E	finite-difference operator		
E_j	effective longitudinal diffusion coefficient in jth phase	L^2T^{-1}	$m^2\ s^{-1}$
E_{Mj}	Murphree efficiency based on jth phase		
F_j	volumetric flow of jth phase	L^3T^{-1}	$m^3\ s^{-1}$
f	frequency	T^{-1}	s^{-1}
f_i	fractional holdup of droplets of size range i		
H_{oj}	height of "true" overall transfer unit based on jth phase $(= U_j/k_{oj}a)$	L	m
h_c	height of compartment	L	m
J	net flow of solute	$ML^{-2}T^{-1}$ †	$kg\ m^{-2}\ s^{-1}$ ‡
J'	net flow of solute	MT^{-1}	$kg\ s^{-1}$
k_1, k_2, k_3	constants		
k_{ox}	overall mass-transfer coefficient based on X phase	LT^{-1}	$m\ s^{-1}$

†Or mol for M

‡Or kmol for kg.

		Dimensions	SI Units
L	length or height of differential extractor	L	m
l	characteristic eddy length	L	m
m	reciprocal slope of equilibrium line, dc_x^*/dc_y		
m'	mass of tracer injected	M	kg
N	total number of actual stages required		
N'	stage number counted from tracer injection point		
N_{ox}	number of "true" overall transfer units based on X phase ($= k_{ox}aL/U_x$)		
N_{oxP}	number of overall plug flow transfer units based on X phase [Eq. (32)]		
N_{oxD}	number of overall dispersion transfer units based on X phase [Eq. (33)]		
N_{ox}^1	number of overall transfer units per actual stage, based on X phase ($= k_{ox}ah_c/U_x$)		
N_p	power number for rotor		
n	rotor speed, rev/s	T^{-1}	s^{-1}
P_j	turbulent Peclet number of jth phase, U_jd_c/E_j		
Q	$= F_x/F_y$		
q	intercept of equilibrium line given by Eq. (3)		
Re_d	$= d_iv_i/\nu_d$, droplet Reynolds number		
S	diameter of stator (internal)	L	m
t	time	T	s
U_j	superficial velocity of phase j	LT^{-1}	$m\ s^{-1}$
V	volume of extractor compartment	L^3	m^3
v_i	velocity in column of droplet(s) of diameter d_i	LT^{-1}	$m\ s^{-1}$
X	dimensionless concentration of X phase [Eq. (6)]		
Y	dimensionless concentration of Y phase [Eq. (7)]		
Z	$= z/L$, fractional length within contactor		
z	length measured from X (feed) = phase inlet	L	m
z'	length measured from tracer injection point	L	m

Greek Symbols

α, β, γ	coefficients as defined by Eqs. (9)–(11) or (41)–(43)		
α', β', γ'	coefficients as defined by Eqs. (19)–(21)		
α_j	backmixing ratio for jth phase, that is, ratio of backflow to F_j		
ϵ	voidage of packing		
λ_i	roots of characteristic equation [Eq. (14)]		
λ_i'	roots of modified characteristic equation [Eq. (18)]		
μ	viscosity	$ML^{-1}T^{-1}$	$kg\ m^{-1}\ s^{-1}$
μ_i	roots of characteristic equation [Eq. (48)]		
ν	kinematic viscosity		
ϕ	volume fraction of holdup		
Ψ	ratio of mass-transfer rates for polydispersions without, to those with, effectively infinite coalescence/breakage rates		

		Dimensions	SI Units
ρ	density	ML^{-3}	$kg\ m^{-3}$
σ	variance		

Subscripts

B	backmixed flow
c	continuous phase
d	dispersed phase
f	feed entry section of divided contactor
i	droplet size group
j	X or Y phase
I	Y-phase inlet end, within contactor
k	at junction of sections of divided contactor
N	last stage (i.e., at solvent inlet end)
n	stage within contactor; number of assumed droplet size groups
P	plug flow
s	solvent entry section of divided contactor
x	X phase
y	Y phase
0	X-phase inlet end, within contactor

Superscripts

I	Y-phase inlet end, external to contactor
$N, N + 1$	Y-phase inlet end, external to contactor
0	overall length (applies to L_f, L_s)
o	X-phase inlet end, external to contactor
*	equilibrium value

REFERENCES

1. W. J. Anderson and H. R. C. Pratt, *Chem. Eng. Sci.* **33,** 995 (1978).
2. V. Rod, *Br. Chem. Eng.* **11,** 483 (1966).
3. J. Yeheskel and E. Kehat, *Chem. Eng. Sci.* **26,** 1223 (1971).
4. J. Yeheskel and E. Kehat, *Chem. Eng. Sci.* **26,** 2037 (1971).
5. T. Miyauchi, University of California Radiation Laboratory Report, UCRL-3911, August 1957.
6. T. Miyauchi and T. Vermeulen, *Ind. Eng. Chem. Fund.* **2,** 113 (1963).
7. C. A. Sleicher, *AIChE J.* **5,** 145 (1959).
8. C. A. Sleicher, *AIChE J.* **6,** 529 (1960).
9. T. Miyauchi and T. Vermeulen, *Ind. Eng. Chem. Fund.* **2,** 304 (1963).
10. S. Hartland and J. C. Mecklenburgh, *Chem. Eng. Sci.* **21,** 1209 (1966).
11. J. F. Wehner and R. H. Wilhelm, *Chem. Eng. Sci.* **6,** 89 (1956).
12. C. J. Geankoplis and A. N. Hixson, *Ind. Eng. Chem.* **42,** 1141 (1950); C. J. Geankoplis, P. J. Wells, and E. L. Hawk, ibid. **43,** 1848 (1951).
13. T. E. Gier and J. O. Hougen, *Ind. Eng. Chem.* **45,** 1362 (1953).
14. H. R. C. Pratt, *Ind. Eng. Chem. Fund.* **10,** 170 (1971).
15. H. R. C. Pratt, *Ind. Eng. Chem. Proc. Des. Devel.* **14,** 74 (1975).
16. N. P. Wilburn, *Ind. Eng. Chem. Fund.* **3,** 189 (1964).
17. N. P. Wilburn and W. L. Nicholson, *AIChE-IChE Symposium Series*, No. 1 (Inst. Chem. Eng., London), 1965, p. 105.
18. C. V. McSwain and L. D. Durbin, *Separation Sci.* **1,** 677 (1966).
19. J. C. Mecklenburgh and S. Hartland, *Ind. Chem.*

Eng. Symp. Series, No. 26, (a) 115, (b) 121, (c) 130 (1967).

20. T. Vermeulen, J. S. Moon, A. Hennico, and T. Miyauchi, *Chem. Eng. Progr.* **62,** 95 (1966).
21. S. Stemerding and F. J. Zuiderweg, *Chem. Eng. (Lond.)*, CE 156 (May 1963).
22. J. S. Watson and H. D. Cochrane, *Ind. Eng. Chem. Proc. Des. Devel.* **10,** 83 (1971).
23. H. R. C. Pratt, *Ind. Eng. Chem. Process Des. Devel.* **15,** 544 (1976).
24. A. K. McMullen, T. Miyauchi, and T. Vermeulen, University of California Radiation Laboratory Report, UCRL-3911 Supplement, 1958.
25. H. R. C. Pratt, *Ind. Eng. Chem. Process Des. Devel.* **15,** 34 (1976).
26. V. Rod, *Br. Chem. Eng.* **9,** 300 (1964).
27. V. Rod, in C. Hanson, Ed., *Recent Advances in Liquid-Liquid Extraction*, Pergamon Press, Oxford, 1971, Chapter 7.
28. V. Rod, *Coll. Czech. Chem. Commun.* **30,** 3822 (1965).
29. J. C. Mecklenburgh and S. Hartland, *Can. J. Chem. Eng.* **47,** 453 (1969).
30. N. L. Ricker, F. Nakashio, and C. J. King, *AIChE J.* **27,** 277 (1981).
31. J. Ingham and I. J. Dunn, *Chem. Eng. (Lond.)* (286), 354 (June 1974).
32. R. B. Olney, *AIChE J.* **10,** 827 (1964).
33. R. H. Chartres and W. J. Korchinsky, *Transact. Inst. Chem. Eng.* **53,** 247 (1975).
34. V. Rod and T. Misek, *Proceedings of the International Solvent Extraction Conference 1971* (ISEC71), Vol. 1 (Soc. Chem. Ind., London), 1971, p. 738.
35. W. J. Korchinsky and S. Azimzadeh-Katyloo, *Chem. Eng. Sci.* **31,** 871 (1976).
36. I. Komasawa, E. Kunugita, and T. Otake, *Kagaku Kogaku* (abridged ed.) **4,** 288 (1966); **5,** 125 (1967); I. Komasawa, S. Hisatani, E. Kunugita, and T. Otake, ibid. **4,** 363 (1966).
37. J. B. Lewis, I. Jones, and H. R. C. Pratt, *Transact. Inst. Chem. Eng.* **29,** 126 (1951).
38. R. Gayler, N. W. Roberts, and H. R. C. Pratt, *Transact. Inst. Chem. Eng.* **31,** 57 (1953).
39. J. A. Hamilton and H. R. C. Pratt, *Proceedings of the International Solvent Extraction Conference,* Liege, Belgium (ISEC80), Vol. 1, Session 3, paper 80-19, 1980.
40. J. A. Hamilton, *Droplet Coalescence and Breakage in a Packed Liquid Extraction Column*, Ph.D. thesis, University of Melbourne (1981).
41. M. O. Garg, *Measurement and Modeling of Droplet Coalescence and Breakage in a Pulsed Plate Liquid Extraction Column*, Ph.D. thesis, University of Melbourne (1982).
42. B. W. Mar and A. L. Babb, *Ind. Eng. Chem.* **51,** 1011 (1959).
43. J. Ingham, *Transact. Inst. Chem. Eng.* **50,** 372 (1972).
44. O. Levenspeil and W. K. Smith, *Chem. Eng. Sci.* **6,** 227 (1957).
45. S. D. Kim and M. H. I. Baird, *Can. J. Chem. Eng.* **54,** 81 (1976).
46. L. D. Smoot and R. L. Babb, *Ind. Eng. Chem. Fund.* **1,** 93 (1962).
47. Z. Ziolkowski and M. Pajak, *Chem. Stosow. Ser. B* **7,** 261, 369 (1970) (through *Chem. Abstr.* **74,** 5020*u* and **89,** 122*q*).
48. C. J. Lim, J. E. Henton, G. Bergeron, and S. D. Cavers, *Proceedings of the International Solvent Extraction Conference*, Toronto (ISEC77), Vol. 1 (Can. Inst. Min. Met.), 1979, p. 248.
49. J. Ingham, in C. Hanson, Ed., *Recent Advances in Liquid-Liquid Extraction*, Pergamon Press, Oxford, 1971, Chapter 8.
50. J. B. Wijffels and K. Rietema, *Transact. Inst. Chem. Eng.* **50,** 224, 233 (1972).
51. A. S. Zheleznyak and A. M. Landau, *Theor. Found. Chem. Eng.* **7,** 525 (1973).
52. G. S. Laddha, T. R. Krishnan, S. Viswanathan, S. Vedaijan, T. E. Degaleesan, and H. E. Hoelscher, *AIChE J.* **22,** 456 (1976).
53. T. Vermeulen, J. S. Moon, A. Hennico, and T. Miyauchi, *Chem. Eng. Progr.* **62** (9), 95 (1966).
54. J. S. Watson and L. E. McNeese, *Proceedings of the International Solvent Extraction Conference*, Lyons, (ISEC74) Vol. 2 (Soc. Chem. Ind. London), 1974, p. 1371.
55. H. R. C. Pratt and W. J. Anderson, *Proceedings of the International Solvent Extraction Conference*, Toronto (ISEC77), Vol. 1 (Can. Inst. Min. Met.), 1979, p. 242.
56. A. M. Rosen and V. S. Krylov, *Chem. Eng. J.* **7,** 85 (1974).
57. R. Gayler and H. R. C. Pratt, *Transact. Inst. Chem. Eng.* **35,** 273 (1957).
58. A. M. Rosen, Y. G. Rubezhnyi, and B. V. Martynov, *Khim. Prom.* **46** (1), 132 (1970).
59. T. Miyauchi and H. Oya, *AIChE J.* **11,** 395 (1965).
60. S. Z. Kagan, B. A. Veisbein, V. G. Trukhanov, and L. A. Muzychenko, *Internatl. Chem. Eng.* **13,** 217 (1972).
61. H. Rouyer, J. Lebouhellec, E. Henry, and P. Michel, *Proceedings of the International Solvent Extraction Conference*, Lyons (ISEC74), Vol. 3 (Soc. Chem. Ind., London), 1974, p. 2339.
62. M. O. Garg and H. R. C. Pratt, *Ind. Eng. Chem. Process Des. Devel.* **20,** 492 (1981).
63. F. W. Woodfield and G. Sege, *Chem. Eng. Progr. Symp. Ser.* (*13*), **50,** 14, 174 (1954).
64. M. Nemecek and J. Prochazka, *Can. J. Chem. Eng.* **52,** 739 (1974).
65. M. M. Hafez, M. H. I. Baird, and I. Nirdosh, *Can. J. Chem. Eng.* **57,** 150 (1979).

66. S. Stemerding, E. C. Lumb, and J. Lips, *Chem. Ing. Tech.* **35,** 844 (1963).

67. T. Misek, *Coll. Czech. Chem. Commun.* **40,** 1686 (1975).

68. I. Komasawa and J. Ingham. *Chem. Eng. Sci.* **33,** 479 (1978).

69. J. Ingham, J. R. Bourne, and A. Mogli, *Proceedings of the International Solvent Extraction Conference*, Lyon (ISEC74), Vol. 2 (Soc. Chem. Ind., London), 1974, p. 1299.

70. A. R. Sheikh, J. Ingham, and C. Hanson, *Transact. Inst. Chem. Eng.* **50,** 199 (1972).

71. V. Rod, *Coll. Czech. Chem. Commun.* **33,** 2855 (1968).

72. S. M. Karpacheva, E. I. Zakharov, V. N. Koshkin, V. S. Dyakov, V. M. Martynov, and V. F. Abramkin, *Zh. Prikl. Khim.* (*Engl. transl.*), **47,** 821 (1974).

73. A. M. Rosen and Y. G. Rubezhnyi, *Teor. Osn. Khim. Tekh.* (*Engl. transl.*) **5,** 766 (1971).

7

UNSTEADY-STATE EXTRACTION

L. Steiner and S. Hartland
Swiss Federal Institute of Technology
Zurich, Switzerland

1. INTRODUCTION

Liquid–liquid extraction processes are usually designed to operate at the steady state over long periods of time without change of the operating variables. Unsteady-state operation is rarely used for production purposes (see Chapter 5, Section 5); nevertheless, there are several good reasons for studying it:

1. The development of automatic and computer control (see Chapter 27.2) requires knowledge of the dynamic response of the equipment to changes in feed rates, compositions, and other operational parameters.
2. Parameters such as holdup, backmixing (see Chapter 6), and mass-transfer coefficients may be obtained experimentally by use of unsteady-state techniques. In particular, eddy diffusion coefficients are best obtained by observing the change in shape of an impulse as it passes through the system.
3. Unsteady-state operation is encountered during start-up and when steady-state operation is disturbed in some way.

Some of the special terms involved are listed in Table 1. To quantitatively describe an unsteady-state process, the behavior of a column is approximated by a mathematical model. This should be sufficiently realistic to be accurate and yet simple enough for the computations to remain practically possible. After the model has been selected, all the necessary parameters are determined, some of them by specially designed unsteady-state experiments. Once this has been achieved, the model realistically represents the process in question and may be used to predict the response of the extractor to external influences and changes in operational parameters.

TABLE 1 GLOSSARY OF TERMS

Dynamic analysis—analysis of the system under changing conditions, usually in response to an artificially applied change in the feed stream

Empirical model—model that describes observed performance of system without regard to mechanism involved

Mechanistic model—model describing system based on mechanism of operation

Parameter determination—determination of parameters describing operation of system (in this case from dynamic analysis)

Response characteristics—response of system to change in feed conditions

Transfer function—ratio of Laplace transform of output signal to input signal

It may also be used in scaling up and design of full-scale equipment from laboratory and pilot plant data, as well as in process control and automatization.

Accordingly, this chapter briefly describes the most important models applicable to countercurrent extraction processes, shows how the parameters are determined by unsteady-state techniques and how the information thus obtained is condensed into the response characteristics.

A list of publications dealing with unsteady-state extraction is given in Table 2. Only published and readily available papers are referred to; theses and research reports are excluded. Literature of this kind may be found in the reviews of Pollock and Johnson [1, 33] or in Souhrada's work [2].

Some selected works, not dealing directly with extraction but giving useful details on the techniques used in the description of unsteady state processes, are given in Table 3. Both these tables are arranged in chronological order, with recent works appearing last.

2. MODELING OF EXTRACTION COLUMNS IN THE UNSTEADY STATE

The actual behavior of any piece of equipment is usually so complicated that it cannot easily be described by a reasonably simple set of mathematical equations. However, mathematical description is necessary if the equipment is to be operated without tedious trial and error. In such a case the actual behavior is approximated by a model that can be expressed by exactly defined equations. In the next paragraph a countercurrent extraction column with its complicated flow pattern is replaced by a series of sections in which perfect mixing is assumed. As material balances can easily be written for perfectly mixed vessels, it is thus possible to write a series of equations that, by simultaneous solution, yield the behavior of a whole column.

As shown in Table 2, the development of models for extraction columns started some 30 years ago and was closely connected with modeling of other kinds of apparatus used in the chemical industry. The development of earlier models is collected in reviews by Pollock and Johnson [1, 33] in which two kinds of basic model were proposed with differential or stagewise contact of the phases. Later on, when the importance of backmixing was recognized, the models were improved by considering this influence. Stagewise models developed from equilibrium stages with perfect mixing to nonideal stages with backflows. The differential and stagewise backmixing models are closely related to each other (see Chapter 6), and any differential process may be approximated by a series of finite steps, provided the number of stages is sufficiently large. It is thus possible to describe a spray column by a stagewise model and a sieve-plate extractor by a differential one. For most purposes, these models are interconvertible, and in any case, finite steps must be used to obtain a numerical solution of the differential equations. In addition, the stagewise model is more versatile and can handle difficult situations (e.g., involving partially miscible solvents, side streams, and the simultaneous transfer of several solutes).

2.1. Stagewise Backflow Model

The historical development of this model may be found in the review of Pollock and Johnson [33]. The model in its complete form with backflow in both phases was postulated by Sleicher [34]; its original application to unsteady-state liquid/liquid extraction is due to Souhrada et al. [2].

The extractor may be divided into a large number of stages that may be purely hypothetical, as in the case of packed or spray columns, or real, as in agitated or pulsed-plate columns. It is assumed that both phases are perfectly mixed in each stage, that there are step changes of concentration between them, and that the stages are seldom at equilibrium. The nonideality of the flows inside the column is expressed by the backflows that are hypothetical streams flowing in the direction opposite to the main flows of the phases and mixed with the contents of the next stage. The classical model assumes that these backflowing streams are proportional to the forward flow rate of the mainstream of the same phase, but other assumptions may also be used. Only the simplest cases with immiscible solvents and linear equilibrium relation may be solved analytically, but this presents no restriction with the advent of powerful computers. Mathematically, the model is described by a set of ordinary differential equations, the number of which is equal to or greater than the number of selected hypothetical stages. They may be solved without much difficulty (e.g., with a modified Runge-Kutta procedure) on a computer. Given sufficient computer capacity, difficult problems with sidestreams entering and leaving the columns, and partially miscible solvents with complicated equilibrium conditions may be simulated successfully. The solutions may be extended to extraction with chemical reaction if information on the mechanism of the process involved is available.

The notation for a column divided into N perfectly mixed stages with backmixing in both phases is shown in Fig. 1. If the solvents are perfectly immiscible, regardless of the concentration of the solute, their flows will be constant along the column, and concentrations are best expressed as relative mass fractions. The holdup ϵ is defined as the volumetric ratio of the dispersed phase in each stage. The mass-transfer rate between the phases is expressed in the usual way as

$$r_n = K_c a(X_n^* - X_n) = K_d a(Y_n - Y_n^*) \tag{1}$$

The equilibrium concentration in the dispersed phase Y^* is given by the equilibrium coefficient m, so that $Y_n^* = mX_n$. This equilibrium coefficient is assumed to be a function of the temperature and the concentration of the solute in one of the phases, namely, $m = f(T, X)$, and must be determined experimentally for each system. The coefficients K_c and K_d in Eq. (1) may be converted by using the familiar relationship $K_c/K_d = m$, which assumes that the equilibrium curve is sufficiently straight to be represented by its slope for concentration changes between X_{n-1} and X_{n+1}. The balance equations for a typical stage are then

$$\frac{dX_n}{dt} = \frac{u_s}{h} \{(1+f)(X_{n+1} - X_n) - f(X_n - X_{n-1})\} - \frac{r_n}{\rho_c(1-\epsilon)} \tag{2}$$

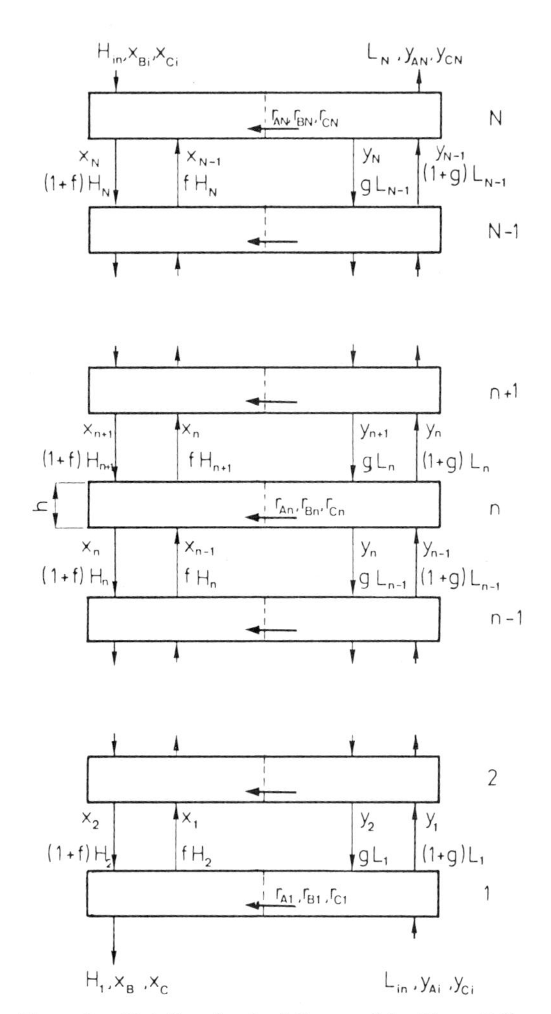

Figure 1. Notation for backflow model with partially miscible solvents.

and

$$\frac{dY_n}{dt} = \frac{v_s}{h} \{(1+g)(Y_{n-1} - Y_n) - g(Y_n - Y_{n+1})\} + \frac{r_n}{\rho_d \epsilon} \tag{3}$$

[see also Chapter 27.2, Eqs. (21) and (22)].

The backmixing is expressed by the backflow coefficients f and g, which are dimensionless numbers indicating the ratio of the backflow inside the column to the feed flow of the same phase outside the column. According to the assumptions of this model, the forward flow of the continuous phase in the column is thus $(1+f)\dot{H}$, whereas $f\dot{H}$ is the backflow. The magnitude of the backflow coefficients is dependent on the number of stages used in the model, and there is

TABLE 2 LITERATURE DEALING WITH UNSTEADY STATE EXTRACTION

Reference	Column Type	Model	Solution[a]	Experimental Data	Observations
3	General	Stagewise counter- and cross-current	A	–	Generalized linear solution of stagewise processes giving response of column to feed disturbances
4	Pulsed plate	Stagewise backflow in one phase	N	Responses to pulse and step forcing	Pulsed column approximated by three-stage model; backmixing was involved in one phase by allowing both phases to pass through the plates by upsurge
5	Rotating disk	Dispersion	A	Pulse testing to obtain backmixing coefficient in continuous phase	Derived empirical equations for backmixing coefficient
6	Pulsed plate	Dispersion approximated by finite differencies	A	Pulse testing to obtain Bode diagrams for different operational conditions	Solution of diffusional equation for linear equilibrium; comparison of calculated Bode diagrams to experimental ones
7	Pulsed plate	Nine different stagewise models from equilibrium stage and uniform mixing to nonequilibrium stage and nonuniform mixing, without backflows.	A, N	Time to approach steady state	Models based on assumption of nonequilibrium and nonuniform mixing predict well steady state behavior and approach to it
8	Single agitated cell	Stirred tank	A, N	Response to step and impulse forcing	Determinations of transfer function, Bode diagram, and mass transfer characteristics

9	Five-stage mixed settler	Cascade of stirred tanks		Pulse testing	Experimental responses to step changes in feed concentrations
10	Packed	Series perfectly mixed cells, no backflows	A	Responses to pulse injection	Model did not adequately describe the frequency response, semiempirical model developed using correcting factors
2	Reciprocating plate	Stagewise backflow	N	Steady-state mass transfer-rate, backmixing coefficients, holdup, approach to steady state for exit concentrations	Models for linear and nonlinear equilibrium, approximative solution for former
11	Countercurrent general	Full stagewise backflow model	N	None	Application of digital simulation to obtain concentration profiles by use of backflow model
12	Multiple mixer	Stagewise backflow	A, N	Approach to steady state, exit concentrations	Computed responses to step and impulse forcing, Bode diagrams for different numbers of stages
13, 14	General	Stagewise backflow	N	None	Methods for calculating mass and backmixing coefficients by impulse testing, computation of concentration profiles, partially miscible solvents
15	Pulsed sieve plate	Stagewise backflow	N	Approach to steady state	Backflow model to simulate approach to steady-state model reduction by transfer function of second order
31	Spray	Dispersion		Axial mixing by pulse testing	Correlating equations for dispersion coefficients

[a] A = analytical; N = numerical

TABLE 3 SOME USEFUL LITERATURE RELATING TO UNSTEADY-STATE PROCESSES

Reference	Topic
16	Programming of unsteady-state stagewise operations on analog computer
17	Pulse testing definitions, residence time distribution
18	Frequency response for perfect mixers, piston flow with longitudinal diffusion, and laminar flow (linear systems)
19	Fitting of nonlinear models to experimental data
20	Computer solution of older stagewise models, review
21	General description of pulse testing methods
22	Analytical solutions of some extraction models
23	Evaluation of frequency characteristics
24	Approximation of complicated models by simpler transfer functions
25	Analytical expressions of different models for residence time distribution
26	Textbook on dynamic behavior of processes
27	Different mixing models involving backmixing, deadspacing, channeling, and mass transfer
28	Textbook on backmixing, detailed description of main models
29	Closed form solution of dispersion model in unsteady state
30	Analytical solution of models with dispersion and stagnant zones

mixing inside the column, even if the backflow coefficients are zero, because of the finite volume of the (well-mixed) stages. The mass-transfer coefficients K_c and K_d used with this model are not identical with those calculated from end concentrations, assuming plug flow conditions inside the column (see Chapter 6). They may be evaluated from concentration profiles along the column axis, both in steady and unsteady stage. Actual solvent velocities are used throughout in the model, so that

$$u_s = \frac{\dot{H}}{(1-\epsilon)\,\rho_c A} \tag{4}$$

$$v_s = \frac{\dot{L}}{\epsilon \rho_d A} \tag{5}$$

Since the backflows do not leave the column, the equations for the first and last stages differ from those for a typical stage. For the first stage the equations become

$$\frac{dX_1}{dt} = \frac{u_s}{h}(1+f)(X_2 - X_1) - \frac{r_1}{\rho_c(1-\epsilon)} \tag{6}$$

$$\frac{dY_1}{dt} = \frac{v_s}{h}\{Y_i + gY_2 - (1+g)\,Y_1\} + \frac{r_1}{\rho_d \epsilon} \tag{7}$$

and for the last stage

$$\frac{dX_N}{dt} = \frac{u_s}{h}\{X_i + fX_{N-1} - (1+f)\,X_N\} - \frac{r_N}{\rho_c(1-\epsilon)} \tag{8}$$

$$\frac{dY_N}{dt} = \frac{v_s}{h}(1+g)(Y_{N-1} - Y_N) + \frac{r_N}{\rho_d \epsilon} \tag{9}$$

The mass-transfer coefficients need not be equal for all the stages, and there may be additional mass transfer in the end stages as a result of the formation and coalescence of drops in these parts of the column. It is also possible for the end-stage volumes to be larger than the rest if this better represents the actual behavior of the column. If there are sidestreams entering or leaving the column, the equations for specific stages may be modified accordingly.

The equations also apply when the stages are well defined, as in a cascade of mixer–settlers, if Eqs. (4) and (5) are rewritten as

$$\frac{u_s}{h} = \frac{\dot{H}}{(1-\epsilon)\,\rho_c V} \tag{4a}$$

$$\frac{v_s}{h} = \frac{\dot{L}}{\epsilon \rho_d V} \tag{5a}$$

where $V = hA$ is the stage volume.

If the mutual solubility of the solvents cannot be neglected, the number of equations increases, as it is now necessary to write the overall balance and the balances for two components for both phases. Denoting the solvents by A and B, the solute by C, and the mass fractions in the aqueous and organic phases by x and y, respectively, the equations are as given in Table 4. It may be seen that relative mass fractions have no advantage here; thus normal mass fractions are used. In addition to the mass-transfer rate for the transfer of the solute from one phase to the other (denoted by r_c), there are two further mass-transfer rates for the solution of the organic phase in water and of the water in the organic phase. The solute equilibrium may again be expressed by using the coefficient m, which is now a function of x_C. The values of x_B^* and y_A^*, which are usually much less important than y_C^*, may be expressed as functions of x_C and y_C, respectively.

The simplest way of applying this model is to use all three mass-transfer rates so that the equations in Table 4 may be directly written into a computer program. Unfortunately, little is known about the rate of mutual solution of the solvents, which is certainly influenced by the presence of the solute. In this case the assumption introduced by Mecklenburgh and Hartland [28] may be used, in which each phase is considered to be always saturated by the other solvent. In unsteady-state operation this is, of course, not always reasonable as infinite transfer rates would be implied. However, if some arbitrary limitation is used, this difficulty may be overcome. The inaccuracy connected with such an assumption is usually low and disappears completely if the solution approaches the steady state. The number of differential equations in the set, which may be as high as six per stage if there are density changes along the column, can be reduced to two if saturation is assumed and the density changes are negligible. By expressing the holdup of the dispersed phase, the drop diameter and mass-transfer rates as functions of, say, x_C, all these parameters can be varied along the column height without introducing further differential equations. In this way a realistic picture of complicated processes may be obtained. Further details on the application of this model are given in Chapter 27.2.

The numerical solution of the model equations is done on a computer without much difficulty. For more complicated cases, the question of computer costs may arise if sets of several hundreds of differential equations are to be solved. In this case the approximation should use as few stages as possible. (The criterion being the amount of backmixing in the column.) The computer cost rises exponentially with the number of stages.

2.2. Other Models

The classical backflow model as described previously is a good tool for general-purpose simulation and a reliable basis for the evaluation of most parameters. However, there are situations when other models have advantages: (1) the concept of well-mixed stages does not correspond to the actual situation; (2) with low backmixing, the differential approach is preferable as the number of stages would be too large otherwise; and (3) the stagewise model becomes too complicated.

2.2.1. Models Dealing with Noncoalescing Drops

The assumption for applicability of the stagewise model is that the contents of all stages are perfectly mixed and that the composition of the backflow is equal to that of the mainstream when these streams leave the same stage. To achieve this, all drops of the dispersed phase should be coalesced, mixed together, and reformed in each stage. The more coalescence and redispersion of the drops taking place in the column, the better is the fit of the backflow model. In actual extractors this condition is sometimes not fulfilled and the drops retain their identities over longer distances along the column axis. In this case the backflow model can be modified by using the backflow concept for the continuous phase only and providing for movement of discrete drops through the cascade of stages according to the specified rules. In the simplest case all drops may be considered to be moving in the same direction and with the same velocity. This case was solved by the present authors [35] and used for the unsteady-stage simulation of spray columns. In another publication [36] cases with drops moving in one direction with different velocities, with different coalescence rate, and with drops moving in different directions were discussed. In the latter two cases the results due Misek [37] and Jiricny and Prochazka [38], respectively, were used. Both these models can be considered as modifications of the classical backflow model, and their application to the unsteady state is readily possible.

2.2.2. The Differential Dispersion Model

This model is commonly used in the literature to describe simpler extraction phenomena. It was

TABLE 4 LIST OF BALANCE EQUATIONS FOR PARTIALLY MISCIBLE SOLVENTS[a]

(a) Overall balances, heavy phase:

$$\frac{d\rho_{x_1}}{dt} = \frac{1}{h}(u_2 - u_1) - \sum_{ABC} r_1 \quad \text{(T.1)}$$

$$\frac{d\rho_{xn}}{dt} = \frac{1}{h}(u_{n+1} - u_n) - \sum r_n \quad \text{(T.2)}$$

$$\frac{d\rho_{xN}}{dt} = \frac{1}{h}(u_{in} - u_N) - \sum r_N \quad \text{(T.3)}$$

(b) Overall balances, light phase:

$$\frac{d\rho_{y_1}}{dt} = \frac{1}{h}(v_{in} - v_1) + \sum r_1 \quad \text{(T.4)}$$

$$\frac{d\rho_{yn}}{dt} = \frac{1}{h}(v_{n-1} - v_n) + \sum r_n \quad \text{(T.5)}$$

$$\frac{d\rho_{yN}}{dt} = \frac{1}{h}(v_{N-1} - v_N) + \sum r_N \quad \text{(T.6)}$$

(c) Solute, heavy phase:

$$\rho_{x_1}\frac{dx_{C_1}}{dt} = \frac{1}{h}\{u_2((1+f)x_{C_2} - fx_{C_1}\} - r_{C_1} \quad \text{(T.7)}$$

$$\rho_{xn}\frac{dx_{Cn}}{dt} = \frac{1}{h}\{u_{n+1}((1+f)x_{C(n-1)} - fx_{Cn}) - u_n((1+f)x_{Cn} - fx_{C(n-1)})\} - r_C \quad \text{(T.8)}$$

$$\rho_{xN}\frac{dx_{CN}}{dt} = \frac{1}{h}\{u_{in}x_{Ci} - u_N((1+f)x_{CN} - fx_{C(N-1)})\} - r_{CN} \quad \text{(T.9)}$$

(d) Solute, light phase:

$$\rho_{y_1}\frac{dy_{C_1}}{dt} = \frac{1}{h}\{v_i y_{Ci} - v_1((1+g)y_{C_1} - gy_{C_2})\} + r_C \quad \text{(T.10)}$$

$$\rho_{yn}\frac{dy_{Cn}}{dt} = \frac{1}{h}\{v_{n-1}((1+g)y_{C(n-1)} - gy_{Cn}) - v_n((1+g)y_{Cn} - gy_{C(n+1)})\} + r_{Cn} \quad \text{(T.11)}$$

$$\rho_{yN}\frac{dy_{CN}}{dt} = \frac{1}{h}\{v_{N-1}((1+g)y_{C(N-1)} - gy_{CN}) - v_n y_{CN}\} + r_{CN} \quad \text{(T.12)}$$

(e) Solvent B, heavy phase: index C in equations (T.7)–(T.9) replaced by B

(f) Solvent A, light phase: index C in equations (T.10)–(T.12) replaced by A

[a]In all sections the balances for first, typical, and final stages are given in order. Explanation of symbols is as follows. Indices: A, heavy solvent (water); B, light solvent (organic); C, solute. Mass fractions: x, in heavy phase; y, in light phase. Densities: ρ_x, heavy phase; ρ_y, light phase. Velocities: u and v are the mass flow rates per unit area of the column (considered as an empty tube).

applied to extraction by Miyauchi and Vermeulen [39] and extended to the unsteady state by Blalock and Clements [29], who obtained an analytical solution, assuming linear equilibrium and immiscible solvents. When notation consistent with the previous sections is used, the partial differential equations describing the dispersion model are

$$\frac{\partial X}{\partial t} = u_s \frac{\partial X}{\partial z} + E_c \frac{\partial^2 X}{\partial z^2} - \frac{r}{\rho_c(1-\epsilon)} \tag{10}$$

$$\frac{\partial Y}{\partial t} = v_s \frac{\partial X}{\partial z} + E_d \frac{\partial^2 Y}{\partial z^2} + \frac{r}{\rho_d \epsilon} \tag{11}$$

the boundary and initial conditions are

$$\left\{\frac{\partial X}{\partial z}\right\}_{z=0} = 0 \tag{12}$$

$$v_s(Y_i - Y_0) + E_d \left\{\frac{\partial Y}{\partial z}\right\}_{z=0} = 0 \tag{13}$$

$$\left\{\frac{\partial Y}{\partial z}\right\}_{z=H} = 0 \tag{14}$$

$$u_s(X_i - X_H) - E_c \left\{\frac{\partial X}{\partial z}\right\}_{z=H} = 0 \tag{15}$$

$$X_{t=0} = Y_{t=0} = 0 \tag{16}$$

This set of equations may be solved numerically by using the well-known difference–differential method, which converts the partial differential equations into a set of ordinary differential equations. This means, in fact, that the differential model is converted into the stagewise one, as follows. By dividing the column into N sections, the derivatives may be approximated, for example, by using three-point Lagrangian formulas, with symmetrical ones for a typical section and unsymmetrical ones for the column ends. Equations (10) and (11) are then replaced by

$$X_0 = \frac{1}{3}(4X_1 - X_2) \tag{17}$$

$$Y_0 = \frac{2hv_s Y_i + E_d(4Y_1 - Y_2)}{2hv_s + 3E_d} \tag{18}$$

$$\frac{dX_n}{dt} = u_s \frac{X_{n+1} - X_{n-1}}{2h} + E_c \frac{X_{n+1} - 2X_n + X_{n-1}}{h^2} - \frac{r_n}{\rho_c(1-\epsilon)} \tag{19}$$

$$\frac{dY_n}{dt} = -v_s \frac{Y_{n+1} - Y_{n-1}}{2h} + E_d \frac{Y_{n+1} - 2Y_n + Y_{n-1}}{h^2} + \frac{r_n}{\rho_d \epsilon} \tag{20}$$

$$X_N = \frac{2hu_s X_i + E_c(4Y_{N-1} - X_{N-2})}{2hu_s + 3E_c} \tag{21}$$

$$Y_N = \frac{1}{3}(4Y_{N-1} - Y_{N-2}) \tag{22}$$

Comparison of Eqs. (19) and (20) with Eqs. (2) and (3) shows that they are identical if the dispersion coefficients E_c and E_d are related to the backflow coefficients f and g by

$$E_c = hu_s\left(\frac{f+1}{2}\right) \tag{23}$$

$$E_d = hv_s\left(\frac{g+1}{2}\right) \tag{24}$$

The dispersion coefficients E_c and E_d do not depend on the number of sections used and hence are best used for reference purposes. It should be emphasized here that the number of stages (or sections, if the dispersion model is being used), should be selected sufficiently large so that for a given amount of backmixing the backflow coefficients remain positive. If not, incorrect results are obtained and it is seldom possible to compute the concentration profiles numerically as the solution tends to oscillate. Further discussion of this model is given in Section 4.4.3(a) of Chapter 27.2.

2.2.3. *Reduced Models for Control Purposes*

The models described previously simulate the actual behavior of the column and may be used for the prediction of concentration profiles and the overall separation. On the other hand, the mathematics involved is complicated, and sometimes, especially in control systems where the column is only one item in a complicated control loop, the solution may be prohibitively expensive. It would be much easier if the column were represented by a simple transfer function like the other items so that methods used in automatic control theory might be used. By observing the calculated or measured responses of extraction columns, one may see that in most cases the shapes are quite simple so that approximation by simpler functions would seem to have a good chance of success. Recently Bauermann and Blass

[15] succeeded in reducing the backflow model for a pulsed sieve-plate column, and the resulting transfer function is second order, with two identical time constants and a time delay. Such a reduction may well be used for control purposes, but the "constants" are in fact complicated functions of the operational variables that must be calculated from experiment or from the complete backflow model. Gibilaro and Lees [24] suggest two model transfer functions, each with three constants that may be used to approximate the behavior of columns in the chemical industry. There is no reason why they should not be applicable in extraction. The proposed transfer functions are

$$G(s) = \frac{\exp(-\tau_1 s)}{(1 + \tau_2 s)(1 + \tau_3 s)} \tag{25}$$

and

$$G(s) = \frac{\exp(-\tau_1 s)}{(1 + \tau_2 s)^k} \tag{26}$$

The coefficients τ_1, τ_2, τ_3, and k may be obtained from response curves by the method of moments, as described in their work. Again, τ_1, τ_2, τ_3, and k are complicated functions of the hydrodynamic parameters and the mass-transfer rate so that for every set of experimental conditions, special values are necessary. Comparison of Eqs. (25) and (26) with the full-scale backflow model has shown that sometimes no solution is found for Eq. (25), whereas Eq. (26) agrees well with the numerical solution over five orders of magnitude as shown in Fig. 2.

3. DETERMINATION OF MODEL PARAMETERS BY UNSTEADY-STATE METHODS

The models described in the previous paragraphs contain parameters that should be determined experimentally for each given item of equipment. In some cases unsteady-state methods, in which a tracer is injected and its path followed through the equipment, are best. The tracer may either be soluble in the other phase or remain in the phase into which it was injected. Most work reported in the literature was carried out by use of the latter method, especially for determining backmixing parameters. The application of soluble tracers is more difficult, but attempts have been made to determine mass-transfer coefficients in this way.

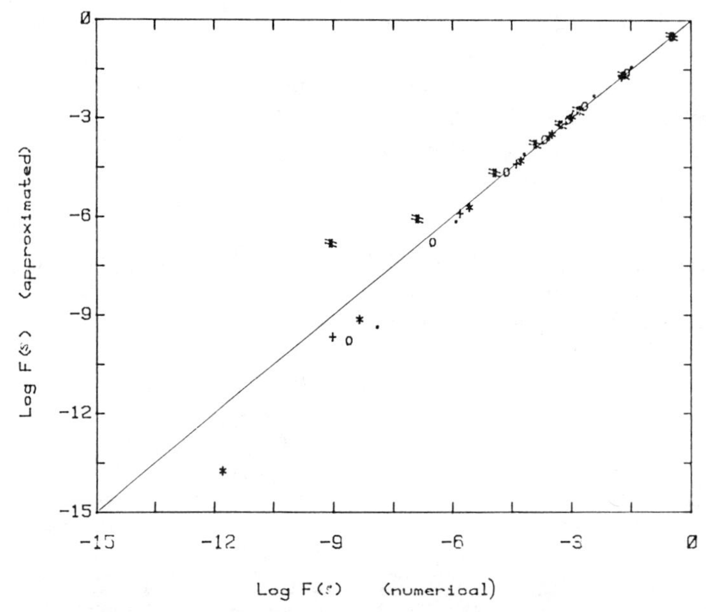

Figure 2. Approximation of transfer function by Eq. (26). For a column represented by the stagewise backflow model, an impulse was injected close to the inlet of dispersed phase and the responses obtained at two points further downstream. Transfer functions were calculated by using Eq. (31) for different operating conditions and correlated using Eq. (26).

3.1. Determination of Backmixing Coefficients

The methods for determining backflow coefficients f and g (Eqs. (2) and (3)] or the dispersion coefficients E_c and E_d [Eqs. (10) and (11)] are based on either steady- or unsteady-state injection of an insoluble tracer into the mainstream of one of the phases and registering the responses in one or two places up or downstream. A survey of experimental work in this area is given in Chapter 6. The theoretical background of all possible variations (using stagewise or differential models, steady- or unsteady-state injections, impulses of different shapes, different boundary conditions, etc.) is given by Mecklenburgh and Hartland [28]. A method tested practically by the present authors will be described here in more detail.

To avoid experimental errors associated with the practical impossibility of producing a perfect Dirac impulse or step change, an imperfect impulse is injected in some part of the apparatus upstream and the responses registered at two points further downstream. As tracer, a saline solution of higher conductivity, a dye, or radioactive material may be used. To obtain coefficients applicable in the models, actual values valid for the inside of the column should be used that are free of end effects. Therefore, both injection and sampling is best done along the column length so the experimental region is "open" at both ends, according to Levenspiel's nomenclature [40]. It is very important to record the responses with utmost accuracy as the tail of the curve has a significant effect on the accuracy of results (e.g., analysis becomes impossible if the signal does not return to its original level after the impulse has passed the measuring point). Earlier problems in this respect can be overcome in most cases by using an on-line process computer. A drift-free digital recording to three significant figures usually ensures sufficient accuracy. Various possibilities exist for the treatment of results; the most important are described in the following paragraphs.

3.1.1. *Method of Moments*

This is the original method described by Levenspiel and Smith [41], which is based on the evaluation of first moments about the origin μ and second moments about the measuring point σ at both positions where the responses are recorded. For the open system, the formula for evaluation of dispersion coefficients was derived by Aris [42]:

$$\mu_2 - \mu_1 = \tau \tag{27}$$

$$\sigma_2^2 - \sigma_1^2 = \frac{2\tau^2}{\mathrm{Pe}} \tag{28}$$

where the numbers 1 and 2 refer to the measuring locations and the first and second moments are defined as follows:

$$\mu = \frac{\int_0^\infty c(t)\, t\, dt}{\int_0^\infty c(t)\, dt} \tag{29}$$

$$\sigma^2 = \frac{\int_0^\infty c(t)\, t^2\, dt}{\int_0^\infty c(t)\, dt} - \mu^2 \tag{30}$$

In this method the tail of the curve is of utmost importance. However, if the measurement is correctly done, good results are achieved and the method may be used without much computing expense.

3.1.2. *Analysis of Transfer Function*

This method was used in solvent extraction by Mixon et al. [43] and generalized by Ostergaard and Michelsen [44]. It assumes strict applicability of the dispersion model. The transfer function $G(s)$ is calculated numerically for several values of the parameter s according to the following formula:

$$G(s) = \frac{\int_0^\infty c_2(t)\, e^{-st}\, dt \Big/ \int_0^\infty c_2(t)\, dt}{\int_0^\infty c_1(t)\, e^{-st}\, dt \Big/ \int_0^\infty c_1(t)\, dt} \tag{31}$$

The values of $G(s)$ for several arbitrarily chosen values of s are then used in

$$\ln \frac{1}{G(s)} = \frac{\tau s}{[\ln(1/G(s))]^2} + \frac{1}{\mathrm{Pe}} \tag{32}$$

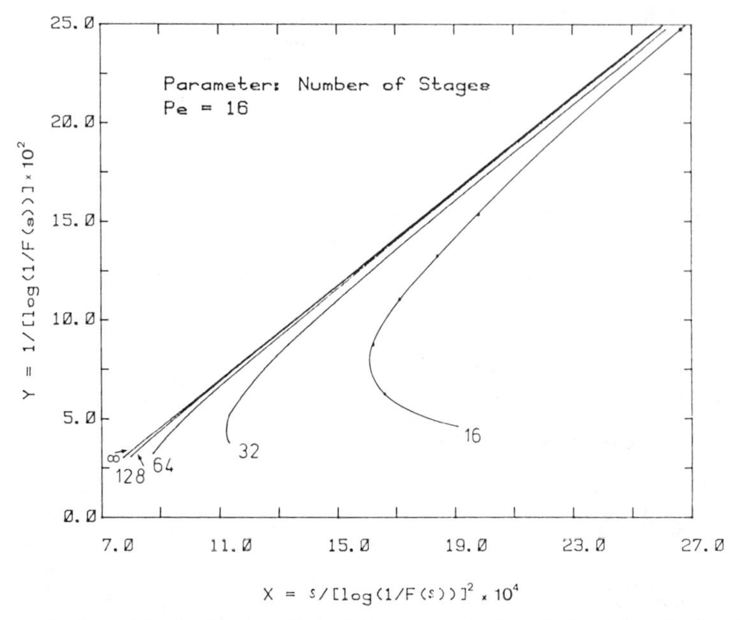

Figure 3. Evaluation of backmixing coefficients from numerically calculated transfer functions [Eqs. (31) and (32)]. The figure shows limits of the method if models with a smaller number of stages are used. Whereas straight lines are obtained for the dispersion model and stagewise models with more than 64 stages, the method fails completely for 16 and less stages.

which may be derived by the Laplace transformation of equation (10) (without mass transfer and for an open system). If the system strictly obeys the differential model, plotting of $\ln[1/G(s)]$ against $s[\ln(1/G(s)]^2$ gives a straight line with slope τ and intercept 1/Pe. In this case "tailing" presents no difficulty, the restricting condition is that a good straight line is actually obtained. For example, if the procedure is used for a stagewise contactor and there are only a few stages between the sampling points (e.g., <10), the line is usually curved and false results are obtained. In our experiments good results were achieved only when the corresponding regression coefficient was better than 0.999. This type of evaluation is shown in Fig. 3.

3.1.3. *Computer Simulation*

This is a rather laborious method. However, the results are insensitive to experimental difficulties, and any model can be used. The responses are obtained in the usual way and fed to the computer, which seeks to reproduce them by varying the backmixing parameters in a model solution until the best fit is obtained. In this way the applicability of the model and results obtained by simpler methods can be checked. The curve is fitted by minimizing the sum of squares of deviations between the experimental and simulated values at a selected number of points. In all methods described here the parameter τ is also evaluated. If the dispersion model is applicable, this parameter gives the time necessary for the impulse to travel between the sampling points in plug flow. As the actual velocity depends on the holdup, the latter is automatically obtained. An example of measured and best-fitting profiles based on the stagewise model is shown in Fig. 4.

3.1.4. *Analysis of Response Curve*

A very practical method of evaluation was derived by Vergnes [45]. The height of the maximum on the response curve c_{max} is multiplied by the time at which the maximum occurs t_{max} and divided by the integral of the area under the curve. (No division is necessary if reduced concentration and reduced time are used.) For an open system, perfect impulse and behavior expressed by the dispersion model area obtained is related to the Peclet number as follows:

$$4\pi R^2 = [\sqrt{1+(\mathrm{Pe})^2} - 1]\exp[\mathrm{Pe} - \sqrt{1+(\mathrm{Pe})^2}] \tag{33}$$

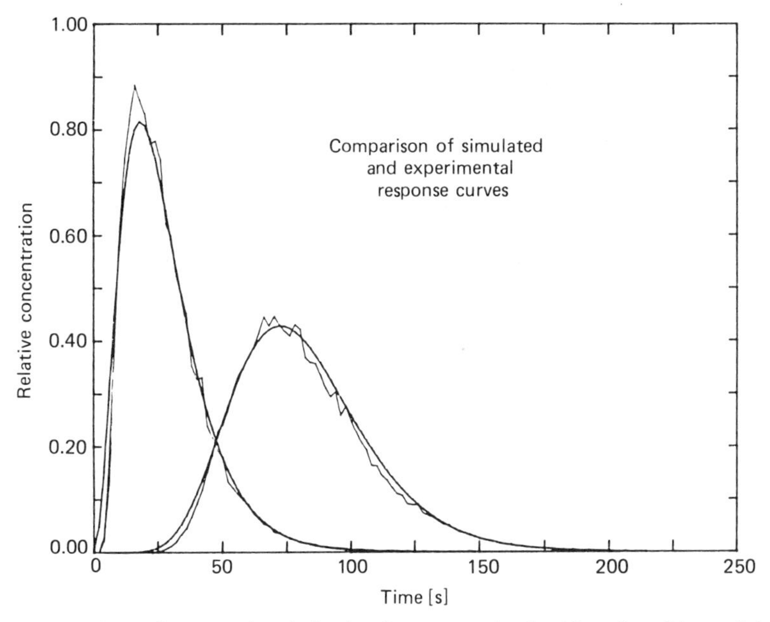

Figure 4. Comparison of measured and simulated responses to short impulse of tracer injected into continuous phase of extraction column. Stagewise backflow model with 32 stages was used to generate the simulated curve; the free parameter is the backflow coefficient g [Eqs. (2), (6), and (8)].

where

$$R = \frac{t_{\max} c_{\max}}{\int_0^\infty c(t)\, dt}$$

For Peclet numbers greater than 10, Eq. (33) may be approximated with an error of less than 1% by

$$\mathrm{Pe} \doteq 4\pi R^2 + 1.5 \tag{34}$$

Surprisingly, this method gives quite accurate results, comparable with the methods listed under Sections 3.1.1 and 3.1.2. Its disadvantage is that a sufficiently short impulse is necessary as only one-point evaluation is possible. If measurements are available at a second point, these may be used independently and the result averaged.

3.2. Determination of Mass-Transfer Rates

If a tracer, partially soluble in the other phase, is injected into an extraction column the mass-transfer rates may be evaluated from the measured response. In this case, as part of the tracer is transferred into the other phase, the area under the response curve decreases with distance while the shape of the response is distorted by axial mixing, as in the case of an insoluble tracer. Simultaneous evaluation by computer simulation of both the mass-transfer and backmixing coefficients is possible, in principle, but the necessary techniques are not yet sufficiently developed. Experiments of this kind have recently been carried out by the present authors. Ostergaard and Michelsen [44] describe how results should be corrected if a partially soluble instead of an insoluble tracer is used to determine backmixing by an impulse method. Steiner and Hartland [13] discuss the possibility of evaluating mass-transfer coefficients from mean concentrations (integrals) of samples collected at two or three points along the column. More should be done in this field before reliable results will be obtained as, more often than not, the hydrodynamic parameters such as drop size and holdup are influenced by the presence of the soluble tracer and change during an experimental run.

4. DYNAMIC ANALYSIS OF BEHAVIOR OF AN EXTRACTION COLUMN

There are several ways of characterizing the dynamic behavior of a piece of equipment. In simpler cases the transfer function is determined experimentally and expressed by a suitable approximate formula. This approach has already been treated here; possible approximations are Eqs. (25) and (26). The transfer function itself may be determined experimentally by evaluation of the shapes of an impulse at the inlet and outlet of a column section by using Eq. (31). Another possibility for characterization of the dynamic behavior is to determine the frequency response. The principles of this method may be found in standard textbooks such as that by Friedly [26], and the application to chemical equipment is described by Clements and Schnelle [21], Staffin and Ju Chin Chu [8], and others. Further discussion is also given in Section 3.2.2 of Chapter 27.2. The most important features are as follows.

If the input to a piece of equipment is varied sinusoidally and the equipment behavior can be described by a linear differential equation with constant coefficients, the output will also be sinusoidal when the steady state is reached. However, the output wave has a different amplitude and is shifted relative to the input wave. The amplitude and the phase shift depend on the frequency, which remains the same for both input and output. The ratio of the input and output amplitudes $M(\omega)$ is called the *magnitude* or *amplitude ratio*, or the *gain*, and the shift is characterized by the phase angle $\phi(\omega)$. The graphs of $\log M(\omega)$ and $\phi(\omega)$ versus $\log \omega$, which are usually plotted together in the same figure, are referred to as the Bode diagram. This diagram contains all the necessary information and is commonly used to characterize the dynamic behavior of systems and pieces of equipment. An example for a typical extraction column is shown in Fig. 5. The advantage of this approach is that a Bode diagram may be determined experimentally and that it makes possible the handling of sometimes complex transfer functions in a simple graphical way. If the corresponding transfer functions are of a lower order, they may be derived directly from the shape of the diagram.

The frequency response may, of course, be determined directly from its definition by introducing a continuous sinusoidal variation at the inlet and measuring the outlet variation at the same time. However, this may be difficult experimentally if, for example, concentration changes in a large column are being investigated. Nevertheless, Gray and Prados [32] used this approach to investigate a packed-column absorber. A more convenient method is to introduce a pulse into the column and measure its shape at the column exit. By use of a mathematical treatment based on the Fourier transformation, the frequency response can be derived from the shapes of the input and output impulses. This procedure is described in detail by Clements and Schnelle [21].

If the shape of the input and output impulses as in Fig. 5 are known and some frequency ω is selected, the following parameters may be ob-

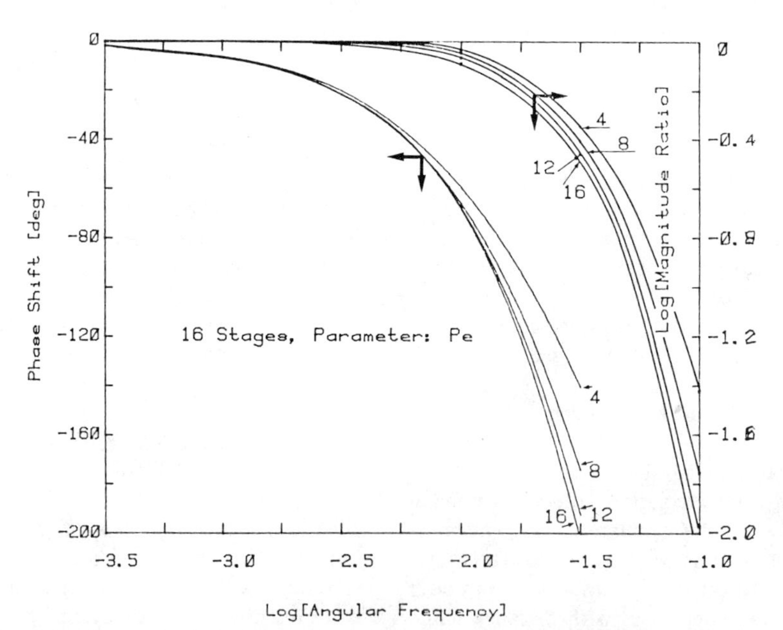

Figure 5. Example of the Bode diagram derived from the stagewise backflow model for one set of operating conditions varying the backflow coefficient (expressed as dimensionless Peclet number).

tained by numerical integration:

$$A = \int_0^{t_e} X_e(t) \cos \omega t \, dt \qquad (35a)$$

$$B = \int_0^{t_e} X_e(t) \sin \omega t \, dt \qquad (35b)$$

$$C = \int_0^{t_i} X_i(t) \cos \omega t \, dt \qquad (35c)$$

$$D = \int_0^{t_i} X_i(t) \sin \omega t \, dt \qquad (35d)$$

These parameters are then combined to obtain

$$\alpha = \frac{AC + BD}{C^2 + D^2} \qquad (36)$$

$$\beta = \frac{AD - BC}{C^2 - D^2} \qquad (37)$$

and the gain and the phase angle are given by

$$M(\omega) = [\alpha^2(\omega) + \beta^2(\omega)]^{1/2} \qquad (38)$$

$$\phi(\omega) = \tan^{-1} \frac{\beta(\omega)}{\alpha(\omega)} \qquad (39)$$

By repeating this procedure for different frequencies, the complete Bode diagram may be constructed.

The numerical integration in Eqs. (35a)–(35d) may be done by using the trapezoidal or Simpson rules, but some difficulties may arise when ω becomes large. A special formula was thus derived by Filon [21] to facilitate evaluation of the integrals without loss of significant digits that cannot be avoided when more usual methods are used.

Comparison of the frequency responses from experimental data and from a simulation procedure is a good test of the applicability of a model in a given situation.

NOTATION

a	interfacial area, m^2/m^3
A	cross section of column, m^2
A, B, C, D	parameters defined by Eq. (35)
c	concentration, kg/m^3
E	dispersion coefficient, m^2/s
f, g	backflow coefficients in continuous and dispersed phases, respectively, (-)
$G(s)$	transfer function
h	height of hypothetical stage, m
H	column height, m
$\dot{H}$	flow rate of continuous phase, kg/s
k	constant in equation (26)
K	mass-transfer coefficient, $kg\, m^{-2}\, s^{-1}$
$\dot{L}$	flow rate of dispersed phase, kg/s
m	equilibrium coefficient, (-)
$M(\omega)$	magnitude ratio [Eq. (38)]
n	number of stages, (-)
Pe	Peclet number (Pe = uH/E) (-)
R	defined in Eq. (33)
r	mass-transfer rate, $kg\, m^{-3}\, s^{-1}$
s	transfer function parameter, (-)
t	time, s
T	temperature, K
u, v	phase velocities defined by Eqs. (4) and (5), m/s
V	stage volume, m^3
x, y	mass fractions in continuous and dispersed phases, respectively, (-)
X, Y	relative mass fractions in continuous and dispersed phases, respectively (solute-free basis), (-)
z	length, m

Greek Symbols

α, β	auxiliary parameters in Eqs. (36) and (37)
ϵ	volumetric holdup of dispersed phase (-)
μ	moment about origin
ω	angular frequency, s^{-1}
ρ	density, kg/m^3
σ	moment about mean value
τ	defined in Eq. (27)
τ_1–τ_3	constants in Eqs. (25) and (26)
$\phi(\omega)$	phase shift [Eq. (39)]

Indices

A, B, C	solvents and solute in partially miscible system
c	continuous phase
d	dispersed phase
e	exit

i	inlet
n	typical stage
N	last stage
s	solvent
*	equilibrium value

REFERENCES

1. G. G. Pollock and A. I. Johnson, *Can. J. Chem. Eng.* **47,** 469 (1969).
2. F. Souhrada, J. Landau, and J. Prochazka, *Can. J. Chem. Eng.* **48,** 322 (1970).
3. L. Lapidus and N. R. Amundson, *Ind. Eng. Chem.* **42,** 1071 (1950).
4. B. A. Di Liddo and R. A. Walsch, *Ind. Eng. Chem.* **53,** 801 (1961).
5. K. R. Westerterp and P. Landsman, *Chem. Eng. Sci.* **17,** 363 (1962).
6. J. W. Watjen and R. M. Hubbard, *AIChE J.* **9,** 614 (1963).
7. J. C. Biery and D. R. Boylan, *Ind. Eng. Chem.* **2,** 44 (1963).
8. H. K. Staffin and J. C. Chu, *AIChE J.* **10,** 98 (1964).
9. P. E. Burns and C. Hanson, *Br. Chem. Eng.* **12,** 75 (1967).
10. J. E. Doninger and W. F. Stevens, *AIChE J.* **14,** 591 (1968).
11. I. J. Dunn and J. Ingham, *Verfahrenstechnik* **6,** 399 (1972).
12. D. A. Jones and W. L. Wilkinson, *Chem. Eng. Sci.* **28,** 539 (1973).
13. L. Steiner and S. Hartland, *Proceedings of the International Solvent Extraction Conference*, Lyons, 1974, (Soc. Chem. Ind., London), 1974, p. 2289.
14. L. Steiner and S. Hartland, *Chemische Rundschau*, **28** (1975).
15. H.-D. Bauermann and E. Blass, *Chem. Ing. Technik.* **49,** 175 (1977).
16. A. Acrivos and N. R. Amundson, *Ind. Eng. Chem.* **45,** 467 (1953).
17. P. V. Danckwerts, *Chem. Eng. Sci.* **2,** 1 (1953).
18. H. Kramers and G. Alberda, *Chem. Eng. Sci.* **2,** 173 (1953).
19. D. W. Marquardt, *Chem. Eng. Progr.* **55** (6), 65 (1959).
20. D. Yesberg and A. I. Johnson, *Can. J. Chem. Eng.* **38,** 49 (1960).
21. W. C. Clements and K. B. Schnelle, *Ind. Chem. Eng. Proc. Des. Devel.* **2,** 94 (1963).
22. V. N. Chernyshev, *Internatl. Chem. Eng.* **6,** 608 (1966).
23. J. R. Hays, W. C. Clements, and T. R. Harris, *AIChE J.* **13,** 374 (1967).
24. L. G. Gibilaro and F. P. Lees, *Chem. Eng. Sci.* **24,** 85 (1969).
25. J. Villermaux and W. P. H. Van Swaaij, *Chem. Eng. Sci.* **24,** 7 (1969).
26. J. C. Friedly, *Dynamic Behaviour of Processes*, Prentice-Hall, Englewood Cliffs, NJ, 1972.
27. A. Burghardt and L. Lipowska, *Internatl. Chem. Eng.* **13,** 227 (1973).
28. J. C. Mecklenburgh and S. Hartland, *The Theory of Backmixing*, Wiley, London, 1975.
29. K. E. Blalock and W. C. Clements, Jr., *Chem. Eng. J.* **9,** 137 (1975).
30. M. Popovic and W. D. Deckwer, *Chem. Eng. J.* **11,** 62 (1976).
31. G. S. Laddha, T. R. Krishnan, S. Viswanathan, S. Vedaiyan, T. E. Degaleesan, and H. E. Hoelscher, *AIChE J.* **22,** 456 (1976).
32. R. I. Gray and J. W. Prados, *AIChE J.* **9,** 211 (1963).
33. G. G. Pollock and A. I. Johnson, The Dynamics of Liquid/Liquid Extraction Processes, in J. A. Marinsky and Y. Marcus, Eds., *Ion Exchange and Solvent Extraction*, Vol. 6, Marcel Dekker, New York, 1974, Chapter 2.
34. C. A. Sleicher, Jr., *AIChE J.* **5,** 145 (1959).
35. L. Steiner, M. Horvath, and S. Hartland, *Ind. Eng. Chem. Proc. Des. Devel.* **17,** 175 (1978).
36. L. Steiner and S. Hartland, *Chem. Ing. Technik.* **52,** 602 (1980).
37. C. Hanson, Ed., *Recent Advances in Liquid/Liquid Extraction*, Pergamon Press, New York, 1971.
38. W. Jiricny and J. Prochazka, Congress CHISA, Prague, 1978.
39. T. Miyauchi and T. Vermeulen, *Ind. Eng. Chem. Fund.* **2,** 113 (1963).
40. O. Levenspiel, *Chemical Reaction Engineering*, Wiley, New York, 1962.
41. O. Levenspiel and W. K. Smith, *Chem. Eng. Sci.* **6,** 227 (1957).
42. R. Aris, *Chem. Eng. Sci.* **9,** 266 (1959).
43. F. O. Mixon, D. R. Whitaker, and J. C. Orcutt, *AIChE J.* **13,** 21 (1967).
44. K. Ostergaard and M. L. Michelson, *Can. J. Chem. Eng.* **47,** 107 (1969).
45. F. Vergnes, *Chem. Eng. Sci.* **31,** 88 (1976).

8

SPECIAL TECHNIQUES

M. H. I. Baird
McMaster University
Hamilton, Canada

1. INTRODUCTION

This short chapter deals with several aspects of extraction that have not been described in Chapters 1–7 and do not fall under the headings of established equipment or processes.

These aspects have however been the subjects of basic research for many years, and considerable progress has been made toward general practical application.

2. LIQUID MEMBRANE EXTRACTION

2.1. Basic Principles

In a conventional solvent extraction process the degree of extraction is limited primarily by the ratio of solvent to feed and the distribution ratio of the solute between the phases. The solute may subsequently be removed from the solvent in a stripping operation, allowing the solvent to be reused, but the extent of solute stripping is again limited by the phase ratio and the distribution ratio.

In liquid membrane extraction, the extraction and stripping operations described previously are combined in a single process and the usage of solvent is greatly reduced. The principle was devised by Li [1] in the 1960s. In the simplest arrangement, the solvent is present as a very thin spherical film or *membrane*, stabilized by a suitable surfactant and interposed between the feed solution and the strip solution (Fig. 1*a*). Preliminary results were given [2, 3] on the separation of aromatic hydrocarbons from mixtures with aliphatic hydrocarbons by permeation through very thin spherical water films stabilized by surfactants. Film diameters were as wide as 5 mm, but the film thickness could not be measured directly. The two important parameters for mass transfer are the *selectivity* of the solvent for the solute (relative to a reference component) and the *permeability* of the film, defined as the ratio of mass flux to concentration driving force. Permeabilities were found to range from 0.1 to 20 μm/s depending on the type of hydrocarbon feed employed and the nature of the surfactant.

The spherical film arrangement, although convenient for laboratory measurements, offers only a relatively low specific interfacial area for mass transfer. In an alternative arrangement [2] the strip solution and the membrane solvent are first emulsified to form extremely fine droplets

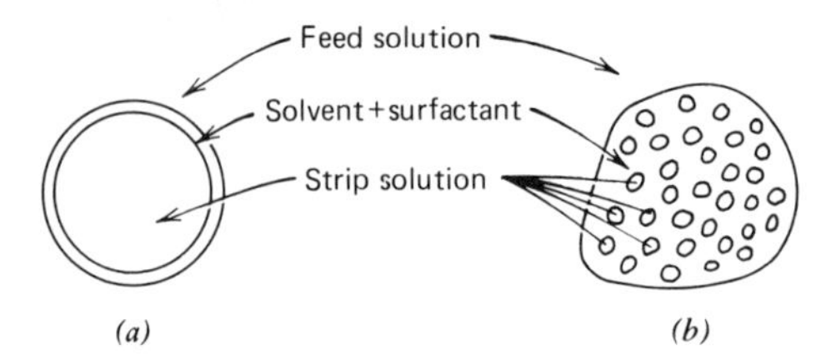

Figure 1. Two modes of liquid membrane extraction: (*a*) spherical film; (*b*) emulsion globule.

(diameter in the order of 1 μm) dispersed in the membrane solvent as shown in Fig. 1*b*. Droplets of this emulsion are then contacted with the feed with moderate agitation followed by conventional settling and separation. The loaded emulsion is then "broken" by centrifugation, electrical methods, heat, or addition of special reagents, so that the membrane solvent may be returned for further use. A complete circuit is shown in Fig. 2.

It should be noted that the places of the feed and the strip solution may be reversed in Figs. 1*a*, and 1*b*, and 2. The choice of which solution is to be continuous and which is to be emulsified will depend on interfacial properties and other variables.

2.2. Applications and Modeling

Liquid membrane extraction is particularly useful when the strip solution contains a reagent that takes up the solute. An early example of this was the extraction of traces of phenol from waste water by an emulsion of aqueous sodium hydroxide droplets in a mixture of surfactant and middle distillate hydrocarbon [4]. Although the reverse permeation of sodium phenolate through the hydrocarbon does not occur, the breakdown of the emulsion droplets can lead to a gradual increase in the phenol concentration in the water after the initial extraction; therefore, the treatment time, surfactant concentration and agitator speed must be carefully set to obtain optimum phenol removal.

The preceding approach has been extended to the removal of metal cations from waste water [5], and particularly to the extraction of copper from waste water or hydrometallurgical leach liquors [6, 7]. In this case chemical reaction occurs in the organic membrane phase, which contains a complexing agent for copper. Extraction from the feed into the liquid membrane occurs as follows:

$$2RH + Cu^{2+} \rightleftharpoons R_2Cu + 2H^+ \quad (1)$$

The copper enters the acid strip solution by the reverse reaction:

$$R_2Cu + 2H^+ \rightleftharpoons 2RH + Cu^{2+} \quad (2)$$

This process of *facilitated transport* is equivalent to a transfer of copper ions to the strip solution balanced by a reverse transfer of hydrogen ions to the feed.

Copper transfer has recently been modeled kinetically [8] in terms of an equivalent spherical membrane, that is, the configuration of Fig. 1*a*. Progress has also been made with models for the more complicated (but more practical) configuration shown in Fig. 1*b*, in which transfer occurs through the globule of membrane solvent to the assembly of micrometer-sized droplets of

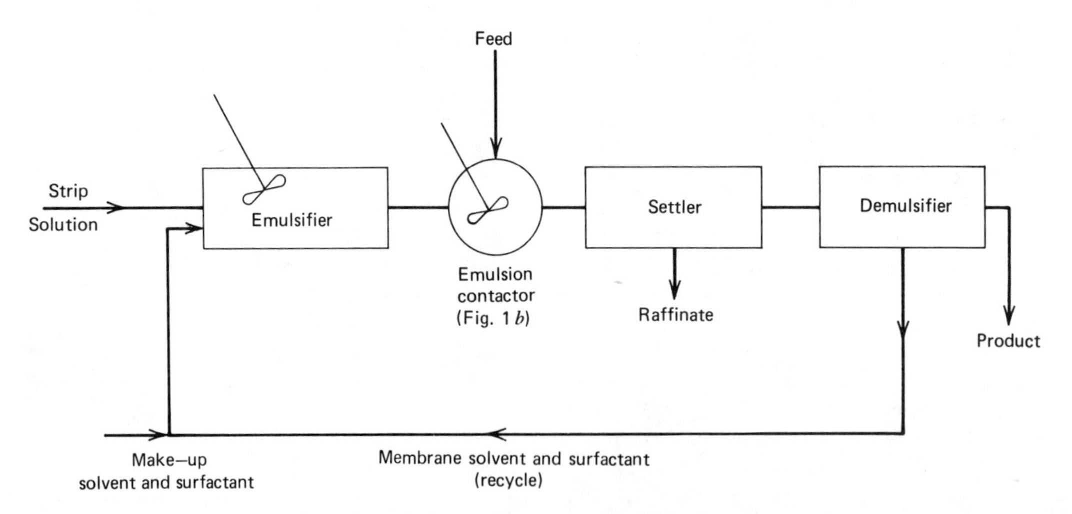

Figure 2. Liquid membrane extraction circuit.

strip solution [9, 10]. It has been concluded [10] that the outer layer of membrane solvent in the globule plays a critical part in determining the overall transfer resistance.

Since its debut in 1968 [1], liquid membrane extraction has been developed at an ever-increasing rate, and a review 10 years later [11] cited 79 references. Although at the time of writing there is no commercial process in which liquid membrane extraction is used, the technique offers widespread industrial potential [12].

3. CYCLIC EXTRACTION PROCESSES

3.1. Controlled Cycling

Controlled cycling can be applied to any countercurrent mass-transfer process; interest in controlled cycling extraction has been slow to develop but is increasing. The concept was introduced by Cannon in 1952 [13] and first applied to extraction by Szabo et al. [14].

Consider a countercurrent extraction column as shown schematically in Fig. 3. In normal continuous operation a steady-state concentration profile is set up in each phase, with mass transfer occurring at a rate depending on the concentration driving force in each stage (see Chapter 5).

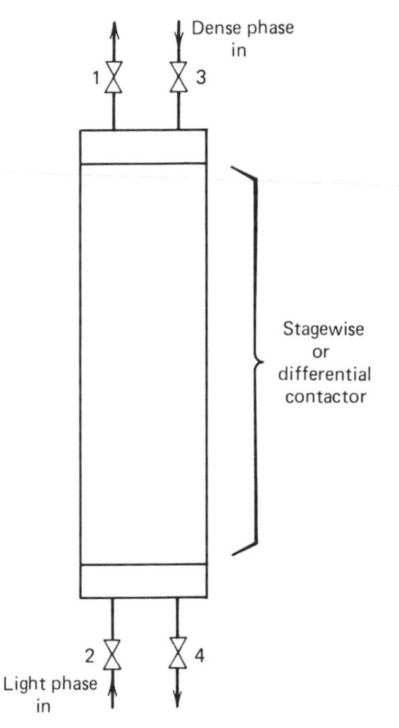

Figure 3. Simplified diagram of controlled cycling (redrawn from Szabo et al. [14]).

In controlled cycling, the conditions during a cycle are *unsteady*, with repetition of the following basic cycle: (1) light phase flowing; valves 1 and 2 open, valves 3 and 4 closed; (2) relaxation period with all valves closed; (3) dense phase flowing; valves 3 and 4 open, valves 1 and 2 closed; and (4) relaxation period with all valves closed.

Szabo et al. [14] carried out exploratory experiments with a sieve-plate column and a packed column, each of 50.8-mm diameter, with total cycle times in the order of 10 s. In the case of the sieve-plate column, the permissible throughput was greatly increased by controlled cycling compared to normal operation; the other major effect of controlled cycling was to render the column efficiency (expressed as the number of equivalent theoretical stages) independent of throughput rate. The main effect of controlled cycling in the packed column was to improve the efficiency, while not greatly affecting the maximum permissible throughput.

Further results were obtained with a 152-mm-diameter sieve-plate column by Belter and Speaker [15]. They went beyond the range of throughputs studied previously [14] and found that column efficiency decreased slightly with increasing throughput, but in other respects the results in the 152-mm column compared closely with those [14] in the 50.8-mm column. Belter and Speaker [15] also developed a mathematical treatment of countercurrent flow under controlled cycling conditions. The analysis showed that under ideal conditions of stagewise contact with N actual stages, the effective number of stages could reach $2N$ with controlled cycling. The analytical modeling technique [15] was limited in application to systems with a linear equilibrium relationship, but the technique has been extended by using a graphical method [16] that can handle nonlinear equilibria. In this case it was also shown that the number of effective stages could be doubled by controlled cycling, provided ideal cycling conditions were met. An important condition for ideality is that the total holdup of each phase should flow between stages in each cycle, without backmixing.

The modeling methods have been improved by Darsi and Feick [17], who checked their results with experimental data in a 101.6-mm column. Boyadzhiev and his co-workers [18, 19] have developed modeling methods to take account of axial mixing effects.

Recently a production model controlled-cycling extractor based on a series of packed columns has been developed for use in the pharmaceutical industry [20], but full performance data are not available. Controlled-cycling extraction has been found to be an energy-saving alternative to distillation in a case of organic chemical recovery [21].

3.2. Parametric Pumping and Cycling Zone Extraction

These cyclic processes must not be confused with controlled cycling. *Parametric pumping* was introduced by Wilhelm et al. [22] in the 1960s and was at first applied to separations involving a fixed bed of solids, for example, adsorption and ion exchange. Parametric pumping was extended to liquid–liquid extraction by Wankat [23]. The basic principle (Fig. 4) is that one phase is held stationary (i.e., within fixed stages) while the second phase is alternately passed to the right and to the left through the cascade. The cycle is sufficiently slow that some process parameter, usually the temperature, can be varied synchronously with the flow direction of the second phase. It can be shown [23] that although there is very little net bulk flow of either phase from one cycle to the next, there is a significant migration ("pumping") of the transferring solute from the cold reservoir to the hot reservoir.

Data were obtained with a rotating horizontal helix contactor and with a discrete staged system by use of centrifuge test tubes. The observed separation was not quite as good as theoretically predicted, but the predicted trends were followed. It was suggested [23] that a staged column capable of 180° inversion or a system with stationary liquid coated on a solid support could also be used for parametric pumping.

Another cyclic technique, known as *cycling zone extraction* [24], is shown in Fig. 5. A solvent phase is retained in m regions of n stages, while a mobile feed phase is fed continuously in one direction. The temperatures in each region are cycled according to the scheme shown. The variation of temperatures causes each region to alternately store and reject solute; proper matching of the cycle period and the time constants of the stages can result in a good overall separation of the solute, building up into strong "waves" of concentration, leaving region m. The modeling technique and small-scale results with the use of centrifuge test tubes have been reported [24].

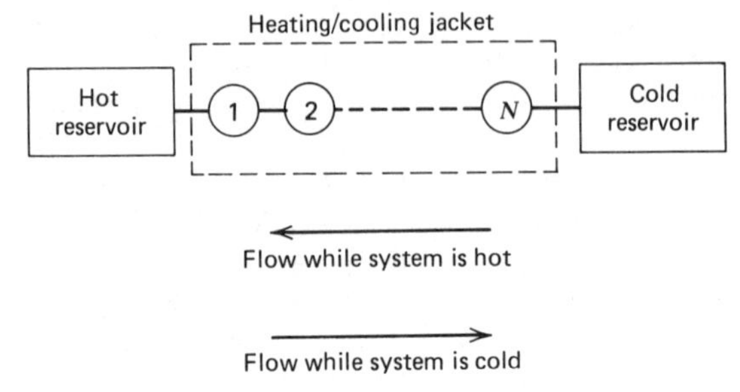

Figure 4. Parametric pumping extraction (redrawn from Wankat [23]).

4. ELECTROSTATIC EFFECTS ON EXTRACTION

4.1. Drop Formation and Characteristics

In conventional extraction equipment the interfacial contact area may be increased by reducing the mean droplet size (Chapter 4); however, this has the effect of reducing the mass-transfer coefficient since the droplet terminal velocity is reduced and since circulation and oscillation are suppressed in very small droplets. The formation

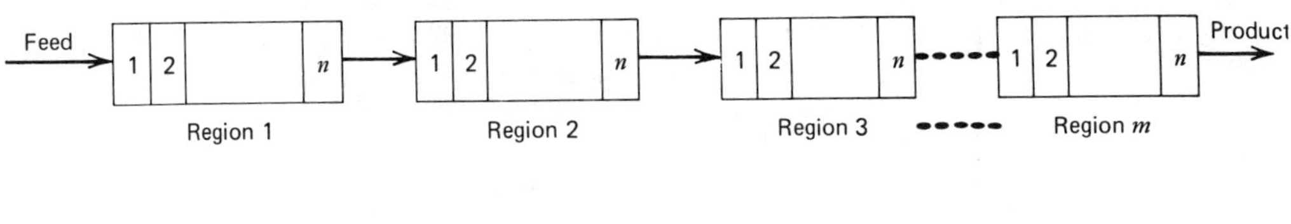

	Regions $m = 1, 3, 5 \ldots$	Regions $m = 2, 4, 6 \ldots$
First part of cycle	Hot	Cold
Second part of cycle	Cold	Hot

Figure 5. Cycling zone extraction.

of droplets in an electrostatic field has been found to result in small droplets that nevertheless exhibit extremely high mass-transfer coefficients.

Thornton and co-workers [25–27] studied single-droplet behavior under these conditions for a parallel-plate electrode geometry. The nominal electrostatic field strength is typically in the region of 0.2 MV/m; and for maintenance of this level, the continuous phase should be of low conductivity, that is, organic. It should be noted that the field strength cannot be calculated simply as (degree of applied voltage per cell length) because of the effect of the finite conductivity of the continuous phase. When a droplet is formed in the electrostatic field, a charge is induced on its surface, which has the effect of (1) making the interface behave as if its interfacial tension were very low and (2) augmenting the gravitational pull on the droplet. As the field strength is increased, the drop formation changes from the *discrete drop regime*, which resembles conventional detachment, to the *electrostatic dispersion regime*, in which a continuous stream of minute, highly charged droplets is shed from the sharp tip of the pendant droplet, which has an elongated conical appearance. The detached droplets were found [26] to move at higher velocities in an electric field than in uncharged conditions, despite their smaller size. They were also found to oscillate.

4.2. Mass-Transfer Effects

Single-drop mass-transfer studies [27] with furfural droplets dispersed in *n*-heptane indicated up to 50% enhancement in the mass-transfer coefficient, depending on the applied voltage and the cell length. The furfural-*n*-heptane system was later used in multidroplet mass-transfer experiments [28] with a modified form of the single-drop cell [27], operating as a continuous countercurrent spray column. The average droplet velocity relative to the continuous phase was enhanced, the mean droplet size was reduced, and the number of dispersed-phase transfer units was increased with increasing applied voltage.

Electrostatically enhanced mass transfer in ternary systems has been reported [29] with the use of an improved basic cell, in which the dispersed phase enters the contact region by falling as jets through a nitrogen layer (Fig. 6). On contact with the continuous phase, the jets of the conducting dispersed phase break up rapidly to form drops of approximately 1-mm diameter. The Murphree stage efficiency of such a unit was enhanced from 50% to over 80% as the applied

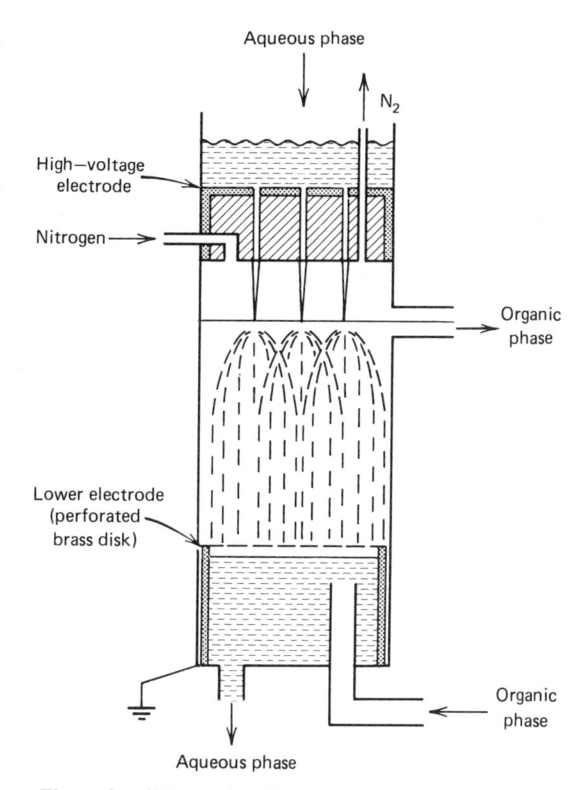

Figure 6. "Three-phase" electrostatic extractor (redrawn from Bailes [29]).

voltage was increased from 0 to 48 kV with a 146-mm depth of continuous phase. The cells of the type shown in Fig. 6 can be vertically stacked in a cascade with a large voltage applied between the top and bottom ends.

The effect of electrostatic fields on metal extraction (nickel and copper) has also been investigated by Bailes [29, 30]. As expected, the rate of extraction was increased, particularly at higher nominal field strengths corresponding to the electrostatic dispersion regime. The direction (polarity) of the field was found to affect the enhancement in some cases.

Although electrostatic fields promote the formation of very fine droplets, they also enhance coalescence [29] (see also, Chapter 4, Section 3.3) and so electrostatic effects continue to be promising where a combination of efficient mass transfer and short interfacial contact time is desired.

5. SONIC AND ULTRASONIC VIBRATION

The operating frequencies of pulsed columns (Chapter 11) and reciprocating-plate columns (Chapter 12) seldom exceed 6 Hz because of limitations of the mechanical drives or pulsing arrangements. However, with the use of electromagnetic and piezoelectric generators, it is possible to vibrate liquid–liquid systems at frequencies in the *sonic* range of about 20 Hz to 16 kHz and in the *ultrasonic* range beyond 16 kHz. The effects of these high frequencies on liquid–liquid extraction form part of what has been termed *sonochemical engineering* [31, 32].

In the low sonic range, it is possible to operate electromagnetic vibrating-plate extractors at amplitudes in the order of a few millimetres. Thus Elenkov et al. [33] reported on a vibrating-plate mixer–settler operating at 17–25 Hz, whereas Nirdosh and Baird [34] devised a laboratory cell for kinetic measurements, in which the flat interface was agitated by 60-Hz vibrations of disk plates located in each phase. The main agitation mechanism was thought to be that of vibration-induced circulation, although some standing-wave interfacial vibrations were sometimes seen to occur.

In general, sonic and ultrasonic extractors do not contain any internals, so that the generators can act on the contents of the extraction vessel without obstruction. Siemens and Franke [35] found that pulsation frequencies of 5–50 Hz could increase extraction rates of acetic acid from carbon tetrachloride drops into water by a factor of 4. Peaks in extraction rate were observed at certain frequencies corresponding to the calculated resonant frequencies for drop shape oscillation. Saint-James and Graham [36] operated a *pulverising pulsed column* at frequencies of 30–100 Hz and found appreciable mass-transfer improvements, and similar results were reported by Schulze et al. [37].

Ultrasonic radiation has been found effective in several solid–liquid extraction processes as summarized by Fogler [31]. For liquid–liquid extraction in 3.8- and 7.5-cm-diameter spray columns, Chen and Chon [38] found that the height of a transfer unit was reduced by up to 20% at frequencies of 100 and 20 kHz. However, if the acoustic power input was increased beyond a certain level, cavitation effects resulted in excessive emulsification in the liquid phases.

It is generally recognized that sonic and ultrasonic vibrations can enhance mass transfer by several distinct mechanisms, notably, induced droplet oscillation, surface waves, acoustic streaming, and cavitation [31]. It is difficult to isolate the effects of these different mechanisms, and thus no general design and scale-up procedures exist for sonically and ultrasonically enhanced extraction, despite the interesting effects that have been observed on the small scale.

REFERENCES

1. N. N. Li, U.S. Patent 3,410,794 (November 12, 1968).
2. N. N. Li, *Ind. Eng. Chem. Proc. Des. Devel.*, **10**, 215 (1971).
3. N. N. Li, *AIChE J.* **17**, 459 (1971).
4. N. N. Li and A. L. Shrier, Liquid Membrane Water Treating in N. N. Li, Ed., *Recent Developments in Separation Science*, Vol. 1, Chemical Rubber Company, Cleveland, 1972.
5. T. Kitagawa, Y. Nishikawa, J. W. Frankenfeld, and N. N. Li, *Env. Sci. Technol.* **11**, 602 (1977).
6. J. W. Frankenfeld and N. N. Li, Liquid Membranc Systems, *Ion Exchange for Pollution Control*, Chemical Rubber Company, Cleveland, 1979.
7. J. W. Frankenfeld, R. P. Cahn, and N. N. Li, *Separation Sci. Technol.* **16**, 385 (1981).
8. T. P. Martin and G. A. Davies, *Proceedings of the International Solvent Extraction Conference*, Liege (ISEC80), Vol. 2, Paper 80-230, 1980.
9. L. Boyadzhiev, T. Sapundzhiev, and E. Benzenshek, *Separation Sci.* **12**, 541 (1977).
10. G. Casamatta, C. Chavarie, and H. Angelino, *AIChE J.* **24**, 945 (1978).
11. W. Halwachs and K. Schugerl, *Chem. Ing. Technik.* **50**, 767 (1978); *Internatl. Chem. Eng.* **20**, 519 (1980).
12. M. M. Hafez, *Proceedings of the International Solvent Extraction Conference*, Liege (ISEC80), Vol. 2, Paper 80-89, 1980.
13. M. R. Cannon, *Oil Gas J.* **51** (12), 268, 271, 318, 320 (1952).
14. T. T. Szabo, W. A. Lloyd, M. R. Cannon, and S. M. Speaker, *Chem. Eng. Progr.* **60** (1), 66 (1964).
15. P. A. Belter and S. M. Speaker, *Ind. Eng. Chem. Proc. Des. Devel.* **6**, 36 (1967).
16. J. Lovland, *Ind. Eng. Chem. Process Des. Devel.* **7**, 65 (1968).
17. C. R. Darsi and J. E. Feick, *Can. J. Chem. Eng.* **49**, 95 (1971).
18. L. Boyadzhiev and G. Angelov, *Izvest. Khim.* **8** (4), 703 (1975).
19. G. Angelov, E. Atanasova, and L. Boyadzhiev, *Izvest. Khim.* **9** (1), 159 (1976).
20. M. E. Breuer, C. Y. Yoon, D. P. Jones, and M. J. Murry, *Chem. Eng. Progr.* **73** (6), 95 (1977).
21. W. M. Langdon, A. Bengali, and G. Gavlin, *Proceedings of the International Solvent Extraction Conference*, Liege (ISEC80), Vol. 3, Paper 80-102 (Universite de Liege), 1980.

22. R. H. Wilhelm, A. W. Rice, and A. R. Bendelius, *Ind. Eng. Chem. Fund.* **5**, 141 (1966).
23. P. C. Wankat, *Ind. Eng. Chem. Fund.* **12**, 372 (1973).
24. P. C. Wankat, *Separation Sci.* **8**, 473 (1973).
25. G. Stewart and J. D. Thornton, *Inst. Chem. Eng. Symp. Ser.* No. 26, 29 (1967).
26. G. Stewart and J. D. Thornton, *Inst. Chem. Eng. Symp. Ser.* No. 26, 37 (1967).
27. P. J. Bailes and J. D. Thornton, *Proceedings of the International Solvent Extraction Conference*, The Hague (ISEC71), Vol. 2 (Soc. Chem. Ind., London), 1971, p. 1431.
28. P. J. Bailes and J. D. Thornton, *Proceedings of the International Solvent Extraction Conference*, Lyons (ISEC74), Vol. 2 (Soc. Chem. Ind., London), 1974, p. 1011.
29. P. J. Bailes, *Proceedings of the International Solvent Extraction Conference*, Toronto (ISEC77), Vol. 1 (Can. Inst. Min. Met., Montreal), 1979, p. 233.
30. P. J. Bailes and I. Wade, *Proceedings of the International Solvent Extraction Conference*, Liege (ISEC80), Paper 80-196 (Universite de Liege), 1980.
31. H. S. Fogler, *Chem. Eng. Progr. Symp. Ser.* **67** (109), 1 (1971).
32. M. J. Ashley, *Chem. Eng.* (*Lond.*), 368 (June (1974).
33. D. Elenkov, L. Boyadzhiev, and I. Krustev, *Isvest. Inst. Obshta Khim., Bulg. Akad. Nauk* **4**, 181 (1966) [through Chem. Abstr. **66**, 67087 (1966)].
34. I. Nirdosh and M. H. I. Baird, *AIChE Symp. Ser.* **74** (173), 107 (1978).
35. W. Siemens and M. Franke, *Chem. Ing. Technik.* **30**, 165 (1958).
36. R. Saint-James and G. P. Graham. *Genie Chimique* **91**, 39 (1964).
37. W. W. F. Schulze, W. Koelle, and K. Haberer, *Fresenius' Z. Anal. Chem.* **230** (1), 7 (1967) [through *Chem. Abstr.* **68**, 24772 (1968)].
38. E. C. Chen and W. Y. Chon, *Chem. Eng. Progr. Symp. Ser.* **63** (77), 44 (1967).

Part II

Industrial Extraction Equipment

9.1

PRINCIPLES OF MIXER–SETTLER DESIGN

J. C. Godfrey and M. J. Slater
University of Bradford
United Kingdom

1. INTRODUCTION

The range of applications of mixer–settlers is now very large, and equipment is available in a wide variety of shapes and sizes. Miniature glass units can be purchased for bench top laboratory studies, whereas at the other end of the scale concrete and steel plant units process hundreds of cubic meters per hour. No column contactor currently available would accept such large flows.

Many designs of mixers have been developed; the most recent innovation has been in large throughput pump–mix units with a trend to multicompartment designs. Recent patents describe a combined mixer–settler that can operate at a controllable volume fraction different from the feed flow ratio. General design expertise has perhaps not kept pace with design innovation. The understanding of extraction performance is reasonably well developed, but scale-up procedures are less so. Other aspects of mixer performance, such as phase inversion, residence time distribution, entrainment generation, and drop size are not yet adequately documented for design purposes. However, both innovation and research on mixer design are continuing, and better mixers and design procedures can be expected.

For most purposes, simple settlers continue to be used, but attention is being paid to more complex arrangements to reduce equipment size and minimize entrainment. The settler should properly be considered as a separator since its job is to allow the two phases to disengage completely before transfer to other mixers. The two distinct processes of *settling* (sedimentation) and *coalescence* of drops with each other and finally with an interface are involved and both need to be better understood for sound design methods to be developed. Chapter 4 reviews the present state of knowledge in this area. In the past decade plant sizes have grown to such an extent that considerable economic benefit results from better design and performance.

2. MIXERS

In a commercial mixer–settler plant the function of the mixer is to provide an adequate combination of agitation and residence time for the desired degree of extraction. Additionally, many large mixers also incorporate a pumping action to avoid the need for separate pumps for interstage liquid transport. Many successful mixer designs have been developed, and some are described in following sections of this chapter. In most designs of liquid–liquid mixers it has been possible to obtain adequate extraction by allowing generous residence times. It may be for this reason, and general lack of need, that design procedures for liquid–liquid mixing tanks have not been developed to the degree encountered in column design.

The usual procedure is to estimate a suitable residence time from small-scale batch extraction tests. Residence time is dependent on impeller speed N, impeller diameter D, and tank size T. Short residence times are possible at high impeller speed, but high speed is associated with the difficulties of high entrainment and possibly slow coalescence in the settler. Once a relationship between residence time and impeller speed has been established for one or more impeller–tank ratios in batch tests, it is possible to make estimates of suitable operating conditions for a full-scale plant.

Approach to equilibrium is the usual basis for calculation:

$$E = \frac{C_0 - C_t}{C_0 - C_e}$$

where C_0 is the initial solute concentration, C_t is the concentration after time t, and C_e is the equilibrium concentration attainable after prolonged contacting. This definition may be used for both batch, E_b, and continuous, E_f, operations, and it is frequently found that the approximation

$$1 - E_b = \exp(-kt_b)$$

may be used to describe batch data. The value of k can then be used to relate values of E_f and t_c for a continuous mixer of the same dimensions:

$$E_f = \frac{kt_c}{1 + kt_c}$$

Thus various combinations of contact time, impeller speed, and diameter may be determined for any value of E_f on the basis of batch data.

There is a problem in estimating the agitation conditions required to give a specified value of E_f for a full-scale plant. The relationships between E and t just described apply only to the agitation conditions at which the experimental data were collected. There is no fundamentally based scale-up procedure that will maintain the same E–t relationships in full-scale plant. A range of scale-up criteria have been used, mostly lying somewhere in effect between the alternatives of constant tip speed and constant power per unit volume. For constant tip speed, the power input per unit volume on the full-scale plant is much less than that used in the laboratory experiments. For constant power input per unit volume, the tip speed on the full-scale plant is much larger than in the laboratory. At the present state of knowledge the importance of these differences is not fully understood.

A better understanding can be obtained by conducting experimental work at the largest pilot scale possible. This need not be a very expensive exercise since large-scale batch data can be very useful.

The procedure discussed previously is the most widely recommended approach to mixer design but does not include some important aspects of mixer performance, such as power consumption, pumping capacity, hold-up characteristics, phase inversion characteristics, drop size, and mass-transfer coefficients. For some impellers, power consumption characteristics are well documented for single-phase systems, and the assumption can be made that similar characteristics will be obtained for two-phase systems. However, in larger mixer–settlers the pump–mix principle is often used. The power and pumping characteristics of most pump–mix impellers are not generally available. For nonpumping impellers, the relationship between $P_o = P/\rho N^3 D^5$ and Re $= D^2 N\rho/\mu$ is all that is required, and often the value of P_o is known and constant in the Re range of interest. For pumping impellers it is necessary to know the relationship between gH/N^2D^2 and Q/ND^3 for pumping performance and $Q\rho H/P$ and Q/ND^3 for power requirements (where P is power, H is head, and Q is flow). In many pump–mix designs there is a provision for clearance adjustment between inlet and impeller that significantly affects the preceding relationships but can allow trial and error adjustments to be made in full-scale plant. Considerable work is necessary if a detailed specification of mixing and pumping is required, but the experiments involved can be easily conducted if a pilot plant is to be built.

In all cases some consideration is given to power and pumping characteristics, but the important topics of holdup and phase inversion are often overlooked. In a continuously operated mixer there is a minimum agitation condition below which the volume fraction in the tank does not correspond to the flow ratio of the two liquid feed streams. This may be attributed to a number of causes depending on operating conditions, including short circuit of light phase in a tank with bottom inlet and top outlet. The problem arises when the impeller speed is too low to produce a near-homogeneous distribution of dispersion in the tank, and this can lead to both poor extraction performance and phase inversion. Phase inversion occurs when the volume fraction of dispersed phase in the tank exceeds an upper limit. This characteristic is usually described by presenting data for volume fraction at inversion as a function of impeller speed. There is an ambivalence region where either phase may be dispersed given a suitable start-up procedure. Laboratory measurements suggest that this region is very wide, expressed in terms of volume fraction and that the effect of impeller speed is of secondary importance. However, plant experience, where inversion problems are not unknown, suggests that there is a more narrow region in large equipment. Correlations are available for both holdup and phase inversion characteristics but are not suitable for the prediction of performance and are of use only in developing data analysis techniques. For design purposes, direct data plotting will be adequate in nearly all cases.

At present a knowledge of drop size is of little direct use in a design exercise but is of considerable interest in understanding the extraction process. There are many correlations, but, as mentioned previously, reliable predictive methods are not yet established. In fact, enormous differences are obtained when various correlations are used to predict for one set of conditions. Similarly, mass-transfer coefficient studies for liquid drops have been very extensive, but the information is not directly applicable to liquid–liquid mixers for a number of reasons. From the computational point of view, the prediction of mass-transfer coefficient is still difficult because of problems in estimating drop surface conditions and internal hydrodynamics. More importantly, many applications of the mixer–settler are for processes where rates of extraction are slow with a significant effect of chemical reaction rate. In such cases calculations based on mass-transfer coefficients significantly underestimate residence time requirements. In other cases, chemical reaction can enhance mass transfer. For further discussion of these effects, see Chapter 2.1.

Over the years the mixer–settler has appeared in many forms. For the majority of designs, the mixer directly or indirectly provides for the transport of liquids from one stage to the next. In some designs mixing and settling occur in the one tank with little or no physical barrier between the two regions. Many variations of the pump–mix theme have been developed since about 1970, and designs have been developed for large flow rates. For large plant, simple designs using a bottom-located turbine in a baffled tank were developed, followed by more sophisticated mixers with a draught tube inlet to allow central location of a turbine impeller in a square unbaffled tank. Other designs use a large pump–mix impeller in a small, shallow tank or use pumping and mixing impellers on the same shaft. The indirect use of impeller energy to provide for interstage liquid transfer is illustrated by the Windscale integral box design, where the density difference between the dispersion and the coalesced light phase is utilized to provide the necessary pressure drop for flow. This leads to a very simple mechanical design but requires deep settlers if large throughputs are to be used.

More recently, multicompartment mixers have been used to provide greater stage efficiency. This approach is based on the homogeneous reactor theory comparison of stirred tanks in series with a single stirred tank. For the same total volume stirred tanks in series will give a better approach to equilibrium by virtue of a closer approach to plug flow.

Various design configurations have been used; it is important to minimize backflows between compartments as backmixing greatly reduces the effectiveness of the stages in series concept. An alternative approach is to use two impellers on one shaft in a tall cylinder divided into two compartments. Some attention has also been given to the use of in-line mixers. A virtually perfect cocurrent plug flow can be obtained by using in-line mixers, and some studies have been published on their use as liquid–liquid contactors. The mixers themselves have no moving parts, but pumps are required for fluid flow and must provide for the pressure drop across the mixer. The energy consumption is comparable to, or less than, that for stirred tanks.

In a recent design a single compartment mixer and settler allows for a holdup value in the impeller region that is significantly different from the flow ratio and can be controlled at a specified value by a simple weir adjustment. This offers considerable advantages for processes where the flow ratio and optimum phase ratio for extraction are different, eliminating the need for recycle streams.

3. SETTLERS

Considerable attention has been given to settler design in recent years, but the performance characteristics are not easily simulated on a small scale. Only gradually are data from large-scale plant—with the attendant difficulties of flow variation, sampling, and dirty liquors—becoming available for interpretation. The large number of variables affecting performance renders the task difficult.

For satisfactory utilization of settler capacity, the dispersion band is usually of near uniform thickness (ΔH) and not so deep that small increases of flow rate lead to large increases in thickness. Dispersion band thickness is usually related to specific dispersed phase flow rate (Q_d/A) by

$$\Delta H = K \left[\frac{Q_d}{A}\right]^y$$

where y can typically be in the range 1–7. Other equations are equally useful, particularly

$$\frac{Q_d}{A} = \frac{k_1 \Delta H}{1 + k_2 \Delta H}$$

which has some theoretical support. The use of batch tests to determine such equation parameters is receiving attention and the question is one of limits of validity rather than one of fundamental incompatibility of batch and flow data.

In general, experimental work on a relatively large scale is required to resolve present design problems; however, with care it is possible to design a settler that works adequately even if not optimum in an economic sense.

The method of introduction of dispersion from the mixer into the settler is important. The use of a dam baffle to create a deep band in the dispersion inlet region helps to achieve even distribution across the settler with little disturbance to flow. Coalescence aids such as removable baskets of packing may be used at the inlet point. These should not suffer a large pressure drop, become blocked frequently, or suffer loss of performance if phase inversion occurs.

The capital cost of a system is often strongly dependent on solvent inventory. Settlers may be made smaller by enhancing coalescence to give greater dispersed phase flow rate per unit horital area. A variety of coalescing devices have been used, including mesh pads (see Chapter 4), packs of horizontal trays and simple plates, and vertical baffles. Any dependence on surface properties is a potential source of failure and the deposition of solids or crud must be considered. Large reductions in settler size are possible, but the economic advantage can be substantially eroded by the cost of such internals, maintenance costs, and possible increases in entrainment losses. All these devices have long histories and few applications. Simple horizontal plate packs deserve more study since they also have the potential to reduce entrainment losses. Throughput is improved by using a multiplicity of thin dispersion bands if dispersion band depth is a strongly nonlinear function of dispersed phase throughput per unit area. Electrical coalescers may also find acceptance in some circumstances since dramatic coalescence rate improvements can be obtained.

A region with little or no dispersion band at the end of a settler may help to reduce entrainment and give a higher margin of safety (avoiding flooding) in case of flow surges or sudden reductions in coalescence rate (due to phase inversion or solids).

Weirs are an important aspect of settler design. Low velocities are required, so full-width weirs are generally used. Ideally, a multiplicity of cross-weirs or side weirs is required to remove disengaged liquid as it is released, if entrainment is a serious problem. However, weirs are expensive to construct, and some compromise is necessary. Air entrainment at weirs and surging should be avoided in order to minimize mixer disturbance.

The operational cost of a settler mainly concerns solvent loss by entrainment. Environmental considerations may limit solvent entrainment in raffinates to less than 10 ml/m^3. Interactions with other processes such as ion exchange, leaching, evaporation, and electrowinning may also be restrictive. Tiny droplets may be found in both phases in a settler. The generation of these droplets may be in the mixer, caused by improper agitation, poor inlet arrangement with premixing of liquids, phase instability (local inversions), and air ingestion. The tiny droplets may also arise from drop–drop coalescences and coalescences at the interface or at solid surfaces. It is speculated that the concentration of drops in the settler exit streams can be reduced by the processes of sedimentation, trapping within a deep dispersion band or filtration by mesh packings placed in the settler. It has been suggested that the concentration in exit streams can be increased by poor inlet distribution to the settler, high vertical velocities fluidizing the dispersion band and elutriating drops, and high local velocities in the separated phases. All these factors concerning entrainment are affected by physical properties and possibly electrical phenomena, so it is not surprising that an understanding of the interactions has not yet been achieved and that a consensus of opinion on the use of devices for improvement of settler performance has not yet been reached.

External treatment to remove entrained material is possible of course, with the use of techniques such as flotation, absorption, or electrostatics, and continuing attention to these methods is desirable.

9.2

SIMPLE BOX-TYPE MIXER–SETTLER

L. Lowes
Seascale
United Kingdom†

1. INTRODUCTION

A simple design of horizontal multistage mixer–settler was developed at Windscale for the United Kingdom Atomic Energy Authority (later British Nuclear Fuels Limited) for the reprocessing of spent nuclear fuels to accomplish the separation of uranium, plutonium, and fission products to very high levels of decontamination from one another [1]. Simplicity in the design was largely dictated by the radioactive nature of the materials to be handled and the consequent need to house the contactors within heavily shielded cells where access for maintenance or adjustment, once the plant was commissioned, would become virtually impossible. The degree of reliability and stability of operation achieved, however, suggests that the system could be applied with advantage in nonradioactive fields. This is particularly true for aqueous metallurgical and other separation and purification processes where many practical stages of solvent extraction are known to be required. Cascades of mixer–settlers involving a series of three extraction and stripping cycles with a total of over 100 stages have operated satisfactorily over long periods with total flows in excess of 15 m^3/hr. Such a system using a 20% tributyl phosphate solution in odorless kerosene, as the extractant for nitric acid solution of uranium, plutonium, and fission products has been employed without modification at Windscale over many years and decontamination factors of the order of 10^7 are achieved.

Some of the many designs of mechanically agitated mixer–settlers [2, 3] introduce complexities of construction or operation such as interstage pumping or adjustable weirs between stages in order to control interfaces in the settling compartments. The simple box type contactor described here is intended for the multistage contacting of two substantially immiscible liquids where all except the final (heavy-phase outflow) interfaces are automatically located within the settler by the hydraulic balances through the various liquid–transfer ports. Methods of calculating interface location in relatively small mixer–settlers have been described [4], and these, together with the present observations, have been confirmed by experience in the nuclear fuel reprocessing field [5] with somewhat larger units.

†Retired, British Nuclear Fuels Ltd., Windscale Works, United Kingdom.

2. GENERAL DESCRIPTION

A typical mixer–settler is illustrated in Fig. 1, which will be seen to have an egg-box type of construction with a double row of mixers and settlers alternating along the box. Each mixing chamber is connected with its associated settler by means of a horizontal mixed phase port situated about midway up the dividing wall, and has a heavy phase port toward the bottom and a light phase port or weir near the top, through which it receives its feeds respectively from the upstream and downstream settlers of adjacent stages. The mixer is agitated by a suitable mechanical impeller, such as a turbine, in such a way that the whole volume of the chamber is adequately mixed so that it can be regarded as having substantially uniform density. There is a baffle opposite the mixed-phase port on the mixer side to minimize the transmission of agitation effects through the port, and if the port is not so large as to allow backflow from the settler, the phase ratio in the mixer will be very nearly the same as the ratio of the feeds to the unit. A system of hydraulic balances through the three ports, taking into account the port pressure drops, may then be used to calculate the equilibrium interface positions. In the separate case of the final (heavy-phase outlet) settler, there is no corresponding mixing chamber against which to balance, and thus a pneumatically controlled weir is provided as shown in Fig. 1 with the principle of operation illustrated in Fig. 2. The mechanical driving force for the transport of the liquids through the system does not derive from any direct pumping action by the impeller, but by the simple overflow from settler into mixer made possible by the level differences established by the differing densities of the mixed and unmixed phases. In practice, for stable operation over a wide range of flow rates, it is desirable to design the mixing impeller to minimize dynamic effects at the liquid-transfer ports.

3. THE MIXER

The volume of the mixer is determined by the residence time required to attain an economic degree of equilibrium. The rate of solute transfer may be determined by batch experiments, and it has been found that small-scale measurements are adequate for defining the residence time required, at least in the nitric tributyl phosphate–kerosene system. The depth of the mixing chamber is set by the hydraulic requirements of the unit as a whole; the cross section may then be determined to give the required volume. In prac-

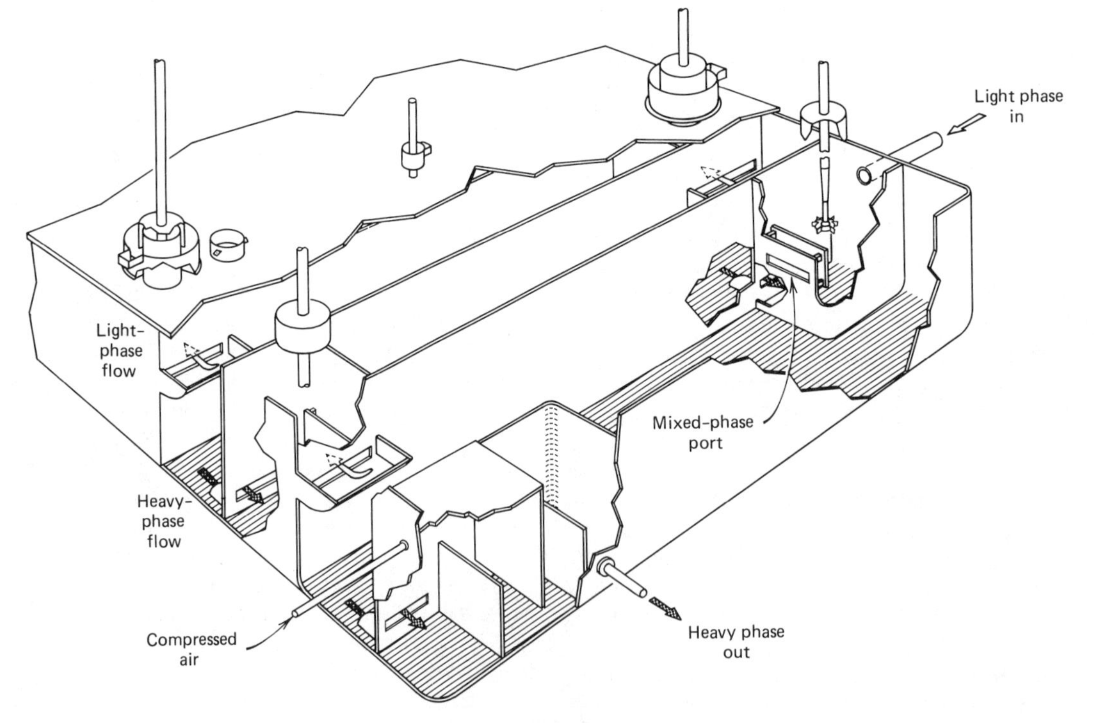

Figure 1. Typical mixer–settler [9].

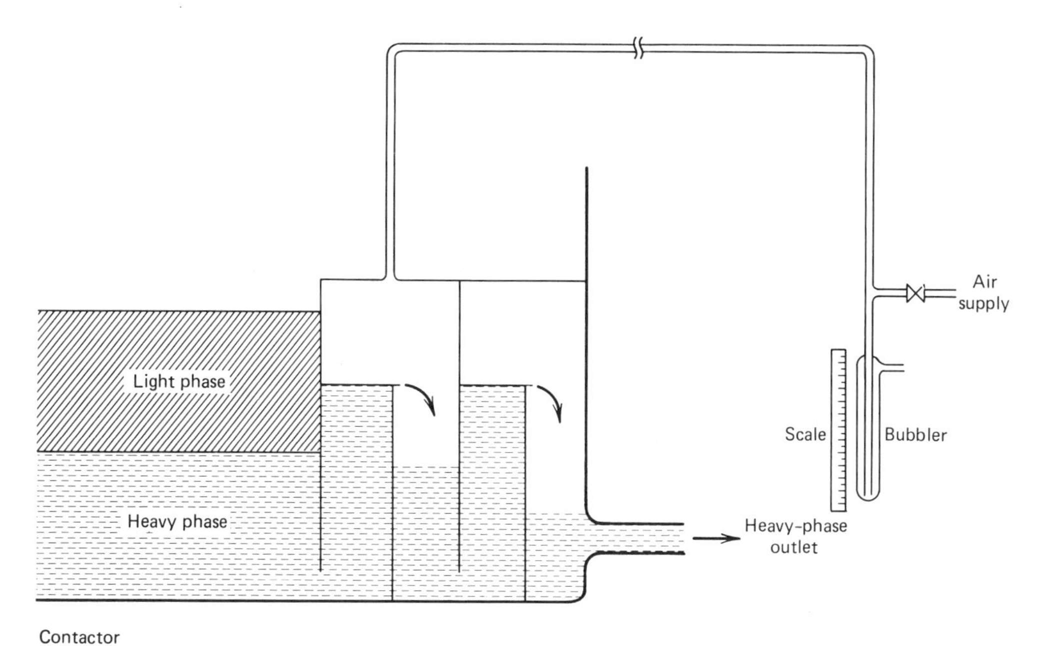

Figure 2. Pneumatic interface controller for final stage (schematic).

tice, an approximately cubical shape is usually found to be suitable. However, the mixing compartments are normally several times smaller than the settling compartments, and in order to obtain a convenient overall design for the mixer–settler unit, the mixer may frequently be oversized. This is of little consequence, except possibly in systems where it is desirable to restrict the contact time on account of, for example, rapid degradation of the solvent. With a turbine-type impeller of a diameter about half the width of the mixer, it is usually found that antiswirl baffles are unnecessary, at least in mixers up to about 610 mm (24 in.) wide, and adequate mixing can be achieved at lower impeller speeds, typically less than 200 rpm, so that there is minimum generation of very fine droplet (haze) that may prove difficult to settle. Baffles fitted on the settler side over both heavy- and light-phase ports serve to prevent mixed phases being thrown back from mixer to settler, which could otherwise cause stage bypassing to occur.

4. THE SETTLER

The time required for a dispersion to coalesce in a settler is usually many times the period needed for mass transfer [6], and in a large settling chamber the droplets of dispersed phase form a parallel bank over the whole surface of the interface. Throughput, phase ratio, temperature, chemical composition, and emulsion type all influence the emulsion band thickness as illustrated for a water-in-oil dispersion of nitric acid in tributyl phosphate–kerosene in Fig. 3. This shows how the specific settling rate, expressed in terms of the total flow rate per unit plan area of the settler ($m^3\ hr^{-1}\ m^{-2}$) varies with emulsion band thickness under different conditions of composition and temperature. It should be noted that the scale-up of settler size is not on a simple volumetric basis since it is the thickness of the emulsion band that characterizes the flow capacity of the settler. Ryon et al. [7] found that the thickness of the band increases exponentially with increasing flow and showed that settler areas can be scaled up by factors of up to 1000 on this basis.

For some time there has been increasing interest in discovering methods of accelerating coalescence in order to reduce the relative size of the settling chamber, and Logsdail [8] has recently reviewed some of these. The present illustration, however, assumes no form of assisted coalescence in the settlers.

5. CALCULATION OF LEVELS IN THE MIXER-SETTLER

Where the light-phase levels are weir controlled to a constant height along the system and the emulsion band is considered to be disposed about

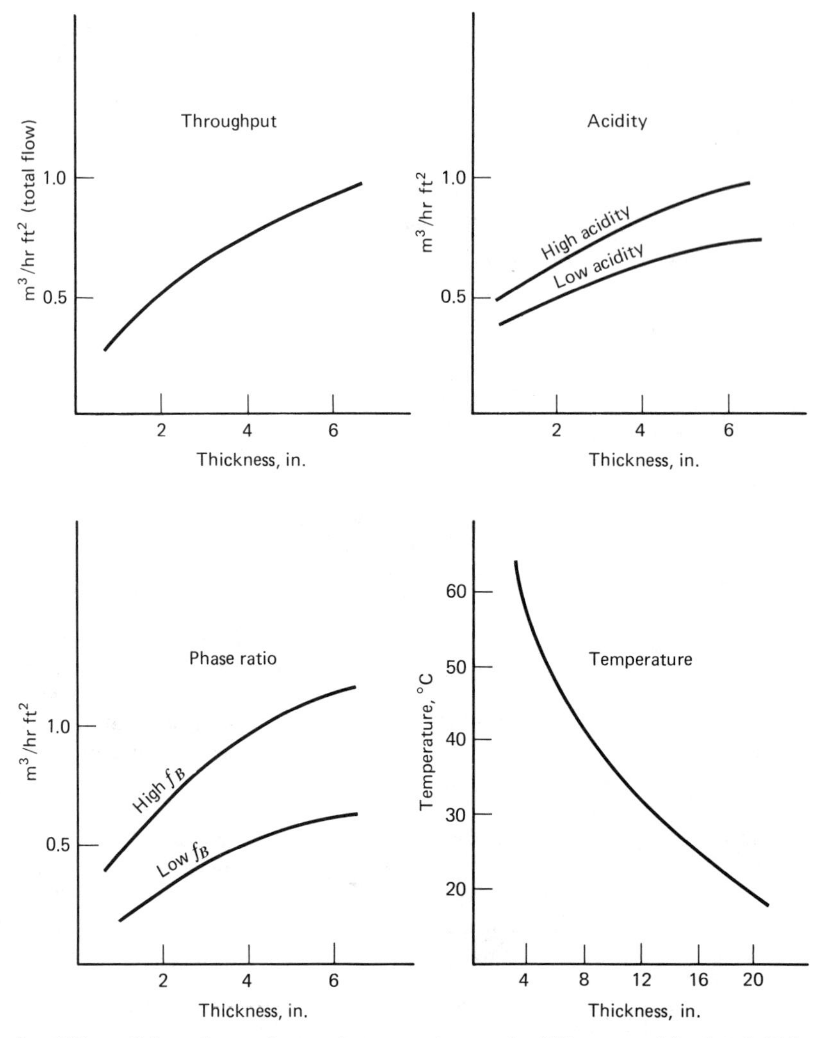

Figure 3. Effect of throughput, phase ratio temperature, and acidity on emulsion band thickness in a mixer–settler [11].

a nominal interface (i.e., the position an interface would have had if there were no emulsion present), its location may be calculated. A compound balance through the heavy-phase and mixed-phase ports in consecutive stages with uniform phase densities (Fig. 4*a*) yields the following result:

$$x = bf_A + \frac{\Delta_M + \Delta_A}{\rho_A - \rho_B} \tag{1}$$

To obtain x less than b, that is, to maintain the interface below the mixed-phase port, we require

$$\frac{\Delta_M + \Delta_A}{\rho_A - \rho_B} < bf_B \tag{2}$$

and this condition must be maintained throughout the unit. With this condition satisfied, the levels of the light phase may be determined by balances through the mixed-phase port:

$$df_A(\rho_A - \rho_B) = H_B\rho_B + h\rho_M + \Delta_M \tag{3}$$

If h is zero or negative, the light phase from the next upstream stage cannot flow freely into the mixer in question and the level of the light phase in that settler will rise until the required head is obtained. It follows, of course, that all upstream stages must now be flooded in this sense, with the light phase weirs progressively more deeply submerged. Thus, to maintain independent light-phase levels, each controlled solely by the overflow weir from settler to mixer, the unit must be

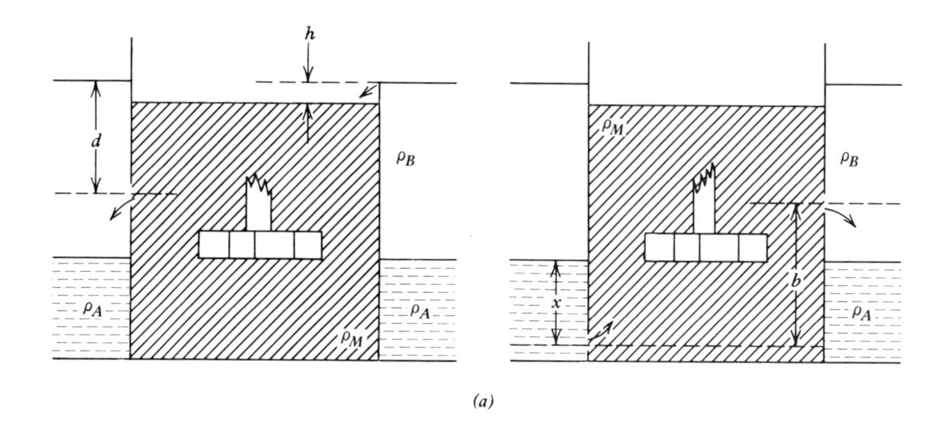

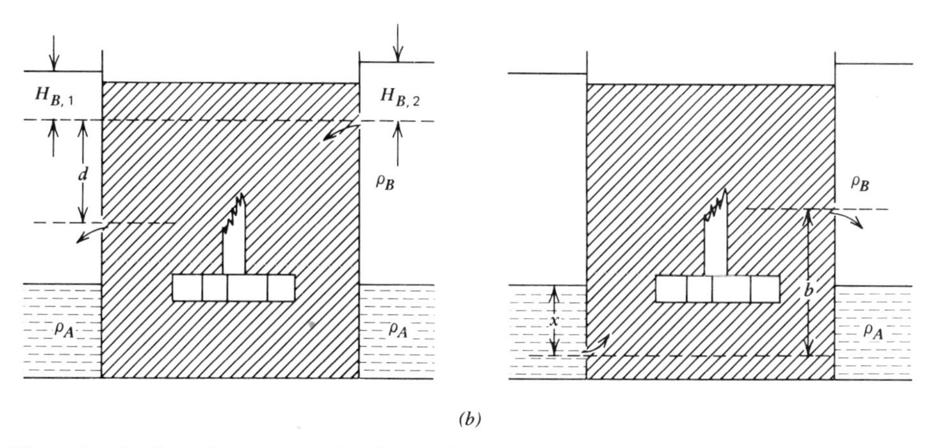

Figure 4. Sections of a mixer–settler showing liquor and port levels [10]: (*a*) nonflooded operation; (*b*) flooded operation.

designed to satisfy the inequality

$$df_A(\rho_A - \rho_B) > H_B\rho_B + \Delta_M \qquad (4)$$

For a given process, the aqueous fraction f_A and the liquid specific gravities ρ_A and ρ_B are fixed by the flowsheet and the design can vary only d, H_B, and Δ_M. The value of d can be increased by increasing the depth of the unit, and H_B and Δ_M can be decreased by increasing the width of the light-phase weir and the mixed-phase port, respectively. These are normally made to the full width available, and thus any increase can be only at the expense of an increase in total area of the unit. This leads to very large units for high throughputs, but the size can be reduced markedly by allowing the level of the light phase to increase from stage to stage, as less additional depth is required than would be required to obtain constant levels by increasing d.

Where the light-phase levels are permitted to step up along the system in this way, we have "flooded operation," but it should be emphasized that the term as used here denotes a stable hydraulic condition with no associated loss in extraction efficiency.

6. LIGHT-PHASE SURFACE LEVELS IN A FLOODED MIXER-SETTLER

In order to control backmixing in a mixer–settler where the mixed-phase level in a mixer rises above the solvent inlet weir, it is desirable to replace the simple overflow weir with a port of similar design to the aqueous and mixed-phase ports. Hydraulic balances may then be derived in a similar manner to give the buildup of light-phase level from stage to stage:

$$(H_{B,1} - H_{B,2})\rho_B = \Delta_M + \Delta_B - df_A(\rho_A - \rho_B) \qquad \text{(Fig. 4}b\text{)} \quad (5)$$

The total buildup light phase over a unit of n stages is then

$$(n - 1)\frac{\Delta_M + \Delta_B - df_A(\rho_A - \rho_B)}{\rho_B} \tag{6}$$

7. INTERFACE LEVELS IN A FLOODED MIXER-SETTLER

If it is assumed in advance that in consecutive stages both downstream and upstream interfaces are below the mixed-phase port, the compound balance by way of the aqueous and mixed-phase ports yields the result

$$X_2 = bf_A + \frac{\rho_B(H_{B,1} - H_{B,2})}{\rho_A - \rho_B} + \frac{\Delta_M + \Delta_A}{\rho_A - \rho_B} \tag{7}$$

where $H_{B,1}$ is the light-phase level above the weir or port in the upstream (light-phase flow) stage and $H_{B,2}$ is the corresponding level in the downstream stage.

The change in the light-phase level between consecutive stages has already been derived:

$$(H_{B,1} - H_{B,2})\rho_B = \Delta_M + \Delta_B - df_A(\rho_A - \rho_B) \tag{8}$$

Consequently,

$$X_2 = bf_A - df_A + \frac{\Delta_B + 2\Delta_M + \Delta_A}{\rho_A - \rho_B} \tag{9}$$

Since X_2 is less than b by assumption, it follows that

$$\frac{\Delta_B + 2\Delta_M + \Delta_A}{\rho_A - \rho_B} < bf_B + df_A \tag{10}$$

This last relationship represents the condition that the interfaces should remain below the mixed-phase port. Thus, granted this condition, and given that the last (externally controlled) interface is below the mixed-phase port, all interfaces will be constant and independent where both phases are of uniform density.

8. EFFECT OF EMULSION BAND

The preceding arguments are based on the assumption of an interface of zero thickness. In practice, each settler contains an emulsion band that may occupy an appreciable proportion of the available depth. If the emulsion composition is denoted by f_B' (not necessarily the same as the volume fraction of light phase in the mixers) and the band thickness by t, there will be within the band a notional interface distance tf_B' from the top boundary of the emulsion. Regardless of whether the mixer–settler is flooded, provided the last (externally controlled) emulsion band is held wholly below the mixed-phase port, there will be a condition in terms of emulsion band thickness that permits a constant and independent band position throughout the mixer–settler. This condition is

$$x + tf_B' < b \tag{11}$$

Hence for a nonflooded box in which

$$x = bf_A + \frac{\Delta_M + \Delta_A}{\rho_A - \rho_B} \tag{12}$$

we have

$$\frac{\Delta_M + \Delta_A}{\rho_A - \rho_B} < bf_B - tf_B' \tag{13}$$

and for a flooded box where

$$x = bf_A - df_A + \frac{\Delta_A + 2\Delta_M + \Delta_B}{\rho_A - \rho_B} \tag{14}$$

we have

$$\frac{\Delta_A + 2\Delta_M + \Delta_B}{\rho_A - \rho_B} < bf_B + df_A - tf_B' \tag{15}$$

If these conditions are observed, interface heights are

$$x = bf_A + \frac{\Delta_M + \Delta_A}{\rho_A - \rho_B} - tf_B' \quad \text{for nonflooded boxes} \tag{16}$$

and

$$x = bf_A + \frac{\Delta_B + 2\Delta_M + \Delta_A}{\rho_A - \rho_B} - tf_B' \quad \text{for flooded boxes} \tag{17}$$

In all the preceding cases it has been assumed that the two liquids maintain uniform densities throughout the system in spite of the transfer of solute that will be taking place between the two phases. This is never true in practice, of course, but where mass-transfer effects are sufficient to alter the densities significantly between one stage and the next, balances may be taken between individual stages, fitting in the appropriate densities and applying them in the same way. When a high degree of separation or purification is required in a long series of stages, it will usually be found that large changes in densities occur only in one or two stages whereas the remainder will be similar to the idealized cases described previously.

The rigorous relationships that should be used where the density is changing from stage to stage are:

Flooded mixer–settler

$$H_1\rho_{B,1} = (H_2 + d)\rho_{B,2} + \Delta_M + \Delta_B - d\rho_{M,2} \quad (18)$$

$$X_2(\rho_A - \rho_{B,2}) = b(\rho_{M,1} - \rho_{B,2}) + d(\rho_{B,1} - \rho_{B,2}) + H_1\rho_{B,1} - H_2\rho_{B,2} + \Delta_M + \Delta_A \quad (19)$$

Nonflooded mixer–settler

$$X_2(\rho_A - \rho_{B,2}) = b(\rho_{M,1} - \rho_{B,2}) + d(\rho_{B,1} - \rho_{B,2}) + \Delta_A + \Delta_M \quad (20)$$

It will be apparent that these contain an additional term, $d(\rho_{B,1} - \rho_{B,2})$, which disappears if the density is constant.

The preceding survey of interface and light-phase level behavior in a mixer–settler shows that as the scale and throughput of a mixer–settler system is increased to the point where light-phase level could only be weir controlled in a very deep unit, a design for stable flooded operation, with light-phase levels in the settlers varying from stage to stage may be applied.

If the design steps are summarized, the essential dimensions of a unit are dictated by the space required for the emulsion band and by the hydraulic requirements outlined previously. The settler area for a chosen emulsion band thickness is first determined and the height of a unit to contain the emulsion band between the mixed-phase port and heavy-phase port assessed. The mixer volume is then set by the residence time required for mass transfer and the width of each stage thus derived. Calculation of port pressure drops and top levels will then indicate whether flooded operation will occur or is required. The impeller should be of simple design and located to operate at low speed to give adequate mixing for mass transfer, but with minimum formation of very fine droplets (haze). There should be minimum dynamic effects on the pressure drops at ports. Experience with simple box-type mixer–settlers over a range of sizes and throughputs has shown them to be capable of prolonged stable operation with consistent extraction performance even when quite sudden flowsheet changes are imposed. Although the relatively large size of the units compared with, say, pulsed columns leads to penalties in space requirement and materials inventory, there is only a slow deviation from equilibrium under fault conditions, thus usually allowing time for corrective action to be taken. Interruptions in operation can be tolerated because equilibrium is largely maintained throughout the system when it is totally shut down, and processing can normally be resumed without run-down and run-up procedures. Solids present in the feeds at least in moderate amounts, do not appear to interfere with contactor operation by choking any of the ports. The impeller in the mixing chamber exerts a scouring action at the heavy-phase port, and solids migrate with one or other of the phases to pass out of the system. Provided the residence time in the mixer is sufficient for the phases to reach substantially complete chemical equilibrium, the number of extraction stages remains constant over the whole range of physical throughput capacity and overall extraction performance can thus be predicted from calculation or laboratory-scale trials.

Haze generation is kept to a minimum by the proper design and speed of the impeller so that unwanted carry-over of dispersed phase in the reverse direction between stages is so small that stage efficiency is virtually unaffected. In the present instance, haze carry-over amounts to about 0.2% of the continuous phase flow, and as an added precaution against contamination of the continuous phase product, the settler of the final stage in the series is made somewhat larger than the normal stages to give additional settling time. Entrainment of the continuous phase is much less than that of the dispersed phase.

ACKNOWLEDGMENT

The work of Mr. M. J. Larkin of BNFL and Mr. M. C. Tanner (UKAEA) in deriving many of the quoted relationships during the development of mixer–settlers at Windscale is gratefully acknowledged.

NOTATION

b	vertical distance between heavy-phase and mixed-phase ports
d	distance of light-phase weir or port above mixed-phase port
H_B, H_1-H_2	head of light phase over weir
h	difference between mixed-phase level and the weir level, taken as negative above and positive below the weir
x	height of an ideal interface above the heavy-phase port
t	thickness of the emulsion band
ρ_A, ρ_B, ρ_M	heavy-, light-, and mixed-phase-specific gravities
f_A	volume fraction of heavy phase in the mixer $[=(1 - f_B)]$
n	number of stages, or nth stage
f_B'	volume fraction of light phase in the emulsion band
Δ_M	total pressure drop developed at the mixed-phase port
Δ_A	total pressure drop developed at the heavy-phase port
Δ_B	total pressure drop developed at the light-phase port

Pressure drops are expressed in terms of height of fluid. In other cases any set of consistent units may be used in which force and mass are not defined independently.

REFERENCES

1. B. F. Warner, *Proceedings of the 3rd UN Conference on Peaceful Uses of Atomic Energy, Geneva*, Vol. 10, 1964, p. 224.
2. T. Verneulen, M. W. Davis, and T. E. Hicks, *Chem. Eng. Progr.* **50,** 188 (1954).
3. B. V. Coplan, J. W. Davidson, and E. L. Zebroski, *Chem. Eng. Progr.* **50,** 403 (1954).
4. J. A. Williams, L. Lowes, and M. C. Tanner, *Transact. Inst. Chem. Eng.* **36,** 6 (1958).
5. L. Lowes and M. J. Larkin, paper published at the Institution of Chemical Engineers Symposium, Newcastle-Upon-Tyne, UK, 1967, series No. 26.
6. R. C. Kintner, T. B. Drew, J. W. Hoopes, Jr., and T. Vermeulen, Eds., *Advances in Chemical Engineering*, Vol. 4, Academic Press, New York, 1964, p. 52.
7. A. D. Ryon, F. L. Daley, and R. S. Lowrie, *Chem. Eng. Progr.* **55,** 10 (1959).
8. D. H. Logsdail, *Process Eng.*, 40 (October 1978).
9. C. Hanson, Ed., *Recent Advances in Liquid-Liquid Extraction*, Pergamon Press, New York, 1971, p. 146.
10. C. Hanson, Ed., *Recent Advances in Liquid-Liquid Extraction*, Pergamon Press, New York, 1971, p. 147.
11. C. Hanson, Ed., *Recent Advances in Liquid-Liquid Extraction*, Pergamon Press, New York, 1971, p. 145.

9.3

THE DAVY McKEE MIXER–SETTLER

G. C. I. Warwick†
British Nuclear Fuels Ltd.
United Kingdom

J. B. Scuffham
Davy McKee Ltd.
United Kingdom

1. DESIGN

The development of reagent systems for the extraction of copper from low-concentration waste and leach liquors brought a need for specially designed liquid–liquid contactors. There were several problems associated with these systems: flow rates of liquor were much higher than those previously encountered in extraction technology, the cost of the reagent was comparatively high, the system was prone to solid and bacterial contamination, and the plants were usually to be located in remote areas. In addition, the process downstream of the solvent extraction plant demanded very low entrainment values in the final aqueous product.

1.1. Fundamental Design Basis

The first decision to be made was the basic choice of the type of contactor. Although column contactors offered potential advantages in space saving and system enclosure, they had never been operated on such a scale (1640 m^3/hr of mixed phase), and their added sophistication was seen to be a source of difficulty in the types of location and the dirtiness of the liquors encountered. It was felt, therefore, that the simplest form of contactor, the mixer–settler, would show advantages in ease of maintenance, whilst the increased space requirements were not a significant factor in non urban locations. Also, the scale-up of mixer–settlers was much less of a problem than that of column systems.

Having made this fundamental decision, it was necessary to address the problems of applying mixer–settler technology to large-scale copper extraction systems. The chemical processes are basically very simple, involving the extraction of copper into an organic solution from which it is subsequently recovered by an aqueous solution of higher acidity (see Chapter 25.1), and the number of extraction and recovery stages is low, but it is still important to minimize the number

†Deceased; also formerly with Davy McKee (Oil and Chemicals), Ltd., United Kingdom.

of stages by maximizing the extraction efficiency of each stage.

In metallurgical systems the composition of the aqueous feed to the solvent extraction plant varies. It is thus more appropriate to measure performance not in terms of degree of extraction, but in terms of the degree of approach to the equilibrium:

$$\text{Percent extraction} = \frac{C_{F,\text{aq}} - C_{R,\text{aq}}}{C_{F,\text{aq}}} \times 100$$

$$\text{percent strip} = \frac{C_{F,\text{org}} - C_{\text{ext}}}{C_{F,\text{org}}} \times 100$$

extraction efficiency (% approach to equilibrium)

$$= \frac{C_{F,\text{aq}} - C_R}{C_{F,\text{aq}} - C_R^*} \times 100$$

strip efficiency (% approach to equilibrium)

$$= \frac{C_{F,\text{org}} - C_{\text{ext}}}{C_{F,\text{org}} - C_{\text{ext}}^*} \times 100$$

where $C_{F,\text{aq}}$ = copper concentration of aqueous feed

C_R = copper concentration of raffinate

$C_{F,\text{org}}$ = copper concentration of organic feed

C_{ext} = copper concentration of the extract

C_R^* = copper concentration of the equilibrium raffinate

C_{ext}^* = copper concentration of the equilibrium extract

which is a true measure of the efficiency of the system. As is well known, the rate of mass transfer is a function of the intensity of mixing, which itself is a function of the power applied to the mixing operation. Unfortunately, as the intensity of mixing increases, so does the difficulty of separating the phases after contact. In a mixer-settler this means an increase in the size of the settler and hence in the volume of organic phase in the contact stage. It was also known that increased mixing rates could lead to increased levels of entrainment of each phase.

Because of these counteracting aspects of mixing intensity, it was clear that mixing could not be considered on its own, but only as part of a combined mixing and settling system and all development work was done on this basis.

The next fundamental decision to be made was the choice of the type of impeller to be used in the mixing operation. It seemed advantageous to consider a pump-mix impeller, that is, one that would enable the two phases to be transferred from stage to stage without the need for interstage pumps. Apart from the clear benefit of eliminating capital equipment, there is a much more fundamental reason for the choice. When the separated phases leave the settler, they still contain small quantities of the opposite phase dispersed as fine droplets. The high shear associated with pumps reduces the size of the droplets still further and leads to their being stabilized against simple methods of entrainment removal.

Several types of impeller were evaluated to establish whether they would be capable of producing the necessary combination of mixing, entrainment minimization, and head development. In this latter respect it should be noted that heads of 0.5–1.0 m are needed in the largest size of plant, where the configuration of the mixer-settler is as shown in Fig. 1. Ultimately a double shrouded impeller with eight swept back vanes was chosen as the type that fitted all requirements best (Fig. 2). In addition, provision was made for the installation of shearing blades on the top and bottom shrouds to increase power input without increasing impeller tip speed.

1.2. Mixer Design

The impeller type having been chosen, the next problem was its position within the mixing chamber. From the work of Miller and Mann [1]

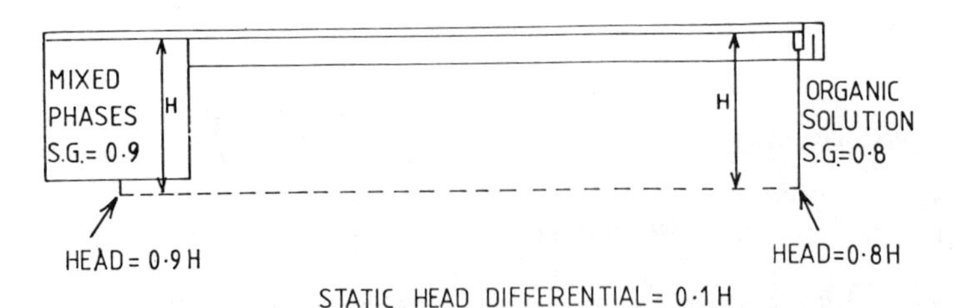

Figure 1. Diagram showing differential head between organic weir and mixer.

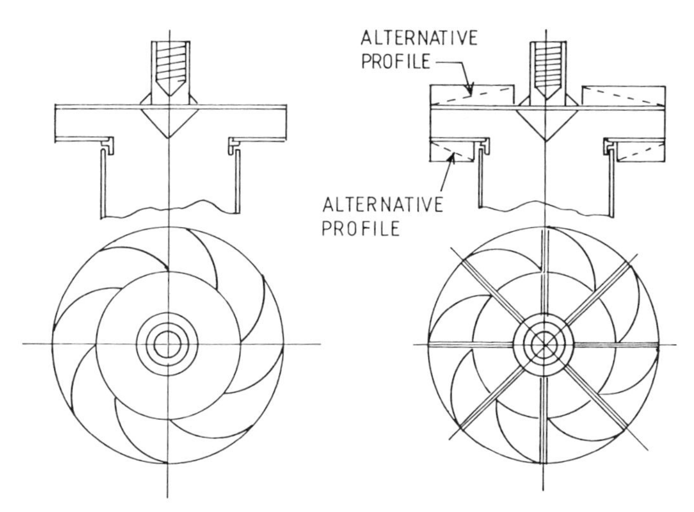

Figure 2. Designs of impeller.

it was predicted that the impeller should operate in the heavy-phase zone of the two-phase system and that the optimum height was either 0.4 times the total liquid depth from the bottom of the chamber or the interface height less the impeller depth, whichever was the lesser. Confirmatory tests were performed on an experimental mixer, and a basic geometric configuration as shown (Fig. 3) was produced. It can be seen that with the impeller located in the center of the mixing chamber, the incoming phases are led to the eye of the impeller by means of a draft tube. The head required can be minimized by leading each phase independently through the draft tube rather than allowing the phases to mix. Figure 3 shows that the mixing box has a cubical form, which was chosen principally because a cubical mixer was the easiest form to construct in the large size of plant considered. There is no fundamental reason why a cylindrical chamber should not be used; the only necessary modification is the addition of baffles at the vessel walls.

To avoid vortexing and also to promote better recirculation within the mixing chamber, a baffle was installed in the top of the chamber (Fig. 4). The baffle serves a secondary purpose of separating the controlled agitation of mixing from the quiescent condition necessary for good settler operation. With this as the fundamental basis of design for the mixer, experimental work was carried out at laboratory and pilot scale to derive the characteristics of the system.

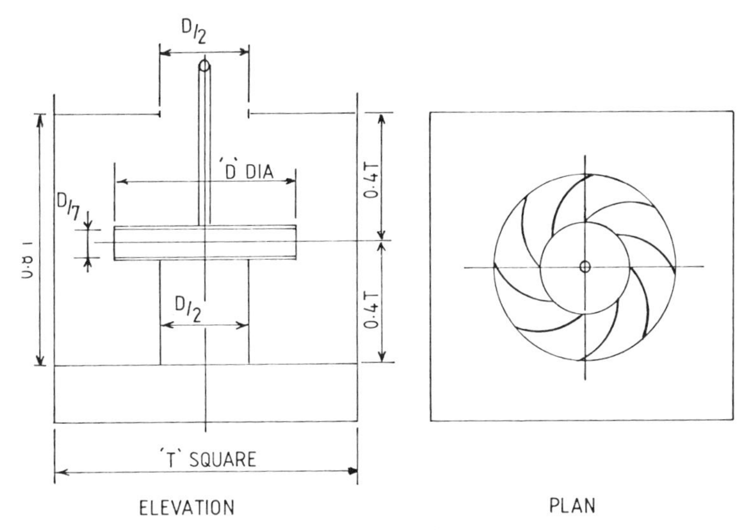

Figure 3. Mixer configuration.

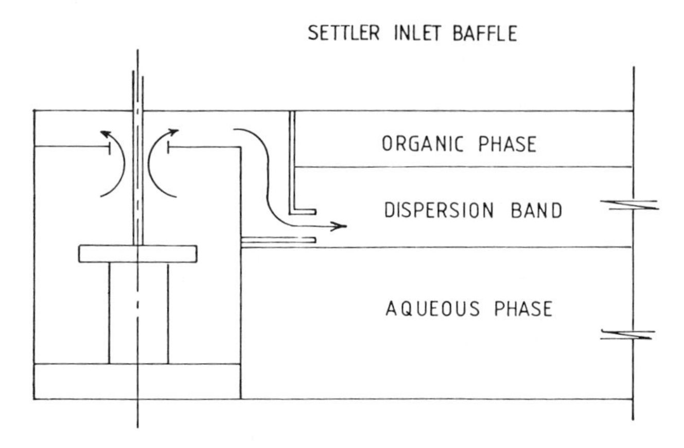

Figure 4. Baffles between mixer and settler.

Typical characteristics are shown in Figs. 5*a* and 5*b*. In addition, the pumping performance can be modified by raising the impeller away from the draft tube. A further parameter that required investigation was the tendency of the impeller to produce entrainment. It was found (Fig. 6) that this was primarily a function of the tip speed of the impeller and led to the establishment of a limit of 6 m/s on the operating tip speed.

The extraction system itself is the main determinant of the residence time within the mixing chamber. For copper–LIX systems, residence times of the order of 2–3 min are called for, whereas in uranium–Alamine or copper–Acorga P5000 systems they may be of the order of a few seconds only. From knowledge of the residence time, the geometric configuration of the mixing chamber can be calculated, and hence the dimensions of the impeller. With this information, application of the characteristics of the impeller enables tip speed and power input calculations to be made. In the first full-scale plants fully variable hydraulic gear boxes were included in the drive mechanism, but it was found subsequently that the system was so stable that speed control was not a serious problem, and simple belt or direct drives were included.

An important aspect of mixer performance is the ability to maintain the continuity of the desired phase. For instance, in the copper–LIX extraction system the final extraction stage should occur with an organic continuous phase, thereby minimizing the entrainment of organic phase in the aqueous raffinate stream. Similarly, it is necessary to prevent the loss of acid by operating the last stage of stripping aqueous continuous. To enable the phase continuity to be set up and maintained, recycle provisions are made where it would be difficult otherwise to arrange this because of an adverse phase ratio. The recycle is taken from the appropriate weir box at the end of the settler.

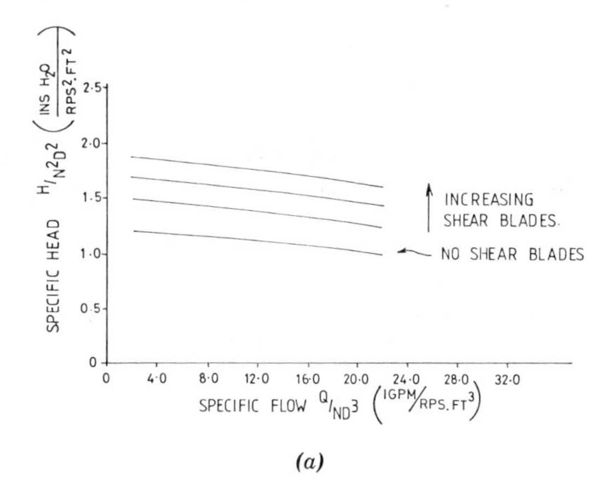

(*a*)

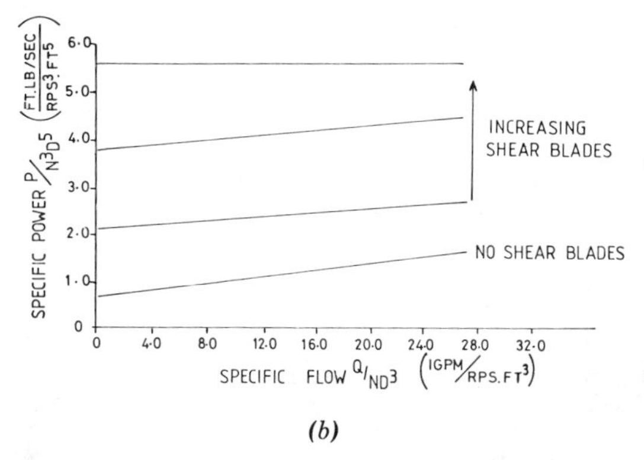

(*b*)

Figure 5. (*a*) Specific head versus specific flow for various impeller configurations. (*b*) Specific power versus specific flow for various impeller configurations.

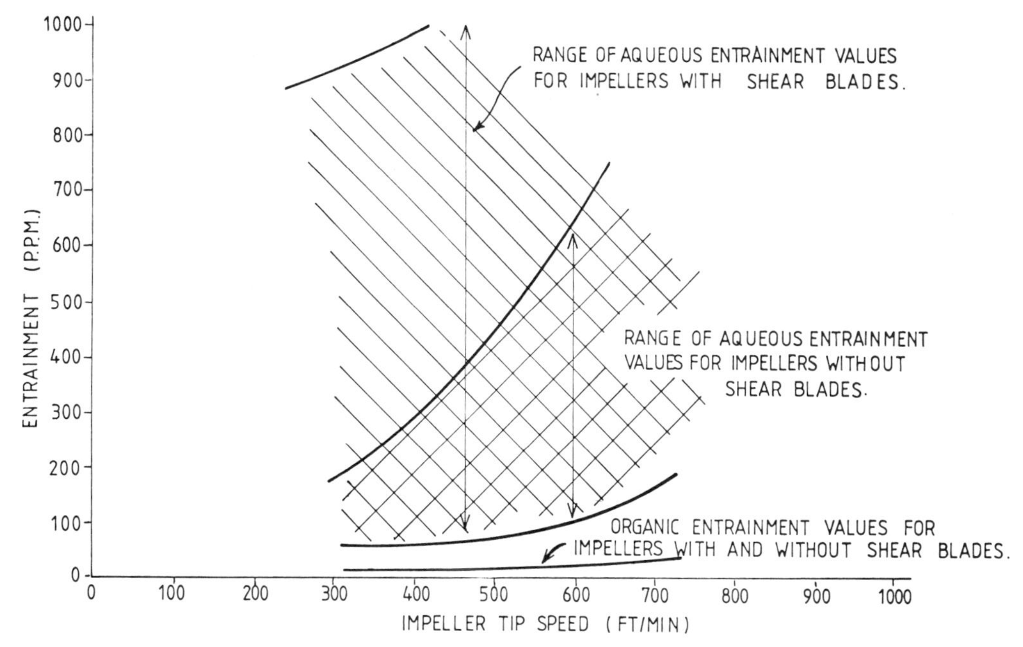

Figure 6. Entrainment versus impeller tip speed for organic continuous operation. Basis specific settler flow 1.4 i.gal/min · ft^2 settled organic depth 2.8 in. (settler), settled aqueous depth 5.2 in., temperature 23°C.

1.3. Settler Design

In the simplest gravity-assisted settler the first consideration is the settling area required. It is not necessary to achieve complete separation of the phases with a dispersion band that diminishes in thickness to zero at the offtake, but rather to produce a stable dispersion band of a thickness that can be accommodated within the settler depth. Figure 7 shows how a dispersion band thickness can vary according to different power inputs to the mixer with different settler areas.

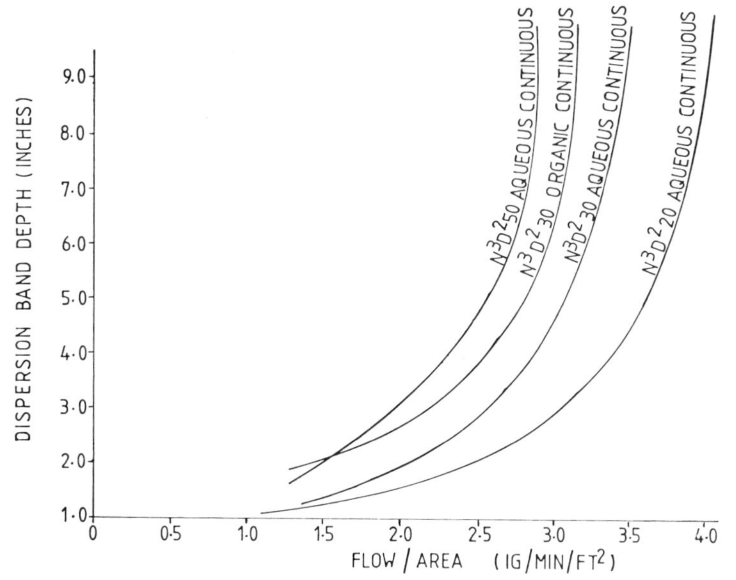

Figure 7. Effect of variation of phase continuity and mixer power on settler dispersion band depth. Basis: aqueous phase 0.05 g/liter of Cu^{2+}, pH 1.5; organic phase 5% v/v LIX64N in Shellsol T; organic : aqueous ratio 1 : 1.

Clearly, the depth of dispersion band must be less than that at the flooding point, and normally 10 cm is chosen. From experimental work of this type the settler area requirement can be calculated for particular mixing conditions.

The next parameter is the velocity of the separated organic phase in the settler. It was known that this affected the stability of the dispersion band and that velocities above certain levels could lead to wave formation in the dispersion band and reentrainment of the dispersion in the separated phases. Although earlier work had indicated that the limiting velocity might be as low as 1 cm/s, further work showed that for the depth of organic considered, ~20 cm, major problems arose only at velocities of the order of 16 cm/s and that design velocities of 10 cm/s were safe. This was an important parameter in determining the shape of the settler, and it enables a limiting width to be established at the required depth of organic phase.

With this information it was possible to match the overall shape of the settler to that of its mixer at any particular scale. It can be seen from Fig. 8 that the arrangements change drastically from the experimental scale to full scale plant (Fig. 8). The particular difficulty that arises is that at full scale, the settler is two to three times wider than the mixer that feeds it; without careful design, this can lead to maldistribution within the settler.

To overcome the problem, the mixed phase leaving the mixer passes into a settler inlet baffle, where it is distributed across the settler and

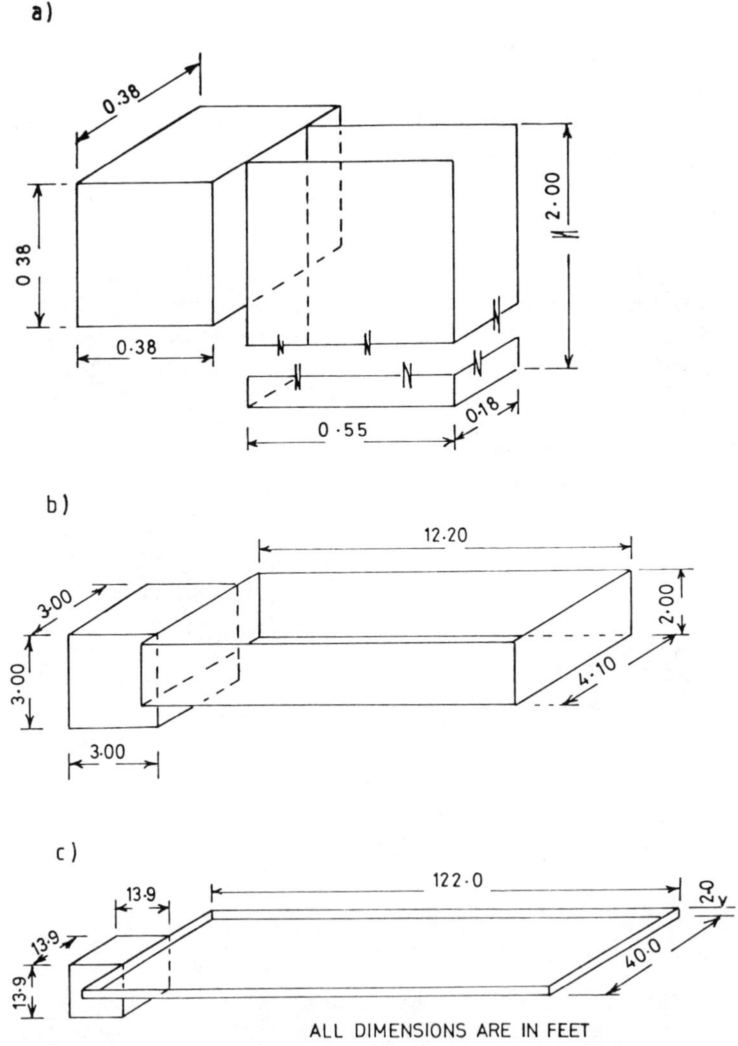

Figure 8. Comparison of three sizes of mixer–settlers to show effect of scale-up: (*a*) bench scale; (*b*) pilot-plant scale; (*c*) main plant scale.

enters the settler in the dispersion band at the coalescing interface. This minimizes turbulence in the dispersion bed (Fig. 4). Next, the settler liquors pass through a picket fence baffle that extends across the whole width of the settler. This baffle consists of two rows of vertical bars, one behind the other, with the bars placed so that those in the second row are located opposite open spaces in the first. Gaps between the bars are approximately one-third of the width of the bars. The baffle evens out the flow distribution and also dampens wave motion in the dispersion band. At the off-take end of the settler, the organic flows over a fixed weir into a full-width collecting box. The downstream side of the weir is profiled to minimize entrainment of air into the organic phase, since entrainment air bubbles encourage bacterial growth. The aqueous stream flows under the organic weir box and then over an adjustable weir into the aqueous weir box. On the larger mixer-settlers the weir is adjustable in 2 m sections so that any leveling necessary can be carried out. The adjustment mechanism is a screw, operated by handwheel.

In the copper extraction system the greatest disparity between aqueous and organic flows occurs in the stripping stages where ratios as low as 1 : 20 can be experienced. Here recycle of aqueous phase is particularly necessary, and a separate offtake from the aqueous weir box serves this purpose.

To prevent loss of organic phase during any maloperation, an overflow weir is located in the settler behind the aqueous weir, set 5-10 cm above the normal operating level of the aqueous weir. It is positioned so that the aqueous phase will tend to overflow at that point before any organic can flow over the settler walls. The total depth of the larger settlers is of the order of 1.0 m, including 0.3 m of freeboard. The installation of this overflow has an additional benefit as part of the fire-prevention system. The use of internal water sprays is considered as a widely applicable method of cooling and extinguishing, and because of the safety overflow system, direct spraying of water onto the organic surface can be used since the water does not displace the organic phase from the settler (see Chapter 30). Settlers of the larger type of plant have dimensions of the order of 12 X 36 m, and since they are often located in regions of high dust levels and/or tropical rain conditions, it is necessary to protect them by covers provided with inspection hatches. The covers are now usually of metallic construction, but, depending on the scale of the plant, other materials such as fire-resistant fabric can be used.

1.4. Construction Materials

Several materials have been proposed and used for the mixer-settler itself. For the first major plant of this design built in Zambia, the structure of the mixer-settler was concrete and this was lined with thin-gauge stainless steel welded onto strips set into the concrete, with the settlers elevated on concrete pillars. On the other hand, at the Anamax Twin Buttes plant, the mixer settlers were built of free-standing stainless steel inside supporting skeletons of carbon steel. The simplicity of the design thus lends itself to a wide variety of construction materials and building techniques. On smaller-scale plants, for instance, polypropylene has been used successfully but such material is limited in applicability because of poor expansion characteristics and lack of resistance to certain chemical systems. On very small scale, Plexiglas® and nylon have been very successful materials, allowing full observation of the performance of the plant.

1.5. Instrumentation

A series of mixer-settlers designed in accordance with the preceding principles is easy to control and requires no direct level control in the weir boxes. The main instrumentation necessary is flow control of the major process streams at the inlet to the plant section. Recycle adjustment is necessary but need not be measured directly. Where phase continuity in the mixer box is of critical importance, such as in the raffinate stage, where organic loss is minimized by operating in an organic continuous mode, conductivity probes are located in the mixing chambers with alarms for any undesired change.

2. PERFORMANCE

The acceptability of a design concept is measured in terms of the number of plants built, the range of systems covered, and the operating performance compared to design. To date, the DM (Davy McKee) design of mixer-settler has been used in four pilot plants and seven full-scale plants and for numerous laboratory and bench-scale systems.

Table 1 shows the design conditions for a selection of plants and illustrates the range of systems to which the design concept has been applied. The concepts, based on general experience prior to 1970, have been developed further in the light of operating performance and demands of different chemical systems.

TABLE 1 SELECTION OF SOLVENT EXTRACTION PLANTS USING THE DM DESIGN OF MIXER-SETTLER

Plants	Startup	Type of Feed	Aqueous Feed		Organic Phase	Number of Streams
			Concentration	Flow Rate (m^3/hr)		
Nchanga Consolidated Copper Mines Ltd. Pilot Plant	1972	0.5% Cu tailings agitation leach	1.8 gpl Cu pH 2.3	7.8	16% LIX64N MSB 210	1
Nchanga Consolidated Copper Mines Ltd. (Stage II) Chingola, Zambia	1974	0.5% Cu tailings agitation leach	3–4 gpl Cu pH 2	3200	24% LIX64N Escaid 100 14% SME529 Escaid 100	4
Anaconda Copper Co., Butte, Montana. (Arbiter Plant).	1974	Ammonia/O_2 leach of sulfide concentration	28–33 gpl Cu 150 gpl NH_3	84	32% LIX65N Nap. 470B.	1
Anamax Mining Co., Twin Buttes, Arizona	1975	Agitation leach of oxide ore	2.5 gpl Cu pH 1.9	1400	12% LIX64N	2
Albright & Wilson Ltd. Marchon Works	1976	Green phosphoric acid	57% P_2O_5	18.7	M.IBK	1
Johannesburg Consolidated Investments Ltd. Millsite	1977	Eluate	7 gpl U_3O_8 100 gpl H_2SO_4	7.5	5% Alamine 3% isodecanol Escaid 100	1

Published information on performance is scarce. One of the few papers describing the DM system from an operator's viewpoint states [2] that the NCCM plant operated satisfactorily, although with stage efficiencies less than expected. Solvent losses were low, about 13 ppm organic in the raffinate and 45 ppm organic in the advance electrolyte. Examination of operating data from other locations has confirmed the low solvent losses experienced at Nchanga but has also shown generally much higher-stage efficiencies than were achieved at NCCM. An inference here is that the NCCM design could be at the limit of scale.

A design of impeller similar to that of DM was developed by McDonald [3] in 1970 and put into effect on a full-scale plant at Harmony Gold Mining Company in South Africa. The plant has operated highly successfully since startup with what are believed to be the lowest organic-in-aqueous entrainments in a uranium plant anywhere in the world. The impeller in this instance is situated at the bottom of the mixer box and has been given the name "MUMP" by Harmony to identify a design concept in which the impeller both mixes and pumps.

2.1. Operation Problems

A particular problem has been described elsewhere, although it has not been noted in any DM plant—namely, phase continuity stability in the mixer box. At certain plants it has not been possible to operate outside a given closely defined range of O/A ratio. Abbas [4] has shown that over a wide range (termed the *ambivalence range*) of O/A ratio, either phase continuity can exist and the mode is stable.

To establish a preferred operating phase continuity in a mixer at start-up, the mixer is filled with the required continuous phase until the impeller is covered. The impeller is then rotated to draw in the dispersed phase. This technique is valid for either phase continuity. A system shock is required to cause a change from one phase continuity to the other. Misrahi and Barnea [5] have shown that the continuity is markedly affected by the presence of solid, and, depending on the wetting characteristics of the surface of the material, one or other continuity is preferred. It is suggested that surfaces with high energy tend to force aqueous continuity on the system. In reality, the majority of

ILLUSTRATING A RANGE OF FEED SOLUTION TYPE, CONCENTRATION, AND THROUGHPUT

Number of Stages			Regen-	Phase Ratio (O/A) Recirculation over Stages not Accounted for		Temperature	Mixer Residence Time	Settler-Specific Flow liter m^{-2} min^{-1} / US gpm/ft^2		Spent Electrolyte Composition (g/liter) and Temperature (°C)					
											Commercial		Starting Sheet		
Extraction	**Scrub**	Strip	eration	Extraction	Strip	of SX (°C)	(min)	Extraction	Strip	Cu	H_2SO_4	T	Cu	H_2SO_4	T
3		2		1:1	3:1	25	3	52/1.3	68/1.7	30	165	25	–	–	–
3		2		1:1	4.5:1	30–35	3	60/1.5	80/2	35	180	45			
2	1	2		E1–1:1 E2–2:1	S2–2:1 S3–1:1	35–40	2.5	70/1.75	70/1.75	20–25	125–130	35–40			
4		3		1.2:1	10:1		3.5	60/1.5	60/1.5	25	135	43			35
2	5	2									Not applicable				
4	3	4	1	1:1	2:1	35	0.5	80/2.0	80/2.0		Not applicable				

solids present in aqueous solvent extraction metallurgical feed liquors are inorganic, typically calcium sulfate and silica. Hence for systems with high solids contents, aqueous continuity is the preferred mode, as the system tends to revert to this regime almost irrespective of the O/A ratio. It has been shown that the nature of the crystalline surface, such as the way in which calcium sulfate was produced, has a strong influence on the rate of phase inversion. Rowden et al. [6] showed that a mixer operates most efficiently in terms of extraction efficiency and entrainment over a narrow range of operating ratios. This limits the influence that an excess of one phase can have in terms of mixer continuity. In addition, when a phase inversion takes place, massive entrainments occur that can result in a high degree of impurity transfer [6].

2.2. Development in Design: Aided Coalescence

In addition to the physical characteristics of the system, the principal factors that influence settling area are phase continuity [9], the droplet size distribution produced in the mixer [10] (on which is superimposed the secondary droplets produced during coalescence [11, 12]), the temperature of the system [14], the flow conditions and the interface control in the settler. In the settler, complete phase disengagement must be achieved if extraction efficiency is to be maximized. Entrainment between stages means loss in efficiency, and the presence of extractant in the outlet streams results in either the loss of solvent or contamination of the product stream [12], both directly affecting operating costs. Provided the settler is correctly designed and operated so that the primary dispersion is separated, entrainment consists only of the secondary dispersion of very small droplets (diameter <50 μm).

Gravity settlers normally operate in the range 3-5 m^3 m^{-2} hr^{-1}. This can be increased to 15-20 m^3 m^{-2} hr^{-1} by using random or regular packing as a coalescence aid. Their performance is governed by the type of dispersion produced. If a packing is designed for one phase continuity and an inversion takes place by design or default, the performance of the coalescer decreases, leading in extreme cases to flooding of the settler.

Davies et al. [12] showed the unique coalescence characteristics of a system when two dissimilar materials are used as a packing medium. A high-surface-energy material, such as metal or glass, in combination with a low-surface-energy material, such as polypropylene or PTFE, gives rise to many point contacts between them. These point contacts, commonly referred to as *junctions*, are regions of rapid coalescence. In the case of a mesh packing, comprising filaments, rapid drainage of the coalesced phases away from the junctions can occur. This allows the capacity of the coalescer to be increased to 50–60 m^3 m^{-2} hr^{-1}, irrespective of the phase continuity of the dispersion flowing to the packing. These effects are also discussed in Chapter 4 (Section 5) of this handbook.

A proprietary product, DC KnitMesh [15, 16], utilizing this junction effect in a regular pattern of knitted mesh has been introduced. The potential of this product was explored and investigated at pilot plant scale by Davy Powergas Ltd. [8], and this work led to a radical change in design concepts taking into account the product's unique characteristics [17, 18]. The predicted design performance was established at SEC Corporation, El Paso, Texas where a KnitMesh coalescer designed to take the full solution flow in the copper extraction section was installed in place of a conventional settler. The unit was operated successfully for 9 months by SEC staff. Extraction efficiencies were reported to be high, entrainments were low, and removal of the KnitMesh pad for periodic cleaning was achieved with little downtime and easy startup [19].

2.3. Concept Application

A measure of the versatility of the DM concept of mixer–settler design can be seen by its application to systems other than copper, for which it was initially developed. For example, in the early 1970s the Marchon Division of Albright and Wilson, Whitehaven, England, developed the chemistry of a process for purification of green phosphoric acid to give a product of high quality. The organic solvent used was MIBK [7] (see also Chapter 26). To establish the basis of design for a full-scale plant and to determine the effect of unknowns such as feed solution clarity and phase separation characteristics (the organic reagent had a mutual solubility in the aqueous phase to the extent of about 2%), product quality, materials of construction, optimum operating procedure, and control, a pilot plant was built at Whitehaven. The plant was constructed in polypropylene and was totally enclosed to take into account the system's volatility. On the basis of results from this pilot plant, a full-scale plant capable of treating 4115 igph (imperial gallons per hour) of green phosphoric acid and producing technical-grade acid of high quality was designed and built in 1977. This plant has now operated successfully for 2 years, and extensions are planned.

2.4. CMS Contactor

Development of a novel new Davy-McKee CMS (combined mixer–settler) has been recently described by Scuffham [20]. It consists of a single vessel in which, under operating conditions, three zones coexist: an upper separation zone comprising a separated organic phase; a lower separation zone comprising a separated aqueous phase; and a central zone comprising the mixed phases or dispersion. A standard DM impeller provides the agitation and pumping duty, whilst the "egg box" baffles positioned at the periphery of the dispersion reduce mass rotation and act as straighteners or turbulence minimizers.

The CMS is a stagewise contactor with most of the advantages of conventional mixer–settlers. However, it also has additional advantages in ease of operation, better control, tolerance to aqueous feed suspended solids and lower cost. These advantages, particularly those of lower capital cost, led to the CMS design being chosen for two commercial uranium recovery plants in the Republic of South Africa, a single stream, 250 m^3/h aqueous feed plant for Western Areas Gold Mining Company Ltd. and a two stream, 700 m^3/h aqueous feed plant for Randfontein Estates Gold Mining Company Ltd. (Fig. 9). Both plants, which include integrated organic recovery sections, were successfully put on-stream in early 1982 and became the first commercial uranium based solvent extraction plants to treat aqueous feeds containing significant concentrations of suspended solids while simultaneously yielding raffinate phase organic levels significantly below those of more conventionally based uranium plants.

The CMS represents a marked departure from traditional mixer–settler design concepts and offers significant advantages over such designs. Although first realized in the uranium solvent extraction process, these advantages are equally applicable to many of the solvent extraction processes where mixer–settlers have traditionally been employed and the use of the concept in such areas is to be expected.

Figure 9. Two stream, 700 m^3/h aqueous feed CMS based uranium plant. (Photograph reproduced by permission of Randfontein Estates Gold Mining Co. Ltd.).

REFERENCES

1. S. A. Miller and C. A. Mann, *Transact. Am. Inst. Chem. Eng.* **40,** 709 (1944).
2. J. A. Holmes, A. D. Deuchar, L. N. Stewart, and J. D. Parker, Design, Construction and Commissioning of the N'Changa Tailings Leach Plant, Chapter in J. C. Yannopoullos and J. C. Agarwal, *Extraction Metallurgy of Copper*, The Metallurgical Society of AIME, 1976, pp. 922-925.
3. A. R. F. McDonald, The Design, Erection and Operation of a Purlex Plant at Buffelsfontein Gold Mining Co. Ltd.–Discussion, *J. South Afr. Inst. Min. Met.* 193 (1971).
4. A. A. Abbas, Dispersion Studies in a Turbine Mixer, Ph.D. thesis, University of Manchester Institute of Science and Technology, UK, 1976.
5. J. Misrahi and F. Barnea, *Br. Chem. Eng.* **15** (4), 497 (1970).
6. G. A. Rowden, J. B. Scuffham, and G. C. I. Warwick, *Proceedings of the International Solvent Extraction Conference*, Lyons (Soc. Chem. Ind., London), 1974, p. 81.
7. T. A. Williams, U.S. Patent 3,914,382 (1975), Purification of Wet Process Phosphoric Acid using MIBK, Albright & Wilson, Warley, England.
8. I. D. Jackson, J. B. Scuffham, G. C. I. Warwick, and G. A. Davies, *Proceedings of the International Solvent Extraction Conference*, Lyons (Soc. Chem. Ind., London), 1974, p. 567.
9. J. B. Lott, G. C. I. Warwick, and J. B. Scuffham, *Transact. Soc. Min. Eng. of AIME*, **252,** 27 (1972).
10. G. A. Rowden, J. B. Scuffham, G. C. I. Warwick, and G. A. Davies, Considerations of Ambivalence Range and Phase Inversion in Hydrometallurgical Solvent Extraction Processes, *Inst. Chem. Eng. Symp. Ser.* No. 42, Paper 17 (1975).
11. G. A. Davies, G. V. Jeffreys, and D. G. Smith, *Can. J. Chem. Eng.* **48,** 328 (1970).
12. G. A. Davies, G. V. Jeffreys, and M. Azfal, *Br. Chem. Eng.* **17,** 708 (1972).
13. G. Eggett, W. R. Hopkins, and J. B. Scuffham, Electrowinning of Copper from Solvent Extraction Electrolytes–Problems and Possibilities, *Second International Symposium on Hydrometallurgy*, AIME, Chicago, 1973, pp. 127-154.
14. G. C. I. Warwick and J. B. Scuffham, The Design of Mixer Settlers for Metallurgical Duties, *International Symposium on Solvent Extraction in Metallurgical Processes*, Tech. Instituut K.VIV, Genootschap Metalkunde, Antwerp, May 4-5, 1972, p. 36.
15. A New Packing for Gas-liquid and Liquid-liquid Separation Processes, KnitMesh Ltd., Greenfield, Holywell, Flintshire, UK.
16. G. A. Davies, G. V. Jeffreys, and D. P. Bayley, UK Patent 1,409,045 (1975), Method and Apparatus for Coalescing Dispersed Droplets.
17. I. D. Jackson, G. M. Newrick, and G. C. I. Warwick, A Recent Development in the Design of Hydrometallurgical Mixer Settlers, *Inst. Chem. Eng. Symp. Ser.* No. 42, Paper 15 (1975).
18. G. M. Newrick, UK Provisional Patent 19266/74 (1974).
19. I. D. Jackson, R. D. Eliasen, and J. H. Cibella, "Solvent Extraction Plant Cost Reduction by Accelerated Coalescence, paper presented at the 1976 Annual Meeting Arizona Section, AIME, December 6, 1976.
20. J. B. Scuffham, *Chem. Eng.*, 328 (July 1981).

9.4

IMI MIXER–SETTLERS

E. Barnea
Edward L. Bateman Ltd.
South Africa†

D. Meyer
IMI Institute for Research and Development
Israel

1. INTRODUCTION

IMI, a pioneer in the industrial application of solvent extraction processes to heavy inorganic chemical industry, has devoted much effort to the development of high-throughput contacting equipment. The concept of equipment design was, fundamentally, absolute hydraulic independence between mixer and settler, and between adjacent units. This completely eliminates back-mixing and preserves concentration profiles on shutdown. To date, IMI liquid–liquid contactors have been installed in 15 plants, with a total of over 400 units. The different configurations described in the following paragraphs have been developed to meet the requirements of a wide variety of liquid–liquid systems.

†Formerly with IMI Institute of Research and Development, Israel.

2. MIXER DESIGN

2.1. The Design Problem

The conflicting demands on a mixer for liquid–liquid contacting service, especially where the pump–mix configuration is employed, are immediately apparent. The contradictions are such that the formulation of an optimal design for universal application is a virtual impossibility, and a working design for the real world must be a compromise fulfilling the most important requirements. The compromise must, however, be based on an integrated concept of mixer–settler design that takes the actual characteristics of the liquid–liquid system into account. These considerations have led IMI to develop two design solutions, namely, the axial pump–mix unit and the turbine pump–mix unit.

2.2. Axial Pump-Mix Unit [1-3]

The IMI axial pump-mix unit features separate mixing and pumping apparatus on the same shaft; mechanical considerations fix the diameter of pumping and mixing impellers, whereas mass transfer and phase separation determine the speed range. The low head required suggests an axial flow unit, but the range of speeds compatible with the physical limitations forces the design outside the normally accepted limits for such pumps. The design solution has produced a unit of reasonable mechanical efficiency at relatively low speeds. The axial flow pump-mix (Fig. 1) consists of a baffled vessel (*a*) divided into a lower, mixing section and an upper delivery section by a partition or deck (*b*). The pump-mix unit is mounted on a centrally located shaft (*c*) and consists of a mixing impeller (*d*) of the upthrust marine type, which assists the pumping function, and an axial flow pump assembly shrouded in a draft tube (*e*). The pump assembly consists of an axial flow impeller (f) and static guide vanes for entry (*g*) and exit (*h*) ends. The incoming liquids are delivered at the level of the mixing impeller (*j*, *k*) and are mixed and lifted through the draft tube by the pumping assembly until they discharge onto the deck, from where they flow by gravity to the settler. The entire assembly of draft tube and shaft may be withdrawn through the deck for maintenance.

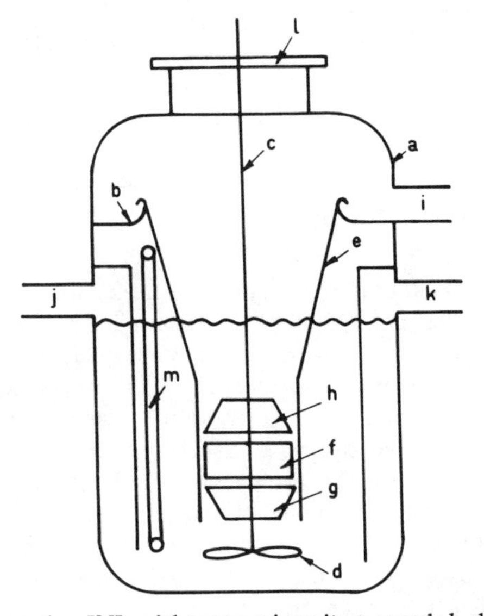

Figure 1. IMI axial pump-mix unit: *a*, vessel; *b*, deck; *c*, shaft; *d*, mixing impeller; *e*, draft tube; *f*, pumping impeller; *g*, entry guide vane; *h*, exit guide vane; *i*, dispersion delivery; *j*, light-phase feed; *k*, heavy-phase feed; *l*, mounting flange; *m*, sight glass.

The axial pump-mix unit is characterized by a small high-shear mixing zone and a relatively large recoalescence zone. Its main advantages are simplicity, low cost, and insensitivity to operating conditions. Flow rate variations of up to 25–30% result in small variations in liquid level in the mixing region, giving a compensating variation in pump delivery. Outside this range, the rotational speed must be adjusted to maintain the pump suction. The unit provides an ideal solution where mass transfer is easy and phase separation is relatively insensitive to mixing parameters and dispersion type.

2.3. Turbine Pump-Mix Unit [4]

This unit was designed for cases where the characteristics of the axial pump-mix limit its applicability. It is based on a turbine impeller that fulfills mixing and pumping functions simultaneously. The turbine pump-mix unit features high mass-transfer efficiency, as a result of the combination of maximum utilization of mixer volume and an intensive coalescence-redispersion cycle induced by a forced recirculation of the dispersion. Application to sensitive liquid-liquid systems [5] and the production of a narrow drop size distribution of large average drop size require limitation of the shearing force; this is achieved by use of a turbine of large diameter relative to the tank diameter, rotating relatively slowly. Sensitivity to the presence of solids is considerably reduced since narrow clearances are avoided.

The turbine pump-mix (Fig. 2) is designed as a large-diameter single turbine (*a*) (70–90% of tank diameter) rotating quite slowly (40–50 rpm for a 1.5-m-diameter turbine in a typical case). The turbine is shrouded and has a large number of blades (*b*). It is located in a closed mixing compartment (*c*) that has a tangential exit (*d*). The mixing compartment is fitted with stator blades (*e*) above the impeller (and below, if required), so designed as to force a flow reversal from the periphery to the turbine suction at the center. The incoming fluids are delivered to the turbine suction by way of a baffled entry (*f*, *g*), which eliminates mixing of feeds before entry into the turbine and accidental formation of an undesired dispersion. Vertical baffles are not required in the mixing compartment since the flow reversal by the stator(s) is sufficient to give the turbulence required for mixing. The turbine action

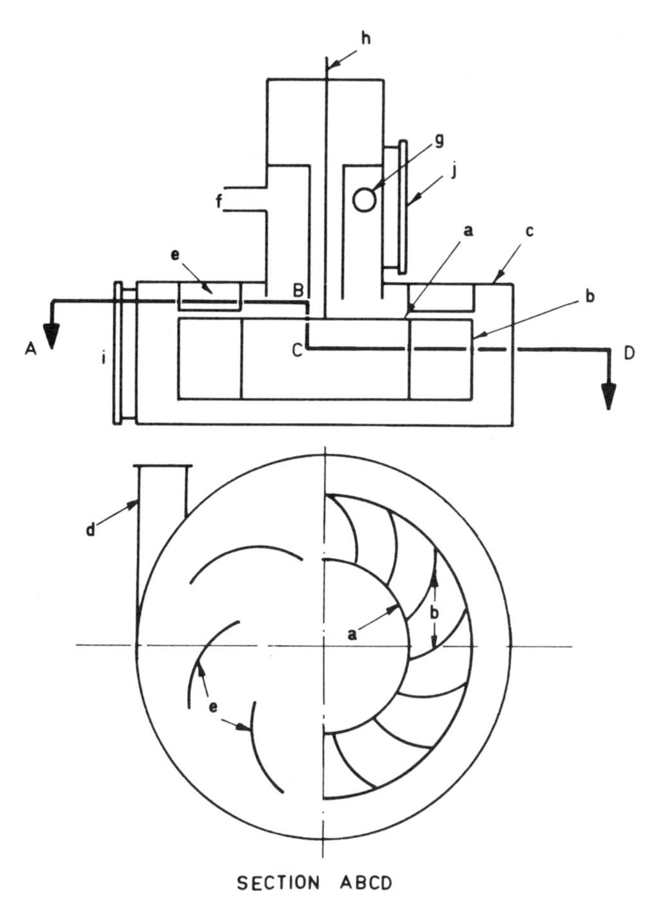

Figure 2. IMI turbine pump-mix unit: *a*, turbine; *b*, turbine blades; *c*, mixing compartment; *d*, tangential dispersion delivery; *e*, stator blades; *f*, light-phase entry; *g*, heavy-phase entry; *h*, shaft; *i*, sight glass; *j*, sight glass.

creates strong radial currents, whose total flow will be several times greater than the net turbine throughput. The relationship between net throughput and recirculating flow depends on the geometry of the stator blades. Thus the dispersion passes through the shearing zone many times; the vortex caused by the circulating flow induces coalescence, so that the overall turbine action ensures the multiple dispersion–recoalescence cycle that is essential for efficient mass transfer at low residence times. The multiple cycle also yields drops having a narrow size distribution. Since the entering liquids are fed directly to the high-shear zone stable operation may be achieved with dispersions containing up to 80% of dispersed phase.

2.4. Design Correlations

Examples of characteristic curves based on dimensional analysis are given in Figs. 3 and 4, for the turbine and axial pumps, respectively. For the turbine pump, Fig. 3 indicates that there is a fairly moderate dependence of pumping efficiency and head developed on pump throughput, thus greatly reducing the requirement for a variable pump speed. The low efficiency for fluid transfer is due to the high recirculation rate that promotes mass transfer. For the axial pump-mix, Fig. 4 shows two regions of sharp dependence of head on throughput, with an unstable region between them. This is characteristic of axial flow pumps, and operation in this region is avoided. The self-compensating character of the axial flow pump is apparent; a change in flow rate will automatically adjust the liquid level in the mixer to give the required head. The contribution of the upthrust mixing impeller to the pumping efficiency is also evident. The characteristics, when used within the space bounded by experimental data, can be reliably used for design of either turbine or axial pumps.

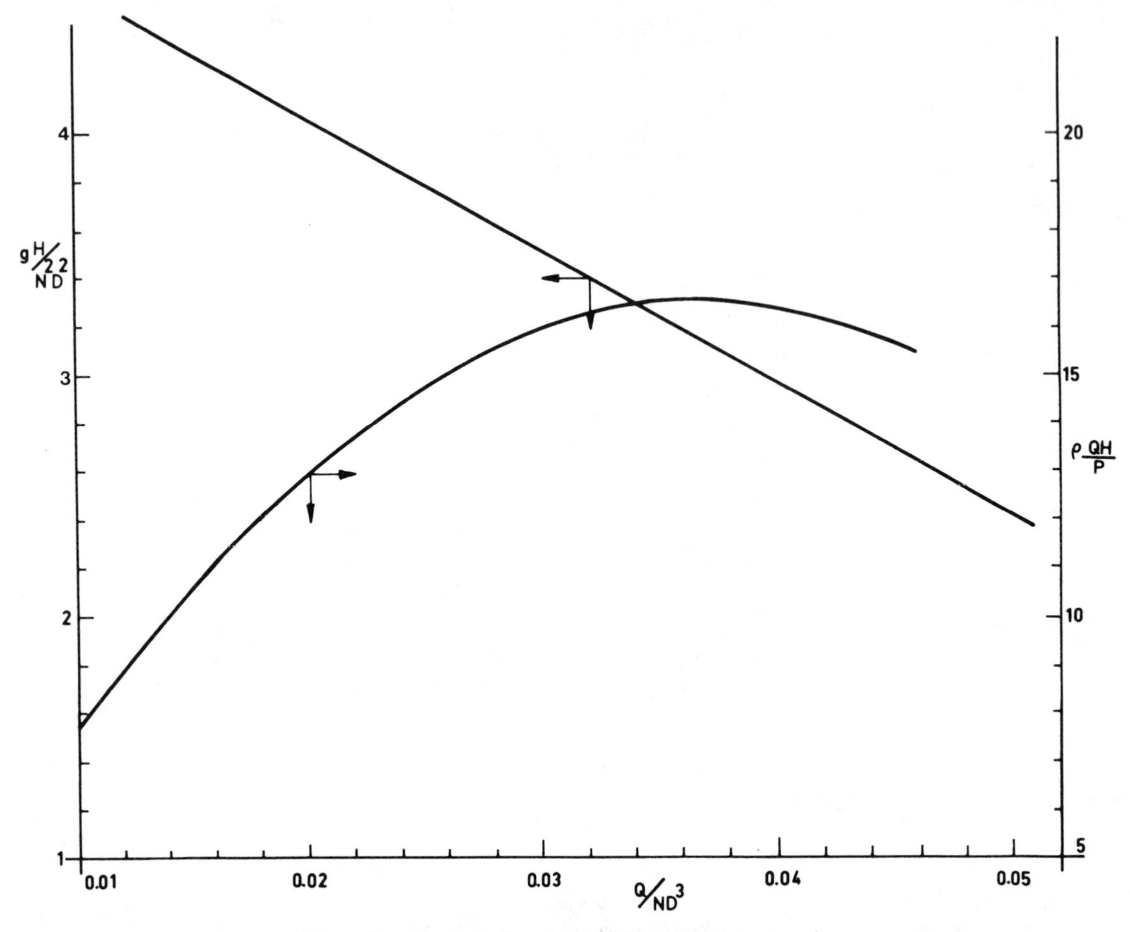

Figure 3. Dimensionless characteristic–Axial pump-mix.

3. SETTLER DESIGN

3.1. Introduction

Despite the greater effect of area compared with height on settler capacity, the IMI design is based on a dispersion band thickness of 30–100 cm for reasons discussed in the following paragraphs. Ryon and co-workers [6] have shown that the relationship between throughput per unit of cross-sectional area and dispersion band thickness is of the form

$$\Delta H = K\left(\frac{Q}{A}\right)^{y} \tag{1}$$

This equation has since been given a theoretical foundation [7] that indicates that $y \leqslant 2.5$ in the general case; $y = 2.5$ represents a particular, yet frequently encountered case and may be used for an optimization exercise, based on a number of simplifying assumptions: (a) the system obeys Eq. (1) with $y = 2.5$; (b) clear layers 30 cm deep are maintained above and below the dispersion band for operational flexibility and reliability; and (c) optimization is for minimum volume, which reflects both minimum solvent holdup and is a function of investment in buildings and plant.

The overall liquid depth H, with a dispersion band thickness ΔH is given by

$$\begin{aligned} H &= \Delta H + (2 \times 30) \\ &= \Delta H + 60 \end{aligned} \tag{2}$$

From Eq. (1)

$$H = K\left(\frac{Q}{A}\right)^{2.5} + 60 \tag{3}$$

or

$$\frac{Q}{A} = K'\,(H - 60)^{0.4} \tag{4}$$

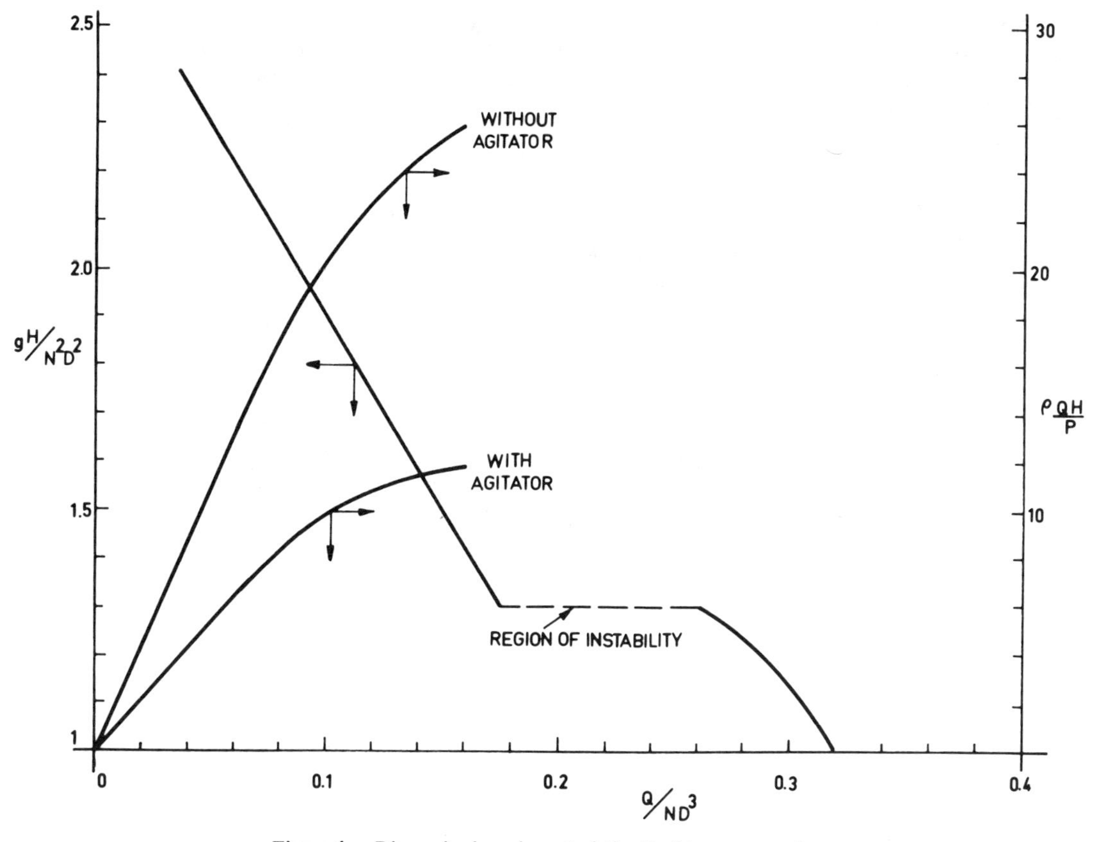

Figure 4. Dimensionless characteristic–Turbine pump-mix.

From assumption c

$$\max\left(\frac{Q}{V}\right) = \frac{Q}{AH} = K' \frac{(H - 60)^{0.4}}{H} \qquad (5)$$

and

$$d\frac{(Q/V)}{dH} = -K' \frac{(H - 60)^{0.4}}{H^2} + \frac{0.4K'}{H(H - 60)^{0.6}} = 0$$

Hence $H - 60 = 0.4\,H$, and

$$H = 100 \text{ cm}$$

$$\Delta H = 40 \text{ cm}$$

Consideration of equipment and building costs tends to increase the optimum value of H, whereas an increase in the value of exponent y will have the opposite effect. Neglect of the need for clear layers drives the optimum value of ΔH toward zero. This explains the tendency to design "swimming pool" settlers, although it is clear that a practical settler cannot operate without substantial layers of separated liquids. IMI deep layer settlers are normally designed for dispersion band heights of 30–100 cm and overall liquid depths of 100–160 cm.

Whereas Eq. (1) characterizes the primary separation, the effects of secondary separation ("haze" or entrainment) have an equal or greater importance. In a deep layer settler, the appearance of haze is to a large extent eliminated. First, feeding the dispersion into the settler close to the coalescence front forces the dispersion to flow through a bed of drops, thus greatly enhancing the probability of collision-coalescence for the smallest droplets generated in the mixer. Second, significant flow velocities normal to the direction of separation are generated in a shallow settler, especially if the settler is end fed; these velocities tend to carry the entrainment to the settler. These flows are minimized in a centrally fed deep-layer settler.

Two sublayers exist in the dispersion band [7]. Close to the coalescence front there is a relatively shallow, dense layer of large droplets. Adjacent to it is a fluidized bed of droplets, which has a dispersed phase holdup slightly less than that of the feed, which constitutes the major part of the dispersion band. Therefore, the "natural" location for feed introduction is between dense layer and the fluidized bed, that is, close to the coalescence front. Feed introduction at another level will lead to gravity induced currents and a loss of separation capacity [7]. The fluidized-bed layer normally determines the capacity of a deep-layer settler; simply speaking, it provides residence time for feed droplets to grow by collision–coalescence until they are sufficiently large to settle against the continuous phase [7].

Settler capacity may thus be increased either by increasing drop size or by reducing the size of drop required for settling; that is, reducing the continuous-phase velocity. These approaches are exploited in IMI mixer–settlers, and are expressed in the IMI diffuser precoalescer and the IMI compact settler, respectively, both of which are described in the following paragraphs.

3.2. Simple Deep-Layer Settler

The effective operation of a deep-layer gravity settler depends mainly on the formation of a well-structured dispersion band, as described previously and minimization of turbulence that disturbs the dispersion band structure. IMI deep-layer settlers minimize turbulence parallel and normal to the direction of the flow in both circular and rectangular configurations.

3.2.1. Construction Details

In the circular settler the dispersion is fed to the settler center with peripheral outflow weirs for light and heavy phases; the dispersion inlet is designed as a diffuser that has a perforated cylindrical partition to ensure even feed distribution. The diffuser entry is fitted with an antiswirl baffle. The perforated area is designed to give a pressure drop of 0.5–1 cm of liquid. The perforations are located at a level that will permit some fluctuation in the level of the coalescence front. Density currents are confined within the diffuser, so that the dispersion flows into the settler body at the optimum level.

The rectangular settler has a similar construction. The dispersion is fed through a channel to a perforated diffuser running along the longitudinal axis of the settler. The diffuser perforations are sized to ensure that the pressure drop through the perforations is similar to the drop along the diffuser length. Light and heavy outflows run along the settler sides and are connected by a cross-channel at the end wall to permit easy removal. Adjacent settlers may thus be contiguous to save space, and may have a common wall if this is convenient.

3.2.2. Auxiliaries

High local velocities at the light-phase overflow are avoided by careful leveling or sawtooth construction. Heavy phase is removed by a standard hydrostatic balancing leg. This may be made adjustable in order to position the dispersion band. The seal must be located at a level that will at all times ensure retention of a heavy-phase "heel" in the settler; thus accidental passage of light phase to the heavy-phase channel is avoided in case the seal leaks.

The dispersion band location is observed by a sight glass. Connection of the sight glass to the settler body at intervals of 150–200 mm is adequate to permit observation of band thickness and identification of the dispersion type.

3.3. Compact Settler [8, 9]

In the IMI compact settler, the primary separation efficiency of shallow dispersion bands is combined with the advantages outlined for a deep-layer settler. The compact settler consists of a number of shallow settlers stacked vertically, one above the other, with efficient drainage to keep the separated phases in thin films, and manifolds for feed distribution and withdrawal of separated phases. The principle can be illustrated by a logarithmic plot of Eq. 1 (Fig. 5) for a particular case where $y = 2.7$. For a simple gravity settler, doubling of the dispersion band height from $\Delta H = b$ to $\Delta H = 2b$ will increase the specific throughput by a factor of $2^{1/2.7}$, that is, by 29%. Two shallow settlers, each of depth b, will have double the throughput, and if these are stacked one above the other, retaining the original area, then for the combined depth of $2b$ the settler capacity will be $2^{1.7/2.7}$, that is, 55% more than the capaicty of a simple settler of depth $2b$.

A number n of such thin settlers stacked vertically, each with a depth of b, will have a maximum throughput that is $(n)^{(y-1)/y}$ times the maximum throughput of a single deep-layer settler having the same area and overall depth ($\Delta H = nb$). Graphically, the combined characteristics of such a set of steelers stacked vertically is expressed as a straight line with slope of 1.0

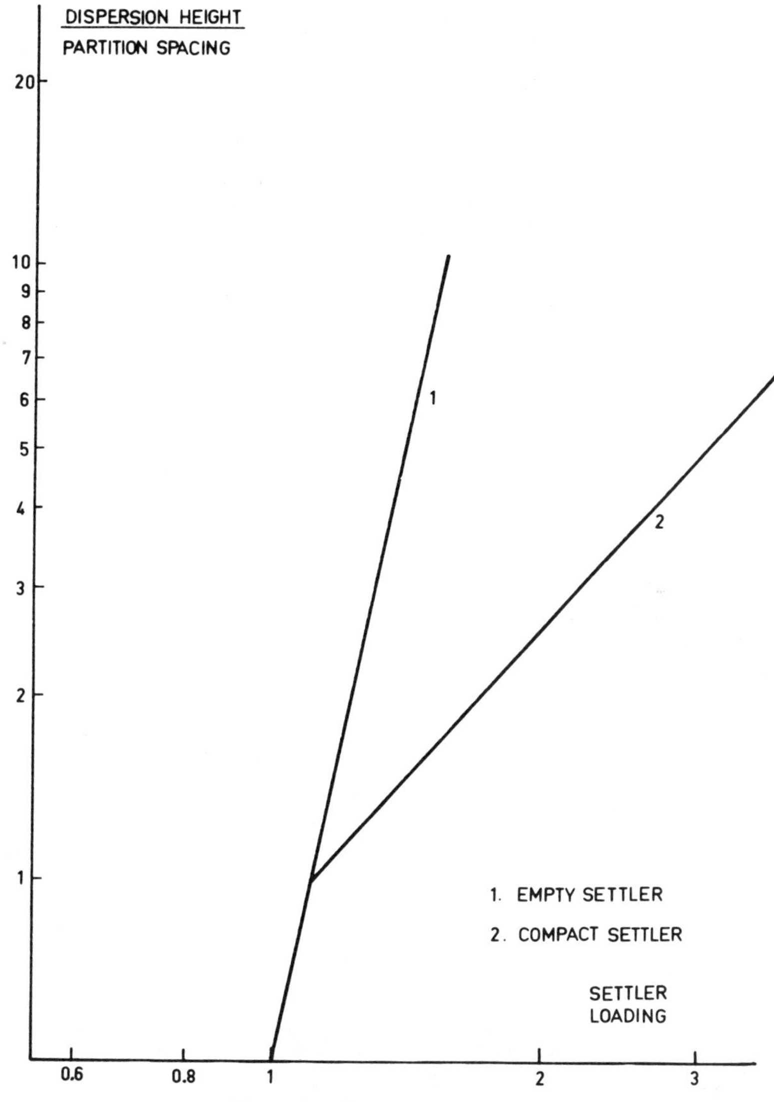

Figure 5. Compact settler effect.

branching from the single settler characteristic at the point $\Delta H = b$.

More fundamentally, the vertical velocity of the draining continuous phase within each such compartment is only $1/n$ of the velocity that would exist if the total feed of n such settlers would be to an empty vessel. Hence, smaller droplets can settle and the mean residence time required for drop-to-drop coalescence is reduced. The combined streams of separated continuous and dispersed phases must, however, be delivered to the corresponding manifolds above and below the set of shallow settlers. This can be done by collecting the individual streams in vertical channels termed "chimneys." The vertical velocity in such chimneys is relatively high, but since no settling takes place there, the separation function is not affected. Obviously, the maximum theoretical effect described previously cannot be realized in full, as (1) part of the settler volume is needed for vertical drainage (the chimneys) and is not available for separation and (2) if the overall depth of a shallow settler is b, the maximum thickness of a dispersion band contained in it is less than b as a result of the physical thickness of the partitions and the thickness of the separated phase films.

The basic guidelines for the IMI design concept are as follows:

Simplicity. The design should be simple, avoiding expensive items such as seals, and based on available standard materials with minimum work.

Flexibility. The design should be compatible with various geometric shapes of settlers, for example cylindrical and rectangular, and should be adaptable to retrofitting to existing deep-layer settlers for capacity increase.

Maintenance. The internal elements should be in relatively small modules that can easily be removed one at a time, for cleaning, without halting the settler.

The set of shallow settlers stacked vertically is simply a rack of partitions slightly inclined toward the horizontal. The slopes of the racks alternate so that alternate chimneys drain the separated light and heavy phases. The chimneys are designed with divergent cross sections to accomodate the accumulating flows of the separated phases.

A schematic section through a basic partition element is shown in Fig. 6. The dispersion is fed in the direction perpendicular to the paper (along the partition length). The space confined between two adjacent partitions defines a shallow settler in which phase separation takes place. Separated heavy and light phases form films at the space boundaries, which drain downward and upward, respectively, along the partition breadth to be collected in the chimneys flowing vertically downward and upward, respectively, which lead to the lower and upper manifolds for separated phases.

Optimization with regard to the vertical spacing between plates, angle of inclination, the proportion of area occupied by chimneys and number of partitions per rack is done on a case-to-case basis, as it depends on the properties of the liquid–liquid system (values of K and y, tendency to entrainment, etc.) as well as by economic and other considerations (cost of solvent, materials of construction, presence of solids, etc.). Typical figures are as follows:

Partition spacing	20–30 mm
Inclination	10–15%
Chimney area	10–20% of settler surface
Number of partitions per rack	25–40

Typical specific throughputs are three to eight times those obtainable by a single deep-layer gravity settler, of the same depth.

The IMI compact settler is thus characterized by high specific throughputs (typically 10–60 $m^3\ hr^{-1}\ m^{-2}$). High localized velocities (of the order of 0.1m/s) may be encountered in the manifolds near the chimneys. If the system tends to produce entrainment, it becomes necessary to spread the vertical movement of both separated phases away from the dispersion band evenly across the whole area. This is achieved by inserting a perforated "false bottom" below and a system of extended overflow weirs (not only around the periphery, but with addition of longitudinal or radial channels) and a perforated plate above the partition racks. The perforations are calculated for a nominal pressure drop of 1 cm of liquid. Typical pilot plant data with various partition spacings are given in Fig. 7.

IMI compact settlers have been operating continuously in industrial plants since 1970, and experience indicates virtually no need for maintenance. Individual racks can easily be taken out for washing of adhering crud; however, this has not been necessary in most cases.

The compact settler not only confers economic advantages in the original design of liquid–liquid contacting plants, but also permits "debottleneck-

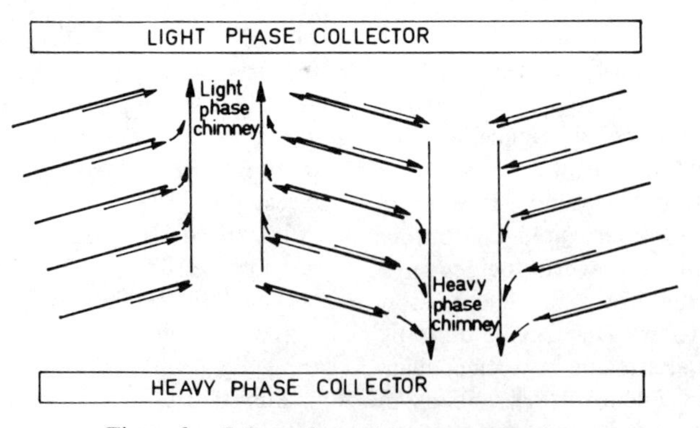

Figure 6. Schematic partition representation.

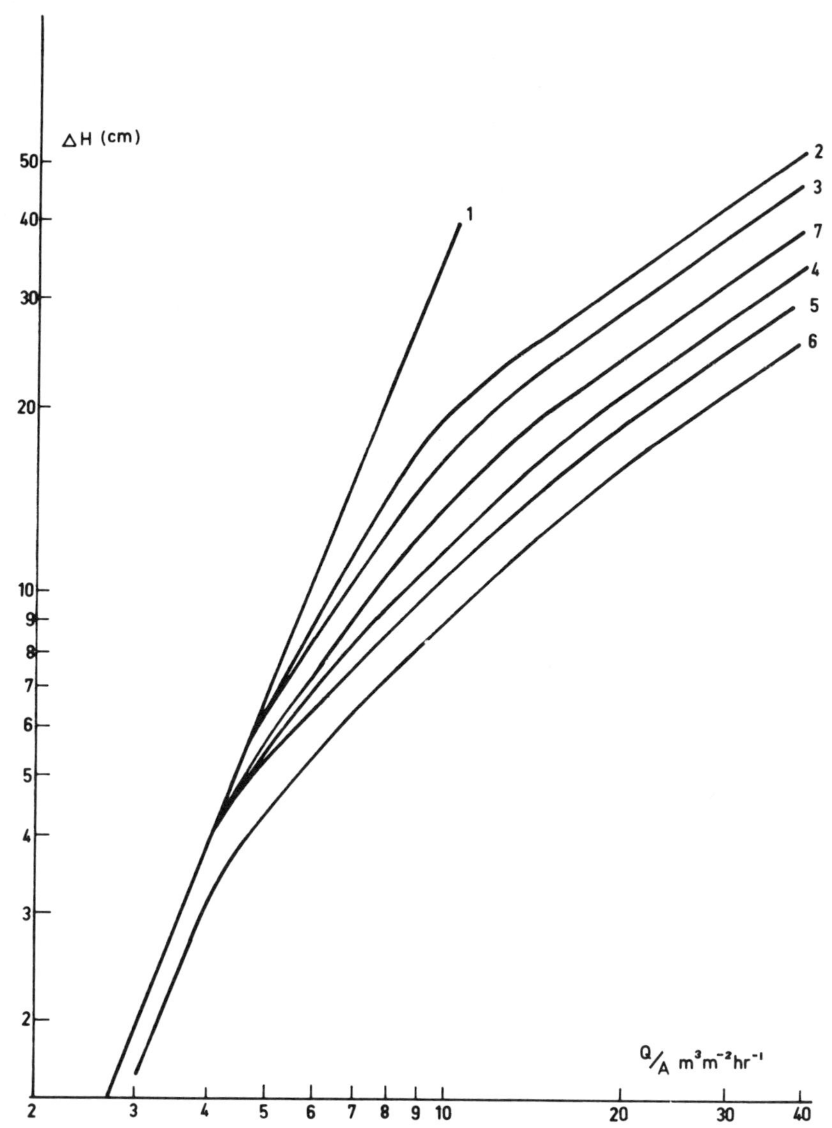

Figure 7. Effect of partition spacing: curve 1, empty settler; curve 2, b = 60 mm; curve 3, b = 40 mm curve 4, b – 28 mm; curve 5, b = 20 mm; curve 6, b = 12 mm; curve 7, b = 8 mm.

ing" or capacity increase of existing equipment, even if this was not originally designed with "compactization" in mind. Moreover, the application of the compact settler principle can be graded to desired capacity increases on a virtually linear scale by increasing the number of partitions. This permits minimization of the design factor for settlers, and a gradual increase in "compactization" intensity can be relied on to cope with required capacity increases without immobilizing capital, and the need for major future expenditure can be avoided.

3.4. IMI Diffuser–Precoalescer [10, 11]

The IMI diffuser-precoalescer is compatible with both the IMI deep-layer settler and the IMI compact settler. The simple diffuser described previously is replaced by a "sandwich" design composed of two perforated surfaces filled randomly with commercially available standard packing elements made of a material that is preferentially wetted by the dispersed phase (e.g., metal or glass for aqueous dispersion and polypropylene or graphite for organic dispersions). The nominal size of packing elements should be 6–10 mm, and the gap between perforated plates should be 5–15 cm.

The IMI diffuser–precoalescer has a dual function: (1) it acts as a more efficient diffuser, ensuring smooth and even introduction of the feed with efficient restriction of turbulence; and (2) induces precoalescence; that is, the outflowing dispersion has a higher mean drop size than that produced by the mixer.

The diffuser–precoalescer effect is additive; separation capacity increases of 20–100% have been observed, with a typical value of 50%. Its use is especially attractive combined with the compact settler since the high separation capacity is further enhanced.

Whereas the compact settler principle changes the slope of the settler characteristic, tending to a slope of 1.0, the effect of the diffuser–precoalescer is expressed by a shift to the right.

Figure 8 compares continuous settling characteristics for a typical liquid–liquid system (based on actual test result) of:

1. IMI deep-layer settler (simple diffuser).
2. IMI deep-layer settler with diffuser–precoalescer.
3. IMI compact settler (simple diffuser)
4. IMI compact settler with diffuser–precoalescer.

The capacities of each of these settlers operating with a disperion band thickness of 100 cm for a particular phase system are given in Table 1. The coarse size of packing elements used ensurcs a

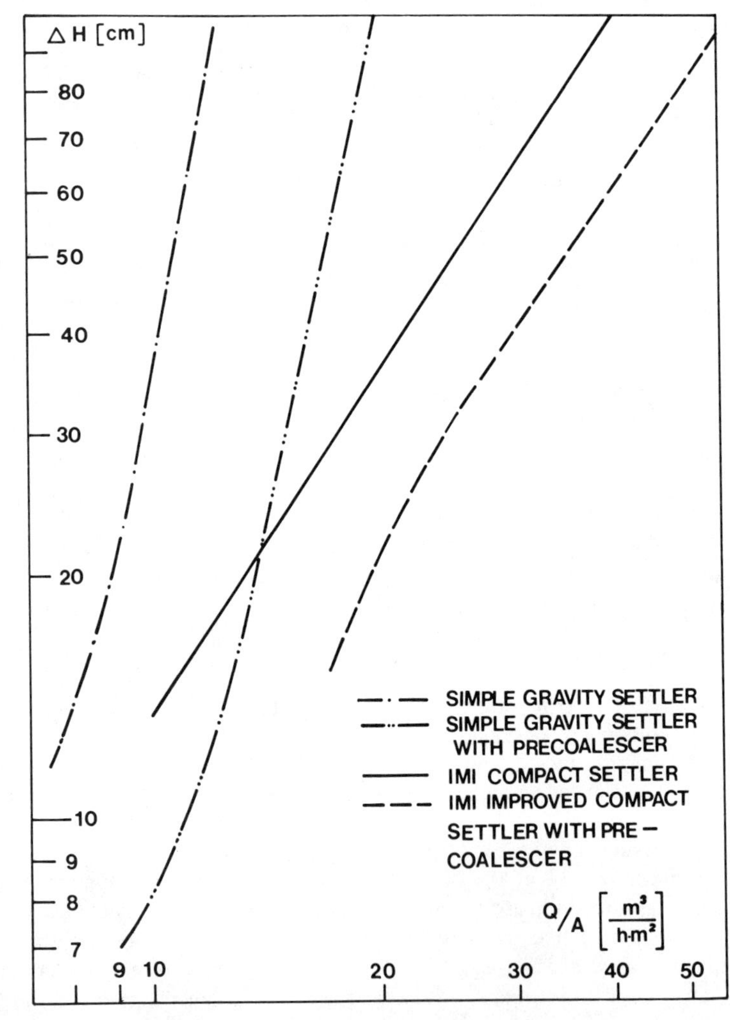

Figure 8. Comparison of IMI settler characteristics.

TABLE 1 COMPARISON OF CAPACITIES OF DIFFERENT CONFIGURATIONS FOR A PARTICULAR CASE

Type of Settler	Specific Throughput, $m^3\ hr^{-1}\ m^{-2}$	Specific Throughput Relative to Deep-Layer Settler	Specific Throughput Relative to a Shallow Settler[a]
1. IMI deep-layer settler (simple diffuser)	11.8	1.0	1.7
2. IMI deep-layer settler with diffuser–precoalescer	18.9	1.6	2.7
3. IMI compact settler (simple diffuser)	38.7	3.3	5.5
4. IMI compact settler with diffuser–precoalescer	54.8	4.6	7.8

[a]Specific throughput for a shallow settler is taken as 7 $m^3\ hr^{-1}\ m^{-2}$ for this particular phase system.

very low pressure drop within the range required in any case for proper operation of the diffuser. It also allows for the presence of some solids. Periodic washing, if necessary, is rather simple.

The diffuser–precoalescer can easily be retrofitted to existing settlers. It can also be used as a substitute for a design factor in an original design, by designing the diffuser with a double layer that can later be filled with packing, if this becomes necessary as a result of changes in the liquid–liquid system or for debottlenecking.

4. SPECIAL ITEMS

4.1. Settler for Three Phases [12]

Third-phase formation, generally considered undesirable in liquid–liquid extraction systems, can be exploited by using the formation of the third phase as a control point since it gives rise to a pseudoinvariant system. Among the reasons why this approach has not been more extensively exploited are the complexity of the three-phase separation mechanism and of the equipment required for accomplishing it. Analysis of dispersion band structure and separation mechanism [7] have indicated a way in which this difficulty can be overcome [12]. The three-phase settler contains the normal diffuser; this is surrounded by the first concentric chamber, which has sufficient area for the disengagement of the third phase (generally the lightest). The chamber wall has passages for the main dispersion and for the separated phases. The main dispersion separates in the second concentric chamber. The settler was developed for the IMI wet process phosphoric acid cleaning process [13] and has been applied on an industrial scale.

4.2. Mixer for High Mass-Transfer Efficiency [12]

In the usually encountered multistage contacting operations there is little incentive to increase mass-transfer efficiency above the 85–95% level normally obtained in mixer–settlers. However, where a difficult phase separation is combined with a low stage requirement, substantial economic benefits can result from a close approach to equilibrium. This cannot be achieved in a simple mixing vessel, even taking into account the natural "plug flow" element, and a series of discrete mixing chambers is required [12]. This requirement is realized in the IMI compartmented turbine, in which two or three mixing chambers are superimposed on each other, separated by annular baffles. A mixing impeller is provided for each chamber, the impellers being mounted on a common shaft. Ths mixing chambers are baffled to eliminate vortices. Incoming liquids are fed to the lowest chamber, and the final dispersion overflows from the highest chamber; flow is cocurrent.

This approach to equilibirum, for cases having a small number of stages, permits abandoning the pump–mix principle without incurring an excessive economic penalty. Interstage transfer is thus achieved by pumping of individual phases, or a combination of gravity flow and pumping.

The compartmented turbine has been implemented on an industrial scale, and mass-transfer efficiencies of 98–99% have been achieved.

NOTATION

A	settler area
D	diameter of pump impeller
g	constant of gravitational acceleration
H	pumping head
ΔH	dispersion band thickness
K	constant of proportionality
N	impeller rotational speed
P	power absorbed in pumping
Q	total flow through settler or pump
y	exponent dependent on system properties
ρ	fluid density

REFERENCES

1. U.S. Patent 3,489,526.
2. IMI Staff Report, *Proc. Eng. Plant and Cont.*, September 1967, 16-7.
3. IMI Patent Application Israel 34096 (1970).
4. U.S. Patent 3,973,759.
5. J. Mizrahi, E. Barnea, and D. Meyer, *International Solvent Extraction Conference,* ISEC74, Lyon, Vol. 1, Society of Chemical Industry, London, 1964, pp. 141–168.
6. A. D., Ryon, F. L. Daley, and B. S. Lowrie, *Chem. Eng. Progr.* **55**(10), 70 (1959).
7. E. Barnea, and J. Mizrahi, *Transact. Inst. Chem. Eng.* (*Lond.*) **53**, Part 1, 61–69; Part 2, 70–74; Part 3, Part 4, 83–92.
8. J. Mizrahi and E. Barnea, *Proc. Eng.* **60** (January 1973).
9. U.S. Patent 3,563,389.
10. Israel Patent 450064 (1973).
11. E. Barnea and J. Mizrahi, *Proceedings of the International Solvent Extraction Conference,* ISEC 77, Toronto, 1977, Vol. 1, Canadian Institute of Mining and Metallurgy, Montreal, Canada, 1979, p. 374.
12. E. Barnea, *Proceedings of the International Solvent Extraction Conference,* ISEC 77, Toronto, 1977, Vol. 1, Canadian Institute of Mining and Metallurgy, Montreal, Canada, 1979, p. 347.
13. IMI Staff Report, IMI Technology for Cleaning of Wet Process Acid, American Chemical Society General Meeting, Chicago, August 1973.

9.5

LURGI MIXER–SETTLERS

E. Müller and H. M. Stönner
Lurgi Kohle and Mineralöltechnik GmbH
West Germany

Mixer–settler extractors are generally used by Lurgi in preference to differential contactors, when high extraction yields are needed (>99.9%) and sufficient mass transfer is possible only after formation of a fresh surface. Various types of mixer–settler have been developed, depending on process requirements. In all cases, however, settler design and scale-up are based on one settling model.

The selection of mixer–settler type takes into account many factors, including plant size and available area, number of stages, solvent ratio, mass-transfer kinetics, type of dispersion, dispersion stability and settling rate, suspended solids, solvent recovery and losses, safety and environmental criteria, and special customer requirements. Lurgi uses horizontal and vertical (column) mixer–settlers that are considered separately in the two main sections of this chapter.

1. HORIZONTAL MIXER–SETTLERS

The mixer has the function of providing sufficient interfacial area for equilibrium to be reached in a reasonable contact time, but not producing such a fine dispersion that the subsequent phase separation is difficult. Consideration must also be given to the transportation of the two liquids between stages in countercurrent flow and to special needs such as organic-phase dispersion or aqueous-phase dispersion; the normal mass-transfer direction is from the dispersed to continuous phase.

To meet these requirements, Lurgi has found

different technical solutions, which are described in the following paragraphs.

1.1. Mixer Types and Phase-Recycle Devices

1.1.1. Mixing Pumps as Applied by Lurgi

Pumps with a special head-discharge characteristic are used for mixing and pumping in cases where equilibrium transfer between the two liquids is reached in a short time and no emulsion problems occur, such as in the *phenosolvan* process. In smaller plants up to about 100 m^3/hr, built-in vertical pumps are used, whereas in bigger plants the pumps are arranged externally.

A specially developed vertical pump with an inverted impeller is shown in detail in Fig. 1. These pumps have a turn-down ratio from 100 to 0%, whereas no upsets are caused by gas or vapor locks. This type of pump is not available commercially for higher loads, so for large extraction plants, LURGI selects horizontal pumps with a flat pumping characteristic curve. To prevent gas entrainment, a low pump head is used in combination with internal recycle of one liquid phase.

1.1.2. Lurgi Phase Mixer

In some applications, especially for metal extraction, it is necessary to obtain a definite dispersion type. This will depend primarily on the phase ratio, but, within certain limits, it can be influenced by the method of mixing. LURGI has developed a phase mixer by which it is possible to run the mixer–settler safely very close to the phase inversion limit so that the recycle flow rate can be minimized.

The phase mixer works as shown in Fig. 2: the desired dispersion type is produced by a premixer, which is situated upstream of a propeller-pump producing an axial flow. The phase to be dispersed is sent through the inner conical, perforated pipe and introduced into the continuous phase. The throughput and mixing intensity can be adjusted by varying the impeller angle and the rotational speed. Phase mixers of this type are in use in several metal extraction plants in Europe and Africa.

When a higher retention time is needed to attain equilibrium, the mixer volume must be enlarged. This is carried out by providing an enlarged pipe between the phase mixer and the settlers. Only in cases where the required retention time is extremely long are conventional mixing tanks with stirrers applied.

1.1.3. Adjustment of Phase Ratio

The phase ratio in the mixer required for the particular dispersion type is very often not identical with the ratio of the flows of the entering liquids in the extraction process. In this case the phase with the lower flow rate must be increased by a stage-internal recycle. For exact phase ratio control, LURGI uses one of the following methods:

1. The total dispersion flow leaving the mixer is controlled by a valve in proportion

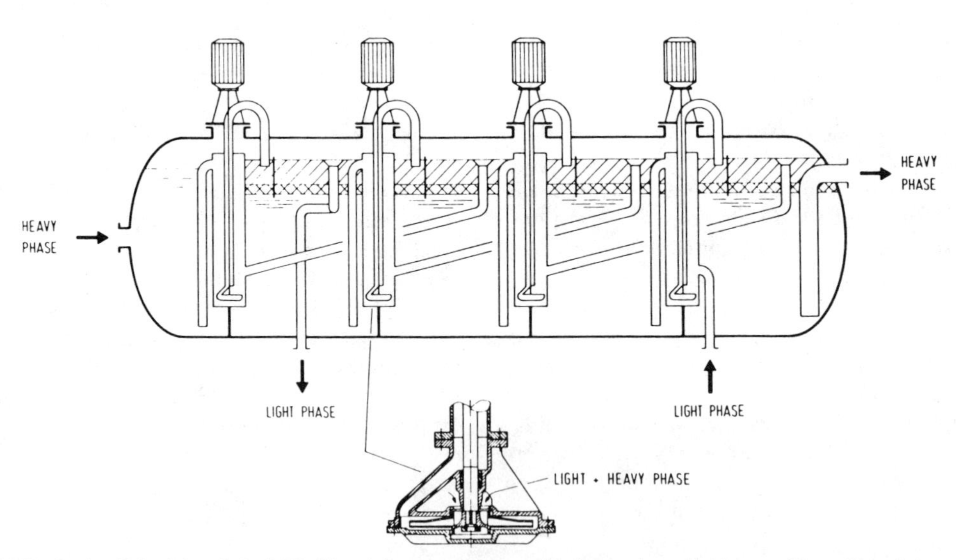

Figure 1. Principle of the LURGI multistage extractor with internal vertical inverted impeller mixing pumps.

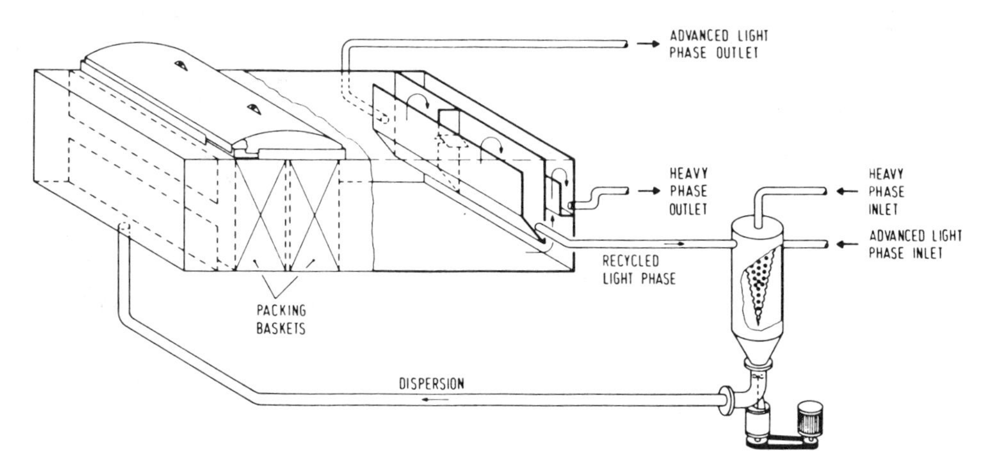

Figure 2. Lurgi phase mixer (dense phase dispersed) combined with a packed settler and divided overflow weir (light phase recycled).

to the feed flow. Thus a constant phase ratio is maintained without disturbing the suction side of the phase mixer. This arrangement can respond quickly to changes in the feed flow rate.

2. For readily emulsifying systems whose separation characteristics are worsened by the throttling effect of a control valve, a divided overflow weir device may be used. This is a simple flow splitter for one phase, allowing control of the ratio between exit and recycle flow rates.

1.2. Settlers

1.2.1. Horizontal Settlers

When it is necessary for the extraction stages to be on ground level for operation, cleaning, and maintenance, horizontal mixer–settlers are applied. The settlers are designed to provide sufficient settling area for a given dispersion flow. For high flows and low settling rates, large areas are required for conventional designs. Sometimes these large settlers have been built as open basins, causing fire hazards and settling disturbances and solvent vapor losses due to wind. To minimize the plant area required and to build enclosed safe extractors, Lurgi has developed different settler types. The design of all these different types is, however, based on settling model 1 as shown in Fig. 3. Equation (1), shown in Fig. 3, describes the dependency of the dispersion band height on dispersion flow rate per unit settler area. Point A characterizes a conventional empty settler, operated at a medium dispersion band thickness and with a safety margin from flooding conditions at point C. Point B characterizes high-load settlers, equipped with trays or packing, in which the dispersion band is kept small [e.g., LMTS (Section 1.2.2) and packed settlers (Section 1.2.3)]. Point C characterizes a settler compartment operated at the flood point, where the emulsion band equals the weir height [used in stratified flow settlers (Section 1.2.4)]. Examples of a conventional empty settler A has been shown in connection with mixing pumps in Fig. 1.

The position of the dispersion band between the two separate liquids is influenced mainly by the height difference between the overflows for light and dense phase and by the different densities of the two liquids.

Empty settlers are simple and easily cleaned but have large plot area requirements and may be prone to flooding due to deterioration of the settling properties of the dispersion, such as by temperature decrease.

1.2.2. Lurgi Multitray Settler

The Lurgi multitray settler (LMTS) was developed for treating high liquid flow rates on small plot areas. The LMTS (Fig. 4) is very flexible regarding upsets in settling behavior or overloads because it operates at a point where the dispersion band thickness rises less steeply than it does near the flooding point.

The LMTS contains a large number of settling trays operating at point A in Fig. 3, where q_0 is nearly proportional to h. This results in a load higher than that in an empty settler, where the maximum q_0 is limited by k_2 according to Eq. (1). Each tray of the LMTS contains side walls

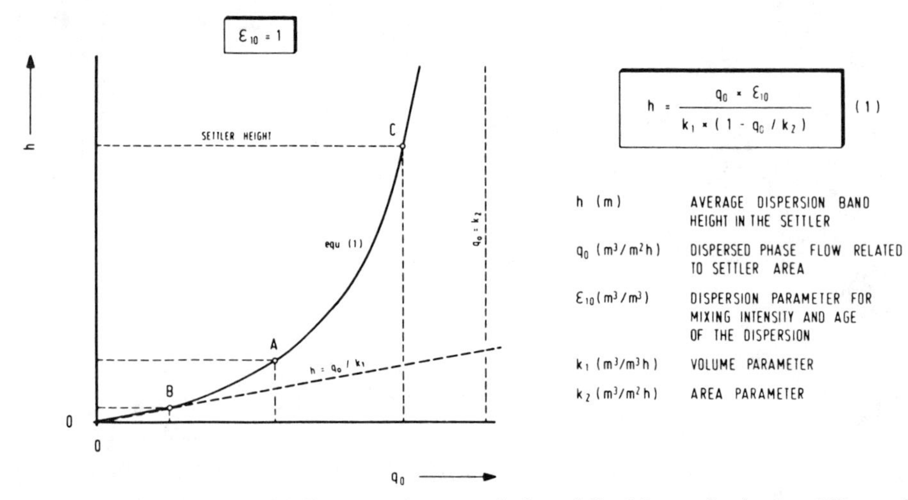

Figure 3. Characterization of different settler types A, B, and C, with use of a simple settling model [Eq. (1)].

and weirs to retain the small dispersion band inside the tray.

The settler surrounding the trays consists of a feed compartment with predistributors for the trays, collecting chambers for the clear phases produced in the trays, and the usual weirs for recovery of light and dense phases. The dispersion is fed by a mixer through the predistribution holes to the single trays. On the trays the dispersion band is limited by the height of an inlet weir. Any surplus mixture enters the next tray. By this simple method, all trays are evenly loaded. One of the settled phases leaves the tray through the open side of the tray into the surrounding settler. The other phase runs at the end of the settling tray through a slot and then over an outlet weir, by which the dispersion band is positioned in the tray. From here, the separated liquid moves in large drops through the surrounding phase to the interface in the collecting chamber, where it joins the streams from the other trays. The trays can be located in either the dense or the light phase of the settler or in both phases. Trays located in the heavy phase are open at the bottom and closed at the top so that light phase and dispersion are trapped while heavy phase is connected with the surrounding liquid and vice versa for trays in the light phase.

It is preferable to position the trays in the heavy phase, especially in cases where the feed to the extractor contains suspended solids and in cases where fungus tends to grow in the light organic phase.

1.2.3. Packed Settler

The active volume of this settler is filled with a packing material enhancing the settler capacity. Its special features are high throughput per unit area, high overload flexibility, and extremely clear settled phases.

This type of settler works at point A (Fig. 3).

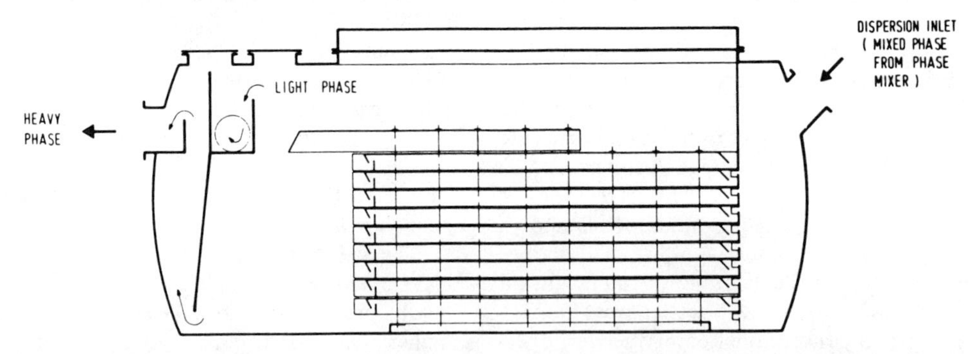

Figure 4. Lurgi multitray settler (LMTS) with trays located in the dense phase and small light phase (solvent) inventory.

The overall settling rate is proportional to the number of packing layers occupied by the dispersion wedge formed as a result of the pressure drop of the packing material. At the same time, laminar flow through the packing is achieved without backmixing, which accounts from the clarity of the settled phases. The path length of the dispersion flows through the packing is optimized so that the settler width is proportional to the throughput. The packing is arranged in baskets, and single baskets can be removed for cleaning while the extractor is in operation.

1.2.4. Stratified Flow Settler

This settler is suitable for dirty feedstocks and resembles a cascade of empty settlers. The inlet compartment operates with a maximum possible dispersion band height, whereas the following compartments work with smaller bands. This has the advantage that while part of the settler is operated at maximum load, the final chamber provides flexibility by working with a small dispersion band. All chambers remain open for inspection and cleaning.

1.2.5. Experience with the Special Settler Types

Whereas smaller plants are still equipped with empty settlers, the special settlers have been introduced in plants with large liquid flows. The LMTS (Section 1.2.2) has been used in 30 settlers handling 140–540 m^3/hr of total flow. The packed settler (Section 1.2.3) has been used in three settlers operating at 440 m^3/hr of total flow. The stratified flow settler (Section 1.2.4) has been used in three settlers operating at 1400 m^3/hr of total flow.

All types of settler are operating with good results. The LMTS and the packed settler require smaller plot areas. For final solvent recovery, coalescers with after settlers are used in most cases. Facilities have been developed for cleaning the trays or packings.

2. LURGI EXTRACTION TOWER

2.1. Horizontal versus Vertical Mixer–Settler Designs

The mixer–settler operates on the principle derived from solvent extraction in a separating funnel. A countercurrent flow can be obtained by arranging the funnels in a sequence, one beside the other, and in two rows, one on top of the other, with the discharge cocks operated by hand (see Chapter 17.1).

The next step in the early development was to handle both phases by means of pumps performing the double duty of mixing and transferring the liquids. For this system to become practicable, it was necessary to develop a method by which the interface level in the settler could be kept constant. Laboratory and test arrangements where small dimensions are involved use the principle of the florence flask for this purpose, whereas larger industrial-scale units often employ a float to operate a control valve instead. The use of pumps also caused another problem: it was necessary to match the pump delivery with the throughput of the extractor. The standard solution for this problem is to have a control valve actuated by way of the liquid level in the receiver, but this can be economically justified only if the units are very large and consist of merely a few stages. For all other applications, pumps with a very low head are placed underneath the settler at sufficient depth to prevent them from taking in air. This is essential because the aspiration of air produces extremely fine dispersions that take a long time to settle. Moreover, if the pump cannot be arranged at a conveniently low height, the suction pipe must be installed at a sufficiently low level.

This makes the disadvantages of the horizontal mixer–settler design obvious: it requires a large area, the interface control system is complicated and may not work smoothly; and special constructions are required for the mixing and transfer pumps. On the other hand, the system has the following advantages: either phase may be partly or completely withdrawn or fed in at each stage. Either phase may be recycled several times at each stage, although additional controls are required for this purpose. Whenever very large settlers with a floor area of more than 50 m^2 are required, lined concrete basins protected by light roofs are more economical than steel tanks.

The major advantage of the extraction *tower* lies in its relatively small floor area—an advantage that is of particular importance when many stages are used.

The interface control system works irrespective of fluctuations in density and quantity (provided the system does not operate close to its maximum load). The extractor is completely filled with liquid so that no pump can draw in air. One disadvantage of the extraction tower is that it is not possible to replace one phase by another if desirable. This can be done only if several extractors, which must be independent of one another, are stacked vertically. Moreover, one of the two phases must be recycled in the extraction tower. If the system operates at low load, it may be necessary to take precautions against phase inversion.

2.2. Pump Delivery, Phase Recycling, and Adjustment of Interface Level

The main problem in any type of mixer–settler is that an interface level must be maintained at each stage and that the pump capacity must be adjusted to a throughput which is not exactly known. In the horizontal arrangement, this problem is overcome by applying the principle of the florence flask and by installing the pumps at a low level. For this system to work, each stage must be open to atmosphere and the effluent phases run off at equal pressure.

By contrast, the extraction tower is completely filled with liquid. The interface levels are maintained by drawing off the liquid from the middle of the chamber (or at the desired interface level). If the interface level in the chamber rises, a larger part of the dense phase will be drawn off, thereby causing the interface level to drop again; thus the interface is always at the level of the draw-off nozzle. The draw-off chamber is connected to the settler itself by means of one of the two phases. For the other phase, a pipe leads to the draw-off chamber of the adjoining stage. Hydrostatic equilibrium normally causes the interface level in the settler to be the same as that in the draw-off chamber. If the throughput rate is very high, however, the interface level will also be influenced by the frictional pressure drop of the phases. In this case it will be easier to control the interface level by means of a control valve (see Chapter 27.3).

The pumps must handle at least that throughput rate that flows from one stage to the next. However, since this rate is not known exactly because of the internal circulation, the pump delivery must be somewhat higher. It must be possible for this surplus to vary without interfering with the normal operation of the extractor, and this is achieved by recycling one phase through the stage. The extraction tower is thus characterized by a recycled and a straight-through phase. Moreover, the pump delivery may be roughly adjusted by means of a "butterfly" valve.

2.3. Extractor Operation with Recycling Light Phase

Figure 5*a* shows a schematic front and side view of the extractor. The mixture enters the settling chamber *A* and separates there. The partition *B* retains the dispersed intermediate layer while the dense phase flows off underneath partition *B* at point *C* to the adjoining draw-off chamber *E*. The light phase flows off above partition *B* at point *D* and then through riser *E* to the draw-off chamber of the stage above. Part of it passes through riser *G* to the draw-off chamber two stages above. Only this part will have moved on by one stage, whereas the rest will be recycled to the stage from which is has originated. The baffle plate *H* prevents the light phase from bypassing the draw-off nozzle *J* and skipping one stage. This is a major risk whenever the density

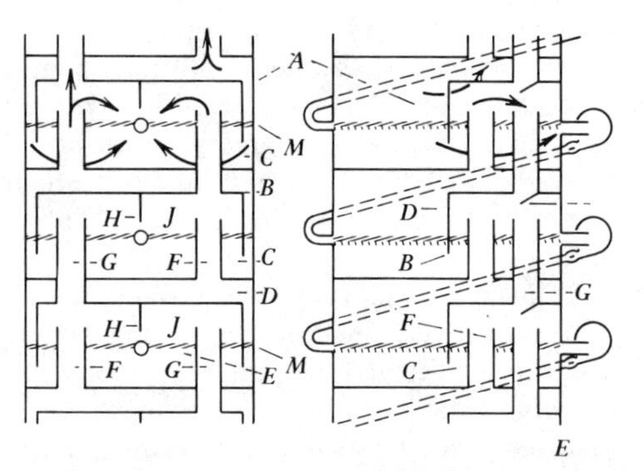

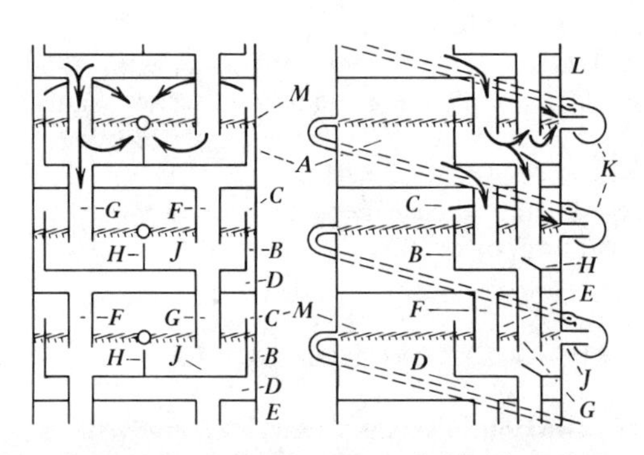

Figure 5. (*a*) Front and side view of extractor if light phase is recycled. (*b*) Front and side view of extractor if heavy phase is recycled. Legend: *A*, settling chamber; *B*, draw-off, partition; *C*, outlet, straight-through phase; *D*, outlet, recycled phase; *E*, draw-off chamber; *F*, pipe for returning the recycled phase; *G*, pipe for straight way of the recycled phase; *H*, bypass preventer; *J*, draw-off nozzle; *K*, pump; *L*, butterfly valve; *M*, phase interface level.

of the light phase increases from bottom to top. Through nozzle *J*, pump *K* withdraws the mixture from the draw-off chamber *E* and transfers it to the settler *A* of the next stage below.

2.4. Extractor Operation with Recycling Dense Phase

Figure 5*b* shows the front and the side views of the extractor when the dense phase is recycled. It corresponds to Fig. 5*a* placed upside down. In this case the light phase overflows partition *B* at point *C* to reach the adjoining draw-off chamber *E*. The dense phase flows off underneath partition *B* at point *D* and passes through downcomer *F* to the draw-off chamber of the stage underneath. Part of the phase is transferred through downcomer *G* to the draw-off chamber two stages down. Only this part has moved on by one stage, whereas the rest is recycled to the stage where it has originated. The baffle plate *H* prevents the dense phase from by-passing the draw-off nozzle *J* and skipping one stage. This is a major risk whenever the density of the dense phase increases from top to bottom. Through nozzle *J*, pump *K* withdraws the mixture from draw-off chamber *E* and transfers it to the settler of the stage above.

2.5. Extractor Operation at Low Load

The operating load of the extractor is characterized by the fact that each pump transfers at a rate that is predetermined by the setting of the butterfly valve. This flow includes the throughputs of both phases and the flow of the recycled phase. If the system operates at underload and the butterfly valves are not readjusted, the dispersion throughput of the settler will remain constant while the proportion of the recycled phase in the extractor will increase. For assessment of the effect of underload operation, the separating behavior must be known as a function of different phase ratios. If, for instance, the holdup of light (hydrocarbon) phase rises from 20 to 45% due to recycling, during extraction of aromatics, the proportion of the light phase in the dispersion will increase further whenever the system operates at underload, and the separation will be improved in this case. Consequently, the efficiency of the extractor is not impaired by operating it at low level. However, increasing the throughput to full load should be done gradually.

When the hydrocarbons obtained in the aromatics extraction process are washed with water to remove the solvent, the original holdup of water can be increased from 5 to 25% by recycling through the stage. If during underload conditions, the proportion of hydrocarbons is reduced to less than half, the dispersion may contain more than 50% water. Then the phases will invert; that is, the water will change from the dispersed to the continuous phase—a type of dispersion that settles much more slowly, so that the settlers may overflow. In this case hydrocarbons that have already been extracted are recycled to the feed so that the holdup is maintained at a constant level. In other cases it may be more suitable to readjust the butterfly valves.

2.6. Practical Applications

The Lurgi extraction tower has so far been used mainly for the extraction of aromatic compounds by the *arosolvan* process (Chapter 18.2.1), in which a selective solvent is used to separate the aromatics from other hydrocarbons in an extractor in which the light phase is recycled. It yields the solvent in a dispersed form, and the mixture settles with a high separating rate. During the subsequent extraction of the hydrocarbons with water to remove residues of the solvent, however, the dense phase (water) is recycled because its holdup in the mixture is only 3 to 10%. The recycling arrangement ensures that both phases will be present in the mixers at any time, even during fluctuations of the quantities or the interface levels.

The Lurgi tower has also been used to extract hydrogen sulfide from liquid propane and to extract phenols from the waste water from coal gasification plants. The largest extractor built so far has a diameter of 8 m, and the dispersion is separated in the settler at a rate of more than 1000 tons/hr. The smallest extractor has a diameter of 0.4 m and has been used for test purposes, such as to develop a process for extracting isoprene. Most extractors have six pumps connected to the same shaft, which is driven by a single motor. If the diameter of the unit exceeds 5 m, each stage has its own motorized in-line pump.

2.7. Conclusions and Recent Developments

The Lurgi extraction tower was originally developed for extraction of aromatics (arosolvan process), but it is also well suited for the extraction of other substances, in particular if the following aspects are important:

A large number of stages on a small floor area.

A high degree of reliability in scaling up the unit with respect to throughput and efficiency.

Handling of widely different phase quantities.

No decrease in efficiency at underload operation.

With a newly developed design in which the number of flanges has been kept very small, operation has also become possible at an overpressure of more than 2 bars. A certain drop in efficiency during underload operation must, however, be taken into account with this design.

10

NONMECHANICALLY AGITATED CONTACTORS

Stuart D. Cavers
The University of British Columbia
Canada

1. GENERAL CHARACTERISTICS

The main advantage of nonmechanically agitated contactors is the absence of moving parts. In the simplest type of contactor, the spray column, drop formation and motion are determined largely by interfacial tension and density difference. The addition of internals (baffles, packing, and perforated plates) has several beneficial effects: reduced axial mixing of the continuous phase; enhanced mass transfer as a result of increased residence time of the dispersed phase; and increased drop coalescence and break-up in the main column. However, the addition of internals reduces column capacity.

Much of the performance data available for nonmechanically agitated contactors has been obtained from laboratory-scale apparatus. Too little is known about the effects of increasing scale and of changing the chemical system. A further complication is that, whereas much of the hydrodynamic data available is for phases at equilibrium, mass transfer can have profound effects. For example, for transfer from the dispersed to the continuous phase, coalescence is much increased over that which occurs for the reverse direction of transfer, resulting in reduced interfacial area and less efficient operation. Interfacial agitation can also be important, but Komosawa and Ingham [1] show that, at least for two systems, the important effects of interfacial instability are those associated with drop size (and hence internal circulation) and not directly with interfacial agitation. Surface-active agents (surfactants) can profoundly affect these kinds of behavior, and it is likely that surfactant contaminants will be present in industrial-scale equipment. In short, the present state of knowledge is such that contactors of industrial scale cannot in general be designed accurately without preliminary pilot plant work [2]. If possible, pilot plants should use the same process streams as the full-scale plant so that the effect of contaminants will be included.

However, the designer should be familiar with the available performance data in order to obtain approximate designs and cost estimates and to estimate the size of the pilot plant.

2. WETTED-WALL COLUMNS AND SIMILAR DEVICES

A wetted-wall column consists of a vertical tube through which one phase flows in the central core, with the second phase flowing countercurrently as a thin film along the pipe wall. Other related devices have been proposed [3, 4] (see also Chapter 3, Section 4). The main advantage is relatively simple hydrodynamics; the interfacial area is known, at least roughly. However, since this area is relatively small, these columns are not likely to be useful in other than laboratory work, and even here operating difficulties cause their usefulness to be less than first expected.

3. SPRAY COLUMNS

3.1. Description

In spray columns, one phase is dispersed as drops that move through the second, or continuous, phase as a result of the difference in phase densities. At the main interface the drops coalesce and leave the column. Figure 1 shows the general layout ordinarily used in experimental work, a design due to Blanding and Elgin [6]. The expanded end sections permit the introduction of

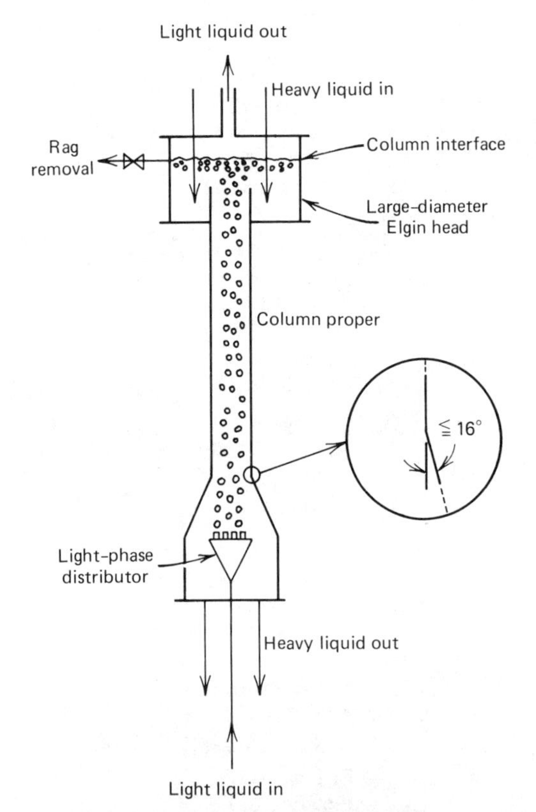

Figure 1. An Elgin spray column, light liquid dispersed.

each phase without undue disturbance and prevent premature flooding.

Drops are produced either at perforations drilled in a flat plate or preferably at the ends of short tubes (nozzle tips). Each of these is chamfered to a sharp edge to reduce any spreading of the dispersed phase across the metal surface, thus providing a controlled drop size. Mayfield and Church [7] suggest the use of a punched plate to approximate this type of distributor. The burr caused by the punching, and extending out from the plate in the direction of drop flow, is left in place [8].

Spray columns have usually been operated at relatively low holdups with a dispersed packing of drops [9]. However, if drops are produced at a rate slightly higher than that at which they coalesce at the main interface, a dense packing spreads from there toward the drop distributor [10]. For operation with a dense packing and with the lighter phase dispersed, the bottom of the column should be straight sided, but the top should be provided with a conical section [10].

As stressed by Steiner et al. [11], it is best to choose the direction of mass transfer so that the drops do not coalesce until the column interface is reached; for most† systems this means that mass transfer should be from the continuous to the dispersed phase [12]. In these circumstances coalescence is slow at the column interface, and residence time and interfacial area are relatively high. It is advantageous, if, by following this rule, the phase dispersed is the one available in larger volume per unit time, since a correspondingly large interfacial area will result. If the phase dispersed is also the one of higher cost and greater flammability, so much the better, since the dispersed phase ordinarily has the lower holdup in the column.

The nozzles used for drop production range in diameter from about 1.3 to 6.4 mm [13]. The drops should be sufficiently small to provide adequate interfacial area in the column, but not so small that premature flooding will occur or that the drops lack internal circulation. Drops appear to be stagnant at diameters smaller than about 1–2 mm [1]. Without circulation, mass transfer within drops is effected by the relatively slow mechanism of molecular diffusion. Hanson [14] points out that, as a result, the beneficial effect of high interfacial area at low drop size may be more than counterbalanced.

†In this chapter the exceptions are ignored (see Chapter 4).

3.2. Applications

There have been relatively few applications of the spray column in industry. Although it can handle high flow rates and is relatively simple and cheap, it suffers from serious backmixing of the continuous phase. Driving forces for mass transfer are reduced, and, as pointed out by Laddha and Degaleesan [15, p. 212], a spray column ordinarily is equivalent to not more than a few theoretical stages. In general, this situation appears to worsen as the diameter of the column is increased. However, Horvath et al. [16] show that, at least at small diameters, much of the effect attributed to diameter may be due to a lack of straightness and verticality. In any case, absence of mechanical complications, ease of cleaning, and the fact that the spray tower can handle suspended solids, make it worth considering where such advantages may be particularly important. Although there has been relatively little experience in operation with dense packing, such operation probably should be considered if the use of a spray column seems attractive.

Spray columns have served as neutralizing scrubbers for organics and as direct-contact heat exchangers [17].

3.3. Performance Data

Methods must be available to the designer for predicting drop size, holdup, superficial velocities corresponding to flooding, mass-transfer coefficients (or heights of transfer units), and (especially for the continuous phase) axial mixing coefficients.

3.3.1. Drop Diameter

Many correlations exist for predicting drop size. For the formation of drops singly, directly on the nozzle tips, Horvath et al. [18] recommend the correlation by de Chazal and Ryan [19]. On the other hand, Bühler et al. [20] use the method of Izard [21]. However, most spray columns operate with the drops being formed at the ends of jets of dispersed-phase fluid extending beyond the nozzle tips.

The following equation due to Scheele and Meister [22] is recommended [18] for calculating the nozzle velocity at which a jet appears:

$$v_j = \left[\frac{3\sigma(1 - d_o/d_e)}{\rho_d d_o}\right]^{0.5} \tag{1}$$

where d_e is the drop size that would be produced by a v_o equal to v_j if a jet did not form. Presumably, Eq. (30) of Ref. 22, Part I should be used for calculating d_e; in this calculation Eq. (6) of Ref. 18 can be useful. As the velocity is increased further a critical value is reached at which the jet is of maximum length and breaks up into drops of approximately uniform size [18]. Also, the drop area produced approximates a maximum [18, 23]. The corresponding nozzle velocity is given by the equation [5, p. 467; 23; 24]

$$v_{om} = 2.69 \left(\frac{d_j}{d_o}\right)^2 \left(\frac{\sigma_2/d_j}{0.514\rho_d + 0.472\rho_c}\right)^{0.5} \tag{2}$$

(Unfortunately, the state of the art is such that it appears that the calculated v_j can be greater than v_{om} in some cases.)

If a third component is present, an effective interfacial tension between σ_2 and that of the extraction system should be used in this equation, even if the phases are at equilibrium [5, p. 468; 23; 25). The ratio of nozzle to jet diameter for use in Eq. (2) is calculated by means of Eq. (8) in Chapter 4 (from Fig. 2, Ref. 23); presumably this same σ is used [25]. The average d_e is 2.1 d_j. [For more recent results, see Eqs. (15)–(17) and Fig. 8 of Ref. 24.] Equation (10) in Chapter 4 also is useful for predicting drop size. Equation (9) in Chapter 4 gives recommended limits for d_o if drops are to be formed by jet breakup. For a suggestion by Treybal [25] with respect to the use of Eq. (2), see Section 3.4.1.

Variations in continuous-phase velocity within the range ordinarily used have no effect on drop size [26, 27]. Also, holdup, and hence presumably drop size, is about the same for transfer from the continuous to the dispersed phase as for the corresponding mutually saturated binary [28]. The equations mentioned here apply to these cases. Section 2.3 in Chapter 4 describes the use of Equation (10) in that chapter for the reverse direction of transfer.

3.3.2. Dispersed-Phase Holdup (Dispersed Packing)

Traditionally holdup has been predicted [5, 13] by using the fact that the ratio v_s/v_t is a unique function of ϕ_d for all fluid–particle systems [29, 30]. The correlation by Zenz [31], for instance, as in Fig. 11.7 of Treybal [5], provides v_s at any ϕ_d for solid particles, with the value of v_s at a ϕ_d of zero as v_t. With v_s/v_t known as a function of ϕ_d for solid particles, v_s for liquid particles can be obtained if v_t for the liquid particles is available, since the function is the same for the liquid–liquid case. A modified version [32, Fig. 10] of the Hu–Kintner [33] correlation [5, Fig. 5.14] can be used to obtain v_t if μ_c is less than 30 cP. Below an ordinate of 1.0, calculation of v_t by Eq. (5.75) of Ref. 5 (limited to μ_c below 2 cP) is preferable [8]. A plot of v_s for the drops against ϕ_d is solved simultaneously with the first equality of Eq. (35) in Section 5.3 (with $\epsilon = 1$ and $v_r = v_s$) to provide the desired ϕ_d.

Alternatively, ϕ_d can be obtained from the second equality of Eq. (35) in Section 5.3 (again with $\epsilon = 1$) used with the first equation in Table 2 of Chapter 4.

Transfer of a solute from the continuous phase to the dispersed results in holdups only slightly higher than those found with no solute present [28]. However, as a result of coalescence and larger drops, ϕ_d is reduced considerably for the reverse direction of transfer.

Thornton [35] obtained Eq. (30) in Chapter 4 for predicting ϕ_{df} and showed that ϕ_{df} reaches a maximum of 0.5 (at $V_c = 0$).

Perrut and Loutaty [34] consider dense-packing conditions.

3.3.3. Flow Rates at Flooding

A slightly simplified form of the correlation by Minard and Johnson [36] is recommended [5, p. 478; 13] (lb_m ft hr units)

$$V_{cf} = \frac{10{,}000\,\Delta\rho^{0.28}}{[0.483\mu_c^{0.075}\rho_c^{0.5} + d_m^{0.056}\rho_d^{0.5}(V_d/V_c)^{0.5}]^2} \tag{3}$$

for "rejection" of drops from a conical end section of a tower; V_{cf} from Eq. (3) will be 7–10% high for a column with straight sides [5, p. 478].

3.3.4. Mass Transfer

Most data are from laboratory equipment of up to 150-mm diameter. In many cases only inlet and outlet concentrations were measured. Therefore, the reported mass-transfer coefficients include end effects due to mass transfer during drop formation and during drop coalescence at the column interface. Furthermore, the coefficients have been calculated by using log-mean driving forces. However, because of axial mixing, the true average driving force is lower than the log mean and the true coefficients tend to be higher than the reported values.

Laddha and Degaleesan [15, p. 233] recommend the following for predicting overall capacity coefficients:

$$K_c a = mK_d a = 0.081 \cdot \frac{\phi_d(1 - \phi_d)\left(\frac{g^3 \Delta\rho^3}{\sigma\rho_c^2}\right)^{0.25}}{(N_{\mathrm{Sc},c})^{0.5} + \frac{1}{m}(N_{\mathrm{Sc},d})^{0.5}}$$

$$\text{with } m = \frac{C_d^*}{C_c} \tag{4}$$

Presumably these include an allowance for end effects, since no separate calculation of the latter appears in an example design [15, p. 249]. However, the example does include a correction for axial mixing, even though Eq. (4) probably is based on data that were not entirely free of axial mixing effects. Therefore, the tower proposed may be too tall. Indeed, mass transfer in spray columns is sufficiently complicated that Eq. (4) may provide only an approximate design. Alternatively, the equations of Tables 1 and 2 can be used; however, a lengthier procedure is required, and, in light of the uncertainties still present, there may not be a commensurate improvement in accuracy.

Table 1 includes data for designing columns in which the drops form at the nozzle tips; Table 2 gives supplementary data for use when the drops form at the ends of jets, which is the usual case. However, since changes of concentration near the bottom of the column can be small for jetting conditions [1, 11, 37], and since data for mass transfer during coalescence of drops at the column interface are sparse, it may be sensible to ignore both drop formation and coalescence in design and calculate solely on the basis of drop rise. A somewhat conservative design may result.

If mass transfer in drop formation is to be allowed for in design, Eq. (19) (Table 2) gives the area of the jets. Equations (20) and (21) provide the mass transfer over this area. Equations (5) and (6) (Table 1) are then used to calculate the mass transfer as the drops break away at the ends of the jets.

The equations given for drop rise in Table 1 are used as if they have been corrected for end effects and axial mixing, although this may not always be the case. Similarly, these equations are assumed to apply to drop swarms, although, as originally proposed, all but Eq. 12 were for other conditions. At first glance this assumption would seem to be risky [5, p. 480], but it has received support elsewhere [11; 15, p. 157] and, in any case, would seem to be better for the dispersed than for the continuous-phase coefficient. The equations used should be appropriate to the drop behavior expected. Drops probably are stagnant below a d_e of about 2 mm [1], but, if larger than this, circulating below an ordinate of about 60 in Fig. 10 of Ref. 32 and oscillating above. Surface-active contaminants can reduce mass-transfer rates markedly by inhibiting oscillation or circulation [5, p. 184]; sometimes even large drops [40] can be stagnant. In this respect Skelland [41] suggests that dispersing the phase present in greater volume may be less successful than dispersing the one with a considerably lower mass-transfer resistance because relatively little harm will be done if the controlling resistance is in the continuous phase.

For all the drop-rise equations tabulated, the surface area to be used is that of a sphere of volume equal to that of a drop. Therefore, the equation

$$a_r = \frac{6\phi_d}{d_e} \tag{26}$$

applies for calculation of the interfacial area.

3.3.5. *Axial Mixing*

The equation due to Zheleznyak and Landau [49]

$$\frac{E_c}{v_c} = 6.5\,(N_{\mathrm{Re}})^{0.987}(\phi_d)^{0.814}(\bar{\mu})^{3.89} \tag{27}$$

is satisfactory for predicting values of E_c for dispersed packing of drops and $D < 10$ cm, if there is no mass transfer, or if transfer is from the continuous to the dispersed phase [18] but should not be used for the reverse direction of transfer. Here E_c is a superficial coefficient. The characteristic length used in N_{Re} is

$$D_h = \frac{D(1 - \phi_d)}{1.5(D\phi_d/d_m) + 1} \tag{28}$$

together with v_c and the velocity v_s [Eq. (26) in Chapter 4]. In Eq. (27):

TABLE 1 EQUATIONS FOR MASS-TRANSFER COEFFICIENTS IN SPRAY AND PERFORATED-PLATE TOWERS AT LOW SOLUTE CONCENTRATIONS AND MASS-TRANSFER RATES

Equation[a, b]	Equation Number	References	Remarks
Drop Formation and Detachment at Nozzle Tips (or at Ends of Jets)			
$k'_{dfa} = 0.0432 \dfrac{d_e}{t_{fa}} \left(\dfrac{\rho_d}{M_d}\right)_{av} \left(\dfrac{v_o^2}{d_e g}\right)^{0.089} \left(\dfrac{d_e^2}{t_{fa} D_{vd}}\right)^{-0.334} \left(\dfrac{\mu_d}{\sqrt{\rho_d d_e \sigma}}\right)^{-0.601}$	(5)	38	Equations (5) and (6) apply to the area of the drops as formed: for one drop = area of sphere of volume equal to that of drop = πd_e^2 if symbol d_e used
$k'_{cfa} = 0.386 \left(\dfrac{\rho_c}{M_c}\right)_{av} \left(\dfrac{D_{vc}}{t_{fa}}\right)^{0.5} \left(\dfrac{\rho_c \sigma}{\Delta\rho g t_{fa} \mu_c}\right)^{0.407} \left(\dfrac{g t_{fa}^2}{d_e}\right)^{0.148}$	(6)	39, 40	Equations (5) and (6) are best for high-σ systems
$t_{fa} = \dfrac{n_o(\pi/6)\, d_e^3}{Q_{de}}$	(7)	40	For use with Eqs. (5), (6), (16), and (17)
$A_{fa} = n_o \pi d_e^2$	(8)	40, 41	For use with Eqs. (5) and (6)
Drop Rise			
$k'_{dr} = -\left(\dfrac{d_e}{6t}\right)\left(\dfrac{\rho_d}{M_d}\right)_{av} \ln\left(1 - \dfrac{\pi D_{vd}^{0.5} t^{0.5}}{0.5 d_e}\right)$	(9)	40–43	For stagnant drops: use d_e for area calculation; fractional extraction should be <0.5
$k'_{cr} = 0.74 \left(\dfrac{D_{vc}}{d_e}\right)\left(\dfrac{\rho_c}{M_c}\right)_{av} \left(\dfrac{d_e v_s \rho_c}{\mu_c}\right)^{0.5} \left(\dfrac{\mu_c}{\rho_c D_{vc}}\right)^{0.333}$	(10)	40, 44	For stagnant drops: single spheres; for single oblate spheroids, see Skelland [41, pp. 285, 407]

$k'_{dr} = 31.4 \frac{D_{vd}}{d_e} \left(\frac{\rho_d}{M_d}\right)_{av} \left(\frac{4D_{vd}t}{d_e^2}\right)^{-0.34} \left(\frac{\mu_d}{\rho_d D_{vd}}\right)^{-0.125} \left(\frac{d_e v_s^2 \rho_c}{\sigma}\right)^{0.37}$	(11)	40, 55	For circulating drops: single drop stream; use d_e for area calculation; for low σ and μ_c
$k'_{cr} = 0.725 \left(\frac{\rho_c}{M_c}\right)_{av} \left(\frac{d_e v_s \rho_c}{\mu_c}\right)^{-0.43} \left(\frac{\mu_c}{\rho_c D_{vc}}\right)^{-0.58} v_s(1 - \phi_d)$	(12)	5, 13, 40, 46	For circulating drops: drop swarm; use d_e for area calculation; for low σ
$k'_{dr} = 0.32 \left(\frac{D_{vd}}{d_e}\right)\left(\frac{\rho_d}{M_d}\right)_{av} \left(\frac{4D_{vd}t}{d_e^2}\right)^{-0.14} \left(\frac{d_e v_s \rho_c}{\mu_c}\right)^{0.68} \left(\frac{\sigma^3 \rho_c^2}{g\mu_c^4 \Delta\rho}\right)^{0.10}$	(13)	40, 45	For oscillating drops: single drop stream; use d_e for area calculation; for low σ and μ_c
$k'_{cr} = \left(\frac{D_{vc}}{d_e}\right)\left(\frac{\rho_c}{M_c}\right)_{av} \left[50 + 0.0085 \left(\frac{d_e v_s \rho_c}{\mu_c}\right)^{1.0} \left(\frac{\mu_c}{\rho_c D_{vc}}\right)^{0.7}\right]$	(14)	40, 47	For oscillating drops: single drop stream; use d_e for area calculation; for low σ
$t = \frac{\text{distance of rise or fall}}{v_s = v_t(1 - \phi_d)}$	(15)	40	For use in Eqs. (9), (11), and (13)
Drop Coalescence			
$k'_{dca} = 0.173 \left(\frac{d_e}{t_{fa}}\right)\left(\frac{\rho_d}{M_d}\right)_{av} \left(\frac{\mu_d}{\rho_d D_{vd}}\right)^{-1.115} \left(\frac{\Delta\rho g d_e^2}{\sigma}\right)^{1.302} \left(\frac{v_s^2 t_{fa}}{D_{vd}}\right)^{0.146}$	(16)	38, 40	Equations (16) and (17) apply to area A_{ca} from Eq. (18); t_{fa} appears because as one drop is formed, one coalesces [48]; best for high-σ systems; at least Eq. (16) includes cases in which the drops did and did not coalesce immediately on arrival at column interface
$k'_{cca} = (5.959)(10^{-4}) \left(\frac{\rho_c}{M_c}\right)_{av} \left(\frac{D_{vc}}{t_{fa}}\right)^{0.5} \left(\frac{\rho_c v_s^3}{g\mu_c}\right)^{0.332} \left(\frac{d_e^2 \rho_d \rho_c v_s^3}{\mu_d \sigma}\right)^{0.525}$	(17)	39, 40	(see comment for Eq. (16))
$A_{ca} = A_C - A_D$	(18)	40	For use with Eqs. (16) and (17); $A_D = 0$ for spray columns

[a] See Skelland and Huang [24, 48] for examples of evaluation of appropriate average concentration differences for use with these mass-transfer coefficients. However, complicated averaging is not always used [41] (see also Sections 3.4.5 and 6.5.8) and probably should be avoided at the present state of knowledge.

[b] Driving force = Δx or Δy; for driving force = ΔC, delete factor $(\rho/M)_{av}$.

TABLE 2 EQUATIONS FOR CALCULATION OF MASS TRANSFER DURING DROP FORMATION UNDER JETTING CONDITIONS[a]

Controlling Resistance in	Equation for Determining	Equation	Equation Number	References	Remarks
	Jet area	$n_o A_j = n_o \pi \left(\frac{d_o + d_j}{2}\right) L_j$	(19)	24, 48	
	Dispersed-phase mass-transfer coefficient[b]	$k'_{dj} = 2\left(\frac{D_{vd}}{\pi t_e}\right)^{0.5}\left(\frac{\rho_d}{M_d}\right)_{av};\ t_e = \frac{L_j}{v'_j}$	(20)	24, 48	Apply to jet (based on penetration theory); apply to driving forces: Δx or Δy
	Continuous-phase mass-transfer coefficient[b]	$k'_{cj} = 2\left(\frac{D_{vc}}{\pi t_e}\right)^{0.5}\left(\frac{\rho_c}{M_c}\right)_{av};\ t_e = \frac{L_j}{v'_j}$	(21)	24	For use with ΔC driving forces delete $(\rho/M)_{av}$
Dispersed phase	Jet length	$\frac{L_j}{d_o} = 5.0767\,(\Delta N_{We})^{0.5499}\left(\frac{\mu_c}{\mu_d}\right)^{0.5245}$	(22)	24, 48	Moderate mass-transfer rate, σ high; single drop stream; valid up to maximum jet length
	Jet contraction	$\frac{d_o}{d_j} = 2.7350\left(\frac{d_o}{\pi\sqrt{\sigma/\Delta\rho g}}\right) + 0.5718$	(23)	24, 48	Moderate mass-transfer rate; single drop stream
Continuous phase	Jet length	$\frac{L_j}{d_o} = 10.8341\,(\Delta N_{we})^{0.5860}\left(\frac{\mu_c}{\mu_d}\right)^{0.5097}$	(24)	24	Low mass-transfer rate, σ high; single drop stream; valid up to maximum jet length
	Jet contraction	$\frac{d_o}{d_j} = 2.6086\left(\frac{d_o}{\pi\sqrt{\sigma/\Delta\rho g}}\right) + 0.8495$	(25)	24	Low mass-transfer rate, σ high; single drop stream

[a]See Eqs. (5) and (6) in Table 1 for mass-transfer coefficients for drop formation and detachment at the ends of the jets.
[b]See footnote *a* to Table 1.

$$\bar{\mu} = \frac{\mu_c + \mu_d}{\frac{2}{3}\mu_c + \mu_d} \tag{29}$$

Division of numerator and denominator of Eq. (28) by D shows that, in contrast to the results due to Wijffels and Rietema [50], this correlation implies that D will have little effect on E_c if D is large. Little directly usable information is available for larger-diameter columns, and extreme caution is needed in scale-up [51].

For the dispersed phase, ordinarily only the large drops need be considered in a population in which small drops at equilibrium with the continuous phase are also present, and, for practical purposes, an assumption that the large drops are in plug flow usually is satisfactory [11, 52, 53] unless transfer is from the dispersed phase and (or) the drops are nonuniform.

For dense packing of drops, E_c is much reduced [54], and the drops again are essentially in plug flow [53].

3.4. Design

In this section it is assumed that the solute concentration in a known volume per hour of feed stream is to be reduced to a specified level by a dispersed, less dense extract phase of which the flow rate has been decided, such as by optimization of the solvent:feed ratio. The procedures refer to dispersed packing; however, dense packing is considered briefly in a final subsection. A straight-line equilibrium relationship is assumed.

3.4.1. Drop Distributor and Drop Size

The value chosen for d_o should exceed 0.05 in. to avoid clogging and be less than 0.25 in. to avoid excessive reduction of interfacial area [13]; a d_o of 0.1 in. is satisfactory for a typical column. Operation with a v_o above v_j [Eq. (1)] at a value v_{om} [Eq. (2)] usually is desirable. Then

$$n_o = \frac{4Q_d}{v_{om}\pi d_o^2} \tag{30}$$

However, when the column diameter is chosen, a check should be made that n_o orifices can be fitted into the tower cross section at the level of the distributor; if not, v_o is increased [5, p. 483], and less than maximum interfacial area is accepted (see Section 3.3.1 for prediction of drop size). However, Treybal [25] indicates that use of Eq. (2) can result in a value of $v_{om} < 0.1$ m/s, the rule-of-thumb value of v_j. He suggests that, if this occurs, v_o be set at 0.1 to 0.15 m/s at least, and d_e estimated from his Fig. 10.46, due to Hayworth and Treybal [55]. [Presumably Eq. (10) in Chapter 4 also would serve.]

3.4.2. Diameter of Column

Equation (3) provides V_{cf}, from which

$$D = \sqrt{\frac{4Q_c}{0.4\pi V_{cf}}} \tag{31}$$

based on use of 40% of V_{cf} [5, p. 479] to allow for uncertainties. Next, Q_c and Q_d are divided by $\pi D^2/4$ to provide V_c and V_d.

3.4.3. Interfacial Area

The dispersed-phase holdup is obtained as suggested in Section 3.3.2. Then Eq. (26) gives a_r.

3.4.4. Height of Column with Allowance for Axial Mixing

Values of $K_c a$ can be calculated by Eq. (4), and the corresponding HTU, by the following equation:

$$(HTU)_{oc} = \frac{V_c}{K_c a} \tag{32}$$

As described later, this is used in Pratt's method [56] for correcting for axial mixing and obtaining H.

Design on the basis of Tables 1 and 2 is more complex. It is assumed that operation is with jets at the spray nozzles, and, initially, that end effects are to be allowed for in design. Equations (20) and (21) (Table 2) provide k_{dj} and k_{cj}. From these, K_{dj} is obtained by means of Eq. (59) in Chapter 3. Equations (5) and (6) (Table 1) are used similarly to provide K_{dfa}. If the mass transfer during jetting, drop formation, and detachment is not likely to be large, the overall driving force to be used with these K values can be calculated from the terminal concentrations at the column bottom [5, p. 484]; otherwise, see, for example, Skelland and Huang [48]. The applicable areas are given by Eqs. (8) and (19). Each K is used with its corresponding driving force and area to provide a mass-transfer rate, calculable also as Q_d times the change in C_d, so that C_d can be determined for the detached drops. The corresponding C_c is obtained by a plug-flow material balance around the bottom of the column. These concentrations can be used in

Pratt's method [56], as if they were terminal concentrations, along with the concentrations of the phases just below the bed of drops coalescing into the column interface. These last concentrations are obtained by calculations similar to those used for the column bottom but based on Eqs. (16)-(18) (Table 1). The uncorrected terminal concentrations are used, of course, if end effects are to be ignored or if the design is to be based on Eq. (4).

Next, values of K_c for drop rise are obtained from phase coefficients from Table 1 appropriate to the kind of drop behavior expected, a_r is obtained from Eq. (26), and $(HTU)_{OC} = (HTU)_{OR}$ is obtained from Eq. (32) ($a = a_r$). An E_c now is predicted, for example, from Eq. (27), and a continuous-phase Peclet number calculated:

$$N'_{\mathrm{Pe},c} = \frac{V_c d_e}{E_c} \tag{33}$$

Pratt's [56] Eq. (58) (based on $N'_{\mathrm{Pe},d}$ = infinity) then gives the length of the column, unless V_c/mV_d is unity, in which case his Eq. (60) applies. (For extension of the length calculation to systems with curved equilibrium lines, see Ref. 57.)

3.4.5. Dense Beds

Care must be taken to match drop formation and coalescence rates to avoid flooding due to the region of dense packing extending to the tower bottom. Croix et al. [58] discuss operation with a dense bed and consider design; however, correlations for mass-transfer coefficients appear to be unavailable.

4. BAFFLE CONTACTORS

The performance of a spray column can be improved by the introduction of baffles, for instance, side to side or center to side [5, Fig. 11.12], and disk and "doughnut" [59]; the latter are used for recovering acetic acid. Few design data are available; Morello and Poffenberger [59] reproduce a chart useful in designing for required flow capacity.

5. PACKED COLUMNS

5.1. Description

A spray column becomes a packed tower if its shell is filled with packing pieces. Packing reduces axial mixing, lengthens residence time, and causes distortion of the drops; the result is increased mass transfer [60]. However, packing reduces the area available for flow, thus lowering capacity. Eckert [60] prefers Intalox saddles, followed by Pall rings and then Raschig rings. The nominal size should be below $0.125D$ to minimize wall effects and channeling; 15 to 38-mm pieces are used industrially [17, 60]. The packing material should be wet preferentially by the continuous phase, and in start-up the column should be filled with continuous phase before the flow of dispersed phase is begun, to ensure that the latter flows through the packing as droplets, rather than in rivulets on the packing surfaces.

The packing support, a bar grid or perforated plate, should provide a maximum of open area without allowing packing to fall through. Drops should be formed at nozzles extending 1-2 in. into the packing [5, p. 487; 61] to avoid premature flooding, which can occur if the drops are produced below the packing support [6]. Maldistribution of flows is reduced by redistributing the phases every 10-15 ft [5, p. 487] or even every 5-7 ft [60].

A tower is packed with the shell filled with water to reduce breakage of packing pieces. Settling of these toward a final level can be encouraged by blowing air through the water-filled tower. The tower is kept free of packing for several feet at the end where the coalesced dispersed phase leaves. The column interface between the coalesced drops and the continuous phase is maintained in this region. In the Elgin design [6] this would be larger in diameter than the packed sections, but increased D is not necessary in the case of the packed tower, provided the continuous phase can enter without undue disturbance [61]. An increase in diameter of more than 30-40% at the dispersed-phase entry is not justifiable [61], and straight-sided towers have been suggested [5, p. 487]. Figure 2 shows a column of this type.

Valves to allow the removal of scum ("rag") usually are provided at the column interface and at similar interfaces below packing supports where the phases are redistributed.

In general, choice of the phase to be dispersed parallels that for spray towers; therefore, the extract phase should be dispersed (assumed here). However, a highly viscous continuous phase can cause difficulties [60].

5.2. Applications

Although packed towers are not suitable for liquids containing suspended solids or for those of

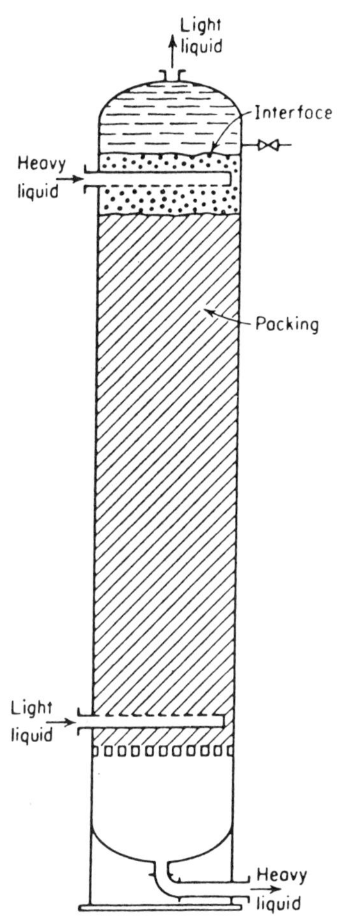

Figure 2. Packed extraction tower, light liquid dispersed. (From *Mass Transfer Operations*, 2nd Ed., by Robert E. Treybal. Copyright © 1955, 1968 by McGraw-Hill, Inc. Used with the permission of McGraw-Hill Book Company.)

high interfacial tension [5, pp. 486, 498], they are used in a wide range of applications. Normally a packed tower is preferred over a spray tower, as the reduced flow capacity is less significant than the improved mass transfer. Packed towers should not be used if the ratio of flow rates is outside the range 0.5–2.0 because of probable flooding when suitable holdup and interfacial area are provided [8].

5.3. Performance Data

5.3.1. Drop Diameter

Packings above the critical size given by Eq. (11) in Chapter 4 always should be used, and nozzle tips should be chosen (Sections 3.3.1 and 3.4.1) to produce drops slightly larger than the size given by [62]

$$\frac{d_{vs}\Delta\rho\sigma}{\mu_c^2} = 1.42\left(\frac{\Delta\rho\sigma^3}{\mu_c^4 g}\right)^{0.475}\left(\frac{v_k\epsilon\phi_d}{V_d}\right) \quad (34)$$

(Table 2 in Chapter 4 gives v_k.) These drops will break down in a short distance to this size, which will be almost independent of the size and type of packing [8] and the dispersed-phase flow rate but will increase slowly with the continuous-phase rate [63]. Interfacial areas will be large relative to those obtained if packings of a size less than or equal to the critical size were used [8].

Equation (34) applies if the phases are mutually saturated and no undistributed solute is present or if solute is transferred from a continuous to a dispersed phase, each of which is saturated with the main component of the other [62]. (However, holdup results [15, p. 265] suggest that drop size may be somewhat smaller for the latter case.)

5.3.2. Dispersed-Phase Holdup

Although static holdup (Section 3, Chapter 4) may have some effect on mass transfer [5, p. 488], practical holdups for use in design can be estimated by using [64]

$$v_r = \frac{V_d}{\epsilon\phi_d} + \frac{V_c}{\epsilon(1 - \phi_d)} = v_k(1 - \phi_d) \quad (35)$$

valid [8] for $d_p > d_{pc}$ up to the second break or transition point [65] in the holdup-V_d plot, together with the equation for v_k in Table 2 in Chapter 4. This replaces an earlier correlation for v_k [8, Fig. 10.9]; the latter was derived from work with mutually saturated, mostly binary, systems [64]. Holdups under flooding conditions can be estimated as described in Section 3.2 in Chapter 4.

Watson et al. [66] believe that, for $\phi_d < 0.10$, the superficial slip velocity [67] is a better correlating parameter than is the characteristic velocity.

5.3.3. Flow Rates at Flooding

Houlihan and Landau [68] tested various relationships and recommend that due to Sakiadis and Johnson [69], rewritten as ($\mu = \text{lb}_m\ \text{hr}^{-1}\ \text{ft}^{-1}$); σ = dynes/cm)

$$1 + 0.835\left(\frac{\rho_d}{\rho_c}\right)^{0.25}\left(\frac{V_d}{V_c}\right)^{0.5} = 1.06c\left[\left(\frac{V_{cf}^2 a_p}{g\epsilon^3}\right)\left(\frac{\rho_c}{\Delta\rho}\right)\mu_c^{0.25}\sigma^{0.25}\right]^{-0.25} \quad (36)$$

The test data were limited to cases in which $D > 3$ in. and $D/d_p \geqq 6$ and were almost all for mutually saturated phases with no mass transfer. Values of c are as follows [68, 69] ($a_p = \text{ft}^{-1}$):

Packing	c
Raschig rings	$0.87\ \epsilon^{0.0068}/a_p^{0.043}$
Berl saddles	$1.2\ \epsilon^{0.78}/a_p^{0.0351}$
Lessing rings	$1.02\ \epsilon^{0.0068}/a_p^{0.043}$
Spheres	$0.95\ \epsilon^{0.0068}/a_p^{0.043}$

Although limits for the correlation are proposed [69, Fig. 8], it appears that few data lie beyond them [68].

It is suggested [68] that, although Eq. (36) is best overall, Eq. (13) in Dell and Pratt [61] is preferable for Raschig rings with liquids of viscosity below 6 cP. This recommendation refers to the original equation [61], and not to the equation as modified in an addendum [70] to Ref. 61.

Watson et al. [66] object to including a_p in correlations when the packing is wetted by the continuous phase. They propose a flooding correlation for Raschig rings based on data covering a wide range of physical properties. However, D/d_p was only 5.3.

5.3.4. *Mass Transfer*

Many published mass-transfer results are from towers of small D, and many are uncorrected for the effect of axial mixing on the driving force, and for end effects. Treybal [13, Table 21-3] gives selected sources of data.

Laddha and Degaleesan [15, p. 297] use the following equation for prediction of $K_c a$; division of the right side by m gives $K_d a$ ($m = dC_d^*/dC_c$):

$$K_c a = \frac{0.06\phi_d(1 - \phi_d)}{[(a_p/\epsilon^3 g)(\rho_c/\Delta\rho)(\sigma/\Delta\rho\, g)]^{0.5}[(N_{\text{Sc},c})^{0.5} + (1/m)(N_{\text{Sc},d})^{0.5}]} \quad (37)$$

In an example problem [15, p. 297] this $K_c a$ is used to calculate the column height for plug flow, with the implication that K_c has been corrected for axial mixing. If this is not completely so, a conservative design should result from using Eq. (37) and allowing for axial mixing subsequently. This correlation is said [15, p. 290] to apply to both directions of transfer; however, ϕ_d changes with direction (Section 3.2.1, Chapter 4).

Alternatively, the tentative correlation

$$\frac{k}{v_k} = c\left(\frac{d_{vs}^0 v_k}{D_v}\right)^{-0.5} \quad (38)$$

due to Pratt and Anderson [71] can be used to obtain k_c ($c = 0.58$) and k_d ($c = 0.50$). Prediction of d_{vs}^0 and v_k is a source of possible error in the k values [71]; however, the former is given reasonably well by [62]

$$\frac{d_{vs}^0 \Delta\rho\sigma}{\mu_c^2} = 1.42\left(\frac{\Delta\rho\sigma^3}{\mu_c^4 g}\right)^{0.475} \quad (39)$$

Use of an interfacial area incorporating a "shielding" correction is recommended [71]:

$$a_e = \frac{2.0}{d_{vs}^0}\left(\frac{V_d}{v_k}\right)^{0.66} \quad (40)$$

5.3.5. *Axial Mixing*

Few data are available. Wen and Fan [72] have recorrelated the data due to Vermeulen et al. [73] and Watson and McNeese [74] and give the following equation for predicting E_c for various packing sizes and shapes:

$$\frac{\epsilon V_c d_p}{E_c} = (1.12)(10^{-2})\, Y^{-0.5} + (7.8)(10^{-3})\, Y^{-0.7} \quad (41)$$

with

$$Y = \left(\frac{\psi\mu_c}{d_p V_c \rho_c}\right)^{0.5}\left(\frac{V_d}{V_c}\right) \quad (42)$$

The Raschig ring data [73] also have been recorrelated with the sphericity omitted [71]. Data for E_d are of doubtful validity, and axial mixing in the dispersed phase can be assumed to be zero pending further study [71].

5.4. Design and Scale-Up

Earlier remarks about the desirability of a pilot plant study preceding the design of the prototype should be taken particularly seriously in the light of the complications inherent in packed towers.

The reader should review the general recommendations related to design given in describing packed columns in Section 5.1.

5.4.1. Packing

The d_p chosen should be above the critical [Eq. (11) in Chapter 4], but less than $0.125D$. The fact that this is true must be checked later, after D has been calculated. Next, a_p and ϵ should be obtained. Published values are available [5, p. 493]. However, these vary with manufacturer and with the way the packing is installed, including the degree of settling achieved. Therefore, values appropriate to the particular installation should be used if possible.

5.4.2. Flooding

The equations recommended in Section 5.3.3 provide V_{cf}; V_{df} is calculated from the known V_d/V_c ratio (see the first paragraph in Section 3.4).

5.4.3. Holdup; Tower Diameter [5, pp. 489, 498; 8; 13; 64]

A ϕ_d in the range 0.1–0.25 is assumed; low values are appropriate if V_d/V_c is below about 0.5. The value of V_c implied by the assumption is calculated next by Eq. (35) (with v_k from Table 2 in Chapter 4). This V_c should not exceed $0.5V_{cf}$, or probably, $0.4V_{cf}$ for systems of high interfacial tension. If necessary, a lower holdup is chosen and the calculations repeated. The tower diameter is given by

$$D = \left(\frac{4Q_c}{\pi V_c}\right)^{0.5} \tag{43}$$

5.4.4. Drop Formation

Eckert [60] suggests that d_o be 5 to 6 mm; below 1–3 mm clogging can be a problem [5, p. 499; 13]. In any case, d_o should be sufficiently large that d_e for the drops as formed (Section 3.3.1) exceeds the value given by Eq. (34). The total number of nozzles is calculated by dividing Q_d by $(v_o)[(\pi/4)\,d_o^2]$. If this number will not fit into the space available in the tower, some adjustment of d_o will be needed.

5.4.5. Interfacial Area

With d_{vs} available from Eq. (34) and ϕ_d chosen, a is calculated from

$$a = \frac{6\epsilon\phi_d}{d_{vs}} \tag{44}$$

or a_e from Eq. (40) is used if appropriate.

5.4.6. Mass-Transfer Coefficients

For a design based on phase coefficients, those chosen should, if possible, be values corrected for end effects and axial mixing. These then are combined [Eq. (58) in Chapter 3] and, with a (or a_e) from Eq. (44) [or (40)], K_ca (or K_ca_e) is calculated. Alternatively, K_ca can be predicted from Eq. (37). The capacity coefficient then is converted to $(HTU)_{oc}$ [Eq. (32)], appropriate to Pratt's [56] method (since transfer from the continuous phase is assumed here). Any mass transfer during drop formation and coalescence not allowed for in the correlations will provide some margin of safety in the design.

5.4.7. Design for Height, with Allowance for Axial Mixing

The methods due to Pratt [56, 57] are recommended. For example, for a linear equilibrium curve [56] and no axial mixing of the dispersed phase, Eq. (58) of Ref. 56 gives H on the basis that transfer is to be from the continuous to the dispersed phase and Pratt's F is not unity. The required $N'_{\text{Pe},c}$ is obtained from Eq. (41) (right side divided by ϵ).

5.4.8. Scale-Up

Scale-up should be approached with caution even if pilot plant studies have been done.

Extrapolation of gas–liquid work by Groenhof [75, 76] to the liquid–liquid case suggests that in scale-up the number of distribution points per square meter of packed area should be kept constant for each phase. Rosen and Krylov [77] stress for scale-up, the importance of insuring uniform distribution, high flow rates, and division of the packing into sections with good mixing of each phase between.

Treybal [5, p. 487] suggests that the prototype should have the same size and type of packing as in the pilot column. One result is that following the 1:8 rule in the pilot tower ensures that the rule will be followed in the full-scale column also; however, use of large (e.g. 38-mm) packing in the prototype then implies a D of at least 30 cm for the pilot tower.

If the packing size is not changed in scaling up, and the other preceding suggestions are followed, the allowable V values, the K values, and a could remain about constant. Then D will increase proportionally to $Q^{0.5}$, and H will stay the same, if no change occurs in axial mixing effects. The final tower, then, will be wide relative to its height [78]. However, if larger packing were to be used in the prototype, the allowable V_c and V_d would be increased. Typically, the effect of such increases approximately cancels the adverse mass-transfer effects to be expected from the larger packing [78], and the scaled-up volume will be the same regardless of whether the packing size is increased. However, with higher V_c and V_d, the prototype can be relatively tall and slender. Jordan [78] believes that, economically, it probably makes sense to go from about 0.5-in. packing in the pilot column to about 1.0 in. in the prototype.

The effect on E_c of increasing D probably will be small if the precautions suggested are followed.

Rosen and Krylov [77, Fig. 4] suggest that in industrial columns the *HTU* is decreased by increasing the flow rates of the phases, whereas the reverse is true in laboratory towers.

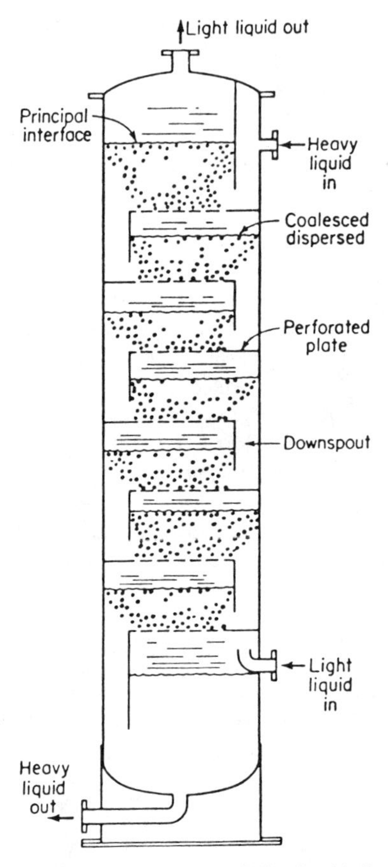

Figure 3. Sieve-tray extractor (light liquid dispersed). (From *Liquid Extraction*, 2nd Ed., by Robert E. Treybal. Copyright © 1963 by the McGraw-Hill Book Company, Inc. Used with the permission of McGraw-Hill Book Company.)

6. PERFORATED-PLATE (SIEVE-PLATE) COLUMNS

6.1. Description

Perforated-plate columns consist of a series of trays, provided with orifices, and placed one above the other in a vertical shell. Drops of dispersed phase form from the orifices, and for $\rho_d < \rho_c$ (assumed throughout this discussion), flow upward through a layer of continuous phase and then coalesce under the next plate of the tower. The continuous phase flows horizontally across a plate, and then through a downspout to the plate below. Figures 3 and 4 show typical arrangements. The second figure shows bypasses for "rag", an emulsionlike material produced by the accumulation of dust and so on. A rag draw-off should be provided also at the main interface of the column [7]. Further discussion will be confined to towers of the type described here; however, other types exist [59]. Advantages of sieve-tray columns include elimination of axial mixing between trays, and whatever mass-transfer benefits accrue from repeated coalescence and redispersion of drops.

As in spray towers, spreading of the drops across the plate during formation should be avoided, such as by using punched holes with the burr left in place. About 55–60% of the plate area is perforated [78]; holes are omitted opposite the downspout from above and near the downspout to the plate below. No weirs are provided; the downspout is flush with the plate from which it carries continuous phase. It extends well below the layer of dispersed phase under the plate.

Holes of d_o from about 8 down to 3 mm are used and are located in square or triangular arrays on about 16-mm centers.

Towers of D up to 12 ft are common [13, 59], with tray spacings of 0.5–2 ft; about 1.5 ft, the minimum for providing entry ports for cleaning [5, p. 503], is a spacing used frequently in large columns [41].

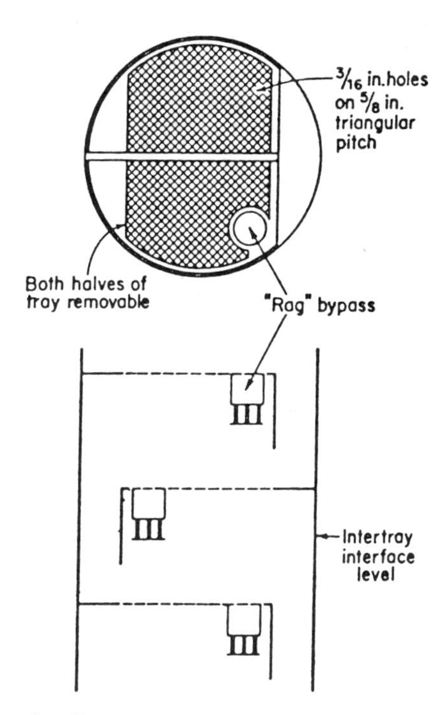

Figure 4. Sieve-tray design. (From *Liquid Extraction*, 2nd Ed., by Robert E. Treybal. Copyright © 1963 by the McGraw-Hill Book Company, Inc. Used with the permission of McGraw-Hill Book Company, and reprinted with permission from *Ind. Eng. Chem.*, **44**, 2253 (1952) by F. D. Mayfield and W. L. Church, Jr. Copyright 1952 by the American Chemical Society.)

Laddha and Degaleesan [15, p. 305] state that the raffinate usually is dispersed, presumably to avoid possible premature flooding due to slowness of coalescence below a plate [37, 79], a situation that might arise if the usual policy of dispersing the extract phase (Section 3.1) were followed. However, evidence is contradictory on the effect of direction of transfer on transfer efficiency [80, 81]. The value of m may have a practical bearing on this effect (see Section 6.5.7).

6.2. Applications

Perforated-plate columns can provide good efficiency and high throughputs, particularly if σ is low [7, 13, 25] and are well established in many different applications, especially in petroleum refining and petrochemical plants.

6.3. Interpretation of Performance Data

Although a sieve-tray tower is a series of stages, each individual tray is a differential-type contactor. Therefore performance data can include both stage efficiencies, and mass-transfer coefficients or *HTU*s. However, it is convenient to convert these last to efficiencies and use a stage-type design. Skelland and coworkers [24, 40, 41, 82] and Treybal [25] describe such conversions.

6.4. Performance Data

6.4.1. Overall Efficiency

The following empirical equation of Treybal [5, p. 507] as modified by Krishnamurty and Rao [13, 83] can be used for estimating values of E_o (ft lb$_m$ hr units):

$$E_o = \left[\frac{(10{,}047)\,(H_c)^{0.5}}{\sigma}\right]\left(\frac{V_d}{V_c}\right)^{0.42}\left(\frac{1}{d_o}\right)^{0.35} \tag{45}$$

The equation should serve reasonably well, at least for preliminary estimates, although many values of σ had to be estimated, and data were for $D < 9$ in. and showed considerable scatter around the correlation. The equation summarizes the effect of various operating parameters. For example, high σ produces low E_o, and increasing H_c increases E_o. However, above an H_c of 16 to 18 in. improvement is small [8], and spacings over 24 in. seldom are used [78].

6.4.2. Holdup and Flooding

The holdup for use in Eq. (26) can be estimated [25] by means of the Zenz correlation (Section 3.3.2) or directly from the second equality of Eq. (35) (with $\epsilon = 1$). For the latter method, v_k is obtained from Table 2 in Chapter 4; Chapter 4 suggests the use of the third equation in Table 2 of that chapter; however, earlier, Laddha and Degaleesan [15, p. 313] proposed the use of the first (spray-column) equation of that table. The value of V_c is zero because the continuous phase flows horizontally, and V_d is based on a cross-sectional area $S(= A_{ca})$, calculated by Eq. (18) (Table 1) [5, p. 508; 40; 41].

Holdup at flooding and the corresponding superficial velocities can be estimated as described in Section 3.2 of Chapter 4.

6.4.3. Mass-Transfer Coefficients and Areas

Tables 1 and 2 list equations giving mass-transfer coefficients, $n_o A_j$, A_{fa}, and A_{ca}; A_r is obtained from the following:

$$A_r = a_r(S)(H_c - h) \qquad (46)$$

with a_r calculated by Eq. (26). Table 2 almost always will be required because of the advantages of operating with the velocity in the perforations above the jetting point [7, 24].

6.4.4. *Axial Mixing*

There is little or no backmixing between compartments. Little information is available with respect to axial mixing coefficients other than preliminary suggestions [77, 84] for prediction equations.

Axial mixing probably is low in the dispersed phase [15, p. 316; 85]. The degree of mixing in the continuous phase must be between certain limits [78], that is, between plug flow across the tray, and thorough backmixing with uniform concentration. It has been stated that the former case may be approximated for large D and the latter for small D [5, p. 506; 15, p. 316]. However, current thinking seems to be that the continuous phase on a tray should be taken as completely mixed [25, 41]. Such an assumption will not always be exact but can aid in making the design for height conservative.

6.5. Design and Scale-Up

6.5.1. *Perforations; Drop Size*

Holes in the range 3–8 mm are usual [25]. A check should be made to see that Eq. (9) in Chapter 4 is satisfied by the d_o chosen, which should be toward the lower limit if σ is high and toward the higher limit if σ is low. The value of v_o to be used is calculated as v_{om} [Eq. (2)], and d_e is obtained as described in Section 3.3.1. However, Treybal [25] suggests checking that $(v_o = v_{om}) > 0.1$ m/s; his recommended procedure if it is less has been given in Section 3.4.1. (Skelland [41] gives 0.15–0.3 m/s as the normal range for v_o for plate columns.)

The number of orifices per plate is given by

$$n_o = \frac{4Q_d}{\pi d_o^2 v_o} \qquad (47)$$

The plate area required to provide n_o perforations is [41], for holes located at the vertices of equilateral triangles

$$A_{pz} = \frac{n_o \pi (\text{pitch})^2}{3.62} \qquad (48)$$

and for holes on a square pitch

$$A_{pz} = \frac{n_o \pi (\text{pitch})^2}{3.14} \qquad (49)$$

Typically a pitch of about 12–20 mm is used [25, 41].

6.5.2. *Downspout Area*

To avoid lowered efficiency due to recycling of dispersed phase and to reduce any tendency of the column to flood, the velocity of continuous phase in the downcomers should equal the terminal velocity of a drop of arbitrarily small diameter (e.g., 0.7 mm) [25] (for v_t, see Section 3.3.2; A_D equals Q_c divided by this v_t.)

6.5.3. *Tower Diameter*

The plate area below a downcomer must be kept free of perforations so as to prevent drops bypassing the plate above. In addition, perforations cannot be provided over the area of the downspout to the plate below and are omitted over a strip [5, p. 508], 1 in. wide beside it to reduce entrainment in the descending continuous phase. The total cross-sectional area of the tower then is $2A_D$ plus the area of this strip; that of a peripheral ring [5, p. 508], 2 in. wide, involved in attaching the tray to the shell; A_{pz}; and the area required for any rag bypass. This total area gives the tower diameter. Skelland and Conger [40, Table I] use a peripheral band width of (0.5) or (1.0) (pitch), depending on the size of the column.

6.5.4. *Depth of Coalesced Dispersed-Phase Liquid under Each Plate* [5, p. 503; 13; 86]

This is calculated by means of Eq. (23) in Chapter 4 and is sufficient to provide the driving force required for the flow of both phases. The first term on the right-hand side (for the continuous phase) does not include friction due to flow in the downspout (usually negligible) but allocates a total of 4.5 velocity heads to contraction and expansion losses and direction changes, as the fluid enters and leaves the downcomer. The second two terms (for the dispersed phase) allow for surface tension and orifice resistance. If the equation shows a coalesced layer of depth under about 50 mm, it is likely that all the perforations will not operate unless the tray is exactly level; in these circumstances a constriction can be used in the bottom of the downspout to increase the continuous-phase term. The effect is calculated as for an orifice [25].

6.5.5. *Flooding; Downcomer Length*

Equation (23) in Chapter 4 shows that increasing the flow rate of either phase will increase the depth of the coalesced layer under each plate. If either or both flows reach too high a rate, the dispersed phase will flow up the downcomers, and the tower will flood. To allow flexibility, each downspout should extend well below the bottom of the coalesced layer (e.g., 0.20 m below for a tray spacing of 0.35 m and a coalesced layer thickness of 0.05 m [25]). The downspout should not end so close to the plate to which it leads that the continuous phase must accelerate in flowing onto the plate; the area for flow between downspout and plate should exceed A_D.

6.5.6. *Distance Between Plates*

Downspout length is one factor influencing plate spacing. In addition, it is necessary that the separation be large enough that (1) jet breakup can occur and the resulting drops rise through a sufficient distance that suitable interfacial area is provided for mass transfer, (2) the velocity of the continuous phase as it flows across the plate is less than the velocity in the downcomer so as to keep entrainment low, and (3) that hand- or manholes can be provided to permit cleaning. The overall result is a spacing of 18–24 in. for commercial-size columns. In small towers (where cleaning may be less important) the spacing should be at least twice the thickness of the coalesced layer [15, p. 312].

6.5.7. *Tower Height: Calculation from Overall Efficiency*

A simple approach is first to use one of the well-known procedures to find the number of theoretical trays required. Then, division by E_o from Eq. (45) converts this to the number of actual trays, and multiplication of the latter by the tray spacing provides the active height of the column.

In practical operations V_E/V_R usually approximates the reciprocal of the distribution coefficient m [78]. Then Eq. (45) predicts a higher E_o for $m < 1$ if the extract is dispersed, and for $m > 1$ if the raffinate is dispersed. The result for raffinate dispersed appears to contradict the expectation that coalescence would reduce efficiency in this case; perhaps the relatively short distance of drop rise for one plate makes the tendency toward coalescence relatively unimportant (here $m = C_E^*/C_R$).

6.5.8. *Tower Height: Conversion of Mass-Transfer Coefficients into Stage Efficiencies*

This approach is more sophisticated than using E_o, but may not be much more accurate because of the tentative nature of present performance data.

The following equation, based on a derivation due to Treybal [25], can be used to obtain the Murphree dispersed-phase efficiency:

$$E_{Mdn} = \frac{K_{dfa}A_{fa} + K_{dr}A_r + K_{dca}A_{ca}}{Q_d + K_{dr}A_r/2 + K_{dca}A_{ca}} \tag{50}$$

In deriving this equation it is assumed that Q_d is unchanged, and C_c uniform over a stage (Section 6.4.4). Also, it is assumed that C_{dn}, the concentration of drops forming on plate n, is the same as that of drops coalescing under plate n and that C_{dn} can be used in calculating the driving force for drop formation and $C_{d,n+1}$ for drop coalescence, for the stage lying between plates n and $(n + 1)$. Mass-transfer coefficients for drop formation and coalescence, as well as for drop rise, must be included in the calculation of E_{Md} because new drops are formed and recoalesced once per tray, instead of once per column, as in a spray tower. The K values are calculated in the usual way from the phase coefficients (Tables 1 and 2). [See Eqs. (58) and (59) in Chapter 3.]

If the drops are formed at the ends of jets, $K_{dfa}A_{fa}$ should be replaced in Eq. (50) by

$$K_{dj}n_oA_j + K_{dfa}A_{fa} \tag{51}$$

in which the first term refers to mass transfer from the jets (Table 2) and the second to drop formation and detachment at the ends of the jets (Table 1) (see Section 3.3.4). Values of A_{fa} and A_{ca} are given in Table 1, and A_j in Table 2; each of these is specifically for use with the corresponding mass-transfer coefficients from these tables. The value of A_r is calculated by means of Eq. (46). Note that when K is used rather than K', the factor $(\rho/M)_{av}$ is omitted from the equations of Tables 1 and 2.

In principle, E_{Md} changes from plate to plate; however, in practice, the correlations used are tentative, and E_{Md} is a rough estimate [25]. Therefore, it usually should be accurate enough to calculate E_{Md} for approximately average conditions and use it throughout the tower to calculate the number of actual trays required, such as by construction of a pseudoequilibrium curve and stepping-off plates.

Skelland and his co-workers [41, 82] offer a similar but somewhat more elaborate procedure based on Tables 1 and 2 to obtain E_{Md}. They provide a "provisional design procedure" in the form of a computer program [40, 41] later modified [24] to allow for drop formation at the ends of jets. The procedure is best for systems of $\sigma > 25$ dynes/cm, [corresponding, e.g., to Eqs. (5), (6), (16), and (17) being for high-σ systems, but notice that many of the drop-rise equations of Table 1 are for low σ]. Skelland and Shah [87] suggest designing conservatively by choosing a system giving oscillating drops, but calculating on the basis of stagnant drops.

If the equilibrium curve and operating line are straight (or nearly so) equations are available to calculate E_o from a constant E_{Md} [5, pp. 403, 509; 8, p. 242; 15, p. 114].

6.5.9. *Scale-Up* [78]

If, as mentioned in Section 6.4.4, it can be assumed that increasing the tower diameter increases the tendency toward plug flow of the continuous phase, large plates should, in general, be more efficient than small. For complete backmixing of the continuous phase (small tray) $E_{Md} = E_{Mdp}$, but for plug flow of the continuous phase (large tray), E_{Md} is related to E_{Mdp} through an equation that includes mV_d/V_c as parameter [5, Eq. (11.43)] ($m = C_d^*/C_c$). By means of these relationships small and large trays of the same E_{Mdp} can be compared on the basis of the suggested assumption about continuous-phase backmixing. If E_{Mdp} is relatively high (e.g., >0.4), a large tray has a higher E_{Md} than a small tray, the percent increase in E_{Md} rising as mV_d/V_c goes from e.g., 0.5–2. For relatively low values of E_{MdP}, (e.g., <0.1), however, the small and the large tray have almost the same E_{Md} whatever the value of mV_d/V_c. Further work on backmixing is needed to clarify these matters.

In scaling-up from a small to a large-diameter tower, there often is an increase in tray spacing (H_c) to 24 in., for example. According to Eq. (45), an increased H_c should result in a higher E_o.

However, expected increases of efficiency with scale may not always be realized. There may, for example, be higher concentrations of surfactants in the larger-scale operation, and a consequent decrease in drop circulation, and/or interfacial turbulence, or there may be an interfacial-barrier effect.

7. MISCELLANEOUS TYPES OF NONMECHANICALLY AGITATED CONTACTORS

Various nonmechanically agitated contactors have been described and tested in addition to those already considered here. These include the spouted mixer–settler designed by Johnston et al. [88]; the air-agitated column by Fernandes and Sharma [89]; the fluid-diode contactor by Thompson [90], which makes use of flow-interference effects to convert an alternating applied pressure into countercurrent motion of the phases; the cocurrent ejector by Acharjee et al. [91]; and, related to the latter, pipeline, and other in-line mixers, including the Kenics Mixer®. In addition, electrically augmented extraction has been investigated by Bailes and Thornton [92].

Godfrey and Slater [93] review previous work of the cocurrent in-line mixer category and summarize some of the research done at Bradford. Pipeline contactors have been investigated by Watkinson and Cavers [94] and Shah and Sharma [95]. In a Kenics Mixer® the liquids are pumped cocurrently through a tube containing a series of elements that cut the flows and bring them into intimate contact as a well-blended dispersion. The elements consist of a series of short, fixed, lengths of sheet material each bent into a helical shape of length : diameter ratio of about 1.5. Right- and left-hand helical elements alternate down the mixer and are welded together in such a way that the trailing edge of each element is offset by 90° from the leading edge of the next succeeding element (Fig. 5). Chen [96], Chen and MacDonald [97], and Tunison and Chapman [98] provide good descriptions of this mixer. The latter reference gives some mass-transfer data, as does Chen [99]. He suggests also the form of a correlation for mass-transfer capacity coefficients. Several other nonmechanically agitated in-line mixers of this general type have been developed. These include the ISG (Interfacial Surface Generator) unit manufactured under license from the Dow Chemical Company, the Koch unit manufactured by Koch Engineering Company, and the LPD (Low Pressure Drop) mixer, like the ISG, developed by Dow. Brief descriptions of these have appeared in the literature [100]. Other examples include the Etoflo mixer manufactured by E. T. Oakes Ltd., Mac-

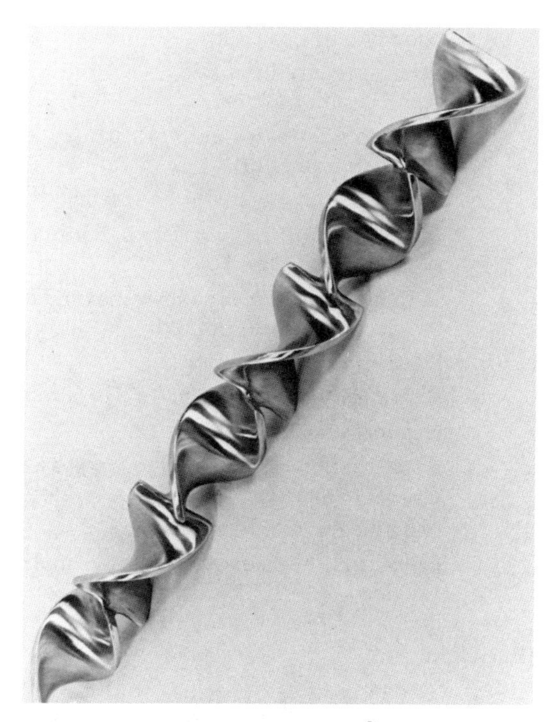

Figure 5. Assembly of Kenics-Mixer® elements. (Courtesy of Kenics Division, Chemineer, Inc.)

clesfield, Cheshire; the Komax mixer of Komax Systems, Inc., Carson, California; and the In-Line Blender produced by Greey Mixing Equipment Ltd., Toronto, Ontario.

NOTATION

A area, m^2 or ft^2

A_C cross-sectional area of the entire column at the cross section of interest, such as at the main, or column or stage interface, where drops coalesce before leaving the column or stage, m^2 or ft^2

A_{ca} total interfacial area in a spray or packed column, or under a tray of a sieve-plate column, associated with drop coalescence at a main interface, m^2 or ft^2

A_D cross-sectional area of one downcomer, m^2 or ft^2

A_j interfacial area between one jet of dispersed phase and the surrounding continuous phase, m^2 or ft^2

A_r total interfacial area associated with drop rise in a spray column, packed column, or one actual stage of a plate column, m^2 or ft^2

A_{pz} area of perforated zone of a plate, m^2 or ft^2

a specific interfacial area, m^2/m^3 or ft^2/ft^3 of active extractor volume

a_e effective value of a

a_p specific interfacial area of packing, m^2/m^3 or ft^2/ft^3 of packed volume

a_r as for a, but based on holdup of drops between where drops form and where drops coalesce; that is, for rising drops, m^2/m^3 or ft^2/ft^3

c a parameter in Eq. (36); a parameter in Eq. (38)

C concentration, $kg \cdot mol/m^3$ or $lb \cdot mol/ft^3$

C^* solute concentration in a liquid phase indicated by subscript, in equilibrium with the other liquid phase in the system, $kg \cdot mol/m^3$ or $lb \cdot mol/ft^3$

ΔC difference in concentration, $kg \cdot mol/m^3$ or $lb \cdot mol/ft^3$

D diameter of mixing vessel or extraction column, m or ft

D_h hydraulic diameter, m or ft [see Eq. (28)]

D_v diffusivity of solute in the appropriate phase, m^2/hr or ft^2/hr

d_e equivalent diameter: diameter of a sphere of volume equal to that of a drop, m or ft

d_j diameter of jet at breakup, m or ft

d_m mean drop diameter, m or ft

d_o nozzle, perforation, or orifice diameter, m or ft

d_p packing size, m or ft

d_{vs} volume/surface, or Sauter mean diameter $= \Sigma n d_e^3 / \Sigma n d_e^2$, m or ft

d_{vs}^0 characteristic value of d_{vs}, that is, the value at $V_c = 0$ and V_d approaching zero [62], m or ft

E superficial axial eddy diffusivity, superficial axial dispersion coefficient; that is, based on empty cross-section of column, m^2/hr or ft^2/hr

E_{Md} Murphree dispersed-phase stage efficiency, fractional

E_{Mdp} Murphree dispersed-phase point efficiency, fractional

E_o overall stage efficiency, fractional

g acceleration of gravity, m/hr^2 or ft/hr^2

H active height of extractor (height of packed section, etc.), m or ft

H_c height of a compartment, or plate spacing, m or ft

h height of the coalesced layer in a compartment, m or ft

HTU height of a transfer unit, m or ft

K overall mass-transfer coefficient, kg · mol hr^{-1} m^{-2} (ΔC), or lb · mol hr^{-1} ft^{-2} (ΔC)

K' as for K, but in the units $\Delta(x$ or $y)$ replaces ΔC

Ka overall mass-transfer capacity coefficient, kg · mol hr^{-1} m^{-3} (ΔC), or lb · mol hr^{-1} ft^{-3} (ΔC)

k individual, or phase, mass-transfer coefficient, kg · mol hr^{-1} m^{-2} (ΔC), or lb moles hr^{-1} ft^{-2} (ΔC)

k' as for k, but in the units $\Delta(x$ or $y)$ replaces ΔC

L_j length of jet, m or ft

ln natural logarithm

M average molecular weight, kg/kg · mol or lb_m/lb · mol

m distribution coefficient, dimensionless, for example, $m = C_E^*/C_R$; slope of equilibrium curve, for example, $m = dC_d^*/dC_c$

n number of, for instance, perforations, dimensionless; tray number, counting from the bottom of the column

N'_{Pe} superficial particle Peclet number for axial mixing = Vd_e/E or Vd_p/E, dimensionless

N_{Re} particle Reynolds number, for example, $d_e v_t \rho_c/\mu_c$, dimensionless; however, see special N_{Re} of Eq. (27), defined by following equations and text

$N_{Sc,c}$ Schmidt number for the continuous phase = $\mu_c/(\rho_c D_{vc})$, dimensionless

$N_{Sc,d}$ Schmidt number for the dispersed phase = $\mu_d/(\rho_d D_{vd})$, dimensionless

N_{We} Weber number, dimensionless

ΔN_{We} $(v_o^2 - v_j^2)\, d_o \rho_d/\sigma$

Q total volumetric flow rate, m^3/hr or ft^3/hr

S cross-sectional area of tower, less area of one downspout, m^2 or ft^2

t time, for example, of rise (or fall) of a drop, hr

t_e time of exposure, hr

t_{fa} time of formation of a single drop, hr

V superficial velocity, m^3 m^{-2} hr^{-1}, or ft^3 ft^{-2} hr^{-1}

v velocity, m/hr or ft/hr

v_j jetting velocity in nozzle, calculated by Eq. (1), m/hr or ft/hr

v'_j velocity in jet based on $(\pi/4)\,[0.5\,(d_o + d_j)]^2$, m/hr or ft/hr

v_k characteristic velocity, m/hr or ft/hr

v_o velocity in orifice, nozzle, or perforation, m/hr or ft/hr

v_{om} as for v_o, but that producing maximum interfacial area, m/hr or ft/hr

v_r relative velocity, m/hr or ft/hr

v_s slip velocity, m/hr or ft/hr

v_t terminal velocity, m/hr or ft/hr

x concentration in continuous phase, mole fraction

Y as defined by Eq. (42)

y concentration in dispersed phase, mole fraction

Greek letters

ϵ voids in a packed section, volume fraction

μ absolute viscosity, kg m^{-1} hr^{-1}, or lb_m ft^{-1} hr^{-1}

$\bar{\mu}$ relative viscosity; see Eqs. (27) and (29)

ν kinematic viscosity, m^2/hr or ft^2/hr

ρ density, kg/m^3 or lb_m/ft^3

$\Delta\rho$ positive difference in density, kg/m^3 or lb_m/ft^3

σ interfacial tension, kg/hr^2, dynes/cm [Eq. (36)], or lb_m/hr^2

σ_2 equilibrium interfacial-tension for a binary liquid mixture, kg/hr^2 or lb_m/hr^2

ϕ volume fraction of a liquid in the void volume of a vessel or an extractor

ψ sphericity of packing, dimensionless

Subscripts

av average

C entire column; see A_C

c continuous phase; or a compartment

ca drop coalescence

D downcomer

d dispersed phase

E extract

e cross section of entry of a stream; effective; equivalent; or exposure

f flooding

fa	drop formation
j	jet
lm	log mean value
m	maximum; with d = mean
n	*n*th stage; sequence of numbering stages is bottom to top
$n + 1$	$(n + 1)$th stage
o	orifice, nozzle, or perforation; or overall value
p	packing; or point value
R	raffinate
r	drop rise
s	slip
t	terminal

REFERENCES

1. I. Komosawa and J. Ingham, *Chem. Eng. Sci.* **33,** 541 (1978).
2. J. C. Godfrey, C. Hanson, and M. J. Slater, *Chem. Ind.* (17), 713 (September 3, 1977).
3. R. K. Warner, *Chem. Eng. Sci.* **3,** 161 (1954).
4. N. F. Murphy, J. E. Lastovica, and A. E. Skrzec, *AIChE J.* **2,** 451 (1956).
5. R. E. Treybal, *Liquid Extraction*, 2nd ed., McGraw-Hill, New York, 1963.
6. F. H. Blanding and J. C. Elgin, *Transact. Am. Inst. Chem. Eng.* **38,** 305 (1942).
7. F. D. Mayfield and W. L. Church, Jr., *Ind. Eng. Chem.* **44,** 2253 (1952).
8. R. E. Treybal, *Mass Transfer Operations*, 2nd ed., McGraw-Hill, New York, 1968.
9. R. Letan and E. Kehat, *AIChE J.* **13,** 443 (1967).
10. E. Kehat and R. Letan, *Ind. Eng. Chem. Proc. Des. Devel.* **7,** 385 (1968).
11. L. Steiner, M. Horvath, and S. Hartland, *Proceedings of the International Solvent Extraction Conference* (ISEC 77), Vol. 1, Canadian Institute of Mining and Metallurgy, Montreal, 1979, p. 366 (C.I.M.M. Special Volume 21).
12. A. R. Smith, J. E. Caswell, P. P. Larson, and S. D. Cavers, *Can. J. Chem. Eng.* **41,** 150 (1963).
13. R. E. Treybal, Liquid–Liquid Systems, in R. H. Perry and C. H. Chilton, Eds., *Chemical Engineers' Handbook*, 5th ed., McGraw-Hill, New York, 1973, pp. 21-3–21-23.
14. C. Hanson, Solvent Extraction: The Current Position, in C. Hanson, Ed., *Recent Advances in Liquid-Liquid Extraction*, Pergamon Press, Oxford, 1971, pp. 1–13.
15. G. S. Laddha and T. E. Degaleesan, *Transport Phenomena in Liquid Extraction*, Tata McGraw-Hill, New Delhi, 1978.
16. M. Horvath, C. Pikios, and S. D. Cavers, paper presented at the AIChE National Meeting, Boston, August 1979; accepted for publication as R&D note by *AIChE J.*
17. K.-H. Reissinger and J. Schröter, *Chem. Eng. (NY)*, **85** (25), 109 (1978)
18. M. Horvath, L. Steiner, and S. Hartland, *Can. J. Chem. Eng.* **56,** 9 (1978).
19. L. E. M. de Chazal and J. T. Ryan, *AIChE J.* **17,** 1226 (1971).
20. B. Bühler, B. Covelli, and F. Widmer, *Chimia* **31,** 307 (1977).
21. J. A. Izard, *AIChE J.* **18,** 634 (1972).
22. G. F. Scheele and B. J. Meister, *AIChE J.* **14,** 9, 15 (1968).
23. R. M. Christiansen and A. N. Hixson, *Ind. Eng. Chem.* **49,** 1017 (1957).
24. A. H. P. Skelland and Y.-F. Huang, *AIChE J.* **25,** 80 (1979).
25. R. E. Treybal, *Mass Transfer Operations*, 3rd ed., McGraw-Hill, New York, 1980.
26. F. W. Keith, Jr. and A. N. Hixson, *Ind. Eng. Chem.* **47,** 258 (1955).
27. S. Vedaiyan, T. E. Degaleesan, G. S. Laddha, and H. E. Hoelscher, *AIChE J.* **18,** 161 (1972).
28. S. Laddha, T. E. Degaleesan, and G. S. Laddha, *Indian Chem. Eng.* **16** (3), T59 (1974).
29. R. E. C. Weaver, L. Lapidus, and J. C. Elgin, *AIChE J.* **5,** 533 (1959).
30. B. O. Beyaert, L. Lapidus, and J. C. Elgin, *AIChE J.* **7,** 46 (1961).
31. F. A. Zenz, *Petrol. Refiner*, **36** (8), 147 (1957).
32. A. I. Johnson and L. Braida, *Can. J. Chem. Eng.* **35,** 165 (1957).
33. S. Hu and R. C. Kintner, *AIChE J.* **1,** 42 (1955).
34. M. Perrut and R. Loutaty, *Chem. Eng. Sci.* **27,** 669 (1972).
35. J. D. Thornton, *Chem. Eng. Sci.* **5,** 201 (1956).
36. G. W. Minard and A. I. Johnson, *Chem. Eng. Progr.* **2,** 62 (1952).
37. S. D. Cavers and J. E. Ewanchyna, *Can. J. Chem. Eng.* **35,** 113 (1957).
38. A. H. P. Skelland and S. S. Minhas, *AIChE J.* **17,** 1316 (1971).
39. A. H. P. Skelland and C. L. Hemler, unpublished work at University of Notre Dame, 1969, as described by Skelland [41].
40. A. H. P. Skelland and W. L. Conger, *Ind. Eng. Chem. Process Des. Devel.* **12,** 448 (1973).
41. A. H. P. Skelland, *Diffusional Mass Transfer*, Wiley-Interscience, New York, 1974, Chapter 8.
42. A. I. Johnson, A. Hamielec, D. Ward, and A. Golding, *Can. J. Chem. Eng.* **36,** 221 (1958).
43. T. Vermeulen, *Ind. Eng. Chem.* **45,** 1664 (1953).

44. A. H. P. Skelland and A. R. H. Cornish, *AIChE J.* **9,** 73 (1963).
45. A. H. P. Skelland and R. M. Wellek, *AIChE J.* **10,** 491 (1964).
46. C. L. Ruby and J. C. Elgin, *Chem. Eng. Progr. Symp. Ser.* **51** (16), 17 (1955).
47. F. H. Garner and M. Tayeban, *An. R. Soc. Esp. Fis. Quim., Ser. B* **56-B,** 479, 491 (1960).
48. A. H. P. Skelland and Y.-F. Huang, *AIChE J.* **23,** 701 (1977).
49. A. S. Zheleznyak and A. M. Landau, *Theor. Found. Chem. Eng.* **7,** 525 (1973).
50. J.-B. Wijffels and K. Rietema, *Transact. Inst. Chem. Eng.* **50,** 224, 233 (1972).
51. M. H. I. Baird, personal communication, 1979.
52. L. Steiner, M. Stemper, and S. Hartland, *Chem.-Ing.-Tech.* **50,** 389 (1978), and the full paper of which this is the synopsis.
53. R. Letan and E. Kehat, *AIChE J.* **15,** 4 (1969).
54. M. Perrut, R. Loutaty, and P. Le Goff, *Chem. Eng. Sci.* **28,** 1541 (1973).
55. C. B. Hayworth and R. E. Treybal, *Ind. Eng. Chem.* **42,** 1174 (1950).
56. H. R. C. Pratt, *Ind. Eng. Chem. Process Des. Devel.* **14,** 74 (1975).
57. H. R. C. Pratt, *Ind. Eng. Chem. Process Des. Devel.* **15,** 34 (1976).
58. J. M. Croix, C. Labroche, C. Lackme, and A. Merle, *Proceedings of the International Solvent Extraction Conference* (ISEC74), Vol. 1, Society of Chemical Industry, London, 1974, p. 913.
59. V. S. Morello and N. Poffenberger, *Ind. Eng. Chem.* **42,** 1021 (1950).
60. J. S. Eckert, *Hydrocarbon Process.* **55** (3), 117 (1976).
61. F. R. Dell and H. R. C. Pratt, *Transact. Inst. Chem. Eng.* **29,** 89 (1951).
62. R. Gayler and H. R. C. Pratt, *Transact. Inst. Chem. Eng.* **31,** 69 (1953).
63. J. B. Lewis, I. Jones, and H. R. C. Pratt, *Transact. Inst. Chem. Eng.* **29,** 126 (1951).
64. R. Gayler, N. W. Roberts, and H. R. C. Pratt, *Transact. Inst. Chem. Eng.* **31,** 57 (1953).
65. R. Gayler and H. R. C. Pratt, *Transact. Inst. Chem. Eng.* **29,** 110 (1951).
66. J. S. Watson, L. E. McNeese, J. Day, and P. A. Carroad, *AIChE J.* **21,** 1080 (1975).
67. J. S. Watson and L. E. McNeese, *AIChE J.* **19,** 230 (1973).
68. R. Houlihan and J. Landau, *Can. J. Chem. Eng.* **52,** 758 (1974).
69. B. C. Sakiadis and A. I. Johnson, *Ind. Eng. Chem.* **46,** 1229 (1954).
70. F. R. Dell and H. R. C. Pratt, *Transact. Inst. Chem. Eng.* **29,** 270 (1951).
71. H. R. C. Pratt and W. J. Anderson, *Proceedings of the International Solvent Extraction Conference* (ISEC77), Vol. 1, Canadian Institute of Mining and Metallurgy, Montreal, 1979, p. 242 (C.I.M.M. Special Volume 21).
72. C. Y. Wen and L. T. Fan, *Models for Flow Systems and Chemical Reactors*, Marcel Dekker, New York, 1975.
73. T. Vermeulen, J. S. Moon, A. Hennico, and T. Miyauchi, *Chem. Eng. Progr.* **62,** 95 (1966).
74. J. S. Watson and L. E. McNeese, *Ind. Eng. Chem. Process Des. Devel.* **11,** 120 (1972).
75. H. C. Groenhof, *Chem. Eng. J.* **14,** 181 (1977).
76. H. C. Groenhof, *Chem. Eng. J.* **14,** 193 (1977).
77. A. M. Rosen and V. S. Krylov, *Chem. Eng. J.* **7,** 85 (1974).
78. D. G. Jordan, *Chemical Process Development, Part 2*, Interscience, New York, 1968.
79. C. J. Geankoplis and A. N. Hixson, *Ind. Eng. Chem.* **42,** 1141 (1950).
80. K. Murali and M. R. Rao, *J. Chem. Eng. Data*, **7,** 468 (1962).
81. F. H. Garner, S. R. M. Ellis, and J. W. Hill, *AIChE J.* **1,** 185 (1955).
82. A. H. P. Skelland and A. R. H. Cornish, *Can. J. Chem. Eng.* **43,** 302 (1965).
83. R. Krishnamurty and C. K. Rao, *Ind. Eng. Chem. Proc. Des. Devel.* **7,** 166 (1968).
84. A. M. Rosen, Yu. G. Rubezhnyi, and B. V. Martynov, *Soc. Chem. Ind.* (*Engl. transl.*) (2), 66 (1970).
85. J. B. Angelo and E. N. Lightfoot, *AIChE J.* **14,** 531 (1968).
86. R. J. Bussolari, S. Schiff, and R. E. Treybal, *Ind. Eng. Chem.* **45,** 2413 (1953).
87. A. H. P. Skelland and A. V. Shah, *Ind. Eng. Chem. Proc. Des. Devel.* **14,** 379 (1975).
88. T. R. Johnston, C. W. Robinson, and N. Epstein, *Can. J. Chem. Eng.* **39,** 1 (1961).
89. J. B. Fernandes and M. M. Sharma, *Chem. Eng. Sci.* **23,** 9 (1968).
90. D. W. Thompson, *Can. J. Chem. Eng.* **48,** 236 (1970).
91. D. K. Acharjee, A. K. Mitra, and A. N. Roy, *Indian J. Technol.* **16,** 262 (1978).
92. P. J. Bailes and J. D. Thornton, *Proceedings of the International Solvent Extraction Conference* (ISEC 74), Vol. 2, Society of Chemical Industry, London, 1974, p. 1011.
93. J. C. Godfrey and M. J. Slater, *Chem. Ind.* (19), 745 (October 7, 1978).
94. A. P. Watkinson and S. D. Cavers, *Can. J. Chem. Eng.* **45,** 258 (1967).
95. A. K. Shah and M. M. Sharma, *Can. J. Chem. Eng.* **49,** 596 (1971).
96. S. J. Chen, *KTEK-1: The Static Mixer® Unit and*

Principles of Operation, Kenics Corporation, North Andover, MA, 1972.

97. S. J. Chen and A. R. MacDonald, *Chem. Eng. (NY)*, **80** (7), 105 (1973).

98. M. E. Tunison and T. W. Chapman, *Chem. Eng. Progr. Symp. Ser.* **74** (173), 112 (1978).

99. S. J. Chen, *KTEK-6: Interphase Heat and Mass Transfer Operations in the Kenics Mixer Unit*, Kenics Corporation, North Andover, MA, revised 1978.

100. Other Motionless Mixers, *Chem. Eng. (NY)*, **80** (7), 111 (1973).

11.1

PULSED PACKED COLUMNS

A. J. F. Simons
DSM/Central Laboratory
Geleen, The Netherlands

1. INTRODUCTION

When a separation problem concerns purification of a liquid or recovery of components from mixtures, the pulsed packed column (PPC) provides an effective and highly satisfactory answer. It is capable of handling large throughputs to any desired degree of efficiency in both small- and large-scale operations.

Mechanical agitation through pulsation produces shear forces in the liquid [1] as a result of which a good contact between the two phases is ensured. The power input needed for good mass-transfer conditions is relatively low. The total liquid holdup is pulsated without any moving parts in the column itself. The power input, regulated by the pulsating velocity (product of amplitude and frequency), is set to strike a balance between throughput and efficiency. DSM has applied the PPC since 1970 because with this contactor (1) the countercurrent principle is utilized, (2) the axial mixing is low, (3) the distribution of the dispersed phase over the cross-sectional area is uniform, (4) the specific mass-transfer area can be controlled, (5) the extraction efficiency has been found independent of the column diameter under hydrodynamically similar conditions, (6) the extraction process is more economical, and (7) mechanical problems (e.g., with the original type of pulsator) have been overcome. To obtain full advantage of these characteristics, the extraction unit and its pulsator should be properly designed. Large-scale PPCs of up to 3.0 m in diameter are already widely used with throughputs in excess of 200 m^3/hr.

2. TECHNICAL DESCRIPTION

The PPC (Fig. 1) comprises a vertical shell filled with packing, specifically, Raschig rings and a pulsator at the column base that imparts an up–down movement to the total liquid contents in the column.

2.1. The Column

Two liquid phases, one in the form of drops, pass through the column countercurrently to each other. The dispersed phase enters through a simple distributor to ensure moderate initial distribution of the dispersed drops over the cross-

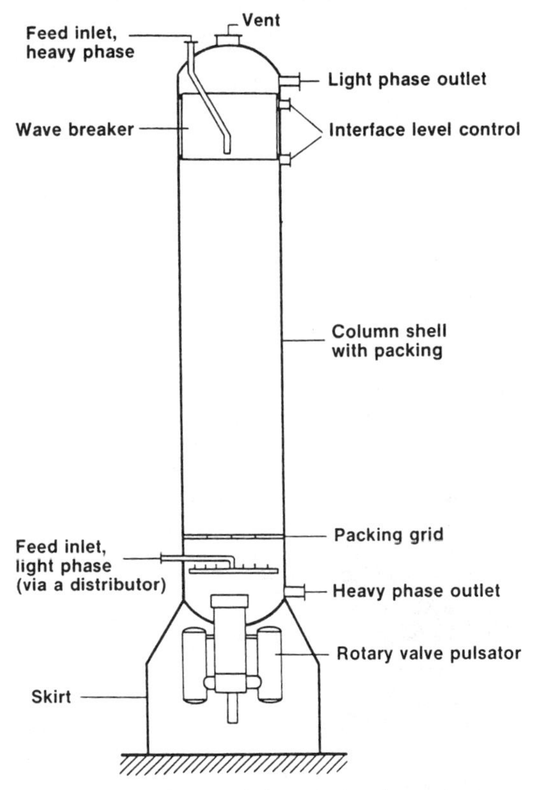

Figure 1. Diagram of the pulsed packed column.

sectional area. This distribution is effectively and properly maintained by the pulsing action, thus rendering redistributors superfluous. In the top or at the base of the column, the dispersed phase coalesces to form an interface layer.

For proper functioning of the contactor, the dispersed phase must be prevented from wetting the packing. If the dispersed phase does wet the packing material, it may flow as a ruptured film over the packing, causing a decrease in throughput. Choice of packing material (ceramic, polyethylene, carbon, steel, etc.) depends on the chemical and physical characteristics of the fluids to be handled. For columns with diameter $D \geqslant 0.1$ m, a packing size of $d_p = 0.025$ m is generally used. The available flow area then comprises over 70% of the total cross-sectional area.

If the density difference between the packing material and the continuous phase is low, it is advisable to provide a mechanical restraint on top of the packing stack to prevent this from moving as a result of pulsation forces. If ceramic material is used, the packing will not move in the normal range of pulsation velocities ($0.005 < A_p \cdot f < 0.0125$ m/s). In general, for a certain pulsation velocity, the pulse amplitude or the frequency can be fixed independently. For industrial columns, DSM has chosen a fixed frequency of abour 5400 cycles per hr.

Often there is a great concentration gradient along the column length, so that both the interfacial tension and/or the density difference between the phases may vary considerably. This calls for a locally variable energy dissipation, which can be realized by modifications such as different packing sizes in the stack, empty column sections, and variations in diameter.

2.2. The Pulsator

Driving an industrial PPC is a problem that should not be underestimated. Often, several tons of liquid must be pulsated, and this not only calls for application of large forces, but also gives rise to considerable pressure variations. This clearly brings out the need for proper insight into (1) the system of forces involved and the stresses these cause in the material, (2) the possibility of resonance of the various column parts, and (3) the civil engineering requirements to be satisfied by the foundation. It will be evident, therefore, that the pulsator is an essential part of a PPC.

The pulsator generally used in industry is the pump pulsator [2], the function of which is apparent from Fig. 2. A rotating basket valve, connected to the base of the extraction column, alternately links the column with a pair of suction vessels and a pair of pressure vessels; this determines the frequency of pulsation. The basket itself has two holes in the wall opposite each other. The casing, in which the basket turns, has four ports, so arranged that each pair of opposite drums can be lined up with the holes in the basket. Two lines are provided, connecting one pair of drums with the suction end and the other pair with the discharge end of a recycling pump; the variable volumetric pump flow rate controls the length of the pulse stroke.

The power requirements for this method of pulsation are reported to be of the same order as other mechanically agitated columns [1]. Depending on process scale and process conditions, other pulsing mechanisms [3–7] may also be applied, such as air pulsing. Pneumatic pulsators are used in columns up to 2 m in diameter and with packed heights of up to 10 m [3, 4].

3. INDUSTRIAL APPLICATIONS

At present PPCs are in operation for research and process development and in industrial plants.

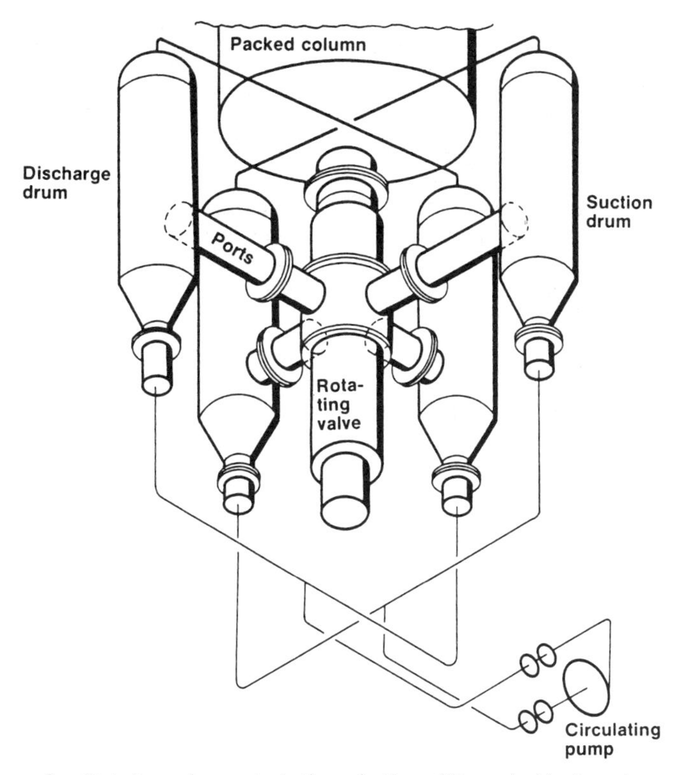

Figure 2. Rotating valve controls the pulsation of liquor inside the column.

Diameters range from 0.03 m for the smallest laboratory-scale applications to 3.0 m in industrial use. Most of the large-scale applications are found at DSM and in Stamicarbon processes (Stamicarbon is a wholly owned subsidiary of DSM). Some examples are given in Table 1. In literature from the Soviet Union, several publications [8–11] have appeared on pulsation equipment; from these it can be concluded that pulsation columns of various diameters (0.6–1.6 m) operate under various conditions. But most probably these columns are equipped with KRIMZ packing [12], which in fact is a stack of sieve plates of special design. It is reported [11] that for $D < 1$ m, the scale-up factor for this stack of plates is 1, and for diameters of up to 2 m, this factor increases to 1.5–1.8.

TABLE 1 PULSED PACKED COLUMNS IN OPERATION

Column Diameter, m	Height of Packing, m	Solute
1.1	6	Caprolactam
1.9	10	Caprolactam
2.7	6	Oxime–anone
2.1	9	Oxime
3.0	9	Oxime
1.0	6	Methanol
1.1	4.5	*N*-Butanol–isopropylalcohol

4. HYDRODYNAMIC PERFORMANCE AND MASS-TRANSFER DATA

Extensive investigations have been carried out to develop a process engineering model for the PPC with useful correlations for the various parameters. Such a model enables a PPC to be designed without, or with only a few, additional experiments. Attempts have been made to adapt the correlations as closely as possible to existing know-how and theories [13].

4.1. Holdup Correlation

The drop size spectrum generated in a PPC is described by the Sauter mean diameter d_m, to which a characteristic velocity V_0, has been attributed:

$$V_r = \frac{V_d}{\varphi \cdot \theta} + \frac{V_c}{(1-\varphi) \cdot \theta} = (1-\varphi) \cdot V_0 \quad (1)$$

This approach was originally suggested for unpulsed columns [14] and is further discussed in Chapter 4, Section 3.1.

The characteristic velocity V_0 is almost independent of the holdup φ of the dispersed phase but does depend on the physical properties of the liquid–liquid system and on the amplitude, frequency, and shape of the pulse. The characteristic velocity V_0 decreases with increasing pulsation velocity. With increasing pulsation velocity the maximum allowable phase velocities, above which flooding occurs, decrease. The characteristic velocity V_0 is determined mainly by the drop diameter d_m and by the difference in density, $\Delta\rho$, between the two phases. Some examples of results obtained with different systems (without mass transfer) are given in Fig. 3. From this figure two distinct regimes can be observed:

Region 1. Small droplet size; V_0 varies linearly with d_m and is almost solely dependent on amplitude and frequency; it is virtually independent of packing size.

Region 2. Large drop size; V_0 is less dependent of d_m. However, it varies with packing diameter d_p.

Theoretical prediction of V_0 from a knowledge of single-drop behavior under pulsing conditions has not yet been satisfactorily achieved [15]. Therefore the following empirical correlations were presented [13] for the two regions:

For region 1

$$V_0 = 6.32 \cdot 10^{-3} \cdot d_m^{0.727} \Delta\rho^{0.815} \cdot (2A_p \cdot f^2)^{-0.254} [S(1 - \theta)]^{-0.184} \cdot \mu^{-0.35} \tag{2}$$

if

$$L = d_m^{0.787} \cdot \Delta\rho^{0.255} \cdot (2A_p \cdot f^2)^{-0.144} \cdot [S(1 - \theta)]^{0.426} \leqslant 0.406 \tag{3}$$

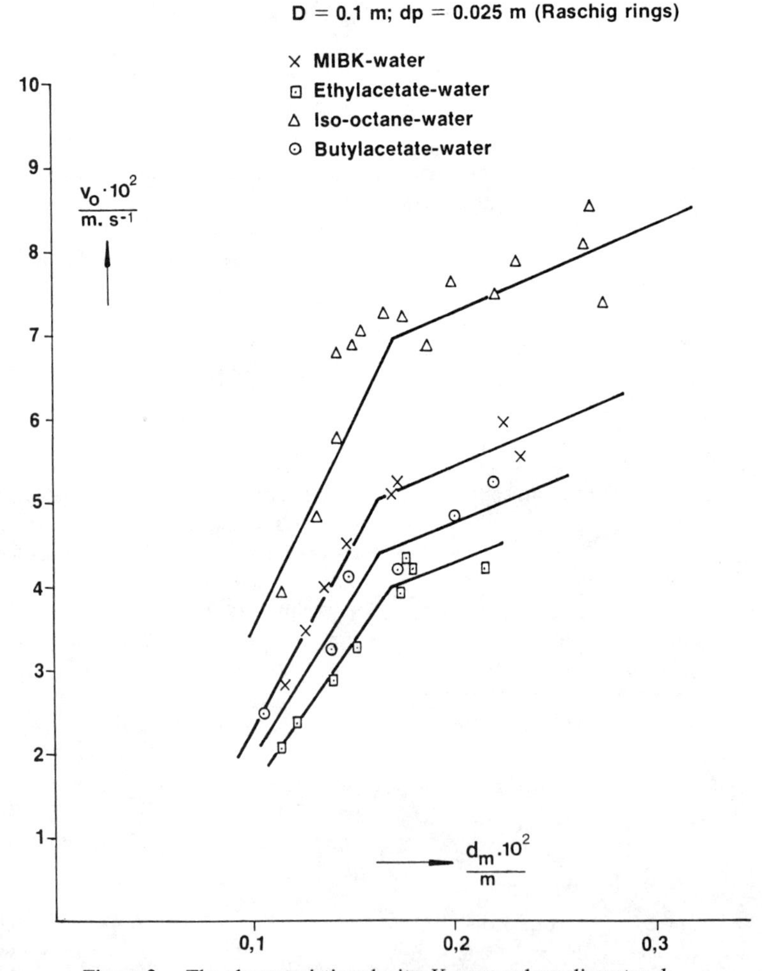

Figure 3. The characteristic velocity V_0 versus drop diameter d_m.

For region 2

$$V_0 = 2.57 \cdot 10^{-3} \cdot d_m^{-0.06} \cdot \Delta\rho^{0.56} \cdot (2A_p \cdot f^2)^{-0.11} [S(1 - \theta)]^{-0.61} \cdot \mu^{-0.35} \quad (4)$$

if

$$L > 0.406$$

In addition, the relationship

$$\frac{d_m}{d_p} \leqslant 1.37 \cdot \Delta\rho^{-0.5} \quad (5)$$

must be satisfied, and it must be emphasized that the preceding correlations hold only for Raschig rings and $0.55 < \theta < 0.80$. It should also be emphasized that Eqs. (2)-(5) are not dimensionless and that all quantities must be given in SI units (International System of Units).

These equations are based on data taken from columns ranging in diameter from 0.05 to 0.75 m and packed heights of 1-5 m. Mutually saturated systems were used covering an extensive range of physical properties, such as the interfacial tension σ varying from 5 to 50 mN/m.

4.2. Flooding Velocities

During countercurrent flow of the phases the velocities of the dispersed and continuous phases V_d and V_c, respectively, can be increased up to specific maxima. Beyond these, flooding occurs. Qualitative relationships between φ and variables such as V_d, V_c and $A_p \cdot f$ are illustrated elsewhere [16]. The literature contains several equations for describing the flooding velocities [17], but practically all of these are based on holdup correlations similar to Eq. (1). Equations applicable for a PPC are

$$V_{d,f} = 2 \cdot \theta \cdot V_0 \cdot \varphi_f^2 (1 - \varphi_f) \quad (6)$$

$$V_{c,f} = \theta \cdot V_0 \cdot (1 - 2 \cdot \varphi_f)(1 - \varphi_f)^2 \quad (7)$$

which have been obtained after differentiation of Eq. (1). In Eqs. (6) and (7), φ_f is a function only of the ratio γ of the linear velocities:

$$\gamma = \frac{V_{d,f}}{V_{c,f}} = \frac{V_d}{V_c} = \frac{Q_d}{Q_c} \quad (8)$$

Mass transfer may have a great influence on V_0, for example, because of interfacial tension gradients (Marangoni effect). The solute concentrations at either end of the column may differ widely, and hence V_0 may change with the height of the column. In Fig. 4 maximum throughputs have been given as a function of the phase ratio, which can be used for a rough estimate of the column diameter. In the region of the flooding point a column cannot be operated; therefore,

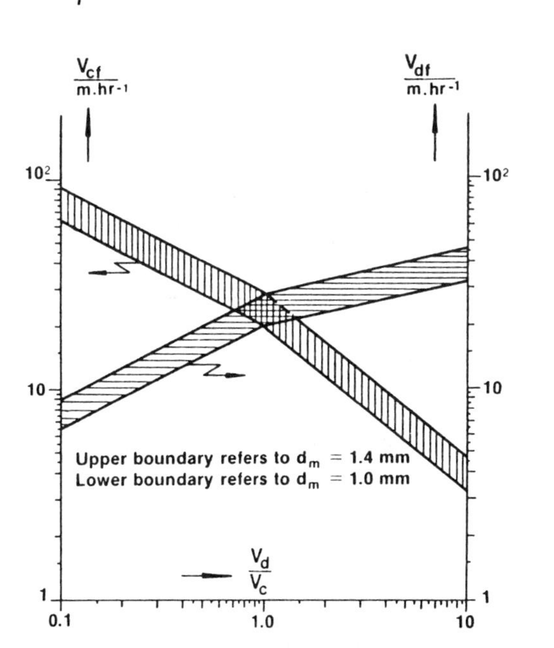

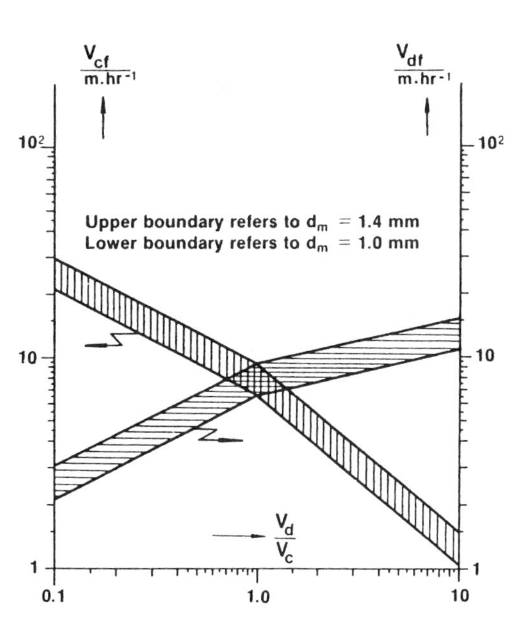

Figure 4. Throughput capacity of a pulsed packed column.

actual velocities are recommended that are 0.65–0.9 times the flooding velocities. The magnitude of this factor to be applied for the PPC depends mainly on the variations in the physical parameter values.

There are only a few other correlations available in the literature from which flooding points or holdup values can be predicted [18–20]. In general, all call for additional experimental investigations with the actual systems.

4.3. Drop Size

Equations (2) and (4) contain the drop diameter d_m and the pulsation velocity $A_p \cdot f$. Within the range of pulse rates normally used in a PPC, the reciprocal drop diameter and the pulse rate are linearly related; that is

$$\frac{1}{d_m} = \frac{1}{d_0} + c_1 \cdot A_p \cdot f \tag{9}$$

The term d_0 denotes a hypothetical drop diameter in the absence of pulsation, but in general its value is larger than the actual diameter without pulsation. It is found to be well described by the expression

$$d_0 = 1.39\left(\frac{\sigma}{\Delta\rho \cdot g}\right)^{0.5} \tag{10}$$

Equation (10) is very similar to the relationship derived for a simple unpulsed packed column when the liquid velocities are small [14].

Mass transfer can have a pronounced effect on the drop diameter, especially if raffinate is the dispersed phase. For determination of the interfacial area a, the influence of mass transfer on the diameter must be incorporated; the interfacial area has been defined as

$$a = \frac{6 \cdot \theta \cdot \varphi}{d_m} \tag{11}$$

The relationship between drop diameter, physical parameters, and working variables in a PPC, as based on the most extensive and larger-scale experimental work [13], is defined by the semiempirical correlation

$$\frac{1}{d_m} - \frac{1}{d_0} = 6700\left(\frac{1-\theta}{\theta}\right)^{0.95}\left(\frac{2\rho_c \cdot A_p \cdot f}{\mu_c \cdot S}\right) \cdot \left(\frac{\mu_c^2 \cdot S}{\sigma \cdot \rho_c}\right)^{0.5}\left(\frac{\sigma \cdot S^2}{\Delta\rho \cdot g}\right)^{0.23} \tag{12}$$

The applicability of Eq. (12) for describing the droplet diameter is shown in Fig. 5 (see Table 2 also). For most of the experimental values, the spread around this correlation is less than 15%. Comparison of available correlations in the literature shows that there are considerable deviations among them [13]; see also Chapter 4, Section 2.4.

4.4. Axial Mixing

Regarding axial mixing in the continuous phase, a distinction must be made between the contributions from the pulsation velocities and the contributions from the phase flow rates.

Axial mixing in the continuous phase, in the absence of a dispersed phase, can be described as follows [13]:

$$E_c = \phi_1 \cdot \frac{2 \cdot A_p}{\theta} \cdot f \cdot d_p + \phi_2 \cdot c_2 \cdot V_c \cdot d_p \tag{13}$$

with

$$\phi_1 = f(A_p, \theta) \tag{14}$$

and

$$\phi_2 = f(A_p, f, \theta, \phi_1) \tag{15}$$

In the region of practical pulsation velocities and phase rates, the presence of the dispersed phase increases the axial mixing coefficient E_c by about 20%. The degree of axial mixing in the dispersed phase is a simple function of the holdup of the dispersed phase [13]; that is

$$E_d = \phi_3 \cdot \theta \cdot \varphi \tag{16}$$

with

$$\phi_3 = f(d_p) \tag{17}$$

To ensure a uniform distribution of drops over the cross-sectional area, the pulsation velocity $A_p \cdot f$ must exceed a given minimum. If the pulsation velocity is below the minimum, axial mixing is increased by hydraulic nonuniformities (see Chapter 6, Section 8.1.3). This minimum depends on the geometry of the internals as well as on the physical parameters of the systems. In general, $(A_p \cdot f)_{\min}$ represents a smaller value than the most economical, or the optimum pulsation rate. It has been observed in tests at DSM that backmixing for a given system in both the continuous and dispersed phases is independent

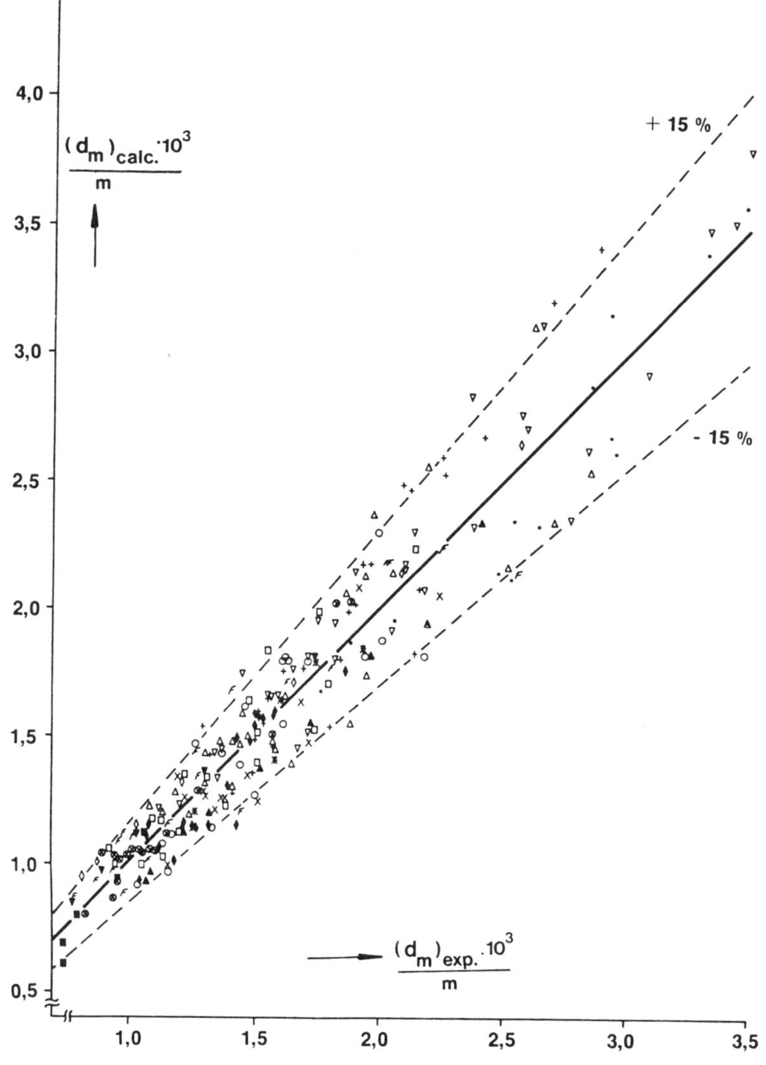

Figure 5. Comparison of calculated and measured drop diameters.

of column diameter, with the value of E_d always being smaller than the value of E_c [13, 21]. The independence of backmixing and diameter is subject to the preservation of hydrodynamic similarity, that is, the same type and size of packing, pulsation velocities, flow rates per unit area, and so on. The principal conclusion from this is that *the scale-up factor for column height is unity*. A large column may thus be treated as a bundle of columns of smaller diameter and the same height whose collective cross-sectional area is equal to that of the large column.

4.5. Mass Transfer

Fundamental mass-transfer studies have been carried out with industrial systems as well as with model systems, for example:

Benzene–caprolactam–water
Toluene–acetone–water
MIBK–acetone–water
Ethylacetate–acetone–water
Butylacetate–acetone–water
MIBK–acetic acid–water

The influence of both pulsation velocity and throughput on the height of a transfer unit (HTU) has been determined. For each set of fixed parameters (C_F, ϵ, d_p) there exists an optimal combination of throughput and pulsation velocity.

From the measured HTU values, true mass-

TABLE 2 DATA FOR FIG. 5

D	d_p	Material of Raschig Ring	Key	System
0.05	0.01	Polyethylene	■	water–hexane
0.05	0.015	Carbon	▲	water–toluene
0.05	0.015	Ceramic	⊗	MIBK–water
			▼	ethylacetate–water
			◇	isooctane–water
			◆	butylacetate–water
0.05	0.016	Polyethylene	⊙	water–toluene
0.10	0.012	Ceramic	*	MIBK–water
			⊡	MIBK–water–acetic acid
0.10	0.025	Ceramic	×	MIBK–water
			□	ethylacetate–water
			△	isooctane–water
			○	butylacetate–water
			●	toluene–water
			▽	toluene–water ($0.0033 < A_p < 0.0165$ m)
			▽	benzene–39% ammonium sulfate solution in water
0.23	0.025	Ceramic	+	toluene–water
0.23	0.035	Ceramic	△	benzene–water

transfer coefficients have been calculated and compared with values predicted by theoretical models. The following procedure was adopted. The influence of axial mixing on the decrease of the driving force can be incorporated in the height of an apparent transfer unit, so

$$H \equiv \text{NTU}_{od} \cdot (\text{HTU}_{od,\,\text{plug}} + \text{HDU}_{od}) \quad (18)$$

with

$$\text{HDU}_o = f\,[(N_{\text{Pe}})_c, (N_{\text{Pe}})_d] \quad (19)$$

Depending on the calculation procedure for HDU_o, in Eq. (18) a clear distinction must be made between HDU_{oR} and HDU_{oE} [28]. The theoretical height of a transfer unit $\text{HTU}_{od,\,\text{plug}}$, based on the dispersed phase, is defined as

$$\text{HTU}_{od,\,\text{plug}} = \frac{V_d}{K_{od} \cdot a} \quad (20)$$

with the overall mass-transfer coefficient given by

$$\frac{1}{K_{od}} = \frac{1}{k_d} + \frac{1}{k_c \cdot m'} \quad (21)$$

If $m' < 100$, as is often the case, the resistance to mass transfer in the continuous phase is negligible. In general, k_c has values in the same order of magnitude as mass-transfer coefficients in the continuous phase of stirred gas–liquid reactors, that is, $k_c \sim 3 \cdot 10^{-4}$ m s^{-1}.

For mass transfer in the dispersed phase, the assumption of rigid drops simplifies the equation of continuity for the solute considerably. Nevertheless, the resulting general expression for k_d is complex. But a mean value for the mass-transfer coefficient approaches a limiting value, depending on both N_{Bi} and ϵ', for greater values of contact time. In Fig. 6 asymptotic values for $(N_{\text{Sh}})_d$ have been given as a function of ϵ' and N_{Bi} [22, 23]. Measured K_{od} values can then be compared with predicted values for $k_{d,\,\text{rigid}}$ [24].

If the drop size increases, circulation inside the drop becomes possible, causing an increase of the mass-transfer coefficient. The circulation inside the drop and hence the mass-transfer coefficient of the dispersed phase can thus be influenced by varying the pulsation velocity. In Fig. 7 a comparison of overall mass-transfer coefficients K_{od} is shown for the system benzene–caprolactam–water, gained in columns with different diameters [24]. Values for other systems are given elsewhere [13]. Only few mass transfer data have been reported [25, 26]. None of these

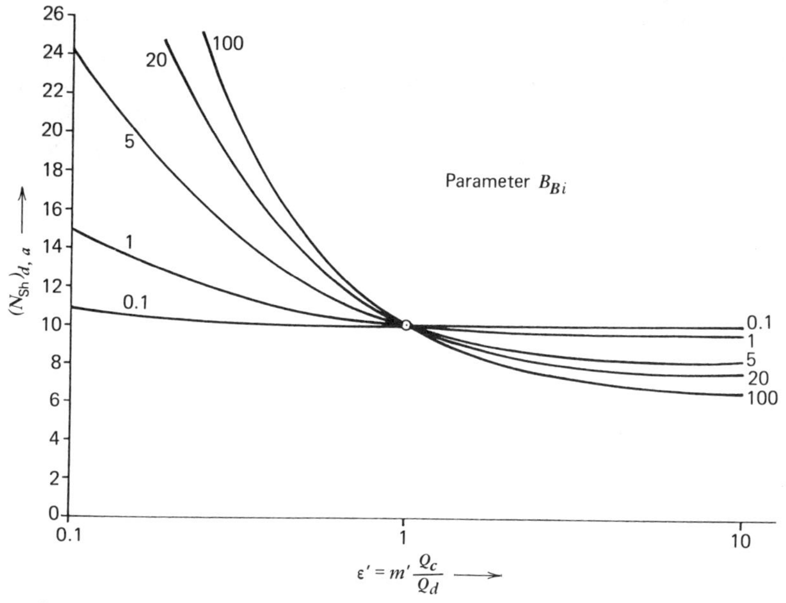

Figure 6. Asymptotic values of $(N_{Sh})_d$ for countercurrent extraction of spherical particles.

reports compare measured values for mass-transfer coefficients with predicted ones. Nevertheless, in one of them the Handlos–Baron circulation model is suggested for calculating the dispersed-phase film coefficient [25].

5. SCALEUP AND DESIGN

Design of a PPC involves (1) chemical engineering expertise to determine the main dimensions of the column (diameter D and column height H)

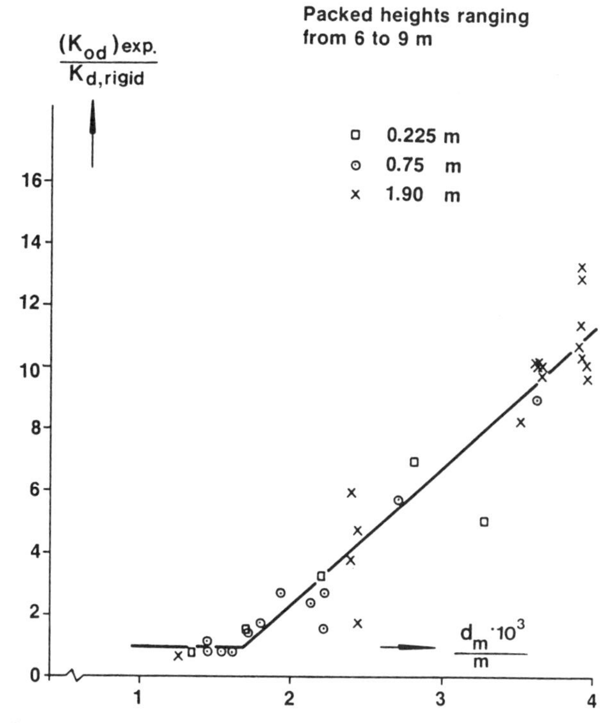

Figure 7. Comparison of overall mass-transfer coefficients K_{od} as calculated from experimental results gained in PPCs with different diameters.

and (2) mechanical engineering expertise to design the column and its pulsator.

For determination of the *diameter*, information is needed about the volumetric flow rates and the maximum allowable linear velocities of the phases; the latter are defined by the characteristic velocity of the drops and the dispersed-phase holdup. The procedure for calculating column diameters then starts with the choice of a drop diameter:

$$d_m = f(A_p \cdot f) \rightarrow V_0 \rightarrow V_d \quad \text{and} \quad V_c \rightarrow D$$

General trends for the column diameter as a function of pulsation velocity and drop diameter are given elsewhere [16, 24].

The column *length* is defined by the required separation efficiency as well as by *basic quantities*, such as distribution coefficients, kinetics of relevant reactions, physical properties of the liquids, and *process conditions*, such as linear velocities of the phases, frequency and amplitude of the pulsation, and packing characteristics. The column length needed for achieving a given extraction efficiency is independent of column diameter, provided the pulsation velocity is adequate and hydrodynamic similarity is maintained. Even if the extraction is combined with a chemical reaction, the PPC can be correctly scaled up on the basis of experimental results obtained in smaller columns. The length of these pilot columns should be chosen according to the efficiency desired; then complete understanding of the transfer processes and reaction kinetics is not needed for designing large-scale equipment. For physical extraction, scaling up can be done on the basis of the process engineering model for the PPC.

Should there be some doubt about the correctness of the process parameters, a few additional (standard) experiments can be carried out. For every drop diameter chosen (as for determination of D), the height of a transfer unit can then be calculated, corrected for axial mixing. Between the several parameters determining HTU_{od} there exists a great interdependence; a complete strategy for the calculation of H has been given elsewhere [24].

NOTATION

A_p	pulse amplitude (distance of half a stroke), m
a	specific interfacial area, m^{-1}
c_1, c_2	constants
C	concentration, kmol m^{-3}
C^*	solute concentration in a liquid phase in equilibrium with other liquid phases in the system, kmol m^{-3}
D	diameter of extraction column, m
d_0	hypothetical drop diameter under unpulsed conditions, m
d_m	mean drop diameter, m
d_p	packing size, m
D_v	diffusivity of solute in dispersed phase, $m^2 s^{-1}$
E	axial dispersion coefficient, $m^2 s^{-1}$
f	pulse frequency, s^{-1}
g	acceleration of gravity, m s^{-2}
H	active height of packed section, m
HDU	height of a diffusion unit, m
HTU	height of a transfer unit, m
K_o	overall mass-transfer coefficient, m s^{-1}
k	mass-transfer coefficient, m s^{-1}
m	distribution coefficient, C_E/C_R^*
m'	distribution coefficient, C_c/C_d^*
N_{Bi}	Biot number = $m' \cdot k_c \cdot d_m/2 \cdot D_v$
NTU	number of transfer units
N_{Pe}	Peclet number for axial mixing = $V \cdot H/E$
N_{Sh}	Sherwood number = $k \cdot d_m/D_v$
Q	volumetric flow rate, $m^3 s^{-1}$
S	surface area per unit volume of packing material, m^{-1}
V	superficial velocity, m s^{-1}
V_0	characteristic velocity, m s^{-1}
γ	ratio of phase flow rates
ϵ	extraction factor = $m \cdot V_E/V_R$
ϵ'	extraction factor = $m' \cdot Q_c/Q_d$
θ	voids in packed section, volume fraction
μ	viscosity, mPa s
ρ	density, kg m^{-3}
$\Delta\rho$	positive difference in density, kg m^{-3}
σ	interfacial tension, N m^{-1}
φ	volume fraction of dispersed phase
$\phi_{1,2,3}$	functions defined by Eqs. (14), (15), and (17), respectively

Subscripts

c	continuous phase
d	dispersed phase
E	extract
F	feed
f	flooding

o	overall, except for V_0 and d_0
plug	plug flow
R	raffinate
r	relative

REFERENCES

1. A. J. F. Simons, in B. H. Lucas, et al., Eds., *Proceedings of the International Solvent Extraction Conference 1977*, Vol. 2 (CIM Special Volume 21), Canadian Institute of Mining and Metallurgy, Montreal, 1979, p. 677.
2. (a) British Patent Specification 1306628; (b) M. Rosenzweig, *Chem. Eng.*, p. 28F. (July 10, 1972).
3. Brochure *Bronswerk Utrecht*, Vlampijpstraat 2, P.O. Box 2435, 3500 GK, Utrecht, The Netherlands.
4. British Patent Specification 1.164710.
5. M. H. I. Baird and G. M. Ritcey, in J. D. Thornton, et al., Eds., *Proceedings of the International Solvent Extraction Conference*, Lyon, 1974, Vol. 2, Society of Chemical Industry, London, 1974, p. 1571.
6. British Patent Specification 1321942.
7. H. W. Brandt, K. H. Reissinger, and J. Schröter, *Verfahrenstechnik* **8**, (8), 1 (1975).
8. S. M. Karpacheva, *Sov. Chem. Ind.* **5** (8), 534 (1973).
9. S. M. Karpacheva, *Chem. Technik* **26** **(9)**, **554** **(1974)**.
10. M. I. Kurochkina, *Zhurnal Prikladnoi Khimii* **49** (1), 192 (1976).
11. S. M. Karpacheva, *Sov. Chem. Ind.* **10** (2), 135 (1978).
12. S. M. Karpacheva, et al., *Sov. Chem. Ind.* **10** (8), 683 (1978).
13. N. M. Spaay, A. J. F. Simons, and G. ten Brink, in J. G. Gregory et al., Eds., *Proceedings of the International Solvent Extraction Conference*, The Hague, 1971, Vol. 1, Society of Chemical Industry, London, 1971, p. 281.
14. R. Gayler, N. W. Roberts, and H. R. C. Pratt, *Transact. Inst. Chem. Eng.* **31**, 57 (1953).
15. R. Clift, J. R. Grace, and M. E. Weber, *Bubbles, Drops and Particles*, Academic Press, New York, 1978, p. 306.
16. A. J. F. Simons, *Chem. Ing. Technik* **48** (5), 487 (1976).
17. J. C. Godfrey, C. Hanson, and M. J. Slater, *Chem. Ind.* 713 (1977).
18. F. Widmer, *Chem. Ing. Technik* **39** (15), 900 (1967).
19. G. V. Potnis, Ph.D. thesis 1970, through G. S. Laddha et al., *Transport Phenomena in Liquid Extraction*, Tata McGraw-Hill, New Delhi, 1976, p. 389.
20. G. K. Dwarakanath, M.Tech. thesis 1973, through G. S. Laddha et al., *Transport Phenomena in Liquid Extraction*, Tata McGraw-Hill, New Delhi, 1976, p. 389.
21. J. S. Moon, A. Hennico, and Th. Vermeulen, UCRL-10928, 1963.
22. M. A. G. Vorstman, and H. A. C. Thijssen, in J. G. Gregory et al., *Proceedings of the International Solvent Extraction Conference*, The Hague, 1971, Vol. 2, Society of Chemical Industry, London, 1971, p. 1071.
23. J. A. M. Spaninks, Ph.D. thesis, Agricultural Research Reports 885, Wageningen, The Netherlands, 1979.
24. A. J. F. Simons, *Chem. Ind.*, 748 (October 7, 1979).
25. E. Bender, et al., *Chem. Ing. Technik* **51** (3), 192 (1979).
26. A. S. Zheleznyak, *Zhurnal Prikladnoi Khimii* **47** (5), 1099 (1974).
27. S. M. Karpacheva et al., *Sov. Chem. Ind.* (2), 129 (1973).
28. G. ten Brink and A. J. F. Simons, *Het ingenieursblad* **41**, 464 (1972).

11.2

PULSED PERFORATED-PLATE COLUMNS

D. H. Logsdail
AERE Harwell
United Kingdom

M. J. Slater
University of Bradford
United Kingdom

1. INTRODUCTION

The pulsed plate column is a differential contactor with the application of mechanical energy and is used for a diverse range of processes. Probably its best known application has been in the nuclear fuel industry. For many process systems, the application of mechanical energy is often required for efficient extraction. The pulsed column has a clear advantage over other mechanical contactors when processing corrosive or radioactive solutions since the pulsing unit can be remote from the column.

2. DESCRIPTION

The pulsed plate column was patented by Van Dijk in 1935 [1]. The major claim of the patent was a series of perforated plates mounted horizontally on a vertical reciprocating shaft in a column. The pulsed column used today is based on a second claim in which the plates are fixed and the liquids are pulsed by the reciprocating motion of a bellows or a piston in a cylinder.

The liquids are fed continuously to the column, flowing countercurrently, and are removed continuously from opposite ends of the column. If the holes in the plates are sufficiently small, the phase to be dispersed, usually the less dense phase, will not flow through the column. If a pulsing action is applied at the base of the column, the dispersed phase is forced through the perforations on the pulse upstroke and the continuous phase flows downward on the downstroke. If the heavy phase is to be dispersed, it is forced through the holes on the downstroke and the continuous light phase flows upward on the upstroke. The liquids are pulsed in a cyclic manner, usually sinusoidally, so that a vertical pumping action is imposed on the countercurrent flow of the two phases. The pulsing action disperses one of the liquids into drops as it is forced through the plate perforations and agitates the continuous phase. In this way the liquids can be agitated without the need for internal moving parts.

The operating characteristics of pulsed plate columns have been described by Sege and Woodfield [2], Geier [3], and Richardson [4], among others. For constant phase flow rates, there are three stable types of operation and two flood mechanisms depending on the pulse frequency and amplitude. Typical frequencies are in the range 1-3 Hz (lower for large-diameter columns), and amplitudes of up to one plate spacing are generally acceptable.

The different regions of operation are shown in Fig. 1. The sum of the flows of both phases is plotted against the frequency-amplitude product. At low pulsation and flow rates the dispersed phase coalesces under the plate (or above, depending on which phase is dispersed). This region is termed the *mixer–settler region* since the liquids are mixed and allowed to settle as in mixer–settler contactors. If the phases are fed to the column at a greater rate than they can pass through the plates by the pumping action of the pulser, each will be discharged from the end of the column at which it enters. This behavior is termed *flooding due to insufficient pulsation.* Although operation is stable in this region, mass-transfer rates are usually poor. If there is no flow when there is no pulsing, then approximately†

$$V_{cf} + V_{df} = fA_p \tag{1}$$

However, for low interfacial tension systems, where the dispersed phase can pass through the holes under the head available between the plates, throughputs higher than those described by Eq. (1) are obtained [5, 6].

As the pulsing and flow rates increase, the inertial and shear forces increase. These forces hinder the coalescence of drops, and a dispersion or emulsion is formed. This region of operation is called the *emulsion zone.* The greater the pulsing, in general, the smaller the drops. As in the mixer–settler zone when one (or both) phase flow rate becomes too great, the column will be flooded. Flooding also results at fixed flow rates if the frequency–amplitude product is sufficiently increased. In this zone there is little change in drop size during the pulse cycle, and it is in this region that the best mass-transfer rates are usually obtained. With some systems, between the emulsion region and flooding the emulsion becomes unstable and coalescence and phase inversion occurs in different parts of the column.

A simple description of the onset of flooding in the emulsion zone is that the pulsing action produces drops that have a terminal settling velocity less than the linear superficial velocity of the continuous phase. These drops accumulate in the disengaging section of the column at the continuous phase outlet and leave the column with the continuous phase. However, the pattern of hydrodynamic behavior is far more complex than this simple explanation implies, and the precise nature of flooding phenomena has defied rigorous analysis so far. Not only are the frequency–amplitude product and phase flow rates variables, other important factors need to be taken into account. These include surface, viscous, inertial, and shearing forces; plate wettability; coalescence rates; and the presence or absence of internal circulation in the drops. Consequently, nearly all flooding data for the emulsion region are correlated empirically, and some flooding correlations are described in the following paragraphs. Maximum flows under favorable conditions depend on the physical properties of the system but can be as high as 200 $m^3\ m^{-2}\ h^{-1}$ (sum of both phases).

During column operation when solute transfer is taking place, the drop size and solute concentration will vary over the column length. Hence the dispersed-phase holdup and coalescence rate will vary and phase inversion can occur. Prediction of inversion is not possible, although a holdup greater than about 30% is usually required before it does occur.

It is usual to disperse the solvent to minimize the inventory and hence costs and also for safety reasons, but sometimes it is advantageous to disperse the aqueous phase, such as when processing radioactive and corrosive solutions.

The effect of different pulse wave forms on column capacity at flooding is small [12].

Concatenated pulsed plate columns can provide the equivalent contacting length of a single tall column with much less head room while a single pulse generator is being used [13].

†A list of symbols used in this and subsequent equations is given in the Notation section at the end of the chapter.

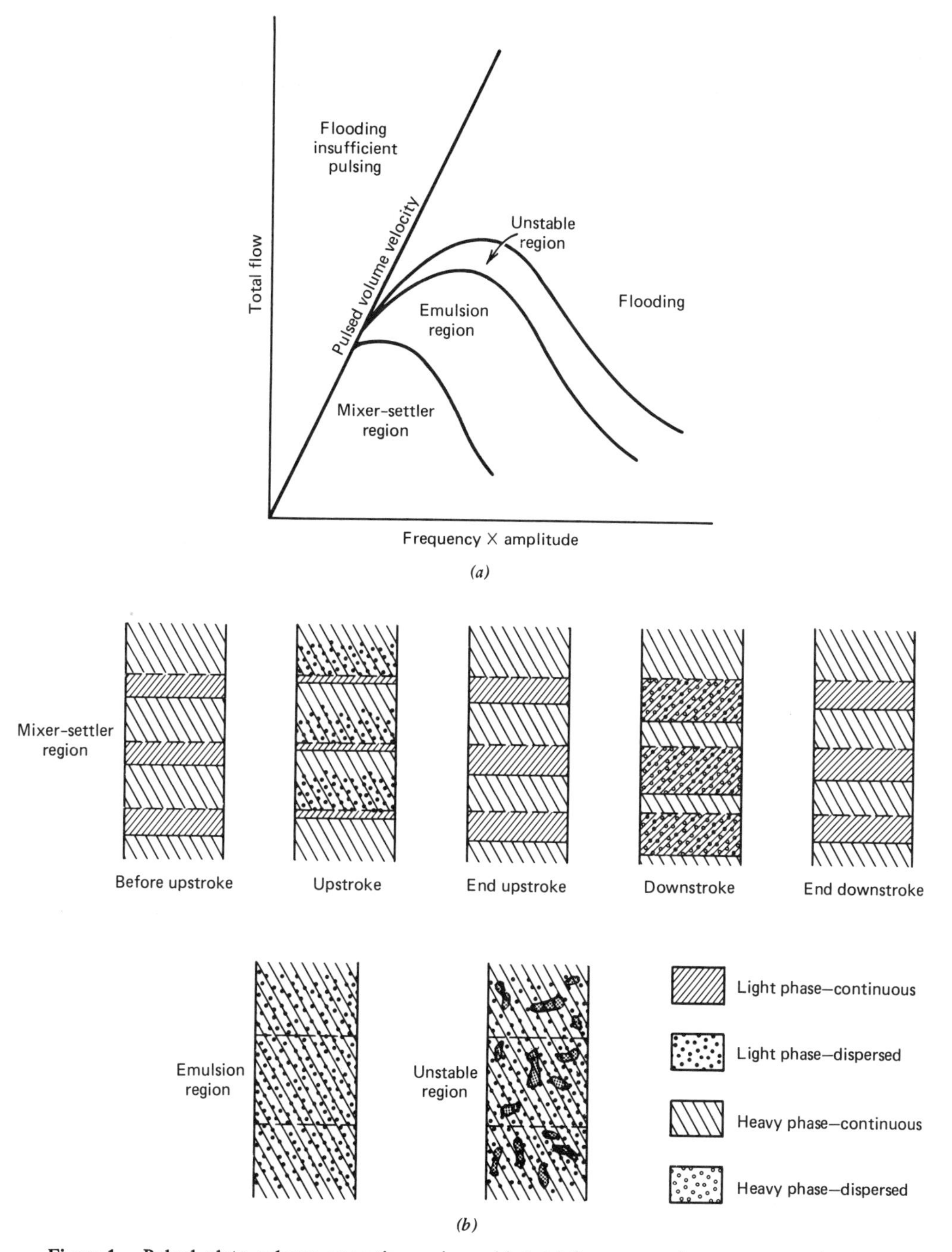

Figure 1. Pulsed plate column operating regions: (*a*) total flow versus frequency × amplitude; (*b*) phase dispersion.

3. PLATE CARTRIDGES

The assembly of plates in a pulsed plate column is called the *plate cartridge.* Generally sieve plates with square-edged orifices are used, but sometimes it is advantageous to use nozzle-shaped orifices or punched holes, particularly if the dispersed phase tends to wet the plates. The range of geometries used is shown in Table 1.

The many studies of plate design and its effect

TABLE 1 RANGE OF PLATE CARTRIDGE GEOMETRIES

Plate diameter	10–2500 mm
Orifice diameter	0.83–4.78 mm
Free area	10–60%
Spacing	12.5–100 mm
Nozzle depth	1.0–1.5 mm

on column performance has led to the use of a "standard geometry" [7]; plates with 3 mm diameter holes on a triangular array to give 23% free area are spaced 50 mm apart in the column (plate thicknesses of up to 2 mm are usual).

For maximum performance, the plate geometry can be tailored for a particular process. Geier [7] used a graded cartridge to give a uniform dispersion over the column length for the system (nitric acid–uranyl nitrate–30% TBP/diluent). Higher throughputs and better column operating stability were obtained than with plates having the same geometry over the column length. Columns having alternate pairs of stainless steel and polytetrafluoroethylene (Teflon®) sieve plates have a much better performance than an all-stainless steel sieve-plate column. Geier also found that the nozzle depth influenced the degree to which the aqueous dispersed phase wetted the plate surface and hence the column performance. Plates with nozzle depths of 0.76 mm gave only marginal improvement in capacity over sieve plates, whereas a depth of 1.02 mm gave a marked improvement in column throughput with little deterioration in mass-transfer efficiency. No additional improvement was obtained if the depth was increased to 6.4 mm.

The clearance between plate and column wall can affect column performance. Clearances of up to 3 mm have little effect for an aqueous continuous system, but the effect is much more marked if the organic phase is continuous [8–11].

Increase of column diameter reduces the mass transfer efficiency of the pulsed plate column [2, 9–11]. By replacing a sieve plate with a louvre redistributor plate at about 1 m intervals, scale-up effects can be reduced significantly when the densities of the solution are greater at the top of the column than at the bottom [2]. In studying the scale-up of small pulsed plate columns, Coggan [28] found that in terms of throughput, columns could be scaled up from 20 mm diameter but that coalescence in a column of less than about 12.5 mm diameter produced inconsistent results.

4. DESIGN

4.1. Holdup and Flooding Data

Thornton [12] and Logsdail and Thornton [11] studied six solvent–water systems, 11 different plate geometries, four column diameters, and three pulse wave forms. All the data were obtained for the emulsion region of column operation and in the absence of mass transfer were well correlated with use of the concept of characteristic drop velocity.

Berger et al. [6] have published data on two of the test systems recommended by the European Federation of Chemical Engineering (EFCE) Working Party on Distillation, Absorption and Extraction and on one other system.

For the high interfacial tension system, toluene–acetone–water (EFCE system) maximum throughput increases slightly as the free area and hole size increase. For large free area (60%) and large holes (6.5 mm), a wide range of frequency and amplitude may be used.

For the medium interfacial tension system, ethyl hexanol–acetic acid–water the results are similar, but the different densities and viscosities restrict the frequency–amplitude range.

For the low interfacial tension system, *n*-butanol–succinic acid–water (EFCE system), a marked difference is noted. The limit on flooding for low pulsation tends to rise as both the free area and hole size increase. The drop size is much smaller than the holes and the column is not dependent on pulsing to make the dispersed phase flow through the plates.

The results also showed that the sum of the superficial velocities at flooding can exceed the pulse velocity fA_p substantially, the more so as interfacial tension decreases. Pulse velocities exceeding 25 mm/s reduce flooding limits markedly. Mass transfer was out of drops.

Other holdup and flooding studies have been reported by:

Bell and Babb [16]
Defives and Schneider [17]
De Witte et al. [18]
Durandet and Talmont [19]
Edwards and Beyer [5]
Foster [20]
Geier [3]
Logsdail and Larner [21]
Miyauchi [22]
Pike and Erickson [23]

Sehmel and Babb [14]
Swift [24]
Cohen and Beyer [15]

Many of these studies were made in the absence of mass transfer.

4.2. Holdup and Flooding Relationships

For the mixer–settler region, Griffith et al. [25] suggest that the holdup x is approximately given by the flow of dispersed phase in the time for one pulse; hence

$$x = \frac{V_d}{fH_c} \tag{2}$$

Values of x tend to be too high in comparison with Sehmel and Babb's [14] data.

The flooding limit due to insufficient pulsation for mixer–settler conditions was correlated by Griffith [25] as

$$V_{cf} + V_{df} = fA_p \left[1 + \frac{V_c}{V_d}\right] \quad \text{for} \quad V_d > V_c \tag{3}$$

$$V_{cf} + V_{df} = fA_p \left[1 + \frac{V_d}{V_c}\right] \quad \text{for} \quad V_d < V_c \tag{4}$$

These equations are not validated by the results due to Logsdail and Larner [21] or by Kagan [26]. In general, the value of $(V_{cf} + V_{df})$ usually equals or exceeds fA_p.

For well-agitated conditions, ($fA_p > 12$ mm/s), Thornton [12] uses the concept of a slip velocity (a relative velocity between phases) experimentally related to a function of holdup as follows:

$$V_{\text{slip}} = \bar{V}_0(1 - x) \tag{5}$$

where $V_{\text{slip}} = V_d/x + V_c/(1 - x)$ and $\bar{V}_0$ is called the *characteristic velocity*.

This approach has been used successfully for a wide variety of columns with the limitations of $x < 0.2$ and no coalescence between drops. This equation has been confirmed for flow ratios of 0.01–100 for several systems. Defining flooding as the point at which drops are substantially rejected through the continuous phase exit Thornton has derived formulas for the superficial velocities at the onset of flooding [12].

The proposed correlation for the characteristic velocity $\bar{V}_0$ is

$$\frac{\bar{V}_0 \mu_c}{\sigma} = 0.6 \left[\frac{\psi \mu_c^5}{\rho_c \sigma^4}\right]^{-0.24} \left[\frac{d_0 \rho_c \sigma}{\mu_c^2}\right]^{0.90} \cdot \left[\frac{\mu_c^4 g}{\Delta\rho\, \sigma^3}\right]^{1.01} \left[\frac{\Delta\rho}{\rho_c}\right]^{1.8} \left[\frac{\mu_d}{\mu_c}\right]^{0.30} \tag{6}$$

where

$$\psi = \frac{\pi^2(1 - S^2)(fA_p)^3}{2S^2 C_o^2 H_c} \quad \text{and} \quad C_o = 0.6 \tag{7}$$

The correlation gives $\bar{V}_0$ values for solvent dispersed, in 75 mm (3 in.), 150 mm (6 in.), and 300 mm (12 in.) diameter columns when no mass transfer is occurring. When mass transfer takes place from solvent drops to the aqueous phase, interfacial effects may encourage coalescence; it has been found necessary to multiple $\bar{V}_0$ by 2.6 in such cases. For mass transfer into solvent drops, a 10% reduction in $\bar{V}_0$ is recommended. For mass transfer into aqueous drops, Coggan [28] found no effect.

The manner in which $\bar{V}_0$ depends on fA_p differs for different investigations. Miyauchi and Oya [22] use a criterion based on

$$\phi = \frac{fA_p}{(\beta_c H_c)^{0.33}} \frac{\mu_d^{0.5}}{(\sigma\, \Delta\rho)^{0.25}} \quad \text{(cgs units)} \tag{8}$$

where

$$\beta_c = \frac{S^2}{(1 - S)(1 - S^2)}$$

When ϕ exceeds 0.21, viscous drag conditions obtain. For ϕ less than 0.21, Thornton's equation for $\bar{V}_0$ agrees with Miyauchi's data.

Thornton's data for six systems show that the dependence of $\bar{V}_0$ on ϕ (and hence on fA_p) changes markedly at about $\phi = 0.25$; the majority of the data have values of ϕ lower than this and $\bar{V}_0 \propto (fA_p)^{-0.72}$. For $\phi > 0.25$, an exponent near zero is required, as confirmed by Defives and Schneider [17]. However, Miyauchi and Oya [22] showed $\bar{V}_0 \propto (fA_p)^{-1.2}$ approximately for $0.2 < \phi < 0.6$. Values of ϕ exceeding 0.25 are not likely to be used industrially.

Flooding limits have been correlated empirically by McAllister [29] for all published data to about 1967, showing that Eqs. (3) and (4) are not always valid. Use of the correlation requires a trial and error procedure and the paper should be consulted. Thornton's equations are said to give generally satisfactory results.

In most cases of industrial relevance it will be impossible to predict the extent of drop coalescence, and reliance on Eq. (5) without practical confirmation could lead to serious overestimation of flooding limits [11, 20].

4.3. Drop Sizes and Drop Motion

Miyauchi and Oya [22] showed that for well-agitated conditions [30]

$$d_m \propto \left(\frac{fA_p}{H_c^{0.33}}\right)^{-1.2} \tag{9}$$

Misek [31] proposes

$$\frac{d_m}{d_0} = 0.439 \left[\frac{\sigma S^{0.5}}{d_0 \rho_c (\pi f A_p + V_c)^2}\right]^{0.6} \tag{10}$$

based on four systems (solvent dispersed; interfacial tension 0.003–0.037 N/m). The drop sizes were not measured but were calculated according to the assumption that the characteristic velocity equals the terminal velocity of a rigid drop. Predicted sizes seem small (Table 2), and Misek states that the equation is valid only for $d_m \ll d_0$. Large values of d_0 (5–35 mm) were used.

Kubica and Zdunkiewicz [32] have developed a correlation for drop size in a column 61 mm in diameter, d_0 3 mm only, based on five systems:

$$\frac{d_m}{D} = 0.0135 \left(\frac{f^2 A_p^2 \rho_c d_0}{\sigma}\right)^{-0.225} \left(\frac{\mu_c}{\mu_d}\right)^{0.0163} \tag{11}$$

For mild agitation, the equation

$$d_m = 0.92 \left(\frac{\sigma}{g\, \Delta\rho}\right)^{0.5} \tag{12}$$

might be used to indicate the maximum drop sizes to be expected. However, if drop sizes are similar to plate hole sizes then the equation is not likely to be valid.

In mixer–settler operating conditions, correlations developed for drop sizes formed at nozzles might be useful [33, 34].

Misek [31] has shown that under well-agitated conditions drops might be expected to move as if they were rigid since drag forces do not act in one direction long enough for circulation patterns to be set up inside drops. The Miyauchi–Oya [22] data support this idea, thus:

$$\bar{V}_0 = V_T = \frac{d_m^2 g\, \Delta\rho}{18\mu_c(1 + 0.15\, \mathrm{Re}^{0.687})} \quad \text{for} \quad 0.2 < \mathrm{Re} < 800 \tag{13}$$

$$\mathrm{Re} = \frac{\rho_c d_m V_T}{\mu_c}$$

Some calculated drop sizes are presented in Table 2 for comparison.

4.4. Axial Mixing

Axial dispersion may be represented by

$$\frac{\partial c}{\partial t} = E \frac{\partial^2 c}{\partial Z^2} - V \frac{\partial c}{\partial Z} \tag{14}$$

Sometimes the definition is based on a true velocity so care is required when interpreting data based on this relationship.

TABLE 2 CALCULATED DROP SIZES[a]

fA_p (mm/s)	Ref. 31 [Eq. (10)], d_m (mm)	Mild Agitation [Eq. (12)], d_m (mm)	Based on $\bar{V}_o$ [Eqs. (6) and (13)], d_m (mm)	Ref. 32 [Eq. (11)], d_m[b] (mm)
20	0.70	2.1	1.9	1.3
30	0.45	2.1	1.4	1.1
40	0.32	2.1	1.1	1.0
60	0.20	2.1	0.85	0.8

[a]System: MIBK–acetic acid–water; solvent dispersed; standard geometry.
[b]$D = 61$ mm.

Many investigations of axial mixing in pulsed plate columns have been carried out [22, 26, 36–46]. Most of these used small-diameter columns (about 50 mm), stainless steel sieve plates, and a continuous water phase.

Mar and Babb [42] used three systems, with the water as the continuous phase in each case, in a 50 mm (2 in.) diameter column. A steady-state tracer technique was used to measure back-mixing coefficients. Some of the results were checked by the pulse injection method, and no significant differences were found.

It is likely that plate wetting can affect axial mixing since Mar and Babb found that the axial mixing coefficient was approximately doubled when the stainless steel plates were replaced with polyethylene plates.

Sehmel and Babb [43] also used a steady state tracer injection method to measure axial mixing coefficients in a 50 mm (2 in.) diameter column with stainless steel plates. The geometry of the plates was not varied. Water was the continuous phase for each of the three systems studied. Maximum axial mixing was found to occur close to the transition frequency, defined as the frequency for minimum dispersed phase holdup [14]. This frequency is also the same as that at which the column can be observed to change from the mixer–settler region to the emulsion region.

Miyauchi and Oya [22] made an extensive investigation of axial mixing in both continuous and dispersed phases. Two column diameters, 32 and 54 mm, were used and water was the continuous phase with methyl isobutyl ketone the dispersed phase. A pulse injection technique was used to measure axial mixing coefficients. The results for the continuous phase for the emulsion region were represented by

$$\frac{E_c}{(1-x)fA_pH_c} = \frac{V_c/(1-x)fA_p}{2\beta - 1/n'} + \frac{1}{\beta} \tag{15}$$

where β is the number of perfectly mixed stages in series between each pair of plates.

Assuming that the geometry of the compartment formed by a pair of plates was the major controlling factor for β, the following empirical equation for β was found to correlate the data well:

$$\beta = \frac{0.57(D^2H_c)^{0.33}S}{d_0} \tag{16}$$

In large-diameter columns β assumes unrealistic values [47].

Specific information on the TBP–water system reported by Rao et al. [49] and Rouyer et al. [48] is useful for nuclear applications. In other cases the use of Miyauchi and Oya's correlation is the only one that can be recommended for aqueous continuous operation with steel plates and emulsion zone operation. Data for organic phase continuous systems with steel nozzle plates (as used in nuclear fuel reprocessing plants) or with plastic plates are lacking.

The effect of column diameter is important but has not been adequately tested. Rouyer et al. [48] have given results for 30% TBP–dodecane dispersed in nitric acid for 45 and 600 mm columns with the same geometry (not given). The results show that for the continuous phase

$$\frac{E_{45}}{E_{600}} \simeq 2 = \left(\frac{D_{45}}{D_{600}}\right)^{-0.27} \tag{17}$$

Miyauchi's correlation would give $E_{45}/E_{600} = 5.5$, Kagan [26] gives $E_{54}/E_{200} = 1.75$.

4.5. Mass Transfer

4.5.1. Introduction

In most cases mass-transfer performance has been reported in terms of overall mass-transfer coefficients or transfer unit heights without correction for axial mixing effects. The height equivalent of a theoretical stage has also been reported but is generally not to be recommended.

Unless each case is studied carefully and film transfer units evaluated, the application of correlations to other systems and to other scales of operation is risky.

The apparent transfer unit height uncorrected for axial mixing (but not the true value) will sometimes show a minimum value in the dispersion region because at high pulse velocities increasing axial mixing may be more important than decreasing drop size, increasing interfacial area, and increasing mass-transfer coefficients [2, 12, 38]. If one phase gives most of the resistance to mass transfer, a minimum may not exist.

Most investigators have used simple nonionic systems with solutes having high diffusivities. In cases of metals extraction, chemical kinetics may be rate controlling and mass transfer much slower. Details of systems, column geometries, and pulsing conditions used by various investigations are given in Table 3.

TABLE 3 DETAILS OF SYSTEMS, COLUMN GEOMETRIES, AND PULSING CONDITIONS USED BY VARIOUS INVESTIGATORS[a]

System	Reference	Column Height × Diameter	Plates Hole Size, Free Area, Spacing	Pulsing Frequency, Amplitude	Comment
Water–acetic acid–MIBK	90	45 × $1\frac{1}{2}$ in.	$\frac{1}{32}$ in., 0.23, 1 in.	0.33–1.33 s^{-1}, $\frac{1}{8}$–2 in.	Aqueous dispersed; fixed flow ratio
	91	800 × 51 mm	3 mm, 0.23, 40 mm	6–40 mm/s	
	38	48 × 2 in.	$\frac{1}{16}$–$\frac{1}{8}$ in., 1.6–2.2 in., stainless steel	0.5–1.5 s^{-1}, 0.5–1.0 in.	Solvent dispersed; mass transfer out of drops
	92	1220 × 40 mm	$\frac{3}{64}$ and $\frac{5}{64}$ in., 3 in.	0.3–0.9 s^{-1}, 1.5–6.0 mm	Corrosion and blocking of small holes
	66	4 × $\frac{3}{4}$ in.	0.04 in., 0.28, 2 in., stainless steel	0.5–3.3 s^{-1}, 4 mm	Study of dynamics
	44	550 × 58 mm	1.5 mm, 0.081, 52 mm, stainless steel	0.5–3.3 s^{-1}, 1.0–8.5 mm	Solvent dispersed
Water–acetone–toluene	12	42 × $2\frac{7}{8}$ in.	$\frac{1}{8}$ and $\frac{1}{16}$ in., 0.13–0.62, 0.5–3.0 in., brass	1.5–7.0 s^{-1} 0.25–0.87 in.	Solvent dispersed
	11	34 × 6 in., 34 × 9 in., 70 × 12 in.	$\frac{1}{8}$ in., 0.25, 2 in.	2–4 s^{-1}, 0.24–0.65 in.	Solvent dispersed
	6	5000 × 80 mm	3.0–6.5 mm, 0.23–0.60, 50 mm	≤3.3 s^{-1}, 8 mm	Solvent dispersed; mass transfer out of drops
Water–benzoic acid–toluene	93	22 × 2 in.	$\frac{1}{16}$ in., 0.085, 2 in.		
	97	152 × 7.62 cm	0.476–0.635 cm, 0.226–0.360, 5–10 cm	0.5–1.5 s^{-1}, 8–16.46 mm	Solvent dispersed; mass transfer out of drops
Water–acetone–1,1,2-trichloroethane	38	48 × 2 in.	$\frac{1}{16}$–$\frac{1}{8}$ in., 1.6–2.2 in.	0.5–1.5 s^{-1}, 0.5–1.0 in.	Solvent dispersed; mass transfer out of drops
Water–succinic acid–*n*-butanol	6	5000 × 80 mm	3.0–6.5 mm, 0.23–0.60, 50 mm	≤3.3 s^{-1}, 8 mm	Solvent dispersed; mass transfer out of drops
Water–acetic acid–ethyl hexanol	6	5000 × 80 mm	3.0–6.5 mm, 0.23–0.60, 50 mm	≤3.3 s^{-1}, 8 mm	Solvent dispersed; mass transfer out of drops
Water–boric acid–isoamyl alcohol	15	20 × 1 in.	0.04 in., 0.09, 2 in.	≤1.2 s^{-1}	

Water–propionic acid–TBP	28	4.8 × $\frac{1}{4}$–2 in.	$\frac{1}{8}$ in., 0.24–0.29, 2 in., stainless steel, nozzles	0.33–3 s^{-1}, 5–50 mm	Aqueous dispersed; wetting problems
Water–acetic acid–ethyl acetate	92	1220 × 40 mm	$\frac{3}{64}$ and $\frac{5}{64}$ in., 3 in.	0.3–0.9 s^{-1}, 1.5–6.0 mm	Corrosion and blocking of small holes
Water–alcohols–hydrocarbon	61	6400 × 49 mm	1 mm, 57 mm	0.83–8.3 s^{-1}, 1–4 mm	
Water–phenol–kerosine	26	400 × 56 mm	2 mm, 0.082, 50 mm, stainless steel		Solvent dispersed
Uranium–H_2SO_4 leach pulp–Alamine 336	71	144 × 10 in., 480 × 2 in.	$\frac{3}{16}$ and $\frac{3}{8}$ in., 0.27 and 0.32, 4 in. or $\frac{3}{16}$ in., 0.27, 2 in.	$\leqslant$0.83 s^{-1}, $\leqslant$2 rise drop	Solvent dispersed; mass transfer into drops
Uranium–thorium–rare earths–H_2SO_4 in pulp–D2EHPA	73	480 × 2 in.	$\frac{3}{16}$ in., 0.26, 2 in., stainless steel 316	0.83 s^{-1}, 1 rise drop	Solvent dispersed
Cu–LIX 63 and D2EHPA Co–Ni–D2EHPA	94	360 × 2 in.	$\frac{3}{16}$ and $\frac{1}{8}$ in., .34, 2 in., stainless steel 316	0.83 s^{-1}, 0.75 rise drop	Extraction–aqueous dispersed; scrub–solvent dispersed
Cu–Ni–Zn–$(NH_4)_2SO_4$–Kelex	94	396 × 2 in.	$\frac{3}{16}$ and $\frac{1}{8}$ in., 0.26 and 0.28, 2 in., stainless steel 316	0.5–1.25 s^{-1}, 50 mm drop rise	Solvent dispersed
Co–Ni chloride–ammonium thiocyanate–MIBK	25	140 × 2 in.	$\frac{1}{32}$ in., 0.23, 1 and 2 in.	0.25–2.1 s^{-1}, 0.25–2.0 in.	Solvent dispersed
Y-91–Pm-147 in nitric acid–D2EHPA	59	48 N 0.75 in., 24 × 0.75 in.	0.032 in., 0.13, 1.5 in.	0.83 s^{-1}, 10 mm	
Ta–Nb–HCl–HF–MIBK	95	144 × 2 in.	$\frac{3}{32}$ in., 0.32, 1.875 in., polythene	0.66 and 1.0 s^{-1}, 1 in.	Polythene plates and column
Zn–Hf–HNO_3–di-heptylsulfoxide, Fe–HCl	96	4 × 0.04 m, 3.75 × 0.037 m	0.40, 50 mm, Teflon	0.66 and 0.83 s^{-1}, 40 mm	Aqueous dispersed; Teflon plates
Zr–Hf–HNO_3–TBP	97	× 51 mm × 152 mm × 254 mm × 559 mm height not given	4.8 and 3.2 mm 50.8 mm 4.8 mm 50.8 mm Stainless steel–Teflon dual plates	0.58 s^{-1}	Aqueous dispersed

[a] All dimensions are as given in the references.

4.5.2. *Correlations*

Smoot and Babb [38] have correlated overall transfer units using MIBK–acetic acid–water, the organic phase dispersed and solute transfer from the dispersed to the continuous phase. The H_{oc} values have been corrected for axial mixing, but the operating conditions were not clearly identifiable (Smoot [51]). The equation is

$$H_{oc} = 504 H_c \left(fA_p \frac{d_0 \rho_d}{\mu_d}\right)^{-0.4} \left(\frac{V_c}{fA_p}\right)^{0.43} \cdot \left(\frac{V_c}{V_d}\right)^{0.56} \left(\frac{d_0}{H_c}\right)^{0.62} \tag{18}$$

The Smoot, et al. [51] correlation for values of H_{oc} uncorrected for axial mixing (in ft lb_m hr units) is

$$H_{oc} = \frac{10.4 V_c^{0.54} D^{0.32} H_c^{0.68} \sigma^{0.097} \Delta\rho^{1.04}}{D_v^{0.865} (fA_p/S)^{0.43} d_0^{0.43} \rho_d^{2.34} \mu_d^{3.27} V_d^{0.64}} \tag{19}$$

based on 285 experiments by various investigators. It applies to the emulsion region and steel sieve plates. It is valid only for mass transfer from the solvent dispersed phase to an aqueous continuous phase and for cases in which the major resistance to mass transfer is in the dispersed phase.

Thornton's [12] correlation for H_{oc} uncorrected for axial mixing is

$$H_{oc} = b \left(\frac{\mu_c^2}{g\rho_c^2}\right)^{0.33} \left(\frac{\mu_c g}{\bar{V}_0^3 (1-x)^3 \rho_c}\right)^{2m/3} \cdot \left(\frac{\Delta\rho}{\rho_c}\right)^{(2/3)(m-1)} \left(\frac{V_d}{V_c}\right)^{0.50} \cdot \left(\frac{V_c^3 \rho_c}{g \mu_c x^3}\right)^{0.33} \tag{20}$$

For toluene–acetone–water and butyl acetate–acetone–water, m was 0.5 for mass transfer from continuous aqueous into dispersed solvent. For the reverse direction of transfer, m was 0.25. For the toluene system, b was approximately 3×10^3 (both directions of transfer) and 1.3×10^3 for the butyl acetate system. The range of fA_p was 10–50 mm/s. Experimental measurements are needed to determine m and b.

Eguchi and Nagata [44] give for the mixer-settler region:

$$H_{oc} = \frac{V_c}{0.019 V_d^{0.5} (fA_p)^{0.2}} \tag{21}$$

The value of H_{oc} is corrected for axial mixing. Flows are cm s^{-1}, A_p cm, and f min^{-1}. The correlations can be compared by using data for the MIBK–acetic acid–water system for water continuous and mass transfer out of drops (Table 4).

True values of H_{oc} can be obtained from correlations based on single drops. By use of equations by Handlos and Baron [52] and Garner et al. [53], sensible values are obtained. The Rose–Kintner [54] equation for K_d gives similar results. These equations also demonstrate that the dispersed phase offers most of the resistance to mass transfer in the MIBK–acetic acid–water system.

Table 5 shows the effect of increasing pulse velocity at fixed flows. A gradual decrease in H_{oc} with increasing agitation and approach to flooding is predicted. Thornton [12] found a very slight increase in H_{oc}; Berger et al. [6] show that the HETS is constant.

Feick and Anderson [55] and Coggan [56] show that increased agitation mainly affects the interfacial area and the effect on mass transfer film coefficients is not great. Berger et al. [6] show that for high-interfacial-tension systems, best results are obtained with small holes and small free area. For low-interfacial-tension systems, plate geometry has no marked effect.

TABLE 4 COMPARISON OF CALCULATED VALUES OF $H_{oc}{}^a$

fA_p (mm/s)	$\bar{V}_0$ (mm/s)	V_d (mm/s)	d_m (mm)	Ref. 51, $H_{oc}{}^b$ (m)	Ref. 38, H_{oc} (m)	Ref. 12, H_{oc} (m)	Ref. 44, H_{oc} (m)	Circulating Drop, H_{oc} (m)	Rigid Drop, H_{oc} (m)
20	69.7	5.65	1.9	1.37	0.42	0.20	1.09	0.28	6.75
30	52.1	4.22	1.4	1.19	0.26	0.20	0.87	0.20	2.77
40	42.3	3.43	1.1	1.08	0.19	0.20	0.74	0.15	1.41
60	31.6	2.56	0.85	0.93	0.12	0.20	0.59	0.11	0.65

[a] $V_d = V_c; x = 0.1$.
[b] $D = 51$ mm.

TABLE 5 VALUES OF H_{oc} WITH INCREASING PULSE VELOCITY AT FIXED FLOWS[a]

fA_p (mm/s)	Ref. 38, H_{oc} (m)	Ref. 51, H_{oc} (m)	V_{df} (mm/s)	Percent Flooding
20	0.34	1.45	10.2	34
30	0.24	1.22	7.6	45
40	0.19	1.08	6.2	55
60	0.14	0.90	4.6	75

[a]Flow rate 3.43 mm/s, $V_d = V_c$.

Sege and Woodfield [2] found no scale-up effect between 75 and 200 mm diameter columns on H_{oc} but reported a fourfold increase when the diameter was increased from 75 to 600 mm. Logsdail and Thornton [11] found a 30% increase on increasing the diameter from 150 to 300 mm and proposed the equation $H_{oc} \propto \exp(D/2)$, but this would seriously overestimate the effect for diameters much greater than 300 mm. Rouyer et al. [48] state that the variation in transfer unit height for diameters of 50–300 mm is not marked. The behavior of different chemical systems and the effects of axial mixing are sufficient to explain these varying observations. Values for H_{oc} will vary within a column if drop size or holdup vary and if the equilibrium line is curved. Plank [57] gives a procedure that accounts for curved equilibrium lines. Khemangorn et al. [58] have studied the influence of the direction of mass transfer on column performance for a single system, but the results cannot be generalized. Data for large columns described by Reissinger [60] may be of interest.

4.6. Optimization and Modeling of Column Performance

The optimum operating condition is not necessarily coincident with minimum transfer unit height. A minimum column volume should be found between the pulse velocity limits at maximum throughput and at minimum transfer unit height. A useful indicator of performance is the ratio $H_{oc}/(V_c + V_d)$ or its reciprocal [61], which provides a combined measure of the transfer efficiency and capacity per unit volume of column. It can be used to compare a pulsed plate column with other contactors, provided the same system and identical feed conditions are used.

Korchinsky [62] has modeled column sizes by using correlations and found that neither the column diameter nor column height depend much on fA_p in the range 10–40 mm/s for the MIBK–acetic acid–water and toluene–acetone–water systems. Comparison with other columns indicates that the minimum volumes for a given duty, of a pulsed packed column, pulsed sieve-plate column, and rotating-disk contactor are much alike. Minimum volumes for simple packed and sieve-plate columns are generally much greater.

Simulation and control of pulsed columns have been studied by Bauermann and Blass [91], Foster et al. [20], Diliddo and Walsh [63], Biery and Boylan [64], Britto [65], and Watjen and Hubbard [66].

5. MODIFIED PULSED COLUMNS

Pulsed plate columns have been modified in a number of ways to improve their performance. A pulsed sieve-plate column with a central rotating stirrer is claimed to be more efficient at lower power input without affecting flooding limits [67] (See also Chapter 14). Churet et al. [68] describe a pulsed column with plates having a segment removed and the plates positioned with the open section alternately one side then the other. No operating data are given. Air injection has been shown to increase flow capacity by Dewitte and Geens [69]. Krimz plates contain rectangular apertures fitted with vanes angled at 20–30°. Alternate plates direct flow clockwise and anticlockwise. The free area is up to 60% and columns up to 1.5 m diameter have been built. Investigation of all aspects of performance have been carried out by Karpacheva and co-workers [70, 87].

6. SOLIDS IN SUSPENSION

To avoid emulsification in systems containing solids in suspension, low pulse velocities are usually required. If the interfacial tension is high in these systems, small plate holes are needed to produce a small drop size. Uranium extraction from ore leach pulps has been investigated thoroughly by Ritcey [71–73]. Plate holes of 9.5 mm ($\frac{3}{8}$ in.) and free area of 27% were used. Smaller holes caused greater solvent loss. Very gentle pulsing action was used for treating the pulps. The pulse frequency and amplitude are adjusted to minimise backmixing of the dispersed phase. The principal difficulty is that of solvent loss by adsorption on solids or by entrainment, but it has been shown that for some ores, the loss is economically acceptable. Pulps with up to 55 wt% fine silica solids have been treated with dispersed solvent at low holdup. Total flow rates of the order of 20–30 $m^3\ m^{-2}\ h^{-1}$ were used. Huppert et al. [74] found that a voluminous precipitate in a zirconium–dibutyl phosphate system could be handled without difficulty in a column having plates with 23% free area and 3 mm nozzle holes. The column was self-cleaning, and there was no accumulation of solids or plugging of holes.

7. INDUSTRIAL APPLICATIONS

Kuylenstierna and Ottertun (Stora Kopparbergs, Sweden) [75] report a nitric acid and hydrofluoric acid extraction from pickling liquor with TBP. A 3.0 m high, 90 mm diameter high-density polythene column was operated with the organic phase continuous. An average HETS of 0.5 m at solvent:aqueous ratios of 2.4 and 1.8 was obtained. A full-scale plant was used between 1974 and 1978 when the steel plant was closed down. The full-scale extraction column was 10 m high and 350 mm in diameter, with a throughput of 25 $m^3\ m^{-2}\ h^{-1}$.

Demarthe et al. [76] have described CEA/Societe le Nickel work on nitric acid recovery from nickel sulphate using 50% TBP. Columns of 45 mm diameter and 4–6 m height with 50 mm plate spacing and 23% free area were used. With continuous solvent phase, the HETS was about 0.5 m. Entrainment losses were less than 10 ml/m^3.

Reissinger and Schroeter [60] give dimensions and constructional details of nine large columns of up to 2500 mm diameter, operated at frequencies of about 1–2 Hz and amplitudes of 3–7 mm. The applications are not given.

The Eries Company (Genas, France) has constructed a number of columns with plates like an RDC. Phenols may be extracted from tar distillates using, for example, dichloroethane or butyl acetate. Caustic soda has been used in columns 450 mm in diameter and 6 m high and 750 mm in diameter and 7 m high. The recovery of phenols with sulfuric acid has been carried out in columns 500 mm in diameter and 4 m high.

Various columns up to 10 m high and 600 mm in diameter have been built for processing organic intermediates. Pharmaceutical applications include stripping of antibiotics from solvent with up to 25% solids in suspension by use of a column 750 mm in diameter and 4 m high. Trials with nickel and cobalt extraction on columns 225 mm in diameter and 6 m high, iron recovery from phosphoric acid, and phosphoric acid production and purification are in progress.

Pulsed columns for nuclear separation processes were first investigated in the United States [77]. This led quickly to the use of pulsed plate columns for uranium recovery from stored waste [78], plutonium recovery [79], and irradiated fuel reprocessing at the large Hanford Purex Plant [2, 3]. They have also been installed in the reprocessing plants at West Valley, New York and Barnwell, South Carolina. In other countries pulsed columns have been or are being installed in the following plants: Trombay, India; Marcoule, France; Eurochemic, Belgium; and Windscale, United Kingdom. The more pertinent investigations of pulsed column application in the nuclear industry are listed in Table 6.

8. PULSE GENERATORS

Pulse generators for pulsed plate columns can be divided into two main groups, those with mechanical and those with fluid operation.

Several types of mechanically driven generator have been described. The conventional type has a piston driven by a crank and connecting rod or a yoke mechanism, which produces a pulse with a sinusoidal wave form. These are similar to positive-displacement pumps, which, with simple modification, can serve as pulse generators. If the displacement can be varied as in metering pumps, the pulse amplitude may be varied. Frequency variation is achieved either by controlling the drive motor speed or by incorporating a variable-speed drive coupling in the gear train.

In addition to the mechanical complexity of these generators, a disadvantage is the leakage of process solution past the piston seal. This can be overcome by having an electromagnetically operated piston at the base of a U tube with the column forming one side of the U and the other side being a balance leg. A diaphragm or bellows can also be used to provide a completely enclosed system. Both can be operated by either a crank-driven rod or a rotating cam. This is a good method for small-diameter columns, and diaphragms have been successfully used for many years at Windscale (UK) in the pulsed packed columns used for plutonium purification [80].

Two types of fluid-operated pulse generator have been described: (1) those that use an incompressible fluid that operates a drive piston coupled to a diaphragm bellows or pulse piston [80]; and (2) those that use a compressible fluid,

TABLE 6 INVESTIGATIONS OF PULSED COLUMN APPLICATION IN THE NUCLEAR INDUSTRY

System	Reference	Comments
Redox-type process (MIBK–uranyl nitrate–nitric acid)	77	One of the earliest studies of operation and performance of a pulsed column; studied effects of plate geometry and pulsing conditions
Redox-type process	99	Used a 5-ft. × 5-in.-diameter column with stainless steel plates; 0.04-in. hole diameter found to be large enough for this system
Purex-type process (TBP–diluent–uranyl nitrate–nitric acid) (TBP-tri-*n*-butyl phosphate)	2	Studied operation and performance of pulsed column, using range of plate geometries and column diameter; recommended use of redistributor plates in larger-diameter columns
Purex-type process	9	Studied operation and performance of pulsed columns in a range of diameters up to 27 in. for the various stages of the Purex process
Purex-type process	19	Studied operation and performance of pulsed columns, including choice of dispersed phase
Purex-type process	21	A detailed study of operation and performance of pulsed columns using stainless steel and Teflon-coated nozzle plates for organic and aqueous-phase dispersed systems
Purex-type process	48	Studied operation and performance of pulsed columns for different geometries and column diameters
Purex-type process	100	Studied operation and performance for organic-phase continuous, nozzle plates, and mixed-plate cartridges
Purex-type process	101	Studied operation and performance of pulsed columns for the Hanford plant
Purex-type process	102	Studied operation and performance, using range of plate geometries and also alternative packings
Purex-type process	103	Studied use of thick and thin organic polymer sieve plates together with stainless steel nozzle plates
Purex-type process	104	Studied the use of mixed-plate cartridges
Purex-type process	106	Measured the effects of elevated temperature on column performance
Purex-type process	107	Studied the effects of operating and column geometry variables on coalescence in pulsed columns
Purex-type process	108	Studied a uranium recovery process in a 25-ft × 3-in.-diameter column
Purex-type process	78	Evaluated a uranium recovery process in a range of column diameters, 3–16 in.
Purex-type process	109	Describes in detail the design and operation of a full scale Purex process plant
Uranium–thorium separation	110	Studied the operation and performance of a 2-in.-diameter column fitted with "standard" geometry plates

for example, air, working on one side of a U tube with the pulsed column forming the other leg [82–84]. Air pulsing has received the greatest attention because air, or a suitable inert gas, can be in direct contact with the process solutions, and the only mechanical parts required are those needed to control the air flow to and from the pulse leg.

A pulse generator has been developed by the Dutch State Mines Company for pulsing large-

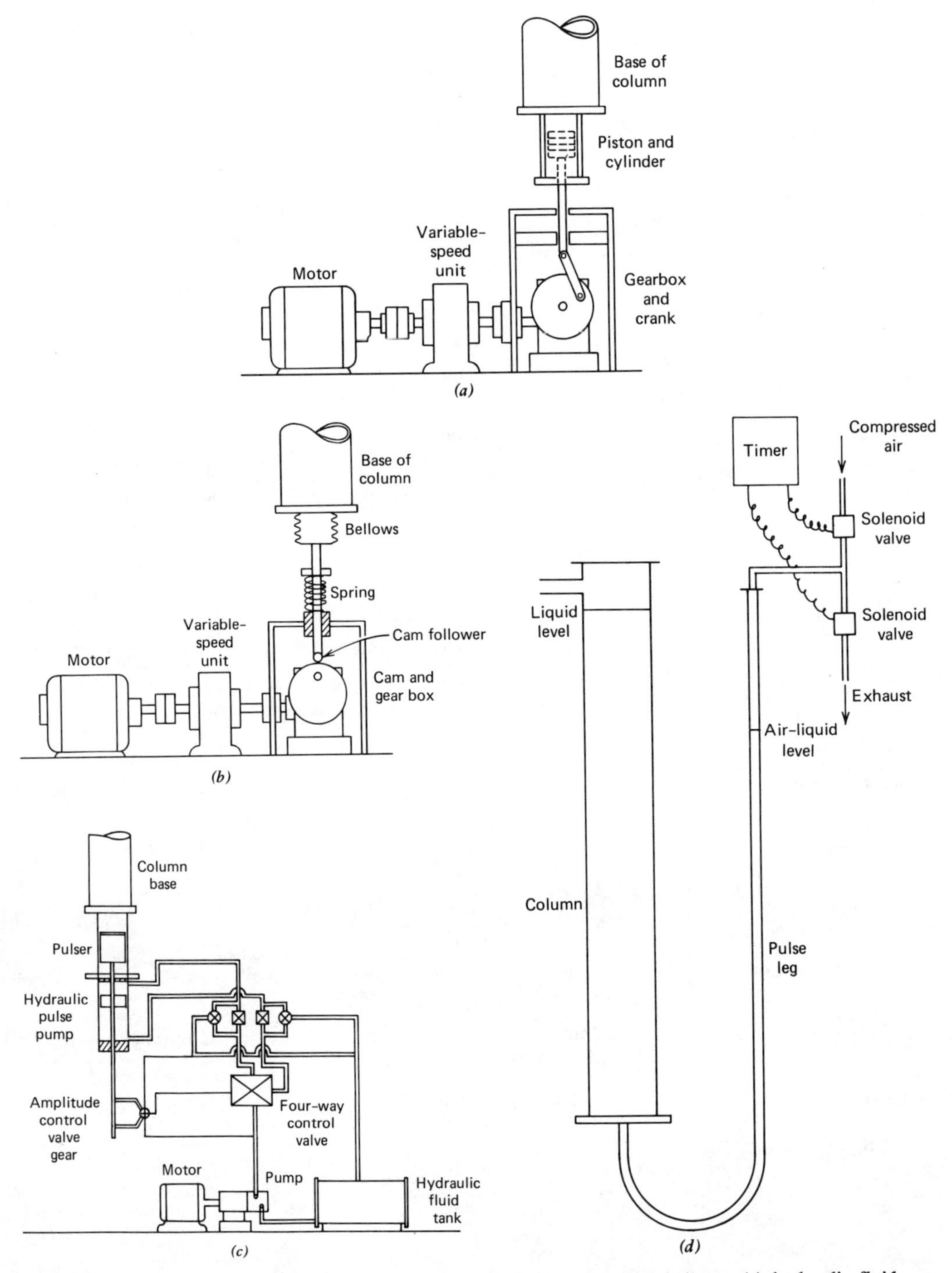

Figure 2. (*a*) Simple reciprocating pulse generator; (*b*) cam-operated bellows; (*c*) hydraulic fluid-operated pulse generator; (*d*) air-operated pulse generator.

diameter columns. A rotary valve connected to the base of the column links the column first with a pair of pressure vessels (upstroke) and then with a pair of suction vessels (downstroke). Liquid is pumped by a centrifugal pump through the pressure vessels and sucks from the suction vessels. The speed of rotation of the valve controls the pulse frequency, and the pumping rate controls the pulse amplitude [111]. Schematic arrangements of various pulse generators are shown in Fig. 2.

Weech and Knight [83] derived the three equations needed to characterize an air pulse system. The equations can be used to calculate the pulse amplitude for a range of frequencies and air pressures and also to show that the amplitude decreases markedly at higher frequencies. Baird and Ritcey [84] considered the case of a column operating at its natural frequency. The advantages for pulsing at the natural frequency are that the energy and air consumption are minimized, a sinusoidal wave form is obtained, and the energy and air requirements are easy to predict [85, 86].

Jealous and Johnson [81] derived the following equation for pulsing power requirements.

$$\text{Power} = S_1 \Bigg[(\rho_1 h_1 - \rho_2 h_3) g + \left(\rho_1 h_1 + \rho_2 h_2 \frac{S_1}{S_2} \right) \frac{d^2 y}{dt^2} + \left\{ \frac{\dfrac{n(1 - S^2)}{0.36 S^2} \rho_1 + \left(\dfrac{S_1}{S_2} - 1 \right)^2 \rho_2}{2} \right\} \cdot \left(\frac{dy}{dt} \right)^2 \Bigg] \frac{dy}{dt} \tag{23}$$

The term y is defined by the motion of the liquid that is derived from the pulse generator. Experimentally determined power requirements gave good agreement with calculated values at moderate frequencies; however, significant differences occurred at higher frequencies.

Pulse air consumption and power dissipation have been measured in pulsed packed and baffle plate columns [88]. The measurements were found to agree approximately with theoretical predictions based on a quasi-steady-state model. Agreement with this model was also found for the time-averaged power consumption in a reciprocating plate extraction column [89].

ACKNOWLEDGMENTS

Correspondence and discussions in the preparation of this chapter with Dr. S. Anderson, Professor A. L. Babb, Mrs. R. Berger, Dr. J. M. de Carvalho, Dr. P. Fremeaux, Dr. T. Misek, Dr. H. Reissinger, and Mr. G. M. Ritcey are gratefully acknowledged.

NOTATION†

A_p	pulse amplitude (stroke)
c	solute concentration
C_0	orifice coefficient
d_m	mean drop diameter
d_0	plate hole diameter
D	column diameter
D_v	diffusivity of solute in dispersed phase
E	axial dispersion coefficient
f	pulse frequency
g	acceleration due to gravity
g_c	gravitational conversion factor
H_c	axial distance between adjacent plates
H_{oc}	overall height of a transfer unit
h_1	effective height of column
h_2	total length of pulse leg to point P
h_3	height of P above bottom of column
K	mass transfer coefficient
n	number of plates in column
n'	number of stages in column
r	V_d/V_c
S	plate free area
S_1	cross-sectional area of column
S_2	cross-sectional area of pulse leg
t	time
T	plate thickness
V	superficial liquid flow rate in column
$\bar{V}_0$	characteristic drop velocity (mean drop velocity when $V_c = 0$ and $V_d \simeq 0$)
V_{slip}	mean drop velocity relative to continuous phase
V_T	terminal drop velocity
x	fractional holdup of dispersed phase
y	linear displacement of liquid in column
Z	axial distance
β	see Eq. (16)

†Any consistent set of units may be used unless stated in the text.

$\Delta\rho$	density difference of phases
μ	viscosity
ρ	phase density
ρ_1	effective density of fluid in column
ρ_2	density of fluid in pulse leg
σ	interfacial tension
ψ	maximum frictional power absorbed per unit mass of fluid by plate holes
ϕ	see Eq. (8)

Subscripts

c	continuous phase
d	dispersed phase
f	value at flooding

REFERENCES

1. W. J. D. Van Dijk, U.S. Patent 201,186 (1935).
2. G. Sege and F. W. Woodfield, *Chem. Eng. Progr.* **50** (8), 396 (1954).
3. R. G. Geier, USAEC Report TID-7534, Book 1, 107, 1957.
4. G. Richardson, USAEC Report HW-SA-2083, 1961.
5. R. B. Edwards and G. H. Beyer, *AIChE J.* **2,** 148 (1956).
6. R. Berger, W. Leuckel, and D. Wolf, *Chem. Ind.* 760 (1978).
7. R. G. Geier, *Proceedings of the 2nd Conference on Peaceful Uses of Atomic Energy*, Vol. 17, Geneva, 1958, Paper 515.
8. R. L. Stevenson and J. G. Bradley, USAEC Report HW-19170, 1951.
9. G. A. Nicholson, USAEC Report HW-40550, Vol. 1, 1956, pp. 2, 79, 109.
10. G. L. Richardson and G. Sege, USAEC Report HW-27171, 1954.
11. D. H. Logsdail and J. D. Thornton, *Transact. Inst. Chem. Eng.* **35,** 331 (1957).
12. J. D. Thornton, *Transact. Inst. Chem. Eng.* **35,** 316 (1957).
13. A. C. Jealous and E. Lieberman, *Chem. Eng. Progr.* **52** (9), 366 (1956).
14. G. A. Sehmel and A. L. Babb, *Ind. Eng. Chem. Process Des. Devel.* **2,** 38 (1963).
15. R. M. Cohen and G. H. Beyer, *Chem. Eng. Progr.* **49,** 279 (1953).
16. R. L. Bell and A. L. Babb, *Ind. Eng. Chem. Process Des. Devel.* **8,** 392 (1969).
17. D. Defives and G. Schneider, *Genie Chim.* **85,** 245 (1961).
18. R. Dewitte, F. Galuppi, and J. A. Fernandez, Eurochemic Report ETR-108, 1961.
19. J. Durandet and X. Talmont, *Bull. Inform. Sci. et Tech. (Paris)* (42), 17 (1960).
20. H. R. Foster, R. E. Mckee, and A. L. Babb, *Ind. Eng. Chem. Process Des. Devel.* 9 (2), 272 (1970).
21. D. H. Logsdail and G. S. Larner, UKAEA Reports AERE R4408 and AERE R4409, 1964.
22. T. Miyauchi and H. Oya, *AIChE J.* **11,** 395 (1965).
23. F. P. Pike and E. E. Erickson, USAEC Report ORO-288, 1955.
24. W. H. Swift, USAEC Report HW-33953, 1954.
25. W. L. Griffith, G. R. Jasny, and H. T. Tupper, USAEC Report AECD-3440, 1952.
26. S. Z. Kagan, M. E. Aerov, V. Lonik, and T. S. Volkova, *Internatl. Chem. Eng.* **5,** 656 (1965).
27. W. S. Groenier, R. A. McAllister, and A. D. Ryon, USAEC Report ORNL 3890, 1966.
28. G. C. Coggan, *Inst. Chem. Eng. Symp. Ser.* (26), 151 (1967).
29. R. A. McAllister, W. S. Groenier, and A. D. Ryon, *Chem. Eng. Sci.* **22,** 931 (1967).
30. K. Endoh and Y. Oyama, *J. Sci. Res. Inst.* No. 1486, **52,** 131 (1958).
31. T. Misek, *Coll. Czech. Chem. Commun.* **28,** 570 (1963); **29,** 1955 (1964).
32. J. Kubica and K. Zdunkiewicz, *Inz. Chem.* **7,** 903 (1977).
33. G. F. Scheele and B. J. Meister, *AIChE J.* **14,** 9 (1968).
34. C. B. Hayworth and R. E. Treybal, *Ind. Eng. Chem.* **42,** 1174 (1950).
35. J. M. Coulson and J. F. Richardson, *Chemical Engineering*, Vol. 2, Pergamon Press, New York, 1960, p. 486.
36. L. L. Burger and W. H. Swift, USAEC Report HW-28867, 1953.
37. A. L. Bell, Ph.D. thesis, University of Washington, Seattle, 1964.
38. L. D. Smoot and A. L. Babb, *Ind. Eng. Chem. Fund.* **1,** 93 (1962).
39. B. E. Clayburgh, Ph.D. thesis, Oklahoma State University, Stillwater, 1961.
40. K. Nishikawa, W. Nakamura, W. Eguchi, and S. Nagata, *Chem. Eng. (Jap.)* **26,** 678 (1962).
41. S. C. Jones, Ph.D. thesis, University of Michigan, Ann Arbor, 1963.
42. B. W. Mar and A. L. Babb, *Ind. Eng. Chem.* **51,** 1011 (1959).
43. G. A. Sehmel and A. L. Babb, *Ind. Eng. Chem. Process Des. Devel.* **3,** 210 (1964).
44. W. Eguchi and S. Nagata, *Kagaku Kogaku* **23,** 146 (1959).

45. G. Kyuchukov and D. Elenkov, *Bulg. Izv. Khim.* 8, 240 (1975).
46. W. Arthayukti, G. Muratet, and H. Angelino, *Chem. Eng. Sci.* **31,** 1193 (1976).
47. P. Novotny, J. Prochazka, and J. Landau, *Can. J. Chem. Eng.* **48,** 405 (1970).
48. H. Rouyer, J. Lebouhellec, E. Henry, and P. Michel, *Proceedings of the International Solvent Extraction Conference, 1974, Lyons, France,* Society of Chemical Industry, London, 1974.
49. K. V. K. Rao, S. A. K. Jeelani, and G. R. Balasubramanian, *Can. J. Chem. Eng.* **56,** 120 (1978).
50. F. W. Woodfield and G. Sege, *Chem. Eng. Progr. Symp. Ser.* **50** (13), 14 (1954).
51. L. D. Smoot, B. W. Mar, and A. L. Babb, *Ind. Eng. Chem.* **51,** 1005 (1959).
52. A. Handlos and T. Baron, *AIChE J.* **3,** 127 (1957).
53. F. Garner, A. Foord, and M. Tayeban, *J. Appl. Chem.* **9,** 315 (1959).
54. P. M. Rose and R. C. Kintner, *AIChE J.* **12,** 530 (1966).
55. G. Feick and H. M. Anderson, *Ind. Eng. Chem.* **44,** 404 (1952).
56. G. C. Coggan, *Inst. Chem. Eng. Symp. Ser.* (26), 138 (1967).
57. C. A. Plank and E. R. Gerhard, *Ind. Eng. Chem. Process Des. Devel.* **1,** 34 (1963).
58. V. Khemangorn, J. Molinier, and H. Angelino, *Chem. Eng. Sci.* **33,** 50 (1978).
59. M. A. Mandil, G. W. Mason, and D. F. Peppard, *Ind. Eng. Chem. Process Des. Devel.* **2,** 106 (1963).
60. K. H. Reissinger and J. Schroeter, *Inst. Chem. Eng. Symp. Ser.* (54) (1978).
61. I. I. Ponikarov, A. M. Nikolaev, and N. M. Zhavoronkov, *Internatl. Chem. Eng.* **2,** 546 (1962).
62. W. J. Korchinsky, *Can. J. Chem. Eng.* **52,** 468 (1974).
63. B. A. Diliddo and T. T. Walsh, *Ind. Eng. Chem.* **53,** 801 (1961).
64. J. C. Biery and D. R. Boylan, *Ind. Eng. Chem. Fund.* **2,** 44 (1963).
65. S. E. J. Britto, M.Sc. thesis, University of Bradford, Bradford, UK, 1968.
66. J. E. Watjen and R. M. Hubbard, *AIChE J.* **9,** 614 (1963).
67. H. Angelino and J. Molinier, *Proceedings of the International Solvent Extraction Conference, 1971, The Hague, Netherlands,* Society of Chemical Industry, London, 1971.
68. G. Churet, P. Fremeaux, E. Henry, and R. Malaterre, French Patent 75 34372, 1975.
69. R. DeWitte and L. Geens, Eurochemic Report ETR-159 and ETR-155, 1963.
70. S. M. Karpacheva, E. I. Zakharov, V. N. Koshkin, O. K. Maimur, V. A. Syakov, and V. F. Abramkin, *J. Appl. Chem. USSR* **47,** 317, 339, 821 (1974).
71. G. M. Ritcey, E. G. Joe, and A. W. Ashbrook, *Transact. AIME* **238,** 330 (1967).
72. G. M. Ritcey, M. J. Slater, and B. H. Lucas, in D. J. I. Evans and R. S. Schoemaker, Eds., *Proceedings of the 2nd International Hydrometallurgy Symposium, Chicago, 1973.*
73. G. M. Ritcey, *Chem. Ind.* 1294 (1971).
74. K. L. Huppert, W. Issel, and W. Knock, *Proceedings of the International Solvent Extraction Conference*, 1974, Lyons, *France*, Society of Chemical Industry, London, 1974.
75. U. Kuylenstierna and H. Ottertun, *Proceedings of the International Solvent Extraction Conference, 1974, Lyons, France,* Society of Chemical Industry, London, 1974.
76. J. M. Demarthe, M. Tarnero, P. Miquel, and J. P. Goumonoly, *Proceedings of the International Solvent Extraction Conference, 1974, Lyons, France,* Society of Chemical Industry, London, 1974.
77. W. A. Burns, C. Groot, and C. M. Slansky, USAEC Report HW-14728, 1949.
78. R. L. Stevenson and J. G. Bradley, USAEC Report HW-19170, 1951.
79. B. F. Judson, *Progress in Nuclear Energy*, Series III, Vol. 2, 302, Pergamon Press, 1958.
80. G. R. Howells, T. G. Hughes, D. R. Makey, and K. Saddington, *Proceedings of the 2nd International Conference on Peaceful Uses of Atomic Energy*, Geneva, 1958, Vol. 17, p. 3.
81. A. C. Jealous and H. F. Johnson, *Ind. Eng. Chem.* **47,** 1159 (1955).
82. J. D. Thornton, *Chem. Eng. Progr. Symp. Ser.* **50** (13), 39 (1954).
83. M. E. Weech and B. E. Knight, *Ind. Eng. Chem. Process Des. Devel.* **6,** 481 (1967).
84. M. H. I. Baird, and G. M. Ritcey, *Proceedings of the International Solvent Extraction Conference, 1974, Lyons, France,* Society of Chemical Industry, London, 1974.
85. M. H. I. Baird, *Proceedings of the Joint AIChE/IChemE Symposium, London,* Vol. 6, 1965, p. 53.
86. W. J. W. Vermijs, *Proceedings of the Joint AIChE/IChemE Symposium, London,* Vol. 6, 1965, p. 98.
87. E. I. Zakharov and S. M. Karpacheva, *Atomnaya Energiya* **45,** 343 (1978).
88. M. H. I. Baird and J. H. Garstang, *Chem. Eng. Sci.* **22,** 1663 (1967).
89. M. M. Hafez and M. H. I. Baird, *Transact. Inst. Chem. Eng.* **56,** 229 (1978).

90. M. W. Belaga and J. E. Bigelow, USAEC Report KT-133, 1952.
91. H. D. Bauermann and E. Blass, *Ger. Chem. Eng.* **1,** 99 (1978).
92. W. A. Chantry, R. L. Von Berg, and H. F. Wiegandt, *Ind. Eng. Chem.* **47,** 1153 (1955).
93. W. H. Li and W. M. Newton, *AIChE J.* **3,** 56 (1957).
94. G. M. Ritcey, *Inst. Chem. Eng. Symp. Hydromet.* **42** (April 1975).
95. J. R. Werning, K. B. Higbie, J. T. Grace, B. F. Speece, and H. L. Gilbert, *Ind. Eng. Chem.* **46,** 644 (1954).
96. G. Laurence, M. T. Chaieb, D. Hubsieger, P. Michel, and J. Talbot, *Proceedings of the International Solvent Extraction Conference, 1974, Lyons, France*, Society of Chemical Industry, London, 1974.
97. G. M. Ritcey and A. W. Ashbrook, Eds., *Solvent Extraction–Principles and Applications to Process Metallurgy*, Elsevier, Amsterdam, 1979.
98. J. C. Mishra and D. K. Dutt, *Chem. Eng. World*, **7** (9), 33 (1972).
99. W. S. Figg and J. E. Bradley, USAEC Report HW-19023, 1951.
100. G. M. Hesson, USAEC Report HW-49181, 1957.
101. R. G. Geier, USAEC Report TID-7534, Vol. 1, 107, 1957.
102. G. Jansen and G. L. Richardson, USAEC Report HW 68846, 1961.
103. R. D. Dierks and N. P. Wilburn, USAEC Report HW 79717, 1963.
104. G. L. Richardson, USAEC Report HW-84473 1964.
105. G. L. Richardson and G. Sege, USAEC Report HW-27171, 1954.
106. L. L. Burger, USAEC Report HW 29001, 1953.
107. W. R. Hamilton, USAEC Report HW 56281, 1959.
108. E. Lieberman and A. C. Jealous, USAEC Report ORNL 1543, 1953.
109. Purex Plant Technical Manual, USAEC Report HW 31000, 1955.
110. A. D. Ryon and R. S. Lowrie, USAEC Report ORNL 3732, 1965.
111. A. J. F. Simons, *Chem. Ind.* 748 (1978).

12

Reciprocating-Plate Extraction Columns

Teh C. Lo
Hoffmann-La Roche, Inc.
United States

Jaroslav Prochazka
Czechoslovak Academy of Sciences
Czechoslovakia

1. INTRODUCTION

Various types of mechanical agitation have been used to improve mass transfer in liquid–liquid contactors (see Chapter 16). The input of pulsating energy is one technique for obtaining a high rate of mass transfer in contactors. The principle of adding pulsating mechanical energy to an extraction column was originated in 1935 by Van Dijck [1] who proposed that extraction efficiency of a perforated-plate column could be improved either by pulsing the liquid contents of the column or by reciprocating the plate. The first idea has been extensively utilized (see Chapters 11.1 and 11.2). The second technique had been relatively little exploited until the late 1950s. In 1959 Karr [2] reported data on a 76-mm (3-in.)-diameter open-type perforated reciprocating-plate column. The column was further developed by Karr and Lo [3] by employing baffles for scale-up. Various columns characterized by reciprocating perforated plates or packing have been reported [4–11]. Interest in research on this type of column has increased in the past two decades. Most of the research data have lent support to the conclusion that reciprocating-plate columns generally have high volumetric efficiencies.

While pulsed columns are open to the criticism that considerable energy is required to pulse the entire liquid content of a column, particularly on a large-scale commercial extractor [12], reciprocation of the plates is an alternative solution to achieving uniform dispersion and similar mixing patterns by use of relatively much less energy. Some units that have been built for industrial use and that have particular research interest are described in the following paragraphs. The general features and industrial applications of this type of column are given in Table 1 in Chapter 16.

Two main types of reciprocating plate column, the "open" type [3] and the "segmental passage" type [46], are widely used in industry. The main difference between the two types lies in the design and function of the plate. In the first design the plates are of open structure with large holes and large free area (~58%). In the second design the plates have small holes and small free area with or without downcomers. These two types of columns have been developed independently, with the former in use mainly in North America and the latter mainly in eastern Europe and the Soviet Union.

2. OPEN-TYPE PERFORATED RECIPROCATING-PLATE COLUMN

2.1. Description of Column

The column developed by Karr and Lo [3] consists of a stack of perforated plates and baffle plates that have a free area of about 58%. The central shaft that supports the plates is reciprocated by means of a reciprocating drive mechanism located at the top of the column. The amplitude is adjustable generally in the range 3–50 mm ($\frac{1}{8}$–2 in.), and the reciprocating speed is variable up to 1000 strokes/min. Figure 1 shows a schematic arrangement of a 0.9-m (36-in.)-diameter reciprocating-plate column. Figure 2*a* shows a typical perforated plate and Fig. 2*b* shows a typical baffle plate. Figure 3 shows the arrangement of perforated plates and baffle plates. The baffle plates are used periodically in the plate stack to minimize axial mixing [3, 14, 15]. Some baffle plates are made of Teflon and are slightly larger in diameter than the stainless steel perforated plates and baffle plates. This prevents metal–metal contact of steel plates and shell. Figure 4 shows a coupling arrangement of a reciprocating-plate column.

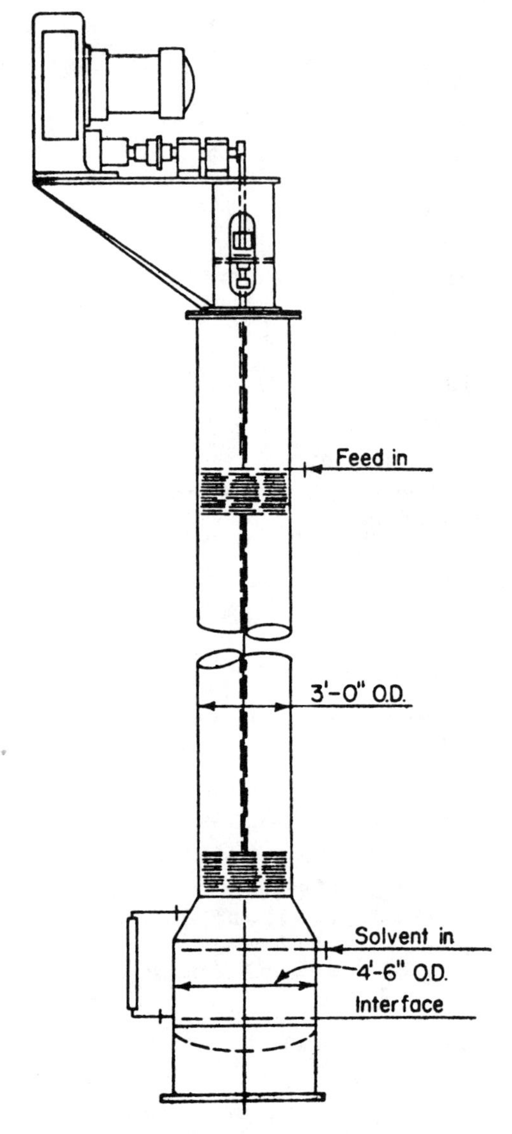

Figure 1. Schematic arrangement of the 914-mm (36-in.) reciprocating-plate column. [Courtesy of *Chem. Eng. Prog.* **72**(9), (1976).]

2.2. Performance Data

Volumetric efficiency as a measure of the effectiveness of a given extraction column is defined as follows:

$$\text{Volumetric efficiency} = \frac{\text{total throughput}}{\text{HETS}} = 1/\text{hr}$$

where total throughput = $m^3\ m^{-2}\ hr^{-1}$ or $ft^3\ ft^{-2}\ hr^{-1}$

HETS = height equivalent of theoretical stage, m (ft)

Performance data on various column sizes, 25, 76, 305, and 914 mm (1, 3, 12, and 36 in.) in diameter, have been reported [2, 3, 13–15]. A minimum HETS of 153 mm (6.12 in.) and a volumetric efficiency of 311 hr^{-1} were achieved

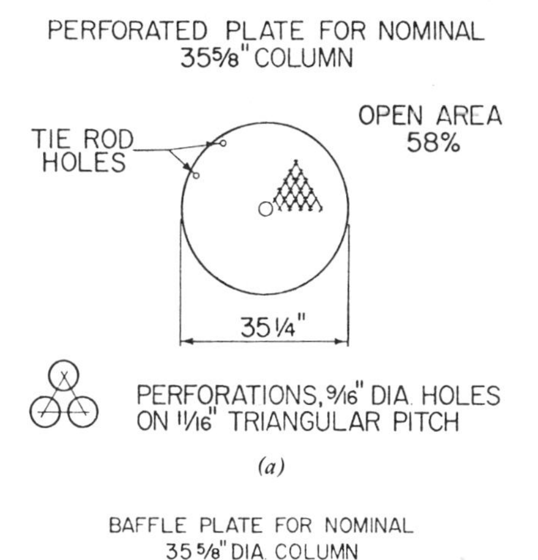

Figure 2. (*a*) Typical perforated plate. (*b*) Typical baffle plate.

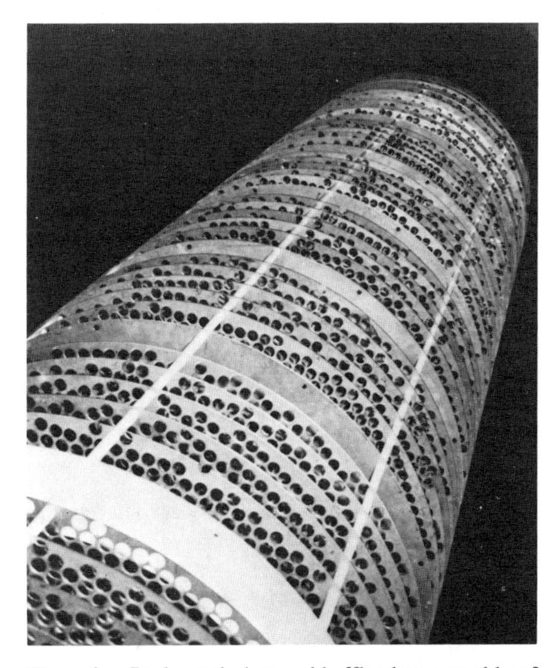

Figure 3. Perforated-plate and baffle-plate assembly of a reciprocating-plate column. (Courtesy of Chem-Pro Equipment Corporation.)

Figure 4. Coupling arrangement of a reciprocating-plate column (Courtesy of Chem-Pro Equipment Corporation.)

in a 300-mm (12-in.)-diameter column by use of the MIBK-acetic acid–water system. A minimum HETS of 508 mm (20 in.) has been measured in a 914-mm (36-in.)-diameter column by use of a relatively difficult extraction system, *o*-xylene-acetic acid–water. A summary of minimum HETS values for various column sizes at specific throughputs is given in Table 1. Figure 5 shows the effect of reciprocating speed on HETS.

2.3. Hydrodynamic Characteristics

Research studies on flooding [16, 17], drop size and holdup [18], power consumption [19], and axial mixing [17, 20] have been reported by Baird and co-workers. In the reciprocating-plate columns the presence of many closely spaced reciprocating impellers leads to conditions that closely approximate the "uniform isotropic turbulence" basis of Kolmogoroff's [21] theory. The uniform distribution of energy dissipation across the cross-sectional area of the column gives relatively uniform droplet size and low axial mixing, resulting in a high mass-transfer rate and low HETS. Based on his studies on a 51-mm (2-in.)-diameter reciprocating-plate column and a 152-mm (6-in.)-diameter pulsed-plate column, Baird has reported that reciprocating-plate columns of small diameter have relatively low axial mixing and that column diameter has an insignificant effect. This is in agreement with the conclusion of Rosen and Krylov [22] that axial dispersion coefficients increase with column diameter only in large columns under two-phase flow conditions that give significant transverse nonuniformity in the flow pattern.

For four different systems in which the interfacial tension was approximately 30 mN/m, the

TABLE 1 SUMMARY OF MINIMUM HETS VALUES AND VOLUMETRIC EFFICIENCIES FOR A RECIPROCATING-PLATE COLUMN[a]

Column Diameter in.	Amplitude, in.	Plate Spacing, in.	Agitator Speed, Strokes/min	Extractant	Dispersed Phase	Minimum HETS	Throughput, gal hr^{-1} ft^{-2}	Volumetric Efficiencies V_t/HETS, h^{-1}
			MIBK-Acetic Acid-Water System					
1	$\frac{1}{2}$	1	360	MIBK	Water	3.1	572	296
			401			2.8	913	523
1	$\frac{1}{2}$	1	278	Water	MIBK	4.2	459	175
			152			8.1	1030	204
3	$\frac{1}{2}$	1	330	MIBK	Water	4.9	600	196
	$\frac{1}{2}$	1	245			6.3	1193	304
	$\frac{1}{2}$	2	355			7.5	1837	393
	$\frac{1}{2}$	1	320	Water	Water	4.3	548	205
	$\frac{1}{2}$	1	230			6.7	1168	280
	$\frac{1}{2}$	2	367	Water	Water	5.0	1172	376
			240			7.75	1707	353
12	$\frac{1}{2}$	1	430	Water	MIBK	5.8	547	151
(with			285			5.7	1167	328
baffle)	$\frac{1}{2}$	1	244	MIBK	MIBK	4.4	599	218
			170			5.6	1193	342
	$\frac{1}{2}$	1	250	MIBK	Water	7.2	602	134
			225			7.2	1200	268
			150			14.0	1821	208
	$\frac{1}{2}$	1	225	Water	Water	7.0	555	127
			200			9.5	1170	197
			150			11.05	1694	246
	$\frac{1}{2}$	1	275	Water	MIBK	9.5	1179	199
	$\frac{1}{2}$	1	200	MIBK	MIBK	7.8	595	123
			150			6.2	1202	311
			Xylene-Acetic Acid-Water System					
3	1	1	267	Water	Water	9.1	424	75
3	$\frac{1}{2}$	1	537	Water	Water	8.2	424	83
3	$\frac{1}{4}$	1	995	Water	Water	7.7	424	88
3	1	2	340	Water	Water	9.1	804	142
36	1	1	168	Water	Water	23.3	425	29[b]
36	1	1	168	Xylene	Water	20.0	442	36[b]

[a]Karr and Lo [2, 3, 13–15].
[b]Because of instrumentation limits, the maximum volumetric efficiencies have not been explored.

following correlation of the flooding rate has been found by Baird et al. [16]:

$$U_d + 0.67U_c = 24.2(Af)^{-1.2}\rho_c^{-1/3} \cdot (\rho_d - \rho_c)^{2/3}\left(\frac{\mu_c}{\mu_{co}}\right)^{-1/3} \quad (1)$$

The length units in this equation are centimeters.

A semitheoretical equation was derived for the general case [16]

$$1.5\,U_d + U_c = 0.0224\left(\frac{\sigma^3}{\psi^2\bar{\rho}}\right)^{0.2}\left(\frac{g^2\Delta\rho^2}{\rho_c\mu_c}\right)^{1/3} \quad (2)$$

Power dissipation per unit volume is expressed as

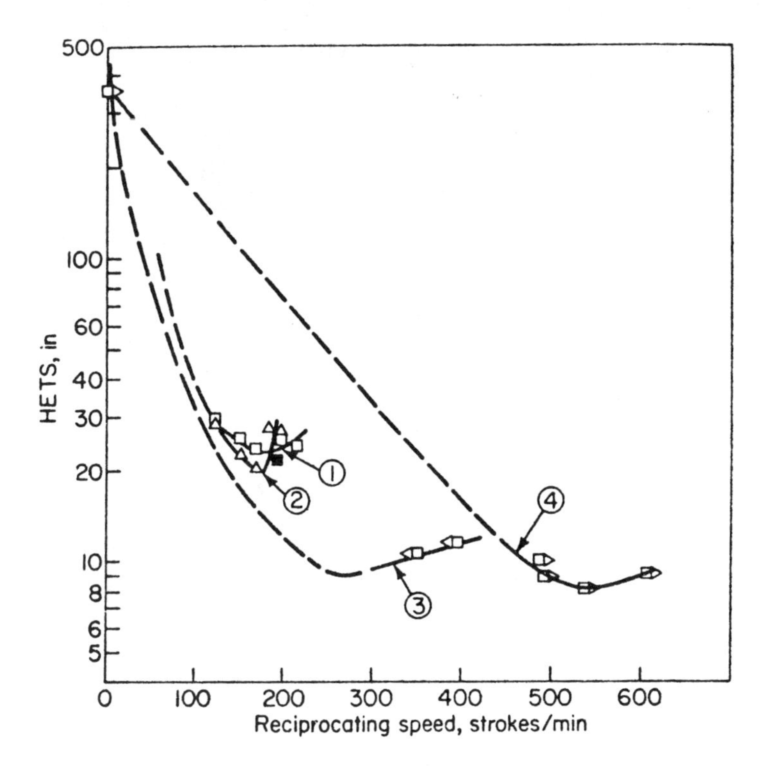

Symbol	Curve No.	Column diam, in	Phase dispersed	Phase extractant	Double amplitude, in	Plate spacing, in	Total throughput, gal/(h)(ft²)
□	1	36	Water	Water	1	1	425
Δ	2	36	Water	Xylene	1	1	442
<□	3	3	Water	Water	1	1	424
□>	4	3	Water	Water	½	1	424
■	Predicted minimum based on exponents of 0.36 in Eq. (4) and 0.14 in Eq. (5).						

Figure 5. Effect of reciprocating speed on HETS, *o*-xylene–acetic acid–water system.

$$\psi = \left(\frac{2\pi^2}{3}\right)\bar{\rho}\left(\frac{1 - A_o^2}{C^2 A_o^2}\right)\frac{(Af)^3}{l} \tag{2a}$$

It can be seen that in cases where physical properties (e.g., density difference, $\Delta\rho$, and interfacial tension σ) vary in different sections of the column, power input or agitation should also be varied to prevent one section of the column from severely limiting throughput. Since agitation speed and amplitude are fixed throughout the column, the only means for variation of power input per unit volume of column (or agitation intensity) is to vary the plate spacing. Since power input per unit volume of column ψ is inversely proportional to the plate spacing l, it can be concluded from Eq. (2) that the optimum relative plate spacing in the different sections of the column is given in terms of the important properties $\Delta\rho$ and σ by the following equation [44]:

$$l \alpha \frac{1}{(\Delta\rho)^{5/3}(\sigma)^{3/2}} \tag{3}$$

This equation is useful in optimization of the degree of agitation throughout the column and hence in maximization of the throughput and extraction efficiency for a particular column.

2.4. Design and Scale-up

2.4.1. Design from Basic Principles

A great deal of work has been done on the design of internally agitated extraction columns from basic principles. A recent attempt to apply basic principles to design the above-mentioned reciprocating-plate column has been made [23]. The design model is based on several fundamental studies of hydrodynamics and mass transfer in reciprocating-plate columns of this type.

Basic correlations and theory are combined for modeling of the reciprocating-plate column. For a given system, flow rates, plate configuration, and the diameter at which flooding occurs can be calculated. If the required degree of extraction is known, the height of the plate stack can also be calculated. Figure 6 shows the information flow diagram for the model. The central box contains basic data that must be known. Calculations are made moving clockwise from ψ, the specific power dissipation, through drop size, to column diameter D selected to be slightly greater than the diameter for flooding; then the holdup ϕ is calculated from the slip velocity and then interfacial area a and mass-transfer coefficient. Finally, axial mixing data are introduced into the calculation, and the "plate stack height" is given for the required separation. The computation can be carried out on a small computer. The difficulty in applying the model lies not so much in the calculations, but in the choice of accurate values of "parameters" such as the axial dispersion coefficient and the mass-transfer coefficient. The model is useful as a general guide for the design, provided such information is available. Because the factors relating to mass transfer and hydrodynamics of a system in an extractor are extremely complex, particularly for mixed solvent and feedstocks of commercial interest, and usually at least some of the basic information is lacking (e.g., axial mixing coefficient and mass-transfer coefficient), one inevitably resorts to pilot tests, coupled with the use of empirical or semiempirical scale-up procedures. At the present state of knowledge, pilot-scale testing remains an almost inevitable preliminary to a full-scale contactor design.

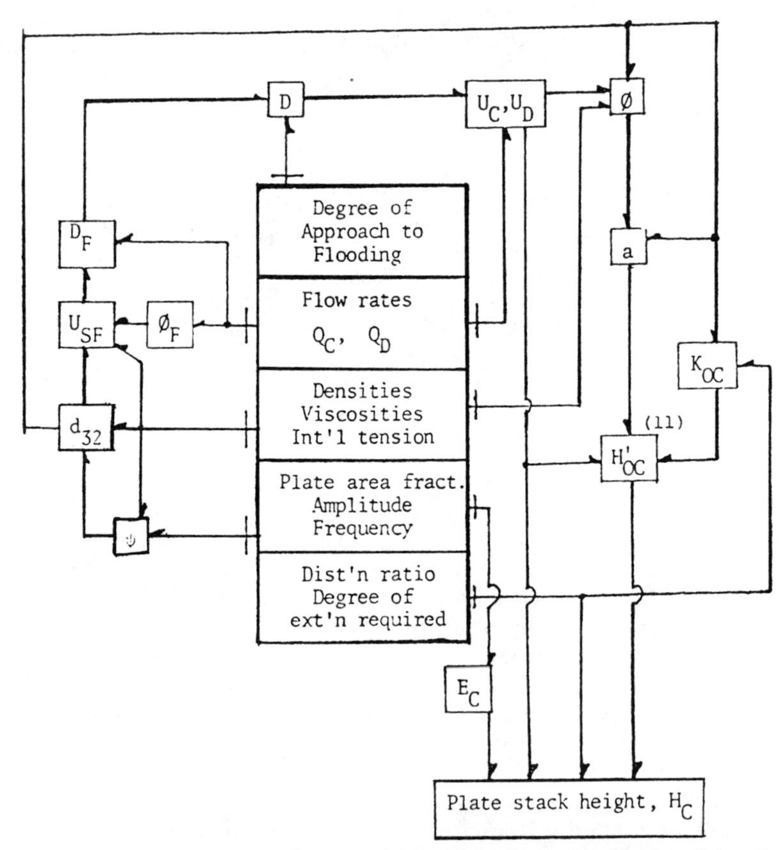

Figure 6. Information flow diagram for a model for designing a reciprocating-plate column [23]. (ψ = power dissipation per volume; d_{32} = sauter mean droplet diameter; ϕ = holdup (ϕ_F at flooding); U_S = slip velocity (U_{SF} at flooding); D = column diameter (D_F at flooding); U_C, U_D = superficial velocities; a = specific interfacial area; H'_{OC} = height of a transfer unit; E_C = axial dispersion coefficient; H_C = plate stack height).

2.4.2. *Pilot-Scale Test*

A successful design of a commercial reciprocating-plate extractor can be achieved only after optimization experiments have been done on a pilot scale. The pilot tests provide the following qualitative and quantitative information for scale-up and design of extractors:

- Total throughput and agitation speed.
- HETS or HTU.
- Stage efficiency.
- Hydrodynamic conditions—droplet dispersion, phase separation, flooding, emulsive-layer formation, and so on.
- Selection of dispersed phase or direction of mass transfer.
- Solvent : feed ratio,
- Material of construction and its wetting characteristics.
- Confirmation of desired separation.

The following procedures for obtaining the pilot-scale data [3, 44] for the reciprocating plate extraction columns are recommended:

1. *Estimation of Throughput, Agitation Speeds, and Test Column Height.* A preliminary estimate of the allowable range of throughput and agitation speed can be approximated from the correlations by Baird [16, 17] or from previous experience. The height of the plate stack selected for the tests will depend on the number of theoretical stages required for the separation. The design model described earlier is useful for this estimation purpose provided the information is available.

2. *Estimation of Plate Spacing Variation.* If the density difference, $\Delta\rho$, and the interfacial tension σ are known to vary from the top of the column to the bottom of the column, it is possible to estimate optimum plate spacing variables from Eq. (3). The minimum plate spacing to be used is 5 cm (2 in.) unless the plate spacing variation is very large, in which case the minimum plate spacing should be 2.5 cm [44].

3. *Minimum HETS versus Throughput.* Experiments are conducted at several different throughputs, and HETS is determined as a function of agitation speed. Since the minimum HETS occurs close to the flooding point in small-diameter columns, it is generally sufficient to determine the agitation speed at which the column will just flood and then reduce the agitation by 5–10%. Thus the HETS values determined will be close to the minimum values. The minimum values of HETS are then plotted against throughput. Generally, HETS increases with increase of throughput.

4. *Volumetric Efficiency versus Throughput.* Maximum volumetric efficiency at each throughput is then plotted against throughput and, if sufficient data are obtained, the volumetric efficiency will pass through a maximum, which is usually close to the optimal design throughput for the scale-up.

5. *Demonstration of Extraction Process.* The extraction process should be demonstrated to confirm that there are sufficient stages in the test column to obtain the desired separation and quality of product. There could be pitfalls in determining the height of an equivalent theoretical stage and then calculating the height of the column on the basis of the number of theoretical stages required for the separation. The possible pitfalls are due to inaccurate distribution data or inadequate *HETS* data obtained from the tests.

Provision should be made for solvent recovery and recycle. The solvent should be recycled for a reasonably long time to ensure that slow buildup of trace contaminants does not affect the performance of the extractor. Finally, the extended runs should be carried out to confirm the optimal process.

2.4.3. *Scale-Up Procedures*

On the basis of performance data obtained in 25-, 76-, 305-, and 914-mm (1-, 3-, 12-, and 36-in.)-diameter columns mentioned earlier, the following empirical equations for scale-up based on an experimental column of D_1 in diameter were presented by Karr and Lo [3, 14, 15]:

$$\frac{(\text{HETS})_2}{(\text{HETS})_1} = \left(\frac{D_2}{D_1}\right)^{0.38} \tag{4}$$

The corresponding reciprocating speed (*SPM*) required for the large-diameter column is given by

$$\frac{(\text{SPM})_2}{(\text{SPM})_1} = \left(\frac{D_1}{D_2}\right)^{0.14} \tag{5}$$

The exponent for HETS as a function of a diameter for a difficult extraction system which requires a higher operating speed and/or amplitude would be expected to be somewhat greater than the exponent for a relatively easy extraction system inasmuch as axial mixing is known to increase with reciprocating speed, especially in large-diameter columns. The exponent 0.38 in Eq. (4) is based on the relatively difficult extraction system of *o*-xylene–acetic acid–water. For MIBK–acetic acid–water, a relatively easy extraction system, the average exponent value is 0.25 although the exponent varies from 0.19 to 0.36, depending on which phase is dispersed and which phase is the extractant. For purposes of a safer design, an exponent value for a relatively difficult extraction system should be chosen.

The following scale-up procedures are recommended by Karr and Lo [3, 14, 15]:

1. Data are obtained on 25, 51, and 76-mm (1-, 2-, 3-in.)-diameter columns that need not have baffle plates [2, 13]; 51- and 76-mm (2- and 3-in.) pilot-plant columns are preferable for scale-up to columns exceeding 610 mm (2 ft) in diameter.
2. The optimum performance of the pilot-scale column is determined. The criterion for optimum performance is maximum volumetric efficiency in a column having optimum plate spacing.
3. The following parameters are kept constant for scaling up from the pilot plant data: (a) total throughput per unit area; (b) plate spacing; and (c) amplitude or stroke length.
4. The expected HETS is calculated by means of Eq. (4). A small upward–downward adjustment of the exponent may be justified if the degree of agitation required differs markedly from that of the system of *o*-xylene–acetic acid–water.
5. The corresponding reciprocating speed required is calculated by use of Eq. (5).
6. Suitably designed baffle plates should be provided at a suitable spacing.

This procedure has been successfully used to scale up several dozen commercial columns to 1.0 m in diameter [15, 43]. It is expected that the scale-up correlation can be extended to columns larger than 1.0 m in diameter.

2.5. Commercial Applications

The reciprocating-plate columns have gained increasing industrial application in the pharmaceutical, petrochemical, chemical, hydrometallurgical, and waste-water treatment industries [15]. They have been found to be suitable for processing mixtures with emulsifying tendencies (e.g., fermentation broth) and for liquids containing suspended solids. The reciprocating-plate column is now being employed extensively in laboratories, pilot plants, and industry around the world. A few applications are described to illustrate specific points.

A 1.0-m (40-in.)-diameter stainless steel column containing 9.75 m (32 ft) of plate stack is used for purification of an organic compound.

A 0.76-m (30-in.)-diameter stainless steel column containing 12.8 m (42 ft) of plate stack is used for fractional liquid extraction of a pharmaceutical product with more than 99.9% recovery efficiency and with high specification purity (Fig. 7).

A 0.61-m (24-in.)-diameter column is operating on a hexane–water solvent system. This system has a very high interfacial tension, approximately 40 mN/m and, therefore, requires a high degree of agitation, specifically, 5.4 Hz at a stroke length of 18 mm (325 strokes/min at $\frac{3}{4}$-in. stroke length).

A 0.3-m (12-in.)-diameter column was used for extraction of a whole fermentation broth containing about 5% unfiltered mycelia. A 0.46-m (18-in.)-diameter column was scaled up from a 25-mm-diameter column for similar application. Emulsions are avoided because of the uniform agitation over the cross-sectional area of the column.

A variety of corrosion-resistant materials of construction can be used in the reciprocating-plate column. A 0.76-m (30-in.)-diameter Teflon plate stack 9.1 m (30 ft) long is used in a glass-lined shell for a corrosive material.

Small-diameter reciprocating-plate extraction columns have been successfully used for countercurrent and fractional liquid extraction in laboratory and pilot-plant process development and scale-up work [13, 45]. A 25-mm (1-in.)-diameter column has proved to be very useful for the process feasibility study in laboratory and process development work [13, 45] and is shown in Fig. 12 of Chapter 17.1. A 51-mm (2-in.)-diameter reciprocating-plate column was used in pilot-plant work leading to the development of tetraethylene glycol (TETRA) as an efficient solvent in the Udex process [24]. Figure 8 shows a typical 51-mm (2-in.)-diameter pilot-scale extraction column [25a].

Applications and performance of the reciprocating-plate column as a cocurrent contactor

Figure 7. An industrial-scale reciprocating-plate column for multistage fractional extraction. (Courtesy of Hoffmann–La Roche, Inc.)

Figure 8. A 51-mm-diameter, pilot-scale reciprocating-plate column. (Courtesy of Julius Montz, GmbH.)

have been recently reported [25b, 26]. The advantages are high throughputs and reduced settling volume. The cocurrent mixer configuration has also been used effectively as a plug-flow reactor for two liquid phases. This type of agitated column is found to be particularly suitable for gas–liquid reaction operations with suspended solid catalysts and is expected to be more widely applied to various mass-transfer operations [27, 28] (e.g., gas absorption, gas–liquid reactions, liquid–liquid mass transfer with simultaneous chemical reaction). A vacuum distillation column utilizing reciprocating-plate contact between liquid and vapor was reported to produce 100% efficient operation [29].

The advantages of the open-type perforated reciprocating-plate extraction column can be summarized as follows: (1) high throughput and high mass transfer (hence, low HETS), and thus high volumetric efficiency; (2) great degree of versatility and flexibility, thus permitting optimization of laboratory or pilot-plant unit; (3) handling of emulsifiable material and liquids containing suspended solids; (4) facility of operation and low maintenance required; (5) simplicity in construction and a wide range of materials of construction.

3. RECIPROCATING COLUMNS–PERFORATED PLATES WITH SEGMENTAL PASSAGES

3.1. Description of Columns

This type of reciprocating-plate column, which is manufactured commercially under the trademark vibrating-plate extractor (VPE), has been developed comparatively recently [10, 30–32]. There are two modifications of the VPE: columns with uniform motion of plates and those with countermotion of plates.

The column with uniform plate motion is depicted in Fig. 9. The shell (1) of the column proper is connected at each end to settling sections of larger diameter. The upper settling section (2) is provided with a distributor for the heavy phase (3) and an overflow for the light phase (4). The lower settling section (5) contains an outlet for the heavy phase (6) and a distributor for the light phase (7). According to the choice of the phase being dispersed, either section may be equipped with an interface control. In the discussion that follows, the lighter phase is assumed to be dispersed.

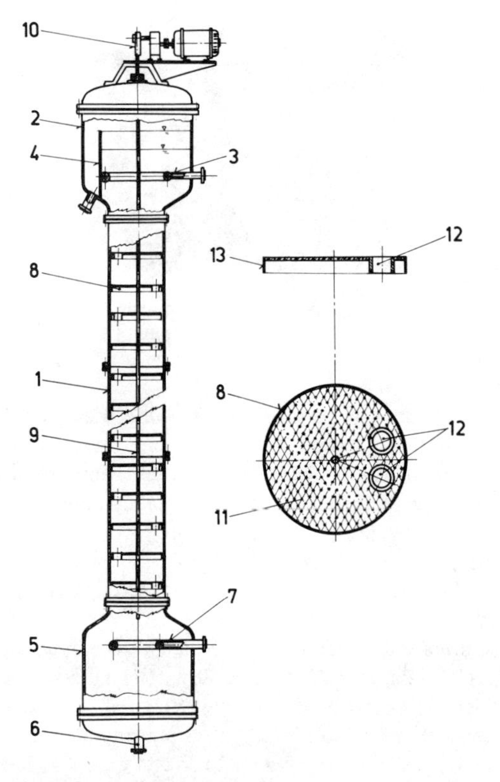

Figure 9. Column with uniform motion of plates.

A stack of perforated plates (8) is situated in the shell, fastened to a rod (9) connected to an eccentric (10) that imparts to the whole stack a vertical harmonic motion, the amplitude and frequency of which are among the important controlling variables. The geometry of the perforated plates is specific for the VPE. The plates are provided with a number of small circular holes for the dispersed phase (11) and with one or more large openings for the continuous phase (12). These openings may either be circular or have various forms suitable from the point of view of plate design. In columns of smaller diameter the passages for continuous phase of neighboring plates are placed on opposite sides of the column axis, so that a crossflow of phases between the plates can occur. On large plates, the distribution of passages is such that several parallel sections with a crossflow of phases are created. As a general rule, the plates are provided with a vertical cylindrical wall (13) on their circumference extending against the flow of the dispersed phase. A similar lining may be applied also to the passages for continuous phase.

This type of reciprocating plate displays several favorable properties. The separation of passages for continuous and dispersed phases ensures high throughputs of the dispersed phase even in case of high flow rates of the continuous phase. It also permits the maintenance of a stable mixer-settler regime over a wide range of amplitude-frequency products. In this regime a pattern with one dispersed phase can be ensured during the entire period of reciprocating motion because the passages for the continuous phase suppress the dispersion of the continuous phase into the layer of dispersed phase on the plate. This is important with respect to the well-known fact that the coalescence rates may depend strongly on the choice of the dispersed phase. In the small perforations the drops are formed or reshaped by a mechanism similar to that of periodic outflow from a nozzle. Splitting under such conditions does not require high velocity, and thus the VPE operates at relatively low amplitudes and frequencies. This implies low mechanical stress and energy consumption.

The overall shape of the column with countermotion of plates does not differ from that of the column with uniform plate motion. In this case, however, the set of plates is divided into inter-

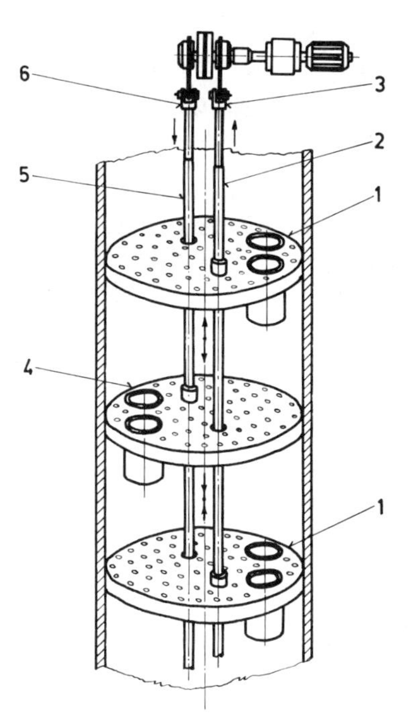

Fig. 10. Column with countermotion of plates.

Figure 11. Mixer–settler regime.

laced stacks, each of which performs its own harmonic motion. In practice, these motions are of the same amplitude and frequency but are shifted in phase by 180° so that the instantaneous velocities of neighboring plates are of equal magnitudes and opposite directions. This principle is represented schematically in Fig. 10. The group of plates (1) is fastened on the rod (2) actuated by the eccentric (3) and is free to move through the plates (4). These are fastened on the rod (5) connected to the eccentric (6). The countermotion of plates ensures high mechanical stability of the apparatus and contributes to its extraction efficiency.

3.2. Capacity and Mass-Transfer Data

The qualitative features of the flow patterns in a VPE column resemble those of the pulsed columns with perforated plates. Thus when the liquid system used is an easily coalescing one and when the velocity of reciprocating motion of plates is low and the phase velocities are high, a mixer–settler regime tends to be established. Under the plates, a coalesced dispersed-phase layer or densely packed drops form, at least during a part of the period of reciprocating motion. An example of this regime is shown in Fig. 11, where caprolactam is extracted from the toluene to the water, with the organic phase being dispersed. The other marginal pattern is called "the emulsion-type regime," which is likely to be established when the liquid system used is a slowly coalescing one. In this case high intensities of reciprocating motion are applied and the velocities of phases are low. These conditions prevail in extraction of caprolactam from a crude water solution (dispersed phase) into toluene (Fig. 12).

The limiting flow rates of a VPE depend on a number of factors, including phase velocity ratio, amplitude and frequency of vibrations, plate geometry, physical properties of liquids, intensity and direction of mass transfer, and the choice of phase being dispersed. The limiting flow rates of the dispersed phase for a VPE plate and a standard perforated plate as currently used in pulsed columns are compared in Fig. 13. Both plates have equal fractional free areas and diameters of perforations. The VPE plate clearly gives higher flow rates, particularly in the mixer–settler regime. In this case the normal perforated plates flood as a result of insufficient pulsation and the limiting flow rates decrease almost proportionally with Af, but the limiting flow rates of the VPE plates are independent of Af.

Figure 12. Emulsion-type regime.

A guideline can be obtained from Fig. 13 for safe prediction of VPE capacity. In the emulsion-type regime the capacity can be calculated according to the relationships derived for a normal sieve plate with holes of the same diameter and equal fractional free area of perforations and of openings for the continuous phase [33]. The maximum value for this plate can be taken as a conservative estimate for the entire mixer–settler region. The continuous phase velocity has little effect on the limiting velocity of the dispersed phase as long as the condition

$$\frac{U_c}{A_{oc}} \leqq 0.3 \left(\frac{2\pi Af}{A_{oc} + A_{od}} \right) \tag{6}$$

is observed for the specific free area of the openings for the continuous phase.

For the prediction of the specific volumetric holdup of the dispersed phase, the relationship proposed by Misek [36]

$$\frac{U_d}{X} + \frac{U_c}{1 - X} = U_o(1 - X) \exp(\mathrm{bX}) \tag{7}$$

may be applied. In Fig. 14 the results for the toluene-water system are given. They confirm the independence of the characteristic velocity U_o from U_d and show the systematic decrease

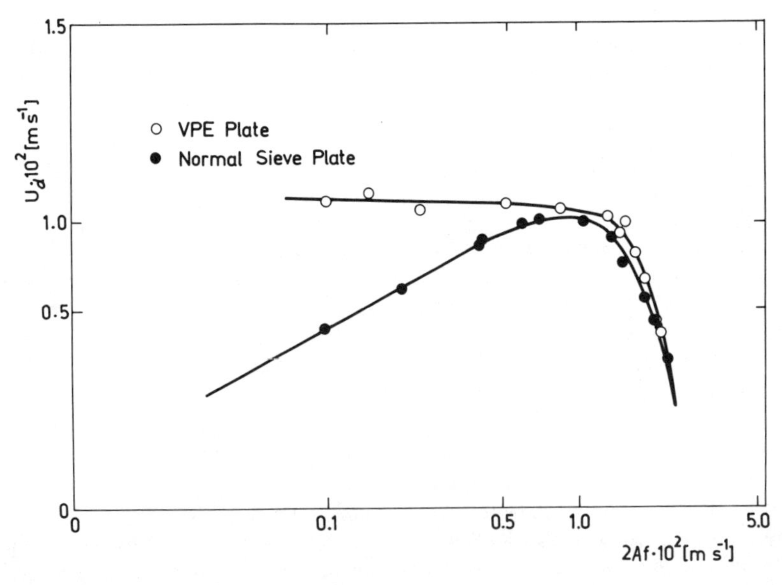

Figure 13. Comparison of limiting velocities of a dispersed phase for VPE plates and normal perforated plates of equal free areas.

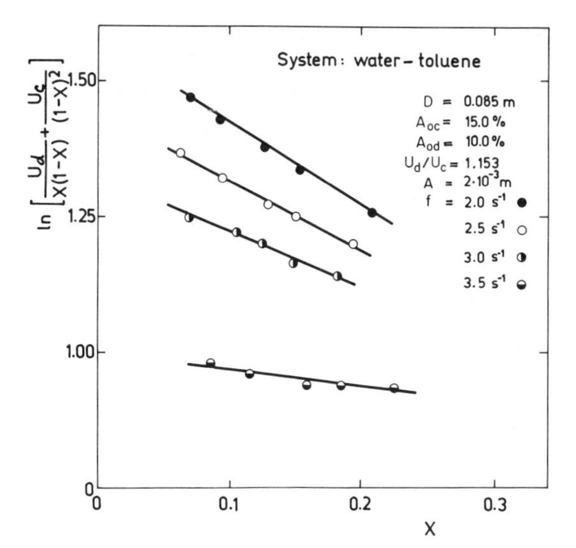

Figure 14. Dependence of holdup on dispersed-phase velocity.

of U_o and b with increasing frequency of vibrations. In practical design, the holdup is kept within 15–25%, which corresponds to 70–80% of flooding throughput. From Fig. 13 it follows that, in the transition regime as well as in the emulsion-type regime, these conditions are easily maintained in column operation for a wide range of throughputs by varying the frequency and the amplitude of vibration.

Important design variables of a VPE are maximum pressure difference across the plate and maximum power consumption. A detailed analysis of the forces acting in a vibrating or a pulsed-plate column and some experimental data have been published [35]. For a column with a total free plate area equal to 20% and with water as the continuous phase, an estimate of these quantities may be obtained from the relationships

$$\Delta p_{\max} = 9.7 \times 10^3\, Af + 1.8 \times 10^6\, (Af)^2 \quad (8)$$

$$N = \frac{2\pi Afn\Delta p_{\max}\pi D^2}{4\eta} \quad (9)$$

The amplitude of pressure pulsations acting on the bottom of the column can be estimated as the sum of the effects of individual plates; thus, for a VPE with uniform motion of plates, it equals $n\,\Delta p_{\max}$. For a column with countermotion of plates, however, these pulsations are negligible, as the effects of neighboring plates cancel.

No general relationships for estimation of the mass-transfer rates have been available until now. In practice, large units are designed on the basis of experiments on pilot-scale models. At column diameters of up to 50 cm, geometric similarity of plates is preserved; plates of larger diameters are constructed as described previously. By observing these conditions and keeping the plate spacing constant, it has been found that limiting phase velocities and HETS are approximately equal in the model and in a production unit. For optimization of the column geometry, the amplitude and frequency of vibrations, and the phase velocities, the volumetric efficiency

$$E_v = \frac{U_c + U_d}{\text{HETS}} \quad (10)$$

has been found to be a suitable objective function. A typical plot of these quantities versus Af is shown in Fig. 15. It represents the results of extraction of caprolactam from water to benzene for throughputs corresponding to 70% of flooding rates, with the use of uniform motion and countermotion of plates, respectively.

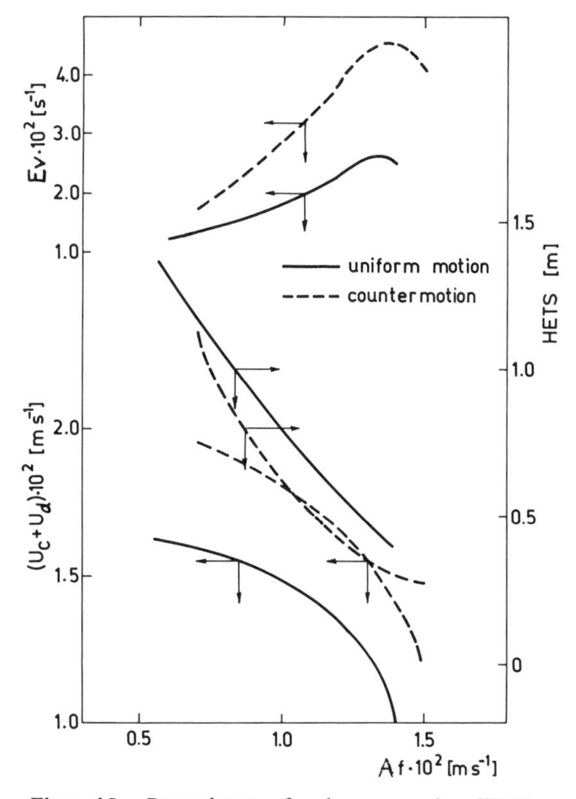

Figure 15. Dependence of column capacity, HETS, and volumetric efficiency on intensity of vibrations.

TABLE 2 APPLICATION OF VPE

System	Direction of Mass Transfer	Dispersed Phase	Diameter, m	Length, m	$(U_c + U_d) \times 10^2$, m/s	Yield, %
Water–Caprolactam–Trichloroethylene	W → O	W	0.500	6.8	1.89	99.8
	O → W	O	0.500	8.0	1.40	99.9
Water–Caprolactam–Benzene	W → O	W	0.085	4.0	1.61	99.7
	O → W	O	0.085	4.0	1.83	98.6
Water–Phenols–Butylacetate	W → O	O	0.085	4.0	2.74	97.3
Water–Phenols–Diisopropylethers	W → O	O	0.050	4.0	1.34	98.8
Fermentation broth–Ketol–Butylacetate	W → O	O	0.600	7.0	1.25	90.0
Fermentation broth–Erythromycine–Butylacetate	W → O	W	0.050	4.0	1.04	98.0
Water–Acrylic acid–Isopropylacetate	W → O	W	0.085	4.0	1.35	99.1

3.3. Applications

In the industrial application of the VPE, advantage is taken of the following properties: high limiting flow rates accompanied by high extraction efficiency; adaptability to liquid systems with a wide range of properties; low energy consumption; simplicity of construction and maintenance; easy adjustment to temporary changes of working conditions; and simple scale-up. Typical data from industrial units as well as experimental models are summarized in Table 2. The largest units for phenol extraction have a diameter of 1.2 m and a plate stack height of 9 m, with a capacity of 80 m^3/hr of phenolic waste water.

4. MISCELLANEOUS RECIPROCATING COLUMNS

Several other types of extraction column containing reciprocating parts of various shapes have been reported. More complex moving and stationary parts, or various modifications of reciprocating motion, have been introduced according to the main objectives pursued.

Wellek et al. [9] have proposed a column with reciprocating wire-mesh packing [36]. In this column reciprocating motion is imparted to a cylindrical body of wire-mesh packing fastened to a vertical shaft. In the experiments described, the void volume of the packing was 95% and the diameter of the mesh wire was 0.27 mm. Harmonic motion was used, with displacements ranging from 4.6 to 23 mm and frequencies of up to 18 Hz.

In the experiments reported, flooding flow rates and extraction performance were studied in two liquid systems: benzene–acetic acid–water and MIBK–acetic acid–water. For both systems, the flooding flow rates were high; the total volumetric flow rate varied from 0.015 to 0.034 m/s. Whereas for the second system the flooding flow rates decreased linearly with increasing intensity of vibrations, for the first system this was true only when the water phase dispersed. For the dispersed organic phase, the flooding flow rate in the system benzene–acetic acid–water was independent of the intensity of vibrations.

The calculated overall HETS was within the range of 0.15–0.6 m. For the benzene system, the following correlations were recommended.

Water dispersed

$$\frac{H}{\mathrm{HTU}_w} = 0.78 + 0.37 \left(\frac{Af}{U_w + U_o} \right) \tag{11}$$

Benzene dispersed

$$\frac{H}{\mathrm{HTU}_o} = 0.98 + 0.33 \left(\frac{Af}{U_w + U_o} \right) \tag{12}$$

Both relationships hold for mass transfer from the aqueous to the organic phase. Generally, for the MIBK system, the HTUs of the reciprocating wire-mesh column have been found to be higher than those of other types of extraction column, but comparably lower for the benzene system. The volumetric efficiencies $E_v = (U_w + U_o)/$ HETS, however, were almost invariably higher than the values reported for other competing types of extractor.

Whereas the reciprocating contactor designs described until now have pursued the aims of both high flow rates and high extraction efficiencies, other proposed types show the tendency to increase the extraction efficiency at the cost of column capacity. Thus Gelperin et al. [7] investigated the effect of separation of perforated plates of a reciprocating-plate column by one or more pairs of stationary rings and disks. The perforated plates were fastened on a common shaft in pairs. The perforations were conical, with the direction of the cones opposite in neighboring plates. Thus a pair of perforated plates together with the corresponding rings and disks formed one stage of a multistage contactor. The systems dibutyl ether–acetic acid–water and isoamylalcohol–boric acid–water were used. The combined flow rates of phases did not exceed 0.0028 m/s, and the HETS ranged from 0.05 to 0.80 m.

The columns described consisted of more or less distinct mixing and settling zones. A step further in the development of a vertical mixer-settler represents the multistage extractor due to Fenske and Long [6]. This apparatus consists of horizontal stages of rather complex form composed of a mixing chamber with a vibrating perforated stirrer and a settling zone. The phases flow cocurrently through each stage, and after separation the lighter phase proceeds to the mixing chamber of the stage above and the heavier phase to the mixing chamber below. This extractor, however, is recommended only for the purpose of measuring multicomponent system equilibria and is thus beyond the scope of this chapter.

Another stagewise reciprocating-plate extractor is the multistage vibrating disk column (MVDC) introduced by Tojo et al. [11, 27, 37–40]. This is a vertical column divided into stages by means of horizontal partition plates with central circular holes of relatively small diameter. Perforated disks of a diameter smaller than that of the column are mounted on a common shaft passing through the holes of the partition plates. These disks are situated at the middle of each compartment and may perform rotation and vertical vibration simultaneously. Various waveforms of disk vibration have been investigated [11, 39]. The MVDC was originally applied for gas–liquid contacting [11, 27, 40]. In this case, the vibrating disks were not perforated.

The last type of reciprocating extractor dealt with here is described by Ziehl [41] and Simonis [42] (see also Chapter 14). It is characterized by a very large free area in the column plate and a complex motion of the mixing elements. These elements have a starlike form and are fastened to a common shaft to which reciprocating motion and rotation are imparted. Specific throughputs amount to 0.02 m/s. Frequencies of reciprocating motion range to 3 Hz, and amplitudes, to 0.05 m. The speed of the rotational motion is in the range 0.15–0.30 Hz.

The largest units have an inner diameter of 0.5 m. Several examples of application of this contactor are reported, including extraction of phenols from waste waters and coal tars and extraction of aniline by use of nitrobenzene and uranium purification. In a column of 0.08-m diameter and 6-m length, the phenol content could be reduced by extraction with butyl acetate from the initial concentration of 2.8 g/liter to the final concentration of 20 mg/liter by using the ratio of volumetric flow rates of solvent to water of 1 : 5.

NOTATION

a	specific interfacial area, m^{-1} or cm^{-1}
A	amplitude (distance of half-stroke), m or cm
A_o	fractional open area, m^2 or cm^2
b	see Eq. (7)
C	orifice coefficient
D	diameter of column, m or cm
E_v	volumetric efficient, s^{-1} or hr^{-1}
f	frequency, Hz
g	acceleration of gravity, cm/s^2
H	packing height, m or cm
HETS	height of an equivalent theoretical stage, m or cm
HTU	height of a transfer unit, m or cm
K	overall mass transfer coefficient, m/s or cm/s
n	number of plates
N	power input, W
U	superficial velocity, m/s or cm/s
U_o	characteristic velocity, m/s or cm/s
X	specific volumetric holdup of dispersed phase
l	plate spacing, m or cm

ρ density, g/cm^3
$\bar{\rho}$ average density
μ viscosity, Pa/s (10 P), $lb\ ft^{-1}\ hr^{-1}$
μ_{co} reference viscosity, 0.01 P
ψ power dissipation per unit mass, W/kg
Δ density difference ($\rho_d - \rho_c$), g/cm^3
σ interfacial tension, N/m or dynes/cm
η mechanical efficiency of vibration

Subscripts

c continuous phase
d dispersed phase
w water phase
o organic phase or overall
f flooding
s slip value

REFERENCES

1. W. J. D. Van Dijck, U.S. Patent 2,011,186 (1935).
2. A. E. Karr, *Am. Inst. Chem. Engr. J.* **5**, 446 (1959).
3. A. E. Karr and T. C. Lo, *Proceedings of the International Solvent Extraction Conference (ISEC) The Hague 1971*, Vol. 1, Society of Chemical Industry, London, 1971.
4. N. Issac and R. L. DeWitte, *Am. Inst. Chem. Eng. J.* **4**, 498 (1958); *Dechema Monogr.* **32**, 218 (1959).
5. A. Guyer, A. Guyer, Jr., and K. Mauli, *Helv. Chim. Acta* **38**, 790, 955 (1955).
6. M. R. Fenske and R. B. Long, *Chem. Eng. Progr.* **51**, 194 (1955).
7. N. I. Gelperin, V. L. Pebalk, and Yu. K. Czechomov, *Khim. Promyshlennos* **41**, 37 (1965).
8. D. Elenkov et al., *Khim. Ind. Sofia*, **4**, 181 (1966).
9. R. Wellek et al., *Ind. Eng. Chem. Process Des. Devel.* **8**, 515 (1969).
10. J. Prochazka, J. Landau, F. Souhrada, and A. Heyberger, *Br. Chem. Eng.* **16**, 42 (1971).
11. K. Miyanami, K. Tojo, and T. Yano, *J. Chem. Eng. Jap.* **6**, 518 (1973).
12. P. J. Bailes, C. Hanson, and M. A. Hughes, *Chem. Eng.* **83**(2), 86 (1976).
13. T. C. Lo and A. E. Karr, *Ind. Eng. Chem. Process Des. Devel.* **11**, 4, 495 (1972); paper presented at Engineering Foundation Conference on Mixing Research, Andover, NH, August 9–13, 1971.
14. A. E. Karr and T. C. Lo, *Chem. Eng. Progr.* **72**, 68 (1976).
15. A. E. Karr and T. C. Lo, *Proceedings of the International Solvent Extraction Conference (ISEC), Toronto 1977*, Vol. 1, Canadian Institute of Mining and Metallurgy, Montreal, 1979, p. 355.
16. M. H. I. Baird, R. G. McGinnis, and G. C. Tan, *Proceedings of the International Solvent Extraction Conference (ISEC), The Hague, 1971*, Vol. 1, Society of Chemical Industry, London, 1971, p. 251.
17. M. M. Hafez, M. H. I. Baird, and I. Nirdosh, *Can. J. Chem. Eng.* **57**, 150 (1979).
18. M. H. I. Baird and S. J. Lane, *Chem. Eng. Sci.* **28**, 947 (1973).
19. M. M. Hafez and M. H. I. Baird, *Transact. Inst. Chem. Eng. (Lond.)* **56**, 229 (1978).
20. S. D. Kim and M. H. I. Baird, *Can. J. Chem. Eng.* **54**, 81 (1976).
21. A. N. Kolmogoroff, *Akad. Nauk USSR* **30**, 301; **31**, 538; **32**, 16 (1941).
22. A. M. Rosen and V. S. Krylov, *Chem. Eng. J.* **7**, 85 (1974).
23. M. M. Hafez, M. H. I. Baird, and I. Nirdosh, *Proceedings of the International Solvent Extraction Conference (ISEC) Liège, 1980*, Association des Ingenieurs Sortis de l'Université de Liège, Belgium, 1980, Paper 80-41.
24. G. S. Somekh, *Proceedings of the International Solvent Extraction Conference (ISEC), The Hague 1971*, Vol. 1, Society of Chemical Industry, London, 1971, p. 323.

25a. Exhibition during *International Solvent Extraction Conference (ISEC), Lyon, 1974.*

25b. A. E. Karr, *Proceedings of the International Solvent Extraction Conference (ISEC) Liège, 1980*, Association des Ingenieurs Sortis de l'Universite de Liège, Belgium, 1980, Paper 80-28.

26. S. H. Noh, Cocurrent Extraction in a Reciprocating Plate Column, M.Eng. thesis, McMaster University, Hamilton, Ontario, 1981.
27. K. Tojo, K. Miyanami, and T. J. Yano, *Chem. Eng. Jap.* **7**, 123 (1974).
28. K. Takeka, reprint of *Proceedings of the 10th General Symposium of the Society of Chemical Engineers*, Japan, 1971, p. 124.
29. R. S. Metcalfe, Ph.D. thesis, Pennsylvania State University, University Park, PA, 1970.
30. J. Prochazka *Dechema Monographien* **65**, 325 (1970).
31. J. Prochazka, J. Landau, and F. Souhrada, U.S. Patent 3,488,037 (1970).
32. J. Landau, J. Prochazka, and F. Souhrada, U.S. Patent 3,583,856 (1971).
33. R. A. McAlister, W. S. Groenier, and A. D. Ryon, *Chem. Eng. Sci.* **22**, 931 (1967).
34. T. Misek, *Coll. Czech. Chem. Commun.* **28**, 570, 1631 (1963).
35. M. M. Hafez and J. Prochazka, *Chem. Eng. Sci.* **29**, 1745, 1755 (1974).
36. J. J. Carr, M.S. Thesis, University of Missouri–Rolla, (1963).
37. K. Tojo, K. Miyanami, and T. Yano, *J. Chem. Eng. Jap.* **8**, 122 (1975).

38. K. Tojo, K. Miyanami, and T. Yano, *J. Chem. Eng. Jap.* **8**, 165 (1975).
39. K. Miyanami, K. Tojo, T. Yano, K. Miyaji, and I. Minami, *Chem. Eng. Sci.* **30**, 1415 (1975).
40. K. Miyanami, K. Tojo, I. Minami, and T. Yano, *Chem. Eng. Sci.* **33**, 601 (1978).
41. L. Ziehl and F. Ploger, *Chemie-Ing. Tech.* **33**, 533 (1961).
42. H. Simonis, *Dechema Monographien* (1168–1192), 315 (1971).
43. Chem-Pro Equipment Bulletin KC-11, *Karr Column,* Fairland, NJ; Julius Montz, GmbH, Hofstrube, West Germany.
44. A. E. Karr, *Separation Sci. Technol.* **15**(4), 877 (1980).
45. L. A. Robbins, *Chem. Eng. Progr.* **75**(9), 45 (1979).
46. J. Prochazka, J. Landau, and F. Souhrada, *Coll. Czech. Chem. Commun.* **29**, 3003 (1964).

13.1

ROTATING-DISK CONTACTOR

W. C. G. Kosters
Shell Internationale Petroleum Maatschappij B. V.
The Netherlands

1. INTRODUCTION

The rotating disk contactor (RDC) belongs to the class of agitated extractors. For this type of extractor, the dispersion required for a high mass-transfer rate is obtained by applying outside mechanical energy.

The RDC was developed by the Royal Dutch/Shell Group at Amsterdam Laboratory during 1948–1952 and has since found numerous applications. Some hundreds of RDCs are at present in use worldwide, ranging in diameter from less than 1 m to 4.5 m. The RDC has been investigated by a large number of universities and research institutes with the object of defining capacity and efficiency on the basis of physical properties and RDC geometry. The resultant theoretical considerations and relationships, however, all indicate a strong dependence on physical properties that under actual mass-transfer conditions are nearly impossible to predict with sufficient accuracy [1–3].

In the following pages a more practical approach is described that is being used by industry in the design of commercial, large-diameter RDCs. This approach starts with pilot plant experiments to determine the behavior of the extraction system in question in the RDC and to determine the capacity and efficiency of the extractor under actual process conditions. By using well-proven relationships and scale-up rules, the commercial RDC can then be designed directly from the pilot plant results. This system has been employed in a large number of applications and has proved reliable even for scaling from a pilot plant diameter of 64 mm direct to diameters of 4–4.5 m.

2. DESCRIPTION OF THE RDC

2.1. Mechanical Description

The RDC (Fig. 1) consists of a vertical shell in which horizontal stator rings are installed. The stator rings are flat plates with a central opening. In the middle of the compartments formed by the stator rings rotor disks are installed. The rotor disks are flat plates fitted on a central shaft driven by an electric motor. The diameter of the rotor disks is smaller than the diameter of the stator opening, thus facilitating easy construction, the complete rotor assembly can be installed as a last step.

Above the top stator ring and below the bottom stator ring, settling compartments are installed. Wide-mesh grids are used between the agitated section and the settling zones to nullify the liquid circular motion, thus ensuring optimum settling conditions. The feed inlets are, in general, arranged tangentially so as not to disturb the flow pattern in the inlet compartments.

2.2. Selection of Type of Dispersion

Generally speaking, the decision as to which process stream to disperse should be based on practical considerations and the results of the pilot plant tests. The phase with the largest volume is often selected at the continuous phase. In principle, however, each of the phases can be

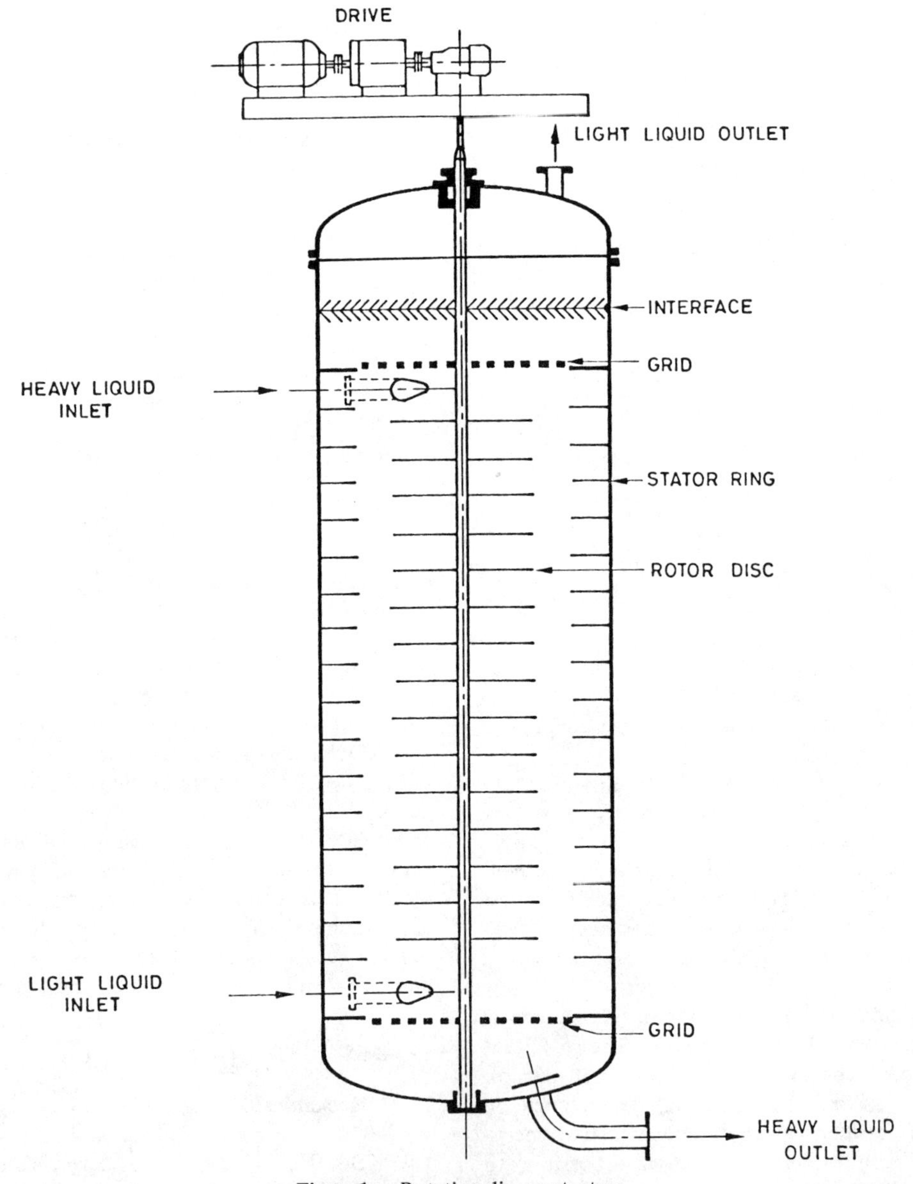

Figure 1. Rotating-disc contactor.

dispersed by setting the interface level in either the top or the bottom settling compartment.

If the interface level is maintained in the top settler, the heavy phase will be continuous; if it is maintained in the bottom settler, the light phase will be continuous.

2.3. Description of Flow Pattern

The agitation pattern (Fig. 2 [4]) obtained by the rotating discs is rather complex. First, the whole liquid contents rotate in the same direction as the rotor. Furthermore, horizontal vortices are created between two adjacent stator rings (Fig. 2). Such a vortex starts from the tip of the rotor disk and flows outwards to the shell, where it is deflected by the shell and the stator ring and returns to the rotor shaft.

The combination of the horizontal rotation and this vortex flow results in a toroidal flow in each compartment. Superimposed on this rather high-flow vortex, the continuous and disperse phases have a vertical flow from their respective inlets to the outlets at the other end.

2.4. Influence of Rotor Speed

The rotor serves as a means to convey the mixing energy to the dispersed phase. The drops of dispersed phase are basically formed at the tip of the rotor. As the flat rotor disks supply a constant, well-distributed energy input to the system; moreover, as no sharp edges are present, the drop size is fairly uniform. The speed of the rotor (and hence the energy input) provides a means for controlling drop size during operation and is, therefore, an additional operating variable. It can be used to advantage in operating the RDC nearly always near to its optimum performance over a wide range of flow (throughput) conditions. In order to obtain such flexibility, the rotor speed should be continuously variable over a wide range with the design rotor speed in the middle of the range.

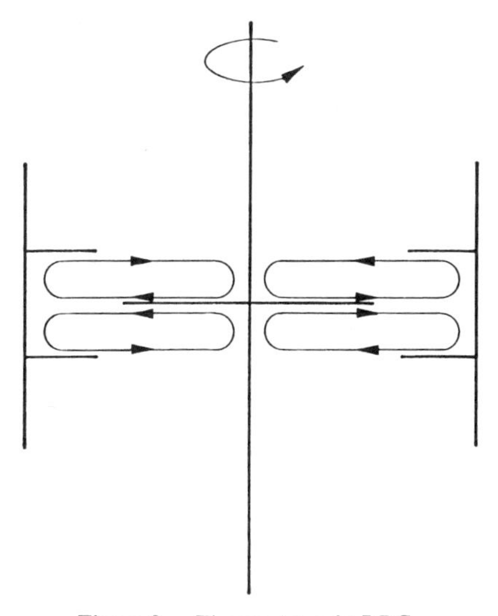

Figure 2. Flow pattern in RDC.

The power input per rotor disk to the find system P is given by the following relationship [4]:

$$\frac{P}{\rho N^3 R^5} = f\left(\frac{\rho N R^2}{\mu}\right)$$

This relationship is depicted in Fig. 3. In most cases the RDC will be operated in the region $(\rho NR^2/\mu) > 10^5$, for which the left-hand part of the equation becomes constant. Thus the power input per rotor disc is then proportional to $\rho N^3 R^5$ and since there is only one rotor disk per RDC compartment, the power input per unit mass is proportional to N^3R^5/HD^2. This latter expression is called the *specific power input group* and is normally used in correlation for RDC performance [4].

2.5. RDC Geometry

As described in Section 2.3, vortices are formed between rotor disks and stator rings. For stabilization of these vortices, a certain baffling action is required from the rotor disks and the stator rings. On the other hand, sufficient free space should be maintained between the rotor and the stator rings to render installation of the rotor possible.

A second consideration is the free area available for vertical flow of the two phases, as this will have an influence on capacity. Obviously, the free area $1 - R^2/D^2$ and S^2/D^2 should be equal in order to ensure optimum flow passage.

The preceding rules have led to the establishment of standard ratios for rotor disk diameter, stator opening diameter and column diameter:

$$\frac{S}{D} = 0.7$$

$$\frac{R}{D} = 0.6$$

These ratios ensure sufficient baffling action, at the same time allowing such an open configura-

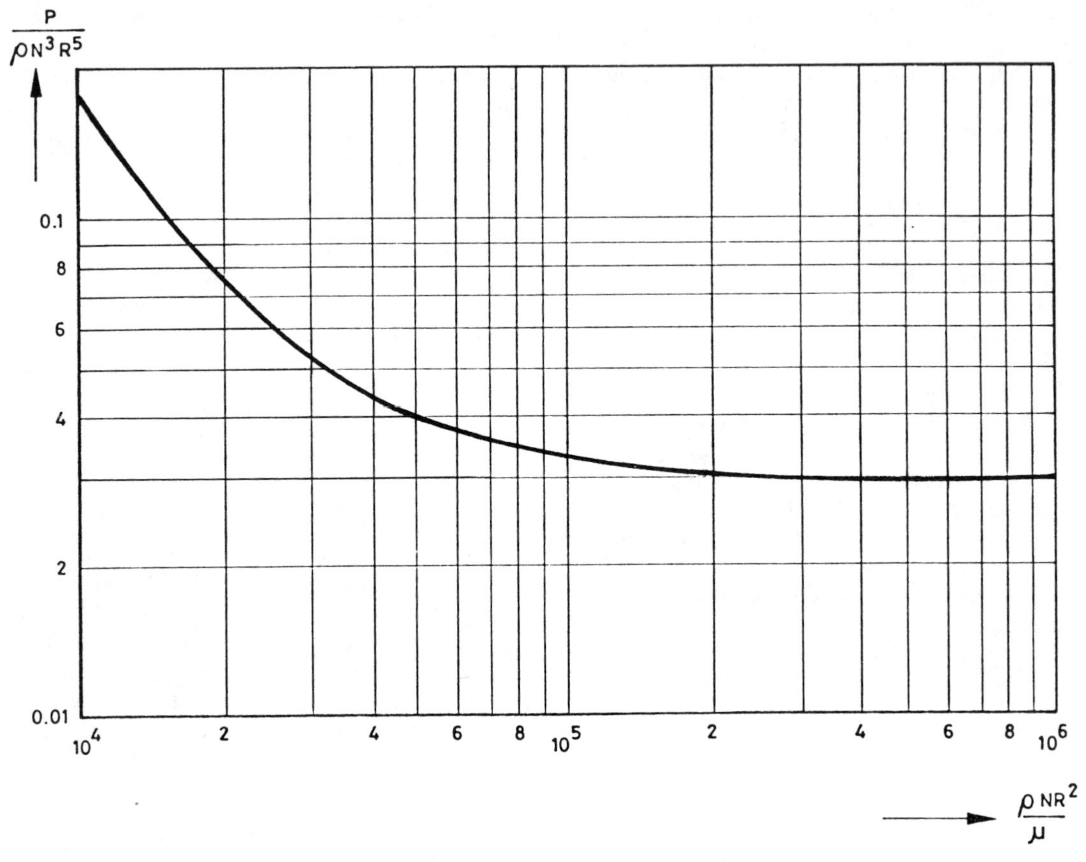

Figure 3. Power input per rotor disk.

tion to be maintained that total liquid flow is not impaired.

Just as the radial path along the rotor disks and stator rings is of importance in stabilizing the vortices, the height dictated by the compartment height is of equal importance. Too high an H/D ratio will result in vortices with large vertical paths, which will increase the chance of instability and also increase flow to the next compartment.

Too small an H/D ratio will result in a rather flat vortex and a higher resistance to vortex flow over the rotor disks and stator rings. The higher resistance may damp the vortex velocity to such an extent that not all the compartment length is used. This means the formation of dead zones along the vessel wall, shorter residence times and, consequently, a lower efficiency.

In practice, the following ratios have been found to be suitable:

RDC diameter D, m	0.5–1.0	1.0–1.5	1.5–2.5	>2.5
H/D	0.15	0.12	0.1	0.08–0.1

It should be realized that in the RDC mass transfer is obtained in a continuous dispersed system; that is, no settling of the phases occurs between adjacent compartments. Consequently only the total stirred height is a measure of efficiency. How many compartments are applied has no direct bearing except on efficiency, except as discussed previously.

2.6. Selection of Building Materials

Process conditions in the RDC may vary considerably; for instance, temperatures from −10 to 200°C and pressures from nearly atmospheric to 50 bar are applied on a commercial scale. Moreover, the process liquids may range from noncorrosive to highly corrosive. The simple mechanical layout of the RDC, however allows the use of nearly any material. Although carbon steel is the most common material applied in the oil and petrochemical industry, stainless steel is often used for applications where high demands are made on product purity (food, pharmaceutical, rare earth extraction applications, etc.) or in corrosive systems. For very corrosive systems RDCs built with glass fiber-reinforced Epikote are in operation on an industrial scale.

3. INDUSTRIAL APPLICATIONS

The RDC can be applied in all those processes in which two liquids (partly or completely immiscible) must be contacted in countercurrent or cocurrent operation. In principle, it can also be used for staged chemical reactions with or without mass transfer. As a consequence of its open structure, the RDC is less prone to fouling. This type of column can, therefore, also be applied for systems with one solid phase and one or two liquid phases (e.g., purification of crystals or washing of solids slurries). The most important commercial applications can be divided into four classes:

I. *Liquid–liquid extraction*

Sulfur dioxide extraction of kerosenes and luboils
Furfural extraction of luboils
Propane–butane deasphalting of crude oil residues
Sulfolane extraction for aromatics production
Caprolactam extraction
Separation of rare earth
Extraction of phenol from waste water
Purification of edible oils
Extraction of synthetic detergents
Separation of chlorinated solvents

II. *Removal of trace components*

Solvent removal by water washing
Solutizer extraction for gasoline

III. *Extraction plus chemical reaction*

Caustic treating
Acid treating
Metals extraction

IV. *Solid–liquid extraction*

Purification of crystalline compounds
Washing of crystals for removal of impurities
Silica gel or molecular sieve treatment in slurry form

The preceding list includes a few examples only and is not an exhaustive review.

Although the RDC is a very versatile piece of equipment, some limitations to its applications should be noted:

1. When the number of equilibrium states is small (two or less), it is in most cases cheaper to use mixer–settlers.
2. For systems with very low interfacial tension and low density differences between the phases, the agitation in the RDC may be too intense, thus leading to rather stable emulsions and/or low capacities. In such cases other extractors may have to be considered.
3. The residence time for the process liquids in the RDC can be high, especially when a large number of equilibrium stages are required. If one or more of the process fluids is not completely stable at the extraction temperature, too high degradation may occur (e.g., as in penicillin extraction). In such a case extractors with a short residence time should be selected.

4. PERFORMANCE DATA

In describing the performance of an RDC, two criteria are of importance, capacity and efficiency (mass transfer). These two are, however, related to each other, since drop diameter is an important factor in both cases. Nevertheless, these two criteria are discussed separately, with cross-references made as appropriate.

4.1. Capacity

The capacity or liquid loading of the RDC is normally expressed as the sum of the superficial velocities of the two phases $V_d + V_c$ [3-5]. For a given system under otherwise constant operating conditions, there is a maximum liquid loading above which, in general, heavy entrainment of the dispersed phase will be noticed (flooding). This maximum loading is determined by the actual settling velocity of the drops under gravitational force.

In describing the capacity, the actual velocities of the phases at the point of the RDCs smallest cross-section should be used. The liquid velocities through this smallest cross section are given by $\overline{V_d}$ and $\overline{V_c}$. However, part of the cross section is occupied by the dispersed phase and part by the continuous phase; the ratio between the two is indicated by the holdup of the dispersed phase in the column h.

The actual local velocities now become

For dispersed phase $\dfrac{\overline{V_d}}{h}$

For continuous phase $\dfrac{\overline{V_c}}{1-h}$

The relative velocity of the two phases, which is called the *slip velocity*, then becomes

$$\overline{V_s} = \frac{\overline{V_d}}{h} + \frac{\overline{V_c}}{1-h} \tag{1}$$

If the holdup approaches zero, this slip velocity equals the settling velocity of the drop for the conditions used. Under actual holdup conditions, however, some hindrance to drop settling will occur, resulting in a lower slip velocity. This effect has been determined experimentally and has been found to be represented adequately by $\exp(-h)$ [6].

The effective slip velocity then becomes

$$\overline{V_e} = \overline{V_s}\ \exp(-h) \tag{2}$$

For a given extraction system the ratio between the phases will be constant and will be given by $\alpha = V_d/V_c$. From this equation it follows that

$$\overline{V_c} = \frac{\overline{V_s}\ \exp(-h)}{\alpha/h + 1/(1-h)} \tag{3}$$

If this relationship, which is valid only for a constant α, is plotted, the ratio $\overline{V_c}/\overline{V_s}$ will pass through a maximum as a function of holdup. The holdup at this maximum indicates the holdup at flooding h_f. This holdup can be calculated, since at this point [6]

$$\frac{d\,\overline{V_c}}{dh} = 0 \qquad \text{at } h = h_f$$

From this it can be calculated that

$$\alpha = \frac{2h_f^2 - h_f^3}{(1-h_f)^3} \tag{4}$$

This relationship is given in Fig. 4, from which the holdup at flooding can be calculated at any phase ratio.

Substitution of h_f into Eq. (3) gives

$$\overline{V_{cf}} = \frac{\overline{V_s}\ \exp(-h_f)}{\alpha/h_f + 1/(1-h_f)} \tag{5}$$

from which the maximum velocity of the continuous phase through the smallest cross section

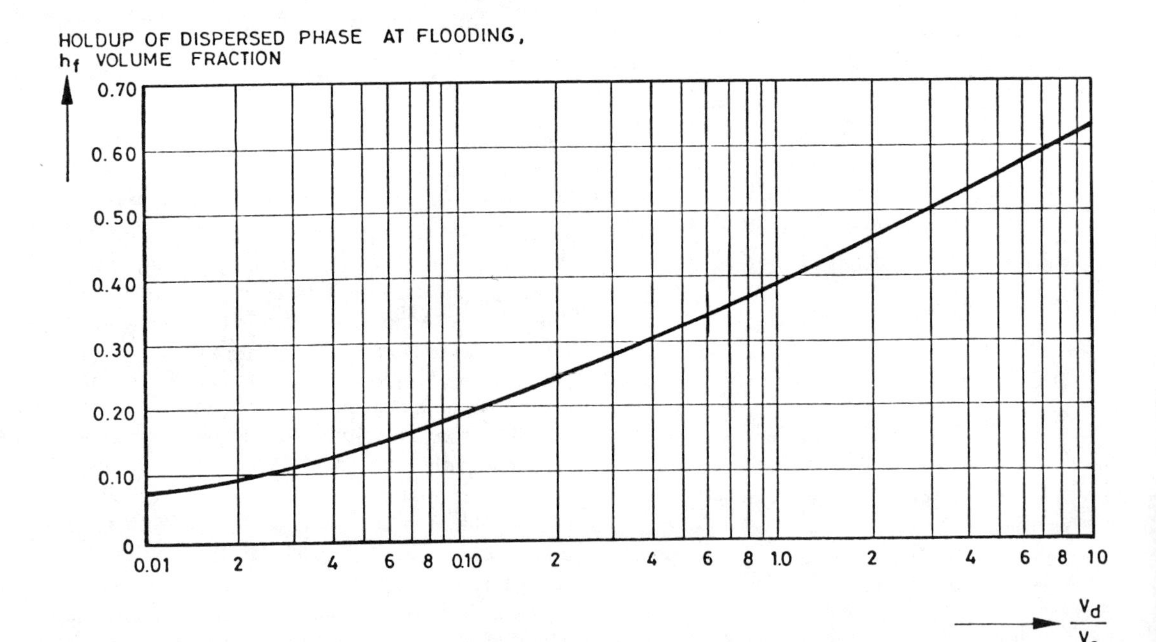

Figure 4. Holdup at flooding.

can be calculated from the effective slip velocity of the system.

For the standard geometry as given in Section 2.5, the stator cross section will always be the smallest and be constant. It can, therefore, be omitted, so that Eq. (5) can be written as

$$V_{c_f} = \frac{V_s \exp(-h_f)}{\alpha/h_f + 1/(1 - h_f)} \tag{6}$$

indicating the superficial velocity of the continuous phase at flooding. In the preceding considerations it has been assumed that the settling velocity of the drop does not change; in other words, a constant drop diameter is assumed.

For a given system, alteration of the rotor speed (and hence the energy input) will have an effect on drop diameter. As stated in Section 2.4, the term N^3R^5/HD^2 is normally used as a measure of energy input. As the power input is increased, the drop diameter decreases, and consequently the slip velocity as well. At low power inputs this effect is rather small; that is, energy is used to establish the normal flow pattern in the RDC. At higher power input the drop size reduction reaches the point where it becomes so large that the slip velocity (and hence the capacity) is substantially reduced.

The effect of power input on capacity is a function of the physical properties of the two phases (density, viscosity, and interfacial tension). Of these the effect of interfacial tension is particularly pronounced. This means that for an accurate prediction of the slip velocity (and hence the capacity), accurate data on the interfacial tension of the system in general and its variations during extraction must be available. Generally speaking, it can be said that these data can easily be obtained for systems in which the total amount of solute transferred is low; that is, solute concentrations in both phases are low, and hence variations in interfacial tension are small.

For a large number of industrial applications, however, the amount of solute transferred is an appreciable fraction of the starting material. Moreover, as solvent : feed ratios are kept to the minimum for economic reasons, the solute concentrations are high, as are their variations in concentration from one end of the extractor to the other.

At the point of high solute concentrations in both phases (i.e., near the feed point) interfacial tension will be small and large variations will be found for small changes in feed composition and/or operating conditions. Moreover, it should be noted that the rate and direction of mass transfer [1, 7, 8] play an important role in drop behavior. The difference in capacity between the two types of dispersion should particularly be noted [1, 7, 9]. From the preceding evidence it is clear that these effects can only be determined with sufficient accuracy for commercial application when the capacity and hence the slip velocity are measured in a pilot plant RDC. In these tests the maximum flow rates can be determined at different rotor speeds. From these data and the data on holdup at flooding obtainable from Fig. 4, the slip velocity can be calculated. If reliable actual holdup measurements can be made under nonflooding conditions, such data can also be used. But under both types of dispersion, conditions should be made to determine the effect on capacity.

In Fig. 5 a generalized correlation is given on the basis of a large number of actual capacity measurements. The curves show clearly the large influence of interfacial tension. It should be noted that these curves can be used only as general yardsticks, as differences in mass transfer rate and direction are not completely accounted for.

4.2. Rate of Mass Transfer

The rate of mass transfer in an RDC can be expressed in general terms as a function of drop size (drop area) and diffusion coefficient. The main difficulty in such relationships is the prediction of average drop diameter under mass-transfer conditions. As already stated in Section 4.1, the drop diameter itself is a function of a number of physical properties such as interfacial tension, density difference, and viscosity as well as power input. Moreover, the mass-transfer rate and direction play a role here [1-3, 7, 10].

Data on average drop diameter and drop diameter distribution under mass-transfer conditions are scarce, and those published mainly concern rather ideal systems, that is, systems such as kerosene–water with a low solute transfer and a relatively high interfacial tension, which varies only slightly during extraction.

In commercial processes the physical data may change considerably during the extraction whereas the diffusion coefficient will also vary over a wide range. Under these conditions the rate of mass transfer will vary considerably over the length of the RDC. As a consequence, any relationship found for mass transfer can be used only to indicate a trend. Pilot plant tests under

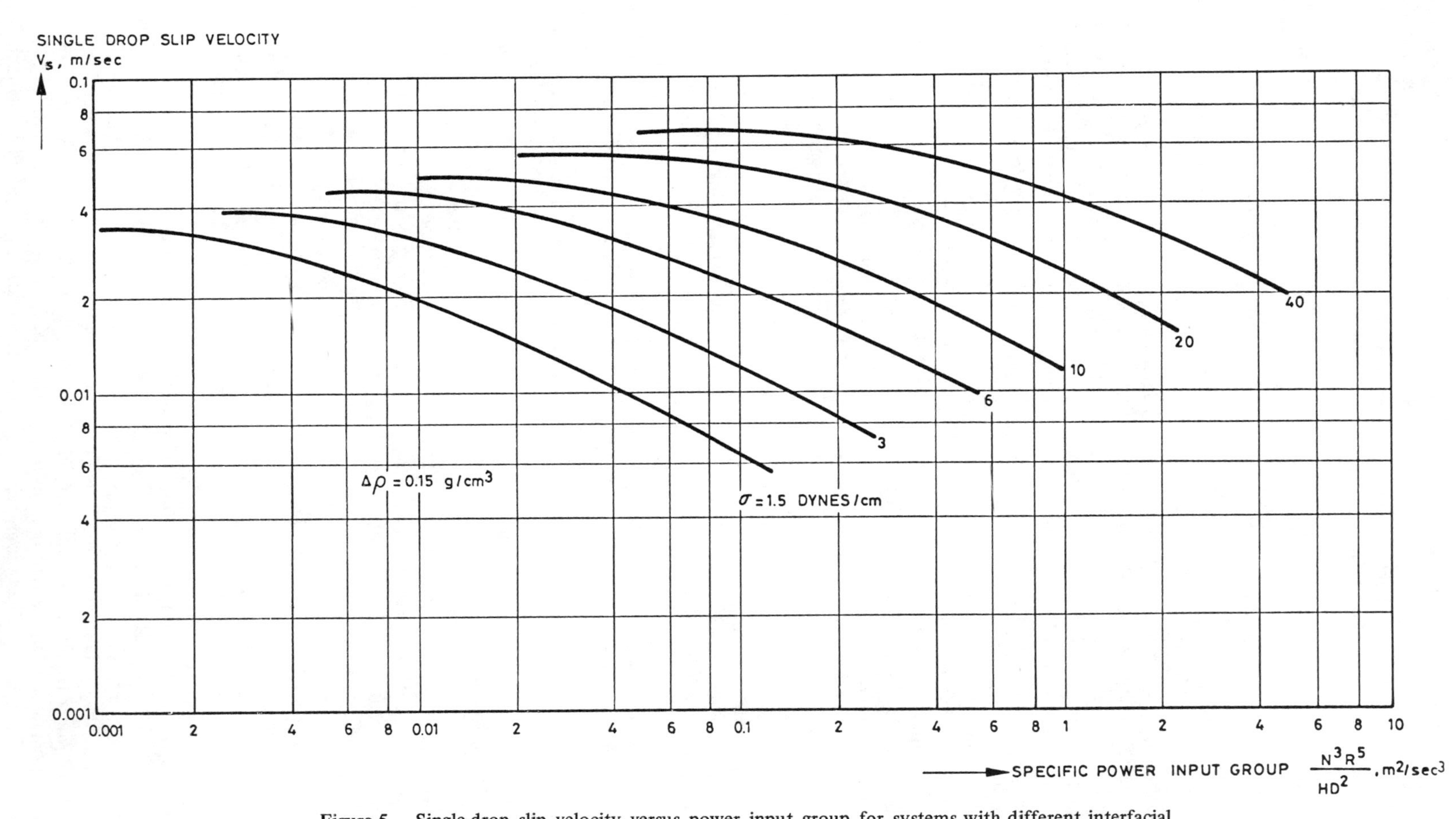

Figure 5. Single-drop slip velocity versus power input group for systems with different interfacial tension.

mass-transfer conditions would have to be performed before accurate data on mass transfer and RDC efficiency could be established.

Rather than use a mass-transfer coefficient, efficiency is expressed as HETS, the height equivalent to a theoretical stage. In pilot plant tests the number of equilibrium stages present in a certain column height can be found by calculation or by comparison with discrete-stage countercurrent extraction results on a laboratory scale. From the number of stages and the total agitated height, used the HETS can be calculated. Repetition of the test at different rotor speeds will give the relationship between HETS and the power input group.

4.3. HTU or HETS

In Section 4.2 it was stated that the efficiency of the RDC can be found by pilot plant tests in terms of HETS. This concept, however, assumes stagewise mixing and settling, whereas the RDC does not operate stagewise but as a continuously operating extractor, so that the concentration profile will also vary continuously. For such systems, the HTU concept (height of a transfer unit) is more appropriate; HETS and HTU are related by

$$\text{HETS} = \frac{\ln F}{F - 1} \cdot \text{HTU} \tag{7}$$

in which F is the extraction coefficient ($F = K \cdot Q$ feed$/Q$ solvent). The HTU as the measure of mass transfer depends on rotor speed. Pilot plant measurements should therefore be carried out at different rotor speeds and for both types of dispersion.

As axial mixing in a pilot plant is very small in relation to the mass-transfer stage height, it can normally be ignored, with HETS alone taken, when required, to represent mass transfer. The extraction coefficient appears in Eq. (7). In simple extractions involving only one solute the average extraction factor for the system can be used. However, in commercial systems (especially in the oil and petrochemical industry) the extraction coefficients for all components being extracted can vary considerably. These coefficients may also change during the extraction. In such a case it is advisable to use the coefficient for the most critical component in the efficiency calculation.

5. SCALE-UP AND DESIGN

5.1. Throughput

From the discussion on column capacity in Section 4.1 it is evident that column capacity is independent of diameter, with the single drop slip velocity for a given system governed only by power input. Since this parameter has a bearing on efficiency, an iterative design method is often required.

The method generally used is as follows: Pilot-plant tests or from other known data yield the specific power input group (N^3R^5/HD^2) at which a good mass-transfer rate is obtained. From a graph relating slip velocity and power input, the slip velocity at the selected power input is obtained for the dispersion selected. From the phase ratio given and the holdup at flooding, the maximum flow rate at flooding for the continuous phase can be calculated by using Eq. (6).

In commercial operation, however, it is customary to install some spare capacity by basing the column design on a certain percentage of the flooding capacity (e.g., 70–80%). This gives the V_c for the design case. From this process the flow of continuous phase Q_c is obtained:

$$\frac{Q_c}{V_c} = \text{column cross section and hence diameter}$$

By use of the standard ratios S/D and R/D as given in Section 2.5, the column diameter can be obtained.

The compartment height can be calculated from the ratios given in Section 2.5, and the rotor speed can be calculated from the selected power input (N^3R^5/HD^2) by inserting the values for R, D, and H calculated. These data plus the values of V_c and V_d can now be used for the calculation of column height. It should be realized that in the column height the axial mixing can play an important role and that this is a function of rotor speed. It may, therefore, be advisable to repeat the calculation for a higher or lower power input to find an optimum for the total column volume.

The same calculation should be carried out by using the reversed dispersion, as this could result in a different column diameter and height. Such an exercise is especially important when an expensive solvent is used. Application of the solvent as the dispersed phase will result in a lower solvent inventory and hence lower solvent costs, even when the total column volume is larger.

5.2. Axial Mixing

In the ideal situation the countercurrent extraction system should proceed as plug flow in both phases. Only in this way will the concentration gradients be maximum and the extraction proceed with maximum efficiency.

In practice, however, the flows of the continuous and dispersed phases will not display true plug flow behavior, but axial mixing will occur; that is, mixing and phase flow in the wrong direction will be found. The axial mixing will result in a decrease in the concentration difference with a resultant reduction in column efficiency. This means that, in order to obtain a given separation, the column height should be increased, or, in other words, the HETS will be larger for cases of severe axial mixing. It is clear that it is important for the designer to calculate the effect of axial mixing on HETS so as to be able to size the column height. To enable both phenomena to be treated separately, the concept of the apparent or effective HTU has been introduced. This concept considers the effective HTU to be composed of two parts; the real mass-transfer HTU and an additional height due to axial mixing, often called the height of a diffusion unit (HDU). Therefore:

$$\mathrm{HTU_{eff}} = \mathrm{HTU} + \mathrm{HDU} \tag{8}$$

As already discussed, the HTU should be determined from pilot plant tests.

The height of the diffusion unit can be calculated from axial diffusivity coefficients (expressed in Peclet numbers) derived from experimental correlations.

For the continuous phase, this comes to

$$\frac{1}{(\mathrm{Pe})_C} = \frac{E_c(1-h)}{HV_c} \tag{9}$$

in which [13]

$$E_c = 0.5\ V_c H + 0.012\ \mathrm{RNH}\ (S/D)^2 \tag{10}$$

For axial mixing in the dispersed phase, much less data are available. It has been found from practical experience that the following relationship gives good results:

$$E_d = E_c \left[\frac{4.2 \times 10^5}{D^2} \left(\frac{V_d}{h}\right)^{3,3}\right] \tag{11}$$

When the correction term is <1, $E_d = E_c$.

The Peclet number for the dispersed phase can now be expressed as follows:

$$\frac{1}{(\mathrm{Pe})_d} = \frac{1}{(\mathrm{Pe})_c} \times \frac{h/V_d}{(1-h)\,V_d} \times \frac{4.2 \times 10^5}{D^2} \left(\frac{V_d}{h}\right)^{3,3} \tag{12}$$

the last term is applied only when >1.

Once the $(\mathrm{Pe})_c$ and $(\mathrm{Pe})_d$ are known, the HDU can be calculated by means of the approximate formula [12]:

$$\mathrm{HDU} = \frac{1}{(\mathrm{Pe})_0/H + 0.8/L \cdot \ln F/F-1} \tag{13}$$

in which

$$(\mathrm{Pe}')_0 = \frac{0.1L/\mathrm{HTU} + 1}{0.1L/\mathrm{HTU} + (\mathrm{Pe}')_1/(\mathrm{Pe}')_2} \cdot (\mathrm{Pe})_1 \tag{14}$$

and

$$\frac{1}{(\mathrm{Pe})_1} = \frac{F}{(\mathrm{Pe})_f} + \frac{1}{(\mathrm{Pe})_s} \tag{15}$$

$$\frac{1}{(\mathrm{Pe})_2} = \frac{1}{(\mathrm{Pe})_f} + \frac{F}{(\mathrm{Pe})_s} \tag{16}$$

If $1 - (\mathrm{Pe})_1/(\mathrm{Pe})_2 \ll 0.1L/\mathrm{HTU}$ and, moreover, $F \approx 1$ and $\mathrm{HDU} \ll L$, the formula can be simplified to

$$\mathrm{HDU} = \frac{H}{(\mathrm{Pe})_d} + \frac{\mathrm{H}}{(\mathrm{Pe})_c} \tag{17}$$

The HTU can now be calculated.

The total agitated height for the RDC now follows from

$$L = (\mathrm{HTU})_{\mathrm{eff}} \frac{\ln F}{F-1} \times N_{\mathrm{eq}} \tag{18}$$

Since L already appears in the formula for HDU, an iterative calculation method should be used. Use Eq. (17) as a first approximation and calculate L. Using this length, recalculate HDU with Eq. (13). In general, two iterations are sufficient to obtain the agitated height L. This method has been used successfully for the design of commercial RDCs.

5.3. Design Example

An example calculation is given for the following case: The feed is an hydrocarbon mixture containing 10% aromatics to be extracted with a polar solvent. From process calculations it has been found that for this extraction a solvent ratio of 5 : 1 wt/wt on feed is required, with the use of 10 equilibrium stages. Physical properties are as follows:

	Feed	Solvent
Specific gravity	0.75	1.2
Viscosity, cP	0.4	1.0

From pilot-plant experiments the slip velocities for solvent continuous and solvent dispersed operation have been determined based on feed and solvent flow. As the pilot-plant RDC had the standard geometry, superficial velocities had been used in the calculation of V_s (Fig. 6). The efficiencies have also been determined from pilot-plant tests and HTUs are shown in Fig. 7. The calculation is carried out in two parts, namely, the solvent continuous phase and the solvent dispersed phase.

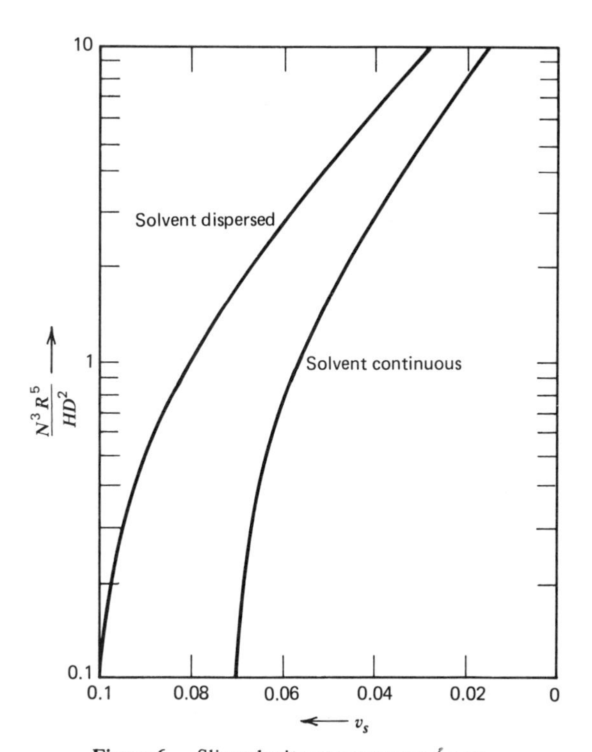

Figure 6. Slip velocity versus power input.

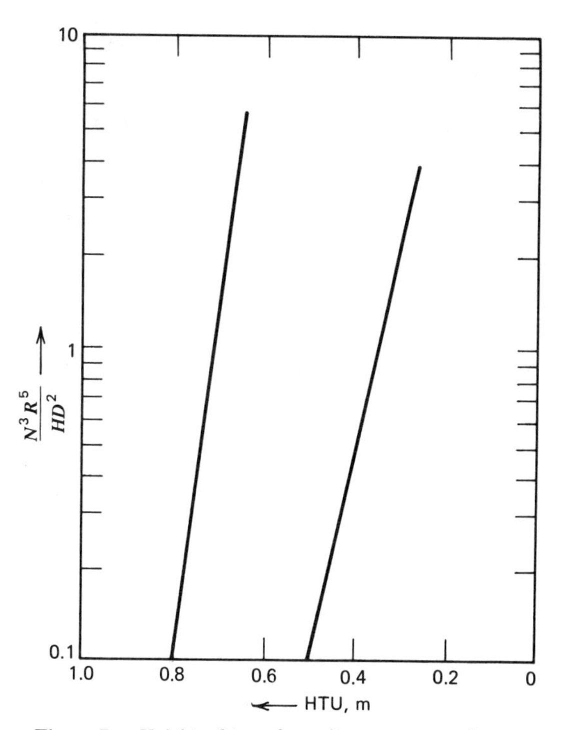

Figure 7. Height of transfer unit versus power input.

5.3.1. *Solvent Continuous Phase*

The first part of the calculation is as follows.

Feed 1000 tons/day, which is

$$\frac{1000}{0.75 \times 86{,}400} = 15.43 \times 10^{-3}\ \text{m}^3/\text{s}$$

Solvent 5000 tons/day, which is

$$\frac{5000}{1.2 \times 86{,}400} = 48.23 \times 10^{-3}\ \text{m}^3/\text{s}$$

From Fig. 6, read for $N^3R^5/HD^2 = 0.4$ that $V_s = 0.065$ m/s. Thus $V_d/V_c = \alpha = 0.32$. From Fig. 4 it follows that $h_f = 0.28$. Hence

$$V_{cf} = \frac{0.065 \exp(-0.28)}{0.32/0.28 + 1/(1 - 0.28)} = 0.0194\ \text{m/s}$$

$$= 19.4 \times 10^{-3}\ \text{m/s}$$

For design, use 75% of flooding: $V_c = 14.55 \cdot 10^{-3}$ m/s; thus

$$V_d = \alpha \cdot V_c = 0.32 \times 14.55 = 4.65\ 10^{-3}\ \text{m/s}$$

The total cross section of RDC = $(48.23 \times 10^{-3})/(14.55 \times 10^{-3}) = 3.31\ \text{m}^2$. The diameter of RDC, D is 2050 mm, say, 2100 mm. From the standard geometry it then follows that

$$S = 1470\ \text{mm}$$

$$R = 1260\ \text{mm}$$

$$H = 210\ \text{mm}$$

From the specific power input group $N^3R^5/HD^2 = 0.4$ it is calculated that $N = 0.5$ rev/sec (30 rpm). To calculate the total agitated height, first calculate the actual holdup by trial and error from the equation, substituting V_c. The result will give $h = 0.11$.

For $N^3R^5/HD^2 = 0.4$ read from Fig. 7: HTU = 0.41 m. Calculate HDU as follows:

$$E_c = 0.5 \times 14.55 \times 10^{-3} \times 0.21 + 0.012 \times 1.26 \times 0.5 \times 0.21 \times \left(\frac{1.47}{2.1}\right)^2 = 2.3 \times 10^{-3}$$

$$E_d = 2.3 \times 10^{-3} \times \frac{4.2 \times 10^5}{2.1^2}\left(\frac{4.65 \times 10^{-3}}{0.11}\right)^{3.3} = 6.4 \times 10^{-3}$$

Then calculate Peclet numbers for the continuous and dispersed phases:

$$\frac{1}{(\text{Pe})_c} = \frac{2.3 \times 10^{-3} \times 0.89}{0.21 \times 14.55 \times 10^{-3}} = 0.67$$

$$\frac{1}{(\text{Pe})_d} = \frac{6.4 \times 10^{-3} \times 0.11}{0.21 \times 4.65 \times 10^{-3}} = 0.72$$

According to the equation,

$$\text{HDU} = 0.21 \times 0.67 + 0.21 \times 0.72 = 0.29$$

$$(\text{HTU})_{\text{eff}} = 0.29 + 0.41 = 0.7$$

The total agitated length will be

$$L = 10 \times 0.7 \times \frac{\ln 0.8}{0.8 - 1} = 7.8\ \text{m}$$

Take L = 8.0 m and start more detailed calculation:

$$\frac{1}{(\text{Pe})_1} = 0.8 \times 0.72 + 0.67 = 1.246$$

$$\frac{1}{(\text{Pe})_c} = 0.72 + 0.8 \times 0.67 = 1.236$$

$$(\text{Pe})_0 = \frac{\dfrac{0.8}{0.41} + 1}{\dfrac{0.8}{0.41} + \dfrac{1.236}{1.246}} \cdot \frac{1}{1.246} = 0.805$$

$$\text{HDU} = \frac{1}{\dfrac{0.805}{0.21} + \dfrac{0.8}{8}\dfrac{\ln 0.8}{0.8 - 1}} = 0.253$$

$$(\text{HTU})_{\text{eff}} = 0.253 + 0.41 = 0.663$$

$$L = 10 \times 0.663 \times \frac{\ln 0.8}{0.8 - 1} = 7.4\ \text{m}$$

This is sufficiently close; take L as 7.5 m. For the sake of safety, to cover small changes in operation, add about 10%; then L = 8.4 m. For H = 0.21 m, this means 40 compartments. The total height of the RDC, assuming a 1-m settling height, now becomes 10.4 m.

Calculate power input. For flexibility, a range of 50–200% of the design rotor speed should be taken. Hence $N_{\text{max}} = 1$ rev/s.

For the continuous phase:

$$\frac{\rho N R^2}{\mu} = \frac{1200 \times 1 \times (1.26)^2}{10^{-3}} \geqslant 10^5$$

Therefore, $P = 0.03$ (from Fig. 4).

The theoretical power consumption is

$$P_{\text{tot}} = 40 \times 0.03 \times 1200 \times 1^3 \times (1.26)^5 = 4.6\ \text{kW}$$

It should be noted that this is only the power consumption of the liquid system. In order to size the actual motor, the resistance in the seals and the efficiency of the gearbox should be taken into account. Tabulation of the results yields the following data:

D = 2100 mm

S = 1470 mm

R = 1260 mm

H = 200 mm

Total height = 10,400 mm

Rotor speed = 15–60 rpm

5.3.2. Solvent Dispersed Phase

The second part of the calculation is as follows.

$$\text{Feed 1000 tons/day} = \frac{1000}{0.75 \times 86{,}400}$$

$$= 15.43 \times 10^{-3}\ \text{m}^3/\text{s}$$

$$\text{Solvent 5000 tons/day} = \frac{5000}{1.2 \times 86{,}400}$$

$$= 48.23 \times 10^{-3}\ \text{m}^3/\text{s}$$

$$\frac{V_d}{V_c} = \frac{48.23}{15.43} = 3.13$$

From Fig. 6 read $h_f = 0.5\, N^3R^5/HD$. From Fig. 4 read for solvent dispersed at 0.4:

$$V_s = 0.0925$$

Hence

$$V_{c_f} = \frac{0.0925 \exp(-0.5)}{3.13/0.5 + 1/0.5} = 6.8\ 10^{-3}\ \text{m/s}$$

For 75% of flood,

$$V_c = 5.1\ 10^{-3}\ \text{m/s}$$

For $\alpha = 3.13$, flow of dispersed phase becomes

$$V_d = 15.9\ 10^{-3}\ \text{m/s}$$

$$\text{Total cross section of RDC} = \frac{15.43 \times 10^{-3}}{6.8 \times 10^{-3}}$$

$$= 2.27\ \text{m}^2$$

Tabulation of the results yields the following data:

Diameter of RDC: $D = 1700$ mm

$S = 1190$ mm

$R = 1020$ mm

$H = 170$ mm

For $N^3R^5/HD^2 = 0.4$, it can be calculated that $N = 0.56$ rev/sec.

To calculate the total agitated height, first calculate the actual holdup by trial and error from the equation, substituting V_c. The result will give $h = 0.25$. For $N^3R^5/HD^2 = 0.4$, read from Fig. 7 that HTU = 0.75. Calculate HDU as follows:

$$E_c = 0.5 \times 5.1 \times 10^{-3} \times 0.17 + 0.012 \times 1.02 \times 0.56 \times 0.17 \times \left(\frac{1.19}{1.7}\right)^2 = 1.0 \times 10^{-3}$$

$$E_d = 1.0 \times 10^{-3} \times \frac{4.2 \times 10^5}{1.7^2}\left(\frac{15.9 \times 10^{-3}}{0.25}\right)^{3.3}$$

$$= 16.35 \times 10^{-3}$$

Calculate Peclet numbers for continuous and dispersed phases:

$$\frac{1}{(\text{Pe})_c} = \frac{1 \times 10^{-3} \times 0.75}{0.17 \times 5.1 \times 10^{-3}} = 0.865$$

$$\frac{1}{(\text{Pe})_d} = \frac{16.35 \times 0.25}{0.17 \times 15.9} = 1.512$$

According to the equation,

$$\text{HDU} = 0.17 \times 0.865 + 0.17 \times 1.512$$

$$= 0.40\ \text{m}$$

$$(\text{HTU})_{\text{eff}} = 0.40 + 0.75 = 1.15\ \text{m}$$

The total agitated height will be

$$L = 10 \times 1.15 \times \frac{\ln 0.8}{0.8 - 1} = 12.8\ \text{m}$$

Take $L = 12.8$ and start more detailed calculation:

$$\frac{1}{(\text{Pe})_1} = 0.8 \times 0.865 + 1.512 = 2.20$$

$$\frac{1}{(\text{Pe})_2} = 0.865 + 0.8 \times 1.512 = 2.07$$

$$(\text{Pe})_0 = \frac{\dfrac{1.28}{0.75} + 1}{\dfrac{1.28}{0.75} + \dfrac{2.07}{2.20}} \cdot \frac{1}{2.20} = 0.465$$

$$\text{HDU} = \frac{1}{\dfrac{0.465}{0.17} + \dfrac{0.8}{12.8}\ \dfrac{\ln 0.8}{0.8 - 1}} = 0.357$$

$$(\text{HTU})_{\text{eff}} = 0.357 + 0.75 = 1.107$$

$$L = 10 \times 1.107 \times \frac{\ln 0.8}{0.8 - 1} = 12.4\ \text{m}$$

This is sufficiently close; take L as 12.50. For the sake of safety, add 10%; then $L = 13.7$ m. For $H = 0.17$, this means 81 compartments. The total height, including 1.0 m settling height on top and bottom will be 15.7 m.

To calculate the power input, take the same margin as for solvent continuous phase operation:

$$N_{max} = 1.12 \text{ rev/s}$$

$$\frac{\rho N R^2}{\mu} > 10^5; \qquad \text{thus } P = 0.03$$

$$P_{tot} = 81 \times 0.03 \times 750 \times 1.12^3 \times 1.02^5 = 2.8 \text{ kW}$$

Tabulation of the results yields the following data:

$D = 1700$ mm

$S = 1190$ mm

$R = 1020$ mm

$H = 170$ mm

Total height = 15,700 mm

Rotor speed = 15–70 rpm

NOTATION

D	diameter
$E_{c,d}$	diffusion coefficient
F	extraction coefficient = $K\left(\frac{Q \text{ feed}}{Q \text{ solvent}}\right)$
H	compartment height
h	holdup
HDU	height of a diffusion unit
HETS	height equivalent to a theoretical stage
HTU	height of a transfer unit
K	distribution coefficient = concentration in feed phase/concentration in solvent phase
N	rotor speed
P	energy input
Pe	Peclet number—Pe_s for solvent phase, Pe_f for feed phase
Q	volumetric flow
R	rotor diameter
S	stator opening diameter
V	superficial velocity
$\overline{V}$	velocity through smallest cross section
V_s	drop slip velocity
α	phase ratio
ρ	density
μ	viscosity

Subscripts

c	continuous phase
d	dispersed phase
eff	effective
f	flooding

REFERENCES

1. D. H. Logsdail et al., *Transact. Inst. Chem. Eng.* **35,** 301 (1957).
2. R. Marr et al., *Verfahrenstechnik* **12,** 139 (1978).
3. C. P. Strand et al., *AIChE J.*, 252 (May 1962).
4. G. H. Reman, *Chim. Ind.–Genie Chim.* **74**(4), 116 (October 1955).
5. S. Stemerding et al., *Erdoel Z.*, 401 (September 1961).
6. G. H. Reman, *De Ingenieur* (45), 128 (November 8, 1957).
7. J. D. Thornton, *Chem. Eng. Sci.* **5,** 201 (August 1956).
8. H. Groothuis et al., *Chem. Eng. Sci.* **12,** 288 (1960).
9. D. B. Broughton, USP 3433735, March 18, 1969.
10. G. S. Laddha et al., *Can. J. Chem. Eng.* **56,** 137 (April 1978).
11. A. E. Handlos, *AIChE J.* **3**(1), 127 (March 1978).
12. S. Stemerding, *Chem. Eng.*, CE 156 (May 1963).
13. S. Stemerding et al., *Chem. Ing. Technik* **35**(12), 844 (1963).

SUGGESTED READING

A. Bahr et al., *Chem. Technik. (Leipzig)* **29**(2), 84–87 (1977).

L. Blazey et al., *Chem. Zvesti* **32**(3), 328–355 (1978).

L. Bock et al., *Chem. Technik. (Berlin)* **16**(2), 85–88 (1964).

T. Borrell et al. *Chem. Eng. Sci.* **29**(5), 1315–1318 (1974).

A. Brink and N. J. Gericke, *S. Afr. Ind. Chem.* **18,** 152–154 (1964).

S. Bruin, *Transact. Inst. Chem. Eng.* **51**(4), 355–360 (1973).

R. H. Chartres and W. J. Korchinsky, *Transact. Inst. Chem. Ing.* **53,** 247–253 (1975).

R. H. Chartres and W. J. Korchinsky, *Transact. Inst. Chem. Eng.* **56,** 91–95 (1978).

A. L. Chvertkin et al., *Khim i Teckhnol, Topl. i Masl.* **5**(10), 30-32 (1970).

G. J. Glazon et al. *Chem. Technol. Fuel Oils* **6**(9-10), 757-760 (1970).

R. Goerz and G. Hoffmann, *Chem. Technik* **16**(2), 80-85 (1964).

W. Häntsch and S. Weiss, *Chem. Technik* **28**(6), 334-337 (1976).

R. Houlihan and J. Landau, *Can. J. Chem. Eng.* **52**(3), 338-344 (1974).

G. Husing, *Inz. Apar. Chem.* **17**(3), 13-17 (1978).

E. Y. Kung and R. B. Beckmann, *AIChE J.* **7**(2), 319-324 (1961).

G. S. Laddha et al., *Can J. Chem. Eng.* **56,** 137-150 (1978).

J. Magiera et al., *Internatl. Chem. Eng.* **16**(4), 744-750 (1976).

J. Magiera and J. Zadlo, *Inz. Chem.* **7**(1), 113-142 (1977).

R. Marr et al., *Chem. Ing. Technol.* **49**(3), 203-212 (1977).

R. Marr et al., *Verfahrenstechnik* **12**(3), 139-144 (1978).

R. Marr, *Chem. Ing. Technol.* **50**(5), 337-344 (1978).

R. Marr and F. Moser, *Chem. Ing. Technol.* **50**(2), 90-100 (1978).

T. Misek and B. Rozkos, *Internatl. Chem. Eng.* **6**(1), 130-137 (1966).

H. G. Mitrofanov et al., *Chem. Technol. Fuel Oils* **10**(5-6) 350-352 (1974).

D. Möhring and S. Weiss, *Chem. Technik* **29**(5) 262-265 (1977).

C. J. Mumford and A. A. A. Al-Hemiri, *Proceedings of the International Solvent Extraction Conference*, Lyons, Society of Chemical Industry, London, 1974, Vol. 2, pp. 1591-1617.

H. Pajah and Z. Ziolkowski, *Chem. Stosow Ser. B* **7**(2), 261-278 (1970).

M. V. R. Rao et al., *Br. Chem. Eng.* **12**(5), 719-721 (1967).

M. V. R. Rao et al., *Chem. Petro-Chem. I* **6**(1), 37-39 (1975).

G. H. Reman and R. B. Olney, *Chem. Eng. Progr.* **51**(3), 141-146 (1955).

W. Spaethe et al., *Verfahrenstecknik* **10**(9), 567-571 (1978).

W. Spielman et al., *Chem. Technik* **29**(2), 84-87 (1977).

T. Sripathi and M. V. R. Rao, *Chem. Eng. World* **10**(4), 101 (1975).

T. Sripathi et al., *Chem. Eng. World* **11**(2), 51-53 (1976).

F. B. Stainthorp et al., *Transact. Inst. Chem. Eng.* **42,** T198-T208 (1964).

V. Shtrobel et al., *Zh. Prikl. Khim.* **36**(12), 2672-2680 (1963).

W. Strobel, *Chem. Technik (Berlin)* **16**(6), 376 (1964).

E. H. Tagiltsewa, *Neft Neftekhim* (6), 42-43 (1971).

J. Temniskov and D. Elenkov, *Dokl. Bolg. Akad. Nauk.* **25**(12), 1673-1676 (1972).

V. B. Thegze et al., *Oil Gas J.* 90-94 (May 8, 1964).

W. J. Thomas, *J. Appl. Chem.* **20,** 261-273 (September 1970).

V. G. Trukhanov et al., *Khim Prom. (Moscow)* **47**(8), 620-624 (1971).

H. J. A. Vermys and H. Kramers, *Chem. Eng. Sci.* **3,** 55-67 (1954).

13.2

ASYMMETRIC ROTATING DISK EXTRACTOR

T. Míšek
Research Institute of Chemical Equipment
Czechoslovakia

J. Marek
Luwa AG
Switzerland

1. INTRODUCTION

The asymmetric rotating-disk (ARD) extractor [1] is a rotary-agitated column developed with emphasis on balanced throughput and separation performance and on reduction of axial mixing effects. It has been in industrial use since 1965. Dependable design and scale-up procedures based on theoretical and experimental background are available for this contactor.

2. DESCRIPTION OF CONTACTOR

Basic arrangement of the ARD extractor is shown in Figs. 1 and 2. The contactor consists of a cylindrical shell, a baffled stator, and a multistage agitator. The shell houses an extraction section and two settling zones. The extraction section is divided by an asymmetrically positioned vertical stator baffle into a contact zone and a transport zone. Both of these are subdivided by horizontal baffles into a series of staggered chambers that communicate with each other through openings on both sides of the vertical baffle. Disk impellers mounted on the agitator shaft and centered in each contact chamber provide mechanical agitation.

2.1. Principle of Operation

The feed and solvent are introduced into the contactor at opposite ends of the extraction section. As a result of density difference, the liquids flow

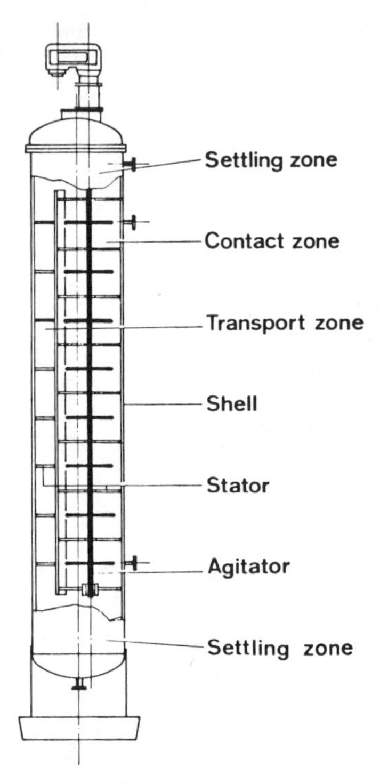

Figure 1. ARD extractor.

countercurrently through this section, one as continuous and the other as dispersed phase, toward the respective settling zones. Here they become separated from the other phase and leave

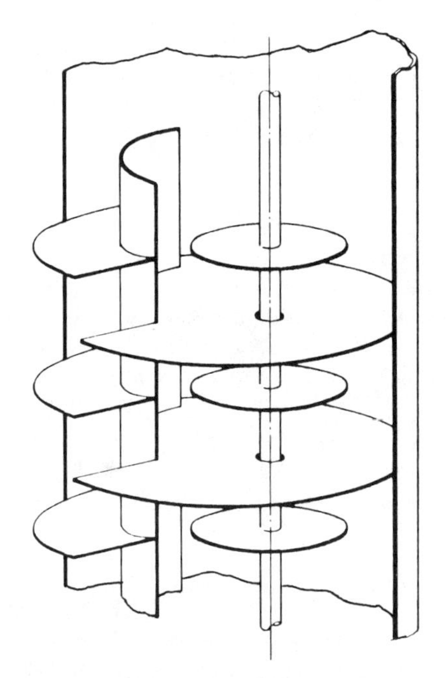

Figure 2. Extraction section. (Courtesy of LUWA AG, Zurich, Switzerland.)

as raffinate and extract. The phase interface and in consequence the phase character can be freely chosen, with an interface in the top (bottom) settling zone leading to heavy (light) liquid continuous. The stator baffling in the extraction section causes both liquids to flow in a helical path alternatively through adjacent contact and transport chambers. In the contact chambers fresh droplets of the dispersed phase are created and both phases are intimately mixed by the disk impellers. In the transport chambers droplets partly recombine, and the phases disengage and flow to the next upper and lower contact chamber, respectively.

The agitator speed influences the droplet size of the dispersed phase, the throughput capacity and the separation effect of the contactor. The repeated dispersion and coalescence of droplets leads to continuous renewal of interphase surface and thus to high mass-transfer rates. The ordered helical flow pattern of both phases is a prerequisite for high throughput and low axial mixing effects, a characteristic feature of the ARD extractor.

2.2. Size and Geometry

The ARD extractor is being manufactured within a broad size range. Diameters of 72–3000 mm and more and lengths of 2–30 m are available.

The internal geometry is standard in most cases, with nominal values (with D_c in millimeters) as follows: free cross section of phase transport 25%; chamber height 1.3 $D_c^{0.67}$; and disk diameter 0.49 D_c. For special applications, different values are used. Thus the free cross section and chamber height can be increased for higher specific throughput. The disk diameter can be varied along the contact zone to compensate for changing physical properties of the liquid dispersion. The agitator speed depends on contactor diameter and liquid properties; a nominal rpm value is $15000/D_c^{0.78}$. A variator is used to adjust the speed for optimal performance.

Mechanical design of the ARD extractor provides the possibility to pull the stator and agitator out of the shell. Sectional [2] or scissor-type [3] baffle arrangements are used on large-diameter contactors for easy insertion of the agitator into the stator and of the stator into the shell. Another development [4] shown in Fig. 3 features a direct pull-out agitator. In this case additional disks mounted on the shaft rotate in the openings of the horizontal baffles and prevent short-circuiting of the liquids between adjacent contact chambers.

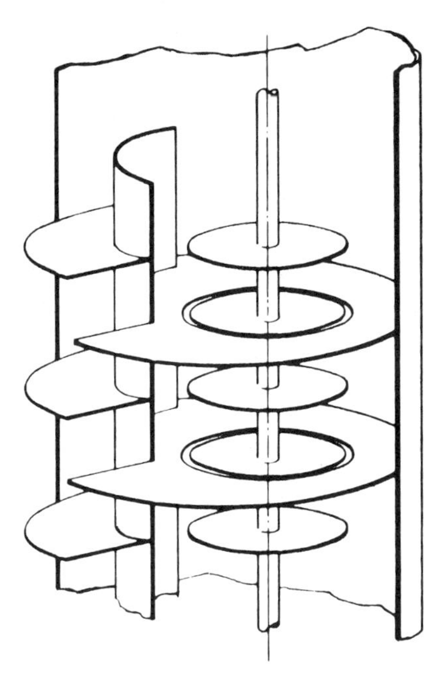

Figure 3. Extraction section with a pull-out agitator. (Courtesy of LUWA AG, Zurich, Switzerland.)

A large-capacity ARD extractor has been developed and tested [5]. This modification, equipped with three agitators, can be built up to a diameter of 6000 mm.

3. INDUSTRIAL APPLICATIONS

The ARD extractor is a general purpose contactor. It is used for processing liquid systems within a broad range of physical properties: density differences from 500 to 20 kg/m^3; continuous-phase viscosity 0.5–1000 mPa s; and interfacial tension 1-50 $\cdot$ 10^{-5} N/cm. It is well suited for treating liquids with emulgating tendency or with suspended solids.

Typical industrial applications are found in the following fields:

1. Organic and petrochemical: acetic acid, acrylic acid, aromatics, caprolactam, cumene, esters, isoprene, pyridine, resorcinol, wool wax, and many others.
2. Inorganic and metallurgical: hydrogen peroxide, metal salts of cobalt, nickel, magnesium, zinc, rare earths, uranium dioxide, phosphoric and other acids, and others.
3. Miscellaneous; glycerine from soap, foodstuff constituents, pharmaceuticals, phenol and other products from liquid wastes, spent catalysts, and so on.

Asymmetric rotating-disk extractors are currently in operation in more than 100 processes.

4. PERFORMANCE

A fairly complete theoretical description of the throughput and mass-transfer performance of the ARD extractor has been developed on the basis of research and industrial measurements.

4.1. Throughput

Hydrodynamic behavior of the agitated liquid system consisting of the dispersed and continuous phase in countercurrent flow determines the throughput capacity of the extractor. The forces acting on the dispersed phase in the extraction section give rise to a polydisperse droplet population. Elementary processes of stochastic character, such as droplet breakup, random movements, and coalescence, take place along the path of the droplet cloud.

4.1.1. Droplet Size and Velocity

The agitation and other conditions prevailing in the contact zone produce different sizes of droplets in the dispersed phase. The droplet size distribution may be conveniently described [6] by the relationship

$$H\left(\frac{d}{d_m}\right) = \frac{1}{6\beta^4}\left(\frac{d}{d_m}\right)^3 \exp\left(-\frac{d}{\beta d_m}\right) \tag{1}$$

where $H(d/d_m)$ is the volumetric size distribution density of droplets with a diameter d, d_m is a maximal droplet diameter, and β is a constant with an average value of 0.13 that characterizes the sharpness of distribution.

The maximal diameter d_m depends on the physical properties of the dispersion and on the geometry of the extractor. The following relationship can be used for the region of intensive agitation [7]:

$$\frac{d_m}{D} = 86.9 \frac{\sigma_i \exp(0.0887\,\Delta D)}{n^2 D^2 \rho_c} \left(\frac{h_m}{D_c}\right)^{0.46} \tag{2}$$

with ΔD in centimeters. For the region of mild agitation where the droplet size is independent of rotor speed, Eq. (3) is valid:

$$d_m = 2.03 \sqrt{\frac{\sigma_i}{\Delta\rho g}} \tag{3}$$

The transition between both regions is characterized [8] by a critical rotor speed

$$n_{cr}^2 = 42.8 \frac{\sqrt{\sigma_i \Delta\rho} \exp(0.0877 \Delta D)}{D^2 \rho_c} \left(\frac{h_m}{D_c}\right)^{0.46} \tag{4}$$

For systems with large viscosity difference between the dispersed and continuous phase, the relationship derived in [9]

$$\frac{d_m}{D} = \frac{427.5}{\mathrm{We}_M} - \frac{15520}{\mathrm{Re}_M} \left(\frac{\mu_c}{\mu_d}\right)^{0.17} \tag{5}$$

gives better results than does Eq. (2). For systems with mass transfer from the continuous into the dispersed-phase droplet size increase of about 25% has been observed.

Droplets move in the extraction section erratically and it is difficult to measure the vertical component of their velocity. Such measurements have, however, been performed [10]. The results for systems within a broad range of interfacial tensions indicate that the vertical droplet velocity approaches the terminal velocity of rigid spheres.

The movement of larger droplets falls into the transition region between turbulent and laminar flow, where the vertical velocity can be determined from

$$u = 0.249 d \left(\frac{\Delta\rho^2 g^2}{\rho_c \mu_c}\right)^{1/3} \tag{6}$$

4.1.2. *Holdup and Flooding*

The holdup, that is, the volumetric fraction of the dispersed phase, determines the limiting flow conditions in the extractor. Taking into account the velocity distribution of droplets of different size, a relationship between the flow distribution density $F(d)$ and size distribution density $H(d)$ has been derived [11]

$$U_d F(d)\, \partial d = \left[\epsilon u X(1 - X) \exp(aX) - \frac{U_c X}{1 - X} \cdot H(d)\, \partial d\right] \tag{7}$$

If the influence of entrainment of small droplets is neglected; Eq. (7) is integrated to

$$\frac{U_d}{X} + \frac{U_c}{1 - X} = \epsilon(1 - X) \exp(aX) \int_0^\infty u\, H(d)\, \partial d \tag{8}$$

where the integral term may be interpreted as a characteristic droplet velocity. For the region of validity of Eq. (6), this velocity is given by

$$u_0 = \int_0^\infty u\, H(d)\, \partial d = 0.249 d_{43} \left(\frac{\Delta\rho^2 g^2}{\rho_c \mu_c}\right)^{1/3} \tag{9}$$

The parameter a in Eq. (3) is a measure of the intensity of droplet coalescence in the liquid system. Its value can be roughly determined from the relationship

$$a = \frac{z}{\alpha} - 4.1 \tag{10}$$

where z is the coefficient of coalescence and α is a velocity exponent. On the basis of the theory of turbulent coalescence [12], the value of z is given [13] by

$$z = 0.0159 \left[\frac{D_c}{\nu_c}\left(\frac{\sigma_i}{\rho_c d_{43}}\right)^{0.5}\right]^{0.5} \tag{11}$$

Equation (11) is valid for pure liquids without mass transfer. For systems containing surface-active impurities and for liquids with mass transfer from the continuous into the dispersed phase, the coalescence is often negligible (i.e., $z = 0$). A test is normally necessary for the determination of z for systems with mass transfer from the dispersed into the continuous phase. The velocity exponent α depends on Reynolds number. It is equal to one within the region of validity of Eq. (6).

When the throughput in the extractor is increased, flooding conditions can be reached whereby the two phases cease to flow countercurrently. The limiting holdup at flooding reaches a value characterized by the conditions $\partial U_d/\partial X = 0$ and $\partial U_c/\partial X = 0$. On solving Eq. (8) for these conditions, the relationship

$$\frac{U_d}{U_c} = \frac{2X_f^2[1 - X_f + a(X_f - X_f^2/2 - 0.5)]}{(1 - X_f)^2 [1 - 2X_f + a(X_f - X_f^2)]} \tag{12}$$

is obtained. Another type of flooding that leads to phase inversion cannot be predicted theoretically. A third type leading to excessive entrainment of one phase in the other can be determined by an appropriate integration of Eq. (7).

Figure 4 is a graphical representation of the flooding relationship [12]. Regions of phase inversion and excessive entrainment are also shown. Figure 4 can be conveniently used for the selection of operating throughput of the ARD extractor. From the given phase flow ratio U_d/U_c, parameter a, and selected value of the tolerable entrainment ratio U_z/U_d, the limiting holdup X is found and the appropriate limiting specific throughput $Q = (U_d + U_c)$ is calculated by solving Eqs. (8), (9), and (12). The operating throughput is selected within 50–80% of the limiting value.

4.1.3. *Agitation*

It has been previously mentioned that the ARD extractor can be operated within a region of intensive or mild agitation whereby the transition point is characterized by the critical rotor speed given by Eq. (4).

Intensive agitation can be used for mixtures of

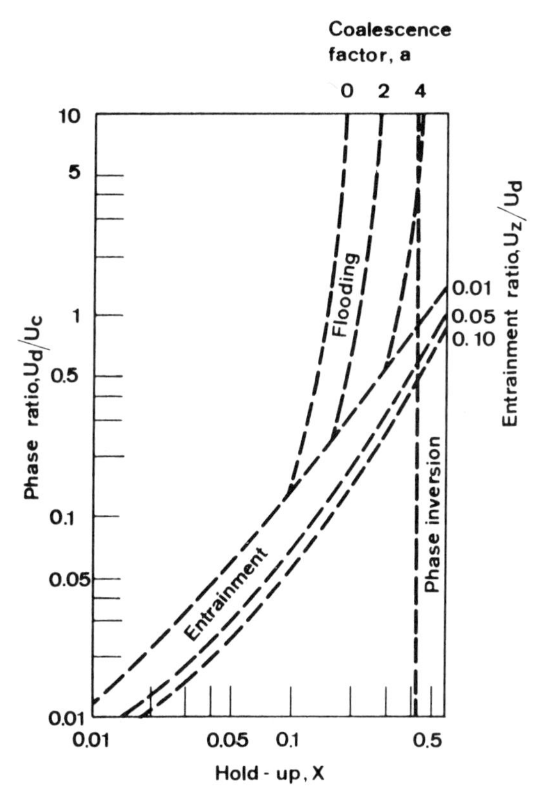

Figure 4. Limiting flow conditions.

pure liquids containing no surface-active impurities and for systems with mass transfer from the dispersed into the continuous phase. These conditions favor droplet coalescence. Equation (2) is used for the selection of rotor speed n for a maximal droplet diameter d_m. The relationship

$$d_m = 0.136 \sqrt[5]{\frac{\sigma_i \mu_c^2}{\rho_c \, \Delta\rho^2}} \tag{13}$$

is useful for the determination of an appropriate droplet size to be used.

Mild agitation is indicated for mixtures of liquids containing impurities and for systems with mass transfer from the continuous into the dispersed phase. Rotor speed is in this case lower than the critical value from Eq. (4). The power input of a disk impeller can be determined from the well-known relationship

$$\frac{P}{n^3 D^5 \rho_c} = k_5 \left(\frac{D^2 n \rho_c}{\mu_c}\right)^{-k_6} \tag{14}$$

Experimentally determined [14] values of the constants are $k_5 = 23.1$, $k_6 = 0.568$ for the laminar region, and $k_5 = 0.261$, $k_6 = 0.155$ for the turbulent region, respectively. The overall power input of the agitator is given by the value [Eq. (14)] multiplied by the number of disks and increased by energy losses in the agitator drive and bearings.

4.2. Mass Transfer

The conditions influencing mass transfer in the agitated liquid system are rather complex. Nevertheless, a set of relationships for diffusional solute transfer and for axial mixing in both phases was found that describes with acceptable accuracy the mass-transfer performance of the contactor.

4.2.1. *Diffusion*

Two mechanisms are involved in diffusional mass transfer of solute inside the droplets of the dispersed phase. In rigid spherical droplets the solute is transported from the interior to the surface by means of molecular diffusion. The mass-transfer coefficient for this case can be determined from a relation [15] modified to

$$\exp\left(-k_{dr} a' \frac{h_c}{U_d}\right) = 1 - \left[1 - \exp\left(\frac{4\pi^2 \mathfrak{D}_d h_c X}{U_d d^2}\right)\right]^{1/2} \tag{15}$$

In oscillating droplets with internal circulation the solute is transfered by eddy diffusion and the mass transfer coefficient [16] is given by

$$k_{do} = \frac{0.00375u}{1 + \mu_d/\mu_c} \tag{16}$$

The conditions in droplets of different sizes are such that both mechanisms must be considered. It has been found [17] that the following relationship represents the diffusional mass-transfer coefficient in the dispersed phase:

$$\frac{1}{k_d} = \frac{A}{k_{do}} + \frac{1 - A}{k_{dr}} \tag{17a}$$

where the transition factor A is a function of rotor speed:

$$A = 2.29 - 0.92 \frac{n}{n_{cr}} \pm 0.22, \qquad A\langle 1;0\rangle \tag{17b}$$

The constant 0.22 expresses the influence of the direction of mass transfer. The plus sign is valid for solute transfer into the droplets and the minus sign, for transfer from the droplets.

A well-known relationship [18] is used to express the mass-transfer coefficient in the continuous phase:

$$k_c = 0.13 \left(\frac{4P}{\pi D_c^2 h_m}\right)^{1/4} \frac{\mathfrak{D}_c^{2/3}}{\rho_c^{1/2} \mu_c^{5/12}} \tag{18}$$

4.2.2. *Axial Mixing*

Axial mixing significantly influences solute transfer in both phases. Both back- and forward-mixing effects have been investigated. Backmixing is understood to be random flow of bulk elements of the phase due to axial concentration gradients. Forward mixing is a similar effect caused by size distribution of the dispersed phase.

From a large number of measurements, the following relationships were derived for the effect of backmixing in the dispersed phase:

$$\frac{\xi_d}{U_d h_m} = 0.5 + \frac{ZX}{U_d} \tag{19}$$

and in the continuous phase

$$\frac{\xi_c}{U_c h_m} = 0.5 + \frac{Z(1 - X) + U_{zw}}{U_c} \tag{20}$$

The backflow parameters Z and U_{zw} are given by [19, 20]

$$Z = \frac{1.82 \cdot 10^{-3}\ \epsilon n D}{h_m^{\ 1/3}} \left(\frac{D}{D_c}\right)^{2/3} \tag{21}$$

$$U_{zw} = 2.71 \cdot 10^{-4}\ U_d (nD)^2 \tag{22}$$

with h_m in centimeters and nD in centimeters per second.

The effect of forward mixing has also been investigated [21]. The resulting dependence of correction factors ϕ on the extraction factor $R_d = U_c/m_d U_d$ is shown in Fig. 5. The distribution coefficient $m_d = x_d^*/x_c$ is defined here as the ratio of equilibrium concentrations in the dispersed and the continuous phase, respectively.

4.2.3. *HTU and HETS*

The relationships for diffusional mass-transfer and forward-mixing effects can be combined in

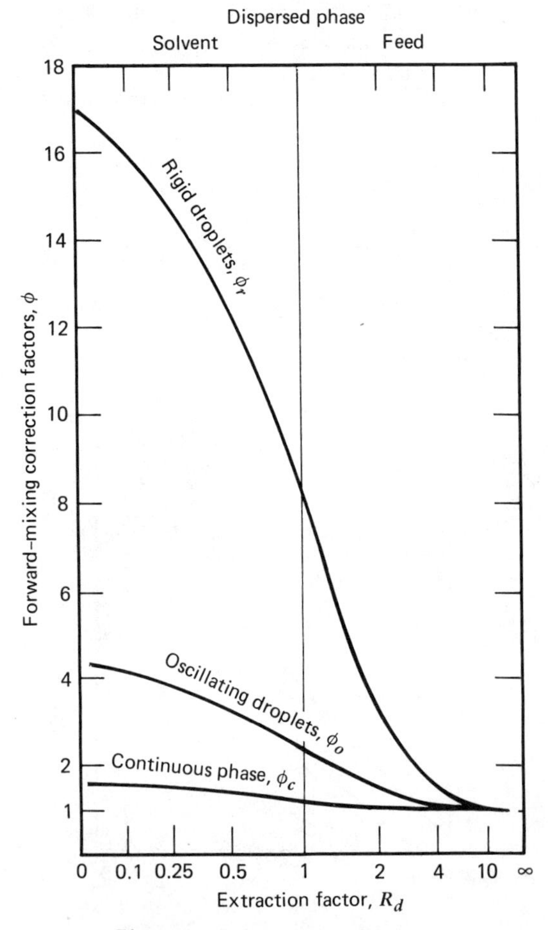

Figure 5. Forward-mixing factors.

accordance with Eq. (17) into an overall coefficient of mass transfer:

$$\frac{1}{K_d} = \frac{A}{k_{do}}\Phi_o + \frac{1-A}{k_{dr}}\Phi_r + \frac{m_d}{k_c}\Phi_c \quad (23)$$

where the coefficients k_{do}, k_{dr}, k_c, and A are calculated from Eqs. (15)-(18) and the correction factors Φ_o, Φ_r, and Φ_c are found from Fig. 5. The corresponding height of a transfer unit is

$$\text{HTU}_{0d} = \frac{U_d d_{32}}{6XK_d} \quad (24)$$

The height of a transfer unit corrected for axial mixing in both phases can be calculated [22] as

$$\text{HTU}_d = \text{HTU}_{0d} + \frac{\xi_d}{U_d}\alpha + \frac{\xi_c}{U_c}R_d\beta \quad (25a)$$

where

$$\alpha = \frac{1}{1 + (1 - R_d)\,\xi_d/(U_d \text{HTU}_d)} \quad (25b)$$

$$\beta = \frac{1}{1 - (1 - R_d)\,\xi_c/(U_c \text{HTU}_d)} \quad (25c)$$

The backmixing factors ξ_d/U_d and ξ_c/U_c are calculated from Eqs. (19) and (20). Equations (25a-c) are solved by iteration. The height equivalent of a theoretical stage is obtained by a transformation of the HTU_d value:

$$\text{HETS} = \text{HTU}_d \frac{R_d(-\ln R_d)}{1 - R_d} \quad (26)$$

The described procedure for HETS determination is strictly valid only for a constant distribution coefficient m_d. For systems where m_d varies with concentration, an average obtained from the values at the feed and raffinate end gives satisfactory results.

5. DESIGN AND SCALE-UP

The relationships presented in Section 4 form a self-contained description of the hydrodynamic and mass-transfer performance of the ARD extractor. They can be used for design purposes and, in simplified form, for scale-up of pilot test data.

5.1. Design Procedure

Direct design of an ARD extractor based on theoretical relationships is possible if basic physical data such as equilibrium distribution of solute, densities, viscosities, interfacial tension, and diffusivities for both phases are known.

The design procedure consists of the following steps:

1. Calculation of mass balance and number of theoretical stages necessary for the specified separation.
2. Selection of dispersed phase.
3. Preliminary selection of extractor diameter based on an approximate correlation of specific throughput.
4. Selection of rotor speed and calculation of power input based on relations in Section 4.1.3.
5. Calculation of hydrodynamic parameters and limiting holdup based on relations in Sections 4.1.1 and 4.1.2.
6. Selection of operating throughput and check of extractor diameter.
7. Calculation of mass-transfer coefficients and axial mixing corrections based on relationships in Sections 4.2.1 and 4.2.2.
8. Calculation of HTU_d and HETS based on relationships in Section 4.2.3.
9. Determination of the height of extraction section from the number of theoretical stages and HETS.

The design procedure is iterative and may best be applied in the form of a computer program. It has been checked by measurements on experimental ARD extractors and on a limited number of plant contactors of diameter up to 2800 mm.

Direct design is, however, applicable with a fair degree of confidence only for simple three-component liquid systems with known physical properties that do not excessively vary over the range of concentrations between the feed and raffinate end.

5.2. Scale-Up Procedure

The majority of industrially important extraction systems does not fulfill the requirements stated in Section 5.1. The design of ARD extractors for such applications is based on scale-up of pilot test data.

In a pilot test the optimal throughput and

rotor speed of the test extractor are found and the separation effect is evaluated from product compositions. Then HETS and HTU_{0d} values are calculated backward by use of the relationships given in Section 4.2.3. The scale-up is based on the assumption that the physical properties of test and plant liquid system and the geometry of test and plant contactor do not significantly differ. In this case a number of parameters in the relationships shown can be taken as constants, and relatively simple scale-up rules result for such variables as specific throughput, rotor speed, and HTU_{0d}.

The scale-up procedure consists of the following steps:

1. Calculation of mass balance and number of theoretical stages necessary for the specified separation.
2. Scale-up of specific throughput based on optimal test value and a simplified expression resulting from the relationships given in Sections 4.1.1 and 4.1.2 and determination of the corresponding extractor diameter.
3. Scale-up of rotor speed based on optimal test value and a simplified expression resulting from the relationships given in Section 4.1.3 and calculation of power input.
4. Scale-up of HTU_{0d} based on test value and a simplified expression resulting from the relationships given in Sections 4.2.1 and 4.2.3.
5. Calculation of axial mixing corrections HTU_d and HETS based on relationships given in Sections 4.2.2 and 4.2.3.
6. Determination of the height of the extraction section from the number of theoretical stages and HETS.

The described scale-up procedure is straightforward and is available as computer program. Its validity was confirmed in a large number of industrial applications. It can be used with confidence up to a throughput scale-up ratio of 1000.

5.3. Design Example

The basic design and scale-up procedure is illustrated in the following example.

An ARD extractor is needed for the extraction of acrylic acid from an aqueous solution in which ethyl acetate is used as solvent. The feed rate is 27,400 kg/hr, and the concentration is 15%. The maximum residual concentration in raffinate is 60 ppm. A computation based on a distribution coefficient of 4.3 indicates that six theoretical stages are required at a volumetric solvent : feed ratio of 0.8.

A pilot test has been performed. The computer evaluation printout of the best test point is shown in Table 1. The values obtained are used as the scale-up basis.

5.3.1. *Extractor Diameter*

As there is only a slight difference between the test and the plant phase ratio, the plant-specific throughput can be taken as equal to the test value:

Specific throughput	20.3 $m^3\ m^{-2}\ hr^{-1}$
Feed rate 27,400/1012	27.1 m^3/hr
Solvent rate 27.1 : 0.8	21.7 m^3/hr
Extractor diameter $\sqrt{48.8/20.3 \cdot 0.785}$	1.75 m
Selected diameter	1800 mm

5.3.2. *Rotor Speed*

For comparable conditions, the relationships shown lead to an approximate proportionality $n \sim D_c^{-1.2}$:

Nominal rotor speed: $500(150/1800)^{1.2}$	25 rpm
Variable speed range	16 ÷ 124 rpm

5.3.3. *HTU and HETS*

The relationships shown lead to an approximate proportionality $HTU_{0d} \sim D_c^{0.2}$:

HTU_{0d}: $158(1800/150)^{0.2}$ 260 mm

Equations (20)–(22) lead to a backmixing factor in the continuous phase:

$$\frac{\xi_c}{U_c} = D_c^{0.667}\left(0.645 + 5.08 \cdot 10^{-5} D_c^{1.166}\frac{n}{U_c}\right)$$

where $U_c = \dfrac{Q}{1 + V_S/V_F}$

Specific throughput Q:
$(27.1 + 21.7)/0.785 \cdot 1.8^2$ 19.2 $m^3\ m^{-2}\ hr^{-1}$

TABLE 1 TEST EVALUATION PRINTOUT[a]

TEST POINT			4.4
	EXTRACTOR		
TYPE			ARD STANDARD
DIAMETER, MM			150
HEIGHT EXTR. SECTION, MM			3000
DISPERSED PHASE			FEED
	MASS BALANCE		
PRODUCT	G, KG/H	V, L/H	CONC., W-PC
FEED	181.1	179.0	15.0
SOLVENT	161.3	179.0	0
RAFFINATE	154.0	152.2	0.0051
EXTRACT	188.4	202.2	14.4
	PHYSICAL PROPERTIES		
DENSITY FEED, G/CM3			1.012
SOLVENT, G/CM3			0.901
DIFFERENCE, G/CM3			0.111
INTERFAC. TENSION, DYNE/CM			9.6
VISCOSITY CONT. PHASE, POISE			0.0048
EQUILIBRIUM CONSTANTS: M			4.3
N			1
	THROUGHPUT, ROTOR SPEED		
THROUGHPUT, L/H			358.0
SPECIFIC, M3/M2.H			20.3
ROTOR SPEED, R.P.M.			500
CRITICAL, R.P.M.			230
	THEORETICAL STAGES, HETS, HTU		
YIELD, PC			99.97
PHASE RATIO, L/L			1.000
EXTR. FACTOR			0.222
NUMBER OF TH. STAGES			5.25
HETS, MM			572
VOL. EXTR. FACTOR			4.857
HTU(D), MM			287
HTU(0D), MM			158

[a]Abbreviations of units: MM, millimeter; KG/H, kg/hr; L/H, liters/hr; W = PC, weight percent; G/CM3, g/cm^3; DYNE/CM, dynes/cm; M3/m2.H, m^3 m^{-2} hr^{-1}; R.P.M., rpm; PC, percent; L/L, liquid/liquid.

Specific flow rate U_c:	
19.2/(1 + 21.7/27.1)	10.7 m^3 m^{-2} hr^{-1}
Factor ξ_c/U_c:	
$1800^{0.667}$ (0.645 + 5.08 $\cdot\, 10^{-5} \cdot 1800^{1.166} \cdot$ 25/10.7)	206 mm
Extraction factor R_d:	
4.3 · 21.7/27.1	3.44

If the axial mixing correction in the dispersed phase is neglected, Eqs. (25a, c) give

$$\mathrm{HTU}_d = 260 + 206 \cdot 3.44\beta$$

where $\beta = 1/(1 + 2.44 \cdot 206/\mathrm{HTU}_d)$.

Iterative calculation starting with $\beta = 0.5$ gives

β	0.500	0.550	0.564	0.569
HTU_d, mm	614	650	660	663
HETS [eq. (26)]:				
$663 \cdot 3.44 \cdot \ln 3.44/2.44$				1155 mm

5.3.4. *Extraction Section*

The following data are obtained:

Number of theoretical stages	6
Height of extraction section	
1155 · 6	6930 mm
Selected	7000 mm
Chamber height	190 mm
Number of chambers:	
7000/190	36

5.4. Case History

In a caprolactam (CL) production process [23] crude reaction product is purified by means of two-step extraction by use of ARD extractors. The feed consisting of CL in an alkaline solution is extracted with toluene and CL is reextracted from toluene solution into water. Caprolactam is thus separated from a number of side products present in the reaction mixture. Toluene solvent is used in closed cycle, including a regeneration step.

Two industrial plants are in operation. The plant designated as "T" contains two contactors with a diameter of 1800 mm, and the plant designated "D" contains four machines with a diameter of 2800 mm arranged in two lines. The ARD extractors for plant T were designed on the basis of tests performed in a pilot unit with a diameter of 200 mm, and the design of contactors for plant D was based on data obtained in plant T. The scale-up methods described previously were applied.

Performance data from both plants were used to check the validity of basic relations and scale-up procedure [24]. Figure 6 is a graphical representation of the height of diffusional transfer unit versus power input for ARD extractors at plant T.

The theoretical dependence was calculated from system properties and contactor dimensions by means of the relationships given in Section 4. The actual values were obtained from measured concentrations by backward calculation of the height equivalent of a theoretical stage, height of a transfer unit, and axial mixing corrections. Figure 7 contains a similar representation of the height equivalent of a theoretical stage versus rotor speed for ARD extractors at plant D. Both figures show a good agreement of theoretical values and measured data and confirm the applicability of the presented relationships to industrial contactors and the dependability of design and scale-up procedures.

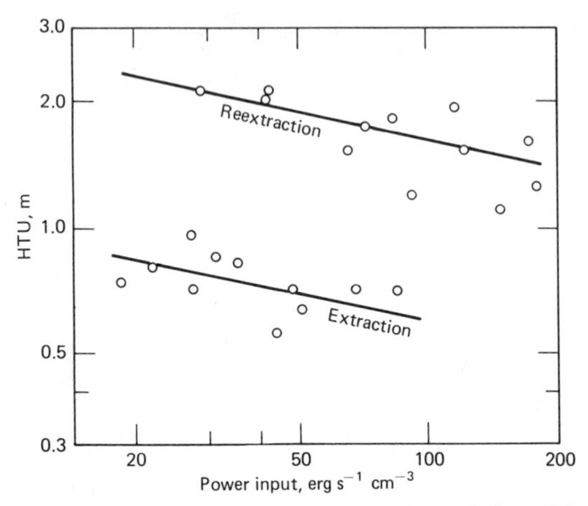

Figure 6. Height of diffusional transfer unit for ARD extractors at plant T.

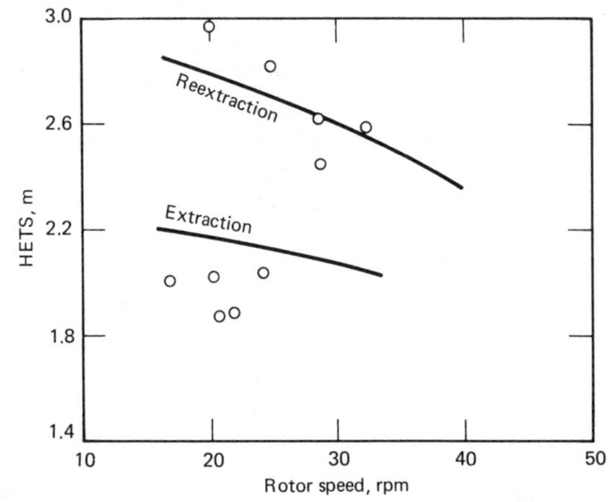

Figure 7. Height equivalent of a theoretical stage for ARD extractors at plant D.

NOTATION†

a	coalescence parameter
a'	specific interfacial area
A	transition factor
d	droplet diameter
D	disk diameter

†Any set of consistent units may be used in the relations if not stated otherwise.

D_c	extractor diameter
ΔD	minimal distance of disk from wall
$\mathfrak{D}$	diffusivity of solute
F	flow distribution density of droplets
g	acceleration due to gravity
h_m	chamber height
H	size distribution density of droplets
HETS	height equivalent of a theoretical stage
HTU_d	height of a transfer unit corrected for axial mixing
HTU_{0d}	height of a transfer unit
k	mass-transfer coefficient
K	overall coefficient of mass transfer
m_d	distribution coefficient
n	rotor speed
P	power input of a disk
Q	specific throughput
R_d	extraction factor
Re_M	Reynolds number
u	vertical droplet velocity
u_0	characteristic droplet velocity
U	flow rate per unit cross section
We_M	Weber number
x	concentration of solute
X	fractional holdup of dispersed phase based on total cross section
z	coefficient of coalescence
Z	backflow per unit cross section
α	velocity exponent
β	distribution constant
Δ	difference
ϵ	fractional free cross section
μ	viscosity
ν	kinematic viscosity
ξ	backmixing factor
ρ	density
σ_i	interfacial tension
Φ	forward-mixing factor

Subscripts

c	continuous; column
cr	critical
d	dispersed
f	flooding
m	maximal
o	oscillating droplets
r	rigid droplets
z	entrainment
*	equilibrium

REFERENCES

1. T. Míšek and J. Marek, *Br. Chem. Eng.* **15**, 202 (1970).
2. B. Rozkoš and T. Míšek, Swiss Patent 446116 (1967).
3. H. Linder, Swiss Patent 548218 (1974).
4. J. Marek, Swiss Patent 604806 (1977).
5. T. Míšek and B. Rozkoš, Report No. 1420, Research Institute of Chemical Equipment, Prague, 1974.
6. J. Přerovská and T. Míšek, paper presented at 3rd CHISA Congress, Mariánské Lázně, (Czechoslovakia), 1969.
7. T. Míšek, *Coll. Czech. Chem. Commun.* **28**, 426 (1963).
8. T. Míšek; Rotating Disk Extractors and Their Design (Czech), SNTL, Prague, 1964.
9. D. Stangl and T. Míšek, paper presented at 6th CHISA Congress, Prague, 1978.
10. T. Míšek, *Coll. Czech. Chem. Commun.* **28**, 570, (1963).
11. T. Míšek, *Coll. Czech. Chem. Commun.* **32**, 4018 (1967).
12. V. B. Levic, *Dokl. Akad. Nauk SSSR* **99**, 809 (1954).
13. T. Míšek, *Coll. Czech. Chem. Commun.* **29**, 2086 (1964).
14. T. Míšek, *Coll. Czech. Chem. Commun.* **29**, 1967 (1964); private communication.
15. A. B. Newmann, Transact. *AIChE* **27**, 310 (1931).
16. A. E. Handlos and T. Baron, *AIChE J.* **3**, 127 (1957).
17. T. Míšek, D. Sc. Thesis, Technical University, Pardubice, (Czechoslovakia), 1972.
18. P. M. Calderbank and M. B. Moo-Young, *Chem. Eng. Sci.* **16**, 39 (1961).
19. T. Míšek, *Coll. Czech. Chem. Commun.* 40, 1686 (1975).
20. J. Haman and T. Míšek, Report No. 1573, Research Institute of Chemical Equipment, Prague, 1975.
21. V. Rod and T. Míšek, paper presented at 2nd CHISA Congress, Mariánské Lánzně (Czechoslovakia), 1965.
22. T. Míšek and V. Rod, in *Advances in Liquid Extraction,* C. Hanson, Ed., Pergamon Press, London, 1971.
23. M. Taverna and M. Chiti, *Hydrocarbon Process.*, 137 (November 1970).
24. T. Míšek, J. Marek, and J. Bergdorf, in *Proceedings of the International Solvent Extraction Conference* (ISEC 77), Canadian Institute of Mining and Metallurgy, Montreal, 1979, p. 701.

13.3

SCHEIBEL COLUMNS

E. G. Scheibel
E. G. Scheibel, Inc.
United States

1. INTRODUCTION

Jantzen [1] in 1932 published the first data on the use of internal agitation to promote the countercurrent mixing of two liquid phases in an extraction column. He used a small multistage glass column to demonstrate that liquid extraction could fractionate coal tar distillates into more pure compounds than could be isolated by fractional distillation. His equipment was not practical for scale-up to commercial application, and other designs were described in the patent literature by Schöneborn [2] in 1934 and Othmer [3] in 1935, and also in 1935 Van Dijck [4] obtained the first of his numerous patents on liquid extraction processes in which he mentioned the use not only of rotation to promote phase mixing, but also of a reciprocating motion.

Multistage liquid extraction processes developed over this period depended on the use of mixer–settlers for implementation, and this imposed an economic limitation on the number of theoretical stages that could be utilized. The full potential of liquid extraction processes could not be realized because of lack of suitable commercial equipment [5]. Even as late as 1947, Fenske et al. [6] stated: "Despite such advantages of liquid-liquid separational processes, the problems of accumulating twenty or more theoretical stages in a small compact and relatively simple countercurrent operation have not yet been wholly solved."

The following year, Scheibel published performance data on the first of his multistage extractor column designs [7] that proved to be highly effective in laboratory, pilot plant, and small commercial-scale processes. Scale-up to large-diameter applications indicated that this initial design was expensive, and in 1956 a modification was described [8], 9] that was subsequently fabricated in diameters of up to 2.1 m. A subsequent modification in 1968 [10] reduced the costs of fabricating and maintaining the larger-diameter columns. The advantages of this third design are not fully realized until diameters exceed 1.5 m, although smaller columns are in commercial operation.

2. DESCRIPTION OF COLUMNS

Simplicity and efficiency are the two key parameters that must be optimized in the design of multistage extraction equipment. Review of the numerous descriptions of such equipment and their performance characteristics in the literature reveals that the simplest designs are usually the least effective whereas the most efficient designs are usually the most complex and, consequently, expensive. Scale-up is complicated by the fact that the most practical design of small-scale equipment for laboratory or pilot plant operation may not be the most economical for large-scale installations, and conversely the most economic commerical equipment may not be practical on a small scale.

The different designs described in the following sections have been developed to satisfy the requirements for the different scales of operation, and cost analysis of columns of 6-m diameter and over indicate that mixer–settlers have an economic advantage that becomes greater in processes requiring relatively small numbers of stages.

2.1. First Scheibel Extractor

In 1946 the only large-scale commercial liquid extraction process utilized seven mixer–settler stages, and it was considered impractical to design processes requiring more theoretical stages. It was obvious that the development of new liquid extraction processes, particularly the fractionation of mixtures as demonstrated in the Craig apparatus [11], required a simple device that could provide a large number of theoretical stages.

Figure 1 shows the first Scheibel extractor design [7] in which the countercurrent liquid phases are contacted in mixing zones with flat-bladed turbine-type agitators and the mixture separates in calming zones. In these zones the light phase passes upward to the mixing stage above and the heavy phase passes to the stage below. The calming zones are filled with knitted wire-mesh packing, which serves two purposes. First, it isolates the agitator flow patterns between adjacent mixing zones to prevent the loss of efficiency due to backmixing. Without this mesh, a high degree of agitation in all the mixing stages would cause complete backmixing and result in only one theoretical stage in the entire column. The height and density of mesh must be greater when a higher degree of agitation is required to mix the two immiscible phases. Second, the wire mesh provides the necessary baffling to remove the rotational motion imparted to the liquid mixture by the agitator. Thus the wire mesh achieves the same effect as the vertical baffles used in the Oldshue–Rushton column described in Chapter 13.4.

Wire mesh is unique in its ability to provide both effects, and it is also self-supporting. Calming could be provided by any conventional tower packing, but the smaller void volume restricts the flow through such a packing and reduces the baffling effect. The use of a packing support that virtually eliminates the baffling effect will necessitate the use of vertical baffles in the mixing zone.

Figure 2 shows the vertical components of the flow pattern in the mixing stage and illustrates

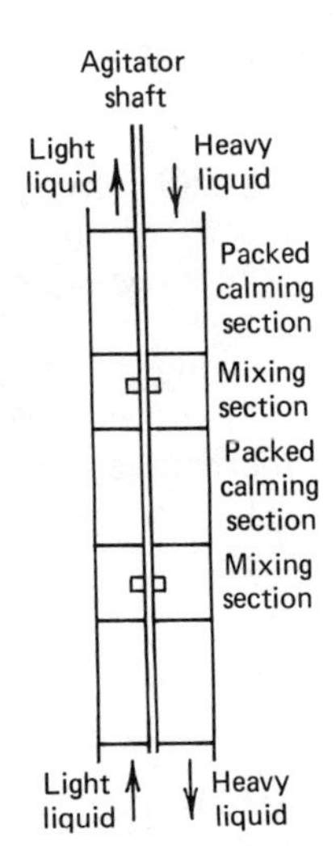

Figure 1. Schematic diagram of first Scheibel extractor design.

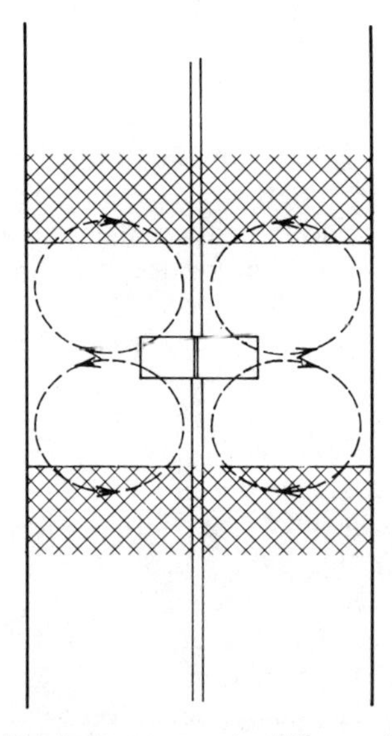

Figure 2. Vertical components of flow pattern in mixing stage of first Scheibel column.

the economic problems involved in scale-up of this design. If the diameter of the column is doubled, the solvent capacity, which is a function of the cross-sectional area, will be increased fourfold, but if the geometry of the flow pattern is retained, the volume of packing required will increase eightfold. Consequently, the wire-mesh packing becomes the most expensive item in the larger-diameter columns, and makes them uneconomical. For this reason, the largest column constructed of this design was 1 m in diameter.

The simplicity makes this design most suitable for laboratory and pilot plant columns of up to 0.3 meters in diameter. It has a major advantage for small-scale laboratory studies because it is available in sizes down to 2.5 cm in diameter.

2.2. Second Scheibel Extractor with Mesh Packing

When the economic limitations in the scale-up of the first design became apparent, a more cost effective method was developed to provide baffling in the mixing zone and at the same time minimize the height of a large-diameter multistage column with a given number of theoretical stages [8]. This modification is shown in Fig. 3, where the vertical flow of the phases in the mixing zone is diverted by horizontal annular baffles at the wall of the column, thereby preventing mixing with the circulation in the adjacent mixing zones.

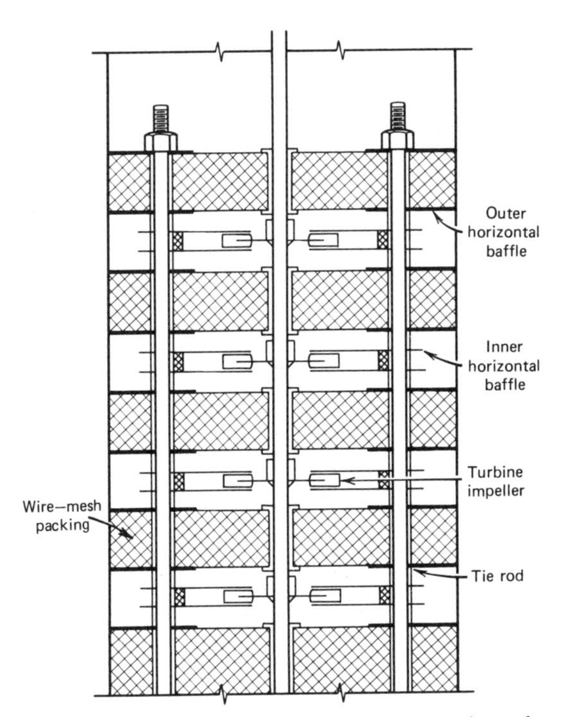

Figure 3. Construction details of 12-in. extraction column with baffled mixing stages and intermediate mesh sections.

To remove the rotational motion and also ensure complete mixing across the column, the liquid mixture is pumped between two inner annular baffles with several layers of wire mesh between them at the discharge of the impeller blades. This mesh has a dual purpose. It not only removes the rotational motion imparted by the impeller, but also serves to break up the liquid droplets, thereby increasing mass transfer between the phases. The amount of this mesh must be adequate to effect the baffling and will be greater when the liquid phases require high agitator speeds for complete mixing.

This design (Fig. 3) reduces the amount of mesh required between mixing stages to a minimum, and for liquid systems with low interfacial tension, it is economical up to 1 m in diameter.

2.3. Second Scheibel Extractor Without Mesh Packing

Performance data on the baffled mixing-stage design [8] indicated that the elimination of the mesh from the calming zone did not always result in a significant reduction in stage efficiency. Furthermore, when baffled mixing stages were installed in the void calming sections, the total number of theoretical stages in a given height was actually increased on systems of high interfacial tension that are difficult to mix and require high agitator speeds.

When the intermediate mesh sections are eliminated, the mesh thickness around the discharge of the impeller will be more critical since this becomes the only means for removing the rotational velocity component imparted by the agitator. In a few cases where the feed streams carried suspended solids, this mesh was omitted to preclude the possibility of plugging. The stage efficiency under these conditions was reduced to half that normally obtained on the same solvent system, presumably as a result of poorer droplet dispersion in the impeller discharge.

Extraction columns with contiguous mixing stages are the most economical for large-diameter columns. The internals are assembled externally on support rods as shown in Fig. 3 and then inserted into the column shelll. Cartridges that are too large to be handled by this technique have been assembled internally, but the cost of providing the access way in the column and positioning the agitators after field erection is greater than that for the external cartridge assembly.

2.4. Third Scheibel Extractor

In the baffled mixing stage shown in Fig. 3 the agitator diameter must be larger than the center inlet of the adjacent annular baffles. This is essential for elimination of bypass of the pumped fluid mixture at low agitator speeds and to ensure a positive radial pressure drop between the two inner baffles to pump the liquid mixture through the mesh layers. It is impossible to remove the agitator shaft after assembly of the cartridge, and the agitators must be positioned accurately on the shaft to provide suitable clearances with the baffles after the shaft is installed in the column. This adjustment is most convenient when cartridges are assembled externally. Internals that are rigidly installed in the column cannot be readily inspected and adjusted.

These problems are solved with the latest design [10], which is shown in Fig. 4. The agitator consists of a closed impeller with essentially the same circumferential area at all radii. It generates a vena contracta at a point beyond its outer diameter and directs the liquid flow between the inner baffles as shown. The agitator has a smaller diameter than the opening in the inner baffle, and the shaft can be removed for inspection and maintenance as required. When column diameters exceed 1.5 m, this removal provides an access way through the column for inspection, cleaning, and repair, if necessary. Columns of smaller diameter have been installed to provide the convenience of inspecting and servicing the moving parts. Cleaning of the internals can be effected through the center opening, but for more extensive repairs, the internal baffles must be removed to reach the problem area. With small diameters, there may be very little advantage over the removal of a complete cartridge for external cleaning and repair.

This third design is preferable for columns of such large diameter that a removable cartridge assembly is impractical.

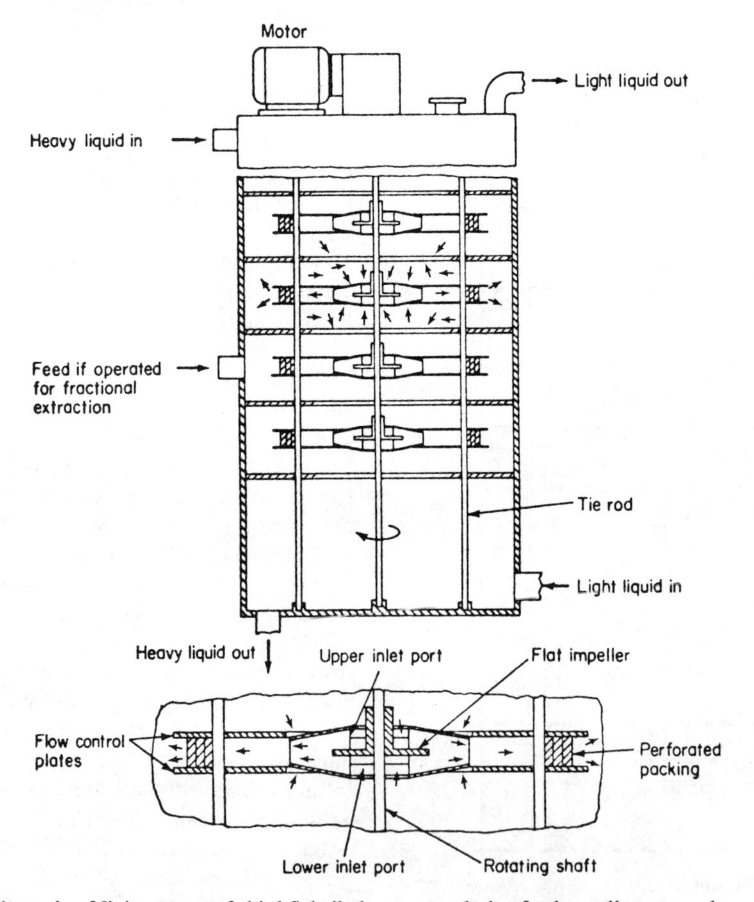

Figure 4. Mixing stages of third Scheibel extractor design for large-diameter columns.

3. PERFORMANCE DATA AND SCALE-UP DESIGN

The operation of all extraction equipment is a compromise between capacity and efficiency. The concept of a theoretical stage is based on mixing the countercurrent phases to achieve equilibrium and then separating them completely. The first step requires a large contacting surface and/or a long contacting time. The smaller the droplet size of the dispersed phase, the greater the rate of mass transfer betweent he phases, but according to the Stokes law relationships, smaller droplets will move more slowly countercurrent to the continuous phase, thus reducing throughput and increasing contact time.

The use of external power for mixing can provide the full range of capability up to stable emulsion formation. Unlike mixer–settlers, where the residence times of the individual phases in the mixing zones can be regulated by the hydrostatic head in these zones, the residence time of the dispersed phase in an agitated column is a function of the power input. High power inputs increase the approach to equilibrium between the phases by increasing (1) the interfacial area for mass transfer and (2) the residence time of the droplets in the mixing and settling zones. This latter effect, however, reduces the capacity of the column, and excessive agitator speeds at any given throughput will result in a "flooding" condition where a second liquid interface appears in the column, either at the end of the column opposite to that at which it is being controlled, or more immediately through a phase inversion in one of the stages that will ultimately result in a second interface at the opposite end of the column.

It is obvious that systems having high interfacial tension will require more power to extend their surfaces by decreasing the droplet size. However, the theoretical power of surface expansion is only a minor fraction of the total power consumed in the mixing zones of an agitated column. It is not sufficient to simply generate such surface. The surface must be uniformly distributed over the cross section of the column if localized effects are to be avoided and equilibrium is to be approached with the total counterflowing continuous phase. Consequently, the kinetic energy imparted to the liquid phases by the agitator is more important than the extension of the surface, particularly in systems where there is a large difference in the volumetric flow rates of the two liquids.

The following sections present general correlations for the capacity and efficiency of the different types of Scheibel column, and since the performance of agitated mixing stages is dependent on surface phenomena, it is extremely sensitive to the presence of foreign materials, both soluble and insoluble. Unexpected problems encountered in full-scale commercial operation can usually be traced to some impurity not present in the original laboratory and pilot plant streams. Simulated tests may not provide meaningful data, and preliminary studies should be conducted on liquid solutions having the closest possible resemblance to the commercial streams. Variable-speed drives are always installed on production columns, so the performance can be optimized under actual conditions. Pilot plant studies should ensure that the final design will be capable of operation in the optimum range of conditions. It is also essential that these studies establish which phase should be dispersed in the column, because it is not always possible to provide a sufficient range of operating speeds in large columns to permit operation with either phase dispersed.

3.1. Column Throughput

The capacity of an agitated extraction column is very sensitive to the interfacial properties. The presence of surface-active agents (surfactants) or insoluble material that collects at the interface will tend to stabilize emulsions and require a longer time for phase separation. Flooding velocities are functions of the power input such that the extractors will have a high efficiency and low throughput at high agitator speeds and a low efficiency and high throughput at low agitator speeds.

For column diameters in excess of 0.5 m, throughputs in the range of 20–40 $m^3\ hr^{-1}\ m^{-2}$ of column cross section will normally result in maximum stage efficiencies. Studies on simulated solvent systems will invariably indicate throughputs higher than those obtainable in practical systems, and particular attention must be paid to the solids that collect at the interface in the pilot plant studies. The solids may be stray foreign matter or may precipitate from solution when a solubilizing component is extracted from the feed. They will normally collect at the interface in one of the two phases. If they favor the light phase and the liquid interface is controlled at the top of the column, they will be carried away in the light phase and probably not present a problem. If they collect below the interface, they will be trapped and eventually tend to form a stable emulsion in the column and require a reduction in either agitator speed, throughput, or both. When the loss in the stage efficiency becomes intolerable, the column must be shut down and cleaned. Consequently, the best solution to the problem of solids collecting below the interface is to control the interface level at the bottom of the column so that they can be readily eliminated in the heavy solvent phase.

Optimum column performance is always obtained by dispersing the phase flowing at the smaller rate, but it may be more practical to sacrifice some capacity and efficiency to operate at constant conditions than to periodically vary the agitator speed between cleanings. Clues to abnormal behavior will frequently be overlooked in preliminary laboratory studies but should be readily apparent in prolonged pilot plant work. Pilot plant operations of more than 6 months are not uncommon for assurance that the full-scale plant can be operated trouble-free for at least this length of time.

Since all scale-ups retain the geometry to some degree, residence times in the mixing zones increase. This makes it possible to achieve the same approach to equilibirum, that is, stage efficiency, at lower power inputs in larger-diameter columns. Conversely, a very fine dispersion is required to achieve a high stage efficiency in stages with diameters and heights of only a few centimeters. Optimum conditions in the smallest column diameters of 2.5 cm for the first Scheibel design and 7.5 cm for the second type with baffled mixing stages are obtained at less than one-third of the superficial flow rates obtained in commercial columns. Capacities of the smallest models of these two designs are normally about 3 and 30 liters/h of total liquids, respectively.

3.2. Extraction Efficiency

The design and operation of agitated extraction columns require knowledge of the conditions for maximum stage efficiency at a reasonable throughput so that equivalent conditions can be related in different column diameters. The basic parameters for mixing-stage efficiency are the same in all column designs. For a given stage efficiency, the power input per volume of solvents flowing through the mixing stage should be proportional to ratio of the volume of the continuous to the dispersed phase [8].

Dispersion of 10% as much solvent uniformly in a continuous phase requires 10 times as much power as required to disperse equal volumes to reach the same approach to equilibrium in a given stage. The additional power is utilized in producing a finer uniform dispersion that then requires a larger settling volume to disengage the phases. This settling problem is handled differently in the various Scheibel column designs.

3.2.1. First Scheibel Extractor

The height of the wire-mesh packing is the key factor in the stage efficiency of this column. Solvent systems with low interfacial tension and low viscosity such as the transfer of acetic acid between water and methyl isobutyl ketone provide appreciable mass transfer in a spray column. Others with high interfacial tension involving hydrocarbons and water provide very little mass transfer in a spray column after the liquid droplets have formed.

If the power input is increased and sufficient packing is utilized to prevent overlapping of the flow patterns in the mixing stages and also provide sufficient volume for complete phase separation between the stages, the "easy extraction system" with low interfacial tension will exhibit considerably more than one theoretical stage for each combination of a mixing and calming zone. The excess efficiency derives from the transfer during countercurrent flow in the calming zone. In the "difficult extraction systems" with high interfacial tension, very little mass transfer occurs in the calming zone, and these systems approach closer to one theoretical stage as the packing height increases [7, 12]. Figure 5 shows the variation of stage efficiency with packing height in a 2.5-cm-diameter extraction column, extracting acetic acid from water with methyl isobutylketone, which is a typical system that has low interfacial tension. Also, shown in this Fig. 5 are the data for the dehydration of ethyl alcohol with the use of both methyl *n*-amyl ketone–glycol and xylene–glycol solvent systems. These are typical high-interfacial-tension systems with further disadvantages due to the high viscosity of the solvent since all data for Fig. 5 were obtained at ambient temperatures.

The packing height required for maximum number of theoretical stages in a given column height occurs at a stage efficiency below the maximum, and this optimum stage efficiency depends on the solvent system as well as the diameter of the column.

Karr and Scheibel [14] studied mass transfer in the mixing zone of a 30-cm column by intro-

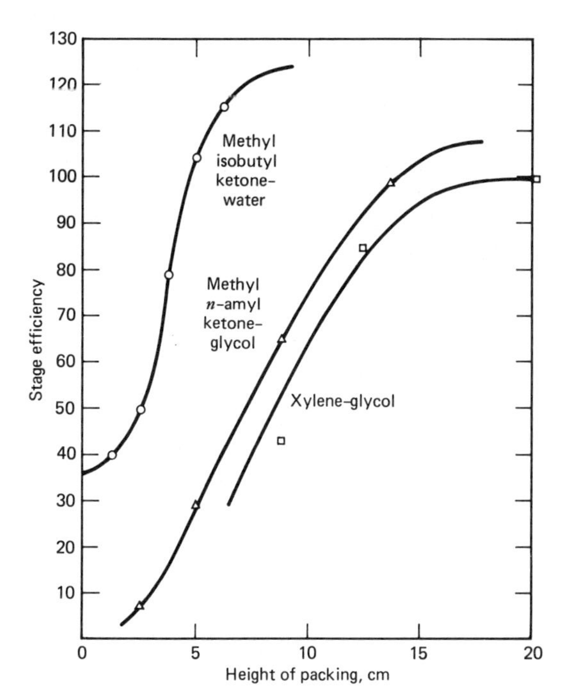

Figure 5. Variation of stage efficiency with packing height in 2.5-cm column for different solvent systems.

ducing the light phase through 48 tubes evenly distributed over the cross-sectional area at the top level of the lower mesh packing and the heavy phase through the same tube arrangement at the bottom of the upper packing. The same solvent systems were used as in previous studies [13] on a three-stage column. Theoretically, this device should establish the transfer due to the mixing zone and make it possible to evaluate the mass transfer in the packing by difference. However, agitator speeds were below those of the previous study, and direct comparison was unreliable.

The discrepancy was attributed to the fact that the solvent introduction system was more efficient than that which occurs during the normal coalescence and phase separation in the packing, and this initial dispersion made it easier to circulate the mixture through the agitator. It was found that the large number of small liquid streams entering the mixing stage provided one-fourth to one-third theoretical stage by acting as a void column at zero agitator speed. The large amount of baffling provided by the tubes in the packing also confined the flow pattern to the mixing stage, thereby reducing the amount of power that could be put into the mixing zone. There could also have been considerable mass transfer in the small amount of liquid mixture that did flow into the densely packed section containing all the tubes so that direct comparison with the data for the normal column design was not valid. The mixing-stage study demonstrated that it was possible to achieve essential equilibrium between the phases in the mixing zones, even though it occurred at a much lower agitator speed than when the multistage column was operated countercurrently. The single-stage data obtained in this study is probably more representative of mixer-settler performance than of countercurrent flow in a multistage column.

However, the data at zero agitator speed demonstrated that the excess efficiency above 100% could be readily attributed to the countercurrent flow through a void column. Earlier observations on the use of wire mesh as an extraction column packing had indicated that it was no more effective than a void column.

Data on a 7.6-cm-diameter mixing stage was also obtained in this study, and a general correlation was developed for the mass transfer in different solvent systems. Mass-transfer coefficients are normally based on concentration difference as the driving force, and the values may differ significantly according to the choice of the solvent phase. When the extraction factor is unity, the number of transfer units will be the same in both phases and also the same as the number of theoretical stages in the observed separation. Under all other operating conditions, the number of transfer units differ in the two phases and the number of theoretical stages is intermediate between the two values. Muphree stage efficiencies also differ when calculated based on concentrations in the different phases.

Thermodynamic definition of activity makes it identical in equilibrium phases. Consequently, stage efficiencies based on activity differences will be the same when evaluated in the different phases, and transfer units defined in terms of activity driving force rather than concentration will be independent of the phase considered. To develop a fundamental correlation of stage efficiencies that could be applied to different solvent systems, Karr and Scheibel calculated the efficiencies of mixing stages based on activities. They found that the mass-transfer correlations depended on the direction of mass transfer when the light phase was dispersed. When the heavy phase was dispersed, it was possible to develop a single correlation of the data for both directions of diffusion presumably because under these conditions the amount of power input required to obtain a reasonable dispersion is so much greater that it becomes the dominant factor.

For solute transfer into the droplets

$$\frac{E_{Md}}{1 - E_{Md}} = 0.0091 \frac{H_c}{d_i} \frac{\delta(\gamma x)}{\delta c_d} \left(\frac{\Delta \rho}{\sigma}\right)^{1.5} (Nd_i)^4$$

where all quantities are in metric units, N is in revolutions per hour, and γx represents the thermodynamic activity, that is, activity coefficient times mole fraction, in the dispersed phase. For the reverse direction of diffusion, the correlation is

$$\frac{E_{Md}}{1 - E_{Md}} = 0.092 \frac{H_c}{d_i V_d} \frac{\delta(\gamma x)}{\delta c_d} \left(\frac{\Delta \rho}{\sigma}\right)^{1.5} (Nd_i)^4$$

For the heavy phase dispersed, the mass-transfer function for both directions of diffusion was correlated as

$$\frac{E_{Md}}{1 - E_{Md}} = 0.0028 \frac{H_c}{d_i^{0.3}} \frac{\delta(\gamma x)}{\delta c_d} \left(\frac{\Delta \rho}{\sigma}\right)^{1.5} (Nd_i)^3$$

As previously noted, these equations had been developed for the efficiency of the mixing stage when there is very little radial flow through the packing. This is an idealized concept of the first Scheibel extractor, in which the solute concentration is constant over the mixing zone, that is, complete mixing, and linear over the packing height. Honekamp and Burkhart [15] studied mass transfer in the packing by measuring the concentration profile over the height of the column and observed that there was no sharp distinction at the ends of the wire-mesh packing. They calculated the mass transfer in the mixing section from the previous correlations and found that the mass transfer in the packing accounted for more than half of the stage efficiency at the higher throughputs and was negligible at very low throughput. The packing contribution to the mass transfer decreased with decreasing agitator speed and throughput, but the effect of the packing appeared to increase at the highest agitator speeds. This latter effect is probably not real and appears because the preceding correlations do not include the mass transfer that takes place as the liquid mixture flows radially through the ends of the packing adjacent to the mixers.

Honekamp and Burkhart noted that when the agitator was operated at very low speed, the column was not as effective as a packed column, thereby confirming earlier studies were the wire-mesh packing was found less effective than other column packings and actually no better than a void column. The prime purpose of the mesh is to eliminate the rotational flow and isolate the indivudal flow patterns in the mixing stages. The excess mass transfer occurs primarily as a result of the coalescence of the droplets in the packing. This effect is most noticeable at the higher agitator speeds, where there is more circulation through the ends of the packing.

Honekamp and Burkhart also found that the dispersed phase holdup in extraction increased with increasing agitator speed and was very nearly proportional to the dispersed phase flow rate at constant continuous phase flow rate.

Jeffreys et al. [16] extended the previous investigation of characteristic droplet size in the wire-mesh packing by studying the droplet break-up and coalescing properties of different wire mesh in the calming section of 3-in Scheibel extractor. They concluded that the droplet size distribution in the packing was more complex than reported by the previous investigators and consisted of three groups: (1) droplets so small that they passed unaffected through the mesh from one mixing stage to the next; (2) droplets that are slightly larger, but smaller than the interstices of the packing, and are only slightly altered in passing between stages; and (3) droplets larger than the mesh size, which collected at the surface of the packing until they coalesce to sufficient size to break away, leaving sufficient liquid on the mesh to resume the coalescence.

Since the first group would eventually coalesce to the larger sizes if it were allowed to pass through sufficient mesh and the third group would under the same conditions eventually break up to smaller sizes, the "characteristic droplet size" occurs in the second group, and for the standard wire mesh design contining 97.5% void volume, the characteristic droplet diameter was close to 2 mm for kerosene in water.

Jeffreys et al. [16] also observed a dramatic effect of void space on the flooding velocity of the column with a kerosene–water system. They did not measure stage efficiencies for transfer between the two liquids, but the interfacial tension classifies this as a "difficult extraction system." When power input to the agitator is increased to provide higher mixing-stage efficiency, the height of the low-density packing required to completely isolate the mixing-stage flow patterns will reduce its apparent advantage, and the most practical packing properties will be intermediate between those studied. Their data clearly illustrate how critical the selection of the mesh packing can be for handling the more difficult extraction systems.

3.2.2. Second Scheibel Extractor

In addition to economic considerations, the baffled mixing zone was developed to overcome some of the problems that had been encountered with the original design. In several applications where the dispersion of a viscous heavy phase was desired, the column could not be used. At low agitator speeds the heavy liquid collected at the bottom of the mesh and large droplets would periodically fall through the mixing stage without being deflected into the liquid flow pattern. When the agitator speed was increased to catch these droplets and carry them into the agitator, the speed was so great that the droplets were immediately broken up into an emulsion that would not separate. It was, therefore, impossible to operate the column under any suitable mixing conditions. The problem was not as noticeable with light-phase dispersions because the normal effect of centrifugal force draws the light phase to the center of the column and into the agitator.

Bypassing of the dispersed heavy phase down the wall of the column was also a problem that was noticeable when the packing did not fit tightly against the wall. Columns operating under these conditions showed abnormally large throughputs and very low efficiencies. A new mixing-stage design was needed that would prevent the dispersed phase from passing through without being caught in the flow pattern, no matter how slow the agitator speed, and also to prevent the heavy phase from bypassing several mixing stages by flowing down the walls after coalescence. The baffled extraction column designs described in Sections 2.2 and 2.3 meet these requirements, and since the annular baffles on the column wall alter the vertical component of the agitator flow pattern, a reduction of packing height and cost results.

Efficiency data were obtained on baffled mixing stages with intermediate mesh sections, and the mesh was removed to observe the effectiveness of the baffles and to obtain data on the mass transfer taking place in the mesh. Conclusions were similar to those deduced from the data on the first design. The packing did not provide much more transfer than a void column. On an easy extraction system, it ensured complete isolation of the mixing flow patterns and provided additional mass transfer. On a difficult extraction system it did not provide appreciable mass transfer, and the most effective use of the column height could be achieved by providing adjacent mixing stages even though the individual stage efficiency was less than obtained with intermediate calming sections.

Figure 6 shows a correlation of stage efficiency of the mixing zone as a function of power input and solvent ratio. The original correlation [8] has been extended to lower power correlation numbers by subsequent studies. Figure 6 shows that the use of excessive power in the mixing stage reduces the efficiency and this is due to the increase in backmixing between stages that results from both the overlapping flow patterns and the formation of small droplets that do not coalesce and are carried into the next stage with the continuous phase. The first effect causes backmixing of the continuous phase, and the second produces backmixing of the dispersed phase.

According to Fig. 6, the optimum power correlation numbers occurs in the range 100–200, and theoretically all extraction columns should be designed for these conditions. The correlation was developed from data on pure solvent systems. In practice, this optimum power input will frequently emulsify the liquid mixture, particularly at low solvent ratios, and the column must be operated at a lower power input and corresponding lower stage efficiency. Performance data have been found to follow this curve, and inability to reach the maximum efficiency can generally be traced to the presence of emulsifying agents in the solutions.

The power input can be calculated from the correlation due to Mack [17], which had been originally checked by power measurements on a 30-cm-diameter column [8] and subsequently verified on larger columns. At normal operating conditions, the agitator speed is in the turbulent region. In the laminar flow region, the agitation is not adequate to break up the droplets sufficiently for effective mass transfer.

4. SCALE-UP AND DESIGN OF COMMERCIAL COLUMNS

Performance data of throughput and efficiency are required on small-scale columns for confident scale-up to large commercial columns. The geometry of the stage designs permit much greater throughputs per unit cross section in the larger diameters. Superficial velocities in 2.5-cm columns of the first design will be two to five

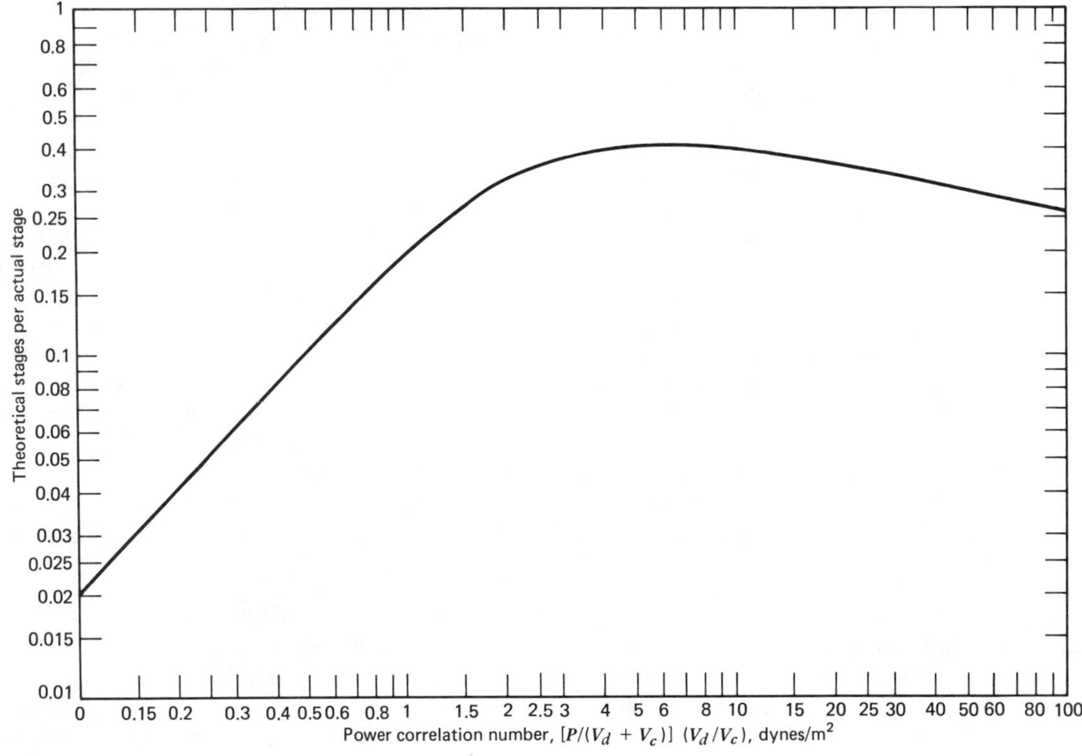

Figure 6. Correlation of mixing-stage efficiency with power input and liquid flow rates.

times greater in 0.5-m or larger columns; and scale-up is empirically based on prior experience. In general, the smaller the capacity of the laboratory column, the greater will be the scale-up factor. The larger laboratory columns will permit smaller scale-up factors; therefore, it is desirable to base the final design on the largest possible test column. When data are obtained on pilot plant columns of 0.5 m diameter, the full-scale unit can be designed to provide essentially the same superficial linear velocities.

The smallest diameter available in the baffled mixing-stage design is 7.5 cm, and the scale-up factors for capacity are essentially the same as for the 2.5-cm columns of the first design. The minimum size is limited by the complexity of the mixing-stage design. A column containing baffled mixing stages 5 cm in diameter with 2-cm stage height was rejected because it provided throughputs and efficiencies only about 10% of the large units, and its performance data cannot be scaled up with suitable confidence.

The normal scale-up procedure is based on providing the same stage efficiency in the larger column as obtained in the test column and adjusting the power input accordingly. If the test column contains sufficient stages, it thus can also be used to confirm the process design as well as provide the necessary scale-up data. This is particularly important in the fractionation of multicomponent mixtures.

The diameter:height ratios in the mixing stage will be less than 3 in the small-diameter columns, and this can be increased to 8 in columns of 3-m diameter. If the same power input per volume of fluids is maintained in the larger-diameter columns, the radial velocities in the mixing stages will be greater in the larger columns. This factor limits the diameter:height ratio in large columns. The longer residence time in the mixing stages permits operation at lower power inputs, and this is the prime factor in providing greater throughput capacity.

Multiple agitators have not been used in the Scheibel extractors because of anticipated problems in obtaining uniform mixing in this type of design and a resultant decrease in the stage efficiency. Economic studies also indicate that when columns larger than 4 m in diameter are required, it is preferable to install multiple units. In exceptionally large installations handling more than 400 m^3/hr of solvents where four or more individual multistage columns may be required, simple mixer-settler arrangements can be more economic when investment costs are compared.

The advantages of multistage extractors lie in their ability to provide a large number of theoretical stages in a single unit. This is particularly desirable for operation under pressure. The break-even point between mixer-settler combinations and multistage columns occurs at smaller throughputs for atmospheric pressure operation and small numbers of theoretical stages. At throughputs of less than 200 m^3/hr (50,000 bbl/day), multistage extractors are indicated. Present commercial applications of liquid extraction processes do not exceed these limits.

NOTATION

- c_d solute concentration in dispersed phase, kgmol/m3
- d_i impeller diameter, m
- E_{Md} Murphree dispersed-phase stage efficiency, fractional
- H_c height of mixing stage, m
- N rotational speed, rph
- V_c superficial velocity of continuous phase
- V_d superficial velocity of dispersed phase
- X solute concentration in dispersed phase, mole fraction
- γ activity coefficient
- $\Delta\rho$ positive difference in density, g/cm^3
- σ interfacial tension, dynes/cm

REFERENCES

1. E. Jantzen, *Dechema Monograph* **5**(48), 81 (1932).
2. H. Schöneborn, U.S. Patent 1,949,496 (March 6, 1934).
3. D. F. Othmer, U.S. Patent 2,000,606 (May 7, 1935).
4. W. J. D. Van Dijck, U.S. Patent 2,011,186 (August 3, 1935).
5. A. J. Frey and E. G. Scheibel, *Jubilee Volume, Emil Barrell, Hoffmann-LaRoche,* Basel, Switzerland, 1946, p. 446.
6. M. R. Fenske, C. S. Carlson, and D. Quiggle, *Ind. Eng. Chem.* **39**, 1932 (1947).
7. E. G. Scheibel, *Chem. Eng. Progr.* **44**, 681 (1948).
8. E. G. Scheibel, *AIChE J.* **2**, 74 (1956).
9. E. G. Scheibel, U.S. Patent 2,850,362 (September 2, 1958).
10. E. G. Scheibel, U.S. Patent 3,389,970 (June 25, 1968).
11. L. C. Craig, *J. Biol. Chem.* **155**, 519 (1944).
12. E. G. Scheibel, *Ind. Eng. Chem.* **42**, 1497 (1950).
13. E. G. Scheibel and A. E. Karr, *Ind. Eng. Chem.* **42**, 1048 (1950).
14. A. E. Karr and E. G. Scheibel, CEP Symposium Series No. 10, 1954, p. 73.
15. J. R. Honekamp and L. E. Burkhart, *Ind Eng. Chem. Process Des. Devel.* **1**, 176 (1962).
16. G. V. Jeffreys, G. A. Davies, and H. B. Piper, *Proceedings of International Solvent Extraction Conference* (The Hague), 1971, p. 680.
17. D. E. Mack, *Chem. Eng.* **58**, No. 3, 109 (March, 1951).

13.4

OLDSHUE-RUSHTON COLUMN

James Y. Oldshue
Mixing Equipment Company, Inc.
United States

1. DESCRIPTION OF CONTACTORS

Agitated liquid–liquid extraction columns have been in use since the early 1950s [5], when three types of column were introduced: the rotating-disk column (RDC); the York–Scheibel column; and the Oldshue-Rushton column. They contrast to the previously available columns that obtained mass transfer by flowing streams across a packing, through perforated plates, through bubble caps, or through spray nozzles.

Three mixer column characteristics are important:

1. Volumetric mass-transfer rates are improved by the use of a controlled mixing regime.
2. The ability for accurate scale-up is possible since the dispersion is produced by rotating impellers with known characteristics. Use of the flowing energy of liquid streams for dispersion can cause problems in liquid distribution, changing wall effects on scale up, and channeling.
3. One or both phases may contain solids. Agitated columns are essentially self-cleaning and can be used on a wide variety of process streams.

The two open-type columns, the RDC and the Oldshue–Rushton column, differ from the original York–Scheibel column. The latter had individual mixing and packed settling stages. This meant that higher mixer power levels could be used and higher mass-transfer rates could be obtained in the mixing zone; however, it was necessary to provide a lower power level and lower mass-transfer rate in the settling zone. The net result was about equal performance in terms of total stage height from either type. The open-type column offers construction simplicity and only one interface to consider. A frequent requirement for alloy construction in large columns further constrains the type of internal complexity that is economically justified.

Another type of system is the mixer–settler. A high mass transfer rate can be obtained in the mixer since a settling tank is provided for separation of the two phases.

In general, comparisons of all types of extractor types must be made on cost estimates since rarely is there the same process operating simul-

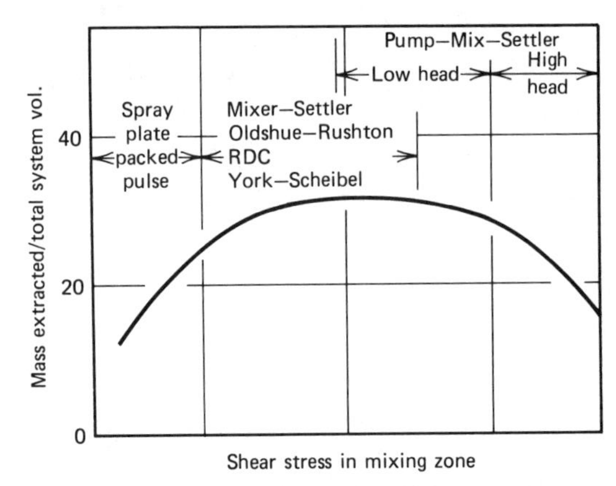

Figure 1. Effect of shear stress in mixing zone on overall extraction performance.

taneously on two different types of equipment. Cost estimates can vary considerably. Usually, however, the total volume in the different systems is quite similar, and the major cost difference revolves around the cost of land, the necessity of multiple column units because of the plant size involved, and whether the mixer–settlers must be enclosed in a structure.

When mixer–settlers become very large and interstage pumping requirements are added to mixing requirements, the increased fluid shear stresses associated with developing high lift heads may bring the mixer regime farther into an area of high mixer horsepower, producing a dispersion that cannot be adequately compensated for by additional settler volume. Figure 1 illustrates this situation.

2. INDUSTRIAL APPLICATION

The main industrial application area is for low and medium-viscosity fluids, generally ranging up to 500 cP and having density differences of at least 0.05. Solids may be suspended in one or both phases.

For countercurrent operations, the column is normally designed as shown schematically in Figs. 2–4. The columns arc normally made of metal, and any metal or alloy can be used that can be machined and welded. In addition, linings or coatings such as rubber, fiberglass, or other materials can be applied to prevent usual types of corrosion. Any reasonable temperature or pressure combination can be used, since the column can be equipped with mechanical seals or stuffing boxes, just as in the case of any fluid mixing application.

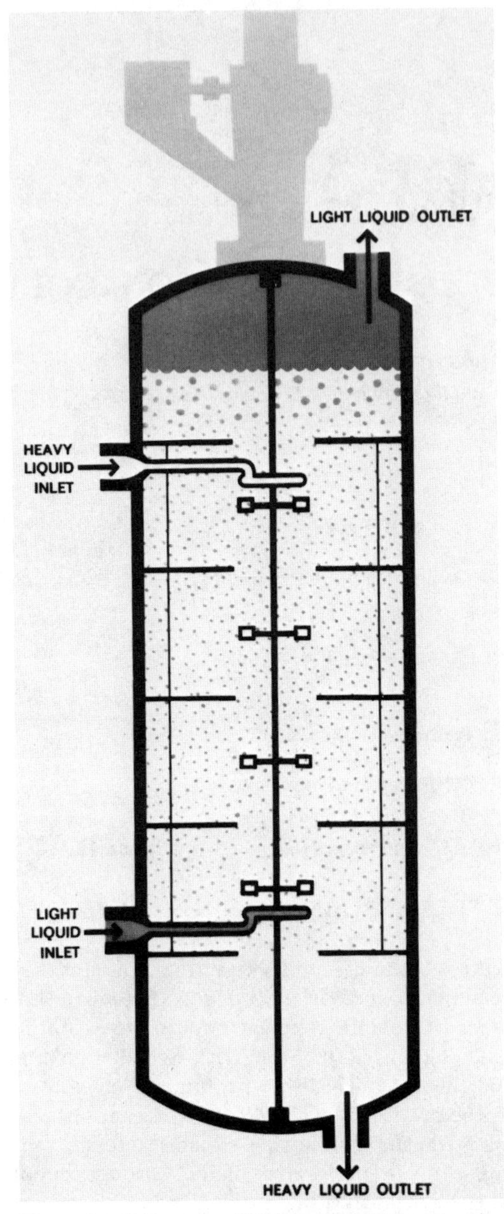

Figure 2. Schematic of Oldshue–Rushton extractor.

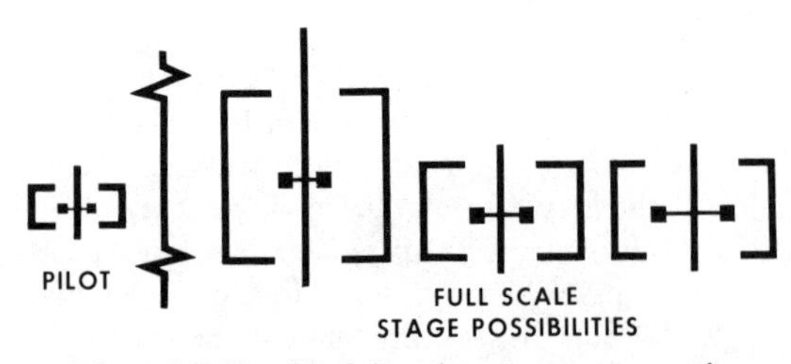

Figure 3. Possible full scale extractor geometries.

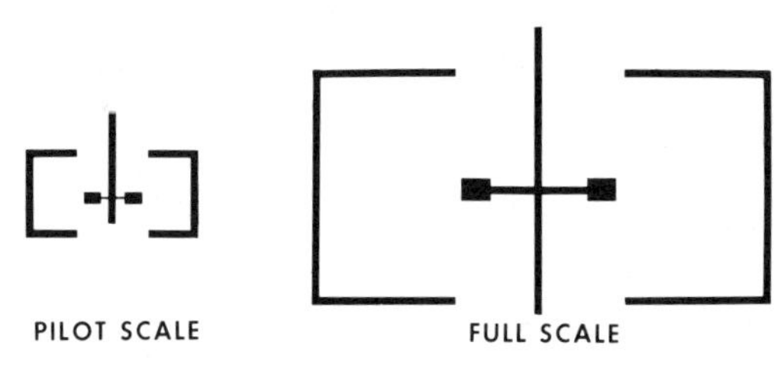

FULL SCALE COMPARED TO PILOT SCALE	
RESIDENCE TIME	HIGHER
BLEND TIME, UNDISPERSED	LONGER
INTERSTAGE MIXING, UNDISPERSED	DIFFERENT
INTERSTAGE MIXING, DISP.	DIFFERENT
CONCENTRATION GRADIENT, DISP.	HIGHER
MAX. IMPELLER ZONE SHEAR RATE	HIGHER
AVE. IMPELLER ZONE SHEAR RATE	LOWER
AVE. TANK ZONE SHEAR RATE	LOWER
TURBULENT SHEAR RATES	DIFFERENT

Figure 4. Mixing factors compared for pilot and full scale.

In addition, the Oldshue–Rushton column can be used in a cocurrent mode for any single- or multiple-phase continuous cocurrent contacting operation, including such diverse functions as leaching, reacting, bleaching, or in processing paperstock, and many other types of cocurrent extraction systems.

3. PERFORMANCE DATA

3.1. Column Throughput

Considerable data are given in the referenced articles [1–8]. In general, for a density difference of 0.2 and with low viscosity aqueous and organic materials with viscosities around 0.001 Pa · s total capacity is a nominal 70 ml/s in a 150-mm-diameter column. Flow rates of the continuous and dispersed phases can vary from 3 : 1 to 1 : 3 with either phase dispersed. When the ratio of one phase to the other is greater than 3 : 1, then the low-volume phase should normally be the dispersed phase.

It is normally necessary to run a 150-mm-diameter pilot scale column to obtain capacity data. If the pilot column is not sufficiently high to produce the desired total theoretical stages required, the performance in the middle and the rich end should be synthesized so that the effect of different viscosities and surface tensions on capacity can be observed.

Even though the impeller diameter and type in each of the stages in the column are usually the same, they can be varied to accommodate different power levels and tip speeds for different requirements at various locations in the column. The column throughput should be measured with actual rates of extraction taking place since the direction and quantity of mass-transfer does effect the capacity of the column. Columns smaller than 150 mm, with impellers smaller than 80 mm in diameter, do not give column throughput relationships that are similar and scalable to full-scale design. These smaller columns can be used for obtaining chemical extraction data and investigating phase ratios and other variables in a relative fashion but are not particularly useful in direct scale-up calculations.

3.2. Stage Efficiency

Stage efficiency on a 150-mm-diameter pilot column, with 75-mm height stages, varies from 1 to 90%. In scale-up, stage efficiencies tend to increase as a result of the increasing effect of residence time as the absolute stage height is increased in a bigger column. This, however, does not mean that the height equivalent to a theoretical stage (HETS) increases; the *HETS* actually decreases in scale-up, but the difference in HETS between systems that are easy and those that are difficult to extract in columns of at least 2-m diameter becomes quite negligible (Fig. 5).

3.3. Interstage Mixing Between Stages

Several investigators have studied the interstage mixing between stages. It can be analyzed as the

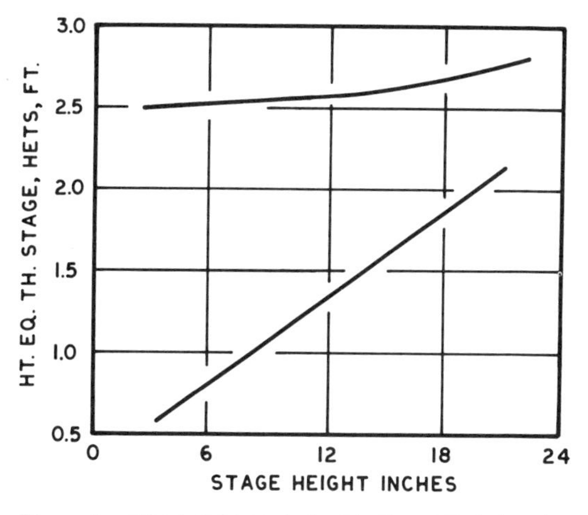

Figure 5. The height equivalent to theoretical stage increases even though there is a marked increase in stage efficiency on scale-up.

effect of the interstage mixing between compartments in terms of the interstage mixing value a compared to the liquid throughput L and its effect on stage efficiency.

Another correlation method assumes the axial diffusion model, which involves a Peclet number, which is made up of the throughput of the phase being considered, the height of the stage compartment, divided by the axial diffusion coefficient, and the fraction of the phase being considered X.

For a continuous phase flowing independently, X_c becomes 1, and it turns out that the same correlation is used for a countercurrent dispersed phase present in which X_c has the actual physical value. Figure 6 shows a typical plot of the reciprocal of the Peclet number versus a group relating the tip speed of the impeller, the superficial velocity of the continuous phase, the diameter of the opening between compartments, and the impeller size, and the fraction of the continuous phase in the column. This typically ranges between 80 and 95%, which can be used for estimation in the absence of actual data.

The axial interstage mixing in the dispersed phase is much more complex. Data are widely scattered, and actual experimental data must be obtained. Accurate numbers are to be obtained on a given actual system. However, Fig. 7 gives an approximate idea of the Peclet number for the disperse phase as a function of the power level in the stage and is useful for estimating various size columns.

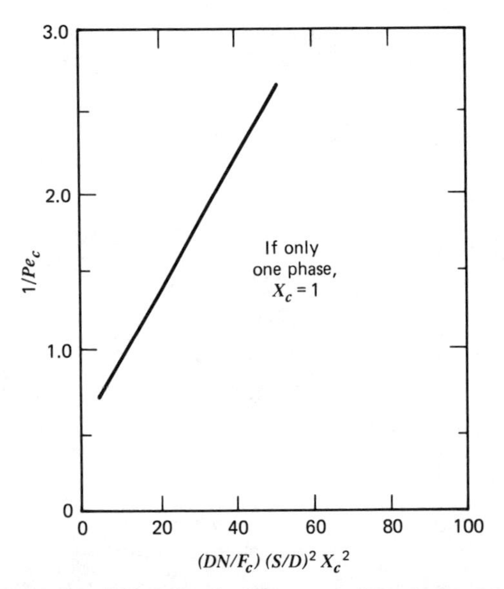

Figure 6. Correlation for interstage mixing in the continuous phase as a function of several mixing variables and stage geometry variables. (Adopted from Komasawa and Ingham [7].)

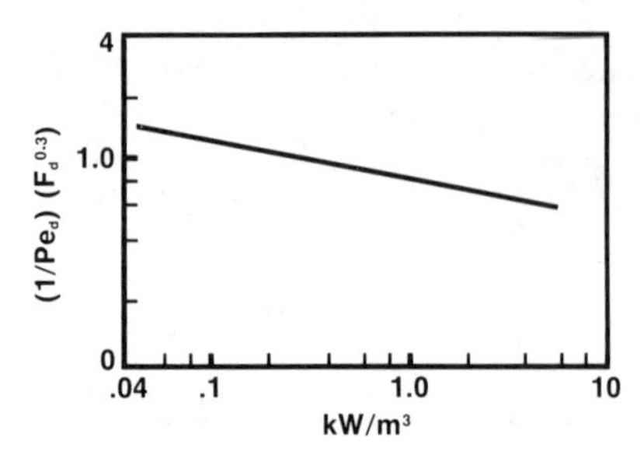

Figure 7. Axial mixing in the disperse phase as a function of power level. (Adopted from Komasawa and Ingham [7].)

4. SCALE-UP AND DESIGN

A large-scale extractor will usually be quite different in several respects from the small pilot plant unit used for study. A 150-mm-diameter column with a 75-mm stage height and a 60-mm-diameter stage opening similar to that shown in Fig. 2, and reported in Ref. 6 is often used to obtain pilot plant data. Ingham [3, 7] contains additional data on these columns.

By separating the several mixing factors involved, it is possible to scale-up the extraction results to obtain any desired geometry, which can then be used to calculate the cost, and then economic comparisons can be made to carry out the overall evaluation. Figures 3 and 4 show some of the properties that change on scale-up. In the pilot plant the usual measurement is the number of theoretical stages obtained in the equipment. It is then possible to calculate the stage efficiency.

The most reliable scale-up technique is to determine the mass-transfer coefficient, $K_L a$, which is normally based on the dispersed phase. In the pilot plant the effect of interstage mixing is evaluated, the effect of the ratio of the slope of operating line to equilibrium equipment line is evaluated, and the $K_L a$ is determined for the residence time in the small-scale equipment. Interstage mixing is a flow induced by the mixer that is in addition to the fluid throughput and tends to reduce concentration gradients. Scale-up can then be made to several different geometries on full scale. For any chosen geometry, the effect of residence time ratio, fluid shear rates, blend time, interstage mixing of the undispersed phase, degree of mixing of the dispersed phase, and power level is considered.

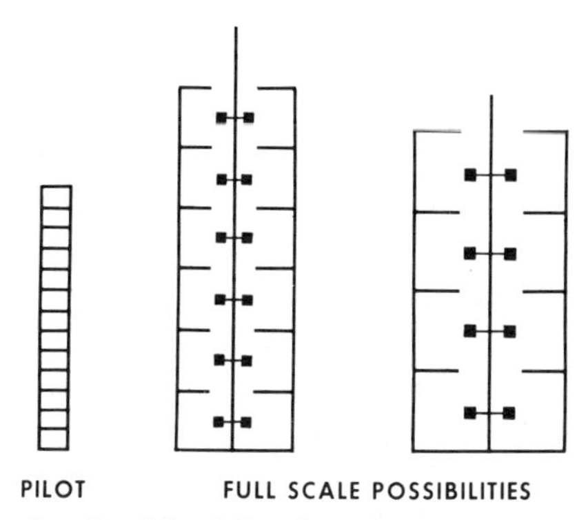

Figure 8. Possible full-scale column geometries for equivalent performance.

Actually, many different combinations of geometry are possible to accomplish the same degree of extraction. For example, a given capacity can be obtained with large openings in a small-diameter column or smaller openings in a larger-diameter column (Fig. 8). Also, different stage heights can be considered, which further vary the number of mechanical stages required. In general, the stage efficiency will be considerably higher on full scale than it is on small scale, as shown in Fig. 9.

4.1. Stage Efficiency

Overall extraction stage efficiency is determined by interstage mixing, [1-3] which is related to the interstage mixing stage efficiency, residence time, and mass-transfer rate. It is also a function of concentration driving force. Any mixing of the dispersed phase tending to reduce its concentration gradient will decrease column performance. It is normally assumed that the undispersed phase is completely mixed and has virtually no concentration gradient. This is less true as

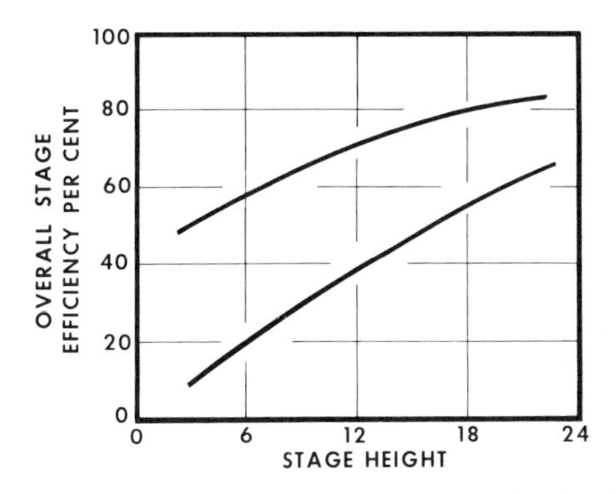

Figure 9. Effect of stage height on stage efficiency for Oldshue-Rushton column.

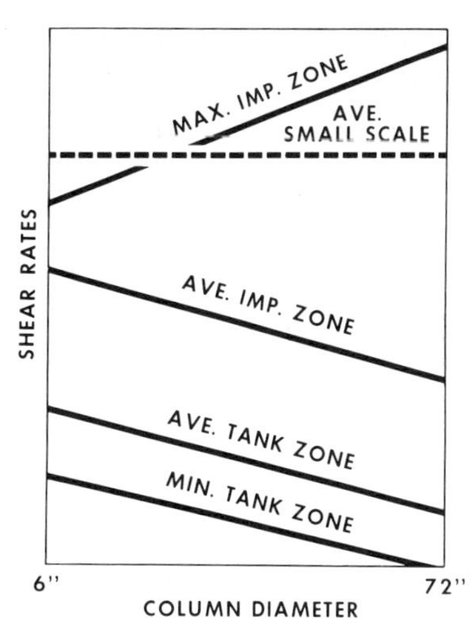

Figure 10. Effect of scale-up on selected shear rates.

column scale-up proceeds to increasingly large systems. The larger the opening between the stages, the lower the stage efficiency and the greater the throughput. There is also the function of the impeller size: tank size ratio and impeller size: opening size ratio. In addition, allowing the radial flow impeller to establish a dual flow pattern in each stage does establish some desirable concentration gradients in the system.

4.2. Fluid Shear Stresses

The role of shear stress is quite important and, among other things, determines what will occur to entrainment on full scale. Figure 10 shows that as we scale up, there is a tendency for the maximum shear rate in the impeller zone to increase while the average shear rate throughout the impeller zone tends to decrease [4]. If bubble size is proportional to shear stress, we will have a greater distribution of bubble sizes from the shear stresses shown in Fig. 10 than we would have in the small-scale unit. We may have the same average mass-transfer coefficient, but there will be a difference in droplet size in the dispersed phase.

4.3. Column Throughput

There is no way at the present time to calculate the throughput capacity of a full size liquid-liquid extraction column without making a small-scale test. The test purpose is to get the capacity through a standard set of geometry conditions so that extrapolation can be made to other geometry options. Table 1 indicates how capacity

TABLE 1 EFFECT OF SIZE OF OPENING BETWEEN COMPARTMENTS[a]

Compartment Opening, mm	Maximum Stage Efficiency	Minimum HETS, mm	Flow Rate, kg s^{-1} m^{-2}
Constant Flow Rate			
0	100	2560	0[b]
54	83	3098	2.9[b]
82	52	4953	2.9
152	38	6731	2.9
At Maximum Efficiency			
0	100	2560	0[b]
54	83	3098	2.9[b]
82	67	3860	5.4[b]
152	38	6731	6.0[b]

[a]Typical data for operation with methyl isobutyl ketone, water, acetic acid; four stages; 101.6-mm stage height, 152-mm-diameter column; extraction, water → ketone.
[b]Optimum flow rate.

changes with different stage openings in a small unit for one chemical system. Normally, the low-volume phase is dispersed and the higher viscosity phase is dispersed. These two requirements may not be possible in every case. One important criterion is to determine the performance under throughput turndown conditions. Figures 11 and 12 show that, depending on the relative mass-transfer characteristics in the system, turndown ratios can either help or hurt the overall column performance. It is essential that these turndown ratios be specified and a suitable test be made in the pilot plant to determine their effect on column performance. At a lower throughput the increased residence time may either be important or insignificant. The increased interstage mixing lowers performance. Normally, the performance of a high-K_La system is lowered with turndown, whereas the performance of a low-K_La system is enhanced.

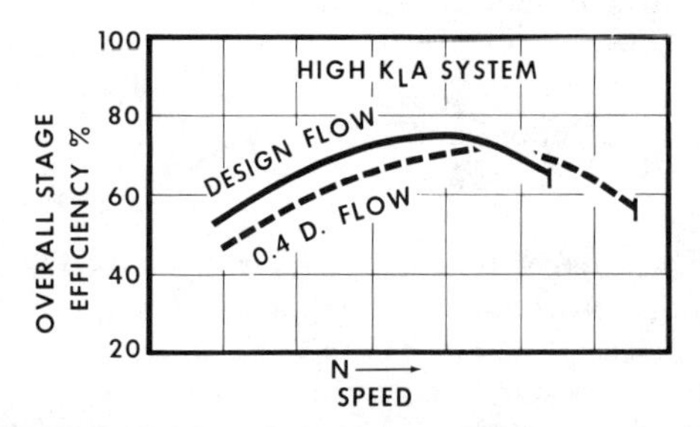

Figure 11. Effect of throughput turndown ratio and stage efficiency for high-K_La systems.

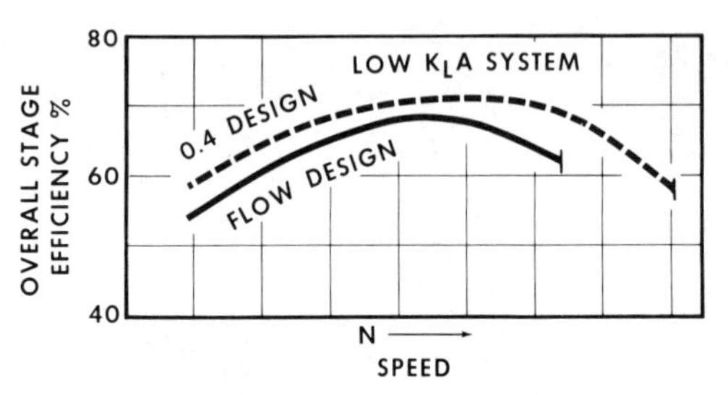

Figure 12. Effect of turndown ratio and stage efficiency for low-K_La systems.

In general, the dispersed phase has less entrainment than the continuous phase. Therefore, special separators must be placed in the continuous-phase withdrawal zone to prevent mixing energy "leaking out" of the last mixing stage.

Many studies have been conducted to attempt to compare break-time studies with capacity of liquid–liquid extraction columns. In a later section examples of some of these break-time studies are illustrated in relation to column capacity.

4.4. Pilot Plant

In the pilot column, normally a minimum of four flow rates are required. At a speed of about 300 rpm, flow rates are adjusted upward until the column just begins to flood. At this point the flow rate is reduced to 80% of this value and extraction tests are performed. Speed is then increased by approximately 20% and the flooding point determined. The flow rate is then reduced to 80% of that value and the extraction test is performed.

The flow is then cut in half, and at the same mixer speed the extraction performance is measured. In addition, the speed is raised to a flood point and is then backed off to about a 7% lower value to determine what effect increased power would have during the turndown condition.

4.5. Batch Studies

It is very helpful in analyzing column performance to have batch extraction tests as well as batch break-time tests. These batch studies should be done with impellers that simulate the shear rate that will be obtained in the full-scale unit. This requires the use of nongeometrically similar impellers as well as special runs designed to duplicate certain shear rate parameters in the system.

Break-time studies involve measurement of the time for coalescence to occur after certain rpm and residence time runs in the batch system

TABLE 2 BREAK-TIME DATA[a]

Revolutions/min (rpm)	Time of Mixing, s				
	5	10	15	30	60
	Break Time, s				
300	38	47	52	58	60
400	59	77	87	99	104
500	88	108	116	118	120
600	110	134	138	136	136
700	137	142	141	138	140
800	142	145	146	147	146
1000	141	124	123	117	126

[a]Specifications:

System	Methyl isobutyl ketone–water–acetic acid
Tank	152-mm diameter Four baffles 13 mm wide 152-mm liquid level
Impeller	51-mm diameter, mounted at interface, stainless steel, four flat blades
Break-point level	6 mm above interface
Interface	76 mm

(Table 2). For these tests, phase ratios are chosen as the feed ratio, even though the column holdup may not be that value.

It is important to make runs with each of the phases dispersed, even though it is customary in the continuous column to have the small-volume phase dispersed. As long as the small volume phase is not less than one-third of the large-volume phase flow rate, it is possible to have the lower flow rate phase be the undispersed phase. Figure 13 indicates methods of predicting which phase will be dispersed in the batch experiments.

Solute concentration can have a large effect on break time. Therefore, break-time studies should be run simulating the rich end of the column as well as conditions simulating the lean end of the column. The approximate volumetric phase ratio in the feed streams should be used

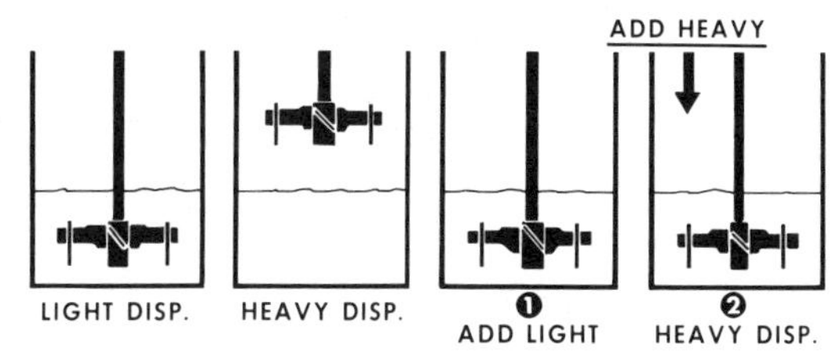

Figure 13. Method of controlling which phase is dispersed in batch extractions.

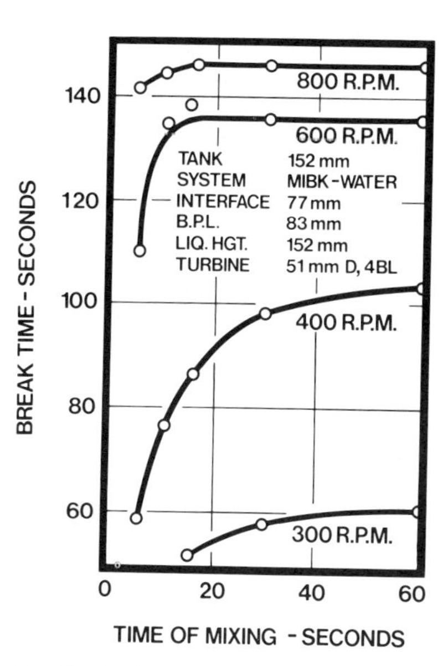

Figure 14. Break time as a function of time of mixing in speed.

but in no case should be greater than 2 : 1. A 150-mm-diameter jar with a 50-mm-diameter turbine and a 150-mm total liquid level is typical. The impeller should be placed so that each phase may be dispersed in turn in successive tests. Several speeds are run, and break time is measured. Observation of break time with various mixer and system variables can give a qualitative picture of the effect on column throughput.

Figure 14 shows the relationships between break time and mixer–impeller speed for the water, acetic acid, methyl isobutyl ketone (MIBK) system. One of the peculiarities found was that under certain conditions there was a decrease in break time with an increase in mixer speed. However, even though the break time was slower, there was a cloudiness in the dispersed phase and, therefore, considerably more entrainment in that phase.

Another observation that illustrates the importance of temperature in column performance is a curve of break-time versus temperature. Figure 15 shows this relationship and indicates that temperatures must be carefully specified for full-scale design and properly controlled during continuous flow performance experiments (see also Table 3).

4.6. Design Example

An example of a scale-up calculation is shown. In a 150-mm-diameter column, a 29% stage efficiency was obtained. This is now corrected for

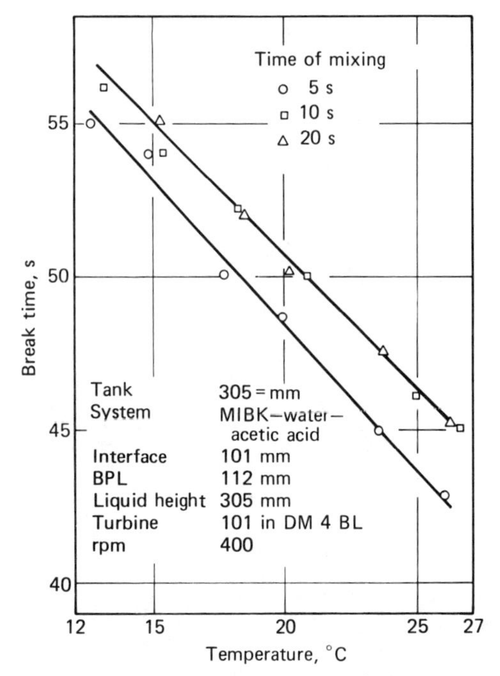

Figure 15. Break time as a function of temperature.

TABLE 3 BREAK-TIME TEMPERATURE DATA[a]

Time of Mixing, s	Break Time, s	Temperature, °C
5	55.5	12
10	58.1	12
20	57.1	12
30	56.5	12
40	55.2	13
5	53.6	14
10	55.6	14
20	54.8	14
30	54.5	14
40	54.0	15
5	50.6	18
10	52.7	18
20	52.8	19
30	52.1	19
40	52.3	19
20	49.9	20
30	49.9	20
40	49.5	21
5	46.9	21
10	49.4	21
5	45.0	23
10	47.4	23
20	47.2	23
30	46.6	24
40	46.5	24
5	42.8	26
10	45.4	26
20	45.4	27
30	45.2	27
40	45.2	27
5	38.9	30
10	42.7	30
20	42.8	30
30	42.5	30
40	42.8	31
5	38.9	32
10	41.5	32
20	41.3	32
30	41.5	32
40	41.5	32

[a]Specifications:

System	methyl isobutyl ketone–water–acetic acid
Tank	305 mm, four baffles, 25 mm wide 305-mm liquid level
Impeller	101-mm diameter, mounted at interface, stainless steel, four flat blades
Break-point level	13 mm above interface
Interface	101 mm
Turbine speed	400 rpm

interstage mixing as well as the ratio of operating line to equilibrium line. A K_La value of 16.2 was obtained (Table 4).

This K_La is now scaled up to a full-sized column 3050 mm in diameter with a 1800-mm-high stage. A 2.5-kW impeller is determined to be compatible with desired capacity relationships. The effect of interstage mixing in both continuous and discontinuous phases is determined, and an overall stage efficiency of 85% is calculated. A suitable design factor is then applied.

TABLE 4 SCALE UP EXAMPLE 1

Parameter	Pilot	Plant
T	150 mm	3050 mm
Z_S	75 mm	1800 mm
P		2.5 kW
K_La		16.2
Efficiency (no D.F.)[a]	29%	85%
Efficiency (D.F.)[a]		55%

[a]D.F. = design factor.

TABLE 5 SCALE UP EXAMPLE 2

Parameter	Pilot	Plant
T	150 mm	3060 mm
Z_S	75 mm	1800 mm
P		3 kW
$K_L a$		4.6
Efficiency (no D.F.)[a]	4%	65%
Efficiency (D.F.)[a]		35%

[a]D.F. = design factor.

Another example starts with 5% stage efficiency on the 150-mm-diameter column. This is now corrected for interstage mixing as well as the ratio of operating line to equilibrium line. A $K_L a$ value of 4.6 was obtained (Table 5).

This $K_L a$ is now scaled up to a full-sized column 3060 mm in diameter with the same 1800 mm-high stage. A 3-kW impeller is required. The overall stage efficiency is now 65%. Again, a suitable design factor can be applied.

NOTATION

D	impeller diameter
DF	design factor
F	superficial velocity of phase being considered
HETS	height equivalent to a theoretical stage
$K_L a$	mass-transfer coefficient
N	impeller speed
P	power
Pe	Peclet number
S	diameter of opening between compartments
X	fraction of phase being considered
Z_s	height of individual stage

Subscripts

c	continuous phase
d	dispersed phase

REFERENCES

1. R. Bibaud and R. Treybal, *AIChE J.* **12** (3), 472 (1966).
2. H. F. Haug, *AIChE J.* **17** (3), 585 (1971).
3. J. Ingham, *Transact. Inst. Chem. Eng.* **50**, 372 (1972).
4. J. Oldshue, *Biotech. Bioeng.*, 8 (1), 3 (1966).
5. J. Oldshue, in Y. Marcus, Ed., *Solvent Extraction Review*, Vol. 1, Marcel Dekker, New York, 1971, pp. 185–213.
6. J. Oldshue and J. Rushton, *Chem. Eng. Progr.* **48** (6), 297 (1952).
7. I. Komasawa and J. Ingham, *Chem. Eng. Sci.* **33**, 479–485 (1978).
8. J. Y. Oldshue, F. Hodgkinson, and J.-C. Pharamond, *Proceedings of the International Solvent Extraction Conference, 1974, Lyons*, Vol. 2, Society of Chemical Industry, London, 1974, p. 1651.

13.5

THE KÜHNI EXTRACTION COLUMN

A. Mögli and U. Bühlmann
Kühni Ltd.
Switzerland

1. INTRODUCTION

One of the most important criteria in the design of a liquid–liquid contactor is adaptability for varying conditions of operation and differing system parameters. During actual column operation it can often be observed that the droplet size and hence the holdup of dispersed phase vary considerably along the column length. The overall column conditions of throughput and mass-transfer efficiency thus represent average values, which may differ considerably from the optimum. To avoid such shortcomings, it is desirable to be able to modify the column design parameters from stage to stage. This is obtained in the Kühni column by the ability to vary the geometry of the compartments, the size of the mixing turbine, and the available free area of flow through the stator plates bounding each compartment. The Kühni contactor can thus be designed to obtain minimal axial mixing, constant holdup of dispersed phase, and maximum capacity for each column section. This design capability thus provides a large degree of flexibility and an overall high efficiency for the apparatus.

2. CONSTRUCTION

The major elements in the construction of the agitated column compartments of the Kühni contactor (Fig. 1) consist of:

1. An assembly of turbine mixers mounted on a central shaft. The turbine mixer is basically a double-entry radial flow impeller which generates the characteristic flow pattern of the Kühni contactor compartments (Fig. 2).
2. Stator plates that serve to separate each compartment from its neighbor in the vertical direction. These are provided with circular holes to allow axial flow of the two phases through the column. The number and diameter of the holes determine the available free area of flow for the stator plates.

For small- and medium-sized columns up to 1.5 m in diameter, the bearings, rotor assembly, and stator plates are assembled as compact units of up to 5 m in length and are then inserted into the column shell, where they are

Figure 1. Compartment of a Kühni contactor; turbine mixer and stator plates.

fixed between flanges (rack mounting). The internals of large columns, however, are mounted individually inside the column shell. For this purpose the turbine mixers and stator plates are inserted in sections through manholes located about every 6–10 m along the column.

Shaft speeds are relatively low, in the range 80–200 rpm for pilot columns and 5–20 rpm for large-production scale contactors. Thus the power consumption of the shaft drive is relatively low, for instance, less than 3 kW for a 2.5 m diameter column. High precision in the fabrication of the rotor shafts is not required; moreover, low shaft speeds result in a long lifetime of the shaft sealing and internal bearings. Maintenance can thus be limited to the normal servicing of the shaft seal and the motor drive.

For columns of diameters greater than 3 m, normally three turbine mixers on parallel axes are employed (Tri-Extractor) in order to avoid nonagitated dead-zone behavior in the column compartments and thus preserve ease of scale-up.

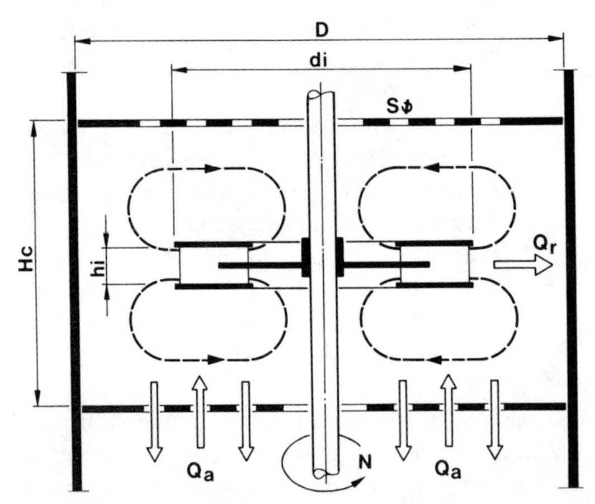

Figure 2. Characteristic flow pattern and geometry of the Kühni contactor.

3. INDUSTRIAL APPLICATIONS

At present over 300 Kühni contactors are in operation for research, process development, and industrial processing. Diameters range from 60 mm for the smallest laboratory-scale column to up to 2.5 m for high-capacity columns. The application of Kühni contactors covers a broad range of the current extraction process technology as shown by the following summary:

Petrochemicals industry	1. Aromatics extraction and stripping (Institut Français du Pétrole process)
	2. Lube oil extraction: purification of crude oil fractions
Pharmaceutical and chemical industry	1. Extraction of vitamins, glycosides, alkaloids, and other organic substances from aqueous solutions
	2. Acetic acid extraction from aqueous solutions
	3. Solvent recovery and purification by extraction: dimethylformamide, pyridine, dimethylsulfoxide, nitrobenzene, dichlorobenzene, and so on
	4. Vanilla extraction
	5. Citric acid recovery from molasses
Heavy-chemical industry	1. Phosphoric acid purification (Rhône-Poulenc process, Phorex process APC, Hoechst process, Albright & Wilson process)
Hydrometallurgy	1. Zinc removal from nickel–cobalt solutions
	2. Uranium extraction from leach liquors and crude phosphoric acid solutions
Chemical waste water treatment	1. Extraction of organic impurities as phenol and nitrobenzene from aqueous effluents
	2. Zinc removal from fiber manufacturing effluents

4. HYDRODYNAMICS AND MASS TRANSFER

A major investigation program has been realized over the last decade to reveal the hydrodynamic behavior and mass-transfer properties of the Kühni column. Because of its adaptable compartment and turbine mixer geometry, the influence of not only different physical properties, but also of geometric parameters had to be considered. Investigations are still proceeding. The following summary gives a review on the present stage of knowledge.

4.1. Mixing Turbine

Power numbers for the mixing turbine have been measured by Fischer [1]. In the range of normal rotational speeds the power number is independent of the impeller Reynolds number, and the influence of viscous shear forces is thus relatively slight in comparison to systems employing rotating-disk rotors.

A further characteristic quantity important for scale-up is the radial discharge flow rate Q_r (Fig. 2). Hody [2, 3] used a hot-film anemometer to determine the effective velocities at the turbine discharge. He derived the following relationship, which has proved valid for a broad range of turbine and compartment sizes:

$$\frac{Q_r}{d_i^3 N} = c\,\frac{h_i}{d_i}\left(\frac{\nu_{H_2O}}{\nu_c}\right)^{0.0267} \qquad (1)$$

4.2. Column Throughput, Flooding

Maximum column throughput or flooding is not well defined. Phase inversion, the formation of longish drops, and high carry-over of dispersed phase all indicate that the upper limit of column capacity has been attained. Phase inversion generally occurs, when the hold-up has reached approximately 50%, a rule of thumb confirmed also for other types of contactors [4].

Column capacity can be varied within a general range of 10-50 $m^3\,m^{-2}\,h^{-1}$ by varying the fractional free cross section of the stator plates and adjusting the rotational speed. The capacity range of a column with a given geometry, however, is smaller and strongly depends on the properties of the liquids involved. Turndown ratios (defined as ratio of maximum to minimum total throughput) are normally in the range 1.5-2, but values as high as 3 have been measured without an appreciable loss of efficiency when the rotational speed has been corrected appropriately.

4.3. Drop Size and Holdup

Fischer [1, 5] measured drop size and holdup for a broad range of operational conditions and a wide variety of liquid-liquid systems. A photographic method has been chosen for the determination of drop sizes to avoid disturbance of the flow pattern despite the extensive work required. It was found that the drop size distribution in the Kühni column is best approximated by a Mugele-Evans relationship.

According to the theory due to Kolmogoroff, two different correlations can be derived for the mean drop size diameter dependent on whether inertial or both inertial and viscous forces prevail. A comparison of theory and experimental results shows that a critical factor must be the interfacial tension of the liquid-liquid system (Fig. 3). Taking the influence of coalescence and the fact of diminishing drop size with increasing number of stages into account by separate factors, Fischer found the following two relationships to apply:

$$\sigma > 0.012\,\frac{\text{kg}}{\text{s}^2}:\ \frac{d_m}{d_i} = c_1 N_{We}^{-0.61}(1.0 + 2.0x_d) \cdot \left(1.0 + \frac{11.0}{n_c^{1.22} N_{We}^{0.5}}\right) \qquad (2)$$

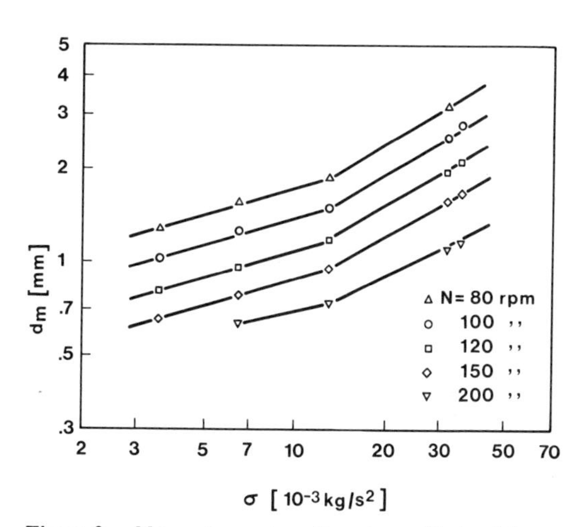

Figure 3. Mean drop size diameter: effect of interfacial tension Q_d/Q_c = 1.0. Reprinted by permission from A. Fischer, Hydrodynamik und Stoffaustausch in einer Flüssig-Flüssig-Rührextraktionskolonne, Ph.D. thesis No. 5016, ETH, Zurich, 1973.

$$\sigma < 0.012 \frac{\text{kg}}{\text{s}^2}: \frac{d_m}{d_i} = c_2 N_{\text{We}}^{-0.3} N_{\text{Re}_i}^{-0.4} (1.0 + 2.0 x_d) \cdot \left(1.0 + \frac{11.0}{n_c^{1.22} N_{\text{We}}^{0.5}}\right) \quad (3)$$

The constant c_1 varies between 0.24 and 0.32, and c_2 between 2.4 and 2.8 according to the fractional free cross section of the stator plates.

Holdup was determined by the same author [1, 5] by use of a manometric technique that allowed determination of not only the mean holdup, but also its variation throughout the column. For correlation of holdup with the drop size and the physical properties of the liquids involved, an expression must be found for the slip velocity $\bar{v}_0$ between the two phases that is defined by

$$\bar{v}_0 = \frac{V_d}{x_d} + \frac{V_c}{(1 - x_d)} \quad (4)$$

Ingham et al. [6] correlated the average holdup measured in a Kühni column on the basis of the slip velocity equation proposed by Misek. Fischer, however, used the chart given by Andersson, describing the behavior of rigid spheres in fluidized beds and corrected the slip velocity for internal circulation:

$$\frac{\bar{v}_0(\text{drops})}{\bar{v}_0(\text{spheres})} = 0.382 \frac{N_{\text{Re}}^{0.25} S_\phi^{0.42}}{\sigma^{0.12}} \left(\frac{\mu_c}{\mu_d}\right)^{0.5} \quad (5)$$

4.4. Axial Mixing

Several investigators [2, 3, 6-9] have been engaged with determining the degree of backmixing in the Kühni column, employing mostly the axial dispersion flow model to correlate their measurements. The experimental technique and the method of evaluation, however, varied considerably. A comparison of the results for single-phase backmixing gives a fairly good agreement (Fig. 4) between the different authors, with the exception of the measurements by Hody [2, 3] that inexplicably level off at low flow rates and high rotational speeds.

Kolmogoroff's theory again yields a good correlating expression [9] for single-phase backmixing:

$$\frac{1}{N_{\text{Pe}_c}} = c_1 + c_2 \left(\frac{h_i}{d_i}\right)^{0.3} \frac{d_i^2 N}{D V_c} \left(\frac{D}{H_c}\right)^{0.33} S\phi^{c_3} \quad (6)$$

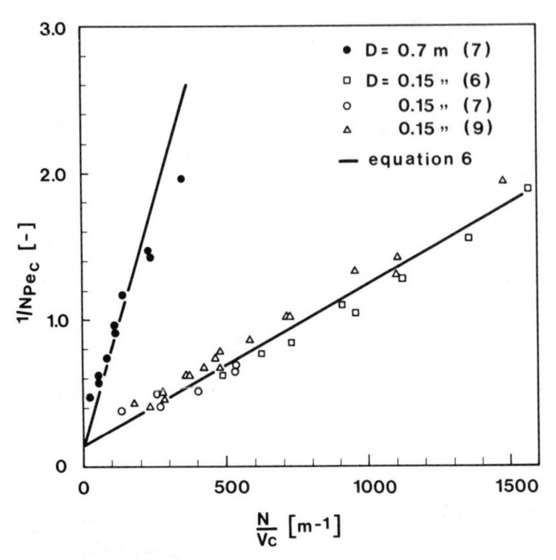

Figure 4. Correlation of single phase backmixing.

The three constants have been determined for differing turbine sizes and column diameters ranging up to 0.7 m. Figure 4 illustrates the reliable nature of the equation for scale-up calculations. In addition, Eq. (6) has proved applicable for the Tri-Extractor, but the constant c_2 must be corrected according to the additional number of rotational shafts.

$$c_{2,\text{Tri}} = c_2 n_R^{0.33} \quad (7)$$

For holdup values up to approximately 7%, Hody [2] found no influence of the dispersed phase on continuous-phase backmixing. Ingham [6], however, found that under normal operating conditions the degree of backmixing was always less than the corresponding single-phase value. He proposed the inclusion of an average holdup term, in the form of correlation previously used for the single-phase results, to take into account the effects of two-phase operation:

$$\frac{1}{N_{\text{Pe}_c}} = c_1 + c_2' \frac{d_i^2 N}{D V_c} (1 - x_d) \quad (8)$$

4.5. Efficiency and Mass Transfer

The first fundamental studies relating to mass transfer efficiency were made by using acetic acid extraction from aqueous solutions with methyl isobutyl ketone (MIBK) and ethyl acetate as solvents [10, 11]. The influence of both rotation speed and throughput on stage efficiency was determined; the stage efficiency was defined as the ratio of the number of theoretical stages to the number of actual stages in the column. It could be clearly shown, by fitting the results to a second-order response surface, that an optimal throughput and rotational speed exists for each set of fixed parameters (geometry of extraction stage, composition of feed streams, phase ratio). Optimal stage efficiencies of up to 80% corresponding to over 10 theoretical stages per meter of contactor height have been measured. Fischer [1, 5] took a further important step. In an extensive experimental study, he measured both drop size and holdup in two-column sections during actual extraction experiments. It was possible, therefore, to calculate the interfacial area and the local mass-transfer coefficient. First, however, the measured HTU_o values had to be corrected for axial mixing. As a first approximation, the following relationship has been adopted:

$$HTU_o = HTU + HDU \tag{9}$$

The evaluation of the experimental results by this method showed that mass-transfer coefficients in the order of magnitude of 10^{-4} m/s are attained. Furthermore, it was shown that the contribution of backmixing to the overall height of a transfer unit HTU_o is not insignificant.

5. SCALE-UP

The operational behavior of the Kühni column has been studied and documented to a great extent. As for other types of contactors a great part of the investigations have been made under ideal conditions and often in the absence of mass transfer in order to separate the influence of specific parameters. During actual extraction processes, however, the physical properties of the two liquid media may vary considerably according to the changing solute concentrations in the two phases. In addition, the effects of other liquid impurities or that of extraneous solid matter must be considered. It is evident that the complex behavior of the liquid–liquid systems involved is not easily predictable. The design of commercial extractors thus requires, in addition to a scale-up procedure, some laboratory and pilot plant investigations with the system concerned.

5.1. Pilot Plant Investigations

Small-scale columns may be designed on the basis of equilibrium data and a series of simple batch mixing tests. In this case, a safe overdesign is often more economical than a series of expensive pilot plant tests.

For large columns, however, pilot plant investigations must be carried out to obtain reliable data for scale-up. Kühni has developed a special experimental technique that allows scale-up from pilot column size. Experience has shown that a column diameter of 100–150 mm is sufficient for this purpose.

One of the most important aims of the pilot tests is the determination of the optimal compartmental geometry and its variation throughout the column. An illustrative, practical example is shown in Table 1, where conditions of uniform geometry in the column are contrasted with those obtained following optimization in the pilot plant.

Of equal importance are the experimental data obtained with respect to column capacity and efficiency, without which a safe scale-up procedure cannot be guaranteed.

5.2. Scale-Up Procedure

The scale-up procedure used for the Kühni column is based on the similarity of geometric parameters and the maintenance of similar hydrodynamic conditions in the pilot and operational scale columns. Similarity in the hydro-

TABLE 1 STAGE GEOMETRY, INFLUENCE ON HYDRODYNAMIC COLUMN PERFORMANCE FOR A SYSTEM WITH VARYING INTERFACIAL TENSION

Geometry	Location	σ, kg/s^2	d_i, mm	$S\phi$, —	N, rpm	d_m, mm	x_d, —	a, m^2/m^3
Uniform	Top	$33 \cdot 10^{-3}$	85	0.4	120	2.3	0.06	157
	Bottom	$6 \cdot 10^{-3}$	85	0.4	120	1.0	0.26	1560
Modified	Top	$33 \cdot 10^{-3}$	100	0.3	160	1.3	0.21	969
	Bottom	$6 \cdot 10^{-3}$	65	0.4	160	1.0	0.26	1560

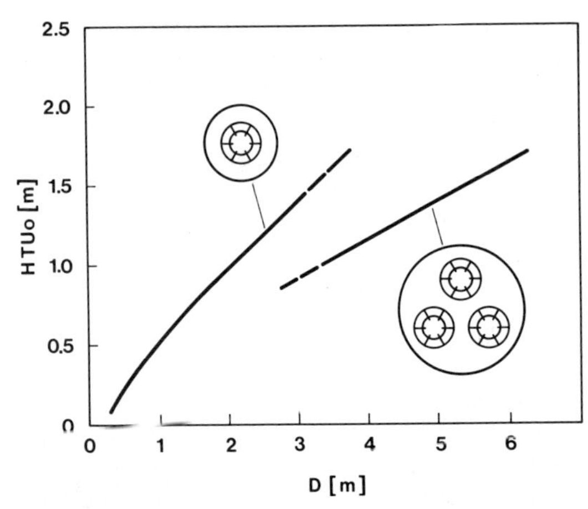

Figure 5. Increase of HTU_o with column diameter.

dynamic behavior in the two columns implies the following relationships:

Droplet diameter	$d_{m_1} = d_{m_2}$
Holdup	$x_{d_1} = x_{d_2}$
Specific throughput	$V_{c_1} = V_{c_2}$
Specific radial discharge flow rate	$\frac{Q_{r_1}}{D_1 H_{c_1}} = \frac{Q_{r_2}}{D_2 H_{c_2}}$

With this set of conditions fulfilled, not only the column capacity but also the rate of mass transfer will be the same for both the model and large-scale columns. To obtain equal overall mass-transfer efficiency, however, one must also take into account the influence of axial mixing. This is done by using Eq. (9), where the overall HTU_o values are split into the separate contributions representing plug flow and axial mixing.

The HDU value and therewith HTU_o can then be corrected to correspond to the conditions in the large-scale column. This procedure has proved to be applicable for all column sizes. Figure 5 shows typical HTU_o values for both the normal Kühni contactor and the Tri-Extractor as a function of the column diameter. This shows clearly the advantage of the latter type of construction for very large column diameters.

NOTATION

a	specific interfacial area, m^2/m^3
c, c_1, c_2, c_3	various constants, dimensionless
D	diameter of extraction column, m
d_i	turbine mixer diameter, m
d_m	Sauter mean drop diameter, m
E	axial dispersion coefficient, m^2/s
H_c	height of compartment, m
h_i	height of turbine mixer, m
HDU	height of an eddy diffusion unit, m
HTU	height of a transfer unit, m
k	mass-transfer coefficient, m/s
n_R	number of rotational shafts, dimensionless
n_c	number of compartments, dimensionless
N	rotational speed, rps
N_{Pe}	Peclet number for axial mixing = VH_c/E, dimensionless
N_{Re_i}	impeller Reynolds number = $d_i^2 N\rho_c/\mu_c$, dimensionless
N_{Re}	Reynolds number = $\bar{v}_0 d_m \rho_c/\mu_c$, dimensionless
N_{We}	Weber number = $N^2 d_i^3 \rho_c/\sigma$, dimensionless
Q	volumetric flow rate, m^3/s
S_ϕ	fractional free cross section of stator plates, dimensionless
$\bar{v}_0$	slip velocity, m/s
V	superficial velocity, m/s
x	holdup, dimensionless
σ	interfacial tension, kg/s^2
ρ	density, kg/m^3
μ	viscosity, kg/ms
ν	kinematic viscosity, m^2/s

Subscripts

a	axial
c	continuous phase
d	dispersed phase
r	radial
o	overall
1	pilot size
2	operational size

REFERENCES

1. A. Fischer, Hydrodynamik und Stoffaustausch in einer Flüssig-Flüssig-Rührextraktionskolonne, Ph.D. thesis No. 5016, ETH, Zürich, 1973.
2. D. Hody, Untersuchung der Rückvermischung in einer Flüssig-Flüssig-Extraktionskolonne mit rotierenden Einbauten, Ph.D. thesis No. 5560, ETH, Zürich, 1975.

3. D. Hody, *Chemische Rundschau* **28,** 9 (1975).

4. S. Weiss and R. Würfel, *Chem. Technik* **27,** 442 (1975).

5. A. Fischer, *Chemische Rundschau* **26,** 1 (1973).

6. J. Ingham, J. R. Bourne, and A. Mögli, *Proceedings of the International Solvent Extraction Conference 1974*, Vol. 2, Society of Chemical Industry, London, 1974, pp. 1299–1317.

7. Kühni AG, LAB 219, unpublished report on backmixing experiments, 1976.

8. J. E. Prenosil, *Verfahrenstechnik* **11,** 470 (1977).

9. R. Bauer, Vermischungseffekte im ARD- und Kühni-Extraktor, 7th seminar on extraction, Graz (Austria), 1978.

10. A. Mögli, *Chem. Ing. Technik* **37,** 210 (1965).

11. A. Mögli and E. Gusset, *Verfahrenstechnik* **1,** 357 (1967).

13.6

THE RTL (FORMERLY GRAESSER RAINING-BUCKET) CONTACTOR

J. Coleby
North Wales
United Kingdom†

1. INTRODUCTION

The RTL contactor was developed in the late 1950s at the chemical works of R. Graesser Ltd., North Wales, UK. Solvent extraction steps were required involving liquids with low density difference, low interfacial tension, and pronounce emulsification tendencies. Agitation levels thu could not be excessive, but the process called for multiple-stage operation in reasonably small apparatus.

The result of the research at Graesser [1-3] was a novel type of contactor that not only solved the original process problem, but has since been applied to many other extraction processes, such as pyrethrum refining, purification of herbicide products, and dearomatization of naphthas. Since 1977 the proprietors of the contactor have been RTL S.A. (117 George St., London W1H 5TB, UK).

2. DESCRIPTION

In its modern form, the RTL contactor consists of a single rotor operating on a substantially horizontal axis in a cylindrical stator, see Fig. 1.

†Formerly of R. Graesser Ltd., North Wales, United Kingdom.

The rotor comprises a series of circular baffles, between which are mounted a number of cylindrical buckets, partly open in the direction of rotation. Under normal operating conditions the contactor is filled with the two liquid phases and the interface level is controlled at the equatorial position. Thus the slow motion of the rotor causes each phase to cascade through the other to provide the contacting. The two liquid phases flow countercurrently through annular gaps between the rotor baffles and the stator walls.

The available sizes range from 100-mm-diameter × 1-m-long pilot scale models in glass or stainless steel to 2-m diameter × 8-m-long production models in mild steel or stainless steel. A typical production unit is shown at the assembly stage in Fig. 2.

3. DESIGN PRINCIPLES

As already noted, the interface level in the RTL contactor is kept at the horizontal axis. The two main design variables for a given contactor are the *rotor speed* and the annular *gap* between the rotor and the stator.

The speed of the rotor is adjusted according to the bucket capacity and the settling properties of the liquid–liquid system. The aim is to provide a similar holdup and drop size distribution regard-

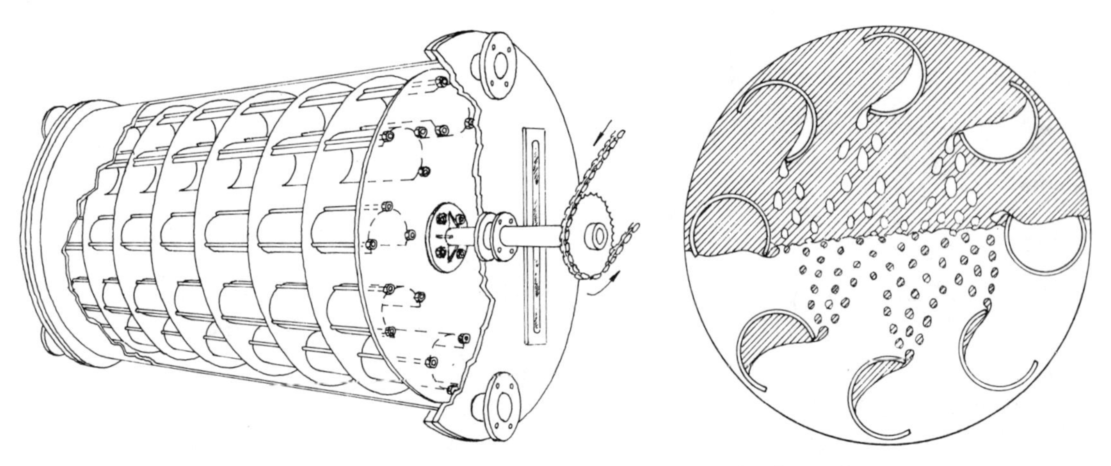

Figure 1. RTL contactor. Sectional views showing mode of operation.

Figure 2. RTL contactor (1.5-m diameter × 5.5 m long) being assembled. (Photograph courtesy of manufactures, Widnes Foundry and Engineering Co., UK.

less of contactor size. The drops are relatively large in comparison to those found in most other contactor types. For this reason, and because of the high frequency of coalescence and redispersion, the area-based mass-transfer coefficients are high. It has been found that scale-up of mass transfer from pilot tests is successful if a basis of constant interfacial area per unit contactor volume is maintained.

The annular gap thickness has little effect on drop size and similar parameters, but it does have a profound effect on back-mixing. A narrow gap leads to reduced backmixing and hence improved countercurrent performance, but the gap must not be so narrow as to significantly hinder the countercurrent flow.

The same type of compromise is reached in the choice of the number of compartments; a large number of compartments reduces axial mixing but increases fluid friction and the capital cost of the contactor.

Interface level in the contactor can be controlled by various means, namely: (1) simple hydrostatic control; (2) internal weir; (3) external weir; (4) pump-separator; and (5) separator-pump. Examples of these are shown on Figs. 3*a*–3*e*, respectively.

Two contactor design procedures are used by RTL. The first of these requires pilot plant results for the system under study, and these are then scaled up to the desired contactor capacity. The second procedure requires only a knowledge of the basic system properties such as interfacial tension, density difference, and partition coefficient. On the basis of extensive past experience it has been possible to estimate the volumetric

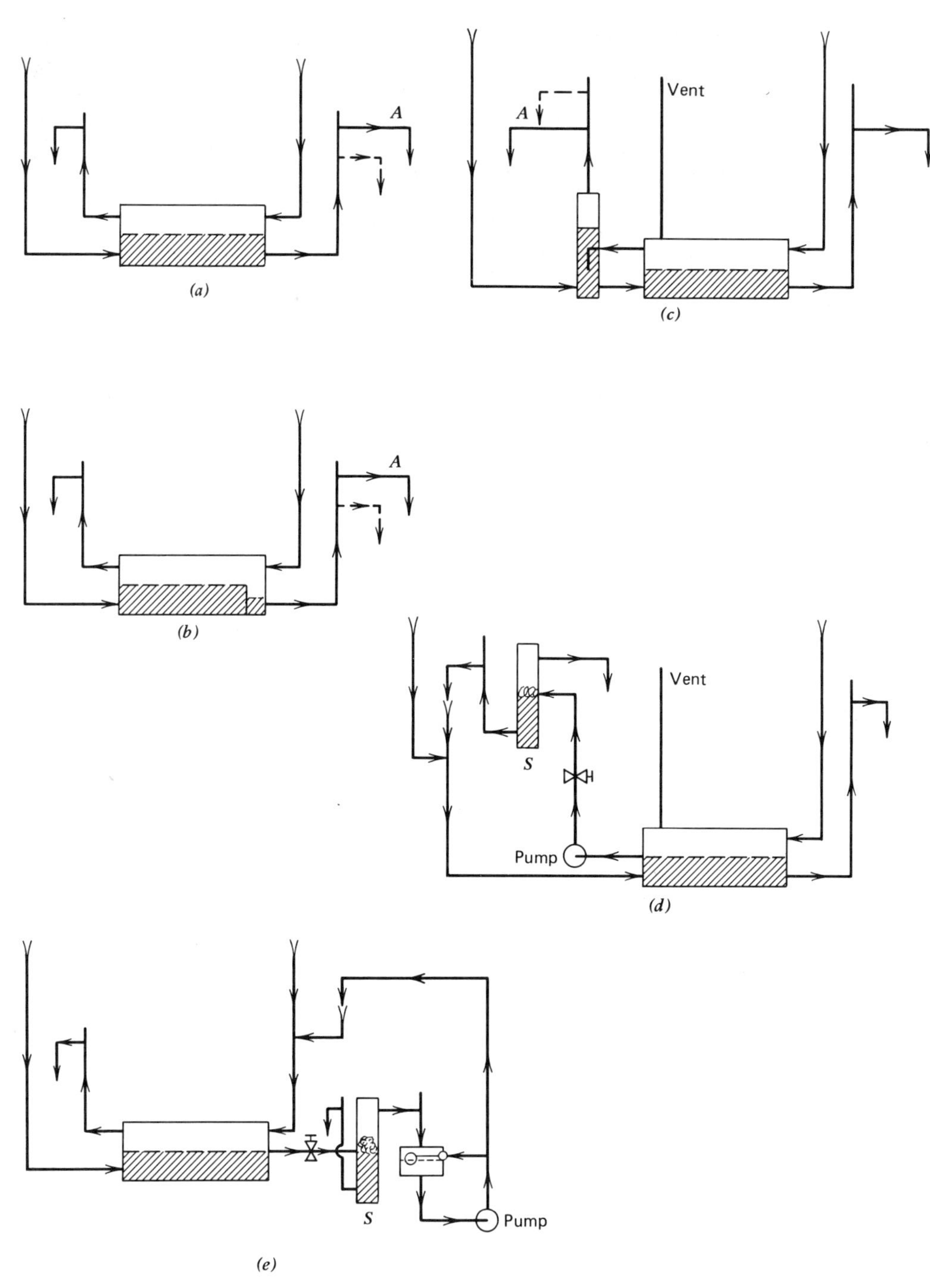

Figure 3. Methods of interface control in RTL contactor: (*a*) simple overflow; (*b*) internal weir (controlling on heavy phase out); (*c*) external weir (controlling on light phase out); (*d*) pump–separator (controlling on light phase out); (*e*) separator-pump (controlling on heavy phase out) (*A*, adjustable head; *S*, separator).

mass-transfer rates and other variables quite accurately from the basic properties. Production contactors are usually designed to give up to 0.3 theoretical stages per compartment (on the basis of an extraction factor of unity). Phase flow ratios of up to 6 : 1 can be readily accommodated, and by special design modifications it has been possible to operate at ratios of up to 20 : 1.

4. CONCLUSIONS AND FUTURE DEVELOPMENT

The RTL contactor differs in some major respects from vertical (tower) type contactors. It has the operating advantages that (1) density changes during extraction will not induce axial circulation since the axis is horizontal and (2) the RTL contactor recovers easily from any sudden shutdown since "drainage" of material from stage to stage cannot occur. These factors of the RTL, in addition to its ability to handle easily emulsified systems, have contributed to its successful application over the past 20 years.

Possibilities exist for the application of the RTL contactor to systems containing a solid phase; thus the apparatus has been used for the treatment of oil-bearing sands, with the use of kerosene to remove bitumen as one product, leaving clean-treated sand as the underflow.

The fundamentals of the operation of the RTL contactor have not been studied as extensively as have those of other commercial contactors, but some preliminary studies have been made [4, 5] on axial mixing and mass transfer in a glass pilot-scale unit (150-mm diameter, 760 mm long). Further work in this area and on other aspects of the contactor is desirable.

REFERENCES

1. J. Coleby, British Patent 860,880 (R. Graesser Ltd.).
2. J. Coleby, British Patent 972,035 (R. Graesser Ltd.).
3. J. Coleby, British Patent 1,037,573 (R. Graesser Ltd.).
4. A. R. Sheikh, J. Ingham, and C. Hanson, *Transact. Inst. Chem. Eng.* **50,** 199 (1972).
5. P. S. M. Wang, J. Ingham, and C. Hanson, *Transact. Inst. Chem. Eng.* **55,** 196 (1977).

14

MISCELLANEOUS ROTARY-AGITATED EXTRACTORS

M. H. I. Baird
McMaster University
Canada

1. INTRODUCTION

This chapter briefly covers those extractors with rotary agitation that have not been described elsewhere. The category is quite large because of the almost infinite possible combinations of rotor and baffle arrangements that can be included in a design and the further possibilities of inclining the contactor axis or introducing pulsation or oscillation. The contactors are classified here according to the *mechanical means of agitation.* It must be noted that whereas some of the contactors have been known for over 20 years, others have only recently been publicized. Because of the great diversity of designs and reported applications of the contactors, this chapter is mainly descriptive, and it is not possible to recommend specific selection criteria for the various types available.

2. EXTRACTORS WITH PURELY ROTARY AGITATION

2.1. Vertical Axis

A useful comparison of the effects of different types of rotor and stator on mass transfer and hydrodynamics in extraction columns (200-mm diameter) is given by Weiss et al. [1]. The performance of a conventional RDC configuration is compared with (1) stator rings and impellers made of round bar steel, (2) radial stationary baffle plates and impellers made of round bar steel, and (3) radial stationary baffle plates and impellers made of flat steel. The HTU and Sauter mean drop diameter dependences on stirrer speed and other variables were found to vary by up to 50% depending on the configurations.

Many rotary column designs are concerned with the prevention of interstage backmixing. One of the most ingenious of these designs is the *Treybal contactor* [2, 3], which was developed from experiments [4] with vertically stacked mixer–settlers. Each compartment contains a central "mixer" section and an outer annular "settler" section. The two-phase dispersion produced in the central section enters the annular section through ports and settling takes place. The settled dense phase flows down to the next lowest mixing section while the settled light phase flows up to the next mixer.

In the *Wirz column* [3, 5] the mixed phases are thrown outward by the impeller through baffles and a gauze coalescer, as shown in Fig. 1 [3]. Leisibach [5] has reported performance data for columns of 100–500-mm diameter; HTU values were found to be in the range 100–300 mm.

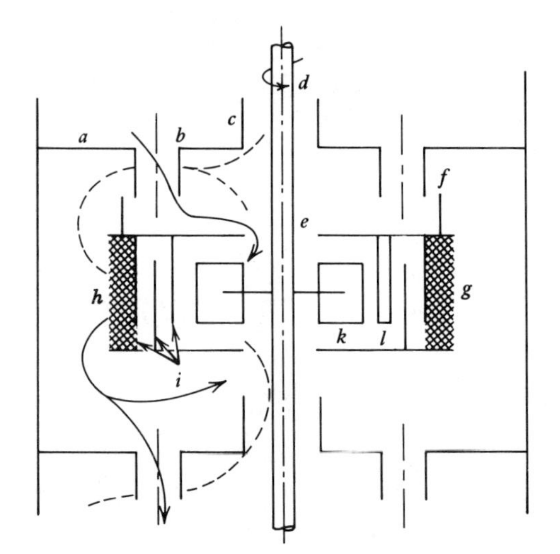

Figure 1. Compartment in a Wirz column. [Reprinted by permission from C. Hanson, *Br. Chem. Eng.* **13**(11), 49 (1968).]

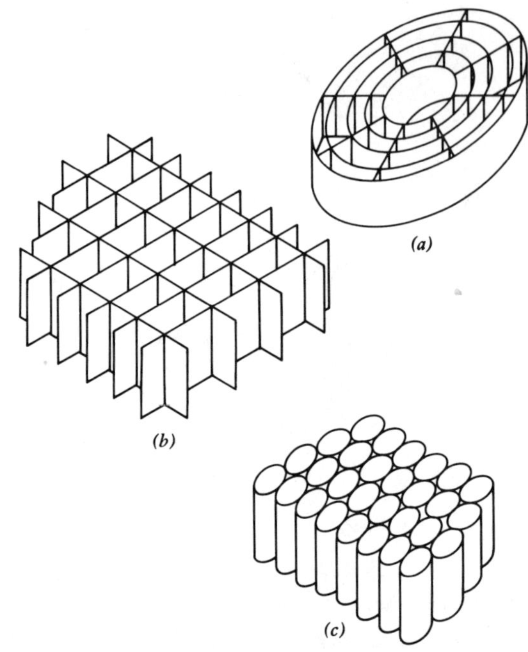

Figure 2. Possible constructions of enhanced coalescence plates. [Reprinted by permission from L. Steiner and S. Hartland, *Separation Sci. Technol.* **15**, 907 (1980).]

Whereas the Wirsz and Treybal columns both depend largely on coalescence for the suppression of backmixing, the *rotary annular column* operates entirely in the *emulsion regime*, with the droplets continually agitated by shear between the cylindrical rotor and the column wall. Rotary annular-type columns have been used for over 50 years on the laboratory scale (see Chapter 17.), but the first detailed study was due to Thornton and Pratt [6], who reported on glass columns of 50–150-mm diameter. Davis and Weber [7] studied mass transfer in the uranyl nitrate-HNO_3–TBP system and also observed droplet motion in the Taylor vortices in the sheared annular space. The relatively small residence times, the low heights of a transfer unit (as low as 60–80 mm [7]), and the ease of cleaning make this type of column particularly attractive to the nuclear industry. A high-speed annular contactor has recently been described [8] in which the phases after contact are allowed to settle in a chamber in the center of the rotor. Settling is accelerated by the centrifugal motion. The contactors can be combined in multistage arrays for laboratory-scale testing.

Steiner and Hartland [9, 10] have developed a special type of stator plate to reduce backmixing in impeller agitated columns. The plates are known as *enhanced coalescence plates* (EC plates) because they are of a material that is preferentially wet by the dispersed phase. Backmixing is further prevented by the honeycomb configuration of the plates, as illustrated in Fig. 2.

Many other variations on the theme of vertical-axis rotary agitated extraction columns have been proposed, and these are summarized below in Table 1.

2.2. Inclined Axis

A recent patent [23] describes an extractor consisting simply of an inclined rotating tube divided into chambers by perforated disks. Countercurrent flow of the liquid phases occurs by gravity, and the action of the disks is to cause some coalescence and redispersion of droplets and to limit backmixing.

A more complicated arrangement [24] involves a *helicoidal* tube rotating about an inclined axis. In this case it is possible to introduce the denser liquid at the bottom of the column and the light liquid at the top since the phase movement depends partly on the direction of helix rotation.

2.3. Horizontal Axis

The first horizontal-axis rotary extractor was introduced by Van Dijck and Ruys [25]. It consists of alternate mixing and settling chambers separated by perforated stators. Mixing is effected

TABLE 1 SUMMARY OF SOME VERTICAL-AXIS ROTARY EXTRACTORS

Reference Source	Special Features
Buhlmann [11]	Complete coalescence occurs between each stage by use of special arrangement
Burova et al. [12]	Slotted rotors are separated by stator rings
Ludwig [13]	Magnetically driven rotating jet ring sprays dispersed phase into continuous phase
Mass [14]	Design can include different sizes and types of impeller at different parts of column
Nakamura and Hiratsuka [15]	Radially supported arc plates are located at 60° angles fixed at rotating axis
Nishizawa [16]	Both the agitator and the housing rotate; there is mixer–settler multistage action with centrifugal phase separation
Postl and Marr [17]	Special baffles encourage separation and coalescence of phases between compartments
Saxon and Kingsbaker [18]	Vertical rotor is formed by concentric walls radially divided into cells for solvent
Schmidiger [19]	Externally operated valves control flow between chambers; retention times can be controlled independently
Sokov and Putilova [20]	Partitions are placed in center of column, and rotors in concentric circles on periphery
Spaethe and Weiss [21]	Backmixing is reduced by radial baffles and specially shaped partition walls
Tudose [22]	Chambers are separated by disks containing square holes with caps

by impellers mounted on a common horizontal shaft. This extractor does not appear to have developed beyond the laboratory stage.

Another horizontal multistage rotary extractor, developed primarily for laboratory use, is described by Signer et al. [26]. It consists of a cylindrical housing, rotating slowly about a horizontal axis and divided longitudinally into compartments. Openings at the center of each end of the housing permit continuous addition and removal of the liquid phases, which form two layers according to density. Countercurrent flow occurs by way of perforations in the compartment walls. Signer et al. [26] modified this design to give a static housing and static compartments, agitated by rotating-disk elements immersed in each compartment.

The horizontal axis rotating-disk principle has been applied on a much larger scale for the treatment of slurries and ore pulps in the *rotary film contactor*. The applicability of this contactor to liquid–liquid systems is discussed by Logsdail and Lowes [27].

In the preceding contactors [26, 27] the disks are only partially immersed in liquids, but a U.S. patent [28] calls for complete immersion of the rotating disk in the two liquids, with V-shaped cutouts provided in the disks to facilitate axial liquid flow. The disk should be preferentially wetted by the phase containing the component to be transferred.

3. ROTARY EXTRACTORS WITH HYDRAULIC PULSATION

A series of studies of the effects of rotary agitation on pulsed column performance has been carried out by Angelino and his co-workers [29-31]. The columns used were of 40- and 50-mm diameter, with conventional fixed perforated plates having 23.6 and 20% open area, respectively. Rotary agitation was provided by disc or turbine agitators positioned between each pair of perforated plates; a typical arrangement is shown in Fig. 3. It was found [29] that rotary agitation combined with pulsation could give a lower HTU than that obtainable by pulsation

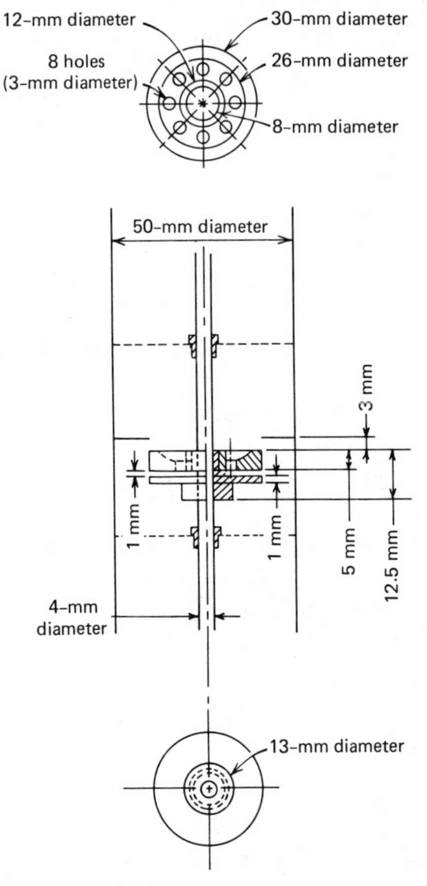

Figure 3. Rotary agitation of a pulsed column. [Reprinted by permission from H. Angelino and J. Molinier, *Proceedings of the International Solvent Extraction Conference, The Hague* (ISEC 71), Vol. 1, Society of Chemical Industry, London, 1971, p. 688.]

alone. It was later concluded [31] that this improvement was due to an increase in dispersed phase holdup between the plates and not to a decrease in axial mixing.

There are some references to rotary-pulsed columns in the Soviet literature [32, 33].

4. EXTRACTORS WITH ROTATIONAL-OSCILLATORY AGITATION

In this class of extractor, energy for agitation is transmitted only through a shaft, but it performs a combination of oscillatory and rotary motion.

The *Humboldt-Ziehl column* (Klockner-Humboldt-Deutz AG, Koln) has achieved widespread industrial application in continental Europe. It consists of a vertical column without fixed internals, containing a central drive shaft that oscillates axially and twists at the same time. Mixing elements made of plastic or metal are attached at intervals to the drive shaft and have the form of sets of radial branches, usually six or eight per element, which lightly touch the inner wall of the column. The combination of the twirling motion and relatively large open area provides uniform agitation but freedom from problems of fouling, such as by traces of solid particles. Detailed design principles and performance data for Humboldt-Ziehl columns are given by Vogeno [34] and Simonis [35, 36].

The *oscillating baffle contactor* due to Thomas [37] also has a central shaft in a cylindrical column with no static internals; but in this case the shaft only twists back and forth, without any axial motion. The mixing elements consist of four longitudinal (cruciform) perforated baffles extending over the agitated length of the column. In operation, the liquids in the column are swept in alternate angular directions and forced back and forth through the perforations. Performance data for a 70-mm-diameter column with a baffle length of 770 mm are reported [37] but as far as is known, the contactor has not been used commercially.

REFERENCES

1. S. Weiss, W. Spaethe, R. Wurfel, and D. Mohring, *Proceedings of the International Solvent Extraction Conference, Lyons* (ISEC 74), Society of Chemical Industry, London, 1974, p. 2315.
2. R. E. Treybal, U.S. Patent 3,325,255 (June 13, 1967).
3. C. Hanson, *Br. Chem. Eng.* **13**(11), 49 (1968).
4. R. E. Treybal, *Chem. Eng. Progr.* **60**(5), 77 (1964).
5. J. Leisibach, *Chem. Ing. Technik* **37,** 205 (1965).
6. J. D. Thornton and H. R. C. Pratt, *Transact. Inst. Chem. Eng.* **31,** 289 (1953).
7. M. W. Davis and E. J. Weber, *Ind. Eng. Chem.* **52,** 929 (1960).
8. R. A. Leonard, G. J. Berstein, A. A. Ziegler, and R. H. Pelto, *Separation Sci. Technol.* **15,** 925 (1980).
9. L. Steiner and S. Hartland, *Separation Sci. Technol.* **15,** 907 (1980).
10. L. Steiner and S. Hartland, *Chem. Eng. Progr.* **76**(12), 60 (1980).
11. U. Buhlmann, *Proceedings of the International Solvent Extraction Conference, Liege* (ISEC 80), Vol. 1, paper 80-213 (University of Liege, Belgium), 1980.

12. L. E. Burova, E. M. Guseinov, and A. N. Planovskii, *Kim. Neft. Mashinostr.* (1), 25 (1971); through *Chem. Abstr.* **74,** 88990*j*.

13. H. Ludwig, German Patent 2,221,554 (November 15, 1973); through *Chem. Abstr.* **80,** 110,443*f*.

14. H. J. Maas, East German Patent 92,222 (September 5, 1972); through *Chem. Abstr.* **79,** 20,768r.

15. A. Nakamura and S. Hiratsuka, *Kagaku Kogaku* **30,** 1003 (1966); through *Chem. Abstr.* **67,** 34,122*q*.

16. H. Nishizawa, German Patent 2,412,183 (October 24, 1974); through *Chem. Abstr.* **82,** 60682*b*.

17. J. Postl and R. Marr, *Proceedings of the International Solvent Extraction Conference, Liege* (ISEC 80), paper 80-148 (University of Liege, Belgium), 1980.

18. A. F. Saxon and C. L. Kingsbaker, British Patent 1,431,802 (April 14, 1976); through *Chem. Abstr.* **85,** 48693*k*.

19. F. Schmidiger, German Patent 2,322,872 (October 31, 1974); through *Chem. Abstr.* **82,** 158,179*t*.

20. Y. F. Sokov and Z. D. Putilova, Soviet Patent 292,346,23 (October 1972); through *Chem. Abstr.* **78,** 86404*e*.

21. W. Spaethe and S. Weiss, German Patent 2,345,261 (June 20, 1974); through *Chem. Abstr.* **82,** 32,798*a*.

22. R. Tudose, *Rev. Chim. (Bucharest)* **12,** 544 (1961); through *Chem. Abstr.* **56,** 12698*i*.

23. Y. Kabasawa and T. Tanimura, German Patent 2,632,149 (January 27, 1977); through *Chem. Abstr.* **87,** 7904*c*.

24. H. F. Wiegandt, U.S. Patent 3,390,963 (July 2, 1968); through *Chem. Abstr.* **69,** 60,254*u*.

25. W. J. D. Van Dijck and J. D. Ruys, *Perfumery Essent, Oil Record*, **28,** 91 (1937).

26. R. Signer, K. Alleman, E. Kohli, W. Lehmann, W. Ritschard, and H. Meyer, *DECHEMA Monogr.* **27,** 32 (1956).

27. D. H. Logsdail and L. Lowes, in C. Hanson, Ed., *Recent Advances in Liquid-Liquid Extraction*, Pergamon Press, Oxford, UK, 1971, pp. 148-150.

28. P. G. Grimes and M. C. Raether, U.S. Patent 3,351,434 (November 7, 1967); through *Chem. Abstr.* **68,** 70,501*s*.

29. H. Angelino, C. Alran, L. Boyadzhiev, and S. P. Mukherjee, *Br. Chem. Eng.* **12,** 627 (1967).

30. H. Angelino and J. Molinier, *Proceedings of the International Solvent Extraction Conference, The Hague* (ISEC 71), Vol. 1, Society of Chemical Industry, London, 1971, p. 688.

31. S. Barame, J. Molinier, and H. Angelino, *Can. J. Chem. Eng.* **51,** 156 (1973).

32. Y. N. Mileshin, Y. A. Shurchkova, and S. Z. Bikmaev, *Khim-Farm. Zh.* **8**(7), 47 (1974); through *Chem. Abstr.* **82,** 21,802*w*.

33. C. D. Murshudli, *Uch. Zap. Azerb. in-t nefti i khimii* Ser. 9 (4), 78 (1974); through *Chem. Abstr.* **83,** 134,066*y*.

34. W. Vogeno, *Riechst., Aromen, Koerperpflegem.* **19**(3), 108 (1969).

35. H. Simonis, *DECHEMA Monogr.* **65,** 315 (1970).

36. H. Simonis, *Proc. Eng.* 110 (November 1972).

37. W. J. Thomas, *Transact. Inst. Chem. Eng.* **47,** T304 (1969).

15

CENTRIFUGAL EXTRACTORS

M. Hafez
Imperial Oil
***Canada*†**

Centrifugal extractors are efficient and compact and have a number of desirable characteristics. They are being successfully used in a variety of chemical process industries.

This chapter gives an overview of the history, development, operation, and performance of various centrifugal extractors. It also discusses their hydrodynamic parameters, although only scarcely studied.

†Formerly of McMaster University, Canada.

1. HISTORICAL

The history and development of centrifugal extractors are closely related to the penicillin extraction process [1]. The problems inherent in large-scale extraction of penicillin [1, 2] played an important role in centrifugal extractors development, namely, the need for large solvent: feed ratios, short residence times, and overcoming the tendency of the system to emulsify.

Initially centrifugal separators were used as cocurrent extractors, but operational problems were numerous and extraction efficiency was low. The first attempts to overcome these difficulties combined high-speed mixers with centrifugal separators to build countercurrent cascades [3]. This approach was successful but complicated by a large number of vessels and sophisticated controls.

These problems indicated that a multistage countercurrent centrifugal extractor was needed or that the mixer and separator should be combined together in a single unit to simplify cascade design. Both approaches were followed and have resulted in the present designs.

1.1. Continuous Countercurrent Centrifugal Extractors

The concept of using centrifugal force to achieve countercurrent mixing and separation was developed in the early 1930s for gas–liquid [4] as well as liquid–liquid [5–7] contact. The principle of this design is to introduce the heavy phase near the axis of a rotating container while the light phase is introduced near the periphery. The applied centrifugal force causes the two phases to move radially in countercurrent mode such that the heavy phase is displaced to the periphery and the light phase, to the axis of rotation. Contact within the extractor may be differential or in a multiple of stages.

Placek [4] and Podbielniak [5] developed a horizontal differential design consisting of a perforated spiral wrapped around a shaft. The spiral was eventually replaced by concentric cylinders [8] and provision for phase introduction and removal through incorporation of radial tubes [9]. By 1944 this extractor was successfully tested for penicillin extraction.

Coutor [6] of Lürgi developed a vertical multistage design, which can be envisaged as a series of mixers–separators mounted on top of one another and suitably connected to perform countercurrent contact. Lürgi and Westfalia, the latter producing centrifugal separators, subsequently developed the Luwesta extractor based on Coutors idea. By 1949 this design was successfully demonstrated for penicillin extraction and acetic acid recovery from cellulose acetate solutions.

These two designs were followed by others. In the first category the Quadronic was developed in 1964 by Liquid Dynamics [12, 13]. In the second category the α-Laval [3, 14], the series Robatel SGN [15], and the unpressurized vertical extractor UPV [17] were developed early in the 1960s.

1.2. Single-Stage Centrifugal Extractors

In this category the mixer and the separator are combined in one unit that mixes and separates by centrifugal action; the two phases may also be externally mixed. The phases enter at the bottom of the unit through a mixing propeller and up to the separating section. In some designs mixing occurs at the top of the extractor through a centripetal pump or an external mixer. Units are connected in series such that phase flow is cocurrent through each unit but countercurrent through the cascade.

Examples of this design include the series Robatel BXP [18] and the Westfalia TA extractors [23], which were developed in the 1960s. The SRL was developed by Savannah River Laboratories in 1961 [19] and a modification of that, the ANL annular extractor, was developed by Argonne National Laboratories in 1973 [21, 22]. In the early 1970s MEAB developed the AKUFVE system, a combination of a static mixer and an H centrifuge. The versatility of the static mixer and the unique design of the H centrifuge largely eliminated the complications of separate mixers.

2. CLASSIFICATION AND CHARACTERISTICS

As shown in Table 1, centrifugal extractors can generally be classified into vertical and horizontal designs in which contact may be differential or in multistages and staging may be internal or external (cascades).

Compared with other extractors, they offer a number of desirable characteristics. Large capacity per unit volume, short residence times, small solvent inventory, and no need for insulation. They handle systems that easily emulsify, have small density difference, and need large solvent to feed ratios. From a mechanical point of view, they are precision equipment and hence expensive.

3. APPLICATIONS

Centrifugal extractors have been and continue to be extensively used in pharmaceutical recovery processes. They are also used in other chemical, petroleum, hydrometallurgical, and atomic energy applications. Table 2 lists some such applications together with the types of extractor used.

TABLE 1 CLASSIFICATION OF CENTRIFUGAL EXTRACTORS

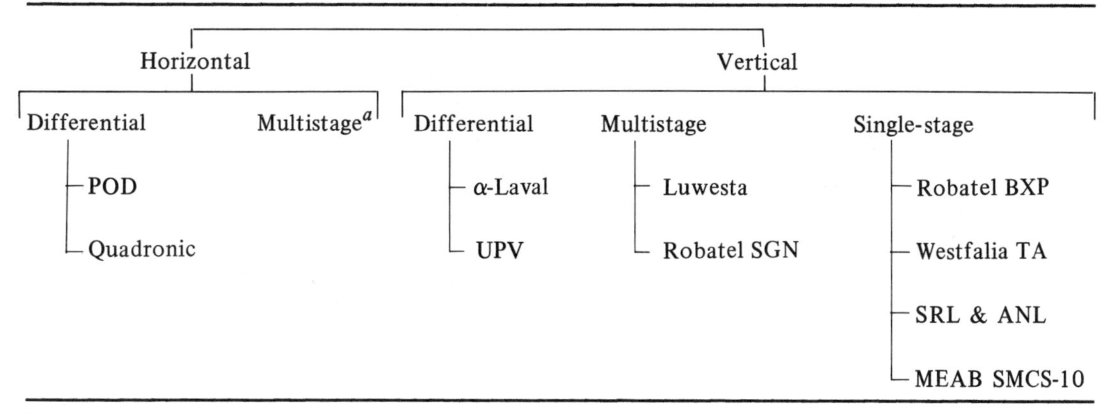

[a]Horizontal multistage designs are not attractive.

TABLE 2 APPLICATIONS OF CENTRIFUGAL EXTRACTORS

Process	Characteristics	Extractor	Feature	Reference
Pharmaceutics				
Antibiotic extraction Vitamines refining	Systems easily degrade	Podbielniak, Westfalia, α-Laval, Quadronic	Sealed, fast	2, 3, 7, 23–25
Atomic energy				
Uranium extraction and stripping	Hazardous	Robatel BXP, SRL, ANL	Small volume, shielding, small solvent inventory	19–22, 31, 32
Lube Processing				
Aromatic removal	Easy emulsifiable, small-density difference	Podbielniak, unpressurized α-Laval	Emulsion handling, fast settling	28–30, 57
Hydrometallurgical				
Metal extraction ion exchange	Easy emulsifiable, corrosive	Podbielniak, α-Laval	Emulsion handling, compact	38
Solid–liquid Processes Perfume extraction	Solid–liquid systems	Robatel SGN, Westfalia	Solid handling	39
Miscellaneous				
Acid treatment	Corrosive	Podbielniak, Westfalia	Compact	23, 38
Ether extraction	Toxic	Podbielniak	Compact, small solvent inventory	38
Waste water	Small concentrations	Podbielniak, Quadronic	Efficient	33, 34
Soap manufacture	Solids, viscous	Quadronic	Solid handling, easy cleaning	36

4. TYPES AND OPERATING PRINCIPLES

Ten different centrifugal extractor designs are presently in common use for a wide range of applications from pharmaceuticals to waste treatment and from lube oil processing to nuclear applications.

The operating principle of all types is essentially the same irrespective of particular design. In the differential contact extractors (Fig. 1*a*) the light phase is introduced near the periphery

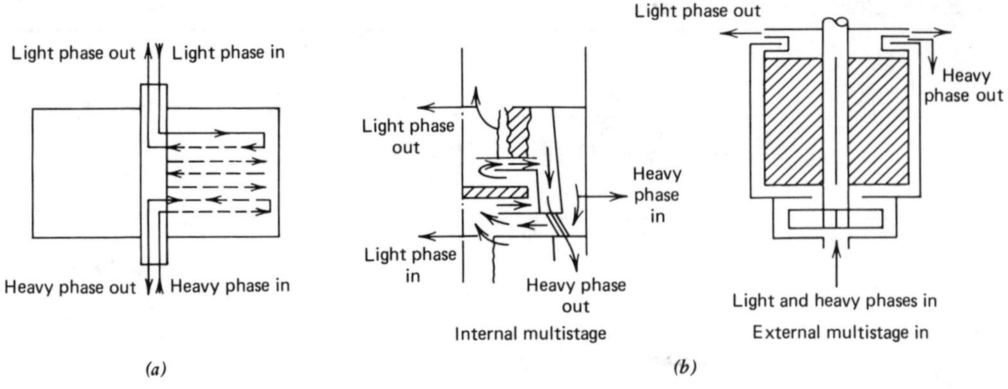

Figure 1. Operating principle of centrifugal extractors: (*a*) differential contact; (*b*) multistage contact.

and the heavy phase, near the shaft. Centrifugal force causes radial countercurrent flow of the heavy phase outward and the light phase, inward, where they are collected at the rim and shaft, respectively. In staged extractors, on the other hand, both phases are introduced into a mixing compartment and out to a settling compartment by centrifugal force (Fig. 1*b*). Countercurrent contact proceeds internally (multistage designs) or externally (cascades of single-stage designs).

4.1. Differential-Contact Centrifugal Extractors

4.1.1. The Podbielniak POD Extractor [38–41]

This is a horizontal countercurrent extractor. It consists of a drum rotating around a shaft, which is equipped with central and annular passages at each end for fluid entrance and exit (Fig. 2). The drum contains concentric perforated cylinders that can be replaced by using removable side plates. Hydrodynamically balanced mechanical seals ensure hermetic operation. Fluids are introduced and removed under pressure by means of removable radial tubes threaded to the rotating shaft and connected to the proper passage. The tubes have orifice-type distributers at the required radial position to collect or introduce the two phases. During operation three zones are formed within the extractor: two narrow zones near the shaft and the rim for phase clarification and a large zone in which extraction takes place (see also Fig. 12).

4.1.2. The Quadronic Extractor [13, 36, 37, 42]

This is also a horizontal countercurrent extractor (Fig. 3). It was introduced with the idea of increasing the versatility of the POD design. To this end, the internal design was modified to enable easy dismantling in order to facilitate introducing and/or removing the phases at arbitrary radial positions, to enable handling of three-phase systems, and to bring about controlled mixing. The latter is achieved by using different orifice sizes at different radial positions and is claimed to significantly affect performance. Three internal designs are being used: the orificed disk column; the perforated strips; and the sectional concentric orificed bands [42]. The external design, connections, and operation of the Quadronic are otherwise similar to those of the POD.

4.1.3. The α-Laval Extractor [3, 14]

The α-Laval is a vertical countercurrent extractor (Fig. 4). Both phases enter under pressure at the bottom of the vertically rotating bowl; the light phase enters near the periphery and the heavy phase, near the center. The two phases move radially and countercurrently up and down concentric channels to their respective exits. The phases mix as they enter or leave each channel through the joining orificed openings. The heavy phase leaves at the top by way of the outermost channel. In a similar manner the light phase leaves at the top through the innermost channel. This design differs from the horizontal one in that an interface is established in every channel and the two phases flow in countercurrent layers except at the mixing ports.

4.1.4. The Unpressurized Vertical UPV Extractor [17, 43, 57]

Although patented in the United States in 1964 [17], this design is not commercially available.

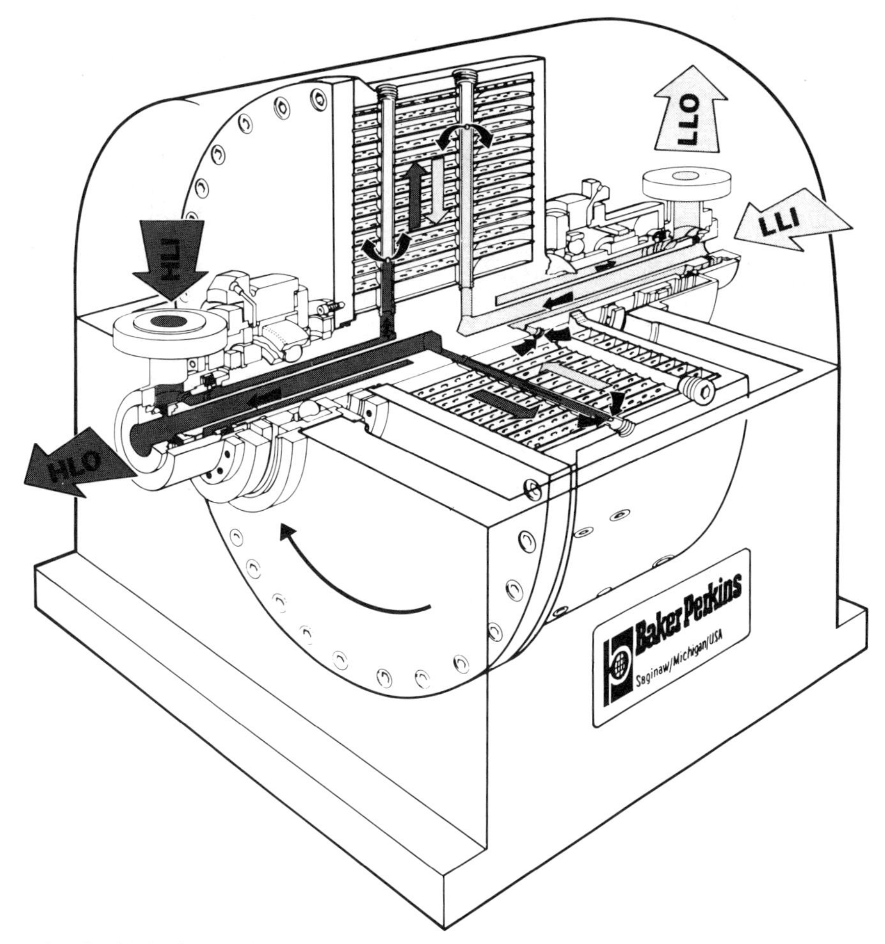

Figure 2. Podbielniak centrifugal extractor POD. (Courtesy of Baker Perkins, Inc., Michigan [38].)

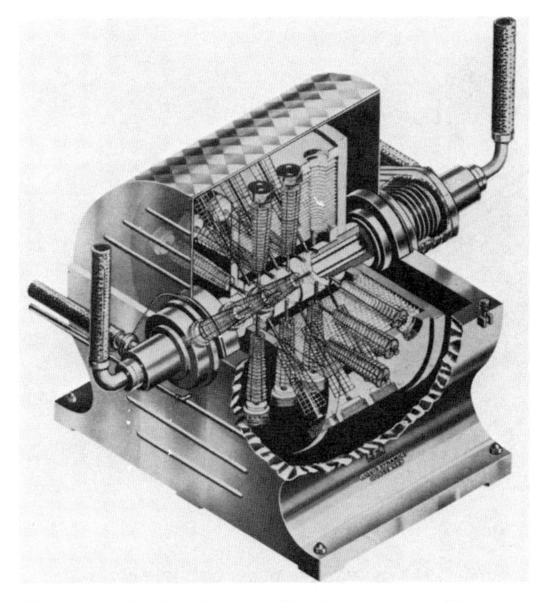

Figure 3. Quadronic centrifugal extractor. (Courtesy of Liquid Dynamics, Illinois [37].)

A modification thereof is, however, commercially used in the Soviet Union [57] and, together with the POD, are the most studied designs. The extractor, schematically shown in Fig. 5, is a vertical differential contactor rotating around a vertical shaft supported on two ball bearings. The two phases are introduced at the top under their own pressure to move radially in countercurrent flow. The heavy phase enters the mixing zone through horizontal orifices and is removed at the bottom near the periphery, while the light phase is removed at the top near the shaft. Various internals have been used, including perforated cylinders [44], corrugated disks [17, 44] and X-shaped packing [43].

4.2. Multistage Centrifugal Extractors

4.2.1. The Luwesta Extractor [6, 7]

This vertical multistage countercurrent extractor is schematically shown in Fig. 6. It consists of a

Figure 4. Alfa-Laval centrifugal extractor. (Courtesy of the de Laval Separator Company, New York [14].)

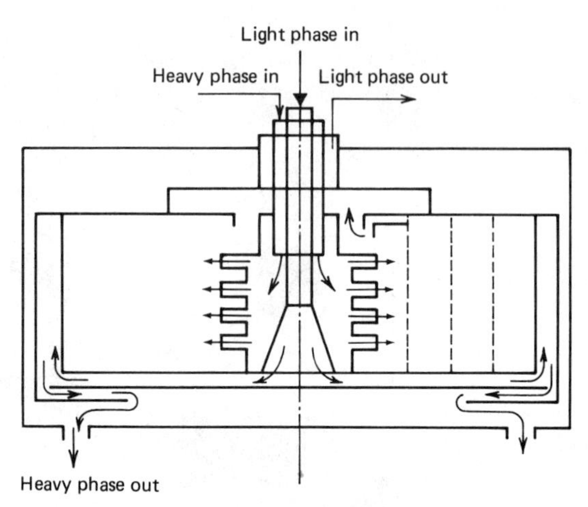

Figure 5. Unpressurized vertical centrifugal extractor UPV (schematic diagram).

stationary casing and a rotating shaft that contains passageways connecting distributer and collector rings mounted on the shaft [7]. The shaft and casing are equipped with inclined disks and baffles, respectively, to centrifuge and pump the two phases, which enter and are removed under pressure at the top. The light phase is introduced to a stage together with the heavy

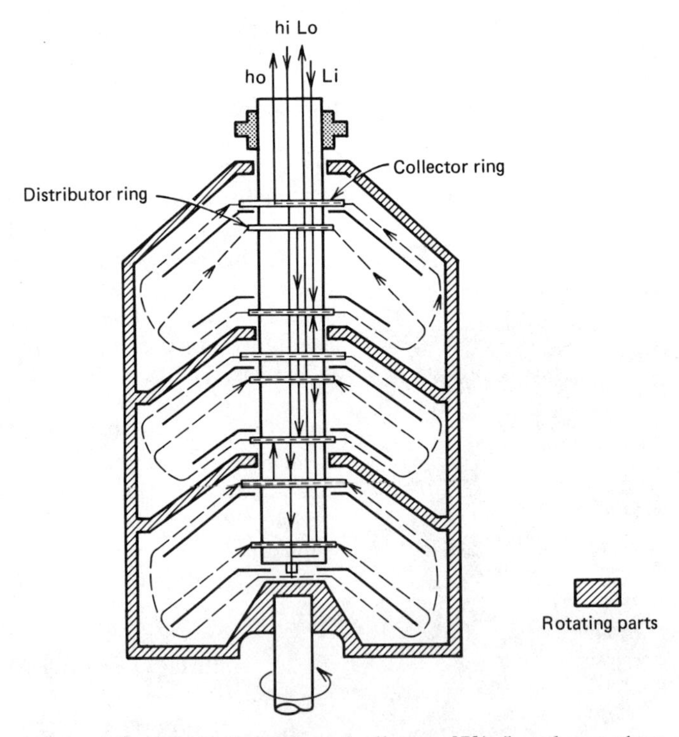

Figure 6. Luwesta centrifugal extractor (schematic diagram [7]) (ho = heavy phase out; hi = heavy phase in; lo = light phase out; li = light phase in).

phase from the stage below through distributor disks. The released mixture moves radially and separates into two phases. Each phase enters its respective collector ring and moves down or up to the next distributer disk, until both phases are finally discharged from the top.

4.2.2. The Robatel SGN Extractor [15, 16, 27]

The design of the series SGN is similar to the original Coutor concept [7]. Mostly used for nuclear processes, the extractor is usually suspended with the motor mounted on top and sheilded from the extractor proper. The extractor itself (Fig. 7) consists of a rotating bowl divided by baffles into horizontal compartments on top of one another (stacked vertically). The stationary central shaft carries mixing disks that run through each compartment and serve to mix and pump the two phases to the settling part of the stage. The latter is separated from the mixing zone by annular plates. Each compartment has connections to lead the settled phases to the previous and following stages. The standard

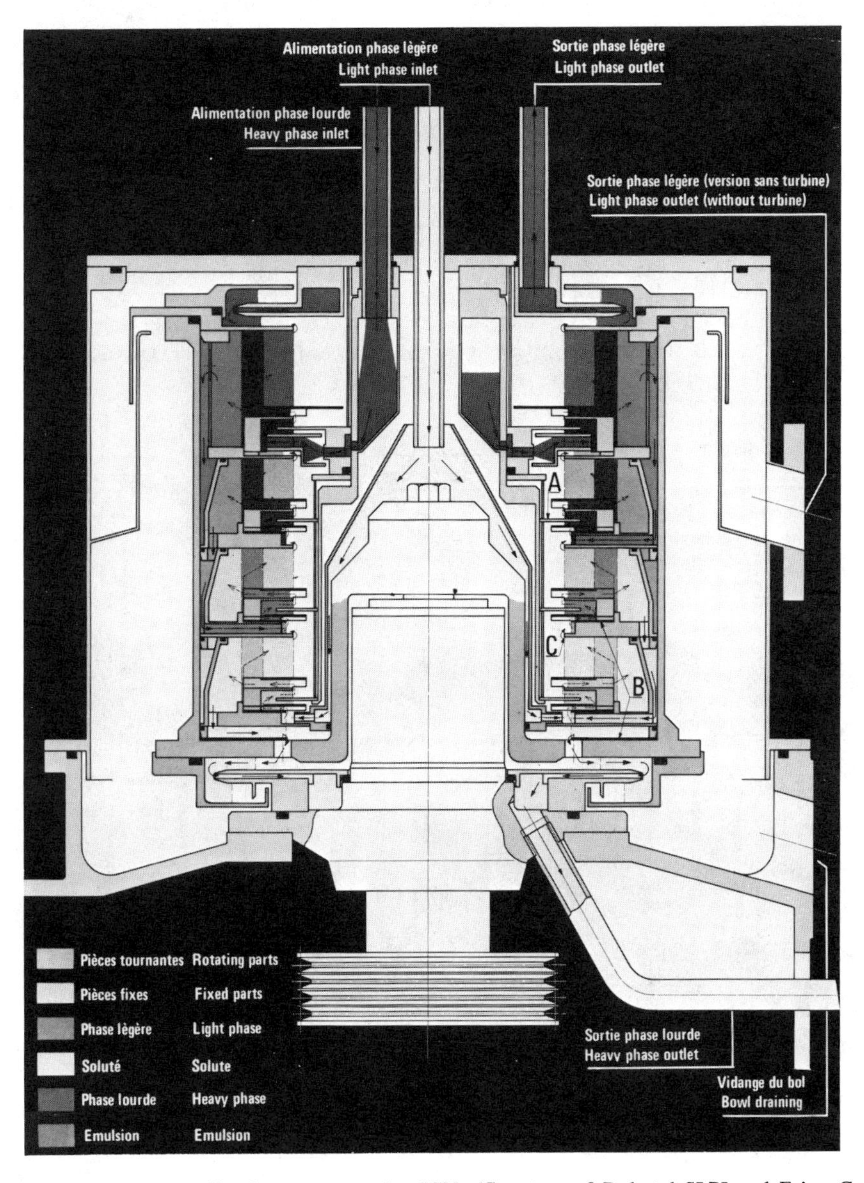

Figure 7. Robatel centrifugal extractor series SGN. (Courtesy of Robatel SLPI and Eries, Genas, France.)

design has three to eight stages. The heavy and light phases, which are introduced near the bottom and top, respectively, flow countercurrently and leave the extractor at opposite ends. Other arrangements have been used, however.

4.3. Single-Stage Extractors

4.3.1. The Robatel BXP Extractor [18]

The series BXP belong to the single stage designs that are used in cascades. A common outer casing contains a battery of units in series with motors mounted on top of each unit (Fig. 8). Turbine mixers at the bottom of the units mix the two phases. The mixture is separated in the rotating bowls and the two phases sent to the respective units by way of overflow weirs which also control the interfaces. The light phase entering the last unit is mixed with the heavy phase overflowing from the previous unit. The heavy phase leaves the cascade, while the light phase continues its flow until finally discharged from the first unit, where the heavy phase is introduced.

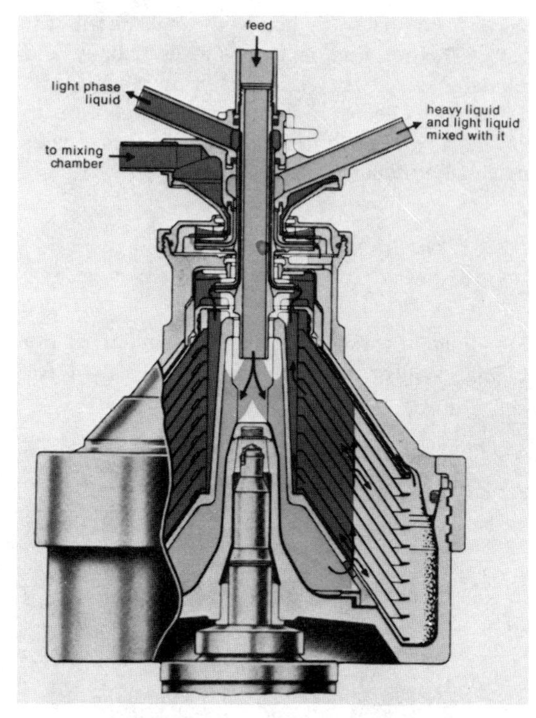

Figure 9. Westfalia TA centrifugal extractor. (Courtesy of Westfalia-Separator, Centrico, Inc., New Jersey [23].)

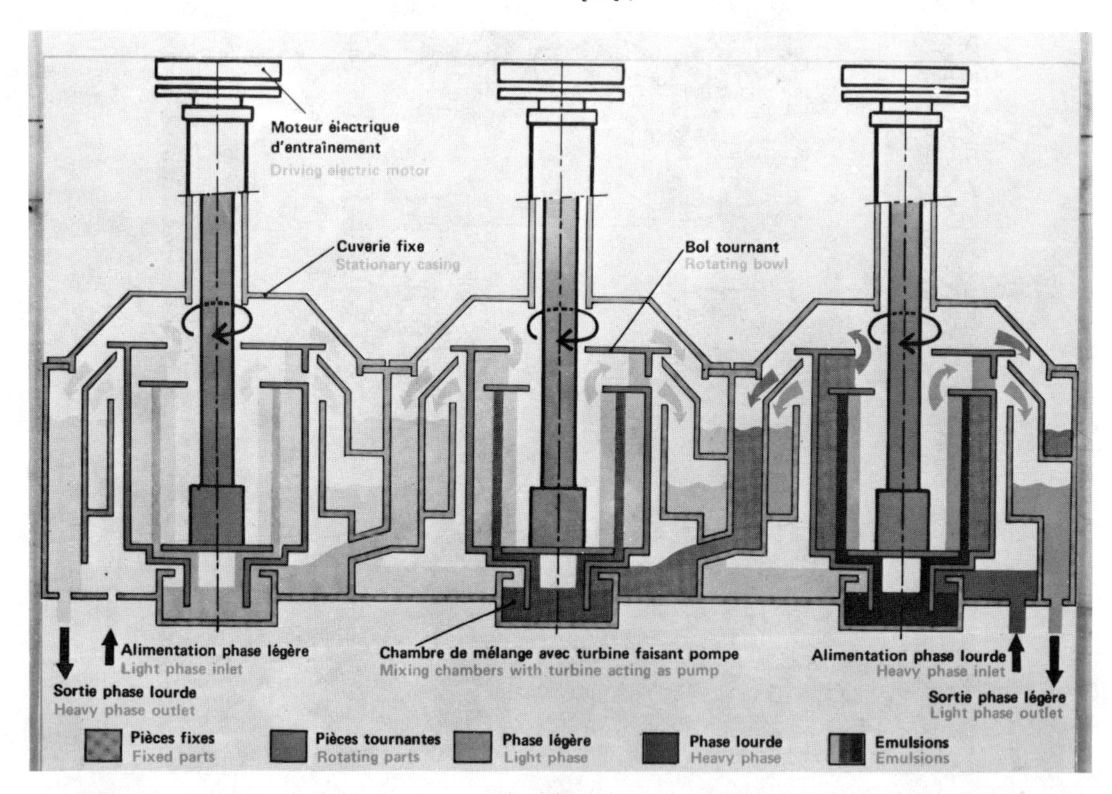

Figure 8. Robatel centrifugal extractors series BXP. (Courtesy of Robatel SLPI and Eries, Genas, France.)

4.3.2. *The Westfalia TA Extractor* [23]

Beside the Luwesta, Westfalia has a single-stage extractor used in cascades (Fig. 9). Like the Luwesta, the TA is a disk-type centrifuge with a stationary bowl and rotating shaft directly mounted on the driving motor. Both phases enter and leave the extractor at the top. However, this design differs from the Robatel in that phases are mixed at the top of any unit and separated in the previous one. Thus in a countercurrent cascade a mixture entering at the top of unit "i" is separated into a light phase that moves on to the top of unit "i + 2". The heavy phase, on the other hand, is mixed on its way out with the light phase from unit "i - 2" by a centrifugal pump at the top of the unit "i" and fed to unit "i - 1". An additional mixing pump is thus required to mix the fresh heavy phase with the spent light phase before entering the last unit.

4.3.3. *The SRL and ANL Extractors* [19-22]

The SRL is a single stage extractor, similar to the BXP design, consisting of a cylindrical bowl suspended inside a stationary casing (Fig. 10). The driving motor is mounted at the top and its shaft extends down the bowl and ends with a four-blade mixing paddle. The two phases entering at the bottom are mixed by the paddle in the mixing zone, which is equipped with dispersion and deflection baffles. In the settling zone, which is provided with radial vanes, the two phases separate and discharge over circular weirs. Air pressure is applied on the light-phase outlet to control the flow and liquid height over weirs.

The ANL extractor (Fig. 11) is a modification of the SRL to ensure safe operation in liquid metal fast-breader reactors. To this end the mixing paddle, the motor shaft, and the radial

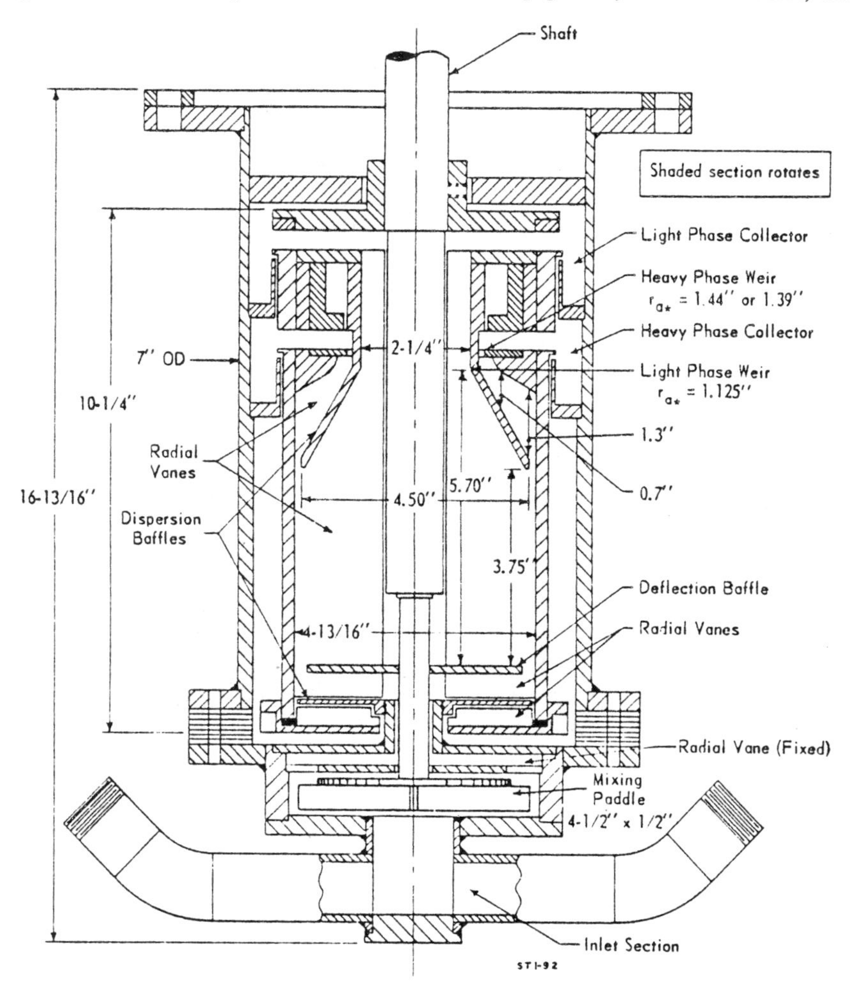

Figure 10. Savannah River Laboratories SRL centrifugal extractor. (Courtesy of E. I. DuPont de Nemours & Company, AED, SRL, South Carolina, [20].)

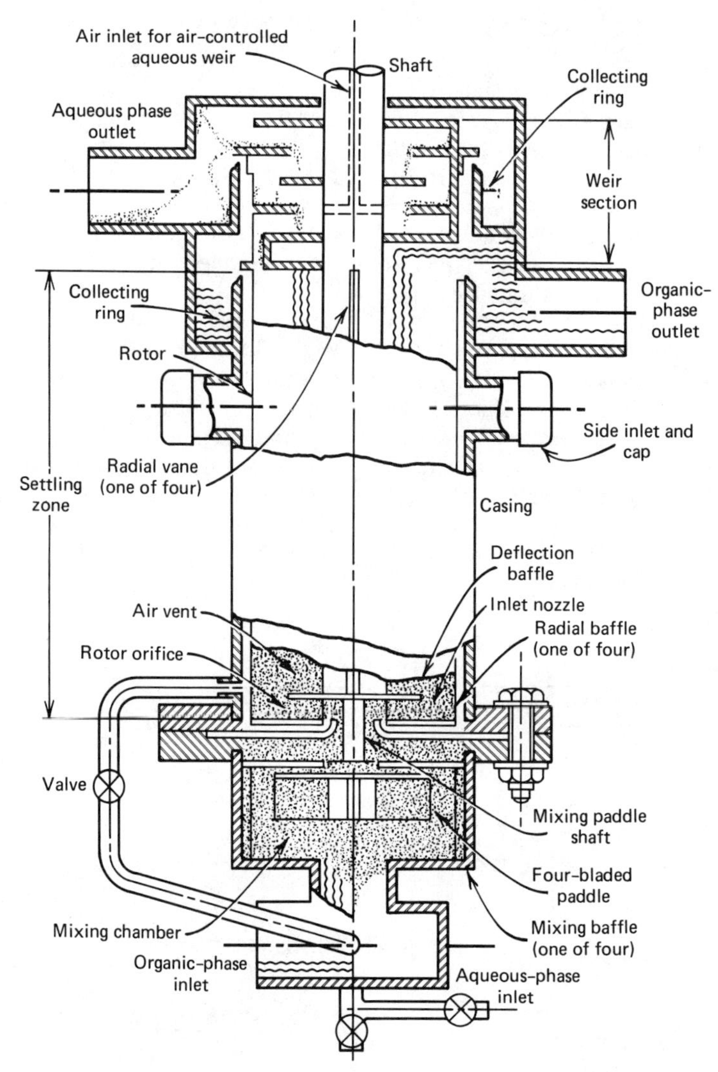

Figure 11. Argonne National Laboratories annular centrifugal extractor ANL. (Courtesy of Argonne National Laboratories, Illinois [21].)

vanes are all removed. Instead, the two phases are tangentially entered at opposite ends of the stationary casing and are mixed in the annular space as they descend by the shearing forces. The mixture enters at the bottom through an annular orifice to be separated and discharged over the weirs. The rotating bowl is not supported at the bottom to minimize corrosion problems; hence, depending on the size and weight of the bowl, there is a critical speed not to be exceeded to ensure balanced operation.

4.3.4. The MEAB SMCS-10 Extractor [58, 59]

This is a typical arrangement of the MEAB units made up of a static mixer and an H centrifuge, both comprising a single-stage extractor. In this arrangement [59] the two phases enter a static mixer directly mounted on top of the centrifuge bowl, which, in turn, is mounted over the motor on ball bearings. The mixture enters centrally into the centrifuge and is accelerated in an inner chamber and forced into a multichamber separation zone. Separated phases are discharged by centrifugal pumps.

4.4. Comparison Between Various Designs

In summary, there is a large number of centrifugal extractor designs to meet the needs of a wide range of processes. The operating conditions of these designs also vary over a wide range, as shown in Table 3. Data reported for the SRL,

TABLE 3 OPERATING CONDITIONS OF CENTRIFUGAL EXTRACTORS[a]

Extractor	Model	Volume, m^3	Capacity, m^3/hr	rpm	Motor Mounting	Motor Power, kW	Diameter, m
Podbielniak	E 48	0.925	113.5	1,600	Side	24	1.2
Quadronic	Hiatchi 4848	0.9	72	1,500	Side	55	1.2
α-Laval	ABE 216	0.07	21	6,000	Top	30	
UPV			6	1,400	Bottom	14	
Luwesta	EG 10006		5	4,500	Bottom		
Robatel SGN	LX6 70NL	0.072	3.5	1,600	Top, side		1.3
Robatel BXP	BXP 800	0.220	50	1,000	Top	15	0.8
Westfalia	TA 15007	0.028	30	3,500	Top	63	0.7
SRL/ANL		0.003	0.05	3,500	Top		0.1
MEAB	SMCS-10	0.00012	0.3	22,000	Bottom		

[a]Operating pressures are in the range 300–1750 kPa; operating temperatures cover a very wide range; operating flow ratios cover the range $\frac{10}{1}$ - $\frac{1}{10}$ easily.

ANL, UPV, and SMCS-10 are for prototypes. No data are available on large sizes. All vertical extractors except the SGN and the UPV operate at high speeds (3000–22,000 rpm); the horizontal extractors run at a maximum of about 1600 rpm.

Vertical extractors, with the exception of the α-Laval are not sealed. Horizontal designs are sealed by necessity. The SRL, ANL, and BXP have suspended rotors and top mounted motors that can be completely shielded from the extractor. Only the Luwesta and the TA designs have stationary bowls and rotating shafts.

Single-stage designs are used in batteries to achieve the required separation. Depending on the process and the number of stages needed, batteries may or may not be superior to continuous extractors.

5. HYDRODYNAMICS

The study of centrifugal extractors hydrodynamics is complicated by lack of visibility and large and changing pressure fields. Except for the UPV [44, 45], which was especially manufactured from transparent plastic, and the POD, little information is available.

Four hydrodynamic parameters are considered—pressures, holdup, flooding, and backmixing. The first parameter is important only in centrifugal extractors.

5.1. Pressures and Controls

The pressure at any point A in a centrifugal force field is given by

$$P_A = P_r + p_A \frac{\omega^2 R_A^2}{2} \tag{1}$$

where the reference point r is usually at the axis of rotation, R_A is the distance between points A and r, and p_A is the density of the liquid between the two points. This equation holds for any design, irrespective of which phase is dispersed.

In horizontal pressurized extractors (Figs. 12*a* and 12*b*) setting of the inlet pressures of the two phases and the outlet pressure of the

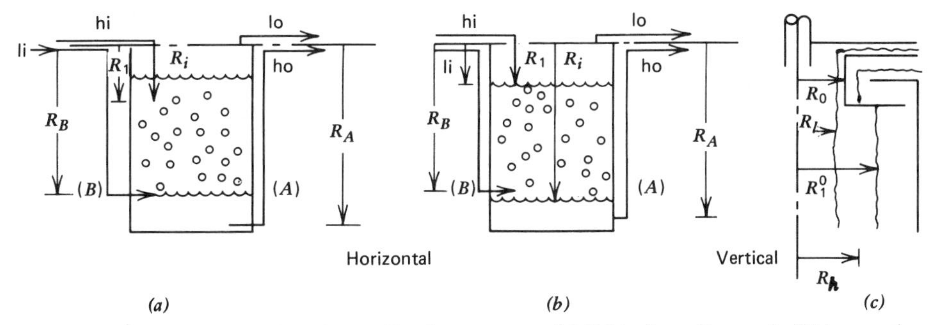

Figure 12. Operational modes of centrifugal extractors: (*a*) light phase dispersed; (*b*) heavy phase dispersed; (*c*) SRL single stage (ho = heavy phase out; hi = heavy phase in; lo = light phase out; li = light phase in).

light phase defines the interface and operation of the extractor. The Appendix is a summary of the governing equations that have been studied by Todd [46], Todd and Davies [47], and Barson and Beyer [48].

Vertical pressurized extractors are described by similar equations that predict pressures necessary to maintain the interface within each compartment.

Equation (1) can also be used for single-stage extractors to define interfaces in the settling zone. Webster et al. [49] included the effect of liquid rise over the discharge weirs and allowed for the air pressure acting on the light-phase outlet in SRL designs (Fig. 12*c*).

The pressures in a typical extractor are adjusted by four control valves V_j. A control model comprising the valve equations, pressure relations, mass balance, and interface displacement [Eqs. (A-1–A-4)] was linearized and solved by Boland and Van der Kaa [50]. They show that this model adequately describes the extractor operation and that the transient variations of the interface R_i are proportional to the reciprocal of its reference point $\overline{R}_i$. The latter point shows that for stable operation, it is desirable to have large $\overline{R}_i$. On the other hand, this reduces the mixing volume and hence the extraction efficiency. An optimum value must be found. A typical control scheme proposed by Boland and Van der Kaa [50] is shown in Fig. 13.

5.2. Holdup

Holdup is defined as the volume fraction of the dispersed phase in the extractor emulsion zone (Fig. 12*a*).

In pressurized extractors holdup is directly proportional to the operating pressures. From the Appendix it can be shown that the following equation results for light-phase dispersed:

$$\phi = 1 - \frac{2\pi b (P_{li} - P_{lo})}{\omega^2 \Delta p V_m} \tag{2}$$

where b is the extractor width and V_m the emulsion volume. For heavy-phase dispersed, Eq. (2) holds, but V_m is slightly different. Jakobson and Beyer [51] measured the holdup in a POD by stopping the flows and displacing the heavy phase. They found the holdup to vary linearly with P_{lo} at constant P_{li} in agreement with Eq. (2).

In unpressurized extractors the general equation (light-phase dispersed):

$$\frac{U_h}{1-\phi} + \frac{U_l}{\phi} = U_R \tag{3}$$

can be used to calculate the holdup [52]. The value U_R is a function of the extractor geometry and system properties [44, 52]:

$$\log\left[\left(\frac{p_l U_R^2}{\Delta p\, \omega^2 R^2 d_e}\right) \cdot \left(\frac{\Delta p\, \omega^2 R^3}{\sigma}\right)^{\alpha} \cdot \left(\frac{h}{R}\right)^{\beta} \cdot \left(\frac{p_l U_l d_e}{\mu_l}\right)^{\zeta} \cdot (\cos \Theta)^{\epsilon}\right] = A - B \cdot \left(\frac{Q_h}{Q_l}\right)^{a} \cdot \left(\frac{p_l}{p_h}\right)^{b} \tag{4}$$

in which only α, β, ζ, a, and b are independent of the internal design whereas ϵ, A, and B are. The

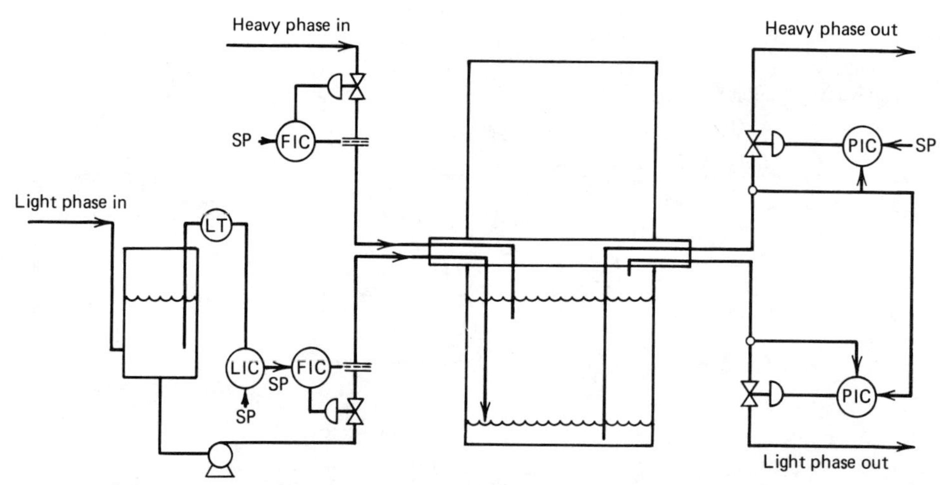

Figure 13. Pressures and control scheme of a pressurized centrifugal extractor.

drop size d_e in Eq. (4) is given in terms of the forming orifice diameter d_0 by [53]

$$d_e = 2.3 d_0 \left(\frac{\sigma}{\Delta p\, \omega^2 r d_0^2} \right)^{0.37} \tag{5}$$

where r is the radius at which the drop is formed. Ponikarov et al. [53] measured the holdup and holdup profile in a UPV by manometric and photographic methods and determined the constants in Eqs. (4) and (5) for different cases. They also found that the holdup passes through a shallow maximum near the middle of the extractor.

5.3. Limiting Capacity "Flooding"

Three types of flooding are possible in centrifugal extractors: (1) light-phase (shaft) flooding, which occurs when the heavy phase is carried with the light phase; (2) heavy-phase (rim) flooding, which occurs when the light phase is carried with the heavy phase; and (3) capacity flooding, which occurs within the extractor by phase inversion and results in contamination of both leaving phases.

In pressurized extractors the first two types are caused by very small or very large light-phase outlet pressure, respectively (Fig. 12*a*). Over the wide range of pressures within which the interface is maintained in the extractor, only capacity flooding occurs and is independent of the outlet light-phase pressure [54].

In the UPV design [52, 53] the first two types of flooding occur at extremely small or extremely large flow ratios. Capacity flooding occurs at the heavy-phase inlet nozzles by phase inversion.

In most designs [19, 20, 44, 45, 47, 54] the total flow rate at capacity increases with diameter and speed of extractor. This is unlike mechanically agitated extractors, where capacity decreases with increasing agitation.

5.4. Backmixing

Backmixing in centrifugal extractors, unlike mechanically agitated ones, decreases with increasing speed of rotation and flow rates. This is due to the fact that centrifugal mixing, as compared wtih pulse or reciprocating mixing, is unidirectional.

Because of the nature of flow in centrifugal extractors (size of stages not necessarily constant; driving forces, velocities and flow fields different at different points), it is not possible to use a Peclet number. Only an overall backflow coefficient can be measured [44, 51]. Todd and Davies [54] measured single-phase backmixing by use of a tracer technique. At a constant speed the backflow coefficient rapidly decreases as the flow rate increases. Similar but slower response was also observed in two-phase flow. At increasing speeds [55] the backflow coefficient decreases, but only at constant percentage of the flooding capacity, which is to be expected.

Backmixing is of no concern in externally staged designs.

5.5. Summary of Hydrodynamics

In summary, very little has been done on centrifugal extractor dynamics because of experimental difficulties. However, a reasonable understanding of the pressures and controls is available. The estimation of holdup from pressure equations or empirically is possible, and holdup profiles at least in one design are similar to other extractors. The capacity and backmixing dependence on the operating variables is also available, and the difference between centrifugal and other designs is recognized. However, need still exists for more studies over a wider range of variables and designs, especially as the importance of centrifugal extractors in process industries continues to increase.

6. PERFORMANCE

Performance of centrifugal extractors depends for a given system on the speed of rotation ω, the total flow rate Q_t, flow ratio R, and holdup ϕ.

Some data are available from small-size extractors, but very little from large units. Once installed these units are run to meet production requirements and their performance over a range of variables is seldom evaluated.

Typical examples of performance data are listed in Table 4, which show that:

1. Stage efficiency of single stage extractors is higher than other designs, but more than one stage are necessary to complete the extraction.
2. Performance largely depends on the system, as can be seen from comparing the "POD" for lube oil versus penicillin extraction. The extractor is twice as efficient for the first system.

TABLE 4 PERFORMANCE OF CENTRIFUGAL EXTRACTORS

Extractor	System	Physical Properties			Operating Variables				Number of Theoretical Stages	Efficiency, %[b]	Reference
		p_h/p_l	h/l		rpm	$R = Q_h/Q_l$	Q_t, m^3/hr	Flooding, %			
Podbielniak											
B-10	Kerosene–NBA[a]–water	998/801	1/1	24	3000	0.5	5.1	73	6–6.5	24	54
D-18	Kerosene–NBA–water	998/801	1/1	24	2000	0.5	11.1	58	5–5.5	15	54
A-1	Oil–aromatics–phenol[b]	1010/877	2/5	3	5000	3.5	0.01–0.02	33–66	5–7.7	28–43	28
9000	Broth–penicillin B–pentacetate				2900	4.4	7.5		1.8	10.7	24
					2900	3.4	7.5		2.04		24
					2900	2.4	7.5		2.21	14.0	24
9500	Some system				2900	3.5	7.5		2.04		25
					2700	3.5	7.5		2.19		25
					2500	3.5	7.5		2.30		25
					2300	3.5	7.5		2.36	15.0	25
	Oil–aromatics–furfural				2000	4.0	12.0	90	3–6	24	29
A-1	IAA[c]–boric acid–water	998/810	1/5	6	5000	1–0.3	0.01–0.03	44–95	3.5–7.7	20–45	48, 56
					3000	1.0	0.01	44	2.3	13.5	48, 56
					4075	1.0	0.01	44	2.8	16.5	48
					4600	1.0	0.01	44	2.96	17.5	48
UPV	Oil–aromatics–phenol[b]	1010/910	2/8	2	1400	0.8–1.2	6	75	2–5.8	33–92	45
Robatel SGN											
LX-168N	Uranyl nitrate–30% TBP				1500	1–0.2	2.1–4.5		7	87	26, 27
LX-324	Some system				3100	1.6	24–63		3.4–3.9	68–80	15
SRL single stage	Uranyl nitrate–Ultrasene				1790	0.5–1.5	6.4–12	33–96	0.92–0.99	92–99	19
ANL single stage	Uranyl nitrate–TBP/dodacane				3500	0.3–4	0.8–1.6	50	0.97–1	97–100	21

[a]Normal butyl amine.
[b]Containing 1.7–5% water.
[c]Isoamyl alcohol.
[d]Number of theoretical and actual stages.

3. Performance of centrifugal extractors is comparable to, if not better than, other extraction equipment [26, 28, 55].

APPENDIX. PRESSURES AND CONTROLS IN CENTRIFUGAL EXTRACTORS

A. The Pressure Equations

The relationship between the various operating pressures in a horizontal centrifugal extractor can be derived by using Fig. 12:

i–Light-phase dispersed (Fig. 12*a*):

$$P_B = P_{li} + \frac{\omega^2 R_B^2 p_l}{2}$$

$$= P_{lo} + \frac{\omega^2 R_i^2 p_h}{2} + \omega^2 (R_B^2 - R_i^2) p_m \quad \text{(A-1)}$$

$$P_A = P_{ho} + \frac{\omega^2 R_A^2 p_h}{2}$$

$$= P_{hi} + \omega^2 (R_B^2 - R_1^2) p_m + \omega^2 (R_A^2 - R_B^2) p_h \quad \text{(A-2)}$$

ii–Heavy-phase dispersed (Fig. 12*b*):

$$P_B = P_{li} + \frac{\omega^2 R_B^2 p_l}{2}$$

$$= P_{lo} + \omega^2 (R_B^2 - R_1^2) p_m + \omega^2 R_1^2 p_l \quad \text{(A-}\bar{1}\text{)}$$

$$P_A = P_{ho} + \frac{\omega^2 R_A^2 p_h}{2}$$

$$= P_{hi} + \omega^2 (R_i^2 - R_1^2) p_m + \omega^2 (R_A^2 - R_i^2) p_h \quad \text{(A-}\bar{2}\text{)}$$

B. Total Mass-Balance Equations

$$(Q_h + Q_l)_{\text{in}} = (Q_h + Q_l)_{\text{out}} = Q_t \quad \text{(A-3)}$$

C. Interface Displacement

$$\frac{d\bar{R}_i}{dt} = \frac{Q_{h\,\text{in}} - Q_{h\,\text{out}}}{2\pi R_i} = \frac{Q_{l\,\text{out}} - Q_{l\,\text{in}}}{2\pi R_i} \quad \text{(A-4)}$$

D. Pressure-Valve Equations

$$Q_j = k \cdot C_{vj} \cdot (V_j) \cdot \left(\frac{P_j}{p_j}\right)^{0.5} \quad \text{(A-5)}$$

REFERENCES

1. W. J. Podbielniak, H. R. Kaiser, and G. J. Ziegenhorn, *Chem. Eng. Progr.* **103,** 45 (1970).
2. D. B. Todd and G. R. Davies, *Filtration Separation* **1,** 1 (1973).
3. H. E. Zurcher, *Chem. Ind.* **21,** 683 (1976).
4. A. Placek, U.S. Patent 1,036,523 (1933).
5. W. J. Podbielniak, U.S. Patent 1,936,524 (1935).
6. C. Coutor, French Patent 769,254 (1934).
7. H. Eisenlhor, *Chem. Ing. Technik.* **23,** 12 (1951).
8. W. J. Podbielniak, U.S. Patent 2,004,001 (1935).
9. A. Placek, U.S. Patent 2,281,616 (1942).
10. N. J. Angelo, U.S. Patent 2,652,975 (1953).
11. W. J. Podbielniak, C. M. Doyle, and W. G. Podbielniak-Doyle, U.S. Patent 2,758,784 (1956).
12. C. M. Doyle, U.S. Patent 3,107,218 (1963).
13. C. M. Doyle, W. G. Podbielniak-Doyle, E. H. Rauch, and C. D. Lowry, *Chem. Eng. Progr.* **64,** 68 (1968).
14. α-ALFA-LAVAL, Report No. PC40044E.
15. M. Tarnero and J. Dollfus, *Genie Chim.* **99,** 1595 (1968).
16. J. Dollfus, paper presented at Solvent Extraction Conference, Imperial College, London, 1969.
17. G. Madany, U.S. Patent 3,133,880 (1964).
18. P. Miachon, *Tech. Mod.* **20,** 102 (1971).
19. A. T. Clark, Jr., Savannah River Laboratories, Aiken, SC, AEC R&D Report, DP371, 1962.
20. D. W. Webster, Savannah River Laboratories, Aiken, SC, AEC R&D Report, DP370, 1962.
21. G. J. Bernstein, D. E. Grossvenor, and J. F. Lane, *Nucl. Technol.* **20,** 200 (1973).
22. G. J. Bernstein, R. A. Leonard, A. A. Ziegler, and M. J. Steindler, paper to the 84th National Meeting of the AIChE, Atlanta, 1978.
23. Westfalia OEH 15007, Report No. 3592c/1067.
24. D. W. Anderson and E. F. Lau, *Chem. Eng. Progr.* **51,** 507 (1955).
25. A. Hall, *Chem. Process.* **40**(6), 78 (1977).
26. C. Bernard, P. Michel, and M. Tarnero, *Proceedings of International Solvent Extraction Conference* (ISEC), The Hague, Paper 102, 1971, p. 1282.
27. M. Tarnero, *BIST Commissariat á l'Energie Atomique* **184,** 35 (1973).
28. O. M. Fox, *Chem. Eng. Progr.* **42,** 133 (1963).
29. S. Ishigaki and H. R. Kaiser, *Petro/Chem. Eng.* **35**(5), 194 (1963).
30. D. B. Todd and F. C. Rao, *Hydrocarbon Process.* **46**(8), 115 (1967).
31. G. J. Bernstein, D. E. Grosvenor, J. F. Lenc, and N. M. Levitz, USAEC Report ANL-7968, 1973.
32. A. A. Kishbaugh, Savannah River Laboratories, AEC R&D Report, DP841 (1963).
33. H. R. Kaiser, *Sewage Ind. Wastes* **27,** 311 (1955).

34. D. B. Todd and C. A. Hooper, *Chem. Eng. Progr.* **67,** 60 (1971).
35. D. B. Todd, R. W. De Cicco, and H. R. Kaiser, *Chem. Eng. Progr.* **61,** 74 (1965).
36. W. J. Beach, *Soap Chem. Spec.* **40,** 169 (1964).
37. Liquid Dynamics, Bulletin No. 15, 1968.
38. Baker Perkins, Inc., Report CM-470, 1968.
39. G. R. Davies, H. R. Kaiser, and D. B. Todd, 3rd Symposium on Hazardous Chemicals Handling and Disposal, Indianapolis, April 1972.
40. D. B. Todd and W. J. Podbielniak, *Chem. Eng. Progr.* **61,** 69 (1965).
41. W. J. Podbielniak, U.S. Patent 2,670,132 (1954).
42. C. M. Doyle, W. G. Podbielniak-Doyle, and E. H. Rauch, paper to the National Meeting of the AIChE, St. Louis, 1968.
43. V. G. Bochkarev and V. G. Maminov, *Kazan. Khim.-Tekhnol. Inst. Trudy* **53,** 55 (1974).
44. I. I. Ponikarov, V. V. Kafarov, and Yu. A. Dulatov, *Zh. Prikl. Khim.* **45,** 2001 (1972).
45. I. I. Ponikarov, V. V. Kafarov, and O. A. Tseitlin, *Zh. Prikl. Khim.* **45,** 560 (1972).
46. D. B. Todd, *Chem. Eng.* **79,** 152 (1972).
47. D. B. Todd and G. R. Davies, paper presented at the International Conference on Solvent Extraction (ISEC), Toronto, No. 8b, 1977.
48. N. Barson and G. H. Beyer, *Chem. Eng. Progr.* **49,** 243 (1953).
49. D. S. Webster, C. L. Williamson, and J. F. Ward, Savannah River Laboratories, Aiken, SC, AEC R&D Report, DP371, 1961.
50. J. J. Boland and J. M. Van der Kaa, Proceedings of 5th IFAC Congress, Paper 1.4, 1972.
51. F. M. Jakobson and G. H. Beyer, *AIChE J.* **2,** 283 (1956).
52. I. I. Ponkarov, V. V. Kafarov, and Yu. A. Dulatov, *Zh. Prikl. Khim.* **46,** 1041 (1973).
53. I. I. Ponikarov, V. V. Kafarov, and Yu. A. Dulatov, *Zhurnal Prikladnoi Khimii* **44,** 2748 (1971).
54. D. B. Todd and G. R. Davies, *Proceedings of International Solvent Extraction Conference* (ISEC), Vol. 3, Lyons, 1974, p. 2380.
55. D. B. Todd, *Chem. Eng. Progr.* **62,** 119 (1966).
56. C. R. Bartels and G. Kleinman, *Chem. Eng. Progr.* **45,** 589 (1949).
57. N. A. Evtyukhin et al., *Khim. Tekhnol. Topl. Masel* **3,** 30 (1975).
58. J. O. Liljenzin, J. Rydberg, and G. Skarnemark, paper presented at the Symposium on Separation Science and Technology for Energy Applications, Session IV, Gatlinburg, TN, November 1979.
59. "MEAB," METALEXTRAKTION AB, Technical Bulletin, 1979.

16

SELECTION, PILOT TESTING, AND SCALE-UP OF COMMERCIAL EXTRACTORS

H. R. C. Pratt
University of Melbourne
Australia

C. Hanson
University of Bradford
United Kingdom

1. SELECTION OF EXTRACTION EQUIPMENT

1.1. Extractor Selection Chart

The selection of a suitable contactor to accomplish a given duty involves complex decisions in the absence of prior experience with the same system on the pilot or full scale, and the probability of subjective judgment is correspondingly high. The difficulty of selection arises from the wide range of contactor types available for consideration and the large number of design variables involved; an early (1950) but useful survey is given by Morello and Poffenburger [1]. In an attempt to rationalize the selection procedure, Pratt [2] in 1954 presented an extractor selection chart in which numerical ratings were given for each contactor type against each design requirement. More recently, Oliver [3] has dis-

TABLE 1 EXTRACTOR SELECTION CHART

Design Requirements		Gravity-Separated Extractors (Category No.)																Centrifugally Separated			
		Continuous Contact					Discontinuous Contact							Mixer–Settlers				Continuous Contact		Mixer–Settler	
							Without Inter-stage Settling: Mechanical			With Settling											
		Nonmechanical			Mechanical					Non-mechanical	Mechanical			Horizontal		Vertical					
Reference	Description	A1 Spray Column	A2 Baffle Plate Column	A3 Packed Column	B1 Pulsed Packed Column	B2 Raining Bucket Contactor	C1 Rotary Agitated Columns	C2 Reciprocating Plate Column	C3 Pulsed Plate Column	D1 Perforated Plate Column	E1 Scheibel Column	E2 ARDC Column	E3 Rotary Film Contactor	F1 Pump-settler	F2 Agitated Mixer-Settlers	G1 Pump-Settler	G2 Agitated Mixer-Settlers	H1 Perforated Plate	H2 Film Flow Type (de Laval)	J1 LUWESTA	J2 ROBATEL
1	Total throughput																				
	<0.25 m^3/hr	3	3	3	3	3	3	3	3	3	3	3	1	0	1	0	1	3	1	0	0
	0.25–2.5 m^3/hr	3	3	3	3	3	3	3	3	3	3	3	3	1	3	1	3	3	3	1	1
	2.5–25 m^3/hr	3	3	3	3	3	3	3	3	3	3	3	3	3	3	3	3	3	3	3	3
	25–250 m^3/hr	3	1	3	3	1	3	1	1	3	1	1	1	3	3	3	1	0[a]	0[a]	0[a]	0[a]
	>250 m^3/hr	1	0	1	1	0	1	0	1	1	0	0	1	5	5	1	1	0[a]	0[a]	0[a]	0[a]
2	*NTS*																				
	$\leqslant 1.0$	5[b]	3	3	3	3	3	3	3	3	3	3	3	3	3	3	3	3	3	3	3
	1–5	1[c]; 0	3	3	3	3	3	3	3	3	3	3	3	3	3	3	3	3	3	3[d]	3
	5–10	0	3	3	3	3	1	3	3	3	3	3	1	3	3	3	3	1[a]	1	0[a]	3
	10–15	0	1	1	3	1	1	3	3	1	3	1	1	3	3	1	1	0[a]	0[a]	0[a]	0[a]
	>15	0	1	1	1	1	1	1	1	1	1	1	0	3	3	1	1	0[a]	0[a]	0[a]	0[a]
3	(a) Physical properties[g]																				
	$(\sigma/\Delta\rho g)^{1/2} > 0.60$	1	1	1	3	1	3	3	3	1	3	3	3	3	3	3	3	5	5	5	5
	(b) Density difference																				
	$0.05 \geqslant (\Delta\rho) \geqslant 0.03$ g/cm^3	3	3	3	0	3	0	0	0	1	0	1	3	1	1	1	1	5	5	5	5
	(c) Viscosity[c]																				
	μ_c and/or $\mu_d > 20$ cP	1	1	1	1	1	1	1	1	1	1	1	1	1	1	1	1	1	1	1	1

4	Slow heterogeneous reaction:																				
	$k_I < 4 \times 10^{-5}$ m/s	0	1	1	3	1	3	3	3	1	3	3	0	0	3	0	3	3	3	3	3
5	Slow homogeneous reaction:																				
	$t_{1/2}$ = 0.5–5 min	1	1	1	1	3	1	1	1	3	1	1	1	3	5	3	3	0	0	1	1
	> 5 min	0	0	0	0	1	0	0	0	1	0	0	0	3	5	3	3	0	0	0	0
6	Extreme phase ratio																				
	$F_d/F_c < 0.2$ or >5	1	1	1	1	3	1	1	1	3	1	1	3	1	5^e	1	5^e	3	3	3	3
7	Short residence time	0	0	0	1	0	1	1	1	0	0	0	0	0	0	0	0	5	5	3	3
8	Ability to handle solids																				
	Trace (<0–.1% in feed)	3	3	1	1	5	3	3	3	1	1	1	5	3	3	3	3	1	1	1	1
	Appreciable (0.1–1% in feed)	1	1	1	1	3	3	3	3	0	0	0	5	1	1	1	1	1	1	1	1
	Heavy (>1% in feed)	1	1	0	0	1	1	1	1	0	0	0	5	1^f	1^f	1^f	1^f	0	0	1	1
9	Tendency to emulsify																				
	Slight	3	3	1	1	3	1	1	1	3	1	1	3	1	1	1	1	5	5	5	5
	Marked	1	1	1	0	1	0	0	0	1	0	0	1	0	0	0	0	3	3	3	3
10	Limited space available																				
	Height	0	1	1	1	5	1	1	1	1	1	1	3	5	5	0	0	5	5	5	5
	Floor	5	5	5	5	0	5	5	5	5	5	5	0	0	0	5	5	5	5	5	5
11	Special materials required																				
	Metals (stainless steel, Ti, etc.)	5	3	5	5	3	3	3	3	3	3	3	3	3	3	3	3	5	5	5	5
	Nonmetals	5	3	5	1	1	0	0	1	1	0	0	1	1	5	1	1	0	0	0	0
12	Radioactivity present																				
	Weak (mainly α, β)	5	5	5	3	1	1	1	3	3	1	1	1	3	5	3	1	1	1	1	1
	Strong γ	5	5	5	3	0	0	0	3	3	0	0	0	3	5	1	0	0	0	0	0
13	Ease of cleaning	5	3	1	1	3	3	3	3	1	1	3	5	3	3	3	3	1	3	3	1
14	Low maintenance	5	5	5	3	3	3	3	3	5	3	3	3	3	3	3	3	1	1	1	1

[a]Multiple units in series and/or parallel can be used.
[b]For immeasurably fast homogeneous reaction.
[c]For diameters ⩽ 15 cm.
[d]Two or three stages only in single machine.
[e]With recirculation of separated phases to mixer.
[f]Requires provision for solids removal from settler.
[g]See text for effect of direction of transfer.

cussed the general criteria for the selection of contactors, and Hanson [4] has given a simple form of chart that provides broad guidance to selection (see also works by Laddha and Degaleesan [5] and Reissinger and Schroter [6]. A useful performance comparison of column contactors has recently been published [45].

In Table 1 a modified and updated version of Pratt's selection chart [2] is presented. Notes are given in the following sections on the various contactor types and design features considered, with an appropriate classification of the former; these are followed by an account of the method of use of the chart and by two worked examples. A summary of features and fields of industrial application of commercial extractors is also given by Lo [46] (Table 2), and a classification of commercial extractors is given in Fig. 1.

It must be emphasized that the contactor features given in the selection chart (Table 1) and the accompanying notes are based on the particular experience and opinions of the authors. It is possible that in some cases the ratings given for proprietary-type equipment may appear to conflict with the manufacturers' claims, in which case due consideration should be given to the latter in making a final assessment.

1.2. Types of Contactor

In the classification adopted in the selection chart (Table 1), extractors are first divided into gravity-separated and centrifugally separated types. Of these the former, which greatly predominate, are subdivided into continuous contact (i.e., differential) types, discontinuous contact types, and true mixer–settlers. These are further subdivided as described in the following paragraphs, using the same reference system as that given in the chart. An alternative classification of major types of commercial extractor has also been presented by Lo [46] as shown in Fig. 1.

1.2.1. Continuous-Contact Gravity-Separated Extractors

The extractors included under this heading consist of those in which the concentration gradients are differentially continuous throughout the con-

TABLE 2 SUMMARY OF FEATURES AND FIELDS OF INDUSTRIAL APPLICATION OF COMMERCIAL EXTRACTORS[a]

Types of Extractor	General Features	Fields of Industrial Application
Unagitated columns	Low capital cost, low operating and maintenance cost, simplicity in construction, handles corrosive material	Petrochemical, chemical
Mixer–settlers	High stage efficiency, handles wide solvent ratios, high capacity, good flexibility, reliable scale-up, handles liquids with high viscosity	Petrochemical, nuclear, fertilizer, metallurgical
Pulsed columns	Low HETS, no internal moving parts, many stages possible	Nuclear, petrochemical, metallurgical
Rotary-agitation columns	Reasonable capacity, reasonable HETS, many stages possible, reasonable construction cost, low operating and maintenance cost	Petrochemical, metallurgical, pharmaceutical, fertilizer
Reciprocating-plate columns	High throughput, low HETS, great versatility and flexibility, simplicity in construction, handles liquids containing suspended solids, handles mixtures with emulsifying tendencies	Pharmaceutical, petrochemical, metallurgical, chemical
Centrifugal extractors	Short contacting time for unstable material, limited space required, handles easily emulsified material, handles systems with little liquid density difference	Pharmaceutical, nuclear, petrochemical

[a]Reprinted by permission from T. C. Lo, Recent Developments in Commercial Extractors, Engineering Foundation Conference on Mixing Research, Rindge, NH, 1975.

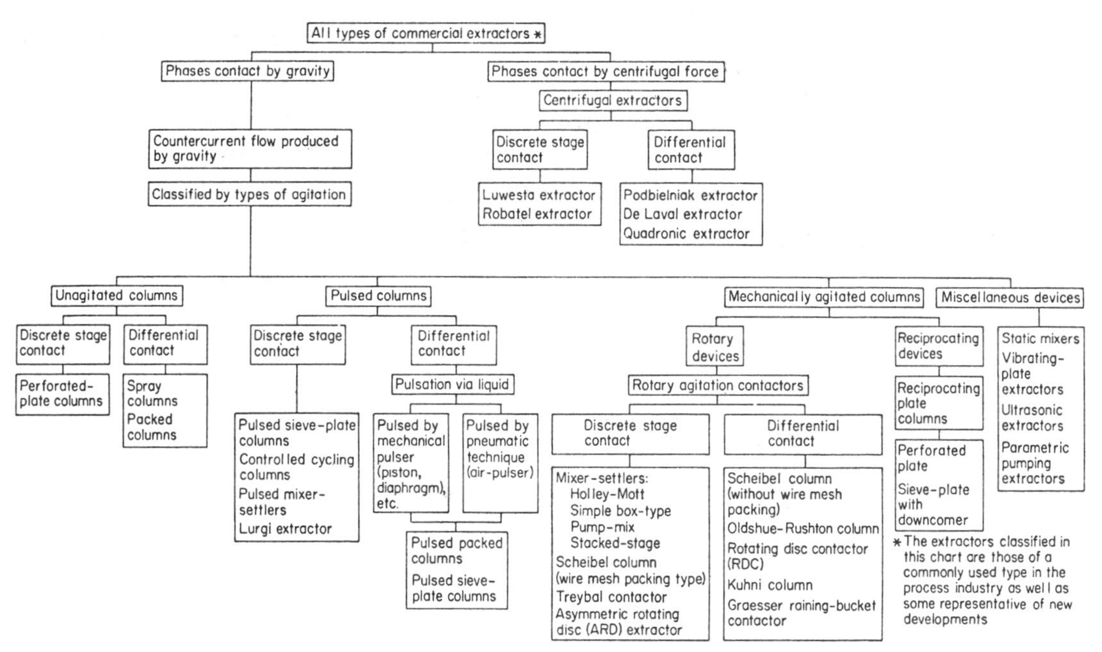

Figure 1. Classification of commercial extractors. (Reprinted by permission from T. C. Lo, "Recent Developments in Commercial Extractors," a paper presented at Engineering Foundation Conference on Mixing Research at Rindge, New Hampshire, 1975.)

tacting section. Both nonmechanical columns and mechanically agitated types are included.

(a) Nonmechanical Columns (Type A)

1. *Spray Column* (Chapter 10, Fig. 1). Although the spray column is the simplest of all contactor types, it suffers from the major disadvantage of a massive entrainment of continuous phase by the dispersed-phase droplets [7, 8]. As a result, in other than very small diameters suitable only for the laboratory, it behaves effectively as a cocurrent contactor with a recirculation of continuous phase in a narrow annulus around the wall. The effective number of theoretical stages under these circumstances cannot exceed unity irrespective of height; thus the use of these columns is restricted to cases where an immeasurably fast irreversible homogeneous reaction occurs in the extract phase, such as in the removal of acid from an organic liquid by scrubbing with aqueous alkali.

2–3. *Baffle-Plate and Packed Columns* (Chapter 10, Figs. 2 and 8). The circulatory flow of continuous phase that is inherent in spray columns can be greatly reduced by the provision of baffle plates to deflect the flow of dispersed phase from side to side or from center to wall and back or by the use of a packing, such as Raschig rings. By such means backmixing of continuous phase can be reduced to small proportions, especially with low dispersed-phase ratios, giving satisfactory multistage operation. With large diameters both types become less favorable, the former because of the need to increase the height to maintain geometric similarity and the latter, because of high packing costs.

(b) Mechanically Agitated Contactors (Type B)

1. *Pulsed Packed Column* (Chapter 11.1, Fig. 1). The performance of packed columns can be materially improved by pulsing of the continuous phase, although at the expense of a reduction in throughput. This results from a decrease in droplet size, with consequent increase in interfacial area, which more than compensates for the increased axial dispersion that is a consequence of the pulsation. A disadvantage with the use of ring packings is that, unless restrained, these tend to orientate with pulsing, forming "chimneys" that increase backmixing and reduce performance.

2. *Raining-Bucket Contactor.* This contactor, shown in Chapter 13.6, Fig. 1 is of horizon-

tal type, requiring little headroom, and is simple to maintain. It has been applied in cases where a moderate duty is required, especially with systems with a tendency to emulsify under moderate agitation.

1.2.2. Discontinuous-Contact Gravity-Separated Extractors

These contactors comprise mainly compartmental columns, mostly with some form of agitation; in some cases partial disengagement of the phases is promoted between compartments. The nonagitated perforated plate column with downcomers or risers for continuous phase, analogous to the gas–liquid bubble plate column, is somewhat arbitrarily included under this heading, although it can also be considered as an unagitated mixer–settler.

(a) Compartmental Contactors Without Interstage Settling (Type C)

1. *Rotary-Agitated Columns.* This group comprises the rotary disk (RDC; Chapter 13.1, Fig. 1), multiturbine (Oldshue–Rushton; Chapter 13.4, Fig. 2), and Kuhni columns (Chapter 13.5, Fig. 1), all of which give comparable performances. These contactors have all been used extensively for nonemulsifying systems in the petroleum and chemical industries. However, they suffer a deterioration in performance on scale-up to large diameters, as a result of increased backmixing and the need to increase compartment height in order to maintain a satisfactory mixing geometry.

2. *Reciprocating Plate Columns.* These [9, 10] employ perforated plates mounted on a vertical shaft that is subjected to a reciprocating motion (Chapter 12, Figs. 1 and 9). Early designs [9] used plates with 2–4-mm holes and 15–25% free area on a spacing of 3–10 cm, which were oscillated at 30–200 cycles/min with an amplitude of 5–30 cm. A more recent design [10] employs plates with about 16-mm holes and 60–65% free area on 25–50-mm spacing, oscillated at 100–500 cycles/min with 10–40-mm amplitude; such columns have been operated satisfactorily with diameters of up to 1.5 m. Both types suffer a deterioration in performance on scaling up the diameter as a result of increased backmixing and channeling, despite maintaining the plate spacing constant.

3. *Pulsed Plate Column.* This column [9] is similar to the first of the foregoing reciprocating-plate types, except that the continuous phase itself is oscillated by external means instead of the plates (Chapter 11.2, Table 1). As with the previous type, the plate spacing is maintained constant on scaling up the diameter, and although the performance deteriorates as a result of increased backmixing, this can be partially overcome by suitable baffling [11].

(b) Nonmechanical Compartmental Contactors with Interstage Settling (Type D)

1. *Perforated Plate Columns.* In simplest form (Chapter 10, Fig. 3) these contactors employ perforated plates with downcomers or risers for the continuous phase (depending on whether it is the heavier or lighter phase); more complex forms such as the Koch "Kaskade" plate have also been used [12]. A disadvantage is the lack of flexibility in operation as a result of the need to maintain a layer of coalesced dispersed phase below or above the plates.

(c) Mechanical Compartmental Contactors with Interstage Settling (Type E)

1. *Scheibel Column.* In this contactor, shown in Chapter 13.3, Figs. 3 and 4, alternate compartments are agitated by means of impellers on a central shaft, and the intermediate ones contain a packing of open-weave wire mesh. In a later design the impeller is enclosed in a shroud, permitting a reduction in the compartment height. The packed sections were originally intended to promote countercurrent settling between stages, although there is doubt as to the extent to which this occurs. However, they appear to lead to increased mass-transfer rates, although at the expense of reduced capacity; with systems of high interfacial tension, the packing is sometimes omitted.

2. *Asymmetric Rotary-Disk Column (ARDC).* This form of rotary-disk column [included in type C1, Section 1.2.2 (a)] employs an off-center rotor with stator baffles designed to promote partial coalescence between compartments [13] (Chapter 13.2, Fig. 1). By this means backmixing is reduced, with correspondingly less deterioration in performance on scaling up in diameter.

3. *Rotary Film Contactor.* This contactor comprises a series of compartments, each con-

taining a number of slowly rotating disks on a horizontal shaft [14]. The phases remain separated within compartments, and contacting is effected by bringing films of one phase on the disks into contact with the other phase. This contactor was developed for use with feedstocks of high solids content, such as ore leach suspensions.

1.2.3. Gravity-Separated Mixer-Settler Cascades

Contactors of this type offer considerable scope for mechanical variation, and the number of designs proposed is correspondingly large [1, 15]. However, for the present purpose, these have been reduced to four basic types, in each of which the mixed phases from each stage pass in cocurrent manner to the settlers and the separated phases pass thence in countercurrent flow to adjacent stages.

An important feature of some mixer–settlers is the ability to control the phase ratio in the mixing compartment independently of the external solvent ratio, to obtain a high stage efficiency. This is achieved by the provision of adjustable ports between mixer and settler, above and below the mixed phase port, to permit recirculation of one or both phases back to the mixer after separation.

No allowance is made in the selection chart (Table 1) for the use of special devices in the settlers, such as baffles or mesh packing, to promote coalescence.

(a) Horizontal Mixer–Settlers (Type F)

1. *Pump-Settler.* In this contactor a centrifugal pump is employed in each stage, both to mix the phases and to transfer the resulting emulsion to the settlers. The contact time is thus very short; nevertheless, stage efficiencies are usually fairly high, except at low dispersed-phase ratios. Various types of nonmechanical in-line mixer are also available [15, 16], but these entail the use of pumps to impart the flows and may be expected to have similar characteristics.

2. *Agitated Mixer–Settlers.* Separate mixing chambers provided with rotary or reciprocating agitators, or even air agitation, are used in this class of extractor. Such chambers are often provided within the individual mixer-settler stages, with the use of suitable partitions that carry the mixed phase and recirculation ports. Many such stages are often arranged compactly in single box contactors by suitable longitudinal partitioning with transverse subpartitions at alternate ends to form the mixing chambers (Chapter 9.1, Fig. 1). Pump–mix agitators are often used in large-scale installations (see Chapters 9.1-9.5).

(b) Vertical Mixer–Settlers (Type G)

Vertical mixer–settlers offer a useful saving in floor area, although they are less suitable for the very largest throughputs. The simplest form is again the pump–settler, which consists of a series of stacked settling chambers with external pumps (Chapter 9.5). A variety of types that use rotary or reciprocating agitators have also been described [1, 17].

1.2.4. Centrifugally Separated Extractors

Contactors that use centrifugal force to effect the phase separation are characterized by extreme compactness, high throughput, and very low residence time and hence are particularly suitable for handling labile materials, such as antibiotic broths. Such contactors fall naturally into two classes: differential types and mixer–settlers.

(a) Differential Contactors (Type H)

1. *Perforated-Plate Types.* The Podbielniak extractor, the earliest centrifugal type, comprises a series of perforated concentric cylinders rotated about a horizontal axis, with rotary seals through which the two phases are introduced and withdrawn [18, 19] (Chapter 15, Fig. 2). A contactor of somewhat similar type is the *Quadronic* [20, 22], in which fixed or adjustable orifices can be inserted to give the desired characteristics.

2. *Film Flow Type.* The de Laval extractor, an underdriven vertical shaft machine, contains a number of concentric cylinders with helical ribs leading to transfer ports located alternately at top and bottom [19-21]. By this means a very long (≤26-m) contact path is provided for the two phases, and chambers are located at the phase outlets to ensure complete separation (Chapter 15, Fig. 4).

(b) Centrifugal Mixer-Settlers (Type J)

1. *Luwesta Extractor* (Chapter 15, Fig. 6). This contactor, the earliest of its class, is an adaption of the underdriven centrifugal cream separa-

tor with two or three stages mounted vertically in a single assembly [19, 23]. Each stage comprises a spray disk mixer and centrifugal separator; the ports and channels for transfer and conveyance of the flows are carried in a stationary central tube.

2. *Robatel Extractor* (Chapter 15, Figs. 7 and 8). This contactor, marketed by Saint-Gobain Techniques Nouvelles, is a more recent type that is available with up to 12 stages [20]. Residence times are quoted as being 5–10 s per stage.

1.3. Design Requirements of Contactors

In the extractor selection chart shown in Table 1 the foregoing contactors are listed across the top and the possible design requirements, down the left-hand margin. Numerical ratings are allocated to each contactor against each design requirement according to the following scheme:

0—unsuitable

1—possibly suitable, but subject to further verification (see Section 1.4)

3—satisfactory

5—outstanding

The bases on which the ratings have been allocated are summarized as follows:

1. *Total Throughput.* The minimum *total* throughput considered, of 0.25 m^3/hr (ca. 50 gal/hr) of feed plus extract phase, corresponds to a column of about 10 cm in diameter for typical systems, that is, that of large laboratory or small pilot-scale units. Capacities of 2.5–250 m^3/hr embrace the majority of industrial processes, as used in the petroleum and chemical industries. Still larger contactors (capacities >250 m^3/hr) have been used particularly in the recovery of copper from dilute mine liquors.

2. *Number of Theoretical Stages.* No allowance is made under this heading for the increase in HETS or HTU with cross section experienced with some types of contactor, since this is taken into account under the previous heading. The heights of most types of vertical contactor tend to become excessive with more than 10–15 stages, necessitating "doubtful" ratings. This limitation does not always arise with horizontal contactors, particularly mixer–settlers, but the increase in floor area occupied is taken into account later.

3. *Physical Properties of System.* For packed columns, the Sauter mean droplet diameter d_{32} at low flow rates is given by [24]

$$d_{32} = 0.92 \left(\frac{\sigma}{\Delta\rho g}\right)^{1/2} \qquad (1)$$

The parameter $(\sigma/\Delta\rho g)^{1/2}$ is probably a controlling factor for all types of nonmechanical extractor; therefore, ratings are given in the selection chart (Table 1) for values above 0.60 cm on the basis that a large drop size will lead to a reduced interfacial area and hence a lower performance. With mechanically agitated contactors, the effect of a large value of this parameter can be overcome by an increase in power input. Contactor performance is often materially influenced by effects arising from interfacial tension gradients accompanying solute transfer, particularly from organic into aqueous phase (Marangoni effect; see Chapter 3, Section 3.3.2); thus special consideration must be given to this possibility. Account is also taken in the selection chart (Table 1) of the effects of density difference $\Delta\rho$ and viscosity. The former is significant in that phase separation can be slow and interface control difficult with some contactors in the range $0.05 \geqslant \Delta\rho \geqslant 0.03$; values below 0.03 can seldom be handled except with centrifugal contactors. A high viscosity of one or both phases (>20 cP) is likely to have a marked effect on performance and phase separation. No simple rule can be given for predicting this; hence each case must be given special consideration.

4. *Slow Heterogeneous Reaction.* The effect of slow interfacial reaction is always to reduce the rate of mass transfer and hence to increase the height of column required, or to reduce the stage efficiency. The effect can be minimized by increasing the interfacial area, that is, by an increased agitation rate, at the expense of a reduced throughput. The ratings given in the selection chart (Table 1) relate to an effective first-order interfacial reaction rate constant k_I of below about 4×10^{-5} m s^{-1} (0.5 ft/hr).

5. *Slow Homogeneous Reaction.* This case, which is relatively rare in liquid extraction, is treated in Chapters 2.1 and 2.2. Slow reactions occur mainly in the bulk of the extract phase, with very little in the boundary layer or "film"; with very slow reactions, unreacted solute may still be present in the final extract. Allowance is made in the chart by relating the residence time

of the contactor to the half-time $t_{1/2}$ of the reaction. Since reactions are normally of higher order than first, a mean value of the *pseudo-half-time* is calculated for the contactor, that is, averaged to allow for the change in reagent concentration. For an nth order reaction in which one solute molecule reacts with $(n - 1)$ reagent molecules, this is given by

$$t_{1/2} = \frac{0.693}{k_n B^{n-1}} \tag{2}$$

where k_n is the nth order velocity constant and B the reagent concentration.

6. *Extreme Phase Ratios.* The effect of a low value of F_d/F_c is to reduce the dispersed-phase holdup and hence the interfacial area; a high value may cause large backmixing, and in both cases the performance suffers. An important exception is that of mixer–settlers in which provision is made for recirculation of the separated phases back to the mixer. Special consideration must be given in cases where large changes in flow ratio occur, such as that resulting from solute transfer.

7. *Short Residence Time.* It is occasionally necessary, particularly when handling labile materials such as penicillin, to minimize the contact time in order to avoid loss of valuable product. Centrifugal contactors are the most favorable in such cases, and other types are assessed according to the approximate residence time per theoretical stage.

8. *Ability to Handle Solids.* Some contactor types are liable to blockage and require dismantling for cleaning; notable exceptions are the pulsed plate and rotary-agitated columns and the raining-bucket contactor, which can usually accept appreciable amounts of solids. Special provision is made for solids removal in the Luwesta centrifugal extractor. The only types that can handle large amounts of solids satisfactorily are the rotary film and, to a lesser extent, the raining-bucket contactors [Sections 1.2.2(c) type E3 and Section 1.2.1(b) type B2 respectively].

9. *Tendency to Emulsify.* Nonmechanical columns and the rotating bucket and rotary film mechanical contactors will often handle feeds containing traces of emulsifying agent satisfactorily, although pilot-scale tests are essential for new systems, especially of biological origin. Mechanical contactors, with the notable exception of the centrifugal types, on the other hand, produce such a fine dispersion that phase separation is usually unsatisfactory. An attempt is made in the selection chart (Table 1) to differentiate between feeds with a slight foaming tendency and those with a marked tendency to emulsify.

10. *Limited Space Available.* Separate entries are given under this heading for limitations on height and floor area required. A high rating under both headings infers overall compactness, as with the centrifugal contactors. The entries assume that at least five theoretical stages are required.

11. *Special Materials Required.* Under corrosive or other particular circumstances special materials of construction are required, and two cases are considered: special metals or alloys (stainless steel, titanium, etc.) and nonmetals (e.g., rubber, glass, or enamel lining, impervious graphite, or plastics). In general, the former are applicable to all contactor types, but the simpler and/or more compact designs are to be preferred on cost grounds. On the other hand, the nonmetals are generally unsuitable for mechanically agitated contactors unless the shafts can be carried in external bearings, as with horizontal mixer–settlers; pulsed columns are also applicable provided a suitable pulsing mechanism can be devised.

12. *Radioactivity Present.* Most contactor types can be used when only weak radioactivity (mainly α and β) is present, although the more complex mechanical types are less suitable if frequent cleaning is required. With very strong γ radiation present, particularly as in nuclear fuel reprocessing, the requirement is for long maintenance-free operation behind heavy shielding. Experience here has shown that nonmechanical columns and mechanically agitated contactors with external drives (i.e., pulsed columns and horizontal mixer–settlers) are the most suitable types.

13. *Ease of Cleaning.* It is assumed that cleaning is required for solids removal rather than for changing of feedstock, which requires only simple washing out.

14. *Low Maintenance.* This heading refers to mechanical maintenance associated with shaft drives and seals, pulsing mechanisms, and so forth.

1.4. Use of Selection Chart

To use the selection chart, it is first necessary to obtain the equilibrium data for the given system, to select a suitable solvent ratio and to calculate the total throughput and number of (plug flow) theoretical stages required. The various design requirements to be taken into account are then listed; these will always include Nos. 1 and 2, together with 3 if applicable.

The next step is to reject any contactors for which a zero appears against any of the design requirements. The remaining contactors are then listed, together with any ratings of 1 or 5. In comparing these contactors, particular weight should be given to ratings of 5; on the other hand, a rating of 1 does not necessarily infer that a contactor is of doubtful or marginal suitability, but that a further detailed study is required, such as one involving laboratory or pilot-scale experimentation, or a detailed economic assessment.

Example 1. It is desired to recover acetic acid from a 20% aqueous solution by extraction with ethyl acetate at a dispersed-phase ratio of 1.25. The contactor is to be located inside a building of limited floor area, but of sufficient height to accommodate a 40-plate column for distillation of the extract.

The design requirements are as follows:

1. Total throughput: 20 m^3/hr
2. NTS: 8
3. Physical properties:
   $(\sigma/\Delta\rho g)^{1/2} = 0.27$ cm†
   $\Delta\rho = 0.11$ g/cm^3†
   μ_c and $\mu_d \ll 20$ cP†

8. Trace of fibrous solids present
10. Limited floor area available.
11. Materials of construction–stainless steel
13. Ease of cleaning
14. Low maintenance

†Not relevant (outside chart range).

Referring to Table 1, types A1, B2, E3, F1, F2 and J1 are unsuitable because of entries of zero against requirements 2 or 10. In addition, the centrifugal types are likely to require more extensive maintenance and will be disregarded. The ratings of the remaining types are as follows (√ = rating of 3, i.e., satisfactory):

Type Ref.	Type of Contactor	Design Requirement 8	11	13	14
A2	Baffle plate	√	√	√	5
A3	Packed	√	5	1	5
B1	Pulsed packed	√	5	1	√
C1–3	Rotary, reciprocating, pulsed plate	√	√	√	√
D1	Perforated plate	√	√	1	5
E1	Scheibel	√	√	1	√
E2	ARDC	√	√	√	√
G1	Pump–settler (vertical)	√	√	√	√
G2	Mixer–settler (vertical)	√	√	√	√

An inspection of this tabular list indicates that the baffle plate column (type A2) followed closely by the mechanically agitated columns (types C1–3 and E2) are likely to prove the most suitable. The packed column (type A3) and perforated-plate column (type D1) are also suitable, apart from some doubt regarding cleaning of deposited solids.

The baffle-plate column has, in fact, been extensively employed in the past for this duty.

Example 2. A contactor is required for the recovery of copper from 450 m^3/hr of mine liquor containing 5.0 kg/m^3 of Cu^{2+} by extraction with an oxime-type reagent dissolved in kerosene. The copper will be recovered as an aqueous solution by acid stripping of the extract in a second contactor.

The design requirements are summarized as follows:

1. Total throughput:
   1000 m^3/hr for extraction; 800 m^3/hr for stripping
2. NTS: 3.5 for extraction; 2 for stripping

3. Physical properties:
   $(\sigma/\Delta\rho g)^{1/2} = 0.25$ cm†
   $\Delta\rho = 0.1$ g/cm^3†
   $\mu_c, \mu_d < 20$ cP†
4. Heterogeneous reaction: $k_I = 10^{-5}$ m/s
6. Dispersed-phase ratio: 1.5 for extraction; 5.0 for stripping
8. Solids present: ca. 0.1%
13. Ease of cleaning
14. Low maintenance

Inspection of the selection chart (Table 1) shows that the ratings for requirements 1, 2, and 4 are as follows:

Type Ref.	Type of Contactor	Design Requirement 1	2	4
C3	Pulsed plate column	1	√	√
D1	Perforated-plate column	1	√	1
E3	Rotary film	1	√	0
F1	Horizontal pump–settler	5	√	0
F2	Horizontal mixer–settler	5	√	√

It is clear that only type F2, the horizontal mixer–settler, can meet all these requirements with certainty; further reference to the chart (Table 1) shows that it can also meet the remaining requirements, although recirculation of the separated aqueous phase would be desirable in the stripper in view of the small aqueous : organic-phase ratio.

The only other contactor that could possibly be considered would be the pulsed plate column, but no experience is available with the very large diameters required, and multiple columns in parallel might be required.

2. PILOT TESTING AND SCALE-UP: MIXER-SETTLERS

2.1. General

This section considers contactors in categories F and G in the selection chart (Table 1). The design of such contactors involves determination of two factors, the throughput (controlled by the coalescence rates in the settlers) and the stage efficiency (a function of mass transfer rates in the mixers). Ideally, the design would be made from first principles by using predicted mass-transfer and coalescence rates; however, this is not yet possible and it is necessary in practice to scale up from pilot-plant tests, preferably supplemented by tests on a single full-scale stage.

The scale-up of mixer–settlers is simpler than that of column extractors in that axial mixing is unimportant. The performance required to meet the flowsheet requirements is thus expressed in terms of the number of ideal (theoretical) plug flow stages required, using the graphical or analytical methods of Chapter 5, Sections 2.5.1 and 2.5.2. These are then related to the number of actual stages by means of the stage efficiency (Chapter 5, Section 4.1). Alternatively, the number of actual stages can be calculated directly, for a linear equilibrium relationship, by means of the expressions for case 7 (backflow model) given in Table 3 of Chapter 6, on putting $n = N$ and solving for N; the value of N_{ox}^1 is related to the Murphree efficiency by Eq. (88) in Chapter 6.

2.2. Pilot-Scale Testing

The design of mixing and settling compartments is approached separately, while recognizing their interdependence: (1) the mixer must interdisperse the two phases, creating an adequate interfacial area and allowing sufficient residence time for the desired stage efficiency to be achieved; and (2) the settler must contain a dispersion band, providing suitable conditions and residence time for sedimentation and coalescence to take place, without undue entrainment.

Pilot-scale tests can be of either batchwise or continuous type, although the former are necessarily limited in scope and are suitable only for providing preliminary data. Continuous testing can be carried out with a single stage, varying the feed conditions to simulate each stage, in turn, of a multistage plant; however, runs with a multistage unit are to be preferred, in order to ensure that hydraulic operation will be satisfactory.

It should be noted that the performance of a solvent extraction plant is temperature dependent. If the production plant is to meet specifica-

†Not relevant (outside chart range).

tions at all times, the pilot-plant data should be obtained for the least favorable temperature likely, which, apart from the possible effect on the equilibrium line, will be the lowest temperature.

2.2.1. *Batch Method*

Batch tests can be useful for checking the effects of different conditions (e.g., temperature), alternative reagents, different impeller types, and so on. However, since the hydrodynamic regime in a batch-mixing vessel is significantly different from that in a corresponding continuous unit, it cannot be expected to give an accurate representation of the performance of the latter.

The tests should be carried out in a vessel of the same geometry as that to be used for the final mixer (fully baffled cylindrical, or square sectioned), with the same type of impeller and the same impeller: vessel diameter ratio. It is recommended that the key dimension (vessel diameter or length of side) should not be less than 15 cm. Unless a pumping impeller is required, use of a turbine with six flat blades [25] is recommended with a diameter one-third that of the key dimension.

The tests consist of introducing the two feed liquids carefully into the vessel in the flowsheet ratio and operating the agitator for a set time, after which a sample is withdrawn and the phases quickly separated and analyzed. This is then repeated for different mixing times. At the end of the test, the rate of phase separation is measured by stopping the agitator and recording the levels of both the sedimentation and coalescence interfaces as a function of time.

It is usually found that a plot of the logarithm of the fractional approach to equilibrium against time approximates to a straight line for a fixed agitation rate. The rate coefficient k is obtained from the gradient using the following relation, applicable to a straight equilibrium line [26]:

$$E_B = 1 - e^{-kt} \qquad (3)$$

The space time required to give the desired fractional approach to equilibrium in a single compartment continuous mixer under the same conditions can then be predicted *approximately* by substituting this value of k in [15]

$$E_H = \frac{k\tau}{1 + k\tau} \qquad (4)$$

where τ is the mean residence time and E_H the Hausen efficiency (for further details, see Slater et al. [26]). The appropriate Murphree efficiency can then be calculated from the relationships given in Table 2 of Chapter 5.

From the settling test, the rate of movement of the interface is determined for the dispersion band depth of interest. Coupling of this with the area of the vessel gives a specific rate of coalescence, which can be used to obtain an approximate measure of the size of continuous settler required (for further details, see Godfrey et al. [27]).

2.2.2. *Continuous Method*

The pilot unit should comprise one or more stages of geometrically similar design to that envisaged for the final plant, with the key linear dimension of the mixing compartment not less than 15 cm or 10% that of the full-scale unit, whichever is the greater. It is useful to provide a vertical movable baffle across the settler to enable the dispersion band depth and effective settler area to be varied. Feed and product tanks must be of adequate size to permit steady-state operation to be established (usually a minimum of three or four mean residence times). Temperature control of the feeds is desirable, and a record of temperature should be kept. If possible, real process liquors should be employed so that the results reflect the effects of any impurities present.

The pilot unit is operated with the same flow ratio and feed concentrations as specified in the flowsheet, and samples of the product phases are withdrawn for analysis. It is important to ensure complete phase separation, since entrainment of one phase in the other can give erroneous results. A solute material balance is calculated and, if it agrees to within 3–4%, the overall stage efficiency is calculated (Chapter 5, Section 4.1). The Murphree efficiency of *each* stage can also be calculated if the compositions of the phases are determined between stages. The total flow should be varied to show the effect of residence time on stage efficiency. The effects of other parameters such as phase continuity, impeller size, type, and speed should also be investigated. For the settler, measurements are made of the dispersion band depth as a function of specific throughput, with variation of the latter by either moving a baffle or changing feed rate. The collapse of the dispersion in the continuous mixer can also be used as a measure of settling characteristics [27].

It should be noted that the linear velocities of the settled phases in a pilot settler will be less than in a production unit of equivalent geometry. Since this is a key parameter for controlling entrainment losses, it may be desirable to provide longitudinal baffles in the pilot settler to give linear velocities equal to those expected at full scale [28].

2.3. Scale-Up

2.3.1. Mixer

The parameters that contribute to performance—drop size, dispersed-phase holdup, and flow pattern—have different scale-up characteristics, so that there can be no simple scale-up rule. In the case of pump–settler units (category F1 Table 1), scale-up is essentially in terms of pumping capacity, that is, total volume pumped per unit time.

For normal agitated mixers, the method usually adopted is to fix the residence time, and hence the mixer volume, from the pilot-plant work and try to provide equivalent agitation. Of the two criteria suggested for the latter, that is, equal tip speed (Nd_I) or equal power input per unit volume ($N^3 d_I^2$), the latter has been shown to approach much more closely to reality [28]. However, since agitation rate is the most important scale-up parameter, it is desirable to make provision for adjusting this on the full scale by variation of agitator speed or diameter, noting that power input $\propto (N^3 d_I^5)$ for the fully turbulent region. This can best be done by prior tests on a single full-scale stage with provision for such variations before the design is finally frozen.

2.3.2. Settler

Experience suggests that settlers can be scaled up fairly reliably on a basis of constant specific throughput at a given dispersion band depth, provided the characteristics of the dispersion are maintained constant. With increase in size, it is important to ensure that the dispersion leaving the mixer is distributed across the whole width of the settler without leaving any excessive dead space. Reference has already been made to the need to keep linear velocities of the settled phases sufficiently low to avoid excessive entrainment. On very large scale units, this may warrant increase in the width:length ratio so as to keep down the total holdup of solvent [26].

3. PILOT TESTING AND SCALE-UP: COLUMNS AND RELATED TYPES

3.1. General

The contactors to be considered here comprise the continuous and discontinuous gravity-separated types listed in the selection chart (Table 1), categories A–E inclusive. Ideally, after such a contactor has been selected for a particular duty, its dimensions would be determined theoretically from first principles as follows:

1. Calculate the flooding or maximum operating rate corresponding to the required solvent ratio using the methods described in the previous chapters; hence select a practical operating value (usually 50–60% of flooding) and determine the contactor cross section.
2. Obtain H_{ox} or N_{ox}^1 for the particular contactor from appropriate correlations of interfacial area and mass-transfer coefficients and also the parameters Pe_j or α_j from correlations of axial dispersion data. Then calculate the contactor length, or the number of compartments required to achieve the desired performance, using the methods described in Chapter 6.

Procedures for predicting maximum throughput are generally sufficiently reliable for engineering design in the absence of interfacial turbulence (Marangoni effect; see Chapter 3, Section 3.3.2). The latter is important in the transfer of many solutes, especially organic from solvent into aqueous phase, and leads to a substantially *increased* throughput, requiring experimental determination.

The prediction of mass transfer rates is more difficult, and methods are available only for packed [29] and perforated-plate columns [30] under conditions of no interfacial turbulence; however, even these require further refinement to make them acceptable for general use. For other types of contactor, particularly those with mechanical agitation, there is a lack of comprehensive data on interfacial area, and also on axial dispersion for commercial sizes, so that it is not possible to obtain true area mass-transfer coefficients from the available performance data. The design of such contactors should, therefore, be based on prior experience, which in the case of new processes must be obtained by pilot-plant experiments using the actual system of interest. A description of the principles involved in conducting such experiments and in scaling up from pilot or small commercial sizes to larger-sized equipment is given in the following paragraphs.

3.2. Pilot-Scale Testing

Column extractors and the like require a more extensive test procedure than do mixer–settlers, for which much useful information can be obtained by using even a single stage, as described in Section 2. This is due mainly to axial dispersion, which is negligible with true mixer–settlers but significant with most other types of contactor.

It is assumed that the aim of the pilot-scale experiments is to obtain design data for a commercial unit, rather than to simulate the performance of the latter on a reduced scale. The experimental program hence must be designed to yield data of a fundamental nature, including, if relevant, the effect of concentration on the controlling parameters. On this basis the following requirements must be kept in mind:

1. The pilot unit should have a throughput of preferably not less than 10% of that of full scale, subject to a lower limit on column diameter of about 8 cm (3 in.) and a length equivalent to not less than about four or five true transfer units (about two or three plug flow transfer units) or, with compartment-type contactors, a minimum of 8–10 compartments.
2. The design of the column internals should be such that the same dispersion characteristics can be maintained in pilot- and full-scale units. Rules for three common types of contactor are as follows:
   a. *Packed Column.* The packing size should be the same in both pilot- and full scale units when using the plug flow model for scale-up (Section 3.3). This restriction is unnecessary when allowing for axial dispersion, so that the packing size can then be reduced in the pilot column provided it is not below the "critical size" [31].
   b. *Reciprocating Columns.* For both oscillating and pulsed plate columns, the plate design (i.e., hole size and free area) and the plate spacing must be maintained constant.
   c. *Rotary-agitated Columns.* For rotary disk and multi-impeller columns the column internals are made geometrically similar except for the compartment height, which is varied as $d_c^{2/3}$ [Section 3.3.2(e)].
3. If the equilibrium relationship is nonlinear, the tests should if possible be conducted over concentration ranges for which it can be approximated by straight lines, to permit the use of analytical methods for scale-up.
4. If interfacial oscillation is likely to occur, tests should be conducted over the smallest practicable concentration ranges to enable the effect of concentration on performance to be assessed. It should be noted in this regard that the effects of interfacial oscillation on mass transfer and flooding rate diminish as equilibrium is approached, that is, toward the low-concentration (extractant inlet) end of the contactor.

Occasions will arise when some of the preceding requirements cannot be met simultaneously, necessitating compromise. This is particularly true if interfacial oscillation is likely, since there is then conflict between the first requirement (for length) and the fourth. It must be emphasized in this connection that interfacial turbulence is the most intractable problem encountered in extractor design, and it is difficult to foresee the development of a satisfactory theoretical method of predicting its effect on performance and throughput.

The test procedure, in which the actual system of interest is used, is as follows:

1. If considered necessary (e.g., because of the likelihood of interfacial oscillation), measurements are made of the flooding rate as a function of throughput and, if relevant, the agitation rate. These are then interpreted in terms of the characteristic velocity [32]. Alternatively, the latter can be obtained from measurements of the dispersed-phase hold up [31, 32] (Chapter 4, Section 2).
2. Mass-transfer runs are then carried out at the same flow ratio and column loading as for the full-scale unit, varying the agitation rate (if relevant) to find the optimum combination of throughput and performance.

3. After the best conditions have been selected, *repeated* runs are carried out under steady-state conditions, taking samples of the inlet and exit streams for analysis. Overall material balances on solute are calculated for each run, and only those that agree to within ±2–3% are accepted.

Some workers have attempted to determine both the backmixing Peclet numbers and H_{ox} from the measured concentration profiles of the two phases along the column [33]. However, it is difficult in practice to obtain uncontaminated samples of the separate phases for analysis (especially the dispersed phase), so that the accuracy is poor; thus the use of this method is not recommended in normal circumstances.

3.3. Scale-Up on Plug Flow Basis

3.3.1. Basis of Method

Although scale-up is simple with use of the plug flow model, unfortunately, axial dispersion has a major effect on the performance of most types of column extractor other than the perforated-plate column (type D1). Nevertheless, such scale-up methods have been proposed for reciprocating and pulsed plate columns (types C2 and 3), using empirical functions of the diameter ratio to allow for the performance loss on scale-up due to increased backmixing. Such methods can be used with reasonable assurance provided care is taken not to extrapolate outside the concentration range used in the pilot-scale tests, especially if interfacial turbulence is likely to occur.

To apply this method, the plug flow *NTS* or *NTU* is first calculated from the pilot plant data by use of one of the methods described in Chapter 5. For this purpose, the plug flow and not the true operating line must be used, even though there may be backmixing (Chapter 6, Section 2.1.6). The resulting value of the *NTS* or *NTU* is then used to calculate the length required for the full-scale column. Recommended procedures for a number of specific types of contactor are given in Section 3.3.2.

3.3.2. Specific Procedures

(a) Perforated-Plate Column [Type D1]. Scale-up of these contactors is based on the assumption that the stage efficiency is independent of diameter. This is normally satisfactory provided that the plates are accurately aligned and the orifices completely sealed by the layer of coalesced dispersed phase; it may in fact be slightly conservative because of the effect of cross-flow of continuous phase (Chapter 5, Section 4.1.2). On this basis, therefore, the number of actual stages required for the full-scale column is given by

$$N_{S,\,\mathrm{act}} = \frac{N_{S,\,\mathrm{theor}}}{E_o} \tag{5}$$

where E_o is the ratio of theoretical to actual stages for the pilot-scale column.

The overall efficiency E_o is subject to variation if the equilibrium line is curved, or if interfacial turbulence occurs. In such cases it is necessary to use the Murphree efficiency and to determine this as a function of concentration by analyzing the continuous phase leaving each stage of the pilot column, locating the concentrations on the operating line and constructing the real stages.

(b) Packed Column [Type A3]. Backmixing in packed columns is confined to the continuous phase and is dependent on the packing size but not the column diameter [34]. Values of H_{ojP} obtained for the pilot column are thus applicable directly to a larger unit provided the same packing is used and the concentration range covered in the two units is approximately the same. Packed columns may exhibit appreciable variations in performance on refilling because of the random nature of the filling.

(c) Reciprocating-Plate Column [Type C2]. The HETS for the open-type reciprocating-plate column [10] (Section 1.2.2) decreases rapidly with increase in frequency at a given flow ratio and amplitude of oscillation until a minimum is reached, beyond which it begins to increase [35]. This minimum is the result of a rapid increase in backmixing, which counteracts the improvement in performance due to the corresponding increase in interfacial area with frequency. It has also been found that the minimum HETS value is higher, and occurs at lower frequencies, for columns of larger diameter with the same plate spacing, showing that the backmixing increases with diameter. These results can be expressed quantitatively as follows for two columns of diameter d_{c1} and d_{c2} [35]:

$$\frac{(\mathrm{HETS})_{d_{c2}}}{(\mathrm{HETS})_{d_{c1}}} = \left(\frac{d_{c2}}{d_{c1}}\right)^{0.38} \tag{6}$$

$$\frac{f_{d_{c2}}}{f_{d_{c1}}} = \left(\frac{d_{c1}}{d_{c2}}\right)^{0.14} \tag{7}$$

A method of scale-up on this basis was proposed [35] (Chapter 12, Section 2.4.3), using a pilot-scale column of diameter preferably 5–8 cm with plates of the same hole size, free area and spacing, and the same amplitude of oscillation as the full scale column.

(d) Pulsed Plate Column [Type C3]. Logsdail and Thornton [36, 37] showed that the flooding rates for pulsed plate columns of diameters of 7.5–30 cm (3–12 in.) are unaffected by diameter, but that H_{ocP} increases according to the following relationship in the "emulsion" region:

$$\frac{(H_{ocP})_{d_{c2}}}{(H_{ocP})_{d_{c1}}} = \exp\left[0.5(d_{c2} - d_{c1})\right] \tag{8}$$

where d_c is expressed in feet. This increase in H_{ocP} was ascribed to an increase in backmixing with diameter. It is not known to what extent this relationship can be extrapolated to larger diameters, but it is of interest to note that it gives a H_{ocP} ratio of 3.08 for columns of 30-in. and 3.0-in. diameter, compared with a HETS ratio of 2.40 from Eq. (6); however, further extrapolation gives unrealistically large values of H_{ocP}.

It is apparent that columns of this type can be scaled up in a similar manner to the reciprocating plate type by using Eq. (8) in place of Eq. (6), provided the frequency, as well as the plate spacing and pulse amplitude, are maintained constant. However, the increase in H_{ocP} with diameter can be largely overcome, at least in diameters of up to 24 in., by the provision of louvre-type baffles that, however, appreciably reduce the throughput [11].

(e) Rotary-Agitated Columns [Types C1]. The scale-up of these columns is more involved than that of the foregoing types since an increase in height is required not only to compensate for increased backmixing of both phases, but also for the greater compartment height needed to maintain a satisfactory mixing geometry. In addition, the agitator speed must be reduced in order to compensate for its increased diameter; therefore, the following scale-up procedure is recommended:

1. Maintain geometric similarity of agitator and stator baffle dimensions in pilot- and full-scale unit, that is, by keeping d_r/d_c, d_s/d_c, and, for impeller agitators, d_w/d_r constant.
2. Increase the compartment height so that $h_c \propto d_c^{2/3}$ [38].
3. Reduce the agitator speed in order to maintain the same droplet characteristic velocity.

On this basis the agitator speed for the full-scale unit is obtained as follows for the rotary disk column [32]:

$$N_2 = N_1 \left(\frac{d_{c1}}{d_{c2}}\right)^{1.9} \left(\frac{h_{c2}}{h_{c1}}\right)^{0.9} \tag{9a}$$

$$= N_1 \left(\frac{d_{c1}}{d_{c2}}\right)^{1.3} \quad \text{when} \quad h_c \propto d_c^{2/3} \tag{9b}$$

Corresponding relationships for the multi-impeller (Oldshue–Rushton) and Kuhni columns are not available.

If this method is used, the flooding rates, and hence the column loadings, are the same for the two columns. However, scale-up of column height on a plug flow basis is not reliable for contactors of this type, and it is recommended that the method taking backmixing into account, described in Section 3.4, should be used.

3.4. Scale-Up Allowing for Axial Dispersion

3.4.1. Procedure

In brief, this method consists in substituting the experimental value of the exit extract composition from the pilot scale unit, together with the corresponding backmixing and other parameters, into the appropriate solution of the diffusion or backflow model and solving for H_{ox} or N_{ox}^1, respectively. This value is then assumed to apply to the full-scale unit and is substituted, together with the backmixing parameters and desired exit extract composition for the latter, into the same model solution to obtain the height or number of (actual) stages required [39].

In more detail, the steps involved in the case of a linear equilibrium relation are as follows:

1. Obtain values of the backmixing parameters Pe_j or α_j for both pilot- and full-scale units from the appropriate correlation or other source.
2. Substitute the experimental value of Y^0 or Y^1 for the pilot-scale unit, together with the backmixing parameters, column

length or number of stages, and **E** into the appropriate solution of the diffusion or backflow model (Chapter 6), and solve by iteration for the value of H_{ox} or N_{ox}^1, respectively.

3. Substitute this value of H_{ox} or N_{ox}^1 together with the *desired* exit extract phase composition, Y^0 or Y^1, the backmixing parameters for the full-scale unit and **E** into the same solution, or the equivalent simplified solution in Table 5 in Chapter 6 and solve for the length or number of actual stages (i.e., compartments) required.

In step 2 the exact model equations, which must be solved by computer, are given as cases 1 to 6 in Tables 1 and 3 in Chapter 6. The procedure for the diffusion model consists in assuming a value for H_{ox} and substituting this, together with the values of Pe_x and/or Pe_y, the extraction factor and the contactor length into the characteristic equation, which is then solved for the roots λ_i; these are then substituted into the Y-profile equation to obtain Y^0. If this does not agree with the measured value, a new values of H_{ox} is assumed and the calculation repeated until convergence is obtained. The solution can be obtained directly by computer, using the secant method, for example, and iterating on $\sqrt{H_{ox}}$ to avoid convergence on negative values [40]. The procedure is similar for the backflow model, with use of the backmixing ratios and number of actual stages in place of Peclet number and length, respectively, solving for the roots μ_i and iterating on $\sqrt{N_{ox}^1}$ to obtain Y^1. If $L > 6$ ft (2 m) or $N_{S,\text{act}} > 6$, the simplified equations given in Table 5 in Chapter 6, suitable for solution by programmable calculator, can be used. The method is illustrated in Example 3 (below) for a typical case.

When the equilibrium relationship cannot be approximated by a single straight line, the preceding equations are not applicable and the method becomes more time consuming. If the equilibrium curve can be represented with sufficient accuracy by two or three straight line segments, the most convenient procedure is to adjust the length of the pilot-scale unit to operate within these concentration ranges so that the values of H_{ox} or N_{ox}^1 can be determined for each by the foregoing procedure. The size of the full-scale contactor can then be calculated by subdivision into sections by use of the method described in Chapter 6, Section 3.2. In extreme cases when this method cannot be used, it is necessary to resort to one of the graphical or, preferably, numerical procedures described in Section 4 in Chapter 6.

Example 3. It is desired to treat 1.30 m^3/hr of toluene containing 50 kg/m^3 of acetone with 1.54 m^3/hr of water in a 0.305 m diameter pulsed plate column at 21°C to reduce the acetone content to 1.0 kg/m^3. Determine the number of plates required, assuming these to have 3.2-mm holes and 24.6% free area, with spacing of 5.0 cm.

Experiments were reported [36] for a 7.4-cm-diameter column provided with 18 plates of the same type, operating with the toluene phase dispersed. The following results were obtained at 21.5°C for a pulse amplitude of 16.0 mm and frequency of 3.0 Hz, with the same relative phase flow rates as the present full-scale unit (i.e., 81.6 and 96.5 liters/hr of toluene and water phases, respectively, and a mass balance error of -1.7%). See Chapter 6 for notation, taking toluene as the X (i.e., feed) phase and water as the Y (i.e., extractant) phase:

$$c_x^0 = 50.0, \qquad c_y^{N+1} = 0.0, \qquad c_y^1 = 36.90 \text{ kg m}^{-3}$$

The equilibrium relationship for this system at 21.5°C, in concentration units of kg/m^3, can be expressed as follows:

$$c_x^* = 0.6370c_y - 0.2922$$

Hence from Eq. (7) in Chapter 6 the value of c_y^1 in dimensionless units is $Y^1 = 0.4674$. From the overall material balance, the exit toluene and aqueous-phase flows are 76.8 and 101.3 liters/hr, respectively, giving $\mathbf{E} = 0.5130$ and 0.5072 at the toluene and water inlet ends, respectively, with a mean of 0.5101.

The estimated value of the backmixing ratio α_y for the continuous phase is 2.39 [Eq. (115) in Chapter 6] and that for the dispersed phase is zero [42]. Substitution in the backflow model equations (case 5 in Table 3 of Chapter 6) and solution by iteration gave $N_{ox}^1 = 0.3812$.

It is assumed that the 0.305-m-diameter column will be operated with the same pulse frequency and amplitude as the small column. The value of N_{ox}^1 will thus be taken as the same for both units (i.e., 0.3812). The estimated value of α_y for the large unit is 7.46 and from the overall material balances $c_y^1 = 39.33$ kg/m^3, that is, $Y^1 = 0.4982$, and $\mathbf{E} = 0.50777$; substitution of these values in the simplified equation for the backflow model, case 5 in Table 5, Chapter 6, gives $N = 58.11$ stages; therefore, on rounding off upward, 59 stages (i.e., plates) are required.

3.4.2. Limitations of Method

The procedure described in Section 3.4.1 assumes that H_{ox} or N_{ox}^1 is constant, irrespective of scale, under similar hydrodynamic conditions; this, in turn, requires that K_{ox} and a in the expressions $H_{ox} = U_x/k_{ox}a$ and $N_{ox}^1 = k_{ox}ah_c/U_x$ are themselves constant. That this is likely to be the case may be deduced from the fact that both the dispersed-phase holdup and the droplet size distribution are unaffected by scale for a given system and column geometry, operating at the same superficial flow velocities and, if relevant, the same degree of agitation. Thus the mass-transfer coefficient and interfacial area, which are controlled by the droplet size and, in the latter case, the holdup as well, should both be constant as required. The parameters H_{ox} and N_{ox}^1 are in fact idealized measures of mass-transfer performance; however, as shown in Chapter 5, Section 4.2, the change in concentration produced by a given number of transfer units is dependent on the flow pattern and is reduced to an extent depending on the degree of backmixing.

A problem that requires further attention when using either plug flow or axial dispersion model is that of entrance effects. These can take three forms: (1) the boundaries of the extractor may not correspond to the feed stream inlets; (2) additional mass transfer occurs during droplet formation and coalescence; and (3) appreciable distances are required along the contactor from the ends for the flow pattern and droplet size distribution to become established. Wilburn [41] has described a modified analytical solution to take account of the first effect, which, however, involves numerical inversion of a 12×12 matrix. Correlations of mass-transfer data to single droplets during formation and coalescence, relevant to the second effect, are available (Chapter 3, Section 4.4.3; Chapter 4, Section 4), but their applicability to droplet swarms in real extractors is not known.

It is clear that complex problems are involved in correcting realistically for individual entrance effects. On the other hand, these can be minimized as a whole by the use of a relatively long pilot-scale contactor, in order to spread the errors over the maximum possible number of transfer units or stages. However, this in itself can involve difficulties, namely, in requiring precise analysis due to the resulting close approach to equilibrium or to a conflict with other requirements given in Section 3.2, so that a careful compromise is necessary.

3.4.3. Application to Specific Column Types

(a) Packed Columns [Type A3]. The performance in this case should be expressed in terms of H_{ox}, that is, by using the diffusion model; the simpler equations of cases 3–6 (Table 3 in Chapter 6) are applicable since there is no dispersed-phase backmixing. The continuous-phase backmixing is controlled by the packing size, not by the column diameter [34], and it is permissible to reduce the packing size in the pilot-scale column [see Section 3.3.2(b)] provided it is not below the "critical size" [31] (Chapter 4, Section 5.2).

(b) Pulsed Packed Column [Type B1]. As for the unpulsed type, backmixing in the dispersed phase is negligible, and that in the continuous phase is unaffected by the column diameter [43]. A method of scale-up based on the use of the backflow model has been described by Simons [43].

(c) Rotary-Agitated Columns [Type C1]. The backflow model should always be used for contactors of this type, with the performance expressed in terms of N_{ox}^1. The assumption that this is the same in both pilot- and full-scale extractors is probably conservative in view of the necessary increase in compartment height with diameter. In other respects the scale-up procedure is the same as that described previously [Section 3.3.2(e)].

(d) Reciprocating and Pulsed-Plate Columns [Types C2–3]. These contactors usually have a small plate spacing, especially in Karr's form of reciprocating-plate column [10]. Either model is thus applicable, but the backflow model is to be preferred unless the number of plates is very large (see Section 3.4.1, Example 3). The simplified equations of cases 3–6 in Table 3 of Chapter 6 can be used for the pulsed plate column since dispersed-phase backmixing is negligible [42] (Chapter 6, Section 8.2). This apparently does not apply to the Karr form of reciprocating plate column [44], for which the corresponding equations for cases 1 or 2 must therefore be used.

NOTATION

		Dimensions	SI Units
A	pulse amplitude (i.e., total excursion of continuous phase or plates)	L	m
B	concentration of reactant B in bulk phase	(mol) L^{-3}	kmol m^{-3}
c_j	concentration in phase j	ML^{-3} (or mol L^{-3})	kg m^{-3} (or kmol m^{-3})
d	characteristic dimension of contactor	L	m
d	diameter; subscript c refers to column, I to impeller, r to rotor, s to inside of stator baffle	L	m
d_w	width of impeller	L	m
d_{32}	Sauter mean (i.e., volume-surface) diameter	L	m
E	extraction factor, mU_x/U_y or mF_x/F_y		
E_{Bj}	efficiency of batch mixer based on phase j		
E_H	Hausen efficiency of single continuous mixer stage		
E_j	effective longitudinal diffusion coefficient in phase j	L^2T^{-1}	m^2s^{-1}
E_o	overall efficiency of stage		
F_j	volumetric flow rate of phase j	L^3T^{-1}	m^3s^{-1}
f	frequency	T^{-1}	s^{-1}
g	acceleration due to gravity	LT^{-2}	m s^{-2}
H_{oj}	height of "true" overall transfer unit based on phase j	L	m
H_{ojP}	height of an overall plug flow transfer unit based on phase j	L	m
HETS	height equivalent to a theoretical stage	L	m
h_c	height of compartment	L	m
k_j	rate constant for mixer based on phase j	T^{-1}	s^{-1}
k_n	nth order reaction rate constant	$L^{3(n-1)}T^{-1}$ (mol)$^{1-n}$	$m^{3(n-1)}T^{-1}$ (kmol)$^{n-1}$
m	reciprocal slope of equilibrium line, dc_x^*/dc_y		
N	agitator speed	T^{-1}	s^{-1}
N_{ox}^1	number of overall transfer units per stage based on X phase		
N_S	number of plug flow stages		
Pe_j	Peclet number for phase j (= U_jd/E_j)		
t	batch mixing time	T	s
$t_{1/2}$	half-time of chemical reaction	T	s
U_j	superficial velocity of phase j	LT^{-1}	m s^{-1}
X	dimensionless concentration of feed phase [Eq. (6) in Chapter 6]		
Y	dimensionless concentration of extractant phase [Eq. (7) in Chapter 6]		

		Dimensions	SI Units
α_j	backmixing ratio of phase j, that is, ratio of backflow rate to F_j		
λ_i	root i of characteristic equation (diffusion model)		
μ	viscosity	$ML^{-1}T^{-1}$	$N\ s\ L^{-2}$
μ_i	root i of characteristic equation (backflow model)		
$\Delta\rho$	density difference of phases	ML^{-3}	$kg\ m^{-3}$
σ	interfacial tension	MT^{-2}	$N\ m^{-1}$
τ	mean residence time	T	s

Subscripts

act	actual (stages)
j	X or Y phase (defined in Chapters 5 and 6); continuous or dispersed phase
theor	theoretical (stages)
1	pilot-scale contactor
2	full-scale contactor

Superscripts

I	exit X or inlet Y phase (diffusion model)†
N	exit X phase (backflow model)†
$N+1$	inlet Y phase (backflow model)†
0	inlet X phase (either model); exit Y phase (diffusion model)†
1	exit Y phase (backflow model)†
*	equilibrium value

†External to column.

REFERENCES

1. V. S. Morello and N. Poffenburger, *Ind. Eng. Chem.* **42,** 1021 (1950).
2. H. R. C. Pratt, *Ind. Chem.* **30,** 475, 597 (1954).
3. E. D. Oliver, *Diffusional Separation Processes*, Wiley, New York, 1966, p. 362.
4. C. Hanson, *Chem. Eng.* **75,** 76 (1968).
5. G. S. Laddha and T. E. Degaleesan, *Transport Phenomena in Chemical Engineering*, Tata-McGraw Hill, New Delhi, 1976.
6. K. H. Reissinger and J. Schroter, *Chem. Eng.* **85,** 109 (1978).
7. W. J. Anderson and H. R. C. Pratt, *Chem. Eng. Sci.* **33,** 995 (1978).
8. J. B. Wijffels and K. Rietema, *Transact. Inst. Chem. Eng.* **50,** 224, 233 (1973).
9. W. J. D. van Dijcke, U.S.P. 2,011, 186 (1935).
10. A. E. Karr, *AIChE J.* **5,** 446 (1959); T. C. Lo and A. E. Karr, *Ind. Eng. Chem. Process Des. Devel.* **11,** 495 (1972).
11. F. W. Woodfield and G. Sege, *Chem. Eng. Progr., Symp. Ser. No. 13*, **50,** 14, 174 (1954).
12. F. C. Koch, U.S.P. 2,176,429; F. D. Fuqua, *Petrol. Process* **3,** 1050 (1948).
13. T. Misek and J. Marek, *Br. Chem. Eng.* **15,** 202 (1970); Anonymous, *Chem. Eng.* **68** (9), 58 (May 1, 1961).
14. A. A. North and R. A. Wells, *Transact. Inst. Min. Met.* **74,** 463 (1964–1965).
15. R. E. Treybal, *Liquid Extraction*, 2nd ed., McGraw-Hill, New York, 1963.
16. J. C. Godfrey et al., *Proceedings of the International Solvent Extraction Conference 1980* (ISEC 80), Vol. 1, Session 4A, Paper No. 80-31 (Association des ingenieurs sortis de l'Université de Liege, Belgium), 1980.
17. R. E. Treybal, U.S.P. 3,325,255 (1967); *Chem. Eng. Progr.* **60** (5), 77 (1964).
18. W. J. Podbielniak, U.S.P. 2,044,966 (1935).
19. D. B. Todd and W. J. Podbielniak, *Chem. Eng. Progr.* **61** (5), 69 (1965).
20. D. H. Logsdail and L. Lowes, in C. Hanson, Ed.,

Recent Advances in Liquid-Liquid Extraction, Pergamon Press, New York, 1971, Chapter 5.

21. F. T. E. Palmqvist and S. Beskow, U.S.P. 3,108,953 (1959).
22. C. M. Doyle, U.S.P. 3,107,218 (1963).
23. H. Eisenlohr, *Ind. Chem.* **27**, 271 (1951); C. Coutor, U.S.P. 2,036,924 (1936).
24. R. Gayler and H. R. C. Pratt, *Transact. Inst. Chem. Eng.* **31**, 69 (1953).
25. J. H. Rushton, E. W. Costick, and H. J. Everett, *Chem. Eng. Progr.* **46**, 395, 467 (1950).
26. M. J. Slater, G. M. Ritcey, and R. F. Pilgrim, *Proceedings of the International Solvent Extraction Conference 1974* (ISEC 74), Vol. 1, Society of Chemical Industry, London, 1974, p. 107.
27. J. C. Godfrey, D. K. Chang-Kakoti, M. J. Slater, and S. Tharmalingam, *Proceedings of the International Solvent Extraction Conference 1977* (ISEC 77), Vol. 1, Canadian Institute of Mining and Metallurgy, Montreal, 1979, p. 406.
28. J. R. Orjans, C. W. Notebaart, J. C. Godfrey, C. Hanson, and M. J. Slater, *Proceedings of the International Solvent Extraction Conference 1977* (ISEC 77), Vol. 1, Canadian Institute of Mining and Metallurgy, Montreal, 1979, p. 340.
29. H. R. C. Pratt and W. J. Anderson, *Proceedings of the International Solvent Extraction Conference 1977* (ISEC 77), Vol. 1, Canadian Institute of Mining and Metallurgy, Montreal, 1979, p. 242.
30. A. H. P. Skelland and W. L. Conger, *Ind. Eng. Chem. Process Des. Devel.* **12**, 445 (1973).
31. R. Gayler, N. W. Roberts, and H. R. C. Pratt, *Transact. Inst. Chem. Eng.* **31**, 57 (1953).
32. D. H. Logsdail, J. D. Thornton, and H. R. C. Pratt, *Transact. Inst. Chem. Eng.* **35**, 300 (1957).
33. J. C. Mecklenburgh and S. Hartland, *Inst. Chem. Eng. Symp. Ser.* No. 26, 115 (1967).
34. R. Gayler and H. R. C. Pratt, *Transact. Inst. Chem. Eng.* **35**, 273 (1957).
35. A. E. Karr and T. C. Lo, *Proceedings of the International Solvent Extraction Conference 1971* (ISEC 71), Vol. 1, Society of Chemical Industry, London, 1971, p. 299; *Chem. Eng. Prog.* **72**, 68 (1976).
36. J. D. Thornton, *Transact. Inst. Chem. Eng.* **35**, 316 (1957).
37. D. H. Logsdail and J. D. Thornton, *Transact. Inst. Chem. Eng.* **35**, 331 (1957).
38. H. R. C. Pratt, *Ind. Chem.* **31**, 505, 552 (1955).
39. H. R. C. Pratt and M. O. Garg, *Ind. Eng. Chem. Process Des. Devel.* **20**, 489 (1981).
40. A. R. Curtis, in L. C. W. Dixon, Ed., *Optimization in Action*, Academic Press, New York, 1976.
41. N. P. Wilburn, *Ind. Eng. Chem. Fund.* **3**, 189 (1964).
42. M. O. Garg and H. R. C. Pratt, *Ind. Eng. Chem. Process Des. Devel.* **20**, 492 (1981).
43. A. J. F. Simons, *Proceedings of the International Solvent Extraction Conference 1977* (ISEC 77), Vol. 2 (Can. Inst. Min. Met., Montreal), 1979, p. 677.
44. M. M. Hafez, M. H. I. Baird and I. Nirdosh, *Proceedings of the International Solvent Extraction Conference 1980* (ISEC 80), Vol. 1, Session 5A, Paper No. 80-41 (Association des ingenieurs sortis de l'Université de Liege, Belgium), 1980.
45. J. Stichlmair, *Chem. Ing. Technik* **52** (3), 253 (1980).
46. T. C. Lo, *Recent Development in Commercial Extractors*, paper presented at Engineering Foundation Conference on Mixing Research at Rindge, NH, 1975, The Engineering Foundation, New York.

17.1

GENERAL LABORATORY-SCALE AND PILOT-PLANT EXTRACTORS

M. H. I. Baird
McMaster University
Canada

Teh C. Lo
Hoffmann-La Roche Inc.
United States

1. INTRODUCTION

The widespread industrial application of solvent extraction is predated by its applications in analytical chemistry and laboratory-scale preparations. Thus many laboratory procedures and types of equipment have not changed significantly over the years, so that some of the older reviews [1–3] in this area are still useful. The recent surveys by Schiebel [4] and Lo [5] are recommended.

A prominent recent addition to the available laboratory extraction equipment is the AKUFVE apparatus, which is dealt with in a separate chapter (Chapter 17.2). Some types of contactor, originally developed in the laboratory or on a pilot scale, have become popular on the industrial scale and are dealt with in a separate manner.

2. DETERMINATION OF PARTITION COEFFICIENTS

In principle, this is not a difficult procedure as it is merely necessary to maintain contact between two liquid phases at a controlled temperature for sufficient time to reach equilibrium. The phases may be shaken or stirred to reduce the time for equilibration. In the absence of slow chemical reactions, the equilibration time can be as low as a few minutes under well-agitated conditions. A suitable cell is shown in Fig. 24 in Chapter 1. Alternatively, the phases may be placed in a sealed tube with an air space, with the tube then repeatedly inverted. Equilibrium is always reached after 50 inversions, and usually far fewer are needed [1]. The equilibrated phases are allowed to settle, and are then each sampled for analysis.

Although the apparatus involved is simple, the determination of many equilibria can be time consuming and the AKUFVE apparatus (Chapter 17.2) provides for partly automatic determination.

3. KINETIC MEASUREMENTS

The need for kinetic data on the laboratory scale has increased with the development of extraction processes involving slow chemical reactions (e.g., metal extractions) (Chapter 2.2). Even in the case of diffusionally limited extraction, mass-transfer rates become important in the design of high-capacity equipment where the "contact time" is very short.

There is a large class of laboratory contactors in which the liquid–liquid contact area is constant and precisely known. This class includes stirred cells such as the Lewis cell, wetted-wall contactors, and single droplet contactors. These devices are described in Section 4 of Chapter 3. They are particularly useful in elucidating the controlling step (diffusion or chemical reaction) in mass transfer, as the diffusional mass-transfer rate per unit area can often be predicted from the hydrodynamic conditions.

Kinetic measurements may also be carried out with droplet dispersions in well-agitated vessels, operated either batchwise or continuously. The AKUFVE apparatus (Chapter 17.2) may be used for this purpose, or a small stirred tank contactor may be preferred (see Chapter 2.1). It must be noted that, although such dispersion contactors resemble industrial equipment more closely than do the constant-area devices, extreme care must be taken in scaling up the kinetic measurements. This is because the measured mass-transfer rate (per unit volume) is in most cases dependent on the specific interfacial area as well as diffusion and reaction rates. The scale-up of interfacial area in agitated systems is a complex function of interfacial tension, specific power input, and other factors, as noted in Chapter 4.

4. EXTRACTION OF A SOLUTE

4.1. Batch Extraction

A simple but effective laboratory method of solute removal is to extract the feed solution repeatedly (*multiple contact*) with fresh batches of solvent. Each treatment usually consists of an equilibration in an agitated vessel, followed by phase separation in a separatory funnel. Handling is reduced if the initial agitation is done by shaking the phases in the same separatory funnel rather than using a different vessel for agitation. The concentration of solute remaining in the raffinate decreases with the number of treatments and may be calculated analytically if the extraction factor **S** is constant (see Section 2.4.2 of Chapter 5). For partially miscible solvents, graphical calculation procedures may be used as noted in Section 5.2 of Chapter 5. It can be shown that multiple treatments are much more effective than a single contact by use of the same total amount of solvent.

The consumption of solvent for a given degree of solute removal and/or the number of contacts required are reduced if a countercurrent arrangement of the contacts is used. This arrangement is sometimes termed *multiple-contact pseudo-countercurrent* extraction [1] and is illustrated in Fig. 1. In this simplified example, it is assumed that the extraction factor is unity; that is, the solute distributes itself in equal amounts between the phases on each contact. The feed is assumed to contain unit mass of solute and in the example it is divided initially into two equal parts, tubes 0 and 1. In transfer number 0, the feed in tube 0 is contacted with fresh solvent in tube 0. Then, in transfer 1, the extract in tube 0 is equilibrated with the fresh feed in tube 1, while fresh solvent in tube 1 is contacted with the raffinate in tube 0. Continuation of this process results in extraction of 62.5% of the solute as shown. This compares favorably with the 50% extraction obtainable in a single stage. It also compares very favorably with multiple contacts with fresh solvent. From Eq. (20) in Chapter 5, it can be seen that for the same degree of extraction of 62.5% (i.e., $X_N = 0.375$), it would be necessary to have no fewer than 25 contacts with the same total amount of fresh solvent (i.e., $S = 0.04$ and $N_S =$

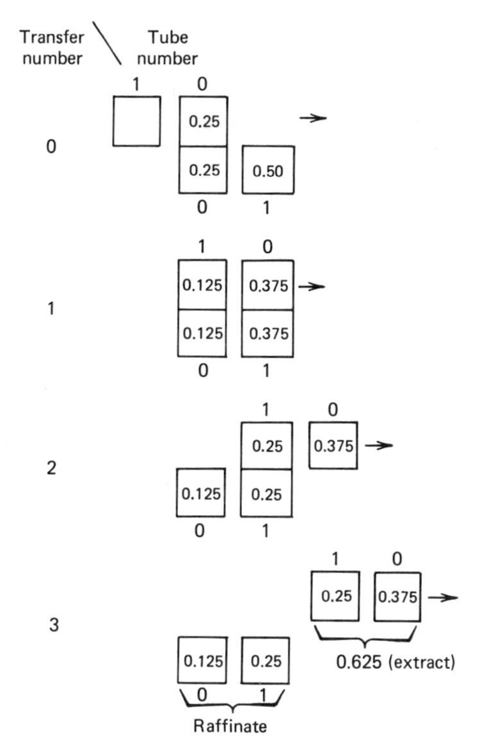

Figure 1. Multiple-contact pseudocountercurrent extraction. [Reprinted by permission from L. C. Craig, D. Craig, and E. G. Scheibel, Laboratory Extraction and Distribution, in A. Weissberger, Ed., *Separation and Purification* (*Techniques of Organic Chemistry*, Vol. 3), 2nd ed., Wiley-Interscience, New York, 1956, p. 149.]

25), whereas the pseudocountercurrent scheme in Fig. 1 requires only four contacts.

4.2. Differential Batch Extraction

In this process solute is gradually removed from a fixed amount of feed solution by continuously passing fresh solvent through it. The calculation procedures are given in Section 5.3 of Chapter 5 and are analogous to those for batch distillation. Some chemistry-oriented texts [1, 2] describe this operation as "continuous extraction," but this is true only in terms of the solvent phase. Many ingenious glassware configurations have been developed [1, 2], two examples of which are given in Fig. 2.

4.3. Continuous Extraction

Here, the flows of *both* phases are continuous, and the calculation procedures are those given in Chapter 5.

On the laboratory scale, countercurrent flow

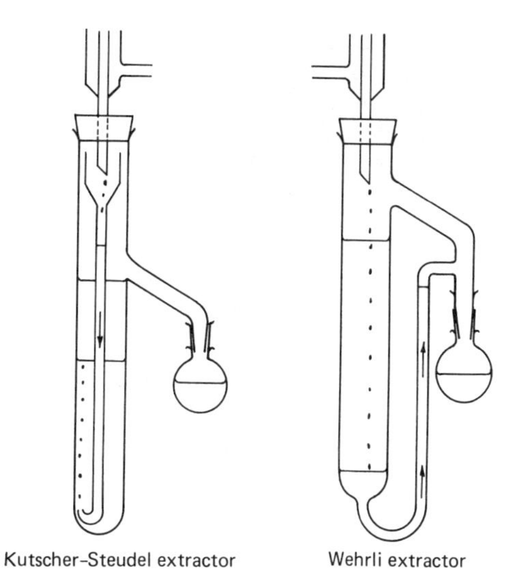

Figure 2. Apparatus for differential batch extraction. [Reprinted by permission from L. C. Craig, D. Craig, and E. G. Scheibel, Laboratory Extraction and Distribution, in A. Weissberger, Ed., *Separation and Purification* (*Techniques of Organic Chemistry*, Vol. 3), 2nd ed., Wiley-Interscience, New York, 1956, p. 149.]

is almost invariably used as it provides optimum solute removal for a given extraction factor. Continuous countercurrent equipment is highly effective for fractionation as well as for simple solute removal, and it will be considered separately in Section 6.

5. BATCH FRACTIONATION

The objective of fractionation by solvent extraction is to separate two or more solute components on the basis of their differing distribution coefficients between two liquid phases. Under laboratory conditions the practical purpose of this separation may be to either (1) isolate the various components of a mixture for analytical purposes, or (2) simulate a large-scale process by using very small amounts of materials.

In either case, batchwise countercurrent operation is popular, as the material demands for batch operation tend to be less than those for continuous operation, and the advantage of countercurrent contact over other modes are well known.

5.1. Basic Principles

These principles were developed independently by Craig [6] and Stene [7] in 1944.

Consider the simple arrangement shown in Fig. 3. At the top of the triangle is a circle representing the initial feed, which, after equilibration with two immiscible solvents, forms two phases; the light product phase is shown entering the circle below and to the right of the feed, while the denser phase is shown moving below and to the left. Fresh amounts of heavy and light solvent are respectively added, and it will be seen that the number of products increases by one as the process is continued from one line to the next.

If p and q are the respective fractions of any particular solute in the light and heavy product phases from any stage, it can easily be shown that

$$p = \frac{\mathbf{S}}{\mathbf{S} + 1} \tag{1}$$

$$q = \frac{1}{\mathbf{S} + 1} \tag{2}$$

Taking the amount of the given solute initially as 1 and assuming that **S** and hence p and q are constant in all stages, the binomial distribution pattern illustrated in Fig. 3 can easily be established. In general, it can be shown that the total amount T of solute in the rth stage from the right (taking the right-hand stage as $r = 0$) in the nth line from the top is given by

$$T_{n,r} = \frac{n!}{r!(n-r)!} \; p^r q^{n-r} \tag{3}$$

It can further be shown that as n is increased, the distribution of solute across the line of stages approaches the normal probability curve, with the maximum occuring at stage r_m, where

$$r_m = \frac{n\mathbf{S}}{\mathbf{S} + 1} \tag{4}$$

For two solutes with differing partition coefficients between the solvents, the values of **S** will also differ; therefore, the solutes can be separated out in "peaks" whose resolution improves as n is increased (see Fig. 4). Further discussion of the calculation methods is given by Craig et al. [1], Treybal [3], and Scheibel [4].

An alternative operating scheme, known as "double withdrawal," develops into rows con-

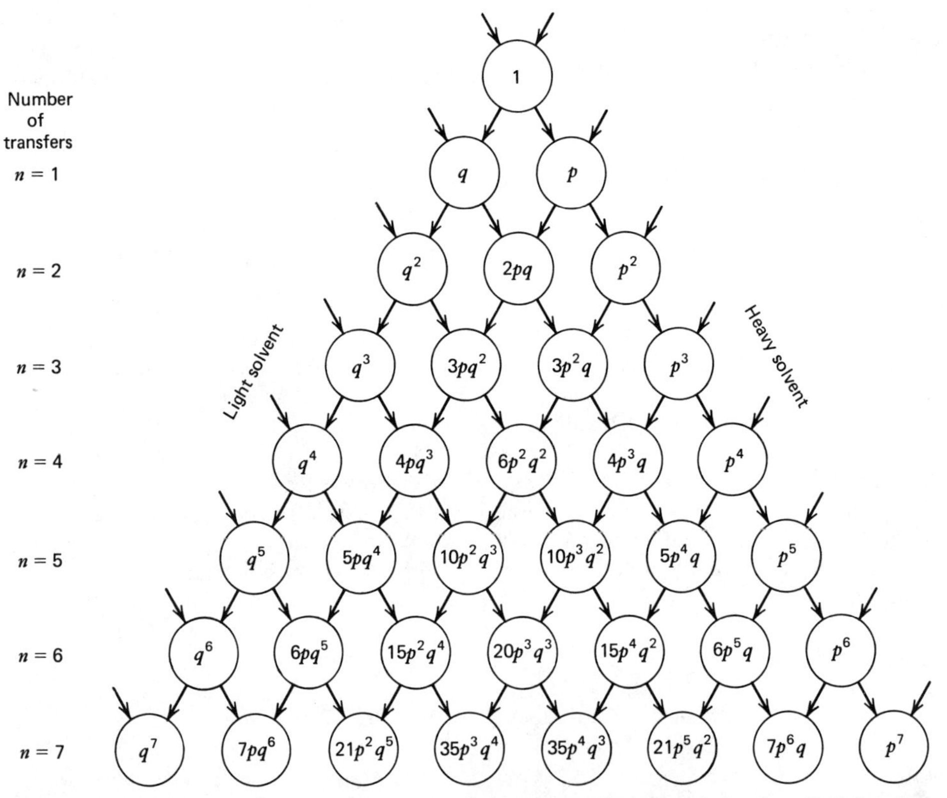

Figure 3. Distribution pattern for batch fractionation. [Reprinted by permission from E. G. Scheibel, Liquid–Liquid Extraction, in E. S. Perry and A. Weissberger, Eds., *Separation and Purification* (*Techniques of Chemistry*, Vol. 12), 3rd ed., Wiley-Interscience, New York, 1978, p. 77.]

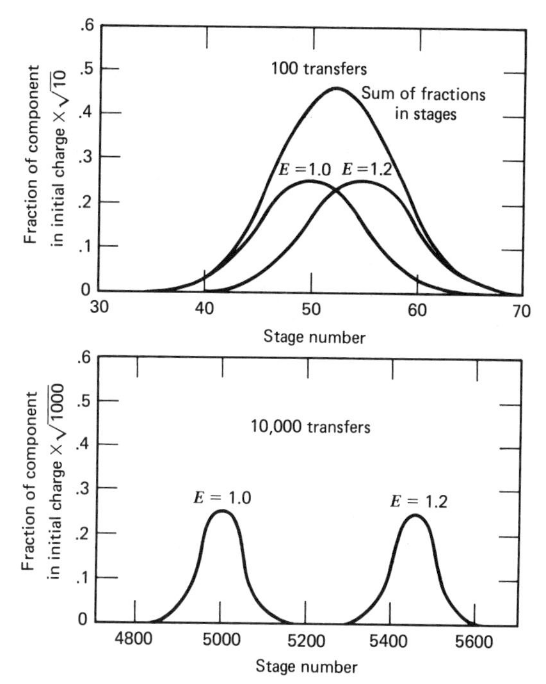

Figure 4. Effect of number of transfers on resolution of compounds having distribution coefficients differing by 20% (E = extraction factor). [Reprinted by permission from E. G. Scheibel, Liquid–Liquid Extraction, in E. S. Perry and A. Weissberger, Eds., *Separation and Purification* (*Techniques of Chemistry*, Vol. 12), 3rd ed., Wiley-Interscience, New York, 1978, p. 77.]

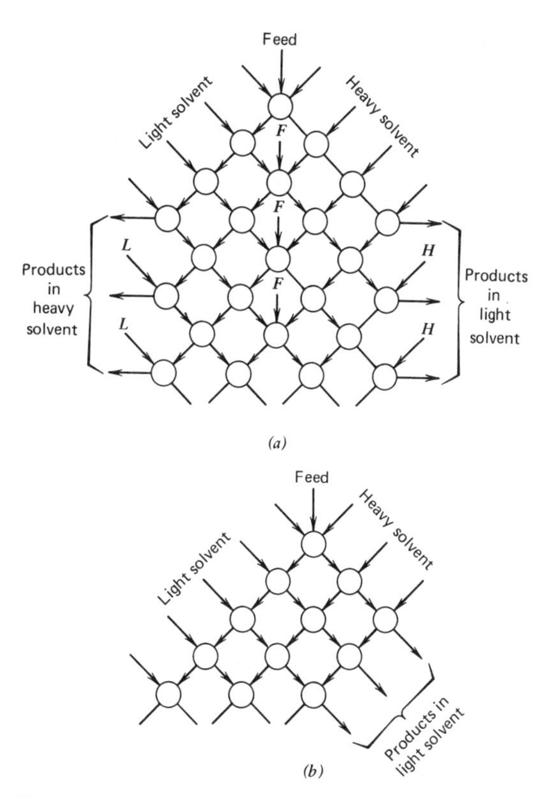

Figure 5. (*a*) Fractionation with "double withdrawal;" (*b*) fractionation with "single withdrawal."

taining a constant number of stages, with phase removal and solvent addition at each end of the row and repeated additions of feed at one of the intermediate stages in each line. This scheme (Fig. 5*a*) is important in that it leads to a batchwise simulation of continuous countercurrent fractional extraction after many rows have been completed. In the "single withdrawal" scheme (Fig. 5*b*) only one solvent stream is withdrawn and the solute(s) emerge from the stages in a peaked distribution. The theory of these schemes has been worked out independently by Compere and Ryland [8] and Scheibel [9].

It should be noted that in all cases involving a very large number of contacts, the analytical calculation methods lose accuracy when there is any partial miscibility between solvents or when the distribution coefficient for one solute is influenced by the presence of another. However, even in this case, the procedures shown in Figs. 3 and 5 are valuable in laboratory studies of new systems.

5.2. The Craig Contactor

The implementation of batchwise countercurrent fractionation schemes requires automatic contacting, not only because of the large number of contacts required, but also because of the necessity to carry out each contact in an identical manner without human error.

The main purposes of Craig contactor are (1) separation of a mixture, (2) proof of purity of substance, (3) identification of substance, and (4) determination of partition coefficients.

Craig's first apparatus [6] consisted of a 150-mm-diameter stainless steel cylinder into which 25 holes were machined. These holes corresponded to equilibration stages, and the cylinder was divided into two parts that could be rotated, allowing the heavy phase from one stage to be brought into contact with the light phase from the adjacent stage. However, this device was soon replaced by an apparatus [1] in which the stages were made entirely of glass and could be rack mounted in any number required.

The apparatus is described in detail in several texts [1, 3, 4], and the mode of operation is shown in Fig. 6. Each stage in the Craig contactor consists of a specially made glass tube, typically with an overall length of about 300 mm and inside diameter of 15 mm in the main section. Equilibrium within a tube is attained

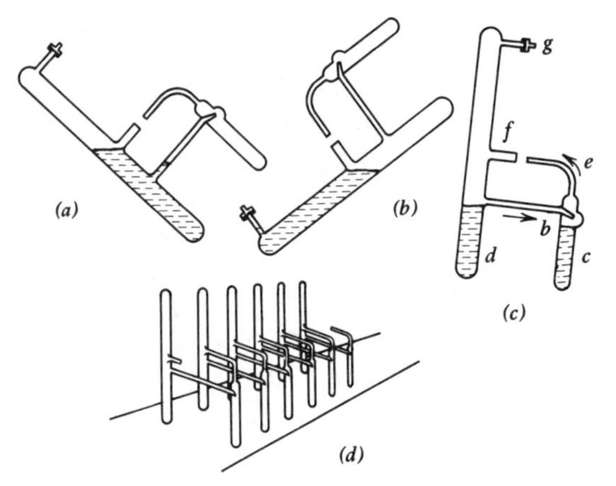

Figure 6. Craig contactor tubes. (Reprinted by permission from R. E. Treybal, *Liquid Extraction*, 2nd ed., McGraw-Hill, New York, 1963, p. 359.)

by tilting it between the positions shown in Fig. 6*a* and 6*b*. The light phase then flows through side arm *b* into chamber *c*, which is an integral part of the adjacent tube. Assemblies of as many as 1000 tubes have been constructed [4], and, with unattended operation, thousands of contacts can be carried out. Figure 7 shows an automatic Craig contactor with 500 contacting stages.

A disadvantage of the Craig contactor is that it cannot readily accommodate changes in phase volume during extraction. In such cases it may be preferable to use a manually operated double rack of separatory funnels [4].

Figure 7. Craig contactor with 500 contacting stages.

6. CONTINUOUS COUNTERCURRENT EXTRACTION

The main purposes of continuous countercurrent extraction on laboratory, minipilot-plant [10], and pilot-plant scales are (1) process feasibility study and development, (2) study of process control, (3) separation of relatively large quantities of solute for product study, and (4) obtaining data for extractor scale up and design. The equipment that has proved useful is described in Section 6.1.

6.1. Mixer–Settlers

6.1.1. *Bench-Scale Multistage Contactor*

A continuous bench-scale multistage countercurrent mixer–settler was developed by Anwar et al. [11]. It consists of a basic stage module as shown in Fig. 8*a*, in which the light (L) and heavy (H) phases are pumped and mixed by an impeller on a motor-driven shaft G. The phases, after contact to approach equilibrium, pass into an annular settling chamber S, from which they leave the module by side arms L and H as shown in the second side elevation (Fig. 8*b*). The modules may be interconnected, as shown in a plan view configuration in Fig. 8*c*, to give continuous countercurrent contact. A 20-stage unit

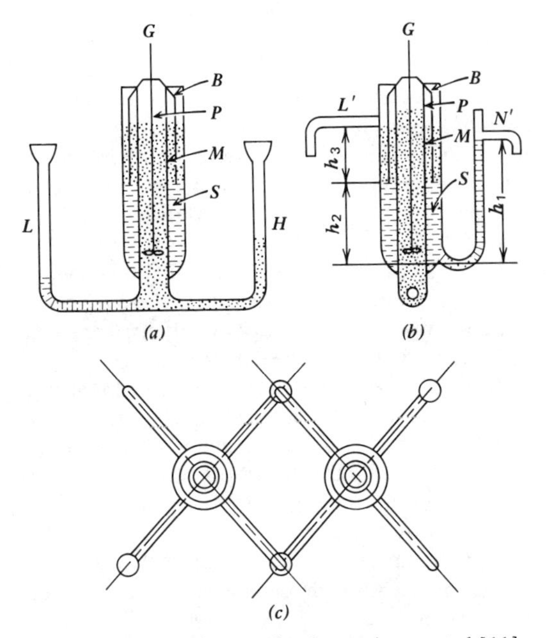

Figure 8. Glass mixer–settler due to Anwar et al [11]: (*a*), (*b*) side elevations; (*c*) plan view. [Reprinted from M. M. Anwar, C. Hanson, and M. W. T. Pratt, *Chem. Ind.*, 1090 (August 9, 1969).]

Figure 9. Bench-scale mixer-settler (Horbury Technical Service Ltd., England).

has been built [11], which is capable of handling flows between 0.1 and 2.0 liters/hr at stage efficiencies of virtually 100%. Figure 9 shows a photograph of a multistage contactor that is commercially available in Pyrex [12]. Because the unit gives a known number of theoretical stages, it is very useful for laboratory work on development and optimization of processes that require confirmation of design flowsheets. It has been successfully used in development work for nuclear fuel processing [13].

6.1.2. Laboratory-Scale Continuous Multistage Extractor

Rahn and Smutz [14] describe a small, continuous multistage, box-type mixer-settler. The mixer-settler uses the pump-mix principle in mixing chambers and individual interface control in settling chambers. The contactor is reported to have high stage efficiency, flexibility of operation, and simplicity of control. The extractor was developed particularly for separation of rare earths.

6.1.3. M-X Processor Laboratory-Scale Mixer–Settler

These mixer–settler units were designed in Sweden [15] and consist of a mixing chamber and a box-type settler. The heavier phase is pumped by a turbine-pump through a bottom inlet tube into the center of the mixing chamber. The lighter phase enters the upper part of the chamber by gravity. Efficient mixing is obtained by impeller wings on the underside of the turbine-pump. The dispersion in the mixer enters the settler through a centrally located port. The lighter phase leaves the settler over a weir. The heavy phase flows under a fixed baffle into a separate compartment, where it is taken out through an adjustable jack leg. The position of the phase boundary in the settler is controlled by adjustment of this jack leg. A separate jack leg is used for recirculation of the heavy phase to the mixing chamber. The size of the mixing chamber ranges from 100 to 1000 ml. The size of the settling chamber varies from 1 to 5 liters. It has been reported [16] that once the unit is set up, it can be operated continuously with little or no attention. New equilibrium can be quickly reestablished when the operation condition is changed, and the unit also can be operated with minimum of solution. Figure 10 shows an arrangement of pilot units for a vanadium recovery process including extraction, scrubbing, stripping, and washing stages.

6.1.4. Davy Power Gas Mixer–Settler

Davy power gas (DPG) pilot units [17] operate on the pump-mix principle (see Chapter 9.3)

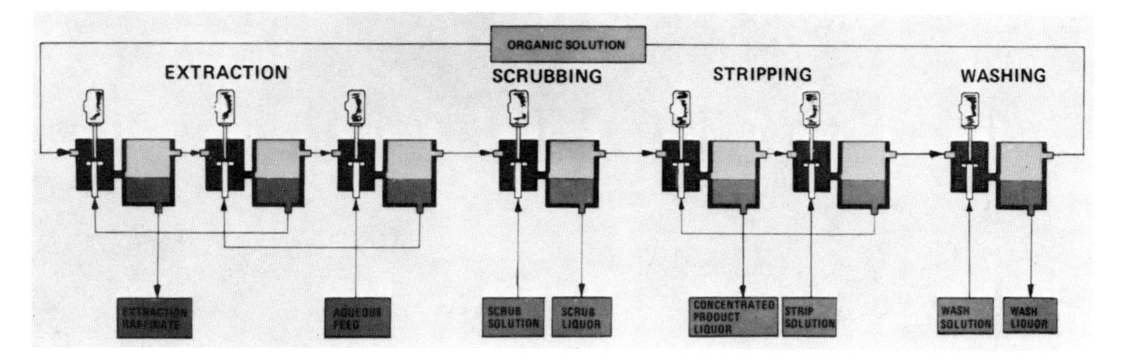

Figure 10. An arrangement of M-X processor laboratory-scale mixer–settler for a vanadium recovery process.

and have a square mixing chamber of about 1 liter and a rectangular settling volume of 4–5 liter. In metallurgical extraction, the DPG unit would be preferred where the process evaluation also required a metal-winning step [16].

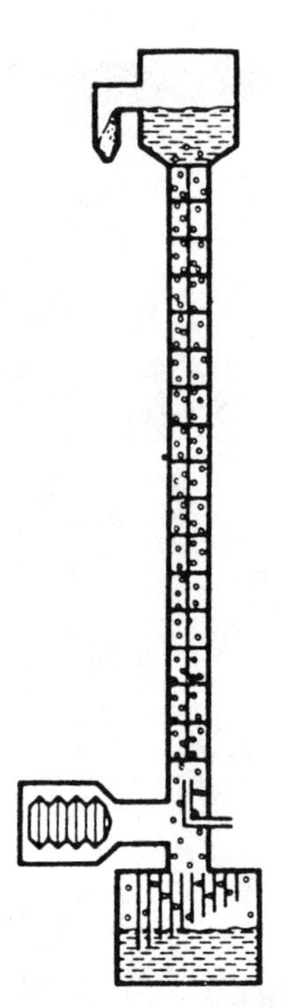

Figure 11. Pulsed perforated-plate column.

6.1.5. Other Types

A single-stage unit with a settler area of 0.09 m^2 and a mixer volume of 7.5 liters is available [18]. This unit is rugged and compact, with dimensions 1.0 × 0.4 × 1.0 m. It is claimed to be suitable for pilot-scale studies or small-scale production.

A laboratory mixer–settler with up to 10 stages [19] is rated for total flows of up to 250 ml/min. The mixer volumes are 180 ml and the settler areas are 61.5 cm^2.

There is an inverse relationship between the size of each stage and the number of stages that may conveniently be incorporated in a continuous mixer–settler system. The French Atomic Energy Commission (CEA) has designed a range of extremely compact miniaturized multistage mixer–settlers [20], the chambers of which are machined cavities in a block of special glass. Standard 16-stage modules are available, and these may be coupled if necessary. Mixer volumes range between 1.65 and 15 ml; total flows range between 0.5 and 13 ml/min. The combination of low retention volumes and chemical passivity makes these units particularly suitable for nuclear research.

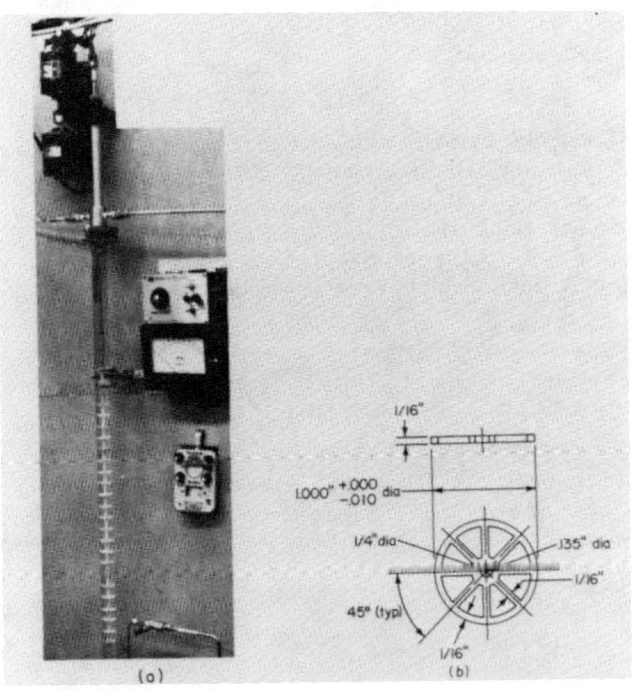

Figure 12. A 25-mm (1-in)-diameter reciprocating-plate extraction column and reciprocating-plate detail.

6.2. Differential Contactors

6.2.1. Pulsed Columns

Laboratory-scale and pilot-plant scale pulsed packed [21] and perforated-plate [22] columns have been reported (see also Chapters 11.1 and 11.2). Pilot test units, 40–50 mm and 80–100 mm in diameter with totally automated pulsed perforated-plate columns, have been reported [23]. Figure 11 shows a schematic of an operating pulsed perforated-plate column developed by the French Atomic Energy Commission (CEA).

6.2.2. Reciprocating-Plate Extraction Column

A 25-mm (1-in.)-diameter open-type perforated reciprocating-plate column has been reported by Lo and Karr [24] (also see Chapter 12). Figure 12 shows column arrangements and reciprocating-plate details. It is simple in construction and versatile for process feasibility study in the laboratory and in the minipilot plant [10]. The column has successfully been used for countercurrent and fractional liquid extraction [24]. A minimum HETS of 70 mm (2.8-in.) and volumetric efficiencies of up to 530 per hour were achieved on an MIBK–acetic acid–water system. The use of 51-mm (2-in.) and 76-mm (3-in.) columns is recommended for pilot-plant study to obtain data for column scale-up and design. Columns of various sizes, constructed of Pyrex with Teflon or stainless steel perforated plates, are commercially available [25].

6.2.3. Rotary-Agitated Columns

Rotating-disk contactors (RDCs) of 76-mm (3-in.) diameter and 150-mm (6-in.) diameter (also see Chapter 13.1) have been used for laboratory and pilot-plant studies [26, 27].

A 25-mm (1-in.)-diameter Scheibel column (see also Chapter 13.3) has been reported for laboratory process feasibility studies, and a 76-mm (3-in.)-diameter column with a Pyrex shell has been widely used for process and scale-up studies in the laboratory as well as in the pilot plant [28]. Data on scale-up and performance of a 1.45 m (4 ft 9 in.)-diameter Scheibel extractor have been reported by Lo [5].

A 152-mm (6-in.)-diameter Oldshue–Rushton column (see also Chapter 13.4) has been used in pilot-plant studies [29] to obtain data for scale-up and design of the columns.

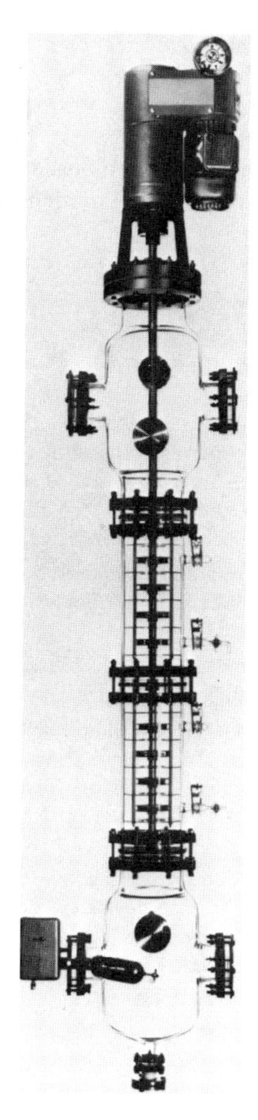

Figure 13. A 60-mm laboratory-scale Kühni column.

A 60-mm (2.35-in.)-diameter Kühni column (see also Chapter 13.5) having a Pyrex shell has been reported for pilot work [30]. Figure 13 shows a photograph of the column.

In another study [31] 101–152-mm (4–6 in.)-diameter RTL (formerly Graesser raining bucket) contactors (also see Chapter 13.6) were used for laboratory and pilot-plant development work.

6.2.4. Centrifugal Extractors

A bench-scale Podbielniak centrifugal extractor [32, 33] (see also Chapter 15) has been described as being useful for laboratory and pilot work. Bench-scale and pilot-plant ROBATEL centrifugal extractors, with two to four stages, have been reported [23] by the French Atomic Energy Commission (CEA).

6.2.5. *Miscellaneous Extractors*

An ultrasonic laboratory extractor has been reported for laboratory work [34] (see also Chapter 8). The extractor is used for processing small quantities of material in many stages.

REFERENCES

1. L. C. Craig, D. Craig, and E. G. Scheibel, Laboratory Extraction and Distribution, in A. Weissberger, Ed., *Separation and Purification* (*Techniques of Organic Chemistry*, Vol. 3), 2nd ed., Wiley-Interscience, New York, 1956, p. 149.
2. G. H. Morrison and H. Freiser, *Solvent Extraction in Organic Chemistry*, Wiley, New York, 1957, p. 79.
3. R. E. Treybal, *Liquid Extraction*, 2nd ed., McGraw-Hill, New York, 1963, p. 359.
4. E. G. Scheibel, Liquid–liquid Extraction, in E. S. Perry and A. Weissberger, Eds., *Separation and Purification* (*Techniques of Chemistry*, Vol. 12), 3rd ed., Wiley-Interscience, New York, 1978, p. 77.
5. T. C. Lo, Commercial Liquid–Liquid Extraction Equipment, in P. Schweitzer, Ed., *Handbook of Separation Techniques for Chemical Engineers*, McGraw-Hill, New York, 1979, Section 1.10.
6. L. C. Craig, *J. Biol. Chem.* **155,** 519 (1944).
7. S. Stene, *Ark. Kem., Miner. Geol.* **A18** (18) (1944).
8. E. L. Compere and A. Ryland, *Ind. Eng. Chem.* **46,** 24 (1954).
9. E. G. Scheibel, *Ind. Eng. Chem.* **46,** 43 (1964).
10. L. A. Robbins, *Chem. Eng. Progr.* **45** (September 1979).
11. M. M. Anwar, C. Hanson, and M. W. T. Pratt, *Chem. Ind.*, 1090 (August 9, 1969).
12. Horbury Technical Services, Ltd. (Yorkshire, England), *Mixer–Settler MKIII–A Continuous Bench-Scale Multistage Countercurrent Liquid–Liquid Contactor.*
13. A. Naylor, W. Baxter, A. Duncan, and A. F. D. Scott, *J. Nucl. Energy* (*Parts A/B*) **18,** 331 (1964).
14. R. W. Rahn and M. Smutz, *Ind. Eng. Chem. Process Des. Devel.* **8,** 289 (1969).
15. M-X Processer Reinhardt and Company AB, *M-X Processer Laboratory Scale Mixer–Settlers*, Molndal, Sweden.
16. D. R. Spink, University of Waterloo, Ontario, Canada, personal communication, 1981.
17. Davy International Ltd., Cleveland, UK, *Davy Power Gas Pilot Mixer–Settlers.*
18. Joy Manufacturing Company, Denver Equipment Division, Colorado Springs, Colorado, Sales Bulletin, 1980.
19. Bell Engineering, Tucson, Arizona, Sales Bulletin, 1980.
20. Establissements Sonal, Argenteuil, France; see CEA Bulletin 14/69, 1969.
21. N. U. Spaay, A. J. F. Simons, and C. P. ten Brink, *Proceedings of the International Solvent Extraction Conference, The Hague, 1971*, Vol. 1, Society of Chemical Industry, London, 1971, p. 281.
22. H. Rouyer, J. Lebouhellec, E. Henry, and D. Michael, *Proceedings of the International Solvent Extraction Conference, Lyons, 1974*, Society of Chemical Industry, London, 1974, p. 2339.
23. ERIES, London (1974) Le Commisariat A L'energie Atomique et La Societé Robatel SIPI (Genas, France), Bulletin on Pilot Test Unit Pulsed Column; Bulletin on Robatel Centrifugal Extractors.
24. T. C. Lo and A. E. Karr, paper presented at the Engineering Foundation Conference on Mixing Research, Andover, N.H. 1971; *Ind. Eng. Chem. Process Des. Devel.* **11** (4), 495 (1972).
25. Chem-Pro Equipment Company Bulletin KC-11, *Karr Column*, Julius Montz, GMbH (Hofstrabe, West Germany).
26. York Process Equipment Company, Parsippany, NJ, Bulletin 43, *The York Multi-Stage Rotating Disc Contactor.*
27. B. V. Escher, Technical Bulletin, *Rotating Disc Contactor for Solvent Extraction Processes*, The Hague, The Netherlands.
28. York Process Equipment Company, Parsippany, NJ, Bulletin 5M/263, *York–Scheibel Solvent Extraction Equipment for Laboratory and Pilot Plant.*
29. Mixing Equipment Company, Inc. (Rochester, NY), Bulletin on *Lightning Pilot-Scale Extractor.*
30. A. G. Kühni (Allschwil, Switzerland), Bulletin on *Laboratory-Scale Kühni Extractor.*
31. RTL Company (117 George St., London WH1 5TB), Bulletin on *RTL Contactor for Liquid–Liquid, Solid–Liquid Solvent Extraction.*
32. Baker Perkins, Inc. (Saginaw, Michigan), Bulletin on *Podbielniak Continuous Centrifugal Contactors, Pilot Plant or Laboratory Scale, Model A-1.*
33. D. B. Todd and G. R. Davis, *Proceedings of the International Solvent Extraction Conference*, Vol 3, Society of Chemical Industry, London, 1974, p. 2379.
34. Personal communication, Atomic Energy of Canada Ltd., Pinawa, Manitoba, Canada.

17.2

THE AKUFVE SOLVENT EXTRACTION SYSTEM

H. Reinhardt
MEAB Metallextraktion AB
Sweden

J. H. A. Rydberg
Chalmers University of Technology
Sweden

1. INTRODUCTION

The acronym AKUFVE is an abbreviation (in Swedish) for "apparatus for continuous study of distribution factors in liquid–liquid extraction." It was developed during 1962–1967 [1–8] to improve accuracy and rapidity in measurement of solvent extraction distribution factors.

The AKUFVE allows continuous measurement of the distribution of a dissolved species between the immiscible liquid phases, usually an organic solvent and an aqueous solution, provided the amount or concentration of the dissolved species can be measured on-line in solution (e.g., for β or γ radioactive elements [3, 9] or optically absorbing organic species [4, 10]). For other substances (e.g., an α radioactive element), sampling for offline measurements is necessary, although the liquid distribution system may run continually [11, 12]. The accuracy achieved in the distribution ratio measurements, which usually is better than 1%, has made the AKUFVE useful for the determination of equilibrium constants [12–14], thermodynamic constants [14–17], kinetics [18–20], and extraction isotherms [21–24]. Under normal conditions one distribution point is obtained each minute. With a data-logging system, points may be sampled even more rapidly, allowing studies of kinetics with half-lives down in the 10-s range [3, 25].

2. TECHNICAL DESCRIPTION

2.1. The H Centrifuge

The heart of the AKUFVE system is the H centrifuge [2, 5]. It is characterized by a comparatively high speed of rotation [10,000–25,000 rpm], short holdup time (0.3–5 s) and extremely high phase-separation efficiency; it usually produces both outgoing phases with a purity better than 99.9%, with respect to entrainment of the other phase [2].

The H centrifuge has been built in several sizes, of which the H-10 and H-33 versions are commercial. In principle, the design is the same [5], although the dimensions (both absolutely

TABLE 1 TECHNICAL DATA FOR H CENTRIFUGES[a]

Centrifuge Type	H-10 tr	H-10 efr	H-33 tr
Flow capacity (benzene/water), liters/hr	100	100	300
Bowl volume, liters	0.015	0.015	0.12
Hold-up time, s	0.5	0.5	1.5
Rotational speed			
No liquid flow, rpm	19,500	24,000	22,000
Maximum liquid flow, rpm	15,000	21,500	14,000
Maximum air consumption (6 bar), m^3/min	0.25	–	0.45
Maximum motor power, W	200	400	500

[a]Phase purity: 0.04% H_2O in benzene; 0.13% benzene in H_2O.

and relatively) vary. Pneumatic and synchronous electric motors are used. Table 1 summarizes dimensions and technical data. The smaller centrifuge (H-10) operates at higher angular velocity, which results in a considerably shorter holdup time of only 0.3 s, compared with 2 s for the larger one.

Tests show that absolute phase separation is achieved by the H centrifuges for most systems: none of the phases leaving the centrifuge contain entrained droplets of the other phase [1, 2, 28]. Such absolute phase separation requires optimal adjustment of the liquid flow parameters.

Figure 1 shows the principle of the centrifuge design. The two-phase mixture enters the centrifuge through the central inlet; the outlets for the two separated phases from the centrifuge are concentrically located. The rotating centrifuge bowl consists of (in order of liquid flow): (1) an acceleration chamber: (2) a separation chamber where the heavier phase is accumulated at the periphery and the lighter phase centrally; (3) pick-up chambers, one for the heavy phase located below and one for the light located above the separation chamber; and (4) stationary turbine-like wheels (scaling disks) that dip down into the rotating liquids in the pick-up chambers. Because of the location of pick-up chambers at each side of the separation chamber, no backmixing can occur. The turbine wheels provide pressure for the outgoing separated phases.

By adjustment of the inflow rate, the counterpressure on the outgoing phases and the speed of rotation, the interface between the light and heavy phases in the separation chamber can be forced to move radially. It is necessary to control this interface properly to obtain optimal phase separation. Different solvent extraction systems require different adjustments because of differences in density, surface tension, and viscosity. It may sometimes be necessary to exchange the turbine wheels for wheels of other dimensions. Since all wetted parts are made of titanium and fluorocarbons (Teflon, Viton, or similar material), the centrifuge is resistant to corrosion by most liquid systems. When additional corrosion resistance is required palladium-stabilized titanium is used.

Control arrangements are required for the centrifuge (see Fig. 2). These are a variable-speed

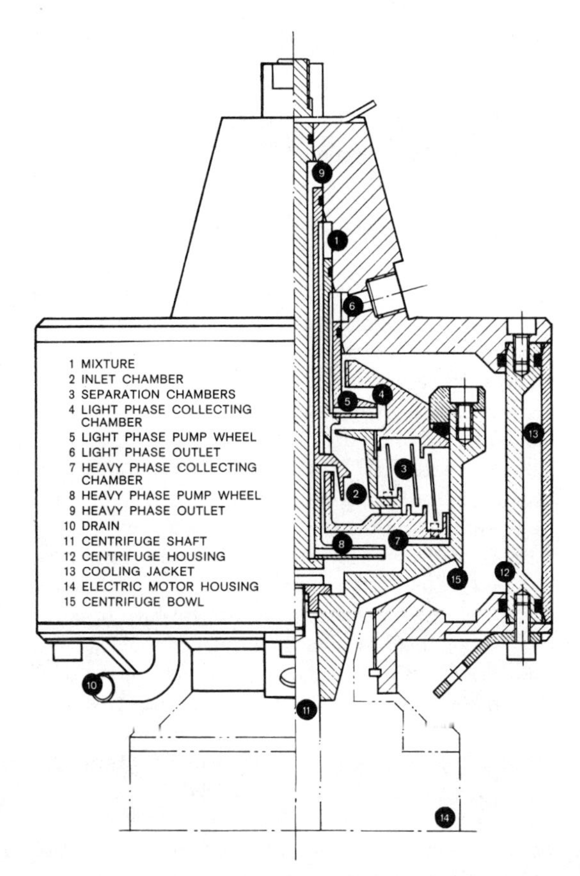

Figure 1. Cutaway of the H centrifuge [28].

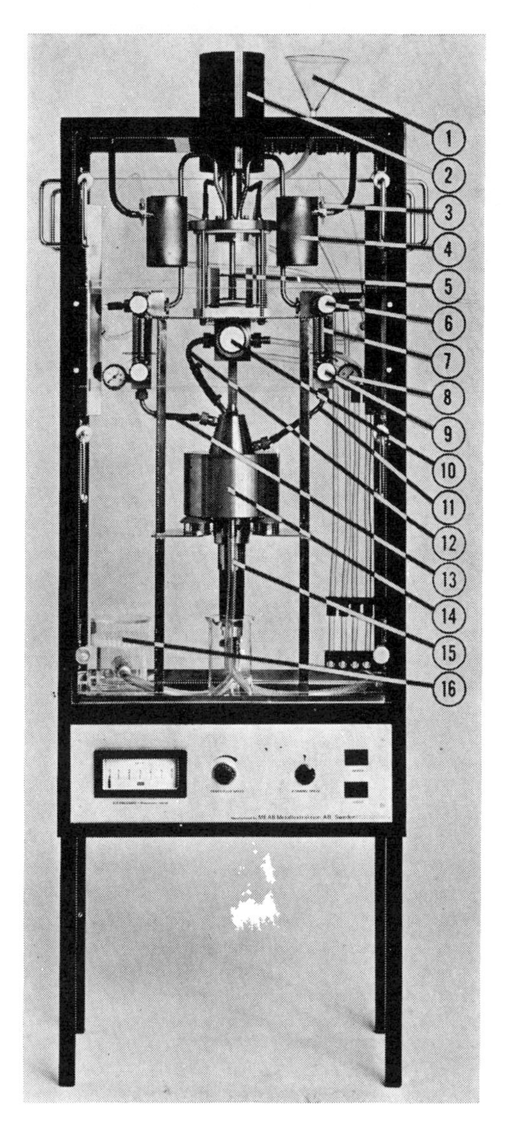

Figure 2. The AKUFVE system with the H-33 centrifuge: (1) main feed inlet; (2) stirrer motor; (3) sampling valve; (4) heat exchanger; (5) mixer; (6) valve "detector or mixer"; (7) flowmeter; (8) pressure gauge; (9) throttling valve; (10) valve "centrifuge-closed-drain"; (11) centrifuge outlet, light phase; (12) centrifuge inlet; (13) centrifuge outlet, heavy phase; (14) centrifuge; (15) centrifuge (air) motor; (16) silencer.

control for the motor, outlet pressure regulating valves, manometers, and flow meters.

2.2. The AKUFVE System

The AKUFVE system (Figs. 2 and 3) is used for rapid obtaining of distribution (Fig. 4) and extraction equilibrium (Fig. 5) curves. The two liquid phases are mixed in the mixing chamber and again separated into two outgoing pure phases in the centrifuge. Each phase passes the flow meter, the sampling valve, and on-line detectors (in an external loop) and is then returned by way of the heat exchanger to the mixing vessel, providing a closed circuit. Chemical reagents are added to the mixing chamber when desired. The mixing chamber is either of the dynamic stirrer type (Fig. 2) or is a static Kenics mixer (25) (Fig. 7). The latter is used for very short mixing times (<1 s).

During sampling, usually 0.1–0.5 ml is withdrawn at the sampling valve. For on-line measurements, with the use of a radioactive tracer of the element, the detector cell may in its simplest version be a flexible tube wound around a scintillation crystal in a lead shield. A special "pocket" is used for on-line pH-measurement with combined glass electrodes.

Heat develops as a result of the acceleration and retardation of the liquids in the centrifuge; thus heat exchangers are necessary for constant-temperature work, especially with the H-33 centrifuge. The total liquid (both phases) volume of an AKUFVE system with the use of the H-33 centrifuge is $\sim$500 ml; the H-10 system (see Fig. 7) requires only $\sim$100 ml. The latter is preferable when the use of chemicals is expensive, dangerous, or otherwise undesirable. On the other hand, the H-33 is easier to handle in routine laboratory work; also, sampling removes a smaller fraction of the total liquid volume.

2.3. Data Collection

The primary purpose of the original AKUFVE system was to obtain accurate distribution data in a minimum of time. For the system shown in Fig. 3, complete mixing equilibrium (but not necessarily chemical equilibrium) is obtained within one minute. In order to measure the concentration of the species of interest in each phase, in addition to pH, temperature, reagents added, and so on as often as once per minute, a data-logger is required [3, 25]. In sampling for external measurements (e.g., by using an external detector combined with an automatic sample changer), an operator can normally collect a sample of each phase in approximately 3 minutes. The data logger can provide outputs in the form of punched tape, magnetically stored information, and so on or can be connected on-line to a computer [28]. In elaborate systems, automatic burettes have been connected to the AKUFVE systems. However, even in simple sampling for off-line measurements, the AKUFVE provides more exact data in a shorter time than seems achievable with simpler techniques.

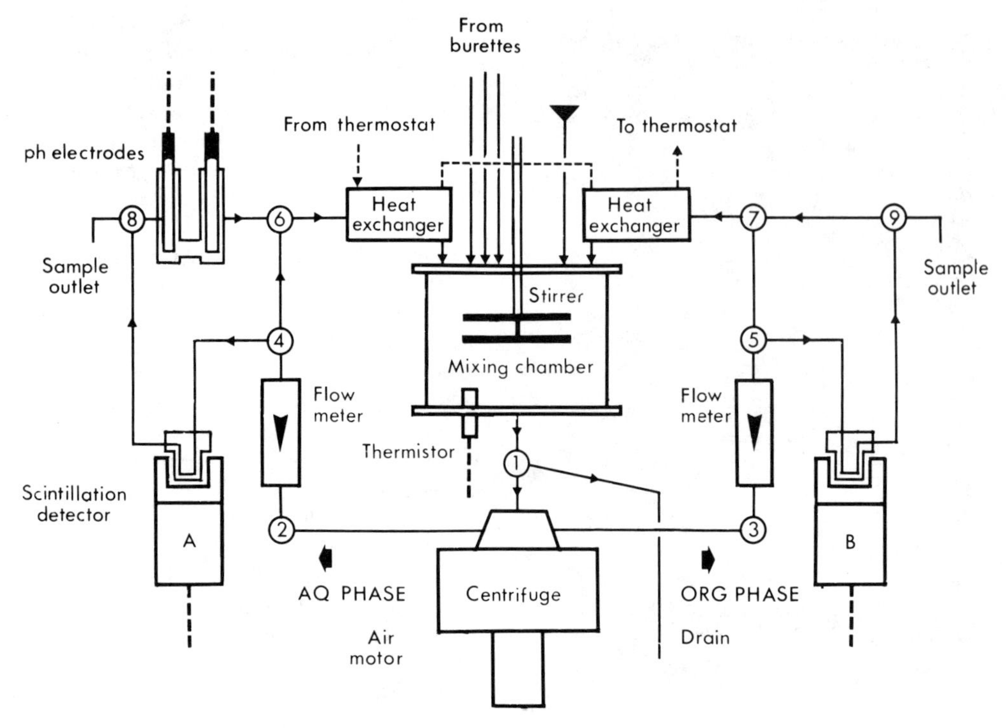

Figure 3. Diagram of the AKUFVE liquid flow system (○ = valve [3]).

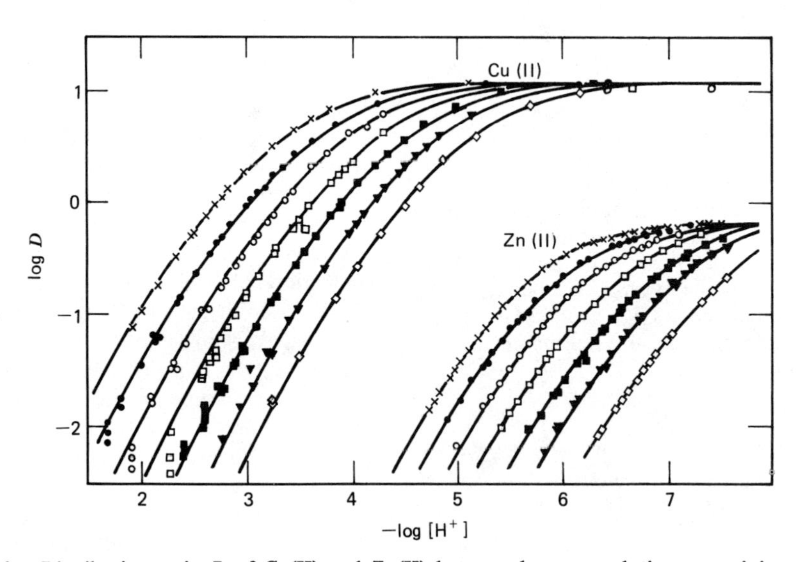

Figure 4. Distribution ratio D of Cu(II) and Zn(II) between benzene solutions containing various amounts of acetylacetone and 1 M $NaClO_4$ as a function of the H^+ concentration [28]. Each curve is obtained in one AKUFVE run lasting 2–3 hr.

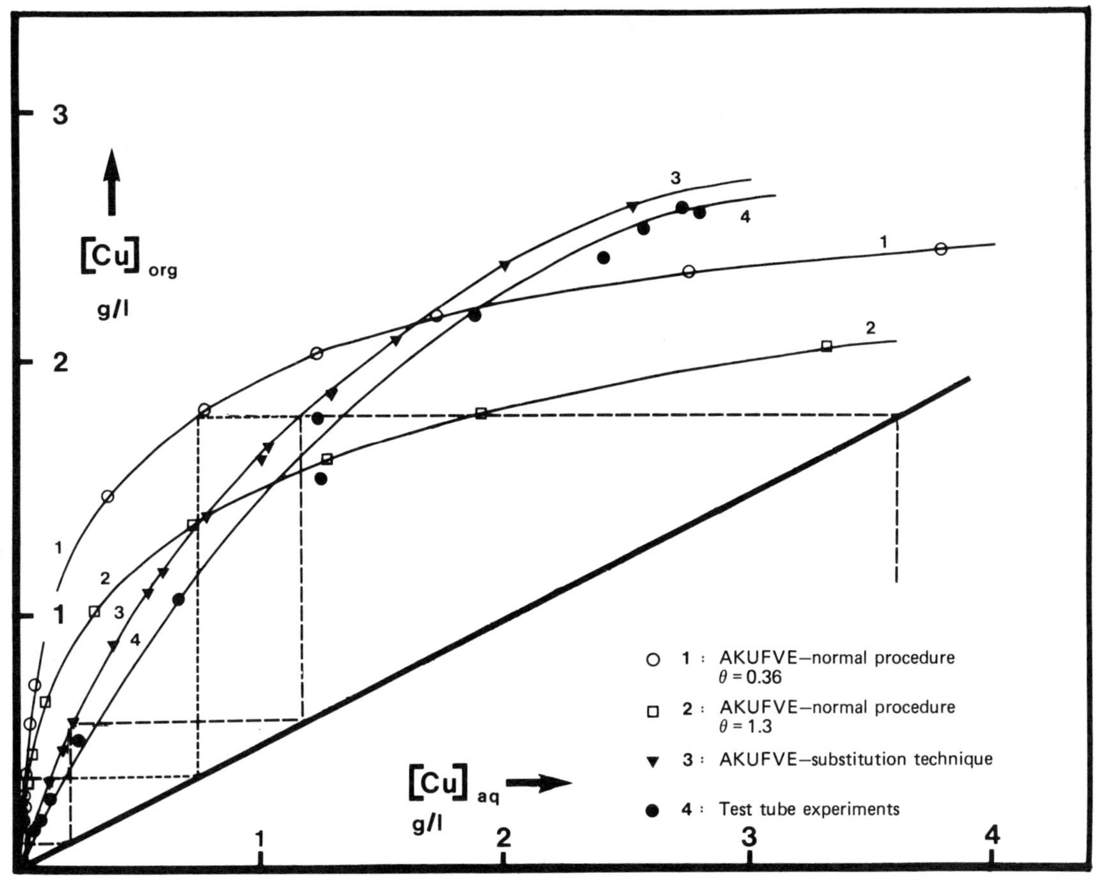

Figure 5. Equilibrium curves for the Cu^{2+}-H_2SO_4-LIX64N in kerosene system. All experiments start with same concentration of copper (~3 g/liter), H_2SO_4 (2.5 g/liter), and LIX (10%) in kerosene.

3. APPLICATIONS

Various applications of the AKUFVE have already been described in review papers [27, 28].

3.1. Basic Solvent Extraction Studies

Basic research on solvent extraction with use of the AKUFVE has been concerned mostly with indentification of the extracted complexes and studies of their interaction with different solvents from measurements of distribution curves (Fig. 4). When possible, these curves are theoretically analyzed and explained in terms of formation constants for various metal complexes and distribution constants for the extractable complex [12–14]. From studies of the distribution of a metal as a function of temperature or of organic solvent composition, thermodynamic constants and specific interaction parameters may be determined [14–17]. Since the AKUFVE is an on-line system, it has also been used for *kinetic studies* of solvent extraction [18–20].

3.2. Extraction Curves for Process Application

McCabe–Thiele diagrams are used for calculation of the number of stages in industrial metal extraction (separation) operations. Figure 5 illustrates such a diagram obtained with the AKUFVE system [22, 23]. To obtain information suitable for application on technical countercurrent processes, the "substitution technique" is used. Curves 1 and 2 are obtained with the AKUFVE system by using the basic D-value procedure (as for Fig. 4), curve 3 with the AKUFVE by using the substitution technique, and curve 4 with a normal test tube procedure.

In curve 1 the phase ratio is 0.36 (aqueous continuous), and in curve 2 the phase ratio (organic : aqueous) is 1.3 (organic continuous). The copper concentration is increased in steps by adding a small amount of a concentrated copper sulfate solution. After each addition, the copper concentrations in the two solutions are measured. In this procedure the sulfate concentration and pH change successively.

In curve 3 an aqueous feed solution containing 3.8 g/liter of Cu in H_2SO_4 and the organic LIX64N solution are added to the AKUFVE (the H-33 system). The concentration of copper in both solutions is measured. A portion (>100 ml) of the organic solution is then withdrawn from the AKUFVE and replaced with the same amount of fresh organic solution. In this way the total sulfate concentration is constant, whereas the

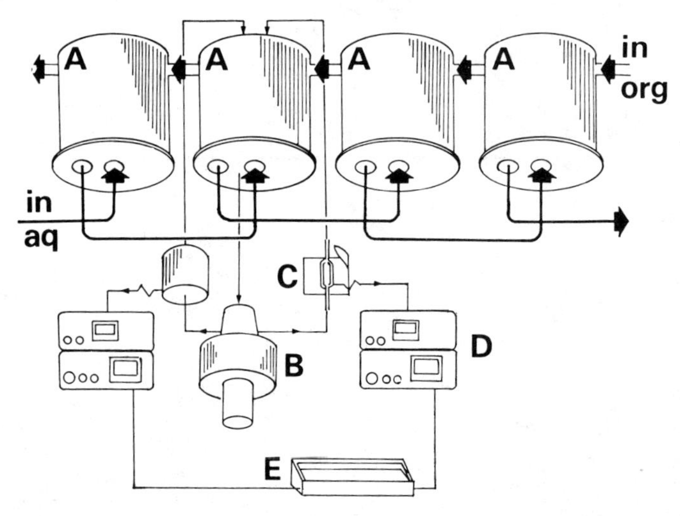

The AKUFVE-system with radiometric detection connected to a four stage mixer-settler unit
A – mixer-settlers; B – H-centrifuge for absolute phase separation; C – on line measuring cells for γ-radiation; D – power supply and rate meter; E – two pen recorder

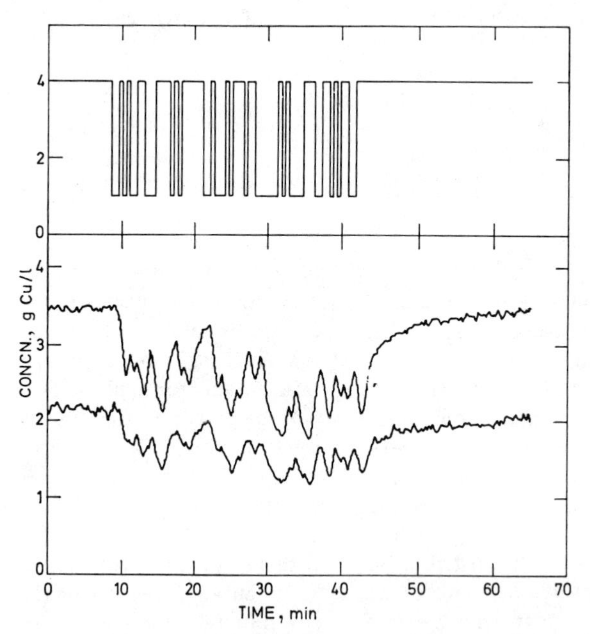

Response curves obtained for concentration PRBS changes in stage 1 of a four stage mixer-settler unit

Figure 6. Example of H centrifuge being used for studying process dynamics [29].

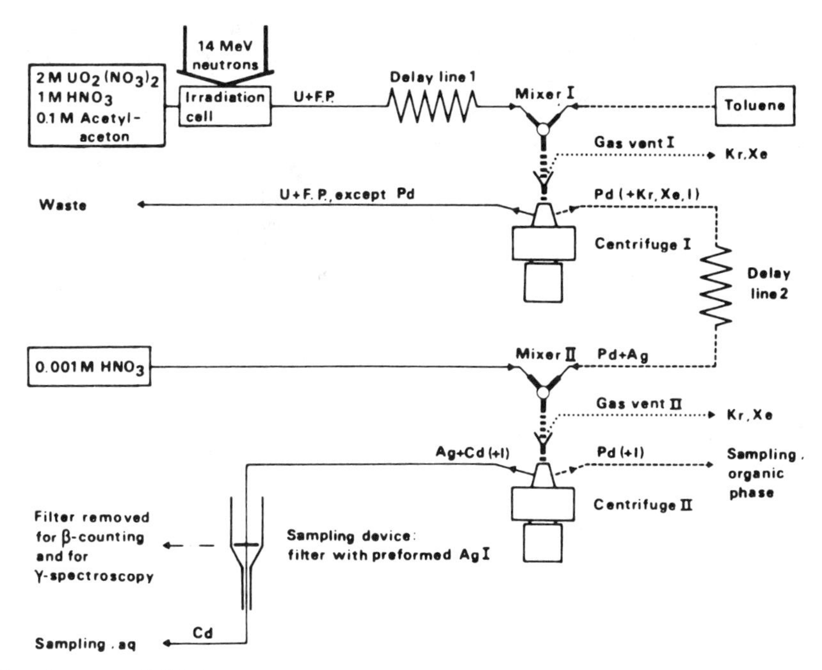

Figure 7. Two-stage SISAK system used for separation and isolation of palladium from fission products. Solid and dashed lines refer to aqueous and organic flows, respectively [30].

total copper concentration in the system is decreased in a manner similar to that in a mixer–settler battery. The H^+ concentration is allowed to vary freely, as in the industrial process. The substitution procedure is repeated to cover the concentration range of the process. If necessary, the total copper concentration can be increased by replacement of the aqueous solution in the AKUFVE [24].

Curve 4 was obtained in normal test tube experiments by measuring the equilibrium values when H_2SO_4 solutions containing copper (as sulfate) were equilibrated with LIX64N in kerosene at different phase ratios (organic : aqueous), varying from 20 : 1 to 1 : 20.

As can be seen from Fig. 5, quite different McCabe–Thiele constructions are obtained if curve 1 or 2 is used instead of curve 3 or 4. However, this discussion holds only for single-metal systems; theoretical calculations and practical evaluations are discussed by Liljenzin [24(a)]. The AKUFVE substitution technique has been frequently used in applied laboratories to obtain data for industrial solvent extraction systems.

3.3. Solvent Extraction Dynamics

In industrial solvent extraction with mixer–settlers, the physical conditions are rarely ideal. Phase transfer and response to external changes may be slow. To investigate about what happens under true *dynamic conditions*, small liquid volumes from an individual large-scale stage have been continually withdrawn to an H centrifuge for direct on-line analyses of the composition of the phases. The use of this technique is illustrated in Fig. 6 [29]. The dynamic response of the H centrifuge itself has been studied by Aronsson [30].

3.4. Separation of Short-Lived Products

The short contact time of the H-10 centrifuge (~0.3 s) and its high separation efficiency makes

Figure 8. Four H-10 centrifuges with static mixers (see also Fig. 7) and control heads built into a four-stage SISAK countercurrent extraction battery [32].

it useful for *rapid separation* of *short-lived products*. In the *SISAK technique* (SISAK = short-lived isotopes studied by the AKUFVE technique) about 20 new fission isotopes with half-lives down to 1 s have been isolated and studied on line [30–32], allowing the determination of detailed decay schemes; see Fig. 7. The SISAK technique uses several AKUFVE units connected in series to make up a multistage separation system. The transfer time through the four-stage battery in Fig. 8 is ~10 s.

ACKNOWLEDGMENT

The AKUFVE was developed with economic support from the Swedish Technical Research Council and Swedish Atomic Energy Commission.

REFERENCES

1. J. Rydberg, *Acta Chem. Scand.* **23,** 647 (1969).
2. H. Reinhardt and J. Rydberg, *Acta Chem. Scand.* **23,** 2773 (1969).
3. C. Andersson, S. Andersson, J. O. Liljenzin, H. Reinhardt, and J. Rydberg, *Acta Chem. Scand.* **23,** 2781 (1969).
4. H. Johansson and J. Rydberg, *Acta Chem. Scand.* **23,** 2797 (1969).
5. J. Rydberg and H. Reinhardt, U.S. Patent 3,442,445 (1969).
6. J. Rydberg and H. Reinhardt, U.S. Patent 3,615,227 (1969).
7. H. Reinhardt, *Chem. Ind.* 363 (1972).
8. J. O. Liljenzin and H. Reinhardt, *Ind. Eng. Chem. Fund.* **9,** 248 (1979).
9. J. F. Desreux, *Proceedings of the International Solvent Extraction Conference*, The Hague (Soc. Chem. Ind.), 1971, p. 1328.
10. S. S. Davies, G. Elson, E. Tomlinson, G. Harrison, and J. C. Dearden, *Chem. Ind.* 677 (1976).
11. M. Cox and D. S. Flett, *Proceedings of the International Solvent Extraction Conference, The Hague 1971*, Society of Chemical Industry, 1971, p. 204.
12. J. O. Liljenzin and J. Stary, *J. Inorg. Nucl. Chem.* **32,** 1357 (1970).
13. D. R. Spink and D. N. Okuhara, *Proceedings of the International Solvent Extraction Conference, Lyons 1974*, Society of Chemical Industry, 1974, p. 2527.
14. B. Allard, S. Johnson, and J. Rydberg, *Proceedings of the International Solvent Extraction Conference, Lyons 1974*, Society of Chemical Industry, 1974, p. 1419.
15. J. O. Liljenzin, J. Stary, and J. Rydberg, in A. Kertes and Y. Marcus, Eds., *Solvent Extraction Research*, Wiley, New York, 1969.
16. D. S. Flett, D. R. Spink, *Proceedings of the International Solvent Extraction Conference, Toronto 1977*, Canadian Institute of Mining and Metallurgy Publication 21, 1977, p. 496.
17. J. Rydberg, *J. Ind. Chem. Soc.* **51,** 15 (1974).
18. J. O. Liljenzin, K. Vadasi, and J. Rydberg, *Transact. Roy. Inst. Technol. (Stockholm)* (280), 407 (1979).
19. E. S. Perez de Ortiz, M. Cox, and D. S. Flett, *Proceedings of the International Solvent Extraction Conference, Toronto 1977*, Canadian Institute of Mining and Metallurgy Publication 21, 1977, p. 198.
20. H. Ottertun and E. Strandell, *Proceedings of the International Solvent Extraction Conference, Toronto, 1977*, Canadian Institute of Mining and Metallurgy Publication 21, 1977, p. 198.
21. A. Aue, L. Skjutare, G. Björling, H. Reinhardt, and J. Rydberg, *Proceedings of the International Solvent Extraction Conference, Lyons 1974*, Society of Chemical Industry, 1974, p. 447.
22. J. Rydberg, H. Reinhardt, B. Lundén, and P. Haglund, *International Symposium on Hydrometallurgy*, Chicago, 1973, AIME Publ., p. 589.
23. H. Reinhardt, H. Ottertun, and T. Troëng, *Inst. Chem. Eng. Symp. Ser.* **41,** WI (1975).
24. (a) J. O. Liljenzin, in *Reinstoffprobleme*, Vol. V, Akademie Verlag, Berlin, 1977, p. 169; (b) J. O. Liljenzin, *Proceedings of the International Solvent Extraction Conference, Toronto, 1977*, Canadian Institute of Mining and Metallurgy Publication 21, 1977, p. 295.
25. S. O. S. Anderson and D. R. Spink, *Can. Res. Devel.*, 16 (Nov./Sec. 1970).
26. T. O. Craig and T. E. Jenkins, *Chem. Ind.* 268 (1978).
27. H. Reinhardt and J. Rydberg, *Het Ingenieursblad* **41,** 453 (1972).
28. J. Rydberg, H. Reinhardt, and J. O. Liljenzin, *Ion Exch. Solv. Extr.* **3,** 111 (1973).
29. G. Aly, Å. Jernqvist, H. Reinhardt, and H. Ottertun, *Chem. Ind.* 1046 (1971).
30. P. O. Aronsson, Dissertation, Chalmers University of Technology, Göteborg, 1974.
31. G. Skarnemark, Dissertation, Chalmers University of Technology, Göteborg, 1977.
32. P. O. Aronsson, B. E. Johansson, J. Rydberg, G. Skarnemark, J. Alstad, E. Beyersen, E. Kvåle, and M. Skarestad, *J. Inorg. Nucl. Chem.* **35,** 2397 (1974).

Part IIIA

Industrial Processes—Organic

18.1

PETROLEUM AND PETROCHEMICALS PROCESSING—INTRODUCTION

P. J. Bailes
University of Bradford
United Kingdom

The need to separate mixtures of aliphatic and aromatic hydrocarbons provided one of the first large-scale applications for solvent extraction. In this connection the particular advantage of solvent extraction was that it represented a physical means for separating groups of components of similar chemical type. Thus a solvent that preferentially dissolved aromatic components from a mixture of hydrocarbons could be used to remove aromatics from kerosenes and lubricating oils. In the former case this treatment improves the clean burning properties of the fuel, whereas separation of the aromatics from the isoparaffins and naphthenes in a lubricating oil improves the viscosity–temperature relationship. In both these examples the end product is a high-quality raffinate; however, solvent extraction is now used equally extensively for the production of high-purity aromatic extracts from catalytic reformates containing some 45–60% aromatics. The benzene, toluene, and xylenes (BTX) produced in this manner are the essential raw material for such products as polystyrene, nylon, and Terylene.

Although great volumes of "broad-cut" raffinates of improved paraffinicity are still required for fuels and lubricants, the development emphasis is on the production of high-purity aromatic extracts. This has resulted in several different commercial processes for the manufacture of BTX, and detailed descriptions of the three principal processes are given in this chapter.

Although the treatment of lubricating oils and BTX extraction are the largest applications of solvent extraction in the organic field, the technique is also used in the manufacture of several important petrochemicals. For example, the production of high-purity fiber-grade caprolactam can be achieved by using vacuum distillation and crystallization from water, but the favored route employs solvent extraction. This need not be confined to the final purification of the caprolactam monomer, however, and it is frequently used earlier in the processing scheme for the recovery of the product intermediate cyclohexanone oxime. Even when the caprolactam is produced from caprolactone and not cyclohexanone oxime, solvent extraction with the process cyclohexane is still used to recover unreacted cyclohexanol and cyclohexanone from aqueous liquors.

A particular instance where solvent extraction provides substantial savings in plant investment and operating costs is the commercial production of anhydrous acetic acid. Here the solvent is used to recover the acid from an aqueous phase and its use greatly reduces the amount of water that must be distilled in order to produce anhydrous acid.

Extraction processes for organic materials are not all based on relatively unspecific physical interactions between the solute and solvent. There can be a definite chemical reaction with the solvent analogous to that which is commonly found in the solvent extraction of metallic species. Thus xylenes react rapidly and reversibly to form complexes with boron trifluoride in the presence of liquid hydrogen fluoride. Under the correct conditions this behavior can be used to separate *m*-xylene from its more marketable isomers because the *m*-xylene complex is much more stable than the others and, furthermore, is preferentially soluble in the excess hydrogen fluoride.

It is clear that solvent extraction in the petroleum and petrochemical industries has proved an economically viable way of improving product purity and reducing energy consumption, despite the fact that solvent recovery invariably involves costly distillation. This is partly the result of careful process integration, but also the development and large-scale use of new types of extractor has been important. The variety and size of column contactors that is to be found in the industry is unparalleled elsewhere. For example, the design of pulsed packed columns and assymetric rotating disk contactors (RDCs) owes much to the experience gained in caprolactam extraction.

These considerations, together with the processes that have been outlined in this introduction, are covered in much greater detail in the chapters that follow.

18.2

AROMATICS–ALIPHATICS SEPARATION

P. J. Bailes
University of Bradford
United Kingdom

The first efficient method of recovering aromatics was the *Udex process*, which used a glycol-based solvent. This was introduced in 1952 and was superior then because, for the first time, a wide-boiling feedstock could be treated directly without expensive prefractionation and yet the BTX products were of unusually high purity. The combination of solvent extraction and extractive stripping that made this possible has since been adopted for other solvents such as sulfolane and *N*-methyl pyrrolidone.

Numerous factors contribute to the success of an aromatics extraction process. For example, the process must be flexible and easy to operate. Flow sheet design must be sound since solvent loss and heat requirements can be greatly reduced by the skillful combination of extraction, extractive distillation, and heat exchange. The contactor designed to use the solvent must be efficient and flexible to allow minimization of capital and operating costs. It is the solvent characteristics, however, that really have the most effect on the process economics.

Many solvents that can be seriously considered for aromatics extraction have been found, but only a few of these have received commercial acceptance. The solvents of principle industrial interest in this field are listed in Table 1, together with some information on physical properties. It should be noted that they all have densities and boiling points that are higher than the material to be extracted and that all are, to a greater or lesser extent, polar molecules and water soluble. The solvent properties of primary importance, however, are selectivity and capacity.

It is necessary that the solvent form a nonideal solution with the mixture to be separated. The solvents used in aromatic–aliphatic separations have a purely physical interaction with the hydrocarbon phase. The hydrocarbons are repelled by the polar solvent to a greater or lesser extent, depending on the type of hydrocarbon; olefins and paraffins have less affinity for the solvent than do aromatics and bicycloparaffins. The relative repulsion that a solvent shows for two different hydrocarbons is a measure of its selectivity. This may be quantified as the ratio of the distribution coefficient for one component between hydrocarbon and solvent phases to that of the other component between the two phases. The selectivity of a solvent for aromatics may be defined as the ratio of the activity coefficient of a particular aliphatic component in the solvent to that of a desired aromatic, if it is assumed that the components behave in a near ideal manner in the hydrocarbon phase. Reference hydrocarbons for testing solvent selectivity might be, for example, *n*-heptane : toluene, hexane : benzene, or methyl cyclopentane : benzene.

The activity coefficients in the solvent phase will depend on temperature and composition. Strictly, therefore, comparison of the selectivity of various solvents is valid only if the coefficients are measured under equivalent conditions. This can be achieved by choosing standard hydrocarbon concentrations or by working with infinite dilution, that is, by limiting activity coefficients. In either event, since activity coefficients may vary with concentration in different ways for different solvents, it is conceivable that the relative performances of solvents will change a little with concentration.

The solvent power or capacity is the next most important property after selectivity, since

TABLE 1 COMMON SOLVENTS FOR BTX AROMATICS EXTRACTION

Solvent	Boiling Point, °C	Specific Gravity at 20°C	Specific Heat 20–30°C
Polyethylene glycol–water mixtures	180–280	1.02–1.13	0.55
Sulfolane	285	1.26	0.35
N-Methyl pyrrolidone (NMP)	200	1.03	0.40
N-Formylmorpholine (NFM)	244	1.12	0.38
Dimethyl sulfoxide (DMSO)	195	1.10	0.49
Tetraethylene glycol	325	1.13	0.63[a]

[a]At 150°C.

this is a measure of the quantity of the material to be separated that can be contained in the solvent phase. It can be defined as the reciprocal of the activity coefficient of a component in the solvent; for convenience, it is usually quantified by the distribution coefficient of an aromatic such as toluene.

Solvents can be readily compared by examining their selectivity–capacity characteristics, provided these are defined according to a common basis. Difficulties can arise as a result of inconsistent definitions of selectivity and capacity and also in cases where the solvent contains an additive such as water. Nevertheless, such comparisons are very helpful for selection between existing commercial solvents and for screening new solvents. Extensive comparative data are available in the literature, and the reader is especially referred to the work by van Aken and Broersen [1] and Asselin and Persak [2].

The ideal solvent should have sufficient capacity to ensure good aromatics recovery without excessive solvent circulation rates. Selectivity for aromatics should, of course, be high over this capacity range. With real solvents, this situation can be only approached since, as the solvent contains progressively more of the material being separated, it gradually loses the properties that distinguish it as a selective solvent. In other words, capacity is achieved at the expense of selectivity. It follows from this, for example, that the changing composition of the solvent as it passes through the extractor will influence selectivity and may impose a minimum solvent rate in order to retain selectivity at the aromatic-rich end of the extractor.

The successful commercial solvents all demonstrate a good compromise between selectivity and capacity. Clearly, the extraction process requires that the hydrocarbon solubility in the solvent be such that two liquid phases can form, and this is not the case if the capacity is too large. Equally, too low a capacity necessitates unacceptable solvent rates. The solvent properties should be such that operation with a high degree of selectivity is possible at an intermediate capacity.

The desired combination of properties can be possessed by a pure solvent, as is the case with sulfolane. More frequently, however, a second polar component will be added to a solvent to achieve the required optimum between selectivity and capacity. In this regard, water, with its good selectivity and very low capacity, is often mixed with solvents to improve their selectivity with some sacrifice of capacity. The advantage of using water is that it is usually already present in the process for washing solvent from the raffinate stream. The disadvantage is the additional heat consumption at the solvent recovery stage. Other solvent mixtures are used, such as ethylene glycol with *N*-methyl pyrrolidone, as is discussed in the section on the Arosolvan process. Perhaps the main benefit of using a mixed solvent is that the properties can be tailored to fit a change in feedstock.

An exchange between selectivity and capacity can also be achieved by altering the temperature. A reduced solvent temperature favors selectivity but diminishes capacity.

The secondary properties of a solvent must also be favorable if it is to form the basis of a commercially viable process. Features such as poor thermal stability or poor density difference between solvent and hydrocarbon phases can easily exclude a solvent with otherwise excellent properties. Other factors must be taken into account, such as melting point, toxicity, price, resistance to chemical attack, and those physical properties that directly affect the cost of solvent recovery. The importance of these solvent characteristics is discussed further in the sections on specific commercial processes.

Although the different processes in BTX aromatics extraction are primarily distinguished by the solvent on which they are based, it is also true that when new solvents have been introduced, they have often been used in novel types of extractor. This equipment then becomes a standard feature of the particular process.

The RDC thus tends to be used with sulfolane; sieve-tray columns are usually employed with the Udex process, and the Kühni column has been applied with dimethylsulfoxide. There are obvious advantages to the process developer and to the user if proprietary knowledge of a solvent can also be extended to the equipment. This is clearly demonstrated with sulfolane, where both the solvent and the extractor were Shell developments. Similarly, the *N*-methyl pyrrolidone processes developed by Lurgi make use of their vertical multistage mixer–settler extractor.

The accumulated experience with the design and operation of the extraction columns has contributed to the success of the Udex, Sulfolane, and *N*-methyl-pyrrolidone processes and is perhaps one of the principal reasons why these three processes dominate the aromatics extraction business at the moment. Process development is not static, of course, and gradual improvements in terms of both quality and recovery of aromatics and also capital and operating costs are continually being made to existing processes. It seems unlikely, however, that a superior process based on a new solvent will be discovered in the foreseeable future.

REFERENCES

1. A. B. van Aken and J. M. Broersen, *Proceedings of the International Solvent Extraction Conference (ISEC)*, Toronto, 1977, Vol. 2, Canadian Institute of Mining and Metallurgy, Montreal, 1979, p. 693.
2. G. F. Asselin and R. A. Persak, *Petrolieri Internazional*, 46 (September 1978).

18.2.1

NMP (AROSOLVAN) PROCESS FOR BTX SEPARATION

E. Müller
Lurgi Khole und Minerelöltechnik GmbH
West Germany

1. FUNDAMENTALS OF THE PROCESS

Extraction is the most important process for the recovery of pure aromatics from hydrocarbon mixtures. It requires a solvent that is not miscible with nonaromatics and has a higher solvency for aromatics than for nonaromatics.

When extracting a hydrocarbon mixture by using a selective solvent, one obtains two streams, both of which do not have the desired purity. The one stream is the raffinate, that is, the nominally aromatics-free nonaromatic hydrocarbons. The residual aromatics content of the raffinate stream can be reduced to an economically feasible level by a great number of extraction stages and by using surplus solvent. The solvent in the raffinate must be recovered and returned to the cycle. The second stream is the aromatics-laden solvent. Apart from the aromatics, the extract also extracts some nonaromatics (see Fig. 1).

The mixture of solvent and hydrocarbons leaving the extractor is charged to the top of a distillation column. The 1.2–2-fold quantity of nonaromatics entrained with the solvent is removed as overhead vapor. The condensed vapor is returned to extraction as countercurrent reflux to the laden solvent. This recycle contains the low-boiling aromatics, chiefly benzene, plus nonaromatics. As the distillation proceeds in the presence of the solvent, it is an extractive distillation without a rectification section. The distillate, therefore, contains a higher proportion of high-boiling nonaromatics than would be the case in the absence of the solvent. On its way through the extractor, the benzene and some of the nonaromatics pass into the solvent phase. The benzene and low-boiling nonaromatics thus become concentrated in this extraction and extractive distillation loop and displace the high-boiling nonaromatics, which cannot be separated in the extractive distillation. The residual nonaromatics content of a few hundred parts per million de-

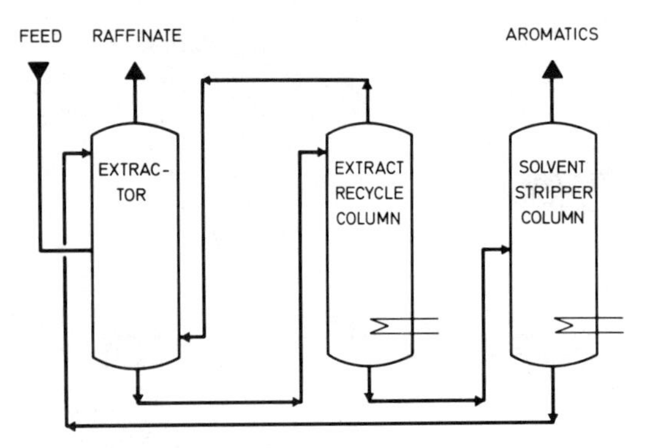

Figure 1. Principle of aromatics extraction plant.

pend mainly on the concentration in the feed and the number of extractor stages between feed and recycle point.

The extractive distillation bottoms product is a mixture of solvent and aromatic hydrocarbons that can easily be separated by distillation. The raffinate is treated by water wash to remove dissolved solvent. The resulting mixture of solvent and water can be separated either in a separate distillation column or in one of the two other distillation columns.

2. ASPECTS OF THE CHOICE OF SOLVENT

Apart from the basic requirements, the characteristics of importance for a solvent for use in a commercial process are selectivity, capacity, aromatics range, solubility in hydrocarbons, density, viscosity, boiling point, thermal stability, solidification point, corrosiveness, toxicity, and price.

Consideration of all major characteristics has led to the selection of *N*-methyl-pyrrolidone (NMP) as the solvent, to which ethylene glycol or water is added to adjust the optimum polarity.

2.1. Boiling Temperature

Separation of the solvent from the extracted substance is generally achieved by distillation, the cost of which is of considerable importance for the economics of the process. For solvents boiling above the extract, the following applies: the higher the boiling point of the solvent, the lower the reflux ratio for distillation and the higher the energy requirements for heating the solvent to its boiling point.

A favorable boiling point for a solvent is 200°C, which is 50°C above xylene and 25°C above the C_9 aromatics. Solvents boiling in that range are preferred.

2.2. Thermal Stability

Apart from the aspect of heat economy, thermal stability is of major importance for the choice of solvent, as it has the following influence on the economics of a process: (1) Polymers would require the provision of regeneration facilities; (2) decomposition products would require postpurification of the aromatics; and (3) corrosive substances would require the use of stainless steel equipment.

To keep the cost for replacement of decomposed solvent at a reasonably low level in terms of overall operating costs, the solvent loss should not be higher than about 0.1 kg/ton of aromatics.

N-Methylpyrrolidone has proved successful in terms of thermal stability in commercial plant operation. It has not been possible so far to prove a thermal decomposition reaction mechanism or velocity for NMP.

Use of a solvent with a solidification point of above ambient temperature requires that all equipment and piping be steam traced. The low solidification points of NMP (-24°C), and of the mixtures of NMP with water or glycol (less than -40°C) are characteristics highly appreciated by users.

2.3. Density and Viscosity

Density and viscosity are additional properties to be considered for the choice of a solvent for extraction. It is desirable to have a good differential density between the two phases to promote phase separation. Low viscosity is preferred because this favors both phase separation and mass transfer. Whereas the NMP lies in the middle of all solvents with regard to density, it has almost the optimum value as far as viscosity is concerned.

2.4. Selectivity and Capacity

Selectivity is the principal aspect normally considered in solvent choice. Capacity tends to be neglected but can play an equally important role.

The capacity of a solvent is defined by the distribution factor, that is, the aromatics concentration ratio in the solvent to the raffinate phase. Selectivity is defined by the ratio of the distribution factor of the aromatics to the distribution factor of the nonaromatics.

To compare solvents in terms of these two

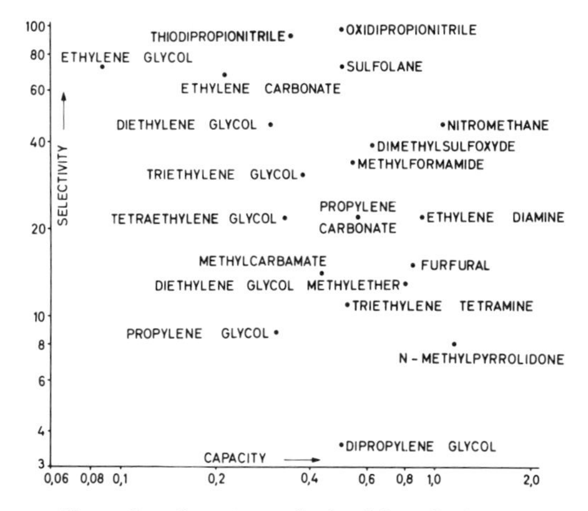

Figure 2. Capacity and selectivity of solvents.

properties, they are plotted on a diagram where the abscissa shows the capacity and the ordinate the selectivity. Figure 2 presents this comparison for those solvents of industrial interest.

As is seen later, most solvents with high capacity have a low selectivity and vice versa. *N*-Methylpyrrolidone has a very high capacity but not a high selectivity.

An economic evaluation shows that the solvent capacity has a much greater influence on the production costs than the selectivity of the solvent. However, the capacity depends not only on the distribution factor of the aromatics, but also on the limits for the formation of a two-phase system. These characteristics are referred to as the *aromatics range* of the solvent.

2.5. Aromatics Range

The aromatics range limits the permissible aromatics content in the feed. When this limit is exceeded, the two phases in the feed stages of the extractor disappear, and the extraction process breaks down.

As countermeasure, the aromatics content in the feed may be reduced by recycling part of the raffinate or the solvent circulating rate may be increased. Another countermeasure that is frequently applied involves increasing the aromatics range by the addition of water or other highly polar compounds to the solvent.

2.6. Addition of Polar Mixing Component

The addition of water has a twofold advantage: it increases the selectivity and decreases the boiling point of the solvent. On the other hand, it has the disadvantage that solvent capacity is reduced and heat consumption in the distillation columns increased because water must be co-evaporated azeotropically during stripping of the aromatics.

N-Methylpyrrolidone is a solvent that can be used without additive only for the extraction of small quantities of aromatics from a mixture, for instance, for the production of aromatics-free solvent for oilseed extraction. When employed for aromatics extraction from reformates and hydrogenated pyrolysis gasoline, a mixing component is added to the NMP. As the NMP has a high capacity and a low selectivity, the mixing component must have a high selectivity and a low capacity.

The use of a mixing component is quite an advantage since, by varying the type and quantity of the mixing component, the solvent can be adapted and optimized for various duties. This is of particular importance for the extraction of feedstocks with different aromatics contents. When the aromatics content in the feed is low, the proportion of the mixing component with high capacity in the solvent mixture is increased and vice versa.

The primary requirement demanded from the mixing component is that it must have a high polarity. On the basis of extensive comparative measurements, the Arosolvan process uses water or monoethylene glycol.

Figure 3 shows the influence of solvent composition on the solvent properties by use of the coordinate system of selectivity and capacity presented in Fig. 2. The lines indicating identical production costs are entered as thin lines. The line of the NMP–water mixture is identical to that for NMP–glycol. This diagram also shows a very important characteristic of this solvent com-

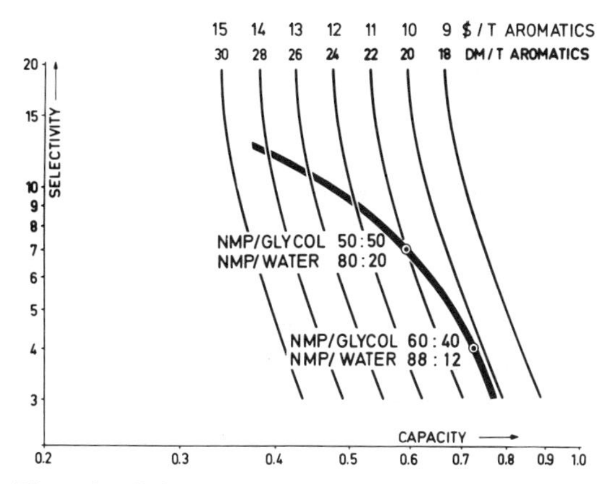

Figure 3. Influence of solvent composition on selectivity and capacity.

bination: the curve does not run on a straight line between the two components but declines toward the right-hand top into the zone of high capacity and high selectivity. Accordingly, the properties of the mixture are better than might be expected from the mixing rule.

2.7. Adaptation of the Process

Each of the two mixing components requires a different process. The major difference between the two mixing components lies in the boiling point and heat requirements for the distillation columns, in particular the solvent stripper. When water is used, the temperatures are low but more heat is consumed because the water goes azeotropically overhead with the aromatics. The extra cost for water evaporation thus increases with increasing proportion of higher boiling aromatics to be recovered. On the other hand, a certain amount of water always must be evaporated to remove solvent from the raffinate. If the low-boiling aromatics represent the major proportion, water evaporation will not cause any additional cost.

Reduction of the column bottom temperature offers the possibility of heat exchange with condensation heat occurring in the distillation units, which will reduce the cost for heating considerably. The economic advantage of such an integrated heat supply system is especially pronounced at high plant capacities.

Another difference between the two processes is the application of a pentane antisolvent with the use of NMP–water. This antisolvent ensures safe maintenance of the two-phase system and a low extract recycle rate. Higher investment costs are thus compensated by lower operating costs. The aromatics purity is the same in both cases.

3. OPERATION OF THE AROSOLVAN PROCESS USING NMP–GLYCOL

The feedstock is charged to the middle of the extractor and the solvent to the extractor top stage. The raffinate, that is, the nonaromatics of the feedstock, leaves the extractor top, saturated with solvent. An extract of solvent and aromatics, saturated with nonaromatics corresponding to the equilibrium conditions, is withdrawn at the bottom. This mixture is fed to the extract recycle column. The column overhead product is recycled to the extractor bottom stage. This recycle contains almost all the nonaromatics.

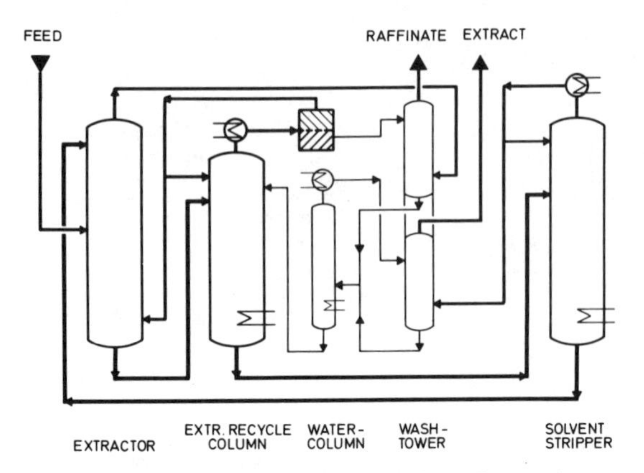

Figure 4. Arosolvan process with NMP–glycol.

The concentration of the remaining nonaromatics boiling within the toluene and xylene range depends on the number of stages, the extract recycle rate, and the concentration in the feedstock. It amounts to a few hundred parts per million. The nonaromatics content in the benzene further depends on the number of trays in the extract recycle column and is about 50 ppm. The aromatics and the solvent are separated by vacuum distillation, where a small proportion of the glycol is distilled overhead azeotropically and separated in the reflux drum as heavy phase. The glycol dissolved in the aromatics is removed by countercurrent extraction with water. Also, the raffinate still contains a few percent of solvent that are washed out by countercurrent contact with water. The bulk of the water in the solvent-laden wash water is removed in a distillation column heated with hot solvent. The remaining water is charged to the extract recycle column, where it is removed overhead azeotropically together with the hydrocarbon vapors, where the solvent is combined with the main solvent stream in the column bottom. Pure aromatics and raffinate are recovered from the top of the two water extractors. This process is illustrated in Fig. 4.

4. OPERATION OF AROSOLVAN PROCESS USING NMP–WATER

Operation of the Arosolvan process by the use of NMP–water (Fig. 5) is essentially the same as that of the NMP–glycol process. The feedstock is again charged to the middle, with the solvent to the top and the extract recycle to the bottom of the extractor. The extraction temperature is about 35°C (cf. 60°C for NMP–glycol). The

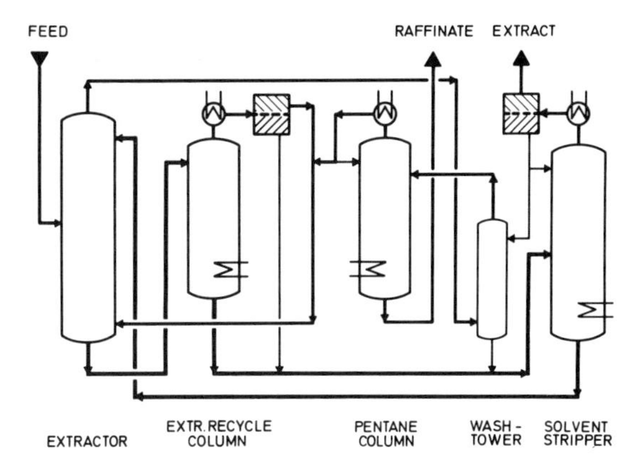

Figure 5. Arosolvan process with NMP–water.

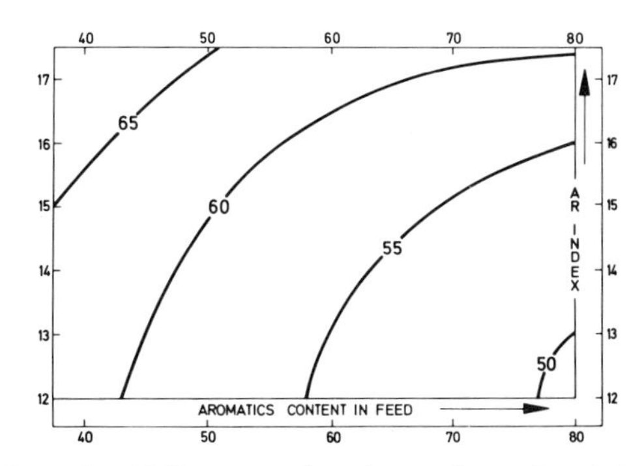

Figure 6. NMP content in solvent mixture in relation to feed properties.

raffinate, after water washing, is supplied to a distillation column where a pentane recycle is produced. This recycle stream flows to the extractor together with the recycle from the extract recycle column. A water distillation column is not necessary in this process as the water from the solvent stripper top is free of solvent and can be used direct for raffinate washing; thus a water wash for the aromatics is not required, either.

5. ASPECTS FOR THE CHOICE OF GLYCOL OR WATER AS MIXING COMPONENT AND COMPOSITION OF SOLVENT MIXTURES

The Arosolvan process with NMP–water as solvent is applied mainly when the advantage of low-temperature distillation can be utilized by heating the columns with the vapors from other distillation processes so that the heat is used twice.

For all other applications, extraction with NMP-glycol solvent is the more economic process. The composition of the solvent is adapted to the properties of the feedstock. The lowest solvent inventory gives the lowest production costs. The lower limit of the solvent inventory is governed by two aspects: (1) when the aromatics content in the feedstock is high, the danger exists that the two phases will disappear; and (2) when the aromatics content in the feedstock is low, the required aromatics yield is not attained. The solvent composition should be chosen so that both limits are reached at the same time. As the composition of the aromatics also plays a role, an aromatics index has been established that is 10 for benzene, 13 for toluene, 18 for xylene, and 30 for C_9 aromatics. The index for mixtures is calculated in accordance with the relative proportion of each individual component. It has been found that mixtures with an identical index require the same solvent composition. A chart showing the solvent composition for various feedstocks is plotted in Fig. 6.

6. HEAT ECONOMY

The sensible heat picked up by the solvent when heated from extractor temperature to its boiling temperature in the stripper bottom, must be recovered as far as possible. This is realized at three different locations (see Fig. 7):

1. The laden solvent leaving the extractor is heated to boiling before entering the extract recycle column.
2. The temperature at the lowermost tray of the extract recycle column is about 30°C below the bottom temperature; thus the effluent from the lowermost tray is heated in a heat exchanger before entering the steam-heated reboilers (see Fig. 8).

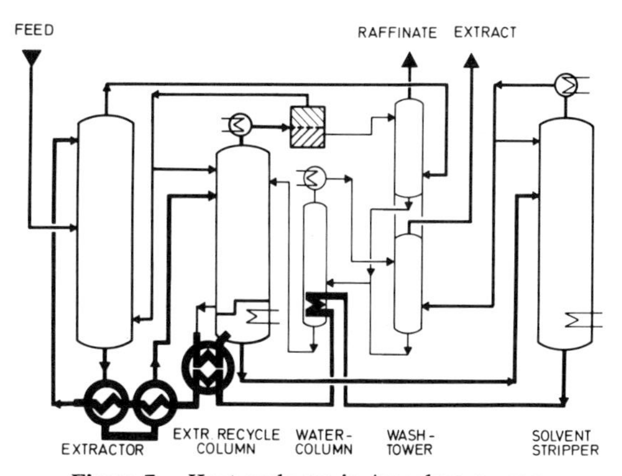

Figure 7. Heat exchange in Arosolvan process.

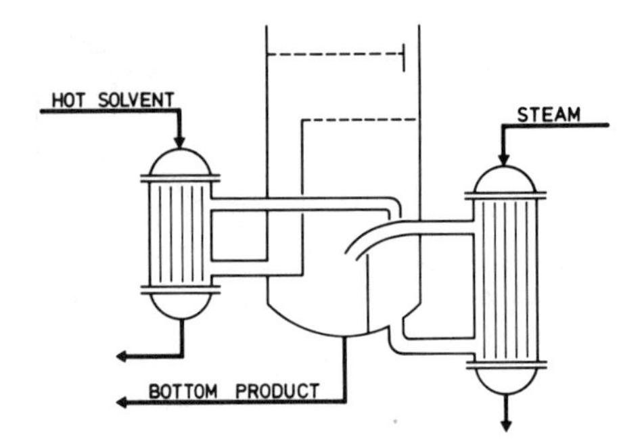

Figure 8. Use of hot solvent for heating extract recycle column.

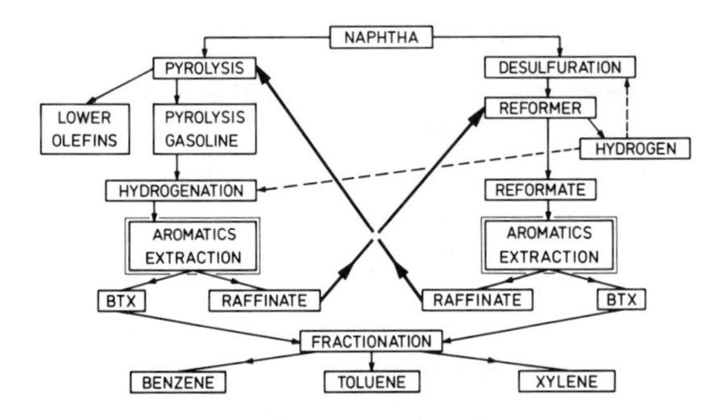

Figure 9. Combination of petrochemical processes for production of aromatics and lower olefins.

3. The water distillation column is also heated with hot solvent from the stripper bottom.

Moreover, heat-saving systems can be incorporated in combination with the aromatics separation units. In some cases the total heat is utilized twice, and no heating steam is required in the plant as all the heat is supplied by direct-fired furnaces. The low bottom temperatures in the NMP–water process permit this favorable heat utilization.

7. AROMATICS REMOVAL AND UTILIZATION OF RAFFINATE

Descriptions of aromatics extraction often deal only with the recovery of aromatics. The raffinate, however, has two characteristics that are demanded for highly priced special boiling point spirits: it should be entirely free of sulfur compounds and largely free of aromatics, particularly benzene.

There can be other undesired properties that must be eliminated: raffinate from pyrolysis gasoline has a high naphthene content that is not desirable for many applications. Raffinate from reformate still contains a few percent of olefins that must be removed by hydrogenation or by treatment with sulfuric acid or clay.

Another valuable application of the raffinates is their use as feedstocks for production of aromatics. Raffinate from pyrolysis gasoline can be charged direct to a reformer, where it furnishes a satisfactory aromatics yield because of its high naphthene content. Raffinate from reformate is a suitable feedstock for pyrolysis because of its low naphthene content (the process scheme of such a plant is illustrated in Fig. 9).

In some cases the raffinate is even the major

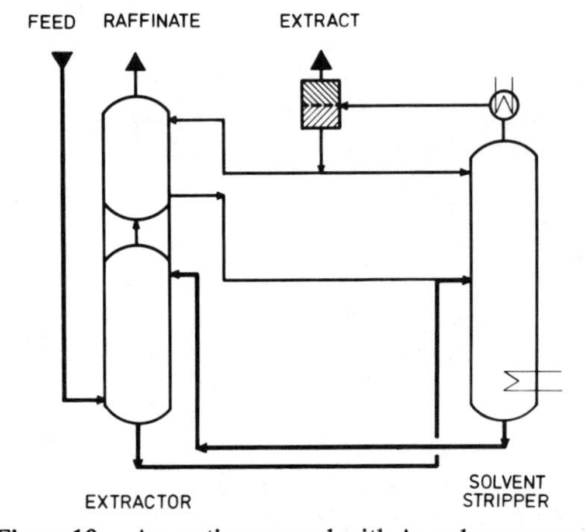

Figure 10. Aromatics removal with Arosolvan process.

product. The Arosolvan process is used for the removal of aromatics from straight-run gasoline for use as a solvent. The preferred solvent is NMP, with 5–10% water. The operating costs range between 0.2 and 0.5 tons of steam per ton of feedstock (a flow diagram of the process is shown in Fig. 10). It differs from extraction with NMP–water in that the pentane column and the extract recycle column are omitted. The process yields an extract with 30–60% aromatics and 60–90% of the feedstock as aromatics-free raffinate. When an extract recycle column is used, extract with more than 98% aromatics can be separated and a correspondingly high yield of aromatics-free raffinate can be obtained.

8. INFLUENCE OF NUMBER OF EXTRACTOR STAGES ON PURITY AND YIELD

In the Arosolvan process, the feedstock is charged to the middle of the extractor. The extractor is designed as a tower, as is the usual practice in the petrochemical industry. The number of extractor stages above the feed tray determines the yield. The plants are generally so designed that the yield of xylene is 95%, in which case the yield of benzene is 99.9%, of toluene 99.5%, and of C_9 aromatics 60%.

In some cases even higher yields are required in order to use the raffinate as aromatics-free solvent. In this case the number of stages above the feed tray is increased. The number of extractor stages below the feed tray determines the purity of the xylene and, in part, also the toluene purity. In these stages nonaromatics in the xylene and toluene boiling range are exchanged for low-boiling nonaromatics that can be removed by distillation in the extract recycle column. With a recycle of 10% of the solvent inventory, the paraffinic nonaromatics in the xylene boiling range are reduced by a factor of 100 and the naphthenic nonaromatics, by 30. The following parameters must be considered in the choice of number of extractor stages: (1) required aromatics purity; (2) nonaromatics content in feedstock; (3) naphthene content in nonaromatics; and (4) extract recycle rate. To adapt the number of extractor stages to the required conditions in an optimum way, the Arosolvan process uses a mixer–settler extractor that has the advantage that the number of stages can be reliably determined in advance and that it can be designed for any number of stages independently of the extractor diameter and without the need to reduce the number of stages at low load operation. A detailed description of the Lurgi tower extractor is given in Chapter 9.4.

18.2.2

UNION CARBIDE TETRA PROCESS

J. A. Vidueira
Union Carbide Corporation
Tarrytown Technical Center
United States

1. UDEX PROCESS

The Udex extraction process for the separation of aromatic hydrocarbons from petroleum fractions was introduced into commercial operation in the early 1950s. The original solvents used in this process were water solutions with diethylene glycol (DEG) or triethylene glycol (TEG) [1]. The structural formulas of DEG and TEG are as follows:

HO CH_2CH_2O CH_2CH_2OH	DEG
HO $\{CH_2CH_2O\}_2$ CH_2CH_2OH	TEG

With the growing demand for petrochemicals and high-octane gasoline, more and more extraction capacity has been and will be needed. The trend in the industry is to employ solvents of increasing solvency to achieve this capacity increase. The industry started with DEG. The liquid–liquid equilibria in the benzene–heptane system with DEG are shown in Fig. 1. The data for the conditions used are 125°C with the solvent containing 8% water. At this water concentration, the solvent has a boiling point of about 140°C. Diethylene glycol solvent is very selective in that it does not dissolve much aliphatics; however, its capacity for benzene is low. The limited solubility with benzene reduces extraction capacity so that benzene distributes in only about a 1:4 extract:raffinate. The distribution coefficients with toluene and C_8 aromatics (which determine the effective capacity of the solvent) are substantially lower.

With TEG at 121.5°C (Fig. 2), the results are better. In this case the solvent contains 5% water, and its boiling point is about 140°C. This solvent is also very selective, and the extract has low heptane solubilities. Benzene distributes in a ratio just under 1:2 between extract and raffinate. Triethylene glycol has nearly twice the extraction capacity of DEG. It is also important to realize that even highly aromatic feeds can be treated.

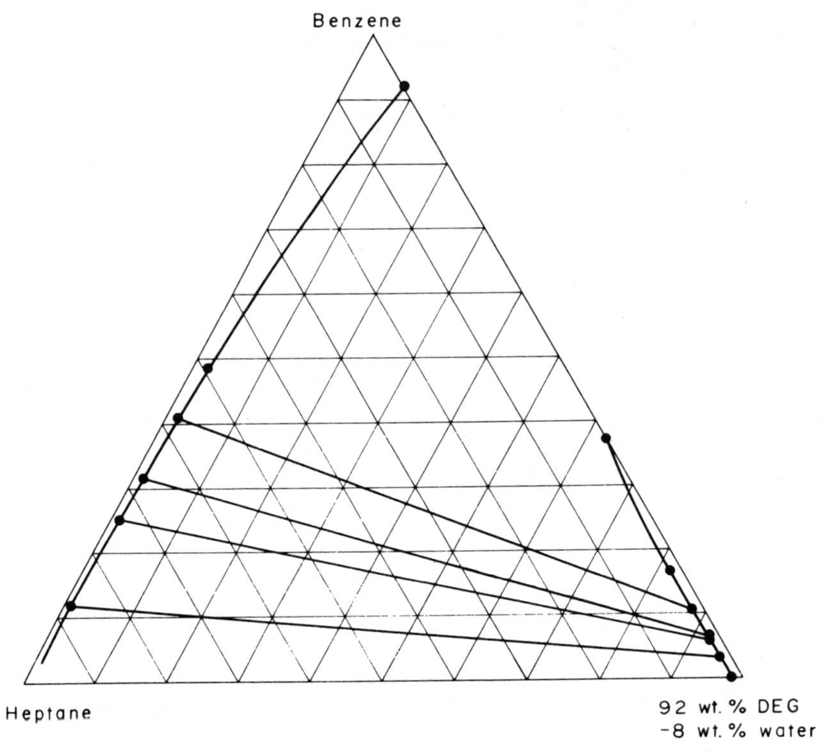

Figure 1. Liquid–liquid equilibria in the benzene–heptane–92 wt. % DEG–8 wt. % water system at 125°C (vol. %).

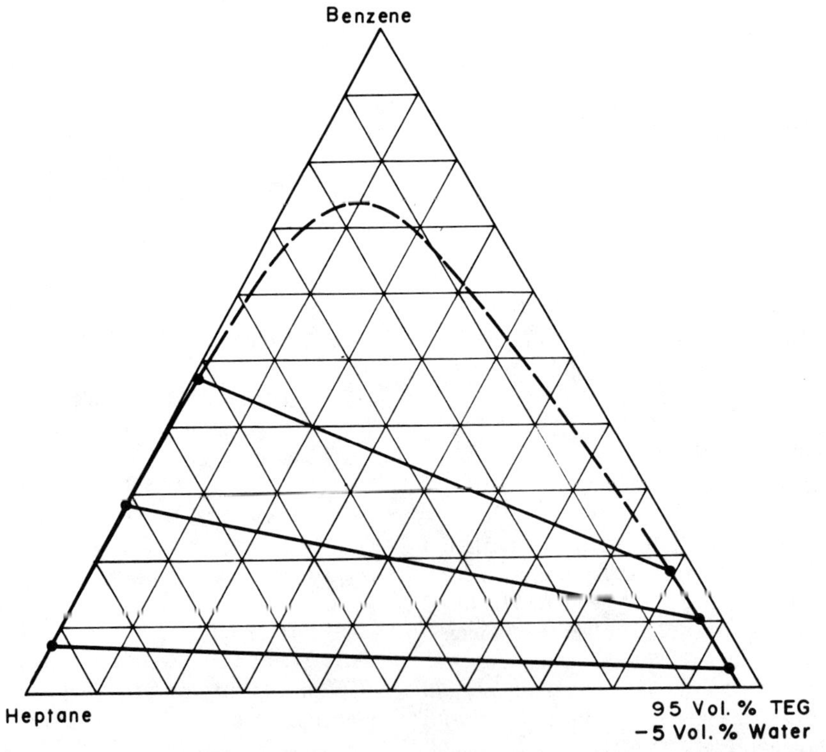

Figure 2. Liquid–liquid equilibria in the benzene–heptane–95 vol. % TEG–5 vol. % water system at 121.5°C (vol. %).

2. TETRA PROCESS

Around 1968, researchers at Union Carbide Corporation realized that a solvent with higher capacity than DEG or TEG was desirable for efficient recovery of aromatics. The main qualities sought were:

- Capacity
- Selectivity
- Cost
- Affinity for hydrocarbons (relative to wash medium)
- Absence of corrosiveness
- Vapor pressure (relative to aromatics)
- Thermal stability
- Viscosity
- Surface tension
- Density
- Toxicity

Although most extraction processes appear to be similar on the surface, there are significant differences in the reliability, operability, and cost that are direct results of the physical characteristics of the solvent as listed here.

Union Carbide researchers experimented with several solvents and finally selected tetraethylene glycol (tetra) as the best one to use in the separation of aromatics from gasoline fractions. Some properties of this solvent are given in Table 1.

TABLE 1 PROPERTIES OF TETRAETHYLENE GLYCOL

Molecular weight	194
Specific gravity, 20°C/20°C	1.1257
Viscosity (at 300°F), cP	1.8
Flash point (open cup), °F	360
Specific heat (at 300°F), BTU lb^{-1} $°F^{-1}$	0.625
Vapor Pressure (at 20°C), mm Hg	0.01
Boiling point (at 300 mm Hg), °F	556
Refractive index (N_D^{20})	1.4598
Price, $/lb	0.46

The liquid–liquid equilibrium data for the system benzene–heptane–tetra–H_2O are shown in Fig. 3 and illustrate why tetra is the best of the solvents studied. The solvent contains 3.9% water and hence has a boiling point of about 140°C. Benzene distributes quite favorably in tetra. The tie lines are rather flat. The data are at 100°C. Highly aromatic feeds can also be treated. The data are summarized in Fig. 4, which depicts the distribution coefficient as a function of benzene concentration in the raffinate. The average distribution coefficient with DEG is about 0.25, and that with TEG is nearly 60% higher, at about 0.40. However, toluene and C_8 aromatics are so much more easily extracted with TEG that the amount of solvent needed is about half that re-

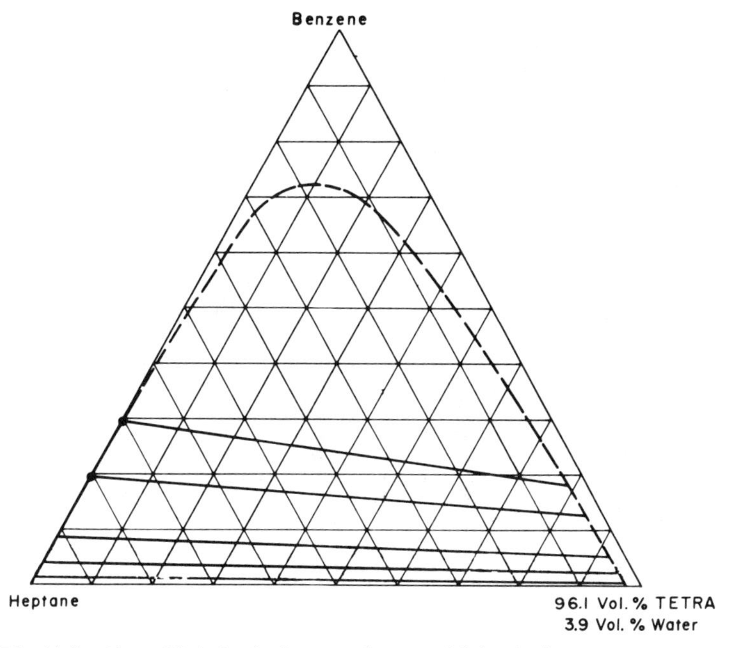

Figure 3. Liquid–liquid equilibria in the benzene–heptane–96.1 vol. % TETRA–3.9 vol. % water system at 100°C (vol. %).

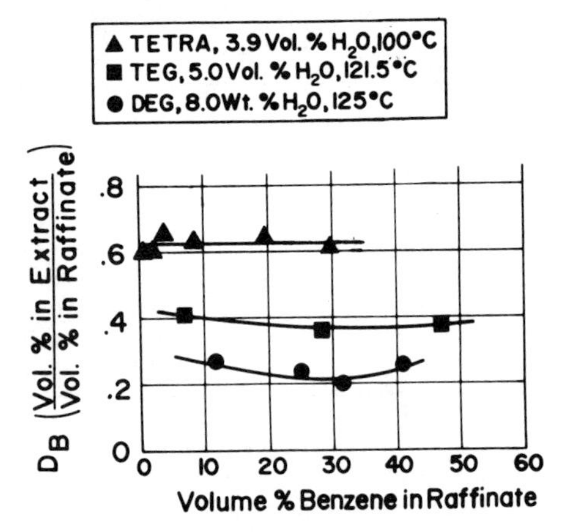

Figure 4. Distribution of benzene between extract and raffinate for various glycols.

quired with DEG. The distribution coefficient with tetra is about 0.6 compared with 0.4 with TEG. Here again, the comparison is that the solvent:feed (S/F) ratio with tetra would be nearly half that with TEG.

After tetra was selected as the solvent, Union Carbide embarked on an experimental program to define the best parameters for extraction of aromatics utilizing this solvent on a commercial scale. A fully integrated, continuous pilot plant was built to determine how the excellent physical properties and favorable capacity and selectivity that tetra shows for aromatics could be utilized. The pilot plant was operated for long periods on various feedstocks, both synthetic and real, and the results confirmed the laboratory bench-scale findings. The superiority of tetra was shown by low S/F ratios, low reflux:feed (R/F) ratios, high recoveries of all aromatics, high purity of benzene and other aromatics, and low operating costs.

In the TETRA process (Fig. 5) feed is charged to the extractor at about the middle of the column. It flows upward and is contacted by the tetra solvent flowing downward. The solvent selectively extracts aromatics. The undissolved aliphatics continue flowing up the column and are removed from the top as the raffinate phase. The part of the extractor above the feed plate serves as the aromatics recovery section; the part below is the purification section where aromatics are purified by back-extraction with a light reflux.

The aromatics-rich extract leaves the extractor bottom and is sent to the top of the stripper. An extractive distillation (further aromatics purification) occurs in the upper part of the stripper. Here light aliphatics are stripped from benzene. Light overhead distillate consisting of hydrocarbons and water is condensed and decanted. The hydrocarbon layer is recycled back to the extractor as reflux. Its composition is approximately 60% aromatics (mostly benzene) and 40% light aliphatics.

The lower part of the column is operated as a steam distillation unit with injection of stripping water to remove aromatics from the solvent. The pure aromatics are withdrawn as a sidestream product from the stripper. The stripper bottom stream is lean tetra solvent, which is recycled back to the top of the extractor. Both the raffinate and aromatics are sent to Union Carbide's

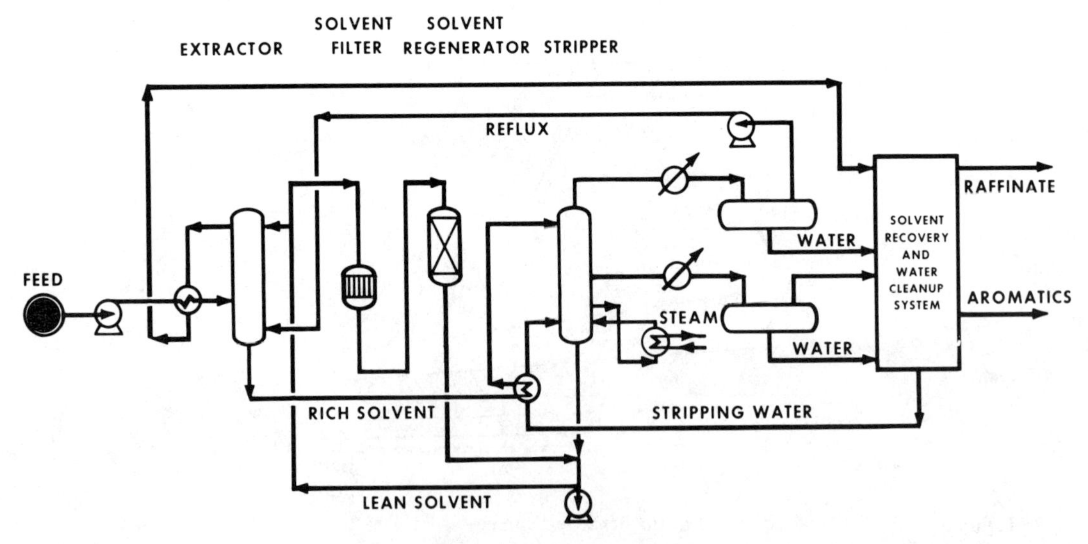

Figure 5. TETRA process.

patented water-wash system, where any dissolved tetra is recovered and returned to the extraction unit.

In addition to a highly selective and high-capacity solvent, TETRA incorporates in its practice the following Union Carbide technology:

- Unique water-wash system to recover the solvent.
- Multiple upcomer (MU) liquid extraction trays.
- Multiple downcomer (MD) distillation trays.
- Solvent regeneration system on a continuous basis.
- Advanced thermodynamic models used to optimize the process designs on an individual basis through computer simulations.
- Elimination of the water still.

Regarding elimination of the water still, stripping water has a very low glycol content as a result of the solvent's high boiling point; thus the water can be used for wash purposes. Other competitive processes require the use of a water still.

3. WATER-WASH SYSTEM

The BTX extract and raffinate streams are washed with water in mixer–settlers before going to storage (Fig. 6). These wash steps recover valuable traces of tetra that would otherwise be lost. Solvent losses in the wash drums are extremely low (⩽20 ppm in both extract and raffinate) because of the high affinity of tetra for water.

Water used for washing the product is part of the closed-loop water cycle that integrates stripping steam with wash water. Water decanted from the stripper overhead decanter is preferable for wash purposes because of its low tetra content. This water stream is first contacted with a small benzene recycle stream in a water clean-up drum to remove any entrained aliphatics from the reflux that would otherwise contaminate the aromatic product. Clean water from the water clean-up drum is injected into the water circulation loop on the aromatic wash drum, and extract washings are subsequently routed to the stripping water accumulator. The raffinate wash system consists of two mixer–settlers with water flowing countercurrently to the hydrocarbon. This arrangement reduces the overall water requirement and maintains low solvent losses.

The spent wash water from the primary raffinate wash drum is routed back to the extractor and is introduced at the feed. As a result, the selectivity of the solvent is increased in the extractor purification section due to the higher water content. Spent water from the aromatic wash drum and from the sidestream decanter are held in the stripping water accumulator before being recycled back to the process as stripping steam. The use of this water-wash system elimi-

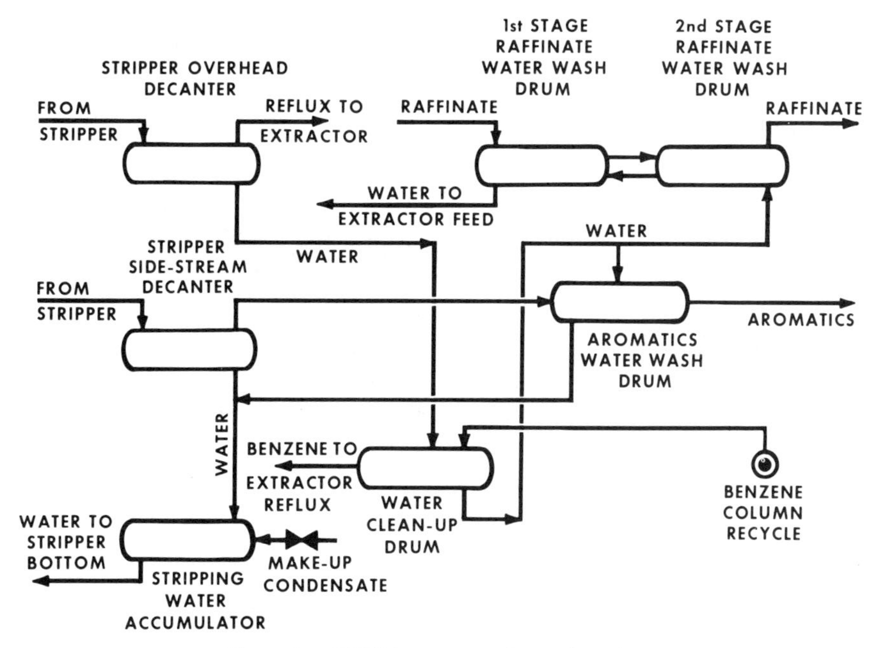

Figure 6. TETRA process water-wash system.

nates the need for a water still column to separate the entrained solvent from the water used to wash the raffinate and BTX products. The savings in equipment costs and heat requirements obtained by the elimination of this column is one of the advantages of using the TETRA process.

4. MULTIPLE-UPCOMER LIQUID EXTRACTION TRAYS

One feature of the TETRA process is the use of MU liquid extraction trays (Fig. 7). These utilize completely new upcomer designs, two completely new sieve-tray designs, and a new idea for tray mechanical support. With the introduction of these innovations, the cross-sectional area of the extractor tower required for a given throughput is significantly reduced. Multiple-upcomer trays are designed for utilization of all surfaces, resulting in much higher extraction efficiency and column capacity. Where the column can be designed to take full advantage of these improvements, column diameters can be reduced sufficiently to yield significant savings in hardware cost and increase throughput–in some instances as much as 40%.

5. MULTIPLE-DOWNCOMER TRAYS

The TETRA process stripper utilizes Union Carbide's multiple downcomer (MD) trays to effect the extractive distillation step required to separate the BTX product from the solvent and purify it. Multiple-downcomer trays offer a substantial capacity increase over conventional multipass trays in handling separations in heavily liquid-loaded distillation systems.

6. SOLVENT REGENERATION

Regeneration of the solvent is on a continuous basis to remove impurities that tend to accumulate in the closed cycle. It is carried out by filtration and adsorption. The regeneration medium is an inexpensive adsorbent and the bed is sized for an expected life of about 6 months, at which time the regenerator can be shut down and the adsorbent replaced without upsetting the extraction process.

Typical impurities removed by adsorption are high-molecular-weight condensation products and low-molecular-weight acids. The use of adsorption eliminates the need for an expensive

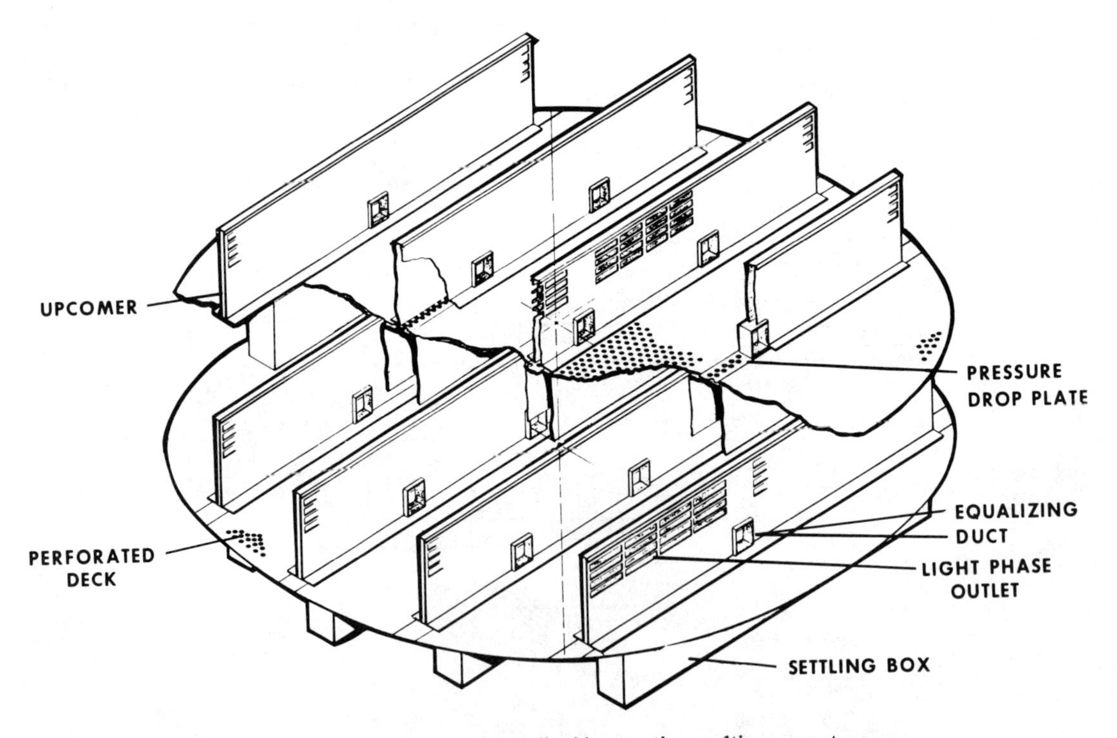

Figure 7. Union Carbide liquid–liquid extraction multiupcomer tray.

and expensive-to-operate high-vacuum distillation column, reduces solvent decomposition (elimination of high temperatures), and eliminates vacuum and air leaks associated with it.

7. THERMODYNAMIC MODEL

Extensive liquid–liquid equilibrium (LLE) data and vapor–liquid equilibrium (VLE) data were obtained experimentally at Union Carbide's research laboratories to develop a model capable of describing the physicochemical interaction of aromatic and nonaromatic hydrocarbons with the solvent–tetra plus water. The data consisted of binary LLE, binary VLE, and multicomponent LLE at temperatures and pressures similar to the operating temperatures and pressures of the TETRA process. These data were correlated with the latest thermodynamic models (NRTL and UNIQUAC) for calculation of activity coefficients to allow simulation of the entire process in conjunction with a Union Carbide proprietary digital computer simulator so that the final design is an optimum one.

8. OPERATING PERFORMANCE SUMMARY

To date there are 35 TETRA units in operation and design. Table 2 summarizes 25 of these units and the accomplishments of each and present feed rates and product specifications. The total combined feed rate is about 350,000 bbl/day (barrels per day). Feed compositions range from 16 to 92% total aromatics. The figures are for a typical day's operation.

The feeds are C_6–C_{12} hydrocarbons derived from low-severity reformates to highly aromatic hydrogenated pyrolysis gasolines. The majority of these units operate to produce high-purity BTX for chemical uses. However, two treat somewhat heavier feedstocks for different objectives. One is used to upgrade octane ratings for motor fuel; the other is concerned with aromatic and aliphatic solvents production. Many of these were Udex units that were converted primarily to achieve the large-capacity increase possible with TETRA. There are others, however, that had little or no additional feed available, and these justified the conversion on the benefits of increased aromatic recovery and reduced utilities. Conversion to TETRA can be justified on the basis of energy conservation alone.

The TETRA conversions that replaced lower glycol systems (Udex) encompassed a wide variety of feedstocks. Substantial capacity increases were realized; two were as high as 80–100%. The values of TETRA conversions range from about \$600,000/year for the smaller units to \$5.10 million for the larger units with large feed rate increases. The cost of heat was assumed to be \$2.00/MM BTU.

9. BENZENE PURITY

It is common in the industry to refer to freezing point as the measure of benzene purity. The impurities in the benzene arise from two sources: aliphatics from the aromatics extraction unit and toluene from the downstream BTX fractionation train. Hence "benzene purities" refer to the final product purity. The current ASTM specification calls for a benzene freezing point of 5.35°C on an anhydrous basis. This corresponds to a benzene purity of about 99.65%.

The TETRA process gives benzene of extremely high purity on a regular basis. Most units produce freeze points of 5.40–5.50°C (99.75–99.94%), with 5.45°C as the average value. It has also demonstrated its ability to consistently produce high-purity toluene and C_8 aromatics. Many TETRA units achieve nonaromatic levels of 500–1000 ppm in both toluene and C_8 aromatics on a regular basis.

10. AROMATICS RECOVERIES

Aromatics recoveries with TETRA are very high. A typical set of aromatics recoveries would be 100% benzene, 99.5% toluene, and 95.0% xylenes. The two units shown in Table 2 that achieve xylene recoveries of only 90.0% could be higher if not for equipment limitations inherent in the original design.

11. OCTANE UPGRADING

One unit (Table 2, column 11) processes a 27,000-bbl/day mixture of heavy reformate and cracked gasoline of about 57% aromatic content. Its purpose is to produce a high-octane concentrate from a medium-octane stream. It is used as a blending stock for gasoline. The total aromatics recovery from the feed is over 99%. The aromatics-free raffinate finds a dual use both as re-

TABLE 2 OPERATING PERFORMANCE SUMMARY

Unit:	1	2	3	4	5	6	7
Location	North America	North America	North America	North America	North America	North America	North America
Feedstock	B	TX	C_9–C_{12}	BTX	BTX C_9	BTX C_9	BTX
Type	Ref[a]	Ref	Heavy napthene	Pg[b]	Ref	Pg	Ref
Aromatics, %	16	46	56	92	54	77	42
Feed rate, bbl/day	2400	5200	3500	11,600	10,800	10,000	16,500
Recoveries, %							
Benzene	99.9	–	–	99.2	100	100	100
Toluene	–	99.0	–	98.6	99.5	99.9	99.9
Xylenes	–	94.0	–	96.4	95.0	90.0	95.0+
C_9 + Aromatics	–	–	96.0	–	68.0	65.0	–
Benzene purity freeze point, °C	5.43	–	–	5.48	5.40	5.45	5.45+

[a]Ref = Reformate.
[b]Pg = Pyrolysis gasoline.

cycled feed to the reformer and as a component of jet fuel.

12. SOLVENT PRODUCTION

Another unit (Table 2, column 3) processes 3500 bbl/day of a C_9–C_{12} kerosene fraction of about 56% aromatic content. This unit is aimed at the efficient production of an aromatic concentrate and an aliphatic concentrate. Each is to be made into a broad range of hydrocarbon solvent fractions. The aromatic extract has a minimum purity of 98%. The overall aromatic recovery from the feed is 96%.

13. ECONOMICS

Typical costs for a unit to process 10,000 bbl/day of feed (50% aromatics) are as follows:

Investment†	
$ per bbl/day	350
Utilities (per barrel of feed)	
Steam (125 psig), lb	125
Water, cooling (20°F rise), gal.	650
Power, kWh	0.03

†Basis: second quarter 1978, including solvent, U.S. Gulf Coast battery limits turnkey plant, but excluding off sites.

14. PERFORMANCE ADVANTAGES OF THE TETRA PROCESS

The Union Carbide TETRA process combines the excellent extraction characteristics of a low-cost solvent with sound engineering practices, resulting in an economical and reliable system for producing high-purity aromatics. The basic advantages of the TETRA process are:

1. TETRA is a highly selective process for separating aromatics from aliphatics: even C_5 and C_6 naphthenes are easily separated.
2. Extraction capacity is high at ordinary temperatures, resulting in a low S/F ratio without risking one-phase formation.
3. The TETRA process extractor is small in diameter as a result of its design.
4. Solvent makeup costs are extremely low as a result of the low cost per pound and minimal loss rate because of the high affinity of tetra for water.
5. In treatment of highly aromatic feeds, tetra maintains its high capacity well below the critical solution temperature of the feed–solvent mixture.
6. Aromatics are easily separated from light saturates and solvent. The TETRA process utilizes a single low-pressure distillation column not requiring any

8	9	10	11	12	13	14	15	16	17
North America	North America	North America	North America	Europe	South America	North America	Europe	North America	North America
BTX	BT	B	C_8–C_{10}	BTX C_9	BTX	BT	BTX C_9	BTX	BT
Ref	Ref	Ref	Ref + Pg	Ref	Ref	Ref	Ref + Pg	Ref	Ref
45	33	30	57	49	40	44	60	38	32
5600	12,500	4250	27,000	11,800	5200	4000	11,000	12,100	7700
100	100	100	–	100	100	100	100	100	100
99.9	99.6	–	–	99.5	98.5	99.0	99.5	99.9	99.0
–	–	–	99.2	95.0	90.0	–	96.0	97.0	–
–	–	–	99.2	90.0	–	–	80.0	–	–
5.40+	5.45	5.40+	–	5.44	5.35	5.40	5.49	5.49	5.52

vacuum to achieve product quality and recovery.

7. The lean solvent temperature in the extractor is generally the same as that of the stripper bottom operating temperature, thus eliminating the need for expensive heat exchange equipment and the unavoidable heat losses that result when stripper temperature is above the extractor temperature.
8. Tetraethylene glycol is relatively nontoxic and biodegradable.
9. With prevention of impurities and with proper pH control, the solvent is nonfoaming and noncorrosive. This eliminates the need for continuous addition of antifoams and for expensive materials of construction. All equipment in TETRA can be constructed of carbon steel. Admiralty condenser tubes might be preferable when water-cooled exchangers are used.
10. The TETRA process employs a simple, effective solvent regenerator. There are no tar disposal or solvent loss problems.

REFERENCE

1. G. S. Somekh and B. I. Friedlander, Tetraethylene Glycol–A Superior Solvent for Aromatics Extraction, *Advances in Chemistry Series*, No. 97, American Chemical Society, 1970.

18.2.3

SULFOLANE EXTRACTION PROCESSES

W. C. G. Kosters
Shell Internationale Petroleum Maatschappij B. V.
The Netherlands

1. INTRODUCTION

The Sulfolane extraction process was developed by the Royal Dutch/Shell group. It was first applied commercially in 1961, since which time a large number of units have been built under license. The process is based on the use of the polar solvent Sulfolane, which has a very high degree of selectivity. Moreover, its high boiling point permits application on feedstocks with boiling ranges of up to 250°C.

2. THE SOLVENT

Sulfolane (tetrahydrothiophene-1,1-dioxide) is a heterocyclic compound. Its main physical properties are given in Table 1.

It should be noted that pure Sulfolane has a high melting point. However, addition of a few percent of water and/or hydrocarbons reduces the melting point considerably. Nevertheless, in cold climates steam tracing of lines containing Sulfolane is necessary.

The thermal and oxidation stabilities of Sulfolane are very good up to temperatures of about 220°C. In practical applications a bulk temperature of 175–185°C is normally used in order to maintain film temperatures below 220°C. At these temperatures some influence of oxygen on solvent stability has been noted; therefore, it is recommended that air be excluded from the feed to the extraction unit. At lower temperatures (e.g., ambient as in storage tanks) hardly any effect has been found.

As a consequence of the good stability of the solvent, corrosion rates are small and Sulfolane extraction plants can be built from carbon steel.

Sulfolane is regarded as only slightly toxic. Its median tolerance level TL_m as measured by ASTM Method D1345-70 is 4800 mg/liter and is thus just below the nontoxic 5000-mg/liter level. It is not irritating to the skin and, as a consequence of its high boiling point, not dangerous by inhalation. It is also nonexplosive.

As Sulfolane is highly water soluble, it can easily be recovered from final products by water washing.

3. DESCRIPTION OF THE PROCESS

A flow scheme of a standard Sulfolane plant is given in Fig. 1 [1, 2]. Feed and solvent are contacted in countercurrent flow in the extractor. The aromatics are extracted by the solvent; the

TABLE 1 SULFOLANE PROPERTIES

Chemical formula	$C_4H_8SO_2$
Molecular weight	120.17
Specific gravity at 30/30°C	1.266
Boiling point	285°C
Freezing point	27.4-27.8°C
Freezing point + 3 wt% water	9°C
Flash point, open cup	350°F
Heat of vaporization	100°C 125 kcal/kg
	200°C 122 kcal/kg
Specific heat of liquid	100°C 0.4 kcal/kg · °C
	200°C 0.46 kcal/kg · °C
Viscosity	
At 100°C	2.5 cP
At 200°C	0.97 cP

nonaromatics leave the extractor at the top in the raffinate. As the latter is saturated with solvent, an additional processing step is required for recovery of this solvent. Since Sulfolane is completely miscible with water, water washing is the preferred method.

The solvent phase that leaves the extractor at the bottom contains all the aromatics from the feed. However, some mainly light—nonaromatics will also be dissolved in the solvent.

The extract phase is routed to the extractive stripping column. In the presence of the solvent the relative volatility of non-aromatics compared with aromatics is enhanced. Consequently, the nonaromatics can easily be distilled from the extract phase. In general, part of the lightest aromatic present will also be evaporated. The overhead vapor is condensed and recirculated to the bottom of the extractor. The aromatics contained in this recycle are again extracted. The nonaromatics will flow countercurrently to the solvent and will eventually leave the system in the raffinate.

In the extraction section between the feed and the recycle inlet, the upflowing hydrocarbon phase will desorb heavy nonaromatics from the solvent phase. As a result, the nonaromatics present in the extract phase leaving the bottom of the extractor will be predominantly the lighter, lower-boiling ones, which facilitates separation in the extractive stripper.

The bottom product of the extractive stripper consists of all the solvent and nearly all the aromatics. This mixture is fed to the recovery column for separation of solvent and extract. The recovery column is operated under vacuum with steam stripping to facilitate the removal of the aromatics. The bottom product, Sulfolane, contains some water and only small amounts of aromatics and is recirculated to the extractor. Some reflux is applied in the top of the recovery column in order to knock back solvent; therefore, the overhead product is free from solvent. After condensation and phase separation, the extract can be transferred directly to the next processing step (in general a fractionation unit to separate the aromatics).

The water phase is first used to wash the raffinate for recovery of dissolved solvent and is then recirculated to the bottom of the recovery

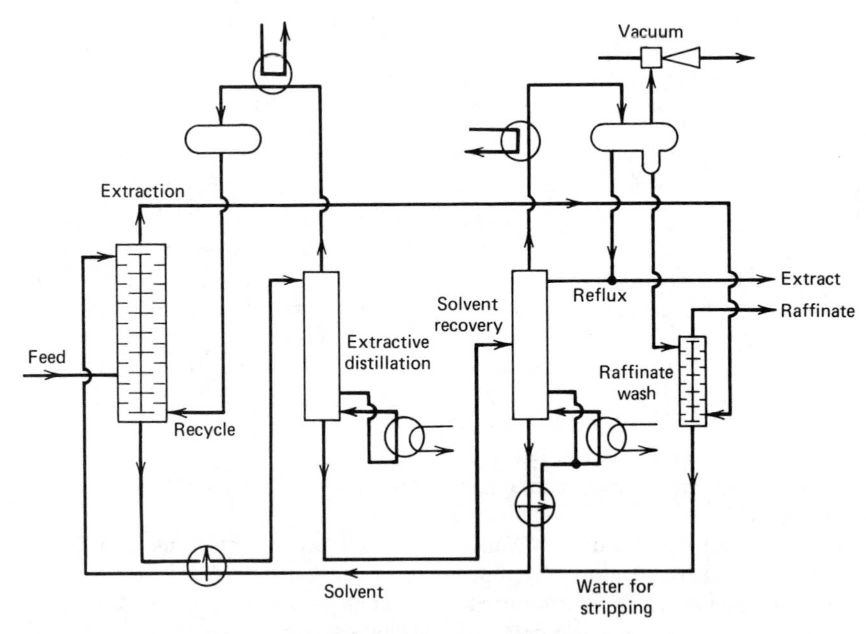

Figure 1. Sulfolane process flow scheme.

column to serve as stripping steam. Steam generation is obtained by heat exchange with the hot solvent.

As can be seen from the flow scheme, the solvent temperature is adapted to the required process conditions solely by heat exchange. As a consequence, heat consumption of the process is nearly independent of solvent circulation.

The normal process scheme for a new Sulfolane extraction plant has been described in the previous paragraphs. It should be noted, however, that it is also possible to replace the glycol-type solvents used in UOPs and the Udex process [3, 4] (see Chapter 18.2.2) by Sulfolane. Such substitution can result in higher plant capacity and lower energy consumption [5].

4. COMMERCIAL APPLICATIONS

4.1. BTX Manufacture

The production of benzene, toluene, and xylenes (BTX) of very high purity is the major application of the Sulfolane extraction process. These aromatics are required with very high purities as feedstock for petrochemicals. Such high purities cannot be obtained by distillation, since the boiling points of the nonaromatics and the aromatics are too close, while some azeotropes are also formed.

Any feedstock with a high aromatic content can be used for BTX manufacture. The feedstocks used in commercial units are mainly from two different sources: (1) the BTX fraction from catalytic reforming of naphtha; and (2) the BTX fraction from the pyrolysis gasoline obtained in cracking naphtha or gas oil for ethylene manufacture. As this fraction as produced in the ethylene cracker is highly unsaturated and often has a high sulfur content, hydrogenation is required as a pretreatment.

These feedstocks normally contain between 50–80%wt aromatics. Of the two, the pyrolysis fraction is the more difficult to extract since the nonaromatics contain a high percentage of naphthenes which are more difficult to separate from the aromatics. Some average data on yield and purities are given in Table 2.

TABLE 2

Feedstock: mixture of catalytic reformate and pyrolysis gasoline

Feed composition: 25 wt.% nonaromatics; 25 wt.% benzene; 25 wt.% toluene; 25 wt.% xylenes

Products		
Benzene	Recovery	99.9%
	Nonaromatic content	300–500 ppm
Toluene	Recovery	98.5%
	Nonaromatic content	500–800 ppm
Xylenes	Recovery	95%
	Nonaromatic content	800–1500 ppm[a]

[a]Dependent on plant operation and feedstock quality.

4.2. Manufacture of Special-Boiling-Point Solvents [6]

Hydrocarbon solvents are fractions which boil in the gasoline range. For the solvents with boiling points below 160°C, the term "SBP" (special-boiling-point) solvent is commonly used. These solvents are employed for a variety of industrial purposes and are often tailored to the needs of the customers. Besides a strict specification on boiling range, a specification on aromatic content often also exists. These specifications can either be a maximum allowable (e.g., for benzene) or ask for a certain range required for its specific use. Such requirements can hardly ever be met by straight-run distillates.

Sulfolane extraction is highly suitable as a processing step for such solvents since it can produce raffinates of very low aromatic content, together with extracts of high aromatic content. The raffinates can be used as such or can be back-blended with suitable extract fractions to meet the aromatic content specification. In this way a highly flexible production scheme can be obtained since any solvent can be supplied to a client in accordance with any specification on aromatic content.

The feedstocks used for the manufacture of SBP solvents are normally hydrotreated light distillate fractions. Two different processing schemes can be adopted: (1) the light feedstocks can be used as produced in the crude distilling unit and the raffinates and extracts produced in the Sulfolane extraction unit fractionated to meet product specifications; and (2) the light distillates can be fractionated first and the products fed in blocked-out operation to the Sulfolane extraction unit. Which of these two schemes is more attractive depends on the types and volumes of product required.

Some results of extraction of light distillates are given in Table 3, in which both wide-boiling-range and narrow-boiling-range feedstocks are covered.

It should be noted that for those feedstocks having a low aromatic content, it is seldom

TABLE 3 EXTRACTION OF LIGHT DISTILLATES

	Feedstock A		Feedstock B	
	Design	Actual	Design	Actual
Feed				
Benzene, %	1.0	0.2	0.5–1.0	0.07
Total aromatics, %	3.0	1.1	12.5	9.2
ASTM boiling range, °C	70–95	70–100	85–160	81–163
Raffinate				
Benzene, ppm	150	<10	150	<5
Total aromatics, %	0.1	<0.01	0.3	0.2
Sulfolane, ppm		<5		5.7
Extract				
Total aromatics, %		55–60[b]	95	95
Sulfolane, ppm		<5		<5

	Feedstock C		Feedstock D		Feedstock E		Feedstock F	
	Design	Actual	Design	Actual	Design	Actual	Design	Actual
Feed								
ASTM distillation, °C	110–145	110–145	100–140	100–140	40–90	30–100	60–90	60–90
Total aromatics, %	77	76	12	10	0.8	1	0.8	0.7
Raffinate								
Total aromatics, %	5–10	7	<1	<1	<0.1	0.08	<0.1	<0.01
Extract								
Total aromatics, %	99.9	99.9[a]		>99	–	–[b]		–[b]

[a]For toluene manufacture (toluene recovery 99.7%).
[b]No requirements for extract purity in view of small amounts.

worthwhile to produce extracts of high aromatic content (>90%). The gain in raffinate yield is rather small, while the increase in heat consumption may be substantial. For such feedstocks, it is also possible to use only an extractor and feed the extract phase directly to the recovery column.

4.3. Extraction of Heavier Feedstocks [7, 8]

Sulfolane, having a high boiling point, can also be applied successfully to feedstocks boiling above the gasoline range, up to a final boiling point of 250°C. In general, the standard process scheme is used. However, for these heavier feedstocks, it is more difficult and sometimes impossible to separate aromatics and Sulfolane in the top part of the recovery column. In those cases the extract and the water phase will contain too much Sulfolane. In order to reduce solvent losses and to render the products solvent free, a separate water-wash circuit may be required. Both raffinate and extract are water washed before being pumped to storage or the next processing step. It is evident that heat requirements and thus operating costs will be higher.

In the boiling range of 150–250°C two types of product can be distinguished: (1) white spirits with a boiling range of 150–200°C; and (2) kerosenes with a boiling range of 150–250°C. White spirits are normally used as solvents with both low and high aromatic content. Using Sulfolane extraction, both types can be produced from a straight-run distillate. As mentioned previously, back blending of raffinate and extract is possible to make products of a specified aromatic content.

Kerosene is in general extracted in order to improve its burning properties (smoke point). Extraction of kerosene with liquid sulfur dioxide was one of the first extraction processes carried out on a commercial scale. The same or better results can be obtained with Sulfolane, because the solvent does not need to be evaporated in the solvent recovery section.

Some operating results on white spirit and kerosene extraction are given in Table 4. From these data it is obvious that very sharp separation can be obtained.

4.4. Other Flow Schemes

The standard process scheme for Sulfolane has been described previously. However, other schemes are also used, based on the excellent properties of Sulfolane as an extraction solvent.

It has already been stated that it is possible

TABLE 4 WHITE SPIRITS AND KEROSENE

Distribution range, °C	150–195	150–225	160–260	200–290
Weight percent aromatics in feed	18	20	20	
Weight percent aromatics in raffinate	1.5–3	2–3	1–2	4–5[a]
Weight percent aromatics in extraction	98	97	90–99	99

[a]Resulting mainly from aromatics remaining in the lean solvent in the recovery column.

to use Sulfolane in an extraction step only, although in that case the extract purity is low. On the other hand, it is also possible to use Sulfolane in an extractive distillation process. Such a process is used when only one aromatic must be produced (e.g., benzene) and for a feedstock with a narrow boiling range.

For wider-boiling feedstock covering more than one carbon number, extractive distillation is in general not suitable, since the highest-boiling nonaromatic will have a volatility close to or lower than that of the most volatile aromatic. In such cases either the recovery of the lowest-boiling aromatic is low or the concentration of nonaromatics in the highest-boiling aromatic is high. It is evident that for such feedstocks, the use of extractive distillation alone is not attractive. An improvement on such a scheme can be obtained by a combination of extractive distillation and extraction, particularly for feedstocks with a high aromatic content. Such feedstocks are fed to the extractive distillation column and the column is operated such that acceptable product purity is obtained. For feedstocks covering more than one aromatic (e.g., benzene–toluene or toluene–benzene), this will mean a great loss of the lowest-boiling aromatic to the top product. An improvement on overall performance can be obtained by feeding the overhead product to an extraction step.

Which of the schemes discussed is the most appropriate depends on the feedstocks available and on aromatics purity and recovery required.

REFERENCES

1. H. Voetter and W. C. G. Kosters, *Proceedings, 6th World Petroleum Congress*, Frankfurt, 1963, Section III, p. 131.
2. D. Broughton and G. F. Asselin, *Proceedings, 7th World Petroleum Congress*, Vol. IV, 1967, p. 65.
3. H. W. Grote, *Chem. Eng. Progr.* **54** (8), 43 (1958).
4. H. W. Grote and D. B. Broughton, *Petrol. Refiner* **38** (6), 161 (1959).
5. F. S. Beardmore and W. C. G. Kosters, *J. Inst. Petrol.* **49** (469), 1 (1963).
6. W. C. G. Kosters, *Proceedings of the International Solvent Extraction Conference (ISEC), The Hague, 1971*, Vol. 1, Society of Chemical Industry, London, 1971, p. 359.
7. H. Voetter and W. C. G. Kosters, *Erdoel und Kohle* **19** (4), 267 (1966).
8. W. G. G. Kosters, *Erdoel und Kohle* **23** (4), 205 (1970).

18.2.4

OTHER EXTRACTION PROCESSES

P. J. Bailes
University of Bradford
United Kingdom

The Sulfolane, Udex, and Arosolvan processes for BTX all rely on solvent extraction coupled with extractive distillation. This arrangement allows the selective properties of the solvent to be used also for the removal of the last traces of nonaromatics in the extractive distillation step. An alternative purification procedure is to use a second nonaromatic solvent.

Such a process has been developed by the Institut Français du Pétrole [1]. Here the primary solvent is dimethylsulfoxide (DMSO) containing several percent water. The selectivity and low viscosity of the solvent allow the extraction to take place entirely at ambient temperature, which is in any case desirable for solvent stability. The process arrangement is essentially as shown in Fig. 1.

In the primary extractor, the feed is contacted counter-currently with the aqueous DMSO, the water content of the solvent being dictated by similar considerations to those described for *N*-methyl pyrrolidone. In the lower section of the first extraction column the solvent stream is treated with a reflux consisting of a mixture of aromatics and paraffins. This improves the aromatics content of the extract and removes the high-boiling nonaromatics. The extract is then treated in a second extractor with a paraffinic solvent, preferably the same as the paraffin used for reflux. The extract from this operation contains the aromatics, which are easily recovered by distillation after a water wash, while the DMSO raffinate is returned to the primary extractor. The paraffin is also continuously recycled, and for successful operation it must be chosen to give simple separation by distillation from the aromatics.

N-formylmorpholine (NFM) is a solvent that is used for the recovery of high-purity BTX in processes offered by Krupp-Koppers [2] and Snamprogetti [3], known respectively as the Aromex and Formex processes. In both cases the NFM contains water, which serves to improve its selectivity but introduces the possibility of hydrolysis for morpholine and formic acid formation. Fortunately, hydrolysis reaches an equilibrium that is strongly biased in favor of the solvent; thus the concentrations of acid and base are very small and do not present corrosion problems with carbon steel equipment. Precautions are taken to avoid solvent losses through decomposition of formic acid and morpholine entrainment.

A feature of the Aromex and Formex processes is the manner in which water content is adjusted between the liquid extraction and extractive distillation steps. For example, in the Aromex process a good selectivity for aromatics in the liquid–liquid extraction step is brought about by water contents of 4–10% in the solvent. The extract then passes to an extractive distillation unit with stripping and rectifying sections. The solubility of nonaromatics that is necessary in the top section of the extractive distillation column is achieved by using a lower water content ($\leqslant 3\%$) in NFM. The overhead product of the extractive distillation has a particularly low aromatics content and is used to advantage as a countersolvent in the sieve plate-liquid extraction column.

In the case of a feed where the benzene content is negligible in comparison with the heavier aromatics, cost savings can result by using an additional liquid extraction column with a

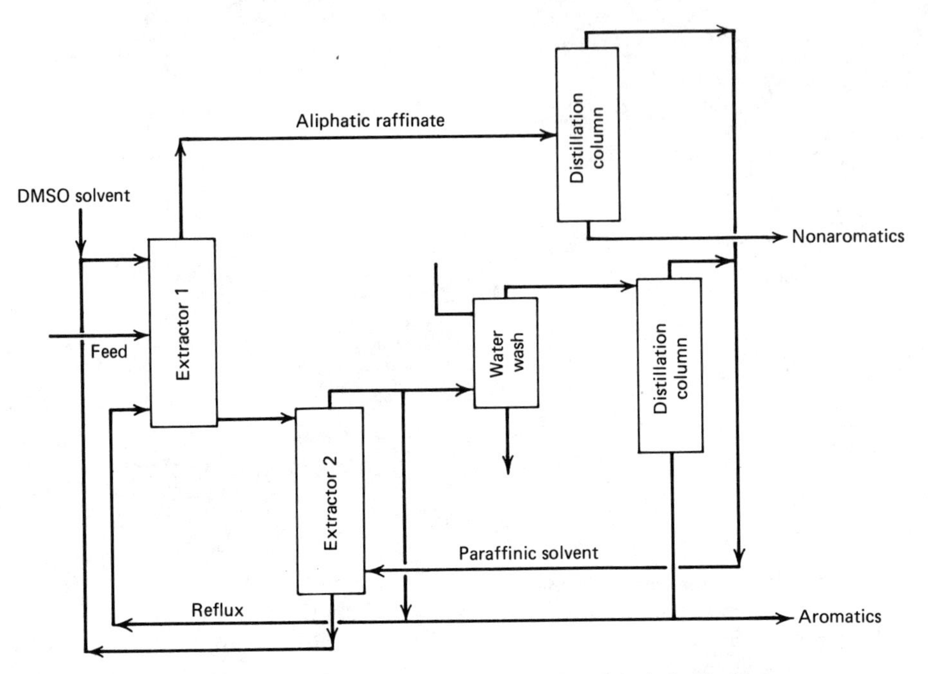

Figure 1. The IFP aromatics extraction using dimethylsulfoxide.

secondary solvent. This scheme is exploited in one variant of the Formex process.

At the present time, the number of installations that use liquid–liquid extraction with NFM is small compared with those based on Sulfolane, NMP, and glycols. However, the first plant was commissioned only in the early 1970s, and this must be compared with Sulfolane, for example, where operation commenced in 1961.

REFERENCES

1. B. Choffe, C. Raimbault, F. P. Navarre, and M. Lucas, *Hydrocarb. Process* **45** (5), 188 (1966).
2. G. Preusser and J. Franzen, in *Proceedings of the International Solvent Extraction Conference (ISEC), Toronto, 1977*, Vol. 2, Canadian Institute of Mining and Metallurgy, Montreal, 1979, p. 707.
3. E. Cinelli, S. Noe, and G. Paret, *Hydrocarb. Process* **51** (4), 141 (1972).

18.3

LUBE OIL EXTRACTION

B. M. Sankey and D. A. Gudelis
Imperial Oil Research Department
Canada

1. HISTORICAL BACKGROUND

Early attempts to utilize crude petroleum for its lubricating properties met with obvious deficiencies such as viscosity, odor, and the hazards associated with high volatility. Distillation permitted the required viscosity to be obtained, but the presence of aromatic, resinous or asphaltic materials gave poor resistance to formation of sludge, acid, and carbon. Filtration through animal charcoal or bone black, a process historically used to purify all types of oil, has been extensively practiced since about 1870. The first effective liquid-phase treating process utilized sulfuric acid; it was applied to petroleum following success in the refining of coal tar products. This process, which involved selective removal of unstable components by means of chemical reaction, was not a true solvent extraction process but did effectively upgrade the oil and was widely used by refineries. Disposal of the acid sludge became a major problem that contributed to the obsolescence of acid treating, although there are a few plants still operating on a version of this process.

To achieve further progress, it was necessary to understand the composition of lubricating oil, at least to the extent of which species needed to be removed. L. Edeleanu carried out a systematic study in the early 1900s and identified the undesirable components as aromatic or unsaturated hydrocarbons. With this knowledge, Edeleanu then set about searching for solvents that would selectively remove these constituents. Liquid sulfur dioxide was discovered to meet the requirements of a selective solvent and was initially and successfully used in the refining of kerosene.

The first plant to exploit this process for lube extraction was constructed at Rouen, France and started operating in 1911. Subsequently, the SO_2 process found worldwide acceptance for many years.

Further developments in automotive and aviation engines required quality levels beyond those attainable with SO_2 because of the relatively low oil solubility in this solvent. Numerous compounds were investigated in the 1930s; this search resulted in the adoption of phenol, furfural, nitrobenzene, and dichlorodiethyl ether for the extraction of lube distillates [1].

These early applications recognized the advantage of countercurrent operation, but this was achieved by separate mixer–settler stages; vertical towers with continuous countercurrent flow later became the standard method of contacting.

Economic and processing factors led to the emergence of furfural and phenol as the two dominant processes for distillate extraction, although significant advantages are claimed for a new process employing *N*-methyl pyrrolidone as solvent.

For utilization of residual feedstocks for lube production, separation of asphaltic material first had to be achieved. Since the boiling point exceeds the decomposition temperature, even under conditions of vacuum distillation, alternate methods were sought to achieve this separation. Extraction with liquid propane was discovered to be very effective, and the first propane deasphalting plant was constructed in 1935. Although many process improvements have been applied to propane deasphalting, including the change from mixer–settler equipment to continuous countercurrent towers, this remains the primary upgrading step for residua prior to further refining.

By combining propane with a polar solvent, the Duo-Sol process can accept both distillate and residual feedstocks. Although plants of this type were widely used between 1935 and 1950, no new ones have been built for well over a decade.

2. PURPOSE OF LUBE OIL EXTRACTION

Lubricating oils are used in a variety of industrial, hydraulic, and crankcase applications, many of which require different characteristics. The wide range of finished-product requirements is matched to the much more limited number of base stocks feasible in large-scale processing by means of blending and additive technology. Since engine oils represent the largest single end use with relatively severe quality requirements, this application is a dominant consideration affecting processing.

Distillate feedstocks used for lube oil manufacture contain a wide spectrum of hydrocarbon types and molecular weights. Hydrocarbon types present include *n*-paraffins, isoparaffins, naphthenes, aromatics, and mixed aliphatic–aromatic ring structures. Nonhydrocarbons typically comprise compounds containing nitrogen, sulfur, and oxygen.

A key requirement for lubricating oil in its operating environment, beyond the need to form a fluid layer between moving metal surfaces, is chemical stability. The components most prone to oxidative and thermal degradation and also having the poorest viscometric properties are, in general, those of highly aromatic or polar character. These are the components most soluble in a polar solvent under conditions giving partial miscibility. The purpose, then, of solvent extraction in the lube oil context is to selectively remove certain aromatic or polar components having poor stability or viscometric properties. Parameters related to chemical composition that are useful for indicating extraction severity include viscosity, viscosity index (a measure of the rate of change of viscosity with temperature), viscosity–gravity constant, specific gravity, and refractive index.

3. GENERAL CONSIDERATIONS

3.1. Lubricating Oil Refining Sequence

The solvent extraction steps used in lube oil refining fit into the overall scheme shown in Fig. 1. Following atmospheric distillation of crude oil, the atmospheric residue is distilled under vacuum, producing a number of distillate cuts and a vacuum residue. High-molecular-weight asphaltic components are removed from the vacuum residue in a deasphalting step. Deasphalted oil and the various distillates are then upgraded in the solvent extraction step by way of "blocked" operation that keeps the different boiling-range streams separate. This is necessary in order to obtain efficient separation in the solvent extraction process and to provide the capability for blending a wide range of finished viscosities after processing is complete. Components removed in solvent extraction, although not wanted in lubricating oils, can be utilized for fuels production or in specialized applications requiring high aromaticity. Following solvent extraction, dewaxing is necessary if wax is present and further finishing steps may be employed to improve color and other properties. Note that

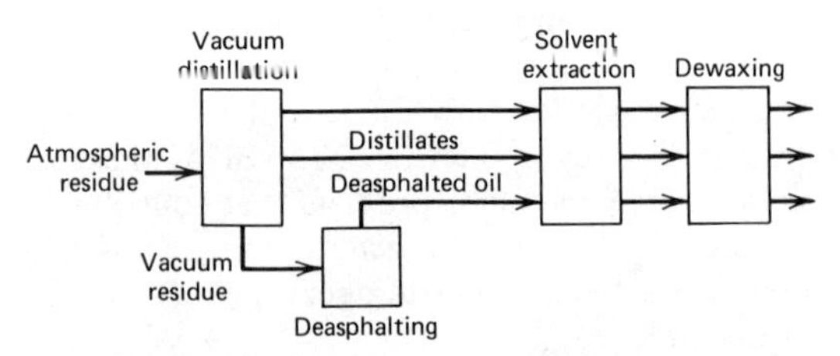

Figure 1. Lubricating oil-refining sequence.

after residual feedstock has been deasphalted to remove high-molecular-weight asphaltic material, the deasphalted oil undergoes the same processing as a distillate feedstock.

3.2. Characteristics of Solvent-Oil Systems

Two general classes of solvent are employed in lubricating oil extraction: (1) polar solvents for selective removal of aromatics and (2) light hydrocarbons for separation of desired lube molecules from high-molecular-weight asphalt. Although both systems must operate in a two-liquid-phase region, overall miscibility properties are quite different in the two cases. Polar solvent-oil systems exhibit an upper critical solution temperature (UCST) so that mutual solubility increases with increasing temperature until complete miscibility occurs for all compositions above the UCST. Mixtures of certain light hydrocarbons with high-molecular-weight lubricating oil components, on the other hand, exhibit a lower critical solution temperature (LCST). Above the LCST, increasing the temperature results in a decreased miscibility of solvent and oil.

Because lubricating oil streams comprise extremely complex mixtures, the classical representation of liquid extraction on a ternary diagram cannot be directly applied. However, a major advance was made by Hunter and Nash [2], who represented oil composition by an additive property plotted along one side of the triangle. In practice, the solvent concentration is shown as a volume fraction, and a property that is additive on a volume basis is used to express the oil composition. Suitable properties are density, refractive index, and viscosity-gravity constant [3].

In the construction of the triangular equilibrium diagram for an oil-solvent system, the solubility curve and tie lines are usually determined from experiments in which the feedstock (e.g., lubricating oil distillate) is mixed with various proportions of the solvent (e.g., phenol). In each case the extract and raffinate layers are separated and analyzed for solvent content and oil quality. If these data are obtained for a sufficient number of mixtures, an equilibrium diagram of the type shown in Fig. 2 may be constructed.

Although such a diagram is useful for representing data and comparing solvents, it is found that the number of theoretical stages predicted from this type of diagram for a given separation is less than the number of theoretical stages needed in practice. The main reason for this difference [4] in a complex mixture such as lubricating oil is that the tie lines determined from single-stage extractions of the feed are not the same as those prevailing in a countercurrent treater.

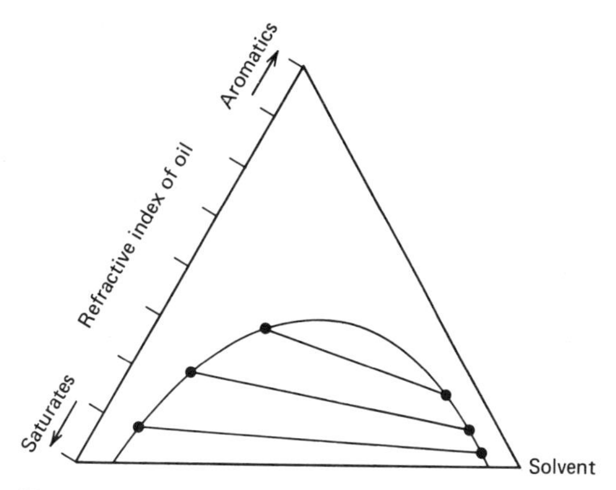

Figure 2. Ternary-phase diagram applied to lube oil-solvent system.

The ideal solvent for lubricating oil extraction would have properties that include the following:

1. High selectivity and high capacity for aromatics in the lube oil molecular-weight range.
2. Good chemical and thermal stability under process conditions.
3. Low viscosity and a density significantly higher or lower than that of the feedstock, factors that aid gravity settling.
4. Residual traces of solvent not deleterious to either oil performance or any downstream processing.
5. Boiling point significantly lower than the lightest oil to be processed.
6. Low latent heat of vaporization.
7. Noncorrosive and nontoxic.
8. Fluid at ambient temperatures.
9. Low cost.

No single solvent meets all these requirements, and some degree of compromise is inevitable. Some key properties of commercially significant solvents are listed in Table 1.

4. SOLVENT EXTRACTION PROCESSES

All lubricating oil extraction plants utilize the same basic concept, namely, countercurrent con-

TABLE 1 PROPERTIES OF COMMERCIALLY USED SOLVENTS

	Sulfur Dioxide (Liquid)	Phenol	Furfural	NMP	Propane (Liquid)
Molecular weight	64	94	96	99	44
Boiling point, °C	-10	182	162	204	-45
Melting point, °C	-76	41	-37	-25	-190
Specific gravity at 20°C	1.45	1.07	1.16	1.03	0.51
Latent heat of vaporization, kJ/kg at boiling point	395	479	450	472	426
Viscosity, cP at 50°C	0.2	3.2	1.2	1.0	0.07

tacting of feedstock and solvent, with the solvent continuously recovered by distillation for reuse. Although early applications of solvent extraction in this area utilized mixer–settler equipment, with some application of centrifugal extractors, virtually all modern processes employ countercurrent treating towers with some form of static or rotating internals. The settling characteristics of the specific system being used are a key factor in designing the optimum internal hardware.

Detailed discussion of specific processes is limited to those of current commercial significance. Thus nitrobenzene, used commercially in the past [5], is no longer considered a feasible process, largely because of the high degree of solvency of the material for lube oil components. This necessitates extraction temperatures at or below ambient, which, in turn, can require refrigeration equipment and that precludes the processing of waxy feeds. Chlorex [6] was used extensively through the 1940s and early 1950s but has now been replaced by more modern processes.

4.1. Sulfur Dioxide Extraction

The Edeleanu process [7] uses liquid sulfur dioxide, which is very selective for aromatics but has low capacity for components in the lubricating oil molecular-weight range. As a single solvent, therefore, the Edeleanu process is most suited to the refining of kerosene, gas oils, and light lubricating oils such as white oils and transformer oils.

A schematic flow diagram is shown in Fig. 3. Charge stock is pumped through a coalescer to remove free water and then through a drier. It is essential to exclude water from the system in order to minimize corrosion and prevent chemical reaction of SO_2 with the hydrocarbons present. The charge is next heat exchanged with raffinate solution prior to entering the lower section of the packed extraction tower. Liquid SO_2 enters the upper section of the extractor

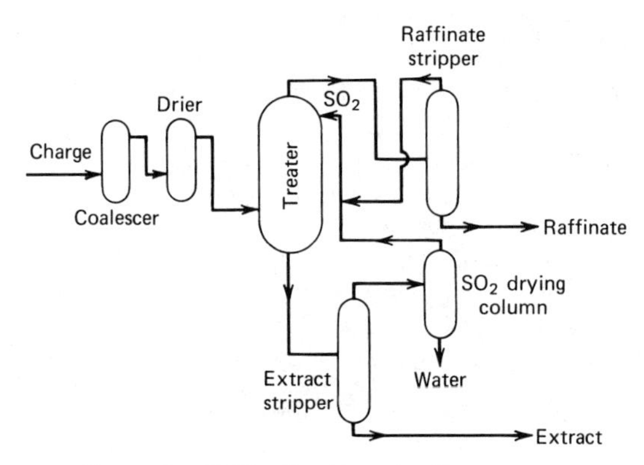

Figure 3. Sulfur dioxide extraction process.

and contacts the rising hydrocarbon phase. Following countercurrent contact, raffinate leaves the extraction tower at the top, while extract phase, which contains most of the SO_2 used for the extraction, accumulates in the base section. Both phases are pumped to their respective recovery systems and the SO_2 recovered for recycle. Part of the sulfur dioxide itself, after recovery, is passed through a drying tower prior to condensation.

Use of sulfur dioxide is associated with relatively high solvent losses of about 0.15 lb per barrel of charge. This economic and environmental penalty, plus concerns over corrosion, have resulted in obsolescence for this process for lubricating oil extraction, although certain old plants continue operation, sometimes processing a range of feedstocks that include lube distillates along with lighter petroleum streams.

4.2. Sulfur Dioxide-Benzene Extraction

Addition of benzene to sulfur dioxide [8] increases the solvent capacity at some loss in selectivity; however, this does permit more paraffinic feeds to be processed. Variation of the percentage of benzene in the solvent mixture makes it possible to select the most advantageous treating conditions for any charge stock so that the desired specification product can be obtained. Design of the plant is essentially the same as in Fig. 3, except for a more elaborate solvent recovery system. The use of benzene involves the incorporation of steam stripping facilities, producing benzene–water–sulfur dioxide mixtures that increase the potential for corrosion. This solvent system is no longer considered a serious contender for new plant design.

4.3. Phenol Extraction

The solvent power of phenol is high, but extraction temperatures are kept within a convenient range (50–100°C) by the use of water, which is completely miscible with phenol above 66°C.

The phenol extraction process [9] was first commercialized in 1930 and has been extensively used ever since, being second only to furfural in current worldwide plant capacity.

A schematic flow diagram is shown in Fig. 4. The feedstock is first contacted with waste vapors (principally water vapor) in an absorber; any phenol present dissolves in the hydrocarbon feedstock and is thus recovered. Feed then enters the lower section of the extraction tower, which may use perforated-plate internals. Solvent is pumped into the treater near the top tray. Water, or a phenolic water stream derived from the solvent recovery system, is introduced near the bottom of the extractor in an amount up to about 10% of the volume of phenol used. This reduces oil solubility in the extract solution, thereby generating reflux by rejecting the least soluble components from the extract phase before it leaves the contacting tower.

After settling at the base of the treater to disengage any oil phase, extract solution is removed and fractionated to remove water as a constant-boiling phenolic water vapor; this stream goes either to the absorber or, after condensation, to the extractor. Phenolic extract is then stripped to remove solvent. Raffinate phase is similarly stripped and recovered solvent combined for recycle.

The solvent losses of a phenol plant are low at about 0.03% on solvent turnover. The main difficulties are associated with the high melting point and toxicity of phenol that necessitate special handling facilities.

4.4. Furfural Extraction

In terms of worldwide installed capacity, furfural [10] is today the most widely used extraction solvent for lubricating oil manufacture. Furfural is an aldehyde made by hydrolysis of agricultural wastes such as oat and rice hulls and other materials. Since furfural is prone to decomposition, certain precautions are necessary to prevent high solvent losses and fouling and corrosion of equipment. The feedstock must be thoroughly deaerated and recovery temperatures held as low as possible, perferably below 220°C. Treating is conducted in a counterflow tower, either packed or by employing a rotating-disk contactor (RDC) with a temperature gradient imposed. Below the feed point, provision is made for cooling and recycling extract phase to provide reflux (Fig. 5). Extract phase from the treater goes to a flash tower, where any water present passes overhead together with some furfural. This latter stream is fed to a water-from-furfural stripper, giving a bottoms product of pure furfural and an overhead that gives two liquid phases on condensation. The water phase is fed to a stripper where the azeotrope takes all the furfural present overhead; the bottoms stream represents the net removal of water from the system.

Attention to details of design, materials of

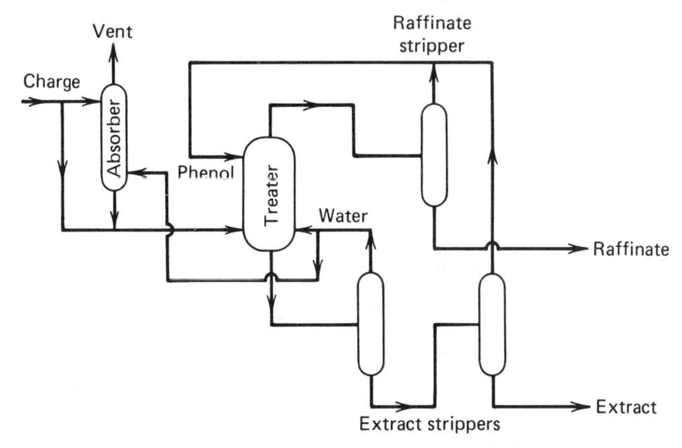

Figure 4. Phenol extraction process.

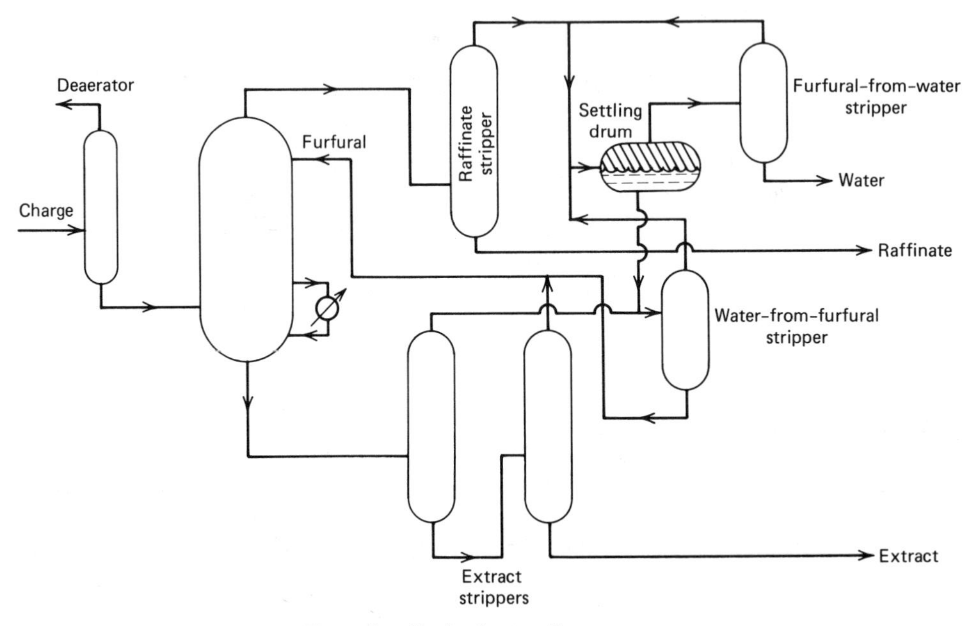

Figure 5. Furfural extraction process.

construction, and care in operation has overcome the relative instability of furfural to give an economical and flexible method of solvent extraction.

4.5. *N*-Methyl Pyrrolidone Extraction

N-Methyl-2-pyrrolidone (NMP) is highly selective for aromatics and was originally used in combination with water or ethylene glycol for the extraction of benzene from nonaromatics in the same boiling range [11] (see Chapter 18.2.1). More recently, this solvent has been applied to lubricating oil refining. *N*-Methyl-2-pyrrolidone has good thermal stability, and solvent losses are quite low; this latter consideration is important in view of the relatively high solvent cost. Control of oil solubility is effected by varying the extraction temperature and water content of the solvent.

No feed pretreatment is required, and water can be removed from the system either by way of an absorber (as for phenol; see Fig. 4) or by distillation. Two commercial processes have been developed [12] that both employ the same basic steps of countercurrent extraction with multiple stages of extract and raffinate stripping. Either steam of nitrogen may be used as the stripping agent.

N-Methyl-2-pyrrolidone has advantages in safety and physical properties over phenol and in stability over furfural. Relative to these two established solvents, the superior selectivity of NMP can be employed either to give increased throughput or to reduce the fuel required per unit of throughput.

4.6. Duo-Sol Extraction

None of the processes hitherto described is capable of processing residual streams containing relatively large amounts of asphaltic materials. The Duo-Sol process [13] employs two practically immiscible solvents, propane and "selecto" (a blend of cresol and phenol). Propane has the prime function of rejecting asphalt, whereas selecto removes aromatic compounds. In essence, the Duo-Sol process effects propane deasphalting and solvent extraction as one operation. The final selecto extract contains the undesirable aromatic components together with the asphalt, whereas the propane raffinate contains the more paraffinic hydrocarbons.

Contacting is carried out in a continuous countercurrent mixer–settler system; each stage comprises a horizontal cylindrical vessel with the necessary mixing equipment. A normal plant employs seven of these extractors. Extraction conditions are 40–65°C and 2 MPa; propane : selecto : feed ratios are typically 3 : 4 : 1 by volume but vary greatly depending on the type of crude and feedstock and the desired product. The phenol content of selecto is 35–40% by volume, and water content is held below 0.15%.

The use of two solvents involves more oper-

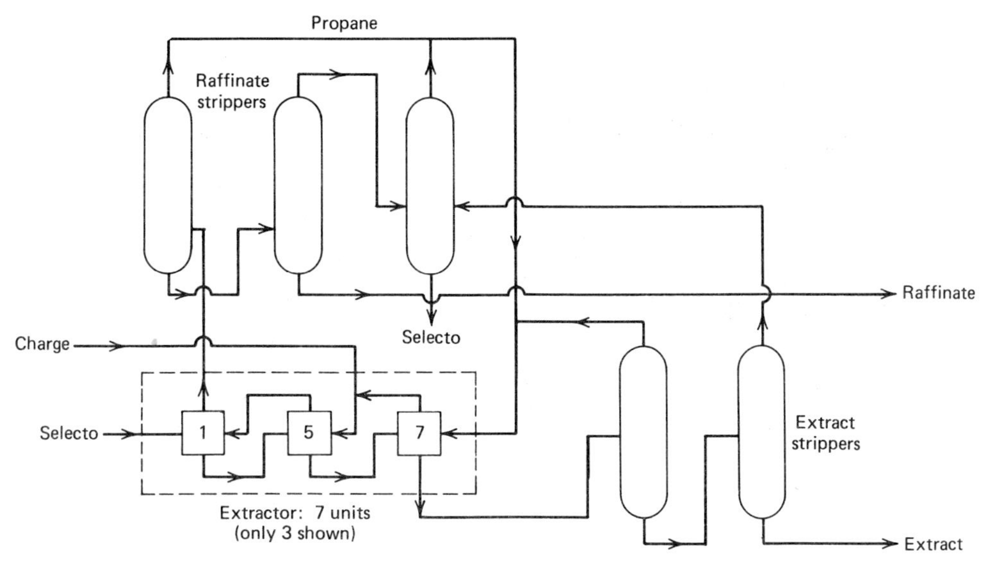

Figure 6. Duo-Sol extraction process.

ating variables than is the case for a single solvent and entails a more complicated recovery design since both solvents must be handled in both recovery systems.

A simplified flow diagram is shown in Fig. 6. Propane and selecto enter the extraction system at opposite ends while feed is introduced near the middle stage.

Terminal raffinate and extract phases each flow through separate towers to remove first propane and then selecto. In addition, final steam or vacuum stripping may be employed.

Investment and operating costs associated with Duo-Sol plants are in general higher than separate deasphalting plus single-solvent plants; the latter approach has thus become the preferred one.

4.7. Propane Deasphalting

The light paraffin hydrocarbons such as propane and butane exhibit a selective solvent action with heavy petroleum fractions. Separation is primarily on the basis of molecular weight, but there is some preferential solubility for paraffinic over aromatic hydrocarbons in any given molecular-weight range.

On contacting a vacuum residue feedstock with liquid propane, therefore, the desired paraffinic components dissolve in the propane-rich (upper) phase while highly aromatic and high-molecular-weight asphaltic material concentrates in a separate liquid phase. By defining a pseudoternary system of propane–oil–asphalt, ternary diagrams can be constructed in the classical manner, with the modifications for a complex mixture described in Section 3.2.

The commercial process [14] employs a countercurrent tower equipped with baffles or rotating-disk internals [15]: tight mesh packing or small clearances must be avoided because of potential plugging from high-viscosity feed or asphalt.

In propane deasphalting the two key variables are temperature and solvent:feed ratio. The extraction tower temperature may be in the range 40–90°C, and solvent:feed ratio is normally within the range 3:1–10:1 (by volume). For maintenance of propane in the liquid phase under these conditions, the operating pressure is about 3 MPa. A temperature gradient is maintained in the tower by steam coils located above the contacting section; this provides reflux for optimum separation.

An overall flow scheme is shown in Fig. 7. Following contacting, the light phase contains the bulk of the propane present, together with the deasphalted oil. Most of the solvent present in this phase is recovered in evaporators; residual solvent is removed by steam stripping. The asphalt phase flows to a flash tower and steam stripper. Overheads from both strippers are combined, cooled to remove water, and the propane is compressed and recycled.

In addition to its application in lubricating oil refining, solvent deasphalting is used to upgrade petroleum streams prior to fluid catalytic cracking or catalytic hydrogenation processes. Solvents other than propane may be used; alkanes in the C_4–C_6 range require progressively

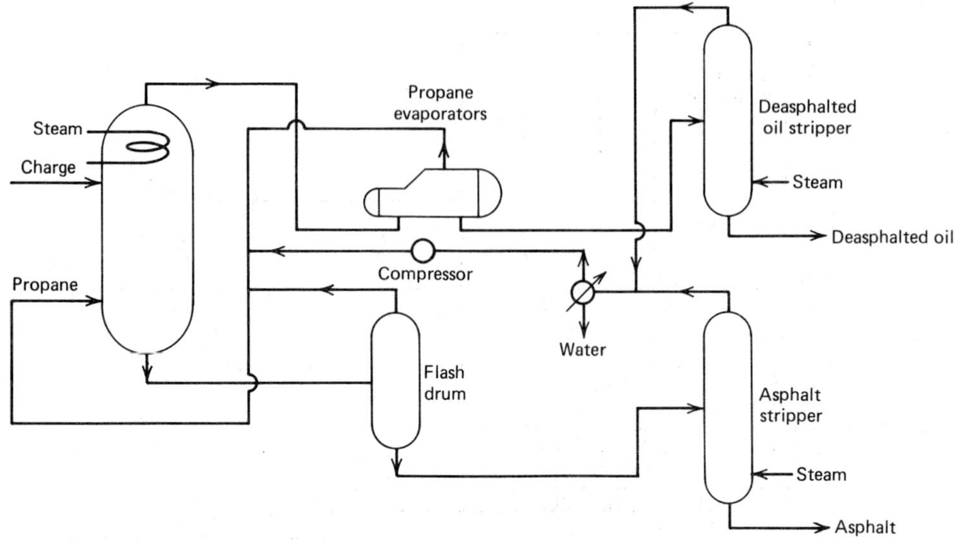

Figure 7. Propane deasphalting process.

higher operating temperatures and give deasphalted oil at a higher yield but with less complete asphalt separation, relative to propane.

4.8. Future Developments

Because of the high investment associated with petroleum refining process units, major changes occur relatively infrequently and only after extensive development. Most advances within a given process are achieved by stepwise improvements to increase throughput, decrease solvent losses, reduce energy consumption, and so on.

Since future energy costs are predicted to account for an ever-increasing fraction of total manufacturing costs, processes requiring minimum energy will become dominant. The prime use of energy in solvent extraction is for solvent recovery by means of distillation. Solvents offering improved selectivity and capacity, and hence reduced treat rates will, therefore, be favored; the use of NMP is a recent indication of this trend. High-efficiency contacting equipment using rotating or oscillating components can similarly reduce solvent treats by increasing the number of theoretical stages available.

The improved process control achievable with computers should reduce solvent losses to very low levels so that solvent cost becomes a less significant factor. Solvents rejected on this basis in the past may, therefore, become much more attractive if they meet other key requirements.

REFERENCES

1. V. A. Kalichevsky and K. A. Kobe, *Petroleum Refining with Chemicals*, Elsevier, Amsterdam, 1956, pp. 319–381.
2. T. G. Hunter and A. W. Nash, *J. Inst. Petrol.* **22,** 49 (1936).
3. J. B. Hill and H. B. Coats, *Ind. Eng. Chem.* **20,** 641 (1928).
4. R. M. Butler, A. E. Spence, A. R. Barncroft, and A. C. Plewes, *Can. J. Technol.* **34,** 340 (1956).
5. S. W. Ferris and W. F. Houghton, *Petrol. Refiner* **11,** 560 (1932).
6. J. M. Page, C. C. Buchler, and S. H. Diggs, *Ind. Eng. Chem.* **25,** 418 (1933).
7. E. J. Dawson, *Third World Pet. Congr.* **3,** 107 (1951).
8. J. C. Albright, *Natl. Petrol. News* **27** (10), 25 (1935).
9. D. W. Kenny and W. B. McCluer, *Oil Gas J.* **39** (36), 48 (1941).
10. L. C. Kemp, G. B. Hamilton, and H. H. Gross, *Ind. Eng. Chem.* **40,** 220 (1948).
11. E. Muller, *Chem. Ind.* **11,** 518 (1973).
12. Anonymous, *Hydrocarbon Processing* **57** (9), 185, 195 (1978).
13. M. B. Miller, *Petrol. Refiner* **26** (4), 138 (1947).
14. J. G. Ditman and R. L. Godino, *Hydrocarbon Processing* **44** (9), 1975 (1965).
15. V. B. Thegze, R. J. Wall, K. E. Train, and R. B. Olney, *Oil Gas J.* **59** (19), 90 (1961).

18.4

EXTRACTION OF CAPROLACTAM

A. J. F. Simons and N. F. Haasen
DSM/Central Laboratory
The Netherlands

1. PRODUCTION OF CAPROLACTAM

1.1. Introduction

Caprolactam is the monomer of Nylon 6 and specifications for fiber-grade material are extremely stringent, calling for extensive purification. Details of properties are available [8]. Of the many known production processes for caprolactam, only a few are used commercially. These are all multistep processes with coproduction of different amounts of ammonium sulfate and organic by-products. About 85% of world caprolactam production is based on the two classical intermediates cyclohexanone and cyclohexanone oxime, yielding caprolactam after Beckmann rearrangement. The main producers include Allied Chemical, BASF, Bayer, DSM, Leuna Werke, Mitsubishi, Montedison, and Ube Industries. The remaining 15% comes from the commercial processes of Snia Viscosa and Toray. The main starting materials for lactam production are benzene and toluene and the first step is the production of cyclohexanone. Surveys of commercial routes are available elsewhere [1, 13, 33].

1.2. Production of Cyclohexanone Oxime and Hydroxylamine

Three alternative processes are used for the technical production of cyclohexanone oxime from cyclohexanone and hydroxylamine ("hyam") [1–7].

1.2.1. Raschig Process [6, 34]

A solution of ammonium nitrite is mixed with aqueous ammonia and sulfur dioxide, yielding the intermediate product hydroxylamine disulfonic acid. This is hydrolyzed in another reactor into hydroxylamine monosulfonic ammonia, and the latter is further hydrolyzed into hyamsulfate. The ammonium bisulfate coproduced is neutralized with aqueous ammonia and used for production of ammonium sulfate. The mixture of hydroxylamine sulfate and ammonium sulfate is sent to the oxime preparation section.

In the oximation of cyclohexanone, hydroxylamine sulfate is reacted with cyclohexanone in a number of constant-flow stirred tank reactors (CSTRs) at a temperature of approximately

85°C. Formation of free sulfuric acid is avoided by the use of aqueous ammonia as neutralizing agent while keeping the pH at 4.5. The overall reaction is as follows:

$$2\ \text{C}_6\text{H}_{10}\text{=O} + (NH_3OH)_2SO_4 + 2NH_4OH \longrightarrow 2\ \text{C}_6\text{H}_{10}\text{=NOH} + (NH_4)_2SO_4 + 4H_2O \quad (1)$$

The exothermic reaction can also be carried out in differential columns [9, 28], such as vibrating-plate columns, rotating-disk contactors (RDCs), and pulsed packed columns (PPCs).

The two liquid phases in the resulting mixture of oxime and ammonium sulfate solution are subsequently separated in a gravity settler. The upper layer, consisting of oxime, is passed to the Beckmann rearrangement unit for conversion into caprolactam. The ammonium sulfate solution is stripped of its traces of organic substances and further processed in a crystallization unit. Application of this route produces about 2.7 tons of ammonium sulfate per ton of caprolactam.

1.2.2. *NO Reduction*

Nitrogen monoxide (Inventa/BASF) [7, 35] is hydrogenated in dilute sulfuric acid in the presence of a platinum catalyst. Oximation of cyclohexanone with the resultant hydroxylammonium salt solution is similar to the oximation in the Raschig route.

Ammonium sulfate is still coproduced as a result of side reactions in NO reduction and neutralization during oxime formation. The sulfate production of this route amounts to 0.5–0.8 tons per ton of oxime.

1.2.3. *HPO Process*

The hydroxylamine phosphate oxime process (Stamicarbon/DSM) [10–12] differs from the conventional processes in that the acid liberated during the oximation of cyclohexanone with hydroxylamine need not be neutralized. This is achieved by having the production of hydroxylamine and the oximation of anone take place in a recirculating buffer solution of phosphoric acid and ammonium dihydrogen phosphate. Thus oxime is produced without coproduction of ammonium sulfate. A simplified diagram of the process is shown in Fig. 1. Hydroxylamine is produced by reduction of nitrate ions with hydrogen [13] in the presence of a noble metal catalyst:

$$\underset{\text{pH} \sim 0}{HNO_3 + H_3PO_4 + H_2PO_4^- + 3H_2} \xrightarrow{\text{Pd/C}} \underset{\text{pH} \sim 2}{NH_3OH^+ + 2H_2PO_4^- + 2H_2O} \quad (2)$$
$$\dashrightarrow NH_4^+, N_2, N_2O$$

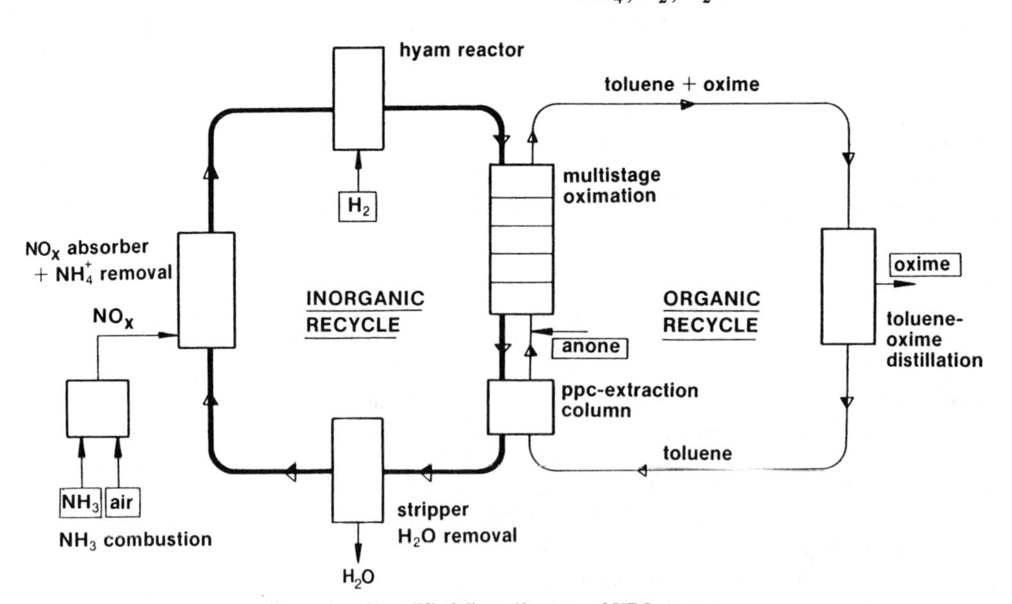

Figure 1. Simplified flow diagram of HPO process.

The reaction product is sent to the oximation reactors. The process liquid, which contains the hydroxylamine, is contacted countercurrently in a cascade of mixer–settlers with a solution of cyclohexanone in toluene. The anone reacts with the hyam to form oxime, and this product is extracted by the toluene phase (extractive reaction). This oximation reaction can also be performed and scaled up in a countercurrent column of the PPC type. The overall reaction can be represented by

$$\text{C}_6\text{H}_{10}{=}\text{O} + NH_3OH^+ + 2H_2PO_4^- \xrightarrow[\text{pH} \sim 2]{} \text{C}_6\text{H}_{10}{=}\text{NOH} + H_2PO_4^- + H_3PO_4 + H_2O \quad (\text{pH} \sim 1) \qquad (3)$$

The toluene phase containing the oxime is washed with water and sent to the distillative toluene–oxime separation. The bottom product of the distillation unit, oxime, is passed to the Beckmann rearrangement section for conversion into caprolactam. The inorganic liquor flowing from the oximation section contains some oxime and anone and must be freed of these components before being recycled to the hyam reactor. This is done by extraction with toluene from the distillative toluene–oxime separation. This step is carried out in a PPC [14, 15]. (For description, see Chapter 11.1.)

The resulting inorganic liquid is stripped for removal of the reaction water and any organic components still present. It is then led through a tower for absorption of the nitrous gases NO/NO_2. During the absorption, nitrate ions are formed that are used in the production of hydroxylamine, while the ammonium ions coproduced in the hyam preparation decompose.

1.3. Beckmann Rearrangement

The cyclohexanone oxime obtained by the method just described is rearranged to caprolactam in an acid medium. The reaction product is a rather viscous mixture of caprolactam and oleum, forming a one-phase layer. Separation of caprolactam from the crude product is described in several patents [1, 3–5]. There are two basic routes:

Direct extraction from the one-phase mixture with an extractant capable of loosening the (weak) chemical bond between caprolactam and sulfuric acid.

Extraction from a highly concentrated caprolactam phase in physical equilibrium with an aqueous ammonium sulfate solution obtained by complete neutralization with ammonia of the rearrangement mixture. About 1.8 tons of sulfate are then produced per ton of caprolactam.

2. PURIFICATION OF CAPROLACTAM

Crude lactam contains various impurities that must be reduced to a very low level [16]. A series of physicochemical purification steps is applied, including extraction, chemical treatment, and final vacuum distillation.

2.1. Direct Extraction from the Rearrangement Mixture

To reduce the amount of by-product ammonium sulfate, several companies have been working on extraction from the rearrangement mixture either with or without partial neutralization of the acid. For reduction of the attraction between the acid and the lactam, most of the patents call for dilution of the rearrangement mixtures with water before extraction. The extractant should thus have a low degree of solubility in water. Additives may be used to improve phase separation. Various extractant patents are summarized in Table 1.

Total elimination of ammonium sulfate was made possible after the development in 1972 of Stamicarbon's sulfuric acid recirculation caprolactam process [17–19]. The process characteristics are:

1. Preparation of a sulfuric acid–caprolactam mixture based on the conventional Beckmann rearrangement with oleum of cyclohexanone oxime, obtained with hydroxylamine produced by DSM's HPO process.
2. Separation of caprolactam by extraction, such as with a chlorinated hydrocarbon solvent from the sulfuric acid–caprolactam mixture, neutralized into ammonium bisulphate. The resulting mixture remains homogeneous.
3. Recovery of oleum from the ammonium bisulfate by converting it by pyrolysis, utilizing natural gas at approximately

TABLE 1 EXTRACTION OF LACTAMS FROM REARRANGEMENT MIXTURES: PATENT EXAMPLES

Patent Specification (Year)	Solvent	Diluent for Rearrangement Mixture	Extraction Temperature, °C
British 1,313,959 (1973)	Chloroform 1,1,2,2-Tetrachloroethane	H_2O	30
U.S. 3,820,972 (1974)	Chloroform	H_2O	30
Belgian 803, 171 (1974)	1,1,2,2-Tetrachloroethane	H_2O	
U.S. 3,850,910 (1974)	Tetrachloroethane	H_2O	150
U.S. 3,859,278 (1975)	Tetrachloroethane	H_2O	150
French 2,190,822 (1974)	Alkylphenols	H_2O	25–30
French 2,190,821 (1974)	Alkylphenols	H_2O	25–30
Japanese 48-4792 (1973)	Alkylphenol, e.g., *o*-isopropylphenol	None	80–150
Japanese 48-10976 (1973)	Phenol, alkylphenols, chlorophenols, nitromethane, alkylchlorides	H_2O (Small amounts)	20–170
Japanese 48-15952 (1973)	Methylcylohexane, kerosene	None	Ambient
U.S. 3,700,669 (1972)	Liquid SO_2	H_2O	-40 to +60
Japanese 47-42688 (1972)	Anisole, *p*-dimethoxybenzene, acetophenone	H_2O	55

1000°C, into sulfur dioxide and nitrogen, and by processing the sulfur dioxide into oleum. Similar procedures have been developed by other companies [7, 20]. The capital investment and the steep rise in energy costs have limited the industrial introduction of these techniques.

2.2. Extraction from Caprolactam Oil

Caprolactam can be separated from the crude product obtained from the Beckmann rearrangement by neutralizing the acid mixture to pH ~4.5 with ammonia. Water is added so as to keep the $(NH_4)_2SO_4$ in solution. As a result of the neutralization, two immiscible phases are formed that can be separated mechanically. The ternary system $(NH_4)_2SO_4$–caprolactam–water at about 50°C is shown in Fig. 2. The two layers are separately processed. The saturated aqueous ammonium sulfate layer [40% $(NH_4)_2SO_4$] contains around 1.5% caprolactam, which is generally subjected to an extraction procedure.

The crude lactam top layer, caprolactam oil, consists of 65–70% caprolactam with organic and inorganic impurities. The impurities can be removed largely by extraction or distillation of the caprolactam oil. In industrial processes, extraction is applied most widely. A general scheme of the extractive purification is shown in Fig. 3.

All impurities that are more soluble in water than in the organic solvent are left behind in the aqueous raffinate. The use of a sufficient quantity of solvent limits caprolactam losses in the aqueous raffinate to a minimum.

The organic extract, containing for most solvents around 20% caprolactam, still contains impurities that dissolve better in the organic solvent than in water. For removal of these impurities, the caprolactam is reextracted with water. The organic solvent is again used for extraction of crude lactam, while part of it is distilled in order to avoid accumulation of impurities.

The aqueous 30% caprolactam solution leaving the reextraction column contains some dissolved solvent. This solvent is stripped off and recirculated. Aqueous caprolactam is then treated further, as mentioned before, to produce caprolactam of an extremely high quality.

3. SOLVENTS FOR CAPROLACTAM EXTRACTION

Several solvents have been proposed for the recovery of caprolactam from both the lactam oil and the aqueous ammonium sulfate liquor, including toluene, benzene, methylene chloride, chloroform, trichloroethylene, nitrobenzene, carbon tetrachloride, dichloroethane, and a mixture of benzene and cyclohexane. The equilibria of the ternary organic solvent–caprolactam–water systems are described [21–23]. Likewise, the equilibrium relationships for some ternary systems of caprolactam and 40% ammonium sulfate solution are published [25, 26]. A Hand diagram

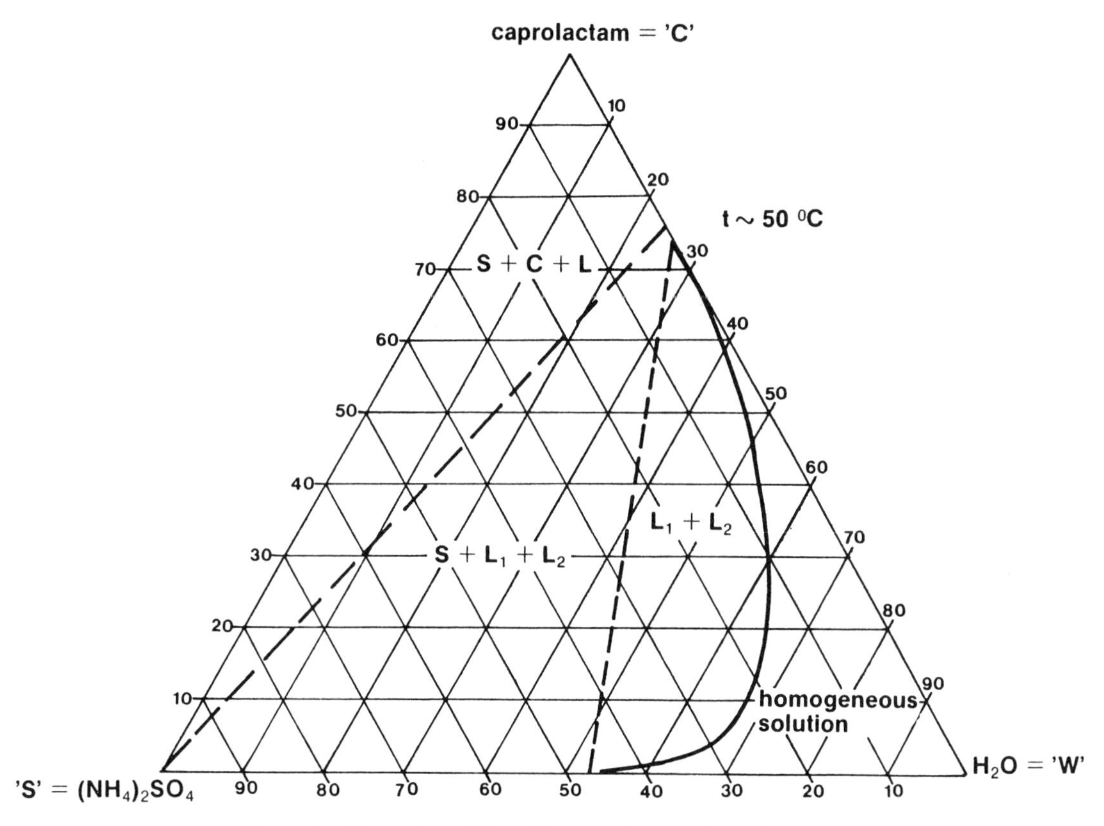

Figure 2. The system $(NH_4)_2SO_4$–caprolactam–H_2O schematic.

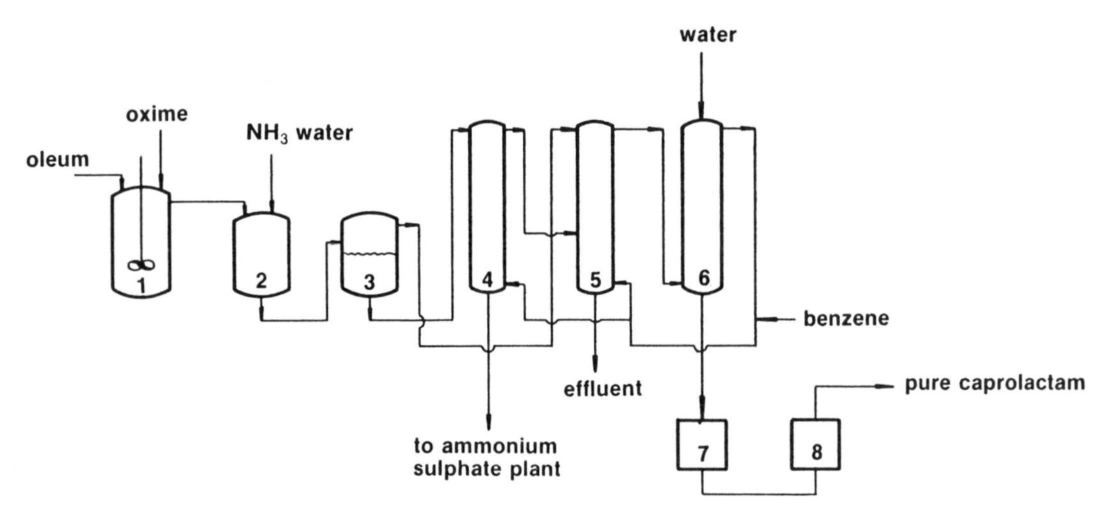

Figure 3. Caprolactam production.

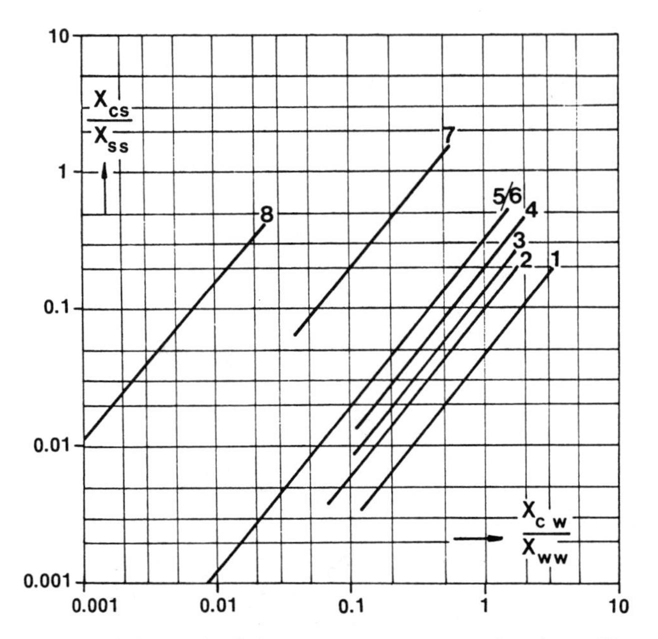

Figure 4. Equilibria (Hand diagram) of the ternary systems organic solvent (s)–caprolactam (c)–water (w). Line 1, tetrachloroethane [21, 23]; line 2, dichloroethane [22, 23]; line 3, nitrobenzene [21]; line 4, benzene [21, 23]; lines 5 and 6, trichloroethylene, methylene chloride [21, 22]; line 7, chloroform [21, 22]; line 8, liquid SO_2 [24].

with smooth curves on the basis of published equilibria is presented in Fig. 4.

It must be emphasized that such equilibria are for pure components and are of only restricted value for an industrial process. Especially at very low caprolactam concentrations, a remarkable difference may exist between pure component distribution coefficients and those determined with actual plant liquids [26]. A ternary diagram for the (pure) benzene–caprolactam–water system ($t = 20°C$) is given in Fig. 5 and the equilibrium line in Fig. 6. The equilibrium for the (pure) system benzene–caprolactam–40% ammonium sulfate can be described with the following modified equation [25]:

$$Y = 6.25X + 30.4X^2 + 30.1X^3 \qquad (4)$$

4. EXTRACTION PROCEDURE

From a chemical engineering point of view, recovery of caprolactam from the lactam oil and the sulfate layer as feedstocks can be done in several ways [15, 26, 27]. An example of a calculation is given for a 50,000 metric ton/year caprolactam plant.

In the diagram shown below, column 1 is for

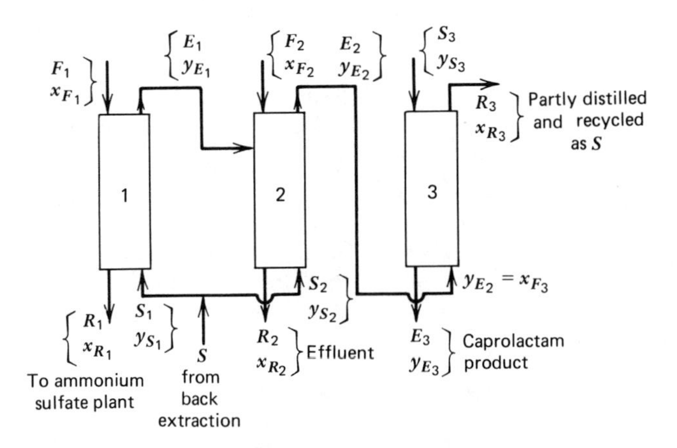

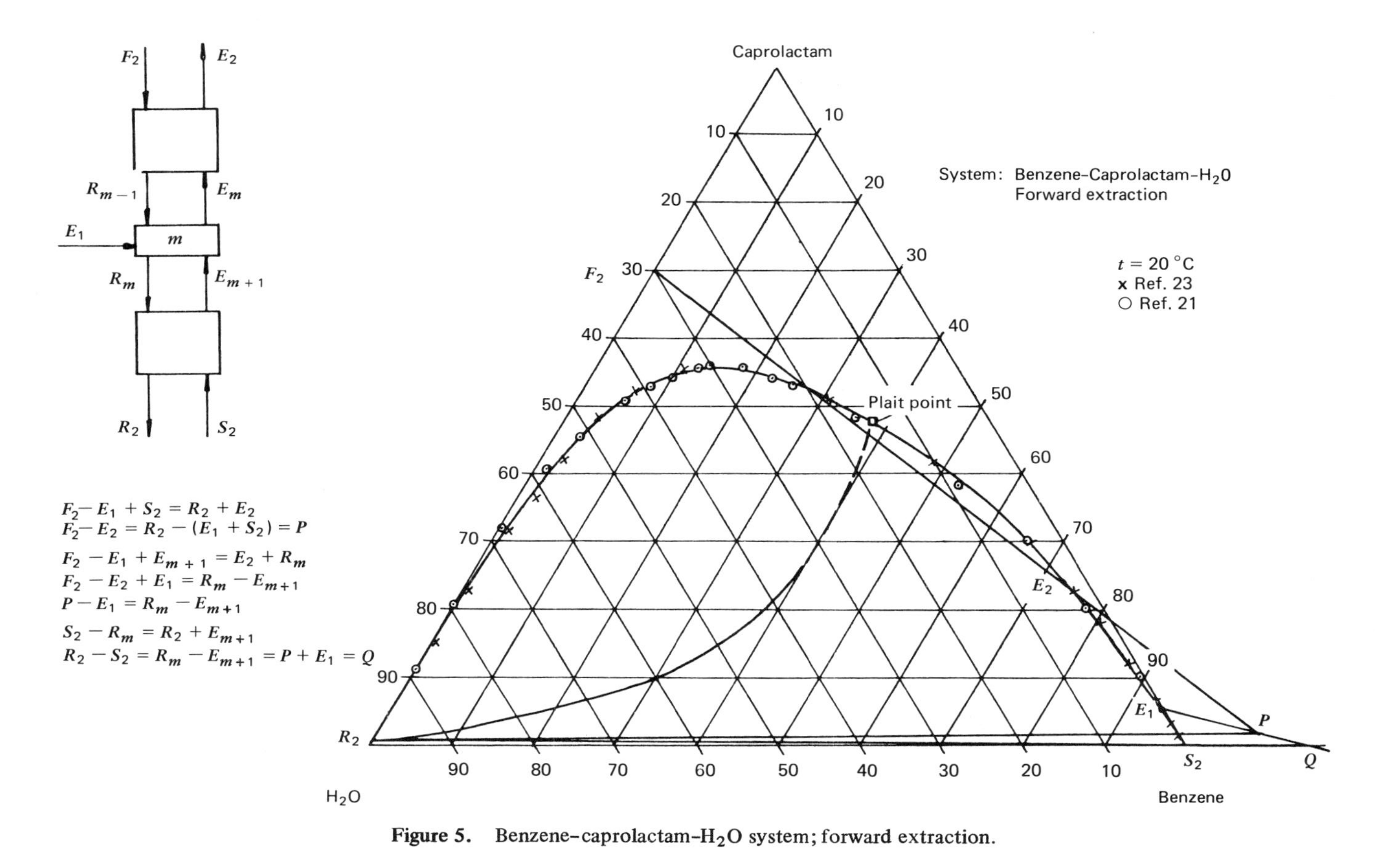

Figure 5. Benzene-caprolactam-H_2O system; forward extraction.

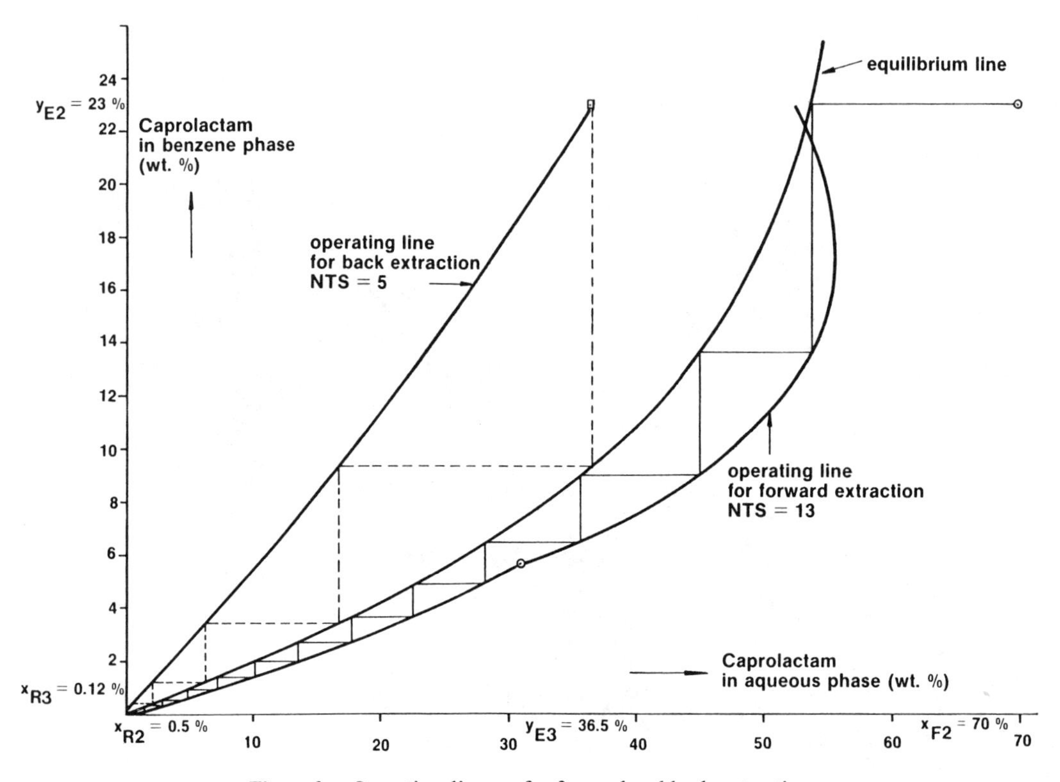

Figure 6. Operating diagram for forward and back extraction.

ammonium sulfate extraction (caprolactam into benzene), column 2 provides forward extraction into benzene, and column 3 is for back extraction into water. The flow scheme, flow rates, and concentrations are based on published data [5] with a slight modification: the total benzene flow to column 1 and 2 is divided into two flows. With "pure" benzene, the caprolactam content in the effluent from column 2 can be decreased to a lower value than with benzene containing the total caprolactam amount from the ammonium sulfate column. Rather arbitrarily the flow rate to column 1 is put at $S_1 = \frac{1}{4}S$. It then follows that

$F_1 = 30.6$ tons/hr	$R_1 = 30.0$ tons/hr
$x_{F_1} = 1.15\%$;	$x_{R_1} = 0.05\%$.
$S_1 = 5.3$ tons/hr	$E_1 = 5.9$ tons/hr
$y_{S_1} = 0.005\%$	$y_{E_1} = 5.7\%$
$F_2 = 8.8$ tons/hr	$R_2 = 2.2$ tons/hr
$x_{F_2} = 70.0\%$	$x_{R_2} = 0.5\%$
$S_2 = 15.7$ tons/hr	$E_2 = 28.2$ tons/hr
$y_{S_2} = 0.005\%$	$y_{E_2} = 23.0\%$
$S_3 = 11.0$ tons/hr	$R_3 = 21.5$ tons/hr
$y_{S_3} = 0$	$x_{R_3} = 0.12\%$
$E_3 = 17.7$ tons/hr	
$y_{E_3} = 36.5\%$	

An operating line has been constructed for both the forward and back extraction on the basis of Fig. 5 [29], as shown in Fig. 6. Stepwise construction between operating lines and equilibrium line then shows the stages and the corresponding raffinate and extract concentrations. As can be seen from Fig. 6, the operating line for forward extraction crosses the equilibrium line, which generally happens if the solute concentration at the plait point is less than the solute concentration in the feed. It is also important to note that, especially for ideal stages, the operating curves in reality exist only at the corner points of the steps on the operating curves. Solutions indicated by coordinates of the curves between such points do not exist. The stepwise construction for both forward and back extraction results in (1) NTS = 13 for forward extraction; this is equivalent to $\text{NTU}_{od} = \text{NTU}_{oR} = 14.0$ for a decrease in caprolactam concentration from $x_R = 53.7\%$ to $x_{R_2} = 0.5\%$ plus one theoretical stage and (2) NTS = 5 for back extraction that is equivalent to $\text{NTU}_{od} = \text{NTU}_{oE} = 3.9$.

The forward extraction is characterized not only by a steep concentration gradient, but also by a strongly changing phase ratio and changing phase rates. A countercurrent column for this extraction, the diameter of which has been based on the flow rates of the solute-rich stream, is therefore underloaded for the greater part of its column length. The great change in mass flow rate of the raffinate phase in the forward extraction and the ratio of the phase flow rates suggest the use of the aqueous raffinate phase as the dispersed one. With the added complication of Marangoni instability, it is very difficult to design this extraction column from first principles, and very often it is designed on the basis of pilot plant data together with experience obtained with industrial columns. The back extraction does not show such a change in mass flow rate in either phase. From material balances around the stages it may be concluded in this case that the extract phase should be dispersed. The design of the column can be approached from first principles.

Both forward and back extraction are carried out in several types of column [5], including PPC [15], RDC [2], and ARD [27, 31, 32] with different systems.

Increase in the temperature of the extraction operation causes a shift of the equilibrium line, which results in a decrease in the number of theoretical stages for the forward extraction but an increase in the number of stages for the back extraction if the same temperature is applied. Whether the samc temperature must be applied for the three columns can be determined only from total mass and energy balances around the purification plant. Solvent rates and solvent : feed ratios also must be optimized [30].

For the ammonium sulfate extraction, which is much simpler from a chemical engineering point of view, a similar McCabe–Thiele approach results in NTS = 3.0; this is equivalent to $NTU_{od} = NTU_{oE} = 1.9$.

5. APPROXIMATE COLUMN DIMENSIONS

General design procedures for various types of column have been given in other chapters. On the basis of the outlined procedures, column dimensions have been estimated for a 50,000 metric ton/year caprolactam plant (extraction with benzene).

The diameters of the three columns are about $D = 1.6$ m if the columns are designed as RDC. The extractive heights then vary between $H = 6$ m for the ammonium sulfate extraction and $H = 22$ m for the forward extraction. If the columns are designed as PPCs, the diameters vary between $D = 1.2$ m and $D = 1.8$ m for the ammonium sulfate and back extraction. Heights are between 3 and 14 m. These column dimensions must be interpreted with care. Often chemical reactions produce coproducts that may behave as surfactants. Surfactants not only have an effect on the throughput capacity ("decrease"), but are also liable to block off the interfacial area, which causes a decrease in the extraction efficiency. The differences in extractive heights for the RDCs and PPCs described here are caused by the difference in scale-up factor, which is 1 for the PPC (see also Chapter 11).

NOTATION

D	diameter of extraction column, m
E	mass flow rate of extract phase, kg hr^{-1}
F	mass flow rate of feed phase, kg hr^{-1}
H	active height of extractor, m
NTS	number of theoretical stages
NTU	number of transfer units
R	mass flow rate of raffinate phase, kg hr^{-1}
S	mass flow rate of solvent phase, kg hr^{-1}
t	temperature, °C
X	concentration (in raffinate phase), weight fraction
x	concentration in raffinate phase, weight percent
Y	concentration in extract phase, weight fraction
y	concentration in extract phase, weight percent

Subscripts

d	dispersed phase
E	extract phase
F	feed
o	overall
R	raffinate phase
S	solvent
1	to column 1
2	to column 2
3	to column 3
$1–m$	stage number

REFERENCES

1. H. C. Ries and R. G. Muller, *Caprolactam*, Part I, 1965, Process Economic Program, Stanford Research Institute, Stanford, CA.
2. *Brochure Caprolactam*, Stamicarbon BV, Geleen, The Netherlands, P.O. Box 10.
3. H. C. Ries and R. G. Muller, *Caprolactam,* Part II, 1965, Process Economic Program, Stanford Research Institute, Stanford, CA.
4. H. C. Ries, *Caprolactam*, Supplement A, 1968, Process Economic Program, Stanford Research Institute, Stanford, CA.
5. K. K. Ushiba, *Caprolactam,* Supplement B, 1976, Process Economic Program, Stanford Research Institute, Stanford, CA.
6. M. Taverna et al., *Hydrocarbon Processing*, 137 (November, 1970).
7. ECN report, in *European Chemical News*, 24 (April 30, 1976).
8. K. Kahr et al., in *Ullmann's Encyklopädie der Technische Chemie*, Vol. 9, Verlag Chemie, GmbH, 1975, Weinheim, West Germany.
9. V. Rod, *Proceedings ISCRE4/ESCRE6*, Heidelberg 1976, Dechema, Frankfurt, West Germany, 1976.
10. J. Damme et al., *Chem. Eng. World* **6,** 44 (1971).
11. S. J. Loyson et al., *Hydrocarbon Processing*, 92 (November 1972).
12. A. H. de Rooy et al., *ACS-AIChE*, General Meeting, New York, April 1976.
13. C. G. M. van de Moesdijk, Ph.D. thesis, DSM/Central Laboratory, Geleen, The Netherlands, 1979.
14. A. J. F. Simons, in B. H. Lucas et al., *Proceedings of the International Solvent Extraction Conference (ISEC), Toronto 1977*, Vol. 2, CIM Special Volume 21, 1979, p. 677.
15. A. J. F. Simons, *Chem. Ind.* 748 (October 7, 1979).
16. L. Polo-Friz et al., *J. Chromatogr.* **39** (3), 253 (1969).
17. Belgian Patent 803,171 (February 4, 1979).
18. *Eur. Chem. News*, 21 (Dec. 22-29, 1972).
19. A. H. de Rooy et al. *Chem. Eng.* 54 (March 18, 1974).
20. A. Heath, *Chem. Eng.* 70 (July 22, 1974).
21. K. Tettamanti et al., *Periodica Polytechnica* **4,** 201 (1960).
22. G. I. Kudryavtseva et al., *J. Appl. Chem. USSR* **26,** 1129 (1953).
23. A. G. Morachevskii et al., *J. Appl. Chem. USSR* **33,** 1755 (1960).
24. Japanese Patent 6815407 (Toyo Rayon Kabushiki Kaisha, Japan).
25. A. G. Kasatkin et al. *Khim. Prom.* 488 (1960).
26. Z. Ziolkowski et al., *Przemysl Chemicsny* **43** (3), 150 (1964); **43** (4), 224 (1964).
27. T. Misek et al., in B. H. Lucas et al., *Proceedings of the International Solvent Extraction Conference (ISEC), Toronto 1977*, Vol. 2, CIM Special Volume 21, 1979, p. 701.
28. Netherlands Patent NL 7601822.
29. R. E. Treybal, *Liquid Extraction*, McGraw-Hill, New York, 1963.
30. S. Hartland, *Counter-Current Extraction*, Pergamon Press, Oxford, 1970.
31. R. Levitanaite et al., *Khim. Prom.* **46,** (4), 2 (1970).
32. W. H. Zehnder et al., *Inform. Chim.* **8,** 81 (March–April, 1970).
33. C. G. M. van de Moesdijk, PT-Procestechniek **36** (3), 147 (1981); **36** (4), 199 (1981).
34. F. Seel, *Fortschr. Chem. Forsch.* **4,** 301 (1963).
35. K. Jockers, *Nitrogen* **50,** 27 (1967).

18.5

ACETIC ACID EXTRACTION

C. Judson King
University of California, Berkeley
United States

1. INTRODUCTION

Recovery of acetic acid from water is one of the oldest applications of solvent extraction, having been proposed by Göring about a century ago [1]. In fact, Göring's patent is remarkable [2] in that it anticipates a number of the more sophisticated developments that have come about in more recent years, including use of countercurrency in the extraction, recovery of residual dissolved solvent from the raffinate water by distillation, use of ethyl acetate as the solvent, and utilization of the azeotropic distillation properties of that solvent to facilitate removal of coextracted water from the recovered acid.

1.1. Applications

Numerous manufacturing processes yield aqueous waste or by-product streams containing acetic acid. These include manufacture of cellulose acetate, aspirin, camphor, and the explosive RDX, as well as semichemical pulping of wood and other processes that use acetic acid as a raw material or a solvent [2–5]. In addition, various manufacturing methods for acetic acid involve separation from water. The principal present-day routes to acetic acid are carbonylation of methanol, liquid-phase oxidation of hydrocarbons such as butane, and oxidation of acetaldehyde [6–9]. Earlier processes, yielding more dilute aqueous solutions and much less used today, include fermentation of alcohol and destructive distillation of wood, yielding pyroligneous acid [3, 10]. Recently, there has been renewed effort to develop inexpensive routes to acetic acid as a chemical feedstock, based on fermentation of biomass and/or wastes. These also lead to dilute aqueous solutions.

1.2. Alternative Separation Processes

Several important fluid separation processes were initially developed in the context of recovery of acetic acid from water. These include liquid–liquid extraction, azeotropic distillation, and extractive distillation [2]. The incentive for separation processes other than simple distillation arises from two factors: (1) the relative volatility between water and acetic acid is not much removed from unity and becomes poorest for dilute solutions of acetic acid in water; and (2) water is the more volatile component, meaning that for dilute solutions, all the water must be vaporized overhead, leading to a large energy cost per unit of acetic acid recovered.

Extractive distillation was used for years in the Suida process for recovery of acetic acid from pyroligneous acid (6–7% acetic acid) [10]. Recycled wood oils were used as the extractive agent. In subsequent plants for synthetic acetic acid, extractive distillation was largely replaced by azeotropic distillation for higher-concentration streams. Methyl and ethyl acetates, diisopropyl ether, and benzene are commonly used entrainers for the azeotropic distillation, although other esters and ethers, ketones, chlorinated hydrocarbons, and alcohols have also been used [4, 7, 8, 12, 13].

Other processes that may be suitable in specific applications include freeze concentration, which has been used for many years as a backyard process for concentration of vinegar; adsorption with carbon or anion exchangers; and chemical derivatization, followed by separation of the derivatives and reconversion, such as was practiced in the calcium acetate process for recovery of acetic acid from pyroligneous acid [10].

Using cost factors appropriate for that time, Brown [12] contrasted extraction with various solvents, azeotropic distillation with benzene or diethyl ketone, and simple fractionation. He found that extraction was the most favored approach except for feeds above about 80% acetic acid content weight for weight (w/w), where azeotropic distillation became preferred. Different conclusions were reached by Eaglesfield et al. [4], who concluded that extraction was preferred for feeds containing up to 35% acetic acid w/w, with azeotropic distillation being more attractive for more concentrated feeds. Brockhaus and Förster [7] support the conclusion of Eaglesfield et al., while also indicating that simple fractionation would be considered only for feeds containing a few percent or less of water. Recent developments in extractive agents and changes in the economic structure may make extractive distillation an attractive choice for feeds in the 30–80% range.

2. CONVENTIONAL EXTRACTION TECHNOLOGY

The basic elements of a typical process for separation of acetic acid and water by extraction are shown in Fig. 1 [12, 14, 15]. In addition to the extractor itself, a method is needed for regeneration of the solvent and recovery of acetic acid from it. Since the solvents used tend to have substantial water solubility themselves, there is usually also some means for removal of residual solvent from the raffinate. In Fig. 1 this is accomplished by distillation or stripping. Heat exchange is used as warranted.

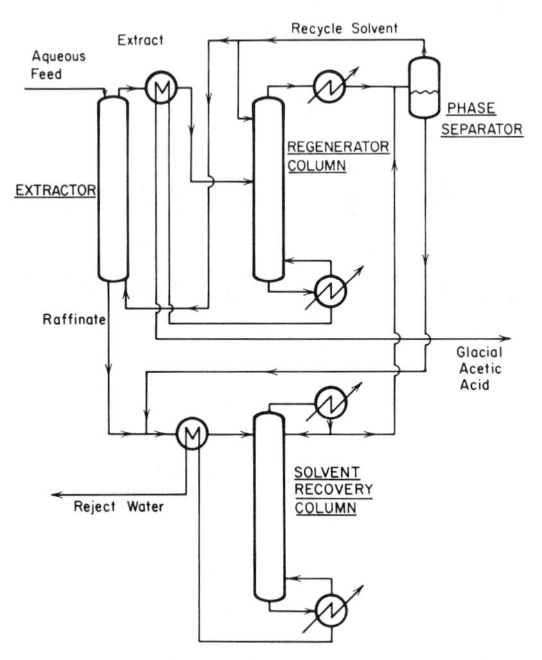

Figure 1. Typical extraction process for recovery of acetic acid from water.

2.1. Regeneration of Solvent and Further Concentration of Acetic Acid

In Fig. 1 solvent regeneration is accomplished by distillation. In principle, solvent regeneration could be accomplished in other ways but seldom is. Back extraction into NaOH solution is one possibility, but in most cases it would not be desirable to degrade the acetic acid to sodium acetate.

Since the desired product is usually glacial acetic acid, and since most solvents coextract a substantial amount of water along with the acetic acid, it is usually necessary to concentrate the extracted acid further. This can be done after solvent regeneration, but again the low volatility of water relative to acetic acid makes it difficult to carry this out by simple distillation. Consequently, azeotropic distillation is the usual means for removing water from the extracted acid. This can be done by adding a separate entrainer and carrying the azeotropic distillation out in a separate tower following solvent regeneration [2, 5, 13]. However, it often makes sense to choose a solvent for extraction that itself can serve effectively as the entrainer for the azeotropic distillation, thus avoiding the addition of yet another chemical to the process [4, 7, 12]. If separate columns are used for solvent regeneration and acid concentration, the acid concentrator must precede the solvent regenerator. Instead, the two operations are usually combined into one column, as shown in Fig. 1, with an overhead phase-separator drum serving to segregate water sufficiently from the recycle solvent. The water layer from this drum is then fed to the solvent-recovery stripping column. In Fig. 1 the split-stream feed of recycle solvent to the top of the regenerator column serves to supply sufficient entrainer (solvent) to build up its concentration adequately in the lower parts of the column to give the azeotropic distillation effect for water removal.

This synergistic coupling of azeotropic distillation with extraction emphasizes the desirability of gaining selective extraction of acetic acid, as opposed to water, in the extractor. In that way the extraction step serves as a preconcentrator of the feed to the azeotropic distillation column. With a higher ratio of acetic acid to water in the feed to the azeotropic distillation, the cost of the azeotropic distillation is reduced, and thus the combination of extraction and azeotropic distillation is cheaper than azeotropic distillation alone.

2.2 High-Boiling Versus Low-Boiling Solvents

One of the more basic choices is in regard to the use of a solvent with a boiling point lower or higher than that of acetic acid. Some of the criteria to be considered are the following. The use of a high-boiling solvent usually leads to a lower steam requirement for the reboiler of the solvent regenerator, since the acetic acid and coextracted water, rather than all the solvent, are taken overhead. With a high-boiling solvent the acetic acid is recovered as a distillate, whereas with a low-boiling solvent (Fig. 1) it comes as a still residue, whch may contain heavy impurities. On the other hand, a high-boiling solvent can collect heavy impurities itself, thereby leading to a need for a continual or periodic solvent purge. Loss of solvent in the purge could be costly, or else some additional processing step is needed to recover solvent from the purge. Use of a high-boiling solvent precludes the combination of water removal and solvent regeneration into a single tower. Furthermore, a high-boiling solvent leads to a higher reboiler temperature in the regenerator, wih possible attendant needs for higher-pressure steam and/or greater thermal decomposition in the boiler. Alternatively, vacuum distillation can be used.

Othmer [2, 3, 5] and Brown [12] have indicated ways in which extraction with a high-boiling solvent may be integrated with azeotropic distillation for solvent regeneration and further concentration of acetic acid. Brown [12] carried out conceptual designs for a high-boiling solvent (isoamyl acetate) and several different low-boiling solvents (diisopropyl ether, diethyl ether, ethyl acetate, chloroform, and ethyl acetate–benzene) and found that the high-boiling solvent led to about 30% higher process costs than the best low-boiling solvents, even though the high-boiling solvent led to about 30% lower steam costs. However, since the time (1963) of his study the ratio of energy costs to equipment costs has escalated, and it may thus be appropriate to give high-boiling conventional solvents serious consideration again. Methylcyclohexanone is another candidate high-boiling solvent, considered in some detail by Eaglesfield et al. [4].

2.3. Solvent Selection

One of the most important solvent properties is the equilibrium distribution coefficient K_D, expressed here as weight fraction acetic acid in the solvent phase divided by weight fraction acetic acid in the aqueous phase at equilibrium. Values of K_D at high dilution and references to more complete data are given by Treybal [16] for extraction of acetic acid by numerous different solvents. Additional data are available elsewhere [4, 17, 18]. The high-dilution K_D data can be summarized as follows for homologous series of compounds with different functional groups:

Family	Range of K_D
n-Alcohols (C_4–C_8)	1.68–0.64
Ketones (C_4–C_{10})	1.20–0.61
Acetates (C_4–C_{10})	0.89–0.17
Ethers (C_4–C_8)	0.63–0.14

The equilibrium distribution coefficient K_D undergoes a continual transition between the extreme values shown. In each case the lowest-molecular-weight member shows the highest value of K_D. Although they give high K_D, alcohols tend to esterify with acetic acid and hence are seldom used. Ketones have the next highest K_D but do not give good azeotroping properties for removal of water by subsequent azeotropic distillation. Hence acetates and ethers are the more commonly used solvents. For acetates one must guard against loss of solvent due to acid hydrolysis of the ester, but this can usually be avoided with proper precautions.

The presence of salts in the aqueous phase, such as in aqueous effluents from chemical pulping, can increase K_D markedly. Othmer [19] has pointed out that for extraction of acetic acid from these effluents, K_D of 4–6 can be achieved with acetone, which becomes only partly miscible because of the high salt content.

The rather low values of K_D realized with conventional solvents lead to a need for relatively high solvent : feed flow ratios, of the order of 1.6–4.0. This follows since K_D(S/W), where S/W is the mass flow ratio of solvent to water, must be greater than 1 for good recovery in a counter-current extractor and is usually set in the range of 1.3–2.5 for optimal and/or reliable operation. Thus one important criterion in solvent selection is realizing a high value of K_D so as to enable the lower expense associated with less solvent flow. A second important criterion is high selectivity for extraction of acetic acid over water so as to reduce the amount of water that must be removed from the acetic acid after extraction. Eaglesfield et al. [4] have examined the trade-off between high-capacity (K_D) and selectivity in some detail for ethyl acetate, mixtures of ethyl acetate and benzene, isopropyl acetate, and methyl cyclohexanone as solvents.

The desire for selective extraction of acetic acid over water enters in another way as well, if the solvent is also to be used as the entrainer in a subsequent azeotropic distillation column for dewatering and solvent recovery. The solvent–entrainer will have a certain maximum carrying capacity for water, which is set by the composition of the binary azeotrope. Thus, for a solvent that is also to be the entrainer, an important criterion is that the selectivity for acetic acid over water be sufficiently high and that the percentage of water in the water–solvent azeotrope be sufficiently high, in combination, that the solvent can entrain all the water overhead in the azeotropic distillation column. This can become a controlling limitation for higher feed concentrations (e.g., $\geqslant$15% w/w acetic acid) because the greater loading of acetic acid brings more water with it into the extract.

Other guidelines for solvent selection that are similar to those for other systems are reviewed by Brown [12]. They include low solubility in water (to reduce the load on the solvent recovery stripper), adequate volatility of the solvent (for a low-boiling solvent, sufficiently high to facilitate distillation of the solvent from acetic acid in the regenerator and stripping of residual solvent from water in the solvent-recovery column), density, interfacial tension, price, stability, toxicity, and compatibility with system contaminants.

On the basis of a number of detailed conceptual designs, Eaglesfield et al. [4] recommended extraction with ethyl acetate for feeds containing up to 16% w/w acetic acid. Above 16% acetic acid in the feed, they found that the limit of the ability of the solvent–entrainer to carry all the coextracted water overhead became controlling, and they thus recommended use of a mixed solvent composed of ethyl acetate and benzene for feeds between about 16 and 25% w/w acetic acid. The benzene is added to the solvent to suppress coextraction of water, and the amount needed in the solvent mixture increases more or less linearly from zero for the 16% feed to 25% w/w for a 25% feed. They noted that isopropyl acetate has entrainer capacity and selectivity properties that offset its lower K_D and concluded that it could be considered as an alternative to ethyl acetate–benzene for feeds above 16% acetic acid, while actually being the preferred solvent for feeds containing over 25% acetic acid. Brown [12] reached similar conclusions, favoring extraction with ethyl acetate or diethyl ether for feeds containing less than about 15% w/w acetic acid and isopropyl ether or a mixture of ethyl acetate and benzene for feeds above 15% and up to about 40% acetic acid. In all these cases the solvent also serves as the entrainer in the subsequent azeotropic distillation. Brown indicated that chloroform could be a preferred solvent for feeds containing about 60% acetic acid, but, as indicated earlier, Eaglesfield et al. and Brockhaus and Förster [7] contend that azeotropic distillation alone, rather than extraction, is preferred for feeds containing more than 35% w/w acetic acid.

Although the papers reaching these conclusions are 17 and 27 years old, it appears that they still reflect current industrial plants well. For new facilities, extractive distillation and extraction with high-boiling solvents may have a more prominent role.

2.4. Extraction Equipment

The choice of contactor for extraction of acetic acid is rather independent of the choices of solvent and methods of regeneration and subsequent concentration. A principal factor influencing equipment choice is generally the desire for a large number of equivalent equilibrium stages or transfer units. This comes from the usual need for high recovery fraction, as well as the economic incentive for keeping solvent flows, and thus K_D(S/W), as low as possible. Other factors affecting equipment selection are the desirability in most cases of high throughput capacity and low cost.

Plate columns match these needs rather well. At least one major manufacturer uses a column with about 100 plates for extraction of acetic acid with isopropyl acetate on a large scale. Rotating-disk contactor (RDC) columns would be a logical alternative; and packed, York–Scheibel, or Karr reciprocating-plate columns might also be considered.

2.5. Process Economics

In 1963 Brown estimated a total operating cost of about U.S. $1.50/m^3 of water ($6/1000 gal) for extraction of feeds containing 2–10% w/w acetic acid, using either ethyl acetate or diethyl ether. Using 1973 economics, Ricker et al. [20, 21] estimate a total operating cost of $3.90/m^3 of water ($15/1000 gal) for extraction of acetic acid from a 5% feed with ethyl acetate. In both cases the solvent serves also as the entrainer. These figures include no credit for recovered acetic acid.

For recovery of acetic acid from dilute aqueous streams, these costs indicate that the lowest feed concentration allowing economically profitable recovery is in the range of 2–4%, depending on the value accorded to recovered acetic acid (1980 market price = U.S. 21¢/lb) and the necessary return on investment.

3. CHEMICALLY COMPLEXING EXTRACTANTS

The best opportunity for reducing the cost of extraction of acetic acid, or for reducing the lower-limit feed concentration for economic recovery, appears to lie in development of solvents giving substantially higher K_D and thereby reducing the large solvent-circulation rates required. Conventional solvents (esters, ethers, and ketones) are limited to K_D of about 1.0 or less for a practical process.

Higher values of K_D can be obtained with chemically complexing extractants—notably strong, organic Lewis bases, which can undergo acid–base complexing with acetic acid. Because of the complexing, such extractants can also be more selective for acetic acid over water. These substances are necessarily high boiling, with the attendant advantages and disadvantages noted previously. They are also more costly than the conventional solvents, which means that solvent losses become more critical. However, a number of strong Lewis-base extractants have been developed and used commercially for some years in hydrometallurgical processing; this gives a backlog of experience with them.

The principal complexing extractants explored to date have been phosphoryl compounds and amines. Tri-*n*-butyl phosphate (TBP) gives K_D of about 2.3 for acetic acid at high dilution [17, 18, 22]. Values of K_D vary approximately linearly when TBP is mixed with hydrocarbon diluents [22]. The 1980 market price of TBP is about U.S. $1.20/lb in contrast to 25 and 33¢/lb for isopropyl acetate and ethyl acetate, respectively. The phosphoryl oxygen becomes a still stronger electron donor with removal of oxygens from the other linkages to the phosphorus atom, giving successively phosphonates, phosphinates, and phosphine oxides [22]. Trioctyl phosphine oxide (TOPO) (1980 market price about $7–$9/lb) has been studied extensively and developed semicommercially as an extractant for acetic acid [23, 24]. As of this writing, it does not appear to have been used yet on a large scale, probably because of a combination of the novelty, the solvent cost, and the difficultilies of buildup of heavies in a high-boiling solvent.

Amines are as strong, or stronger, extractants for acetic acid than are phosphine oxides and are less costly [20–22, 25]. Tertiary amines are preferable [21, 25] since primary amines are too soluble in water and secondary amines are prone to irreversible amide formation with acetic acid during regeneration. Tertiary amines are available in the Alamine (Henkel, Inc.) and Adogen (Sherex, Ashland Chemicals) series at 1980 market prices in the range of $1.00–$1.50/lb.

Both TOPO and amines should be used in solvent mixtures with one or more diluents, for several reasons:

1. The diluent is helpful for reducing viscosity and controlling both the density difference between phases and the interfacial tension.
2. The phosphine oxides and amines suitable for use as extractants are quite high boiling, and the diluent thereby sets the reboiler temperature for the regeneration column and provides most of the vapor boilup at the bottom of the column.

3. The extractants themselves are relatively poor solvents for the acid–base complexes that are formed with acetic acid. The diluent can thus be an important solvating medium for the complex. This means that values of K_D can be much higher for intermediate solvent compositions than for either pure extractant or pure diluent. Quite marked behavior of this sort has been observed [21, 22, 25].

For amines, Ricker et al. [21, 25] found that alcohols as diluents gave the highest K_D values but were again subject to the problem of esterification with acetic acid. Ketones appeared to be next best as diluents from the standpoint of high K_D, although chloroform would be much superior, if it can be used. With TOPO, alcohols were inferior even to aromatics as diluents, presumably because the alcohols hydrogen bond with the phosphoryl oxygen, blocking acetic acid. Ketones gave the highest K_D values of the diluents examined for TOPO.

Complexing extractants give some coextraction of water—for example, about 16% w/w water and 84% acetic acid in the extract, on a solvent-free basis for extraction from a feed containing 6.6% acetic acid with a solvent mixture composed of Alamine 336 and diisobutyl ketone (DIBK) [21, 25]. This water can be removed in a dehydrating extractive distillation column before a solvent-regeneration column [20, 21]. The complexing extractant would act as solvent in the extractive distillation in a way analogous to the use of the same solvent for extraction and for azeotropic distillation (Fig. 1). In some cases the economic incentive to avoid coextraction of water may be such that it is worth using a diluent or a modifier (e.g., a hydrocarbon) in the solvent mixture that gives lower K_D for extraction but greater selectivity for acetic acid over water.

In these extractant–diluent systems, the optimal molecular weight for the extractant reflects a compromise between high K_D (low molecular weight) and excessive solubility of the extractant or of the complex in the aqueous phase (high molecular weight). Trioctylphosphine oxide and Alamine 336 (tri-octyl/decyl amines) appear to be near optimal in this sense. The solubility of Alamine 336 in water is less than 10 ppm [21, 25], and that of TOPO is about 1 ppm [23]. The optimal diluent molecular weight reflects a compromise between high K_D (low molecular weight) and either low solubility in water or sufficiently low volatility relative to acetic acid for solvent regeneration to be able to take place by distillation in the presence of the extractant, which suppresses the volatility of acetic acid (high molecular weight). Among the ketones, DIBK seems to be an effective compromise in that regard, with the relative volatility of acetic acid to DIBK being about 3.5 for a solvent containing 40% Alamine 336 in DIBK [20]. Values of K_D for extraction of acetic acid (1% w/w in raffinate) are about 2.5 for both 50% Alamine 336 in DIBK and TOPO in 2-heptanone [25].

Because of the stoichiometry of the complexing reaction and the law of mass action, K_D values with the complexing extractants become substantially less at high solute loadings, and for that reason these extractants are probably of most interest for relatively dilute feeds (e.g., $\leqslant$10% acetic acid).

Ricker et al. [20, 21] estimate total operating costs of U.S. $1.90 and $2.78 (1978 basis) per cubic meter of water ($7.20 and $10.50 per 1000 gal) for extraction of acetic acid from 5% w/w aqueous solution into solvent mixtures of 50% Alamine 336 in DIBK and 40% TOPO in 2-heptanone, respectively. This would lower the economically recoverable feed concentration by up to a factor of 2 in comparison with use of ethyl acetate. Among the more critical assumptions was that an average loss of 20 ppm of extractant could be achieved over time, based on experience with similar systems. The difference in operating costs for TOPO and the amine reflects primarily the higher cost for TOPO itself.

In cases of separations made in particularly complex systems, such as fermentation broths, where contamination or loss would be a major problem, it may be worthwhile to consider use of one of the complexing extractants impregnated into membranes, with an appropriate diluent. In cases where multiple carboxylic acids are present and should be fractionated, use of dissociation extraction with a stoichiometric deficient quantity of one of the Lewis-base extractants is a possibility [26].

REFERENCES

1. German Patent 28,064 (December 18, 1883), Th. Göring.
2. D. F. Othmer, *Chem. Eng. Progr.* **54** (7), 48 (1958).
3. E. LeMonnier, Ethanoic Acid, in *Kirk-Othmer Encyclopedia of Chemical Technology,* 2nd ed., Vol. 8, Wiley-Interscience, New York, 1965, pp. 386–404.

4. P. Eaglesfield, B. K. Kelly, and J. F. Short, *Ind. Chem.* 29, 147, 243 (1953).
5. D. F. Othmer, *Ind. Eng. Chem.* **50** (3), 60A (1958).
6. F. S. Wagner, Jr., Acetic Acid, in *Kirk-Othmer Encyclopedia of Chemical Technology,* 3rd ed., Vol. 1, Wiley-Interscience, New York, 1978, pp. 124–147.
7. R. Brockhaus and F. Förster, Essigsäure, in *Ullman's Encyklopädie der technischen Chemie,* 4th ed., Vol. 11, Verlag Chemie, Weinheim, 1972, pp. 57–74.
8. K. S. McMahon, Acetic Acid, in J. J. McKetta and W. A. Cunningham, Eds., *Encyclopedia of Chemical Processing and Design,* Vol. 1, Marcel Dekker, New York, 1976, pp. 216–240.
9. R. P. Lowry and A. Aguilo, *Hydrocarb. Process* **53** (11), 103 (1974).
10. W. F. Schurig, Acetic Acid, in *Kirk-Othmer Encyclopedia of Chemical Technology,* 1st ed., Vol. 1, Interscience, New York, 1947, pp. 56–78.
11. J. A. Gerster, Distillation, Section 13 in R. H. Perry, C. H. Chilton, and S. D. Kirkpatrick, Eds., *Chemical Engineers' Handbook,* 4th ed., McGraw-Hill, New York, 1963, p. 13–14.
12. W. V. Brown, *Chem. Eng. Progr.* **59** (10), 65–68 (1963).
13. D. G. Weaver and W. A. Biggs, Jr., *Ind. Eng. Chem.* **53**, 773 (1961).
14. L. A. Robbins, Liquid–Liquid Extraction, Section 1.9 in P. A. Schweitzer, Ed., *Handbook of Separation Techniques for Chemical Engineers,* McGraw-Hill, New York, 1979, pp. 1-255–1-282.
15. J. Coleby, Industrial Organic Processes, in C. Hanson, Ed., *Recent Advances in Liquid-Liquid Extraction,* Pergamon, London, 1971, Chapter 4.
16. R. E. Treybal, Liquid Extraction, Section 15 in R. H. Perry and C. H. Chilton, Eds., *Chemical Engineers' Handbook,* 5th ed., McGraw-Hill, New York, 1973, p. 15-8.
17. K. W. Won, Ph.D. dissertation, University of California, Berkeley, 1974.
18. J. M. Wardell, M.S. Thesis, University of California, Berkeley, 1976.
19. D. F. Othmer, paper presented at Anaheim, CA meeting, American Chemical Society, March 1978.
20. N. L. Ricker, E. F. Pittman, and C. J. King, *J. Separation Process Technol.* **1** (2), 23 (1980).
21. N. L. Ricker and C. J. King, Report No. EPA-600/2-80-064, U.S. Environmental Protection Agency, April 1980; see also N. L. Ricker, Ph.D. dissertation, University of California, Berkeley, 1978.
22. J. M. Wardell and C. J. King, *J. Chem. Eng. Data* **23**, 144 (1978).
23. R. W. Helsel, *Chem. Eng. Progr.* **73** (5), 55 (1977).
24. U.S. Patent 3,816,524 (June 11, 1974), R. R. Grinstead (assigned to Dow Chemical Company).
25. N. L. Ricker, J. N. Michaels, and C. J. King, *J. Separation Process Technol.* **1** (1), 36 (1979).
26. G. C. Jagiradar and M. M. Sharma, *J. Separation Process Technol.* **1**(2), 40 (1980).

18.6

MGC XYLENE EXTRACTION PROCESS BY USE OF $HF-BF_3$

Tamotsu Ueno
Mitsubishi Gas Chemical Company
Japan

1. INTRODUCTION

Currently, xylenes are prepared mainly by aromatics extraction from reformate. A minor portion is also available from thermal cracking of gasoline (TCG) and from coal. The proportions of the isomers depend on the origin; typical figures are shown in Table 1. PX, OX, and EB are the raw materials for terephthalic acid, phthalic anhydride, and styrene monomer, respectively; MX is the raw material for isophthalic acid and some specialty chemicals but has a smaller market than the other three isomers. There is thus an incentive to convert MX to other isomers in order to give maximum production of the other isomers. By utilizing the difference of physical properties shown in Table 2, OX and EB can be separated by superfractionation in two columns, whereas PX is recovered by either crystallization or adsorption on zeolite.

The MGC process for MX extraction and its isomerization was developed as the key technology for the production of four isomers. Extraction of the least required isomer prior to any other separation may seem a roundabout route but was shown to be feasible since elimination of the most abundant and otherwise difficult-to-separate components makes a significant contribution to the ease of separation of the other three isomers, both with respect to increase in concentration and decrease in quantity [1]. When maximum production of OX and PX is required, it is advantageous for $HF-BF_3$ as solvent that the MX thus extracted is readily converted to OX and PX with the use of $HF-BF_3$ as a catalyst.

The chemistry of xylene extraction and its isomerization with the use of $HF-BF_3$ has been passably studied [2–4]. However, the possibility of *perfect* MX separation and of solvent recovery was vague, whereas materials of construction were not known. After these problems had been solved, MGC concentrated its effort on maximizing the efficiency of the combined extraction–isomerization process. The basic research started in 1961, with demonstration by a pilot plant of 1 metric ton/day in 1965, and the first commercial plant was launched in 1968. The technology is characterized by two distinct features that originate from the nature of $HF-BF_3$.

First, extraction occurs with the formation of HF-soluble protonated complexes of moderate stability. The order of the stabilities is MX $\gg$ OX $>$ PX $>$ EB. Hydrogen fluoride by itself is a

TABLE 1 COMPOSITION OF XYLENES (%)

	EB	PX	OX	MX
Reformate	15	20	20	45
TCG	30	15	15	40
Coal	10	20	20	50

poor solvent of xylenes, as it has no selectivity among the isomers; hence the addition of BF_3 is indispensable in the extraction of any isomers.

Second, at an elevated temperature, HF–BF_3 acts as a good isomerization catalyst for MX, PX, and OX, thus making it possible to establish a suitable process combining MX extraction and isomerization.

2. MECHANISM OF COUNTERCURRENT MX EXTRACTION

Figure 1 illustrates a typical example of the commercial process. Both HF(l) and BF_3(g) are mixed in mixer M_1 with the hydrocarbon mixture withdrawn from extraction column E_1. Complex formation takes place according to

$$HF(l) + BF_3(g) + OsX(l) = OsXHBF_4(HF) \quad (1)$$

Hydrogen fluoride is used in some excess; the amount is determined from the vapor pressure of the system, heat requirements, and so on. The heat of complex formation is approximately 7 kcal mol^{-1} $liter^{-1}$. The complex of OsX thus formed in HF undergoes the complex exchange [Eq. (3)], which is put between two elementary steps [Eqs. (2) and (4)], while it contacts with the MX-rich hydrocarbons that flow up in columns E_1 and E_2:

$$MX(l) = MX(HF) \quad (2)$$

$$MX(HF) + OsXHBF_4(HF) = MXHBF_4(HF) + OsX(HF) \quad (3)$$

$$OsX(HF) = OsX(l) \quad (4)$$

Equations (2)–(4) are regarded as the essential process for the extraction of MX. The heat of reaction (3) is approximately 3 kcal mol^{-1} $liter^{-1}$.

When OsX is specified to an isomer, the equilibrium constant of reaction 3 is called *relative basicity*, as expressed by Eq. (5) (see Table 2):

$$K_r = \frac{[MXHBF_4(HF)]\,[OX(HF)]}{[OXHBF_4(HF)]\,[MX(HF)]} \quad (5)$$

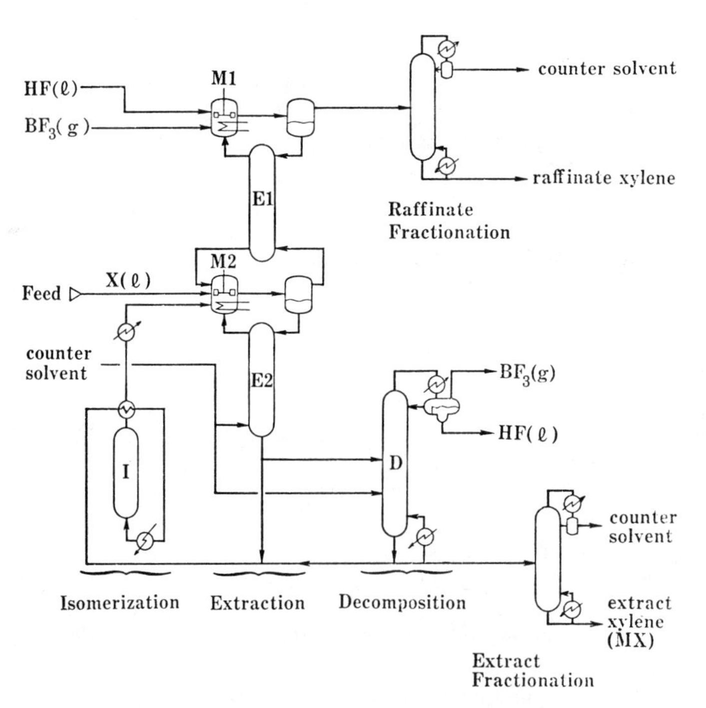

Figure 1. Flow diagram of extraction–isomerization.

TABLE 2 PHYSICAL AND RELATED PROPERTIES OF XYLENES

	EB	PX	OX	MX	Remarks
Freezing point (°C)	-94.98	13.26	-25.18	-47.87	Ref. 6
Boiling point (°C)	136.19	138.35	144.41	139.10	Ref. 6
Relative basicity	0.14	1	2	ca. 100	MGC's data
Separation factor		1		3–100	MGC's data

A conventional but more useful property is given by Eq. (6):

$$\alpha_e = \frac{[MXHBF_4(HF) + MX(HF)]\ [OsX(l)]}{[OsXHBF_4(HF) + OsX(HF)]\ [MX(l)]} \quad (6)$$

where α_e corresponds to relative volatility of vapor–liquid equilibrium. As is obvious from Fig. 2, α_e increases either when the system is poor in MX or when a countersolvent is added. The augmentation of α_e in the preceding cases is attributable to the decrease of nonselective physically dissolving xylenes that are denoted MX(HF) or OsX(HF). It is a nature of the MX complex ($MXHBF_4$) that gives the nonselective solvent power to HF.

A C_5–C_6 saturated hydrocarbon mixture is introduced to the bottom of column E_2 as a countersolvent. Reflux of MX is not necessary. The temperature in the extraction section is kept between -10 and 10°C. Excluding reaction (3), isomerization of xylenes [reaction (8)] or any other types of chemical reaction occuring in the HF phase are kept substantially *frozen* at this temperature.

After the extract is withdrawn from the bottom of column E_2, it is fed into the decomposer D, where the degradation of the complex $MXHBF_4$ takes place [reaction (7)] along with vaporization of the whole HF. Heat is given by the vapor of C_5–C_6 hydrocarbons, which are used in common with the countersolvent. Thus MX(l) is obtained in the form of a solution with saturated hydrocarbons:

$$MXHBF_4(HF) = MX(l) + HF(l) + BF_3(g) \quad (7)$$

The degradation is conducted in such a short time that no chemical reactions, including isomerization of xylene, occur to any detectable

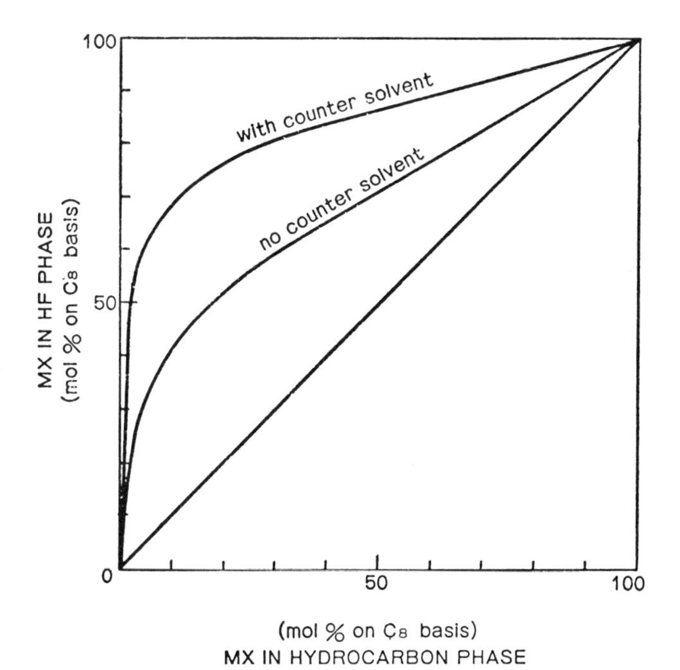

Figure 2. X–Y diagram of MX extraction.

extent. After cooling of the top gas, HF(l) and BF_3(g) are obtained. Since the decomposer D is operated under pressure, BF_3(g) is recycled to the extraction section without any additional compression. After topping the saturated hydrocarbons from the effluent of the decomposer D and from the effluent of the mixer M_1, over 99.5% extract MX and raffinate xylenes of less than 0.05% MX (both on C_8 basis) are currently produced. Fluorine concentration in both products is less than 1 ppm.

Considering the roles of HF and BF_3 from the technological point of view, HF could be considered merely as a *second phase* and BF_3 as the medium or extractant driving the substance to be extracted from the hydrocarbon to the HF phase.

3. COMBINATION OF MX EXTRACTION AND ITS ISOMERIZATION

When isomerization of MX is needed, a portion of the extract is mixed with the desired portion of decomposer effluent and fed into the isomerizer I, where the two liquids are contacted at 80–120°C. The composition of xylenes (C_8 basis) in the hydrocarbon layer changes toward the thermodynamic equilibrium [5], excluding EB, the formation of which is inhibited by a kinetic reason:

$$\begin{array}{ccccc} PX(l) & & MX(l) & & OX(l) \\ \upharpoonleft\downharpoonright & & \upharpoonleft\downharpoonright & & \upharpoonleft\downharpoonright \\ PX(HF) & & MX(HF) & & OX(HF) \\ \upharpoonleft\downharpoonright & & \upharpoonleft\downharpoonright & & \upharpoonleft\downharpoonright \\ PXHBF_4(HF) & \rightleftharpoons & MXHBF_4(HF) & \rightleftharpoons & OXBF_4(HF) \end{array} \qquad (8)$$

The ultimate composition of xylenes in the hydrocarbon layer becomes approximately 24% PX, 18% OX, and 58% MX. The composition of xylenes in HF phase (catalyst layer) is in the range 70–90% MX; the balance is OX and PX. The isomerizate, a mixture of two liquids, is cooled and fed into mixer M_2. An alternative, that is, separating the two layers, feeding the hydrocarbon layer into the mixer M_2 and recycling the catalyst layer through the isomerizer is also possible. However, since the catalyst layer contains xylenes of the preceding composition, it is more advantageously recycled through the lower part of the extractors in order to accomplish maximum recovery of PX and OX produced in the isomerizer. The MX splitting is not affected by the newly formed loop of isomerization–extraction.

This combination of MX extraction and isomerization could be classified as an example of an extractive reaction.

4. VARIATIONS

The MGC process has a wide variety of applications. The cases discussed in Sections 4.1 and 4.2 have been developed as process improvements.

4.1. Partial Decomposition-Isomerization

In contrast to the decomposition stated above, partial decomposition and isomerization (PDI) is an operation in which reaction (7) is made to occur (suppressing the vaporization of HF to a minimum level) along with isomerization of MX [reaction (8)] by controlling the residence time of the extract in a unit referred to as a *partial decomposer–isomerizer.* A saving on both erection cost and utilities expenses is accomplished. Needless to mention, pure MX cannot be obtained through this unit.

4.2. Twin Extractors

A twin extraction system is recommended when the EB concentration is high in the feed xylenes and a large production of PX and OX is required. The feed xylenes are fed to the first extraction unit, which gives a raffinate containing the EB. The second unit treats only the isomerizate, which is free of EB, thus avoiding useless mutual dilution of two types of raffinate. The first raffinate goes to the PX–OX splitting section after separation of EB; the second raffinate is fed directly to the PX–OX splitting section. The load reduction in the EB splitting section is sufficient to compensate for the twin extractors.

5. COMMERCIAL PLANTS

Table 3 shows brief specifications of existing plants (1979).

NOTATION

α_e	separation factor
K_r	relative basicity
EB	ethylbenzene
PX	*p*-xylene; similarly, MX for *m*-xylene and OX for *o*-xylene
OsX	mixture of three xylenes other than *m*-xylene

TABLE 3 COMMERCIAL PLANTS

Country	Number of Plants	Capacity, Metric Ton/Year	Extraction	Breakdown (Decomposition and Isomerization)[a]
Japan	1	300,000	Twins	One D, one I, one PDI
Italy	1	200,000	Twins	One D, one I, one PDI
United States	1	200,000	Single	One D

[a]D, decomposor; I, isomerizer; PDI, partial decomposor–isomerizer.

MX(l) *m*-xylene in the liquid state; a symbol in parentheses indicates the state or phase where the substance exists—l for liquid state, g for gaseous state [e.g., HF(l) for HF liquid phase]

REFERENCES

1. T. Ueno and T. Nakano, in *The Proceedings of the Eighth World Petroleum Congress*, Vol. 4, Applied Science Publishers, Barking, UK, 1971, p. 187.
2. D. A. MacCaulay, B. H. Shoemaker, and A. P. Lien, *Ind. Eng. Chem.* **42**, 2103 (1950).
3. D. A. MacCaulay and A. P. Lien, *J. Am. Chem. Soc.* **74**, 6246 (1952).
4. D. A. MacCaulay, in G. A. Olah, Ed., *Friedel-Crafts and Related Reactions*, Vol. II, Part 2, Interscience, New York, 1964, p. 1049.
5. W. J. Taylor, *J. Res NBS* **37**, 116 (1946).
6. F. D. Rossini, K. S. Pitzer, R. L. Arnett, R. M. Braun, and G. C. Pimentel, *Selected Values of Physical and Thermodynamic Properties of Hydrocarbons and Related Compounds*, Carnegie Press, Pittsburgh, 1953, p. 71.

18.7

MISCELLANEOUS PROCESSES

P. J. Bailes
University of Bradford
United Kingdom

The role of solvent extraction in the production of high-grade raw materials for the manufacture of Terylene and nylon has already been indicated. Another example, also from the synthetic fiber industry, occurs in the manufacture of acrylic fibers, where it can be more economic to recover acrylic acid from aqueous solution by use of solvent extraction [1, 2] rather than distillation. This requires a solvent that is sufficiently specific for acrylic acid to avoid excessive solvent usage and recovery problems. The more specific solvents tend to be too soluble in the aqueous phase or give unwanted side reactions, and hence mixed solvents such as benzene and methyl ethyl ketone have been considered.

Styrene is manufactured from ethylbenzene, and its separation from the unreacted ethylbenzene presents a difficult problem primarily because of their close similarity in volatility. Current practice is to use vacuum distillation for the separation, but a large number of theoretical plates is required, and there can be problems with styrene polymerization. Solvent extraction with diethylene glycol is feasible but not economic. However, it has now been found that styrene can be separated and recovered in high yield and purity by using a new two-phase solvent system [3]. The separation is accomplished by contacting the styrene–ethylbenzene mixture under anhydrous conditions with a specially prepared concentrated cuprous nitrate–propionitrile polar phase and a paraffinic phase. The styrene is selectively extracted into the polar phase by formation of a styrene–cuprous complex, whereas the ethylbenzene distributes predominantly into the paraffinic phase. The two phases are then separated.

In a multistage column contactor the polar phase, the paraffinic phase, and the styrene–ethylbenzene mixture can be charged respectively to the top, bottom, and middle of the extractor. The propionitrile solution containing the styrene–cuprous complex and the paraffin containing the ethylbenzene can be continuously removed as the bottom and top products, respectively. The styrene can be recovered from the polar phase either by decomposing the complex or by displacing the styrene with a different olefin in another liquid–liquid contacting step.

The separation of olefins and diolefins from hydrocarbons of lesser unsaturation with solutions of cuprous salts has been tried before. In fact, reaction with ammoniacal copper acetate solvents in mixer–settlers has been used industrially for butadiene purification, although the preferred method is now extractive distillation. Nevertheless, the new styrene process could have a future since present processes based on distillation are not without their problems.

There are other extractions based on chemical interactions that feature in several well-established refinery processes, for example, mercaptan removal and sulfonation. As mercaptans are easily oxidized and acidic, they can be readily removed by chemical means in processes that include solvent extraction with aqueous alkaline solution.

REFERENCES

1. British Patents 995,471, 995,472, 997,888, and 1,055,532 (1964).
2. German Patent 2,161,525 (1973).
3. G. C. Blytas, U.S. Patent 3,801,664 (1974).

19

USE OF SOLVENT EXTRACTION IN PHARMACEUTICAL MANUFACTURING PROCESSES

K. Ridgway
University of London
United Kingdom

E. E. Thorpe
Beecham Pharmaceuticals, Ltd.
United Kingdom

1. INTRODUCTION

The first major and still by far the most important use of solvent extraction in the pharmaceutical industry is in the manufacture of penicillin. It is also the case that in this and most other pharmaceutical manufacturing processes there is a strong economic advantage in not publishing process improvements so the information available in the literature is relatively sparse.

Solvent extraction is employed in the manufacture of a number of other antibiotics and also for the production of some nonantibiotic pharmaceuticals of animal or vegetable origin.

There are also a number of pharmaceutical compounds that are synthetic in nature and that as part of their chemical synthesis employ solvent extraction techniques.

2. PENICILLIN

2.1. Nature

Penicillin is an umbrella term covering a range of compounds having the basic structure

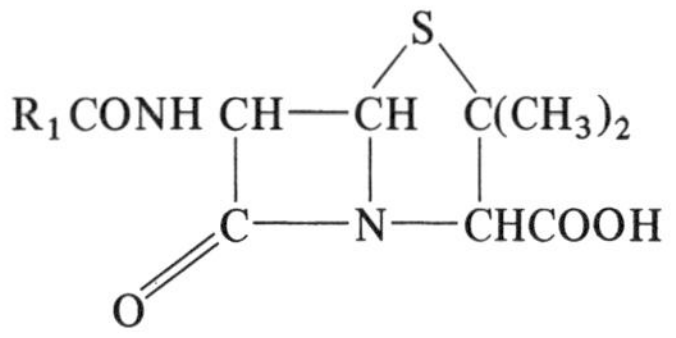

```
                        S
                       / \
R1CONH CH——CH   C(CH3)2
            |        |        |
           C——N——CHCOOH
          //
         O
```

A number of natural penicillins differing only in the composition of the R_1 group are produced by strains of the mold *Penicillium chrysogenum*. Of these only penicillin G (R_1 = benzyl) and penicillin V (R_1 = phenoxymethyl) are important commercially.

In 1959 Batchelor and his colleagues [1] announced the isolation of the penicillin nucleus 6-amino penicillanic acid ($R_1CO = H$), thus opening the way for the chemical manufacture of penicillin analogs not found in nature. The nucleus is produced on an industrial scale either by chemical treatment of the naturally occurring penicillins or by enzymatic removal of the side chain by use of penicillin acylases. A variety of "semisynthetic" penicillins can then be produced by acylation of the 6-amino group. Examples from this range, which are in current use, are ampicillin and amoxicillin (Beecham trademarks Totacillin and Amoxil, respectively).

2.2. The Process

A report of the early development of a practicable industrial process for penicillin manufacture is given by Souders et al. [2]. Toward the end of World War II it was possible to produce a *Penicillium* culture broth with an active penicillin content of 20–60 ppm, and a total solids content of about 3%, or 1000 times the penicillin concentration. The method of extraction at that time was by the adsorption of the penicillin onto activated charcoal, followed by elution with chloroform. Only about onc third of thc penicillin was recovered, and that at only 20% purity.

Rapid process improvements took place: (1) new strains of the *Penicillium* organism were developed so that the penicillin content of the broth rose by a factor of several hundred; and (2) the fact that the penicillins are acidic compounds, more soluble in many organics than in water whereas their ions and salts are more soluble in water than in organics, was exploited. Thus the addition of a mineral acid to the broth will produce the acid form of penicillin, so that it can be extracted into an organic solvent such as amyl or butyl acetate or methyl isobutyl ketone.

If the penicillin is required as an intermediate for chemical or biochemical modification, it can be precipitated directly from the organic solvent by the addition of a concentrated acetate or phosphate buffer. The raising of pH results in the formation of the appropriate salt that, because there is little water present in the buffer, comes out of the two-phase mixture of solvent and buffer. The separated solid is solvent washed to remove impurities and is then dried.

Alternatively, the organic solution can be contacted with a somewhat more dilute buffer. The penicillin salt then passes to the aqueous phase. This aqueous extract is again acidified so that the penicillin can be reextracted back into organic solvent from which the pure penicillin can be crystallized as the sodium, potassium, or organic base salt. The repeated transfer from one solvent to another, besides allowing most of the available penicillin to be removed from the broth, has the advantage of selectively rejecting most of the unwanted impurities.

The need to transfer the penicillin more than once arises because both the impurities and the penicillins have distribution ratios that vary with pH in any one solvent (Fig. 1) and also with the solvent (Fig. 2), and in most cases these distribution ratio lines cross, giving the analog of an azeotropic point in a distillation operation.

The ionization equilibrium constant for an organic monobasic acid RCOOH is defined as

$$k_i = \frac{\{H^+\}\{RCOO^-\}}{\{RCOOH\}}$$

The un-ionized acid will distribute itself at equilibrium between an organic solvent and an aqueous phase according to its distribution ratio

$$k_d = \frac{\{RCOOH\}_{org}}{\{RCOOH\}_{aq}}$$

The observed and macroscopically important overall distribution ratio K will be given by the

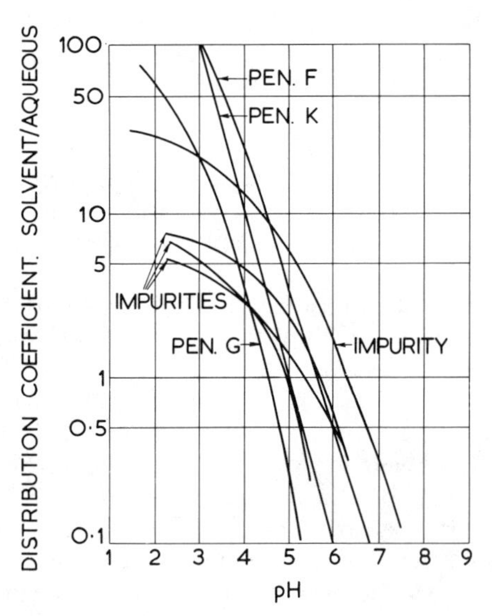

Figure 1. Overall distribution coefficients for some penicillins and impurities plotted against pH [2].

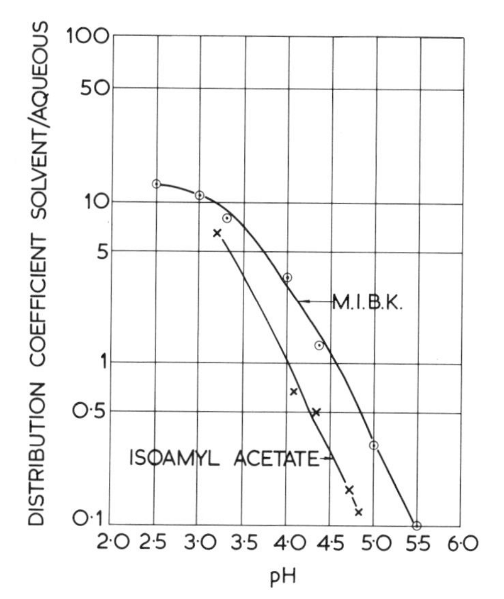

Figure 2. Distribution coefficient of Penicillin G between solvent and water for isoamylacetate and methyl isobutylketone (MIBK) [3].

ratio of the acid concentration in the organic layer to the total acid concentration in the aqueous layer, which will be the sum of the concentrations of the un-ionized and ionized species, that is

$$K = \frac{\{\mathrm{RCOOH}\}_{\mathrm{org}}}{\{\mathrm{RCOOH}\}_{\mathrm{aq}} + \{\mathrm{RCOO}^-\}_{\mathrm{aq}}}$$

and since

$$\{\mathrm{RCOO}^-\}_{\mathrm{aq}} = \frac{k_i\,\{\mathrm{RCOOH}\}_{\mathrm{aq}}}{\{\mathrm{H}^+\}}$$

it follows that

$$\begin{aligned} K &= \frac{\{\mathrm{RCOOH}\}_{\mathrm{org}}}{\{\mathrm{RCOOH}\}_{\mathrm{aq}} + k_i\,\{\mathrm{RCOOH}\}_{\mathrm{aq}}/\{\mathrm{H}^+\}} \\ &= \frac{\{\mathrm{H}^+\}\,\{\mathrm{RCOOH}\}_{\mathrm{org}}}{\{\mathrm{H}^+\}\,\{\mathrm{RCOOH}\}_{\mathrm{aq}} + k_i\,\{\mathrm{RCOOH}\}_{\mathrm{aq}}} \\ &= \frac{\{\mathrm{RCOOH}\}_{\mathrm{org}}}{\{\mathrm{RCOOH}\}_{\mathrm{aq}}} \cdot \frac{\{\mathrm{H}^+\}}{\{\mathrm{H}^+\} + k_i} \\ &= k_o\,\frac{\{\mathrm{H}^+\}}{\{\mathrm{H}^+\} + k_i} \end{aligned}$$

Thus plots of overall distribution ratio against hydrogen ion concentration (i.e., pH) will be curves that at low pH values asymptotically approach the k_o value for the particular compound and solvent. This is illustrated in Fig. 1, where overall distribution ratios are shown for several penicillins in a common solvent. Most of the impurities have distribution ratios such that extraction into an organic solvent at low pH tends to leave them behind in the broth, whereas transfer to an aqueous solution at higher pH leaves them behind in the organic solvent. Repetition of the transfer under a similar pH regime leads to further purification. The alpha value for two compounds, the analog of their relative volatility in distillation, is the ratio of their distribution ratios. There is clearly an analogy with dissociation extraction (see Chapter 21).

2.3. Industrial Plant

A flow diagram of a penicillin plant is shown in Fig. 3. The charge to the fermenter is corn-steep liquor, a waste product in the wet milling of corn, or any similar cheap nutrient solution. Additional nutrients, metabolism modifiers to persuade the *Penicillium* organism to produce the required product, and antifoaming agents may also be added. The pH is adjusted and the fermenter contents are steam sterilized. After cooling to 25°C or so, the fermenter is inoculated with a pure high-yielding mutant strain of *Penicillium* and air sparged with sterile air. Fermentation is continued for 6–8 days, with the periodic addition of extra nutrients and antifoaming agents, and the use of turbine agitation to provide a maximum air–medium interface.

At the end of the fermentation period, the contents of the fermentation vessel are passed to a rotary vacuum drum filter to remove the mycelium, leaving the penicillin in the aqueous solution. The pH is reduced to about 2–2.5 with sulfuric acid, and the solution is contacted with butyl or amyl acetate, into which the penicillin is extracted. Typically a 3 : 1 or 4 : 1 broth : solvent ratio is used. This organic solution is contacted with an aqueous buffer solution at pH 6 or so, producing a penicillin-rich aqueous solution. This is then reacidified and contacted either with the same organic solvent or with a different one such as chloroform or methyl isobutyl ketone to prevent the transfer of any ester-soluble impurities. Subsequent treatment yields the penicillin in solid form.

The foul organic phase and spent broth streams, which contain dissolved and dispersed solvent, are stream stripped to recover the solvent, which is then recycled to the process.

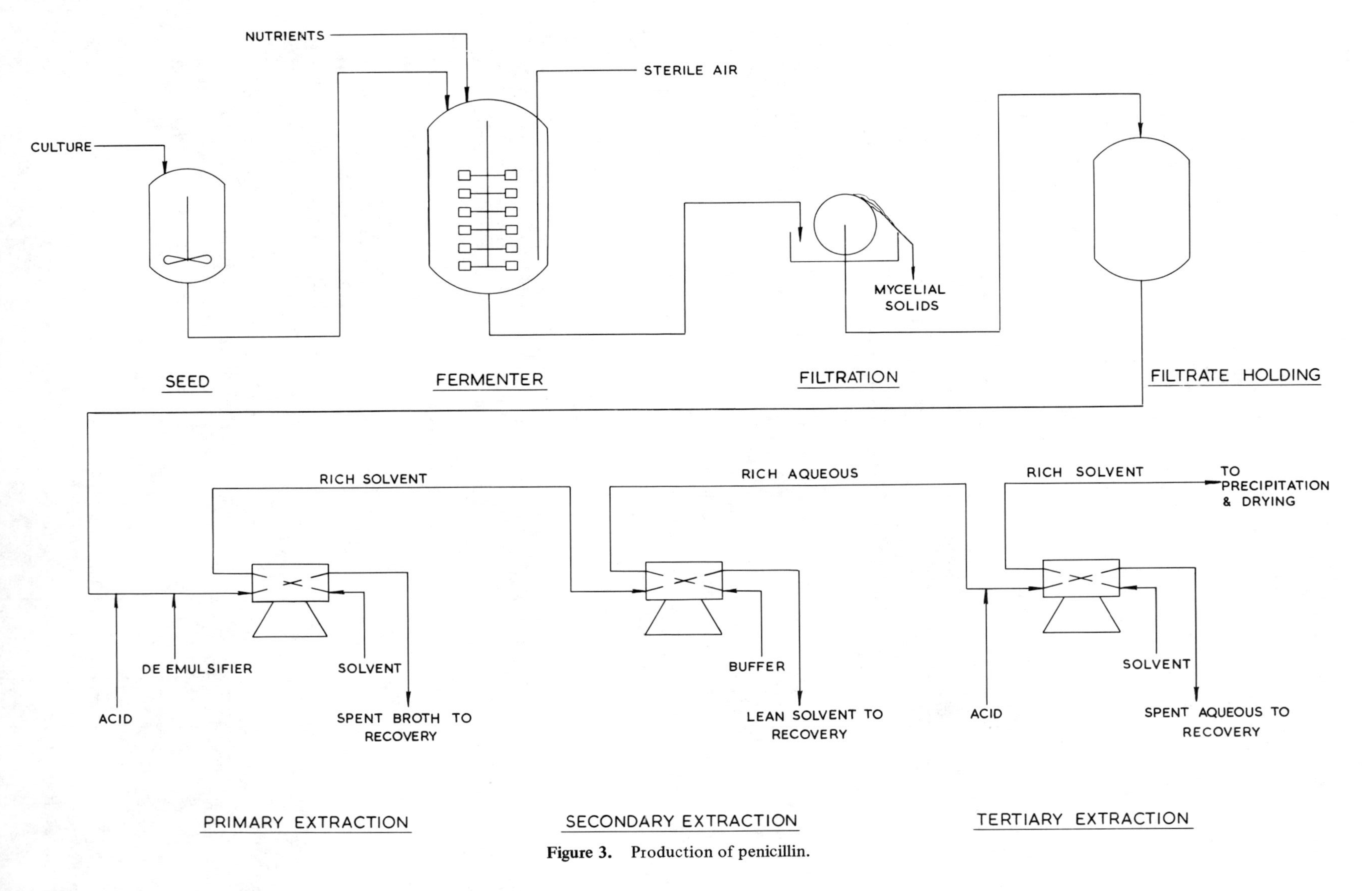

Figure 3. Production of penicillin.

TABLE 1 HALF-LIFE[a] OF SODIUM PENICILLIN G IN AQUEOUS SOLUTION AS A FUNCTION OF pH AND TEMPERATURE [4]

pH	Half-Life, hours		
	0°C	10°C	24°C
2.0	4.25	1.30	0.3
3.0	24	7.6	1.7
4.0	197	52	12
5.0	2000	341	92

[a]Half-life is defined as time in hours required to deactivate 50%.

TABLE 2 DISTRIBUTION COEFFICIENTS FOR PENICILLIN AT pH 4 BETWEEN AQUEOUS SOLUTIONS AND THE LISTED SOLVENTS

Methyl cyclohexanone	180
Dimethyl cyclohexanone	160
Methyl cyclohexanol	80
2-Chloro-2^1-methoxy diethyl ether	57
Cyclohexyl acetate	62
Furfuryl acetate	44
Methyl isobutyl ketone	33
Dimethyl phthalate	30
2-Ethyl hexanol	26
Amyl acetate	20
Diethyl oxalate	20

2.4. Type of Extractor

The difficulty in penicillin extraction is that decomposition is relatively rapid at low pH and high temperature, as can be seen from Table 1, where the half-life of sodium penicillin G is plotted as a function of pH at various temperatures.

The need for rapid extraction, particularly in the low-pH part of the process, favors the use of centrifugal contactors, where the residence time is typically of the order of seconds. Of the machines available, the Podbielniak extractor, a horizontal-axis machine, is perhaps the most widely employed. It is used extensively in the United States and in Eastern Block countries. The Alfa-Laval extractor, originating from Sweden, is more widely found in Western Europe. This is a vertical axis machine operating at a somewhat higher *g* force to the Podbielniak. Other machines that have been used are the Liquid Dynamics Quadronic extractor, the Robatel extractor, and machines from the Westfalia Company of West Germany.

2.5. Process Improvements

Suggestions for improvement have appeared in small numbers in the literature over the years.

One of the earliest examinations of the purification and concentration of penicillin by solvent extraction was made by Rowley et al. as early as 1946 [5]. They reported the distribution coefficients for undissociated penicillin between an aqueous solution at pH 4 and a range of organic solvents. Their values are given in Table 2.

Most of these solvents were more effective than amyl acetate, which is perhaps the most widely used, and they are all efficient at pH 4; the selectivity increases as the pH is reduced. In particular, the selectivity of methylcyclohexanone approaches that of chloroform, which is recognized as a highly selective solvent for penicillin.

The change of partition coefficient with pH for the extraction of benzyl and heptylpenicillins into butyl acetate was measured by Kansawa et al. [6]. Using a phosphate buffer and a Whitmore tower as a continuous extractor, they were able to obtain a transfer of 70–85% of the available penicillin from the broth through the organic phase to the buffer solution.

Eisenlohr and Scharlau [7] described the use of a chain of Luwesta extractors to carry out the extraction mentioned in the previous paragraph. Sulfuric acid was added to the filtered medium and penicillin extracted into butyl acetate, with the use of added wetting agents to prevent emulsification. The butyl acetate was extracted with an aqueous basic phosphate buffer, and after reacidification it was transferred into butyl acetate before penicillin G was isolated as the salt of an organic base.

Zhukovskaya [8] compared the performance of a number of contactors, constructing a nomogram for each that gave the relationship between the amount of penicillin remaining unextracted, the concentration of the aqueous phase, and the number of theoretical contact stages. Overall he decided that the optimum for benzylpenicillin extraction was the operation of a Podbielniak separator, extracting the antibiotic at pH 3 into butyl acetate.

Later, the same author [9] with other coauthors reported on the performance of an Alfa-Laval countercurrent differential contact centrifugal extractor in the extraction of benzylpenicillin into butyl acetate, and out again into sodium bicarbonate solution at pH 7. With a phase ratio of 6 : 1, 1.4–1.8 contact stages were obtained for the transfer from liquor to butyl acetate, but at the same throughput at the re-extraction stage the performance was equivalent to less than 0.7 stages, even though the overall yield of the transfer operation was better than 85%.

Emulsion formation is a common problem in the extraction of penicillin from the fermentation liquor. Current practice to overcome this is to add a precise quantity of deemulsifying agent to the aqueous stream just prior to extraction. The Boots Company suggested [10] the addition of strong aqueous calcium chloride solution to remove by precipitation, as a calcium soap, any surface-active compounds that might cause emulsification. A 40 wt.% solution is added, at just over the 1% level, to the liquor at the end of the fermentation. After stirring for half an hour, filtration is carried out with a rotary vacuum filter, the cake is washed, and the filtrate is acidified with phosphoric acid to pH 2 before being contacted with butyl acetate.

The traditional process for the manufacture of penicillin requires that the broth be filtered to remove mycelium prior to solvent extraction. Direct extraction of the unfiltered fermentation broth has the potential of increasing the extraction efficiency by up to 10% since penicillin adsorbed onto the mycelial solids can also be extracted. In addition, there are the advantages that filter aid is not required and the broth is not diluted with wash water from the filtration stage. Anderson and Lau [11] of the Parke Davis Company reported the successful use of Podbielniak extractors to extract chloromycetin (a nonpenicillin antibiotic) from unfiltered broth. However, they were less successful with penicillin because of the problems of emulsion formation and solids buildup in the extractor. Workers at Hoffman-La Roche [12] described the extraction of a whole fermentation broth with the use of butanol as the solvent. The nature of the active component was not disclosed. A reciprocating-plate column was used and, by operating at a low agitator speed, emulsions were avoided and a good extraction efficiency was obtained. Since the date of publication (1955) of the Anderson–Lau paper, it is understood that Podbielniaks have been used for the commercial extraction of unfiltered broth [13]. A mixture of lipophilic deemulsifying agent and water-soluble wetting agent is used to give selective cleaning actions on both the solvent and broth streams. The precipitated solvents are dispersed so they mainly flow out with the spent broth instead of accumulating in the contactor. Periodic cleanup is carried out by "cleaning in place" systems.

2.6. Penicillin Ester Production

Penicillin esters, where the acidic carboxyl group on the nucleus has been esterified to give $COOR_2$, have certain advantages over unesterified penicillins, such as increased effectiveness of adsorption when taken orally, and reduced side effects.

An interesting application of solvent extraction through the use of phase-transfer catalysis for the manufacture of the ampicillin ester talampicillin is described in a British patent [14]. The phase-transfer catalyst, a quaternary ammonium salt (Q^+X^-), increases the organic solubility and hence extracts an intermediate ($RCOO^-Na^+$) from aqueous solution by means of the ion pair ($RCOO^-Q^+$) into the water-immiscible solvent methylene dichloride. Here it reacts with an alkyl halide (R_2X) to give the desired product ($RCOOR_2$) in high yield. This is shown pictorially as follows:

$$\begin{array}{lccc} \text{Aqueous} & RCOO^-Na^+ + Q^+X^- & \rightleftharpoons & RCOO^-Q^+ + Na^+X^- \\ \text{Interface} & \cdots\cdots \Updownarrow \cdots\cdots & & \cdots\cdots \Updownarrow \cdots\cdots \\ \text{Organic} & RCOOR_2 + QX & \leftrightharpoons & [RCOO^-Q^+] + R_2X \end{array}$$

Subsequent extraction back into aqueous solution and further treatment releases the talampicillin.

3. OTHER ANTIBIOTICS

Fermentation followed by solvent extraction has become the route for the preparation of many antibiotics so the general outline of the plant is as described previously.

3.1. Erythromycin

This is an organic base extracted under basic rather than the acidic conditions required for penicillin.

A range of solvents, including butyl acetate, amyl acetate, ethylene dichloride, chloroform, and methylene dichloride were tested by Kostareva et al. [15]. The nature of the extractant used had no effect on the purity of the final product, but the acetate esters had the highest extracting capacity. It is apparent that butyl acetate is the favored solvent for the extraction because Kostareva et al. [16] in a later publication described the use of hydroxy acids to isolate erythromycin from solutions in butyl acetate. Lactic, citric, malic, and tartaric acids were tried; with the best results, a yield of 82%, being obtained with lactic acid.

The overall degree of purification obtained by butyl acetate extraction is affected by both pH and temperature, and Rusin et al. [17] have measured their effect on the distribution between water and butyl acetate and suggested an optimal extraction regime. Product quality, it was claimed by Kostareva et al. [18], could be improved by the addition of potassium sulfate to a buffered aqueous extract, transferring the separated precipitate into butyl acetate, and bringing the pH of the erythromycin solution to 6.5.

3.2. Tetracyclines

These are compounds with amphoteric properties and are soluble in both aqueous acid and aqueous base. The acid salts tend to be soluble in organic solvents. A U.S. patent [19] describes the extraction of oxytetracycline from a fermentation broth with *n*-butanol.

In some processes it is used in a reverse manner, holding the compound in the organic phase while washing with water to remove water-soluble impurities. For example, Pereira da Luz [20] suggests that the tetracycline-containing broth from the fermenter be acidified to pH 3 or less with sulfuric acid. It is then filtered and rendered alkaline to separate the calcium and magnesium salts of tetracycline, which are then suspended rather than dissolved in as little as 10% of the original volume of an organic solvent, which may be butyl alcohol, butyl acetate, isoamyl acetate, or methyl isobutyl ketone. Sodium chloride is added to the suspension, and the organic phase is washed with water at pH 10. The aqueous extract is treated with activated charcoal and a little oxalic acid to remove calcium. After reacidification and bringing back to pH 4 with sodium hydroxide, the product is allowed to crystallize, with a yield of 92%.

A slightly modified method [21] is to acidify the culture medium with sulfuric acid to pH 1.8 and add 0.5% potassium ferrocyanide. This is again filtered and extracted with butyl acetate containing 3% of a complexing agent, which is a mixture of two-thirds dodecyl trimethyl ammonium chloride and one-third olein. The pH is adjusted to 9.5 with sodium bicarbonate solution and the liquid contacted with 5% oxalic acid solution. The butyl acetate layer is then a concentrate of tetracycline, from which a crystalline product is obtained by adding ammonium hydroxide to bring the pH to 4.1, and the crystals are washed with methylene chloride.

3.3. Bacitracin

Bacitracin can be removed from fermentation broth [22] by adjusting the pH to 7 with sodium hydroxide and extracting with *n*-butanol. The antibiotic is removed from this into aqueous phosphoric acid, to which calcium hydroxide is added to precipitate calcium phosphate. The filtrate is concentrated by evaporation under vacuum, when the addition of zinc chloride precipitates the bacitracin.

3.4. Cephalosporins

These are a range of antibiotics similar in chemical structure to the penicillins. Cephalosporin C, which is produced by fermentation, differs from the naturally occurring penicillins in that it is amphoteric and is too highly soluble in water for extraction with organic solvents in the usual way. Consequently, one way of isolating it from aqueous solution is by adsorption onto a resin bed and subsequent elution. However, there are potential yield and purity advantages in carrying out a solvent extraction prior to the aforementioned isolation stages. For these reasons, chemical conversion of the cephalosporin C while still in the broth, in order to render it more lipophilic, has been proposed.

A paper by Andrisano et al. [23] describes the conversion of cephalosporin C to the *N*-tosyl and *N*-*p*-cumyl sulfonyl derivatives by use of sulfonyl chloride. It is then possible to extract the resulting compound by use of ethyl acetate, methyl isobutyl ketone, or *n*-butanol. The compound MIBK was found to be the best of the solvents used. The derivatives were further purified by ion-exchange chromatography.

Bywood and his co-workers [24] give details of a process they call *extractive esterification* whereby cephalosporin C and a closely related compound deacetyl cephalosporin C are extracted from fermentation liquors. The method involves blocking of the basic group on the antibiotic side chain with an acylating reagent. This is carried out in aqueous solution at slightly alkaline pH and is then followed by treatment at acid pH with a diazo alkane in a water-immiscible solvent such as methylene dichloride. A mixture of the bis esters of the *N*-blocked cephalosporins thus accumulates in the organic phase. The methylene dichloride solution can then be used as the input to additional processes, giving a wide range of cephalosporin antibiotics.

4. NON-ANTIBIOTICS

Like the antibiotics, a number of other pharmaceuticals are produced in the aqueous liquors of fermentation broths and hence are obtainable by solvent extraction. Vitamins A, B_{12}, and C have been recovered in this way. Prednisolone, a steroid compound, can be extracted from a bacterial fermentation by use of chloroform [25]. Dactinomycin, used in cancer chemotherapy, is obtained from a fermentation broth by use of ether as the organic phase [26].

Solvent extraction is also employed after solids from medicinal plants or animals have first been percolated to obtain the active ingredient. Reserpine, a tranquillizer and antihypertensive, is extracted by ethylene dichloride from a methanol percolate of the African root *Rauwolfia vomitoria* [27]. In the case of morphine production [28], poppy straw is first milled and percolated. The solution obtained is contaminated with plant particles and is polished in a self-cleaning centrifuge. It is then fed to the solvent extraction process where the morphine base is picked up by an organic solvent; a 96–98% extraction yield in a three-stage process is achievable. Reextraction by use of an acid solution to recover the morphine is then made. Other alkaloids such as caffeine, codeine, quinine, and strychnine can be obtained in a similar manner.

REFERENCES

1. F. R. Batchelor, F. P. Doyle, J. H. C. Nayler, and G. N. Rolinson, *Nature (Lond.)* **183**, 257 (1959).
2. M. Souders, G. J. Pierotti, and C. L. Dunn, The Recovery of Penicillin by Extraction with a pH Gradient, in *The History of Penicillin Production*, Chemical Engineering Progress Symposium Series, Vol. 66, No. 100, 1970, Chapter V.
3. R. L. Feder, Recovery of Penicillin–Distribution Coefficients and Vapour Liquid Equilibria, M.S. thesis, Polytechnic Institute of Brooklyn, 1947; through W. J. Podbielniak, H. R. Kaiser, and G. J. Zeigenhorn, Centrifugal Solvent Extraction, Ref. 2, Chapter VI.
4. R. G. Benedict, W. H. Schmidt, and R. Coghill, *J. Bacteriol.* **3**, 51 (1946).
5. D. Rowley, H. Steiner, and E. Zimkin, *J. Soc. Chem. Ind.* **65**, 237 (1946).
6. T. Kansawa, H. Kawai, and T. Ito, *Ann. Rep. Takeda Res. Lab.* **9**, 57 (1950).
7. H. Eisenlohr and A. Scharlau, *Pharmaz. Ind.* **17**, 207 (1955).
8. S. A. Zhukovskaya, *Khim.-Farm. Zh.* **6** (6), 42 (1972).
9. E. Sh. Radzinskaya, V. N. Rusin, M. F. Skobin, S. A. Zhukovskaya, and R. Kh. Izmailov, *Khim.-Farm. Zh.* **11** (12), 77 (1977).
10. S. W. Stroud and H. M. P. Ransley, British Patent 760,351 (October 10, 1956); assigned to Boots Pure Drug Company Ltd.
11. D. W. Anderson and E. F. Lau, *Chem. Eng. Progr.* **51**, 507 (1955).
12. A. E. Karr and T. C. Lo, in *Proceedings of the International Solvent Extraction Conference (ISEC), Toronto, 1977*, Canadian Institute of Mining and Metallurgy, Montreal, 1979, p. 355.
13. Baker Perkins Ltd., unpublished communication.
14. Beecham Group Ltd., British Patent 1,565,656 (December 13, 1975).
15. M. G. Kostareva, K. K. Eremina, A. N. Fedorova, G. A. Bityukskaya, M. I. Vyatkina, and T. L. Vasil'era, *Antibiotiki (Moscow)* **16** (2), 124 (1971).
16. M. G. Kostareva, N. P. Bednyagina, and G. A. Bityutskaya, *Khim.-Farm. Zh.* **8** (8), 44 (1974).
17. V. N. Rusin, S. A. Zhukovskaya, and V. L. Pebalk, *Antibiotiki (Moscow)* **20** (3), 213 (1975).
18. M. G. Kostareva and N. I. Vyatkina, U.S.S.R. Patent 306,669 (October 25, 1977).
19. B. A. Sobin, A. C. Finlay, and J. H. Kane, U.S. Patent 2,516,080 (July 18, 1950); assigned to Charles Pfizer & Company, Inc.
20. A. Pereira da Luz, British Patent 1,140,517 (January 22, 1969).
21. C. Pal, M. Gavrilescu, S. User, D. Urescu, and L. Natalia-Margineau, Romanian Patent 53,908 (November 9, 1971).
22. G. M. Miescher, U.S. Patent 3,795, 663 (March 5, 1974).
23. R. Andrisano, G. Guerra, and G. Mascellani, *J. Appl. Chem. Biotechnol.* **26**, 459 (1976).
24. R. Bywood, C. Robinson, H. C. Staples, D. Walker, and E. M. Wilson, in J. Elks, Ed., *Recent Advances*

in the Chemistry of β-Lactam Antibiotics, The American Chemical Society, 1977, p. 139.

25. A. Nobile, U.S. Patent 2,837,464 (June 3, 1958); assigned to Schering Corporation.
26. S. A. Waksman and H. B. Woodruff, U.S. Patent 2,378,876 (June 19, 1945); assigned to Merck & Company, Inc. through M. Sittig, *Pharmaceutical Manufacturing Encyclopedia*, Noyes Data Corporation, 1979, p. 166.
27. *Chem. Eng.* **64** (4), 230 (1957); through R. N. Shreve and J. A. Brink, *Chemical Process Industries*, McGraw-Hill, New York, 1977, p. 784.
28. Westfalia Separator Ltd., unpublished communication.

20

LIQUID-LIQUID EXTRACTION IN THE FOOD INDUSTRY

W. Hamm
Unilever Research
United Kingdom

1. INTRODUCTION

The application of liquid extraction in the food industry long predates modern technology. In the eighth century the use of aqueous alcohol as an extraction solvent was known in the Arab world, and before then extraction with water was used to recover aroma-bearing extracts. Insofar as the food industry is usually working with materials of plant or animal origin, a division between liquid–solid and liquid–liquid extraction is not always immediately discernible, and it may sometimes be considered a matter of semantics whether a particular operation should be classified under one of these headings or the other.

Despite this long-standing application of extraction in the food industry, which for our purposes is taken to include the production of all types of beverage, modern diffusional separation processes in the industry are still in an early phase of development. This may be attributed in part to the influence of tradition on processing techniques used, but also to the nature of the material being processed. In particular, the prevalence of large molecules gives rise to two problems that have certainly impeded the transfer of technology from the technologically more advanced process industries—the lack, until very recently, of comprehensive analytical techniques capable of the detailed type of analysis required and the difficulty of finding solvents for the macromolecules often encountered. On the other hand, a factor in favor of liquid–liquid extraction is the ability to work at or near ambient temperature, an important advantage when dealing with thermally sensitive materials.

As the sophistication of the industry in its use of raw materials grows, it can be expected that opportunities for the application of liquid–liquid extraction will increase, but such growth will depend on the range of solvents available. The availability of solvents is governed, inter alia, by legislation covering the levels of solvent residues permitted in foodstuffs, a subject dealt with later.

The removal of undesirable constituents from a raw material to be used in the food industry is obviously an operation that can be carried out by liquid–liquid extraction in the same way as the extraction of a valuable constituent.

As in the chemical process industries, liquid–liquid extraction competes with other separation processes in the development of new processes or the improvement of existing ones, and a detailed knowledge of the relevant physical properties is a prerequisite to making the correct choice of process. Unfortunately, the appropriate data are often difficult to find in the literature, and process developers often must begin by generating at least some of their own data. In this respect the advances in the methodology of analytical chemistry provide powerful support to the development of new separation processes, but the difficulty of correlating composition with quality criteria such as taste, flavor, and keepability is a recurring feature of process development in the industry. This aspect particularly affects scale-up, as flexibility is needed to adjust the process in the light of experience.

2. APPLICATIONS

2.1. Extraction of Lipids

Lipids constitute a very varied class of edible materials based on esterification products of glycerol and fatty acids. Triglycerides, in which all three hydroxyl groups of the glycerol molecule have been esterified by one or other of the naturally occurring fatty acids, form the major constituent of this class and the major component of the fats and oils found in a vast array of food products. Monoglycerides, obtained as the reaction product of glycerol and fats or fatty acids, are important because of their interface behavior in a lipophilic-hydrophilic system, but diglycerides are normally undesirable by-products of the above-mentioned glycerolysis reaction or of natural deterioration of fats. A minor member of this class in terms of quantity but one that plays an important role in the food industry is lecithin, in which one of the hydroxyl groups of the glycerol molecule has been esterified by phosphoric acid [1]. Lipids provide many opportunities for liquid–liquid extraction, although in practice the number of applications is very limited.

Edible oils and fats in their crude state contain a mixture primarily of mixed triglycerides and fatty acids, with mono- and diglycerides also present at minor levels. The removal of the fatty acids is a major step in the process of purification (refining) of oils and fats, primarily because the yield of neutral oil in this operation has a very significant effect on the economics of refining. Deacidification has been carried out predominantly by neutralization of the free fatty acids with aqueous alkali, with the soaps formed (which are virtually insoluble in the oil at the operating temperature) transferring to the aqueous phase, but deacidification by steam stripping has gained in importance in recent years.

Liquid–liquid extraction offers an alternative route for deacidification that becomes of interest when the free fatty acid (FFA) content of the crude oil is relatively high and when the temperatures normally applied in steam stripping (220–270°C) are unacceptable. The reason for the potential advantage of extraction is to be found in the fact that losses of neutral oil into the extract phase in an extraction process may be considerably lower than losses of neutral oil experienced in alkali refining. As the value of the fatty acid by-product is very much lower than that of neutral oil, this advantage, against which the cost of operating a solvent-based process must be set, becomes significant when the (alkali) refining losses begin to rise sharply as the free fatty acid content exceeds, say, 5%.

Solvent choice for the extraction of fatty acids is governed by the difference in polarity between fatty acids (polar) and triglycerides (apolar). Thus a polar solvent such as acetone or one of the shorter-chain alcohols is capable of giving extracts containing low levels of triglycerides. The addition of water to the solvent enables the user to optimize with respect to solvent selectivity and capacity, since such addition reduces the capacity of the solvent for triglycerides but to a lesser extent also for fatty acids. The neutral triglycerides constitute the second solvent.

Aqueous acetone containing 10–15% water is discussed as a deacidification solvent by Foresti and Giuffrida [2], who claim a number of advantages for a process based on this solvent. In particular, they point to the low losses of neutral oil into the extract phase. Rigamonti and Botto [3] have also studied the system oil–acetone–water and provide useful ternary data and at the same time demonstrate the applicability of the Hand and Othmer–Tobias correlation methods to their data. Rigamonti and co-workers [4] also published data on the use of methanol and ethanol to extract fatty acids from triglyceride–fatty acid mixtures and in particular studied the use of anhydrous and aqueous methanol for this purpose. Figure 1 shows their results with extraction at 20°C.

A smoothed Hand plot (Fig. 2) demonstrates the effect of the addition of water to the solvent (methanol) on the fatty acid partitioning: the slope of the Hand correlation ($\log x^E_{FFA}/x^E_{solv}$ with $\log x^R_{FFA}/x^R_{oil}$) remains unchanged, but the position of the correlation is shifted toward the raffinate axis. The data thus enable the user to interpolate raffinate and extract compositions for intermediate solvent water contents. Where extraction must be carried out at higher temperatures, as is the case when saturated fatty acids are present, a similar relationship is to be expected, although in this case there will be a displacement

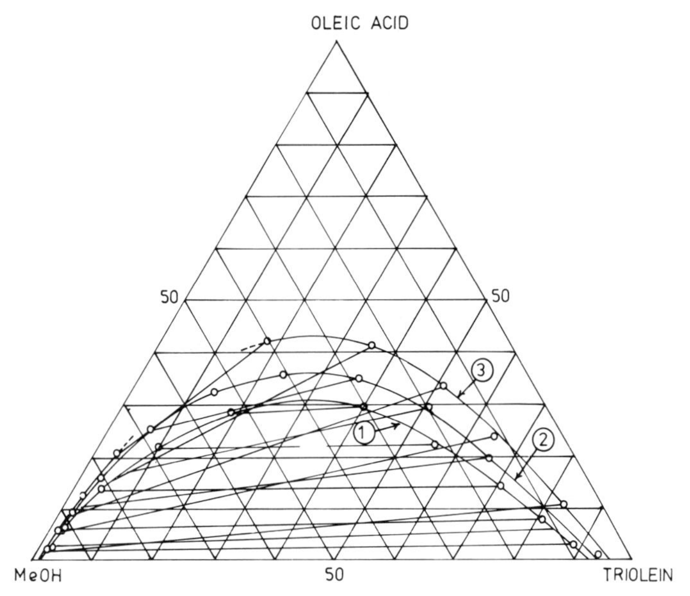

Figure 1. Miscibility curve at 20°C: (1) triolein–oleic acid–methanol (anhydrous) system; (2) triolein–oleic acid–methanol (5% H_2O) system; (3) triolein–oleic acid–methanol (10% H_2O) system.

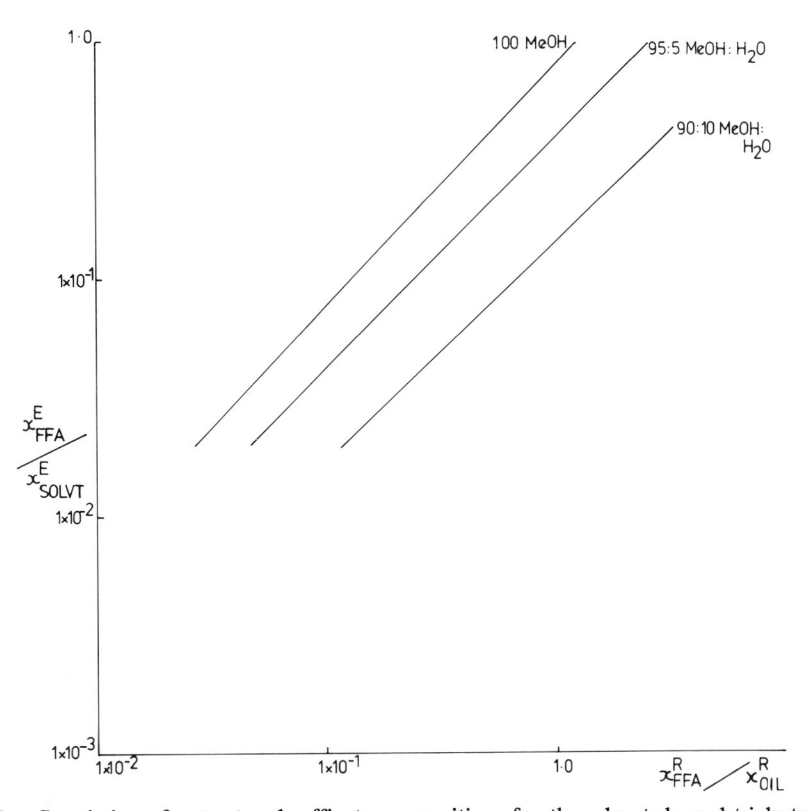

Figure 2. Correlation of extract and raffinate compositions for the solvent-glycerol trioleate–oleic acid system at 20°C solvent: methanol–water (composition as indicated).

of the correlation line due to the effect of temperature. The effect of increased temperature will be in the same direction as that of reduced water content of the methanol.

The fatty acid–triglyceride–solvent system is conventionally represented on a ternary diagram by use of solvent water content as parameter; however, the system in practice is more complex because of the presence of diglycerides at a significant level in oils containing high levels of free fatty acids. Diglycerides affect mainly the slope of the tie lines, leading to an increased separating power requirement for a given deacidification duty.

The separating power required will, of course, be a function of initial and final free fatty acid contents of the oil to be processed. It has generally been found uneconomical to design for reduction in FFA content to below 1%, and the data available suggest that a contactor capable of giving the equivalent of 5–10 equilibrium stages would normally be required. A rotating-disk contactor (RDC) or other mechanically agitated device can be expected to give the required separating power, but care must be taken to minimize the energy dissipation in the contactor as interfacial tension falls to very low levels as the free fatty acid content of the raffinate is reduced. Low density differences between raffinate and extract (<100 kg/m^3) and the low interfacial tension ($1\text{-}3 \times 10^{-3}$ N/m) lead to low contactor capacities ($V_c + V_d \approx 2 \times 10^{-3}$ m/s).

Since edible oils of vegetable origin are frequently recovered from oilseeds by means of hexane extraction, deacidification by extraction of the hexane miscella with a second solvent has periodically been advocated. Vandervoort [5] patented a process for miscella refining in which alkali neutralization of free fatty acids is carried out in a hexane–isopropanol–water solvent system and the field has recently been reviewed by Cavanagh [6]. Thomopoulos [7] discusses deacidification in a double-solvent system in which hexane is the apolar solvent while the polar solvent is either an alcohol or a ketone. He concludes that 96% ethanol is the most suitable of the polar solvents examined and, for very high FFA contents in the crude oils, shows that refining losses are far below those to be expected in alkali refining.

The partitioning of triglycerides into fractions of differing degrees of unsaturation has been a subject of interest since the early 1940s, when processes were developed for production of highly unsaturated fractions from oil rich in such glycerides, for use in surface coatings. The partitioning effect is obtained by reason of the greater polarity of the unsaturated glycerides.

The Solexol process [8] makes use of the solvent properties of liquid propane in the region of its critical point ($T_c = 368$ K, $P_c = 43$ bars). The propane–triglyceride mixture, homogeneous at ambient temperature, forms a two-phase system as the temperature is raised toward the critical value and partitioning of the triglycerides on the basis of unsaturation occurs. Extraction is thus carried out in the range 60–90°C in countercurrent manner, with the use of control of the temperature gradient to provide reflux. Critical solution temperatures and equilibrium data for several systems comprising propane, fatty acids, and triglycerides were reported by Drew and Hixson [9]. Dickinson and Meyers [10] reviewed the scope and function of the plants installed in the early 1950s, and Moore [11] gave details of a plant used in a typical application. In this case a propane:oil ratio of 17:1 was used with the plant operating at 40 bars. The process has also been used to extract a fraction rich in vitamin A.

The use of furfural in edible oil processing was also developed for the purpose of extracting fractions rich in unsaturated triglycerides, and a number of plants were built. Kenyon et al. [12] reviewed the potential of the process and provided details of a plant processing soybean oil. Ruthruff and Wilcock [13] had earlier documented equilibrium and solubility data for this system, as well as demonstrating the advantages of using an apolar solvent such as naphtha for back extracting the oil fractions from the extract and raffinate phases formed during furfural extraction.

Operation of the propane and furfural extraction plants for triglyceride partitioning was discontinued after a relatively short process life as a result of the evolution of new products superior to those produced by extraction. Interest in the use of furfural for liquid extraction of edible oils has persisted, however, as is shown by more recent work on marine oils [14] and on the equilibrium data [15]. The first of these papers indicates that useful separations in terms of degree of unsaturation can be obtained even with low furfural:oil ratios. The authors of the latter work, using model systems of the triglycerides glycerol trioleate (triolein) and glycerol trilinoleate (trilinolein) with the solvents furfural and heptane, show the differences in phase equilibria as unsaturation of the triglycerides increases. The authors also calculate the separating power required for a mixed triglyceride system and conclude that separation of relatively pure triglycerides from such systems requires a large number of stages and extremely high solvent ratios (300–450 kg of solvent per kilogram of oil). A comparison of the two reports demonstrates the difficulty of attempting a separation of specific components, whereas the separation based on a mean degree of unsaturation, as indicated by measurements such as that of the Iodine Value, is far less demanding. The use of liquid extraction for production of fractions enriched in unsaturated triglycerides remains a subject of interest, as is shown by a recent patent [16] in which a solvent comprising *N*-methyl pyrrolidone together with minor additions of a polar solvent, such as ethylene glycol, is advocated. The partition coefficient for this separation is relatively poor, and an appreciable enrichment of unsaturated glycerides requires high solvent ratios and a large number of stages, leading to high capital and operating costs.

Liquid extraction has also been proposed [17] for the enrichment of monoglycerides as a step in the production of monoglyceride concentrates used in the food industry as emulsifiers, and in this case the solvent system proposed is a mixture of *n*-octane and ethanol. Again the partition coefficient is such that a large number of stages (>10) is required to give a concentrate containing more than 90% monoglycerides. For the capacities involved, the cost of operating such a process cannot compete with one based on molecular distillation of the crude monoglyceride reaction product. The use of liquid–liquid extraction to upgrade lecithin, a by-product of edible oil refining and a valuable food additive, is discussed by Liebing and Lau [18]. In this case an alcoholic solvent can be used to extract the lecithin from the lecithin–oil mixture produced in oil refining. Extraction of lipids from protein concentrates, such as fish protein concentrates, is a problem in the field of liquid–solid rather than liquid–liquid extraction, but the equilibrium data for the system lipid–solvent–water are of wider interest. McPhee et al. [19] report data for a number of marine oil–alcohol–water systems, and their work shows the importance of the water content of the system in guiding the choice of solvent.

2.2. Decaffeination

The removal of caffeine (trimethylxanthine) from coffee is an operation of long-established commercial importance. The development of a process for liquid–liquid extraction of caffeine from an aqueous extract of the soluble solids of coffee [20] avoids contact of the beans with a nonaqueous solvent. In this process trichlorethylene is the preferred solvent, and Sivetz [21] gives some details of the method of operation. Countercurrent contactors are used for the extraction, with the organic solvent as the continuous phase, and the solvent aqueous-phase ratios vary between 8 : 1 for decaffeination of the extract from roasted beans to 40 : 1 for the extract from green beans. The trichlorethylene solution of caffeine is then reverse-extracted into water after concentration to 1% caffeine strength by changing the phase ratio and reducing the temperature of extraction (Fig. 3). Partition coefficients for caffeine between trichlorethylene and water are given as 0.20 and 0.35 at 25 and 70°C, respectively. It can be estimated that the

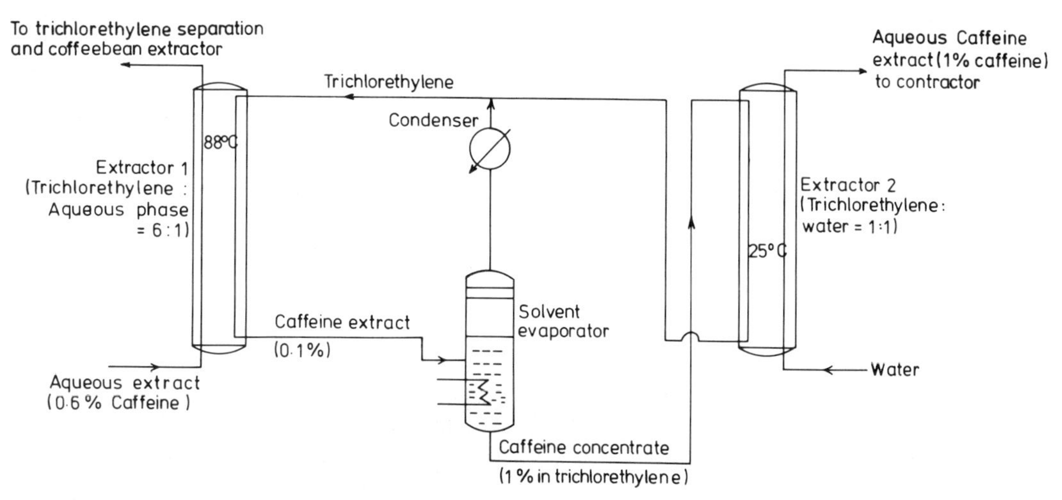

Figure 3. Decaffeination of aqueous extract with trichlorethylene. (Based on U.S. Patent 2,309,092.)

transfer of caffeine to the trichlorethylene requires approximately five theoretical stages for a 98% removal of caffeine, and a mechanically agitated contactor such as the RDC may be used for this purpose. Due to the risk of emulsification in the extractor, energy dissipation during extraction must be minimized.

More recently the use of triglycerides of fatty acids for the extraction of caffeine from the aqueous extract has been advocated [22] and, from the information available, the partition coefficient for caffeine between triglycerides and water appears to be significantly lower than those quoted above. As the latter results were obtained when working with relatively concentrated solutions, it is to be expected that values somewhat closer to those quoted for the trichlorethylene-water system would be obtained when working with lower concentrations. Use of triglycerides for this extraction has the advantage of replacing a potentially questionable solvent by one completely acceptable to the industry but at the same time also raises problems concerning fat-water separation after extraction. To minimize the risk of entrainment of fat droplets during the separation, the coffee extract (from the bean) is concentrated to a total solids content of 45-55% before caffeine is extracted.

A further development in this field requires the use of liquid carbon dioxide just below its critical point as an extracting solvent for the caffeine in the extract coming from the liquid-solid extractors. In a recent patent [23] a technique is described for contacting an aqueous raw coffee extract with carbon dioxide by use of temperature and pressure conditions immediately below the critical point of the gas. The caffeine thus removed is subsequently recovered on a suitable sorbent. Still another development has been the extraction of caffeine by means of supercritical gas extraction from the bean [24], thus causing the process to move from liquid-liquid to liquid-solid extraction.

Decaffeination is also of interest in the production of tea and the techniques described previously can be applied.

2.3. Extraction of Flavors and Aromas

Although of necessity a small-scale operation, the extraction of flavors and aromas has attracted a considerable amount of attention, and a number of applications are known. Alcoholic solvents are commonly used, but ethylene chloride has also been used and offers the advantage of forming a heterogeneous azeotrope with water, thus facilitating solvent recovery.

Citrus oils such as lemon and orange oils are important sources of flavors for the food industry. These oils contain high proportions of hydrocarbons of the terpene and sesquiterpene type—more than 90% in the case of lemon oil and above 95% in orange oil—whereas the valuable oxygenated component citral may be present at a level of 4-5%. Solubility and stability criteria require that the hydrocarbon content of the oil be reduced substantially before the oil is used as a food additive, particularly in beverages. Of the methods available for this purpose, liquid extraction, either with a single solvent or with a polar-apolar solvent pair, is the most suitable. Single-solvent extraction by use of a polar solvent, generally an aliphatic alcohol, is suitable only if the consequent need for large solvent : oil ratios is not disadvantageous, as may be the case if the citral extract is to be used in the dilute form.

Extraction of citral with a solvent pair, sometimes referred to as *Nardenization*, generally entails the use of solvent pairs such as an aqueous alcohol, for example methanol and a short-chain hydrocarbon (e.g., pentane). Reman [25] reported on this separation as carried out in an early experimental RDC of 1500-mm height and 82-mm diameter. An alcoholic solvent and gasoline were respectively fed to the top and bottom of the contactor, the feed point for the lemon oil (containing 4.2% citral) being located in the lower half of the column. The alcohol : oil ratio was held at 4.0, and the gasoline : oil ratio at approximately 1.4. By this means Reman was able to achieve a citral recovery of over 95% and a citral concentration in the extract of approximately 51%. He estimated that the extraction required the equivalent of 10 equilibrium stages. The sensitivity of the process to emulsification is shown by the low overall velocity of just under 2 mm/s in the contactor. Polyhydric alcohols can also be used in conjunction with a hydrocarbon [26].

The use of liquid carbon dioxide as extraction solvent, already discussed in the context of caffeine extraction, has also been advocated for aroma recovery. Randall et al. [27] describe laboratory work on the extraction of a range of components representative of the aromas of fruit juices. As in the case of caffeine extraction, extraction is carried out in the immediately subcritical range, and an important advantage associated with the use of liquid carbon dioxide is the very low solubility of water (0.1%) and the insolubility of components such as sugars and proteins in the solvent. This work was subsequently scaled up to pilot-plant scale [28] and a 10-stage Scheibel column used for the extraction, in which the solvent formed the dispersed phase. A solvent to aqueous feed ratio of approximately 0.8 : 1 was used. In an earlier paper Schultz and Randall [29] compare the activity coefficients of important aroma components as determined experimentally and when calculated according to well-established empirical rules and demonstrate the relationship between the number of carbon atoms in the molecule and its partition coefficient between liquid carbon dioxide and water at 16.7°C.

The use of liquid–solid extraction to produce a hop extract from partially dried whole hops has established itself firmly in the brewing industry. The extract produced when using methylene chloride, methanol, or hexane as solvent contains the hard and soft resins primarily responsible for the bitter flavor and also contains some of the seed fat. The use of this extract in wort boiling is considered to confer a number of advantages on the process [30], amongst which more efficient use of the bittering substances is of considerable importance. Further improvements in the efficiency of utilization of the flavoring and aromatic constituents of hops involved [31] the separation of the hop extract by liquid–liquid extraction into a number of fractions (α-acids, β-acids, other resins, and seed fat). For this purpose, a polar–apolar solvent system is used in which the polar solvent is aqueous methanol and in which water content and pH of the solvent are used to maximize separation. The separated α-acids (humulones) are subsequently isomerized in order to enhance their flavor contribution.

An alternative pathway to isomerized α-acid production, claimed [32] to be more economical and scaled up successfully, comprises a number of liquid–liquid extraction stages in which the α-acids of the hop extract are progressively separated from the other components by making use of the solubility of the α-acids in weak alkali. In a first stage, for which a mechanically agitated contactor may be used, the hop extract dissolved in a water-immiscible solvent is contacted with a weak potassium carbonate solution that extracts the α-acids and thus separates them from the other constituents of the hop extract. The alkaline solution is subsequently washed with fresh petroleum spirit to remove remaining impurities and is then counterextracted with *n*-butanol to give a 10% concentration of the α-acids in the alcohol. This solution proceeds to the isomerization stage after replacement of the butanol by water and thereafter to final purification (see Fig. 4). The separating power and conditions required in the first stage will, of course, depend

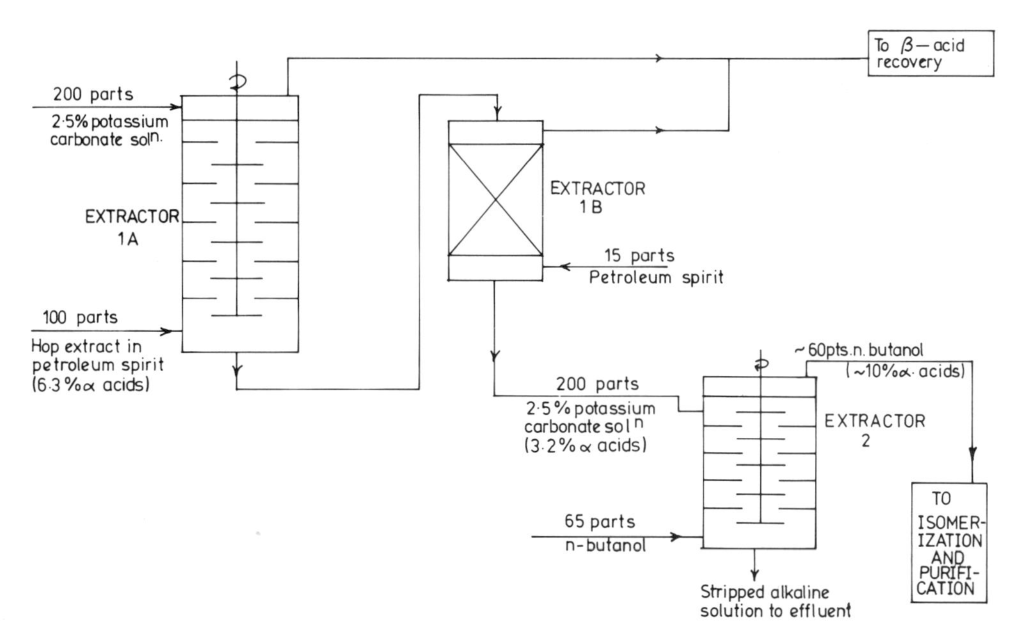

Figure 4. Separation of α-acids from hop extract by liquid–liquid extraction.

on the selectivity obtained in the previous liquid-solid extraction process, and claims for greater selectivity when using liquid carbon dioxide for extraction of the hops [33] suggest that extract produced by that method will simplify the subsequent separation as described previously.

3. PHYSICAL PROPERTIES

Design and operation of liquid-liquid extraction processes in the food industry require the availability of the appropriate physical properties of components, mixtures, and systems, but the literature reveals considerable gaps in the various data banks, and in many cases data must be determined experimentally. The use of predictive techniques is fraught with uncertainty since the multicomponent nature of natural products makes their reduction to pseudobinary or pseudoternary systems for purposes of estimation or calculation a potentially misleading operation. Furthermore, the presence of minor or even trace components, which can affect certain physical properties significantly, may vary with the season or the location. For these reasons, it is generally advisable to carry out the necessary determinations on the actual system to be processed rather than on an apparently analogous model system.

Solubility data are required for purposes of solvent choice and also provide guidance on possible concentration ranges to be employed. Of the lipids the triglycerides, which are relatively apolar, are freely soluble in solvents such as hydrocarbons and acetone, with the latter also acting as a solvent for the more polar fatty acid. As more polar groups are introduced into the molecule, the solubility in solvents such as the aliphatic alcohols increases whereas that of the triglycerides in alcohols is very limited (except in the case of castor oil, because of the presence of a hydroxyl group in the principal fatty acid constituent). The presence of water in the alcohol depresses the solubility of all lipids, none of which show any appreciable solubility in water at temperatures of 100°C or below. The solubility of water in fatty acids is reported to be rather greater than that of fatty acids in water [1]. Skau [34] reports a method of predicting solubility data for fatty acids and their derivatives in related solvents. For this purpose, he uses the ideal relationship between the freezing points and heats of solutions of two homologous compounds, one of which serves as reference substance. The solubilities of the principal fatty acids in various common solvents are given in plotted form.

Partition coefficients for fats and fatty acids between various solvent pairs are reported by a number of authors [3, 15, 35, 36], but where these are derived from measurements on actual systems the data must be assessed critically in view of the general lack of information on the possible presence of minor components. Selectivity values within a given class of lipids are normally close to 1.0 as a result of the dominating influence of molecular size. The random distribution of fatty acids in natural triglycerides also makes partitioning of triglycerides on a component basis very difficult.

Sivetz [21] reports solubility data for caffeine in a number of solvents and the data are useful in designing caffeine extraction processes.

Flavor reproduction depends on reproducing the profile of flavor components present in the raw material, and thus solubility data on individual components are not very useful. In addition, concentrations are often extremely low, sometimes at the level of parts per billion, and analytical techniques may not be adequate for quantitative determination of concentration. Subjective assessment may thus be required for determination of whether extraction from the raw material has achieved the required objective. The short-chain aliphatic alcohols, in some cases in aqueous admixture, are frequently used as extracting solvents. Randall et al. [27] quote data for partitioning of various alcohols and esters between liquid carbon dioxide and water. The flavor components of hops are soluble in solvents of varying polarity, but the principal flavor component (α-acids) is separated by transfer to a more polar phase.

Other data required pertain to conditions in the extractor. Density, viscosity, and interfacial tension all affect the process conditions, and in each case the values required are those of the extract and raffinate phases rather than those of the pure components. Density differences between the phases can be expected to be below 250 kg/m^3 and in certain cases below 100 kg/m^3, although the use of halogenated hydrocarbons and of liquid carbon dioxide will give higher density differences. Since low density differences increase the risk of entrainment, care must be taken to ascertain the minimum density difference in operation.

Viscosity is seldom a serious problem in food industry extraction processes as the material to be extracted is in most cases dissolved in a sol-

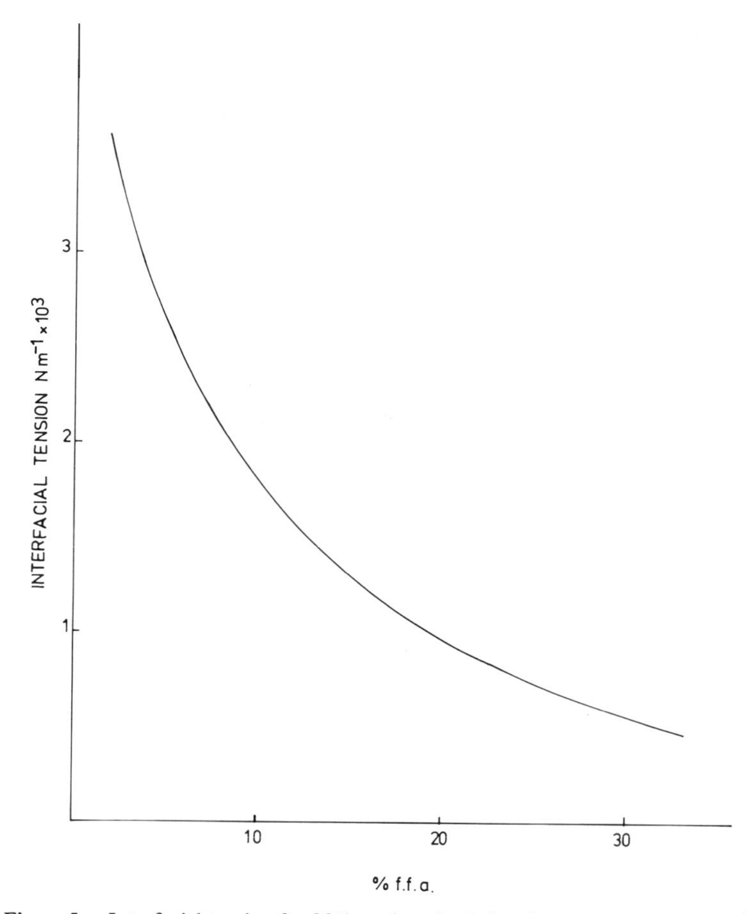

Figure 5. Interfacial tension for 98% methanol–oil–free fatty acid system at 50°C.

vent of low viscosity. An exception is to be found in some lipid extraction processes since the triglycerides then sometimes act as solvents. Their viscosities vary over a considerable range but can be expected to be between 10 and 20 × 10^{-3} Ns/m^2 in the temperature range 40–70°C. These relatively high viscosities lead process designers to opt for the fat phase forming the disperse phase in a contactor.

Interfacial tension data relate to phases rather than components, and although very few data have been reported, it is known that low interfacial tension values are often experienced in practice, giving rise to a limitation on energy dissipation in the contactor if emulsification is to be avoided. The presence of trace components having surface-active characteristics can have a sharply adverse effect on the interfacial tension for a given system. This effect can obviously also occur with components present in larger amounts, as shown in Fig. 5.

4. SOLVENT SELECTION

In the chemical process industries the choice of solvent for an extraction process depends primarily on evaluation of readily quantifiable parameters such as solvent capacity, selectivity, and recovery costs and the less easily quantified question of operating hazards attributable to use of the solvent. Although the food industry must consider all these criteria, it is also obliged to assess the toxicologic hazards associated with the use of a given solvent. Furthermore, the possibility of thermal damage to the product during solvent recovery means that solvents having high boiling points may also be unacceptable. Obviously, the problem of exposure to the vapors of hazardous solvents in a working atmosphere is common to all process industries.

A growing awareness that certain solvents can endanger health, mainly because of carcinogenic tendencies, has led to increased toxicologic testing and, in turn, to greater efforts by regulatory authorities to control the range of solvents used in the processing of foodstuffs [37]. The aromatic solvents have largely been excluded from the processing of edible materials for this reason, and halogenated hydrocarbons have come under close scrutiny. The use of trichlorethylene for the extraction of edible materials has recently been severely restricted in the United States [38], and concern about the attitude of the authorities to other halogenated solvents has led to moves to replace other solvents of this group in use in the food industry. A number of the shorter-chain hydrocarbons (C_5–C_8) are freely used in liquid extraction, but in the case of industrial hexane, which is also extensively used in liquid–solid extraction in the food industry, the polycyclic hydrocarbon content is subject to strict control [39].

Where a "permitted" solvent is used in extraction, the residual levels allowed in the product are fixed by the regulatory authorities, and the effect of these measures has been to ensure rigorous and carefully monitored desolventizing. Consequently there is a strong incentive to use solvents that present no problems during solvent recovery. Although the alcohols of lower molecular weight are generally considered to be in this category, care must be exercised when using them to prevent the occurrence of undesired esterification reactions. The use of aqueous alcohols effectively serves this purpose. The extensive efforts to use liquid carbon dioxide as an extraction solvent, which have been discussed earlier, also reflect the search for a solvent unlikely to encounter toxicologic difficulties.

Where macromolecules such as proteins are to be separated, the relatively simple solvents already discussed are inapplicable because of lack of solvency and possible effect on molecular structure. A pointer to the more complex solvent systems required is provided by recent Swedish work [40], which has shown that for protein fractionation, an aqueous dual-polymer system (in the case referred to, dextran and polyethylene glycol were used) can be designed to give a biphasic system displaying protein partitioning characteristics.

ACKNOWLEDGMENTS

The author wishes to thank Dr. A. Brench of Unilever Research, Sharnbrook, Bedfordshire, and Mr. J. M. Fincher of White Stevenson Ltd., Reigate, Surrey, for valuable discussions in the course of preparing this text.

NOMENCLATURE

x weight fraction of component indicated in subscript (FFA = free fatty acids)

Superscripts

E refers to extract phase

R refers to raffinate phase

REFERENCES

1. D. Swern, Ed., *Bailey's Industrial Oil and Fat Products*, 4th ed., Vol. 1, Wiley-Interscience, New York, 1979.
2. B. Foresti and A. Giuffrida, *Oleagineux* **13** (1), 131 (1958).
3. R. Rigamonti and G. Botto, *Oleagineux* **13** (1), 199 (1958).
4. R. Rigamonti et al. *Chim. Ind.* **33,** 619 (1951).
5. German Federal Republic Patent 961,380 to Extraction Continue de Smet SA (1957).
6. G. C. Cavanagh, *J. Am. Oil Chem. Soc.* **53** (6), 361 (1976).
7. C. Thomopoulos, *Rev Franc des Corps Gras* **18** (3), 143 (1971).
8. H. J. Passino, *Ind. Eng. Chem.* **41** (2), 280 (1949).
9. D. A. Drew and A. N. Hixson, *Transact. AIChE* **34,** 675 (1944).
10. N. L. Dickinson and J. M. Meyers, *J. Am. Oil Chem. Soc.* **29** (6), 235 (1952).
11. E. B. Moore, *J. Am. Oil Chem. Soc.* **27** (3), 75 (1950).
12. R. L. Kenyon et al., *Ind. Eng. Chem.* **40,** 1162 (1948).
13. R. F. Ruthruff and D. F. Wilcock, *Transact. AIChE* **34,** 675 (1944).
14. R. Oscar Contreras et al., *J. Am. Oil Chem. Soc.* **48** (3), 98 (1971).
15. P. L. Chueh and S. W. Briggs, *J. Chem. Eng. Data* **9** (2), 207 (1964).
16. British Patent 1,444,551 to Unilever Ltd (1976).
17. J. A. Monick and R. E. Treybal, *J. Am. Oil Chem. Soc.* **33** (5), 193 (1956).
18. H. Liebing and J. Lau, *Fette, Seifen, Anstrichmittel* **78** (3), 123 (1976).
19. A. D. McPhee et al. *J. Chem. Eng. Data* **17** (2), 244 (1972).
20. U.S. Patent 2,309,092 to General Foods Corporation (1943).

21. M. Sivetz, *Coffee Processing Technology*, Vol. II, AVI Publishing Company, 1963.
22. German Federal Republic Patent Application 2,721,765 to Soc des Produits Nestle SA (1977).
23. German Federal Republic Patent Application 2,357,590 to HAG AG (1975).
24. British Patent 1,346,134 to HAG AG (1974).
25. G. H. Reman, *Proceedings of the Third World Petroleum Congress*, Section III, The Hague, 1951.
26. Swiss Patent 548,743 to Soc. des Produits Nestle SA (1974).
27. J. M. Randall, W. G. Schultz, and A. I. Morgan, *Confructa* **16** (1), 10 (1971).
28. W. G. Schultz, T. H. Schultz, R. A. Coulson, and J. S. Hudson, *Food Technol.* **28** (6), 32 (1974).
29. W. G. Schultz and J. M. Randall, *Food Technol.* **24** (11), 94 (1970).
30. F. Schur and H. Pfenninger, *Brewers Guardian* **107** (9), 37 (1978).
31. Canadian Patent 619,563 to Canadian Breweries, Ltd. (1961).
32. British Patent 1,274,678 to Bush Boake Allen, Ltd. (1972).
33. D. R. J. Laws et al., *J. Inst. Brewing* **83** (1), 39 (1977).
34. E. L. Skau, *J. Am. Oil Chem. Soc.* **47** (7), 233 (1970).
35. K. Sreenivasan and D. S. Viswanath, *Indian J. Technol.* **11** (2), 83 (1973).
36. E. Bak and C. J. Geankoplis, *Ind. Eng. Chem. Data Ser.* **3** (2), 256 (1958).
37. Food Additives and Contaminants Committee on the Review of Solvents in Food SAC/REP/25, HMSO, London, 1978.
38. Fed. Reg **42**, 49465 (1977).
39. *Specifications for the Identity and Purity of Some Extraction Solvents and Certain Other Substances*, FAO Nutrition Meeting Report Series 48B, WHO/FOOD ADD/70, 40, FAO, New York, 1971.
40. G. Johansson, *J. Chromatogr.* **150,** 63 (1978).

21

MISCELLANEOUS ORGANIC PROCESSES FOR CHEMICALS FROM COAL AND ISOMER SEPARATIONS

M. W. T. Pratt
University of Bradford
United Kingdom

1. SOLVENT EXTRACTION OF COAL

An extensive literature on the solvent extraction of coal has been reviewed by Dryden in a book edited by Lowry [1] and also summarized by Beyer [2].

In recent years, several processes to depolymerize, desulfurize, and deash coal by solvent extraction have been developed and operated up to large pilot-plant scale. These have included process variations such as shallow extraction, deep extraction, and separate catalytic hydrogenation of the solvent.

A typical process for solvent-refined coal is illustrated in Fig. 1. This is operated in the United States by Pittsburgh-Midway at Tacoma, Washington (50 tons/day) and by Southern and Catalytic at Wilsonville, Alabama (6 tons/day). Typically, a slurry of coal in a coal-derived solvent is mixed with hydrogen, preheated to 450°C at 1500-psi pressure, and digested adiabatically. Most of the coal is dissolved and depolymerized, with sulfur removed as hydrogen sulfide gas. In the separation section, unused hydrogen is removed and recycled. A mineral residue is filtered, and a vacuum column then recovers solvent for recycle. The column bottom product is solvent refined coal, having low ash and sulfur content. However, some technical disadvantages to the process are excessive hydrogen consumption, long residence times in the dissolver, and slow filtration rates.

The increasing cost of petroleum and its relative scarcity have given new importance to coal as an alternative, both as a fuel and also as a raw material source for the chemical industry. Although the production of olefins, the key building blocks of the petrochemicals industry, could be achieved from alternative higher boiling sources to the increasingly demanded naphtha (e.g., kerosene and gas oil), these are not so attractive for the production of the main aromatic primary petrochemicals—benzene, toluene, and xylenes. Coal provides an alternative source, but not by the traditional carbonization processes, which yield only about 3 gal/ton of coal. New methods have been developed for the liquefaction of coal employing solvent extraction techniques to provide a good supply of aromatics. Two such processes have been described by Davies [3] covering work at the Coal Research Establishment in Britain. One uses a supercritical gas as a solvent, the other a coal derived liquid. Although the extracts differ, they are

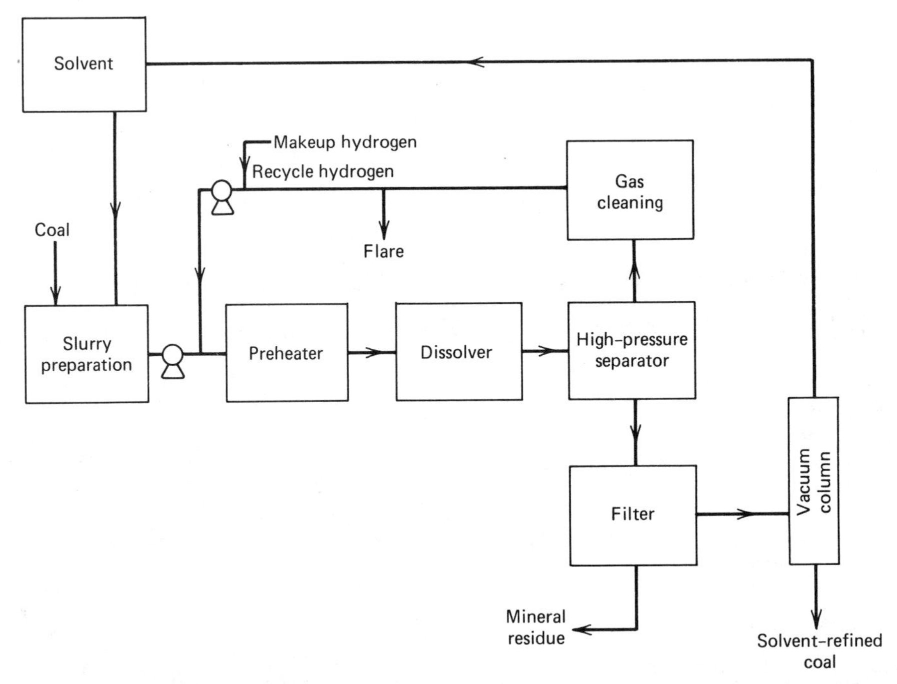

Figure 1. Solvent-refined coal process.

both converted to liquid distillates by catalytic hydrotreatment.

Gases at high pressure above their critical temperature may have especially good solvent properties. Coal decomposes at about 400°C to give liquids that are too involatile to distill. They can, however, be dissolved by such solvents as toluene and naphtha, which have critical temperatures close to 400°C. The process flow sheet is outlined in Fig. 2. The extract, which represents up to 40% of the weight of bituminous coal, has a relatively low molecular weight and a higher hydrogen:carbon ratio (1.0) than the coal (0.7). The residue char is suitable for combustion and gasification to produce the energy and hydrogen that is needed for liquefaction.

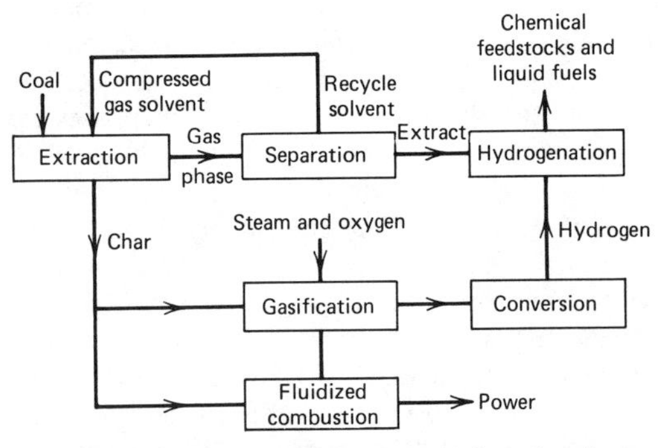

Figure 2. Supercritical gas extraction of coal.

The extract, which contains small aromatic structures linked by ether or methylene bridges, is hydrogenated to produce a distillable oil almost free of heteroatoms and rich in cycloparaffins, alkyl benzenes, hydroaromatics, and higher aromatics. A typical analysis quoted by Davies is shown in Table 1.

A more conventional method of extraction uses a liquid solvent, such as anthracene oil, at a digestion temperature of 370–450°C (3 parts oil to 1 part coal). The digest is filtered at 250°C to give a clean coal solution containing less than

TABLE 1 ANALYSIS OF OIL FROM HYDROTREATMENT OF SUPERCRITICAL GAS EXTRACT OF COAL[a]

	Fraction Distillation Range, °C				
Chemical Type, Weight Percent	<170	170–250	250–300	300–350	350–420
Paraffins	17	4	3	6	8
Cycloparaffins	36	30	17	12	13
Alkyl benzenes	39	20	5	1	0
Hydroaromatics	0	43	46	34	6
Higher aromatics	0	3	26	46	73
Unidentified	8	0	3	1	0
Yield wt. %	27	22	20	15	10

[a] Adapted from Davies [3].

0.1% ash. This may be used to produce a high-grade coke from which graphite electrodes for steelmaking arc furnaces may be made, or concentrated to a pitchlike material for the manufacture of carbon fibers.

After filtration and evaporation of the solvent for recycle, the extract may be hydrogenated and fractionated to yield low-boiling hydrocarbon distillates that have high potential value as a feedstock for the manufacture of primary aromatics, or as a gasoline blending stock. Again, the undigested coal residue may be burned to raise steam and gasified to yield part of the hydrogen required in the hydrogenation reactors. It is proposed in this process that some solvent be fed with the coal extract to the hydrocracker, so that an equivalent amount of solvent must be recovered from the hydrocracked products to maintain a solvent balance. All hydrogenated oils have a high solvent power for coal as a result of the presence of polynuclear hydroaromatics having a hydrogen-donor effect that promotes a mild hydrogenation of the coal during digestion. Extracts made with hydrogenated oils differ somewhat from those made from anthracene oil as they are richer in hydrogen and have a lower average molecular weight.

Considerable progress has been made with the development of these processes in fundamental and pilot-plant studies covering a wide range of coals, solvents, and operating conditions of digestion and hydrogenation. These result in a variety of extract compositions and hydrogenated distillates.

2. EXTRACTION OF PHENOLS FROM COAL TARS AND LIQUORS

In the branch of the chemical industry concerned with the recovery of chemicals from coal, solvent extraction techniques have long been of interest, chiefly for the extraction and separation of phenolic tar acids.

Coal carbonization to form coke or smokeless fuel continues to be an important industrial process, although during the 1950s and 1960s coke was virtually completely replaced in many countries by petroleum and natural gas as raw material for the chemical synthesis of organic compounds, ammonia, and town gas. However, coke continues to be required for iron production, and smokeless fuel has tended to displace coal for domestic consumption. Also, the growing expense and shortage of petroleum compared with coal will surely lead to a resurrection of the importance of coke in the chemical industry.

The coal tar formed as a by-product in coke production and the rather different liquid coal oil obtained in relatively low temperature carbonization giving smokeless fuels, are initially distilled to yield light, middle, and heavy oil fractions and a residue of pitch. The light and middle oils are rich in phenolic compounds and these are separated, in effect, by a solvent extraction process in which the oils are contacted with aqueous caustic soda solution. The acidic phenolic compounds react to form their sodium salts in the aqueous phase; the sodium salt solution is washed with a light neutral oil to remove the heavy neutral oil and then is steam treated to remove the last traces of neutral oils and bases. Undissociated phenols are "sprung" from the aqueous phase by treatment with carbon dioxide (using, e.g., flue gas containing 12% CO_2) and separated from the resulting sodium carbonate solution, which is treated with lime to regenerate caustic soda.

A number of alternatives to the caustic soda extraction of phenols have been proposed involving extraction with selective solvents. Some of these processes have been used, but they tend to extract the phenolic series efficiently only as far as the cresols and some of the xylenols. Higher compounds are not completely extracted, and the purification of the extract to separate the impurities of pyridine bases, alcohols, and hydrocarbons from the high-boiling phenols is difficult. However, the light liquor produced at the rate of 40 gal/ton in the low-temperature carbonization of coal to form smokeless fuel contains valuable phenols of which 95–96% may be extracted by countercurrent contact with isobutyl acetate [4]. The extract is distilled to recover the solvent for recycle, and the concentrated extract is distilled to give monohydric and dihydric phenol fractions, which pass through further separation procedures.

Among the processes for separation of phenol homologues, one mentioned by Coleby [5] involving solvent extraction techniques is dissociation extraction, described in detail in Section 3.

3. DISSOCIATION EXTRACTION

3.1. The "Classical" Process

Dissociation extraction is a technique for the separation of mixtures of organic acids or bases that depends on differences in the dissociation constants of the components of the mixture. It can be applied to closely related or isomeric compounds, the separation of which may be difficult by the more common methods of distillation, solvent extraction, or fractional crystallization.

To illustrate the principle of dissociation extraction, consider, as an example, a mixture of the weak organic acids *m*- and *p*-cresol. These have normal boiling points differing by only a fraction of a degree and have closely similar solubilities in many solvents. The difference in molecular arrangement of the two isomers does result, however, in an appreciable difference in their strengths as acids, shown by their relative dissociation constants, which are 9.8×10^{-11} for *m*-cresol and 6.7×10^{-11} for *p*-cresol. This difference may be exploited as the basis of a practicable separation by dissociation extraction. Thus, if the mixture is partially neutralized by strong base, such as aqueous caustic soda, there will be competition between the two organic acids for reaction with the base. The stronger organic acid, which has the larger dissociation constant (*m*-cresol), will react preferentially with the base to form in the aqueous solution an ionized cresylate salt, insoluble in organic solvents. The weaker acid (*p*-cresol) will remain predominantly in its undissociated form, soluble in organic solvents. If the partial neutralization is followed by a conventional extraction with an organic solvent, the weaker acid will be concentrated in the solvent and the stronger, as its dissociated salt, in the aqueous phase. If this process of dissociation extraction is applied in a multistage system with countercurrent flow of organic solvent and aqueous caustic soda, the feed mixture of cresols may be separated into its component isomers with high purity. *m*-Cresol may be regenerated from its salt in the aqueous phase by treatment with a strong mineral acid. Clearly, a mixture of organic bases may be separated similarly by treatment with a stoichiometric deficiency of aqueous strong acid.

Although the basic principle of dissociation extraction has been known at least since the work of Warnes [6] in 1924, the theory of the technique has only recently been adequately developed. Its first practical applications were merely in laboratory separations of particular mixtures. It was later applied on a commercial basis by use of multistage contactors, first considered by Twigg [7] in 1949.

Application of dissociation extraction has been considered chiefly for the separation of organic acids and bases occurring in coal tar, particularly of *m*- and *p*- cresols [8–10], xylenols [11], and the 2.6-lutidine, 3- and 4-picoline system [12, 13]. It has been applied commercially for the separation of dichlorophenols, penicillin acids, and benzoic acid derivatives.

In early work on dissociation extraction by Anwar, Hanson, and Pratt, the theory of the process was developed for those systems in which the organic compounds to be separated have a reasonably high affinity for the aqueous phase, so that the reaction in that phase with the mineral acid or base goes to completion [14]. As will be explained later, this is not the case for the cresol isomers and many other organic compounds. The "complete reaction" system studied experimentally was the separation of 3- and 4-picoline, with a deficiency of hydrochloric acid used as reagent, and chloroform and benzene as alternative organic solvents. The assumption of complete reaction in the aqueous phase of the reagent with a stoichiometric amount of the organic isomers leads to a relatively simple theoretical treatment. An expression was derived for the separation factor α_{AB} for bases A and B, having dissociation constants K_A and K_B ($K_A < K_B$) and distribution coefficients of the undissociated bases (organic phase concentration/aqueous phase concentration): D_A and D_B. If N is the molar concentration of reagent (HCl) originally present in the aqueous phase, and terms in square brackets represent concentrations in the organic phase, the expression deduced was

$$\alpha_{AB} = \frac{D_A}{D_B} \cdot \frac{D_B N + [B] + (K_A/K_B) \cdot (D_B[A]/D_A)}{D_B N(K_A/K_B) + [B] + (K_A/K_B) \cdot (D_B[A]/D_A)}$$

If the total concentration of bases in the organic phase is represented by T, where $T = [A] + [B]$, and the ratio of the bases in this phase by δ, where $\delta = [A]/[B]$, the equation may be rearranged to

$$\alpha_{AB} = \frac{D_A K_B}{D_B K_A} \left[\frac{N(\delta + 1) + T\{(1/D_B) + (K_A \delta / K_B D_A)\}}{N(\delta + 1) + T\{(K_B/K_A D_B) + (\delta / D_A)\}} \right]$$

By means of this equation, it is possible to calculate separation factors in terms of important process parameters (total concentration and concentration ratio of the organic isomers and their dissociation constants and distribution coefficients, together with the feed concentration of reagent). Note that the equation does not involve pH. This parameter was included in earlier theoretical treatments of dissociation extraction [15] but is very difficult to control in a practical plant, and its inclusion distracts attention from the key process variables.

As the total concentration of bases decreases, T tends to zero, and the preceding equation shows that α approaches the value $D_A K_B / D_B K_A$. This sets the maximum of α and indicates that the separation is greater if the isomer that is the stronger acid or base (having the greater value of dissociation constant) has the lower value of D, that is, the smaller tendency in the undissociated state to distribute to the organic phase. It is desirable to choose an organic solvent that gives as high as possible a ratio of D_A/D_B, as this will reinforce the difference in dissociation constants to achieve a high separation factor. In contrast, when T becomes very high and the proportion of dissociated bases is small, α tends to D_A/D_B. This represents the minimum value of separation factor.

As the concentration of extraction reagent N increases from zero, α increases from D_A/D_B and may pass through a maximum or become fairly constant at concentrations that depend on the physical parameters of the system.

Experimental results [14] for the picoline system demonstrated the effect of changes in these process variables and showed good agreement between measured and calculated separation factors.

The effect of choice of solvent was illustrated by Wadekar and Sharma [16] in the separation of *o*-cresol and 6-chloro-*o*-cresol, having, respectively, normal boiling points of 191.2° and 191.3°C and pK_a values of 10.28 and 8.69 at 25°C. With benzene as solvent, practically no separation occurred, despite the large difference in dissociation constants, but with di-*n*-butyl ether as solvent, separation factors of 3–4 were achieved.

Later work by Anwar et al. [17] applied the new theory to multistage processes for separations of two-component mixtures by dissociation extraction. A procedure was established by which the number of equilibrium stages required for a given separation could be determined by stagewise calculation or by an approximate graphical method, analogous to the McCabe–Thiele diagram. Again, good agreement was obtained between experiment and theory for the picolines separation. A new type of laboratory-scale mixer–settler was developed [18] for these tests.

The theory of dissociation extraction may be extended to cover the separation of mixtures of more than two organic compounds, which are commonly experienced. Kafarov et al. [19] have developed a generalized theory for multicomponent systems with a strong extracting agent. In unpublished work, Wadekar and Sharma have gone on to consider multicomponent mixtures containing multifunctional groups, such as substituted dicarboxylic and tricarboxylic aromatic acids, and bifunctional aromatic amines.

3.2. Modified Process Involving Weak Reagents

Although dissociation extraction processes have been applied commercially, their application on a wide scale has been hampered by the continuous consumption of strong alkali and strong acid. This operating cost has limited the application of the "classic" dissociation extraction process, described earlier, to the separation of compounds of high intrinsic value in comparison with the mineral acid and alkali used in their recovery.

This high operating cost is always involved in separation processes where there is a strong interaction between the component to be separated and the reagent used to achieve the separation (e.g., the strong reaction between mineral acids or alkalis and organic bases or acids or between a metal ion and a strong complexing agent). This is due to the difficulty in breaking down the product formed in order to free the purified component and regenerate the separation reagent. In the ideal separation process, there will be only a weak interaction between the separating reagent and the component to be separated, sufficiently strong to form the basis of a practical separation process but weak enough to be broken down without the expenditure of a large amount of chemical or thermal energy.

The basis of the modified process of dissociation extraction described by Anwar et al. [20] is the use, in place of strong mineral acids or alkalis, of only weakly acidic or basic reagents so that the reaction product between them and the organic components may be broken down simply by contact with a solvent having a strong affinity for the separated component. The reagent is thereby regenerated and can be used again. This avoids the continuous consumption of materials, and the significant reduction in operating costs that results should make dissociation extraction a cheaper and, therefore, more versatile process.

Consider again the separation of *m*- and *p*-cresols. If the cresols, either alone or dissolved in an organic solvent that has only a moderate affinity for cresols, such as a mixture of 70% by volume *n*-hexane and 30% benzene, are contacted with an aqueous solution of a weak base such as sodium phosphate, Na_3PO_4, and the proportion of phosphate is less than stoichiometric, the competition between the cresols for reaction with the phosphate to form sodium cresylate will enable a normal separation by dissociation extraction to be achieved. The reaction, which will not proceed to completion in the aqueous phase because of the weakly basic nature of trisodium phosphate, can be summarized as follows:

$$\begin{array}{ll} \text{Organic phase} & C_6H_4(CH_3)OH \\ & \quad\Updownarrow \\ \text{Aqueous phase} & C_6H_4(CH_3)OH + PO_4^{3-} \rightleftharpoons C_6H_4(CH_3)O^{1-} + HPO_4^{2-} \end{array}$$

The more acidic *m*-cresol reacts preferentially, forming a dissociated salt soluble in the aqueous phase, and the less acidic *p*-cresol remains predominantly in its undissociated form, soluble in the organic phase. In this way the isomers are partially separated and a high degree of separation may be achieved in a multistage process. The aqueous phase containing disodium hydrogen phosphate and sodium *m*-cresylate is subsequently contacted with an organic solvent having strong affinity for cresols, such as benzene or chloroform. The reaction is reversed and recyclable trisodium phosphate is regenerated. The cresol is recovered from the organic solvent by distillation.

The distribution equilibria for the separation and recovery stages were measured [20] for this cresols separation system and the process was mathematically modelled to permit optimization of design [21]. It was found that the efficiency of the *m*-cresol recovery was much improved if a small proportion of phosphoric acid was added to the aqueous phosphate salt of *m*-cresol fed to the recovery contactor. This free acid was neutralized by caustic soda before recycle of the regenerated sodium phosphate stream. The formation of extra sodium phosphate resulting from this enables a purge stream of aqueous sodium phosphate to be constantly rejected, preventing the buildup of impurities in the recycled reagent. The simplified flow sheet for the proposed separation process is shown in Fig. 3.

Although the experimental results confirmed the viability of the process, the low separation factor and aqueous-phase loadings of cresols necessitated liquid–liquid contactors with many stages and capable of handling large flow volumes. The result of these process complications was that the overall economics of the process were less advantageous than had been initially expected. No attempt has been made to optimize the choice of reagent or solvents, and it is probable that alternative separation systems would be more attractive economically.

The technique was also applied to the separation of the coal tar bases, 2-6-lutidine with 3- and 4-picoline. In the process proposed [22], the picoline isomers were separated by competitive reaction with an aqueous solution of the weakly acidic salt sodium dihydrogen phosphate.

The weakly acidic or basic reagent used in the modified dissociation extraction process need not necessarily be an inorganic compound. Weak organic acids or bases can be considered, provided they are preferentially soluble in the aqueous phase, and these may have practical advantages. Further work [23] established that the organic base monoethanolamine could be used successfully to separate mixtures of 2-3 and 2-6-dichlorophenols, a process of commercial interest. In a multistage contactor, the dichlorophenols, dissolved in toluene, were passed countercurrent to a stream of 4% aqueous monoethanolamine. The 2-6-dichlorophenol, concentrated as its salt in the aqueous phase, was then treated with ethyl acetate, leading to an efficient extraction of the purified dichlorophenol into

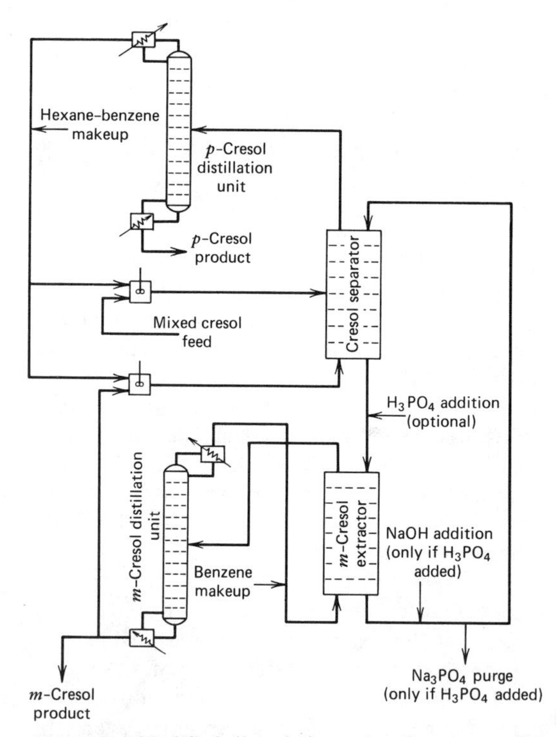

Figure 3. Modified dissociation extraction process for the separation of *m*- and *p*-cresols.

the organic solvent and regeneration of the monoethanolamine for recycle. It was also shown [24] that monoethanolamine could be used as reagent for cresols separation.

The modified dissociation extraction process may be used not only to separate isomers, but also to remove a particular class of compound from a mixture. It could be used, for example, to remove the phenolic tar acids from the middle oil resulting from coal carbonization and also to extract the phenolic impurities formed during aromatic nitration [25].

As the reaction in the aqueous phase between the organic compounds being separated and the weak reagent used in the modified dissociation extraction process is likely to be incomplete, that is, the reagent is not wholly used up even though it is present in stoichiometric deficiency, the theoretical treatment of the process, developed for the case of complete reaction, was not appropriate and so a new approach was made [22, 23] to cover incomplete reaction.

Consider, as an example, the separation of two organic acids HA and HB by a basic organic reagent C. The equilibria established are thus

Organic phase	HA HB	
	⇅ ⇅	
		$HB + C \rightleftharpoons B^- + CH^+$
Aqueous phase	HA +	$C \rightleftharpoons A^- + CH^+$

The dissociation constants of the acids HA and HB in an aqueous solution are respectively

$$K_A = \frac{[H^+][A^-]}{[HA]} \tag{1}$$

$$K_B = \frac{[H^+][B^-]}{[HB]} \tag{2}$$

and the acid dissociation constant of the reagent C is given by

$$K_C = \frac{[H^+][C]}{[CH^+]} \tag{3}$$

Now one mole A^- or B^- is formed for every mole of C that reacts; thus by mass balance

$$R = [C]_{eq} + [A^-] + [B^-] \tag{4}$$

and

$$[CH^+] = [A^-] + [B^-] \tag{5}$$

where $[C]_{eq}$ is the concentration of undissociated reagent remaining at equilibrium and R is the initial concentration of reagent C. Equation (4) gives

$$[C]_{eq} = R - ([A^-] + [B^-]) \tag{6}$$

Where subscript aq refers to the aqueous phase, the equilibrium constants for the reactions in the aqueous phase are

$$K_1 = \frac{[CH^+][B^-]}{[HB]_{aq}[C]_{eq}} = \frac{K_B}{K_C} \tag{7}$$

$$K_2 = \frac{[CH^+][A^-]}{[HA]_{aq}[C]_{eq}} = \frac{K_A}{K_C} \tag{8}$$

and for the exchange reaction between the two acids

$$HB + A^- \rightleftharpoons HA + B^-$$

$$K_3 = \frac{[HA]_{aq}[B^-]}{[HA]_{aq}[A^-]} = \frac{K_B}{K_A} \tag{9}$$

Thus from Eqs. (5)–(7), eliminating $[CH^+]$ and $[C]_{eq}$:

$$K_1 = \frac{[B^-]([A^-] + [B^-])}{[HB]_{aq} \cdot (R - [A^-] - [B^-])} \tag{10}$$

Similarly

$$K_2 = \frac{[A^-]([A^-] + [B^-])}{[HA]_{aq} \cdot (R - [A^-] - [B^-])} \tag{11}$$

Also, where subscript o refers to the organic solvent phase, the distribution coefficients of the undissociated isomers are

$$D_A = \frac{[HA]_o}{[HA]_{aq}} \tag{12}$$

$$D_B = \frac{[HB]_o}{[HB]_{aq}} \tag{13}$$

and $[HB]_{aq}$ and $[HA]_{aq}$ can also be eliminated from Eqs. (10) and (11). It is thus necessary to know the values of D at the concentrations under consideration, but methods for determining D have been found to be unreliable. However, since the total amounts of each acid in the aqueous phase can be considered as process variables, a simple substitution in Eqs. (9)–(11) eliminates the need for knowledge of the distribution coefficients.

Let total acid in aqueous phase $[HA]_{aq} + [A^-] = M$ and $[HB]_{aq} + [B^-] = N$. Also, let $[A^-] = x$ and $[B^-] = y$. Then $[HA]_{aq} = M - x$ and $[HB]_{aq} = N - y$.

By solution of any two of the Eqs. (9)–(11) simultaneously, a cubic equation of the form

$$Ay^3 + By^2 + Cy + D = 0 \qquad (14)$$

is generated where

$$A = K_1 + K_2 - K_1K_3 - 1$$

$$B = M(K_1 - 1) + N(2K_1K_3 - K_1 - K_3) + R(K_1K_3 - K_1)$$

$$C = K_1N(R - M) - K_1K_3(N + 2R)$$

$$D = K_1K_3N^2R$$

The separation factor is defined as the ratio of the overall distribution coefficients D'_A and D'_B, hence

$$\alpha_{AB} = \frac{D'_A}{D'_B} = \frac{[HB]_o([HB]_{aq} + [B^-])}{[HB]_o([HA]_{aq} + [A^-])} \qquad (15)$$

Thus substitution for $[HA]_o$ and $[HB]_o$ from Eqs. (12) and (13) yields

$$\alpha_{AB} = \frac{D_A}{D_B} \cdot \frac{[HA]_{aq}}{[HB]_{aq}} \cdot \frac{([HA]_{aq} + [A^-])}{([HB]_{aq} + [B^-])} \qquad (16)$$

and, using the same notation as before:

$$\alpha_{AB} = \frac{D_A}{D_B} \cdot \frac{(M - x)}{(N - y)} \cdot \frac{N}{M} \qquad (17)$$

By combination of Eqs. (9) and (17), the expression for separation factor is further simplified to

$$\alpha_{AB} = \frac{D_A}{D_B} \cdot \frac{K_B}{K_A} \cdot \frac{x}{y} \cdot \frac{N}{M} \qquad (18)$$

Provided the ratio of the distribution coefficients can be determined, the separation factor can be calculated once values of x and y have been found from Eq. (14). Distribution coefficients must be found experimentally, and values measured for separate isomers are not valid when both isomers are present together. However, the ratio of the distribution coefficients was found experimentally to be fairly constant over a wide range of concentrations. By use of an average experimental value of the distribution coefficient ratio, the estimated separation factors calculated from Eq. (18) were found to be close to experimental measurements of this quantity, and thus equilibrium data can be predicted by the relationship with a reasonable degree of accuracy.

3.3. Recent Process Developments: Use of Strong Reagents with Recovery by Solvent Extraction

A limitation of the dissociation processes using weakly acidic or basic reagents, as was found for the cresols–phosphate system, is that separation factors and aqueous phase loadings may be low, so that the number of contact stages required and the volume of the equipment used must be relatively high. These disadvantages must be set against the advantage of elimination of consumption of reagent chemicals achieved by this type of process in comparison with the "classic" process.

Recent work [24] has shown that it is possible to combine the advantages of the classical and weak-reagent types of dissociation extraction process. The classical strong reagent, aqueous caustic soda, was used to separate m- and p-cresols, but the organic solvents in the separation and recovery sections had different affinities for cresols. This enabled the separation reaction to be reversed and purified m-cresol recovered without the consumption of chemicals. Chemists have traditionally believed that reaction between a strong base (e.g., caustic soda) and weak acids (cresols) proceeds to completion, although the pH will not be 7 when stoichiometrically equivalent quantities of the two are mixed. This is true for homogeneous systems, but if two phases are present, reaction in the aqueous phase may be far from complete. In such two-phase systems cresol will face two competing tendencies: to react with caustic soda in the aqueous phase to form a salt, and also to concentrate in the organic solvent phase, in which it will be preferentially soluble. These competing attractions limit the completion of reaction in the aqueous phase, and the extent of reaction is influenced by the nature of the organic solvent and its affinity for cresols. With a solvent of relatively low affinity, such as toluene or carbon tetrachloride, reaction between cresols and caustic soda in the aqueous phase proceeds more toward completion, and, in a multistage process, the cresol isomers may be separated. The aqueous phase, containing the purified m-cresol salt, is then contacted with an organic solvent of high affinity for the cresol, such as octan-1-ol; the aqueous phase reaction is partially reversed and m-cresol extracted into the organic solvent from which it may be separated

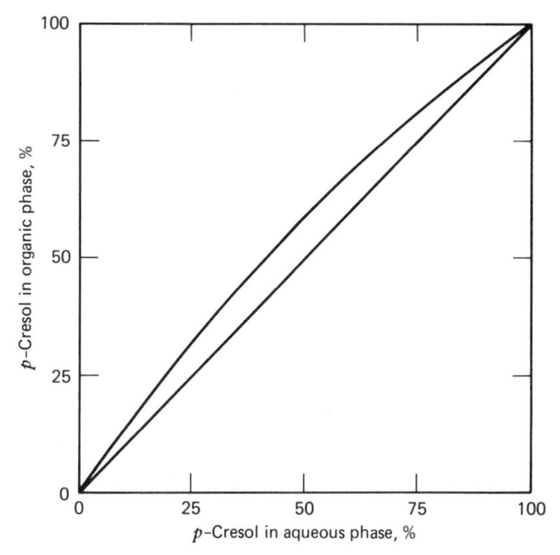

Figure 4. Separation of *m*- and *p*-cresols distributed between 0.5 *N* caustic soda and toluene (300 g of total cresols in 1 liter of each phase).

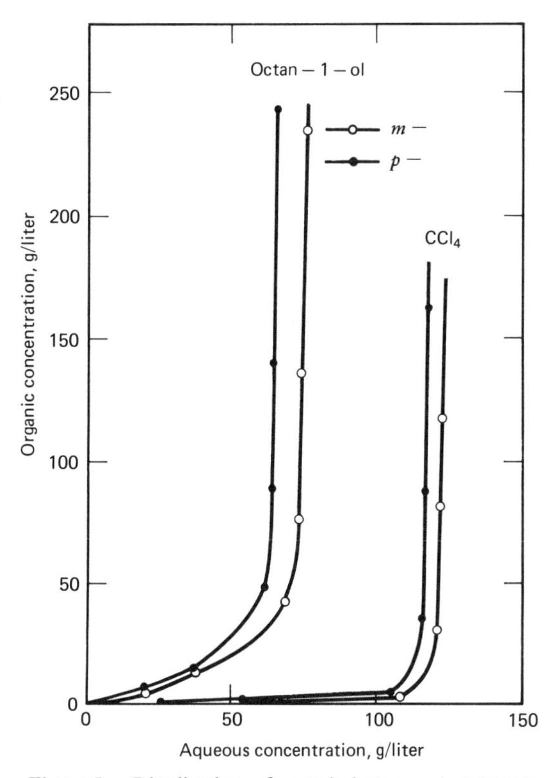

Figure 5. Distribution of cresols between 1 *N* NaOH and the solvents carbon tetrachloride and octan-1-ol.

by distillation. The regenerated caustic soda in the aqueous phase may then be recycled to the cresol separation stage.

The separation of the cresol isomers distributed between 0.5 *N* caustic soda and toluene is illustrated in Fig. 4. The distribution of cresols between 1 *N* caustic soda and the solvents carbon tetrachloride and octan-1-ol is shown in Fig. 5, which illustrates the differences in extent of reaction in the presence of these two solvent. Modeling of the cresol recovery stage with a multistage contactor demonstrated that 76% of *m*-cresol was recovered from an aqueous phase containing 108 g/liter of *m*-cresol in 1 *N* caustic soda in five

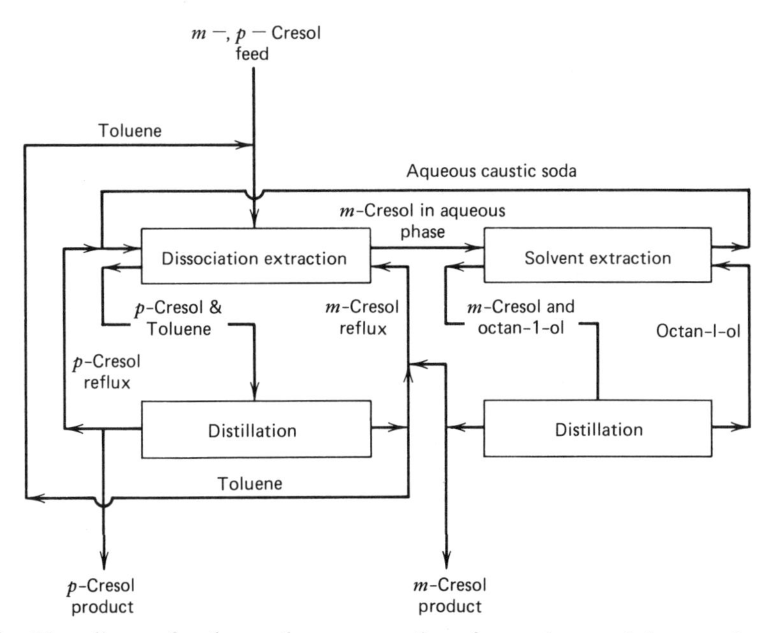

Figure 6. Flow diagram for the continuous separation of *m*- and *p*-cresols by use of caustic soda reagent.

stages. The simplified overall flowsheet proposed for this cresols separation process is shown in Fig. 6.

The use of caustic soda instead of trisodium phosphate as reagent results in a higher separation factor (1.5 instead of 1.4) and aqueous phase loading of cresols (ca. 70 g/liter compared with 50 g/liter). Since recovery from the aqueous phase is achieved without the continuous consumption of chemicals, it is probable that the overall process costs are lower than in the systems previously considered for separation of cresols by dissociation extraction. This latest type of dissociation extraction process, with the use of strong reagents but with recovery from the aqueous phase achieved simply by contact with organic solvent, may also prove to be practical and economically attractive for the separation of other mixtures of acidic or basic organic isomers or chemically closely related compounds. Dissociation extraction is an example of mass transfer with simultaneous chemical reaction. Rate aspects are covered in Chapter 2.1.

REFERENCES

1. H. H. Lowry, Ed., *Chemistry of Coal Utilization*, supplementary volume, Wiley, New York, 1963, pp. 237–249.
2. G. H. Beyer, *Proceedings of the International Solvent Extraction Conference (ISEC), Toronto, 1977*, Vol. 2, Canadian Institute of Mining and Metallurgy, Montreal, 1979, p. 715.
3. G. O. Davies, *Chem. Ind.*, 560 (1978).
4. J. G. M. Thorne, *Chem. Process.* (March 1970).
5. J. Coleby, in C. Hanson, Ed., *Recent Advances in Solvent Extraction*, Pergamon Press, London, 1971, pp. 121–127.
6. A. R. Warnes, *Coal Tar Distillation*, 3rd ed., Benn Brothers, London, 1924, p. 228.
7. G. H. Twigg, *Nature* **163**, 1006 (1949).
8. S. R. M. Ellis and J. D. Gibbon, in J. M. Pirie, Ed., *The Less Common Means of Separation*, Institution of Chemical Engineers, London, 1964, p. 119.
9. C. A. Walker, *Ind. Eng. Chem.* **42**, 1226 (1950).
10. V. A. Kostyuk, G. S. Mickhailova, S. M. Grigorev, E. Ya. Chernomordik, and G. A. Marbas, *Coke Chem. USSR*, **12**, 48 (1968).
11. J. Coleby, *Symposium on Liquid-Liquid Extraction*, Institution of Chemical Engineers, Newcastle-upon-Tyne, April 1967.
12. A. Yamamoto, Japanese Patent 1517 (1956).
13. A. E. Karr and E. G. Scheibel, *Ind. Eng. Chem.* **46**, 1583 (1954).
14. M. M. Anwar, C. Hanson, and M. W. T. Pratt, *Transact. Inst. Chem. Eng.* **49**, 95 (1971).
15. W. S. Wise and D. E. Williams, in J. M. Pirie, Ed., *The Less Common Means of Separation*, Institution of Chemical Engineers, London, 1964, p. 112.
16. V. V. Wadekar and M. M. Sharma, *J. Separ. Proc. Technol.* **2**, 1 (1981).
17. M. M. Anwar, C. Hanson, A. N. Patel, and M. W. T. Pratt, *Transact. Inst. Chem. Eng.* **51**, 151 (1973).
18. M. M. Anwar, C. Hanson, and M. W. T. Pratt, *Chem. Ind.* 1090 (1969).
19. V. V. Kafarov, V. G. Vygon, A. I. Chulok, and V. A. Kostyuk, *Russ. J. Phys. Chem.* **50**, 1618 (1976).
20. M. M. Anwar, C. Hanson, and M. W. T. Pratt, *Proceedings of the International Solvent Extraction Conference (ISEC), The Hague, 1971*, Vol. 2, Society of Chemical Industry, London, 1971, p. 911.
21. M. W. T. Pratt and J. Spokes, *Proceedings of the International Solvent Extraction Conference (ISEC), Toronto, 1977*, Vol. 2, Canadian Institute of Mining and Metallurgy, Montreal, 1979, p. 723.
22. M. M. Anwar, S. T. M. Cook, C. Hanson, and M. W. T. Pratt, *Proceedings of the International Solvent Extraction Conference (ISEC), Lyons, 1974*, Vol. 1, Society of Chemical Industry, London, 1974, p. 895.
23. M. M. Anwar, S. T. M. Cook, C. Hanson, and M. W. T. Pratt, *Proceedings of the International Solvent Extraction Conference (ISEC), Toronto, 1977*, Vol. 2, Canadian Institute of Mining and Metallurgy, Montreal, 1979, p. 671.
24. M. M. Anwar, M. W. T. Pratt, and M. Y. Shaheen, *Proceedings of the International Solvent Extraction Conference (ISEC), Liège, 1980*, Vol. 2, Association des Ingénieurs sortis de l'Université de Liège, 1980, Session 8, Paper 80-64.
25. C. Hanson, T. Kaghazchi, and M. W. T. Pratt, in L. F. Albright and C. Hanson, Eds., *Industrial and Laboratory Nitrations*, American Chemical Society Symposium Series No. 22, American Chemical Society, Washington, DC, 1976, pp. 151–154.

22

EXTRACTIVE REACTION PROCESSES

C. Hanson
University of Bradford
United Kingdom

1. INTRODUCTION

The primary objective of solvent extraction is conventionally taken to be *separation*. This may be accompanied by a chemical reaction, which usually has an effect (enhancement or retardation) on extraction rates relative to those effects that would be expected solely from diffusional considerations. The general principles of such interactions are discussed in Chapter 2.1.

The use of a chemical reaction for separation purposes implies that the reaction can be reversed by suitable means (e.g., change of pH) so that the original species can be recovered, such as in a stripping operation. This is characteristic of the extraction processes for metal ions (Chapter 2.2). The term "extractive reaction process" used in this chapter is taken to mean an essentially *irreversible* reaction in which a desired *new species* is produced in a heterogeneous liquid–liquid dispersion. Some typical unit processes involving heterogeneous liquid phases are alkylation, esterification, hydrolysis of esters, nitration, saponification, sulfation, and sulfonation.

Such processes are of considerable industrial importance, yet until quite recently they have been treated from the point of view of reactor design rather than liquid–liquid extraction. There is now increasing recognition of the common features of extractive reaction processes and liquid–liquid extraction operations; an obvious example is the hydrodynamics of phase interdispersion.

The accurate design of a reactor for a heterogeneous liquid-phase process demands accurate rate data and their correct interpretation. A full interpretation depends on knowledge of the *key reacting species* and the *locale of the reaction*. The reaction may occur homogeneously throughout either bulk phase, at the interface or in a reaction zone adjacent to (or spanning) the interface. At the two extremes, the process may involve a slow reaction with fast diffusion, in which case it will be kinetically controlled and diffusional resistances will be unimportant, or it may involve an instantaneous reaction so that the rate is controlled entirely by diffusion. Most real cases are intermediate. The rate-controlling mechanism will affect the response of the process to changes in temperature, concentration and/or concentration profile, agitation intensity, phase continuity, and the presence of surface-active impurities. It should be recognized that the rate-controlling step may itself vary with temperature, agitation, the concentrations of key reactants, and other parameters and that it is not necessarily the same under all conditions.

The rates of such processes have been studied with a variety of contacting devices, the most important of which are simple agitated vessels, constant interfacial area agitated cells of the Lewis type, and laminar jets (see Chapters 2.1 and 3). The first have the disadvantage of not providing an easy measure of interfacial area. Some of the problems encountered are discussed in the following sections on specific processes.

2. AROMATIC NITRATION

The nitration of aromatic hydrocarbons is an important process in the manufacture of many explosives, dyestuffs, plastics, and other materials. It is usually carried out in the liquid phase by contacting the hydrocarbon with nitrating acid. The latter is a mixture of nitric and sulfuric acids with water, with the composition chosen to suit the reactivity of the aromatic substrate. The two phases are essentially immiscible. The chemistry of aromatic nitration has been extensively studied as it offers an ideal example of electrophilic substitution. The nitrating species is generally believed to be the nitronium ion NO_2^+ formed by the reaction

$$HNO_3 + 2H_2SO_4 \rightleftharpoons NO_2^+ + H_3O^+ + 2HSO_4^- \quad (1)$$

The sulfuric acid thus acts as an ionizing medium. The nitronium ion then attacks the aromatic substrate:

$$ArH + NO_2^+ \longrightarrow ArNO_2 + H^+ \quad (2)$$

Kinetic studies played an important part in the elucidation of this mechanism. However, in these studies, problems of interphase transfer were avoided by the choice of solvents to allow homogenous nitration, so the results are not directly applicable to the industrial situation.

Albright [1] and Hanson et al. [2] independently reviewed the literature on heterogeneous nitration kinetics in the mid-1960s. They both cast doubt on the assumption of all previous workers that mass-transfer resistance was eliminated by use of a high level of agitation, permitting the use of a simple kinetic model. Although the rate of nitration in a continuous-flow stirred tank reactor becomes virtually independent of agitation at high agitation rates, this itself does not prove the absence of any mass-transfer resistance. Once very high rates have been established, further increase in agitation may not bring about any further significant decrease in drop size (and hence increase in interfacial area per unit volume). Alternatively, very high rates of agitation may produce a decrease in mass-transfer coefficient to counterbalance any increase in interfacial area that does occur. This could arise through suppression of internal circulation in the drops as their size falls and also from suppression of drop interaction [3].

Hanson et al. [4] studied the nitration of toluene in a miniature CFSTR (constant-flow stirred tank reactor) in light of the above-mentioned effects and applied a number of criteria. They found that the reaction rate increased rapidly with agitation at modest levels but then flattened out, as seen by previous workers. However, Hanson et al. [4] pointed out the importance of considering rate per unit volume of *aqueous phase* since it is generally agreed that reaction takes place in that phase. Other workers had used rate per unit volume of dispersion. The difference does not matter provided the phase ratio in the reactor remains unchanged, but Hanson et al. [4] found a significant change in the relative holdup of aqueous phase in the reactor with change in agitation even though the feed phase ratio was maintained constant. Results quoted per unit volume of dispersion thus included a component due to changing aqueous phase fractional holdup at low agitation levels.

The question of relative phase holdup in continuous-flow agitated tanks has been studied in more detail by Godfrey and Grilc [5]. It is an important parameter in determing the degree of agitation required in the mixer of a mixer–settler since inadequate agitation can lead to stratification, lowered efficiencies, poor phase stability, and high entrainment losses. This is considered further in Chapter 9.

Temperature was also used as a possible criterion by Hanson et al. Earlier workers had quoted a doubling of the rate constant per 10°C increase as evidence of kinetic control. However, although temperature will affect the true reaction velocity constant, it was pointed out that it will also change the solubility of the aromatic in the aqueous phase (a key parameter as the reaction was known to take place in that phase) and the concentration of nitronium ions [by the effect of temperature on the equilibrium constant for reaction (1)]. The observed overall temperature coefficient is thus not easy to interpret.

The most convincing evidence was based on the effect of phase continuity. Inversion produced a different rate under otherwise identical conditions, and this could be explained only by the presence of a diffusional resistance. Inversion will not change residence times, but it will vary interfacial area through its effect on drop size.

Giles et al. [6] repeated the preceding work, using a larger reactor fitted with probes, whereby the dispersion could be photographed to provide a measure of drop size and interfacial area. Incidentally, they showed [7] the effect of reaction on drop size and the difficulty of using established correlations (see Chapter 4) that were obtained with systems at equilibrium. They developed a model for toluene nitration on the assumption of a fast reaction taking place in a zone in the aqueous phase adjacent to the interface. The parameters of distribution coefficient and toluene diffusivity required by the model were measured, but the rates predicted by the model did not agree with those measured experimentally. This was attributed to the fact that molecular diffusivities were being used for the prediction, whereas conditions in the stirred tank were highly turbulent. Hanson and Ismail [8] subsequently carried out toluene nitration in a laminar jet and found that in this case the model did predict the observed rate. This emphasizes that a rate model will give useful results only if the values of diffusivity employed are applicable to the hydrodynamic conditions existing in the reactor.

Strachan and co-workers [9] have made a major contribution to the understanding of aromatic nitration under heterogeneous conditions, demonstrating inter alia the existence of three regimes for rate determination, primarily controlled by composition of the nitrating acid. At low concentrations of sulfuric acid, toluene nitration does exhibit a true kinetic regime. With increase in sulfuric acid concentration, it passes through an intermediate regime before exhibiting fast-reaction characteristics under typical industrial conditions, as observed by Hanson and co-workers. This demonstrates the dependence of the rate-controlling mechanism on external conditions.

All the work described previously was centered on toluene. Similar studies have been reported with other hydrocarbons, including benzene [10] and chlorobenzene [11, 12]. Strachan et al. have also worked on dinitration.

3. AROMATIC SULFONATION

Many sulfonation reactions are now carried out with liquid SO_3. However, some can involve reaction between the hydrocarbon and concentrated sulfuric acid. Such systems are heterogeneous and have many similarities with nitration processes, discussed earlier, except for the fact that the products are water soluble and thus remain in the aqueous phase rather than having to diffuse back into the organic phase, as with nitration.

Early work with benzene and other hydrocarbons [13], using the same approach that Hanson et al. [4] employed for nitration, showed the presence of diffusional resistances. Grosjean and Sawistowski [14] have carried out a more systematic investigation of toluene sulfonation by using a constant interfacial area cell, again showing the existence of different kinetic regimes depending on the sulfuric acid concentration.

4. ALKYLATION REACTIONS

Alkylation is used in the petroleum industry as a means of increasing octane number without addition of lead tetraethyl. It has been shown [15] that the type of emulsion is an important parameter during the alkylation of isobutane with light olefins and that the quality of the alkylate produced under acid-continuous conditions is significantly higher than when the hydrocarbon phase is continuous. Coupled with the sensitivity of such reactions to the degree of agitation, this observation suggests the involvement of a mass-transfer-controlled mechanism. Alkylation has been the subject of a recent symposium [16].

5. HYDROLYSIS OF FATS

The hydrolyses of animal and vegetable fats for production of glycerine and fatty acids are important operations and are classic examples of mass transfer with chemical reaction under heterogeneous liquid-phase conditions. In modern plants the hydrolysis is carried out continuously by contacting with hot water in a countercurrent column under high pressure. Jeffreys et al. [17] have studied such a system at fullscale. They have analyzed the column performance in terms of both transfer units and theoretical stages, using models assuming mass transfer with simultaneous chemical reaction. The results indicate the importance of mass-transfer resistance and also the effect of the chemical reaction.

6. REACTOR DESIGN

Unit processes such as those discussed previously have often been carried out industrially in a batchwise manner. A full understanding of mechanism is not then critical, except for possible aspects such as by-product formation. It is when the processes are carried out continuously that the information on the mechanism is needed and the potential applicability of liquid extraction technology is greatest. Current equipment types are described in standard reference works [18] and in Part II of this handbook.

Countercurrent contacting has obvious attractions in terms of maximizing the driving force. If a process involves a homogeneous reaction in one phase, there may be a case for increasing the holdup of that phase over that determined by the overall flow by using some form of recycle or by choice of dispersed phase. If reaction takes place at the interface or in a zone adjacent to the interface, a large interfacial area is likely to be beneficial. Countercurrent operation may also have advantages in terms of spreading the heat release in the case of exothermic reactions, thereby facilitating temperature control.

In general, the design principles for industrial liquid extraction contactors given in Part II are applicable to extractive reaction processes. Because of the nature of many of the reactions involved (examples of which are listed in the second paragraph in Section 1), special consideration must often be given to temperature control and safety.

REFERENCES

1. L. F. Albright, *Chem. Eng.* **73** (9), 169 (1966).
2. C. Hanson, J. G. Marsland and G. Wilson, *Chem. Ind.* 675 (1966).
3. C. Hanson, in, *Recent Advances in Liquid-Liquid Extraction*, C. Hanson, Ed., Pergamon Press Oxford, 1971, Chapter 12.
4. C. Hanson, J. G. Marsland, and G. Wilson, *Chem. Eng. Sci.* **26**, 1513 (1971).
5. J. C. Godfrey, and V. Grilc, *Proceedings of the 3rd European Conference on Mixing* (BHRA), York, UK, 1979.
6. J. Giles, C. Hanson, and H. A. M. Ismail, in *Industrial and Laboratory Nitrations,* ACS Symposium Series No. 22, American Chemical Society, Washington, DC, 1976, Chapter 12.
7. C. Hanson, J. G. Marsland, and J. Giles, *Proceedings of the International Solvent Extraction Conference* (ISEC71), Vol. 1, (Soc. Chem. Ind.), London, 1971, p. 94.
8. C. Hanson, and H. A. M. Ismail, *Chem. Eng. Sci.* **32**, 775 (1977).
9. A. N. Strachan, in *Industrial and Laboratory Nitrations,* ACS Symposium Series No. 22, American Chemical Society, Washington, DC, 1976, Chapter 13.
10. D. F. Schiefferle, C. Hanson, and L. F. Albright, in *Industrial and Laboratory Nitrations,* ACS Symposium Series No. 22, American Chemical Society, Washington, DC, 1976, Chapter 11.
11. C. Hanson, J. G. Marsland, and M. A. Naz, *Chem. Eng. Sci.* **29**, 297 (1974).
12. J. W. Chapman and A. N. Strachen, in *Industrial and Laboratory Nitrations,* ACS Symposium Series No. 22, American Chemical Society, Washington, DC, 1976.
13. M. Sohrabi, T. Kaghazchi, and C. Hanson, *J. Appl. Chem. Biotechnol.* **27**, 453 (1977).
14. P. R. L. Grosjean and H. Sawistowski, *Proceedings of the International Solvent Extraction Conference* (ISEC) Liege, 1980, Association des Ingenieurs Sortis de l'Université de Liège, 1980, Paper 16.
15. K. W. Li, R. E. Eckert, and L. F. Albright, *Ind. Eng. Chem. Process Des. Devel.* **9**, 434 (1970).
16. L. F. Albright and A. R. Goldsby, *Industrial and Laboratory Alkylations,* ACS Symposium Series 55, American Chemical Society, Washington, DC, 1977.
17. G. V. Jeffreys, V. G. Jensen, and F. R. Miles, *Transact. Inst. Chem. Eng.* **39**, 389 (1961).
18. T. C. Lo, Commercial Liquid-Liquid Extraction Equipment, in, *Handbook of Separation Techques for Chemical Engineers,* P. A. Schweitzer, Ed., McGraw-Hill, New York, 1979, Section 1.10.

23

INDUSTRIAL EFFLUENT TREATMENT (Nonmetals)

Donald Mackay
University of Toronto
Canada

Magda Medir
Universitat de Barcelona
Spain

1. INTRODUCTION

Although solvent extraction has many features that suggest that it may have broad applicability to the removal of organic solutes from waste water, its application has been relatively limited, principally to phenol removal. The advantages of solvent extraction are that it is a conservative process, recovering rather than destroying the solute; thus there is potential for economic credit to the operation from the sale of the solute, and it is not subject to the toxicity instabilites encountered with biotreatment units. The main disadvantage is that even with solvents that are only slightly water soluble the high flow rate of waste water normally results in a high solvent loss. In addition, the solvent may have adverse environmental effects.

Solvent extraction is most suitable for (1) removal of solutes that may be present in high concentrations, especially when they are sufficiently high to interfere with biotreatment and (2) in cases in which the solute has a high partition coefficient between the organic solvent and water. Increasing environmental concerns about toxic organic compounds such as phenols and hydrocarbons and their halogenated derivatives, organonitrogen and organosulfur compounds, coupled to a possible increase in coal conversion technology that generates these compounds, suggests that solvent extraction may play an increasing role in waste water effulent treatment. Reviews of this operation and its applicability have been compiled by several groups of investigators [1-6].

It is first useful to review briefly the physicochemical features of water–solvent–solute systems, since these features are usually quite different in character from those of other solvent extraction systems and because they play a key role in solvent selection.

2. PHASE EQUILIBRIUM AND SOLVENT SELECTION

The partition coefficient K_x or ratio of solute concentration (x mole fraction) in the organic phase o and in the aqueous phase w is determined by the activity coefficient γ as follows:

$$K_x = \frac{x_o}{x_w} = \frac{\gamma_w}{\gamma_o}$$

In dilute systems (<1%) of nonassociating systems the activity coefficient is often fairly constant, and its concentration dependence can be ignored. Partition coefficients are frequently defined as K_M, the ratio of concentration in molarity (mol/liter) or mass concentration (g/liter). The relationship between K_x and K_M at low solute concentrations and for nearly immiscible solvents in water is

$$K_M = K_x \frac{v_w}{v_o}$$

where v is the molar volume (liters/mol) of water and solvent phases. In applying solvent extraction to treatment of a given solute in water, the value of γ_w is defined and for sparingly soluble solutes is approximately equal to the reciprocal of the mole fraction solubility of the solute in water. The solvent selection process is essentially that of determining a suitable γ_o. High values of K_x or K_M are desirable since they permit a high water:solvent flow ratio; thus low values of γ_o are preferred.

This is illustrated in Table 1 for phenol partition between selected solvents [1]. It illustrates the unfortunate trend that high partition coefficients are achieved at the expense of high solvent solubilities in water. At a given water flow rate the water:solvent volume flow rate ratio can be assumed to approximately equal K_M. Clearly, the solvent losses for more soluble solvents become unacceptably high. This results in a tradeoff between (1) high partition coefficient, low solvent rate, and high solvent loss and (2) low partition coefficient, high solvent rate, and low solvent loss. Ultimately, the optimum is determined by economic considerations, especially the cost of solvent recovery and the value of the solvent.

TABLE 1 DATA FOR PHENOL PARTITION BETWEEN WATER AND SELECTED SOLVENTS AT 25°C [1]

Solvent	Solubility in Water, Wt. %	Phenol K Values: $\frac{\text{mole fraction}}{\text{mole fraction}}$	Phenol K Values: $\frac{\text{mg/liter}}{\text{mg/liter}}$
Diethylketone	3.2	556	94.5
Isoamyl alcohol	2.4	225	37.2
n-Butyl acetate	1.2	525	71.
Methylcyclohexanol	1.0	353	51.6
Isopropyl ether	0.9	227	29.
1, 2-Dichloroethane	0.82	19.3	4.38
n-Hexanol	0.56	345	49.6
Tetrachloroethane	0.32	16.2	2.76
Benzene	0.178	11.4	2.3
Toluene	0.05	11.8	1.97
Trichloroethylene	0.1	5.2	1.03
Carbontetrachloride	0.083	2.6	0.477
n-Octanol	0.054	347	39.6
Pentachloroethane	0.05	7.7	1.15
n-Decanol	0.02	299	27.4
m-Xylene	0.02	10.48	1.53
Tetrachloroethylene	0.015	2.5	0.405
Cyclohexane	0.0055	0.96	0.159
n-Hexane	0.00095	0.96	0.132

[a]The activity coefficient of phenol in water is approximately 50.

The most attractive solvents are hydrocarbons (especially aromatic hydrocarbons) and polar but sparingly soluble organics such as higher alcohols, esters and ethers, and a number of commercial solvents of unpublished composition. An interesting possibility is to use mixtures of such solvents in which the attempt is made to retain a high partition coefficient but reduce solubility.

Reviews of literature data on partition equilibrium of solvents between water and organic solvents relevant to solvent extraction have been given in the previous references cited. Data for mixed or dual-solvent systems have been given by Mackay and Medir [7], and correlations for activity coefficient with molecular properties have been reported by Kiezyk and Mackay [8].

3. PROCESS CONFIGURATIONS

Several process configurations are possible in which the water is subjected to extraction and solvent removal and the solvent stripped of solute prior to recycle. The sequence and nature of the separation process depends on the physico-chemical properties of the solute–solvent–water system. The simplest configuration is that illustrated in Fig. 1, in which several possibilities exist for each operation, as is discussed in the paragraphs that follow.

3.1. Solvent Extraction

A conventional solvent extraction unit is used with the denser fluid (usually water) introduced at the top and with the water phase usually dispersed in order to reduce backmixing, assuming that a column contactor is employed. Conventional design procedures are applicable, although it should be recognized that such units employ

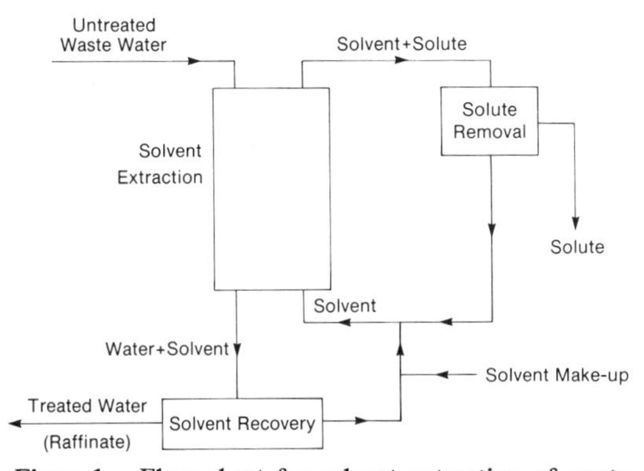

Figure 1. Flow sheet for solvent extraction of waste water.

unusually large water:solvent ratios. The hydraulics, mass-transfer characteristics, and stage design procedures are described elswhere in the handbook. It can be assumed that the exit water, depleted of solute, is saturated with solvent and that the exit solvent is saturated with water.

3.2. Solute Recovery

Several methods are available, including a second solvent extraction, distillation, chemical reaction, or precipitation. Solvent extraction is used for the removal of phenols from aromatic solvents using aqueous alkaline solutions as solvent. Distillation or evaporation is normally employed to separate a volatile solvent (e.g., a low-molecular-weight hydrocarbon) from a less volatile solute. If the solute were volatile, it would probably be removed from the waste water by gas stripping rather than by solvent extraction.

3.3. Solvent Recovery

Economic considerations and effluent restrictions usually dictate that the solvent be removed to some extent from the effluent water. The options for this separation include gas stripping, adsorption, and a second solvent extraction. Gas stripping, although possibly expensive in energy consumption, may be advantageous since the characteristics of a solvent that result in low solubility (i.e., high activity coefficient) also induce high relative volatility with respect to water. This operation is particularly attractive if it can be accomplished at near-ambient temperatures without appreciable heating of the water. This may involve low-pressure operation and selection of a volatile solvent. The commonest removal process (as distinct from recovery) is biotreatment, a process that is often necessary in any event to treat other plant effluents. Solvent extraction can thus be regarded as a companion to biotreatment in that solvent extraction removes and recovers high concentrations of solute that have economic value and could interfere with biotreatment and leaves a low residual concentration for biotreatment that cannot be recovered economically and is less toxic. In such cases it is debatable whether solvent extraction is an effluent treatment or an in-plant treatment.

3.4. Dual-Solvent Systems

Earhart et al. [5] have suggested novel configurations (Figs. 2 and 3) in which a volatile solvent such as isobutane is used in conjunction with the principal polar extracting solvent, such as an

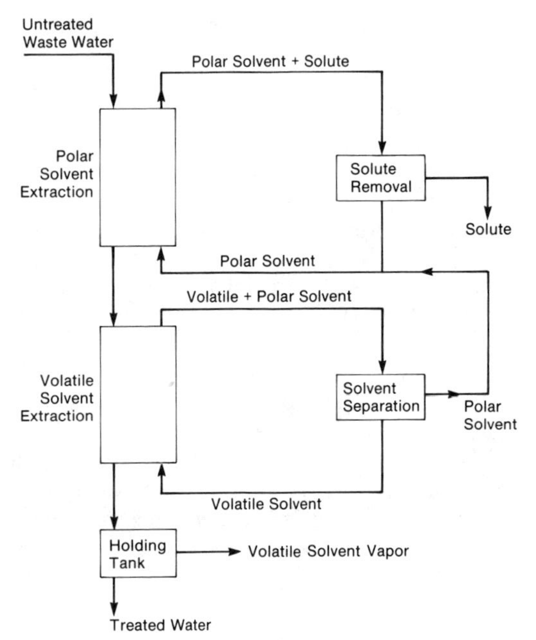

Figure 2. Dual-solvent flow sheet with separate solvent cycles.

ester or an ether. Figure 2 is essentially a modification of the process illustrated in Fig. 1, in which the solvent removal stage is a second solvent extraction with a volatile solvent. The volatile and polar solvent extract mixture is then separated (probably by distillation) and the individual streams recycled. The volatile solvent is readily removed from the final water effluent as vapor. The second approach (Fig. 3) is to combine the extraction columns, feeding the polar

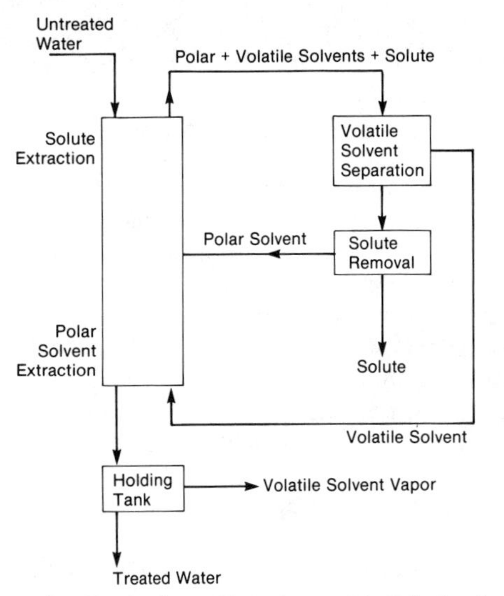

Figure 3. Dual-solvent flow sheet with linked solvent cycles.

solvent to the middle and the volatile solvent to the bottom. The advantage is that only one column is required instead of two, but the presence of the volatile solvent in the upper half of the column serves to dilute the polar solvent and reduce the partition coefficient.

A third option is to use a mixture of a volatile and a polar solvent which has a sufficiently high partition coefficient but has a low solubility in water. The properties of some mixed solvents have been reported by Mackay and Medir [7].

4. PROCESSES

Solvent extraction, as an industrial operation, has been applied principally to the recovery of phenolic compounds from water effluents since World War II. Removal of other organic compounds by solvent extraction has been carried out only to a limited extent, usually on a pilot or small scale.

4.1. Phenol Recovery Processes

Phenols occur as major contaminants in the effluent waters of industrial processes such as oil refining, coke and coal processing, and phenolic resin manufacture and in several other chemical and metallurgical operations. This has led to the development of several phenol recovery processes by use of solvent extraction. The treatment of these water effluents by solvent extraction alone seldom yields a phenol content in the extracted water sufficiently low to satisfy discharge levels established by environmental regulatory agencies; thus this method can be regarded as a first treatment for concentrated phenolic wastes prior to biological oxidation.

Solvent extraction is usually applicable to effluents exceeding approximately 0.2 m^3/min and phenol concentrations exceeding 0.1 g/liter. The configuration of the solvent extraction process depends on the origin of the effluent to be treated and the solvent used. Commonly used solvents are crude oil, light oil, benzene, toluene, and Benzol; and more selective solvents are isopropyl ether, tricresylphosphate, methyl isobutyl ketone, butyl acetate, and methylene chloride. Descriptions of some of the more important processes are given in the following paragraphs.

4.1.1. *Benzene-Caustic Process* [9]

Phenols are extracted from the effluent water with benzene, and the extract is then treated with the caustic soda solution (aqueous sodium hydroxide) that converts the phenols to sodium phenolates. With the use of steam, carbon dioxide, or other chemicals, the phenols are separated from the aqueous phase. The equipment used for the extraction stage is usually (1) a countercurrent flow packed column in which phenol recovery efficiencies are 92–93%, (2) a Podbielniak centrifuge with an efficiency of 95%, or (3) a pulsed packed column with an efficiency of 98–99%. The benzene–caustic process has been extensively used in the dephenolization of coke oven effluents, which contain generally 1–3 g/liter of phenols. Its application to dephenolization of effluents from low-temperature carbonization of brown coal and hydrogenation plants presents problems due to the large quantities of multivalent phenols, chiefly pyrocatechols, that are not removed because of their low partition coefficient into the benzene [10].

4.1.2. *Tricresylphosphate Process*

This process was developed by IG Farbenindustrie AG for the dephenolization of brown coal low-temperature carbonization and hydrogenation effluents. Tricresylphosphate (Triphos) is a high-boiling point solvent that permits the recovery of phenol by vacuum distillation with a very high purity. However, the multivalent phenols contained in the effluent tend to build up in the recirculating solvent, causing regeneration difficulties and increasing the cost of the operation [10, 11].

4.1.3. *Phenosolvan Process*

This process developed by Lurgi [12–14] initially used butyl acetate as solvent, but more recently isopropyl ether has been used. Figure 4 is a flow diagram of the process. The phenolic effluent, after removal of suspended matter by filtration and cooling, is treated with the solvent in a multistage countercurrent extractor, which is often a train of either horizontal or vertical mixer–settlers (*A*). The extract is separated into crude phenol and phenol-free solvent in a distillation column (*B*), and the solvent is recycled to the extractor. The extracted water, which is free from phenols but still contains some solvent, is preheated and freed of solvent in a desorption column (*C*) by circulating stripping gas. The solvent is recovered from the stripping gas by washing with crude phenol (*D*) and passed to the extract distillation column (*B*). The crude phenol can be sold or processed to commercial-grade pure product.

Isopropyl ether has a lower partition coefficient ($K_M = 20$) for phenol than butyl acetate ($K_M = 50$), thus requiring a greater number of extraction stages. However, it is not subject to hydrolysis, and the separation of solvent and phenol by distillation is easier and less costly because of the low solvent boiling point. Peroxide formation, which is the principal hazard encountered with ethers, is prevented because phenols act as oxidation inhibitiors and the plant operates under an inert atmosphere.

The residual phenol content in the dephenolized effluent is usually in the range 5–20 mg/liter. Steam volatile phenols are typically extracted to the extent of 99%, and other organic substances (multivalent phenols, neutral oils, or pyridine bases) are removed to the extent of approximately 70%

The Phenosolvan process is widely used in dephenolization of effluents from plants involving phenol synthesis, coke ovens, coal gasification, low-pressure carbonization, and plastics manufacture.

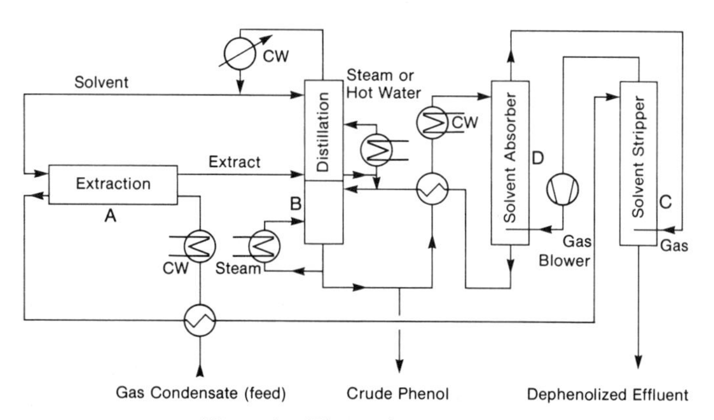

Figure 4. Phenosolvan process.

4.1.4. *Phenex Process*

This process was developed by the Humble Oil and Refining Company and is offered by Howe-Baker Engineers, Inc. [15].

Catalytic cracking operations produce significant amounts of phenol compounds, and the condensate distillate water often contains more than 300 mg/liter of phenols. A light catalytic cycle oil available from the same process is used to extract the phenols from the water.

The distillate water and the cycle oil are passed through a "butterfly" mixing valve to assure good phase contact and mass transfer to an electrostatic treating drum operating at 20,000–40,000-V potential between the electrodes inside the drum. The water/oil interface is controlled externally and allows a continuous flow of the treated water to the refinery's drainage system. The catalytic cycle oil flows overhead to treating facilities.

Phenol recovery ranges from 75% to over 90% in a single-stage extraction process for 5 : 10 oil: water volume ratios. The cycle oil also removes much of the oil present in the distillate water. The phenols present in the cycle oil after extraction act as oxidation inhibitors, improving color stability and reducing sediment formation, which is a side benefit since cycle oil is blended into distillate fuels.

4.1.5. *Barret Process*

Waste waters from tar distillation plants are dephenolized by extracting with a high-specific-gravity solvent in a rotating-disk contactor (RDC). The solvent is extracted with sodium hydroxide and the phenol is removed as sodium phenolate. The solvent has low solubility in water, limited volatility, low freezing point, and high partition coefficient and is effective in processing waste phenolic liquors with a pH as high as 10. The process is claimed to produce concentrations as low as 1.6 mg/liter [6].

4.1.6. *Ifawol Process*

A high-boiling-point solvent is used in a countercurrent packed column. Phenols are recovered from the solvent by vacuum distillation. The removal efficiency is approximately 99% [9].

4.1.7. *Holley–Mott Process*

A mixer–settler is used to contact coke oven gas liquor and the solvent. The solvents employed are benzene or light creosote oil fortified with tar bases such as quinoline and isoquinoline to raise the partition coefficient [16].

4.1.8. *Distillers Company Ltd. Process*

In the cumene process for the manufacture of phenols, there can be substantial losses of phenol in waste waters. A mixture of 80% cumene and 20% mesityl oxide (a by-product of the process) is used for the extraction of phenols from the effluent waters [17].

4.2. Other Recovery Processes

These include extraction of salicyclic and other hydroxyaromatic acids from waste water with methyl isobutyl ketone as solvent [18], extraction of thiazole-based chemicals from a rubber processing effluent with benzene [18], and recovery of fatty acids from effluent waters produced in wool scouring. The effluent waters contain a wool grease that is separated by addition of acid. The grease is then extracted with an alcohol, to obtain the free fatty acids [19].

A process for deoiling quench water has been developed by Gulf Oil Corporation, in which the oil quench water containing about 6000 mg/liter of dissolved and emulsified oil is extracted with a light aromatic oil and the extract recycled for refinery processing. The water is treated to coalesce any remaining oil in a sand filter. It is then sufficiently clean for plant reuse, containing about 20 mg/liter of oil [1].

Acetic acid recovery has been described with the use of solvent extraction in neutral sulfite semichemical pulping plants and cellulose acetate plants. Ketones, ethers, and esters have been found to be effective solvents [1] (see also chapter 18.5).

4.3. Processes Under Development

Several research efforts in the field of solvent extraction for recovery of organic pollutants from industrial waste waters have been reported.

Earhart et al. [5, 20] have investigated the recovery of organic pollutants from industrial waste waters by use of volatile solvents such as isobutane or isobutylene with extraction under pressure. The use of a volatile solvent in combination with a conventional polar solvent has been studied in process configurations with both separate and linked solvent cycles, as illustrated in Figs. 2 and 3 and as described earlier. Experiments with waste waters from petroleum refining and petrochemical manufacture plants have shown that the dual-solvent process may be particularly attractive in cases where phenolic substances are major contaminants.

Union Carbide Corporation has investigated the potential for the extraction of caprolactam, acrylonitrile and acetonitrile from manufacturing plant waste waters [4].

Investigations are reported to be under way [4] on the removal of caprolactam, acrylonitrile, and acetonitrile from chemical plant waste waters, on the removal of lower-molecular-weight halogenated hydrocarbons (by use of a hydrocarbon), sulfonic acid dyes (by use of an amine), and carboxylic acid and alkyl amines.

In general, solvent extraction has greatest potential applicability when the organic solute has a low solubility in water, partitions readily into a solvent of low water solubility, and is unusually noxious or valuable. Since many of the organic compounds of greatest environmental concern are so because of their tendency to partition into and accumulate in the lipid phases of biota, it seems likely that solvent extraction, which operates industrially on the same partitioning principle, may find wider future application in the treatment of water effluents containing such compounds.

REFERENCES

1. P. R. Keizyk and D. Mackay, *Can. J. Chem. Eng.* **49**, 747 (1971).
2. C. Hanson, Solvent Extraction, *Chem. Eng.* **75** (18), 76 (1968).
3. D. Mackay and M. Medir, in *Proceedings of the International Solvent Extraction Conference, (ISEC), Toronto, 1977,* Vol. 2, Canadian Institue of Mining and Metallurgy, Montreal, 1979, p. 791.
4. A. D. Little, Inc., *Physical, Chemical and Biological Treatment Techniques for Industrial Waste,* report to U.S. EPA Office of Solid Waste Management, Washington, DC, 1977, No. C-78950.
5. J. P. Earhart, K. W. Won, C. J. King, and M. M. Prausnitz, *Extraction of Chemical Pollutants from Industrial Waste Waters with Volatile Solvents,* USEPA-600/2-76-220, December 1976.
6. A. N. Heller, E. W. Clarke, and W. M. Reiter, *Proceedings of the Industrial Waste Conference*, Purdue University, Lafayette, IN, May 1957.
7. D. Mackay and M. Medir, *Can. J. Chem. Eng.* **53**, 274 (1975).
8. P. R. Keizyk and D. Mackay, *Can. J. Chem. Eng.* **51**, 741 (1973).
9. K. M. Lanouette, *Chem. Eng.* (Deskbook issue), **84** (22), 99 (1977).
10. R. Jauernik, *Erdol und Kohle* **13**, 252–275 (1960).
11. W. W. Hodge, *Industrial Waste,* American Chemical Society, Monograph Series No. 118, Reinhold, New York, Chapter 14.
12. F. Wohler, *Removal and Recovery of Phenol and Ammonia from Gas Liquor,* 1978, Südafrika, England, Lurgi, Kohle and Mineraloltechnik GmbH.
13. R. Jauernik (transl.), *Erdol und Kohle* p. 252–277 (1960).
14. Upgrading of Solid Fuels, Handbook, 1971, Lurgi Kohle Mineraloltechnik GmbH.
15. W. L. Lewis and W. L. Martin, *Hydrocarb. Process.* **46** (2), 131 (1967).
16. D. G. Murdoch and M. Cuckney, *Transact. Inst. Chem. Eng.* **24**, 90 (1946).
17. P. A. Witt Jr., and M. C. Forbes, *Chem. Eng. Progr.* **67** (10), (1971).
18. *Petrochemical Effluents Treatment Practices,* report by the Federal Water Pollution Control Administration, U.S. Department of the Interior, February 1970, Report No. 120120.
19. J. Coleby, Industrial Organic Processes, in C. Hanson, Ed., *Recent Advances in Liquid-Liquid Extraction,* Pergamon Press, Oxford, 1971, Chapter 4.
20. J. P. Earhart, K. W. Won, M. Y. Wong, J. M. Prausnitz, and C. J. King, *Chem. Eng. Prog.* **73** (5), 67 (1977).
21. D. B. Todd, and C. A. Hopper, paper presented at the AIChE Meeting, Houston, TX, March 1971.

Part IIIB

Industrial Processes—Inorganic

24

COMMERCIAL SOLVENT SYSTEMS FOR INORGANIC PROCESSES

Douglas S. Flett and John Melling
Warren Spring Laboratory
United Kingdom

Michael Cox
Hatfield Polytechnic
United Kingdom

1. INTRODUCTION

Solvent extraction is a well-established process within the inorganic chemical and hydrometallurgical industries. The solvent, the initial organic phase, is usually a mixture of components. The extractant is commonly mixed with a diluent, mainly to confer appropriate flow properties on the resultant solvent. Other components may be added such as modifiers to alter physical characteristics, synergists to increase extraction, or catalysts to increase the extraction rate. It is now well known that diluents and modifiers are not neutral in their effects on equilibria and kinetics; these factors were discussed in Chapter 2.2.

Examples of extractants are listed in Table 1, with emphasis on those in commercial use. The classification is that employed in Chapter 2.2. Apart from the carboxylic acids and acid chelating extractants, many of the extractants listed (and a large number of similar compounds) were evaluated in the late 1940s and early 1950s for solvent extraction processes required by the nuclear programme. A history of the development of reagents for nuclear fuel reprocessing and uranium extraction from leach liquors has been given recently by Coleman [1]. It is fascinating to find that tributyl phosphate (TBP), now so widely used in nuclear fuel reprocessing, was rejected originally for this application because of the novelty of using an admixture with a hydrocarbon diluent. Later, in the early and mid-1960s, a number of chelating extractants were produced. The LIX series of oxime extractants from General Mills Chemicals, Inc. (now Henkel Corporation) were the first in the field. Their development and that of oxime extractants by other companies have been described by Flett [2] and Swanson [3]. The Kelex extractants based on 8-hydroxyquinoline were developed by Ashland Chemical Company (now Sherex Chemical Company) almost contemporaneously with the LIX oximes. In the 1970s chelating extractants with diketone and quinoline sulfonamide groupings have been produced, but commercial exploitation of these is still at an early stage. Carboxylic acids, originally proposed as cheaper alternatives to alkylphosphoric acids, have never achieved widespread commercial use.

In the early years of commercial application, fuel kerosenes were used as diluents, but following better understanding of the role of the diluent in the overall process and an appreciation of the potential of the market, specialist products have become available from the major oil companies. A variety of modifiers, synergists, and catalysts are also available.

In the face of this rather bewildering array of solvent components, how is the choice of optimum solvent to be made? A large number of factors—both chemical and physical, technical and economic—must be considered. These include solvent strength, selectivity, loading, ease of stripping, rates of extraction and stripping, chemical stability, aqueous-phase solubility of solvent components, organic-phase density and viscosity, interfacial properties of the solvent–aqueous system, volatility and flammability of the solvent, toxicity of the solvent within the working area and the outside environment, and solvent cost. The choice of solvent components favoring any of the properties will frequently result in a poorer performance in other respects; thus compromises must be made if the optimum solvent for a specific application is to be produced.

2. SELECTION OF PROCESS OPTION

When considering potential solvent extraction processes, the use of commonly available extractants should be considered first. Obviously, comparison of requirements against existing processes will be of considerable assistance. A thorough knowledge of the chemistry of both the aqueous phase to be treated and the various solvent systems available is essential. The chemistry of the solvent extraction of reactive solutes was considered in Chapter 2. More detailed information regarding the extraction properties and behavior of specific compounds available commercially is usually scattered through the literature, and direct comparisons of reagent performance are the exception rather than the rule. Specific data on particular compounds are usually available from the manufacturers and are application orientated. Literature data, on the other hand, may well be rather academic or of relevance in the area of analytic application. Several commercial extractants have been based on successful analytic solvent extraction reagents, including Kelex and the many oxime reagents related to salicylaldoxime.

To combat problems of data availability, Warren Spring Laboratory and the Mineral Industry Research Organisation have produced a Solvent Extraction Data Service that provides indexed tabulations of published distribution data for most metals and phosphoric acid, with all the currently available commercial extractants. This source permits ready selection of solvents for potential process requirements and definition of parameters needed for process operation and plant design. Much data on the physical properties of reagents and diluents are also included.

The major criticism of literature data is that, whereas extraction behavior and chemistry have usually been studied in depth, in many cases little or no information is given with respect to stripping behavior.

2.1. Interface Between Solvent Extraction and Other Unit Processes

The solvent extraction step must interface with both upstream and downstream processes, which can impose constraints.

TABLE 1 SOLVENT EXTRACTION REAGENTS FOR HYDROMETALLURGY

Class	Type	Examples	Manufacturers	Commercial Uses
Acid extractants	Carboxylic acids	Naphthenic acids	Shell Chemical Co.	Copper–nickel separation
	Alkyl phosphoric acids	Di-2-ethylhexylphosphoric acid (D2EHPA)	Union Carbide	Yttrium recovery, europium extraction, nickel–cobalt separation
		Octylphenylphosphoric acid (OPPA)	Mobil Oil Co.	Uranium extraction
	Aryl sulfonic acids	SYNEX 1051	King Industries, Inc.	Magnesium extraction
Acid chelating extractants	Hydroxyoximes	LIX63, LIX64N, LIX65N, LIX70	Henkel Corporation	Copper and nickel extraction
		SME 529	Shell Chemical Co.	Copper extraction
		P5000 series	Acorga Ltd.	Copper extraction
	Oxine derivatives	Kelex 100	Sherex Chemical Co.[a]	Proposed for copper extraction
	β-Diketones	Hostarex DK16	Farbwerke Hoechst AG	Proposed for copper extraction from ammoniacal solution
		LIX54	Henkel Corporation	Copper extraction from ammoniacal extraction
		XI51	Henkel Corporation	Proposed for cobalt extraction from ammoniacal solution
	Alkarylsulfonamide	LIX34	Henkel Corporation	Proposed for copper extraction from acidic leach liquors
	Polyols		Dow Chemical Co.	Boron extraction
Anion exchangers	Primary amines	Primene JMT	Rohm and Haas	No known commercial use
	Secondary amines	LA-2	Rohm and Haas	Zinc and uranium extraction
		Adogen 283	Sherex Chemical Co.[a]	Zinc and tungsten extraction
	Tertiary amines	Various Alamines; in particular Alamine 336	Henkel Corporation	Widely used; cobalt, tungsten, vanadium, uranium extractions etc.
		Various Adogens; in particular Adogen 364, Adogen 381, Adogen 382	Sherex Chemical Co.[a]	Cobalt, vanadium, and uranium extractions
	Quaternary amines	Aliquat 336	Henkel Corporation	Vanadium extraction; other possible uses are chromium, tungsten, and uranium extraction
		Adogen 464	Sherex Chemical Co.[a]	Similar to Aliquat 336
Solvating extractants	Phosphoric, phosphonic, and phosphinic acid esters	Tributyl phosphate (TBP)	Union Carbide, Albright and Wilson	Nuclear fuel reprocessing, U_3O_8 refining, iron extraction, zirconium–hafnium separation, niobium–tantalum separation, rare earth separations, acid extraction
		Phosphonic acid esters	Farbwerke Hoechst AG	No known commercial use
		Trioctylphosphine oxide (TOPO)	Henkel Corporation Cyanamid	Recovery of uranium from wet process phosphoric acid liquors (with D2EHPA)
	Various alcohols, ethers, ketones	Butanol–pentanol	Various	Phosphoric acid extraction
		Diisopropyl ether	Various	Phosphoric acid extraction
		Methyl isobutyl ketone (MIBK)	Various	Niobium-tantalum separation, zirconium-hafnium separation
	Alkyl sulfides	Di-*n*-hexyl sulfide		Palladium extraction

[a]Previously Ashland Chemical Co.

2.1.1. *Feed Solution*

Little quantitative information is available on the allowable solids content of the aqueous feed to solvent extraction, but less than 10 ppm of solids is recommended in copper extraction from acid sulfate solution [4], less than 20 ppm of solids for South African Purlex plants using amines [5], and up to 300 ppm for uranium plants using alkyl phosphates [6].

Organic molecules may also be present in the aqueous feed solution to solvent extraction as a result of either the presence of organic molecules in the original arising (common if the solid is mined from a surface deposit), or by deliberate addition of surface-active agents (surfactants) such as flotation or flocculating or electroplating agents. The effects of these may be quite drastic, but unfortunately, very little quantitative information is available.

It is important that the feed solutions do not cause excessive loss of extractive capacity in the solvent by chemical degradation. Particular problems in this area are discussed below.

2.1.2. *Raffinate Solution*

In a number of processes, such as copper recovery by leach–solvent extraction–electrowinning, the raffinate from extraction is returned to leaching. It will contain organic reagents, both soluble and entrained. Efforts are made to recover this organic material, but any not recovered may be deleterious to the leaching circuit. For example, the organic reagents may be toxic to the bacteria involved in a bacterial leach, although fears on this score have yet to be substantiated. If the raffinate is returned to leaching, it may not always be desirable to reduce the solute value to very low levels during solvent extraction. An example is the recovery of copper from scrap by ammonia leaching [7].

The presence of organic phase in a raffinate solution that is returned to the countercurrent decantation section of a plant causes problems in this section by absorption on the solid particles and subsequent flotation. This increases the size of plant required and usually results in a higher solids content in the decanted product [8].

2.1.3. *Stripping and Other Extraction Processes*

In many processes stripping of metal values from the loaded solvent simultaneously converts the solvent to the form required for extraction, but this is not always the case, especially with anion exchangers.

It is not common in commercial practice for aqueous solutions from one solvent extraction circuit to be further processed by use of a second solvent. Special problems are posed here, as any cross-contamination will result in a mixed organic phase and the properties of the two extractants may oppose one another. One process that has operated successfully with TBP followed subsequently by tertiary amine extraction is the Falconbridge Matte Leach process [9]. The TBP is claimed not to be harmful to the amine extraction. This is perhaps not surprising as TBP can be used as a modifier for amine extractants. Two new plants use amine extraction followed by di-2-ethylhexylphosphoric acid (D2EHPA) extraction: the Zincex process [10] for zinc and a plant at Moab, Utah, where Atlas Mineral are recovering uranium with tertiary amine followed by vanadium recovery with D2EHPA [11]. The consequences of cross-contamination in these plants could be very serious, and it is not clear what steps are taken to prevent it.

2.1.4. *Solute Recovery Processes*

In some cases, such as acid extraction, the solution produced is a satisfactory product. In general, however, further processing is required to convert the solute to saleable solid form as metal, oxide, or salt. A particularly interesting method for the recovery of metal value direct from the loaded organic phase with consequent regeneration of the solvent is reduction to metal with hydrogen in an autoclave [12, 13]. Thus far, this has been carried out only on the laboratory scale. Oxime extractants are degraded by the treatment, but carboxylic acids are suitable extractants for the process.

Electrowinning as part of a flow sheet involving solvent extraction is mainly of importance in the copper industry, and to some extent for nickel and zinc. With nickel and zinc the stripping conditions are arranged to give electrolytes very similar to those used in conventional electrowinning. With copper there has been sufficient incentive to adapt the conventional process to high acid strip solutions as described in Chapter 25.1.

2.1.5. *Effluent Treatment*

The presence of organic reagents in aqueous effluents can cause a further problem with solvent extraction. It is usually necessary to install equipment to recover organic solvent for reuse in order to minimize processing costs. Traces of organic may still remain and must be removed in accordance with local legislation.

Liming plants are effective for the removal of organic reagents by adsorption on the precipitates. Their success in the removal of tertiary amines from uranium plant raffinates has been demonstrated [14]. Liming is a common method of effluent treatment, and it would seem likely that, where it is employed, organic solvents will be removed at the same time. However, liming is essentially a method of removing heavy metals from solution and hence is not applicable to all effluents.

The effects of organic solvent on the environment is a matter of considerable importance, but there have been few definitive studies in this field.

2.1.6. Construction Materials

The use of both organic and aggressive aqueous phases in the same equipment can pose difficulties in materials selection, but it is unlikely that solvent choice will be influenced by materials problems.

3. SELECTION OF SOLVENTS

3.1. Structures of Extractants, Diluents, and Modifiers

The structures and some physical properties of the common commercially available extractants are shown in Tables 2–5. The formulation and physical properties of some proprietary diluents are given in Table 6 and the properties of common modifiers, in Table 7. Compounds used as synergists or catalysts are included in Tables 2–5. The proprietary extractants are freely available throughout the Western world. However, the availability of some of the diluents listed is limited geographically as they are products of particular refineries; thus diluent availability in the region where a process is to operate must be checked before a solvent evaluation program is implemented.

For the extractants, the molar costs of active reagent can be determined as a basis for comparison from the molecular weights and purities plus the price. Impurities present may remain from the synthesis and can be deleterious to the process. Diluents are often added to lower the pour point, whereas other additives may be purposely included in the formulation.

3.2. Chemistry of Extraction

3.2.1. Acid Extractants

The chemistry of these has been described in Chapter 2.2. Carboxylic acids are relatively weak extractants and are fairly soluble in the aqueous phase at acidities where significant extraction occurs. The sulfonic acids are much stronger extractants. A major problem is their very high surface activity, which can lead to difficulties in phase separation. Di-2-ethylhexylphosphoric acid (D2EHPA) was selected from a number of phosphoric acids for use in the DAPEX process for the extraction of uranium from leach liquors. Subsequently, a number of other applications including rare earth separations, zinc extraction and cobalt–nickel separation have resulted. A particularly interesting aspect of the latter application is the remarkable increase in separation factor on raising the temperature of the system from ambient to 50°C or above [15]. A major current use for this extractant is the extraction of traces of uranium from phosphoric acid liquors in an admixture with trioctylphosphine oxide (TOPO). Octylphenylphosphoric acid (OPPA) is a competitor in this application, but the process using D2EHPA/TOPO for extraction from a previously oxidized feed liquor followed by reductive stripping is currently preferred to the treatment of the original liquor with OPPA followed by oxidative stripping [16]. This appears to be due in part to doubts about the solvent solubility and stability in contrast to that of D2EHPA/TOPO. In a number of applications D2EHPA is used as the sodium salt to simplify pH control. In these cases, a modifier must be present to maintain a single organic phase when converting the free-acid form, from stripping, to the sodium form.

3.2.2. Acid Chelating Extractants

The development of commercial acid chelating extractants has stemmed from the need to provide alternative process options for the treatment of dilute copper bearing acidic leach liquors. Fewer complications are found in the extraction chemistry than with the acid extractants. The hydroxyoximes extract copper selectively from dilute acid leach liquors; however, they also extract other metal ions. Ferric iron is extracted, but at the pH values of the leach liquors (~2.5) the separation factor is about 200. The only metal ion likely to interfere with copper under

TABLE 2 STRUCTURE AND PROPERTIES OF ACIDIC EXTRACTANTS

Name	Formula	Molecular Weight of Active Extractant	As Received Extractant		
			Active Extractant, wt. %	Specific Gravity	Flash Point, °C
Versatic 10	$R_1-C(R_2)(R_3)-COOH$; $R_1 + R_2 + R_3 = C_8$	175	99.6	0.91	129
Di-2-ethylhexyl-phosphoric acid	$(C_4H_9CH(C_2H_5)CH_2O)_2POOH$	322	100	0.98	
Octylphenyl-phosphoric acid	$ROPO(OH)_2 + (RO)_2POOH$; $R = (CH_3)_3CCH_2C(CH_3)_2C_6H_5$	–	–	–	
SYNEX 1051	Dinonylnaphthalenesulfonic acid structure (SO_3H, two R groups); $R = C_9H_{19}$	458	50	0.92	

these conditions is molybdenum, and problems have been encountered at a plant in South America, although treatment of the organic phase with alkali or ammonia can remove the molybdenum. Under ammoniacal conditions, however, both copper and nickel are easily extracted, as is cobalt(II); although cobalt tends to oxidize to the cobalt(III) state, which cannot easily be stripped from the organic phase. Cobalt(III), if present in the aqueous feed, is nonextractable. Copper–nickel separations from ammoniacal solution are possible, although the preferred technique is one of total extraction and selective stripping [17]. Thus the hydroxyoximes are quite versatile.

As seen from Table 3, there are a variety of oximes to choose from. The chloro compound, LIX70, is the most powerful extractant, but as a consequence, a much stronger acid strip solution is required for stripping. Interfacing directly with an electrowinning plant for copper thus poses problems as there are limits to the degree of acidity that can be tolerated. The chloro compound LIX70 would be very difficult to operate in circuit with an electrowinning plant. Although LIX65N has the right selectivity properties required for copper extraction, its rate of extraction is too slow in acid conditions. Addition of LIX63 accelerates copper extraction without prejudicing selectivity. As the extraction rate increases with increasing pH, the need for LIX63 addition does not arise when treatment of ammoniacal solution is contemplated. The compound SME529 has an extraction rate similar to that of LIX64N but, because of its lower molecular weight, SME529 will extract more copper per mole than will LIX64N. The aromatic β-hydroxyoxime P1 is a very fast, quite strong extractant. The P5000 series was formulated by addition of nonyl phenol, an impurity present in all the oximes from the synthesis process, which interacts with the free oxime reducing the extractant strength and improving the strip performance.

In ammoniacal solution the hydroxyoximes tend to extract ammonia, although this ammonia is transferred to the aqueous phase on metal loading. Nevertheless, ammoniacal circuits must make provision for washing the metal-loaded organic phase to avoid ammonia carry-over into the strip circuit [18].

Kelex 100 will extract a much larger number of aqueous species than the oxime extractants. Although developed specifically to compete with the hydroxyoxime reagents for copper extraction, it suffers from the drawback that the quinoline nitrogen is much more basic than the oximic nitrogen. This results in extraction of acid into the strip circuit that severely affects the overall acid balance. The selectivity for copper over ferric iron is kinetic in nature for steric reasons, and the same cobalt oxidation problem exists as with hydroxyoximes. The acid uptake property of this reagent has prevented commercial application and it is unlikely that it will ever be used for copper extraction. However, its use on a small scale for the recovery of arsenic and gallium is expected.

For the β-diketone reagents there is an obvious difference in pH functionality between the fluorinated and nonfluorinated compounds. The cobalt oxidation problem also occurs with these reagents. They are much weaker than the hydroxyoximes and hence will not compete with the hydroxyoximes for metal extraction from acidic liquors. They are aimed, in fact, at the treatment of copper containing ammonical liquors as they extract less ammonia than do the hydroxyoximes. The first commercial use of LIX54 may be in the treatment of copper drosses from the Imperial Smelting process, for which there are three plants under construction or planned [19]. In contrast to copper extraction, ammonia is coextracted with nickel in a 1 : 1 ammonia : nickel mole ratio [18], and this would appear to limit the usefulness of LIX54 to nickel-free solutions.

The newest chelating extractant is LIX34. The behavior of this quinoline sulfonamide in copper extraction is similar to that of LIX64N, but it does not extract ferric iron at all. It is considered to be only a matter of time before this reagent finds commercial application. The extractant LIX34 also extracts nickel, cobalt, zinc, cadmium, and lead, but at pH values much higher than copper. Thus it may have a versatile future as an extractant for nonferrous metals. Polyols have found a very specific application to the extraction of boron from brines.

3.2.3. *Anion Exchangers*

For metal extraction, consideration must be given to anionic metal complexes formed between the metal cation and either the anion already present (sulfate, chloride, nitrate, etc.) or water-soluble complexing agents deliberately added to the system. To optimize such systems, it is necessary to know which of the range of complexes formed in the aqueous phase is extracted. Generally, if the ligand concentration is increased, the degree of extraction is also increased, although maxima may appear in plots of degree of extraction versus ligand concentration as a result of competition with acidic species or formation of inextractable or less extractable multivalent species (see Chapter 2.2.) Increasing acid concentration almost always results in competition between the metal complex and the acid for the amine. As a general rule, the relative order of metal complex extraction from chloride solution is primary amines < secondary < tertiary, whereas the reverse order is true for sulfate complexes. Selectivity can be varied by the concentration and type of the complexing anion. Clearly, some metal ions may exist in solution partly as cationic and partly as anionic species, and the relative amounts are determined by the stability constants of the complexing or ion pair reaction and the concentration of the ligand. The uranyl ion is a good example, as from sulfate solution it is possible to extract species such as $UO_2(SO_4)_2^{2-}$ with amines or the uranyl cation with acidic extractants such as D2EHPA.

To achieve a high solvency of the amine in the organic phase and prevent third-phase formation in alkaline stripping, a modifier is usually added to the solvent. Alcohol modifiers inhibit vanadium extraction, and so Adogen 382, which has excellent solvency in kerosene diluents, is used for vanadium extraction from acid leach liquors.

3.2.4. *Solvating Extractants*

Similar comments can be made for this class as for the alkylamines, although here neutral metal complexes rather than complex anions are extracted. The extractants are all weakly basic, so they can either solvate the central metal ion of the neutral species, or at moderate to high acidities they can solvate the proton giving rise to the extraction of simple or complex acids. Where phosphorus-containing compounds are involved, the extractability increases with the number of C–P bonds in the extractant, specifically, phosphine oxides > phosphonates > phosphates. Tributyl phosphate is the most widely used solvating extractant. It is preferred over amine extractants in processes for recovering acids from pickling liquors because of higher loadings and better phase separation behavior. It is also a competitor with methyl isobutyl ketone for niobium–tantalum and zirconium–hafnium separations, as it has the advantages of lower losses and a much lower flammability.

Alcohol extractants are used for phosphoric acid from hydrochloric acid leach liquors, and

TABLE 3 STRUCTURE AND PROPERTIES OF CHELATING EXTRACTANTS

Name	Formula	Molecular Weight of Active Extractant	As Received Extractant		
			Active Extractant, wt. %	Specific Gravity	Flash Point, °C
	Aromatic β-Hydroxyoximes				
	[Structure: 2-hydroxyphenyl oxime with R_1 on C, R_2 ortho to OH, R_3 para; C=N—OH] R_1 / R_2 / R_3				
LIX65N	ϕ / H / C_9H_{19}	339	39.2	0.88	85
LIX64N	LIX65N + LIX63	339	40.1	0.88	85
LIX70	ϕ / Cl / C_9H_{19}	375	40.4	0.90	–
SME529	CH_3 / H / C_9H_{19}	276	50	0.93	74
P1	H / H / C_9H_{19}	232	–	–	–
P5100	P1 + nonyl phenol	262	47.5	0.96	78
P5300	P1 + nonyl phenol	262	25.0	0.95	96
OME	C_4–C_5 / H / C_9H_{19}	328	91.3	0.94	124
	Aliphatic α-Hydroxyoximes				
LIX63	$C_4H_9CH(C_2H_5)CH(OH)C(NOH)CH(C_2H_5)C_4H_9$	257	–	–	–
	Oxine Derivatives				
Kelex 100	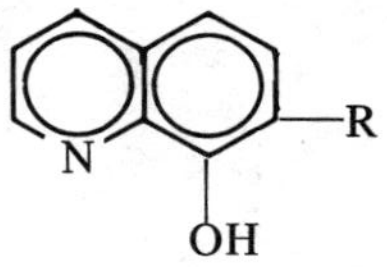 [Structure: 8-hydroxyquinoline with R at position 7] R = dodecenyl	311	74–80	0.99	–

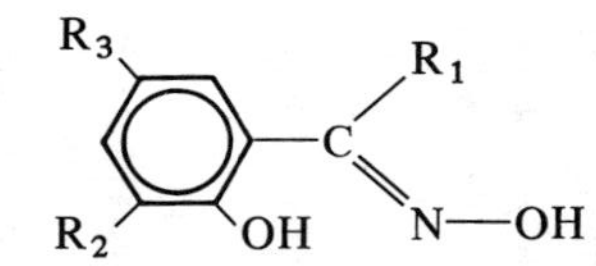

	β-Diketone Derivatives					
	R_1–C_6H_4–$COCH_2COR_2$					
	R_1	R_2				
LIX54	*p*-Dodecyl	CH_3	330	–	–	–
XI51	*p*-Dodecyl	CF_3 or C_2F_5	–	–	–	–
DK16	Various alkyl or alkenyl groups	$CH_{0-3}Cl_{0-3}F_{0-3}$	–	–	0.93	–
	Alkarylsulfonamides					
LIX34	8-quinolyl–$NHSO_2R$; R = *p*-Dodecylbenzene		438	–	–	
	Polyols					
	Phenol ring bearing OH, CH_2OH, *x* and R substituents					

TABLE 4 STRUCTURE AND PROPERTIES OF ANION EXCHANGERS

Name	Formula	Molecular Weight of Active Extractant	As Received Extractant: Active Extractant, wt. %	As Received Extractant: Specific Gravity	As Received Extractant: Flash Point, °C
	Primary Amines (RNH_2)				
Primene JMT	$R = (CH_3)_3C(CH_2C(CH_3)_2)_4$—	269–325	100	0.84	–
	Secondary Amines (R_2NH)				
LA-2	$R = C_{12}–C_{13}$	351–393	100	0.83	180
Adogen 283	$R = C_{13}$	385	92	0.83	–
	Tertiary Amines (R_3N)				
Adogen 364	$R = C_8–C_{10}$	~380	96.2	0.81	–
Alamine 336	$R = C_8–C_{10}$	~392	95	0.81	168
Adogen 368	$R = C_8–C_{12}$	–	95.8	0.82	–
Hostarex A327	$R = C_8–C_{12}$	~395	–	0.81	203
Adogen 381	R = isooctyl	353	95.1	0.82	–
Alamine 308	R = isooctyl	–	–	–	–
Hostarex A324	R = isooctyl	~363	–	0.81	166
Adogen 382	R = isodecyl	437	95.1	0.82	–
	Quaternary Ammonium Compounds ($R_3N(CH_3)^+Cl^-$)				
Adogen 464	$R = C_8–C_{10}$	~431	90	0.84	–
Aliquat 336	$R = C_8–C_{10}$	~442	>88	0.88	132

ether extractants for phosphoric acid from sulfuric acid solutions.

The sulfide and sulfoxide reagents are relative newcomers to this group of extractants. However, considerable Russian work has been carried out on these compounds [20]. The only likely commercial application in the near future is the use of di-*n*-hexyl sulfide for palladium recovery in platinum group metal processing in South Africa [21].

3.2.5. *Diluents*

As the viscosities of many reagents are much too high for direct use in solvent extraction equipment, it is common practice to dissolve them in an organic diluent. The diluents considered for metal and inorganic processing are usually hydrocarbons selected on the basis of a flash point above 60°C, to minimize evaporation loss and the risk of fire, and with a specific gravity of about 0.8 to aid phase separation. Chlorinated hydrocarbons, which are denser than water and dilute aqueous solutions, have been suggested for commercial use, but no applications are known. In general, the diluents used contain a mixture of paraffinic, aromatic, and naphthenic hydrocarbons. Recently the major oil companies have developed ranges of diluents with a metallurgical rather than a fuel specification specially for use in solvent extraction. The Escaid range from Exxon, MSB210 from Shell, and Chevron Ion Exchange Solvent are examples of this trend.

In metal extraction the use of aromatic diluents for extractants such as the aryl hydroxyoximes results in slower kinetics in extraction and stripping and in weaker solvents [22, 23]. However, the more aromatic the diluent, the higher the solubility of the metal complex. The properties of the diluent also influence the phase separation characteristics of the system, and a balance between the chemical and physical behavior must be attained. Test methods for the selection and characterization of diluents have been discussed recently [24].

The use of solubility parameters for correlation of solvent properties has not proved an especially useful tool for the selection of solvents for inorganic species. However, the behavior of

TABLE 5 STRUCTURE AND PROPERTIES OF SOLVATING EXTRACTANTS

Name	Formula	Molecular Weight	Specific Gravity	Flash Point, °C	Viscosity (25°C), cP
	Carbon-Oxygen-Bonded Donors				
Ethers (R_1OR_1) or ($R_2OCH_2CH_2OR_2$)					
Diisopropyl ether	$R_1 = (CH_3)_2CH$	102	0.726	-25	0.38
Dibutylcellosolve	$R_2 = C_4H_9$	174	0.837	–	1.34
Alcohols (ROH)					
n-Butanol	$R = C_4H_9$	74	0.81	32	2.46
n-Pentanol	$R = C_5H_{11}$	88	0.82	33	3.31
Ketones (R_1COR_2)					
Methyl isobutyl ketone	$R_1 = CH_3$, $R_2 = (CH_3)_2CHCH_2$	100	0.804	14	0.55
	Phosphorus-Oxygen-Bonded Donors $\mathbf{R_1R_2R_3PO}$				
Phosphoric acid esters					
Tri-*n*-butylphosphate	$R_1 = R_2 = R_3 = C_4H_9O$	266	0.97	193	3.56
Phosphonic acid esters					
Hostarex PO212	$R_1 = R_2 = C_4H_9O$, $R_3 = C_4H_9$	250	0.94	—	5
Hostarex PO224	$R_1 = R_2 = C_8H_{17}O$, $R_3 = C_8H_{17}$	418	0.91	—	16
Phosphine oxides					
Trioctylphosphine oxide	$R_1 = R_2 = R_3 = C_8H_{17}$	386		(solid)	
	Sulfur-Containing Extractants				
Sulfides (RSR)					
Dihexyl sulfide	$R = C_6H_{13}$	202	–	–	–

TABLE 6 PROPERTIES OF SOME PROPRIETARY DILUENTS FOR SOLVENT EXTRACTION

	Composition, %			Specific Gravity	Flash Point, °C	Viscosity (25°C), cP
	Paraffins	Naphthenes	Aromatics			
Amsco Odorless Mineral Spirits[a]	85	15	0	0.76	53	–
Chevron Ion-Exchange Solvent[b]	52.3	33.3	14.4	0.80	91	1.70
CycloSol 63[c]	1.5		98.5	0.89	66	–
Escaid 100[d]	80		20	0.8	78	1.52
Escaid 110[d]	99.7		0.3	0.79	74	1.52
Escaid 350 (formerly Solvesso 150)[d]	3.0	0	97.0	0.89	66	1.20
Kermac 470B (formerly Napoleum 470)[e]	48.6	39.7	11.7	0.81	79	2.10
MSB210[c]	97.5		2.5	0.78	74	–
Shell 140[c]	45	49	6.0	0.79	61	–
Shellsol R[c]	17.5		82.5	0.89	79	1.71

[a]Union Oil Co.
[b]Chevron Oil.
[c]Shell.
[d]Esso or Exxon.
[e]Kerr McGee.

Kelex 100 in various diluents has been described in terms of diluent solubility parameters [25].

3.2.6. *Modifiers and Other Additives*

Because of solubility limitations of the metal complex, a phenomenon known as *third-phase separation* can occur. Invariably it is the organic phase that splits into two to yield a metal complex-rich phase at the aqueous interface and a diluent-rich phase above. To overcome this problem, a third component, known as a *diluent modifier*, is added to restore these organic phases to a single phase. Such modifiers are invariably solvating reagents such as tributyl phosphate, isodecanol, or nonyl phenol. Addition of such modifiers can also improve the phase separation characteristics of the system, and additions for this purpose are known. Like the diluent, the modifier is not inert and will affect the equilibrium distribution of the metal species and rate of extraction.

Addition of other components to enhance equilibrium distribution (synergists) or to increase the rate of extraction (catalysts) are also fairly common. Probably the best known example of synergism is the UO_2^{2+}–D2EHPA–TBP system, whereas the best example of catalysis is the Cu^{2+}–LIX65N–LIX63 system.

3.3. Solvent Prices

The cost of solvent can form a high proportion of the capital cost of a plant, and so the price of the components will play a role in solvent choice. Up-to-data information cannot be quoted here, and the reader is recommended to obtain this from the manufacturers. However, it is perhaps worth noting that prices for a number of reagents have remained fairly constant in U.S. dollar terms over a number of years. In general, prices for as received reagents decrease in the following order: acid chelating extractants $\approx$ di-2-ethylhexylphosphoric acid $>$ basic extractants $\approx$ P-bonded solvating extractants $>$ acid extractants (excluding D2EHPA) $\approx$ C-bonded solvating extractants $\approx$ modifiers $>$ diluents. The extractant content of the commercial products must be considered when prices are compared.

TABLE 7 PROPERTIES OF MODIFIERS

Name	Specific Gravity	Flash Point, °C
2-Ethylhexanol	0.833	85
Isodecanol	0.841	104
Nonyl phenol	0.95	140
Tri-*n*-butyl phosphate	0.973	193

3.4. Density

A density difference between the solvent and aqueous phases of at least 0.1 g/cm^3 must be maintained if rapid phase separation is to occur. Specific gravities of the various solvent components have been included in Tables 2–7. Most solvents are formulated to be less dense than the aqueous phase, and the density of most diluents is around 0.8 g/cm^3 to ensure this. The use of tri-*n*-butyl phosphate ($\rho = 0.97$ g/cm^3) can pose problems if concentrated solutions with resultant high metal loadings are used.

3.5. Viscosity

For successful operation of a liquid–liquid extraction circuit, it is essential that the viscosity of the solutions be sufficiently low to permit (1) ready flow of the solutions through the equipment, (2) the use of low-power agitators for phase dispersion, and (3) ready phase separation. In many applications the viscosity of the aqueous phase is sufficiently low, but the solvent phase must be selected to have the desired low viscosity, preferably less than 2 cP. This will be particularly critical for highly loaded organic phases in some systems where polymerization results in increased viscosity. One way of reducing the viscosity is to operate at a higher temperature. Available viscosity data are given in the following paragraphs.

3.5.1. *Acid Extractants*

Little data are available on the viscosities of these extractants under practical conditions, that is, with the use of other than pure compounds as diluents. The designer should be aware of the possibility that large increases in viscosity can take place with metal loading of the solvent. Measurements of the viscosity of solutions of the sodium salt of D2EHPA in toluene at 28°C showed an increase from 0.89 cP for a 5-v/o NaD2EHPA solution of up to 16.2 cP for a 30-v/o solution [26]. The effects of nickel loading on a D2EHPA solution (20-v/o in toluene, 28°C) that had been previously 39% converted to the sodium form, with a resultant increase in viscosity from 1.02–1.32 cP, were determined. The viscosity rose from 0.65 cP at 2 g/liter of nickel to 2.01 cP at 19.3 g/liter of nickel.

3.5.2. Acid Chelating Extractants

Information on the viscosity of these extractants in proprietary diluents is presented in Fig. 1 as a function of extractant, concentration, and temperature [27–29]. Although information on the effect of metal loading is scanty, the presence of copper leads to only a small increase in viscosity.

3.5.3. Anion Exchangers

There appears to be little quantitative information on the viscosities of these extractants that relates to commercial practice. The quaternary ammonium compounds are more viscous than the free amines and amine salts.

3.5.4. Solvating Extractants

The viscosities of several solvating extractants are given in Table 5. The lower viscosity of the ethers and methyl isobutyl ketone allows application of these undiluted, whereas the higher viscosity of tri-*n*-butyl phosphate and the alcohols normally requires dilution with a low-viscosity diluent. The viscosity of TBP can be lowered [30] by using a high-temperature (η = 1.74 cP at 40°C and 0.92 cP at 60°C) or by dilution (η for 20-v/o TBP in kerosene = 1.70 cP at 25°C and 0.95 cP at 40°C). With addition of nitric acid, the viscosity of TBP reaches a maximum at about 1.5 *M* HNO_3 (η = 5.7 cP at 25°C) and decreases to close to the original value at 5 *M* HNO_3.

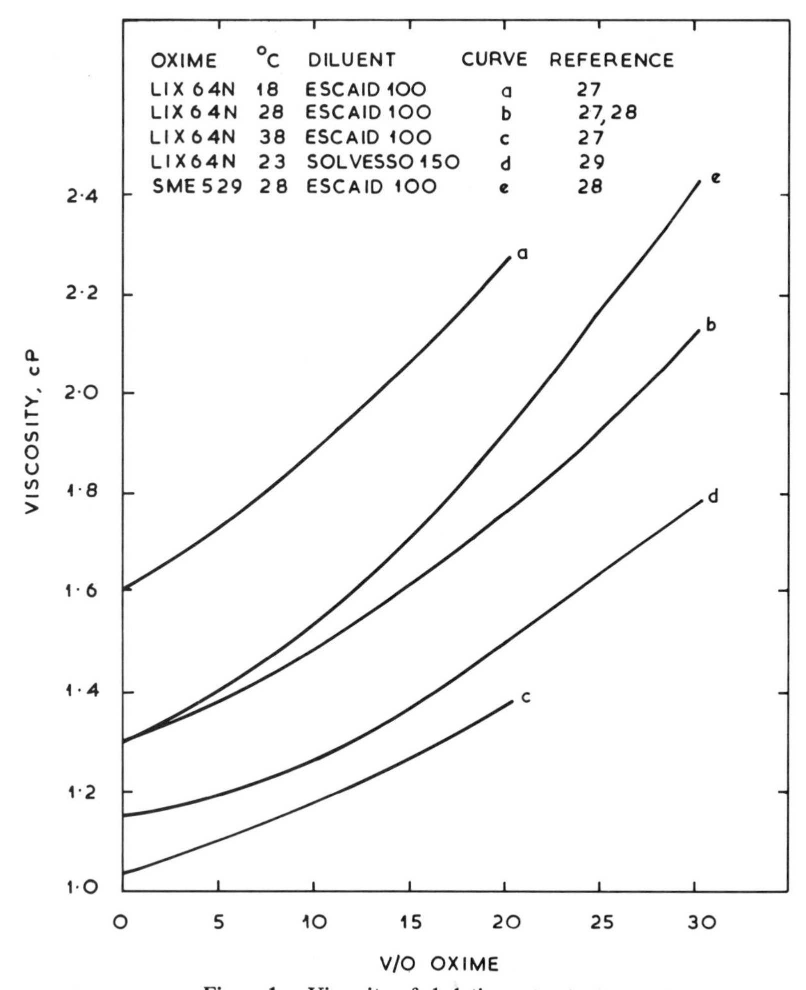

Figure 1. Viscosity of chelating extractants.

TABLE 8 INTERFACIAL TENSIONS OF WATER-ALKANE, WATER-ALCOHOL, AND WATER-KETONE INTERFACES AT 20°C [33]

Number of Carbon Atoms	Alkane	σ, mN/m	Alcohol	σ, mN/m	Ketone	σ, mN/m
4			*n*-Butanol	1.8		
5			*n*-Pentanol	4.4		
6	Hexane	51.1	*n*-Hexanol	6.8		
7	Heptane	50.2	*n*-Heptanol	7.7	Methylpentylketone	12.4
8	Octane	50.8	*n*-Octanol	8.5	Methylhexylketone	14.1

3.6. Interfacial Properties

The importance of interfacial properties in metal extraction has been reviewed recently [32]. Along with other properties of the system, the interfacial properties influence the extraction rate and the characteristics of the dispersion. With highly interfacially active solvents, there is a strong tendency to form dispersions that are slow to coalesce. Interfacial tension is the simplest interfacial property to measure but, as data for practical systems are scanty, the results for pure systems must be used as a guide. With water as the component, data for alkanes, alcohols, and ketones are given in Table 8 and for other hydrocarbons and chlorinated hydrocarbons, in Table 9.

Literature data [35–41] for several extractants in ternary system with pure diluents are shown in Fig. 2, in which the interfacial pressure (i.e., the difference between the interfacial tension in the absence and presence of extractant) is plotted as a function of extractant concentration in the organic phase. Figure 2 shows a very wide variation in interfacial activity between the extractants and also that the effect of diluent can be significant.

TABLE 9 INTERFACIAL TENSIONS OF WATER-HYDROCARBON AND WATER-CHLORINATED HYDROCARBON INTERFACES AT 20°C

Hydrocarbon	σ, mN/m	Reference
Benzene	35.0	33
Toluene	36.1	33
m-Xylene	37.9	33
Escaid 100	29.1[a]	34
Escaid 350	36.3[a]	34
Tetrachloroethylene	47.5	33
Dichloroethane	31.0	33

[a]Aqueous phase 1 *M* sodium sulfate.

Interfacial tension is also affected by the molecular nature of the interfacial species; for example, the values for carboxylic acids such as lauric acid are constant with respect to the pH of the aqueous phase in the acid–neutral region but fall considerably in alkaline solution as dissociation of the acid occurs. The interfacial tension may be reduced by the presence in commercial extractants of interfacially active impurities. For the hydroxyoxime extractants, there has been a considerable effort to design molecules with the optimum interfacial geometry with respect to the rate of extraction. Comparisons between extractants have proved difficult because of the effects of impurities in the commercial products [42].

3.7. Solvent Stability

The chemical stability of the various solvent components is important for two reasons. Degradation by reaction with the aqueous phase is one way in which loss of solvent components may occur. If a differential loss of one component occurs, the problem of analytical control of the solvent composition is increased. Moreover, the products of degradation may be deleterious to the physical or chemical performance of the process. Available information on solvent stability is summarized in the paragraphs that follow. The effect on solvent choice is seen clearly from the restriction on additives for oximes and the choice of TBP rather than amines for the extraction of chromic acid.

3.7.1. Acid and Acid Chelating Extractants

Information on the stability of these classes of extractant is available only for the oxime extractants.

Oximes undergo acid hydrolysis with the formation of the parent ketone. Another possible

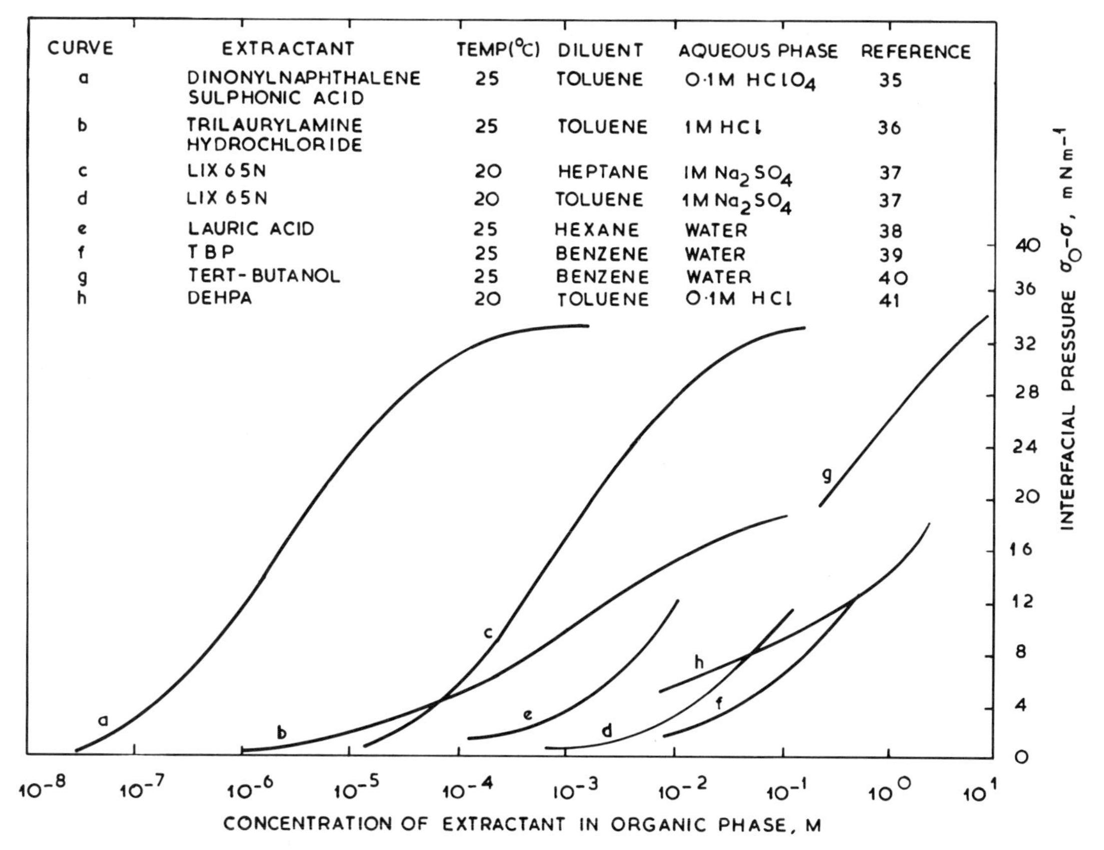

Figure 2. Interfacial pressure measurements for various extractants.

degradation route for arylhydroxy oximes is by the Beckmann Rearrangement to yield an *N*-phenolated amide.

Evidence of both degradation routes has been obtained recently [43]. The rate of loss of loading capacity for members of the P5000 series in contact with 150 g/liter of sulfuric acid has been measured over 90 days. Values of 0.005%/hr in Escaid 100 for 15-v/o P5100 and 0.007%/hr for 30-v/o P5300 were found [44]. The degradation products are claimed to have no effect on extractant performance.

Oxime degradation has recently been shown to be catalyzed by sulfonic and phosphoric acids [43]. Of three acids tested, degradation increased in the order di-2-ethylhexylphosphoric acid $<$ Dowfax 2AO (a sulfonic acid) $\ll$ dinonylnaphthalenesulfonic acid. Very large losses of extractive capacity occurred during 5 weeks of circuit operation with 20-v/o LIX65N and 2-v/o added acid in Escaid 100. These large losses appear to exclude the use of D2EHPA and DNNS as catalysts in LIX circuits and the use of Dowfax 2AO as an antifoaming agent in the associated electrowinning operation. The contamination of the LIX65N solution by Dowfax 2AO was also deleterious in phase separation.

3.7.2. Anion Exchangers

Originally the Bufflex process for uranium extraction was operated with an ammonium nitrate–nitric acid strip solution. Degradation of the amine occurred, leading to a loss of capacity and emulsion problems. Degradation did not occur to a significant extent in the absence of nitrate, and so an ammoniacal–ammonium sulfate strip was adopted [45]. Amines are much less stable than TBP in the presence of chromium(VI) [46]. The stability of quaternary ammonium salts has been investigated in closed-loop operation [47]. On exposure to base, an amine and an alkene are produced by the Hofmann elimination reaction. Quaternary compounds can also be oxidized by atmospheric oxygen in the presence of strong base, but the reaction is extremely slow in the absence of an aqueous phase or the presence of strong acid.

3.7.3. Solvating Extractants

(a) Tri-*n*-butyl Phosphate This extractant is attacked only slowly by 70% nitric acid at room temperature, but hydrochloric acid and zirconium cause fairly rapid degradation [48]. The hydrolytic products are di- and monobutyl phosphoric acid and phosphoric acid:

$$(BuO)_3P{=}O \xrightarrow{k_1} (HO)(BuO)_2P{=}O \xrightarrow{k_2} (HO)_2(BuO)P{=}O \xrightarrow{k_3} (HO)_3P{=}O$$

The substituted phosphoric acids are extractants in their own right and may well have a deleterious effect on the selectivity of the process. First-order rate constants (k_1) are given in Table 10. The first-order rate constants for the subsequent steps (k_2 and k_3) at 76°C in the aqueous phase are 3.4×10^{-3}/hr and 1.6×10^{-3}/hr, respectively. If a 30% solution of TBP in kerosene is contacted with 1 *M* nitric acid at 25°C, the rate of production of dibutylphosphoric acid is 1.5×10^{-4} g/liter hr. The rate is increased by a factor of two at 5 *M* acid and of 10 at 50°C. Work on the hydrolysis of TBP has been reviewed [49].

Tri-*n*-butylphosphate is also fairly stable in the presence of chromic acid [46]. Pure TPB loaded with 23.6 g/liter of chromium(VI) from 4 *M* acid was stored in the dark for 24 hr. Only 0.3% of the chromium was reduced to chromium(III).

(b) Other Solvating Extractants Other common extractants are not stable in the presence of nitric acid [50]. Isopentyl alcohol, diisopropyl ether, dibutyl carbitol, and methyl isobutyl ketone are reportedly especially unstable at high acid loadings. The reaction between ethers and the acid is violent. The chief oxidation products of dibutyl carbitol are butanol and glyoxal [$(CHO)_2$], but oxalic acid, acetic acid, and butyric acid are also formed. The alcohols formed from ethers and ketones may affect the selectivity of the solvent, and oxalic acid may react with the metal ion.

TABLE 10 RATE CONSTANTS FOR TRI-*n*-BUTYLPHOSPHATE HYDROLYSIS

Temperature, °C	k_1, hr^{-1} per Mole of Nitric Acid	
	Aqueous Phase	Organic Phase
105	1.2×10^{-2}	5×10^{-3}
76	2.5×10^{-3}	8×10^{-6}
25	1.7×10^{-4}	8×10^{-7}

3.8. Solubility of Solvent Components

Loss by dissolution in the aqueous phase is inevitable and may require recovery depending on the amount lost. Information on the solubility of the various extractants, diluents, and modifiers is summarized in the text that follows. Extractants such as the chelating acids and amines listed in Tables 3 and 4 have been designed to have low aqueous solubilities. The solvating extractants such as ethers and ketones that have been used in the past for metal separations have become less popular, partly through the development of these less soluble alternatives. The acid extractants are limited to the treatment of relatively low pH solutions because of the high solubility of the sodium and ammonium salts. The solubility is influenced by temperature and the nature of the aqueous phase. It is considerably reduced by the presence of added electrolyte, the well-known "salting-out effect," and this is illustrated by some of the examples quoted in the following paragraphs, which are presented as a guide to trends in solubility behavior.

3.8.1. Acid Extractants

Measurements have been made of the solubility of Versatic 911, which is similar to Versatic 10, but mainly in the regions where commercial use of the reagent is unlikely. At 20°C the solubility in water has been quoted [51] as 0.3 g/kg solution, but at pH 4.0 the solubility is only 45 ppm [52]. Ashbrook [53] investigated the effects of pH and ionic strength on the solubility and found that a rapid increase in solubility, above about 200 ppm, occurs beyond threshold pH values of 6.32, 6.59, 7.14, 7.43, and 7.75 for ammonium sulfate solutions of molarity 0.5, 1.0, 2.0, 3.0, and 4.0, respectively.

The solubility of di-2-ethylhexylphosphoric acid follows the same trends [34, 54], with 19 ppm dissolving in a 0.1 *M* perchlorate solution at 20°C from a 3.1-v/o solution in *n*-octane at pH 3.5 but 122 ppm at pH 4.3; and from a 1.7-v/o solution in kerosene containing 5-v/o isodecanol, 956 ppm dissolves in water, and 304 ppm in 0.1 *M* sodium sulfate, yet only 12 ppm in 1 *M* sodium sulfate solution.

3.8.2. *Acid Chelating Extractants and Amines*

The solubilities of these extractants are very much lower than those mentioned previously and are generally less than 10 ppm and often less than 5 ppm, although higher values have been given by some workers. Comparative values for pure oximes have been measured recently [55] and are given in Table 11. Values for P1 were also determined as a function of pH and temperature, with values of less than 2 ppm at 25°C in the pH range 1.0–5.8, and at pH 1.78 (25°C) in the temperature range 15–45°C. Above pH 8 a rapid increase in solubility occurred with values of 2.48 ppm at pH 9.5 and 37 ppm at pH 11.5.

3.8.3. *Solvating Extractants*

Not only are solvating extractants appreciably soluble in water (the carbon-bonded extractants are much more so than the phosphorus-bonded reagents) but the organic phase also dissolves substantial quantities of water. Mutual solubilities (50) are shown in Table 12. There is an extensive literature on the solubility of TBP in aqueous solution as a function of concentration, diluent, electrolyte concentration, temperature, and acidity [56, 57].

3.8.4. *Diluents and Modifiers*

Little information is available on the solubilities of proprietary diluents. The initial solubility of Escaid 100 is about 100 ppm [58]. Most of the data on modifier solubility relate to the DAPEX process [59] in which uranium is extracted from acid sulfate solution with di-2-ethylhexylphosphoric acid. From a solvent containing 2.6-v/o TBP, 10–15 ppm TBP is reported in raffinates of various sulfate concentrations over the pH range 0.4–1.8. About 100 ppm of 2-ethylhexanol is reported in similar aqueous solutions from an organic phase containing 2-v/o 2-ethylhexanol. The solubility of pure nonyl phenol under the conditions of Table 11 is 7.8 ppm.

TABLE 11 AQUEOUS SOLUBILITY[a] OF PURE OXIMES AT 25°C [55] ppm

P1	1.78
SME529	0.84
LIX65N	0.33
LIX63	4.2

[a]Aqueous phase 0.05 *M* sodium sulfate at pH 1.78.

TABLE 12 MUTUAL SOLUBILITIES OF SOLVATING EXTRACTANTS AND WATER AT 25°C [50]

Extractant	Aqueous Phase, wt. % S	Organic Phase, wt. % H_2O
Diisopropyl ether	1.2	0.63
Dibutylcellosolve	0.2	0.6
Methyl isobutyl ketone	1.7	1.9
n-Butanol	7.31	20.4
n-Pentanol	2.19	7.5
	(g/liter S)	(g/liter H_2O)
Tri-*n*-butyl phosphate	0.39–0.42	64

3.9. Hazards

Ideally, the solvent chosen should be nonflammable and nontoxic, both within the working area and in the outside environment. Unfortunately, most solvent extraction plants contain large volumes of flammable liquid, and the risk of fire must be minimized both by choice of high-flash-point solvents and careful plant design. Toxic hazards within the working environment do not appear to be a major restriction on solvent selection. Potential environmental effects, especially in the long term, have not yet been quantified. A particular problem here is the effect on a bacterial leaching operation of traces of solvent in the barren raffinate returned to the leach. Safety is discussed in more detail in Chapter 30.

4. CONCLUSIONS

Solvents capable of a wide range of duty can be formulated from existing reagents, for which a large amount of laboratory and operating plant data are available. Development of both new reagents and of new applications for existing reagents will doubtless continue. The performance of a solvent must be evaluated in test work using the actual aqueous liquor to be processed. Important areas where knowledge is lacking are solvent stability in contact with novel aqueous solutions and the possible long-term environmental effects of any solvent components discharged to natural waters.

REFERENCES

1. C. F. Coleman, *J. Tenn. Acad. Sci.* **53,** 102, (1978).

2. D. S. Flett, *Transact. Inst. Min. Met. Sect. C* **83,** 30, (1974).
3. R. Swanson, *Proceedings of the International Solvent Extraction Conference (ISEC), Toronto, 1977,* Vol. 1, Canadian Institute of Mining and Metallurgy, Montreal, 1979, p. 3.
4. J. P. Evans, *Min. Mag.* **133,** 271, (1975).
5. A. Faure and T. H. Tunley, *The Recovery of Uranium, Symposium,* Sao Paulo, 1970, International Atomic Energy Agency, Vienna, 1971, pp. 241–251.
6. R. C. Merritt, *The Extractive Metallurgy of Uranium,* Colorado School of Mines Research Institute, Golden, 1971.
7. C. R. Merigold, D. W. Agers, and J. E. House, *Proceedings of the International Solvent Extraction Conference (ISEC), The Hague, 1971,* Society of Chemical Industry, London, 1971, p. 1351–1355.
8. G. A. Rowden and G. Collins, paper presented at *Symposium on Solvent Extraction,* University of Newcastle-upon-Tyne, September 1976.
9. E. Wigstol and K. Froyland, *Ingeniursblad* **41,** 476, (1972).
10. E. Diaz Nogueira and J. M. Regife Vega, *Quim. Ind.* **23,** 693, (1977).
11. L. White, *Eng. Min. J.* **177** (1), 87, (1976).
12. A. R. Burkin, paper presented at Richardson Conference, Imperial College, London, July 1973.
13. A. R. Burkin and J. E. A. Burgess, *Ingeniursblad* **41,** 459, (1972).
14. R. K. Ryan and P. G. Alfredson, *Austral. Inst. Min. Met. Process* **253,** 25, (1975).
15. D. S. Flett and D. W. West, *Complex Metallurgy 78,* Institution of Mining and Metallurgy, London, 1978, pp. 49–57.
16. F. J. Hurst, *Transact. Soc. Min. Eng. AIME,* **262,** 240 (1977).
17. J. C. Agarwal, N. Beecher, G. L. Hubred, D. L. Natwig, and R. R. Skarbo, *Eng. Min. J.* **177** (12), 74 (1976).
18. D. S. Flett and J. Melling, *Hydrometallurgy* **4,** 135 (1979).
19. A. O. Adami, G. R. Firkin, and A. W. Robson, *Complex Metallurgy 78,* Institution of Mining and Metallurgy, London, 1978, pp. 36–42.
20. V. A. Mikhailov, *Proceedings of the International Solvent Extraction Conference (ISEC), Toronto, 1977,* Vol. 1, Canadian Institute of Mining and Metallurgy, Montreal, 1979, p. 52.
21. R. I. Edwards, *Proceedings of the International Solvent Extraction Conference (ISEC), Toronto, 1977,* Vol. 1, Canadian Institute of Mining and Metallurgy, Montreal, 1979, p. 24.
22. K. J. Murray and C. J. Bouboulis, *Eng. Min. J.* **174** (7) 74, (1973).
23. A. J. van der Zeeuw, in G. A. Davies and J. B. Scuffham, Eds., *Hydrometallurgy,* Institution of Chemical Engineers, London, Symposium Series No. 42, 1975, paper 16.
24. H. A. Jung, *Chem. Ind.* **1976,** 170 (1976).
25. D. R. Spink and D. N. Okuhara, *Proceedings of the International Solvent Extraction Conference (ISEC), Lyons, 1974,* Society of Chemical Industry, London, 1974, pp. 2527–2540.
26. K. Durrani, M.Sc. Thesis, Bradford University, UK, 1976.
27. M. A. Hughes, J. S. Preston, and R. J. Whewell, *J. Inorg. Nucl. Chem.* **38,** 2067 (1976).
28. M. A. Hughes, P. D. Middlebrook, and R. J. Whewell, *J. Inorg. Nucl. Chem.* **39,** 1679 (1977).
29. T. A. Tunley and J. C. Paynter, *Recovery of Copper from Sulfate Leach Liquors by Liquid Ion Exchange with LIX64N,* National Institute for Metallurgy, Johannesburg, Report 964, 1970.
30. C. J. Hardy, D. Fairhurst, H. A. C. McKay, and A. M. Willson, *Transact. Faraday Soc.* **60,** 1626 (1964).
31. D. G. Tuck, *Transact. Faraday Soc.* **57,** 1297 (1961).
32. M. Cox and D. S. Flett, *Proceedings of the International Solvent Extraction Conference (ISEC), Toronto, 1977,* Canadian Institute of Mining and Metallurgy, Montreal, 1979, p. 63.
33. A. A. Abramzon, *Kolloid Zhur.* **29,** 467 (1967).
34. Warren Spring Laboratory, unpublished data.
35. R. Chiarizia, P. R. Danesi, G. D'Alessandro, and B. Scuppa, *J. Inorg. Nucl. Chem.* **38,** 1367 (1976).
36. M. Pizzichini, R. Chiarizia, and P. R. Danesi, *J. Inorg. Nucl. Chem.* **40,** 669 (1978).
37. D. S. Flett, *Acc. Chem. Res.* **10,** 99 (1977).
38. A. F. H. Ward and L. Tordai, *Recueil* **71,** 396 (1952).
39. M. Tomoaia, Z. Andrei, and E. Chifu, *Rev. Roum. Chim.* **18,** 1547 (1973).
40. E. Sada, S. Kito, and M. Yamashita, *J. Chem. Eng. Data* **20,** 376 (1975).
41. S. J. Lyle and D. B. Smith, *J. Colloid Interf. Sci.* **61,** 405 (1977).
42. D. S. Flett, paper presented AIME Annual Meeting, New Orleans, February 1979.
43. A. J. Oliver and V. A. Ettel, *Can. Met. Quart.* **15,** 383 (1976).
44. Acorga Ltd., technical bulletin.
45. A. Faure, S. Finney, H. P. Hart, C. L. Jordaan, P. J. Lloyd, R. E. Robinson, D. van Heerden, and E. B. Viljoen, *J. South Afr. Inst. Min. Met.* **66,** 319 (1966).
46. J. P. Cuer, W. Stuckens, and N. Texier, *Proceedings of the International Solvent Extraction Conference (ISEC), Lyons, 1974,* Society of Chemical Industry, London, 1974, pp. 1185–1200.
47. C. E. O'Neill, V. A. Ettel, A. J. Oliver, and I. J. Itzkovitch, *Can. Inst. Min. Met. Bull.* **69** (7), 86 (1976).

48. H. A. C. McKay, in D. Dyrssen, J. O. Liljensin, and J. Rydberg, Eds., *Solvent Extraction Chemistry*, North-Holland, Amsterdam, 1967, pp. 185–194.

49. P. Moszkowicz, *Contribution a l'Etude des Transferts de Masse Interfaciaux. Extraction de l'Uranium et du Plutonium*, CEA CEN, Fountenay-aux-Roses, CEA-R-4735 (1976).

50. Y. Marcus and A. S. Kertes, *Ion Exchange and Solvent Extraction of Metal Complexes*, Wiley-Interscience, London, 1969.

51. A. W. Fletcher, D. S. Flett, J. L. Pegler, D. P. Pepworth, and J. C. Wilson, *Advances in Extractive Metallurgy*, Institution of Mining and Metallurgy, London, 1965, pp. 686–711.

52. A. W. Fletcher and D. S. Flett, in H. A. C. McKay, T. V. Healy, I. L. Jenkins, and A. Naylor, Eds., *Solvent Extraction Chemistry of Metals*, Macmillan, London, 1965, pp. 359–375.

53. A. W. Ashbrook, *J. Inorg. Nucl. Chem.* **34,** 1721 (1972).

54. Z. Kolarik, *Coll. Czech. Chem. Commun.* **32,** 311 (1967).

55. H. J. Foakes, J. S. Preston, and R. J. Whewell, *Anal. Chim. Acta* **97,** 349 (1978).

56. K. Alcock, S. S. Grimley, T. V. Healy, J. Kennedy, and H. A. C. McKay, *Transact. Faraday Soc.* **52,** 39 (1956).

57. E. A. Belousov and L. Yu. Zakahrova, *Russ. J. Phys. Chem.* **49,** 1695 (1975).

58. W. Manfroy and T. Gunkler, *Proceedings of the International Solvent Extraction Conference (ISEC), Lyons, 1974*, Society of Chemical Industry, London, 1974, pp. 726–745.

59. C. A. Blake, D. J. Crouse, C. F. Coleman, K. B. Brown, and A. D. Kelmers, *Progress Report: Further Studies on the Dialkylphosphoric Acid Extraction (DAPEX) Process for Uranium*, ORNL–2172, 1956.

25.1

COMMERCIAL PROCESSES FOR COPPER

J. F. C. Fisher† *and C. W. Notebaart*‡
Mining Industry Technical Services
Zambia

1. INTRODUCTION

Hydrometallurgical extraction of copper from aqueous solutions, with the use of cementation on iron, was practiced in Spain as early as the middle of the eighteenth century [1]. In the nineteenth century leaching of oxidized copper either by natural or downward-percolating acid water or by artificially produced sulfuric acid was practiced in Russia and the United States, with the copper being recovered from solution by cementation. This method was generally found to be suitable for the lower-grade oxidized copper ores and residues that could not be extracted economically by direct smelting. It did depend, however, on the availability of adequate and inexpensive sources of acid and of iron scrap. Sulfuric acid has almost always been favored, as it can usually be produced easily from the sulfur of sulfide ore minerals. The advent of electrowinning for recovery of copper from its acid sulfate solutions led to significant economies compared with cementation due to elimination of the necessity for iron scrap, regeneration of a part of the sulfuric acid for leaching, and the much greater value of the final product, cathodes, com-

†Now with Anglo American Corporation, Republic of South Africa.
‡Now with Billiton Research B.V., The Netherlands. Also, formerly with Nchanga Consolidated Copper Mines Ltd. and Roan Consolidated Mines Ltd., Zambia.

pared with copper cement powder. Despite these advantages, electrowinning only replaced cementation in a few places because of the severe limitations on electrowinning as regards solution quality.

The development of solvent extraction reagents specific for copper in the mid-1960s allowed the improved economics of electrowinning to be applied to the low-copper-tenor, impure solutions derived from many sulfuric acid leaching operations. Solvent extraction has also been applied to the recovery of copper from ammoniacal solutions produced by the ammonia leaching of chalcopyrite (Arbiter process [2]), scrap metal [3], and lead refinery drosses [41].

Solvent extraction in the two leach systems is based on the following exchange reactions:

Sulfuric acid leaching:

$$2HX_{(org)} + Cu^{2+}_{(aq)} \rightleftharpoons CuX_{2(org)} + 2H^{+}_{(aq)} \quad (1)$$

Ammonia leaching:

$$Cu(NH_3)^{2+}_{4(aq)} + 2OH^{-} + 2H_2O + 2HX_{(org)} \rightleftharpoons CuX_{2(org)} + 4NH_4OH_{(aq)} \quad (2)$$

In copper extraction at low H^+ concentrations, both reactions proceed to the right, and thus the leaching medium—sulfuric acid or ammonia—is regenerated. The raffinates from the copper solvent extraction step are thus recycled to the leaching circuit.

The loaded organic in both cases is stripped with a solution high in sulfuric acid and copper, to reverse reaction (1), to form the feed to electrolysis. In the ammonia system contamination of the sulfuric acid electrolyte should be avoided and the loaded organic is thus washed thoroughly prior to stripping.

A typical schematic flow diagram for a copper solvent extraction plant is shown in Fig.1, in which the left-hand circle represents the leach circuit, the center circle the solvent extraction circuit, and the right-hand one the electrowinning circuit.

The three main opeations—leaching, solvent extraction, and electrowinning—need to be carefully integrated such that a maximum overall profitability is achieved. This needs detailed evaluation of the interaction effects, and where conflicting requirements exist, determination of the best compromise. A brief outline of these interfacing problems is given in the following two sections.

2. CONSTRAINTS ON COPPER SOLVENT EXTRACTION PLANTS

2.1. Leaching Operations

Acid leaching operations can be divided into two principal groups:

1. Percolation leaching, in which the solids are contacted by allowing leaching solution to flow through a static bed of solids.
2. Agitation leaching, in which a finely ground suspension of solids is agitated, more or less violently, together with the leaching solution.

2.1.1. Percolation Leaching

In percolation leaching, acidified liquor is allowed to permeate and percolate through a static

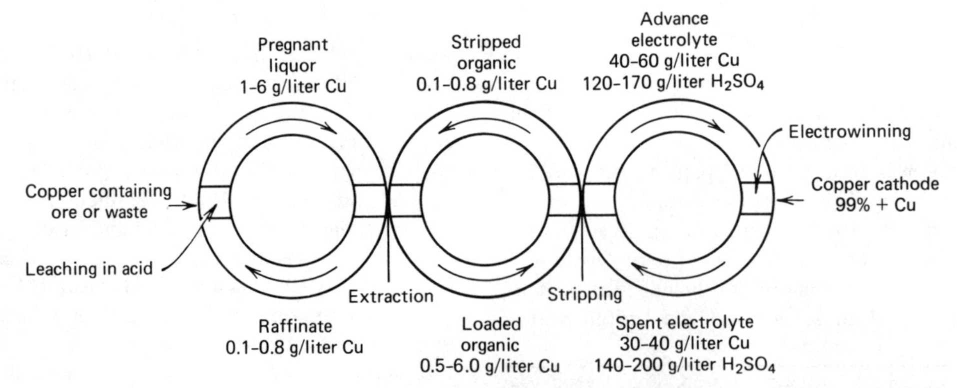

Figure 1. Diagrammatic representation of copper extraction by use of solvent extraction. The indicated copper tenors are typical of sulfuric acid leach circuits. For the ammonia leach system, the circuit would be similar, but the copper concentrations in the leaching solution circulation are much higher.

bed of ore, which may be contained within vats (as at the Kennecott Ray silicate leach plant), may be piled in an open heap (Ranchers Exploration Co. [4] and Cyprus Bagdad [5]), or may be the ore-body itself in place (in situ leaching). Oxidized copper minerals are leached directly by acid; some sulfide minerals are leached by ferric sulfate, and by bacteria that oxidize iron and sulfur. Copper and acid tenors of the pregnant liquor depend on many factors such as grade and type of mineralization, acid tenor of feed liquor, iron tenors, and so on. Copper tenors of pregnant liquor are generally in the range 0.5–5 g/liter, which is satisfactory for subsequent solvent extraction.

Where bacterial action is an important part of the leaching process, both metal and acid tenors in solution must be limited to avoid poisoning the bacteria; these limitations serve to restrict the free acid content of pregnant liquor to concentrations which are acceptable for solvent extraction. Toxicity of entrained organic in raffinate towards the leaching bacteria has been of concern, but there is no evidence that the levels of entrainment which are regarded as normal have any deleterious effects.

Where bacterial action is insignificant, high acid concentrations in the leach liquor will generally lead to higher copper recoveries, or faster leach kinetics. However, high free-acid concentrations in pregnant liquors should be avoided, as unfavorable solvent extraction equilibrium conditions may be created, leading to higher equipment and working costs; in the worst case, neutralization of pregnant liquor might be required. However, this problem can usually be avoided by suitable design of the leach circuitry, and by ensuring that high acid liquors are largely neutralized by contact with fresh ore, in order to yield pregnant liquors suitable for solvent extraction.

In contrast with agitation leaching, the pregnant solutions from a percolation leach plant have very low suspended solids contents as a result of the filtering action of the heap of crushed ore; the problems of solution clarification and crud removal which characterize agitation leach/solvent extraction plants are therefore avoided.

2.1.2. Agitation Leaching

In agitation leaching, finely ground copper ore particles are agitated in the acid leach solution; by contrast with percolation leaching, recoveries of copper to solution are much greater, but there is the added problem of solid–liquid separation and the losses caused by liquor entrainment. In both of the two important plants using agitation leaching, the NCCM tailings leach plant in Zambia [6] and the Anamax Twin Buttes plant in Arizona [7], the first stage in solid–liquid separation is countercurrent washing in thickeners. This is comparatively cheap, but has the disadvantage that the resulting liquor may still contain some 50–500 ppm of suspended solids. Removal of these suspended solids either before solvent extraction by special clarifiers, or during solvent extraction by special crud-handling facilities, poses a number of difficulties; these difficulties are frequently increased by supersaturation of the solution with calcium sulfate, derived from dolomite or calcite, which are often associated with oxidized copper ores.

The cost of agitation leaching plant is very largely a function of the volumetric flow in the plant. To reduce flows in the wash circuit, the incoming liquor flowrate is minimized by maximizing the density of the feed pulp. The wash flow rate is then selected such that the best compromise is reached between plant size and costs and overall copper recovery. A smaller washflow reduces leach–countercurrent decantation (CCD) equipment size but, for a given solvent extraction recovery, results in higher losses from the CCD circuit and thus lower overall recoveries. Reduced washflow also implies higher copper tenor in the pregnant liquor, resulting in high-acid tenors of interstage aqueous and raffinate flows. This requires increased extractant tenors and possibly higher phase ratios, which increases the mixer–settler size and organic inventories. For very high pregnant liquor tenors, interstage neutralization could even be required. Obviously for an existing plant, maximization of washflow within the constraints of the equipment gives maximum recoveries both in the wash circuit and in solvent extraction.

2.2. Electrowinning

Copper electrowinning is carried out in an electrowinning cell containing inert lead alloy anodes and either copper starting sheets or rigid blanks of titanium as the cathodes. The electrolyte is a solution of copper sulfate in sulfuric acid. At the cathode, copper ions are reduced to copper metal, whereas oxygen is evolved at the anode.

The most important constraints are those pertaining to copper and acid content of the electrolyte. For satisfactory electrowinning, the copper content of the spent electrolyte should be not less than 25 g/liter and preferably in the range 30–35 g/liter. To avoid any risk of copper sulfate crystallization, the copper tenor in the advance electrolyte should not exceed 60 g/liter. To ensure maximum conductivity of the electrolyte, the free sulfuric acid content should not be less than 80 g/liter. The acid content should be held at the minimum possible above this figure in order to minimize acid attack on the lead anodes. Experience has shown that the maximum practical acid concentration lies in the range 160–200 g/liter.

To obtain maximum advantage from the process, the copper deposited should be as pure as possible. Many factors effect this purity, the most important of these are:

(a) Density of Deposit A dense deposit with minimum entrained electrolyte requires copper tenors as indicated in the preceding paragraphs, a cathode current density of not more than 450 A/m^2, addition of small quantities of chloride ion, and of an organic surface active agent to depress nodular growth. The presence of an organic film on the surface of the electrolyte, such as arises from entrained organic from solvent extraction, has been shown to cause a loose powdery deposit, and is to be avoided if at all possible.

(b) Impurity Content of the Electrolyte Various impurities in the electrolyte affect the copper deposit in different ways. Bismuth and selenium are to some extent codeposited with copper. A high concentration of metal cations other than copper, such as magnesium or aluminum, reduces mobility of the copper ion and reduces the density of the deposit. Solid particles in the electrolyte can be incorporated in the copper deposit by physical entrapment in pore spaces. Also, they may be attracted to the cathode by their positive zeta potential.

(c) Anode Degradation Acid attack on the anode releases flakes of lead sulfate. The larger flakes may be physically incorporated in the cathode and are generally discovered during a quality inspection; these grossly contaminated cathodes are normally rejected. Fine flakes may be trapped in pore spaces and are subsequently overgrown with copper or are incorporated as a result of the electrokinetic phenomenon described previously.

Two other factors in electrowinning have an important bearing on the solvent extraction process:

1. **Iron Content of the Electrolyte.** Iron in the electrolyte causes a reduction in current efficiency and its content must be kept at a minimum, preferably below 2 g/liter. Good selectivity of the copper extractant against iron is thus an important requirement.
2. **Acid Mist Evolution.** The evolution of oxygen at the anode produces a fine acid mist. The most effective method of reducing this mist is to add a surface–active agent to the electrolyte to induce a foam cover on the cell. Unfortunately, practically all such reagents either do not form a froth as a result of solvation by the entrained organic phase or reduce the organic/aqueous interfacial tension to the point where dispersion bands increase and emulsion may be formed in the solvent extraction circuit. However, the proprietary reagent 7C 100, marketed by the 3M Company, has been used with some success at a number of operations in the United States. The alternative method of using single or multiple layers of polythene balls on the surface of the cell is much less effective.

2.3. Size of Copper Solvent Extraction Operations

Copper solvent extraction plants are an order of magnitude larger than any other type of solvent extraction plant presently in operation, as the following statistics will show:

Annual tonnage of metal produced in a single operation	5000–80,000 ton/year
Flow rate of pregnant liquor	200–3000 m^3/hr
Surface area of solvent extraction plant	$\leqslant 50{,}000\ m^2$

The large size of the plants bring particular problems, including (1) design of the solvent extraction equipment, typically mixer–settlers, (2) provision of adequate fire-fighting facilities, (3) provision of facilities for preventing accidental discharge to streams or rivers, and (4) high capital cost of the organic inventory of the plant.

3. REAGENTS FOR COPPER SOLVENT EXTRACTION

3.1. Extractants

Of the reagents currently available and under consideration for copper extraction, most belong

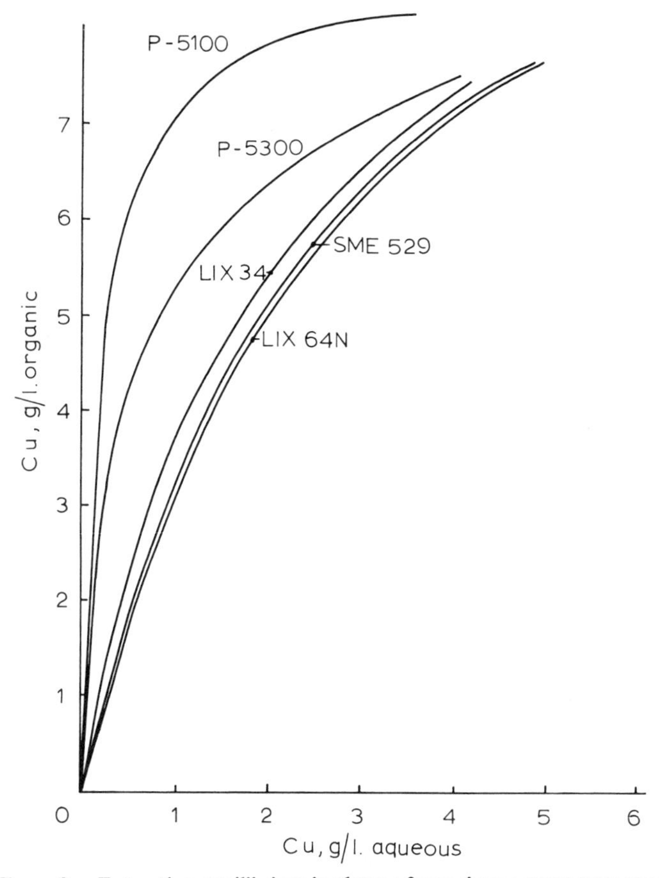

Figure 2. Extraction equilibrium isotherms for various copper extractants.

to the family of hydroxyoximes; exceptions being the Kelex reagents, based on 8-hydroxyquinoline, and LIX 34, which is 8-(alkarylsulfonamido)-quinoline. [18] The various reagents are listed in Table 1 with an indication of properties, uses, and limitations.

The Henkel reagents LIX 64, LIX 63 and LIX 64N were the first to achieve commercial acceptance. During the early 1970s, the reagent LIX 64N became the standard extractant for copper solvent extraction. It consists of a mixture of LIX 65N, a hydroxybenzophenone oxime, and LIX 63, an aliphatic hydroxyoxime, which acts as a kinetic modifier. LIX 64N has a loading capacity of 0.25 g Cu per liter/vol.% in kerosene at pH 2.0. LIX 64N tenors in kerosene up to 30% v/v can be used at temperatures below 40°C with satisfactory phase disengagement. Extraction isotherms of LIX 64N for a typical acid leach liquor are compared with other extractants in Fig. 2.

TABLE 1 REAGENTS AVAILABLE FOR COPPER EXTRACTION

Trade Name	Manufacturer[a]	Composition	Uses and Properties	Limitations
LIX63	1	5,8-Diethyl-7-hydroxy-6-dodecanone oxime	Kinetic promoter for other hydroxy oxime reagents	Low loading capacity on its own; degrades faster than other LIX reagents
LIX64	1	2-Hydroxy-5-dodecyl benzophenone oxime with added LIX63	Low tenor, low-acid copper solutions; replaced by LIX64N	Maximum concentration in kerosene of 10%; efficient for extracting from copper tenors exceeding 3 g/liter

TABLE 1 (*Continued*)

Trade Name	Manufacturer[a]	Composition	Uses and Properties	Limitations
LIX64N	1	LIX65N, with added LIX63	The standard reagent for low–medium copper tenor, acid leach solutions; also used for ammoniacal solutions	Slow-medium kinetics in acid solutions, poisoned by Co^{2+} in ammoniacal solutions
LIX65N	1	2-Hydroxy-5-nonyl benzophenone oxime	High-temperature extraction of copper	Too slow without kinetic modifier for normal operations
LIX70	1	Chlorinated LIX65N with added LIX63	Strong chelating agent for high-acid, high-copper solutions	Difficult to strip requiring acid tenors around 300 g/liter H_2SO_4
LIX71	1	Mixture of LIX64N and LIX70	Intermediate between LIX64N and LIX70	–
LIX73	1	Mixture of LIX64N and LIX70	Intermediate between LIX64N and LIX70	–
LIX34	1	8-(Alkaryl sulfonamido)quinoline	Similar to LIX64N but higher selectivity against Fe; may be stripped with somewhat lower-strength acid solutions	Comparatively high cost
LIX54	1	See Table 3 Chapter 24	Extraction from ammoniacal solutions, high copper loadings, good selectivity, loads no ammonia. Rapid kinetics. Strips with relatively weak H_2SO_4	Needs to be operated aqueous phase continuous. Fe loads in weak acid solution and is difficult to strip with H_2SO_4.
LIX622	1	–	Strong chelate, high maximum loading, rapid kinetics	Requires high acid for stripping.
LIX6022	1	–	Weakened form of LIX622, similar to LIX64N but with rapid kinetics	–
SME 529	2	2-Hydroxy-5-nonyl acetophenone oxime	Higher maximum loading capacity than LIX64N; extraction similar to LIX64N, stripping easier	Slow kinetics, significantly slower than LIX64N

TABLE 1 (*Continued*)

Trade Name	Manufacturer[a]	Composition	Uses and Properties	Limitations
SME 530	2	SME 529 with kinetic modifier	Similar to SME 529 but kinetics faster–similar to LIX64N	–
P50	3	Substituted salicyl aldoxime	Basis for reagents P5100, P5200, P5300, and PT 5050	–
P5100	3	P50 diluted with an equal part nonyl phenol	Strong chelate, high maximum loading copper, useful for high acid, high copper tenor solutions; rapid kinetics	Requires strong acid electrolyte for stripping; not compatible with rubber linings
P5300	3	One part P50 with three parts nonyl phenol	Chelating power weakened by nonyl phenol, maximum loading similar to LIX64N, favorable extraction isotherm but more difficult stripping as compared with LIX64N; faster kinetics than LIX64N	Not compatible with rubber linings
P5200	3	One part P50 with two parts nonyl phenol	Intermediate between P5100 and P5300	Not compatible with rubber linings
PT5050	3	Two parts P50 with one part tridecanol	Similar to P5100 but compatible with rubber linings	–
P17	3	Phenyl benzyl ketone oxime	Similar to LIX64N	Low solubility of copper complex in organic phase
Kelex 100	4	Substituted 8-hydroxy quinoline with isodecanol modifier	Extracts from high-copper-tenor, medium acid solutions, stripped by lower acid solution than most other reagents	Acid carry-over from strip solution, requiring acid wash stages; modifier required to assist phase disengagement
Kelex 120	4	Similar to Kelex 100 but with approximately 20% isodecanol	As Kelex 100 but better phase disengagement	As Kelex 100

[a]Manufacturer: (1) Henkel Corporation; (2) Shell Chemical Company; (3) Acorga Ltd.; (4) Sherex Chemical Company (previously Ashland Chemical Company).

LIX 64N is currently in use for concentrating and purifying acid leach solutions containing up to 6–7 g Cu per liter. For higher copper concentrations an alternative, stronger reagent would be recommended (e.g., LIX 70 series, or P5100) because of the effect of increasing hydrogen ion concentration in the later extraction stages.

In the ammoniacal leach system LIX 64N has very favorable extraction properties for copper. The loading capacity is higher than for sulfate solutions (0.3–0.4 g Cu per liter vol.%) and kinetics are fast [8]. Because of the uptake of free NH_3, the loaded organic should be washed prior to stripping to avoid contamination of the electrolyte. Nickel, cobalt (Co^{2+}), and zinc are coextracted with copper. Nickel and copper can be subsequently separated by selective stripping or by copper cementation from the nickel strip liquors using nickel metal. Coextraction of Co^{2+} should be avoided as the Co^{2+} oxidizes to Co^{3+} in the organic phase, forming a complex which is very difficult to strip and consequently builds up in the loaded organic. Cobalt in the pregnant liquor should therefore be mainly in the cobaltic form, which is not extracted by LIX 64N. LIX 64N can also be used in the extraction of copper from chloride [9, 10] and nitrate [11] solutions.

Since 1975, various other reagents have achieved wide commercial acceptance. The most important of these are the Acorga P-5000 [13, 14] series including the reagent PT-5050. These Acorga reagents are all based on the reagent P-50, which is a substituted salicyl aldoxime. P-50 is a strong chelating agent, with a high maximum loading for copper and relatively fast kinetics. It is modified by the addition of nonyl phenol to yield the commercial reagents P-5100, P-5200, and P-5300. While satisfactory in most circumstances, the nonyl phenol modifier in these reagents causes degradation of natural and synthetic rubbers; consequently, they are unsuitable for use in plants which have any significant amount of rubber-lined construction. For such cases, the alternative reagent PT-5050 is available in which the nonyl phenol modifier is replaced by tridecanol. In other respects PT-5050 is similar to P-5100 in performance characteristics.

Henkel has also developed a number of reagents with stronger chelating power and more rapid kinetics. The LIX 70 series [15, 16] were developed first; they have strong chelating powers and extract copper efficiently from relatively high acid solutions. Subsequently the reagents LIX 622 and 6022 were marketed and are now in commercial use. They are similar in performance characteristics to the Acorga reagents P-5100 and P-5300, respectively. The strong chelating reagents, of which LIX 622 and P-5100 are typical, can be used for treatment of relatively high copper tenor liquors or for accomplishing extraction from low tenor liquors in fewer stages. They have the disadvantage, however, of requiring increased acid concentrations in the stripping solutions, and if these are higher than 180–200 g/liter serious problems may arise in electrowinning due to more rapid attack on the lead anodes and to increased acid mist generation.

The improved kinetics of the newer reagents, compared with LIX 64N, permit satisfactory operation down to temperatures below 10°C. Previously a number of plants in the United States had to use some auxiliary heating in winter to maintain adequate kinetics.

The only other reagent that has achieved commercial recognition is the Shell Chemical Company SME 529 [12], which is similar to LIX 64N but with rather higher maximum loading capacity, balanced by somewhat slower kinetics. Shell Chemical Company also markets a reagent SME 530, which is similar to SME 529 but with an added kinetic modifier.

Mention must also be made of the Kelex series of reagents that were developed by the Ashland Chemical Company [17]. They are somewhat different from the hydroxyoxime family of reagents in that, while they extract particularly well from relatively high-copper, high-acid pregnant liquor, they do not require a very high acid concentration in the strip liquor. This could have been very useful were it not that these reagents also take up acid from the stripping acid solution and transfer it to the last extraction stage. Acid carry over may be prevented by adding wash stages, but this obviously complicates the circuit. Some problems of insoluble complex formation have also been noted with the Kelex reagents. So far they have not found a commercial application.

3.2. Diluents

All copper solvent extraction plants presently in operation use kerosene with an intermediate to low aromatics content (0–25%) as a diluent for the extractant.

The chemical composition of a diluent may affect the following extractant properties: loading capacity; equilibrium loading; kinetics; stability; selectivity; and phase disengagement of the organic phase. Variation in these properties have been related to the solvency power of the diluent [19], which itself is related to the aromaticity. An increasing aromaticity may result in [19–25] (1) a reduced maximum loading capacity, but only at high aromatics content, (2) a decrease in equilibrium loading, (3) a reduced pH functionality at high aromaticity, (4) a decrease in kinetics, (5) faster phase separation, (6) a decrease in stripping efficiency, (7) reduced selectivity of copper over iron, (8) improved copper chelate solubility, and (9) increased extractant stability. These changes should be regarded only as trends. Not all extractants show changes in each of these properties, and not to the same extent. In fact, variation in aromaticity among diluents currently used in solvent extraction would appear to have very little effect on extraction properties.

The use of perchloroethylene as a diluent for LIX64N has been investigated [26]. Comparison with Escaid 100 with the use of 10% v/v LIX64N showed considerably faster phase disengagement, improved iron rejection, a better stripping efficiency, and improved kinetics. Another advantage is that perchloroethylene is not flammable. Perchloroethylene has a higher solubility in water and evaporates faster than the normal kerosene diluent. As perchloroethylene is approximately twice as expensive as normal kerosene, a diluent recovery system is required. Perchloroethylene has been used as a diluent in the Metal–Chem Plant, Mesa, Arizona, which is now closed. Settler specific flows of 387 liters min^{-1} m^{-2} could be achieved [3] that is, four to six times as high as those for which most other plants were designed.

4. COPPER SOLVENT EXTRACTION EQUIPMENT DESIGN

All large-scale copper solvent extraction plants treating acid leach liquors with low to intermediate copper tenors use mixer–settlers, arranged for countercurrent extraction and stripping. The main reasons for choosing mixer–settlers instead of other contacting equipment are the high feed flow rates and the relatively slow kinetics of the copper extractants for low-pH sulfate solutions, requiring contacting times of 2–4 min per stage. The copper solvent extraction plants presently in operation on low copper tenor acid leach liquors have two to four extraction and two to three stripping stages.

Because of the nature of the hydrometallurgical processes treating low-grade copper ores or tailings, a development toward very large scale operations has taken place, with inevitable uncertainties in the scale-up of mixer–settlers. After the successful scale-up of the Ranchers plant [4] from pilot-plant data, decisions for construction of much larger plants were taken with more confidence. The first of these very large plants, treating low tenor leach liquors was built at Chingola, Zambia in 1974 [6]. This plant operated successfully and was followed by the large Twin Buttes plant [7] in 1976. However, years of operating experience on these plants highlighted a number of areas in which improvements in mixer and settler design could be made and led to changes in design concepts and scale-up parameters.

4.1. Mixer Design

The first commercial copper solvent extraction plant at the Bluebird mine in Arizona [4] uses circular mixer boxes with straight-bladed, top-shrouded mixing impellers. Dispersion is pumped from the mixer by an axial flow pump to the far end of the settler, and the separated liquors gravitate to the appropriate mixers. With the exception of Ray Mines, all subsequent plants used the pump–mix systems [27], in which mixing and pumping is done by the same impeller. The advantage of this system is that no external pumps are required. Recently, certain disadvantages of the pump–mix system were recognized and a new mixer was developed in which pumping and mixing are again separated [28].

In mixer design a compromise must be reached between the objectives of achieving the highest possible mass transfer and keeping entrainments as low as possible.

The main mixer types and configurations that are being used in operating copper solvent extraction circuits or that have been developed for planned treatment plants are described in Sections 4.1.1–4.1.4.

4.1.1. Davy McKee Mixer

The Davy McKee pump–mix impeller [29-30] (See Chapter 9.2) is double shrouded with backswept blades, which were designed for minimum entrainment. It is centrally located in the mixer and the clear phases are introduced through a bottom feed box and a draft tube into the eye of the impeller. Recirculation, and thus mass transfer (Fig. 3), is increased by setting the gap between the impeller and the draft tube at a maximum so that the pumping requirements are just met. The mass transfer may be further increased by attaching spoiler blades at the top and/or bottom of the impeller, although this effect is

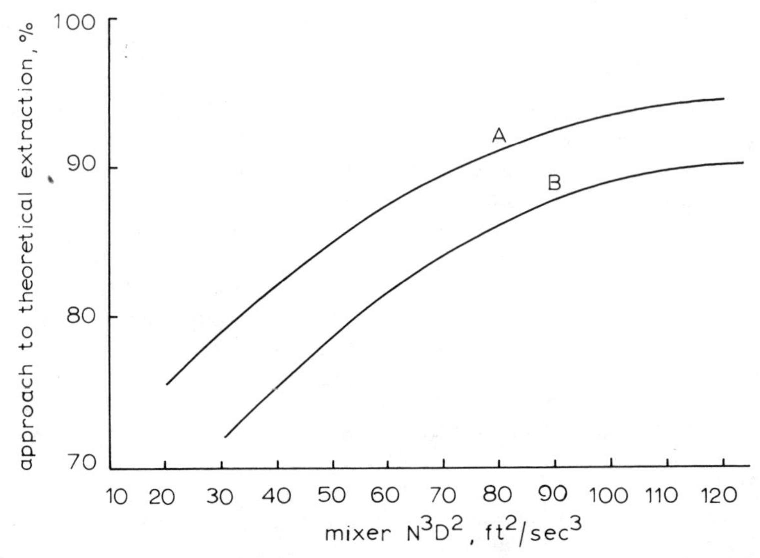

Figure 3. Effect of recirculation of dispersion through the mixing impeller zone (from Warwick and Scuffham [30]) (*A* = recirculation in mixer; *B* = no recirculation).

generally small. This increases the power input considerably and the drive mechanism should be designed to cope with this. Mixing vessels fitted with these impellers in the Chingola Tailings Leach Plant and the Twin Buttes Plant have a square cross section and do not have vertical baffles. Vortex formation and air entrainment are prevented by a top plate with a central hole through which the drive shaft passes and the dispersion is discharged.

4.1.2. *Holmes–Narver Mixer*

In the Holmes–Narver mixer design the pump-mix impeller is located at the bottom of the mixing vessels where the clear phases are introduced [31]. The impeller is top shrouded and has six straight blades. Single-stage mixers using this impeller type have been installed in the Bagdad and Cyprus Johnson plants.

In this mixer design, recirculation, and thus redispersion and coalescence, does not occur. The bulk of the mass transfer takes place during the very short passage of the two phases through the impeller. The later Holmes–Narver design, the "low-profile" mixer [32], consists of three or more mixing compartments in series (Fig. 4). The first compartment contains their standard mixing impeller, located near the bottom. One or more additional mixing impellers are installed in the other compartments. The "low-profile" mixer offers the following advantages:

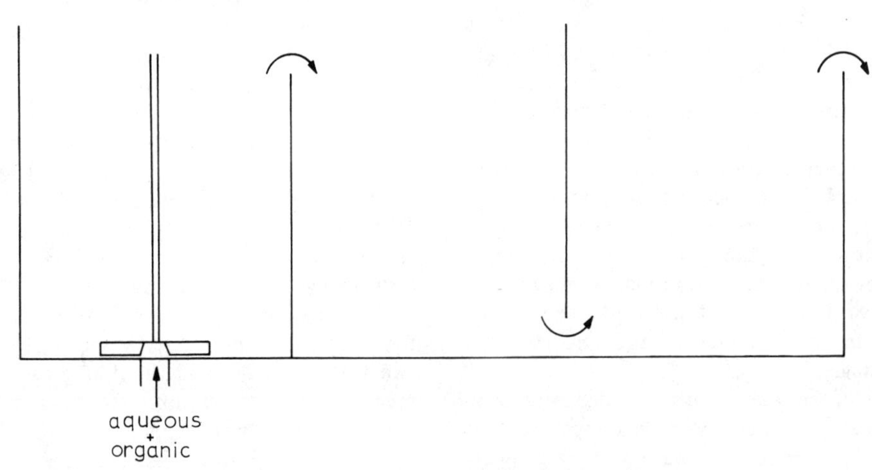

Figure 4. Holmes–Narver low-profile mixer.

1. As compared to the single stage design, for a given mixer, residence time and agitation intensity, mass transfer will be increased as a result of a greater plug flow component and coalescence and redispersion by the additional mixing impellers.
2. A lower pumping head is required.
3. The lower elevation of the mixer and settler will reduce the cost of a support structure. It should be noted that room for piping is required and that if the mixer–settler is on ground level, some excavation work and a special drainage system may be required.

The low-profile mixer has operated satisfactorily in the Cities Service plant [32]. The stage efficiencies in the extraction circuit were 60% with agitation in one compartment and 75% with agitation in two compartments. It is understood that additional agitator mechanisms will be installed to further increase mass transfer.

4.1.3. NCCM Mixer

The NCCM mixer [28] was designed for the expansion of the existing tailings leach plant at Chingola and consists of three mixing compartments, each of which contains a turbine impeller with six straight blades (Fig. 5) located at half height. Two vertical baffles in each compartment located on opposite walls prevent rotation of the dispersion and vortex formation, and thus no top baffle is required. The clear phases are separately introduced into the mixer, the organic phase through a central inlet, and the aqueous phase through a side inlet. The aqueous phase gravitates from one stage to the next, whereas the organic, is pumped with external, axial flow pumps.

Two main considerations led to this mixer design:

1. Recirculation of dispersion through the impeller zone and thus redispersion and coalescence in the two previous mixer designs were considered inadequate. The centrally located Rushton impeller in the NCCM mixer gives a high recirculation rate of dispersion throughout the mixer volume.
2. Organic continuous dispersion proved highly unstable in the Davy McKee mixers in the Chingola plant, even at organic-to-aqueous (O/A) phase ratios as high as O/A = 1.6 : 1 in extraction and O/A = 4 : 1 in stripping. It is believed that this was due to:
   a. The short residence time of the dispersion in the impeller zone. A sudden decrease in organic flow rate will result in an immediate deficit of this phase in the impeller zone and cause a local phase inversion from organic-to-aqueous continuity, which then may quickly spread throughout the mixer.
   b. Accumulation of aqueous phase in the bottom part of the mixer. The gap between the double-shrouded impeller and the draught tube is probably not wide enough to permit sufficient recirculation. Also, the power input per unit volume is probably insufficient (see section on mixer scale-up).

In the new mixer design, separation of aqueous phase is prevented by the much improved mixing in the lower half of the mixer and by the increased power input per unit volume as compared to the existing mixer. Phase stability is further improved by the separate introduction of

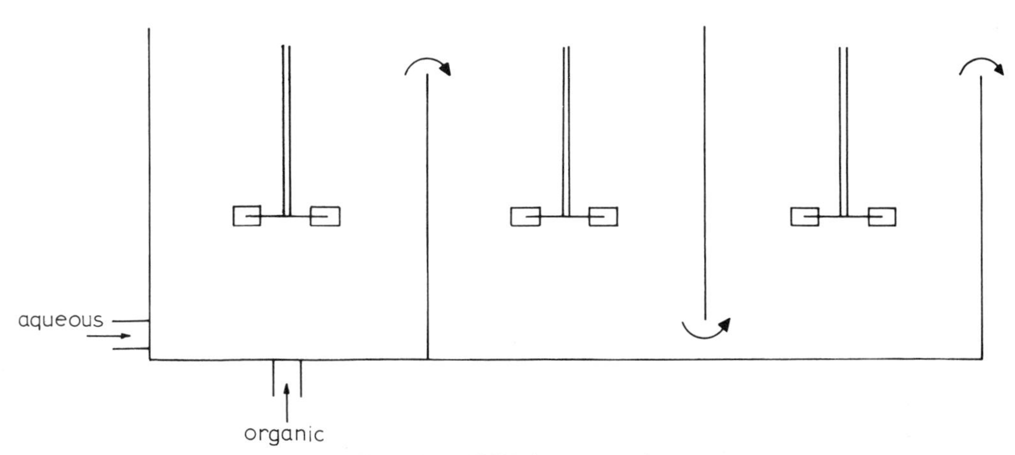

Figure 5. NCCM three-stage mixer.

the phases well away from the high shear impeller zone, so that the whole mixer volume acts as a buffer against surges in the continuous-phase flow rate.

A large-scale test mixer based on this design has operated successfully in the first extraction stage of one of the streams in the Chingola Tailings Leach Plant treating a dispersion flow of 1400 m^3/hr (phase ratio 1 : 1). Average stage efficiencies in the first extraction stage were 95% at a total residence time of 3 min and a power input of 1 hp/m^3. Organic phase continuity could be maintained at phase ratios as low as O/A = 0.7 : 1. Entrainment levels were comparable to those of the Davy McKee mixer.

4.1.4. *Kenics Mixer*

The only known planned application of plug flow, static mixers is located in Zaire for the treatment of impure leach liquors at Luilu and Shituru. This type of mixer is possible because of the high temperature of the pregnant liquor (60°C) produced in a leach circuit, in which the main feed constituent is roaster calcine. Kinetics of LIX 65N at this temperature are very fast. In view of the potential fire hazard at high temperature, a completely sealed mixer–settler system was chosen. Gecamines tested the Kenics static mixer successfully on a pilot scale. This mixer is essentially a pipe of which certain sections contain alternate right- and left-handed helical blades. These helical segments are positioned such that at their contact the edges are at 90°, so that splitting of flows and changes in rotation direction of the liquid flow occur. This generates a dispersion with a narrow drop size distribution and ensures efficient radial mixing.

4.2. Mixer Scale-Up

In the design of mixers for very large copper solvent extraction plants the two criteria—constant tip speed and constant power input per unit volume—have both been used for scale-up based on geometric similarity. The Davy McKee mixers in the Chingola and Twin Buttes plants were scaled up by use of constant impeller tip speed. In Chingola this proved not entirely satisfactory as stage efficiencies were some 5–10% lower than obtained on the pilot plant. In subsequent scale-up test work for the design of the NCCM mixer, it was found that constant power input per unit volume gave better results. On this basis the large-scale test mixer was designed (see section NCCM mixer). Average stage efficiencies in this mixer differed very little from those obtained in a pilot-scale unit.

Apart from scale-up with respect to mass transfer, entrainment and phase stability must also be considered. If constant power input per unit volume is used as the scale-up criterion for obtaining a particular stage efficiency, it might result in excessive entrainment, although this did not prove the case for the NCCM mixer at Chingola. On the other hand, if constant tip speed is used for scale-up, it should be realized that the power input per unit volume decreases with increasing mixer size to the point at which segregation of the phases occurs. This has been found to occur in a pilot-scale mixer at 0.2 hp/m^3. The power input per unit volume in the Chingola mixers is close to this critical value; this may contribute to the phase stability problems encountered.

It was recently shown that for a given mixer size but different impeller diameters, the stage efficiencies remain unchanged if the tip speed is kept constant. Organic-in-aqueous entrainments varied and reached a minimum at a tank width : impeller diameter ratio of approximately 3 : 1.

The following procedure for mixer design and scale-up has thus been proposed [28] : (1) the tip speed for the required stage efficiency is determined in a laboratory mixer; (2) at this selected tip speed the optimum impeller–tank diameter ratio with respect to entrainment is determined in the same vessel under continuous flow conditions; and (3) the scale-up from the laboratory mixer to commercial size is carried out by use of geometric similarity and equal power input per unit volume.

4.3. Settler Design

Settler area requirements are generally determined from pilot-plant data for given mixing conditions by use of the scale-up criterion of constant flow of dispersed phase per unit settler area. Dispersion band depths are determined for various specific settler flows, which can be conveniently done by changing the settler area with a vertical dam baffle. Important factors to be considered are extractant tenor, diluent type, phase continuity, temperature, and suspended solids in the pregnant liquor. As the settling area required for organic continuous dispersions is considerably greater than that for aqueous continuous dispersions, settlers are designed for organic phase continuity, even when several stages operate aqueous continuous, to avoid flooding when phase inversion occurs.

Temperature has a considerable effect on dispersion bands: in pilot-plant studies in Chingola the dispersion band volume was halved when the temperature increased from 25 to 33°C. It is thus very important to take seasonal temperature changes into consideration.

The functional relationship between dispersion band thickness and specific settler flow in the copper–LIX64N system is generally exponential and can be adequately described by the function $H = k(Q/A)^n$, where H = dispersion band thickness, Q = dispersion flow rate, A = settler area, k = constant, and n is an exponent that varies from 1.5 to 7. If a high exponent is found, low specific settler flows are selected with an ample safety margin to avoid the steep part of the curve and thus flooding of the settler. For a low exponent, the curve can be approximated by a straight line, and the selection of the specific flow is less critical. It should be pointed out that for a given extraction process, the exponent may be affected by a number of factors, including size, quantity, and type of suspended solids. These suspended solids generally result in a more rapid rate of coalescence, although masking of the interface may occur at high concentrations.

Several coalescing aids have been proposed to reduce the required settler area, thereby reducing construction costs and organic inventory costs:

1. **Vertical Baffles.** Vertical baffles cause an increase in dispersion band depth and reduce the velocity of dispersion. Thick dispersion bands also tend to reduce aqueous in organic entrainment significantly as compared to thin, wedge-shaped dispersion bands. A settler design incorporating several vertical baffles, each with short horizontal baffles to increase the area available for coalescence [33], has been proposed for the copper solvent extraction plant at Panguna, Papua New Guinea, In this design most of the coalescence takes place in the first one third of the settler, whereas the remainder of the settler serves to reduce entrainment in the separated phases. This settler gave a very marked reduction in aqueous entrainment as compared to an unbaffled settler.
2. **Vertical Screens.**
3. **Perforated Plates.**
4. **Vertical Parallel Rods.**
5. **Knit-mesh Packing** [34]. This consists of interwoven, stainless steel and polypropylene filaments. Coalescence of organic droplets is enhanced by the high-surface-energy, organic wetted polypropylene, whereas coalexcence of aqueous droplets is promoted by the low-surface-energy, aqueous wetted stainless steel. The presence of both materials ensures enhancement of coalescence regardless of the phase continuity.
6. **Precoalescence by a Packing with Rashig Rings [35].**
7. **Horizontal Baffles.** Horizontal or near-horizontal baffles or trays extending over a major part of the settler area divide the incoming dispersion into what can be visualized as a series of superimposed settlers and, therefore, allow a very significant reduction in total settler area required. Examples are the Lurgi multitray settler [36] and the IMI compact settler [37] (see Chapter 9).

These coalescence aids certainly do reduce settler area, but they also complicate crud and scale removal, and it is thought that they have as yet not been installed on any operating plant.

Settlers installed in copper solvent extraction plants all have a rectangular shape, although there is no obvious reason why circular settlers with a center feed should not be considered. The latter would, in fact, have the advantage of low linear velocities of the separated phases at the peripheral weirs, thus minimizing entrainment.

The length:width ratio is determined by the opposing requirements of minimizing the organic inventory and maintaining acceptable entrainment levels. The entrainment levels in the separated phases are a function of the linear velocities in these phases at the discharge weirs. For each phase, a critical linear velocity exists above which entrainment rises sharply until eventually gross flooding occurs as the dispersion band is dragged over the organic weirs or under the organic weir into the aqueous weir. Generally, sufficient aqueous depth is allowed to be well below the critical velocity of the clear aqueous, so that only the critical velocity in the organic phase needs to be considered. This critical linear velocity of the organic phase typically is in the region 5–10 cm/s (Fig 6). For a given settler area, the organic inventory becomes proportional to the depth of the clear organic phase. To avoid exceeding the critical linear velocity, the clear organic depth can be decreased only by simultaneously increasing the settler width. The settler width is limited by the problems of even distribution of the dispersion from mixer to settler and of maintaining uniform flow within the settler. The latter could be overcome by installing lengthwise flow straighteners. At very shallow organic depths disturbances in the upper surface of the dispersion band could cause a sharp rise in aqueous entrainments.

In most operations the settler length:width ratio is in the region 2–4:1. The first radical change from this aspect ratio was made in the design of settlers for the Cities Service plant, which have a length to width ratio of 1:1. For the Chingola Tailings Leach Plant expansion, a settler length:width ratio of 0.93:1 is proposed to reduce organic inventories [28].

4.4. The Davy McKee CMS Contactor

Over recent years Davy McKee have developed a combined mixer–settler [39, 40], consisting of a large tank with a centrally placed mixer, in which separation of the organic and aqueous phases takes place in the top and bottom sections, respectively. The mixer is characterized by a relatively large mixing volume, with low impeller tip speeds; mixer kinetics and mass transfer are not good by comparison with other mixers presently in use for copper solvent extraction, but the "settler" area is reduced by a factor of 8–10. The overall equipment cost and organic inventory are thereby considerably reduced. The combined mixer–settler has found application already in the uranium solvent extraction industry; with some modifications its application is likely to be extended to copper solvent extraction, in association with reagents characterized by relatively rapid kinetics.

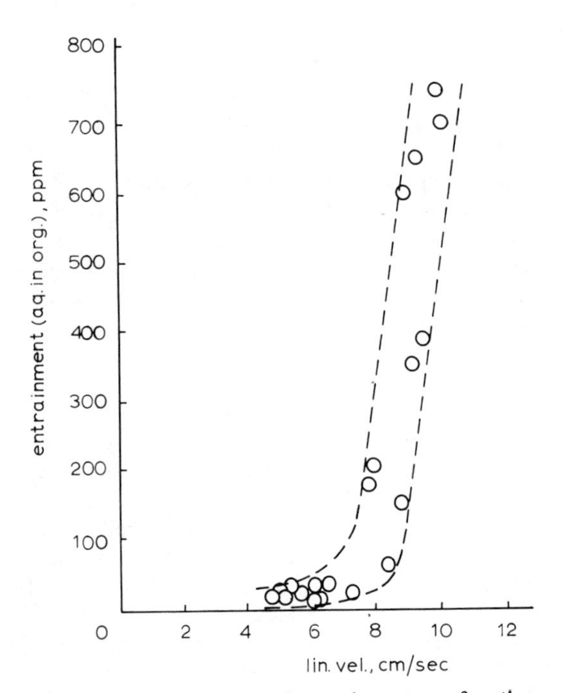

Figure 6. Aqueous inorganic entrainment as a function of linear velocity (from Orjans et al. [28]).

5. PROBLEM AREAS IN DESIGN AND OPERATION OF COPPER SOLVENT EXTRACTION PLANTS

5.1. Entrainment

Both organic-in-aqueous and aqueous-in-organic entrainment are partially generated in the mixer and partially in the settler. Factors that may affect entrainment are described in Sections 5.1.1–5.1.4.

5.1.1. Mixing Power Input

High mixing power input may increase both aqueous and organic entrainment.

5.1.2. Phase Ratio

The phase ratio has a marked effect on entrainment [38]. For organic continuous dispersions, the aqueous entrainment increases considerably at phase ratios higher than approximately O/A = 1.5–2:1 in both extraction and stripping (Fig 7). In the extraction circuits of copper solvent extraction plants the phase ratio is generally less than 2:1. In the stripping circuit high phase ratios are usually applied, and thus it is necessary to recycle the aqueous phase to maintain a mixer phase ratio of 1:1. A further reason for applying aqueous recycle is that at high phase ratios the mass-transfer rate and thus stage efficiencies are reduced, possibly as a result of a decreased rate of coalescence and redispersion of the dispersed phase in the mixer. The organic entrainment appears to be virtually independent of the phase ratio (Fig. 7).

For aqueous-phase continuity, the aqueous-in-organic entrainment reaches a minimum at a phase ratio of around O/A = 1.5:1, although the exact location of this minimum is not certain. The organic entrainment increases sharply at phase ratios less than 1:1 (Fig. 7).

5.1.3. Phase Continuity

Although a relationship exists between entrainment level and phase continuity, this is not well defined and may be different for particular conditions, as indicated by different operating practice in a number of plants. In the Cities Service plant all mixers operate aqueous continuous ex-

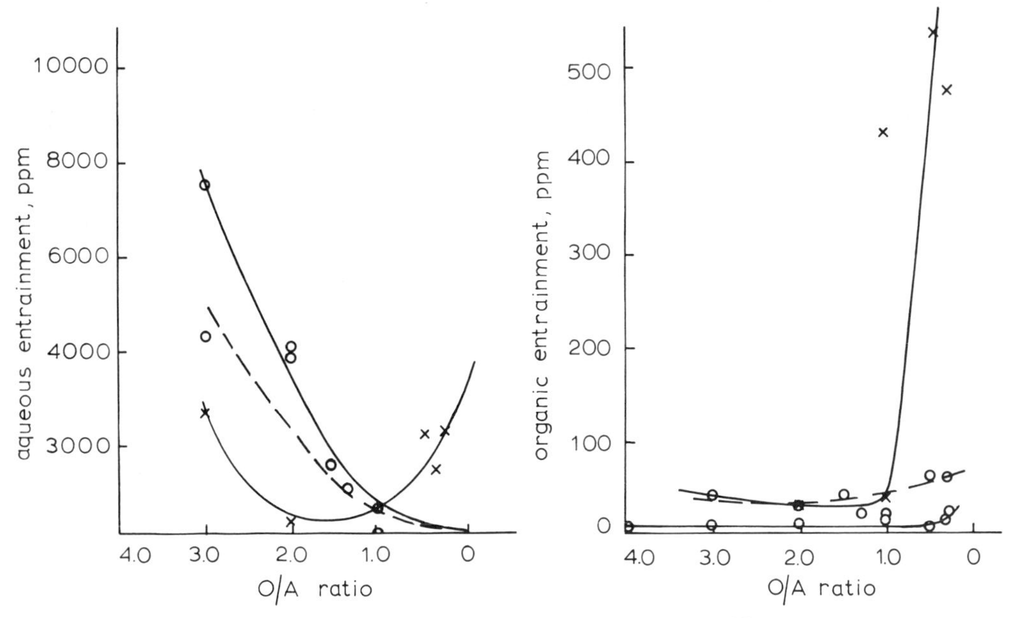

Figure 7. Effect of phase ratio on entrainment (from Rowden et al. [28]) (solid line, extraction; dashed line, strip; 0–0, organic continuity; X–X, aqueous continuity).

cept for the last extraction mixer. Here, organic continuity has been found to reduce organic loss in raffinate. However, less organic in electrolyte entrainment has been observed when operating the second strip mixer aqueous continuous. In the Anamax Twin Buttes plant the first extraction mixer operates aqueous continuous, which reportedly gives lower aqueous entrainment in the loaded organic. In the Cyprus Johnson plant all mixers operate aqueous continuous because of lower aqueous-in-organic entrainment. The Bagdad plant operates with organic-phase continuity in the last extraction stage and the second stripping stage to reduce organic entrainment.

Entrainment studies in the Chingola plant showed that the organic entrainment is virtually the same for both phase continuities; stage efficiencies also seem to be independent of phase continuity. Although normal aqueous-in-organic entrainment is low during operation in the aqueous continuous mode, the flotation of dispersion band fragments causes relatively high overall aqueous carry-over. This is generally difficult to quantify with the normal method of aqueous entrainment determination, which uses centrifuge tubes with a calibrated capillary at the lower end to measure the aqueous volume. The volume of these tubes is relatively small. Much larger sample volumes would be required for estimation of the aqueous carry-over due to the dispersion fragments.

5.1.4. *Air Entrainment*

Entrained air causes a marked increase in the entrainment of organic which is generated in the mixer. It will also enhance the flotation of dispersion fragments in the settler when operating in the aqueous continuous mode. Vortex formation in the mixer should thus be avoided, and all weir profiles should be carefully designed to minimize turbulence.

5.2. Solids and Crud Handling

Suspended solids in the form of gangue minerals and gypsum precipitated from supersaturated leach liquors accumulate on the settler bottom and in the dispersion band at the coalescing front and are generally referred to as *crud*. The dispersion band crud contains, when removed, considerable quantities of organic phase that is generally recovered by centrifuging. Removal of interfacial crud should be carried out regularly to prevent buildup and subsequent breakthrough at the weirs.

A crud problem frequently exists in leaching of ground ore or tailings with subsequent solids–liquid separation in thickeners. Dump or *in situ* leach operations usually have a very low level of suspended solids in the pregnant liquor.

Prevention of crud formation by an efficient

clarification–filtration system is evidently an area to which considerable attention should be devoted. Overflow from the first washing thickener is generally filtered in sand filters, either directly or following an intermediate clarification step, which serves to remove the bulk of the coarser suspended solids and ensures a more uniform feed solids level to the sand filter.

Considerable problems may result from calcium supersaturation of the pregnant liquor, which results in gypsum precipitation in the feed nozzles and in the bed of the sand filter, thus reducing filter capacity. Gypsum scale formation in the pipework of the solvent extraction circuit results in a reduction of flow rates and necessitates expensive descaling campaigns. Calcium supersaturation can be reduced by contacting the pregnant liquor with gypsum seed crystals in a sludge-bed clarifier. A critical pH exists below which good clarities cannot be maintained for long periods with a gypsum sludge bed unless portions of the bed are regularly replaced with freshly precipitated gypsum, produced by partially neutralizing some raffinate.

5.3. Organic Additives

Flocculants, smoothing agents, and mist suppressants for electrowinning must be carefully tested to ensure compatability with solvent extraction operations. Cationic flocculants are generally not compatable with LIX64N in kerosene, but several nonionic and anionic polyacrylamides and guars have been found to be satisfactory. Nonionic polyacrylamide flocculants are currently in use in the leaching circuit of the Chingola Tailings Leach Plant. Both nonionic polyacrylamides and guars are used for inhibition of nodular growth in electrowinning. Satisfactory foaming agents for mist suppression in electrowinning have proved difficult to find since the structural properties that lead to a lowering of air–water surface tension, and foam formation, also lead to low surface tensions between aqueous and organic

TABLE 2 DETAILS OF COPPER SOLVENT EXTRACTION PLANTS COMMISSIONED UP TO 1979

Company and Location	Commenced	Type of Feed	Rate of Production Tons/Year	Pregnant Aqueous Feed: Flow Rate, m^3/hr	g/liter Cu	pH
Ranchers' Exploration Bluebird Mine, Arizona	1968	Heap leach liquor, sulfate	7,500	460	1.8–2.3	1.8–1.9
Cyprus Bagdad Copper Co. Bagdad, Arizona	1970	Heap leach liquor, sulfate	7,000	550	1.4–1.6	2.0–2.3
Nchanga Consolidated Copper Mines, Chingola, Zambia	1974	Agitation leach liquor, sulfate	80,000	3200	3–4	2.0
Arbiter Plant Anaconda Co., Anaconda, Montana	1974	Ammonia leach liquor from sulfide concentrate	36,000	84	28–33	180 g/liter NH_3
Anamax Mining Co., Twin Buttes, Arizona	1975	Agitation leach liquor, sulfate	33,000	1400	2.9	1.9
Capital Wire and Cable, Casa Grande, Arizona	1970	Ammonia leach liquor from scrap copper	2,500	11.5	30	3g/liter NH_3
Cyprus Johnson Copper Co., Johnson Camp, Arizona	1975	Heap leach liquor, sulfate	5,000	570	1.3–1.4	2.1
Cities Service Co., Miami, Arizona	1976	*In situ* leach liquor, sulfate	5,500	680	1.0	1.0 g/liter H_2SO_4
Cerro Verde, Minero Peru, Arequipa, Peru	1977	Heap leach liquor, sulfate	33,000	1400	5	1.5–1.8
Inspiration Consolidated Copper Company, Arizona	1979	Heap leach liquor, sulfate	26,000	908.4	1.4	4.0 g/liter H_2SO_4

phases, with a consequent tendency to formation of stable emulsions. However, a patented reagent has recently been under test at the Inspiration Copper Corporation plant in Arizona with apparently successful results.

5.4. Materials of Construction

The combination of copper and high acid, together with kerosene, complicates the choice of materials for copper solvent extraction plants. Stainless steel, either free standing or as a lining for concrete, is the most favored material for withstanding corrosive attack. Glass-fiber-reinforced plastics have been used for piping. Plasticized PVC has been proposed as a settler lining material, but leaching of the plasticizer by kerosene can occur, and this may affect solvent extraction. Rubber-lined mild steel has been used for construction of tanks and storage vessels. Nitrile rubber must be used as natural rubbers are rapidly attacked by kerosene. Lead lining cannot be used for mixer settlers as entrained organic droplets cause rapid pitting corrosion.

6. DESCRIPTION OF TYPICAL PLANTS

6.1. NCCM Tailings Leach Plant, Chingola, Zambia

In the Tailings Leach Plant [6] at Chingola (see Fig. 8 for flow sheet) the feed material is fine tailings from the concentrator. This material contains a number of acid soluble minerals, malachite, chrysocolla, cupriferous mica, cuprite, azurite, and pseudomalachite. Leaching is carried out in a countercurrent two-stage process in pachucas. The secondary leach takes place at relatively high acid strength to achieve maximum copper extraction. Solution from the second-stage leach is contacted with fresh tailings in the primary leach to yield a low-acid pregnant solution for solvent extraction. Solids–liquid separation is

Extraction Agent	Diluent	Number of Streams	Flow per Stream, m^3/hr	Number of Stages		Phase Ratio (O/A)		Temperature of Solvent Extraction °C
				Extraction	Strip	Extraction	Strip	
12% LIX64N	Nap 470B	1	460	3	2	1.15:1	3:1	25–30
8% LIX64N	Chevron, Nap 470B	4	140	4	3	1:1	20:1	16–25
24% LIX64N 14% SME529	Escaid 100	4	800	3	2	1:1	4.5:1	30–35
32% LIX64N	Nap 470B	1	84	2 + 1 wash	2	–	–	35–40
12% LIX64N	Chevron Kerosene	2	700	4	2	1.2.1	12:1	23°C minimum
20% LIX64N	–	1	11.5	1 + 1 wash	2	6:1	3:1	–
6% LIX64N	Nap 470B	5	120	3	2	1:1	20:1	10–25
6.2% LIX64N	Nap 470B	2	340	3	2	1:1	–	22–26
30% LIX64N	–	4	385	3	2	1:1	–	15
7% Acorga P5300	Chevron ion-exchange solvent	1	908.4	2	2	1.1:1	10:1	19

Continued on page 666

TABLE 2 (*Continued*)

Company and Location	Fire-Fighting Facilities	Materials of Construction: Mixer–Settlers	Pipes	Tanks	Mixer Type
Ranchers Exploration Bluebird Mine, Arizona	Water posts	Stainless steel and fiberglass-lined concrete	FRP	FRP- and RFP-lined steel	Circular tanks, turbine mixer
Cyprus Bagdad Copper Co. Bagdad, Arizona	Foam sprinklers in settlers	Stainless steel supported by mild steel	PVC- and PVC-lined steel	PVC-lined steel or concrete	Holmes–Narver pump–mix
Nchanga Consolidated Copper Mines, Chingola, Zambia	Foam sprinklers in settlers	Stainless steel lined concrete	PVC-lined FRP	Rubber-, FRP-lined mild steel	Davy McKee pump–mix
Arbiter Plant Anaconda Co., Anaconda, Montana	–	304L stainless steel	Stainless steel and PVC	Stainless steel	Davy McKee pump–mix
Anamax Mining Co., Twin Buttes, Arizona	Water sprinklers in settlers	Stainless steel	Stainless steel, PVC in tank-house	Rubber-lined steel, stainless steel	Davy McKee pump–mix
Capital Wire and Cable, Casa Grande, Arizona					
Cyprus Johnson Copper Co., Johnson Camp, Arizona	Water sprinklers in settlers	Fiber-reinforced plastic	Fiber-reinforced plastic	Fiber-reinforced plastic	Holmes–Narver, pump–mix
Cities Service Co., Miami, Arizona	Water posts	Stainless-steel-lined concrete	Stainless steel	Stainless steel	Holmes–Narver, low profile
Cerro Verde, Minero Peru, Arequipa, Peru	Heat-activated sprinklers in settlers	Stainless steel	FRP and stainless steel	Plastic-lined concrete, stainless steel	Wright Engineers design, cylindrical tank
Inspiration Consolidated Copper Company, Arizona	Automatic foam system to mixers, settlers, and pipe trenches, actuated by fusible link detectors	316L stainless steel	Stainless steel or high-density polyethylene	Mainly 316L stainless steel	Two-stage mixers, pump-mixer followed by a radial flow mixer

carried out in countercurrent wash thickeners in four to five stages, with barren raffinate being used as the wash liquor.

The overflow from the first washing thickener was previously filtered in 11 pressure sand filters. A major difficulty was the precipitation of gypsum in the sand bed, which caused serious reduction in capacity and consequently in overall efficiency of the filter plant. Filtration in pressure sand filters has now been abandoned. Pending construction of an alternative clarification plant, the CCD1 overflow is now being clarified in one of the wash thickeners. The solids content of the pregnant liquor to solvent extraction frequently exceeds 200 ppm; thus crud buildup is extremely rapid.

Pregnant liquor, generally containing 3–6 g/liter Cu at pH 1.9–2.0, is treated in four parallel solvent extraction streams. Three of these streams use a 22-vol % solution of LIX64N in ESSO kerosene Escaid 100. The fourth stream uses a 14% solution of the Shell extractant SME 529, also in Escaid 100. Consideration is being given currently to use of Acorga reagent PT 5050. Each stream consists of three extraction and two strip stages and treats approximately 700 m^3/h of pregnant liquor.

Mixer–settlers are made of stainless steel lined concrete. The square box mixers use double-shrouded pump–mix turbines with backswept blades. The top of the mixer has a full-width horizontal baffle to prevent vortex formation

	Mixers		Settlers				
Residence Time, min	Depth, m	Width, m	Length, m	Width, m	Depth, m	Specific Settler Flow, liter/m^2 min	Crud-Handling Facilities
2	–	–	–	–	–	80	Filtered in plate/ frame
2	–	1.2 diameter	19.8	4.3	1.14	80	Centrifuge
extr. 3 strip 3	3.7 3.2	5.9 4.0	36.5 26.2	12.2 12.2	0.76 0.76	60/80	Centrifuge
2.5	–	–	–	–	–	70	–
3.5	3.7	4.6	E1, E2: 51 E3, E4: 45.6 S: 37.3	13.7		60	Centrifuge
3	–	–	–	–	–	80	Centrifuge
3 except 1st stage strip, which has 6	1.0	–	15.2	11.9	–	75	Centrifuge
3	3.6	3.25 diameter	21.3	9.1	–	53	–
2 in first compartment, 1 in second	3.3	3.6	25	17	1.21	61	Movable hand-held suction pump to storage and then to Delaval centrifuge

and short circuiting. The mixer discharges through a 1.7-m outlet in the vortex baffle, concentric with the center shaft, by way of a flared launder into a full-width entry slot into the settler at the level of the dispersion band. Each settler is fitted with two picket fence baffles. Important dimensions of the mixer–settler units are as follows:

Apparatus	Parameter	Extraction	Strip
Mixers	Width	5.9 m	4.0 m
	Depth	3.7 m	3.2 m
Mixing impeller	Diameter	2.7 m	2.3 m
Settlers	Width	12.2 m	12.2 m
	Length	36.5 m	26.2 m
	Depth	0.76 m	0.76 m
First picket fence	Distance from inlet	13.1 m	9.1 m
Second picket fence	Distance from inlet	30.5 m	21.9 m
Specific settler flow		60 liters/min m^2	80 liters/min m^2

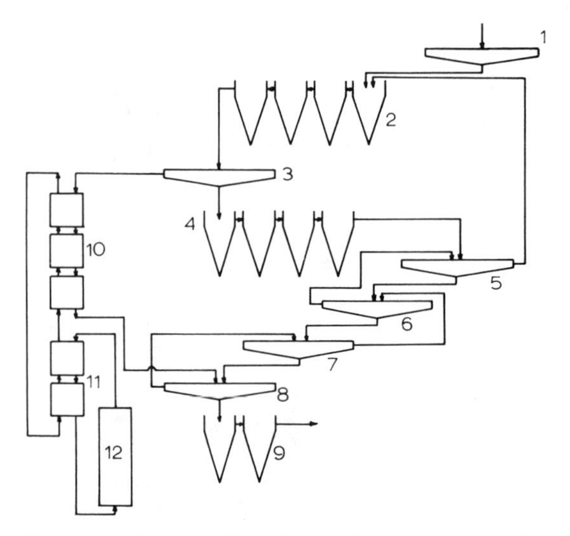

Figure 8. Nchanga Consolidated Copper Mines Ltd. Tailings Leach Plant at Chingola, Zambia: (1) preleach thickener; (2) primary leach (eight pachucas); (3) CCD-1 washing thickener; (4) secondary leach (four pachucas); (5-8) CCD2-5 washing thickener; (9) neutralization (four pachucas); (10) solvent extraction circuit (three stages); (11) stripping circuit (two stages); (12) tankhouse.

Phase ratios are 1:1 in extraction and 4:1 in strip, with aqueous phase recycle around each mixer-settler unit on the strip side, to give a 1:1 phase ratio in the mixers. Because of reasons outlined in previous sections, it has proved impossible to operate the mixers in any but the aqueous continuous mode. Extraction stage efficiencies of 85–90% and stripping stage efficiencies of 75–85% are generally achieved, yielding raffinates of 0.4–0.5 g/liter Cu. However, a major problem with this plant is the variable nature of the feed material, giving rise to pregnant liquors with copper contents varying from 3 to 6 g/liter. At the upper end of this range, extraction efficiencies naturally fall off because of approach to maximum loading capacity of the extractant and the generation of excess acid from the extraction reaction.

Crud formation results in excessive organic loss, estimated at 170 ppm based on overall raffinate flow. Crud is removed from both extraction and strip stages several times per year by dumping into ground sumps and is then separated in pusher centrifuges. The high rate of crud formation was not foreseen at the design stage, and future plants of this type will probably include more advanced technical crud removal and handling facilities.

The electrowinning circuit is operated at an average current density of 320 A/m^2; it may be noted that current densities of up to 450 A/m^2 were found practical in pilot-scale and plant test work. This is a considerable increase over most electrowinning operations and is made feasible by the improved electrolyte purity resulting from solvent extraction. Initial problems with filtration led to carry-over of fine solids that adversely affected cathode quality. Problems with entrainment of organic in electrolyte, causing a loose powdery deposit at the top edge of the cathode, were also encountered. Subsequent operation has shown that high quality cathodes can be produced at the exceptionally high current densities in use at Chingola.

6.2. Cities Service Company, Miami, Arizona

This relatively new plant [32] was commissioned in 1976. Pregnant liquor for solvent extraction is generated by percolating acid solution through an oxidized and weathered cap overlying a mined-out copper sulfide orebody. Copper in the oxidized cap occurs mainly as chrysocolla, wad, and cupriferous mica, with smaller amounts of sulfide copper. In addition to the normal acid leaching of the oxide copper minerals, bacterial leaching of the sulfide is taking place. It is notable that this is unaffected by traces of solvent extraction reagents. As the acid solution percolates through the orebody, the acid content becomes neutralized and the copper content builds up. The pregnant liquor contains 1.0 g/liter Cu, about 0.5 g/liter H_2SO_4, and very low suspended solids, generally less than 25 ppm, and thus no filtration is required.

Pregnant liquor is treated in two parallel solvent extraction streams, where it is contacted with 6% LIX64N in kerosene. The plant features the Holmes–Narver low-profile mixer with an effective liquor depth of about 1 m. Each separate compartment is now provided with a turbine. The first strip mixer has six compartments, giving a total residence time of 6 min. All other mixers have three compartments, giving a total residence time of 3 min. Linear velocity in the settlers has been reduced by making them almost square in shape. This coupled with the low rate of crud formation, and the effective distribution of the feed into the settler by means of a picket fence baffle at the settler feed end has reduced the organic losses to about 30 ppm based on raffinate flow. Crud is removed through a pipe near the organic weir and separated in a centrifuge.

Stripping of the loaded organic is carried out

TABLE 3 DETAILS OF COPPER SOLVENT EXTRACTION PLANTS COMMISSIONED 1979–1981

Company and Location	Commenced	Type of Feed	Rate of Production, Tons/Year	Pregnant Aqueous Feed Flow Rate m^3/hr	Pregnant Aqueous Feed g/liter Cu	Pregnant Aqueous Feed pH	Extraction Agent	Diluent	Number of Streams	Flow per Stream, m^3/hr	Number of Stages Extraction	Number of Stages Strip
Duval Corp., Battle Mountain	1979	Heap leach liquor sulfate	4,000	400	1.2	2.0	7% LIX64N	Chevron	1	400	3	2
Kennecott Corp., Ray Mine, Arizona	1980	Vat leach liquor sulfate	38,000	1400	3.5	1.8	ACORGA P5100	Escaid 200 Kermac	2	700	2	2
Sociedad Minora Pudahuel, Lo Aguirre, Chile	1980	Tailings leach liquor sulfate	15,000	–	5.0	1.9	30% LIX64N	–	–	–	2	2
CIA Minera de Cananea, Sonora, Mexico	1980	Heap leach liquor sulfate	15,000	900	2.5	1.9	ACORGA P5100	Chevron	1	450	2	2
Minero Peru, Cerro De Pasco, Peru	1980	Mine water liquor sulfate	6,000	–	1.0	1.5	ACORGA P5100	–	–	–	3	2
Noranda Lakeshore Mine, Arizona	1981	Vat leach liquor sulfate	25,000	454	6–7	1.6	ACORGA P5100	Chevron	1	454	3	2
Cities Service Co., Pinto Valley, Arizona	1981	Dump leach liquor sulfate	3,500	1400	0.35	2.2	1.7% LIX622	Chevron	2	700	2	1

with spent electrolyte containing 150 g/liter Cu and 30–33 g/liter H_2SO_4, at a phase ratio of approximately 20:1. A phase ratio of 1:1 is achieved in individual stages by internal recirculation of aqueous.

The absence of solids and of organic entrainment in the advance electrolyte permits production of high-quality cathode in electrowinning at the relatively high current density of 270 A/m^2.

Further details on this plant and other operating copper solvent extraction plants are given in Tables 2 and 3.

REFERENCES

1. W. E. Greenawalt, *The Hydrometallurgy of Copper,* McGraw-Hill, New York, 1912, pp. 205–215.
2. M. C. Kuhn, N. Arbiter, and H. Kling, *Can. Min. Met. Bull.* 67(742), 62–73 (1974).
3. D. S. Flett, *Transact. Inst. Min. Met. Sect. C* 83, C30–C38 (1974).
4. K. L. Power, *Proceedings of the International Solvent Extraction Conference (ISEC), The Hague, 1971*, Society of Chemical Industry, London, 1971, pp. 1409–1415.
5. H. J. McGarr, *Chem. Eng.* 77 (17), 82–84 (1970).
6. J. A. Holmes, A. D. Deuchar, L. N. Stewart, and J. D. Parker, *International Symposium on Copper Extraction and Refining*, Las Vegas, 1976, Extractive Metallurgy of Copper, AIME, New York, 1976, pp. 907–925.
7. G. Rossiter, Anamax Twin Buttes oxide plant operating experience–first year. Paper presented at the Arizona Section of AIME Hydrometallurgical Division Spring Meeting, 1976.
8. C. R. Merigold, D. W. Agers, and J. E. House, *Proceedings of the International Solvent Extraction Conference (ISEC), The Hague, 1971*, Society of Chemical Industry, London, 1971, pp. 1351–1455.
9. P. G. Christie, V. I. Lakshmanan, and G. J. Lawson, *Proceedings of the International Solvent Extraction Conference (ISEC), Lyons, 1974*, Society of Chemical Industry, London, 1974, pp. 685–697.
10. J. C. Paynter, *J. South Afr. Inst. Min. Met.* **74**, 158–170 (November 1973).
11. P. B. Queneau, and J. B. Prater, Nitric acid process for recovering metal values from sulphide ore materials containing iron sulphides. U.S. Patent 3793428.
12. A. J. Van der Zeeuw, *International Symposium on Copper Extraction and Refining*, Las Vegas, 1976, Extractive Metallurgy of Copper, AIME, New York, 1976, pp. 1039–1055.
13. J. A. J. Tumilty, R. F. Dalton, and J. P. Massam, *Adv. Extract. Met.* 123–131 (1977).
14. J. A. J. Tumilty, J. P. Massam, and G. W. Seward, *Proceedings of the International Solvent Extraction Conference (ISEC), Toronto, 1977*, Canadian Institute of Mining and Metallurgy, Montreal, 1979, p. 542–551.
15. General Mills Inc., LIX70-a major advance in liquid ion exchange technology. Paper presented at Centennial Meeting AIME, New York, 1971.
16. D. W. Agers and E. R. de Ment, *Igenieursblad*, **41**, 433–441 (1972).
17. G. M. Ritcey and B. H. Lucas, Some aspects of the extraction of metals from acidic solutions by Kelex 100. Paper presented at CIM Annual Meeting, Quebec, 1973.
18. M. J. Virnig, *Proceedings of the International Solvent Extraction Conference, (ISEC), Toronto, 1977*, Canadian Institute of Mining and Metallurgy, Montreal, 1979, p. 535–542.
19. G. M. Ritcey, and B. H. Lucas, *Proceedings of the International Solvent Extraction Conference (ISEC), Lyons, 1974*, Society of Chemical Industry, London, 1974, pp. 2437–2481.
20. Nchanga Consolidated Copper Mines Limited, *The Standard Laboratory Procedure for the Initial Evaluation of Extractants and Diluents*, unpublished company report, 1971.
21. K. J. Murray and C. J. Bouboulis, *Eng. Min. J.* **174,** 74–77 (July 1973).
22. A. J. Van der Zeeuw, *Symposium on Hydrometallurgy*, Manchester, 1975, Institution of Chemical Engineers Symposium Series No. 42, pp. 16.1–16.11.
23. S. Dobson, and A. J. Van der Zeeuw, *Chem. Ind.*, 175 (1976).
24. R. F. Dalton, F. Hauxwell, and J. A. J. Tumilty, *Chem. Ind.*, 181 (1976).
25. E. Hogfeldt, *Chem. Ind.*, 184 (1976).
26. W. Manfroy and T. Gunkler, *Proceedings of the International Solvent Extraction Conference (ISEC), Lyons, 1974*, Society of Chemical Industry, London, 1974, pp. 726–745.
27. B. V. Coplan, J. K. Davidson, and E. L. Zebroski, *Chem. Eng. Progr.* **50,** 403–408 (1954).
28. J. R. Orjans, C. W. Notebaart, J. C. Godfrey, C. Hanson, and M. J. Slater, *Proceedings of the International Solvent Extraction Conference (ISEC), Toronto, 1977*, Canadian Institute of Mining and Metallurgy, Montreal, 1979, p. 340–346.
29. G. C. I. Warwick, J. B. Scuffham, and J. B. Lott, *Proceedings of the International Solvent Extraction Conference (ISEC), The Hague, 1971*, Society of Chemical Industry, London, 1971, pp. 1373–1385.
30. G. C. I. Warwick, J. B. Scuffham, *Ingenieursblad* **41,** 442–449 (August 1972).
31. P. Paige, *Some Ideas on Solvent Extraction*, Holmes and Narver, Inc. publication.
32. A. D. Kennedy and C. L. Pfalzgraff, *Proceedings of the International Solvent Extraction Confer-*

ence (ISEC), Toronto, 1977, Canadian Institute of Mining and Metallurgy, Montreal, 1979, p. 333–339.

33. I. E. Lewis, *Proceedings of the International Solvent Extraction Conference (ISEC), Toronto, 1977*, Canadian Institute of Mining and Metallurgy, Montreal, 1979, p. 325–333.

34. I. D. Jackson, J. B. Scuffham, G. C. I. Warwick, and G. A. Davies, *Proceedings of the International Solvent Extraction Conference (ISEC), Lyons, 1974*, Society of Chemical Industry, London, 1974, pp. 567–598.

35. J. Mizrahi, F. Barnea, and D. Meyer, *Proceeaings of the International Solvent Extraction Conference (ISEC), Lyons, 1974*, Society of Chemical Industry, London, 1974, pp. 141–168.

36. H. M. Stonner and F. Wohler, *Symposium on Hydrometallurgy*, Manchester, 1975, Institution of Chemical Engineers Symposium Series No. 42, pp. 14.1–14.12.

37. J. Mizrahi and E. Barnea, *Process Eng.* 60–65 (January 1973).

38. G. A. Rowden, J. B. Scuffham, G. C. I. Warwick, and G. A. Davies, *Symposium of Hydrometallurgy*, Manchester, 1975, Institution of Chemical Engineers Symposium Series No. 42, pp. 17.1–17.17.

39. G. A. Rowden, M. Dilley, C. F. Bonney and G. A. Gillett, Davy McKee's CMS Contactor—Its development and applicability, *Symposium on Hydrometallurgy*, Manchester, 1981, Society of Chemical Industry, London, 1981.

40. J. B. Scuffham, *The Chem. Eng.* (370), 328 (1981).

41. W. Hopkin, et al, *Erzmetall*, 35, 192–195 (1982); also paper to be presented at AIME Symposium on Hydrometallurgy, Atlanta, 1983.

25.2

COMMERCIAL PROCESSES FOR NICKEL AND COBALT

G. M. Ritcey

CANMET
Department of Energy, Mines and Resources
Ottawa, Canada

1. INTRODUCTION

Cobalt–nickel ores and concentrates can be leached with ammonia [1–5] or acid [6–8]. In the ammoniacal leaching process the separation of nickel and cobalt, from an ammonium carbonate solution containing the ammines of these metals, can be achieved either by distillation [1, 2], by precipitation as the basic carbonate [2, 9, 10], or by electrolysis [2]. Nickel and cobalt have been recovered separately as metals from ammonium sulfate solutions by hydrogen reduction under pressure [4, 5]. Solvent extraction has been shown to be suitable for the difficult separation of the two metals from alkaline media [11, 12], and many processes have been proposed for separation from acidic solutions. The following is a summarized version of a more comprehensive treatise by Ritcey and Ashbrook [13].

2. SULFURIC ACID SYSTEM

The separation of nickel and cobalt has been a problem for metallurgists for many years, particularly from sulfuric acid medium. With respect to such systems, the extractants available for extraction of nickel and/or cobalt perform best in the pH 4–6 range. Thus in the case of cationic and the chelating extractants, copper and iron would have to be first removed before nickel or cobalt could be extracted. Also any iron present would require prior removal. With the chelating extractants, cobalt is difficult to strip, unless extracted and maintained in the cobaltous form.

A process for separation of cobalt and nickel from sulfate liquors has been proposed by Magner of Dow Chemical Company [14]. The flow sheet is shown in Fig. 1. Copper is removed by precipitation with H_2S, and at the same time Fe^{3+} is reduced to Fe^{2+}, which is not extractable in the next step. Cobalt and nickel are coextracted with dinonylnaphthelene sulfonic acid as a 10% solution in kerosene, which loads about 2 g/liter of each metal. Both metals are stripped by use of 1–6 *M* HCl, which also regenerates the solvent for recycle, yielding a strip solution containing about 5 g/liter of each metal. The acid concentration used for stripping is governed by the subsequent purification step, which requires 5–6 *M* HCl for the selective extraction and separation of cobalt from nickel. Triisooctylamine (TIOA) is the extractant in either an aromatic diluent such as toluene, or an aliphatic kerosone with TBP as modifier. Cobalt is recovered from the loaded solvent by water stripping, yielding a solution containing about 50 g Co/liter. High-purity cobalt can be produced from the strip solution by electrolysis or precipitation with an

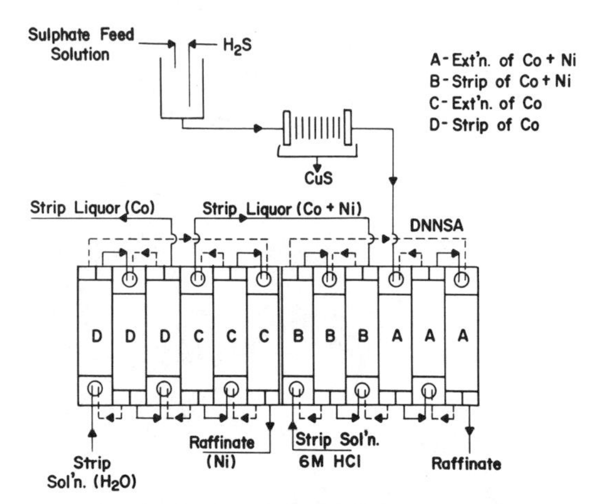

Figure 1. Separation of cobalt–nickel from sulfate and chloride systems.

alkali. By adding sulfuric acid to the strip solution, equivalent to the hydrochloric acid, followed by distillation of the HCl, the solution is converted to cobalt sulfate, followed by electrolysis to produce cobalt metal. Cobalt products are reported to contain less than 1 part of nickel per 1000 parts of cobalt.

Nickel is produced by the addition of sulfuric acid to the raffinate, distillation of the hydrochloric acid followed by electrolysis of the nickel sulfate to produce nickel metal containing less than 3 parts of cobalt per 1000 parts of nickel. Hydrochloric acid regenerated by distillation is recycled to the front end of the circuit to strip the cobalt–nickel mixture from the dinonylnaphthelene sulfonic acid extractant. Sulfuric acid formed in the electrolytic cells is also recycled to the process; thus reagent consumptions are kept to a minimum.

Investigations have been reported into the use of carboxylic acids for extraction [15, 16] and a process described for separation of copper and nickel using a carboxylic acid and naphthenic acid, after prior treatment of the ore by flotation and roasting, followed by leaching and impurity removal. This has not been used commercially.

Gindin et al. have published several papers on the use of C_7–C_9 fatty acids for the extraction of divalent metals [17–20]. An order of extractability, at a given pH, is given as $Sn^{4+} > Bi^{3+} > Fe^{3+} > Pb^{2+} > Al^{3+} > Cu^{2+} > Cd^{2+} > Zn^{2+} > Ni^{2+} > Co^{2+} > Mn^{2+} > Mg^{2+} > Na^{+}$. The authors showed that the technique of exchange extraction could be applied to the separation of two metals with similar extraction properties, for example, nickel from cobalt.

A process was developed by the Israel Mining Industries by use of a carboxylic acid, α-bromolauric acid [21], which is also pH dependent. The order of extraction, from lowest to highest pH, was $Fe^{3+} > Al > Cu > Ni > Zn > Co >> Mn > Ca$. Extraction at 50°C improved phase disengagement. Using limestone for pH adjustment, the Fe^{3+} is extracted at pH 2.1, followed by stripping with sulfuric acid. The raffinate from extraction, containing copper and nickel, is contacted at pH 3.1 with the α-bromolauric acid in kerosene mixture to extract and separate copper from nickel. Copper is recovered by stripping with H_2SO_4.

Eldorado Nuclear piloted a process developed by Ritcey and Ashbrook [22] for the recovery of copper, nickel, cobalt, and silver at their Port Hope Refinery. Feed material consisted of residues, dump chemical precipitates, ores, concentrates, as well as high-grade cobalt arsenide speiss containing appreciable amounts of silver and other metals. Leaching was achieved with sulfuric or a mixed nitric–sulfuric acid medium. For the extraction and separation of cobalt from nickel, D2EHPA was the best of several extractants evaluated. Usually, 10–30% D2EHPA in an aliphatic diluent with either TBP or isodecanol as modifier composed the solvent. The extraction with use of D2EHPA is pH dependent, and the order of extraction of some of these metals is $Fe^{3+} > Zn^{2+} > Cu^{2+} > Co^{2+} > Ni^{2+} > Mn^{2+} > Mg^{2+} > Ca^{2+}$. Because the extraction is pH dependent and the extraction characteristics of cobalt and nickel are very similar, the equilibrium pH during extraction must be closely maintained. Earlier work [23] on maintaining the equilibrium pH of an ammoniacal (pH 11) extraction system showed that the use of an alkali salt of the D2EHPA extractant could maintain the desired pH for optimum extraction and separation of cobalt from nickel.

The process [22, 24] is probably different from most other processes for the following three reasons: (1) the solvent is preequilibrated with an alkali to provide the desired pH buffering during extraction; (2) nickel impurity is removed by scrubbing with a solution containing a high concentration of cobalt; and (3) differential column contactors are used.

Formation of the sodium salt of a D2EHPA-kerosene solution can be accomplished by addition of a stoichiometric amount of 50 wt. % sodium hydroxide solution. Similarly, addition of the stoichiometric amount of either ammonium hydroxide solution (28 wt. %) or anhydrous am-

TABLE 1 EFFECT OF NH_4OH QUANTITY USED TO PREEQUILIBRATE THE SOLVENT ON THE RAFFINATE pH (FEED pH 4.0)

Ammonia Used to Preequilibrate Solvent (lb NH_4OH per 100 Imperial Gallons)	Raffinate pH
15	3.29
25	3.90
35	4.45
45	5.27
55	6.18

monia, also results in a single-phase system. The effect of the amount of NH_4OH used to preequilibrate the solvent on the resulting raffinate is shown in Table 1.

The diagrammatic flow sheet for the process is shown in Fig. 2. After contact of the feed solution at about pH 5.0–5.5 with 20% D2EHPA previously equilibrated with NH_4OH, the loaded solvent is scrubbed with a cobalt sulfate solution at a pH of 5.0–5.5. A bleed of the scrub recycle stream is returned to the extraction circuit as the nickel content is increased as a result of scrubbing. The cobalt in the scrubbed solvent can be stripped by sulfuric, nitric, or hydrochloric acid. If sulfuric acid stripping is used, the resultant cobalt sulfate solution can be reduced by hydrogen in the conventional manner [8]. A portion of the cobalt sulfate solution is returned to the scrub circuit. If nitric acid is used for stripping, the cobalt nitrate strip solution is then evaporated, denitrated to the cobalt oxide, and finally reduced to cobalt metal under a hydrogen atmosphere. The nitrous oxide from denitration is recovered in a gas scrub circuit and converted to nitric acid for return to the stripping circuit.

Increased Co–Ni separation was achieved at elevated temperatures and by using 20% D2EHPA rather than 30% D2EHPA. The choice of continuous phase during extraction, scrubbing, or stripping can play an important role in determining maximum Co–Ni separation. The organic continuous system is to be preferred when equal quantities of cobalt and nickel are present in the extraction circuit. In treatment of a feed solution containing a high Co:Ni ratio, it was found that either organic or aqueous continuous operations were equally effective. At high Ni:Co ratios, only the organic continuous system was effective for the extraction and separation of cobalt from nickel. A typical column profile from an organic continuous extraction run is shown in Fig. 3. Aqueous continuous is preferable in the scrubbing stage. Tank stripping, either aqueous or organic continuous, is used. The loss of D2EHPA to the aqueous raffinate appears to be pH dependent, of the order of 30 ppm at an equilibrium pH of 6.0.

The solvent extraction separation of cobalt and nickel with the use of D2EHPA as described previously is under licence to Nippon Mining in Japan and International Nickel Company of Canada, where it is incorporated in their new refinery in Sudbury. The Nippon circuit went onstream late in 1975 [25, 26], whereas the Inco circuit is not yet in operation.

The D2EHPA separation process was also used by the same authors [27] for the recovery of cobalt from solutions arising from the hydrometallurgical treatment of superalloy scrap material.

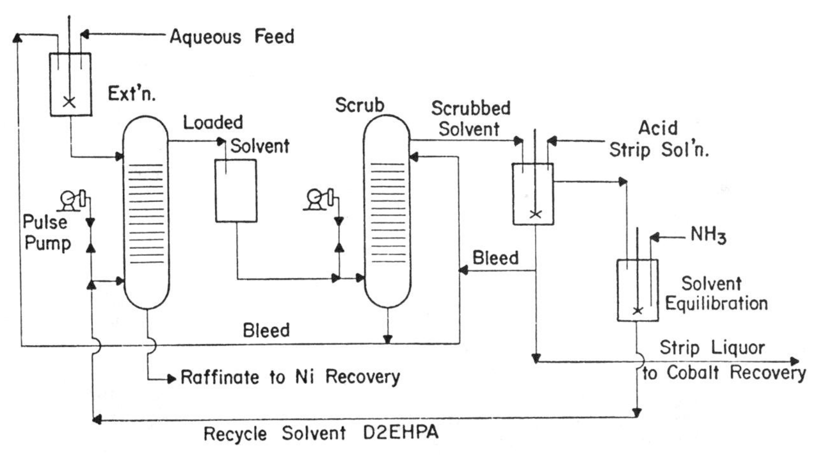

Figure 2. Cobalt–nickel separation by use of D2EHPA as extractant.

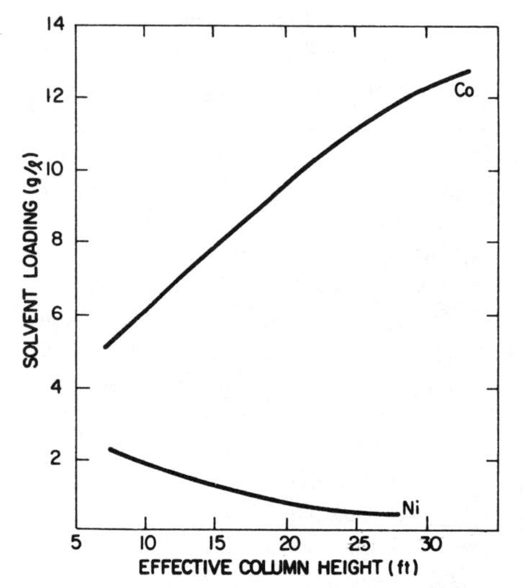

Figure 3. Column profile–organic continuous extraction.

Pyrites Company (United States) used the process of solvent extraction for the removal of impurities in their feed solutions prior to recovery of cobalt by precipitation as hydroxide or carbonate [28]. Feed stock consisted of pyrite cinder and residues such as spent catalysts and sludges that were leached in sulfuric acid. An alkali salt of D2EHPA was used for pH control as earlier described by Ritcey and Ashbrook [22] for the separation of cobalt and nickel. The feed to the solvent extraction process, after removal of copper and iron, had the composition (in grams per liter) 0.13 Cu, 0.03 Fe, 0.10 Mn, 11.0 Zn, and 11.0 Co. Plant operation consisted of 13 stages of mixer–settlers: 5 for extraction, 3 for scrubbing, 2 for stripping, 1 for solvent treatment, and 2 for solvent wash (see also Chapter 25.9).

Several workers [29–33] have reported on the possible use of mixtures of oximes and carboxylic acids for separation of nickel and cobalt.

3. CHLORIDE SYSTEMS

One of the early plants using the chloride system is that at Kristiansand, Norway, which has an annual production of 20,000,000 lb of nickel and 10,000 lb of cobalt [18, 36]. The flow sheet for the extraction is shown in Fig. 4 [34, 35].

The filtrate from leaching of the matte is oxidized with O_2 at 65°C to convert iron to ferric, sulfide to elemental sulfur, and Cu^+ to Cu^{2+}. The elemental sulphur is filtered and washed to remove any entrained copper, before sending to waste. The filtrate is cooled to room temperature before going to solvent extraction. Feed to the solvent extraction operation contains about 120 g Ni/liter, 165 g HCl/liter, and 2 g/liter of each of iron, cobalt, and copper. Iron is first extracted, after oxidation with oxygen, in two stages, by use of 4% TBP, followed by three-stage water stripping. Cobalt and copper are extracted with Ashland 383 (10% TIOA) in three stages, followed by three-stage water stripping, first for cobalt at O/A 30/1, and then for copper at O/A 20/1. Any remaining iron is removed in an additional water stripping stage. Solvesso 100 is the diluent throughout the plant. The copper strip is

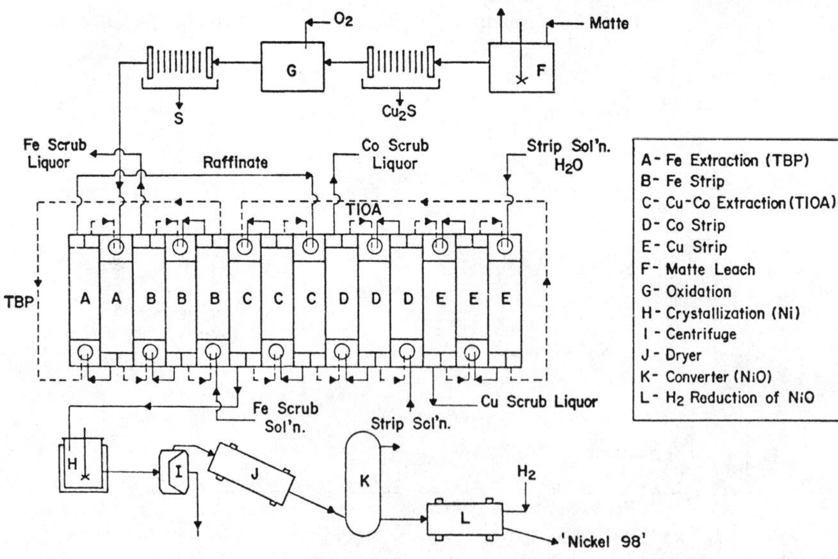

Figure 4. Falconbridge Nikkelverk A/S matte leach process.

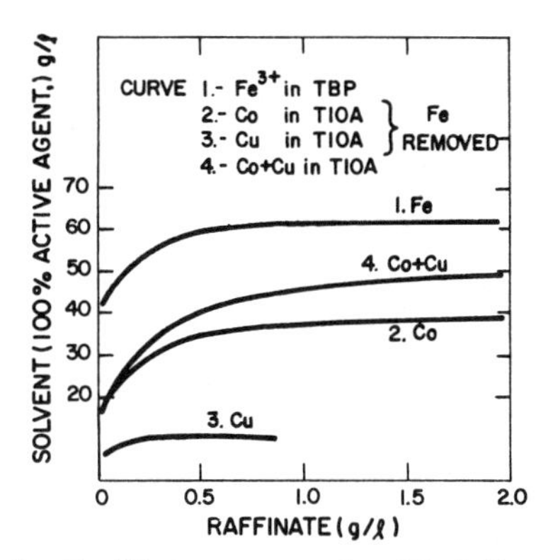

Figure 5. Equilibrium curves for 5% TBP and 10% TIOA.

recycled to the main refinery circuit, and the cobalt strip is further refined in the cobalt refinery, while the nickel chloride in the raffinate from extraction is crystallized and subsequently reduced to nickel powder. All streams are returned to a scrubber system for HCl removal at an efficiency of 99.8% recovery.

Equilibrium loading curves for iron, copper, and cobalt in the Falconbridge process are shown in Fig. 5.

A flow sheet has been proposed by Itzkovitch et al. [36] for the purification of nickel anolyte at Inco's refinery at Thomson, Manitoba.

The U.S. Bureau of Mines have done considerable research on production of high-purity nickel and cobalt from crude nickel metal and high-grade ferronickel [37]. The process, operated on a pilot scale, comprises the application of solvent extraction to extract and separate cobalt and iron from nickel, and the electrodeposition of pure nickel and cobalt from purified chloride electrolytes by use of diaphragm cells. A typical chloride-based electrolyte analyzed (in grams per liter) as 100 Ni, 160 Cl, 60 SO_4, and 50 Na. Iron was extracted with the secondary amine, Amberlite LA-1. A contact time of 3 min at 50°C was required for extraction, with the use of an aromatic diluent. After the LA-1 was cycled in the process, the initial high solubility of 2 mol % was reduced to 10 ppm. Because of better efficiency in the water stripping stage, 0.25 *M* amine was used instead of 0.5 *M* amine. Zinc, if present in the system, would be coextracted with the iron.

Cobalt was extracted using 0.5 *M* TIOA in an aromatic diluent, followed by water stripping. Copper and manganese, if present, were also extracted, as was most of any lead in the feed. Cobalt, manganese, and lead were completely stripped in three stages of water stripping, whereas only 5% of the copper was removed, which would build up on the solvent if not all removed before extraction. To produce both a concentrated cobalt chloride strip liquor and solvent free of both the copper and cobalt, the loaded solvent was scrubbed with water to strip the cobalt. A loaded solvent containing 2 g Co/liter and 0.01 g Cu/liter was stripped in six stages plus one water scrub stage, at an O/A of 25, to yield a cobalt strip containing 50 g Co/liter and 0.01 g Cu/liter; and a cobalt free liquor containing 0.24 g Cu/liter.

Cobalt was recovered from the strip liquor by either electrorefining or as a precipitate and conversion to the oxide. Cobalt chloride hexahydrate was recovered from the chloride strip liquor by evaporation of the water and crystallization of the salt. By reextraction from the chloride solution into TIOA, nickel was virtually eliminated, resulting in a cobalt chloride product containing 40 ppm of nickel. Cobaltous carbonate or cobaltous hydroxide could be precipitated at pH 8 by using either sodium carbonate or sodium hydroxide, and calcination to 900°C resulted in cobalt oxide. Another alternate method for a cobalt production was to precipitate cobaltic hydroxide by using chlorine and sodium carbonate.

In another process developed by the U.S. Bureau of Mines, Brooks et al. [38] described a process for recovering nickel, cobalt, molybdenum and chromium from superalloy waste grindings. The process includes preparation of the scrap, dissolution in chloride solution, carbon adsorption, three successive solvent extraction separations, and selective precipitations. A flow sheet for one version appears as Fig. 6 in Chapter 25.9.

The Gullspång Electrochemical Company in Sweden have developed a process for the separation and recovery of iron, cobalt, and nickel from scrap alloy [39]. The process involves first a pyrometallurgical pretreatment, followed by electrolytic dissolution in a chloride media, separation through solvent extraction, and electrolytic recovery of metals. The flow sheet [40] is shown diagrammatically in Fig. 6.

Ferric iron is first extracted at low chloride concentrations, ranging between 50 and 100 g/liter, in three stages of mixer–settlers by use of a solvent consisting of 25% Alamine 336 and 15% dodecanol in a kerosene diluent. To obtain extraction of the other metals, the chloride concentration is increased by controlled evaporation. The extraction curves in Fig. 7 illustrate the selectivity of metal extraction depending on the

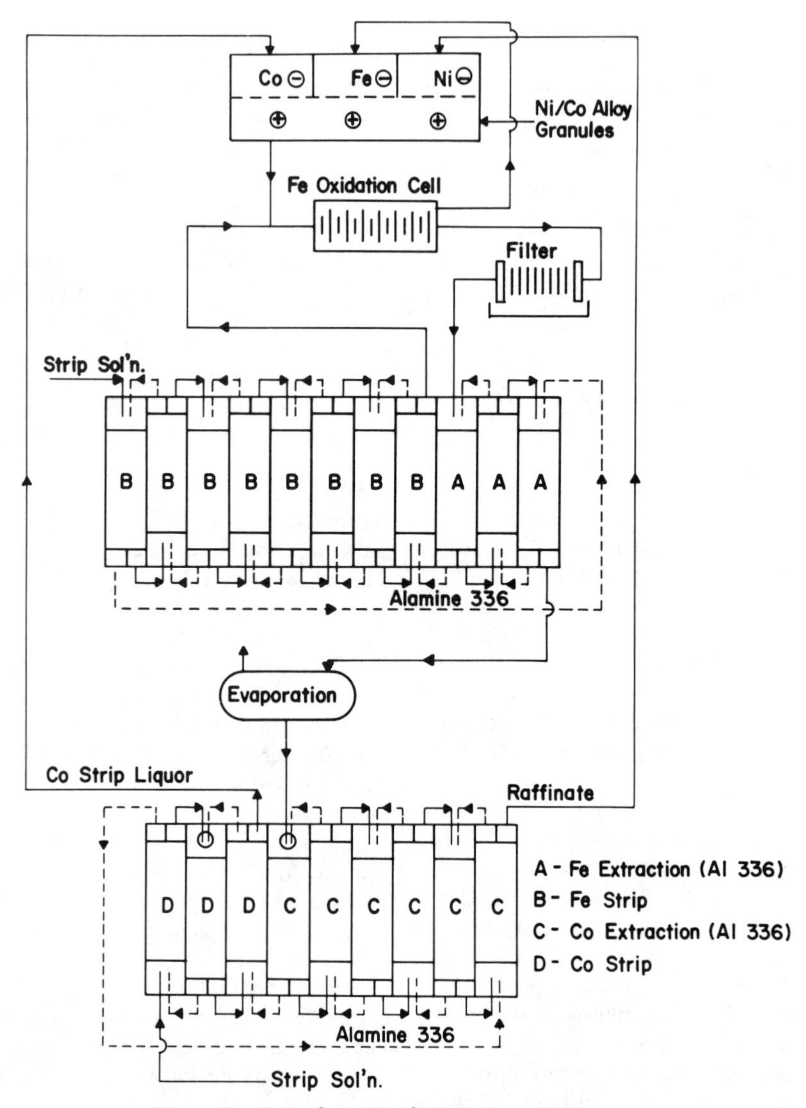

Figure 6. Gullspång nickel/cobalt alloy scrap process.

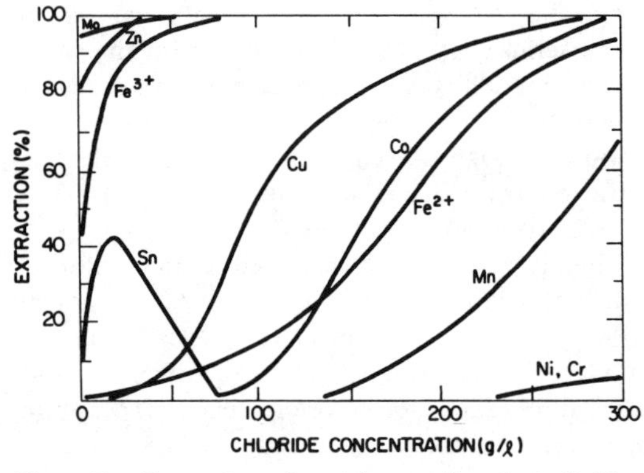

Figure 7. Extraction of metals as a function of chloride ion concentration.

total chloride content [39]. Evaporation of the raffinate to a chloride content of greater than 200 g/liter enables the cobalt to be extracted in six stages of mixer–settlers. The condensate from evaporation is used for water stripping—eight stages for iron recovery, whereas cobalt is stripped in three stages with slightly acid (pH 2–3) condensate. The raffinate, containing the nickel, is diluted with condensate to a chloride concentration of 102–150 g/liter. The three strip solutions of iron, cobalt, and nickel are recovered electrolytically in separate compartments of the dissolution cell. It has been reported that the quality of the cobalt and nickel is suitable for use by the company in alloy steel manufacture [40].

At Metallurgie Hoboken, Belgium, nickel-cobalt alloy scrap is dissolved in a chloride leach solution, followed by oxidation and lime addition to precipitate iron and chrome [40] (see also Chapter 25.9). Copper is removed by cementation on cobalt metal, followed by lead and manganese precipitation with chromic acid and cobalt oxide, respectively. Two successive solvent extraction steps are employed by use of a tertiary amine, one to remove zinc at a low (5–10 g HCl/liter) chloride ion concentration and the other to extract cobalt at a high chloride concentration (110 g HCl/liter). The zinc is recovered from the loaded solvent as $Zn(OH)_2$ by stripping-precipitation with sodium hydroxide, followed by filtration of the organic phase. Cobalt and nickel powders as well as cobalt oxide and nickel powder are produced by several reduction routes noted in the flow sheet shown as Fig. 8.

Baggott et al. [41] have described a process for the recovery of cobalt and nickel from chloride solutions of superalloy scrap, followed by solvent extraction for the nickel–cobalt separation (see Chapter 25.9, Fig. 7).

Société Le Nickel in France have described a process for the recovery of nickel and cobalt from metallurgical wastes [42] at their Le Havre [43] plant. The slag feed material is purified by

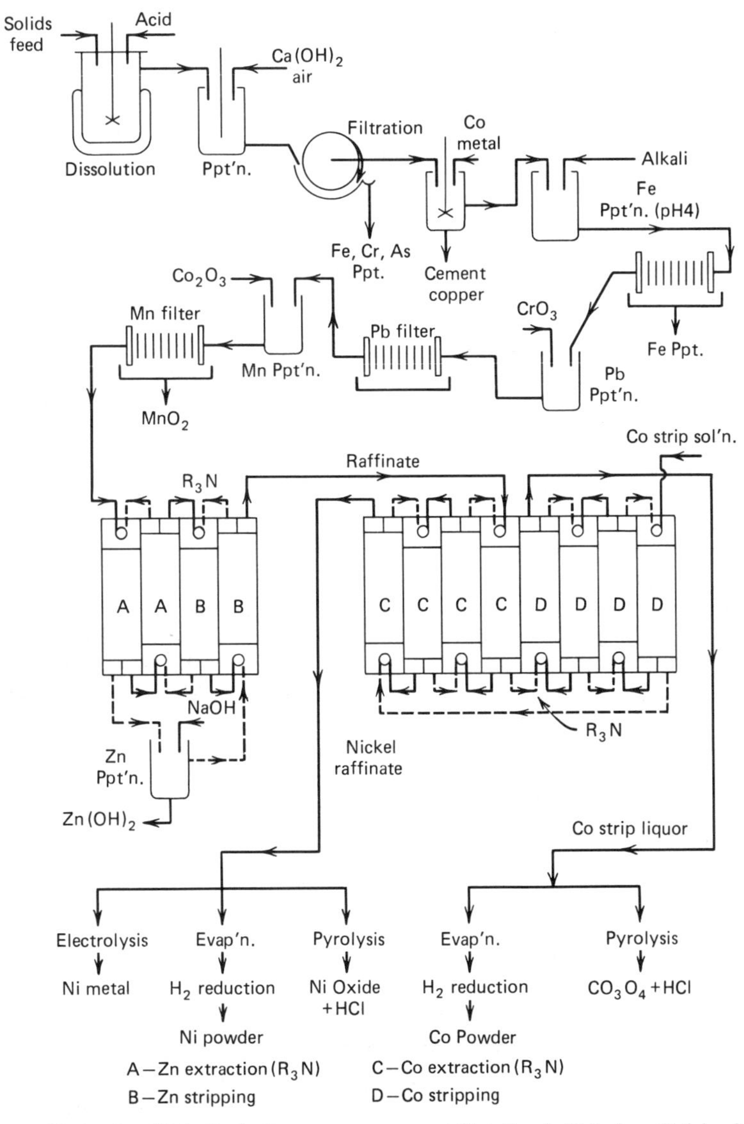

Figure 8. Nickel/cobalt recovery process at Metallurgie Hoboken (Belgium).

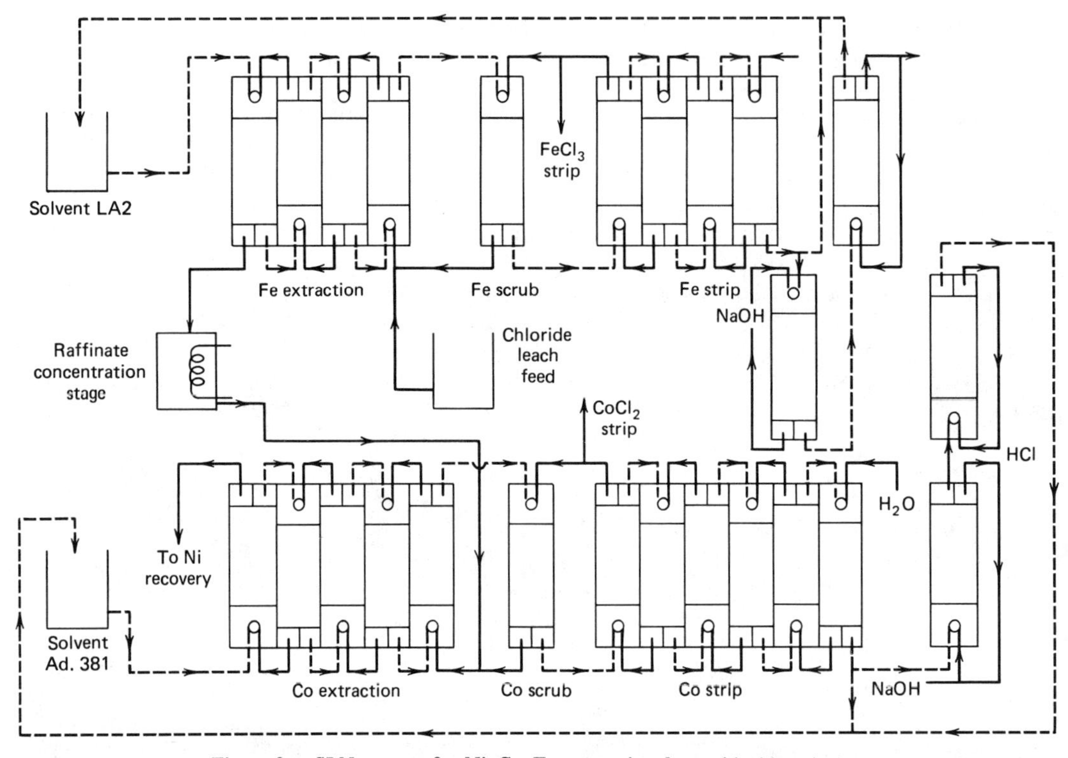

Figure 9. SLN process for Ni–Co–Fe separation from chloride solution.

injecting gaseous chlorine into semirefined molten matte under a layer of sodium chloride. The resultant sodium chloride slag contains the three metallic chlorides of iron, cobalt, and nickel. Water leaching the slag solubilizes the metal chlorides, which are fed to the solvent extraction plant. The leach solution composition (in grams per liter) is 56 Ni, 21.6 Co, 36.4 Fe, 0.15 Cu, and 55 Na and containing 7.3 N Cl^- and 0.2 N H^+. The flow sheet for the process is shown in Fig. 9. Amberlite LA-2 in Naphtha 90/160 (99% aromatic content) is first used to extract the iron. For the cobalt–nickel separation 0.30 M Adogen 381 in Naphtha 90/100, and containing 3% by volume of Shell octylol, resulted in a loading capacity of 8 g Co/liter. The raffinate from iron extraction must be concentrated to increase the chloride level to greater than 4.5 or 5 M. Four stages of extraction are required for iron removal, followed by a single-stage scrubbing and four stages of water stripping. Evaporation of the raffinate produces the desired chloride level for feed to five extraction stages for cobalt removal (plus copper). Following a single stage of scrubbing, the cobalt is removed by six stages of water stripping. Prior to recycle of the stripped solvents, in both circuits, back to extraction, a clean-up treatment is necessary, initially with NaOH and then with HCl. Soluble losses amount to 3 mg/liter in the iron strip liquor (1.3 N HCl), 3.5 mg/liter in the cobalt strip liquor (2.7 N HCl) and 3.0 mg/liter in the purified nickel raffinate (4.9 N HCl). This is equivalent to 330 mg amines per kilogram of nickel produced.

Deepsea Ventures have done considerable work on the treatment of ocean nodules by chloride processing [44–52]. Manganese, nickel, copper, and cobalt are selectively converted to water-soluble chlorides by use of HCl gas at a temperature of 200–400°C. Nickel, cobalt, and copper are separated and recovered by solvent extraction followed by electrowinning. The manganese is recovered from the raffinate by crystallization as the manganese chloride, followed by reduction to the metal.

In another Deepsea Ventures patent, using a mixture of coal, chlorine, and the nodules, the metal chlorides can be volatilized [46] at 800–1000°C. Again iron is converted to the oxide with steam. The condensed chlorides are water leached, the gangue removed, and the metals subsequently recovered by solvent extraction by use of the Kelex 100 reagent. Copper is first extracted. The raffinate is then contacted again with Kelex at pH 3.5 to coextract cobalt and nickel, which are then separated by stripping the nickel with spent electrolyte from the nickel electrowinning cell. Cobalt is stripped with HCl and purified by extraction with TlOA, following by stripping and electrolysis [46, 49]. Iron is extracted from a chloride solution with a tertiary amine, if not previously converted to the iron oxide [50].

4. ALKALINE AMMONIUM SULFATE OR CARBONATE SYSTEMS

The S. E. C. Corporation recovers copper and nickel from a crystallizer discharge stream operated by Phelps Dodge, a local copper refinery [53, 54]. The copper is first recovered as cathodes by solvent extraction and electrowinning from the acidic solution. The raffinate is neutralized and filtered to remove impurities and the pH adjusted to the alkaline side before recovery of zinc and nickel. LIX64N is used in both the copper and nickel circuits. The flow sheet for the plant is shown in Fig. 10.

Feed solution from Phelps Dodge contains about 70 g Cu/liter, 20 g Ni/liter, 1–3 g/liter of Fe + Al, and a trace of Zn, at a pH of 1–2. To prevent hydrolysis, the solution is diluted fourfold and is fed into extraction, in contact with 30% LIX64N in Napoleum 470 at a flow of 56 gal/min. The circuit is operated at 50% loading capacity (3.75 g Cu/liter) plus pH control with NH_3 in order to achieve a low raffinate (with respect to copper) to be subsequently fed to nickel recovery. The copper is stripped from the solvent by using 155–170 g/liter of free H_2SO_4, to attain an electrolyte of 35–45 g Cu/liter. Raffinate from the copper circuit is adjusted with anhydrous NH_3 to pH 8.2, which precipitates the iron and aluminum. The decant is filtered through a precoated filter.

The order of preferential loading is reported as Cu > Ni > Zn. Because zinc is present and could coextract with the nickel and subsequently produce serious electrowinning problems, it is removed from the solution by extraction with an unnamed extractant. The raffinate is then contacted with 9% LIX64N in two stages of mixer-settlers, with 60% utilization of the LIX64N capacity to extract the nickel. The pH is controlled at 8.5 and 9.5–10 in the two stages of extraction.

The nickel-loaded solvent is washed, in one

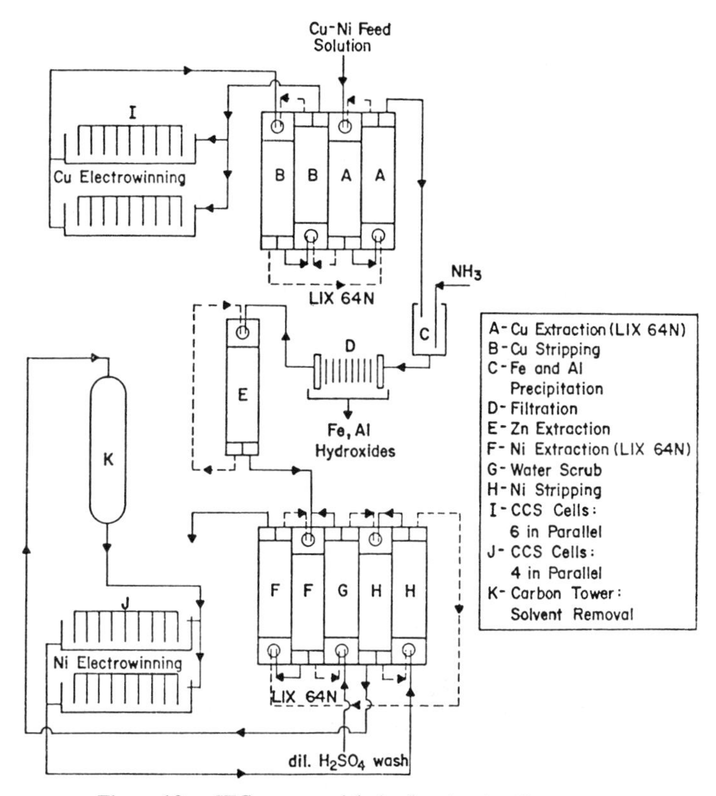

Figure 10. SEC copper–nickel solvent extraction process.

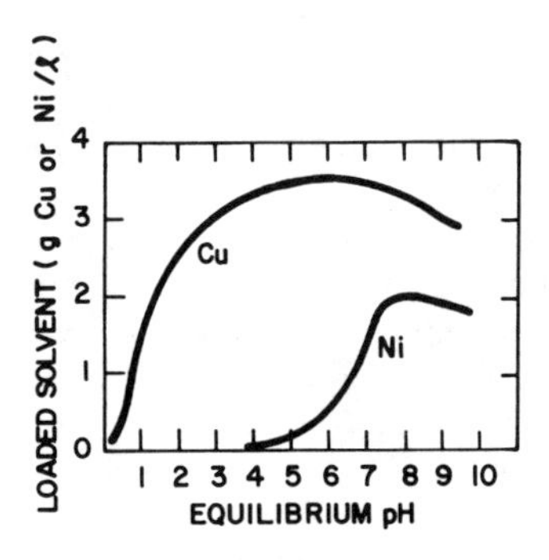

Figure 11. Copper-nickel loading on LIX64N versus equilibrium pH.

stage, with H_2SO_4-actified water (pH 3.5–5.0) to remove any entrained NH_3, which also removed about 3–5% of the nickel. If the ammonia were not removed, nickel ammonium sulfate would be precipitated in the stripping circuit. Two stages of stripping, controlled at pH 1.6 and 2.2 with H_2SO_4, recovers the nickel from the solvent.

Other work on the use of LIX64N for extraction from ammoniacal ammonium carbonate solutions has been reported by Merigold and Sudderth [55]. In their investigation of the nickel–copper–ammonia–carbonate system, containing 34 g Ni/liter, 2.9 g Cu/liter, 100 g NH_3/liter, and 100 g CO_3/liter, it is reported that (1) LIX64N loads copper preferentially to nickel, (2) strong ammonia-ammonium carbonate solutions, barren of copper and nickel, will scrub the nickel before stripping the solvent for copper recovery, (3) copper requires relatively strong acid for stripping, whereas nickel requires weak acid, and (4) copper present with the nickel in the electrolyte can be selectively removed by subsequent extraction with LIX64N (this is demonstrated in Fig. 11).

Stripping of nickel is accomplished by use of a feed containing 50 g Ni/liter and 37 g H_2SO_4/liter. Because of the presence of copper in the nickel electrolyte, a second extraction circuit for removal of copper is necessary prior to electrowinning of the nickel. The loaded solvent is scrubbed with a hot solution containing 10 g/liter of Na_2SO_4 to remove the ammonia that otherwise would be transferred to the strip circuit. The circuit proposed for recovery of copper and nickel from ammoniacal ammonium carbonate solutions is shown in Fig. 12. Subsequent to the above work on ammoniacal solution, treatment of a New Caledonian nickel laterite has been evaluated [56].

In addition to LIX64N for the extraction of nickel from ammoniacal sulfate or carbonate sys-

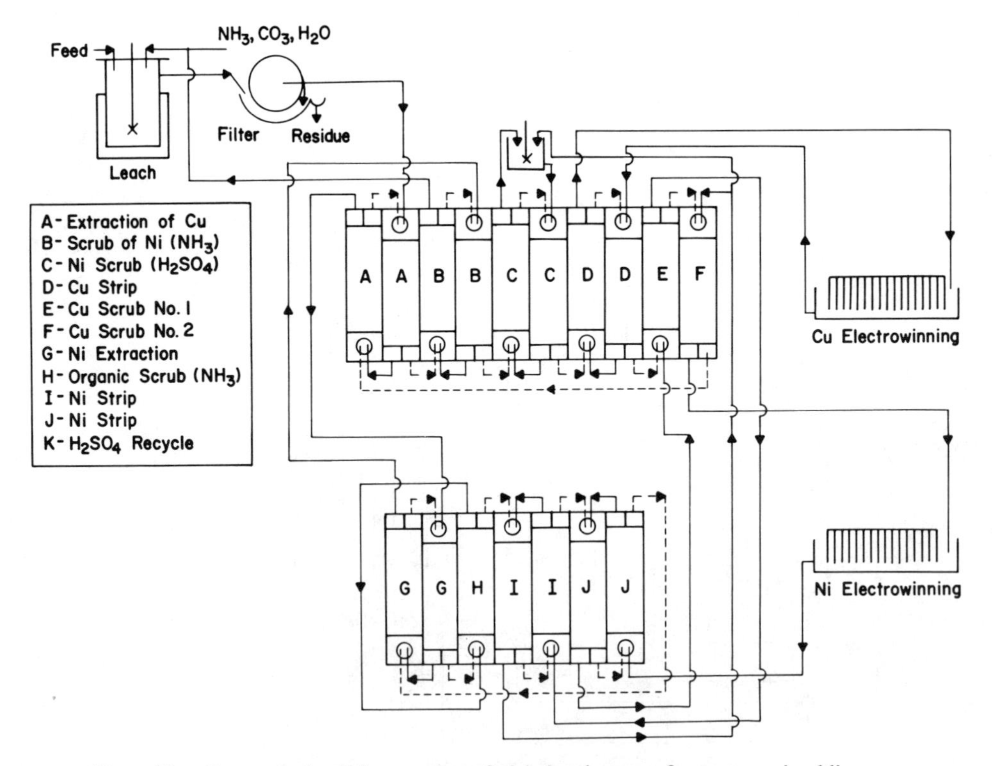

Figure 12. Proposed circuit for recovery of nickel and copper from ammoniacal liquors.

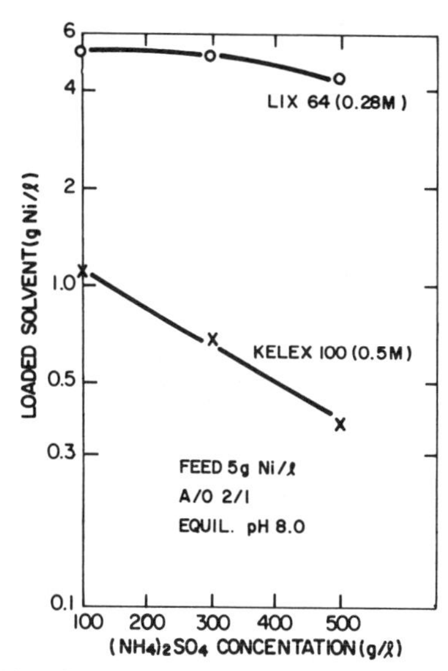

Figure 13. Comparison of LIX64N and Kelex 100 for nickel extraction at various $(NH_4)_2SO_4$ levels.

tems, Kelex has also been reported as suitable. Although less effective with respect to equilibrium loading than LIX64N, it has better selectivity for copper in the presence of nickel [57]. This is shown in Fig. 13 comparing the loading capability of 0.28 *M* LIX64N with 0.5 *M* Kelex 100, at an equilibrium pH 8.0, and varying concentrations of $(NH_4)_2SO_4$. The general order of extraction is Cu > Zn > Co > Ni. The order may be altered by interactions of solvent concentration, temperature, contact-time, anion concentration, and equilibrium pH.

Cobalt, as Co^{3+}, is readily extracted by Kelex 100 from ammoniacal solution, and the extractability increased with increasing pH from 7 to 9 and increasing $(NH_4)_2SO_4$ concentration from 50 to 500 g/liter. Repeated stagewise scrubbing with a Cu-Ni-$(NH_4)_2SO_4$ solution was unsuccessful in removing the Co^{3+}, and instead, the copper was scrubbed and replaced by the cobalt. If cobalt is extracted in the Co^{2+} state, it is readily removed from the solvent by stripping with 10% H_2SO_4. However, if the loaded solvent is exposed to air, the Co^{2+} is oxidized to Co^{3+}, resulting in a rapid decrease in the stripping of cobalt (Fig. 14) [58].

A process for the separation of cobalt and nickel from ammoniacal solution has been described by Ritcey and Ashbrook with the use of D2EHPA [11]. For optimum conditions of extraction of cobalt and separation from nickel, the solution must be oxidized to ensure that the cobalt is in the Co^{3+} state, that the alkalinity is pH 11.0 to 11.5, and that the solution contains not

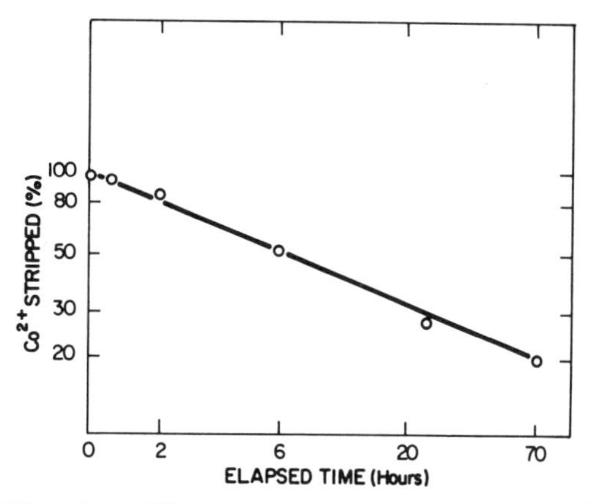

Figure 14. Effect of time on the oxidation of Co^{2+} and stripping.

greater than 40 g SO_4/liter (Fig. 15). In an alkaline ammoniacal nitrate solution, up to 100 g NO_3/liter can be tolerated.

Shell SME529, although designed for the extraction of copper from acidic solutions [59], is also capable of extraction from alkaline solutions [59, 60]. In acidic solutions nickel, cobalt, and zinc are not chelated, but Mo^{6+} forms a strong complex. From alkaline solutions, copper, nickel, and cobalt are strongly extracted. Like the other chelating extractants, the cobalt must be either Co^{3+}, and not extracted, or as Co^{2+} with controlled stripping conditions [60]. Zinc is weakly extracted and, in the presence of copper and nickel, is rejected. Small amounts of NH_3 are coextracted with the metal, which is removed in a two-stage scrub by use of 1 *M* $(NH_4)HCO_3$.

A system was described by van der Zeeuw [59], who used a 35% volume solution of SME529 in Shell MSB210 diluent. The aqueous

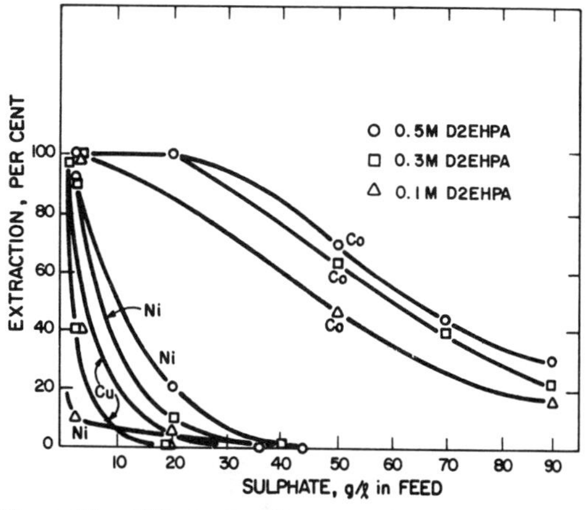

Figure 15. Effect of sulfate on extraction of copper, nickel, and cobalt with D2EHPA.

solution had the composition (in grams per liter) 53.5 Cu, 7.5 Ni, 52.7 CO_3, and 76.4 total NH_3. Extraction and stripping were at 40°C. One stage was required for extraction. For removal of any coextracted nickel, a neutralized tankhouse bleed solution containing (in grams per liter) 29.7 Cu, 9.7 Ni, and 14.7 free H_2SO_4 was used for two-stage scrubbing. Copper was removed from the scrubbed solvent by stripping with spent electrolyte containing 30 g Cu/liter and 150 g H_2SO_4/liter in two stages.

Leach solutions arising from alkaline leaching of sulfide concentrates or solutions from leaching of ores with ammonium carbonate contain high concentrations of either $(NH_4)_2SO_4$ or $(NH_4)_2CO_3$. Processes for the recovery of copper from such leach solutions [61], as well as cobalt [12], have been described by Ritcey and Lucas. The alkaline leach produced by ammonium leaching of nickel sulfide ore, for example, by the Sherritt process, can contain high concentrations of nickel and copper and a trace of cobalt. By use of LIX63 (61), LIX 64N (55), Kelex 100 (57), or SME529 (60), the copper is first selectively extracted and separated from the nickel and cobalt. In the conventional Sherritt Gordon process, the nickel is recovered by precipitation by using hydrogen reduction at 350°F [4, 62]. The resulting solution from nickel precipitation contains approximately 1 g/liter each of cobalt and nickel and 500 g/liter of $(NH_4)_2SO_4$, which is contacted with H_2S at atmospheric pressure in order to precipitate the residual cobalt and nickel. The cobalt is recovered from this mixed sulfide in several stages [4]. To eliminate the many process steps and equipment, an alternative process using solvent extraction was developed as shown in Fig. 16.

After contact of the leach feed solution at pH 8.0 with Versatic 911 or Versatic 10, the loaded solvent is purified by scrubbing with a cobalt ammine sulfate solution at pH 8 to remove small amounts of coextracted nickel. A bleed of the scrub recycle stream is returned to the extraction circuit as the nickel content increases due to scrubbing. The scrubbed solvent containing the cobalt can be stripped by various mineral acids, and sulfuric acid stripping is shown in Fig. 16. The resultant cobalt sulfate solution is reduced by hydrogen in the conventional manner [5], and the residual ammonium sulfate is recovered by crystallization. A portion of the cobalt sulfate solution is returned to the scrub circuit to maintain the necessary cobalt balance in that circuit.

The nickel contained in the raffinate can be recovered by subsequent extraction into Versatic 911, LIX64N, SME529, or Kelex 100, followed by stripping with sulfuric acid. This adjusted strip solution would be amenable to hydrogen reduction to produce nickel powder [5]. Another alternative would be recovery by precipitation of the sulfide.

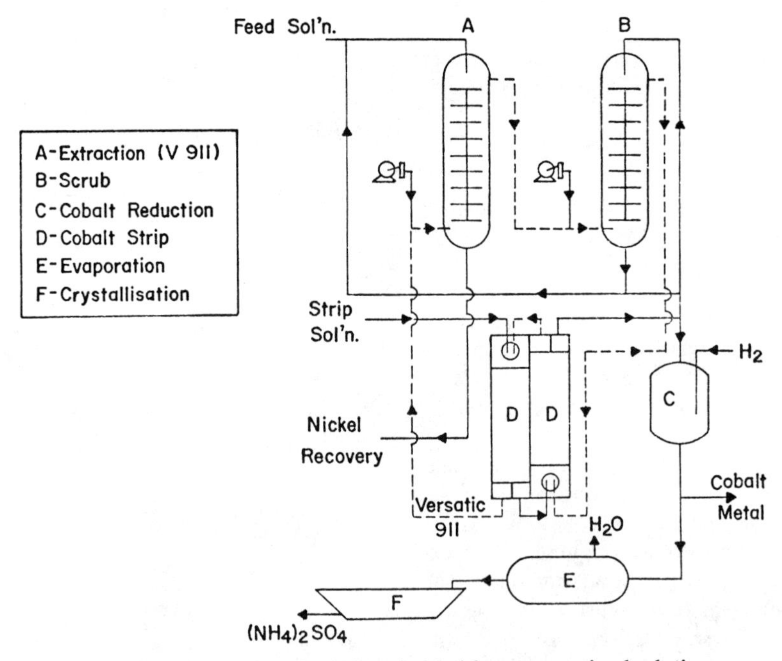

Figure 16. Separation of cobalt and nickel from ammoniacal solution.

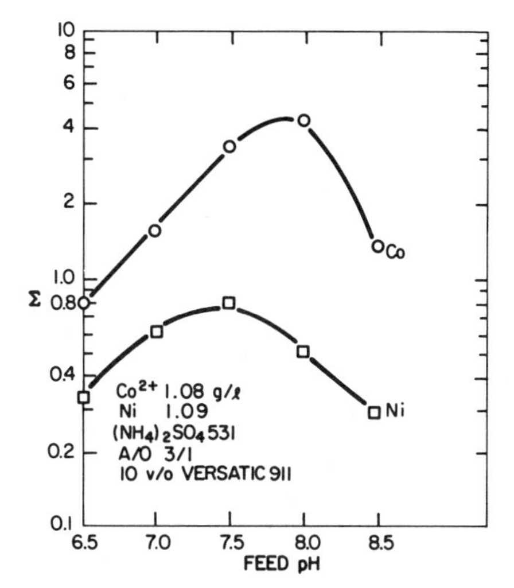

Figure 17. Effect of feed pH on separation of cobalt and nickel.

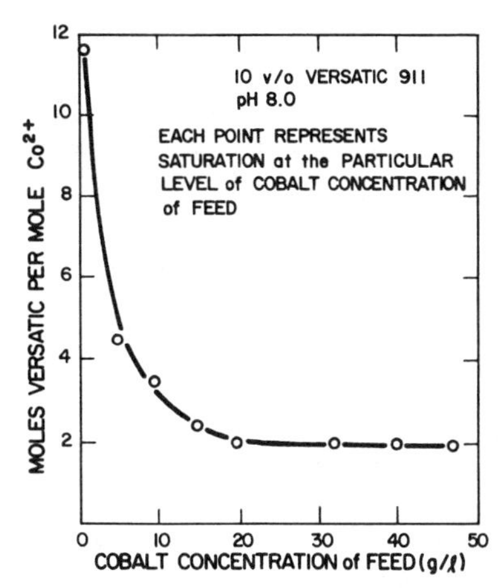

Figure 19. Effect of cobalt concentration in feed on cobalt loading of Versatic 911.

The extraction of cobalt and separation from nickel using Versatic 911 is very pH dependent, as well as being sensitive to the $(NH_4)_2SO_4$ concentration as shown in Figs. 17 and 18. Also it is prefereable for the cobalt to be in the Co^{2+} state, although Co^{3+} will also extract, but to a lesser extent. Saturation of the Versatic 911 is achieved only with feed solution containing 20 g/liter, indicating that two moles of Versatic 911 are required per mole of cobaltous cobalt at saturation (Fig. 19). The loaded solvent can be readily scrubbed to remove any coextracted nickel. The cobalt is readily recovered from the loaded solvent with 10% H_2SO_4, and a 50 g Co/liter strip solution can be produced in one stage of stripping with an efficiency of 90%.

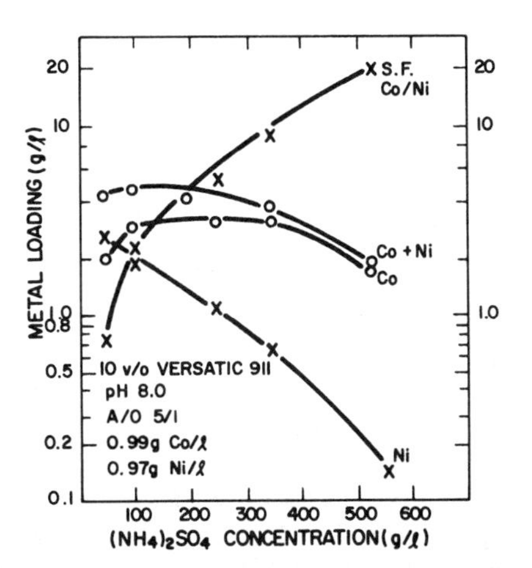

Figure 18. Effect of $(NH_4)_2SO_4$ concentration on separation of cobalt from nickel.

Extraction and separation of cobalt from nickel can occur from several types of solutions, with the following orders:

Order of Co–Ni separation

$$NO_3^- > Cl^- > SO_4^= > CO_3^=$$

Order of total metal loading

$$NO_3^- > SO_4^= > CO_3^= > Cl^-$$

Soluble losses in the system are dependent on the equilibrium pH and the $(NH_4)_2SO_4$ concentration. Increased solubility results from increase in pH above 8.0 and decrease in $(NH_4)_2SO_4$ concentration below 4 *M*. At the conditions in the flow sheet, at pH 8.0, and 4 *M* $(NH_4)_2SO_4$, the loss is of the order of 10 ppm Versatic 911.

In recent years, work has been described in the literature on the direct reduction of the loaded solvent to produce metal powders [63, 64]. This route might be applicable to the recovery of both the cobalt and the nickel from the metal–carboxylate without the requirement of a stripping circuit.

REFERENCES

1. M. H. Caron, *Transact. Am. Inst. Min. Engr.* **188**, 67–90 (1950).
2. M. H. Caron, *Transact. Am. Inst. Min. Eng.* **188**, 91–103 (1950).
3. Nicaro Expands Nickel Capacity, *Eng. Min. J.* **158**, 82–89 (September 1947).
4. W. Kunda, J. P. Warner, and V. N. Mackiw, *Transact. Can. Inst. Min. Met.* **65**, 21–25 (1962).

5. R. Stauffer and S. Lindsay, *Transact. Can. Inst. Min. Met.* **70**, 161–166 (1967).
6. R. B. Schaal, U.S. Patent 2,379,659 (1945).
7. J. S. Mitchell, *J. Metals NY* **9**, 1945 (1957).
8. H. Veltman, D. J. Evans, and V. N. Mackiw, *Bull. Can. Inst. Min. Met.* **57**, 1281–1287 (1964).
9. M. H. Caron, *De Ingenieur* **69** (36), M19–M32 (1957).
10. M. H. Caron, *De Ingenieur* **68** (18), M8–M12 (1956).
11. G. M. Ritcey and A. W. Ashbrook, *Transact. Inst. Min. Met. C* **78** (751), 57–63 (June 1969); U.S. Patent 3,438,768 (April 1969).
12. G. M. Ritcey and B. H. Lucas, in *Proceedings of the International Solvent Extraction Conference (ISEC), The Hague, 1971*, Society of Chemical Industry, London, 1971, pp. 463–475; also U.S. Patent 3,718,458 (1973).
13. G. M. Ritcey and A. W. Ashbrook, in *Solvent Extraction–Principles and Applications to Process Metallurgy*, Vol. 2, Elsevier, Amsterdam, 1979.
14. J. E. Magner, in P. Queneau, Ed., *Extractive Metallurgy of Copper, Nickel and Cobalt*, Proceedings of the International Symposium, TMS/AIME, New York, 1960.
15. A. W. Fletcher and K. D. Hester, *Transact. Soc. Min. Eng. AIME* **229**, 282–291 (1964).
16. A. W. Fletcher and D. S. Flett, The Separation of Nickel from Cobalt by Solvent Extraction with a Carboxylic Acid, paper presented at Annual AIME Meeting, Washington DC, 1969.
17. L. M. Gindin, P. I. Bobikov, E. F. Kouba, and A. V. Bugeava, *Russ. J. Inorg. Chem.* **5** (8), 906–910 (August 1960).
18. L. M. Gindin, P. I. Bobikov, E. F. Kouba, and A. V. Bugaeva, *Russ. J. Inorg. Chem.* **5** (10), 1146–1149 (October 1960).
19. L. M. Gindin, P. I. Bobikov, E. F. Kouba, and A. V. Bugaeva, *Russ. J. Inorg. Chem.* **6** (12), 1412–1416 (December 1961).
20. P. I. Bobikov and L. M. Gindin, *Internatl. Chem. Eng.* **3** (1), 133–138 (January 1963).
21. R. Blumberg and P. Melzer, *Proceedings of International Mineral Processing Congress*, 1964, Vol. 1, Gordon & Breach, New York, 1964, pp. 139–145.
22. G. M. Ritcey and A. W. Ashbrook, U.S. Patent 3,399,055 (August 1968).
23. G. M. Ritcey and A. W. Ashbrook, *Transact. Inst. Min. Met. C* **78** (751), 57–63 (June 1969); also U.S. Patent 3,438,768 (April 1969).
24. G. M. Ritcey, A. W. Ashbrook, and B. H. Lucas, Development of a Solvent Extraction Process for the Separation of Cobalt from Nickel, presented at Annual Meeting, AIME, San Francisco, February 1972, CIM Bulletin, January 1975.
25. Private communication to G. M. Ritcey and A. W. Ashbrook.
26. M. Ando, S. Emoto, M. Takahashi, T. Kasai, and N. Nakayama, Nickel and Cobalt Refinery of Nippon Mining Co., paper presented at 7th Annual Hydrometallurgical Meeting, CIM, Vancouver, August 1977.
27. A. W. Ashbrook and G. M. Ritcey, Hydrometallurgical Processing of Superalloy Scrap for the Recovery of Cobalt and Nickel, paper presented at the Conference of Metallurgists, CIMM, Montreal, 1971.
28. L. F. Cook and W. W. Szmokaluk, in *Proceedings of the International Solvent Extraction Conference (ISEC), The Hague, 1971*, Society of Chemical Industry, London, 1971, pp. 451–462.
29. B. G. Nyman and L. Hummelstedt, in *Proceedings of the International Solvent Extraction Conference (ISEC), Lyons, 1974*, Vol. 1, Society of Chemical Industry, London, pp. 669–684.
30. D. S. Flett, M. Cox, and J. D. Heels, *Proceedings of the International Solvent Extraction Conference (ISEC), Lyons, 1974*, Vol. 3, Society of Chemical Industry, London, pp. 2560–2575.
31. L. Hummelstedt, H. E. Sund, J. Karjaluoto, L. O. Berts, and B. G. Nyman, *Proceedings of the International Solvent Extraction Conference (ISEC), Lyons, 1974*, Vol. 1, Society of Chemical Industry, London, pp. 829–848.
32. G. M. Ritcey and B. H. Lucas, *CIM Bull.* (February 1974).
33. S. O. Fekete, G. A. Meyer, and G. R. Wicker, The Selective Extraction of Nickel and Cobalt from Acid Leach Solutions Using a Mixed Solvent System, paper presented at 1977 AIME, Atlanta, TMS Paper No. A77-95.
34. P. G. Thornhill, E. Wigstol, and G. VanWeert, *J. Met.* 13–18 (July 1971).
35. E. Wigstol and K. Froyland, in *Proceedings of Solvent Extraction in Metallurgical Processes*, Antwerp, May 4–5, 1972, Technologische Instituut K. VIV, Antwerp, 1972, pp. 71–81.
36. I. J. Itzkovitch, V. A. Ettel, and A. S. Gendron, *CIM Transact.* **77**, 58–62 (1974).
37. P. T. Brooks and G. M. Potter, U.S. Bureau of Mines, Report of Investigation, RI 74-2, 1970.
38. P. T. Brooks, G. M. Potter, and D. A. Martin, U.S. Bureau of Mines, Report of Investigation, RI 7316, 1969.
39. A. Aue, L. Skjutare, G. Björling, H. Reinhardt, and J. Rydberg, in *Proceedings of the International Solvent Extraction Conference (ISEC), The Hague, 1971*, Society of Chemical Industry, London, 1971, pp. 447–450.
40. A. W. Fletcher, *Chem. Ind.* 414–419 (May 1973).
41. E. R. Baggott, A. W. Fletcher, and T. A. W. Kirkwood, Recovery of Valuable Metals from Nickel–

Cobalt Alloy Scrap, paper presented at Ninth Commonwealth Mining and Metallurgical Congress, London, May 1969.

42. C. Bozec, J. M. Demarthe, and L. Gandon, in *Proceedings of the International Solvent Extraction Conference (ISEC), Lyons, 1974*, Vol. 2, Society of Chemical Industry, London, pp. 1201–1230.
43. D. S. Flett and D. R. Spink, *Hydrometallurgy*, **1** 207 (1976).
44. P. H. Cardwell, *Min. Congr. J.* 38–43 (November 1973).
45. W. S. Kane and P. H. Cardwell, U.S. Patent 3,752,745 (1973).
46. W. S. Kane and P. H. Cardwell, U.S. Patent 3,773,635 (1973).
47. W. S. Kane and P. H. Cardwell, U.S. Patent 3,795,596 (1974).
48. W. S. Kane and P. H. Cardwell, U.S. Patent 3,809,624 (1974).
49. W. S. Kane and P. H. Cardwell, U.S. Patent 3,810,827 (1974).
50. W. S. Kane and P. H. Cardwell, U.S. Patent 3,832,165 (1974).
51. H. L. McCutchen and P. H. Cardwell, U.S. Patent 3,855,059 (1974).
52. R. Sisselman, *E/MJ*, 75–86 (April 1975).
53. R. D. Eliasen, The Operation of a Nickel Solvent Extraction and Electrowinning Circuit, *Proceedings of Symposium on Solvent Ion Exchange*, AIChE, Tucson, Arizona, May 1973.
54. R. D. Eliasen and E. Edmunds, Jr., *CIM Bull.* 82–86 (February 1974).
55. C. L. Merigold and R. B. Sudderth, *Proceedings of the International Symposium on Hydrometallurgy*, Chicago, February 1973, AIME, New York, 1973, pp. 552–587.
56. C. R. Merigold and W. H. Jensen, in *Proceedings of the International Solvent Extraction Conference (ISEC), Lyons, 1974*, Vol. 2, Society of Chemical Industry, London, pp. 1231–1262.
57. G. M. Ritcey and B. H. Lucas, *CIM Bull.* (February 1975).
58. G. M. Ritcey, unpublished data.
59. A. J. Van der Zeeuw, in *Extractive Metallurgy of Copper–Hydrometallurgy and Electrowinning*, Vol. II, AIME, 1976, Port City Press, Baltimore, 1976, pp. 1039–1055.
60. G. M. Ritcey and B. H. Lucas, unpublished data.
61. G. M. Ritcey and B. H. Lucas, *CIM Transact.* **L75**, 82–86, 1972; also U.S. Patent 3,761,249 (1973).
62. V. N. Mackiw, W. C. Lin, and W. Kunda, *J. Met. Transact.* **209**, 786 (1957).
63. A. R. Burkin, *Powder Met.* **12**, 243–250 (1969).
64. F. D. Richardson, *Jernkont, Ann.* **153**, 359–370 (1969).

25.3

COMMERCIAL PROCESSES FOR TUNGSTEN AND MOLYBDENUM

M. B. MacInnis and T. K. Kim
GTE Products Corporation
United States

1. INTRODUCTION

Tungsten and molybdenum are very similar chemically. Both are transition metals in group VI of the periodic table. They form compounds with oxidation states of 0, +2, +3, +4, +5, and +6; the last is the most stable.

Tungsten has a higher melting point (3387°C) than any other element. As a consequence, it is widely used for components and structural parts that require very high temperature resistance. At temperatures above 1650°C, tungsten has the highest tensile strength of all metals. Its use as filaments in incandescent lamps is the best known. However, cutting and wear-resisting materials, welding and hard-facing rods, tool and die steels, superalloys, and nonferrous alloys account for 80% of recent consumption.

Molybdenum has a high melting point (2610°C) and a high resistance to corrosion. As a consequence, molybdenum has been widely used for components and structural parts that require hot strength and corrosion resistance. The largest quantities are used in alloy steel, which requires a combination of wear resistance, strength, and toughness.

Tungsten ore concentrates are usually processed by either an acid process or an alkali process. In the alkali process the tungsten is converted to sodium tungstate at atmospheric or above-atmospheric pressure. The resulting sodium tungstate is first converted to calcium tungstate and/or tungstic acid. The tungstic acid is finally converted to ammonium tungstate, which is the precursor of the commercial product, ammonium paratungstate. In the acid process the tungstic acid is the first product prepared, and further processing is similar to that in the alkali process. Tungsten chemicals (ammonium paratungstate, tungstic acid, calcium tungstate, and sodium tungstate) are produced as coproducts at some tungsten-processing plants and as primary products in others. Usually, the tungsten plant must select the kind and quality of concentrates that can be processed.

Molybdenum ore concentrates are generally processed by a method in which the initial step is roasting of the sulfide to molybdenum oxide. The technical-grade molybdenum oxide is then sublimed in air to produce pure molybdenum trioxide. As an alternate method, the technical-grade molybdenum oxide is upgraded by chemical leaching.

Tungsten and molybdenum compounds polymerize at lower pH to form high-molecular-weight isopolyanions. Many tungstate species have been proposed to exist in aqueous solution. Generally, it has been known that a series of polytungstate anions predominate and rarely does one find cationic species [1-4]. Similarly, many molybdate species have been proposed to exist in aqueous solution. It is known that a series of polymolybdate anions predominate in the region of pH greater than 2. However, unlike tungsten, molybdenum forms cationic species [4-6] at a lower pH.

The introduction of solvent extraction allows one to take advantage of this polymerization and at the same time permits flexibility in the processing of molybdenum and tungsten from any source to obtain high quality in high yields.

2. TUNGSTEN SOLVENT EXTRACTION

2.1. Tungsten Sources

Tungsten is found and produced on nearly all continents and occurs in the minerals scheelite, $CaWO_4$; ferberite, $FeWO_4$; hubnerite, $MnWO_4$; and wolframite, $(Fe, Mn)WO_4$. Since the ores contain at best only 2-3% tungsten, a concentrate must be produced first. The objective is to produce a concentrate containing at least 60% WO_3, although concentrates as low as 15% WO_3 are used. The tungsten concentrate contains various concentrations of different impurities depending on the source of the ore. The usual impurities are tin, copper, arsenic, antimony, bismuth, phosphorus, molybdenum, calcium, iron, manganese, lead, sulfur, and silicon.

Another significant source of tungsten is tungsten scraps, and the contained impurities are again dependent on the source.

2.2. Conversion of Tungsten to Sodium Tungstate Solution

Tungsten concentrates are converted to sodium tungstate by pressure leaching or fusion with sodium carbonate or by baking with sodium hydroxide at a temperature of 90-150°C followed by water leaching to dissolve the sodium tungstate [7]. Tungsten scraps can also be converted to the water-soluble sodium tungstate by fusion with sodium nitrate.

2.3. Preparation of Tungsten Solvent Extraction Feed Solution

The feed solution for the solvent extraction process is usually prepared from an impure sodium tungstate solution. During the extraction process, undesirable impurities such as silicon, arsenic, phosphorus, and molybdenum are coextracted with tungsten. These impurities should be removed from the solvent extraction feed solution so that they are not transferred to the tungsten product. Silicon can be removed by filtration as SiO_2 after adjustment of the pH of the sodium tungstate solution to 9.5. Its presence in tungsten metal products causes brittleness. Arsenic, phosphorus, molybdenum, and similar impurities can be removed from sodium tungstate solutions. The addition of excess sodium hydrogen sulfide to a solution with a pH at least 8 causes the thio complexes to form. These impurities precipitate as the sulfides when the pH is adjusted to about 3 and are then separated by filtration. The major impurity is usually molybdenum [8], and its presence in tungsten is very detrimental to the life of incandescent filaments. After the separation of impurities such as silicon, phosphorus, arsenic, and molybdenum from the sodium tungstate solution, the excess sodium hydrogen sulfide is converted to sodium sulfate by oxidation. The pH of the sodium tungstate solution is adjusted to 2, and it is used as the tungsten solvent extraction feed solution. Tungsten concentration of the feed solution is 150-200 g WO_3/liter.

2.4. Tungsten Solvent Extraction System

The solvent extraction system consists of a high-molecular-weight tertiary alkyl amine, tricaprylyl amine; tri-*n*-butyl phosphate; and kerosene. Tri-*n*-butyl phosphate is the modifier. The optimum composition with respect to loading, rate of extraction, and phase separation for processing the sodium tungstate solution prepared as described in Section 2.3 is 12% by volume tricaprylyl amine, 12% by volume tri-*n*-butyl phosphate, and 76% by volume kerosene [9]. The tricaprylyl amine is converted to the sulfate salt prior to the tungsten extraction and is prepared by being contacted with 1.5 *N* H_2SO_4. The mechanism involved in the tungsten extraction is an anion exchange and was described previously [2].

2.4.1. *Extraction Isotherm*

The extraction isotherm was plotted and a McCabe–Thiele diagram was prepared for estimating the number of theoretical stages. The optimum contact time or equilibration time was determined prior to the construction of the extraction isotherm and found to be 3 min. Figure 1 is the tungsten extraction isotherm. The plot shows that two stages of extraction are adequate for complete extraction of tungsten when the organic:aqueous phase ratio is 1.3 and about 90% of the tungsten is extracted in the first stage.

2.4.2. *Stripping Isotherm*

The stripping isotherm was plotted and a McCabe–Thiele diagram was prepared for estimating the number of theoretical stages. The optimum contact time or equilibration time was determined prior to construction of the stripping isotherm and found to be two minutes. Figure 2 is the tungsten-stripping isotherm. The plot shows that one stage of stripping is sufficient for complete stripping of tungsten. The stripping agent is a dilute ammonium tungstage solution fortified with ammonia gas to a pH of 11.

2.4.3. *Continuous Extraction System*

A laboratory-scale mixer–settler system (Fig. 3) was operated. The design of the solvent extraction system was based on the data obtained from the extraction and stripping isotherms. The extraction isotherm showed that two stages of extraction are adequate. However, three stages were used as a precautionary measure against unexpected tungsten losses. The water washing of the

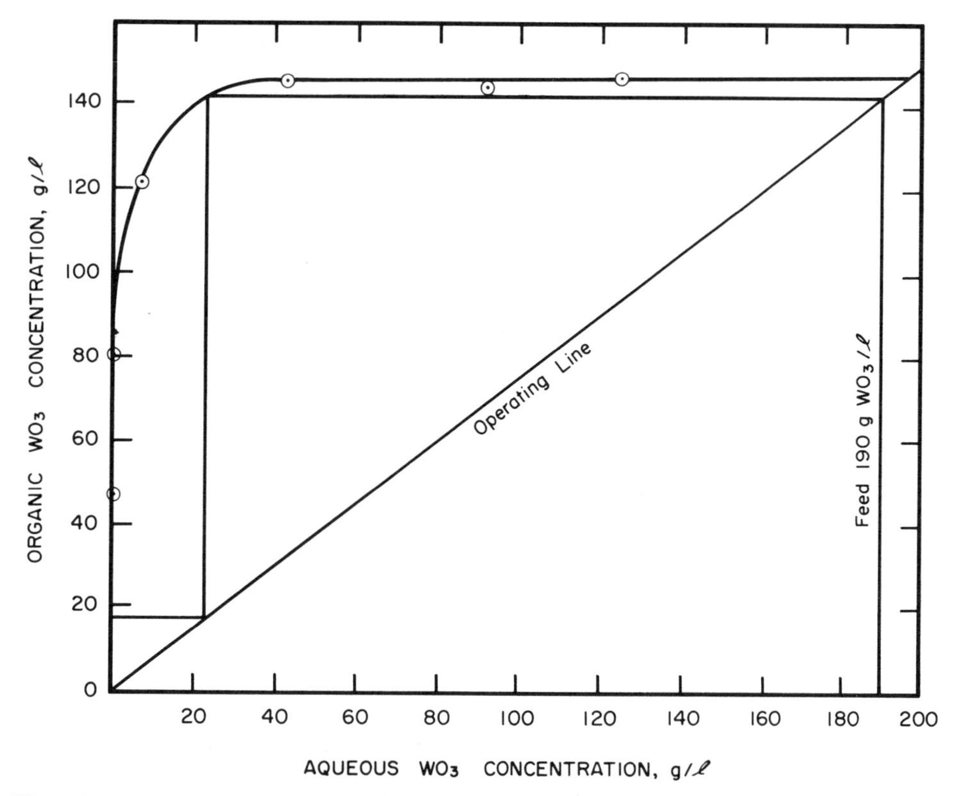

Figure 1. Tungsten extraction isotherm. Organic solution: 12% by vol. Alamine 336–12% by vol. TBP–76% by vol. kerosene. Aqueous solution: 190 g WO_3/liter, pH 2.0.

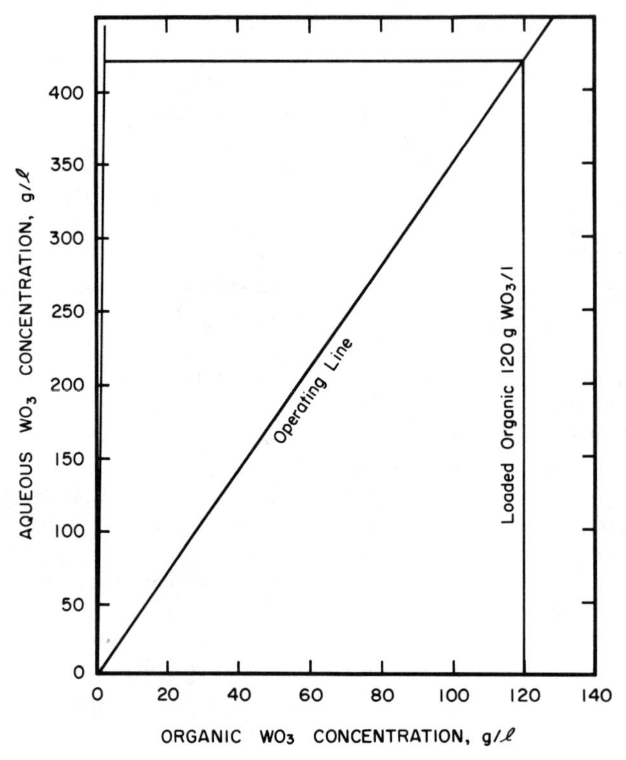

Figure 2. Tungsten stripping isotherm. Stripping solution: ammonium tungstate solution (30 g WO_3/liter) fortified with ammonia gas, pH 11.

organic following the extraction circuit is used to minimize aqueous-phase entrainment. In the stripping circuit a column contactor having a length : diameter ratio of 4.5 was used. The stripping solution consisted of a dilute ammonium tungstage solution fortified with ammonia gas to a pH of 11. The ammonium tungstate solution obtained from the mixer-settler system had a concentration of 350 g WO_3/liter. The product was then isolated by the crystallization of ammo-

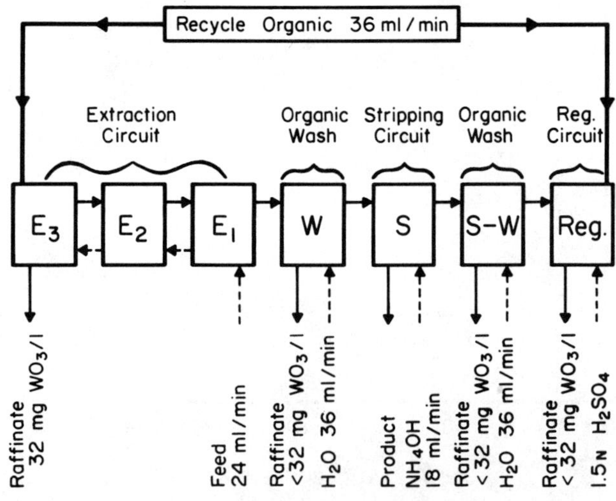

Figure 3. Continuous laboratory mixer-settler system. Organic solution: 12% by vol. Alamine 336–12% by vol. TBP–76% by vol. kerosene. Aqueous solution: 190 g WO_3/liter, pH 2.0.

nium paratungstate, which is the precursor of pure WO_3. The system was then upscaled and operated as a manufacturing plant.

3. MOLYBDENUM SOLVENT EXTRACTION

3.1. Molybdenum Sources

Molybdenum is predominantly found in the earth's crust in the form of molybdenite (MoS_2). The concentration in the ore is about 0.3–0.6% by weight. The ore is usually concentrated to 60–90% MoS_2 by weight. Copper porphyries are also an important source of molybdenum, although the molybdenite content of these ores is considerably less than the molybdenum porphyries and runs anywhere from 0.005 to 0.1% MoS_2. The usual impurities found in the concentrated ore are copper, iron, potassium, chromium, silicon, aluminum, nickel, and calcium.

3.2. Conversion of Molybdenum to Sodium Molybdate Solution

Molybdenum concentrates are converted to sodium molybdate by fusion, at 700–950°C, in an oxidizing atmosphere with at least 15% in excess of the theoretical amount of sodium carbonate required to convert the molybdenum and sulfide present in the concentrate to water-soluble molybdenum and sulfur compounds. A sodium molybdate solution is obtained by leaching the fusion mass with water [10].

3.3. Preparation of Molybdenum Solvent Extraction Feed Solution

The presence of sulfide ion and silica give rise to phase-disengagement problems during the solvent extraction process. The addition of hydrogen peroxide in excess of the amount theoretically required converts the sulfide to sulfate when it is added at a pH of greater than 7. The solution is then heated to destroy the excess hydrogen peroxide, and the pH of the solution is maintained at about 9 for sufficient time to allow the precipitation of silica, which is then removed by filtration [11]. The sodium molybdate solution is adjusted to a pH of 4.5 and is used as the molybdenum solvent extraction feed solution. The concentration of the feed solution is 100 g MoO_3/liter.

3.4. Molybdenum Solvent Extraction System

The solvent extraction system consisted of tricaprylyl amine (a high-molecular-weight tertiary alkyl amine) and an aromatic petroleum solvent (SC#28 of Solvents and Chemical Co.). The optimum composition with respect to loading, rate of extraction, and phase separation for processing the sodium molybdate solution prepared in Section 3.3 is 15% by volume tricaprylyl amine and 85% by volume SC#28 [11]. Tricaprylyl amine is converted to the sulfate salt prior to molybdenum extraction by being contacted with 1.5 N H_2SO_4. Molybdenum forms both anionic and cationic types of polymolybdate and the mechanism involved in extraction is a function of pH. At higher pH, the mechanism of extraction is anion exchange, whereas at lower pH the mechanism is an adduct type [12, 13].

3.4.1. Extraction Isotherm

The extraction isotherm was plotted and a McCabe–Thiele diagram was prepared for estimating the number of theoretical stages. The optimum contact time or equilibration time was determined prior to the construction of the extraction isotherm and found to be less than 1 min. Figure 4 is the molybdenum extraction isotherm. The plot shows that two stages of extraction are adequate for complete extraction of the molybdenum when the organic:aqueous

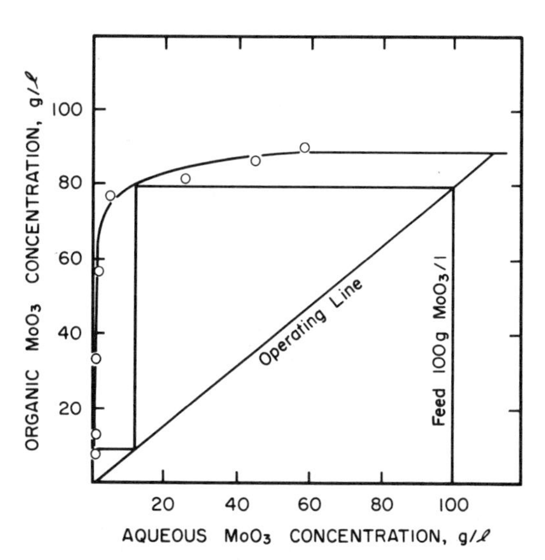

Figure 4. Molybdenum extraction isotherm. Organic solution: 15% by vol. Alamine 336–85% SC#28. Aqueous solution: 100 g MoO_3/liter, pH 4.5.

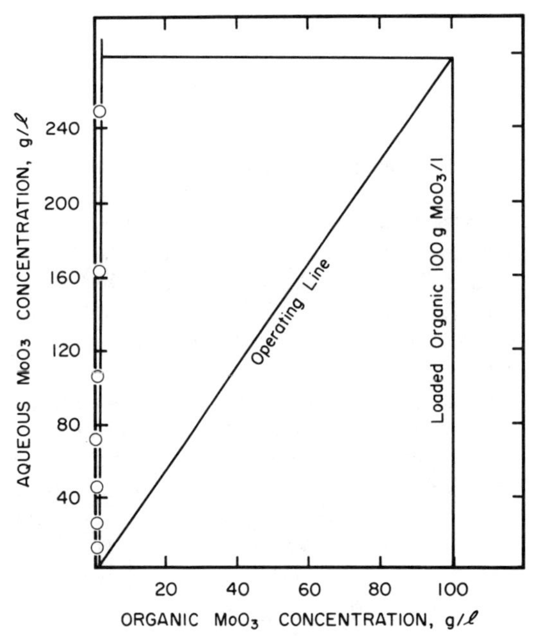

Figure 5. Molybdenum stripping isotherm. Stripping agent: concentrated ammonium hydroxide.

phase ratio is 1.2, and about 90% of the molybdenum is extracted in the first stage.

3.4.2. *Stripping Isotherm*

The stripping isotherm was plotted and a McCabe–Thiele diagram was prepared for estimating the number of theoretical stages. The optimum contact time or equilibration time was determined prior to the construction of the stripping isotherm and found to be less than 1 min. Figure 5 is the molybdenum stripping isotherm. The plot shows that one stage is sufficient for almost 100% stripping of molybdenum when the stripping solution is concentrated ammonium hydroxide.

3.4.3. *Continuous Extraction System*

The pilot-scale mixer–settler system shown in Fig. 6 was operated. The system was based on the data obtained from the extraction and stripping isotherms. The extraction isotherm showed that two stages of extraction are sufficient. However, three stages were used as a precautionary measure against unexpected molybdenum losses. The pH of the aqueous solution after the first extraction stage but before the subsequent two extraction stages was maintained by continuous addition of dilute sodium hydroxide solution. Otherwise, undesirable sulfate anions would be transferred to the product as a result of the adduct mechanism [12, 13]. The water washing of the organic following the extraction circuit is used to minimize aqueous-phase entrainment. The stripping solution was 14 *M* NH_4OH and complete stripping was obtained in one stage. The ammonium molybdate solution from the mixer–settler system had a concentration of 175 g MoO_3/liter. The product was then isolated by the crystallization of ammonium paramolybdate, which is the precursor of pure MoO_3.

REFERENCES

1. D. L. Kepert, *Progr. Inorg. Chem.* **4,** 199 (1962).
2. T. K. Kim, R. W. Mooney, and V. Chiola, *Separation Sci.* **3,** 467 (1968).

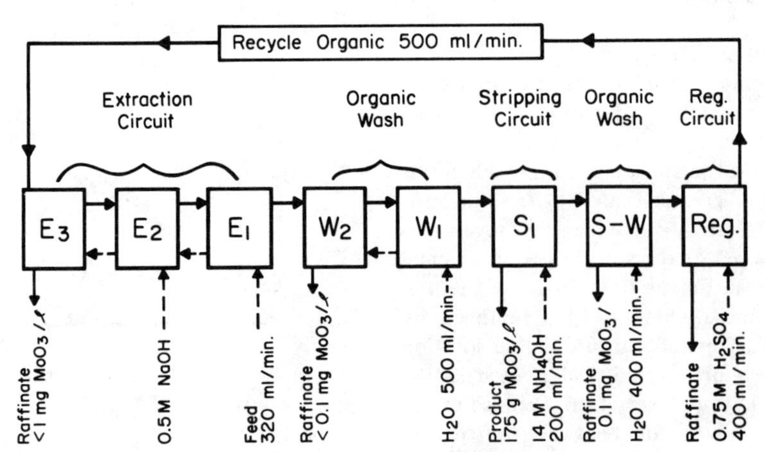

Figure 6. Continuous pilot-plant mixer–settler system. Organic solution: 15% by vol. Alamine 336–85% by vol. SC#28. Aqueous solution: 100 g MoO_3/liter, pH 4.5.

3. M. B. MacInnis, in H. F. Mark, J. J. McKetta, and D. F. Othmer, Eds., *Kirk-Othmer Encyclopedia of Chemical Technology*, Vol. 22, Wiley, New York, 1969, p. 346.
4. K. H. Tytko and O. Glemser, in H. J. Emeleus and A. G. Sharpe, Eds., *Advance in Inorganic Chemistry and Radiochemistry*, Vol. 19, Academic Press, New York, 1976, p. 239.
5. B. N. Laskorin, V. S. Ul'yanov, and R. A. Sviridova, *Zh. Priklad, Khim.* **35,** 2409 (1962).
6. L. G. Anokhina, N. A. Agrinskaya, and V. I. Petrashen, *Zh. Neorgan. Khim.* **15,** 155 (1970).
7. B. E. Martin and J. E. Ritsko, U.S. Patent 3911077 (1975).
8. C. R. Kurtak, U.S. Patent 3158438 (1964).
9. V. Chiola and F. W. Liedtke, British Patent 1240524 (1971).
10. B. E. Martin and M. B. MacInnis, U.S. Patent 3725524 (1973).
11. T. K. Kim, L. R. Pagnozzi, M. B. MacInnis, and J. M. Laferty, U.S. Patent 3770869 (1973).
12. M. B. MacInnis, T. K. Kim, and J. M. Laferty, *J. Less-Common Met.* **36,** 111 (1974).
13. T. K. Kim and M. B. MacInnis, in D. L. Horrocks and C. T. Peng, Eds., *Proceedings of the International Conference on Organic Scintillators and Liquid Scintillation Counting*, Academic Press, New York, 1971, p. 925.

25.4

COMMERCIAL PROCESSES FOR CHROMIUM AND VANADIUM

N. M. Rice
University of Leeds
United Kingdom

1. AQUEOUS CHEMISTRY

The solvent extraction chemistry and processing of chromium and vanadium are dominated by their aqueous-phase behavior and especially the multiplicity of oxidation states that occur [1–3]. Furthermore, there are complications as a result of the hydrolysis and polymerization of anionic species in the highest stable oxidation states as a function of pH and concentration [1–9] and of complex formation, such as with sulfate and chloride [4]. All affect the extractability of the metals.

The behavior of chromium(VI) is comparatively simple and well established [1]. Below pH 1–2, it exists as un-ionized H_2CrO_4. Between pH values of about 2 and 7, it exists as $Cr_2O_7^{2-}$ above 0.02 mol/dm^3 (~1 g/liter), whereas $HCrO_4^-$ predominates below this concentration independently of pH. At pH values above about 7, the CrO_4^{2-} ion is stable.

The behavior of vanadium(V) is less well characterized. The general pattern is that at pH values less than about 1 the VO_2^+ cation predominates; above this V_2O_5 precipitates between pH 1 and 4 at concentrations over about 5×10^{-4} mol/dm^3. At higher pH values polymerization occurs up to about pH 10–12, above which the VO_4^{3-} cation exists. At very low concentrations the "metavanadate" ion ($H_2VO_4^-$) predominates from pH 3 to about 9. Various sequencies of polymerization reactions have been proposed to account for behavior between pH 3 and 10. Proposed species include $V_6O_{17}^{4-}$, $V_3O_9^{3-}$, $HV_2O_7^{3-}$, and $V_{10}O_{28}^{6-}$ as well as their protonated analogs [5–9]; some evidence is available from solvent extraction loading data [5–7].

1.1. Precipitation

Precipitation is important in the recovery of vanadium from strip product liquors. Table 1 shows solubility data for various species. Unhydrated V_2O_4, V_2O_5, and Cr_2O_3 are often specified [1] as these are the thermodynamically stable species, although their rates of formation

TABLE 1 SOLUBILITY AND SOLUBILITY PRODUCTS AT 298K [2]

Oxidation Number	Species	K_s	pH Limit[a]	Species	K_s	pH Limit[a]
2+	$V(OH)_2$	4×10^{-16}	8.3	$Cr(OH)_2$	1×10^{-17}	7.5
3+	$V(OH)_3$	4×10^{-35}	3.9	$Cr(OH)_3$	6×10^{-31}	5.3
4+	$VO(OH)_2$	3×10^{-24}	4.2	–	–	–
5+	NH_4VO_3	2.6×10^{-3}	>4	–	–	–
		0.6[b]				
	V_2O_5	0.07[b]	1.4–4.1	–	–	–
	$Na_2H_2V_6O_{17}$	0.03[b]	2–3	–	–	–
6+	–	–	–	$Na_2Cr_2O_7 2H_2O$	185[b]	–
	–	–	–	$Na_2CrO_4 6H_2O$	85[b]	–

[a]The pH at which the hydroxide is in equilibrium with 10^{-4} mol/dm^3 metal ion concentration and above which it becomes increasingly stable.
[b]Solubility in grams per 100 g of H_2O at 298 K.

may be slow relative to the hydroxides or hydrated oxides. The precipitation of vanadium(V) occurs only slowly at ambient temperature but is accelerated by heating to 70–90°C. Usually "acid red cake" (HVO_3) precipitates, but red cake or $Na_2H_2V_6O_{17}$ is precipitated industrially by prolonged boiling (1–6 hr) of sodium metavanadate ($NaVO_3$) solutions at a pH of 2–3 in the presence of sulfuric acid [10]. This product, which contains 86% V_2O_5, usually requires redissolution and reprecipitation as ammonium metavanadate (AMV), NH_4VO_3, from which ammonia may be removed by heating to 200°C [2]. Ammonium metavanadate can also be precipitated directly from alkaline media by addition of ammonia or during the precipitation stripping of organic extracts with ammonium salts [7].

Chromium(VI) does not form insoluble species and must be recovered from solution by crystallization of the hydrated chromate or dichromate or by electrolysis.

2. EXTRACTION CHEMISTRY

2.1. Lower Oxidation States

Virtually no data exist for Cr(II) or V(II), which are unstable in aqueous solution, although V^{2+} is said to be extracted at pH values between Cu^{2+} and Ni^{2+} by naphthenic acid [11].

Although no data exist for the extraction of V^{3+} by liquid cation exchangers, it should be sufficiently stable. It can form complexes with thio-β-diketone ligands [12] and is slightly extracted from solutions containing less than 20 g/liter chloride by amine at pH 2 [13].

2.2. Chromium(III)

Chromium(III) can be extracted as an ion pair from perchlorate solution by TBP [14]. It is extracted at pH values between those of Fe(III) and Cu(II) by carboxylic acids [11] and as $Cr(OH)^{2+}$ by di-2-ethylhexyl phosphoric acid (D2EHPA) from sulfate media [15]. Both thenoyltrifluoracetone (TTA) and long-chain amines extract Cr(III) from chloride [13, 16] and sulfate media [16]. Extraction from chloride media is relatively low, even at 10 *M* Cl^- in LiCl solution [17] but is appreciable from sulfate solution. A detailed study [18] of the extraction of Cr(III) from sulfate medium by Primene JM-T followed by stripping with 3 *M* HCl established that, even at 80°C, the inert nature of Cr(III) sulfate complexes caused extraction kinetics to be too slow for the development of a practicable process.

2.3. Vanadium(IV)

It has been shown that VO^{2+} cations can be extracted from sulfate [16, 17] and chloride [20–22] solutions containing 0.01–1.0 mol/dm^3 acid by D2EHPA in kerosene. Polymerization of the extracted species occurs [23]. Studies of the extraction kinetics have recently been published [23, 24].

Neutral $VOCl_2$ species are extracted from chloride solution by TBP and TOPO [25], and the anionic species $VOCl_3^-$, $VO(SCN)_4^{2-}$, and $VO(SCN)_5^{3-}$ are extracted by amines from chloride and thiocyanate media, respectively [26–28].

No synergistic effect has been observed on the addition of TBP modifier in D2EHPA as for UO_2^{2+}

[10]; in fact, a small antagonistic effect is observed [23].

2.4. Chromium(VI)

In sulfuric acid media (1–4 mol/dm^3 H_2SO_4) H_2CrO_4 is extracted by solvating reagents such as TBP [29–31] or TOPO [32].

Amines (e.g., Alamine 336, TOA, and Amberlite LA-2) can extract Cr(VI) anions [33–37] from sulfate solutions. Competition between chromate and sulfate is strong [30]. Chromium(VI) can also be extracted from chloride media containing 6 *M* HCl by tribenzyl and triisooctyl amines [34].

Both Alamine 336 and Aliquat 336 extract Cr(VI) from sulfate media of pH 1–7, and the latter reagent also extracts it from carbonate solutions of pH 7–11 with pH 8 reported as optimum [35, 36]. The dichromate ion is extracted from acid media [37].

Mathematical models have been fitted to binary equilibrium data for the extraction of Cr(VI) by Aliquat 336 from chloride media of pH 2–13 as well as for exchange with vanadate (taken as VO_3^-) and OH^- ions [35]. Empirical constants were found, but no extraction equilibrium constants are available in the literature [38]. Water is coextracted [5].

Synergism occurs in the extraction of Cr(VI) by mixtures of trioctylamine and alkylphosphates, including D2EHPA, from various acids, although TBP shows no such effect [39].

Oxidation of the organic phase by Cr(VI) is always a possibility, but reaction rates are negligible if the organic phase is kept dark for alkaline media [35, 39]. No amine reagent stable to acid chromate media over long periods is known [5].

Kinetics for extraction with Aliquat 336 are rapid, and equilibrium established within 3 min of contact time. Aliquat 336 can be stripped effectively with mixtures of 1.5 *M* NH_3 and NH_4Cl or $(NH_4)_2SO_4$. The compound NH_4NO_3 causes a third phase to form [35].

2.5. Vanadium(V)

There appear to be few data for the extraction of VO_2^+ cations by cation exchangers, but vanadium(V) is said to follow UO_2^{2+} during extraction with D2EHPA at pH 0.4–1 [10] at one U.S. mill. Neutral chlorocomplexes of vanadium(V) can be extracted by TBP in CCl_4 [41], and the α-hydroxyoxime XI-8A (Henkel Corporation) can extract vanadium(V) from sulfate media [42], probably as the VO_2HSO_4 complex. The latter species is also extracted by long-chain alcohols from 1–10 *M* H_2SO_4 [43] and is polymerized above 0.01 *M*.

Above pH values of about 2, anionic vanadium(V) species can be extracted by amines [5–7, 19, 36, 40–48]. There is uncertainty as to the exact nature of the extracted species. Aliquat 336 extracts vanadium(V) strongly from alkaline as well as acid media [5, 7, 36, 45, 48].

Mathematical models have been fitted on the basis of RVO_3 and R_2HVO_4 extraction from chloride and chromate media by this extractant [5]. Extraction kinetics were found to be rapid, and stripping with NH_3/NH_4Cl or $NH_3/(NH_4)_2SO_4$ was possible. The most efficient stripping agents were 1.5 *M* NH_4Cl or NH_4NO_3 with 0.9–1.8 *M* NH_3. Single ammonium salts were not effective. Ammonium metavanadate (AMV) precipitation occurs during stripping. This is assisted by temperatures of 50–60°C [5, 7].

Oxidation of amines by vanadium(V) has been said to necessitate the use of centrifugal contactors to minimize contact time [47, 49], but no information on this phenomenon is available.

3. ANALYTICAL CHEMISTRY

Solvent extraction has been used for the analytical separation of chromium and vanadium from other metals [50].

4. SEPARATION OF METALS FROM ASSOCIATED ELEMENTS

The major industrial separation problems that can be solved by solvent extraction are (a) uranium and vanadium from acid sulfate leach liquors (as well as from iron and aluminum), (b) chromium and vanadium (and also aluminum) from titaniferous iron ores, and (c) chromium and sulfate from waste solutions.

4.1. Uranium–Vanadium Separation

Both carbonate and sulfuric acid leaching of uranium ores dissolves some vanadium, but its recovery is low, except at higher than normal acid concentrations or unless a salt roasting–water leaching step is included [10].

There are two main solvent extraction methods for the recovery of uranium from acid leach liquors where it is present as UO_2^{2+}, based on either amine extractants (e.g., the AMEX and PURLEX processes) or on D2EHPA (the DAPEX process) [10] see Chapter 25.11).

1. *Amine Processes.* Uranium is strongly extracted by tertiary amines at pH 1. At this pH the distribution coefficient of vanadium(V) is less than 1 and cationic vanadium(IV) is not extracted [10]. At pH 2 or more, however, the distribution coefficient of vanadium(V) rises to over 20. Thus there are two possible variants [51]: (1) simultaneous extraction of both metals at pH 2 followed by selective stripping first of vanadium by dilute acid followed by uranium in the conventional way; or (2) successive extraction of uranium at pH 1 followed by extraction of vanadium from the raffinate either as vanadium(V) or vanadium(IV), formed deliberately by reduction with iron or SO_2 [10].

Method (2) is commonly used with vanadium recovered from uranium extraction raffinates [8]. Interactions can occur between uranium and vanadium as well as vanadium and molybdenum (which may be present in some ores) [51].

2. *D2EHPA Processes.* It has been established that VO^{2+} is less strongly extracted by D2EHPA than UO_2^{2+} and requires a higher aqueous pH. Two extraction routes—both used in practice—have been developed [10] (see Table 2): (a) sequential extraction of uranium at pH values of less than 1.8 (followed by Na_2CO_3 stripping) and of vanadium(IV) at pH values of 1.8–2.2 (followed by stripping with 1 *M* sulfuric acid); and (b) simultaneous extraction of UO_2^{2+} and VO^{2+} at a pH of approximately 2 followed by selective stripping of vanadium with 1 *M* sulfuric acid and then uranium by sodium carbonate. A flow sheet for such a process is shown in Fig. 1. Iron(III) is strongly extracted by D2EHPA at these pH values, so it is necessary to reduce it

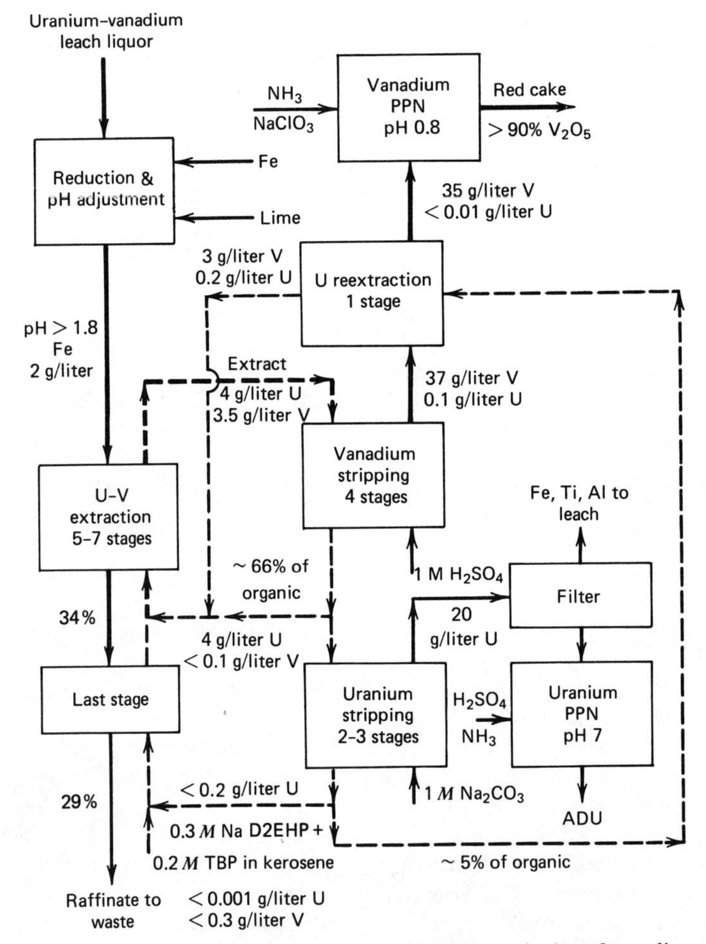

Figure 1. DAPEX process modified for extraction and selective stripping of vanadium and uranium.

TABLE 2 COMMERCIAL VANADIUM EXTRACTION PROCESSES[a]

Feed			Solvent Extraction Process Parameters											
				Load			Scrub			Strip				
Type and Source [Reference]	Oxidation Number	Aqueous Phase Composition[b]	Solvent Composition[b]	Stages	O/A Ratio	Raffinate Concentration[b]	Reagent Concentration[b]	Stages	O/A Ratio	Reagent Concentration[b]	Stages	O/A Ratio	Product[b] and Recovery Method	Status[c]
Uranium D2EHPA solvent extraction raffinate (DAPEX) [10, 47, 49]	IV	Sulfate, pH 1.8–2.2, 3.5 V_2O_5	7.5% D2EHPA, 3% TBP, kerosene	5	1/1	0.15 V	–	–	–	30 H_2SO_4	4	30/1	100–120 V, oxidize and and precipitate "acid red cake"	C
Uranium resin-in-pulp ion-exchange eluates, vanadium–iron ore salt roast leach liquors, uranium tailings and lignite ash leach liquors [10, 47, 49]	IV	Sulfate, pH 1.8–1.9	3–10% D2EHPA, 2–5% TBP, kerosene	6	2.2	<0.1 V	10 Na_2CO_3	2[d] ?[e]	?[e]	14 H_2SO_4	4 (50°C)	8/1	55–60 V_2O_5, oxidize and precipitate "red cake" at pH 0.8 and 70°C	C
Sulfuric acid leach of dolomitic shale [47, 53]	V	Sulfate, pH 2.5, 3 V_2O_5	0.075 *M* Di (tridecyl)amine amine in kerosene	3	4	0.1 V	–	–	–	1.5 Na_2CO_3	2	8/1	Add NH_3 to precipitate AMV, organic phase 12 g/liter V	P
Salt roast/water leach of uranium–vanadium ore [47, 49]	V	Sulfate–chloride pH 3	Alamine 336, kerosene	3	?[e]	?[e]	–	–	–	Soda ash	?[e]	?[e]	Add ammonia to precipitate AMV	CD
Tailings from carbonate leach of uranium ore: Leached with HCl and H_2SO_4 and oxidized with $KMnO_4$ [10, 47, 49]		Sulfate–chloride pH 1.7, Zn, Fe, Al, Cl, and V	4.5% Tertiary amine, 1.7% D2EHPA, 1.4% HDPA[f] 1.3% Decanol, kerosene	?[e]	?[e]	?[e]	5 H_2SO_4[g] 3 NaCl[g]	?[e]	?[e]	10 Na_2CO_3	?[e]	?[e]	Precipitate sodium uranyl vanadate	CD

[a]All plants in the United States of America.
[b]Concentrations in g/liter except % = % w/v.
[c]C = commercial, P = pilot, CD = defunct.
[d]Scrubbing of 10% of solvent flow after vanadium stripping to remove uranium.
[e]No data available.
[f]Heptadecyl phosphoric acid.
[g]To remove zinc.

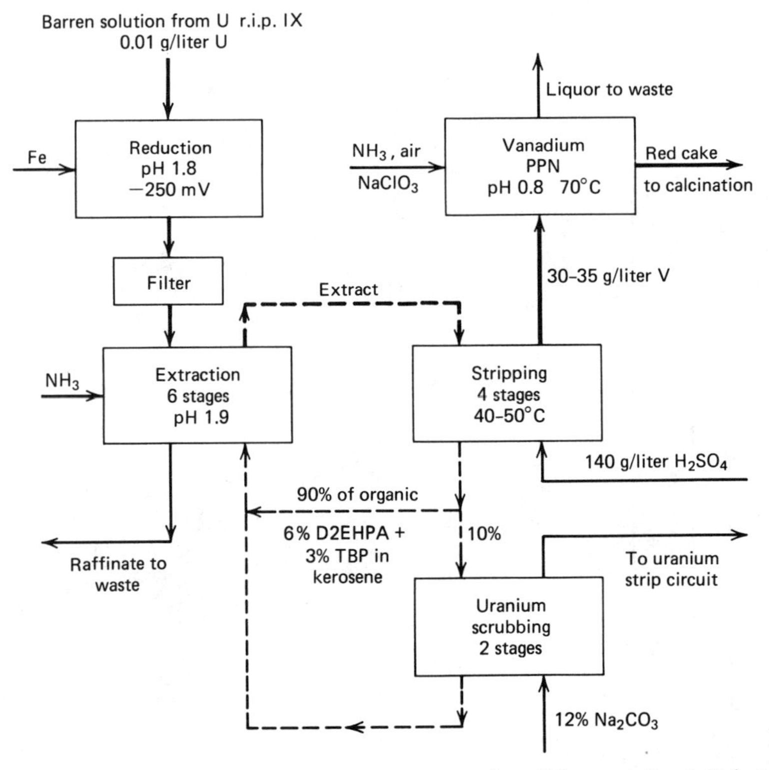

Figure 2. Vanadium recovery plant of Mines Development, Inc., Edgemont, South Dakota treating uranium resin-in-pulp ion exchange barren solutions.

to Fe(II) by use of iron or SO_2 [23]. This also has the advantage that vanadium(V) is reduced to vanadium(IV). Figure 2 shows a flow sheet [48] of a typical plant.

3. *Mixed Solvent Process.* All the DAPEX processes use TBP as a modifier to prevent precipitation of the NaEHP during stripping and as a uranium synergist. In one plant [10, 49] a solvent containing 3–4.5% tertiary amine, 0.25–1.7% D2EHPA, 0.75–1.4% hexadecylphosphoric acid, and 1–1.3% *n*-decanol in kerosene was used to extract UO_2^{2+} and vanadium(V) simultaneoulsy from an aqueous phase at pH 1.7 after oxidation with $KMnO_4$. The feed was from a salt roast process and had a high Cl^- content, and Zn^{2+} was coextracted and could be removed with 5% H_2SO_4–NaCl after Na_2CO_3 stripping of uranium and vanadium.

4.2. Chromium-Vanadium Separation

Chromium(VI) is more strongly extracted from acid media (pH 5) by quaternary ammonium extractants, whereas vanadium(V) is more strongly extracted from alkaline media [5, 35, 36, 48] of pH 9, and chromium is again more strongly extracted above pH 12. Figure 3 shows the simplified flow sheet of a process for treating the caustic liquor from the leaching of titaniferous magnetite concentrate [47, 48]. Chromium is extracted first by using Adogen 464 (0.2 *M*) in Shellsol 140/5% isodecanol at 50°C, followed by scrubbing with Na_2CrO_4 solution to remove coextracted vanadium and aluminum and then stripping with NaCl (1.5 *M*). Vanadium is recovered by a similar process by using a 5-g/liter $NaVO_3$ scrub solution and precipitation stripping of AMV with 10 g/liter of NH_3 + 140 g/liter of NH_4Cl.

Vanadium can be preferentially extracted at pH 9 or the metals coextracted and chromium removed by scrubbing with $NaVO_3$ solution [36]. However, coextraction of AlO_2^- and its hydrolysis can cause crud formation (see Table 2).

Recently a process has been developed in China [46] (Fig. 4) on the basis of the preferential extraction of vanadium(V) from acid media of pH 4.4–7.5 by the primary amine N-1923 (10% in kerosene). By careful control of acid

TABLE 3 PROPOSED EXTRACTION PROCESSES FOR CHROMIUM AND VANADIUM

Feed				Solvent Extraction Process Details											
					Load			Scrub			Strip				
Source and Composition [Reference]	Pretreatment	Metal and Oxidation Number	Aqueous Phase Composition[a]	Solvent Composition[a]	Stages	O/A Ratio	Organic Concentration[a]	Reagent Concentration[a]	Stages	O/A Ratio	Reagent Concentration[a]	Stages	O/A Ratio	Product and Recovery Method Raffinate Concentration[a]	Location and Status[c]
Titaniferous iron ore 52% Fe, 7% Ti, 0.65% Si, 0.2% V, 5% Al, 2% Cr [47, 48]	Na_2CO_3 roast, NaOH leach	Cr(VI)	NaOH, pH 13.5, 0.4 V, 3 Cr, 6.9 Al	0.2 *M* Aliquat 336 or Adogen 464 in Shell 140/10% isodecanol preequilibrated with Na_2CO_3	7 (50°C)	1.03/1	2.9 Cr	5 Na_2CrO_4	4	10/1	5 *M* NaCl	4	7/1	66 g/liter Na_2CrO_4, crystallize, raffinate 0.02 Cr, solvent loss 50 ppm	Canada (P)
		V(V)	Raffinate of above 0.4 V, >0.02 Cr, pH 13.4	As above	4	1.2/1	0.3 V	5 $NaVO_3$, pH 12	4	9/1	10 NH_3, 140 NH_4Cl	3	2/1	Precipitation strip to AMV, raffinate 0.006 V, 7 Al	
Titaniferous iron ore [46]	Na_2SO_4 roast, 1250–1280°C, alkali leach precipitate SiO_2 with H_2SO_4, add 0.2 *N* H_2SO_4 during extraction	V(V)	5–6 V, 10 Cr, 1 Al, 1 Si, pH 8.6	10% N-1923 (primary amine) in kerosene	1 (20°C)	1/1	6 V 0.1 Cr	10 $NaVO_3$, pH 7.8	1 13°C	10/1	0.8–1.5 *M* Na_2CO_3 or $(NH_4)_2CO_3$	1 (55°C)	20/1	$NaVO_3$ or AMV, filter and calcine, raffinate 0.002 V	China (P)
		Cr(VI)	Raffinate of above pH 7.4, >0.2 V, add acid, H/Cr = 3.2, pH 2.2	As above	2 (20°C)	1/1	10 Cr	–	–	–	0.5–1.5 *M* Na_2CO_3 or $(NH_4)_2CO_3$	1	10/1	$Na_2Cr_2O_7$ or $(NH_4)_2Cr_2O_7$, crystallize and calcine to Cr_2O_3, raffinate 0.007 Cr	
Oil power station fly ash, 5% V, 2000 tons/year [23]	H_2SO_4 leach, SO_2 reduction	V(IV)	25 V; pH 1.9; 25 Na, $MgSO_4$; 3–5 Fe(II)	20% D2EHPA, 15% TBP, kerosene washed with water after strip	3 (25°C)	4/1	6 V	6 *M* H_2SO_4[b]	1	?[e]	1 *M* H_2SO_4[d]	3	10/1	Precipitate $VO(OH)_2$, Recycle raffinate to leach	Sweden (P)

[a]Concentrations in g/liter or mole/liter (*M*), % = % w/v.
[b]After stripping vanadium.
[c]Pilot plant (P).
[d]Combine with stripping agent solution for vanadium.
[e]No data available.

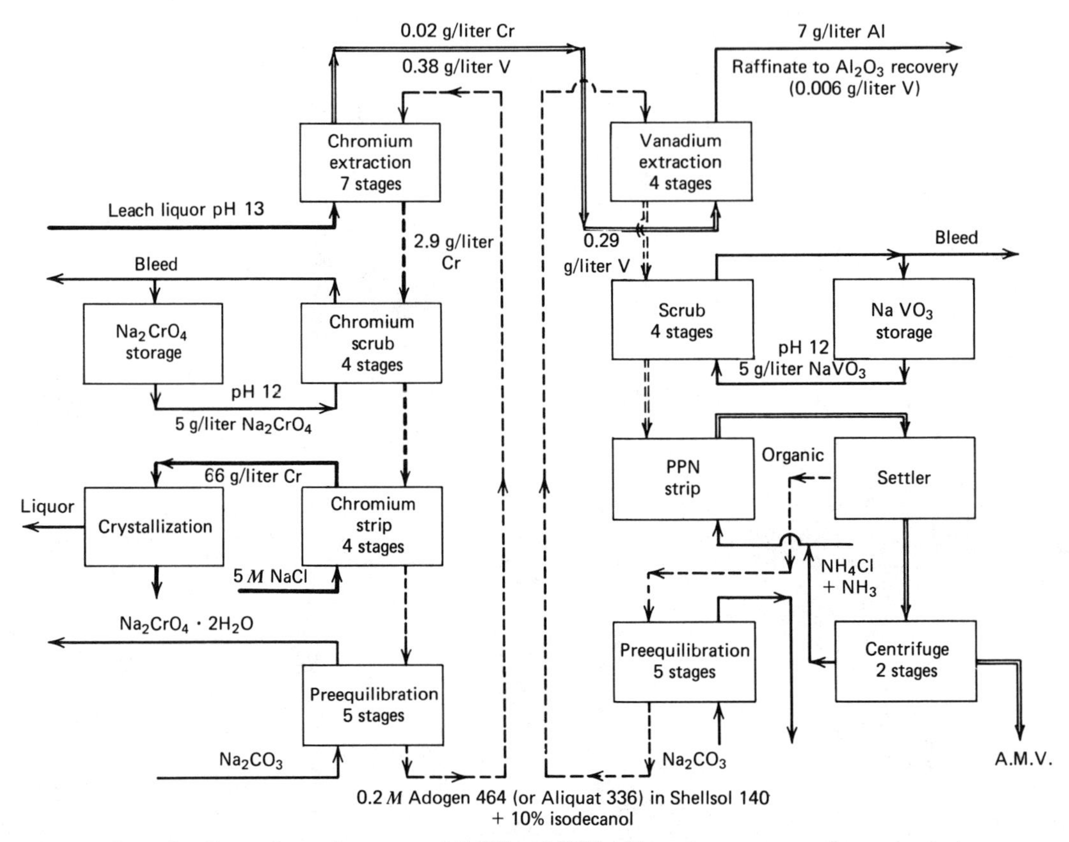

Figure 3. Separation and recovery of Cr(VI) and V(V) with quaternary ammonium extractants.

addition to give H^+:V ratio of 1, very little chromium(VI) is coextracted. The presence of chromium(VI) is necessary, however, for good vanadium extraction. Separation is best at 5–10°C and poor at 40°C. Scrubbing with $NaVO_3$ removes coextracted Cr from the solvent, which can be stripped with Na_2CO_3, or $(NH_4)_2CO_3$. Cr(VI) can then be recovered from the vanadium raffinate (see Fig. 4). An interesting feature is the use of regenerated solvent to "scavenge" traces of chromium from the raffinate from chromium recovery prior to returning it to the chromium extraction step (see Table 2).

4.3. Chromium(VI)-Sulfuric Acid Separation

This type of separation is important in the treatment of wastes from CrO_3 plants and plating effluents [30]. There are two possible types of extractant, TBP and amines, but the latter extract sulfate in preference to Cr(VI) and their capacity for chromium falls with rising acidity, whereas that of undiluted TBP rises from 6 to 7 g/liter of Cr(VI) with 2 N H_2SO_4 to 55 g/liter of Cr(VI) for 4 N acid [28]. Amines are superior for dilute acid feeds. The selectivity of TBP for Cr(VI) over sulfate and other metals ions is excellent. Extracts are kept dark to minimize oxidation of TBP by Cr(VI). Stripping may be carried out (1) by using caustic soda with Na : Cr = 2 : 1, (2) with water at pH 4, or (3) with $NaCrO_4$ solution that regenerates $Na_2CrO_2O_7$. A commercial process is in operation by Product Chimique Ugine Kuhlman in France [30, 31], but details are sparse. The strip liquor contains more than 200 g/liter of Cr(VI) and less than 20 mg/liter of Cr(III).

5. COMMERCIAL SOLVENT EXTRACTION PROCESSES

Eleven vanadium plants in the United States have used solvent extraction; nine of these were associated with uranium mills, and five were active in 1971 [10]. Of the two plants recovering vanadium only, one was operating in 1973 (Pyrites Co., Wilmington, Delaware [69]) with a feed from the hydrolysis liquor of TiO_2 pigment precipitation. The status of the other at Wilson Springs, Arkansas [49] is uncertain, but an

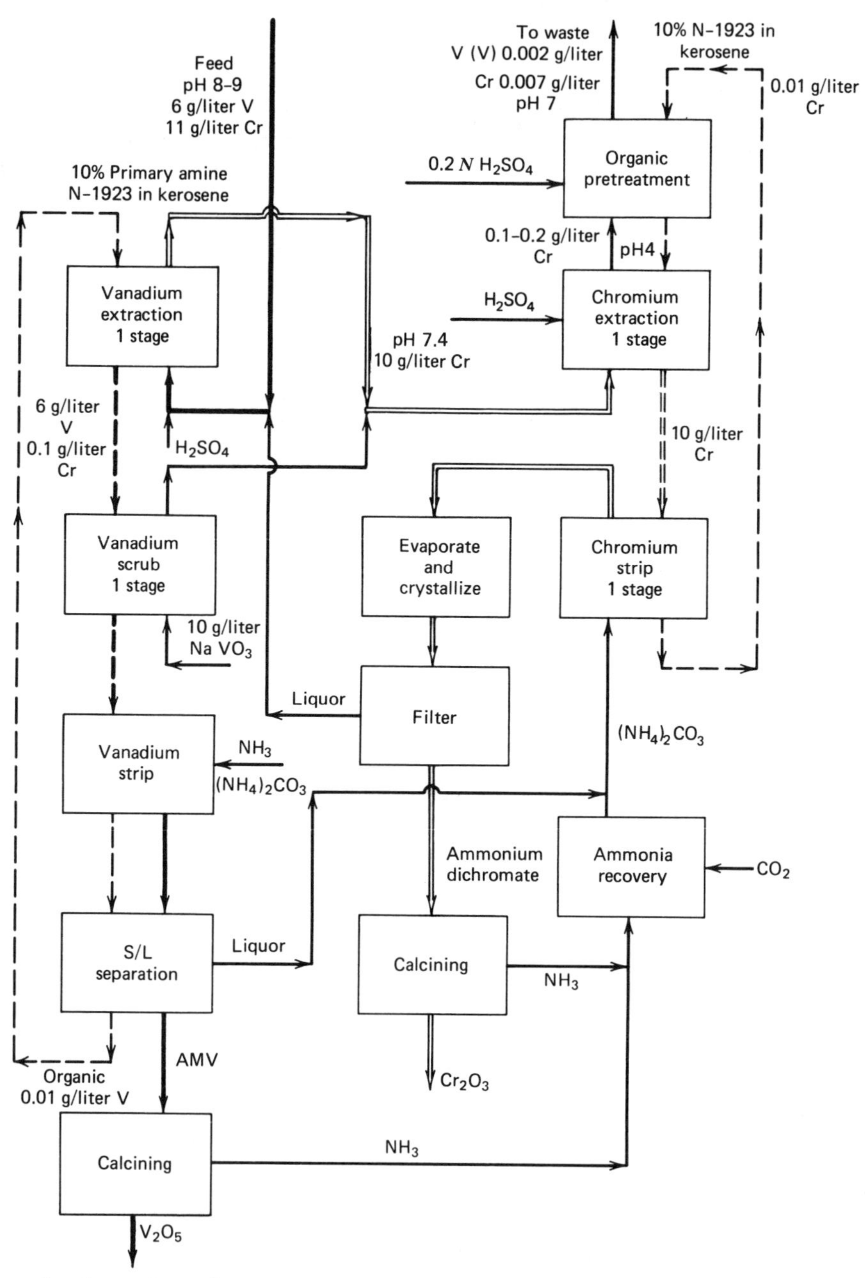

Figure 4. Separation and recovery of V(V) and Cr(VI) by extraction with primary amine N-1923.

amine at pH 3 was used to extract vanadium in Podbielniak contractors. No commercial vanadium plants are known to be fully operational outside the United States, although a D2EHPA circuit is to be incorporated in a proposed plant of LKAB at Ranstad, Sweden [23].

Only one commercial Cr(VI) extraction plant has operated in France. No data have been published [30, 31].

Table 2 shows operating data from plants where sufficient detail has been reported. Little information on amine circuits has been published. Data for defunct plants are incorporated for interest, as is a proposed circuit for recovering vanadium from dolomitic shales [53]. The first plant listed in Table 2 formerly used the simultaneous extraction–selective stripping version of the DAPEX process illustrated in Fig. 1. Figure 2 illustrates the flow sheet for the second plant. No cost data are available.

6. PROPOSED PROCESSES

A full-scale plant for extracting vanadium from oil-fired power station ash is to be built in Sweden by SOTEX AB [23].

Pilot plants for recovering vanadium and chromium from titaniferous magnetite iron ores have been operated in Canada [48] and China [46] (see Figs. 3 and 4, respectively) and in South Africa [54], whereas by-product vanadium recovery from uranium ore processing has been studied on a pilot scale in Brazil [51] and France [55]. The last two processes also involve molybdenum separation.

The Canadian process is well documented, but there is insufficient information regarding the others to be very specific. A number of other proposed processes [56, 57] include recovery of vanadium from wet process phosphoric acid streams with alkyl phosphates or fluorinated β-diketone extractants [58]; fly ash or smelter slags [54, 59]; and chromium from wastes and ores, including the use of a mixture of D2EHPA and dinonyl naphthalene sulfonate [60]. The removal of Cr(VI) from waste water by means of a resin impregnate containing TOPO has recently been suggested [61]. Few details of these processes have appeared outside patents and reports.

REFERENCES

1. M. Pourbaix, *Atlas of Electrochemical Equilibria*, Oxford Pergamon, Brussels, Cebelcor 1966 (J. A. Franklin, Transl.), pp. 23–45 (Vanadium) and 236–271 (Chromium).
2. G. H. Aylward and J. J. V. Findlay, *S.I. Chemical Data*, 2nd ed., Wiley, Sydney, 1974.
3. W. M. Latimer and J. H. Hildebrand, *Reference Book of Inorganic Chemistry*, 3rd ed., Macmillan, New York, 1952.
4. L. G. Sillen and A. E. Martell, *Stability Constants of Metal-Ion Complexes*, Special Publications Nos. 17 and 25, Royal Society of Chemistry, London, 1966 and 1971 (supplement).
5. M. A. Hughes and T. M. Lever, *Proceedings of the International Solvent Extraction Conference (ISEC), Lyons, 1974*, Society of Chemical Industry, London, 1974, p. 1147.
6. A. R. Burkin, *Proceedings of the Conference on Unit Processes in Hydrometallurgy*, Dallas, 1963, AIME, New York, p. 529.
7. D. W. Agers, J. L. Drobnick, and C. J. Lewis, Recovery of Vanadium from Acidic Solutions by Liquid Ion Exchange, paper presented at AIME Annual Meeting, New York, February 1962.
8. F. J. Rosotti and H. Rosotti, *Acta Chem. Scand.* **10,** 957 (1956).
9. W. P. Griffith and J. D. Wickins, *J. Chem. Soc.* A 1087 (1966).
10. C. Merritt, *Extractive Metallurgy of Uranium*, Colorado School of Mines Research Institute/U.S. Atomic Energy Commission, Denver, 1971.
11. D. S. Flett and M. Jaycock, in Y. Marcus and J. Marinsky, Eds., *Ion Exchange and Solvent Extraction*, Vol. 3, Marcel Dekker, New York, 1973, pp. 1–50.
12. M. Cox and J. Darken, *Co-ord Revs.* **I,** 27 (1971).
13. A. Aue, L. Skjutare, and G. Bjorling, *Proceedings of the International Solvent Extraction Conference (ISEC), The Hague*, 1971, Society of Chemical Industry, London, 1971, p. 447.
14. J. Agett and D. J. Udy, *J. Inorg. Nucl. Chem.* **32,** 2802 (1970).
15. K. Kimura, *Bull. Chem. Soc. Jap.* **33,** 1038 (1960).
16. T. Ishimori and E. Nakamura, *JAERI* Report No. 1047 (1963).
17. T. M. Florence and Y. J. Farrar, *Aust. J. Chem.* **22** (2), 473 (1969).
18. D. S. Flett and D. W. West, *Transact. SME/AIME* **247** (4), 288 (1970).
19. C. A. Blake, Jr., et al., *Proceedings of the U.N. Conference on Peaceful Uses of Atomic Energy*, Vol. 28, Geneva, 1958, p. 289.
20. T. Sato and T. Tukeda, *J. Inorg. Nucl. Chem.* **32** (10), 3387 (1970).
21. T. Sato et al. *Proceedings of the International Solvent Extraction Conference (ISEC), Toronto 1977*, Canadian Institute of Mining and Metallurgy, Montreal, 1979, p. 159.
22. T. Rigg and J. O. Garner, *J. Inorg. Nucl. Chem.* **29,** 2019 (1967).
23. H. Ottertun and E. Strandell, *Proceedings of the International Solvent Extraction Conference (ISEC), Toronto, 1977*, Canadian Institute of Mining and Metallurgy, Montreal, 1979, p. 501.
24. F. Islam and R. K. Biswas, *J. Inorg. Nucl. Chem.* **42,** 415, 421 (1980).
25. T. Sato, S. Ikoma, and T. Nakamura, *Hydrometallurgy* **6** (1, 2), 13 (1980).
26. T. Sato et al., *J. Inorg. Nucl. Chem.* **39,** 395, 401 (1977).
27. T. Sato and T. Nakamura, *Nippon, Kaj. Kaishi* **9,** 1367 (1980) (in Japanese).
28. T. Sato, S. Kotani, and O. Terrao, *Proceedings of the International Solvent Extraction Conference (ISEC), Lyons, 1974*, Society of Chemical Industry, London, 1974, p. 2249.
29. D. G. Tuck, *J. Chem. Soc.* 111 (1963).
30. J. P. Cuer, W. Stuckens, and N. Texier, *Proceedings of the International Solvent Extraction Conference (ISEC), Lyons, 1974*, Society of Chemical Industry, London, 1974, p. 1185.

31. J. P. Cuer et al., French Patent 72,13318 (1972).
32. S. C. White and W. J. Ross, *Extraction of Chromium with TOPO*, U.S. Atomic Energy Commission Report No. ORNL-2326, 1955.
33. E. L. Smith and J. E. Page, *J. Soc. Chem. Ind. (Lond.)* **67**, 48 (1948).
34. F. L. Moore, *Anal. Chem.* **29**, 1660 (1957); **30**, 908 (1958).
35. Henkel Corporation (General Mills, Inc.), *Chromium*, Technical Bulletin S1–6, 1961.
36. R. R. Swanson, H. N. Dunning, and J. E. Holise, *Eng. Min. J.* (10) (1961).
37. C. Deptula, *J. Inorg. Chem.* **30** (5), 1309 (1968).
38. A. S. Kertes, Y. Marcus, and E. Yanir, *Equilibrium Constants of Liquid-Liquid Distribution Reactions*, Part III, *Alkylammonium Salt Extractants*, IUPAC/Butterworths, London, 1974.
39. C. Deptula, *Proceedings of the International Solvent Extraction Conference (ISEC), The Hague, 1971*, Society of Chemical Industry, London, 1971, p. 638.
40. C. F. Coleman, K. B. Brown, J. G. Moore, and K. A. Allen, *Proceedings of the U.N. Conference on Peaceful Uses of Atomic Energy*, Vol. 3, Geneva, 1958, p. 472.
41. P. H. Tedesco and V. B. de Rumi, *J. Inorg. Nucl. Chem.* **42**, 269 (1980).
42. Henkel Corporation (formerly General Mills, Inc), *Xl-8A as a Vanadium Extractant*, Preliminary Evaluation Report, 1973.
43. S. Kopacz and L. Paidovski, *Russ. J. Inorg. Chem.* **16**, 236 (1970).
44. C. F. Coleman, K. B. Brown, J. G. Moore, and D. J. Crouse, *Ind. Eng. Chem.* **50**, 1756 (1958).
45. Henkel Corporation (formerly General Mills, Inc.) *Vanadium*, Technical Bulletin CDS 3-60, 1960.
46. Yu Shu-Chiou et al., *Proceedings of the International Solvent Extraction Conference (ISEC), Liège, 1980,* Université de Liège, 1980.
47. G. M. Ritcey and A. W. Ashbrook, *Solvent Extraction Principles and Applications to Process Metallurgy*, Part II, Elsevier, Amsterdam, 1979.
48. G. M. Ritcey and B. H. Lucas, *Proceedings of the International Solvent Extraction Conference (ISEC), Toronto, 1977*, Canadian Institute of Mining and Metallurgy, Montreal, 1979, p. 520.
49. J. B. Rosenbaum, D. R. George, and J. T. May, *Metallurgical Applications of Solvent Extraction*, Part 2, *Practice and Trends*, USBM Information Circular IC 8502, January 1971, p. 13.
50. A. K. De, S. M. Khopkar, and R. A. Chalmers, *Solvent Extraction of Metals*, Van Nostrand-Reinhold, London, 1970.
51. B. Floh, B. Abrao, and E. Calman Cooke, *Proceedings of the IAEA Symposium on the Recovery of Uranium*, Sao Paulo, Brazil, 1970, p. 267.
52. N. M. Rice, *Visits to North American Hydromet Plants and Research Centres in 1973*, IMM Report, p. 54.
53. P. T. Brooks and G. M. Potter, USBM, RI, 7932, 1974.
54. A. Faure et al., NIM Report No. 1620.
55. E. Saliano et al., *Proceedings of the International Solvent Extraction Conference (ISEC), Toronto, 1977*, Canadian Institute of Mining and Metallurgy, Montreal, 1979, p. 592.
56. D. S. Flett and D. R. Spink, *Hydrometallurgy* **2**, 207 (1976).
57. D. S. Flett and D. R. Spink, WSL Report No. LR254ME, 1977.
58. M. F. Lucid et al., U.S. Patent Numbers 3,700,415; 3,734,469; and 3,764,274.
59. B. Judd and R. Kemp, *New Zealand DSIR*, Report CD 2186, 1974.
60. C. P. Bruen and C. A. Wamser, U.S. Patent No. 3,787,555.
61. E. J. Fuller, German Patent 2,950,567 (1980).

25.5

COMMERCIAL PROCESSES FOR CADMIUM AND ZINC

Gunnar Thorsen
University of Trondheim
Norway

1. INTRODUCTION

Commercial solvent extraction processes for zinc and cadmium actually in operation are limited, and, in fact, the only true direct solvent extraction process for the production of zinc as the main product on a commercial scale appears to be the *Espindesa process* developed by Technical Reunidas in Spain. However, there is considerable interest and activity with a number of potential applications within production routes for zinc and cadmium.

Developments have been reviewed by Flett and Spink [1] for the period 1972–1974 and by Flett [2] for 1975–1976. A recent discussion on the extraction of zinc in complex systems [3] is also very useful.

In general terms, one can say that the organophosphoric acid D2EHPA is by far the most useful and widespread extractant for zinc and cadmium from complex solutions. The pH functionalities for these metals are well suited in that both zinc and cadmium are extracted in the lower acid region of pH $\approx$ 1–2. This may be compared with the solvent extraction systems of carboxylic acids such as naphthenic acids and Versatic acids, where the equilibrium distribution curves for zinc and cadmium normally are in the region of pH $\approx$ 5–6.

One practical consequence of this difference in pH functionality is that in the presence of copper the order of extraction is changed. In a carboxylic acid system, copper will be extracted before zinc and cadmium, whereas the order of extraction is reversed in the D2EHPA system. This is further illustrated later.

In the present chapter the possibilities of using solvent extraction techniques in the commercial production of zinc and cadmium are outlined by presenting a few selected process developments carried out at the author's laboratory. In addition to the Espindesa process, these examples should be of interest to those concerned with the treatment of zinc- and cadmium-bearing sources.†

2. THE ESPINDESA PROCESS

The flow sheet of the Espindesa process is shown in outline according to Flett [4] in Fig. 1. The

†The author is indebted to NORZINK AS of Norway for kind permission to publish some of the data from the bench-scale pilot-plant operations to be presented.

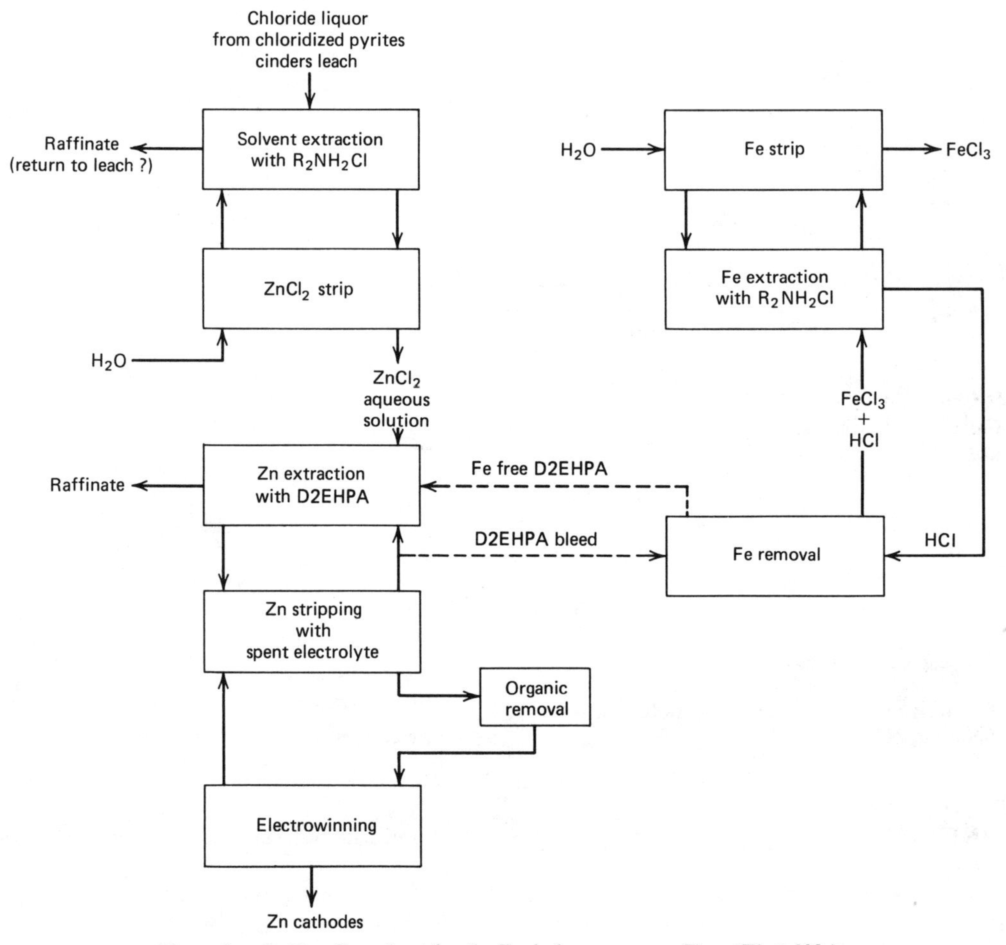

Figure 1. Outline flow sheet for the Espindesa process. (From Flett [2].)

process has been used at the Bilbao plant of Metalquimica del Nervion in Spain since late 1976 for the production of 8000 tons/year of zinc extracted from a leach liquor obtained by leaching chloridized pyrites cinders. A second plant is under construction for startup in early 1980 in Portugal with a design capacity of 11,000 tons/year [5].

The Espindesa process [6] involves a two-stage solvent extraction operation, with the first stage using a secondary amine for extracting pure zinc chloride from an impure feed liquor. In the second stage D2EHPA is used to extract zinc cations from the aqueous zinc chloride strip solution resulting from the first stage. By stripping with sulfuric acid in the second stage, the zinc is converted into a pure zinc sulfate solution from the original zinc chloride.

Any ferric iron present will be coextracted and must be kept at a low level. However, iron actually extracted into D2EHPA is separated from the zinc by selective stripping of zinc with the sulfate solution. The ferric iron forms a very strong complex with D2EHPA in the organic solution. When zinc is stripped with the spent electrolyte from the zinc electrowinning plant, iron will not be removed and may circulate in the system. The necessary bleed of iron is taken from a continuous closed-loop side stream, stripping with concentrated hydrochloric acid. From the resulting strong aqueous chloride solution, ferric iron is once more extracted with the secondary amine and eventually stripped with water from the amine solution. Thus the iron will end up in a dilute effluent stream as an aqueous ferric chloride solution. Accordingly, there will be a consumption of hydrochloric acid equivalent to the ferric chloride bleed.

The extraction of zinc in the second extraction circuit needs careful pH control, with the use of ammonia or lime, which means that alkali must be provided and probably used more or less equivalent to the zinc extracted. If ammonia were used, one could think of evaporating the resulting ammonium chloride solution and in some way recycling the chloride and/or ammonia. Although this operation undoubtedly would be uneconomic, some sort of treatment of the effluents seems necessary to make the Espindesa process generally acceptable from environmental reasons.

3. RECOVERY OF ZINC FROM SPENT ELECTROLYTE ACID SOLUTION

In the electrowinning circuit in the hydrometallurgical production of zinc, buildup of impurities such as magnesium may require a bleed of the return acid from the tankhouse. By electrolytic deposition, the zinc in this bleed stream may be brought down to a level of say 3 g/liter of zinc. A typical sulfuric acid concentration will then be about 200 g/liter of H_2SO_4.

From this strong sulfuric acid solution, the zinc may be recovered by a two stage extraction approach similar to the Espindesa process by adding chloride ions to the bleed solution. From an early process development assigned to NORZINK AS for recovery of zinc from galvanizing zinc ashes [7], it has been found that zinc chloride can be extracted from an acid sulfate solution by triisooctyl amine (TIOA). On this basis a process has been worked out as shown in Fig. 2.

It was found that some excess of chloride ions is necessary to obtain a satisfactory extraction of zinc. Figure 3 illustrates the point, showing the need for a minimum of 10 g/liter of chloride excess. With this excess, a four-stage extraction would reduce 3 g/liter of zinc down to about 0.2 g/liter with an aqueous:organic ratio of unity.

4. SOLVENT EXTRACTION OF ZINC AND CADMIUM WITH ACID EXTRACTANTS

The two main groups of cation extracting systems, apart from the chelating agents, are carboxylic acids and organophosphoric acids. Among these, Versatic 911 or Versatic 10 (Shell), and D2EHPA extractants are by far the most interesting for potential commercial use.

The following is an example of a solvent ex-

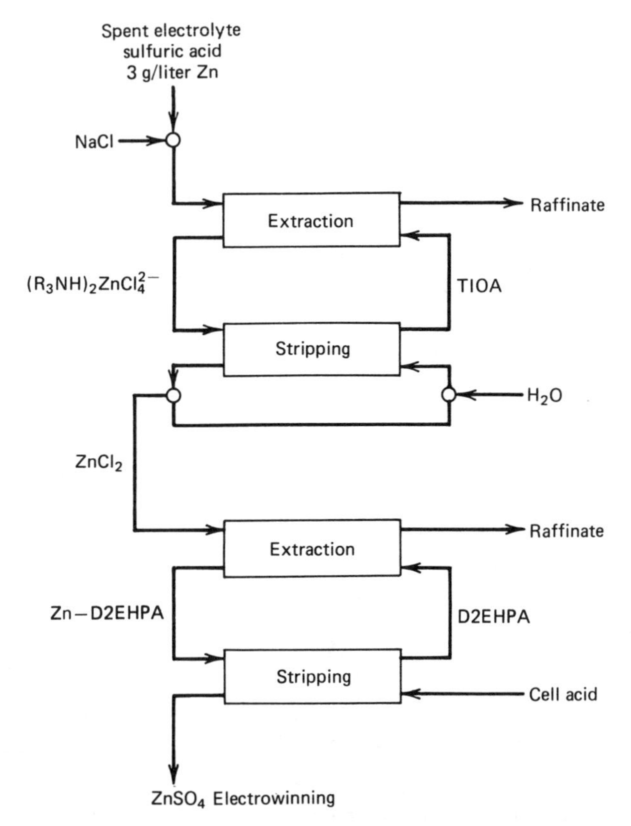

Figure 2. Schematic flow sheet for solvent extraction of zinc from spent electrolyte acid solution.

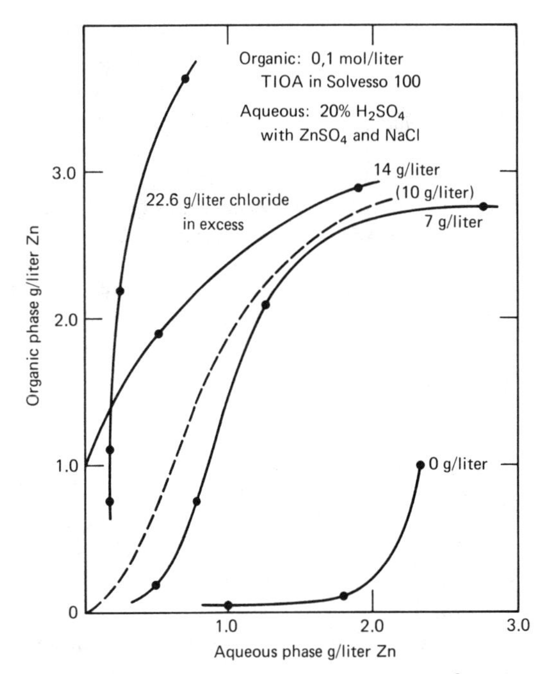

Figure 3. Equilibrium extraction of $ZnCl_4^{2-}$ with TIOA from spent electrolyte sulfuric acid solution.

traction process where both zinc and cadmium appear as metal values to be recovered and separated as product streams. The solution of the mixed metals comes from dissolution of a cementation residue from the purification of zinc sulfate electrolyte by zinc dust cementation. This consists of the metals copper, zinc, cadmium, nickel, cobalt, arsenic, and antimony. Copper, zinc, and cadmium are the main constituents for recovery, the remaining metals being impurities in minor quantities. It is to be noted, however, that it is of vital importance to take care of trace metals because of their detrimental effect in the electrowinning process for zinc.

For consideration of potential solvent extraction systems, the distribution curves for Versatic 911 and D2EHPA are given in Figs. 4 and 5. The pH was adjusted by adding sodium hydroxide [8]. With the main object of separating copper, zinc, and cadmium, and also considering the cobalt and nickel present, the D2EHPA appears more attractive than the Versatic system. Further developments were thus based on D2EHPA as extracting agent.

The process itself starts with an oxidation and direct dissolution of the cementation product in a mixture of water and 30% D2EHPA in Shellsol. This operation results in a distribution between the aqueous and organic phases of the seven metals present as shown in Table 1.

The distribution will depend on factors such as the degree of oxidation of the cementation product as well as the presence of sulfate ions for the formation of the aqueous salt solution. Other important factors are the relative amounts of the metals present and the loading capacity of the organic extractant.

The organic and aqueous phases obtained from the leaching are both feed streams to a multistage solvent extraction process where copper, zinc, and cadmium leave as separate product streams of their pure sulfates ready for electrowinning or crystallization.

It should be noted that the distributions of the various metals given in Table 1 do not fit the curves given in Fig. 5. It must be remembered, however, that the distribution data were measured for single ideal metal solutions at 20°C.

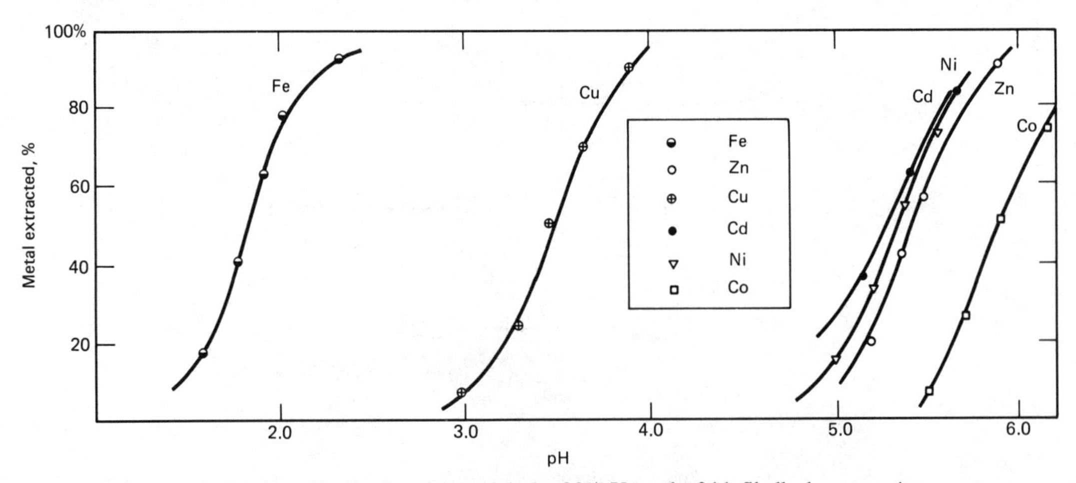

Figure 4. Equilibrium distribution for metals in 30% Versatic 911–Shellsol system (temperature 20°C). Extraction from single metal sulfate solutions of 5 g/liter metal [8].

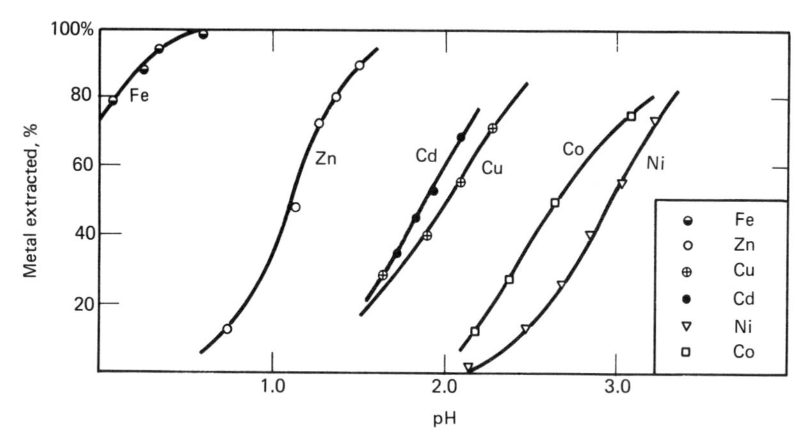

Figure 5. Equilibrium distribution for metals in 30% D2EHPA-Shellsol system (temperature 20°C). Extraction from single metal sulfate solutions of 5 g/liter metal [8].

TABLE 1 EQUILIBRIUM CONCENTRATIONS (IN GRAMS PER LITER) FROM DISSOLVING CEMENT COPPER IN A MIXTURE OF WATER AND 30% D2EHPA IN SHELLSOL (pH = 2.4, $T = 55°C$)

Metals Present	Cu	Zn	Cd	Co	Ni	As	Sb
D2EHPA phase	5.1	12.8	1.2	0.003	0.001	0.0015	0.015
Aqueous phase	13.6	0.15	7.3	0.11	0.11	0.13	0.17

The diagram serves only as a guideline at the start of a development program. Real process parameters have to be worked out on actual process streams, as in the present project.

To achieve the specified purities of the product streams, sulfuric acid and caustic soda are added to control the pH at various stages within the overall solvent extraction circuit.

A very important aspect of the process economics comes from dissolution of the solid cementation product by direct leaching with a mixture of water and the organic solution of D2EHPA. The amount of alkali needed for the separation of the metal ions by the cation extracting system is greatly reduced when a major part of the solid raw material is directly dissolved into the organic phase [9]. This principle of integrating the leaching and solvent extraction steps is further illustrated and discussed in the succeeding example of iron extraction from zinc leach liquors.

Figure 6 shows in a block diagram the process

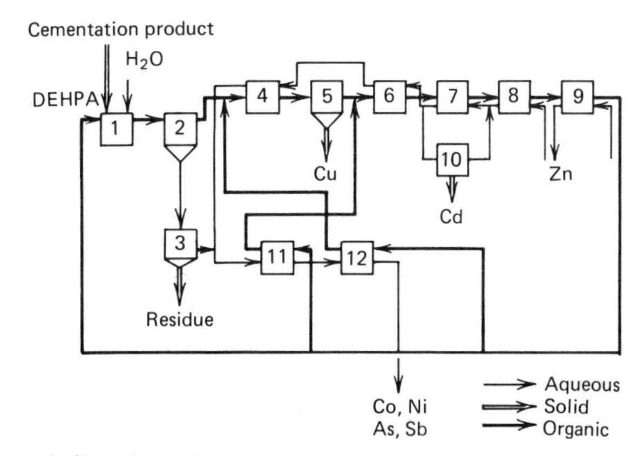

Figure 6. Schematic flow sheet of solvent extraction process for zinc dust cementation product.

flow sheet principles. Starting with the leaching step [1], the two phases from Table 1 are separated [2] in two streams. The aqueous phase meets an organic solution of D2EHPA for extracting cadmium and zinc in one section [11] and copper in the next [12], leaving cobalt, nickel, arsenic, and antimony in the raffinate phase. The organic-phase mainstream from the leaching step [1] and from the copper extraction [12] is scrubbed [4] for removing the last traces of cobalt, nickel, arsenic, and antimony prior to stripping of copper [5] as a copper sulfate solution or directly as copper crystals.

Downstream from the copper stripping [5], the organic solution is enriched with the cadmium organic flow from the extraction step [11]. The last trace of copper is scrubbed [6] prior to the cadmium stripping [7] with sulfuric acid from the cadmium electrowinning [10]. A scrubbing stage for cadmium [8] precedes the final complete stripping of zinc [9]. The stripping stream of zinc is returned to the main zinc sulfate circuit for electrowinning.

Each step in the block diagram may consist of several stages that must be adjusted by adding acid or alkali for pH control to meet the particular specifications. This is illustrated in Fig. 7, where the steady-state profile of the pH and concentrations are shown for the three stages in the cadmium stripping section [7], where a pure cadmium sulfate solution suited to electrowinning is produced by stripping with sulphuric acid. The zinc concentration must be kept low in the strip solution.

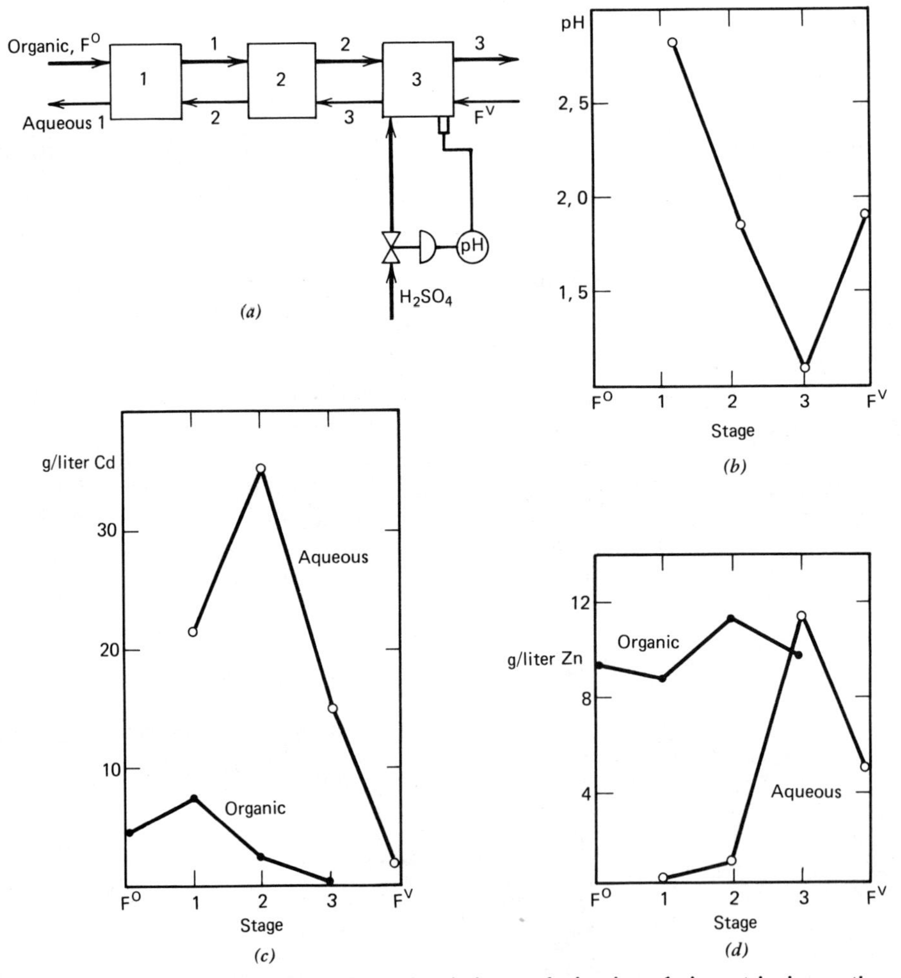

Figure 7. Concentration and pH profiles of cadmium and zinc in cadmium stripping section of D2EHPA solvent extraction process for zinc dust cementation product: (*a*) three-stage stripping of cadmium; (*b*) pH profile in aqueous phase; (*c*) Cd profile; (*d*) Zn profile.

The steady-state concentrations given at a particular stage number are for the streams leaving the stage. It will be noted that a maximum concentration of cadmium (35 g/liter) is obtained in stage 2, whereas the stream leaving the stripping section contains 21 g/liter. This is to be compared with the aqueous profile of zinc, which shows that its concentration is at minimum in the exit stream.

5. SOLVENT EXTRACTION OF IRON IN ZINC HYDROMETALLURGY

The recovery of zinc and cadmium from the cementation product referred to above is typical of a process where both metals are extracted into the organic phase and subsequently separated. In zinc hydrometallurgy, however, there is also a challenge for solvent extraction in purifying the zinc sulfate leach liquor. In particular, the removal of iron is a major problem.

Present practice in the zinc industry is to leach the calcine with sulfuric acid in two stages. In the first, gentle leaching with spent electrolyte dissolves some 80–85% of the zinc, leaving the less soluble zinc ferrites behind. These are dissolved in the second leaching stage, where hot and highly concentrated sulfuric acid is used. The latter increases the overall zinc yield but, at the same time, the iron is brought into solution and then must be separated from the zinc sulfate solution. This problem has been met by various process routes for precipitation of iron as jarosites, goethite, or hematite [10]. However, these do exhibit disadvantages.

An obvious alternative would be to apply a suitable solvent extraction system for extracting the iron from the acid zinc sulfate solution. Possibilities appear to be the carboxylic acid cation exchangers such as Versatic acids [9]. The efficient operation of these extractants is demonstrated by the distribution curves for Versatic 911 in Fig. 4, where ferric ions are shown to be extracted quite readily at a low pH. It has usually been quoted as a drawback of acid cation exchangers that the liberated proton in the extraction equation has to be neutralized by an equivalent amount of alkali [11]. This is economically prohibitive, at least for extraction of the appreciable amount of iron present in the leach liquors.

To overcome this problem, the prohibitive feature of adding alkali to the processing circuit can be avoided by a technique outlined in patents by the present author [9]. The crucial point of the process principle for application in zinc hydrometallurgy is to take advantage of the alkaline property of the zinc calcine itself. By a direct leaching of the calcine with the organic acid (in this case Versatic acid is to be preferred), the organic salt of zinc–Versatic is formed quite readily in the organic phase by the reaction:

$$\mathrm{ZnO(solid)} + \mathrm{2HR(org)} \longrightarrow \mathrm{ZnR_2(org)} + \mathrm{H_2O} \tag{1}$$

As will be seen from Fig. 4, there is a considerable difference in the pH values between the extraction curves for zinc and iron in the Versatic system. By contacting the iron-bearing leach liquor from the hot acid leach of the zinc ferrites, the exchange reaction will take place:

$$\mathrm{2Fe^{3+}(aq)} + \mathrm{3ZnR_2(org)} \rightleftharpoons \mathrm{3Zn^{2+}(aq)} + \mathrm{2FeR_3(org)} \tag{2}$$

Prior to reaction (2), any free acid in the leach liquor will be neutralized by the reaction

$$\mathrm{2H^{+}(aq)} + \mathrm{ZnR_2(org)} \rightleftharpoons \mathrm{Zn^{2+}(aq)} + \mathrm{2HR(org)} \tag{3}$$

which will adjust the pH upward into regions where the exchange reaction (2) will take place. The preceding exchange reaction for purification of zinc calcine leach solutions has also been discussed by Van der Zeeuw [12].

The iron loaded organic phase may readily be stripped by any mineral acid. If hydrochloric acid is used, the ferric chloride produced may be split into a salable iron oxide, whereas the hydrochloric acid is recovered for recirculation. A more promising stripping route, however, appears to be precipitation of the iron directly from the organic phase by the recent technique called hydrolytic stripping [13]. By heating the iron-loaded organic phase in an autoclave in the presence of water in the temperature region of 180–200°C, the following simple hydrolysis reaction will take place:

$$\mathrm{2FeR_3(org)} + \mathrm{3H_2O} \longrightarrow \mathrm{Fe_2O_3(solid)} + \mathrm{6HR(org)} \tag{4}$$

The Versatic acid will be regenerated without any noticeable degradation at the temperature employed. The simplified process flow sheet for the zinc process as above is shown in Fig. 8.

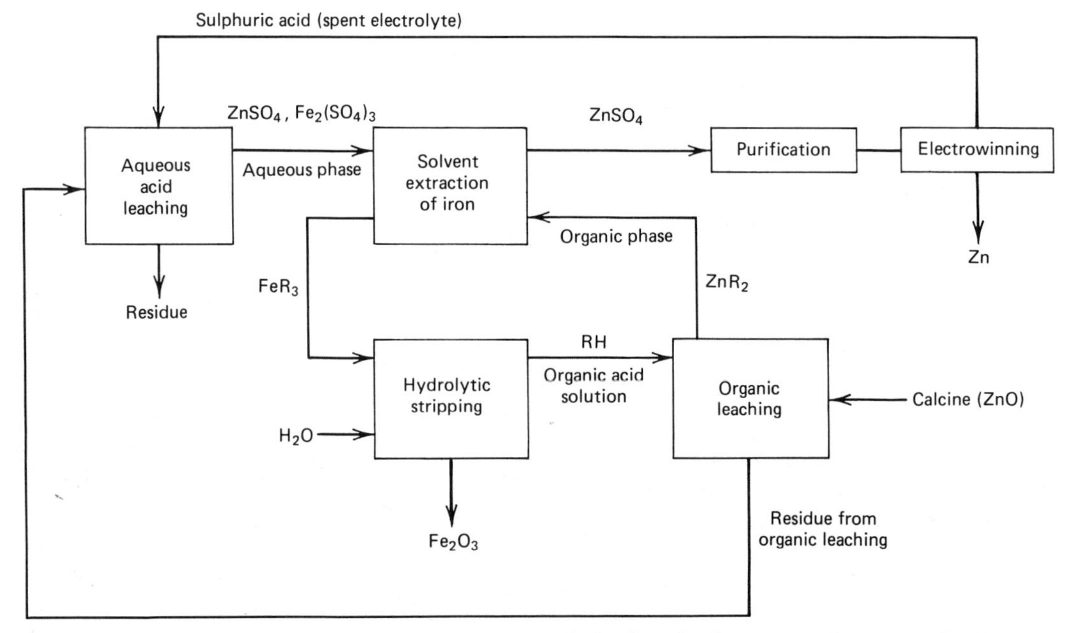

Figure 8. Schematic flow sheet of integrated organic leach and solvent extraction process in zinc hydrometallurgy [9, 13].

REFERENCES

1. D. S. Flett and D. R. Spink, *Hydrometallurgy* **1**, 207–240 (1976).
2. D. S. Flett, Solvent extraction of non-ferrous metals: A review 1975–1976. Warren Spring Laboratory LR254(ME), 1977.
3. G. Barthel, H. Fischer, and U. Scheffler, Separation and recovery of copper and zinc from sulphuric acid solutions of metal salts by solvent extraction, in *Complex metallurgy '78*, IMM, London, 1978, pp. 121–133.
4. D. S. Flett, Solvent Extraction in Scrap and Waste Processing, paper presented at the Joint IMM/SCI Conference on Impact of Solvent Extraction and Ion Exchange on Hydrometallurgy, University of Salford, March 21–22, 1978, 20 pp.
5. Anonymous, *Eng. Min. J.* **179** (11), 133 (1978).
6. Davy Powergas Ltd., *Espindesa Zinc Solvent Extraction Process*, brochure, February 1978.
7. G. Steintveit, F. Dyvik, G. Thorsen, and A. Hjemås, Process for Treating Chloride-Containing Zinc Waste, British Patent 1,366,380 (1971).
8. A. Grislingås, Dr. Ing. thesis, The Norwegian Institute of Technology, University of Trondheim, 1976.
9. G. Thorsen, Extraction and Separation of Metals from Solids Using Liquid Cation Exchangers, U.S. Patent 4,008,134; British Patent 1,474,944.
10. A. R. Gordon, Improved Use of Raw Material, Human and Energy Resources in the Extraction of Zinc, in *Advances in Extractive Metallurgy 1977* IMM, London, 1977, pp. 153–160.
11. E. L. T. M. Spitzer, The Use of Organic Chemicals for the Selective Liquid/Liquid Extraction of Metals, *International Symposium on Solvent Extraction in Metallurgical Processes*, Antwerp, May 4–5, 1972, Technologisch Instituut K. VIV, Antwerp, 1972, pp. 14–18.
12. A. J. Van der Zeeuw, *Hydrometallurgy* **2**, 275–284 (1977).
13. G. Thorsen and A. J. Monhemius, Precipitation of Metal Oxides from Loaded Carboxylic Acid Extractants by Hydrolytic Stripping, paper presented at the 108th AIME Annual Meeting, New Orleans, February 18–22, 1979.

25.6

COMMERCIAL PROCESSES FOR RARE EARTHS AND THORIUM

L. Sherrington

Wokingham
United Kingdom†

1. INTRODUCTION

The separation of rare earths by liquid–liquid extraction depends mainly on four well-known extractants: (1) TBP; (2) quaternary ammonium compounds; (3) tertiary carboxylic acids (Versatic acid); and (4) D2EHPA. Reagents 1 and 2 extract rare earth nitrates from aqueous solution; reagents 3 and 4 form compounds with rare earth metals. Exchange of rare earths between organic and aqueous phases occurs readily in all cases.

Separation factors are typically 1.5–2.5 for neighboring members of the rare earth series. The preparation of high-purity products requires 30–60 stages of separation.

Mixer–settlers have been used almost exclusively for these separations. Stainless steel equipment is preferred when the aqueous phase contains rare earth nitrates. Versatic acid and D2EHPA processes may use rare earth chloride solutions in equipment coated with epoxy resin. All the extractants are diluted with trimethyl benzenes (Shellsol A) or aliphatic hydrocarbons. A typical plant will contain (1) solvent loading section, (2) countercurrent separation section, (3) backextraction section, and (4) further purification of extractant for recycling when necessary.

Most stages are required for the separation section. This consists of about 50 stages with a feed to stage 25 or thereabouts consisting of aqueous rare earth nitrates (chlorides). Extractant loaded with rare earths enters at the raffinate end (stage 1) and is transferred to the back-extraction section after contact in the 50 stages. In difficult separations considerable reflux at each end is necessary; up to 90% of the rare earths in the aqueous raffinate flow may be used to load extractant in the solvent loading section. Also, the aqueous rare earths recovered in the back-extraction section are concentrated if necessary and up to 90% of them returned to stage 50.

The design and control of the separation section is greatly facilitated by having constant molar concentrations of rare earths in the organic phases of each stage. Feed, aqueous layers, and recycled back-extract should also be at the

†Formerly of RTZ Ltd., United Kingdom.

same concentration, but not necessarily the same as the organic phase. For the separation of two rare earths, the standard methods described in Chapter 5 may be used to predict relative mass flows of rare earths in feed, organic phase, and aqueous phase; they also predict reflux ratios and the number of stages required. Multicomponent separations can be predicted only by a computer program simulating the extraction process.

Process control by atomic absorption, recording spectrophotometer or x-ray fluorescence is necessary at the feed point and about five stages each side of it. Product analysis when sufficiently rapid results can be obtained also helps control.

2. LANTHANUM-TBP PROCESS

Separation of light rare earths (La to Sm) is carried out by using the extractant TBP diluted with 50% by volume of Shellsol A. Tributylphosphate forms the compound $La(NO_3)_3 \cdot 3TBP$ and extracts a maximum of 100 g/liter of lanthanum oxide when in equilibrium with aqueous lanthanum nitrate at 200 g/liter of lanthanum oxide. Separation factors increase with rare earth concentration in the aqueous layer. The process uses an aqueous layer at 450 g/liter of rare earth oxides in equilibrium with the extractant containing 100 g/liter of rare earth oxides. Separation factors for neighboring rare earths La–Ce, Ce–Pr, Pr–Nd, and Nd–Sm are 2, 1.5, 1.5, and 2, respectively. Rare earths heavier than Sm cannot reasonably be separated by this system.

The simplest process for pure lanthanum uses a feed containing 98% La, 1% Pr, and 1% Nd. This mixture as carbonates is added to 4 M HNO_3 in stainless steel vessels until a small excess of carbonate is present. After boiling for 1 hr, the solution is cooled and clarified on a precoated rotary vacuum filter. The clear product is evaporated to 450 g/liter rare earth oxides. This feed is free from cerium, and the most difficult separation is La–Pr, with a separation factor of 3.

Pure lanthanum accumulates at the raffinate end of the plant, where the reflux ratio requires about 70% of the rare earths flowing along the aqueous extract section to be transferred to the ingoing solute-free extractant. Figure 1 is a flow diagram of the process, and Table 1 gives the relative mass flows of rare earths in the process and the liquid flows. The high feed flow in this separation results in solvent loading occurring in stages 1–3 at the raffinate end; stages 4–40 operate at 450 g/liter and 100 g/liter of rare earths in aqueous and organic phases, respectively.

Stainless steel equipment is used for this process, and the liquid flows in Table 1 are for 90-liter mixer–settlers; these are built in groups of 10, having pump–mixers with impellers operated by direct drive at 750 rpm. Prolonged settling is needed in the back-extraction section, and this is achieved by doubling the length of the settling chambers. The back-extraction evaporator is steam heated with a stainless steel coil. The distillate may be used for back extraction and topped up with deionized water. Rare earths Pr and Nd accumulate in the evaporator and holding tanks and can be removed occasionally. Products can be converted to oxide by precipitation as

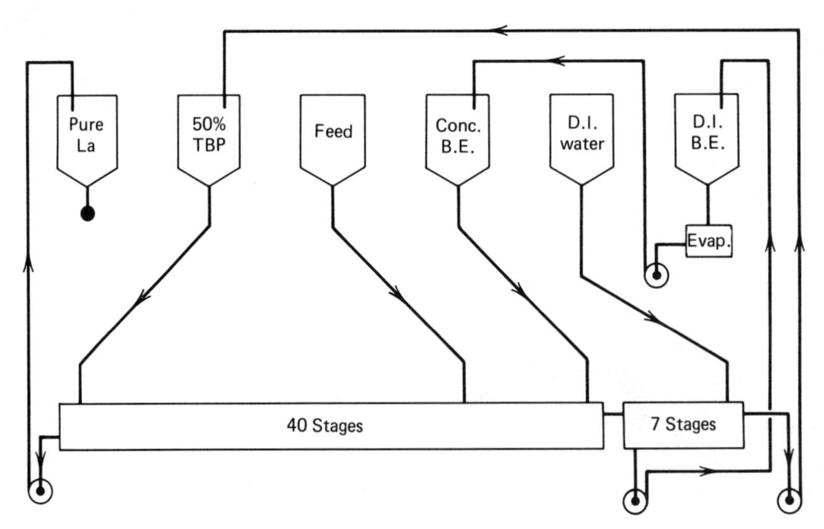

Figure 1. Lanthanum purification by TBP. (B.E. = back extract.)

TABLE 1 LANTHANUM TBP PROCESS–SX FLOWS

	Relative Mass Flow of Rare Earths	Liquid Flows, liters/hr
Organic phase	1.0	225
Feed	0.31	21
Scrub	0.99	50
Raffinate	0.30[a]	71
Back-extract	–	225

[a]Mass flow after loading organic phase.

oxalates from 100 g/liter solutions and igniting the filtered product at 900°C in trays in a tunnel kiln.

This process is based on a cerium-free feed, and this simplifies the separation. The TBP process can be used to separate La and Ce; the lower separation factor of 2 reduces feed flow and increases reflux. An example of this type of process is described later under the section on Pr–Nd separation. The major costs in TBP processes are steam for evaporation required to maintain reflux, and rare earth losses.

3. LANTHANUM-VERSATIC ACID PROCESS

The extractant is Versatic acid 911 containing a mixture of tertiary carboxylic acids with 9–11 carbon atoms. When diluted with 50% Shellsol A, it dissolves about 90 g/liter of rare earth oxide; higher loadings produce precipitates that inhibit settling. Higher separation factors of 3 and 1.8 are obtained for La-Ce and Ce–Pr, respectively; factors for Pr–Nd and so on are similar to the TBP system.

Loaded solvent is in equilibrium with rare earth nitrate or chloride solutions in the pH range 3.5–6. Solvent loading is achieved by mixing barren organic phase with aqueous rare earth nitrates and adding base until pH 6 is reached. This represents substantially complete tranference to the organic phase; a small loss of water-soluble Versatic acid occurs at pH 6, but this disappears at lower pH levels.

A typical process uses a separate solvent loading unit consisting of two mixers in series; flows of 4 M NH_3 and 1.33 M $La(NO_3)_3$ are adjusted to give an aqueous layer of pH 6. The separation section operates with feed and all aqueous layers at 1.33 M rare earth nitrate; solvent is constant at 0.34 M rare earths. Back extraction is achieved by 4 M HNO_3 and uses six stages with double mixers to increase contact times.

TABLE 2 LANTHANUM VERSATIC PROCESS–SX FLOWS

	Relative Mass Flow of Rare Earths	Liquid Flows, liters/hr
Organic phase	1.0	425
Feed	0.31	45
Back-extract flow	0.99	144
Pr and similar	0.01	1.3
Raffinate	1.3	188
Solvent loading (rare earths)	1.0	144
4 M NH_3	–	144

Lanthanum purification including Ce removal uses reflux ratios similar to those for the TBP process. Mass flows and liquid flows are given in Table 2. If cerium removal is not required, the process is designed on the La–Pr separation factor of 5; a reflux ratio at the raffinate end of 1 : 1 may be used. This process can also be adapted for removal of lanthanum from a crude feed containing 65% La, 9% Pr, and 26% Nd. Relative mass flows of rare earths are given in Table 3.

Versatic acid separations are subject to interference from iron impurities as rare earths suppress the solubility of the iron compound in the extractant. The last stage of the separation section and the first back-extraction stage are coupled directly and iron impurities can be removed by syphoning off a small amount of the aqueous phase where the change from acid to neutral con-

TABLE 3 LANTHANUM STRIPPING BY VERSATIC ACID

	Relative Mass Flows of Rare Earths	Liquid Flows, liters/hr
Organic phase	1.0	450
Feed	0.77	173
Pure lanthanum product	0.5	120
Pr + Nd product	0.27	53
Back-extract recycled	0.73	176
Raffinate for loading organic phase	1.0	240
Back-extraction 4 M NHO_3	–	240
4 M NH_3	–	240

ditions occur; the same syphon can collect product from the back-extraction end of the plant. Flows are controlled to keep conditions in the syphon stage constant.

4. PRASEODYMIUM-NEODYMIUM SEPARATIONS: CONTINUOUS

Rare earth Pr–Nd separation factors are about 1.5 for most extractants. Tributylphosphate is selected as processing costs are cheaper. Because of the difficult separation, the plant and flow sheet differ from the lanthanum–TBP process in that the feed is lower, the reflux is much higher, and special solvent loading equipment is necessary. Sales of Pr and Nd are for 95-98% purity; higher purities would require many more stages.

A plant of 60 stages, center fed with Pr–Nd mixture containing 25% Pr, uses the relative mass flows of rare earths given in Table 4. The liquid flows in Table 4 are figures for a separation based on 300 liter mixer–settlers. The raffinate for this type of process must be evaporated to 450 g/liter and recycled to double the aqueous flow at stage 3; thus stages 1–3 act as solvent loading section, ensuring TBP saturation with rare earths in stages 4–60. All aqueous phases in these stages are at 450 g/liter rare earth oxides. Products accumulate in evaporated raffinate and back-extract holding tanks.

5. PRASEODYMIUM-NEODYMIUM SEPARATIONS: TOTAL REFLUX

The continuous separation of Pr–Nd describes the flow conditions and products when the plant has reached equilibrium. For start-up from a new plant, mixer–settlers, evaporators, and so on would be filled with mixed rare earths and operated until products reach purity. It is more useful to stop the feed and run the plant under total reflux; in other words, the relative mass flows of rare earths in solvent and aqueous are equal. This can be continued until the raffinate recycle and back-extract recycle reach the required purity. The Pr and Nd separate as two bands at each end of the mixer–settlers with a "crossover" section in the middle changing from 95% Pr to 98% Nd in about 18 stages. Given enough stages, a mixture of three rare earths; for example, La, Pr, and Nd would form three bands of purified La, Pr, and Nd under total reflux. This principle was adopted by Thorium Ltd. to produce more than two products from one SX plant [1, 2].

The basic plant is identical to that described for continuous Pr–Nd separation. Sixty-five stages of separation are used at 300 liter each. Feed is prepared from rare earth carbonate as before and contains 38% La, 10% Pr, 35% Nd, 17% Sm, and so on. After filling with rare earths, the plant runs continuously for 5–6 days to separate into bands. Then 99.6% La accumulates in the evaporated raffinate stock tank and evaporator; Pr and Nd bands form in the mixer–settlers, and a crude Sm, and so on, concentrate in evaporated back-extract stock tank. After analysis, products are removed from various sections and recovered. To ensure reasonable size of Pr and Nd bands, extra feed is introduced at stage 18 at startup and flows at 160 liters/hr for 5 days (see Table 5). During this forced feed period, crude lanthanum and samarium concentrates accumulate from the two evaporators. The forced feed is stopped and the process operated under total reflux for 5 days to complete purification of the four products. The products are withdrawn from mixer–settlers and holding tanks. The empty stages are half-filled with feed, and then solvent flow and forced feed flow are commenced to restart the cycle. Each cycle produces 770 kg 95% Pr, 1360 kg 99% Nd, and 2500 kg 99.6% La.

TABLE 4 CONTINUOUS PRASEODYMIUM-NEODYMIUM SEPARATIONS

	Relative Mass Flows of Rare Earths	Liquid Flows, liters/hr
Extractant	1.0	1000
Feed	0.27	54
Returned back-extract	0.8	160
Aqueous in extraction section	1.07	214
Raffinate recycle	1.0	200
Nd product	0.2	40
Pr product	0.07	14
Back-extract	–	1000

TABLE 5 TOTAL REFLUX FOR La, Pr, AND Nd SEPARATION

	Relative Mass Flows of Rare Earths	
	Forced Feed	Total Reflux
Extractant	1.0	1.0
Feed (stage 15)	0.8	–
Returned back-extract	0.6	1.0
Raffinate recycle (stage 3)	–	1.0

6. YTTRIUM PURIFICATION

This element behaves like a rare earth but is erratic. Thus, with the use of Versatic acid, it separates with terbium. With D2EHPA it separates near erbium. Most yttrium purification methods use two such processes to ease the separation. Some use extraction of rare earth–yttrium thiocyanates into quaternery ammonium extractant or TBP. One patent claims that extraction of thiocyanate at 0°C gives a process that is selective for yttrium. Thorium Ltd.'s process uses two plants of 56 stages operating with forced feed and total reflux as described previously above [3]. The first stage has Versatic acid as solvent. Feed is prepared from a 50% yttrium concentrate with forced feed into box 15 for 2 days followed by 3 days under total reflux. Light rare earths accumulate at the raffinate end and are removed. A slow syphon at the back-extraction end operates during total reflux to remove accumulated heavy rare earths and iron. Product is unloaded from 16 center boxes at 95% Y and less than 10 ppm Nd and 50 ppm Er. Final purification is made in a typical TBP total reflux plant of 56 stages. Operating conditions are different; feed and aqueous concentrations in settlers are kept constant at 90 g/liter (solvent 21 g/liter). Raffinate recycle is run at a concentration of 130 g/liter to keep solvent loading constant. The remaining impurities (mainly Sm, Gd, Tb, Dy, and Ho) accumulate in the back-extraction concentrate; this is also maintained at 90 g/liter. Separation factors for Y–Gd and so on are about 2 in this process. After analysis, product is unloaded from about 20–25 stages at 99.9%.

The company MCI-Megon A/S operates a small plant producing 30 tons/year of high-purity yttrium oxide in Norway [4]. The raw material used is a 60% Y_2O_3 concentrate extracted from xenotime. After dissolution in nitric acid, this is upgraded to 99.999% Y_2O_3 by a solvent extraction process. Quaternary ammonium compounds dissolved in an aromatic diluent are used as extractants. Separation is based on the different extraction behavior of the lanthanides in nitrate and thiocyanate systems and on the fact that yttrium extracts like a "heavy" in the nitrate system and like a "light" in the thiocyanate system (see Fig. 2). It is thus possible to extract away the lights in the nitrate process and subsequently the heavies, using quaternary ammonium nitrate and thiocyanate; yttrium collects in the aqueous raffinate. In addition, yttrium is extracted from the aqueous phase to separate it from non-rare-earth (RE) impurities. Yttrium

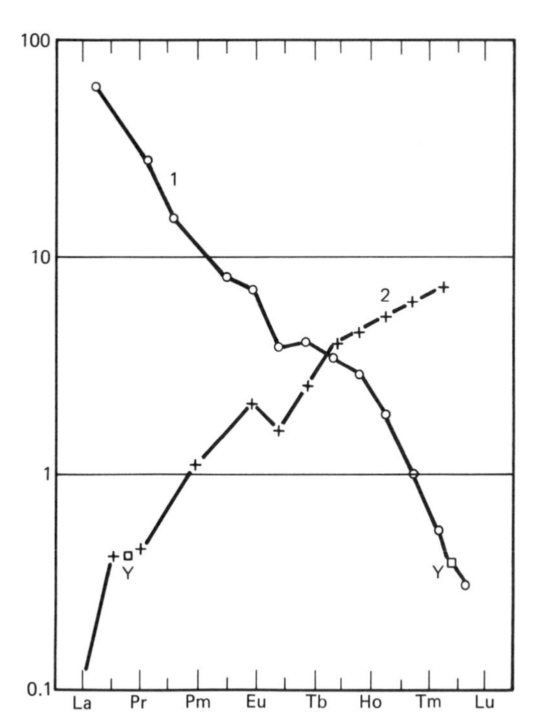

Figure 2. Distribution coefficients for yttrium and rare earths: (1) aliquat 336–solvesso, nitrate system; (2) aliquat 336–xylene, thiocyanate system.

nitrate is then stripped from the extractant with very pure water, and yttrium oxalate is precipitated and ignited to Y_2O_3. The process has been patented and described briefly in previous publications. A simplified flow diagram is shown in Fig. 3. The extraction processes are carried out in about 100 stainless steel mixer–settlers. Maintenance of correct flow ratios is a key factor in controlling the process. Yttrium concentrations in various process streams are monitored by an x-ray fluorescence based on-stream analyzer.

Patent information is available on the yttrium process of the Molybdenum Corporation [5]. This is a two-step process with the use of a continuous Versatic acid-type process to separate the heavy earths and a quaternary amine–nitrate process to separate light rare earths.

7. D2EHPA PROCESSES

This reagent gives the highest separation factors in the region Sm–Lu and is probably widely used for preparing europium concentrates where partial separation from samarium is essential. The Molybdenum Corporation describe their process in a publication by the Denver Equipment Company [6].

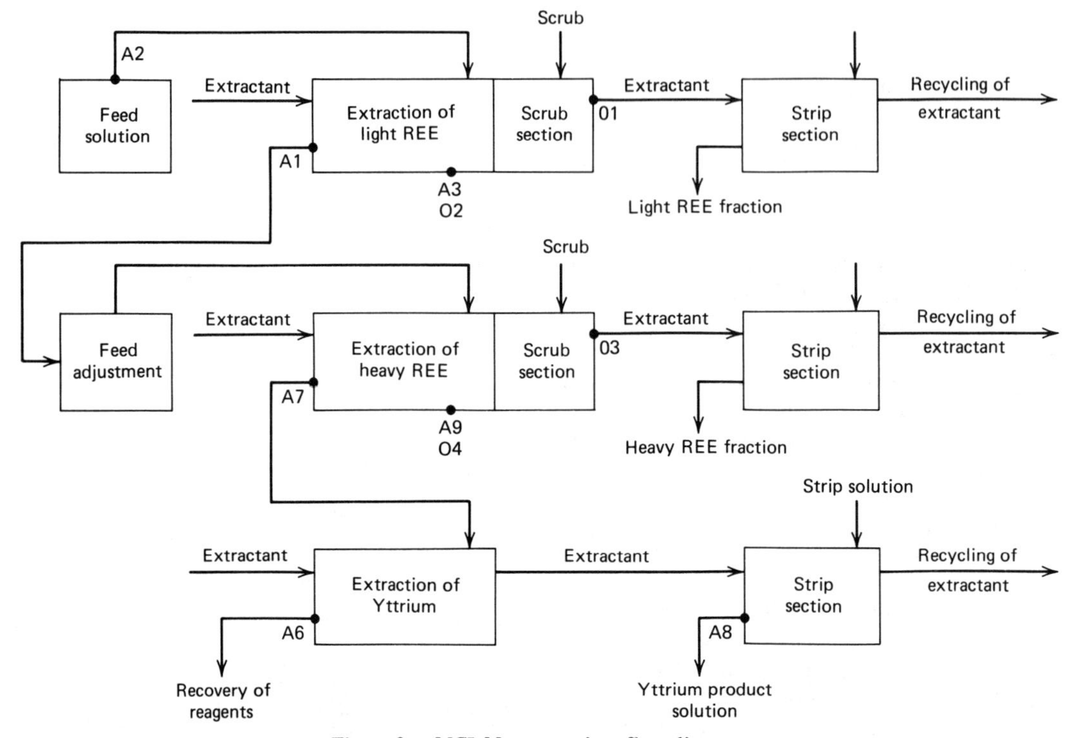

Figure 3. MCI–Megon yttrium flow diagram.

This extractant is used in a system that is similar to Versatic processes; the reagent forms dimers in normal diluents such as H_2R_2 or $RE(HR_2)_3$. It is a strongly acidic reagent and extracts rare earths from acid solutions having 2–0.2 *M* excess acid. Complete back-extraction may need high flows of 5 *M* acid to eliminate heavy earths. Further information and details of Sm and Gd preparation have been described [7].

8. RHÔNE-POULENC RARE EARTH SEPARATIONS

Detailed information on this company's processes are not available, but they have outlined a comprehensive manufacturing procedure for processing monazite that uses solvent extraction separations for all rare earth and thorium purification. Monazite is decomposed to give thorium and rare earth hydroxides that are converted to acid nitrate solution. Successive continuous extraction processes—probably using TBP—give Th, high-purity La, Ce, Pr, and Nd. The residue of Sm, Y, and so on is then separated by a variety of extractants.

9. COMPUTER PROGRAMS

W. P. Kemp of Thorium Ltd. developed programs based on batchwise equilibrium of organic and aqueous phases for each stage, assuming constant concentrations of total rare earths in each phase and fixed separation factors. This is readily applicable to TBP and Versatic acid systems. It was adapted to D2EHPA systems by assuming constant total concentrations of rare earths plus hydrogen ion in each phase. It also assumed constant separation factors for rare earths from each other and a more complex equilibrium between hydrogen ion and rare earths. A 50-stage process required about 250 calculations of stage transfers to reach a steady state. B. Gaudernack et al. [8] developed a comprehensive program for calculating equilibrium conditions for their yttrium process. This has been published in detail elsewhere.

10. THORIUM

Purification of thorium by solvent extraction was first established at Thorium Ltd. in 1958 [9]. It was used again in 1972 with some minor modifications. The process involves extraction of tho-

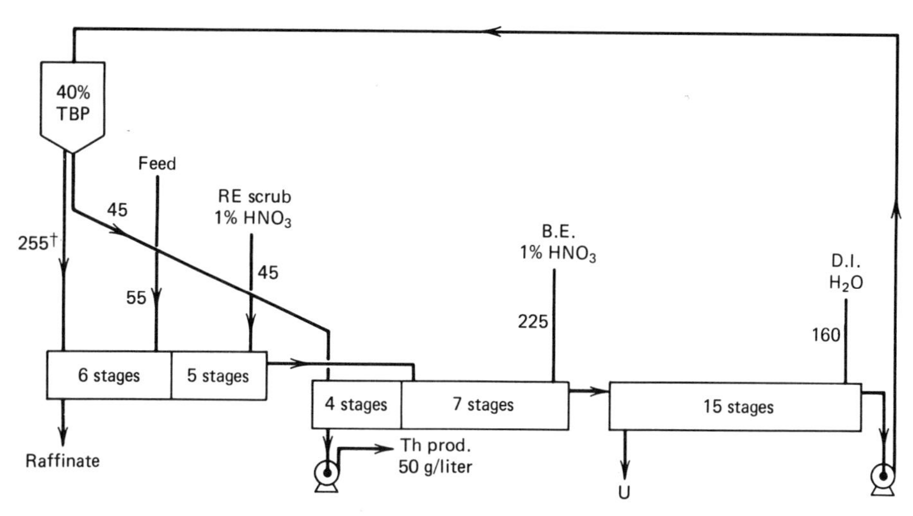

Figure 4. Thorium process. †Flows in liters/hr for 90 liter mixer-settlers.

rium and uranium from nitrate solution with TBP–Shellsol A, scrubbing with water to remove residual rare earths, and back-extraction with 1% HNO_3. Feed contains 180 g/liter of ThO_2, 10 g/liter of P_2O_5, 150 g/liter of Ln_2O_3, and 0.06–0.3 g/liter of U. Uranium is left in the organic phase and is eliminated by water washing. The thorium back-extract is scrubbed with organic phase to remove uranium from the product (see Fig. 4). The thorium nitrate contains less than 1 ppm of uranium, and 5 ppm of rare earths is obtainable.

REFERENCES

1. L. G. Sherrington and W. P. Kemp, British Patent 1,026,791 (1966) (amended 1969).
2. C. G. Brown, British Patent 1,262,469 (1972).
3. L. G. Sherrington and W. P. Kemp, British Patent 1,180,922 (1970).
4. B. Gaudernack, U.S. Patent 3,751,553 (1973).
5. C. M. Trimble and D. B. Strott, U.S. Patent 3,640,678 (1972).
6. Denver Equipment Company (USA), Bulletin No. M4-B167, 1967.
7. C. G. Brown and L. G. Sherrington, *J. Chem. Tech. Biotechnol.* **29**, 193–209 (1979).
8. B. Gaudernack et al., *Proceedings of the International Solvent Extraction Conference (ISEC), Lyons, 1974*, Vol. 3, Society of Chemical Industry, London, 1974, p. 2631.
9. R. J. Callow, *The Industrial Chemistry of Lanthanom, Yttrium, Thorium and Uranium*, Pergamon Press, New York, 1967, pp. 108–119.

25.7

COMMERCIAL PROCESSES FOR PRECIOUS METALS

R. I. Edwards
University of Natal
South Africa†

W. A. M. te Riele
Council for Mineral Technology
South Africa

1. INTRODUCTION

The precious metals—silver, gold, and the platinum metals—are recovered from a wide variety of sources that present metallurgical problems of a widely differing nature.

In the case of gold, the bulk of production is derived from relatively simple ores by the elegant and highly successful cyanide process. Silver is produced mainly as a by-product of base metal (lead and copper) refining operations, but its recovery involves a relatively simple combination of pyrometallurgical and electrochemical operations.

Although solvent extraction techniques have sometimes been proposed in these two cases [1–4], they are unlikely to become significant and are not considered further here.

In many other instances, and especially for the platinum metals, production is derived from difficult and complex materials in which all the precious metals occur together. It is in this context that solvent extraction technology is likely to play a useful role, and the following discussion considers this application.

The source materials to which solvent extraction procedures would seem appropriate include copper–nickel matte leach residues, copper anode slimes, silver anode slimes, and recycled scrap. Such materials are invariably treated by hydrometallurgical techniques, although sometimes other processes are first employed for the removal of the bulk of the gold and silver.

The hydrometallurgical processes are normally carried out in chloride media, and the application of solvent extraction techniques to the processing is thus largely concerned with the behavior of such solutions.

2. PROCESS CHEMISTRY

The aqueous chemistry of these metals is extremely complex, and reference should be made

†Formerly of National Institute for Metallurgy, South Africa; and Research and Development Manager, Engelhard Industries Ltd., United Kingdom.

to several excellent texts [5, 6] for details. In addition, a review [7] of the more pertinent aspects of the chloride chemistry is available. Only a brief summary of the more important aspects of the behavior of the metals in chloride medium can be given here.

The most prominent feature of the chemistry is the very great tendency of the metals to form complexes in solution. In chloride media they are typically present as anionic or neutral, chloro- or mixed aquo-chloro complexes. Only in special circumstances are cationic species present in the absence of strong neutral ligands such as thiourea and ammonia. In the presence of the latter, a great number of cationic species can be formed; this is often useful for back extraction from an organic extractant. The structure and charge of the complex formed varies considerably from metal to metal and also depends on the oxidation state of the metal. In addition, the stabilities of the full chlorocomplexes are widely different. This wide range in properties of the chlorocomplexes can be utilized to effect separations; differing solubilities of ammonium salts are classically used, but separation by liquid anion exchange is also largely dependent on these differences. Table 1 lists some typical complexes formed by the metals in chloride medium.

The stabilities of the chlorocomplexes listed in the preceding paragraph varies widely; for gold, platinum, and palladium, the complexes are extremely stable, whereas for the other metals, especially in the tervalent oxidation state, aquated complexes are more stable. In addition, rhodium, iridium, osmium, and ruthenium tend to form polynuclear chloride- and oxygen-bridged complexes of great number.

The kinetics of reaction of these metal complexes is also an extremely important facet of their chemistry; in general, these kinetics are very much slower than those for corresponding base metal complexes. This means that even separations based on thermodynamic differences can be affected by kinetic factors, often adversely. Typically, a clear-cut separation between two metals based on differences in properties of two anionic chlorocomplexes is blurred because of the presence in solution of nonequilibrium concentrations of different complexes of both metals.

On the other hand, the reactivities of the metal complexes are widely different; this fact could be exploited to achieve separations based on kinetic differences. Where the technique employed involves reaction between two phases, however, the utility of the method is restricted to those metals showing reasonably high reactivity, that is, Ag, Au, Pd(II), and possibly Pt(II), with the other metals reacting too slowly for a

TABLE 1 PRECIOUS METAL CHLOROCOMPLEXES

Metal	Oxidation Number	Coordination Number	Anion Charge		
			-1	-2	-3
Ag	I	2	$AgCl_2^-$		
Au	III	4	$AuCl_4^-$		
Pt	II	4		$PtCl_4^{2-}$	
	IV	6		$PtCl_6^{2-}$	
Pd	II	4		$PdCl_4^{2-}$	
	IV	6		$PdCl_6^{2-}$	
Ru	(II)	6		$RuNOCl_5^{2-}$	
	III	6			$RuCl_6^{3-}$
	IV	6		$RuCl_6^{2-}$	
Os	III	6			$OsCl_6^{3-}$
	IV	6		$OsCl_6^{2-}$	
Rh	III	6			$RhCl_6^{3-}$
Ir	III	6			$IrCl_6^{3-}$
	IV	6		$IrCl_6^{2-}$	

TABLE 2 PSEUDO-FIRST-ORDER RATE CONSTANTS FOR THE REACTION $MCl_x + L \longrightarrow MCl_{x-1}L + Cl^-$ [a,b]

Metal Species	k, s^{-1}	Metal Species	k, s^{-1}
Ag(I), Au(I)	$\sim 10^7$	Rh(III)	$\sim 2 \times 10^{-3}$
Pd(II)	$\sim 10^3$	Ir(III)	$\sim 10^{-5}$
Au(III)	$\sim 10^2$	Os(III)	$\sim 4 \times 10^{-6}$
Pt(II)	~ 10	Pt(IV)	$< 10^{-10}$
Ru(III)	~ 1		

[a] This reaction is written for L, a "soft" water-soluble ligand, such as CN^-.
[b] For octahedral (six-coordinate) complexes the rate of reaction is not greatly dependent on the nature of the entering ligand. For linear and square-planar (four-coordinate) complexes, the rate is highly dependent on the entering ligand. The rate constants quoted for Ag(I) → Pt(II) are thus correct only to within an order of magnitude.

convenient extent of reaction to take place in a reasonable time.

The order of reactivity for these metal complexes is Ag(I), Au(I) >> Pd(II) > Au(III) > Pt(II) > Ru(III) >> Rh(III) > Ir(III) > Os(III) >> Ir(IV), Pt(IV); this is illustrated in Table 2.

3. SOLVENT EXTRACTION SYSTEMS

3.1. Anion-Exchange Systems

By far the largest number of systems investigated so far are based on anion-exchange mechanisms, although the literature shows a great deal of confusion between ion-pair formation (i.e., anion-exchange) reactions and true complex formation.

Two effects are used to achieve separation by anion exchange: (1) the differing stabilities of chlorocomplex anions and (2) differences in structure, charge, and size between different chlorocomplexes. The first effect is most useful in separating the precious metals from base metals; in moderately weak chloride medium most base metals will be present in solution in cationic forms, whereas the precious metal chlorocomplexes are stable (or inert to substitution) at fairly low chloride concentrations. A number of references illustrate this type of separation, where amines are generally used as the anion-exchange solvent [8–11].

The second effect is very useful in achieving both group and individual precious metal separations. The degree to which a particular anion is extracted by a liquid anion exchanger is controlled largely by the charge:size ratio of the anion; the smaller this ratio, the higher the extraction coefficient.

Reference to Table 1 shows that for precious metals chlorocomplexes, a large spread in this ratio is evident, with anionic charges ranging from −1 to −3. This may be correlated with the extraction coefficient data published by Gindin et al. [12] for a QAB system and by Casey et al. [13] for the TBP system. As expected, published data [14, 15] show extremely high coefficients for gold extraction by anion-exchange systems; in fact, the anionic chlorocomplex can in this case be extracted by reagents (ethers, alcohols, ketones, etc.) not normally considered to be anion-exchangers [16–19].

There are many published examples of specific separations. Those of purely analytical interest are not considered here, as an excellent review of these may be found in the standard texts on analysis by Beamish and van Loon [20] and Ginzburg et al. [21].

Systems of possible commercial application are as follows:

1. Separation of "primary" platinum metals (platinum and palladium) from "secondary" (ruthenium, rhodium, iridium, and osmium) can be accomplished by using amine systems (secondary, tertiary, and quaternary ammonium salts), when the secondary metals are reduced to their tervalent oxidation states [12, 22, 23]. In some instances this effect is reinforced by "aquating" the secondary metal complexes before solvent extraction to decrease the extraction coefficient to the minimum [23]. More limited separations (e.g., platinum from iridium and rhodium) have been published [24, 25], and these, of course, are of the same type.
2. Individual separations of some of the platinum metals can also be accomplished. Among the published possibilities are the separation of rhodium from iridium by use of TBP [26] or amines [27], and purification of rhodium from other metals such as ruthenium [28]. Even the separation of platinum from palladium is possible by use of the rather small differences in charge:size ratios between the $PtCl_6^{2-}$ anion and the $PdCl_4^{2-}$ in the TBP system [13].
3. The separation of gold is a special case

and, as mentioned previously, very weak base anion exchangers can be used.

The very wide range of separations that may be accomplished by anion exchange appears to offer a simple method for their refining. This versatility is, however, offset to a great extent by the difficulty of achieving an efficient and simple method of recovery of the metals from the organic phase. This problem can be severe and a wide variety of methods have been employed in attempts to overcome it:

1. Reversal of the anion-exchange equilibrium, that is, stripping by contact with an aqueous solution containing an anion of comparable extraction coefficient. Where Cl^- is used as the competing anion, only systems of fairly low extraction coefficient are useful, such as primary and secondary amines, and in one case a special tertiary amine [29–31]. Perchlorate stripping can be used [32] but poses several problems with regard to the removal of the ClO_4^- ion from the system.
2. Deprotonation, that is, conversion of the anion exchanger to the free-base form. This is possible through the use of water for extractants such as TBP and alcohols and of alkalies for amines [33]. The usefulness of this approach is also restricted by the aqueous chemistry of the metals and by the fact that the free-base extractant can usually form coordination complexes with the extracted metal, rendering extraction irreversible [34, 35].
3. Formation of a neutral or cationic aqueous-soluble complex of the extracted metal. With base metals, this is done very simply by lowering the chloride concentration to the point where the aquo ion is formed, thus removing the metal from the organic phase. This is not possible with the precious metals, for which strong complexing agents must be used. A wide variety of such reagents are proposed for analytical purposes; for process application, a very much smaller choice of reagents is possible, including NH_3 [36], hydrazine [33], thiourea [32] as typical examples.
4. Alteration of the oxidation state of the extracted metal. Iridium, for example, may be back extracted if the highly extractable $IrCl_6^{2-}$ complex can be reduced to the relatively nonextractable complex $IrCl_6^{3-}$ [37].
5. Direct reduction to metal. This has been used for platinum and palladium [38, 39] and is commonly used for gold [40] but could not be used for the other metals because of the extremely slow reduction kinetics.

3.2. Ligand-Exchange Systems

Ligand-exchange reactions of the type

$$MCl_n^{-(n-m)} + n{-}mL \longrightarrow MCl_m L_{n-m} + (n-m)Cl^-$$

where L is a organic-soluble ligand can also be used to effect separations. However, the speed of such reactions is such that only silver, gold, and palladium can undergo sufficient extent of reaction to be extracted to any significant extent.

The basis for separation in this case is almost always kinetics; it is unlikely that an economic ligand could be found with sufficient thermodynamic selectivity to achieve successful separations. However, the differences in kinetics that do occur are so large that an extremely high degree of selectivity can be obtained.

The rate order in which the metals will be extracted is normally Ag(I) >> Au(III) > Pd(II) >> Pt(II). The reversal between Pd(II) and Au(III) over that shown in Table 2 is due to the fact that two ligands must be exchanged in the case of Pd(II), whereas only one must be exchanged in the case of Au(III) to form a neutral complex.

The ligands proposed for such applications contain nitrogen [41, 42] or sulfur [43–46] as donor atoms; carbon- and phosphorus-containing ligands could be useful as well.

Two possible methods of metal stripping can be used in such systems: (1) where L can be protonated, the complex may be destroyed and the metal stripped with strong acid solutions; in this case, however, the protonated ligand HL^+ can behave as an anion exchanger; or (2) the metal can otherwise only be stripped by a further ligand-exchange reaction to form an aqueous-soluble charged complex.

It should be pointed out, however, that although many of the published separations are based on clearly defined and simple principles, in practice, the separation that is achieved is less effective than could be expected. A typical example is the separation of iridium. Very often the attainment of a solution containing the pure

$IrCl_6^{2-}$ anionic species is difficult; separation of this species is easy, but normally other, non-extractable iridium complexes are present in solution.

4. PROCESS APPLICATIONS

In the past few years the number of applications of solvent extraction in large-scale refining operations has increased rapidly, and it seems likely that this trend will continue in the future. At present, the technique is used mainly as an aid in the classic separation processes, but there are a few examples of processes based mainly on solvent extraction techniques; these include both patented processes not yet in operation and also full-scale operations.

The minor applications include examples of group separations, individual metal separations, and uses as final purification steps before pure metal production. Notable among these are the applications described in Sections 4.1–4.3.

4.1. Group Separations

Soviet workers have been particularly active in this field, and several reports of pilot-plant operation have been published. Dolgikh et al. [8] have described the use of primary amines for concentration of the precious metals from a base-metal-rich chloride solution derived from the chlorination of Cu–Ni slimes. This work has been extended to a two-stage process in which the "primary" platinum metals are first extracted selectively with a tertiary amine and the "secondary" platinum metals thereafter with a primary amine [21]. Thus separation into three groups is obtained. It is not known whether this work has since been extended to full-scale plant operation.

A similar example of such a separation is the process at the Hoboken refinery of Metallurgie Hoboken Overpelt, as reported by Tougarinoff et al. [39].

At this refinery, gold and platinum metals are recovered from silver anode slimes by a nitric acid route that separates the components into a gold–platinum-rich stream and a palladium–rhodium-rich stream. The latter is treated by solvent extraction with a tertiary amine to remove platinum, palladium, and gold from the aqueous phase. The precious metals are stripped from the organic phase by reduction to metals; the mixed metal concentrate thus produced is treated further by conventional techniques for their separation.

The process has since been altered and the role of the solvent extraction circuit reduced to the minor one of recovery of low concentrations of platinum, palladium, and gold from various process streams.

4.2. Individual Metal Separations

The most prominent example is the separation of gold by use of very weak base extractants. The dibutyl carbitol system, first described by Morris and Kahn [16] has been in use at the Acton Refinery of the International Nickel Company since 1971. The industrial application has been described by Rimmer [40].

The system is used to remove gold from a precious metal leach liquor, derived from a Cu–Ni matte concentrate, in which gold is a minor constituent and the primary platinum metals are the major constituent.

As both selectivity and distribution coefficient are extremely high, batch contacting is employed; in fact, a crossflow system is employed to build up gold in the organic phase to ±25 g/liter. The aqueous concentration of gold is reduced from ±5 g/liter to $<$10 ppm.

Base metals that form singly charged or doubly charged anions under the conditions of extraction (4 *M* HCl) are coextracted and are scrubbed back with weak (1.5 *M*) hydrochloric acid.

As the gold cannot be stripped easily by conventional methods, it is reduced directly to metal with an oxalic acid solution to produce a coarse gold sand of high purity ($>$99.95%).

The operation appears to be simple and effective; the only major problem is the high solubility of the extractant (0.3%) in the aqueous phase and consequent high loss (±4% per cycle).

The selective recovery of palladium has also been suggested as a candidate for solvent extraction. The basis of the processes published thus far have been ligand exchange; the solvents used include sulfur compounds [43–46], oximes [41], and various industrial products [47, 48] that could contain carbon–carbon double bonds. It is not clear whether any of these processes have yet reached commercial operation.

4.3. Purification Processes

In this case the extraction process is used to remove other noble metals and base metal contaminants from a particular metal as part of the purification train. A typical example is provided by rhodium; solvent extraction techniques are used for the removal of particularly iridium, but also other contaminants from rhodium-rich solutions. Tributyl phosphate [26, 49] and various amines [27] have been proposed for use in this application, but the most successful commercial system appears to be the TBP one.

This is used for example at the Hanau refinery of Degussa. A rhodium-rich process stream results from the selective removal by classical techniques of the major components, specifically, platinum, palladium, gold, and base metals. Iridium, platinum, and palladium are, however, still present in low levels and are removed in a two-stage TBP extraction. In the first stage the primary metals are extracted, leaving Ir(III) and Rh(III) in the raffinate. This is then oxidized to produce Ir(IV), which is extracted selectively from the rhodium. Rhodium is then further refined by classical procedures; iridium is stripped with water and the solution accumulated until sufficient metal is obtained to enable further purification to take place.

It is believed that similar systems are in operation at other precious metals refineries.

Information about processes in which solvent extraction procedures play the dominant role is available in only three cases; in only one of these is it known for certain that full-scale operation has been achieved. Patents [28, 33] issued to PGP Industries (Inc.) could form the basis for a platinum–metals refining process, but it is not known whether the operations as described in the patents have been tested on a large scale.

The basis for the separation process includes the following steps:

1. The primary metals are coextracted with a secondary amine, and palladium is preferentially stripped by complexation with hydrazine or other ligand. Platinum is then stripped by deprotonation of the amine with sodium bicarbonate.
2. Ruthenium and iridium are oxidized to the +4 oxidation state and extracted with a tertiary amine, leaving rhodium and base metals in the aqueous phase. These are stripped together by a series of steps that involve the destruction of the chlorocomplexes with alkali and their redissolution in weak acid under reducing conditions to form presumably nonextractable aquated complexes.

Further separations and purifications could be accomplished by conventional means.

Patents issued to Matthey Rustenburg Refining (UK) [25, 37, 41] are believed to form the basis for an extensive pilot-plant operation at the Royston refinery. Although the patents allow for a number of variations in the basic process, the most probable sequence of steps is as follows:

1. Gold, silver, ruthenium, and osmium are removed from solution before the solvent extraction process by conventional means, at least in the case of the latter three.
2. Palladium is extracted from a weakly acidic medium by use of an oxime-type solvent (LIX64), stripped with concentrated acid, and recovered by an ion-exchange technique.
3. The solution is reduced selectively and Pt(IV) is removed by solvent extraction with a long-chain amine, either secondary or tertiary, and stripped with alkali.
4. The solution is then oxidized and Ir(IV) extracted with a similar amine and stripped by reduction to Ir(III).

The individual metal streams arising from the solvent extraction separations are then processed further to recover the separated metals in pure form.

A series of patents issued to the National Institute for Metallurgy, South Africa provides the most complete description of a solvent-extraction process yet published [31, 43, 49, 50]. Some variation is allowed in the process, but the most general combination of steps is as follows:

1. Gold and silver are removed by conventional or other means.
2. Palladium is extracted with an alkyl sulfide (a ligand-exchange process) and stripped with ammonia and precipitated as the insoluble complex $Pd(NH_3)_2Cl_2$. The process steps have been described recently [51].
3. Platinum is extracted from the reduced solution with a highly selective long-chain amino acid (an anion-exchange process), stripped with hydrochloric acid, and processed further by conventional means [51].
4. Osmium is removed by distillation of the volatile tetroxide before the ruthenium is extracted as the pentachloronitrozyl complex $RuNOCl_5^{2-}$ with a tertiary amine. This is stripped with an alkali and con-

verted by oxidation to RuO_4^{2-} and finally precipitated as RuO_2.

5. Iridium is then removed from solution by resin anion exchange and eluted as an Ir(III) complex. This is oxidized and the resulting solution purified by use of the TBP system.

With some minor modifications, this process is now being operated in South Africa at two platinum metal refineries. The modifications consist basically of omission of step 2 so that palladium is coextracted with platinum and a mixed strip solution is produced.

Solvent extraction techniques have gained a firm foothold in the industry, and their usage is likely to increase greatly in the next few years.

REFERENCES

1. G. N. Shivrin et al., *Tsvetnye Metally* **39** (12), 15-18 (1966).
2. N. R. Das and S. N. Bhattacharyya, *Talanta* **23** (7), 535-540 (1976).
3. A. I. Sinel'nikova et al., Extraction of Gold and Silver from Cyanide Solutions, Soviet Patent 144028 (1962).
4. D. C. Madigan, *AMDEL Bull.* **6,** 1-6 (1968).
5. W. P. Griffith, *The Chemistry of the Rarer Platinum Metals*, Wiley-Interscience, London, 1967.
6. F. R. Hartley, *The Chemistry of Platinum and Palladium*, Applied Science Publishers, London, 1973.
7. R. I. Edwards, *J. Metals* **28** (8), 4-9 (1976).
8. V. I. Dolgikh et al., *Tr. Vses Nauchn.-Tekhn. Sovesch Protsessy Zhidkostnoi Ekstraktsii i Khemosorbtsii*, 2nd, Leningrad, 1964, pp. 307-311; *Chem. Abstr.* **65:** 11423*d* (1966).
9. C. Pohlandt, *Natl. Inst. Met. Report No. 1881*, 1977 (South Africa).
10. V. I. Dolgikh et al., *Tsvetnye Metally* **40** (2), 27-32, 1967.
11. V. F. Borbat and E. F. Kouba, *Tr. Vses. Nauchn-Tekhn. Soveshch. Protsessy Zhidkostnoi Ekstraktsii i Khemosorbtsii*, 2nd, Leningrad, 1964, pp. 299-306; *Chem. Abstr.* **65,** 11423*c* (1966).
12. L. M. Gindin et al. *Isvest. Sib. Otd. Akad. Nauk SSSR, Serv. Khim. Nauk* **1,** 89-96 (1967); *Chem. Abstr.* **67,** 68140*q* (1967).
13. A. T. Casey et al., *Solvent Extraction Chemistry, Proceedings of the International Conference*, Goteborg, 1966, North Holland, Amsterdam, 1967, pp. 327-334.
14. Z. B. Maksimovic, *Proceedings of the International Solvent Extraction Conference (ISEC), Lyons 1974*, Vol. 2, Society of Chemical Industry, London, 1974, pp. 1937-1947.
15. A. A. Yadav et al., *Separation Sci.* **5** (5), 637-643 (1970).
16. D. F. Morris and M. A. Khan, *Talanta* **15** (11), 1301-1305 (1968).
17. G. S. Lopatin and I. N. Plaksin, *Tsvetnye Metally* **4** (4), 87-90 (1961).
18. C. R. Boswell and R. R. Brooks, *Mikrochim. Ichnoanal. Acta* **5** (6), 814-821 (1965).
19. M. Fieberg et al., *Natl. Inst. Met. Report No. 1996*, **1978** (South Africa).
20. F. E. Beamish and J. C. Van Loon, *Recent Advances in the Analytical Chemistry of the Noble Metals*, 1st ed., Pergamon Press, Oxford, 1972, pp. 34-58.
21. S. I. Ginzburg et al., *Analytical Chemistry of the Platinum Metals*, Wiley, New York, 1975, pp. 434-458.
22. V. F. Borbat and E. F. Kouba, *Tsvetnye Metally* **12** (4), 60-65 (1969).
23. L. M. Gindin et al., *Sin. Ochistka Anal. Neorg. Mater., Tr. Konf. "Nauka-Proizvod"* (1965), published 1971, pp. 41-51; *Chem. Abstr.* **77,** 51318*g* (1972).
24. L. M. Gindin and S. N. Ivanova, *Izv. Sibirsk. Otd. Akad. Nauk SSSR, Ser. Khim Nauk* **2,** 28-34 (1964); *Chem. Abstr.* **62,** 4681*h* (1965).
25. J. J. MacGregor, Improvements in and Relating to the Refining of Metals, British Patent 5682673 (1973).
26. R. B. Wilson and W. D. Jacobs, *Anal. Chem.* **33,** 1650-1652 (1961).
27. M. Ziegler, Methods of Separating Iridium from Rhodium, German Patent 2144151.0 (1971).
28. J. Baltz and E. Coltrinari, Process for the Separation of Platinum Group Metals, U.S. Patent 4,012,481 (1977).
29. V. F. Borbat and E. F. Kouba, *Protsessy Zhidkostnoi Ekstraktsii i Khemosorbtsii*, Moskva-Leningrad, Khimiya, Moscow, 1966, p. 299.
30. M. A. Khattak and R. J. Magee, *Recl. Trav. Chim. Pays-Bas* **88** (6), 584-591 (1969).
31. R. I. Edwards and M. J. Nattrass, The Separation and Purification of Platinum and Palladium, South African Patent 76/3679 (1977).
32. A. Warshawsky, Separating and Purifying Platinum Metals and Gold, South African Patent 72/0308 (1972).
33. J. Baltz and E. Coltrinari, Separation and Selective Recovery of Platinum and Palladium by Solvent Extraction, U.S. Patent 4,041,126 (1977).
34. L. M. Gindin et al., in *Solvent Extraction Chemistry, Proceedings of the International Conference, Goteborg, 1966*, North Holland, Amsterdam, 1967, pp. 433-438.
35. V. I. Dolgikh et al., *Tsvetnye Metally* **4,** 97 (1964).

36. W. A. Te Riele et al., *Natl. Inst. Met., Report No. 1637*, 1974 (South Africa).

37. J. J. MacGregor, Refining Noble Group Metals, British Patent 57793/73 (1973).

38. V. I. Dolgikh et al., *Tsvetnye Metally* **11** (1963).

39. B. Tougarinoff et al., Advances in Extractive Metallurgy, in *Proceedings of Symposium*, London, 1967, IMM, London, 1968, pp. 741–758.

40. B. F. Rimmer, *Chem. Ind.* **2,** 63–66 (1974).

41. J. B. Payne, Improvements in and Relating to the Separation of Metals, British Patent 5779673 (1973).

42. U. B. Talwar and B. C. Haldar, *Indian J. Chem.* 7 (8), 803–805 (1969).

43. R. I. Edwards, The Separation of Platinum Group Metals and Gold, SA Patent 745109 (1974).

44. B. P. Blednov et al., *Sb. Nauch Tr., Krasnoyarskaya Inst. Tsvetnye Metally* **4,** 180–186 (1971); *Chem. Abstr.* **79,** 95135*r* (1973).

45. V. M. Shul'man and T. V. Zagorskaya, *Isv. Sib. Otd. Akad. Nauk SSSR, Ser. Khim Nauk* **2,** 142–144 (1971); *Chem. Abstr.* **76,** 104448*n* (1972).

46. V. A. Pronin et al., *Russ. J. Inorg. Chem.* **18** (17), 1016–1018 (1973).

47. V. T. Athavale et al., *Indian J. Chem.* **5** (11), 585 (1967).

48. H. Nohe, Verfahren zur Abtrennung von Palladiumhaltigen Verbindungen durch Extraktion aus Wasserigen Lösungen, German Patent 1,222,485 (1966).

49. A. P. Evers et al., The Recovery and Purification of Iridium, SA Patent 763681 (1977).

50. R. I. Edwards and M. M. Fieberg, The Separation and Purification of Ruthenium, SA Patent 763680 (1977).

51. R. I. Edwards, *Proceedings of the International Solvent Extraction Conference (ISEC), Toronto 1977*, Canadian Institute of Mining and Metallurgy, Montreal, 1979.

25.8

COMMERCIAL PROCESSES FOR OTHER METALS

C. Hanson
University of Bradford
United Kingdom

1. INTRODUCTION

The purpose of this short chapter is to cover those metals for which solvent extraction is of current or potential commercial interest but that have not been dealt with in other chapters. More general coverage of the fundamental extraction chemistry of metals is available in standard texts [1, 2]. Reference should also be made to useful reviews of the solvent extraction of nonferrous metals by Flett et al. [3, 4], of nuclear applications by Jenkins [5], and of commercial processes by Blumberg [6] and, more recently, by Ritcey and Ashbrook [7].

2. ALUMINUM

The potential application of solvent extraction for the recovery or purification of aluminum has not aroused much commercial interest, although a number of investigational projects have been reported [7]. Thus the U.S. Bureau of Mines has used 0.4 *M* monododecyl phosphoric acid to extract aluminum from waste solutions. Gypsum is precipitated during the extraction and must be removed.

A possible process for separation of iron from aluminum by use of amines in chloride media has been claimed [8]. More recently, work has been published [9] on a process for purification of aluminum derived from nonbauxitic materials. The latter are first leached with a mineral acid; hydrochloric or sulfuric acid is suggested. Several systems were investigated for removal of iron from the resultant solution: carboxylic acids; amines; and TBP. All achieved their objective, although no economic comparison is available. Ritcey and Ashbrook [7] report a similar process in the United States with the use of Alamine 336. Cocco et al. [10] recommend the use of TBP to separate iron from aluminum in hydrochloric acid solution. Removal of iron, by solvent extraction with an amine, features in a process recently suggested for recovery of aluminum from coal combustion ash [30].

There is no evidence that any of the preceding processes having been applied on a large scale.

3. ANTIMONY-BISMUTH SEPARATION

A process has been developed and operated on a pilot scale (600 kg of feed per hour) for antimony-bismuth separation during processing of siderite-sulfide ore [11]. Pyrometallurgical treatment of this gives a flue dust containing antimony, bismuth, and arsenic. This is leached with hydrochloric acid and any mercury reduced by metallic antimony. The solution is contacted with 40% TBP in a kerosene-type diluent, which extracts antimony preferentially. The extract is stripped with acidified ammonium chloride solution. Bismuth is precipitated with ammonia from the raffinate. The precipitate is subsequently redissolved in hydrochloric acid and the bismuth purified by extraction into 40% TBP. The process was operated in bench-scale mixer-settlers and gave products of high purity. It has since been used in a pilot plant with RDC extractors.

4. BERYLLIUM

There is commercial interst in the use of solvent extraction in the recovery and purification of beryllium, and at least two companies have built pilot plants for the purpose [12, 13]. That operated by Vitro Chemical Company was designed to handle the clarified liquor from sulfuric acid leaching of a low-grade ore. In each case an organophosphorous extractant was employed. The U.S. Bureau of Mines has also described a process [14]. Iron is first extracted from the leach liquor by using Primene JM-T in kerosene, and the beryllium is then extracted with 0.5 *M* D2EHPA in kerosene with an alcohol modifier.

5. BORON

Reports have been published [6, 7] of the commercial use of polyhydric alcohols to recover boron, such as from low-grade brines, but the detail available is limited.

Su and et al. [15] have described work on the development of a process for treating the mineral ascharite. This is first decomposed with hydrochloric acid at elevated temperature. The boric acid in the clarified liquor is then extracted by use of 50% 2-ethylhexanol in kerosene. The extract is stripped in two steps: first with water, giving boric acid; and finally with sodium hydroxide solution, giving borax. Mixer-settlers are suggested as contactors (six stages for extraction, three for water stripping, and two for the alkaline strip). It is recommended that both extraction and stripping be carried out at 80°C. The aqueous product streams are passed to crystallizers. The raffinate is treated for removal of iron, aluminium, and calcium. It then comprises essentially pure magnesium chloride. This is spray dried and then decomposed, giving a product of MgO and hydrogen chloride, which is absorbed and recycled to leaching. The essential steps in the proposed flow sheet are shown in Fig. 1. See also Chapter 26 (Section 2.8).

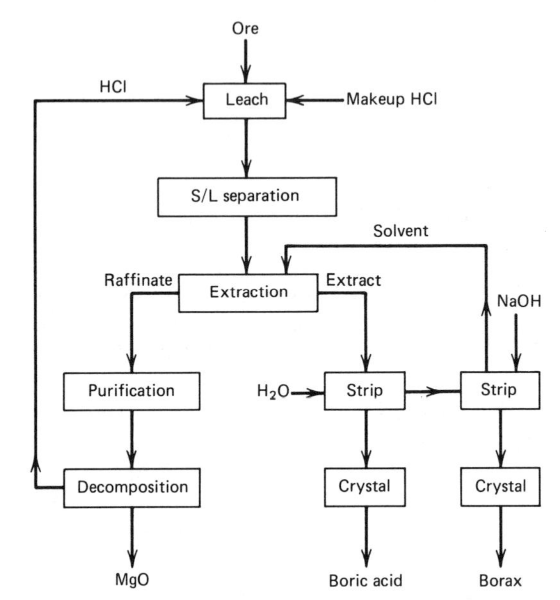

Figure 1. Outline flow sheet for production of boric acid and borax from ascharite [15].

6. GALLIUM

Ritcey and Ashbrook [7] list three cases of solvent extraction being considered for gallium, although it seems doubtful whether any of these are in commercial use. In addition, reference should be made to the work of Herak and Jagodic [16] on the separation of gallium from zinc by aminophosphoric acids.

De Schepper [17] has reported on a process for gallium extraction that is linked to a clear industrial application. This arises as a solution in sulfuric and hydrochloric acids (8–9 M H^+), also containing iron, zinc, arsenic, and other metallic impurities during manufacture of germanium from certain concentrates. Tributylphosphate was considered the most likely extractant and Ga–Fe(III), the critical separation. Measurement of distribution coefficients showed Fe(III) to be preferentially extracted by pure TBP but, with 10% TBP in Escaid 110 (plus 25% isodecanol modifier), gallium has the higher value of D. The distribution coefficients vary with HCl molarity, and this can be exploited in scrubbing. For enhanced iron removal, the proposed process first involves reduction of most of the Fe(III) to Fe(II) by addition of iron powder [Fe(II) is not extracted]. The extract is first scrubbed and then stripped with HCl. Gallium is finally precipitated by addition of NaOH. Continuous tests with countercurrent contacting showed the process to be capable of recovering 99.5% of the gallium, giving a concentrate with a Ga:Fe ratio of 50. Further development could doubtless be made, but it is not known whether the process has been used on a commercial basis.

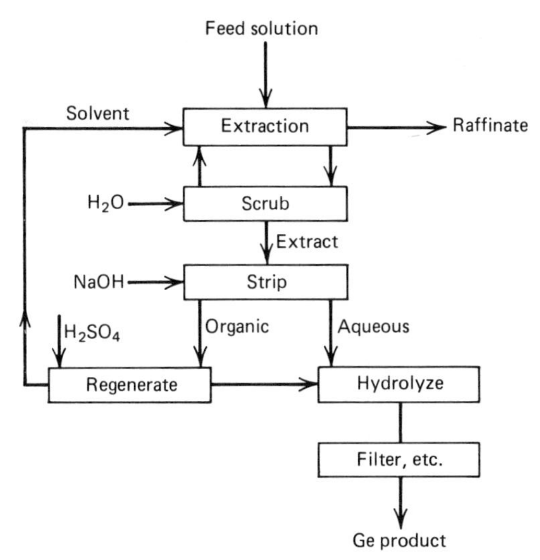

Figure 2. Proposed flow sheet for extraction of germanium with LIX63 [19].

7. GERMANIUM

There are a number of early references to the possible use of solvent extraction for the recovery or purification of germanium, such as extraction of $GeCl_4$ with carbon tetrachloride as a recovery process during transistor manufacture [18]. Although there is no evidence of these suggestions having been used, solvent extraction is now used commercially for the concentration and purification of germanium by Metallurgy Hoboken-Overpelt in Belgium, and the process has been described by de Schepper [19].

The extractant chosen is the α-hydroxyoxime LIX63. This was evaluated with feeds derived from both sulfuric and hydrochloric acid leaching and mixtures of the two. The extractant is very selective for germanium at higher acidities. Four stages of extraction were employed in the test work. The extract was scrubbed with water and then stripped by use of aqueous sodium hydroxide solution. The water scrub reduces the acid content of the organic phase and hence the sodium hydroxide consumption in stripping. The solvent phase finally must be regenerated by contact with acid to remove sodium from the oxime. The flow sheet is shown in outline form in Fig. 2. The kinetics of extraction are slow. The process yields a good germanium concentrate and is claimed to be more economic than traditional alternatives.

The separation of germanium from acidic leach liquors is important in zinc production, as a good yield during electrolysis is only obtained when germanium has been completely eliminated from the electrolyte. Solvent extraction has been evaluated for this purpose [20]. Ten extractants were examined and Kelex 100 chosen as the most promising for further study. Diluents were also screened and kerosene with 10% octanol was chosen for the process. The germanium extraction isotherm depends on aqueous phase acidity, and its form indicates the formation of two different species depending on the acidity. The rate of extraction is quite high at low pH but decreases rapidly with decrease in acidity. The process has been tested on a bench scale.

8. IRON

There is no economic justification for the use of solvent extraction in the production of iron. The removal of iron as a contaminant from solutions of other metal values is usually achieved by precipitation. However, there are a few processes in which solvent extraction is used for this purpose (e.g., the Falconbridge process for nickel; see Fig. 4 of Chapter 25.2). Solven extraction has been suggested [21] for the preparation of high-purity iron compounds, but there is no evidence of its commercial use. The proposed process uses aliphatic monocarboxylic acids as extractants.

9. LITHIUM

Interest has been expressed in the possible use of solvent extraction for the recovery and purification of this element. Gabra and Torma [22] report work with the use of aliphatic alcohols as extractants. They conclude that *n*-butanol is the most promising and present data for both lithium chloride distribution and its separation from sodium chloride. These suggest the possibility of obtaining a high-quality product in a comparatively small number of stages.

More recently, Epstein et al. [23] have studied the extraction of lithium from the Dead Sea. They point out that demand for the metal could increase markedly if lithium-based fuel cells come into widespread use. In the proposed process, aluminum chloride is first dissolved in the brine. The pH is then raised to precipitate aluminum hydroxide. The latter precipitate contains a high proportion of the lithium present in the original brine. After separation, the solids are dissolved in aqueous hydrochloric acid. The lithium and aluminum chlorides are then separated by solvent extraction: *n*-hexanol, 2-ethylhexanol, and methyl isobutyl ketone are suggested as suitable solvents. The lithium-depleted aluminum chloride solution is recycled to the precipitation stage. An approximate economic evaluation suggests a production cost by this route roughly equal to the current selling price of lithium metal. Although apparently discouraging, the situation would change significantly if demand for lithium increases. It is also relevant to note that this was achieved with a feed brine containing only 40 ppm of lithium.

10. MAGNESIUM

Consideration has been given to the possible use of solvent extraction for recovery of magnesium chloride from bitterns and other brines. This is interesting in that the most promising system appears to be a mixture of cationic and anionic extractants, such as an equimolar mixture of Aliquat 336 and Acid 810 [24]. This coextracts both the cation and the anion, and the large excess of chloride ions in the feed displaces the equilibrium to the right:

$$Mg^{2+}_{aq} + 2Cl^{-}_{aq} + 2R_4N \cdot R'COO_{org} \rightleftharpoons 2R_4NCl_{org} + Mg(R'COO)_{2\ org}$$

Magnesium chloride can be stripped from the extract by contacting with water. A flow sheet has been proposed with preliminary economic assessment [25]. As in the case of lithium considered in Section 9, this is not very promising, as the cost of production is comparable with the current value of the product. The cost of solvent losses in raffinate and product streams is a severe limitation with a low-value product of this type, even though the feed is available at no cost. The proposed flow sheet is interesting as it shows the need to estimate an optimum concentration of metal to leave in the solvent recycled from stripping so as to minimize total capital cost.

11. MANGANESE

Ritcey and Ashbrook [7] summarize work carried out in Canada on the use of solvent extraction to recover metal values from manganiferous shales. A flow sheet is proposed involving successive extraction of copper, zinc, and manganese. It is not clear whether it has been used commercially.

Manganese nodules from the sea bed have excited a great deal of interest, not so much as a commercial source of manganese as for their content of other metals such as copper and cobalt. Much developmental work has been undertaken, which has recently been reviewed by Monhemius [26]. Solvent extraction appears in a number of the proposed processing routes.

12. MERCURY

Solvent extraction with triisoctyl amine has been reported [27] as a process for the recovery of mercury from a chloride plant effluent.

13. POTASSIUM

The recovery of potassium chloride from Dead Sea brines has attracted considerable attention, and a number of possible processes have been proposed. Epstein et al. [28] have suggested one involving solvent extraction. The inital separation is achieved by precipitation of the potassium as its perchlorate by addition of calcium perchlorate. An aqueous slurry of the potassium perchlorate is then contacted countercurrently with a tertiary amine in the form of its hydrochloride (in a suitable diluent). Anion exchange takes place, giving an aqueous solution of potassium chloride. The amine leaves in the perchlorate form. It is reacted with lime to produce the calcium perchlorate required for the initial precipitation and is then reconverted to the hydrochloride by treatment with hydrochloric acid. Although this flow sheet was shown to be operable, it has been suggested recently [29] that is is not economic because of the consumption of reagents. An alternative version, still based on a perchlorate cycle, is suggested, but an economic analysis is not yet available.

REFERENCES

1. Y. Marcus and A. S. Kertes, *Ion Exchange and Solvent Extraction of Metal Complexes,* Wiley-Interscience, London, 1967.
2. A. S. Kertes, Chapter 2 in C. Hanson, Ed., *Recent Advances in Liquid-Liquid Extraction,* Pergamon Press, Oxford, 1971.
3. D. S. Flett and D. R. Spink, *Hydrometallurgy* **1**, 207 (1976).
4. D. S. Flett, *Hydrometallurgy* **3**, 199 (1978).
5. I. L. Jenkins, *Hydrometallurgy* **4**, 1 (1979).
6. R. Blumberg, Chapter 3 in C. Hanson, Ed., *Recent Advances in Liquid-Liquid Extraction*, Pergamon Press, Oxford, 1971.
7. G. M. Ritcey and A. W. Ashbrook, *Solvent Extraction, Principles and Applications to Process Metallurgy, Part II,* Elsevier, Amsterdam, 1979.
8. British Patent 933,233 (1963).
9. P. Mühl, K. Gloe,, C. Fischer, G. Ziegenbalg, and H. Hoffman, *Hydrometallurgy* **5**, 161 (1980).
10. A. Cocco, I. Colussi, I. Kikic, and S. Meriani, *Proceedings of the International Solvent Extraction Conference (ISEC), Liège 1980*, Association des Ingénieurs sortis de l'Université de Liège, 1980, Paper 186.
11. V. Bumbalek, K. Carmak, and J. Haman, ibid., Paper 177.
12. J. D. Moore, and L. D. Lash, *Mining Congr. J.* 49(7), 44 (1963).
13. *Chem. Eng. News* 43, 70 (1965).
14. L. Crocker, et al., U.S. Bureau of Mines, RI 6173 (1963).
15. Y. F. Su, D. Y. Yu, and S. D. Chen, *Proceedings of the International Solvent Extraction Conference (ISEC), Liège 1980*, Association des Ingénieurs sortis de l'Université de Liège, 1980, Paper 57.
16. M. J. Herak and V. Jagodić, *Proceedings of the International Solvent Extraction Conference (ISEC), The Hague 1971*, Society of Chemical Industry, London, 1971, p. 656.
17. A. de Schepper, *Hydrometallurgy* **4**, 285 (1979).
18. C. Hanson and D. A. Kaye, *Metallurgia* 182 (1965).
19. A. de Schepper, *Hydrometallurgy* **1**, 291 (1976).
20. G. Cote and D. Bauer, *Hydrometallurgy* **5**, 149 (1980).
21. P. Mühl, et al., *Hydrometallurgy* **1**, 113 (1975).
22. G. G. Gabra and A. E. Torma, *Hydrometallurgy* **3**, 23 (1978).
23. J. A. Epstein, E. M. Feist, Y. Marcus, and J. Zmora, *Hydrometallurgy* **6**, 269 (1981).
24. C. Hanson, M. A. Hughes, and S. L. N. Murthy, *J. Inorg. Nucl. Chem.* **37**, 191 (1975).
25. C. Hanson and S. L. N. Murthy, *Proceedings of the International Solvent Extraction Conference (ISEC), Lyons 1974*, Vol. 1, Society of Chemical Industry, London, 1974, p. 779.
26. A. J. Monhemius, in A. R. Burkin, Ed., *Topics in Non-Ferrous Extractive Metallurgy,* Blackwell, Oxford, 1980.
27. R. Caban and T. W. Chapman, *AIChE J.* **18**, 904 (1972).
28. J. A. Epstein, D. Altaras, E. M. Feist, and J. Rosenweig, *Hydrometallurgy* **1**, 39 (1975).
29. E. Barnea, J. E. Gai, G. Harel, and J. Metcalf, *Proceedings of the International Solvent Extraction Conference (ISEC), Liège 1980*, Association des Ingénieurs sortis de l'Université de Liège, 1980, Paper 56.
30. F. G. Seeley, W. J. McDowell, L. K. Felker, A. D. Kelmers, and B. Z. Egan, *Hydrometallurgy* **6**, 277 (1981).

25.9

SECONDARY METALS AND METALS RECOVERY FROM SOLID WASTES

D. S. Flett
Warren Spring Laboratory
United Kingdom

1. INTRODUCTION

With the upsurge in hydrometallurgical processing, solvent extraction has become one of the major unit processes for solution purification, metal separation, and concentration. The availability and uses of extractants for this purpose and their commercial application are described in this handbook and elsewhere [1].

With the success of hydrometallurgical processing in the primary sector, it is natural that application to the secondary sector should be considered, and there is every sign that it will be successful because of its flexibility, capital cost savings, and scale advantages in that small plants can be built and operated with economy. The field has been extensively reviewed [2–7].

Application of solvent extraction in the processing of scrap and wastes can involve different constraints from its application in primary extractive metallurgy. In the main, the differences relate to the need to treat combinations of metals not encountered in primary processing and the scale of operations, which will affect capital and labor-related costs. Although many processes involving solvent extraction have been devised for recovering metals from scrap and wastes, relatively few have yet achieved commercial application.

2. COPPER SCRAP AND WASTE

Scrap and waste treatment plants built or under construction for copper recovery are summarized in Table 1. Copper scrap and wastes arise in many forms such as flotation tailings, metal scrap, bearing metal, brass scrap, furnace drosses, brass fume, and so on. Although the amount of copper produced by the NCCM tailings leach plant in Zambia is equivalent to that of a small copper smelter and really is more akin to a primary producer than a scrap processor, flotation tailings of this sort are, strictly speaking, a waste material. However, this plant has been well documented [8] and is not discussed further (see also Chapter 25.1). Table 1 shows that the other commercial applications involve ammoniacal leaching of wastes. Two scrap operations have been located in Casa Grande, Arizona, and both have closed. A third plant operated by Metal Chem in Mesa, Arizona treated comminuted copper scrap by ammoniacal leaching, solvent extraction and

TABLE 1 SOLVENT EXTRACTION PLANTS FOR COPPER RECOVERY FROM SCRAP AND WASTES

Plant	Type of Feed	Comments	Reference
NCCM, Zambia	Acid liquor from flotation tailings leach	World's largest copper solvent extraction plant; final product–copper cathodes.	8
Criterion Corp., Casa Grande, Arizona, United States	Ammoniacal liquor from cement and scrap leaching	Operated in old Capital Wire & Cable plant (recently closed down); product was copper cathodes	7
Hachinohe Smelting Co. Ltd., Hachinohe, Japan	Ammoniacal liquor from ISF copper dross leaching	Extractant is LIX64N; product will be copper cathode	9
ConZinc Rio Tinto, Cockle Creek, Australia	Ammoniacal liquor from ISF copper dross leaching	Extractant choice not known; product will be $CuSO_4$	9

electrowinning. That this plant too has closed is evidence of the vagaries of the scrap business.

The processing of copper dross from the Imperial Smelting Furnace (ISF) shows how solvent extraction can be used to treat a waste material and provide an added degree of flexibility to existing processes. The flow sheet is shown in Fig. 1. The organic-phase scrub requirement arises from the fact that the hydroxyoxime reagents commonly used for copper extraction from ammoniacal liquors extract ammonia. The ammonia content of the organic phase decreases with copper loading, and there is no evidence to suggest that any ammonia is directly associated with the copper. Dilute acid scrubs of the loaded organic phases are necessary to avoid ammonia carry-over into the strip liquor. Cognizant of this fact, Henkel Corp. have brought out a new reagent LIX54, a β-diketone, which does not extract ammonia. It has other advantages for copper recovery from ammoniacal liquors in that it has a much higher copper transfer capacity than

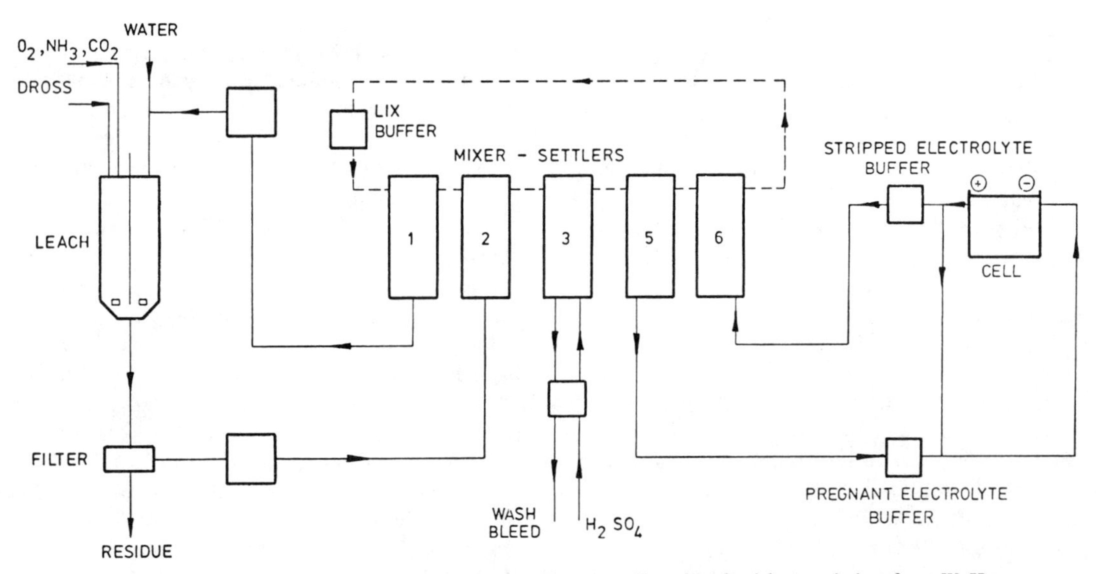

Figure 1. Pilot-plant flow sheet for ISF dross processing. (Reprinted with permission from W. Hopkins, Process for Treatment of Zinc–Lead and Lead Blast Furnace Copper Drosses, paper presented at 105th AIME Annual Meeting, Las Vegas, 1976.)

the hydroxyoximes and, as it is a weaker extractant, very strong strip liquors can be produced by use of dilute acid [1, 7]. The potential reductions in plant size and attendant capital cost are considerable. In one case [11] tests showed that a reduction in plant size by a factor of 10 was possible by the use of LIX54 in place of the hydroxyoxime reagent. A similar reagent, Hostarex DK16, has been developed by Hoechst for copper recovery from ammoniacal liquors [1].

Use of ammoniacal leaching and solvent extraction for copper recovery from scrap has distinct advantages as, theoretically, copper metal is converted to copper sulfate and all reagents are recycled within the process without loss. The process chemistry is described by the following equations:

$$Cu^0 + Cu(NH_3)_4^{2+} + 4NH_4OH \longrightarrow 2\,Cu(NH_3)_4^+ + 4H_2O \quad (1)$$

$$4Cu(NH_3)_4^+ + O_2 + H_2O \longrightarrow 4\,Cu(NH_3)_4^{2+} + 4OH^- \quad (2)$$

$$Cu(NH_3)_4^{2+} + 2\overline{RH} + 2H_2O \longrightarrow \overline{CuR_2} + 4NH_4OH \quad (3)$$

$$\overline{CuR_2} + H_2SO_4 \longrightarrow CuSO_4 + 2\overline{RH} \quad (4)$$

Although the technology is sound, successful commercial operation depends on process economics, such as scrap availability, energy costs, and copper product price as witnessed by the history of Arizona plants.

Several developments have taken place and await commercial adoption. Thus Warren Spring Laboratory has developed a process for recovering copper from clad materials [12]. Bearing scrap was leached with ammonia–ammonium carbonate in the presence of air in a rotating inclined drum (Fig. 2). The ammoniacal copper solution was extracted with LIX64N and copper finally recovered by electrowinning. The process economics appear to offer the greatest opportunities where copper-based alloys are employed, such as in lead–bronzes or similar bearing material. The clean steel backing produced by this method of treatment of bearing metal can be an additional source of revenue.

A process for treatment of copper–zinc flue dust has been proposed by MX Processer as part of their H-MAR (Metals and Acid Recovery) concept [13]. The flow sheet is shown in Fig. 3. The flue dust is leached with sulfuric acid and copper extracted by LIX64N. Copper metal is recovered by electrowinning. The solvent extraction raffinate is then used to leach zinc flue dusts, purified by zinc cementation and zinc recovered by solvent extraction with di-2-ethylhexyl phos-

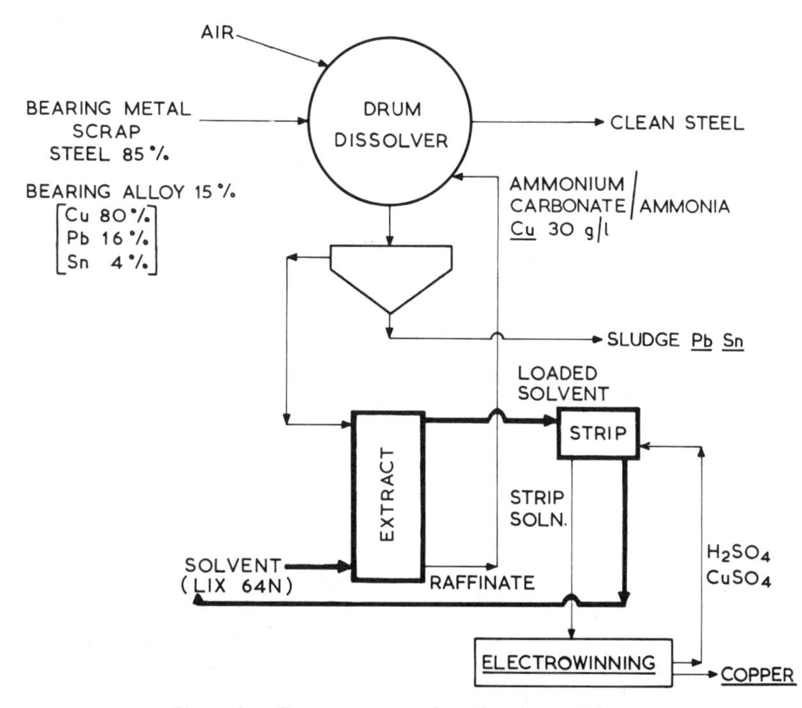

Figure 2. Copper recovery from bearing metal scrap.

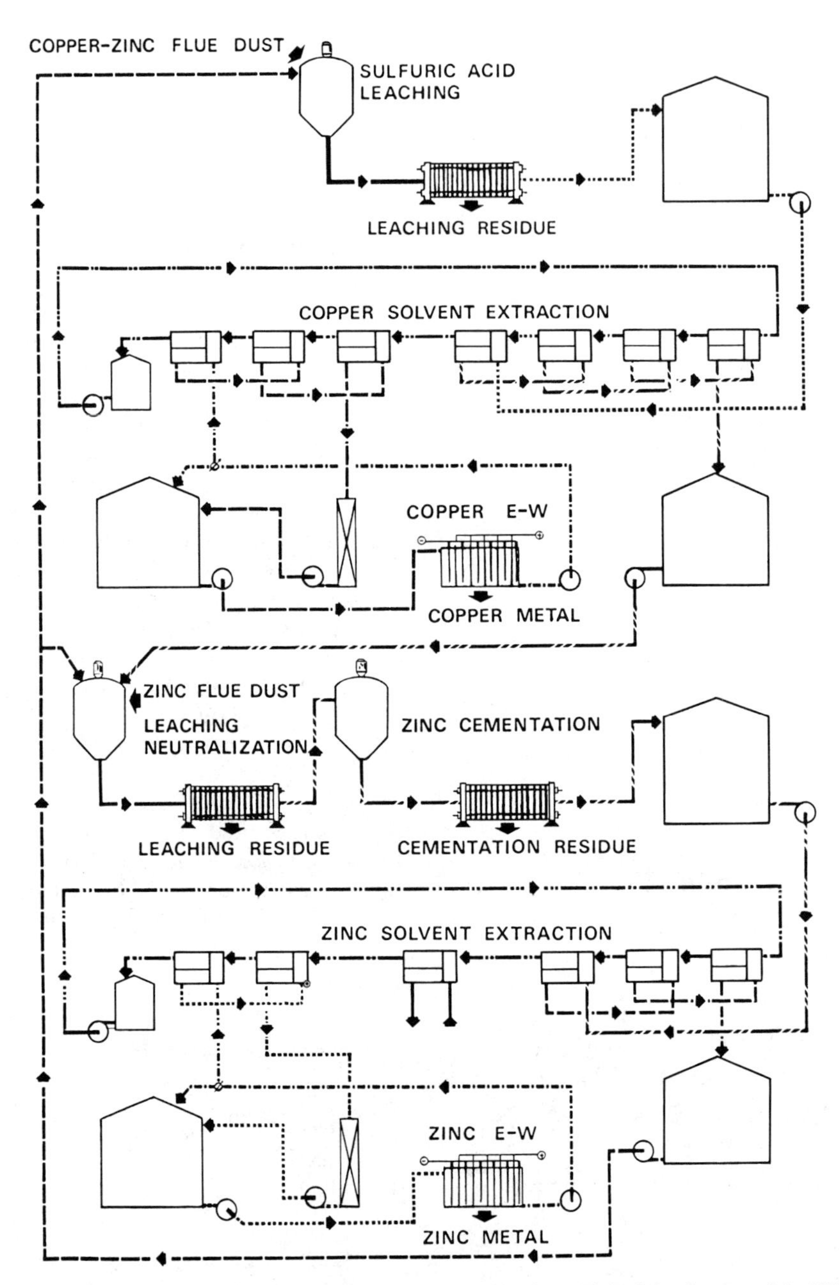

Figure 3. Acidic MAR process. (Reprinted with permission from H. Reinhardt, copyright MX Processer AB, Kråketorpsgaten 20, 431 33 Mölndal, Sweden.)

phoric acid and electrowinning. The final raffinate is returned to copper zinc flue dust leaching. (See also Chapter 25.10 for use of the H-MAR process.)

Thus there are various ways in which solvent extraction can be employed to recover copper from scrap. Many other process applications exist, as an examination of the patent literature will show. The reagent requirement is clearly seen to be the hydroxyoxime reagents for acidic liquors and, whereas they are equally applicable for ammoniacal liquors, ammonia transfer problems and capital cost savings indicate that the best reagent choice for ammoniacal liquor processing should be a β-diketone reagent such as LIX54 or Hostarex DK16.

3. NICKEL AND COBALT SCRAP AND WASTE

The use of solvent extraction for these materials is well established in principle, although few plants are operating in practice. A summary of known commercial operations is given in Table 2. The best known is that operated by Metallurgie Hoboken in Belgium [2]. This is a very flexible process, as is shown in Fig. 4. The circuit is chloride based, and lime is used to precipitate iron. The extractant is Alamine 336, and use is made of the differences in extractability of complex metal chlorides with differing chloride concentrations. Hence zinc is extracted at relatively low chloride concentrations, whereas cobalt requires higher chloride concentrations for extraction. This process has operated successfully for several years. It is not known just what types of feed are handled but, as indicated in the flowsheet, a variety of types can be catered for (see also Chapter 25.2).

Slags may be classed as a residue, and in this context the plant of Société Le Nickel [14] at Le Havre deserves mention. Here slags produced by chlorine blowing of nickel matte are leached, iron is removed by extraction with Amberlite LA-2, and cobalt is recovered by extraction with Adogen 381 to provide a pure nickel chloride solution for nickel recovery by electrowinning. The chlorine recovered at the anode can be recycled to the chlorine blow. The flow sheet appears as Fig. 9 in Chapter 25.2.

In the United Kingdom, solvent extraction has been used to clean up a nickel liquor arising from sulfuric acid dissolution of sludges [7]. Operated for a time by Wimborne Hydrometallurgicals Ltd., the process used solvent extraction with D2EHPA to remove zinc and copper from the nickel liquor. The plant, although technically successful, has now closed down.

The Pyrites Inc. process [15] (the Pyrites plant has also closed down) also employed D2EHPA to clean up a cobalt liquor. Iron, manganese, and zinc were removed, and a novel feature was the removal of iron from the organic phase by reductive stripping with an alkaline sugar solution (see Chapter 25.2).

When D2EHPA is used as the preferred solvent, iron is always a problem. This extractant has a very high affinity for Fe^{3+} to the extent that oxidative extraction of Fe^{2+} takes place. As D2EHPA has considerable potential in the secondary field as an extraction agent, this high affinity for ferric iron is a considerable disadvantage. Thus it is usually essential to remove iron to as great an extent as possible at an early stage in the process to avoid problems in the solvent extraction step.

Thus the INCO process for nickel–cobalt sepa-

TABLE 2 SOLVENT EXTRACTION PLANTS FOR NICKEL AND COBALT RECOVERY FROM SCRAP AND WASTES

Plant	Type of Feed	Comments
Metallurgie Hoboken, Belgium	Various Ni, Co chloride liquors from scrap and waste leaching	Alamine 336 used widely as extractant; metal oxides produced by pyrohydrolysis
Société Le Nickel, Le Havre, France	Chloride liquors from leaching slag from chlorine blowing of nickel matte	Iron extracted by LA-2; cobalt recovered by Adogen 381 to provide pure $NiCl_2$ for electrolysis
INCO, Copper Cliff, Canada	Sulfate liquor from leaching IPC plant residue	Iron removed by hydrolysis prior to cobalt SX with D2EHPA in a column contactor at 80°C; plant not yet operated.
Wimborne Hydrometallurgicals Ltd., United Kingdom	Acid leach of plating bath sludges Ni, Zn (Cu)	D2EHPA SX for Zn (Cu) removal (plant closed down)
Pyrites, Inc., United States	H_2SO_4 leach of pyrite cinders	D2EHPA SX for Fe, Zn, Mn removal from cobalt (plant closed down)

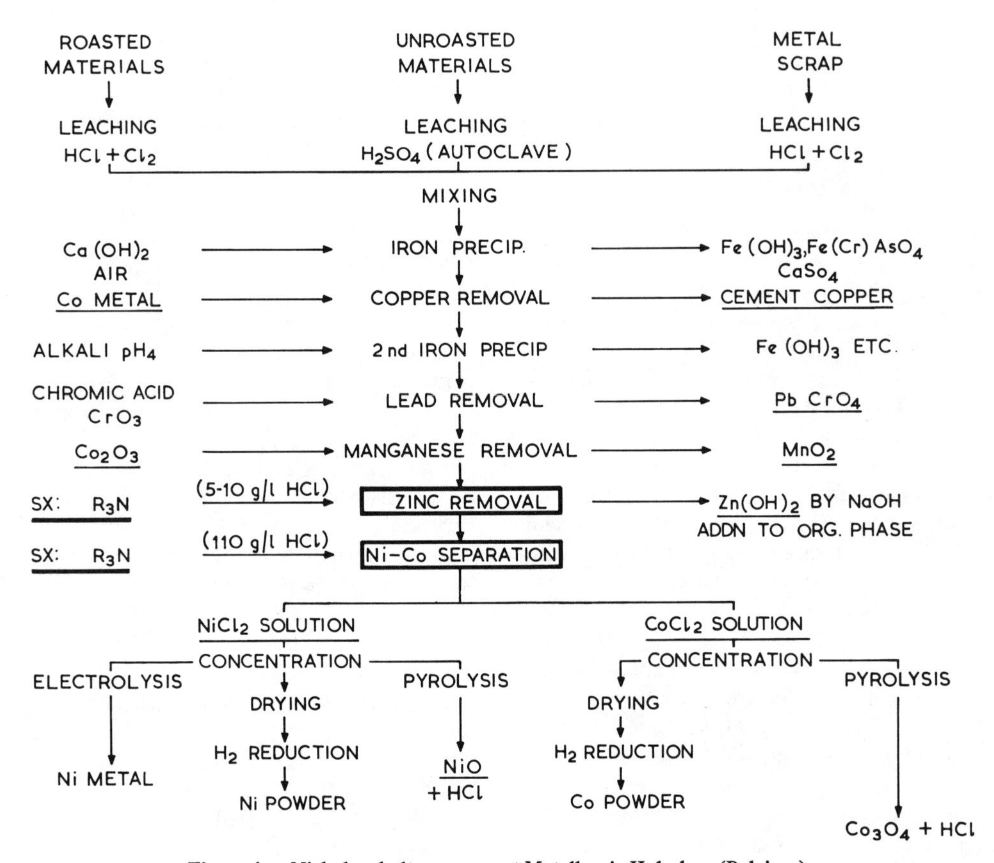

Figure 4. Nickel–cobalt recovery at Metallurgie Hoboken (Belgium).

ration in treatment of IPC plant residues requires complete iron removal by hydrolysis prior to separation of nickel and cobalt by solvent extraction with D2EHPA. The separation will be carried out at elevated temperatures in a column contactor [7]. The process, which has not yet been operated, is based on the work of Ritcey et al. [16]. The chemistry of the process and continuous countercurrent mixer–settler data have been reported [17].

In the Swedish (Gullspång) process for nickel scrap (Fig. 5), iron is kept essentially in the ferrous state throughout and cannot be separated from the nickel by solvent extraction. The solution is evaporated to yield a chloride concentration sufficient to ensure successful separation of cobalt from nickel by extraction with a tertiary amine. It is important to note how material and electrolytic balances are maintained, and that, whereas the formation of carbides theoretically prevents the dissolution of chromium, some does dissolve. Hence the products are impure; the nickel is contaminated with iron and chromium, and the cobalt is contaminated with iron. These products are suitable, however, for in-house use for alloy steel manufacture by Gullspångs Elec-trochemiska AB, who have developed the process.

Flow sheets for two other processes are shown in Figs. 6 and 7. The USBM process has several solvent extraction steps that rely on different reagents rather than variations in chloride concentration to remove molybdenum, iron, and cobalt (see also Chapter 25.2). The Warren Spring Laboratory process, on the other hand, has only one operation, employing a tertiary amine for cobalt recovery. Separation from the small amount of iron remaining after precipitation as hydroxide is carried out by controlling the chloride concentration during stripping, in this case by adjusting the organic:aqueous phase ratio. Thus with a small amount of water the chloride concentration is too high for iron stripping to occur, and a cobalt strip solution is obtained. Iron is subsequently stripped with a much larger volume of water. Metal hydroxides are precipitated with magnesia and the hydro-

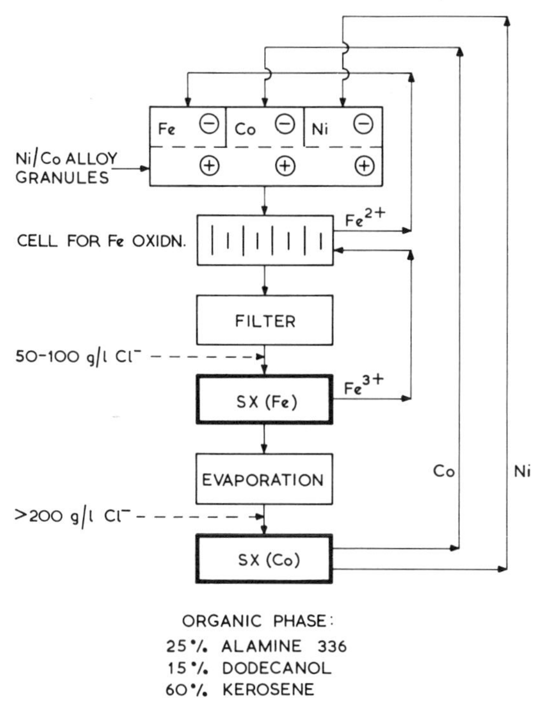

Figure 5. Nickel-cobalt alloy scrap process (Gullspång Electrochemical Co., Sweden.)

chloric acid is regenerated by spray evaporation and pyrohydrolysis of the magnesium chloride solutions.

The large number of operations employed is common to all the processes, and this inevitably leads to high capital charges and the need to operate on a large scale. The uncertainty surrounding availability of scrap arisings in sufficient quantities to feed the processes has prevented their full-scale adoption.

4. ZINC SCRAP AND WASTE

Recovery of zinc from flue dusts has already been mentioned in the context of copper–zinc recovery by the acid leach MAR process. Other work on treatment of zinc flue dusts from ISF sinter plants for cadmium recovery has involved acid leaching with H_2SO_4, ion exchange to convert the cadmium and zinc to chlorides, and solvent extraction separation by use of TBP or naphthenic acid (a carboxylic acid; Shell Chemicals Ltd). The aim was to produce a pure cadmium solution for production of cadmium chemicals. Both processes were successfully pilot planted, and it was concluded that the TBP process was preferred [18]. The use of D2EHPA was not considered, but the selectivity shown toward zinc by D2EHPA should warrent its investigation should a similar requirement arise.

A new commercial process for zinc recovery from pyrite cinders has been developed by Technicas Reunidas in Spain. Called the *Zincex process*, this involves the recovery of zinc from liquors produced by leaching chloridised pyrite cinders by means of two solvent extraction steps with final metal production by electrowinning

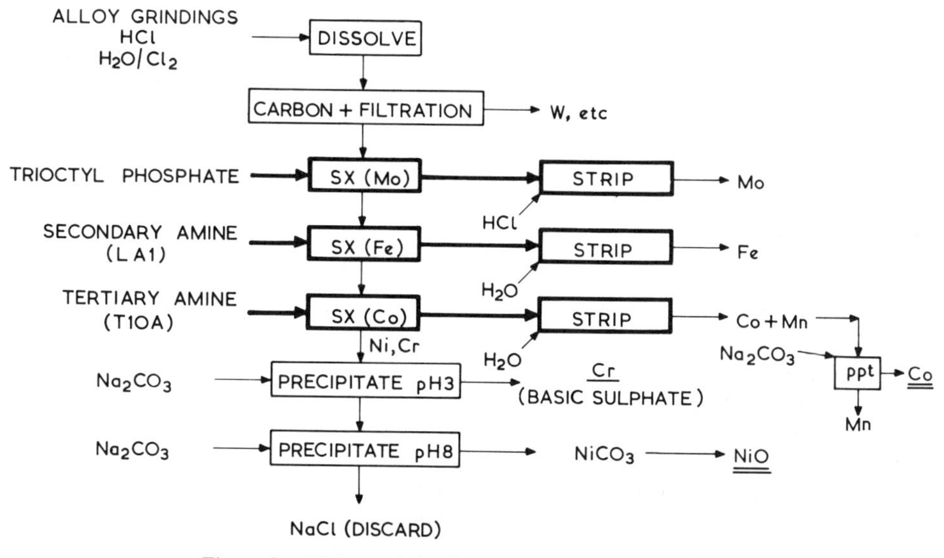

Figure 6. Nickel–cobalt alloy scrap process (USBM, USA).

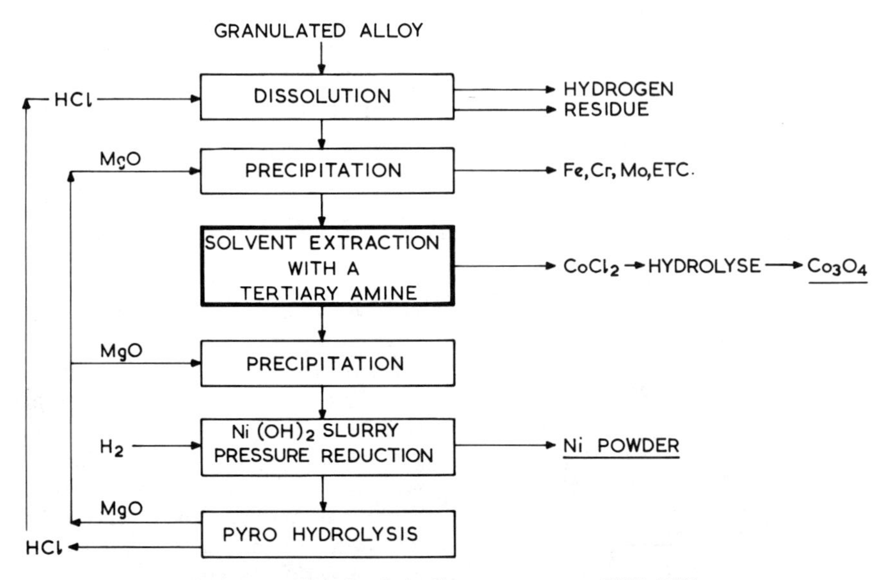

Figure 7. Nickel–cobalt alloy scrap process (WSL, UK).

[7, 19]. This is discussed in more detail in Chapter 25.5, and the flow sheet appears as Fig. 1 in that chapter.

Removal and recovery of zinc from steelmaking dusts provides a potential application for solvent extraction in the reprocessing of wastes. The problem arises because of increased recycle of scrap steel with a high nonferrous content and hence production of dusts rich in nonferrous metals that cannot be safely dumped and present a toxic hazard in handling. Technical feasibility of development of zinc removal and recovery by leach, solvent extraction, and electrowinning processes has been shown at Warren Spring Laboratory [29]. Processes for treating Waelz oxides, produced by fuming off the zinc from steel dusts, also using solvent extraction of zinc with D2EHPA are also under examination. Similar interests are found in Sweden, where MX Processer AB are developing similar processes involving solvent extraction [13] (see Chapter 25.10).

5. GENERAL SCRAP AND WASTE

The recovery of metals from spent catalysts has received considerable attention, although no commercial application of solvent extraction is known. Warren Spring Laboratory [21] developed a process for recovery of uranium and nickel from catalysts using HNO_3 leaching and uranium extraction with TBP. Elsewhere, for example, processes have been patented for treatment of spent catalysts used in processing petroleum [22, 23]. Sulfuric acid leaching is followed by amine extraction of molybdenum and rhenium, or, as an alternative, rhenium is recovered by ion exchange and the rhenium chloride eluate is extracted with TBP to produce a pure perrhenic acid free from iron and molybdenum. Processes using solvent extraction for recovery of rhenium from copper smelters flue dusts have been proposed (see also Chapter 25.10).

Metals recovery from hydroxide sludges and nickel–cadmium battery scrap by ammoniacal leaching and solvent extraction has been described in the Am-MAR process [13], which is also applicable to the copper–zinc flue dusts discussed earlier. The flow sheet is shown in Fig. 8. Ammoniacal carbonate leaching leaves iron and chromium in the residue, whereas valuable metals such as copper, zinc, nickel, and cadmium report in the leach liquor. The flow sheet calls for solvent extraction of copper and nickel with hydroxyoximes or β-diketones. Although the MX Processer flow sheet shows differential extraction of copper and nickel to produce cathode copper and nickel sulfate, a coextraction of copper and nickel followed by selective stripping as recommended for manganese nodule processing would be preferable [1]. Zinc or cadmium are then precipitated as carbonates after the ammonia is removed and recovered by thermal stripping. The ammonia is absorbed and recycled as shown in Fig. 8. Although the technical feasibility of this flow sheet is not in doubt, the economics are

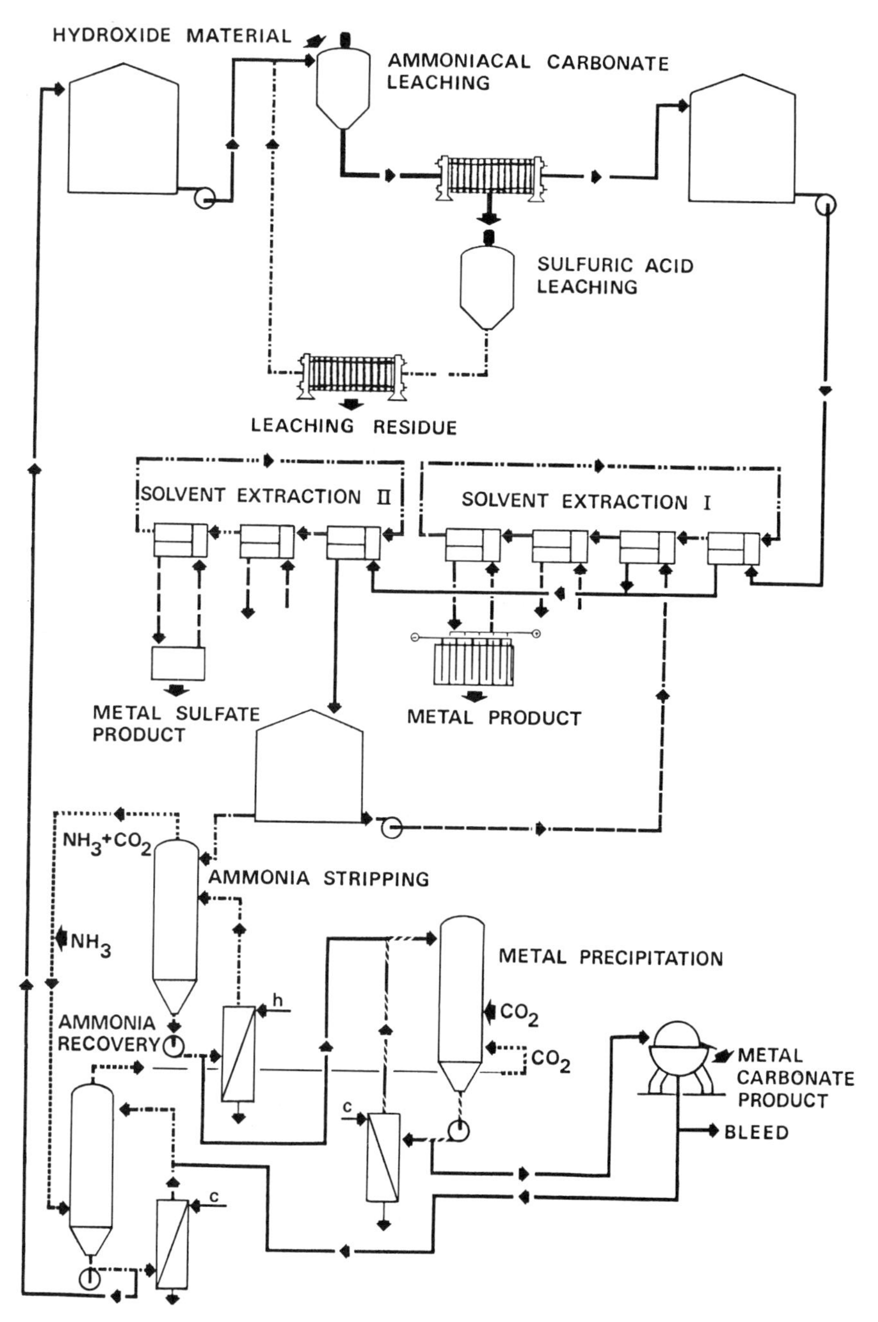

Figure 8. Ammoniacal MAR process. (Reprinted with permission from H. Reinhardt, copyright MX-Processer AB, Kråketorpsgaten 20, 431 33 Mölndal, Sweden.)

very scale dependent, and thus successful application is attendant on identification of a suitable and continuing source of arisings. A summary of derived cost data [13] is given in Table 3.

Vanadium recovery from chromium-bearing or phosphorus-bearing blast furnace slags has been proposed by use of solvent extraction [24]. The process involves salt roasting of slag, and the resultant material is leached with water and vanadium extracted with an alkyl amine.

A process for the recovery of vanadium and nickel from soot and ash produced by power sta-

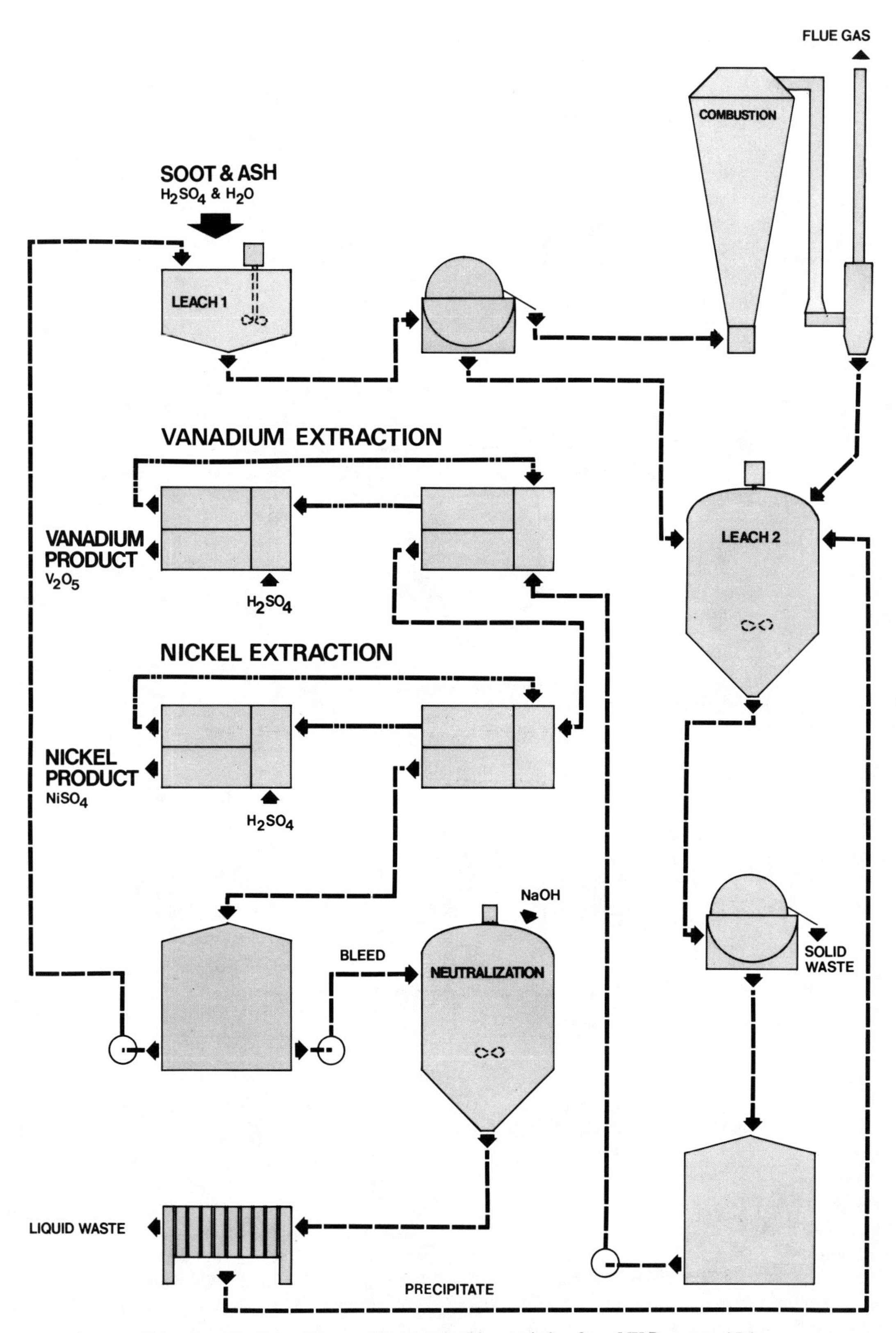

Figure 9. The Sotex Process. (Reprinted with permission from MX-Processer AB.)

TABLE 3 SUMMARY OF PROCESS AND ECONOMIC DATA FOR THE MAR PROCESSES

	H-MAR Zn-Dust	H-MAR Cu-Zn-Dust	H-MAR Cu-Zn-Dust	Am-MAR Cu-Zn-Dust	Am-MAR Cu-Zn-Dust	AM-MAR Cu-Zn-Ni-Sludge	Am-MAR Ni-Cd-Scrap
Feed material, tons	52,500[a]	6,200[b]	12,400[b]	6,200[b]	12,400[b]	12,000[c]	2,400[d]
Product Cu, tons	–	480	960	530	1,060	260	–
Product Zn, tons	10,000	2,400	4,800	2,500[e]	5,000[e]	300[e]	–
Product Ni, tons	–	–	–	–	–	180[f]	300[f]
Product Cd, tons	–	–	–	–	–	–	200[g]
Direct investment U.S. $1000	12,600	5,100	7,900	3,200	5,300	3,600	3,600
Operating cost U.S. $1000	5,000	1,500	2,600	1,500	2,700	1,100	1,100
Product value U.S. $1000	7,800	2,600	5,100	2,700	5,300	1,400	2,500
Return on assets, %	10	8	17	20	32	–	24

[a]Five percent zinc content (20% in ferrites).
[b]Forty percent copper rich dust and 60% zinc-rich dust.
[c]Dry substance 30%.
[d]Inclusive of electrode frames, leachable material 60%.
[e]As carbonate or oxide.
[f]As nickel sulfate.
[g]As cadmium carbonate.

tions fired with fossil fuels has been developed by MX Processor AB [25, 26]. The flow sheet is shown in Fig. 9. In this process, called the *Sotex process*, a mixture of D2EHPA and TBP is recommended for extraction of vanadium. Again, iron can be a problem, and its extraction is minimized by SO_2 treatment of the solvent extraction feed to reduce both iron and vanadium to the di- and tetravalent states, respectively. Iron is scrubbed from the organic phase with 6 *M* H_2SO_4. Then $VO_2 \cdot nH_2O$ is precipitated from the organic-phase strip liquor by addition of soda. The extractant for nickel is not revealed but the final product is $NiSO_4$ crystals.

Finally, the treatment of waste dumps from the gold mines in South Africa for the recovery of contained uranium is an activity of current importance. Here solvent extraction has a key role to play, regardless of whether it is by direct extraction of uranium from the leach liquor [27] or by extraction from continuous ion-exchange eluates [28].

5. CONCLUSIONS

There is considerable interest in the application of solvent extraction to scrap and waste processing. However, few such processes are in commercial operation. Solvent extraction in scrap and waste processing tends to be used as a sophisticated separation process for production of fairly high purity metals commanding premium prices. For low-value materials, it may be regarded as a capital intensive process and thus require a certain level of arisings to ensure profitability and certainty of supply to ensure satisfactory return on investment in the long run.

REFERENCES

1. D. S. Flett, *Chem. Ind.* (17), 706 (1977).
2. A. W. Fletcher, *Chem. Ind.* (9), 414 (1973).
3. H. Reinhardt, *Chem. Ind.* (5), 210 (1973).
4. M. A. Hughes, *Chem. Ind.* (24), 1042 (1975).
5. D. S. Flett and D. R. Spink, *Hydrometallurgy* **1**, 207 (1976).
6. D. S. Flett, *Warren Spring Laboratory Report LR254(ME)*, 1977.
7. D. S. Flett, *J. Chem. Tech. Biotechnol.* **29**, 258 (1979).
8. J. A. Holmes, A. D. Deuchar, L. N. Stewart, and J. D. Parker, *Extractive Metallurgy of Copper*, Vol. II, AIME, New York, 1976, p. 907.
9. A. O. Adami, G. R. Firkin, and A. W. Robson, *Complex Metallurgy '78*, The Institution of Mining and Metallurgy, London 1978, p. 36.
10. D. S. Flett and J. Melling. *Hydrometallurgy* **4**, 135 (1979).

11. W. Hopkin, paper presented at 105th AIME Annual Meeting, American Institute of Mining, Metallurgical and Petroleum Engineers, Las Vegas, Nevada, 1976.
12. D. Pearson, *Chem. Ind.* (6), 220 (1977).
13. S. O. S. Andersson and H. Reinhardt, *Proceedings of the International Solvent Extraction Conference (ISEC), Toronto 1977*, Vol. 2, Canadian Institute of Mining and Metallurgy, Montreal, 1979, p. 798.
14. C. Bozec, J. M. Demarthe, and L. Gandon, *Proceedings of the International Solvent Extraction Conference (ISEC), Lyons 1974*, Vol. 2, Society of the Chemistry Industry, London, 1974, p. 1201.
15. L. F. Cooke and W. W. Szmokaluk, *Proceedings of the International Solvent Extraction Conference (ISEC), The Hague 1971*, Vol. 1, Society of Chemistry Industry, London, 1971, p. 451.
16. G. M. Ritcey, A. W. Ashbrook, and B. H. Lucas, *Can. Min. Met. Bull.* **68** (753), 111 (1975).
17. D. S. Flett and D. W. West, *Complex Metallurgy '78*, The Institution of Mining and Metallurgy, London, 1978, p. 49.
18. A. W. Fletcher, D. S. Flett, J. L. Pegler, D. P. Pepworth, and J. C. Wilson, *Advances in Extractive Metallurgy*, 1967, The Institution of Mining and Metallurgy, London, 1968, p. 686.
19. E. Diaz Nogueira and J. M. Regife Vega, *Quim. Ind. (Madrid)* **23,** (10) 693, (1977).
20. Anonymous, *Chem. Eng.* **85** (4), 6809 (1978).
21. S. Rao, *Warren Spring Laboratory Report No. LR201(ME)*, 1974.
22. A. J. Derosset and K. A. Morgan, U.S. Patent 3,855,385 (1974).
23. K. A. Morgan, U.S. Patent 3,932,579 (1976).
24. R. B. Coleman and G. W. Clevenger, *Intermountain Assoc. Petrol. Geol. Ann. Field Conf. Guidebook* (15), 241 (1967).
25. A. B. Sotex, British Patent 522,459 (1978).
26. H. Ottertun and E. Strandell, *Proceedings of the International Solvent Extraction Conference (ISEC), Toronto 1977*, Vol. 2, Canadian Institute of Mining and Metallurgy, Montreal, 1979, p. 501.
27. W. T. Ruhmer, F. Botha, and J. S. Adams, *J. South Afr. Inst. Min. Met.* **77,** 134 (1977).
28. E. B. Viljoen and D. W. Boydell, *J. South Afr. Inst. Min. Met.* **78,** 199 (1978).
29. Pearson, D., Recovery of Zinc from Metallurgical Dust and Fumes, in *Process and Fundamental Considerations of Selected Hydrometallurgical Systems*, Martin C. Kuhn, Ed., 1981, American Inst. of Mining, Metallurgical and Petroleum Engineers, pp. 153–168.

25.10

RECOVERY OF METALS FROM LIQUID EFFLUENTS

S. O. S. Andersson
P.R. Processutveckling AB
Sweden

H. Reinhardt
MEAB Metallextraktion AB
Sweden

1. GENERAL

The various methods available for recovery of metals from effluents have been reviewed and discussed in recent publications [1–3]. Successful use of solvent extraction as a production procedure for metals in the mining industry has directed attention to the use of this method for recovery of metals from liquid effluents. This is also demonstrated in several review papers on effluent treatment [4–8]. The recent general reviews on solvent extraction by Flett and Spink [9] and by Flett [10, 11] also contain much information on applications of solvent extraction in this field.

Solvent extraction can remove metals from an

effluent to give environmentally acceptable levels. Contamination of the liquid effluent with potentially undesirable organic compounds must be avoided. This can be solved by the use of suitable absorbents such as carbon. The final effluent control is often best performed with a lime neutralization unit; any organic entrainment in the effluent will be reduced by adsorption on the gypsum sludge produced.

On the basis of the different solvent extraction processes developed or in use for recovery of metals from liquid effluents, two principal cases can be defined, as illustrated in Fig. 1. Case 1 in Fig. 1 comprises high-concentration effluents. These usually involve actual process solutions that must be discarded because of unsuitable composition, or, more commonly, because of the buildup of harmful impurities. Examples of solutions belonging to this case are used pickling and plating baths and process bleed streams.

The purpose of recovery is to purify the solution by solvent extraction of the impurities and return of the raffinate as a usable solution to the original process. It is not always necessary to obtain complete removal of impurities when the solution is recirculated. The solvent extraction unit can in this case be regarded as a "kidney" operating on the process stream. The process costs must be balanced against the value of the recovered process solution and, less often, against the value of the extracted material.

A subcase to case 1 in Fig. 1 is when the concentration of a component in a process solution is increased during the processing, as in pickling and etching operations. This component can be partly or fully extracted, thus regenerating the solution which, after makeup, can be reused.

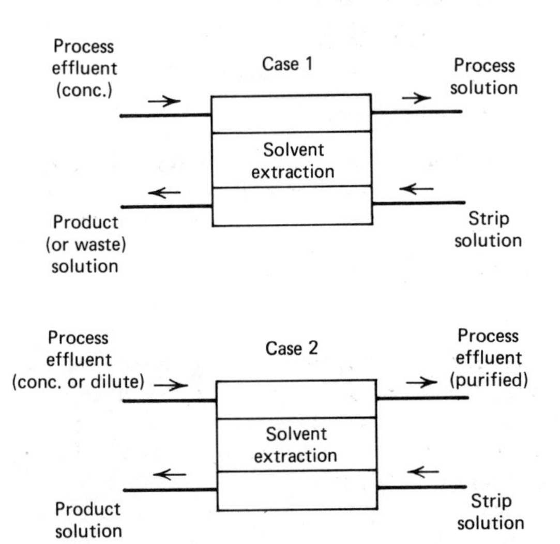

Figure 1. Principal cases of solvent extraction for metals recovery from effluents.

Case 2 in Fig. 1 represents the situation where the liquid effluent, dilute or concentrated, is discarded after the solvent extraction treatment. Low-metal-concentration effluents, often with large volume flows, occur as rinse waters to remove dragout from pickling and plating baths, for example, or from washing of filter cakes. Mine waters, wet scrubber solutions and drainage waters from dumps and so on are also included in case 2. Low-metal-concentration effluents may also occur in waste process solutions as a result of inefficient recovery.

The recovered metals (valuable or harmful) are obtained in a product solution. The treated effluent may be environmentally acceptable for discharge to the recipient or may need further treatment. The process costs must be balanced against the value of the recovered metals and the environmental advantage of the effluent treatment.

A survey of the literature and contacts with people involved in hydrometallurgy have shown that a large number of processes involving solvent extraction for recovery of metals from liquid effluents have been proposed. Some of the processes have been piloted, but very few have been carried through to commercial operation. Thus very little information is available on practical operation and economic feasibility.

2. RECOVERY OF ZINC

2.1. Recovery of Zinc from Spent Hydrochloric Acid Pickle Liquors

Pickling in the galvanizing industry is commonly performed with hydrochloric acid. The fresh solution contains about 18% free HCl. Pickling increases the iron content up to 100–130 g/liter. The solution will also contain zinc 20–120 g/liter, depending on the amount of rejected galvanized material that is pickled. The iron and zinc concentrations balance, so that a high zinc concentration will be accompanied by a low iron concentration. The spent liquor contains 1–2% free HCl. The zinc content prohibits the conventional treatment of the liquor by thermal decomposition to iron oxide and hydrochloric acid.

2.1.1. *Metsep Process* [12, 13]

One method for zinc recovery is the Metsep process, illustrated in Fig. 2. The zinc is separated

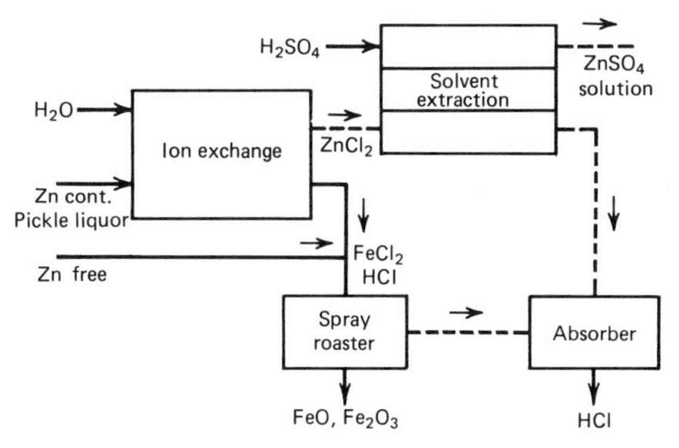

Figure 2. The Metsep process for zinc recovery from spent pickle liquor.

from the iron chloride solution by continuous resin ion exchange. Anionic zinc–chloro complexes are absorbed on a strong-base ion-exchange resin. In the pickling, only iron in the ferrous state is produced; this is the basis for good separation. Before the resin can be eluted, occluded pickle liquor must be washed from the resin. This is done by recirculation of a part of the eluate, which is then mixed with the feed.

The resin is eluted with water to give an eluate of zinc chloride. This is converted to sulfate medium by solvent extraction with D2EHPA and stripping with sulfuric acid. The product is a zinc sulfate solution suitable for electrowinning. The chloride-bearing effluent from ion exchange is pyrolyzed (spray roaster) to produce hydrochloric acid and iron oxide. The solvent extraction raffinate is used in the hydrochloric acid production.

Typical assays from plant operation give a barren solution from ion exchange containing 22 g/liter of HCl, 80 g/liter of Fe, and <1 g/liter of Zn. The eluate contains 1.3 g/liter of HCl, 1.5 g/liter of Fe, and 15 g/liter of Zn. The solvent extraction produced a raffinate containing 15.5 g/liter of HCl, 1.5 g/liter of Fe, and 2.3 g/liter of Zn and a strip solution containing 95 g/liter of Zn, 300 g/liter H_2SO_4, and 5 ppm of chloride.

The Metsep process was developed by the National Institute of Metallurgy (NIM) in South Africa and operated commercially by Woodall Duckham, Ltd. in Johannesburg, South Africa. The zinc sulfate solution was sold to an electrolytic zinc refinery. Changes in market forces have caused this plant to close down.

2.1.2. *MeS Process* [8]

An alternative to the Metsep process, named the *MeS process*, has been developed for the recovery of zinc from pickle liquors. The flow sheet of this process is given in Fig. 3. It uses a solvent extraction circuit for the initial separation of zinc from iron in the pickle liquor.

Zinc is extracted as a zinc chloride complex with TBP as extractant. To avoid extraction of iron, this must be in the ferrous state. The preferential extraction of zinc over iron is somewhat less with TBP than with the best amine extractant, but this is balanced by operational advantages such as higher loading.

Zinc is stripped from the organic solution with water or dilute sulfuric acid. The zinc chloride strip solution is mixed with sulfuric acid mother liquor in a boiler, thus evaporating off hydrochloric acid and crystallizing out zinc sulfate. The zinc sulfate is separated by centrifugation from the cooled mother liquor. By adjusting the conditions in the boiler, chloride-free zinc sulfate suitable for electrowinning can be produced. The distilled hydrochloric acid–water mixture can be further rectified to obtain 6 molar HCl, which can be directly returned to the pickling operation. The iron chloride raffinate can be treated in a pyrolysis plant to produce iron oxide and hydrochloric acid, or it can be used for the production of flocculation chemicals for sewage water treatment.

The process was developed by MX-Processer AB in Sweden and has been piloted in Holland at a galvanizing plant with encouraging results. Building of a full-scale plant is presently under evaluation. The raffinate that was produced contained less than 100 ppm of zinc, which makes the $FeCl_2$ solution acceptable for water treatment. Residues of extractant were removed by adsorption on active carbon.

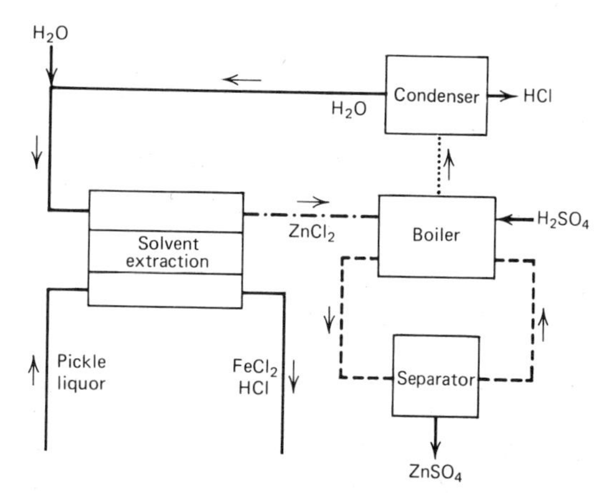

Figure 3. The MeS process for zinc recovery from spent pickle liquor.

2.1.3. *Espindesa Process* [14, 16]

It is interesting to compare the Metsep and MeS processes with the Espindesa (Zincex) process, which was developed by Technicas Reunidas in Spain. This process was developed for the recovery of zinc from a leach liquor resulting from chloride roasting of pyrite cinders. The basic composition of such a leach liquor is similar to the pickle liquors.

The Espindesa process is very similar to the Metsep process but uses a primary solvent extraction circuit, with an amine extractant, for the separation of zinc from iron. (For more information on the Espindesa process see Chapters 25.5 and 25.9.)

2.2. Recovery of Zinc from Rayon Manufacturing Effluents

In the manufacture of rayon, rinse waters and other zinc-containing liquid effluents are produced. The total liquid effluent in a rayon plant may amount to several m^3/min, and the zinc concentration may be 0.1–1 g/liter. In addition to zinc, the effluent contains sodium sulfate and surface-active agents (surfactants), and the pH of the effluent is normally about 1.5–2.

Recovery of zinc has been successfully accomplished by use of the Vålberg process [17], outlined in Fig. 4. Zinc is extracted from the effluent with D2EHPA in kerosene. At an effluent pH of 2, more than 95% of the zinc can be extracted in two stages. Sulfuric acid is used for stripping. By adjusting the net flow rate of the stripping solution, the concentration of zinc in the sulfuric acid may be increased to 50 g/liter or more. Thus the zinc–sulfuric acid solution can be returned directly to the rayon spinning bath.

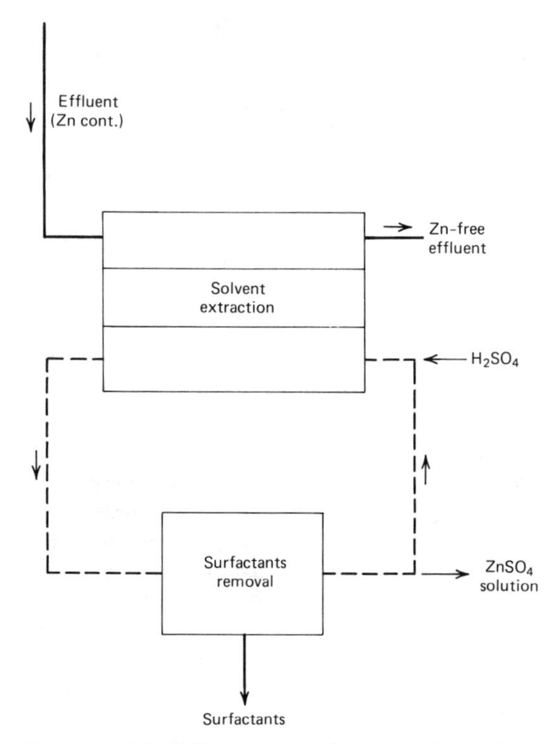

Figure 4. The Vålberg process for zinc recovery from rayon manufacturing effluents.

A full-scale plant was built in 1975 at the factory of Svenska Rayon AB in Vålberg, Sweden. The plant comprises two extraction stages and one stripping stage. The recovered zinc is returned directly to the factory. After the extraction of zinc, the final effluent treatment is lime neutralization before discharge.

2.3. Recovery of Zinc and Cyanide from Zinc Electroplating Rinse Waters

Zinc electroplating is carried out from alkaline zinc cyanide solutions. Because of the toxicity of cyanide, rinse waters from zinc plating must be treated before discharge, and cyanide in particular must be reduced to very low levels.

Zinc cyanide can be efficiently extracted from alkaline solutions by quaternary amines. This property has been used in a solvent extraction process for the removal of zinc and cyanide from zinc electroplating rinse waters developed for the Union Carbide Corporation (USA) [18, 19].

Figure 5 shows the flow sheet. The process is based on the simultaneous extraction of both zinc and cyanide. The decontaminated raffinate is recycled as fresh rinse water. The amine extractant is regenerated by stripping with sodium hy-

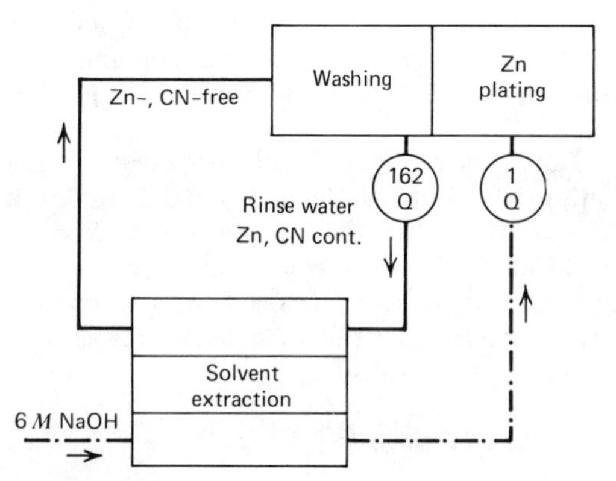

Figure 5. Process for recovery of zinc and cyanide from plating rinse water.

droxide, whereas recovered zinc and cyanide are recycled to the plating bath.

A typical composition of the contaminated rinse water was 40 ppm of cyanide and 23 ppm of zinc. Solvent extraction reduced these values to 0.4 ppm of cyanide and 0.07 ppm of zinc. Active carbon treatment reduced these levels further and at the same time removed entrained and dissolved amine, down to 0.1 ppm. A ratio of feed to strip solution of 162 : 1 produced a strip solution containing 3–4 g/liter of zinc. The same procedure has also been demonstrated for cadmium cyanide plating rinse waters. It is not known whether the process has been operated commercially.

2.4. Recovery of Zinc (and Copper) From Mine Waters

Solvent extraction of copper from mine waters can be carried out with the well-established technique for production of copper from dump leach liquors, practiced in Arizona and other parts of the world (see Chapter 25.1). The solvent extraction technique for recovery of zinc has been outlined by Hazen and Henrickson [20] but has not been practiced commercially. A process for the recovery of both copper and zinc from mine waters is based on the H-MAR flow sheet [21], developed in Sweden (Fig. 6).

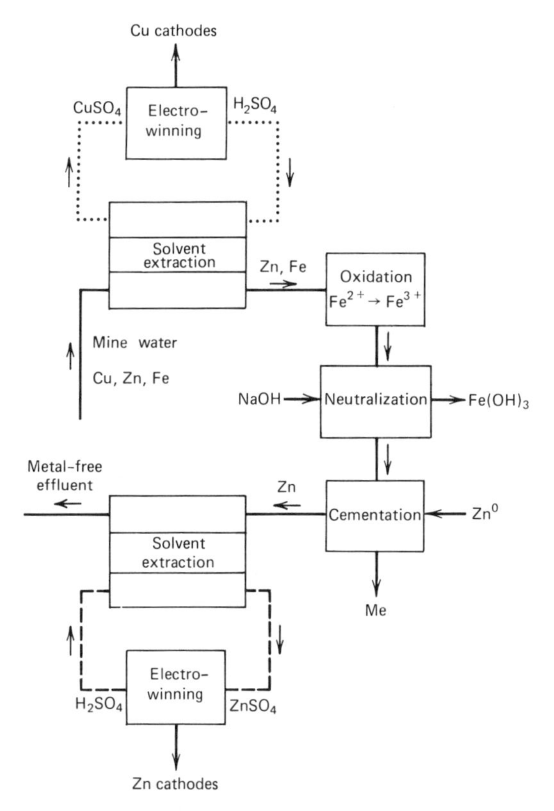

Figure 6. The H-MAR process for recovery of zinc and copper from mine water.

Copper is first recovered by solvent extraction by use of the extractant LIX64N. Iron is then selectively precipitated with calcium carbonate or sodium hydroxide, with air used for oxidation. Zinc is finally extracted from the filtrate in a second solvent extraction circuit with D2EHPA. The process has been tested in pilot-plant operation but has not yet been used commercially.

3. RECOVERY OF COPPER

The higher value of copper compared to zinc and the simplicity of producing copper metal electrolytically makes recovery of copper generally more favorable than recovery of zinc. The availability of copper-selective extractants with a high affinity for copper from weakly acidic and ammoniacal solutions and with a simultaneous rejection of, particularly, ferric iron present a favorable technical basis. The procedures and their feasibility have also been well established in the existing hydrometallurgical copper industry.

3.1. Copper Recovery from Silver Refining Electrolytes [22, 23]

Copper is almost always a significant impurity in unrefined silver, whether derived from scrap or from ore. In the electrolytic refining of silver it is necessary to limit the copper buildup to such a level that deposition and occlusion of copper salts at the cathode are negligible. For a conventional silver nitrate bath, operation at a pH of about 1.5–2.5, and with a silver concentration in the range of 60–160 g/liter, a copper concentration of 60 g/liter is considered the highest permissible. A solvent extraction process for the continuous removal of copper (Fig. 7) has been developed and built by Brookside Metal Company Ltd. in the United Kingdom. Electrolyte is withdrawn from the silver cells and copper is extracted from the solution by use of the extractant SME529. Flow rates are adjusted so that only a portion of the copper is removed from the electrolyte. In the Brookside plant the flow rate was reported to be 4 liters/min and the feed concentration was 15–20 g/liter of copper, resulting in a raffinate containing 10–15 g/liter of copper. After a water scrub, the loaded solvent is stripped

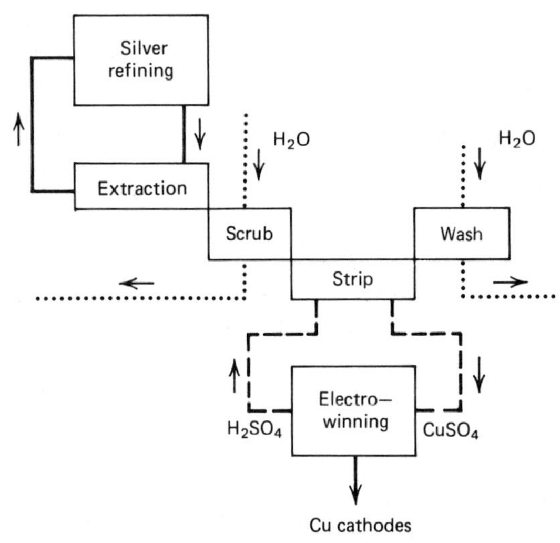

Figure 7. Process for recovery of copper from silver refining electrolyte.

with the barren copper sulfate electrolyte from a copper electrowinning unit.

One problem in this process is the degradation of the organic solvent by the acidic nitrate solution. Despite indications from laboratory tests, some nitration occurs. The result is organic compounds that form much more stable complexes with copper than can be handled. To overcome this effect, the raffinate is limited to a pH of not lower than 1.5.

3.2. Copper from Ammoniacal Etching Solutions

Etching of copper, with the use of an ammoniacal solution, is a common procedure in the manufacture of printed circuit boards for the electronics industry. The ammoniacal etching solution contains free ammonia, one or more ammoniacal salts, and oxidants. During etching, the copper concentration in the solution is increased. Maximum etching efficiency is obtained when the ammoniacal solution contains 110–130 g/liter of copper, and it gradually disappears when the copper concentration is 150–170 g/liter. Thus, to keep the etching efficiency constant and optimal, the etching solution must be continuously removed and regenerated, or replaced with fresh solution.

3.2.1. Process by Criterion Corporation [24]

The process principle involves complete removal of copper from the spent etchant to produce a fresh etchant that, after makeup, can be sold again to the printed circuit board producers.

The process, which is outlined in Fig. 8, has been patented by the Criterion Corporation in the United States. It is based on LIX64N and is reported to involve the selective extraction of copper and chloride ions.

It is reported that a process of this type has been operated in the UK by Proteus Reclamation Ltd. [6], recovering 300 kg/day of copper with Acorga P5100 used as extractant.

3.2.2. MECER Process [18]

By withdrawing etchant directly from the etching line and recirculating it through a solvent extraction circuit, the copper concentration in the etchant can be maintained within the optimal range of 110–130 g/liter. This is practiced in the MECER process outlined in Fig. 9. Treatment of the rinse water obtained from rinsing the circuit boards after etching is also integrated into the process.

In the first extraction stage the etchant is mixed with an organic solution containing LIX54 in kerosene. The copper concentration in the etchant, initially approximately 130 g/liter, is reduced to approximately 90 g/liter. The regenerated etchant is returned to the etching line after careful removal of entrained organic solvent. The copper in the rinse water is extracted

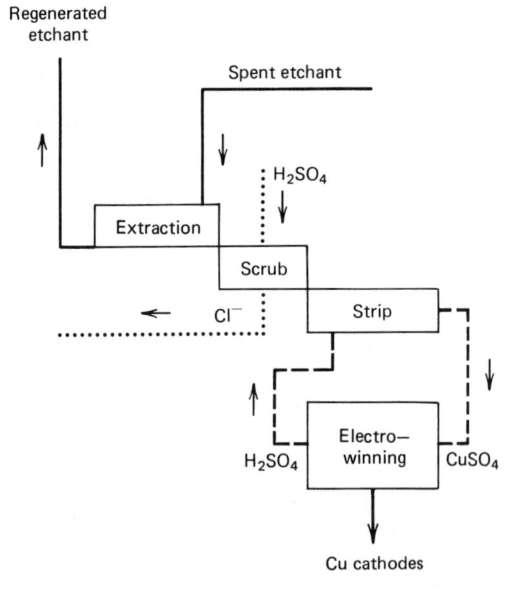

Figure 8. Process for recovery of copper from spent ammoniacal chloride etchant.

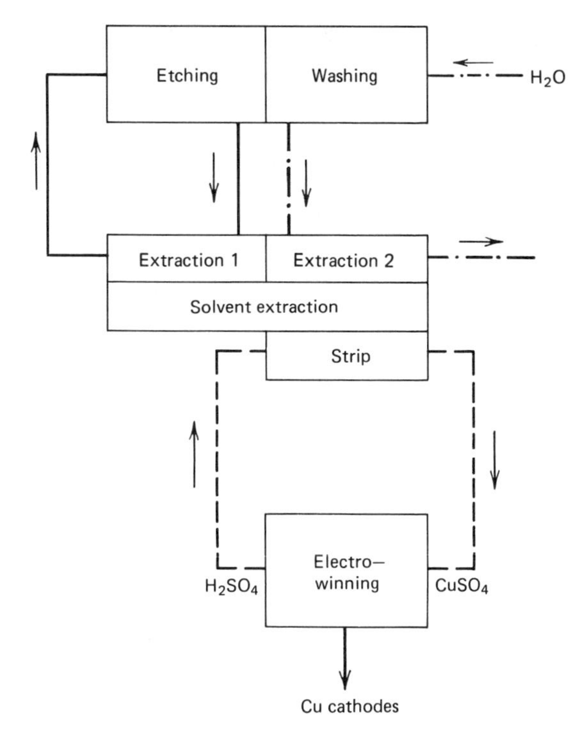

Figure 9. The MECER process for on-line regeneration and copper recovery from ammoniacal etchant.

in the second extraction stage. At the same time any entrainment of etchant from the first extraction stage is washed out. Copper is stripped from the organic solvent with barren copper electrolyte in the stripping stage. The solvent is returned to extraction. Copper metal is produced by electrowinning the titanium cathodes.

The process is in operation in two prototype installations, with good results, and commercial units are now marketed by P.R. Processutveckling AB, Sweden. Etchant makeup has been reduced by 95% and no negative influence on product quality has been encountered.

3.3. Other Examples of Copper Recovery

Other examples of copper recovery from liquid effluents are known, but they differ in nature very little from the commercial recovery of copper from copper sulfate leach solutions, that is, extraction with LIX64N, stripping with spent electrolyte, and production of copper cathodes, as described in Chapter 25.1. Examples are the recovery of copper from sulfuric acid pickling of copper [3], from mine and smelter waters [25] and from copper tankhouse bleed streams [26].

4. RECOVERY OF NICKEL

4.1. Nickel from Plating Baths and Rinse Waters

A method for recovery of nickel from rinse waters has been proposed and investigated by Flett and Pearson [3]. The flow sheet is outlined in Fig. 10. The process is based on extraction of nickel with D2EHPA in its sodium form, to avoid pH changes during extraction:

$$2NaD2EHP_{(org)} + Ni^{2+} \longrightarrow Ni(D2EHP)_{2\,(org)} + 2Na^{+}$$

A typical feed solution may contain 1–2 g/liter of nickel. Laboratory tests showed that nickel could be effectively removed in two extraction steps. The resulting raffinate contained 4 mg/liter of nickel. By loading the solvent with nickel, the transfer of sodium to the strip solution was minimized. Nickel was stripped from the solvent with dilute sulfuric acid. Recovery of nickel from the strip solution is best achieved by electrowinning. The main problem encountered was the high loss of extractant, which amounted to as much as 0.5 g/liter. This could be considerably reduced by an increased sodium sulfate concentration in the aqueous phase (55 mg/liter of D2EHPA at 0.5 molar sodium sulfate).

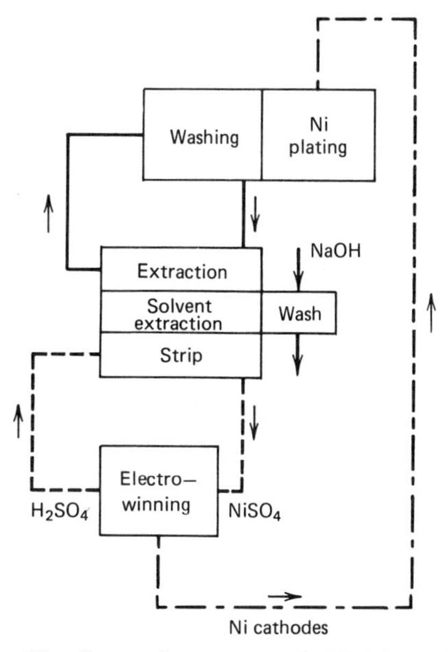

Figure 10. Process for recovery of nickel from plating rinse water.

5. RECOVERY OF CHROMIUM

In chromium plating, a buildup of impurities such as Fe(III), Cr(III), Ni, Cu, and Zn gradually takes place, making the bath unusable for plating. A rinsing step is always used after plating to produce a rinse water effluent containing Cr(VI) and the metals mentioned. Two alternative solvent extraction procedures are available: extraction of Cr(VI) or extraction of impurities.

5.1. Extraction of Chromium(VI)

It is well known that chromate and dichromate ions can be extracted from acid solutions by use of TBP or an amine extractant. Cuer et al. [27] have investigated the applicability of the extractants Alamine 336, LA-2, and TBP. They report excellent stability for TBP and have designed a process for the recovery of 99.5% of chromium(VI) from industrial effluents. The particular cases described refer to waste liquors originating from the production of chromium anhydride (CrO_3) and from processes using this product (chromium plating and metal treatment). The extracted chromic acid is recovered as sodium chromate by stripping with sodium hydroxide solution. Concentrations of 200 g/liter of Cr(VI) in the strip solution are reached. A similar process is reported in a Japanese patent [28], which includes the use of a chromic acid wash of the organic solvent to remove extracted impurities such as iron and chloride.

Rothmann et al [29] have proposed the use of an amine extractant for recovery of chromium, together with vanadium, in effluents from the processing of vanadium- and chromium-containing slags from steelmaking. Difficulties with the precipitation of silica, which interfered with the phase separation, were reported.

5.2. Extraction of Impurities from Chromium Plating Baths

The approach taken by MX Processer AB, Sweden [30] is to extract the impurities from the spent chromium plating bath, thus regenerating the solution for further use in the plating process. The flow sheet of the process is outlined in Fig. 11.

The extraction is carried out by use of a mixture of HDNNS and TBP. The acidity of the plating bath limits the extraction efficiency. Dilution of the plating bath with water significantly improves the operation; water balance is partly maintained by the natural evaporation during plating. Small amounts of chromic acid may be extracted but can be selectively scrubbed with water. The scrubbing solution is used for dilution of the plating bath before extraction; 5 molar

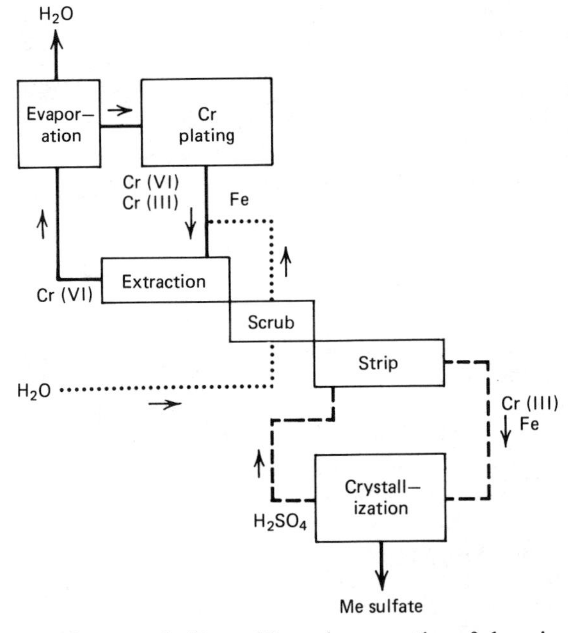

Figure 11. Process for removal of impurities and regeneration of chromium plating bath.

H_2SO_4 or HCl is used for stripping, giving a metal concentration in the strip solution of more than 60 g/liter.

6. RECOVERY OF MERCURY

6.1. Recovery of Mercury from Chlor–Alkali Plant Effluent

The release of mercury with the brine effluents from chlor–alkali plants has long been regarded as environmentally unacceptable. The mercury contamination in brine effluents is typically 10 ppm, and this must be reduced to the parts-per-billion level to be acceptable for discharge.

A solvent extraction method has been proposed by Gronier [31]. It is based on the ready and rapid extraction of mercury from chloride medium by high-molecular-weight tertiary and quaternery amines. With a mercury contamination of 10 ppm in the effluent, a concentration factor 2500 is claimed to be obtainable, leaving a strip solution of 25 g/liter of mercury. The main problem in this system is the loss of organic material to the brine effluent. The amount of mercury that is extractable in a particular case depends, to some extent, on the pH of the brine. With tertiary amines, better than 99% is achieved at pH 3, but there is a decrease at higher pH.

A closely related process based on the amine solvent extraction system has been patented by Chapman and Caban [32]. It is not known whether any commercial application has been considered.

7. RECOVERY OF OTHER METALS

7.1. Noble Metals

Work on the recovery of noble metals, mainly from plating solutions, has been reported. Gold and silver are extracted from cyanide solutions with quaternary amines [33, 34]. Gold can be stripped from the organic solvent with alkaline potassium cyanide solutions.

Extraction of platinum and palladium from waste solutions, by use of an amine extractant, has also been reported [35]. Commercial applications may well exist but have not been reported in the literature.

7.2. Molybdenum and Rhenium

A flow sheet for the recovery and separation of molybdenum and rhenium from ore roaster scrubber liquors has been patented by Continental Ore Corporation [36]. Molybdenum and rhenium are extracted from the scrubber liquor with a tertiary amine. The raffinate is recirculated to the scrubber, and the metals are recovered from the extract by ammonia stripping. The strip liquor is treated with magnesium oxide to precipitate extracted impurities. Sulfuric acid is then added, and pure ammonium tetramolybdate is crystallized. The mother liquor is further treated by amine extraction, followed by a pyridine extraction, to produce a crude rhenium product. It is not known if any commercial application has been attempted.

8. MISCELLANEOUS

8.1. Recovery of Nitric and Hydrofluoric Acids in Stainless Steel Pickling

Stainless steel is pickled with a mixture of HNO_3 and HF. As the metal concentrations increase, the pickling action decreases and finally the high-concentration solution must be discarded. A solvent extraction procedure for recovery of the acids and metals (Fig. 12) was developed [37, 38] and put into operation some years ago. The process, named the *AX process*, utilizes the capacity of undissociated acids, such as HNO_3 and HF, to form adduct complexes with certain extractants such as TBP.

The same technique, with TBP extraction, has been used in a pilot-plant operation in France for recovery of nitric acid from nickel sulfate solutions [39]. A process for the recovery of hydrofluoric acid with the use of amine extraction has also been proposed [40].

A process closely related to the AX process has been patented in Japan [41]. The acids are

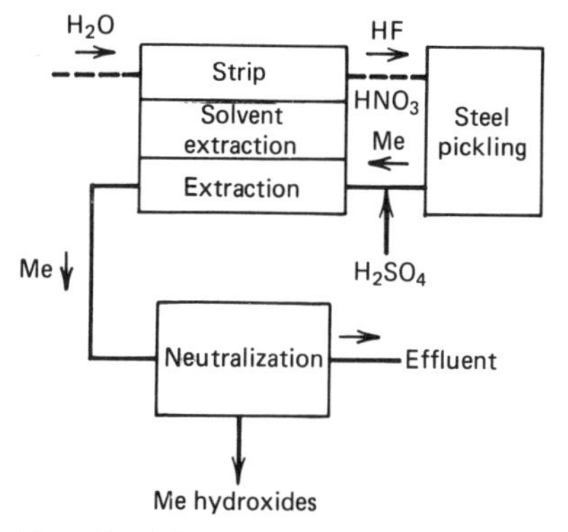

Figure 12. The AX process for regeneration and metals recovery from spent stainless steel pickle liquor.

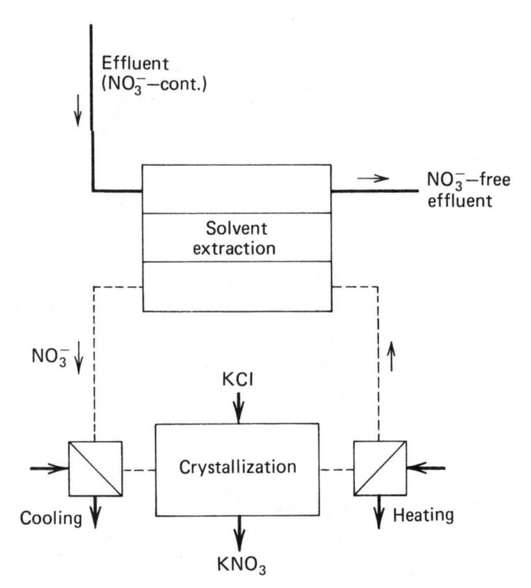

Figure 13. Process for the removal and recovery of nitrate from effluents.

extracted with TBP and then separated by scrubbing the organic solvent with dilute HNO_3 to remove HF.

8.2. Recovery of Nitrate from Effluents

Pilot-plant tests of nitrate removal from cellulose nitrate plant waste water have been undertaken in Finland [42, 43]. The process (Fig. 13) is claimed to be superior to established nitrate-removal methods such as ion exchange and chemical or biological decomposition, although recovery is not as high. The recovered nitrates are obtained as a potassium nitrate product.

The nitrate is extracted with a solvent containing 5–10% of a secondary amine in aliphatic kerosene. During stripping, nitrate is removed from the organic solvent by a salt solution containing about 15% KNO_3 and 25% KCl (by weight). Potassium nitrate is crystallized after cooling, and the mother liquor is recirculated to stripping.

Nitrate recovery from titanium leach liquor effluents has been proposed by USBM [44] by use of TBP extraction. The flow sheet is nearly the same as for the AX process. Nitrate is recovered as nitric acid at concentrations suitable for reuse in the titanium process.

REFERENCES

1. A. W. Fletcher, Metal Recovery from Effluents, *The Application of Chemical Engineering to the Treatment of Sewage and Industrial Liquid Effluents*, Institution of Chemical Engineers Symposium Series No. 41, London, 1975.
2. T. W. Cadman and R. W. Dellinger, *Chem. Eng.*, 8, 79 (1974).
3. D. S. Flett and D. Pearson, *Chem. Ind.*, 639 (1975).
4. P. R. Kiezyk and D. Mackay, *Can. J. Chem. Eng.* **49,** 747 (1971).
5. M. A. Hughes, *Chem. Ind.*, 1042 (1975).
6. D. S. Flett, Solvent Extraction in Scrap and Waste Processing, paper presented at IMM/SCI Conference on Impact of Solvent Extraction and Ion Exchange on Hydrometallurgy, University of Salford, 1978.
7. H. Reinhardt, *Chem. Ind.*, 210 (1975).
8. H. Reinhardt, Some Hydrometallurgical Processes for the Reclamation of Metal Waste, paper presented at IWTU Conference, Waterloo, 1978.
9. D. S. Flett and D. R. Spink, *Hydrometallurgy* **1,** 207 (1976).
10. D. S. Flett, Solvent Extraction of Non-ferrous Metals: A Review 1975–1976, *Warren Spring Laboratory Report LR257 ME.*
11. D. S. Flett, *Rept. Progr. Appl. Chem.* **57,** 37 (1974).
12. A. K. Haines, T. H. Tunley, W. A. M. TeRiele, F. L. D. Cloete, and T. D. Sampson, *J. South Afr. Inst. Min. Met.* **74,** 149 (1973).
13. T. H. Tunley, P. Kohler, and T. D. Sampson, *J. South Afr. Inst. Min. Met.* **77,** 423 (1976).
14. R. Wood, *Process Eng.* **6,** 6 (1977).
15. Davy Powergas Ltd., London, private communication.
16. J. M. Vega and E. D. Nogusira, *Quimica e Industria* **23,** 694 (1977).
17. H. Reinhardt, H. Ottertun, and T. Troeng, Solvent Extraction Process for Recovery of Zinc from Weakly Acidic Effluent, *The Application of Chemical Engineering to the Treatment of Sewage and Industrial Liquid Effluents*, Institution of Chemical Engineers Symposium Series No. 41, London, 1975.
18. F. L. Moore and W. S. Groemier, *Plating Surface Finishing* 26 (August 1976).
19. F. L. Moore, *Separation Sci.* **10** (4) 489 (1975).
20. W. C. Hazen and A. V. Henrickson, Process for Concentrating Copper and Zinc Values Present in Aqueous Solution, U.S. Patent 2,992,894 (1961).
21. S. O. S. Andersson and H. Reinhardt, *Proceedings of International Solvent Extraction Conference (ISEC), Toronto 1977*, Vol. 2, Canadian Institute of Mining and Metallurgy, Montreal, 1979, p. 798.
22. W. Hunter, Electrolytic Refining, U.S. Patent 3,975,244 (1976).
23. W. Hunter, The Use of Solvent Extraction for Purification of Silver Nitrate Electrolyte, paper

presented at the Second International Conference on Precious Metals, New York, May 1978.

24. W. D. Hamby and M. D. Slade, Process for Regenerating and for Recovering Metallic Copper from Chloride-Containing Etching Solutions, U.S. Patent 4,083,758 (1978).

25. G. Barthel, *J. Metals* 7 (July 1978).

26. R. B. Sudderth and W. H. Jensen, Utilization of LIX63 in Some Liquid Ion Exchange Systems, paper presented at Canadian Institute of Mining and Metallurgy, 76th General Meeting, Montreal, Quebec, April 1974.

27. J. P. Cuer, W. Stukens, and N. Texier, *Proceedings of International Solvent Extraction Conference (ISEC), Lyons 1974*, Vol. 2, Society of Chemical Industry, London, 1974, p. 1185.

28. S. Nishimura and M. Watanabe, *Japan Kokai* **76,** 90,999 (1976).

29. H. Rothmann, G. Bauer, A. Stuhr, and H. J. Retelsdorf, *Metall. (Berlin)* **30** (8), 737 (1976).

30. H. Reinhardt and H. D. Ottertun, Swedish Patent Application 76-13686-0.

31. W. S. Gronier, Application of Modern Solvent Extraction Techniques to the Removal of Trace Quantities of Toxic Substances from Industrial Effluents, Oak Ridge National Laboratories, Tennessee, Report ORNL-TM-4209, 1973.

32. T. W. Chapman and R. Caban, Extraction of Mercuric Chloride from Dilute Solution and Recovery, U.S. Patent 3,899,570 (1975).

33. M. D. Ivanoviskii, M. A. Meretukov, V. D. Potekhin, and L. S. Strizhko, *Izvest. Vyssh. Ucheb. Zadev. Tsvetnye Metally* **17** (2), 36 (1974).

34. B. N. Laskorin, A. I. Shilin, V. V. Shatalov, M. I. Abramova, and V. Yu. Smirnov, *Prib. Sist. Upr.* (4), 50 (1974).

35. E. Nerkova, K. Vasilev, and D. Denev, *Metalurgiya (Sofia)* (7) 31 (1972).

36. Continental Ore Corporation, New York, Improvements in or Relating to the Recovery of Molybdenum and Rhenium Values, British Patent 1,364,933 (1974).

37. U. Kuylenstierna and H. Ottertun, *Proceedings of the International Solvent Extraction Conference (ISEC), Lyons 1974*, Vol. 3, Society of the Chemistry Industry, London, 1974, p. 2803.

38. J. H. Dempster and P. Björklund, Operating Experience in Recovery and Recycling of Nitric and Hydrofluoric Acids from Waste Liquors, *Proceedings of the 4th Annual Meeting of the Hydrometallurgy Section of the Metallurgical Society, CIM,* Toronto, August 1974, p. 68.

39. J. M. Demarthe, M. Tarnero, P. Miguel, and J. P. Goumondy, *Proceedings of the International Solvent Extraction Conference (ISEC), Lyons 1974*, Vol. 2, Society of Chemical Industry, London, 1974, p. 1275.

40. W. H. Hardwick and P. F. Wace, *Chem. Process. Eng.* 283 (June 1965).

41. S. Nishimura and M. Watanabe, *Jap. Kokai* **76,** 18, 993 (1976).

42. R. Remirez, *Chem. Eng.* **83** (14), 36C (1976).

43. T. K. Mattila and T. K. Lehto, *Ind. Eng. Chem. Process Des. Devel.* **16** (4), 469 (1977).

44. J. R. Ross, S. R. Borrowmann, and D. R. George, *Solvent Extraction of Nitrate from Titanium Leacher Effluent*, U.S. Bureau of Mines Report RI 7733, 1973.

25.11

COMMERCIAL PROCESSES FOR URANIUM FROM ORE

P. J. D. Lloyd
Chamber of Mines Research Organisation
South Africa

1. INTRODUCTION

The purpose of solvent extraction in uranium ore processing is primarily to concentrate the uranium to the point where it can be recovered from solution economically. A secondary but hardly less important purpose is to reject impurities. It is probable that more uranium has been won from its ores by solvent extraction than by any other process.

The reasons for the general predominance of solvent extraction are as follows:

1. A variety of solvents are available. Even though different ores require different lixiviants, an extractant can usually be found that will permit both concentration and purification of uranium from the resultant solution.
2. High concentration ratios are available. The forward-distribution ratio (O/A) is usually $>>1$, and the back-distribution ratio (O/A) is often $<<1$.
3. A variety of highly selective solvents is available, so excellent purification of uranium is possible.
4. The process is usually quite insensitive to the inevitable changes in feed composition and can be engineered in such a way that malfunctions or errors in operation do not lead to drastic loss of efficiency.

2. CHOICE OF EXTRACTANT

The choice of extractant is determined largely by the choice of lixiviant, which in turn is determined by the composition of the ore. In Table 1 the various types of ore and the preferred

TABLE 1 ORE TYPES, LIXIVIANTS, AND EXTRACTANTS

Ore Type	Examples	Lixiviant, Leaching Conditions	Extractant	Remarks
Highly basic (>10–15% $CaCO_3$ equiv.) containing U^{6+}	Ambrosia Lake–Smith Lake ores, Grants, New Mexico	Na_2CO_3; elevated temperatures and sometimes pressures; O_2; ore sometimes roasted before leaching, or sulfide concentration separated	Nil (quaternary ammonium compound feasible, but not used)	Concentration and purification achieved by selective leaching and precipitation
Weakly basic to acidic ores, containing uranium as uraninite, coffinite, uranothorite, brannerite, etc.; sulfide and other flotation concentrates	Many Canadian, Witwatersrand, Australian, and U.S. ores	H_2SO_4, oxidant, mild temperature and pressure except for refractory ores	Dialkyl orthophosphoric acids; secondary or tertiary alkyl amines	Acids have slower kinetics, require stronger strip than amines, sensitive to Fe^{3+}; amines more sensitive to V, Mo, anions and sus-suspended solids, and are usually slower settling
Phosphates	Florida, United States	H_2SO_4–H_3PO_4	Dialkyl pyrophosphoric acids; synergistic combinations such as dialkyl orthophosphoric acids–trialkyl phosphine oxides	Uranium extracted as U^{4+}; multiple cycles may be necessary for adequate purification
High-grade ores or concentrates	Uranothorianite concentrate, Palabora, South Africa	HNO_3	Tri-*n*-butyl phosphate	Ethers were used—abandoned because hazardous

routes for their treatment are summarized. Most commercial uranium is produced from ores of the second type, that is, the weakly basic to acidic types, from which the uranium can be recovered by sulfuric acid leaching in the presence of an oxidant.

In 1978 over 34,000 tons per day (or 80%) of the installed ore processing capacity in the United States relied on sulfuric acid leaching and solvent extraction; just over 4000 tons per day (or 9% of the total) employed acid leaching and ion exchange; and nearly 5000 tons per day (11% of the total) used carbonate leaching and precipitation. Ion exchange was preferred in all cases for the treatment of liquors from *in situ* leaching, and solvent extraction for recovery from phosphate solutions [1]. A similar predominance of sulfuric acid leaching and solvent extraction occurs elsewhere in the world.

In general, the amine extractants have been preferred for solvent extraction from sulfate solutions. Table 2 compares the two classes of extractant.

A variety of extractants have been used to extract uranium from phosphoric acid solutions. In an early process, U^{4+} was extracted with dialkyl pyrophosphoric acids, but the need to reduce the uranium to the tetravalent state and the tendency for the pyrophosphoric acid extractant to hydrolyze at the P–O–P bond led to excessive costs [2].

Relatively high concentration solutions of long-chain dialkyl orthophosphoric acids will also extract U^{4+}, but a synergistic combination of di(2-ethyl-hexyl) phosphoric acid (D2EHPA) and tri-*n*-octyl phosphine oxide (TOPO) will extract UO_2^{2+}. Because it is not necessary to reduce the whole feed stream, processes based on this combination have been preferred in recent years. In 1979 it was estimated that production from Florida phosphate ores would amount to over 2000 tons/year before 1982.

The only other leaching agent of any commercial importance is HNO_3. This has been applied to leach uranium from the high-grade pitchblende ores of Zaire, Canada, and Czechoslovakia. Currently it is used for the recovery of uranium from a uranothorianite concentrate produced at Palabora, Republic of South Africa. A variety of solvents may be used to extract UO_2^{2+} from HNO_3 solutions. Ethers were used initially but have been generally abandoned in favor of TBP on grounds of the reduced fire hazard.

The TBP extraction of uranium from nitrate solutions is fully covered in Chapters 25.12 and 25.13. A description of the Palabora plant is given by Tunley and Nel [3].

TABLE 2 COMPARISON OF AMINE AND ALKYL PHOSPHORIC ACID EXTRACTANTS FOR RECOVERY OF URANIUM(VI) FROM SULFATE SOLUTIONS

Amine	Alkyl Phosphoric Acid
Very rapid extraction	Relatively slow extraction because of replacement of anions in primary co-ordination sphere of uranyl ion by extractant radical
Sensitive to presence of vanadates; molybdates; and anions such as HSO_4^-, NO_3^-, Cl^-, CNS^-, $(S_nO_{n+4})^{2-}$	Sensitive to the presence of Fe^{3+}, Th^{4+}, V^{4+}, Ti^{4+}, rare earths, molybdenum
Stable extractants of low stability	Slightly unstable extractants but generally cheaper than amines
Tendency to form third phases with cheap aliphatic diluents forces use of modifiers such as long-chain alcohols or an aromatic diluent	Compatible with many diluents, but sodium salts form third phases that require addition of tributyl phosphate (TBP) for compatibility
Tend to be surface active and thus may be affected by presence of suspended solids (particularly siliceous), flotation reagents, oils, tars, etc.	Relatively insensitive to the presence of suspended solids
Relatively slow phase separation	Relatively rapid phase separation
Simple to strip under relatively mild conditions	Stripping usually with alkali carbonates, which necessitates addition of TBP to prevent third-phase formation; can be stripped with strong acids (e.g., HCl)

3. AMINE SOLVENT EXTRACTION OF URANIUM FROM SULFATE SOLUTIONS (AMEX)

3.1. Pretreatment of Leach Solution

Because the amine extractants are sensitive to a variety of contaminants in the feed to the extraction step, various pretreatments of the feed may be necessary.

3.1.1. Solids

Filtration of the feed is usually necessary to ensure that the suspended solids concentrations are less than about 20 ppm. Fixed-bed sand filters, either atmospheric or pressure, are usually used for this purpose. In one case the use of an anthracite "sand" was found to reduce molybdenum contamination by adsorption of the molybdenum on the anthracite as well as removing suspended solids.

3.1.2. Eluex

Where the uranium concentration in the feed solution is very low, or where the leach solution contains fine solids that are either difficult to remove by filtration or present in too great a concentration, it may be desirable to precede solvent extraction by an ion-exchange step. This combined process is known as an *Eluex* or Bufflex process.

In these processes advantage is taken of the ability of basic ion-exchange resins to be eluted with moderately strong (typically 10%) sulfuric acid solutions. Uranium is then extracted from the eluate with tertiary amines or quaternary alkyl ammonium salts in the normal manner. The extractability of uranium by the solvent is reduced under the strongly acidic conditions, mainly because of competition between the uranyl sulfate complexes and bisulfate anions, but not to such an extent that solvent extraction is impossible.

The prior ion-exchange step may be very effective in removing solids, particularly if a continuous contactor of the fluidized-bed type is used. In addition, the use of two stages of concentration and purification in series can lead to a high-purity product, even though a similar chemistry is employed in each stage.

3.1.3. Interfering Anions [4]

The two anions of greatest concern are those of V^{5+} and Mo^{6+}. The effect of vanadium present in the ore is best overcome by leaching at or adjusting the leach solution to an oxidation potential of less than about −0.8 V, when most of the vanadium is present in the nonextracted V^{4+} state. (More information on the behavior of vanadium is given in Chapter 25.4.) The effect of molybdenum may be reduced by pretreatment of the feed solution. Such pretreatments include partial neutralization, particularly at temperatures of 100°C or higher, when the molybdenum is coprecipitated with iron, a two-stage countercurrent leach with a high final pH (about pH 3.2), and adsorption of the molybdenum on activated carbon.

More common anions that interfere, such as NO_3^- and Cl^-, do not present such severe problems as molybdenum and vanadium. Nevertheless, care must be taken to keep their concentrations as low as possible.

3.1.4. Temperature

It is often found advantageous to leach at temperatures above ambient. As uranium extractability falls off markedly with increasing temperature, it is necessary to cool the leach solution before solvent extraction. This also reduces the fire hazard.

3.2. Choice of Solvent

3.2.1. Amine

Although a range of amines has been available and, indeed, used on various plants at various times, it has generally been found that the tertiary alkylamines, with alkyl groups in the range of C_8–C_{12}, offer the best combination of acceptable performance, reasonable first cost, and low loss as a result of solubility or degradation. Such amines are widely available under trade names such as Alamine 336 and Adogen 363 (see Chapter 24).

3.2.2. Diluent

The choice of diluent is not critical from the process point of view. A range of aliphatic (kerosenes) and aromatic solvents has been used satisfactorily. More important are operational parameters. For safety, the flash point must be as high as possible, and it must be remembered that the actual flash point reduces with pressure, so that the hazard increases with altitude. To assist a rapid phase disengagement, both the viscosity and the density of the solvent must be as low as possible. It is difficult to replace a low-density diluent with a higher density one because interfaces in settlers are set by fixed weirs.

3.2.3. Choice of Modifier

The modifiers that have been found satisfactory, in that they prevent third-phase formation without affecting uranium extraction or phase disengagement, are effective at relatively low concentrations and have a low solubility, are long-chain alcohols, typically nonanol and decanol. These are added to amine solution in highly paraffinic diluents but are not needed in highly aromatic diluents. Alternatively, a highly aromatic diluent may be mixed with a paraffinic diluent, in which case no modifier is needed.

3.2.4. Choice of Concentrations

The prime choice is that of the concentration of amine in the solvent mixture. Both the distribution ratio and the capacity of the solvent for uranium vary linearly with amine concentration over practical ranges. Reduction of the concentration increases the number of stages necessary to ensure essentially complete uranium extraction, the size of the stripping stages, and the organic flow rates. Increasing the concentration of the extractant increases the viscosity and the density of the solvent mixture, both of which increase the settler area needed for a given duty, which in turn increases the solvent inventory unnecessarily. Experience has shown that an amine concentration in the range 2.5–10% is usually satisfactory.

For paraffinic diluents, addition of a long-chain alcohol in a concentration similar to or slightly less than that of the amine usually serves to avoid third-phase formation. Thus typical amine-alcohol-diluent combinations are 5–4–91%, 2.5–3–94.5%, and 10–9–81%. For amine-aromatic-paraffinic diluent mixtures, typical combinations are 5–95–0%, 4–36–60%, and 6–50–44%.

3.3. Choice of Extraction Conditions

3.3.1. Sulfate and Sulfuric Acid Concentrations

There is a wide range of sulfate and acid concentrations over which extraction is effective. Between about 30 and 150 g SO_4^{2-}/liter the distribution ratio D falls according to $D = k_1(SO_4^{2-})^{-2}$, at constant pH and total amine concentration. Similarly, at constant sulfate and amine concentrations, the distribution ratio falls approximately according to $D = k_2(H^+)^{-0.5}$ over the range pH 0.5–2.5. A single measurement of the distribution ratio at any one condition can be extrapolated to other conditions with sufficient accuracy to check on the likely effect of changes in leaching or pretreatment conditions.

3.3.2. Uranium Concentrations

A Langmuir adsorption model is sufficient to account for the effect of changes in aqueous or organic uranium concentrations:

$$D = \frac{U_0}{U_a} = k_3(C - k_4 U_0)$$

where C is the extractant concentration in the organic phase. Figure 1 shows typical isotherms at three different sulfuric acid concentrations. If replotted as $1/U_a$ versus $1/U_0$, straight lines are obtained, as required by the above equation.

It is possible to calculate the parameters of the equation from their slopes and intercepts. As an example, Table 3 gives the calculation of the parameters in this specific case.

The utility of this approach is that k_3 describes the distribution ratio for unit concentrations of solvent at given aqueous phase conditions and infinitely dilute uranium concentration; the effects of changes in the aqueous phase are thus immediately apparent. In this case, concentrations exceeding 125 g/liter have a bad effect on extraction. On the other hand, k_4 is inversely proportional to the maximum loading of the solvent, and it is apparent that for this system there is a definite advantage in exceeding 125 g/liter of acid at high aqueous uranium concentrations. The linear dependence on the amine concentration C makes it possible to determine readily the total effect of changing this concentration.

3.3.3. Effect of Various Anions

Interfering anions have the effect of complexing the solvent, which usually has the effect of reducing the distribution ratio. The equilibria become complex, but the Langmuir approach outlined previously remains a reasonable approximation. The effect of increased chloride concentration at fixed sulfuric acid concentration is shown in Fig. 2. Again, the continuous curves are those of the "best-fit" equations to the isotherms.

3.3.4. Phase Ratio

The phase ratio should be such that organic-continuous emulsions are formed if the loss of

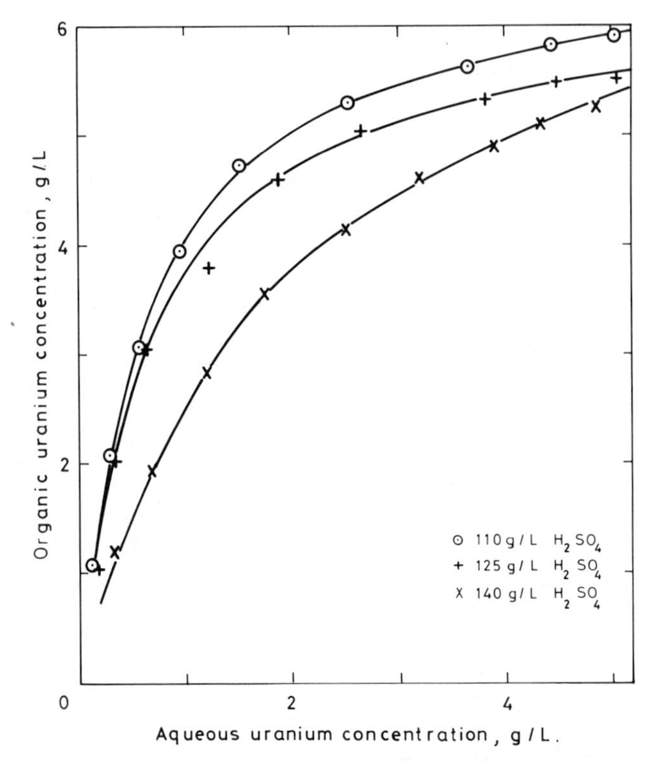

Figure 1. Isotherms for extraction of uranium from sulfuric acid by 5% Alamine 336 in kerosene–2% isodecanol at 20°C.

solvent by entrainment is to be minimised. Stable organic-continuous emulsions cannot be maintained for long periods at overall phase ratios much above 1.6 aqueous: 1 organic (1.6 A: 1 O). However, the uranium concentration in typical leach solutions is of the order of 1 g/liter, whereas the organic phase can readily be loaded to 5 g/liter or higher. The nominal phase ratio is thus 5 A: 1 O or higher. Therefore, within any extraction stage it is necessary to recycle organic phase to ensure low enough phase ratios for organic-continuous conditions. Typical ratios used are 1 A: 1.5 O.

In introducing the aqueous phase into the emulsion, it is essential to ensure that it is dispersed and cannot coalesce into large drops. If there are large drops present, the emulsion may be dispersed within these drops, causing a local inversion in the phases, so defeating the whole object of running organic-continuous.

3.3.5. Mixing [5]

The kinetics of the amine extraction of uranium are sufficiently fast that no particular accent need be placed on mixing conditions. Where con-

TABLE 3 CALCULATION OF ISOTHERM PARAMETERS FOR URANIUM EXTRACTION FROM SULFURIC ACID SOLUTIONS BY 5% ALAMINE 336 IN KEROSENE–2% ISODECANOL AT 20°C

Sulfuric Acid Concentration, g/liter	Slope = k_3C	Intercept at $1/U_a = 0 = k_4/k_3C$ liters/g	k_3 (1/percent)	k_4 liters/g
110	10.03	0.149	2.01	1.49
125	9.20	0.158	1.84	1.46
140	3.79	0.134	0.758	0.507

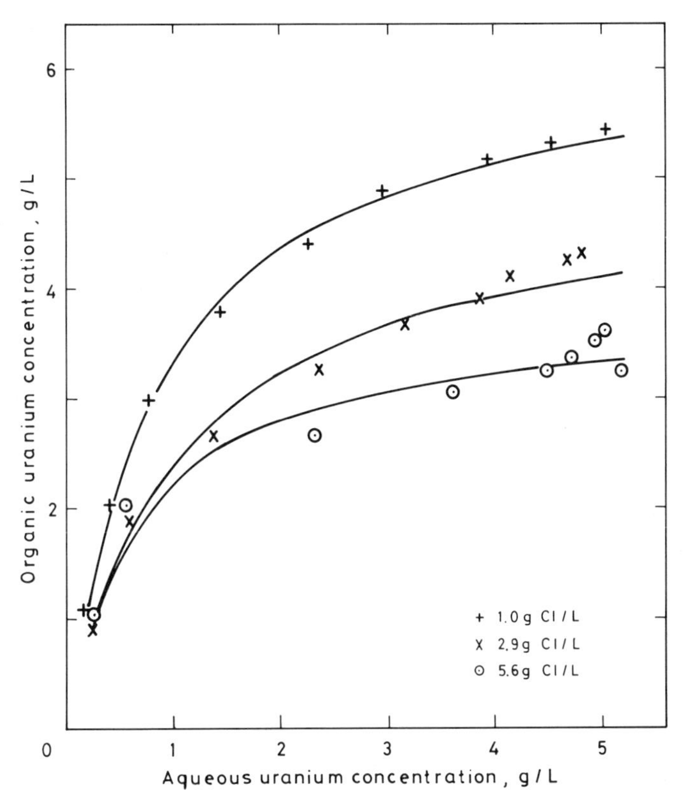

Figure 2. Isotherms for extraction of uranium from sulfuric acid in the presence of chloride ions by Alamine 336 in kerosene–isodecanol.

ventional Treybal-type mixers are employed, power inputs are typically 1–2 kW/m^3. Proprietary types of mixer may have different power characteristics, particularly if they are of the pump–mix type. The general aim should be to avoid high rates of shear that would lead to the generation of fine dispersions. In turbine-type agitators, tip speeds should not exceed about 5 m/s. Mixers are typically sized to provide <1 min of residence time.

3.4. Scrubbing or Washing

Where a high-purity product is desired, a "scrub" or "wash" of the loaded solvent may be employed. Both terms have been used to describe this step. Typically the step takes the form of several stages of scrub with 10% sulfuric acid at 1 A:50 O phase ratio, possibly followed by several stages with an alkaline $(NH_4)_2SO_4$ solution, or clean water scrub at a similar phase ratio.

In the acidic section impurities such as ferric iron and other cations with a lower distribution ratio than uranium are removed. In the alkaline section some of the uranium transfers, but any entrained aqueous phase is also removed. Any amine bisulfate $R_3NH_2SO_4$ is neutralized in this step, which simplifies pH control in subsequent stages. The presence of excessive alkalinity in the scrub feed promotes the formation of a siliceous "crud."

It should be noted that the product of the overall solvent extraction process is invariably less than nuclear grade; at best, scrubbing can contribute only to a higher-than-normal grade of final concentrate.

3.5. Choice of Stripping Conditions

Various strip reagents are available. They include acid chloride and nitrate solutions, neutral ammonium sulfate, and alkaline carbonate solutions. There is little to choose between these reagents on technical grounds; there is some evidence for solvent loss in the presence of strong nitrate solutions, and alkaline carbonate stripping has advantages when the solvent contains molybdenum or vanadium in addition to uranium. The choice of reagent is thus largely determined by economic considerations. The precipitation of the uranium from the strip solution and the subsequent handling and treatment of the precipitate also play a small part.

3.5.1. *Acidic Stripping*

The stripping reaction is a pure ion-exchange one:

$$\overline{(R_3NH)_4UO_2(SO_4)_3} + 4HX \rightleftharpoons \overline{4R_3NHX} + UO_2SO_4 + 2H_2SO_4$$

but it is immediately obvious that this leaves the solvent in a salt form which may not be suitable for uranium extraction. In fact, the amine chlorides are quite readily converted to the amine sulfates, and the amine chlorides are extractants for uranium in their own right. Thus, after chloride stripping, the solvent could be recycled directly. In the case of nitrate stripping, however, the nitrate ion is not readily displaced by sulfate, and the amine nitrate is a poor extractant for uranium from sulfate solution. The amine nitrate must thus be regenerated to the free base form. The cost of regeneration of the full solvent stream is relatively high; therefore, nitrate stripping has not found wide favor.

Chloride stripping typically requires 1.0–1.5 *M* solutions, acidified to about pH 2 to prevent hydrolysis and precipitation. Overall phase ratios of about 1 A:15 O are used, although some aqueous may be recycled from the settler to the mixer to improve both mixing and settling. Typically four mixer–settler stages are used, with mixing power inputs of the order of 2–5 kW/m^3 with residence times of about 1 min in the mixer.

Much of the molybdenum and phosphate in the loaded solvent is not removed in chloride stripping. At low concentrations of these impurities, the regeneration of a portion of the recycle solvent stream by contact with an alkaline chloride is necessary to prevent buildup of these impurities. Moderate concentrations of these impurities may lead to the formation of polymeric phosphomolybdate–amine complexes in the organic phase, with resultant loss of amine.

Uranium is recovered from the strip solution by precipitation as ammonium or sodium diuranate (ADU and SDU), as a crude "yellow cake" by the addition of magnesia, or as uranium peroxide $UO_4 \cdot 2H_2O$ by the addition of H_2O_2. The separation from molybdenum, vanadium, and phosphorous impurities is best with this last precipitant.

3.5.2. *Neutral Stripping* [6]

This route is probably the most widely employed. The loaded solvent is contacted with concentrated $(NH_4)_2SO_4$ solution, and the pH is slowly raised with ammonia. The uranyl complex with the solvent is hydrolyzed, and the uranium transfers to the aqueous phase, where the high sulfate concentration maintains it in solution. Most of the amine sulfate is also hydrolysed, so the quantity of ammonia required is close to that required by the amine content of the organic phase.

Typical strip conditions require four stages with the pH gradually increasing from the region of pH 3.5 in the first mixer to the region of pH 5 in the last. The final pH varies from circuit to circuit, depending on process parameters. The ammonium sulfate concentration is not critical and typically lies in the range 100–150 g $(NH_4)_2SO_4$/liter. The flow ratio is typically 1 A:5 O, although the aqueous phase may be recycled to maintain aqueous-phase continuous conditions in some stages to simplify pH control. Mixing conditions are typically a retention time of 5 min per stage with a power input of 5–10 kW/m^3 in each stage.

The strip section is usually run at temperatures above ambient (typically 30°C), partly because this speeds stripping and partly because the aqueous phase is heated during uranium precipitation.

Uranium is recovered from the strip solution by addition of ammonia to pH $>$ 7.0. The precipitation is usually arranged to be as rapid as possible by fast stirring, the use of NH_3 gas for pH control, and increase of temperature sometimes to as much as 85°C, to minimize the formation of basic ammonium uranyl sulfates, which are undesirable in many products.

A bleed of ammonium sulfate is taken from the recycled aqueous stream after precipitation. Disposal of this bleed may present environmental problems because of the traces of uranium and solvent, as well as macro amounts of ammonia. Typically the volume flow rate of bleed is about 10% of the solvent flow rate.

3.5.3. *Alkaline Stripping*

Contact of the loaded solvent with sodium or ammonium carbonate solutions brings about total hydrolysis of the amine salts in the organic phase. The soluble uranyl carbonate complexes are formed in the aqueous phase.

Typically the aqueous phase contains 80–120

g Na_2CO_3/liter, and the flow rate is adjusted to maintain a pH >7.5 in the strip circuit. The flow ratio is of the order 1 A:20 O or higher, depending on the amine content of the organic phase.

Three or four stages are sufficient to provide adequate stripping. In each stage the mixer residence time is 1 min or less and the installed power, about 2 kW/m^3. When fresh solvent is being treated, traces of organic acids may accumulate at the interface during contact with the alkaline solution and interfere with phase separation. With extended operation, this problem disappears.

Uranium is recovered from the strip liquor by precipitation as ADU or SDU, and the strip solution is regenerated by addition of fresh NaOH or NH_4OH and recarbonation. Molybdenum is not precipitated with the uranium and can be recovered from the recycle solution by, for instance, extraction with quaternary alkyl ammonium salts or by acidification.

The recycled solvent is in the free-base form, of course, and care should be taken to ensure that there is sufficient free acid in the feed to resulfate the solvent without raising the pH level unduly.

3.6. Regeneration

A regeneration stage is often used to clean a portion of the recycle solvent following chloride or neutral stripping in order to prevent complexes not stripped along with uranium from building up in the solvent. This can be done periodically, but most large-scale operators prefer continuous regeneration. A portion of the stream is contacted with sodium carbonate or ammonium carbonate as in alkaline stripping. If this is done after neutral stripping, the amine is already largely in the free-base form, and the flow rate of alkaline solution is thus very low. On many plants the whole of the aqueous phase is recycled for several days, and then a portion is replaced. The wash solution may be returned to the leaching section to recover traces of uranium present.

3.7. Amine Extraction Circuits [4, 5]

Examples of circuits using amine solvent extraction to recover uranium are given in Figs. 3 and 4. There is little to choose between the various circuits on an operational basis, and costs are not widely different. Table 4 compares reagent consumptions on two of these circuits. Note that, because the Kerr–McGee ore is far richer, consumptions of reagents on a per ton basis appear far higher, but on a per kilogram U_3O_8 basis they appear far lower than the Buffelsfontein consumptions.

4. ALKYL PHOSPHORIC ACID EXTRACTION OF URANIUM FROM SULFATE SOLUTIONS (DAPEX) [7]

4.1. Pretreatment of Leach Solution

The demand for clarification of leach solutions is not as extreme in the acid system as it is in the amine system. Concentrations of suspended solids of the order of 100 ppm are perfectly tolerable. However, the acids are less selective than the amines. Tetravalent vanadium, molybdenum, rare earths, Fe^{3+}, Ti^{4+} and Th^{4+} are all strongly extracted. Reduction with the use of scrap iron, Na_2S, NaHS, or SO_2 is a necessary pretreatment, in particular to overcome Fe^{3+} extraction. However, V^{5+} is usually reduced to the extractable V^{4+} state. An electromotive force (EMF) of about −300 mV is the usual target before extraction. Some precipitates may form during reduction, and it may be necessary to clean reductant beds carefully from time to time to prevent large quantities of these precipitates from entering the solvent extraction circuit.

4.2. Choice of Solvent

4.2.1. Phosphate

A variety of substituted phosphoric acids, phosphonates, phosphinates, and phosphine oxides have been tested at various times, but the only solvent to find commercial use in this application has been di (2-ethyl hexyl) phosphoric acid (D2EHPA).

4.2.2. Diluent

Remarks that may be made here about the diluent are very similar to those made in Section 3.2.2, and, again, the choice is determined more by operational than by chemical parameters.

4.2.3. Modifiers

A prime role of the modifier is to prevent third-phase formation during stripping, because in stripping under alkaline conditions the alkali salt that forms has a limited solubility in most diluents. The usual choice of modifier is tri-*n*-

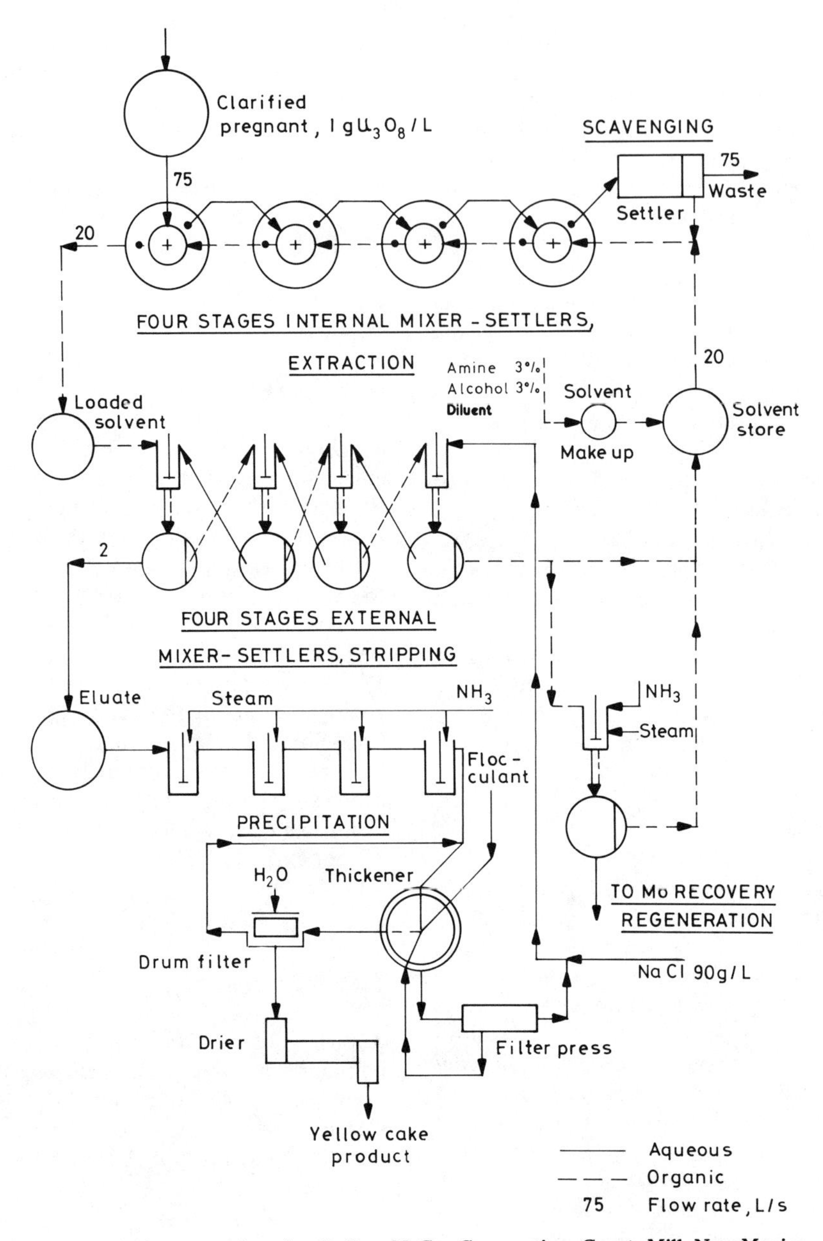

Figure 3. Amine extraction circuit, Kerr-McGee Corporation, Grants Mill, New Mexico.

butyl phosphate (TBP), the use of which has the advantage that there is a synergistic effect during extraction. Possible modifiers such as tri-*n*-octyl phosphine oxide (TOPO) show an even greater synergistic effect but have not found commercial use in this application, partly on grounds of higher cost and partly because the peak in their synergistic effect occurs at too low a concentration for them to act effectively as modifiers in the true sense.

4.2.4. Choice of Concentrations

The distribution ratio for uranium(VI) increases as the second power of the D2EHPA concentration. However, the distribution ratios for a number of possible contaminants of concern, such as Fe(III), V(IV) and Al(III), rise as fast or faster. Above 10% D2EHPA (approximately 0.3 *M*), significant extraction of these contaminants may take place. The concentration of

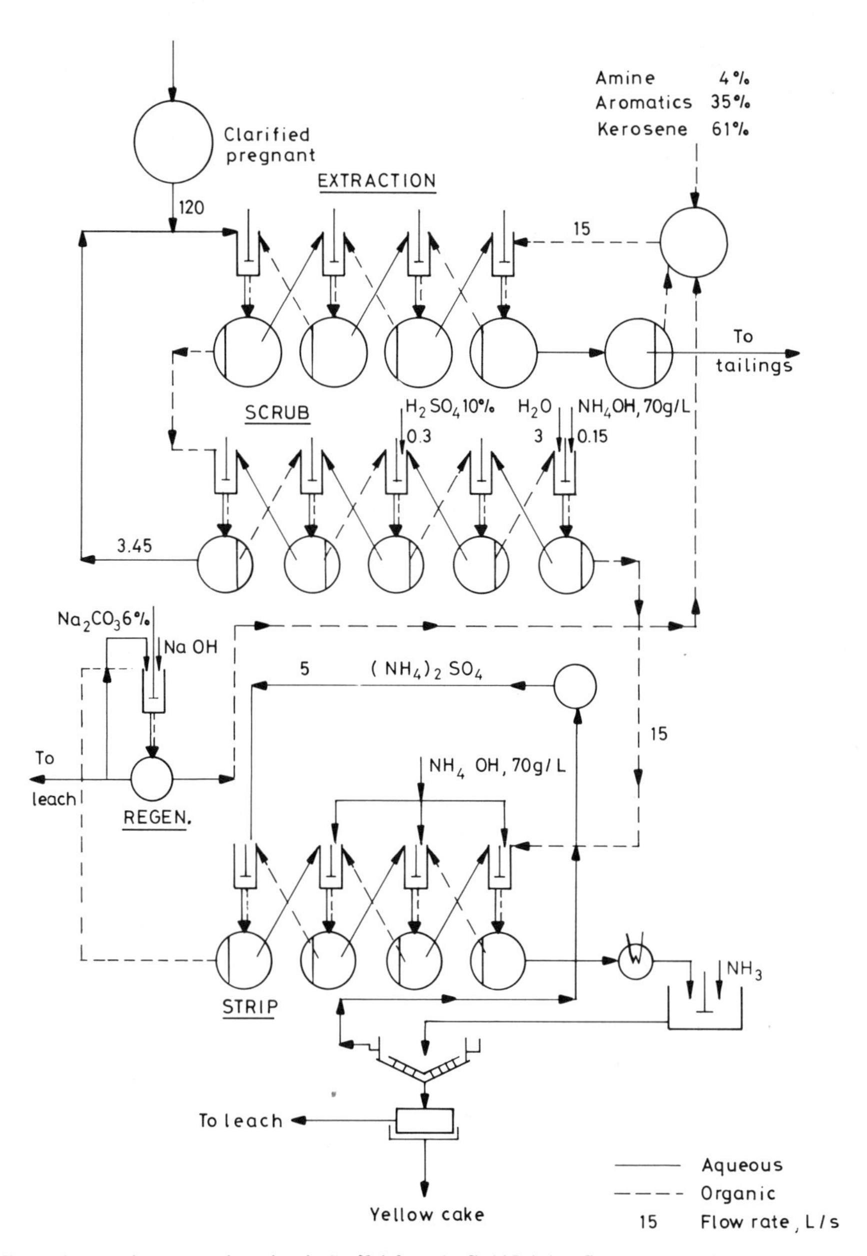

Figure 4. Amine extraction circuit, Buffelsfontein Gold Mining Company, Stilfontein, Transvaal.

D2EHPA employed is thus usually in the range 4–10%.

The modifier prevents third-phase formation when close to equimolar. The TBP concentration is thus typically in the range 3–8%.

4.3. Choice of Extraction Conditions

In the pH range of about 0–2, the log (distribution ratio) for U(VI) rises with the first power of the pH, but those of Fe(III) and V(IV) rise as the second power. It is thus advantageous to maintain the solution acidity such that pH < 1, unless these contaminants are present at low concentration.

As the D2EHPA extracts U(VI) as the UO_2^{2+} cation directly, the formation of uranylsulfate complexes will reduce the extraction as the sulfate concentration increases. Similarly, it is desirable to keep the concentration of other anions that form strong complexes with the uranyl

TABLE 4 COMPARISON OF REAGENT CONSUMPTIONS OF VARIOUS AMINE SOLVENT EXTRACTION CIRCUITS

	Consumption			
	Kerr–McGee		Buffelsfontein	
Reagent	kg/ton feed	kg/kg U_3O_8	kg/ton feed	kg/kg U_3O_8
Amine	0.014	0.007	0.010	0.032
Modifier	0.032	0.016	0.015	0.048
Kerosene	0.38	0.19	0.19	0.60
Ammonia	0.63	0.315	0.14	0.44
Flocculant	0.02	0.01	–	–
Filter aid	0.09	0.045	–	–
NaCl	3.2	1.6	–	–
NaOH	–	–	0.026	0.083
Na_2CO_3	–	–	0.018	0.057

cation, such as chloride and phosphate, as low as possible.

The effect of increasing the uranium concentration in the organic phase is not as simply correlated as in the case of the amine system (see Section 3.3.2). This is partly because of the very complex chemistry of the synergistic system and partly because of the possible extraction of contaminating cations. However, very often it is found that, to a reasonable approximation, much of the isotherm can be linearized by setting

$$D = k_3 (C - k_4 U_0)^2$$

As previously (Section 3.3.4), it is often desirable to arrange a recycle of the organic phase within each stage to ensure organic-continuous emulsions. Similar precautions should be taken to disperse the aqueous phase into the emulsion.

Unlike the amine system, the D2EHPA system is a relatively slow extractant of uranium from sulfate solution. Residence times in mixers of at least two min are necessary to ensure a reasonable approach to equilibrium, and the power input required is accordingly increased to at least 6 kW/m^3.

4.4. Scrubbing

Advantage may be taken of the rapid drop in the distribution ratio of vanadium with drop in pH to scrub any vanadium that has been extracted. Typically 5–10% sulfuric acid is used at 1 aq.: 20 org.

4.5. Stripping

Either strong acids, particularly those that have anions that complex uranium strongly, such as HCl, or alkali carbonates may be used to strip uranium. Sodium carbonate is preferred on economic grounds. Any iron or titanium present precipitates, and the strip solution must thus be filtered before uranium precipitation. The sodium carbonate concentration in the strip solution is typically 10–15%. The stripping reaction is

$$\overline{((RO)_2PO_2)_2UO_2} + Na_2CO_3 \rightleftharpoons \overline{2(RO)_2PO_2Na} + UO_2(CO_3)_2$$

with the uranium solubilized in the aqueous phase as the tricarbonato complex by the presence of excess carbonate. The sodium carried to the extraction section will cause an increase in pH that must be taken into account there. Phase ratios in stripping are typically about 1 A:10 O. Stripping is rather slow and requires a retention time of the order of 3 min with a power input of about 6 kW/m^3.

There are a variety of ways for recovering uranium from the strip solution after filtration. Acidification to destroy the carbonate followed by precipitation of the uranium as yellow cake with alkalis, alkaline earth, or ammonia is the usual route. Neutralisation followed by precipitation as uranyl peroxide by addition of H_2O_2 has been employed. This has the advantage of not precipitating vanadium or molybdenum with the uranium to any significant extent.

4.6. D2EHPA Extraction Circuits

As in the case of the amine system, there are a number of minor variants on individual circuits. Figure 5 shows a typical circuit treating ordinary leach liquors, and Fig. 6 shows an Eluex circuit, where solvent extraction is preceded by ion exchange. In the latter case, a higher concentration of D2EHPA is necessary in the organic phase for effective extraction of uranium from stronger sulfuric acid solution than the typical leach liquor. This increases the phase ratio (A/O) in extraction.

In Table 5 typical reagent consumptions are given for Dapex processes treating high-grade solutions.

5. SOLVENT EXTRACTION OF URANIUM FROM PHOSPHORIC ACID SOLUTIONS

5.1. Pretreatment of Acid Feed [2, 8]

Where the ore has been roasted before leaching to extract the phosphoric acid (and uranium), the levels of dissolved and suspended organic matter in the solution are low, and the solution is referred to as "green acid." Where the ore has been leached without any pretreatment, the resultant solution is laden with organic matter and carbon and is known as "black acid." The organic content of the black acid gives rise to emulsions with many extractants, and humic acids present in the acid will contaminate the solvent

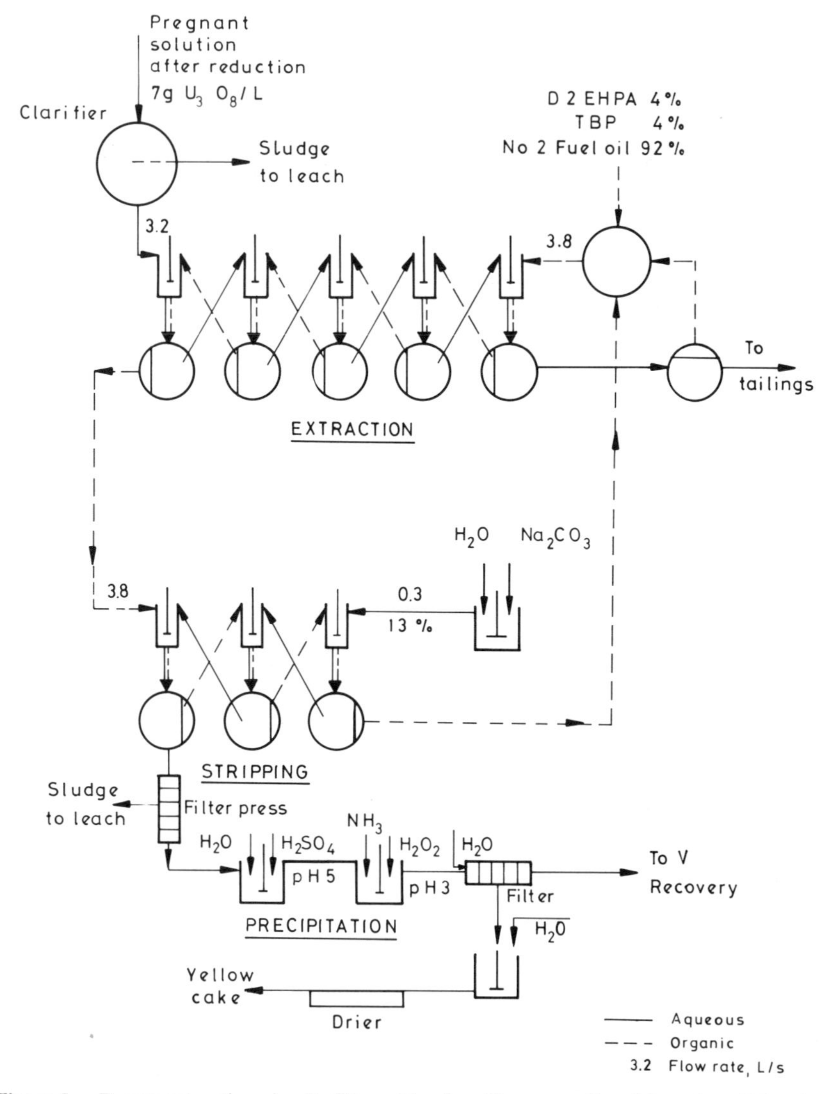

Figure 5. Dapex extraction circuit, Climax Uranium Company, Grand Junction, Colorado.

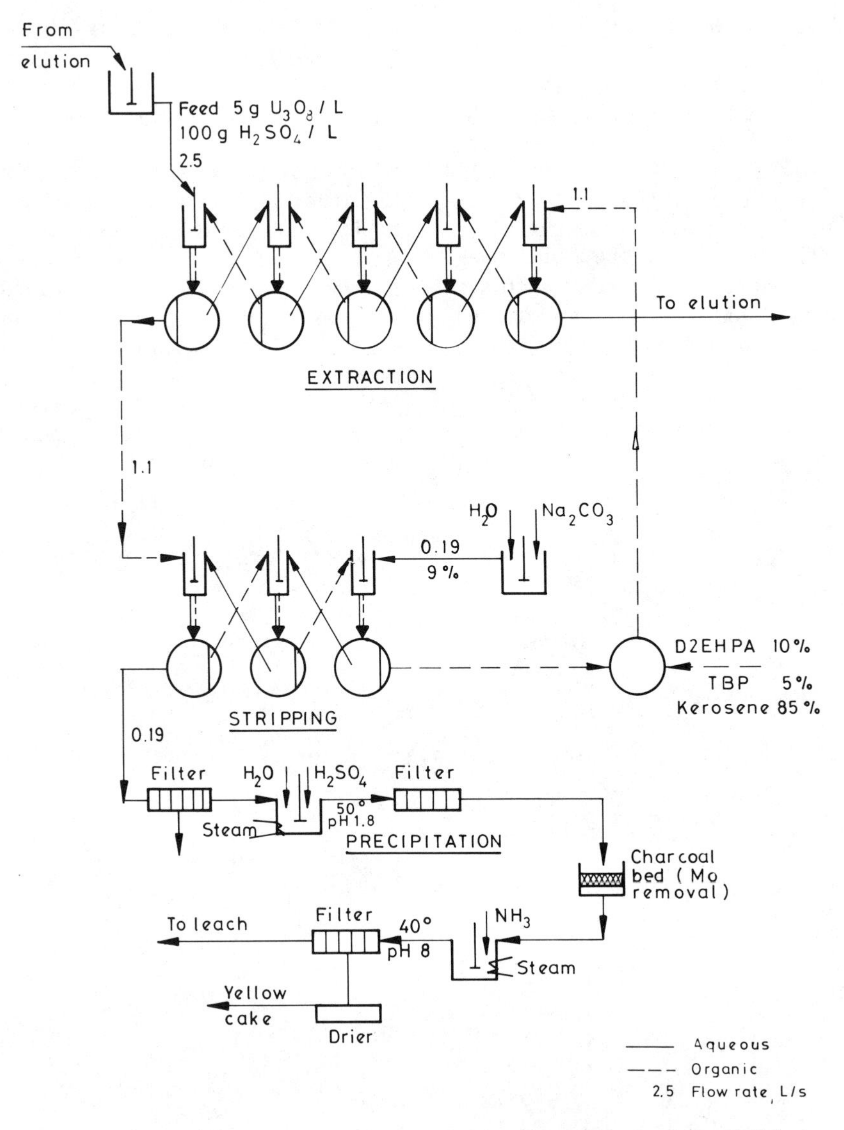

Figure 6. Eluex extraction circuit, Mines Development, Inc., Edgemont, South Dakota.

TABLE 5 APPROXIMATE REAGENT CONSUMPTIONS IN DAPEX CIRCUITS FOR 5 g U_3O_8/LITER FEED SOLUTIONS

Reagent	Consumption kg/kg U_3O_8
D2EHPA	0.012
TBP	0.025
Diluent	0.18
Na_2CO_3	1.3
H_2SO_4	2.2
NH_3	0.3

and may react with the extractant itself. Satisfactory recovery of uranium from this source thus requires removal of the organic matter before solvent extraction.

The usual methods of treatment involve wet oxidation by addition of chlorate, flocculation, and clarification, followed by passage of the acid through beds of activated carbon. Solvent extraction without the presence of an extractant for uranium has also been tested. The emulsion that forms is removed and treated separately for recovery of the solvent.

The uranium is usually present in the acid in the U^{4+} state. It may be necessary to oxidize it to the U^{6+} state (e.g., by addition of $NaClO_3$ or H_2O_2) if the extractant used is effective only for U^{6+}, or to reduce it fully (e.g., by addition of scrap iron) if the extractant is effective only for U^{4+}. Typically the EMF (vs. standard calomel) should be under 300 mV to ensure U^{4+} and over 420 mV to ensure U^{6+}. It may be noted that the presence of fluoride at about 1 *M* concentration in the acid catalyzes the $U^{4+} \rightleftharpoons U^{6+}$ equilibrium.

The final pretreatment necessary is usually to reduce the temperature, typically to less than 45°C. Lower temperatures would be far more conducive to efficient extraction and would be safer in that the solvent would be further from its flash point. However, the acid is usually saturated with $CaSO_4$ at about 60°C. It can be held reasonably satisfactorily in supersaturated solution at about 45°C, but at less than this level crystallizes out slowly. The precipitate not only causes great operational problems, but also removes some uranium as a coprecipitate.

5.2. Choice of Solvent System

Three solvent systems are known to be reasonably effective for extracting uranium from typical phosphoric acid process solutions containing over 30% P_2O_5. They are compared in Table 6. At present it is impossible to say which system is to be preferred on economic grounds.

Most of the common diluents are satisfactory in these systems. No modifier is necessary because conditions that might give rise to third phases are not present in either extraction or stripping. In the D2EHPA–TOPO system the optimum molar ratio appears to be about 4 (D2EHPA) per TOPO.

5.3. Choice of Extraction Conditions

There is little to be done to modify the aqueous-phase conditions other than what has been outlined in the previous section. The high concentration of phosphate and the high acidity are the inevitable consequences of superphosphate production, and considerable cost would be incurred if the aqueous phase were to be diluted to enhance uranium recovery by solvent extraction. This sort of one-way compromise is the inevitable result whenever one product is seen as the by-product of another, as in this case.

During the primary extraction phase the solvent is not highly loaded. This is a consequence of the low concentration of uranium in the original acid (typically $\leqslant$200 ppm) and the low distribution ratio for uranium. In consequence, the concentration of uranium in the organic phase varies linearly with that in the aqueous phase. The slope of this line, the distribution ratio, has maximum values of 10 for 0.5 *M* D2EHPA/ 0.12 *M* TOPO, 20 for 20% octylphenyl phosphoric acid, and 40 for 5% tetradecyl pyrophosphoric acid. The distribution ratio varies as about the 1.5 power of the D2EHPA concentration, at constant D2EHPA:TOPO ratio, and as approximately the second power of the other acid concentrations.

There are no published data on the kinetics

TABLE 6 COMPARISON OF EXTRACTANTS FOR RECOVERING URANIUM COMMERCIALLY FROM PHOSPHORIC ACID SOLUTIONS

Extractant	Advantages	Disadvantages
Tetraoctyl pyrophosphoric acid	Cheap extractant, high distribution ratio for uranium as U^{4+}, single cycle of operation	Extractant unstable (hydrolysis at –P–O–P–linkage), difficult to strip (15% HF)
Octyl phenyl orthophosphoric acid (mono and di)	Extracts uranium as U^{4+}, relatively cheap extractant, and relatively high distribution ratio	Difficult to strip (54% P_2O_5 solutions); some phase separation problems; little commercial experience to date
D2EHPA–TOPO	Easily stripped, relative to other extractants; stable, well-known solvent system	Requires oxidation of uranium to U^{6+} before extraction; relatively expensive extractant

of the extraction of uranium from phosphates in these systems, but it seems likely, by analogy with the sulfate system, that the rate of extraction will be fairly slow, and power inputs of the order of 6kW/m^3 with residence times of the order of 2 min will be necessary to achieve a reasonable approach to equilibrium.

5.4. Scrubbing

A scrub with 25% H_2SO_4 is used in the pyrophosphate system to remove extracted calcium. Scrubbing is also used to reduce the carry-over of phosphoric acid from the extractions stage to the final stage of stripping (see Section 5.6). The pyrophosphoric acid is further scrubbed with 25% H_2SO_4 after stripping, to remove HF before recycling. Finally, as the phosphoric acid in the raffinate finds industrial use, it is essential to recover all the entrained solvent from it. The raffinate thus not only passes through an after-settler, but also may be scrubbed in the air-flotation device before passing down a packed tower containing an adsorbent such as carbon for the solvent.

5.5. Stripping

In the pyrophosphoric acid system the uranium is removed in the tetravalent state and simultaneously precipitated as UF_4 by stripping with 25% sulfuric acid containing 15–20% HF.

The primary stripping process in the D2EHPA and phenyl phosphoric systems is reversal of the extraction step by a change of oxidation state of the uranium. For instance, in the phenyl phosphoric acid system, the uranium is stripped from the organic phase by concentrated (>50% P_2O_5) phosphoric acid containing an oxidant such as $NaClO_3$ to oxidize the uranium to the (inextractable) U^{6+} state. Similarly, the D2EHPA-TOPO systems are stripped by normal-strength H_3PO_4 containing enough Fe^{2+} to reduce the extracted uranium to the U^{4+} state. The stripping reaction is very slow because of this interphase oxidation–reduction. Mixer residence times should be of the order of 5 min and power inputs, of the order of 10 kW/m^3.

In the second cycle of purification (see Section 5.6.) the uranium can be stripped economically by sodium or ammonium carbonate and precipitated as the ammonium uranyl tricarbonate (AUT) salt, which forms a dense precipitate with large crystallites that thus settles readily and is easily filtered.

5.6. Circuits for Uranium Extraction from Phosphoric Acid

Because the initial concentration of uranium in the phosphoric acid is low (<200 mg/liter typically) and because the solvents offer low distribution ratios for uranium, the concentration that can be achieved in a single extraction–stripping cycle is low. Uranium concentrations sufficient for economic recovery from solution by precipitation are not available.

It is thus usually necessary to follow the first extraction–stripping cycle with a second similar cycle. The uranium in the strip solution from the first cycle is oxidized or reduced to make it extractable again. In the second cycle concentrations of the order of 20 g U_3O_8/liter can be reached at which point precipitation as AUT or ADU becomes practicable.

Figure 7 shows a design for a circuit employing D2EHPA–TOPO in both cycles. It has been suggested that there may be merit in the use of phenyl phosphoric acids in the first cycle and D2EHPA–TOPO in the second, but the risk of cross-contamination of the two cycles may be excessive for industrial use.

Figure 8 shows a circuit employed for the recovery of uranium by use of tetradecyl pyrophosphoric acid. In this case the initial UF_4 product is redissolved in nitric acid, AlF_3 is precipitated to remove most of the fluoride ions, and the uranium is reextracted by use of TBP, and finally precipitated as ADU [9].

Comparison of Fig. 7 or 8 and Fig. 3 indicates the greater complexity of winning by-product uranium from phosphate solutions. Nevertheless, this is cost-effective because the product does not have to bear any of the mining or milling costs. Moreover, there is a relative simplification whereby several "first-cycle" plants can be operated at different phosphate producers and the first-cycle strip solution can be piped or sent by tank car to a central plant where a relatively large-scale second-cycle plant operates. Where several phosphate producers are in close proximity, this obviously makes sense [10].

Purification of phosphoric acid is further considered in Chapter 26.

6. OVERALL CIRCUIT DESIGN

6.1. Number of Stages Needed

Most full-scale operations have flow rates that exceed those for the satisfactory operation of

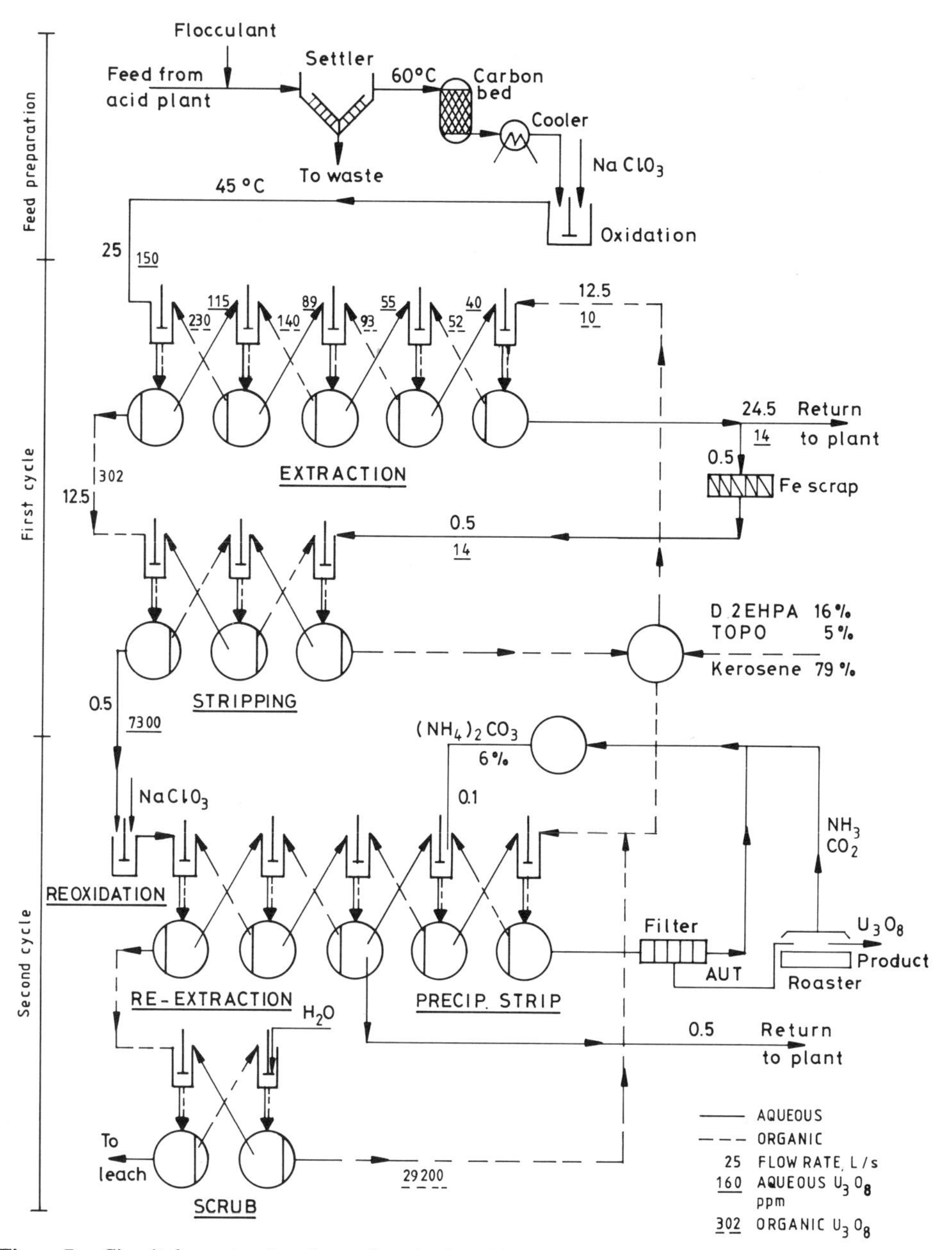

Figure 7. Circuit for extraction from phosphoric acid solutions by use of D2EHPA-TOPO mixtures, International Minerals Corporation, New Wales, Florida.

countercurrent towers, and mixer–settlers are usually employed.

The number of ideal stages required can be found by the methods given in Chapter 5. The stage efficiency will be a function primarily of mixing intensity and kinetics of extraction. The power inputs and residence times given previously should be sufficient to give a stage efficiency of the order of 90%.

6.2. Phase Ratio

The choice of phase ratio is fairly readily made from the known feed concentration and the desired recovery. In addition, a decision may be taken to load the organic phase as highly as possible, either to "squeeze out" coextracted impurities or, for example, to minimize the amount of acid transferred to the strip section.

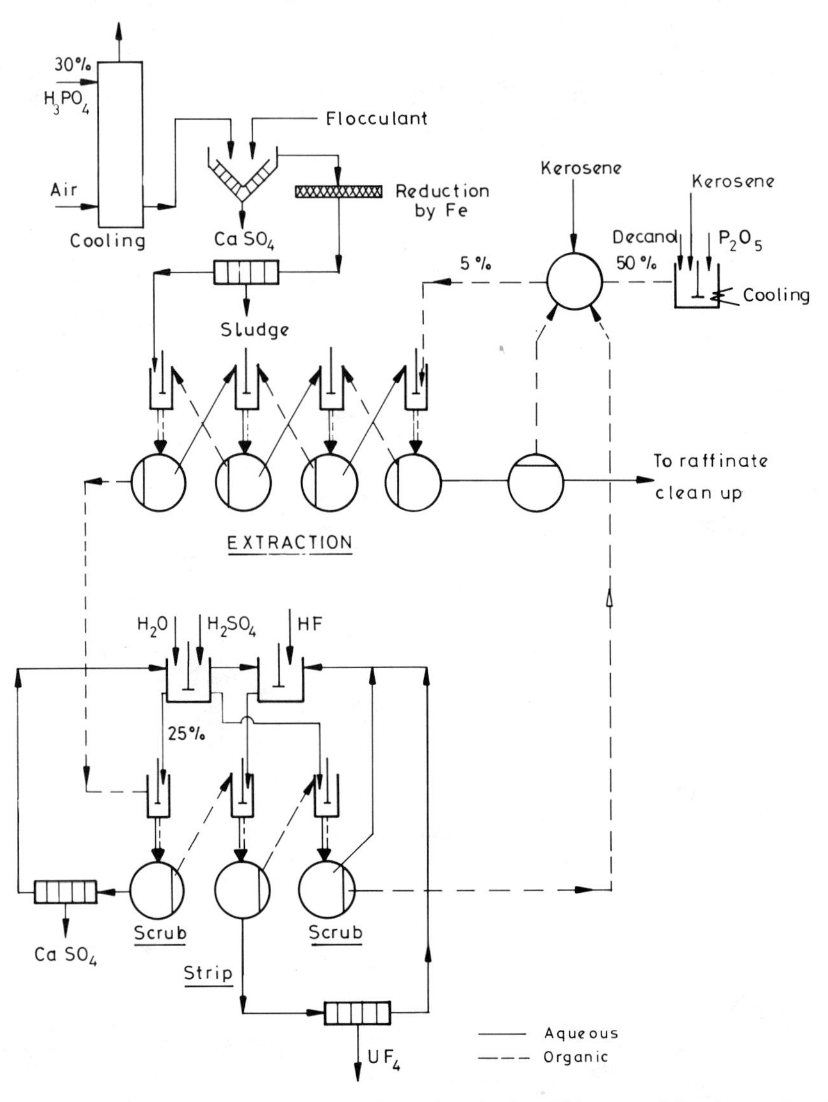

Figure 8. Circuit for extraction of uranium from phosphoric acid by use of decyl pyrophosphoric acid, Gardinier, Inc., Tampa, Florida.

It is not often recognized how sensitive the process becomes when driven hard to load the organic phase fully. For instance, on one plant the phase ratio was chosen so that the organic phase was loaded to within 16% of its maximum value. When the recycle solvent contained 0.007 g U_3O_8/liter, the final raffinate value was 4.9 mg U_3O_8/liter. When the recycle solvent increased to 0.020 g U_3O_8/liter, the raffinate value rose to about 0.040 g U_3O_8/liter over a period of 12 hrs. This may seem like a small increase but, on the plant in question, treating about 500 m^3/hr of solution, the loss of uranium during this test amounted to about 105 kg U_3O_8. Increasing the organic flow rate rapidly restored normal raffinate values. Figure 9 indicates how the total cost of an amine extraction circuit may vary with solvent loading (which is directly related to phase ratio), and indicates, as is found in many cases in practice, that the optimal loading is in fact over a fairly narrow range.

7. ENGINEERING ASPECTS

7.1. Type of Equipment

In the extraction section the large flows usually influence the choice of equipment in the direction of a cascade of mixer–settlers. Methods for

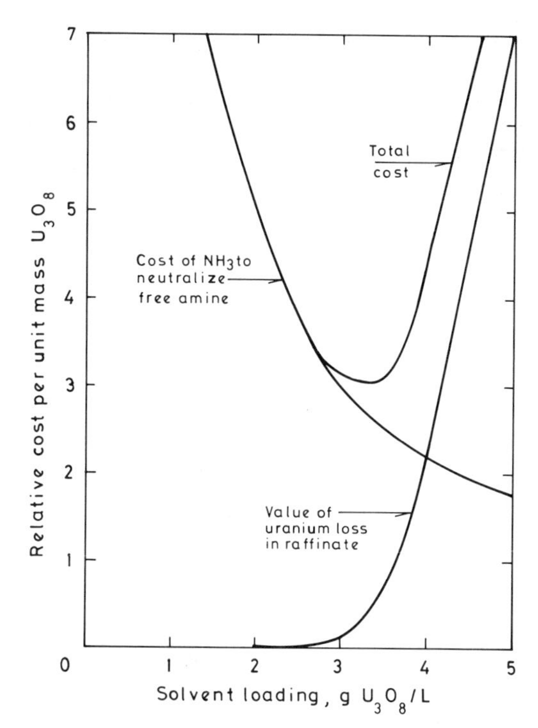

Figure 9. Variation in relative cost of stripping as a function of loading of uranium in an amine solvent.

the design of these are covered in Chapter 9. In scrub and strip sections a wider choice is possible. Again, however, for operational reasons, mixer–settlers are usually chosen. One stage can be readily bypassed for maintenance if required, with little loss in production. The handling of "crud" by means of interface drains is relatively simple. The only real drawback is the solvent inventory, and although this can be a nuisance or even a hazard, its total cost is not excessive. A variety of columns and even centrifugal contactors have been used from time to time, but their use is not widespread.

7.2. Materials of Construction

In the sulfate system Type 316 stainless steel is satisfactory for most purposes where metals are essential. The combination of dissolved chlorides and low oxygen availability must be avoided if pitting corrosion is to be prevented.

Glass-reinforced polyester is an excellent constructional material in these plants [11]. Particularly when used as a liner within concrete structures, it offers reasonable fire resistance. Saran linings have also been used successfully in phosphoric acid plants, where the risk of polyester failures is significant.

Structural steel is difficult to protect in the plant environment, and it must be protected carefully if it is to have a reasonable life. A high build coating in several layers is desirable, with excellent adhesion to the underlying metal. The coating should be readily patched to permit repair. Long-term resistance to organics is not normally necessary, except at the floor level.

Piping in unplasticized PVC is generally satisfactory, provided it is properly supported. It is beneficial to protect it from light and extremes of temperature as far as possible.

Gasketing of Viton is usually satisfactory. Fluorocarbons are satisfactory as a material of ultimate resort but need careful use under plant conditions.

In the phosphate systems there are virtually no structural metals of any use. The principal constructional material is glass-reinforced polyester, in which care must be taken to seal the glass reinforcement (because of the presence of F^- in high concentrations in the aqueous phase). Pump and similar parts may be cast from high-Cr alloys such as Alloy 20, but, again, organics are to be preferred.

7.3. Safety

7.3.1. Fire [12]

As two uranium plants have thus far been destroyed by fire, the designer should be aware of the potential problem and take appropriate action to minimize loss if fire should occur.

Primary parameters in reducing the fire risk are:

1. Selection of a diluent with a high flash point (noting that flash points decrease as altitude above sea level increases).
2. Reduction of the temperature of extraction as far as is practical.
3. Reducing potential sources of ignition to the minimum (possible sources often overlooked include lightning and static electricity that can build up when liquids of low conductivity, such as diluents, are moved).
4. Preventing accumulations of flammable vapor as far as possible.
5. Avoiding spillage of solvent as far as possible (and this includes spillage under conditions of fire).

Among other precautions, the designer should attempt:

1. To install pipe runs that are well fire-protected and that will not lead to excessive loss of organic if they do fail.
2. To isolate all major pieces of equipment containing organics as far as possible.
3. To provide secondary containment for all large volumes of organics.
4. To arrange for access to fire-fighting equipment.
5. To keep air spaces above settlers well ventilated and cool.
6. To clear and handle all spillage away from the major vessels.

Fire-fighting arrangements should be designed primarily to cool and maintain the integrity of vessels. Water sprays over open areas of solvent appear to be the best solution. Adequate freeboard must be provided for the water used for fire fighting to escape without spilling the organic phase. Further discussion of fire prevention is given in Chapter 31.

7.3.2. *Environment*

The major environmental hazard is that posed by the release of the extractants to streams or rivers. The amines in particular, but also the phosphates and the diluents, are known to be toxic to bacteria, algae, and fish at concentrations readily found in raffinates [13].

Fortunately, the more toxic substances are also fairly surface active and adsorb well on surfaces. It is thus possible to immobilize these reagents by adsorption. Neutralization of raffinates with lime will, in the sulfate system, usually generate sufficient surface to scavenge the extractants effectively. Alternatively, if the raffinates are used to repulp leach residues to permit the residues to be pumped to slimes dams, the extractants will adsorb on the mineral surfaces. The long-term fate of such adsorbed residues is not known, however.

7.3.3. *Radiation Safety*

Because the uranium will carry its daughter products with it through much of the process for the recovery of the uranium, some note should be taken of possible radiation hazards.

To date no undue concentration of activity has been noted at any section of the solvent extraction system anywhere in the world. The preparation of the final concentrate is not without its hazards, and care should be taken in this area to ensure cleanliness and high standards of hygiene, as well as restricting access to these areas. In the main solvent extraction plant, however, radiation hazards are low. Regular urine analysis should be employed to check that ingestion of uranium is not excessive.

REFERENCES

1. Anonymous, *World Mining* **32** (2), 48–51 (1979).
2. B. F. Greek, O. W. Allen, and D. E. Tynan, *Ind. Eng. Chem.* **49** (4), 628–638 (1957).
3. T. H. Tunley and V. W. Nel, *Proceedings of the International Solvent Extraction Conference (ISEC), Lyons 1974,* Society of Chemical Industry London, 1974, pp. 1519–1533.
4. R. C. Merritt, *The Extractive Metallurgy of Uranium,* Colorado School of Mines Research Institute, 1971.
5. B. G. Meyburgh, *J. South Afr. Inst. Min. Met.* **71**(3), 55–66 (1970).
6. D. J. Crouse, U.S. Atomic Energy Commission Report ORNL-2941, Oak Ridge National Laboratory, TN, 1960.
7. C. A. Blake *et al., Ind. Eng. Chem.* **50**(12), 1763–1767 (1958).
8. F. J. Hurst, *Transact. Soc. Min. Eng. AIME* **262**(3), 240–248 (1977).
9. A. P. Kouloheris, *Chem. Eng.* 82–84 (August 11, 1980).
10. R. C. Ross, *Eng. Min. J.* **176**(12), 80–85 (1975).
11. A. Walz, *Corrosion* (Rueil Malmaison, Fr.) **20** (2), 155–158 (1972).
12. G. Collins, J. H. Cooper, and M. R. Brandy, *Eng. Min. J.* **179**(12), 58–64 (1978).
13. G. Dave, H. Blanck, and K. Gustaffson, *J. Chem. Tech. Biotechnol.* **29**, 249–257 (1979).

25.12

RECOVERY OF URANIUM AND PLUTONIUM FROM IRRADIATED NUCLEAR FUEL

A. Naylor and P. D. Wilson
British Nuclear Fuels Ltd.
United Kingdom

1. INTRODUCTION

Nuclear power has been in use for one purpose or another for over three decades [1, 2]. For most of that time it has been considered essential to reprocess the fuel in order to recover valuable materials, initially for military purposes and more recently to make the best use of finite energy resources [3] and to facilitate fission product waste management [4]. Solvent extraction quickly established itself as the dominant chemical process, and the purpose of this chapter is to describe the various forms it takes in the recovery of uranium and plutonium from irradiated fuel.

The nuclear industry is at present based almost exclusively on the thermal fission of uranium; a ^{235}U atom, in nature one in about 140 of the total, is split by a slow neutron to yield energy, fission products, and several more neutrons of which one—slowed by passage through a moderator—causes a further fission. Other neutrons are absorbed by structural materials and other absorbers, by shielding, or by ^{238}U that is then converted by successive nuclear reactions to plutonium. Part of this itself undergoes fission before the net consumption of fissile material, the accumulation of neutron-absorbing fission products, and the structural deterioration of the fuel element require it to be discharged. Most of the uranium remains unchanged (although with a reduced proportion of the fissile isotope ^{235}U) and can be purified for further use; the plutonium is also worth separating for use either as a substitute for ^{235}U in thermal reactors or more effectively in fast neutron reactors as an intermediate in the net conversion of relatively abundant but nonfissile ^{238}U to energy and fission products [5].

There is an alternative cycle based on ^{232}Th and ^{233}U, but it has not been widely adopted [6], and, although it may become more prominent, particularly if supplies of uranium approach exhaustion, it is not considered here. Neither is much attention paid to the highly enriched uranium (often alloy) fuel of research or material testing reactors. Instead, this chapter concentrates on those fuels that have provided, or are planned to provide in the future, nuclear energy on a large scale; these are (1) uranium metal, of natural isotopic composition, as used in the first commercial generation of British and French power stations, (2) uranium oxide, with the proportion of fissile ^{235}U generally enriched to a few percent, from various other types of thermal reactor, and (3) mixed uranium and plutonium oxides from fast reactors.

2. FEATURES OF FUEL REPROCESSING

The reprocessing of nuclear fuel is unusual among industrial activities in the number and stringency of requirements to be met:

1. Recovery of uranium and plutonium must be high, usually about 99%, not only because of the intrinsic value of these materials, but because their presence in raffinate streams exacerbates the problems of waste management [7].
2. They are initially accompanied by a range of fission products, from germanium to the middle rare earth elements, representing every group of the periodic table [8, 9]; some of these are gaseous or volatile, and most include unstable isotopes emitting ionizing radiation. This radiation is eventually degraded to a considerable heat output.
3. Present techniques of refabrication require that uranium and plutonium be decontaminated from fission products by factors in the region 10^6–10^8 [7].
4. Technical reliability is essential since repair or replacement of equipment is difficult and expensive once the plant has been committed to active use.
5. Because of the intense radioactivity, the earlier stages, at least, must be operated remotely with minimal maintenance.
6. Besides the usual demands of industrial safety, radiation doses to the work force and the public must be kept as low as reasonably practicable and must not in any case exceed rather restrictive limits [10], while the fissile material handled must not be allowed to adopt a configuration in which a self-sustaining nuclear chain reaction ("criticality") might occur [11] (see also Chapter 31).
7. Any waste streams, not sufficiently innocuous to be discharged immediately to the environment under current or foreseeable regulations, must be in a form suitable for storage either for the indefinite future or until the harmful constituents have decayed to insignificant levels.

3. BASIC PROCESS

3.1. General Outline

Solvent extraction has emerged as the favored separative process, essentially because of the fortunate coincidence that uranium and plutonium can both be extracted readily from nitric acid at convenient concentrations, whereas most fission products are not [1, 12]. Furthermore, the cleanness of separation can easily be improved by concatenation of stages in fairly simple equipment amenable to remote control and continuous operation. This last point is important in allowing plant of moderate size to attain the required throughput, although it does require special conditions at the start and end of a campaign since a steady state takes some time to reach.

It must be emphasized, however, that this chemical separation is only one of the processes to which irradiated fuel is subjected. Postirradiation storage, transport, disassembly, dissolution in nitric acid, and feed clarification are important and elaborate operations in their own right, as are product finishing and waste treatment processes. Nevertheless, they are outside the scope of this chapter, which is concerned only with the conversion of uranium, plutonium, and fission products from a mixed aqueous solution into separate aqueous streams.

The feed solution generally contains uranium at a concentration in the region of 1 M in about 3 M nitric acid. Levels of other constituents vary according to the nature of the fuel, with concentrations of inbred plutonium and of fission products increasing with irradiation, as shown in Table 1. The essence of the process is quite sim-

TABLE 1 FEED COMPOSITIONS FROM VARIOUS REACTOR FUELS [13-15]

Fuel Type	Magnox	Thermal Oxide	Fast Reactor
Irradiation, MW day/metric ton	3500	37,000	52,000 (mean)
Cooling time, months	6	12–60	6
Uranium molarity	1.25	1.25	1
Percentage of ^{235}U (approx.)	0.3	1–2	<0.2
Plutonium molarity (approx.)	0.002	0.01	0.16
Zirconium molarity $\times 10^4$	13	130	124
Ruthenium molarity $\times 10^4$	6.5	71	125
Cesium molarity $\times 10^4$	6.4	62	125
Strontium molarity $\times 10^4$	3.3	32	24
Iodine molarity $\times 10^4$	0.63	6.7	15
Total fission products Ci/liter	100	150–820	ca. 3000
Neptunium molarity $\times 10^4$	0.22	5.8	1.5
Americium molarity $\times 10^4$	0.12	2.9	17
Curium molarity $\times 10^4$	0.0014	0.66	0.53

ple, as shown in Fig. 1 [16]. In practice, several cycles of extraction and backwashing (stripping) are needed for adequate purification of the products.

The solution of fuel in nitric acid (the least corrosive to stainless steel of the common mineral acids) is extracted with a suitable organic solvent, which removes over 99.9% of the uranium and plutonium, leaving behind usually more than 99.9% of the fission products, and so on, which are concentrated for indefinite storage. The small proportion extracted by the solvent consists chiefly of ruthenium, zirconium, niobium, cerium, and iodine, together with some transuranic elements, in particular neptunium. The extracted fission products are among those with relatively high fission yields and are sufficiently long-lived to require subsequent removal from the uranium and plutonium streams. Cerium-144 is of minor importance, as a large decontamination factor (10^4–10^5) is achieved in the first extraction, while the ^{131}I isotope (with a half-life of 8.05 days) is less significant than others provided that the fuel has been stored (or "cooled") for many months. Although the ^{95}Zr and ^{95}Nb isotopes are relatively short-lived, with half-lives of 65 and 35 days, respectively, they are present in sufficient quantities to affect uranium and plutonium product qualities unless the fuel has been cooled for several years before reprocessing. Ruthenium-106 (half-life 1 year), ^{129}I (half-life 1.6×10^7 years), and ^{237}Np (half-

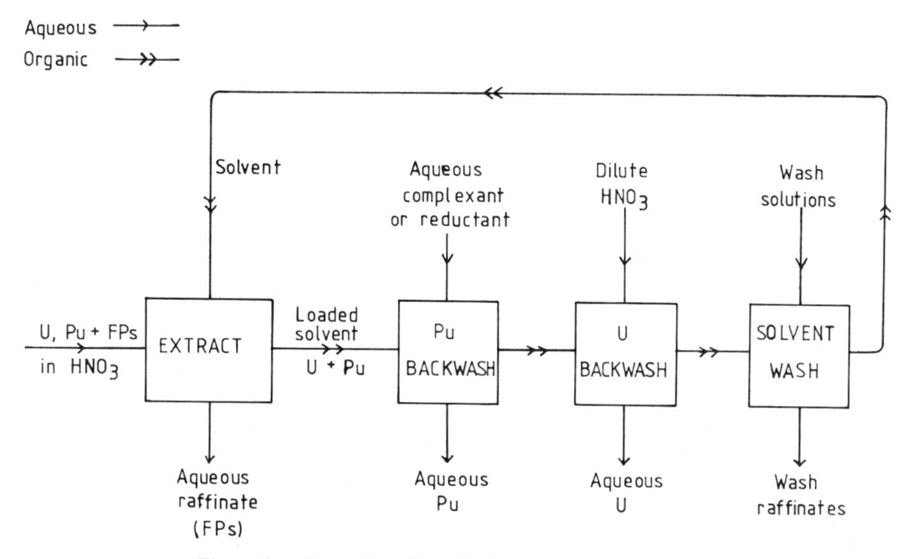

Figure 1. General outline of solvent extraction process.

life 2.1 × 10^6 years) are the long-lived isotopes with sufficient effects on the reprocessing cycles, product qualities, and environmental discharge limitations to warrant particular attention during solvent extraction. The problem with ^{129}I is confined mainly to the first solvent extraction cycle and the waste streams arising from it; a good proportion of the iodine is frequently removed during or immediately after the dissolution stage, so that relatively small quantities come into contact with the organic solvent, while the fraction that does remain in the dissolver feed solution is usually extracted into the first-cycle organic solvent and retained in it until solvent purification. Ruthenium-106 and ^{237}Np are the isotopes that persist throughout the series of solvent extraction cycles and require specific reagents and flow sheet conditions for their removal.

In theory, the difference in extraction behavior between uranium and plutonium might be sufficient to permit separation without further complications, but in practice it would require a tighter control over process conditions than it is convenient to guarantee, and this difference is invariably supplemented by some more positive separating technique [17]. The most common is to reduce plutonium from the extractable tetra- or hexavalent state to the virtually inextractable trivalent form, which passes out with the reducing aqueous stream while the uranium remains in the solvent. From this it is backwashed separately later on, leaving the solvent to be purified and recycled. The uranium and plutonium streams are then purified further as required.

Many features of a process affect its performance, but two essential requirements are a suitable solvent and appropriate contacting equipment. The historical development of these choices is thus discussed as part of the basic process. This is illustrated by a specific example, the Magnox reprocessing plant at Windscale, United Kingdom, and variants are considered later.

3.2. Choice of Solvent

The solvent must meet many requirements, most common to all comparable industrial separative processes, but a few specific to the nuclear industry, as follow:

1. It must selectively extract the valuable elements (uranium and plutonium) under easily accessible conditions, leaving the impurities behind.
2. The distribution ratio and attainable solvent loading must be sufficiently high for complete recovery of uranium and plutonium with moderate solvent: aqueous flow ratios, but back extraction must be possible without too great a change in conditions.
3. The solvent must be stable against chemical or, more particularly, radiolytic attack, and any degradation that may nevertheless occur should not interfere significantly with the process before purification.
4. Density, viscosity, and surface tension must allow ready mixing with the aqueous phase and rapid disengagement afterward.
5. Volatility and solubility in water should be low to prevent both undue losses and possible side reactions in the aqueous or vapor phases.
6. The flash point should be well above any temperature likely to arise by design or accident. This is particularly important in that a fire, besides the customary hazards, might create difficulties in controlling the spread of radioactive material.
7. The solvent should not be particularly toxic. This is perhaps less important than in processes subject to less rigid control, but there would be no sense in adding unnecessarily to handling problems.
8. The solvent should be conveniently available and not too expensive in the quantities required, generally tens to hundreds of tons for a full-scale modern plant.

The first solvent to be used for large-scale reprocessing was Hexone, or methyl isobutyl ketone (Hanford, USA, 1951–1960) [18], which was unsatisfactory as it had an aqueous solubility of 20 g/liter with inconvenient degradation products, reacted vigorously with nitric acid at concentrations greater than 3 *M*, and required 1–2 *M* aluminum nitrate to salt uranium and plutonium out of less acid aqueous phases. Because of this salt, the fission product solution remaining after extraction could not be evaporated to small bulk, and the resulting volumes of waste have yet to be equaled by subsequent reprocessing activities throughout the western world.

The triple ether Butex, 2,2′-dibutoxy diethylether, was used at Windscale, United Kingdom, from 1952 to 1964 for Magnox fuel [19] and on a limited scale for oxide fuel [20], during 1969–1973. This solvent was expensive and inconveniently soluble in and reactive with aqueous nitric acid but had the great advantage of needing no extra salting agent. Various long-chain amines have a similar virtue, but backwashing can then be difficult, and there are some problems of stability to radiation [21]. Tri-*n*-butyl phosphate (TBP) is cheaper and more stable and gives better separations than does either Butex or Hexone and, at least in the first cycles, is now universally used in a system generally known as the *Purex process* [22]. It is diluted with a hydrocarbon that reduces the density, viscosity, and extractive power to convenient levels: the proportion of TBP in the mixture is usually 5–30% by volume.

The first large-scale reprocessing plants that used TBP were in the United States at Savannah River in 1954 and Hanford in 1956 [23, 24]. In Britain it was first used in 1956 for recovering plutonium and enriched uranium and then in 1960 for uranium purification cycles [25]. In the Second Separation Plant, which started operating in 1964 at Windscale, United Kingdom, the extracting solvent throughout the process is 20% TBP in odorless kerosene (OK) [26]. At Marcoule, France, 40% TBP was initially used in the highly active cycle [27], but more recently 30% TBP has been used throughout the process [28]. The diluent in both instances was hydrogenated propylene tetramer (HPT). At La Hague, which started in 1967, the solvent is 30% TBP–HPT [29].

At Dounreay, United Kingdom, 6–30% TBP–OK has been used in the reprocessing of fast reactor fuels since 1962 [30]. In summary, TBP in some diluent is now adopted at least as first cycle solvent in all fuel reprocessing plants.

3.3. Choice of Contactor

The need for remote routine operation and the difficulties of maintenance in case of breakdown mean that equipment should be as simple as is compatible with other requirements, with as few moving components as possible in contact with radioactive materials. Multistage operation is necessary to achieve the required separations, and transfer between the stages should, if possible, be automatic—for instance, under gravity, supplemented where necessary by mechanically simple devices such as air lifts.

The first solvent extraction installations at Windscale used packed columns [19], which could hardly be simpler but suffered from inconvenient height and restricted throughput per unit cross section, besides a risk of channeling if the packing should orientate itself in a regular pattern. The second separating plant at Windscale uses banks of mixer–settlers [L. Lowes (Chapter 9.2) and Ref. 26] which take up more floor space but only a single story in height. They require stirrer units, but these are driven through seals by motors mounted outside the shielding and thus accessible for maintenance. One great virtue is that each physical stage is an almost ideal equilibrium stage, so that it is possible to develop process conditions and test possible variations on a very small scale, with quantities of fissile material and radioactive fission products manageable without undue difficulty in suitable laboratory conditions, in confidence that the results will be directly applicable to the plant-sized equipment [31]. This is particularly valuable in investigations regarding which variations in process conditions can be tolerated without risking, for instance, a criticality incident [26].

The avoidance of such an incident in a plant of adequate dimensions to handle several metric tons of fuel per day depends largely on concentration limitation, and although this can be assured with the relatively low plutonium and ^{235}U content of Magnox fuel, the margins would become smaller with light-water-reactor (LWR) fuels, where these proportions would be up to five times higher (Table 1). Furthermore, the separation of phases after mixing at each stage implies a long residence time, of the order of an hour in, say, a twenty-stage bank. Apart from requiring long run-up and run-down times, this is no great disadvantage in the processing of Magnox fuel, but with fission product concentrations up to 10 times higher in LWR processing solutions, radiolytic degradation of the solvent would become troublesome during such exposure times [32].

Accordingly, for these fuels, contactors must be designed with favorable geometry (or with fixed neutron absorbers to prevent criticality) and short residence times. Pulsed columns are most commonly used, starting with Idaho Falls in 1953 [33] and Hanford in 1956 [34], but centrifugal contactors (essentially mixer–settlers with greatly accelerated settling) have also been installed at Savannah River, United States [35] and Cap La Hague, France [36]. This type of contactor is advantageous in reducing residence times and holdup volumes of process liquors but is mechanically complex and sensitive to plugging by suspended solids.

4. SPECIMEN PROCESS

The process operated for Magnox fuel in the Second Separation Plant at Windscale is now de-

TABLE 2 PRINCIPAL OPERATING REPROCESSING PLANTS FOR POWER REACTOR FUELS

		First Cycle Conditions		Subsequent Cycle Conditions	
Country	Location	Solvent	Equipment	Solvent	Equipment
France	Marcoule	30% TBP–HPT	Mixer–settlers	30% TBP–HPT	Pu cycles: pulsed columns; U Cycles: mixer–settlers
France	Cap La Hague	30% TBP–HPT	First contactor, centrifugal; remainder, mixer–settlers	30% TBP–HPT	Mixer–settlers
Germany	Karlsruhe	30% TBP–*n*-dodecane	Mixer–settlers	30% TBP–*n*-dodecane	Mixer–settlers
					Final purification: U, silica gel; Pu, anion exchange
India	Trombay	30% TBP–Shellsol-T	Pulsed columns	30% TBP–Shellsol-T	Pulsed columns
					Final Pu purification, anion exchange
Japan	Tokai-Mura	30% TBP–dodecane	Mixer–settlers	30% TBP–dodecane	Mixer–settlers
United Kingdom	Windscale	20% TBP–OK	Mixer–settlers	20% TBP–OK	Mixer–settlers
United States	Savannah River	30% TBP–*n*-paraffin	First contactor, centrifugal; remainder, mixer–settlers	30% TBP–*n*-paraffin	Mixer–settlers
					Final purification: U, silica gel; Pu, anion exchange
United States	Hanford	30% TBP–*n*-paraffin	Pulsed columns	30% TBP–*n*-paraffin	Pulsed columns
					Final purification: U, silica gel; Pu, anion exchange
United States	Idaho Falls	30% TBP–paraffin	Pulsed columns	Hexone	Pulsed columns

scribed [26, 37]. Variations adopted elsewhere are noted later (Table 2).

4.1. First Cycle

The feed solution contains 300 g U/liter as uranyl nitrate in 3 *M* nitric acid, together with Pu(IV) nitrate, fission products, and transuranic elements (Table 1). It also contains small proportions of magnesium from the cladding, iron and aluminum originally alloyed for metallurgical reasons with the uranium, and corrosion products from the dissolver. The solution is fed to an intermediate point in a multistage battery of mixer–settlers where it is brought into countercurrent contact with an approximately fivefold flow of 20% TBP–OK, which also extracts small proportions of a few fission products such as ruthenium (a notorious chemical chameleon) and zirconium, together with rather larger proportions of the neptunium present. To reduce the level of these contaminants, particularly zirconium, the loaded solvent is scrubbed with fresh 3 *M* nitric acid, which inevitably also removes some of the uranium and plutonium and is therefore combined with the active feed.

The loaded and scrubbed solvent now passes to a backwash contactor where plutonium is first backwashed with 0.4 *M* HNO_3, to prevent hydrolysis, and then uranium with more dilute acid to allow for its greater extractability. The two aqueous streams are not actually separate, but the aqueous uranium solution is acidified before passing to the plutonium backwash contactor. The solvent, thus stripped of its uranium and plutonium load, is purified and recycled (Fig. 2).

4.2. Solvent Degradation

The solvent inevitably suffers some radiolytic degradation, chiefly through contact with the fission product solution during extraction of uranium and plutonium, but to a slight extent (liable to increase with more highly irradiated fuels) as a result of the extracted material. There is also some purely chemical attack by nitric acid, but this is generally insignificant under process conditions [38].

Solvent degradation takes two major forms [39]:

1. Tri-*n*-butylphosphate is in effect hydrolysed to dibutylphosphoric acid (HDBP), monobutylphosphoric acid (H_2MBP), and eventually orthophosphoric acid in progressively decreasing yields. There may also be some chain extension by the attachment of hydrocarbon fragments.
2. The diluent is partly converted to nitrate esters, nitrocompounds, ketones, carboxylic acids, and other compounds. The extent of attack depends largely on the nature of the hydrocarbon diluent: *n*-paraffins are the most resistant and branched-chain paraffins less so, whereas unsaturated or cyclic hydrocarbons are particularly susceptible. It has thus been argued that the *n*-paraffins are the most suitable diluents, and there has been a gradual trend in other plants to substitute them for the cheaper but less uniformly stable distillate cuts (e.g., kerosene) used previously. On the other hand, it appears that the very lability of these diluents may enable them to act as energy sinks, diverting attack away from TBP itself [31].

The effects of solvent degradation can be serious. Even at concentrations as low as 10^{-5} *M*, HDBP can significantly enhance the distribution ratios of trace uranium, plutonium, and fission products such as zirconium. It may thus depress decontamination factors, especially from zirconium and niobium, and increase the minute proportions

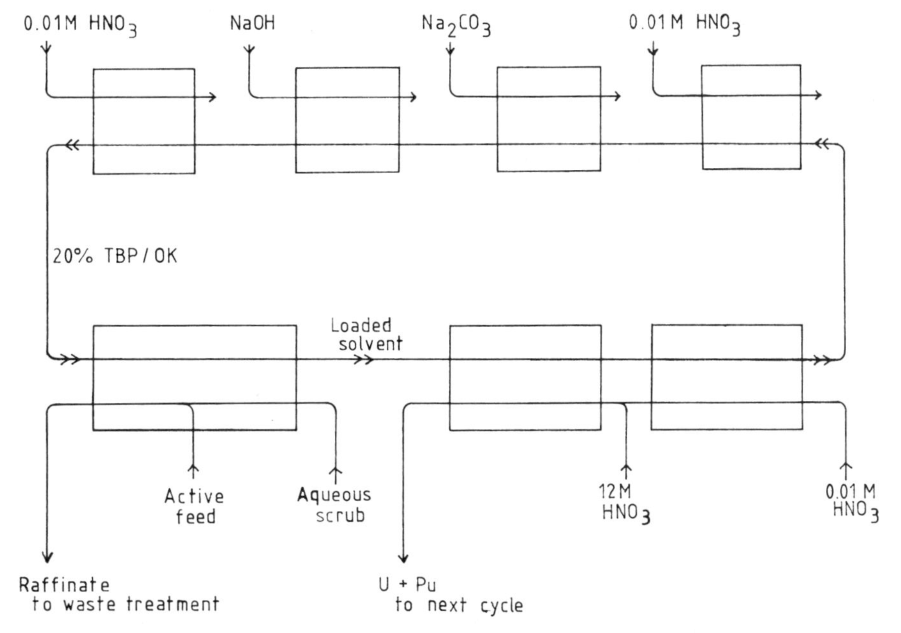

Figure 2. Purex Highly Active (HA) cycle (Magnox plant, Windscale).

of uranium and plutonium remaining in the backwashed solvent. At higher concentrations, salts such as zirconium dibutylphosphate may be precipitated as troublesome "cruds" that elsewhere have been known to interfere seriously with operation [32, 40, 41]. Degradation products of the diluent may also detract from decontamination, although the components responsible have yet to be positively identified; for a long time they were thought to be hydroxamic acids derived by further reaction of nitroparaffins [42], but synthetic hydroxamic acids have failed to reproduce the observed behavior [43].

4.3. Solvent Purification

For reduction of the effect of degradation, the solvent must be purified before being used again. In the Windscale Magnox plant, this is achieved by successive washes with dilute sodium carbonate (in which uranium and plutonium form soluble complexes) and sodium hydroxide solutions [26]. For assurance of adequate residence times and appropriate phase ratios without excessive flows of aqueous solution (which cannot be greatly concentrated for storage), recirculating contactors of the Holley–Mott type are used.

The alkaline washes remove acidic components such as HDBP and carboxylic acids and to some extent moderately water-soluble impurities such as butanol and ketones. About 95% of inorganic impurities such as fission products (notably ^{106}Ru, ^{95}Zr, and ^{95}Nb) are also removed; hence the need to restrict the volume of wash solutions to minimize storage problems. Neutral degradation products such as nitrates or nitro compounds are not significantly removed and so build up to a limit (a few percent) set by the slight but inevitable losses of solvent from the system.

4.4. Separation and Purification of Uranium and Plutonium

The uranium and plutonium in the aqueous stream from the first or highly active cycle pass to the second cycle, where they are again extracted as before. This time, however, the plutonium is backwashed with an aqueous solution of ferrous sulfamate [17] [the ferrous ion to reduce Pu(IV) to the virtually inextractable Pu(III), the sulfamate to scavenge nitrite that would reverse the reduction), and any uranium partitioning into the aqueous phase is scrubbed out with fresh solvent (Fig. 3). The uranium is then backwashed separately and receives a further cycle of purification, as does the plutonium after reoxidation to the extractable tetravalent state. Thus each product receives three cycles of purification.

5. MAJOR PROCESS VARIATIONS

The main reprocessing plants, with important distinctive features, are listed in Table 2. Regardless of the form of flow sheet and the conditions chosen for reprocessing nuclear fuel, the main considerations are the safety, reliability, and stability of the process; high recovery of the uranium and plutonium; and adequate purification from fission products (notably ruthenium, zirconium, niobium, and iodine), the transuranic element neptunium, and from any other deleterious impurities such as corrosion products.

5.1. Flow Sheet Parameters

The chief parameters affecting extraction efficiency of uranium and plutonium, along with high

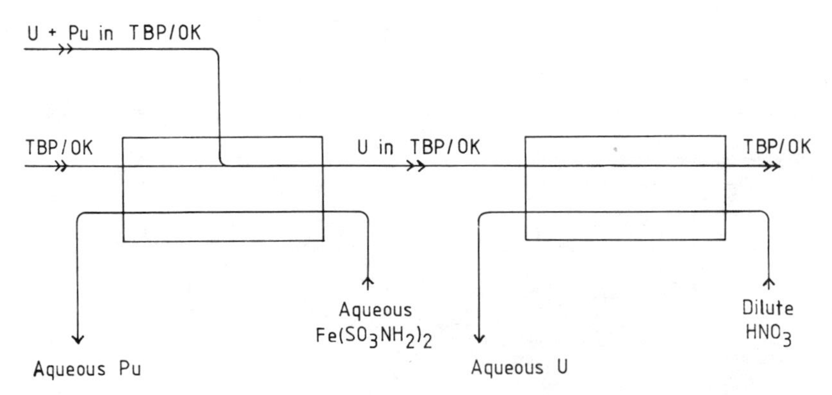

Figure 3. Uranium–plutonium separation, Windscale Magnox Reprocessing Plant.

removal factors for fission products and transuranics, are nitric acid concentration (in the extraction and scrub contactors), flow ratio, temperature, saturation (or loading of the uranium and plutonium in the organic phase), solvent quality and degradation, and the residence times of the phases in the contactors.

Flow sheet conditions are chosen to optimize the process requirements as outlined. To achieve these objectives, a vast amount of distribution, kinetic, and process data have been obtained for ruthenium, zirconium, niobium, and neptunium as well as for uranium and plutonium. The removal of these other elements from the uranium and plutonium streams usually requires two or three cycles of solvent extraction. The fission products zirconium, niobium, and, in particular, ruthenium have a number of nitrate and nitro species with different extractabilities and rates of conversion [44] : to minimize extraction of these unwanted elements, a compromise must be reached between the sometimes conflicting flow sheet requirements for their efficient removal. For example, decontamination from ruthenium is favored by high temperature and acidity during scrubbing, but zirconium and niobium are more effectively scrubbed out at a low acidity. There have thus been suggestions for combining both regimes in different parts of the process. Again intercycle conditioning is sometimes used in an attempt to convert the extractable species selected by the preceding cycle into an equilibrium mixture with proportions similar to those in the initial feed solution.

For neptunium removal, numerous flow sheets have been developed, taking into account the fact that neptunium adopts various valencies in a nitrate medium and that the nitrato species so formed have different extractabilities: Np(IV) and Np(VI) are extractable and Np(V) is virtually inextractable [45]. At the Savannah River plant neptunium is diverted to the aqueous raffinate of the first highly active extraction contactor by conversion to Np(V) with nitrous acid at the lower stages of the extraction section, and by high uranium loadings in the 30% TBP–OK phase [46]. At Hanford, neptunium was separated by the use of a reflux and recycle flow sheet in the second cycle of the solvent extraction process [47]. In the United Kingdom, on the Magnox reprocessing plant, neptunium is mainly removed in the uranium purification cycle, the feed to which is heated at low acidity before extraction of the uranium, while ferrous sulfamate is used as a reducing agent in the scrub sections of the contactor. On the proposed new oxide plant at Windscale, it is proposed to remove the neptunium in the two uranium purification cycles by the use of reducing agents such as hydroxylamine and by low acid conditions [14].

High solvent loadings generally improve decontamination by reducing the concentrations of extractant available to the impurities. Nevertheless, there is an interesting exception in technetium which forms an anion (presumably TcO_4^-) tending to associate with extracted uranium [48]. In any case the solvent loading must not be increased to the extent of risking instability in the process or unacceptable losses of uranium and plutonium to raffinates.

5.2. Early or Late Split

The choice of the second cycle as location for the uranium–plutonium separation ("late split") is a consequence of the intractability of the oxidation products of ferrous sulfamate (ferric ion and sulfuric acid) which emerge in the raffinate from the plutonium purification cycle. Their presence is a nuisance both in limiting the extent to which concentration for storage might be possible and in aggravating the corrosion of vessels. A late rather than an early split reduces the activity associated with this stream and limits the range of fission products present; the principal contaminant is ^{106}Ru, with a half-life of 1 year, so that after a suitable period of delay storage it is possible to discharge these wastes to the environment without approaching the authorized limits.

On the other hand, the late split means that the whole of the first cycle and much of the second must be accommodated in equipment sufficiently large for the full uranium throughput, yet protected against any possibility of criticality while taking the whole plutonium throughput. This can be achieved reliably by concentration limitation alone with the small plutonium content of Magnox fuel or with the limited quantities of oxide fuel such as have also been passed through the same solvent extraction equipment after pretreatment in a "head-end" plant.

With the amounts of fully irradiated oxide fuel that are to be processed in the projected Thermal Oxide Reprocessing Plant (THORP) at Windscale, concentrations of plutonium under flow sheet conditions would still be less than could lead to criticality, but to allow for the remote possibility of accumulation following a particular type of maloperation, the coprocessing contactors may be fitted with fixed neutron absorbers, such as hafnium plates or boron-loaded stainless steel packings [14]. This is an expensive precaution, and the number of contactors to be so poisoned is thus to be minimized by adoption of an early split.

5.3. Plutonium Backwashing Agents

Although some flow sheets have depended on selective complex formation, such as by sulfuric acid, for the separation of plutonium from uranium, reduction to the almost inextractable Pu(III) is the more common method, although there is some variation in choice of reductant [49]. The classic reagent is ferrous sulfamate, but this has the disadvantages already mentioned. They are exacerbated by the need for several times the stoichiometric quantity of reductant to maintain the required redox potential and to allow for some cycling of plutonium within the contactor [17]. Even Pu(III) has a significant, although slight, solubility in the solvent phase, where there is no ferrous ion or nitrite scavenger. Nitrous acid, on the other hand, has a high distribution ratio and autocatalyzes the oxidation of Pu(III) by nitric acid. Thus some plutonium is always extracted into the scrub solvent (necessary to remove uranium from the plutonium stream) and carried toward the plutonium backwashing section, where it joins the freshly arriving flow (Fig. 4).

In the presence of excessive quantities of nitrous acid or insufficient reductant the recycling of plutonium could build up so far as to vitiate the separation. In equipment without geometric limitation or fixed neutron absorbers it might, if unsuppressed, lead to criticality; this was a possibility that had to be examined especially carefully in relation to the Windscale mixer–settlers, where the large physical dimensions require concentrations to be kept below the "infinite sea" minimum level for criticality. Fortunately, these very dimensions imply a volume capable of absorbing a large mass without approaching such concentrations, and computation confirmed by small-scale trials showed that if a buildup were to occur, there would be ample time for corrective action [37].

With the increasing amounts of plutonium in the more highly irradiated fuels, the quantity of reductant required increases at least in proportion. If ferrous sulfamate were retained for the purpose, this increase, together with the concomitant rise in fission product levels, probably exacerbated by an early split, would create great problems in waste management, particularly as limits on discharges to the environment are likely to become if anything more restrictive than at present. There is thus a general tendency to replace ferrous sulfamate by reductants leaving no intractable residues. Methods that have received special attention are reduction with U(IV) (which ultimately follows the uranium product stream), with hydroxylamine (decomposing to gaseous products), or by electrolysis that introduces no foreign material in itself—although as with the others, hydrazine is commonly used as nitrite scavenger [49].

Neither of the chemical reductants is ideal. Hydroxylamine is relatively slow acting, requires a rather high temperature to be effective within

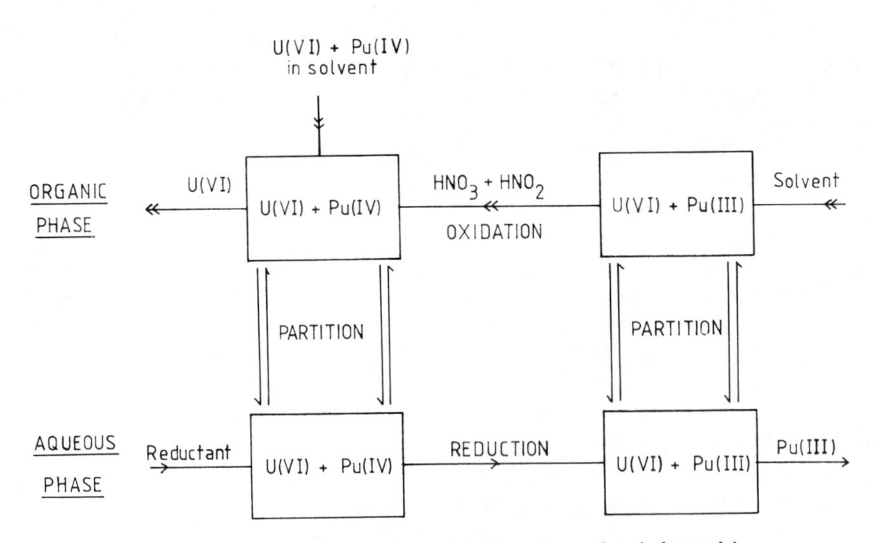

Figure 4. Recycling uranium–plutonium separation (schematic).

the residence time of a pulsed column, and is very sensitive to acid; the rate of reaction is inversely related to the fourth power of the acidity [50], and where the acidity of the separation contactor is dominated by that of the loaded solvent feed, this can lead to considerable problems in control. Furthermore, the gaseous oxidation products have aroused some concern about their possible effects on the operation of a pulsed column, although in practice they seem to cause little, if any, difficulty.

Uranous nitrate suffers from fewer disadvantages. With an inverse-square dependence on acidity [51, 52], it is less sensitive than hydroxylamine (although more so than ferrous sulfamate), and its oxidation product is quite innocuous, particularly if the reductant is generated from the uranium product of the current campaign in order to avoid degradation of enrichment bands. The reductant is itself extractable into the solvent so that it can be eliminated from the plutonium product by the solvent scrub. Whether this extractability aids the reductive backwashing is uncertain; it brings the reductant into direct contact with the bulk of the Pu(IV) feed, but there is a possibility that it may join in the autocatalytic reactions between Pu(III) and nitric acid [17]. Because of the extractability of U(IV), a suitable concentration profile along the contactor, providing the higher level near the loaded solvent feed position where the bulk of the reduction occurs, can be achieved only by means of a divided feed with the greater part going to this position. To attain the correct profile quickly at startup, additional temporary feeds at other points may also be needed. However, these arrangements are easily made; reduction with uranous nitrate has been employed successfully at Eurochemic [53] and Cap La Hague [54] and is planned for the Thermal Oxide Reprocessing Plant at Windscale [14].

Electrolytic reduction might appear the ideal method in that the reductive potential can be controlled directly and the capacity supplied exactly where it is needed. It is true that such a process has operated successfully in small mixer-settlers [49] and minature pulsed columns [55], but there is little experience of using it on a larger scale. Stray currents could cause corrosion problems, so materials more expensive and intractable than the conventional stainless steel may have to be used, while side reactions with nitric acid complicate control of the chemistry. Nevertheless, electrolytic reduction has been incorporated in the Barnwell plant, South Carolina [56] and is proposed for a major German reprocessing plant [57].

5.4. Other Process Variations

5.4.1. Extractant

Although TBP is the standard solvent for the first cycle, amines have sometimes been used for later cycles, particularly for plutonium purification in which their selective affinity for plutonium is put to good use [29, 58, 59]. Amines can also be used when it is desired to extract uranium [60]. A difficulty that has apparently led to a decline in amine processes, at least for irradiated fuel, is that the extraction of valuable materials is not readily reversed without the use of complexing agents, such as sulfuric or formic acid, in the aqueous phase [61]. These give rise to problems of waste management. Furthermore, to maintain satisfactory qualities, amines must be used with a diluent, generally an aromatic hydrocarbon, in which the solubility of metal–amine–nitrate adducts is limited. As a result, there is a tendency (also present although less marked in TBP systems [62]) for separation of the organic phase into two [63]. The resulting "third phase," comprising the adduct with a relatively small proportion of dissolved diluent, is intermediate in density between aqueous and lighter organic phases, and its presence usually renders a solvent extraction process inoperable. To prevent its formation, a "solvent modifier," such as a long-chain alcohol, may be added, but this generally detracts from extractive properties [64].

5.4.2. Solvent Concentration

Tributylphosphate is diluted to reduce its density, viscosity, and extractive power (which if unimpaired would be too high for convenient backwashing) and sometimes to limit the concentration of fissile material in the solvent phase for the sake of ensuring nuclear safety. The actual extent of dilution varies to some extent. The most widely used proportions of TBP are 20 or 30% (Section 3.2), although concentrations of 4–8% are not unknown for particular purposes.

The advantage of a high concentration is that high loadings are obtainable. Thirty percent TBP is slightly more than a molar solution, so that concentrations of uranium or plutonium (which both form predominantly 1:2 complexes with TPB [65]) could in theory approach 0.5 molar, although in practice degrees of saturation exceeding about 70–80% generally lead to instability: maximum loadings with more dilute solvent are correspondingly reduced. Thus in a plant limited by solvent flow rate rather than consistency, the use of the highest practicable concentration maximizes the throughput obtainable during extraction in equipment of given size or, conversely, minimizes the dimensions for a required throughput: however, this advantage may be countered by the increased aqueous flow required for effective backwashing.

Small dimensions are desirable to simplfy criticality control, but this is sometimes achieved by concentration limitation ensured ironically by a very low TBP concentration, but of course this assumes no undue accumulation in the aqueous phase. In mixer–settlers, where flow rates may be limited by the rate of phase disengagement between stages, the improvement in hydraulic properties following a reduction in TBP content allows an increase in flow rate sufficient to compensate for the reduced concentration [30]. It may also lead to improved decontamination from fission products and an easier uranium–plutonium separation.

5.4.3. Process Configuration

It is usual for uranium and plutonium to progress steadily forward throughout the system, but there are exceptions in the "reflux" flow sheets sometimes used, for instance, to produce a more concentrated plutonium product solution than is possible without undue losses in the direct flow sheet. In one example [66] the plutonium backwashed from the uranium–plutonium separation system is conditioned and fed back to the forward-extraction column, so adding to the plutonium input until the product concentration reaches the required level, whereupon an appropriate proportion is bled off for further processing. A problem with this type of flow sheet is that impurities may be recycled as well as plutonium, leading to reduced decontamination factors, so that reflux has largely fallen out of favor.

6. DEVELOPMENT OF PROCESSES

6.1. Need for Development

Recovery of uranium and plutonium by solvent extraction from irradiated nuclear fuel is now a well-established technique of nearly three decades' standing, but development is still needed for several reasons:

1. There is constant pressure to reduce effluent discharges and improve the management of stored wastes.
2. The tendency toward increased burnup of reactor fuels, from 3–4000 MW day/metric ton in the British Magnox reactors to about 40,000 MW day/metric ton in pressurized water reactors (PWRs) of American basic design, requires that correspondingly greater quantities of plutonium and fission products be handled. This means changes in process chemistry to match changes in feed composition (see Table 1) and direct or consequential alterations to the equipment to prevent undue solvent degradation or risk of criticality.
3. Problems or new demands arising on existing plant often require a certain amount of development work for the most effective resolution.

6.2. Process of Development

The established method of development in the reprocessing industry for the solvent extraction processes is the use of small-scale fully active laboratory and pilot-plant techniques in conjunction with large-scale nonactive engineering equipment. The methods used in the United Kingdom for Magnox fuel reprocessing have been described in detail previously [31], and similar techniques are to be followed for the new oxide reprocessing plant [14].

The development work progresses along two simultaneous and complementary lines:

1. Full-scale trace-active testing of equipment, such as mixer–settlers and pulsed columns, with uranium, nitric acid, and the relevant organic solvent. Hydrodynamic and uranium transfer data are obtained, and the effects of varying solvent quality, solids, surfactants, and simulated fission products are investigated.
2. Small-scale (1 : 1500 to 1 : 6000) miniature fully radioactive plant studies (i.e., uranium, plutonium, fission products, nitric acid, and relevant solvent) of the solvent extraction flow sheets [67].

The purpose of the minature pilot plant is to investigate details of the chemical process under radioactive countercurrent conditions, whereas problems of hydraulics, mass transfer, process control, and instrumentation are studied on the full-scale pilot plant. Stage for stage, chemical conditions in miniature and full-scale contactors can be expected to correspond closely, so that observations on minor components in the miniature equipment apply to the full-scale plant. The data from the miniature pilot plant are related to those from the large-scale equipment by direct comparison of the extraction, scrub, and backwash efficiencies of the key component, uranium, in both types of equipment. The behavior of real and simulated fission products is also correlated. The distribution behavior and extraction efficiencies of the other components, such as plutonium, genuine fission products, and transuranics, are determined only in the miniature equipment and then by means of the correlations developed between the scales, extrapolated to the large-scale plant performance. This development system has proved to be successful for the solvent extraction processes used throughout the western world and is being used for future plants for the reprocessing of thermal oxide and fast reactor fuels.

6.3. Future Development

Because of the experience accumulated over the last 30 or so years and the need to ensure that any departure from established technology can be achieved without sacrifice of safety or reliability, reprocessing tends to progress by evolution rather than revolution. Many of the developments that may be expected to occur during the next few decades are in areas such as fuel storage, breakdown and dissolution at one end of the process, and the immobilization of wastes at the other. These are vitally important but not directly related to solvent extraction. Several of the probable developments in solvent extraction were adumbrated earlier.

Among the most important are those required to ensure that with the processing of high-burnup oxide fuels and the consequent increase in fission product activity, plutonium, and other transuranic elements in the feed solution to the solvent extraction process, environmental discharge levels are not increased and, in fact, are reduced to the minimum consonant with the technology available at the time. To this end, much of the effort on flow sheet refinement has been directed toward eliminating the use of chemical reagents that limit the flexibility of waste management. The idea would be to segregate from the long-lived radionuclides those relatively short-lived wastes that would add appreciably to the cost or problems of storing them, to minimze the volumes of all stored wastes, and to avoid constraints on combining other wastes destined for permanent storage with the highly active aqueous raffinate stream from the first contactor of the extraction cycle.

Three particular areas of interest have been (1) the choice of reducing agent for the separation of plutonium from uranium, (2) the chemical conditioning step preparing plutonium for extraction in the purification cycles, and (3) the use of alternative washing reagents other than sodium carbonate and hydroxide for the purification of the recycled TBP–diluent. The first of these was considered in some detail previously, but it is noteworthy that for the plutonium reduction steps at the uranium–plutonium separation stage, the reagent ferrous sulfamate has generally been replaced in new plants by tetravalent uranium [14, 29], and other large-scale plants have considered electrolytic reduction [56, 57]. In general, electrolytic reduction and chemical methods such as the use of organic reducing agents (instead of hydrazine and hydroxylamine) are at the laboratory or pilot–plant stage. Recently the use of di-*tert*-pentylhydroquinone or 2,3-dichloro-1,4-naphthaquinone as organic-soluble reductants for plutonium (and neptunium) has been suggested [68], but a holding reductant such as hydroxylamine nitrate is necessary in the aqueous phase to prevent reoxidation of plutonium(III).

Uranium(IV) is itself produced in an ancillary process by electrolytic reduction of uranyl nitrate, but there is a possibility of replacing this by photolytic reduction [69].

After the separation of plutonium from the uranium, the trivalent plutonium in the aqueous phase is reconverted to the extractable tetravalent plutonium before extraction in the plutonium purification cycles. The reagent widely used in most Purex processes is sodium nitrite [26, 29], but in the search to eliminate salts from effluent streams, alternatives have been investigated. At Marcoule, France [40] the large-scale reprocessing plant uses mixed $NO-NO_2$ gases, generated by the action of nitric acid on sodium nitrite, and removes the excess by air sparging before extraction. A similar process is envisaged for the new oxide reprocessing plant at Windscale [14], but other techniques such as electrolytic oxidation, as suggested for the German reprocessing plant [57], are being studied. The possibility of photolytic oxidation has also been mooted [69].

For solvent purification, the traditional method of alkali washing is generally effective but leaves a contaminated "medium active" effluent loaded with nonvolatile solutes that severely limit the degree of concentration attainable for storage. To alleviate this problem, there have been two suggested approaches: to replace sodium carbonate and hydroxide by an alkali such as hydrazine with the excess destroyed electrolytically [57] or to dispense with washing altogether and remove impurities either by a distillation process [70] or by a solid sorbent such as an anion exchange resin [71]. Such a substitution would raise further problems of waste disposal.

Another area of development work associated with environmental discharge levels is the minimizing of the actinide contents of waste streams. A number of establishments are investigating the use of organic solvents other than TBP–diluent to remove plutonium, americium, and neptunium from aqueous effluents. At Hanford (USA) dibutyl butylphosphonate has been used for a number of years, but it is of interest to note that Schulz and McIsaac were investigating the use of bidentate ligand dihexyl-*N*, *N*-diethylcarbamylmethylene phosphate for waste streams at both Hanford and Idaho Falls plants [72]. The reagent was highly selective for actinides but had some disadvantages in being hard to purify. Substituted hydroxamic group solvents, patented and called *HX-70,* were being investigated by Grossi and Gasparini for the removal of plutonium from aqueous effluent streams [73]. These solvents showed potential for some waste treatments but were susceptible to radiolysis and attack by nitrous acid.

Although the main purpose of these developments is to simplify treatment and disposal of the wastes by removing the long-lived components, it is also expected that large-scale reprocessing plants will provide processes for recovering and recycling at least the plutonium from waste streams.

The Barnwell plant (USA) employs a "saveall" pulsed column extraction system to recover uranium and plutonium from selected aqueous raffinate streams from the main-line process [74]. For recovery of plutonium from such waste streams in the reprocessing of thermal oxide fuels, the new oxide reprocessing plant at Windscale is likely to use a mixer–settler extraction and backwash process [14]. These additional systems allow for any process upsets and avoid any need to recycle streams during the main campaigns.

The thermal oxide fuel to be reprocessed in the new plant contains uranium with a proportion of ^{235}U still higher than natural and, therefore, eminently worth recycling to reactors after re-enrichment. However, it is then necessary to avoid contamination of the enrichment plant by volatile transuranic fluorides, such as those formed by neptunium. This means that these elements must be reduced to very low proportions in the purified uranium product from the solvent extraction process.

To achieve these requirements, the flow sheets for neptunium removal in the uranium purification cycles are being investigated so that the maximum flexibility can be achieved.

As discussed earlier (Section 3.3.), the increased levels of fission product activity to be handled in the reprocessing of thermal and fast-reactor oxide fuels require serious attention to the limitation of radiolytic damage to the organic solvent in the contacting equipment. Modification of the solvent (probably not to the extent of a fundamental departure from the well-proven TBP-diluent system) has been considered, but a more effective approach has been to reduce the contactor residence times. In the highly radioactive sections of the process the mixer–settler equipment has been replaced by pulsed columns [14] or centrifugal contactors [35, 36]. However, away from the highly radioactive sections of the plant, there is still a tendency to keep the longer residence time equipment or at most install pulsed column contactors. In any choice of contacting equipment, improvement in process performance by reduction in residence times must be weighed against the complexity and maintenance requirements of the new equipment.

The initial reprocessing plants, particularly those treating uranium metal fuel, have separated uranium and plutonium from each other with very high efficiencies, such as removal factors of 10^6 for plutonium from uranium and for uranium from plutonium [26]. As discussed earlier, the uranium stream must be purified thoroughly to prevent contamination of the enrichment plant, but in the future, where plutonium is to be used for fast reactor fuel, complete removal of uranium from it will not be necessary. Furthermore, for the transport and storage of plutonium products, there are some advantages in safety and containment to be gained from the presence of a proportion of uranium. Flow sheets will thus be adapted, and in the plutonium stream probably simplified, in accordance with the altered balance of requirements.

In summary, development of the Purex process continues to refine and consolidate the basic solvent extraction flow sheets and to ensure that future highly irradiated fuels with higher plutonium contents can be successfully and effectively processed with stringent environmental discharge standards.

REFERENCES

To avoid undue repetition, the titles of certain compendia containing several references have been abbreviated as follows:

2nd Internatl. Conf.
Peaceful Uses of Atomic Energy, Proceedings of the Second International Conference, Geneva, 1958, United Nations, Geneva, 1958.

3rd Internatl. Conf.
Proceedings of the Third International Conference on the Peaceful Uses of Atomic Energy, Geneva, 1964, United Nations, New York, 1965.

4th Internatl. Conf.
Peaceful Uses of Atomic Energy, Proceedings of the Fourth International Conference, Geneva, 1971, International Atomic Energy Agency/ United Nations, New York and Vienna, 1972.

KR 126
Reprocessing of Fuel from Present and Future Power Reactors, Kjeller Report 126, Institutt for Atomenergi, Kjeller, Norway, 1967.

Long
J. T. Long, *Engineering for Nuclear Fuel Reprocessing,* Gordon and Breach, New York, 1967.

ISEC 71
Proceedings of the International Solvent Extraction Conference (ISEC), The Hague 1971, Society of Chemical Industry, London, 1971.

ISEC 74
Proceedings of the International Solvent Extraction Conference (ISEC), Lyons 1974, Society of Chemical Industry, London, 1974.

ISEC 77
Proceedings of the International Solvent Extraction Conference (ISEC), Toronto 1977, Canadian Institute of Mining and Metallurgy, Montreal, 1979.

1. Long, pp. 7–10.
2. ATOM, U. K. Atomic Energy Authority Publication No. 138, April 1968.
3. A. Ferrari et al., in *Nuclear Power and its Fuel Cycle,* Vol. 3, International Atomic Energy Agency, Vienna, 1977, pp. 249–257.
4. N. L. Franklin, Irradiated Fuel Cycle, in *Nuclear Energy Maturity Proceedings of the European Nuclear Conference,* Vol 8, Paris, April 1975, pp. 1–16.
5. H. Kronberger and N. L. Franklin, *J. Br. Nucl. Energy Soc.* 4, 37 (1965).
6. J. T. Roberts, KR126, p. 279.
7. Long, p. 24.
8. S. Katcoff, *Nucleonics* 18, 201 (1960).
9. Long, Appendix A.
10. *Radiation Protection,* ICRP Publication 9, Pergamon Press, Oxford, 1966.
11. J. T. Daniels, *J. Br. Nucl. Energy Soc.* 4, 23 (1965).
12. R. L. Stevenson and P. E. Smith, in S. M. Stoller and R. B. Richards, Eds., *Reactor Handbook,* Volume II, *Fuel Reprocessing,* 2nd ed., Interscience, New York, 1961, Chapter 4.
13. A. Duncan and A. Naylor, U.K. Atomic Energy Authority PG Report 812(W), 1967.
14. B. F. Warner, *Proof of Evidence to Windscale Inquiry,* Whitehaven, 1977.
15. U. S. Atomic Energy Commission Report ORNL-4436, 1970, Appendixes A and B.
16. Long, Chapter 3.
17. A. Naylor, KR126, pp. 172–200.
18. S. Lawroski and M. Levenson, in *Symposium on the Reprocessing of Irradiated Fuels, Brussels, 1957,* Book 1, U. S. Atomic Energy Commission Report TID-7534, pp 45–68.
19. G. R. Howells et al., 2nd Internatl. Conf., Vol. 17, pp. 3–24.
20. T. G. Hughes et al., 4th Internatl. Conf., Vol. 8, pp. 367–373.
21. I. D. Eubanks, *Atomic Energy Rev.* 7, 49 (1969).
22. V. R. Copper and M. T. Walling, Jr., 2nd Internatl. Conf., Vol. 17, pp. 291–323.
23. A. W. Joyce, L. C. Perry, and E. B. Sheldon, *Chem. Eng. Progr. Symp. Ser. 56,* 28, 21 (1960).
24. E. R. Irish, U.S. Atomic Energy Commission Report HW-60116 1959.
25. T. G. Hughes and D. W. Clelland, *Aqueous Reprocessing Chemistry for Irradiated Fuels,* Brussels Symposium 1963, OECD European Nuclear Energy Agency, pp. 189–204.
26. B. F. Warner et al., 3rd Internatl. Conf., Vol. 10, pp. 224–230.
27. C. Jonannaud, 3rd Internatl. Conf., Vol. 10, pp. 215–221.
28. P. Patigny et al., ISEC 74, Vol. 3, pp. 2019–2033, Fig. 1.
29. M. Duboz, *Energie Nucleaire* 7, 228 (1965).
30. A. L. Mills and E. Lillyman, ISEC 74, Vol. 2, pp. 1499–1517.
31. A. Naylor and M. Larkin, ISEC 71, Vol. 2, pp. 1356–1372.
32. G. Koch et al., *Kerntechnik* 18, 253 (1976).
33. R. D. Modrow, G. F. Offutt, and B. R. Wheeler, ISEC 71, Vol. 1, pp. 556–564.
34. G. L. Richardson and A. M. Platt, *Progress in Nu-*

clear Energy, Series IV, Vol. 4, Pergamon Press, London, 1961, pp. 279–307.

35. D. A. Orth, J. M. McKibben, and W. C. Scotten, ISEC 71, Vol. 1, pp. 514–533.
36. P. Auchapt et al., 4th Internatl. Conf., Vol. 8, pp. 547–557.
37. B. F. Warner, *Design and Development of the Windscale Reprocessing Plant*, U.K. Atomic Energy Authority Production Group, Warrington; adapted from Kerntechnik 9, 249 (1967).
38. L. L. Burger, in F. R. Bruce et al., Ed. *Progress in Nuclear Energy, Series III, Process Chemistry,* Vol. 2, Pergamon Press, London, 1958, pp. 307–319.
39. A. Naylor, KR126, pp. 120–142.
40. C. Breschet and P. Miquel, ISEC 71, Vol. 1, pp. 565–576.
41. K. L. Huppert, W. Issel, and W. Knoch, ISEC 74, Vol. 3, pp. 2063–2074.
42. A. J. Huggard and B. F. Warner, *Nucl. Sci. Eng.* **17**, 638 (1963).
43. T. V. Healy and A. Pilbeam, ISEC 74, Vol. 1, pp. 459–468.
44. A. Duncan, A. Naylor, and B. F. Warner in H. A. C. McKay, Ed., *Solvent Extraction Chemistry of Metals,* Macmillan, London, 1965, pp. 3–26.
45. T. Tsuboya et al., ISEC 74, Vol. 3, pp. 1985–1992.
46. W. L. Poe, A. W. Joyce, and R. I. Martens, *Ind. Eng. Chem. Process. Des. Devel.* **3**, 314 (1964).
47. R. E. Isaacson and B. F. Judson, *Ind. Eng. Chem. Process. Des. Devel.* **3**, 296 (1964).
48. F. Macasek, *Radiochem. Radioanal. Lett.* **22**, 175 (1975).
49. H. Schmieder et al., ISEC 74, Vol. 3, pp. 1997–2108.
50. G. S. Barney, *J. Inorg. Nucl. Chem.* **38**, 1677 (1976).
51. T. W. Newton, *J. Phys. Chem.* **63**, 1493 (1959).
52. W. Baxter and A. Naylor, in H. A. C. McKay, Ed., *Solvent Extraction Chemistry of Metals,* Macmillan, London, 1965, pp. 117–132.
53. J. van Geel et al., ISEC 71, Vol. 1, pp. 577–592.
54. J. Couture, *Chem. Eng. Progr. Symp. Ser. No. 94,* **65**, 26 (1969).
55. A. Naylor et al., British provisional patent application 55943/70.
56. R. I. Newman, *Nuclear Engineering International* (November 1972), p. 938.
57. F. Baumgaertner et al., ISEC 77, Vol. 2, pp. 599–604.
58. J. N. C. van Geel, Eurochemic Report ERT 188, 1968.
59. A. V. Hultgren, KR126, pp. 201–228.
60. S. Cao, H. Dworschak, and A. Hall, ISEC 74, Vol. 2, pp. 1453–1480.
61. G. Koch, J. Schoen, and G. Franz, KFK Report 893, 1970.
62. A. L. Mills and W. R. Logan, in D. Dyrssen et al., Ed., *Solvent Extraction Chemistry,* North-Holland, Amsterdam, 1967, pp. 322–326.
63. J.-C. Saey, CEA Report 3478, 1968.
64. F. Baroncelli, G. Scibona, and M. Zifferero, *J. Inorg. Nucl. Chem.* **25**, 205 (1963).
65. T. V. Healy and H. A. C. McKay, *Transact. Faraday Soc.* **52**, 633 (1956).
66. P. Auchapt, J. - P. Giraud, and X. Talmont, *Energie Nucleaire* **10**, 181 (1968).
67. B. F. Warner et al., ISEC 74, Vol. 2, pp. 1481–1497.
68. G. Grossi, ISEC 77, Vol. 2, pp. 634–639.
69. M. Goldstein, J. J. Barker, and T. Gangwer, *Nuclear Engineering International* (September 1977), p. 69.
70. H. J. Clark and G. S. Nichols, U.S. Atomic Energy Commission Report DP 849, 1965.
71. W. W. Schulz, ISEC 71, Vol. 1, pp. 174–185.
72. W. W. Schulz and L. D. McIsaac, ISEC 77, Vol. 2, pp. 619–630.
73. F. Baroncelli, G. Grossi, and G. M. Gasparini, ISEC 77, Vol. 2, pp. 640–644.
74. U.S. Atomic Energy Commission Docket 50332-42, pp. 4.79–4.82.

25.13

URANIUM PURIFICATION

A. W. Ashbrook and V. I. Lakshmanan
Eldorado Nuclear Ltd.
Canada

1. INTRODUCTION

Reviews on the uses of solvent extraction in the nuclear industry for the extraction of uranium and thorium have been published [1-3], and the subject is treated comprehensively in various books and conferences [4-12]. Hence no attempt is made here to discuss the various types of extraction either fundamentally or in their application. The subjects on uranium purification involving solvent extraction dealt with in this chapter concern those with relevance to the production of nuclear-grade uranium from concentrates by various refineries around the world. Application of the technique of solvent extraction for processing uranium ores in yellow cake production is dealt with separately in Chapter 25.11.

To prepare the yellow cake concentrates produced by various mills for use as reactor fuel, they need to be further purified to meet certain specifications [13]. Currently the most widely used wet method for the purification of yellow cake to meet such specifications is by solvent extraction employing tri-*n*-butyl phosphate (TBP). Other extractants, such as phosphine oxides, phosphonates, phosphinates, phosphinic acids, sulfoxides, ketones, and ethers have either been tested or used in pilot-plant studies [14-21]. Reagents such as organophosphorous compounds mentioned previously are more polar than TBP, which may result in inefficient uranium stripping and thus its purification. Tertiary amines (Alamine 336) have been successfully employed in uranium purification on a small scale, such as in the PNC process [22]. Use of TBP as an extractant for nitric acid leach solutions from thorium- and uranium-containing ores has also been tested [23].

2. EXTRACTION CHEMISTRY

Except for the Japanese PNC process and the Allied process (USA), all other refinery operations employing solvent extraction in the United Kingdom, United States, France and Canada dissolve the yellow cake directly, or after calcination, into nitric acid prior to extraction of the uranium into TBP, as illustrated by the following reactions:

$$(NH_4)_2U_2O_7 + 6HNO_3 \longrightarrow 2UO_2(NO_3)_2 + 2NH_4NO_3 + 3H_2O \quad (1)$$

$$U_3O_8 + 8HNO_3 \longrightarrow 3UO_2(NO_3)_2 + 2NO_2\uparrow + 4H_2O \quad (2)$$

Extraction of uranyl nitrate with TBP in various diluents from nitric acid media has been reported by several authors [24, 25]. In the system TBP · HNO_3 · H_2O it appears fairly certain that nitric acid displaces water from the solvent hydrate (TBP · H_2O) to form TBP · HNO_3 (except at very low concentrations where some TBP · H_2O · HNO_3 may form), with a strong hydrogen bond.

Extraction or uranium from nitric acid solution with TBP in a diluent can be represented by the following reaction:

$$UO^{2+}_{2\,aq} + 2NO^{-}_{3\,aq} + 2TBP_{org} \rightleftharpoons UO_2(NO_3)_2 \cdot 2TBP_{org} \quad (3)$$

Neglecting activity coefficients, the equilibrium constant may be written as

$$K = \frac{[UO_2(NO_3)_2 \cdot 2TBP]_{org}}{[UO_2^{2+}]_{aq}[NO_3^-]^2_{aq}[TBP]^2_{org}} \quad (4)$$

Equation (4) clearly indicates the dependence of uranium extraction on the aqueous uranium, nitrate, and TBP concentrations. The reaction is exothermic, and hence the effect of increased temperature in decreasing uranium extraction is apparent.

Extraction isotherms for uranyl nitrate with the use of 20 vol. % solution of TBP in kerosene are shown in Fig. 1. Under extreme conditions of nitric acid and TBP concentrations, direct solvation of the uranyl ion by TBP leads to the formation of the proton-solvated trinitrato uranyl acid, $(HTBP)_2UO_2(NO_3)_3$, forming a third phase [26].

Extraction of uranium into the organic phase from nitrate media with TBP can be suppressed by complexing uranyl ion with anions such as F^-, SO_4^{2-}, PO_4^{3-}, and CO_3^{2-}, and these effects can usually be reduced by the addition of cations that form more stable complexes with these anions than does uranium. An example is the addition of aluminum to reduce the effects of fluoride ion.

Under the conditions normally employed in the purification of uranium by solvent extraction, elements with distribution coefficients >0.01 are Mo^{6+}, Ce^{4+}, and Th^{4+}; however, Ce^{4+} is not met with under normal conditions, and Th^{4+} in high concentrations is separated as a result of saturation of the TBP by uranium. With 80–90% saturation of the organic phase by uranium, the distribution coefficient of thorium becomes less than 0.01 [27]. Molybdenum is by far the major problem in the solvent extraction purification. This is normally reduced in the refining process by suitable blending of the yellow cake feeds.

3. COMMERCIAL PRACTICE FOR URANIUM PURIFICATION

The refinery operations using wet purification methods in the United Kingdom (British Nuclear

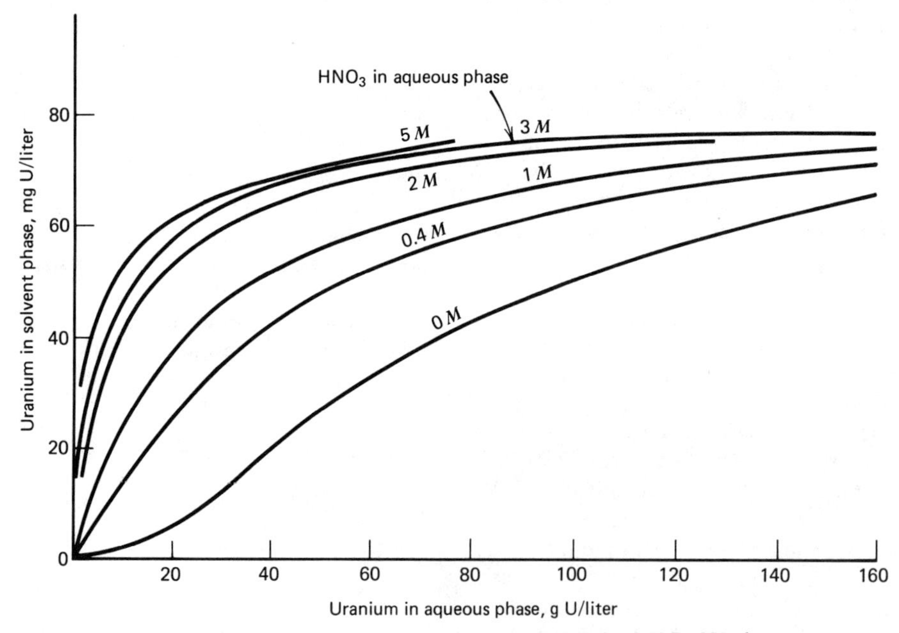

Figure 1. Partition of uranium between aqueous and 20% TBP–OK phases.

Fuels Ltd., Springfields), France (Comurhex, Malvesi), the United States (Kerr-McGee, Oklahoma), and Canada (Eldorado Nuclear Ltd., Port Hope) employ basically similar processes. The flow sheet employed by one of the refinery operators (Eldorado Nuclear Ltd., Canada) is shown in Fig. 2.

Essentially the solvent extraction process for

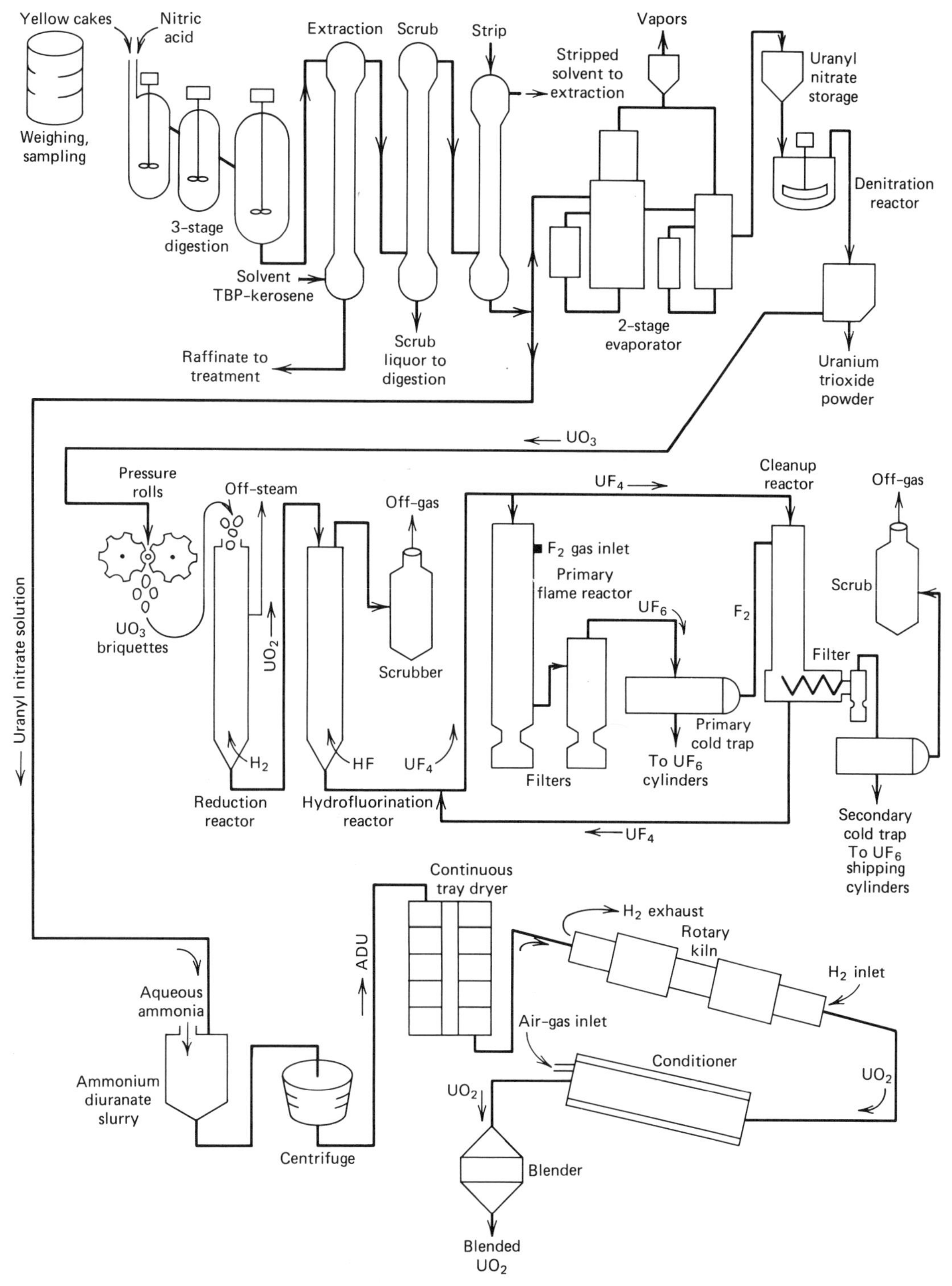

Figure 2. Uranium purification flow sheet used by Eldorado Nuclear Ltd., Canada.

TABLE 1 COMMERCIAL CONVERSION PLANTS IN THE WESTERN WORLD

Location	Owner	Present Capacity, Metric Tons of Uranium per Year	Planned Expansion, Metric Tons of Uranium per Year
Canada			
Port Hope, Ontario	Eldorado Nuclear Ltd.	5,500	14,500 (1983) (second facility)
France			
Malvesi (to UF_4)			
Pierrelatte (UF_4–UF_6)	Comurhex	12,000	14,000 (?)
United Kingdom			
Springfields, Lancs.	British Nuclear Fuels Ltd.	9,500	11,500 (?)
United States			
Gore, Oklahoma	Kerr–McGee	9,090	

uranium purification employed by the United Kingdom, France, or United States is similar to Eldorado's practice, except that the contactors used by BNFL are mixer–settlers; at Kerr–McGee they are pumper decanters, and Comurhex uses agitated columns. Tri-*n*-butylphosphate concentrations are maintained at 20, 30, and 40 vol. %, respectively. The free-acid concentration in the feed to extraction is maintained at about 0.5 *M*.

Present capabilities of conversion plants in the western world using wet purification processes are shown in Table 1.

Earlier operations in the United States, such as Fernald and Weldon Springs, either no longer process yellow cake or have ceased operations. Details of these processes have been summarized elsewhere [29, 30].

4. ELDORADO NUCLEAR LTD., PORT HOPE, CANADA

The refinery accepts feeds such as ammonium diuranate (ADU), calcined ammonium diuranate, magnesium diuranate, and sodium diuranate from Canadian, U.S., and other mills.

Mill concentrates are usually −100 mesh. The feeds are weighed, sampled, and blended to produce a suitable feed [31]. After calcination if ADU is used, the blended feed is conveyed to digestion circuits where it is dissolved in nitric acid at 80°C to form uranyl nitrate. At Port Hope the digestion circuit consists of four batch digesters, a combination that solubilizes more than 99.5% of the uranium in the feed. The product is an acidic slurry containing suspended solids, some free acid (2 *M*) and about 350 kg/m^3 uranium as uranyl nitrate. In contrast to operations in the United Kingdom and France, the feed is not filtered prior to entering the solvent extraction stage [32, 33].

From the digestion circuit the feed is pumped to the extraction stage, and the flow is controlled by a rotary dip feeder. The contactors used in the solvent extraction circuit consist of a 76-cm-diameter Mixco column and three 56-cm-diameter pulse columns. The Mixco column is used for extraction and the pulse columns, for scrubbing and stripping.

Eldorado uses 25 vol. % TBP in EXXON DX 3641 as the solvent. The loaded organic phase (90–100 kg/m^3 of uranium) after extraction is discharged from the top of the Mixco column and cascades into the bottom of the scrub column. Impurities and some uranium are discharged from the bottom of the extraction column as the raffinate, which is regularly monitored. The raffinate, containing 0.1–0.5 kg/m^3 of uranium, and thorium, vanadium, silica and radium, and so on is sent to the raffinate treatment circuit.

In the scrub column the loaded solvent is contacted with a small flow of pure water or dilute uranyl nitrate hydrate (UNH) to remove impurities and free acid. The aqueous discharge from the bottom of the column is recycled back to the digestion system. Scrubbed solvent from the top of the scrub column flows into the bottom of the strip column, where it is contacted with a relatively larger flow of pure water. The stripped solvent is then treated with Na_2CO_3, centrifuged, and recycled back to the extraction column. The

OK liquor (130 kg/m^3 of uranium) stream is discharged for boildown and then denitrated for conversion to UO_3. The UO_3 thus produced is used as feed for UF_6 production. Some of the UNH produced by Eldorado is also used to produce ceramic UO_2 for the CANDU reactors.

5. CONCLUSIONS

In the wet method of uranium purification, TBP has thus far been the extractant of choice. Nevertheless, studies on alternative extractants could prove useful.

Environmental restrictions on the use of ammonia and nitrate could warrant alternative concentrate production and dissolution agents other than nitric acid. Extraction of uranyl nitrate with TBP takes place through a solvation process, and studies on drop size and coalescence with uranium extraction efficiency would be productive. One mining company has recently found in-line mixing to be more efficient than mixer–settlers in uranium extraction with a tertiary amine. If proved, in-line mixers for the uranium–TBP system could reduce emulsion formation, increase throughput, and handle feeds containing solids.

If fuel reprocessing is acceptable and breeder reactors become more popular, depletion of natural uranium reserves may warrant studies on the extraction and purification of $UO_2 \cdot ThO_2$ and $UO_2 \cdot PuO_2$ mixed oxide systems.

ACKNOWLEDGMENTS

The authors appreciate the assistance of Messrs. R. J. McClure, W. J. Stinson, and R. G. Wilkinson of Eldorado Nuclear Ltd. for providing us with useful discussions and Mr. R. J. Macdonald of CANMET for carrying out the CISTI literature search on this subject.

REFERENCES

1. I. L. Jenkins, *Hydrometallurgy* **4**, 1 (1979).
2. E. A. Newland, Australian Atomic Energy Commission, Report AAEC/LIB/Bib-361, 1972.
3. R. Derry, Warren Spring Laboratories, UK, Mineral Dressing Information Note No. 7, 1972.
4. Y. Marcus and A. S. Kertes, *Ion Exchange and Solvent Extraction of Metal Complexes,* Wiley-Interscience, London, 1969.
5. C. Hanson, *Recent Advances in Liquid–Liquid Extraction,* Pergamon Press, Oxford, 1971.
6. G. M. Ritcey and A. W. Ashbrook, *Solvent Extraction: Principles and Applications to Process Metallurgy,* Elsevier, Amsterdam, 1979.
7. J. Stary, *The Solvent Extraction of Metal Chelates,* Macmillan, New York, 1964.
8. D. Dyrssen, J. O. Liljenzin, and J. Rydberg, *Solvent Extraction Chemistry,* North-Holland, Amsterdam, 1967.
9. H. A. C. McKay, T. V. Healy, I. L. Jenkins, and A. Naylor, *Solvent Extraction Chemistry of Metals,* Macmillan, London, 1965.
10. *Proceedings of the International Solvent Extraction Conference (ISEC), The Hague 1971*, Society of Chemical Industry, London, 1971.
11. *Proceedings of the International Solvent Extraction Conference (ISEC), Lyons 1974,* Society of Chemical Industry, London, 1974.
12. *Proceedings of the International Solvent Extraction Conference (ISEC), Toronto 1977,* Canadian Institute of Mining and Metallurgy, Montreal, 1979.
13. W. J. Stinson, *Proceedings of the Canadian Uranium Producers' Metallurgical Committee,* CANMET Report ERP/MSL77-208, 1977.
14. N. P. Galkin and B. N. Sudarikov, *Technology of Uranium,* Israeli Program for Scientific Translations, Jerusalem, 1966.
15. G. J. Laurence, M. T. Chaieb, and J. Talbot, *Proceedings of the International Solvent Extraction Conference (ISEC), The Hague 1971,* Society of Chemical Industry, London, 1971, p. 1150.
16. S. R. Mohanty and A. J. Reddy, *J. Inorg. Nucl. Chem.* **37**, 1791 (1975).
17. S. R. Mohanty and A. J. Reddy, *Proc. Chem. Symp., Dept. Atomic Energy, India* **2**, 193 (1975).
18. M. Certui, G. Marcu, and M. Diaconesa, *Stud. Univ. Babes. Bolyai Ser. Chem.* **21**, 63 (1976).
19. B. Tamkhina and M. J. Herak, *J. Inorg. Nucl. Chem.* **38**, 1505 (1976).
20. M. F. Lucid, U.S. Patent 3,821,351 (1974).
21. J. P. Shukla, U. K. Agarwal, and K. Bhatt, *Separation Sci.* **8**, 383 (1973).
22. S. Takada, T. Amanuma, and G. Fukuda, *Proceedings on Recovery of Uranium,* IAEA, Sao Paulo, 1970, p. 97.
23. T. H. Tunley and V. W. Nel, *Proceedings of the International Solvent Extraction Conference (ISEC), Lyons 1974*, Society of Chemical Industry, London, 1974, p. 1519.
24. Y. Marcus, *Chem. Rev.* **63**, 139 (1963).
25. H. A. C. McKay and T. V. Healy, *Progr. Nucl. Chem. Ser. III, Process Chem.* **2**, 546 (1958).
26. A. S. Solovkin, N. S. Povitskii, and K. P. Lunichkina, *Zhur. Neorg. Khim.* **5**, 2115 (1960).
27. V. S. Yemel'yanov and A. I. Yevstynukhim, *The Metallurgy of Nuclear Fuel,* Pergamon Press, London, 1969.

28. W. J. Stinson, *Extractive Metallurgy of Uranium,* course organized by the University of Toronto and the CIM, 1978.

29. C. D. Harrington and A. Ruehle, *Uranium Production Technology,* Van Nostrand, Princeton, NJ, 1959.

30. J. H. Gittus, *Uranium,* Butterworths, London, 1963.

31. R. M. Berry, paper presented at the Annual General CIM Meeting, Montreal, 1969.

32. P. G. Alfredson, B. G. Charlton, R. K. Ryan, and V. K. Vilkaitis, Australian Atomic Energy Commission Report AAEC/E-344, 1975.

33. H. Page, paper presented at Solvent Extraction and Ion Exchange Group of Society of Chemical Industry, Springfields Works, May 23, 1974.

25.14

PROCESSES FOR ZIRCONIUM–HAFNIUM AND NIOBIUM–TANTALUM

G. A. Yagodin and O. A. Sinegribova
The Mendeleev Institute of Chemical Technology
Soviet Union

1. ZIRCONIUM-HAFNIUM

The hafnium content of zirconium used for reactor engineering should not exceed 0.05%. At present extraction methods are widely used for the separation of zirconium and hafnium. The multiplicity of forms of these elements in solution leads to the fact that zirconium and hafnium can be separated by extraction from practically any medium by selecting corresponding extractants: neutral; acid; or basic ones [1].

1.1. Extraction with Neutral Oxygen-Containing Compounds

1.1.1. Extraction in the Form of Thiocyanate Complexes

The first extraction method for separation of zirconium and hafnium was the separation of the thiocyanate complexes by extraction into diethyl ether, described in 1947–1948 by Fisher et al. [2, 3]. Zirconium (hafnium) sulfates have been used as initial salts. However, Fisher et al. subsequently discarded the use of sulfates because of a spontaneous precipitation and suggested the use of chlorocompounds of zirconium and hafnium as the initial forms [4].

Thiocyanate complexes of zirconium and hafnium are extractable not only into diethyl ether but also into other polar organic substances containing an electronegative atom of oxygen [5]. Hafnium thiocyanate is preferably extracted into the organic phase. The extraction is usually carried out from hydrochloric or sulfuric acid solutions. Although increase of the H_2SO_4 concentration results in decreased extractability of zirconium and hafnium, the separation between them from sulfuric acid media is higher (Table 1), probably as a result of the preferable complexing of less readily extractable zirconium with sulfate ion in an aqueous phase [9].

In most of the extraction systems with thio-

TABLE 1 COMPARISON OF SELECTIVITY OF ZIRCONIUM AND HAFNIUM THIOCYANATE EXTRACTION FROM CHLORIDE AND SULFATE MEDIA

	Separation Factor			
Extractant	From Chloride Media	Reference	From Sulfate Media	Reference
Diethyl ether	10	4	⩽75	2
			32	4
Hexone	9	4	80	10
Tri-*n*-butylphosphate	4	11	21.6	8
Diisoamylmethylphosphonate	4	9	24.9	8
Acetophenone	5	6	33.3	6
n-Butyl acetate	6–13	4	–	–
Ethyl acetate	–	–	30	6
Cyclohexanone	–	–	60–140	7

cyanate, hydrolized thiocyanate compounds of zirconium and hafnium pass into the organic phase. Cl^- and SO_4^{2-} anions with zirconium and hafnium are not extractable. For instance, in DAMPA and TBP compounds, $M(OH)_2(NCS)_2 \cdot 2S$ are extractable, where S = DAMPA or TBP [12, 13].

When performing a counterflow extraction to extract hafnium thiocyanate with TBP from sulfuric acid media, the authors discovered that from stage to stage the separation gets poorer, which indicates the possibility of extracting hydrolised copolymers of zirconium and hafnium [14]. The extraction of zirconium and hafnium thiocyanates into TBP in the form of hydrolized and polymerized compounds has been confirmed also in reports by Zajtcev et al. [15].

Tetrathiocyanate complexes of zirconium and hafnium are extractable only from solutions with a high ratio of SCN^-: M^{4+}, as the extractability of hydrolized compounds is higher than that of nonhydrolized [13-16] ones. The thiocyanate group is coordinated with metal through a nitrogen atom [15, 17].

The separation of thiocyanate complexes of zirconium and hafnium by extraction into hexone has been accomplished on a commercial scale [18]. The initial form is zirconium tetrachloride or agglomerate, obtained following the alkali winning of ore, dissolved in HCl. The aqueous phase containing 100–120 g/liter zirconium is treated with 1-3 *M* NH_4SCN. The extraction is carried out with hexone containing 2.7–2.8 *M* HSCN. Hafnium is reextracted with 5 *M* H_2SO_4 and precipitated as the hydroxide. An additional purification is achieved by reprecipitation of hafnium in the form of salicilate, phthalate or basic sulfate. Zirconium is also precipitated for an additional purification as sulfate, which then can be treated with ammonia and calcined to the oxide form. To stabilize the HSCN, thioglycolic acid is added, or a mixture of hexone with butylacetate or methylethyl ketone [19]; it is advisable to remove iron from the aqueous phase [20].

The extraction of hafnium thiocyanate from highly concentrated solutions of zirconium sulfate into TBP (S = 6–7) was used to extract hafnium simultaneously when producing pure sulfatezirconates [21].

1.1.2. Extraction with Neutral Organophosphorus Compounds (NOPC)

The extraction of zirconium and hafnium salts from HNO_3, HCl, $HClO_4$, and H_2SO_4 into NOPC proceeds through a solvation mechanism. Compounds in the form of $Zr(Hf)A_4 \cdot qS$, where A = inorganic anion and S = extractant, pass into the organic phase. When extracting from nitric acid solutions, the quantity S is usually equal to 2 [22–24]; however, the existence of zirconium (hafnium) monosolvate is possible along with disolvate [24–27].

An increase in the acid concentration contributes to the extraction of zirconium and hafnium into NOPC [28–30]. Korovin et al. [31] have shown that at an ionic strength I = const = 4.0, zirconium and hafnium are extractable from nitric acid solutions into TBP according to the reaction

$$M(OH)_2^{2+} + 2H^+ + 4NO_3^- + 2TBP \rightleftharpoons M(NO_3)_4 \cdot 2TBP + 2\,H_2O$$

An analogous reaction is postulated for the extraction of zirconium into TBPO from 1.5 to 8 M HNO_3 [32]. Solovkin [33] has concluded that the extraction mechanism is

$$Zr(OH)_i^{4-i} + (4 - i)NO_3^- + 2\,TBP \rightleftharpoons M(OH)_i(NO_3)_{4-i} \cdot 2TBP$$

where $i = 0$–2.

Changes in the distribution coefficients of zirconium and hafnium at higher pH values are related to suppression of hydrolysis of these elements. Since the activities a_{H^+} and $a_{Zr^{4+}}$ are determined by the solution composition, the dependence of $\log D$ on $\log[H^+]$ can be different at various salt levels in the aqueous phase [25].

An abrupt enhancement of zirconium and hafnium extraction with increase in pH in the high acidity range (4–6 M HNO_3), when the hydrolysis of zirconium and hafnium is suppressed, is explained by the interaction of zirconium (hafnium) nitrate with adducts $(HNO_3)_m \cdot TBP$ [34]:

$$M^{4+} + 4\,NO_3^- + n(HNO_3)_m \cdot TBP \rightleftharpoons M(NO_3)_4 \cdot n(HNO_3)_m \cdot TBP$$

where $n = 4$ at $m = 1$ and $n = 2$ at $m = 2$.

The change in the mechanism of zirconium and hafnium extraction into NOPC in the high acid concentration range has been confirmed [32, 35–36].

Because of a high tendency of zirconium and hafnium to complex, hydrolyze and polymerize, the reaction kinetics in the extraction of micro- and macroquantities of the elements can vary substantially. When extracting macroquantities of hydrolized compounds of zirconium and hafnium, they may polymerize in the organic phase [37], leading to a poorer separation of zirconium from hafnium (Table 2) [38].

The trend of zirconium and hafnium to hydrolytic polymerization results in the fact that, under certain conditions (low acidity, high metal concentration, and higher temperatures), these elements form extremely stable polymers that are not in equilibrium with the monomeric forms. The properties of these stable multinuclear compounds (SMC) have been reported [39]. The base of the SMC structure is a tetrameric polyion in which the atoms of zirconium

TABLE 2 DISTRIBUTION AND SEPARATION OF ZIRCONIUM AND HAFNIUM EXTRACTED INTO TBP OF VARIOUS CONCENTRATIONS[a]

TBP, vol. %	D_{Zr}	D_{Hf}	S
40	0.12	0.04	3.00
50	0.21	0.10	2.10
60	0.34	0.27	1.26
80	0.60	0.64	0.94

[a] The initial aqueous solution is 1.6 M $ZrOCl_2 \cdot 8\,H_2O$ + 4 M $Ca(NO_3)_2 \cdot 4\,H_2O$ [38].

(hafnium) are completely or partially bound not by hydroxyl groups, but by oxygen bridges. The presence of such compounds makes it impossible, even in a multistage process, to transport all the metal available in the aqueous phase into the organic phase. A complete decomposition of the SMC of zirconium and hafnium occurs only following continuous heating in concentrated nitric acid [40].

In general, one should bear in mind that the extractability of zirconium nitrate is strongly dependent on the nature of the initial compound [41]. The authors recommend that, for investigations into the extraction of the monomeric nitrate or oxinitrate of zirconium, the initial species should be α-oxinitrate of zirconium $[Zr_4(OH)_8(H_2O)_x(NO_3)_8]\,(H_2O)_{16-x}$ [41] or, at low acidity in the presence of notable quantities of NO_3^- ion, α-oxichloride of zirconium $[Zr_4(OH)_8(H_2O)_{16}]Cl$ [42]. A mathematical model for the extraction of zirconium nitrate into TBP is available [43, 44].

Zirconium and hafnium are extractable from hydrochloric acid solutions into TBP in the form of $Zr(Hf)Cl_4 \cdot 2$ TBP [31, 45]. No solvates of the type $H_2MCl_6 \cdot 2$ TBP have been found, although it is pointed out [46] that the hafnium extraction mechanism changes in the high HCl concentration range.

Extraction systems containing TBP, DAMPA, and chloride compounds of zirconium in macroquantities are unstable when heated; the extractant decomposes with cleavage of alkylchloride and formation of polymeric zirconium salts of alkylphosphoric acids [47, 48].

Tributylphosphate extracts hydroxoperchlorates of zirconium and hafnium from solutions deficient in $HClO_4$, their extractability increasing with the hydrolysis rate [49].

Tributylphosphate extracts zirconium from sulfuric acid solutions only at high concentrations of H_2SO_4 [50–51]. The compound extracted contains sulfuric acid. The extractability of zirconium compounds varies in the series nitrate > chloride > sulfate > bromide [51].

From mixtures of inorganic acids, NOPC extracts zirconium and hafnium with separation factors considerably exceeding the corresponding values for extraction from solutions of individual acids [51–56] (see, e.g., Fig. 1). It has been found that mixed nitrate–fluoride [52], nitrate–chloride [53] and nitrate–perchlorate [54] complexes have an enhanced extractability. However, the change in the distribution coefficients has been caused mainly by change in the activity coefficients in the extraction system [31, 54, 55]. A sharp enhancement of zirconium extractability from HCl with the addition of H_2SO_4 has been caused by the dehydration action of sulfuric acid [51].

When extracted with NOPC, zirconium is preferentially transported into the organic phase. In the extraction from nitric acid media into TBP, $S = 3$–30 [57]. A more selective extractant is DAMPA ($S = 40$) [58]. However, the selectivity of the extraction process is determined to a great extent by the aqueous-phase composition. At 4 *M* HNO_3 the maximum selectivity (Fig. 2) is evidently related to the transition from hydrolized forms to nonhydrolized ones in the aqueous phase [9]. A still higher separation is obtained when using mixtures of acids, especially those of HNO_3 and HCl (Fig. 3) [31]. The

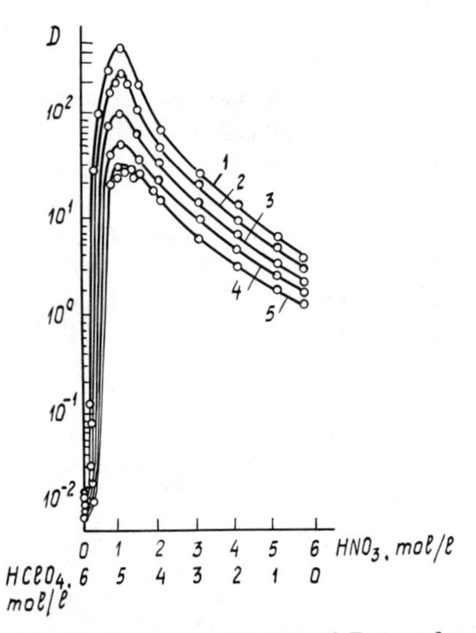

Figure 1. Distribution coefficient of Zr as a function of the HNO_3 and $HClO_4$ concentration in extraction from solutions with $I_H = 6$ mol/liter [55]: (1) 15 g/liter ZrO_2; (2) 20 g/liter ZrO_2; (3) 30 g/liter ZrO_2; (4) 40 g/liter ZrO_2; (5) 50 g/liter ZrO_2.

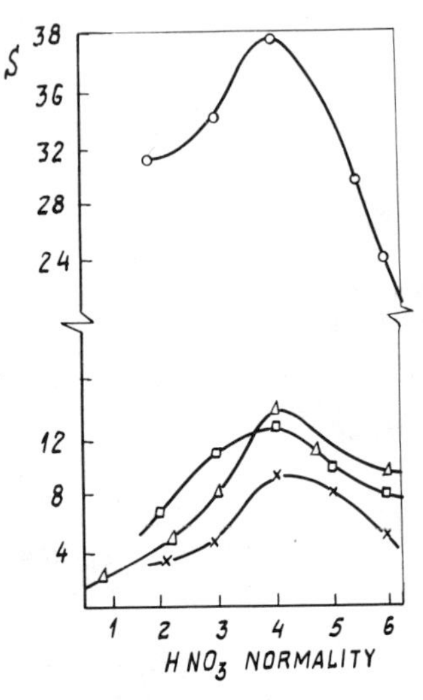

Figure 2. Separation factor Zr/Hf as a function of nitric acid concentration in aqueous phase [9]. Initial metal concentration = 10 g/liter (×–40% TBP; ○–10% DAMPA; △–10% DAMPA, 99% Hf; □–6.5% TOPO).

change in the total concentration of metal in the aqueous phase is almost without effect on the selectivity of the extraction process [9], but a relative increase in the hafnium concentration in the solution results in a poorer separation of zirconium from hafnium (Table 3) [59].

Several versions have been developed of the process for purifying zirconium from hafnium by extraction into TBP [18, 60]. Most often, the initial solution is obtained by dissolving an agglomerate of zirconium with sodium hydroxide in HNO_3. A solution with a high concentration of HNO_3 or salting-out agent containing up to 60 g/liter of zirconium, is delivered for the extraction. This solution must be freed from silicic acid whose presence results in the formation of

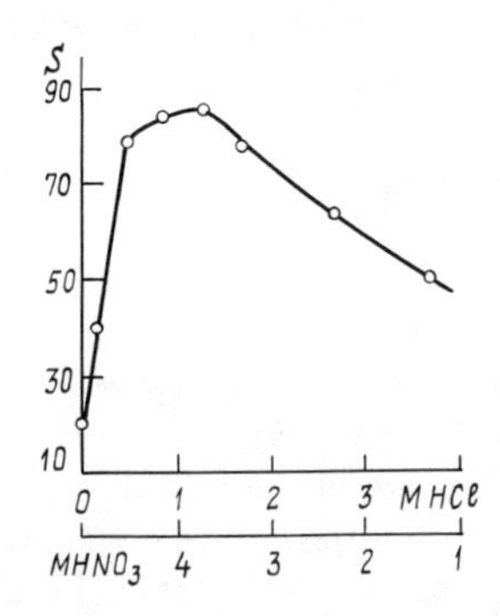

Figure 3. Separation factor Zr/Hf as a function of the ratio of nitric and hydrochloric acid concentrations in aqueous solution when extracting with sulfoxides [31].

TABLE 3 DISTRIBUTION OF ZIRCONIUM AND HAFNIUM AT VARIOUS CONCENTRATIONS IN THE INITIAL SOLUTION[a]

HfO_2 in Initial Form, %	Concentration $\sum Zr(Hf)O_2$, g/liter: Aqueous Phase	Organic Phase	Content HfO_2 in $\sum Zr(Hf)O_2$, %: Aqueous Phase	Organic Phase	Distribution Factors: $D_{\sum Zr(Hf)}$	D_{Zr}	D_{Hf}	Separation Coefficient, S
2.4	39.3	18.4	3.0	0.14	0.47	0.48	0.023	20.9
16.2	35.2	17.2	23.9	1.8	0.49	0.63	0.040	15.6
50.0	37.2	14.6	69.5	11.7	0.38	1.23	0.072	16.9
95.8	34.0	6.2	97.5	85.9	0.18	1.0	0.16	6.23

[a]Extractant–50% TBP in *o*-xylene; $V_{org}:V_{aq} = 1:1$ [59].

interphase films and poorly laminating emulsions [18]. Silicic acid is removed by chemical methods such as precipitation with gelatin [61], by dehydration with the formation of a readily filtered sediment of SiO_2 [62], or by mechanical techniques using vibration [63]. Tributylphosphate is used in the form of a 20–60% solution in aliphatic or aromatic diluents. The extract is washed with 5 M NHO_3, and is reextracted with water. Zirconium from the reextracted product is precipitated in the form of hydroxide and sulfate. The best refining of the hafnium product is achieved by ion-exchange methods [64–66]. Presented in Table 4 are the conditions for the separation of zirconium from hafnium by extraction with TBP from nitrate solution.

1.1.3. Extraction with Neutral Extractants Containing no Phosphorus

The separation of zirconium from hafnium by extraction with alcohols, esters, ketones, and so on is usually not high [67]. More attention has recently been paid to a new class of extractants–dialkylsulfoxides and petroleum sulfoxides (PSO). The zirconium compound extractable with dialkylsulfoxides has the composition $ZrA_4 \cdot 2S$ [68, 69]. The separation factor from 2 M HNO_3 is equal to ~10 [70]. Petroleum sulfoxides obtained by the oxidation of a fraction of petroleum sulfides with the boiling temperature of 200–250°C extract zirconium and hafnium from solutions containing up to 6.7 M HCl or 6 M HNO_3; at a higher acidity PSO would be destroyed [71].

When nitric acid is partly replaced by hydrochloric acid, the distribution factor of zirconium passes through a maximum, whereas hafnium extraction becomes poorer and the separation factor passes through a maximum (Table 5) [72].

1.2. Extraction with Acidic Extractants

Zirconium and hafnium could be extracted by ion exchange with acid extractants from solutions of HNO_3, HCl, $HClO_4$ and H_2SO_4, of a concentration such that these elements are in the form of simple or complex cations.

When extracted into organophosphoric acids (HA), zirconium and hafnium form with an anion of organophosphoric acid a compound $Zr(Hf)A_4$; when HA is excessive, the compound is solvated with one or two molecules or organophosphoric acid. Inorganic anions are sometimes in the compound extracted. This is especially characteristic of the nitrate ion. Compounds containing Cl^- and ClO_4^- are extractable from low-acidity solutions [5]. Under certain conditions a compound containing sulfate ion could pass into the organic phase [73, 74]. The extraction of hydrolized compounds is possible [5].

At a higher aqueous phase acidity the extraction mechanism changes to solvation. In the organic phase saturated with metal, formation of polymeric compounds is possible [5].

Zirconium and hafnium form extremely stable compounds with organophosphoric acids. It is possible to reextract these elements only by using readily complexing species, for example, NH_4F, $(NH_4)_2C_2O_4$, or $NaHCO_3$ for reextracting from D2EHPA [75].

Zirconium and hafnium compounds with organophosphoric acids are under certain conditions stronger extractants for some elements than organophosphoric acids themselves [76]. For instance, with europium [77, 78] the compound $Zr(Hf)A_4 \cdot 2HA$ is an extractant having an enhanced extractability that extracts europium through the ion-exchange mechanism:

$$2Eu^{3+} + 3\,H_2MA_6 \rightleftharpoons Eu_2(MA_6)_3 + 6H^+$$

TABLE 4 CONDITIONS FOR THE SEPARATION OF ZIRCONIUM AND HAFNIUM BY SOLVENT EXTRACTION FROM TBP NITRATE SOLUTIONS [60]

Extractant–TBP, vol. %	60	20	50	60	60	50
Diluent	White spirit	Kerosene	Xylene	Heptane	Dibutyl alcohol	Heptane
Working solution: Zr(Hf) concentration g/liter	22	15	80	92	51	90
Hafnium content, %	2.4	2.0	2.0	2.4	2.1	2.0
HNO_3, mole/liter	3.0	6.5	3.0	5.0	2.6	8.0
$NaNO_3$, mole/liter	3.5	–	–	–	–	–
HCl, mole/liter	–	–	–	–	1.74	–
$CaCl_2$, mole/liter	–	–	–	–	0.7	–
Number of separation stages	9	10	10	14	7	Not mentioned
Separation factor	10	10	10	3–30	25–40	Not mentioned
Relationship of flows: working solution : washing solution : extractant	0.48 : 0.48 : 1	2.3 : 1 : 10	0.4 : 0.2 : 1	1 : 1.18 : 4.85	1 : 1 : 1	1 : 0.6 : 3.6
Extracted products: Zr(Hf) concentration g/liter	10	4.5	35	17	36	Not mentioned
Hafnium content, %	0.02	0.04	0.01	0.01	0.017	–
Refined product Zr(Hf) g/liter	1.0	1.0	4.4	2.7	0.7	–
Hafnium content, %	20–40	⩽50	33	43.8	⩽96	–

The enhancement is especially high in the case of hafnium (Fig. 4). The increase in the distribution factors is also observed in the extraction of actinide elements [78]. The enhancement of the extractive properties of organophosphoric acids is caused mainly by redistribution of the electron density in the HA molecules, solvating $Zr(Hf)A_4$ and resulting in change in the acid properties.

TABLE 5 DISTRIBUTION OF ZIRCONIUM AND HAFNIUM BETWEEN MIXED NITRIC-HYDROCHLORIC ACID SOLUTIONS AND 50% PSO SOLUTION [72]

Cl^- Concentration in Aqueous Phase g-ion/liter	Equilibrium Concentration of $Zr(Hf)O_2$, g/liter		Concentration in Equilibrium Phases of HfO_2, %		Distribution Factor		Separation Factor
	Aqueous Phase	Organic Phase	Aqueous Phase	Organic Phase	D_{Zr}	D_{Hf}	S
6	5.4	1.2	27.4	0.91	0.304	0.0074	41.1
4.5	4.97	33.5	27.6	0.55	9.26	0.134	69.1
3	5.12	33.9	27.3	0.78	9.05	0.189	47.7
1.5	5.14	26.9	25.4	0.95	6.97	0.195	35.8
0	5.3	25.8	25.6	1.1	6.43	0.211	30.7

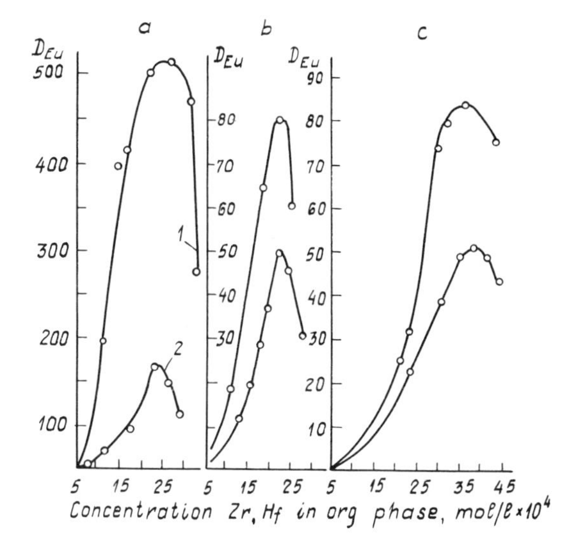

Figure 4. Europium distribution coefficient as a function of Zr(Hf) content in organic acid [77]: (1) Hf; (2) Zr. (*a*) Organic phase, 0.15 *M* DAPA in octane; aqueous phase, 1.54 *M* HNO_3. (*b*) Organic phase, 0.01 *M* DNPA in decane; aqueous phase, 0.36 *M* HNO_3. (c) Organic phase, 0.018 *M* 2EHPPA in octane; aqueous phase, 0.05 *M* HNO_3.

1.3. Extraction with Basic Extractants

Zirconium and hafnium are extracted by basic extractants (primary, secondary, ternary amines and quaternary ammonium bases) from solutions in which the elements are present in the anionic form. Cerrai and Testa [79] have studied the extraction of zirconium and hafnium from hydrochloric acid solutions with primary, secondary, and ternary amines and found that ternary amines have the maximum selectivity. [The value of S, when using tri-*n*-octyl-amine (TOA), reached 29]. The extraction of chlorides into TOA has been described in [80]. For the extractive separation of zirconium and hafnium it has been proposed to use tri-phenylamine [81], tri-*n*-benzylamine [82, 83], dibutylamine, and tri-*n*-butylamine [83]. The separation factor for the extraction into tri-benzylamine under optimal conditions is 26, into tri-*n*-butylamine 7–10, into dibutylamine 2. As reported [81], hafnium preferably passes into the triphenylamine-in-kerosene phase; however, zirconium is more readily extractable in most of the other extraction systems with amines.

When extracting zirconium and hafnium from hydrochloric acid media into tricaprylmonomethyl ammonium chloride (Aliquat 336), the distribution coefficients increase sharply with HCl concentration. At an acidity of 7.8 *M* HCl, $S = 72$ [84].

The mechanism of zirconium and hafnium sulfate extraction with amines is very intricate. The problem of its elucidation is aggravated by the complicated chemical behavior of both zirconium and hafnium in sulfate solutions.

Extraction with tri-*n*-octylamine solutions has been studied more thoroughly. As a rule, the investigators point out the difference in the compounds of zirconium (hafnium) extractable from weakly acid or strongly acid solutions into TOA [85, 86] and other amines [87].

Extraction of zirconium and hafnium sulfates into TOA has been studied in detail [88–93]. On the basis of extraction equilibrium and kinetic data, and studies of the sulfate-group exchange rate and infrared spectroscopy of the extracted products, it has been found that at a high acidity nonhydrolized complexes are extractable and at a low acidity hydrolized compounds pass into the organic phase; the extraction mechanism of both types of complexes is complicated and the extraction reaction is a multistage one.

Judging from the equality of the effective activation energies of the sulfuric acid reextraction from tri-*n*-octylamine (TOA) bisulfate solution and the extraction of nonhydrolized hafnium sulfate with TOA bisulfate [88], and also data on the SO_4^{2-}-ion exchange kinetics between the aqueous and organic phases [89], the following extraction mechanism is proposed for nonhydrolized sulfates [90]:

$$M(SO_4)_3^{2-} + 2\,(NR_3H)\,HSO_4 \rightleftharpoons M(SO_4)_3(NR_3H)_2 + 2HSO_4^- \quad (1)$$

$$2(NR_3H)HSO_4 \rightleftharpoons (NR_3H)_2SO_4 + H_2SO_4 \quad (2)$$

$$M(SO_4)_3(NR_3H)_2 + (NR_3H)_2SO_4 \rightleftharpoons M(SO_4)_4(NR_3H)_4 \quad (3)$$

In this case the second and third stages of the process are accomplished at the interface, where the TOA sulfate concentration exceeds the equilibrium value because of a higher surface activity of sulfate [94].

In the extraction of nonhydrolized sulfate complexes of zirconium and hafnium with TOA, the separation factor is equal to 8–10 [92].

In the extraction of hydrolized sulfates the destruction of polymers takes place with the formation of monomeric compounds, for example, $Zr(OH)_2(SO_4)_2$. At the moment of formation, the coordination sphere of such a complex is not saturated; therefore, their extraction at the first stage follows the addition mechanism:

$$Zr(OH)_2(HSO_4)_2 + (NR_3H)_2SO_4 \rightleftharpoons Zr(OH)_2(SO_4)_2(HSO_4)_2(NR_3H)_2 \quad (4)$$

Then a slow decrease in the rate of zirconium hydrolysis in the organic phase occurs:

$$Zr(OH)_2(SO_4)(HSO_4)_2(NR_3H)_2 \rightleftharpoons Zr(SO_4)_3(NR_3H)_2 + H_2O \quad (5)$$

The final stage of the process is similar to that accomplished in the extraction of nonhydrolized complexes:

$$Zr(SO_4)_3(NR_3H)_2 + (NR_3H)_2SO_4 \rightleftharpoons Zr(SO_4)_4(NR_3H)_4 \quad (6)$$

Since at the first stage of the process a compound is formed with the ratio Zr:TOA = 1:2, but at the final stage the ratio Zr:TOA = 1:4, the metal concentration in the organic phase varies anomalously: initially the amount of metal passing into the organic phase apparently exceeds the equilibrium value [91].

Stable hydrolized polymers are also capable of passing into the organic phase following the exhaustion of hydrolized monomers and destruction of unstable multinuclear complexes in conformity with the extractability series of zirconium sulfates with TOA:nonhydrolized complexes > hydrolized monomers > stable hydrolized polymers [92].

The dependence of the separation factor of zirconium and hafnium sulfates on the concentration of free acid has been determined for the first time [93] (Fig. 5).

The extraction of zirconium and hafnium sulfates with amines could be used for the separation of zirconium from hafnium [95] and for purifying these elements [96].

Zirconium and hafnium are extracted by TOA sulfate from fluorometallate solutions as complexes ZrF_6^{2-} and HfF_6^{2-} through the anion-exchange reaction [97]:

$$MF_6^{2-} + 2(NR_3H)HSO_4 \rightleftharpoons MF_6(NR_3H)_2 + 2\ HSO_4^-$$

The extraction of zirconium and hafnium by Aliquat 336 solutions from sulfuric acid media is

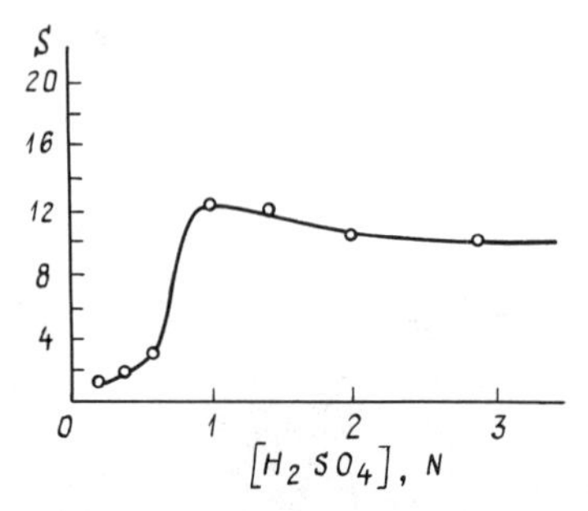

Figure 5. Separation factor Zr/HF as a function of H_2SO_4 concentration in the extraction with TOA [92]: (1) 0.2 *M* TOA in xylene; [Zr] = 8.9 g/liter; Hf/Zr content = 0.8.

reported [98]. The maximum extractability is observed in the range of 0.02–0.05 *M* H_2SO_4, the maximum separation ($S = 80.2$) at 0.8 *M* H_2SO_4.

The use of acid mixtures as the initial aqueous phases leads to improved selectivity in a number of cases [80].

2. NIOBIUM-TANTALUM

Niobium and tantalum are separated in order to obtain pure niobium and its alloys for nuclear power engineering, aviation, and space engineering, and pure tantalum for chemical plant construction. The separation uses solvent extraction techniques. Tantalum and niobium are stable in solution only as tantalates, niobates, and complex ions, such as those with fluoride, thiocyanate, and various complexing ions [99]. Most often, extraction systems with HF or with mixtures of HF and H_2SO_4 or other acids are used [100].

2.1. Extraction with Neutral Oxygen-Containing Extractants

2.1.1. Extraction with Ketones

The distribution of niobium and tantalum when extracted with ketones such as diisopropyl ketone, methylisobutyl ketone, and cyclohexanone has been studied in detail. Tantalum is extractable with diisopropyl ketone from mixtures of 0.4 *M* HF with HCl, HNO_3, H_2SO_4, and $HClO_4$ much more readily than niobium. The maximum separation factor for tantalum equal to 880 was obtained with extraction from a solution of 0.4 *M* HF + 3.92 *M* HNO_3 [101]. Niobium is readily extracted in the absence of HF, such as from 10 *M* HCl [102].

In analytical practice [103-106], as well as in engineering [107], the separation of niobium and tantalum by solvent extraction with methylisobutyl ketone (hexone) is widely used. Werning et al. [108] have shown that the extraction of niobium and tantalum with hexone increases at higher concentrations of HCl and HF in the aqueous phase. However, at low concentrations of acids tantalum passes into the organic phase much better than niobium. The maximum extraction of tantalum is observed at 4-6 *M* HF; niobium at 4 *M* concentration is essentially not extracted. The separation factor in such systems exceeds 700. Tantalum is readily extracted also from mixtures of 0.2 *M* HF + 4 *M* HNO_3 and 0.2 *M* HF + 6 *M* H_2SO_4 [109]. The compounds H_2TaF_7 and H_2NbF_7 are extractable into hexone [110, 111]. The preferential extraction of tantalum is explained by the greater stability of the H_2TaF_7 complex in aqueous solutions as compared to H_2NbF_7, which is easily hydrolized to form H_2NbOF_5 but not extractable with hexone. The mixture of hexone with diisobutyl ketone [112] and diisopropyl ketone [113] was used to separate niobium and tantalum at the stage of reextraction with hydrochloric acid from an organic phase containing water-free $TaCl_5$ and $NbCl_5$. The compound 12 *N* HCl reextracts tantalum preferentially, whereas 6 *N* HCl favors Nb.

The separation of niobium and tantalum by solvent extraction with hexone is commercially used [107]. The ore is extracted by hydrofluoric acid, and the resulting extract is filtered; then sulfuric acid is added to the clear solution; niobium and tantalum are extracted together with hexone. Niobium is removed at the stage of washing the organic phase with water. Further metal extraction from the depleted ores is suggested with the use of hexone [114].

It is said that the use of cyclohexanone results in pure niobium, whereas the use of hexone gives pure tantalum [115]. A fluoride complex of tantalum is much better extracted into an organic phase than that of niobium. In the absence of HF, excess niobium is almost not extractable with cyclohexanone; however, increase in HF or other mineral acid concentration contributes to a better extractability of Nb, especially evident when using HCl and HF (Fig. 6 [115]). At HCl concentrations of 170-180 g/liter, formation of nonflocculating emulsions was observed. Cyclohexanone could be used also as a collective extractant to extract niobium and tantalum from a solution containing 100 g/liter HF + 250 g/liter H_2SO_4 with subsequent reextraction of niobium with water. Goroshchenko et al. [116] suggest

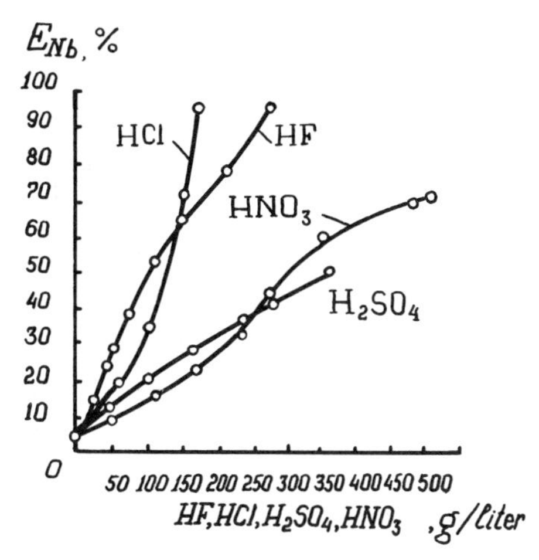

Figure 6. Extraction of niobium into cyclohexanone as a function of the HF, HCl, H_2SO_4, and HNO_3 concentration in the initial solution [115].

that cyclohexanone extracts the fluoride complex of tantalum from sulfate-ammonium solutions better than do all other extractants: even small additions of NH_4F to a solution containing 470 g/liter of H_2SO_4 and 215 g/liter of $(NH_4)_2SO_4$ are sufficient for a complete transport of tantalum into the cyclohexanone phase. Niobium is totally reextractable from cyclohexanone containing fluorides of niobium and tantalum by a mixture of 100g/liter H_2SO_4 + 50 g/liter $(NH_4)_2SO_4$, because of a high tendency of the niobium fluorocomplex to hydrolyse.

The compound extractable into cyclohexanone has a composition $HMF_6 \cdot qH_2O \cdot n\text{CHex}$ where $q \approx 4$ and $n < 4$ [117]. A high selectivity for extraction by cyclohexanone is explained by the difference in the extraction constants (6.6×10^6 for niobium and 8.7×10^{10} for tantalum). To determine the optimal conditions for separating niobium and tantalum with cyclohexanone ($S \approx 10^3$-10^4), the Box-Wilson method of hill-climbing [118, 119] has been used.

Tantalum is extractable with cyclohexanone only as fluoride complexes; niobium is also extractable in the form of thiocyanate complexes, for instance, extraction from a solution with an H_2SO_4 concentration above 2 *M* containing 4.9×10^{-1} *M* SCN^- gives the niobium compound $[H(H_2O)_n(\text{CHex})_m]$ $[Nb(NCS)_4]$ [120].

Figure 7 presents a continuous scheme of niobium and tantalum separation by extraction with cyclohexanone [121]. This permits processing of solutions obtained in the sulfuric acid winning of titanium-niobium-tantalum raw materials with subsequent precipitation of the main bulk of the titanium by salting out a double sulfate of titanyl

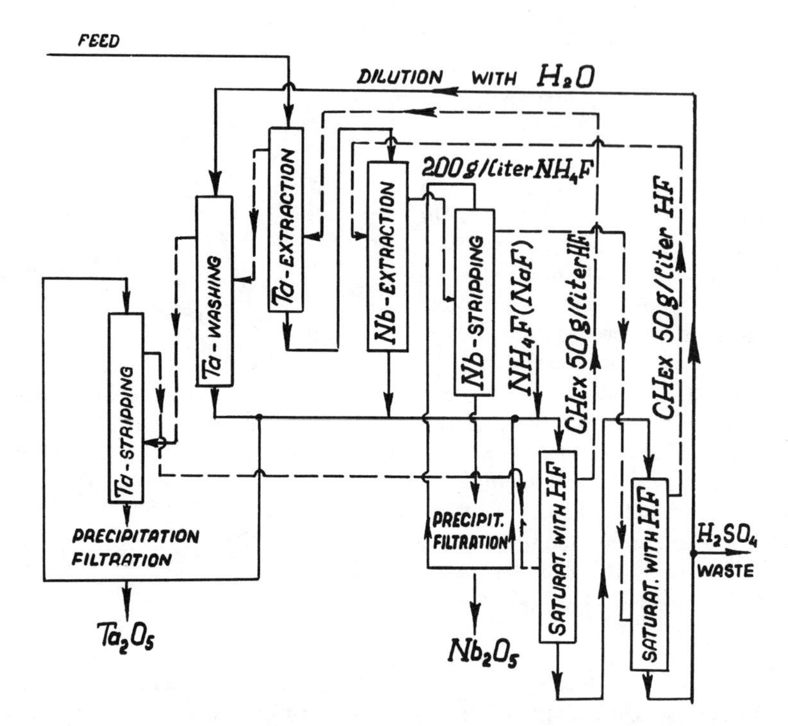

Figure 7. Continuous scheme for Nb and Ta separation by extraction with cyclohexanone [100].

and ammonium $(NH_4)_2TiO(SO_4)_2 \cdot H_2O$. The solutions obtained contain 340–400 g/liter of H_2SO_4, 180–200 g/liter $(NH_4)_2SO_4$, 7–15 g/liter Nb_2O_5, 0.5–1.5 g/liter Ta_2O_5, and 3–4 g/liter TiO_2. In column 1, tantalum is extracted from the sulfuric acid solution; in column 2, the extract is washed to separate niobium; in column 3, tantalum is reextracted with NH_4F; in column 4, niobium is extracted from sulfuric acid solution; in column 5, niobium is reextracted with NH_4F. In column 6, cyclohexanone from the tantalum circuit becomes saturated with HF from the spent sulfuric acid solution and is returned to the cycle. In column 7, cyclohexanone from the niobium circuit is saturated with hydrofluoric acid. Thus HF is introduced into the process as a solution in cyclohexanone. The losses of HF are replenished by the addition of NaF, NH_4F, or technical-grade HF to the spent sulfuric acid solution prior to delivery into columns 6 and 7. This scheme gave the final products containing: in Ta_2O_5, TiO_2 0.15%; SiO_2 1.0%; Fe_2O_3 0.25%; SO_3 0.40%; Nb_2O_5 0.75%; in Nb_2O_5, TiO_2 0.30%; SiO_2 0.55%; Fe_2O_3 0.25%; SO_3 0.15%; Ta_2O_5 0.02%.

One should bear in mind that the extractability of cyclohexanone becomes poorer as a result of continuous contact with acid solutions, as a result of the accumulation of the cyclohexanone self-condensation product–cyclohexylene cyclohexanone [122].

The extraction of ^{95}Nb with oxygen-containing solvents from nitric acid media has been studied [123, 124]. In the analytical chemistry of tantalum and niobium, the extractions of thiocyanate complexes into ethylacetate [125, 126], into ester [127–129] and into *n*-butanol [130] have been used.

2.1.2. Extraction with Tributylphosphate (TBP) and Trioctylphosphine Oxide (TOPO)

Tantalum is readily extractable with TBP from dilute solutions of HF, whereas extraction of niobium is appreciable only in moderately acid solutions and increases abruptly with HF concentration above 8 *M* [131]. Nisimura et al. [132] have shown that, when extracting with 15% TBP, the tantalum distribution coefficient decreases from 13.03 to 3.12 with HF concentration growth from 1.28 to 15.9 *M*, whereas the niobium distribution coefficient under the same conditions increases from 0.1 to 28.8.

Extraction proceeds through the solvation mechanism. Tantalum is extractable in the form of $HTaF_6$ [133, 134] or (at $[HF] > 10\,M$) in the form of H_2TaF_7 [133, 135]. The solvation number is equal to 3 [133, 134]. Niobium is extractable in the form of $HNbF_6 \cdot 3$ TBP or $H_2NbF_7 \cdot 3$ TBP; from solutions saturated with niobium the extraction of $HNbOF_4 \cdot 3$ TBP or $H_2NbF_5 \cdot 3$ TBP [133] is possible. Giganov and Ponomarev [133] are of the opinion that the separation of tantalum and niobium, when extracted from weakly acid solutions, is due to formation of poorly hydrolized niobium compounds. However, Korovin and Kol'tsov [136] suggest that both of these elements are extractable as hydrolized compounds, but that tantalum at the low acidity is extracted through the solvation mechanism as $Ta(OH)F_4 \cdot n$TBP and $TaF_5 \cdot n$TBP, but at the high acidity, through the hydration-solvation mechanism, whereas niobium is extractable only through the hydration-solvation mechanism, so that niobium at the low acidity is essentially not transported into the organic phase.

The difference in the extractabilities at low concentrations of HF permits efficient separation of niobium and tantalum at both the extraction and reextraction stages. The HF concentration should be below 3 *M* [132]. To obtain pure tantalum and niobium from the ore concentrate containing ~61% Ta_2O_5; ~14% Nb_2O_5; ~12% Fe; and also impurities MnO_2, TiO_2, SnO_2, SiO_2, the concentrate was dissolved in HF, and then tantalum and niobium were extracted with TBP from 15 to 16 *M* HF. Then both the elements were reextracted with ammonia; this was followed by selective reextraction of tantalum from 1.15 *M* HF. The purification was considerably better than in the extraction with hexone. Ryabchikov and Volynets [137] have extracted niobium and tantalum from 5 to 7 *M* HF into 80% TBP in kerosene and then have separated the elements at the stage of washing out with 1 *M* HF. In the aqueous phase, HF could partially be replaced with H_2SO_4. From purely sulfuric acid solutions, niobium and tantalum are extractable with TBP, the extractability increasing with the H_2SO_4 concentration. Niobium is preferentially transported into the organic phase. From H_2SO_4 of 7.5–11.35 *M*, niobium is extractable as a complex $m HNbO(SO_4)_2 \cdot n$TBP [138].

However, in the presence of fluoride ion, sulfuric acid acts as a salting-out agent; $HNbF_6 \cdot 3$ TBP compound is extracted and, possibly, $H_2NbF_7 \cdot 3$ TPB and $HTaF_6 \cdot 3$TBP, $H_2TaF_7 \cdot 3$TBP [139–141]. Kaplan and Baram [142] have shown that the isotherms in the tantalum and niobium extraction from a sulfate–fluoride medium are close; although the mass concentration of tantalum in the organic phase is higher, the niobium in it is more abundant in terms of molar concentration.

Two technological schemes have been suggested for the separation of niobium and tantalum. If the solution contains much tantalum, the latter could be extracted from low acid solutions (1 *M* HF + 0.5 *M* H_2SO_4), with niobium left in the aqueous phase; if niobium prevails, both metals could be transported into the organic phase and separated at the reextraction stage. For the coextraction of niobium and tantalum, it is necessary to have a high acidity in the aqueous solution (6 *M* HF + 3 *M* H_2SO_4 [139]; 6 *M* HF + 8 *N* H_2SO_4 [140]; 12–13 *M* HF + 3–4 *N* H_2SO_4 [133]). Niobium is reextractable with water or 0.5 HF and tantalum, with 5% solution of soda and NH_4F solution [139, 140]. Both metals could be reextracted with ammonia, then tantalum could be reextracted from the solution of 1–1.5 *M* HF + 1–1.5 *M* H_2SO_4 [133].

Tantalum and niobium are extractable from chloride solutions at a high concentration of HCl (10–11 *M*) [143] (Fig. 8) or salting-out agents $CaCl_2$, H_2SO_4 [144]. Figure 8 shows the distribution coefficient dependence on the HCl concentration in the aqueous solution [145]. The extract contains the hydrolized compounds $H_2[NbO(OH)Cl_4]$ [146], $HNbOCl_3$, H_2NbOCl_5 [147, 148], $NbOCl_3$ and $TaOCl_3$ [144]. As reported [145], these compounds contain OH groups: $HNb(OH)Cl_5$, $HNb(OH)_2Cl_4$, $Nb(OH)Cl_4$, $Nb(OH)_2Cl_3$, $Nb(OH)_3Cl_2$. Extraction of $NbCl_5$ is possible. Extraction of niobium and tantalum in the form of chlorine-containing complex acids up to 11 *M* HCl concentration is scarcely probable [144]. The solvation numbers

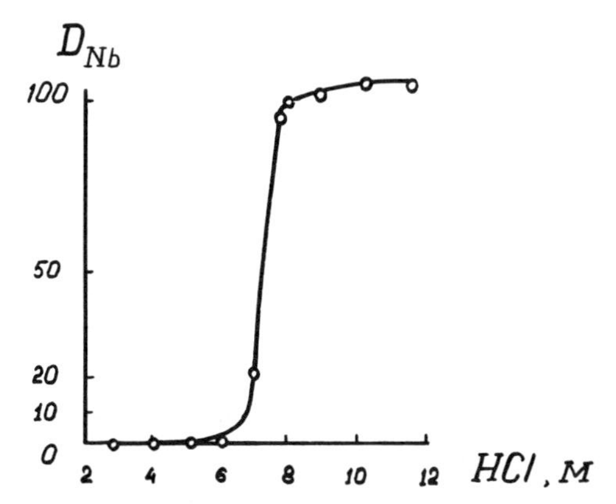

Figure 8. Distribution coefficient of Nb as a function of hydrochloric acid concentration when extracted with TBP [145]. Extractant, 100% TBP; $[Nb]_{init} = 0.02\,M$.

determined in various studies were: for Ta, 1 [143], 1 and 2 [144, 149], for Nb, 1 [145], 1 and 2 [143, 144, 149], 3 [147, 148]. A method is suggested for separation of niobium and tantalum, consisting of dissolving solid commercial chlorides containing 3% tantalum in TBP with subsequent washing out with NH_4F (100 g/liter) solution in counterflow [150]. In that case the refined product is pure niobium. Niobium is also reextractable with sulfuric acid and tantalum, with sulfuric acid in the presence of KSCN [151].

A comparison has been made between the extraction of tantalum and niobium from HCl with various extractants. Tantalum is extractable from 10 *M* HCl only by TBP, and to a lesser degree with acetophenone. The extractability of niobium from 9 *M* HCl by TBP is comparable with that by acetophenone, benzaldehyde, alkylacetates, and certain alcohols [152]. Microquantities of niobium are extractable from nitric acid solutions by TBP at high concentrations of HNO_3 [153]. The compound $LiTaF_6$ is extractable with TBP from aqueous solutions containing K_2TaF_7 and also considerable quantities of lithium chloride, nitrate, or sulfate [154]. Niobium in the compound $[H \cdot mH_2O \cdot TBP] \cdot [NbO(NCS)_4] \cdot n$HSCN is extractable from solutions of HCl and H_2SO_4 in the presence of 0.2 *M* NaSCN [155]. The composition of a tantalum complex has been determined when extracted from 3 *M* $HClO_4$ in the presence of HF into TOPO [156]; $TaF_4(ClO_4) \cdot q$ TOPO [for $q = 2$, $K = (1.00 - 0.26) \times 10^4$]:

$$TaF_4^+ + ClO_4^- + q\ \text{TOPO} \rightleftharpoons TaF_4(ClO_4) \cdot q\ \text{TOPO}$$

2.1.3. *Extraction with N-Oxides*

To extract and separate niobium and tantalum, a study has been made using *N*-oxides such as 2-*n*-nonylpyridine-*N*-oxide (2-NPO) [157, 158], trialkylamine oxide (TAAO) [158, 159], and 4-(5-nonyl)-pyridine-*N*-oxide [159] that under certain conditions have proven to be selective for tantalum. The optimal acidity is 0.01 *M* HCl, HNO_3, or H_2SO_4. The following reactions for the extraction into 4-(5-nonyl)pyridine-*N*-oxide (NPyOx) are suggested:

$$Ta(OH)_nA_{5-n} + 3\text{NPyOx} \rightleftharpoons Ta(OH)_nA_{5-n} \cdot 3\text{NPyOx}$$

where $n \leqslant 5$, A = Cl or NO_3 and

$$Ta(OH)_n(SO_4)_{(5-n)/2} + 3\ \text{NPyOx} \rightleftharpoons Ta(OH)_n\,(SO_4)_{(5-n)/2} \cdot 3\ \text{NPyOx}$$

The reactions with TAAO are similar [159].

When using fluoride solutions containing 62.3 g/liter of $Ta(Nb)_2O_5$ and 1.6–4.3 *M* HF, the extraction of Ta is almost complete and decreases only negligibly at very high concentrations of HF. The maximum separation of tantalum and niobium ($S \sim 10^3$) is observed with solutions having compositions close to stoichiometric ones. In the presence of excess (4 *M*) HF and 3 *M* H_2SO_4, *N*-oxides prove to be collective extractants for both tantalum and niobium. The compound NH_4NO_3 is the most selective reactant for the separation at the reextraction stage. In that case practically all tantalum remains in the organic phase. The compounds NH_4F, KOH, and $(NH_4)_2CO_3$ are used for its reextraction.

2.1.4. *Extraction with Sulfoxides*

For the extraction of tantalum and niobium from hydrochloric acid solutions, the selectivity of alkylsulfoxides is much better than that of TPB [160] (Fig. 9). Petroleum sulfoxides could be used for the extraction of tantalum and niobium. Rozen et al. [161] have extracted tantalum and niobium from fluoride–sulfate solutions

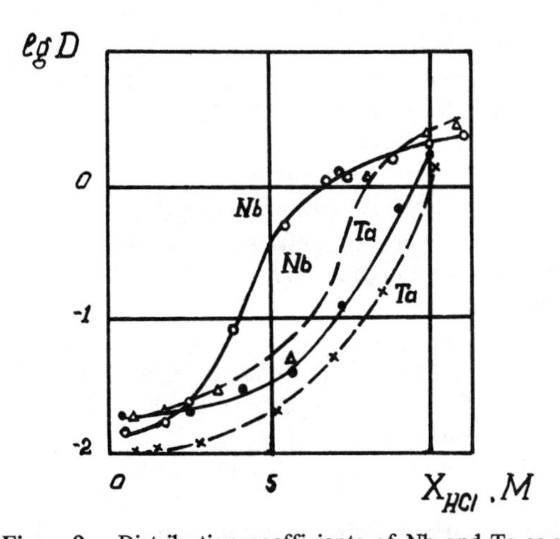

Figure 9. Distribution coefficients of Nb and Ta as a function of hydrochloric acid concentration when extracted with sulfoxides [160]. Extractant, 0.5 *M* DOSO in benzene (solid lines); 0.5 *M* TBP in benzene (dashed lines); [Nb] = $5.2 \cdot 10^{-2}$ gr · atom/liter; [Ta] = $5.0 \cdot 10^{-2}$ gr · atom/liter.

($5\ M$ HF-$5\ N\ H_2SO_4$) with a mixture of sulfoxides obtained by the oxidation of petroleum sulfides extracted from a petroleum fraction (boiling temperature 260–360°C). The extraction was carried out by 0.7 *M* (SO group) solution of sulfoxides in *o*-xylene. The following compounds are extractable.

$$TaF_6^- + H^+ + 3S \rightleftharpoons HTaF_6 \cdot 3S$$

$$TaF_7^- + 2H^+ + 3S \rightleftharpoons H_2TaF_7 \cdot 3S$$

$$NbOF_4^- + H^+ + 3S \rightleftharpoons HNbOF_4 \cdot 3S$$

where S is an extractant.

Golovacheva et al. [162] have studied the dependence of tantalum and niobium extraction on the HF concentration at varying concentrations of H_2SO_4 in the initial solution. Sulfoxides for this purpose are obtained by the oxidation of petroleum sulfides from a petroleum fraction boiling between 240 and 390°C. The petroleum sulfoxides have an average molecular weight of 260 and are dissolved as a 65% solution in benzene. The results are shown in Fig. 10. Petroleum sulfoxides surpass TBP so far as efficiency is concerned. Table 6 presents the relative efficiency values in terms of the distribution coefficients in the extraction with 1 *M* extractants from solutions containing 0.23 mole/liter of niobium (tantalum) at a 1 : 1 phase volume ratio.

Chernyak et al. [163] have extracted tantalum and niobium from a solution containing 115 g/liter Nb_2O_5; 8.7 g/liter Ta_2O_5, 40 g/liter free HF and 300 g/liter H_2SO_4 with petroleum sulfoxides obtained from a petroleum fraction boiling at 280–360°C. The sulfoxide sulfur content was 7.1%. A 50% solution of petroleum sulfoxide in C_6–C_8 alcohols was used to avoid the formation of a third phase. The extracted volume contained 140 g/liter of Nb_2O_5 and 15 g/liter Ta_2O_5, which exceeded the volume of undiluted TBP. Niobium could be reextracted by water and tantalum by NH_4F solution (50 g/liter [162]). The usage of solutions obtained from depleted ores has been described [164].

The disadvantage of sulfoxides is their high viscosity, which necessitates the use of diluted solutions. Nikolaev et al. [165, 166] have studied the extraction of niobium and tantalum from fluoride–sulfate media into diluted and undiluted sulfoxides. In both cases the addition of HF to the solution containing ~300 g/liter of H_2SO_4 and ~200 g/liter of $(NH_4)_2SO_4$ caused an abrupt increase in the tantalum extraction (D_{max} at ~30 g/liter HF). In spite of the troubles caused by the high viscosity of the extractant, the undiluted petroleum sulfoxides can extract large quantities of tantalum and niobium into the organic phase (more than 200 g/liter of Ta_2O_5 and more than 150 g/liter of Nb_2O_5).

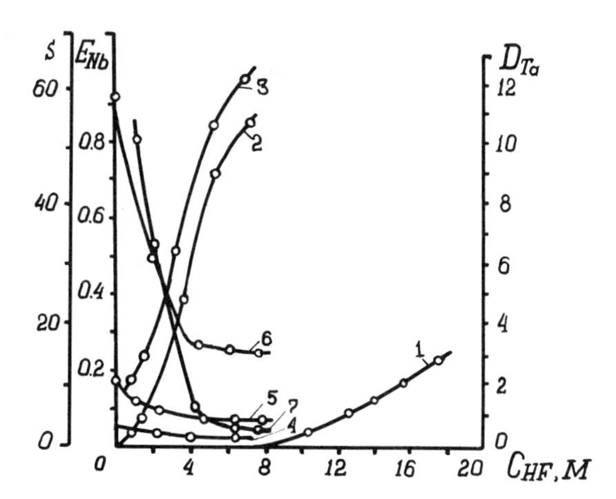

Figure 10. Distribution coefficients of Nb and Ta and their separation factor (*S*) as a function of HF concentration in conjugate aqueous phase for various H_2SO_4 contents of initial solution [165]. Extractant: 1 *M* PSO in benzene

Curves 1–3 refer to Nb
Curves 4–6 refer to Ta
Curve 7 gives separation factor
Curves 1, 4: no H_2SO_4
Curves 2, 5: 1.0 mol/liter H_2SO_4
Curves 3, 6, 7: 3.0 mol/liter H_2SO_4

2.2. Extraction with Amine Extractants

The extraction of niobium and tantalum from solutions with TOA-fluoride occurs as a result of the ion-exchange reaction [167]:

For Nb:

$$2R_3NHF + NbOF_5^{2-} \rightleftharpoons (R_3NH)_2NbOF_5 + 2F^-$$

or, in case of a high degree of organic phase saturation:

$$(R_3NH)_2NbOF_5 + H_2NbOF_5 \rightleftharpoons 2R_3NH_2NbOF_5$$

For Ta:

$$2R_3NHF + TaF_7^{2-} \rightleftharpoons (R_3NH)_2TaF_7 + 2F^-$$

TABLE 6 RELATIVE EFFICIENCY (RE) OF PETROLEUM SULFOXIDES AND TBP IN EXTRACTION OF NIOBUM AND TANTALUM [162]

Free Concentration of Acids in Initial Solution, mol/liter		Distribution Factor of Metal		RE
HF	H_2SO_4	D_{PSO}	D_{TBP}	D_{PSO}/D_{TBP}
		Niobium		
2	1	0.04	0.04	1.0
2	2	0.05	0.05	1.0
2	3	0.18	0.13	1.38
2	4	0.43	0.29	1.48
0.9	3	0.13	0.10	1.3
1.75	3	0.17	0.13	1.31
3.5	3	0.22	0.16	1.37
5.25	3	0.40	0.28	1.42
7.0	3	0.54	0.27	2.00
		Tantalum		
1	1	0.90	0.82	1.10
1	2	2.90	2.45	1.18
1	3	6.00	4.58	1.30
1	4	29.40	12.60	2.33
0	3	7.00	5.10	1.37
2	3	3.23	2.10	1.54
4	3	1.43	0.85	1.68
6	3	1.33	0.70	1.90

and

$$2R_3NHF + H_2Ta_4O_5F_{14}^{2-} \rightleftharpoons (R_3NH)_2H_2Ta_4O_5F_{14} + 2F^-$$

The distribution coefficient of tantalum is constant (~200) in the 1-5 *M* HF concentration range, whereas that of niobium decreases sharply as the HF concentration increases to 2 *M*. Extraction with TOA can give a separation of $S \sim 400$, but TOA can also be used as a collective extractant with a subsequent reextraction of niobium with 7% HCl, 6-10% HNO_3 or 14% NH_4Cl. Tantalum is reextractable only with concentrated HNO_3 (600-800 g/liter) or 25% NH_4OH.

Niobium, in contrast to tantalum, is readily extractable by amines from hydrochloric acid solutions of high acidity (>4-6 *M*) [168-172] or at a high concentration of chloride-ions [168]. However, the extractability is strongly dependent on the diluent used: when using $CHCl_3$ it is much lower than in the case of CCl_4. The following niobium compounds pass into the tribenzylamine phase in chloroform [172]: TBA · H · $Nb(OH)_2Cl_4$ · TBA · HCl and TBA · H · Nb · $(OH)_2Cl_4$ · 2TBA · HCl. Tantalum is almost completely extractable into the organic phase from hydrochloric acid solutions in the presence of 0.25 *M* HF. It should be noticed that the addition of HF to any mineral acid improves the extraction of tantalum with amines [173]. Tantalum fluoride is extractable from sulfuric acid solutions into Amberlite LA-2 through the reaction [134]:

$$HTaF_6 + nR_3NH^+ \rightleftharpoons TaF_6 \cdot nR_3NH + nH^+$$

The extraction of tartrate and oxalate complexes of niobium and tantalum by tri-*n*-octylamine has been studied [174-176] and so has that of citrate complexes into Aliquat 336 [177].

2.3. Extraction with Acidic Extractants

The use of acid extractants is not consistent with the extraction chemistry of niobium and tantalum. Dialkylphosphoric acids extract neither niobium nor tantalum [134, 178]. The extraction of tantalum fluoride into D2EHPA proceeds through the solvation mechanism [134]. However, niobium is readily extractable with monoalkylphosphoric acids [178] from sulfuric acid solutions in the form of hydrolized or polymerized compounds. Dibutylphosphate extracts cation complexes of niobium and tantalum formed in H_2SO_4 solutions in the presence of H_2O_2 [179, 180]. The extraction of tantalum and niobium with di-*n*-octyl-methyl-bis-phosphoric acid has been described [181]. In the analytical chemistry of niobium, and to a lesser degree of tantalum, extraction of intracomplex compounds is used, such as with TTA [182, 183], diantipyrilmethane [184, 185], 8-oxiquinoline [184, 185], PMBP [186, 187], and cupferron [188–192].

GLOSSARY OF CHEMICAL TERMS

Hexone	methylisobutyl ketone
CHex	cyclohexanone
NOPC	neutral organophosphorus compounds
TBP	orthophosphoric acid tributyl ester (tri-*n*-butyl phosphate)
DAMPA	methylphosphonic acid diisoamyl ester (diisoamyl methylphosphonate)
TBPO	tributylphosphine oxide
TOPO	trioctylphosphine oxide
SMC	stable multinuclear compounds of zirconium (hafnium)
DOSO	dioctylsulfoxide
PSO	petroleum sulfoxide
2-NPO	2-*n*-nonylpyridine-*N*-oxide
TAAO	trialkylamine oxide
NPyOx	4-(5-nonyl)pyridine-*N*-oxide
TTA	thenoyltrifluroacetone
D2EHPA	di-2-ethylhexylphosphoric acid
DAPA	diamylphosphoric acid
DNPA	dinonylphosphoric acid
2EHPPA	2-ethylhexylphenylphosphoric acid
TOA	trioctylamine
TBA	tribenzylamine
Amberlite LA-2	trialkylmethylamine
Aliquat 336	tricaprylmonomethyl ammonium chloride
PMBP	1-phenyl-3-methyl-4-benzoylpyrazolone-5
BPHA	*N*-benzoyl-*N*-phenylhydroxylamine.

REFERENCES

1. T. Sekine and Y. Hasegawa, *Solvent Extraction Chemistry*, Dekker, New York 1977.
2. W. Fisher and W. Chalybaeus, *Z. anorgan. Chem.* **225**, 79 (1947).
3. W. Fisher, W. Chalybaeus, and M. Zumbusch, *Z. anorgan. Chem.* **225** 277 (1948).
4. West German Patent 10110061 (1957).
5. A. S. Solovkin and G. A. Yagodin, Extractsionnaja Chimija Zirconia i Hafnia, in *Itogi Nauki. Neorganicheskaja Chimija*, Part III, VINITI, Moscow, 1976, p. 83.
6. I. V. Vinarov, E. I. Kovaleva, and I. I. Byk, *Ukr. Chim. Zhur.* **34**, 62 (1968).
7. D. L. Motov and T. G. Leshtaeva, *Izvest. Vuzov. Tcvetnaja Metallurgija* (6), 113 (1962); (1), 121 (1963).
8. G. A. Yagodin and O. A. Mostovaja, *Zh. Prikl. Chim.* **33**, 2459 (1960).
9. G. A. Yagodin, O. A. Sinegribova, and A. M. Chekmarev, *Proceedings of the International Solvent Extraction Conference (ISEC), The Hague, 1971*, Vol. 2, Society of Chemical Industry, London, 1971, p. 1124.
10. B. Lustman and F. Kerze, Jr., *The Metallurgy of Zirconium*, McGraw-Hill, New York, 1955.
11. N. Isaak and R. Witte, *Energia Nucl.* **1** (2), 71 (1957).
12. O. A. Sinegribova and G. A. Yagodin, *Zh. Neorgan. Chim.* **10**, 1250 (1965).
13. A. M. Golub and V. M. Sergunkin, *Zh. Prikl. Chim.* **43**, 1203 (1970).
14. E. A. Pepeliaeva, O. A. Sinegribova, N. A. Shostenko, G. A. Yagodin, *Zh. Prikl. Chim.* **46**, 1301 (1973).
15. L. M. Zajtcev, V. M. Kluchnikov et al., *Zh. Neorgan. Chim.* **17**, 2761, 2766 (1972).
16. N. A. Shostenko, L. G. Nehamkin, and G. A. Yagodin, *Zh. Neorgan. Chim.* **18**, 1336 (1973).
17. O. A. Sinegribova and G. A. Yagodin, *Zh. Neorgan. Chim.* **16**, 2237, 2760 (1971).
18. W. D. Jamrack, *Rare Metal Extraction by Chemcal Engineering Technology*, Pergamon Press, Oxford, 1963.
19. Japanese Patent 681974 (1966).
20. U. S. Patent 3006719 (1961).
21. E. A. Pepeliaeva, N. A. Shostenko, and G. A. Yagodin, *Nauchn. tr. N.-i. i Proektn. in-t Redkomet. Prom. (Moscow)*, **44**, 113 (1972).
22. K. Alcock, F. C. Bedford, W. H. Hardwick, H. A. C. McKay, *J. Inorg. Nucl. Chem.* **4**, 100 (1957).

23. N. M. Adamski, S. M. Karpacheva, I. N. Melnikov, and A. M. Rozen, *Radiochimija* **2**, 400 (1960).
24. A. S. Solovkin, *Zh. Neorgan. Chim.* **2**, 611 (1957).
25. G. F. Egorov, V. V. Fomin, Yu. G. Frolov, and G. A. Yagodin, *Zh. Neorgan. Chim.* **5**, 1044 (1960).
26. E. N. Lebedeva, S. S. Korovin, and A. M. Rozen, *Zh. Neorgan. Chim.* **9**, 1744 (1964).
27. G. P. Nikitina and M. F. Pushlenkov, *Radiochimija* **4**, 137 (1962).
28. S. Siekierski, *J. Inorgan. Nucl. Chem.* **12**, 129 (1959).
29. E. H. Scadden and N. E. Ballow, *Anal. Chem.* **25**, 1602 (1953).
30. M. J. Hure and R. Saint-James, *Chimia jadernogo goriuchego (Moscow)*, 511 (1956).
31. S. S. Korovin, A. M. Reznick, E. N. Lebedeva, and I. A. Apraksin, in *Organicheskie Reagenti v Analiticheskoi Chimii Zirkonia,* Nauka, Moscow, 1970, p. 63.
32. G. P. Nikitina and M. F. Pushlenkov, *Radiochimija* **5**, 436, 456; **6** 347 (1964).
33. A. S. Solovkin, *J. Radioanal. Chem.* **21**, 15 (1974).
34. Z. N. Tsvetkova, A. S. Solovkin, N. S. Povitski, and I. P. Davydov, *Zh. Neorgan. Chim.* **6**, 489 (1961).
35. I. A. Sheka and S. A. Kacherova, *Ukr. Chim. Zhur.* **28**, 38 (1962).
36. O. A. Mostovaja, T. V. Momot, and G. A. Yagodin, *Zh. Neorgan. Chim.* **9**, 1280 (1964).
37. V. M. Kluchnikov, L. M. Zajtcev, S. S. Korovin, E. S. Solovieva, and I. A. Apraksin, *Zh. Neorgan. Chim.* **17**, 780, 3030 (1972).
38. G. A. Yagodin, S. V. Chizhevskaja, and O. A. Sinegribova, *Zh. Neorgan. Chim.* **20**, 189 (1975).
39. G. A. Yagodin and V. G. Kazack, in *Chimija Processov Extraktcii,* Nauka, Moscow, 1972, p. 182.
40. G. A. Yagodin, V. V. Taracov, and V. G. Kazack, *Radiochimija* **15**, 118 (1973).
41. A. Sarsenov, D. T. Lien, O. A. Sinegribova, G. A. Yagodin, *Zh. Neorgan. Chim.* **19**, 2519 (1974).
42. L. M. Zajtcev, V. M. Kluchknikov, I. A. Apraksin, *Zh. Neorgan. Chim.* **17**, 2761 (1972).
43. Yu. V. Granovski et al., *Zav. Lab.* **29**(1), 60; (3), 321; (10), 1220 (1963).
44. E. P. Nikitina, Yu. V. Granovski, L. N. Komissarova, et al., *Zav. Lab.* **40** 566 (1974).
45. A. E. Levitt and H. Freund, *J. Am. Chem. Soc.* **78**, 1545 (1956).
46. N. Ishinose, *Talanta* **19**, 1644 (1972).
47. J. Moffat and R. D. Thompson, *J. Inorg. Nucl. Chem.* **16**, 365 (1961).
48. O. A. Sinegribova and G. A. Yagodin, *Zh. Neorgan. Chim.* **14**, 2469 (1969).
49. I. M. Gavrilova, G. N. Vronskaja, V. M. Kluchnikov, et al., *Zh. Neorgan. Chim.* **19**, 478 (1974).
50. F. G. Zharovski et al., *Ukr. Chim. Zhurn.* **32**, 747 (1966); **37**, 694, 939 (1971); **38**, 1055 (1972).
51. R. Shabana and F. Hafez, *J. Radioanal. Chem.* **29**, 99 (1976).
52. G. A. Yagodin, G. E. Kaplan, O. A. Mostovaja, et al., *Zh. Neorgan. Chim.* 8, 1973 (1963).
53. A. M. Reznick, A. M. Rozen, S. S. Korovin, and I. A. Apraksin, *Dokl. AN SSSR*, **143**, 1413 (1962); *Radiochimija* **5**, 49 (1963).
54. S. S. Korovin, K. Dedich, E. N. Lebedeva, and A. M. Reznick, *Zh. Neorgan. Chim.* **7**, 2475 (1962).
55. S. S. Korovin, E. N. Lebedeva, A. M. Rozen, et al., *Zh. Neorgan. Chim.* **12** 1006 (1967).
56. A. M. Rozen, *Radiochimija* **20**, 273 (1968).
57. R. P. Cox, H. C. Peterson, and G. H. Beyer, *Ind. Eng. Chem.* **50**, 141 (1958).
58. G. A. Yagodin, O. A. Mostovaja and A. M. Chekmarev, *Izvest. VUZov. Chimija i Chim. Technoogija* **3**(1), 135 (1960).
59. S. S. Korovin, E. N. Lebedeva, and A. M. Reznick, *Izvest VUZov. Chimija i Chim. Technologija* **5**(2), 231 (1962).
60. A. S. Solovkin and G. A. Yagodin, Extractcionnaja Chimija Zirconia i Hafnia in *Itogi Nauki. Neorganicheskaja Chimija* Part II. VINITI, Moscow, 1970, p. 5.
61. H. C. Coi, *Can. Min. Met. Bull.* **58**, 193 (1956).
62. British Patent 976050 (1964).
63. Avt. svid. USSR 487651 (1975).
64. British Patent 986294 (1965).
65. French Patent 1425429 (1966).
66. U. S. Patent 3346330 (1967).
67. V. I. Tihomirov and V. P. Ionov, *Radiochimija* **14**, 754 (1972).
68. D. C. Kennedy and J. Fritz, *Talanta* **17**, 837 (1970).
69. A. M. Rozen, Yu. I. Murinov, Yu. E. Nikitin, *Radiochimija* **14**, 513, 754 (1972).
70. Shanker Ramendra and K. S. Venkateswarlu, *J. Inorg. Nucl. Chem.* **32**, 2369 (1970).
71. Yu. A. Tcylov, A. M. Reznick, N. A. Shostenko, and Z. S. Vasiljeva, *Izvest. VUZov. Chimija i Chim. Technologija* **19**, 269 (1976).
72. Yu. A. Tcylov, A. M. Reznick, and A. N. Turanov, *Izvest VUZov. Chimija i Chim. Technologija* **19**, 1079 (1976).
73. G. A. Yagodin, T. V. Nikiforova, and A. M. Chekmarev, *Tr. MCh. TI im. Mendeleeva (Moscow)* **67**, 133 (1970).
74. H. D. Liem and O. A. Sinegribova, *Acta Chem. Scand.* **25**, 301 (1971).
75. E. A. Pepeliaeva, N. A. Shostenko, N. B. Onisimova, and T. I. Popova, *N. Tr. N.-i i Proekt. in-t Redkomet. Prom.* **44**, 113 (1972).
76. B. Wiver, *J. Inorg. Nucl. Chem.* **30**, 2223 (1968).

77. O. A. Sinegribova, G. A. Yagodin, et al., *Zn. Neorgan. Chim.* **20**, 189, 2536 (1975); **22**, 1640 (1977).
78. N. A. Plesskaja, O. A. Sinegribova, I. K. Shvetcov, and E. G. Chudinov, *Thezici Dokl. Vsesojuznoi Konf. po Chimii Extrakcii*, Novosibirsk, 1978, p. 199.
79. E. Cerrai and C. Testa, *Energia Nucl.* **6**(11), 707; (12), 768 (1959).
80. T. Sato and H. Watanabe, *Anal. Chim. Acta* **54**, 439 (1971).
81. SRR Patent 52459 (1970).
82. West German Patent 1188179 (1970).
83. Canadian Patent 639404 (1962).
84. F. Bonifati, E. Cerrai, and G. Chersini, *Energia Nucl.* **14**, 578 (1967).
85. Z. Malek, D. Schrotterova, V. Yedinakova, et al., *Proceedings of the International Solvent Extraction Conference (ISEC), Lyons 1974*, Vol 1, Society of Chemical Industry, London, 1974, p. 477.
86. A. Cornea and T. Vellea, *Metallurgia (SRR)* **25**, 1973 (1973).
87. H. F. Aly, A. El-Haggan, and A. A. Adbel-Roussoul, *Zh. Anorg. Allg. Chem.* **378**, 315 (1970).
88. V. G. Chernishov, V. V. Tarasov, and G. A. Yagodin, *Tr. MCh. TI im. Mendeleeva (Moscow)* **67**, 140 (1970).
89. G. A. Yagodin and A. M. Chekmarev, *Tr. MCh. TI im. Mendeleeva (Moscow)* **67**, 129 (1970).
90. A. M. Chekmarev and G. A. Yagodin, *Tr. MCh. TI im. Mendeleeva (Moscow)*, 89, 29 (1975).
91. G. A. Yagodin, A. M. Chekmarev, and L. M. Vladimirova, *Zh. Neorgan. Chim.* **14**, 1603 (1969).
92. G. E. Kaplan, G. A. Yagodin, S. D. Moiseev, et al., in *Razdelenije Blizkih po Svoistvam Redkih Metallov*, Metallurgizdat, Moscow 1962, p. 28.
93. A. M. Chekmarev, L. G. Molokanova, and G. A. Yagodin, *Dep. VINITI*, N 252-277 (1976).
94. W. J. McDowell and C. F. Coleman, *J. Inor. Nucl. Chem.* **29**, 1325 (1967).
95. Japanese Patent 10617 (1970).
96. A. M. Chekmarev, V. G. Chuprinko, and G. A. Yagodin, *Tr. MChTI im. Mendeleeva (Moscow)* **69**, 147 (1972).
97. G. A. Yagodin and A. M. Chekmarev, in *Ekstraktcija*, Vol. 2, Atomizdat, Moscow, 1962, p. 141.
98. I. S. El-Yamani, M. Y. Farah, and F. A. Abd. El-Alim, *J. Radioanal. Chem.* **45**, 125 (1978).
99. F. Cotton and J. Wilkinson, *Sovremennaya neorganicheskaya Khimia*, Mir, Moscow **3**, (1969).
100. Ya. G. Goroshchenko, *Khimiya niobiya i tantala*, Naukova Dumka, Kiev, (1965).
101. P. S. Stevenson and H. G. Hicks, *Anal. Chem.* **25**, 1517 (1953).
102. H. G. Hicks and R. S. Gilbert, *Anal. Chem.* **26** 1205 (1954).
103. G. W. C. Milner, G. A. Barnet, and A. A. Smales, *Analyst* **80** (950), 380 (1955).
104. M. L. Theodore, *Anal. Chem.* **30**(4), Part I, 465 (1958).
105. P. Senise and L. Sant-Agostino, *Anal. Chem. Acta* **22**(3), 296 (1960).
106. R. Münchow, *Materialprüfung* **2**(5), 171 (1960).
107. R. Kiffer and H. Braun, *Vanadij Niobij*, Metallurgiya, Moscow 1968.
108. J. R. Werning, K. B. Higbie, J. T. Grace, B. F. Speece, and H. L. Gilbert, *Ind. Eng. Chem.* **46**, 644 (1954).
109. S. Nisimura and H. Irokawa, *Jap. Analyst* **12**(10), 933 (1963).
110. E. L. Koerner, M. Smutz, and H. A. Wilhelm, *Chem. Eng. Progr.* **56**(9), 492 (1958).
111. D. Y. Soisson, *Ind. Eng. Chem.* **53**(11), 861 (1961).
112. J. R. Werning and K. B. Higbie, *Ind. Eng. Chem.* **46**(12), 2491 (1954).
113. T. R. Bhat, Y. W. Gokhale, and B. S. Matnur, *Indian J. Technol.* **1**(4), 165 (1963).
114. West German Patent 2214817 (1976), FRG, cl. c22B34I24.
115. T. F. Zhitkova, P. S. Kindyakov and A. I. Vajsenberg, *Nauchn. Tr. Giredmet. (Moscow)* **1**, 623 (1959).
116. Ya. G. Goroshchenko, M. I. Andreeva, and A. G. Babkin, *Zh. Prikl. Khim.* **32**, 1904 (1959).
117. A. G. Babkin and Ya. G. Goroshchenko, *Dokl. Akad. Nauk SSSR* **173**(4), 873 (1967).
118. O. M. Petrukhin and I. P. Alimarin, *Zav. Lab.* **32**(10), 1239 (1966).
119. M. Yu. Medvedev, V. G. Majorov, A. G. Babkin, A. I. Nikolaev, and N. M. Kondratovich, in *Planirovanie Ehksperimenta*, Moscow, 1966, p. 280.
120. A. M. Sych and E. N. Stradomskaya, *Zh. Neorg. Khim.* **21** 2479 (1976).
121. Ya. G. Goroshchenko, A. G. Babkin, V. G. Majorov, and S. A. Fedyushkina, *Zh. Prikl. Khim.* **34**, 48 (1961).
122. V. G. Majorov, Yu. I. Balabanov, A. G. Babkin, V. K. Kopkov, and A. M. Nikolaev, in *Khimicheskaya tekhnologiya Pererabotki Redkometal'nogo syr'ya Kol'skogo Poluostrova*, L., Nauka, Leningrad, 1972, p. 36.
123. V. G. Timoshev, K. A. Petrov, A. V. Rodionov, V. V. Balandina, A. A. Volkova, A. V. El'kina, and Z. I. Nagnibeda, *Radiokhimiya* **2**, 419 (1960).
124. V. M. Vdovenko, A. S. Krivokhatskij, and Yu. K. Gusev, *Radiokhimiya* **2**, 531 (1960).
125. R. Vanossi, *An. Asoc. Quim. Argent* **41**, 127 (1953).
126. R. Vanossi, *An. Asoc. Quim. Argent.* **42**, 59 (1954).

127. L. I. Zemtsova, *Metody Khimicheskogo Analiza Mineral'nogo Syr'ya*, Nedra, Moscow, 1965, p. 118.
128. G. V. Rozovskaya, *Metody Khimicheskogo Analiza Mineral'nogo Syr'ya,* Nedra, Moscow, 1965, p. 129.
129. D. C. Canada, *Anal. Chem.* **39**(3), 381 (1967).
130. A. M. Golub and A. M. Sych, *Zh. Prikl. Khim.* **39**, 2400 (1966).
131. R. P. Giganov and V. D. Ponomarev, *Tsvetnaya Metallurgiya,* Alma-Ata, Izd. AN Kaz. SSR, 1962, pp. 115, 125.
132. S. Nisimura, D. Morijama, and I. Kusima, *J. Jap. Inst. Metals* **26** (1), 60 (1962).
133. I. P. Giganov and V. D. Ponomarev, *Tr. Inst. Metallurgii i Obogashcheniya AN Kaz. SSR* **5**, 125 (1962).
134. W. Sanad, A. Haggas, and N. Todros, *J. Radioanal. Chem.* **42**, 59 (1978).
135. S. Nisimura D. Morijama, and I. Kusima, *J. Jap. Inst. Metals* **26**(1) 52 (1962).
136. S. S. Korovin and Yu. I. Kol'tsov, *Zh. Neorg. Khim.* **14**, 1062 (1969).
137. D. I. Ryabchikov and M. P. Volynets, *Zh. Anal. Khim.* **14**, 700 (1959).
138. N. N. Kuznetsova, *Zh. Anal. Khim.* **23**, 1485 (1968).
139. G. P. Giganov, V. D. Ponomarev, and O. A. Khan, *Razdelenie Blizkikh po svojstvam Redkikh Metallov,* Metallurgizdat, Moscow, 1962, p. 79.
140. G. P. Giganov, V. D. Ponomarev, and O. A. Khan, *Izv. AN Kaz. SSR, Ser. Metallurg. Obogashch. i. Ogneupor.* **3**(6), 73 (1959).
141. I. I. Baram, G. E. Kaplan, and B. N. Laskorin, *Zh. Neorg. Khim.* **10**, 507 (1965).
142. G. E. Kaplan and I. I. Baram, *Zh. Neorg. Khim.* **10**, 703 (1965).
143. I. M. Gibalo and D. S. Al'bardi, *Vestn. Mosk. Univ., Khimiya* (3), 111 (1967).
144. Yu. I. Kol'tsov, S. S. Korovin, and K. I. Petrov, *Zh. Neorg. Khim.* **14,** 1065 (1969).
145. Yu. I. Kol'tsov and S. S. Korovin, *Izv. VUZov, Tsvetnye. Metally* (1), 69 (1967).
146. V. N. Startsev, Yu. I. Sannikov, S. S. Stroganov, and E. I. Krylov, *Zh. Neorg. Khim.* **13**, 1608 (1968).
147. V. N. Startsev and E. I. Krylov, *Zh. Neorg. Khim.* **11**, 2820 (1966).
148. V. N. Startsev and E. I. Krylov, *Zh. Neorg. Khim.* **12**, 526 (1967).
149. Yu. I. Kol'tsov, I. A. Apraksin, S. S. Korovin, V. G. Pervykh, and K. I. Petrov, *Zh. Neorg. Khim.* **11,** 667 (1966).
150. I. P. Giganov and V. T. Mel'nik, *Tr. Vses. Nauchn. Issl. Gorno-Metallurg. Inst. Tsvetnye. Metally.* (10), 372 (1967).
151. A. M. Golub and A. M. Sych, *Zh. Prikl. Khim.* **39**, 2658 (1966).
152. I. M. Gibalo, D. S. Al'bardi, and G. A. Eremina, *Zh. Anal. Khim.* **22**, 816 (1967).
153. N. E. Brezhneva, V. I. Levin, G. V. Korpusov, et al., *Trudy Vtoroj Mezhdunarodnoj Konferentsii po Mirnomu Ispol'zovaniyu Atomnoj Ehnergii, Geneva, 1958,* Atomizdat, Moscow, 1959, p. 57.
154. M. A. Zakharov, S. S. Korovin, V. M. Klyuchnikov, and I. A. Apraksin, *Zh. Neorg. Khim.* **18**, 1916 (1973).
155. A. M. Sych, A. F. Alekseev, and E. N. Stradomskaya, *Zh. Neorg. Khim.* **19**, 1046 (1974).
156. L. P. Varaga, W. D. Walkey, L. S. Nicolson, M. L. Madden, and J. Patterson, *Anal. Chem.* **37** (8), Part 1, 1003 (1965).
157. A. S. Chernyak, V. G. Torgov, G. I. Druzhina, et al., *Izvest. Sib. Otd. Akad. Nauk SSSR, Ser. Khim. Nauk.* **14** (5), 118 (1967).
158. A. S. Chernyak, G. I. Druzhina, V. A. Mikhajlov, and V. G. Torgov, *Izvlechenie Zolota, Almazov, Redkikh i Tsvetnykh Metallov iz Rud,* Nedra, Moscow, 1970, p. 331.
159. I. M. Ejaz and D. I. Carswell, *J. Radioanal. Chem.* **29**, 259 (1976).
160. V. A. Mikhajlov, V. G. Torgov, E. N. Gil'bert, et al., *Proceedings of the International Solvent Extraction Conference (ISEC), The Hague, 1971,* Vol. 2, Society of Chemical Industry, London, 1971, p. 1112.
161. A. M. Rozen, Yu. I. Murinov, Yu. E. Nikitin, and A. A. Abramova, *Radiokhimiya* **14**, 752 (1972).
162. T. S. Golovacheva, V. N. Startsev, V. D. Novokshonova, Yu. E. Nikitin, and N. V. Yankovskaya, *Izvest. Sib. Otd. Adad. Nauk SSSR, Ser. Khim. Nauk.* 7(3), 36 (1973).
163. A. S. Chernyak, G. I. Smirnov, A. S. Bobrova, V. A. Mikhajlov, V. G. Torgov, G. Ya. Druzhina, and O. N. Kostromina, *Izvest. Sib. Otd. Akad. Nauk SSSR, Ser. Khim. Nauk* 7(3) 42 (1973).
164. A. I. Nikolaev and A. G. Babkin, *Khimiya i Khimicheskaya Tekhnologiya Mineral'nogo Syr'ya*, Apatity, 1975, p. 60.
165. T. S. Golovatcheva, V. N. Startsev, V. D. Novokshonova, Yu. E. Nikitin, and N. V. Yankovskaya, *Izvest. Sib. Otd. Akad. Nauk SSSR, Ser. Khim. Nauk* 7(3), 36 (1973).
166. A. I. Nikolaev, A. G. Babkin, V. G. Mikhajlov, and B. E. Chistyakov, *Izvest. Sib. Otd. Akad. Nauk SSSR, Ser. Khim. Nauk* 7(3), 32 (1973).
167. B. N. Laskorin, G. E. Kaplan, T. A. Uspenskaya, and R. I. Barushkova, *Razdelenie Blizkikh po Svojstvam Redkikh Metallov,* M., Metallurgizdat, Moscow, 1962, p. 71.
168. T. Sato and S. Kikushi, *Z. Anorg. Allgem. Chem. Bd.* **365**(5–6), 330 (1969).
169. N. A. Ivanov, I. P. Alimarin, N. M. Gibalo, et al.,

Izvest. Ahad. Nauk SSSR, Ser. Khim. (12), 2664 (1970).

170. G. Lidicotte and F. Moore, *J. Am. Chem. Soc.* **74** (6), 1618 (1952).
171. I. P. Alimarin, N. A. Ivanov, and I. M. Gibalo, *Zh. Analit. Khim.* **24**, 1521 (1969).
172. T. Omori and N. Suzuki, *Bull. Chem. Soc. Jap.* **35**(10), 1633 (1962).
173. H. Marchart and F. Hecht, *Michrochim. Acta* (6), 1152 (1962).
174. A. N. Nevzorov and L. A. Bychkov, *Zh. Anal. Khim.* **19** 1336 (1964).
175. B. I. Nabivanets and E. A. Mazurov, *Ukr. Khim. Zh.* **32**(7), 739 (1966).
176. C. Djordjevic, H. Gorican, and S. L. Tan, *J. Less-Common Metals* **11**(5), 372 (1966).
177. L. Rigali and P. Barbano, *Energia Nucleare* **14**(8), 168 (1967).
178. V. N. Nikolaevskij, A. F. Morgunov, and Yu. G. Frolov., *Tr. Mosk. Khim.-Tekhnol. Inst. im. Mendeleeva, (Moscow)* (7), 133 (1972).
179. A. K. Babko and V. F. Gorlach, *Zh. Neorg. Khim* **11**, 2835 (1966).
180. V. F. Gorlach, *Zh. Anal. Khim.* **26**, 2372 (1971).
181. C. Djordjevic, H. Gorcian, and S. L. Tan, *J. Inorg. Nucl. Chem.* **28**(6–7), 1451 (1966); **29**(6), 1505 (1967).
182. R. Guillaumont, J. C. Frank, and R. Muxart, *Radiochem. Radioanal. Lett.* **4**, 73 (1970).
183. A. Jurriaanse and F. L. Moore, *Anal. Chem.* **39** (4), 494 (1967).
184. I. P. Alimarin and I. M. Gibalo, *Vestn. Mosk. Univ., Ser. Mat. Mekh. Astron. Fiz. Khim.* **11**, 185 (1956).
185. I. P. Alimarin and G. N. Bilimovich, *Tsuj Syan'-khan, Zh. Neorg. Khim.* **7**, 2728 (1962).
186. O. D. Savrova, I. M. Gibalo, S. S. Spiridonova, and F. I. Lobanov, *Zh. Anal. Khim.* **28** 817 (1973).
187. B. F. Myasoedov, N. P. Molochnikova, and P. N. Palej, *Radiokhimiya* **12**, 829 (1970).
188. I. P. Alimarin and I. M. Gibalo, *Dokl. Adad, Nauk SSSR* **109**(6), 1137 (1956).
189. J. F. Reed, *Talanta* **10**(4), 347 (1963).
190. K. V. Troitskij, in *Primenenie Mechenykh Atomov v AnaliticheskojKhimii*, Inst. Geokhimii i Analit. Khimii AN SSSR, 1955, p. 148.
191. T. Kiba, S. Ohasi, and T. Maeda, *Bull. Chem. Soc. Jap.* **33**, 818 (1960).
192. K. F. Karlysheva, L. A. Malinko, and I. A. Sheka, *Zh. Neorg. Khim.* **11**, 540 (1966).

26

MISCELLANEOUS INORGANIC PROCESSES

R. Blumberg
Miles Israel Ltd.
Israel†

1. INTRODUCTION, CLASSIFICATION, AND SOLVENT SELECTION

When one excludes the recovery of metals from the definition of inorganic processes, one finds that solvent extraction has been applied in a very limited number of cases. Furthermore, although considerable effort has been expended in synthesizing tailor-made reagents for metal recovery, whether these are cation exchangers or complexants, little attention has been paid in other cases to specific solvents. A striking exception, perhaps, is the extensive work done on modifying special classes of solvents for water extraction in desalination. On the other hand, the interaction between components of solvent phases has been studied to some extent with the aim of exploitation for specific purposes [1]. In the main, however, one can say that only a very limited number of solvents has been applied thus far in inorganic processing for a variety of ends.

If one accepts the concept that solvent extraction is essentially a separation procedure and also the fact that in inorganic processes one phase of the extraction pair will always be an aqueous phase, then one should perhaps list the types of separation aimed at, before considering examples of processes where these separations are exploited:

1. *Separation of Water from Inorganic Compounds.* In certain cases this may be the separation desired, such as in water desalination or in water transfer for concentrating aqueous solutions; in other cases there may be unavoidable transfer of water between phases, accompanying the transfer of hydrated species, such as acids or salts. For solvent selection, the mutual miscibility of water and solvent, and the influence of other components of the projected liquid–liquid extraction system on this mutual miscibility,

†Also with Miles Laboratories, Inc., United States.

takes on a considerable significance. The range of mutual miscibilities in systems–water–organic liquid–is very broad; in certain cases a solvent may dissolve more than its own weight of water, which is an enormous water:solvent ratio on a molar basis. It becomes difficult then to define the second phase as a solvent phase, and at most one can talk of the heavy or the light member of the pair or of the water-rich and solvent-rich phases, respectively.

2. *Separation Between Similar Compounds.* This is exemplified by separation between acids or between salts. Again, there are processes where these separations are the main objective, but there are also processes that include such separation steps inevitably by the nature of the components in the aqueous phase. As a first approach to solvent selection for separations between similar materials, it may be useful to know the solubility of the separate inorganic components, especially of the hydrated species, in various organic solvents. As a second approach, a knowledge of the distribution of solute between an aqueous phase containing the single solute and the organic solvent may be helpful.

3. *Separation Between Dissimilar Compounds.* This is exemplified by purification operations or the recovery of values from waste streams.

In all cases, regardless of the separation sought, solvent selection may entail a very sophisticated consideration of thermodynamic factors or a wholly empirical approach to practical constraints; usually it will be a combination of both.

2. PROCESSES

In inorganic processing probably the largest effort to introduce solvent extraction as a separation tool has been spent on phosphoric acid [2]. Initially the effort was directed toward separating the acid by direct extraction from reaction mixtures obtained by acid attack on phosphate rock. Subsequently the main effort has been directed toward purifying what is conventionally called "wet process" phosphoric acid. The basic transfer operations are similar in these processes and have been described in principle [3, 4].

Apart from phosphoric acid, one of the greatest successes has been attained in directing reversible metathetic reactions between acids and salts towards the desired product [5]. In this framework several reversible processes of the type $MA + HB \rightleftharpoons MB + HA$ have been brought to pilot scale, although only one has thus far been implemented industrially. In most cases the nonimplementation can be ascribed to economic considerations, since the price differentials for large-tonnage inorganic chemicals are generally small, and also because solvent extraction operations entail certain minimum limit of costs per unit of volume handled.

Much effort has been expended on the separation of salts from brines; here one is in direct competition with evaporative crystallizations, and again economics are all determining.

Liquid anion exchange has been applied in a process proposed for converting sodium chloride into sodium bicarbonate. This, too, has been taken to the stage of piloting, but has not been implemented because of doubtful economics related mainly to ancillary operations [6].

The striking aspect of the processes mentioned previously is the generality of their character and of the technological thinking involved. They rely almost wholly on separation factors, that is, the ratio of distribution coefficients, and on the effect of additional nonextracted components in the aqueous phase to enhance the transfer into the solvent; thus transfer back to an aqueous phase is also promoted, once these additional components have been removed with the raffinate.

There have, of course, also been various developments based on single, unique separations. Among such, borate recovery from brines, K^+ recovery by crown ethers, hydrogen peroxide separation can be listed. These are specific processes developed for defined purposes.

Inorganic processes are generally characterized by the limited number of transfer stages required and hence by the very sharp concentration gradients generated. As a consequence, stage efficiency is of primary importance, since backflow of even small quantities of one phase at a high concentration can completely eliminate the concentration gradient across a battery of stages and thus reduce the efficiency completely. This aspect may have considerable bearing on equipment selection.

The processes described here have all been proved technologically, at least on pilot scale. Not all have yet turned out economically favorable, however, usually because of small margins between raw materials and product costs or because of costs of ancillary operations. The solvent extraction process itself is seldom energy intensive, so one can anticipate that there will be more inorganic processes utilizing solvent extraction as energy costs continue to rise.

2.1. Phosphoric Acid Production

2.1.1. *By Means of Hydrochloric Acid*

In special circumstances it may be desirable to use hydrochloric acid instead of sulfuric acid for attack on phosphate rock. This poses a problem, however, since the separation of the H_3PO_4 from the highly water soluble $CaCl_2$ cannot follow the procedure utilized in the more common case of sulfuric acid, where $CaSO_4$ precipitation is the main separation tool; nor indeed can another separation procedure such as evaporation be used. Solvent extraction, however, can be considered if a solvent can be selected that permits separation between hydrogen ions and other cations. This is indeed so, and a number of readily available solvents, which are only partially miscible with water (thus satisfying the first requirement for a liquid–liquid extraction regime), show considerable preference for extracting acids over salts. Conceptually it is immediately possible, therefore, to consider a process for recovery of H_3PO_4 from the dissolution liquor resulting from the attack of hydrochloric acid on phosphate rock:

$$Ca_3(PO_4)_2 + 6HCl \rightleftharpoons 3CaCl_2 + 2H_3PO_4 \quad (1)$$

For a solvent to be considered for this process, it must be a weakly basic solvating solvent that will readily give up the acid to water again. Such solvents are alcohols, amides, and tributyl phosphate in practice, as well as ketones, esters, and ethers in principle. The process always consists of two main steps, extraction and washing, with auxiliary operations related to recovery of residual HCl from the calcium chloride raffinate brine and separation of trace cations in a purification extraction step [7].

No attempt is made to separate the HCl from the H_3PO_4 by extraction, but instead evaporation, used for concentrating the phosphoric acid, serves also to drive out the volatile HCl. Since this HCl, although accompanied by water, must be returned to the process, another mass-transfer operation between phases is utilized for promoting separation, so as to retain the HCl in the system, while rejecting a part of the water by transfer of water from solvent to the calcium chloride brine. The brine, therefore, serves two important purposes here, acting both as a salting-out agent for the acids and as a water retainer.

As can be seen from the flow diagram in Fig. 1, liquid–liquid extraction is the central operation in this process. However, the auxiliary operations of (1) dissolution liquor preparation and solid–liquid separation and (2) concentration of the product acid by evaporation are technologically no less important and economically probably more significant than the solvent extraction operation itself [8].

The concentration of the dissolution liquor will be determined by the quality of the phosphate rock, by the concentration of the hydrochloric acid available, and by the extent to which washing of insoluble solids is necessary for preventing losses of P_2O_5 values; the last item, in turn, is dependent on the characteristics of the insoluble solids and the procedure used for solid–liquid separation and washing. The concentration of the dissolution liquor determines the concentration of the extract and hence also of the aqueous acids obtained by washing the extract; these aqueous acids constitute the product going to evaporation.

The mass transfer in the liquid extraction operation is rapid and reversible. All inorganic anions present will extract as acids, distributing between the aqueous and solvent phases; the relative separations achievable are dependent in principle on relative distribution coefficients and concentrations of anions and on total acidity. The success of the solvent extraction operation depends, therefore, on the manipulation of the component streams and the separations achieved. Only Cl^- and PO_4^{3-} anions are important on the macroscale, with F^- a significant minor component. However, separation between Cl^- and PO_4^{3-} can be attained only by evaporation. On the other hand, since the solvents selected have a high preference for extraction of H^+ over other cations, a simple, interposed backwash (purification) of extract permits control of cation transfer.

An important aspect for equipment selection in the liquid extraction operation will be the necessity to maintain the profile of cation concentrations in the purification battery. Thus the concentration range of Ca^{2+} across the multistage purification battery will cover two orders of magnitude. Any entrainment or backflow may thus completely annul the effectiveness of this operation.

This process has been successfully implemented in a number of cases; however, the availability of hydrochloric acid for this purpose is not widespread, and hence application of the process is limited, notwithstanding the high quality of the phosphoric acid obtained from it.

2.1.2. *By Means of Nitric Acid*

A process in similar vein, but utilizing nitric acid, has also been proposed [9]. Since calcium nitrate

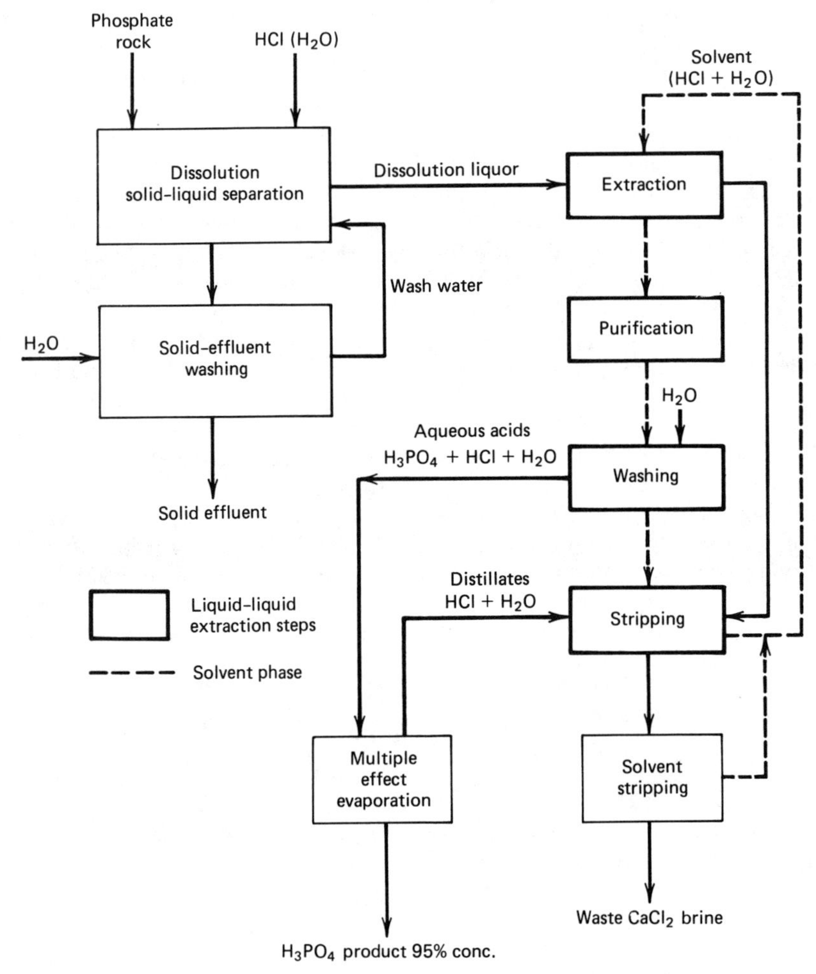

Figure 1. Phosphoric acid by way of hydrochloric acid.

itself is a valuable by-product, it is recovered by crystallisation, thus eliminating the necessity to dispose of a waste brine.

2.2. "Wet Process" Phosphoric Acid Purification

2.2.1. Derived from Hydrochloric Acid Process

There are two direct derivations from the hydrochloric acid process described in Section 2.1:

1. One process utilizes a calcium chloride brine in closed cycle as a means of purifying wet process phosphoric acid [10, 11]. In this way the advantages of selective separation of H_3PO_4 from cationic impurities by solvent extraction can be exploited without the need for hydrochloric acid for attacking the rock. It is necessary, however, to precipitate impurities from the calcium chloride brine and to reconcentrate it, usually in a multiple effect evaporator, before it can be recycled. Figure 2 presents the main process steps.

2. The second direct derivation relates to the use of the same type of solvent, especially alcohols, for extracting H_3PO_4 directly from wet process phosphoric acid, but without the addition of the calcium chloride brine as salting-out agent. As a result, the cationic impurities remain in a dilute waste acid stream that is diverted to fertilizer production [12]. This group is very similar to the processes described in Section 2.2.2.

2.2.2. Other Processes

In specific applications of solvent extraction to inorganic processing, the pecularities of phase diagrams have been exploited to an unusual de-

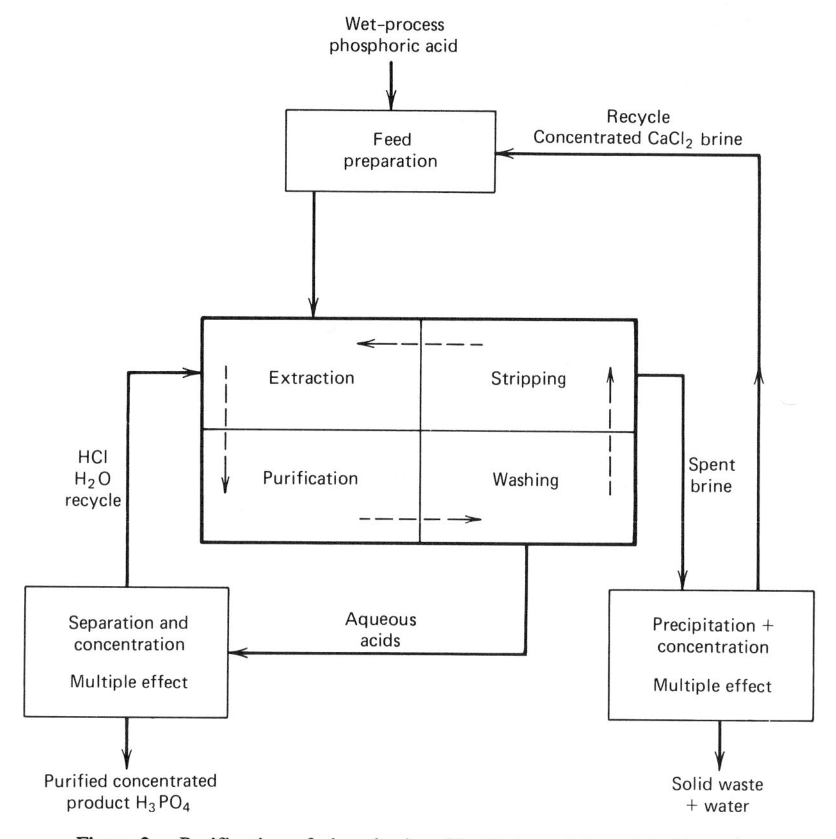

Figure 2. Purification of phosphoric acid utilizing calcium chloride brine.

gree. This is nowhere more striking than in the process for the cleaning of wet phosphoric acid, where a group of solvent types have been applied [13-15].

The essential definition of the group of solvent types is that they should not extract acid substantially below some threshold concentration and that there should be a marked effect of temperature on the distribution coefficient of the acid. Ketones, ethers, and esters are of the group considered. The differences between them is a matter of degree, and which one is the selected solvent in a specific case depends on local circumstance.

These processes are characterized by the disproportion of acid and impurities leading to a larger quantity of purer product and a smaller quantity of a more impure residual acid. The rejection of cationic impurities present in normal wet process phosphoric acid is marked; however, inorganic anions such as sulfate and fluoride, which are always present in wet phosphoric acid, are rejected to a much lesser extent. The flow sheets in all cases are quite similar; a typical example is presented in Fig. 3. In some circumstances it is possible to make use also of specific aspects of phase diagrams, such as the appearance of a third liquid phase, as a special control feature, or to admix a second modifying solvent to change the ratio of clean to residual acid [16].

When phosphoric acid is extracted by solvents of the classes mentioned, with the characteristics specified, there is a marked energetic effect since the extraction is exothermic whereas the acid release is endothermic. In practice it is usually desirable to handle the heat balance by intent, that is, by extracting at a lower temperature with heat removal and releasing the acid at a higher temperature with heat input. When ethers are selected as solvent, the advantage gained by separating between the two temperatures as far as is economic and practicable may be very pronounced; with ketones this is less important, and the system may even be left to adjust itself by internal heat transfer.

Since this process is based on a disproportionating extraction, the residual acid will contain the bulk of the impurities; hence it is relatively viscous and tends also to form solid precipitates, thus requiring an original approach to equipment

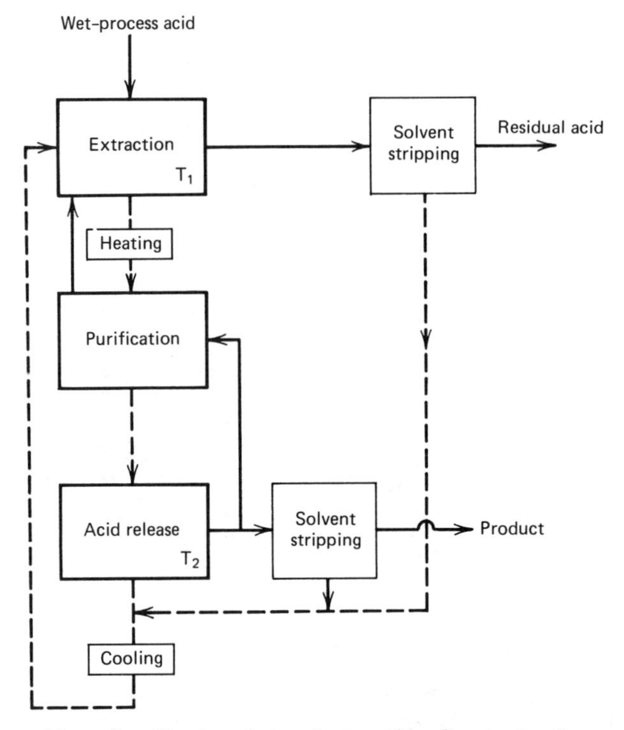

Figure 3. Cleaning of phosphoric acid by direct extraction.

selection, both for the liquid–liquid contacting and separation and for the heat-transfer steps.

In all cases extraction and acid release entail a very limited number of theoretical stages, two to three at most. Usually a backwash or scrub step for purifying the extract is interposed between the two main operations. This backwash or scrub deals mainly with cationic impurities that are coextracted, although only to a limited extent, in the extraction step.

The economics of the process are controlled in some degree by the use to which the residual acid can be put. These processes have been exploited on industrial scale in a number of locations using ethers as well as ketones.

2.3. Metathetic Salt–Acid Reactions [17, 18]

2.3.1. Potassium Nitrate [19]

One of the most elegant and unusual processes utilizing liquid extraction for attaining an inorganic transformation is that for converting potassium chloride and nitric acid into potassium nitrate and hydrochloric acid:

$$KCl + HNO_3 \rightleftharpoons KNO_3 + HCl \qquad (2)$$

Since this reaction is reversible, separation between products and starting materials is not easy; even though it is possible to separate crystalline KNO_3 by proper selection of aqueous phase composition, it would not be possible to recycle the mother liquor, since the second product—HCl—cannot be recovered by simple means because distillation removes both HNO_3 and HCl, which then interact to give NOCl. Compared to these possibilities, solvent extraction presents a very attractive alternative.

The process, as shown schematically in Fig. 4, has two main separation cycles, and the first of these relates to removal of a quantity of HCl commensurate with the quantity of KNO_3 produced, so that steady state can be maintained in the metathetic conversion. This conversion is performed at a pseudo-invariant point, at fixed temperature and acidity, in the presence of solid product KNO_3 and a vanishingly small quantity of solid KCl; hence this is a single-stage contact equilibrium system, containing two liquid and two solid phases.

The second solvent extraction cycle is needed for separating between HNO_3 and HCl, since the solvent (C_4 or C_5 alcohols) will extract HNO_3 as well as HCl in a ratio related to their distribution coefficients.

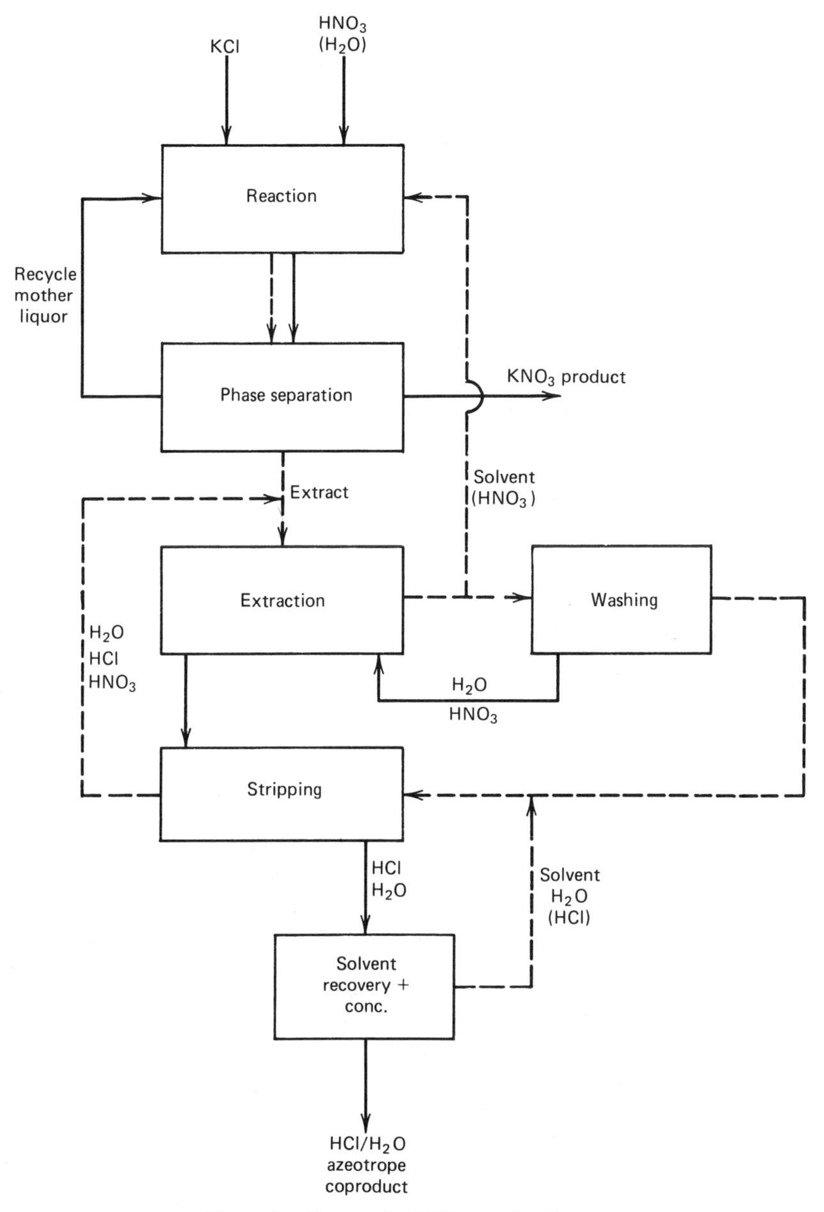

Figure 4. Process for KNO_3 production.

Finally, in principle, it is necessary to balance the water transfer and to control the water entering and leaving so as to remain in steady state with regard to the volume of aqueous phase in the reaction system.

From the flow diagram in Fig. 4 it is clear that liquid–liquid extraction here constitutes the process itself, contrary to many other cases where extraction serves only as a separation or purification step interposed between other nonextraction operations. The process has been developed in such a manner as to attend first to producing the potassium nitrate product, essentially free of potassium chloride contamination, and subsequently to separate between the acids.

The conversion reaction is performed in a multicomponent, multiphase system; this influences the selection of process conditions as well as the selection and design of the contacting and separation equipment. Since mass transfer between the liquid phases is rapid, dissolution of the feed potassium chloride and crystallization of the product potassium nitrate are rate controlling. The separation of the by-product hydrochloric acid from nitric acid, which itself has a higher distribution coefficient, is brought about by liquid–liquid extraction only, utilizing the difference in distribution coefficients, much in the same way as differences in relative volatility are utilized for attaining separations in distillation.

2.3.2. *Other Alkali Salts*

Since alkali chlorides are the basic starting points for other alkali salts, it is natural to consider the possibility of producing such products by metathetic reactions with the intervention of solvent extraction. Thus alkali phosphates and sulfates can be considered:

$$\text{Alkali chlorides} + H_3PO_4 \rightleftharpoons \text{alkali phosphates} + HCl \quad (3)$$

$$\text{Alkali chlorides} + H_2SO_4 \rightleftharpoons \text{alkali sulfates} + HCl \quad (4)$$

These reactions are complicated by the fact that acid salts of phosphate and sulfate are formed at higher acidity levels which are more favorable to hydrochloric acid extraction; these acid salts must then be decomposed by a second solvent extraction cycle. On the other hand, the HCl byproduct is volatile relative to H_3PO_4 or H_2SO_4, and hence can be separated by direct evaporation, thus obviating the need for the learned type of liquid–liquid extraction/separation utilized in the case of potassium nitrate.

Although these transformations have been fully demonstrated from the physicochemical point of view and their technological feasibility has been shown on pilot scale, they have not been implemented industrially, for reasons of low cost margins between raw materials and products.

2.4. Separation and Recovery of Halide Salts from Brines

Halide brines, apart from providing potassium salts, are also the source of bromide ion for bromine production and of magnesium chloride for production of magnesium metal. Classic processes for recovery of salts from brines are based on crystallization; phase diagrams are exploited for separating successive solid phases, for instance, common salt–NaCl, then carnallite–$KCl \cdot MgCl_2 \cdot 6H_2O$, and finally bishoffite–$MgCl_2 \cdot 6H_2O$. During these crystallization steps the bromide ion concentrates in the mother liquor, which can then be used for bromine recovery by oxidation with elementary chlorine.

An interesting liquid–liquid extraction process has been developed for separating magnesium bromide and magnesium chloride from halide brine mother liquors, containing low concentrations of K^+ and Na^+ after potassium chloride recovery but fairly high concentrations of Ca^{2+} [20, 21]. *A priori* the application of nonspecific solvents in solvent extraction for separating desired components of such essentially similar types must be based on thermodynamic differences. Furthermore, one can anticipate that separations of this type will not be straightforward; in other words, it will not be possible to achieve the desired purity of any one product in a single extraction battery, irrespective of the number of contact stages. Figure 5 presents the highly integrated flow sheet that permits these separations, using six solvent extraction batteries. Interactions between Cl^- and Br^- are unexpectedly great, and indeed permit unusual concentrations to be attained in Br^- [22].

Academically the separations achieved in this process and the strong interactions between components should provide a fertile field for a thermodynamic study. Very little is known about the species present in the various streams, only very limited measurements of ΔH of extraction have been attempted [23], and activity coefficients are not available; yet it has been possible, on the basis of distributions, calculated separation factors, and material balances alone to develop this sophisticated flow sheet. All transfers are rapid and reversible, and the whole separation/selectivity is controlled by thermodynamics, not by kinetics.

The brine itself is passed through only one solvent extraction battery, the extraction step, to produce the extract. When magnesium bromide is the only end product desired, an intermediate chloride wash is utilized for scrubbing out the coextracted chloride, and incidentally also any coextracted Ca^{2+}, followed by a stripping or washing of the purified extract. The chloride wash can be utilized as a reservoir for accumulating Br^-, so that a manyfold concentration of Br^- in the end product is attained [22]. If pure magnesium chloride is also a desired product, an additional extraction step is interposed on the aqueous product from the chloride wash.

In this process, where a partially water-miscible solvent, such as a C_4 or C_5 alcohol, is used, and where all species extracted are highly hydrated, water transfer between the phases takes on considerable significance. A part of the raffinate

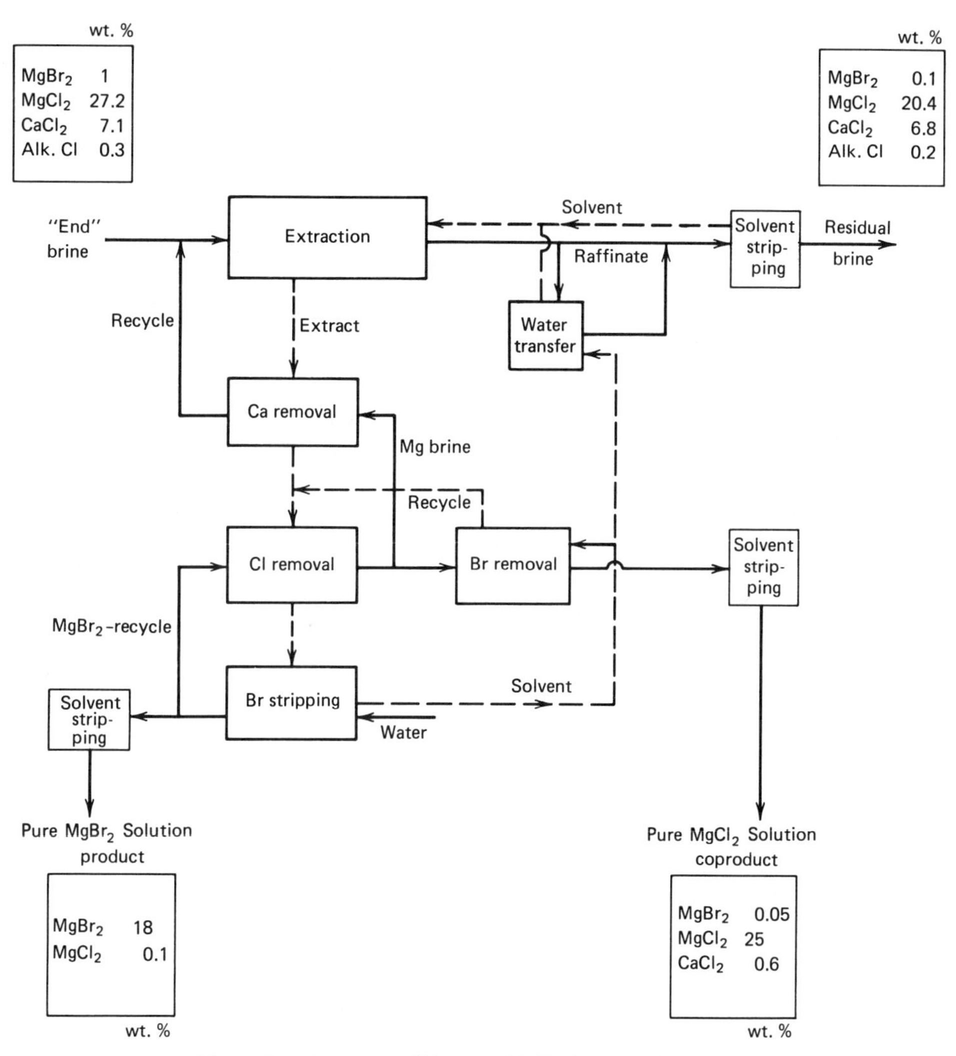

Figure 5. Recovery of $MgBr_2 + MgCl_2$ from halide brines.

brine from extraction is utilized as a water acceptor from the washed solvent, thus preventing undue dilution within the extraction itself.

As in other inorganic separations of a similar type, where the number of theoretical stages for achieving a particular subordinate separation is small but concentration gradients are steep, the selection of equipment for the phase separation step may be critical. Entrainment or backflow may completely annul the effectiveness of a particular step.

Economically, full recovery of solvent from raffinate and product(s) is required. Since large volumes of liquids are to be handled, the "capacity" of the solvent for the extracted species is important: even though the alcohols have skew mutual miscibility curves with water, thus favoring their choice, their solubility in aqueous brines is nevertheless too high to be ignored. In the present state of high energy costs, this aspect may be controlling in regard to implementation. This process has been demonstrated very successfully on pilot scale but has not been implemented industrially.

2.5. Anion Exchange: Production of Sodium Bicarbonate from Sodium Chloride and Carbon Dioxide

A process has been proposed for the use of solvent extraction, in the form of anion exchange, as a means of producing sodium bicarbonate from sodium chloride and carbon dioxide giving hydrochloric acid as by-product [24].

It consists of two anion exchange steps as shown in Eqs. (5) and (6) and a thermal operation represented by Eq. (7) to complete the cycle:

$$\mathrm{NaCl + CO_2 + H_2O + \overline{Am} \longrightarrow \overline{AmH^+Cl} + NaHCO_3} \quad (5)$$

$$\mathrm{\tfrac{1}{2}MgO + \overline{AmH^+Cl} \longrightarrow \overline{Am} + \tfrac{1}{2}MgCl_2 + \tfrac{1}{2}H_2O} \quad (6)$$

$$\mathrm{\tfrac{1}{2}MgCl_2 + \tfrac{1}{2}H_2O \xrightarrow{500-600^\circ C} \tfrac{1}{2}MgO + HCl} \quad (7)$$

Am denotes amine and a bar over a formula indicates organic phase. The process itself serves as a model utilizing interactive effects in fitting a compound solvent system to very carefully balanced reactions [6]; not only is it necessary that the basicity of the amine in its environment permits sodium bicarbonate to be present as solid phase, but it must also be sufficiently weak for magnesium oxide to react with the amine chloride, even in the presence of aqueous magnesium chloride.

Conceptually the flow sheet is simple, as can be seen from Fig. 6. As is common also with solid ion exchangers, there are loading and regenerating cycles. However, as with liquid cation exchange, phase separation between the two liquids is sharp, so no "sweetening-off," washing, and "sweetening-on" operations are required, thus avoiding the dilution effects so characteristic of solid exchangers. Unlike most liquid cation exchange systems, these are multiphase (more than two) multicomponent systems, where the extraction comprises two liquids, one solid and a gas phase, whereas the regeneration comprises two liquid phases and one solid phase.

Because of the multiphase nature of the extraction operations, equipment design has required considerable ingenuity. From the industrial point of view, it seems unlikely at this time that this process can be implemented because of the energy-intensive thermal step.

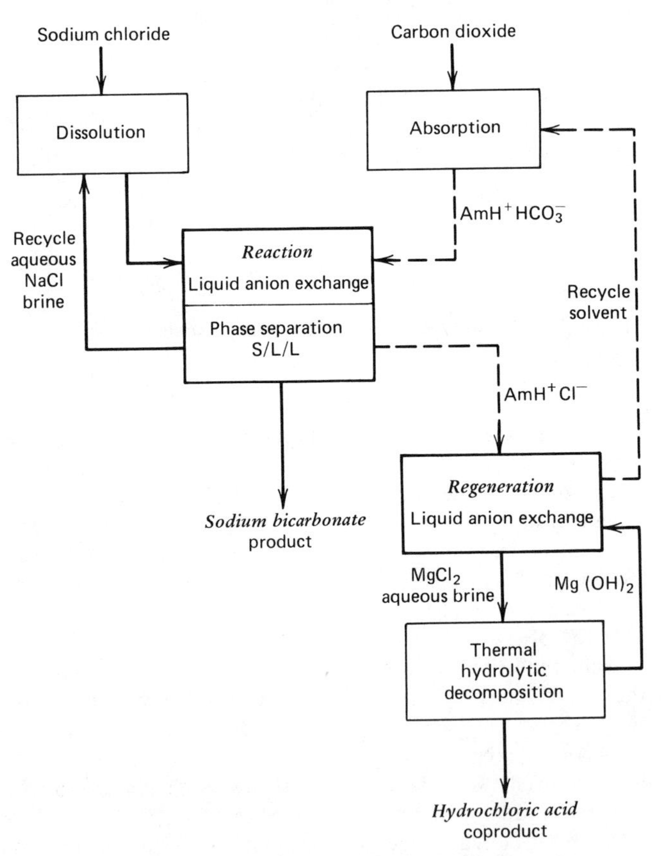

Figure 6. Process for $NaHCO_3$ production.

2.6. Water Transfer

2.6.1. Desalination

The selective separation and recovery of soft water from brackish waters by solvent extraction of the water was studied intensively in the 1950s and 1960s in the framework of water desalination. The approach was first one of solvent screening; then synthetic modifications were sought, leading to the expressed aims, namely, high selectivity toward water, even at higher salt content in the aqueous phase, a high capacity for water dissolution, and a strong temperature dependency of the water distribution [25]. The flowsheet postulated, using a solvent with these properties, was simple in concept, as can be seen in Fig. 7, consisting of water transfer by extraction followed by water release by raising the temperature of the hydrated extract to bring about the separation of water. This was an extremely sophisticated exploitation of solvent/water interactions as a function of temperature. Several solvents, mainly tertiary or secondary aliphatic amines, were found promising, and the process was proven technologically on a pilot scale.

The economics of the process are dependent on the efficiency of heat transfer and possibly on the quantity of water transferred per unit of solvent. Other factors, such as residual traces of solvent in the desalinated water and solvent losses in the rejected brine, are naturally of considerable weight in assessing the process. It appears that this project, a model in process development, did not reach implementation because a more economically competitive process, also based on low energy requirements, namely, *reverse osmosis*, was being developed too during the same decade and seemed more promising.

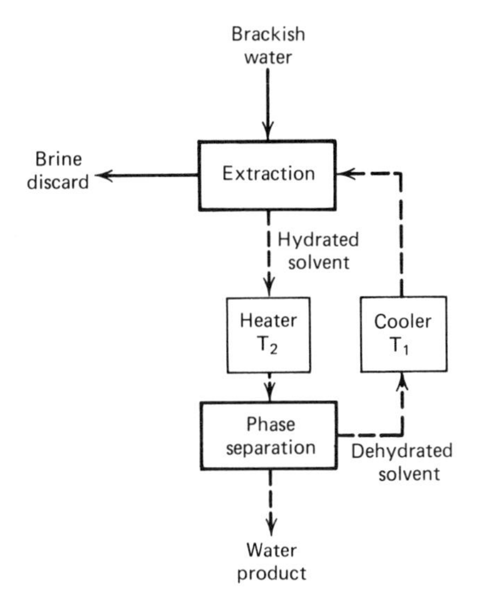

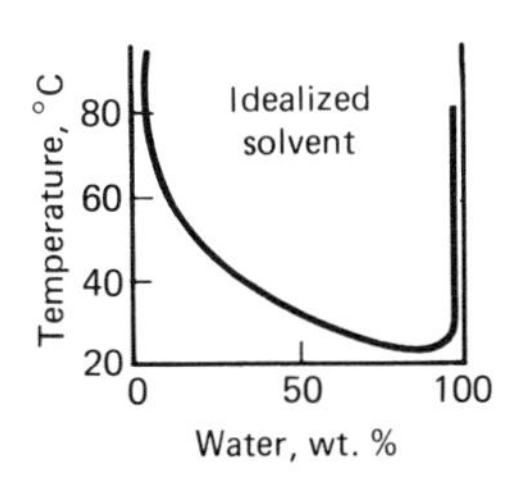

Figure 7. Process for Desalination.

2.6.2. Concentration of Aqueous Solutions

An entirely different purpose in water transfer was studied at about the same time to attain concentration without the high energy input of water evaporation [26]. The primary proposal related to the utilization of differences in partial vapor pressure of water of various solutions to induce transfer of water from one solution to another through a partially miscible solvent acting as a water carrier.

The proposal was tested on pilot scale for concentrating a dilute solution of copper sulfate to the point of crystallization by utilizing *n*-butanol as the medium for transfer of water to sodium chloride [27].

The system is classic in its simplicity; as can be seen from the flow sheet in Fig. 8, it consists of a solvent hydration step and a solvent dehydration step. The level of water in the butanol phase entering the hydration step is determined by the water vapor pressure of sodium chloride solutions at equilibrium in the dehydration step. Since saturated aqueous sodium chloride solution has a lower partial water vapor pressure than does saturated aqueous copper sulfate, the water-transfer operation can remove all the water, except for the water of hydration of the crystallizing copper sulfate pentahydrate.

This is a typical example of the utilization of differences in partial water-vapor pressure as the driving force for water transfer. In fact however, the same approach can be used here as in water desalination, by exploiting a temperature dependency of water distribution to solvent, instead of the water-vapor pressure differences of two aqueous solutions.

Process feasibility was demonstrated in a large number of cases, including concentration of

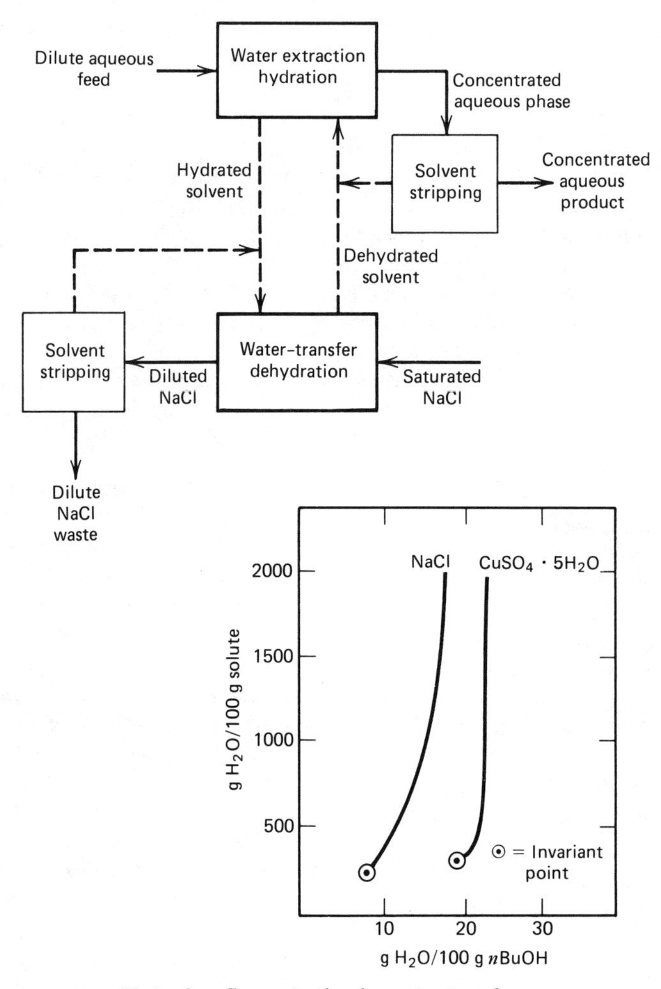

Figure 8. Concentration by water transfer.

aqueous wastes from various industries [28]. However, economics always prevented implementation. Now, however, with the high cost of energy, this approach to water transfer for solution concentration may be worthy of review.

2.7. Boric Acid Recovery from Aqueous Solutions

A process has been described [29] for recovering boric acid from borax by neutralizing with aqeous sulfuric acid and then extracting the boric acid by an aliphatic water-immiscible alcohol such as 2-ethyl hexanol, in an inert diluent such as a C_{12}–C_{17} long-chain hydrocarbon or a water-immiscible glycol ether.

The process consists of an extraction step and a stripping step, both at 80°C. The boric acid is recovered from the strip solution by cooling, and sodium sulfate is separated from the raffinate, also in a cooler–crystallizer. This process has apparently not been implemented industrially [30]. See also Chapter 25.8.

2.8. Miscellaneous

A process for purification of electrolyte in copper winning or refining by solvent extraction of a bleed stream of electrolyte has been in industrial use since 1973 [31]. Actual operating details are not available. However, the process, as described in one of the relevant patents [32], entails the extraction of arsenic and acid from the impure electrolyte bleed into a trialkyl or triaryl phosphate, at a temperature of up to 60°C.

Solvent extraction has been used for a number of years in the manufacture of hydrogen peroxide [33]. In this case, contrary to all the examples given in this section, the organic phase is the feed stream, and water is the extractant; the organic phase constitutes the medium for the quinone–quinol cyclic operation, which produces the hydrogen peroxide from H_2 and O_2 as a net reaction. The solvent medium must be selected to keep both the oxidized and reduced components in solution and must not add to process hazards, hence mixed solvents are usually used. These are selected from a variety of components, such as alcohols, ketones, esters, chlorinated hydrocarbons, alkyl benzenes, and alkylated naphthalenes, and the mass transfer takes place in the mixed-solvent–peroxide–water system.

In inorganic chemical processing there are many cases in which solvent extraction operations have been interposed for specific purifications. Usually the details of such operations are not published, as they are maintained as "in-house" know-how. Such purification steps may rely on specific extractions, or on liquid ion exchange—which falls outside the scope of the present chapter. The patent literature is a good source of information regarding the nature of operations and solvents that have been considered, although patents do not necessarily carry details of preferred procedures, nor does a valid patent imply implementation. Among the solvent types considered are alkyl or aryl phosphate esters such as tributyl or triphenyl phosphate; amines, particularly tertiary alkyl amines; amine salts; ketones such as methyl isobutyl ketones; and various esters.

REFERENCES

1. R. Blumberg and J. E. Gai, *Proceedings of the International Solvent Extraction Conference (ISEC), Toronto, 1977*, Vol. 1, Canadian Institute of Mining & Metallurgy, Montreal), 1979, p. 9.
2. R. Blumberg, *Solvent Extraction Rev.* 1(1), 93 (1971).
3. A. Baniel and R. Blumberg, in A. V. Slack, Ed., *Fertilizer Science and Technology Series*, Vol. 1, *Phosphoric Acid* Part 2, Marcel Dekker, New York, 1968, Chapter 11, p. 889.
4. A. Baniel and R. Blumberg, in A. V. Slack, Ed., *Fertilizer Science and Technology Series*, Vol. 1, *Phosphoric Acid*, Part 2, Marcel Dekker, New York, 1968, Chapter 8, p. 709.
5. A. Baniel and R. Blumberg, *Chimie et Industrie* 78(4), 327 (1957).
6. R. Blumberg, J. E. Gai, and K. Hajdu, *Proceedings of the International Solvent Extraction Conference (ISEC), Lyons, 1974*, Vol. 3, Society of Chemical Industry, London, 1974, p. 2787.
7. A. Baniel and R. Blumberg, *Dechema Monographien* 33(477–502), 57 (1959).
8. IMI Staff Report, *Proceedings of the International Solvent Extraction Conference (ISEC), The Hague 1971*, Vol. 2, Society of Chemical Industry, London, 1971, p. 1386.
9. R. J. Piepers, in A. V. Slack, Ed., *Fertilizer Science and Technology Series*, Vol. 1, *Phosphoric Acid*, Part 2, Marcel Dekker, New York, 1968, Chapter 11, p. 913.
10. Toyo Soda Manufacturing Company Ltd., Japanese Patent 7,753 (1964).
11. A. Baniel, R. Blumberg, and A. Alon, French Patent 1,396,077 (1965).
12. A. V. Slack, in A. V. Slack, Ed., *Fertilizer Science and Technology Series*, Vol. 1, *Phosphoric Acid*, Part 2, Marcel Dekker, New York, 1968, Chapter 8, p. 721.
13. A. Baniel and R. Blumberg, Belgian Patent 661,743 (1965).
14. R. Blumberg and A. Baniel, U.S. Patent 3,903,247 (1975).
15. Albright and Wilson Ltd., German Patent Application 2,320,877 (1973).
16. R. Blumberg, Newer Developments in Cleaning Wet Process Phosphoric Acid, The Fertilizer Society of London, Proceedings No. 151, 1975.
17. A Baniel and R. Blumberg, U.S. Patent 2,894,813 (1959).
18. A. Baniel and R. Blumberg, U.S. Patent 2,902,341 (1959).
19. Y. Araten, A. Baniel, and R. Blumberg, Potassium Nitrate, The Fertilizer Society of London Proceedings No. 99, 1967.
20. A. Baniel and R. Blumberg, Israeli Patent 23,760 (1968).
21. A. Baniel and R. Blumberg, Israeli Patent 41,225 (1976).
22. J. E. Gai, *Proceedings of the International Solvent Extraction Conference (ISEC), Toronto 1977*, Vol. 1, Canadian Institute of Mining and Metallurgy, Montreal, 1979, p. 316.
23. R. Blumberg, J. E. Gai, and J. Moscovici, Enthalphy Effects in the Extraction of Magnesium Halides by *n*-Octanol, paper presented to the Israel Chemical Society 43rd Annual Meeting, October 1975.
24. A. Baniel et al., Israeli Patents 33,551 and 33,552 (1968).
25. Texas A & M Research Foundation, *Saline Water R & D Progress Report No. 35*, U.S. Office of Saline Water, February 1960.
26. A. Baniel, *J. Appl. Chem.* 8, 611 (1958); 9, 521 (1959).

27. R. Blumberg, E. Cejtlin, and F. Fuchs, *J. Appl. Chem.* **10**, 407 (1960).

28. A. Baniel and R. Blumberg, *Ind. Chemist* **39**, 460 (1963).

29. C. G. Brown and B. R. Sanderson, U.S. Patent 4,058,588 (November 15, 1977).

30. B. R. Sanderson, Borax Holdings Ltd., private communication.

31. Metallurgie Hoboken-Overpelt, Brussels, Belgium, Personal communication.

32. A. de Schepper and A. Van Peteghem, U.S. Patent 4,061,564 (1977).

33. Laporte Process, *Br. Chem. Eng.* **4**, 88 (1959).

Part IV

Cost and Engineering

27.1

COMPUTATION AND MODELING TECHNIQUES

A. L. Mills
Atomic Energy Research Establishment
Harwell
United Kingdom

1. INTRODUCTION

With the present availability of small computers and powerful hand-held calculators, the engineer is now able to make many detailed computations in respect of process design and indeed can model the plant or process in a theoretical manner before even constructing or operating a pilot plant. This situation has brought about a need not only to understand both computational and modeling techniques in respect of the process being designed or studied, but also the reasons for computation and modeling, the data requirement for such procedures, the information to be gained by using such procedures together, and the application of such data.

In the present chapter the computational method is considered first so that a process flow sheet might be developed; these methods are then extended to modeling techniques. There is obviously no one general method of modeling a solvent extraction process since the final requirements will vary, but a selection of methods is presented that should enable the reader to appreciate the technique involved.

2. COMPUTATIONAL PROCEDURES

The basic technique for single solute multistage solvent extraction processes was described by Varteressian and Fenske [1]. This technique is still of value and is frequently used since it is simple and requires easily obtained distribution data for its use. Furthermore, it is amenable to "adjustments" to allow for variations in process parameters such as backmixing, incomplete settling, salting out, and other phenomena that might not be amenable to direct measurement.

Figures 1*a* and 1*b* show the familiar Fenske diagrams (also known as McCabe–Thiele diagrams) for the extraction and scrub sections of a hypothetical process. The strip section is taken to be the inverse of extraction. The number of theoretical stages for each process is stepped off as shown. This technique gives initially only the ideal system; any allowance for backmixing or other plant features must be made either by "estimating" the number of extra stages required or by adjusting the equilibrium and/or operating lines from the ideal values as shown in Fig. 1 to the true values as shown in Fig. 2. Figure 2 shows

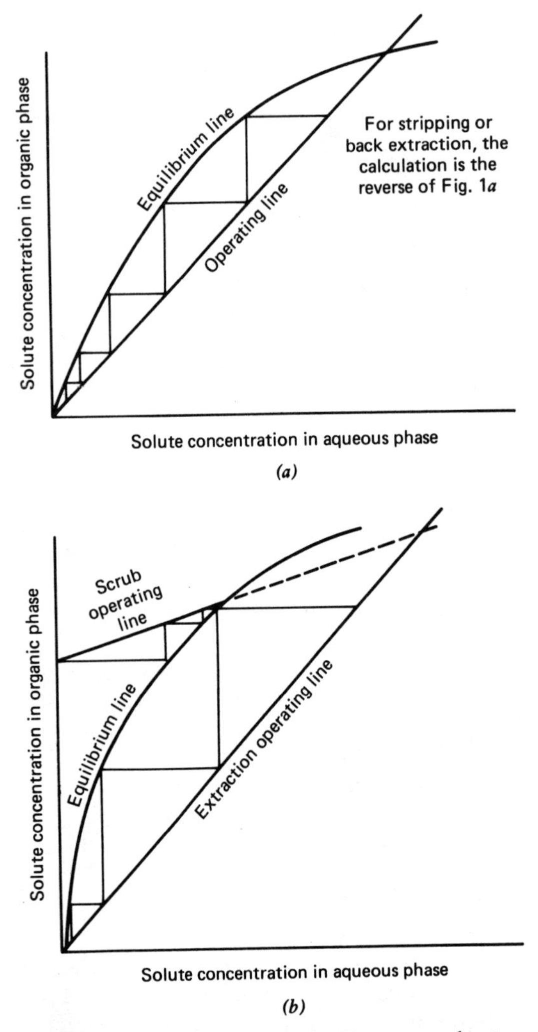

Figure 1. (*a*) Extraction system; (*b*) compound extraction–scrub system.

an adjusted equilibrium line where incomplete settling, phase carry-over, and backmixing (see Chapter 6) have been included as "lumped parameters," that is, not uniquely determined, and a "plant-devised" equilibrium line is now used. This gives a requirement for an increased number of stages in the process and the calculated result is a better approximation to actual requirements. Clearly, in order to draw Fig. 2, it is necessary to have sufficient knowledge of the plant and process such that the plant-derived equilibrium line can be drawn. This assumes either that the engineer has sufficient data with which to modify the true equilibrium line or has access say to a pilot plant from which the equilibrium line can be obtained by sampling. The engineer then relies on knowledge of scale-up to ensure that the final plant is in agreement with the calculated design. This retrospective technique is of general application since it is not unusual for there to be unknown features in the plant that cannot be fully allowed for at the design stage.

Any graphical technique such as that just described can be written in algebraic terms to coincide with the most sophisticated computational procedures available. It is not part of this chapter to discuss computation per se but the point is noted. Since the results of the calculation are to have a practical application, it is desirable that the stages of the calculation and the input to the calculation itself also have a practical meaning.

Accepting the simple graphical method and retrofitting technique described briefly here, the chapter now moves on to a more complex, general system for a multistage process.

Let us consider first the basis of liquid–liquid extraction. A necessary condition for two immiscible phases to be in equilibrium is that the chemical potentials for each phase are equal (see Chapter 1):

$$\mu_{\text{org}} = \mu_{\text{aq}} \tag{1}$$

Furthermore, for a species x in each phase

$$\mu_{xo} = \mu_{xa} \tag{2}$$

where subscripts o and a refer to organic and aqueous phase, respectively, but in Eq. (1)

$$\mu_{\text{org}} = \sum \mu_{xo} \quad \text{and} \quad \mu_{\text{aq}} = \sum \mu_{xa} \tag{3}$$

where the summations refer to all the species in the organic or aqueous phases. However, in

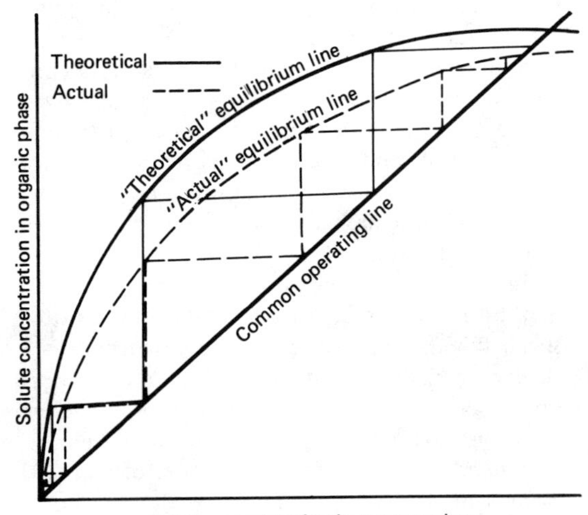

Figure 2. Modified extraction diagram (cf. Fig. 1*a*).

a multisolute system it should be noted that the rates of mass transfer for each species could well be different and that some species may act as salting-in or salting-out agents for other species; thus Eq. (1) might be fulfilled rapidly, but the condition implied in Eq. (3) need not coincide with the conditions of Eq. (1). That is to say that the chemical potentials of the alternate phases may be equal and remain constant after a time t_1, but the components of the chemical potentials of the alternate phases [Eq. (3)] might not be equal and constant until time t_n where $t_n > t_1$, as a result of the different rates of mass transfer of species, salting-in and salting-out effects, and other variables.

This is shown diagrammatically in Fig. 3, where μ the chemical potential in the system is plotted as a function of time for three solutes. The total potential equilibrium (total μ) is attained rapidly at time t_m because of the enhanced extraction of species 1, which achieves pseudoequilibrium because of its rapid extraction rate and the salting-out effects of the other species (2 and 3). With time, however, the requirement for $\mu_{2_o} = \mu_{2_a}$ and $\mu_{3_o} = \mu_{3_a}$ is achieved, and the value of μ_{1_o} falls to give $\mu_{1_o} = \mu_{1_a}$ at a time t_n without alteration of the overall condition $\Sigma\mu_o = \Sigma\mu_a$. Thus, unless the time for this final condition as described in Eq. (2) for all species is realized, a nonequilibrium system is being dealt with.

In the paragraphs that follow this is taken as the general case. The implications of this on the calculation–modeling procedure are that (1) the type of contactor will affect the model, (2) rates of mass transfer are required to be built in, (3) other influencing data such as backmixing are required, and (4) factors affecting the thermodynamic energy distribution in the system such as salting-out agents, complex formation, and so on must be included. Clearly, there are many factors that will now affect in some way the mathematical description of the process. A number of these may not be fully definable (in a practical sense), which is why recourse to "retrofitting" might be required when the model is applied to a given plant.

Let us now consider contactor type in the light of the preceding discussion; the simplest contactor type for calculational purposes is the mixer–settler operating under conditions of static or steady state equilibrium, and the pulsed column operating under conditions of dynamic equilibrium is probably the most complex system. Centrifugal contactors operating in either the steady state or dynamic equilibrium conditions are intermediate between these two contactors since, like mixer–settlers, their stage boundaries can be geometrically defined and are fixed, even though the stage efficiency (due, e.g., to kinetic factors) may not be the theoretical stage efficiency.

Mixer–settlers have finite boundaries, and depending on the residence times in both the mixer and the settler, distribution equilibrium of all the species present may be achieved. If the contactor is operated such that this is the case, for each

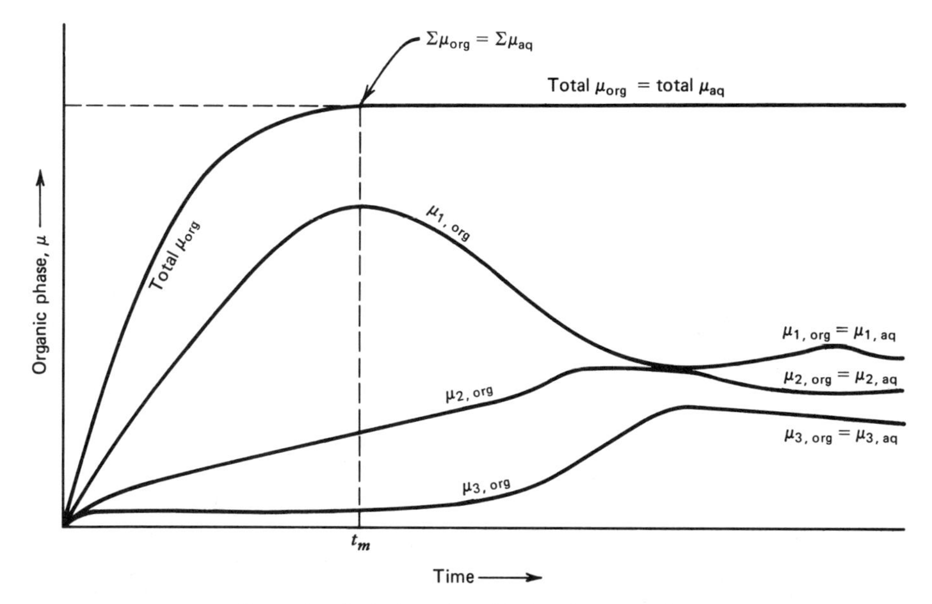

Figure 3. Variation of μ with time for extraction.

species we can write D_∞ = org/aq, where D_∞ is the equilibrium value of the distribution coefficient at infinite time. But for time t if $D_t \neq D_\infty$, then $D_t = kD_\infty$, where k is a constant and is different for each species. It is possible, therefore, to adjust the equilibrium values for each species in any stage of a mixer–settler to allow for nonattainment of equilibrium for either kinetic reasons, imperfect mixing, lack of settling, or alternate phase carry-over by this empirical means and thus make a theoretical model of the system even though some of the factors affecting the practical system may not be uniquely known or amenable to measurement (see Chapter 5, Section 4). Hence for any mixer–settler system the basic stage of operation of IN = OUT can be written for n stages. For a pulsed column operating without the mixer–settler region, that is, under conditions of dynamic equilibrium, the situation is very different.

With a single solute, the stage height varies along the length of the column as the solute concentration varies. Additionally, backmixing will require an increase in stage height in order to approach the theoretical efficiency per stage and backmixing is a function of not only the physical properties of the liquids in the column, but also the pulse amplitude and frequency. With a single solute, mass-transfer rate affects only the overall length of the column, or for a fixed length column the solute residence time. However, for a multisolute system, the system becomes complex.

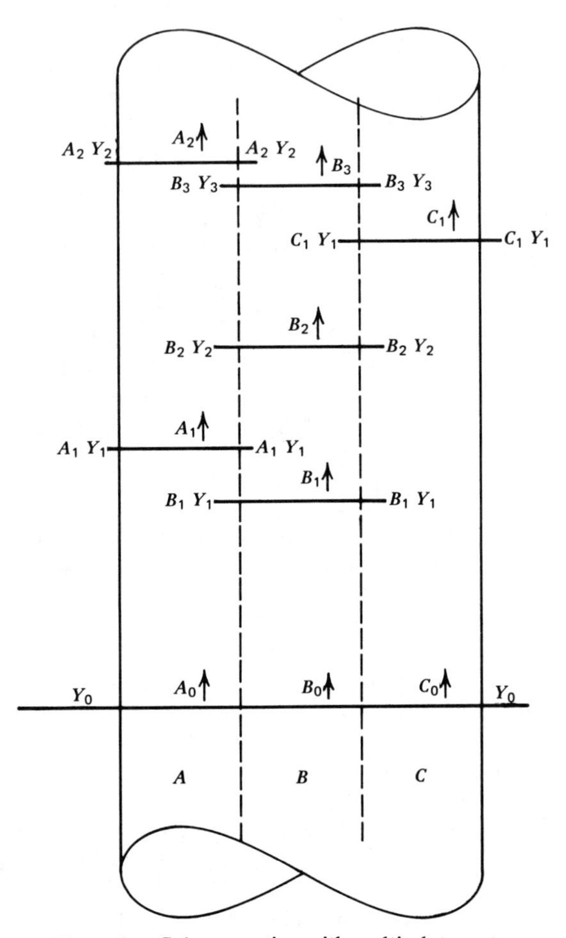

Figure 4. Column section with multisolute system.

Consider the arbitrary column section shown in Fig. 4 and three extracting solutes, A, B, and C where the rates of extraction are $A > B > C$. The column has been split into three vertical sections for convenience as shown. In the column length we have two extraction stages for A, with an entrant concentration of A_0 and an exit concentration A_2; three stages for B, entrant and exit concentrations B_0 and B_3; and one stage for C, entrant and exit concentrations C_0 and C_1, respectively. It can be seen from Fig. 4 that in stage 1 for A—that is, for the change from A_0 to A_1, B changes from B_0 to a value intermediate to B_1 and B_2 whereas C changes from a value intermediate to C_0 and C_1 since the stage heights for the components are not equal. There is then no practical equivalent of a stage for all the components, and clearly there are "edge" effects at the boundary of each "stage" for any one extracting species as a result of the effects of the other solutes. In calculating the overall height of a column operating under conditions of dynamic equilibrium, for example, it is necessary thus to consider the least extractable species or the slowest extracting species and to design the column to accommodate this species. Clearly, any changes in feed composition (i.e., the ratio of solutes or the gross concentration) will affect column operation and flow sheet modifications might well be required if the column is to operate satisfactorily. In addition to allowing for variations to distribution coefficients and rates of extraction, the effect of salting in/out must be allowed for, particularly at the stage "fringe" regions where nonstandard concentration conditions apply.

A further difficulty in computing stage heights is due to the nonphysical static equilibrium of the column. The column itself is assumed to be pulsed and hence its contents are oscillating; thus even given well-defined stage heights at (physical) equilibrium, there is a "smearing" of stage boundaries as a result of pulsing, which further complicates analysis of the pulsed column system. To some extent this effect can be included in the backmixing coefficients for the system, and in practice this is usually adequate.

For centrifugal contactors we have the situation that either full or partial stage efficiency is obtained in one centrifugal unit. Thus in the extreme n units in series are effectively one stage, if the series is operated in the mixer–settler mode. At the other extreme a series of centrifugal contactors can be made to represent and operate in a pulsed column dynamic equilibrium type mode, without the superimposition of effects due to pulsing, of course.

3. MODEL REQUIREMENTS

The physical problems associated with modeling a given type of contactor have been noted briefly in Section 2. It is now necessary to consider particular requirements of the model and the data available for modeling. At the outset it is accepted that the model has to be "real," that is, that it is required for a real plant and that experimentally derived data, not necessarily complete, are available. Basic data can be either theoretically or experimentally derived.

3.1. Theoretically Derived Data

Various attempts have been made to obtain distribution data theoretically, notably by Rozen et al. [2], but most thermodynamic methods fail when dealing with real concentrated solutions since it is either impossible or very difficult to make the necessary thermodynamic measurements to confirm the theory. Most models are based on experimentally determined distribution data, together with a theoretical approach in order to enable interpolation and extrapolation of data to be carried out (see Chapter 1). This technique is useful when the number of variables in a system is reasonably small and where the chemistry of the system is understood.

3.2. Experimentally Derived Data

With the recent increase in solvent extraction processes and the availability of machines such as the AKUFVE [3] (see Chapter 17.1) for the determination of distribution data, it has become both necessary and possible to determine distribution data for a complete range of solute and solvent concentrations—at various temperatures if required—rapidly, reproducibly and accurately. In these cases analysis of the data is achieved by a variety of mathematical techniques to obtain statistical correlations that are often simplified or assisted by recourse to the chemistry of the system.

Whether the theoretical or experimental approach is used care must be taken to ensure the statistical accuracy of the analysis, and sufficient tests on interpolated values, together with a determination of the limits of extrapolation, must be carried out; otherwise the data might be responsible for considerable errors in the model.

The distribution data calculational procedure will form a subroutine for the whole model and is applicable to both steady-state and non-steady-state systems.

3.3. Mixer–Settlers

Considering now the stage concept of IN = OUT, that is, what is fed to a stage—mass or volume, must leave the stage, albeit sometimes in modified form because of chemistry, mutual solubility of solvents, or other factors. Two sets of equations are required to describe the system:

1. *Mass and Volume Balance Equations.* These are straightforward and selfexplanatory. These may be modified as required to account for features such as incomplete phase separation, mutual solubility of solvents, volume changes due to the transfer of solutes, and so on. These data will be obtained either from laboratory or in-plant measurements or may indeed come from a "retrofitting" exercise as described earlier.
2. *Hydrodynamic and Hydraulic Equations.* To some extent these might be subsumed or incorporated into the relationship dealing with volume balances, but the hydrodynamic relationships should be kept separate since they can be used to describe flow patterns in the mixer–settler bank. An example is described later in the SEPHIS procedure.

As is seen later, there are then sufficient equations to describe the model. For a single- or multisolute system with zero or minimal interaction between solutes, a simple calculation to obtain overall mass balance consistent with the inputs and the boundary conditions of the system is sufficient.

Thus if the inputs (feeds) to the system and the product–raffinate levels are defined, it can be seen that it is a simple matter to calculate the number of stages required for extraction, say, with each stage being assumed to be in static equilibrium. Clearly, given any two of the three sets of variables–inputs, products–raffinates, or number of physical stages–the third variable can be calculated. Optimization can be carried out by feed plate matching, and departures from 100% efficiency for any reason whatsoever can be allowed for by modifying the stagewise distribution data.

For a multisolute system where each solute is extractable and each solute mutually interacts with other solutes, an iterative process is useful (Fig. 5). After each iteration the results are compared with the previous iteration, and if the error or differences is within the preset limits, the calculation is deemed to have converged and is complete.

An advantage of the iterative procedure is that if the interactions are known, or if the mass-transfer rates are known, for example, these can be superimposed on the calculation and the approach to the equilibrium can be studied.

In both the simple (single or noninteractive) systems and the interactive systems, perturbations to the system can be studied as a function

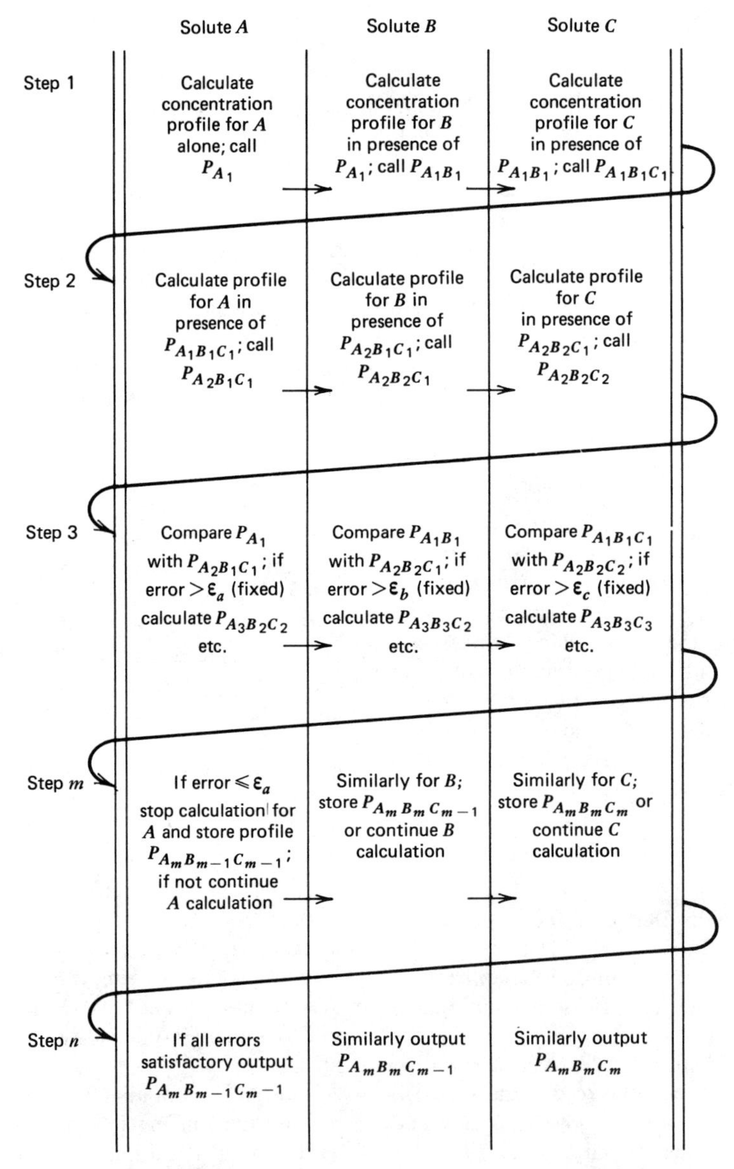

Figure 5. Iterative process for three solute interactive system.

of time, such as changes in feed flow rates, provided the model is constructed in such a way that feed flow rates and mixer–settler volumes can be related.

Some calculational procedures that have been described in the literature are discussed briefly in the following list, which although not exhaustive, is intended to be representative of more recently published procedures and should form a background for further reading. It will be appreciated that it is not possible to give a "general" method for the computation of solvent extraction systems since the physical and chemical processes vary widely.

1. Recently Hoh and Bautista [4] presented a method for the prediction of distribution coefficients based on the thermodynamic equations for extraction of the species, using chemically based models. The model requires the following data to be available; (a) the appropriate stability constants of the ligands involved; (b) the equilibrated ligand concentrations in the aqueous phase; (c) the equilibrated organic concentration; and (d) the distribution coefficients for a given system. The development of the model equation (as follows) requires calculation of the degree of formation of the extracting species and an estimate of the effective equilibrium constant. By use of a linear regression technique, the predicted distribution coefficients can then be obtained. From thermodynamic considerations, these authors derive the expression $K_d = K_1 \alpha_o [NO_3^-]_A^4 [TBP]_o^2$ for the systems Pu(IV) and Np(IV) nitrate–nitric acid/TBP–diluent, where K_d is the distribution coefficient, K_1 is an effective equilibrium constant for the extraction equilibrium, and α_o is the degree of formation of the extracting species, and subscripts A and o refer to aqueous and organic phases, respectively. For the systems examined, the authors predicted distribution data at different nitrate ion and TBP concentrations and obtained a correlation coefficient $R^2 = 0.996$ between experimentally derived and calculated data. These authors have also used the method for prediction of distribution coefficients in the Cu–LIX65N and Cu–Kelex 100 systems [5].

2. Gai [6] describes the modeling of the IMI process for the separation and recovery of $MgBr_2$ and $MgCl_2$ and applies the model-derived data to the process. The equilibrium correlations use density correlations for both the aqueous and organic phases and a semiempirical–semitheoretical "salinity" function for the aqueous phase. Solute-solute interaction requires both the density correlations and the salinity function to be combined so that systems for the two salts can be described. With the use of a regression analysis technique, the constants in the correlations were determined and the correlations reproduced the experimental results to ±2%.

Having obtained the required equilibrium correlations and by defining the feed compositions, Gai was able to calculate the performance of his mixer–settler battery by using material balances as the criterion for convergence. Use of an IBM370 the iterations can be completed in about 2 s.

3. Tierney [7] uses matrix methods to obtain an algorithm for a three-component liquid-liquid extraction system. Experimentally determined distribution data are used, and the calculational procedure requires iteration on only one composition variable per stage. Data are presented for the extraction of acetone from water to trichloroethane and for the separation of styrene and ethylbenzene by use of diethylene gylcol. In both cases the distribution data are available as ternary diagrams. The calculational procedure devised takes advantage of the fact that the distribution ratios (as expressed on a ternary diagram) can take the form of the equation $\Theta^{\dot{\gamma}} = \Theta^{\dot{\gamma}}(X^1)$, $1 \leqslant \dot{\gamma} \leqslant 3$, where Θ is the distribution coefficient matrix, $\Theta^{\dot{\gamma}}$ is the distribution ratio vector for component, and X^1 is the composition vector for the reference component in the reference phase. If the composition and amounts of all feeds, the number of stages, the interstage flow patterns and the distribution ratios are given, the process can be fully completed using an iterative procedure. With the use of a DEC1055 computer system programmed in BASIC, a five-stage extractor problem converged in about 3 s. The author notes that computing time would increase as the square of the number of stages and that for extraction with 25 or more stages the use of a high-level language such as FORTRAN would be advantageous.

4. Gaudernack et al. [8] have examined the simulation of rare earth extraction under dynamic and static conditions and have present algorithms for the process together with a comparison of experimental and theoretical data. In dealing with multicomponent systems such as the lanthanides, the correlation and prediction of distribution data is simplified because of the similarity of the extraction behavior of all species in the system. Thus for one particular solute, Gaudernack et al. were able to write $D_j = D_{mj} O$, where D_{mj} is a function of parameters

specific for the extraction system but independent of the rare earth and yttrium concentration and O is a correction factor derived from equilibrium data by means of nonlinear regression analysis.

Under static conditions, mass-balance relationships may be used to calculate the solute concentration profiles in the system. Where incomplete mass transfer or some other feature causing departure from ideality is considered, the mass-balance relationships are modified accordingly prior to the calculation.

The dynamic model is described for both complete and incomplete mass transfer, but because of lack of measurements on the input side of the experimental mixer–settler units, the author comments that it was not possible to obtain good agreement between the model and experiment.

4. SIMTEX [9] AND SEPHIS [10]

Although written independently in the United Kingdom and the United States, these two procedures have much in common. Both will handle the extraction of four macro solutes, namely (1) uranium, plutonium(IV), plutonium(VI), and nitric acid (SIMTEX); and (2) uranium, plutonium(IV), plutonium(III), and nitric acid (SEPHIS MOD 4), together with inextractable nitrate salts in each system. Both procedures are for mixer–settler plants operating under static equilibrium conditions, although both can be used to calculate start-up–shut-down behavior and both can examine the effects of applied perturbations.

Both procedures use an idealized model for a mixer–settler in the general balance equation, specifically

Quantities IN (to a given stage)

= quantities OUT (from a given stage)

where any modifications due to mutual solubility say of solvents, imperfect mixing–settling and so on can be allowed for in the stagewise calculations. In SEMPHIS MOD 4 each settler is split into three zones; SIMTEX considers the settler as an entity. (A British modification of SIMTEX, QUANTEX, splits each settler into three zones and also enables calculations to be made under non-steady-state conditions.)

The following is a general description of SEPHIS and SIMTEX. Details of SEPHIS have been published and are, unlike the details of SIMTEX, generally available. For published information on the procedure, the reader is referred to the reference given.

Differential equations (SEPHIS) describing the mixers and settlers are formulated thus:

$$\frac{d(V_{maj}X_y + V_{moj}Y_{ij})}{dt} = A_{j-1}X_{i,j-1} + O_{j+1}Y_{i,j+1} + A_{fj}X_{fi,j} + O_{fj}Y_{fi,j} - (A_j + A_{pj})X_{ij} - (O_j + O_{pj})Y_{ij} \quad (4)$$

where A = aqueous flow, liters/min
O = organic flow, liters/min
V = volume, liters
X = organic phase, solute concentration
Y = organic phase, solute concentration

subscripts m = mixer
a = aqueous phase
o = organic phase
j = stage number
i = solute number
f = feed stream

Now if the volumes and flow rates are assumed constant and the solutes in the mixer are at equilibrium, then $D_i = Y_{i,j}/X_{i,j}$, where D is the distribution coefficient. Hence we can write Eq. (4) in the form

$$\frac{dx_{i,j}}{dt} = \{A_{j-1}x_{i,j-1} + O_{j+1}y_{i,j+1} + A_{fj}x_{fi,j} + O_{fj}y_{fi,j} - [A_j + A_{pj} + D_i(O_j + O_{pj})] \cdot x_{i,j}\}/(V_{maj} + D_iV_{moj}). \quad (5)$$

where subscript p refers to product stream. The value of $X_{i,j}$ is obtained by the Runge–Kutta method, then D_i can be calculated. By assuming that $X_{i,j+1}$ will vary only by small amounts during the time interval.

$$X_{i,j-1} = \frac{(X_{i,j-1,t} + X_{i,j-1}, t + \Delta t)}{2}$$

and

$$Y_{i,j+1} = \frac{(Y_{i,j+1,t} + Y_{i,j+1}, t + \Delta t)}{2}$$

can be assumed, and hence the differential can be solved by an iteration procedure. The iteration procedure is rather similar to that already described earlier in this chapter.

Clearly, the equations for settlers are similar to those for the mixer given earlier and are written in the same form.

Equations for describing distribution coefficients are based on a semitheoretical–semiempirical approach by the use of experimentally derived data. In both SEPHIS and SIMTEX tri-*n*-butyl phosphate (TBP) is the extractant, but the correlations used to fit the experimental data are somewhat different; for example, in SEPHIS, for uranyl nitrate extraction,,

$$K_u = \frac{[UO_2(NO_3)_2 \cdot 2TBP]}{[UO_2^{2+}]\,[TBP]^2} = K_u^1[NO_3^-]^2$$

where K_u = pseudo-mass-equilibrium constant for uranyl nitrate and K_u^1 = pseudo-mass-action equilibrium constant for uranyl nitrate. Similar "SEPHIS" equations may be written for the plutonium nitrates and nitric acid.

In SIMTEX the concept of free or uncomplexed TBP is used and the correlation equations are of the form $K = F(H)\,B(b)\,f^2G^2$, where K is the distribution coefficient, H is the aqueous-phase acidity, f is the fraction of uncomplexed TBP, and G is a saturation function.

SEPHIS MOD 4 enables chemical reactions in the mixer and the settler to be considered. The reduction of Pu(IV) to Pu(III) by three different mechanisms is available.

Although uranium–plutonium systems are considered in both procedures, clearly they can be modified to cope with any analogous extraction and at the present time perhaps form the most generally applicable computational procedures for mixer–settlers available. Typical SIMTEX concentrations profiles are given in Fig. 6.

As shown in Fig. 3, different rates of mass transfer effectively mean an increase in the number of HTUs or NTUs to achieve equilibrium, and thus the simplest model could vary the HTU or NTU for each component in the system. To obtain the "best" model, it is considered that concentration profiles from an operating system would be required to be analyzed in order to provide data for a general model.

Correlation of equilibrium distribution data for A and B could take the form of methods already described and would be a subroutine for the computational procedure.

Equations of the type used in SEPHIS or SIMTEX could be used to obtain stage equilibrium data, but such data would have to be modified for backmixing.

4.1. Columns

Modeling of a pulsed column is considerably more difficult than modeling of a mixer–settler system, and it is probably correct to state, at least for the published literature, that no full model of a column exists. The present discussion thus deals briefly with the problems rather than their solutions, which are not yet available.

An early and still acceptable model of a pulsed column is that due to DiLiddo and Walsh [11], who assume immiscibility of phases, aqueous-phase continuous systems, and a sinusoidal pulse form in the column itself that operates in the mixer–settler region. The column is considered as a series of geometrically defined stages, that restricts the model at the outset to essentially the mixer–settler region. Nonetheless, it gives a good approximation to the actual performance of a 4-in. column and has been used successfully to predict transient behavior. It is perhaps true to regard this particular paper as a link between mixer–settlers and columns proper.

Other authors, for example, Miyauchi and Vermeulen [12], have considered specific aspects of columns, but none have examined the problem as a whole.

Therefore, in lieu of any readily available models, this section is confined to the requirements for a model of a pulsed column with references as appropriate to available data.

4.1.1. Distribution Data Correlation

It has been seen that there are several methods available for the correlation and prediction of distribution data. The actual method to be used will depend on the type of data being handled. However, assuming that a multisolute interactive system is being examined and that the extractant is dissolved in a diluent, then for three solutes, for instance, a variable extractant concentration and a variable aqueous-phase acidity, there are at least five variables or dimensions to be considered simultaneously if we wish to correlate all the data in one model. This excludes external degrees of freedom such as temperature. In an interactive system, reduction of the number of dimensions is not possible, although with the correct experimental design, the number of experimental measurements to be made can be minimized.

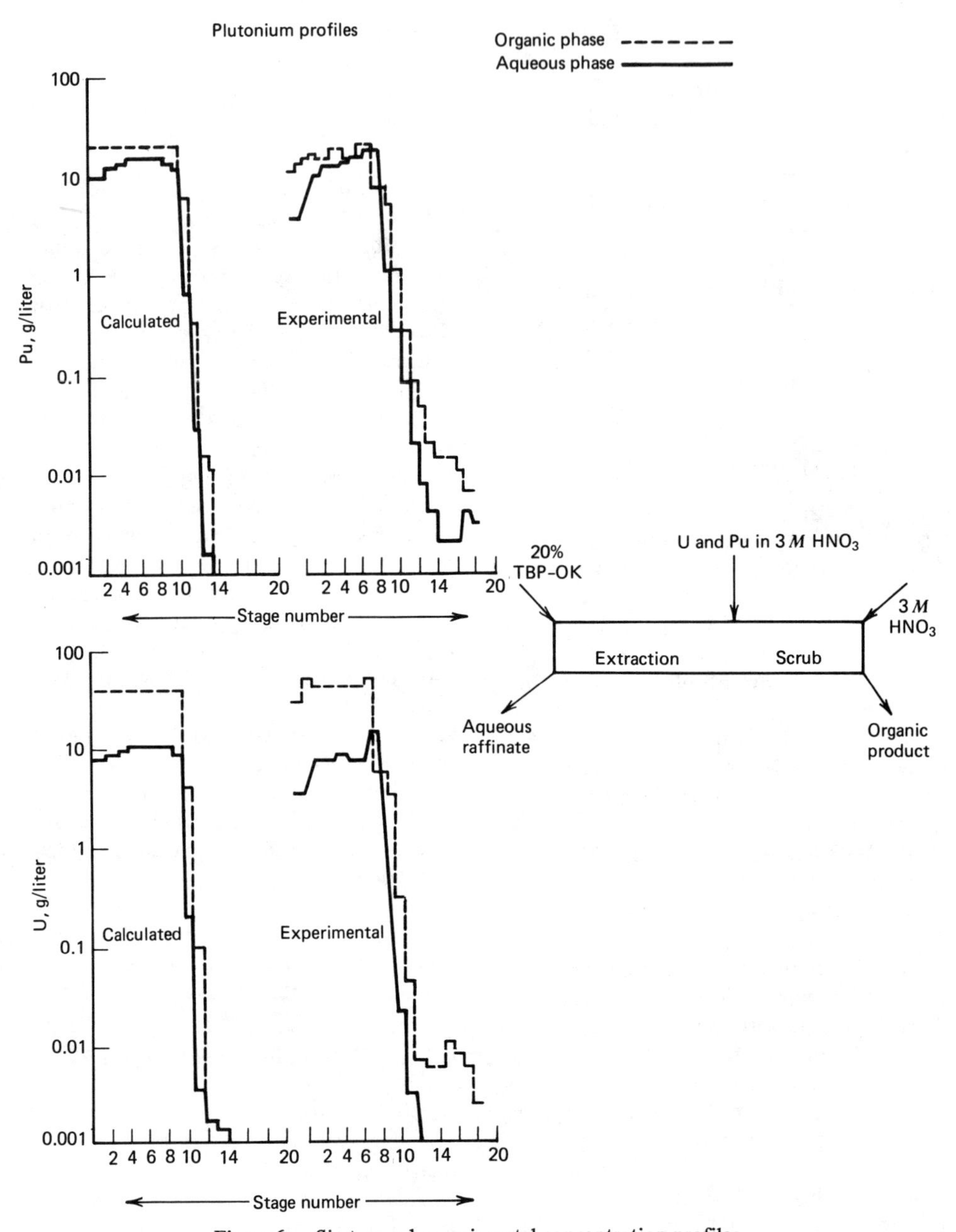

Figure 6. Simtex and experimental concentration profiles.

4.1.2. Frequency

In a pulsed column two major considerations must be observed: (1) at low frequencies the column can be considered as a series of n stages operating in the mixer–settler mode; but (2) as the pulse frequency increases, the degree of backmixing in either or both phases is increased. Thus in its general form a pulsed column can be regarded as a series of n stages operating at a level of backmixing depending on pulse amplitude, frequency and shape, and relative flow rates of the liquid through the column in addition to column geometry.

At low liquid flow rates and suitable pulse characteristics, the column operates in the mixer–settler mode, where each of the n stages can be uniquely defined and true distribution equilibria for all solutes is obtained in each stage.

At high flow rates, for an interactive multisolute system irrespective of pulse characteristics the behavior as described in Fig. 4 is obtained. Under these conditions it is essential that both backmixing and mass-transfer effects be built into the model if it is to be effective.

A number of authors have considered the effects of backmixing in varying degrees of mathematical complexity [e.g., 13, 14]. Pratt, however, has considerably simplified and extended the work of these and other authors, and his [15–17] is described in detail in Chapter 6.

It should be commented that all the work referred to is reasonably mathematical, and the reader should examine the original papers for the detailed arguments. None of the authors considers in detail the rate of mass transfer in their arguments. This has yet to be done.

The complexity of interactive multisolute systems is such that the methods due to Pratt cannot be directly used since the equilibrium lines are not amenable to his simplifying treatment. It would seem, however, that for a single solute system, the rate of extraction could be incorporated into the model by breaking the equilibrium line into a sufficient number of units to simulate each HTU. Although Pratt has not reported in his papers his approach to the analysis of column and column design, this approach could be easily transferred to a computer, and both modeling and control of a column could be examined with the effect of backmixing as one of the parameters to be studied.

5. SUMMARY AND CONCLUSIONS

It has been seen from the foregoing that there is no general approach to modeling of a solvent extraction system. The type of model required depends on the contactor to be modeled and whether the process is to be operated under static or dynamic equilibrium conditions. For multisolute systems, solute–solute interaction must be taken into account, and the operating characteristics of the contactor must then be superimposed on these interactions.

If modeling is approached from a purely theoretical aspect, it is unlikely that all the required data are available from such considerations. Alternatively, a purely empirical approach requires too great a knowledge of the behavior to be successful unless an iterative approach to the problem is taken by using data taken from the process itself.

Most authors, for various types of process, have used a semiempirical–semitheoretical approach that appears to give satisfactory results in practice.

Clearly, because of their operating characteristics, mixer–settlers are the simplest type of contactor to model, even for interactive systems. Column contactors are modeled to take into account backmixing but not solute–solute interactions or rates of solute mass transfer to the same extent as mixer–settler models. Undoubtedly this will be studied over the next decade.

REFERENCES

1. K. A. Varteressian and M. R. Fenske, *Ind. Eng. Chem.* **28**, 928 (1936).
2. A. M. Rozen, *Solvent Extraction Chemistry*, North-Holland, Amsterdam, 1967, p. 195ff.
3. J. O. Liljinzin, *Proceedings of the International Solvent Extraction Conference (ISEC), Toronto 1977*, Canadian Institute of Mining & Metallurgy, Montreal, 1979, p. 295.
4. Y.-C. Hoh and R. G. Bautista, *Proceedings of the International Solvent Extraction Conference (ISEC), Toronto 1977*, Canadian Institute of Mining & Metallurgy, Montreal, 1979, p. 273.
5. Y. C. Hoh and R. G. Bautista, *Met. Transact. (AIME) B*, **9B**, 69–72 (1978).
6. J. E. Gai, *Proceedings of the International Solvent Extraction Conference (ISEC), Toronto 1977*, Canadian Institute of Mining & Metallurgy, Montreal, 1979, p. 316.
7. J. W. Tierney, *Proceedings of the International Solvent Extraction Conference (ISEC), Toronto 1977*, Canadian Institute of Mining & Metallurgy, Montreal, 1979, p. 289.
8. B. Gaudernack et al., *Proceedings of the International Solvent Extraction Conference (ISEC), Lyons 1974*, Society of Chemical Industry, London, 1974, p. 2631.
9. W. R. Burton A. L. Mills, *Nucl. Eng. (Lond.)*, **8**, 248 (1963).
10. See, for example, D. E. Horner, *A Mathematical Model and a Computer Program for Estimating Distribution Coefficients for Plutonium, Uranium and Nitric Acid in Extractions with Tri-n-butylphosphate*, Oak Ridge National Laboratories ORNL/TM 2711 (available from Technical Information Center, P.O. Box 62, Oak Ridge, TN 37830, USA); W. C. Scotten, *Solvex: A Computer Program for the Simulation of Solvent Extraction Processes*, USA Report DP 1391; September 1975 (available as above); G. B. Watson and R. H. Rainey, *Modifications of the SEPHIS Computer*

Code for Calculating the Purex Solvent Extraction System, USA Report ORNL/TM 5123, December 1975 (available as above); W. S. Gronier et al., *Fast Reactor Fuel Reprocessing*, proceedings of a symposium, Society of Chemical Industry, London, 1980, p. 195.

11. B. A. DiLiddo, and T. J. Walsh, *Ind. Eng. Chem.* **53**, 801, (1961).
12. T. Miyauchi and T. Vermeulen, *Ind. Eng. Chem. Fund.* **2**, 113, (1963).
13. E. A. Schleicher, *AIChE J.* **6**, 529 (1960).
14. S. Hartland and J. C. Mecklenburgh, *J. Chem. Eng. Sci.* **21**, 1209 (1966).
15. H. R. C. Pratt, *Ind. Eng. Chem. Process. Des. Devel.* **14**(1), 1975.
16. H. R. C. Pratt, *Ind. Eng. Chem. Process. Des. Devel.* **15**(1), 1976.
17. H. R. C. Pratt, *Ind. Eng. Chem. Process. Des. Devel.* **15**(4), 1976.

27.2

DYNAMIC BEHAVIOR AND CONTROL

W. L. Wilkinson† *and J. Ingham*
University of Bradford
United Kingdom

1. INTRODUCTION

Much of the work with which chemical engineers are involved concerns only the *steady-state behavior* of processes, that is, the relationship between certain input variables $\theta_{i1}, \theta_{i2}, \ldots$ and the corresponding output variables $\theta_{o1}, \theta_{o2}, \ldots$ when the variables are constant and do not change with time. The dependence of the stage concentrations in a mixer–settler unit on flow rates, feed concentrations, temperature, and other parameters is an example. The *dynamic behavior* of processes, on the other hand, involves the relationships between time-dependent inputs $\theta_{i1}(t), \theta_{i2}(t), \ldots$ and the output variables $\theta_{o1}(t), \theta_{o2}(t), \ldots$, which will also vary with time. For example, how will the stage concentrations change with time following a sudden step change in the feed concentration?

The unsteady-state characteristics of processes are important in two classes of problem: (1)

†Present address: British Nuclear Fuels Ltd., United Kingdom.

startup problems, in which it is desired to predict the rate at which a process proceeds to equilibrium from a given initial state, such as the rate at which a mixer–settler, initially filled with pure solvent and solute free aqueous phase, approaches equilibrium after the feeds have been put on and (2) *control problems*, in which information is required on the open-loop response of the controlled variables in a process to uncontrolled disturbances in certain input streams, deliberate changes in load, and changes in the corrective actions that it is proposed to use in the control of the process. This is the important problem.

In considering the control of solvent extraction processes, we must take into account that, in general, there are many different types of process, a wide range of contactors and many possible types of control from simple manual systems to sophisticated schemes involving on-line computers.

However, in every case we should ask three questions:

1. What do we want to achieve by control?
2. Why do we want to achieve these objectives;
3. How can we most simply achieve these objectives?

There are many possible answers to these questions.

For example, the *objectives* could be to maintain a desired concentration profile or to maintain desired interface levels, and these would have to be achieved in spite of disturbances, load changes and set point changes. Possible reasons for these objectives could be to (1) maintain product quality, (2) avoid losses, (3) maximize throughput, (4) minimize operating costs, and (5) ensure safe operation.

The various *control strategies* available are (1) manual control, (2) traditional automatic control, and (3) computer control.

The various types of automatic control are considered qualitatively in turn and the techniques available for control system design are then discussed.

2. CONTROL STRATEGIES

2.1. Environmental Control

This is illustrated in Fig. 1 for the case of a typical mixer–settler. If we are given the inputs F, X_F, and X_A and the required outputs Y_E and X_R are specified, we can calculate the flow rates S and A and control them at these values. With this mode of control, we thus protect the process by controlling its environment. This would be satisfactory in a perfect world, but in practice we cannot control the environment completely, and unsuspected disturbances would upset the process.

2.2. Feedback Control

The problem with environmental control is that we do not measure continuously the parameter we are trying to control. However, with feedback control (Fig. 2), the controlled variables is measured, and if it is in error, we feed the error back as a corrective action. For example, we could maintain the solvent saturation approximately constant by regulation of the feed rate as shown in Fig. 2.

The duty of the controller is thus to maintain the error near to zero in spite of disturbances. This is termed *regulator operation*, in which the objective of the control system is to keep the controlled condition constant in spite of distur-

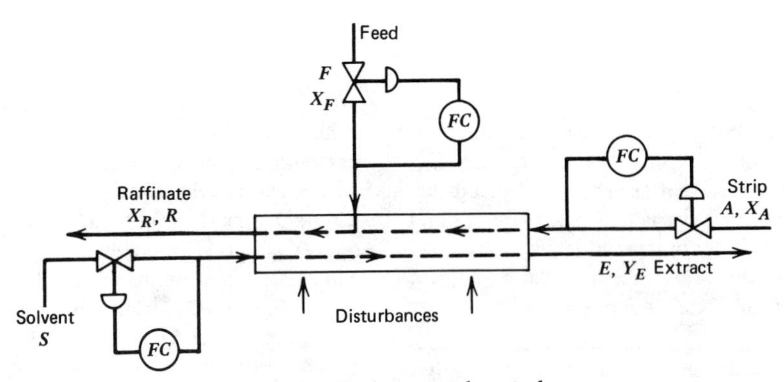

Figure 1. Environmental control.

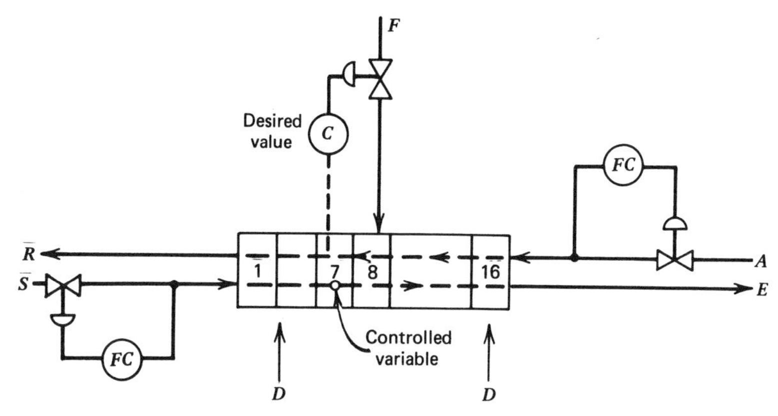

Figure 2. Feedback control.

bances and load changes by means of the control action. Alternatively, there is *servo operation*, in which the control system attempts to follow changes in the desired value of the controlled variable and keep the controlled variable close to this value by varying the control action.

2.3. Feed-Forward Control

Although feedback control is ideal in principle, it has some serious limitations in practice: the main point is that it can respond only to a disturbance when it has made its presence felt by producing an error. With fast processes, such as pressure and flow control, this does not really matter, but with slow processes this can lead to a very unsatisfactory response and solvent extraction processes are normally very sluggish.

With feed-forward control, this problem is avoided since the disturbance is measured and a correction immediately applied to counteract it, that is, before the process has been upset by it (see Fig. 3).

The problem is that it is obviously difficult to anticipate *all* the disturbances, and even if we could, it is unlikely that they could be measured *accurately* and the necessary corrective actions predicted *exactly*.

If we fail in any of these, an error will result and in feed-forward control we do not measure the error.

2.4. Feed-Forward–Feedback Control

This combined system (Fig. 4) gives us the best of both worlds. The feed-forward control system measures the main disturbances and provides fast corrective action before the process is seriously affected. The feedback system acts as a trim. It measures the controlled variable and attempts to eliminate any error, albeit slowly, due to possible shortcomings of the feed-forward system.

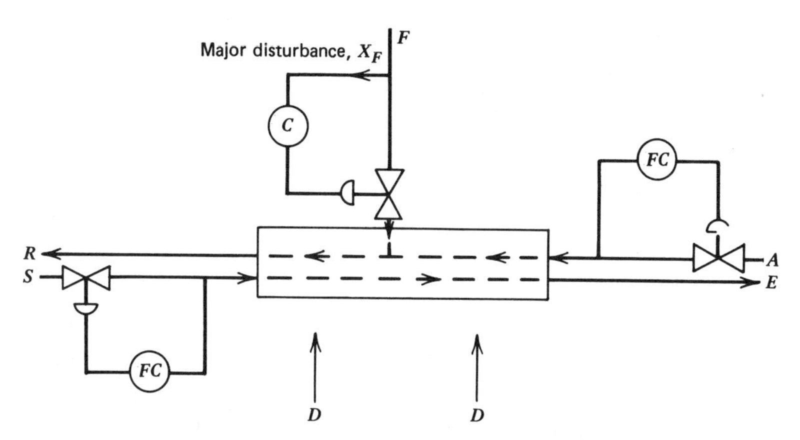

Figure 3. Feed-forward control.

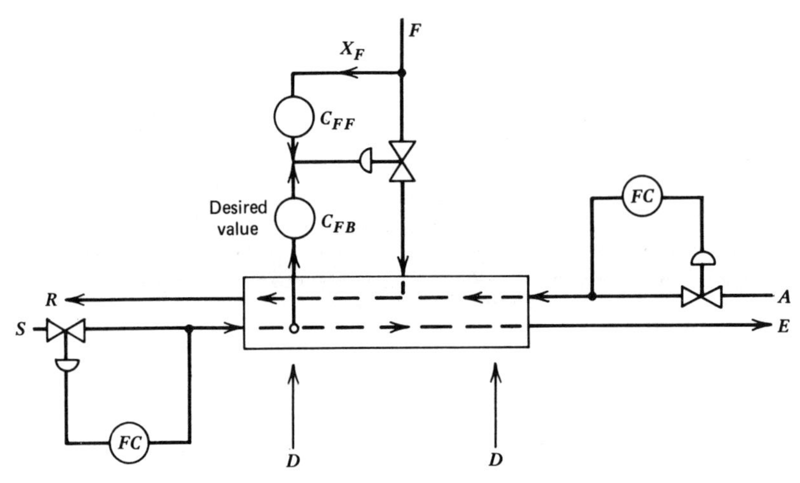

Figure 4. Feed-forward–feedback control.

This is likely to be a preferred method for many solvent extraction plants.

3. CONTROL SYSTEM DESIGN

Empirical and theoretical methods of control system design are considered in turn.

3.1. Empirical Methods

Most control systems are still designed empirically as follows: (1) design the plant without considering how it will be controlled; (2) when the design is complete, devise a logical control scheme based on past experience or a similar plant; and (3) build the plant, install the control system, and tune it by established methods.

3.1.1. Reaction Curve Method

This relies on the fact that the response of the controlled variable of most processes to a step change in the corrective action is sigmoidal in shape as in Fig. 5 and can be represented approximately by a delay time t_d and an exponential response with a time constant of τ.

The results of a theoretical analysis on this basis can be generalized to give satisfactory control. For proportional, integral, and derivative control, the control parameters K_c (proportional gain), I (integral action time), and D (derivative action time) are related to τ and t_d as follows:

$$K_c = \frac{1}{K}\frac{\tau}{\tau_d}\left(1.333 + \frac{t_d}{4\tau}\right) \tag{1}$$

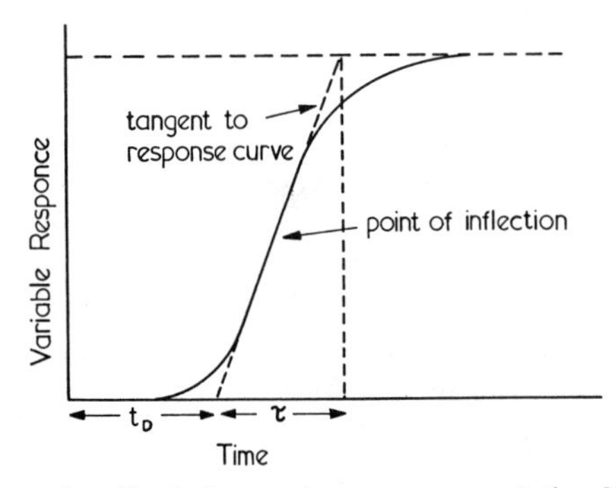

Figure 5. Typical response curve representation [1].

$$I = t_d\,\frac{32 + 6t_d/\tau}{13 + 8t_d/\tau} \tag{2}$$

$$D = t_d\,\frac{4}{11 + 2t_d/\tau} \tag{3}$$

In these equations K is the overall process gain, that is, the ratio of the magnitudes of the output and the input signals.

The application of these results saves the effort of using a trial-and-error approach to find the controller settings that best satisfy the performance criteria. Experience has shown that the controller settings obtained in this way give a fairly good closed-loop response. When the reaction curve is determined experimentally, all the elements in the feedback loop should be included in the test.

3.1.2. Loop Tuning

Alternatively, after the control system has been installed, trial-and-error methods can be used to find suitable controller parameters.

One approach is to first set the integral time of a three-mode controller to infinity and the derivative time to zero, or the lowest possible value and then introduce a number of step changes into the closed-loop system with the controller gain set to various values. In this way we can find the gain that corresponds to the generation of continuous oscillations—that is, the limit of stability. This gain is the *ultimate gain* K_u, and the period of the oscillations is the *ultimate period* P_u. Ziegler and Nichols [1] suggested the use of these values to calculate the controller settings according to Eqs. (4)–(6) for proportional plus integral plus derivative:

$$K_c = 0.6K_u \tag{4}$$

$$I = \frac{P_u}{2} \tag{5}$$

$$D = \frac{P_u}{8} \tag{6}$$

This is quite satisfactory for many processes for example, flow, pressure and temperature control, and also for more complex processes where considerable practical expertise has been established, such as distillation processes. However, there are many processes for which this empirical method could lead to costly mistakes, and *some* solvent extraction processes may well be in this category. A more theoretical approach would then be of value.

3.2. Theoretical Methods

To design a process control system we need, in the simplest terms, the dynamic relationship between (1) the controlled variable and disturbances and load changes, (2) the controlled variable and the control actions, and (3) the dynamic behavior of the other components in the system, (controller, valves, measuring instruments, etc.).

In other words, we need to know the *transfer functions* of all the components in the loop shown in block diagram form for the simple feedback control of the solvent extraction plant shown in Fig. 6. In addition to the dynamic behavior of the controller, represented by the transfer function G_c, the measuring unit G_m, and the valve G_v, we need to know G_{p1}, the transfer function relating to the open-loop response of the solvent concentration Y_n to the feed F (other things being constant); G_{p2}, the transfer function relating Y_n and X_F, the feed concentration and G_{p3}, the transfer function relating Y_n and V, the solvent flow rate. The corresponding block diagram for the feed forward system is given in Fig. 7.

The information on the dynamic characteristics can be obtained by (1) developing mathematical models based on the physics and chemistry of the process (this approach is discussed in detail later) and (2) experimentally, by injecting known perturbing signals and measuring the response. Note that method 1 can be applied to *new* processes whereas method 2 can be used only on *existing* plants.

3.2.1. Mathematical Models

Mathematical models can be developed that predict the necessary dynamic relationships for control system design for both stagewise and differential contactors.

This topic is discussed fully in Section 4. Suffice it to say here that such mathematical models can be derived to furnish the dynamic relation-

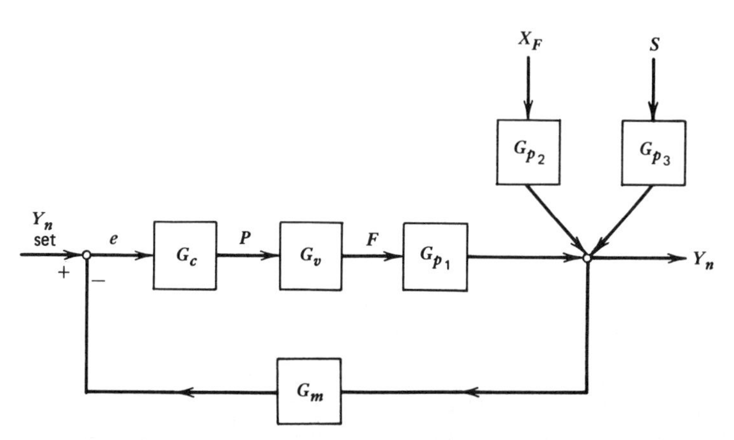

Figure 6. Block diagram for feedback control.

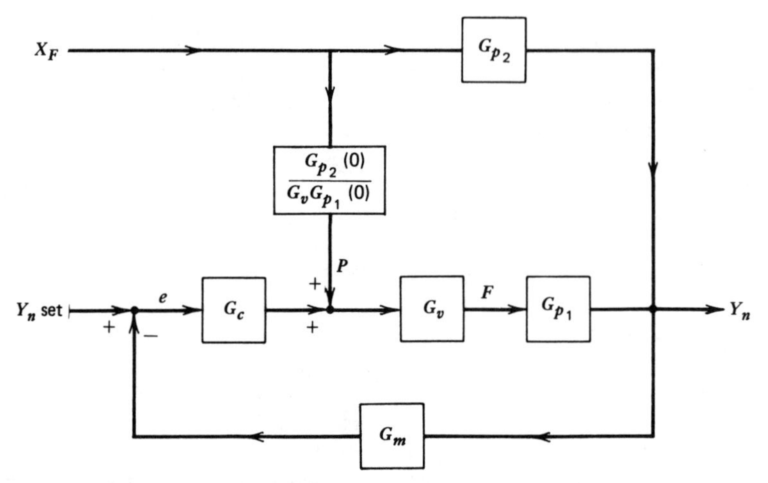

Figure 7. Block diagram for feed-forward–feedback control.

ships or transfer functions listed previously as necessary for the design of the control system for solvent extraction plants.

3.2.2. Experimental Techniques for Obtaining Dynamic Characteristics of Processes

If the plant exists, the dynamic characteristics can be obtained by imposing perturbing signals and observing the response. A variety of signals can be used: (1) steps, (2) impulses, (3) sinusoidal signals, and (4) pseudo-random binary signals. These inputs, together with the corresponding forms of the outputs, are illustrated in Fig. 8.

The choice depends on many factors. Steps and impulses are easy to apply, but the accuracy of the information obtained is rarely very good. Sinusoidal signals (frequency response) and pseudo-random binary signals (PRBS) require more expensive equipment, but the data obtained are of high accuracy, and moreover these signals can be made so small that they do not upset the process.

The experimental data can be used directly for control system design, and well-established techniques are available, particularly for dealing with frequency response data [1]. Alternatively, the results can be approximated by analytical forms and numerical solutions obtained for a range of possible control schemes.

3.2.3. Control of a Large Mixer–Settler

Simple transient response experiments have been carried out on a large 16-stage mixer–settler used for the purification of uranyl nitrate. The unit is

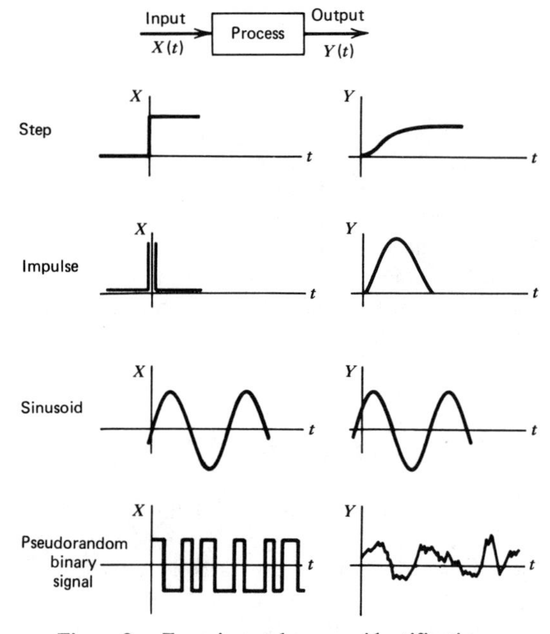

Figure 8. Experimental process identification.

as shown in Fig. 2. The object of the work was to find the response of the uranium concentration in the solvent phase at stage 7 to a step change in the feed flow to stage 8. It was intended to maintain the solvent loading at stage 7 constant by regulating the feed rate to counteract the effect of feed concentration disturbances.

The open loop response to a 10% increase in feed rate, imposed when the unit was at equilibrium was approximately first order and the transfer function relating changes in uranium concentration y_7 at stage 7 to feed flow F may be represented by

$$G_{p_1} = \frac{\bar{y}_7(s)}{F(s)} = \frac{K_p}{1 + \tau_1 s} \tag{7}$$

The time constant τ_1 is approximately 3 hr, and K_p was found to be 0.01% uranium/liter^{-1} hr^{-1}.

The block diagram of the control loop is identical to that in Fig. 6 and the closed-loop response to changes in feed concentration is then given by

$$\frac{\bar{y}_7(s)}{\bar{x}_F(s)} = \frac{G_{p_2}}{1 + G_c G_v G_{p_1} G_m} \tag{8}$$

It is assumed that $G_m \sim 1$; that is, the measurement response is fast compared with the plant response and similarly $G_v = K_v$, where K_v is the valve constant. For the three-term controller used, G_c was given by

$$G_c = K_c \left[1 + \frac{D}{I} + Ds + \frac{1}{Is} \right] \tag{9}$$

where K_c is the controller gain; D and I are the derivative and integral action times, respectively; and s is the Laplace transform variable.

Equation (8) then becomes

$$\frac{\bar{y}_7(s)}{\bar{x}_F(s)} = \frac{G_{p_2}[1 + \tau_1 s]}{1 + \tau_1 s + K_c K_v K_p [1 + D/I + Ds + 1/Is]} \tag{10}$$

The form of the response is determined by the denominator in Eq. (10), that is, whether the system is overdamped or oscillatory and if, oscillatory the period and subsidence ratio.

For example, if we choose a subsidence ratio of $1/\sqrt{2}$ and a period of 1 hr, it can be shown analytically that the controller parameters should be as follows:

Gain	6.2 (16% proportional bandwidth)
Integral action	6 min
Derivative action	3 min

The controller was set with these values initially and then tuned further by trial and error. Operation was satisfactory. A purely ad hoc method of setting up the controller without a preliminary experimental and theoretical analysis would probably have been unsuccessful.

4. DYNAMIC CHARACTERISTICS OF SOLVENT EXTRACTION CONTACTORS

4.1. Differential and Stagewise Contactors

The relative geometries and operational characteristics of the many types of solvent extraction contacting equipment that are now available have been discussed in previous chapters. In this chapter, solvent extraction equipment is considered to consist of two basic forms: (1) stagewise contactors, of which mixer–settlers are taken as the prime example, and (2) differential column contactors of many differing forms. These two basic types exhibit markedly different operational characteristics, and this applies especially to their dynamic behavior and to the related control problem.

Mixer–settlers are characterized by relatively low throughputs and large holdup volumes, and thus by relatively sluggish but stable responses. The effects of any disturbances tend to be transmitted relatively slowly through the contactor, and thus their correction by controller action tends to be a rather lengthy process, such that the resultant variations in the contactor outlet stream also tend to be of long duration. In the absence of any special kinetic effects, mixer–settlers nearly always tend to operate at close to equilibrium conditions. Thus the resultant steady-state effects of any controller action are easily predictable, since the main result is simply to change the slope of the operating line (feed : solvent mass flow rate ratio). Most mixer–settler cascades include a fairly large degree of overdesign. Therefore any such controller effects are unlikely to pose a serious problem during plant operation. Because of their equilibrium nature, mixer–settler cascades present no special difficulties following shutdown or during intermittent operation since the plant conditions remain at the same state as they achieved prior to feed or solvent stoppage. The relative stability of mixer–settlers, is, of course, one of the prime reasons for their preference in many industrial applications.

On the other hand, differential column contactors are generally characterized by high throughputs and low holdup volumes and thus by relatively rapid response characteristics. Optimum operational efficiencies are usually obtained over a relatively narrow range of operating conditions, and hence such contactors tend to be relatively unstable where conditions depart substantially from those corresponding to design. Since the disturbances are transmitted rapidly through the column, the controller action must be correspondingly fast, thus requiring high controller gains, which will themselves tend to increase instability. Column contactors usually operate under non-equilibrium conditions, in which the rate of mass transfer is related directly to the area available for interfacial mass transfer and thus to the volumetric holdup of dispersed phase. Any large controller action can have a substantial effect on the hydrodynamic conditions within the column, either by reducing the holdup and hence the effective rate of mass transfer or by increasing holdup with attendant dangers of column failure by flooding or by phase inversion. Thus the resultant effects of controller action are not so easily predictable, the column will not have so much overdesign capacity, and the controller action must be strictly limited to far narrower practicable ranges of operation than for mixer–settlers.

4.2. Dynamic Modeling

The mathematical modeling approach has the advantage that it can be applied to new processes. This is important since the need to include a consideration of probable process dynamics even at the very earliest plant design stage is being recognized increasingly. The task of achieving a realistic dynamic model for the process represents a very considerable accomplishment since it is much more difficult than that required to develop a steady-state model, and with our present state of knowledge, even a satisfactory steady-state model of the process is often difficult to achieve. It must be accepted that the actual hydrodynamic and mass-transfer processes in a solvent extraction contactor are exceedingly complex, and it is impossible to achieve an exact representation of all the various phenomena that occur. Generally speaking, it is a sound modeling approach to begin with the simplest possible representation of the system and then to increase the complexity combined with a continuous process of validation of basic theory, mathematical relations, parameter values, and numerical techniques until the simplest and most robust model is achieved that gives a reasonable accuracy of solution. It should be remembered that the accuracy of the final results depends on many factors, including the relevance of the basic theory and its basic assumptions, errors in the translation of the theory into mathematical equations, errors in the translation of the equations into a form suitable for solution, and errors in the actual solution technique. The repeated validity checking required in the development of a dynamic model has the advantage of forcing the process engineer to think more thoroughly about the nature of the actual cause and effect sequences of the process and often exposes considerable areas of ignorance. The accuracy obtained must, however, be balanced against the increased effort required.

4.3. Solution Techniques

The earliest modeling approaches, as represented by the pioneering work of Marshall and Pigford [2], were based on classical mathematical techniques. As shown in later sections, the analysis of stagewise or lumped parameter extraction devices leads to forms of solution that may be expressed eventually as sets of simultaneous differential equations. Likewise, the analysis of truly differential or distributed parameter extraction devices leads to a mathematical description in terms of partial differential equations. Assuming constant coefficients and linear equilibrium relationships, these can be solved by Laplace transformation to eliminate the derivative terms and to reduce the models to a series of simultaneous linear algebraic equations that can be more easily solved by use of appropriate boundary conditions. The Laplace transformation method may also be employed for nonlinear cases if the model behavior is restricted to small perturbations about some steady-state value and the model equations are linearized by use of the first-order Taylor series approximation. The application of the analytical method is limited to simplified linearized systems based on appropriate balance equations with constant coefficients. Thus it is impossible to include many real-life effects such as hydrodynamic factors into the model. Once an analytical solution has been obtained, however, this may prove to be a very useful check on the more realistic numerical methods. Owing to the very high speeds of solutions obtained, many workers have been attracted by the technique of analog computation. One of the major disadvantages of this technique, however, is the large number of components required for realistic problems. With the rapid growth of digital computational capability over the last decade, the advent of mini and micro computers and software, the use of analog computing even in hybrid form has been largely superseded by digital means, except perhaps in very specialist applications. By use of the digital computer, the chemical engineer is now able to tackle and solve problems, which only 10 years ago would have seemed to be unthinkable. This is made possible by the growth in the capacities and speeds of the machine and by use of sophisticated numerical integration routines that are now available as standard packages. Special digital simulation languages such as MIMIC and CSMP have been developed specifically to provide direct simple means of solution to sets of simultaneous differential and algebraic equations and are especially convenient for the noncomputing specialists as suggested by Ingham and Dunn [3, 4]. Otherwise, problems can be solved by using standard languages such as FORTRAN, using simple subroutine calls to effect the integration. In digital computation, all previous restraints are removed and the model can be completely generalized to include variable-parameter terms, nonlinear terms, multisolute and multifeed extraction, hydrodynamic terms, and closed-loop controller expressions as required.

One difficulty that may be encountered in the use of numerical integration techniques is that of equation stiffness. This occurs when, for example, there are large differences in the time constants for the various differential equations. Thus a minor component with a fast time constant may require very small integration step sizes and may thereby come to dominate the overall speed of solution which can be substantially reduced. Special numerical routines specifically designed to handle stiff systems have been developed, therefore. A comparison of the relative running times for a five-stage nonequilibrium model carried out by Wilkinson and McDonald [5] using of differing integration routines gave the following results:

Method	Relative Running Time
Nardsieck	4.6
Krogh	1.45
Runge-Kutta-Merson	1.21
Gear	1.0

The method due to Gear is specially formulated for the solution of stiff equations, whereas that proposed by Nardsieck is known to be notoriously difficult when applied to stiff systems. The method of Runge-Kutta-Merson, which is, however, generally considered suspect, came out surprisingly well. Most advanced digital simulation languages, however, provide the user with a choice of integration routines, and the most appropriate can be selected for a particular problem on the basis of experience.

4.4. Dynamic Models

4.4.1. Stagewise Cascades

The simplest modeling approach is that adopted by Franks [6], in which the cascade is represented by a series of perfectly mixed equilibrium stages of constant volume (see Fig. 9).

For any stage n, the dynamical component balance equation may be expressed by

$$\begin{array}{c}\text{Rate of change of}\\ \text{solute holdup}\end{array} = \begin{array}{c}\text{Inflow rate}\\ \text{of solute}\end{array} - \begin{array}{c}\text{Outflow rate}\\ \text{of solute}\end{array}$$

or

$$\begin{aligned}\frac{d}{dt}(H_x X_n + H_y Y_n) &= LX_{n-1} + VY_{n+1} - LX_n - VY_n \\ &= L(X_{n-1} - X_n) + V(Y_{n+1} - Y_n) \quad (11)\end{aligned}$$

where H_x and H_y are the total phase holdup volumes in each stage, L and V are the phase volumetric throughput rates, and X_n and Y_n are the solute concentration in the two phases leaving stage n.

The equilibrium stage condition is satisfied by the condition that

$$Y_n = \int_{\text{eq}} (X_n) \quad (12)$$

For the case of a linear equilibrium relation where

$$Y_n = mX_n \quad (13)$$

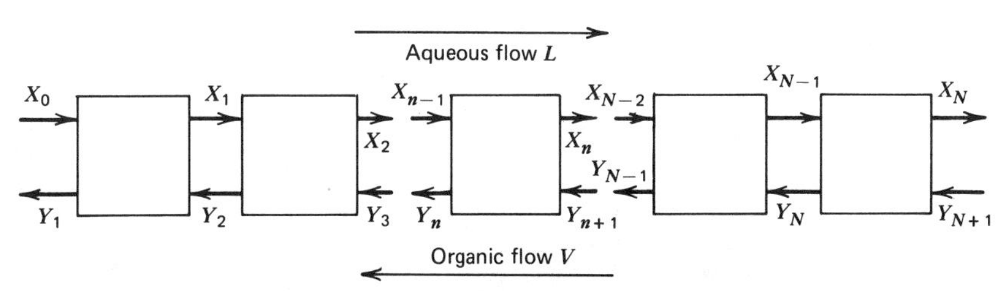

Figure 9. Countercurrent stagewise cascade [6].

and m is the distribution coefficient, the dynamic behavior of the cascade can be solved analytically [7]. Thus substituting for Y_n in Eq. (11) gives the relationship

$$(H_X + mH_Y)\frac{dX_n}{dt} = mV(X_{n+1} - X_n) + L(X_{n-1} - X_n) \quad (14)$$

or

$$(\theta_X + \theta_Y)\frac{dX_n}{dt} = X_{n-1} - (1+\lambda)X_n + \lambda X_{n+1} \quad (15)$$

where

$$\theta_X = \frac{H_X}{L}, \qquad \theta_Y = \frac{H_Y}{V}, \quad \text{and} \quad \lambda = \frac{mV}{L} \quad (16)$$

Thus

$$\frac{dX_n}{dt} = \frac{X_{n-1} - (1+\lambda)X_n + \lambda X_{n+1}}{\theta_X + \theta_Y} \quad (17)$$

where the resulting set of N simultaneous linear difference differential equations, assuming constant values of θ_X, θ_Y, and λ, can be solved by classical means to give concentrations X_n and hence Y_n as a function of time for, say, a step change forcing disturbance in the feed and solvent inlet concentrations X_0 and Y_{N+1} (7).

For nonlinear equilibrium concentrations, a numerical solution approach must be adopted. The approach suggested by Franks [6] is illustrated by the information flow diagram shown in Fig. 10.

In view of the preceding observations, it is suggested that the procedure be used to solve for Y_n or X_n depending on whichever is the largest.

This poses difficulties in the solution; however, since the equilibrium relationship must be able to be differentiated to obtain dX_n/dt, for example, where $Y_n = mX_n$, $dY_n/dt = mdX_n/dt$, and this procedure may not always be mathematically convenient, especially in the case of multicomponent extraction where the different solutes exhibit mutually interacting equilibrium relationships. Otherwise, considerable computational effort must be spent in iterating between the component balance equation [Eq. (11)] and the equilibrium relationship [Eq. (12)].

An alternative approach follows the natural cause and effect sequence of the equilibrium process by incorporating an expression for the mass-transfer rate Q_n into the dynamic model. Thus the component continuity equations for the solute in each phase of any stage n may be expressed by

Rate of accumulation of solute in phase i

= inflow rate of solute in phase i

- outflow rate of solute in phase i

± rate of solute mass transfer

Thus for phase X

$$H_X\frac{dX_n}{dt} = L(X_{n-1} - X_n) - Q_n \quad (18)$$

and for phase Y

$$Y_Y\frac{dY_n}{dt} = V(Y_{n+1} - Y_n) + Q_n \quad (19)$$

where

$$Q_n = K_n a_n (X_n - X_n^*) \quad (20)$$

where K_n is the overall coefficient for mass transfer in stage n, a_n is the total interfacial area for

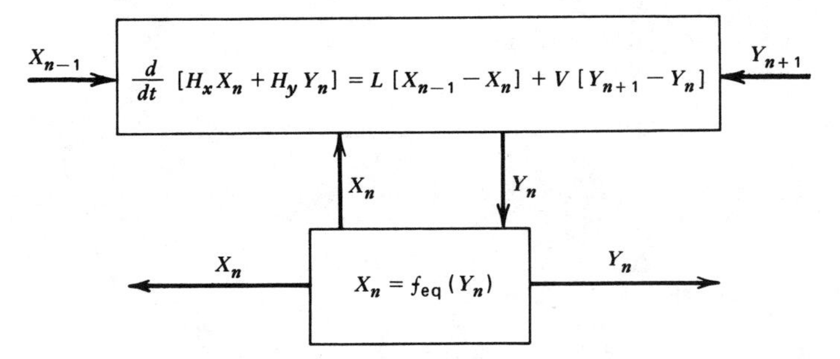

Figure 10. Information flow diagram for stage n [6].

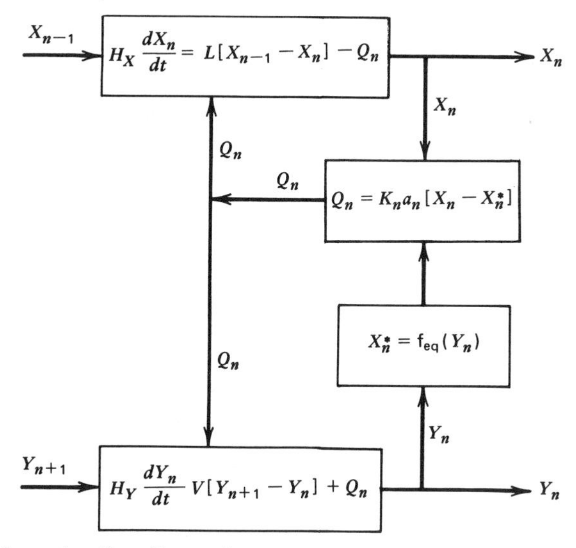

Figure 11. Information flow diagram for stage n based on rate and equilibrium relationships.

mass transfer in stage n, and X_n^* is the equilibrium solute concentration in stage n.

The equilibrium condition for the stage can be satisfied by using an arbitrarily high value for the term $K_n a_n$ since for a finite rate of transfer Q_n, a high value of $K_n a_n$ forces the driving-force term $(X_n - X_n^*)$ to be very low and hence the value of X_n^* to approach very closely to X_n. The arbitrary value taken for $K_n a_n$ thus acts as a high-gain factor to force conditions toward equilibrium. Although this technique offers a more inherently satisfactory means to solution, it does tend to be rather more expensive in computer time. An information flow diagram for this solution technique is shown in Fig. 11, in which the two component continuity equations for the two phases are shown to be linked by the equilibrium and rate relationships.

Because of the use of the rate equation in the preceding approach, this method is also representative of the general approach to the problem of stagewise cascades operating under non-equilibrium conditions by use of a finite and realistic value for the mass transfer capacity coefficient term $K_n a_n$ in the model. This again appears more intrinsically useful than the alternative practice of representing non-equilibrium conditions by means of a Murphree stage efficiency, which, unlike the $K_n a_n$ term, is less likely to be constant during plant transients. In addition, the general form of the equations is now also applicable to the case of non-equilibrium stagewise models for differential contactors as discussed in Section 4.4.3.

The extension of the modeling approach to allow for backmixing between stages, cascades with sidestreams or multiple feeds, or extraction combined with chemical reaction is also easily accomplished by appropriate modification of the inflow and outflow terms or by inclusion of a reaction rate term to work in conjunction with the mass transfer rate term. For example, for a cascade with constant backmixing flows L_B and V_B as shown in Fig. 12, the component continuity equations are very easily modified. With a backmixing flow L_B in the aqueous phase, the resultant forward flow along the cascade is thus $L + L_B$ since the backmixing does not appear exterior to the column. Similarly, with backmixing flow V_B, the forward flow for the organic phase is modified to $V + V_B$.

Allowance for the backmixing flow contributions in the inflow and outflow terms of the continuity equation for each phase in stage n gives

$$H_X \frac{dX_n}{dt} = (L + L_B)(X_{n-1} - X_n) + L_B(X_{n+1} - X_n) - Q_n \quad (21)$$

$$H_Y \frac{dY_n}{dt} = (V + V_B)(Y_{n+1} - Y_n) + V_B(Y_{n-1} - Y_n) + Q_n \quad (22)$$

Figure 13 shows a digital simulation program for a five-stage cascade with backmixing, written in the BEDSOCS programming language [8]. The simplicity and directness of this solution approach are readily apparent. Statements 30-210 set the constants, initial conditions, and graphical

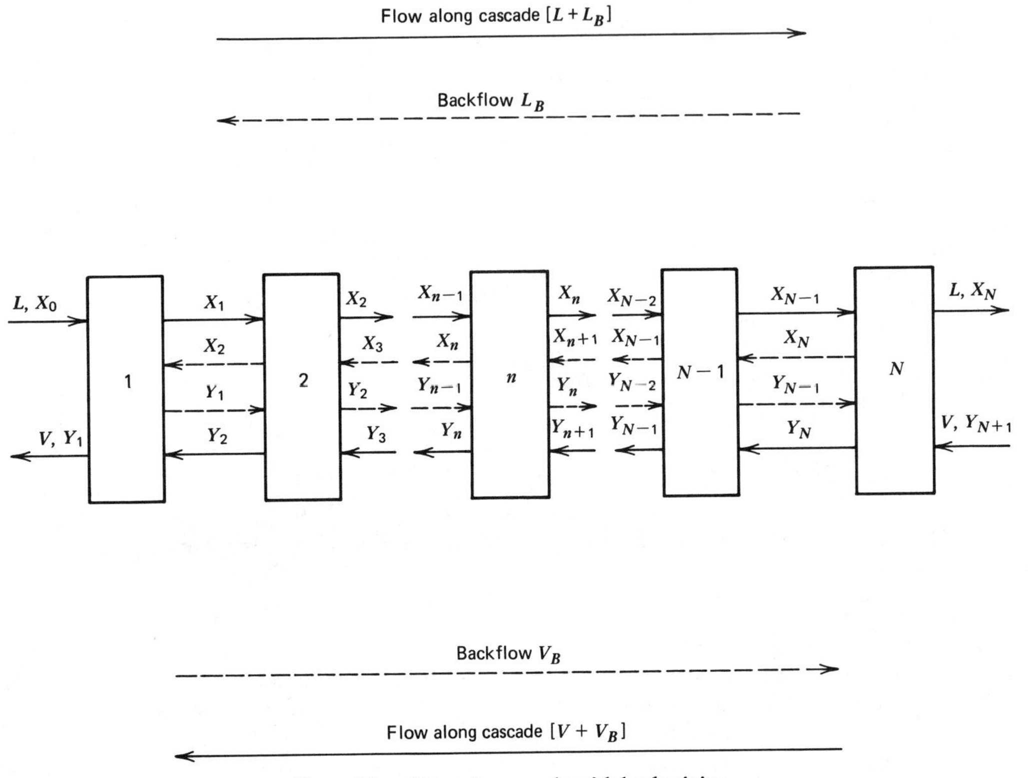

Figure 12. Stagewise cascade with backmixing.

display parameters for the problem. The model equations are represented by statements 260–420 located in the DYNAMIC REGION of the program, where the equations are solved in parallel by use of a Runge–Kutta–Merson integration procedure initiated by the DER statements. The direct relationship of the program statements representing the model and Eqs. (21) and (22) is readily apparent. Note that slightly different balance equations are used for the end stages because of the absence of backmixing flow contributions exterior to the column. Figure 14 shows the time-dependent variations in the feed phase concentrations X_n as a function of time, starting from initially zero solute concentrations along the cascade and following a unit step change in the feed concentration X_0. The effects of the backmixing in increasing the inlet-phase solute concentration jump, the flattening of the concentration profiles, the reduced extraction efficiency, and an increase in the speed of response are readily apparent. Note that although the program is formulated for simplicity in terms of a linear equilibrium relationship, this does not represent a restrictive condition in the solution procedure.

4.4.2. Mixer-Settlers

A realistic description of the process dynamics, however, also involves a consideration of the hydrodynamic characteristics of both the mixer and settler compartments and possibly also their hydraulic interactions (see Fig. 15).

There is a large volume of theoretical work in the field of open-loop dynamic behavior of mixer–settlers but little experimental information on their response. Difficulties of analytical solution and computer size limitations have prevented the development of a truly representative model for the description of a mixer–settler. This has resulted in many assumptions, the most common of which are (1) perfect mixing in the mixer, (2) immiscibility of the two liquid phases, (3) constant flow rates of both phases, (4) no mass transfer in the settler, (5) equilibrium achieved in the mixer, and (6) plug flow in each phase of the settler.

(a) Mixers. Most of the evidence suggests that the mixers, which must be well stirred in order to achieve efficient mass transfer, can be modeled as perfectly mixed tanks, in a manner analogous to

```
10  REM             *********** PROGRAM 1 **********
20  REM EQUILIBRIUM FIVE STAGE EXTRACTION CASCADE WITH BACKMIXING
30  REM PHASE VOLUMES IN EACH STAGE
40  LET H1=H2=2
50  REM VOLUMETRIC PHASE FLOWS
60  LET L=1
70  LET V=3
80  REM INLET CONCENTRATIONS
90  LET X0=1
100 LET Y6=0
110 REM EQUILIBRIUM DISTRIBUTION COEFFICIENT
120 LET M=.8
130 REM MASS TRANSFER GAIN FACTOR FOR EQUILIBRIUM CONDITION
140 LET K=100
150 REM BACKMIXING FLOWS
160 LET L1=5*L
170 LET V1=5*V
180 REM DISPLAY CHARACTERS
190 LET @1=@(1)=10
200 LET @(2)=@(4)=0
210 LET @(3)=.5
220 DYNAMIC
230   EQUATIONS
240     INDVAR T
250     REM STAGE 1
260     DER X1=(L*X0+L1*X2-(L+L1)*X1-Q1)/H1
270     DER Y1=((V+V1)*(Y2-Y1)+Q1)/H2
280     Q1=K*(X1-Y1/M)
290     REM STAGES 2-4
300     DER X2=((L+L1)*(X1-X2)+L1*(X3-X2)-Q2)/H1
310     DER X3=((L+L1)*(X2-X3)+L1*(X4-X3)-Q3)/H1
320     DER X4=((L+L1)*(X3-X4)+L1*(X5-X4)-Q4)/H1
330     DER Y2=((V+V1)*(Y3-Y2)+V1*(Y1-Y2)+Q2)/H2
340     DER Y3=((V+V1)*(Y4-Y3)+V1*(Y2-Y3)+Q3)/H2
350     DER Y4=((V+V1)*(Y5-Y4)+V1*(Y3-Y4)+Q4)/H2
360     Q2=K*(X2-Y2/M)
370     Q3=K*(X3-Y3/M)
380     Q4=K*(X4-Y4/M)
390     REM STAGE 5
400     DER X5=((L+L1)*(X4-X5)-Q5)/H1
410     DER Y5=(V*Y6+V1*Y4-(V+V1)*Y5+Q5)/H2
420     Q5=K*(X5-Y5/M)
430     DISPLAY X1,X2,X3,X4,X5
440   EQUEND
450   IF T >= @1 THEN  STOP
460 DYNEND
```

Figure 13. Digital simulation program for a 5-stage cascade with backmixing.

the treatment described in Section 4.4.1. Under conditions of large variations in the aqueous and organic phases, the total mixed volume may, however, vary. This can be modeled by a total continuity equation where

Rate of change in total mass in mixer

= mass inflow rate

- mass outflow rate

that is

$$\frac{d}{dt}(H_m \rho_m) = L_F \rho_{V_F} + V_F \rho_{V_F} - (L_m + V_m)\rho_m \qquad (23)$$

where H_m is the total holdup in the stage $= H_L + H_V$, ρ_m is the mean density in the mixer, subscripts F refer to inlet conditions, and m refers to the mixer outlet.

The total outlet flow from the mixer may be related to the difference in the driving head between the levels in the mixer h_m and that in the settler h_s, where $h_m = H_m/A$ and A is the cross-

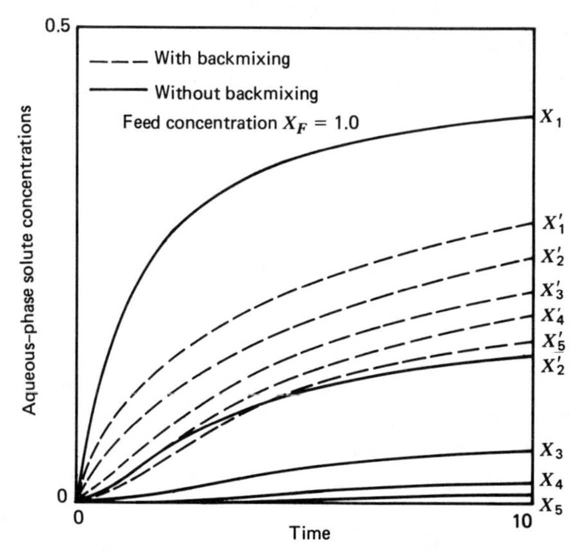

Figure 14. Transient solute concentration profiles for a five-stage equilibrium cascade with backmixing on startup.

sectional area of the mixing tank, such that

$$L_m + V_m \propto \sqrt{h_m\rho_m - h_s\rho_s} \tag{24}$$

where ρ_s is the mean density of the material in the settler and ρ_s and ρ_m are related to the relative phase densities. This hydraulic effect has, however, generally been neglected in most modeling exercises, and it is most commonly assumed that the mixers and settlers operate under constant total volume conditions. It may, however, be important in the related modeling of the settler that the individual variations of the holdups in the two phases in the mixer should be considered. At high levels of agitation it is reasonable to assume that the ratio of the two phase holdup volumes is in direct proportion to the inlet volumetric flow rates.

The aqueous outlet flow L_m will be given by

$$L_m = \frac{(L_F + V_F)\,H_L}{H_L + H_V} \tag{25}$$

Thus variations in the aqueous holdup in the mixer H_L will be given by

$$\frac{dH_L}{dt} = L_F - \frac{(L_F + V_F)\,H_L}{H_L + H_V} \tag{26}$$

where H_V is the solvent phase holdup in the mixer.

The corresponding equations for the organic-phase holdup will be

$$V_m = \frac{(L_F + V_F)\,H_V}{H_L + H_V} \tag{27}$$

and

$$\frac{d(H_V)}{dt} = V_F - \frac{(L_F + V_F)\,H_V}{H_L + H_V} \tag{28}$$

A solute balance in the aqueous phase gives

$$\frac{d}{dt}(H_L X_m) = L_F X_F - L_m X_m - Q_m \tag{29}$$

where Q_m is the rate of transfer of solute from the aqueous to the solvent phase in the mixer, as shown in Fig. 15.

Similarly, for the organic phase

$$\frac{d}{dt}(H_V Y_m) = V_F Y_F - V_m Y_m + Q_m \tag{30}$$

Addition of these relationships gives

$$\frac{d}{dt}(H_L X_m + H_V Y_m) = L_F X_F + V_F Y_F - L_m X_m - V_m Y_m \tag{31}$$

which is similar to Eq. (11) derived previously.

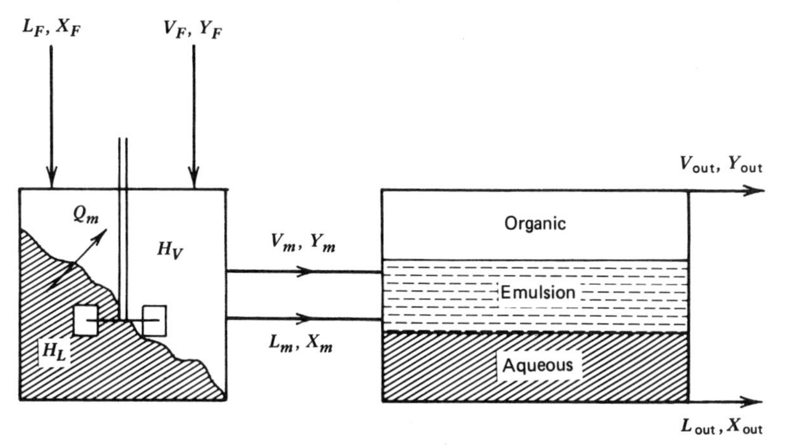

Figure 15. Mixer–settler unit.

Assuming an equilibrium expression of the form

$$Y_m = A_1 + A_2 X_m + A_3 X_m^2 \qquad (32)$$

$$\frac{dY_m}{dt} = A_2 \frac{dX_m}{dt} + 2A_3 X_m \frac{dX_m}{dt} \qquad (33)$$

so that

$$[H_L + H_V(A_2 + 2A_3 X_m)] \frac{dX_m}{dt} = L_F X_F + V_F Y_F - L_m X_m - V_m Y_m \qquad (34)$$

which in conjunction with Eqs. (26), (28), and (30) is solvable for both X_m and Y_m and for any mixer in a cascade, knowing the appropriate input quantities L_F, V_F, X_F, and Y_F.

The combination of the dynamic equations for both mixer and settler can, however, lead to problems of stiffness in the numerical integration since the time constant for the settler equations is often many times greater than that of the mixer. In such circumstances it may be possible to neglect the mixer dynamics in comparison with those of the settler and to replace the dynamic differential equations for the mixer by steady-state algebraic relationships. This is the basis of several proprietary computer programs as in Refs. (9)–(11).

(b) Settlers. The simplest approach is to assume that both phases in the settler are in plug flow and that the settler concentration response for each phase is simply that leaving the mixer delayed by the time required to pass through the appropriate phase in the settler. This is very easy to represent on a digital computer by setting up an array in which present values are stored and are read from the array after delay time t but is practically impossible to represent on an analog computer. The method proposed by Franks [12] by use of a circular array in which the values retain a fixed place in the array and input and output pointers move around the array is especially convenient. In practice, however, there is considerable evidence of appreciable mixing processes in the settler and especially in the emulsion band, such that various means—including combinations of plug and mixed flow, series of perfectly mixed stages, and recirculation streams—have been used to represent the settler dynamics. This type of representation is, however, rather arbitrary, and the particular flow scheme and its characteristic parameters are certain to alter according to the particular situation considered. What is needed, therefore, is a more fundamental approach based on observations on what is actually happening in the settler. A more basic approach due to Wilkinson [13] is as follows. Considering a conventional box mixer–settler as in Fig. 15, it may be observed that as the flow rate increases the length of the emulsion wedge increases but with high throughput corresponding to a correct sizing of the settler, the emulsion band thickness becomes uniform. It is known that the rate of coalescence in an emulsion band is a function of the band depth. Once the depth becomes uniform, it is thus reasonable to assume that the rate of coalescence will be uniform over the entire area of the settler. This means that the liquid flux into both phases will be uniform, and on this basis the flow patterns in the two phases can be predicted. This, in turn, allows a dynamic model of the settler to be developed.

The Emulsion Band. Treating the band as a differential process, this is divided into N finite difference elements as shown in Fig. 16 such that the composition of each phase in each element is assumed to be uniform. It is assumed that there is no mass transfer in the emulsion band, that is, that the phases are in equilibrium at the entrance to the settler and that the aqueous phase is dispersed.

Consider the nth stage. Since there is a constant flux of both phases from the emulsion band the flows displacing the solute along the band will be as follows:

Aqueous phase in $$L_{n-1} = (N + 1 - n)\frac{L_m}{N} \quad (35)$$

Aqueous phase out $$L_n = (N - n)\frac{L_m}{N} \quad (36)$$

Organic phase in $$V_{n-1} = (N + 1 - n)\frac{V_m}{N} \quad (37)$$

Organic phase out $$V_n = (N - n)\frac{V_m}{N} \quad (38)$$

since for a uniform flux out of the band, the flow along the band is proportional to the number of elements.

If we assume that the flows across a section boundary are at the means of the concentration of the adjacent sections for both phases, the concentrations of the phases entering the sections are

$$Y_{n-1} = \tfrac{1}{2}(Y_{n-1} + Y_n) \quad (39)$$

$$X_{n-1} = \tfrac{1}{2}(X_{n-1} + X_n) \quad (40)$$

The change in the holdups of each phase in the emulsion band H_L and H_V are given by

$$\frac{dH_L}{dt} = \frac{L_m}{N} - L' \quad (41)$$

$$\frac{dH_V}{dt} + \frac{V_m}{N} - V' \quad (42)$$

where L' and V', respectively, are the aqueous and solvent fluxes from each element of the emulsion band.

The variation in the height of the emulsion band l_e is given by

$$A_s\frac{dl_e}{dt} = L_m + V_m - N(L' + V') \quad (43)$$

but since l_e is not always dependent on V' if the aqueous phase is dispersed, then $V_m = NV'$, and Eq. (43) reduces to

$$A_s\frac{dl_e}{dt} = L_m - NL' \quad (44)$$

The aqueous flux V' is a function of l_e where the relationship may take several forms. For

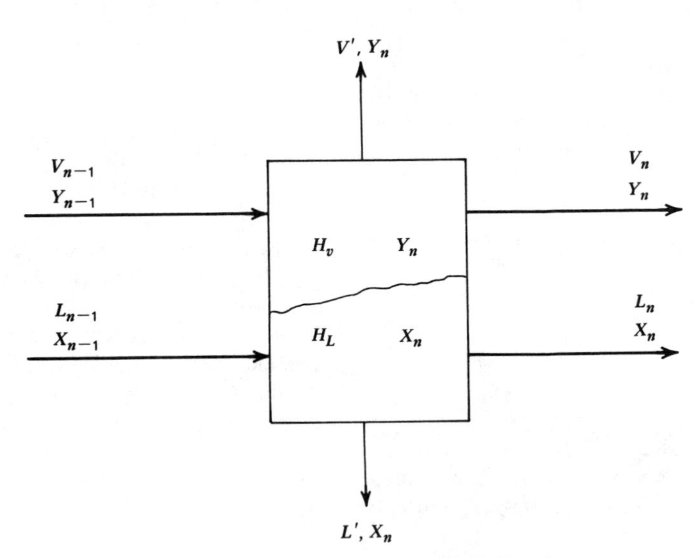

Figure 16. Section n of the emulsion band.

example,

$$V' = k_e l_e^n \tag{45}$$

where n is observed to lie between 0.5 and 1.0.

A solute balance in the aqueous phase for section n of the emulsion band gives

$$\begin{aligned}\frac{d}{dt}(H_L X_n) &= H_L \frac{dX_n}{dt} + X_n \frac{dH_L}{dt} \\ &= \frac{(N+1-n)\, L_m (X_n + X_{n-1})}{2N} \\ &\quad - \frac{(N-n)\, L_m (X_n + X_{n+1})}{2N} - V'X_n\end{aligned} \tag{46}$$

substituting for dH_L/dt from Eq. (41) gives

$$H_L \frac{dX_n}{dt} = \frac{L_m}{2N}[(N+1-n)\, X_{n-1} - X_n - (N-n)\, X_{n+1}] \tag{47}$$

and similarly for the organic phase

$$H_V \frac{dY_n}{dt} = \frac{V_m}{2N}[(N+1-n)\, Y_{n-1} - Y_n - (N-n)\, Y_{n+1}] \tag{48}$$

this representing the dynamic behavior of the emulsion band.

The Continuous Phase. The flow situation in the continuous phase is represented in Fig. 17 where the length of the settler is l_S and the areas between the streamlines are A_1, A_2, and so on. The flow field in the two continuous phases may be solved by using the potential flow equation

$$\frac{\partial^2 \phi}{\partial x^2} + \frac{\partial^2 \phi}{\partial y^2} = 0 \tag{49}$$

where ϕ is the stream function.

The length of the settler is divided into N equal sections with ϕ having values of l to N. Equation (51) is then solved numerically using Thompson's algorithm for a square mesh grid. Values of the flow area A_1 to A_{10} are thus calculated and the dead space obtained by linear interpolation of the areas.

The residence time τ in each "flow channel" can then be found from

$$\tau_n = \frac{A_n W}{F} \tag{50}$$

where $F = L'$ or V' depending on the phase to be considered and W is the width of the settler.

This procedure is, however, rather time consuming, even for a single stage, and in order to render the results useful in the development of a dynamic model for a multistage mixer–settler, the computed results for a single stage may be approximated by a delay time τ_3 in conjunction with a second-order system with time constants τ_1 and τ_2, by use of a transfer function of the

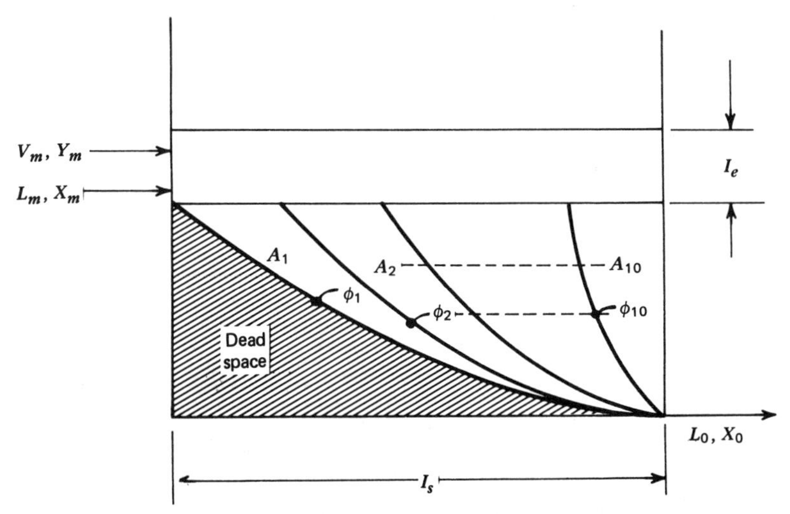

Figure 17. Streamlines in the aqueous phase in the settler.

form

$$\frac{\theta_o(s)}{\theta_i(s)} = \frac{K_{\exp}(-\tau_3 s)}{(1+\tau_1 s)(1+\tau_2 s)} \tag{51}$$

This form of transfer function was found to describe the response of a single stage with reasonable accuracy and is easily extended to the case of multiple mixer–settler cascades by use of conventional control engineering techniques and has the further advantage that the various time constants are predictable from basic theory. Most of the basic assumptions in the model are confirmed by experimental observation [13], but it was observed that the horizontal velocity component of the emulsion band plays an important role on the height of the emulsion band. In the settler the emulsion band absorbs momentum from both liquid phases, thus decreasing its velocity with distance from the mixed-phase port. This deceleration, in turn, results in increased coalescence rates–which, in turn, decrease the height of the emulsion band.

4.4.3. Differential Column Contactors

As with mixer–settlers, there has been a large volume of theoretical work on the modeling of the open-loop dynamics of liquid extraction columns but relatively little experimental confirmation and few closed-loop studies. Differential contactors can be treated either as distributed parameter models described by partial differential equations or by nonequilibrium stagewise models described by sets of difference–differential equations. At high degrees of dispersion these two types of model simply represent different approaches to the same problem and in the limit give equivalent results. In treating differential contactors, the following assumptions are commonly made: (1) constant physical properties; (2) uniform radial concentrations; and (3) equilibrium conditions at the interface only.

(a) Distributed Parameter Models for Differential Contactors. Considering a simple counterflow extraction column, unsteady-state component balances carried out over a differential element of column length allowing for backmixing in both phases lead to the following pair of coupled partial differential equations:

$$h_L \frac{\partial X}{\partial t} = -L \frac{\partial X}{\partial Z} + D_L \frac{\partial^2 X}{\partial Z^2} - Ka\,[X - X^*] \tag{52}$$

$$h_V \frac{\partial Y}{\partial t} = V \frac{\partial Y}{\partial Z} + D_V \frac{\partial^2 Y}{\partial Z^2} + Ka\,[X - X^*] \tag{53}$$

where h_L, h_V are the respective fractional holdup volumes for the two phases, X and Y are the respective solute concentrations, D_L and D_V are the backmixing eddy diffusivities, t is time, Z is the distance along the extractor measured from the L-phase inlet, K is the overall mass-transfer coefficient, a is the specific interfacial area/unit volume of contactor, X^* is the equilibrium solute concentration, and L and V are the superficial flow velocities for the aqueous and organic phases.

Early formulations of this case tended to neglect the backmixing terms, but this is now recognized to be an essential component in a realistic representation of extraction column behavior. The preceding equations can be solved analytically if in addition to the earlier assumptions, it is assumed that (1) flow velocities, fractional holdups, mass-transfer coefficients, and eddy diffusivities are constant and (2) the equilibrium relationship is linear.

It is apparent that the model equations can be solved to give the solute compositions X and Y as functions of both time t and position Z and may be solved for either of the two independent variables. Thus different solution approaches have been developed, although the formulation of the problem remains valid for either variable. The technique of Laplace transform taken with respect to time, as used by Burge and Clements [14], represents probably the most satisfactory approach. Laplace transformation yields two simultaneous algebraic equations that can be combined to yield a single characteristic equation of fourth order whose roots determine the nature of the response, such that the transformed feed phase concentration is given by

$$\bar{X}_{(z,s)} = P_1 e^{p_1} + P_2 e^{p_2} + P_3 e^{p_3} + P_4 e^{p_4} \tag{54}$$

where P_1, P_2, P_3, and P_4 are the characteristic roots and the coefficients p_1, p_2, p_3, and p_4 are determined by consideration of the Wehner-Wilhelm boundary conditions [15]:

$$\bar{X}_{in}(0^-, s) = \bar{X}(0^+, s) - \frac{1}{P_L} \frac{d\bar{X}(0^+, s)}{dZ} \tag{55}$$

$$\frac{d\bar{X}(1^+, s)}{dZ} = 0 \tag{56}$$

$$\bar{Y}_{in}(1^-, s) = \bar{Y}(1^-, s) + \frac{1}{P_V} \frac{d\bar{Y}(1^-, s)}{dZ} \tag{57}$$

$$\frac{d\bar{Y}(0^-, s)}{dZ} = 0 \tag{58}$$

assuming that dispersion only occurs between the dimensionless distances $Z = 0^+$ and $Z = 1^-$ as represented by the Peclet number P_L and P_V and that external to the column $Z = 0^-$ and $Z = 1^+$, plug flow conditions prevail.

If a numerical means of solution is applied to the model equations, the restrictive conditions of (1) and (2) (in the preceding paragraph) no longer apply, and the dynamic model can be generalized to a much greater degree. Thus if desired Ka, h_L, h_V, D_L, D_V, L, and V could be assumed to be functions of column height, time, and indeed solute concentrations, thus allowing hydrodynamic variations (if known) to be incorporated into the model. Naturally the limitation to a linear equilibrium relationship could also be relaxed. For numerical solutions, it is necessary to eliminate one of the independent variables (t or Z) by the use of a finite-difference approximation. Thus, for example, the time derivative may be replaced by

$$\frac{\partial X}{\partial t} = \frac{X_{m+1,n} - X_{m,n}}{\Delta t} \tag{59}$$

$$= \frac{X_{m,n} - X_{m-1,n}}{\Delta t} \tag{60}$$

$$= \frac{1}{2} \frac{X_{m+1,n} - X_{m-1,n}}{\Delta t} \tag{61}$$

representing first-order, forward, backward, and central difference approximations, respectively; where $X_{m,n}$ is the solute concentration in element n at time $t = m\,\Delta t$.

As in the analytical solution approach, it is more usual to approximate the distance derivative terms, thus for a finite difference element n of length Δz, the model Eqs. (52) and (53) may be reformulated as

$$h_{L_n} \frac{dX_n}{dt} = \frac{L_n}{2} \left(\frac{X_{n+1} - X_{n-1}}{\Delta Z} \right) + D_{L_n} \left(\frac{X_{n+1} - 2X_n + X_{n-1}}{\Delta Z^2} \right) - K_n a_n (X_n - X_n^*) \tag{62}$$

$$h_{V_n} \frac{dY_n}{dt} = \frac{V_n}{2} \left(\frac{Y_{n+1} - Y_{n-1}}{\Delta Z} \right) + D_{V_n} \left(\frac{Y_{n+1} - 2Y_n + Y_{n-1}}{\Delta Z^2} \right) + K_n a_n (X_n - X_n^*) \tag{63}$$

using first-order central difference approximations.

Note that the overall mass transfer coefficient K_n, when employed in the case of a nonlinear equilibrium, will vary as a function of the solute concentration X_n, even when the film coefficients for the respective liquid phases remain constant, since the overall resistance to mass transfer is then dependent on the local slope of the equilibrium curve. For nonlinear equilibrium instances, therefore, the corresponding sets of difference-differential equations (62) and (63) should be accompanied by additional relations, relating K_n to X_n.

Alternatively, more complex second-order terms may be used to reduce the error of the approximation. As long as the integration step length and the length of the finite difference element satisfy the so called stability criteria, increased accuracy can also be obtained by increasing the number of elements N but this must be balanced against the cost of the additional computational effort. For computational ease, the lengths of the finite difference elements are usually taken as equal, but in parts of the column where conditions are changing very rapidly (i.e., toward the phase inlets), it may be advantageous to reduce ΔZ and to employ correspondingly greater values in the more central regions of the column where conditions change slowly.

Thus by use of finite differencing techniques the original two partial differential equations are transformed for N finite-difference elements into $2N$ difference–differential equations that can be solved numerically in conjunction with the appropriate finite-difference approximations of Eqs. (56)–(58) for the end stages or by specially formulated expressions that take into account the relative locations of the phase inlets and outlets and perhaps the interface.

(b) Stagewise Models for Differential Contactors. At the prevailing high levels of dispersion normally encountered in extraction columns, it is also possible to represent the behavior of differential contactors by the use of a nonequilibrium stagewise model as represented in Fig. 18. Allow-

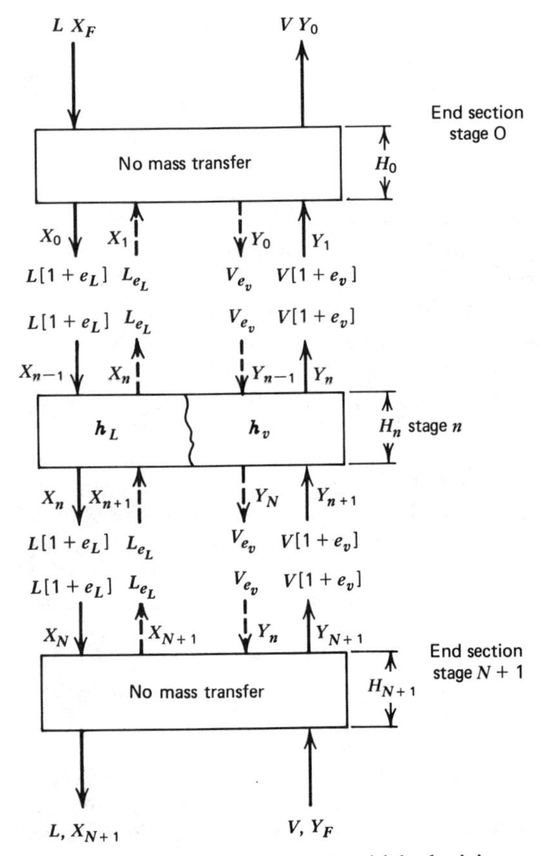

Figure 18. Stagewise contactor with backmixing.

ing for backmixing in each phase, unsteady-state material balances for the two phases in any stage n of the contactor yield the following difference–differential equations

$$Hh_L \frac{dX_n}{dt} = L(1 + e_L)X_{n-1} - L(1 + 2e_L)X_n + Le_L X_{n+1} - Ka(X_n - X_n^*)H \quad (64)$$

$$Hh_V \frac{dY_n}{dt} = V(1 + e_V)Y_{n+1} - V(1 + 2e_V)Y_n + Ve_V Y_{n-1} + Ka(X_n - X_n^*)H \quad (65)$$

where L and V are the superficial phase velocities through the column, H is the height of stage, h_L and h_V are the fractional holdups of each phase, K is the overall mass-transfer coefficient, a is the specific interfacial area for mass transfer, X and Y are the phase solute concentrations, e_L and e_V are the backmixing factors for each phase and subscript n refers to stage n of the extractor.

Equations (64) and (65) are, of course, similar to those derived to the generalized case of a stagewise contactor in Section 4.4.1 [Eqs. (21) and (22)].

The main assumptions made in this model are (1) perfect mixing in the stages, (2) constant stage volume, (3) constant physical properties and (4) equilibrium conditions at the interface.

The resulting model relationships have been solved analytically by Jones and Wilkinson [16], by use of the following additional assumptions: (5) flow rates, holdups, backmixing, and mass-transfer capacity coefficients are constant; (6) the equilibrium relationship is linear and further (7) the column is equipped with end sections in which there is negligible mass transfer.

The solution approach is based on a linearization of the model relations in terms of small perturbations X_n' and Y_n' about some steady-state value. Thus by defining

$$X_n' = X_n - \bar{X}_n \quad (66)$$

$$Y_n' = Y_n - \bar{Y}_n \quad (67)$$

$$X_n^{*\prime} = X_n^* - \bar{X}_n^* \quad (68)$$

where the bar superscript relates to steady state and using the steady-state condition

$$\frac{d\bar{X}_n}{dt} = \frac{d\bar{Y}_n}{dt} = 0$$

one may rewrite the model equations as

$$Hh_L \frac{dX_n'}{dt} = L(1 + e_L)X_{n-1}' + Le_L X_{n+1}' - L(1 + 2e_L)X_n' - Ka(X_n' - X_n'^*)H \quad (69)$$

$$Hh_V \frac{dY_n'}{dt} = V(1 + e_V)Y_{n+1}' + Ve_V Y_{n-1}' - V(1 + 2e_V)Y_n' + Ka(X_n' - X_n'^*)H \quad (70)$$

Taking Laplace transforms and rearranging leads to a fourth-order characteristic difference equation for X_n' with a solution of the form

$$X_n' = A_1 r_1^n + A_2 r_2^n + A_3 r_3^n + A_4 r_4^n \quad (71)$$

where r_1, r_2, r_3, and r_4 are the characteristic roots of the equation and A_1, A_2, A_3, and A_4 are coefficients to be determined from the boundary conditions.

The boundary conditions are obtained from steady-state solute balances over the end stages such that

$$X_0' = \frac{X_F' + e_L X_1'}{1 + e_L} \tag{72}$$

$$Y_0' = Y_1' \tag{73}$$

$$X_{N+1}' = X_N' \tag{74}$$

$$Y_N' = \frac{Y_F' + e_V Y_{N-1}'}{1 + e_V} \tag{75}$$

Assuming a pure solvent feed, these relations provide four simultaneous equations for determination of A_1, A_2, A_3, and A_4.

The transfer function of the system to changes in the inlet aqueous feed concentration is then given by

$$\frac{X_n'(s)}{X_F'(s)} = A_1 R_1^n + A_2 R_2^n + A_3 R_3^n + A_4 R_4^n \tag{76}$$

where R_1, R_2, R_3, and R_4 are the appropriate roots.

A similar treatment gives the transfer function of the system to changes in the feed rate, although this neglects any hydrodynamic effects.

The model equations can also be solved numerically in which case the restrictive conditions of 2, 3, 5, 6, and 7 [preceding Eq. (66)] no longer apply. Thus the steady-state boundary conditions for the end stages may be replaced by dynamic relationships in which a finite rate of solute transfer may be allowed. Thus for end stage 0:

$$H_0 h_{L_0} \frac{dX_0}{dt} = LX_F - L(1 + e_L) X_0 + Le_L X_1 - K_0 a_0 (X_0 - X_0^*) H_0 \tag{77}$$

$$H_0 h_{V_0} \frac{dY_0}{dt} = V(1 + e_V)(Y_1 - Y_0) + K_0 a_0 (X_0 - X_0^*) H_0 \tag{78}$$

and for end stage $N + 1$

$$H_{N+1} h_{L_{N+1}} \frac{dX_{N+1}}{dt} = L(1 + e_L)(X_N - X_{N+1}) - K_{N+1} a_{N+1} (X_{N+1} - X_{n+1}^*) H_{N+1} \tag{79}$$

$$H_{N+1} h_{V_{N+1}} \frac{dY_{N+1}}{dt} = VY_F - V(1 + e_V) Y_{N+1} + Ve_V Y_N + K_{N+1} a_{N+1} (X_{N+1} - X_{N+1}^*) H_{N+1} \tag{80}$$

where subscripts 0 and $N + 1$ relate to specific conditions in the end stage.

The correct modeling of the end stages is of obvious importance in a realistic dynamical modeling exercise, and Prenosil [18] has given useful advice in this respect based on the analysis of a Kühni column extractor.

Note that for nonlinear equilibrium cases, the overall mass transfer coefficients K_n will vary as functions of X_n as explained previously in the case of differential contactors.

(c) Hydrodynamics of Differential Column Contactors. Under controller closed-loop conditions, it is desirable to include some consideration of the hydrodynamic changes in the phase flow rates. Under such conditions the fractional holdup volumes h_{L_n} and h_{V_n} will vary with time and from stage to stage, with consequent variations in the phase flow rates L_n and V_n. This can have a considerable effect on the column response characteristics, especially since variations in holdup will effect the solute transfer rate terms Q_n by consequent variations in the specific interfacial area term a_n where $Q_n = K_n a_n (X_n - X_n^*)$ and a_n is related to h_{V_n}.

The transient holdup behavior of a pulsed sieve-plate solvent extraction column has been investigated by Foster et al. [19], and the following approach is based on a slight modification of his method. Under transitory hydrodynamic conditions, the phase superficial velocities through the column L and V will vary from stage to stage of the column as a function of time such that the representation of the column as in Fig. 18 must be changed to that of Fig. 19. For completeness, it is also assumed that the backmixing flow contributions $e_L L$ and $e_V V$ of Fig. 18 also vary with time and may be represented for stage n as the backmixing flows L_{Bn} and V_{Bn}, respectively.

A dynamic balance for the dispersed-phase holdup in stage n gives

$$H_n \frac{dh_{V_n}}{dt} = V_{n+1} + V_{Bn+1} + V_{Bn-1} - V_n - 2V_{Bn} \tag{81}$$

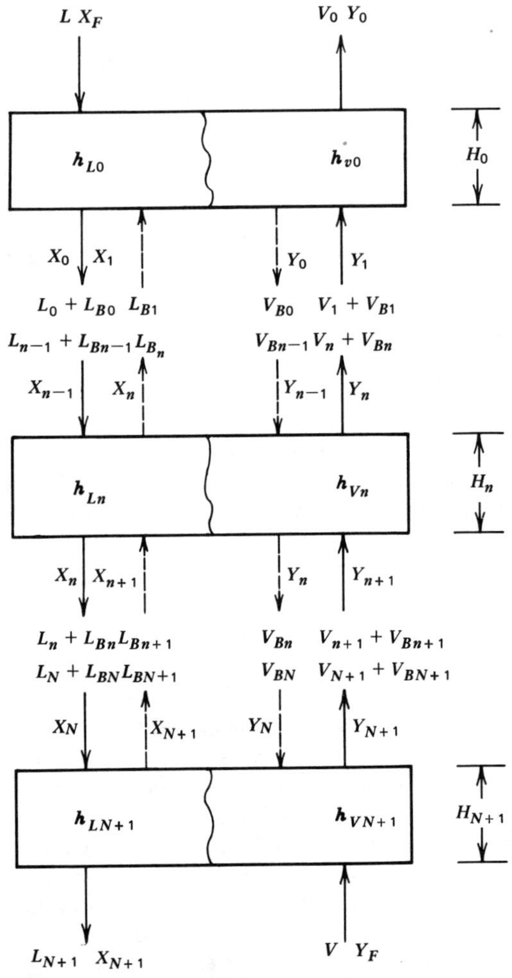

Figure 19. Stagewise contactor with backmixing under varying hydrodynamic conditions.

and since the two liquid phases are incompressible, for stage n

$$V_{n+1} + V_{Bn+1} + V_{Bn-1} + L_{n-1} + L_{Bn-1} + L_{Bn+1} = V_n + 2V_{Bn} + L_n + 2V_{Ln} \quad (82)$$

for the overall column

$$L + V = L_{N+1} + V_0 \quad (83)$$

and for the top section of the column

$$L + L_{Bn+1} + V_{n+1} + V_{Bn+1} = L_n + L_{Bn} + V_{Bn} + V_0 \quad (84)$$

In practice, however, it is unlikely that the present state of knowledge will provide for variations in V_{Bn} and L_{Bn}. Thus, assuming these to be small, Eqs. (81), (82), and (84) reduce to

$$H_n \frac{dh_{V_n}}{dt} = V_{n+1} - V_n \quad (85)$$

$$V_{n+1} + L_{n-1} = V_n + L_n \quad (86)$$

$$L + V_{n+1} = L_n + V_0 \quad (87)$$

Foster assumed that h_{V_n} obeys a characteristic velocity equation such that

$$\frac{L_n}{h_{L_n}} = \frac{V_n}{h_{V_n}} = V_{K_n}[1 - h_{V_n}]\, f(h_{V_n}) \quad (88)$$

where $h_{L_n} + h_{V_n} = 1$ and V_{K_n} is the characteristic velocity for the dispersed-phase droplets in stage n.

For the special case, considered by Foster where V_{N+1} and L_N are fixed and $V_{N+1} - L_N = \Delta$ = const, this leads to the straightforward solution of $N + 1$ nonlinear differential equations that are readily solved by numerical integration. Excellent agreement was found by Foster between experiment and the theoretical predictions of the model, although the model solution led the experimental data slightly. This was attributed to the slowness of coalescence in the dispersed phase, and when the model was modified to include a corrective lag term in the characteristic velocity equation, better agreement was obtained. Under normal operating conditions, however, either V_{N+1} or L_N will be variable quantities, and under such circumstances difficulty in computation arises, since the combination of Eqs. (87) and (88) represents an IMPLICIT algebraic loop, which must be solved by a convergence procedure at every integration step in order to determine the appropriate values of L_n or V_n satisfying the value of h_{v_n} generated in Eq. (85), which can obviously be a very time consuming procedure. An alternative approach may be based on experimental data relating h_v as a function of, say, V for constant ranges of L. These may be expressed as polynomial expressions where the constant quantities are functions of the particular value of L. A simplified block information diagram, ignoring the backmixing terms, is shown in Fig. 20. The resultant calculation procedure is then straightforward. A computer simulation of the resulting dynamical changes in the holdup profile for a five-stage column with end stages following a step change in solvent flow rate is shown in Fig. 21.

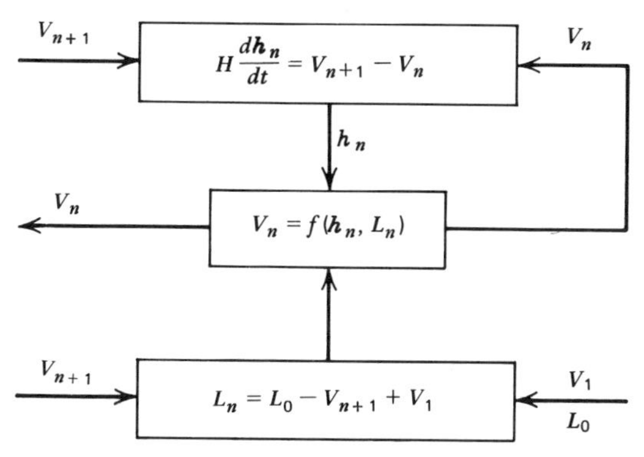

Figure 20. Information flow diagram for simplified hydrodynamic model.

(d) Multicomponent Systems. The previous treatments have been confined to the cases of single-solute extraction between immiscible liquid phases. Assuming that the liquid phases remain immiscible, the modeling approach for multicomponent systems remains the same except that owing to possible complications in the form of the equilibrium relationships, it is preferable to base this on the use of the rate equation for nonequilibrium stages and perhaps the use of a high-gain factor to force near equilibrium conditions in the case of mixer–settler cascades. The hydrodynamic relationships, of course, remain the same.

For each component, it is necessary to write two continuity equations for the two phases. Thus for a four-component mixture where $i = 1, 2, 3, 4$ applied to a stagewise differential column

$$H_n h_{L_n} \frac{dX_{in}}{dt} = L_n(1 + e_{L_n}) X_{in-1} - L_n(1 + 2e_{L_n}) X_{in} + L_n e_{L_n} X_{in+1} - K_{in} a_{in} (X_{in} - X_{in}^*) H_n \tag{89}$$

$$H_n h_{V_n} \frac{dY_{in}}{dt} = G_n(1 + e_{V_n}) Y_{in+1} - V_n(1 + 2e_{V_n}) Y_{in} + V_n e_{V_n} Y_{in-1} + K_{in} a_{in} (X_{in} - X_{in}^*) H_n \tag{90}$$

where, for example, the equilibrium concentrations X_{in}^* may be expressed as functions of all four solute concentrations

$$X_{in}^* = \int_{\text{eq}} (Y_1, Y_2, Y_3, Y_4) \tag{91}$$

This obviously represents a considerable increase in computational effort, but apart from this, no major additional difficulty.

(e) Simulation of Closed-Loop Behavior. The generalized controller equations for proportional and proportional–integral control may be expressed by

$$G = \bar{G} + K_c \epsilon \quad \text{proportional} \tag{92}$$

$$G = \bar{G} + K_c \epsilon + \frac{K_c}{\tau_I} \int \epsilon \, dt \quad \text{proportional and integral} \tag{93}$$

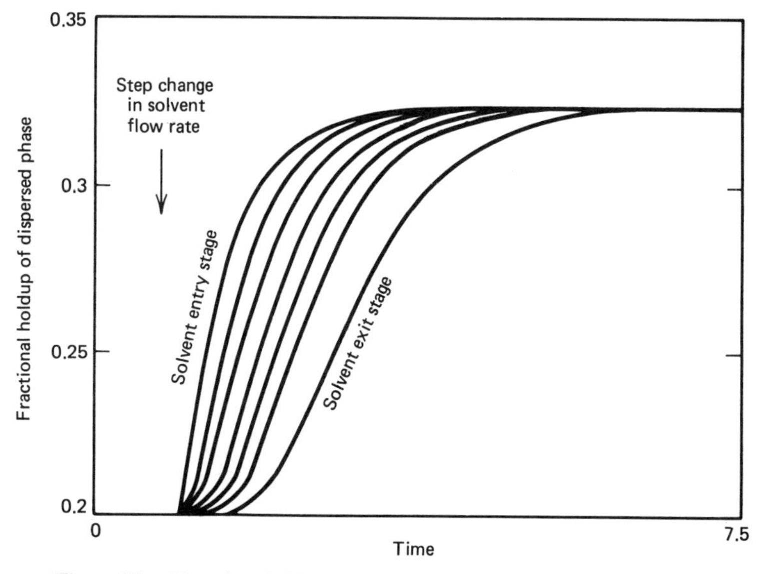

Figure 21. Transient holdup profiles for a five-stage column contactor.

where G is the manipulated variable, $\bar{G}$ is the base value for G, K_c is the proportional gain, τ_I is the integral action time, and ϵ is the error.

Thus, for example, manipulation of the solvent flow rate V to maintain constant raffinate outlet concentration X_{N+1} for simple proportional and feedback action may be incorporated into a digital simulation program by the use of direct statements equivalent to

$$E = X_{N+1} - X_{N+1\,(\text{set})} \tag{94}$$

$$\frac{dE_1}{dt} = E \tag{95}$$

$$V = \bar{V} + K_c E + K_I E_1 \tag{96}$$

where E_1 is the integral of the error and K_c and K_I are controller constants.

If required, it is also possible to include the dynamic effects of the measuring elements, measurement transmitters, and the control valve characteristics into the simulation program as discussed by Franks [6, 12].

(f) Generalized Model. The model equations represented in Sections 4.4.3(b)–4.4.3(d) are easily combined to form a generalized model in which variations of mass-transfer coefficient, interfacial area, backmixing, and holdup along the length of the column or as a function of solute concentration (if known) may be included, by defining the appropriate values of K, a, e_L and e_V, and h_L and h_V for each particular stage in terms of the other related variables. One such model, involving the closed-loop response of a five-stage column with end stages, two-solute extraction with simple mutually interacting nonlinear equilibrium relationships, and incorporation of column hydrodynamic changes was solved by the authors by use of a BEDSOCS digital simulation program on a Hewlett Packard 2100 minicomputer with 16K core in a time of approximately 2 min.

5. PREVIOUS WORK

The literature prior to 1968 is excellently reviewed by Pollock and Johnson [20] and is not repeated here.

5.1. Mixer–Settlers

Hanson and Sharif [21], in a continuation of earlier work, studied the hydrodynamics of mixer–settlers. They found conclusive evidence of mixing in the settler, showing that perfect mixing was a better assumption than plug flow.

Cadman and Hsu [22] developed a nonlinear dynamic representation of multistage mixer–settlers in an attempt to estimate the system transfer function, which could then be used for control purposes. They assumed perfect mixing in both mixer and settler and, by means of instantaneous mass balances and empirical holdup relationships, derived the nonlinear equations for the system. They made no assumptions regarding constant flow rate, immiscibility of phases, or linear equilibrium. A specific solution technique is presented for a quadratic equilibrium relationship. The equations are linearized and Laplace transformed in order to develop the system transfer function. The transfer function [23] for a multistage cascade is obtained by expanding the transfer function for a two-stage system as determined by Cadman and Hsu [22].

Kikiudai et al. [24] represented an attempt to simulate mixer–settler response with an analog computer. The experimental work was carried out in a single-stage, laboratory-scale mixer–settler unit, using the system nitric acid–uranium–TBP. They assumed a hybrid model for the settler, which gave quite adequate results. Their main problem was, however, the analog computation associated with highly complex equilibrium relationships. It was suggested that for large-scale equipment, hybrid computers should be used.

Rouyer, Chazal, and Demarthe, Kikiudai, and Hoclet [25–31] have presented parallel studies of the hydrodynamics of mixer–settlers for the system water–nitric acid–uranium–30% TBP. Rouyer explained the mixing phenomena in the settler and obtained experimental results for upward steps in aqueous feed concentration. Demarthe investigated both upward and downward steps, but his work is too complex to be used for computer control. Rouyer et al. [32] have also recently reported on the practical application of modal control theory to a liquid–liquid extraction column.

Beetner et al. [33] presented a simulation technique for the equipment described in the paper by Casto et al. [34], which was a 10-stage, box-type pump–mix mixer–settler. They divided the equipment volume into 20 regions in order to simulate the organic phase in the settler. The aqueous phase was described in terms of four regions, perfectly mixed, with two recycle mixing areas. The transient time period was finite differenced, and there was interstage flow at the end of each time increment only. Mass transfer takes place only when interstage flow is not taking place, thus allowing the accumulation terms to be omitted from the model.

Aly et al. [35–37] followed Apelblat and Faraggi [38] and introduced in one of their models a Murphree plate efficiency. They also used a different method to identify the system, solving the state space representation of their models. He obtained experimental results for a four-stage unit operating on a copper extraction process and applied perturbations by use of step changes and pseudo-random binary sequence (PRBS) signals and the AKUFVE system for continuous sampling of the mixer. It is possible to conclude from their results that the use of efficiencies derived from steady-state data is not valid and that the more simple models, making use of the assumptions already outlined, represent the experimental results better than does the complex non-equilibrium stage model.

Drown and Thomson [39] investigated the hydrodynamic characteristics of horizontal liquid–liquid settlers. They found that the settlers could be characterized by three distinct regions: the entrance region; the emulsion band region; and the exit region. The experimental measurements included studies of the expanding jet in the entrance region, of the velocity distributions (by use of laser velocimetry and cinephotographs) of the emulsion band region, and vortex flow patterns in the exit region. They found that the velocity distribution varied considerably from plug flow and that the emulsion band zone was capable of absorbing momentum from both bulk phases. This caused a deceleration of the emulsion band, which, in turn, enhanced the dynamic coalescence rates. Some success was obtained in an attempt to describe the observed phenomena with a simplified mathematical model.

Bobrow et al. [40] employed the DYNSIS modular computer program to study the dynamics of a multistage butadiene extraction plant. This program is similar to the well-known steady-state modular flow sheeting programs such as PACER, GEMCS, and FLOWPACK, but whereas these programs are concerned with iterating towards steady state, each sequence in computation in DYNSIS represents a step in time. Unlike the solution approach adopted by digital simulation languages such as MIMIC and CSMP, the modular computing blocks are items of chemical equipment rather than sets of differential equations.

Figure 22 shows a typical mixer–settler stage and the corresponding information flow diagram,

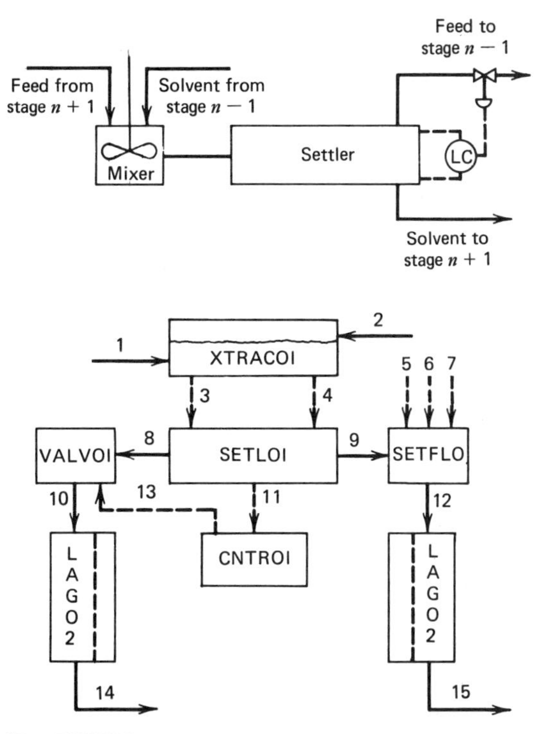

Figure 22. DYNSIS representation of a typical mixer–settler stage [40].

where the subroutines XTRAC01 represents the transfer of solute in a perfectly mixed tank; LAG02 represents the time delays for the two phases in the settler, allowing for some fractional bypassing; and SETL01 represents the variation in level in the settler, which information is passed to the level controller, subroutine CNTRL01, and hence to the control valve, subroutine VALV01, which controls the hydrocarbon phase flow from SETL01.

Libhaber et al. [41] used the MIMIC digital simulation language to model the extraction and purification stages of the IMI phosphoric acid process, using an equilibrium stagewise model in which the separate behavior of each mixer and settler were treated as one constant-volume equilibrium stage. El-Rifai [42] derived linear dynamic models for countercurrent mixer–settler cascades involving simultaneous solvent extraction and first-order chemical reaction in the organic phase. Typical results demonstrating the effects of chemical rate constant, extraction factor, residence time, and number of stages on the frequency response of the two outlet streams are presented.

Barnea et al. [43] have derived a dynamic model for continuously mixed chemically reacting liquid–liquid dispersions, allowing for simultaneous interfacial heat- and mass-transfer effects. It was found that there is often a dynamic dependence of drop size on phase composition, temperature and/or impeller speed and that the lag between drop size and system variables often causes oscillatory behavior in the concentration and temperature of the reactor, which can lead the reactor into undesirable operating conditions.

Ochsenfeld et al. [44] have successfully employed a multistage, multicomponent equilibrium stage to confirm the dynamic behavior of a laboratory-scale hot cell mixer–settler cascade used for the extraction of fast breeder reactor fuel. Although the dynamic simulation confirmed the basic flow sheet calculations for the process, it highlighted important effects that could be caused by relatively small changes (7–13%) in the flow rate of the active feed, where in the concentration ranges near to the organic saturation limit, relatively large concentration changes along the cascade, with attendant danger of breakthrough in the exit raffinate stream, were shown. These changes, however, occurred so slowly that corrective action was possible.

5.2. Column Contactors

In addition to the critical review cited previously [20], Pollock and Johnson carried out a major study of the dynamic characteristics of a three-stage multimixer column using the system water–acetic acid–methyl isobutyl ketone. In a study of the relative advantages of various testing techniques, it was concluded that step testing was equally reliable as pulse methods [45]. Techniques for determining column concentration profiles during both steady-state and transient conditions were investigated [46], and frequency and time domain techniques were used for the comparison of various mathematical models, which, however, excluded the effects of backmixing [47]. It was concluded that a realistic mathematical model for an extraction column requires allowance for the variation in mass-transfer coefficient with concentration and column length and proper consideration of the end effects. Further work by Pang and Johnson [48] discussed the determination of the frequency response characteristics and the modeling of a liquid–liquid extraction column in the frequency domain [49]. A minimum line operating policy for an extraction column was also discussed [50, 51].

Souhrada et al. [52] were among the first to use a stagewise model with backmixing and employing stage efficiency terms to represent the approach to equilibrium, which was solved numerically by use of the Runge–Kutta–Merson numerical procedure.

Landau et al. [63] modeled the dynamic behavior of reciprocating-plate columns, using a one-dimensional two-parameter model in which an extraction stage is considered to consist of three distinct regions: two regions of perfect mixing adjacent to the plates, caused by the pumping action of the plates, connected by a region of axially dispersed plug flow. The two parameters of the model thus consist of the eddy diffusivity E_B in the axially dispersed flow region and the length α of one perfectly mixed region. This model leading to a system of differential equations with time dependent coefficients was simplified to allow parameter estimation in the Laplace domain.

Wang et al. [54] used the MIMIC digital simulation language to model the dynamic concentration changes in a Graesser raining-bucket contactor by means of both non-equilibrium stagewise model and finite differenced axial dispersion flow model with backmixing. Both models were found to give equivalent results but, although representing the outlet responses to changes in feed concentration with some success, failed to adequately model the internal dynamic changes with any accuracy. This was attributed to failures in describing the behavior of the end sections and to possible variations in local mass-transfer and hydrodynamic changes along the length of the contactor.

Steiner et al. [55] modeled the dynamic behavior of a spray column by use of a modification of the nonequilibrium stagewise model, in which the dispersed phase was assumed to consist of noncoalescing drops. It is claimed that this approach represents a more realistic approach for some types of column and also reduces the computational time.

Bauermann and Blass [56] modeled the behavior of a pulsed sieve-plate column by use of the non-equilibrium-stage backmixing model and successfully reduced their results to a second-order transfer function with two identical time constants and a time delay. Such an approach, as suggested in Section 4.4.2(d), in which the constants are obtained from an analysis of the theoretical mode, is probably more useful in practice than the use of the full model for control implementation.

6. CONTROL STUDIES

6.1. Mixer–Settlers

Mills et al. [57, 58] have described problems in the control of uranium extraction in a mixer–settler plant. Uranyl nitrate is extracted in tributylphosphate as UO_2 $(NO_3)_2$-2TBP, and ideally all the molecules of the solvent TBP should react with uranyl nitrate such that the solvent phase is fully saturated. A typical steady-state solvent profile is shown in Fig. 23, together with a block diagram of the control scheme. In stages 0 to n–m a solvent saturation of 60% is obtained, and at stage n–m all the solute has been extracted from the aqueous phase and m stages of the cascade are redundant. If the solvent saturation could be increased to 90% as shown by the dashed lines, the whole curve is raised and more stages can be utilized for the separation, and ideally there should be no solute present in the last stage. Most solvent extraction processes are manually controlled and operate at low saturation, thus providing spare stages to allow for some degree of plant maloperation compared with the latter situation where there is little spare capacity and automatic control must be used. Concentration measurements taken in stages corresponding to the plateau will fail to determine the position of the concentration profile and a measurement at the end fails to establish the degree of saturation along the cascade. Thus it is necessary to base control on measurements at the points where the slope of the profile changes most rapidly, thus defining both the position and the degree of saturation to the profile. The control feed was controlled by a concentration analyzer located in the sensitive region of the profile and the set points for the two end feed flow controllers set by calculated results obtained from the steady-state SIMTEX computer program [59]. By use of proportional control and a pump output response curve, satisfactory control was achieved, since by selection of the correct gain for the servo loop, the pump response can be high outside and low inside the dead band region, which corresponds to ±1% of the desired concentration value. The approach to equilibrium under controlled and manual conditions was similar in both cases, but control was accurate to ±0.5% of equilibrium, whereas manual operation gave no better than 10% for this highly nonlinear plant

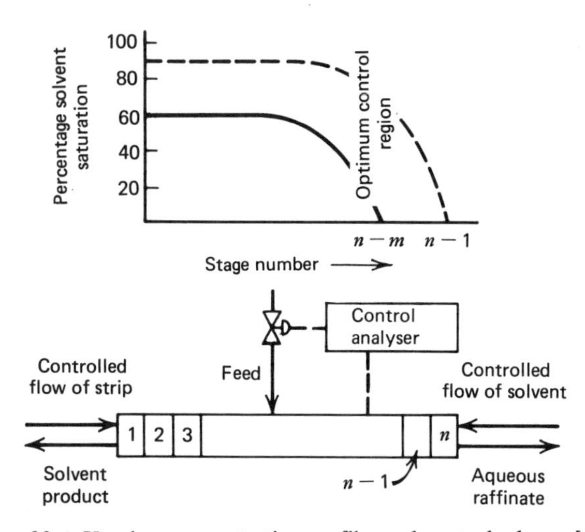

Figure 23. Uranium concentration profiles and control scheme [57].

containing long delay times. The controlled response was also greatly superior for changes in the solvent and strip feed flows because of the time delay between measurement of product quality and the input of the transient condition.

6.2. Extraction Columns

Studies in the automatic control of solvent extraction columns have been reported by McDonald and Wilkinson [5, 60], in which the theoretical model equations for a nonequilibrium stagewise extractor with backmixing, as described earlier, were used to investigate possible control strategies designed to maintain constant raffinate composition, following changes in the inlet feed concentration, by regulation of the feed flow rate. These included studies of simple feedback control, feed-forward control, feed-forward control, and their combinations. The theoretical predictions were compared with the results of experiments on a 23-stage, 150-mm-diameter Oldshue–Rushton column by use of the nitric acid–water–TBP system.

6.2.1. Open-Loop Results

The open-loop response of the controlled variable x_{raf} following step changes in the feed concentration x_F and feed flow rate L were compared with the corresponding theoretical predictions. Figure 24 shows the response of x_{raf} to downward and upward steps of equal magnitude in L. The agreement is reasonable, but the experimental response is slower than that predicted as dynamic effects are neglected in the model. Figure 25 shows the response to step changes in x_F, where the agreement is improved.

6.2.2. Closed-Loop Results

(a) Single Feedback Loop. The system was examined with proportional and proportional plus integral controller actions, and the experimental results are shown in Fig. 26, where K_{pr} is the feedback proportional gain and K_{ir} is the feedback integral constant $=K_{pr}/\tau_I$.

The introduction of proportional control improves the raffinate response considerably in

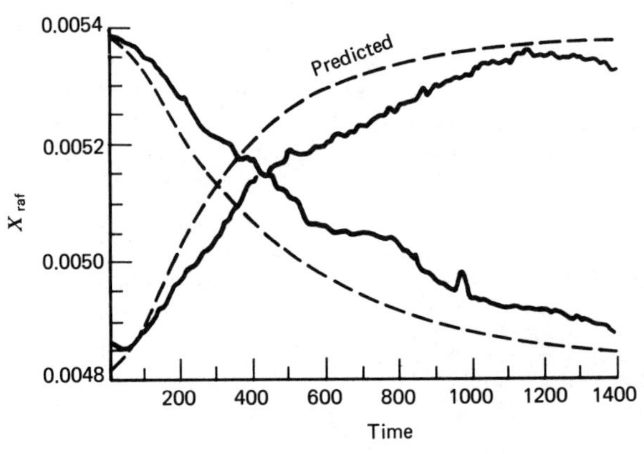

Figure 24. Open-loop response to step changes in feed rates.

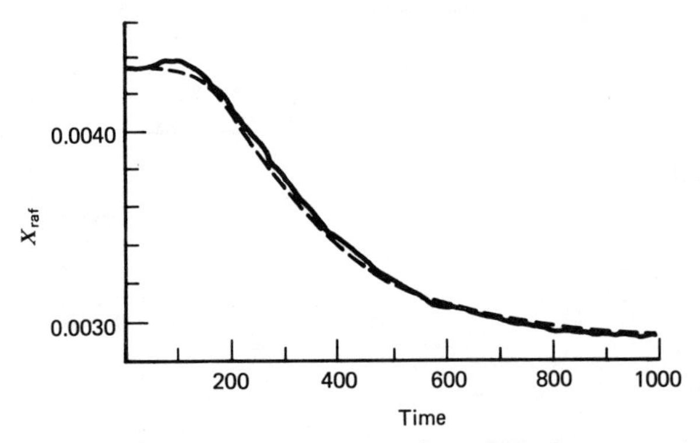

Figure 25. Open-loop responses to step change in feed concentration.

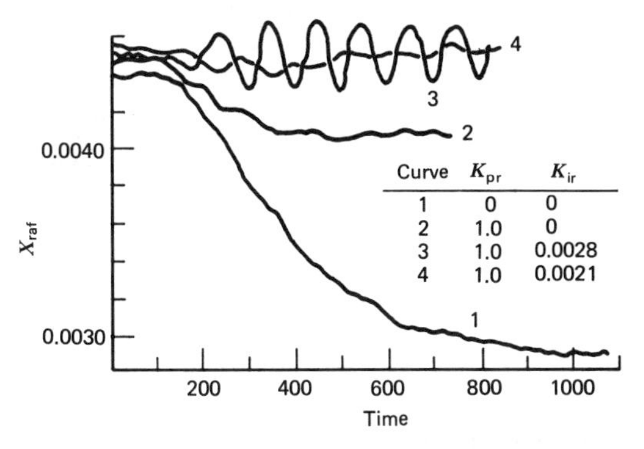

Figure 26. Results of P and $P + I$ feedback control.

comparison with the open-loop case but some offset remains. Increase in the gain reduces this offset until instability sets in. A small amount of integral action eliminates offset as expected, but the response becomes slightly less stable. Increase in integral action eventually worsens the control by causing oscillations.

Typical examples of comparisons between experimental and predicted responses are given in Fig. 27 for proportional-only control and also for proportional plus integral control. In the case of proportional control, the time delay and the approach to steady state is well predicted, but the predicted offset was somewhat lower than that found experimentally. With proportional plus integral control, the zero offset and the initial time delay are satisfactorily predicted, but the oscillatory nature of the experimental response is not predicted as this is due to additional hydrodynamic factors not included in the model.

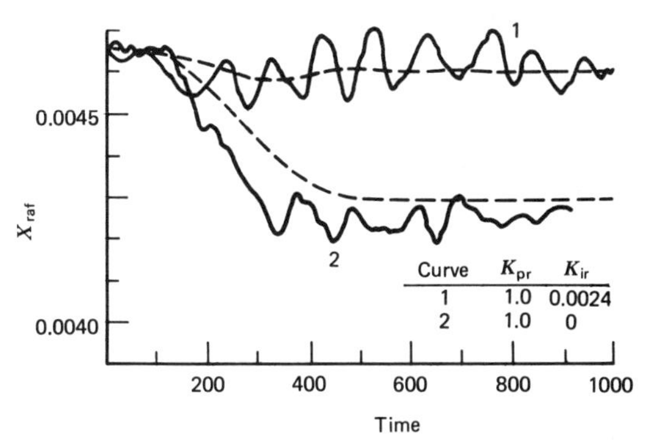

Figure 27. Comparison of experimental and predicted results for feedback control.

6.2.3. *Feedback–Feed-Forward Control*

Figure 28 shows the results of a sequence of tests comparing an open-loop response with the corresponding feedback-only and feedback–feed-forward controlled results, where K_{pf} is the feed-forward proportional gain. The addition of a feedforward component to the control scheme has the unexpected effect of increasing x_{raf}, at least initially. The response eventually peaks and the expected downward trend is introduced. For $K_{pf} = 2.6$, the response passes through an oscillatory period before settling down at a steady-state value very near to the zero offset point. This is a fortuitous result, as the size of the feed-forward action is just sufficient to overcome the inherent offset at this particular value of the feedback controller gain. For $K_{pf} = 5.2$, the peak is higher and the final steady state settles at a negative offset value. The response is fully damped.

This behavior is a result of the difference in the response times for L and x_F. The downward step in x_F is detected even before it enters the column and corrective action is applied immediately through a corresponding upward step in L, the size of which depends on the feedforward controller gain. This upward step has the imme-

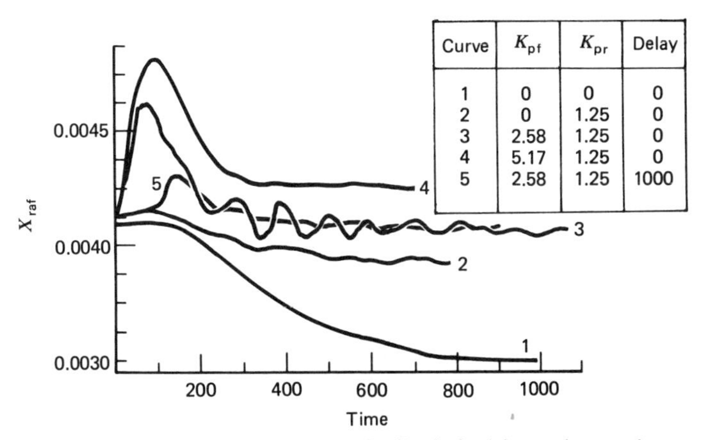

Figure 28. Results of open-loop feedback–feed-forward control.

diate effect of increasing x_{raf} and continues until the effects of the step in x_F itself have had the time to pass down the column and into the raffinate stream. At this point the raffinate concentration begins to fall toward the final steady state, governed by the combined effects of the feed-forward and feedback control actions. The severity of the control due to the feed-forward element is sufficient to induce instability, as can be seen for $K_{pf} = 2.6$. The instability would have been more marked for the higher value of K_{pf} if the increase in L would not have been sufficiently large to saturate the control system.

Since this unexpected behavior is due to the feed-forward control action being implemented before the disturbance has reached the raffinate stream, an experiment was carried out in which the feed-forward element of the control was delayed for 1000 s, an approximation to the dead time associated with the disturbances in x_F. This run was carried out for $K_{pf} = 2.6$, and, since the final value of L was the same as for the equivalent run without a delay, the final values of x_{raf} were also identical. However, as the implementation of the feed-forward control was delayed, the magnitude of the peak was greatly reduced. Furthermore, by minimizing the transient offset in x_{raf}, the oscillations caused due to the corrective efforts of the feedback system can be avoided, giving a more stable overall response.

The simulation of the feed-forward results was hampered by the inability to correctly simulate the raffinate response due to variations in L. Since L varies in a stepwise manner in this case, this effect is even more noticeable. Figure 29 shows a typical result. The failure to correct for the rapid increase in L following feed-forward control action resulted in the high experimental peak value, whereas the difference in the final steady-state values is similar to that noted for feedback-only control.

It can be seen from these results that, rather than preventing the development of erroneous intermediate concentration profiles during the early part of the run, the simple forms of feed-forward control used here actually induces this behavior. However, this form of control could be improved by suitable dynamic compensation.

7. COMPUTER CONTROL

Despite the large sales volume of process control computers, there have been very few actual reports of direct computer control (DDC) specifically related to solvent extraction applications. For example, Flett et al. [61], in discussing the proposed application of computer control to a projected rare metal separation process, list only four reported applications. These relate to two preliminary studies using now rather out of date types of computer controlling simple extraction systems and reference to two industrial applications for the extraction of boric acid and cobalt–nickel. Direct digital control of the Falconbridge Matte Leach Plant in Norway for nickel–cobalt separation has now been described in more detail [64]. This computer system, based on the use of a DEC PDP 8 minicomputer, has been in operation for 10 years, and only 7 days have been lost by computer failure. This system, which was designed completely without analog back-up controllers and involving relatively little cost associated with the instrumentation, has been shown to run economically reliably and to the overall benefit of the process. Although it is realized

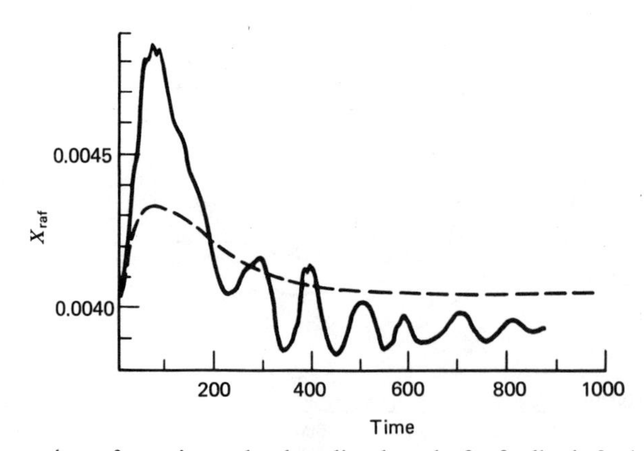

Figure 29. Comparison of experimental and predicted results for feedback–feed-forward control.

that improvement in stream analyzers and better computer resident models would increase process efficiency, it is felt that the further benefits would be small compared to the additional investment and programming effort required. A computer based on stream analytical system has also recently been described for the control of a rare earth separation process [65]. Although the introduction of the analyzer has brought about substantial operational advantages, no decision has yet been made concerning its application to computer closed-loop control, although this may be possible on the basis of the use of an established mathematical model once the analyzer is believed to be operating at its full potential and capacity. A further important application of DDC is the CALUS TAMARA project [62], at the Kernforschungszentrum, Karlsruhe, where a uranium extraction pilot plant was run with the aim of demonstrating the feasibility of DDC to nuclear reprocessing applications. Although the viability of the computer for off-line and in-line studies for process monitoring, logging of data, supervision of alarm systems, and so on was confirmed, application of the computer to actual online process control studies remained to be fully verified.

In general, the first spate of computer control studies was not generally commercially successful. These early systems were designed as digital equivalents of their conventional analog controller counterparts and apart from some additional flexibility were costly, unreliable, and created additional interface problems between the computer, the plant, and the operator and were also generally too ambitious. With the development of cheap and reliable minicomputers, less ambitious schemes employing direct digital control were more successful, but there remained a fear of the vulnerability of a single computer to failure. This had led to a trend toward supervisory control in which the conventional analog controllers are retained for critical control loops but which have their set points supervised by the computer [63].

The increase in the size of the schemes has resulted in a trend toward a devolution of computing power that has been encouraged by the advent of the microprocessor. Thus the DDC is at the lowest level of command, being responsible for only one plant. Above this is the supervisory computer, responsible for several individual plants; and above this is the computer responsible for coordinating a complex of several plants; and at the top is the management information computer. With this scheme the fear of a single computer failure is overcome, since each process control computer should have sufficient capacity to take over control of other sections of the plant or overall supervisory control and because communication between the digital computers is far superior to that between the computer and its analog systems. The outstanding ability of the computer is its ability to acquire, analyze, and disseminate large amounts of information with high speed, accuracy, and flexibility, and it is much more sensible to use the computer in its prime function as a high-speed calculation device rather than use it as a counterpart to the analog conventional controller. A typical computer system will have a disk store for storing models, storage for plant data on floppy disks, a core store for carrying out control algorithm calculations, an analog to digital conversion unit controlled by a multiplexer that takes process variables in the form of electrical signals and feeds them into the computer store and a digital-to-analog or digital-to-digital/analog output interface for transmitting control commands to the plant, as indicated in Fig. 30.

Thus the conventional three-term controller equation can be replaced by a sampled data control algorithm represented by

$$M_n = K_c e_n + \frac{K_c T}{T_i} \sum_i^n e_i + \frac{T_d K_c}{T} (e_n - e_{n-1}) + M_R \tag{97}$$

where M_n is the value of the manipulated variable at the nth sampling instant, e_n is the error at the nth sampling instant, T is the sampling time, K_c is the proportional gain, T_I is the integral time, and T_d is the derivative time, and M_R is the base (midrange) valuc for the manipulated variable. However, it is doubtful that more than two terms would actually be required for DDC. Since it is unlikely that the plant model can give an exact representation of plant behavior, especially where the plant has highly nonlinear characteristics, and where these characteristics may change with time; by use of modern control identification techniques it is possible to utilize the computer as a self-optimizing and model reference adaptive control system (Fig. 31) to automatically compensate for variations in the system dynamics by continuous adjustments in the plant model and controller characteristics aimed

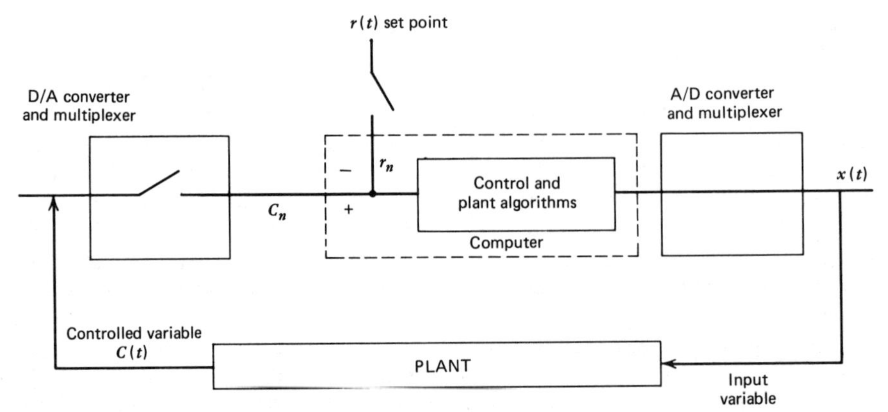

Figure 30. Typical computer control system.

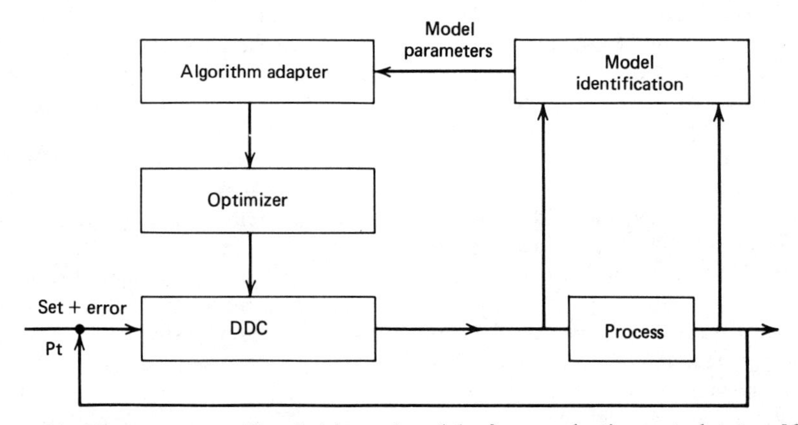

Figure 31. Computer as a self-optimizing and model reference adaptive control system [63].

at optimum plant performance. Thus in the identification stage the present status of the process is determined by identifying the current values of the dependent and independent model parameters by use of a weighting function to automatically discard old data. By this means the current values of the model parameters can be calculated by way of linear regression techniques and the process and control algorithms continuously modified. These data and the present situation of the process are then fed into the optimizer to determine the necessary corrective action required for resultant optimum behavior.

It has been claimed that solvent extraction processes are not themselves worthy of DDC since, apart from specialist applications (e.g., nuclear reprocessing), their operation is seldom critical and the time base for mixer–settler plants is so long as to obviate critical control problems. In economic terms there is little financial gain to be obtained by straightforward replacement of conventional analog controllers by DDC. The solvent extraction plant is, however, only part of the overall plant where DDC may be fully justified and indeed may, in future, be indispensable. With the flexibility and power capability provided by present microprocessor techniques, it makes obvious sense to extend the overall plant DDC system to the solvent extraction sectors. This, combined with the use of plant optimization routines and model reference adaptive system with the use of modern techniques of model identification techniques for sampled data systems, should play a very important part in future developments, and it would be highly unrealistic to base future projections of the use of DDC on the unsatisfactory and limited experiences of the past.

REFERENCES

1. J. M. Douglas, *Process Dynamics and Control*, Vol. 2, Prentice-Hall, Englewood Cliffs, NJ, 1972.

2. W. R. Marshall and R. L. Pigford, *The Application of Differential Equations to Chemical Engineering Problems*, University of Delaware (1947).
3. I. J. Dunn and J. Ingham, *Chem. Eng. Sci.* **27**, 1751 (1972).
4. J. Ingham and I. J. Dunn, *Chem. Eng.* 354 (1972).
5. C. R. McDonald and W. L. Wilkinson, *Proceedings of the International Solvent Extraction Conference (ISEC), Lyons 1974*, Vol. 3, Society of Chemical Industry, London, 1974, p. 2608.
6. R. G. E. Franks, *Mathematical Modelling in Chemical Engineering*, Wiley, New York, 1966.
7. U. Lelli, *Annali di Chimica* **56**, 113 (1966).
8. R. J. Ord-Smith and J. Stephenson, *Computer Simulation of Continuous Systems*, Cambridge University Press, 1975.
9. W. S. Groenier, Report ORNL-4746 UC-10, 1972.
10. S. B. Watson and R. H. Rainey, Report ORNL-TM-5-23 UC-77, 1975.
11. W. C. Scotten, Report DP-1391, 1975, E. I. du Pont de Nemours and Company.
12. R. G. E. Franks, *Modelling and Simulation in Chemical Engineering*, Wiley-Interscience, New York, 1972.
13. W. L. Wilkinson and F. Medina Gomez, The Dynamic Behaviour of a Mixer–Settler for Solvent Extraction, *Chem. Eng. Sci.* (in press).
14. D. A. Burge and W. C. Clements, *Chem. Eng. Sci.* **27**, 1537 (1972).
15. J. F. Wehner and R. H. Wilhelm, *Chem. Eng. Sci.* **6**, 89 (1956).
16. D. A. Jones and W. L. Wilkinson, *Chem. Eng. Sci.* **23**, 537 (1973).
17. D. A. Jones and W. L. Wilkinson, *Chem. Eng. Sci.* **23**, 1577 (1973).
18. J. E. Prenosil, *Verfahrenstechnik* **11**, 470 (1977).
19. H. R. Foster, R. E. McKee, and A. L. Babb, *Ind. Eng. Chem. Process. Des. Devel.* **9**, 272 (1970).
20. G. C. Pollock and A. I. Johnson, *Can. J. Chem. Eng.* **47**, 469 (1969).
21. C. Hanson and M. Sharif, *Can. J. Chem. Eng.* **43**, 132 (1970).
22. T. W. Cadman and C. K. Hsu, *Transact. Inst. Chem. Eng.* **48**, 432 (1973).
23. C. K. Hsu, *Transact. Inst. Chem. Eng.* **49**, 251 (1971).
24. T. Kikiudai, P. Michel, H. Rouyer, and G. Chazal, paper presented at CEA Conference, 1967, p. 1129.
25. H. Rouyer and J.-M. Demarthe, paper presented at CEN Commission Energie Atomique (1971).
26. H. Rouyer and J.-M. Demarthe, *Bull. Inform. Sci. Techn. Comm. Energ. Atomique* **134**, 49 (1973).
27. H. Rouyer and T. Kikiudai, *Comptes Rend. Acad. Sci. Ser. C.* **270**, 271 (1970).
28. H. Rouyer, *Comptes Rend. Acad. Sci. Ser. C* **270**, 580 (1970).
29. G. Chazal, O. Hoclet, and H. Rouyer, CEA Conference, 1970, p. 1579.
30. G. Chazal, paper presented at CEN Commission Energy Atomique, 1971.
31. J.-M. Demarthe, CE Report A-R-4350, 1973.
32. P. Bonnefoi, A. Poujol, G. Zwingelstein, C. Darzler, and H. Rouyer, paper presented at CEA Conference, 1977, p. 4023.
33. G. A. Beetner, A. L. Frey, and R. G. Bautista, *Transact. Soc. Mining Eng. AIME* **254**, 349 (1973).
34. M. G. Casto, Y.-C. Hoh, M. Smutz, and R. G. Bautista, *Ind. Eng. Chem. Process. Des. Devel.* **12**, 432 (1973).
35. G. Aly, A. Jernqvist, H. Reinhardt, and H. Ottertun, *Chem. Ind.* 1046 (1971).
36. G. Aly and B. Wittenmark, *J. Appl. Chem. Biotechnol.* **22**, 1165 (1972).
37. G. Aly and H. Ottertun, *J. Appl. Chem. Biotechnol.* **23**, 643 (1973).
38. A. Apelblat and M. Faraggi, *J. Nucl. Energy* **20**, 953 (1966).
39. D. C. Drown and W. J. Thomson, *Ind. Eng. Chem. Process. Des. Devel.* **12**, 197 (1977).
40. S. Bobrow, J. W. Ponton, and A. I. Johnson, *A. I. Can. J. Chem. Eng.* **49**, 391 (1971).
41. M. Libhaber, P. Blumberg, and E. Kehat, *Ind. Eng. Chem. Process. Des. Devel.* **13**, 39 (1974).
42. M. A. El-Rifai, *Chem. Eng. Sci.* **30**, 79 (1975).
43. D. Barnea, M. S. Hoffer, and W. Resnick, *Chem. Eng. Sci.* **33**, 205, 219 (1978).
44. W. Ochsenfeld, H-J. Bleyl, D. Ertel, F. Heil, and G. Petrich, paper presented at the Fast Reactor Fuel Reprocessing Conference, Dounreay, May 1977.
45. G. C. Pollock and A. I. Johnson, *Can. J. Chem. Eng.* **47**, 565 (1969).
46. G. C. Pollock and A. I. Johnson, *Can. J. Chem. Eng.* **48**, 64 (1970).
47. G. C. Pollock and A. I. Johnson, *Can. J. Chem. Eng.* **48**, 711 (1970).
48. K. H. Pang and A. I. Johnson, *Can. J. Chem. Eng.* **47**, 477 (1969).
49. K. H. Pang and A. I. Johnson, *Can. J. Chem. Eng.* **49**, 837 (1971).
50. K. H. Pang and A. I. Johnson, *Can. J. Chem. Eng.* **49**, 398 (1971).
51. K. H. Pang and A. I. Johnson, *Can. J. Chem. Eng.* **49**, 529 (1971).
52. F. Souhrada, J. Laudau, and J. Prochazka, *Can. J. Chem. Eng.* **48**, 322 (1970).
53. J. Laudau, A. Dim, and L. W. Shemilt, *Can. J. Chem. Eng.* **53**, 9 (1975).
54. P. S. M. Wang, J. Ingham, and C. Hanson, *Transact. Inst. Chem. Eng.* **55**, 196 (1977).

55. L. Steiner, M. Horvath, and S. Hartland, *Ind. Eng. Chem. Process. Des. Devel.* **17**, 175 (1978).

56. H.-D. Bauermann and E. Blass, *Germ. Chem. Eng.* **1**, 99 (1978).

57. A. L. Mills, A. J. Oliphant, and M. Parkinson, *Atom* **237** (1967).

58. A. L. Mills and P. C. Bell, *Proceedings of the International Solvent Extraction Conference* (ISEC71), The Hague, 1971.

59. W. R. Burton and A. L. Mills, *Nucl. Eng.* 248 (1963).

60. C. R. McDonald and W. L. Wilkinson, *Proceedings of the International Solvent Extraction Conference (ISEC), Toronto 1977*, Special Vol. 21, Canadian Institute of Mining and Metallurgy, Montreal, 1979, p. 309.

61. D. S. Flett, G. W. Cutting, and P. Carey, paper presented at IMPC 10th Congress, (1973).

62. H. R. Mache and K. Landmark, Report, Kernforschungszentrum Karlsruhe, 1978.

63. C. L. Smith, *Digital Computer Process Control, Intext. Educational Publication*, 1972.

64. P. E. Vembe, *Proceedings of the International Solvent Extraction Conference (ISEC), Toronto 1977*, Special Vol. 21, Canadian Institute of Mining and Metallurgy, Montreal, 1979, p. 761.

65. B. Gaudernack and O. B. Michelsen, *Proceedings of the International Solvent Extraction Conference (ISEC), Toronto 1977*, Special Vol. 21, Canadian Institute of Mining and Metallurgy, Montreal, 1979, p. 754.

27.3

INSTRUMENTATION AND CONTROL

S. Plonsky
Hoffmann-LaRoche Inc.
United States

1. SCOPE AND INTENT OF CHAPTER

Chapter 27.2 gave a comprehensive account of fundamentals of the control of solvent extraction, with emphasis on modeling and dynamic characteristics. This chapter is of a practical nature. The intent is to assist the reader in the details of the design of workable and reliable extractor control systems. The general subject of automatic process control is not covered in any depth; many excellent sources of information are readily available [1–4, 7, 8].

Equipment of the type covered here is essential to the implementation of the control strategies described in Chapter 27.2. The emphasis is on examples of the control of extraction columns, but much of the instrumentation is also used in the control of mixer–settlers. The reader is referred to the relevant chapters (Chapters 9.1–9.5) for examples.

The control functions encountered in this chapter present few problems from the standpoints of rapid response and control loop stability. System synthesis and equipment selection, however, may contain the seeds of difficulties in the field. It is the author's hope that something of value will be offered the reader in facilitating recognition of areas of the design requiring special consideration.

2. CONSTITUENTS OF AN EXTRACTOR CONTROL SYSTEM

In contrast to the control of distillation columns, where one is concerned with both material and energy balance controls to ensure the desired separations, an extraction column control system requires only material balance control. Thus the control systems are inherently simpler.

The minimum requirements are means to control or monitor solvent and feed streams to the extractor so that they are within the permissible operating ranges and inventory control accomplished by maintaining the position of the light-phase level and interface level.

A schematic representation of a simple system for a column with one feed and two solvent streams, a vented settling section, and pumped light and heavy-phase drawoff is shown in Fig. 1. The flow and level control systems may be rearranged in a variety of ways to accommodate physical layout and process conditions. For instance, some streams may come from other processing units up and downstream of the extractor and cannot be controlled independently of their source as in Fig. 1. The column may be designed to run flooded under pressure; one or both of the drawoffs may be removed by gravity. Some of these possibilities are examined in the following paragraphs.

2.1. Level and Interface Level Control Arrangements

In Fig. 1 the interface level and a discrete light-phase level are maintained in the upper portion of the column by controls LC-1 and LC-2, respectively. (For a guide to standard instrumentation symbols, see Ref. 5). The location of the drawoff connection for the light phase should always be below the lower light-phase level instrument connection to ensure a flooded drawoff line for all possible controller setpoints in the range 0–100% of instrument range.

Figure 2*a* illustrates instrument connections where the process design requires the interface to be controlled at the bottom of the column.

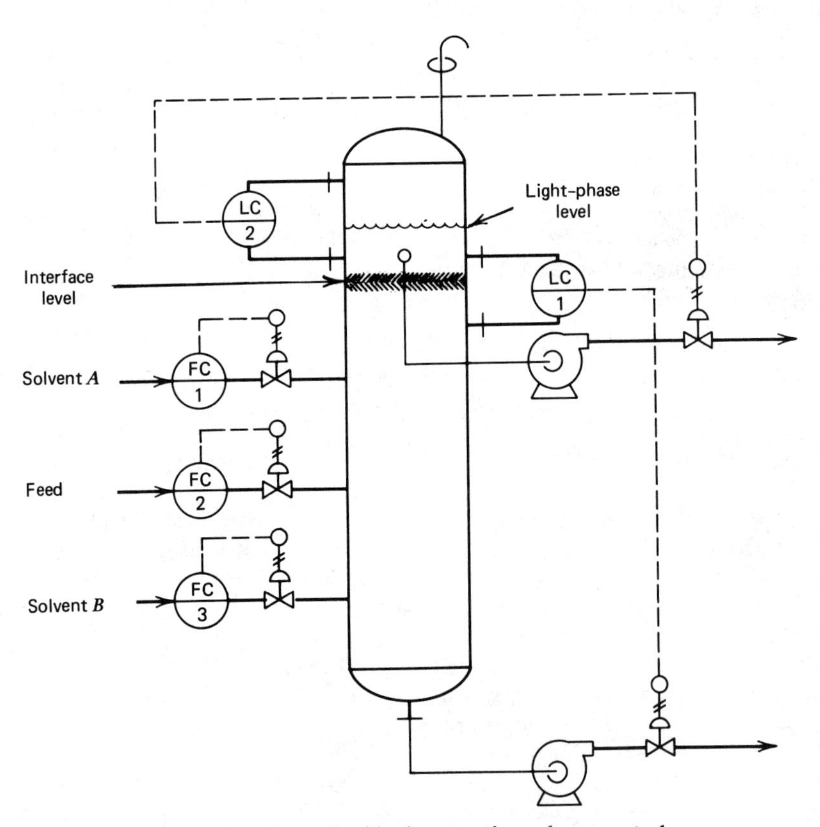

Figure 1. Generalized basic extraction column control.

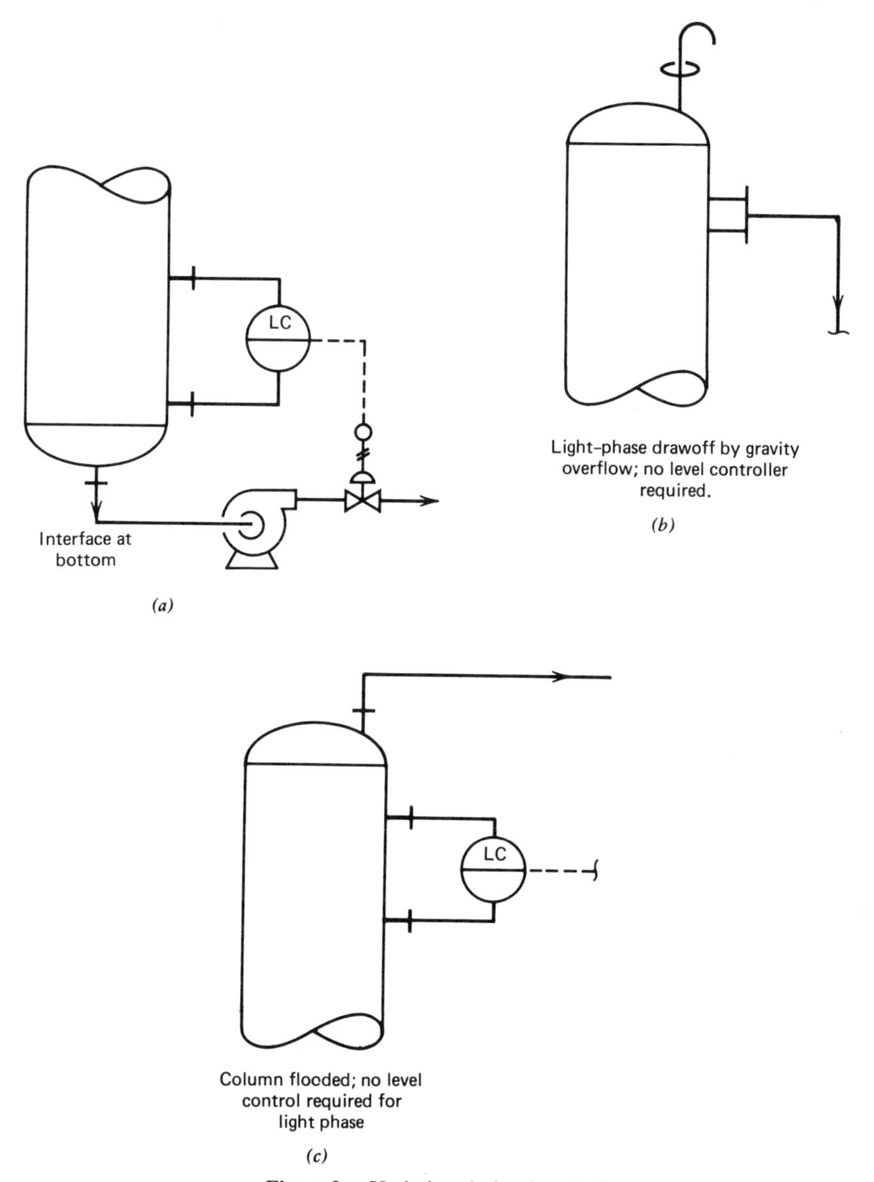

Figure 2. Variations in level control.

Functionally, the operation of controller LC-1 is no different from that in Fig. 1. Figure 2*b* illustrates the fact that the light-phase level controller can be dispensed with entirely if it is possible to overflow by gravity to a receiver at an elevation lower than that of the column overflow nozzle. Figure 2*c* illustrates an arrangement with the column being operated flooded, under pressure. Since there is no discrete light-phase level, no controller is required. The inventory of separated light phase is maintained by the positioning of the interface by interface level controller LC-1 as previously.

2.2. Flow and Flow Ratio Control

Figure 1 illustrates a flow control configuration that is simple and suitable if the column is to be run at nearly constant rate for long periods of time. For example, the sources of the feed and solvent streams may be large surge tanks, each equipped with a column feed pump.

If, for example, the feed stream originates directly from one or more continuous process units, its flow rate will normally be adjusted by their automatic control systems. Accordingly, it will vary with respect to time to some degree in

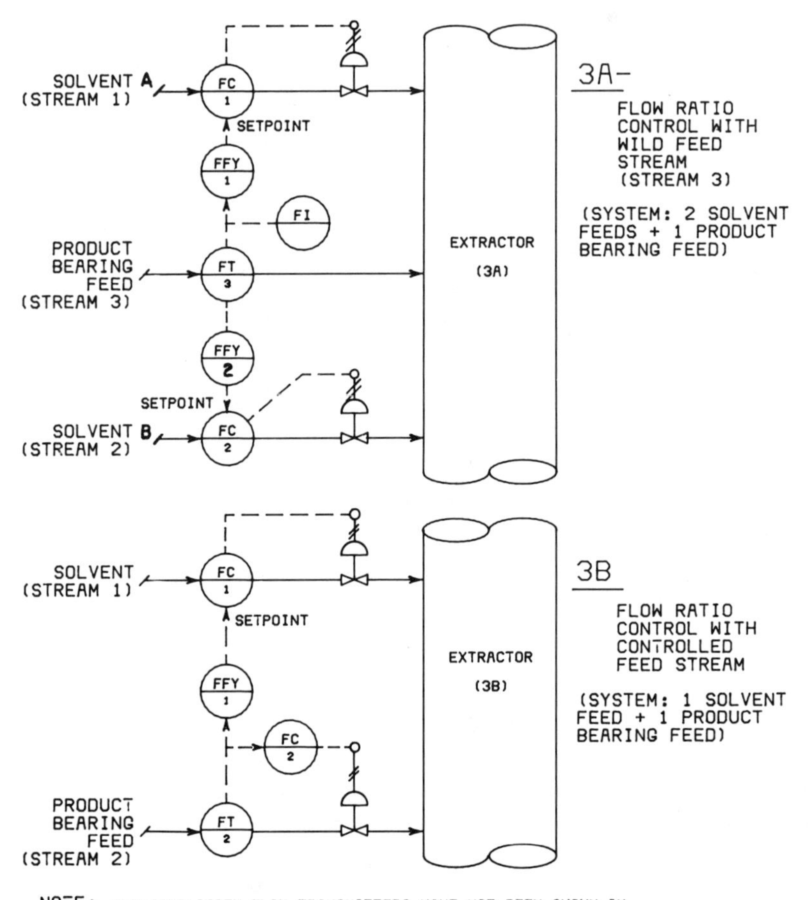

Figure 3. Flow ratio control.

composition and volumetric rate, reflecting variations in feed to the source units and the quality of their control systems. In a well-designed plant, the variations will be within permissible performance limits for the extractor.

For volume-varying feed streams, the solvent : feed ratios may be kept constant by a feedforward ratio control system such as that shown in Fig. 3*a*. The wild stream from other units is measured and converted to a proportional signal by flow transmitter FT-3. This signal is sent to (manually) adjustable ratio device FFY, which multiplies the FT-3 flow signal by the preset value. The modified value becomes the setpoint or instruction to flow controller FC-1 and (FC-2, if two solvent streams are employed). Since a flow control loop response time is very rapid (in the order of seconds), the solvent streams will vary in proportion to the wild stream in a fixed ratio. Commercially, FFY–FC combination units are available as flow ratio controllers.

Another variation of the flow control scheme given in Fig. 1 is shown in Fig. 3*b*. This permits a single throughput adjustment by the operator and is a great convenience if frequent adjustment of the unit feed rate must be made. With adjustment of the setpoint of controller FC-2, the other streams will follow automatically in proportion. For instance, if the feed stream is withdrawn from a surge tank that, in turn, is fed by periodic discharges from a batch process, adjustment is needed to keep the continuous outflow in balance with the intermittent inflow. An automated approach to this problem is discussed briefly in Section 2.4.

2.3. Analysis Instrumentation

The application of in-process analytical measuring and control equipment should be approached with the realization that this class of instrumentation will be quite expensive, will be more fragile than the usual equipment, and further, will require more frequent calibration and maintenance.

Some analysis instrumentation has been time tested and can be applied with a high degree of confidence. For instance, one might consider automatically correcting a feed stream of variable concentration by using the liquid density as an inferential measure of concentration. U-Tube densitometers are available with a long-term accuracy of ± 1 g/cm^3.

Continuous in-line pH control of a feed stream is certainly practical in consideration of currently available equipment and techniques [6]. In this case, calibration should be checked by manual sample once per shift. Figure 4 illustrates one way each of these controls may be accomplished.

The reader should approach in-process analysis of feed, raffinate, and extract streams with somewhat more caution if the objective is chromatography, ultraviolet or infrared absorption, refractive index, or other measurements of similar complexity. Equipment is available that can be applied successfully if one is prepared to bear the initial cost and to provide the required care and attention. If changes in these streams are relatively slow, as is commonly the case, it might be wise to consider the traditional manual sample-laboratory analysis method as an alternate way of monitoring extractor performance.

2.4. System Configurations: Information Transmission and Display

For the basic continuous process control functions required for extraction, one could employ pneumatic analog controllers, electronic analog controllers, or a microprocessor-based system capable of performing the same tasks. For more complex functions, microprocessors are necessary.

If, for example, the extraction unit (or units) is essentially a stand-alone operation with the operating control center no more than 300 ft (90 m) from the plant, one might choose a pneumatically powered system for controlling, record-

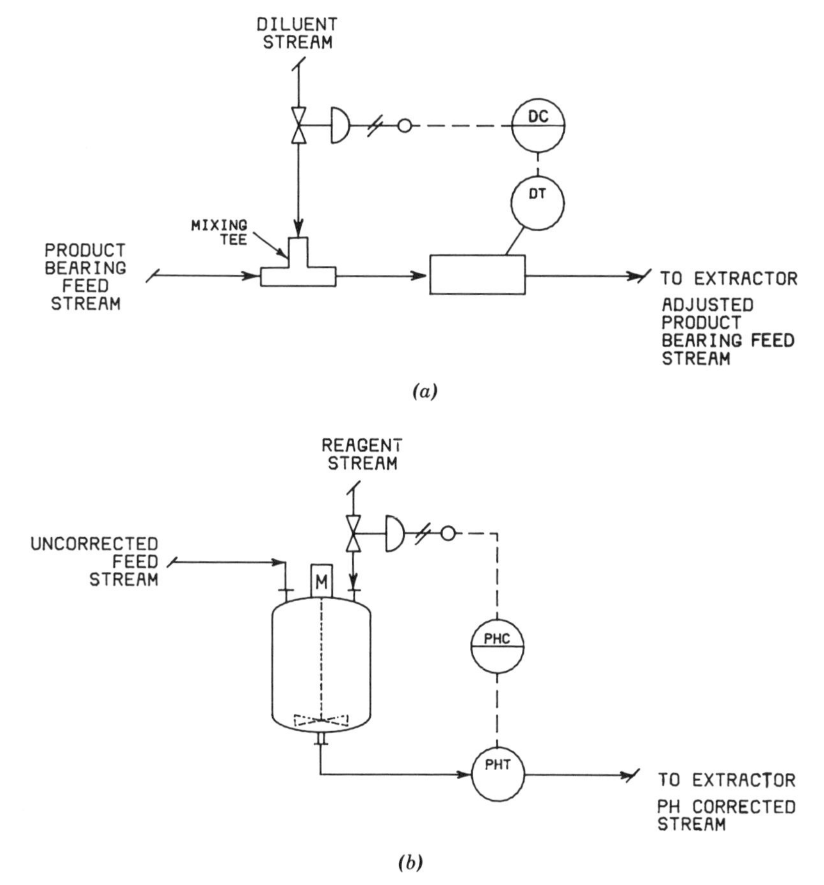

Figure 4. Density and pH control.

ing, and indicating continuous process variables. Normally, controllers, recorders, and indicators will be mounted on a control panel. If the plant is indoors and the panel is in the manufacturing area, pneumatic control is often the best choice from an economic and functional standpoint. The cases are self-purged with clean air. The pneumatic units inherently are without safety hazard, but electrical or electronic components must be explosionproof in flammable solvent areas.

If a central control room is planned and the extraction unit operations are only part of the process train, usually electronic systems will be chosen, if only because of rapid signal transmission over relatively long distances. Electrical components in the control room need not be explosionproof. If the field-mounted control components are part of an intrinsically safe design, the signal transmission cables need not be carried in expensive heavy-walled conduit systems, and the transmitters themselves need not be provided with heavy metal housings in a hazardous area. (Intrinsically safe systems are designed so that under any condition of failure, it is not possible to produce a spark of sufficient energy to ignite an explosive mixture of the solvent systems being handled in the area. They have passed this criterion in the opinion of independent agencies such as the Factory Mutual group or the Underwriters' Laboratories and bear a label so marked (see also Chapter 30).

Whether one chooses a conventional analog control system in the control room with discrete indicators, chart recorders, and controllers dedicated in individual control loops or a microprocessor-based system with shared hardware and software, the field transmitters and current-to-pneumatic converters for control valves will be the same and will operate on a 4–20 mA signal range.

Microprocessor-based systems today are cost-competitive on an initial capital outlay basis, with the older analog systems. Today's designs are modular. Each module will handle the function of from one to 16 analog loops with nonconventional enhancements available, such as nonlinear gain, preprogrammed excursion limits, and dead bands, to name a few. Control of calculated variables is more readily implemented. Operator communication and information display through cathode ray tube (CRT) terminals is standard with these systems. Data reduction and organization for production records are also possible.

Computer control of extraction is still at an evolutionary stage as noted in Chapter 27.2 (Section 7), but its use can be expected to increase to meet the requirements of complex control functions. A stand-alone minicomputer or host minicomputer–distributed microprocessor system may have to be employed to implement control strategies that exceed the logical capabilities of standard commercial microcomputer-based products.

Referring to the batch process–continuous process interface control method discussed in Section 2.2, the logical capabilities of a microprocessor-based system could be employed to automate the setpoint adjustment of a flow controller being fed from an intermittently filled tank. With the use of the same logical method that an operator would, the estimated time of arrival of the next batch could be calculated by determining batch status. The tank level desired at that time would be compared with present level and withdrawal rate. If they do not match, a correction would be made. This procedure would be repeated periodically. If the preprogrammed logic could not achieve the desired results within equipment constraints, the operator would be alerted.

3. AUXILIARY INSTRUMENTATION

3.1. Detection and Alarm of Abnormal Conditions

An operator's duties in a control center will encompass the surveillance of several unit operations, which from time to time will experience process upsets and equipment malfunction. Some of the malfunctions may indeed originate with the field-mounted sensors in the process. They often operate at extremes of pressure and temperature; are exposed to the elements; and may be subject to physical damage, in contrast to the control equipment, which is installed in the climate-controlled atmosphere of the control center.

Auxiliary detectors are often installed to warn the operator of process malfunctions. In some cases automatic shutdown may be initiated if the consequence of the detected fault warrants this action. The detection devices should preferably be independent of the primary control system for reasons stated previously. The simplest and most reliable device that will serve the purpose will be the best choice (see Fig. 5).

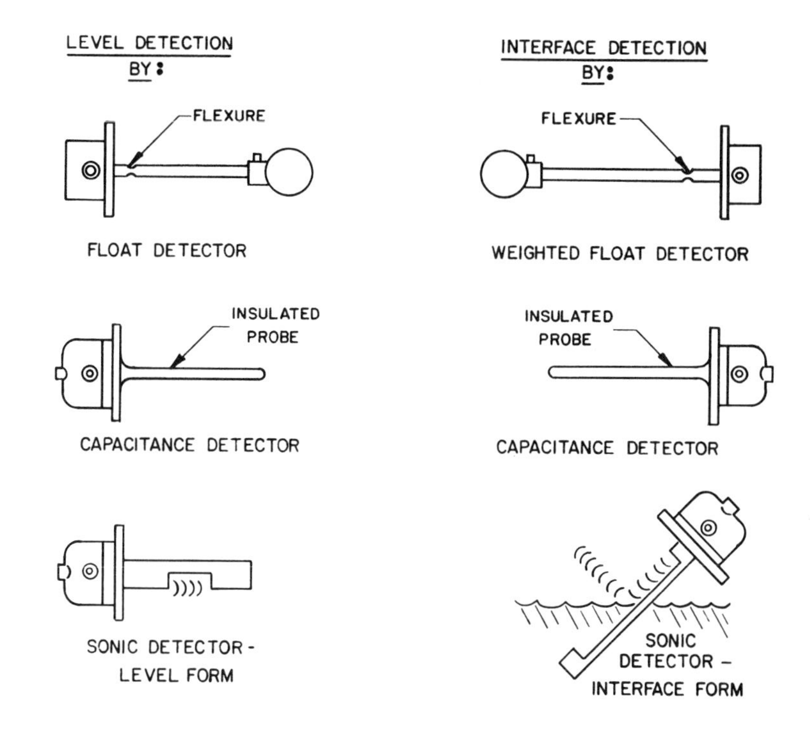

Figure 5. Level and flow detectors.

3.1.1. Level Location

In configurations where a discrete light-phase level exists, a simple float-operated switch will often do the job of abnormal level detection. Capacitance and sonic signal activated switches are also commonly employed.

3.1.2. Interface Location

Abnormal location of the column interface may be detected by the same type of device as employed for the light-phase level described earlier.

They are modified by the manufacturer for interface application. For example, a hollow float can be weighted so that it sinks in the light phase but will be bouyant in the heavy phase.

3.1.3. Emulsification

Emulsification detection is not difficult in many cases. If the normal specific gravity difference between the phases is 0.2 or greater, emulsification can be sensed easily be a weighted float immersed in the light-phase space. If the two phases have a normal difference in dielectric constant of 2.0 or more, emulsification can reliably be detected by a capacitance sensing device in either the light-phase space within the extractor or the light-phase piping.

3.1.4. Flow Detection

Before evaluation of means of flow detection other than the primary flow measuring device, the potential causes of difficulty and their consequences should be examined. A pump may be accidentally shut down, in which case simple electrical detection at the motor control circuit may suffice, or detection of loss of feed line pressure may be employed.

Another method of detecting a gross change in flow is the use of a paddle-type flow switch in the line, which is activated by the impact force due to the velocity of the flowing liquid. This could detect an inoperative pump, closed block valve, plugged strainer, or failure in the flow control loop, for instance.

Feed loss or off-normal feed may be inferred from the positions of the control valves in the interface and level control loops relative to their normal states, which, in turn, are proportional to the output of their respective controllers. If we wish to detect deviation of flow from a normal value within close limits, the initiating signal can be derived from the flow transmitter in the control loop. This method is suitable for detection of malfunction other than the flow transmitter itself.

3.2. Temperature Control of Feed Streams

An often used auxiliary control is the temperature control of feed streams, since solubility and viscosity are temperature dependent. This method, which is simpler than temperature control of the extraction apparatus itself, is an inferential means of controlling the temperature of the fluid at the points of contact. If the exact temperature within the column is critical, a cascade control system as illustrated may be provided (see Fig. 6).

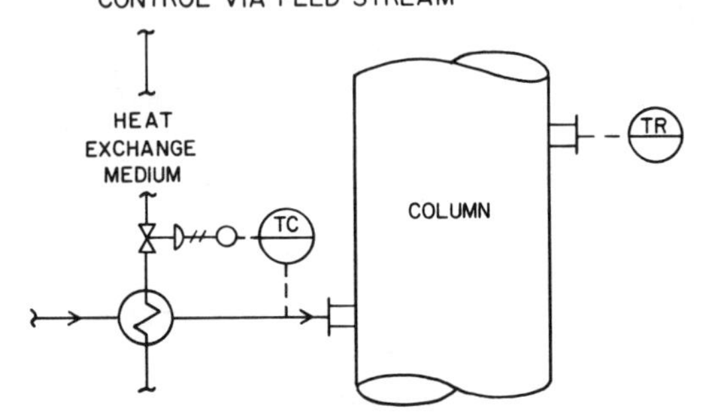

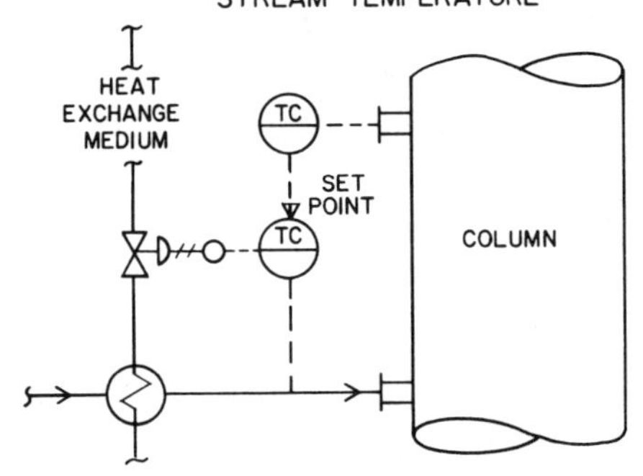

Figure 6. Column temperature control.

3.3. System Security

The detection means outlined in Sections 3.1.1–3.1.4 are usually incorporated into an alarm system that alerts the operator by visual and/or audible means to the particular problems. As previously mentioned, they may be incorporated into a loss prevention system that will automatically shut down the operation.

The temptation to a beginner when developing a concept for such an automatic system is to mentally mix hardware and logic together in formulation of the basic design. This almost always impedes the development of a sound design. The first step is to set down in clear language what one wishes to accomplish, without regard to the equipment necessary for the task. This goes against the grain for an engineer trained in practical possibilities. But the chances are that the means will be available at reasonable cost. If not, the concept can be modified to accommodate available technology as the detail design proceeds. The next recommended step is to depict the operation by means of a logic diagram. The logic diagram with the verbal description closely defines the system and is in a form that can more readily be understood by plant operating personnel and the originator than, say, an electrical schematic. Figure 7 illustrates a simple system that uses the logic notation of Instrument Society of America Standard S5.2.

4. DETAILS OF SYSTEM SYNTHESIS

4.1. Materials of Construction

For instrumentation parts wetted by process fluids, the view taken in selection of materials of construction must be somewhat difficult from the one used for process equipment. Commonly available corrosion data for metals are based on tests of samples whose surfaces are not subject to very high shearing forces. Alloys such as the stainless steels and the Hastelloys, which might

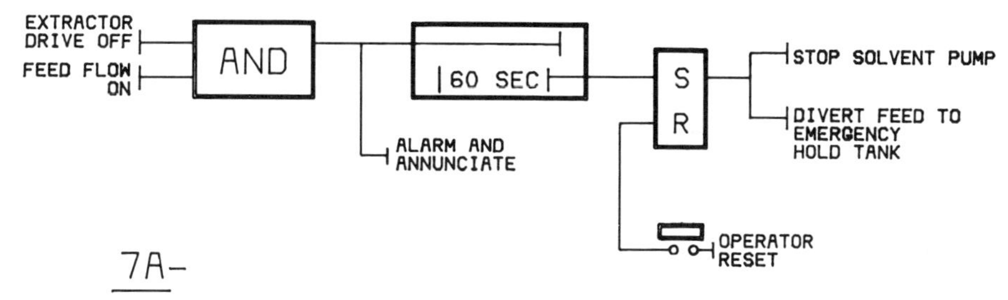

"SHOULD THE EXTRACTOR DRIVE STOP DURING PROCESSING, THE OPERATOR SHALL BE WARNED. IF THE CONDITION PERSISTS FOR 60 SECONDS, THE SOLVENT PUMP SHALL BE STOPPED AND THE FEED DIVERTED TO THE EMERGENCY HOLD TANK."

(a)

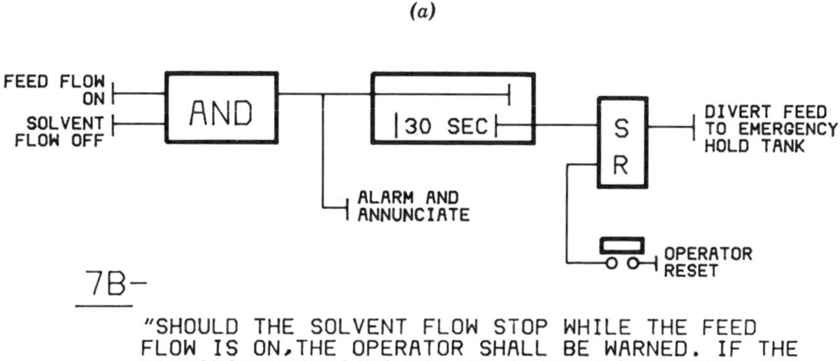

"SHOULD THE SOLVENT FLOW STOP WHILE THE FEED FLOW IS ON, THE OPERATOR SHALL BE WARNED. IF THE CONDITION PERSISTS FOR 30 SECONDS, THE FEED SHALL BE DIVERTED TO THE EMERGENCY HOLD TANK."

(b)

Figure 7. Text and logic diagrams.

be expected to have a reasonable service life, may sometimes corrode at a very rapid rate when used for control valve trim. The reason is that the velocity between the valve plug and seat is so high that the passivating layer, which normally protects the metal beneath, strips off as rapidly as it is formed.

One procedure that should be followed is the determination of the thickness of the wetted parts. For instance, a metal diaphragm in a level transmitter may be 5 mils (0.125 mm) in thickness. A corrosion rate of 2 mils (0.05 mm) per year—which is acceptable for process equipment—would, of course, be unsuitable for the level instrument.

4.2. Special Problems

4.2.1. Solids

At times, finely divided solids may be present in the product-bearing stream. This presents a problem only when the piping connecting the instrumentation to process lines or equipment terminates in a nonflowing volume below the point of connection. The solids will slowly accumulate and plug up these spaces, rendering the device inoperative. In an application such as this the instrument should be mounted above the connection or a substitute flow-through device or arrangement selected.

4.2.2. Tars and Rags at Interface

In some applications tarry residues or other foreign materials in the product stream will agglomerate and float at the interface. This is rarely a problem if a sidestream connection is provided to draw these accumulations off periodically. Tars will tend to foul float-type interface detectors. If it is known that they will exist, another choice should be made.

4.2.3. Degradation in Stagnant Pockets

Sensitive materials may form degradation products in stagnant pockets that may exist in a sensor such as a differential pressure transmitter. In these problem applications the use of a flow-through flow transmitter and differential pressure devices fitted with flush-mounted diaphragm seals will avoid the problem. Float-type level sensors should be mounted inside the extractor settling section instead of in chambers external to it.

4.2.4. Sterilization and Cleanliness Requirements

For pharmaceutical and food applications, the entire assemblage, including the instrumentation in contact with the process fluids, must often be steam sterilized and sometimes easily demountable. First, design of configurations with no pockets makes sterilization simple. Control valves with angle body design are self-draining and suitable for this application. If demountability for washing is required, the range of available instrumentation is somewhat restricted, but an adequate selection of flowmeters, control valves, and diaphragm-type level devices are available with quick disassembly food-industry standard connections. For steam sterilizing, the temperature rating of the equipment being specified should be checked for adequacy.

4.3. Field Instrumentation

A brief survey of field instrumentation is given here. For the reader interested in detail, the handbook by Considine [4] and manufacturers' technical brochures are recommended.

4.3.1. Flow Measuring Devices

The oldest and still widely used method of measuring flow in a pipeline is the orifice plate and differential-pressure transmitter. Its assets are simplicity and relatively low cost. It has a limited range of operation of 4 : 1, with accuracy falling off rapidly below 3 : 1. Flow is proportional to the square root of the transmitted signal; hence a nonlinear scale is required. For flow ratio control, the signal must be linearized with an auxiliary square-root extractor. This is recommended for general applications. For situations described in Sections 4.2.1, 4.2.3, and 4.2.4, this is not the best choice.

Magnetic flowmeters are obstructionless and appropriately employed with electrically conductive fluids. For streams that may coat the inner surface, an ultrasonic cleaning auxiliary is available. This meter produces a linear signal over a 10 : 1 range when properly sized and will operate successfully with corrosive media. For instance, one could specify a Teflon liner with platinum alloy electrodes.

Rotamater transmitters are suitable for clean liquid service for flow rates of 2–75 gal/min (8.5–320 liters/min). They are moderate in cost and produce a linear signal over a 10 : 1 range. For process safety, only the metal tube variety should be specified. If they are carefully used, many years of good service can be expected. Teflon-lined units are available for corrosive fluids. Damage can occur to the float assembly if it is subject to sudden high surges of flow.

Turbine meters are a suitable choice for clean, noncorrosive service. Although they are relatively high in cost, they have the advantage of excellent accuracy and can be coupled with totalizing devices with ease. A pulse output and a linear signal over as much as a 20 : 1 range can be obtained. A disadvantage is maintenance cost due to turbine-bearing wear.

Vortex shedding meters are a suitable choice where a pocket-free flow-through device is desired. A large open area is presented to the flow path, and a linear signal with a 10 : 1 range is produced. Vortex frequency detectors of several designs are commercially available. For solids-bearing streams, a sealed detector should be specified.

4.3.2. Level and Interface Measuring Devices

For extraction column level and interface service, the three most widely used devices are vertical tubular float transmitters (which do not truly float but change in net weight by Archimedes' principle), differential pressure (or hydrostatic head) transmitters, and electrical capacitance transmitters.

The most commonly specified tubular float transmitters range in diameter from 3 to 4 in. (75–100 mm) and are 14–32 inches (350–800 mm) in length. Essentially, the measurement is one of weight. A signal proportional to the average buoyant force over the length of the float is produced. One may view both level and liquid interface measurements as essentially the same; the level measurement is a liquid/air interface.

For liquid/liquid interface, the float is weighted with lead shot to produce a zero signal when totally immersed in the light phase. Density differences between light and heavy phases as small as 0.05 g/cm^3 are readily accommodated. When applying this type of transmitter to a pulsed or vertically agitated column, one must mount it externally in a separate chamber connected at top and bottom to the settling section. By inserting a wire screen mesh between the flanged connections, the oscillations communicated to the float will be greatly diminished.

Differential-pressure transmitters used as level devices measure the difference in hydrostatic head between two fixed taps in the side of the settling section. The device is usually mounted so that its high-pressure connection is at the same elevation as its bottom tap. For interface measurement, the upper tap and its piping is flooded with the light phase. For level measurement, the upper tap communicates with the vapor space, but the piping down to the instrument is usually filled with the light-phase liquid.

In each case the instrument is adjusted to read zero when it "sees" the light phase at its bottom tap.

The failing with this arrangement for interface measurement is that if the interface accidentally rises above the upper tap for that measurement, the heavy phase will pour into the connecting piping, displacing the lighter fluid. It will remain there even after the interface has descended, giving a false reading. A solution to this problem is a pneumatic "repeater" at the upper tap instead of a direct connection. This diaphragm-activated device produces a pneumatic pressure on a 1 : 1 basis equal to the hydraulic pressure on its diaphragm. The pressure thus duplicated is connected to the low side of the differential device.

For practical purposes, one should not specify an instrument with less than a 5-in. (125-mm) water column span. For example, in interface service if the difference in density between the phases were 0.2 g/cm^3, the instrument taps would have to be 5/0.2 or 25 in. (625 mm) apart.

An electrical capacitance transmitter produces a signal proportional to the average dielectric constant of the medium along its length.

In the usual construction one electrode is a vertical stainless steel rod of small diameter covered with a thin skin of Teflon. The second electrode is either the wall of the extractor, or in the case of small dielectric constant difference between the phases, a concentric perforated metal sheath. In either level or interface service, the instrument is adjusted to produce a zero signal when the light phase is at the lower end of the electrode. An ideal application is the interface between organic and aqueous phases, because of the very large dielectric constant difference between the two. For reliable service, the configuration should produce a change of not less than 10 pF from zero to 100% of range.

4.3.3. Final Control Elements

In general, a final control element in an extraction column control system will control the flow of a process fluid or a heating or cooling medium.

For liquids being delivered by a dedicated pump, control of the pump speed may be employed to vary the flow rate. Electronic motor speed controls are now available at reasonable cost for use with ordinary alternating-current (AC) induction motors. With energy costs ever on the rise, this method should be considered when the flow is large and energy saving may warrant the extra investment.

A control valve is the other alternative and operates on the principle of introducing a variable hydraulic resistance in the flow path, lowering the potential energy of the fluid entering it. For detailed expositions, the reader is referred to Refs. 3 and 4. The most common construction in the smaller sizes, 3 in. (75 mm) and under, is that of a globe valve. For reasons of economy, one should consider butterfly and rotating plug designs for larger sizes. Most piping specifications call for flanged connections. Considerable savings can be realized by utilizing between-flange flangeless body designs, in which the valve body is secured between two pipe flanges with replacement of the flange bolts by long studs.

Correct sizing will preclude difficulties during operation. The valve will function adequately if it is assigned no more than 15% of the total friction drop of the system. A reasonable rule is to size the valve for 110% of maximum required flow or 160% of normal flow, whichever is greater.

5. SYSTEM INSTALLATION, CALIBRATION, AND STARTUP SYSTEM MAINTENANCE

An adequate set of specifications and installation drawings should be provided to the contractor who will do the installation and prestart-up checkout and calibration of the instrumentation. If these cannot be provided with the technical talent available, responsible instrumentation contractors can provide engineering for the job as well.

Of prime importance are the drawings necessary for instrumentation piping. One such detail for a flowmeter is illustrated in Fig. 8.

Avoidance of damage to the instrumentation during construction and testing is another concern that must be addressed. A safe storage area

ORIFICE STEAM FLOW MEASUREMENT HORIZONTAL.	APPROVED: STANDARD DESIGN PRACTICE
	DATE JUNE 1979 PAGE 1 OF 1

SUPERSEDES IN-100

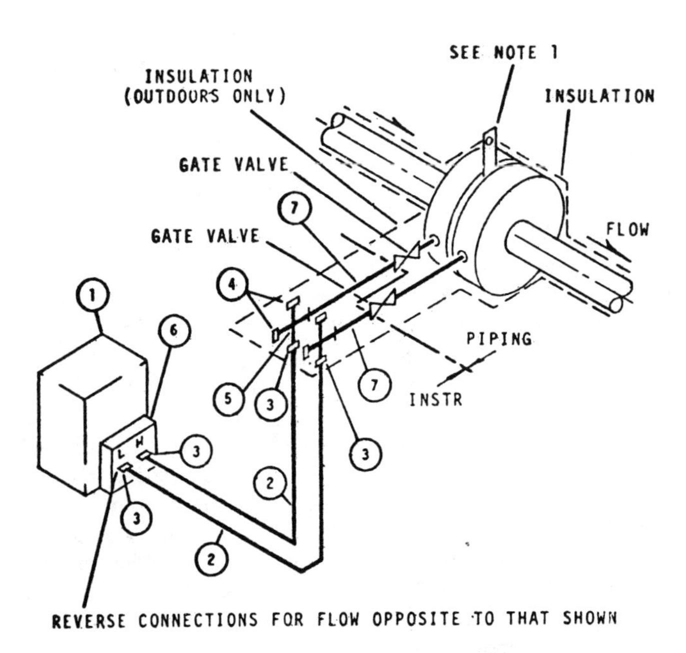

[1]

NOTES

1. INSTALL ORIFICE PLATE WITH PADDLE IDENTIFICATION MARKING (SHARP EDGE) ON UPSTREAM (HIGH PRESSURE) SIDE.

2. FOR OUTDOOR INSTALLATION FILL LINES FROM CROSS TO CELL WITH DIBUTYL PHTHALATE. IF THIS FILL IS NOT AVAILABLE, USE 75% ETHYLENE GLYCOL AND 25% WATER MIX. REPLACE EACH FALL.

3. REFER TO INSTRUMENT LISTING AND 07 301 FOR PIPE AND TUBING MATERIAL SPECIFICATION REFERENCES SIGNIFIED BELOW BY - XXX.

ITEM	MATL CODE	QUANTITY	DESCRIPTION
1	–	1	DIFFERENTIAL PRESSURE INSTRUMENT
2	203-XXX	AS REQD	1/2" (12 mm) OD TUBING
3	208-XXX	4	1/2" (12 mm) OD TUBING X 1/2" (DN 15) MPT FITTING
4	107-XXX	4	1/2" (DN 15) PIPE PLUG
5	105-XXX	2	1/2" (DN 15) PIPE CROSS
6	100-XXX	1	3 VALVE MANIFOLD (SUPPLIED WITH INSTR)
7	103-XXX	AS REQD	1/2" (DN 15) PIPE

Figure 8. Flowmeter piping.

should be provided, physically separate from that used for piping and equipment. During pressure testing, line flushing, and cleaning, instrument isolation valves should be shut. In-line devices such as turbine meters that could be damaged should be removed from the line and replaced with spool pieces. Orifice plates are normally installed after line cleaning.

Even though the instrument system had been checked out and bench calibrated at mechanical completion, a thorough functional check should be made again by instrument technicians prior to startup. At this time, seal fluids are added. Pre-start-up zero checks and preliminary controller settings are made in accordance with tabulated data provided to the technician. During startup, technicians should be available for controller tuning, zero and calibration checks with actual process fluids, and troubleshooting, should malfunction occur. Periodic maintenance during plant turnaround periods is highly recommended. If in-house service is not possible, contract maintenance should be provided.

NOTATION

DC	density controller
DT	density transmitter
FC	flow controller
FFY	flow : flow ratio device
FT	flow transmitter
LC	level controller
M	motor
pHT	pH transmitter
pHC	pH controller
S	set
R	reset
TC	temperature controller
TR	temperature recorder

REFERENCES

1. P. S. Buckley, *Techniques of Process Control*, Wiley, New York, 1964.
2. D. Campbell, *Process Dynamics*, Wiley, New York, 1958.
3. J. W. Hutchison, Ed., *ISA Handbook of Control Valves*, Instrument Society of America, 1971.
4. D. Considine, Ed., *Process Instruments and Controls Handbook*, McGraw-Hill, New York, 1957.
5. *Standard Instrumentation Symbols and Identification*, ISA-S5.1, R1981, Instrument Society of America.
6. F. G. Shinskey, *pH and pIon Control in Process and Waste Streams*, Wiley, New York, 1973.
7. D. Considine and S. Ross, Eds., *Handbook of Applied Instrumentation*, McGraw-Hill, New York, 1964.
8. B. Liptak, Ed., *Instrument Engineer's Handbook*, Chilton, Philadelphia, 1969.

28

ENGINEERING DESIGN CONSIDERATIONS FOR AN EXTRACTION PLANT

Harold H. Bieber and Robert Kern
Hoffmann-La Roche Inc.
United States

1. INTRODUCTION

In this chapter we outline the general principles and procedures used to develop the detail engineering for an extraction process plant. Generally, there are three sets of parameters that define an extraction processing system: (1) batch versus continuous processing; (2) single- versus multiple-stage separation; and (3) solid–liquid, liquid–liquid, or gas–liquid process. In recent publications [1, 2] the authors graphically depicted the alternatives with the above three parameters. Once the process concept is chosen, plant specifics, operating criteria, and economics determine the extractor to be used. For the most part, design considerations for an extraction plant are similar and not unduly sensitive to the combination of the three parameters just outlined. One exception is a plant that handles solids (leaching, decantation, digestion, etc.). However, this is a specialized and separate subject.

In the organic chemical industries, the most often employed extraction process is one involving two or more liquids or liquid–liquid extraction. For liquid–liquid extraction operations, the three most commonly used extractors are the mixer–settler; the multistage, mechanically agitated or pulsed column; and the high-speed rotating contactor of the Westphalia or Podbielniak type. Batch or continuous processing and staged separations can be accomplished in any of these units. Development of engineering design in the sections to follow concentrates on liquid–liquid extraction processing employing these extractor types.

2. PROCESS DESIGN

2.1. Process Selection

As with any design, the process selected should be the simplest and most reliable operation that can be built and operated at minimum cost. Generally, liquid–liquid extraction is used when a multicompound feedstock must be separated into two or more components, one of which is the valuable product. Extraction becomes increasingly attractive as the difficulty of the separation increases. Liquid–liquid extraction operations are usually carried out in relatively dilute solutions with two or more solvents. Solvent selection greatly influences the degree of separation and thus must be thoroughly investigated if the process and overall economics are to be optimized. Capital costs, as well as the utility costs, are sensitive to the solvents employed. Higher solvent selectivities for the valuable solute result in smaller process equipment trains and lower operating costs. Although our knowledge and understanding of extraction operations has increased significantly in recent years, the designer is still faced with two constraints: (1) the process must be thoroughly studied in the laboratory and pilot plant; and (2) to ensure operability, a complete understanding of system stability is required and must be developed. For example, the presence of trace impurities could markedly change the selectivity of the system and/or the hydraulics that affect equipment performance. The designer who knows the effect of impurities on the process can design operations that are reliable. In addition, the designer must fully understand the factors influencing scale-up from laboratory and pilot plant data. Otherwise, the commercial plant could be plagued with start-up problems and will not meet expected performance. In short, design must be based on well-conceived and executed experiments coupled with the designer's understanding of process variability and its effect on plant performance.

2.2. Process Parameters

The proper selection of the solvent system to achieve the desired separation and to ensure that the process is relatively stable to normal daily variations experienced on a commercial scale is the key to success. Laboratory data also are used to develop the system sensitivity to variations in process parameters such as flow, temperature, pressure, pH, and concentrations.

The effect of trace impurities and minor variations in concentration on the distribution coefficients are important to the analysis of the stability and flexibility of an extraction process. Thus the uniformity of the feedstock and the "cleanliness" of the solvents and the feed are critical to the operation. If daily variations in interfacial tension and concentration are experienced, the degree of separation and the throughput could become unreliable.

2.3. Process Equipment

Compatible materials of construction for the extractors as well as the auxiliaries in the plant are essential. Trace metal contaminants have been known to alter the extraction operation significantly. Care also must be taken so that no foreign solvents or liquids from pump seals, gearboxes, grease fittings, or other components bleed into the system, as these impurities also could seriously impair the operation.

2.4. Process Control

The control of the feed and solvents flow rates, coupled with the ability to monitor the compositions of all streams entering and leaving the extractor, are of paramount importance to the extraction operation. In a countercurrent fractional liquid extraction (with feed introduced into the middle of a column and solvents introduced at both ends), all column feeds usually are controlled by flow. Ratio controlling the solvents or the solvent to the feed are alternatives. If pH control is important to the separation, care must be taken to select a system that ensures a steady flow of the pH controlling reagent. In the separation of complex organic compounds, the ability to analyze both the incoming and outgoing extract and raffinate streams quickly and accurately enhances control of the operation. The designer must strive for a simple and rapid way of determining the degree of separation in the extraction operation. On-line analyzers, if feasible and reliable, replace the laborious task of frequently sampling and analyzing the streams to determine performance. Column hydraulic stability is dependent on the reliability of the interface controller. If the solvent phases are relatively clean, a conventional interface controller such as a float, conductivity or resistance probe, or a simple differential pressure (dp) cell will prove adequate. If there is a tendency to build up an emulsion or "rag" at the interface, a side stream withdrawal from the interface must be provided. This stream could be clarified and returned to the column, discarded if it is of sufficiently small quantity or recycled to the appropriate solvent recovery unit.

3. PLANT LAYOUT

An extraction plant is comprised of the following modules: (1) extraction equipment; (2) solvent and products recovery; (3) product purification; (4) auxiliaries; and (5) storage facilities for feed, solvents, chemicals, and products.

The designer who develops the extraction plant equipment and piping arrangement must give equal attention to all process, utility, and auxiliary equipment. The extractor itself (important as it is) is only one link in the process train. An extraction process flow scheme is shown in Fig. 1.

Basic layout principles are considered as they apply to the components of an extraction plant and as they relate to the requirements of process, operation, safety, maintenance, and construction. A conceptual treatment is given here, and more specific details are available in the literature [3, 4].

3.1. Extractors

From a plant layout standpoint, extractors can be grouped as follows: (1) horizontal centrifugal types; (2) vertical centrifugal types; (3) extraction columns (packed columns and columns with mechanical agitation); and (4) mixer–settlers. Space, handling, and access requirements differ for each type and are considered separately.

3.1.1. Horizontal Centrifugal Extractors

Centrifugal extractors of the Podbielniak and Quadronic type require overhead space for cover removal. The cover is rather large and heavy. Since this type of rotating equipment usually is located in a sheltered area, an overhead trolley beam running over the extractor center line and hoist can be provided. The minimum vertical space from floor to underside of the trolley beam should be roughly twice the height of extractor. However, this overall dimension should be carefully verified so that the vertical dimensions of hoist, hook, sling, and cover removal clear the machine. The trolley beam usually is designed so that it extends the full length of the extractor and over the laydown space, where the cover can be stored.

Piping to the extractors should be designed so as not to interfere with cover and internals removal. The valves at each machine inlet and outlet should be accessible. Break flanges should be provided where the piping must be removed to provide maintenance access.

3.1.2. Vertical Centrifugal Extractors

The bearings of vertical centrifugal extractors are located in the bottom of the machine (Westphalia, Alfa-Laval). To replace bearings, the internals usually are removed as one unit from the outer shell. Therefore, piping connected to the top cover must be removed. For easy internals removal, a trolley beam should be provided over the machine.

When two or three extractors are arranged in a row, each should have a separate trolley beam (Fig. 2*b*) with removal space in front of the machine. A common trolley beam over a line of extractors is not practical (Fig. 2*a*) since lifting height is increased, and piping and valve removal is necessary at the adjacent machine. Permanent hoists are not necessary because this type of maintenance is not frequent.

3.1.3. Column-Type Extractors

The piping to column-type extractors is similar to distillation column arrangements, which are discussed in a later section.

Spray, sieve-tray, and packed-column extractors have no moving parts. For these types, manholes are provided for removal of the internals. Packed columns require space in the layout for filling and removing the packing. Top filling and bottom removal necessitates lug or leg supports that provide open space at a bottom manhole. Side packing removal should be over a platform or at floor level. For adding the packing through the top of the column or manhole, a floor level or a 360° platform is recommended.

Agitated column extractors that have rotating (Scheibel, Oldshue-Rushton, Kuehni, and Rotating Disk) or reciprocating internals. A motor is located at the top or bottom, driving the shaft at the center of the column through a gearbox. In a top drive arrangement, the motor is supported on the column. A floor or 360° platform

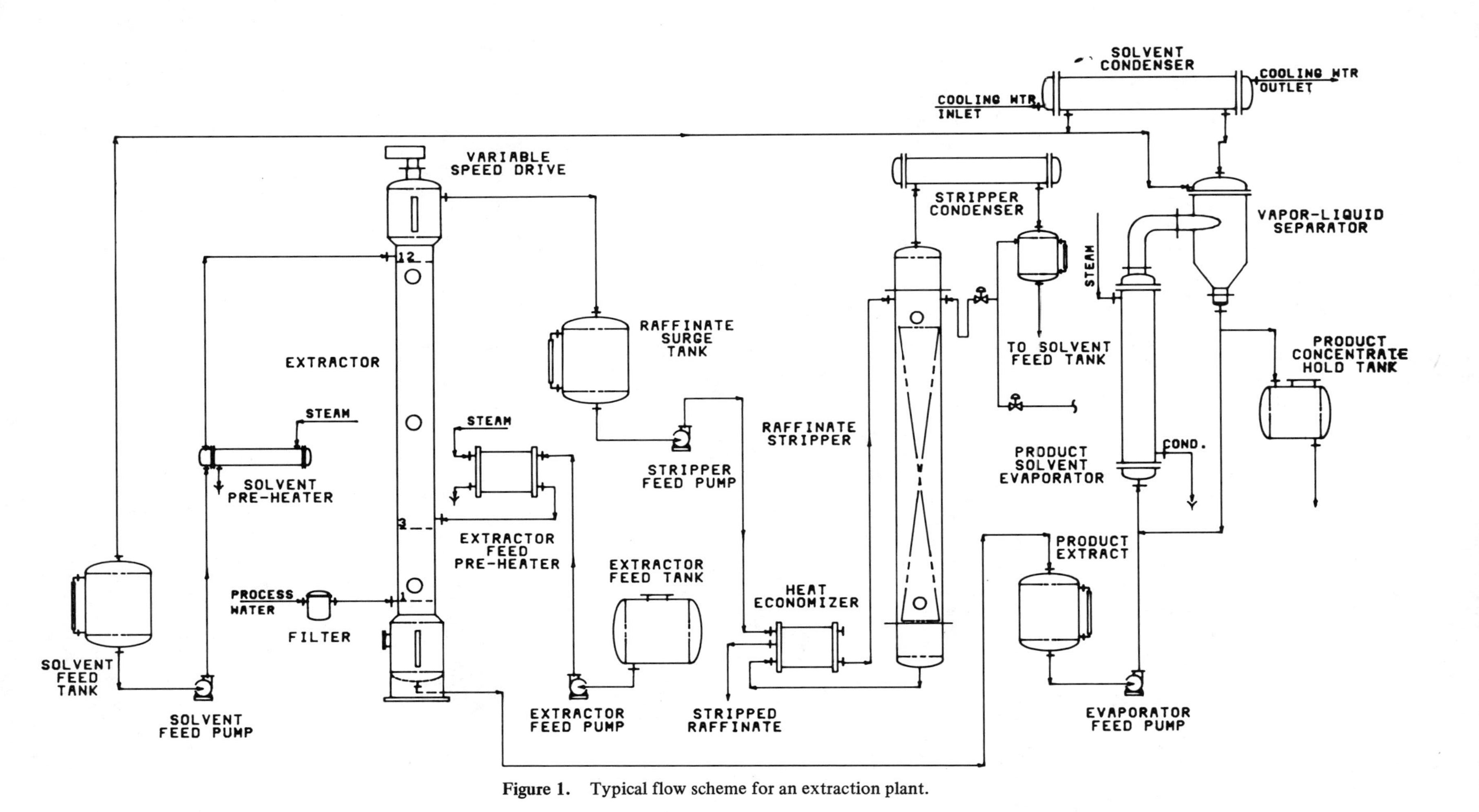

Figure 1. Typical flow scheme for an extraction plant.

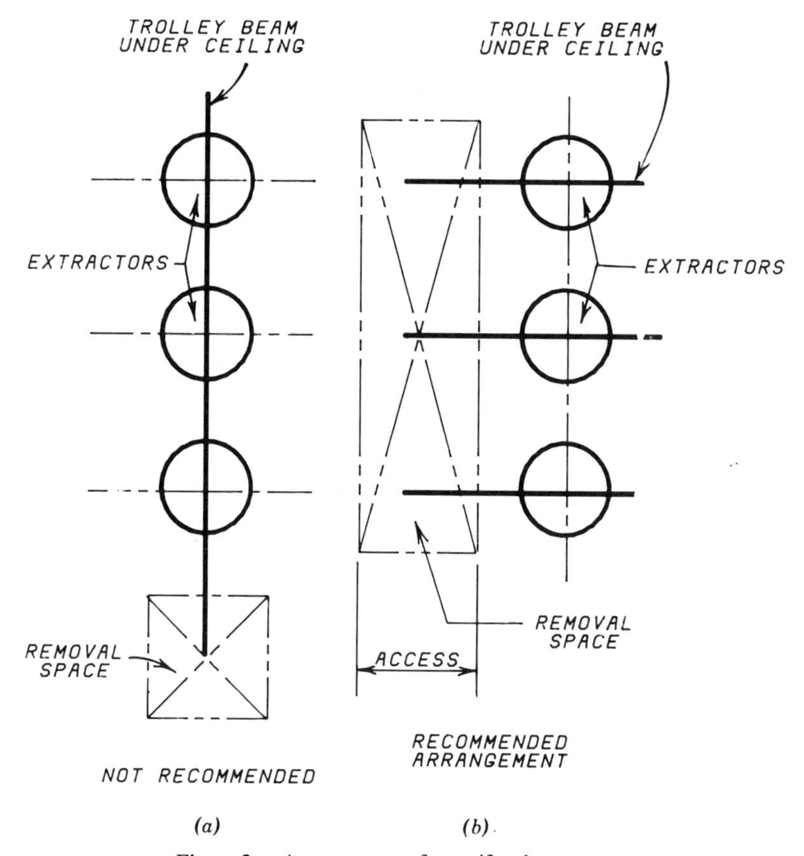

Figure 2. Arrangement of centrifugal extractors.

should be provided for ease of inspection and maintenance.

Extractors with pulsating agitation should be supported at grade, separate from building steel and footings. This will avoid the transmittal of undesirable vibration to the floor, surrounding steel, and adjacent piping. Space is required above the column for shaft removal. If the building height permits, a hitching point and hoist can accomplish agitator shaft removal. If the column is located close to the ceilings, a hatch is needed with handling provisions to accommodate an A-frame hoist on the floor or roof above. If mobile crane access is planned to the building, outside clearance, boom angle, and lifting heights must be considered.

3.1.4. Mixer–Settlers

Single and multistage settler units are considered from layout viewpoint. In the one stage, single-unit horizontal mixer–settler, mixing takes place at one end, the dispersed liquids flow horizontally, and separation takes place in the designed calming zone, which usually is located at the opposite end of the vessel. For in-place maintenance, open space must be left for cover removals and facilities provided for mixer removal. A trolley beam parallel and over the center line of this unit normally is adequate. If the settling zone is fitted with internals to improve phase separation (wire mesh), removal space for internals must be provided in the plant arrangement.

The multistage mixer–settler has several units in line. Vertical removal space should be provided under a trolley beam located over the mixer motors. Care must be taken to arrange convenient access to associated pumps, instrument panels, valves, and local instruments. Removal space, recommended by the manufacturers, should be provided.

3.2. Solvent and Product Recovery

In the conventional liquid–liquid extraction process, two streams leave the extractor, a raffinate containing dissolved impurities and/or by-

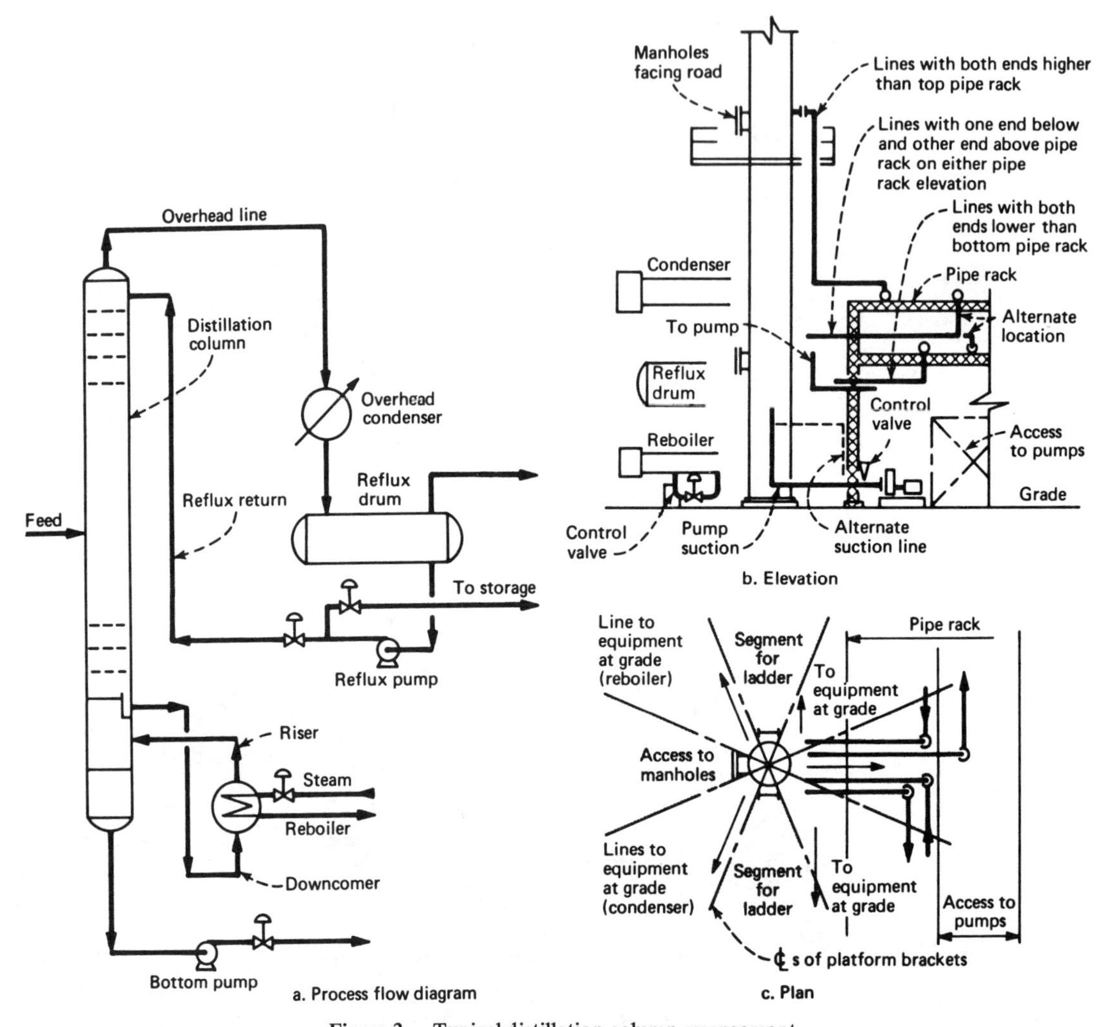

Figure 3. Typical distillation column arrangement.

products and an extract stream containing the desired product dissolved in the selected solvent. The processing steps following the extraction usually are concerned with one or more of the following operations: (1) chemical treatment; (2) solvent evaporation; (3) solvent stripping; (4) solvent purification; (5) product purification (fractional distillation; high vacuum distillation; and crystallization, isolation, and drying). Layout concepts and principles for these equipment trains are not unique to extraction plants but are common to most all chemical manufacturing plants. Short abstracts of layout considerations for three of the most frequently employed unit operation process trains (distillation, evaporation, and continuous vacuum distillation) are discussed.

3.2.1. Distillation

Probably the most common unit operation is the concentration and recovery of solvents and solutes by continuous distillation or stripping. Typical arrangements of the column and its associated condenser and reboiler are shown in Fig. 3. Layout concepts for condensers, reboilers, and other heat-exchange equipment are discussed in Section 3.2.2.

The principal features, manholes, tower platforms, and pipe runs of a tower are shown schematically in elevation in Fig. 3*b*. Nozzle elevations are determined by process requirements; and manhole elevations by maintenance requirements. For economy and easy support, piping should drop or rise immediately on leaving the nozzle and should run parallel, and as close as possible, to the tower itself. The horizontal elevation for piping (after the lines leave their vertical runs) is governed by the main pipe rack elevations. Lines that run directly to equipment at grade, more or less in the direction of the main pipe rack, often have the same elevation as the pipe rack.

Reboiler-line elevations are determined by the drawoff and return nozzles, and their orientation is influenced by thermal-flexibility considerations. Reboiler lines (and overhead lines) should be as simple and as direct as possible.

The plan view (Fig. 3*c*) of a tower shows the segments of its circumference allotted to piping, nozzles, manholes, platform brackets, and ladders in a pattern that usually leads to a well-designed layout. The entire circumference of a tower theoretically is available for arranging these components.

An elevated condenser is more convenient from the standpoint of tower piping layout because the large overhead line leaves the tower at a high level and crosses directly to the condenser. This opens a segment at lower elevations for piping or for a ladder from grade to the first platform. The availability of a nearby supporting structure or the ability to mount the condenser on the column usually dictate the elevated location. Whether the condenser is at grade or at an elevated level, flexibility, thermal load, and support connected with the large-diameter overhead lines must be considered.

The relief valve protecting the tower usually is connected to the overhead line. A relief valve discharging to atmosphere should be located on the highest tower platform. In a closed relief-line system, the relief valve should be located on the lowest tower platform above the relief system header. This will result in the shortest relief valve discharge leads. The entire relief-line system should be self-draining.

3.2.2. Heat Exchangers

Reboilers associated with distillation columns, liquid–liquid exchangers arranged in flow sequence with the process train, and evaporators (rising- and falling-film, vertical, shell-and-tube units) are the principal types of heat exchange equipment encountered in extraction plants. Layout and piping considerations unique to these pieces of equipment are described.

(a) Reboilers. In thermosyphon reboilers liquid flows from the tower trapout boot through a downcomer pipe into the bottom of the exchanger. The liquid is heated, leaves the reboiler in the return piping as a vapor–liquid mixture, and flows back to the tower.

In vertical reboilers heating usually occurs on the shell side. In horizontal reboilers, heating is on the tube side. Piping to reboilers is designed as simply and directly as possible within the limitations of thermal-expansion forces. Symmetrical arrangements between the drawoff and reboiler-inlet nozzles, as well as between the reboiler outlet and return connection on the tower, are preferred to have equal flow in the reboiler circuit. A nonsymmetrical piping configuration also is acceptable for a more economical or more flexible piping design.

Most reboilers are at grade next to the tower, with center line elevations of about 3–5.5 ft above ground level for exchangers 24–36 in. in diameter. Exchangers at grade provide economical arrangements—valves and instruments are accessible, tube-bundle handling is convenient, and maintenance is easy. In this arrangement the static heads are well determined between the exchanger's centerline and the drawoff and return nozzles on the tower. Vertical reboilers usually are supported on the distillation column itself.

(b) Condensers. A gravity-flow condenser is located above the level of the terminating point of the condenser's outlet line (see raffinate stripper condenser in Fig. 1). The static head pressure difference, between the vertical overhead line and the condenser's outlet line, must be equal to or greater than the sum of (1) the pipe system resistance, (2) exchanger pressure drop, and (3) required pressure difference across a control valve.

The required distance between fractionator inlet and exchanger center line (or bottom of exchanger) gives the required static head pressure differential.

Control valves in gravity-flow condenser systems should be located at a low point of the return line and product stream. A seal loop is provided to prevent a reversed flow of vapor in the condenser's outlet line.

In a pumped reflux configuration (Fig. 3*a*) the condenser is located close to grade. In layout design, usually the reflux drum elevation is established first in accordance with the required net positive suction head (NPSH) of the reflux pump. The condenser then is located to satisfy

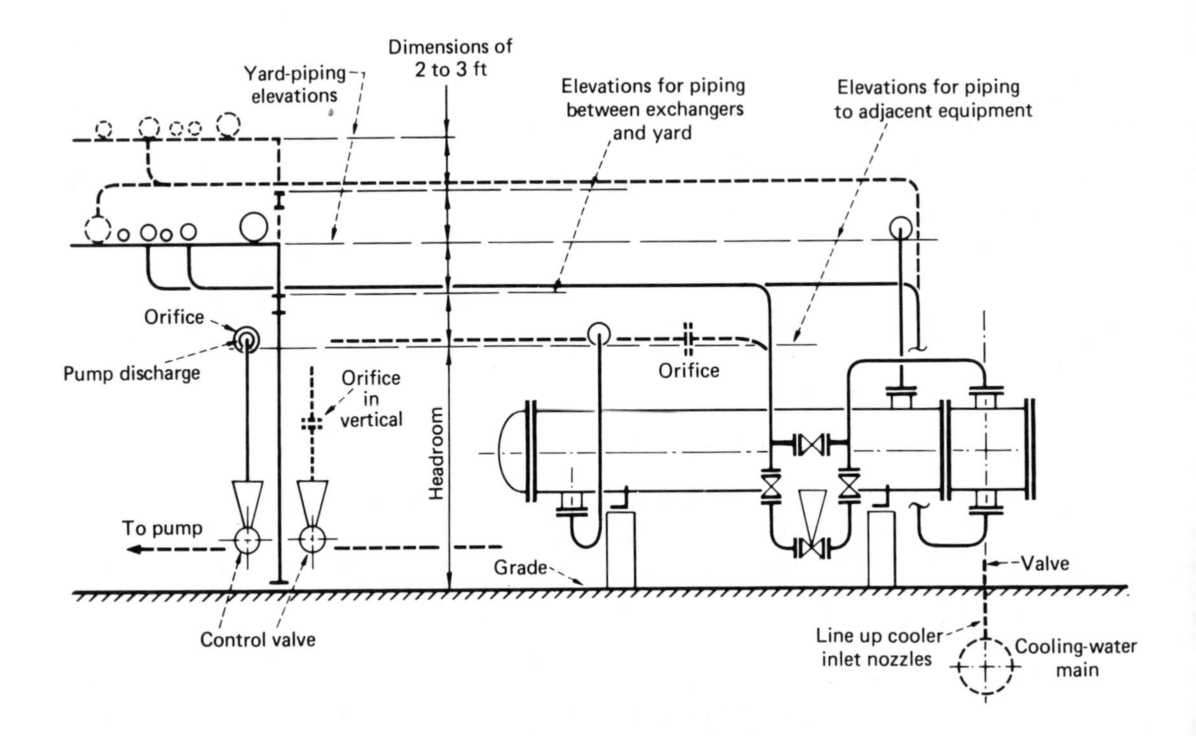

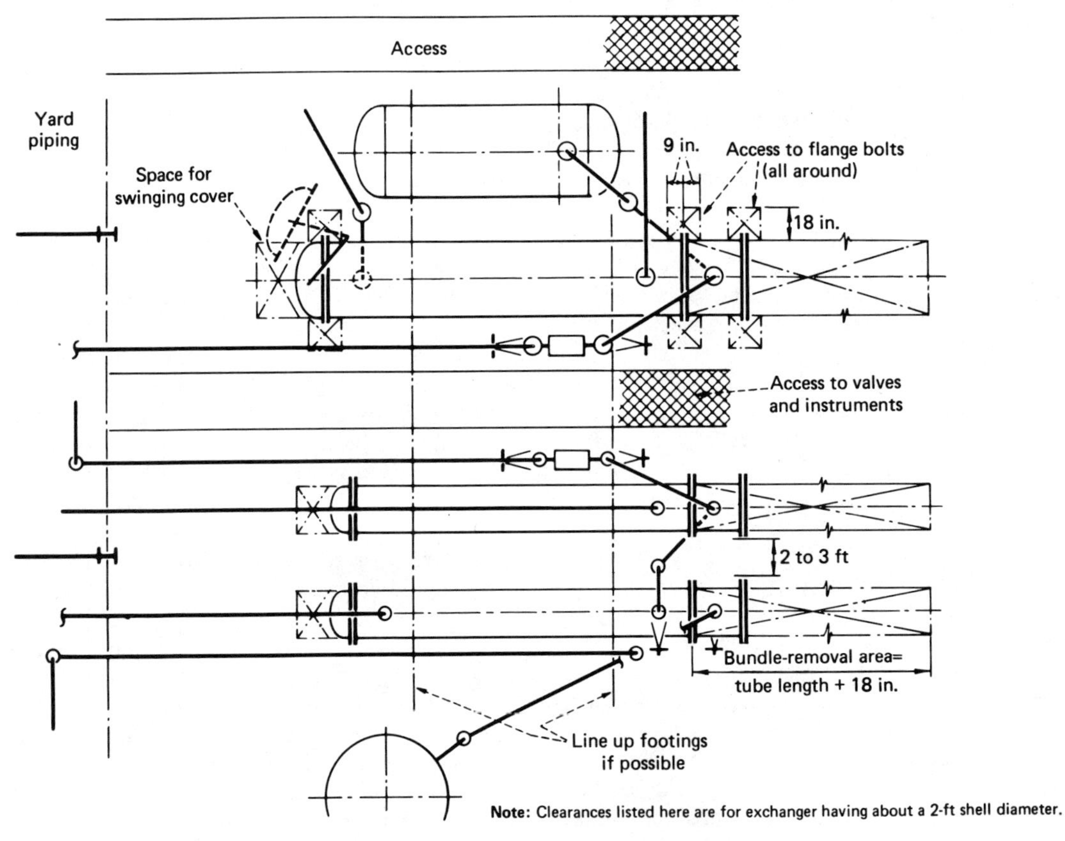

Figure 4. Typical exchangers and piping arrangement.

hydraulic conditions and to optimize piping and structural costs.

(c) Liquid–Liquid Heat Exchangers. Figure 4 shows a horizontal heat-exchanger arrangement in plan and elevation. The overall layout of the process equipment consistent with process flow usually determines the location of the exchanger and influences the arrangement of exchanger piping and access. The channel ends of exchangers face the main access. The shell cover usually faces the pipe rack. Service piping to and from the unit should be direct, with access to valves and instruments.

Locally mounted pressure and temperature indicators on exchanger nozzles, on the shell, or on process lines should be visible from the access aisles. Similarly, gauge glasses and other auxiliaries on exchangers should be visible from this aisle.

Excessive piping strains on exchanger nozzles from the actual weight of pipe and fittings and from forces of thermal expansion should be avoided.

(d) Evaporators. Rising-film, vertical, and shell-and-tube evaporators commonly are designed for first-cut solvent recovery operations in extraction plants. The evaporator is close coupled with a vapor–liquid separator, condenser, and solvent concentrate pumps. Where the solutes are heat sensitive, a falling-film evaporator is employed. Figure 5 illustrates these two evaporator configurations. Plant arrangement configurations are similar to that described in Section 3.2.3.

3.2.3. Continuous Vacuum Evaporation

In those cases where the final product is heat sensitive, viscous, and/or has high boiling points, a wiped-film vacuum evaporator of the Pfaudler design (spring loaded wall wipers) or the Luwa type (fixed wall to blade clearance) may be employed.

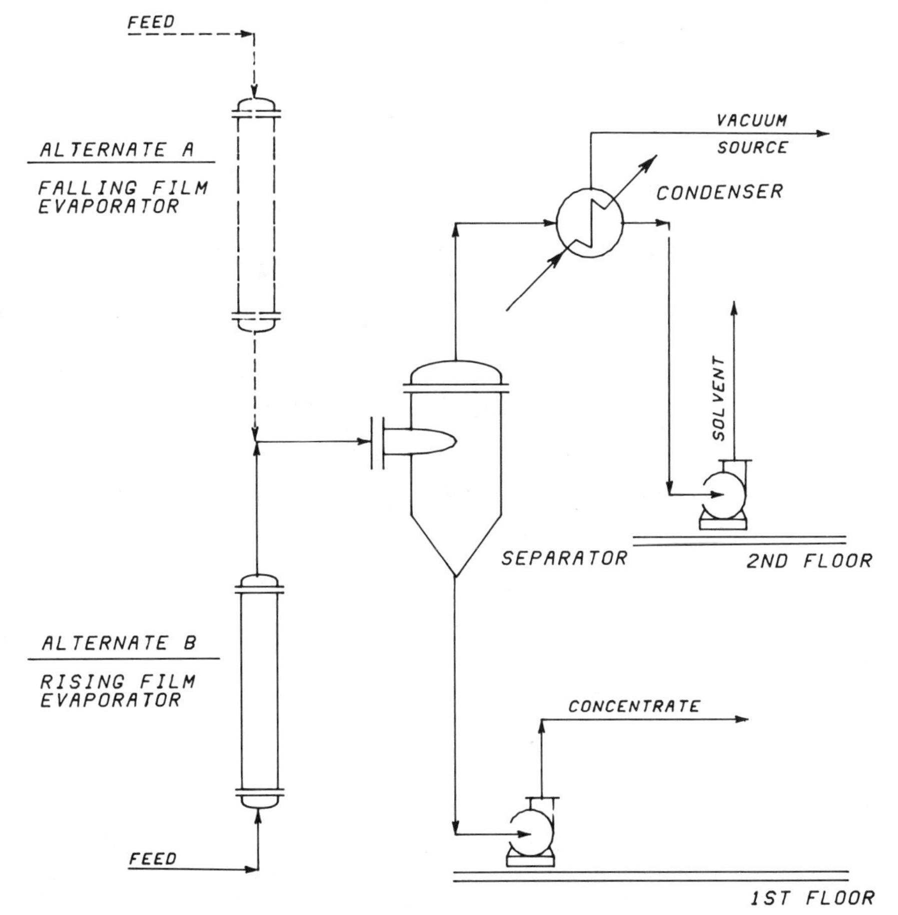

Figure 5. Solvent recovery arrangements.

Figure 6. Typical wiped film chemical processing system.

The Pfaudler thin-film, wiped-film evaporator (see Fig. 6*a*), consists of a jacketed cylinder for heating, usually four spring-loaded wiper blades 90° apart, and a louvered entrainment separator concentrically arranged to the cylinder wall. Vapors pass through the entrainment separator and are cooled with a U-tube bundle located at the lower extended section in or under the evaporator shell. The low-speed wiper rotor is driven through gears and shaft by a relatively small motor.

Figure 6*b* shows a typical wiped-film evaporator installation with its associated auxiliaries and residue pumps. Steam jet vacuum pumps are shown as the vacuum source but could be replaced by an equivalent mechanical pumping system.

A conceptual vertical arrangement also is indicated on Fig. 6*b*. The wiped-film evaporator is placed through the floor, supported on lugs resting on floor beams. Static head elevation requirements to the distillate and residue pumps, plus the wiped-film evaporator's vertical dimensions, will set the minimum second floor elevation. Re-

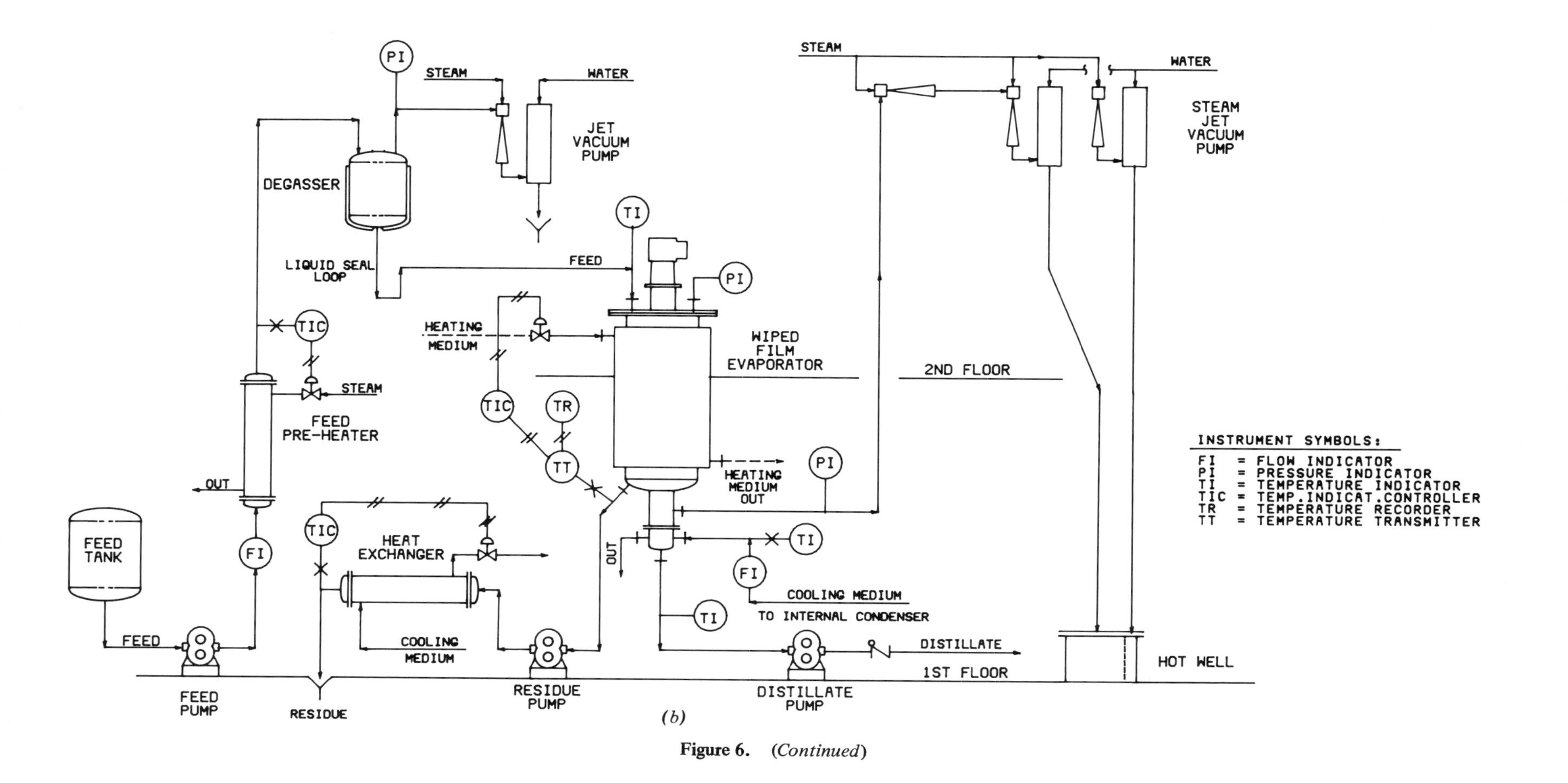

(*b*)

Figure 6. (*Continued*)

moval space and clearances for the bottom condenser on the first floor and for the wiper rotor removal on the second floor must be provided. A trolley beam or hitching point above the wiped-film evaporator eliminates expensive future rigging for rotor removal.

The feed line from the degasser to the wiped-film evaporator is usually a gravity-flow line. Piping configuration sets the minimum elevation difference.

A minimum of 34-ft elevation is required from the hot well at grade to the bottom of barometric condensers in the steam jet vacuum pump assembly.

Accessibility from an operating and maintenance viewpoint must be considered as with any well-designed processing unit.

3.3. Feed, Solvents, and Product Storage

Storage tanks (vertical or horizontal) are normally grouped. Nonflammables are located in paved and curbed areas, whereas flammable chemicals in maximum 5-ft-high dikes. The dike volume is at least 10% greater than the enclosed tank volumes. Specifications and safety standards determine dike requirements. Dike and curbed areas are sloped to one corner and drained through a foot valve into a suitable collection system. Ladders are provided over dike walls.

Discharge or drawoff valves are located close to the tanks. Transfer pumps are either within the curbed area or in the case of flammable solvents, outside the dike with suction lines penetrating the dike wall. Generally, pumps (and tanks) are lined up for convenient access to pump valves, control valves, instruments, and motor starters. A pipe rack usually is required over or parallel to the pumps for process and utility lines, electrical cables, and instrument lines.

3.4. Overall Plant Arrangement

Whether storage, extraction equipment, and recovery and purification equipment are grouped together or are separated by a distance, a pipe rack connects the pieces. The pipe rack contains process lines in and out of the battery limits; all interconnecting process piping; and utility headers supplying steam, air, gas, and water to process equipment. Instrument lines and electric supply conduits are also in the pipe rack. If convenient, heating, ventilating, air conditioning (HVAC) ductwork also can be supported on the building pipe rack.

Most of the time the pipe rack is overhead, and usually access aisles are provided under the pipe rack. The economy of piping primarily depends on the length of the pipe rack. This necessitates a compact process sequence equipment arrangement. Equipment arranged on one side of a pipe rack will increase the pipe rack cost. Equipment arrangement with a central pipe rack will require less pipe rack length than a one-sided arrangement. A comparison is illustrated in Fig. 7.

The heavy lines around the process equipment

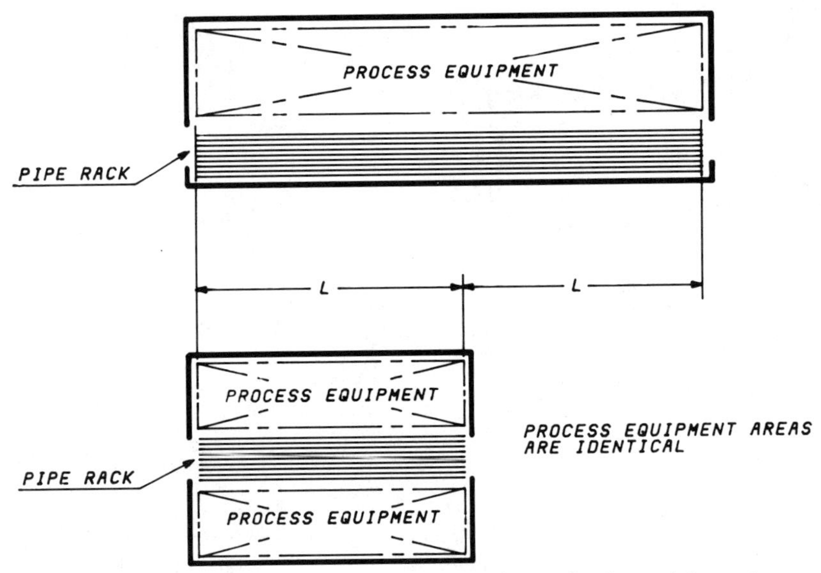

Figure 7. Compact equipment arrangements (bottom) reduces piping cost.

and pipe rack shown in Fig. 7 indicate the floor space if the whole extraction plant is housed. Completely enclosed buildings are expensive because ventilation and possible air conditioning are required. This represents substantial capital cost, and if climatic conditions permit, open-sided structures should be considered to reduce building costs. Storage, workup, and recovery equipment can be arranged outside, housing or roofing only the extraction equipment. This is desirable because of the occasional maintenance required for this equipment. If practical, equipment arranged outdoors at grade usually provides the most economic plant arrangement.

4. PROJECT EXECUTION

The design, construction, and startup of an extraction plant for the most part is identical to the procedures and methods employed for any chemical manufacturing facility. What is specific is that we are looking at a narrower and more specific type of plant, namely, one that is concerned primarily with unit operations rather than unit processes and with the handling of liquids. The phases of an extraction plant project are (1) process engineering, (2) concept development and arrangement studies, (3) detail engineering, (4) construction and mechanical checkout, and (5) startup. Figure 8 shows the overall project design activities diagram.

4.1. Process Engineering

The successful operation of an extraction plant is dependent on a complete understanding of the process and its sensitivity to upset as well as the ability to monitor and control the operation to maintain steady-state conditions. Therefore, it is essential that the operating and process departments work diligently to develop a thorough understanding of the process and to establish a firm basis of design. Although the aforementioned is the case for any process design, it is of paramount importance for an extraction process in that system performance almost always is based on laboratory and pilot-plant data. Thus a thorough understanding of the experimental data and their reliability can often be the difference between success and failure. With this word of caution, the information flow from the client to the process engineer and the design information produced and distributed to the design disciplines is outlined in the sections that follow.

4.1.1. Basic Process Design

Conceptual process design establishes a design basis and fixes the scope of the process plant. During this phase of work, the chemistry is reviewed and established, studies of alternatives to minimize capital and operating costs are carried out, operating and design philosophies are established, and all pertinent technical data are collated and reviewed. Typical activities during this phase of work are as follows:

1. List the approved process yields.
2. Determine plant capacity, operating basis, and future expansion capabilities.
3. Develop operating control philosophy.
4. Tabulate raw material requirements, specifications, availability, and storage requirements.
5. Tabulate intermediates, product and by-product specifications, and storage requirements.
6. Develop physical property data manual.
7. Study and analyze unusual process safety considerations including a review of all chemicals being handled in the process. Publish guidelines for the project.
8. Obtain utility design information.
9. Establish process design safety factors.
10. List materials of construction for raw materials, intermediates, and final product (piping, equipment). Review and establish piping specifications for the project.
11. Obtain guidelines for disposal of effluents and vent discharges.

4.1.2. Detail Process Design

The resultant of the process engineering activities outlined in Section 4.1.1 is the development of a conceptual process design package that includes process flow diagrams, equipment lists, utility requirements, effluent and vent tabulations, and layout requirements. After review and approval, detail design of the piping and instrumentation flow sheet (P&I) commences. These P&I flow sheets are the governing document for detail engineering and completely define the process plant. Line sizes are given, and locations of size and specification changes are indicated. All valves (including type) are shown with proper

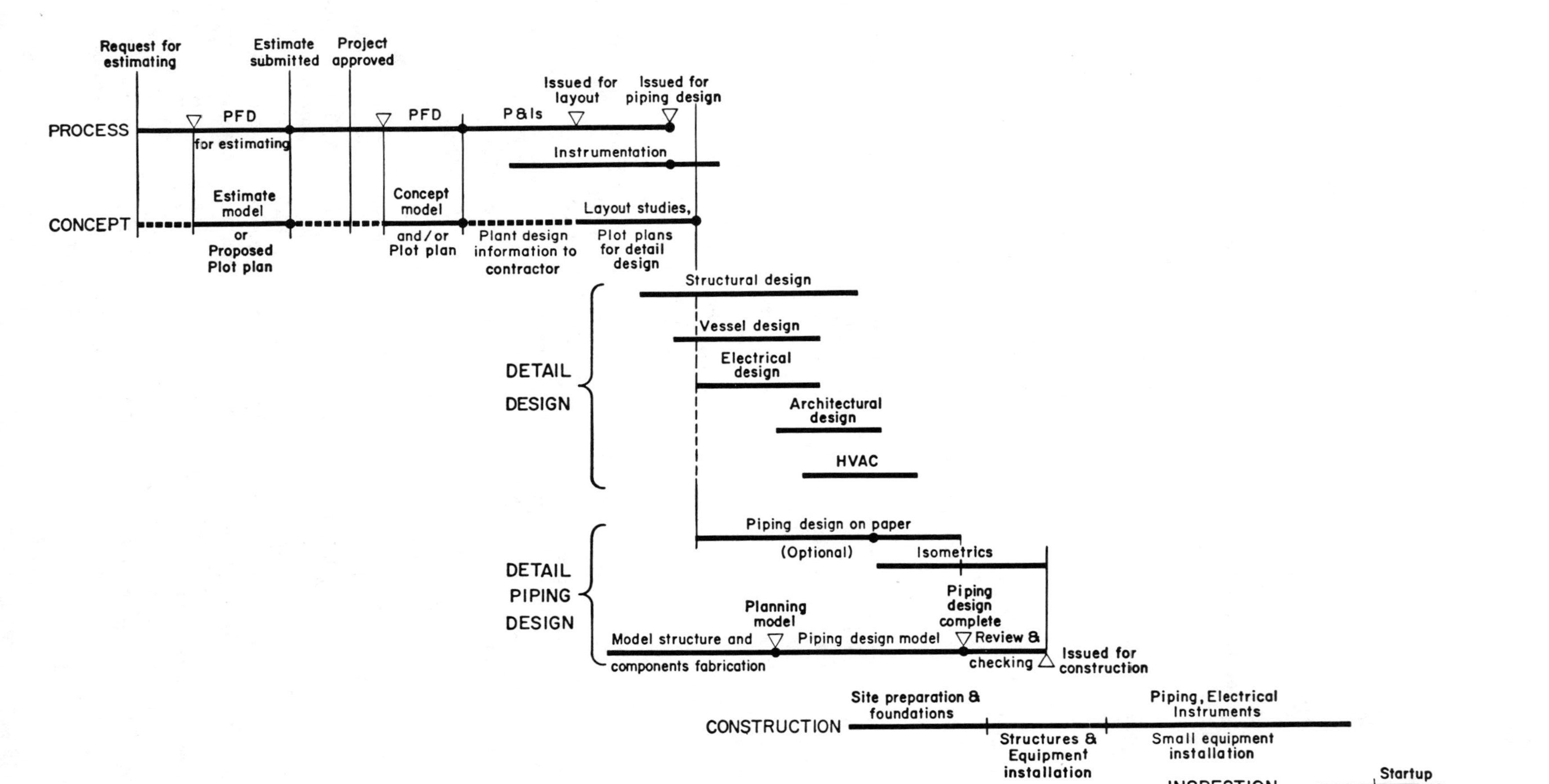

Figure 8. Overall project design activities diagram. Legend: PFD = process flow diagram; P&I = piping and instrumentation flow diagram; HVAC = heating, ventilating, and air conditioning; ● = formal approvals of basic designs.

locations in the pipelines. This is especially important for extraction plant operations, since flexibility is important to be able to cope with daily variability. Thus P&Is should note and detail:

1. Minimum-length lines or circuits.
2. Gravity-flow lines. Most of these lines are readily recognized. Dimensions of seal loops in gravity-flow lines, sloping lines, and pocket-free, gravity-flow lines should be indicated clearly on P&Is.
3. Special pipelines. Some pipelines have special features and limitations that must be taken into account during plant layout and piping design. Examples are glass piping having special fittings and support requirements; glass-lined, Teflon-lined, or other specialty piping that usually have flanged fittings and limitations for line length; and jacketed and steam-traced lines.
4. Location of critical instruments requiring frequent maintenance or calibration.
5. Location of sampling points.

As detail process engineering progresses, it is important to begin involving the concept and other design disciplines. Providing these people with the basic process concepts and design philosophy at an early stage allows them to crystallize their design thoughts for discussion at project review meetings throughout the course of the job.

4.2. Concept Design

The three major steps of mechanical-plant design development are shown in Fig. 8: (1) plant-design concept (concept design); (2) arrangement studies (design planning); and (3) detail design. These are not always sharply divided areas. Concept design is often absorbed in the planning of arrangement studies. Although equipment arrangements can be made in combination with piping layouts, these activities normally become separate and well defined when dealing with large and complex plants.

Extensive conceptual design is needed for the usually one-of-a-kind extraction plants. A great deal of coordination is required at this stage among the process, project, design, maintenance, safety, and operating disciplines. Essential process and plant-design requirements are established, an overall agreement is worked out, and design principles are determined.

Input data for concept design are (1) plant location, available space, layout constraints, and requirements, (2) operating and control requirements, control room and motor control center, and specific considerations, (3) process flow diagrams, including process equipment sizes and data, process flow concepts, and special piping–control requirements, (4) process descriptions and safety considerations, (5) laboratory, storage, and handling, (6) personnel and architectural requirements, and (7) project design data.

With this information, a layout concept is conceived on a model or plot plan. Horizontal and vertical equipment relationships are generated. A system of access, emergency-escape routes, and material-handling provisions is determined. Operating and control procedures, ease of maintenance, and constructability are investigated. Space allocation is provided for personnel activities (laboratories, offices, locker rooms, storage, etc.); control rooms, instrument panels, motor-control centers are planned; and plant site, structures and the building size are determined.

4.3. Arrangement Studies

Arrangement studies or planning studies are more detailed developments of concept design. These drawings are used for approval by all disciplines for final design. Planning studies are prepared so that all design sections will be able to work simultaneously and independently for producing the detail construction drawings. These studies also coordinate the requirements of specifications, operations, maintenance, and safety as well as the process and project data information to produce construction drawings.

A general outline of the information requirements needed for a typical extraction plant to produce layout studies is:

1. A good preliminary issue of the P&Is, specifications, and instrumentation data.
2. Vessel drawings with elevations, pipe connections, and insulation requirements.
3. Instrument connection details to vessels.
4. Heat-exchanger details.
5. Pump and mechanical equipment details and NPSH requirements
6. Weight and size of equipment that require handling.
7. Electrical requirements: motor-control center; main cable runs; junction boxes; and starters.

8. Instrument requirements: control house; panels; main instrument-line tray locations; and transmitter locations.
9. Building sizes and requirements for personnel facilities.

4.4. Detail Engineering

Pipeline and instrumentation flow sheets and approved arrangement studies are the two basic documents on which detail engineering is based. Whereas the P&Is specify the chemical or process requirement, the final arrangement study with associated mechanical standards and equipment drawings defines the basis for the plants mechanical design. Dimensions are given for equipment location; overall dimensions for structures; and routing of piping, electrical, and instrument lines. Generally, all equipment and piping components are shown that occupy floor space. Access, removal space, and handling facilities are also defined.

The piping department is supplied with the location of all equipment and nozzles, instruments, control valves, controllers and transmitters, valves and manifolds, orifice runs, and main pipe with supporting steel.

The structural department will receive overall structural dimensions and elevations, equipment weights, floor-loading design criteria, the location of all equipment footings, roadways and paving, trenches and culverts; pipe loads affecting structural design; platform, staircase, and ladder requirements; and fireproofing and equipment-handling requirements.

The vessel department receives the types of all vessel support; all nozzle, manhole, handhole, and platform-bracket locations; loads affecting vessel design; heat exchanger support; and davit positions on exchangers.

For the electrical and instrument departments, the instrument leads, cable runs in trenches or overhead, and switchboard and instrument locations are shown. The mechanical department will receive the location of HVAC, refrigeration air compressor, blowers, and ductwork locations.

The architects will have all the building requirements planned for detailing all the architectural features.

The output from all the disciplines are drawings for construction.

4.5. Construction and Mechanical Checkout

The principal functions of design engineering dis ciplines during the construction phase are to provide the necessary technical information and clarifications to keep the field work moving and to monitor the construction to ensure that the design intent is being followed. Mechanical checkout is usually a contractor responsibility but is followed closely by engineering. When the plant is judged to be mechanically complete, the following activities will have been accomplished:

1. Installation checked against P&Is. Startup strainers installed.
2. Piping water and/or pressure tested and judged to be tight. All lines flushed, drained, and dried where necessary.
3. All vacuum systems tight and meeting leak-rate criteria. Documentation on file.
4. Equipment inspected, cleaned, drained, and dried wherever necessary. All equipment internals in place and properly installed.
5. Mechanical equipment run in. (Optional; may be done as part of water runs.)
6. Electrical systems operational. Alarm settings established, rotating equipment checked for rotation, and interlocks reviewed and approved.
7. Instrument systems operational. Valves stroked and action checked, instruments calibrated, control settings reviewed and preset, interlocks set and reviewed for startup, and programmers stepped.
8. Utility systems activated (steam, cooling tower water, compressed air, hot oil, brime, process water, nitrogen, etc.).

4.6. Startup

The startup of a chemical manufacturing facility may be divided into four distinct phases as follows:

1. **Prestartup.** All activities prior to mechanical completion of the plant including planning and organization of the startup, training and participation in mechanical completion activities, development of operating instructions and associated control procedures.
2. **Water and Solvent Runs.** All activities relating to mechanical checkout and readying of the plant for the start of initial chemical trial runs.
3. **Initial Production Trials.** Starts with the first introduction of feed stocks. Con-

tinues stepwise as the plant is brought up to full design rate and held at the point for a predetermined length of time and/or batches.

4. **Closeout.** Monitoring–critiquing during steady-state production until major problems are identified and/or resolved and a summary report written.

4.6.1. Water and Solvent Runs

Water and solvent runs are the first on-line flow checks of the completed extraction plant. The major steps of this phase of the shakedown are as follows:

1. Calibrate pumps, weigh scales, and tanks. As previously discussed, constant flow rates to the extractor are essential to monitoring and controlling extraction performance.
2. Confirm pumping and other transfer operations.
3. Check agitators under load; mechanically run in rotating equipment.
4. Simulate operating cycles wherever possible and as completely as possible, including transfers, heating, cooling, vacuum operations, and pumpouts. Review safety interlock systems.
5. Test evaporations, distillations, and so on, using water wherever possible.
6. Review control systems and settings under water run operations.
7. Flush through, drain, dry, and inspect tested units. Begin removing temporary strainers. Cleanliness is essential for extraction plants.
8. Simulate upset and emergency situations.
9. Review and modify operating instructions, if necessary.

At the end of the water runs, confirmation of the mechanical integrity of the plant is obtained, all utility and instrument systems are operational, mechanical failures have been uncovered and corrected, and "minor" process and installation modifications deemed necessary for production have been made.

Following successful water testing, plant operations are simulated with the process solvents. By simulating process conditions, designs can be evaluated for the first time; system hydraulics verified; operating temperatures and pressures checked; and control systems set up, checked, and fine tuned. Flowing solvents through the extractor allows the designer and operator to observe the liquid dispersion and to verify the scale-up of the mixing. Startup, shutdown, and emergency situations also can be practiced under near "real" situations.

At the end of the solvent runs the system is drained, dried, and readied for product trials. The solvents should be repurified and an assessment made that there is no in-leakage of foreign substances (oil, grease, seal fluids, etc.) that could upset the extraction.

4.6.2. Initial Production Trials

The order and manner in which the first product trial runs are made usually are developed by the startup team. Product trial runs, wherever possible, are made with materials obtained from an existing plant. It is usual to start up at lower than design rates and to step up to design throughput in discrete increments.

At the end of the trial run period all equipment should have been run under operating conditions and major problems resolved, control systems should be operational and for the most part fine tuned, and operating personnel should be fully familiar with all phases of the plant's operations. Extractor performance and scale-up are verified during this phase of the startup.

In the final phase of startup all materials being processed in the new plant are synthesized in the new installation. When enough confidence is obtained that the new installation can meet production commitments, the new facility is brought up to design rates for steady-state trial production runs of 1–2 weeks. The plant is judged to be onstream when it is capable of producing specification product at the design rate.

REFERENCES

1. T. C. Lo, in P. Schweitzer, Ed., *Handbook of Separation Techniques for Chemical Engineers*, McGraw Hill, New York, 1979, Section 1.10.
2. K. H. Reissinger and J. Schroter, *Chem. Eng.* (November 6, 1979).
3. R. Kern, *Plant Layout*, Chemical Engineering Reprint No. 004-1978.
4. R. Kern, *Practical Piping Design*, Chemical Engineering Reprint No. 238-1977.

29.1

COST OF EQUIPMENT

Donald R. Woods
McMaster University
Canada

1. INTRODUCTION

Many different technically feasible alternatives exist to solve any particular problem. To select the "best" technically feasible solution and to optimize a process, we need the results of three economic calculations: (1) the financial attractiveness or the return on the capital investment; (2) the total amount of money required to build the plant or the fixed capital investment; and (3) the annual operating cost, including depreciation as an expense. The calculations are related because the financial attractiveness requires that we know the capital and the operating cost, and the operating cost requires the capital cost as one of its components.

Many books review how to calculate the different financial attractiveness criterion, including discounted cash flow (DCF), net present value (NPV), equivalent annual cost (EAC), simple return on investment (ROI), and payback time [1-4]. This is not discussed here; rather, the focus is on methods and data for estimating the capital and operating costs that are specific to solvent extraction.

Different Methods Needed for Different Stages in the Project

Many different estimates of the capital and operating costs need to be made from the beginning of a project until the "best" plant is in operation. Near the beginning of a project the estimates of these are order of magnitude and identify the general areas where we should be directing our attention. The errors in the estimates may be as large as +100% or −50%. Later as the feasibility of the project is explored further and alternatives are narrowed down to several, the cost and economic calculations should be within ±30%. Figure 1 illustrates how different methods are needed to estimate the capital requirement and implications of a project at different stages in the development.

2. CAPITAL COST ESTIMATION

The capital cost is the amount of money spent to build the plant. This includes the following items:

1. The cost of purchasing and installing all the necessary equipment into such a condition that it is operable. This includes the FOB (free on board the supplier) cost of all fabricated equipment; delivery costs, sales taxes, and duties; installed concrete; structural steel; electrical hookups; instruments; piping and insulation; and painting and all installation materials and labor.
2. The contractor's fees.

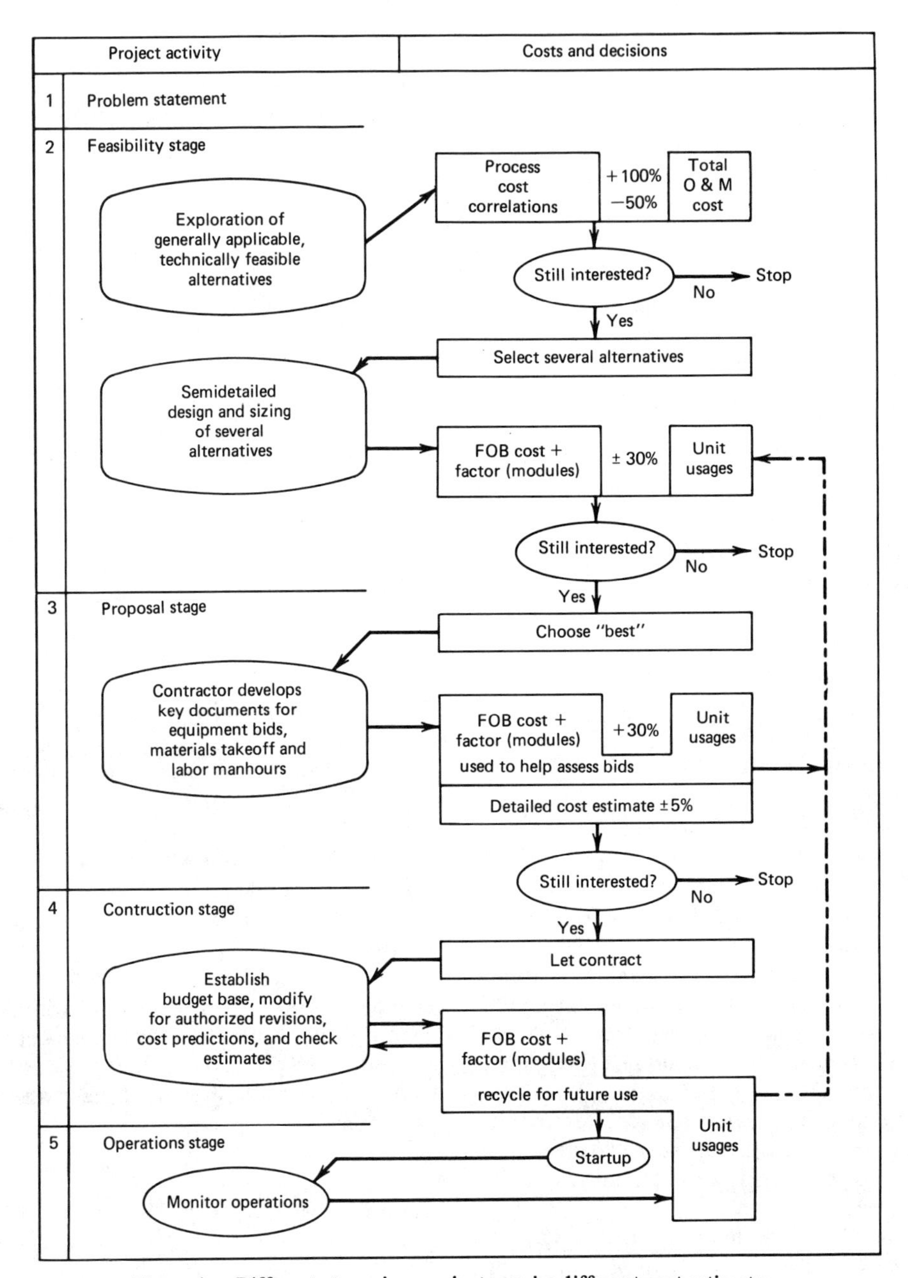

Figure 1. Different stages in a project require different cost estimates.

3. The engineering expense to design and prepare the drawings and for all home office expenses related to the project.
4. The field expense for supervision of construction and for construction facilities and security thereof.
5. The legal fees related to the project and its administration and financing.

Other items that need to be accounted for, but not necessarily in the preceding cost, are:

6. The buildings and the land.
7. The site preparation and yardwork.
8. The interest on the capital borrowed during the construction stage of the project.

9. The startup costs.
10. A contingency to allow for unexpected delays, strikes, or difficulties encountered during the construction. Naturally, a contingency factor exists only for estimates of uncompleted projects. Once a plant has been constructed, any contingency that was necessary would be allocated to the source of the expense.
11. Special architectural requirements.
12. Special land needs such as blasting, piling, or drainage.

The working capital is the money required to purchase the initial supply of on-hand chemicals, solvents, catalysts, or packings. Because this money is usually completely recoverable when the plant becomes obsolete and is shut down, this working capital is not considered to be fixed capital investment. However, because the total initial amount of capital required is the fixed capital investment plus the working capital, and because the solvent inventory can be expensive, the working capital should always be estimated. A general breakdown of these capital cost components is given in Table 1.

All equipment cost data must refer to a given time. Many different inflation indexes can be used to account for the effects of inflation from one time to another. Details are given elsewhere [4]. The Marshall–Stevens (or Marshall-Swift) (MS) index is probably the most reliable index to use for solvent extraction processing of all the indexes published in the open literature. Current values are reported in such magazines as *Chemical Engineering.* Such indexes can be used to update previous data to within about 3 months of the present time. To account for inflation in the future, an estimate is made of the anticipated annual rate of inflation. In the 1960s it was about 4% per annum; in 1974 it was 18% per annum; and in the late 1970s it was between 6 and 8% per annum in North America.

To use the MS inflation index to obtain the value at a "new" time, multiply the value at the given time by the ratio of the "new" to "given" values of the inflation index. For example, to estimate the cost of installation in 1976 of a plant that cost $100,000 in 1960, we multiply by the ratio of the MS in 1976 to the MS in 1960:

$$\$100{,}000 \times \frac{485}{238} = \$204{,}000$$

Although the MS index and U.S. dollars are used in this book, costs in other currencies and other times can be found by either (1) working in U.S. dollars and with the MS index to obtain costs at the *time* of interest and then using the currency exchange rates pertinent to that time or (2) converting U.S. dollars to the currency of interest and working with the inflation index pertinent to construction in that country and in that currency. Inflation indexes for Canada, the United Kingdom, the Netherlands, and West Germany are available [5, 6].

2.1. Order of Magnitude Estimates

Sometimes the capital cost is available for the complete process based on the cost of previously installed plants. Such costs normally include engineering, field expense, and contractors' fees. The costs normally refer to a battery limits (BL) installation in that the following costs are excluded: offsites such as steam raising; cooling water; utilities; perhaps waste water treatment; storage; and shipping and receiving. The costs include the solvent extraction and solvent recovery. The costs are correlated in terms of the feed or product flow rates by an equation of the form

$$\text{Capital cost} = \text{reference cost}\left(\frac{\text{capacity of interest}}{\text{reference capacity}}\right)^n \quad (1)$$

Table 2 summarizes the various data needed in Eq. (1) to estimate the BL cost for some extraction processes. All data refer to an inflation index value MS = 600. The costs in Table 2 are for only the solvent extraction section of a process.

The operating rate is the ratio of the number of 24 hr-days of actual operation per annum to the 365 calendar days. The daily stream–day capacity is converted to annual by multiplying by the product of the operating rate and 365.

The reference capacity is always in units of thousands of metric tons per annum, except where the capacity is stated otherwise. Then the column entry refers only to the stated units, and *not* to thousands of such units. In the n column, if the costs are available for only one size of plant, a best estimate of the exponent n is given in brackets.

Example. Estimate the BL capital cost at MS = 600 of an SX plant to recover 5000 Mg/a of crude naphthenic acid from spent caustic.

TABLE 1 TYPICAL COST COMPONENTS FOR THE INSTALLATION OF A CARBON STEEL SX BL UNIT

	Total Dollars, $	Labor Dollars, $	Material Dollars, $	
1. FOB equipment	1000		1000	
2. Freight, tax, duties	(80)			80 is neither L nor M
3. Subtotal "delivered" or purchased	1080			
4. Uncrate, inspect, haul to site, and hook up	450	450		
5. Subtotal "installed"	1530			
6. Piping	867	272	595	
7. Instruments	154	40	114	
8. Insulation	110	30	80	May not be needed
9. Painting	33	20	13	
10. Electrical	69	20	49	Low if mechanical agitation used
11. Concrete	158	60	98	
12. Structural steel	158	80	78	
13. Buildings	–			
14. Offsites	–			
15. Subtotal "labor plus materials" excluding item 2	2999	972	2027	L/M = 0.48; L + M = 2.99
2. Freight, tax, duties	80			
16. Subtotal "physical plant cost"	3079			
17. Engineering home office expense	277	9% of L + M		Values increase for small-sized projects
18. Field expense; supervision, temporary facilities, rental and services	616	20% of L + M		
19. Subtotal direct plant costs (or BM)	3972			
20. Contractors' fees	120	3–5% of 19		
21. Contingencies	400	10–15% of 19		
22. Subtotal fixed capital investment	4492			
23. Royalty				
24. Land	45	1–2% of 22		
25. Spare parts				
26. Interest during construction	270	6% of 22		
27. Legal fees	45	1% of 22		
28. Working capital: solvent inventory	450	10% of 22		
29. Other working capital: inventory of necessary feeds and products, etc.	450	10% of 22		
30. Subtotal total investment	5752			
31. Startup expenses	450	8–10% of 22		
32. Total turnkey Cost	6202			

Solution. From Table 2, the cost correlation is

$$\text{Cost} = 0.38 \times 10^6 \left(\frac{5 \times 10^3}{10 \times 10^3}\right)^{0.70}$$

$$= \$230{,}000 \text{ (U.S. MS = 600)}$$

2.2. Feasibility Study Estimates

Estimation of the fixed capital investment to within about ±30% requires that a flow diagram be developed and that all the main plant items (MPI) be sized. The procedure is:

1. Estimate the FOB cost of the MPI.

TABLE 2 ESTIMATION OF BATTERY LIMIT COST FOR VARIOUS EXTRACTION PROCESSES

Product	Process	Capacity, 10^3 Mg/a	Cost, $\$10^6$	Classification	Range	n	Error, %	Operating Rate, %	Comments
Copper	From acidic dump leach liquor containing 2 g/liter of Cu including 313 stainless steel mixer settlers, three extraction two stripping stages; including engineering, field expense, contractors fees; excluding inventory, offsites								
	(capacity: input leach liquor flow rate in liter/s)	100[a]	3.8	BL	60–900	0.70	10		
	As above at 1 g/liter × 1.2					(0.70)	30	90	
	As above at 30 g/liter	7[a]	0.4			(0.70)			
	(Capacity: output copper at 90% recovery for 2 g/liter feed)	10	50	BL	3.5–40	0.67	10	90	
	for 1 g/liter feed (capacity: output)	10	9.8	BL	2–10	0.67	30	90	
	From copper ore Australian conditions yielding 3 g/liter								
	(capacity: input leach liquor flow rate in liter/s)	100[a]	10.2	BL	60–300	0.57		90	
	Electrowinning only at 250–300 A/m^2; including cells, transformers, rectifiers, and electrical distribution; excluding starting sheets								
	(capacity: output copper at 90% recovery)	10	6.0	BL	2–60	1.00	30	90	
	From dump leach liquor–solvent extraction plus electrowinning: liquor concentration 1 g/liter								
	(capacity: output copper) At 0.5 g/liter × 1.5 At 2 g/liter × 0.75 At 4 g/liter × 0.62 From sea nodules × 0.2 Cu recovery only From sea nodules × 0.3 Cu and Ni recovery From 34 g/liter × 0.045	10	11	GR	5–30	0.75	30	90	

TABLE 2 *(Continued)*

Product	Process	Capacity, 10^3 Mg/a	Cost, $\$10^6$	Classification	Range	n	Error, %	Operating Rate, %	Comments
Chronium–vanadium	From titaniferous magnetive via alkaline roast–leach; includes SX only, feed storage–chromium circuit, vanadium circuit; excludes solvent								
	(capacity: ore) or	5000	67	BL	–	(0.70)			Solvent: 57% I_F
	(Capacity: aqueous leach liquor flow, liters/s)	695[a]							
	(Capacity: daily chromium production as $Na_2CrO_4 \cdot 4H_2O$, Mg/day	912[a]							
Nickel	From laterite via solvent extraction; includes disengagement pumps, piping, instrumentation; excludes flotation, electrowinning, and activated carbon adsorption and excluding copper recovery circuit, excluding solvent								
	(capacity: output nickel)	12.5	1.8			(0.70)			
	Complete system including heat exchange, flotation, electrowin and carbon adsorption × 5.2								
	Integrated copper recovery circuit with electrowin, cement leach, SX and product electrowin. SX only and excluding solvent.								
	(capacity: output copper) including full circuit × 4.7	0.062	0.070					99	
Uranium	From acid leach liquor from 0.1% U_3O_8 ores; solvent extraction section only; excludes solvent inventory, crushing, grinding, leaching, filtering, precipitation, drying, packaging, reagent handling, tailings, and effluent treatment								
	(capacity: feed uranium-rich ore)	990	4.0	BL	40–3000	0.73	40	90	$TM = 4.75$
(As UO_3)	As above but including Th recovery circuit to recover 0.60 Mg thorium sulfate per Mg U_3O_8								
	(capacity: feed ore)	990	5.0	BL		(0.73)			

(As UF_4)	Pechiney PNC process from strong acid leach liquor; including solvent extraction, chloride conversion and separation, hydrofluorination, filtration, and dehydration; excluding electrolytic reduction, tailing, and effluent treatment ore preparation							
	(capacity: UF_4 product) To include electrolytic reduction × 2	1.75	4.6	BL		(0.70)	95	
	From spent fuel rods; including solvent extraction, building, piping, shielding, and instrumentation; excluding engineering, design, and field expense							
	(capacity: uranium processed)	1.65	2.1			(0.70)	90	
Aromatics, BTX mixture	From mixed hydrocarbon stream via Lurgi Arosolvan process, for about 60% of the feed as BTX (LURGI tower extractor); excluding product purification							
	(capacity: product BTX)	100	1.2	BL	100–500	0.78	90	Solvent: 10%
	From C_6–C_8 reformate cut via IFP, DSMO process (RDC extractor); excluding product purification							
	(capacity: product BTX)	100	1.8	BL	100–300	0.60	95	Solvent: 10%
	From C_6–C_8 reformate via Shell sulfolane (RDC extractor); excluding product purification							
	(capacity: product BTX)	300	6.0	BL	130–450	1.60	95	
	From gas oil, via furfural extraction, (RDC extractor)							
	(capacity: product BTX)	100	5.4	BL		(0.70)	95	
	From gas oil, via phenol extraction (centrifugal extractor)							
	(capacity: gas oil feed)	260	5.0	BL		(0.70)	95	
Crude naphthenic acid	From spent caustic	5	0.38	BL		(0.70)	82	

[a]Column title does not apply. The units in brackets apply. Example: for copper from acidic dump leach liquor, the capacity is 100 liter/s.

TABLE 3 FABRICATED EQUIPMENT COSTS FOR SOLVENT EXTRACTORS

	Size	Unit	Cost, 10^3 $	Range	n	Error %	L + M	L/M	BM	Comments
RDC carbon steel vertical pressure vessel; including internals, rotor, drive, and explosion-proof motor; FOB	10	(height, m) (diameter, m)$^{1.5}$	26	0.5–75	0.61	40				MS = 600
Agitator column extractor (Oldshoe–Rushton, Scheibel), 304 stainless steel; FOB; including motor and variable-speed drive	4	(height, m) (diameter, m)	44	0.1–200	0.85					
	or 10	(height, m) (diameter, m)$^{1.5}$	90	0.5–55	0.66					
Pulsed plate column, 316 stainless steel tower and plates at 100-mm separation; including pulsing mechanism and necessary auxiliaries; excluding pump, piping, foundation; FOB	10	(height, m) (diameter, m)$^{1.5}$	100	0.5–100	0.81					
Reciprocating-plate column, 316 stainless steel tower and plates; including pulsing mechanism, motor, and drive; excluding pump, piping, foundation, FOB	10	(height, m) (diameter, m)$^{1.5}$	100	0.5–100	0.75					
Continuous centrifugal extractor, 316 stainless steel; FOB; including flexible connectors, explosionproof motor, variable-speed driver, ammeter, tachometer; excluding pumps, starter, flowmeters, control valves	2.2	Aqueous feed rated capacity, liter/s	51	0.03–2.2 2.2–36	0.25 0.58		1.70			
Continuous centrifugal extractor, Alloy 20; FOB as above.	2.2	Aqueous feed rated capacity, liter/s	84	1.9–20	0.48		1.45			
Hydrocyclone, carbon steel FOB unit only	25	Nominal body diameter, cm	1.4	2–200	1.07					
Hydrocyclone, carbon steel; installed, hydrocyclone only; excluding piping, pump.	50	Total flow capacity, liters/s	9	9–1300	0.35	60				
Hydrocyclone, carbon steel; FOB, unit only, for aqueous: oil (A:O) ratios of 1:1 Factors for A:O ratios: 1:2 or 2:1 × 1.3 1:5 or 5:1 × 2.6 1:9 or 9:1 × 6.5	10	$\frac{(\Sigma, \text{m}^2)}{(\Delta p, \text{kPa})^{0.8}}$	0.86	3–20	2.19	40				
Mixer–settler, carbon steel; FOB; separate mixer and settler; including explosion-proof motor, drive: cost per stage; excluding inter-	10	Flow of aqueous feed stream to be treated, liters/s	7.4	1–10 10–100	0.22 0.60	50	1.52			L + M if include interstage pumps and piping; L + M = 2.51

connecting pumps and piping; tankage and crud removal system; assume A:O = 1:1	or									
Factors (for flows less than 40 liter/s):	20	Total mixer settler volume, m^3	7.4	1–20	0.27					
With no nozzles, costs are approx. × 0.9				20–300	0.61					
With nozzles and internals × 1.2										
Factors (for materials of construction): carbon steel × 1.0; rubber-lined × 2.1; PVC × 1.75; PVC + fiber-glass-reinforced poly-ester opanol lined × 2.4; 304 s/s × 3.00 × 2.0				Excluding piping			1.52	0.25	2.1	Factors for FOB above correlation to include interstage piping; $TM = 4.0$
Including interstage piping and pumps × 1.54–1.75							2.51		3.5	
excluding tankage, crud removal system × 2.0										
Mixer–settler, carbon steel, L + M installed	1.5	Aqueous feed capacity to one stage, liters/s	7	0.1–1.5	0		incl.		1.4	
Separate mixer–settler, including explosion-proof motor, drive, and "within-module" piping, concrete, steel, instruments, electrical, insulation and paint, and necessary labor	1.5		7	1.5–10	0.40					
	10		14.8	$10–10^3$	0.70	40				
Cost per stage but excluding interconnecting piping and pumps, tankage, and crud removal system										
Factors (for materials of construction): carbon steel × 1.00; concrete × 0.70; 316 stainless steel × 2.00; rubber-lined × 1.4										MS = 600
Including interstage piping and pump × 1.54–1.75										
Including tankage and crud removal system × 2.00										
Units for other designs and materials of construction can be estimated by sizing the units separately and developing the costs from prices of tanks and mixers, pumps, and piping where needed										

2. Multiply the FOB cost by experience factors that convert this cost into the BL cost.
3. Multiply this BL cost by experience factors to obtain the offsite costs and hence the total fixed capital investment.
4. Add the contributions for working capital, startup and financing where applicable.
5. Inflate to the time of interest.

The FOB costs for the MPI needed in a solvent extraction process are usually correlated in terms of the capacity or size of equipment. The correlating equation is of the same general form as Eq. (1) and is summarized in Table 3.

The experience factors must account for all the cost contributions listed in Table 1. Although many different types of factor can be used [4, 6–8], the bare module factors are summarized here. For this method, the FOB cost is assigned a value of 1.00. We visualize a module as being everything within about 3 m of the outer boundary of the MPI in question. This is illustrated in Fig. 2 for a pulsed column. Fig. 2(*a*) shows the FOB unit and Fig. 2(*b*) the working unit. *All* costs required to convert the FOB unit into the working unit are accounted for by factors. Thus the $L + M$ factor is one that when multiplied by the FOB carbon steel cost, yields a total cost value that accounts for all the labor costs, material costs, and the FOB equipment costs within the module. In Table 1 the total costs to be accounted for are $972 for labor, $1027 for module materials, and $1000 for the FOB equipment, totaling $2999. This is an L + M factor that, when multiplied by the FOB cost, yields the same cost value; $2999/$1000 = 2.99. The L/M factor is the ratio of the labor cost to total materials cost within the module. In Table 1 this is $972/(1027 + 1000) = 0.48. The BM factor is one that, when multiplied by the FOB carbon steel cost, yields a total cost value that accounts for the labor, all the materials plus freight, duties, tax, engineering home office expenses, and field supervision expense. In Table 1 this factor is 3.97 to account for 3.97 ($1000) = $3972 cost. The TM factor accounts for the BM cost plus contractor's fees and contingency. In Table 1 this is 4.49 to account for 4.49 ($1000) = $4492 cost. The allowances for royalties, land, spare parts, interest during construction, legal fees, working capital, solvent inventory, and startup expenses are estimated separately in addition to the TM module cost. Some percentage

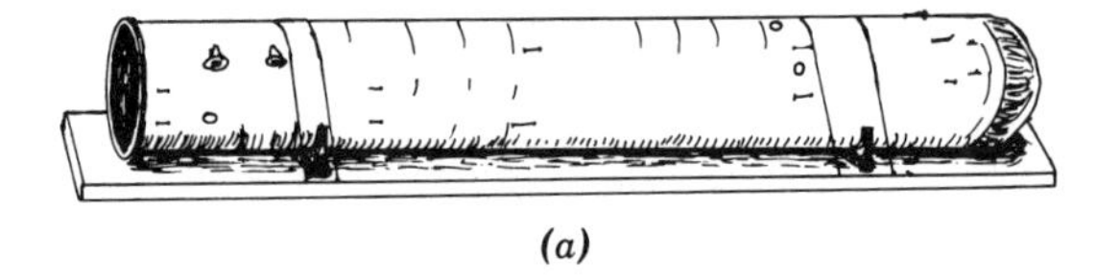

(*a*)

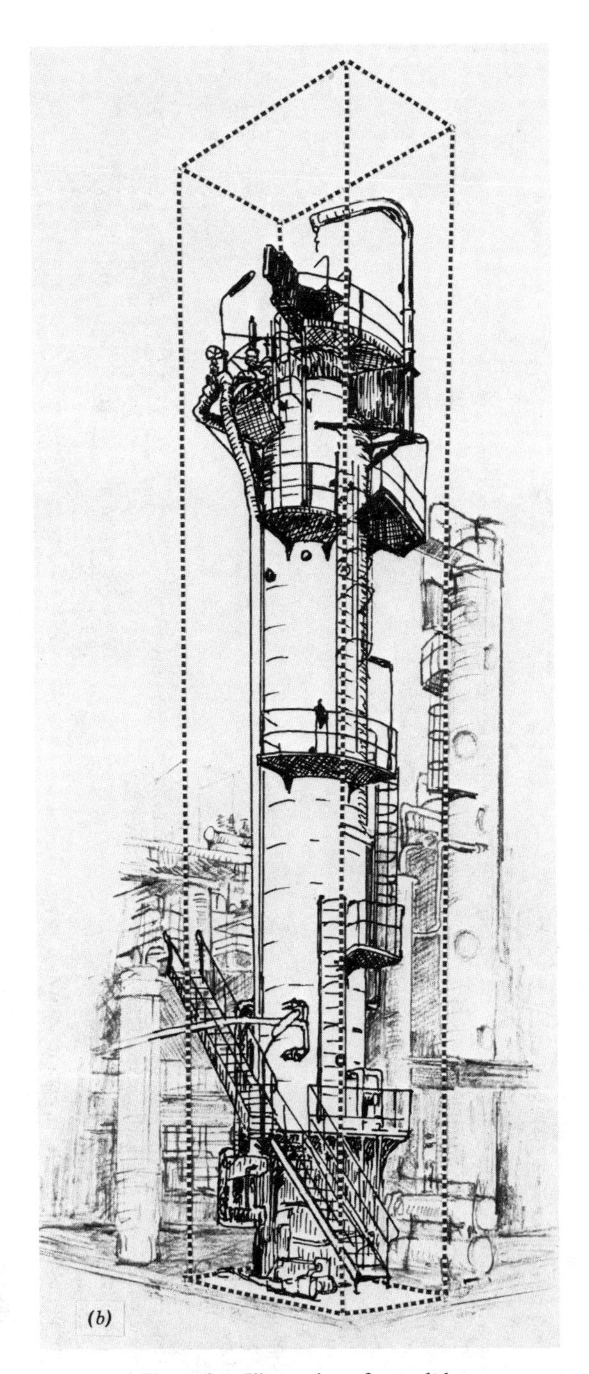

(*b*)

Figure 2. Illustration of a module.

values are given in Table 1 for general conditions. Most of the L + M, L/M, and BM factors given in Table 3 apply to carbon steel construction. If the SX unit is comprised of other materials of construction, then cost first an "imaginary" carbon steel unit, apply the factors given in Table 3, and add the FOB equipment alloy increment to the total cost.

Example: What is the total battery limit cost for three mixer settlers, to handle an aqueous flow of 50 liters/s? The materials of construction are rubber-lined carbon steel. The design is such that no interstage piping and pumps are necessary. The tankage and crud removal system is already available. The costs should refer to MS = 600 conditions.

Solution. From Table 3, we cost first a carbon steel unit, determine the total module cost, and then add the alloy increment. The carbon steel FOB cost is

$$7.4 \times 10^3 \left(\frac{50 \text{ liters/s}}{10}\right)^{0.6} = \$19.4 \times 10^3 \text{ U.S.}$$

The L + M factor is 1.52, and the BM factor is 2.1. The carbon steel BM cost is

$$19.4 \times 10^3 \times 2.1 = \$40{,}740 \text{ (U.S.)}$$

To this must now be added the alloy allowance. From Table 3, the factor for a rubber-lined unit is 2.1. Hence the FOB cost of this is $40,740 or an alloy increment of 40,740 - 19,400 = $21,340 (U.S.). Hence the BM cost is the carbon steel BM cost plus the alloy increment is (40,740 + 21,340) = $62,080 U.S. To this must be added the contractor's fees and perhaps a contingency allowance. From Table 1, select 5% for the former and 10% for the latter to yield 62,080 × 1.15 = $71,392 U.S. BL per unit. For three units, the BL cost is $214,176 for the fixed capital investment. The error is ±50%, so the fixed BL cost is between $110,000 and $320,000. An estimate of the solvent cost is approximately $20,000.

The working capital expense for solvent extraction processes can be significant because of the cost of the initial charge of solvent. For estimating purposes (when the solvent inventory has not been calculated) the cost of the solvent can be approximated to be 10% of the fixed capital investment. Hence, the total working capital for product and raw material inventory might be estimated as 15–20%. The startup costs range between 5 and 17% of the fixed capital investment. More details are given by Derrick and Sutor [9].

3. SUMMARY

Capital cost estimates of different levels of accuracy must be made at different stages of the project development. Here cost correlation methods, accurate to within ±40%, were described. Module factor methods combined with FOB equipment estimates give improved accuracy, approximately ±15–30%. Data were given for SX processes, equipment, and installation. Techniques of accounting for inflation are described. Detailed cost estimation procedures were not discussed.

REFERENCES

1. F. C. Jelen, ed. *Cost and Optimization Engineering*, McGraw-Hill, New York (1970).
2. J. H. Kempster, *Financial Analysis to Guide Capital Expenditure Decisions*, Research Report 43, National Association of Accountants, New York, 1967.
3. C. G. Edge, *A Practical Manual on the Appraisal of Capital Expenditure*, The Society of Industrial and Cost Accountants of Canada, Hamilton, Ontario, 1960.
4. D. R. Woods, *Financial Decision-Making in the Process Industry*, Prentice-Hall, Englewood Cliffs, NJ, 1975.
5. American Association of Cost Engineers, *Cost Engineer's Notebook: Construction Indices for Canada, U.K., Netherlands and West Germany*, AACE, 1977.
6. O. P. Kharbanda, *Process Plant and Equipment Cost Estimation*, Sevak Publications, Bombay, India, 1977.
7. K. M. Guthrie, *Process Plant Estimating, Evaluation and Control*, Craftsman Book Company, Solana Beach, CA, 1974.
8. C. A. Miller, *AACE Bull.* **7** (3), 92 (1965).
9. G. C. Derrick and W. L. Sutor, *AACE Bull.* **19** (3) p. 1 (1977).

29.2

COST OF PROCESS

M. W. T. Pratt
University of Bradford
United Kingdom

1. OPERATING COSTS

Following the general convention of dividing process costs into capital costs—the expense of obtaining a working plant—and operating costs—the charges involved in running the plant after it has been provided—then the general types of operating cost, directly and regularly incurred by the typical solvent extraction plant are those described in the following paragraphs. Most of these costs have been reviewed in detail by Ritcey and Ashbrook [1].

1.1. Cost of Feed Solution

This could be the main raw material cost of the process, but solvent extraction processes differ from the norm in the chemical industry since the feed may be available quite cheaply or may even be free. This is because the feed for solvent extraction may be regarded as a waste or effluent stream that would otherwise have little value. In a dump leaching operation, for example, the solvent extraction plant may escape any share of the cost of mining, separation, and conveying the material forming the dump, and its costs may be considered to commence only at the point in processing at which the leaching solution is fed to the dump.

The feed stream may arise as a by-product from some other chemical processing or metallurgical operation, in which case the price level at which the feed is to be charged to the solvent extraction section will be settled by the manufacturing firm's policy on internal transfer pricing. It may be costed simply in terms of the value of the best available alternative use, such as the thermal energy value of a hydrocarbon stream;

alternatively, the feed may be required to bear a proportion of the processing costs of the plant that provides it or may even include a contribution toward profit. Clearly, it is of great economic advantage toward the commercial viability of a solvent extraction process if the feed containing the recoverable and valuable product is obtained at a low price.

1.2. Cost of Preconditioning the Feed Stream or Extractant

If it is necessary to adjust the pH of the feed to the optimum required for the chosen extractant, this will involve not only capital costs for the plant equipment necessary for this stage of the process, but also extra operating costs. The most important of these is likely to be the cost of the chemical conditioning agent, such as acid or alkali, which will be continuously consumed during plant operation. In addition, an extra contribution will be incurred to the operating costs of labor, power, and maintenance for the plant. Other possible pretreatment operations, such as filtration of the feed liquor, sometimes preceded by the precipitation of an unwanted metallic impurity, will similarly result in additional expenditure for equipment and operating costs for filtration materials and chemicals, together with the others mentioned previously.

In some solvent extraction processes significant operating costs are incurred for treatment of the extractant. For example, in a TBP–HNO_3 process for the recovery of zirconium, preequilibration of the extractant by nitric acid costs [1] about 20¢/lb of zirconium produced; however, in the extraction of uranium from sulfate solutions by tertiary amines, the preequilibration of the solvent by sulfuric acid prior to extraction costs less than 1¢/lb of U_3O_8 product, and in some other processes the costs for pretreatment of the extractant are negligible.

1.3. Scrubbing and Stripping Costs

Metals extraction processes may involve the stripping of the laden extractant by a solution of a chemical treatment agent, commonly an acid or base.

Sometimes, a prior scrubbing operation is required to remove an impurity and this again may involve contact with a treatment solution. In either or both of these processes, operating costs are involved for the provision of the treatment chemicals, either to replace losses in recycled process streams or to provide chemicals that are consumed by reaction in the normal process operation. Obviously, if water alone may be used for the stripping operation, these costs are minimized, but if chemical agents are involved, costs may be significant. In the recovery of uranium from amines by treatment with sodium carbonate solution, for example, costs incurred are 1–2¢/lb of U_3O_8 recovered.

Scrubbing and stripping will, of course, necessitate extra operating costs for labor, power, and maintenance associated with these stages of the plant.

1.4. Distillation Costs of Laden Extractant

In organic separation processes it is common for the laden extractant to be separated from the recovered solute by distillation. The major operating expense will then be the thermal energy or steam costs, although again a contribution will be made to the overall labor, maintenance, and general power requirements.

The choice of organic : aqueous feed flow ratio in this type of process provides a classic exercise in cost optimization: a higher solvent treat may achieve the wanted separation in fewer equilibrium stages, thus saving capital costs for the liquid–liquid contactor, but producing a greater flow rate of laden solvent to the distillation column, requiring higher thermal energy operating costs. The capital and operating costs must be compared to discover the minimum overall process costs, and the optimum choice of operating conditions will, of course, change with variation in the interest rate on capital and other cost factors.

1.5. Solvent Losses

Although the costs of the initial stock of the extractant used in the solvent extraction process may be regarded as a capital cost, the charges for the "makeup" quantities of extractant necessary for the regular replenishment of the losses that tend to occur are an operating cost. Losses of the solvent, or any component of the extractant phase including diluent and modifier, may be due to one or more of the causes described in Sections 1.5.1–1.5.6.

1.5.1. Solubility of Extractant in Raffinate Phase

Constituents of the two phases, regarded as mutually insoluble, do have a finite, even if low, equilibrium solubility, and this results in regular losses of extractant in the aqueous process stream unless special recovery steps by distillation or adsorption are included.

1.5.2. Entrainment of Extractant in Raffinate Phase

This occurs as small droplets, which do not separate in the settler or phase separation section of the liquid–liquid contactor, are carried along and may be discharged from the plant in the aqueous waste stream. In that case the entrainment may result in environmental pollution, as well as causing increased operating costs through loss of extractant. The extent of entrainment is influenced by the choice of phase to be dispersed in the solvent extraction equipment and also may be a function of mixer design and operation, especially impeller type and speed, of other factors affecting the drop size in the contacting equipment, and also of suspended solids content. It may be minimized by design modifications. Modern large-scale mixer–settlers have organic entrainment values of 13 ppm in comparison with values of 120–160 ppm obtained with earlier plants [2].

The direct costs of organic entrainment for replacement of the extractant lost can readily be calculated once the concentration of entrained organic phase is known, and some estimates for copper extraction processes have been given by Rowden and Collins [2].

Entrainment not only results in direct operating costs for replacement of extractant, but may also involve increased capital costs: the organic material may have a corrosive effect on common materials of construction and necessitate the use of more expensive materials. Organic entrainment in an acid leach solvent extraction plant for copper production can result in the phenomenon of "organic burn" in the electrowinning unit, which may result in the affected copper cathodes having a lower market value. Aqueous entrainment in the organic phase also may result in costly losses [2].

1.5.3. Degradation of Extractant

This may be a consequence of chemical attack or atmospheric oxidation and is especially likely to be significant if pH or operating concentration levels are allowed to reach the extreme ends of the normal range.

1.5.4. Crud Formation

"Crud" is the term describing the pollutant phase containing mineral or biological solids that tends to build up at the phase interfaces in solvent extraction plant. The solids forming the basis for crud formation may be suspended in the aqueous stream entering the plant, be precipitated as a result of supersaturation or pH changes, or originate from airborne dust. The volume of crud may be considerably augmented by entrainment of organic and aqueous phases to form a flocculant mass. This will be periodically removed from the plant, and if it is discarded without further treatment, it can result [2] in substantial costs through loss of the extractant. Extra labor costs and other charges incurred from repeated plant shutdown for cleaning may also result from "crud" formation.

1.5.5. Evaporation of Extractant

Loss of one or more components of the extractant may occur due to evaporation, a loss that is likely to be more severe if the plant is exposed to the open air, winds, and solar radiation, and if operating temperature is high.

1.5.6. Physical Losses Due to Leaks or Spillages

These are likely to occur from time to time, even in the best regulated plants. Good engineering, safety devices, and an efficient draining system help to minimize these losses.

1.6. Labor Costs

The costs of process labor, together with the supervisory labor, is usually quite a minor operating cost for a solvent extraction process and may cost only a few cents per pound of metal or other product recovered. In the estimation of labor costs for a new plant, all normal overhead charges for pension funds, insurance, holiday pay, and so on should be added on to the direct labor cost representing the wages or share of salary of the operatives involved. Typically, only from one to three workers per shift would be required for a solvent extraction plant that has been reasonably designed and automated.

1.7. Maintenance Costs

These may be estimated either from experience gained in the operation of similar plants; or the annual costs taken as a certain percentage of fixed capital costs, for example; or on any other rule-of-thumb basis.

1.8. Energy Costs

Power will be required for operation of the mixers and pumps of the plant, together with heating and lighting requirements, although these costs are normally a very small proportion of total costs.

1.9. Cost of Product Lost in Raffinate

The loss of wanted solute in the raffinate results in an operating cost of the value of the unrecovered material if the costing of the plant is based on the treatment of a given quantity of feed of specific composition. Like depreciation, it is not a cost that actually must be paid, but represents an inefficiency in the recovery process that could possibly be avoided by technical improvements. If the plant were to be costed on the alternative basis of the quoted production rate of product, such recovery inefficiencies would be represented, not directly as a separate operating cost, but by the larger throughput of the feed stream (which may or may not have to be paid for) and also the larger size of plant necessary to achieve the given output of product.

1.10. Cost of Treatment of Raffinate

If the raffinate must be distilled to recover the solvent it contains, which may be required in organic separations, additional operating costs will be incurred. Similarly, if the raffinate must be further treated by charcoal adsorption or otherwise to remove entrained or dissolved solvent for environmental considerations, that is, to achieve a plant effluent that will not cause pollution problems, extra costs will result. Important factors determining costs are volume of effluent stream, its solvent content, ease of separation (e.g., relative volatility), and type and design of the recovery plant.

Besides the direct operating costs, considered in the preceding paragraphs, indirect allocation for depreciation of the capital value of the plant will annually be made, and also indirect overhead charges will be assigned to cover the provision of central administration and services. These general cost factors, involved in the operation of all chemical plant, are not considered here in any further detail.

1.11. Relative Importance of Operating Costs

In comparing the relative importance of the various operating costs, it is generally agreed that some cost items are minor, and labor, maintenance, and energy costs (other than for distillation) are normally among these. Solvent losses were commonly regarded as the most significant operating cost of solvent extraction processes, but modern experience based on detailed breakdown of costs has shown that other costs, in particular the costs of solvent recovery, have been greater. For example, Warner [3] quoted the following ranges for costs ($) in hydrometallurgical separations per ton of metal processed:

Capital cost of equipment	5–140
Solvent losses	3–30
Solvent recovery	85–335

These costs are now out of date, but the relative magnitude should still be valid.

Precise operating costs clearly depend on the process, scale of operation, location and date, and some particular examples are quoted in the following section.

1.12. Practical Examples of Costings

Extensive information concerning the detailed costings of modern solvent extraction plants has not yet been published in the general scientific literature. In assessment of such cost information that is available there is always the problem that existing solvent extraction processes tend to differ in type and concentration of feed solution and in details of the processing route used. These differences, together with variations in costs between countries and localities, greatly complicate the comparison and estimation of costs for solvent extraction processes, although the published information may give a "feel" for the order of costs to be expected.

A very useful example of cost data in the metals extraction area is the recent survey by Whittaker and Gray [4] on the interaction of metallurgical and economic factors in the hydrometallurgical processing of copper ores. In this paper, costs are compared for the processing of copper ores not amenable to physical concentration, containing 0.2–5% or more copper in oxide ores in which a leaching operation by dump, vat, or agitated tank is followed by solvent extraction and electrowinning. This type of processing accounts for 10–15% of the world's copper production, and the proportion could increase.

The late 1977 costs data were derived from actual plants that Davy Powergas Ltd. had constructed and were based on an ore treatment rate of 10,000 tons/day. Major variables affecting process economics were identified as ore grade, recovery, leaching rate, and acid consumption. Compared with these, other variables such as equipment selection, power costs, and labor rates were considered to have only a relatively minor effect. In accordance with standard accounting policy, it was assumed that the unconcentratable ore used in the hydrometallurgical process is obtained as a by-product from the preparation of concentratable ore, with all the mining and comminution costs borne by the latter.

The key operating costs, assumed constant in the processes surveyed, were taken (in U.S. currency values) as:

Labor	\$20,000/worker-year
Electrical power	4.0¢/kWhr
Maintenance spares	3% of delivered equipment cost per year
Water	5.6¢/m^3
Sulfuric acid	\$0–\$30/ton
Contingency	\$200,000/year

It was assumed that site services, such as workshops, laboratories, offices, and mobile equipment, were provided by the mine and concentrator parts of the operation, and the production costs for the hydrometallurgical stage did not cover these services.

Production costs were given as the sum of all direct operating costs and capital charges (interest on outstanding loan debts, interest on equity capital, repayments of loan capital, taxes, and profit). It was assumed that the plant would be financed on a 3 : 1 debt : equity ratio with interest rates of 8% on long-term debt and 15% on the equity. A repayment of debt in 10 years from commissioning was assumed. Figures 1–3 give the cost information presented for each of the three leaching modes. Production costs are divided into three sections: leaching; solvent extraction; and electrowinning. Costs in each section are split between capital charges and direct operating costs. The cost of leaching acid is a main variable (except in dump leaching) and is shown separately for acid costs in the range \$0–\$30/ton.

In the cases examined, production costs vary from about 22¢/lb to more than 137¢/lb, chiefly because the volume of liquor that must be processed per ton of copper varies from 107 m^3 to more than 4300 m^3. It is to be noted particularly that general statements about the cost of hydrometallurgical processing are bound to be imprecise unless they are closely related to ore grade and leach efficiency.

For the solvent extraction stage, investment and operating costs are related primarily to the volume of pregnant liquor treated. Costs per ton of copper are chiefly dependent on the copper loading of the pregnant liquor, which may vary by a factor of at least 10 : 1 in the approximate range 0.5–5.0 g/liter. Direct operating costs of the solvent extraction stage are 13–21¢/m^3 for dump leaching, 26–41¢/m^3 for vat leaching, and 17–31¢/m^3 for agitated leaching operation. In terms of costs per pound of cathode copper, the operating costs vary from over 25¢/lb to 2¢/lb, as Figs. 1–3 indicate, with the minimum for the vat leaching operation.

The reason vat leaching gives the most favorable processing costs is because it enables the volume of liquor per ton of ore to be controlled most closely and minimized.

Whittaker and Gray conclude that, although hydrometallurgical processing of copper ores that cannot be concentrated by physical methods can be a highly profitable operation, the economics can clearly vary greatly with the ore mineralization and the selection of processing route. The latter should be chosen so as to minimize the volume of liquor generated per ton of copper produced.

Another publication [5], relating to the same firm's experience in the construction of copper extraction plant, details the operating costs at 1974 pre-oil crisis levels for a plant to produce 100 tons/day of copper. On the basis of the costs:

Solvent	\$2.75/lb
Diluent	\$0.33/U.S. gal
Power	8¢/kWhr
Water	90¢/1000 gal

and salaries:

Supervisor	\$500/month
Operator	\$400/month
Chemist	\$350/month

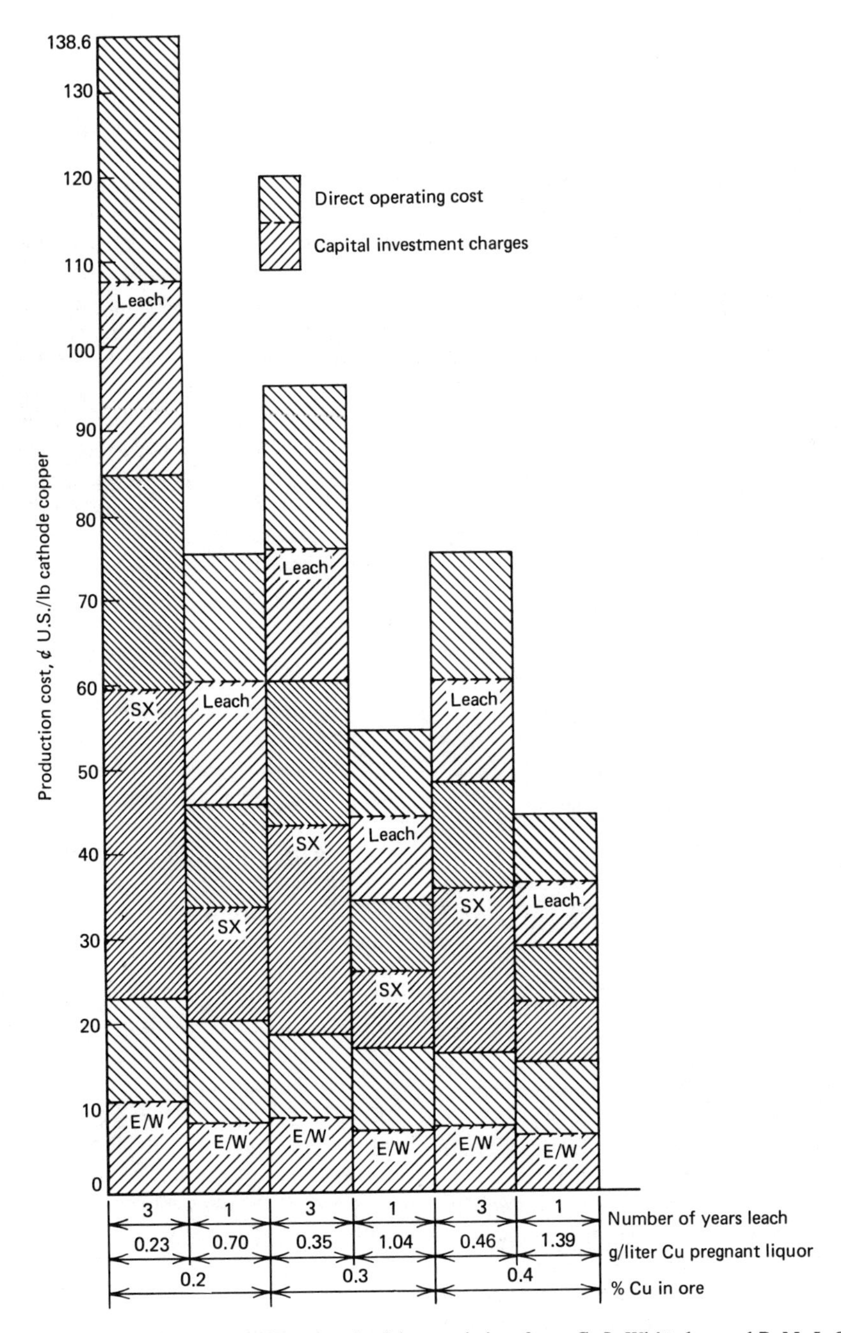

Figure 1. Dump leaching costs. [Reprinted with permission from C. J. Whittaker and P. M. J. Gray, *Proceedings of the 11th Commonwealth Mining and Metallurgical Congress, Hong Kong, 1978* (Institute of Mining and Metallurgy, London), p. 4.]

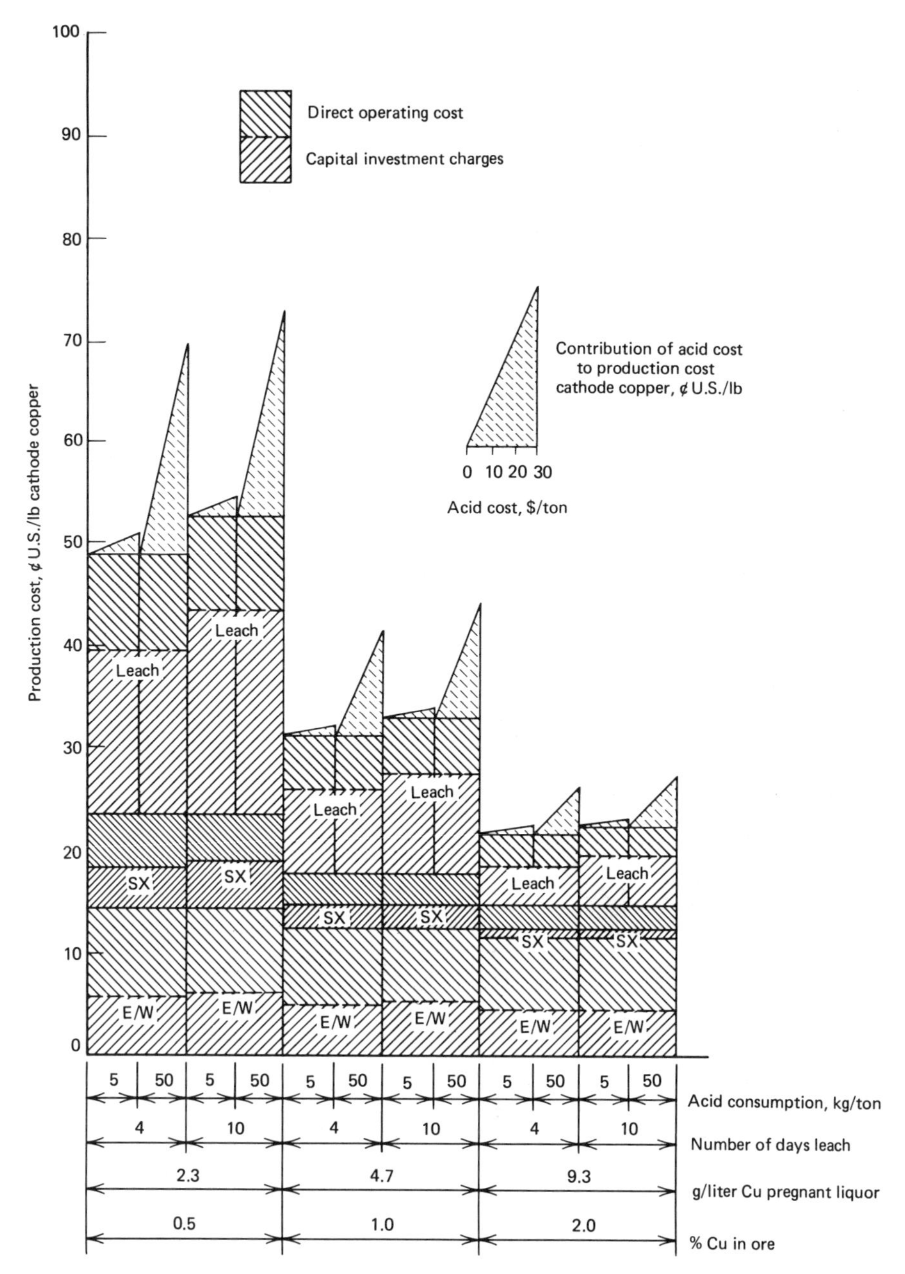

Figure 2. Vat leaching costs. [Reprinted with permission from C. J. Whittaker and P. M. J. Gray, *Proceedings of the 11th Commonwealth Mining and Metallurgical Congress, Hong Kong, 1978* (Institute of Mining and Metallurgy, London), p. 4.]

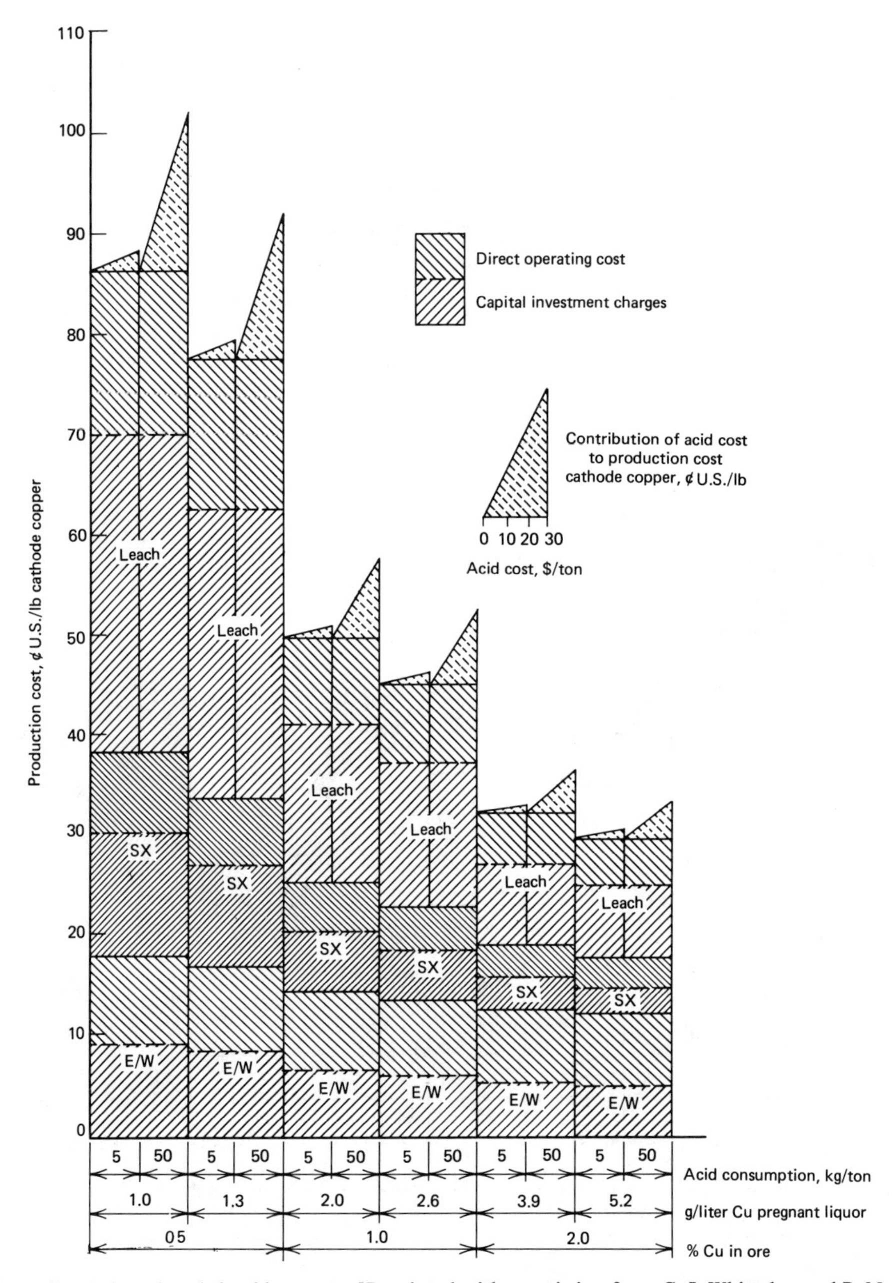

Figure 3. Agitated tank leaching costs. [Reprinted with permission from C. J. Whittaker and P. M. J. Gray, *Proceedings of the 11th Commonwealth Mining and Metallurgical Congress, Hong Kong, 1978* (Institute of Mining and Metallurgy, London), p. 4.]

Individual operating costs were:

	Copper, ¢/lb
Solution makeup	
Solvent 0.99 m^3/day,	
diluent 1.48 m^3/day	2.45
Power	
505 kW	0.40
Water	
74 gal/min	0.04
Labor	
One supervisor (daytime only)	
One chemist (daytime only)	
One control-room attendant (Shift-work (42 hr/week))	
One operator (Shift-work (42 hr/week))	0.11
One assistant operator (Shift-work (42 hr/week))	
Half-time maintenance worker (Shift-work (42 hr/week))	
Direct operating costs = 2.45 + 0.40 + 0.04 + 0.11	3.00
Take maintenance	1.00
Operating cost SX plant	4.00
Overall Operating Costs	
Mining (open pit)	4.0
Leaching	9.0
Electrowinning	8.0
Solvent extraction	4.0
Total	25¢/lb of copper

The economics of extraction and separation of two coextractable metals, such as nickel and cobalt, is more complex than for the extraction of a single metal. Generally, more stages are required, accurate control of pH is essential, the extraction equipment can become more complex, and choice of solvent system can be a very significant economic factor. Ritcey and Ashbrook [1] present the cost information given in Table 1 covering the use for cobalt and nickel extraction and separation from sulfuric acid solution of either carboxylic acid or alkyl phosphoric acid (D2EHPA) extractants. Although the solubility of Versatic 911 is about 100 ppm (compared with 30 ppm for D2EHPA), the relative costs of the extractants are about 30¢/lb for the carboxylic acid and $1.25 for D2EHPA. Solvent losses are, however, responsible for the largest cost differences between the processes.

Ritcey [6] has given comparative solvent extraction processing costs for metal systems, shown in Table 2, based on a feed solution flow of 100 gal/min. Typically, a range of feed concentrations and variety of specialized treatments results in a wide spread in cost per pound of product (3.8–81.0¢/lb).

In a new process developed by O'Neill et al. [7] for the separation of cobalt, copper, and zinc from nickel sulfate process streams, the extractant was a quaternary ammonium thiocyanate dissolved in aromatic diluent. It was estimated that 1 kg of cobalt product required the use of about $800 worth of the organic reagent: 30 v/o of Aliquat 336 in Solvesso 150. The mass balance for Aliquat recovery was over 99.5%, and the

TABLE 1 ESTIMATED PROCESSING COSTS FOR THE EXTRACTION OF COBALT OR NICKEL USING NAPHTHENIC, VERSATIC 911, AND DI(2-ETHYLHEXYL)PHOSPHORIC ACIDS AS EXTRACTANTS[a]

Process	Metal, ¢/lb		
	Naphthenic Acid (1 *M*)	Versatic 911 (1 *M*)	D2EHPA (0.6 *M*)
Solvent loss	1.80	2.70	0.38
Alkali requirements (pH control)	3.00	3.00	3.00
Scrubbing (recycle strip liquor)	0.20	0.20	0.20
Stripping (10 v/o H_2SO_4)	1.80	1.80	1.80
Maintenance and labor	1.30	1.30	1.30
Depreciation (equipment and instrumentation)	1.74	1.74	1.39
Total estimated costs, ¢/lb metal	9.8	10.7	8.1

[a]100 gal/min aqueous feed flow; 10 g of cobalt and nickel per liter in feed; pH 6.5.

TABLE 2 APPROXIMATE SOLVENT EXTRACTION COSTS FOR RECOVERY OF SOME METALS (Flow Rate 100 gpm)

						Processing Costs, ¢/lb Metal Produced							
Metal	Feed Concentration (g/liter)	Aqueous System	Organic Extractant	A/O Ratio	Equipment Type	Acidification Adjustment	Equipment and Instrumentation	Solvent Treatment	Scrubbing	Stripping	Estimated Solvent Losses	Labor and Maintenance	Total SX Costs
Co–Ni	Co 10, Ni 10	H_2SO_4, pH 5–6	20% D2EHPA	1/1	Columns	0.2	1.4	3.1	0.2	1.8	0.6	1.3	8.6
Cu	Cu 2	H_2SO_4, pH 2	10% LIX64N	2/1	Mixer–settlers	–	3.3	–	–	1.5	1.5	7.0	13.3
Cu	Cu 15	H_2SO_4, pH 0.5–1.5	20% Kelex 100	1/1	Mixer–settlers	–	0.3	–	–	1.5	0.9	1.1	3.8
RE & Y	Y 0.10	H_2SO_4, pH 1.5–2.0	3.3% D2EHPA	20/1	Centrifuge and columns	18.0	0.6	–	–	30.4	7.0	25.0	81.0
U	U 2.0	H_2SO_4, pH 1.5–2.0	5% Adogen 364 or Alamine 336	2.5/1	Mixer–settlers	–	2.5	0.5	–	1.2	3.0	2.5	9.7
Zr–Hf	Zr 95, Hf 1.5	HNO_3, 7.5 *N*	50% TBP	1/3	Columns	16.0	0.20	20.0	4.0	–	0.8	0.03	41.0
Hf–Zr	Zr 100, Hf 2	HCl–HCNS (1 *M*) (3 *M*)	MIBK	1/2	Columns	6.0	4.6	2.8	3.9	5.2	3.9	30.0	56.4

overall chemical reagent costs for the cobalt-nickel separation, including effluent cleanup, were less than \$1.00/kg cobalt.

In the IMI phosphoric acid process, the requirement for solvent makeup has been reported [8] as 4 kg of the organic alcohol solvent per ton of P_2O_5 produced, representing only 4% of the total transformation costs. In this process the utilities operating costs, involved mainly in the sections of the plant for recovery of solvent by vacuum steam stripping of the raffinate aqueous phase and for concentration of the aqueous phosphoric acid product, totaled \$8.82/ton P_2O_5, of which steam costs (\$/ton) were 7.18, electricity 0.49, cooling water 0.49, process water 0.14, and boiler water 0.52. Direct operating man-hours varying between 2 and 5 operators per shift were in the range 0.2–1.5/ton of P_2O_5, and doubling hourly wage rates to allow for supervision and overheads resulted in a labor cost of \$1.2–\$9/ton.

These examples are typical of recent published cost information concerning solvent extraction processes but do not represent a comprehensive survey of available data.

2. OPTIMAL COST DESIGN FOR EXTRACTION PLANT

2.1. General Principles

Because of practical constraints in time and budget allowed for process development, many solvent extraction processes have been brought into operation commercially with little prior attempt at detailed optimization of process design. Yet such optimization is necessary for maximization of profitability and assurance that a proposed novel process gains the best chance of implementation. Such a quantitative assessment requires a full mathematical modeling of the process, which may then be optimized. Many standard texts are devoted to the relevant process optimization techniques, for which a full account is beyond the scope of this book.

The process optimization is based on the following framework:

1. *Identification of ALL the System Varibles.* These are divided into the types of (a) *environmentally fixed variables*, fixed by factors external to the process and including the composition of the raw material feed stream and the purity specification of the wanted product, (b) *design variables*, the setting of which may be freely chosen by the engineer over a range, which may be limited by internal constraints of the process, and (c) *state variables*, which become fixed when discrete values are assigned to the design variables. The allocation of variables into these categories is not precise, and there may be an arbitrary choice as to whether to consider a particular variable as, for example, a design or state variable. The design variables for each unit in the process are listed together with the relationships that connect them. The system design relationships governing the behavior of each unit are obtained from theoretical equations, experimental data fitted by mathematical models, manufacturer's recommendations, plant scale-up data, and other sources of fact or intuition.

2. The flow of design information through the system is identified. The system is acyclic if the information flow through the series of design equations involves no recycle of information. In such systems a change in design of a given component affects only those process units that are downstream, and the acyclic system is optimized when its downstream components are suboptimized with respect to the feed they receive from the units upstream.

3. The output parameters for each unit are modeled mathematically in terms of the input parameters.

4. Each unit is then suboptimized in turn for a particular objective function by use of a parametric survey.

5. An acyclic system may then be optimized by use of dynamic programming, in which the most downstream unit is suboptimized first, then the last two units, and so on until the whole process is included. Mathematically, the optimization is concerned with finding the maximum or minimum of a mathematical function that is deterministic, having a definite value when the continuously variable independent variables, such as concentration or flow rate, are fixed within the various constraints or restrictions limiting their permissible range of values. Linear functions subject to linear constraints are dealt with by linear programming techniques, but in general the relationships concerned with the operation of solvent extraction plant will be nonlinear and require a dynamic programming approach.

The objective function to be optimized may involve the maximum yield of a single product or, in the case of an economic evaluation, the production cost, which should be minimized (or the financial profit to be maximized). Methods of seeking the optimum by iterative techniques, in which the objective function is calculated at one point for given values of the independent variables, and then a sequence of further points generated that represent ever-improved approximations to the optimum, include direct search and gradient methods.

2.1.1. *Direct Search Methods*

These methods do not require the solution of the partial-differential equations of the objective function. Instead, the objective function is calculated directly for particular settings of the variables. It is recalculated for different positions of the variables, and the new objective function is compared with its original value. As an increasing number of points are evaluated, the design parameters are progressively varied in the direction that minimizes the objective function.

Some developments of this method in effect use the various function values to derive approximations to the derivatives of the objective function or to fit polynomials or surfaces through selected points.

2.1.2. *Gradient Search Methods*

These methods involve the evaluation of the partial derivatives of the objective function with respect to the independent (design) variables. From these, the best direction of change of the variables is found, which is the one that gives the greatest slope of descent in the value of the objective function. By progressively altering the setting of the variables in this direction, the minimum should be most quickly reached.

A major problem with these iterative processes is convergence. The change in function value for successive iterations may be erratic, and even if it is a small change, it is not certain that the vicinity of the minimum has been reached. Also, it cannot be guaranteed that the global, rather than a local, minimum will be found for a non-uni-modal function, which has more than one minimum. The constraints limiting the allowable range of the independent variables may also cause problems in some cases.

2.2. Examples of Optimization of Solvent Extraction Processes

The simple organic product type of solvent extraction system involving distillation both for solvent recovery and raffinate purification was analyzed by Treybal [9]. Subsequently, Jenson and Jeffries [10] considered the hourly profit from such a process as a function of the cost of the contactor (operating and fixed costs, together with depreciation) and of the solvent (recovery, makeup, and capital charges); the relative magnitude of the two are functions of the solvent treat. Jenson and Jeffries gave charts having contour lines showing how the cost varied with the choice of operating parameters. From these, the conditions for maximum profit for a particlar application could be determined.

Jeffries et al. [11] extended Treybal's treatment, considering the annual cost of operation of the process. From the partial derivatives of the total annual cost, with respect to the design parameters, expressions were given from which the optimum design could be selected if the appropriate costs and physical data were available. Jeffries and co-workers went on to describe a computer programme to achieve the optimization by the steepest gradient method proposed by Hooke and Jeeves.

The work of Jeffries and co-workers confirmed that, for a typical organic extraction process, the costs of solvent recovery by distillation had the major influence.

Robinson and Paynter [12] have described a procedure for the optimization of the design of a countercurrent liquid–liquid extraction plant for copper using LIX64N. In this typical metals extraction process an increase in the concentration of active agent in the organic phase results in higher loadings in the organic phase, leading to reduced stripping costs, higher efficiency for the same number of stages or fewer stages for the same extraction efficiency, increased operating costs for solvent losses, and increase or decrease in working capital associated with the solvent inventory. The optimum number of extraction stages is a function of the concentration of active agent, the flow ratios, the cost of unextracted metals, and the interest rate on capital.

For optimization of the process, it was necessary to:

1. Model the equilibrium curve as a function of concentration of active agent and factors affecting equilibrium distribution, including pH, temperature, and effect of interfering ions. On the basis of experimental equilibrium data, a curvilinear regression was used to model the equilibrium lines at different concentrations of acid.

2. Model the operating line and relate it to the equilibrium curve to obtain the number of stages necessary and the amount of metal lost in the barren stream leaving the plant.
3. Compensate for departure from 100% stage efficiency due to kinetic effects of interphase mass transfer.
4. Generate a series of cost functions that include the effect on operating and capital costs of the solvent losses, solvent holdup, metal losses, and other variables.

A nonderivate search technique was used to minimize the objective function and determine the optimum plant configuration.

The optimization procedure of Robinson and Paynter is a model that should be capable of adaptation to many similar metals extraction systems.

Process modeling of solvent extraction systems for uranium–plutonium separation has been reported by Wood and Williams [13] and Burton and Mills [14]. Hughes [15] and co-workers have described techniques for the modeling of equilibrium curves for metals where the family of curves obtained at various pH levels are described mathematically in terms of an equilibrium surface. This is a useful basis for the development of a mathematical model of a complete process.

The mathematical modeling and optimization has been described by Pratt and Spokes [16] of a plant for the separation by dissociation extraction of 400 tons/year of a mixture of 40% metacresol and 60% paracresol into the separate isomers of purity ranging from 85 to 99%. The process involves partial neutralization of the feed cresols, dissolved in organic solvent, by an aqueous solution of trisodium phosphate in a multistage countercurrent contactor. The aqueous phase leaving this contactor, containing the salt of purified metacresol, meets a countercurrent stream of another organic solvent in a secondary contactor. The weak dissociation reaction is reversed and metacresol extracted into the organic solvent, thereby regenerating the phosphate salt, which is recycled. The process thus avoids the continuous consumption of chemicals involved in "classic" dissociation extraction processes using strong aqueous alkalis and acids as reagents.

The performance of the primary separation and the back-extraction contactors were mathematically modeled from the experimental data for separation factor and cresols distribution between the phases. The costs of these two stages were estimated, together with those of the associated distillation steps to separate the product cresols from their solvent streams. An optimum process configuration and corresponding cost for various product purities was obtained by a combination of dynamic programming and direct search optimization techniques.

With several engineering unit processes, such as distillation and heat transfer, cost estimation techniques have achieved the ultimate "goal" through the development of a module that permits the prediction of overall costs of a particular process simply from knowledge of desired throughput, product specification, and the relevant physical properties of the chemical system to be treated. Such modules depend on the comparison of a reasonable quantity of achieved cost data derived from working processes, from which the necessary correlations may be derived, and do not require detailed design of proposed new processes for which a cost estimation is desired. However, the number of commercial solvent extraction processes is so small at present, relative to the rich variety of feed streams used, that the achievement of a similar module for solvent extraction will probably not be realized for some time, if at all.

REFERENCES

1. G. M. Ritcey and A. W. Ashbrook, *Process Metallurgy I, Solvent Extraction: Principles and Applications to Process Metallurgy, Part II*, Elsevier, Amsterdam, 1979.
2. G. A. Rowden and G. Collins, *Solvent Extraction*, symposium held at the University of Newcastle-upon-Tyne, September 7–9, 1976, Society of Chemical Industry and Institute of Chemical Engineers, London.
3. B. F. Warner, in D. Dryssen et al., Ed., *Solvent Extraction Chemistry*, North-Holland, Amsterdam, 1967.
4. C. J. Whittaker and P. M. J. Gray, *Proceedings of the 11th Commonwealth Mining and Metallurgical Congress, Hong-Kong, 1978* (Institute of Mining and Metallurgy, London).
5. I. O. Nichols and D. Wilson, *Solvent Extraction*, symposium held at the University of Newcastle-upon-Tyne, September 7–9, 1976, Society of Chemical Industry and Institute of Chemical Engineers, London).
6. G. M. Ritcey, *CIM Bull.* 68, 85 (1975).
7. C. E. O'Neill, V. A. Ettel, A. J. Oliver, and I. J. Itzkovitch, *CIM Bull.* 69, 86 (1976).
8. IMI Staff Report, *Proceedings of the International Solvent Extraction Conference (ISEC), The Hague 1971*, Vol. 2, Society of Chemical Industry, London, 1971, p. 1386.

9. R. E. Treybal, *Liquid Extraction*, 2nd ed., McGraw-Hill, New York, 1963, pp. 541–544.

10. V. G. Jenson and G. V. Jeffries, *Br. Chem. Eng.* **6**, 676, 1961.

11. G. V. Jeffries, C. J. Mumford, and M. H. Herridge, *J. Appl. Chem. Biotechnol.* **22**, 319, 1972.

12. C. G. Robinson and J. C. Paynter, *Proceedings of the International Solvent Extraction Conference (ISEC), The Hague 1971*, Vol. 2, Society of Chemical Industry, London, 1971, p. 1416.

13. J. Wood and J. A. Williams, *Transact. Inst. Chem. Eng.* **36**, 382 (1958).

14. W. R. Burton and A. L. Mills, *Nucl. Eng.* **8**, 248 (1963).

15. M. A. Hughes, S. Andersson, and C. Forrest, *Internatl. J. Mineral Process.* **2**, 267 (1975).

16. M. W. T. Pratt and J. Spokes, *Proceedings of the International Solvent Extraction Conference (ISEC), Toronto 1977*, Vol. 2, Canadian Institute of Mining and Metallurgy, Montreal, 1979, p. 723.

30

SAFETY AND ENVIRONMENTAL CONSIDERATIONS (Nonnuclear Operation)

J. B. Scuffham and G. A. Rowden
Davy McKee Ltd.
United Kingdom

1. INTRODUCTION

It is generally recognized that there are three guiding principles to be followed in the successful design and safe operation of a solvent extraction plant [1]:

1. The plant must be designed and constructed to ensure that safety features are incorporated efficiently without sacrificing the overlying economic considerations.
2. The plant must be managed by people who understand the operation of the plant and have set up efficient operation and control procedures.
3. The plant must be operated by staff dedicated to safe practice and trained in safety procedures, including the safe handling of flammable liquids and gases.

Statistical records show [2] that most accidents in the solvent extraction industry (as in all industry) are caused not so much by technological failure as by human error.

About 10% of accidents are caused by technological failure and the remainder by some human being acting in a careless and thoughtless way. This may suggest that designers and operators are together doing enough to ensure safety, but since injury or death commonly results from accidents in the solvent extraction industry, we must always lean toward the overcautious in engineering considerations.

Few technical details of accidents in the solvent extraction industry are given in the literature, and it is largely through the assistance of several colleagues and associates in the industry that we can include their experiences to the benefit of all. However, since accidents involve insurance, which, in turn, involves responsibility, we have referred directly only to those events that have been reported openly and do not refer to specifics, although the reader can be assured that the experiences are real.

The legal requirements of safety regulations and codes of practice, such as the NFPA [3], must be followed. However, planning and design cannot be effective if their sole purpose is simply compliance with this or that regulation; rather they should be taken as the *minimum requirements* for safety.

Of all the health hazards, fire is the most sensational, but toxicity can in certain circumstances be as disastrous in terms of human life. Care must thus be taken in relation to fire and explosions, static electricity, toxicity, exposure, and other hazards as for example, from the discharge of liquors into a watercourse.

2. DESIGN

2.1. The Process

The starting point in a design of solvent extraction plant is the process flow sheet. The more detailed the knowledge of the steps involved, the more complete is the design of the plant and hence the probable safety of operation. In addition to the temperature, pressure, and flow of each process step for normal operating conditions, it is important to have information on (1) the physical and chemical properties of the system components, (2) the ranges of flows, pressures, and temperatures, (3) the extremes of environmental or local conditions, (4) the critical operating levels, (5) the effects of maloperation or failure of associated equipment, and (6) the extent of formation of undesirable products, for instance, precipitates commonly referred to in the industry as "crud" (a word owes its origins to Chalk River unidentified deposit).

2.1.1. System Components

In liquid–liquid extraction (sometimes referred to as *solvent extraction* or SX) the organic phase generally consists of an active species, the reagent, dissolved in a carrier, the diluent. Thus the physical property requirements extend to such information as boiling point and range, flammability, flash point, dielectric constant (static hazard), spontaneous heating, toxicity, and corrosion characteristics. It is important to recognize that detailed and realistic data on the preceding factors are essential to the design of a safe solvent extraction system, as fire is the most common hazard.

As process requirements often call for diluent/reagent combinations that are not covered by the data sheets prepared by the manufacturers/vendors, these data should be obtained by experiment. It could be dangerous to assume a number in a "best-guess" manner. Reagent and diluent manufacturers are usually most helpful and prepared to generate data requested, even to the extent of suggesting the most appropriate combination for a given duty. If this is not possible, Chemical Safety Reference Data Sheet 485, prepared by the Chemical Section of the National Safety Council (NSC), can be a useful starting point.

2.1.2. Flammabilities

The flammability of a liquid is defined in terms of its *flash point*, that is, the lowest temperature at which the vapor mixture formed above the liquid with air will sustain combustion when ignited by an external source. This corresponds to the lower flammable limit when a flammable vapor mixture is formed. Lowering of pressure causes reduction of the flash point of a liquid; therefore, care must be exercised when SX plants are to be built at altitude.

Under the NFPA code, flammable liquids are defined as having flash points below 140°F and a vapor pressure not exceeding 40 psig at 100°F. NFPA Class I liquids have flash points below 100°F (38°C); Class II liquids have flash points in the range 100–140°F, and Class III liquids have flash points above 140°F (>60°C). Plants are in operation with extractants or dilutents that have flash points in the range 59–175°F (15–79°C).

Typical examples of extractants and diluents are given in Table 1. The choice of operating temperature is thus a compromise between the influence of the circumscribing unit operations, the local geographical conditions, and the economic optimum physical and chemical characteristics of the solvent extraction circuit.

In general, the higher the operating temperature, the better the rate of mass transfer between the phases but more importantly, the better the rate of settling or separation of the phases. In most process flow sheets the temperature of the pregnant liquor reporting to the SX plant is within the range 20–35°C, depending on upstream unit operations. The cost of heating the liquors above the equilibrium temperature is generally not justified, except where cold conditions

TABLE 1 TYPICAL EXAMPLES OF EXTRACTANTS AND DILUENTS

Component	Chemical or Trade Name	Flash Point, °F (°C)
Extractant	Methyl isobutyl ketone (MIBK)	59 (15)
Diluent	Solvesso 100 (Esso)	108 (42)
Diluent	Illuminating paraffin	118 (48)
Diluent	Napoleum 470 (Kerr-McGee)	175 (79)
Diluent	MSB 210 (Shell)	162.5 (72)

prevail or the cost associated with liquid–liquid contactor design for the extremes of seasonal temperature is in excess of that for heating requirements.

However, when a combination of continuous ion exchange (CIX) and SX is used, such as in a Bufflex uranium circuit, the pregnant liquor feed to the SX plant is the eluate from the CIX plant that could be operating at 50°C. A heat exchanger between eluate to SX and raffinate from SX is thus often needed to maintain a reasonable operating temperature in the SX plant. In this case the choice of operating temperature is a compromise between the greater safety at lower temperatures and the improved operating characteristics at higher temperatures, which implies smaller, less costly equipment for the same duty [4]. A further example of a cooling requirement is the stripping circuit of a copper SX plant. The strip liquor is the spent electrolyte flowing from the tankhouse where the cell temperature is commonly 50–60°C. A heat exchanger between the strip and advance electrolyte may be needed to maintain the operating temperature of the solvent extraction plant at an appropriate value.

Another influence on the choice of SX operating temperature is the solubility of the metal–organic complex in the diluent. If the solubility increases markedly with temperature, there is an incentive to operate at higher temperatures employing the reagent to its full capacity, thereby reducing both plant and inventory costs.

2.1.3. *Static Electricity*

The ability to dissipate static charges and the time it takes to do so are important in controlling hazards from static electricity.

The dielectric constant ϵ and conductivity K of a liquid affect the half value time, measured in seconds, to the relationship [5]

$$t_{1/2} = \frac{\epsilon(6.5)}{K}$$

(units of K - picomhos/m)

which is the time taken for the static charge to decay to half its original value. The relaxation time T, which is a measure of how long it takes for a material to lose its static charge, is related to the half value time by the simple equation [5]

$$T = t_{1/2} \times 1.44$$

One rule of thumb is that the safe upper limit of the half value time is 0.012 s [6].

Data related to static hazards are given in Table 2 for a number of common solvents.

TABLE 2 DATA RELATED TO STATIC HAZARDS FOR A NUMBER OF COMMON SOLVENTS

Liquid	Specific Conductivity, mho/cm	Dielectric Constant	Relaxation Time, s	Half Value Time, s
Water	5.1×10^{-6}	80.4	1.4×10^{-6}	1×10^{-6}
Ethanol	1.4×10^{-8}	27.5	1.6×10^{-4}	1.1×10^{-4}
Heptane	1.0×10^{-12}	2.0	0.18	0.13
Toluene	1.0×10^{-14}	2.4	21.00	15.00
Benzene	1.0×10^{-13}	2.3	0.2	0.14
Escaid 100 (ESSO)	6.0×10^{-15}	2.1	31.0	21.5
Solvesso 260 (ESSO)	1.5×10^{-13}	2.4	1.4	0.98

All metal equipment for processing, handling, and storing flammable liquids should be grounded. However, grounding and bonding (the joining with a wire of two pieces of equipment that are close enough that a spark could jump from one to another) cannot eliminate all hazards. A steel tank, for instance, could be grounded, but if the liquid in it were a poor conductor, having long relaxation times, a static charge could build up on the liquids' surface and grounding will do little to prevent it. Agitation can cause liquids of low conductivity to develop static sparks on the surface [7].

The lining of tanks by epoxy and other plastic coatings may electrically isolate the liquid. Although thin spots in the coating can cause grounding, it is usually preferable to install an insert or a probe in the vessel.

2.1.4. *Toxicity*

The hazards associated with the use of toxic materials in solvent extraction plants can be regarded best as a combination of the toxicity of the substance and the type, duration, and conditions of exposure [10].

The toxicity of any substance is a relative measure of its poisonous nature with respect to a specific animal or plant under specified exposure conditions. Toxicity may be *acute*, with the effects being apparent after a short exposure period, or *chronic*, with the effects apparent after a long exposure period. Acute toxicity is usually measured by the LD_{50} (medium lethal dose) notation where the associated value is the quantity of compound (milligram or gram per kilogram of animal body weight) that proves fatal to half the animals in the test group. There is no such simple notation for chronic toxicity. Also classified under acute toxicity are irritation, including dermatitis and sensitization. The latter term refers to an increase in adverse reactions after repeated contact and can be similar in physiological mechanism to allergies such as hay fever.

The toxic effects of vapor inhalation are usually expressed in terms of a maximum allowable concentration [9]: "the upper limit of a contaminant in the air which will not cause injury to an individual exposed continuously during the working day and for indefinite periods of time." This maximum allowable level, sometimes known as the threshold limit value (TLV), is based on comfort as well as toxicity data and is usually measured in parts per million in air.

For lists of compounds and their toxicity, the reader is referred to standard tables, such as the E. R. Plunkett, *Handbook of Industrial Toxicology*, Chemical Publishing Company, New York, 1976, or the reader may apply directly to the manufacturer or to the American Petroleum Institute or the Manufacturing Chemists Association. The latter has issued some excellent publications and safety data sheets. However, it must be emphasized that no two situations are identical; therefore, all available information must be obtained and the plant designed and operated accordingly.

Toxicity in relation to solvent extraction plant design and operation is usually associated with the safety of the operating personnel, but there is a growing awareness of the toxic effects of solvent extraction plant effluents on the environment, and both aspects of toxicity are discussed here.

(a) Hazards to Operating Personnel. Hazards to operating personnel can be classified as *oral*, *dermal* (skin–eye contact), and *inhalation*.

Although common solvent extraction chemicals are not particularly toxic [11], many of the aqueous phases are, but the prevention of oral exposure is readily achieved by an awareness of the workers of the potential hazard coupled with good personal hygiene.

Dermal exposure can be very common and must be considered a potential hazard, more particularly with aqueous phase contacts. Common solvent extraction chemicals are not particularly toxic, irritating, or sensitizing [11], but repeated and prolonged exposure can lead to adverse effects. Again, however, the awareness of the worker of the hazard and the provision of adequate protective clothing, goggles, and other protective paraphernalia and washing and showering facilities will reduce the exposure and hence the hazard.

The avoidance of toxic effects due to inhalation of vapor is normally achieved by maintaining the concentration of the toxic compound in the working atmosphere below the TLV value, or failing that, by the supply and enforcement of the wearing of personal respirators. Normally the TLV values of common solvent extraction chemicals are such that good ventilation/extraction equipment for plants enclosed in buildings, or unenclosed locations, are sufficient to prevent TLV values being exceeded for the general plant environment, and atmospheric monitoring should be used as a regular safety check.

Inhalation hazards can vary considerably with location within the plant, however. Any enclosed, unventilated space, such as under settler roofs, particularly in hot environments, should be regarded as a potentially hazardous area. Similarly, the dense nature of hydrocarbon vapors could result in their accumulation in plant sumps. Entry of personnel into such areas should be preceded by monitoring of atmospheric contaminant levels, and, where necessary, short residence times, ventilation, or personal respirators should be employed to reduce exposure.

(b) Hazards to the Environment. The effect of SX plant effluents on the local environment is of increasing concern, particularly in developed countries. Solvent extraction organic phase compounds and/or their degradation products can adversely affect aquatic life [12, 13]. However, the techniques of lime neutralization that are widely applied to the effluents from many plants for inorganic contaminant removal would be expected to remove entrained organic drops by way of adsorption of such drops on the solid precipitates formed. The soluble loss could also be decreased through a similar mechanism, although no data are currently available to confirm this. Removal of soluble organic materials by means of adsorption on active carbon has also been proposed [14] and is used commercially.

2.1.5. Corrosion Characteristics

The procedures in selecting materials to safely contain solutions in a solvent extraction plant are obviously very important in plant design but are not discussed in detail here as they differ little from most kinds of chemical plant. However, it is well worth remembering that combinations of organic and aqueous phases sometimes have corrosion properties that are not held by either of the individual solutions and in cases of doubt experimental evidence should be produced before material choice is finalized.

2.2. The Equipment

Equipment design must follow the needs of the process flow sheet, satisfy the requirements of the safety and design codes and standards pertaining, match the constraint of geographical location and climatic conditions, and be within the commercial limits imposed. No plant will be built if it does not satisfy the legal requirements as well as being a sound commercial venture. Engineers in many countries are fast becoming legally responsible for their design philosophies.

2.2.1. Vessels and Tanks

Vessels are designed according to the process requirement. Tanks should be of a size sufficient for appropriate storage and surge capacity. To minimize organic losses, "after-settlers" are frequently installed. These allow primary dispersion droplets to coalesce and settle from the raffinate, thereby preventing discharge of excess organic into rivers, dumps, or leach vats where they may cause environmental or process problems. A typical example is the discharge of raffinate into a lake or fjord popular with the fishermen. Although a discharge is within the toxic limit, it might cloud or discolor the water, resulting in protests from environmentalists, fishermen, and others. Care is especially necessary when bacterial leaching is practiced, as the organic phase or its decomposition products could be toxic to the bacterium in the dumps and thus slow down or hinder the rate of dissolution.

As a general rule, vessels and tanks are covered in order to keep out dust or debris and allow controlled ventilation where necessary, but the retention of the vapors enhances the risk of fire or explosion. When employing liquids with very low flash points, nitrogen is sometimes used as a blanket [15]. Floating roofs are occasionally used when highly inflammable liquids are employed as solvents. The roof should have very low electrical resistivity and be well grounded. Overhead filling with consequent splashing should be avoided. Additionally, where the vapors are contained, exposure of the operator to fumes containing a high level of toxicant or irritant follows. In the uranium industry ammonia vapor can issue from the strip settlers, especially when inspection hatches are opened for observation or sampling.

If mixer–settlers are used, the settler should be designed with a weir system and sufficient freeboard above the normal liquid level to ensure that in the event of abnormal conditions, organic solution will not overflow the unit. This can readily be achieved by incorporating an overflow weir (Fig. 1) that allows the aqueous phase to overflow preferentially before the organic phase reaches the top of the freeboard allowed [4]. This consideration is of special importance when water is used for fire fighting as it can be applied directly into the mixer–settler without

Figure 1. Emergency overflow weir on copper solvent extraction plants. (Photograph courtesy of Nchanga Consolidated Copper Mines Ltd.)

fear of spreading the fire by overflowing organic phase [4].

Depending on process requirements, a variety of materials of construction can be employed, ranging from stainless-steel-lined concrete [16] through epoxy resin [17] to Haveg materials [19]. In large tanks, means of access or exit for workers in cases of emergency should be provided [19].

2.2.2. Pipes

The design of piping determines the extent and degree of pressure losses, such as in the case of flow between stages with a pump–mix impeller system. The generation of static electricity is proportional to the velocity in the pipe and must, therefore, be restricted to safe values. The API code [20] gives a value of 6 ft/s where the discharge is always beneath the tank liquid level. However, it is recommended that a velocity of 3 ft/s not be exceeded at any point in the solvent extraction pipe system [2].

Fiberglass-reinforced plastic pipes and tanks (FRP/GRP) present a problem because they cannot be grounded, but antistatic agents have been incorporated with varying degrees of success to prevent charges building up. Bonding and grounding of pipes must be carried out effectively. Painting the outside of FRP/GRP pipes with a conducting paint to dissipate any static has been suggested.

It has been suggested that piping be designed with a maximum of mechanical joints in order to minimize maintenance welding [22].

Finally, the location of the pipe runs are important. Wherever possible, organic solution should not be carried overhead in plastic or FRP/GRP pipes. If fires occur beneath, the pipes could melt, allowing the contents to shower over the area, thereby spreading the fire risk.

2.2.3. Motors and Pumps

The specification of electric motors to be used in potentially hazardous areas is laid down in regulations according to the country of use (e.g., [21]). However, for most commercial solvent extraction plants, based on kerosene-type diluents, the conditions necessary for a spark-induced explosion or fire are not present under normal operation, in which case spark or explosion-proof equipment can be specified. Systems in which conditions for a spark-induced explosion or fire are always present require flameproof equipment, and such equipment may also be necessary in specific hazardous sections in an otherwise nonhazardous area.

The choice of pump type is normally based on characteristics, client experience, and cost. However, environmental considerations related to leakage and production of difficult-to-separate organic/aqueous entrainment or emulsions also affect pump selection and design.

2.2.4. Instruments

Instruments can be divided into two categories: those that are essential for operation and control of solvent extraction plant and those that warn of malfunctions. In each case, minimum requirements are established by the code or standard in force (e.g., [21]). Care must be taken where automatic fire signals are installed [2]. All conductors immersed in organic liquids should be well grounded. This applies to such equipment as floats and conductivity probes.

2.2.5. Sumps

The containment of deliberate or accidental spills is one of the most controversial aspects of design, especially the location of drains, gullies, bund walls, dikes, tanks, or pits. As a general rule, storage tank areas should be diked and process sumps be designed to contain 110% of the contents discharged. In the case of mixer–settlers, it is usually 110% of the solution volume in a single mixer–settler. If the material is toxic and persistent in the environment, the dike bottoms should be impervious.

The run of rain and process water gulleys should be such that if spills of organic find their way to the gulleys and are accidentally ignited, the flames will not spread on the surface of the water under or along the full length of the plant.

There is a scarcity of reported information on fires that have occurred in solvent extraction plants, including the most recent one at the site of Rossing Uranium Ltd. in Namibia [18]. One of the fullest descriptions is that by R. Hoey-Petersen [19], who describes the changes in the reconstructed plant at Kristiansand, Norway. One of the most important features was the quench drainage system with a collecting tank in concrete buried in the ground outside the building, as shown in Fig. 2.

The practice of removing the spillage return sumps, or collecting to a location away from the main solvent extraction plant has been followed after fires at other locations.

2.2.6. *Fire Protection*

Fires involving high-flash-point products are generally extinguished by cooling, whereas fires involving low-flash-point products are extinguished by smothering. The dividing line between high- and low-flash-point products is a *fire point* of 45°C, where the fire point is the temperature at which the bulk liquid phase can be made to ignite and burn for at least 5 s.

A properly applied water fog or spray is a very suitable method of protecting solvent extraction settlers that have built-in safety facilities for dealing with overflow. The spray system may also be used to control rise in temperature in the vapor space and may thus even prevent fire. Storage tanks containing flammable liquids should not be

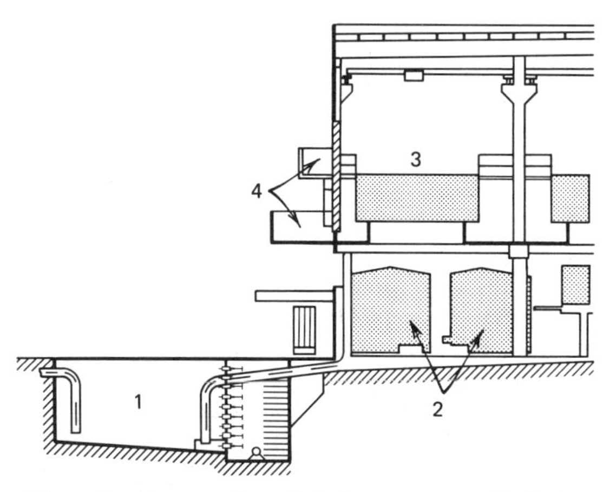

Figure 2. Cross-section of drainage system: (1) drainage tank; (2) surge tanks for solvent; (3) mixer–settlers; (4) escape balconies. (Diagram courtesy of Falconbridge Nicklewerk AS.)

treated in this way because of the wide variation in working level.

Spillage fires require a different approach. Water will not extinguish a fire from a free spillage, which it will usually spread, whereas foam will tend to smother with minimal spread and would be effective on relatively small spills. However, if flammable product continues to discharge, such as from a burst pipe, the area that it could cover could be extensive and require the application of large quantities of foam. Therefore, the design should ensure that any major spills are automatically channelled away from hazardous areas by a safe route to a safe location where they can be contained for extinguishment and/or burnout. In these circumstances fire water hoses provide a very effective means of helping the burning liquid on its way. However, if so much water is used that the sump overflows, burning liquid will again be spread; therefore, the location of the sump is important.

The choice between water and foam must be made for the main hazards, but small fires can be treated by a variety of media, including dry powder which is said to be effective in flame knockdown.

At one time the use of foam was criticized because contamination could potentially arise. More recently the philosophy of plant operators has changed to comparing the cost of losing a charge of inventory with the cost of losing the whole plant through fire. Thus foam is finding more favor than in earlier days.

However, water is more generally available than foam or other extinguishants that may have to be imported into some countries and thus remains the more common first line of defense [23]. Carbon dioxide has been used on occasions as an inert medium within the gas space of liquid–liquid contactors.

A fire detection system is essential. A high-temperature alarm is advisable in mixer–settlers. This should not only control the water spray to the unit from which the signal was received, but should also spray the adjacent units until the cause of the temperature rise has been established and should stop all process flows. If foam is the chosen means of fire protection, the alarm should control only the supply to the individual unit. However, there is a trend toward simultaneous blanketing beneath all the mixer–settlers.

Dangerous conditions signalled from areas other than the mixer–settlers should be dealt with by use of manual equipment, including portable adjustable spray monitors.

If foam is used as a primary or secondary line

of defense, the "light-water" type [4] is said to have the least contaminating effect.

2.3. Layout

To secure maximum fire prevention and protection, the three ingredients of fire–fuel, oxygen, and a source of ignition–must be kept separate to the greatest extent consistent with the design.

Local or national standards control the final choice of location of the vessels, tanks, and pipe runs, but the NFPA standard [3] offers a useful guide, as does API RP500. The main points in the standards are summarized in the following paragraphs.

No sources of ignition may be installed or brought within 15 m of the extraction processes unless the process is purged of solvent. There may be no sources of ignition within 30 m of vapor travel; vapor barriers of noncombustible material may be used to satisfy this requirement. The more safety conscious operators are known to have isolated all the equipment in separate rooms with separate fireproof doors even to the extent of an individual pump room.

The fill spout of any solvent storage tank and the tank, if above ground, must be 7.5 m from the process building. A fence to restrict access must be built 15 m from the process. Apart from the security aspect, it allows control of the accidental source of ignition of the cigarette or pipe smoker or the ignorant operator who requires a "quick light" to read a gauge.

Many of the larger solvent extraction plants are built in the open. If plants are built in a confined area, adequate ventilation is required. In certain instances, buildings are put under a positive pressure.

Emergency exits must be clearly identifiable and well lit (by emergency lighting if necessary), leading to balconies and escape ladders. Operators are often reluctant to use alternative escape routes. It must be remembered that solvent fires spread very rapidly, giving off smoke and acrid fumes, immobilizing the operators very quickly unless they can move freely to a source of fresh air. Personnel must have easy access to multiple exits.

The location of pipes, vessels, and pumps has been described earlier. However, it must be remembered that the tendency to arrange mixer-settlers together so as to minimize pressure drops could mean an increase in fire risk coupled with a restriction to easy access necessary in cases of emergency. Walkways over pipe tracks minimize the need to walk directly on the pipes and lessen the risk of damage.

Sample points must be conveniently placed; it is better to err in placement of excess points rather than too few. The inclusion of a temporary sample point during commissioning or troubleshooting can be an unnecessary added risk. Covers that are to be used as walkways must be designed as such; those that are not must be clearly identified as unsafe.

Side windows can serve as a useful aid to operation. However, they do need to be carefully located, be of leakproof construction, and be accessible to ready cleaning. Installation of skylights on settler roofs may avoid the need to lift inspection hatches that expose the operator to additional risk.

All vessels, pipes, and so on must be well grounded. However, ground fault protection devices should not cause unnecessary power failure; short circuits produced in lighting by rain, snow, and process solution spills can cause safety hazards in the affected areas ranging from personnel being left in total darkness to sump areas and basements being filled with hazardous liquids. One solution is to specify 110 V lighting systems. Short circuits in 110 V systems are separated from substation ground fault breakers by transformers and usually trip individual circuit breakers. These short circuits do not disturb unaffected lighting circuits or process equipment. Short circuits in 440 V lighting systems can cause total facility blackout by tripping an entire substation. Sealing of light fixtures and their supplies is recommended.

Fires can be caused by lightning. Where necessary, it is usual to arrange grounding conductors at high points of the equipment to discharge any strike (see Fig. 3). In areas prone to thunderstorms, a system of grounded electric wires above the hazardous areas is required [20].

3. OPERATION

The key to successful and safe operation is the availability of adequate operating procedures plus their safe implementation. The designer must write clear *operating instructions* for startup, routine running, emergency procedures, and shutdown that can be readily understood by the technical person and easily transmitted to the nontechnical person.

3.1. Safety

Plant operators must follow the operating procedures, and safety should be regarded as a convenience and not a chore. Operators must con-

Figure 3. Settler-mounted lightning conductors. (Photograph courtesy of Nchanga Consolidated Copper Mines Ltd.)

scientiously practice safety awareness during routine duties. Appropriate safety clothes and equipment must be used, and precautions must be followed at all times [24–28]. The results of failure to comply should be readily understood by all. Staff should be aware that by not following necessary instructions, they may be endangering their colleagues as well as themselves.

Routine training and retraining could be especially beneficial where process changes have been introduced. Regular fire drill is recommended. Familiarity with emergency equipment could save lives.

3.2. Hazard Areas

In addition to the solvent extraction plant area, other services or supplies are provided from separate locations, including liquid ammonia storage on uranium plants. Adequate safety measures must be included in their design and location, and operators must be equally familiar with their characteristics.

3.3. Welding

Welding is acknowledged to be the most dangerous activity to be carried out in the solvent extraction plant areas. Some plants remain on-line during weld repairing of mixer and settler bottoms and sides. The area around the welds is fan ventilated to remove any vapor accumulation. The system fluid is said to act as a heat sink. However, it is more usual to take the plant off-line.

3.4. Organic Content of Effluent Streams

The organic content of effluent streams is composed of two parts: the soluble or dissolved part and the entrained part. The amount of dissolved organic is a function of the aqueous system characteristics (temperature, pH, etc.) as well as the nature and characteristic of the organic phase itself, whereas the amount of entrained organic is a function of operating influences. As a general rule, the organic content of an aqueous phase leaving a solvent extraction plant can be assumed to be the solubility in parts per million plus 50 ppm.

There are various ways of minimizing the amount of organic phase lost in the outgoing aqueous streams, including flotation and coalescence [29], which are effective in reducing entrained organic losses, and solid adsorbents such as active carbon [14], which are usually employed to recover soluble losses. Although economic and unit operation interfacing problems have been the main driving forces for the inclusion of organic recovery equipment in the past, additional pressure is expected from the growing awareness of the environmental effects.

REFERENCES

1. J. E. Heilman, *J. Am. Oil Chem. Soc.* **53**, 293 (1976).
2. C. E. M. Critchfield, *J. Am. Oil Chem. Soc.* **53**, 295 (1976).
3. National Fire Protection Association No. 36, *Recommended Practice on Solvent Extraction Plants*, 470 Atlantic Avenue, Boston, MA 02210.
4. G. C. Collins, J. M. Cooper, and M. R. Bandy, Solvent Extraction Plant Design–Safety Aspects, paper presented at the SCI Symposium on Safety in Solvent Extraction Plants, Fire Research Station, Borehamwood, Hertfordshire, UK, April 26, 1977.
5. Electrical Safety Procedures, ISA Monograph No. 110, Instrument Society of America (1965).
6. J. S. Dorsey, *Chem. Eng.* **83**, 203 (September 13, 1976).
7. A. Klinkenberg, Laboratory and Plant Scale-Up Experiments on the Generation and Prevention of Static Electricity, paper presented at the 37th Annual Meeting of the American Petroleum Institute, Chicago, IL, November 12, 1957.
8. F. G. Eichel, *Chem. Eng.* **74**, 153 (March 13, 1967).
9. Anonymous, *Ind. Eng. Chem.* **19**, 742 (1943).
10. M. J. Wallace, *Chem. Eng.* **85**, 73 (April 24, 1978).
11. Anonymous, General Mills, Inc., Chemical Division, Kankakee, IL, USA, *LIX64N and Alamine 336 Toxicity Studies.*
12. A. W. Ashbrook, I. J. Itzkovitch, and W. Sowa, *Proceedings of the International Solvent Extrac-*

tion Conference (ISEC), Toronto 1977, Vol. 2, Canadian Institute of Mining and Metallurgy, Montreal, 1979, p. 781.

13. G. Dave, H. Blanck, and K. Gustafsson, Biological Effects of Solvent Extraction Chemicals on Aquatic Organisms, paper presented at Society of Chemical Industry and Inst. Min. & Met. Symposium on Impact of Solvent Extraction and Ion Exchange on Hydrometallurgy, Salford University, UK, March 21-22, 1978.
14. G. M. Ritcey, B. H. Lucas, and A. W. Ashbrook, *Proceedings of the International Solvent Extraction Conference (ISEC), Lyon 1974*, Vol. 3, London, 1974, p. 2873.
15. Private communication, November 1978.
16. J. A. Holmes, A. D. Deuchar, L. N. Stewart, and J. D. Parker, Design, Construction & Commissioning of the Nchanga Tailings Leach Plant, in Y. C. Yannopoulus and J. C. Agarwal, Eds., *Extractive Metallurgy of Copper*, Vol. II, TMS (AIME), 1976.
17. K. Power, Operation of the first Commercial Liquid Ion Exchange and Electrowinning Plant, *Proceedings of the Extraction Metallurgical Division AIME Symposium*, Denver, 1970.
18. R. J. M. Wyllie, *World Min.* **32** (1) (1979).
19. R. Hoey-Petersen, Fire Prevention in Solvent Extraction Plants, *Proceedings of the First International Loss Prevention Symposium*, The Hague/Delft, The Netherlands, May 28-30, 1974.
20. *Recommended Practice for Protection against Ignitions Arising out of Static, Lightning and Stray Currents*, API Report No. 2003, October 1974.
21. *National Electrical Code Article 500*, Chapter 5, *Hazardous (Classified) Locations*, 70.347.
22. R. J. Klotzbach, private communication, Union Carbide Corp., New York.
23. P. Nash, *Fire Prevent. Sci. Technol.* (10) (December 1974).
24. D. M. Barber and A. M. Tibbetts, *Chem. Eng.* **83,** 2 (February 2, 1976).
25. L. K. Herrick, *Chem. Eng.* **83,** 147 (October 18, 1976).
26. R. D. Jonathan, *Chem. Eng.* **82,** 147 (September 15, 1975).
27. J. R. Gauerke, *Chem. Eng.* **79,** 108 (April 3, 1972).
28. W. I. Morton, *Chem. Eng.* **83,** 127 (October 18, 1976).
29. G. C. I. Warwick and J. B. Scuffham, The Design of Mixer-Settlers for Metallurgical Duties, paper presented at the International Symposium on Solvent Extraction in Metallurgical Processes, Technologisch Instituut KVIV, Antwerp, Belgium, May 4-5, 1972.

31

SAFETY DESIGN FOR NUCLEAR EXTRACTION

J. A. Williams and W. J. Bowers
British Nuclear Fuels Ltd.
United Kingdom

I. INTRODUCTION

Solvent extraction is used in many parts of the nuclear fuel cycle. It is used to extract pure uranium from partly purified ore for the manufacture of nuclear fuel for reactors and in the reprocessing of nuclear fuel after its irradiation in a nuclear reactor to separate valuable plutonium and uranium from the fission products. It is also used for the purification or recovery of plutonium and the recovery of valuable by-products such as neptunium. Many types of fuel are produced for use in nuclear reactors, including (1) natural uranium metal or oxide fuels for MAGNOX or CANDU reactors, (2) uranium oxide fuels enriched in the isotope ^{235}U to about 3% for light-water-moderated reactors (BWR, PWR) or for advanced gas-cooled reactors (AGR), (3) mixed uranium–plutonium oxide fuels enriched up to 30% in PuO_2 for fast breeder reactors, and (4) uranium–aluminum metal alloy fuels containing uranium enriched up to 90% in ^{235}U for material testing or high-flux reactors.

The safety requirements for the processing of nuclear fuel depend on the part of the fuel cycle being considered, together with the type of fuel being treated. Safety considerations for nuclear chemical plants will apply to the whole of the plant, not only to solvent extraction as a unit operation. The instrument lines, feed lines, ancillary tanks, and pumps must all be taken into consideration. Good engineering design practice must be applied, of course, but in addition, the processing of nuclear fuel introduces special factors that are considered here.

Throughout the fuel cycle, nuclear fuel contains radioactive materials. The plant operators and the public must be protected from these materials, which may affect them by external irradiation or by internal radiation caused by inhalation or ingestion of radioactive materials. The latter may occur by inhalation or by ingestion of the material by means of a food chain that might in itself concentrate specific radioactive isotopes. Some elements are retained by the body and collect in particular organs.

Several types of radiation must be taken into consideration. *Alpha particles* are helium nuclei that are emitted in the energy range of 4–9 meV. Most alpha emitters are elements of high atomic weight (>200); ^{238}U, ^{239}Pu, and ^{241}Am are important alpha emitters. The range of an alpha particle in air is only a few inches. *Beta particles* are electrons emitted at high speed from the nucleus of the radioactive atom. The particles have ranges up to many feet in air and will penetrate a few millimeters of body tissue, some of the energy loss giving rise to x rays. Most of the fission products formed in a nuclear reactor give rise to beta emissions. *Gamma rays* are electromagnetic radiation of very short wavelength and are very penetrating. The rays often accompany alpha- and beta-particle emission and are emitted with discrete energies, indicating the transition of an excited nucleus to a more stable energy level. *Neutrons* are uncharged particles of weight similar to a proton. Neutrons are emitted when certain heavy elements undergo spontaneous fission and also by the interaction of alpha particles with light elements such as aluminum, oxygen, or fluorine (see Table 1).

The safety problems associated with the treatment of nuclear fuel are fourfold. *Containment* is the need to confine radioactive materials under both normal operation and accident conditions. The concentration of radioactive elements must be limited in the air that the operators or the public breathe and in the liquid effluent that is discharged from the plant (see section 2). *Shielding* is the protection of operators from external effects of radioactivity (see Section 3). *Criticality* is reached when fissile material in sufficient quantity gives rise to a fission chain reaction with the consequent generation of a high neutron flux and considerable heat. This situation must be avoided (see Section 4). *Long-term plant reliability* is needed under radioactive conditions because normal maintenance may be difficult or even impossible. Special plant designs or maintenance facilities must be used (see Section 5). Instrumentation is needed for control of these nuclear aspects of the plant (see Section 6). The plant must be designed to give an extremely small probability of the accidental release of radioactive material. In many countries special legislation lays down conditions under which nuclear processing plants may be built or operated. Techniques have been developed in the nuclear and aerospace industries that help in the assessment and design of "safe" plants; these are described with the legal aspects in Section 7. Some additional safety matters are noted in Section 8, whereas practical examples of plant design are quoted in Section 9.

The need to apply the design precautions outlined in this chapter will depend on the material being handled; not all the precautions may need to be taken in every case. Unirradiated natural uranium can be handled with a minimum of criticality precautions and with only limited shielding and containment. Irradiated fuels of all types must be handled behind heavy shielding and with full regard to containment and criticality precautions. Unirradiated plutonium containing alpha emitters may be processed in lightly shielded or unshielded plant depending on the amount to be handled and on the presence of higher plutonium isotopes (^{240}Pu) or decay products (^{241}Am), but it must be rigorously contained at all times to protect the operators from the dangers of inhalation or ingestion of the material.

TABLE 1. NEUTRON EMISSION

Spontaneous Fission		α-n reactions	
Isotope	Neutrons, $s^{-1}\,g^{-1}$	Compound	Neutrons, $s^{-1}\,g^{-1}$
^{238}Pu	2.3×10^3	$^{239}Pu\ ^4F$	4.3×10^3
^{240}Pu	9×10^2	$^{239}Pu\ ^2O$	45
^{244}Cm	1.2×10^7	$^{240}Pu\ ^4F$	1.6×10^4

2. CONTAINMENT

The object of containment is to minimize the release of radioactivity into the working environment or the public domain; normal conditions and accident conditions must be taken into account. The permissible levels of release are derived from the permissible levels of dose and ingestion recommended by the International Commission on Radiological Protection (ICRP). These recommendations are adopted in legislation by the national bodies responsible. The need for confinement is illustrated by the recommended maximum permissible concentration (MPC) in air and the body burden values for workers in the nuclear industry:

Nuclide	MPC, μCi/cm³	Body Burden, μCi (μg)
^{239}Pu	2×10^{-12}	0.04 (0.65)
Natural uranium	7×10^{-11}	0.005 (7400)

In solvent extraction equipment it is unlikely that a significant gaseous or vapor-carried release of radioactivity can occur under normal conditions. Release is possible under accident conditions such as a fire, major leakage, or a criticality accident. The primary containment is the process equipment, which must be designed with this in mind. Where the process or equipment might give rise to a suspension of activity in a gaseous phase, such as in an airlift pump or a steam ejector, means of removal of the acitivity should be included. The secondary containment is often a cell in which the plant is placed and where limited access or no access is permitted. The construction of the secondary containment will depend on the material to be processed; short-range activity such as alpha emission can be contained within transparent plastic cells with integral flexible gloves (glove boxes), whereas for penetrating radiation such as gamma-emitting material, a thick concrete wall may be required. Operation of the plant is normally external to the secondary containment and today is as far as is possible carried out from a remote control room. The need to reduce the radiation exposure of operators and maintenance personnel to levels that are as low as reasonably achievable must be borne in mind by the plant designers.

Ventilation of active plants is important and is described in detail in the *Regulatory Guides* [12]. The confinement of activity is by means of multiple bounded zones; zone 1 will be the plant cell or glove box, zone 2 will be an area surrounding the cell and where entry into the cell or removal of materials may give rise to a break in the zone 1 containment, and zone 3 will be the operating area. Pressure differentials should be maintained between building containment zones and the outside atmosphere to ensure that airflow is from less to more active areas.

Before release to the atmosphere, the air should be passed through one or several banks of efficient filters. It may be necessary to dry the air first to protect the filters. The level of activity in the air must be monitored before release to atmosphere. Large sand filters or the recycle of gas to the plant have also been used and may offer a more reliable system with lower effluent discharge in accident conditions.

It is necessary to have a standby electrical supply and means of continuous monitoring of activity in active areas. All materials of construction must as far as possible be fireproof and capable of being decontaminated from radioactivity. Access for maintenance and for the entry or removal of contaminated solid materials or equipment requires specially constructed facilities to minimize the possibility of adventitious release of activity.

3. SHIELDING

On leaving a reactor, a fuel rod will typically contain a few percent of fission products and other radioactive substances, known as *actinides,* which are formed by neutron absorption and subsequent decay. Radioactive decay results in the emission of alpha, beta, gamma, and neutron radiation. Alpha and beta radiation need not be considered in shield design because of their low penetrating power. The penetrating gamma radiation is due primarily to the fission product isotopes, and since most of these reach equilibrium in a reactor over a time short compared with the irradiation, the source strengths are dependent largely on the fuel rating (MW/metric ton) and cooling time. With the trend toward higher fuel burnup (MW day^{-1} metric ton^{-1} U), the actinide isotopes that approach equilibrium concentrations more slowly are causing additional neutron shielding problems. The principal sources, ^{238}Pu, ^{239}Pu, ^{240}Pu, ^{242}Pu, ^{241}Am, ^{242}Cm, and ^{244}Cm, gave rise to neutrons from both spontaneous fission and (α, n) reactions with light elements.

The energy range of interest for gamma radiation from irradiated fuel extends from the low-keV range to several MeV. Attenuation by shielding mainly occurs through the photoelectric effect, Compton scattering, and pair production. Each process results in some or all of the energy of the incident photon being deposited at the site of the interaction, and the last two cause emission at a lower energy in in a new direction. The probability p that a photon can penetrate a shield of thickness x (cm) without suffering absorption or deflection is $p = \exp(-\mu x)$; that is, the attenuation is exponential. It is possible to compute the linear attenuation coefficient μ (cm^{-1}) from the sum of the microscopic cross sections for the preceding processes. Because these processes all involve electrons, the attenuation coefficient and hence the rate at which gamma rays are attenuated are determined primarily by the number of electrons per unit volume. Thus high-density, high-atomic-number materials such as lead make the most efficient gamma shields. The order of attentuation obtained is illustrated by Figs. 1 and 2 [1], which show broad-beam transmission of gamma rays from ^{137}Cs by some common shield materials. It should be noted that such curves alone cannot be used to do shielding calculations since other factors, such as the geometry of a source-shield arrangement, are also important.

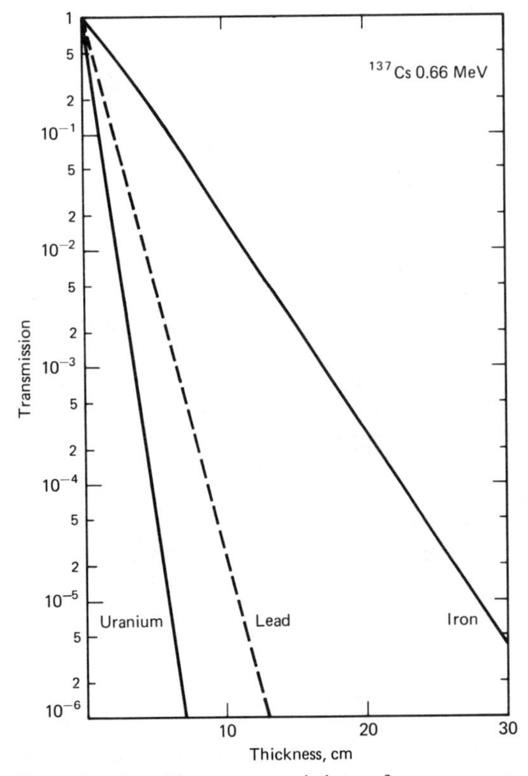

Figure 1. Broad-beam transmission of gamma rays from ^{137}Cs through iron, lead, and uranium.

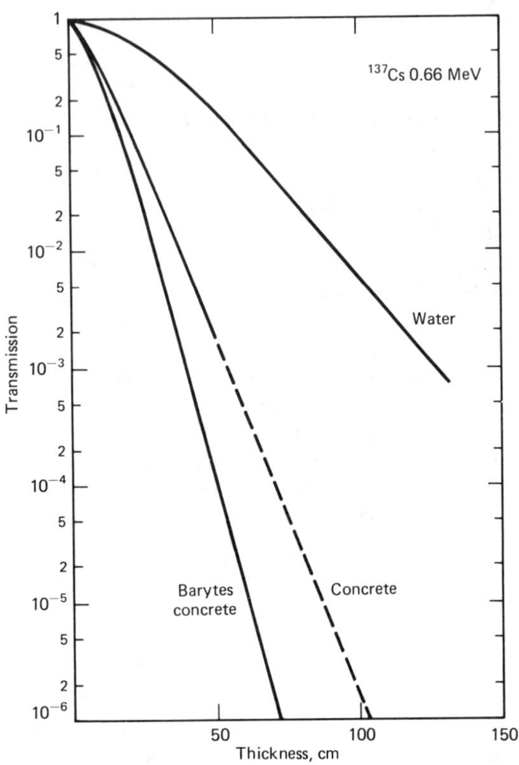

Figure 2. Broad-beam transmission of gamma rays from ^{137}Cs through water, concrete, and barytes concrete.

All gamma ray interactions of interest involve electrons, but neutron reactions involve nuclei, and because such reactions are more complicated and in many cases not well understood, most of the cross sections must be measured experimentally. However, a parameter similar to that used for gamma-ray calculations can be defined as representing the average attenuation of a great many neutrons suffering numerous interactions of various kinds. Normally known as a *removal cross section,* it is applied only to fast neutrons and can be justified by the argument that some collisions effectively ('remove') neutrons either through a large energy loss or change in direction. One of the most important interactions is elastic scattering of the neutron. Since the neutron can lose more energy in such an event if the nucleus of a light element is involved, the most efficient neutron shields have a large hydrogenous content.

Requirements for methods used in the assessment of chemical plant shielding are (1) computational efficiency, with a quick turn-around being essential for production shield design, (2) ability to model the complex arrangements of sources and shields typically encountered, and (3) accuracy consistent with other errors resulting from the definition of the source term and the basic nuclear data available. For the majority of cases involving the bulk shielding of gamma sources, these requirements are met by *kernel methods,* whereby a distributed source is considered as a large number of smaller sources and their contributions are evaluated and summed. The effect of scattered radiation is taken into account by the introduction of an empirically derived factor, known as the *buildup factor.* The technique can be used for simplified analytic solutions and for direct numerical integration. A recent useful tabulation of methods is the *Engineering Compendium on Radiation Shielding* [2].

Discrete-ordinate methods involve the solution of the integrodifferential form of the Boltzman equation, which is the fundamental equation describing the transport of radiation through a medium, by making the variables in the equation discrete. Because of the cumbersome nature of the three-dimensional (3-D) Boltzmann equation, codes based on this approach are generally applied in the situations where, because of symmetry and other considerations, one- or two-dimensional (1-D or 2-D) represenations are adequate. The coupled neutron–gamma ray libraries generally used with such codes [e.g., 3] allow a single calculation, which may also include the effect of a secondary gamma source, where otherwise separate neutron and gamma-ray calculations would have to be performed.

The *Monte Carlo codes* [e.g., 4] simulate the transport of particles through a medium and thereby effectively solve the Boltzmann equation by creating and tracking a large number of fictitious particles by use of random choice and the application of probabilities for possible interactions (e.g., scattering and absorption). Although the method is rigorous and allows a general 3-D description of the problem, a large amount of computing is often necessary to obtain a value with a sufficiently small standard deviation. Monte Carlo codes are thus generally used only when other approaches would not be appropriate (i.e., where scattering is important or the geometry rather complex).

4. CRITICALITY CONTROL IN SOLVENT EXTRACTION PLANT

4.1. General Outline of Criticality

The nuclear industry exists because of the ability of certain very heavy so-called fissile nuclei to split apart following the absorption of a neutron, to give fission product fragments and further neutrons. Each of these nuclear fissions liberates roughly 160 MeV of kinetic energy, the binding energy previously required to hold the fragments together, and two or three "fast" high-energy neutrons. If one of these secondary neutrons happens to collide with another fissile nucleus, a self-sustaining "chain reaction" may be set up that may "diverge" if the number of fissions per second increases with time or "converge" if the number of fissions per second decreases with time. A special case occurs if the number of fissions remains constant; that is, just enough neutrons collide with fissile nuclei to maintain the chain reaction. The system is "critical" under these conditions and "supercritical" or "subcritical" if the fission rate increases or decreases. The ratio of neutrons produced in successive generations is called the *multiplication factor k*.

The conditions required to bring about a critical system are numerous and interdependent. Under ideal conditions less than 500 g of plutonium may be critical. whereas under other conditions tons of plutonium may be safely kept in the same store. It is clear that if anything more than very small quantities of fissile nuclei are present in a system, the possibility of criticality exists. For a fission to occur, there must first be sufficient target nuclei present for there to be a good chance of a neutron colliding with a fissile nucleus. The probability that the neutron will then be absorbed and cause fission also depends on the energy of the incident neutron. The nuclear "cross section" of the target nucleus is a measure of the probability of collision, and this generally increases as neutron energies decrease, that is, as the incident neutrons become progressively "slower." For example, the ^{235}U fission cross section at slow "thermal" neutron energies is roughly 400 times that at the "fast" energies typical of neutrons released in a previous fission. A system is known as a "fast system" if a chain reaction is maintained by fast neutrons alone and a "thermal system" if the chain reaction can only be maintained if the neutrons are slowed down or "moderated" to lower energies. Neutrons slow down most effectively by colliding with nonfissile "moderator" nuclei, and the energy lost per collision depends on the moderating material. Light nuclei such as hydrogen are particluarly good moderators and can slow down a fission neutron to thermal energy in only 18 collisions (on average), whereas 18,000 collisions would be needed if the moderator were lead, for example. The residual "thermal" energy of a slow neutron results from thermal equilibrium with the moderator atoms. There is a chance that a neutron may not just "bounce off" a collision with a nucleus but may be absorbed in a "parasitic absorption" reaction and lost from the system. The probability of this depends on the "absorption cross section" of the nuclei (which is also dependent on the neutron energy), and certain materials are used as neutron "poisons" because they have very large absorption cross sections. At thermal energies, boron, for example, has a absorption cross section of 760 barns, compared to 0.332 barns for hydrogen and 0.17 barns for lead. (1 barn = 10^{-24} cm^2).

Hydrogen acts both as a moderator and an absorber, and these are usually competing effects in terms of effect on nuclear reactivity. Figure 3 shows how the minimum critical mass of uranium and plutonium varies with the hydrogen : fissile ratio. With no hydrogen present (i.e., completely dry, fully dense) the system is fast fissile only. As a moderator is added to the system (i.e., the H: fissile ratio increases) after an initial rise due to increased parasitic absorption, less and less fissile material is necessary to maintain the chain reaction. An optimum point is reached where the critical mass is a minimum, and beyond this point the system is "over moderated." Addition of further moderator does not slow the neutrons down more effectively, but increases the parasitic absorption in the system. Finally, so much para-

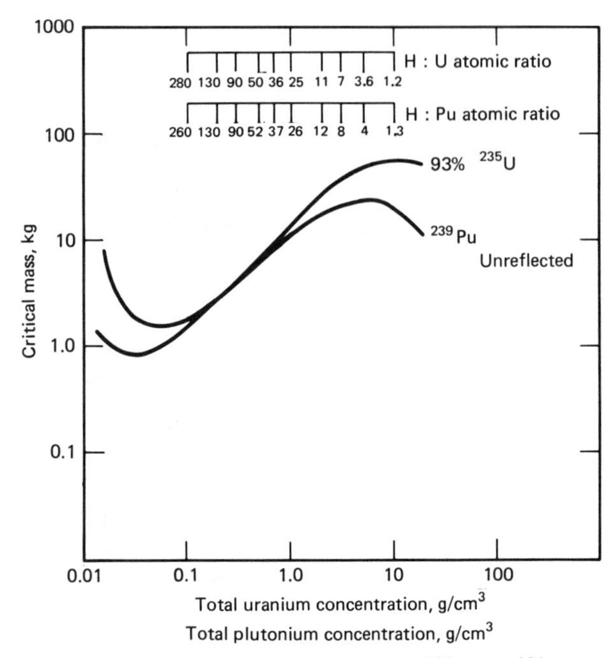

Figure 3. Smallest critical mass against concentration for 93% ^{235}U and ^{239}Pu metal–water mixtures.

sitic absorption takes place that the system cannot be made critical even with an infinite fissile mass. This is the so-called infinite-sea condition. The other possible fate for a neutron is for it to escape from the system entirely, before either parasitic absorption or fission. The amount of this neutron "leakage" depends on the density and total mass of material present and the size and shape of the system. Unless the fissile material is surrounded by a vacuum, however, there is a possibility that the neutron may be reflected back into the system to cause a further fission. The "reflector" material outside the fissile region also influences the nuclear behavior of the system. Figure 4 shows the effect of water reflection on the critical masses of Fig. 3.

We can now see which conditions influence the nuclear behavior of a system, and how they are interrelated:

1. *Neutron leakage* depends on total system mass, material density, and the geometric shape of the system (particularly surface:volume ratio). Thus a critical sphere may be subcritical if expanded to a lower material density, or rearranged as a thin slab or long cylinder, merely because more neutrons can escape. Leakage will be reduced in the presence of a reflector, depending on the reflector thickness and material.
2. *Neutron absorption* is a function of the material composition and particularly the fissile:parasitic absorber nuclei ratio. Competition for neutrons results in the nuclei with the largest value of (cross section times number of nuclei) absorbing the largest number of neutrons. Thus a critical system can be made subcritical by introducing a neutron poison that absorbs so many neutrons that the chain reaction converges.
3. *Neutron moderation* is also determined by the material composition. The introduction or removal of moderation will change the distribution of neutron energies (the neutron "spectrum") and hence all the cross sections and consequently the entire neutron balance of the system.
4. *Neutron interaction* is important in that any neutron lost through leakage may enter another fissile system and affect its neutron economy and vice versa. Two systems that are subcritical in isolation may, if brought together, reduce the total neutron leakage by exchanging neutrons and become critical. The degree of interaction obviously depends on the physical separation and orientation of the systems and the absorption and reflection properties of any material between them.

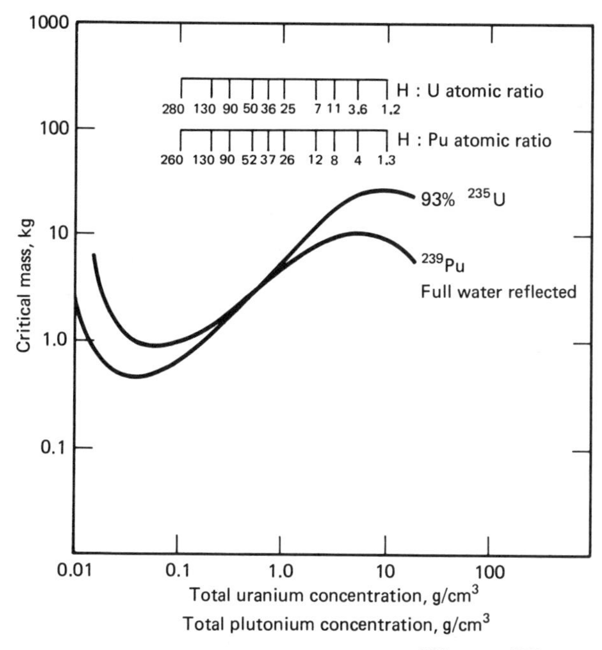

Figure 4. Smallest critical mass against concentration for 93% ^{235}U and ^{239}Pu metal–water mixtures.

4.2. Methods of Criticality Control

Once the underlying mechanisms are understood, the various methods of ensuring that criticality does not occur are largely self-evident. The most usually methods are given in the following list:

1. *Geometry control,* that is, the design such that there is sufficient neutron leakage for criticality to be impossible for any concentration, moderation or possible reflection conditions, leads to "ever-safe geometry" designs, including slab tanks, harp tanks, and columns. Usually, a geometry is not really "ever-safe," but safe only for a range of concentration, moderation, and so on. Operational control is essential to maintain the material in this range (see item 2 below).
2. *Mass control* involves the restriction of an isolated system to less than the known minimum critical mass under optimum conditions of moderation and reflection, so that there is always sufficient neutron leakage to guarantee that criticality cannot occur. Mass limits are most suited to small-scale batch processes.
3. *Composition–concentration control* keeps the fissile concentration always low enough to guarantee sufficient neutron leakage or parasitic absorption in other materials. This is quite often combined with geometry controls to give a system that is geometrically safe for a controlled range of fissile concentration.
4. *Moderation control* involves exclusion of moderating materials so as to considerably increase the minimum critical mass, allowing the safe accumulation of larger amounts of material. Moderation control alone is not sufficient to guarantee safety.
5. *Isolation* minimizes the neutron leakage between systems. If many fissile units are gathered together (e.g., a fissile storage array) each unit may be isolated to a greater or lesser extent by geometric spacing or the use of interstitial shielding–absorbing materials.
6. *Deliberate poisoning,* that is, the addition of a material with a large parasitic absorption cross section, can be very effective in reducing the system reactivity. Stringent operational controls are essential to guarantee that the correct concentration of poison is always present.

In practice, the safety of a given section of plant will often rely on a combination of the possible controls, requiring strict operational limits on combinations of mass, volume, concentration, composition, and other parameters.

4.3. Calculational Techniques

For criticality controls to be effective, the critical mass of fissile materials or the effective value k_{eff} of the multiplication factor k for a given system must be known. Experimental data for simple arrangements of fissile material are compiled in criticality handbooks [5]. This allows a fairly easy evaluation of systems with simple geometries and compositions by comparison with the tabulated critical parameters. Once the reactivity of an isolated unit has been evaluated, interaction effects can be estimated approximately [6].

For complex systems, a rigorous mathematical analysis is necessary for evaluation of the neutron balance and resulting k_{eff} value. These usually require the use of a computer code, and a full treatment is quite a formidable task. Three such codes are KENO [7], MORET [8], and MONK [9]. These use Monte Carlo techniques, which basically set up a mathematical model of the system; inject neutrons; and simulate the subsequent collisions, absorption, and losses. As an increasing number of neutrons are tracked, the estimates of k_{eff} improve. Because of the statistical nature of the code, there will always be some residual uncertainty, unless an infinite number of neutrons are tracked. It is important to remember that this is only one component of the overall uncertainty of the analysis. Other components will arise as a result of uncertainties in material composition, basic nuclear data, approximations in geometry representation, and so on.

4.4. Mechanism and Consequences of a Critical Excursion

If the k_{eff} value of a system slightly exceeds unity, the fission rate will increase exponentially with time at an approximate rate exp $\{[(k_{eff} - 1)/l]\ t\}$. The parameter l is the average time between successive nuclear fissions. Not all neutrons are released instantly following fission; however, a fraction β of typically 0.7% are delayed by roughly 0.1 s (on average). If k_{eff} is less than $1 + \beta$ (i.e., about 1.007), the chain reaction depends on these delayed neutrons. This condition is known as *delayed critical*, and rates of power rise are relatively slow. For example, a system with a $k_{eff} = 1.001$ would typically double in fission rate every minute or so. If k_{eff} is greater than $1 + \beta$, the chain reaction can be maintained by the "prompt" neutrons instantly released on fission. As the system does not now have to wait for the delayed neutrons, the fission rate increases much more quickly. For example, if $k_{eff} = 1.01$, the fission rate would typically double every 10 ms. This condition is known as "prompt critical" and clearly can give rise to large releases of energy and radiation.

Normally the physical changes in the system caused by an exponential power excursion lead to negative reactivity feedback effects. The density or concentration of the fissile material (giving greater neutron leakage) is reduced, and in certain systems the parasitic absorption cross section increases with increased temperature (giving greater neutron absorption). After an initial "power burst," these effects shut down the chain reaction. If the initial conditions that gave the supercritical reactivity persist, the reaction may start up again and oscillate its fission rate for a time until finally settling down to a "just-critical" level of $k_{eff} = 1$ where the negative feedback exactly balances the inital excess reactivity. Experiments to investigate these effects have been described [10].

Total energy yields from observed excursions normally fall in the range 10^{17}–10^{19} fissions, and the energy release from 10^{18} fissions is roughly equivalent to the energy release from exploding about 10 kg of TNT. However, the disruptive effects of such an excursion will not necessarily be so severe, depending on the size of the initial power burst, which, in turn, depends on the maximum superreactivity achieved. For a prompt critical excursion, the power burst can be very disruptive, although in liquid systems this is often not the case unless large pressures build up. For a very slow approach to critical followed by a delayed critical excursion (e.g., starting up a nuclear reactor), there may be no initial power burst, and the energy release may be spread out over many seconds or even minutes. The major hazard thus will seldom be from an explosive shock wave, but from the very large radiation. These effects have been discussed for several historic cases [11]. A more recent excursion occured at a plant reprocessing highly enriched uranium fuel, using a pulsed column solvent extraction system. A dissolver solution containing high-enriched uranium was fed to a pulsed column with a countercurrent TBP flow. Aluminum nitrate at 0.7 *M* was normally used to take the uranium into the solvent phase. The strength of this nitrate feed fell to 0.07 *M*, leading to a concentration of uranium around the feed plate in excess of the controlling maximum safe concentration level. A critical configuration was achieved, and an excursion of 10^{18}–10^{19} fissions occurred. The excursion took place in a shielded cell, so there was no operator dose uptake. The approach to critical must have been quite slow, and the system was probably delayed critical, as there was no observable damage to the exterior of the column. The large total energy release must have caused large pressure transients in the ventilation system, and short-lived fission products from the event were discharged from the site stack. This is a classic example of criticality control by operational procedures (fissile concentration control by the strength of the aluminum nitrate solution) that failed when the procedure was not maintained.

5. LONG-TERM RELIABILITY

Nuclear processing plant is built inside containment and because of the presence of radioactive materials there is limited access for operation or maintenance.

Plants containing only alpha emitters can be contained in glove boxes and the maintenance and operation carried out by use of gloves or simple tools that penetrate the containment wall. In the presence of gamma-emitting materials, however, thick concrete shielding may be necessary and operation is carried out remotely.

The *remote maintenance* approach has been applied at the Hanford, Savannah River and Idaho plants among others in the United States. The plant is constructed behind heavy shielding and essential to the system is the "jumper" system for making and breaking pipe joints remotely. After the joints have been broken, the plant item is removed by means of remotely operated cranes and master–slave manipulators, to a cell where the item can be maintained and/or cleaned. In the meantime the plant item is replaced by a new unit, and the "jumper" joints are remade. The system requires precision jigging and manufacture and must be combined with reliable hoists and manipulators, and is thus expensive. It has the advantage that repairs can be carried out comparatively quickly and it is possible to introduce plant modifications after the plant has been in operation. The system has operated successfully on very large plants for many years.

The *direct maintenance* system used in the Windscale plant in the United Kingdom, and in continental Europe, relies on the plant being extremely reliable. The plant design is kept as simple as possible, moving parts are avoided and all joints are welded and of high quality. Normally all parts of the plant are constructed in corrosion resistant material. Where moving parts cannot be avoided, they are kept, as far as is possible, away from the activity and are preferably brought outside the shield wall. Where this cannot be done the moving part is put into an isolated shielded area (a bulge) to minimize the radiation present and to facilitate semiremote access.

When major maintenance is needed, the plant is decontaminated over a long period in order to reduce the activity levels so that workers can enter the cell. Major plant changes have been carried out in this way at Windscale and Dounreay in the United Kingdom and at Marcoule in France.

The remote maintenance system makes possible the use of more complex solvent extraction equipment, simple drives on mixer–settlers, or the use of centrifugal settling systems. The direct maintenance system requires the use of simpler equipment; box-type mixer–settlers with the bearings and drive taken outside the shielding by way of a long shaft or of air-pulsed columns or mixer–settlers.

6. INSTRUMENTATION AND CONTROL

It is conceivable that radioactive material could be released either in airborne or liquid-borne substances discharged through the normal effluent channels or as the result of breach of plant containment by a fire, explosion, criticality incident, missile impact, or other occurrence. (A criticality incident can be defined as an uncontrolled and divergent fission chain reaction in a system that is designed to operate remote from critical (see Section 4.)

Complete protection against hazards arising from incidents generated internally is based on the plant being operated at the designed conditions. Controls and instruments must be provided for the various plant parameters and to warn the operator of unacceptable divergencies from normal conditions. Failure of this equipment must be included in the overall fault analysis, but it is worthwhile outlining the design philosophy that might be applied in relation to a nuclear extraction plant.

The instrumentation may be considered in two broad categories: (1) that which monitors the performance of the plant for the purposes of plant control by either the operator or automatic equipment linked to the instrumentation and (2) that which is provided solely for personnel protection; generally, this warns the operator when an abnormal condition has been reached. For all instrumentation, it is necessary to assess the performance of the complete system in relation to the demands set by the plant. This frequently shows the need for diversity of measurement and redundancy of equipment. The possible failure modes of the instrumentation and control systems must be determined and the equipment designed to be ('fail-safe') rather than ('fail-to-danger.')

In addition to the normal process control instrumentation, certain instrumentation is specific to nuclear plants. This is reviewed in Sections 6.1–6.4.

6.1. Critical Incident Protection

To ensure that solutions containing fissile material are left at safe, subcritical concentrations, it is necessary to control the strength and rates of flow of the inactive feeds.

For plutonium, the concentration of fissile material in solution can be monitored by scintillation counters, measuring the emitted alpha particles or 384-keV gamma photons or by neutron counters using boron trifluoride counting tubes. The neutron counters make use of the alpha–neutron reaction that occurs in the plutonium solution. Because large numbers of counting tubes are used, scanning systems and automatic data processing are employed. The response of alpha-particle counters to background gamma radiation must be taken into account, as well as the possibility of other alpha-emitting elements being present in the plutonium-bearing stream.

6.2. Critical Incident Detection

Despite the design safeguards against a criticality incident, the possibility of such an event cannot be ruled out. To enable the plant operators to evacuate the building as quickly as possible, it is essential for an unmistakable warning to be broadcast throughout the plant within less than a second of the incident. Detectors are used that respond to gamma radiation; care must be taken to ensure that adequate monitoring is provided for all parts of the plant with the potential for a criticality incident. Information from these instruments is sometimes presented in a criticality control center, remote from the plant. The gamma flux at a detector is inversely proportional to the square of the distance from the point where the critical incident occurs and is also dependent on the attenuation of the radiation due to pipes and other materials. It is necessary, therefore, that these limitations on the effective sensitivity of the detector be considered in relation to their siting.

6.3. Airborne Radioactive Contamination

The plant ventilation system is designed to ensure that under normal operating conditions air always flows away from inactive areas. With a continuous flow of air into the plant, it is necessary for air to be continuously exhausted to the environment, usually by way of a gas clean-up system and a stack. The effluent air must be continuously monitored for the level of alpha and beta particulate activity.

Within the plant it is necessary to measure the level of airborne particulate activity to warn the operators of any possible hazard arising from a breach in the plant containment. This type of monitoring is usually achieved by trapping particulate matter on a filter that is then exposed to a radiation detector.

6.4. Contaminated Liquid Effluents

All liquid effluents from the plant must be monitored to assess the quantity of radioactivity discharged. In addition, it may also be necessary to continuously measure the specific activity of the effluent so that action can be initiated if it exceeds a predetermined level.

7. LEGISLATION AND SAFETY ANALYSIS

In all countries operating nuclear fuel manufacturing or reprocessing facilities, some kind of regulatory control is exercised, and this usually takes the form of licensing by a government ministry or a federal agency. The precise details obviously vary throughout the world, but they all have a common aim, to ensure that the plant does not present an unacceptable risk to members of the public or to its operators.

In most cases regulatory control is exercised in relation to (1) the siting of the facility, (2) the design, (3) the commissioning, and (4) the operation (including the discharge of gaseous and liquid effluent and the disposal of solid radioactive waste).

Approval is given on the basis of documents that must be prepared by the designer, constructor, and operator of the plant (usually the licensee) and submitted to the appropriate regulatory authority. USNRC Regulatory Guide 3.26

[12] indicates typical requirements for this documentation. In some countries, such as the United Kingdom, the regulatory control is exercised by more than one government ministry. The responsibility for authorizing the discharge of effluents, for instance, may rest with a ministry different from that which is responsible for approving the operation of the plant.

It is the duty of the regulatory authority to assess, as far as is reasonably practicable, that the plant does not present an undue risk under (1) normal operating conditions and (2) postulated fault conditions caused by occurrences within the plant and by credible external events.

There is a worldwide move toward the quantitative assessment of the probability and consequences of fault conditions [13]. It is generally considered that the level of risk to which a citizen is subjected by industry should be considerably less than the level of risk which that citizen accepts voluntarily, such as that involved in traveling by automobile. Efforts have been made also to distinguish between the level of risk that is acceptable to individual members of the public and the level of risk that will be accepted by society in the cases where an accident could cause a considerable number of deaths [14].

7.1. Siting

Clearly, if an accident could result in a release of radioactive substances to the environment, the risk to the public will depend to a great extent on the location of the plant in relation to centers of population. The probability of some accident conditions could also depend on the seismology, geology, hydrology, and so on of the site and its proximity to other industry, military establishments, and airfields. For these reasons, siting of of the plant is always subject to regulatory control.

7.2. Design

The design must be assessed to ensure that under normal conditions there is no accumulation of excessive radiation doses by the personnel involved. The recommendations of ICRP are that radiation doses should be as low as reasonably achievable, with social and economic factors taken into account [15]. To demonstrate compliance with this recommendation, it is necessary to make a detailed assessment of the radiation doses incurred in all foreseeable tasks and to show that further reduction in these doses could be achieved only at a cost greater than that which could be justified in relation to the benefit obtained.

Equipment reliability is of key importance in the nuclear industry for the following reasons:

1. Maintenance of radioactive plant, in general, is achieved only at the expense of some radiation dose uptake by the personnel involved. Also, it frequently generates radioactive waste that must be stored or disposed of. Thus maintenance requirements must be kept to a minum.
2. The safety of the plant is usually dependent on the correct functioning of a number of engineered safeguards such as containment barriers, safety interlocks, emergency electrical supplies, instrumentation, and so on.

Therefore, to justify the adequacy of the plant design, it is usually necessary to assess essential equipment quantitatively. The accuracy of these assessments depends on available data on factors such as failure rates and repair times for plant items. In many cases it is necessary to estimate these by using data relating to similar equipment and applying weighting factors to allow for differences in operating environment, maintenance or testing requirements, and the like, or from a detailed analysis of the components.

7.3. Fault Analysis

This demonstrates, as far as is possible at the design stage, that unplanned releases of radioactive material caused by plant faults and by external hazards such as earthquakes, gas cloud explosions, missile impacts, floods, and high winds do not expose the operators or the public to an unacceptable level of risk; therefore, it is necessary to perform a detailed fault analysis.

The object of such analyses is to determine the fault sequences that could occur and their probabilities and the resulting consequences. As the consequences increase in severity, the probability of occurrence must decrease [16]. This concept is equally applicable to nuclear extraction plants or to nonnuclear chemical plants. In the identification and analysis of fault conditions, increasing use is being made of event and "fault tree" techniques pioneered by the aerospace industry and developed further in the study on the safety of water reactors in the United States headed by Norman C. Rasmussen [17]. These techniques impose a structured deductive discipline on the safety assessor and provide the logical framework necessary for the quantification of accident probabilities and consequences. The application of the event tree–fault tree technique is a complex and often interactive process.

8. ADDITIONAL SAFETY MATTERS

There are a number of specific areas in the solvent extraction field that must be considered:

1. The presence of "zirconium" fines may constitute a problem. Light-water fuel is canned in Zircaloy and can give rise to the presence of a proportion of fine metal when the fuel is broken up and dissolved. Zircaloy fines may react violently with water when in a moist state (2–15% w/w H_2O) or under certain conditions with nitric acid. The conditions under which reaction can take place are not known precisely; therefore, it is prudent to remove Zircaloy fines prior to solvent extraction.

2. In the presence of high fluxes of radioactivity, organic solvents can break down. This is known as *solvent degradation.* Tributyl phosphate, for example, will break down to mono- and dibutylphosphates that will strongly bind some of the fission products or plutonium and contaminate effluent streams or later stages of

TABLE 2. REPROCESSING PLANTS AND THEIR SOLVENT EXTRACTION EQUIPMENT

Plant	Startup	Contactor	References
Savannah River (USA)	1954	Mixer–settlers	26
SRP (1966)	1966	Centrifugal contactor for highly active cycle	18
Hanford (USA)			
Uranium recovery plant	1950	Pulsed plate columns	19
Purex reprocessing plant	1956	Pulsed plate columns	20
Marcoule (France)	1960	Mixer–settler; pulsed plate column	
Idaho (USA)		Pulsed column	
Windscale (UK)			
Butex reprocessing	1961	Packed column	21
Purex process	1964	Mixer–settler	22
Dounreay (UK)			
MTR plant	1958	Stirred mixer–settler	23
Fast reactor plant		Pulsed mixer–settler	
Trombay (India)	1965	Pulsed column	
West Valley (USA)	1966	Pulsed column	
Cap La Hague (France)	1966	Mixer–settler; Robatel centrifugal mixer–settler	24
Mol, Eurochemic (Belgium)	1966–1972	Pulsed column	25
Tokai-Mura (Japan)	1977	Mixer–settler	
Barnwell (USA)	Completed 1976	Pulsed column; centrifugal contactor	
Morris (GEC; USA)	Completed 1975 (abandoned)	Pulsed column	
Karlsruhe (West Germany)	1970	Mixer–settler	
ITREC (Italy)		Mixer–settler	
EUREX (Italy)		Mixer–settler	

the process. When the breakdown becomes severe, the organic phase may form emulsions with the aqueous phase and the process of solvent extraction becomes inefficient or impossible. The presence of radioactive solids (crud) that collect at interfaces can enhance solvent degradation. To avoid solvent degradation, the fuel can be "cooled" for a long time to allow activity levels to fall, or the residence time of the organic phase in the extraction can be minimized. A pulsed column or centrifugal extractor will be favored over a mixer–settler in the presence of high activity levels.

3. In the presence of high nitric acid concentrations and at high plutonium and uranium concentrations in the solvent phase, it is possible to form a third phase of density intermediate between organic and aqueous phases. This third phase can be rich in plutonium [i.e., $Pu(NO_3)_4 \cdot 2(TBP)$], and it can collect in one part of the contactor; therefore, it is a criticality hazard as well as an interference with the hydraulic operation of the equipment. It is important to avoid conditions that can produce a third phase.

4. Plutonium will precipitate in water, sometimes forming a colloidal suspension that will collect at an interface. It is important to keep the system acid (ca. 0.01 *N* minimum) under all operating conditions.

9. SOLVENT EXTRACTION EQUIPMENT IN REPROCESSING PLANTS

Many reprocessing plants have operated throughout the world. Table 2 lists these plants with references; it also shows the solvent extractor type used in the plant.

REFERENCES AND SUGGESTED READING

This brief chapter is a basic introduction to the subject, but more detailed study is a necessity for those contemplating the design of nuclear solvent extraction plant. The numbered references are referred to in the text.

General

C. Hanson, *Chem. Process Eng.*, 445–456 (October 1960).

M. Benedict and T. H. Pigford, *Nuclear Chemical Engineering*, McGraw-Hill, New York, 1957.

Containment

C. A. Burchsted, A. B. Fuller, and J. E. Kahn, *Nuclear Air Cleaning Handbook,* ERDA 76-21.

14th U.S. Energy Research and Development Administration Air Cleaning Conference: 1976 Conf.-760822.

13th U.S. Atomic Energy Commission Air Cleaning Conference: 1974 Conf.-740807.

12th U.S. Atomic Energy Commission Air Cleaning Conference: 1972 Conf.-720823.

11th U.S. Atomic Energy Commission Air Cleaning Conference: 1970 Conf.-700816.

10th U.S. Atomic Energy Commission Air Cleaning Conference: 1968 Conf.-680821.

9th U.S. Atomic Energy Commission Air Cleaning 1966 Conf.-660904.

8th U.S. Atomic Energy Commission Air Cleaning Conference: 1963 TID-7677.

7th U.S. Atomic Energy Commission Air Cleaning Conference: 1961 TID-7627.

Shielding

1. *Handbook of Radiation Protection,* Part 1, HMSO, 1971.
2. R. G. Jaeger, Ed., *Engineering Compendium on Radiation Shielding,* Vol. 1, Springer-Verlag, New York, 1968.
3. W. W. Engle, *ANISN: A One Dimensional Discrete Ordinate Code with Anisotropic Scattering,* K. 1693.
4. D. E. Bendall, *MCBEND: A Code for Solving Gamma Ray and Neutron Transport Problems,* AEEW-R1237.

Criticality

5. R. D. Carter, et al., *ARH-600 Criticality Handbook,* Vols. I–III, Atlantic Richfield, Hanford, CA, 1968.
6. J. F. Flagg, Ed., *Chemical Processing of Reactor Fuels*, Academic Press, New York, 1961, Chapter IX.
7. G. E. Whitesides and N. F. Cross, *KENO: A Multigroup Monte Carlo Criticality Program*, CTC-5, 1969.
8. R. Caizergues, et al., *MORET: A Rapid Monte Carlo Code to Calculate K_{eff} of Fissile Material in Complex Geometries*, CEA-N-1645, 1973.
9. V. S. W. Sherrifts, *MONK: A General Purpose Monte Carlo Neutronics Program,* SRD-R-86, 1978.
10. P. Lecorche and R. L. Seale, *A Review of the Experiments Performed to Determine to Radiological Consequences of a Criticality Accident,* Y-CDC-12 UC-46, 1973.

11. OECD Book, *Criticality Control in Nuclear Processing Plant.*

Legislation and Safety

12. Regulatory Guide 3.26, USNRC Guide Series, US Nuclear Regulatory Commission.

13. *Individual Risk–A Compilation of Recent British Data. Safety and Reliability Directorate,* United Kingdom Atomic Energy Authority.

14. J. H. Bowen, Individual Risk vs. Public Risk Criteria, *Chem. Eng. Progr.*, p. 63 (February 1976).

15. *Annals of the ICRP,* ICRP Publication 26.

16. F. R. Farmer, Siting Criteria–A New Approach, paper presented at Conference on Periodic Inspection of Pressure Vessels, May 9–11 1972.

17. Reactor Safety Study–An Assessment of Accident Risks in U.S. Commercial Nuclear Power Plants, Report No. WASH 1400.

Reprocessing Plants

18. D. A. Orth and J. M. McKibben, *Transact. Am. Nucl. Soc.* **12,** 28 (1969).

19. R. L. Stevenson and B. G. Bradley, HW 19170, 1951.

20. R. G. Geien, HW 49542, 1957.

21. G. R. Howells, et al., *Progress in Nuclear Energy Series III*, Pergamon Press, New York, 1961.

22. B. F. Warner et al., *Proceedings of the 3rd International Conference on the Peaceful Uses of Atomic Energy, Geneva,* Vol. 10, 1964, p. 224.

23. C. Birch, et al., *Proceedings of the 2nd International Conference on the Peaceful Uses of Atomic Energy, Geneva,* Vol. 17, 1958, p. 23.

24. C. Bernard, et al., *Proceedings of the International Solvent Extraction Conference (ISEC)*, The Hague 1971, p. 1282, Society of Chemical Industry, London, 1971.

25. C. J. Joseph et al., *Proceedings of the International Solvent Extraction Conference, (ISEC), The Hague 1971*, p. 593, Society of Chemical Industry, London, 1971.

26. R. B. Lemon et al., *Proceedings of the International Conference on the Peaceful Uses of Atomic Energy,* Vol. 9, 1955, p. 532.

INDEX